Page Numbers of Some Important Tables

Physical Constants

Constant	Symbol	Value
Atomic mass unit	u	1.66054×10^{-27} kg
Avogadro's number	N	6.02214×10^{23} mol^{-1}
Bohr radius	a_0	5.292×10^{-11} m
Boltzmann constant	k	1.38066×10^{-23} J/K
Charge of an electron	e	1.60218×10^{-19} C
Faraday constant	F	96,485 C/mol
Gas constant	R	8.31451 J/K · mol 0.08206 L · atm/K · mol
Mass of an electron	m_e	9.10939×10^{-31} kg 5.48580×10^{-4} u
Mass of a neutron	m_n	1.67493×10^{-27} kg 1.00866 u
Mass of a proton	m_p	1.67262×10^{-27} kg 1.00728 u
Planck's constant	h	6.62608×10^{-34} J · s
Speed of light	c	2.99792458×10^8 m/s

SI Units and Conversion Factors

Length

SI unit: meter (m)

1 meter = 1.0936 yards
1 centimeter = 0.39370 inch
1 inch = 2.54 centimeters (exactly)
1 kilometer = 0.62137 mile
1 mile = 5280 feet
= 1.6093 kilometers
1 angstrom = 10^{-10} meter
= 100 picometers

Mass

SI unit: kilogram (kg)

1 kilogram = 1000 grams
= 2.2046 pounds
1 pound = 453.59 grams
= 0.45359 kilogram
= 16 ounces
1 ton = 2000 pounds
= 907.185 kilograms
1 metric ton = 1000 kilograms
= 2204.6 pounds
1 atomic mass unit = 1.66056×10^{-27} kilograms

Volume

SI unit: cubic meter (m³)

1 liter = 10^{-3} m^3
= 1 dm^3
= 1.0567 quarts
1 gallon = 4 quarts
= 8 pints
= 3.7854 liters
1 quart = 32 fluid ounces
= 0.94633 liter

Temperature

SI unit: kelvin (K)

0 K = $-273.15°$C
= $-459.67°$F
K = °C + 273.15
$°C = \dfrac{5}{9}(°F - 32)$
$°F = \dfrac{9}{5}(°C) + 32$

Energy

SI unit: joule (J)

1 joule = 1 kg · m^2/s^2
= 0.23901 calorie
= 9.4781×10^{-4} btu
(British thermal unit)
1 calorie = 4.184 joules
= 3.965×10^{-3} btu
1 btu = 1055.06 joules
= 252.2 calories

Pressure

SI unit: pascal (Pa)

1 pascal = 1 N/m^2
= 1 kg/m · s^2
1 atmosphere = 101.325 kilopascals
= 760 torr (mmHg)
= 14.70 pounds per square inch
1 bar = 10^5 pascals

Chemistry

AN ATOMS FIRST APPROACH

Second Edition

Steven S. Zumdahl
University of Illinois

Susan A. Zumdahl
University of Illinois

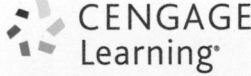

CENGAGE
Learning

Australia • Brazil • Mexico • Singapore • United Kingdom • United States

CENGAGE
Learning

Chemistry: An Atoms First Approach,
Second Edition
Steven S. Zumdahl, Susan A. Zumdahl

Product Director: Mary Finch

Product Manager: Lisa Lockwood

Content Developer: Thomas Martin

Product Assistant: Morgan Carney

Media Developer: Brendan Killion

Marketing Manager: Julie Schuster

Content Project Manager: Teresa L. Trego

Art Director: Cate Rickard Barr

Manufacturing Planner: Judy Inouye

Production Service: Graphic World

Photo Researcher: Sharon Donahue

Copy Editor: Graphic World

Text Designer: Ellen Pettengell Design

Cover Designer: Irene Morris

Cover Image: Laurie Knight/Nikon Instruments

Compositor: Graphic World

For product information and technology assistance, contact us at
Cengage Learning Customer & Sales Support, 1-800-354-9706.
For permission to use material from this text or product,
submit all requests online at **www.cengage.com/permissions.**
Further permissions questions can be e-mailed to
permissionrequest@cengage.com.

Library of Congress Control Number: 2014942489

Student Edition:
ISBN: 978-1-305-07924-3

Loose-leaf Edition:
ISBN: 978-1-305-63267-7

Cengage Learning
20 Channel Center Street
Boston, MA 02210
USA

Cengage Learning is a leading provider of customized learning solutions with office locations around the globe, including Singapore, the United Kingdom, Australia, Mexico, Brazil, and Japan. Locate your local office at **www.cengage.com/global.**

Cengage Learning products are represented in Canada by Nelson Education, Ltd.

To learn more about Cengage Learning Solutions, visit **www.cengage.com.**

Purchase any of our products at your local college store or at our preferred online store **www.CengageBrain.com.**

Printed in the United States of America
Print Number: 08   Print Year: 2017

Contents

Manex Catalapiedra/Getty Images

Milos Jokic/Getty Images

Mark Mawson/Stone/Getty Images

mitzy/Shutterstock.com

Features of *Chemistry: An Atoms First Approach*

Conceptual learning and problem solving have been fundamental to our approach in *Chemistry* through eight successful editions. Our philosophy is to help students learn to think like chemists so they can apply the process of problem solving to all aspects of their lives. In recent years a significant number of instructors has decided to engage their students by covering atoms, bonding, and structures of molecules at the beginning of their general chemistry courses, an approach often called "atoms first." In order to provide proper context within the reorganized topics, we wrote the first edition of *Chemistry: An Atoms First Approach*. In that text, we covered atoms, molecules, energy, gases, liquids, and solids before stoichiometry. While we believe that instructors using the "atoms first" approach want their students to learn the theory of atomic structure and bonding early, feedback from users of the first edition indicated strongly that they want stoichiometry covered earlier in the text to ensure that students have sufficient background to complete basic measurement and stoichiometry experiments early in the course. We have responded to these suggestions in producing the second edition of *Chemistry: An Atoms First Approach*.

We found that users of the first edition strongly support our approach to general chemistry. We have always written with a student-first approach. That is, in writing every page we ask ourselves: "How can we explain the material in a way that will be most clear to the students?" We always develop concepts in accord with the scientific method. That is, we always consider the observed properties of substances first. We then look for common threads among these properties (formulate laws). Finally we help the students understand why and how the theories of chemistry developed. In describing theories we always make clear that models are works in progress. We expect to find areas where the models fail and, in fact, that this occurrence often leads to the greatest progress in our understanding of how nature operates.

One of the main goals of our treatment of chemistry is to help students learn to be effective problem solvers. We want to go beyond memorized steps to help students think their way through the problems. To do this we take a "think like a chemist" approach. In solving problems we ask students several questions to guide them through the process: Where are we going?, What do we know?, and How do we get there? Our goal is to foster creative, concept-based problem solving,

which will serve the students in their lives and careers beyond the general chemistry course.

Over the years, thousands of students and instructors who have used our books have found that these approaches work. Thus in the second edition of the text *Chemistry: An Atoms First Approach* we have returned to our strength: a text that effectively explains chemistry to the students and helps them to learn to be creative problem solvers. The most significant difference from our widely used *Chemistry* textbook is that we present atoms, bonding, and molecules at the beginning of the text and have made sure that the rest of the topics flow smoothly from this starting point, while moving stoichiometry to a position earlier in the text.

To strengthen the atoms first approach we have emphasized at every opportunity throughout the text the importance and advantages of thinking about chemistry from an atomic/molecular perspective.

What's New

Considerable effort went into making the second edition of *Chemistry: An Atoms First Approach* even more focused on the atom. To maintain this emphasis, we added new content (with an atoms first focus) throughout the textbook by revising the core text and adding margin notes, critical thinking questions, and boxed features:

- **Core Text**—Throughout the textbook, various chapters were updated with new content that "speaks" to the atoms first approach. Many introductions were updated to help students focus on an atomic/molecular point of view before delving into the chapter. Many sections were also revised as needed to help enhance the atoms first emphasis throughout the textbook.

- **Margin Notes**—New margin notes were added throughout the textbook to provide insight and expand upon principles being taught from an atoms first approach.

- **Critical Thinking Questions**—New Critical Thinking questions were added throughout the textbook to help students think about what they learned from an atomic/molecular point of view.

- **Connecting to Atoms**—A new boxed feature, "Connecting to Atoms," contains summaries that cover key atoms first concepts at point-of-use. They have visuals and text to help students understand how an atomic/molecular approach clarifies chemical concepts.

■ **ChemWork Problems**—These multiconcept problems (and additional ones) are found interactively online with the same type of assistance a student would get from an instructor.

In addition to the new material described above, we continue to present atoms, bonding, and molecules at the beginning of the text. The biggest change in the TOC involved moving stoichiometry to a position early in the text to where users prefer it. Additional changes from the first edition of *Chemistry: An Atoms First Approach* to the second edition are listed below:

■ **Stoichiometry**—In the first edition of *Chemistry: An Atoms First Approach* we delayed stoichiometry until later in the text. However, users reported that this caused difficulty with the laboratory schedule. Therefore in the second edition we cover stoichiometry much earlier (Chapter 5). The new stoichiometry chapter provides a complete, coherent coverage of chemical stoichiometry.

■ **Chemical Reactions and Solution Stoichiometry**—In keeping with the movement of stoichiometry to Chapter 5, we have presented chemical reactions and solution stoichiometry in Chapter 6.

■ **Naming Simple Compounds**—The section on naming compounds was moved to the end of Chapter 3 (Bonding). Here we are taking advantage of an atoms first approach. Naming compounds makes more sense after bonding has been discussed.

■ **Learning to Solve Problems**—The material on learning to solve problems has been moved to a position much earlier in the text (now in Chapter 5). We also provided a brief introduction to problem solving in Chapter R. It is never too early to start students thinking about solving problems in a conceptual manner.

Hallmarks of *Chemistry: An Atoms First* Approach

■ *Chemistry: An Atoms First Approach* contains numerous discussions, illustrations, and exercises aimed at *overcoming misconceptions*. It has become increasingly clear from our own teaching experience that students often struggle with chemistry because they misunderstand many of the fundamental concepts. In this text, we have gone to great lengths to provide illustrations and explanations aimed at giving students a more accurate picture of the fundamental ideas of chemistry. In particular, we have attempted to represent the microscopic world of chemistry so that students have a picture in their minds of "what the atoms and molecules are doing." The art program along with the animations emphasize this goal. We have also placed a larger emphasis on the qualitative understanding of concepts before quantitative problems are considered. Because using

an algorithm to correctly solve a problem often masks misunderstanding—when students assume they understand the material because they got the right "answer"—it is important to probe their understanding in other ways. In this vein, the text includes a number of *Active Learning Questions* at the end of each chapter that are intended for group discussion. It is our experience that students often learn the most when they teach each other. Students are forced to recognize their own lack of understanding when they try and fail to explain a concept to another student.

With a strong problem-solving orientation, this text talks to students about how to approach and solve chemical problems. We emphasize a thoughtful, logical approach rather than simply memorizing procedures. This approach is thoroughly described in Section 5.3 (Learning to Solve Problems), which promotes the importance of thoughtful, creative problem solving. This section emphasizes to students that thinking through a problem produces more long-term, meaningful learning that can be applied to "real life" than memorizing steps that apply only to a particular type of problem. To help students adopt this way of thinking we have organized the problem-solving process in terms of:

■ Where are we going?

■ What do we know?

■ How do we get there?

■ Reality check, which prompts students to check whether their answer makes sense

As we proceed in the text, we gradually shift more responsibility to the students to think through the examples so that they do not become overly dependent on our help.

One of the characteristics of this text is an innovative method for dealing with acid–base equilibria, the material the typical student finds most difficult and frustrating. The key to this approach involves first deciding what species are present in solution, then thinking about the chemical properties of these species. This method provides a general framework for approaching all types of solution equilibria.

■ The text contains almost *250 Examples,* with more given in the text discussions, to illustrate general problem-solving strategies. When a specific strategy is presented, it is summarized in a Problem-Solving Strategy box and the *Example* that follows it reinforces the use of the strategy to solve the problem. In general, we emphasize the use of conceptual understanding to solve problems rather than an algorithm-based approach. This approach is strongly reinforced by the inclusion of 204 *Interactive Examples,* which encourage students to thoughtfully consider the examples step-by-step.

■ We have presented a thorough *treatment of reactions* that occur in solution, including acid–base reactions. This material appears in Chapter 6, "Types of Chemical Reactions and Solution Stoichiometry," directly after the chapter on chemical stoichiometry, to emphasize the connection between solution reactions and chemical reactions in general. Chapter 6 also includes oxidation–reduction reactions and balancing by oxidation state, because a large number of interesting and important chemical reactions involve redox processes.

- *Descriptive chemistry* and chemical principles are thoroughly integrated in this text. Chemical models may appear sterile and confusing without the observations that stimulated their invention. On the other hand, facts without organizing principles may seem overwhelming. A combination of observation and models can make chemistry both interesting and understandable. In the chapter on the chemistry of the elements we have used tables and charts to show how properties and models correlate. Descriptive chemistry is presented in a variety of ways—as applications of principles in separate sections, in *Examples* and exercises, in photographs, and in *Chemical Connections*.

- Throughout the book a strong *emphasis on models* prevails. Coverage includes how they are constructed, how they are tested, and what we learn when they inevitably fail. Models are developed naturally, with pertinent observation always presented first to show why a particular model was invented.

- *Chemical Connections* boxes present applications of chemistry in various fields and in our daily lives. Margin notes in the *Instructor's Annotated Edition* also highlight many more *Chemical Connections* available on the student Web site.

- We offer end-of-chapter exercises for every type of student and for every kind of homework assignment: questions that promote group learning, exercises that reinforce student understanding, and problems that present the ultimate challenge with increased rigor and by integrating multiple concepts. To further encourage this approach we have included a selection of ChemWork Problems in the text. These multiconcept problems (and additional ones) are found interactively online with the same type of assistance a student would get from an instructor. We have also included biochemistry problems to make the connection for students in the course who are not chemistry majors.

- Judging from the favorable comments of instructors and students who have used our books, the text seems to work very well in a variety of courses. We are especially pleased that *readability* is cited as a key strength when students are asked to assess our textbooks.

Supporting Materials

Please visit http://www.cengage.com/chemistry/zumdahl/atomsfirst2e for information about student and instructor resources for this book and about custom versions.

Acknowledgments

This book represents the efforts of many talented and dedicated people. We particularly want to thank Charles Hartford, Product Manager, for his vision and oversight of the project. Charlie's knowledge of the field and enthusiasm has contributed immensely to the success of our textbooks. We also want to thank Teresa Trego, Senior Content Project Manager, who did an excellent job of coordinating the production of a very complex project. We also especially appreciate the dedication of Tom Martin, Content Developer, who managed the development process in a very supportive and organized manner. He contributed in many important ways to the successful completion of this edition, keeping the details in order and managing many different people with grace and good humor. We are also grateful for the excellent work of Brendan Killion, Media Developer. In addition, Morgan Carney, Product Assistant, did an outstanding job of communicating with us and coordinating the print supplements.

Don DeCoste, University of Illinois at Urbana–Champaign, provided invaluable contributions to this text. In particular he helped us focus on emphasizing to students the importance of thinking from an atomic/molecular point of view. We are especially grateful to Tom Hummel, University of Illinois at Urbana–Champaign, who managed the end-of-chapter problems and the solutions manuals. We greatly appreciate the outstanding work of Gretchen Adams, University of Illinois at Urbana–Champaign, who created the *Interactive Examples*. We are deeply indebted to these three colleagues who contribute in so many ways to our various projects. We are also very grateful to Jim Hall, University of Massachusetts–Lowell, for creating the excellent laboratory manual for the text. We are extremely appreciative of Jim's creative and accurate work over many years on many of our projects. We owe a large debt of appreciation to Professor Douglas Mulford of Emory University for providing a great deal of useful insight as we prepared the second edition. We followed many of Professor Mulford's suggestions as we revised the text.

Thanks to others who supplied valuable assistance for this text: Cate Ricard Barr, Art Director; Sharon Donahue, Photo Researcher; Janet Del Mundo, Marketing Manager; Lorraine Towle, Marketing Coordinator; Christina Brown, Digital Project Manager; and John Sarantakis, Permissions Project Manager.

We are especially thankful to all of the reviewers who participated in different aspects of the development process from reviewing chapters to providing feedback on the development of new features. We sincerely appreciate all of these suggestions.

Reviewers

Second Edition Reviewers

David Boatright, *University of West Georgia*

Nancy Breen, *Roger Williams University*

Elzbieta Cook, *Louisiana State University*

Tracy McGill, *Emory University*

Douglas Mulford, *Emory University*

Edith Osborne, *Angelo State University*

Yasmin Patell, *Kansas State University*

Jerry Poteat, *Georgia Perimeter College*

Michael Sakuta, *Georgia Perimeter College*

Jeff Statler, *University of Utah*

Shane Street, *University of Alabama*

Dennis Taylor, *Clemson University*

Joshua Wallach, *Old Dominion University*

First Edition Reviewers

David Boatright, *University of West Georgia*

Gene Carlisle, *West Texas A&M University*

Tsun-Mei Chang, *University of Wisconsin–Parkside*

Charles Cox, *Georgia Institute of Technology*

Michelle Driessen, *University of Minnesota*

Amina El-Ashmawy, *Collin College*

Deborah Exton, *University of Oregon*

David Geiger, *SUNY-College at Geneseo*

Jon Hardesty, *Collin County Community College*

Jennie Mayer, *Bellevue College*

Katherine Parks, *Motlow State Community College*

Casey Raymond, *SUNY Oswego*

David Smith, *New Mexico State University*

Lydia Tien, *Monroe Community College*

Wayne Wesolowski, *University of Arizona*

David Zax, *Cornell University*

To the Student

As you jump into the study of chemistry we hope that you will find our text helpful and interesting. Our job is to present the concepts and ideas of chemistry in a way you can understand. We hope to encourage you in your studies and to help you learn to solve problems in ways you can apply in all areas of your professional and personal lives.

Our main goal is to help you learn to become a truly creative problem solver. Our world badly needs people who can "think outside the box." Our focus is to help you learn to think like a chemist. Why would you want to do that? Chemists are great problem solvers. They use logic, trial and error, and intuition—along with lots of patience—to work through complex problems. Chemists make mistakes, as we all do in our lives. The important thing that a chemist does is to learn from the mistakes and to try again. This "can do" attitude is useful in all careers.

In this book we develop the concepts in a natural way: The observations come first and then we develop models to explain the observed behavior. Models help us to understand and explain our world. They are central to scientific thinking. Models are very useful, but they also have limitations, which we will point out. By understanding the basic concepts in chemistry we lay the foundation for solving problems.

Our main goal is to help you learn a thoughtful method of problem solving. True learning is more than memorizing facts. Truly educated people use their factual knowledge as a starting point—a basis for creative problem solving. Our strategy for solving problems is explained in Section 5.3. To solve a problem we ask ourselves questions, which help us think through the problem. We let the problem guide us to the solution. This process can be applied to all types of problems in all areas of life.

As you study the text, use the *Examples* and the problem-solving strategies to help you. The strategies are boxed to highlight them for you, and the *Examples* show how these strategies are applied.

After you have read and studied each chapter of the text you'll need to practice your problem-solving skills. To do this we have provided plenty of review questions and end-of-chapter exercises. Your instructor may assign these on paper or online; in either case, you'll want to work with your fellow students. One of the most effective ways to learn chemistry is through the exchange of ideas that comes from helping one another. The online homework assignments will give you instant feedback, and, in print, we have provided answers to some of the exercises in the back of the text. In all cases, your main goal is not just to get the correct answer, but to understand the process for getting the answer. Memorizing solutions for specific problems is not a very good way to prepare for an exam (or to solve problems in the real world!).

To become a great problem solver you'll need these skills:

1. Look within the problem for the solution. (Let the problem guide you.)
2. Use the concepts you have learned along with a systematic, logical approach to find the solution.
3. Solve the problem by asking questions and learn to trust yourself to think it out.

You will make mistakes, but the important thing is to learn from these errors. The only way to gain confidence is to practice, practice, practice and to use your mistakes to find your weaknesses. Be patient with yourself and work hard to understand rather than simply memorize.

We hope you'll have an interesting and successful year learning to think like a chemist!

Steve and Susan Zumdahl

A GUIDE TO *Chemistry, An Atoms First Approach* SECOND EDITION

Connecting To Atoms This new boxed feature contains atoms first summaries that cover key atoms first concepts at point-of-use.

Chemical Connections Interesting applications of modern chemistry show students the relevance of chemistry to the real world.

Connecting To Atoms is a new boxed feature that contains summaries that cover key atoms first concepts at point-of-use. They have visuals and text to help students understand how an atomic/molecular approach clarifies chemical concepts. Examples of topics include Atoms, Ions, and Isotopes: A Pictorial Summary, Lewis Structures and the Periodic Table, Polar Molecules—It's All About Symmetry, and Entropy and Positional Probability.

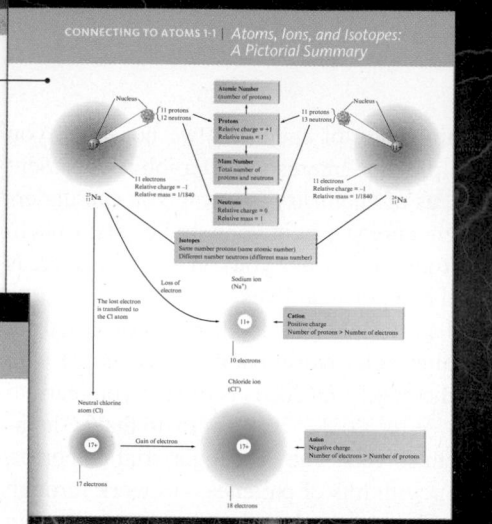

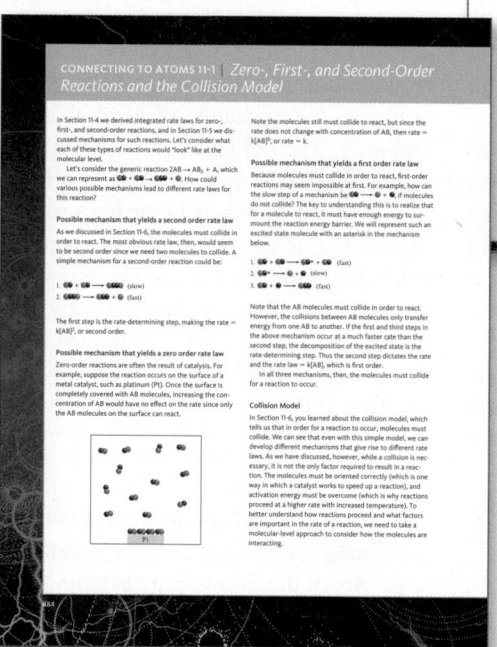

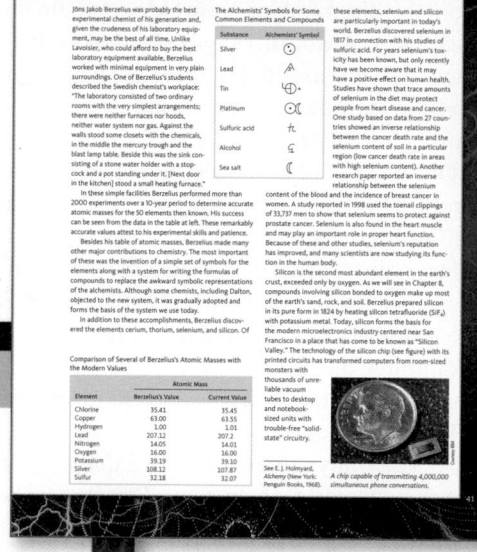

Chemical Connections describe current applications of chemistry. These special-interest boxes cover such topics as the invention of Post-it Notes, farming the wind, and the use of iron metal to clean up contaminated groundwater. Additional *Chemical Connections* are available on the student Web site.

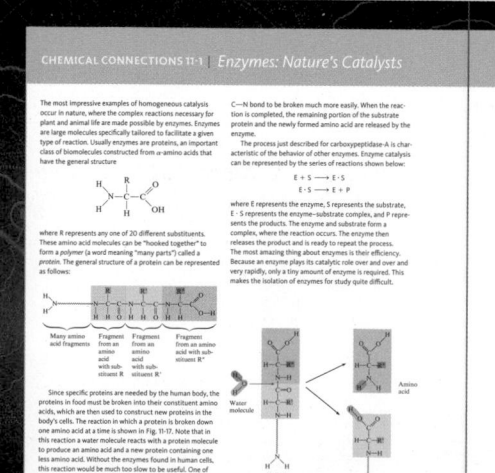

Conceptual Understanding Conceptual learning and problem solving are fundamental to the approach of *Chemistry*. The text gives students the tools to become critical thinkers: to ask questions, to apply rules and models, and to evaluate the outcome.

*"The **first** principles of the universe are **atoms** and empty space; everything else is merely thought to exist."*

—DEMOCRITUS

The authors' **emphasis on modeling** (or chemical theories) throughout the text addresses the problem of rote memorization by helping students better understand and appreciate the process of scientific thinking. By stressing the limitations and uses of scientific models, the authors show students how chemists think and work.

> ### 4-1 | Molecular Structure: The VSEPR Model
> The structures of molecules play a very important role in determining their chemical properties. As we will see later, this is particularly important for biological molecules; a slight change in the structure of a large biomolecule can completely destroy its usefulness to a cell or may even change the cell from a normal one to a cancerous one.
> Many accurate methods now exist for determining **molecular structure**, the three-dimensional arrangement of the atoms in a molecule. These methods must be used if

> **Critical Thinking**
> We now have evidence that electron energy levels in the atoms are quantized. Some of this evidence is discussed in this chapter. What if energy levels in atoms were not quantized? What are some differences we would notice?

The text includes a number of open-ended **Critical Thinking** questions that emphasize the importance of conceptual learning. These questions are particularly useful for generating group discussion.

> **LET'S REVIEW** | *A Summary of the Hydrogen Atom*
> - In the quantum (wave) mechanical model, the electron is viewed as a standing wave. This representation leads to a series of wave functions (orbitals) that describe the possible energies and spatial distributions available to the electron.
> - In agreement with the Heisenberg uncertainty principle, the model cannot specify the detailed electron motions. Instead, the square of the wave function represents the probability distribution of the electron in that orbital. This allows us to picture orbitals in terms of probability distributions, or electron density maps.
> - The size of an orbital is arbitrarily defined as the surface that contains 90% of the total electron probability.
> - The hydrogen atom has many types of orbitals. In the ground state, the single electron resides in the 1s orbital. The electron can be excited to higher-energy orbitals if energy is put into the atom.

Let's Review boxes help students organize their thinking about the crucial chemical concepts that they encounter.

Active Learning Questions
These questions are designed to be used by groups of students in class.

1. Consider the following apparatus: a test tube covered with a nonpermeable elastic membrane inside a container that is closed with a cork. A syringe goes through the cork.

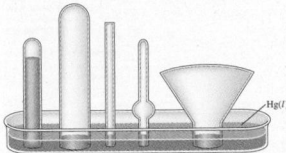

— Syringe

— Cork

— Membrane

   a. As you push down on the syringe, how does the membrane covering the test tube change?

   b. You stop pushing the syringe but continue to hold it down. In a few seconds, what happens to the membrane?

2. Figure 8-2 shows a picture of a barometer. Which of the following statements is the best explanation of how this barometer works?

   a. Air pressure outside the tube causes the mercury to move in the tube until the air pressure inside and outside the tube is equal.

   b. Air pressure inside the tube causes the mercury to move in the tube until the air pressure inside and outside the tube is equal.

   c. Air pressure outside the tube counterbalances the weight of the mercury in the tube.

   d. Capillary action of the mercury causes the mercury to go up the tube.

   e. The vacuum that is formed at the top of the tube holds up the mercury.

Justify your choice, and for the choices you did not pick, explain what is wrong with them. Pictures help!

3. The barometer below shows the level of mercury at a given atmospheric pressure. Fill all the other barometers with mercury for that same atmospheric pressure. Explain your answer.

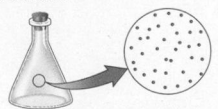

Hg(*l*)

4. As you increase the temperature of a gas in a sealed, rigid container, what happens to the density of the gas? Would the results be the same if you did the same experiment in a container with a piston at constant pressure? (See Fig. 8-17.)

5. A diagram in a chemistry book shows a magnified view of a flask of air as follows:

362

The text includes a number of **Active Learning Questions** at the end of each chapter that are intended for group discussion, as students often learn the most when they teach each other.

Problem Solving This text talks to the student about how to approach and solve chemical problems, as one of the main goals of general chemistry is to help students become creative problem solvers. The authors emphasize a thoughtful, logical approach rather than simply memorizing procedures.

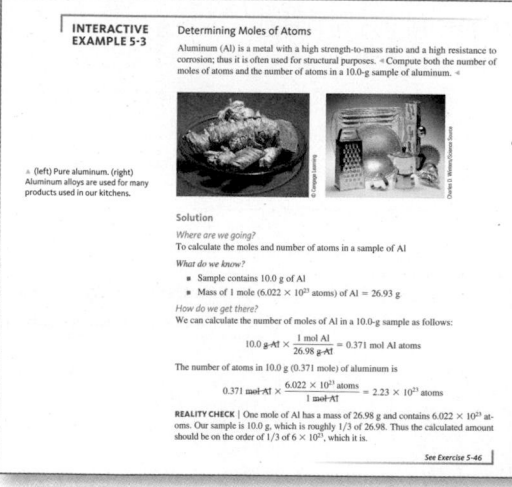

5-3 | Learning to Solve Problems

One of the great rewards of studying chemistry is to become a good problem-solver. Being able to solve complex problems is a talent that will serve you well in all walks of life. It is our purpose in this text to help you learn to solve problems in a flexible, creative way based on understanding the fundamental ideas of chemistry. We call this approach **conceptual problem solving**.

The ultimate goal is to be able to solve new problems (that is, problems you have not seen before) on your own. In this text we will provide problems and offer solutions by explaining how to think about the problems. While the answers to these problems are important, it is perhaps even more important to understand the process—the thinking necessary to get the answer. Although at first we will be solving the problem for you, do not take a passive role. While studying the solution, it is crucial that you interactively think through the problem with us. Do not skip the discussion and jump to the answer. Usually, the solution will involve asking a series of questions. Make sure that you understand each step in the process. This active approach should apply to problems outside of chemistry as well. For example, imagine riding with someone in a car to an unfamiliar destination. If your goal is simply to have the other person get you to that destination, you will probably not pay much attention to how to get there (passive), and if you have to find this same place in the future on your own, you will probably not be able to do it. If, however, your goal is to learn how to get there, you would pay attention to distances, signs, and turns (active). This is how you should read the solutions in the text (and the text in general).

While actively studying our solutions to problems is helpful, at some point you will need to know how to solve problems on your own. If we help you too much as you solve a problem, you won't really learn effectively. If we always "drive," you won't interact as meaningfully with the material. Eventually you need to learn to drive yourself. We will provide more help at the beginning of the text and less as we proceed to later chapters.

There are two fundamentally different ways you might use to approach a problem. One way emphasizes memorization. We might call this the "pigeonholing method." In this approach, the first step is to label the problem—to decide in which pigeonhole it fits. The pigeonholing method requires that we provide you with a set of steps that you memorize and store in the appropriate slot for each different problem you encounter. The difficulty with this method is that it requires a new pigeonhole each time a problem is changed by even a small amount.

Consider the driving analogy again. Suppose you have memorized how to drive from your house to the grocery store. Do you know how to drive back from the grocery store to your house? Not necessarily. If you have only memorized the directions and do not understand fundamental principles such as "I traveled north to get to the store, so my house is south of the store," you may find yourself stranded. In a more complicated example, suppose you know how to get from your house to the store (and back) and from your house to the library (and back). Can you get from the library to the store without having to go back home? Probably not if you have only memorized directions and you do not have a "big picture" of where your house, the store, and the library are relative to one another.

▲ Pigeon holes can be used for sorting and classifying objects like mail.

In **Chapter 5**, "Stoichiometry," the authors dedicate a section, **Learning to Solve Problems,** that emphasizes the importance of problem solving. This section helps students understand that thinking their way through a problem produces more long-term, meaningful learning than simply memorizing steps, which are soon forgotten.

Chapters 1–8 introduce a series of questions into the in-chapter **Examples** to engage students in the process of problem solving, such as **Where are we going?** and **How do we get there?** This more active approach helps students think their way through the solution to the problem.

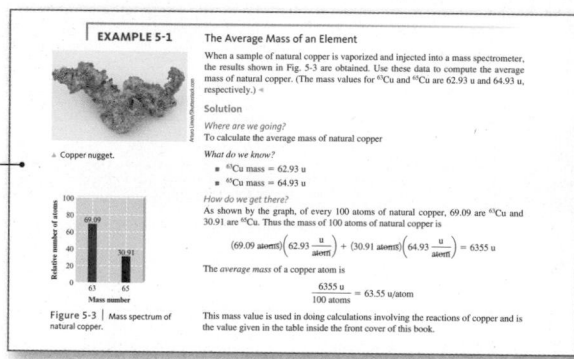

EXAMPLE 5-1 The Average Mass of an Element

When a sample of natural copper is vaporized and injected into a mass spectrometer, the results shown in Fig. 5-3 are obtained. Use these data to compute the average mass of natural copper. (The mass values for ^{63}Cu and ^{65}Cu are 62.93 u and 64.93 u, respectively.)

Solution

Where are we going?
To calculate the average mass of natural copper

What do we know?
- ^{63}Cu mass = 62.93 u
- ^{65}Cu mass = 64.93 u

How do we get there?
As shown by the graph, of every 100 atoms of natural copper, 69.09 are ^{63}Cu and 30.91 are ^{65}Cu. Thus the mass of 100 atoms of natural copper is

$$(69.09 \text{ atoms})\left(62.93 \frac{u}{\text{atom}}\right) + (30.91 \text{ atoms})\left(64.93 \frac{u}{\text{atom}}\right) = 6355 \text{ u}$$

The *average* mass of a copper atom is

$$\frac{6355 \text{ u}}{100 \text{ atoms}} = 63.55 \text{ u/atom}$$

This mass value is used in doing calculations involving the reactions of copper and is the value given in the table inside the front cover of this book.

▲ Copper nugget.

Figure 5-3 | Mass spectrum of natural copper.

INTERACTIVE EXAMPLE 5-3

Determining Moles of Atoms

Aluminum (Al) is a metal with a high strength-to-mass ratio and a high resistance to corrosion; thus it is often used for structural purposes. Compute both the number of moles of atoms and the number of atoms in a 10.0-g sample of aluminum.

▲ (left) Pure aluminum. (right) Aluminum alloys are used for many products used in our kitchens.

Solution

Where are we going?
To calculate the moles and number of atoms in a sample of Al

What do we know?
- Sample contains 10.0 g of Al
- Mass of 1 mole (6.022×10^{23} atoms) of Al = 26.93 g

How do we get there?
We can calculate the number of moles of Al in a 10.0-g sample as follows:

$$10.0 \text{ g Al} \times \frac{1 \text{ mol Al}}{26.98 \text{ g Al}} = 0.371 \text{ mol Al atoms}$$

The number of atoms in 10.0 g (0.371 mole) of aluminum is

$$0.371 \text{ mol Al} \times \frac{6.022 \times 10^{23} \text{ atoms}}{1 \text{ mol Al}} = 2.23 \times 10^{23} \text{ atoms}$$

REALITY CHECK | One mole of Al has a mass of 26.98 g and contains 6.022×10^{23} atoms. Our sample is 10.0 g, which is roughly 1/3 of 26.98. Thus the calculated amount should be on the order of 1/3 of 6×10^{23}, which it is.

See Exercise 5-46

Interactive Examples engage students in the problem-solving process by requiring them to think through the example step-by-step rather than simply scanning the written example in the text as many students do.

Problem-Solving Strategy boxes focus students' attention on the very important process of problem solving.

PROBLEM-SOLVING STRATEGY

Steps to Apply the VSEPR Model

1. Draw the Lewis structure for the molecule.
2. Count the electron pairs and arrange them in the way that minimizes repulsion (that is, put the pairs as far apart as possible).
3. Determine the positions of the atoms from the *way the electron pairs are shared.*
4. Determine the name of the molecular structure from the *positions of the atoms.*

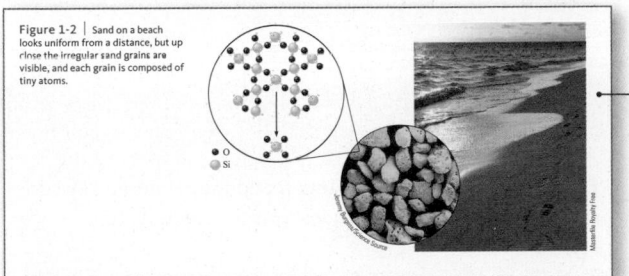

Figure 1-2 | Sand on a beach looks uniform from a distance, but up close the irregular sand grains are visible, and each grain is composed of tiny atoms.

O
Si

The art program emphasizes molecular-level interactions that help students visualize the "micro/macro" connection.

Realistic drawings of glassware and instrumentation found in the lab help students make real connections.

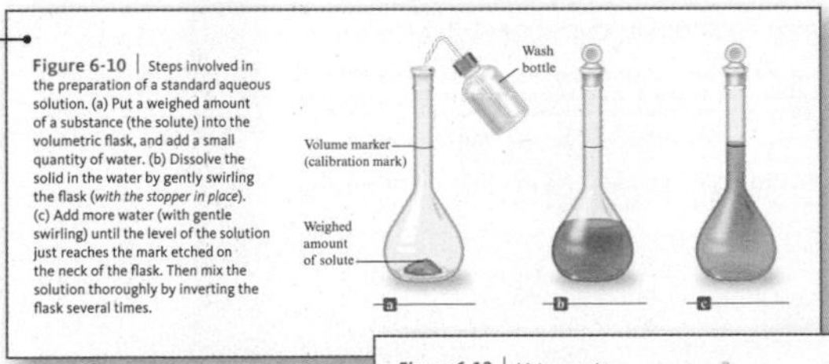

Figure 6-10 | Steps involved in the preparation of a standard aqueous solution. (a) Put a weighed amount of a substance (the solute) into the volumetric flask, and add a small quantity of water. (b) Dissolve the solid in the water by gently swirling the flask (*with the stopper in place*). (c) Add more water (with gentle swirling) until the level of the solution just reaches the mark etched on the neck of the flask. Then mix the solution thoroughly by inverting the flask several times.

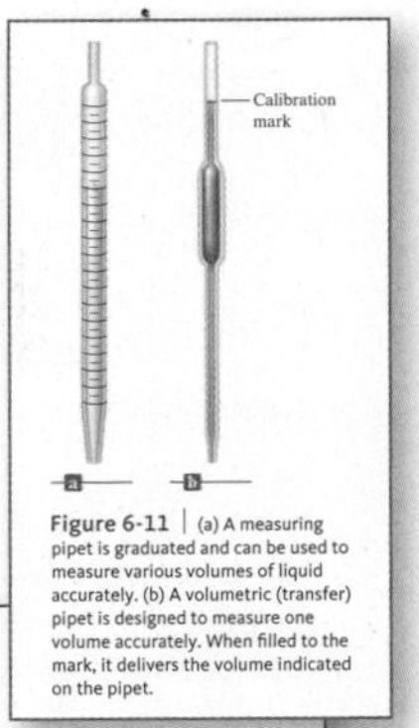

Wash bottle

Volume marker (calibration mark)

Weighed amount of solute

Calibration mark

Figure 6-11 | (a) A measuring pipet is graduated and can be used to measure various volumes of liquid accurately. (b) A volumetric (transfer) pipet is designed to measure one volume accurately. When filled to the mark, it delivers the volume indicated on the pipet.

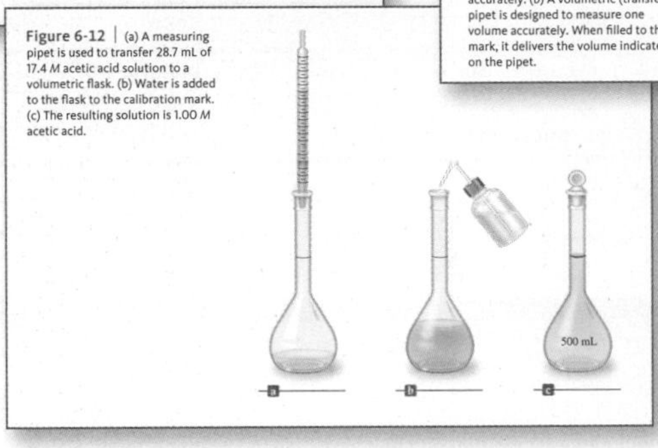

Figure 6-12 | (a) A measuring pipet is used to transfer 28.7 mL of 17.4 M acetic acid solution to a volumetric flask. (b) Water is added to the flask to the calibration mark. (c) The resulting solution is 1.00 M acetic acid.

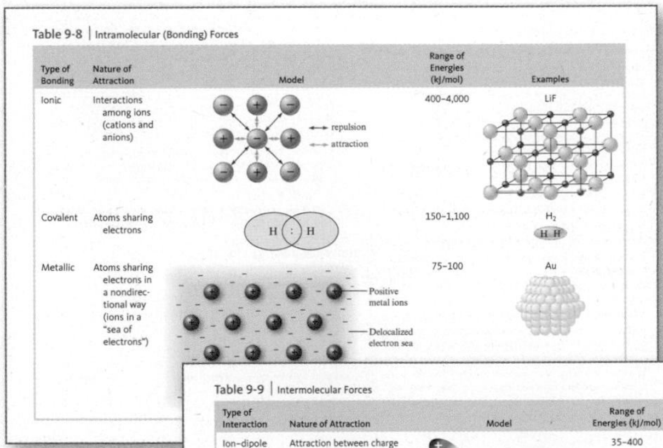

500 mL.

Table 9-8 | Intramolecular (Bonding) Forces

Type of Bonding	Nature of Attraction	Model	Range of Energies (kJ/mol)	Examples
Ionic	Interactions among ions (cations and anions)	repulsion / attraction	400–4,000	LiF
Covalent	Atoms sharing electrons	H : H	150–1,100	H_2 / H H
Metallic	Atoms sharing electrons in a nondirectional way (ions in a "sea of electrons")	Positive metal ions / Delocalized electron sea	75–100	Au

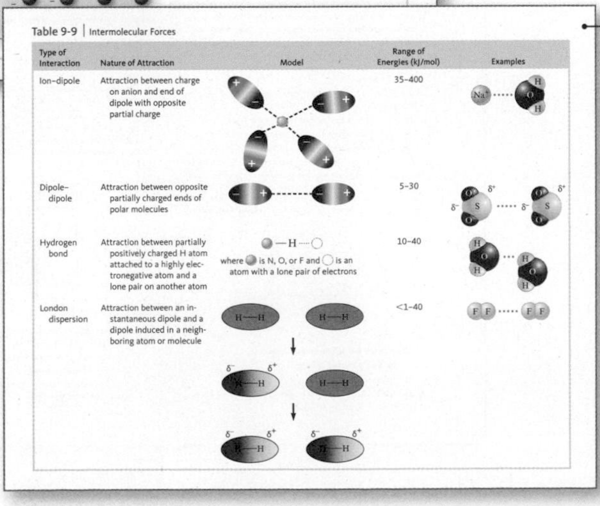

Tables containing visuals help students understand and compare intramolecular and intermolecular forces between ions, atoms, and molecules.

Table 9-9 | Intermolecular Forces

Type of Interaction	Nature of Attraction	Model	Range of Energies (kJ/mol)	Examples
Ion–dipole	Attraction between charge on anion and end of dipole with opposite partial charge		35–400	Na+ ⋯ O H H
Dipole–dipole	Attraction between opposite partially charged ends of polar molecules		5–30	
Hydrogen bond	Attraction between partially positively charged H atom attached to a highly electronegative atom and a lone pair on another atom	—H— where is N, O, or F and is an atom with a lone pair of electrons	10–40	
London dispersion	Attraction between an instantaneous dipole and a dipole induced in a neighboring atom or molecule	H—H H—H	<1–40	F F ⋯ F F

Comprehensive End-of-Chapter Practice and Review
We offer end-of-chapter exercises for every type of student and for every kind of homework assignment.

FOR REVIEW

Key Terms

Section 14-1
common ion
common ion effect

Section 14-2
buffered solution
Henderson–Hasselbalch
equation

Section 14-3
buffering capacity

Section 14-4
pH curve (titration curve)
millimole (mmol)
equivalence point
(stoichiometric point)

Section 14-5
acid–base indicator
phenolphthalein

Buffered Solutions

- Contains a weak acid (HA) and its salt (NaA) or a weak base (B) and its salt (BHCl)
- Resists a change in its pH when H^+ or OH^- is added
- For a buffered solution containing HA and A^-
  - The Henderson–Hasselbalch equation is useful:

$$pH = pK_a + \log\left(\frac{[A^-]}{[HA]}\right)$$

  - The capacity of the buffered solution depends on the amounts of HA and A^- present
- The most efficient buffering occurs when the $\frac{[A^-]}{[HA]}$ ratio is close to 1
- Buffering works because the amounts of HA (which reacts with added OH^-) and A^- (which reacts with added H^+) are large enough that the $\frac{[A^-]}{[HA]}$ ratio does not change significantly when strong acids or bases are added

Acid–Base Titrations

- The progress of a titration is represented by plotting the pH of the solution versus the volume of added titrant; the resulting graph is called a pH curve or titration curve
- Strong acid–strong base titrations show a sharp change in pH near the equivalence point
- The shape of the pH curve for a strong base–strong acid titration before the equivalence point is quite different from the shape of the pH curve for a strong base–weak acid titration
  - The strong base–weak acid pH curve shows the effects of buffering before the equivalence point
  - For a strong base–weak acid titration, the pH is greater than 7 at the equivalence point because of the basic properties of A^-
- Indicators are sometimes used to mark the equivalence point of an acid–base titration
  - The end point is where the indicator changes color
  - The goal is to have the end point and the equivalence point be as close as possible

Review Questions Answers to the Review Questions can be found on the Student Web site (accessible from www.cengagebrain.com).

1. What is meant by the presence of a common ion? How does the presence of a common ion affect an equilibrium such as

$$HNO_2(aq) \rightleftharpoons H^+(aq) + NO_2^-(aq)$$

What is an acid–base solution called that contains a common ion?

2. Define a buffer solution. What makes up a buffer solution? How do buffers absorb added H^+ or OH^- with little pH change?

Is it necessary that the concentrations of the weak acid and the weak base in a buffered solution be equal? Explain. What is the pH of a buffer when the weak acid and conjugate base concentrations are equal?

A buffer generally contains a weak acid and its weak conjugate base, or a weak base and its weak conjugate acid, in water. You can solve for the pH by setting up the equilibrium problem using the K_a reaction of the weak acid or the K_b reaction of

613

Each chapter has a **For Review** section to reinforce key concepts and includes review questions for students to practice independently.

Active Learning Questions

These questions are designed to be used by groups of students in class.

1. Paracelsus, a sixteenth-century alchemist and healer, adopted as his slogan: "The patients are your textbook, the sickbed is your study." Is this view consistent with using the scientific method?

2. What is wrong with the following statement? "The results of the experiment do not agree with the theory. Something must be wrong with the experiment."

3. Which of the following is true about an individual atom? Explain.
   a. An individual atom should be considered to be a solid.
   b. An individual atom should be considered to be a liquid.
   c. An individual atom should be considered to be a gas.
   d. The state of the atom depends on which element it is.
   e. An individual atom cannot be considered to be a solid, liquid, or gas.
   Justify your choice, and for choices you did not pick, explain what is wrong with them.

4. These questions concern the work of J. J. Thomson.
   a. From Thomson's work, which particles do you think he would feel are most important for the formation of compounds (chemical changes), and why?
   b. Of the remaining two subatomic particles, which do you place second in importance for forming compounds, and why?
   c. Propose three models that explain Thomson's findings and evaluate them. To be complete you should include Thomson's findings.

5. Which of the following explain how an ion is formed? Explain your answer.
   a. adding or subtracting protons to/from an atom
   b. adding or subtracting neutrons to/from an atom
   c. adding or subtracting electrons to/from an atom

6. You have a chemical in a sealed glass container filled with air. The setup is sitting on a balance as shown below. The chemical is ignited by means of a magnifying glass focusing sunlight on the reactant. After the chemical has completely burned, which of the following is true? Explain your answer.

250.0 g

   a. The balance will read less than 250.0 g.
   b. The balance will read 250.0 g.
   c. The balance will read greater than 250.0 g.
   d. The scale's reading cannot be determined without knowing the identity of the chemical.

7. You may have noticed that when water boils, you can see bubbles that rise to the surface of the water. Which of the following is inside these bubbles? Explain.
   a. air
   b. hydrogen and oxygen gas
   c. oxygen gas
   d. water vapor
   e. carbon dioxide gas

8. One of the best indications of a useful theory is that it raises more questions for further experimentation than it originally answered. Does this apply to Dalton's atomic theory? Give examples.

9. Dalton assumed that all atoms of the same element were identical in all their properties. Explain why this assumption is not valid.

10. Which (if any) of the following can be determined by knowing the number of protons in a neutral element? Explain your answer.
    a. the number of neutrons in the neutral element
    b. the number of electrons in the neutral element
    c. the name of the element

A blue question or exercise number indicates that the answer to that question or exercise appears at the back of this book and a solution appears in the *Student Solutions Manual*.

Questions

11. The difference between a *law* and a *theory* is the difference between *what* and *why*. Explain.

12. As part of a science project, you study traffic patterns in your city at an intersection in the middle of downtown. You set up a device that counts the cars passing through this intersection for a 24-hour period during a weekday. The graph of hourly traffic looks like this.

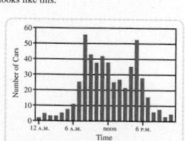

   a. At what time(s) does the highest number of cars pass through the intersection?
   b. At what time(s) does the lowest number of cars pass through the intersection?
   c. Briefly describe the trend in numbers of cars over the course of the day.

Active Learning Questions are designed to promote discussion among groups of students in class.

xviii

Comprehensive End-of-Chapter Practice and Review

A blue question or exercise number indicates that the answer to that question or exercise appears at the back of this book and a solution appears in the *Student Solutions Manual*.

Questions

Questions are homework problems directed at concepts within the chapter and in general don't require calculation.

17. Compare and contrast the bonding found in the $H_2(g)$ and $HF(g)$ molecules with that found in $NaF(s)$.
18. The following electrostatic potential diagrams represent H_2, HCl, or NaCl. Label each and explain your choices.

    a.

    b.

    c.

19. Describe the type of bonding that exists in the $Cl_2(g)$ molecule. How does this type of bonding differ from that found in the $HCl(g)$ molecule? How is it similar?
20. Some of the important properties of ionic compounds are as follows:
    i. low electrical conductivity as solids and high conductivity in solution or when molten
    ii. relatively high melting and boiling points
    iii. brittleness
    iv. solubility in polar solvents
    How does the concept of ionic bonding discussed in this chapter account for these properties?
21. Label the type of bonding for each of the following.

    a.    b.

22. Distinguish between the following terms.
    a. molecule versus ion
    b. covalent bonding versus ionic bonding

197a   Chapter 4   Molecular Structure and Orbitals

b. The bond angle in SO_2 should be similar to the bond angle in CS_2 or SCl_2.
c. Of the compounds CF_4, KrF_4, and SeF_4, only SeF_4 exhibits an overall dipole moment (is polar).
d. Central atoms in a molecule adopt a geometry of the bonded atoms and lone pairs about the central atom in order to maximize electron repulsions.

10. Give one example of a compound having a linear molecular structure that has an overall dipole moment (is polar) and one example that does not have an overall dipole moment (is nonpolar). Do the same for molecules that have trigonal planar and tetrahedral molecular structures.
11. In the hybrid orbital model, compare and contrast σ bonds with π bonds. What orbitals form the σ bonds and what orbitals form the π bonds? Assume the z-axis is the internuclear axis.
12. Write the name of each of the following molecular structures.

    a.
    b.
    c.
    d.
    e.

    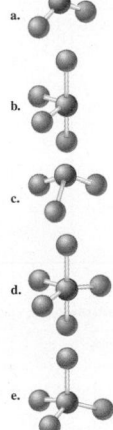

There are numerous **Exercises** to reinforce students' understanding of each section. These problems are paired and organized by topic so that instructors can review them in class and assign them for homework.

13. Give the expected hybridization for the molecular structures illustrated in the previous question.
14. In the molecular orbital model, compare and contrast σ bonds with π bonds. What orbitals form the σ bonds and what orbitals form the π bonds? Assume the z-axis is the internuclear axis.
15. Why are d orbitals sometimes used to form hybrid orbitals? Which period of elements does not use d orbitals for hybridization? If necessary, which d orbitals ($3d$, $4d$, $5d$, or $6d$) would sulfur use to form hybrid orbitals requiring d atomic orbitals? Answer the same question for arsenic and for iodine.

16. The atoms in a single bond can rotate about the internuclear axis without breaking the bond. The atoms in a double and triple bond cannot rotate about the internuclear axis unless the bond is broken. Why?
17. Compare and contrast bonding molecular orbitals with antibonding molecular orbitals.
18. What modification to the molecular orbital model was made from the experimental evidence that B_2 is paramagnetic?
19. Why does the molecular orbital model do a better job in explaining the bonding in NO⁻ and NO than the hybrid orbital model?
20. The three NO bonds in NO_3^- are all equivalent in length and strength. How is this explained even though any valid Lewis structure for NO_3^- has one double bond and two single bonds to nitrogen?

Exercises

In this section, similar exercises are paired.

Molecular Structure and Polarity

21. Predict the molecular structure (including bond angles) for each of the following.
    a. SeO_3
    b. SeO_2
22. Predict the molecular structure (including bond angles) for each of the following.
    a. PCl_3
    b. SCl_2
    c. SiF_4
23. Predict the molecular structure and bond angles for each molecule or ion in Exercises 81 and 87 from Chapter 3.
24. Predict the molecular structure and bond angles for each molecule or ion in Exercises 82 and 88 from Chapter 3.
25. There are several molecular structures based on the trigonal bipyramid geometry (see Table 4-3). Three such structures are

    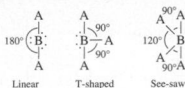

    Linear   T-shaped   See-saw

    Which of the compounds/ions Br_3^-, ClF_3, XeF_4, SF_4, PF_5, ClF_5, and SF_6 have these molecular structures?
26. Two variations of the octahedral geometry (see Table 4-1) are illustrated below.

    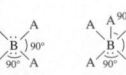

    Square planar   Square pyramid

    Which of the compounds/ions Br_3^-, ClF_3, XeF_4, SF_4, PF_5, ClF_5, and SF_6 have these molecular structures?

Wealth of End-of-Chapter Problems
The text offers an unparalleled variety of end-of-chapter content with problems that increase in rigor and integrate multiple concepts.

New **ChemWork** end-of-chapter multiconcept problems, with many additional problems, are found interactively online with the same type of assistance a student would get from an instructor. These problems are available to assign online for more practice.

ChemWork Problems

These multiconcept problems (and additional ones) are found interactively online with the same type of assistance a student would get from an instructor.

60. Complete the following table, including the mass number and the atomic number with the symbol for the isotope.

Number of Protons	Number of Neutrons	Symbol
9	10	
13	14	
53	74	
34	45	
16	16	

61. Complete the following table.

Atoms	Number of Protons	Number of Neutrons
$^{4}_{2}$He		
$^{20}_{10}$Ne		
$^{48}_{22}$Ti		
$^{190}_{76}$Os		
$^{50}_{27}$Co		

Challenge Problems

89. Consider the following complex ion, where A and B represent ligands.

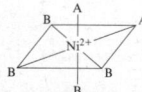

The complex is known to be diamagnetic. Do A and B produce very similar or very different crystal fields? Explain.

90. Consider the pseudo-octahedral complex ion of Cr^{3+}, where A and B represent ligands.

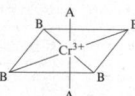

Ligand A produces a stronger crystal field than ligand B. Draw an appropriate crystal field diagram for this complex ion (assume the A ligands are on the z-axis).

91. Consider the following data:

$$Co^{3+} + e^- \longrightarrow Co^{2+} \qquad \mathscr{E}° = 1.82 \text{ V}$$
$$Co^{2+} + 3en \longrightarrow Co(en)_3^{2+} \qquad K = 1.5 \times 10^{12}$$
$$Co^{3+} + 3en \longrightarrow Co(en)_3^{3+} \qquad K = 2.0 \times 10^{47}$$

where en = ethylenediamine.

a. Calculate $\mathscr{E}°$ for the half-reaction

$$Co(en)_3^{3+} + e^- \longrightarrow Co(en)_3^{2+}$$

b. Based on your answer to part a, which is the stronger oxidizing agent, Co^{3+} or $Co(en)_3^{3+}$?

c. Use the crystal field model to rationalize the result in part b.

92. Henry Taube, 1983 Nobel Prize winner in chemistry, has studied the mechanisms of the oxidation–reduction reactions of transition metal complexes. In one experiment he and his students studied the following reaction:

$$Cr(H_2O)_6^{2+}(aq) + Co(NH_3)_5Cl^{2+}(aq)$$
$$\longrightarrow Cr(III) \text{ complexes} + Co(II) \text{ complexes}$$

Challenge Problems take students one step further and challenge them more rigorously than the Additional Exercises.

Integrative Problems

These problems require the integration of multiple concepts to find the solutions.

118. Some nonelectrolyte solute (molar mass = 142 g/mol) was dissolved in 150. mL of a solvent (density = 0.879 g/cm³). The elevated boiling point of the solution was 355.4 K. What mass of solute was dissolved in the solvent? For the solvent, the enthalpy of vaporization is 33.90 kJ/mol, the entropy of vaporization is 95.95 J/K · mol, and the boiling-point elevation constant is 2.5 K · kg/mol.

119. For the equilibrium

$$A(g) + 2B(g) \rightleftharpoons C(g)$$

the initial concentrations are [A] = [B] = [C] = 0.100 atm. Once equilibrium has been established, it is found that [C] = 0.040 atm. What is $\Delta G°$ for this reaction at 25°C?

120. What is the pH of a 0.125-M solution of the weak base B if $\Delta H° = -28.0$ kJ and $\Delta S° = -175$ J/K for the following equilibrium reaction at 25°C?

$$B(aq) + H_2O(l) \rightleftharpoons BH^+(aq) + OH^-(aq)$$

Integrative Problems combine concepts from multiple chapters.

Marathon Problem

This problem is designed to incorporate several concepts and techniques into one situation.

116. Consider the following reaction:

$$CH_3X + Y \longrightarrow CH_3Y + X$$

At 25°C, the following two experiments were run, yielding the following data:

Experiment 1: $[Y]_0 = 3.0\ M$

$[CH_3X]$ (mol/L)	Time (h)
7.08×10^{-3}	1.0
4.52×10^{-3}	1.5
2.23×10^{-3}	2.3
4.76×10^{-4}	4.0
8.44×10^{-5}	5.7
2.75×10^{-5}	7.0

Marathon Problems also combine concepts from multiple chapters; they are the most challenging problems in the end-of-chapter material.

About the Authors

Steven S. Zumdahl earned a B.S. in Chemistry from Wheaton College (IL) and a Ph.D. from the University of Illinois, Urbana–Champaign. He has been a faculty member at the University of Colorado–Boulder, Parkland College (IL), and the University of Illinois at Urbana–Champaign (UIUC), where he is Professor Emeritus. He has received numerous awards, including the National Catalyst Award for Excellence in Chemical Education, the University of Illinois Teaching Award, the UIUC Liberal Arts and Sciences Award for Excellence in Teaching, UIUC Liberal Arts and Sciences Advising Award, and the School of Chemical Sciences Teaching Award (five times). He is the author of several chemistry textbooks. In his leisure time he enjoys traveling and collecting classic cars.

Susan A. Zumdahl earned a B.S. and M.A. in Chemistry at California State University–Fullerton. She has taught science and mathematics at all levels, including middle school, high school, community college, and university. At the University of Illinois at Urbana–Champaign, she developed a program for increasing the retention of minorities and women in science and engineering. This program focused on using active learning and peer teaching to encourage students to excel in the sciences. She has coordinated and led workshops and programs for science teachers from elementary through college levels. These programs encourage and support active learning and creative techniques for teaching science. For several years she was director of an Institute for Chemical Education (ICE) field center in Southern California, and she has authored several chemistry textbooks. Susan spearheaded the development of a sophisticated Web-based electronic homework system for teaching chemistry. She enjoys traveling, classic cars, and gardening in her spare time—when she is not playing with her grandchildren.

REVIEW

Measurement and Calculations in Chemistry

Glassware used to measure volumes of liquids. (Alexander Raths/Shutterstock.com)

1

Making observations is fundamental to all science. These observations can be qualitative or quantitative. A quantitative observation is called a **measurement**, which always has two parts: a **number** and a scale (called a **unit**). Both parts must be present for a measurement to be meaningful. A qualitative observation does not involve a number. Examples of qualitative observations are "the substance is blue" and "the sun is very hot."

In this chapter we will discuss measurements in detail and explain the various properties of the numbers and units associated with measurements. This material should be familiar to you from previous science courses, but we include it here to provide a review of these topics that are critical to the operations of chemistry.

R-1 | Units of Measurement

In our study of chemistry we will use measurements of mass, length, time, temperature, electric current, and the amount of a substance, among others. Scientists recognized long ago that standard systems of units had to be adopted if measurements were to be useful. If every scientist had a different set of units, complete chaos would result. Unfortunately, different standards were adopted in different parts of the world. The two major systems are the *English system* used in the United States and the *metric system* used by most of the rest of the industrialized world. This duality causes a good deal of trouble; for example, parts as simple as bolts are not interchangeable between machines built according to the two systems. As a result, the United States has begun to adopt the metric system.

Most scientists in all countries have for many years used the metric system. In 1960, an international agreement set up a system of units called the *International System* (*le Système International* in French), or the **SI system**. This system is based on the metric system and units derived from the metric system. The fundamental SI units are listed in Table R-1. We will discuss how to manipulate these units later in this chapter. ◄

Because the fundamental units are not always convenient (expressing the mass of a pin in kilograms is awkward), prefixes are used to change the size of the unit. These are listed in Table R-2. Some common objects and their measurements in SI units are listed in Table R-3.

One physical quantity that is very important in chemistry is *volume*, which is not a fundamental SI unit but is derived from length. ◄ ⓘ A cube that measures 1 meter (m) on each edge is represented in Fig. R-1. This cube has a volume of $(1 \text{ m})^3 = 1 \text{ m}^3$. Because there are 10 decimeters (dm) in a meter, the volume of this cube is $(1 \text{ m})^3 = (10 \text{ dm})^3 = 1000 \text{ dm}^3$. A cubic decimeter, that is $(1 \text{ dm})^3$, is commonly called a *liter (L)*, which is a unit of volume slightly larger than a quart. As shown in Fig. R-1,

▲ Soda is commonly sold in 2-liter bottles—an example of the use of SI units in everyday life.

ⓘ A unit such as volume that is based on a fundamental unit is called a "derived unit."

Table R-1 | The Fundamental SI Units

Physical Quantity	Name of Unit	Abbreviation
Mass	kilogram	kg
Length	meter	m
Time	second	s
Temperature	kelvin	K
Electric current	ampere	A
Amount of substance	mole	mol
Luminous intensity	candela	cd

Table R-2 | The Prefixes Used in the SI System (Those most commonly encountered are shown in blue.)

Prefix	Symbol	Meaning	Exponential Notation*
exa	E	1,000,000,000,000,000,000	10^{18}
peta	P	1,000,000,000,000,000	10^{15}
tera	T	1,000,000,000,000	10^{12}
giga	G	1,000,000,000	10^9
mega	M	1,000,000	10^6
kilo	k	1,000	10^3
hecto	h	100	10^2
deka	da	10	10^1
—	—	1	10^0
deci	d	0.1	10^{-1}
centi	c	0.01	10^{-2}
milli	m	0.001	10^{-3}
micro	μ	0.000001	10^{-6}
nano	n	0.000000001	10^{-9}
pico	p	0.000000000001	10^{-12}
femto	f	0.000000000000001	10^{-15}
atto	a	0.000000000000000001	10^{-18}

*See Appendix 1.1 if you need a review of exponential notation.

Table R-3 | Some Examples of Commonly Used Units

Unit	Example
Length	A dime is 1 mm thick. A quarter is 2.5 cm in diameter. The average height of an adult man is 1.8 m.
Mass	A nickel has a mass of about 5 g. A 120-lb person has a mass of about 55 kg.
Volume	A 12-oz can of soda has a volume of about 360 mL.

1000 liters are contained in a cube with a volume of 1 cubic meter. Similarly, since 1 decimeter equals 10 centimeters (cm), the liter can be divided into 1000 cubes each with a volume of 1 cubic centimeter:

$$1 \text{ liter} = (1 \text{ dm})^3 = (10 \text{ cm})^3 = 1000 \text{ cm}^3$$

Also, since $1 \text{ cm}^3 = 1$ milliliter (mL),

$$1 \text{ liter} = 1000 \text{ cm}^3 = 1000 \text{ mL}$$

Thus 1 liter contains 1000 cubic centimeters, or 1000 milliliters.

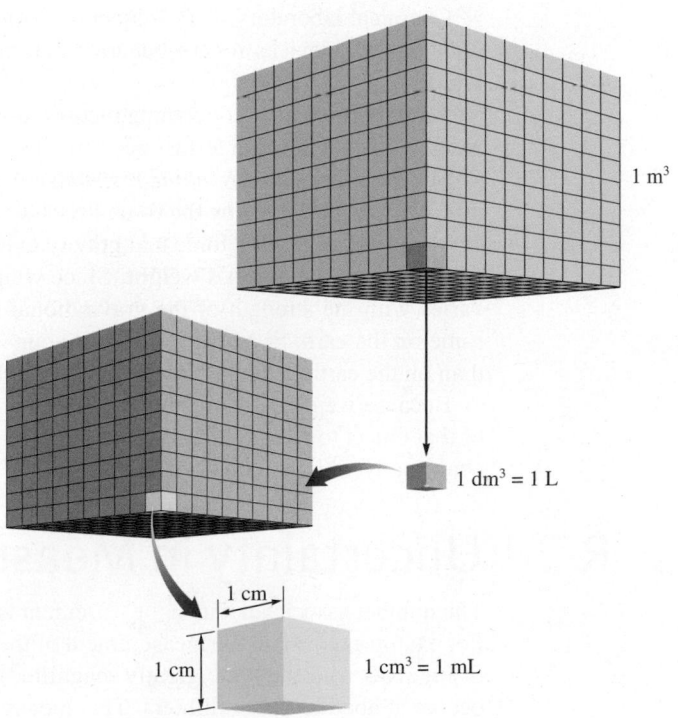

Figure R-1 | The largest cube has sides 1 m in length and a volume of 1 m³. The middle-sized cube has sides 1 dm in length and a volume of 1 dm³, or 1 L. The smallest cube has sides 1 cm in length and a volume of 1 cm³, or 1 mL.

How important are conversions from one unit to another? If you ask the National Aeronautics and Space Administration (NASA), very important! In 1999, NASA lost a $125 million Mars Climate Orbiter because of a failure to convert from English to metric units.

The problem arose because two teams working on the Mars mission were using different sets of units. NASA's scientists at the Jet Propulsion Laboratory in Pasadena, California, assumed that the thrust data for the rockets on the Orbiter they received from Lockheed Martin Astronautics in Denver, which built the spacecraft, were in metric units. In reality, the units were English. As a result the Orbiter dipped 100 kilometers lower into the Mars atmosphere than planned, and the friction from the atmosphere caused the craft to burn up.

NASA's mistake refueled the controversy over whether Congress should require the United States to switch to the metric system. About 95% of the world now uses the metric system, and the United States is slowly switching from English to metric. For example, the automobile industry has adopted metric fasteners, and we buy our soda in two-liter bottles.

Units can be very important. In fact, they can mean the difference between life and death on some occasions. In 1983, for example, a Canadian jetliner almost ran out of fuel when someone pumped 22,300 pounds of fuel into the aircraft instead of 22,300 kilograms. Remember to watch your units!

NASA/JPL

Artist's conception of the lost Mars Climate Orbiter.

Chemical laboratory work frequently requires measurement of the volumes of liquids. Several devices for the accurate determination of liquid volume are shown in Fig. R-2.

An important point concerning measurements is the relationship between mass and weight. Although these terms are sometimes used interchangeably, they are *not* the same. **Mass** *is a measure of the resistance of an object to a change in its state of motion.* Mass is measured by the force necessary to give an object a certain acceleration. On the earth we use the force that gravity exerts on an object to measure its mass. We call this force the object's **weight.** Since weight is the response of mass to gravity, it varies with the strength of the gravitational field. Therefore, your body mass is the same on the earth or on the moon, but your weight would be much less on the moon than on the earth because of the moon's smaller gravitational field.

Because weighing something on a chemical balance involves comparing the mass of that object to a standard mass, the terms *weight* and *mass* are sometimes used interchangeably, although this is incorrect.

R-2 | Uncertainty in Measurement

The number associated with a measurement is obtained using some measuring device. For example, consider the measurement of the volume of a liquid using a buret (shown in Fig. R-3 with the scale greatly magnified). Notice that the meniscus of the liquid occurs at about 20.15 milliliters. This means that about 20.15 mL of liquid has been

Figure R-2 | Common types of laboratory equipment used to measure liquid volume.

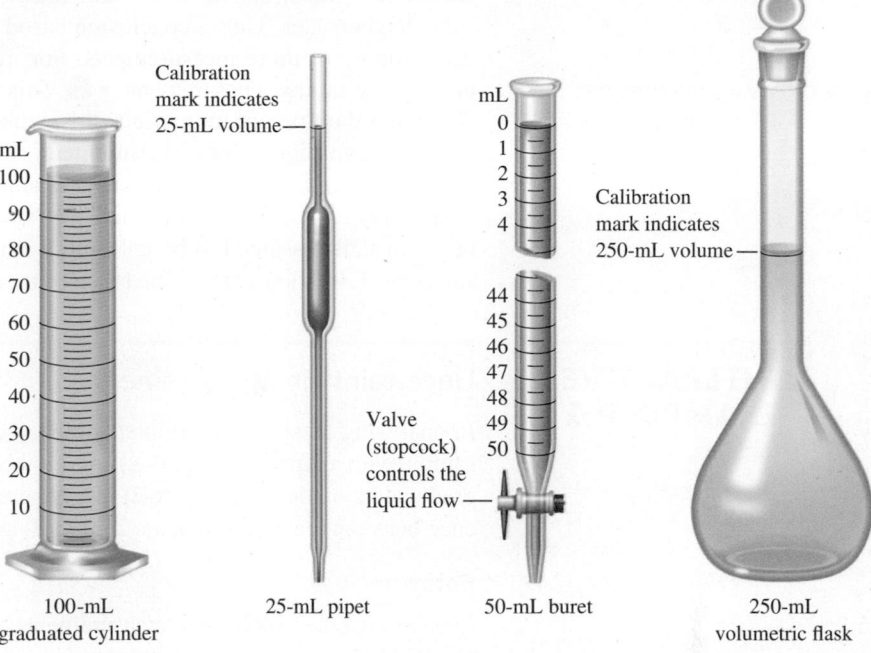

Calibration mark indicates 25-mL volume

mL
100
90
80
70
60
50
40
30
20
10

100-mL graduated cylinder

25-mL pipet

Valve (stopcock) controls the liquid flow

mL
0
1
2
3
4

44
45
46
47
48
49
50

50-mL buret

Calibration mark indicates 250-mL volume

250-mL volumetric flask

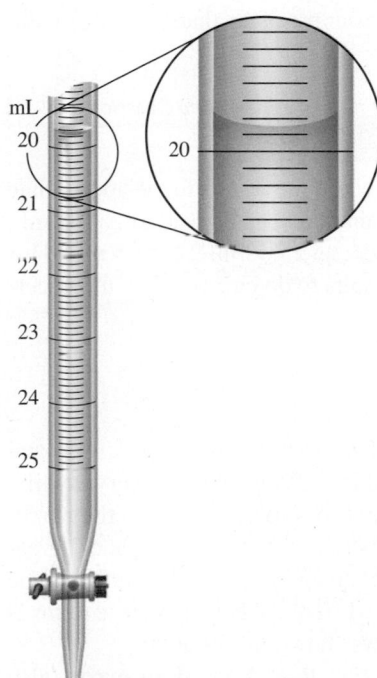

mL
20
21
22
23
24
25

20

Figure R-3 | Measurement of volume using a buret. The volume is read at the bottom of the liquid curve (called the meniscus).

i A measurement always has some degree of uncertainty.

delivered from the buret (if the initial position of the liquid meniscus was 0.00 mL). Note that we must estimate the last number of the volume reading by interpolating between the 0.1-mL marks. Since the last number is estimated, its value may be different if another person makes the same measurement. If five different people read the same volume, the results might be as follows:

Person	Results of Measurement
1	20.15 mL
2	20.14 mL
3	20.16 mL
4	20.17 mL
5	20.16 mL

These results show that the first three numbers (20.1) remain the same regardless of who makes the measurement; these are called *certain* digits. However, the digit to the right of the 1 must be estimated and therefore varies; it is called an *uncertain* digit. We customarily report a measurement by recording all the certain digits plus the *first* uncertain digit. In our example it would not make any sense to try to record the volume of thousandths of a milliliter, because the value for hundredths of a milliliter must be estimated when using the buret.

It is very important to realize that a *measurement always has some degree of* **uncertainty**. The uncertainty of a measurement depends on the precision of the measuring device. For example, using a bathroom scale, you might estimate the mass of a grapefruit to be approximately 1.5 pounds. Weighing the same grapefruit on a highly precise balance might produce a result of 1.476 pounds. In the first case, the uncertainty occurs in the tenths of a pound place; in the second case, the uncertainty occurs in the thousandths of a pound place. ◄ *i* Suppose we weigh two similar grapefruits on the two devices and obtain the following results:

	Bathroom Scale	Balance
Grapefruit 1	1.5 lb	1.476 lb
Grapefruit 2	1.5 lb	1.518 lb

Do the two grapefruits have the same mass? The answer depends on which set of results you consider. Thus a conclusion based on a series of measurements depends on the certainty of those measurements. For this reason, it is important to indicate the uncertainty in any measurement. ◀ ⓘ This is done by always recording the certain digits and the first uncertain digit (the estimated number). These numbers are called the **significant figures** of a measurement.

ⓘ Uncertainty in measurement is discussed in more detail in Appendix 1-5.

The convention of significant figures automatically indicates something about the uncertainty in a measurement. The uncertainty in the last number (the estimated number) is usually assumed to be ±1 unless otherwise indicated. For example, the measurement 1.86 kilograms can be taken to mean 1.86 ± 0.01 kilograms.

INTERACTIVE EXAMPLE R-1

Uncertainty in Measurement

In analyzing a sample of polluted water, a chemist measured out a 25.00-mL water sample with a pipet (see Fig. R-2). At another point in the analysis, the chemist used a graduated cylinder (see Fig. R-2) to measure 25 mL of a solution. What is the difference between the measurements 25.00 mL and 25 mL?

Solution

Even though the two volume measurements appear to be equal, they really convey different information. The quantity 25 mL means that the volume is between 24 mL and 26 mL, whereas the quantity 25.00 mL means that the volume is between 24.99 mL and 25.01 mL. The pipet measures volume with much greater precision than does the graduated cylinder.

See Question R-21

When making a measurement, it is important to record the results to the appropriate number of significant figures. For example, if a certain buret can be read to ±0.01 mL, you should record a reading of twenty-five milliliters as 25.00 mL, not 25 mL. This way, at some later time when you are using your results to do calculations, the uncertainty in the measurement will be known to you.

Precision and Accuracy

Two terms often used to describe the reliability of measurements are *precision* and *accuracy*. Although these words are frequently used interchangeably in everyday life, they have different meanings in the scientific context. **Accuracy** refers to the agreement of a particular value with the true value. **Precision** refers to the degree of agreement among several measurements of the same quantity. Precision reflects the *reproducibility* of a given type of measurement. The difference between these terms is illustrated by the results of three different dart throws shown in Fig. R-4.

Two different types of errors are illustrated in Fig. R-4. A **random error** (also called an *indeterminate error*) means that a measurement has an equal probability of being high or low. This type of error occurs in estimating the value of the last digit of a measurement. The second type of error is called **systematic error** (or *determinate error*). This type of error occurs in the same direction each time; it is either always high or always low. Fig. R-4(a) indicates large random errors (poor technique). Fig. R-4(b) indicates small random errors but a large systematic error, and Fig. R-4(c) indicates small random errors and no systematic error.

In quantitative work, precision is often used as an indication of accuracy; we assume that the *average* of a series of precise measurements (which should "average out" the random errors because of their equal probability of being high or low) is accurate, or close to the "true" value. However, this assumption is valid only if systematic errors are

Figure R-4 | The results of several dart throws show the difference between precise and accurate. (a) Neither accurate nor precise (large random errors). (b) Precise but not accurate (small random errors, large systematic error). (c) Bull's-eye! Both precise and accurate (small random errors, no systematic error).

absent. Suppose we weigh a piece of brass five times on a very precise balance and obtain the following results:

Weighing	Result
1	2.486 g
2	2.487 g
3	2.485 g
4	2.484 g
5	2.488 g

Normally, we would assume that the true mass of the piece of brass is very close to 2.486 grams, which is the average of the five results:

$$\frac{2.486 \text{ g} + 2.487 \text{ g} + 2.485 \text{ g} + 2.484 \text{ g} + 2.488 \text{ g}}{5} = 2.486 \text{ g}$$

However, if the balance has a defect causing it to give a result that is consistently 1.000 gram too high (a systematic error of +1.000 gram), then the measured value of 2.486 grams would be seriously in error. The point here is that high precision among several measurements is an indication of accuracy *only* if systematic errors are absent.

EXAMPLE R-2   Precision and Accuracy

To check the accuracy of a graduated cylinder, a student filled the cylinder to the 25-mL mark using water delivered from a buret (see Fig. R-2) and then read the volume delivered. Following are the results of five trials:

Trial	Volume Shown by Graduated Cylinder	Volume Shown by the Buret
1	25 mL	26.54 mL
2	25 mL	26.51 mL
3	25 mL	26.60 mL
4	25 mL	26.49 mL
5	25 mL	26.57 mL
Average	25 mL	26.54 mL

Is the graduated cylinder accurate?

Solution

The results of the trials show very good precision (for a graduated cylinder). ◄ ⓘ The student has good technique. However, note that the average value measured using the buret is significantly different from 25 mL. Thus this graduated cylinder is not very accurate. It produces a systematic error (in this case, the indicated result is low for each measurement)

ⓘ Precision is an indication of accuracy only if there are no systematic errors.

See Question R-2

R-3 | Significant Figures and Calculations

Calculating the final result for an experiment usually involves adding, subtracting, multiplying, or dividing the results of various types of measurements. Since it is very important that the uncertainty in the final result is known correctly, we have developed rules for counting the significant figures in each number and for determining the correct number of significant figures in the final result.

Rules for Counting Significant Figures

1. *Nonzero integers.* Nonzero integers always count as significant figures.

2. *Zeros.* There are three classes of zeros:

   a. *Leading zeros* are zeros that *precede* all the nonzero digits. These do not count as significant figures. In the number 0.0025, the three zeros simply indicate the position of the decimal point. This number has only two significant figures. ◄ ⓘ

   b. *Captive zeros* are zeros *between* nonzero digits. These always count as significant figures. The number 1.008 has four significant figures. ◄ ⓘ

   c. *Trailing zeros* are zeros at the *right end* of the number. They are significant only if the number contains a decimal point. The number 100 has only one significant figure, whereas the number 1.00×10^2 has three significant figures. The number one hundred written as 100. also has three significant figures. ◄ ⓘ

3. *Exact numbers.* Many times calculations involve numbers that were not obtained using measuring devices but were determined by counting: 10 experiments, 3 apples, 8 molecules. Such numbers are called *exact numbers.* They can be assumed to have an infinite number of significant figures. Other examples of exact numbers are the 2 in $2\pi r$ (the circumference of a circle) and the 4 and the 3 in $\frac{4}{3}\pi r^3$ (the volume of a sphere). Exact numbers also can arise from definitions. For example, one inch is defined as *exactly* 2.54 centimeters. Thus, in the statement 1 in = 2.54 cm, neither the 2.54 nor the 1 limits the number of significant figures when used in a calculation. ◄ ⓘ

ⓘ Leading zeros are never significant figures.

ⓘ Captive zeros are always significant figures.

ⓘ Trailing zeros are sometimes significant figures.

ⓘ Exact numbers never limit the number of significant figures in a calculation.

ⓘ Exponential notation is reviewed in Appendix 1-1.

Note that the number 1.00×10^2 above is written in **exponential notation**. ◄ ⓘ This type of notation has at least two advantages: the number of significant figures can be easily indicated, and fewer zeros are needed to write a very large or very small number. For example, the number 0.000060 is much more conveniently represented as 6.0×10^{-5}. (The number has two significant figures.)

INTERACTIVE EXAMPLE R-3

Significant Figures

Give the number of significant figures for each of the following results.

a. A student's extraction procedure on tea yields 0.0105 g of caffeine.

b. A chemist records a mass of 0.050080 g in an analysis.

c. In an experiment a span of time is determined to be 8.050×10^{-3} s.

Solution

a. The number contains three significant figures. The zeros to the left of the 1 are leading zeros and are not significant, but the remaining zero (a captive zero) is significant.

b. The number contains five significant figures. The leading zeros (to the left of the 5) are not significant. The captive zeros between the 5 and the 8 are significant, and the trailing zero to the right of the 8 is significant because the number contains a decimal point.

c. This number has four significant figures. Both zeros are significant.

See Exercises R-15 through R-18

To this point we have learned to count the significant figures in a given number. Next, we must consider how uncertainty accumulates as calculations are carried out. The detailed analysis of the accumulation of uncertainties depends on the type of calculation involved and can be complex. However, in this textbook we will employ the following simple rules that have been developed for determining the appropriate number of significant figures in the result of a calculation.

Rules for Significant Figures in Mathematical Operations

1. *For multiplication or division*, the number of significant figures in the result is the same as the number in the least precise measurement used in the calculation. For example, consider the calculation

$$4.56 \times 1.4 = 6.38 \xrightarrow{\text{Corrected}} 6.4$$

$\uparrow$ Limiting term has two significant figures $\uparrow$ Two significant figures

The product should have only two significant figures, since 1.4 has two significant figures.

2. *For addition or subtraction*, the result has the same number of decimal places as the least precise measurement used in the calculation. For example, consider the sum

$$
\begin{array}{r}
12.11 \\
18.0 \quad \leftarrow \text{Limiting term has one decimal place}\\
\underline{1.013} \\
31.123 \xrightarrow{\text{Corrected}} 31.1
\end{array}
$$

$\uparrow$ One decimal place

The correct result is 31.1, since 18.0 has only one decimal place.

For multiplication and division: significant figures are counted. For addition and subtraction: decimal places are counted.

Note that for multiplication and division, significant figures are counted. For addition and subtraction, the decimal places are counted. ◄ ⓘ

In most calculations you will need to round numbers to obtain the correct number of significant figures. The following rules should be applied when rounding.

Rules for Rounding

1. In a series of calculations, carry the extra digits through to the final result, *then* round.

2. If the digit to be removed

   a. is less than 5, the preceding digit stays the same. For example, 1.33 rounds to 1.3.

   b. is equal to or greater than 5, the preceding digit is increased by 1. For example, 1.36 rounds to 1.4. ◄ ⓘ

ⓘ Rule 2 is consistent with the operation of electronic calculators.

Although rounding is generally straightforward, one point requires special emphasis. As an illustration, suppose that the number 4.348 needs to be rounded to two significant figures. In doing this, we look *only* at the *first number* to the right of the 3:

$$4.348$$
$$\uparrow$$

Look at this number to
round to two significant figures.

> *i* Do not round sequentially. The number 6.8347 rounded to three significant figures is 6.83, not 6.84.

The number is rounded to 4.3 because 4 is less than 5. It is incorrect to round sequentially. ◄ *i* For example, do *not* round the 4 to 5 to give 4.35 and then round the 3 to 4 to give 4.4.

When rounding, *use only the first number to the right of the last significant figure.*

It is important to note that Rule 1 above usually will not be followed in the Examples in this text because we want to show the correct number of significant figures in *each step* of a problem. This same practice is followed for the detailed solutions given in the *Solutions Guide*. However, when you are doing problems, you should carry extra digits throughout a series of calculations and round to the correct number of significant figures only at the end. This is the practice you should follow. The fact that your rounding procedures are different from those used in this text must be taken into account when you check your answer with the one given at the end of the book or in the *Solutions Guide*. Your answer (based on rounding only at the end of a calculation) may differ in the last place from that given here as the "correct" answer because we have rounded after each step. To help you understand the difference between these rounding procedures, we will consider them further in Example R-4.

INTERACTIVE EXAMPLE R-4

Significant Figures in Mathematical Operations

Carry out the following mathematical operations, and give each result with the correct number of significant figures.

a. $1.05 \times 10^{-3} \div 6.135$

b. $21 - 13.8$

c. As part of a lab assignment to determine the value of the gas constant (R), a student measured the pressure (P), volume (V), and temperature (T) for a sample of gas, where

$$R = \frac{PV}{T}$$

The following values were obtained: $P = 2.560$, $T = 275.15$, and $V = 8.8$. (Gases will be discussed in detail in Chapter 8; we will not be concerned at this time about the units for these quantities.) Calculate R to the correct number of significant figures.

Solution

a. The result is 1.71×10^{-4}, which has three significant figures because the term with the least precision (1.05×10^{-3}) has three significant figures.

b. The result is 7 with no decimal point because the number with the least number of decimal places (21) has none.

c.
$$R = \frac{PV}{T} = \frac{(2.560)(8.8)}{275.15}$$

The correct procedure for obtaining the final result can be represented as follows:

> ▲ When this number is rounded to two significant figures the result is 0.082.

$$\frac{(2.560)(8.8)}{275.15} = \frac{22.528}{275.15} = 0.0818753$$

$$= 0.082 = 8.2 \times 10^{-2} = R \triangleleft$$

© Cengage Learning

The final result must be rounded to two significant figures because 8.8 (the least precise measurement) has two significant figures. To show the effects of rounding at intermediate steps, we will carry out the calculation as follows:

$$\frac{(2.560)(8.8)}{275.15} = \frac{22.528}{275.15} = \overset{\text{Rounded to two}}{\underset{}{\frac{23}{275.15}}}$$

Now we proceed with the next calculation:

$$\frac{23}{275.15} = 0.0835908$$

Rounded to two significant figures, this result is

$$0.084 = 8.4 \times 10^{-2}$$

Note that intermediate rounding gives a significantly different result than was obtained by rounding only at the end. Again, we must reemphasize that in *your* calculations you should round *only at the end*. However, because rounding is carried out at intermediate steps in this text (to always show the correct number of significant figures), the final answer given in the text may differ slightly from the one you obtain (rounding only at the end).

See Exercises R-23 through R-26

There is a useful lesson to be learned from part c of Example R-4. The student measured the pressure and temperature to greater precision than the volume. A more precise value of R (one with more significant figures) could have been obtained if a more precise measurement of V had been made. As it is, the efforts expended to measure P and T very precisely were wasted. Remember that a series of measurements to obtain some final result should all be done to about the same precision.

R-4 | Learning to Solve Problems Systematically

One of the challenges of mastering chemistry is learning to solve the problems that are associated with the concepts of chemistry. In this text, we will first explore the various concepts and then consider problems that test your true understanding of these concepts. It turns out that this process is excellent preparation for "real life" where you will need to think creatively about the situations you encounter in your work and your personal life. Therefore, in this text we will help you learn to approach problems in a thoughtful creative way rather than to rely on rote memorization. A little later in the text when the problems get more complicated, we will consider this process in more detail. However, at this early point in our study of chemistry we will consider some questions that will help you get started being a thoughtful problem solver:

1. What is my goal? Or you might phrase it as: *Where am I going?*
2. Where am I starting? Or you might phrase it as: *What do I know?*
3. How do I proceed from where I start to where I want to go? Or you might say: *How do I get there?*

We will use these ideas as we consider unit conversions in this chapter. Then we will have much more to say about problem solving in Chapter 5, where we will start to consider more complex problems.

R-5 | Dimensional Analysis

It is often necessary to convert a given result from one system of units to another. The best way to do this is by a method called the **unit factor method** or, more commonly, **dimensional analysis**. To illustrate the use of this method, we will consider several unit conversions. Some equivalents in the English and metric systems are listed in Table R-4. A more complete list of conversion factors given to more significant figures appears in Appendix 6.

Consider a pin measuring 2.85 cm in length. What is its length in inches? To accomplish this conversion, we must use the equivalence statement

$$2.54 \text{ cm} = 1 \text{ in}$$

If we divide both sides of this equation by 2.54 cm, we get

$$1 = \frac{1 \text{ in}}{2.54 \text{ cm}}$$

This expression is called a *unit factor*. Since 1 inch and 2.54 cm are exactly equivalent, multiplying any expression by this unit factor will not change its *value*.

The pin has a length of 2.85 cm. Multiplying this length by the appropriate unit factor gives

$$2.85 \text{ cm} \times \frac{1 \text{ in}}{2.54 \text{ cm}} = \frac{2.85}{2.54} \text{ in} = 1.12 \text{ in}$$

Note that the centimeter units cancel to give inches for the result. This is exactly what we wanted to accomplish. Note also that the result has three significant figures, as required by the number 2.85. Recall that the 1 and 2.54 in the conversion factor are exact numbers by definition.

Table R-4 | English–Metric Equivalents

Unit	Example
Length	1 m = 1.094 yd
	2.54 cm = 1 in
Mass	1 kg = 2.205 lb
	453.6 g = 1 lb
Volume	1 L = 1.06 qt
	1 ft³ = 28.32 L

INTERACTIVE EXAMPLE R-5

Unit Conversions I

A pencil is 7.00 in long. What is its length in centimeters?

Solution

Where are we going?
To convert the length of the pencil from inches to centimeters

What do we know?
- The pencil is 7.00 in long.

How do we get there?
Since we want to convert from inches to centimeters, we need the equivalence statement 2.54 cm = 1 in. The correct unit factor in this case is $\dfrac{2.54 \text{ cm}}{1 \text{ in}}$:

$$7.00 \text{ in} \times \frac{2.54 \text{ cm}}{1 \text{ in}} = (7.00)(2.54) \text{ cm} = 17.8 \text{ cm}$$

Here the inch units cancel, leaving centimeters, as requested.

See Exercises R-29 and R-30

Note that two unit factors can be derived from each equivalence statement. For example, from the equivalence statement 2.54 cm = 1 in, the two unit factors are

$$\frac{2.54 \text{ cm}}{1 \text{ in}} \quad \text{and} \quad \frac{1 \text{ in}}{2.54 \text{ cm}}$$

Consider the direction of the required change to select the correct unit factor.

How do you choose which one to use in a given situation? Simply look at the *direction* of the required change. To change from inches to centimeters, the inches must cancel. Thus the factor 2.54 cm/1 in is used. To change from centimeters to inches, centimeters must cancel, and the factor 1 in/2.54 cm is appropriate. ◄ *i*

> ## PROBLEM-SOLVING STRATEGY
> ### Converting from One Unit to Another
> - To convert from one unit to another, use the equivalence statement that relates the two units.
> - Derive the appropriate unit factor by looking at the direction of the required change (to cancel the unwanted units).
> - Multiply the quantity to be converted by the unit factor to give the quantity with the desired units.

INTERACTIVE EXAMPLE R-6

Unit Conversions II

You want to order a bicycle with a 25.5-in frame, but the sizes in the catalog are given only in centimeters. What size should you order?

Solution

Where are we going?
To convert from inches to centimeters

What do we know?
- The size needed is 25.5 in.

How do we get there?
Since we want to convert from inches to centimeters, we need the equivalence statement 2.54 cm = 1 in. The correct unit factor in this case is $\dfrac{2.54 \text{ cm}}{1 \text{ in}}$:

$$25.5 \text{ in} \times \frac{2.54 \text{ cm}}{1 \text{ in}} = 64.8 \text{ cm}$$

See Exercises R-29 and R-30

To ensure that the conversion procedure is clear, a multistep problem is considered in Example R-7.

INTERACTIVE EXAMPLE R-7

Unit Conversions III

A student has entered a 10.0-km run. How long is the run in miles?

Solution

Where are we going?
To convert from kilometers to miles

What do we know?
- The run is 10.00 km long.

How do we get there?
This conversion can be accomplished in several different ways. Since we have the equivalence statement 1 m = 1.094 yd, we will proceed by a path that uses this fact.

Before we start any calculations, let us consider our strategy. We have kilometers, which we want to change to miles. We can do this by the following route:

To proceed in this way, we need the following equivalence statements:

$$1 \text{ km} = 1000 \text{ m}$$
$$1 \text{ m} = 1.094 \text{ yd}$$
$$1760 \text{ yd} = 1 \text{ mi}$$

To make sure the process is clear, we will proceed step by step:

Kilometers to Meters

$$10.0 \text{ km} \times \frac{1000 \text{ m}}{1 \text{ km}} = 1.00 \times 10^4 \text{ m}$$

Meters to Yards

$$1.00 \times 10^4 \text{ m} \times \frac{1.094 \text{ yd}}{1 \text{ m}} = 1.094 \times 10^4 \text{ yd}$$

> 🛈 In the text we round to the correct number of significant figures after each step to show the correct significant figures for each calculation. However, since you use a calculator and combine steps on it, you should round only at the end.

Note that we should have only three significant figures in the result. However, since this is an intermediate result, we will carry the extra digit. Remember, round off only the final result. ◀ 🛈

Yards to Miles

$$1.094 \times 10^4 \text{ yd} \times \frac{1 \text{ mi}}{1760 \text{ yd}} = 6.216 \text{ mi}$$

Note in this case that 1 mi equals exactly 1760 yd *by designation*. Thus 1760 is an exact number.

Since the distance was originally given as 10.0 km, the result can have only three significant figures and should be rounded to 6.22 mi. Thus

$$10.0 \text{ km} = 6.22 \text{ mi}$$

Alternatively, we can combine the steps:

$$10.0 \text{ km} \times \frac{1000 \text{ m}}{1 \text{ km}} \times \frac{1.094 \text{ yd}}{1 \text{ m}} \times \frac{1 \text{ mi}}{1760 \text{ yd}} = 6.22 \text{ mi}$$

See Exercises R-29 and R-30

In using dimensional analysis, your verification that everything has been done correctly is that you end up with the correct units. In doing chemistry problems, you should always include the units for the quantities used. Always check to see that the units cancel to give the correct units for the final result. This provides a very valuable check, especially for complicated problems.

Study the procedures for unit conversions in the following examples.

INTERACTIVE EXAMPLE R-8

Unit Conversions IV

The speed limit on many highways in the United States is 55 mi/h. What number would be posted in kilometers per hour?

Solution

Where are we going?
To convert the speed limit from 55 miles per hour to kilometers per hour

What do we know?

■ The speed limit is 55 mi/h.

How do we get there?

We use the following unit factors to make the required conversion:

Result obtained by
rounding only at the
end of the calculation

$$\frac{55 \text{ mi}}{\text{h}} \times \frac{1760 \text{ yd}}{1 \text{ mi}} \times \frac{1 \text{ m}}{1.094 \text{ yd}} \times \frac{1 \text{ km}}{1000 \text{ m}} = 88 \text{ km/h}$$

Note that all units cancel except the desired kilometers per hour.

See Exercises R-37 through R-39

INTERACTIVE EXAMPLE R-9

Unit Conversions V

A Japanese car is advertised as having a gas mileage of 15 km/L. Convert this rating to miles per gallon.

Solution

Where are we going?

To convert gas mileage from 15 kilometers per liter to miles per gallon

What do we know?

■ The gas mileage is 15 km/L.

How do we get there?

We use the following unit factors to make the required conversion:

Result obtained by
rounding only at the
end of the calculation

$$\frac{15 \text{ km}}{\text{L}} \times \frac{1000 \text{ m}}{1 \text{ km}} \times \frac{1.094 \text{ yd}}{1 \text{ m}} \times \frac{1 \text{ mi}}{1760 \text{ yd}} \times \frac{1 \text{ L}}{1.06 \text{ qt}} \times \frac{4 \text{ qt}}{1 \text{ gal}} = 35 \text{ mi/gal}$$

See Exercise R-40

INTERACTIVE EXAMPLE R-10

Unit Conversions VI

The latest model Corvette has an engine with a displacement of 6.20 L. What is the displacement in units of cubic inches?

Solution

Where are we going?

To convert the engine displacement from liters to cubic inches

What do we know?

■ The displacement is 6.20 L.

How do we get there?

We use the following unit factors to make the required conversion:

$$6.20 \text{ L} \times \frac{1 \text{ ft}^3}{28.32 \text{ L}} \times \frac{(12 \text{ in})^3}{(1 \text{ ft})^3} = 378 \text{ in}^3$$

Note that the unit factor for conversion of feet to inches must be cubed to accommodate the conversion of ft^3 to in^3.

See Exercise R-44

R-6 | Temperature

Three systems for measuring temperature are widely used: the Celsius scale, the Kelvin scale, and the Fahrenheit scale. The first two temperature systems are used in the physical sciences, and the third is used in many of the engineering sciences. Our purpose here is to define the three temperature scales and show how conversions from one scale to another can be performed. Although these conversions can be carried out routinely on most calculators, we will consider the process in some detail here to illustrate methods of problem solving.

The three temperature scales are defined and compared in Fig. R-5. Note that the size of the temperature unit (the *degree*) is the same for the Kelvin and Celsius scales. The fundamental difference between these two temperature scales is their zero points. Conversion between these two scales simply requires an adjustment for the different zero points.

$$\text{Temperature (Kelvin)} = \text{temperature (Celsius)} + 273.15$$
$$T_K = T_C + 273.15$$

or

$$\text{Temperature (Celsius)} = \text{temperature (Kelvin)} - 273.15$$
$$T_C = T_K - 273.15$$

For example, to convert 300.00 K to the Celsius scale, we do the following calculation:

$$300.00 - 273.15 = 26.85°C$$

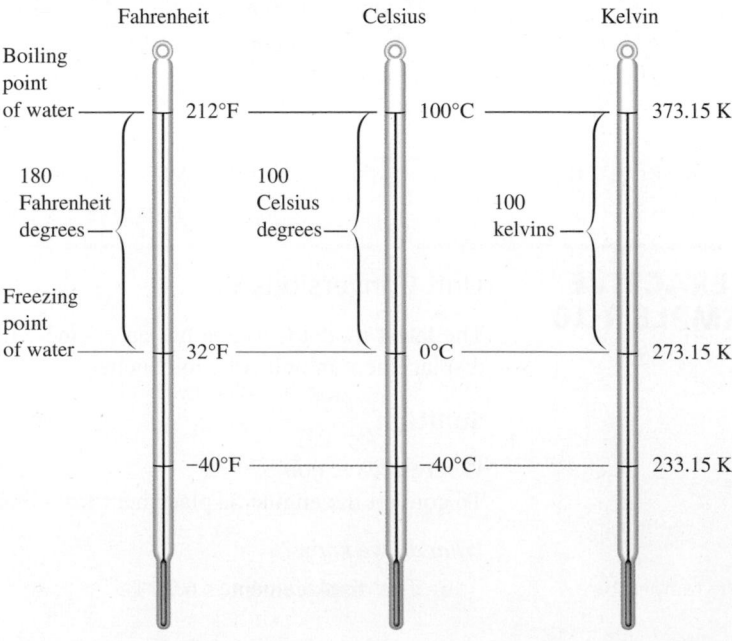

Figure R-5 | The three major temperature scales.

Note that in expressing temperature in Celsius units, the designation °C is used. The degree symbol is not used when writing temperature in terms of the Kelvin scale. The unit of temperature on this scale is called a *kelvin* and is symbolized by the letter K.

Converting between the Fahrenheit and Celsius scales is somewhat more complicated because both the degree sizes and the zero points are different. Thus we need to consider two adjustments: one for degree size and one for the zero point. First, we must account for the difference in degree size. This can be done by reconsidering Fig. R-5. Notice that since 212°F = 100°C and 32°F = 0°C,

$$212 - 32 = 180 \text{ Fahrenheit degrees} = 100 - 0 = 100 \text{ Celsius degrees}$$

Thus 180° on the Fahrenheit scale is equivalent to 100° on the Celsius scale, and the unit factor is

$$\frac{180°\text{F}}{100°\text{C}} \quad \text{or} \quad \frac{9°\text{F}}{5°\text{C}}$$

or the reciprocal, depending on the direction in which we need to go.

Next, we must consider the different zero points. Since 32°F = 0°C, we obtain the corresponding Celsius temperature by first subtracting 32 from the Fahrenheit temperature to account for the different zero points. Then the unit factor is applied to adjust for the difference in the degree size. This process is summarized by the equation

$$(T_\text{F} - 32°\text{F})\frac{5°\text{C}}{9°\text{F}} = T_\text{C} \tag{R-1}$$

where T_F and T_C represent a given temperature on the Fahrenheit and Celsius scales, respectively. In the opposite conversion, we first correct for degree size and then correct for the different zero point. This process can be summarized in the following general equation:

$$T_\text{F} = T_\text{C} \times \frac{9°\text{F}}{5°\text{C}} + 32°\text{F} \tag{R-2}$$

Equations (R-1) and (R-2) are really the same equation in different forms. See if you can obtain Equation (R-2) by starting with Equation (R-1) and rearranging.

At this point it is worthwhile to weigh the two alternatives for learning to do temperature conversions: You can simply memorize the equations, or you can take the time to learn the differences between the temperature scales and to understand the processes involved in converting from one scale to another. The latter approach may take a little more effort, but the understanding you gain will stick with you much longer than the memorized formulas. This choice also will apply to many of the other chemical concepts. Try to think things through!

INTERACTIVE EXAMPLE R-11

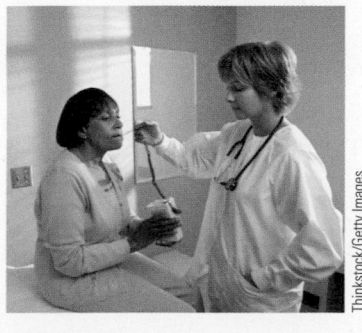

▲ A nurse taking the temperature of a patient.

Thinkstock/Getty Images

Temperature Conversions I

Normal body temperature is 98.6°F. ◄ Convert this temperature to the Celsius and Kelvin scales.

Solution

Where are we going?
To convert the body temperature from degrees Fahrenheit to degrees Celsius and to kelvins.

What do we know?

■ The body temperature is 98.6°F.

How do we get there?
Rather than simply using the formulas to solve this problem, we will proceed by thinking it through. The situation is diagrammed in Fig. R-6. First, we want to convert 98.6°F

Figure R-6 | Normal body temperature on the Fahrenheit, Celsius, and Kelvin scales.

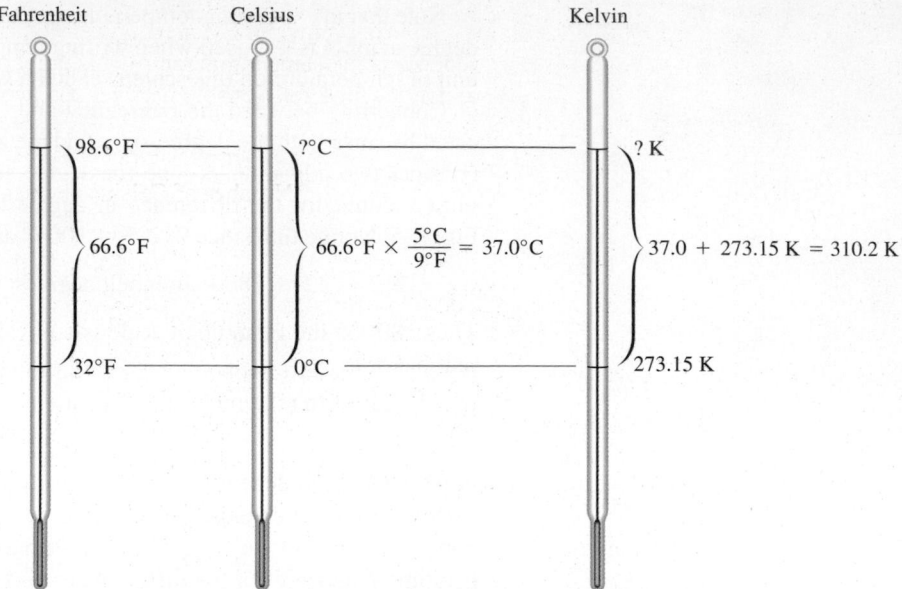

to the Celsius scale. The number of Fahrenheit degrees between 32.0°F and 98.6°F is 66.6°F. We must convert this difference to Celsius degrees:

$$66.6°F \times \frac{5°C}{9°F} = 37.0°C$$

Thus 98.6°F corresponds to 37.0°C.

Now we can convert to the Kelvin scale:

$$T_K = T_C + 273.15 = 37.0 + 273.15 = 310.2 \text{ K}$$

Note that the final answer has only one decimal place (37.0 is limiting).

See Exercises R-45, R-47, and R-48

EXAMPLE R-12   Temperature Conversions II

One interesting feature of the Celsius and Fahrenheit scales is that −40°C and −40°F represent the same temperature, as shown in Fig. R-5. Verify that this is true.

Solution

Where are we going?
To show that −40°C = −40°F

What do we know?

- The relationship between the Celsius and Fahrenheit scales

How do we get there?
The difference between 32°F and −40°F is 72°F. The difference between 0°C and −40°C is 40°C. The ratio of these is

$$\frac{72°F}{40°C} = \frac{8 \times 9°F}{8 \times 5°C} = \frac{9°F}{5°C}$$

as required. Thus −40°C is equivalent to −40°F.

See Exercise R-49

Since, as shown in Example R-12, $-40°$ on both the Fahrenheit and Celsius scales represents the same temperature, this point can be used as a reference point (like 0°C and 32°F) for a relationship between the two scales:

$$\frac{\text{Number of Fahrenheit degrees}}{\text{Number of Celsius degrees}} = \frac{T_F - (-40)}{T_C - (-40)} = \frac{9°F}{5°C}$$

$$\frac{T_F + 40}{T_C + 40} = \frac{9°F}{5°C} \qquad \text{(R-3)}$$

where T_F and T_C represent the same temperature (but not the same number). This equation can be used to convert Fahrenheit temperatures to Celsius, and vice versa, and may be easier to remember than Equations (R-1) and (R-2).

INTERACTIVE EXAMPLE R-13

Richard Megna/Fundamental Photographs © Cengage Learning

▲ Liquid nitrogen is so cold that water condenses out of the surrounding air, forming a cloud as the nitrogen is poured.

Temperature Conversions III

Liquid nitrogen, which is often used as a coolant for low-temperature experiments, has a boiling point of 77 K. ◄ What is this temperature on the Fahrenheit scale?

Solution

Where are we going?
To convert 77 K to the Fahrenheit scale

What do we know?
- The relationship between the Kelvin and Fahrenheit scales

How do we get there?
We will first convert 77 K to the Celsius scale:

$$T_C = T_K - 273.15 = 77 - 273.15 = -196°C$$

To convert to the Fahrenheit scale, we will use Equation (R-3):

$$\frac{T_F + 40}{T_C + 40} = \frac{9°F}{5°C}$$

$$\frac{T_F + 40}{-196°C + 40} = \frac{T_F + 40}{-156°C} = \frac{9°F}{5°C}$$

$$T_F + 40 = \frac{9°F}{5°C}(-156°C) = -281°F$$

$$T_F = -281°F - 40 = -321°F$$

See Exercises R-45, R-47, and R-48

R-7 | Density

A property of matter that is often used by chemists as an "identification tag" for a substance is **density**, the mass of substance per unit volume of the substance:

$$\text{Density} = \frac{\text{mass}}{\text{volume}}$$

The density of a liquid can be determined easily by weighing an accurately known volume of liquid. This procedure is illustrated in Example R-14.

INTERACTIVE EXAMPLE R-14

Determining Density

A chemist, trying to identify an unknown liquid, finds that 25.00 cm³ of the substance has a mass of 19.625 g at 20°C. The following are the names and densities of the compounds that might be the liquid:

Compound	Density in g/cm³ at 20°C
Chloroform	1.492
Diethyl ether	0.714
Ethanol	0.789
Isopropyl alcohol	0.785
Toluene	0.867

Which of these compounds is the most likely to be the unknown liquid?

Solution

Where are we going?
To calculate the density of the unknown liquid

What do we know?

- The mass of a given volume of the liquid.

How do we get there?
To identify the unknown substance, we must determine its density. This can be done by using the definition of density:

$$\text{Density} = \frac{\text{mass}}{\text{volume}} = \frac{19.625 \text{ g}}{25.00 \text{ cm}^3} = 0.7850 \text{ g/cm}^3 \blacktriangleleft \ \textit{i}$$

This density corresponds exactly to that of isopropyl alcohol, which therefore most likely is the unknown liquid. However, note that the density of ethanol is also very close. To be sure that the compound is isopropyl alcohol, we should run several more density experiments. (In the modern laboratory, many other types of tests could be done to distinguish between these two liquids.)

See Exercises R-55 and R-56

> *i* There are two ways of indicating units that occur in the denominator. For example, we can write g/cm³ or g cm⁻³. Although we will use the former system here, the other system is widely used.

Besides being a tool for the identification of substances, density has many other uses. For example, the liquid in your car's lead storage battery (a solution of sulfuric acid) changes density because the sulfuric acid is consumed as the battery discharges. In a fully charged battery, the density of the solution is about 1.30 g/cm³. If the density falls below 1.20 g/cm³, the battery will have to be recharged. Density measurement is also used to determine the amount of antifreeze, and thus the level of protection against freezing, in the cooling system of a car.

The densities of various common substances are given in Table R-5.

R-8 | Classification of Matter

Before we can hope to understand the changes we see going on around us—the growth of plants, the rusting of steel, the aging of people, the acidification of rain—we must find out how matter is organized. **Matter**, best defined as anything occupying space and having mass, is the material of the universe. Matter is complex and has many levels of organization. In this section we will introduce basic ideas about the structure of matter and its behavior.

Table R-5 │ Densities of Various Common Substances* at 20°C

Substance	Physical State	Density (g/cm³)
Oxygen	Gas	0.00133
Hydrogen	Gas	0.000084
Ethanol	Liquid	0.789
Benzene	Liquid	0.880
Water	Liquid	0.9982
Magnesium	Solid	1.74
Salt (sodium chloride)	Solid	2.16
Aluminum	Solid	2.70
Iron	Solid	7.87
Copper	Solid	8.96
Silver	Solid	10.5
Lead	Solid	11.34
Mercury	Liquid	13.6
Gold	Solid	19.32

*At 1 atmosphere pressure.

We will start by considering the definitions of the fundamental properties of matter. Matter exists in three **states**: solid, liquid, and gas. A *solid* is rigid; it has a fixed volume and shape. A *liquid* has a definite volume but no specific shape; it assumes the shape of its container. A *gas* has no fixed volume or shape; it takes on the shape and volume of its container. In contrast to liquids and solids, which are only slightly compressible, gases are highly compressible; it is relatively easy to decrease the volume of a gas. Molecular-level pictures of the three states of water are given in Fig. R-7. The different properties of ice, liquid water, and steam are determined by the different arrangements of the molecules in these substances. Table R-5 gives the states of some common substances at 20°C and 1 atmosphere pressure.

Most of the matter around us consists of **mixtures** of pure substances. Wood, gasoline, wine, soil, and air all are mixtures. The main characteristic of a mixture is that it has *variable composition*. For example, wood is a mixture of many substances, the

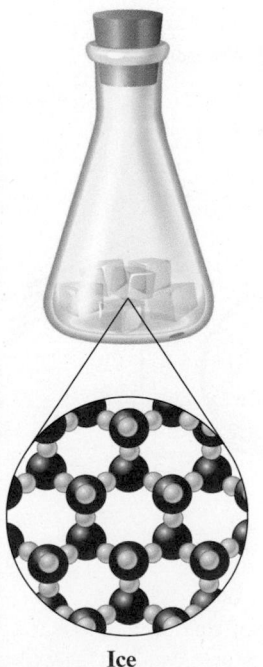

Ice

Solid: The water molecules are locked into rigid positions and are close together.

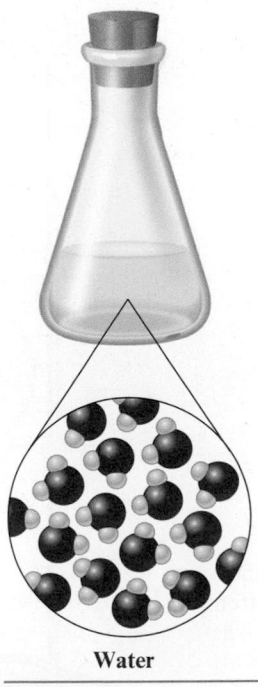

Water

Liquid: The water molecules are still close together but can move around to some extent.

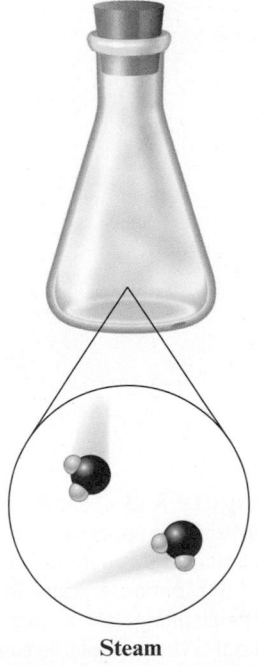

Steam

Gas: The water molecules are far apart and move randomly.

Figure R-7 │ The three states of water (where red spheres represent oxygen atoms and blue spheres represent hydrogen atoms).

proportions of which vary depending on the type of wood and where it grows. Mixtures can be classified as **homogeneous** (having visibly indistinguishable parts) or **heterogeneous** (having visibly distinguishable parts).

A homogeneous mixture is called a **solution**. Air is a solution consisting of a mixture of gases. Wine is a complex liquid solution. Brass is a solid solution of copper and zinc. Sand in water and iced tea with ice cubes are examples of heterogeneous mixtures. Heterogeneous mixtures usually can be separated into two or more homogeneous mixtures or pure substances (for example, the ice cubes can be separated from the tea).

Mixtures can be separated into pure substances by physical methods. A **pure substance** is one with constant composition. Water is a good illustration of these ideas. As we will discuss in detail later, pure water is composed solely of H_2O molecules, but the water found in nature (groundwater or the water in a lake or ocean) is really a mixture. Seawater, for example, contains large amounts of dissolved minerals. Boiling seawater produces steam, which can be condensed to pure water, leaving the minerals behind as solids. The dissolved minerals in seawater also can be separated out by freezing the mixture, since pure water freezes out. The processes of boiling and freezing are **physical changes**. When water freezes or boils, it changes its state but remains water; it is still composed of H_2O molecules. A physical change is a change in the form of a substance, not in its chemical composition. A physical change can be used to separate a mixture into pure compounds, but it will not break compounds into elements.

One of the most important methods for separating the components of a mixture is **distillation**, a process that depends on differences in the volatility (how readily substances become gases) of the components. In simple distillation, a mixture is heated in a device such as that shown in Fig. R-8. The most volatile component vaporizes at the lowest temperature, and the vapor passes through a cooled tube (a condenser), where it condenses back into its liquid state.

The simple, one-stage distillation apparatus shown in Fig. R-8 works very well when only one component of the mixture is volatile. ◀ ⓘ For example, a mixture of water and sand is easily separated by boiling off the water. Water containing dissolved

> ⓘ The term *volatile* refers to the ease with which a substance can be changed to its vapor.

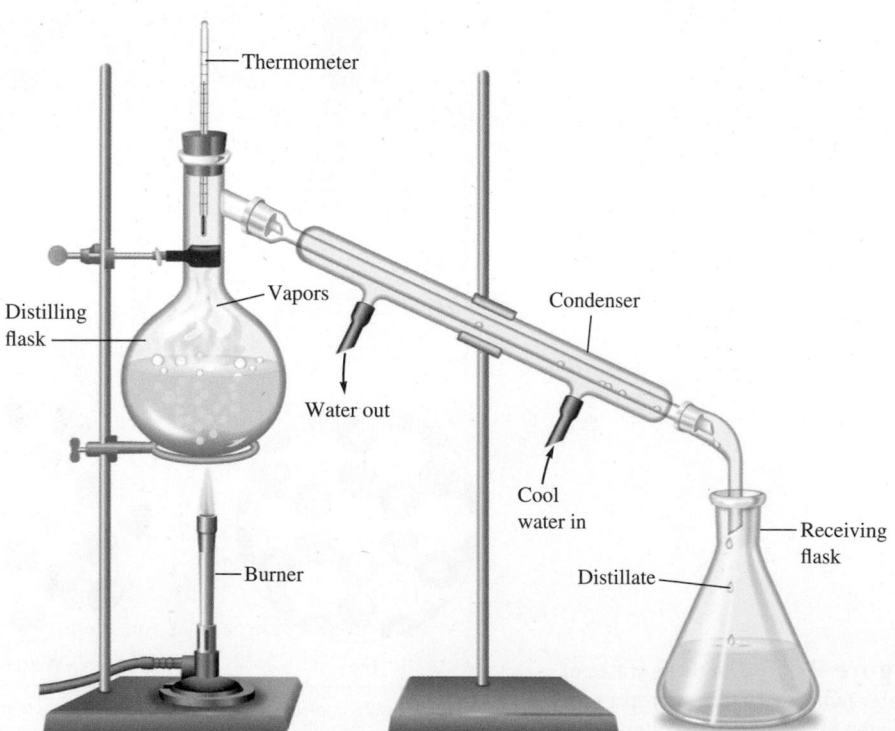

Figure R-8 │ Simple laboratory distillation apparatus. Cool water circulates through the outer portion of the condenser, causing vapors from the distilling flask to condense into a liquid. The nonvolatile component of the mixture remains in the distilling flask.

Figure R-9 | Paper chromatography of ink. (a) A dot of the mixture to be separated is placed at one end of a sheet of porous paper. (b) The paper acts as a wick to draw up the liquid.

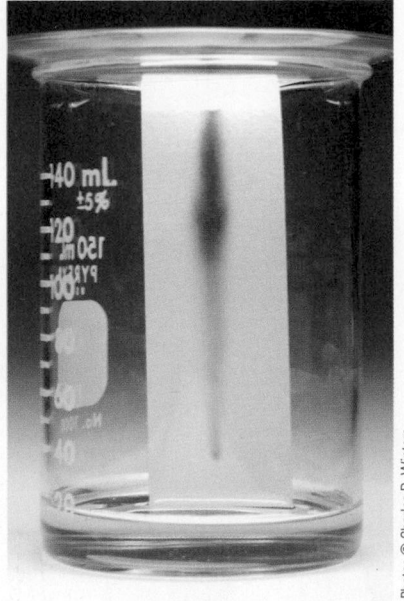

minerals behaves in much the same way. As the water is boiled off, the minerals remain behind as nonvolatile solids. Simple distillation of seawater using the sun as the heat source is an excellent way to desalinate (remove the minerals from) seawater.

However, when a mixture contains several volatile components, the one-step distillation does not give a pure substance in the receiving flask, and more elaborate methods are required.

Another method of separation is simple **filtration**, which is used when a mixture consists of a solid and a liquid. The mixture is poured onto a mesh, such as filter paper, which passes the liquid and leaves the solid behind.

A third method of separation is **chromatography**. Chromatography is the general name applied to a series of methods that use a system with two *phases* (states) of matter: a mobile phase and a stationary phase. The *stationary phase* is a solid, and the *mobile phase* is either a liquid or a gas. The separation process occurs because the components of the mixture have different affinities for the two phases and thus move through the system at different rates. A component with a high affinity for the mobile phase moves relatively quickly through the chromatographic system, whereas one with a high affinity for the solid phase moves more slowly.

One simple type of chromatography, **paper chromatography**, uses a strip of porous paper, such as filter paper, for the stationary phase. A drop of the mixture to be separated is placed on the paper, which is then dipped into a liquid (the mobile phase) that travels up the paper as though it were a wick (Fig. R-9). This method of separating a mixture is often used by biochemists, who study the chemistry of living systems.

It should be noted that when a mixture is separated, the absolute purity of the separated components is an ideal. Because water, for example, inevitably comes into contact with other materials when it is synthesized or separated from a mixture, it is never absolutely pure. With great care, however, substances can be obtained in very nearly pure form.

Pure substances are either compounds (combinations of elements) or free elements. A **compound** is a substance with *constant composition* that can be broken down into elements by chemical processes. An example of a chemical process is the electrolysis of water, in which an electric current is passed through water to break it down into the

Figure R-10 | The organization of matter.

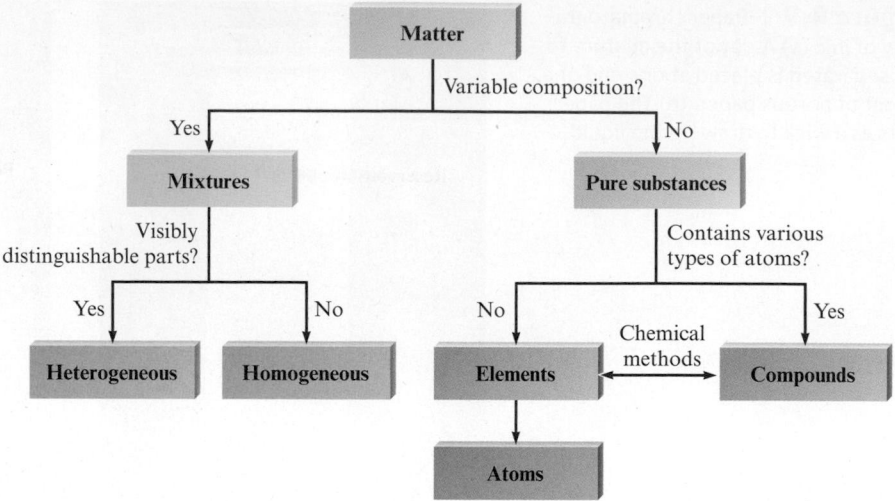

▲ The element mercury (top left) combines with the element iodine (top right) to form the compound mercuric iodide (bottom). This is an example of a chemical change.

free elements hydrogen and oxygen. This process produces a chemical change because the water molecules have been broken down. The water is gone, and in its place we have the free elements hydrogen and oxygen. A **chemical change** is one in which a given substance becomes a new substance or substances with different properties and different composition. ◄ **Elements** are substances that cannot be decomposed into simpler substances by chemical or physical means.

We have seen that the matter around us has various levels of organization. The most fundamental substances we have discussed so far are elements. As we will see in later chapters, elements also have structure: They are composed of atoms, which in turn are composed of nuclei and electrons. Even the nucleus has structure: It is composed of protons and neutrons. And even these can be broken down further, into elementary particles called *quarks*. However, we need not concern ourselves with such details at this point. Fig. R-10 summarizes our discussion of the organization of matter.

R-9 | Energy

Although energy is a familiar concept, it is difficult to define precisely. For our purposes we will define **energy** as the ability to do work or produce heat. **Work** is defined as force acting over a distance. Heat is best defined as energy that flows from one object to another because of a temperature difference between the two objects.

One of the most important characteristics of energy is that it can be converted from one form to another, and in such a process no energy is created or destroyed. This principle is summarized by the Law of Conservation of Energy, which states that energy can be converted from one form to another, but the total quantity of energy is not changed; that is, the energy content of the universe is constant.

Energy can be divided into two categories: kinetic energy (KE) and potential energy (PE). **Kinetic energy** is the energy of motion. A baseball thrown by a pitcher at 90 miles per hour has more kinetic energy than one thrown at 80 miles per hour. **Potential energy** can be defined as stored energy due to position. For example, the water stored behind a dam has potential energy that can be converted to electrical energy, as shown in Fig. R-11. In this case the water behind the dam is at a higher level than the outlet water in the river and thus has a higher PE. This PE can be changed to KE as it runs through the generator turbine, which produces electricity.

Another type of PE important in chemical processes is that due to the attraction of opposite charges. For example, consider two relative positions of a positively charged proton and a negatively charged electron, shown in Fig. R-12. The potential energies of the two cases are different. Does the situation shown in Fig. R-12(a) have a higher or lower PE than that shown in Fig. R-12(b)? We can answer this question by devising

Figure R-11 | Using a dam to convert potential energy to kinetic energy.

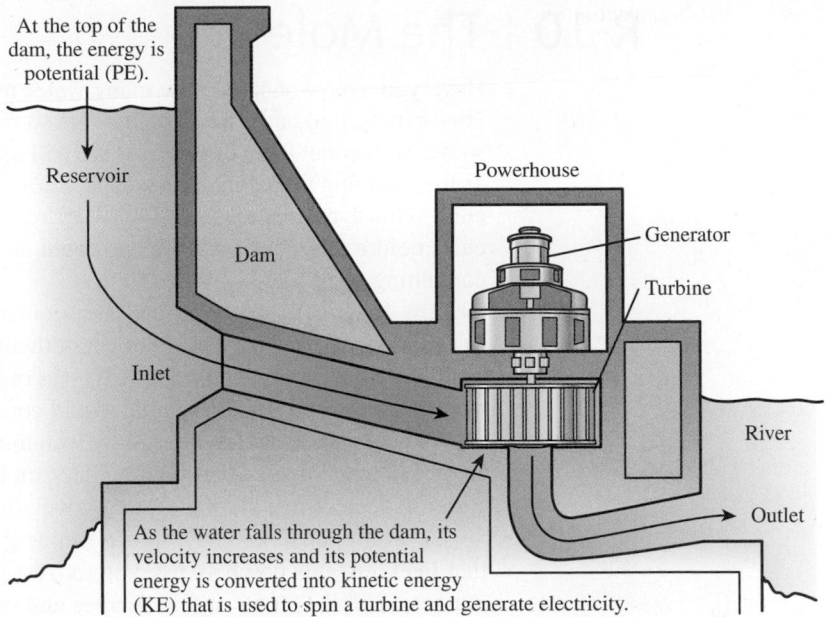

At the top of the dam, the energy is potential (PE).

Reservoir

Dam

Inlet

Powerhouse

Generator

Turbine

River

Outlet

As the water falls through the dam, its velocity increases and its potential energy is converted into kinetic energy (KE) that is used to spin a turbine and generate electricity.

an imaginary tiny machine, as illustrated in Fig. R-13. Work is done by this machine as the negative charge moves toward the positive charge. Which situation [Fig. R-12(a) or Fig. R-12(b)] would produce the most work (the greatest lifting of the weight)? We can see that the situation in Fig. R-12(b), with the charges furthest apart, can do the most work because the negative charge moves the greatest distance and thus lifts the weight the greatest distance. Thus, because Fig. R-12(b) has more stored energy than Fig. R-12(a), Fig. R-12(b) has a higher PE. In general, the farther apart two opposite charges are, the greater the stored energy and the greater the PE.

The significance of charged-based PE in chemistry has to do with the energies of electrons in atoms and the energies of chemical reactions, as we will see in Chapters 4 and 7, respectively.

The fundamental unit of energy in science is the joule (pronounced jewel). A **joule** is defined as a kilogram meter squared per second squared:

$$1 \text{ joule} = \frac{\text{kg} \cdot \text{m}^2}{\text{s}^2}.$$

Often the kilojoule (kJ) is used to describe energies.

$$1 \text{ kJ} = 10^3 \text{ J}$$

We will cover energy in much more detail in Chapter 7.

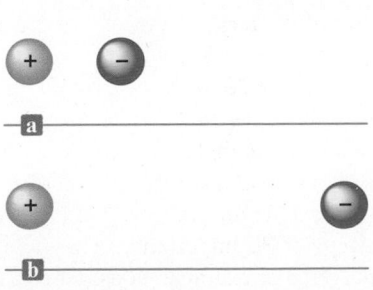

Figure R-12 | Two positions of opposite charges.

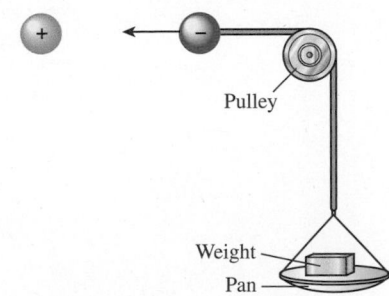

Pulley

Weight

Pan

Figure R-13 | Work is done as the charges move together and the weight is lifted.

R-10 | The Mole

Have you ever wondered how many water molecules are in a glass of water (H_2O)? How can we find out? The molecules are so tiny we can't see them with the naked eye, so we can't count them directly. As we will see in Section 1-1, recent technology, such as the scanning tunneling microscope, allows us to "see" individual atoms and molecules. However, this technology is very recent. How did we count atoms and molecules before this? The answer is we count atoms and molecules by weighing samples containing them.

How do we count by weighing? Consider a sample of jelly beans. What do we need to know about these jelly beans to count them by weighing samples of them? The answer is we need to know the average mass of a jelly bean. For example, if the average mass is 5.0 g, then a 500 g sample would contain 100 jelly beans.

We count atoms and molecules by weighing in the same way we count jelly beans. If we know the average mass of a water molecule, we can determine the number of water molecules in a glass of water by weighing the sample of water in the glass.

We won't detail here how we determine the mass of a water molecule (we will do that later). At this point we ask you to trust us that we have determined the average masses of all the atoms and molecules and can use these masses to count atoms and molecules by weighing samples of them.

When we count the water molecules in a glass of water (by weighing the water) we find that the number present is unimaginably large. The number of molecules in a glass of water is much larger than the age of the earth in seconds (4.32×10^{17} seconds) and larger than the number of milliliters of water in the earth's oceans (1.3×10^{24} mL). Therefore, because of the huge numbers of atoms and molecules in normal-sized samples of matter, we need to invent a unit for describing the number of atoms and molecules present. This unit has to be very large to be convenient. The unit of one dozen (12) works fine for eggs, but it wouldn't help much for describing numbers of atoms or molecules. The unit we have chosen is called the **mole**, which for our purposes we will define as 6.022×10^{23}. Thus, a glass of water that contains 9.0×10^{24} water molecules contains

$$9.0 \times 10^{24} \text{ } H_2O \text{ molecules} \times \frac{1 \text{ mol}}{6.022 \times 10^{23} \text{ molecules}} = 15 \text{ mol } H_2O \text{ molecules.}$$

Note that 15 moles is much more convenient than 9.0×10^{24} water molecules.

We will discuss the mole and its use in chemistry in much more detail in Chapter 5. For the present we simply need to know that the mole is a unit for counting atoms and molecules. For example, when energy terms are given, they are usually given per mole. Examples are the bond energies we will consider in Chapter 3. The bond energy for the hydrogen molecule (H_2) is 432 kJ/mol (where mol is the abbreviation for mole). This means that 432 kJ of energy is required to break one mole (6.022×10^{23}) of H—H bonds. Just remember that when you encounter "per mole" on a unit, it means that 6.022×10^{23} events are represented by the quantity in question.

FOR REVIEW

Key Terms

Section R-1
SI system
mass
weight

Section R-2
uncertainty
significant figures
accuracy
precision
random error
systematic error

Section R-3
exponential notation

Section R-5
unit factor method
dimensional analysis

Section R-7
density

Section R-8
matter
states (of matter)
homogeneous mixture
heterogeneous mixture
solution
pure substance
physical change
distillation
filtration
chromatography
paper chromatography
compound
chemical change
element

Section R-9
energy
work
kinetic energy
potential energy
joule

Section R-10
mole

Quantitative Observations Are Called Measurements.

- Consist of a number and a unit
- Involve some uncertainty
- Uncertainty is indicated by using significant figures
  - Rules to determine significant figures
  - Calculations using significant figures
- Preferred system is SI

Solving Problems Systematically

Use these questions:

- *Where am I going?*
- *What do I know?*
- *How do I get there?*

Temperature Conversions

- $T_K = T_C + 273$
- $T_C = (T_F - 32°F)\left(\frac{5°C}{9°F}\right)$
- $T_F = T_C\left(\frac{9°F}{5°C}\right) + 32°F$

Density

- $\text{Density} = \frac{\text{mass}}{\text{volume}}$

Matter Can Exist in Three States:

- Solid
- Liquid
- Gas

Mixtures Can Be Separated by Methods Involving Only Physical Changes:

- Distillation
- Filtration
- Chromatography

Compounds Can Be Decomposed to Elements Only Through Chemical Changes.

Energy

- The ability to do work or release heat.
  - Kinetic—energy of motion
  - Potential—stored energy of position
  - Joule—SI unit

Mole

- 6.022×10^{23} units

A blue question or exercise number indicates that the answer to that question or exercise appears at the back of this book and a solution appears in the *Student Solutions Manual.*

Questions

1. What is the difference between random error and systematic error?

2. To determine the volume of a cube, a student measured one of the dimensions of the cube several times. If the true dimension of the cube is 10.62 cm, give an example of four sets of measurements that would illustrate the following.

   a. imprecise and inaccurate data

   b. precise but inaccurate data

   c. precise and accurate data

   Give a possible explanation as to why data can be imprecise or inaccurate. What is wrong with saying a set of measurements is imprecise but accurate?

3. A student performed an analysis of a sample for its calcium content and got the following results:

   14.92%   14.91%   14.88%   14.91%

   The actual amount of calcium in the sample is 15.70%. What conclusions can you draw about the accuracy and precision of these results?

4. For each of the following pieces of glassware, provide a sample measurement and discuss the number of significant figures and uncertainty.

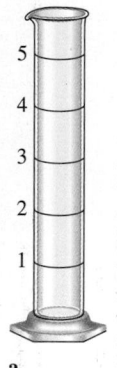  

a.   b.   c.

5. What are significant figures? Show how to indicate the number one thousand to 1 significant figure, 2 significant figures, 3 significant figures, and 4 significant figures. Why is the answer, to the correct number of significant figures, not 1.0 for the following calculation?

$$\frac{1.5 - 1.0}{0.50} =$$

6. Compare and contrast the multiplication/division significant figure rule to the significant figure rule applied for addition/subtraction mathematical operations.

7. A rule of thumb in designing experiments is to avoid using a result that is the small difference between two large measured quantities. In terms of uncertainties in measurement, why is this good advice?

8. Explain how density can be used as a conversion factor to convert the volume of an object to the mass of the object, and vice versa.

9. When the temperature in degrees Fahrenheit (T_F) is plotted versus the temperature in degrees Celsius (T_C), a straight-line plot results. A straight-line plot also results when T_C is plotted versus T_K (the temperature in kelvins). Reference Appendix A1.3 and determine the slope and *y*-intercept of each of these two plots.

10. On which temperature scale (°F, °C, or K) does 1 degree represent the smallest change in temperature?

11. Give four examples illustrating each of the following terms.

    a. homogeneous mixture

    b. heterogeneous mixture

    c. compound

    d. element

    e. physical change

    f. chemical change

12. Use molecular-level (microscopic) drawings for each of the following.

    a. Show the differences between a gaseous mixture that is a homogeneous mixture of two different compounds, and a gaseous mixture that is a homogeneous mixture of a compound and an element.

    b. Show the differences among a gaseous element, a liquid element, and a solid element.

13. What is the Law of Conservation of Energy? Differentiate between kinetic energy and potential energy.

14. Explain the concept of counting by weighing. What is a mole and why do we use the mole unit?

Exercises

In this section, similar exercises are paired.

Significant Figures and Unit Conversions

15. Which of the following are exact numbers?

    a. There are *100* cm in 1 m.

    b. One meter equals *1.094* yards.

    c. We can use the equation

    $$°F = \tfrac{9}{5}°C + 32$$

    to convert from Celsius to Fahrenheit temperature. Are the numbers $\tfrac{9}{5}$ and *32* exact or inexact?

    d. $\pi = 3.1415927$.

16. Indicate the number of significant figures in each of the following:

    a. This book contains more than 1000 pages.

    b. A mile is about 5300 ft.

    c. A liter is equivalent to 1.059 qt.

    d. The population of the United States in 2015 is 3.1 × 10^2 million.

e. A kilogram is 1000 g.

f. The Boeing 747 cruises at around 600 mi/h.

17. How many significant figures are there in each of the following values?

   a. 6.07×10^{-15}

   b. 0.003840

   c. 17.00

   d. 8×10^{8}

   e. 463.8052

   f. 300

   g. 301

   h. 300.

18. How many significant figures are in each of the following?

   a. 100

   b. 1.0×10^{2}

   c. 1.00×10^{3}

   d. 100.

   e. 0.0048

   f. 0.00480

   g. 4.80×10^{-3}

   h. 4.800×10^{-3}

19. Round off each of the following numbers to the indicated number of significant digits and write the answer in standard scientific notation.

   a. 0.00034159 to three digits

   b. 103.351×10^{2} to four digits

   c. 17.9915 to five digits

   d. 3.365×10^{5} to three digits

20. Use exponential notation to express the number 385,500 to

   a. one significant figure.

   b. two significant figures.

   c. three significant figures.

   d. five significant figures.

21. You have water in each graduated cylinder shown:

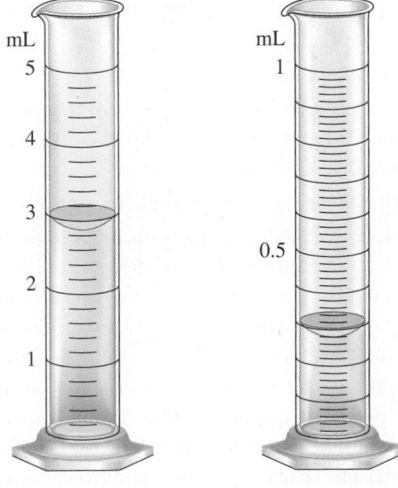

You then add both samples to a beaker. How would you write the number describing the total volume? What limits the precision of this number?

22. The beakers shown below have different precisions.

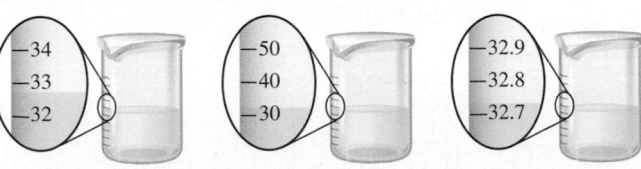

   a. Label the amount of water in each of the three beakers to the correct number of significant figures.

   b. Is it possible for each of the three beakers to contain the exact same amount of water? If no, why not? If yes, did you report the volumes as the same in part a? Explain.

   c. Suppose you pour the water from these three beakers into one container. What should be the volume in the container reported to the correct number of significant figures?

23. Evaluate each of the following and write the answer to the appropriate number of significant figures.

   a. $212.2 + 26.7 + 402.09$

   b. $1.0028 + 0.221 + 0.10337$

   c. $52.331 + 26.01 - 0.9981$

   d. $2.01 \times 10^{2} + 3.014 \times 10^{3}$

   e. $7.255 - 6.8350$

24. Perform the following mathematical operations, and express each result to the correct number of significant figures.

   a. $\dfrac{0.102 \times 0.0821 \times 273}{1.01}$

   b. $0.14 \times 6.022 \times 10^{23}$

   c. $4.0 \times 10^{4} \times 5.021 \times 10^{-3} \times 7.34993 \times 10^{2}$

   d. $\dfrac{2.00 \times 10^{6}}{3.00 \times 10^{-7}}$

25. Perform the following mathematical operations and express the result to the correct number of significant figures.

   a. $\dfrac{2.526}{3.1} + \dfrac{0.470}{0.623} + \dfrac{80.705}{0.4326}$

   b. $(6.404 \times 2.91)/(18.7 - 17.1)$

   c. $6.071 \times 10^{-5} - 8.2 \times 10^{-6} - 0.521 \times 10^{-4}$

   d. $(3.8 \times 10^{-12} + 4.0 \times 10^{-13})/(4 \times 10^{12} + 6.3 \times 10^{13})$

   e. $\dfrac{9.5 + 4.1 + 2.8 + 3.175}{4}$

   (Assume that this operation is taking the average of four numbers. Thus 4 in the denominator is exact.)

   f. $\dfrac{8.925 - 8.905}{8.925} \times 100$

   (This type of calculation is done many times in calculating a percentage error. Assume that this example is such a calculation; thus 100 can be considered to be an exact number.)

26. Perform the following mathematical operations, and express the result to the correct number of significant figures.

   a. $6.022 \times 10^{23} \times 1.05 \times 10^2$

   b. $\dfrac{6.6262 \times 10^{-34} \times 2.998 \times 10^8}{2.54 \times 10^{-9}}$

   c. $1.285 \times 10^{-2} + 1.24 \times 10^{-3} + 1.879 \times 10^{-1}$

   d. $\dfrac{(1.00866 - 1.00728)}{6.02205 \times 10^{23}}$

   e. $\dfrac{9.875 \times 10^2 - 9.795 \times 10^2}{9.875 \times 10^2} \times 100$ (100 is exact)

   f. $\dfrac{9.42 \times 10^2 + 8.234 \times 10^2 + 1.625 \times 10^3}{3}$ (3 is exact)

27. Perform each of the following conversions.

   a. 8.43 cm to millimeters

   b. 2.41×10^2 cm to meters

   c. 294.5 nm to centimeters

   d. 1.445×10^4 m to kilometers

   e. 235.3 m to millimeters

   f. 903.3 nm to micrometers

28. a. How many kilograms are in one teragram?

   b. How many nanometers are in 6.50×10^2 terameters?

   c. How many kilograms are in 25 femtograms?

   d. How many liters are in 8.0 cubic decimeters?

   e. How many microliters are in one milliliter?

   f. How many picograms are in one microgram?

29. Perform the following unit conversions.

   a. Congratulations! You and your spouse are the proud parents of a new baby, born while you are studying in a country that uses the metric system. The nurse has informed you that the baby weighs 3.91 kg and measures 51.4 cm. Convert your baby's weight to pounds and ounces and her length to inches (rounded to the nearest quarter inch).

   b. The circumference of the earth is 25,000 mi at the equator. What is the circumference in kilometers? in meters?

   c. A rectangular solid measures 1.0 m by 5.6 cm by 2.1 dm. Express its volume in cubic meters, liters, cubic inches, and cubic feet.

30. Perform the following unit conversions.

   a. 908 oz to kilograms

   b. 12.8 L to gallons

   c. 125 mL to quarts

   d. 2.89 gal to milliliters

   e. 4.48 lb to grams

   f. 550 mL to quarts

31. Use the following exact conversion factors to perform the stated calculations:

$$5\tfrac{1}{2} \text{ yards} = 1 \text{ rod}$$
$$40 \text{ rods} = 1 \text{ furlong}$$
$$8 \text{ furlongs} = 1 \text{ mile}$$

   a. The Kentucky Derby race is 1.25 miles. How long is the race in rods, furlongs, meters, and kilometers?

   b. A marathon race is 26 miles, 385 yards. What is this distance in rods, furlongs, meters, and kilometers?

32. Although the preferred SI unit of area is the square meter, land is often measured in the metric system in hectares (ha). One hectare is equal to 10,000 m^2. In the English system, land is often measured in acres (1 acre = 160 rod^2). Use the exact conversions and those given in Exercise 31 to calculate the following.

   a. 1 ha = _____ km^2.

   b. The area of a 5.5-acre plot of land in hectares, square meters, and square kilometers.

   c. A lot with dimensions 120 ft by 75 ft is to be sold for $6500. What is the price per acre? What is the price per hectare?

33. Precious metals and gems are measured in troy weights in the English system:

$$24 \text{ grains} = 1 \text{ pennyweight (exact)}$$
$$20 \text{ pennyweight} = 1 \text{ troy ounce (exact)}$$
$$12 \text{ troy ounces} = 1 \text{ troy pound (exact)}$$
$$1 \text{ grain} = 0.0648 \text{ gram}$$
$$1 \text{ carat} = 0.200 \text{ gram}$$

   a. The most common English unit of mass is the pound avoirdupois. What is one troy pound in kilograms and in pounds?

   b. What is the mass of a troy ounce of gold in grams and in carats?

   c. The density of gold is 19.3 g/cm^3. What is the volume of a troy pound of gold?

34. Apothecaries (druggists) use the following set of measures in the English system:

$$20 \text{ grains ap} = 1 \text{ scruple (exact)}$$
$$3 \text{ scruples} = 1 \text{ dram ap (exact)}$$
$$8 \text{ dram ap} = 1 \text{ oz ap (exact)}$$
$$1 \text{ dram ap} = 3.888 \text{ g}$$

   a. Is an apothecary grain the same as a troy grain? (See Exercise 33.)

   b. 1 oz ap = _____ oz troy.

   c. An aspirin tablet contains 5.00×10^2 mg of active ingredient. What mass in grains ap of active ingredient does it contain? What mass in scruples?

   d. What is the mass of 1 scruple in grams?

35. For a pharmacist dispensing pills or capsules, it is often easier to weigh the medication to be dispensed than to count the individual pills. If a single antibiotic capsule weighs 0.65 g, and a pharmacist weighs out 15.6 g of capsules, how many capsules have been dispensed?

36. A children's pain relief elixir contains 80. mg acetaminophen per 0.50 teaspoon. The dosage recommended for a child who weighs between 24 and 35 lb is 1.5 teaspoons. What is the range of acetaminophen dosages, expressed in mg acetaminophen/kg body weight, for children who weigh between 24 and 35 lb?

37. Science fiction often uses nautical analogies to describe space travel. If the starship *U.S.S. Enterprise* is traveling at warp factor 1.71, what is its speed in knots and in miles per hour? (Warp 1.71 = 5.00 times the speed of light; speed of light = 3.00×10^8 m/s; 1 knot = 2000 yd/h, exactly.)

38. The world record for the hundred meter dash is 9.58 s. What is the corresponding average speed in units of m/s, km/h, ft/s, and mi/h? At this speed, how long would it take to run 1.00×10^2 yards?

39. Would a car traveling at a constant speed of 65 km/h violate a 40 mi/h speed limit?

40. You pass a road sign saying "New York 112 km." If you drive at a constant speed of 65 mi/h, how long should it take you to reach New York? If your car gets 28 miles to the gallon, how many liters of gasoline are necessary to travel 112 km?

41. You are in Paris, and you want to buy some peaches for lunch. The sign in the fruit stand indicates that peaches cost 2.45 euros per kilogram. Given that 1 euro is equivalent to approximately $1.32, calculate what a pound of peaches will cost in dollars.

42. In recent years, there has been a large push for an increase in the use of renewable resources to produce the energy we need to power our vehicles. One of the newer fuels that has become more widely available is E85, a mixture of 85% ethanol and 15% gasoline. Despite being more environmentally friendly, one of the potential drawbacks of E85 fuel is that it produces less energy than conventional gasoline. Assume a car gets 28.0 mi/gal using gasoline at $3.50/gal and 22.5 mi/gal using E85 at $2.85/gal. How much will it cost to drive 500. miles using each fuel?

43. Mercury poisoning is a debilitating disease that is often fatal. In the human body, mercury reacts with essential enzymes leading to irreversible inactivity of these enzymes. If the amount of mercury in a polluted lake is 0.4 μg Hg/mL, what is the total mass in kilograms of mercury in the lake? (The lake has a surface area of 100 mi^2 and an average depth of 20 ft.)

44. Carbon monoxide (CO) detectors sound an alarm when peak levels of carbon monoxide reach 100 parts per million (ppm). This level roughly corresponds to a composition of air that contains 400,000 μg carbon monoxide per cubic meter of air (400,000 $μg/m^3$). Assuming the dimensions of a room are 18 ft × 12 ft × 8 ft, estimate the mass of carbon monoxide in the room that would register 100 ppm on a carbon monoxide detector.

Temperature

45. Convert the following Fahrenheit temperatures to the Celsius and Kelvin scales.
    a. −459°F, an extremely low temperature
    b. −40.°F, the answer to a trivia question
    c. 68°F, room temperature
    d. 7×10^7 °F, temperature required to initiate fusion reactions in the sun

46. A thermometer gives a reading of 96.1°F ± 0.2°F. What is the temperature in °C? What is the uncertainty?

47. Convert the following Celsius temperatures to Kelvin and to Fahrenheit degrees.
    a. the temperature of someone with a fever, 39.2°C
    b. a cold wintery day, −25°C
    c. the lowest possible temperature, −273°C
    d. the melting-point temperature of sodium chloride, 801°C

48. Convert the following Kelvin temperatures to Celsius and Fahrenheit degrees.
    a. the temperature that registers the same value on both the Fahrenheit and Celsius scales, 233 K
    b. the boiling point of helium, 4 K
    c. the temperature at which many chemical quantities are determined, 298 K
    d. the melting point of tungsten, 3680 K

49. At what temperature is the temperature in degrees Fahrenheit equal to twice the temperature in degrees Celsius?

50. The average daytime temperatures on the earth and Jupiter are 72°F and 313 K, respectively. Calculate the difference in temperature, in °C, between these two planets.

51. Ethylene glycol is the main component in automobile antifreeze. To monitor the temperature of an auto cooling system, you intend to use a meter that reads from 0 to 100. You devise a new temperature scale based on the approximate melting and boiling points of a typical antifreeze solution (−45°C and 115°C). You wish these points to correspond to 0°A and 100°A, respectively.
    a. Derive an expression for converting between °A and °C.
    b. Derive an expression for converting between °F and °A.
    c. At what temperature would your thermometer and a Celsius thermometer give the same numerical reading?
    d. Your thermometer reads 86°A. What is the temperature in °C and in °F?
    e. What is a temperature of 45°C in °A?

52. Use the figure below to answer the following questions.

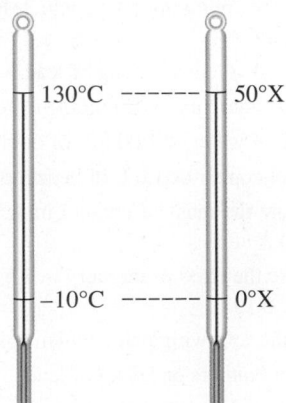

    a. Derive the relationship between °C and °X.
    b. If the temperature outside is 22.0°C, what is the temperature in units of °X?
    c. Convert 58.0°X to units of °C, K, and °F.

Density

53. A material will float on the surface of a liquid if the material has a density less than that of the liquid. Given that the density of water is approximately 1.0 g/mL, will a block of material having a volume of 1.2×10^4 in^3 and weighing 350 lb float or sink when placed in a reservoir of water?

54. For a material to float on the surface of water, the material must have a density less than that of water (1.0 g/mL) and must not react with the water or dissolve in it. A spherical ball has a radius of 0.50 cm and weighs 2.0 g. Will this ball float or sink when placed in water? (*Note*: Volume of a sphere = $\frac{4}{3}\pi r^3$.)

55. A star is estimated to have a mass of 2×10^{36} kg. Assuming it to be a sphere of average radius 7.0×10^5 km, calculate the average density of the star in units of grams per cubic centimeter.

56. A rectangular block has dimensions 2.9 cm $\times$ 3.5 cm $\times$ 10.0 cm. The mass of the block is 615.0 g. What are the volume and density of the block?

57. Diamonds are measured in carats, and 1 carat = 0.200 g. The density of diamond is 3.51 g/cm^3.

    a. What is the volume of a 5.0-carat diamond?

    b. What is the mass in carats of a diamond measuring 2.8 mL?

58. Ethanol and benzene dissolve in each other. When 100. mL of ethanol is dissolved in 1.00 L of benzene, what is the mass of the mixture? (See Table R-5.)

59. A sample containing 33.42 g of metal pellets is poured into a graduated cylinder initially containing 12.7 mL of water, causing the water level in the cylinder to rise to 21.6 mL. Calculate the density of the metal.

60. The density of pure silver is 10.5 g/cm^3 at 20°C. If 5.25 g of pure silver pellets is added to a graduated cylinder containing 11.2 mL of water, to what volume level will the water in the cylinder rise?

61. In each of the following pairs, which has the greater mass? (See Table R-5.)

    a. 1.0 kg of feathers or 1.0 kg of lead

    b. 1.0 mL of mercury or 1.0 mL of water

    c. 19.3 mL of water or 1.00 mL of gold

    d. 75 mL of copper or 1.0 L of benzene

62. a. Calculate the mass of ethanol in 1.50 qt of ethanol. (See Table R-5.)

    b. Calculate the mass of mercury in 3.5 in^3 of mercury. (See Table R-5.)

63. In each of the following pairs, which has the greater volume?

    a. 1.0 kg of feathers or 1.0 kg of lead

    b. 100 g of gold or 100 g of water

    c. 1.0 L of copper or 1.0 L of mercury

64. Using Table R-5, calculate the volume of 25.0 g of each of the following substances at 1 atm.

    a. hydrogen gas

    b. water

    c. iron

    Chapter 8 discusses the properties of gases. One property unique to gases is that they contain mostly empty space. Explain, using the results of your calculations.

65. The density of osmium (the densest metal) is 22.57 g/cm^3. If a 1.00-kg rectangular block of osmium has two dimensions of 4.00 cm $\times$ 4.00 cm, calculate the third dimension of the block.

66. A copper wire (density = 8.96 g/cm^3) has a diameter of 0.25 mm. If a sample of this copper wire has a mass of 22 g, how long is the wire?

Classification and Separation of Matter

67. Match each description below with the following microscopic pictures. More than one picture may fit each description. A picture may be used more than once or not used at all.

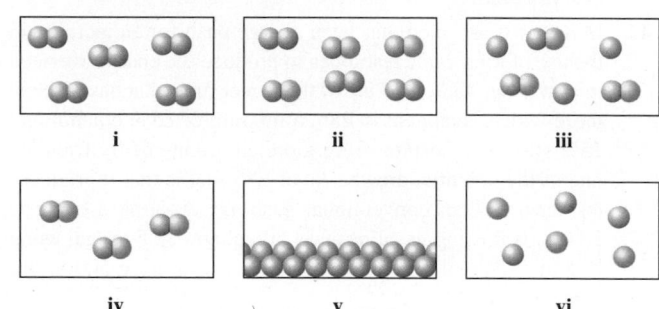

    a. a gaseous compound

    b. a mixture of two gaseous elements

    c. a solid element

    d. a mixture of a gaseous element and a gaseous compound

68. Define the following terms: solid, liquid, gas, pure substance, element, compound, homogeneous mixture, heterogeneous mixture, solution, chemical change, physical change.

69. What is the difference between homogeneous and heterogeneous matter? Classify each of the following as homogeneous or heterogeneous.

    a. a door

    b. the air you breathe

    c. a cup of coffee (black)

    d. the water you drink

    e. salsa

    f. your lab partner

70. Classify the following mixtures as homogeneous or as heterogeneous.

    a. potting soil

    b. white wine

    c. your sock drawer

    d. window glass

    e. granite

71. Classify each of the following as a mixture or a pure substance.

    a. water   f. uranium

    b. blood   g. wine

    c. the oceans   h. leather

    d. iron   i. table salt

    e. brass

    Of the pure substances, which are elements and which are compounds?

72. Suppose a teaspoon of magnesium filings and a teaspoon of powdered sulfur are placed together in a metal beaker. Would this constitute a mixture or a pure substance? Suppose the magnesium filings and sulfur are heated so they react with each other, forming magnesium sulfide. Would this still be a "mixture"? Why or why not?

73. If a piece of hard white blackboard chalk is heated strongly in a flame, the mass of the piece of chalk will decrease, and eventually the chalk will crumble into a fine white dust. Does this change suggest that the chalk is composed of an element or a compound?

74. During a very cold winter, the temperature may remain below freezing for extended periods. However, fallen snow can still disappear, even though it cannot melt. This is possible because a solid can vaporize directly, without passing through the liquid state. Is this process (sublimation) a physical or a chemical change?

75. Classify the following as physical or chemical changes.

    a. Moth balls gradually vaporize in a closet.

    b. Hydrofluoric acid attacks glass and is used to etch calibration marks on glass laboratory utensils.

    c. A French chef making a sauce with brandy is able to boil off the alcohol from the brandy, leaving just the brandy flavoring.

    d. Chemistry majors sometimes get holes in the cotton jeans they wear to lab because of acid spills.

76. The properties of a mixture are typically averages of the properties of its components. The properties of a compound may differ dramatically from the properties of the elements that combine to produce the compound. For each process described below, state whether the material being discussed is most likely a mixture or a compound, and state whether the process is a chemical change or a physical change.

    a. An orange liquid is distilled, resulting in the collection of a yellow liquid and a red solid.

    b. A colorless, crystalline solid is decomposed, yielding a pale yellow-green gas and a soft, shiny metal.

    c. A cup of tea becomes sweeter as sugar is added to it.

Additional Exercises

77. If you had a mole of U.S. dollar bills and equally distributed the money to all of the people of the world, how rich would every person be? Assume a world population of 7 billion.

78. The active ingredient of aspirin tablets is acetylsalicylic acid, which has a density of 1.4 g/cm^3. In a lab class, a student used paper chromatography to isolate another common ingredient of headache remedies. The isolated sample had a mass of 0.384 g and a volume of 0.32 cm^3. Given the data in the following table, what was the other ingredient in the headache remedy?

Density Values for Potential Headache Remedies	
Compound	Density (g/cm^3)
White table sugar	0.70
Caffeine	1.2
Acetylsalicylic acid	1.4
Sodium chloride	2.2

79. Lipitor, a pharmaceutical drug that has been shown to lower "bad" cholesterol while raising "good" cholesterol levels in patients taking the drug, had over $11 billion in sales in 2006. Assuming one 2.5-g pill contains 4.0% of the active ingredient by mass, what mass in kg of active ingredient is present in one bottle of 100 pills?

80. Which of the following are chemical changes? Which are physical changes?

    a. the cutting of food

    b. interaction of food with saliva and digestive enzymes

    c. proteins being broken down into amino acids

    d. complex sugars being broken down into simple sugars

    e. making maple syrup by heating maple sap to remove water through evaporation

    f. DNA unwinding

81. The contents of one 40-lb bag of topsoil will cover 10. square feet of ground to a depth of 1.0 inch. How many bags are needed to cover a plot that measures 200. by 300. m to a depth of 4.0 cm?

82. In the opening scenes of the movie *Raiders of the Lost Ark*, Indiana Jones tries to remove a gold idol from a booby-trapped pedestal. He replaces the idol with a bag of sand of approximately equal volume. (Density of gold = 19.32 g/cm^3; density of sand $\approx 2 \text{ g/cm}^3$.)

    a. Did he have a reasonable chance of not activating the mass-sensitive booby trap?

    b. In a later scene he and an unscrupulous guide play catch with the idol. Assume that the volume of the idol is about 1.0 L. If it were solid gold, what mass would the idol have? Is playing catch with it plausible?

83. A parsec is an astronomical unit of distance where 1 parsec = 3.26 light years (1 light year equals the distance traveled by light in one year). If the speed of light is 186,000 mi/s, calculate the distance in meters of an object that travels 9.6 parsecs.

84. You are driving 65 mi/h and take your eyes off the road for "just a second." What distance (in feet) do you travel in this time?

85. A column of liquid is found to expand linearly on heating 5.25 cm for a 10.0°F rise in temperature. If the initial temperature of the liquid is 98.6°F, what will the final temperature be in °C if the liquid has expanded by 18.5 cm?

86. A 25.00-g sample of a solid is placed in a graduated cylinder and then the cylinder is filled to the 50.0-mL mark with benzene. The mass of benzene and solid together is 58.80 g. Assuming that the solid is insoluble in benzene and that the density of benzene is 0.880 g/cm³, calculate the density of the solid.

87. For each of the following, decide whether the orange block is more dense, the blue block is more dense, or it cannot be determined. Explain your answers.

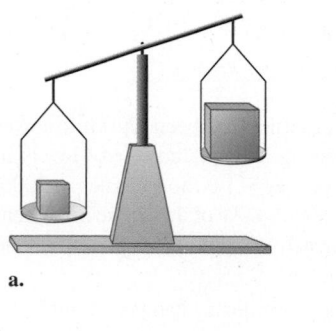
a.

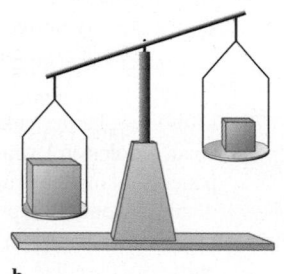
b.

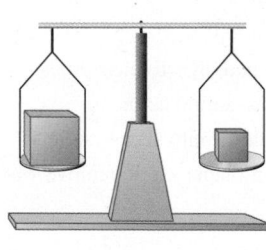
c.

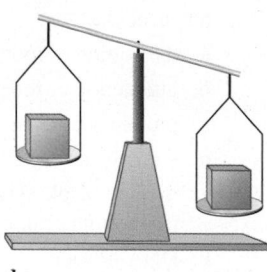
d.

88. According to the *Official Rules of Baseball*, a baseball must have a circumference not more than 9.25 in or less than 9.00 in and a mass not more than 5.25 oz or less than 5.00 oz. What range of densities can a baseball be expected to have? Express this range as a single number with an accompanying uncertainty limit.

89. The density of an irregularly shaped object was determined as follows. The mass of the object was found to be 28.90 g ± 0.03 g. A graduated cylinder was partially filled with water. The reading of the level of the water was 6.4 cm³ ± 0.1 cm³. The object was dropped in the cylinder, and the level of the water rose to 9.8 cm³ ± 0.1 cm³. What is the density of the object, with appropriate error limits? (See Appendix 1.5.)

90. The chemist in Example R-14 did some further experiments. She found that the pipet used to measure the volume of the cleaner is accurate to ±0.03 cm³. The mass measurement is accurate to ±0.002 g. Are these measurements sufficiently precise for the chemist to distinguish between isopropyl alcohol and ethanol?

91. Sterling silver is a solid solution of silver and copper. If a piece of a sterling silver necklace has a mass of 105.0 g and a volume of 10.12 mL, calculate the mass percent of copper in the piece of necklace. Assume that the volume of silver present plus the volume of copper present equals the total volume. Refer to Table R-5.

$$\text{Mass percent of copper} = \frac{\text{mass of copper}}{\text{total mass}} \times 100$$

92. Draw a picture showing the markings (graduations) on glassware that would allow you to make each of the following volume measurements of water, and explain your answers (the numbers given are as precise as possible).

  a. 128.7 mL   **b.** 18 mL   **c.** 23.45 mL

  If you made these measurements for three samples of water and then poured all of the water together in one container, what total volume of water should you report? Support your answer.

93. Consider the accompanying diagram. Ball A is allowed to fall and strike ball B. Assume that all of ball A's energy is transferred to ball B, at point I, and that there is no loss of energy to other sources. What are the kinetic energy and the potential energy of ball B at point II? The potential energy is given by $PE = mgz$, where m is the mass in kilograms, g is the gravitational constant (9.81 m/s²), and z is the distance in meters.

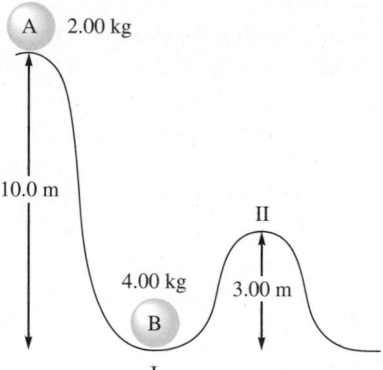

94. An experiment was performed in which an empty 100-mL graduated cylinder was weighed. It was weighed once again after it had been filled to the 10.0-mL mark with dry sand. A 10-mL pipet was used to transfer 10.00 mL of methanol to the cylinder. The sand–methanol mixture was stirred until bubbles no longer emerged from the mixture and the sand looked uniformly wet. The cylinder was then weighed again. Use the data obtained from this experiment (and displayed at the end of this problem) to find the density of the dry sand, the density of methanol, and the density of sand particles. Does the bubbling that occurs when the methanol is added to the dry sand indicate that the sand and methanol are reacting?

Mass of cylinder plus wet sand	45.2613 g
Mass of cylinder plus dry sand	37.3488 g
Mass of empty cylinder	22.8317 g
Volume of dry sand	10.0 mL
Volume of sand plus methanol	17.6 mL
Volume of methanol	10.00 mL

ChemWork Problems

These multiconcept problems (and additional ones) are found interactively online with the same type of assistance a student would get from an instructor.

95. The longest river in the world is the Nile River with a length of 4145 mi. How long is the Nile in cable lengths, meters, and nautical miles?

    Use these exact conversions to help solve the problem:

    $$6 \text{ ft} = 1 \text{ fathom}$$
    $$100 \text{ fathoms} = 1 \text{ cable length}$$
    $$10 \text{ cable lengths} = 1 \text{ nautical mile}$$
    $$3 \text{ nautical miles} = 1 \text{ league}$$

96. Secretariat is known as the horse with the fastest run in the Kentucky Derby. If Secretariat's record 1.25-mi run lasted 1 minute 59.2 seconds, what was his average speed in m/s?

97. The hottest temperature recorded in the United States is 134°F in Greenland Ranch, CA. The melting point of phosphorus is 44°C. At this temperature, would phosphorus be a liquid or a solid?

98. The radius of a neon atom is 69 pm, and its mass is 3.35×10^{-23} g. What is the density of the atom in grams per cubic centimeter (g/cm^3)? Assume the nucleus is a sphere with volume $= \frac{4}{3}\pi r^3$.

99. Which of the following statements is(are) *true*?
    a. A spoonful of sugar is a mixture.
    b. Only elements are pure substances.
    c. Air is a mixture of gases.
    d. Gasoline is a pure substance.
    e. Compounds can be broken down only by chemical means.

100. Which of the following describes a chemical property?
    a. The density of iron is 7.87 g/cm^3.
    b. A platinum wire glows red when heated.
    c. An iron bar rusts.
    d. Aluminum is a silver-colored metal.

Chemical Foundations

Chemistry is involved in the structure, tires, and fuel of a high-performance race car. (HZ/PIXATHLON/SIPA/ Newcom)

When you start your car, do you think about chemistry? Probably not, but you should. The power to start your car is furnished by a lead storage battery. How does this battery work, and what does it contain? When a battery goes dead, what does that mean? If you use a friend's car to "jump start" your car, did you know that your battery could explode? How can you avoid such an unpleasant possibility? What is in the gasoline that you put in your tank, and how does it furnish the energy to drive to school? What is the vapor that comes out of the exhaust pipe, and why does it cause air pollution? Your car's air conditioner might have a substance in it that is leading to the destruction of the ozone layer in the upper atmosphere. What are we doing about that? And why is the ozone layer important anyway?

All these questions can be answered by understanding some chemistry. In fact, we'll consider the answers to all these questions in this text.

Chemistry is around you all the time. You are able to read and understand this sentence because chemical reactions are occurring in your brain. The food you ate for breakfast or lunch is now furnishing energy through chemical reactions. Trees and grass grow because of chemical changes.

Chemistry also crops up in some unexpected places. When archaeologist Luis Alvarez was studying in college, he probably didn't realize that the chemical elements iridium and niobium would make him very famous when they helped him solve the problem of the disappearing dinosaurs. For decades scientists had wrestled with the mystery of why the dinosaurs, after ruling the earth for millions of years, suddenly became extinct 65 million years ago. In studying core samples of rocks dating back to that period, Alvarez and his coworkers recognized unusual levels of iridium and niobium in these samples—levels much more characteristic of extraterrestrial bodies than of the earth. On the basis of these observations, Alvarez hypothesized that a large meteor hit the earth 65 million years ago, changing atmospheric conditions so much that the dinosaurs' food couldn't grow, and they died—almost instantly in the geologic timeframe.

Chemistry is also important to historians. Did you realize that lead poisoning probably was a significant contributing factor to the decline of the Roman Empire? The Romans had high exposure to lead from lead-glazed pottery, lead water pipes, and a sweetening syrup called *sapa* that was prepared by boiling down grape juice in lead-lined vessels. It turns out that one reason for sapa's sweetness was lead acetate ("sugar of lead") that formed as the juice was cooked down. Lead poisoning, with its symptoms of lethargy and mental malfunctions, certainly could have contributed to the demise of the Roman society.

Chemistry is also apparently very important in determining a person's behavior. Various studies have shown that many personality disorders can be linked directly to imbalances of trace elements in the body. For example, studies on the inmates at Stateville Prison in Illinois have linked low cobalt levels with violent behavior. Lithium salts have been shown to be very effective in controlling the effects of manic depressive disease, and you've probably at some time in your life felt a special "chemistry" for another person. Studies suggest there is literally chemistry going on between two people who are attracted to each other. "Falling in love" apparently causes changes in the chemistry of the brain; chemicals are produced that give that "high" associated with a new relationship. Unfortunately, these chemical effects seem to wear off over time, even if the relationship persists and grows.

The importance of chemistry in the interactions of people should not really surprise us, since we know that insects communicate by emitting and receiving chemical signals via molecules called *pheromones*. For example, ants have a very complicated set of chemical signals to signify food sources, danger, and so forth. Also, various female sex attractants have been isolated and used to lure males into traps to control insect populations. It would not be surprising if humans also emitted chemical signals that we were not aware of on a conscious level. Thus chemistry is pretty interesting and pretty important. The main goal of this text is to help you understand the concepts of chemistry so that you can better appreciate the world around you and can be more effective in whatever career you choose.

1-1 | Chemistry: An Atoms-First Approach

In this chapter we give a brief overview of chemistry and of the nature of science. The main purpose of this chapter is to discuss the early development of the model of the atom, setting the stage for a more detailed discussion of modern atomic theory in Chapter 2.

Why spend time considering atoms, and why describe them so early? Observations do not make it immediately obvious that all matter is made of atoms. We cannot see them directly, for example. What do we gain by an understanding of the particulate nature of matter?

Science can be viewed as a cyclical process of observing, questioning, and theorizing. We develop theories and models, such as the atomic model of matter, in order to better explain what we observe.

If, for example, someone nearby opens a bottle of fingernail polish remover, you become aware of it quickly. Why? Because you can smell it. But that leads to an interesting question—why can you smell it? One explanation is that "pieces" of the liquid fingernail polish remover, too small to see, are traveling through the air and interacting with sensors in your nose. These "pieces" must all be alike in some way because the fingernail polish remover always smells the same. Viewing the liquid fingernail polish remover as a collection of molecules (named acetone), and that these molecules can escape into the vapor phase and diffuse through the air, helps us explain what we observe. Thus, to explain what we observe in the macroscopic world (the world we experience), we need to understand what occurs in the microscopic world (the world of atoms and molecules).

But on further consideration we encounter more questions, which is always the case in science. Fingernail polish remover is a collection of acetone molecules, and a molecule, as chemists have learned, is a collection of atoms (in this case, the molecules are composed of carbon, hydrogen, and oxygen atoms). But *why* do molecules exist? That is, why do atoms "stick" together to form a molecule? And why do the acetone molecules "stick" together to form a liquid when other molecules, such as methane (natural gas used to heat our homes), do not stick together at room temperature and thus form a gas? Of course the acetone must evaporate (turn into a vapor) relatively easily in order that we may smell it, and we observe that it evaporates at a much greater rate than water, for example. Why is this? It also turns out that acetone is flammable in air; that is, it will burn with the oxygen in the air, and it forms carbon dioxide and water. How does this happen?

We will see in this chapter that John Dalton developed a theory of atoms and molecules that was relatively simple. His model, for example, does not answer some of the questions we have been asking about acetone. This is acceptable because models in science are always simplifications, and we use different models for different purposes. Albert Einstein is quoted as saying "make everything as simple as possible, but not simpler." As we will see in later chapters, we can use Dalton's relatively simple theory to discuss a chemical reaction as a reorganization of atoms or to understand how a gas exerts pressure. However, to answer more complicated questions, such as "Why do molecules exist?" or "Why do some substances exist at normal temperatures as solids, some as liquids, and others as gases?," we need a more complicated model. This model will be explored in Chapter 2, and its ramifications will be discussed in Chapters 3 and 4. The significance of this model will form the background of much of what we discuss in the text.

As evidenced by the title of this text, here we take an "atoms first" approach to learning chemistry. This is because understanding atoms is crucial to understanding chemistry. The science of chemistry is completely dependent on the properties of atoms; that is, chemistry is really about how atoms behave. To understand our observations we need to think about what the atoms or molecules are "doing"—that is, how they are interacting. Sometimes this will only require a relatively simple model, and sometimes the model will need to be more complex. The questions that we need to answer will guide us.

The point is, when trying to answer our questions (Why can we smell acetone? Why does water evaporate? Why does iron rust?, etc.) we must realize that the understanding comes from thinking of the atoms first. But, of course, this assumes that we have made observations and asked questions. So, the chapter order of the text reflects that we consider the structure of the atoms early and in detail. However, within the chapters we will often begin with observations and questions. We will then use these as a springboard to develop an understanding of what we see by considering the atoms and molecules.

Since the time of the ancient Greeks, people have wondered about the answer to the question, "What is matter made of?" For a long time humans have believed that matter is composed of atoms, and in the previous three centuries we have collected much indirect evidence to support this belief. Very recently, something exciting has happened— for the first time we can "see" individual atoms. Of course, we cannot see atoms with the naked eye but must use a special microscope called a *scanning tunneling microscope* (STM). Although we will not consider the details of its operation here, the STM uses an electron current from a tiny needle to probe the surface of a substance. The STM pictures of several substances are shown in Fig. 1-1. Notice how the atoms are connected to one another by "bridges," which, as we will see, represent the electrons that interconnect atoms.

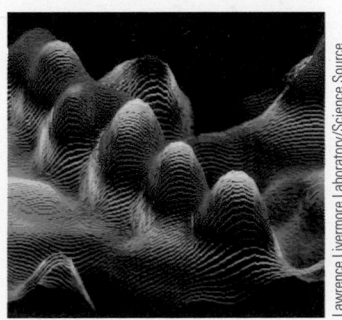

Figure 1-1 | Scanning tunneling micrograph of gold atoms on a graphite substrate.

An image showing the individual carbon atoms in a sheet of graphene.

Scanning tunneling microscope image of DNA.

Philippe Plailly/Science Source

Lawrence Livermore Laboratory/Science Source

Figure 1-2 | Sand on a beach looks uniform from a distance, but up close the irregular sand grains are visible, and each grain is composed of tiny atoms.

So, at this point, we are fairly sure that matter consists of individual atoms. The nature of these atoms is quite complex, and the components of atoms don't behave much like the objects we see in the world of our experience. We call this world the *macroscopic world*—the world of cars, tables, baseballs, rocks, oceans, and so forth. One of the main jobs of a scientist is to delve into the macroscopic world and discover its "parts." For example, when you view a beach from a distance, it looks like a continuous solid substance. As you get closer, you see that the beach is really made up of individual grains of sand. As we examine these grains of sand, we find they are composed of silicon and oxygen atoms connected to each other to form intricate shapes (Fig. 1-2). One of the main challenges of chemistry is to understand the connection between the macroscopic world that we experience and the *microscopic world* of atoms and molecules. To truly understand chemistry, you must learn to think on the atomic level. We will spend much time in this text helping you learn to do that.

Critical Thinking

The scanning tunneling microscope allows us to "see" atoms. What if you were sent back in time before the invention of the scanning tunneling microscope? What evidence could you give to support the theory that all matter is made of atoms and molecules?

One of the amazing things about our universe is that the tremendous variety of substances we find there results from only about 100 different kinds of atoms. You can think of these approximately 100 atoms as the letters in an alphabet, out of which all the "words" in the universe are made. It is the way the atoms are organized in a given substance that determines the properties of that substance. For example, water, one of the most common and important substances on the earth, is composed of two types of atoms: hydrogen and oxygen. There are two hydrogen atoms and one oxygen atom bound together to form the water molecule:

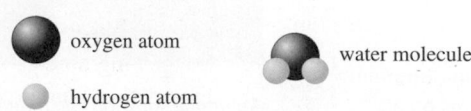

When an electric current passes through it, water is decomposed to hydrogen and oxygen. These *chemical elements* themselves exist naturally as diatomic (two-atom) molecules:

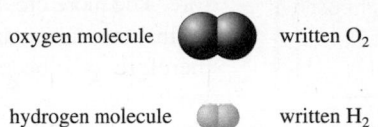

We can represent the decomposition of water to its component elements, hydrogen and oxygen, as follows:

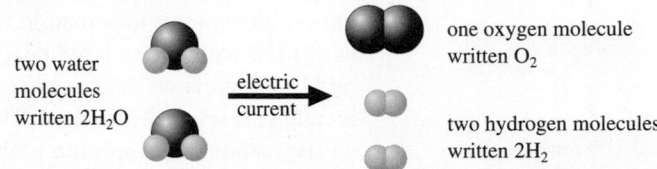

By thinking of the atoms and molecules we can determine the relative numbers of molecules involved in a chemical reaction. We will consider this idea in more detail in Chapter 5.

Notice that it takes two molecules of water to furnish the right number of oxygen and hydrogen atoms to allow for the formation of the two-atom molecules. ◄ *ⓘ* This reaction explains why the battery in your car can explode if you jump start it improperly. When you hook up the jumper cables, current flows through the dead battery, which contains water (and other things), and causes hydrogen and oxygen to form by decomposition of some of the water. A spark can cause this accumulated hydrogen and oxygen to explode, forming water again.

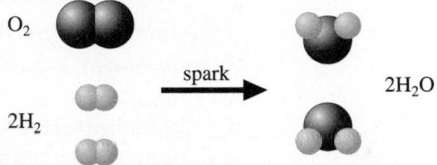

This example illustrates two of the fundamental concepts of chemistry: (1) matter is composed of various types of atoms, and (2) one substance changes to another by reorganizing the way the atoms are attached to each other.

These are core ideas of chemistry, and we will have much more to say about them.

1-2 | The Scientific Method

How do you tackle the problems that confront you in real life? Think about your trip to school. If you live in a city, traffic is undoubtedly a problem you confront daily. How do you decide the best way to drive to school? If you are new in town, you first get a map and look at the possible ways to make the trip. Then you might collect information from people who know the area about the advantages and disadvantages of various routes. On the basis of this information, you probably try to predict the best route. However, you can find the best route only by trying several of them and comparing the results. After a few experiments with the various possibilities, you probably will be able to select the best way. What you are doing in solving this everyday problem is applying the same process that scientists use to study nature. The first thing you did was collect relevant data. Then you made a prediction, and then you tested it by trying it out. This process contains the fundamental elements of science:

1. Making observations (collecting data)
2. Suggesting a possible explanation (formulating a hypothesis)
3. Doing experiments to test the possible explanation (testing the hypothesis)

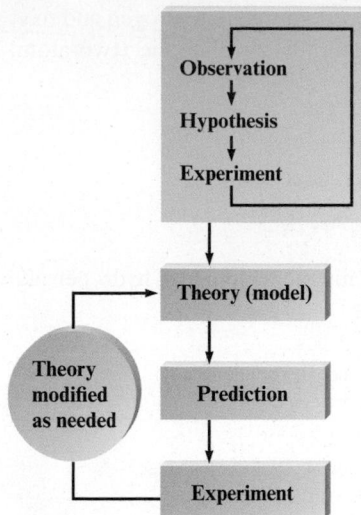

Figure 1-3 | The fundamental steps of the scientific method.

Scientists call this process the *scientific method*. One of life's most important activities is solving problems—not "plug and chug" exercises, but real problems—problems that have new facets to them, that involve things you may have never confronted before. The more creative you are at solving these problems, the more effective you will be in your career and your personal life. Part of the reason for learning chemistry, therefore, is to become a better problem solver. Chemists are usually excellent problem solvers, because to master chemistry, you have to master the scientific approach. Chemical problems are frequently very complicated—there is usually no neat and tidy solution. Often it is difficult to know where to begin.

Science is a framework for gaining and organizing knowledge. Science is not simply a set of facts but also a plan of action—a *procedure* for processing and understanding certain types of information. Scientific thinking is useful in all aspects of life, but in this text we will use it to understand how the chemical world operates. As we have said in our previous discussion, the process that lies at the center of scientific inquiry is called the **scientific method**. There are actually many scientific methods, depending on the nature of the specific problem under study and on the particular investigator involved. However, it is useful to consider the following general framework for a generic scientific method (Fig. 1-3).

Steps in the Scientific Method

1. *Making observations.* Observations may be *qualitative* (the sky is blue; water is a liquid) or *quantitative* (water boils at 100°C; a certain chemistry book weighs 2 kilograms). A qualitative observation does not involve a number. A quantitative observation (called a **measurement**) involves both a number and a unit.

2. *Formulating hypotheses.* A **hypothesis** is a *possible* explanation for an observation.

3. *Performing experiments.* An experiment is carried out to test a hypothesis. This involves gathering new information that enables a scientist to decide whether the hypothesis is valid—that is, whether it is supported by the new information learned from the experiment. Experiments always produce new observations, and this brings the process back to the beginning again.

To understand a given phenomenon, an investigator repeats these steps many times, gradually accumulating the knowledge necessary to provide a possible explanation of the phenomenon.

Scientific Models

Once a set of hypotheses that agree with the various observations is obtained, the hypotheses are assembled into a theory. A **theory**, which is often called a **model**, is a set of tested hypotheses that gives an overall explanation of some natural phenomenon.

It is very important to distinguish between observations and theories. An observation is something that is witnessed and can be recorded. A theory is an *interpretation*—a possible explanation of *why* nature behaves in a particular way. Theories inevitably change as more information becomes available. For example, the motions of the sun and stars have remained virtually the same over the thousands of years during which humans have been observing them, but our explanations—our theories—for these motions have changed greatly since ancient times.

The point is that scientists do not stop asking questions just because a given theory seems to account satisfactorily for some aspect of natural behavior. They continue doing experiments to refine or replace the existing theories. This is generally done by using the currently accepted theory to make a prediction and then performing an experiment (making a new observation) to see whether the results bear out this prediction.

Post-it Notes, a product of the 3M Corporation, revolution-ized casual written communications and personal reminders. Introduced in the United States in 1980, these sticky-but-not-too-sticky notes have now found countless uses in offices, cars, and homes throughout the world.

The invention of sticky notes occurred over a period of about 10 years and involved a great deal of serendipity. The adhesive for Post-it Notes was discovered by Dr. Spencer F. Silver of 3M in 1968. Silver found that when an acrylate polymer material was made in a particular way, it formed cross-linked microspheres. When suspended in a solvent and sprayed on a sheet of paper, this substance formed a "sparse monolayer" of adhesive after the solvent evaporated. Scan-ning electron microscope images of the adhesive show that it has an irregular surface, a little like the surface of a gravel road. In contrast, the adhesive on cellophane tape looks smooth and uniform, like a superhighway. The bumpy surface of Silver's adhesive caused it to be sticky but not so sticky to produce permanent adhesion, because the number of contact points between the binding surfaces was limited.

When he invented this adhesive, Silver had no specific ideas for its use, so he spread the word of his discovery to his fellow employees at 3M to see if anyone had an application for it. In addition, over the next several years development was carried out to improve the adhesive's properties. It was not until 1974 that the idea for Post-it Notes popped up. One Sunday, Art Fry, a chemical engineer for 3M, was singing in his church choir when he became annoyed that the bookmark in his hymnal kept falling out. He thought to himself that it would be nice if the bookmark were sticky enough to stay in place but not so sticky that it couldn't be moved. Luckily, he remembered Silver's glue—and the Post-it Note was born.

For the next three years Fry worked to overcome the man-ufacturing obstacles associated with the product. By 1977 enough Post-it Notes were being produced to supply 3M's corporate headquarters, where the employees quickly

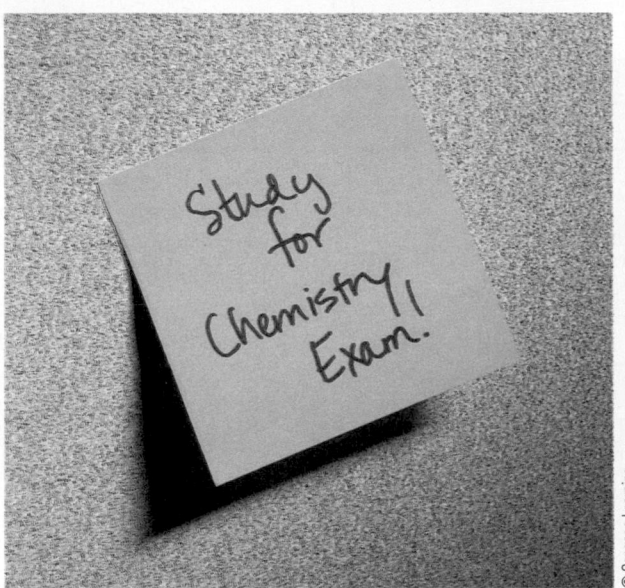

© Cengage Learning

became addicted to their many uses. Post-it Notes are now available in 62 colors and 25 shapes.

In the years since their introduction, 3M has heard some remarkable stories connected to the use of these notes. For example, a Post it Note was applied to the nose of a corpo-rate jet, where it was intended to be read by the plane's Las Vegas ground crew. Someone forgot to remove it, however. The note was still on the nose of the plane when it landed in Minneapolis, having survived a take-off and landing and speeds of 500 miles per hour at temperatures as low as −56°F. Stories on the 3M Web site also describe how a Post-it Note on the front door of a home survived the 140-mile-per-hour winds of Hurricane Hugo and how a foreign official accepted Post-it Notes in lieu of cash when a small bribe was needed to cut through bureaucratic hassles.

Post-it Notes have definitely changed the way we commu-nicate and remember things.

Always remember that theories (models) are human inventions. They represent at-tempts to explain observed natural behavior in terms of human experiences. A theory is actually an educated guess. We must continue to do experiments and to refine our theories (making them consistent with new knowledge) if we hope to approach a more nearly complete understanding of nature.

As scientists observe nature, they often see that the same observation applies to many different systems. For example, studies of innumerable chemical changes have shown that the total observed mass of the materials involved is the same before and

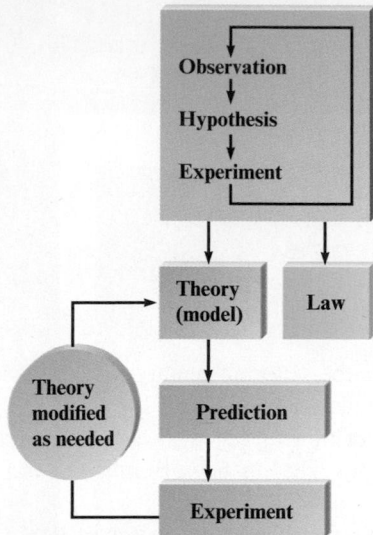

Figure 1-4 | The various parts of the scientific method.

after the change. Such generally observed behavior is formulated into a statement called a **natural law**. For example, the observation that the total mass of materials is not affected by a chemical change in those materials is called the **law of conservation of mass**.

Note the difference between a natural law and a theory. A natural law is a summary of observed (measurable) behavior, whereas a theory is an explanation of behavior. A law summarizes what happens; a theory (model) is an attempt to explain why it happens.

In this section we have described the scientific method as it might ideally be applied (Fig. 1-4). However, it is important to remember that science does not always progress smoothly and efficiently. For one thing, hypotheses and observations are not totally independent of each other, as we have assumed in the description of the idealized scientific method. The coupling of observations and hypotheses occurs because once we begin to proceed down a given theoretical path, our hypotheses are unavoidably couched in the language of that theory. In other words, we tend to see what we expect to see and often fail to notice things that we do not expect. Thus the theory we are testing helps us because it focuses our questions. However, at the very same time, this focusing process may limit our ability to see other possible explanations.

It is also important to keep in mind that scientists are human. They have prejudices; they misinterpret data; they become emotionally attached to their theories and thus lose objectivity; and they play politics. Science is affected by profit motives, budgets, fads, wars, and religious beliefs. Galileo, for example, was forced to recant his astronomical observations in the face of strong religious resistance. Lavoisier, the father of modern chemistry, was beheaded because of his political affiliations. Great progress in the chemistry of nitrogen fertilizers resulted from the desire to produce explosives to fight wars. The progress of science is often affected more by the frailties of humans and their institutions than by the limitations of scientific measuring devices. The scientific methods are only as effective as the humans using them. They do not automatically lead to progress.

1-3 | The Early History of Chemistry

Chemistry has been important since ancient times. The processing of natural ores to produce metals for ornaments and weapons and the use of embalming fluids are just two applications of chemical phenomena that were utilized prior to 1000 B.C.

The Greeks were the first to try to explain why chemical changes occur. By about 400 B.C. they had proposed that all matter was composed of four fundamental substances: fire, earth, water, and air. The Greeks also considered the question of whether matter is continuous, and thus infinitely divisible into smaller pieces, or composed of small, indivisible particles. Supporters of the latter position were Demokritos* of Abdera (*c.* 460–*c.* 370 B.C.) and Leucippos, who used the term *atomos* (which later became *atoms*) to describe these ultimate particles. However, because the Greeks had no experiments to test their ideas, no definitive conclusion could be reached about the divisibility of matter.

The next 2000 years of chemical history were dominated by a pseudoscience called *alchemy*. Some alchemists were mystics and fakes who were obsessed with the idea of turning cheap metals into gold. However, many alchemists were serious scientists, and this period saw important advances: the alchemists discovered several elements and learned to prepare the mineral acids.

---

*Democritus is an alternate spelling.

▲ Robert Boyle (1627–1691) was born in Ireland. He became especially interested in experiments involving air and developed an air pump with which he produced evacuated cylinders. He used these cylinders to show that a feather and a lump of lead fall at the same rate in the absence of air resistance and that sound cannot be produced in a vacuum. His most famous experiments involved careful measurements of the volume of a gas as a function of pressure. In his book *The Skeptical Chymist*, Boyle urged that the ancient view of elements as mystical substances should be abandoned and that an element should instead be defined as anything that cannot be broken down into simpler substances. This conception was an important step in the development of modern chemistry.

Bridgeman Art Library, New York

The foundations of modern chemistry were laid in the sixteenth century with the development of systematic metallurgy (extraction of metals from ores) by a German, Georg Bauer (1494–1555), and the medicinal application of minerals by a Swiss alchemist/physician known as Paracelsus (full name: Philippus Theophrastus Bombastus von Hohenheim [1493–1541]).

The first "chemist" to perform truly quantitative experiments was Robert Boyle (1627–1691), who carefully measured the relationship between the pressure and volume of air. ◄ When Boyle published his book *The Skeptical Chymist* in 1661, the quantitative sciences of physics and chemistry were born. In addition to his findings on the quantitative behavior of gases, Boyle's other major contribution to chemistry consisted of his ideas about the chemical elements. Boyle held no preconceived notion about the number of elements. In his view, a substance was an element unless it could be broken down into two or more simpler substances. As Boyle's experimental definition of an element became generally accepted, the list of known elements began to grow, and the Greek system of four elements finally died. Although Boyle was an excellent scientist, he was not always right. For example, he clung to the alchemists' views that metals were not true elements and that a way would eventually be found to change one metal into another.

The phenomenon of combustion evoked intense interest in the seventeenth and eighteenth centuries. The German chemist Georg Stahl (1660–1734) suggested that a substance he called "phlogiston" flowed out of burning material. Stahl postulated that a substance burning in a closed container eventually stopped burning because the air in the container became saturated with phlogiston. Oxygen gas, discovered by Joseph Priestley (1733–1804),* an English clergyman and scientist (Fig. 1-5), was found to support vigorous combustion and was thus supposed to be low in phlogiston. In fact, oxygen was originally called "dephlogisticated air."

Roald Hoffman

Figure 1-5 | The Priestley Medal is the highest honor given by the American Chemical Society. It is named for Joseph Priestley, who was born in England on March 13, 1733. He performed many important scientific experiments, and among his discoveries was a gas, later identified as carbon dioxide, that could be dissolved in water to produce *seltzer*. Also, as a result of meeting Benjamin Franklin in London in 1766, Priestley became interested in electricity and was the first to observe that graphite was an electrical conductor. However, his greatest discovery occurred in 1774, when he isolated oxygen by heating mercuric oxide.

Because of his nonconformist political views, Priestley was forced to leave England. He died in the United States in 1804.

---

*Oxygen gas was actually first observed by the Swedish chemist Karl W. Scheele (1742–1786), but because his results were published after Priestley's, the latter is commonly credited with the discovery of oxygen.

1-4 | Fundamental Chemical Laws

By the late eighteenth century, combustion had been studied extensively; the gases carbon dioxide, nitrogen, hydrogen, and oxygen had been discovered; and the list of elements continued to grow. However, it was Antoine Lavoisier (1743–1794), a French chemist (Fig. 1-6), who finally explained the true nature of combustion, thus clearing the way for the tremendous progress that was made near the end of the eighteenth century. Lavoisier, like Boyle, regarded measurement as the essential operation of chemistry. His experiments, in which he carefully weighed the reactants and products of various reactions, suggested that *mass is neither created nor destroyed.* Lavoisier's verification of this **law of conservation of mass** was the basis for the developments in chemistry in the nineteenth century. Mass is neither created nor destroyed in a chemical reaction.

Lavoisier's quantitative experiments showed that combustion involved oxygen (which Lavoisier named), not phlogiston. ◄ ❶ He discovered that life was supported by a process that also involved oxygen and was similar in many ways to combustion. In 1789 Lavoisier published the first modern chemistry textbook, *Elementary Treatise on Chemistry*, in which he presented a unified picture of the chemical knowledge assembled up to that time. Unfortunately, in the same year the text was published, the French Revolution broke out. Lavoisier, who had been associated with collecting taxes for the government, was executed on the guillotine as an enemy of the people in 1794.

After 1800, chemistry was dominated by scientists who, following Lavoisier's lead, performed careful weighing experiments to study the course of chemical reactions and to determine the composition of various chemical compounds. One of these chemists, a Frenchman, Joseph Proust (1754–1826), showed that *a given compound always contains exactly the same proportion of elements by mass.* For example, Proust found that the substance copper carbonate is always 5.3 parts copper to 4 parts oxygen to 1 part carbon (by mass). The principle of the constant composition of compounds, originally called "Proust's law," is now known as the **law of definite proportion**. A given compound always contains exactly the same proportion of elements by mass.

❶ Oxygen is from the French *oxygène*, meaning "generator of acid," because it was initially considered to be an integral part of all acids.

Figure 1-6 | Antoine Lavoisier was born in Paris on August 26, 1743. Although Lavoisier's father wanted his son to follow him into the legal profession, young Lavoisier was fascinated by science. From the beginning of his scientific career, Lavoisier recognized the importance of accurate measurements. His careful weighings showed that mass is conserved in chemical reactions and that combustion involves reaction with oxygen. Also, he wrote the first modern chemistry textbook. It is not surprising that Lavoisier is often called the father of modern chemistry.

To help support his scientific work, Lavoisier invested in a private tax-collecting firm and married the daughter of one of the company executives. His connection to the tax collectors proved fatal, for radical French revolutionaries demanded his execution, which occurred on May 8, 1794.

Manchester Literary and Philosophical Society

Figure 1-7 | John Dalton (1766–1844), an Englishman, began teaching at a Quaker school when he was 12. His fascination with science included an intense interest in meteorology, which led to an interest in the gases of the air and their ultimate components, atoms. Dalton is best known for his atomic theory, in which he postulated that the fundamental differences among atoms are their masses. He was the first to prepare a table of relative atomic weights.

Dalton was a humble man with several apparent disabilities: He was not articulate and he was color-blind, a terrible problem for a chemist. Despite these disadvantages, he helped to revolutionize the science of chemistry.

Proust's discovery stimulated John Dalton (1766–1844), an English schoolteacher (Fig. 1-7), to think about atoms as the particles that might compose elements. Dalton reasoned that if elements were composed of tiny individual particles, a given compound should always contain the same combination of these atoms. This concept explained why the same relative masses of elements were always found in a given compound.

But Dalton discovered another principle that convinced him even more of the existence of atoms. He noted, for example, that carbon and oxygen form two different compounds that contain different relative amounts of carbon and oxygen, as shown by the following data:

	Mass of Oxygen That Combines with 1 g of Carbon
Compound I	1.33 g
Compound II	2.66 g

Dalton noted that compound II contains twice as much oxygen per gram of carbon as compound I, a fact that could easily be explained in terms of atoms. Compound I might be CO, and compound II might be CO_2.* This principle, which was found to apply to compounds of other elements as well, became known as the **law of multiple proportions**: When two elements form a series of compounds, the ratios of the masses of the second element that combine with 1 gram of the first element can always be reduced to small whole numbers.

To make sure the significance of this observation is clear, in Example 1-1 we will consider data for a series of compounds consisting of nitrogen and oxygen.

EXAMPLE 1-1   Illustrating the Law of Multiple Proportions

The following data were collected for several compounds of nitrogen and oxygen:

	Mass of Nitrogen That Combines with 1 g of Oxygen
Compound A	1.750 g
Compound B	0.8750 g
Compound C	0.4375 g

Show how these data illustrate the law of multiple proportions.

---

*Subscripts are used to show the numbers of atoms present. The number 1 is understood (not written). The symbols for the elements and the writing of chemical formulas will be illustrated later in the text.

Solution

For the law of multiple proportions to hold, the ratios of the masses of nitrogen combining with 1 gram of oxygen in each pair of compounds should be small whole numbers. We therefore compute the ratios as follows:

$$\frac{A}{B} = \frac{1.750}{0.8750} = \frac{2}{1}$$

$$\frac{B}{C} = \frac{0.8750}{0.4375} = \frac{2}{1}$$

$$\frac{A}{C} = \frac{1.750}{0.4375} = \frac{4}{1}$$

i By assuming the existence of atoms, and that atoms make up molecules, we can explain Dalton's data. This is how science is done—the observation (data) comes first, and then the theory (in this case, the existence of atoms and molecules) is used to explain the observation.

These results support the law of multiple proportions. ◄ *i*

See Exercises 1-29 and 1-30

What was the significance of the Law of Multiple Proportions to Dalton? It made him realize that compounds are made up of atoms and that a given pair of atoms can form a variety of compounds.

1-5 | Dalton's Atomic Theory

In 1808 Dalton published *A New System of Chemical Philosophy*, in which he presented his theory of atoms.

i These statements are a modern paraphrase of Dalton's ideas.

Dalton's Atomic Theory ◄ *i*

1. Each element is made up of tiny particles called atoms.
2. The atoms of a given element are identical; the atoms of different elements are different in some fundamental way or ways.
3. Chemical compounds are formed when atoms of different elements combine with each other. A given compound always has the same relative numbers and types of atoms.
4. Chemical reactions involve reorganization of the atoms—changes in the way they are bound together. The atoms themselves are not changed in a chemical reaction.

It is instructive to consider Dalton's reasoning on the relative masses of the atoms of the various elements. In Dalton's time water was known to be composed of the elements hydrogen and oxygen, with 8 grams of oxygen present for every 1 gram of hydrogen. Dalton made a fundamental assumption: He decided that nature would be as simple as possible. This assumption led him to conclude that the formula for water should be OH. He thus assigned hydrogen a mass of 1 and oxygen a mass of 8.

Using similar reasoning for other compounds, Dalton prepared the first table of **atomic masses** (sometimes called **atomic weights** by chemists, since mass is often determined by comparison to a standard mass—a process called *weighing*). Many of the masses were later proved to be wrong because of Dalton's incorrect assumptions about the formulas of certain compounds, but the construction of a table of masses was an important step forward.

Although not recognized as such for many years, the keys to determining absolute formulas for compounds were provided in the experimental work of the French chemist Joseph Gay-Lussac (1778–1850) and by the hypothesis of an Italian chemist named Amadeo Avogadro (1776–1856). In 1809 Gay-Lussac performed experiments in

Jöns Jakob Berzelius was probably the best experimental chemist of his generation and, given the crudeness of his laboratory equipment, may be the best of all time. Unlike Lavoisier, who could afford to buy the best laboratory equipment available, Berzelius worked with minimal equipment in very plain surroundings. One of Berzelius's students described the Swedish chemist's workplace: "The laboratory consisted of two ordinary rooms with the very simplest arrangements; there were neither furnaces nor hoods, neither water system nor gas. Against the walls stood some closets with the chemicals, in the middle the mercury trough and the blast lamp table. Beside this was the sink consisting of a stone water holder with a stopcock and a pot standing under it. [Next door in the kitchen] stood a small heating furnace."

In these simple facilities Berzelius performed more than 2000 experiments over a 10-year period to determine accurate atomic masses for the 50 elements then known. His success can be seen from the data in the table at left. These remarkably accurate values attest to his experimental skills and patience.

Besides his table of atomic masses, Berzelius made many other major contributions to chemistry. The most important of these was the invention of a simple set of symbols for the elements along with a system for writing the formulas of compounds to replace the awkward symbolic representations of the alchemists. Although some chemists, including Dalton, objected to the new system, it was gradually adopted and forms the basis of the system we use today.

In addition to these accomplishments, Berzelius discovered the elements cerium, thorium, selenium, and silicon. Of

The Alchemists' Symbols for Some Common Elements and Compounds

Substance	Alchemists' Symbol
Silver	☉
Lead	♄
Tin	♃+
Platinum	☾
Sulfuric acid	𝄞
Alcohol	G
Sea salt	☽

these elements, selenium and silicon are particularly important in today's world. Berzelius discovered selenium in 1817 in connection with his studies of sulfuric acid. For years selenium's toxicity has been known, but only recently have we become aware that it may have a positive effect on human health. Studies have shown that trace amounts of selenium in the diet may protect people from heart disease and cancer. One study based on data from 27 countries showed an inverse relationship between the cancer death rate and the selenium content of soil in a particular region (low cancer death rate in areas with high selenium content). Another research paper reported an inverse relationship between the selenium content of the blood and the incidence of breast cancer in women. A study reported in 1998 used the toenail clippings of 33,737 men to show that selenium seems to protect against prostate cancer. Selenium is also found in the heart muscle and may play an important role in proper heart function. Because of these and other studies, selenium's reputation has improved, and many scientists are now studying its function in the human body.

Silicon is the second most abundant element in the earth's crust, exceeded only by oxygen. As we will see in Chapter 8, compounds involving silicon bonded to oxygen make up most of the earth's sand, rock, and soil. Berzelius prepared silicon in its pure form in 1824 by heating silicon tetrafluoride (SiF_4) with potassium metal. Today, silicon forms the basis for the modern microelectronics industry centered near San Francisco in a place that has come to be known as "Silicon Valley." The technology of the silicon chip (see figure) with its printed circuits has transformed computers from room-sized monsters with thousands of unreliable vacuum tubes to desktop and notebook-sized units with trouble-free "solid-state" circuitry.

Comparison of Several of Berzelius's Atomic Masses with the Modern Values

Element	Atomic Mass	
	Berzelius's Value	Current Value
Chlorine	35.41	35.45
Copper	63.00	63.55
Hydrogen	1.00	1.01
Lead	207.12	207.2
Nitrogen	14.05	14.01
Oxygen	16.00	16.00
Potassium	39.19	39.10
Silver	108.12	107.87
Sulfur	32.18	32.07

See E. J. Holmyard, *Alchemy* (New York: Penguin Books, 1968).

Courtesy IBM

A chip capable of transmitting 4,000,000 simultaneous phone conversations.

The Granger Collection, New York

▲ Joseph Louis Gay-Lussac, a French physicist and chemist, was remarkably versatile. Although he is now known primarily for his studies on the combining of volumes of gases, Gay-Lussac was instrumental in the studies of many of the other properties of gases. Some of Gay-Lussac's motivation to learn about gases arose from his passion for ballooning. In fact, he made ascents to heights of over 4 miles to collect air samples, setting altitude records that stood for about 50 years. Gay-Lussac also was the codiscoverer of boron and the developer of a process for manufacturing sulfuric acid. As chief assayer of the French mint, Gay-Lussac developed many techniques for chemical analysis and invented many types of glassware now used routinely in labs. Gay-Lussac spent his last 20 years as a lawmaker in the French government.

which he measured (under the same conditions of temperature and pressure) the volumes of gases that reacted with each other. ◄ For example, Gay-Lussac found that 2 volumes of hydrogen react with 1 volume of oxygen to form 2 volumes of gaseous water and that 1 volume of hydrogen reacts with 1 volume of chlorine to form 2 volumes of hydrogen chloride. These results are represented schematically in Fig. 1-8.

In 1811 Avogadro interpreted these results by proposing that *at the same temperature and pressure, equal volumes of different gases contain the same number of particles*. This assumption (called **Avogadro's hypothesis**) makes sense if the distances between the particles in a gas are very great compared with the sizes of the particles. Under these conditions, the volume of a gas is determined by the number of molecules present, not by the size of the individual particles.

If Avogadro's hypothesis is correct, Gay-Lussac's result,

2 volumes of hydrogen react with 1 volume of oxygen $\longrightarrow$ 2 volumes of water vapor

can be expressed as follows:

2 molecules* of hydrogen react with 1 molecule of oxygen $\longrightarrow$ 2 molecules of water

These observations can best be explained by assuming that gaseous hydrogen, oxygen, and chlorine are all composed of diatomic (two-atom) molecules: H_2, O_2, and Cl_2, respectively. ◄ *ⓘ* Gay-Lussac's results can then be represented as shown in Fig. 1-9. (Note that this reasoning suggests that the formula for water is H_2O, not OH as Dalton believed.) ◄ *ⓘ*

Unfortunately, Avogadro's interpretations were not accepted by most chemists, and a half-century of confusion followed, in which many different assumptions were made about formulas and atomic masses.

During the nineteenth century, painstaking measurements were made of the masses of various elements that combined to form compounds. From these experiments a list of relative atomic masses could be determined. One of the chemists involved in

ⓘ There are seven elements that occur as diatomic molecules:

$$H_2, N_2, O_2, F_2, Cl_2, Br_2, I_2$$

ⓘ Compare Figures 1-8 and 1-9. As we have seen, the macroscopic observations are explained by the microscopic world of atoms and molecules.

2 volumes hydrogen   combines with   1 volume oxygen   to form   2 volumes gaseous water

1 volume hydrogen   combines with   1 volume chlorine   to form   2 volumes hydrogen chloride

Figure 1-8 | A representation of some of Gay-Lussac's experimental results on combining gas volumes.

*A *molecule* is a collection of atoms.

Figure 1-9 | A representation of combining gases at the molecular level. The spheres represent atoms in the molecules.

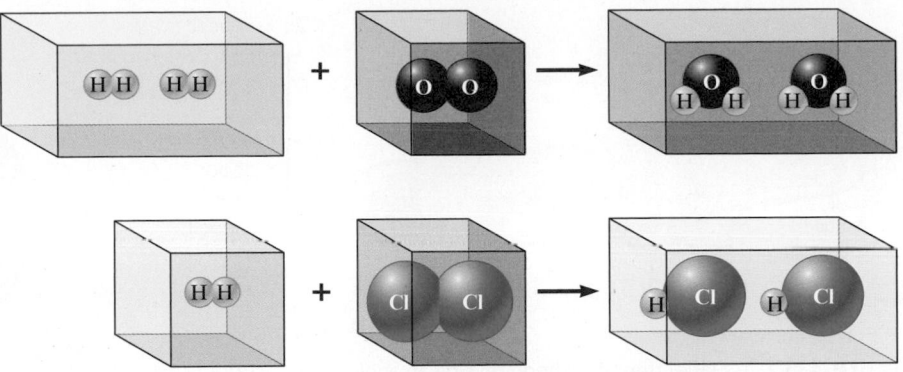

contributing to this list was a Swede named Jöns Jakob Berzelius (1779–1848), who discovered the elements cerium, selenium, silicon, and thorium and developed the modern symbols for the elements used in writing the formulas of compounds.

1-6 | Early Experiments to Characterize the Atom

On the basis of the work of Dalton, Gay-Lussac, Avogadro, and others, chemistry was beginning to make sense. The concept of atoms was clearly a good idea. Inevitably, scientists began to wonder about the nature of the atom. What is an atom made of, and how do the atoms of the various elements differ?

The Electron

The first important experiments that led to an understanding of the composition of the atom were done by the English physicist J. J. Thomson (Fig. 1-10), who studied electrical discharges in partially evacuated tubes called **cathode-ray tubes** (Fig. 1-11) during the period from 1898 to 1903. Thomson found that when high voltage was applied to the tube, a "ray" he called a *cathode ray* (because it emanated from the negative electrode, or cathode) was produced. Because this ray was produced at the negative electrode and was repelled by the negative pole of an applied electric field (Fig. 1-12), Thomson postulated that the ray was a stream of negatively charged particles, now called **electrons**. From experiments in which he measured the deflection of the beam

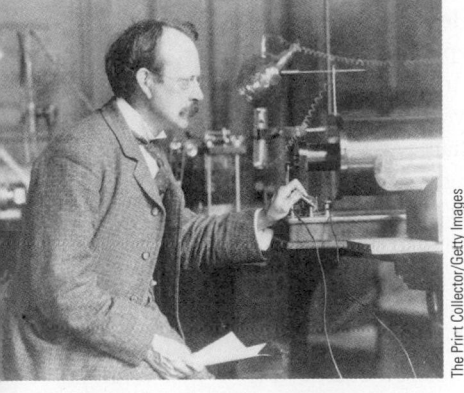

Figure 1-10 | J. J. Thomson (1856–1940) was an English physicist at Cambridge University. He received the Nobel Prize in physics in 1906.

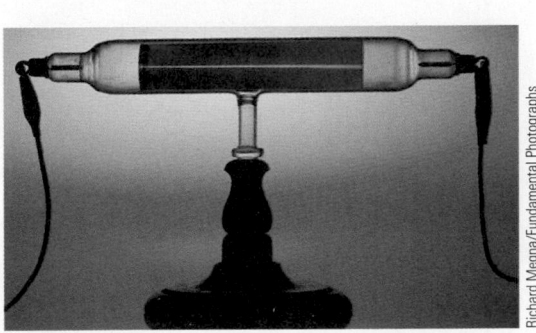

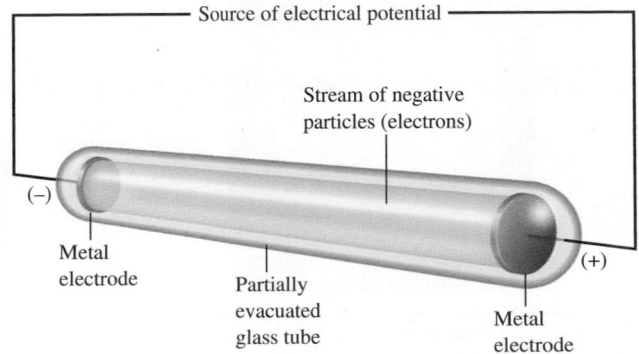

Figure 1-11 | A cathode-ray tube. The fast-moving electrons excite the gas in the tube, causing a glow between the electrodes. The green color in the photo is due to the response of the screen (coated with zinc sulfide) to the electron beam.

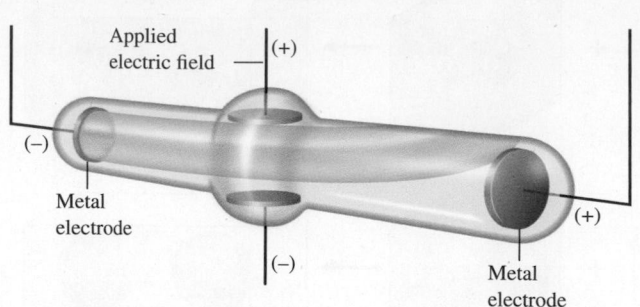

Figure 1-12 | Deflection of cathode rays by an applied electric field.

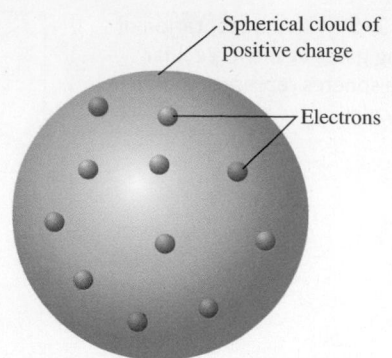

Figure 1-13 | The plum pudding model of the atom.

of electrons in a magnetic field, Thomson determined the *charge-to-mass ratio* of an electron:

$$\frac{e}{m} = -1.76 \times 10^8 \text{ C/g}$$

where *e* represents the charge on the electron in coulombs (C) and *m* represents the electron mass in grams.

One of Thomson's primary goals in his cathode-ray tube experiments was to gain an understanding of the structure of the atom. He reasoned that since electrons could be produced from electrodes made of various types of metals, *all* atoms must contain electrons. Since atoms were known to be electrically neutral, Thomson further assumed that atoms also must contain some positive charge. Thomson postulated that an atom consisted of a diffuse cloud of positive charge with the negative electrons embedded randomly in it. This model, shown in Fig. 1-13, is often called the *plum pudding model* because the electrons are like raisins dispersed in a pudding (the positive charge cloud), as in plum pudding, a favorite English dessert. ◀ A more familiar example of this might be a Jell-O dessert containing pieces of fruit.

In 1909 Robert Millikan (1868–1953), working at the University of Chicago, performed very clever experiments involving charged oil drops. These experiments allowed him to determine the magnitude of the electron charge (Fig. 1-14). With this

▲ A classic English plum pudding, in which the raisins represent the distribution of electrons in the atom.

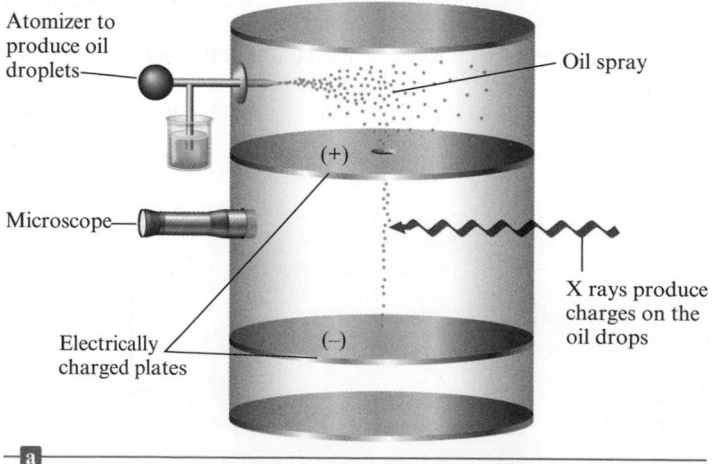

Figure 1-14 | (a) A schematic representation of the apparatus Millikan used to determine the charge on the electron. The fall of charged oil droplets due to gravity can be halted by adjusting the voltage across the two plates. This voltage and the mass of the oil drop can then be used to calculate the charge on the oil drop. Millikan's experiments showed that the charge on an oil drop is always a whole-number multiple of the electron charge. (b) Robert Millikan using his apparatus.

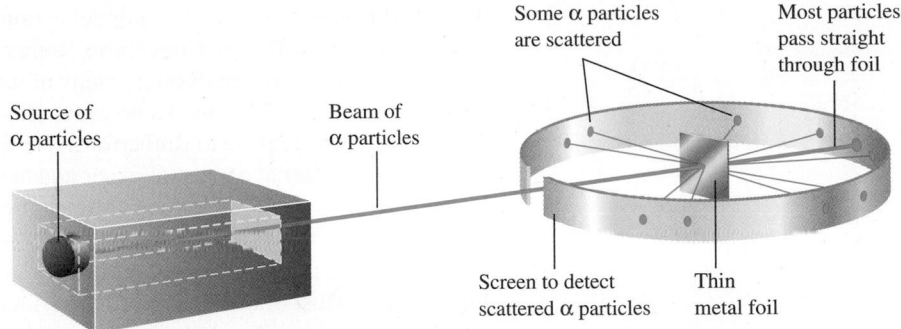

Figure 1-16 | Rutherford's experiment on α-particle bombardment of metal foil.

Figure 1-15 | Ernest Rutherford (1871–1937) was born on a farm in New Zealand. In 1895 he placed second in a scholarship competition to attend Cambridge University but was awarded the scholarship when the winner decided to stay home and get married. As a scientist in England, Rutherford did much of the early work on characterizing radioactivity. He named the α and β particles and the γ ray and coined the term *half-life* to describe an important attribute of radioactive elements. His experiments on the behavior of α particles striking thin metal foils led him to postulate the nuclear atom. He also invented the name *proton* for the nucleus of the hydrogen atom. He received the Nobel Prize in chemistry in 1908.

ⓘ It is interesting that these experiments were conducted by Hans Geiger and Ernest Marsden, Rutherford's students.

value and the charge-to-mass ratio determined by Thomson, Millikan was able to calculate the mass of the electron as 9.11×10^{-31} kilogram.

Radioactivity

In the late nineteenth century scientists discovered that certain elements produce high-energy radiation. For example, in 1896 the French scientist Henri Becquerel found accidentally that a piece of a mineral containing uranium could produce its image on a photographic plate in the absence of light. He attributed this phenomenon to a spontaneous emission of radiation by the uranium, which he called **radioactivity**. Studies in the early twentieth century demonstrated three types of radioactive emission: gamma (γ) rays, beta (β) particles, and alpha (α) particles. A γ ray is high-energy "light"; a β particle is a high-speed electron; and an α particle has a $2+$ charge, that is, a charge twice that of the electron and with the opposite sign. The mass of an α particle is 7300 times that of the electron. More modes of radioactivity are now known, and we will discuss them in Chapter 18. Here we will consider only α particles because they were used in some crucial early experiments.

The Nuclear Atom

In 1911 Ernest Rutherford (Fig. 1-15), who performed many of the pioneering experiments to explore radioactivity, carried out an experiment to test Thomson's plum pudding model. ◄ *ⓘ* The experiment involved directing α particles at a thin sheet of metal foil, as illustrated in Fig. 1-16. Rutherford reasoned that if Thomson's model were accurate, the massive α particles should crash through the thin foil like cannonballs through gauze, as shown in Fig. 1-17(a). He expected the α particles to travel through

Figure 1-17 | (a) The expected results of the metal foil experiment if Thomson's model were correct. (b) Actual results.

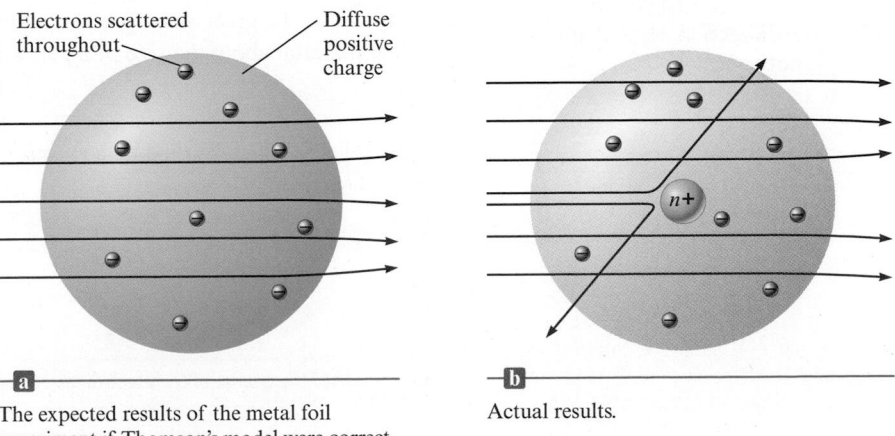

The expected results of the metal foil experiment if Thomson's model were correct.

Actual results.

the foil with, at the most, very minor deflections in their paths. The results of the experiment were very different from those Rutherford anticipated. Although most of the α particles passed straight through, many of the particles were deflected at large angles, as shown in Fig. 1-17(b), and some were reflected, never hitting the detector. This outcome was a great surprise to Rutherford. (He wrote that this result was comparable with shooting a howitzer at a piece of paper and having the shell reflected back.)

Rutherford knew from these results that the plum pudding model for the atom could not be correct. The large deflections of the α particles could be caused only by a center of concentrated positive charge that contains most of the atom's mass, as illustrated in Fig. 1-17(b). Most of the α particles pass directly through the foil because the atom is mostly open space. The deflected α particles are those that had a "close encounter" with the massive positive center of the atom, and the few reflected α particles are those that made a "direct hit" on the much more massive positive center.

In Rutherford's mind these results could be explained only in terms of a **nuclear atom**—an atom with a dense center of positive charge (the **nucleus**) with electrons moving around the nucleus at a distance that is large relative to the nuclear radius. ◄ *ⓘ*

ⓘ Notice how the model of the atom changed over time due to new observations. The newer model builds on the older model.

> ### Critical Thinking
>
> You have learned about three different models of the atom: Dalton's model, Thomson's model, and Rutherford's model. What if Dalton was correct? What would Rutherford have expected from his experiments with gold foil? What if Thomson was correct? What would Rutherford have expected from his experiments with gold foil?

1-7 | The Modern View of Atomic Structure: An Introduction

In the years since Thomson and Rutherford, a great deal has been learned about atomic structure. Because much of this material will be covered in detail in later chapters, only an introduction will be given here. The simplest view of the atom is that it consists of a tiny nucleus (with a diameter of about 10^{-13} cm) and electrons that move about the nucleus at an average distance of about 10^{-8} cm from it (Fig. 1-18).

As we will see later, the chemistry of an atom mainly results from its electrons. For this reason, chemists can be satisfied with a relatively crude nuclear model. The nucleus is assumed to contain **protons**, which have a positive charge equal in magnitude to the electron's negative charge, and **neutrons**, which have virtually the same mass as a proton but no charge. The masses and charges of the electron, proton, and neutron are shown in Table 1-1.

Two striking things about the nucleus are its small size compared with the overall size of the atom and its extremely high density. The tiny nucleus accounts for almost all the atom's mass. Its great density is dramatically demonstrated by the fact that a piece of nuclear material about the size of a pea would have a mass of 250 million tons! ◄

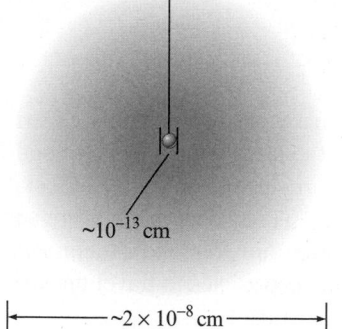

Nucleus

~10^{-13} cm

|———— ~2×10^{-8} cm ————|

Figure 1-18 | A nuclear atom viewed in cross section. Note that this drawing is not to scale.

Table 1-1 | The Mass and Charge of the Electron, Proton, and Neutron

Particle	Mass	Charge*
Electron	9.109×10^{-31} kg	1−
Proton	1.673×10^{-27} kg	1+
Neutron	1.675×10^{-27} kg	None

*The magnitude of the charge of the electron and the proton is 1.60×10^{-19} C.

▲ If the atomic nucleus were the size of this ball bearing, a typical atom would be the size of this stadium.

i Note that a question drives us to consider a more complicated model.

i The *chemistry* of an atom arises from its electrons.

An important question to consider at this point is, *"If all atoms are composed of these same components, why do different atoms have different chemical properties?"* ◄ *i* The answer to this question lies in the number and the arrangement of the electrons. The electrons constitute most of the atomic volume and thus are the parts that "intermingle" when atoms combine to form molecules. Therefore, the number of electrons possessed by a given atom greatly affects its ability to interact with other atoms. As a result, the atoms of different elements, which have different numbers of protons and electrons, show different chemical behavior. ◄ *i*

A sodium atom has 11 protons in its nucleus. Since atoms have no net charge, the number of electrons must equal the number of protons. Therefore, a sodium atom has 11 electrons moving around its nucleus. It is *always* true that a sodium atom has 11 protons and 11 electrons. However, each sodium atom also has neutrons in its nucleus, and different types of sodium atoms exist that have different numbers of neutrons. For example, consider the sodium atoms represented in Fig. 1-19. These two atoms are **isotopes**, or *atoms with the same number of protons but different numbers of neutrons*. Note that the symbol for one particular type of sodium atom is written

$$\text{Mass number} \longrightarrow \ _{11}^{23}\text{Na} \longleftarrow \text{Element symbol}$$
$$\text{Atomic number} \nearrow$$

i Mass number

$$_{Z}^{A}\text{X} \longleftarrow \text{Element symbol}$$

Atomic number

where the **atomic number** Z (number of protons) is written as a subscript, and the **mass number** A (the total number of protons and neutrons) is written as a superscript. (The particular atom represented here is called "sodium twenty-three." It has 11 electrons, 11 protons, and 12 neutrons.) ◄ *i* Because the chemistry of an atom is due to

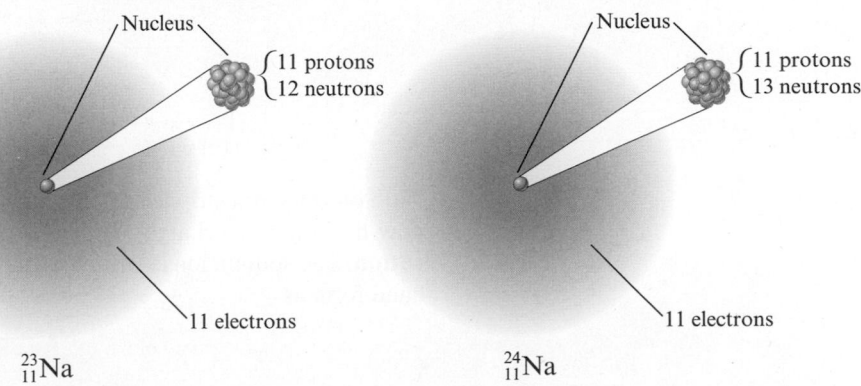

Figure 1-19 | Two isotopes of sodium. Both have 11 protons and 11 electrons, but they differ in the number of neutrons in their nuclei.

its electrons, isotopes show almost identical chemical properties. In nature most elements contain mixtures of isotopes.

Critical Thinking

The average diameter of an atom is 2×10^{-10} m. What if the average diameter of an atom were 1 cm? How tall would you be?

INTERACTIVE EXAMPLE 1-2

Writing the Symbols for Atoms

Write the symbol for the atom that has an atomic number of 9 and a mass number of 19. How many electrons and how many neutrons does this atom have?

Solution

The atomic number 9 means the atom has 9 protons. This element is called *fluorine*, symbolized by F. The atom is represented as

$$^{19}_{9}\text{F}$$

and is called *fluorine nineteen*. Since the atom has 9 protons, it also must have 9 electrons to achieve electrical neutrality. The mass number gives the total number of protons and neutrons, which means that this atom has 10 neutrons.

See Exercises 1-43 through 1-46

Atoms are electrically neutral. That is, they have no net charge because they have equal numbers of protons (+ charge) and electrons (− charge). When one or more electrons are removed or added to a neutral atom, a species with a net charge is formed. These charged species are called **ions**.

To see how the ions are formed, consider what happens when an electron is transferred from a sodium atom to a chlorine atom (the neutrons in the nuclei will be ignored):

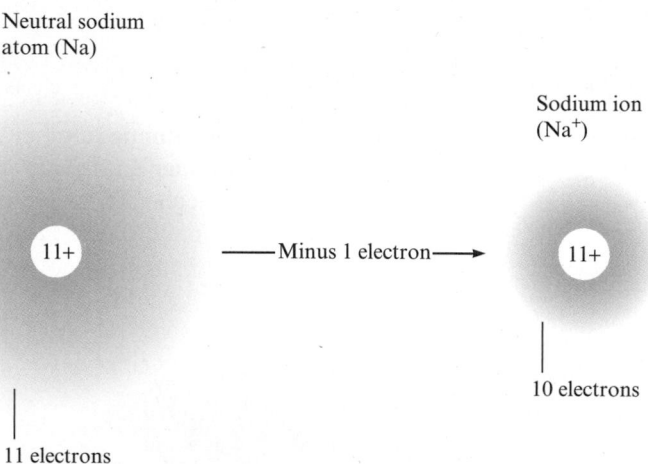

Neutral sodium atom (Na)

Sodium ion (Na⁺)

Minus 1 electron

11+   11+

11 electrons

10 electrons

With one electron stripped off, the sodium, with its 11 protons and only 10 electrons, now has a net 1+ charge—it has become a *positive ion*. A positive ion is called a **cation**. The sodium ion is written as Na^+, and the process can be represented in shorthand form as

$$\text{Na} \longrightarrow \text{Na}^+ + e^-$$

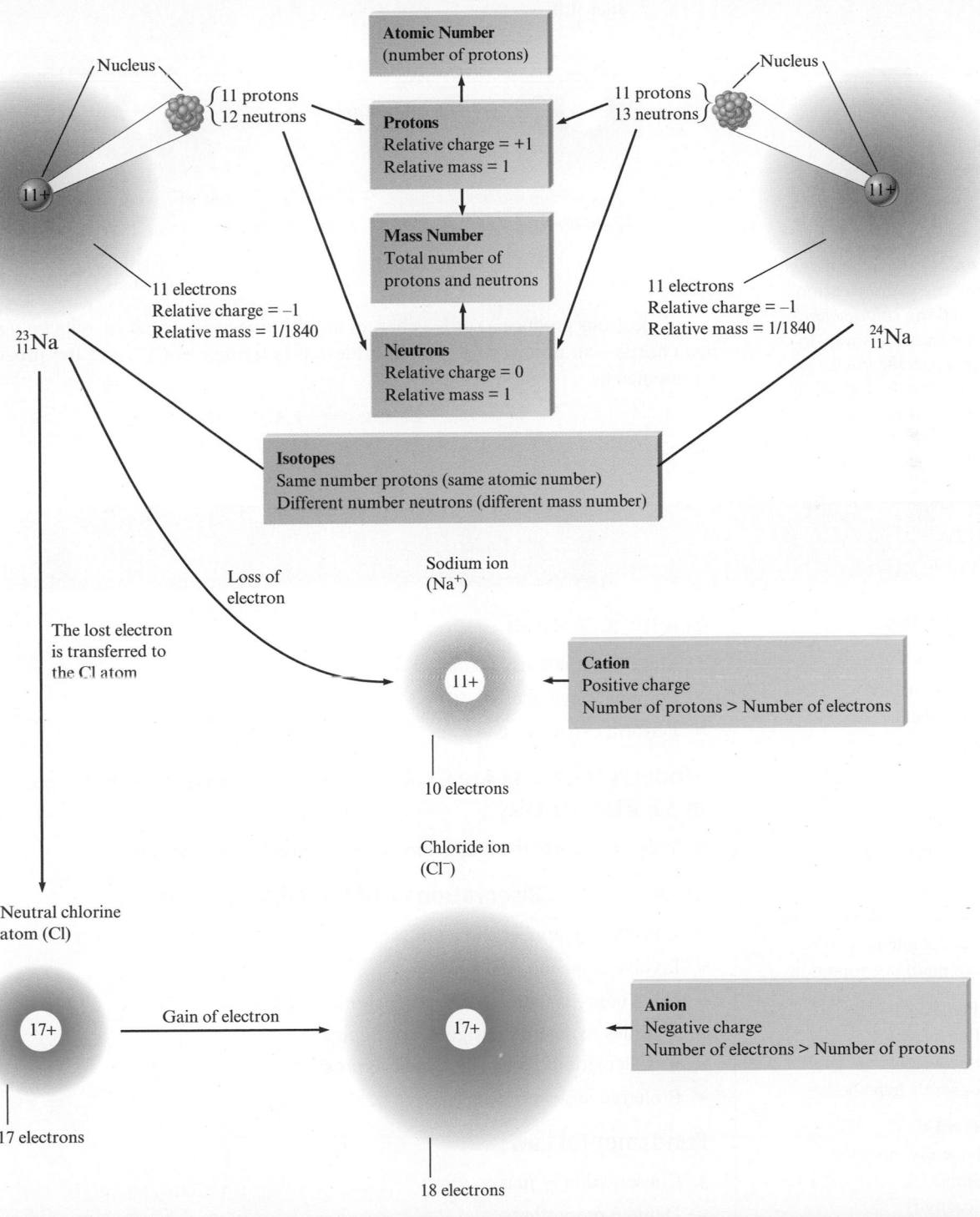

If an electron is added to chlorine,

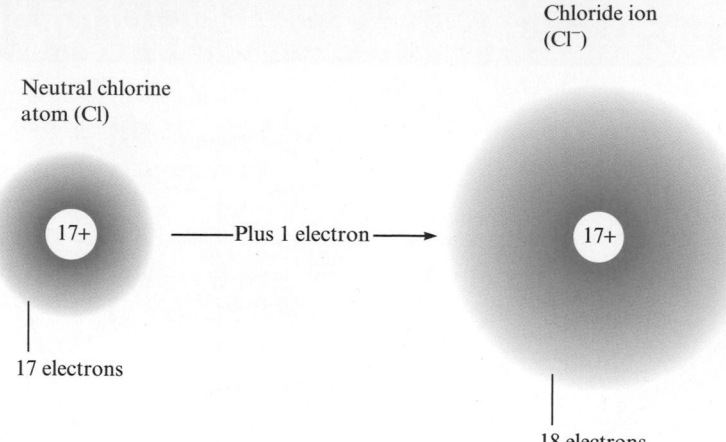

Neutral chlorine
atom (Cl)

Chloride ion
(Cl⁻)

17 electrons

18 electrons

ℹ Na⁺ is usually called the *sodium ion* rather than the sodium cation. Also Cl⁻ is called the *chloride ion* rather than the chloride anion. In general, when a specific ion is referred to, the word *ion* rather than *cation* or *anion* is used.

the 18 electrons produce a net 1− charge; the chlorine has become an *ion with a negative charge*—an **anion**. ◄ *ℹ* The chloride ion is written as Cl⁻, and the process is represented as

$$Cl + e^- \longrightarrow Cl^-$$

FOR REVIEW

Key Terms

Section 1-2
scientific method
measurement
hypothesis
theory
model
natural law
law of conservation of mass

Section 1-4
law of conservation of mass
law of definite proportion
law of multiple proportions

Section 1-5
atomic masses
atomic weights
Avogadro's hypothesis

Section 1-6
cathode-ray tubes
electrons
radioactivity
nuclear atom
nucleus

Scientific Method
- Make observations
- Formulate hypotheses
- Perform experiments

Models (Theories) Are Explanations of Why Nature Behaves in a Particular Way.
- They are subject to modification over time and sometimes fail.

Quantitative Observations Are Called Measurements.
- Consist of a number and a unit
- Involve some uncertainty
- Uncertainty is indicated by using significant figures
  - Rules to determine significant figures
  - Calculations using significant figures
- Preferred system is SI

Fundamental Laws
- Conservation of mass
- Definite proportion
- Multiple proportions

Key Terms

Dalton's Atomic Theory

- All elements are composed of atoms.
- All atoms of a given element are identical.
- Chemical compounds are formed when atoms combine.
- Atoms are not changed in chemical reactions, but the way they are bound together changes.

Early Atomic Experiments and Models

- Thomson model
- Millikan experiment
- Rutherford experiment
- Nuclear model

Atomic Structure

- Small dense nucleus contains protons and neutrons.
  - Protons—positive charge
  - Neutrons—no charge
- Electrons reside outside the nucleus in the relatively large remaining atomic volume.
  - Electrons—negative charge, small mass (1/1840 of proton)
- Isotopes have the same atomic number but different mass numbers.

Review Questions   Answers to the Review Questions can be found on the Student Web site (accessible from **www.cengagebrain.com**).

1. Define and explain the differences between the following terms.
   a. law and theory
   b. theory and experiment
   c. qualitative and quantitative
   d. hypothesis and theory

2. Is the scientific method suitable for solving problems only in the sciences? Explain.

3. Use Dalton's atomic theory to account for each of the following.
   a. the law of conservation of mass
   b. the law of definite proportion
   c. the law of multiple proportions

4. What evidence led to the conclusion that cathode rays had a negative charge?

5. What discoveries were made by J. J. Thomson, Henri Becquerel, and Lord Rutherford? How did Dalton's model of the atom have to be modified to account for these discoveries?

6. Consider Ernest Rutherford's α-particle bombardment experiment illustrated in Fig. 1-16. How did the results of this experiment lead Rutherford away from the plum pudding model of the atom to propose the nuclear model of the atom?

7. Do the proton and the neutron have exactly the same mass? How do the masses of the proton and neutron compare to the mass of the electron? Which particles make the greatest contribution to the mass of an atom? Which particles make the greatest contribution to the chemical properties of an atom?

8. What is the distinction between atomic number and mass number? Between mass number and atomic mass?

Active Learning Questions

These questions are designed to be used by groups of students in class.

1. Paracelsus, a sixteenth-century alchemist and healer, adopted as his slogan: "The patients are your textbook, the sickbed is your study." Is this view consistent with using the scientific method?

2. What is wrong with the following statement? "The results of the experiment do not agree with the theory. Something must be wrong with the experiment."

3. Which of the following is true about an individual atom? Explain.

   a. An individual atom should be considered to be a solid.

   b. An individual atom should be considered to be a liquid.

   c. An individual atom should be considered to be a gas.

   d. The state of the atom depends on which element it is.

   e. An individual atom cannot be considered to be a solid, liquid, or gas.

   Justify your choice, and for choices you did not pick, explain what is wrong with them.

4. These questions concern the work of J. J. Thomson.

   a. From Thomson's work, which particles do you think he would feel are most important for the formation of compounds (chemical changes), and why?

   b. Of the remaining two subatomic particles, which do you place second in importance for forming compounds, and why?

   c. Propose three models that explain Thomson's findings and evaluate them. To be complete you should include Thomson's findings.

5. Which of the following explain how an ion is formed? Explain your answer.

   a. adding or subtracting protons to/from an atom

   b. adding or subtracting neutrons to/from an atom

   c. adding or subtracting electrons to/from an atom

6. You have a chemical in a sealed glass container filled with air. The setup is sitting on a balance as shown below. The chemical is ignited by means of a magnifying glass focusing sunlight on the reactant. After the chemical has completely burned, which of the following is true? Explain your answer.

   a. The balance will read less than 250.0 g.

   b. The balance will read 250.0 g.

   c. The balance will read greater than 250.0 g.

   d. The scale's reading cannot be determined without knowing the identity of the chemical.

7. You may have noticed that when water boils, you can see bubbles that rise to the surface of the water. Which of the following is inside these bubbles? Explain.

   a. air

   b. hydrogen and oxygen gas

   c. oxygen gas

   d. water vapor

   e. carbon dioxide gas

8. One of the best indications of a useful theory is that it raises more questions for further experimentation than it originally answered. Does this apply to Dalton's atomic theory? Give examples.

9. Dalton assumed that all atoms of the same element were identical in all their properties. Explain why this assumption is not valid.

10. Which (if any) of the following can be determined by knowing the number of protons in a neutral element? Explain your answer.

    a. the number of neutrons in the neutral element

    b. the number of electrons in the neutral element

    c. the name of the element

A blue question or exercise number indicates that the answer to that question or exercise appears at the back of this book and a solution appears in the *Student Solutions Manual*.

Questions

11. The difference between a *law* and a *theory* is the difference between *what* and *why*. Explain.

12. As part of a science project, you study traffic patterns in your city at an intersection in the middle of downtown. You set up a device that counts the cars passing through this intersection for a 24-hour period during a weekday. The graph of hourly traffic looks like this.

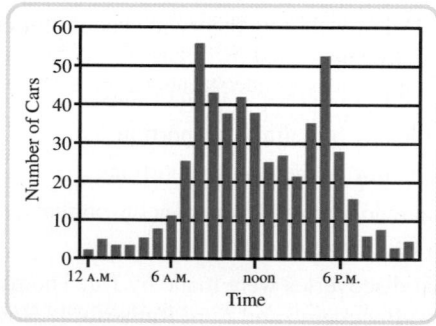

    a. At what time(s) does the highest number of cars pass through the intersection?

    b. At what time(s) does the lowest number of cars pass through the intersection?

    c. Briefly describe the trend in numbers of cars over the course of the day.

d. Provide a hypothesis explaining the trend in numbers of cars over the course of the day.

e. Provide a possible experiment that could test your hypothesis.

13. Explain the fundamental steps of the scientific method. The scientific method is a dynamic process. What does this mean?

14. When hydrogen is burned in oxygen to form water, the composition of water formed does not depend on the amount of oxygen reacted. Interpret this in terms of the law of definite proportion.

15. Explain the law of conservation of mass, the law of definite proportion, and the law of multiple proportions.

16. Chlorine has two natural isotopes: $^{37}_{17}Cl$ and $^{35}_{17}Cl$. Hydrogen reacts with chlorine to form the compound HCl. Would a given amount of hydrogen react with different masses of the two chlorine isotopes? Does this conflict with the law of definite proportion? Why or why not?

17. The vitamin niacin (nicotinic acid, $C_6H_5NO_2$) can be isolated from a variety of natural sources such as liver, yeast, milk, and whole grain. It also can be synthesized from commercially available materials. From a nutritional point of view, which source of nicotinic acid is best for use in a multivitamin tablet? Why?

18. Section 1-5 describes the postulates of Dalton's atomic theory. With some modifications, these postulates hold up very well regarding how we view elements, compounds, and chemical reactions today. Answer the following questions concerning Dalton's atomic theory and the modifications made today.

a. The atom can be broken down into smaller parts. What are the smaller parts?

b. How are atoms of hydrogen identical to each other and how can they be different from each other?

c. How are atoms of hydrogen different from atoms of helium? How can H atoms be similar to He atoms?

d. How is water different from hydrogen peroxide (H_2O_2) even though both compounds are composed of only hydrogen and oxygen?

e. What happens in a chemical reaction and why is mass conserved in a chemical reaction?

19. The contributions of J. J. Thomson and Ernest Rutherford led the way to today's understanding of the structure of the atom. What were their contributions?

20. What is the modern view of the structure of the atom?

21. The number of protons in an atom determines the identity of the atom. What do the number and arrangement of the electrons in an atom determine? What does the number of neutrons in an atom determine?

22. If the volume of a proton is similar to the volume of an electron, how will the densities of these two particles compare to each other?

23. For lighter, stable isotopes, the ratio of the mass number to the atomic number is close to a certain value. What is the value? What happens to the value of the mass number to atomic number ratio as stable isotopes become heavier?

24. What refinements had to be made in Dalton's atomic theory to account for Gay-Lussac's results on the combining volumes of gases?

Exercises

In this section, similar exercises are paired.

Development of the Atomic Theory

25. When mixtures of gaseous H_2 and gaseous Cl_2 react, a product forms that has the same properties regardless of the relative amounts of H_2 and Cl_2 used.

a. How is this result interpreted in terms of the law of definite proportion?

b. When a volume of H_2 reacts with an equal volume of Cl_2 at the same temperature and pressure, what volume of product having the formula HCl is formed?

26. Observations of the reaction between nitrogen gas and hydrogen gas show us that 1 volume of nitrogen reacts with 3 volumes of hydrogen to make 2 volumes of gaseous product, as shown below:

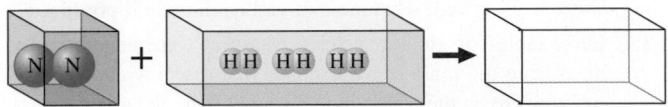

Determine the formula of the product and justify your answer.

27. A sample of chloroform is found to contain 12.0 g of carbon, 106.4 g of chlorine, and 1.01 g of hydrogen. If a second sample of chloroform is found to contain 30.0 g of carbon, what is the total mass of chloroform in the second sample?

28. A sample of H_2SO_4 contains 2.02 g of hydrogen, 32.07 g of sulfur, and 64.00 g of oxygen. How many grams of sulfur and grams of oxygen are present in a second sample of H_2SO_4 containing 7.27 g of hydrogen?

29. Hydrazine, ammonia, and hydrogen azide all contain only nitrogen and hydrogen. The mass of hydrogen that combines with 1.00 g of nitrogen for each compound is 1.44×10^{-1} g, 2.16×10^{-1} g, and 2.40×10^{-2} g, respectively. Show how these data illustrate the law of multiple proportions.

30. Consider 100.0-g samples of two different compounds consisting only of carbon and oxygen. One compound contains 27.2 g of carbon and the other has 42.9 g of carbon. How can these data support the law of multiple proportions if 42.9 is not a multiple of 27.2? Show that these data support the law of multiple proportions.

31. The three most stable oxides of carbon are carbon monoxide (CO), carbon dioxide (CO_2), and carbon suboxide (C_3O_2). The molecules can be represented as

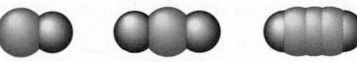

Explain how these molecules illustrate the law of multiple proportions.

32. Two elements, R and Q, combine to form two binary compounds. In the first compound, 14.0 g of R combines with 3.00 g of Q. In the second compound, 7.00 g of R combines with 4.50 g of Q. Show that these data are in accord with the

law of multiple proportions. If the formula of the second compound is RQ, what is the formula of the first compound?

33. In Section 1-1 of the text, the concept of a chemical reaction was introduced with the example of the decomposition of water, represented as follows:

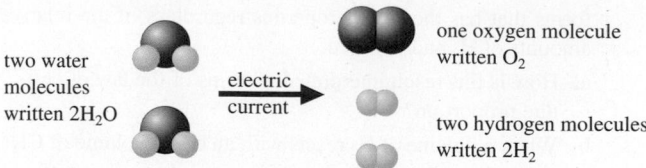

two water molecules written $2H_2O$ electric current   one oxygen molecule written O_2 two hydrogen molecules written $2H_2$

Use ideas from Dalton's atomic theory to explain how the above representation illustrates the law of conservation of mass.

34. In a combustion reaction, 46.0 g of ethanol reacts with 96.0 g of oxygen to produce water and carbon dioxide. If 54.0 g of water is produced, what mass of carbon dioxide is produced?

35. Early tables of atomic weights (masses) were generated by measuring the mass of a substance that reacts with 1.00 g of oxygen. Given the following data and taking the atomic mass of hydrogen as 1.00, generate a table of relative atomic masses for oxygen, sodium, and magnesium.

Element	Mass That Combines with 1.00 g Oxygen	Assumed Formula
Hydrogen	0.126 g	HO
Sodium	2.875 g	NaO
Magnesium	1.500 g	MgO

36. Indium oxide contains 4.784 g of indium for every 1.000 g of oxygen. In 1869, when Mendeleev first presented his version of the periodic table, he proposed the formula In_2O_3 for indium oxide. Before that time it was thought that the formula was InO. What values for the atomic mass of indium are obtained using these two formulas? Assume that oxygen has an atomic mass of 16.00.

The Nature of the Atom

37. From the information in this chapter on the mass of the proton, the mass of the electron, and the sizes of the nucleus and the atom, calculate the densities of a hydrogen nucleus and a hydrogen atom.

38. If you wanted to make an accurate scale model of the hydrogen atom and decided that the nucleus would have a diameter of 1 mm, what would be the diameter of the entire model?

39. In an experiment it was found that the total charge on an oil drop was 5.93×10^{-18} C. How many negative charges does the drop contain?

40. A chemist in a galaxy far, far away performed the Millikan oil drop experiment and got the following results for the charges on various drops. Use these data to calculate the charge of the electron in zirkombs.

2.56×10^{-12} zirkombs 7.68×10^{-12} zirkombs
3.84×10^{-12} zirkombs 6.40×10^{-13} zirkombs

41. Write the symbol of each atom using the $_Z^A X$ format.

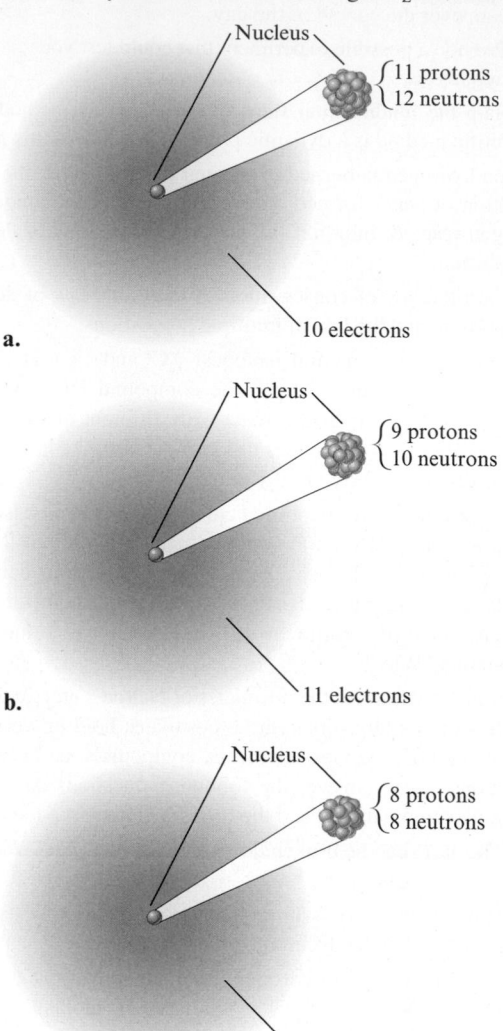

a.
Nucleus
11 protons
12 neutrons
10 electrons

b.
Nucleus
9 protons
10 neutrons
11 electrons

c.
Nucleus
8 protons
8 neutrons
8 electrons

42. For carbon-14 and carbon-12, how many protons and neutrons are in each nucleus? Assuming neutral atoms, how many electrons are present in an atom of carbon-14 and in an atom of carbon-12?

43. How many protons and neutrons are in the nucleus of each of the following atoms? In a neutral atom of each element, how many electrons are present?

a. ^{79}Br d. ^{133}Cs
b. ^{81}Br e. ^{3}H
c. ^{239}Pu f. ^{56}Fe

44. What number of protons and neutrons is contained in the nucleus of each of the following atoms? Assuming each atom is uncharged, what number of electrons is present?

a. $_{92}^{235}U$ d. $_{82}^{208}Pb$
b. $_{13}^{27}Al$ e. $_{37}^{86}Rb$
c. $_{26}^{57}Fe$ f. $_{20}^{41}Ca$

45. Write the atomic symbol ($_Z^A$X) for each of the following isotopes.

    a. $Z = 8$, number of neutrons = 9

    b. the isotope of chlorine in which $A = 37$

    c. $Z = 27$, $A = 60$

    d. number of protons = 26, number of neutrons = 31

    e. the isotope of I with a mass number of 131

    f. $Z = 3$, number of neutrons = 4

46. Write the atomic symbol ($_Z^A$X) for each of the isotopes described below.

    a. number of protons = 27, number of neutrons = 31

    b. the isotope of boron with mass number 10

    c. $Z = 12$, $A = 23$

    d. atomic number 53, number of neutrons = 79

    e. $Z = 20$, number of neutrons = 27

    f. number of protons = 29, mass number 65

47. For each of the following ions, indicate the number of protons and electrons the ion contains.

    a. Ba^{2+} **e.** Co^{3+}

    b. Zn^{2+} **f.** Te^{2-}

    c. N^{3-} **g.** Br^-

    d. Rb^+

48. How many protons, neutrons, and electrons are in each of the following atoms or ions?

    a. $_{12}^{24}$Mg   **d.** $_{27}^{59}$Co^{3+} **g.** $_{34}^{79}$Se^{2-}

    b. $_{12}^{24}$Mg^{2+} **e.** $_{27}^{59}$Co   **h.** $_{28}^{63}$Ni

    c. $_{27}^{59}$Co^{2+} **f.** $_{34}^{79}$Se   **i.** $_{28}^{59}$Ni^{2+}

49. What is the symbol for an ion with 63 protons, 60 electrons, and 88 neutrons? If an ion contains 50 protons, 68 neutrons, and 48 electrons, what is its symbol?

50. What is the symbol of an ion with 16 protons, 18 neutrons, and 18 electrons? What is the symbol for an ion that has 16 protons, 16 neutrons, and 18 electrons?

51. Complete the following table:

Symbol	Number of Protons in Nucleus	Number of Neutrons in Nucleus	Number of Electrons	Net Charge
$_{92}^{238}$U				
	20	20		2+
	23	28	20	
$_{39}^{89}$Y				
	35	44	36	
	15	16		3−

52. Complete the following table:

Symbol	Number of Protons in Nucleus	Number of Neutrons in Nucleus	Number of Electrons	Net Charge
$_{26}^{53}$Fe^{2+}				
	26	33		3+
	85	125	86	
	13	14	10	
		76	54	2−

Additional Exercises

53. Four Fe^{2+} ions are key components of hemoglobin, the protein that transports oxygen in the blood. Assuming that these ions are $^{53}Fe^{2+}$, how many protons and neutrons are present in each nucleus, and how many electrons are present in each ion?

54. Which of the following statements is/are *true*? For the false statements, correct them.

    a. All particles in the nucleus of an atom are charged.

    b. The atom is best described as a uniform sphere of matter in which electrons are embedded.

    c. The mass of the nucleus is only a very small fraction of the mass of the entire atom.

    d. The volume of the nucleus is only a very small fraction of the total volume of the atom.

    e. The number of neutrons in a neutral atom must equal the number of electrons.

55. Identify each of the following elements. Give the number of protons and neutrons in each nucleus.

    a. $_{15}^{31}$X   **c.** $_{19}^{39}$X

    b. $_{53}^{127}$X   **d.** $_{70}^{173}$X

56. The isotope of an unknown element, X, has a mass number of 79. The most stable ion of this isotope has 36 electrons and has a 2− charge. Which of the following statements is(are) *true*? For the false statements, correct them.

    a. This ion has more electrons than protons in the nucleus.

    b. The isotope of X contains 38 protons.

    c. The isotope of X contains 41 neutrons.

    d. The identity of X is strontium, Sr.

57. An element's most stable ion has a 2+ charge. If the ion of element X has a mass number of 230 and has 86 electrons, what is the identity of the element, and how many neutrons does it have?

58. The early alchemists used to do an experiment in which water was boiled for several days in a sealed glass container. Eventually, some solid residue would appear in the bottom of the flask, which was interpreted to mean that some of the water in the flask had been converted into "earth." When Lavoisier repeated this experiment, he found that the water weighed the same before and after heating and the mass of the flask plus

the solid residue equaled the original mass of the flask. Were the alchemists correct? Explain what happened. (This experiment is described in the article by A. F. Scott in *Scientific American*, January 1984.)

59. In a reaction, 34.0 g of chromium(III) oxide reacts with 12.1 g of aluminum to produce chromium and aluminum oxide. If 23.3 g of chromium is produced, what mass of aluminum oxide is produced?

ChemWork Problems

These multiconcept problems (and additional ones) are found interactively online with the same type of assistance a student would get from an instructor.

60. Complete the following table, including the mass number and the atomic number with the symbol for the isotope.

Number of Protons	Number of Neutrons	Symbol
9	10	
13	14	
53	74	
34	45	
16	16	

61. Complete the following table.

Atoms	Number of Protons	Number of Neutrons
$^{4}_{2}$He		
$^{20}_{10}$Ne		
$^{48}_{22}$Ti		
$^{190}_{76}$Os		
$^{50}_{27}$Co		

62. Complete the following table.

Atom/Ion	Protons	Neutrons	Electrons
$^{120}_{50}$Sn			
$^{25}_{12}$Mg^{2+}			
$^{56}_{26}$Fe^{2+}			
$^{79}_{34}$Se			
$^{35}_{17}$Cl			
$^{63}_{29}$Cu			

63. Which of the following is(are) correct?
    a. ^{40}Ca^{2+} contains 20 protons and 18 electrons.
    b. Rutherford created the cathode-ray tube and was the founder of the charge-to-mass ratio of an electron.
    c. An electron is heavier than a proton.
    d. The nucleus contains protons, neutrons, and electrons.

Challenge Problems

64. Confronted with the box shown in the diagram, you wish to discover something about its internal workings. You have no tools and cannot open the box. You pull on rope B, and it moves rather freely. When you pull on rope A, rope C appears to be pulled slightly into the box. When you pull on rope C, rope A almost disappears into the box.

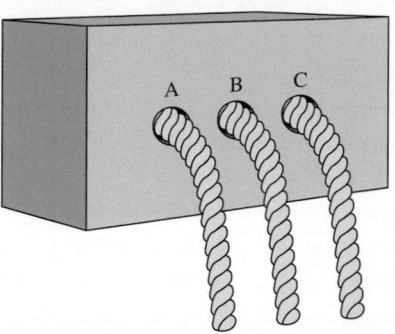

    a. Based on these observations, construct a model for the interior mechanism of the box.
    b. What further experiments could you do to refine your model?

65. Each of the following statements is true, but Dalton might have had trouble explaining some of them with his atomic theory. Give explanations for the following statements.
    a. The space-filling models for ethyl alcohol and dimethyl ether are shown below.

● C
● O
● H

    These two compounds have the same composition by mass (52% carbon, 13% hydrogen, and 35% oxygen), yet the two have different melting points, boiling points, and solubilities in water.
    b. Burning wood leaves an ash that is only a small fraction of the mass of the original wood.
    c. Atoms can be broken down into smaller particles.
    d. One sample of lithium hydride is 87.4% lithium by mass, while another sample of lithium hydride is 74.9% lithium by mass. However, the two samples have the same chemical properties.

66. Reaction of 2.0 L of hydrogen gas with 1.0 L of oxygen gas yields 2.0 L of water vapor. All gases are at the same temperature and pressure. Show how these data support the idea that oxygen gas is a diatomic molecule. Must we consider hydrogen to be a diatomic molecule to explain these results?

67. A combustion reaction involves the reaction of a substance with oxygen gas. The complete combustion of any hydrocarbon (binary compound of carbon and hydrogen) produces carbon dioxide and water as the only products. Octane is a hydrocarbon that is found in gasoline. Complete combustion of octane produces 8 liters of carbon dioxide for every 9 liters of water vapor (both measured at the same temperature and

pressure). What is the ratio of carbon atoms to hydrogen atoms in a molecule of octane?

68. A chemistry instructor makes the following claim: "Consider that if the nucleus were the size of a grape, the electrons would be about 1 *mile* away on average." Is this claim reasonably accurate? Provide mathematical support.

69. You have two distinct gaseous compounds made from element X and element Y. The mass percents are as follows:

Compound I: 30.43% X, 69.57% Y
Compound II: 63.64% X, 36.36% Y

In their natural standard states, element X and element Y exist as gases. (Monatomic? Diatomic? Triatomic? That is for you to determine.) When you react "gas X" with "gas Y" to make the products, you get the following data (all at the same pressure and temperature):

1 volume "gas X" + 2 volumes "gas Y" $\longrightarrow$
2 volumes compound I

2 volumes "gas X" + 1 volume "gas Y" $\longrightarrow$
2 volumes compound II

Assume the simplest possible formulas for reactants and products in the chemical equations above. Then, determine the relative atomic masses of element X and element Y.

Integrative Problems

70. Using the information in Table 1-1, answer the following questions. In an ion with an unknown charge, the total mass of all the electrons was determined to be 2.55×10^{-26} g, while the total mass of its protons was 5.34×10^{-23} g. What is the identity and charge of this ion? What is the symbol and mass number of a neutral atom whose total mass of its electrons is 3.92×10^{-26} g, while its neutrons have a mass of 9.35×10^{-23} g?

71. A single molecule has a mass of 7.31×10^{-23} g. Provide an example of a real molecule that can have this mass. Assume the elements that make up the molecule are made of light isotopes where the number of protons equals the number of neutrons in the nucleus of each element.

Marathon Problem

This problem is designed to incorporate several concepts and techniques into one situation.

72. You have gone back in time and are working with Dalton on a table of relative masses. Following are his data.

0.602 g gas A reacts with 0.295 g gas B
0.172 g gas B reacts with 0.401 g gas C
0.320 g gas A reacts with 0.374 g gas C

a. Assuming simplest formulas (AB, BC, and AC), construct a table of relative masses for Dalton.

b. Knowing some history of chemistry, you tell Dalton that if he determines the volumes of the gases reacted at constant temperature and pressure, he need not assume simplest formulas. You collect the following data:

6 volumes gas A + 1 volume gas B → 4 volumes product
1 volume gas B + 4 volumes gas C → 4 volumes product
3 volumes gas A + 2 volumes gas C → 6 volumes product

Write the simplest balanced equations, and find the actual relative masses of the elements. Explain your reasoning.

C H A P T E R   T W O

Atomic Structure and Periodicity

The line spectra of a number of gases. Each spectrum is unique and allows the identification of the elements. (Ted Kinsman/Science Source)

n the past 200 years, a great deal of experimental evidence has accumulated to support the atomic model. This theory has proved to be both extremely useful and physically reasonable. When atoms were first suggested by the Greek philosophers Democritus and Leucippus about 400 B.C., the concept was based mostly on intuition. In fact, for the following 20 centuries, no convincing experimental evidence was available to support the existence of atoms. The first real scientific data were gathered by Lavoisier and others from quantitative measurements of chemical reactions. The results of these stoichiometric experiments led John Dalton to propose the first systematic atomic theory. Dalton's theory, although crude, has stood the test of time extremely well.

While Dalton's theory is useful in many ways, it fails to explain some relatively common observations. Consider, for example, fireworks. The colors are due to different metal ions in the salts used in the fireworks. But this leads us to ask questions such as "why do these salts emit colors instead of 'white light'?," and "why are there different colors for different salts?" These types of questions require us to consider the nature of light and the nature of the atom. To answer these questions we need a more complicated model. In Chapter 1 we considered some of the experiments most important for shedding light on the nature of the atom. Now we will see how the atomic theory has evolved to its present state.

One of the most striking things about the chemistry of the elements is the periodic repetition of properties. There are several groups of elements that show great similarities in chemical behavior. In fact, these similarities led to the development of the periodic table of the elements as we will discuss in this chapter. That is, the periodic table was developed before atomic structure was known—it was based on observations and properties. As we will see, there is a relationship among the properties of the elements in a column (also known as a group). In this chapter we will see that the modern theory of atomic structure accounts for periodicity in terms of the electron arrangements in atoms.

However, before we examine atomic structure, we must consider the revolution that took place in physics in the first 30 years of the twentieth century. During that time, experiments were carried out, the results of which could not be explained by the theories of classical physics developed by Isaac Newton and many others who followed him. A radical new theory called *quantum mechanics* was developed to account for the behavior of light and atoms. This "new physics" provides many surprises for humans who are used to the macroscopic world, but it seems to account flawlessly (within the bounds of necessary approximations) for the behavior of matter.

As the first step in our exploration of this revolution in science we will consider the properties of light, more properly called *electromagnetic radiation*.

2-1 | Electromagnetic Radiation

One of the ways that energy travels through space is by **electromagnetic radiation**. The light from the sun, the energy used to cook food in a microwave oven, the X rays used by dentists, and the radiant heat from a fireplace are all examples of electromagnetic radiation. Although these forms of radiant energy seem quite different, they all exhibit the same type of wavelike behavior and travel at the speed of light in a vacuum.

Waves have three primary characteristics: wavelength, frequency, and speed. **Wavelength** (symbolized by the lowercase Greek letter lambda, λ) is the *distance*

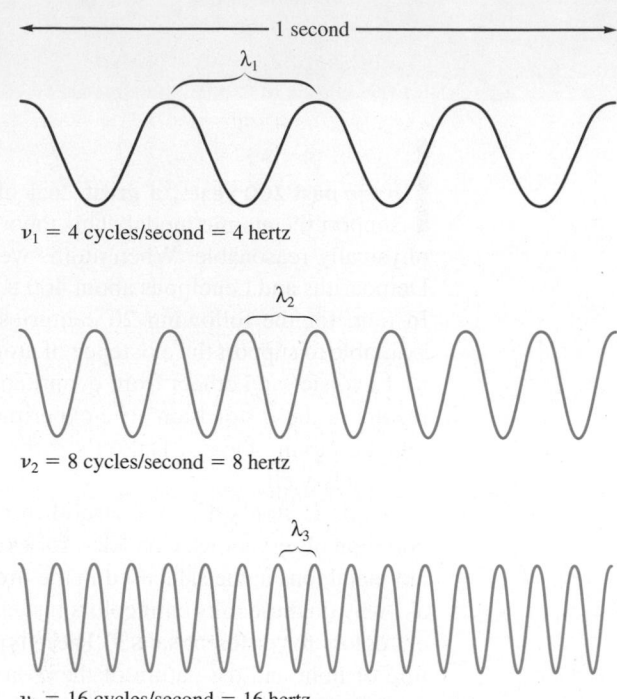

$\nu_1 = 4$ cycles/second = 4 hertz

$\nu_2 = 8$ cycles/second = 8 hertz

$\nu_3 = 16$ cycles/second = 16 hertz

Figure 2-1 | The nature of waves. Many of the properties of ocean waves are the same as those of light waves. Note that the radiation with the shortest wavelength has the highest frequency.

Wavelength λ and frequency ν are inversely related.

c = speed of light
= 2.9979×10^8 m/s

between two consecutive peaks or troughs in a wave, as shown in Fig. 2-1. The **frequency** (symbolized by the lowercase Greek letter nu, ν) is defined as the *number of waves (cycles) per second that pass a given point in space.* Since all types of electromagnetic radiation travel at the speed of light, short-wavelength radiation must have a high frequency. You can see this in Fig. 2-1, where three waves are shown traveling between two points at constant speed. Note that the wave with the shortest wavelength (λ_3) has the highest frequency and the wave with the longest wavelength (λ_1) has the lowest frequency. ◀ ⓘ This implies an inverse relationship between wavelength and frequency, that is, $\lambda \propto 1/\nu$, or

$$\lambda\nu = c$$

where λ is the wavelength in meters, ν is the frequency in cycles per second, and c is the speed of light (2.9979×10^8 m/s). ◀ ⓘ In the SI system, *cycles* is understood, and the unit per second becomes 1/s, or s^{-1}, which is called the *hertz* (abbreviated Hz).

Electromagnetic radiation is classified as shown in Fig. 2-2. Radiation provides an important means of energy transfer. For example, the energy from the sun reaches the

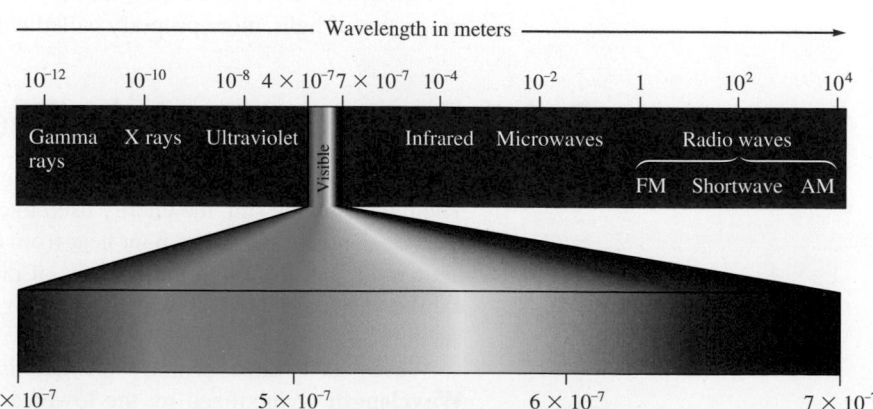

Figure 2-2 | Classification of electromagnetic radiation.

earth mainly in the form of visible and ultraviolet radiation, whereas the glowing coals of a fireplace transmit heat energy by infrared radiation. In a microwave oven the water molecules in food absorb microwave radiation, which increases their motions. This energy is then transferred to other types of molecules via collisions, causing an increase in the food's temperature. As we proceed in the study of chemistry, we will consider many of the classes of electromagnetic radiation and the ways in which they affect matter.

INTERACTIVE EXAMPLE 2-1

ⓘ Consider these observations. Why do we see colors when a metallic salt is dissolved in methanol and ignited or when a pickle is electrified? Why is the light not white? Why do different metals (such as strontium and sodium) produce different colors? We can answer these questions by better understanding atomic structure.

Frequency of Electromagnetic Radiation

The brilliant red colors seen in fireworks are due to the emission of light with wavelengths around 650 nm when strontium salts such as $Sr(NO_3)_2$ and $SrCO_3$ are heated. ◄ ◄ ⓘ (This can be easily demonstrated in the lab by dissolving one of these salts in methanol that contains a little water and igniting the mixture in an evaporating dish.) Calculate the frequency of red light of wavelength 6.50×10^2 nm.

Solution

We can convert wavelength to frequency using the equation

$$\lambda \nu = c \quad \text{or} \quad \nu = \frac{c}{\lambda}$$

where $c = 2.9979 \times 10^8$ m/s. In this case $\lambda = 6.50 \times 10^2$ nm. Changing the wavelength to meters, we have

$$6.50 \times 10^2 \; \text{nm} \times \frac{1 \; \text{m}}{10^9 \; \text{nm}} = 6.50 \times 10^{-7} \; \text{m}$$

and

$$\nu = \frac{c}{\lambda} = \frac{2.9979 \times 10^8 \; \text{m/s}}{6.50 \times 10^{-7} \; \text{m}} = 4.61 \times 10^{14} \; \text{s}^{-1} = 4.61 \times 10^{14} \; \text{Hz}$$

See Exercises 2-39 and 2-40

▲ When a strontium salt is dissolved in methanol (with a little water) and ignited, it gives a brilliant red flame. The red color is produced by emission of light when electrons, excited by the energy of the burning methanol, fall back to their ground states.

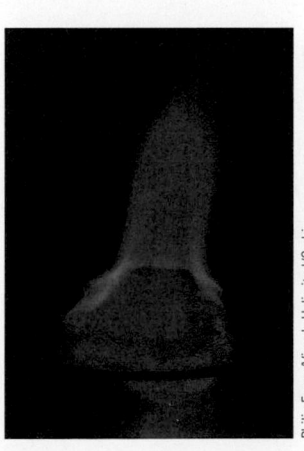

Philip Evans/Visuals Unlimited/Corbis

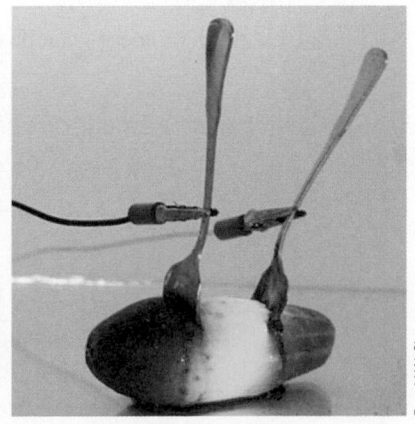

Donald W. Clegg

▲ When alternating current at 110 volts is applied to a dill pickle, a glowing discharge occurs. The current flowing between the electrodes (forks), which is supported by the Na^+ and Cl^- ions present, apparently causes some sodium atoms to form in an excited state. When these atoms relax to the ground state, they emit visible light at 589 nm, producing the yellow glow reminiscent of sodium vapor lamps.

2-2 | The Nature of Matter

It is probably fair to say that at the end of the nineteenth century, physicists were feeling rather smug. Theories could explain phenomena as diverse as the motions of the planets and the dispersion of visible light by a prism. Rumor has it that students were being discouraged from pursuing physics as a career because it was felt that all the major problems had been solved, or at least described in terms of the current physical theories.

At the end of the nineteenth century, the idea prevailed that matter and energy were distinct. Matter was thought to consist of particles, whereas energy in the form of light (electromagnetic radiation) was described as a wave. Particles were things that had mass and whose position in space could be specified. Waves were described as massless and delocalized; that is, their position in space could not be specified. It also was assumed that there was no intermingling of matter and light. Everything known before 1900 seemed to fit neatly into this view.

At the beginning of the twentieth century, however, certain experimental results suggested that this picture was incorrect. The first important advance came in 1900 from the German physicist Max Planck (1858–1947). Studying the radiation profiles emitted by solid bodies heated to incandescence, Planck found that the results could not be explained in terms of the physics of his day, which held that matter could absorb or emit any quantity of energy. Planck could account for these observations only by postulating that energy can be gained or lost only in *whole-number multiples* of the quantity $h\nu$, where h is a constant called **Planck's constant**, determined by experiment to have the value 6.626×10^{-34} J · s. That is, the change in energy for a system ΔE can be represented by the equation

$$\Delta E = nh\nu$$

where n is an integer (1, 2, 3, . . .), h is Planck's constant, and ν is the frequency of the electromagnetic radiation absorbed or emitted. ◀ *i*

Planck's result was a real surprise. It had always been assumed that the energy of matter was continuous, which meant that the transfer of any quantity of energy was possible. Now it seemed clear that energy is in fact **quantized** and can occur only in discrete units of size $h\nu$. Each of these small "packets" of energy is called a *quantum*. A system can transfer energy only in whole quanta. Thus energy seems to have particulate properties. ◀

i Energy can be gained or lost only in integer multiples of $h\nu$.

Planck's constant = 6.626×10^{-34} J · s.

INTERACTIVE EXAMPLE 2-2

The Energy of a Photon

The blue color in fireworks is often achieved by heating copper(I) chloride (CuCl) to about 1200°C. Then the compound emits blue light having a wavelength of 450 nm. What is the increment of energy (the quantum) that is emitted at 4.50×10^2 nm by CuCl?

Solution

The quantum of energy can be calculated from the equation

$$\Delta E = h\nu$$

The frequency ν for this case can be calculated as follows:

$$\nu = \frac{c}{\lambda} = \frac{2.9979 \times 10^8 \text{ m/s}}{4.50 \times 10^{-7} \text{ m}} = 6.66 \times 10^{14} \text{ s}^{-1}$$

So

$$\Delta E = h\nu = (6.626 \times 10^{-34} \text{ J} \cdot \text{s})(6.66 \times 10^{14} \text{ s}^{-1}) = 4.41 \times 10^{-19} \text{ J}$$

A sample of CuCl emitting light at 450 nm can lose energy only in increments of 4.41×10^{-19} J, the size of the quantum in this case.

See Exercises 2-41 and 2-42

Science Source/Getty Images

Figure 2-3 | Albert Einstein (1879–1955) was born in Germany. Nothing in his early development suggested genius; even at the age of 9 he did not speak clearly, and his parents feared that he might be handicapped. When asked what profession Einstein should follow, his school principal replied, "It doesn't matter; he'll never make a success of anything." When he was 10, Einstein entered the Luitpold Gymnasium (high school), which was typical of German schools of that time in being harshly disciplinarian. There he developed a deep suspicion of authority and a skepticism that encouraged him to question and doubt—valuable qualities in a scientist. In 1905, while a patent clerk in Switzerland, Einstein published a paper explaining the photoelectric effect via the quantum theory. For this revolutionary thinking he received a Nobel Prize in 1921. Highly regarded by this time, he worked in Germany until 1933, when Hitler's persecution of the Jews forced him to come to the United States. He worked at the Institute for Advanced Studies in Princeton, New Jersey, until his death in 1955.

Einstein was undoubtedly the greatest physicist of our age. Even if someone else had derived the theory of relativity, his other work would have ensured his ranking as the second greatest physicist of his time. Our concepts of space and time were radically changed by ideas he first proposed when he was 26 years old. From then until the end of his life, he attempted unsuccessfully to find a single unifying theory that would explain all physical events.

The next important development in the knowledge of atomic structure came when Albert Einstein (Fig. 2-3) proposed that electromagnetic radiation is itself quantized. Einstein suggested that electromagnetic radiation can be viewed as a stream of "particles" called **photons**. The energy of each photon is given by the expression

$$E_{\text{photon}} = h\nu = \frac{hc}{\lambda}$$

where h is Planck's constant, ν is the frequency of the radiation, and λ is the wavelength of the radiation.

The Photoelectric Effect

Einstein arrived at this conclusion through his analysis of the **photoelectric effect** (for which he later was awarded the Nobel Prize). The photoelectric effect refers to the phenomenon in which electrons are emitted from the surface of a metal when light strikes it. The following observations characterize the photoelectric effect.

1. Studies in which the frequency of the light is varied show that no electrons are emitted by a given metal below a specific threshold frequency ν_0.
2. For light with frequency lower than the threshold frequency, no electrons are emitted, regardless of the intensity of the light.
3. For light with frequency greater than the threshold frequency, the number of electrons emitted increases with the intensity of the light.
4. For light with frequency greater than the threshold frequency, the kinetic energy of the emitted electrons increases linearly with the frequency of the light.

These observations can be explained by assuming that electromagnetic radiation is quantized (consists of photons) and that the threshold frequency represents the minimum energy required to remove the electron from the metal's surface.

Minimum energy required to remove an electron $= E_0 = h\nu_0$

Because a photon with energy less than E_0 ($\nu < \nu_0$) cannot remove an electron, light with a frequency less than the threshold frequency produces no electrons (Fig. 2-4).

Low-frequency (low-energy) light

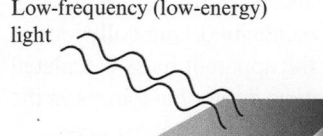

Metal surface

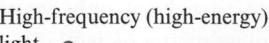
a

High-frequency (high-energy) light

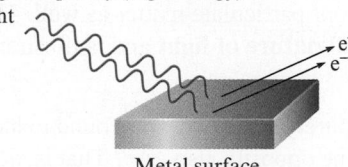

Metal surface

b

Figure 2-4 | The photoelectric effect. (a) Light with frequency less than the threshold frequency produces no electrons. (b) Light with frequency higher than the threshold frequency causes electrons to be emitted from the metal.

On the other hand, for light where $\nu > \nu_0$, the energy in excess of that required to remove the electron is given to the electron as kinetic energy (KE):

$$\mathrm{KE}_{\text{electron}} = \tfrac{1}{2}mv^2 = h\nu - h\nu_0$$

Mass of   Velocity   Energy of   Energy required
electron   of   incident   to remove electron
electron   photon   from metal's surface

Because in this picture the intensity of light is a measure of the number of photons present in a given part of the beam, a greater intensity means that more photons are available to release electrons (as long as $\nu > \nu_0$ for the radiation).

In a related development, Einstein derived the famous equation

$$E = mc^2$$

in his *special theory of relativity* published in 1905. The main significance of this equation is that *energy has mass*. This is more apparent if we rearrange the equation in the following form:

$$m = \frac{E}{c^2} \leftarrow \text{Energy}$$

Mass   Speed of light

Using this form of the equation, we can calculate the mass associated with a given quantity of energy. For example, we can calculate the *apparent* mass of a photon. ◄ ⓘ For electromagnetic radiation of wavelength λ, the energy of each photon is given by the expression

ⓘ Note that the *apparent mass* of a photon depends on its wavelength. A photon does not have mass in a classical sense.

$$E_{\text{photon}} = \frac{hc}{\lambda}$$

Then the apparent mass of a photon of light with wavelength λ is given by

$$m = \frac{E}{c^2} = \frac{hc/\lambda}{c^2} = \frac{h}{\lambda c}$$

The photon behaves as if it has mass under certain circumstances. In 1922 American physicist Arthur Compton (1892–1962) performed experiments involving collisions of X rays and electrons that showed that photons do exhibit the apparent mass calculated from the preceding equation. However, it is clear that photons do not have mass in the classical sense. A photon has mass only in a relativistic sense—it has no rest mass.

We can summarize the important conclusions from the work of Planck and Einstein as follows:

Energy is quantized. It can occur only in discrete units called quanta.

Electromagnetic radiation, which was previously thought to exhibit only wave properties, seems to show certain characteristics of particulate matter as well. This phenomenon is sometimes referred to as the **dual nature of light** and is illustrated in Fig. 2-5.

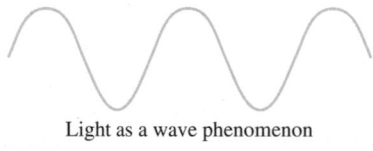

Light as a wave phenomenon

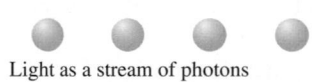

Light as a stream of photons

Figure 2-5 | Electromagnetic radiation exhibits wave properties and particulate properties. The energy of each photon of the radiation is related to the wavelength and frequency by the equation $E_{\text{photon}} = h\nu = hc/\lambda$.

Thus light, which previously was thought to be purely wavelike, was found to have certain characteristics of particulate matter. But is the opposite also true? That is, does matter that is normally assumed to be particulate exhibit wave properties? This question was raised in 1923 by a young French physicist named Louis de Broglie (1892–1987). To see how de Broglie supplied the answer to this question, recall that the relationship between mass and wavelength for electromagnetic radiation is $m = h/\lambda c$. For a particle with velocity v, the corresponding expression is

$$m = \frac{h}{\lambda v}$$

In the animal world, the ability to see at night provides predators with a distinct advantage over their prey. The same advantage can be gained by military forces and law enforcement agencies around the world through the use of recent advances in night vision technology.

All types of night vision equipment are electro-optical devices that amplify existing light. A lens collects light and focuses it on an image intensifier. The image intensifier is based on the photoelectric effect—materials that give off electrons when light is shined on them. Night vision intensifiers use semiconductor-based materials to produce large numbers of electrons for a given input of photons. The emitted electrons are then directed onto a screen covered with compounds that phosphoresce (glow when struck by electrons). While television tubes use various phosphors to produce color pictures, night vision devices use phosphors that appear green, because the human eye can distinguish more shades of green than any other color. The viewing screen shows an image that otherwise would be invisible to the naked eye during nighttime viewing.

Current night vision devices use gallium arsenide (GaAs)–based intensifiers that can amplify input light as much as 50,000 times. These devices are so sensitive they can use starlight to produce an image. It is also now possible to use light (infrared) that cannot be sensed with the human eye to create an image.

This technology, while developed originally for military and law enforcement applications, is now becoming available to the consumer. For example, Cadillac has recently included night vision as an option on its cars. As night-imaging technology improves and costs become less prohibitive, a whole new world is opening up for the technophile—after the sun goes down.

Pilots using their night vision goggles to guide their approach to the runway on a training mission

U.S. Air Force photo/Shannon McKay

Rearranging to solve for λ, we have

$$\lambda = \frac{h}{mv}$$

This equation, called *de Broglie's equation*, allows us to calculate the wavelength for a particle, as shown in Example 2-3. ◄ 🛈

🛈 Do not confuse ν (frequency) with v (velocity).

INTERACTIVE EXAMPLE 2-3

Calculations of Wavelength

Compare the wavelength for an electron (mass = 9.11×10^{-31} kg) traveling at a speed of 1.0×10^7 m/s with that for a ball (mass = 0.10 kg) traveling at 35 m/s.

Solution

We use the equation $\lambda = h/mv$, where

$$h = 6.626 \times 10^{-34} \text{ J} \cdot \text{s} \quad \text{or} \quad 6.626 \times 10^{-34} \text{ kg} \cdot \text{m}^2/\text{s}$$

since

$$1 \text{ J} = 1 \text{ kg} \cdot \text{m}^2/\text{s}^2$$

For the electron,

$$\lambda_e = \frac{6.626 \times 10^{-34} \dfrac{\text{kg} \cdot \cancel{\text{m}} \cdot \text{m}}{\cancel{\text{s}}}}{(9.11 \times 10^{-31} \text{ kg})(1.0 \times 10^7 \, \cancel{\text{m/s}})} = 7.27 \times 10^{-11} \text{ m}$$

For the ball,

$$\lambda_b = \frac{6.626 \times 10^{-34} \dfrac{\text{kg} \cdot \cancel{\text{m}} \cdot \text{m}}{\cancel{\text{s}}}}{(0.10 \text{ kg})(35 \, \cancel{\text{m/s}})} = 1.9 \times 10^{-34} \text{ m}$$

See Exercises 2-53 through 2-56

Notice from Example 2-3 that the wavelength associated with the ball is incredibly short. On the other hand, the wavelength of the electron, although still quite small, happens to be on the same order as the spacing between the atoms in a typical crystal. This is important because, as we will see presently, it provides a means for testing de Broglie's equation.

Diffraction results when light is scattered from a regular array of points or lines. You may have noticed the diffraction of light from the ridges and grooves of a compact disc. The colors result because the various wavelengths of visible light are not all scattered in the same way. The colors are "separated," giving the same effect as light passing through a prism. Just as a regular arrangement of ridges and grooves produces diffraction, so does a regular array of atoms or ions in a crystal, as shown in the photograph below. For example, when X rays are directed onto a crystal of a particular nickel/titanium alloy, the scattered radiation produces a **diffraction pattern** of bright spots and dark areas on a photographic plate, as shown in Fig. 2-6. This occurs because the scattered light can interfere constructively (the peaks and troughs of the beams are in phase) to produce a bright area [Fig. 2-6(a)] or destructively (the peaks and troughs are out of phase) to produce a dark spot [Fig. 2-6(b)].

A diffraction pattern can be explained only in terms of waves. Thus this phenomenon provides a test for the postulate that particles such as electrons have wavelengths. As we saw in Example 2-3, an electron with a velocity of 10^7 m/s (easily achieved by

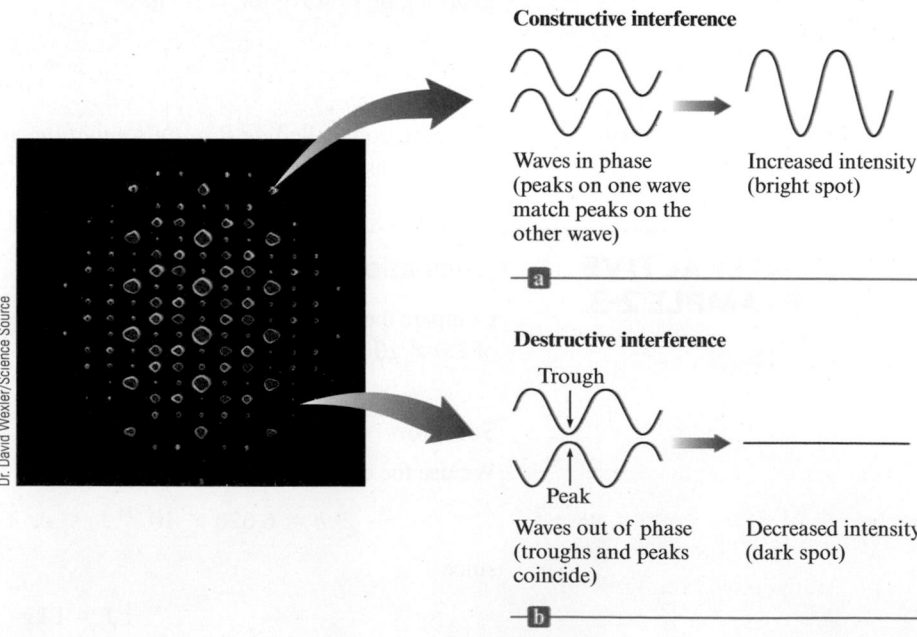

Figure 2-6 | A diffraction pattern of a beryl crystal. (a) A light area results from constructive interference of the light waves. (b) A dark area arises from destructive interference of the waves.

Dr. David Wexler/Science Source

Constructive interference

Waves in phase (peaks on one wave match peaks on the other wave)

Increased intensity (bright spot)

a

Destructive interference

Trough

Peak

Waves out of phase (troughs and peaks coincide)

Decreased intensity (dark spot)

b

acceleration of the electron in an electric field) has a wavelength of about 10^{-10} m, which is roughly the distance between the ions in a crystal such as sodium chloride. This is important because diffraction occurs most efficiently when the spacing between the scattering points is about the same as the wavelength of the wave being diffracted. Thus, if electrons really do have an associated wavelength, a crystal should diffract electrons. An experiment to test this idea was carried out in 1927 by C. J. Davisson and L. H. Germer at Bell Laboratories. When they directed a beam of electrons at a nickel crystal, they observed a diffraction pattern similar to that seen from the diffraction of X rays. This result verified de Broglie's relationship, at least for electrons. Larger chunks of matter, such as balls, have such small wavelengths (see Example 2-3) that they are impossible to verify experimentally. However, we believe that all matter obeys de Broglie's equation.

Now we have come full circle. Electromagnetic radiation, which at the turn of the twentieth century was thought to be a pure waveform, was found to possess particulate properties. Conversely, electrons, which were thought to be particles, were found to have a wavelength associated with them. The significance of these results is that matter and energy are not distinct. Energy is really a form of matter, and all matter shows the same types of properties. That is, *all matter exhibits both particulate and wave properties*. Large pieces of matter, such as baseballs, exhibit predominantly particulate properties. The associated wavelength is so small that it is not observed. Very small "bits of matter," such as photons, while showing some particulate properties, exhibit predominantly wave properties. Pieces of matter with intermediate mass, such as electrons, show clearly both the particulate and wave properties of matter.

2-3 | The Atomic Spectrum of Hydrogen

As we saw in Chapter 1, key information about the atom came from several experiments carried out in the early twentieth century, in particular Thomson's discovery of the electron and Rutherford's discovery of the nucleus. Another important experiment was the study of the emission of light by excited hydrogen atoms. When a sample of hydrogen gas receives a high-energy spark, the H_2 molecules absorb energy, and some of the H—H bonds are broken. The resulting hydrogen atoms are *excited*; that is, they contain excess energy, which they release by emitting light of various wavelengths to produce what is called the *emission spectrum* of the hydrogen atom.

To understand the significance of the hydrogen emission spectrum, we must first describe the **continuous spectrum** that results when white light is passed through a prism, as shown in Fig. 2-7(a). This spectrum, like the rainbow produced when sunlight is dispersed by raindrops, contains *all* the wavelengths of visible light. ◀ In contrast, when the hydrogen emission spectrum in the visible region is passed through a prism, as shown in Fig. 2-7(b), we see only a few lines, each of which corresponds to a discrete wavelength. The hydrogen emission spectrum is called a **line spectrum**. ◀ ⓘ

What is the significance of the line spectrum of hydrogen? It indicates that *only certain energies are allowed for the electron in the hydrogen atom*. In other words, the energy of the electron in the hydrogen atom is *quantized*. This observation ties in perfectly with the postulates of Max Planck. Changes in energy between discrete energy levels in hydrogen will produce only certain wavelengths of emitted light, as shown in Fig. 2-8. For example, a given change in energy from a high to a lower level would give a wavelength of light that can be calculated from Planck's equation:

$$\Delta E = h\nu = \frac{hc}{\lambda}$$

Change in energy   Frequency of light emitted   Wavelength of light emitted

▲ A beautiful rainbow.

Jupiter Images/Getty Images

ⓘ The line spectrum of hydrogen (which we can see) is explained by quantized energy levels (which we cannot see).

The art of using mixtures of chemicals to produce explosives is an ancient one. Black powder—a mixture of potassium nitrate, charcoal, and sulfur—was being used in China well before 1000 A.D. and has been used subsequently through the centuries in military explosives, in construction blasting, and for fireworks. The DuPont Company, now a major chemical manufacturer, started out as a manufacturer of black powder. In fact, the founder, Eleuthère duPont, learned the manufacturing technique from none other than Lavoisier.

Before the nineteenth century, fireworks were confined mainly to rockets and loud bangs. Orange and yellow colors came from the presence of charcoal and iron filings. However, with the great advances in chemistry in the nineteenth century, new compounds found their way into fireworks. Salts of copper, strontium, and barium added brilliant colors. Magnesium and aluminum metals gave a dazzling white light. Fireworks, in fact, have changed very little since then.

How do fireworks produce their brilliant colors and loud bangs? Actually, only a handful of different chemicals are responsible for most of the spectacular effects. To produce the noise and flashes, an oxidizer (an oxidizing agent) and a fuel (a reducing agent) are used. A common mixture involves potassium perchlorate ($KClO_4$) as the oxidizer and aluminum and sulfur as the fuel. The perchlorate oxidizes the fuel in a very exothermic reaction, which produces a brilliant flash, due to the aluminum, and a loud report from the rapidly expanding gases produced. For a color effect, an element with a colored emission spectrum is included. Recall that the electrons in atoms can be raised to higher-energy orbitals when the atoms absorb energy. The excited atoms can then release this excess energy by emitting light of specific wavelengths, often in the visible region. In fireworks, the energy to excite the electrons comes from the reaction between the oxidizer and fuel.

Yellow colors in fireworks are due to the 589-nm emission of sodium ions. Red colors come from strontium salts emitting at 606 nm and from 636 to 688 nm. This red color is familiar from highway safety flares. Barium salts give a green color in fireworks, due to a series of emission lines between 505 and 535 nm. A really good blue color, however, is hard to

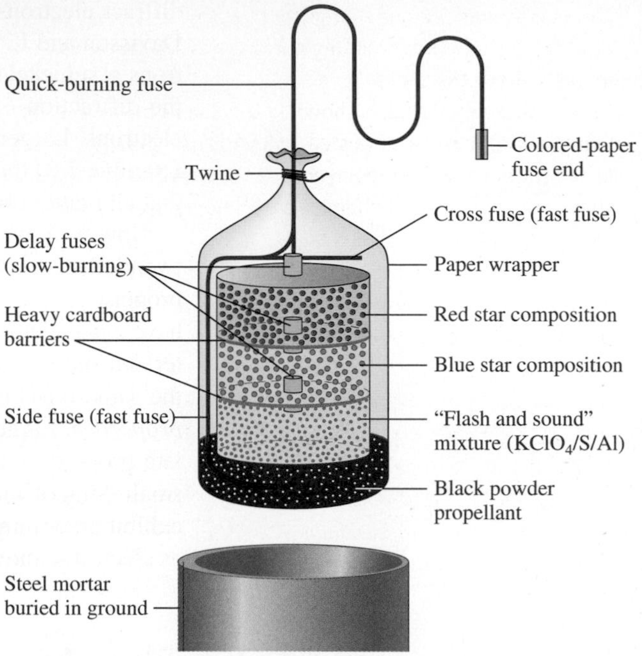

A typical aerial shell used in fireworks displays. Time-delayed fuses cause a shell to explode in stages. In this case a red starburst occurs first, followed by a blue starburst, and finally a flash and loud report.
(Based on *Chemical & Engineering News*, June 29, 1981, p. 24.)

obtain. Copper salts give a blue color, emitting in the 420- to 460-nm region. But difficulties occur because the oxidizing agent, potassium chlorate ($KClO_3$), reacts with copper salts to form copper chlorate, a highly explosive compound that is dangerous to store. (The use of $KClO_3$ in fireworks has been largely abandoned because of its explosive hazards.) Paris green, a copper salt containing arsenic, was once used extensively but is now considered to be too toxic.

In recent years the colors produced by fireworks have become more intense because of the formation of metal chlorides during the burning process. These gaseous metal chloride molecules produce colors much more brilliant than do the metal atoms by themselves. For example, strontium chloride produces a much brighter red than do strontium atoms. Thus chlorine-donating compounds are now included in many fireworks shells.

A typical aerial shell is shown in the diagram. The shell is launched from a mortar (a steel cylinder) using black powder as the propellant. Time-delayed fuses are used to fire the shell in stages. A list of chemicals commonly used in fireworks is given in the table.

Although you might think that the chemistry of fireworks is simple, the achievement of the vivid white flashes and the brilliant colors requires complex combinations of chemicals. For example, because the white flashes produce high flame temperatures, the colors tend to wash out. Thus oxidizers such as $KClO_4$ are commonly used with fuels that produce relatively low flame temperatures. An added difficulty, however, is that perchlorates are very sensitive to accidental ignition and are therefore quite hazardous. Another problem arises from the use of sodium salts. Because sodium produces an extremely bright yellow emission, sodium salts cannot be used when other colors are desired. Carbon-based fuels also give a yellow flame that masks other colors, and this limits the use of organic compounds as fuels. You can see that the manufacture of fireworks that produce the desired effects and are also safe to handle requires careful selection of chemicals. And, of course, there is still the dream of a deep blue flame.

Lance Gerrard

Reflection of fireworks on a lake in France

Chemicals Commonly Used in the Manufacture of Fireworks

Oxidizers	Fuels	Special Effects
Potassium nitrate	Aluminum	Red flame: strontium nitrate, strontium carbonate
Potassium chlorate	Magnesium	Green flame: barium nitrate, barium chlorate
Potassium perchlorate	Titanium	Blue flame: copper carbonate, copper sulfate, copper oxide
Ammonium perchlorate	Charcoal	Yellow flame: sodium oxalate, cryolite (Na_3AlF_6)
Barium nitrate	Sulfur	White flame: magnesium, aluminum
Barium chlorate	Antimony sulfide	Gold sparks: iron filings, charcoal
Strontium nitrate	Dextrin	White sparks: aluminum, magnesium, aluminum–magnesium alloy, titanium
	Red gum	Whistle effect: potassium benzoate or sodium salicylate
	Polyvinyl chloride	White smoke: mixture of potassium nitrate and sulfur
		Colored smoke: mixture of potassium chlorate, sulfur, and organic dye

Figure 2-7 | (a) A continuous spectrum containing all wavelengths of visible light (indicated by the initial letters of the colors of the rainbow). (b) The hydrogen line spectrum contains only a few discrete wavelengths.

Spectrum based on C. W. Keenan, D. C. Kleinfelter, and J. H. Wood, *General College Chemistry*, Sixth Edition (New York: Harper & Row, 1980).

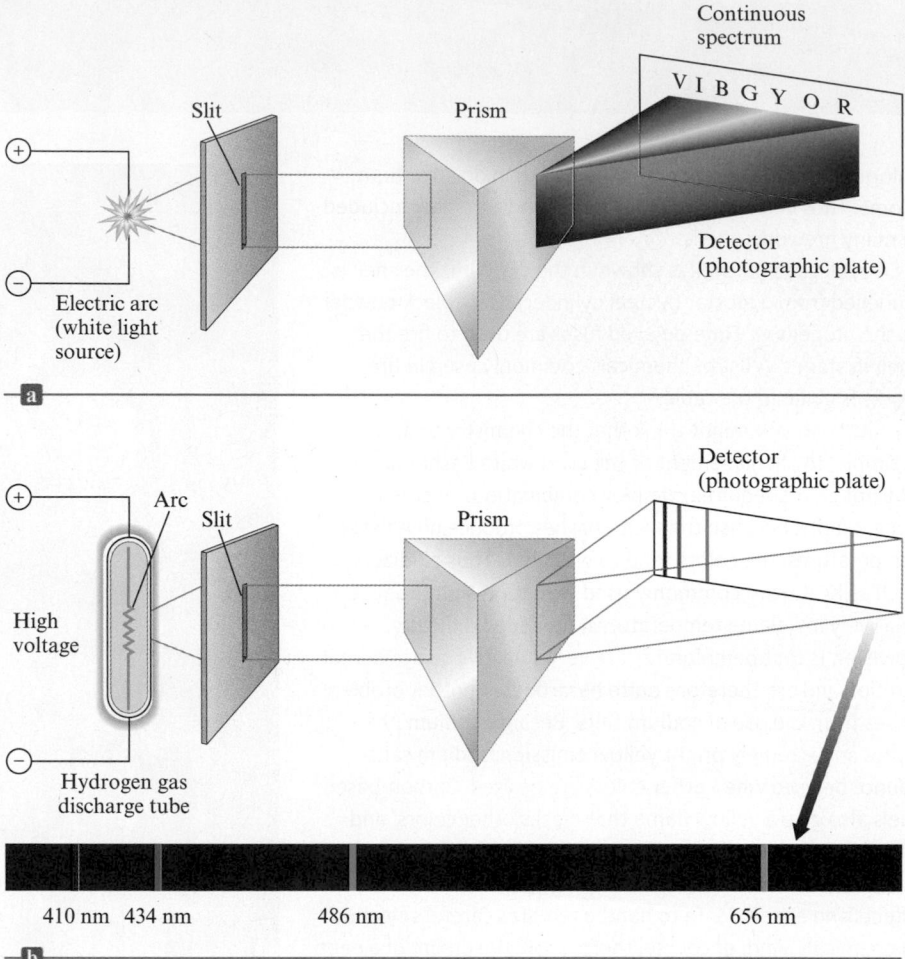

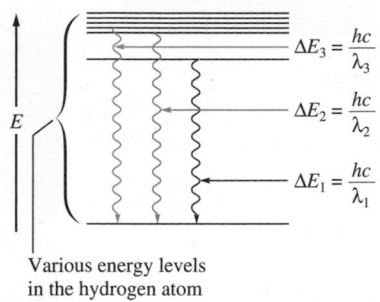

Various energy levels in the hydrogen atom

Figure 2-8 | A change between two discrete energy levels emits a photon of light.

The discrete line spectrum of hydrogen shows that only certain energies are possible; that is, the electron energy levels are quantized. In contrast, if any energy level were allowed, the emission spectrum would be continuous.

Critical Thinking

We now have evidence that electron energy levels in the atoms are quantized. Some of this evidence is discussed in this chapter. What if energy levels in atoms were not quantized? What are some differences we would notice?

2-4 | The Bohr Model

In 1913, a Danish physicist named Niels Bohr (1885–1962), aware of the experimental results we have just discussed, developed a **quantum model** for the hydrogen atom. Bohr proposed that the *electron in a hydrogen atom moves around the nucleus only in certain allowed circular orbits*. He calculated the radii for these allowed orbits by using the theories of classical physics and by making some new assumptions. ◄

From classical physics Bohr knew that a particle in motion tends to move in a straight line and can be made to travel in a circle only by application of a force toward the center of the circle. Thus Bohr reasoned that the tendency of the revolving electron

▲ Niels Hendrik David Bohr (1885–1962) as a boy lived in the shadow of his younger brother Harald, who played on the 1908 Danish Olympic Soccer Team and later became a distinguished mathematician. In school, Bohr received his poorest marks in composition and struggled with writing during his entire life. In fact, he wrote so poorly that he was forced to dictate his Ph.D. thesis to his mother. Nevertheless, Bohr was a brilliant physicist. After receiving his Ph.D. in Denmark, he constructed a quantum model for the hydrogen atom by the time he was 27. Even though his model later proved to be incorrect, Bohr remained a central figure in the drive to understand the atom. He was awarded the Nobel Prize in physics in 1922.

to fly off the atom must be just balanced by its attraction for the positively charged nucleus. But classical physics also decreed that a charged particle under acceleration should radiate energy. Since an electron revolving around the nucleus constantly changes its direction, it is constantly accelerating. Therefore, the electron should emit light and lose energy—and thus be drawn into the nucleus. This, of course, does not correlate with the existence of stable atoms.

Clearly, an atomic model based solely on the theories of classical physics was untenable. Bohr also knew that the correct model had to account for the experimental spectrum of hydrogen, which showed that only certain electron energies were allowed. The experimental data were absolutely clear on this point. Bohr found that his model would fit the experimental results if he assumed that the angular momentum of the electron (angular momentum equals the product of mass, velocity, and orbital radius) could occur only in certain increments. It was not clear why this should be true, but with this assumption, Bohr's model gave hydrogen atom energy levels consistent with the hydrogen emission spectrum. The model is represented pictorially in Fig. 2-9.

Although we will not show the derivation here, the most important equation to come from Bohr's model is the expression for the *energy levels available to the electron in the hydrogen atom*:

$$E = -2.178 \times 10^{-18} \text{ J} \left(\frac{Z^2}{n^2} \right) \qquad (2\text{-}1)$$

Figure 2-9 | Electronic transitions in the Bohr model for the hydrogen atom. (a) An energy-level diagram for the first three electronic transitions. (b) An orbit-transition diagram, which accounts for the experimental spectrum. (Note that the orbits shown are schematic. They are not drawn to scale.) (c) The resulting line spectrum on a photographic plate is shown. Note that the lines in the visible region of the spectrum correspond to transitions from higher levels to the $n = 2$ level.

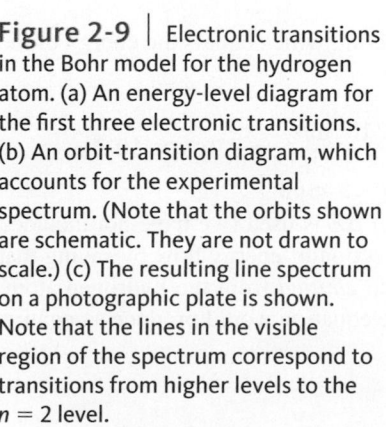

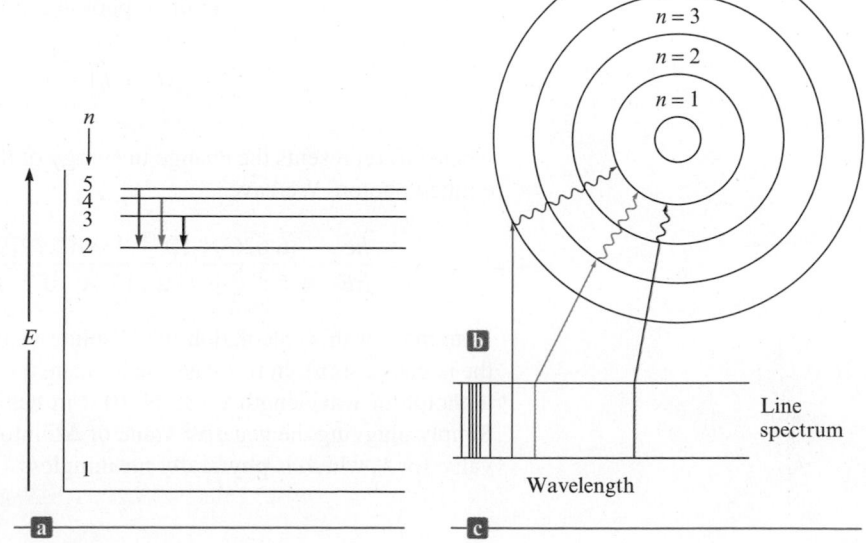

i The J in Equation (2-1) stands for joules.

in which n is an integer (the larger the value of n, the larger the orbit radius) and Z is the nuclear charge. ◄ *i* Using Equation (2-1), Bohr was able to calculate hydrogen atom energy levels that exactly matched the values obtained by experiment.

The negative sign in Equation (2-1) simply means that the energy of the electron bound to the nucleus is lower than it would be if the electron were at an infinite distance ($n = \infty$) from the nucleus, where there is no interaction and the energy is zero:

$$E = -2.178 \times 10^{-18} \, \text{J} \left(\frac{Z^2}{\infty} \right) = 0$$

The energy of the electron in any orbit is negative relative to this reference state.

Equation (2-1) can be used to calculate the change in energy of an electron when the electron changes orbits. For example, suppose an electron in level $n = 6$ of an excited hydrogen atom falls back to level $n = 1$ as the hydrogen atom returns to its lowest possible energy state, its **ground state**. We use Equation (2-1) with $Z = 1$, since the hydrogen nucleus contains a single proton. The energies corresponding to the two states are as follows:

$$\text{For } n = 6: \quad E_6 = -2.178 \times 10^{-18} \, \text{J} \left(\frac{1^2}{6^2} \right) = -6.050 \times 10^{-20} \, \text{J}$$

$$\text{For } n = 1: \quad E_1 = -2.178 \times 10^{-18} \, \text{J} \left(\frac{1^2}{1^2} \right) = -2.178 \times 10^{-18} \, \text{J}$$

Note that for $n = 1$ the electron has a more negative energy than it does for $n = 6$, which means that the electron is more tightly bound in the smallest allowed orbit.

The change in energy ΔE when the electron falls from $n = 6$ to $n = 1$ is

$$\Delta E = \text{energy of final state} - \text{energy of initial state}$$
$$= E_1 - E_6 = (-2.178 \times 10^{-18} \, \text{J}) - (-6.050 \times 10^{-20} \, \text{J})$$
$$= -2.117 \times 10^{-18} \, \text{J}$$

The negative sign for the *change* in energy indicates that the atom has *lost* energy and is now in a more stable state.

The fact that the atom loses energy when the electron changes from $n = 6$ to $n = 1$ makes sense, because in this transition the electron moves closer to the nucleus. Recall from Section R-9 that as opposite charges move closer together, the potential energy decreases. The energy lost is carried away from the atom by the production (emission) of a photon.

The wavelength of the emitted photon can be calculated from the equation

$$\Delta E = h \left(\frac{c}{\lambda} \right) \quad \text{or} \quad \lambda = \frac{hc}{\Delta E}$$

where ΔE represents the change in energy of the atom, which equals the energy of the emitted photon. We have

$$\lambda = \frac{hc}{\Delta E} = \frac{(6.626 \times 10^{-34} \, \text{J} \cdot \text{s})(2.9979 \times 10^8 \, \text{m/s})}{2.117 \times 10^{-18} \, \text{J}} = 9.383 \times 10^{-8} \, \text{m}$$

Note that for this calculation the absolute value of ΔE is used (we have not included the negative sign). In this case we indicate the direction of energy flow by saying that a photon of wavelength 9.383×10^{-8} m has been *emitted* from the hydrogen atom. Simply plugging the negative value of ΔE into the equation would produce a negative value for λ, which is physically meaningless.

INTERACTIVE EXAMPLE 2-4

i Note from Fig. 2-2 that the light required to produce the transition from the $n = 1$ to $n = 2$ level in hydrogen lies in the ultraviolet region.

Energy Quantization in Hydrogen

Calculate the energy required to excite the hydrogen electron from level $n = 1$ to level $n = 2$. ◄ *i* Also calculate the wavelength of light that must be absorbed by a hydrogen atom in its ground state to reach this excited state.

Solution

Using Equation (2-1) with $Z = 1$, we have

$$E_1 = -2.178 \times 10^{-18} \text{ J}\left(\frac{1^2}{1^2}\right) = -2.178 \times 10^{-18} \text{ J}$$

$$E_2 = -2.178 \times 10^{-18} \text{ J}\left(\frac{1^2}{2^2}\right) = -5.445 \times 10^{-19} \text{ J}$$

$$\Delta E = E_2 - E_1 = (-5.445 \times 10^{-19} \text{ J}) - (-2.178 \times 10^{-18} \text{ J}) = 1.633 \times 10^{-18} \text{ J}$$

The positive value for ΔE indicates that the system has gained energy. The wavelength of light that must be *absorbed* to produce this change is

$$\lambda = \frac{hc}{\Delta E} = \frac{(6.626 \times 10^{-34} \text{ J} \cdot \text{s})(2.9979 \times 10^8 \text{ m/s})}{1.633 \times 10^{-18} \text{ J}}$$

$$= 1.216 \times 10^{-7} \text{ m}$$

See Exercises 2-57 and 2-58

At this time we must emphasize two important points about the Bohr model:

1. The model correctly fits the quantized energy levels of the hydrogen atom and postulates only certain allowed circular orbits for the electron.
2. As the electron becomes more tightly bound, its energy becomes more negative relative to the zero-energy reference state (corresponding to the electron being at infinite distance from the nucleus). As the electron is brought closer to the nucleus, energy is released from the system.

Using Equation (2-1), we can derive a general equation for the electron moving from one level ($n_{initial}$) to another level (n_{final}):

$$\begin{aligned}
\Delta E &= \text{energy of level } n_{final} - \text{energy of level } n_{initial} \\
&= E_{final} - E_{initial} \\
&= (-2.178 \times 10^{-18} \text{ J})\left(\frac{1^2}{n_{final}^2}\right) - (-2.178 \times 10^{-18} \text{ J})\left(\frac{1^2}{n_{initial}^2}\right) \\
&= -2.178 \times 10^{-18} \text{ J}\left(\frac{1}{n_{final}^2} - \frac{1}{n_{initial}^2}\right)
\end{aligned} \tag{2-2}$$

Equation (2-2) can be used to calculate the energy change between *any* two energy levels in a hydrogen atom, as shown in Example 2-5.

The long-accepted value for the radius of the proton is 0.8768 fm (fm = 10^{-15} m). However, new experiments suggest that this value might be about 4% too large. This doesn't sound like much, but if proven correct it will raise havoc in the scientific community.

The possible problem with the currently accepted value of the proton radius results from work done by Randolf Pohl and his coworkers at the Max Planck Institute of Quantum Optics in Garching, Germany. These scientists created an exotic form of hydrogen in which they replaced the hydrogen electron by a muon. A muon is a particle that has the same charge as an electron but is 200 times more massive than the electron. Because of its greater mass, the muon has orbitals with smaller average radii than those of an electron. The German team attempted to excite the muon to higher energy levels using laser pulses. They set their laser to detect muon transitions assuming the proton radius was in the range from 0.87 to

0.91 femtometers (fm), the value expected from current theories. After years of failing to see the expected energy transitions, the German scientists were within weeks of shutting down the experiment when they decided to assume a smaller value for the proton radius. They saw the expected transition when they tuned their laser at a value corresponding to a proton radius of 0.84184 fm. Although this value doesn't seem too different from 0.8768 fm, it has tremendous implications. A change in the radius of the proton will affect the value of the charge density of the proton, a value that affects the values of all of the fundamental physical constants since all of these constants are interrelated. Thus, more experiments are needed to determine the correct value of the proton radius. If the new, smaller value proves to be correct, it has important implications for the fundamental theory (quantum electrodynamics) of matter. Perhaps this will lead to a better understanding of nature.

EXAMPLE 2-5   Electron Energies

Calculate the energy required to remove the electron from a hydrogen atom in its ground state.

Solution

Removing the electron from a hydrogen atom in its ground state corresponds to taking the electron from $n_{initial} = 1$ to $n_{final} = \infty$. Thus

$$\Delta E = -2.178 \times 10^{-18} \text{ J} \left(\frac{1}{n_{final}^2} - \frac{1}{n_{initial}^2} \right)$$

$$= -2.178 \times 10^{-18} \text{ J} \left(\frac{1}{\infty} - \frac{1}{1^2} \right)$$

$$= -2.178 \times 10^{-18} \text{ J}(0 - 1) = 2.178 \times 10^{-18} \text{ J}$$

The energy required to remove the electron from a hydrogen atom in its ground state is 2.178×10^{-18} J.

See Exercises 2-65 and 2-66

ⓘ Although Bohr's model fits the energy levels for hydrogen, it is a fundamentally incorrect model for the hydrogen atom.

At first Bohr's model appeared to be very promising. The energy levels calculated by Bohr closely agreed with the values obtained from the hydrogen emission spectrum. However, when Bohr's model was applied to atoms other than hydrogen, it did not work at all. Although some attempts were made to adapt the model using elliptical orbits, it was concluded that Bohr's model is fundamentally incorrect. ◄ ⓘ The model is, however, very important historically, because it showed that the observed quantization of energy in atoms could be explained by making rather simple assumptions. Bohr's model paved the way for later theories. It is important to realize, however, that the current theory of atomic structure is in no way derived from the Bohr model.

Electrons do *not* move around the nucleus in circular orbits, as we shall see later in this chapter.

2-5 | The Quantum Mechanical Model of the Atom

By the mid-1920s it had become apparent that the Bohr model could not be made to work. A totally new approach was needed. Three physicists were at the forefront of this effort: Werner Heisenberg (1901–1976), Louis de Broglie (1892–1987), and Erwin Schrödinger (1887–1961). The approach they developed became known as *wave mechanics* or, more commonly, *quantum mechanics*. As we have already seen, de Broglie originated the idea that the electron, previously considered to be a particle, also shows wave properties. Pursuing this line of reasoning, Schrödinger, an Austrian physicist, decided to attack the problem of atomic structure by giving emphasis to the wave properties of the electron. To Schrödinger and de Broglie, the electron bound to the nucleus seemed similar to a **standing wave**, and they began research on a wave mechanical description of the atom.

The most familiar example of standing waves occurs in association with musical instruments such as guitars or violins, where a string attached at both ends vibrates to produce a musical tone. The waves are described as "standing" because they are stationary; the waves do not travel along the length of the string. The motions of the string can be explained as a combination of simple waves of the type shown in Fig. 2-10. The dots in this figure indicate the nodes, or points of zero vertical displacement, for a given wave. Note that there are limitations on the allowed wavelengths of the standing wave. Each end of the string is fixed, so there is always a node at each end. This means that there must be a whole number of *half* wavelengths in any of the allowed motions of the string (see Fig. 2-10). Standing waves can be illustrated using the wave generator shown in the photo. ◄

A similar situation results when the electron in the hydrogen atom is imagined to be a standing wave. As shown in Fig. 2-11, only certain circular orbits have a

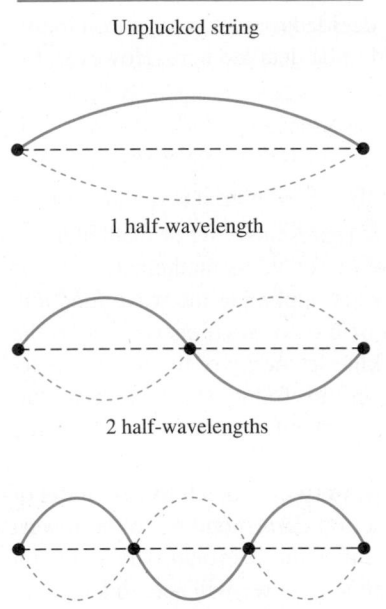

Unplucked string

1 half-wavelength

2 half-wavelengths

3 half-wavelengths

Figure 2-10 | The standing waves caused by the vibration of a guitar string fastened at both ends. Each dot represents a node (a point of zero displacement).

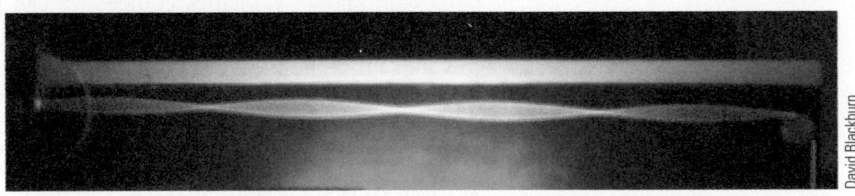

David Blackburn

▲ Wave-generating apparatus.

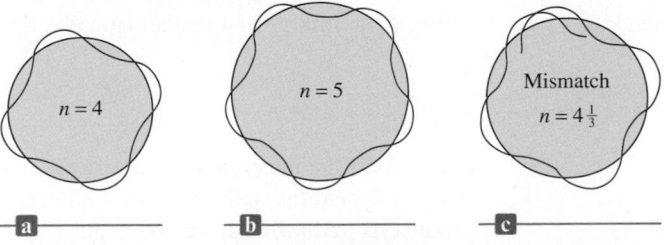

$n = 4$ $n = 5$ Mismatch $n = 4\frac{1}{3}$

a   b   c

Figure 2-11 | The hydrogen electron visualized as a standing wave around the nucleus. The circumference of a particular circular orbit would have to correspond to a whole number of wavelengths, as shown in (a) and (b), or else destructive interference occurs, as shown in (c). This is consistent with the fact that only certain electron energies are allowed; the atom is quantized. (Although this idea encouraged scientists to use a wave theory, it does not mean that the electron really travels in circular orbits.)

circumference into which a whole number of wavelengths of the standing electron wave will "fit." All other orbits would produce destructive interference of the standing electron wave and are not allowed. This seemed like a possible explanation for the observed quantization of the hydrogen atom, so Schrödinger worked out a model for the hydrogen atom in which the electron was assumed to behave as a standing wave.

It is important to recognize that Schrödinger could not be sure that this idea would work. The test had to be whether or not the model would correctly fit the experimental data on hydrogen and other atoms. The physical principles for describing standing waves were well known in 1925 when Schrödinger decided to treat the electron in this way. His mathematical treatment is too complicated to be detailed here. However, the form of Schrödinger's equation is

$$\hat{H}\psi = E\psi$$

where ψ, called the **wave function**, is a function of the coordinates (x, y, and z) of the electron's position in three-dimensional space and $\hat{H}$ represents a set of mathematical instructions called an *operator*. In this case, the operator contains mathematical terms that produce the total energy of the atom when they are applied to the wave function. E represents the total energy of the atom (the sum of the potential energy due to the attraction between the proton and electron and the kinetic energy of the moving electron). When this equation is analyzed, many solutions are found. Each solution consists of a wave function ψ that is characterized by a particular value of E. A specific wave function is often called an **orbital**.

To illustrate the most important ideas of the **quantum (wave) mechanical model** of the atom, we will first concentrate on the wave function corresponding to the lowest energy for the hydrogen atom. This wave function is called the 1s orbital. The first point of interest is to explore the meaning of the word *orbital*. As we will see, this is not a trivial matter. One thing is clear: An orbital is *not* a Bohr orbit. The electron in the hydrogen 1s orbital is not moving around the nucleus in a circular orbit. How, then, is the electron moving? The answer is quite surprising: *We do not know*. The wave function gives us no information about the detailed pathway of the electron. This is somewhat disturbing. When we solve problems involving the motions of particles in the macroscopic world, we are able to predict their pathways. For example, when two billiard balls with known velocities collide, we can predict their motions after the collision. However, we cannot predict the electron's motion from the 1s orbital function. Does this mean that the theory is wrong? Not necessarily: We have already learned that an electron does not behave much like a billiard ball, so we must examine the situation closely before we discard the theory.

To help us understand the nature of an orbital, we need to consider a principle discovered by Werner Heisenberg, one of the primary developers of quantum mechanics. Heisenberg's mathematical analysis led him to a surprising conclusion: *There is a fundamental limitation to just how precisely we can know both the position and momentum of a particle at a given time*. This is a statement of the **Heisenberg uncertainty principle**. Stated mathematically, the uncertainty principle is

$$\Delta x \cdot \Delta(mv) \geq \frac{h}{4\pi}$$

where Δx is the uncertainty in a particle's position, $\Delta(mv)$ is the uncertainty in a particle's momentum, and h is Planck's constant. Thus the minimum uncertainty in the product $\Delta x \cdot \Delta(mv)$ is $h/4\pi$. What this equation really says is that the more accurately we know a particle's position, the less accurately we can know its momentum, and vice versa. This limitation is so small for large particles such as baseballs or billiard balls that it is unnoticed. However, for a small particle such as the electron, the limitation becomes quite important. Applied to the electron, the uncertainty principle implies that we cannot know the exact motion of the electron as it moves around the nucleus. It is therefore not appropriate to assume that the electron is moving around the nucleus in a well-defined orbit, as in the Bohr model.

The Physical Meaning of a Wave Function

Given the limitations indicated by the uncertainty principle, what then is the physical meaning of a wave function for an electron? That is, what is an atomic orbital? Although the wave function itself has no easily visualized meaning, the square of the function does have a definite physical significance. *The square of the function indicates the probability of finding an electron near a particular point in space.* ◀ ⓘ For example, suppose we have two positions in space, one defined by the coordinates x_1, y_1, and z_1 and the other by the coordinates x_2, y_2, and z_2. The relative probability of finding the electron at positions 1 and 2 is given by substituting the values of x, y, and z for the two positions into the wave function, squaring the function value, and computing the following ratio:

$$\frac{[\psi(x_1, y_1, z_1)]^2}{[\psi(x_2, y_2, z_2)]^2} = \frac{N_1}{N_2}$$

The quotient N_1/N_2 is the ratio of the probabilities of finding the electron at positions 1 and 2. For example, if the value of the ratio N_1/N_2 is 100, the electron is 100 times more likely to be found at position 1 than at position 2. The model gives no information concerning when the electron will be at either position or how it moves between the positions. This vagueness is consistent with the concept of the Heisenberg uncertainty principle.

The square of the wave function is most conveniently represented as a **probability distribution**, in which the intensity of color is used to indicate the probability value near a given point in space. The probability distribution for the hydrogen $1s$ wave function (orbital) is shown in Fig. 2-12(a). The best way to think about this diagram is as a three-dimensional time exposure with the electron as a tiny moving light. The more times the electron visits a particular point, the darker the negative becomes. Thus the darkness of a point indicates the probability of finding an electron at that position. This diagram is also known as an *electron density map; electron density* and *electron probability* mean the same thing. When a chemist uses the term *atomic orbital*, he or she is probably picturing an electron density map of this type.

Another way of representing the electron probability distribution for the $1s$ wave function is to calculate the probability at points along a line drawn outward in any direction from the nucleus. The result is shown in Fig. 2-12(b). Note that the probability of finding the electron at a particular position is greatest close to the nucleus and drops off rapidly as the distance from the nucleus increases. We are also interested in knowing the *total* probability of finding the electron in the hydrogen atom at a particular *distance* from the nucleus. Imagine that the space around the hydrogen nucleus is made up of a series of thin spherical shells (rather like layers in an onion), as shown in Fig. 2-13(a). When the total probability of finding the electron in each spherical shell

ⓘ Probability is the likelihood, or odds, that something will occur.

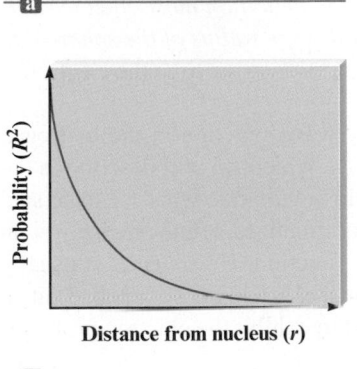

a

b

Figure 2-12 | (a) The probability distribution for the hydrogen $1s$ orbital in three-dimensional space. (b) The probability of finding the electron at points along a line drawn from the nucleus outward in any direction for the hydrogen $1s$ orbital.

(y-axis: Probability (R^2); x-axis: Distance from nucleus (r))

Figure 2-13 | (a) Cross section of the hydrogen $1s$ orbital probability distribution divided into successive thin spherical shells. (b) The radial probability distribution. A plot of the total probability of finding the electron in each thin spherical shell as a function of distance from the nucleus.

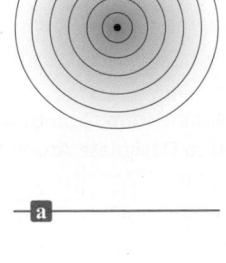

a

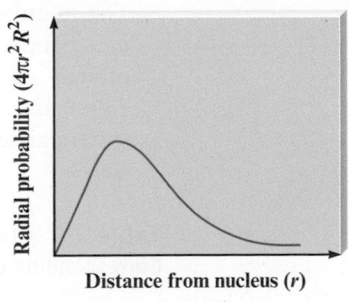

b

(y-axis: Radial probability ($4\pi r^2 R^2$); x-axis: Distance from nucleus (r))

is plotted versus the distance from the nucleus, the plot in Fig. 2-13(b) is obtained. This graph is called the **radial probability distribution**.

The maximum in the curve occurs because of two opposing effects. The probability of finding an electron at a particular position is greatest near the nucleus, but the volume of the spherical shell increases with distance from the nucleus. Therefore, as we move away from the nucleus, the probability of finding the electron at a given position decreases, but we are summing more positions. Thus the total probability increases to a certain radius and then decreases as the electron probability at each position becomes very small. For the hydrogen 1s orbital, the maximum radial probability (the distance at which the electron is most likely to be found) occurs at a distance of 5.29×10^{-2} nm or 0.529 Å from the nucleus. ◄ ⓘ Interestingly, this is exactly the radius of the innermost orbit in the Bohr model. Note that in Bohr's model the electron is assumed to have a circular path and so is *always* found at this distance. In the quantum mechanical model, the specific electron motions are unknown, and this is the *most probable* distance at which the electron is found.

One more characteristic of the hydrogen 1s orbital that we must consider is its size. As we can see from Fig. 2-12, the size of this orbital cannot be defined precisely, since the probability never becomes zero (although it drops to an extremely small value at large values of r). So, in fact, the hydrogen 1s orbital has no distinct size. However, it is useful to have a definition of relative orbital size. *The definition most often used by chemists to describe the size of the hydrogen 1s orbital is the radius of the sphere that encloses 90% of the total electron probability.* That is, 90% of the time the electron is inside this sphere.

So far we have described only the lowest-energy wave function in the hydrogen atom, the 1s orbital. Hydrogen has many other orbitals, which we will describe in the next section. However, before we proceed, we should summarize what we have said about the meaning of an atomic orbital. An orbital is difficult to define precisely at an introductory level. Technically, an orbital is a wave function. However, it is usually most helpful to picture an orbital as a three-dimensional electron density map. That is, an electron "in" a particular atomic orbital is assumed to exhibit the electron probability indicated by the orbital map.

ⓘ 1 Å = 10^{-10} m; the angstrom is most often used as the unit for atomic radius because of its convenient size. Another convenient unit is the picometer:

$$1 \text{ pm} = 10^{-12} \text{ m}$$

2-6 | Quantum Numbers

When we solve the Schrödinger equation for the hydrogen atom, we find many wave functions (orbitals) that satisfy it. Each of these orbitals is characterized by a series of numbers called **quantum numbers**, which describe various properties of the orbital:

The **principal quantum number** (n) has integral values: 1, 2, 3, ... The principal quantum number is related to the size and energy of the orbital. As n increases, the orbital becomes larger and the electron spends more time farther from the nucleus. An increase in n also means higher energy, because the electron is less tightly bound to the nucleus, and the energy is less negative.

The **angular momentum quantum number** (ℓ) has integral values from 0 to $n - 1$ for each value of n. This quantum number is related to the shape of atomic orbitals. The value of ℓ for a particular orbital is commonly assigned a letter: $\ell = 0$ is called s; $\ell = 1$ is called p; $\ell = 2$ is called d; $\ell = 3$ is called f. This system arises from early spectral studies and is summarized in Table 2-1.

Table 2-1 | The Angular Momentum Quantum Numbers and Corresponding Letters Used to Designate Atomic Orbitals

Value of ℓ	0	1	2	3	4
Letter Used	s	p	d	f	g

Table 2-2 | Quantum Numbers for the First Four Levels of Orbitals in the Hydrogen Atom

n	ℓ	Sublevel Designation	m_ℓ	Number of Orbitals
1	0	1s	0	1
2	0	2s	0	1
	1	2p	−1, 0, +1	3
3	0	3s	0	1
	1	3p	−1, 0, 1	3
	2	3d	−2, −1, 0, 1, 2	5
4	0	4s	0	1
	1	4p	−1, 0, 1	3
	2	4d	−2, −1, 0, 1, 2	5
	3	4f	−3, −2, −1, 0, 1, 2, 3	7

Number of Orbitals per Subshell
$s = 1$
$p = 3$
$d = 5$
$f = 7$
$g = 9$

$n = 1, 2, 3, \ldots$
$\ell = 0, 1, \ldots (n - 1)$
$m_\ell = -\ell, \ldots 0, \ldots +\ell$

The **magnetic quantum number** (m_ℓ) has integral values between ℓ and $-\ell$, including zero. The value of m_ℓ is related to the orientation of the orbital in space relative to the other orbitals in the atom.

The first four levels of orbitals in the hydrogen atom are listed with their quantum numbers in Table 2-2. Note that each set of orbitals with a given value of ℓ (sometimes called a **subshell**) is designated by giving the value of n and the letter for ℓ. Thus an orbital where $n = 2$ and $\ell = 1$ is symbolized as 2p. There are three 2p orbitals, which have different orientations in space. We will describe these orbitals in the next section.

INTERACTIVE EXAMPLE 2-6

Electron Subshells

For principal quantum level $n = 5$, determine the number of allowed subshells (different values of ℓ), and give the designation of each.

Solution

For $n = 5$, the allowed values of ℓ run from 0 to 4 ($n - 1 = 5 - 1$). Thus the subshells and their designations are

$\ell = 0 \quad \ell = 1 \quad \ell = 2 \quad \ell = 3 \quad \ell = 4$
$5s \qquad 5p \qquad 5d \qquad 5f \qquad 5g$

See Exercises 2-71 and 2-73

2-7 | Orbital Shapes and Energies

We have seen that the meaning of an orbital is represented most clearly by a probability distribution. Each orbital in the hydrogen atom has a unique probability distribution. We also saw that another means of representing an orbital is by the surface that surrounds 90% of the total electron probability. These two types of representations for the hydrogen 1s, 2s, and 3s orbitals are shown in Fig. 2-14. Note the characteristic spherical shape of each of the s orbitals. Note also that the 2s and 3s orbitals contain areas of high probability separated by areas of zero probability. These latter areas are called **nodal surfaces**, or simply **nodes**. The number of nodes increases as n increases. For s orbitals, the number of nodes is given by $n - 1$. For our purposes, however, we will think of s orbitals only in terms of their overall spherical shape, which becomes larger as the value of n increases. Note that it is difficult to picture how the 1s, 2s, and 3s orbitals might overlap in space. The important thing is that they are distinct mathematically.

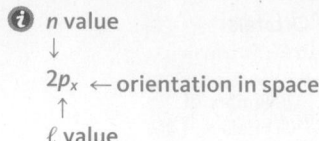

n value
↓
2*p*$_x$ ← orientation in space
↑
ℓ value

The two types of representations for the 2*p* orbitals (there are no 1*p* orbitals) are shown in Fig. 2-15. Note that the *p* orbitals are not spherical like *s* orbitals but have two *lobes* separated by a node at the nucleus. The *p* orbitals are labeled according to the axis of the *xyz* coordinate system along which the lobes lie. For example, the 2*p* orbital with lobes centered along the *x* axis is called the 2*p*$_x$ orbital. ◄ ⓘ

Figure 2-14 | Three representations of the hydrogen 1*s*, 2*s*, and 3*s* orbitals. (a) The square of the wave function. (b) "Slices" of the three-dimensional electron density. (c) The surfaces that contain 90% of the total electron probability (the "sizes" of the orbitals).

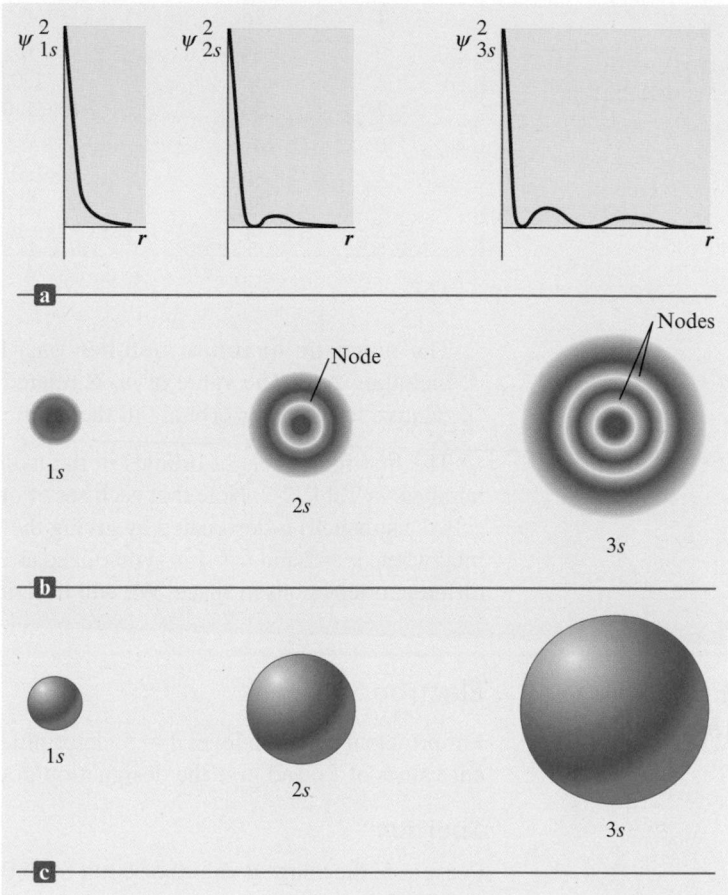

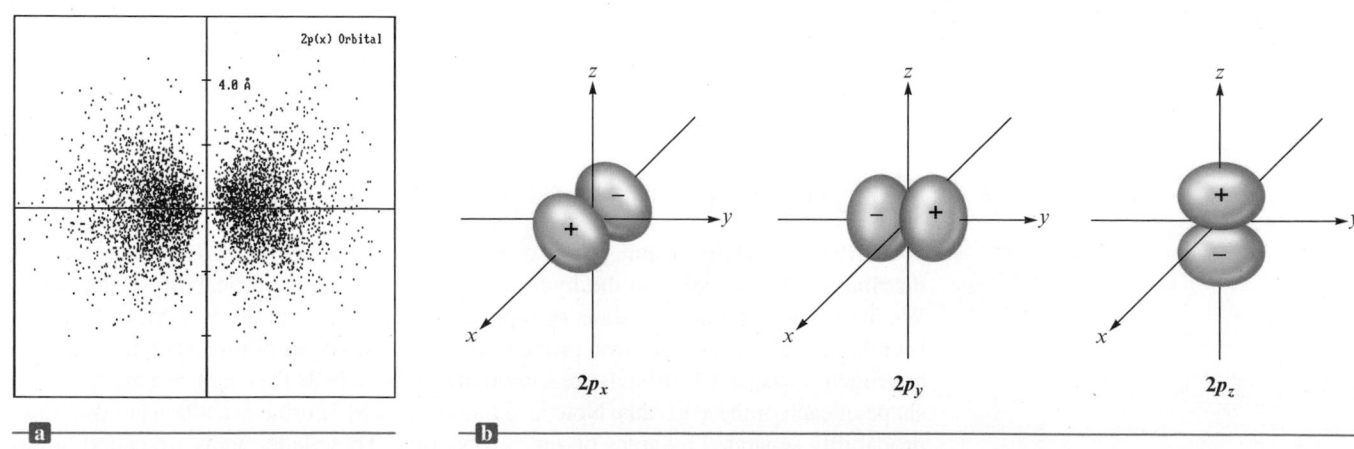

Figure 2-15 | Representation of the 2*p* orbitals. (a) The electron probability distribution for a 2*p* orbital. (Generated from a program by Robert Allendoerfer on Project SERAPHIM disk PC 2402; reprinted with permission.) (b) The boundary surface representations of all three 2*p* orbitals. Note that the signs inside the surface indicate the phases (signs) of the orbital in that region of space.

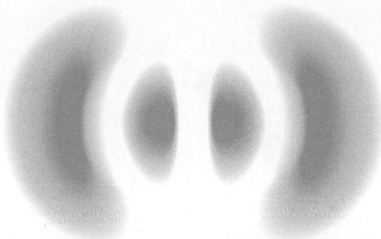

Figure 2-16 | A cross-section of the electron probability distribution for a 3p orbital.

At this point it is useful to remember that mathematical functions have signs. For example, a simple sine wave (see Fig. 2-1) oscillates from positive to negative and repeats this pattern. Atomic orbital functions also have signs. The functions for s orbitals are positive everywhere in three-dimensional space. That is, when the s orbital function is evaluated at any point in space, it results in a positive number. In contrast, the p orbital functions have different signs in different regions of space. For example, the p_z orbital has a positive sign in all the regions of space in which z is positive and has a negative sign when z is negative. This behavior is indicated in Fig. 2-15(b) by the positive and negative signs inside their boundary surfaces. It is important to understand that these are mathematical signs, not charges. Just as a sine wave has alternating positive and negative phases, so too do p orbitals. The phases of the p_x, p_y, and p_z orbitals are indicated in Fig. 2-15(b).

As you might expect from our discussion of the s orbitals, the 3p orbitals have a more complex probability distribution than that of the 2p orbitals (see Fig. 2-16), but they can still be represented by the same boundary surface shapes. The surfaces just grow larger as the value of n increases.

There are no d orbitals that correspond to principal quantum levels n = 1 and n = 2. The d orbitals (ℓ = 2) first occur in level n = 3. The five 3d orbitals have the shapes shown in Fig. 2-17. The d orbitals have two different fundamental shapes. Four of the orbitals (d_{xz}, d_{yz}, d_{xy}, and $d_{x^2-y^2}$) have four lobes centered in the plane indicated in the orbital label. Note that d_{xy} and $d_{x^2-y^2}$ are both centered in the xy plane; however, the lobes of $d_{x^2-y^2}$ lie *along* the x and y axes, while the lobes of d_{xy} lie *between* the axes. The fifth orbital, d_{z^2}, has a unique shape with two lobes along the z axis and a belt

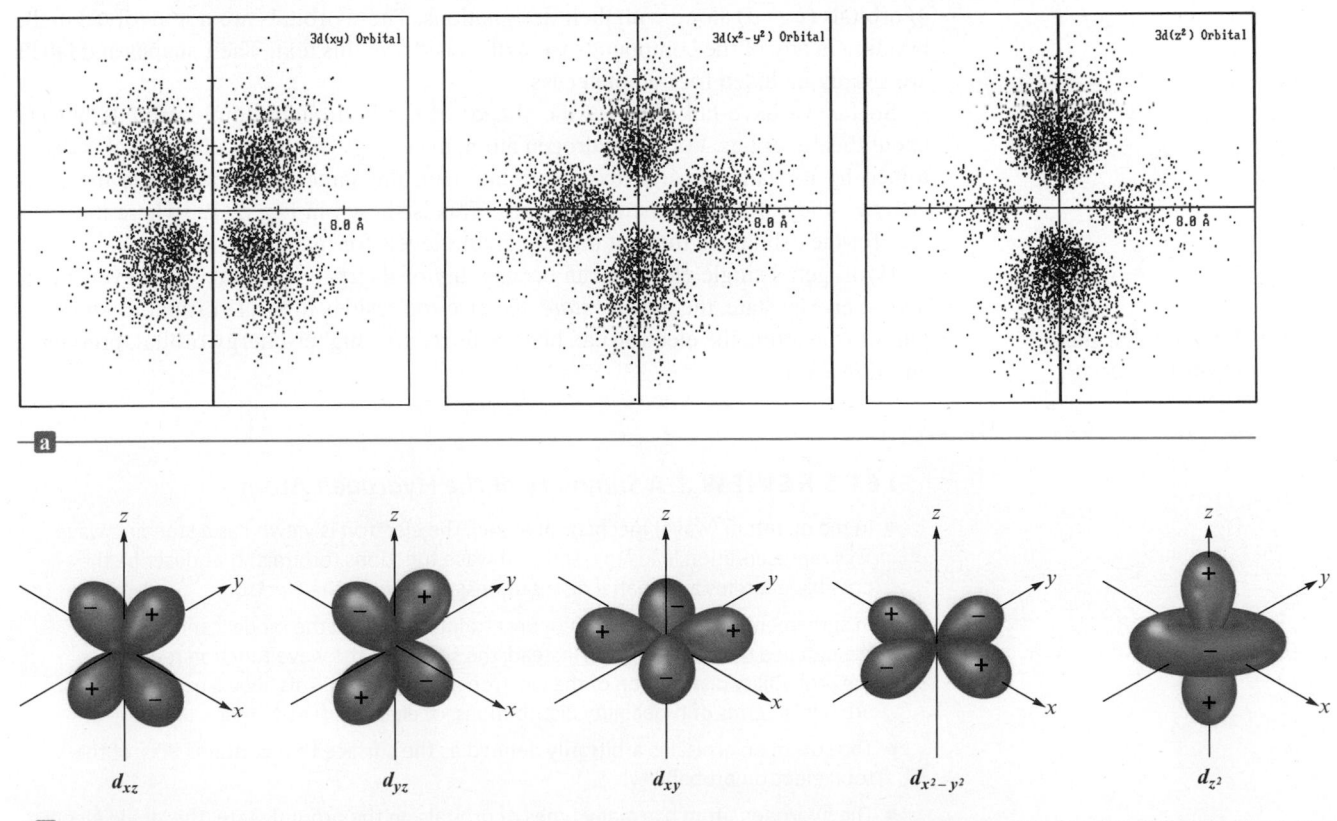

Figure 2-17 | Representation of the 3d orbitals. (a) Electron density plots of selected 3d orbitals. (Generated from a program by Robert Allendoerfer on Project SERAPHIM disk PC 2402; reprinted with permission.) (b) The boundary surfaces of all five 3d orbitals, with the signs (phases) indicated.

Figure 2-18 | Representation of the 4f orbitals in terms of their boundary surfaces.

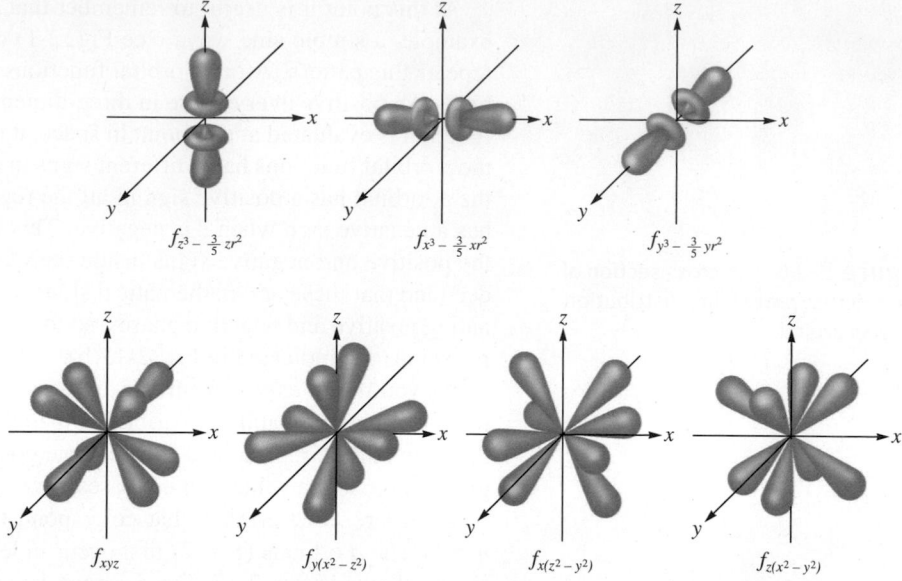

$f_{z^3 - \frac{3}{5}zr^2}$ $f_{x^3 - \frac{3}{5}xr^2}$ $f_{y^3 - \frac{3}{5}yr^2}$

f_{xyz} $f_{y(x^2 - z^2)}$ $f_{x(z^2 - y^2)}$ $f_{z(x^2 - y^2)}$

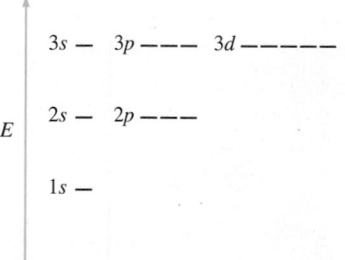

$3s$ — $3p$ ——— $3d$ —————

$2s$ — $2p$ ———

E

$1s$ —

Figure 2-19 | Orbital energy levels for the hydrogen atom.

centered in the *xy* plane. The *d* orbitals for levels *n* > 3 look like the 3*d* orbitals but have larger lobes.

The *f* orbitals first occur in level *n* = 4, and as might be expected, they have shapes even more complex than those of the *d* orbitals. Fig. 2-18 shows representations of the 4*f* orbitals (ℓ = 3) along with their designations. These orbitals are not involved in the bonding in any of the compounds we will consider in this text. Their shapes and labels are simply included for completeness.

So far we have talked about the shapes of the hydrogen atomic orbitals but not about their energies. For the hydrogen atom, the energy of a particular orbital is determined by its value of *n*. Thus *all* orbitals with the same value of *n* have the *same energy*—they are said to be **degenerate**. This is shown in Fig. 2-19, where the energies for the orbitals in the first three quantum levels for hydrogen are shown.

Hydrogen's single electron can occupy any of its atomic orbitals. However, in the lowest energy state, the *ground state*, the electron resides in the 1*s* orbital. If energy is put into the atom, the electron can be transferred to a higher-energy orbital, producing an *excited state*.

LET'S REVIEW | *A Summary of the Hydrogen Atom*

■ In the quantum (wave) mechanical model, the electron is viewed as a standing wave. This representation leads to a series of wave functions (orbitals) that describe the possible energies and spatial distributions available to the electron.

■ In agreement with the Heisenberg uncertainty principle, the model cannot specify the detailed electron motions. Instead, the square of the wave function represents the probability distribution of the electron in that orbital. This allows us to picture orbitals in terms of probability distributions, or electron density maps.

■ The size of an orbital is arbitrarily defined as the surface that contains 90% of the total electron probability.

■ The hydrogen atom has many types of orbitals. In the ground state, the single electron resides in the 1*s* orbital. The electron can be excited to higher-energy orbitals if energy is put into the atom.

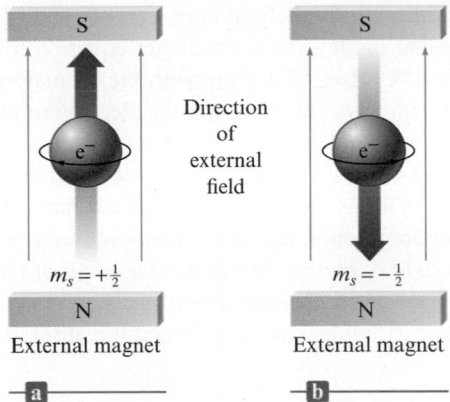

Figure 2-20 | A picture of the spinning electron. Spinning in one direction, the electron produces the magnetic field oriented as shown in (a). Spinning in the opposite direction, it gives a magnetic field of the opposite orientation, as shown in (b).

2-8 | Electron Spin and the Pauli Principle

The concept of **electron spin** was developed by Samuel Goudsmit and George Uhlenbeck while they were graduate students at the University of Leyden in the Netherlands. They found that a fourth quantum number (in addition to n, ℓ, and m_ℓ) was necessary to account for the details of the emission spectra of atoms. The spectral data indicate that the electron has a magnetic moment with two possible orientations when the atom is placed in an external magnetic field. Since they knew from classical physics that a spinning charge produces a magnetic moment, it seemed reasonable to assume that the electron could have two spin states, thus producing the two oppositely directed magnetic moments (Fig. 2-20). The new quantum number adopted to describe this phenomenon, called the **electron spin quantum number** (m_s), can have only one of two values, $+\frac{1}{2}$ and $-\frac{1}{2}$. We can interpret this to mean that the electron can spin in one of two opposite directions, although other interpretations also have been suggested.

For our purposes, the main significance of electron spin is connected with the postulate of Austrian physicist Wolfgang Pauli (1900–1958): In a given atom no two electrons can have the same set of four quantum numbers (n, ℓ, m_ℓ, and m_s). This is called the **Pauli exclusion principle**. Since electrons in the same orbital have the same values of n, ℓ, and m_ℓ, this postulate says that they must have different values of m_s. Then, since only two values of m_s are allowed, *an orbital can hold only two electrons, and they must have opposite spins*. This principle will have important consequences as we use the atomic model to account for the electron arrangements of the atoms in the periodic table.

2-9 | Polyelectronic Atoms

The quantum mechanical model gives a description of the hydrogen atom that agrees very well with experimental data. However, the model would not be very useful if it did not account for the properties of all the other atoms as well.

To see how the model applies to **polyelectronic atoms**, that is, atoms with more than one electron, let's consider helium, which has two protons in its nucleus and two electrons:

Three energy contributions must be considered in the description of the helium atom: (1) the kinetic energy of the electrons as they move around the nucleus, (2) the potential energy of attraction between the nucleus and the electrons, and (3) the potential energy of repulsion between the two electrons.

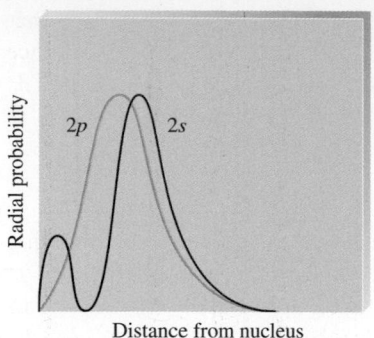

Figure 2-21 | A comparison of the radial probability distributions of the 2s and 2p orbitals.

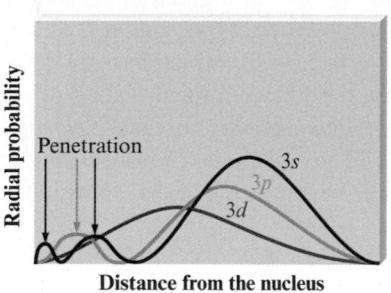

Figure 2-22 | (a) The radial probability distribution for an electron in a 3s orbital. Although a 3s electron is mostly found far from the nucleus, there is a small but significant probability (shown by the arrows) of its being found close to the nucleus. The 3s electron penetrates the shield of inner electrons. (b) The radial probability distribution for the 3s, 3p, and 3d orbitals. The arrows indicate that the s orbital (red arrow) allows greater electron penetration than the p orbital (yellow arrow) does; the d orbital allows minimal electron penetration.

Although the helium atom can be readily described in terms of the quantum mechanical model, the Schrödinger equation that results cannot be solved exactly. The difficulty arises in dealing with the repulsions between the electrons. Since the electron pathways are unknown, the electron repulsions cannot be calculated exactly. This is called the *electron correlation problem*.

The electron correlation problem occurs with all polyelectronic atoms. To treat these systems using the quantum mechanical model, we must make approximations. Most commonly, the approximation used is to treat each electron as if it were moving in a *field of charge that is the net result of the nuclear attraction and the average repulsions of all the other electrons*.

For example, consider the sodium atom, which has 11 electrons:

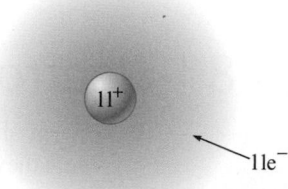

Now let's single out the outermost electron and consider the forces this electron feels. The electron clearly is attracted to the highly charged nucleus. However, the electron also feels the repulsions caused by the other 10 electrons. The net effect is that the electron is not bound nearly as tightly to the nucleus as it would be if the other electrons were not present. We say that the electron is *screened* or *shielded* from the nuclear charge by the repulsions of the other electrons.

This picture of polyelectronic atoms leads to *hydrogenlike orbitals* for these atoms. They have the same general shapes as the orbitals for hydrogen, but their sizes and energies are different. The differences occur because of the interplay between nuclear attraction and the electron repulsions.

One especially important difference between polyelectronic atoms and the hydrogen atom is that for hydrogen all the orbitals in a given principal quantum level have the same energy (they are said to be *degenerate*). This is not the case for polyelectronic atoms, where we find that for a given principal quantum level the orbitals vary in energy as follows:

$$E_{ns} < E_{np} < E_{nd} < E_{nf}$$

In other words, when electrons are placed in a particular quantum level, they "prefer" the orbitals in the order s, p, d, and then f. Why does this happen? Although the concept of orbital energies is a complicated matter, we can qualitatively understand why the 2s orbital has a lower energy than the 2p orbital in a polyelectronic atom by looking at the probability profiles of these orbitals (Fig. 2-21). Notice that the 2p orbital has its maximum probability closer to the nucleus than for the 2s. This might lead us to predict that the 2p would be preferable (lower energy) to the 2s orbital. However, notice the small hump of electron density that occurs in the 2s profile very near the nucleus. This means that although an electron in the 2s orbital spends most of its time a little farther from the nucleus than does an electron in the 2p orbital, it spends a small but very significant amount of time very near the nucleus. We say that the 2s electron *penetrates* to the nucleus more than one in the 2p orbital. This *penetration effect* causes an electron in a 2s orbital to be attracted to the nucleus more strongly than an electron in a 2p orbital. That is, the 2s orbital is lower in energy than the 2p orbitals in a polyelectronic atom.

The same thing happens in the other principal quantum levels as well. Fig. 2-22 shows the radial probability profiles for the 3s, 3p, and 3d orbitals. Note again the

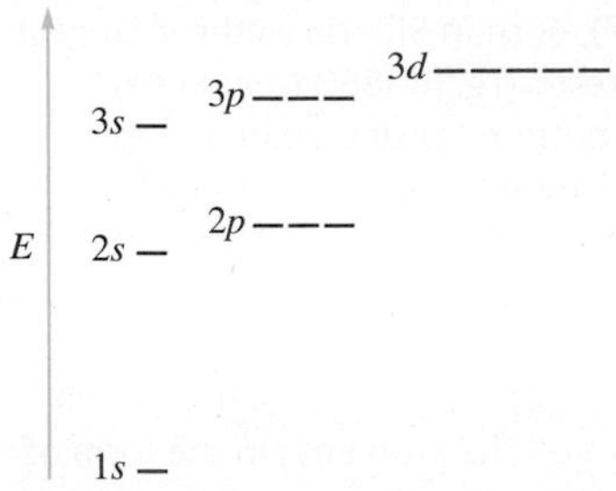

Figure 2-23 | The orders of the energies of the orbitals in the first three levels of polyelectronic atoms.

hump in the 3s profile very near the nucleus. The innermost hump for the 3p is farther out, which causes the energy of the 3s orbital to be lower than that of the 3p. Notice that the 3d orbital has its maximum probability closer to the nucleus than either the 3s or 3p does, but its absence of probability near the nucleus causes it to be highest in energy of the three orbitals. The relative energies of the orbitals for $n = 3$ are

$$E_{3s} < E_{3p} < E_{3d}$$

In general, the more effectively an orbital allows its electron to penetrate the shielding electrons to be close to the nuclear charge, the lower the energy of that orbital.

A summary diagram of the orders of the orbital energies for polyelectronic atoms is represented in Fig. 2-23. We will use these orbitals in Section 2-11 to show how the electrons are arranged in polyelectronic atoms.

Critical Thinking

What if Bohr's model was correct? How would this affect the radial probability profiles as shown in Figs. 2-21 and 2-22?

2-10 | The History of the Periodic Table

The modern periodic table contains a tremendous amount of useful information. In this section we will discuss the origin of this valuable tool; later we will see how the quantum mechanical model for the atom explains the periodicity of chemical properties. Certainly the greatest triumph of the quantum mechanical model is its ability to account for the arrangement of the elements in the periodic table.

The periodic table was originally constructed to represent the patterns observed in the chemical properties of the elements. ◄ ⓘ As chemistry progressed during the eighteenth and nineteenth centuries, it became evident that the earth is composed of a great many elements with very different properties. Things are much more complicated than thesimple model of earth, air, fire, and water suggested by the ancients. At first, the array of elements and properties was bewildering. Gradually, however, patterns were noticed.

The first chemist to recognize patterns was Johann Dobereiner (1780–1849), who found several groups of three elements that have similar properties, for example, chlorine, bromine, and iodine. However, as Dobereiner attempted to expand this model of *triads* (as he called them) to the rest of the known elements, it became clear that it was severely limited.

The next notable attempt was made by the English chemist John Newlands, who in 1864 suggested that elements should be arranged in *octaves*, based on the idea that certain properties seemed to repeat for every eighth element in a way similar to the musical scale, which repeats for every eighth tone. Even though this model managed to group several elements with similar properties, it was not generally successful.

The present form of the periodic table was conceived independently by two chemists: the German Julius Lothar Meyer (1830–1895) and Dmitri Ivanovich Mendeleev (1834–1907), a Russian (Fig. 2-24). Usually Mendeleev is given most of the credit, because it was he who emphasized how useful the table could be in predicting the existence and properties of still unknown elements. For example, in 1872 when Mendeleev first published his table (Fig. 2-25), the elements gallium, scandium, and germanium were unknown. Mendeleev correctly predicted the existence and properties of these elements from gaps in his periodic table. The data for germanium (which Mendeleev called *ekasilicon*) are shown in Table 2-3. Note the excellent agreement between the actual values and Mendeleev's predictions, which were based on the properties of other members in the group of elements similar to germanium.

ⓘ The periodic table was developed according to observed chemical properties of the elements and before the discovery of subatomic particles. As we will see in the remainder of the chapter, our model of the electrons in the atom will account for various trends in the periodic table. This shows us of the importance of valence electrons in the chemical properties we observe.

Art Resource

Figure 2-24 | Dmitri Ivanovich Mendeleev (1834–1907), born in Siberia as the youngest of 17 children, taught chemistry at the University of St. Petersburg. In 1860 Mendeleev heard the Italian chemist Cannizzaro lecture on a reliable method for determining the correct atomic masses of the elements. This important development paved the way for Mendeleev's own brilliant contribution to chemistry—the periodic table. In 1861 Mendeleev returned to St. Petersburg, where he wrote a book on organic chemistry. Later Mendeleev also wrote a book on inorganic chemistry, and he was struck by the fact that the systematic approach characterizing organic chemistry was lacking in inorganic chemistry. In attempting to systematize inorganic chemistry, he eventually arranged the elements in the form of the periodic table.

Mendeleev was a versatile genius who was interested in many fields of science. He worked on many problems associated with Russia's natural resources, such as coal, salt, and various metals. Being particularly interested in the petroleum industry, he visited the United States in 1876 to study the Pennsylvania oil fields. His interests also included meteorology and hot-air balloons. In 1887 he made an ascent in a balloon to study a total eclipse of the sun.

Using his table, Mendeleev also was able to correct several values for atomic masses. For example, the original atomic mass of 76 for indium was based on the assumption that indium oxide had the formula InO. This atomic mass placed indium, which has metallic properties, among the nonmetals. Mendeleev assumed the atomic mass was probably incorrect and proposed that the formula of indium oxide was really In_2O_3. Based on this correct formula, indium has an atomic mass of approximately 113, placing the element among the metals. Mendeleev also corrected the atomic masses of beryllium and uranium.

Because of its obvious usefulness, Mendeleev's periodic table was almost universally adopted, and it remains one of the most valuable tools at the chemist's disposal. For example, it is still used to predict the properties of elements recently discovered, as shown in Table 2-4.

TABELLE II

REIHEN	GRUPPE I. R^2O	GRUPPE II. RO	GRUPPE III. R^2O^3	GRUPPE IV. RH^4 RO^2	GRUPPE V. RH^3 R^2O^5	GRUPPE VI. RH^2 RO^3	GRUPPE VII. RH R^2O^7	GRUPPE VIII. RO^4
1	H =1							
2	Li = 7	Be = 9,4	B = 11	C = 12	N =14	O = 16	F = 19	
3	Na = 23	Mg =24	Al = 27,3	Si = 28	P = 31	S = 32	Cl = 35,5	
4	K = 39	Ca = 40	= 44	Ti = 48	V = 51	Cr = 52	Mn = 55	Fe = 56, Co = 59, Ni = 59, Cu = 63.
5	(Cu = 63)	Zn = 65	= 68	= 72	As = 75	Se = 78	Br = 80	
6	Rb = 85	Sr = 87	?Yt = 88	Zr = 90	Nb = 94	Mo = 96	= 100	Ru = 104, Rh = 104, Pd = 106, Ag = 108.
7	(Ag = 108)	Cd = 112	In = 113	Sn = 118	Sb = 122	Te = 125	J = 127	
8	Cs = 133	Ba = 137	?Di = 138	?Ce = 140	—	—	—	— — —
9	(—)	—	—	—				
10	—	—	?Er = 178	?La = 180	Ta = 182	W = 184	—	Os = 195, Ir = 197, Pt = 198, Au = 199.
11	(Au = 199)	Hg = 200	Tl = 204	Pb = 207	Bi = 208	—	—	
12	—	—	—	Th = 231	—	U = 240	—	— — —

Figure 2-25 | Mendeleev's early periodic table, published in 1872. Note the spaces left for missing elements with atomic masses 44, 68, 72, and 100.

(From *Annalen der Chemie und Pharmacie*, VIII, Supplementary Volume for 1872, page 511.)

Table 2-3 | Comparison of the Properties of Germanium as Predicted by Mendeleev and as Actually Observed

Properties of Germanium	Predicted in 1871	Observed in 1886
Atomic weight	72	72.3
Density	5.5 g/cm^3	5.47 g/cm^3
Specific heat	0.31 J/(°C · g)	0.32 J/(°C · g)
Melting point	Very high	960°C
Oxide formula	RO$_2$	GeO$_2$
Oxide density	4.7 g/cm^3	4.70 g/cm^3
Chloride formula	RCl$_4$	GeCl$_4$
Boiling point of chloride	100°C	86°C

Table 2-4 | Predicted Properties of Elements 113 and 114

Property	Element 113	Element 114
Chemically like	Thallium	Lead
Atomic mass	297	298
Density	16 g/mL	14 g/mL
Melting point	430°C	70°C
Boiling point	1100°C	150°C

A current version of the periodic table is shown in Fig. 2-26. The only fundamental difference between this table and that of Mendeleev is that it lists the elements in order by atomic number rather than by atomic mass. The reason for this will become clear later in this chapter as we explore the electron arrangements of the atom. Another recent format of the table is discussed in the following section.

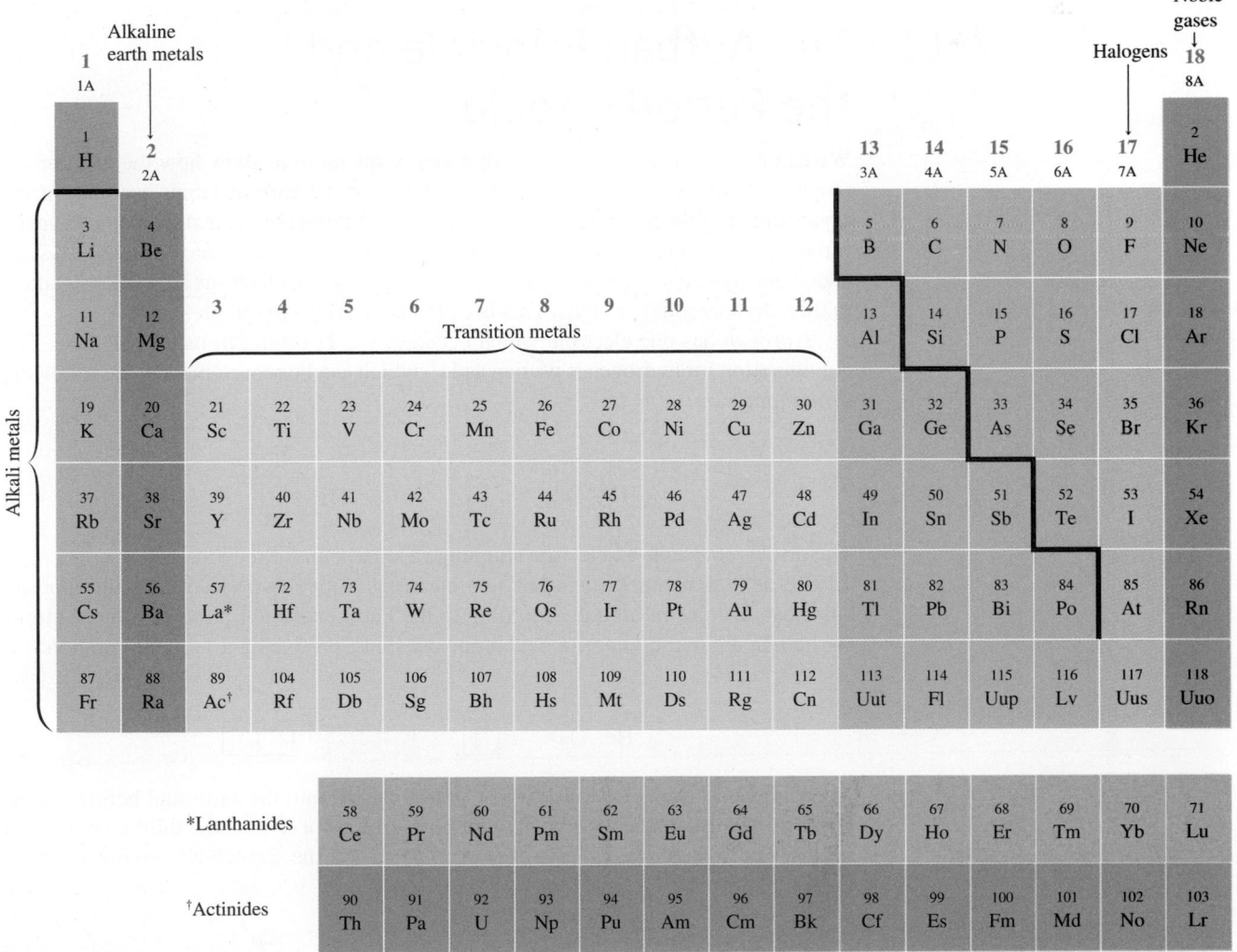

Figure 2-26 | The periodic table.

Although element 112 is not exactly a newborn (it was discovered in 1996), it has just been given a name. The International Union of Pure and Applied Chemistry announced on February 19, 2010, that the official name of 112 is Copernicium (pronounced koh-pur-NEE-see-um) and has the symbol Cn. The name honors Nicolaus Copernicus, who is best known for first recognizing that the planetary system is sun-centered rather than earth-centered.

Now scientists from the Paul Sherrer Institute in Switzerland and the Joint Institute for Nuclear Research in Russia have studied the properties of Cn. By firing a beam of ^{48}Ca projectiles at a target of ^{242}Pu (doped with Nd_2O_3), these

scientists have produced just two atoms of Cn. Given the relatively long lifetime of Cn atoms (a few seconds), the researchers have been able to compare the properties of Cn, Hg, and Ar atoms. When these atoms were injected into a series of gold-covered detectors, it was found that, similar to Hg (and in contrast to Ar), Cn atoms were "mildly" volatile and readily formed bonds with gold. Thus, as predicted by the periodic table, the element Cn, which has an expected electron configuration of $[Rn]7s^26d^{10}$, fits with the Zn, Cd, and Hg group of elements, all of which have the $ns^2(n-1)d^{10}$ configuration.

2-11 | The Aufbau Principle and the Periodic Table

We can use the quantum mechanical model of the atom to show how the electron arrangements in the hydrogenlike atomic orbitals of the various atoms account for the organization of the periodic table. Our main assumption here is that all atoms have the same type of orbitals as have been described for the hydrogen atom. As protons are added one by one to the nucleus to build up the elements, electrons are similarly added to these hydrogenlike orbitals. This is called the **aufbau principle**. ◄ ⓘ

Hydrogen has one electron, which occupies the $1s$ orbital in its ground state. The configuration for hydrogen is written as $1s^1$, which can be represented by the following *orbital diagram:*

> ⓘ *Aufbau* is German for "building up."
>
> H ($Z = 1$)
> He ($Z = 2$)
> Li ($Z = 3$)
> Be ($Z = 4$)
> B ($Z = 5$)
> etc.
> (Z = atomic number)

$$\text{H: } 1s^1 \quad \begin{array}{ccc} 1s & 2s & 2p \\ \boxed{\uparrow} & \square & \square\square\square \end{array}$$

The arrow represents an electron spinning in a particular direction.

The next element, *helium*, has two electrons. Since two electrons with opposite spins can occupy an orbital, according to the Pauli exclusion principle, the electrons for helium are in the $1s$ orbital with opposite spins, producing a $1s^2$ configuration:

$$\text{He: } 1s^2 \quad \begin{array}{ccc} 1s & 2s & 2p \\ \boxed{\uparrow\downarrow} & \square & \square\square\square \end{array}$$

Lithium has three electrons, two of which can go into the $1s$ orbital before the orbital is filled. Since the $1s$ orbital is the only orbital for $n = 1$, the third electron will occupy the lowest-energy orbital with $n = 2$, or the $2s$ orbital, giving a $1s^22s^1$ configuration:

$$\text{Li: } 1s^22s^1 \quad \begin{array}{ccc} 1s & 2s & 2p \\ \boxed{\uparrow\downarrow} & \boxed{\uparrow} & \square\square\square \end{array}$$

The next element, *beryllium*, has four electrons, which occupy the $1s$ and $2s$ orbitals:

$$\text{Be: } 1s^2 2s^2 \qquad \begin{array}{ccc} 1s & 2s & 2p \\ [\uparrow\downarrow] & [\uparrow\downarrow] & [\ ][\ ][\ ] \end{array}$$

Boron has five electrons, four of which occupy the $1s$ and $2s$ orbitals. The fifth electron goes into the second type of orbital with $n = 2$, the $2p$ orbitals:

$$\text{B: } 1s^2 2s^2 2p^1 \qquad \begin{array}{ccc} 1s & 2s & 2p \\ [\uparrow\downarrow] & [\uparrow\downarrow] & [\uparrow][\ ][\ ] \end{array}$$

Since all the $2p$ orbitals have the same energy (are degenerate), it does not matter which $2p$ orbital the electron occupies.

Carbon is the next element and has six electrons. Two electrons occupy the $1s$ orbital, two occupy the $2s$ orbital, and two occupy $2p$ orbitals. Since there are three $2p$ orbitals with the same energy, the mutually repulsive electrons will occupy *separate* $2p$ orbitals. This behavior is summarized by **Hund's rule** (named for the German physicist F. H. Hund): The lowest energy configuration for an atom is the one having the maximum number of unpaired electrons allowed by the Pauli principle in a particular set of degenerate orbitals. By convention, the unpaired electrons are represented as having parallel spins (with spin "up"). ◄

The configuration for carbon could be written $1s^2 2s^2 2p^1 2p^1$ to indicate that the electrons occupy separate $2p$ orbitals. However, the configuration is usually given as $1s^2 2s^2 2p^2$, and it is understood that the electrons are in different $2p$ orbitals. The orbital diagram for carbon is

$$\text{C: } 1s^2 2s^2 2p^2 \qquad \begin{array}{ccc} 1s & 2s & 2p \\ [\uparrow\downarrow] & [\uparrow\downarrow] & [\uparrow][\uparrow][\ ] \end{array}$$

Note that the unpaired electrons in the $2p$ orbitals are shown with parallel spins.

The configuration for *nitrogen*, which has seven electrons, is $1s^2 2s^2 2p^3$. The three electrons in the $2p$ orbitals occupy separate orbitals with parallel spins:

$$\text{N: } 1s^2 2s^2 2p^3 \qquad \begin{array}{ccc} 1s & 2s & 2p \\ [\uparrow\downarrow] & [\uparrow\downarrow] & [\uparrow][\uparrow][\uparrow] \end{array}$$

The configuration for *oxygen*, which has eight electrons, is $1s^2 2s^2 2p^4$. One of the $2p$ orbitals is now occupied by a pair of electrons with opposite spins, as required by the Pauli exclusion principle:

$$\text{O: } 1s^2 2s^2 2p^4 \qquad \begin{array}{ccc} 1s & 2s & 2p \\ [\uparrow\downarrow] & [\uparrow\downarrow] & [\uparrow\downarrow][\uparrow][\uparrow] \end{array}$$

The orbital diagrams and electron configurations for *fluorine* (nine electrons) and *neon* (ten electrons) are as follows:

$$\begin{array}{lccc} & & 1s & 2s \qquad\qquad 2p \\ \text{F: } & 1s^2 2s^2 2p^5 & [\uparrow\downarrow] & [\uparrow\downarrow] \quad [\uparrow\downarrow][\uparrow\downarrow][\uparrow] \\ \text{Ne: } & 1s^2 2s^2 2p^6 & [\uparrow\downarrow] & [\uparrow\downarrow] \quad [\uparrow\downarrow][\uparrow\downarrow][\uparrow\downarrow] \end{array}$$

With neon, the orbitals with $n = 1$ and $n = 2$ are now completely filled.

For *sodium*, the first ten electrons occupy the $1s$, $2s$, and $2p$ orbitals, and the eleventh electron must occupy the first orbital with $n = 3$, the $3s$ orbital. ◄ The electron configuration for sodium is $1s^2 2s^2 2p^6 3s^1$. To avoid writing the inner-level electrons,

i For an atom with unfilled sub-shells, the lowest energy is achieved by electrons occupying separate orbitals with parallel spins, as far as allowed by the Pauli exclusion principle.

© Cengage Learning

▲ Sodium metal is so reactive that it is stored under kerosene to protect it from the oxygen in the air.

Figure 2-27 | The electron configurations in the type of orbital occupied last for the first 18 elements.

H $1s^1$									He $1s^2$
Li $2s^1$	Be $2s^2$		B $2p^1$	C $2p^2$	N $2p^3$	O $2p^4$	F $2p^5$	Ne $2p^6$
Na $3s^1$	Mg $3s^2$		Al $3p^1$	Si $3p^2$	P $3p^3$	S $3p^4$	Cl $3p^5$	Ar $3p^6$

ⓘ [Ne] is shorthand for $1s^22s^22p^6$.

this configuration is often abbreviated as [Ne]$3s^1$, where [Ne] represents the electron configuration of neon, $1s^22s^22p^6$. ◀ ⓘ

The next element, *magnesium*, has the configuration $1s^22s^22p^63s^2$, or [Ne]$3s^2$. Then the next six elements, *aluminum* through *argon*, have configurations obtained by filling the 3p orbitals one electron at a time. Fig. 2-27 summarizes the electron configurations of the first 18 elements by giving the number of electrons in the type of orbital occupied last.

At this point it is useful to introduce the concept of **valence electrons**, *the electrons in the outermost principal quantum level of a main group atom.* The valence electrons of the nitrogen atom, for example, are the 2s and 2p electrons. For the sodium atom, the valence electron is the electron in the 3s orbital, and so on. Valence electrons are the most important electrons to chemists because they are involved in bonding, as we will see in the next two chapters. The inner electrons are known as **core electrons**.

Note in Fig. 2-27 that a very important pattern is developing: The elements in the same group (vertical column of the periodic table) have the same valence electron configuration. Remember that Mendeleev originally placed the elements in groups based on similarities in chemical properties. Now we understand the reason behind these groupings. Elements with the same valence electron configuration show similar chemical behavior.

▲ A vial containing potassium metal. The sealed vial contains an inert gas to protect the potassium from reacting with oxygen.

ⓘ Note that our observations of the similar chemical reactivity of sodium and potassium lead us to an understanding of the relative energies of the orbitals.

The element after argon is *potassium.* ◀ Since the 3p orbitals are fully occupied in argon, we might expect the next electron to go into a 3d orbital (recall that for $n = 3$ the orbitals are 3s, 3p, and 3d). However, the chemistry of potassium is clearly very similar to that of lithium and sodium, indicating that the last electron in potassium occupies the 4s orbital instead of one of the 3d orbitals, a conclusion confirmed by many types of experiments. ◀ ⓘ The electron configuration of potassium is

$$\text{K:}\quad 1s^22s^22p^63s^23p^64s^1 \quad \text{or} \quad [\text{Ar}]4s^1$$

The next element is *calcium:* ◀

$$\text{Ca:}\quad [\text{Ar}]4s^2$$

The next element, *scandium*, begins a series of 10 elements (scandium through zinc) called the **transition metals**, whose configurations are obtained by adding electrons to the five 3d orbitals. The configurations of the first three transition elements are:

$$\text{Sc:}\quad [\text{Ar}]4s^23d^1$$

$$\text{Ti:}\quad [\text{Ar}]4s^23d^2$$

$$\text{V:}\quad [\text{Ar}]4s^23d^3$$

Chromium is the next element. ◀ The expected configuration is [Ar]$4s^23d^4$. However, the observed configuration is

$$\text{Cr:}\quad [\text{Ar}]4s^13d^5$$

▲ Calcium metal.

▲ Chromium is often used to plate bumpers and hood ornaments, such as this statue of Mercury found on a 1929 Buick.

The explanation for this configuration of chromium is beyond the scope of this book. In fact, chemists are still disagreeing over the exact cause of this anomaly. Note, however, that the observed configuration has both the $4s$ and $3d$ orbitals half-filled. This is a good way to remember the correct configuration.

The next four elements, *manganese* through *nickel*, have the expected configurations:

$$\text{Mn: } [\text{Ar}]4s^23d^5 \quad \text{Co: } [\text{Ar}]4s^23d^7$$

$$\text{Fe: } [\text{Ar}]4s^23d^6 \quad \text{Ni: } [\text{Ar}]4s^23d^8$$

The configuration for *copper* is expected to be $[\text{Ar}]4s^23d^9$. However, the observed configuration is

$$\text{Cu: } [\text{Ar}]4s^13d^{10}$$

In this case, a half-filled $4s$ orbital and a filled set of $3d$ orbitals characterize the actual configuration.

Zinc has the expected configuration:

$$\text{Zn: } [\text{Ar}]4s^23d^{10}$$

The configurations of the transition metals are shown in Fig. 2-28. After that, the next six elements, *gallium* through *krypton*, have configurations that correspond to filling the $4p$ orbitals (Fig. 2-28).

The entire periodic table is represented in Fig. 2-29 in terms of which orbitals are being filled. The valence electron configurations are given in Fig. 2-30. From these two figures, note the following additional points:

ⓘ The $(n + 1)s$ orbitals fill before the nd orbitals.

1. The $(n + 1)s$ orbitals always fill before the nd orbitals. ◄ **ⓘ** For example, the $5s$ orbitals fill in rubidium and strontium before the $4d$ orbitals fill in the second row of transition metals (yttrium through cadmium). This early filling of the s orbitals can be explained by the penetration effect. For example, the $4s$ orbital allows for so much more penetration to the vicinity of the nucleus that it becomes lower in energy than the $3d$ orbital. Thus the $4s$ fills before the $3d$. The same things can be said about the $5s$ and $4d$, the $6s$ and $5d$, and the $7s$ and $6d$ orbitals.

ⓘ Lanthanides are elements in which the $4f$ orbitals are being filled.

2. After lanthanum, which has the configuration $[\text{Xe}]6s^25d^1$, a group of 14 elements called the **lanthanide series**, or the **lanthanides**, occurs. This series of elements corresponds to the filling of the seven $4f$ orbitals. ◄ **ⓘ** Note that sometimes an electron occupies a $5d$ orbital instead of a $4f$ orbital. This occurs because the energies of the $4f$ and $5d$ orbitals are very similar.

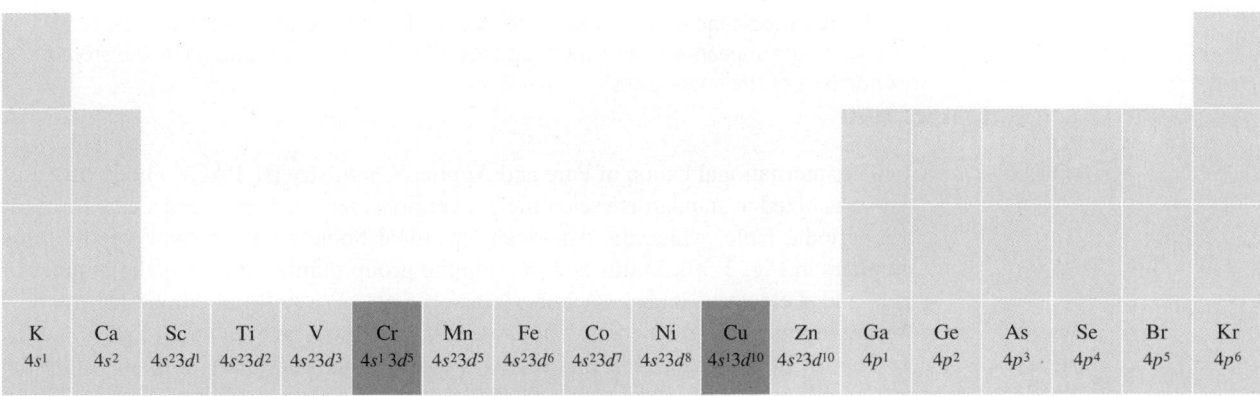

Figure 2-28 | Valence electron configurations for potassium through krypton. The transition metals (scandium through zinc) have the general configuration $[\text{Ar}]4s^23d^n$, except for chromium and copper.

Figure 2-29 | The orbitals being filled for elements in various parts of the periodic table. Note that in going along a horizontal row (a period), the $(n + 1)s$ orbital fills before the nd orbital. The group labels indicate the number of valence electrons (ns plus np electrons) for the elements in each group.

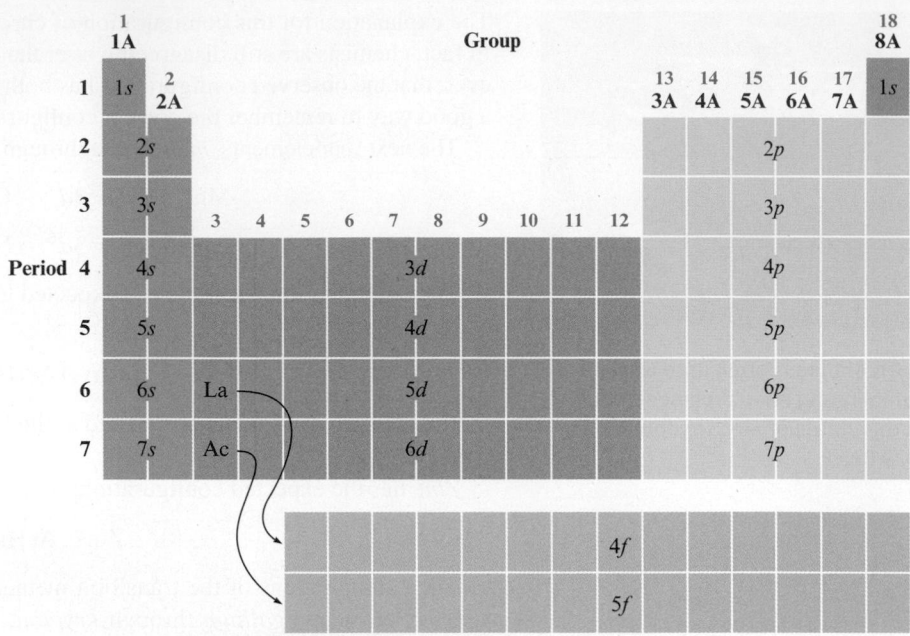

ⓘ Actinides are elements in which the $5f$ orbitals are being filled.

ⓘ The group label tells the total number of valence electrons for that group.

3. After actinium, which has the configuration $[Rn]7s^2 6d^1$, a group of 14 elements called the **actinide series**, or the **actinides**, occurs. This series corresponds to the filling of the seven $5f$ orbitals. ◄ **ⓘ** Note that sometimes one or two electrons occupy the $6d$ orbitals instead of the $5f$ orbitals, because these orbitals have very similar energies.

4. The group labels for Groups 1A, 2A, 3A, 4A, 5A, 6A, 7A, and 8A indicate the *total number* of valence electrons for the atoms in these groups. ◄ **ⓘ** For example, all the elements in Group 5A have the configuration $ns^2 np^3$. (The d electrons fill one period late and are usually not counted as valence electrons.) The meaning of the group labels for the transition metals is not as clear as for the Group A elements, and these will not be used in this text.

5. The groups labeled 1A, 2A, 3A, 4A, 5A, 6A, 7A, and 8A are often called the **main-group**, or **representative**, **elements**. Every member of these groups has the same valence electron configuration.

Critical Thinking

You have learned that each orbital is allowed 2 electrons and this pattern is evident on the periodic table. What if each orbital was allowed 3 electrons? How would this change the appearance of the periodic table? For example, what would be the atomic numbers of the noble gases?

The International Union of Pure and Applied Chemistry (IUPAC), a body of scientists organized to standardize scientific conventions, has recommended a new form for the periodic table, which the American Chemical Society has adopted (see the blue numbers in Fig. 2-30). In this new version the group number indicates the number of s, p, and d electrons added since the last noble gas. We will not emphasize the new format in this book. As a compromise we will usually show both numbering systems.

The results considered in this section are very important. We have seen that the quantum mechanical model can be used to explain the arrangement of the elements in the periodic table. This model allows us to understand that the similar chemistry exhibited by the members of a given group arises from the fact that they all have the

Transition metals

Representative Elements | *d*-Transition Elements | Representative Elements | Noble Gases

Period number, highest occupied electron level

Period	1 1A ns^1	2 2A ns^2	3	4	5	6	7	8	9	10	11	12	13 3A ns^2np^1	14 4A ns^2np^2	15 5A ns^2np^3	16 6A ns^2np^4	17 7A ns^2np^5	18 8A ns^2np^6
1	1 H $1s^1$																	2 He $1s^2$
2	3 Li $2s^1$	4 Be $2s^2$											5 B $2s^22p^1$	6 C $2s^22p^2$	7 N $2s^22p^3$	8 O $2s^22p^4$	9 F $2s^22p^5$	10 Ne $2s^22p^6$
3	11 Na $3s^1$	12 Mg $3s^2$											13 Al $3s^23p^1$	14 Si $3s^23p^2$	15 P $3s^23p^3$	16 S $3s^23p^4$	17 Cl $3s^23p^5$	18 Ar $3s^23p^6$
4	19 K $4s^1$	20 Ca $4s^2$	21 Sc $4s^23d^1$	22 Ti $4s^23d^2$	23 V $4s^23d^3$	24 Cr $4s^13d^5$	25 Mn $4s^23d^5$	26 Fe $4s^23d^6$	27 Co $4s^23d^7$	28 Ni $4s^23d^8$	29 Cu $4s^13d^{10}$	30 Zn $4s^23d^{10}$	31 Ga $4s^24p^1$	32 Ge $4s^24p^2$	33 As $4s^24p^3$	34 Se $4s^24p^4$	35 Br $4s^24p^5$	36 Kr $4s^24p^6$
5	37 Rb $5s^1$	38 Sr $5s^2$	39 Y $5s^24d^1$	40 Zr $5s^24d^2$	41 Nb $5s^14d^4$	42 Mo $5s^14d^5$	43 Tc $5s^14d^6$	44 Ru $5s^14d^7$	45 Rh $5s^14d^8$	46 Pd $4d^{10}$	47 Ag $5s^14d^{10}$	48 Cd $5s^24d^{10}$	49 In $5s^25p^1$	50 Sn $5s^25p^2$	51 Sb $5s^25p^3$	52 Te $5s^25p^4$	53 I $5s^25p^5$	54 Xe $5s^25p^6$
6	55 Cs $6s^1$	56 Ba $6s^2$	57 La* $6s^25d^1$	72 Hf $4f^{14}6s^25d^2$	73 Ta $6s^25d^3$	74 W $6s^25d^4$	75 Re $6s^25d^5$	76 Os $6s^25d^6$	77 Ir $6s^25d^7$	78 Pt $6s^15d^9$	79 Au $6s^15d^{10}$	80 Hg $6s^25d^{10}$	81 Tl $6s^26p^1$	82 Pb $6s^26p^2$	83 Bi $6s^26p^3$	84 Po $6s^26p^4$	85 At $6s^26p^5$	86 Rn $6s^26p^6$
7	87 Fr $7s^1$	88 Ra $7s^2$	89 Ac** $7s^26d^1$	104 Rf $7s^26d^2$	105 Db $7s^26d^3$	106 Sg $7s^26d^4$	107 Bh $7s^26d^5$	108 Hs $7s^26d^6$	109 Mt $7s^26d^7$	110 Ds $7s^26d^8$	111 Rg $7s^16d^{10}$	112 Cn $7s^26d^{10}$	113 Uut $7s^27p^1$	114 Fl $7s^27p^2$	115 Uup $7s^27p^3$	116 Lv $7s^27p^4$	117 Uus $7s^27p^5$	118 Uuo $7s^27p^6$

f-Transition Elements

*Lanthanides	58 Ce $6s^24f^15d^1$	59 Pr $6s^24f^35d^0$	60 Nd $6s^24f^45d^0$	61 Pm $6s^24f^55d^0$	62 Sm $6s^24f^65d^0$	63 Eu $6s^24f^75d^0$	64 Gd $6s^24f^75d^1$	65 Tb $6s^24f^95d^0$	66 Dy $6s^24f^{10}5d^0$	67 Ho $6s^24f^{11}5d^0$	68 Er $6s^24f^{12}5d^0$	69 Tm $6s^24f^{13}5d^0$	70 Yb $6s^24f^{14}5d^0$	71 Lu $6s^24f^{14}5d^1$
**Actinides	90 Th $7s^25f^06d^2$	91 Pa $7s^25f^26d^1$	92 U $7s^25f^36d^1$	93 Np $7s^25f^46d^1$	94 Pu $7s^25f^66d^0$	95 Am $7s^25f^76d^0$	96 Cm $7s^25f^76d^1$	97 Bk $7s^25f^96d^0$	98 Cf $7s^25f^{10}6d^0$	99 Es $7s^25f^{11}6d^0$	100 Fm $7s^25f^{12}6d^0$	101 Md $7s^25f^{13}6d^0$	102 No $7s^25f^{14}6d^0$	103 Lr $7s^25f^{14}6d^1$

Figure 2-30 | The periodic table with atomic symbols, atomic numbers, and partial electron configurations.

same valence electron configuration. Only the principal quantum number of the valence orbitals changes in going down a particular group.

It is important to be able to give the electron configuration for each of the main-group elements. This is most easily done by using the periodic table. If you understand how the table is organized, it is not necessary to memorize the order in which the orbitals fill. ◄ Review Figs. 2-29 and 2-30 to make sure that you understand the correspondence between the orbitals and the periods and groups.

When an electron configuration is given in this text, the orbitals are listed in the order in which they fill.

Predicting the configurations of the transition metals (3*d*, 4*d*, and 5*d* elements), the lanthanides (4*f* elements), and the actinides (5*f* elements) is somewhat more difficult because there are many exceptions of the type encountered in the first-row transition metals (the 3*d* elements). You should memorize the configurations of chromium and copper, the two exceptions in the first-row transition metals, since these elements are often encountered. ◄

Cr: $[Ar]4s^13d^5$
Cu: $[Ar]4s^13d^{10}$

Figure 2-31 | The positions of the elements considered in Example 2-7.

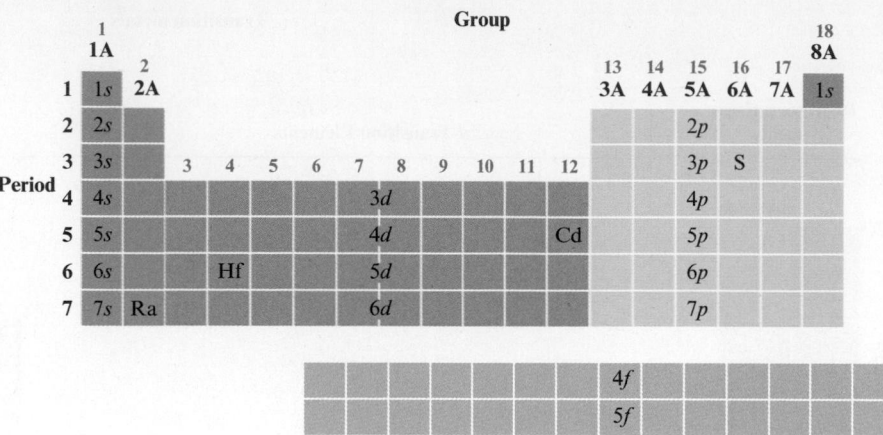

INTERACTIVE EXAMPLE 2-7

Electron Configurations

Give the electron configurations for sulfur (S), cadmium (Cd), hafnium (Hf), and radium (Ra) using the periodic table inside the front cover of this book.

Solution

Sulfur is element 16 and resides in Period 3, where the $3p$ orbitals are being filled (Fig. 2-31). Since sulfur is the fourth among the "$3p$ elements," it must have four $3p$ electrons. Its configuration is

$$S: \quad 1s^2 2s^2 2p^6 3s^2 3p^4 \quad \text{or} \quad [\text{Ne}]3s^2 3p^4$$

Cadmium is element 48 and is located in Period 5 at the end of the $4d$ transition metals, as shown in Fig. 2-30. It is the tenth element in the series and thus has 10 electrons in the $4d$ orbitals, in addition to the 2 electrons in the $5s$ orbital. The configuration is

$$Cd: \quad 1s^2 2s^2 2p^6 3s^2 3p^6 4s^2 3d^{10} 4p^6 5s^2 4d^{10} \quad \text{or} \quad [\text{Kr}]5s^2 4d^{10}$$

Hafnium is element 72 and is found in Period 6, as shown in Fig. 2-31. Note that it occurs just after the lanthanide series. Thus the $4f$ orbitals are already filled. Hafnium is the second member of the $5d$ transition series and has two $5d$ electrons. The configuration is

$$Hf: \quad 1s^2 2s^2 2p^6 3s^2 3p^6 4s^2 3d^{10} 4p^6 5s^2 4d^{10} 5p^6 6s^2 4f^{14} 5d^2 \quad \text{or} \quad [\text{Xe}]6s^2 4f^{14} 5d^2$$

Radium is element 88 and is in Period 7 (and Group 2A), as shown in Fig. 2-31. Thus radium has two electrons in the $7s$ orbital, and the configuration is

$$Ra: \quad 1s^2 2s^2 2p^6 3s^2 3p^6 4s^2 3d^{10} 4p^6 5s^2 4d^{10} 5p^6 6s^2 4f^{14} 5d^{10} 6p^6 7s^2 \quad \text{or} \quad [\text{Rn}]7s^2$$

See Exercises 2-85 through 2-88

2-12 | Periodic Trends in Atomic Properties

We have developed a fairly complete picture of polyelectronic atoms. Although the model is rather crude because the nuclear attractions and electron repulsions are simply lumped together, it is very successful in accounting for the periodic table of elements. We will next use the model to account for the observed trends in several important atomic properties: atomic radius, ionization energy, and electron affinity.

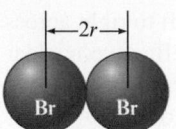

Figure 2-32 | The radius of an atom (*r*) is defined as half the distance between the nuclei in a molecule consisting of identical atoms.

Atomic Radius

Just as the size of an orbital cannot be specified exactly, neither can the size of an atom. We must make some arbitrary choices to obtain values for **atomic radii**. These values can be obtained by measuring the distances between atoms in chemical compounds. For example, in the bromine molecule, the distance between the two nuclei is known to be 228 pm. The bromine atomic radius is assumed to be half this distance, or 114 pm, as shown in Fig. 2-32. These radii are often called *covalent atomic radii* because of the way they are determined (from the distances between atoms in covalent bonds).

For nonmetallic atoms that do not form diatomic molecules, the atomic radii are estimated from their various covalent compounds. The radii for metal atoms (called *metallic radii*) are obtained from half the distance between metal atoms in solid metal crystals.

The values of the atomic radii for the representative elements are shown in Fig. 2-33. Note that these values are significantly smaller than might be expected from the 90% electron density volumes of isolated atoms, because when atoms form bonds, their electron "clouds" interpenetrate. However, these values form a self-consistent data set that can be used to discuss the trends in atomic radii.

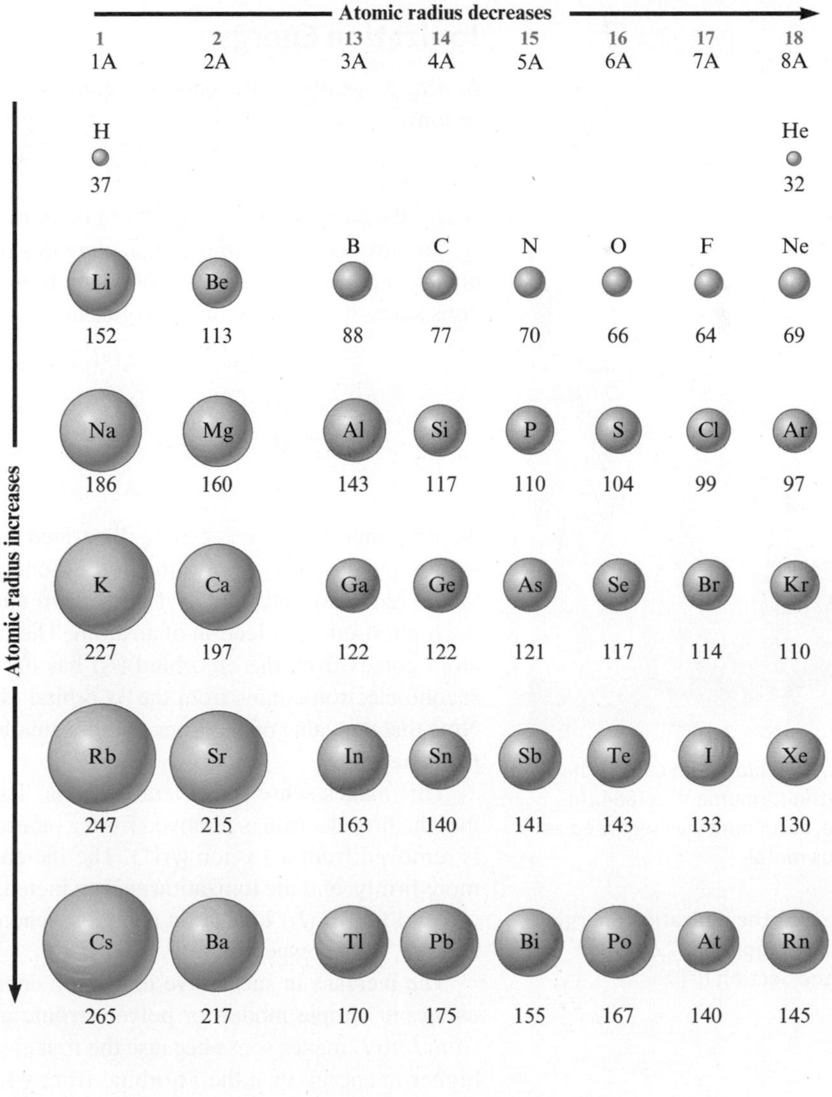

Figure 2-33 | Atomic radii (in picometers) for selected atoms. Note that atomic radius decreases going across a period and increases going down a group. The values for the noble gases are estimated, because data from bonded atoms are lacking.

Atomic radius decreases →

Atomic radius increases ↓

1 1A	2 2A	13 3A	14 4A	15 5A	16 6A	17 7A	18 8A
H 37							He 32
Li 152	Be 113	B 88	C 77	N 70	O 66	F 64	Ne 69
Na 186	Mg 160	Al 143	Si 117	P 110	S 104	Cl 99	Ar 97
K 227	Ca 197	Ga 122	Ge 122	As 121	Se 117	Br 114	Kr 110
Rb 247	Sr 215	In 163	Sn 140	Sb 141	Te 143	I 133	Xe 130
Cs 265	Ba 217	Tl 170	Pb 175	Bi 155	Po 167	At 140	Rn 145

Note from Fig. 2-33 that the atomic radii decrease in going from left to right across a period. This decrease can be explained in terms of the increasing effective nuclear charge (decreasing shielding) in going from left to right. This means that the valence electrons are drawn closer to the nucleus, decreasing the size of the atom.

Atomic radius increases down a group, because of the increases in the orbital sizes in successive principal quantum levels.

INTERACTIVE EXAMPLE 2-8

Trends in Radii

Predict the trend in radius for the following ions: Be^{2+}, Mg^{2+}, Ca^{2+}, and Sr^{2+}.

Solution

All these ions are formed by removing two electrons from an atom of a Group 2A element. In going from beryllium to strontium, we are going down the group, so the sizes increase:

$$Be^{2+} < Mg^{2+} < Ca^{2+} < Sr^{2+}$$

↑ Smallest radius   ↑ Largest radius

See Exercises 2-107, 2-108, and 2-111

Ionization Energy

Ionization energy is the energy required to remove an electron from a gaseous atom or ion:

$$X(g) \longrightarrow X^{+}(g) + e^{-}$$

*ⓘ Ionization energy results in the formation of a **positive** ion.*

where the atom or ion is assumed to be in its ground state. ◄ ⓘ

To introduce some of the characteristics of ionization energy, we will consider the energy required to remove several electrons in succession from aluminum in the gaseous state. ◄ The ionization energies are

$$Al(g) \longrightarrow Al^{+}(g) + e^{-} \qquad I_1 = 580 \text{ kJ/mol}$$
$$Al^{+}(g) \longrightarrow Al^{2+}(g) + e^{-} \qquad I_2 = 1815 \text{ kJ/mol}$$
$$Al^{2+}(g) \longrightarrow Al^{3+}(g) + e^{-} \qquad I_3 = 2740 \text{ kJ/mol}$$
$$Al^{3+}(g) \longrightarrow Al^{4+}(g) + e^{-} \qquad I_4 = 11,600 \text{ kJ/mol}$$

Several important points can be illustrated from these results. ◄ ⓘ In a stepwise ionization process, it is always the highest-energy electron (the one bound least tightly) that is removed first. The **first ionization energy** I_1 is the energy required to remove the highest-energy electron of an atom. The first electron removed from the aluminum atom comes from the $3p$ orbital (Al has the electron configuration $[Ne]3s^2 3p^1$). The second electron comes from the $3s$ orbital (since Al^{+} has the configuration $[Ne]3s^2$). Note that the value of I_1 is considerably smaller than the value of I_2, the **second ionization energy**.

This makes sense for several reasons. The primary factor is simply charge. Note that the first electron is removed from a neutral atom (Al), whereas the second electron is removed from a $1+$ ion (Al^{+}). The increase in positive charge binds the electrons more firmly, and the ionization energy increases. The same trend shows up in the third (I_3) and fourth (I_4) ionization energies, where the electron is removed from the Al^{2+} and Al^{3+} ions, respectively.

The increase in successive ionization energies for an atom also can be interpreted using our simple model for polyelectronic atoms. The increase in ionization energy from I_1 to I_2 makes sense because the first electron is removed from a $3p$ orbital that is higher in energy than the $3s$ orbital from which the second electron is removed. The

▲ Setting the aluminum cap on the Washington Monument in 1884. At that time, aluminum was regarded as a precious metal.

ⓘ Note that the ionization energies are given per mole (6.022×10^{23}) events (see Section R-10).

The Granger Collection, New York

Table 2-5 | Successive Ionization Energies in Kilojoules per Mole for the Elements in Period 3

Element	I_1	I_2	I_3	I_4	I_5	I_6	I_7
Na	495	4560					
Mg	735	1445	7730	Core electrons*			
Al	580	1815	2740	11,600			
Si	780	1575	3220	4350	16,100		
P	1060	1890	2905	4950	6270	21,200	
S	1005	2260	3375	4565	6950	8490	27,000
Cl	1255	2295	3850	5160	6560	9360	11,000
Ar	1527	2665	3945	5770	7230	8780	12,000

← General decrease

*Note the large jump in ionization energy in going from removal of valence electrons to removal of core electrons.

——————— General increase ———————→

Table 2-6 | First Ionization Energies for the Alkali Metals and Noble Gases

Atom	I_1(kJ/mol)
Group 1A	
Li	520
Na	495
K	419
Rb	409
Cs	382
Group 8A	
He	2377
Ne	2088
Ar	1527
Kr	1356
Xe	1176
Rn	1042

largest jump in ionization energy by far occurs in going from the third ionization energy (I_3) to the fourth (I_4). This is so because I_4 corresponds to removing a core electron (Al^{3+} has the configuration $1s^22s^22p^6$), and core electrons are bound much more tightly than valence electrons.

Table 2-5 gives the values of ionization energies for all the Period 3 elements. Note the large jump in energy in each case in going from removal of valence electrons to removal of core electrons.

The values of the first ionization energies for the elements in the first six periods of the periodic table are graphed in Fig. 2-34. Note that in general as we go *across a period from left to right, the first ionization energy increases*. This is consistent with the idea that electrons added in the same principal quantum level do not completely shield the increasing nuclear charge caused by the added protons. Thus electrons in the same principal quantum level are generally more strongly bound as we move to the right on the periodic table, and there is a general increase in ionization energy values as electrons are added to a given principal quantum level.

On the other hand, *first ionization energy decreases in going down a group*. This can be seen most clearly by focusing on the Group 1A (1) elements (the alkali metals) and the Group 8A (18) elements (the noble gases), as shown in Table 2-6. The main reason for the decrease in ionization energy in going down a group is that the electrons

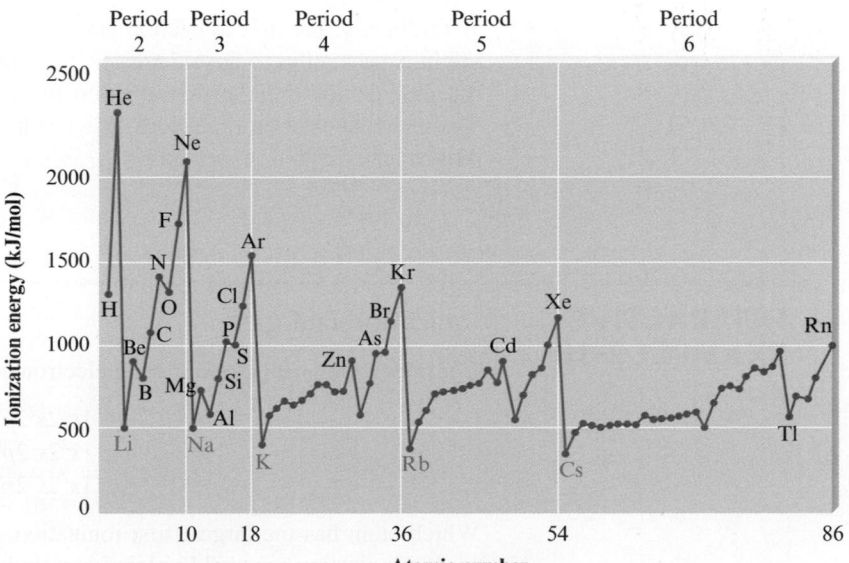

Figure 2-34 | The values of first ionization energy for the elements in the first six periods. In general, ionization energy decreases in going down a group. For example, note the decrease in values for Group 1A and Group 8A. In general, ionization energy increases in going left to right across a period. For example, note the sharp increase going across Period 2 from lithium through neon.

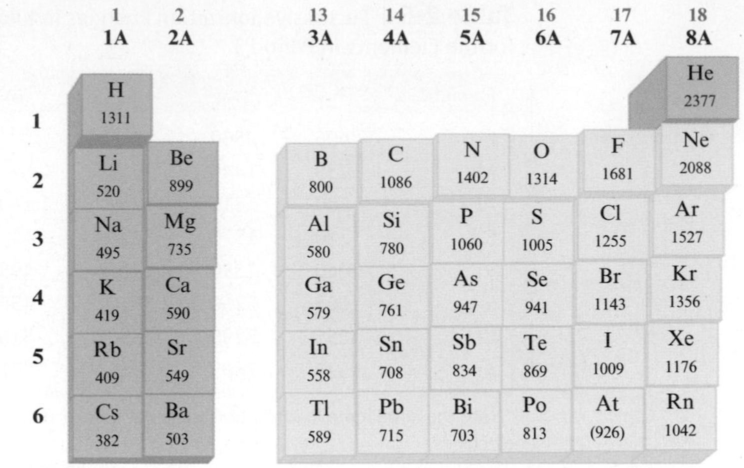

Figure 2-35 | Trends in ionization energies (kJ/mol) for the representative elements.

	1 1A	2 2A		13 3A	14 4A	15 5A	16 6A	17 7A	18 8A
1	H 1311								He 2377
2	Li 520	Be 899		B 800	C 1086	N 1402	O 1314	F 1681	Ne 2088
3	Na 495	Mg 735		Al 580	Si 780	P 1060	S 1005	Cl 1255	Ar 1527
4	K 419	Ca 590		Ga 579	Ge 761	As 947	Se 941	Br 1143	Kr 1356
5	Rb 409	Sr 549		In 558	Sn 708	Sb 834	Te 869	I 1009	Xe 1176
6	Cs 382	Ba 503		Tl 589	Pb 715	Bi 703	Po 813	At (926)	Rn 1042

being removed are, on average, farther from the nucleus. As n increases, the size of the orbital increases, and the electron is easier to remove.

In Fig. 2-34 we see that there are some discontinuities in ionization energy in going across a period. For example, for Period 2, discontinuities occur in going from beryllium to boron and from nitrogen to oxygen. These exceptions to the normal trend can be explained in terms of electron repulsions. The decrease in ionization energy in going from beryllium to boron reflects the fact that the electrons in the filled $2s$ orbital provide some shielding for electrons in the $2p$ orbital from the nuclear charge. The decrease in ionization energy in going from nitrogen to oxygen reflects the extra electron repulsions in the doubly occupied oxygen $2p$ orbital.

The ionization energies for the representative elements are summarized in Fig. 2-35. ◄ ⓘ

ⓘ First ionization energy increases across a period and decreases down a group.

| EXAMPLE 2-9 | Trends in Ionization Energies |

The first ionization energy for phosphorus is 1060 kJ/mol, and that for sulfur is 1005 kJ/mol. Why?

Solution

Phosphorus and sulfur are neighboring elements in Period 3 of the periodic table and have the following valence electron configurations: Phosphorus is $3s^2 3p^3$, and sulfur is $3s^2 3p^4$.

Ordinarily, the first ionization energy increases as we go across a period, so we might expect sulfur to have a greater ionization energy than phosphorus. However, in this case the fourth p electron in sulfur must be placed in an already occupied orbital. The electron–electron repulsions that result cause this electron to be more easily removed than might be expected.

See Exercises 2-115 and 2-116

| INTERACTIVE
EXAMPLE 2-10 | Ionization Energies |

Consider atoms with the following electron configurations:

$$1s^2 2s^2 2p^6$$
$$1s^2 2s^2 2p^6 3s^1$$
$$1s^2 2s^2 2p^6 3s^2$$

Which atom has the largest first ionization energy, and which one has the smallest second ionization energy? Explain your choices.

Solution

The atom with the largest value of I_1 is the one with the configuration $1s^2 2s^2 2p^6$ (this is the neon atom), because this element is found at the right end of Period 2. Since the $2p$ electrons do not shield each other very effectively, I_1 will be relatively large. The other configurations given include $3s$ electrons. These electrons are effectively shielded by the core electrons and are farther from the nucleus than the $2p$ electrons in neon. Thus I_1 for these atoms will be smaller than for neon.

The atom with the smallest value of I_2 is the one with the configuration $1s^2 2s^2 2p^6 3s^2$ (the magnesium atom). For magnesium, both I_1 and I_2 involve valence electrons. For the atom with the configuration $1s^2 2s^2 2p^6 3s^1$ (sodium), the second electron lost (corresponding to I_2) is a core electron (from a $2p$ orbital).

See Exercises 2-117 and 2-118

Electron Affinity

Electron affinity is associated with the production of a **negative** ion.

Electron affinity *is the energy change associated with the addition of an electron to a gaseous atom:* ◀ ⓘ

$$X(g) + e^- \longrightarrow X^-(g)$$

Because two different conventions have been used, there is a good deal of confusion in the chemical literature about the signs for electron affinity values. Electron affinity has been defined in many textbooks as the energy *released* when an electron is added to a gaseous atom. In this book we define electron affinity as a *change* in energy, which means that if the addition of the electron results in a lower energy, then the corresponding value for electron affinity will carry a negative sign.

Fig. 2-36 shows the electron affinity values for the atoms among the first 20 elements that form stable, isolated negative ions—that is, the atoms that undergo the addition of an electron as shown above. As expected, all these elements have negative (exothermic) electron affinities. Note that the *more negative* the energy, the greater the quantity of energy released. Although electron affinities generally become more negative from left to right across a period, there are several exceptions to this rule in each period. The dependence of electron affinity on atomic number can be explained by considering the changes in electron repulsions as a function of electron configurations. For example, the fact that the nitrogen atom does not form a stable, isolated $N^-(g)$ ion, whereas carbon forms $C^-(g)$, reflects the difference in the electron configurations of these atoms. An electron added to nitrogen ($1s^2 2s^2 2p^3$) to form the $N^-(g)$ ion ($1s^2 2s^2 2p^4$) would have to occupy a $2p$ orbital that already contains one electron. The extra repulsion between the electrons in this doubly occupied orbital causes $N^-(g)$ to be unstable. When an electron is added to carbon ($1s^2 2s^2 2p^2$) to form the $C^-(g)$ ion ($1s^2 2s^2 2p^3$), no such extra repulsions occur.

In contrast to the nitrogen atom, the oxygen atom can add one electron to form the stable $O^-(g)$ ion. Presumably oxygen's greater nuclear charge compared with that of nitrogen is sufficient to overcome the repulsion associated with putting a second

Figure 2-36 | The electron affinity values for atoms among the first 20 elements that form stable, isolated X^- ions. The lines shown connect adjacent elements. The absence of a line indicates missing elements (He, Be, N, Ne, Mg, and Ar) whose atoms do not add an electron exothermically and thus do not form stable, isolated X^- ions.

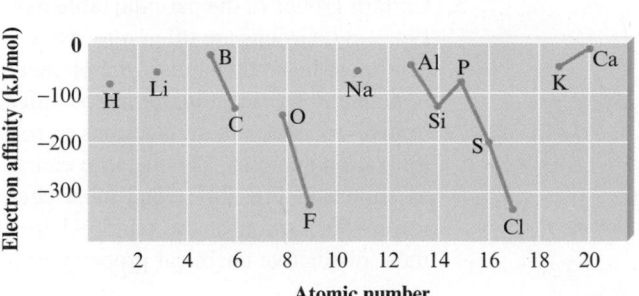

Table 2-7 | Electron Affinities of the Halogens

Atom	Electron Affinity (kJ/mol)
F	−327.8
Cl	−348.7
Br	−324.5
I	−295.2

electron into an already occupied $2p$ orbital. However, it should be noted that a second electron *cannot* be added to an oxygen atom $[\text{O}^-(g) + \text{e}^- \not\rightarrow \text{O}^{2-}(g)]$ to form an isolated oxide ion. This outcome seems strange in view of the many stable oxide compounds (MgO, Fe_2O_3, and so on) that are known. As we will discuss in detail in Chapter 3, the O^{2-} ion is stabilized in ionic compounds by the large attractions that occur among the positive ions and the oxide ions.

When we go down a group, electron affinity should become more positive (less energy released), since the electron is added at increasing distances from the nucleus. Although this is generally the case, the changes in electron affinity in going down most groups are relatively small, and numerous exceptions occur. This behavior is demonstrated by the electron affinities of the Group 7A (17) elements (the halogens) shown in Table 2-7. Note that the range of values is quite small compared with the changes that typically occur across a period. Also note that although chlorine, bromine, and iodine show the expected trend, the energy released when an electron is added to fluorine is smaller than might be expected. This smaller energy release has been attributed to the small size of the $2p$ orbitals. Because the electrons must be very close together in these orbitals, there are unusually large electron–electron repulsions. In the other halogens with their larger orbitals, the repulsions are not as severe.

2-13 | The Properties of a Group: The Alkali Metals

We have seen that the periodic table originated as a way to portray the systematic properties of the elements. Mendeleev was primarily responsible for first showing its usefulness in correlating and predicting the elemental properties. In this section we will summarize much of the information available from the table. We also will illustrate the usefulness of the table by discussing the properties of a representative group, the alkali metals.

Information Contained in the Periodic Table

1. The essence of the periodic table is that the groups of representative elements exhibit similar chemical properties that change in a regular way. The quantum mechanical model of the atom has allowed us to understand the basis for the similarity of properties in a group—that each group member of the representative elements has the same valence electron configuration. *It is the number and type of valence electrons that primarily determine an atom's chemistry.*

2. One of the most valuable types of information available from the periodic table is the electron configuration of any representative element. If you understand the organization of the table, you will not need to memorize electron configurations for these elements. Although the predicted electron configurations for transition metals are sometimes incorrect, this is not a serious problem. You should, however, memorize the configurations of two exceptions, chromium and copper, since these $3d$ transition elements are found in many important compounds.

3. Certain groups in the periodic table have special names. These are summarized in Fig. 2-37. Groups are often referred to by these names, so you should learn them.

4. The most basic division of the elements in the periodic table is into metals and nonmetals. The most important chemical property of a metal atom is the tendency to give up one or more electrons to form a positive ion; metals tend to have low ionization energies. The metallic elements are found on the left side of the table, as shown in Fig. 2-37. The most chemically reactive metals are found on the lower-left portion of the table, where the ionization energies are smallest. The most distinctive chemical property of a nonmetal atom is the ability to gain one

Figure 2-37 | Special names for groups in the periodic table.

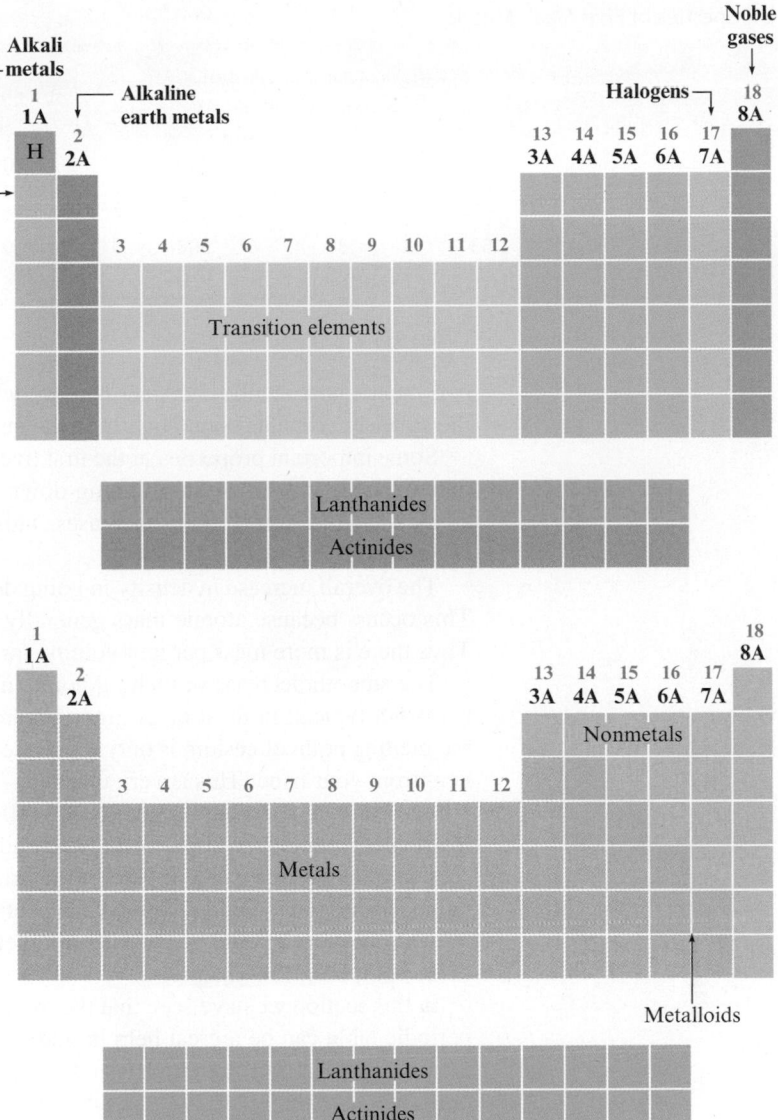

or more electrons to form an anion when reacting with a metal. Thus nonmetals are elements with large ionization energies and the most negative electron affinities. The nonmetals are found on the right side of the table, with the most reactive ones in the upper-right corner, except for the noble gas elements, which are quite unreactive. The division into metals and nonmetals shown in Fig. 2-37 is only approximate. Many elements along the division line exhibit both metallic and nonmetallic properties under certain circumstances. These elements, often called **metalloids**, or sometimes **semimetals**, are: Si, Ge, As, Sb, Te, Po, and At.

The Alkali Metals

The metals of Group 1A (1), the alkali metals, illustrate very well the relationships among the properties of the elements in a group. Lithium, sodium, potassium, rubidium, cesium, and francium are the most chemically reactive of the metals. ◄ We will not discuss francium here because it occurs in nature in only very small quantities. Although hydrogen is found in Group 1A (1) of the periodic table, it behaves as a nonmetal, in contrast to the other members of that group. The fundamental reason for

▲ Potassium reacts vigorously with water.

© Cengage Learning

Table 2-8 | Properties of Five Alkali Metals

Element	Valence Electron Configuration	Density at 25°C (g/cm³)	mp (°C)	bp (°C)	First Ionization Energy (kJ/mol)	Atomic (Covalent) Radius (pm)	Ionic (M⁺) Radius (pm)
Li	$2s^1$	0.53	180	1330	520	152	60
Na	$3s^1$	0.97	98	892	495	186	95
K	$4s^1$	0.86	64	760	419	227	133
Rb	$5s^1$	1.53	39	668	409	247	148
Cs	$6s^1$	1.87	29	690	382	265	169

hydrogen's nonmetallic character is its very small size (see Fig. 2-36). The electron in the small $1s$ orbital is bound tightly to the nucleus.

Some important properties of the first five alkali metals are shown in Table 2-8. The data in Table 2-8 show that in going down the group, the first ionization energy decreases and the atomic radius increases. This agrees with the general trends discussed in Section 2-12.

The overall increase in density in going down Group 1A (1) is typical of all groups. This occurs because atomic mass generally increases more rapidly than atomic size. Thus there is more mass per unit volume for each succeeding element.

The smooth decrease in melting point and boiling point in going down Group 1A (1) is not typical; in most other groups more complicated behavior occurs. Note that the melting point of cesium is only 29°C. Cesium can be melted readily using only the heat from your hand. This is very unusual—metals typically have rather high melting points. For example, tungsten melts at 3410°C. The only other metals with low melting points are mercury (mp −38°C) and gallium (mp 30°C).

The chemical property most characteristic of a metal is the ability to lose its valence electrons. The Group 1A (1) elements are very reactive. They have low ionization energies and react with nonmetals to form ionic solids. We'll discuss these compounds in Chapter 3.

In this section we have seen that the trends in atomic properties summarized by the periodic table can be a great help in understanding the chemical behavior of the elements. This fact will be emphasized over and over as we proceed in our study of chemistry.

Potassium is widely recognized as an essential element. In fact, our daily requirement for potassium is more than twice that for sodium. Because most foods contain potassium, serious deficiency of this element in humans is rare. However, potassium deficiency can be caused by kidney malfunction or by the use of certain diuretics. Potassium deficiency leads to muscle weakness, irregular heartbeat, and depression.

Potassium is found in the fluids of the body as the K^+ ion, and its presence is essential to the operation of our nervous system. The passage of impulses along the nerves requires the flow of K^+ (and Na^+) through channels in the membranes of the nerve cells. Failure of this ion flow prevents nerve transmissions and results in death. For example, the black mamba snake kills its victims by injecting a venom that blocks the potassium channels in the nerve cells.

Although a steady intake of potassium is essential to preserve life, ironically, too much potassium can be lethal. In fact, the deadly ingredient in the drug mixture used for executing criminals is potassium chloride. Injection of a large amount of a potassium chloride solution produces an excess of K^+ ion in the fluids surrounding the cells and prevents the essential flow of K^+ out the cells to allow nerve impulses to occur. This causes the heart to stop beating. Unlike other forms of execution, death by lethal injection of potassium

chloride does not harm the organs of the body. Thus condemned criminals who are executed in this manner could potentially donate their organs for transplants. However, this idea is very controversial.

© imageBROKER/Alamy

The black mamba snake's venom kills by blocking the potassium channels in the nerve cells of victims.

FOR REVIEW

Key Terms

Section 2-1
electromagnetic radiation
wavelength
frequency

Section 2-2
Planck's constant
quantization
photons
photoelectric effect
$E = mc^2$
dual nature of light
diffraction
diffraction pattern

Electromagnetic radiation

- Characterized by its wavelength (λ), frequency (ν), and speed ($c = 2.9979 \times 10^8$ m/s)

$$\lambda\nu = c$$

- Can be viewed as a stream of "particles" called photons, each with energy $h\nu$, where h is Planck's constant (6.626×10^{-34} J · s)

Photoelectric Effect

- When light strikes a metal surface, electrons are emitted
- Analysis of the kinetic energy and numbers of the emitted electrons led Einstein to suggest that electromagnetic radiation can be viewed as a stream of photons

Key Terms

Section 2-3
continuous spectrum
line spectrum

Section 2-4
quantum model
ground state

Section 2-5
standing wave
wave function
orbital
quantum (wave) mechanical model
Heisenberg uncertainty principle
probability distribution
radial probability distribution

Section 2-6
quantum numbers
principal quantum number (n)
angular momentum quantum number (ℓ)
magnetic quantum number (m_ℓ)
subshell

Section 2-7
nodal surfaces (nodes)
degenerate orbital

Section 2-8
electron spin
electron spin quantum number
Pauli exclusion principle

Section 2-9
polyelectronic atoms

Section 2-11
aufbau principle
Hund's rule
valence electrons
core electrons
transition metals
lanthanide series (lanthanides)
actinide series (actinides)
main-group elements (representative elements)

Section 2-12
atomic radii
first ionization energy
second ionization energy
electron affinity

Section 2-13
metalloids (semimetals)

Hydrogen Spectrum

- The emission spectrum of hydrogen shows discrete wavelengths
- Indicates that hydrogen has discrete energy levels

Bohr Model of the Hydrogen Atom

- Using the data from the hydrogen spectrum and assuming angular momentum to be quantized, Bohr devised a model in which the electron traveled in circular orbits
- Although an important pioneering effort, this model proved to be entirely incorrect

Wave (Quantum) Mechanical Model

- An electron is described as a standing wave
- The square of the wave function (often called an orbital) gives a probability distribution for the electron position
- The exact position of the electron is never known, which is consistent with the Heisenberg uncertainty principle: it is impossible to know accurately both the position and the momentum of a particle simultaneously
- Probability maps are used to define orbital shapes
- Orbitals are characterized by the quantum numbers n, ℓ, and m_ℓ

Electron Spin

- Described by the spin quantum number m_s, which can have values of $\pm\frac{1}{2}$
- Pauli exclusion principle: no two electrons in a given atom can have the same set of quantum numbers n, ℓ, m_ℓ, and m_s
- Only two electrons with opposite spins can occupy a given orbital

Periodic Table

- By populating the orbitals from the wave mechanical model (the aufbau principle), the form of the periodic table can be explained
- According to the wave mechanical model, atoms in a given group have the same valence (outer) electron configuration
- The trends in properties such as ionization energies and atomic radii can be explained in terms of the concepts of nuclear attraction, electron repulsions, shielding, and penetration

Review Questions

Answers to the Review Questions can be found on the Student Web site (accessible from **www.cengagebrain.com**).

1. Four types of electromagnetic radiation (EMR) are ultraviolet, microwaves, gamma rays, and visible. All of these types of EMR can be characterized by wavelength, frequency, photon energy, and speed of travel. Define these terms and rank the four types of electromagnetic radiation in order of increasing wavelength, frequency, photon energy, and speed.

2. Characterize the Bohr model of the atom. In the Bohr model, what do we mean when we say something is quantized? How does the Bohr model of the hydrogen atom explain the hydrogen emission spectrum? Why is the Bohr model fundamentally incorrect?

3. What experimental evidence supports the quantum theory of light? Explain the wave-particle duality of all matter. For what size particles must one consider both the wave and the particle properties?

4. List the most important ideas of the quantum mechanical model of the atom. Include in your discussion the terms or names *wave function, orbital, Heisenberg uncertainty principle, de Broglie, Schrödinger,* and *probability distribution.*

5. What are quantum numbers? What information do we get from the quantum numbers n, ℓ, and m_ℓ? We define a spin quantum number (m_s), but do we know that an electron literally spins?

6. How do 2p orbitals differ from each other? How do 2p and 3p orbitals differ from each other? What is a nodal surface in an atomic orbital? What is wrong with 1p, 1d, 2d, 1f, 2f, and 3f orbitals? Explain what we mean when we say that a 4s electron is more penetrating than a 3d electron.

7. Four blocks of elements in a periodic table refer to various atomic orbitals being filled. What are the four blocks and the corresponding orbitals? How do you get the energy ordering of the atomic orbitals from the periodic table? What is the aufbau principle? Hund's rule? The Pauli exclusion principle? There are two common exceptions to the ground-state electron configuration for elements 1–36 as predicted by the periodic table. What are they?

8. What is the difference between core electrons and valence electrons? Why do we emphasize the valence electrons in an atom when discussing atomic properties? What is the relationship between valence electrons and elements in the same group of the periodic table?

9. Using the element phosphorus as an example, write the equation for a process in which the energy change will correspond to the ionization energy and to the electron affinity.

Explain why the first ionization energy tends to increase as one proceeds from left to right across a period. Why is the first ionization energy of aluminum lower than that of magnesium and the first ionization energy of sulfur lower than that of phosphorus?

Why do the successive ionization energies of an atom always increase? Note the successive ionization energies for silicon given in Table 2-5. Would you expect to see any large jumps between successive ionization energies of silicon as you removed all the electrons, one by one, beyond those shown in the table?

10. The radius trend and the ionization energy trend are exact opposites. Does this make sense?

Active Learning Questions

These questions are designed to be used by groups of students in class.

1. What does it mean for something to have wavelike properties? Particulate properties? Electromagnetic radiation can be discussed in terms of both particles and waves. Explain the experimental verification for each of these views.

2. Defend and criticize Bohr's model. Why was it reasonable that such a model was proposed, and what evidence was there that it "works"? Why do we no longer "believe" in it?

3. The first four ionization energies for the elements X and Y are shown below. The units are not kJ/mol.

	X	Y
First	170	200
Second	350	400
Third	1800	3500
Fourth	2500	5000

Identify the elements X and Y. There may be more than one correct answer, so explain completely.

4. Compare the first ionization energy of helium to its second ionization energy, remembering that both electrons come from the $1s$ orbital. Explain the difference without using actual numbers from the text.

5. Which has the larger second ionization energy, lithium or beryllium? Why?

6. Explain why a graph of ionization energy versus atomic number (across a row) is not linear. Where are the exceptions? Why are there exceptions?

7. Without referring to your text, predict the trend of second ionization energies for the elements sodium through argon. Compare your answer with Table 2-5. Explain any differences.

8. Account for the fact that the line that separates the metals from the nonmetals on the periodic table is diagonal downward to the right instead of horizontal or vertical.

9. Explain *electron* from a quantum mechanical perspective, including a discussion of atomic radii, probabilities, and orbitals.

10. Choose the best response for the following. The ionization energy for the chlorine atom is equal in magnitude to the electron affinity for

    a. the Cl atom.   **d.** the F atom.

    b. the Cl^- ion.   **e.** none of these.

    c. the Cl^+ ion.

    Explain each choice. Justify your choice, and for the choices you did not select, explain what is incorrect about them.

11. Consider the following statement: "The ionization energy for the potassium atom is negative, because when K loses an electron to become K^+, it achieves a noble gas electron configuration." Indicate everything that is correct in this statement. Indicate everything that is incorrect. Correct the incorrect information and explain.

12. In going across a row of the periodic table, electrons are added and ionization energy generally increases. In going down a column of the periodic table, electrons are also being added but ionization energy decreases. Explain.

13. How does probability fit into the description of the atom?

14. What is meant by an *orbital?*

15. Explain the difference between the probability density distribution for an orbital and its radial probability.

16. Is the following statement true or false? The hydrogen atom has a $3s$ orbital. Explain.

17. Which is higher in energy, the $2s$ or $2p$ orbital, in hydrogen? Is this also true for helium? Explain.

18. Prove mathematically that it is more energetically favorable for a fluorine atom to take an electron from a sodium atom than for a fluorine atom to take an electron from another fluorine atom.

Questions

19. What type of relationship (direct or inverse) exists between wavelength, frequency, and photon energy? What does a photon energy unit of a joule equal?

20. What do we mean by the *frequency* of electromagnetic radiation? Is the frequency the same as the *speed* of the electromagnetic radiation?

21. Explain the photoelectric effect.

22. Describe briefly why the study of electromagnetic radiation has been important to our understanding of the arrangement of electrons in atoms.

23. How does the wavelength of a fast-pitched baseball compare to the wavelength of an electron traveling at $\frac{1}{10}$ the speed of light? What is the significance of this comparison? See Example 2-3.

24. The following is an energy-level diagram for electronic transitions in the Bohr hydrogen atom.

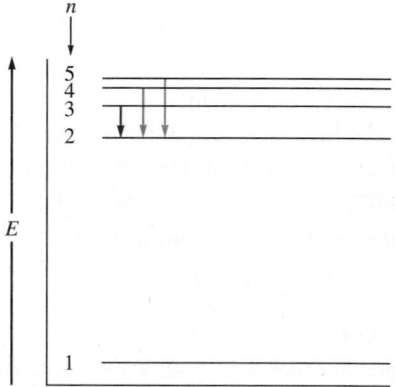

    a. Explain why the energy levels get closer together as they increase. Provide mathematical support for this.

    b. Verify that the colors given in the diagram are correct. Provide mathematical support.

25. The Bohr model works for only one electron species. Why do we discuss it in this text (what's good about it)?

26. We can represent both probability and radial probability versus distance from the nucleus for a hydrogen $1s$ orbital as depicted below.

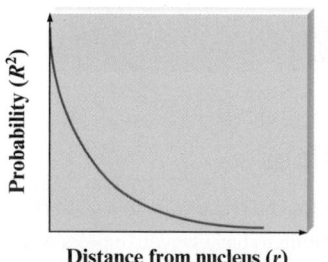 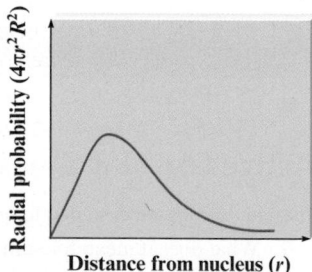

    What does each graph tell us about the electron in a hydrogen $1s$ orbital?

27. Consider the representations of the p and d atomic orbitals in Figs. 2-15 and 2-17. What do the $+$ and $-$ signs indicate?

28. The periodic table consists of four blocks of elements that correspond to s, p, d, and f orbitals being filled. After f orbitals come g and h orbitals. In theory, if a g block and an h block of elements existed, how long would the rows of g and h elements be in this theoretical periodic table?

29. Many times the claim is made that subshells half-filled with electrons are particularly stable. Can you suggest a possible physical basis for this claim?

30. Diagonal relationships in the periodic table exist as well as the vertical relationships. For example, Be and Al are similar in some of their properties, as are B and Si. Rationalize why these diagonal relationships hold for properties such as size, ionization energy, and electron affinity.

31. Elements with very large ionization energies also tend to have highly negative (favorable) electron affinities. Explain. Which group of elements would you expect to be an exception to this statement?

32. The changes in electron affinity as one goes down a group in the periodic table are not nearly as large as the variations in ionization energies. Why?

33. Why is it much harder to explain the line spectra of polyelectronic atoms and ions than it is to explain the line spectra of hydrogen and hydrogenlike ions?

34. Scientists use emission spectra to confirm the presence of an element in materials of unknown composition. Why is this possible?

35. Does the minimization of electron–electron repulsions correlate with Hund's rule?

36. In the hydrogen atom, what is the physical significance of the state for which $n = \infty$ and $E = 0$?

37. The work function is the energy required to remove an electron from an atom on the surface of a metal. How does this definition differ from that for ionization energy?

38. Many more anhydrous lithium salts are hygroscopic (readily absorb water) than are those of the other alkali metals. Explain.

Exercises

In this section, similar exercises are paired.

Light and Matter

39. The laser in an audio CD player uses light with a wavelength of 7.80×10^2 nm. Calculate the frequency of this light.

40. An FM radio station broadcasts at 99.5 MHz. Calculate the wavelength of the corresponding radio waves.

41. Microwave radiation has a wavelength on the order of 1.0 cm. Calculate the frequency and the energy of a single photon of this radiation.

42. A photon of ultraviolet (UV) light possesses enough energy to mutate a strand of human DNA. What is the energy of a single UV photon and 1 mole of UV photons having a wavelength of 25 nm? 1 mol UV photons = 6.022×10^{23} UV photons.

43. Octyl methoxycinnamate and oxybenzone are common ingredients in sunscreen applications. These compounds work by absorbing ultraviolet (UV) B light (wavelength 280–320 nm), the UV light most associated with sunburn symptoms. What frequency range of light do these compounds absorb?

44. Human color vision is "produced" by the nervous system based on how three different cone receptors interact with photons of light in the eye. These three different types of cones interact with photons of different frequency light, as indicated in the following chart:

Cone Type	Range of Light Frequency Detected
S	6.00–7.49×10^{14} s^{-1}
M	4.76–6.62×10^{14} s^{-1}
L	4.28–6.00×10^{14} s^{-1}

What wavelength ranges (and corresponding colors) do the three types of cones detect?

45. Consider the following waves representing electromagnetic radiation:

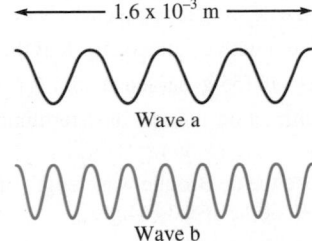

Which wave has the longer wavelength? Calculate the wavelength. Which wave has the higher frequency and larger photon energy? Calculate these values. Which wave has the greater velocity? What type of electromagnetic radiation does each wave represent?

46. One type of electromagnetic radiation has a frequency of 107.1 MHz, another type has a wavelength of 2.12×10^{-10} m, and another type of electromagnetic radiation has photons with energy equal to 3.97×10^{-19} J/photon. Identify each type of electromagnetic radiation and place them in order of increasing photon energy and increasing frequency.

47. Carbon absorbs energy at a wavelength of 150. nm. The total amount of energy emitted by a carbon sample is 1.98×10^5 J. Calculate the number of carbon atoms present in the sample, assuming that each atom emits one photon.

48. X rays have wavelengths on the order of 1×10^{-10} m. Calculate the energy of 1.0×10^{-10} m X rays in units of kilojoules per mole of X rays. (1 mol X rays = 6.022×10^{23} X rays.) AM radio waves have wavelengths on the order of 1×10^4 m. Calculate the energy of 1.0×10^4 m radio waves in units of kilojoules per mole of radio waves. Consider that the bond energy of a carbon–carbon single bond found in organic compounds is 347 kJ/mol. Would X rays and/or radio waves be able to disrupt organic compounds by breaking carbon–carbon single bonds?

49. The work function of an element is the energy required to remove an electron from the surface of the solid element. The work function for lithium is 279.7 kJ/mol (that is, it takes 279.7 kJ of energy to remove 1 mole of electrons from 1 mole of Li atoms on the surface of Li metal; 1 mol Li = 6.022×10^{23} atoms Li). What is the maximum wavelength of light that can remove an electron from an atom on the surface of lithium metal?

50. It takes 208.4 kJ of energy to remove 1 mole of electrons from an atom on the surface of rubidium metal. (1 mol electrons = 6.022×10^{23} electrons.) How much energy does it take to remove a single electron from an atom on the surface of solid rubidium? What is the maximum wavelength of light capable of doing this?

51. It takes 7.21×10^{-19} J of energy to remove an electron from an iron atom. What is the maximum wavelength of light that can do this?

52. Ionization energy is the energy required to remove an electron from an atom in the gas phase. The ionization energy of gold is 890.1 kJ/mol. Is light with a wavelength of 225 nm capable of ionizing a gold atom (removing an electron) in the gas phase? (1 mol gold = 6.022×10^{23} atoms gold.)

53. Calculate the de Broglie wavelength for each of the following.
    a. an electron with a velocity 10.% of the speed of light
    b. a tennis ball (55 g) served at 35 m/s (~80 mi/h)

54. Neutron diffraction is used in determining the structures of molecules.
    a. Calculate the de Broglie wavelength of a neutron moving at 1.00% of the speed of light.
    b. Calculate the velocity of a neutron with a wavelength of 75 pm (1 pm = 10^{-12} m).

55. A particle has a velocity that is 90.% of the speed of light. If the wavelength of the particle is 1.5×10^{-15} m, what is the mass of the particle?

56. Calculate the velocities of electrons with de Broglie wavelengths of 1.0×10^2 nm and 1.0 nm.

Hydrogen Atom: The Bohr Model

57. Calculate the wavelength of light emitted when each of the following transitions occur in the hydrogen atom. What type of electromagnetic radiation is emitted in each transition?
    a. $n = 3 \rightarrow n = 2$
    b. $n = 4 \rightarrow n = 2$
    c. $n = 2 \rightarrow n = 1$

58. Calculate the wavelength of light emitted when each of the following transitions occur in the hydrogen atom. What type of electromagnetic radiation is emitted in each transition?
    a. $n = 4 \rightarrow n = 3$
    b. $n = 5 \rightarrow n = 4$
    c. $n = 5 \rightarrow n = 3$

59. Using vertical lines, indicate the transitions from Exercise 57 on an energy-level diagram for the hydrogen atom (see Fig. 2-9).

60. Using vertical lines, indicate the transitions from Exercise 58 on an energy-level diagram for the hydrogen atom (see Fig. 2-9).

61. Calculate the longest and shortest wavelengths of light emitted by electrons in the hydrogen atom that begin in the $n = 6$ state and then fall to states with smaller values of n.

62. Assume that a hydrogen atom's electron has been excited to the $n = 5$ level. How many different wavelengths of light can be emitted as this excited atom loses energy?

63. Does a photon of visible light ($\lambda \approx 400$ to 700 nm) have sufficient energy to excite an electron in a hydrogen atom from the $n = 1$ to the $n = 5$ energy state? From the $n = 2$ to the $n = 6$ energy state?

64. An electron is excited from the $n = 1$ ground state to the $n = 3$ state in a hydrogen atom. Which of the following statements is/are true? Correct the false statements to make them true.
    a. It takes more energy to ionize (completely remove) the electron from $n = 3$ than from the ground state.
    b. The electron is farther from the nucleus on average in the $n = 3$ state than in the $n = 1$ state.
    c. The wavelength of light emitted if the electron drops from $n = 3$ to $n = 2$ will be shorter than the wavelength of light emitted if the electron falls from $n = 3$ to $n = 1$.
    d. The wavelength of light emitted when the electron returns to the ground state from $n = 3$ will be the same as the wavelength of light absorbed to go from $n = 1$ to $n = 3$.
    e. For $n = 3$, the electron is in the first excited state.

65. Calculate the maximum wavelength of light capable of removing an electron for a hydrogen atom from the energy state characterized by $n = 1$, by $n = 2$.

66. Consider an electron for a hydrogen atom in an excited state. The maximum wavelength of electromagnetic radiation that can completely remove (ionize) the electron from the H atom is 1460 nm. What is the initial excited state for the electron ($n = ?$)?

67. An excited hydrogen atom with an electron in the $n = 5$ state emits light having a frequency of 6.90×10^{14} s^{-1}. Determine the principal quantum level for the final state in this electronic transition.

68. An excited hydrogen atom emits light with a wavelength of 397.2 nm to reach the energy level for which $n = 2$. In which principal quantum level did the electron begin?

Quantum Mechanics, Quantum Numbers, and Orbitals

69. Using the Heisenberg uncertainty principle, calculate Δx for each of the following.
    a. an electron with $\Delta v = 0.100$ m/s
    b. a baseball (mass = 145 g) with $\Delta v = 0.100$ m/s

    How does the answer in part a compare with the size of a hydrogen atom? How does the answer in part b correspond to the size of a baseball?

70. The Heisenberg uncertainty principle can be expressed in the form

$$\Delta E \cdot \Delta t \geq \frac{h}{4\pi}$$

    where E represents energy and t represents time. Show that the units for this form are the same as the units for the form used in this chapter:

$$\Delta x \cdot \Delta(mv) \geq \frac{h}{4\pi}$$

71. What are the possible values for the quantum numbers n, ℓ, and m_ℓ?

72. Identify each of the following orbitals and determine the n and l quantum numbers. Explain your answers.

a.

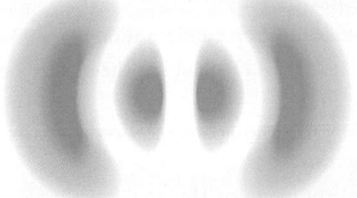

b.   Node   **c.**

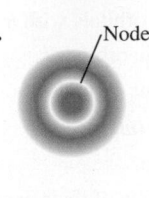   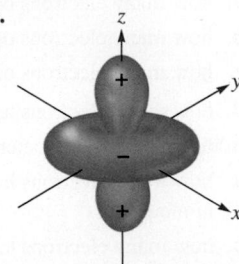

73. Which of the following sets of quantum numbers are not allowed in the hydrogen atom? For the sets of quantum numbers that are incorrect, state what is wrong in each set.

a. $n = 3, l = 2, m_\ell = 2$ **c.** $n = 0, l = 0, m_\ell = 0$
b. $n = 4, l = 3, m_\ell = 4$ **d.** $n = 2, l = -1, m_\ell = 1$

74. Which of the following sets of quantum numbers are not allowed? For each incorrect set, state why it is incorrect.

a. $n = 3, l = 3, m_\ell = 0, m_s = -\frac{1}{2}$
b. $n = 4, l = 3, m_\ell = 2, m_s = -\frac{1}{2}$
c. $n = 4, l = 1, m_\ell = 1, m_s = +\frac{1}{2}$
d. $n = 2, l = 1, m_\ell = -1, m_s - -1$
e. $n = 5, l = -4, m_\ell = 2, m_s = +\frac{1}{2}$
f. $n = 3, l = 1, m_\ell = 2, m_s = -\frac{1}{2}$

75. What is the physical significance of the value of ψ^2 at a particular point in an atomic orbital?

76. In defining the sizes of orbitals, why must we use an arbitrary value, such as 90% of the probability of finding an electron in that region?

Polyelectronic Atoms

77. Total radial probability distributions for the helium, neon, and argon atoms are shown in the following graph. How can one interpret the shapes of these curves in terms of electron configurations, quantum numbers, and nuclear charges?

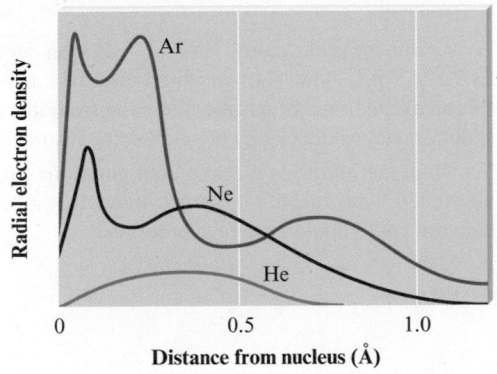

78. The relative orbital levels for the hydrogen atom can be represented as

$$
\begin{array}{l}
3s - \quad 3p --- \quad 3d ----- \\[2mm]
E \quad 2s - \quad 2p --- \\[2mm]
1s -
\end{array}
$$

Draw the relative orbital energy levels for atoms with more than one electron and explain your answer. Also explain how the following radial probability distributions support your answer.

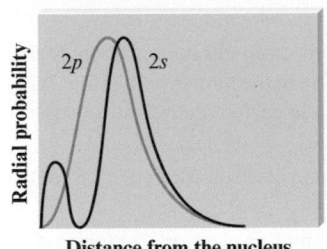   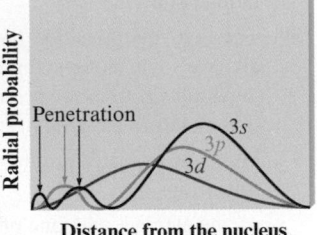

79. How many orbitals in an atom can have the designation $5p$, $3d_{z^2}$, $4d$, $n = 5$, $n = 4$?

80. How many electrons in an atom can have the designation $1p$, $6d_{x^2 - y^2}$, $4f$, $7p_y$, $2s$, $n = 3$?

81. Give the maximum number of electrons in an atom that can have these quantum numbers:

a. $n = 4$
b. $n = 5, m_\ell = +1$
c. $n = 5, m_s = +\frac{1}{2}$
d. $n = 3, l = 2$
e. $n = 2, l = 1$

82. Give the maximum number of electrons in an atom that can have these quantum numbers:

a. $n = 0, l = 0, m_\ell = 0$
b. $n = 2, l = 1, m_\ell = -1, m_s = -\frac{1}{2}$
c. $n = 3, m_s = +\frac{1}{2}$
d. $n = 2, l = 2$
e. $n = 1, l = 0, m_\ell = 0$

83. Draw atomic orbital diagrams representing the ground-state electron configuration for each of the following elements.

a. Na
b. Co
c. Kr

How many unpaired electrons are present in each element?

84. For elements 1–36, there are two exceptions to the filling order as predicted from the periodic table. Draw the atomic orbital diagrams for the two exceptions and indicate how many unpaired electrons are present.

85. The elements Si, Ga, As, Ge, Al, Cd, S, and Se are all used in the manufacture of various semiconductor devices. Write the expected electron configuration for these atoms.

86. The elements Cu, O, La, Y, Ba, Tl, and Bi are all found in high-temperature ceramic superconductors. Write the expected electron configuration for these atoms.

87. Write the expected electron configurations for each of the following atoms: Sc, Fe, P, Cs, Eu, Pt, Xe, Br.

88. Write the expected electron configurations for each of the following atoms: Cl, Sb, Sr, W, Pb, Cf.

89. The four most abundant elements by mass in the human body are oxygen, carbon, hydrogen, and nitrogen. These four elements make up about 96% of the human body. The next four most abundant elements are calcium, phosphorus, magnesium, and potassium. Write the expected ground-state electron configurations for these eight most abundant elements in the human body.

90. The first-row transition metals from chromium through zinc all have some biologic function in the human body. How many unpaired electrons are present in each of these first-row transition metals in the ground state?

91. Write the expected ground-state electron configuration for the following.

    a. the element with one unpaired $5p$ electron that forms a covalent compound with fluorine

    b. the (as yet undiscovered) alkaline earth metal after radium

    c. the noble gas with electrons occupying $4f$ orbitals

    d. the first-row transition metal with the most unpaired electrons

92. Using only the periodic table inside the front cover of the text, write the expected ground-state electron configurations for

    a. the third element in Group 5A.

    b. element number 116.

    c. an element with three unpaired $5d$ electrons.

    d. the halogen with electrons in the $6p$ atomic orbitals.

93. Given the valence electron orbital level diagram and the description, identify the element or ion.

    a. A ground state atom

$$3s \quad\quad 3p$$
$$\boxed{\uparrow\downarrow} \quad \boxed{\uparrow\downarrow}\,\boxed{\uparrow}\,\boxed{\uparrow}$$

    b. An atom in an excited state (assume two electrons occupy the $1s$ orbital)

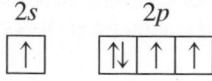

$$2s \quad\quad 2p$$
$$\boxed{\uparrow} \quad \boxed{\uparrow\downarrow}\,\boxed{\uparrow}\,\boxed{\uparrow}$$

    c. A ground state ion with a charge of -1

$$4s \quad\quad 4p$$
$$\boxed{\uparrow\downarrow} \quad \boxed{\uparrow\downarrow}\,\boxed{\uparrow\downarrow}\,\boxed{\uparrow}$$

94. Identify the following elements.

    a. An excited state of this element has the electron configuration $1s^2 2s^2 2p^5 3s^1$.

    b. The ground-state electron configuration is $[Ne]3s^2 3p^4$.

    c. An excited state of this element has the electron configuration $[Kr]5s^2 4d^6 5p^2 6s^1$.

    d. The ground-state electron configuration contains three unpaired $6p$ electrons.

95. In the ground state of mercury, Hg,

    a. how many electrons occupy atomic orbitals with $n = 3$?

    b. how many electrons occupy d atomic orbitals?

    c. how many electrons occupy p_z atomic orbitals?

    d. how many electrons have spin "up" ($m_s = +\frac{1}{2}$)?

96. In the ground state of element 115, Uup,

    a. how many electrons have $n = 5$ as one of their quantum numbers?

    b. how many electrons have $\ell = 3$ as one of their quantum numbers?

    c. how many electrons have $m_\ell = 1$ as one of their quantum numbers?

    d. how many electrons have $m_s = -\frac{1}{2}$ as one of their quantum numbers?

97. Give a possible set of values of the four quantum numbers for all the electrons in a boron atom and a nitrogen atom if each is in the ground state.

98. Give a possible set of values of the four quantum numbers for the $4s$ and $3d$ electrons in titanium.

99. Valence electrons are those electrons in the outermost principal quantum level (highest n level) of an atom in its ground state. Groups 1A to 8A have from 1 to 8 valence electrons. For each group of the representative elements (1A–8A), give the number of valence electrons, the general valence electron configuration, a sample element in that group, and the specific valence electron configuration for that element.

100. How many valence electrons do each of the following elements have, and what are the specific valence electrons for each element?

    a. Ca

    b. O

    c. element 117

    d. In

    e. Ar

    f. Bi

101. A certain oxygen atom has the electron configuration $1s^2 2s^2 2p_x^2 2p_y^2$. How many unpaired electrons are present? Is this an excited state of oxygen? In going from this state to the ground state, would energy be released or absorbed?

102. Which of the following electron configurations correspond to an excited state? Identify the atoms and write the ground-state electron configuration where appropriate.

    a. $1s^2 2s^2 3p^1$

    b. $1s^2 2s^2 2p^6$

c. $1s^2 2s^2 2p^4 3s^1$

d. $[Ar]4s^2 3d^5 4p^1$

How many unpaired electrons are present in each of these species?

103. Which of elements 1–36 have two unpaired electrons in the ground state?

104. Which of elements 1–36 have one unpaired electron in the ground state?

105. One bit of evidence that the quantum mechanical model is "correct" lies in the magnetic properties of matter. Atoms with unpaired electrons are attracted by magnetic fields and thus are said to exhibit *paramagnetism*. The degree to which this effect is observed is directly related to the number of unpaired electrons present in the atom. Consider the ground-state electron configurations for Li, N, Ni, Te, Ba, and Hg. Which of these atoms would be expected to be paramagnetic, and how many unpaired electrons are present in each paramagnetic atom?

106. Identify how many unpaired electrons are present in each of the following in the ground state: O, O^+, O^-, Os, Zr, S, F, Ar.

The Periodic Table and Periodic Properties

107. Arrange the following groups of atoms in order of increasing size.

a. Te, S, Se

b. K, Br, Ni

c. Ba, Si, F

108. Arrange the following groups of atoms in order of increasing size.

a. Rb, Na, Be

b. Sr, Se, Ne

c. Fe, P, O

109. Arrange the atoms in Exercise 107 in order of increasing first ionization energy.

110. Arrange the atoms in Exercise 108 in order of increasing first ionization energy.

111. In each of the following sets, which atom or ion has the smallest radius?

a. H, He

b. Cl, In, Se

c. element 120, element 119, element 116

d. Nb, Zn, Si

e. Na^-, Na, Na^+

112. In each of the following sets, which atom or ion has the smallest ionization energy?

a. Ca, Sr, Ba

b. K, Mn, Ga

c. N, O, F

d. S^{2-}, S, S^{2+}

e. Cs, Ge, Ar

113. Element 106 has been named seaborgium, Sg, in honor of Glenn Seaborg, discoverer of the first transuranium element.

a. Write the expected electron configuration for element 106.

b. What other element would be most like element 106 in its properties?

114. Predict some of the properties of element 117 (the symbol is Uus, following conventions proposed by the International Union of Pure and Applied Chemistry, or IUPAC).

a. What will be its electron configuration?

b. What element will it most resemble chemically?

115. The first ionization energies of As and Se are 0.947 and 0.941 MJ/mol, respectively. Rationalize these values in terms of electron configurations.

116. Rank the elements Be, B, C, N, and O in order of increasing first ionization energy. Explain your reasoning.

117. Consider the following ionization energies for aluminum:

$$Al(g) \longrightarrow Al^+(g) + e^- \qquad I_1 = 580 \text{ kJ/mol}$$
$$Al^+(g) \longrightarrow Al^{2+}(g) + e^- \qquad I_2 = 1815 \text{ kJ/mol}$$
$$Al^{2+}(g) \longrightarrow Al^{3+}(g) + e^- \qquad I_3 = 2740 \text{ kJ/mol}$$
$$Al^{3+}(g) \longrightarrow Al^{4+}(g) + e^- \qquad I_4 = 11,600 \text{ kJ/mol}$$

a. Account for the trend in the values of the ionization energies.

b. Explain the large increase between I_3 and I_4.

118. The following graph plots the first, second, and third ionization energies for Mg, Al, and Si.

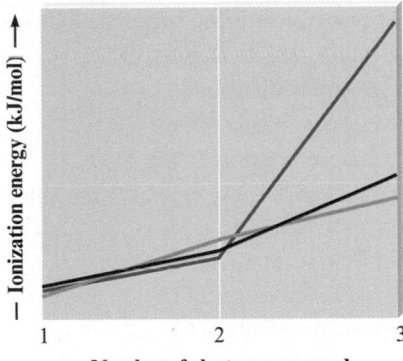

Without referencing the text, which plot corresponds to which element? In one of the plots, there is a huge jump in energy between I_2 and I_3, unlike in the other two plots. Explain this phenomenon.

119. For each of the following pairs of elements

$$(\text{C and N}) \qquad (\text{Ar and Br})$$

pick the atom with

a. more favorable (more negative) electron affinity.

b. higher ionization energy.

c. larger size.

120. For each of the following pairs of elements

(Mg and K)   (F and Cl)

pick the atom with

   a. more favorable (more negative) electron affinity.

   b. higher ionization energy.

   c. larger size.

121. The electron affinities of the elements from aluminum to chlorine are -44, -120, -74, -200.4, and -384.7 kJ/mol, respectively. Rationalize the trend in these values.

122. In the second row of the periodic table, Be, N, and Ne all have positive (unfavorable) electron affinities, whereas the other second-row elements have negative (favorable) electron affinities. Rationalize why Be, N, and Ne have unfavorable electron affinities.

123. Order the atoms in each of the following sets from the least negative electron affinity to the most.

   a. S, Se

   b. F, Cl, Br, I

124. Order the atoms in each of the following sets from the least negative electron affinity to the most.

   a. N, O, F

   b. Al, Si, P

125. The electron affinity for sulfur is more negative than that for oxygen. How do you account for this?

126. Which has the more negative electron affinity, the oxygen atom or the O^- ion? Explain your answer.

127. Write equations corresponding to the following:

   a. the fourth ionization energy of Se

   b. the electron affinity of S^-

   c. the electron affinity of Fe^{3+}

   d. the ionization energy of Mg

128. Using data from the text, determine the following values (justify your answer):

   a. the electron affinity of Mg^{2+}

   b. the ionization energy of Cl^-

   c. the electron affinity of Cl^+

   d. the ionization energy of Mg^- (electron affinity of Mg = 230 kJ/mol)

Alkali Metals

129. Cesium was discovered in natural mineral waters in 1860 by R. W. Bunsen and G. R. Kirchhoff, using the spectroscope they invented in 1859. The name came from the Latin *caesius* ("sky blue") because of the prominent blue line observed for this element at 455.5 nm. Calculate the frequency and energy of a photon of this light.

130. The bright yellow light emitted by a sodium vapor lamp consists of two emission lines at 589.0 and 589.6 nm. What are the frequency and the energy of a photon of light at each of these wavelengths? What are the energies in kJ/mol?

131. Does the information on alkali metals in Table 2-8 of the text confirm the general periodic trends in ionization energy and atomic radius? Explain.

132. Predict the atomic number of the next alkali metal after francium and give its ground-state electron configuration.

133. Complete and balance the equations for the following reactions.

   a. $Li(s) + N_2(g) \rightarrow$

   b. $Rb(s) + S(s) \rightarrow$

134. Complete and balance the equations for the following reactions.

   a. $Cs(s) + H_2O(l) \rightarrow$

   b. $Na(s) + Cl_2(g) \rightarrow$

Additional Exercises

135. "Lithium" is often prescribed as a mood-stabilizing drug. Do you think the "lithium" prescribed is in the elemental form? What is the more likely form of lithium to be prescribed as a drug?

136. A carbon–oxygen double bond in a certain organic molecule absorbs radiation that has a frequency of 6.0×10^{13} s^{-1}.

   a. What is the wavelength of this radiation?

   b. To what region of the spectrum does this radiation belong?

   c. What is the energy of this radiation per photon?

   d. A carbon–oxygen bond in a different molecule absorbs radiation with frequency equal to 5.4×10^{13} s^{-1}. Is this radiation more or less energetic?

137. Mars is roughly 60 million km from the earth. How long does it take for a radio signal originating from the earth to reach Mars?

138. Consider the following approximate visible light spectrum:

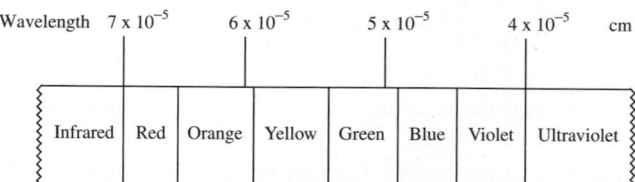

Barium emits light in the visible region of the spectrum. If each photon of light emitted from barium has an energy of 3.59×10^{-19} J, what color of visible light is emitted?

139. One of the visible lines in the hydrogen emission spectrum corresponds to the $n = 6$ to $n = 2$ electronic transition. What color light is this transition? See Exercise 138.

140. Using Fig. 2-30, list the elements (ignore the lanthanides and actinides) that have ground-state electron configurations that differ from those we would expect from their positions in the periodic table.

141. Are the following statements true for the hydrogen atom only, true for all atoms, or not true for any atoms?

   a. The principal quantum number completely determines the energy of a given electron.

   b. The angular momentum quantum number, ℓ, determines the shapes of the atomic orbitals.

   c. The magnetic quantum number, m_ℓ, determines the direction that the atomic orbitals point in space.

142. Although no currently known elements contain electrons in g orbitals in the ground state, it is possible that these elements will be found or that electrons in excited states of known elements could be in g orbitals. For g orbitals, the value of ℓ is 4. What is the lowest value of n for which g orbitals could exist? What are the possible values of m_ℓ? How many electrons could a set of g orbitals hold?

143. The four most abundant elements by mass in the human body are oxygen, carbon, hydrogen, and nitrogen. These four elements make up about 96% of the human body. The next four most abundant elements are calcium, phosphorus, magnesium, and potassium. Excluding hydrogen, which of these elements would have the smallest size? largest size? smallest first ionization energy? largest first ionization energy?

144. Which of the following orbital designations are incorrect: $1s$, $1p$, $7d$, $9s$, $3f$, $4f$, $2d$?

145. The successive ionization energies for an unknown element are

$I_1 = 896$ kJ/mol
$I_2 = 1752$ kJ/mol
$I_3 = 14{,}807$ kJ/mol
$I_4 = 17{,}948$ kJ/mol

To which family in the periodic table does the unknown element most likely belong?

146. An unknown element is a nonmetal and has a valence electron configuration of ns^2np^4.

a. How many valence electrons does this element have?

b. What are some possible identities for this element?

c. What is the formula of the compound this element would form with potassium?

d. Would this element have a larger or smaller radius than barium?

e. Would this element have a greater or smaller ionization energy than fluorine?

147. While Mendeleev predicted the existence of several undiscovered elements, he did not predict the existence of the noble gases, the lanthanides, or the actinides. Propose reasons why Mendeleev was not able to predict the existence of the noble gases.

148. Photosynthesis uses 660-nm light to convert CO_2 and H_2O into glucose and O_2. Calculate the frequency of this light.

149. Photogray lenses incorporate small amounts of silver chloride in the glass of the lens. When light hits the AgCl particles, the following reaction occurs:

$$AgCl \xrightarrow{h\nu} Ag + Cl$$

The silver metal that is formed causes the lenses to darken. The energy change for this reaction is 3.10×10^2 kJ/mol. Assuming all this energy must be supplied by light, what is the maximum wavelength of light that can cause this reaction?

ChemWork Problems

These multiconcept problems (and additional ones) are found interactively online with the same type of assistance a student would get from an instructor.

150. It takes 476 kJ to remove 1 mole of electrons from the atoms at the surface of a solid metal. How much energy (in kJ) does it take to remove a single electron from an atom at the surface of this solid metal?

151. Calculate, to four significant figures, the longest and shortest wavelengths of light emitted by electrons in the hydrogen atom that begin in the $n = 5$ state and then fall to states with smaller values of n.

152. Assume that a hydrogen atom's electron has been excited to the $n = 6$ level. How many different wavelengths of light can be emitted as this excited atom loses energy?

153. Determine the maximum number of electrons that can have each of the following designations: $2f$, $2d_{xy}$, $3p$, $5d_{yz}$, and $4p$.

154. Consider the ground state of arsenic, As. How many electrons have $\ell = 1$ as one of their quantum numbers? How many electrons have $m_\ell = 0$? How many electrons have $m_\ell = +1$?

155. Which of the following statements is(are) *true*?

a. The $2s$ orbital in the hydrogen atom is larger than the $3s$ orbital also in the hydrogen atom.

b. The Bohr model of the hydrogen atom has been found to be incorrect.

c. The hydrogen atom has quantized energy levels.

d. An orbital is the same as a Bohr orbit.

e. The third energy level has three sublevels, the s, p, and d sublevels.

156. Identify the following three elements.

a. The ground-state electron configuration is $[Kr]5s^24d^{10}5p^4$.

b. The ground-state electron configuration is $[Ar]4s^23d^{10}4p^2$.

c. An excited state of this element has the electron configuration $1s^22s^22p^43s^1$.

157. For each of the following pairs of elements, choose the one that correctly completes the following table.

	K and Cs	Te and Br	Ge and Se
The more favorable (negative) electron affinity	_____	_____	
The higher ionization energy	_____	_____	_____
The larger size (atomic radius)	_____	_____	_____

158. Which of the following statements is(are) *true*?

a. F has a larger first ionization energy than does Li.

b. Cations are larger than their parent atoms.

c. The removal of the first electron from a lithium atom (electron configuration is $1s^22s^1$) is exothermic—that is, removing this electron gives off energy.

d. The He atom is larger than the H^+ ion.

e. The Al atom is smaller than the Li atom.

159. Three elements have the electron configurations $1s^2 2s^2 2p^6 3s^2$, $1s^2 2s^2 2p^6 3s^2 3p^4$, and $1s^2 2s^2 2p^6 3s^2 3p^6 4s^2$. The first ionization energies of these elements (not in the same order) are 0.590, 0.999, and 0.738 MJ/mol. The atomic radii are 104, 160, and 197 pm. Identify the three elements, and match the appropriate values of ionization energy and atomic radius to each configuration. Complete the following table with the correct information.

Electron Configuration	Element Symbol	First Ionization Energy (MJ/mol)	Atomic Radius (pm)
$1s^2 2s^2 2p^6 3s^2$	_____	_____	_____
$1s^2 2s^2 2p^6 3s^2 3p^4$	_____	_____	_____
$1s^2 2s^2 2p^6 3s^2 3p^6 4s^2$	_____	_____	_____

Challenge Problems

160. One of the emission spectral lines for Be^{3+} has a wavelength of 253.4 nm for an electronic transition that begins in the state with $n = 5$. What is the principal quantum number of the lower-energy state corresponding to this emission? (*Hint:* The Bohr model can be applied to one-electron ions. Don't forget the Z factor: Z = nuclear charge = atomic number.)

161. The figure below represents part of the emission spectrum for a one-electron ion in the gas phase. All the lines result from electronic transitions from excited states to the $n = 3$ state. (See Exercise 160.)

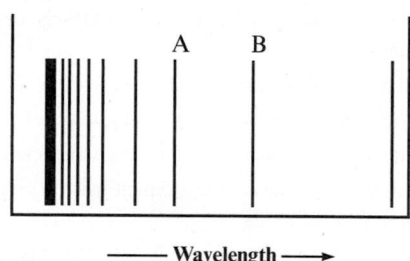

            ——— **Wavelength** ⟶

   a. What electronic transitions correspond to lines A and B?

   b. If the wavelength of line B is 142.5 nm, calculate the wavelength of line A.

162. When the excited electron in a hydrogen atom falls from $n = 5$ to $n = 2$, a photon of blue light is emitted. If an excited electron in He^+ falls from $n = 4$, to which energy level must it fall so that a similar blue light (as with the hydrogen) is emitted? Prove it. (See Exercise 160.)

163. The ground state ionization energy for the one-electron ion X^{m+} is 4.72×10^4 kJ/mol. Identify X and m. (See Exercise 160.)

164. For hydrogen atoms, the wave function for the state $n = 3$, $\ell = 0$, $m_\ell = 0$ is

$$\psi_{300} = \frac{1}{81\sqrt{3\pi}} \left(\frac{1}{a_0}\right)^{3/2} (27 - 18\sigma + 2\sigma^2)e^{-\sigma/3}$$

where $\sigma = r/a_0$ and a_0 is the Bohr radius (5.29×10^{-11} m). Calculate the position of the nodes for this wave function.

165. The wave function for the $2p_z$ orbital in the hydrogen atom is

$$\psi_{2p_z} = \frac{1}{4\sqrt{2\pi}} \left(\frac{Z}{a_0}\right)^{3/2} \sigma e^{-\sigma/2} \cos\theta$$

where a_0 is the value for the radius of the first Bohr orbit in meters (5.29×10^{-11}), σ is $Z(r/a_0)$, r is the value for the distance from the nucleus in meters, and θ is an angle. Calculate the value of $\psi_{2p_z}^2$ at $r = a_0$ for $\theta = 0°$ (z axis) and for $\theta = 90°$ (xy plane).

166. Answer the following questions, assuming that m_s could have three values rather than two and that the rules for n, ℓ, and m_ℓ are the normal ones.

   a. How many electrons would an orbital be able to hold?

   b. How many elements would the first and second periods in the periodic table contain?

   c. How many elements would be contained in the first transition metal series?

   d. How many electrons would the set of $4f$ orbitals be able to hold?

167. Assume that we are in another universe with different physical laws. Electrons in this universe are described by four quantum numbers with meanings similar to those we use. We will call these quantum numbers p, q, r, and s. The rules for these quantum numbers are as follows:

$p = 1, 2, 3, 4, 5, \ldots$.
q takes on positive odd integers and $q \le p$.
r takes on all even integer values from $-q$ to $+q$. (Zero is considered an even number.)
$s = +\frac{1}{2}$ or $-\frac{1}{2}$

   a. Sketch what the first four periods of the periodic table will look like in this universe.

   b. What are the atomic numbers of the first four elements you would expect to be least reactive?

   c. Give an example, using elements in the first four rows, of ionic compounds with the formulas XY, XY_2, X_2Y, XY_3, and X_2Y_3.

   d. How many electrons can have $p = 4$, $q = 3$?

   e. How many electrons can have $p = 3$, $q = 0$, $r = 0$?

   f. How many electrons can have $p = 6$?

168. Without looking at data in the text, sketch a qualitative graph of the third ionization energy versus atomic number for the elements Na through Ar, and explain your graph.

169. The following numbers are the ratios of second ionization energy to first ionization energy:

Na: 9.2
Mg: 2.0
Al: 3.1
Si: 2.0
P: 1.8
S: 2.3
Cl: 1.8
Ar: 1.8

Explain these relative numbers.

170. We expect the atomic radius to increase going down a group in the periodic table. Can you suggest why the atomic radius of hafnium breaks this rule? (See data below.)

Atomic Radii, in pm			
Sc	157	Ti	147.7
Y	169.3	Zr	159.3
La	191.5	Hf	147.6

171. The ionization energy for a $1s$ electron in a silver atom is 2.462×10^6 kJ/mol.

    a. Determine an approximate value for Z_{eff} for the Ag $1s$ electron. Assume the Bohr model applies to the $1s$ electron. Z_{eff} is the apparent nuclear charge experienced by the electrons.

    b. How does Z_{eff} from part a compare to Z for Ag? Rationalize the relative numbers.

Integrative Problems

These problems require the integration of multiple concepts to find the solutions.

172. Answer the following questions based on the given electron configurations and identify the elements.

    a. Arrange these atoms in order of increasing size: $[Kr]5s^24d^{10}5p^6$; $[Kr]5s^24d^{10}5p^1$; $[Kr]5s^24d^{10}5p^3$.

    b. Arrange these atoms in order of decreasing first ionization energy: $[Ne]3s^23p^5$; $[Ar]4s^23d^{10}4p^3$; $[Ar]4s^23d^{10}4p^5$.

173. As the weapons officer aboard the *Starship Chemistry*, it is your duty to configure a photon torpedo to remove an electron from the outer hull of an enemy vessel. You know that the work function (the binding energy of the electron) of the hull of the enemy ship is 7.52×10^{-19} J.

    a. What wavelength does your photon torpedo need to be to eject an electron?

    b. You find an extra photon torpedo with a wavelength of 259 nm and fire it at the enemy vessel. Does this photon torpedo do any damage to the ship (does it eject an electron)?

    c. If the hull of the enemy vessel is made of the element with an electron configuration of $[Ar]4s^13d^{10}$, what metal is this?

Bonding: General Concepts

The nudibranch uses a particular molecule (called an allomone) to defend itself against predators. (Manex Catalapiedra/Getty Images)

As we examine the world around us, we find it to be composed almost entirely of compounds and mixtures of compounds: Rocks, coal, soil, petroleum, trees, and human bodies are all complex mixtures of chemical compounds in which different kinds of atoms are bound together. Substances composed of unbound atoms do exist in nature, but they are very rare. Examples are the argon in the atmosphere and the helium mixed with natural gas reserves.

The manner in which atoms are bound together has a profound effect on chemical and physical properties. For example, graphite is a soft, slippery material used as a lubricant in locks, and diamond is one of the hardest materials known, valuable both as a gemstone and in industrial cutting tools. Why do these materials, both composed solely of carbon atoms, have such different properties? The answer, as we will see, lies in the bonding in these substances.

Silicon and carbon are next to each other in Group 4A of the periodic table. From our knowledge of periodic trends, we might expect SiO_2 and CO_2 to be very similar. But SiO_2 is the empirical formula of silica, which is found in sand and quartz, and carbon dioxide is a gas, a product of respiration. ◄ Why are they so different? We will be able to answer this question after we have developed models for bonding.

Molecular bonding and structure play the central role in determining the course of all chemical reactions, many of which are vital to our survival. Later in this book we will demonstrate their importance by showing how enzymes facilitate complex chemical reactions, how genetic characteristics are transferred, and how hemoglobin in the blood carries oxygen throughout the body. All of these fundamental biological reactions hinge on the geometric structures of molecules, sometimes depending on very subtle differences in molecular shape to channel the chemical reaction one way rather than another.

Many of the world's current problems require fundamentally chemical answers: disease and pollution control, the search for new energy sources, the development of new fertilizers to increase crop yields, the improvement of the protein content in various staple grains, and many more. To understand the behavior of natural materials, we must understand the nature of chemical bonding and the factors that control the structures of compounds. In this chapter we will present various classes of compounds that illustrate the different types of bonds and then develop models to describe the structure and bonding that characterize materials found in nature. Later these models will be useful in understanding chemical reactions.

In Chapters 1 and 2, we were concerned with the structure and properties of atoms. But as we stated earlier, free atoms in nature are rare—natural substances are almost always atoms bonded together. Thus, we will now turn our attention to how the atoms combine to form molecules. Why is it that atoms bond together in a molecule? How do they do this? Once we understand this, we will be able to make sense of the structure of molecules (Chapter 4) and we will use this knowledge to better understand the forces that hold molecules together in liquids and solids (Chapter 9).

© Cengage Learning

▲ Quartz grows in beautiful, regular crystals.

3-1 | Types of Chemical Bonds

From a chemist's viewpoint, the most interesting characteristic of an atom is its ability to combine with other atoms to form compounds. It was John Dalton who first recognized that chemical compounds are collections of atoms, but he could not determine the structure of atoms or their means for binding to each other. During the twentieth century we learned that atoms have electrons and that these electrons participate in bonding one atom to another.

The forces that hold atoms together in compounds are called **chemical bonds**. One way that atoms can form bonds is by *sharing electrons*. These bonds are called **covalent bonds**, and the resulting collection of atoms is called a **molecule**. Molecules can be represented in several different ways. The simplest method is the **chemical formula**, in which the symbols for the elements are used to indicate the types of atoms present and subscripts are used to indicate the relative numbers of atoms. For example, the formula for carbon dioxide is CO_2, meaning that each molecule contains 1 atom of carbon and 2 atoms of oxygen.

Examples of molecules that contain covalent bonds are hydrogen (H_2), water (H_2O), oxygen (O_2), ammonia (NH_3), and methane (CH_4). More information about a molecule is given by its **structural formula**, in which the individual bonds are shown (indicated by lines). Structural formulas may or may not indicate the actual shape of the molecule. For example, water might be represented as

$$H-O-H \quad \text{or} \quad \overset{O}{\underset{H \quad H}{\diagdown}}$$

The structure on the right shows the actual shape of the water molecule. Scientists know from experimental evidence that the molecule looks like this. (We will study the shapes of molecules further in Chapter 4.)

The structural formula for ammonia is

$$\underset{H \quad H \quad H}{\overset{N}{\wedge}}$$

Ammonia.

Note that atoms connected to the central atom by dashed lines are behind the plane of the paper, and atoms connected to the central atom by wedges are in front of the plane of the paper.

In a compound composed of molecules, the individual molecules move around as independent units. For example, a molecule of methane gas can be represented in several ways. The structural formula for methane (CH_4) is shown in Fig. 3-1(a). The **space-filling model** of methane, which shows the relative sizes of the atoms as well as their relative orientation in the molecule, is given in Fig. 3-1(b). **Ball-and-stick models** are also used to represent molecules. The ball-and-stick structure of methane is shown in Fig. 3-1(c).

What is a chemical bond? There is no simple and yet complete answer to this question. For our purposes we will define bonds as forces that hold groups of atoms together and make them function as a unit.

Figure 3-1 | (a) The structural formula for methane. (b) Space-filling model of methane. This type of model shows both the relative sizes of the atoms in the molecule and their spatial relationships. (c) Ball-and-stick model of methane.

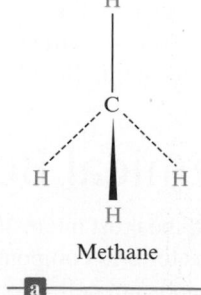

Methane

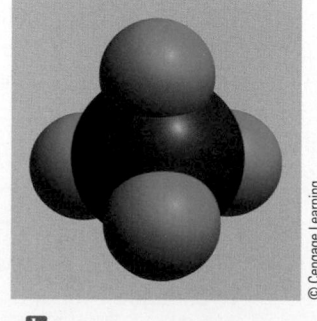

© Cengage Learning

Ken O'Donoghue © Cengage Learning

Did you ever wonder why the part of a pencil that makes the mark is called the "lead"? Pencils have no lead in them now—and they never have. Apparently the association between writing and the element lead arose during the Roman Empire, when lead rods were used as writing utensils because they leave a gray mark on paper. Many centuries later, in 1564, a deposit of a black substance found to be very useful for writing was discovered in Borrowdale, England. This substance, originally called "black lead," was shown in 1879 by Swedish chemist Carl Scheele to be a form of carbon and was subsequently named graphite (after the Greek *graphein*, meaning "to write").

Originally, chunks of graphite from Borrowdale, called marking stones, were used as writing instruments. Later, sticks of graphite were used. Because graphite is brittle, the sticks needed reinforcement. At first they were wrapped in string, which was unwound as the core wore down. Eventually, graphite rods were tied between two wooden slats or inserted into hollowed-out wooden sticks to form the first crude pencils.

Although Borrowdale graphite was pure enough to use directly, most graphite must be mixed with other materials to be useful for writing instruments. In 1795, the French chemist Nicolas-Jaques Conté invented a process in which graphite is mixed with clay and water to produce pencil "lead," a recipe that is still used today. In modern pencil manufacture, graphite and clay are mixed and crushed into a fine powder to which water is added. After the gray sludge is blended for several days, it is dried, ground up again, and mixed with more water to give a gray paste. The paste is extruded through a metal tube to form thin rods, which are then cut into pencil-length pieces called "leads." These leads are heated in an oven to 1000°C until they are smooth and hard. The ratio of clay to graphite is adjusted to vary the hardness of the lead—the more clay in the mix, the harder the lead and the lighter the line it makes.

Pencils are made from a slat of wood with several grooves cut in it to hold the leads. A similar grooved slat is then placed on top and glued to form a "sandwich" from which individual pencils are cut, sanded smooth, and painted. Although many types of wood have been used over the years to make pencils, the current favorite is incense cedar from the Sierra Nevada Mountains of California.

Modern pencils are simple but amazing instruments. The average pencil can write approximately 45,000 words, which is equivalent to a line 35 miles long. The graphite in a pencil is easily transferred to paper because graphite contains layers of carbon atoms bound together in a "chicken-wire" structure. Although the bonding *within* each layer is very strong, the bonding *between* layers is weak, giving graphite its slippery, soft nature. In this way, graphite is much different from diamond, the other common elemental form of carbon. In diamond the carbon atoms are bound tightly in all three dimensions, making it extremely hard—the hardest natural substance.

Pencils are very useful—especially for doing chemistry problems—because we can erase our mistakes. Most pencils used in the United States have erasers (first attached to pencils in 1858), although most European pencils do not. Laid end to end, the number of pencils made in the United States each year would circle the earth about 15 times. Pencils illustrate how useful a simple substance like graphite can be.

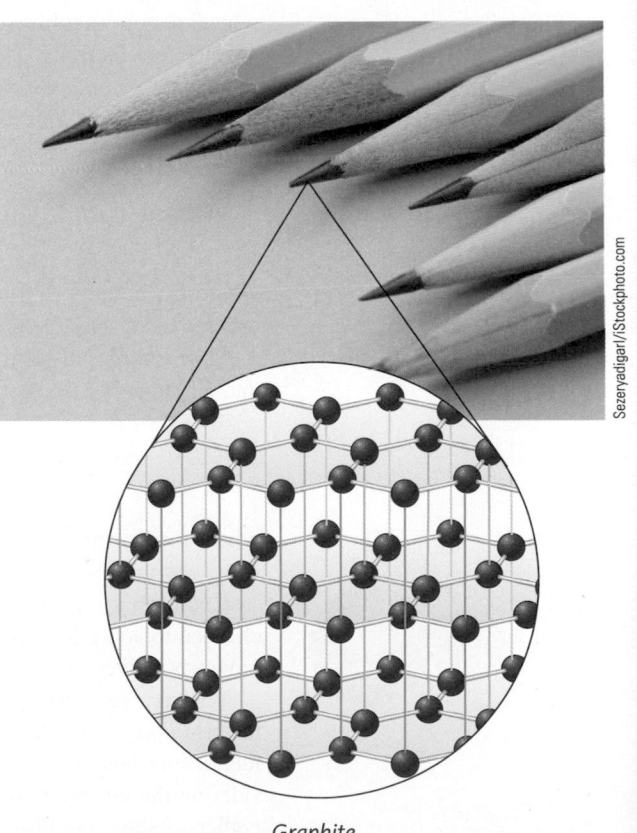

Graphite

Sezeryadigar/iStockphoto.com

There are many types of experiments we can perform to determine the fundamental nature of materials. For example, we can study physical properties such as melting point, hardness, and electrical and thermal conductivity. We can also study solubility characteristics and the properties of the resulting solutions. To determine the charge distribution in a molecule, we can study its behavior in an electric field. We can obtain information about the strength of a bonding interaction by measuring the **bond energy**, which is the energy required to break the bond.

There are several ways in which atoms can interact with one another to form aggregates. We will consider several specific examples to illustrate the various types of chemical bonds.

We observe that when solid sodium chloride is dissolved in water, the resulting solution conducts electricity, a fact that helps to convince us that sodium chloride is composed of Na^+ and Cl^- ions. Therefore, when sodium and chlorine react to form sodium chloride, electrons are transferred from the sodium atoms to the chlorine atoms to form Na^+ and Cl^- ions, which then aggregate to form solid sodium chloride. Why does this happen? The best simple answer is that *the system can achieve the lowest possible energy by behaving in this way.* The attraction of a chlorine atom for the extra electron and the very strong mutual attractions of the oppositely charged ions provide the driving forces for the process. Note that in an ionic substance such as sodium chloride, there are no individual $Na^+\cdots Cl^-$ bonds but many simultaneous interactions among the ions. The resulting solid sodium chloride is a very sturdy material; it has a melting point of approximately 800°C. The bonding forces that produce this great thermal stability result from the electrostatic attractions of the closely packed, oppositely charged ions. This is an example of **ionic bonding**. Ionic substances are formed when an atom that loses electrons relatively easily reacts with an atom that has a high affinity for electrons. That is, an **ionic compound** results when a metal reacts with a nonmetal.

The energy of interaction between a pair of ions can be calculated using **Coulomb's law** in the form

$$E = (2.31 \times 10^{-19}\,\text{J}\cdot\text{nm})\left(\frac{Q_1Q_2}{r}\right)$$

where E has units of joules, r is the distance between the ion centers in nanometers, and Q_1 and Q_2 are the numerical ion charges.

For example, in solid sodium chloride the distance between the centers of the Na^+ and Cl^- ions is 2.76 Å (0.276 nm), and the ionic energy per pair of ions is

$$E = (2.31 \times 10^{-19}\,\text{J}\cdot\text{nm})\left[\frac{(+1)(-1)}{0.276\,\text{nm}}\right] = -8.37 \times 10^{-19}\,\text{J}$$

where the negative sign indicates an attractive force. That is, the *ion pair has lower energy than the separated ions.* ◄

Coulomb's law also can be used to calculate the repulsive energy when two like-charged ions are brought together. In this case the calculated value of the energy will have a positive sign.

We have seen that a bonding force develops when two different types of atoms react to form oppositely charged ions. But how does a bonding force develop between two identical atoms? Let's explore this situation from a very simple point of view by considering the energy terms that result when two hydrogen atoms are brought close together, as shown in Fig. 3-2(a). When hydrogen atoms are brought close together, there are two unfavorable potential energy terms, proton–proton repulsion and electron–electron repulsion, and one favorable term, proton–electron attraction. Under what conditions will the H_2 molecule be favored over the separated hydrogen atoms? That is, what conditions will favor bond formation? The answer lies in the strong tendency in nature for any system to achieve the lowest possible energy. ◄ *ⓘ* A bond will form

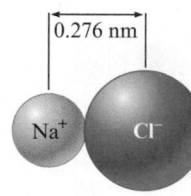

0.276 nm

Na^+ Cl^-

ⓘ A bond will form if the energy of the aggregate is lower than that of the separated atoms.

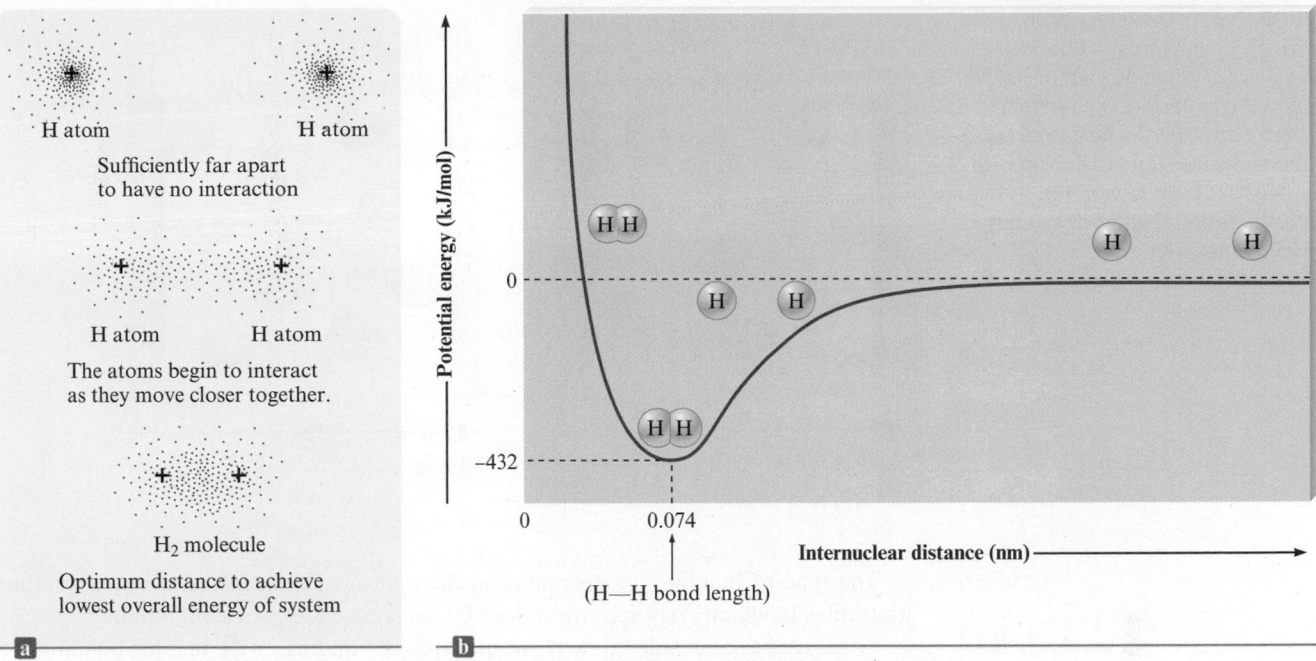

Figure 3-2 | (a) The interaction of two hydrogen atoms. (b) Energy profile as a function of the distance between the nuclei of the hydrogen atoms. As the atoms approach each other (right side of graph), the energy decreases until the distance reaches 0.074 nm (74 pm) and then begins to increase again due to repulsions.

(that is, the two hydrogen atoms will exist as a molecular unit) if the system can lower its total energy in the process.

In this case, then, the hydrogen atoms will position themselves so that the system will achieve the lowest possible energy; the system will act to minimize the sum of the positive (repulsive) energy terms and the negative (attractive) energy term. The distance where the energy is minimal is called the **bond length**. The total energy of this system as a function of distance between the hydrogen nuclei is shown in Fig. 3-2(b). Note several important features of this diagram:

The energy terms involved are the net potential energy that results from the attractions and repulsions among the charged particles and the kinetic energy due to the motions of the electrons.

The zero point of energy is defined with the atoms at infinite separation.

At very short distances the energy rises steeply because of the importance of the repulsive forces when the atoms are very close together.

The bond length is the distance at which the system has minimum energy.

In the H_2 molecule, the electrons reside primarily in the space between the two nuclei, where they are attracted simultaneously by both protons. This positioning is precisely what leads to the stability of the H_2 molecule compared with two separated hydrogen atoms. The potential energy of each electron is lowered because of the increased attractive forces in this area. When we say that a bond is formed between the hydrogen atoms, we mean that the H_2 molecule is more stable than two separated hydrogen atoms by a certain quantity of energy (the bond energy).

We can also think of a bond in terms of forces. The simultaneous attraction of each electron by the protons generates a force that pulls the protons toward each other and that just balances the proton–proton and electron–electron repulsive forces at the distance corresponding to the bond length. ◀ 𝒊

𝒊 Notice that to understand why a bond forms, we need to know the structure of the atoms.

Figure 3-3 | The effect of an electric field on hydrogen fluoride molecules. (a) When no electric field is present, the molecules are randomly oriented. (b) When the field is turned on, the molecules tend to line up with their negative ends toward the positive pole and their positive ends toward the negative pole.

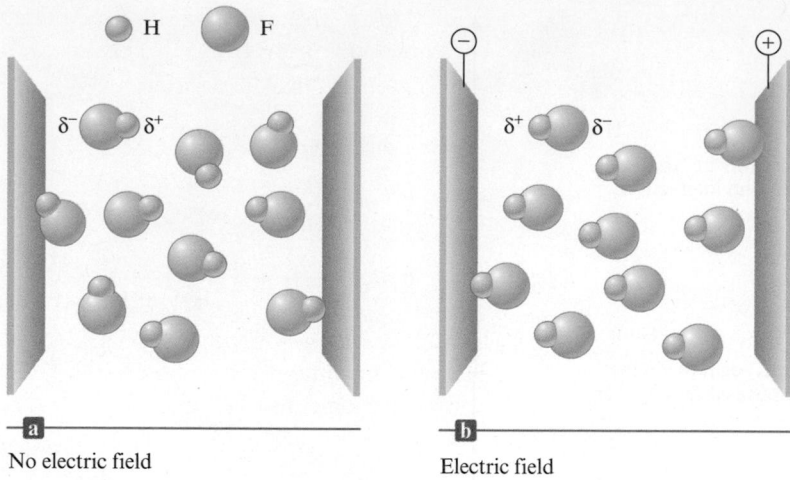

No electric field   Electric field

i Ionic and covalent bonds are the extreme bond types.

The type of bonding we encounter in the hydrogen molecule and in many other molecules in which *electrons are shared by nuclei* is called covalent bonding.

So far we have considered two extreme types of bonding. ◄ *i* In ionic bonding the participating atoms are so different that one or more electrons are transferred to form oppositely charged ions, which then attract each other. In covalent bonding two identical atoms share electrons equally. The bonding results from the mutual attraction of the two nuclei for the shared electrons. Between these extremes are intermediate cases in which the atoms are not so different that electrons are completely transferred but are different enough that unequal sharing results, forming what is called a **polar covalent bond**. An example of this type of bond occurs in the hydrogen fluoride (HF) molecule. When a sample of hydrogen fluoride gas is placed in an electric field, the molecules tend to orient themselves as shown in Fig. 3-3, with the fluoride end closest to the positive pole and the hydrogen end closest to the negative pole. This result implies that the HF molecule has the following charge distribution:

$$H \text{—} F$$
$$\delta^+ \quad \delta^-$$

where δ (lowercase delta) is used to indicate a fractional charge.

The most logical explanation for the development of the partial positive and negative charges on the atoms (bond polarity) in such molecules as HF and H_2O is that the electrons in the bonds are not shared equally. For example, we can account for the polarity of the HF molecule by assuming that the fluorine atom has a stronger attraction for the shared electrons than the hydrogen atom. Likewise, in the H_2O molecule the oxygen atom appears to attract the shared electrons more strongly than the hydrogen atoms do. Because bond polarity has important chemical implications, we find it useful to quantify the ability of an atom to attract shared electrons. In the next section we show how this is done.

3-2 | Electronegativity

The different affinities of atoms for the electrons in a bond are described by a property called **electronegativity**: *the ability of an atom in a molecule to attract shared electrons to itself.*

The most widely accepted method for determining values of electronegativity is that of Linus Pauling (1901–1995), an American scientist who won the Nobel Prizes for both chemistry and peace. To understand Pauling's model, consider a hypothetical molecule HX. The relative electronegativities of the H and X atoms are determined by comparing the measured H—X bond energy with the "expected" H—X bond energy, which is an average of the H—H and X—X bond energies:

$$\text{Expected H—X bond energy} = \frac{\text{H—H bond energy} + \text{X—X bond energy}}{2}$$

The difference (Δ) between the actual (measured) and expected bond energies is

$$\Delta = (\text{H—X})_{act} - (\text{H—X})_{exp}$$

If H and X have identical electronegativities, $(\text{H—X})_{act}$ and $(\text{H—X})_{exp}$ are the same, and Δ is 0. On the other hand, if X has a greater electronegativity than H, the shared electron(s) will tend to be closer to the X atom. The molecule will be polar, with the following charge distribution:

$$\underset{\delta^+ \quad \delta^-}{\text{H—X}}$$

Note that this bond can be viewed as having an ionic as well as a covalent component. The attraction between the partially (and oppositely) charged H and X atoms will lead to a greater bond strength. Thus $(\text{H—X})_{act}$ will be larger than $(\text{H—X})_{exp}$. The greater is the difference in the electronegativities of the atoms, the greater is the ionic component of the bond and the greater is the value of Δ. Thus the relative electronegativities of H and X can be assigned from the Δ values.

Electronegativity values have been determined by this process for virtually all the elements; the results are given in Fig. 3-4. Note that electronegativity generally increases going from left to right across a period and decreases going down a group for the representative elements. The range of electronegativity values is from 4.0 for fluorine to 0.7 for cesium.

The relationship between electronegativity and bond type is shown in Table 3-1. For identical atoms (an electronegativity difference of zero), the electrons in the bond are shared equally, and no polarity develops. When two atoms with very different electronegativities interact, electron transfer can occur to form the ions that make up an ionic substance. Intermediate cases give polar covalent bonds with unequal electron sharing.

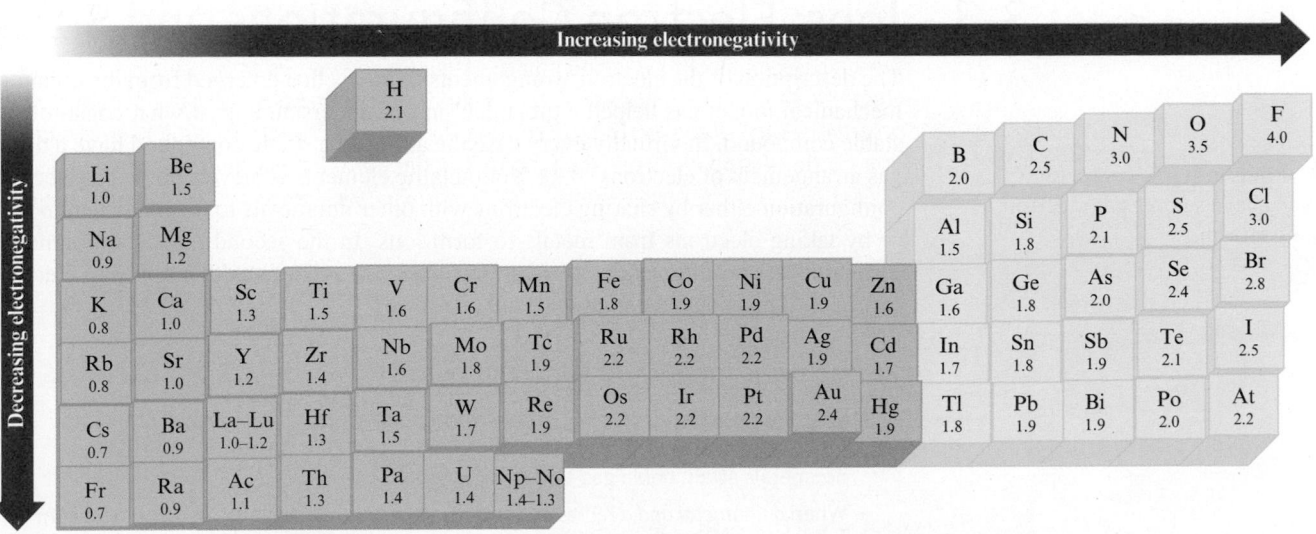

Figure 3-4 | The Pauling electronegativity values. Electronegativity generally increases across a period and decreases down a group.

Table 3-1 | The Relationship Between Electronegativity and Bond Type

Electronegativity Difference in the Bonding Atoms	Bond Type	
Zero	Covalent	Covalent character
↓	↓	
Intermediate	Polar covalent	
↓	↓	
Large	Ionic	Ionic character

Relative Bond Polarities

Order the following bonds according to polarity: H—H, O—H, Cl—H, S—H, and F—H.

Solution

The polarity of the bond increases as the difference in electronegativity increases. From the electronegativity values in Fig. 3-4, the following variation in bond polarity is expected (the electronegativity value appears in parentheses below each element): ◄ ⓘ

$$H\text{—}H < S\text{—}H < Cl\text{—}H < O\text{—}H < F\text{—}H$$
$$(2.1)(2.1) \quad (2.5)(2.1) \quad (3.0)(2.1) \quad (3.5)(2.1) \quad (4.0)(2.1)$$

Electronegativity difference

0	0.4	0.9	1.4	1.9

Covalent bond ——————→ Polar covalent bond
Polarity increases

See Exercises 3-41 and 3-42

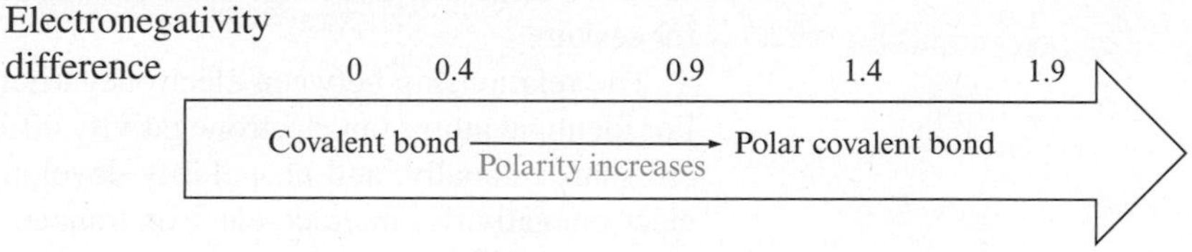

ⓘ Notice that the general trends are consistent with those of atomic radii. In general, smaller atoms across a row tend to have larger electronegativity values. This makes sense because the greater nuclear charge that causes the atom to be smaller will also have more attraction for a shared electron. As we move down a column, the atoms are larger and the shared electrons are further from the nucleus, thus they are not "held" as tightly. Note that electronegativity values generally decrease down a column.

3-3 | Ions: Electron Configurations and Sizes

The description of the electron arrangements in atoms that emerged from the quantum mechanical model has helped a great deal in our understanding of what constitutes a stable compound. In virtually every case the atoms in a stable compound have a noble gas arrangement of electrons. ◄ ⓘ Nonmetallic elements achieve a noble gas electron configuration either by sharing electrons with other nonmetals to form covalent bonds or by taking electrons from metals to form ions. In the second case, the nonmetals form anions, and the metals form cations. The generalizations that apply to electron configurations in stable compounds are as follows:

ⓘ Atoms in stable compounds usually have a noble gas electron configuration.

Electron Configuration of Compounds

- When *two nonmetals* react to form a covalent bond, they share electrons in a way that completes the valence electron configurations of both atoms. That is, both nonmetals attain noble gas electron configurations.

- When *a nonmetal and a representative-group metal* react to form a binary ionic compound, the ions form so that the valence electron configuration of the nonmetal achieves the electron configuration of the next noble gas atom and the valence orbitals of the metal are emptied. In this way both ions achieve noble gas electron configurations.

These generalizations apply to the vast majority of compounds and are important to remember. We will deal with covalent bonds more thoroughly later, but now we will consider what implications these rules hold for ionic compounds.

Predicting Formulas of Ionic Compounds

At the beginning of this discussion it should be emphasized that when chemists use the term *ionic compound*, they are usually referring to the solid state of that compound. In the solid state the ions are close together. That is, solid ionic compounds contain a large collection of positive and negative ions packed together in a way that minimizes the $\ominus \cdot \cdot \ominus$ and $\oplus \cdot \cdot \oplus$ repulsions and maximizes the $\oplus \cdot \cdot \ominus$ attractions. ◄ ⓘ This situation stands in contrast to the gas phase of an ionic substance, where the ions are quite far apart on average. In the gas phase, a pair of ions may get close enough to interact, but large collections of ions do not exist. ◄ ⓘ Thus, when we speak in this text of the stability of an ionic compound, we are referring to the solid state, where the large attractive forces present among oppositely charged ions tend to stabilize (favor the formation of) the ions. For example, as we mentioned in the preceding chapter, the O^{2-} ion is not stable as an isolated, gas-phase species but, of course, is very stable in many solid ionic compounds. That is, $MgO(s)$, which contains Mg^{2+} and O^{2-} ions, is very stable, but the isolated, gas-phase ion pair $Mg^{2+} \cdot \cdot O^{2-}$ is not energetically favorable in comparison with the separate neutral gaseous atoms. Thus you should keep in mind that in this section, and in most other cases where we are describing the nature of ionic compounds, the discussion usually refers to the solid state, where many ions are simultaneously interacting.

To illustrate the principles of electron configurations in stable, solid ionic compounds, we will consider the formation of an ionic compound from calcium and oxygen. We can predict what compound will form by considering the valence electron configurations of the two atoms:

$$Ca \quad [Ar]4s^2$$
$$O \quad [He]2s^22p^4$$

From Fig. 3-4 we see that the electronegativity of oxygen (3.5) is much greater than that of calcium (1.0). Because of this large difference, electrons will be transferred from calcium to oxygen to form oxygen anions and calcium cations in the compound. How many electrons are transferred? We can base our prediction on the observation that noble gas configurations are generally the most stable. ◄ ⓘ Note that oxygen needs two electrons to fill its $2s$ and $2p$ valence orbitals and to achieve the configuration of neon ($1s^22s^22p^6$). And by losing two electrons, calcium can achieve the configuration of argon. Two electrons therefore transferred:

$$Ca + O \longrightarrow Ca^{2+} + O^{2-}$$
$$\underbrace{}_{2e^-}$$

To predict the formula of the ionic compound, we simply recognize that chemical compounds are always electrically neutral—they have the same quantities of positive and negative charges. In this case we have equal numbers of Ca^{2+} and O^{2-} ions, and the empirical formula of the compound is CaO.

The same principles can be applied to many other cases. For example, consider the compound formed between aluminum and oxygen. Because aluminum has the configuration $[Ne]3s^23p^1$, it loses three electrons to form the Al^{3+} ion and thus achieves the neon configuration. Therefore, the Al^{3+} and O^{2-} ions form in this case. Since the compound must be electrically neutral, there must be three O^{2-} ions for every two Al^{3+} ions, and the compound has the empirical formula Al_2O_3. ◄

Table 3-2 shows common elements that form ions with noble gas electron configurations in ionic compounds. In losing electrons to form cations, metals in Group 1A (1) lose one electron, those in Group 2A (2) lose two electrons, and those in Group 3A (3)

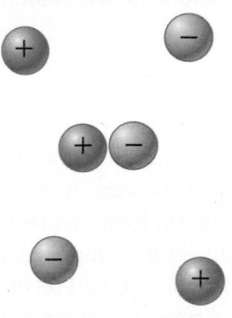

←→ repulsion

←→ attraction

ⓘ In the solid state of an ionic compound the ions are relatively close together, and many ions are simultaneously interacting:

ⓘ In the gas phase of an ionic substance the ions would be relatively far apart and would not contain large groups of ions:

ⓘ While our observations lead us to see that ions in compounds generally have noble gas configurations, this does not mean that the isolated ion is more stable than the parent atom. We will discuss this in more detail at the end of Section 3.5.

Courtesy Aluminum Company of America

▲ A bauxite mine. Bauxite contains Al_2O_3, the main source of aluminum.

Table 3-2 | Common Ions with Noble Gas Configurations in Ionic Compounds

Group 1A	Group 2A	Group 3A	Group 6A	Group 7A	Electron Configuration
H^-, Li^+	Be^{2+}				[He]
Na^+	Mg^{2+}	Al^{3+}	O^{2-}	F^-	[Ne]
K^+	Ca^{2+}		S^{2-}	Cl^-	[Ar]
Rb^+	Sr^{2+}		Se^{2-}	Br^-	[Kr]
Cs^+	Ba^{2+}		Te^{2-}	I^-	[Xe]

lose three electrons. In gaining electrons to form anions, nonmetals in Group 7A (17; the halogens) gain one electron, and those in Group 6A (16) gain two electrons. Hydrogen typically behaves as a nonmetal and can gain one electron to form the hydride ion (H^-), which has the electron configuration of helium.

There are some important exceptions to the rules discussed here. For example, tin forms both Sn^{2+} and Sn^{4+} ions, and lead forms both Pb^{2+} and Pb^{4+} ions. Also, bismuth forms Bi^{3+} and Bi^{5+} ions, and thallium forms Tl^+ and Tl^{3+} ions. There are no simple explanations for the behavior of these ions. For now, just note them as exceptions to the very useful rule that ions generally adopt noble gas electron configurations in ionic compounds. Our discussion here refers to representative metals. The transition metals exhibit more complicated behavior, forming a variety of ions.

Sizes of Ions

Ion size plays an important role in determining the structure and stability of ionic solids, the properties of ions in aqueous solution, and the biologic effects of ions. As with atoms, it is impossible to define precisely the sizes of ions. Most often, ionic radii are determined from the measured distances between ion centers in ionic compounds. This method, of course, involves an assumption about how the distance should be divided up between the two ions. Thus you will note considerable disagreement among ionic sizes given in various sources. Here we are mainly interested in trends and will be less concerned with absolute ion sizes.

Various factors influence ionic size. We will first consider the relative sizes of an ion and its parent atom. Since a positive ion is formed by removing one or more electrons from a neutral atom, the resulting cation is smaller than its parent atom. The opposite is true for negative ions; the addition of electrons to a neutral atom produces an anion significantly larger than its parent atom.

It is also important to know how the sizes of ions vary, depending on the positions of the parent elements in the periodic table. Fig. 3-5 shows the sizes of the most important ions (each with a noble gas configuration) and their positions in the periodic table. Note that ion size increases down a group. The changes that occur horizontally are complicated because of the change from predominantly metals on the left side of the periodic table to nonmetals on the right side. A given period thus contains both elements that give up electrons to form cations and ones that accept electrons to form anions.

One trend worth noting involves the relative sizes of a set of **isoelectronic ions**—*ions containing the same number of electrons*. Consider the ions O^{2-}, F^-, Na^+, Mg^{2+}, and Al^{3+}. Each of these ions has the neon electron configuration. How do the sizes of these ions vary? In general, there are two important facts to consider in predicting the relative sizes of ions: the number of electrons and the number of protons. Since these ions are isoelectronic, the number of electrons is 10 in each case. Electron repulsions therefore should be about the same in all cases. However, the number of protons increases from 8 to 13 as we go from the O^{2-} ion to the Al^{3+} ion. Thus, in going from O^{2-} to Al^{3+}, the 10 electrons experience a greater attraction as the positive charge on the nucleus increases. This causes the ions to become smaller. You can confirm this by

Figure 3-5 | Sizes of ions related to positions of the elements on the periodic table. Note that size generally increases down a group. Also note that in a series of isoelectronic ions, size decreases with increasing atomic number. The ionic radii are given in units of picometers.

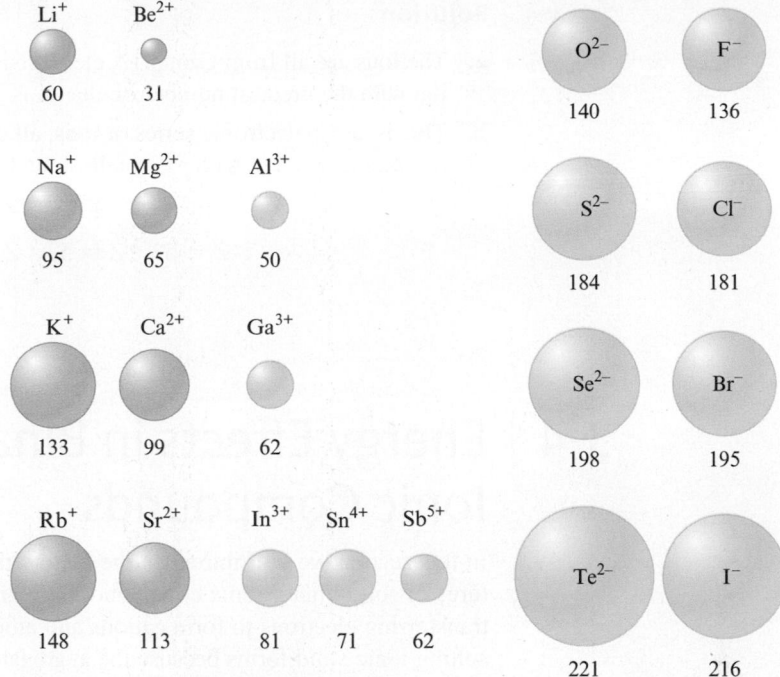

For isoelectronic ions, size decreases as Z increases.

looking at Fig. 3-5. In general, for a series of isoelectronic ions, the size decreases as the nuclear charge Z increases. ◀

Critical Thinking

Ions have different radii than their parent atoms. What if ions stayed the same size as their parent atoms? How would this affect ionic bonding in compounds?

INTERACTIVE EXAMPLE 3-2

Relative Ion Size I

Arrange the ions Se^{2-}, Br^-, Rb^+, and Sr^{2+} in order of decreasing size.

Solution

This is an isoelectronic series of ions with the krypton electron configuration. Since these ions all have the same number of electrons, their sizes will depend on the nuclear charge. The Z values are 34 for Se^{2-}, 35 for Br^-, 37 for Rb^+, and 38 for Sr^{2+}. Since the nuclear charge is greatest for Sr^{2+}, it is the smallest of these ions. The Se^{2-} ion is largest:

$$Se^{2-} > Br^- > Rb^+ > Sr^{2+}$$

　　　↑　　　　　　　　↑
　　Largest　　　　　Smallest

See Exercises 3-55 and 3-56

INTERACTIVE EXAMPLE 3-3

Relative Ion Size II

Choose the largest ion in each of the following groups.

a. Li^+, Na^+, K^+, Rb^+, Cs^+

b. Ba^{2+}, Cs^+, I^-, Te^{2-}

Solution

a. The ions are all from Group 1A elements. Since size increases down a group (the ion with the greatest number of electrons is largest), Cs^+ is the largest ion.

b. This is an isoelectronic series of ions, all of which have the xenon electron configuration. The ion with the smallest nuclear charge is largest:

$$Te^{2-} \; > \; I^- \; > \; Cs^+ \; > \; Ba^{2+}$$
$$Z = 52 \quad Z = 53 \quad Z = 55 \quad Z = 56$$

See Exercises 3-57 and 3-58

3-4 | Energy Effects in Binary Ionic Compounds

In this section we will introduce the factors that influence the stability and the structures of solid binary ionic compounds. We know that metals and nonmetals react by transferring electrons to form cations and anions that are mutually attractive. The resulting ionic solid forms because the aggregated oppositely charged ions have a lower energy than the original elements. Just how strongly the ions attract each other in the solid state is indicated by the **lattice energy**—*the change in energy that takes place when separated gaseous ions are packed together to form an ionic solid:*

$$M^+(g) + X^-(g) \longrightarrow MX(s)$$

 The structures of ionic solids will be discussed in detail in Chapter 9.

The lattice energy is often defined as the energy *released* when an ionic solid forms from its ions. ◄  However, in this book the sign of an energy term is always determined from the change in energy that occurs in the process: negative if the process results in a lower energy, positive if the energy increases. Thus, using this convention, the lattice energy has a negative sign, because when the ions are brought together the energy decreases.

We can illustrate the energy changes involved in the formation of an ionic solid by considering the formation of solid lithium fluoride from its elements: ◄

$$Li(s) + \tfrac{1}{2}F_2(g) \longrightarrow LiF(s)$$

To see the energy terms associated with this process, we break this reaction into steps, the sum of which gives the overall reaction.

▲ Lithium fluoride.

1. Sublimation of solid lithium. Sublimation involves taking a substance from the solid state to the gaseous state:

$$Li(s) \longrightarrow Li(g)$$

The energy of sublimation for $Li(s)$ is 161 kJ/mol.

2. Ionization of lithium atoms to form Li^+ ions in the gas phase:

$$Li(g) \longrightarrow Li^+(g) + e^-$$

This process corresponds to the first ionization energy for lithium, which is 520 kJ/mol.

3. Dissociation of fluorine molecules. We need to form a mole of fluorine atoms by breaking the F—F bonds in a half mole of F_2 molecules:

$$\tfrac{1}{2}F_2(g) \longrightarrow F(g)$$

The energy required to break this bond is 154 kJ/mol. In this case we are breaking the bonds in a half mole of fluorine, so the energy required for this step is (154 kJ)/2, or 77 kJ.

Figure 3-6 | The energy changes involved in the formation of solid lithium fluoride from its elements. The numbers in parentheses refer to the reaction steps discussed in the text.

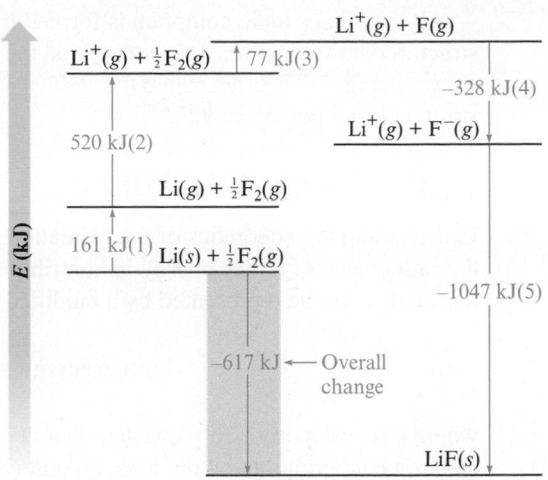

i Remember: when the energy decreases in a process, the sign of the energy change (ΔE) is negative.

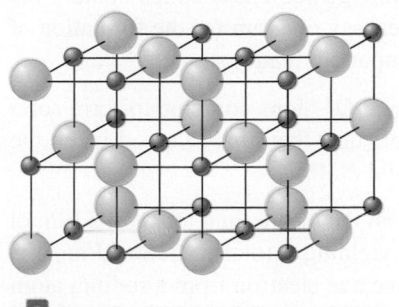

a

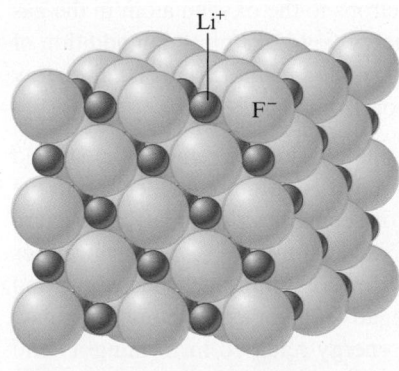

Li$^+$

F$^-$

b

Figure 3-7 | The structure of lithium fluoride. (a) Represented by ball-and-stick model. Note that each Li$^+$ ion is surrounded by six F$^-$ ions, and each F$^-$ ion is surrounded by six Li$^+$ ions. (b) Represented with the ions shown as spheres. The structure is determined by packing the spherical ions in a way that both maximizes the ionic attractions and minimizes the ionic repulsions.

4. Formation of F$^-$ ions from fluorine atoms in the gas phase:

$$F(g) + e^- \longrightarrow F^-(g)$$

The energy change for this process corresponds to the electron affinity of fluorine, which is -328 kJ/mol. ◀ i

5. Formation of solid lithium fluoride from the gaseous Li$^+$ and F$^-$ ions:

$$Li^+(g) + F^-(g) \longrightarrow LiF(s)$$

This corresponds to the lattice energy for LiF, which is -1047 kJ/mol.

Since the sum of these five processes yields the desired overall reaction, the sum of the individual energy changes gives the overall energy change:

Process	Energy Change (kJ)
Li(s) → Li(g)	161
Li(g) → Li$^+$(g) + e$^-$	520
$\frac{1}{2}$F$_2$(g) → F(g)	77
F(g) + e$^-$ → F$^-$(g)	−328
Li$^+$(g) + F$^-$(g) → LiF(s)	−1047
Overall: Li(s) + $\frac{1}{2}$F$_2$(g) → LiF(s)	−617 kJ (per mole of LiF)

This process is summarized by the energy diagram in Fig. 3-6. Note that the formation of solid lithium fluoride from its elements corresponds to a large decrease in energy, mainly because of the very large negative lattice energy. A great deal of energy is released when the ions combine to form the solid. In fact, note that the energy released when an electron is added to a fluorine atom to form the F$^-$ ion (328 kJ/mol) is not enough to remove an electron from lithium (520 kJ/mol). That is, when a metallic lithium atom reacts with a nonmetallic fluorine atom to form *separated* ions,

$$Li(g) + F(g) \longrightarrow Li^+(g) + F^-(g)$$

the process requires energy and is thus unfavorable. Clearly, then, the main impetus for the formation of an ionic compound rather than a covalent compound results from the strong mutual attractions among the Li$^+$ and F$^-$ ions in the solid. The lattice energy is the dominant energy term.

The structure of solid lithium fluoride is represented in Fig. 3-7. Note the alternating arrangement of the Li$^+$ and F$^-$ ions. Also note that each Li$^+$ is surrounded by six F$^-$ ions, and each F$^-$ ion is surrounded by six Li$^+$ ions. This structure can be rationalized by assuming that the ions behave as hard spheres that pack together in a way that both maximizes the attractions among the oppositely charged ions and minimizes the repulsions among the identically charged ions.

All the binary ionic compounds formed by an alkali metal and a halogen have the structure shown in Fig. 3-7, except for the cesium salts. The arrangement of ions shown in Fig. 3-7 is often called the *sodium chloride structure*, after the most common substance that possesses it.

Lattice Energy Calculations

In discussing the energetics of the formation of solid lithium fluoride, we emphasized the importance of lattice energy in contributing to the stability of the ionic solid. Lattice energy can be represented by a modified form of Coulomb's law:

$$\text{Lattice energy} = k\left(\frac{Q_1 Q_2}{r}\right)$$

where k is a proportionality constant that depends on the structure of the solid and the electron configurations of the ions, Q_1 and Q_2 are the charges on the ions, and r is the shortest distance between the centers of the cations and anions. Note that the lattice energy has a negative sign when Q_1 and Q_2 have opposite signs. This result is expected, since bringing cations and anions together releases energy. Also note that the process releases more energy as the ionic charges increase and as the distances between the ions in the solid decrease.

The importance of the charges in ionic solids can be illustrated by comparing the energies involved in the formation of NaF(*s*) and MgO(*s*). These solids contain the isoelectronic ions Na^+, F^-, Mg^{2+}, and O^{2-}. The energy diagram for the formation of the two solids is given in Fig. 3-8. Note several important features:

The energy released when the gaseous Mg^{2+} and O^{2-} ions combine to form solid MgO is much greater (more than four times greater) than that released when the gaseous Na^+ and F^- ions combine to form solid NaF.

The energy required to remove two electrons from the magnesium atom (735 kJ/mol for the first and 1445 kJ/mol for the second, yielding a total of 2180 kJ/mol) is much greater than the energy required to remove one electron from a sodium atom (495 kJ/mol).

Energy (737 kJ/mol) is required to add two electrons to the oxygen atom in the gas phase. Addition of the first electron releases energy (-141 kJ/mol), but addition of the second electron requires a great deal of energy (878 kJ/mol). This latter energy must be obtained indirectly, since the $O^{2-}(g)$ ion is not stable.

ℹ️ Since the equation for lattice energy contains the product $Q_1 Q_2$, the lattice energy for a solid with 2+ and 2− ions should be four times that for a solid with 1+ and 1− ions. That is,

$$\frac{(+2)(-2)}{(+1)(-1)} = 4$$

For MgO and NaF, the observed ratio of lattice energies (see Fig. 3-8) is

$$\frac{-3916 \text{ kJ}}{-923 \text{ kJ}} = 4.24$$

In view of the facts that twice as much energy is required to remove the second electron from magnesium as to remove the first and that addition of an electron to the gaseous O^- ion requires a great deal of energy, it seems puzzling that magnesium oxide contains Mg^{2+} and O^{2-} ions rather than Mg^+ and O^- ions. ◄ ℹ️ The answer lies in the lattice energy. Note that the lattice energy for combining gaseous Mg^{2+} and O^{2-} ions to form MgO(*s*) is 3000 kJ/mol more negative than that for combining gaseous Na^+ and F^- ions to form NaF(*s*). Thus the energy released in forming a solid containing Mg^{2+} and O^{2-} ions rather than Mg^+ and O^- ions more than compensates for the energies required for the processes that produce the Mg^{2+} and O^{2-} ions.

If there is so much lattice energy to be gained in going from singly charged to doubly charged ions in the case of magnesium oxide, why then does solid sodium fluoride contain Na^+ and F^- ions rather than Na^{2+} and F^{2-} ions? We can answer this question by recognizing that both Na^+ and F^- ions have the neon electron configuration. Removal of an electron from Na^+ requires an extremely large quantity of energy (4560 kJ/mol) because a $2p$ electron must be removed. Conversely, the addition of an electron to F^- would require use of the relatively high-energy $3s$ orbital, which is also an unfavorable process. Thus we can say that for sodium fluoride the extra energy required to form the doubly charged ions is greater than the gain in lattice energy that would result.

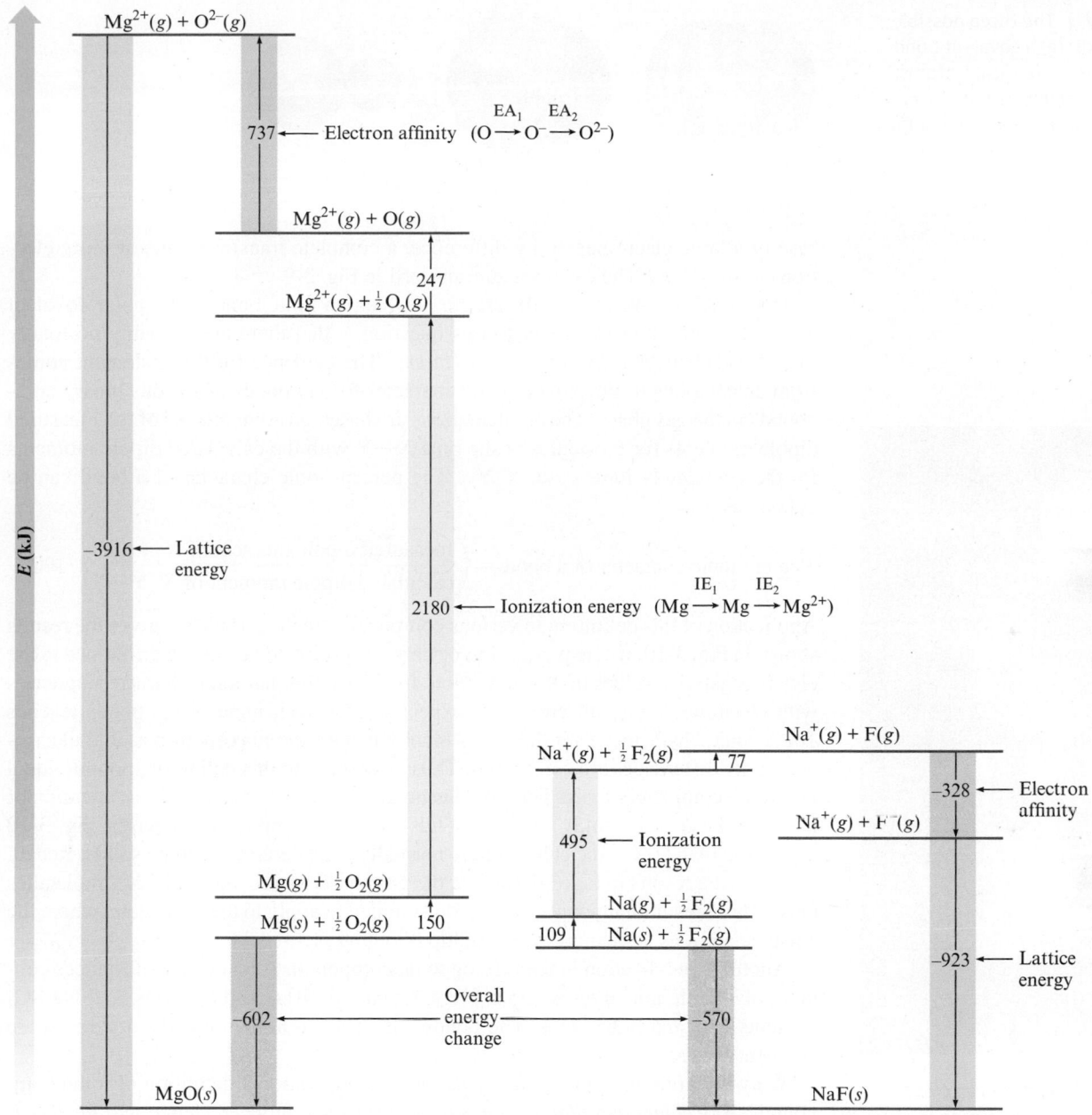

Figure 3-8 | Comparison of the energy changes involved in the formation of solid sodium fluoride and solid magnesium oxide. Note the large lattice energy for magnesium oxide (where doubly charged ions are combining) compared with that for sodium fluoride (where singly charged ions are combining).

This discussion of the energies involved in the formation of solid ionic compounds illustrates that a variety of factors operate to determine the composition and structure of these compounds. The most important of these factors involve the balancing of the energies required to form highly charged ions and the energy released when highly charged ions combine to form the solid.

3-5 | Partial Ionic Character of Covalent Bonds

Recall that when atoms with different electronegativities react to form molecules, the electrons are not shared equally. The possible result is a polar covalent bond or, in the

Figure 3-9 | The three possible types of bonds: (a) a covalent bond formed between identical F atoms; (b) the polar covalent bond of HF, with both ionic and covalent components; and (c) an ionic bond with no electron sharing.

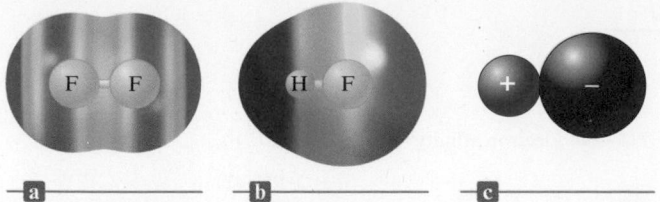

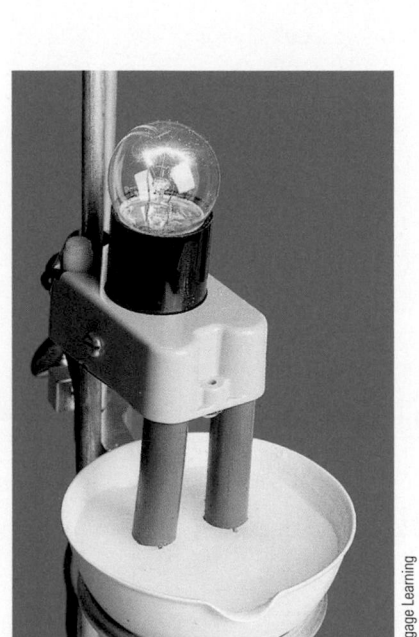

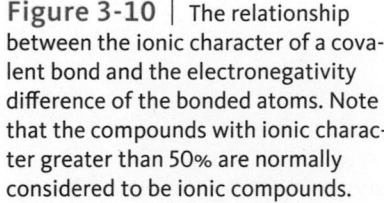

▲ Molten NaCl conducts an electric current, indicating the presence of mobile Na^+ and Cl^- ions.

case of a large electronegativity difference, a complete transfer of one or more electrons to form ions. The cases are summarized in Fig. 3-9.

How well can we tell the difference between an ionic bond and a polar covalent bond? The only honest answer to this question is that there are probably no totally ionic bonds between *discrete pairs of atoms*. The evidence for this statement comes from calculations of the percent ionic character for the bonds of various binary compounds in the gas phase. These calculations are based on comparisons of the measured dipole moments for molecules of the type X—Y with the calculated dipole moments for the completely ionic case, X^+Y^-. The percent ionic character of a bond can be defined as

$$\text{Percent ionic character of a bond} = \left(\frac{\text{measured dipole moment of X—Y}}{\text{calculated dipole moment of } X^+Y^-} \right) \times 100\%$$

Application of this definition to various compounds (in the gas phase) gives the results shown in Fig. 3-10, where percent ionic character is plotted versus the difference in the electronegativity values of X and Y. Note from this plot that ionic character increases with electronegativity difference, as expected. However, none of the bonds reaches 100% ionic character, even though compounds with the maximum possible electronegativity differences are considered. Thus, according to this definition, no individual bonds are completely ionic. This conclusion is in contrast to the usual classification of many of these compounds (as ionic solids). All the compounds shown in Fig. 3-10 with more than 50% ionic character are normally considered to be ionic solids. Recall, however, the results in Fig. 3-10 are for the gas phase, where individual XY molecules exist. These results cannot necessarily be assumed to apply to the solid state, where the existence of ions is favored by the multiple ion interactions.

Another complication in identifying ionic compounds is that many substances contain polyatomic ions. For example, NH_4Cl contains NH_4^+ and Cl^- ions, and Na_2SO_4 contains Na^+ and SO_4^{2-} ions. The bonds within the ammonium and sulfate ions are covalent bonds.

We will avoid these problems by adopting an operational definition of ionic compounds: *Any compound that conducts an electric current when melted will be classified as ionic.* ◄

Figure 3-10 | The relationship between the ionic character of a covalent bond and the electronegativity difference of the bonded atoms. Note that the compounds with ionic character greater than 50% are normally considered to be ionic compounds.

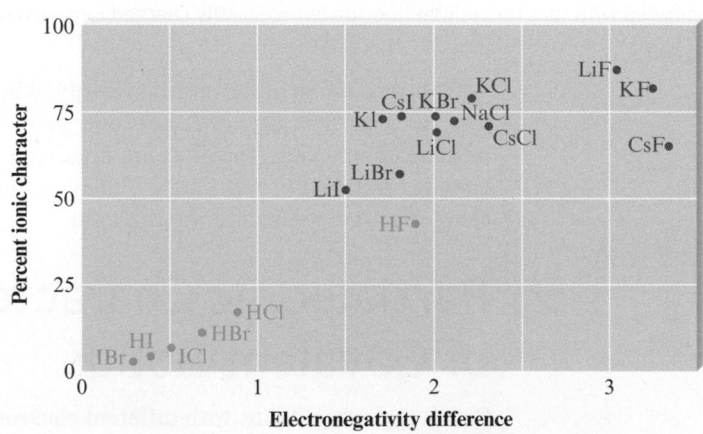

3-6 | The Covalent Chemical Bond: A Model

Before we develop specific models for covalent chemical bonding, it will be helpful to summarize some of the concepts introduced in this chapter.

What is a chemical bond? Chemical bonds can be viewed as forces that cause a group of atoms to behave as a unit.

Why do chemical bonds occur? There is no principle of nature that states that bonds are favored or disfavored. Bonds are neither inherently "good" nor inherently "bad" as far as nature is concerned; *bonds result from the tendency of a system to seek its lowest possible energy*. From a simplistic point of view, bonds occur when collections of atoms are more stable (lower in energy) than the separate atoms. For example, approximately 1652 kJ of energy is required to break a mole of methane (CH_4) molecules into separate C and H atoms. Or, from the opposite view, 1652 kJ of energy is released when 1 mole of methane is formed from 1 mole of gaseous C atoms and 4 moles of gaseous H atoms. Thus we can say that 1 mole of CH_4 molecules in the gas phase is 1652 kJ lower in energy than 1 mole of carbon atoms plus 4 moles of hydrogen atoms. Methane is therefore a stable molecule relative to its separated atoms.

We find it useful to interpret molecular stability in terms of a model called a *chemical bond*. To understand why this model was invented, let's continue with methane, which consists of four hydrogen atoms arranged at the corners of a tetrahedron around a carbon atom. ◄ *i*

i A tetrahedron has four equal triangular faces.

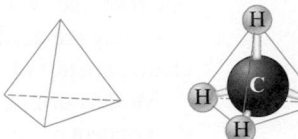

Given this structure, it is natural to envision four individual C—H interactions (we call them *bonds*). The energy of stabilization of CH_4 is divided equally among the four bonds to give an average C—H bond energy per mole of C—H bonds:

$$\frac{1652 \text{ kJ/mol}}{4} = 413 \text{ kJ/mol}$$

Next, consider methyl chloride, which consists of CH_3Cl molecules having the structure

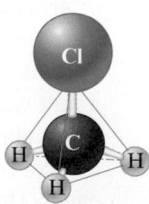

Experiments have shown that approximately 1578 kJ of energy is required to break down 1 mole of gaseous CH_3Cl molecules into gaseous carbon, chlorine, and hydrogen atoms. The reverse process can be represented as

$$C(g) + Cl(g) + 3H(g) \longrightarrow CH_3Cl(g) + 1578 \text{ kJ/mol}$$

A mole of gaseous methyl chloride is lower in energy by 1578 kJ than its separate gaseous atoms. Thus a mole of methyl chloride is held together by 1578 kJ of energy. Again, it is very useful to divide this energy into individual bonds. Methyl chloride can be visualized as containing one C—Cl bond and three C—H bonds. If we assume arbitrarily that a C—H interaction represents the same quantity of energy in any situation (that is, that the strength of a C—H bond is independent of its molecular environment), we can do the following bookkeeping:

$$1 \text{ mol C—Cl bonds plus 3 mol C—H bonds} = 1578 \text{ kJ}$$

$$\text{C—Cl bond energy} + 3(\text{average C—H bond energy}) = 1578 \text{ kJ}$$

$$\text{C—Cl bond energy} + 3(413 \text{ kJ/mol}) = 1578 \text{ kJ}$$

$$\text{C—Cl bond energy} = 1578 - 1239 = 339 \text{ kJ/mol}$$

These assumptions allow us to associate given quantities of energy with C—H and C—Cl bonds.

It is important to note that the bond concept is a human invention. Bonds provide a method for dividing up the energy evolved when a stable molecule is formed from its component atoms. Thus in this context a bond represents a quantity of energy obtained from the overall molecular energy of stabilization in a rather arbitrary way. This is not to say that the concept of individual bonds is a bad idea. In fact, the modern concept of the chemical bond, conceived by the American chemists G. N. Lewis and Linus Pauling, is one of the most useful ideas chemists have ever developed.

Models: An Overview

The framework of chemistry, like that of any science, consists of *models*—attempts to explain how nature operates on the microscopic level, based on experiences in the macroscopic world. To understand chemistry, one must understand its models and how they are used. We will use the concept of bonding to reemphasize the important characteristics of models, including their origin, structure, and uses.

Models originate from our observations of the properties of nature. For example, the concept of bonds arose from the observations that most chemical processes involve collections of atoms and that chemical reactions involve rearrangements of the ways the atoms are grouped. Therefore, to understand reactions, we must understand the forces that bind atoms together. ◄ ⓘ

ⓘ Bonding is a model proposed to explain molecular stability.

In natural processes there is a tendency toward lower energy. Collections of atoms therefore occur because the aggregated state has lower energy than the separated atoms. Why? As we saw earlier in this chapter, the best explanations for the energy change involve atoms sharing electrons or atoms transferring electrons to become ions. In the case of electron sharing, we find it convenient to assume that individual bonds occur between pairs of atoms. Let's explore the validity of this assumption and see how it is useful.

In a diatomic molecule such as H_2, it is natural to assume that a bond exists between the atoms, holding them together. It is also useful to assume that individual bonds are present in polyatomic molecules such as CH_4. Therefore, instead of thinking of CH_4 as a unit with a stabilization energy of 1652 kJ per mole, we choose to think of CH_4 as containing four C—H bonds, each worth 413 kJ of energy per mole of bonds. Without this concept of individual bonds in molecules, chemistry would be hopelessly complicated. There are millions of different chemical compounds, and if each of these compounds had to be considered as an entirely new entity, the task of understanding chemical behavior would be overwhelming.

The bonding model provides a framework to systematize chemical behavior by enabling us to think of molecules as collections of common fundamental components. For example, a typical biomolecule, such as a protein, contains hundreds of atoms and might seem discouragingly complex. ◄ However, if we think of a protein as constructed of individual bonds, C—C, C—H, C—N, C—O, N—H, and so on, it helps tremendously in predicting and understanding the protein's behavior. The essential idea is that we expect a given bond to behave about the same in any molecular environment. Used in this way, the model of the chemical bond has helped chemists to systematize the reactions of the millions of existing compounds.

▲ The concept of individual bonds makes it much easier to deal with complex molecules such as DNA. A small segment of a DNA molecule is shown here.

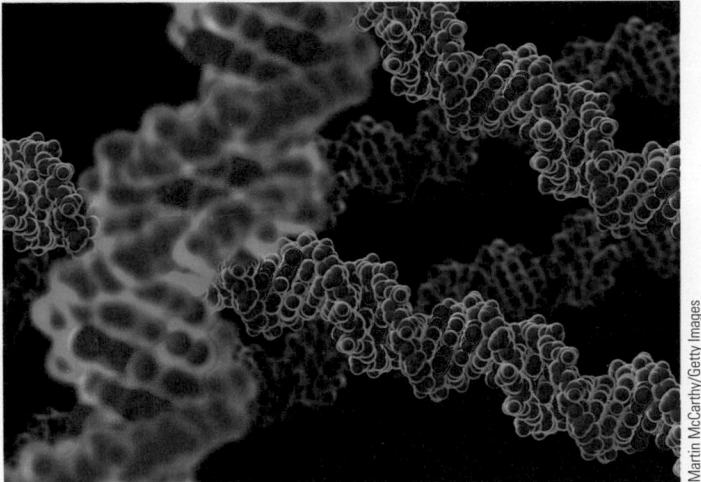

Martin McCarthy/Getty Images

In addition to being useful, the bonding model is physically sensible. It makes sense that atoms can form stable groups by sharing electrons; shared electrons give a lower energy state because they are simultaneously attracted by two nuclei.

Also, as we will see in the next section, bond energy data support the existence of discrete bonds that are relatively independent of the molecular environment. It is very important to remember, however, that the chemical bond is only a model. Although our concept of discrete bonds in molecules agrees with many of our observations, some molecular properties require that we think of a molecule as a whole, with the electrons free to move through the entire molecule. This is called *delocalization* of the electrons, a concept that will be discussed more completely in the next chapter.

LET'S REVIEW | *Fundamental Properties of Models*

- Models are human inventions, always based on an incomplete understanding of how nature works. *A model does not equal reality.*

- Models are often wrong. This property derives from the first property. Models are based on speculation and are always oversimplifications.

- Models tend to become more complicated as they age. As flaws are discovered in our models, we "patch" them and thus add more detail.

- It is very important to understand the assumptions inherent in a particular model before you use it to interpret observations or to make predictions. Simple models usually involve very restrictive assumptions and can be expected to yield only qualitative information. Asking for a sophisticated explanation from a simple model is like expecting to get an accurate mass for a diamond using a bathroom scale.

  For a model to be used effectively, we must understand its strengths and weaknesses and ask only appropriate questions. An illustration of this point is the simple aufbau principle used to account for the electron configurations of the elements. Although this model correctly predicts the configuration for most atoms, chromium and copper, for example, do not agree with the predictions. Detailed studies show that the configurations of chromium and copper result from complex electron interactions that are not taken into account in the simple model. However, this does not mean that we should discard the simple model that is so useful for most atoms. Instead, we must apply it with caution and not expect it to be correct in every case.

- When a model is wrong, we often learn much more than when it is right. If a model makes a wrong prediction, it usually means we do not understand some fundamental characteristics of nature. We often learn by making mistakes. (Try to remember this when you get back your next chemistry test.)

3-7 | Covalent Bond Energies and Chemical Reactions

In this section we will consider the energies associated with various types of bonds and see how the bonding concept is useful in dealing with the energies of chemical reactions. One important consideration is to establish the sensitivity of a particular type of bond to its molecular environment. For example, consider the stepwise decomposition of methane:

Process	Energy Required (kJ/mol)
$CH_4(g) \rightarrow CH_3(g) + H(g)$	435
$CH_3(g) \rightarrow CH_2(g) + H(g)$	453
$CH_2(g) \rightarrow CH(g) + H(g)$	425
$CH(g) \rightarrow C(g) + H(g)$	339

$$\text{Total} = 1652$$

$$\text{Average} = \frac{1652}{4} = 413$$

Although a C—H bond is broken in each case, the energy required varies in a non-systematic way. This example shows that the C—H bond is somewhat sensitive to its environment. We use the *average* of these individual bond dissociation energies even though this quantity only approximates the energy associated with a C—H bond in a particular molecule. The degree of sensitivity of a bond to its environment also can be seen from experimental measurements of the energy required to break the C—H bond in the following molecules:

Molecule	Measured C—H Bond Energy (kJ/mol)
$HCBr_3$	380
$HCCl_3$	380
HCF_3	430
C_2H_6	410

These data show that the C—H bond strength varies significantly with its environment, but the concept of an average C—H bond strength remains useful to chemists. The average values of bond energies for various types of bonds are listed in Table 3-3.

So far we have discussed bonds in which one pair of electrons is shared. This type of bond is called a **single bond**. As we will see in more detail later, atoms sometimes share two pairs of electrons, forming a **double bond**, or share three pairs of electrons, forming a **triple bond**. Notice that the bond energy of a double bond is not twice the bond energy of a single bond. Similarly, the bond energy of a triple bond is not three times the single bond energy. The bond energies for these *multiple bonds* are also given in Table 3-3.

A relationship also exists between the number of shared electron pairs and the bond length. As the number of shared electrons increases, the bond length shortens. This relationship is shown for selected bonds in Table 3-4.

Table 3-3 | Average Bond Energies (kJ/mol)

Single Bonds								Multiple Bonds	
H—H	432	N—H	391	I—I	149			C=C	614
H—F	565	N—N	160	I—Cl	208			C≡C	839
H—Cl	427	N—F	272	I—Br	175			O=O	495
H—Br	363	N—Cl	200					C=O*	745
H—I	295	N—Br	243			S—H	347	C≡O	1072
		N—O	201			S—F	327	N=O	607
C—H	413	O—H	467			S—Cl	253	N=N	418
C—C	347	O—O	146			S—Br	218	N≡N	941
C—N	305	O—F	190			S—S	266	C≡N	891
C—O	358	O—Cl	203					C=N	615
C—F	485	O—I	234			Si—Si	340		
C—Cl	339					Si—H	393		
C—Br	276	F—F	154			Si—C	360		
C—I	240	F—Cl	253			Si—O	452		
C—S	259	F—Br	237						
		Cl—Cl	239						
		Cl—Br	218						
		Br—Br	193						

*C=O(in CO_2) = 799

Bond Energies

Bond energy values can be used to calculate approximate energies for reactions. To illustrate how this is done, we will calculate the change in energy that accompanies the following reaction:

$$H_2(g) + F_2(g) \longrightarrow 2HF(g)$$

This reaction involves breaking one H—H and one F—F bond and forming two H—F bonds. For bonds to be broken, energy must be *added* to the system—a positive ΔE. Consequently, the energy terms associated with bond breaking have *positive* signs. The formation of a bond *releases* energy, so the energy terms associated with bond making carry a *negative* sign. We can write the energy change for a reaction as follows:

$\Delta E =$ sum of the energies required to break old bonds (positive signs) plus the sum of the energies released in the formation of new bonds (negative signs)

This leads to the expression

$$\Delta E = \underbrace{\Sigma n \times D \text{ (bonds broken)}}_{\text{Energy required}} - \underbrace{\Sigma n \times D \text{ (bonds formed)}}_{\text{Energy released}}$$

Table 3-4 | Bond Lengths for Selected Bonds

Bond	Bond Type	Bond Length (pm)	Bond Dissociation Energy (kJ/mol)
C—C	Single	154	347
C=C	Double	134	614
C≡C	Triple	120	839
C—O	Single	143	358
C=O	Double	123	745
C—N	Single	143	305
C=N	Double	138	615
C≡N	Triple	116	891

where Σ represents the sum of terms, D represents the bond dissociation energy per mole of bonds (D *always* has a positive sign), and n represents the moles of a particular type of bond.

In the case of the formation of HF,

$$\Delta E = D_{\text{H}-\text{H}} + D_{\text{F}-\text{F}} - 2D_{\text{H}-\text{F}}$$

$$= 1 \text{ mol} \times \frac{432 \text{ kJ}}{\text{mol}} + 1 \text{ mol} \times \frac{154 \text{ kJ}}{\text{mol}} - 2 \text{ mol} \times \frac{565 \text{ kJ}}{\text{mol}}$$

$$= -544 \text{ kJ}$$

Thus, when 1 mole of $H_2(g)$ and 1 mole of $F_2(g)$ react to form 2 moles of HF(g), 544 kJ of energy should be released.

INTERACTIVE EXAMPLE 3-4

ΔE from Bond Energies

Using the bond energies listed in Table 3-3, calculate ΔE for the reaction of methane with chlorine and fluorine to give Freon-12 (CF_2Cl_2).

$$CH_4(g) + 2Cl_2(g) + 2F_2(g) \longrightarrow CF_2Cl_2(g) + 2HF(g) + 2HCl(g)$$

Solution

The idea here is to break the bonds in the gaseous reactants to give individual atoms and then assemble these atoms into the gaseous products by forming new bonds:

$$\text{Reactants} \xrightarrow[\text{required}]{\text{Energy}} \text{atoms} \xrightarrow[\text{released}]{\text{Energy}} \text{products}$$

We then combine the energy changes to calculate ΔE:

$\Delta E =$ energy required to break bonds $-$ energy released when bonds form

where the minus sign gives the correct sign to the energy terms for the exothermic processes.

Reactant Bonds Broken:

$$CH_4: \quad 4 \text{ mol C}-\text{H} \qquad 4 \text{ mol} \times \frac{413 \text{ kJ}}{\text{mol}} = 1652 \text{ kJ}$$

$$2Cl_2: \quad 2 \text{ mol Cl}-\text{Cl} \qquad 2 \text{ mol} \times \frac{239 \text{ kJ}}{\text{mol}} = 478 \text{ kJ}$$

$$2F_2: \quad 2 \text{ mol F}-\text{F} \qquad 2 \text{ mol} \times \frac{154 \text{ kJ}}{\text{mol}} = 308 \text{ kJ}$$

$$\text{Total energy required} = 2438 \text{ kJ}$$

Product Bonds Formed:

$$CF_2Cl_2: \quad 2 \text{ mol C}-\text{F} \qquad 2 \text{ mol} \times \frac{485 \text{ kJ}}{\text{mol}} = 970 \text{ kJ}$$

$$\text{and}$$

$$2 \text{ mol C}-\text{Cl} \qquad 2 \text{ mol} \times \frac{339 \text{ kJ}}{\text{mol}} = 678 \text{ kJ}$$

$$HF: \quad 2 \text{ mol H}-\text{F} \qquad 2 \text{ mol} \times \frac{565 \text{ kJ}}{\text{mol}} = 1130 \text{ kJ}$$

$$HCl: \quad 2 \text{ mol H}-\text{Cl} \qquad 2 \text{ mol} \times \frac{427 \text{ kJ}}{\text{mol}} = 854 \text{ kJ}$$

$$\text{Total energy released} = 3632 \text{ kJ}$$

We now can calculate ΔE:

$$\Delta E = \text{energy required to break bonds } - \text{ energy released when bonds form}$$
$$= 2438 \text{ kJ} - 3632 \text{ kJ}$$
$$= -1194 \text{ kJ}$$

Since the sign of the value for the energy change is negative, this means that 1194 kJ of energy is released per mole of CF_2Cl_2 formed.

See Exercises 3-69 through 3-76

3-8 | The Localized Electron Bonding Model

So far we have discussed the general characteristics of the chemical bonding model and have seen that properties such as bond strength and polarity can be assigned to individual bonds. In this section we introduce a specific model used to describe covalent bonds. We need a simple model that can be applied easily even to very complicated molecules and that can be used routinely by chemists to interpret and organize the wide variety of chemical phenomena. The model that serves this purpose is the **localized electron (LE) model**, which assumes that *a molecule is composed of atoms that are bound together by sharing pairs of electrons using the atomic orbitals of the bound atoms.* Electron pairs in the molecule are assumed to be localized on a particular atom or in the space between two atoms. Those pairs of electrons localized on an atom are called **lone pairs**, and those found in the space between the atoms are called **bonding pairs**.

As we will apply it, the LE model has three parts:

1. Description of the valence electron arrangement in the molecule using Lewis structures (will be discussed in the next section).
2. Prediction of the geometry of the molecule using the valence shell electron-pair repulsion (VSEPR) model (will be discussed in Chapter 4).
3. Description of the type of atomic orbitals used by the atoms to share electrons or hold lone pairs (will be discussed in Chapter 4).

As we discussed earlier, we use models that are as simple as possible. The quantum mechanical model discussed in Chapter 2 views electron distribution in terms of probability. We now will consider a simpler model in which the electrons exist between the atoms in a molecule.

3-9 | Lewis Structures

The **Lewis structure** of a molecule shows how the valence electrons are arranged among the atoms in the molecule. These representations are named after G. N. Lewis (Fig. 3-11). The rules for writing Lewis structures are based on observations of

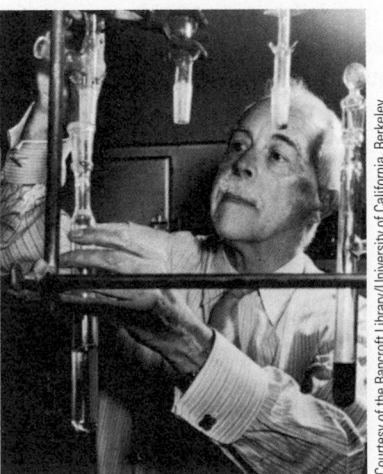

Figure 3-11 | G. N. Lewis (1875–1946).

 While the phrase "atoms want to be like a noble gas" is a way to remember charges for ions and to determine Lewis structures, it does not tell us why this occurs.

ⓘ Lewis structures show only valence electrons.

K⁺   Br⁻

thousands of molecules. From experiments, chemists have learned that the *most important guidelines for the formation of a stable compound is that the atoms achieve noble gas electron configurations.* ◄ ⓘ

We have already seen that when metals and nonmetals react to form binary ionic compounds, electrons are transferred and the resulting ions typically have noble gas electron configurations. An example is the formation of KBr, where the K⁺ ion has the [Ar] electron configuration and the Br⁻ ion has the [Kr] electron configuration. In writing Lewis structures, the rule is that *only the valence electrons are included.* ◄ ⓘ Using dots to represent electrons, the Lewis structure for KBr is

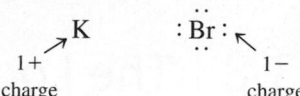

No dots are shown on the K⁺ ion because it has no valence electrons. The Br⁻ ion is shown with eight electrons because it has a filled valence shell.

Next we will consider Lewis structures for molecules with covalent bonds, involving elements in the first and second periods. The principle of achieving a noble gas electron configuration applies to these elements as follows:

- Hydrogen forms stable molecules where it shares two electrons. That is, it follows a **duet rule**. For example, when two hydrogen atoms, each with one electron, combine to form the H_2 molecule, we have

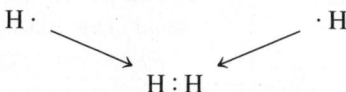

By sharing electrons, each hydrogen in H_2, in effect, has two electrons; that is, each hydrogen has a filled valence shell.

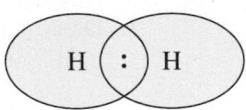

- Helium does not form bonds because its valence orbital is already filled; it is a noble gas. Helium has the electron configuration $1s^2$ and can be represented by the Lewis structure

He :

ⓘ Carbon, nitrogen, oxygen, and fluorine always obey the octet rule in stable molecules.

- The second-row nonmetals carbon through fluorine form stable molecules when they are surrounded by enough electrons to fill the valence orbitals, that is, the $2s$ and the three $2p$ orbitals. Since eight electrons are required to fill these orbitals, these elements typically obey the **octet rule**; they are surrounded by eight electrons. ◄ ⓘ An example is the F_2 molecule, which has the following Lewis structure:

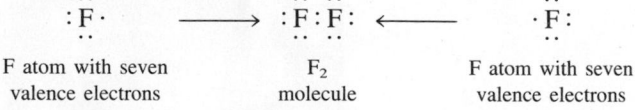

Note that each fluorine atom in F_2 is, in effect, surrounded by eight electrons, two of which are shared with the other atom. This is a *bonding pair* of electrons, as discussed earlier. Each fluorine atom also has three pairs of electrons not involved in bonding. These are the *lone pairs*.

- Neon does not form bonds because it already has an octet of valence electrons (it is a noble gas). The Lewis structure is

$$: \overset{\displaystyle ..}{\underset{\displaystyle ..}{Ne}} :$$

Note that only the valence electrons of the neon atom ($2s^2 2p^6$) are represented by the Lewis structure. The $1s^2$ electrons are core electrons and are not shown.

From the preceding discussion we can formulate the following rules for writing the Lewis structures of molecules containing atoms from the first two periods.

PROBLEM-SOLVING STRATEGY

Steps for Writing Lewis Structures

1. Sum the valence electrons from all the atoms. Do not worry about keeping track of which electrons come from which atoms. It is the *total* number of electrons that is important.

2. Use a pair of electrons to form a bond between each pair of bound atoms.

3. Arrange the remaining electrons to satisfy the duet rule for hydrogen and the octet rule for the second-row elements.

To see how these steps are applied, we will draw the Lewis structures of a few molecules. We will first consider the water molecule and follow the previous steps.

1. We sum the *valence* electrons for H_2O as shown:

$$1 + 1 + 6 = 8 \text{ valence electrons}$$
$$\nearrow \quad \nearrow \quad \nearrow$$
$$H \quad H \quad O$$

2. Using a pair of electrons per bond, we draw in the two O—H single bonds:

$$H-O-H$$

Note that *a line instead of a pair of dots is used to indicate each pair of bonding electrons*. This is the standard notation.

3. We distribute the remaining electrons to achieve a noble gas electron configuration for each atom. Since four electrons have been used in forming the two bonds, four electrons (8 − 4) remain to be distributed. Hydrogen is satisfied with two electrons (duet rule), but oxygen needs eight electrons to have a noble gas configuration. Thus the remaining four electrons are added to oxygen as two lone pairs. Dots are used to represent the lone pairs: ◄

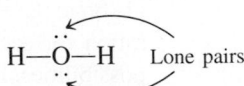

This is the correct Lewis structure for the water molecule. Each hydrogen has two electrons and the oxygen has eight, as shown below:

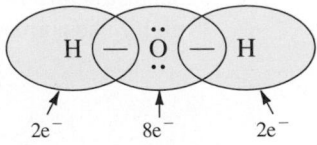

As a second example, let's write the Lewis structure for carbon dioxide.

1. Summing the valence electrons gives

$$4 + 6 + 6 = 16$$
$$$$
C   O   O

2. Parts of electrons are used to form a bond between the carbon and each oxygen,

$$\text{O—C—O}$$

and the remaining electrons are distributed to achieve noble gas configurations on each atom. In this case we have 12 electrons (16 − 4) remaining after the bonds are drawn. The distribution of these electrons is determined by a trial-and-error process. We have 6 pairs of electrons to distribute. Suppose we try 3 pairs on each oxygen to give

$$:\overset{..}{\text{O}}\text{—C—}\overset{..}{\text{O}}:$$

3. Is this correct? To answer this question, we need to check two things:

1. The total number of electrons. There are 16 valence electrons in this structure, which is the correct number.
2. The octet rule for each atom. Each oxygen has 8 electrons, but the carbon has only 4. This cannot be the correct Lewis structure.

How can we arrange the 16 available electrons to achieve an octet for each atom? Suppose there are 2 shared pairs between the carbon and each oxygen: ◀ 𝒊

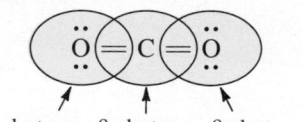

8 electrons   8 electrons   8 electrons

Now each atom is surrounded by 8 electrons, and the total number of electrons is 16, as required. This is the correct Lewis structure for carbon dioxide, which has two double bonds and four lone pairs.

Finally, let's consider the Lewis structure of the CN⁻ (cyanide) ion. Summing the valence electrons, we have

$$\text{CN}^-$$
$$4 + 5 + 1 = 10$$

Note that the negative charge means an extra electron is present. After drawing a single bond (C—N), we distribute the remaining electrons to achieve a noble gas configuration for each atom. Eight electrons remain to be distributed. We can try various possibilities, for example:

$$\overset{..}{\text{C}}\text{—}\overset{..}{\text{N}}$$

This structure is incorrect because C and N have only six electrons each instead of eight. The correct arrangement is

$$[:\text{C}\equiv\text{N}:]^-$$

(Satisfy yourself that both carbon and nitrogen have eight electrons.)

𝒊 $\ddot{\text{O}}=\text{C}=\ddot{\text{O}}$ represents

$$\ddot{\text{O}}:\; :\ddot{\text{C}}:\; :\ddot{\text{O}}$$

INTERACTIVE EXAMPLE 3-5

Writing Lewis Structures

Give the Lewis structure for each of the following.

a. HF   **d.** CH_4

b. N_2 **e.** CF_4

c. NH_3 **f.** NO^+

Solution

In each case we apply the three steps for writing Lewis structures. Recall that lines are used to indicate shared electron pairs and that dots are used to indicate nonbonding pairs (lone pairs). We have the following tabulated results:

	Total Valence Electrons	Draw Single Bonds	Calculate Number of Electrons Remaining	Use Remaining Electrons to Achieve Noble Gas Configurations	Check Number of Electrons
a. HF	$1 + 7 = 8$	H—F	6	H—F̈:	H, 2 F, 8
b. N_2	$5 + 5 = 10$	N—N	8	:N≡N:	N, 8

c. NH_3	$5 + 3(1) = 8$	H—N—H
		      |
		      H
			2	H—N̈—H
				      |
				      H
					H, 2
					 N, 8

d. CH_4	$4 + 4(1) = 8$	H | H—C—H | H	0	H | H—C—H | H	H, 2 C, 8
e. CF_4	$4 + 4(7) = 32$	F | F—C—F | F	24	:F̈: | :F̈—C—F̈: | :F̈:	F, 8 C, 8
f. NO^+	$5 + 6 - 1 = 10$	N—O	8	$[:N≡O:]^+$	N, 8 O, 8

See Exercises 3-79 through 3-82

When writing Lewis structures, do not worry about which electrons come from which atoms in a molecule. The best way to look at a molecule is to regard it as a new entity that uses all the available valence electrons of the atoms to achieve the lowest possible energy.* The valence electrons belong to the molecule, rather than to the individual atoms. Simply distribute all valence electrons so that the various rules are satisfied, without regard for the origin of each particular electron.

---

*In a sense this approach corrects for the fact that the localized electron model overemphasizes that a molecule is simply a sum of its parts—that is, that the atoms retain their individual identities in the molecule.

The periodic table was originally organized to represent patterns in chemical properties of the elements. We now see that there are patterns of valence electron arrangements in the table, which makes sense since the electrons are responsible for how elements react. We have already discussed trends of atomic radius, ionization energy, electron affinity, and electronegativity across rows and down columns of the periodic table.

We can also see periodic trends with respect to Lewis structures. For example, consider the diatomic molecules N_2, O_2, and F_2 as seen below:

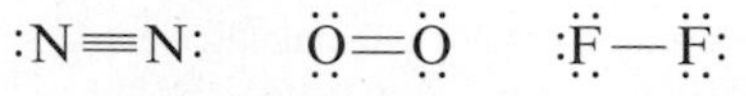

$$:N \equiv N: \quad \ddot{O} = \ddot{O} \quad :\ddot{F} - \ddot{F}:$$

Note that the nitrogen atoms (Group 5A) each have three bonds and one lone pair, the oxygen atoms (Group 6A) have two bonds and two lone pairs, and the fluorine atoms (Group 7A) have one bond and three lone pairs. We can also see a pattern in the compounds formed with hydrogen by the elements in the second row:

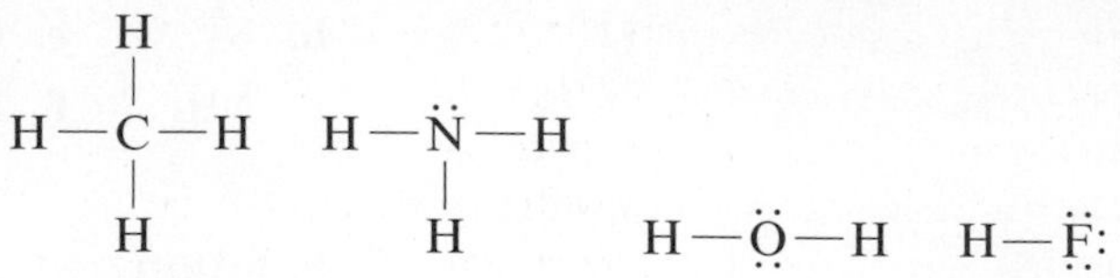

H–C(–H)(–H)–H (methane)   H–N(–H)–H (ammonia)   H–Ö–H (water)   H–F̈: (hydrogen fluoride)

Note the pattern across the row in terms of number of bonds and number of lone pairs, including Group 4A (four bonds and no lone pairs). Of course, neon (Group 8A) does not form a compound with hydrogen, meaning there are 0 bonds.

C	N	O	F	Ne
4 bonds	3 bonds	2 bonds	1 bond	0 bonds
0 lone pairs	1 lone pair	2 lone pairs	3 lone pairs	

It is not the case that all Lewis structures with these elements will follow these patterns exactly, but these patterns give us a good starting point.

3-10 | Exceptions to the Octet Rule

The localized electron model is a simple but very successful model, and the rules we have used for Lewis structures apply to most molecules. However, with such a simple model, some exceptions are inevitable. Boron, for example, tends to form compounds in which the boron atom has fewer than eight electrons around it—it does not have a complete octet. Boron trifluoride (BF_3), a gas at normal temperatures and pressures, reacts very energetically with molecules such as water and ammonia that have available electron pairs (lone pairs). The violent reactivity of BF_3 with electron-rich molecules arises because the boron atom is electron-deficient. Boron trifluoride has 24 valence electrons. The Lewis structure often drawn for BF_3 is

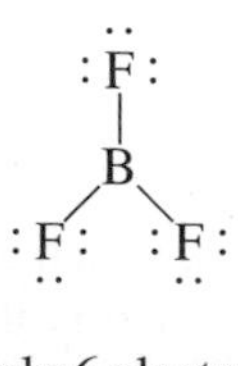

Note that in this structure boron has only 6 electrons around it. The octet rule for boron can be satisfied by drawing a structure with a double bond, such as

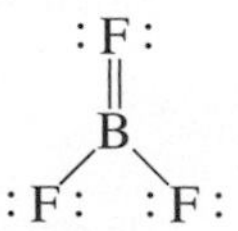

Recent studies indicate that double bonding may be important in BF_3. However, the boron atom in BF_3 certainly behaves as if it is electron-deficient, as indicated by

The element nitrogen exists at normal temperatures and pressures as a gas containing N_2, a molecule with a very strong triple bond. In the gas phase the diatomic molecules move around independently with almost no tendency to associate with each other. Under intense pressure, however, nitrogen changes to a dramatically different form. This conclusion was reached at the Carnegie Institution in Washington, D.C., by Mikhail Erements and his colleagues, who subjected nitrogen to a pressure of 2.4 million atmospheres in a special diamond anvil press. Under this tremendous pressure the bonds of the N_2 molecules break and a substance containing an aggregate of nitrogen atoms forms. In other words, under great pressure elemental nitrogen changes from a substance containing diatomic molecules to one containing many nitrogen atoms bonded to each other. Interestingly, this substance remains intact even after the pressure is released—as long as the temperature remains at 100 K. This new form of nitrogen has a very high potential energy relative to N_2. Thus this substance would be an extraordinarily powerful propellant or explosive if enough of it could be made. This new form of nitrogen is also a semiconductor for electricity; normal nitrogen gas is an insulator.

The newly discovered form of nitrogen is significant for several reasons. For one thing, it may help us understand the nature of the interiors of the giant gas planets such as Jupiter. Also, their success in changing nitrogen to an atomic

solid encourages high-pressure scientists who are trying to accomplish the same goal with hydrogen. It is surprising that nitrogen, which has diatomic molecules containing bonds more than twice as strong as those in hydrogen, will form an atomic solid at these pressures but hydrogen does not. Hydrogen remains a molecular solid at far greater pressures than nitrogen can endure.

A diamond anvil cell used to study materials at very high pressures.

Stever D. Jacobsen/Northwestern University

the reactivity of BF_3 toward electron-rich molecules, for example, toward NH_3 to form H_3NBF_3:

In this stable compound, boron has an octet of electrons.

It is characteristic of boron to form molecules in which the boron atom is electron-deficient. On the other hand, carbon, nitrogen, oxygen, and fluorine can be counted on to obey the octet rule.

Some atoms exceed the octet rule. This behavior is observed only for those elements in Period 3 of the periodic table and beyond. To see how this arises, we will consider the Lewis structure for sulfur hexafluoride (SF_6), a well-known and very stable molecule. The sum of the valence electrons is

$$6 + 6(7) = 48 \text{ electrons}$$

Indicating the single bonds gives the structure on the left below:

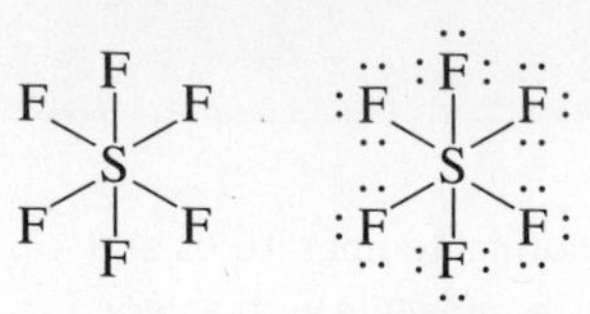

$$
\begin{array}{cc}
\underset{F}{\overset{F}{\underset{|}{\overset{|}{F-\!\!\!\underset{\displaystyle\diagup}{S}\!\!\!-F}}}} & \quad
\end{array}
$$

We have used 12 electrons to form the S—F bonds, which leaves 36 electrons. Since fluorine always follows the octet rule, we complete the six fluorine octets to give the structure on the right above. This structure uses all 48 valence electrons for SF_6, but sulfur has 12 electrons around it; that is, sulfur *exceeds* the octet rule. How can this happen?

The answer to this question is very complicated. Originally, it was assumed that because the sulfur atom has empty *d* orbitals

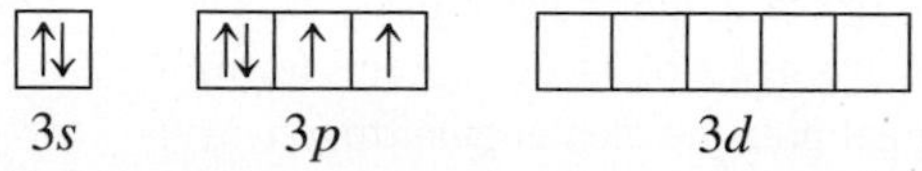

that these orbitals would be used to hold the extra electrons. Recent evidence indicates that this reasoning is not correct. However, the new explanation is beyond the scope of this text. ◀ ⓘ

ⓘ Whether the atoms that exceed the octet rule actually place the extra electrons in their *d* orbitals is a matter of controversy among theoretical chemists. We will not consider this issue in this text.

INTERACTIVE EXAMPLE 3-6

Lewis Structures for Molecules That Violate the Octet Rule I

Write the Lewis structure for PCl_5.

Solution

We can follow the same stepwise procedure we used previously for sulfur hexafluoride.

1. Sum the valence electrons.

$$5 + 5(7) = 40 \text{ electrons}$$
$$\underset{P}{\uparrow} \qquad \underset{Cl}{\uparrow}$$

2. Indicate single bonds between bound atoms.

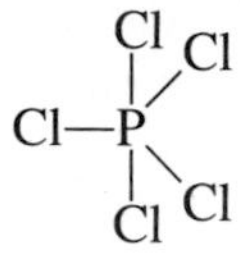

3. Distribute the remaining electrons. In this case, 30 electrons (40 − 10) remain. These are used to satisfy the octet rule for each chlorine atom. The final Lewis structure is

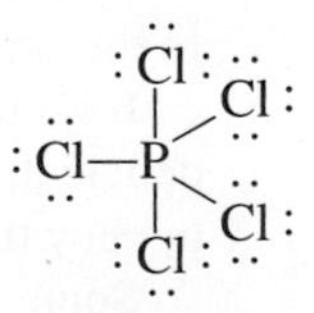

Note that phosphorus, which is a third-row element, has exceeded the octet rule by two electrons.

See Exercises 3-85 and 3-86

> ### LET'S REVIEW | *Lewis Structures: Comments About the Octet Rule*
>
> - The second-row elements C, N, O, and F should always be assumed to obey the octet rule.
> - The second-row elements B and Be often have fewer than eight electrons around them in their compounds. These electron-deficient compounds are very reactive.
> - The second-row elements never exceed the octet rule, since their valence orbitals (2s and 2p) can accommodate only eight electrons.
> - Third-row and heavier elements often satisfy the octet rule but can exceed the octet rule.
> - When writing the Lewis structure for a molecule, satisfy the octet rule for the atoms first. If electrons remain after the octet rule has been satisfied, then place them on any element from Period 3 or beyond.

In the PCl_5 and SF_6 molecules, the central atoms (P and S, respectively) must have the extra electrons. However, in molecules having more than one atom that can exceed the octet rule, it is not always clear which atom should have the extra electrons. Consider the Lewis structure for the triiodide ion (I_3^-), which has

$$3(7) + 1 = 22 \text{ valence electrons}$$

$$\uparrow \qquad \uparrow$$
$$I \qquad -1 \text{ charge}$$

Indicating the single bonds gives I—I—I. At this point, 18 electrons ($22 - 4$) remain. Trial and error will convince you that one of the iodine atoms must exceed the octet rule, but *which* one?

The rule we will follow is that *when it is necessary to exceed the octet rule for one of several third-row (or higher) elements, assume that the extra electrons should be placed on the central atom.*

Thus for I_3^- the Lewis structure is

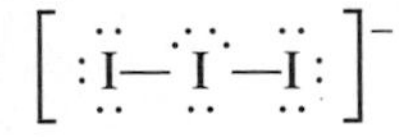

where the central iodine exceeds the octet rule. This structure agrees with known properties of I_3^-.

INTERACTIVE EXAMPLE 3-7

Lewis Structures for Molecules That Violate the Octet Rule II

Write the Lewis structure for each molecule or ion.

a. ClF_3 **b.** XeO_3 **c.** $RnCl_2$ **d.** $BeCl_2$ **e.** ICl_4^-

Solution

a. The chlorine atom (third row) accepts the extra electrons.

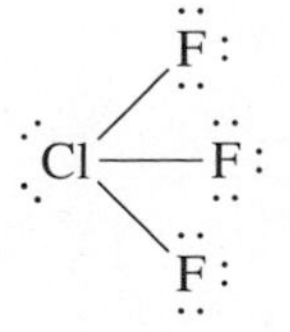

b. All atoms obey the octet rule.

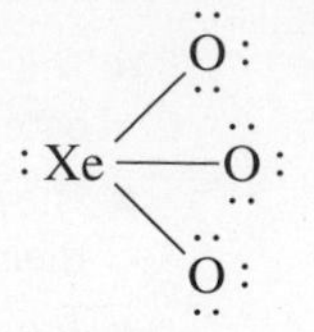

$$: \overset{\displaystyle .\,.}{\underset{}{Xe}} \overset{\displaystyle \overset{.\,.}{O} :}{\diagdown}$$

(Lewis structure: Xe bonded to three O atoms, all with octets)

c. Radon, a noble gas in Period 6, accepts the extra electrons.

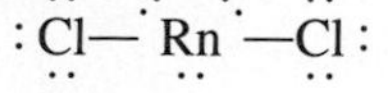

$$: \overset{.\,.}{\underset{.\,.}{Cl}} - \overset{.}{\underset{.}{Rn}} - \overset{.\,.}{\underset{.\,.}{Cl}} :$$

d. Beryllium is electron-deficient.

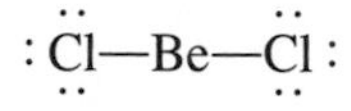

$$: \overset{.\,.}{\underset{.\,.}{Cl}} - Be - \overset{.\,.}{\underset{.\,.}{Cl}} :$$

e. Iodine exceeds the octet rule.

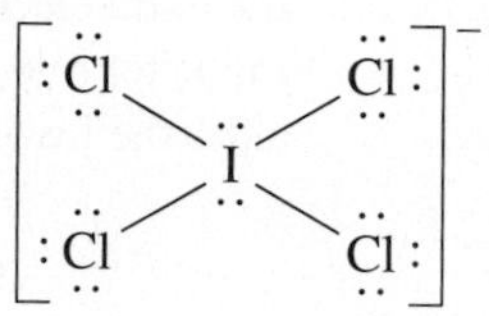

(Lewis structure: I bonded to four Cl atoms in brackets with a negative charge)

See Exercises 3-83, 3-85, and 3-86

3-11 | Resonance

Sometimes more than one valid Lewis structure (one that obeys the rules we have outlined) is possible for a given molecule. Consider the Lewis structure for the nitrate ion (NO_3^-), which has 24 valence electrons. To achieve an octet of electrons around each atom, a structure like this is required:

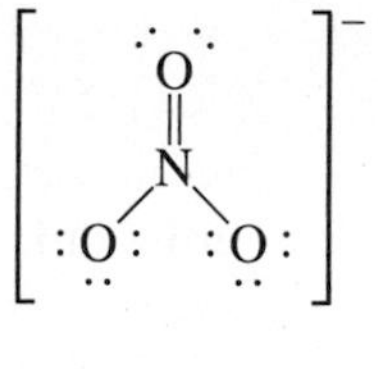

(Lewis structure of NO_3^- showing N double-bonded to one O and single-bonded to two O, in brackets with negative charge)

If this structure accurately represents the bonding in NO_3^-, there should be two types of N—O bonds observed in the molecule: one shorter bond (the double bond) and two identical longer ones (the two single bonds). However, experiments clearly show that NO_3^- exhibits only *one* type of N—O bond with a length and strength *between* those expected for a single bond and a double bond. Thus, although the structure we have shown above is a valid Lewis structure, it does *not* correctly represent the bonding in NO_3^-. This is a serious problem, and it means that the model must be modified.

Look again at the proposed Lewis structure for NO_3^-. There is no reason for choosing a particular oxygen atom to have the double bond. There are really three valid Lewis structures:

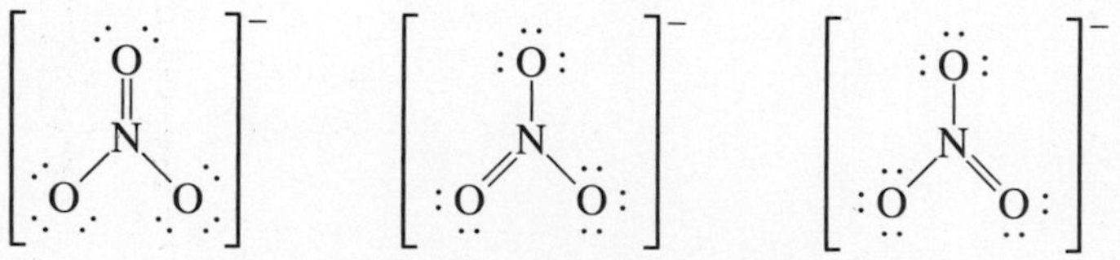

(Three resonance Lewis structures of NO_3^-, each in brackets with a negative charge, showing the double bond to a different oxygen atom in each)

Is any of these structures a correct description of the bonding in NO_3^-? No, because NO_3^- does not have one double and two single bonds—it has three equivalent bonds. We can solve this problem by making the following assumption: The correct description of NO_3^- is *not given by any one* of the three Lewis structures but is given only by the *superposition of all three*.

The nitrate ion does not exist as any of the three extreme structures but exists as an average of all three. **Resonance** is the condition in which more than one valid Lewis structure can be written for a particular molecule (in which the atoms are connected in the same order). The resulting electron structure of the molecule is given by the average of these **resonance structures**. This situation is usually represented by double-headed arrows as follows:

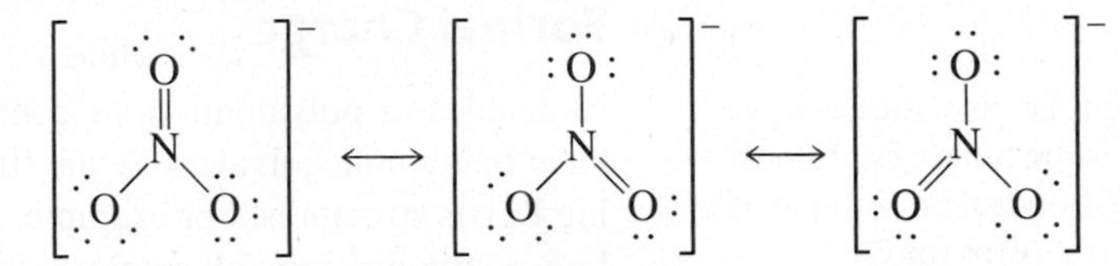

Note that in all these resonance structures the arrangement of the nuclei is the same. Only the placement of the electrons differs. The arrows do not mean that the molecule "flips" from one resonance to another. They simply show that the *actual structure is an average of the three resonance structures*.

The concept of resonance is necessary because the localized electron model postulates that electrons are localized between a given pair of atoms. However, nature does not really operate this way. Electrons are really delocalized—they can move around the entire molecule. The valence electrons in the NO_3^- molecule distribute themselves to provide equivalent N—O bonds. Resonance is necessary to compensate for the defective assumption of the localized electron model. However, this model is so useful that we retain the concept of localized electrons and add resonance to allow the model to treat species such as NO_3^-. ◄ ⓘ

ⓘ The concept of resonance serves as a reminder that all models are simplifications. Electrons are not localized within a molecule, and we will discuss more complex models in Chapter 4.

EXAMPLE 3-8   Resonance Structures

Describe the electron arrangement in the nitrite anion (NO_2^-), using the localized electron model.

Solution

We will follow the usual procedure for obtaining the Lewis structure for the NO_2^- ion. In NO_2^- there are $5 + 2(6) + 1 = 18$ valence electrons. Indicating the single bonds gives the structure

$$O—N—O$$

The remaining 14 electrons $(18 - 4)$ can be distributed to produce these structures:

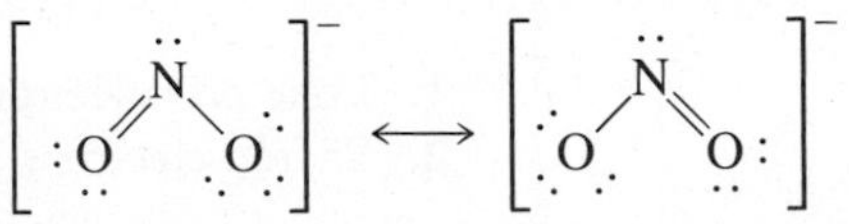

This is a resonance situation. Two equivalent Lewis structures can be drawn. *The electronic structure of the molecule is correctly represented not by either resonance structure but by the average of the two.* There are two equivalent N—O bonds, each one intermediate between a single and a double bond.

See Exercises 3-87 through 3-92

Odd-Electron Molecules

Relatively few molecules formed from nonmetals contain odd numbers of electrons. One common example is nitric oxide (NO), which is formed when nitrogen and oxygen gases react at high temperatures in automobile engines. Nitric oxide is emitted into the air, where it immediately reacts with oxygen to form gaseous nitrogen dioxide (NO_2), another odd-electron molecule.

Since the localized electron model is based on pairs of electrons, it does not handle odd-electron cases in a natural way. To treat odd-electron molecules, a more sophisticated model is needed.

Formal Charge

i Equivalent Lewis structures contain the same numbers of single and multiple bonds. For example, the resonance structures for O_3

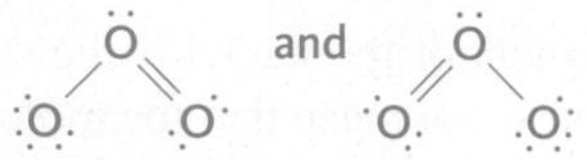

and

are equivalent Lewis structures. These are equally important in describing the bonding in O_3. Nonequivalent Lewis structures contain different numbers of single and multiple bonds.

Molecules or polyatomic ions containing atoms that can exceed the octet rule often have many nonequivalent Lewis structures ◀ *i*, all of which obey the rules for writing Lewis structures. For example, as we will see in detail below, the sulfate ion has a Lewis structure with all single bonds and several Lewis structures that contain double bonds. How do we decide which of the many possible Lewis structures best describes the actual bonding in sulfate? One method is to estimate the charge on each atom in the various possible Lewis structures and use these charges to select the most appropriate structure(s). We will see below how this is done, but first we must decide on a method to assign atomic charges in molecules.

We will now consider a concept called *formal charge* for estimating the charge of an individual atom in a molecule. We will use the formal charges of atoms to evaluate Lewis structures. As we will see below, the **formal charge** of an atom in a molecule is *the difference between the number of valence electrons on the free atom and the number of valence electrons assigned to the atom in the molecule*.

Therefore, to determine the formal charge of a given atom in a molecule, we need to know two things:

1. The number of valence electrons on the free neutral atom (which has zero net charge because the number of electrons equals the number of protons)
2. The number of valence electrons "belonging" to the atom in a molecule

We then compare these numbers. If in the molecule the atom has the same number of valence electrons as it does in the free state, the positive and negative charges just balance, and it has a formal charge of zero. If the atom has one more valence electron in a molecule than it has as a free atom, it has a formal charge of -1, and so on. Thus the formal charge on an atom in a molecule is defined as

Formal charge = (number of valence electrons on free atom)
   − (number of valence electrons assigned to the atom in the molecule)

To compute the formal charge of an atom in a molecule, we assign the valence electrons in the molecule to the various atoms, making the following assumptions:

1. Lone pair electrons belong entirely to the atom in question.
2. Shared electrons are *divided equally* between the two sharing atoms.

Thus the number of valence electrons assigned to a given atom is calculated as follows:

$(\text{Valence electrons})_{\text{assigned}}$ = (number of lone pair electrons)
   $+ \frac{1}{2}$ (number of shared electrons)

We will illustrate the procedure for calculating formal charges by considering two of the possible Lewis structures for the sulfate ion, which has 32 valence electrons. For the Lewis structure

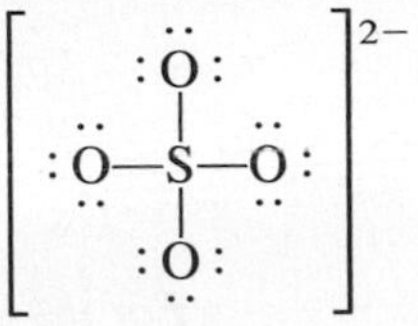

$$
\left[\begin{array}{c} \ddot{\text{O}} \\ | \\ \ddot{\text{O}} - \text{S} - \ddot{\text{O}} \\ | \\ \ddot{\text{O}} \end{array} \right]^{2-}
$$

each oxygen atom has 6 lone pair electrons and shares 2 electrons with the sulfur atom. Thus, using the preceding assumptions, each oxygen is assigned 7 valence electrons.

$$\text{Valence electrons assigned to each oxygen} = 6 \; plus \; \tfrac{1}{2}(2) = 7$$

Lone   Shared
pair   electrons
electrons

$$\text{Formal charge on oxygen} = 6 \; minus \; 7 = -1$$

Valence electrons
on a free O atom
Valence electrons
assigned to each O
in SO_4^{2-}

The formal charge on each oxygen is -1.

For the sulfur atom there are no lone pair electrons, and eight electrons are shared with the oxygen atoms. Thus, for sulfur,

$$\text{Valence electrons assigned to sulfur} = 0 \; plus \; \tfrac{1}{2}(8) = 4$$

Lone   Shared
pair   electrons
electrons

$$\text{Formal charge on sulfur} = 6 \; minus \; 4 = 2$$

Valence electrons
on a free S atom
Valence electrons
assigned to S
in SO_4^{2-}

A second possible Lewis structure is

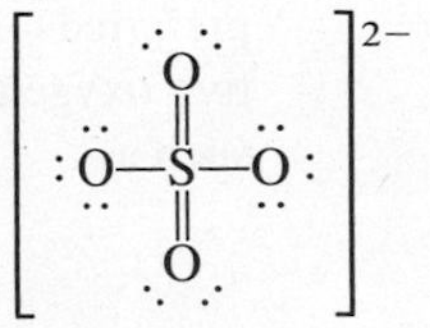

$$
\left[\begin{array}{c} \ddot{\text{O}} \\ \| \\ \ddot{\text{O}} - \text{S} - \ddot{\text{O}} \\ \| \\ \ddot{\text{O}} \end{array} \right]^{2-}
$$

In this case the formal charges are as follows:
For oxygen atoms with single bonds:

$$\text{Valence electrons assigned} = 6 + \tfrac{1}{2}(2) = 7$$
$$\text{Formal charge} = 6 - 7 = -1$$

For oxygen atoms with double bonds:

$$\text{Valence electrons assigned} = 4 + \tfrac{1}{2}(4) = 6$$

Each double
bond has 4
electrons

$$\text{Formal charge} = 6 - 6 = 0$$

For the sulfur atom:

$$\text{Valence electrons assigned} = 0 + \tfrac{1}{2}(12) = 6$$
$$\text{Formal charge} = 6 - 6 = 0$$

We will use two fundamental assumptions about formal charges to evaluate Lewis structures:

1. Atoms in molecules try to achieve formal charges as close to zero as possible.
2. Any negative formal charges are expected to reside on the most electronegative atoms.

We can use these principles to evaluate the two nonequivalent Lewis structures for sulfate given previously. Notice that in the structure with only single bonds, each oxygen has a formal charge of -1, while the sulfur has a formal charge of $+2$.

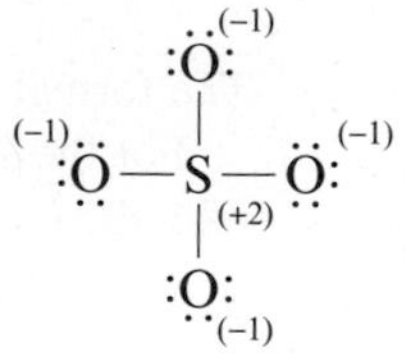

In contrast, in the structure with two double bonds and two single bonds, the sulfur and two oxygen atoms have a formal charge of 0, while two oxygens have a formal charge of -1.

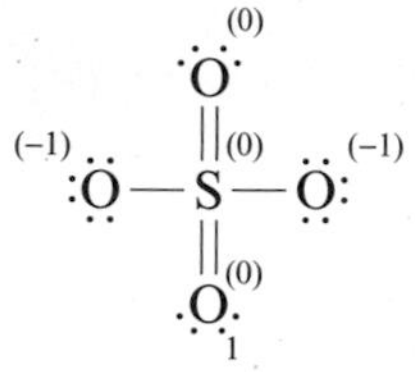

Based on the assumptions given previously, the structure with two double bonds is preferred—it has lower formal charges and the -1 formal charges are on electronegative oxygen atoms. Thus, for the sulfate ion, we might expect resonance structures such as

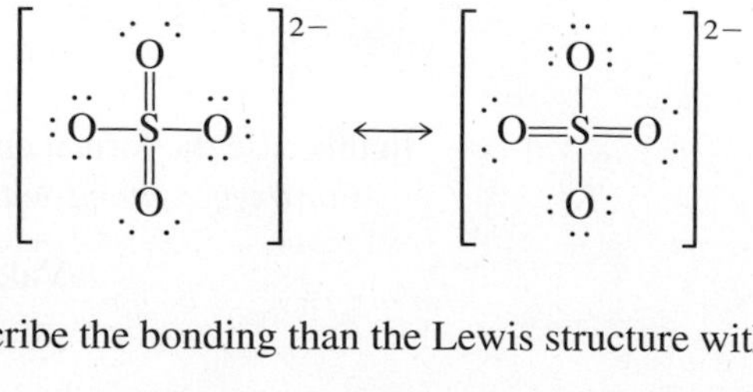

to more closely describe the bonding than the Lewis structure with only single bonds.

Rules Governing Formal Charge

■ To calculate the formal charge on an atom:

1. Take the sum of the lone pair electrons and one-half the shared electrons. This is the number of valence electrons assigned to the atom in the molecule.

2. Subtract the number of assigned electrons from the number of valence electrons on the free, neutral atom to obtain the formal charge.

■ The sum of the formal charges of all atoms in a given molecule or ion must equal the overall charge on that species.

■ If nonequivalent Lewis structures exist for a species, those with formal charges closest to zero and with any negative formal charges on the most electronegative atoms are considered to best describe the bonding in the molecule or ion.

EXAMPLE 3-9

Formal Charges

Give possible Lewis structures for XeO_3, an explosive compound of xenon. Which Lewis structure or structures are most appropriate according to the formal charges?

Solution

For XeO_3 (26 valence electrons) we can draw the following possible Lewis structures (formal charges are indicated in parentheses):

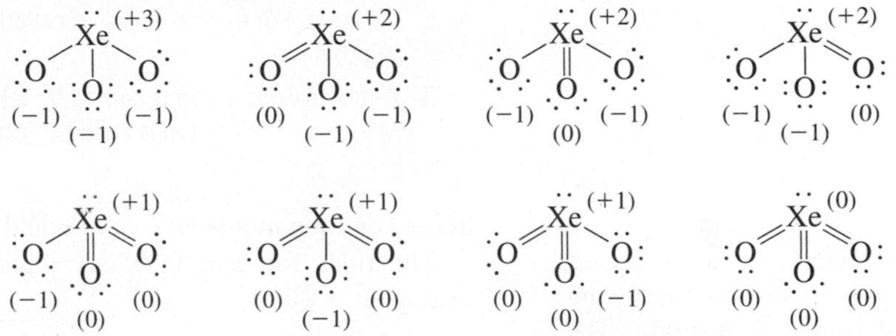

Based on the ideas of formal charge, we would predict that the Lewis structures with the lower values of formal charge would be most appropriate for describing the bonding in XeO_3.

See Exercises 3-99 and 3-100

As a final note, there are a couple of cautions about formal charge to keep in mind. First, although formal charges are closer to actual atomic charges in molecules than are oxidation states, formal charges still provide only *estimates* of charge—they should not be taken as actual atomic charges. Second, the evaluation of Lewis structures using formal charge ideas can lead to erroneous predictions. Tests based on experiments must be used to make the final decisions on the correct description of the bonding in a molecule or polyatomic ion.

3-12 | Naming Simple Compounds

When chemistry was an infant science, there was no system for naming compounds. Names such as sugar of lead, blue vitrol, quicklime, Epsom salts, milk of magnesia, gypsum, and laughing gas were coined by early chemists. Such names are called *common names.* As chemistry grew, it became clear that using common names for compounds would lead to unacceptable chaos. Nearly 5 million chemical compounds are currently known. Memorizing common names for these compounds would be an impossible task.

The solution, of course, is to adopt a *system* for naming compounds in which the name tells something about the composition of the compound. After learning the system, a chemist given a formula should be able to name the compound or, given a name, should be able to construct the compound's formula. In this section we will specify the most important rules for naming compounds other than organic compounds (those based on chains of carbon atoms).

We will begin with the systems for naming inorganic **binary compounds**—compounds composed of two elements—which we classify into various types for easier recognition. We will consider both ionic and covalent compounds.

Binary Ionic Compounds (Type I)

Binary ionic compounds contain a positive ion (cation) always written first in the formula and a negative ion (anion). In naming these compounds, the following rules apply.

Naming Type I Binary Compounds

1. The cation is always named first and the anion second.

2. A monatomic (meaning "one-atom") cation takes its name from the name of the element. For example, Na^+ is called sodium in the names of compounds containing this ion.

3. A monatomic anion is named by taking the root of the element name and adding *-ide*. Thus the Cl^- ion is called chloride.

Some common monatomic cations and anions and their names are given in Table 3-5.

The rules for naming binary ionic compounds are illustrated by the following examples: ◀ *i*

i In formulas of ionic compounds, simple ions are represented by the element symbol: Cl means Cl^-, Na means Na^+, and so on.

Compound	Ions Present	Name
NaCl	Na^+, Cl^-	Sodium chloride
KI	K^+, I^-	Potassium iodide
CaS	Ca^{2+}, S^{2-}	Calcium sulfide
Li_3N	Li^+, N^{3-}	Lithium nitride
CsBr	Cs^+, Br^-	Cesium bromide
MgO	Mg^{2+}, O^{2-}	Magnesium oxide

Table 3-5 | Common Monatomic Cations and Anions

Cation	Name	Anion	Name
H^+	Hydrogen	H^-	Hydride
Li^+	Lithium	F^-	Fluoride
Na^+	Sodium	Cl^-	Chloride
K^+	Potassium	Br^-	Bromide
Cs^+	Cesium	I^-	Iodide
Be^{2+}	Beryllium	O^{2-}	Oxide
Mg^{2+}	Magnesium	S^{2-}	Sulfide
Ca^{2+}	Calcium	N^{3-}	Nitride
Ba^{2+}	Barium	P^{3-}	Phosphide
Al^{3+}	Aluminum		

INTERACTIVE EXAMPLE 3-10

Naming Type I Binary Compounds

Name each binary compound.

a. CsF   **b.** $AlCl_3$ **c.** LiH

Solution

a. CsF is cesium fluoride.

b. $AlCl_3$ is aluminum chloride.

c. LiH is lithium hydride.

Notice that, in each case, the cation is named first, and then the anion is named.

See Exercise 3-105

Formulas from Names

So far we have started with the chemical formula of a compound and decided on its systematic name. The reverse process is also important. For example, given the name calcium chloride, we can write the formula as $CaCl_2$ because we know that calcium forms only Ca^{2+} ions and that, since chloride is Cl^-, two of these anions will be required to give a neutral compound.

INTERACTIVE EXAMPLE 3-11

Formulas from Names for Type I Binary Compounds

Given the following systematic names, write the formula for each compound:

a. potassium iodide

b. calcium oxide

c. gallium bromide

Solution

	Name	Formula	Comments
a.	potassium iodide	KI	Contains K^+ and I^-.
b.	calcium oxide	CaO	Contains Ca^{2+} and O^{2-}.
c.	gallium bromide	$GaBr_3$	Contains Ga^{3+} and Br^-.
			Must have $3Br^-$ to balance charge of Ga^{3+}.

See Exercise 3-105

Binary Ionic Compounds (Type II)

In the binary ionic compounds considered earlier (Type I), the metal present forms only a single type of cation. That is, sodium forms only Na^+, calcium forms only Ca^{2+}, and so on. However, as we will see in more detail later in the text, there are many metals that form more than one type of positive ion and thus form more than one type of ionic compound with a given anion. For example, the compound $FeCl_2$ contains Fe^{2+} ions, and the compound $FeCl_3$ contains Fe^{3+} ions. In a case such as this, the *charge on the metal ion must be specified*. The systematic names for these two iron compounds are iron(II) chloride and iron(III) chloride, respectively, where *the Roman numeral indicates the charge of the cation*.

Another system for naming these ionic compounds that is seen in the older literature was used for metals that form only two ions. *The ion with the higher charge has a name ending in -ic, and the one with the lower charge has a name ending in -ous.* In this system, for example, Fe^{3+} is called the ferric ion, and Fe^{2+} is called the ferrous ion. The names for $FeCl_3$ and $FeCl_2$ are then ferric chloride and ferrous chloride, respectively. In this text we will use the system that employs Roman numerals. Table 3-6 lists the systematic names for many common Type II cations.

INTERACTIVE EXAMPLE 3-12

Naming Type II Binary Compounds

1. Give the systematic name for each of the following compounds:

   a. CuCl   b. HgO   c. Fe_2O_3

2. Given the following systematic names, write the formula for each compound:

   a. Manganese(IV) oxide

   b. Lead(II) chloride

ⓘ Type II binary ionic compounds contain a metal that can form more than one type of cation.

ⓘ A compound must be electrically neutral.

Solution

All of these compounds include a metal that can form more than one type of cation. ◄ ⓘ Thus we must first determine the charge on each cation. This can be done by recognizing that a compound must be electrically neutral; that is, the positive and negative charges must exactly balance. ◄ ⓘ

1.

	Formula	Name	Comments
a.	CuCl	Copper(I) chloride	Because the anion is Cl^-, the cation must be Cu^+ (for charge balance), which requires a Roman numeral I.
b.	HgO	Mercury(II) oxide	Because the anion is O^{2-}, the cation must be Hg^{2+} [mercury(II)]. ◄
c.	Fe_2O_3	Iron(III) oxide	The three O^{2-} ions carry a total charge of 6−, so two Fe^{3+} ions [iron(III)] are needed to give a 6+ charge.

▲ Mercury(II) oxide.

2.

	Name	Formula	Comments
a.	Manganese(IV) oxide	MnO_2	Two O^{2-} ions (total charge 4−) are required by the Mn^{4+} ion [manganese(IV)].
b.	Lead(II) chloride	$PbCl_2$	Two Cl^- ions are required by the Pb^{2+} ion [lead(II)] for charge balance.

See Exercise 3-106

ⓘ A compound containing a transition metal usually requires a Roman numeral in its name.

Note that the use of a Roman numeral in a systematic name is required only in cases where more than one ionic compound forms between a given pair of elements. ◄ ⓘ This case most commonly occurs for compounds containing transition metals, which often form more than one cation. Elements that form only one cation do not need to be identified by a Roman numeral. Common metals that do not require Roman numerals are the Group 1A (1) elements, which form only 1+ ions; the Group 2A (2) elements, which form only 2+ ions; and aluminum, which forms only Al^{3+}. The element silver deserves special mention at this point. In virtually all its compounds silver is found as

Figure 3-12 | Flowchart for naming binary ionic compounds.

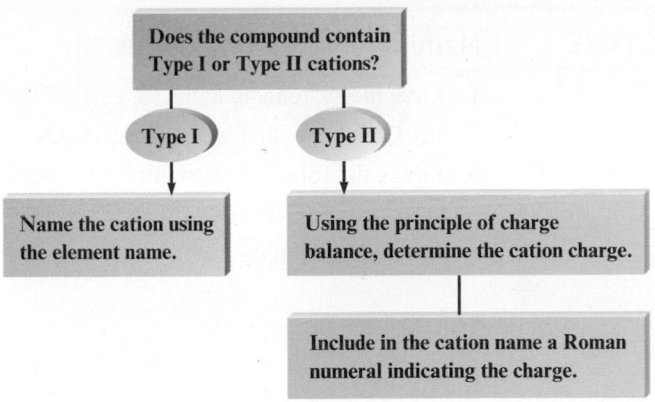

Table 3-6 | Common Type II Cations

Ion	Systematic Name
Fe^{3+}	Iron(III)
Fe^{2+}	Iron(II)
Cu^{2+}	Copper(II)
Cu^{+}	Copper(I)
Co^{3+}	Cobalt(III)
Co^{2+}	Cobalt(II)
Sn^{4+}	Tin(IV)
Sn^{2+}	Tin(II)
Pb^{4+}	Lead(IV)
Pb^{2+}	Lead(II)
Hg^{2+}	Mercury(II)
Hg_2^{2+}*	Mercury(I)
Ag^{+}	Silver[†]
Zn^{2+}	Zinc[†]
Cd^{2+}	Cadmium[†]

*Note that mercury(I) ions always occur bound together to form Hg_2^{2+} ions.
[†]Although these are transition metals, they form only one type of ion, and a Roman numeral is not used.

the Ag^+ ion. Therefore, although silver is a transition metal (and can potentially form ions other than Ag^+), silver compounds are usually named without a Roman numeral. Thus AgCl is typically called silver chloride rather than silver(I) chloride, although the latter name is technically correct. Also, a Roman numeral is not used for zinc compounds, since zinc forms only the Zn^{2+} ion.

As shown in Example 3-12, when a metal ion is present that forms more than one type of cation, the charge on the metal ion must be determined by balancing the positive and negative charges of the compound. To do this you must be able to recognize the common cations and anions and know their charges (see Tables 3-5 and 3-7). The procedure for naming binary ionic compounds is summarized in Fig. 3-12.

Critical Thinking

We can use the periodic table to tell us something about the stable ions formed by many atoms. For example, the atoms in column 1 always form 1+ ions. The transition metals, however, can form more than one type of stable ion. What if each transition metal ion had only one possible charge? How would the naming of compounds be different?

Table 3-7 | Common Polyatomic Ions

Ion	Name	Ion	Name
Hg_2^{2+}	Mercury(I)	NCS^- or SCN^-	Thiocyanate
NH_4^+	Ammonium	CO_3^{2-}	Carbonate
NO_2^-	Nitrite	HCO_3^-	Hydrogen carbonate (bicarbonate is a widely used common name)
NO_3^-	Nitrate		
SO_3^{2-}	Sulfite		
SO_4^{2-}	Sulfate	ClO^- or OCl^-	Hypochlorite
HSO_4^-	Hydrogen sulfate (bisulfate is a widely used common name)	ClO_2^-	Chlorite
		ClO_3^-	Chlorate
		ClO_4^-	Perchlorate
OH^-	Hydroxide	$C_2H_3O_2^-$	Acetate
CN^-	Cyanide	MnO_4^-	Permanganate
PO_4^{3-}	Phosphate	$Cr_2O_7^{2-}$	Dichromate
HPO_4^{2-}	Hydrogen phosphate	CrO_4^{2-}	Chromate
$H_2PO_4^-$	Dihydrogen phosphate	O_2^{2-}	Peroxide
		$C_2O_4^{2-}$	Oxalate
		$S_2O_3^{2-}$	Thiosulfate

INTERACTIVE EXAMPLE 3-13

Naming Binary Compounds

1. Give the systematic name for each of the following compounds:
   a. $CoBr_2$ **b.** $CaCl_2$ **c.** Al_2O_3
2. Given the following systematic names, write the formula for each compound:
   a. Chromium(III) chloride ◄
   b. Gallium iodide

Solution

1.

	Formula	Name	Comments
a.	$CoBr_2$	Cobalt(II) bromide	Cobalt is a transition metal; the compound name must have a Roman numeral. The two Br^- ions must be balanced by a Co^{2+} ion.
b.	$CaCl_2$	Calcium chloride	Calcium, an alkaline earth metal, forms only the Ca^{2+} ion. A Roman numeral is not necessary.
c.	Al_2O_3	Aluminum oxide	Aluminum forms only the Al^{3+} ion. A Roman numeral is not necessary.

2.

	Name	Formula	Comments
a.	Chromium(III) chloride	$CrCl_3$	Chromium(III) indicates that Cr^{3+} is present, so 3 Cl^- ions are needed for charge balance.
b.	Gallium iodide	GaI_3	Gallium always forms 3+ ions, so 3 I^- ions are required for charge balance.

See Exercises 3-107 and 3-108

Richard Megna/Fundamental Photographs © Cengage Learning

▲ Various chromium compounds dissolved in water. From left to right: $CrCl_2$, $K_2Cr_2O_7$, $Cr(NO_3)_3$, $CrCl_3$, K_2CrO_4.

The common Type I and Type II ions are summarized in Fig. 3-13. Also shown in Fig. 3-13 are the common monatomic ions.

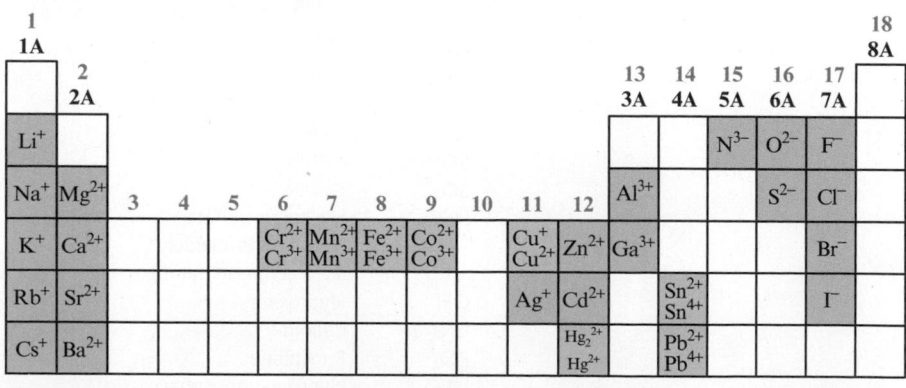

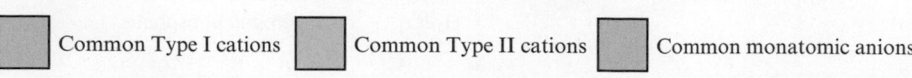

Figure 3-13 | The common cations and anions.

Ionic Compounds with Polyatomic Ions

We have not yet considered ionic compounds that contain polyatomic ions. For example, the compound ammonium nitrate, NH_4NO_3, contains the polyatomic ions NH_4^+ and NO_3^-. Polyatomic ions are assigned special names that *must be memorized* to name the compounds containing them. The most important polyatomic ions and their names are listed in Table 3-7.

Note in Table 3-7 that several series of anions contain an atom of a given element and different numbers of oxygen atoms. These anions are called **oxyanions**. When there are two members in such a series, the name of the one with the smaller number of oxygen atoms ends in *-ite* and the name of the one with the larger number ends in *-ate*—for example, sulfite (SO_3^{2-}) and sulfate (SO_4^{2-}). When more than two oxyanions make up a series, *hypo-* (less than) and *per-* (more than) are used as prefixes to name the members of the series with the fewest and the most oxygen atoms, respectively. The best example involves the oxyanions containing chlorine, as shown in Table 3-7.

| **INTERACTIVE EXAMPLE 3-14** | **Naming Compounds Containing Polyatomic Ions** |

1. Give the systematic name for each of the following compounds:

    a. Na_2SO_4

    b. KH_2PO_4

    c. $Fe(NO_3)_3$

    d. $Mn(OH)_2$

    e. Na_2SO_3

    f. Na_2CO_3

2. Given the following systematic names, write the formula for each compound:

    a. Sodium hydrogen carbonate

    b. Cesium perchlorate

    c. Sodium hypochlorite

    d. Sodium selenate

    e. Potassium bromate

Solution

1.

	Formula	**Name**	**Comments**
a.	Na_2SO_4	Sodium sulfate	
b.	KH_2PO_4	Potassium dihydrogen phosphate	
c.	$Fe(NO_3)_3$	Iron(III) nitrate	Transition metal—name must contain a Roman numeral. The Fe^{3+} ion balances three NO_3^- ions.
d.	$Mn(OH)_2$	Manganese(II) hydroxide	Transition metal—name must contain a Roman numeral. The Mn^{2+} ion balances three OH^- ions.
e.	Na_2SO_3	Sodium sulfite	
f.	Na_2CO_3	Sodium carbonate	

2.

	Name	Formula	Comments
a.	Sodium hydrogen carbonate	$NaHCO_3$	Often called sodium bicarbonate.
b.	Cesium perchlorate	$CsClO_4$	
c.	Sodium hypochlorite	$NaOCl$	
d.	Sodium selenate	Na_2SeO_4	Atoms in the same group, like sulfur and selenium, often form similar ions that are named similarly. Thus SeO_4^{2-} is selenate, like SO_4^{2-} (sulfate).
e.	Potassium bromate	$KBrO_3$	As above, BrO_3^- is bromate, like ClO_3^- (chlorate).

See Exercises 3-109 and 3-110

Binary Covalent Compounds (Type III)

Binary covalent compounds are formed between *two nonmetals*. Although these compounds do not contain ions, they are named very similarly to binary ionic compounds.

Table 3-8 | Prefixes Used to Indicate Number in Chemical Names

Prefix	Number Indicated
mono-	1
di-	2
tri-	3
tetra-	4
penta-	5
hexa-	6
hepta-	7
octa-	8
nona-	9
deca-	10

Naming Binary Covalent Compounds

1. The first element in the formula is named first, using the full element name.
2. The second element is named as if it were an anion.
3. Prefixes are used to denote the numbers of atoms present. These prefixes are given in Table 3-8.
4. The prefix *mono-* is never used for naming the first element. For example, CO is called carbon monoxide, *not* monocarbon monoxide.

To see how these rules apply, we will now consider the names of the several covalent compounds formed by nitrogen and oxygen:

Compound	Systematic Name	Common Name
N_2O	Dinitrogen monoxide	Nitrous oxide
NO	Nitrogen monoxide	Nitric oxide
NO_2	Nitrogen dioxide	
N_2O_3	Dinitrogen trioxide	
N_2O_4	Dinitrogen tetroxide	
N_2O_5	Dinitrogen pentoxide	

Notice from the preceding examples that to avoid awkward pronunciations, we often drop the final *o* or *a* of the prefix when the element begins with a vowel. For example, N_2O_4 is called dinitrogen tetroxide, *not* dinitrogen tetr*a*oxide, and CO is called carbon monoxide, *not* carbon mon*o*oxide.

Some compounds are always referred to by their common names. Three examples are water, ammonia, and hydrogen peroxide. The systematic names for H_2O, NH_3, and H_2O_2 are never used.

INTERACTIVE EXAMPLE 3-15

Naming Type III Binary Compounds

1. Name each of the following compounds:

    a. PCl_5 **b.** PCl_3 **c.** SO_2

2. From the following systematic names, write the formula for each compound:

    a. Sulfur hexafluoride   **b.** Sulfur trioxide   **c.** Carbon dioxide

Solution

1.

Formula	Name
a. PCl_5	Phosphorus pentachloride
b. PCl_3	Phosphorus trichloride
c. SO_2	Sulfur dioxide

2.

Name	Formula
a. Sulfur hexafluoride	SF_6
b. Sulfur trioxide	SO_3
c. Carbon dioxide	CO_2

See Exercises 3-111 and 3-112

The rules for naming binary compounds are summarized in Fig. 3-14. Prefixes to indicate the number of atoms are used only in Type III binary compounds (those containing two nonmetals). An overall strategy for naming compounds is given in Fig. 3-15.

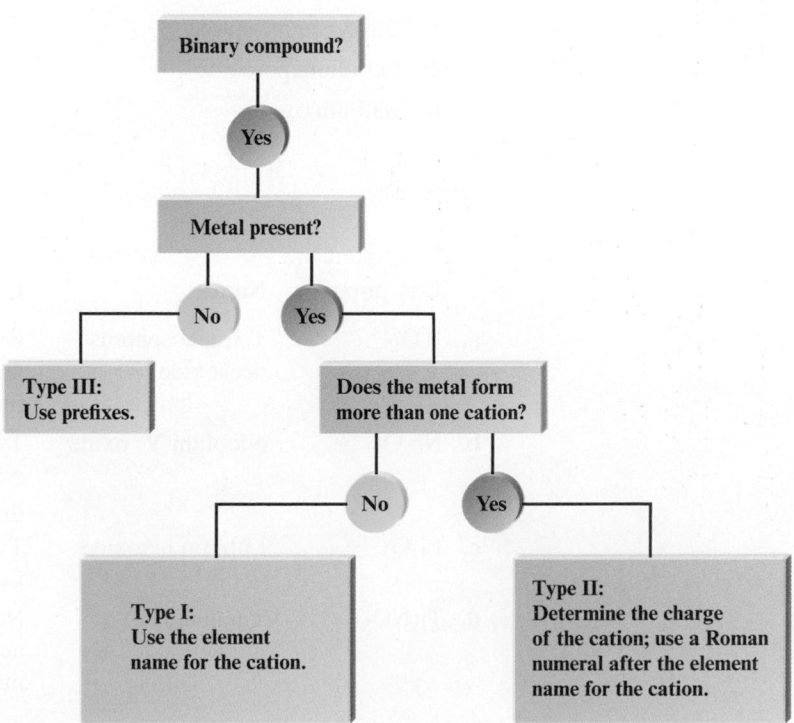

Figure 3-14 | A flowchart for naming binary compounds.

Figure 3-15 | Overall strategy for naming chemical compounds.

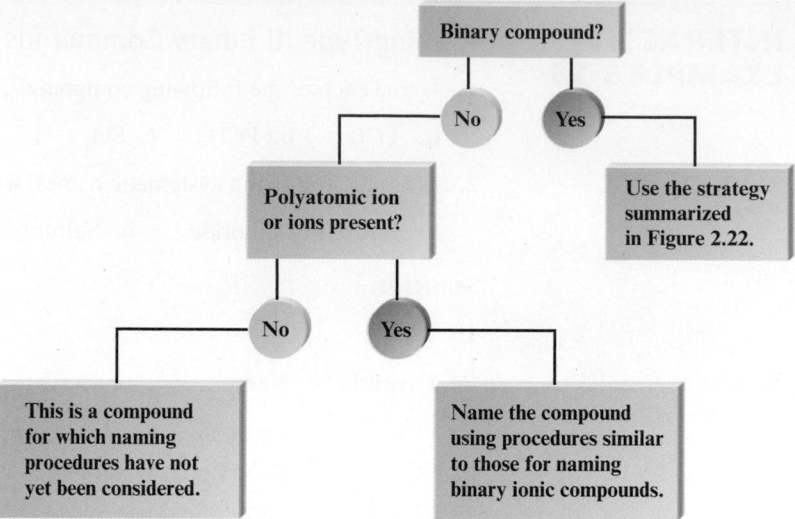

INTERACTIVE EXAMPLE 3-16

Naming Various Types of Compounds

1. Give the systematic name for each of the following compounds:

   a. P_4O_{10}

   b. Nb_2O_5

   c. Li_2O_2

   d. $Ti(NO_3)_4$

2. Given the following systematic names, write the formula for each compound:

   a. Vanadium(V) fluoride

   b. Dioxygen difluoride

   c. Rubidium peroxide

   d. Gallium oxide

Solution

1.

	Compound	Name	Comment
a.	P_4O_{10}	Tetraphosphorus decaoxide	Binary covalent compound (Type III), so prefixes are used. The *a* in *deca-* is sometimes dropped.
b.	Nb_2O_5	Niobium(V) oxide	Type II binary compound containing Nb^{5+} and O^{2-} ions. Niobium is a transition metal and requires a Roman numeral.
c.	Li_2O_2	Lithium peroxide	Type I binary compound containing the Li^+ and O_2^{2-} (peroxide) ions.
d.	$Ti(NO_3)_4$	Titanium(IV) nitrate	Not a binary compound. Contains the Ti^{4+} and NO_3^- ions. Titanium is a transition metal and requires a Roman numeral.

2.

	Name	Chemical Formula	Comment
a.	Vanadium(V) fluoride	VF_5	The compound contains V^{5+} ions and requires five F^- ions for charge balance.
b.	Dioxygen difluoride	O_2F_2	The prefix *di-* indicates two of each atom.
c.	Rubidium peroxide	Rb_2O_2	Because rubidium is in Group 1A, it forms only 1+ ions. Thus two Rb^+ ions are needed to balance the 2− charge on the peroxide ion (O_2^{2-}).
d.	Gallium oxide	Ga_2O_3	Because gallium is in Group 3A, like aluminum, it forms only 3+ ions. Two Ga^{3+} ions are required to balance the charge on three O^{2-} ions.

See Exercises 3-113, 3-117, and 3-118

Acids

ⓘ Acids can be recognized by the hydrogen that appears first in the formula.

When dissolved in water, certain molecules produce a solution containing free H^+ ions (protons). ◂ ⓘ These substances, **acids**, will be discussed in detail later in the text. Here we will simply present the rules for naming acids.

An acid is a molecule in which one or more H^+ ions are attached to an anion. The rules for naming acids depend on whether the anion contains oxygen. If the name of the *anion ends in -ide*, the acid is named with the prefix *hydro-* and the suffix *-ic*. For example, when gaseous HCl is dissolved in water, it forms hydrochloric acid. Similarly, HCN and H_2S dissolved in water are called hydrocyanic and hydrosulfuric acids, respectively.

When the *anion contains oxygen*, the acidic name is formed from the root name of the anion with a suffix of *-ic* or *-ous*, depending on the name of the anion.

Table 3-9 | Names of Acids* That Do Not Contain Oxygen

Acid	Name
HF	Hydrofluoric acid
HCl	Hydrochloric acid
HBr	Hydrobromic acid
HI	Hydroiodic acid
HCN	Hydrocyanic acid
H_2S	Hydrosulfuric acid

*Note that these acids are aqueous solutions containing these substances.

1. If the anion name ends in *-ate*, the suffix *-ic* is added to the root name. For example, H_2SO_4 contains the sulfate anion (SO_4^{2-}) and is called sulfuric acid; H_3PO_4 contains the phosphate anion (PO_4^{3-}) and is called phosphoric acid; and $HC_2H_3O_2$ contains the acetate ion ($C_2H_3O_2^-$) and is called acetic acid.
2. If the anion has an *-ite* ending, the *-ite* is replaced by *-ous*. For example, H_2SO_3, which contains sulfite (SO_3^{2-}), is named sulfurous acid; and HNO_2, which contains nitrite (NO_2^-), is named nitrous acid.

The application of these rules can be seen in the names of the acids of the oxyanions of chlorine:

Acid	Anion	Name
$HClO_4$	Perchlor*ate*	Perchlor*ic* acid
$HClO_3$	Chlor*ate*	Chlor*ic* acid
$HClO_2$	Chlor*ite*	Chlor*ous* acid
HClO	Hypochlor*ite*	Hypochlor*ous* acid

Table 3-10 | Names of Some Oxygen-Containing Acids

Acid	Name
HNO_3	Nitric acid
HNO_2	Nitrous acid
H_2SO_4	Sulfuric acid
H_2SO_3	Sulfurous acid
H_3PO_4	Phosphoric acid
$HC_2H_3O_2$	Acetic acid

The names of the most important acids are given in Tables 3-9 and 3-10. An overall strategy for naming acids is shown in Fig. 3-16.

Figure 3-16 | A flowchart for naming acids. An acid is best considered as one or more H^+ ions attached to an anion.

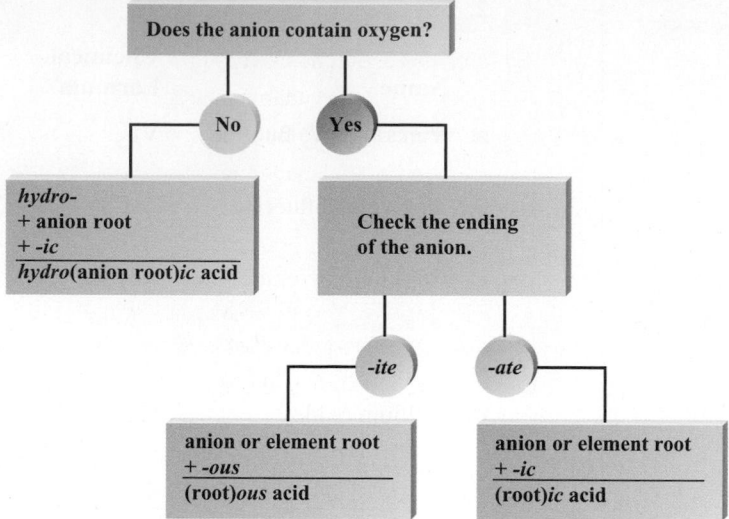

Critical Thinking

In this chapter, you have learned a systematic way to name chemical compounds. What if all compounds had only common names? What problems would this cause?

FOR REVIEW

Key Terms

Section 3-1
chemical bonds
covalent bonds
molecule
chemical formula
structural formula
space-filling model
ball-and-stick models
bond energy
ionic bonding
ionic compound
Coulomb's law
bond length
polar covalent bond

Section 3-2
electronegativity

Section 3-3
isoelectronic ions

Section 3-4
lattice energy

Chemical Bonds

- Hold groups of atoms together
- Occur when a group of atoms can lower its total energy by aggregating
- Types of chemical bonds
    - Ionic: electrons are transferred to form ions
    - Covalent: equal sharing of electrons
    - Polar covalent: unequal electron sharing
- Percent ionic character of a bond X—Y

$$\frac{\text{Measured dipole moment of X} - \text{Y}}{\text{Calculated dipole moment for X}^+ \text{Y}^-} \times 100\%$$

- Electronegativity: the relative ability of an atom to attract shared electrons
    - The polarity of a bond depends on the electronegativity difference of the bonded atoms
- The spatial arrangement of polar bonds in a molecule determines whether the molecule has a dipole moment

Key Terms

Section 3-7
single bond
double bond
triple bond

Section 3-8
localized electron (LE) model
lone pairs
bonding pairs

Section 3-9
Lewis structure
duet rule
octet rule

Section 3-11
resonance
resonance structures
formal charge

Section 3-12
binary compounds
binary ionic compounds
oxyanions
binary covalent compounds
acids

Ionic Bonding

- An ion has a different size than its parent atom
  - An anion is larger than its parent ion
  - A cation is smaller than its parent atom
- Lattice energy: the change in energy when ions are packed together to form an ionic solid

Bond Energy

- The energy necessary to break a covalent bond
- Increases as the number of shared pairs increases
- Can be used to estimate the energy change for a chemical reaction

Lewis Structures

- Show how the valence electron pairs are arranged among the atoms in a molecule or polyatomic ion
- Stable molecules usually contain atoms that have their valence orbitals filled
  - Leads to a duet rule for hydrogen
  - Leads to an octet rule for second-row elements
  - The atoms of elements in the third row and beyond can exceed the octet rule
- Several equivalent Lewis structures can be drawn for some molecules, a concept called resonance
- When several nonequivalent Lewis structures can be drawn for a molecule, formal charge is often used to choose the most appropriate structure(s)

Compounds Are Named Using a System of Rules Depending on the Type of Compound.

- Binary compounds
  - Type I—contain a metal that always forms the same cation
  - Type II—contain a metal that can form more than one cation
  - Type III—contain two nonmetals
- Compounds containing a polyatomic ion

Review Questions

Answers to the Review Questions can be found on the Student Web site (accessible from **www.cengagebrain.com**).

1. Distinguish between the terms *electronegativity* versus *electron affinity*, *covalent bond* versus *ionic bond*, and *pure covalent bond* versus *polar covalent bond*. Characterize the types of bonds in terms of electronegativity difference. Energetically, why do ionic and covalent bonds form?
2. When metals react with nonmetals, an ionic compound generally results. What is the predicted general formula for the compound formed between an alkali metal and sulfur? Between an alkaline earth metal and nitrogen? Between aluminum and a halogen?
3. When an element forms an anion, what happens to the radius? When an element forms a cation, what happens to the radius? Why? Define the term *isoelectronic*. When comparing sizes of ions, which ion has the largest radius and which ion has the smallest radius in an isoelectronic series? Why?
4. Define the term *lattice energy*. Why, energetically, do ionic compounds form? Fig. 3-8 illustrates the energy changes involved in the formation of $MgO(s)$ and $NaF(s)$. Why is the lattice energy of $MgO(s)$ so different from that of $NaF(s)$? The magnesium oxide is composed of Mg^{2+} and O^{2-}

ions. Energetically, why does $Mg^{2+}O^{2-}$ form and not Mg^+O^-? Why doesn't $Mg^{3+}O^{3-}$ form?

5. Explain how bond energies can be used to estimate ΔE for a reaction. Why is this an estimate of ΔE? How do the product bond strengths compare to the reactant bond strengths for a reaction that releases energy? A reaction that gains energy? What is the relationship between the number of bonds between two atoms and bond strength? Bond length?

6. Give a rationale for the octet rule and the duet rule for H in terms of orbitals. Give the steps for drawing a Lewis structure for a molecule or ion.

7. In general, molecules and ions always follow the octet rule unless it is impossible. The three types of exceptions are molecules/ions with too few electrons, molecules/ions with an odd number of electrons, and molecules/ions with too many electrons. Which atoms sometimes have fewer than 8 electrons around them? Give an example. Which atoms sometimes have more than 8 electrons around them? Give some examples. Why are odd-electron species generally very reactive and uncommon? Give an example of an odd-electron molecule.

8. Explain the terms *resonance* and *delocalized electrons*. When a substance exhibits resonance, we say that none of the individual Lewis structures accurately portrays the bonding in the substance. Why do we draw resonance structures?

9. Define formal charge and explain how to calculate it. What is the purpose of the formal charge? Organic compounds are composed mostly of carbon and hydrogen but also may have oxygen, nitrogen, and/or halogens in the formula. Formal charge arguments work very well for organic compounds when drawing the best Lewis structure. How do C, H, N, O, and Cl satisfy the octet rule in organic compounds so as to have a formula charge of zero?

10. The compounds $AlCl_3$, $CrCl_3$, and ICl_3 have similar formulas, yet each follows a different set of rules to name it. Name these compounds, and then compare and contrast the nomenclature rules used in each case.

11. How would you name $HBrO_4$, KIO_3, $NaBrO_2$, and HIO? Refer to Table 3-7 and the acid nomenclature discussion in the text.

Active Learning Questions

These questions are designed to be used by groups of students in class.

1. Explain the electronegativity trends across a row and down a column of the periodic table. Compare these trends with those of ionization energies and atomic radii. How are they related?

2. The ionic compound AB is formed. The charges on the ions may be $+1$, -1; $+2$, -2; $+3$, -3; or even larger. What are the factors that determine the charge for an ion in an ionic compound?

3. Using only the periodic table, predict the most stable ion for Na, Mg, Al, S, Cl, K, Ca, and Ga. Arrange these from largest to smallest radius, and explain why the radius varies as it does. Compare your predictions with Fig. 3-5.

4. The bond energy for a C—H bond is about 413 kJ/mol in CH_4 but 380 kJ/mol in $CHBr_3$. Although these values are relatively close in magnitude, they are different. Explain why they are different. Does the fact that the bond energy is lower in $CHBr_3$ make any sense? Why?

5. Consider the following statement: "Because oxygen wants to have a negative two charge, the second electron affinity is more negative than the first." Indicate everything that is correct in this statement. Indicate everything that is incorrect. Correct the incorrect statements and explain.

6. Which has the greater bond lengths: NO_2^- or NO_3^-? Explain.

7. The following ions are best described with resonance structures. Draw the resonance structures, and using formal charge arguments, predict the best Lewis structure for each ion.
   a. NCO^-
   b. CNO^-

8. The second electron affinity values for both oxygen and sulfur are unfavorable (positive). Explain.

9. What is meant by a chemical bond? Why do atoms form bonds with each other? Why do some elements exist as molecules in nature instead of as free atoms?

10. Why are some bonds ionic and some covalent?

11. How does a bond between Na and Cl differ from a bond between C and O? What about a bond between N and N?

12. Does a Lewis structure tell which electrons come from which atoms? Explain.

13. Evaluate each of the following as an acceptable name for water:
   a. dihydrogen oxide
   b. hydroxide hydride
   c. hydrogen hydroxide
   d. oxygen dihydride

14. Why do we call $Ba(NO_3)_2$ barium nitrate, but we call $Fe(NO_3)_2$ iron(II) nitrate?

33. Without using Fig. 3-4, predict which bond in each of the following groups will be the most polar.
    a. C—F, Si—F, Ge—F
    b. P—Cl or S—Cl
    c. S—F, S—Cl, S—Br
    d. Ti—Cl, Si—Cl, Ge—Cl

34. Without using Fig. 3-4, predict which bond in each of the following groups will be the most polar.
    a. C—H, Si—H, Sn—H
    b. Al—Br, Ga—Br, In—Br, Tl—Br
    c. C—O or Si—O
    d. O—F or O—Cl

35. Repeat Exercises 31 and 33, this time using the values for the electronegativities of the elements given in Fig. 3-4. Are there differences in your answers?

36. Repeat Exercises 32 and 34, this time using the values for the electronegativities of the elements given in Fig. 3-4. Are there differences in your answers?

37. Which of the following incorrectly shows the bond polarity? Show the correct bond polarity for those that are incorrect.
    a. $^{\delta+}$H—F$^{\delta-}$ **d.** $^{\delta+}$Br—Br$^{\delta-}$
    b. $^{\delta+}$Cl—I$^{\delta-}$ **e.** $^{\delta+}$O—P$^{\delta-}$
    c. $^{\delta+}$Si—S$^{\delta-}$

38. Indicate the bond polarity (show the partial positive and partial negative ends) in the following bonds.
    a. C—O   **d.** Br—Te
    b. P—H   **e.** Se—S
    c. H—Cl

39. Predict the type of bond (ionic, covalent, or polar covalent) one would expect to form between the following pairs of elements.
    a. Rb and Cl   **d.** Ba and S
    b. S and S   **e.** N and P
    c. C and F   **f.** B and H

40. List all the possible bonds that can occur between the elements P, Cs, O, and H. Predict the type of bond (ionic, covalent, or polar covalent) one would expect to form for each bond.

41. Hydrogen has an electronegativity value between boron and carbon and identical to phosphorus. With this in mind, rank the following bonds in order of decreasing polarity: P—H, O—H, N—H, F—H, C—H.

42. Rank the following bonds in order of increasing ionic character: N—O, Ca—O, C—F, Br—Br, K—F.

Ions and Ionic Compounds

43. Would you expect each of the following atoms to gain or lose electrons when forming ions? What ion is the most likely in each case?
    a. Ra   **c.** P   **e.** Br
    b. In   **d.** Te   **f.** Rb

44. For each of the following atomic numbers, use the periodic table to write the formula (including the charge) for the simple *ion* that the element is most likely to form in ionic compounds.
    a. 13   **c.** 56   **e.** 87
    b. 34   **d.** 7   **f.** 35

45. Write electron configurations for the most stable ion formed by each of the elements Al, Ba, Se, and I (when in stable ionic compounds).

46. Write electron configurations for the most stable ion formed by each of the elements Te, Cl, Sr, and Li (when in stable ionic compounds).

47. Predict the empirical formulas of the ionic compounds formed from the following pairs of elements. Name each compound.
    a. Li and N   **c.** Rb and Cl
    b. Ga and O   **d.** Ba and S

48. Predict the empirical formulas of the ionic compounds formed from the following pairs of elements. Name each compound.
    a. Al and Cl   **c.** Sr and F
    b. Na and O   **d.** Ca and Se

49. Write electron configurations for
    a. the cations Mg^{2+}, K^+, and Al^{3+}.
    b. the anions N^{3-}, O^{2-}, F^-, and Te^{2-}.

50. Write electron configurations for
    a. the cations Sr^{2+}, Cs^+, In^+, and Pb^{2+}.
    b. the anions P^{3-}, S^{2-}, and Br^-.

51. Which of the following ions have noble gas electron configurations?
    a. Fe^{2+}, Fe^{3+}, Sc^{3+}, Co^{3+}
    b. Tl^+, Te^{2-}, Cr^{3+}
    c. Pu^{4+}, Ce^{4+}, Ti^{4+}
    d. Ba^{2+}, Pt^{2+}, Mn^{2+}

52. What noble gas has the same electron configuration as each of the ions in the following compounds?
    a. cesium sulfide
    b. strontium fluoride
    c. calcium nitride
    d. aluminum bromide

53. Give the formula of a *negative* ion that would have the same number of electrons as each of the following *positive* ions.
    a. Na^+ **c.** Al^{3+}
    b. Ca^{2+} **d.** Rb^+

54. Give an example of an ionic compound where both the anion and the cation are isoelectronic with each of the following noble gases.
    a. Ne   **c.** Kr
    b. Ar   **d.** Xe

55. Give three ions that are isoelectronic with neon. Place these ions in order of increasing size.

56. Consider the ions Sc^{3+}, Cl^-, K^+, Ca^{2+}, and S^{2-}. Match these ions to the following pictures that represent the relative sizes of the ions.

15. Why is calcium dichloride not the correct systematic name for $CaCl_2$?

16. The common name for NH_3 is ammonia. What would be the systematic name for NH_3? Support your answer.

A blue question or exercise number indicates that the answer to that question or exercise appears at the back of this book and a solution appears in the *Student Solutions Manual*.

Questions

17. Compare and contrast the bonding found in the $H_2(g)$ and $HF(g)$ molecules with that found in $NaF(s)$.

18. The following electrostatic potential diagrams represent H_2, HCl, or NaCl. Label each and explain your choices.

a.

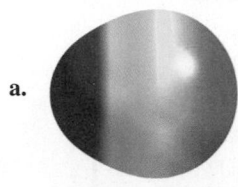

b.

c.

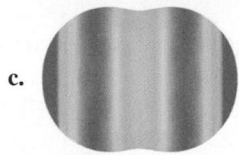

19. Describe the type of bonding that exists in the $Cl_2(g)$ molecule. How does this type of bonding differ from that found in the $HCl(g)$ molecule? How is it similar?

20. Some of the important properties of ionic compounds are as follows:

i. low electrical conductivity as solids and high conductivity in solution or when molten

ii. relatively high melting and boiling points

iii. brittleness

iv. solubility in polar solvents

How does the concept of ionic bonding discussed in this chapter account for these properties?

21. Label the type of bonding for each of the following.

a. 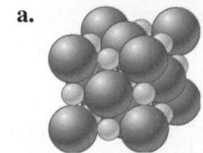   **b.**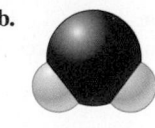

22. Distinguish between the following terms.

a. molecule versus ion

b. covalent bonding versus ionic bonding

c. molecule versus compound

d. anion versus cation

23. What is the electronegativity trend? Where does hydrogen fit into the electronegativity trend for the other elements in the periodic table?

24. When comparing the size of different ions, the general radii trend discussed in Chapter 2 is usually not very useful. What do you concentrate on when comparing sizes of ions to each other or when comparing the size of an ion to its neutral atom?

25. In general, the higher the charge on the ions in an ionic compound, the more favorable the lattice energy. Why do some stable ionic compounds have $+1$ charged ions even though $+4$, $+5$, and $+6$ charged ions would have a more favorable lattice energy?

26. Combustion reactions of fossil fuels provide most of the energy needs of the world. Why do the combustion reactions of fossil fuels produce so much energy?

27. Which of the following statements is/are true? Correct the false statements.

a. It is impossible to satisfy the octet rule for all atoms in XeF_2.

b. Because SF_4 exists, OF_4 should also exist, because oxygen is in the same family as sulfur.

c. The bond in NO^+ should be stronger than the bond in NO^-.

d. As predicted from the two Lewis structures for ozone, one oxygen–oxygen bond is stronger than the other oxygen–oxygen bond.

28. Three resonance structures can be drawn for CO_2. Which resonance structure is best from a formal charge standpoint?

29. By analogy with phosphorus compounds, name the following: Na_3AsO_4, H_3AsO_4, $Mg_3(SbO_4)_2$.

30. Each of the following compounds has three possible names listed for it. For each compound, what is the correct name and why aren't the other names used?

a. N_2O: nitrogen oxide, nitrogen(I) oxide, dinitrogen monoxide

b. Cu_2O: copper oxide, copper(I) oxide, dicopper monoxide

c. Li_2O: lithium oxide, lithium(I) oxide, dilithium monoxide

Exercises

In this section, similar exercises are paired.

Chemical Bonds and Electronegativity

31. Without using Fig. 3-4, predict the order of increasing electronegativity in each of the following groups of elements.

a. C, N, O   **c.** Si, Ge, Sn

b. S, Se, Cl   **d.** Tl, S, Ge

32. Without using Fig. 3-4, predict the order of increasing electronegativity in each of the following groups of elements.

a. Na, K, Rb   **c.** F, Cl, Br

b. B, O, Ga   **d.** S, O, F

57. For each of the following groups, place the atoms and/or ions in order of decreasing size.
   a. Cu, Cu^+, Cu^{2+}
   b. Ni^{2+}, Pd^{2+}, Pt^{2+}
   c. O, O^-, O^{2-}
   d. La^{3+}, Eu^{3+}, Gd^{3+}, Yb^{3+}
   e. Te^{2-}, I^-, Cs^+, Ba^{2+}, La^{3+}

58. For each of the following groups, place the atoms and/or ions in order of decreasing size.
   a. V, V^{2+}, V^{3+}, V^{5+}
   b. Na^+, K^+, Rb^+, Cs^+
   c. Te^{2-}, I^-, Cs^+, Ba^{2+}
   d. P, P^-, P^{2-}, P^{3-}
   e. O^{2-}, S^{2-}, Se^{2-}, Te^{2-}

59. Which compound in each of the following pairs of ionic substances has the most negative lattice energy? Justify your answers.
   a. NaCl, KCl
   b. LiF, LiCl
   c. $Mg(OH)_2$, MgO
   d. $Fe(OH)_2$, $Fe(OH)_3$
   e. NaCl, Na_2O
   f. MgO, BaS

60. Which compound in each of the following pairs of ionic substances has the most negative lattice energy? Justify your answers.
   a. LiF, CsF
   b. NaBr, NaI
   c. $BaCl_2$, BaO
   d. Na_2SO_4, $CaSO_4$
   e. KF, K_2O
   f. Li_2O, Na_2S

61. Use the following data for potassium chloride to estimate ΔE for the reaction:

$$K(s) + \tfrac{1}{2}Cl_2(g) \longrightarrow KCl(s) \qquad \Delta E = ?$$

Lattice energy	−690. kJ/mol
Ionization energy for K	419 kJ/mol
Electron affinity of Cl	−349 kJ/mol
Bond energy of Cl_2	239 kJ/mol
Energy of sublimation for K	90. kJ/mol

62. Use the following data for magnesium fluoride to estimate ΔE for the reaction:

$$Mg(s) + F_2(g) \longrightarrow MgF_2(s) \qquad \Delta E = ?$$

Lattice energy	−2913 kJ/mol
First ionization energy of Mg	735 kJ/mol
Second ionization energy of Mg	1445 kJ/mol
Electron affinity of F	−328 kJ/mol
Bond energy of F_2	154 kJ/mol
Energy of sublimation for Mg	150. kJ/mol

63. Consider the following energy changes:

	ΔE (kJ/mol)
$Mg(g) \rightarrow Mg^+(g) + e^-$	735
$Mg^+(g) \rightarrow Mg^{2+}(g) + e^-$	1445
$O(g) + e^- \rightarrow O^-(g)$	−141
$O^-(g) + e^- \rightarrow O^{2-}(g)$	878

Magnesium oxide exists as $Mg^{2+}O^{2-}$ and not as Mg^+O^-. Explain.

64. Compare the electron affinity of fluorine to the ionization energy of sodium. Does the process of an electron being "pulled" from the sodium atom to the fluorine atom have a negative or a positive ΔE? Why is NaF a stable compound? Does the overall formation of NaF have a negative or a positive ΔE? How can this be?

65. Consider the following:

$$Li(s) + \tfrac{1}{2}I_2(s) \longrightarrow LiI(s) \qquad \Delta E = -272 \text{ kJ/mol}$$

LiI(s) has a lattice energy of −753 kJ/mol. The ionization energy of Li(g) is 520. kJ/mol, the bond energy of $I_2(g)$ is 151 kJ/mol, and the electron affinity of I(g) is −295 kJ/mol. Use these data to determine the energy of sublimation of Li(s).

66. Use the following data (in kJ/mol) to estimate ΔE for the reaction $S^-(g) + e^- \rightarrow S^{2-}(g)$. Include an estimate of uncertainty. ΔE_{sub} is the energy of sublimation.

	Lattice Energy	I.E. of M	ΔE_{sub} of M
Na_2S	−2203	495	109
K_2S	−2052	419	90
Rb_2S	−1949	409	82
Cs_2S	−1850	382	78

$2 Na(s) + S(s) \longrightarrow Na_2S(s)$	$\Delta E = -365$ kJ/mol
$2 K(s) + S(s) \longrightarrow K_2S(s)$	$\Delta E = -381$ kJ/mol
$2 Rb(s) + S(s) \longrightarrow Rb_2S(s)$	$\Delta E = -361$ kJ/mol
$2 Cs(s) + S(s) \longrightarrow Cs_2S(s)$	$\Delta E = -360$ kJ/mol
$S(s) \longrightarrow S(g)$	$\Delta E = 277$ kJ/mol
$S(g) + e^- \longrightarrow S^-(g)$	$\Delta E = -200$ kJ/mol

Assume that all values are known to ±1 kJ/mol.

67. Rationalize the following lattice energy values:

Compound	Lattice Energy (kJ/mol)
CaSe	−2862
Na_2Se	−2130
CaTe	−2721
Na_2Te	−2095

68. The lattice energies of $FeCl_3$, $FeCl_2$, and Fe_2O_3 are (in no particular order) −2631, −5359, and −14,774 kJ/mol. Match the appropriate formula to each lattice energy. Explain.

Bond Energies

69. Use bond energy values (Table 3-3) to estimate ΔE for each of the following reactions in the gas phase.

a. $H_2 + Cl_2 \rightarrow 2HCl$ **b.** $N \equiv N + 3H_2 \rightarrow 2NH_3$

70. Use bond energy values (Table 3-3) to estimate ΔE for each of the following reactions.

a.

$$H-C \equiv N(g) + 2H_2(g) \longrightarrow H-\overset{\overset{\displaystyle H}{|}}{\underset{\underset{\displaystyle H}{|}}{C}}-\overset{\overset{\displaystyle H}{|}}{\underset{\underset{\displaystyle H}{|}}{N}}(g)$$

b.

$$\overset{\displaystyle H}{\underset{\displaystyle H}{>}}N-N\overset{\displaystyle H}{\underset{\displaystyle H}{<}}(g) + 2F_2(g) \longrightarrow N \equiv N(g) + 4HF(g)$$

71. Use bond energies (Table 3-3) to predict ΔE for the isomerization of methyl isocyanide to acetonitrile:

$$CH_3N \equiv C(g) \longrightarrow CH_3C \equiv N(g)$$

72. Acetic acid is responsible for the sour taste of vinegar. It can be manufactured using the following reaction:

$$CH_3OH(g) + C \equiv O(g) \longrightarrow CH_3\overset{\overset{\displaystyle O}{\|}}{C}-OH(l)$$

Use tabulated values of bond energies (Table 3-3) to estimate ΔE for this reaction.

73. Use bond energies to predict ΔE for the following reaction:

$$H_2S(g) + 3F_2(g) \longrightarrow SF_4(g) + 2HF(g)$$

74. The major industrial source of hydrogen gas is by the following reaction:

$$CH_4(g) + H_2O(g) \longrightarrow CO(g) + 3H_2(g)$$

Use bond energies to predict ΔE for this reaction.

75. Use bond energies to estimate ΔE for the combustion of one mole of acetylene:

$$C_2H_2(g) + \tfrac{5}{2}O_2(g) \longrightarrow 2CO_2(g) + H_2O(g)$$

76. Use data from Table 3-3 to determine the energy of reaction for the combustion of methane (CH_4), as shown below:

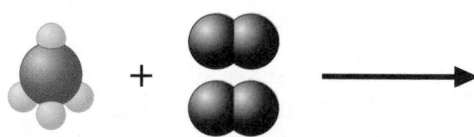

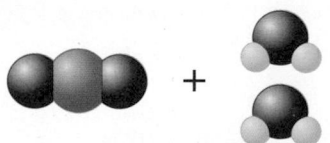

77. Consider the following reaction:

$$\overset{\displaystyle H}{\underset{\displaystyle H}{>}}C=C\overset{\displaystyle H}{\underset{\displaystyle H}{<}}(g) + F_2(g) \longrightarrow$$

$$H-\overset{\overset{\displaystyle F}{|}}{\underset{\underset{\displaystyle H}{|}}{C}}-\overset{\overset{\displaystyle F}{|}}{\underset{\underset{\displaystyle H}{|}}{C}}-H(g) \quad \Delta E = -549 \text{ kJ}$$

Estimate the carbon–fluorine bond energy, given that the C—C bond energy is 347 kJ/mol, the C=C bond energy is 614 kJ/mol, and the F—F bond energy is 154 kJ/mol.

78. Consider the following reaction:

$$A_2 + B_2 \longrightarrow 2AB \quad \Delta E = -285 \text{ kJ}$$

The bond energy for A_2 is one-half the amount of the AB bond energy. The bond energy of $B_2 = 432$ kJ/mol. What is the bond energy of A_2?

Lewis Structures and Resonance

79. Write Lewis structures that obey the octet rule (duet rule for H) for each of the following molecules. Carbon is the central atom in CH_4, nitrogen is the central atom in NH_3, and oxygen is the central atom in H_2O.

a. F_2 **e.** NH_3
b. O_2 **f.** H_2O
c. CO   **g.** HF
d. CH_4

80. Write Lewis structures that obey the octet rule (duet rule for H) for each of the following molecules.

a. H_2CO
b. CO_2
c. HCN

Carbon is the central atom in all of these molecules.

81. Write Lewis structures that obey the octet rule for each of the following molecules.

a. CCl_4 **c.** $SeCl_2$
b. NCl_3 **d.** ICl

In each case, the atom listed first is the central atom.

82. Write Lewis structures that obey the octet rule for each of the following molecules and ions. (In each case the first atom listed is the central atom.)

a. $POCl_3$, SO_4^{2-}, XeO_4, PO_4^{3-}, ClO_4^-
b. NF_3, SO_3^{2-}, PO_3^{3-}, ClO_3^-
c. ClO_2^-, SCl_2, PCl_2^-
d. Considering your answers to parts a, b, and c, what conclusions can you draw concerning the structures of species containing the same number of atoms and the same number of valence electrons?

83. One type of exception to the octet rule are compounds with central atoms having fewer than eight electrons around them. BeH_2 and BH_3 are examples of this type of exception. Draw the Lewis structures for BeH_2 and BH_3.

84. Lewis structures can be used to understand why some molecules react in certain ways. Write the Lewis structures for the reactants and products in the reactions described below.

   a. Nitrogen dioxide dimerizes to produce dinitrogen tetroxide.

   b. Boron trihydride accepts a pair of electrons from ammonia, forming BH_3NH_3.

   Give a possible explanation for why these two reactions occur.

85. The most common exceptions to the octet rule are compounds or ions with central atoms having more than eight electrons around them. PF_5, SF_4, ClF_3, and Br_3^- are examples of this type of exception. Draw the Lewis structure for these compounds or ions. Which elements, when they have to, can have more than eight electrons around them? How is this rationalized?

86. SF_6, ClF_5, and XeF_4 are three compounds whose central atoms do not follow the octet rule. Draw Lewis structures for these compounds.

87. Write Lewis structures for the following. Show all resonance structures where applicable.

   a. NO_2^-, NO_3^-, N_2O_4 (N_2O_4 exists as O_2N-NO_2.)

   b. OCN^-, SCN^-, N_3^- (Carbon is the central atom in OCN^- and SCN^-.)

88. Some of the important pollutants in the atmosphere are ozone (O_3), sulfur dioxide, and sulfur trioxide. Write Lewis structures for these three molecules. Show all resonance structures where applicable.

89. Benzene (C_6H_6) consists of a six-membered ring of carbon atoms with one hydrogen bonded to each carbon. Write Lewis structures for benzene, including resonance structures.

90. Borazine ($B_3N_3H_6$) has often been called "inorganic" benzene. Write Lewis structures for borazine. Borazine contains a six-membered ring of alternating boron and nitrogen atoms with one hydrogen bonded to each boron and nitrogen.

91. An important observation supporting the concept of resonance in the localized electron model was that there are only three different structures of dichlorobenzene ($C_6H_4Cl_2$). How does this fact support the concept of resonance (see Exercise 89)?

92. Consider the following bond lengths:

   C—O   143 pm   C=O   123 pm   C≡O   109 pm

   In the CO_3^{2-} ion, all three C—O bonds have identical bond lengths of 136 pm. Why?

93. A toxic cloud covered Bhopal, India, in December 1984 when water leaked into a tank of methyl isocyanate, and the product escaped into the atmosphere. Methyl isocyanate is used in the production of many pesticides. Draw the Lewis structures for methyl isocyanate, CH_3NCO, including resonance forms. The skeletal structure is

$$
\begin{array}{c}
H \\
| \\
H-C-N-C-O \\
| \\
H
\end{array}
$$

94. Peroxyacetyl nitrate, or PAN, is present in photochemical smog. Draw the Lewis structures (including resonance forms) for PAN. The skeletal structure is

$$
\begin{array}{ccccc}
H & O & & & O \\
| & | & & & \diagup \\
H-C-C-O-O-N & & \\
| & & & & \diagdown \\
H & & & & O
\end{array}
$$

95. Order the following species with respect to carbon–oxygen bond length (longest to shortest).

   CO, CO_2, CO_3^{2-}, CH_3OH

   What is the order from the weakest to the strongest carbon–oxygen bond? (CH_3OH exists as H_3C-OH.)

96. Place the species below in order of the shortest to the longest nitrogen–oxygen bond.

   H_2NOH, N_2O, NO^+, NO_2^-, NO_3^-

   (H_2NOH exists as H_2N-OH.)

Formal Charge

97. Use the formal charge arguments to rationalize why BF_3 would not follow the octet rule.

98. Use formal charge arguments to explain why CO has a less polar bond than expected on the basis of electronegativity.

99. Write Lewis structures that obey the octet rule for the following species. Assign the formal charge for each central atom.

   a. $POCl_3$ **e.** SO_2Cl_2

   b. SO_4^{2-} **f.** XeO_4

   c. ClO_4^- **g.** ClO_3^-

   d. PO_4^{3-} **h.** NO_4^{3-}

100. Write Lewis structures for the species in Exercise 99 that involve minimum formal charges.

101. A common trait of simple organic compounds is to have Lewis structures where all atoms have a formal charge of zero. Consider the following incomplete Lewis structure for an organic compound called *methyl cyanoacrylate*, the main ingredient in Super Glue.

$$
\begin{array}{ccccccc}
H & C-N & & & H \\
| & | & & & | \\
H-C-C-C-O-C-H \\
| & & & | \\
H & O & & H
\end{array}
$$

   Draw a complete Lewis structure for methyl cyanoacrylate in which all atoms have a formal charge of zero.

102. Benzoic acid is a food preservative. The space-filling model for benzoic acid is shown below.

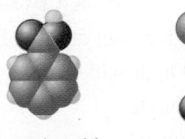

Benzoic acid
($C_6H_5CO_2H$)

Draw the Lewis structure for benzoic acid, including all resonance structures in which all atoms have a formal charge of zero.

103. Oxidation of the cyanide ion produces the stable cyanate ion, OCN^-. The fulminate ion, CNO^-, on the other hand, is very unstable. Fulminate salts explode when struck; $Hg(CNO)_2$ is used in blasting caps. Write the Lewis structures and assign formal charges for the cyanate and fulminate ions. Why is the fulminate ion so unstable? (C is the central atom in OCN^- and N is the central atom in CNO^-.)

104. Nitrous oxide (N_2O) has three possible Lewis structures:

$$\ddot{N}{=}N{=}\ddot{O}\colon \longleftrightarrow \colon N{\equiv}N{-}\ddot{O}\colon \longleftrightarrow \colon\ddot{N}{-}N{\equiv}O\colon$$

Given the following bond lengths,

N—N	167 pm	N=O	115 pm
N=N	120 pm	N—O	147 pm
N≡N	110 pm		

rationalize the observations that the N—N bond length in N_2O is 112 pm and that the N—O bond length is 119 pm. Assign formal charges to the resonance structures for N_2O. Can you eliminate any of the resonance structures on the basis of formal charges? Is this consistent with observation?

Nomenclature

105. Name the compounds in parts a–d and write the formulas for the compounds in parts e–h.
 a. NaBr
 b. Rb_2O
 c. CaS
 d. AlI_3
 e. strontium fluoride
 f. aluminum selenide
 g. potassium nitride
 h. magnesium phosphide

106. Name the compounds in parts a–d and write the formulas for the compounds in parts e–h.
 a. Hg_2O
 b. $FeBr_3$
 c. CoS
 d. $TiCl_4$
 e. tin(II) nitride
 f. cobalt(III) iodide
 g. mercury(II) oxide
 h. chromium(VI) sulfide

107. Name each of the following compounds:
 a. CsF
 b. Li_3N
 c. Ag_2S
 d. MnO_2
 e. TiO_2
 f. Sr_3P_2

108. Write the formula for each of the following compounds:
 a. zinc chloride
 b. tin(IV) fluoride
 c. calcium nitride
 d. aluminum sulfide
 e. mercury(I) selenide
 f. silver iodide

109. Name each of the following compounds:
 a. $BaSO_3$
 b. $NaNO_2$
 c. $KMnO_4$
 d. $K_2Cr_2O_7$

110. Write the formula for each of the following compounds:
 a. chromium(III) hydroxide
 b. magnesium cyanide
 c. lead(IV) carbonate
 d. ammonium acetate

111. Name each of the following compounds:

 a.
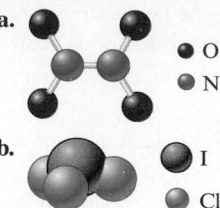

 b.

 O
 N
 I
 Cl

 c. SO_2
 d. P_2S_5

112. Write the formula for each of the following compounds:
 a. diboron trioxide
 b. arsenic pentafluoride
 c. dinitrogen monoxide
 d. sulfur hexachloride

113. Name each of the following compounds:
 a. CuI
 b. CuI_2
 c. CoI_2
 d. Na_2CO_3
 e. $NaHCO_3$
 f. S_4N_4
 g. SF_4
 h. NaOCl
 i. $BaCrO_4$
 j. NH_4NO_3

114. Name each of the following compounds. Assume the acids are dissolved in water.
 a. $HC_2H_3O_2$
 b. NH_4NO_2
 c. Co_2S_3
 d. ICl
 e. $Pb_3(PO_4)_2$
 f. $KClO_3$
 g. H_2SO_4
 h. Sr_3N_2
 i. $Al_2(SO_3)_3$
 j. SnO_2
 k. Na_2CrO_4
 l. HClO

115. Elements in the same family often form oxyanions of the same general formula. The anions are named in a similar fashion. Give the names of the oxyanions of selenium and tellurium: SeO_4^{2-}, SeO_3^{2-}, TeO_4^{2-}, TeO_3^{2-}.

116. Knowing the names of similar chlorine oxyanions and acids, deduce the names of the following: IO^-, IO_2^-, IO_3^-, IO_4^-, HIO, HIO_2, HIO_3, HIO_4.

117. Write the formula for each of the following compounds:
 a. sulfur difluoride
 b. sulfur hexafluoride
 c. sodium dihydrogen phosphate
 d. lithium nitride
 e. chromium(III) carbonate
 f. tin(II) fluoride
 g. ammonium acetate
 h. ammonium hydrogen sulfate
 i. cobalt(III) nitrate
 j. mercury(I) chloride
 k. potassium chlorate
 l. sodium hydride

118. Write the formula for each of the following compounds:
 a. chromium(VI) oxide
 b. disulfur dichloride

c. nickel(II) fluoride

d. potassium hydrogen phosphate

e. aluminum nitride

f. ammonia

g. manganese(IV) sulfide

h. sodium dichromate

i. ammonium sulfite

j. carbon tetraiodide

119. Write the formula for each of the following compounds:

a. sodium oxide

b. sodium peroxide

c. potassium cyanide

d. copper(II) nitrate

e. selenium tetrabromide

f. iodous acid

g. lead(IV) sulfide

h. copper(I) chloride

i. gallium arsenide

j. cadmium selenide

k. zinc sulfide

l. nitrous acid

m. diphosphorus pentoxide

120. Write the formula for each of the following compounds:

a. ammonium hydrogen phosphate

b. mercury(I) sulfide

c. silicon dioxide

d. sodium sulfite

e. aluminum hydrogen sulfate

f. nitrogen trichloride

g. hydrobromic acid

h. bromous acid

i. perbromic acid

j. potassium hydrogen sulfide

k. calcium iodide

l. cesium perchlorate

121. Name the acids illustrated below.

 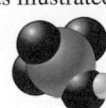 

a.   b.   c.

   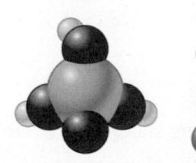 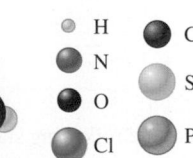

H	C
N	S
O	P
Cl	

d.   e.

122. Each of the following compounds is incorrectly named. What is wrong with each name, and what is the correct name for each compound?

a. FeCl$_3$, iron chloride

b. NO$_2$, nitrogen(IV) oxide

c. CaO, calcium(II) monoxide

d. Al$_2$S$_3$, dialuminum trisulfide

e. Mg(C$_2$H$_3$O$_2$)$_2$, manganese diacetate

f. FePO$_4$, iron(II) phosphide

g. P$_2$S$_5$, phosphorus sulfide

h. Na$_2$O$_2$, sodium oxide

i. HNO$_3$, nitrate acid

j. H$_2$S, sulfuric acid

Additional Exercises

123. Arrange the following in order of increasing radius and increasing ionization energy.

a. N$^+$, N, N$^-$

b. Se, Se$^-$, Cl, Cl$^+$

c. Br$^-$, Rb$^+$, Sr^{2+}

124. For each of the following, write an equation that corresponds to the energy given.

a. lattice energy of NaCl

b. lattice energy of NH$_4$Br

c. lattice energy of MgS

d. O=O double bond energy beginning with O$_2$(g) as a reactant

125. Use bond energies (Table 3-3), values of electron affinities (Table 2-7), and the ionization energy of hydrogen (1312 kJ/mol) to estimate ΔE for each of the following reactions.

a. HF(g) → H$^+$(g) + F$^-$(g)

b. HCl(g) → H$^+$(g) + Cl$^-$(g)

c. HI(g) → H$^+$(g) + I$^-$(g)

d. H$_2$O(g) → H$^+$(g) + OH$^-$(g)

(Electron affinity of OH(g) = −180. kJ/mol.)

126. Write Lewis structures for CO$_3^{2-}$, HCO$_3^-$, and H$_2$CO$_3$. When acid is added to an aqueous solution containing carbonate or bicarbonate ions, carbon dioxide gas is formed. We generally say that carbonic acid (H$_2$CO$_3$) is unstable. Use bond energies to estimate ΔE for the reaction (in the gas phase)

$$H_2CO_3 \longrightarrow CO_2 + H_2O$$

Specify a possible cause for the instability of carbonic acid.

127. Which member of the following pairs would you expect to be more energetically stable? Justify each choice.

a. NaBr or NaBr$_2$

b. ClO$_4$ or ClO$_4^-$

c. SO$_4$ or XeO$_4$

d. OF$_4$ or SeF$_4$

128. What do each of the following sets of compounds/ions have in common?

a. SO$_3$, NO$_3^-$, CO$_3^{2-}$

b. O$_3$, SO$_2$, NO$_2^-$

129. Although both Br$_3^-$ and I$_3^-$ ions are known, the F$_3^-$ ion has not been observed. Explain.

130. Look up the energies for the bonds in CO and N$_2$. Although the bond in CO is stronger, CO is considerably more reactive than N$_2$. Give a possible explanation.

131. The formulas and common names for several substances are given below. Give the systematic names for these substances.

a. sugar of lead   Pb(C$_2$H$_3$O$_2$)$_2$

b. blue vitrol   CuSO$_4$

c. quicklime   CaO

d. Epsom salts   MgSO$_4$

e. milk of magnesia $Mg(OH)_2$
f. gypsum $CaSO_4$
g. laughing gas N_2O

132. Identify each of the following elements:

a. a member of the same family as oxygen whose most stable ion contains 54 electrons

b. a member of the alkali metal family whose most stable ion contains 36 electrons

c. a noble gas with 18 protons in the nucleus

d. a halogen with 85 protons and 85 electrons

133. A certain element has only two naturally occurring isotopes: one with 18 neutrons and the other with 20 neutrons. The element forms $1-$ charged ions when in ionic compounds. Predict the identity of the element. What number of electrons does the $1-$ charged ion have?

134. The designations 1A through 8A used for certain families of the periodic table are helpful for predicting the charges on ions in binary ionic compounds. In these compounds, the metals generally take on a positive charge equal to the family number, while the nonmetals take on a negative charge equal to the family number minus 8. Thus the compound between sodium and chlorine contains Na^+ ions and Cl^- ions and has the formula NaCl. Predict the formula and the name of the binary compound formed from the following pairs of elements.

a. Ca and N   e. Ba and I
b. K and O   f. Al and Se
c. Rb and F   g. Cs and P
d. Mg and S   h. In and Br

135. When molten sulfur reacts with chlorine gas, a vile-smelling orange liquid forms that has a formula of S_2Cl_2. The Lewis structure of this compound has a formal charge of zero on all elements in the compound. Draw the Lewis structure for the vile-smelling orange liquid.

136. The study of carbon-containing compounds and their properties is called *organic chemistry*. Besides carbon atoms, organic compounds also can contain hydrogen, oxygen, and nitrogen atoms (as well as other types of atoms). A common trait of simple organic compounds is to have Lewis structures where all atoms have a formal charge of zero. Consider the following incomplete Lewis structure for an organic compound called *histidine* (an amino acid), which is one of the building blocks of proteins found in our bodies:

Draw a complete Lewis structure for histidine in which all atoms have a formal charge of zero.

ChemWork Problems

These multiconcept problems (and additional ones) are found interactively online with the same type of assistance a student would get from an instructor.

137. What are the formulas of the compounds that correspond to the names given in the following table?

Compound Name	Formula
Carbon tetrabromide	
Cobalt(II) phosphate	
Magnesium chloride	
Nickel(II) acetate	
Calcium nitrate	

138. What are the names of the compounds that correspond to the formulas given in the following table?

Formula	Compound Name
$Co(NO_2)_2$	
AsF_5	
$LiCN$	
K_2SO_3	
Li_3N	
$PbCrO_4$	

139. Complete the following table to predict whether the given atom will gain or lose electrons in forming the ion most likely to form when in ionic compounds.

Atom	Gain (G) or Lose (L) Electrons	Ion Formed
K		
Cs		
Br		
S		
Se		

140. Which of the following statements is(are) *correct*?

a. The symbols for the elements magnesium, aluminum, and xenon are Mn, Al, and Xe, respectively.

b. The elements P, As, and Bi are in the same family on the periodic table.

c. All of the following elements are expected to gain electrons to form ions in ionic compounds: Ga, Se, and Br.

d. The elements Co, Ni, and Hg are all transition elements.

e. The correct name for TiO_2 is titanium dioxide.

141. Classify the bonding in each of the following molecules as ionic, polar covalent, or nonpolar covalent.

   a. H_2
   b. K_3P
   c. NaI
   d. SO_2
   e. HF
   f. CCl_4
   g. CF_4
   h. K_2S

142. List the bonds P—Cl, P—F, O—F, and Si—F from least polar to most polar.

143. Arrange the atoms and/or ions in the following groups in order of decreasing size.

   a. O, O^-, O^{2-}
   b. $Fe^{2+}, Ni^{2+}, Zn^{2+}$
   c. Ca^{2+}, K^+, Cl^-

144. Use the following data to estimate ΔE for the reaction:

$$Ba(s) + Br_2(g) \longrightarrow BaBr_2(s) \qquad \Delta E = ?$$

Lattice energy	−1985 kJ/mol
First ionization energy of Ba	503 kJ/mol
Second ionization energy of Ba	965 kJ/mol
Electron affinity of Br	−325 kJ/mol
Bond energy of Br_2	193 kJ/mol
Enthalpy of sublimation of Ba	178 kJ/mol

145. Use bond energy values to estimate ΔE for the following gas phase reaction:

$$C_2H_4 + H_2O_2 \longrightarrow CH_2OHCH_2OH$$

146. Which of the following compounds or ions exhibit resonance?

   a. O_3
   b. CNO^-
   c. AsI_3
   d. CO_3^{2-}
   e. AsF_3

Challenge Problems

147. Use Coulomb's law,

$$V = \frac{Q_1Q_2}{4\pi\epsilon_0 r} = 2.31 \times 10^{-19}\, J \cdot nm\left(\frac{Q_1Q_2}{r}\right)$$

to calculate the energy of interaction, V, for the following two arrangements of charges, each having a magnitude equal to the electron charge.

   a.

   b.

148. An alternative definition of electronegativity is

$$\text{Electronegativity} = \text{constant (I.E.} - \text{E.A.)}$$

where I.E. is the ionization energy and E.A. is the electron affinity using the sign conventions of this book. Use data in Chapter 2 to calculate the (I.E. − E.A.) term for F, Cl, Br, and I. Do these values show the same trend as the electronegativity values given in this chapter? The first ionization energies of the halogens are 1678, 1255, 1138, and 1007 kJ/mol, respectively. (*Hint:* Choose a constant so that the electronegativity of fluorine equals 4.0. Using this constant, calculate relative electronegativities for the other halogens and compare to values given in the text.)

149. Given the following information:

Energy of sublimation of Li(s)	= 166 kJ/mol
Bond energy of HCl	= 427 kJ/mol
Ionization energy of Li(g)	= 520. kJ/mol
Electron affinity of Cl(g)	= −349 kJ/mol
Lattice energy of LiCl(s)	= −829 kJ/mol
Bond energy of H_2	= 432 kJ/mol

Calculate the net change in energy for the following reaction:

$$2Li(s) + 2HCl(g) \longrightarrow 2LiCl(s) + H_2(g)$$

150. Think of forming an ionic compound as three steps (this is a simplification, as with all models): (1) removing an electron from the metal; (2) adding an electron to the nonmetal; and (3) allowing the metal cation and nonmetal anion to come together.

   a. What is the sign of the energy change for each of these three processes?
   b. In general, what is the sign of the sum of the first two processes? Use examples to support your answer.
   c. What must be the sign of the sum of the three processes?
   d. Given your answer to part c, why do ionic bonds occur?
   e. Given your above explanations, why is NaCl stable but not Na_2Cl? $NaCl_2$? What about MgO compared to MgO_2? Mg_2O?

151. Use data in this chapter (and Chapter 2) to discuss why MgO is an ionic compound but CO is not an ionic compound.

152. Three processes that have been used for the industrial manufacture of acrylonitrile (CH_2CHCN), an important chemical used in the manufacture of plastics, synthetic rubber, and fibers, are shown below. Use bond energy values (Table 3-3) to estimate ΔE for each of the reactions.

   a.

   b.

The nitrogen–oxygen bond energy in nitric oxide (NO) is 630. kJ/mol.

   c.

153. The compound hexaazaisowurtzitane is one of the highest-energy explosives known (*C & E News*, Jan. 17, 1994, p. 26). The compound, also known as CL-20, was first synthesized in 1987. The method of synthesis and detailed performance data are still classified because of CL-20's potential military application in rocket boosters and in warheads of "smart" weapons. The structure of CL-20 is

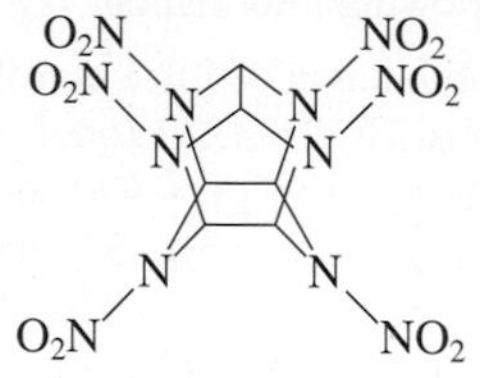

CL-20

In such shorthand structures, each point where lines meet represents a carbon atom. In addition, the hydrogens attached to the carbon atoms are omitted; each of the six carbon atoms has one hydrogen atom attached. Finally, assume that the two O atoms in the NO_2 groups are attached to N with one single bond and one double bond.

Three possible reactions for the explosive decomposition of CL-20 are

i. $C_6H_6N_{12}O_{12}(s) \rightarrow 6CO(g) + 6N_2(g) + 3H_2O(g) + \frac{3}{2}O_2(g)$

ii. $C_6H_6N_{12}O_{12}(s) \rightarrow 3CO(g) + 3CO_2(g) + 6N_2(g) + 3H_2O(g)$

iii. $C_6H_6N_{12}O_{12}(s) \rightarrow 6CO_2(g) + 6N_2(g) + 3H_2(g)$

a. Use bond energies to estimate ΔE for these three reactions.

b. Which of the above reactions releases the largest amount of energy per kilogram of CL-20?

154. Many times extra stability is characteristic of a molecule or ion in which resonance is possible. How could this be used to explain the acidities of the following compounds? (The acidic hydrogen is marked by an asterisk.) Part c shows resonance in the C_6H_5 ring.

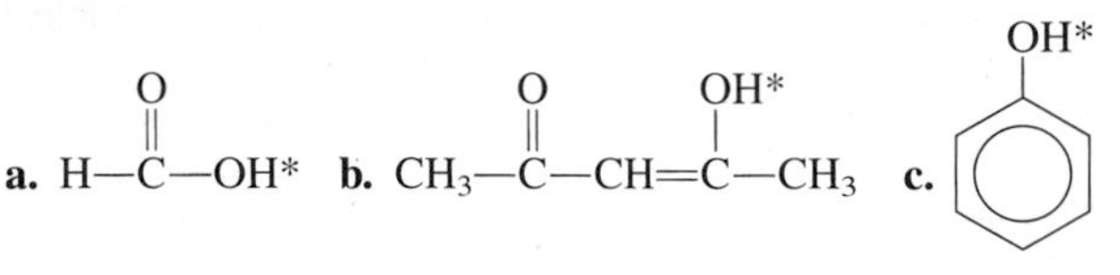

a. H—C(=O)—OH* **b.** CH_3—C(=O)—CH=C—CH_3 with OH* **c.** C_6H_5—OH*

155. Draw a Lewis structure for the *N,N*-dimethylformamide molecule. The skeletal structure is

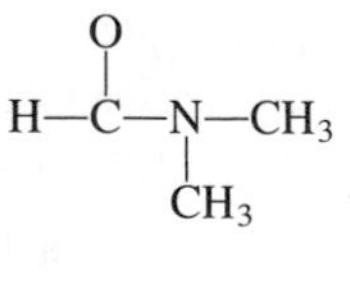

H—C(=O)—N—CH_3 with CH_3

Various types of evidence lead to the conclusion that there is some double bond character to the C—N bond. Draw one or more resonance structures that support this observation.

156. Cholesterol ($C_{27}H_{46}O$) has the following structure:

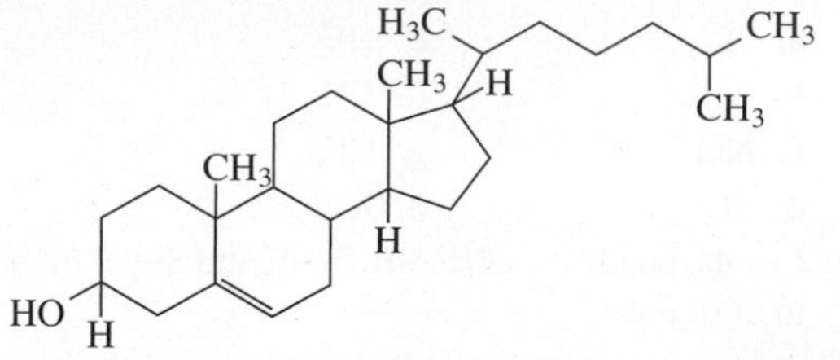

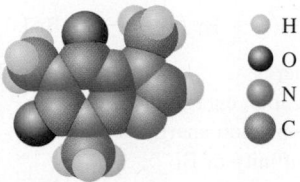

In such shorthand structures, each point where lines meet represents a carbon atom, and most H atoms are not shown. Draw the complete structure showing all carbon and hydrogen atoms. (There will be four bonds to each carbon atom.)

157. Consider the following computer-generated model of caffeine.

Draw a Lewis structure for caffeine in which all atoms have a formal charge of zero.

Integrative Problems

These problems require the integration of multiple concepts to find the solutions.

158. For each of the following ions, indicate the total number of protons and electrons in the ion. For the positive ions in the list, predict the formula of the simplest compound formed between each positive ion and the oxide ion. Name the compounds. For the negative ions in the list, predict the formula of the simplest compound formed between each negative ion and the aluminum ion. Name the compounds.

a. Fe^{2+} **e.** S^{2-}

b. Fe^{3+} **f.** P^{3-}

c. Ba^{2+} **g.** Br^-

d. Cs^+ **h.** N^{3-}

159. Identify the following elements based on their electron configurations and rank them in order of increasing electronegativity: $[Ar]4s^13d^5$; $[Ne]3s^23p^3$; $[Ar]4s^23d^{10}4p^3$; $[Ne]3s^23p^5$.

160. A polyatomic ion is composed of C, N, and an unknown element X. The skeletal Lewis structure of this polyatomic ion is $[X—C—N]^-$. The ion X^{2-} has an electron configuration of $[Ar]4s^23d^{10}4p^6$. What is element X? Knowing the identity of X, complete the Lewis structure of the polyatomic ion, including all important resonance structures.

C H A P T E R   F O U R

Molecular Structure and Orbitals

Chili peppers taste hot because of capsaicin, a complex molecule containing atoms with many different hybridizations. (Image Source)

In Chapter 2 we considered the structure of the atom. We needed this knowledge to be able to discuss chemical bonds and better understand how and why atoms come together to form molecules. In Chapter 3 we used a relatively simple model, the localized electron model, to draw Lewis structures for molecules.

As chemists we are interested in the answers to questions like the following: "Why is water a liquid at room conditions?," and "Why don't water and oil mix?" To answer questions such as these we need to understand the three-dimensional structure of the molecules and how they interact with one another. Lewis structures do not tell us directly about molecular structure. In this chapter, we will focus on molecular structure and more complex bonding models.

4-1 | Molecular Structure: The VSEPR Model

The structures of molecules play a very important role in determining their chemical properties. As we will see later, this is particularly important for biological molecules; a slight change in the structure of a large biomolecule can completely destroy its usefulness to a cell or may even change the cell from a normal one to a cancerous one.

Many accurate methods now exist for determining **molecular structure**, the three-dimensional arrangement of the atoms in a molecule. These methods must be used if precise information about structure is required. However, it is often useful to be able to predict the approximate molecular structure of a molecule. In this section we consider a simple model that allows us to do this. This model, called the **valence shell electron-pair repulsion (VSEPR) model**, is useful in predicting the geometries of molecules formed from nonmetals. The main postulate of this model is that *the structure around a given atom is determined principally by minimizing electron-pair repulsions*. The idea here is that the bonding and nonbonding pairs around a given atom will be positioned as far apart as possible. To see how this model works, we will first consider the molecule $BeCl_2$, which has the Lewis structure

$$: \overset{..}{Cl}—Be—\overset{..}{Cl} :$$

BeCl₂ has only four electrons around Be and is expected to be very reactive with electron-pair donors.

Note that there are two pairs of electrons around the beryllium atom. ◄ ⓘ What arrangement of these electron pairs allows them to be as far apart as possible to minimize the repulsions? Clearly, the best arrangement places the pairs on opposite sides of the beryllium atom at 180 degrees from each other:

This is the maximum possible separation for two electron pairs. Once we have determined the optimal arrangement of the electron pairs around the central atom, we can specify the molecular structure of $BeCl_2$, that is, the positions of the atoms. Since each electron pair on beryllium is shared with a chlorine atom, the molecule has a **linear structure** with a 180-degree bond angle:

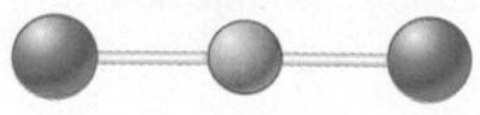

Next, let's consider BF_3, which has the Lewis structure

$$: \ddot{F} :$$
$$|$$
$$: \ddot{F} — B — \ddot{F} :$$

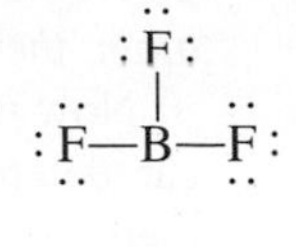

Here the boron atom is surrounded by three pairs of electrons. What arrangement will minimize the repulsions? The electron pairs are farthest apart at angles of 120 degrees:

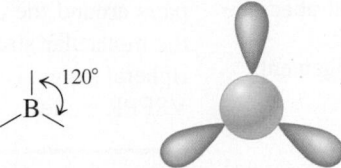

Since each of the electron pairs is shared with a fluorine atom, the molecular structure will be

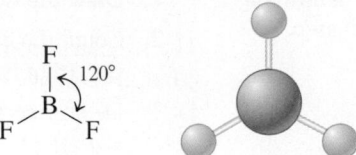

This is a planar (flat) and triangular molecule, which is commonly described as a **trigonal planar structure**.

Next, let's consider the methane molecule, which has the Lewis structure

$$H$$
$$|$$
$$H—C—H$$
$$|$$
$$H$$

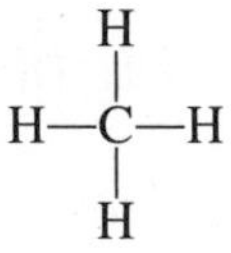

There are four pairs of electrons around the central carbon atom. What arrangement of these electron pairs best minimizes the repulsions? First, let's try a square planar arrangement:

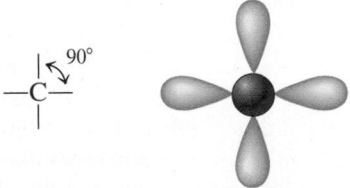

The carbon atom and the electron pairs are centered in the plane of the paper, and the angles between the pairs are all 90 degrees.

Is there another arrangement with angles greater than 90 degrees that would put the electron pairs even farther away from each other? The answer is yes. The **tetrahedral structure** has angles of 109.5 degrees:

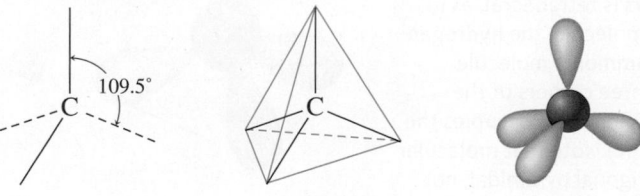

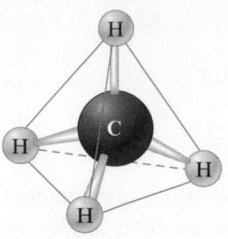

Figure 4-1 | The molecular structure of methane. The tetrahedral arrangement of electron pairs produces a tetrahedral arrangement of hydrogen atoms.

ⓘ Notice that we need to determine the Lewis structure as a first step.

▲ When four uniform balloons are tied together, they naturally form a tetrahedral shape.

Figure 4-2 | (a) The tetrahedral arrangement of electron pairs around the nitrogen atom in the ammonia molecule. (b) Three of the electron pairs around nitrogen are shared with hydrogen atoms, as shown, and one is a lone pair. Although the arrangement of *electron pairs* is tetrahedral, as in the methane molecule, the hydrogen atoms in the ammonia molecule occupy only three corners of the tetrahedron. A lone pair occupies the fourth corner. (c) Note that molecular geometry is trigonal pyramidal, not tetrahedral.

It can be shown that this is the maximum possible separation of four pairs around a given atom. This means that *whenever four pairs of electrons are present around an atom, they should always be arranged tetrahedrally.*

Now that we have the electron-pair arrangement that gives the least repulsion, we can determine the positions of the atoms and thus the molecular structure of CH_4. In methane, each of the four electron pairs is shared between the carbon atom and a hydrogen atom. Thus the hydrogen atoms are placed as in Fig. 4-1, and the molecule has a tetrahedral structure with the carbon atom at the center.

Recall that the main idea of the VSEPR model is to find the arrangement of electron pairs around the central atom that minimizes the repulsions. Then we can determine the molecular structure from knowing how the electron pairs are shared with the peripheral atoms. Use the following steps to predict the structure of a molecule using the VSEPR model.

PROBLEM-SOLVING STRATEGY

Steps to Apply the VSEPR Model

1. Draw the Lewis structure for the molecule. ◀ *ⓘ*
2. Count the electron pairs and arrange them in the way that minimizes repulsion (that is, put the pairs as far apart as possible).
3. Determine the positions of the atoms from the *way the electron pairs are shared*.
4. Determine the name of the molecular structure from the *positions of the atoms*.

We will predict the structure of ammonia (NH_3) using this stepwise approach.

1. Draw the Lewis structure:

$$H\!-\!\overset{\displaystyle ..}{N}\!-\!H$$
$$|$$
$$H$$

2. Count the pairs of electrons and arrange them to minimize repulsions. The NH_3 molecule has four pairs of electrons: three bonding pairs and one nonbonding pair. From the discussion of the methane molecule, we know that the best arrangement of four electron pairs is a tetrahedral array, as shown in Fig. 4-2(a). ◀
3. Determine the positions of the atoms. The three H atoms share electron pairs, as shown in Fig. 4-2(b).
4. Name the molecular structure. It is very important to recognize that the *name* of the molecular structure is always based on the *positions of the atoms*. The placement of the electron pairs determines the structure, but the name is based on the

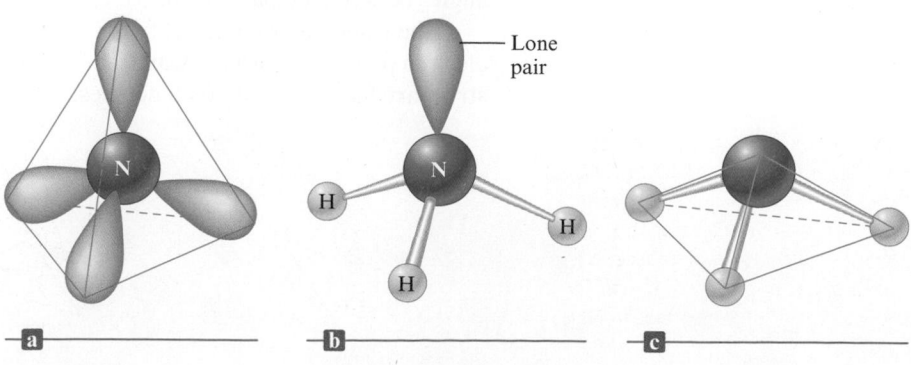

Lone pair

a   *b*   *c*

© Cengage Learning

positions of the atoms. Thus it is incorrect to say that the NH_3 molecule is tetra-hedral. It has a tetrahedral arrangement of electron pairs but not a tetrahedral ar-rangement of atoms. The molecular structure of ammonia is a **trigonal pyramid** (one side is different from the other three) rather than a tetrahedron, as shown in Fig. 4-2(c).

EXAMPLE 4-1   Prediction of Molecular Structure I

Describe the molecular structure of the water molecule.

Solution

The Lewis structure for water is

$$H - \ddot{O} - H$$

There are four pairs of electrons: two bonding pairs and two nonbonding pairs. To minimize repulsions, these are best arranged in a tetrahedral array, as shown in Fig. 4-3(a). Although H_2O has a tetrahedral arrangement of electron pairs, it is not a tetrahedral molecule. The atoms in the H_2O molecule form a V shape, as shown in Fig. 4-3(b) and (c).

See Exercises 4-21 and 4-22

From Example 4-1 we see that the H_2O molecule is V-shaped, or bent, because of the presence of the lone pairs. If no lone pairs were present, the molecule would be linear, the polar bonds would cancel, and the molecule would have no dipole moment. This would make water very different from the polar substance so familiar to us.

From the previous discussion we would predict that the H—X—H bond angle (where X is the central atom) in CH_4, NH_3, and H_2O should be the tetrahedral angle of 109.5 degrees. Experimental studies, however, show that the actual bond angles are those given in Fig. 4-4. What significance do these results have for the VSEPR model?

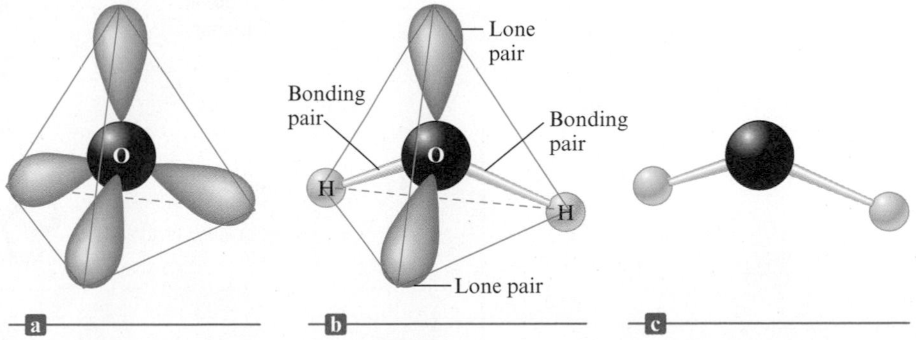

Figure 4-3 | (a) The tetrahedral arrangement of the four electron pairs around oxygen in the water molecule. (b) Two of the electron pairs are shared between oxygen and the hydrogen atoms and two are lone pairs. (c) The V-shaped molecular structure of the water molecule.

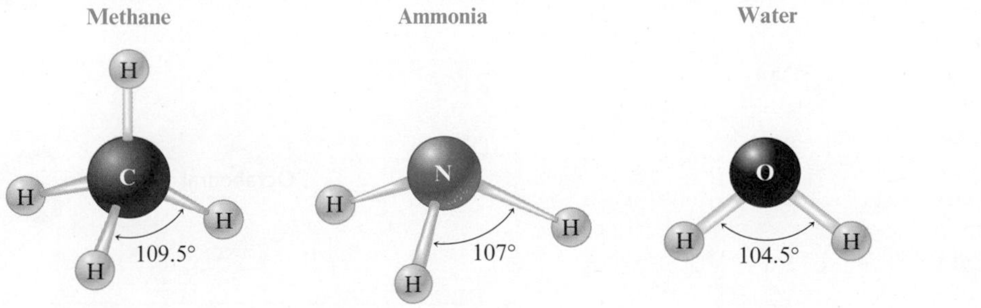

Figure 4-4 | The bond angles in the CH_4, NH_3, and H_2O molecules. Note that the bond angle between bonding pairs decreases as the number of lone pairs increases. Note that all of the angles in CH_4 are 109.5 degrees and all of the angles in NH_3 are 107 degrees.

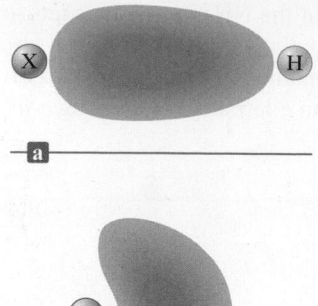

Figure 4-5 | (a) In a bonding pair of electrons, the electrons are shared by two nuclei. (b) In a lone pair, both electrons must be close to a single nucleus and tend to take up more of the space around that atom.

One possible point of view is that we should be pleased to have the observed angles so close to the tetrahedral angle. The opposite view is that the deviations are significant enough to require modification of the simple model so that it can more accurately handle similar cases. We will take the latter view.

Let us examine the following data:

	CH_4	NH_3	H_2O
Number of Lone Pairs	0	1	2
Bond Angle	109.5°	107°	104.5°

One interpretation of the trend observed here is that lone pairs require more space than bonding pairs; in other words, as the number of lone pairs increases, the bonding pairs are increasingly squeezed together.

This interpretation seems to make physical sense if we think in the following terms. A bonding pair is shared between two nuclei, and the electrons can be close to either nucleus. They are relatively confined between the two nuclei. A lone pair is localized on only one nucleus, and both electrons will be close only to that nucleus, as shown schematically in Fig. 4-5. These pictures help us understand why a lone pair may require more space near an atom than a bonding pair.

As a result of these observations, we make the following addition to the original postulate of the VSEPR model: Lone pairs require more room than bonding pairs and tend to compress the angles between the bonding pairs.

So far we have considered cases with two, three, and four electron pairs around the central atom. These are summarized in Table 4-1.

Table 4-1 | Arrangements of Electron Pairs Around an Atom Yielding Minimum Repulsion

Number of Electron Pairs	Arrangement of Electron Pairs	Example
2	Linear	
3	Trigonal planar	
4	Tetrahedral	
5	Trigonal bipyramidal	
6	Octahedral	

Ken O'Donoghue © Cengage Learning

Table 4-2 summarizes the structures possible for molecules in which there are **four** electron pairs around the central atom, with various numbers of atoms bonded to it. Note that molecules with four pairs of electrons around the central atom can be tetrahedral (AB_4), trigonal pyramidal (AB_3), and V-shaped (AB_2).

For five pairs of electrons, there are several possible choices. The one that produces minimum repulsion is a **trigonal bipyramid**. Note from Table 4-1 that this arrangement has two different angles, 90 degrees and 120 degrees. As the name suggests, the structure formed by this arrangement of pairs consists of two trigonal-based pyramids that share a common base. Table 4-3 summarizes the structures possible for molecules in which there are five electron pairs around the central atom, with various numbers of atoms bonded to it. Note that molecules with five pairs of electrons around the central atom can be trigonal bipyramidal (AB_5), see-saw (AB_4), T-shaped (AB_3), and linear (AB_2).

Six pairs of electrons can best be arranged around a given atom with 90-degree angles to form an **octahedral structure**, as shown in Table 4-1.

To use the VSEPR model to determine the geometric structures of molecules, you should memorize the relationships between the number of electron pairs and their best arrangements.

Table 4-2 | Structures of Molecules That Have Four Electron Pairs Around the Central Atom

Table 4-3 | Structures of Molecules with Five Electron Pairs Around the Central Atom

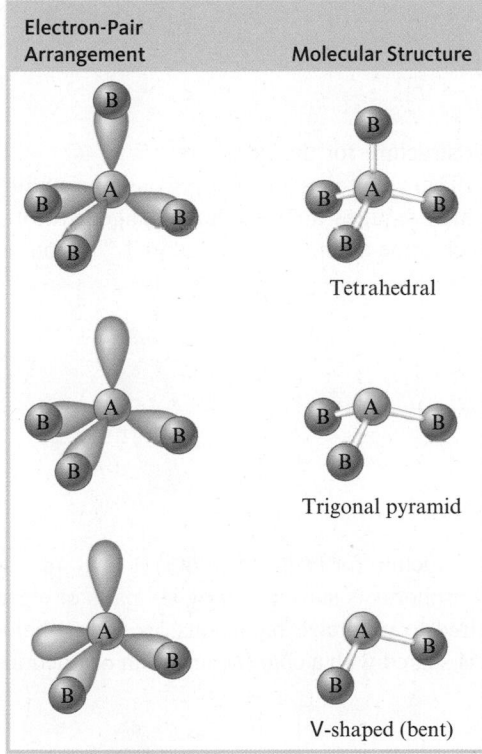

Critical Thinking

You and a friend are studying for a chemistry exam. What if your friend tells you that all molecules with polar bonds are polar molecules? How would you explain to your friend that this is not correct? Provide two examples to support your answer.

INTERACTIVE EXAMPLE 4-2

Prediction of Molecular Structure II

When phosphorus reacts with excess chlorine gas, the compound phosphorus pentachloride (PCl_5) is formed. In the gaseous and liquid states, this substance consists of PCl_5 molecules, but in the solid state it consists of a 1:1 mixture of PCl_4^+ and PCl_6^- ions. Predict the geometric structures of PCl_5, PCl_4^+, and PCl_6^-.

Solution

The Lewis structure for PCl_5 is shown. Five pairs of electrons around the phosphorus atom require a trigonal bipyramidal arrangement (see Table 4-1). When the chlorine atoms are included, a trigonal bipyramidal molecule results:

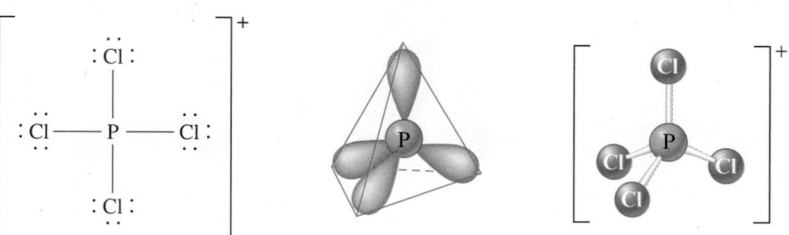

The Lewis structure for the PCl_4^+ ion [$5 + 4(7) - 1 = 32$ valence electrons] is shown below. There are four pairs of electrons surrounding the phosphorus atom in the PCl_4^+ ion, which requires a tetrahedral arrangement of the pairs. Since each pair is shared with a chlorine atom, a tetrahedral PCl_4^+ cation results.

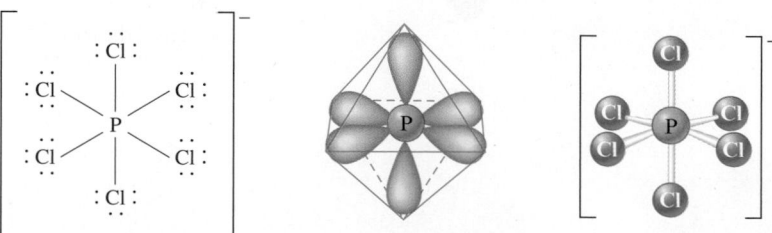

The Lewis structure for PCl_6^- [$5 + 6(7) + 1 = 48$ valence electrons] is shown below. Since phosphorus is surrounded by six pairs of electrons, an octahedral arrangement is required to minimize repulsions, as shown below in the center. Since each electron pair is shared with a chlorine atom, an octahedral PCl_6^- anion is predicted.

See Exercises 4-25 through 4-28

INTERACTIVE EXAMPLE 4-3

Prediction of Molecular Structure III

Because the noble gases have filled *s* and *p* valence orbitals, they were not expected to be chemically reactive. In fact, for many years these elements were called *inert gases* because of this supposed inability to form any compounds. However, in the early 1960s several compounds of krypton, xenon, and radon were synthesized. For example, a team at the Argonne National Laboratory produced the stable colorless compound xenon tetrafluoride (XeF_4). Predict its structure and whether it has a dipole moment.

Solution

The Lewis structure for XeF_4 is

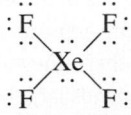

The xenon atom in this molecule is surrounded by six pairs of electrons, which means an octahedral arrangement.

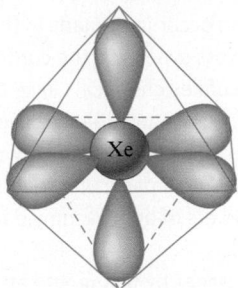

The structure predicted for this molecule will depend on how the lone pairs and bonding pairs are arranged. Consider the two possibilities shown in Fig. 4-6. The bonding pairs are indicated by the presence of the fluorine atoms. Since the structure predicted differs in the two cases, we must decide which of these arrangements is preferable. The key is to look at the lone pairs. In the structure in part (a), the lone pair–lone pair angle is 90 degrees; in the structure in part (b), the lone pairs are

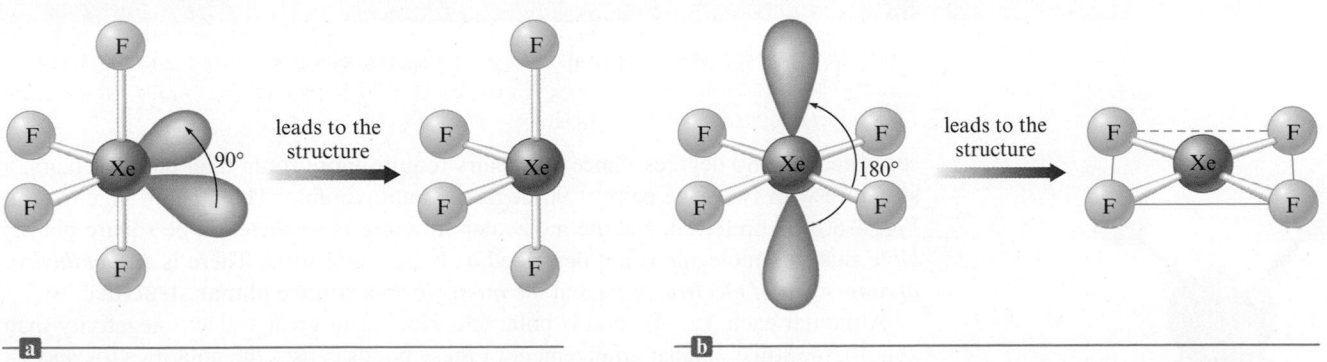

Figure 4-6 | Possible electron-pair arrangements for XeF_4. Since arrangement (a) has lone pairs at 90 degrees from each other, it is less favorable than arrangement (b), where the lone pairs are at 180 degrees.

Chemical Structure and Communication: Semiochemicals

In this chapter we have stressed the importance of being able to predict the three-dimensional structure of a molecule. Molecular structure is important because of its effect on chemical reactivity. This is especially true in biological systems, where reactions must be efficient and highly specific. Among the hundreds of types of molecules in the fluids of a typical biological system, the appropriate reactants must find and react only with each other—they must be very discriminating. This specificity depends largely on structure. The molecules are constructed so that only the appropriate partners can approach each other in a way that allows reaction.

Another area where molecular structure is central is in the use of molecules as a means of communication. Examples of a chemical communication occur in humans in the conduction of nerve impulses across synapses, the control of the manufacture and storage of key chemicals in cells, and the senses of smell and taste. Plants and animals also use chemical communication. For example, ants lay down a chemical trail so that other ants can find a particular food supply. Ants also warn their fellow workers of approaching danger by emitting certain chemicals.

Molecules convey messages by fitting into appropriate receptor sites in a very specific way, which is determined by their structure. When a molecule occupies a receptor site, chemical processes are stimulated that produce the appropriate response. Sometimes receptors can be fooled, as in the use of artificial sweeteners—molecules fit the sites on the taste buds that stimulate a "sweet" response in the brain, but they are not metabolized in the same way as natural sugars. Similar deception is useful in insect control. If an area is

sprayed with synthetic female sex attractant molecules, the males of that species become so confused that mating does not occur.

A *semiochemical* is a molecule that delivers a message between members of the same or different species of plant or animal. There are three groups of these chemical messengers: allomones, kairomones, and pheromones. Each is of great ecological importance.

An *allomone* is defined as a chemical that somehow gives adaptive advantage to the producer. For example, leaves of the black walnut tree contain a herbicide, juglone, that appears after the leaves fall to the ground. Juglone is not toxic to grass or certain grains, but it is effective against plants such as apple trees that would compete for the available water and food supplies.

Antibiotics are also allomones, since the microorganisms produce them to inhibit other species from growing near them.

Many plants produce bad-tasting chemicals to protect themselves from plant-eating insects and animals. The familiar compound nicotine deters animals from eating the tobacco plant. The millipede sends an unmistakable "back off" message by squirting a predator with benzaldehyde and hydrogen cyanide.

Defense is not the only use of allomones, however. Flowers use scent as a way to attract pollinating insects. Honeybees, for instance, are guided to alfalfa and flowers by a series of sweet-scented compounds.

Kairomones are chemical messengers that bring advantageous news to the receiver, and the floral scents are kairomones from the honeybees' viewpoint. Many predators are

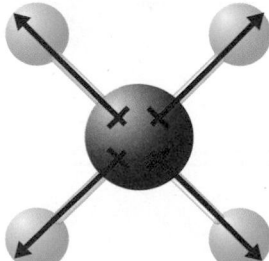

separated by 180 degrees. Since lone pairs require more room than bonding pairs, a structure with two lone pairs at 90 degrees is unfavorable. Thus the arrangement in Fig. 4-6(b) is preferred, and the molecular structure is predicted to be square planar. Note that this molecule is *not* described as being octahedral. There is an *octahedral arrangement of electron pairs*, but the *atoms* form a **square planar structure**. ◄

Although each Xe—F bond is polar (fluorine has a greater electronegativity than xenon), the square planar arrangement of these bonds causes the polarities to cancel.

Thus XeF_4 has no dipole moment, as shown in the margin.

See Exercises 4-31 and 4-32

Close-up of trail ants using pheromones as signals

Alex Wild

Alarm pheromones are highly volatile compounds (ones easily changed to a gas) released to warn of danger. Honeybees produce isoamyl acetate ($C_7H_{14}O_2$) in their sting glands. Because of its high volatility, this compound does not linger after the state of alert is over. Social behavior in insects is characterized by the use of *trail pheromones*, which are used to indicate a food source. Social insects such as bees, ants, wasps, and termites use these substances. Since trail pheromones are less volatile compounds, the indicators persist for some time.

guided by kairomones emitted by their food. For example, apple skins exude a chemical that attracts the codling moth larva. In some cases kairomones help the underdog. Certain marine mollusks can pick up the "scent" of their predators, the sea stars, and make their escape.

Pheromones are chemicals that affect receptors of the same species as the donor. That is, they are specific within a species. *Releaser pheromones* cause an immediate reaction in the receptor, and *primer pheromones* cause long-term effects. Examples of releaser pheromones are sex attractants of insects, generated in some species by the males and in others by the females. Sex pheromones also have been found in plants and mammals.

Primer pheromones, which cause long-term behavioral changes, are harder to isolate and identify. One example, however, is the "queen substance" produced by queen honeybees. All the eggs in a colony are laid by one queen bee. If she is removed from the hive or dies, the worker bees are activated by the absence of the queen substance and begin to feed royal jelly to bee larvae so as to raise a new queen. The queen substance also prevents the development of the workers' ovaries so that only the queen herself can produce eggs.

Many studies of insect pheromones are now under way in the hope that they will provide a method of controlling insects that is more efficient and safer than the current chemical pesticides.

We can further illustrate the use of the VSEPR model for molecules or ions with lone pairs by considering the triiodide ion (I_3^-).

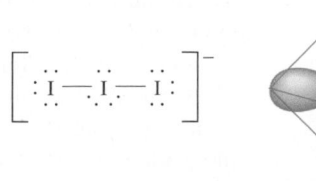 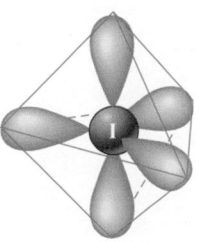

Figure 4-7 | Three possible arrangements of the electron pairs in the I_3^- ion. Arrangement (c) is preferred because there are no 90-degree lone pair–lone pair interactions.

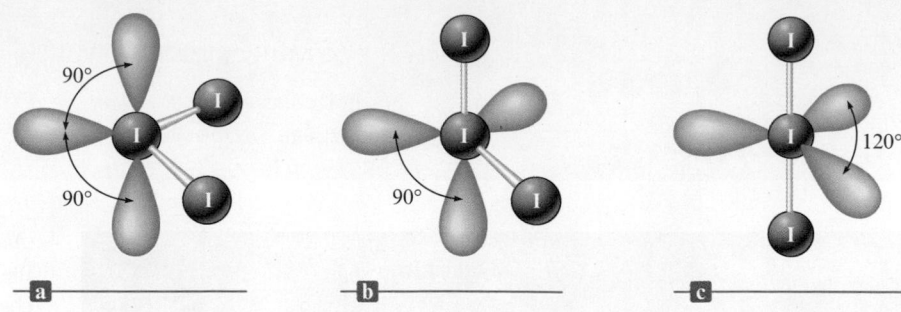

The central iodine atom has five pairs around it, which requires a trigonal bipyramidal arrangement. Several possible arrangements of lone pairs are shown in Fig. 4-7. Note that structures (a) and (b) have lone pairs at 90 degrees, whereas in (c) all lone pairs are at 120 degrees. Thus structure (c) is preferred. The resulting molecular structure for I_3^- is linear:

$$[I-I-I]^-$$

The VSEPR Model and Multiple Bonds

So far in our treatment of the VSEPR model we have not considered any molecules with multiple bonds. To see how these molecules are handled by this model, let's consider the NO_3^- ion, which requires three resonance structures to describe its electronic structure:

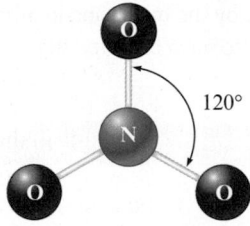

The NO_3^- ion is known to be planar with 120-degree bond angles:

This planar structure is the one expected for three pairs of electrons around a central atom, which means that *a double bond should be counted as one effective pair* in using the VSEPR model. This makes sense because the two pairs of electrons involved in the double bond are *not* independent pairs. Both the electron pairs must be in the space between the nuclei of the two atoms to form the double bond. In other words, the double bond acts as one center of electron density to repel the other pairs of electrons. The same holds true for triple bonds. This leads us to another general rule: For the VSEPR model, multiple bonds count as one effective electron pair.

The molecular structure of nitrate also shows us one more important point: When a molecule exhibits resonance, any one of the resonance structures can be used to predict the molecular structure using the VSEPR model. These rules are illustrated in Example 4-4.

INTERACTIVE EXAMPLE 4-4

Structures of Molecules with Multiple Bonds

Predict the molecular structure of the sulfur dioxide molecule. Is this molecule expected to have a dipole moment?

Solution

First, we must determine the Lewis structure for the SO_2 molecule, which has 18 valence electrons. The expected resonance structures are

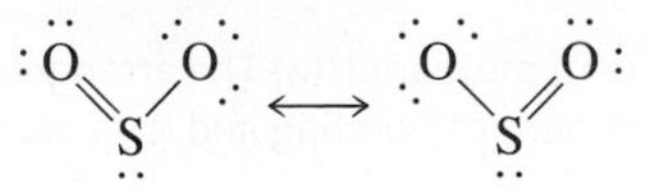

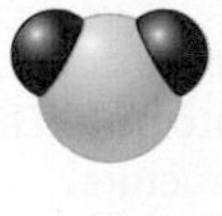

To determine the molecular structure, we must count the electron pairs around the sulfur atom. In each resonance structure the sulfur has one lone pair, one pair in a single bond, and one double bond. Counting the double bond as one pair yields three effective pairs around the sulfur. According to Table 4-1, a trigonal planar arrangement is required, which yields a V-shaped molecule:

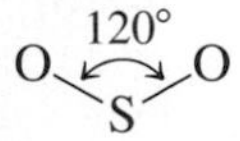

Thus the structure of the SO_2 molecule is expected to be V-shaped, with a 120-degree bond angle. The molecule has a dipole moment directed as shown:

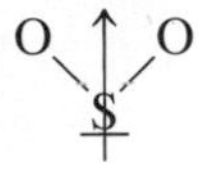

Since the molecule is V-shaped, the polar bonds do not cancel.

See Exercises 4-33 and 4-34

It should be noted at this point that lone pairs that are oriented at least 120 degrees from other pairs do not produce significant distortions of bond angles. For example, the angle in the SO_2 molecule is actually quite close to 120 degrees. We will follow the general principle that *a 120-degree angle provides lone pairs with enough space so that distortions do not occur. Angles less than 120 degrees are distorted when lone pairs are present.*

Molecules Containing No Single Central Atom

So far we have considered molecules consisting of one central atom surrounded by other atoms. The VSEPR model can be readily extended to more complicated molecules, such as methanol (CH_3OH). This molecule is represented by the following Lewis structure:

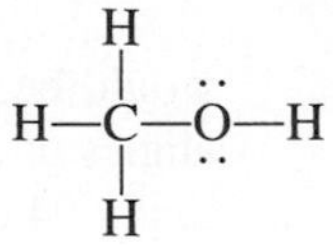

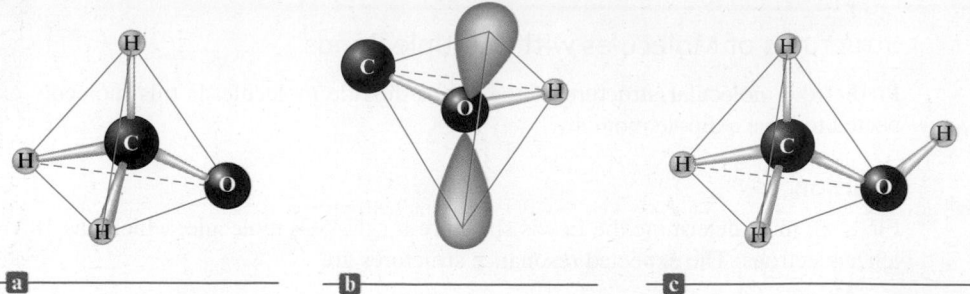

Figure 4-8 | The molecular structure of methanol. (a) The arrangement of electron pairs and atoms around the carbon atom. (b) The arrangement of bonding and lone pairs around the oxygen atom. (c) The molecular structure.

The molecular structure can be predicted from the arrangement of pairs around the carbon and oxygen atoms. Note that there are four pairs of electrons around the carbon, which requires a tetrahedral arrangement, as shown in Fig. 4-8(a). The oxygen also has four pairs, which requires a tetrahedral arrangement. However, in this case the tetrahedron will be slightly distorted by the space requirements of the lone pairs [Fig. 4-8(b)]. The overall geometric arrangement for the molecule is shown in Fig. 4-8(c).

> **LET'S REVIEW** | *Summary of the VSEPR Model*
>
> The rules for using the VSEPR model to predict molecular structure follow:
>
> - Determine the Lewis structure(s) for the molecule.
> - For molecules with resonance structures, use any of the structures to predict the molecular structure.
> - Sum the electron pairs around the central atom.
> - In counting pairs, count each multiple bond as a single effective pair.
> - The arrangement of the pairs is determined by minimizing electron-pair repulsions. These arrangements are shown in Table 4-1.
> - Lone pairs require more space than bonding pairs do. Choose an arrangement that gives the lone pairs as much room as possible. Recognize that the lone pairs may produce a slight distortion of the structure at angles less than 120 degrees.

The VSEPR Model—How Well Does It Work?

The VSEPR model is very simple. There are only a few rules to remember, yet the model correctly predicts the molecular structures of most molecules formed from non-metallic elements. Molecules of any size can be treated by applying the VSEPR model to each appropriate atom (those bonded to at least two other atoms) in the molecule. Thus we can use this model to predict the structures of molecules with hundreds of atoms. It does, however, fail in a few instances. For example, phosphine (PH_3), which has a Lewis structure analogous to that of ammonia,

$$H\!-\!\overset{\displaystyle ..}{P}\!-\!H \qquad\qquad H\!-\!\overset{\displaystyle ..}{N}\!-\!H$$
$$\overset{\displaystyle |}{H} \qquad\qquad\qquad \overset{\displaystyle |}{H}$$

would be predicted to have a molecular structure similar to that for NH_3, with bond angles of approximately 107 degrees. However, the bond angles of phosphine are actually 94 degrees. There are ways of explaining this structure, but more rules have to be added to the model.

CONNECTING TO ATOMS 4-1 | *Polar Molecules—It's All About Symmetry*

In Chapter 3 we learned about polar bonds, which result from the unequal sharing of electrons between the atoms in a chemical bond. For example, in the molecule HF, the electron density is greater around the fluorine atom than the hydrogen atom due to the lone pairs around the fluorine and the greater electronegativity of fluorine compared to hydrogen:

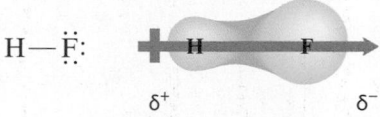

$$\delta^+ \qquad\qquad \delta^-$$

To determine the polarity of more complicated molecules, however, we need to consider the shapes of the molecules, using the rules we have learned from the VSEPR theory.

Consider, for example, methane, ammonia, and water as shown in the following table:

We can see that while each molecule has a tetrahedral geometry of electron pairs around the central atom, ammonia and water each has a dipole moment; that is, each molecule is polar. Methane, however, has slightly polar bonds, but, due to the symmetry of the shape, these bond polarities cancel and the molecule has no dipole moment; methane is a nonpolar molecule.

In determining if a molecule is polar, we need to first draw the Lewis structure, determine the electron pair arrangement, determine the shape, and then decide if the molecule is symmetrical. If the molecule is symmetrical, any polar bonds will cancel resulting in a nonpolar molecule overall.

In Chapter 9 we will discuss intermolecular forces resulting from polar molecules and how the strength of these forces influence molecular properties such as boiling point and surface tension of liquids.

Molecule	Lewis Structure	Electron Pair Arrangement	Electrostatic Potential Diagram
CH_4	H—C—H with H above and H below	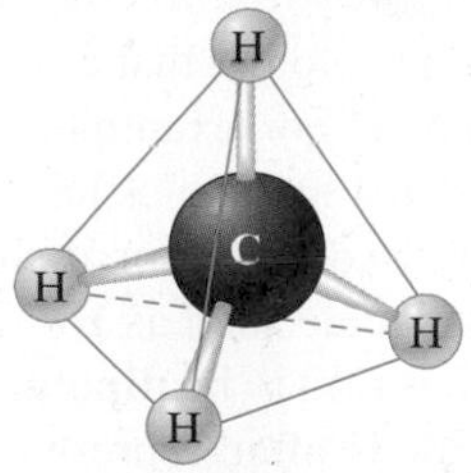 (tetrahedral H–C–H arrangement)	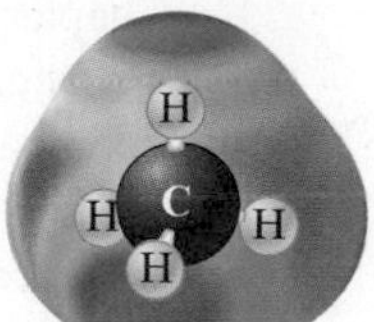
NH_3	H—N—H with H below	Lone pair	
H_2O	H—O—H	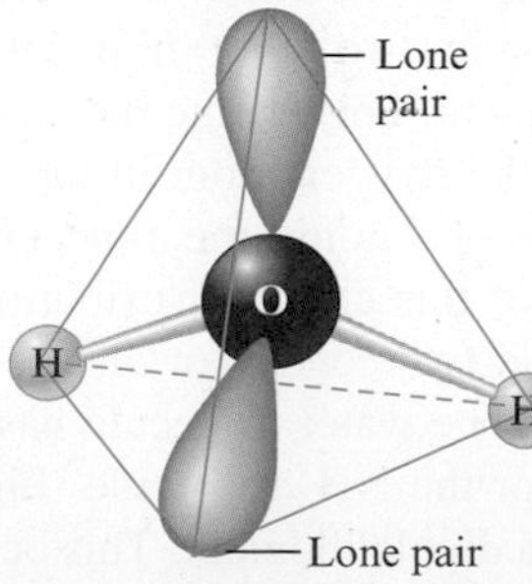 Lone pair / Lone pair	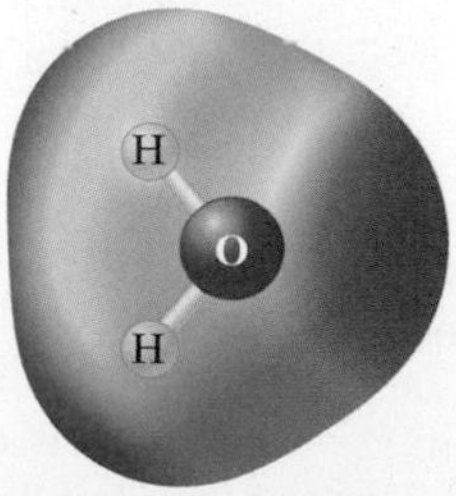

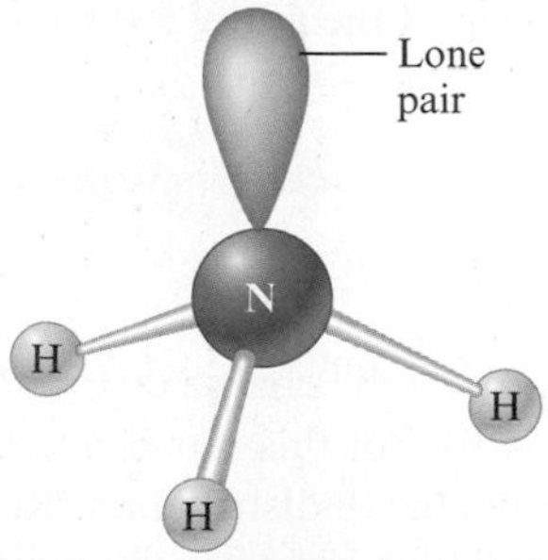

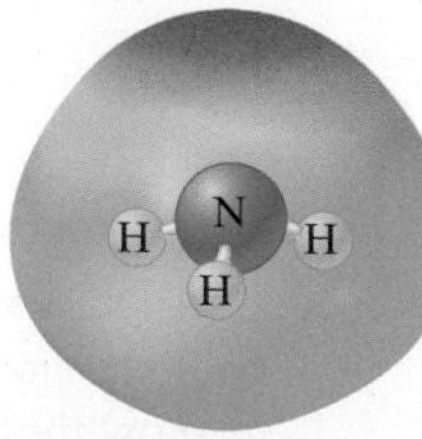

This again illustrates the point that simple models are bound to have exceptions. In introductory chemistry we want to use simple models that fit the majority of cases; we are willing to accept a few failures rather than complicate the model. The amazing thing about the VSEPR model is that such a simple model predicts correctly the structures of so many molecules.

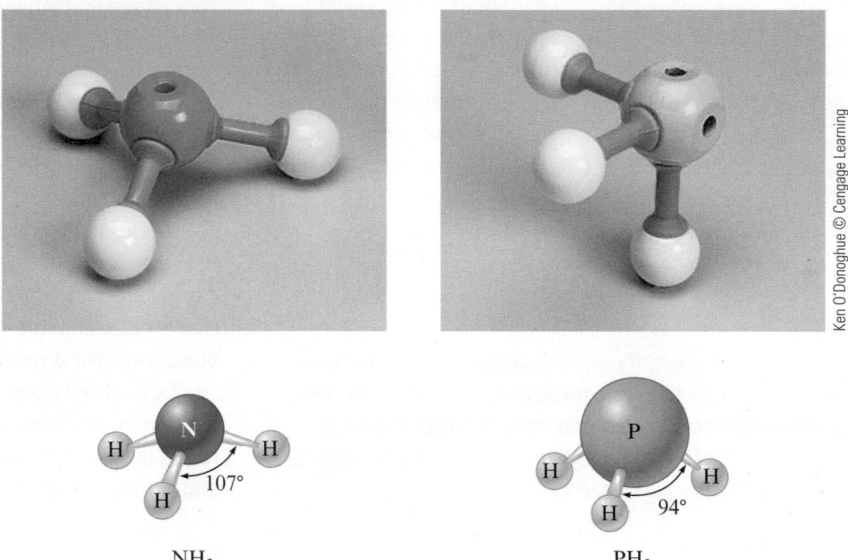

NH₃   PH₃

4-2 | Bond Polarity and Dipole Moments

In Chapter 3 (Section 3-2) we found that covalent bonds between atoms with different electronegativities are polar. For example, we saw that when hydrogen fluoride is placed in an electric field, the molecules have a preferential orientation (see Fig. 3-3). This follows from the charge distribution in the HF molecule, which has a positive end and a negative end. A molecule such as HF that has a center of positive charge and a center of negative charge is said to be **dipolar**, or to have a **dipole moment**. The dipolar character of a molecule is often represented by an arrow pointing to the negative charge center with the tail of the arrow indicating the positive center of charge:

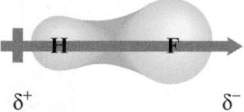

δ^+ δ^-

Another way to represent the charge distribution in HF is by an electrostatic potential diagram (see Fig. 4-9). For this representation the colors of visible light are used to show the variation in charge distribution. Red indicates the most electron-rich region of the molecule and blue indicates the most electron-poor region.

Of course, any diatomic (two-atom) molecule that has a polar bond also will show a molecular dipole moment. Polyatomic molecules also can exhibit dipolar behavior. For example, because the oxygen atom in the water molecule has a greater electronegativity than the hydrogen atoms, the molecular charge distribution is that shown in Fig. 4-10(a). Because of this charge distribution, the water molecule behaves in an electric field as if it had two centers of charge—one positive and one negative—as shown in Fig. 4-10(b). The water molecule has a dipole moment. The same type of behavior is observed for the NH₃ molecule (Fig. 4-11). Some molecules have polar bonds but do not have a dipole moment. This occurs when the individual bond polari-

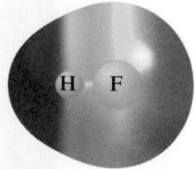

Figure 4-9 | An electrostatic potential map of HF. Red indicates the most electron-rich area (the fluorine atom) and blue indicates the most electron-poor region (the hydrogen atom).

Figure 4-10 | (a) The charge distribution in the water molecule. (b) The water molecule in an electric field. (c) The electrostatic potential diagram of the water molecule.

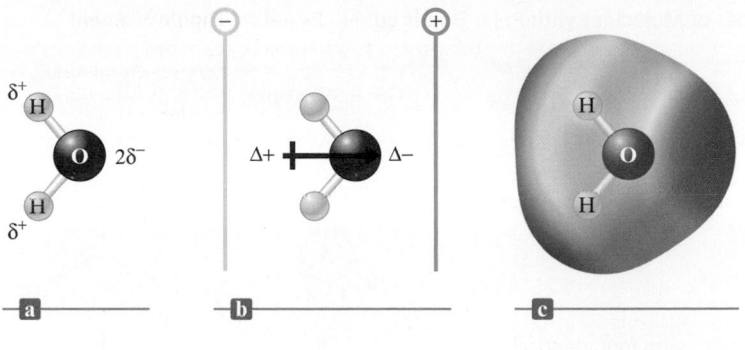

Figure 4-11 | (a) The structure and charge distribution of the ammonia molecule. The polarity of the N—H bonds occurs because nitrogen has a greater electronegativity than hydrogen. (b) The dipole moment of the ammonia molecule oriented in an electric field. (c) The electrostatic potential diagram for ammonia.

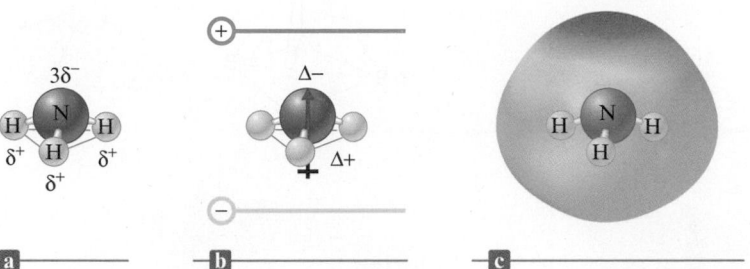

Figure 4-12 | (a) The carbon dioxide molecule. (b) The opposed bond polarities cancel out, and the carbon dioxide molecule has no dipole moment. (c) The electrostatic potential diagram for carbon dioxide.

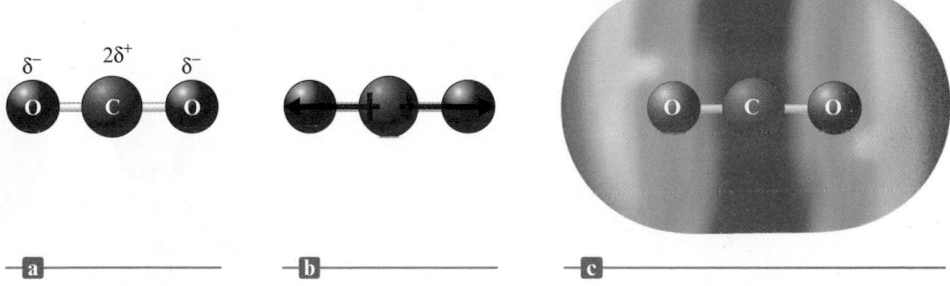

ties are arranged in such a way that they cancel each other out. An example is the CO_2 molecule, which is a linear molecule that has the charge distribution shown in Fig. 4-12. In this case the opposing bond polarities cancel out, and the carbon dioxide molecule does not have a dipole moment. There is no preferential way for this molecule to line up in an electric field. (Try to find a preferred orientation to make sure you understand this concept.)

There are many cases besides that of carbon dioxide where the bond polarities oppose and exactly cancel each other. Some common types of molecules with polar bonds but no dipole moment are shown in Table 4-4.

EXAMPLE 4-5   Bond Polarity and Dipole Moment

For each of the following molecules, show the direction of the bond polarities and indicate which ones have a dipole moment: HCl, Cl_2, SO_3 (a planar molecule with the oxygen atoms spaced evenly around the central sulfur atom), CH_4 [tetrahedral (see Table 4-4) with the carbon atom at the center], and H_2S (V-shaped with the sulfur atom at the point).

Table 4-4 | Types of Molecules with Polar Bonds but No Resulting Dipole Moment

Type	General Example	Cancellation of Polar Bonds	Specific Example	Ball-and-Stick Model
Linear molecules with two identical bonds	B—A—B	←—+ +—→	CO_2	
Planar molecules with three identical bonds 120 degrees apart	B | A B⤴ ⤵B 120°		SO_3	
Tetrahedral molecules with four identical bonds 109.5 degrees apart	B A B B B		CCl_4	

© Ken O'Donoghue
© Cengage Learning

Solution

ⓘ Recall that a blue color is an electron-poor region and red is an electron-rich region.

The HCl molecule: In Fig. 3-3, we see that the electronegativity of chlorine (3.0) is greater than that of hydrogen (2.1). Thus the chlorine will be partially negative, and the hydrogen will be partially positive. ◄ ⓘ The HCl molecule has a dipole moment:

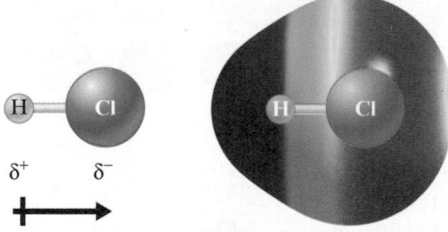

$$\delta^+ \qquad \delta^-$$

$$\longrightarrow$$

The Cl_2 molecule: The two chlorine atoms share the electrons equally. No bond polarity occurs, and the Cl_2 molecule has no dipole moment.

The SO_3 molecule: The electronegativity of oxygen (3.5) is greater than that of sulfur (2.5). This means that each oxygen will have a partial negative charge, and the sulfur will have a partial positive charge:

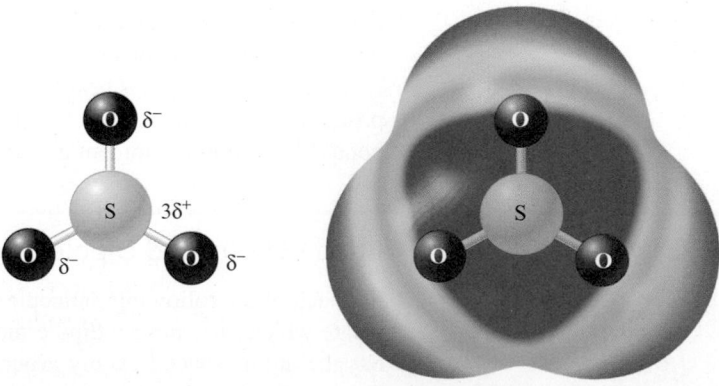

ⓘ The presence of polar bonds does not always yield a polar molecule.

The bond polarities arranged symmetrically as shown cancel, and the molecule has no dipole moment. ◄ ⓘ This molecule is the second type shown in Table 4-4.

The CH₄ molecule: Carbon has a slightly higher electronegativity (2.5) than does hydrogen (2.1). This leads to small partial positive charges on the hydrogen atoms and a small partial negative charge on the carbon:

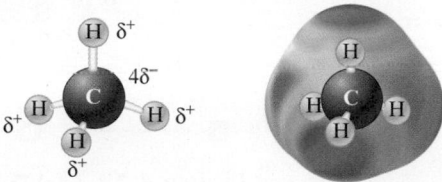

This case is similar to the third type in Table 4-4, and the bond polarities cancel. The molecule has no dipole moment.

The H₂S molecule: Since the electronegativity of sulfur (2.5) is slightly greater than that of hydrogen (2.1), the sulfur will have a partial negative charge, and the hydrogen atoms will have a partial positive charge, which can be represented as

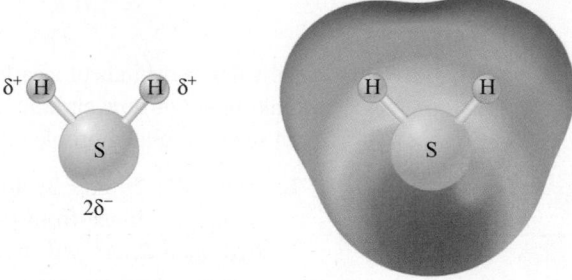

This case is analogous to the water molecule, and the polar bonds result in a dipole moment oriented as shown:

See Exercises 4-29 and 4-30

4-3 | Hybridization and the Localized Electron Model

As we saw in Chapter 3, the localized electron model views a molecule as a collection of atoms bound together by sharing electrons between their atomic orbitals. The arrangement of valence electrons is represented by the Lewis structure (or structures, where resonance occurs), and the molecular geometry can be predicted from the VSEPR model. In this section we will describe the atomic orbitals used to share electrons and hence to form the bonds.

sp³ Hybridization

Let us reconsider the bonding in methane, which has the Lewis structure and molecular geometry shown in Fig. 4-13. In general, we assume that bonding involves only the valence orbitals. ◄ ⓘ This means that the hydrogen atoms in methane use 1*s* orbitals.

ⓘ The valence orbitals are the orbitals associated with the highest principal quantum level that contains electrons on a given atom.

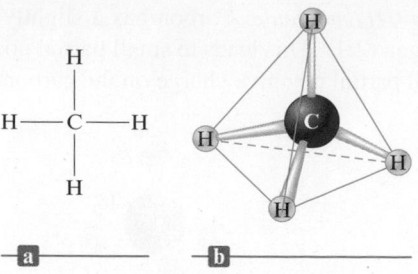

Figure 4-13 | (a) The Lewis structure of the methane molecule. (b) The tetrahedral molecular geometry of the methane molecule.

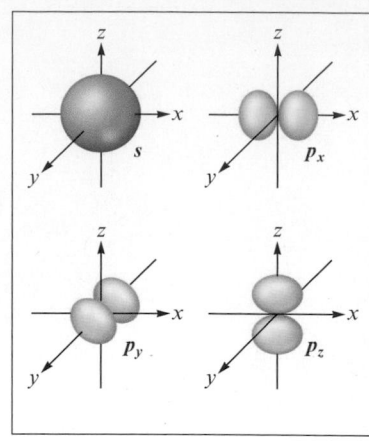

Figure 4-14 | The valence orbitals on a free carbon atom: $2s$, $2p_x$, $2p_y$, and $2p_z$.

The valence orbitals of a carbon atom are the $2s$ and $2p$ orbitals shown in Fig. 4-14. In thinking about how carbon can use these orbitals to bond to the hydrogen atoms, we can see two related problems:

1. Using the $2p$ and $2s$ atomic orbitals will lead to two different types of C—H bonds: (a) those from the overlap of a $2p$ orbital of carbon and a $1s$ orbital of hydrogen (there will be three of these) and (b) those from the overlap of a $2s$ orbital of carbon and a $1s$ orbital of hydrogen (there will be one of these). This is a problem because methane is known to have four identical C—H bonds.

2. Since the carbon $2p$ orbitals are mutually perpendicular, we might expect the three C—H bonds formed with these orbitals to be oriented at 90-degree angles:

However, the methane molecule is known by experiment to be tetrahedral, with bond angles of 109.5 degrees.

This analysis leads to one of two conclusions: Either the simple localized electron model is wrong or carbon adopts a set of atomic orbitals other than its "native" $2s$ and $2p$ orbitals to bond to the hydrogen atoms in forming the methane molecule. The second conclusion seems more reasonable. The $2s$ and $2p$ orbitals present on an *isolated* carbon atom may not be the best set of orbitals for bonding; a new set of atomic orbitals might better serve the carbon atom in forming molecules. To account for the known structure of methane, it makes sense to assume that the carbon atom has four equivalent atomic orbitals, arranged tetrahedrally. In fact, such a set of orbitals can be obtained quite readily by combining the carbon $2s$ and $2p$ orbitals, as shown schematically in Fig. 4-15. This mixing of the native atomic orbitals to form special orbitals for bonding is called **hybridization**. ◄ ⓘ The four new orbitals are called sp^3 orbitals because they are formed from one $2s$ and three $2p$ orbitals (s^1p^3). We say that the carbon atom undergoes sp^3 **hybridization** or is sp^3 **hybridized**. The four sp^3 orbitals are identical in shape, each one having a large lobe and a small lobe (Fig. 4-16). The four orbitals are oriented in space so that the large lobes form a tetrahedral arrangement, as shown in Fig. 4-15. ◄ ⓘ ◄ ⓘ

The hybridization of the carbon $2s$ and $2p$ orbitals also can be represented by an orbital energy-level diagram, as shown in Fig. 4-17. Note that electrons have been

ⓘ Hybridization is a modification of the localized electron model to account for the observation that atoms often seem to use special atomic orbitals in forming molecules. Recall that all models are simplifications and we make models more complex to answer new questions or to explain more observations.

ⓘ sp^3 hybridization gives a tetrahedral set of orbitals.

ⓘ Note that the shapes of the hybrid orbitals shown in Figure 4-15 are "mushroom-like," calculated from wave functions.

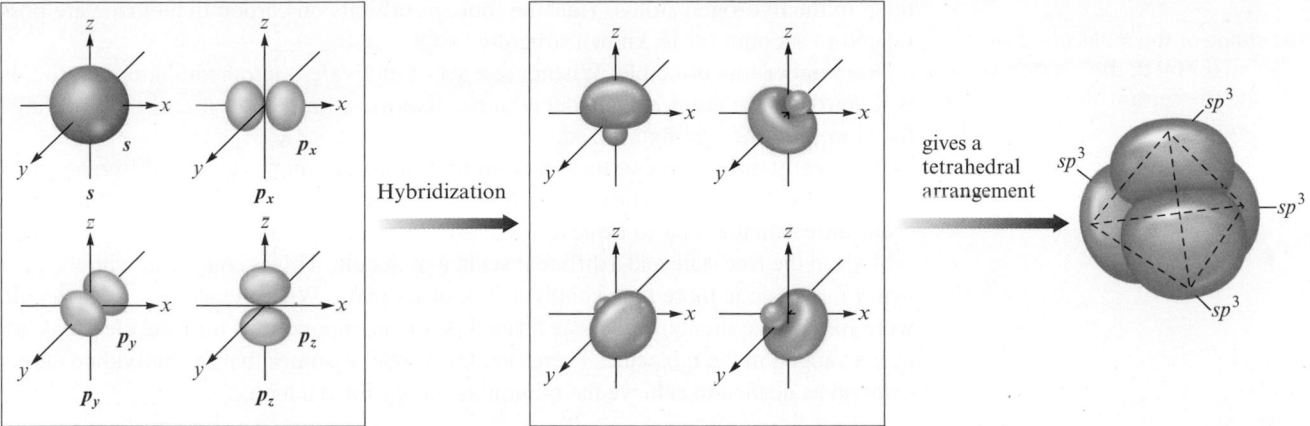

Figure 4-15 | The "native" $2s$ and three $2p$ atomic orbitals characteristic of a free carbon atom are combined to form a new set of four sp^3 orbitals. The small lobes of the orbitals are usually omitted from diagrams for clarity.

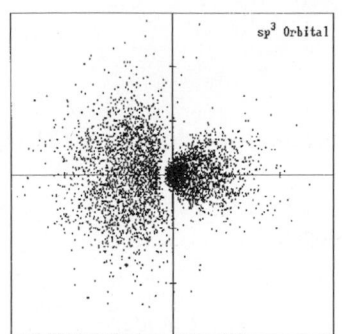

Figure 4-16 | Cross section of an sp^3 orbital. This shows a "slice" of the electron density of the sp^3 orbitals illustrated in the center diagram of Fig. 4-15.

(Generated from a program by Robert Allendoerfer on Project SERAPHIM disk PC2402; printed with permission.)

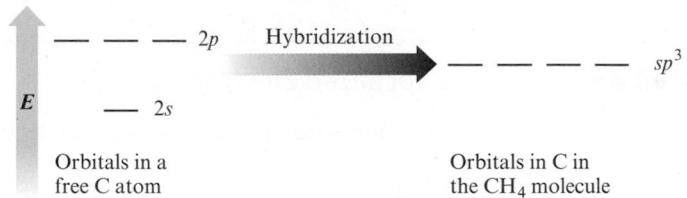

Figure 4-17 | An energy-level diagram showing the formation of four sp^3 orbitals.

omitted because we do not need to be concerned with the electron arrangements on the individual atoms—it is the total number of electrons and the arrangement of these electrons in the *molecule* that are important. We are assuming that carbon's atomic orbitals are rearranged to accommodate the best electron arrangement for the molecule as a whole. The new sp^3 atomic orbitals on carbon are used to share electron pairs with the $1s$ orbitals from the four hydrogen atoms, as shown in Fig. 4-18.

At this point let's summarize the bonding in the methane molecule. The experimentally known structure of this molecule can be explained if we assume that the carbon atom adopts a special set of atomic orbitals. These new orbitals are obtained by combining the $2s$ and the three $2p$ orbitals of the carbon atom to produce four identically shaped orbitals that are oriented toward the corners of a tetrahedron and are used to

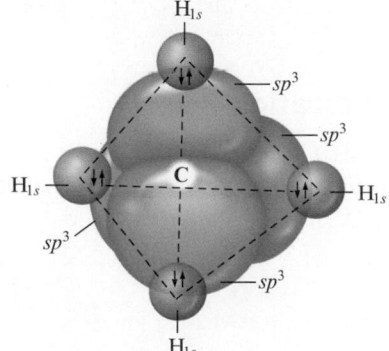

Figure 4-18 | The tetrahedral set of four sp^3 orbitals of the carbon atom are used to share electron pairs with the four $1s$ orbitals of the hydrogen atoms to form the four equivalent C—H bonds. This accounts for the known tetrahedral structure of the CH_4 molecule.

bond to the hydrogen atoms. Thus the four sp^3 orbitals on carbon in methane are postulated to account for its known structure. ◄ ℹ️

Remember this principle: Whenever a set of equivalent tetrahedral atomic orbitals is required by an atom, this model assumes that the atom adopts a set of sp^3 orbitals; the atom becomes sp^3 hybridized.

It is really not surprising that an atom in a molecule might adopt a different set of atomic orbitals (called **hybrid orbitals**) from those it has in the free state. It does not seem unreasonable that to achieve minimum energy, an atom uses one set of atomic orbitals in the free state and a different set in a molecule. This is consistent with the idea that a molecule is more than simply a sum of its parts. What the atoms in a molecule were like before the molecule was formed is not as important as how the electrons are best arranged in the molecule. Therefore, this model assumes that the individual atoms respond as needed to achieve the minimum energy for the molecule.

> ### Critical Thinking
> What if the sp^3 hybrid orbitals were higher in energy than the p orbitals in the free atom? How would this affect our model of bonding?

EXAMPLE 4-6

The Localized Electron Model I

Describe the bonding in the ammonia molecule, using the localized electron model.

Solution

A complete description of the bonding involves three steps:

1. Writing the Lewis structure
2. Determining the arrangement of electron pairs, using the VSEPR model
3. Determining the hybrid atomic orbitals needed to describe the bonding in the molecule

The Lewis structure for NH_3 is

H—N̈—H
|
H

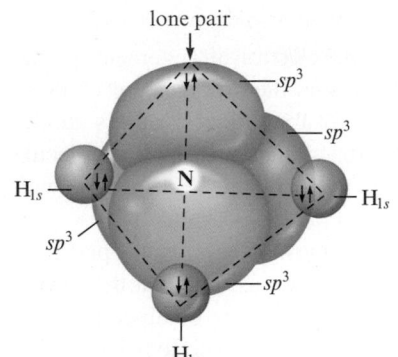

Figure 4-19 | The nitrogen atom in ammonia is sp^3 hybridized.

The four electron pairs around the nitrogen atom require a tetrahedral arrangement to minimize repulsions. We have seen that a tetrahedral set of sp^3 hybrid orbitals is obtained by combining the $2s$ and three $2p$ orbitals. In the NH_3 molecule, three of the sp^3 orbitals are used to form bonds to the three hydrogen atoms, and the fourth sp^3 orbital holds the lone pair, as shown in Fig. 4-19.

See Exercises 4-39 and 4-40

sp^2 Hybridization

Ethylene (C_2H_4) is an important starting material in the manufacture of plastics. The C_2H_4 molecule has 12 valence electrons and the following Lewis structure:

H   H
\   /
C=C
/   \
H   H

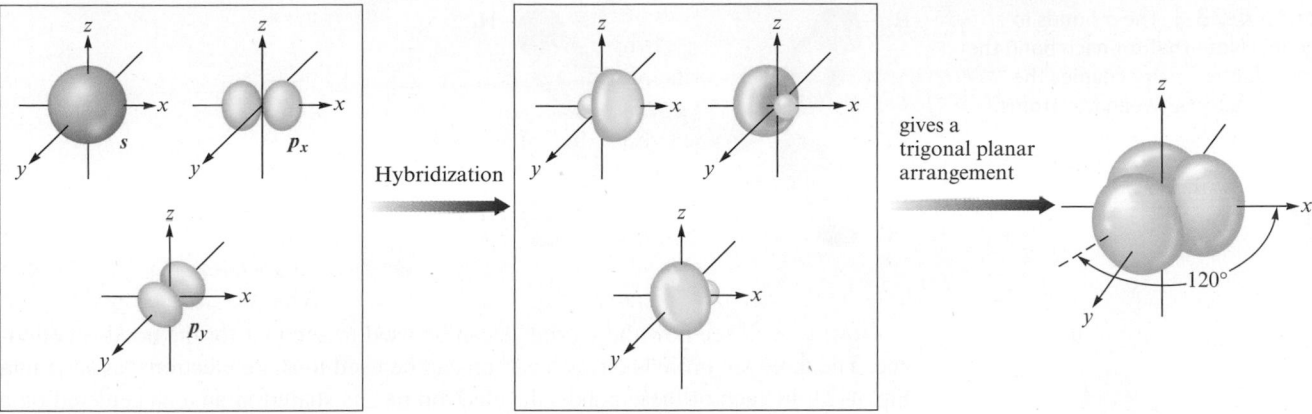

Figure 4-20 | The hybridization of the s, p_x, and p_y atomic orbitals results in the formation of three sp^2 orbitals centered in the xy plane. The large lobes of the orbitals lie in the plane at angles of 120 degrees and point toward the corners of a triangle.

ⓘ A double bond acts as one effective electron pair.

ⓘ sp^2 hybridization gives a trigonal planar arrangement of atomic orbitals.

ⓘ Note in Fig. 4-22 and the figures that follow that the orbital lobes are artificially narrowed to more clearly show their relative orientations.

We saw in Section 4-1 that a double bond acts as one effective pair, so in the ethylene molecule each carbon is surrounded by three effective pairs. ◀ ⓘ This requires a trigonal planar arrangement with bond angles of 120 degrees. What orbitals do the carbon atoms in this molecule employ? The molecular geometry requires a set of orbitals in one plane at angles of 120 degrees. Since the $2s$ and $2p$ valence orbitals of carbon do not have the required arrangement, we need a set of hybrid orbitals.

The sp^3 orbitals we have just considered will not work because they are at angles of 109.5 degrees rather than the required 120 degrees. In ethylene the carbon atom must hybridize in a different manner. A set of three orbitals arranged at 120-degree angles in the same plane can be obtained by combining one s orbital and two p orbitals, as shown in Fig. 4-20. ◀ ⓘ The orbital energy-level diagram for this arrangement is shown in Fig. 4-21. Since one $2s$ and two $2p$ orbitals are used to form these hybrid orbitals, this is called **sp^2 hybridization**. Note from Fig. 4-20 that the plane of the sp^2 hybridized orbitals is determined by which p orbitals are used. Since in this case we have arbitrarily decided to use the p_x and p_y orbitals, the hybrid orbitals are centered in the xy plane.

In forming the sp^2 orbitals, one $2p$ orbital on carbon has not been used. This remaining p orbital (p_z) is oriented perpendicular to the plane of the sp^2 orbitals, as shown in Fig. 4-22. ◀ ⓘ

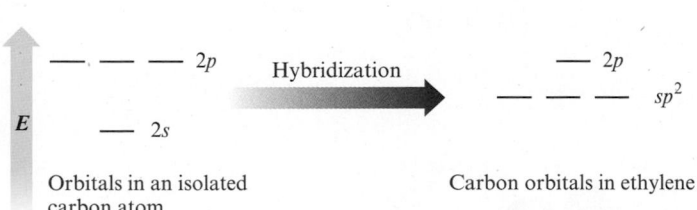

Figure 4-21 | An orbital energy-level diagram for sp^2 hybridization. Note that one p orbital remains unchanged.

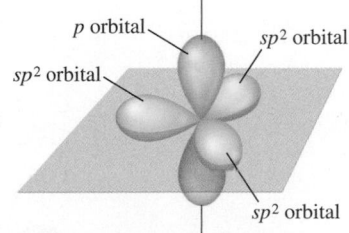

Figure 4-22 | When an s and two p orbitals are mixed to form a set of three sp^2 orbitals, one p orbital remains unchanged and is perpendicular to the plane of the hybrid orbitals. Note that in this figure and those that follow, the orbitals are drawn with narrowed lobes to show their orientations more clearly.

Figure 4-23 | The σ bonds in ethylene. Note that for each bond the shared electron pair occupies the region directly between the atoms.

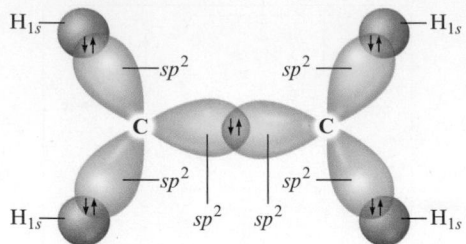

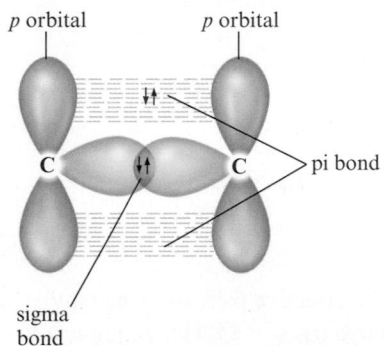

Figure 4-24 | A carbon–carbon double bond consists of a σ bond and a π bond. In the σ bond the shared electrons occupy the space directly between the atoms. The π bond is formed from the unhybridized p orbitals on the two carbon atoms. In a π bond the shared electron pair occupies the space above and below a line joining the atoms.

Now we will see how these orbitals can be used to account for the bonds in ethylene. The three sp^2 orbitals on each carbon can be used to share electrons, as shown in Fig. 4-23. In each of these bonds, the electron pair is shared in an area centered on a line running between the atoms. This type of covalent bond is called a **sigma (σ) bond**. In the ethylene molecule, the σ bonds are formed using sp^2 orbitals on each carbon atom and the $1s$ orbital on each hydrogen atom.

How can we explain the double bond between the carbon atoms? In the σ bond the electron pair occupies the space between the carbon atoms. The second bond must therefore result from sharing an electron pair in the space *above and below* the σ bond. This type of bond can be formed by using the $2p$ orbital perpendicular to the sp^2 hybrid orbitals on each carbon atom (refer to Fig. 4-22). These parallel p orbitals can share an electron pair, which occupies the space above and below a line joining the atoms, to form a **pi (π) bond**, as shown in Fig. 4-24.

Note that σ bonds are formed from orbitals whose lobes point toward each other, but π bonds result from parallel orbitals. A *double bond always consists of one σ bond*, where the electron pair is located directly between the atoms, *and one π bond*, where the shared pair occupies the space above and below the σ bond.

We can now completely specify the orbitals that this model assumes are used to form the bonds in the ethylene molecule. As shown in Fig. 4-25, the carbon atoms use sp^2 hybrid orbitals to form the σ bonds to the hydrogen atoms and to each other, and they use p to form the π bond with each other. Note that we have accounted fully for the Lewis structure of ethylene with its carbon–carbon double bond and carbon–hydrogen single bonds.

This example illustrates an important general principle of this model: Whenever an atom is surrounded by three effective pairs, a set of sp^2 hybrid orbitals is required.

sp Hybridization

Another type of hybridization occurs in carbon dioxide, which has the following Lewis structure:

$$:\!\ddot{O}\!=\!C\!=\!\ddot{O}\!:$$

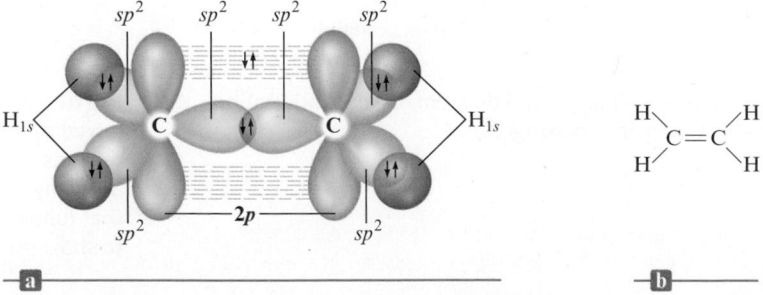

Figure 4-25 | (a) The orbitals used to form the bonds in ethylene. (b) The Lewis structure for ethylene.

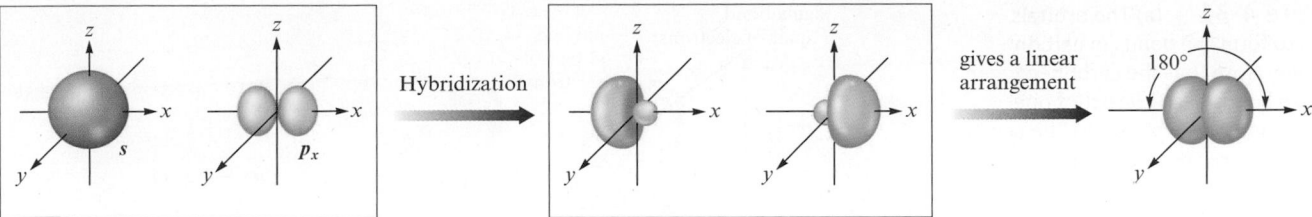

Figure 4-26 | When one *s* orbital and one *p* orbital are hybridized, a set of two *sp* orbitals oriented at 180 degrees results.

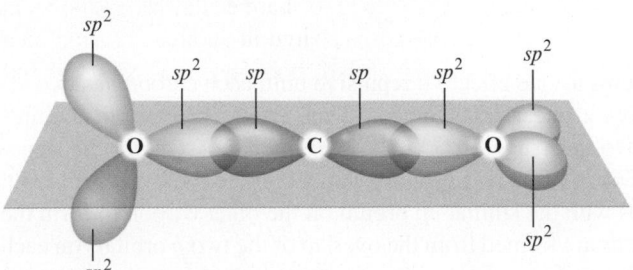

Figure 4-27 | The hybrid orbitals in the CO₂ molecule.

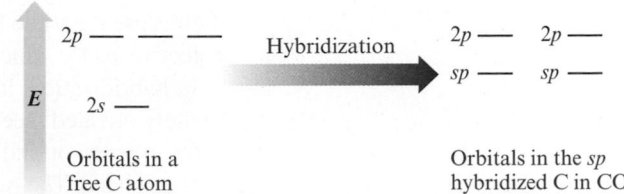

Figure 4-28 | The orbital energy-level diagram for the formation of *sp* hybrid orbitals on carbon.

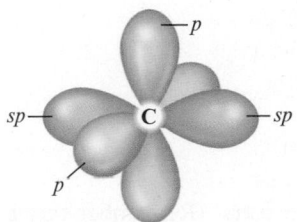

Figure 4-29 | The orbitals of an *sp* hybridized carbon atom.

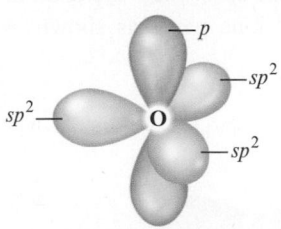

Figure 4-30 | The orbital arrangement for an *sp²* hybridized oxygen atom.

ⓘ More rigorous theoretical models of CO₂ indicate that each of the oxygen atoms uses two *p* orbitals simultaneously to form the pi bonds to the carbon atom, thus leading to unusually strong C=O bonds.

In the CO_2 molecule the carbon atom has two effective pairs that will be arranged at an angle of 180 degrees. We therefore need a pair of atomic orbitals oriented in opposite directions. This requires a new type of hybridization, since neither *sp³* nor *sp²* hybrid orbitals will fit this case. To obtain two hybrid orbitals arranged at 180 degrees requires **sp hybridization**, involving one *s* orbital and one *p* orbital, as shown in Fig. 4-26.

In terms of this model, two effective pairs around an atom will always require *sp* hybridization of that atom. The *sp* orbitals of carbon in carbon dioxide can be seen in Fig. 4-27, and the corresponding orbital energy-level diagram for their formation is given in Fig. 4-28. These *sp* hybrid orbitals are used to form the σ bonds between the carbon and the oxygen atoms. Note that two 2*p* orbitals remain unchanged on the *sp* hybridized carbon. These are used to form the π bonds with the oxygen atoms.

In the CO_2 molecule each oxygen atom* has three effective pairs around it, requiring a trigonal planar arrangement of the pairs. Since a trigonal set of hybrid orbitals requires *sp²* hybridization, each oxygen atom is *sp²* hybridized. One *p* orbital on each oxygen is unchanged and is used for the π bond with the carbon atom.

Now we are ready to use our model to describe the bonding in carbon dioxide. The *sp* orbitals on carbon form σ bonds with the *sp²* orbitals on the two oxygen atoms (see Fig. 4-27). The remaining *sp²* orbitals on the oxygen atoms hold lone pairs. The π bonds between the carbon atom and each oxygen atom are formed by the overlap of parallel 2*p* orbitals. The *sp* hybridized carbon atom has two unhybridized *p* orbitals, pictured in Fig. 4-29. Each of these *p* orbitals is used to form a π bond with an oxygen atom (see Fig. 4-30). The total bonding picture for the CO_2 molecule is shown in Fig. 4-31. Note that this picture of the bonding neatly explains the arrangement of electrons predicted by the Lewis structure. ◀ *ⓘ*

Another molecule whose bonding can be described by *sp* hybridization is acetylene (C_2H_2), which has the systematic name ethyne. The Lewis structure for acetylene is

$$H—C≡C—H$$

---

*We will assume that minimizing electron repulsions also is important for the peripheral atoms in a molecule and apply the VSEPR model to these atoms as well. There is no direct way to test whether or not this assumption is correct.

Figure 4-31 | (a) The orbitals used to form the bonds in carbon dioxide. Note that the carbon–oxygen double bonds each consist of one σ bond and one π bond. (b) The Lewis structure for carbon dioxide.

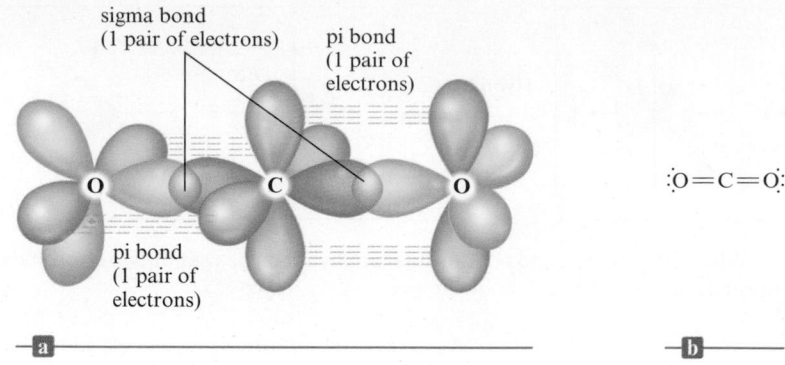

sigma bond
(1 pair of electrons)

pi bond
(1 pair of electrons)

pi bond
(1 pair of electrons)

$:\!\ddot{O}\!=\!C\!=\!\ddot{O}\!:$

Because the triple bond counts as one effective repulsive unit, each carbon has two effective pairs, which requires a linear arrangement. Thus each carbon atom requires *sp* hybridization, leaving two unchanged *p* orbitals (see Fig. 4-28). One of the oppositely oriented (see Fig. 4-26) *sp* orbitals is used to form a bond to the hydrogen atom; the other *sp* orbital overlaps with the similar *sp* orbital on the other carbon to form the sigma bond. The two pi bonds are formed from the overlap of the two *p* orbitals on each carbon. This accounts for the triple bond (one sigma and two pi bonds) in acetylene.

EXAMPLE 4-7

The Localized Electron Model II

Describe the bonding in the N_2 molecule.

Solution

The Lewis structure for the nitrogen molecule is

$$: N \equiv N :$$

where each nitrogen atom is surrounded by two effective pairs. (Remember that a multiple bond counts as one effective pair.) This gives a linear arrangement (180 degrees) requiring a pair of oppositely directed orbitals. This situation requires *sp* hybridization. Each nitrogen atom in the nitrogen molecule has two *sp* hybrid orbitals and two unchanged *p* orbitals, as shown in Fig. 4-32(a). The *sp* orbitals are used to form the σ bond between the nitrogen atoms and to hold lone pairs, as shown in

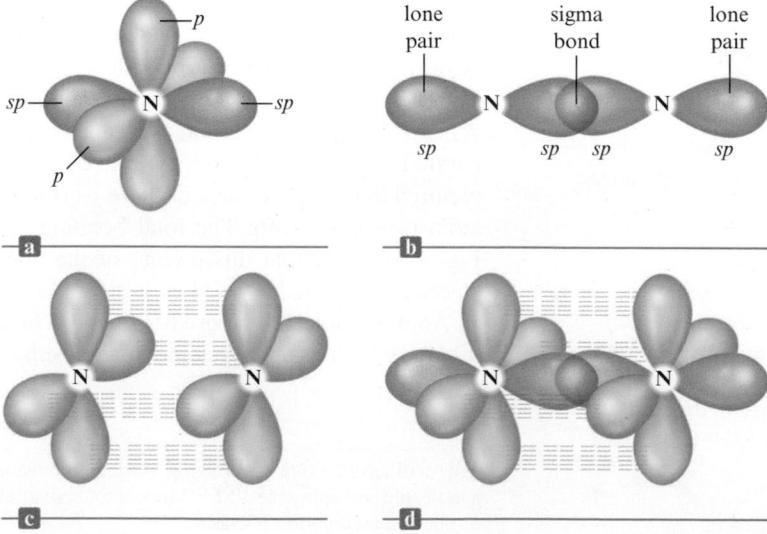

Figure 4-32 | (a) An *sp* hybridized nitrogen atom. There are two *sp* hybrid orbitals and two unhybridized *p* orbitals. (b) The σ bond in the N_2 molecule. (c) The two π bonds in N_2 are formed when electron pairs are shared between two sets of parallel *p* orbitals. (d) The total bonding picture for N_2.

Fig. 4-32(b). The *p* orbitals are used to form the two π bonds [see Fig. 4-32(c)]; each pair of overlapping parallel *p* orbitals holds one electron pair. Such bonding accounts for the electron arrangement given by the Lewis structure. The triple bond consists of a σ bond (overlap of two *sp* orbitals) and two π bonds (each one from an overlap of two *p* orbitals). In addition, a lone pair occupies an *sp* orbital on each nitrogen atom.

See Exercises 4-41 and 4-42

dsp³ Hybridization

To illustrate the treatment of a molecule in which the central atom exceeds the octet rule, consider the bonding in the phosphorus pentachloride molecule (PCl_5). The Lewis structure

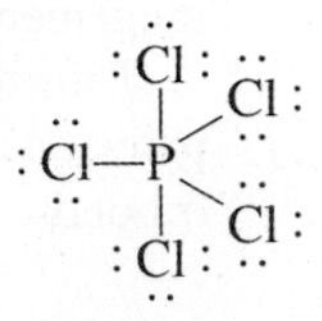

$$:\!\overset{\displaystyle :Cl:}{\underset{\displaystyle :Cl:}{\overset{\displaystyle |}{\underset{\displaystyle |}{Cl\!-\!P}}}}\!\!\overset{\displaystyle Cl:}{\underset{\displaystyle Cl:}{}}$$

shows that the phosphorus atom is surrounded by five electron pairs. Since five pairs require a trigonal bipyramidal arrangement, we need a trigonal bipyramidal set of atomic orbitals on phosphorus. Such a set of orbitals is formed by *dsp³* **hybridization** of one *d* orbital, one *s* orbital, and three *p* orbitals, as shown in Fig. 4-33.

The *dsp³* hybridized phosphorus atom in the PCl_5 molecule uses its five *dsp³* orbitals to share electrons with the five chlorine atoms. Note that a set of five effective pairs around a given atom always requires a trigonal bipyramidal arrangement, which in turn requires *dsp³* hybridization of that atom. Although this model involves the use of *d* orbitals, there is considerable doubt about whether the *d* orbitals are involved in the bonding in these molecules. However, this matter is beyond the scope of this text.

The Lewis structure for PCl_5 shows that each chlorine atom is surrounded by four electron pairs. This requires a tetrahedral arrangement, which in turn requires a set of four *sp³* orbitals on each chlorine atom.

Now we can describe the bonding in the PCl_5 molecule shown in Fig. 4-34. The five P—Cl σ bonds are formed by sharing electrons between a *dsp³* orbital on the phosphorus atom and an orbital on each chlorine.

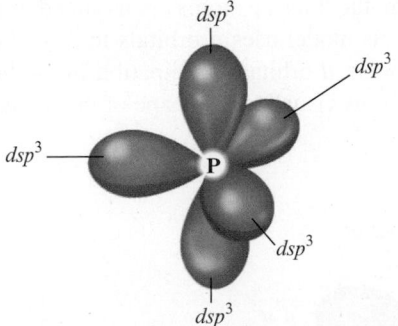

Figure 4-33 | A set of *dsp³* hybrid orbitals on a phosphorus atom. Note that the set of five *dsp³* orbitals has a trigonal bipyramidal arrangement. (Each *dsp³* orbital also has a small lobe that is not shown in this diagram.)

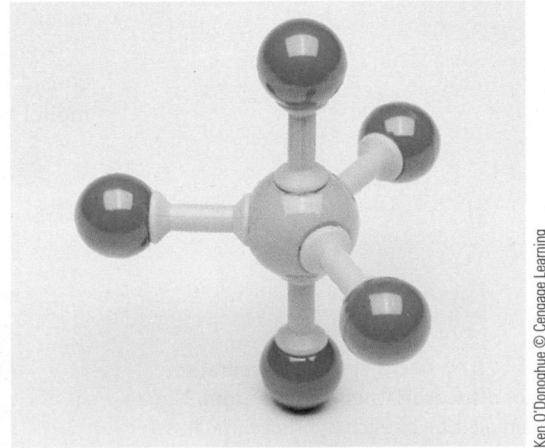

Figure 4-34 | The structure of the PCl_5 molecule.

Ken O'Donoghue © Cengage Learning

EXAMPLE 4-8

The Localized Electron Model III

Describe the bonding in the triiodide ion (I_3^-).

Solution

The Lewis structure for I_3^-

$$\left[:\ddot{\underset{..}{I}}-\ddot{\underset{..}{I}}-\ddot{\underset{..}{I}}: \right]^-$$

shows that the central iodine atom has five pairs of electrons. A set of five pairs requires a trigonal bipyramidal arrangement, which in turn requires a set of dsp^3 orbitals. The outer iodine atoms have four pairs of electrons, which calls for a tetrahedral arrangement and sp^3 hybridization.

Thus the central iodine is dsp^3 hybridized. Three of these hybrid orbitals hold lone pairs, and two of them overlap with sp^3 orbitals of the other two iodine atoms to form σ bonds.

See Exercise 4-48

d^2sp^3 Hybridization

Some molecules have six pairs of electrons around a central atom; an example is sulfur hexafluoride (SF_6), which has the Lewis structure

$$
\begin{array}{ccc}
 & :\ddot{F}: & \\
:\ddot{F} & | & \ddot{F}: \\
 & \diagdown | \diagup & \\
 & S & \\
 & \diagup | \diagdown & \\
:\ddot{F} & | & \ddot{F}: \\
 & :\ddot{F}: & \\
\end{array}
$$

This requires an octahedral arrangement of pairs and in turn an octahedral set of six hybrid orbitals, or d^2sp^3 **hybridization**, in which two d orbitals, one s orbital, and three p orbitals are combined (Fig. 4-35). Six electron pairs around an atom are always arranged octahedrally and require d^2sp^3 hybridization of the atom. Each of the d^2sp^3 orbitals on the sulfur atom is used to bond to a fluorine atom. Since there are four pairs on each fluorine atom, the fluorine atoms are assumed to be sp^3 hybridized.

Again, although this model uses d orbitals to describe the bonding here, there is considerable doubt about d orbital participation in the bonding. We use this simple model because it helps us visualize the shape of the molecule.

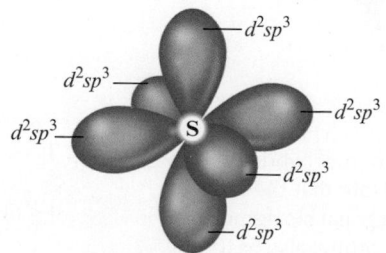

Figure 4-35 | An octahedral set of d^2sp^3 orbitals on a sulfur atom. The small lobe of each hybrid orbital has been omitted for clarity.

INTERACTIVE EXAMPLE 4-9

The Localized Electron Model IV

How is the xenon atom in XeF_4 hybridized?

Solution

As seen in Example 4-3, XeF_4 has six pairs of electrons around xenon that are arranged octahedrally to minimize repulsions. An octahedral set of six atomic orbitals is required to hold these electrons, and the xenon atom is d^2sp^3 hybridized.

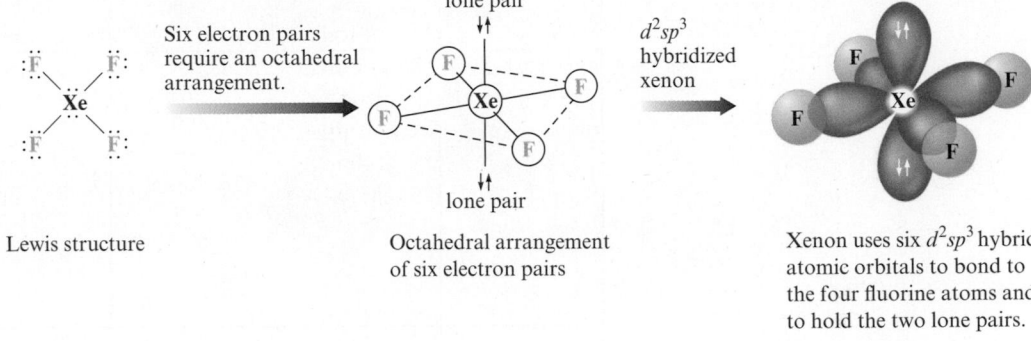

Lewis structure   Octahedral arrangement of six electron pairs   Xenon uses six d^2sp^3 hybrid atomic orbitals to bond to the four fluorine atoms and to hold the two lone pairs.

See Exercise 4-48

The Localized Electron Model: A Summary

The description of a molecule using the localized electron model involves three distinct steps.

PROBLEM-SOLVING STRATEGY

Using the Localized Electron Model

1. Draw the Lewis structure(s).
2. Determine the arrangement of electron pairs, using the VSEPR model.
3. Specify the hybrid orbitals needed to accommodate the electron pairs.

It is important to do the steps in this order. For a model to be successful, it must follow nature's priorities. In the case of bonding, it seems clear that the tendency for a molecule to minimize its energy is more important than the maintenance of the characteristics of atoms as they exist in the free state. The atoms adjust to meet the "needs" of the molecule. When considering the bonding in a particular molecule, therefore, we always start with the molecule rather than the component atoms. In the molecule the electrons will be arranged to give each atom a noble gas configuration, where possible, and to minimize electron-pair repulsions. We then assume that the atoms adjust their orbitals by hybridization to allow the molecule to adopt the structure that gives the minimum energy.

In applying the localized electron model, we must remember not to overemphasize the characteristics of the separate atoms. It is not where the valence electrons originate that is important; it is where they are needed in the molecule to achieve stability. In the same vein, it is not the orbitals in the isolated atom that matter, but which orbitals the molecule requires for minimum energy.

Figure 4-36 | The relationship of the number of effective pairs, their spatial arrangement, and the hybrid orbital set required.

Number of Effective Pairs	Arrangement of Pairs	Hybridization Required
2	—— Linear	sp 180°
3	Trigonal planar	sp^2 120°
4	Tetrahedral	sp^3 109.5°
5	Trigonal bipyramidal	dsp^3 90° 120°
6	Octahedral	d^2sp^3 90° 90°

Octahedral fluorite crystals

The requirements for the various types of hybridization are summarized in Fig. 4-36.

INTERACTIVE EXAMPLE 4-10

The Localized Electron Model V

For each of the following molecules or ions, predict the hybridization of each atom, and describe the molecular structure.

a. CO   **b.** BF_4^- **c.** XeF_2

Solution

a. The CO molecule has 10 valence electrons, and its Lewis structure is

$$: C \equiv O :$$

Each atom has two effective pairs, which means that both are sp hybridized. The triple bond consists of a σ bond produced by overlap of an sp orbital from each atom and two π bonds produced by overlap of $2p$ orbitals from each atom. The lone pairs are in sp orbitals. Since the CO molecule has only two atoms, it must be linear.

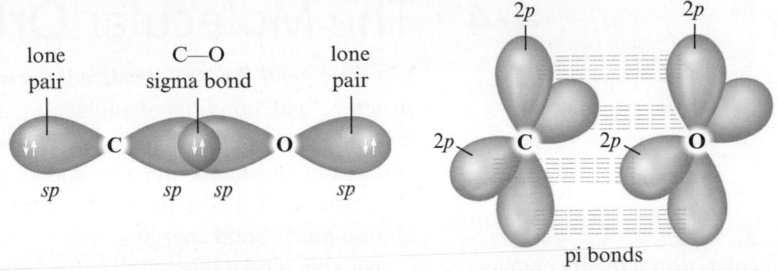

b. The BF_4^- ion has 32 valence electrons. The Lewis structure shows four pairs of electrons around the boron atom, which means a tetrahedral arrangement:

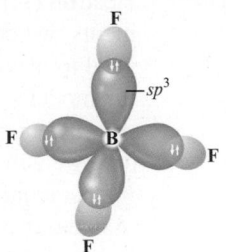

This requires sp^3 hybridization of the boron atom. Each fluorine atom also has four electron pairs and can be assumed to be sp^3 hybridized (only one sp^3 orbital is shown for each fluorine atom). The BF_4^- ion's molecular structure is tetrahedral.

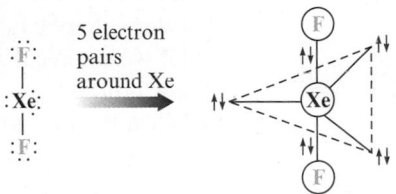

c. The XeF_2 molecule has 22 valence electrons. The Lewis structure shows five electron pairs on the xenon atom, which requires a trigonal bipyramidal arrangement:

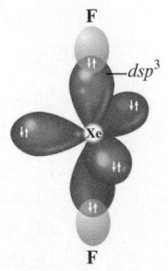

Note that the lone pairs are placed in the plane where they are 120 degrees apart. To accommodate five pairs at the vertices of a trigonal bipyramid requires that the xenon atom adopt a set of five dsp^3 orbitals. Each fluorine atom has four electron pairs and can be assumed to be sp^3 hybridized. The XeF_2 molecule has a linear arrangement of atoms.

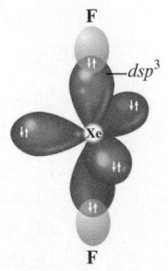

See Exercises 4-49 and 4-50

4-4 | The Molecular Orbital Model

We have seen that the localized electron model is of great value in interpreting the structure and bonding of molecules. However, there are some problems with this model. For example, it incorrectly assumes that electrons are localized, and so the concept of resonance must be added. Also, the model does not deal effectively with molecules containing unpaired electrons. And finally, the model gives no direct information about bond energies.

Molecular orbital theory parallels the atomic theory discussed in Chapter 2.

Another model often used to describe bonding is the **molecular orbital model**. ◄ ⓘ To introduce the assumptions, methods, and results of this model, we will consider the simplest of all molecules, H_2, which consists of two protons and two electrons. A very stable molecule, H_2 is lower in energy than the separated hydrogen atoms by 432 kJ/mol.

Since the hydrogen molecule consists of protons and electrons, the same components found in separated hydrogen atoms, it seems reasonable to use a theory similar to the atomic theory discussed in Chapter 2, which assumes that the electrons in an atom exist in orbitals of a given energy. Can we apply this same type of model to the hydrogen molecule? Yes. In fact, describing the H_2 molecule in terms of quantum mechanics is quite straightforward.

However, even though it is formulated rather easily, this problem cannot be solved exactly. The difficulty is the same as that in dealing with polyelectronic atoms—the electron correlation problem. Since we do not know the details of the electron movements, we cannot deal with the electron–electron interactions in a specific way. We need to make approximations that allow a solution of the problem but do not destroy the model's physical integrity. The success of these approximations can be measured only by comparing predictions based on theory with experimental observations. In this case we will see that the simplified model works very well.

Just as atomic orbitals are solutions to the quantum mechanical treatment of atoms, **molecular orbitals (MOs)** are solutions to the molecular problem. Molecular orbitals have many of the same characteristics as atomic orbitals. Two of the most important are that they can hold two electrons with opposite spins and that the square of the molecular orbital wave function indicates electron probability.

We will now describe the bonding in the hydrogen molecule, using this model. The first step is to obtain the hydrogen molecule's orbitals, a process that is greatly simplified if we assume that the molecular orbitals can be constructed from the hydrogen $1s$ atomic orbitals.

When the quantum mechanical equations for the hydrogen molecule are solved, two molecular orbitals result, which can be represented as

$$MO_1 = 1s_A + 1s_B$$
$$MO_2 = 1s_A - 1s_B$$

where $1s_A$ and $1s_B$ represent the $1s$ orbitals from the two separated hydrogen atoms. This process is shown schematically in Fig. 4-37.

The orbital properties of most interest are size, shape (described by the electron probability distribution), and energy. These properties for the hydrogen molecular orbitals are represented in Fig. 4-38. From Fig. 4-38 we can note several important points:

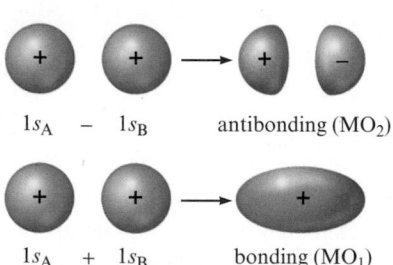

$1s_A$ － $1s_B$ antibonding (MO_2)

$1s_A$ ＋ $1s_B$ bonding (MO_1)

Figure 4-37 | The combination of hydrogen $1s$ atomic orbitals to form MOs. The phases of the orbitals are shown by signs inside the boundary surfaces. When the orbitals are added, the matching phases produce constructive interference, which gives enhanced electron probability between the nuclei. This results in a bonding molecular orbital. When one orbital is subtracted from the other, destructive interference occurs between the opposite phases, leading to a node between the nuclei. This is an antibonding MO.

1. The electron probability of both molecular orbitals is centered along the line passing through the two nuclei. For MO_1 the greatest electron probability is *between* the nuclei, and for MO_2 it is on *either side* of the nuclei. This type of electron distribution is described as *sigma* (σ), as in the localized electron model. Accordingly, we refer to MO_1 and MO_2 as **sigma (σ) molecular orbitals**.

2. In the molecule only the molecular orbitals are available for occupation by electrons. The $1s$ atomic orbitals of the hydrogen atoms no longer exist, because the H_2 molecule—a new entity—has its own set of new orbitals.

Figure 4-38 | (a) The MO energy-level diagram for the H_2 molecule. (b) The shapes of the MOs are obtained by squaring the wave functions for MO_1 and MO_2. The positions of the nuclei are indicated by •.

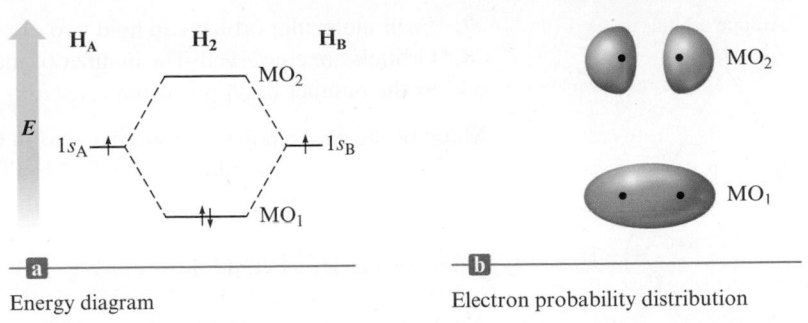

Energy diagram

Electron probability distribution

3. MO_1 is lower in energy than the $1s$ orbitals of free hydrogen atoms, while MO_2 is higher in energy than the $1s$ orbitals. This fact has very important implications for the stability of the H_2 molecule, since if the two electrons (one from each hydrogen atom) occupy the lower-energy MO_1, they will have lower energy than they do in the two separate hydrogen atoms. This situation favors molecule formation, because nature tends to seek the lowest energy state. That is, the driving force for molecule formation is that the molecular orbital available to the two electrons has lower energy than the atomic orbitals these electrons occupy in the separated atoms. This situation is favorable to bonding, or *probonding*.

On the other hand, if the two electrons were forced to occupy the higher-energy MO_2, they would be definitely *antibonding*. In this case, these electrons would have lower energy in the separated atoms than in the molecule, and the separated state would be favored. Of course, since the lower-energy MO_1 *is* available, the two electrons occupy that MO and the molecule is stable.

We have seen that the molecular orbitals of the hydrogen molecule fall into two classes: bonding and antibonding. A **bonding molecular orbital** is *lower in energy than the atomic orbitals of which it is composed*. Electrons in this type of orbital will favor the molecule; that is, they will favor bonding. An **antibonding molecular orbital** is *higher in energy than the atomic orbitals of which it is composed*. Electrons in this type of orbital will favor the separated atoms (they are antibonding). Fig. 4-39 illustrates these ideas.

4. Fig. 4-38 shows that for the bonding molecular orbital in the H_2 molecule the electrons have the greatest probability of being between the nuclei. This is exactly what we would expect, since the electrons can lower their energies by being simultaneously attracted by both nuclei. On the other hand, the electron distribution for the antibonding molecular orbital is such that the electrons are mainly outside the space between the nuclei. This type of distribution is not expected to provide any bonding force. In fact, it causes the electrons to be higher in energy than in the separated atoms. Thus the molecular orbital model produces electron distributions and energies that agree with our basic ideas of bonding. This fact reassures us that the model is physically reasonable. ◄ ⓘ

5. The labels on molecular orbitals indicate their symmetry (shape), the parent atomic orbitals, and whether they are bonding or antibonding. Antibonding character is indicated by an asterisk. For the H_2 molecule, both MOs have σ symmetry, and both are constructed from hydrogen $1s$ atomic orbitals. The molecular orbitals for H_2 are therefore labeled as follows:

$$MO_1 = \sigma_{1s}$$
$$MO_2 = \sigma_{1s}{}^*$$

6. Molecular electron configurations can be written in much the same way as atomic (electron) configurations. Since the H_2 molecule has two electrons in the σ_{1s} molecular orbital, the electron configuration is $\sigma_{1s}{}^2$.

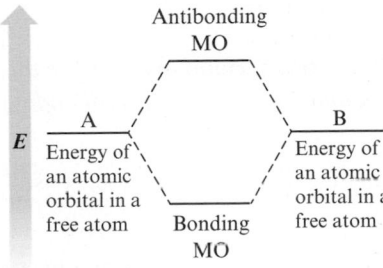

Figure 4-39 | Bonding and antibonding molecular orbitals (MOs).

ⓘ Bonding will result if the molecule has lower energy than the separated atoms.

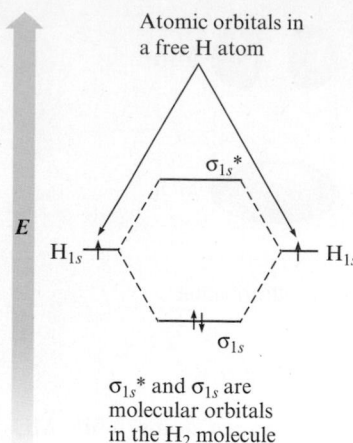

Atomic orbitals in
a free H atom

σ_{1s}^* and σ_{1s} are
molecular orbitals
in the H_2 molecule

Figure 4-40 | A molecular
orbital energy-level diagram for the
H_2 molecule.

ⓘ Although the model predicts that
H_2^- should be stable, this ion has
never been observed, again emphasizing the perils of simple models.

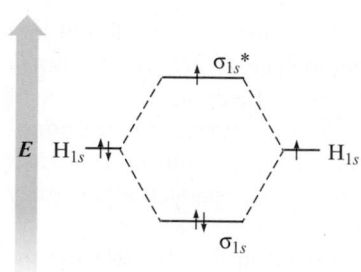

Figure 4-41 | The molecular
orbital energy-level diagram for the
H_2^- ion.

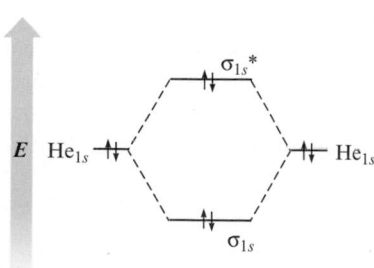

Figure 4-42 | The molecular
orbital energy-level diagram for the
He_2 molecule.

7. Each molecular orbital can hold two electrons, but the spins must be opposite.
8. Orbitals are conserved. The number of molecular orbitals will always be the same as the number of atomic orbitals used to construct them.

Many of the above points are summarized in Fig. 4-40.

Now suppose we could form the H_2^- ion from a hydride ion (H^-) and a hydrogen atom. Would this species be stable? Since the H^- ion has the configuration $1s^2$ and the H atom has a $1s^1$ configuration, we will use $1s$ atomic orbitals to construct the MO diagram for the H_2^- ion, as shown in Fig. 4-41. The electron configuration for H_2^- is $(\sigma_{1s})^2(\sigma_{1s}^*)^1$.

The key idea is that the H_2^- ion will be stable if it has a lower energy than its separated parts. From Fig. 4-41 we see that in going from the separated H^- ion and H atom to the H_2^- ion, the model predicts that two electrons are lowered in energy and one electron is raised in energy. In other words, two electrons are bonding and one electron is antibonding. Since more electrons favor bonding, H_2^- is predicted to be a stable entity—a bond has formed. But how would we expect the bond strengths in the molecules of H_2 and H_2^- to compare?

In the formation of the H_2 molecule, two electrons are lowered in energy and no electrons are raised in energy compared with the parent atoms. When H_2^- is formed, two electrons are lowered in energy and one is raised, producing *a net lowering of the energy of only one electron*. Thus the model predicts that H_2 is *twice as stable* as H_2^- with respect to their separated components. In other words, the bond in the H_2 molecule is predicted to be about twice as strong as the bond in the H_2^- ion. ◄ ⓘ

Bond Order

To indicate bond strength, we use the concept of bond order. **Bond order** is *the difference between the number of bonding electrons and the number of antibonding electrons divided by 2.*

$$\text{Bond order} = \frac{\text{number of bonding electrons} - \text{number of antibonding electrons}}{2}$$

We divide by 2 because, from the localized electron model, we are used to thinking of bonds in terms of *pairs* of electrons.

Since the H_2 molecule has two bonding electrons and no antibonding electrons, the bond order is

$$\text{Bond order} = \frac{2 - 0}{2} = 1$$

The H_2^- ion has two bonding electrons and one antibonding electron; the bond order is

$$\text{Bond order} = \frac{2 - 1}{2} = \frac{1}{2}$$

Bond order is an indication of bond strength because it reflects the difference between the number of bonding electrons and the number of antibonding electrons. *Larger bond order means greater bond strength.*

We will now apply the molecular orbital model to the helium molecule (He_2). Does this model predict that this molecule will be stable? Since the He atom has a $1s^2$ configuration, $1s$ orbitals are used to construct the molecular orbitals, and the molecule will have four electrons. From the diagram shown in Fig. 4-42 it is apparent that two electrons are raised in energy and two are lowered in energy. Thus the bond order is zero:

$$\frac{2 - 2}{2} = 0$$

This implies that the He_2 molecule is *not* stable with respect to the two free He atoms, which agrees with the observation that helium gas consists of individual He atoms.

4-5 | Bonding in Homonuclear Diatomic Molecules

In this section we consider *homonuclear diatomic molecules* (those composed of two identical atoms) of elements in Period 2 of the periodic table. Since the lithium atom has a $1s^2 2s^1$ electron configuration, it would seem that we should use the Li $1s$ and $2s$ orbitals to form the molecular orbitals of the Li_2 molecule. However, the $1s$ orbitals on the lithium atoms are much smaller than the $2s$ orbitals and therefore do not overlap in space to any appreciable extent (Fig. 4-43). Thus the two electrons in each $1s$ orbital can be assumed to be localized and not to participate in the bonding. *To participate in molecular orbitals, atomic orbitals must overlap in space.* This means that only the valence orbitals of the atoms contribute significantly to the molecular orbitals of a particular molecule.

The molecular orbital diagram of the Li_2 molecule and the shapes of its bonding and antibonding MOs are shown in Fig. 4-44. The electron configuration for Li_2 (valence electrons only) is $\sigma_{2s}^{\ 2}$, and the bond order is

$$\frac{2 - 0}{2} = 1$$

Li_2 is a stable molecule (has lower energy than two separated lithium atoms). However, this does not mean that Li_2 is the most stable form of elemental lithium. In fact, at normal temperature and pressure, lithium exists as a solid containing many lithium atoms bound to each other.

For the beryllium molecule (Be_2), the bonding and antibonding orbitals both contain two electrons. In this case the bond order is $(2 - 2)/2 = 0$, and since Be_2 is not more stable than two separated Be atoms, no molecule forms. However, beryllium metal contains many beryllium atoms bonded to each other and is stable for reasons we will discuss in Chapter 9.

Since the boron atom has a $1s^2 2s^2 2p^1$ configuration, we describe the B_2 molecule by considering how p atomic orbitals combine to form molecular orbitals. Recall that p orbitals have two lobes and that they occur in sets of three mutually perpendicular orbitals [see Fig. 4-45(a)]. When two B atoms approach each other, two pairs of p orbitals can overlap in a parallel fashion [Fig. 4-45(b) and (c)] and one pair can overlap head-on [Fig. 4-45(d)].

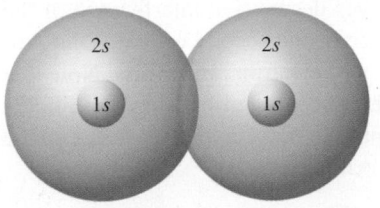

Figure 4-43 | The relative sizes of the lithium $1s$ and $2s$ atomic orbitals.

Beryllium metal.

Ken O'Donoghue © Cengage Learning

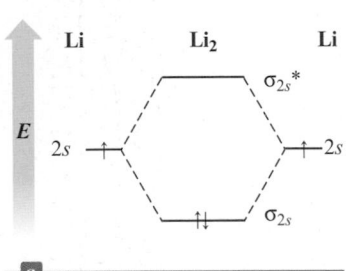

Energy diagram

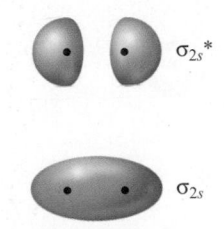

MO Shapes

Figure 4-44 | The molecular orbital energy-level diagram for the Li_2 molecule.

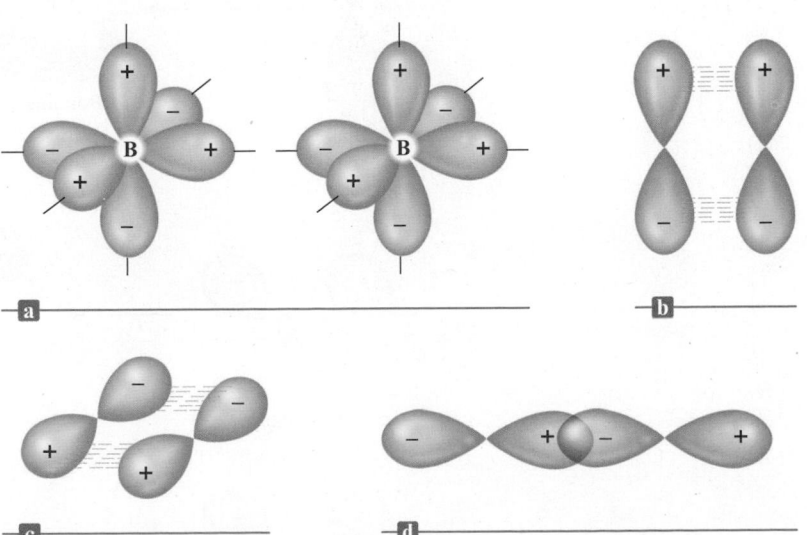

Figure 4-45 | (a) The three mutually perpendicular $2p$ orbitals on two adjacent boron atoms. The signs indicate the orbital phases. Two pairs of parallel p orbitals can overlap, as shown in (b) and (c), and the third pair can overlap head-on, as shown in (d).

First, let's consider the molecular orbitals from the head-on overlap. The bonding orbital is formed by reversing the sign of the right orbital so the positive phases of both orbitals match between the nuclei to produce constructive interference. This leads to enhanced electron probability between the nuclei. The antibonding orbital is formed by the direct combination of the orbitals, which gives destructive interference of the positive phase of one orbital with the negative phase of the second orbital. This produces a node between the nuclei, which gives decreased electron probability. Note that the electrons in the bonding MO are, as expected, concentrated between the nuclei, and the electrons in the antibonding MO are concentrated outside the area between the two nuclei. Also, both these MOs are σ molecular orbitals.

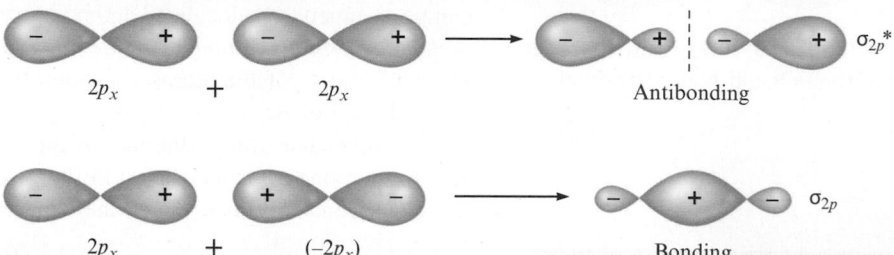

When the parallel p orbitals are combined with the positive and negative phases matched, constructive interference occurs, giving a bonding π orbital. When the orbitals have opposite phases (the signs of one orbital are reversed), destructive interference occurs, resulting in an antibonding π orbital.

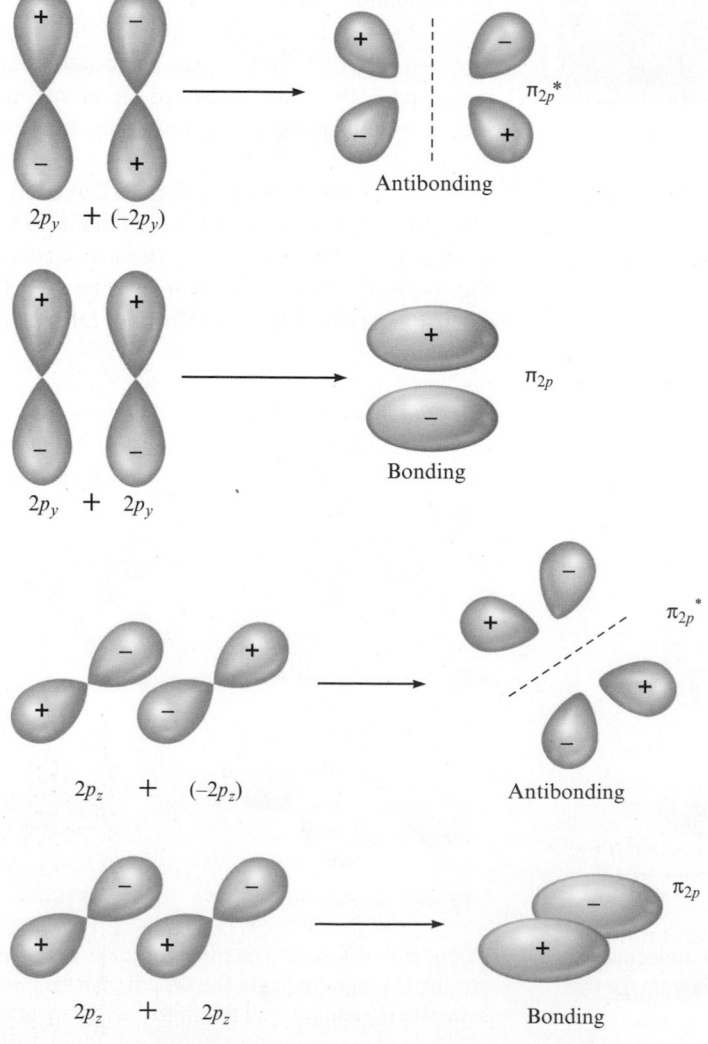

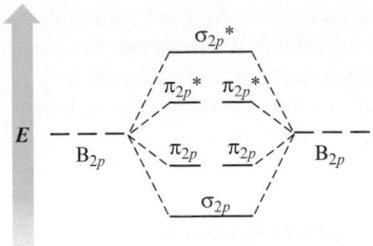

Figure 4-46 | The *expected* molecular orbital energy-level diagram resulting from the combination of the 2p orbitals on two boron atoms.

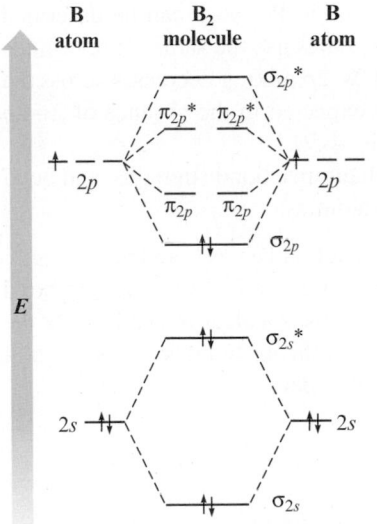

Figure 4-47 | The *expected* molecular orbital energy-level diagram for the B_2 molecule.

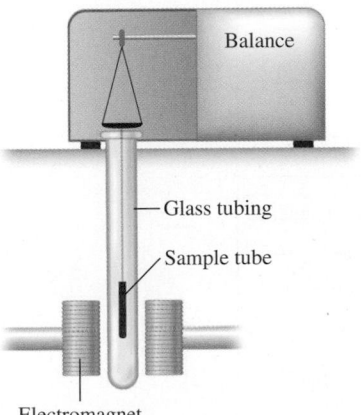

Figure 4-48 | Diagram of the kind of apparatus used to measure the paramagnetism of a sample. A paramagnetic sample will appear heavier when the electromagnet is turned on because the sample is attracted into the inducing magnetic field.

Since the electron probability lies above and below the line between the nuclei, both the orbitals are **pi (π) molecular orbitals**. They are designated as π_{2p} for the bonding MO and $\pi_{2p}{}^*$ for the antibonding MO. A similar set of π molecular orbitals is formed from overlap of the parallel p_z atomic orbitals.

Let's try to make an educated guess about the relative energies of the σ and π molecular orbitals formed from the 2p atomic orbitals. Would we expect the electrons to prefer the σ bonding orbital (where the electron probability is concentrated in the area between the nuclei) or the π bonding orbital? The σ orbital would seem to have the lower energy, since the electrons are closest to the two nuclei. This agrees with the observation that σ interactions are stronger than π interactions.

Fig. 4-46 gives the molecular orbital energy-level diagram *expected* when the two sets of 2p orbitals on the boron atoms combine to form molecular orbitals. Note that there are two π bonding orbitals at the same energy (degenerate orbitals) formed from the two pairs of parallel p orbitals, and there are two degenerate π antibonding orbitals. The energy of the π_{2p} orbitals is expected to be higher than that of the σ_{2p} orbital because σ interactions are generally stronger than π interactions.

To construct the total molecular orbital diagram for the B_2 molecule, we make the assumption that the 2s and 2p orbitals combine separately (in other words, there is no 2s–2p mixing). The resulting diagram is shown in Fig. 4-47. Note that B_2 has six *valence* electrons. (Remember the 1s orbitals and electrons are assumed not to participate in the bonding.) This diagram predicts the bond order:

$$\frac{4 - 2}{2} = 1$$

Therefore, B_2 should be a stable molecule.

Paramagnetism

At this point we need to discuss an additional molecular property—magnetism. Most materials have no magnetism until they are placed in a magnetic field. However, in the presence of such a field, magnetism of two types can be induced. **Paramagnetism** causes the substance to be attracted into the inducing magnetic field. **Diamagnetism** causes the substance to be repelled from the inducing magnetic field. Fig. 4-48 illustrates how paramagnetism is measured. The sample is weighed with the electromagnet turned off and then weighed again with the electromagnet turned on. An increase in weight when the field is turned on indicates the sample is paramagnetic. Studies have shown that *paramagnetism is associated with unpaired electrons and diamagnetism is associated with paired electrons*. Any substance that has both paired and unpaired electrons will exhibit a net paramagnetism, since the effect of paramagnetism is much stronger than that of diamagnetism.

The molecular orbital energy-level diagram represented in Fig. 4-47 predicts that the B_2 molecule will be diamagnetic, since the MOs contain only paired electrons. However, experiments show that B_2 is actually paramagnetic with two unpaired electrons. Why does the model yield the wrong prediction? This is yet another illustration of how models are developed and used. In general, we try to use the simplest possible model that accounts for all the important observations. In this case, although the simplest model successfully describes the diatomic molecules up to B_2, it certainly is suspect if it cannot describe the B_2 molecule correctly. This means we must either discard the model or find a way to modify it.

Let's consider one assumption that we made. In our treatment of B_2, we have assumed that the s and p orbitals combine separately to form molecular orbitals. Calculations show that when the s and p orbitals are allowed to mix in the same molecular orbital, a different energy-level diagram results for B_2 (Fig. 4-49). Note that even though the s and p contributions to the MOs are no longer separate, we retain the simple orbital designations. The energies of π_{2p} and σ_{2p} orbitals are reversed by p–s mixing,

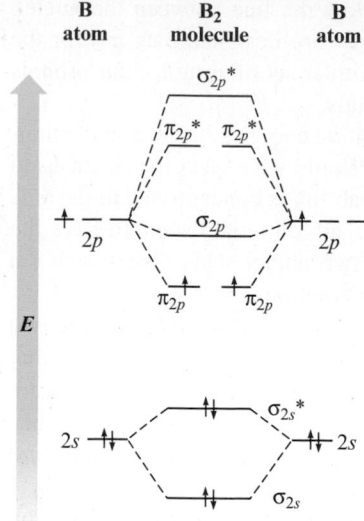

Figure 4-49 | The *correct* molecular orbital energy-level diagram for the B_2 molecule. When *p–s* mixing is allowed, the energies of the σ_{2p} and π_{2p} orbitals are reversed. The two electrons from the B $2p$ orbitals now occupy separate, degenerate π_{2p} molecular orbitals and thus have parallel spins. Therefore, this diagram explains the observed paramagnetism of B_2.

and the σ_{2s} and the σ_{2s}^* orbitals are no longer equally spaced relative to the energy of the free $2s$ orbital.

When the six valence electrons for the B_2 molecule are placed in the modified energy-level diagram, each of the last two electrons goes into one of the degenerate π_{2p} orbitals. This produces a paramagnetic molecule in agreement with experimental results. Thus, when the model is extended to allow *p–s* mixing in molecular orbitals, it predicts the correct magnetism. Note that the bond order is $(4 - 2)/2 = 1$, as before.

The remaining diatomic molecules of the elements in Period 2 can be described using similar ideas. For example, the C_2 and N_2 molecules use the same set of orbitals as for B_2 (see Fig. 4-49). Because the importance of $2s$–$2p$ mixing decreases across the period, the σ_{2p} and π_{2p} orbitals revert to the order expected in the absence of $2s$–$2p$ mixing for the molecules O_2 and F_2, as shown in Fig. 4-50.

Several significant points arise from the orbital diagrams, bond strengths, and bond lengths summarized in Fig. 4-50 for the Period 2 diatomics:

1. There are definite correlations between bond order, bond energy, and bond length. As the bond order predicted by the molecular orbital model increases, the bond energy increases and the bond length decreases. This is a clear indication that the bond order predicted by the model accurately reflects bond strength, and it strongly supports the reasonableness of the MO model.

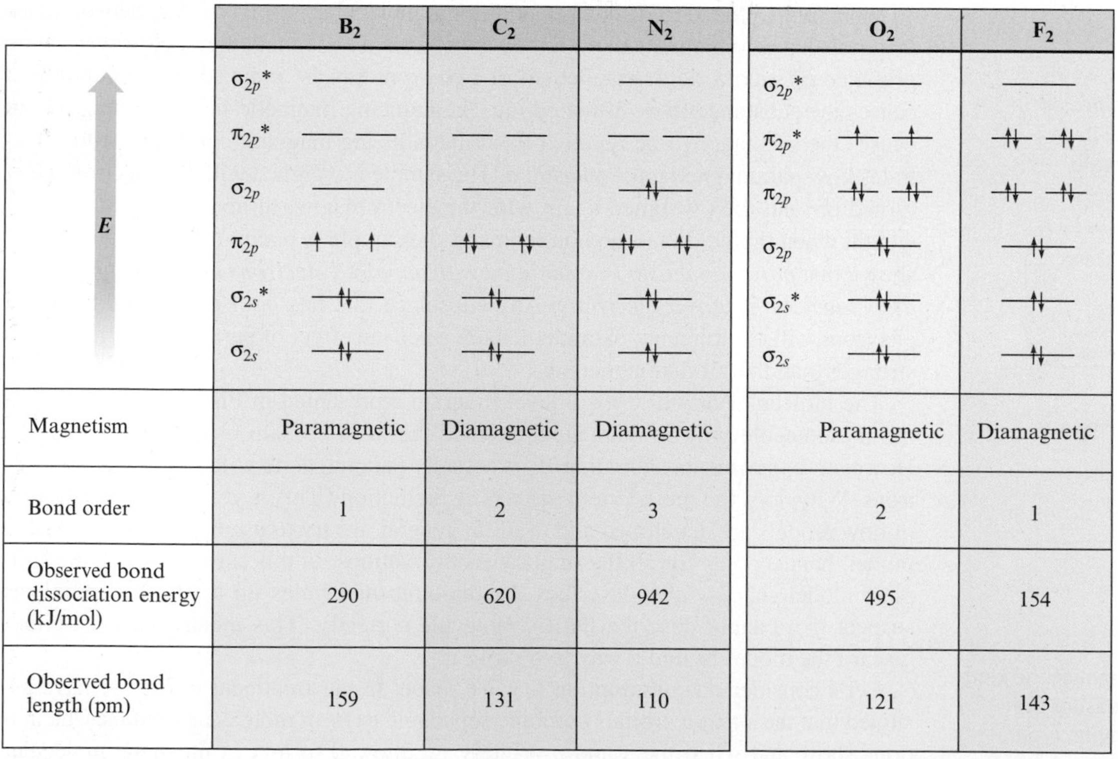

ⓘ For N_2, O_2, and F_2, notice how the bond order matches the number of bonds in the Lewis structure. The molecular orbital model does not contradict the simpler localized electron model.

	B_2	C_2	N_2		O_2	F_2
σ_{2p}^*	——	——	——	σ_{2p}^*	——	——
π_{2p}^*	— —	— —	— —	π_{2p}^*	↿ ↿	⇅ ⇅
σ_{2p}	——	——	⇅	π_{2p}	⇅ ⇅	⇅ ⇅
π_{2p}	↿ ↿	⇅ ⇅	⇅ ⇅	σ_{2p}	⇅	⇅
σ_{2s}^*	⇅	⇅	⇅	σ_{2s}^*	⇅	⇅
σ_{2s}	⇅	⇅	⇅	σ_{2s}	⇅	⇅
Magnetism	Paramagnetic	Diamagnetic	Diamagnetic		Paramagnetic	Diamagnetic
Bond order	1	2	3		2	1
Observed bond dissociation energy (kJ/mol)	290	620	942		495	154
Observed bond length (pm)	159	131	110		121	143

Figure 4-50 | The molecular orbital energy-level diagrams, bond orders, bond energies, and bond lengths for the diatomic molecules B_2 through F_2. Note that for O_2 and F_2 the σ_{2p} orbital is lower in energy than the π_{2p} orbitals. ◄ ⓘ

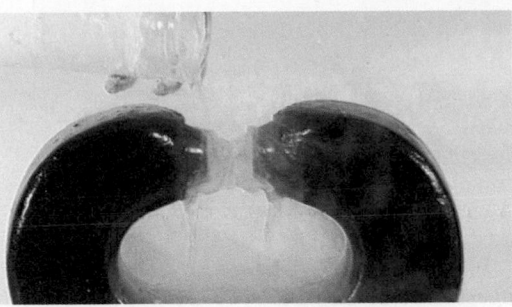
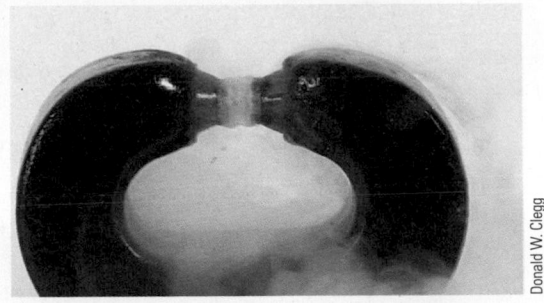

Donald W. Clegg

Figure 4-51 | When liquid oxygen is poured into the space between the poles of a strong magnet, it remains there until it boils away. This attraction of liquid oxygen for the magnetic field demonstrates the paramagnetism of the O_2 molecule.

2. Comparison of the bond energies of the B_2 and F_2 molecules indicates that bond order cannot automatically be associated with a particular bond energy. Although both molecules have a bond order of 1, the bond in B_2 appears to be about twice as strong as the bond in F_2. As we will see in our later discussion of the halogens, F_2 has an unusually weak single bond due to larger than usual electron–electron repulsions (there are 14 valence electrons on the small F_2 molecule).

3. Note the very large bond energy associated with the N_2 molecule, which the molecular orbital model predicts will have a bond order of 3, a triple bond. The very strong bond in N_2 is the principal reason that many nitrogen-containing compounds are used as high explosives. The reactions involving these explosives give the very stable N_2 molecule as a product, thus releasing large quantities of energy.

4. The O_2 molecule is known to be paramagnetic. This can be very convincingly demonstrated by pouring liquid oxygen between the poles of a strong magnet, as shown in Fig. 4-51. The oxygen remains there until it evaporates. Significantly, the molecular orbital model correctly predicts oxygen's paramagnetism, while the localized electron model predicts a diamagnetic molecule.

INTERACTIVE EXAMPLE 4-11

The Molecular Orbital Model I

For the species O_2, O_2^+, and O_2^-, give the electron configuration and the bond order for each. Which has the strongest bond?

Solution

The O_2 molecule has 12 valence electrons $(6 + 6)$; O_2^+ has 11 valence electrons $(6 + 6 - 1)$; and O_2^- has 13 valence electrons $(6 + 6 + 1)$. We will assume that the ions can be treated with the same molecular orbital diagram as for the neutral diatomic molecule:

	O_2	**O_2^+**	**O_2^-**
σ_{2p}^*	—	—	—
π_{2p}^*	↑ ↑	↑ —	↑↓ ↑
π_{2p}	↑↓ ↑↓	↑↓ ↑↓	↑↓ ↑↓
σ_{2p}	↑↓	↑↓	↑↓
σ_{2s}^*	↑↓	↑↓	↑↓
σ_{2s}	↑↓	↑↓	↑↓

The electron configuration for each species can then be taken from the diagram:

$$O_2: \quad (\sigma_{2s})^2(\sigma_{2s}^*)^2(\sigma_{2p})^2(\pi_{2p})^4(\pi_{2p}^*)^2$$
$$O_2^+: \quad (\sigma_{2s})^2(\sigma_{2s}^*)^2(\sigma_{2p})^2(\pi_{2p})^4(\pi_{2p}^*)^1$$
$$O_2^-: \quad (\sigma_{2s})^2(\sigma_{2s}^*)^2(\sigma_{2p})^2(\pi_{2p})^4(\pi_{2p}^*)^3$$

The bond orders are:

$$\text{For } O_2: \quad \frac{8-4}{2} = 2$$

$$\text{For } O_2^{+}: \quad \frac{8-3}{2} = 2.5$$

$$\text{For } O_2^{-}: \quad \frac{8-5}{2} = 1.5$$

Thus O_2^{+} is expected to have the strongest bond of the three species.

See Exercises 4-67 and 4-68

INTERACTIVE EXAMPLE 4-12

The Molecular Orbital Model II

Use the molecular orbital model to predict the bond order and magnetism of each of the following molecules.

a. Ne_2 **b.** P_2

Solution

a. The valence orbitals for Ne are $2s$ and $2p$. Thus we can use the molecular orbitals we have already constructed for the diatomic molecules of the Period 2 elements. The Ne_2 molecule has 16 valence electrons (8 from each atom). Placing these electrons in the appropriate molecular orbitals produces the following diagram:

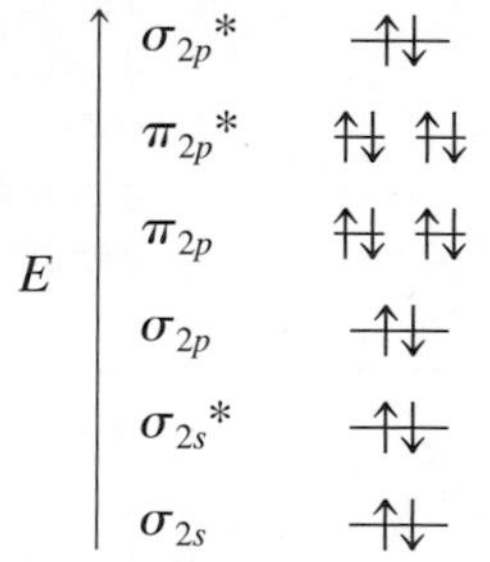

The bond order is $(8 - 8)/2 = 0$, and Ne_2 does not exist.

b. The P_2 molecule contains phosphorus atoms from the third row of the periodic table. We will assume that the diatomic molecules of the Period 3 elements can be treated in a way very similar to that which we have used so far. Thus we will draw the MO diagram for P_2 analogous to that for N_2. The only change will be that the molecular orbitals will be formed from $3s$ and $3p$ atomic orbitals. The P_2 model has 10 valence electrons (5 from each phosphorus atom). The resulting molecular orbital diagram is

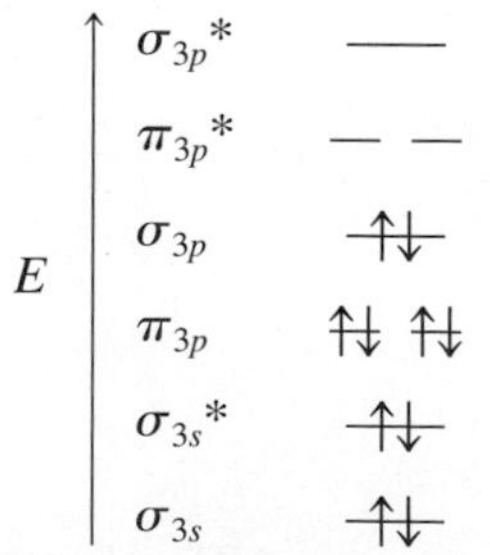

The molecule has a bond order of 3 and is expected to be diamagnetic.

See Exercises 4-65 through 4-70

4-6 | Bonding in Heteronuclear Diatomic Molecules

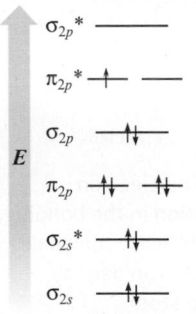

Figure 4-52 | The molecular orbital energy-level diagram for the NO molecule. We assume that orbital order is the same as that for N_2. The bond order is 2.5.

In this section we will deal with selected examples of **heteronuclear** (different atoms) **diatomic molecules**. A special case involves molecules containing atoms adjacent to each other in the periodic table. Since the atoms involved in such a molecule are so similar, we can use the molecular orbital diagram for homonuclear molecules. For example, we can predict the bond order and magnetism of nitric oxide (NO) by placing its 11 valence electrons (5 from nitrogen and 6 from oxygen) in the molecular orbital energy-level diagram shown in Fig. 4-52. The molecule should be paramagnetic and has a bond order of

$$\frac{8-3}{2} = 2.5$$

Experimentally, nitric oxide is indeed found to be paramagnetic. Notice that this odd-electron molecule is described very naturally by the MO model. In contrast, the localized electron model, in the simple form used in this text, cannot be used readily to treat such molecules.

INTERACTIVE EXAMPLE 4-13

The Molecular Orbital Model III

Use the molecular orbital model to predict the magnetism and bond order of the NO^+ and CN^- ions.

Solution

The NO^+ ion has 10 valence electrons $(5 + 6 - 1)$. The CN^- ion also has 10 valence electrons $(4 + 5 + 1)$. Both ions are therefore diamagnetic and have a bond order derived from the equation

$$\frac{8-2}{2} = 3$$

The molecular orbital diagram for these two ions is the same (Fig. 4-53).

See Exercises 4-71 and 4-72

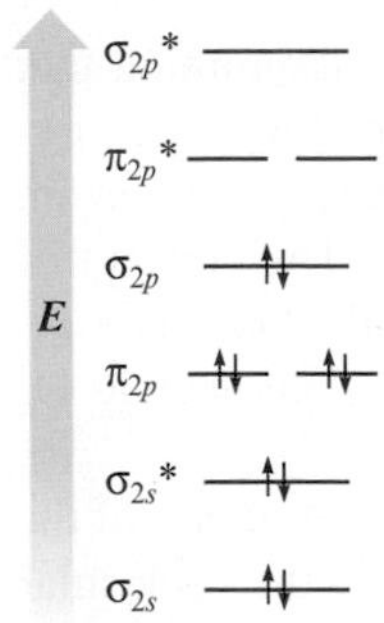

Figure 4-53 | The molecular orbital energy-level diagram for both the NO^+ and CN^- ions.

When the two atoms of a diatomic molecule are very different, the energy-level diagram for homonuclear molecules can no longer be used. A new diagram must be devised for each molecule. We will illustrate this case by considering the hydrogen fluoride (HF) molecule. The electron configurations of the hydrogen and fluorine atoms are $1s^1$ and $1s^22s^22p^5$, respectively. To keep things as simple as possible, we will assume that fluorine uses only one of its $2p$ orbitals to bond to hydrogen. Thus the molecular orbitals for HF will be composed of fluorine $2p$ and hydrogen $1s$ orbitals. Fig. 4-54 gives the partial molecular orbital energy-level diagram for HF, focusing only on the orbitals involved in the bonding. We are assuming that fluorine's other valence electrons remain localized on the fluorine. The $2p$ orbital of fluorine is shown at a lower energy than the $1s$ orbital of hydrogen on the diagram because fluorine binds its valence electrons more tightly. Thus the $2p$ electron on a free fluorine atom is at lower energy than the $1s$ electron on a free hydrogen atom. The diagram predicts that the HF molecule should be stable because both electrons are lowered in energy relative to their energy in the free hydrogen and fluorine atoms, which is the driving force for bond formation.

Because the fluorine $2p$ orbital is lower in energy than the hydrogen $1s$ orbital, the electrons prefer to be closer to the fluorine atom. That is, the σ molecular orbital

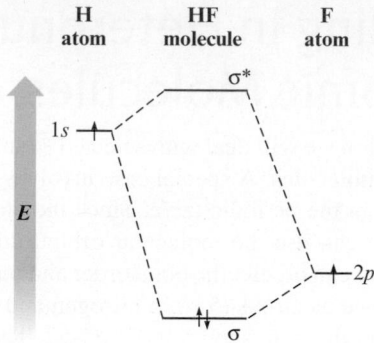

Figure 4-54 │ A partial molecular orbital energy-level diagram for the HF molecule.

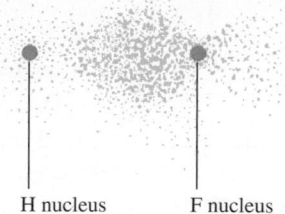

Figure 4-55 │ The electron probability distribution in the bonding molecular orbital of the HF molecule. Note the greater electron density close to the fluorine atom.

ⓘ The molecular orbital model is consistent with our predictions of bond polarity using electronegativity values from Chapter 3.

containing the bonding electron pair shows greater electron probability close to the fluorine (Fig. 4-55). The electron pair is not shared equally. This causes the fluorine atom to have a slight excess of negative charge and leaves the hydrogen atom partially positive. This is *exactly* the bond polarity observed for HF. Thus the molecular orbital model accounts in a straightforward way for the different electronegativities of hydrogen and fluorine and the resulting unequal charge distribution. ◄ ⓘ

4-7 │ Combining the Localized Electron and Molecular Orbital Models

One of the main difficulties with the localized electron model is its assumption that electrons are localized. This problem is most apparent with molecules for which several valid Lewis structures can be drawn. It is clear that none of these structures taken alone adequately describes the electronic structure of the molecule. The concept of resonance was invented to solve this problem. However, even with resonance included, the localized electron model does not describe molecules and ions such as O_3 and NO_3^- in a very satisfying way.

It would seem that the ideal bonding model would be one with the simplicity of the localized electron model but with the delocalization characteristic of the molecular orbital model. We can achieve this by combining the two models to describe molecules that require resonance. Note that for species such as O_3 and NO_3^- the double bond changes position in the resonance structures (Fig. 4-56). Since a double bond involves one σ and one π bond, there is a σ bond between all bound atoms in each resonance structure. It is really the π bond that has different locations in the various resonance structures.

ⓘ In molecules that require resonance, it is the π bonding that is most clearly delocalized.

Therefore, we conclude that the σ bonds in a molecule can be described as being localized with no apparent problems. It is the π bonding that must be treated as being delocalized. ◄ ⓘ Thus, for molecules that require resonance, we will use the localized

Figure 4-56 │ The resonance structures for O_3 and NO_3^-. Note that it is the double bond that occupies various positions in the resonance structures.

One of the best things about New Mexico is the food. Authentic New Mexican cuisine employs liberal amounts of green and red chilies—often called *chili peppers*. Chilies apparently originated in parts of South America and were spread north by birds. When Columbus came to North America, which he originally thought was India, he observed the natives using chilies for spicing foods. When he took chilies back to Europe, Columbus mistakenly called them peppers and the name stuck.

The spicy payload of chilies is delivered mainly by the chemical capsaicin, which has the following structure:

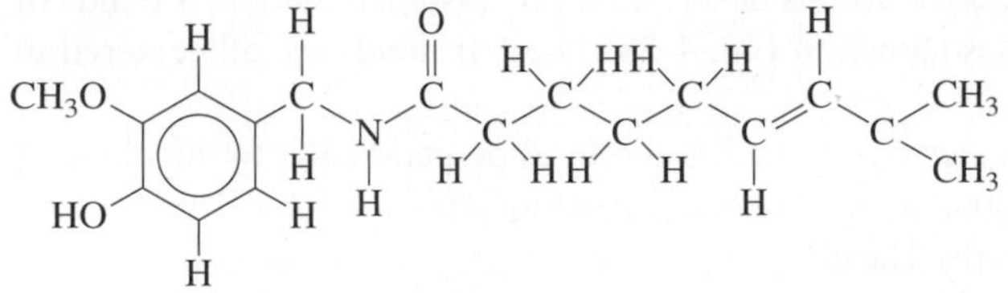

Capsaicin was isolated as a pure substance by L. T. Thresh in 1846. Since then substituted capsaicins have also been found

in chilies. The spicy power of chilies derives mostly from capsaicin and dihydrocapsaicin.

The man best known for explaining the "heat" of chilies is Wilbur Scoville, who defined the Scoville unit for measuring chili power. He arbitrarily established the hotness of pure capsaicin as 16 million. On this scale a typical green or red chili has a rating of about 2500 Scoville units. You may have had an encounter with habanero chilies that left you looking for a firehose to put out the blaze in your mouth—habaneros have a Scoville rating of about 500,000!

Capsaicin has found many uses outside of cooking. It is used in pepper sprays and repellant sprays for many garden pests, although birds are unaffected by capsaicin. Capsaicin also stimulates the body's circulation and causes pain receptors to release endorphins, similar to the effect produced by intense exercise. Instead of jogging you may want to sit on the couch eating chilies. Either way, you are going to sweat.

electron model to describe the σ bonding and the molecular orbital model to describe the π bonding. This allows us to keep the bonding model as simple as possible and yet give a more physically accurate description of such molecules.

We will illustrate the general method by considering the bonding in benzene, an important industrial chemical that must be handled carefully because it is a known carcinogen. The benzene molecule (C_6H_6) consists of a planar hexagon of carbon atoms with one hydrogen atom bound to each carbon atom, as shown in Fig. 4-57(a). In the molecule all six C—C bonds are known to be equivalent. To explain this fact, the localized electron model must invoke resonance [see Fig. 4-57(b)].

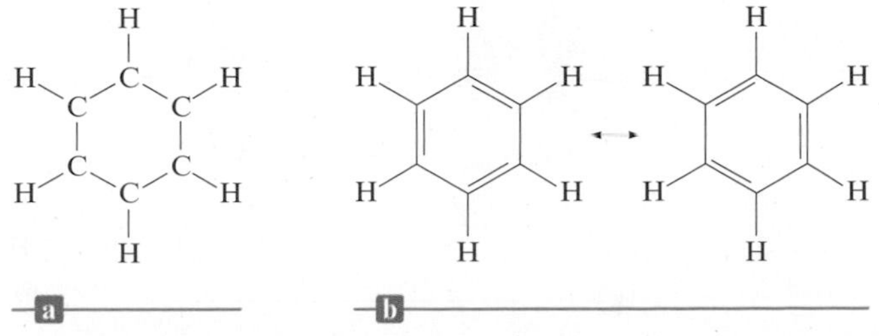

Figure 4-57 | (a) The benzene molecule consists of a ring of six carbon atoms with one hydrogen atom bound to each carbon; all atoms are in the same plane. All the C—C bonds are known to be equivalent. (b) Two of the resonance structures for the benzene molecule. The localized electron model must invoke resonance to account for the six equal C—C bonds.

Figure 4-58 | The σ bonding system in the benzene molecule.

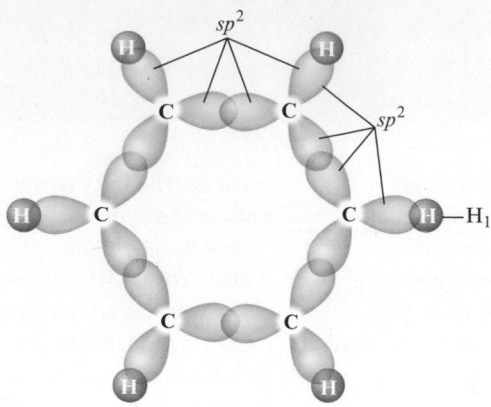

A better description of the bonding in benzene results when we use a combination of the models, as described above. In this description it is assumed that the σ bonds of carbon involve sp^2 orbitals, as shown in Fig. 4-58. These σ bonds are all centered in the plane of the molecule.

Since each carbon atom is sp^2 hybridized, a p orbital perpendicular to the plane of the ring remains on each carbon atom. These six p orbitals [Fig. 4-59(a)] can be used to form six π molecular orbitals. The electrons in the resulting π molecular orbitals are delocalized above and below the plane of the ring [Fig. 4-59(b)]. This gives six equivalent C—C bonds, as required by the known structure of the benzene molecule. The benzene structure is often shown as

to indicate the **delocalized π bonding** in the molecule.

Very similar treatments can be applied to other planar molecules for which resonance is required by the localized electron model. For example, the NO_3^- ion can be described using the π molecular orbital system shown in Fig. 4-60. In this molecule each atom is assumed to be sp^2 hybridized, which leaves one p orbital on each atom perpendicular to the plane of the ion. These p orbitals can combine to form the π molecular orbital system.

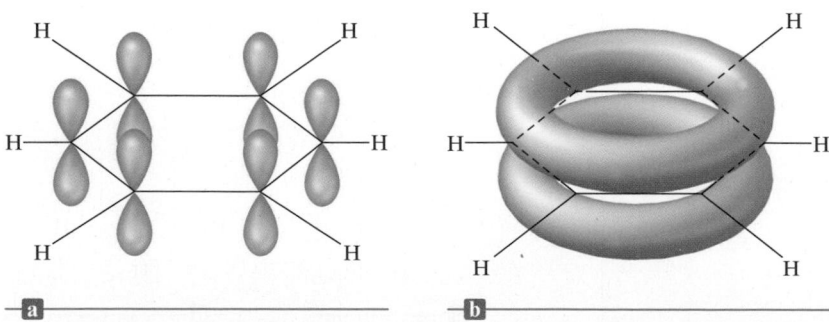

Figure 4-59 | (a) The π molecular orbital system in benzene is formed by combining the six p orbitals from the six sp^2 hybridized carbon atoms. (b) The electrons in the resulting π molecular orbitals are delocalized over the entire ring of carbon atoms, giving six equivalent bonds. A composite of these orbitals is represented here.

Figure 4-60 | (a) The *p* orbitals used to form the π bonding system in the NO_3^- ion. (b) A representation of the delocalization of the electrons in the π molecular orbital system of the NO_3^- ion.

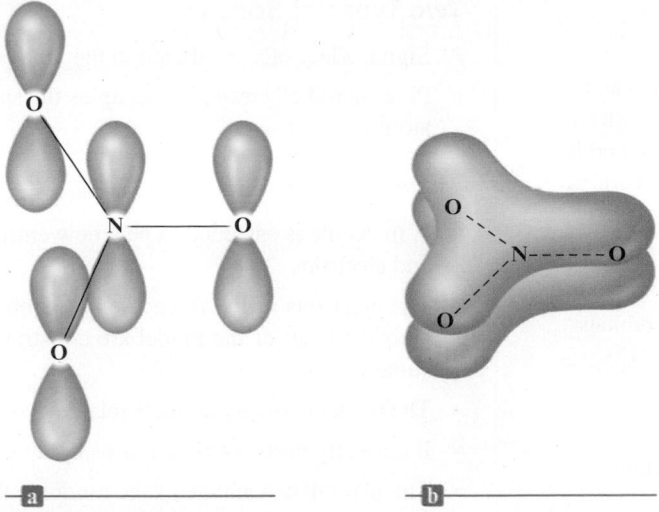

FOR REVIEW

Key Terms

Section 4-1
molecular structure
valence shell electron-pair
    repulsion (VSEPR) model
linear structure
trigonal planar structure
tetrahedral structure
trigonal pyramid
trigonal bipyramid
octahedral structure
square planar structure

Section 4-2
dipolar
dipole moment

Section 4-3
hybridization
sp^3 hybridization
hybrid orbitals
sp^2 hybridization
sigma (σ) bond
pi (π) bond
sp hybridization
dsp^3 hybridization
d^2sp^3 hybridization

VSEPR Model

- Based on the idea that electron pairs will be arranged around a central atom in a way that minimizes the electron repulsions
- Can be used to predict the geometric structure of most molecules

Dipole Moment

- The spatial arrangement of polar bonds in a molecule determines whether the molecule has a dipole moment

Two Widely Used Bonding Models

- Localized electron model
- Molecular orbital model

Localized Electron Model

- Molecule is pictured as a group of atoms sharing electron pairs between atomic orbitals
- Hybrid orbitals, which are combinations of the "native" atomic orbitals, are often required to account for the molecular structure
    - Six electron pairs (octahedral arrangement) require d^2sp^3 orbitals
    - Five electron pairs (trigonal bipyramidal arrangement) require dsp^3 orbitals
    - Four electron pairs (tetrahedral arrangement) require sp^3 orbitals
    - Three electron pairs (trigonal planar arrangement) require sp^2 orbitals
    - Two electron pairs (linear arrangement) require *sp* orbitals

Key Terms

Section 4-4
molecular orbital model
molecular orbitals (MOs)
sigma (σ) molecular orbital
bonding molecular orbital
antibonding molecular orbital
bond order

Section 4-5
pi (π) molecular orbitals
paramagnetism
diamagnetism

Section 4-6
heteronuclear diatomic molecules

Section 4-7
delocalized π bonding

Two Types of Bonds

- Sigma: electrons are shared in the area centered on a line joining the atoms
- Pi: a shared electron pair occupies the space above and below the line joining the atoms

Molecular Orbital Model

- A molecule is assumed to be a new entity consisting of positively charged nuclei and electrons
- The electrons in the molecule are contained in molecular orbitals, which in the simplest form of the model are constructed from the atomic orbitals of the constituent atoms
- The model correctly predicts relative bond strength, magnetism, and bond polarity
- It correctly portrays electrons as being delocalized in polyatomic molecules
- The main disadvantage of the model is that it is difficult to apply qualitatively to polyatomic molecules

Molecular Orbitals Are Classified in Two Ways: Energy and Shape

- Energy
  - A bonding MO is lower in energy than the atomic orbitals from which it is constructed. Electrons in this type of MO are lower in energy in the molecule than in the separated atoms and thus favor molecule formation.
  - An antibonding MO is higher in energy than the atomic orbitals from which it is constructed. Electrons in this type of MO are higher in energy in the molecule than in the separated atoms and thus do not favor molecule formation.
- Shape (symmetry)
  - Sigma (σ) MOs have their electron probability centered on a line passing through the nuclei
  - Pi (π) MOs have their electron probability concentrated above and below the line connecting the nuclei

Bond Order Is an Index of Bond Strength

$$\text{Bond order} = \frac{\text{number of bonding electrons} - \text{number of antibonding electrons}}{2}$$

Molecules That Require the Concept of Resonance in the Localized Electron Model Can Be More Accurately Described by Combining the Localized Electron and Molecular Orbital Models

- The σ bonds are localized
- The π bonds are delocalized

Review Questions

Answers to the Review Questions can be found on the Student Web site (accessible from **www.cengagebrain.com**).

1. Explain the main postulate of the VSEPR model. List the five base geometries (along with bond angles) that most molecules or ions adopt to minimize electron-pair repulsions. Why are bond angles sometimes slightly less than predicted in actual molecules as compared to what is predicted by the VSEPR model?
2. Explain why CF_4 and XeF_4 are nonpolar compounds (have no net dipole moments) while SF_4 is polar (has a net dipole moment). Is CO_2 polar? What about COS?
3. Consider the following compounds: CO_2, SO_2, KrF_2, SO_3, NF_3, IF_3, CF_4, SF_4, XeF_4, PF_5, IF_5, and SCl_6. These 12 compounds are all examples of

different molecular structures. Draw the Lewis structures for each and predict the molecular structures. Predict the bond angles and the polarity of each. (A polar molecule has a net dipole moment, while a nonpolar molecule does not.) See Exercises 25 and 26 for the molecular structures based on the trigonal bipyramid and the octahedral geometries.

4. Why do we hybridize atomic orbitals to explain the bonding in covalent compounds? What type of bonds form from hybrid orbitals: sigma or pi? Explain.

5. What hybridization is required for central atoms that have a tetrahedral arrangement of electron pairs? A trigonal planar arrangement of electron pairs? A linear arrangement of electron pairs? How many unhybridized p atomic orbitals are present when a central atom exhibits tetrahedral geometry? Trigonal planar geometry? Linear geometry? What are the unhybridized p atomic orbitals used for? Describe the bonding in H_2S, CH_4, H_2CO, and HCN, using the localized electron model.

6. What hybridization is required for central atoms exhibiting trigonal bipyramidal geometry? Octahedral geometry? Describe the bonding of PF_5, SF_4, SF_6, and IF_5, using the localized electron model.

7. Electrons in σ bonding molecular orbitals are most likely to be found in the region between the two bonded atoms. Why does this arrangement favor bonding? In a σ antibonding orbital, where are the electrons most likely to be found in relation to the nuclei in a bond?

8. Show how $2s$ orbitals combine to form σ bonding and σ antibonding molecular orbitals. Show how $2p$ orbitals overlap to form σ bonding, π bonding, π antibonding, and σ antibonding molecular orbitals.

9. What are the relationships among bond order, bond energy, and bond length? Which of these can be measured? Distinguish between the terms *paramagnetic* and *diamagnetic*. What type of experiment can be done to determine if a material is paramagnetic?

10. How does molecular orbital theory explain the following observations?

   a. H_2 is stable, while He_2 is unstable.

   b. B_2 and O_2 are paramagnetic, while C_2, N_2, and F_2 are diamagnetic.

   c. N_2 has a very large bond energy associated with it.

   d. NO^+ is more stable than NO^-.

11. Consider the heteronuclear diatomic molecule HF. Explain in detail how molecular orbital theory is applied to describe the bonding in HF.

12. What is delocalized π bonding and what does it explain? Explain the delocalized π bonding system in C_6H_6 (benzene) and O_3 (ozone).

Active Learning Questions

These questions are designed to be used by groups of students in class.

1. What are molecular orbitals? How do they compare with atomic orbitals? Can you tell by the shape of the bonding and antibonding orbitals which is lower in energy? Explain.

2. Explain the difference between the σ and π MOs for homonuclear diatomic molecules. How are bonding and antibonding orbitals different? Why are there two π MOs and one σ MO? Why are the π MOs degenerate?

3. Compare Figs. 4-47 and 4-49. Why are they different? Because B_2 is known to be paramagnetic, the π_{2p} and σ_{2p} molecular orbitals must be switched from the first prediction. What is the rationale for this? Why might one expect the σ_{2p} to be lower in energy than the π_{2p}? Why can't we use diatomic oxygen to help us decide whether the σ_{2p} or π_{2p} is lower in energy?

4. Which of the following would you expect to be more favorable energetically? Explain.

   a. an H_2 molecule in which enough energy is added to excite one electron from the bonding to the antibonding MO

   b. two separate H atoms

5. Arrange the following molecules from most to least polar and explain your order: CH_4, CF_2Cl_2, CF_2H_2, CCl_4, and CCl_2H_2.

6. Which is the more correct statement: "The methane molecule (CH_4) is a tetrahedral molecule because it is sp^3 hybridized" or "The methane molecule (CH_4) is sp^3 hybridized because it is a tetrahedral molecule"? What, if anything, is the difference between these two statements?

7. Compare and contrast the MO model with the local electron model. When is each useful?

8. What are the relationships among bond order, bond energy, and bond length? Which of these quantities can be measured?

A blue question or exercise number indicates that the answer to that question or exercise appears at the back of this book and a solution appears in the *Student Solutions Manual*.

Questions

9. Which of the following statements is/are *true*? Correct the false statements.

   a. The molecules SeS_3, SeS_2, PCl_5, $TeCl_4$, ICl_3, and $XeCl_2$ all exhibit at least one bond angle which is approximately 120°.

197

b. The bond angle in SO_2 should be similar to the bond angle in CS_2 or SCl_2.

c. Of the compounds CF_4, KrF_4, and SeF_4, only SeF_4 exhibits an overall dipole moment (is polar).

d. Central atoms in a molecule adopt a geometry of the bonded atoms and lone pairs about the central atom in order to maximize electron repulsions.

10. Give one example of a compound having a linear molecular structure that has an overall dipole moment (is polar) and one example that does not have an overall dipole moment (is non-polar). Do the same for molecules that have trigonal planar and tetrahedral molecular structures.

11. In the hybrid orbital model, compare and contrast σ bonds with π bonds. What orbitals form the σ bonds and what orbitals form the π bonds? Assume the z-axis is the internuclear axis.

12. Write the name of each of the following molecular structures.

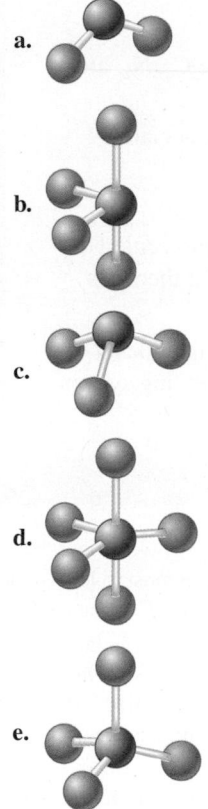

a.

b.

c.

d.

e.

13. Give the expected hybridization for the molecular structures illustrated in the previous question.

14. In the molecular orbital model, compare and contrast σ bonds with π bonds. What orbitals form the σ bonds and what orbitals form the π bonds? Assume the z-axis is the internuclear axis.

15. Why are d orbitals sometimes used to form hybrid orbitals? Which period of elements does not use d orbitals for hybridization? If necessary, which d orbitals ($3d$, $4d$, $5d$, or $6d$) would sulfur use to form hybrid orbitals requiring d atomic orbitals? Answer the same question for arsenic and for iodine.

16. The atoms in a single bond can rotate about the internuclear axis without breaking the bond. The atoms in a double and triple bond cannot rotate about the internuclear axis unless the bond is broken. Why?

17. Compare and contrast bonding molecular orbitals with anti-bonding molecular orbitals.

18. What modification to the molecular orbital model was made from the experimental evidence that B_2 is paramagnetic?

19. Why does the molecular orbital model do a better job in explaining the bonding in NO^- and NO than the hybrid orbital model?

20. The three NO bonds in NO_3^- are all equivalent in length and strength. How is this explained even though any valid Lewis structure for NO_3^- has one double bond and two single bonds to nitrogen?

Exercises

In this section, similar exercises are paired.

Molecular Structure and Polarity

21. Predict the molecular structure (including bond angles) for each of the following.

a. SeO_3

b. SeO_2

22. Predict the molecular structure (including bond angles) for each of the following.

a. PCl_3

b. SCl_2

c. SiF_4

23. Predict the molecular structure and bond angles for each molecule or ion in Exercises 81 and 87 from Chapter 3.

24. Predict the molecular structure and bond angles for each molecule or ion in Exercises 82 and 88 from Chapter 3.

25. There are several molecular structures based on the trigonal bipyramid geometry (see Table 4-3). Three such structures are

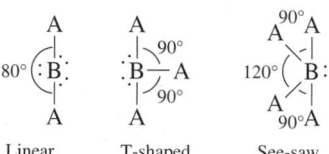

Linear   T-shaped   See-saw

Which of the compounds/ions Br_3^-, ClF_3, XeF_4, SF_4, PF_5, ClF_5, and SF_6 have these molecular structures?

26. Two variations of the octahedral geometry (see Table 4-1) are illustrated below.

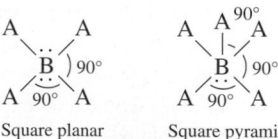

Square planar   Square pyramid

Which of the compounds/ions Br_3^-, ClF_3, XeF_4, SF_4, PF_5, ClF_5, and SF_6 have these molecular structures?

27. Predict the molecular structure (including bond angles) for each of the following. (See Exercises 25 and 26.)
    a. $XeCl_2$
    b. ICl_3
    c. TeF_4
    d. PCl_5
28. Predict the molecular structure (including bond angles) for each of the following. (See Exercises 25 and 26.)
    a. ICl_5
    b. $XeCl_4$
    c. $SeCl_6$
29. State whether or not each of the following has a permanent dipole moment.

a.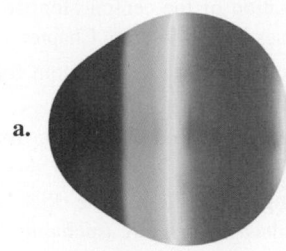

b.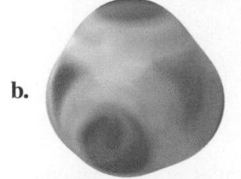

c.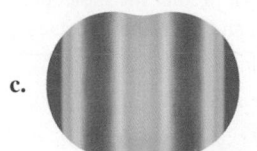

d.

e.

f.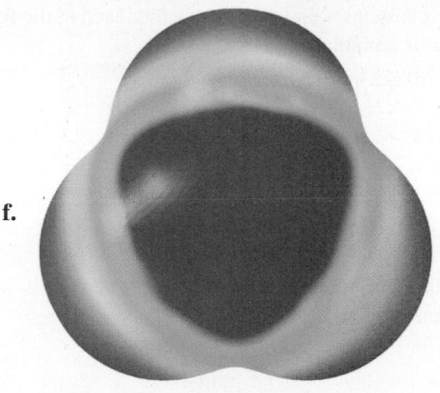

30. The following electrostatic potential diagrams represent CH_4, NH_3, or H_2O. Label each and explain your choices.

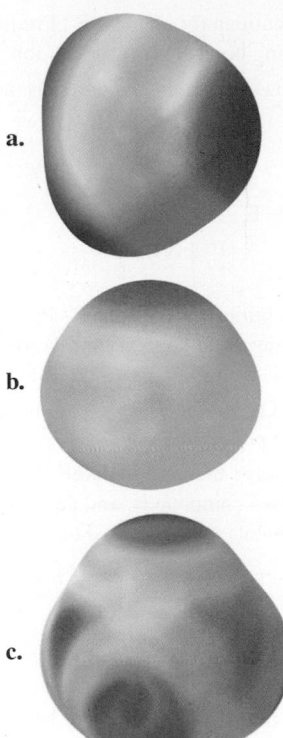

a.

b.

c.

31. Which of the molecules in Exercises 21 and 22 have net dipole moments (are polar)?
32. Which of the molecules in Exercises 27 and 28 have net dipole moments (are polar)?
33. Write Lewis structures and predict the molecular structures of the following. (See Exercises 25 and 26.)
    a. OCl_2, KrF_2, BeH_2, SO_2
    b. SO_3, NF_3, IF_3
    c. CF_4, SeF_4, KrF_4
    d. IF_5, AsF_5
    Which of these compounds are polar?

34. Write Lewis structures and predict whether each of the following is polar or nonpolar.

    a. HOCN (exists as HO—CN)

    b. COS

    c. XeF_2

    d. CF_2Cl_2

    e. SeF_6

    f. H_2CO (C is the central atom.)

35. Consider the following Lewis structure where E is an unknown element:

$$\left[:\overset{\cdot\cdot}{O} - E - \overset{\cdot\cdot}{\underset{\cdot\cdot}{O}}: \right]^{-}$$
$$\quad\quad\quad | $$
$$\quad\quad :\overset{}{\underset{\cdot\cdot}{O}}:$$

What are some possible identities for element E? Predict the molecular structure (including bond angles) for this ion.

36. Consider the following Lewis structure where E is an unknown element:

$$\left[:\overset{\cdot\cdot}{F} - E \overset{\displaystyle :\overset{\cdot\cdot}{O}:}{\underset{\displaystyle \overset{\cdot\cdot}{F}:}{\diagdown}} \right]^{2-}$$

What are some possible identities for element E? Predict the molecular structure (including bond angles) for this ion. (See Exercises 25 and 26.)

37. The molecules BF_3, CF_4, CO_2, PF_5, and SF_6 are all nonpolar, even though they all contain polar bonds. Why?

38. Two different compounds have the formula XeF_2Cl_2. Write Lewis structures for these two compounds, and describe how measurement of dipole moments might be used to distinguish between them.

The Localized Electron Model and Hybrid Orbitals

39. Use the localized electron model to describe the bonding in H_2O.

40. Use the localized electron model to describe the bonding in CCl_4.

41. Use the localized electron model to describe the bonding in H_2CO (carbon is the central atom).

42. Use the localized electron model to describe the bonding in C_2H_2 (exists as HCCH).

43. The space-filling models of ethane and ethanol are shown below.

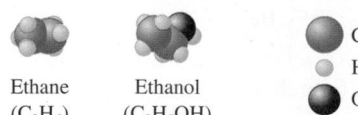

Ethane   Ethanol   ● C
(C_2H_6)   (C_2H_5OH)   ○ H
                         ● O

Use the localized electron model to describe the bonding in ethane and ethanol.

44. The space-filling models of hydrogen cyanide and phosgene are shown below.

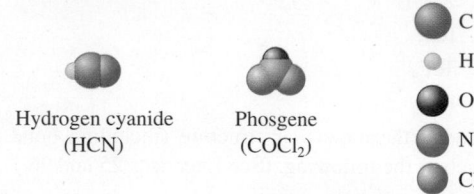

Hydrogen cyanide   Phosgene
(HCN)   ($COCl_2$)

● C
○ H
● O
● N
● Cl

Use the localized electron model to describe the bonding in hydrogen cyanide and phosgene.

45. Give the expected hybridization of the central atom for the molecules or ions in Exercises 81 and 87 from Chapter 3.

46. Give the expected hybridization of the central atom for the molecules or ions in Exercises 82 and 88 from Chapter 3.

47. Give the expected hybridization of the central atom for the molecules in Exercises 21 and 22.

48. Give the expected hybridization of the central atom for the molecules in Exercises 27 and 28.

49. For each of the following molecules, write the Lewis structure(s), predict the molecular structure (including bond angles), give the expected hybrid orbitals of the central atom, and predict the overall polarity.

    a. CF_4 **e.** BeH_2 **i.** KrF_4

    b. NF_3 **f.** TeF_4 **j.** SeF_6

    c. OF_2 **g.** AsF_5 **k.** IF_5

    d. BF_3 **h.** KrF_2 **l.** IF_3

50. For each of the following molecules or ions that contain sulfur, write the Lewis structure(s), predict the molecular structure (including bond angles), and give the expected hybrid orbitals for sulfur.

    a. SO_2

    b. SO_3

    c. $S_2O_3{}^{2-}$ $\left[\begin{array}{c} O \\ | \\ S - S - O \\ | \\ O \end{array} \right]^{2-}$

    d. $S_2O_8{}^{2-}$ $\left[\begin{array}{c} O \quad\quad O \\ || \quad\quad || \\ O - S - O - O - S - O \\ || \quad\quad || \\ O \quad\quad O \end{array} \right]^{2-}$

    e. $SO_3{}^{2-}$

    f. $SO_4{}^{2-}$

    g. SF_2

    h. SF_4

    i. SF_6

    j. $F_3S—SF$

    k. $SF_5{}^{+}$

51. Why must all six atoms in C_2H_4 lie in the same plane?

52. The allene molecule has the following Lewis structure:

Must all hydrogen atoms lie the same plane? If not, what is their spatial relationship? Explain.

53. Indigo is the dye used in coloring blue jeans. The term *navy blue* is derived from the use of indigo to dye British naval uniforms in the eighteenth century. The structure of the indigo molecule is

a. How many σ bonds and π bonds exist in the molecule?

b. What hybrid orbitals are used by the carbon atoms in the indigo molecule?

54. Urea, a compound formed in the liver, is one of the ways humans excrete nitrogen. The Lewis structure for urea is

Using hybrid orbitals for carbon, nitrogen, and oxygen, determine which orbitals overlap to form the various bonds in urea.

55. Biacetyl and acetoin are added to margarine to make it taste more like butter.

Biacetyl   Acetoin

Complete the Lewis structures, predict values for all C—C—O bond angles, and give the hybridization of the carbon atoms in these two compounds. Must the four carbon atoms and two oxygen atoms in biacetyl lie in the same plane? How many σ bonds and how many π bonds are there in biacetyl and acetoin?

56. Many important compounds in the chemical industry are derivatives of ethylene (C_2H_4). Two of them are acrylonitrile and methyl methacrylate.

Acrylonitrile   Methyl methacrylate

Complete the Lewis structures, showing all lone pairs. Give approximate values for bond angles *a* through *f*. Give the hybridization of all carbon atoms. In acrylonitrile, how many of the atoms in the molecule must lie in the same plane? How many σ bonds and how many π bonds are there in methyl methacrylate and acrylonitrile?

57. Two molecules used in the polymer industry are azodicarbonamide and methyl cyanoacrylate. Their structures are

Azodicarbonamide   Methyl cyanoacrylate

Azodicarbonamide is used in forming polystyrene. When added to the molten plastic, it decomposes to nitrogen, carbon monoxide, and ammonia gases, which are captured as bubbles in the molten polymer. Methyl cyanoacrylate is the main ingredient in super glue. As the glue sets, methyl cyanoacrylate polymerizes across the carbon–carbon double bond. (See Chapter 21.)

a. Complete the Lewis structures showing all lone pairs of electrons.

b. Which hybrid orbitals are used by the carbon atoms in each molecule and the nitrogen atom in azodicarbonamide?

c. How many π bonds are present in each molecule?

d. Give approximate values for the bond angles marked *a* through *h* in the previous structures.

58. Hot and spicy foods contain molecules that stimulate pain-detecting nerve endings. Two such molecules are piperine and capsaicin:

Piperine

Capsaicin

Piperine is the active compound in white and black pepper, and capsaicin is the active compound in chili peppers. The ring structures in piperine and capsaicin are shorthand notation. Each point where lines meet represents a carbon atom.

a. Complete the Lewis structure for piperine and capsaicin, showing all lone pairs of electrons.

b. How many carbon atoms are sp, sp^2, and sp^3 hybridized in each molecule?

c. Which hybrid orbitals are used by the nitrogen atoms in each molecule?

d. Give approximate values for the bond angles marked *a* through *l* in the above structures.

59. One of the first drugs to be approved for use in treatment of acquired immune deficiency syndrome (AIDS) was azidothymidine (AZT). Complete the Lewis structure for AZT.

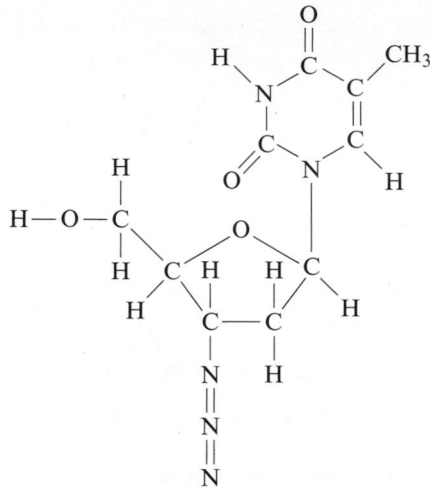

a. How many carbon atoms are sp^3 hybridized?

b. How many carbon atoms are sp^2 hybridized?

c. Which atom is sp hybridized?

d. How many σ bonds are in the molecule?

e. How many π bonds are in the molecule?

f. What is the N=N=N bond angle in the azide ($-N_3$) group?

g. What is the H—O—C bond angle in the side group attached to the five-membered ring?

h. What is the hybridization of the oxygen atom in the $-CH_2OH$ group?

60. The antibiotic thiarubin-A was discovered by studying the feeding habits of wild chimpanzees in Tanzania. The structure for thiarubin-A is

a. Complete the Lewis structure, showing all lone pairs of electrons.

b. Indicate the hybrid orbitals used by the carbon and sulfur atoms in thiarubin-A.

c. How many σ and π bonds are present in this molecule?

The Molecular Orbital Model

61. Consider the following molecular orbitals formed from the combination of two hydrogen $1s$ orbitals:

a. Which is the bonding molecular orbital and which is the antibonding molecular orbital? Explain how you can tell by looking at their shapes.

b. Which of the two molecular orbitals is lower in energy? Why is this true?

62. Sketch the molecular orbital and label its type (σ or π; bonding or antibonding) that would be formed when the following atomic orbitals overlap. Explain your labels.

a.
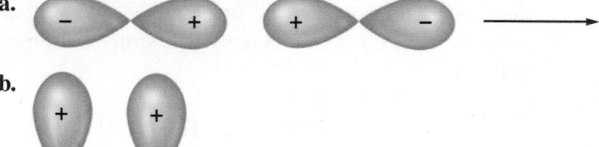

b.
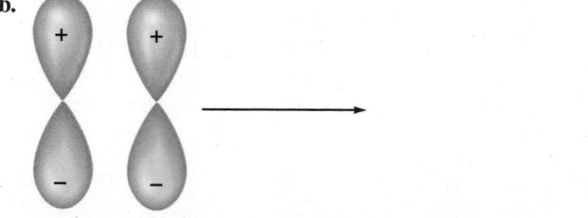

c.

d.
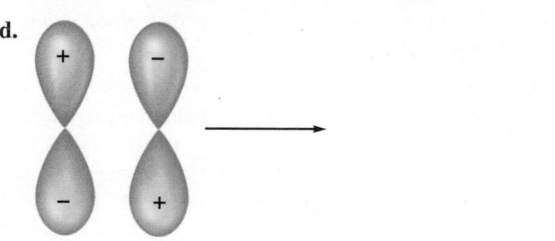

63. Which of the following are predicted by the molecular orbital model to be stable diatomic species?

a. H_2^+, H_2, H_2^-, H_2^{2-}

b. He_2^{2+}, He_2^+, He_2

64. Which of the following are predicted by the molecular orbital model to be stable diatomic species?

a. N_2^{2-}, O_2^{2-}, F_2^{2-}

b. Be_2, B_2, Ne_2

65. Using the molecular orbital model, write electron configurations for the following diatomic species and calculate the bond orders. Which ones are paramagnetic?

a. Li_2 **b.** C_2 **c.** S_2

66. Consider the following electron configuration:

$$(\sigma_{3s})^2(\sigma_{3s}*)^2(\sigma_{3p})^2(\pi_{3p})^4(\pi_{3p}*)^4$$

Give four species that, in theory, would have this electron configuration.

67. Using molecular orbital theory, explain why the removal of one electron in O_2 strengthens bonding, while the removal of one electron in N_2 weakens bonding.

68. Using the molecular orbital model to describe the bonding in F_2^+, F_2, and F_2^-, predict the bond orders and the relative bond lengths for these three species. How many unpaired electrons are present in each species?

69. The transport of O_2 in the blood is carried out by hemoglobin. Carbon monoxide can interfere with oxygen transport because hemoglobin has a stronger affinity for CO than for O_2. If CO is present, normal uptake of O_2 is prevented, depriving the body of needed oxygen. Using the molecular orbital model, write the electron configurations for CO and for O_2. From your configurations, give two property differences between CO and O_2.

70. A Lewis structure obeying the octet rule can be drawn for O_2 as follows:

$$\ddot{\text{O}}{=}\dot{\text{O}}{:}$$

Use the molecular orbital energy-level diagram for O_2 to show that the above Lewis structure corresponds to an excited state.

71. Using the molecular orbital model, write electron configurations for the following diatomic species and calculate the bond orders. Which ones are paramagnetic? Place the species in order of increasing bond length and bond energy.

   a. CO   **b.** CO^+ **c.** CO^{2+}

72. Using the molecular orbital model, write electron configurations for the following diatomic species and calculate the bond orders. Which ones are paramagnetic? Place the species in order of increasing bond length and bond energy.

   a. CN^+ **b.** CN   **c.** CN^-

73. In which of the following diatomic molecules would the bond strength be expected to weaken as an electron is removed?

   a. H_2 **c.** C_2^{2-}

   b. B_2 **d.** OF

74. In terms of the molecular orbital model, which species in each of the following two pairs will most likely be the one to gain an electron? Explain.

   a. CN or NO

   b. O_2^{2+} or N_2^{2+}

75. Show how two $2p$ atomic orbitals can combine to form a σ or a π molecular orbital.

76. Show how a hydrogen $1s$ atomic orbital and a fluorine $2p$ atomic orbital overlap to form bonding and antibonding molecular orbitals in the hydrogen fluoride molecule. Are these molecular orbitals σ or π molecular orbitals?

77. Use Figs. 4-54 and 4-55 to answer the following questions.

   a. Would the bonding molecular orbital in HF place greater electron density near the H or the F atom? Why?

   b. Would the bonding molecular orbital have greater fluorine $2p$ character, greater hydrogen $1s$ character, or an equal contribution from both? Why?

   c. Answer the previous two questions for the antibonding molecular orbital in HF.

78. The diatomic molecule OH exists in the gas phase. The bond length and bond energy have been measured to be 97.06 pm and 424.7 kJ/mol, respectively. Assume that the OH molecule is analogous to the HF molecule discussed in the chapter and that molecular orbitals result from the overlap of a lower-energy p_z orbital from oxygen with the higher-energy $1s$ orbital of hydrogen (the O—H bond lies along the z-axis).

   a. Which of the two molecular orbitals will have the greater hydrogen $1s$ character?

   b. Can the $2p_x$ orbital of oxygen form molecular orbitals with the $1s$ orbital of hydrogen? Explain.

   c. Knowing that only the $2p$ orbitals of oxygen will interact significantly with the $1s$ orbital of hydrogen, complete the molecular orbital energy-level diagram for OH. Place the correct number of electrons in the energy levels.

   d. Estimate the bond order for OH.

   e. Predict whether the bond order of OH^+ will be greater than, less than, or the same as that of OH. Explain.

79. Acetylene (C_2H_2) can be produced from the reaction of calcium carbide (CaC_2) with water. Use both the localized electron and molecular orbital models to describe the bonding in the acetylide anion (C_2^{2-}).

80. Describe the bonding in NO^+, NO^-, and NO, using both the localized electron and molecular orbital models. Account for any discrepancies between the two models.

81. Describe the bonding in the O_3 molecule and the NO_2^- ion, using the localized electron model. How would the molecular orbital model describe the π bonding in these two species?

82. Describe the bonding in the CO_3^{2-} ion using the localized electron model. How would the molecular orbital model describe the π bonding in this species?

Additional Exercises

83. Draw the Lewis structures, predict the molecular structures, and describe the bonding (in terms of the hybrid orbitals for the central atom) for the following.

   a. XeO_3 **d.** $XeOF_2$

   b. XeO_4 **e.** XeO_3F_2

   c. $XeOF_4$

84. Vitamin B_6 is an organic compound whose deficiency in the human body can cause apathy, irritability, and an increased susceptibility to infections. Below is an incomplete Lewis structure for vitamin B_6. Complete the Lewis structure and answer the following questions. *Hint:* Vitamin B_6 can be classified as an organic compound (a compound based on carbon atoms). The majority of Lewis structures for simple organic compounds have all atoms with a formal charge of zero. Therefore, add lone pairs and multiple bonds to the structure below to give each atom a formal charge of zero.

   a. How many σ bonds and π bonds exist in vitamin B_6?

   b. Give approximate values for the bond angles marked a through g in the structure.

   c. How many carbon atoms are sp^2 hybridized?

   d. How many carbon, oxygen, and nitrogen atoms are sp^3 hybridized?

   e. Does vitamin B_6 exhibit delocalized π bonding? Explain.

85. Two structures can be drawn for cyanuric acid:

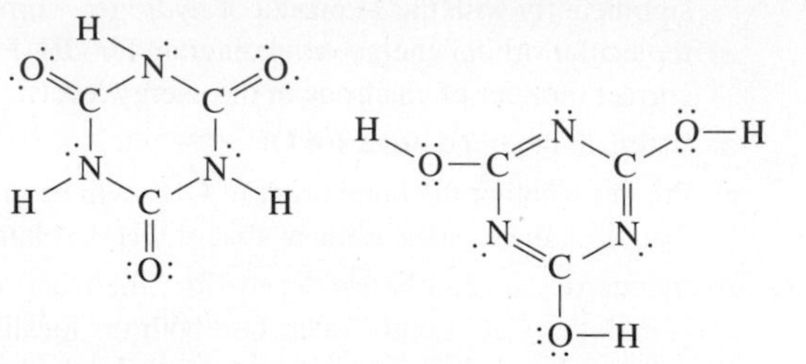

a. Are these two structures the same molecule? Explain.

b. Give the hybridization of the carbon and nitrogen atoms in each structure.

c. Use bond energies (Table 3-3) to predict which form is more stable; that is, which contains the strongest bonds?

86. What do each of the following sets of compounds/ions have in common with each other?

a. SO_3, NO_3^-, CO_3^{2-}

b. O_3, SO_2, NO_2^-

87. What do each of the following sets of compounds/ions have in common with each other?

a. $XeCl_4$, $XeCl_2$

b. ICl_5, TeF_4, ICl_3, PCl_3, SCl_2, SeO_2

88. Aspartame is an artificial sweetener marketed under the name Nutra-Sweet. A partial Lewis structure for aspartame is shown below.

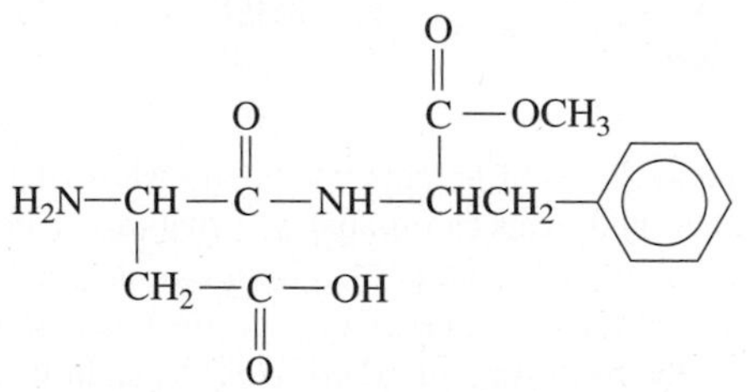

Aspartame can be classified as an organic compound (a compound based on carbon atoms). The majority of Lewis structures for simple organic compounds have all atoms with a formal charge of zero. Therefore, add lone pairs and multiple bonds to the structure above to give each atom a formal charge of zero when drawing the Lewis structure. Also note that the six-sided ring is shorthand notation for a benzene ring ($-C_6H_5$). Benzene is discussed in Section 4-7. Complete the Lewis structure for aspartame. How many C and N atoms exhibit sp^2 hybridization? How many C and O atoms exhibit sp^3 hybridization? How many σ and π bonds are in aspartame?

89. Using bond energies from Table 3-3, estimate the barrier to rotation about a carbon–carbon double bond. To do this, consider what must happen to go from

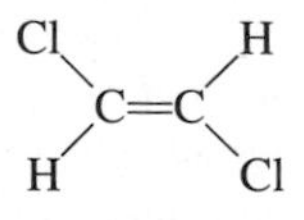

to

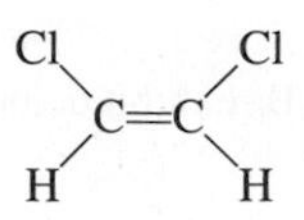

in terms of making and breaking chemical bonds; that is, what must happen in terms of the π bond?

90. The three most stable oxides of carbon are carbon monoxide (CO), carbon dioxide (CO_2), and carbon suboxide (C_3O_2). The space-filling models for these three compounds are

For each oxide, draw the Lewis structure, predict the molecular structure, and describe the bonding (in terms of the hybrid orbitals for the carbon atoms).

91. Refer back to the formal charge Exercises 99 and 100 from Chapter 3. Would you make the same prediction for the molecular structure for each case using the Lewis structure obtained in Exercise 99 as compared with the one obtained in Exercise 100?

92. Which of the following molecules have net dipole moments? For the molecules that are polar, indicate the polarity of each bond and the direction of the net dipole moment of the molecule.

a. CH_2Cl_2, $CHCl_3$, CCl_4

b. CO_2, N_2O

c. PH_3, NH_3

93. The structure of TeF_5^- is

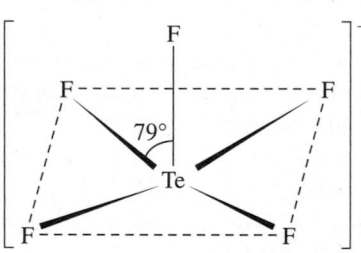

Draw a complete Lewis structure for TeF_5^-, and explain the distortion from the ideal square pyramidal structure. (See Exercise 26.)

94. Complete the following resonance structures for $POCl_3$.

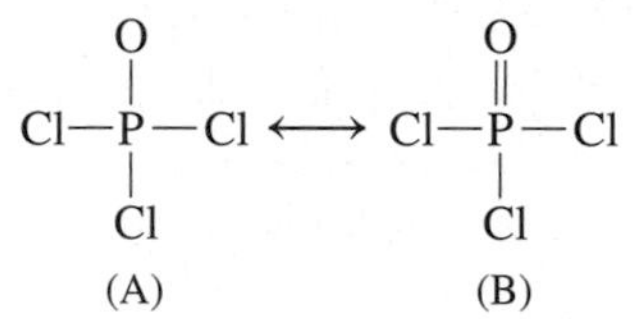

a. Would you predict the same molecular structure from each resonance structure?

b. What is the hybridization of P in each structure?

c. What orbitals can the P atom use to form the π bond in structure B?

d. Which resonance structure would be favored on the basis of formal charges?

95. The N_2O molecule is linear and polar.

a. On the basis of this experimental evidence, which arrangement, NNO or NON, is correct? Explain your answer.

b. On the basis of your answer to part a, write the Lewis structure of N_2O (including resonance forms). Give the formal charge on each atom and the hybridization of the central atom.

c. How would the multiple bonding in : N≡N—Ö : be described in terms of orbitals?

96. Describe the bonding in the first excited state of N_2 (the one closest in energy to the ground state) using the molecular orbital model. What differences do you expect in the properties of the molecule in the ground state as compared to the first excited state? (An excited state of a molecule corresponds to an electron arrangement other than that giving the lowest possible energy.)

97. Using an MO energy-level diagram, would you expect F_2 to have a lower or higher first ionization energy than atomic fluorine? Why?

98. Show how a d_{xz} atomic orbital and a p_z atomic orbital combine to form a bonding molecular orbital. Assume the x-axis is the internuclear axis. Is a σ or a π molecular orbital formed? Explain.

99. What type of molecular orbital would result from the in-phase combination of the two d_{xz} atomic orbitals shown below? Assume the x-axis is the internuclear axis.

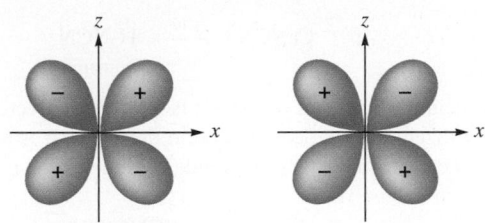

100. Consider three molecules: A, B, and C. Molecule A has a hybridization of sp^3. Molecule B has two more effective pairs (electron pairs around the central atom) than molecule A. Molecule C consists of two σ bonds and two π bonds. Give the molecular structure, hybridization, bond angles, and an example for each molecule.

ChemWork Problems

These multiconcept problems (and additional ones) are found interactively online with the same type of assistance a student would get from an instructor.

101. The formulas of several chemical substances are given in the table below. For each substance in the table, give its chemical name and predict its molecular structure.

Formula	Compound Name	Molecular Structure
CO_2		
NH_3		
SO_3		
H_2O		
ClO_4^-		

102. Predict the molecular structure, bond angles, and polarity (has a net dipole moment or has no net dipole moment) for each of the following compounds.
   a. $SeCl_4$ d. CBr_4
   b. SF_2 e. IF_3
   c. KrF_4 f. ClF_5

103. Draw the Lewis structures for SO_2, PCl_3, NNO, COS, and PF_3. Which of the compounds are polar? Which of the compounds exhibit at least one bond angle that is approximately 120 degrees? Which of the compounds exhibit sp^3 hybridization by the central atom? Which of the compounds have a linear molecular structure?

104. Draw the Lewis structures for $TeCl_4$, ICl_5, PCl_5, $KrCl_4$, and $XeCl_2$. Which of the compounds exhibit at least one bond angle that is approximately 120 degrees? Which of the compounds exhibit d^2sp^3 hybridization? Which of the compounds have a square planar molecular structure? Which of the compounds are polar?

105. A variety of chlorine oxide fluorides and related cations and anions are known. They tend to be powerful oxidizing and fluorinating agents. $FClO_3$ is the most stable of this group of compounds and has been studied as an oxidizing component in rocket propellants. Draw a Lewis structure for F_3ClO, $F_2ClO_2^+$, and F_3ClO_2. What is the molecular structure for each species, and what is the expected hybridization of the central chlorine atom in each compound or ion?

106. Pelargondin is the molecule responsible for the red color of the geranium flower. It also contributes to the color of ripe strawberries and raspberries. The structure of pelargondin is:

How many σ and π bonds exist in pelargondin? What is the hybridization of the carbon atoms marked 1–4?

107. Complete a Lewis structure for the compound shown below, then answer the following questions. What are the predicted bond angles about the carbon and nitrogen atoms? How many lone pairs of electrons are present in the Lewis structure? How many double bonds are present?

108. Which of the following statements concerning SO_2 is(are) *true*?
   a. The central sulfur atom is sp^2 hybridized.
   b. One of the sulfur–oxygen bonds is longer than the other(s).
   c. The bond angles about the central sulfur atom are about 120 degrees.

d. There are two σ bonds in SO_2.

e. There are no resonance structures for SO_2.

109. Consider the molecular orbital electron configurations for N_2, N_2^+, and N_2^-. For each compound or ion, fill in the table below with the correct number of electrons in each molecular orbital.

MO	N_2	N_2^+	N_2^-
σ_{2p}^*			
π_{2p}^*			
σ_{2p}			
π_{2p}			
σ_{2s}^*			
σ_{2s}			

110. Place the species B_2^+, B_2, and B_2^- in order of increasing bond length and increasing bond energy.

Challenge Problems

111. The compound NF_3 is quite stable, but NCl_3 is very unstable (NCl_3 was first synthesized in 1811 by P. L. Dulong, who lost three fingers and an eye studying its properties). The compounds NBr_3 and NI_3 are unknown, although the explosive compound $NI_3 \cdot NH_3$ is known. Account for the instability of these halides of nitrogen.

112. Predict the molecular structure for each of the following. (See Exercises 25 and 26.)

a. $BrFI_2$

b. XeO_2F_2

c. $TeF_2Cl_3^-$

For each formula there are at least two different structures that can be drawn using the same central atom. Draw all possible structures for each formula.

113. Consider the following computer-generated model of caffeine:

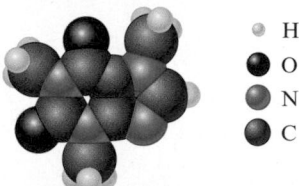

H
O
N
C

Complete a Lewis structure for caffeine in which all atoms have a formal charge of zero (as is typical with most organic compounds). How many C and N atoms are sp^2 hybridized? How many C and N atoms are sp^3 hybridized? sp hybridized? How many σ and π bonds are there?

114. Cholesterol ($C_{27}H_{46}O$) has the following structure:

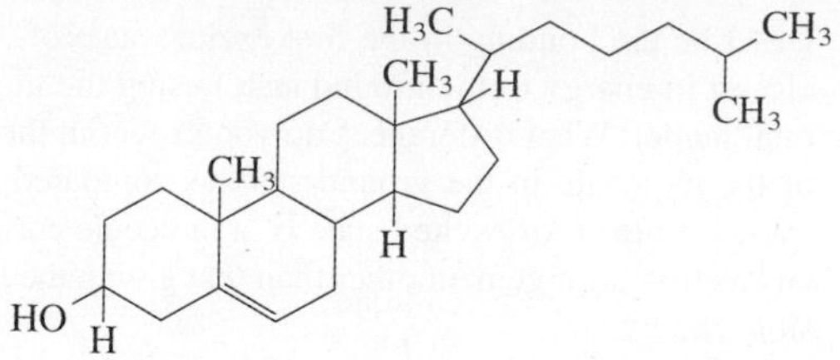

H₃C   CH₃
CH₃ H   CH₃
CH₃
H
HO H

In such shorthand structures, each point where lines meet represents a carbon atom and most H atoms are not shown. Draw the complete structure showing all carbon and hydrogen atoms. (There will be four bonds to each carbon atom.) Indicate which carbon atoms use sp^2 or sp^3 hybrid orbitals. Are all carbon atoms in the same plane, as implied by the structure?

115. Cyanamide (H_2NCN), an important industrial chemical, is produced by the following steps:

$$CaC_2 + N_2 \longrightarrow CaNCN + C$$

$$CaNCN \xrightarrow{\text{Acid}} \underset{\text{Cyanamide}}{H_2NCN}$$

Calcium cyanamide (CaNCN) is used as a direct-application fertilizer, weed killer, and cotton defoliant. It is also used to make cyanamide, dicyandiamide, and melamine plastics:

$$H_2NCN \xrightarrow{\text{Acid}} \underset{\text{Dicyandiamide}}{NCNC(NH_2)_2}$$

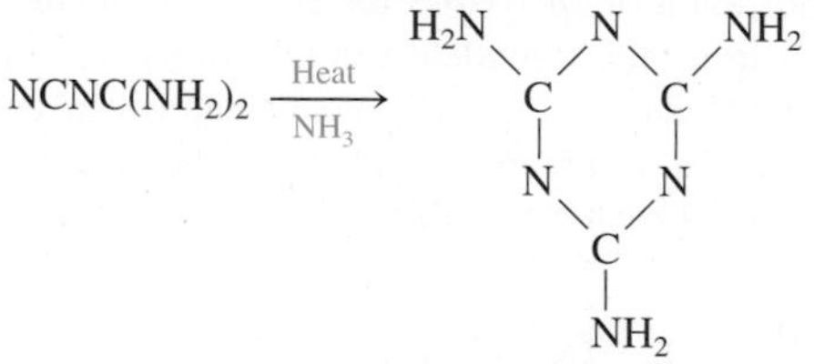

$$NCNC(NH_2)_2 \xrightarrow[NH_3]{\text{Heat}}$$

H₂N   N   NH₂
     C   C
     N   N
          C
          NH₂
     Melamine
     (π bonds
     not shown)

a. Write Lewis structures for NCN^{2-}, H_2NCN, dicyandiamide, and melamine, including resonance structures where appropriate.

b. Give the hybridization of the C and N atoms in each species.

c. How many σ bonds and how many π bonds are in each species?

d. Is the ring in melamine planar?

e. There are three different C—N bond distances in dicyandiamide, $NCNC(NH_2)_2$, and the molecule is nonlinear. Of all the resonance structures you drew for this molecule, predict which should be the most important.

116. As compared with CO and O_2, CS and S_2 are very unstable molecules. Give an explanation based on the relative abilities of the sulfur and oxygen atoms to form π bonds.

117. Values of measured bond energies may vary greatly depending on the molecule studied. Consider the following reactions:

$$NCl_3(g) \longrightarrow NCl_2(g) + Cl(g) \quad \Delta E = 375 \text{ kJ/mol}$$
$$ONCl(g) \longrightarrow NO(g) + Cl(g) \quad \Delta E = 158 \text{ kJ/mol}$$

Rationalize the difference in the values of ΔE for these reactions, even though each reaction appears to involve only the breaking of one N—Cl bond. (*Hint:* Consider the bond order of the NO bond in ONCl and in NO.)

118. Use the MO model to explain the bonding in BeH_2. When constructing the MO energy-level diagram, assume that the Be's $1s$ electrons are not involved in bond formation.

119. *Bond energy* has been defined in the text as the amount of energy required to break a chemical bond, so we have come to think of the addition of energy as breaking bonds. However, in some cases the addition of energy can cause the formation of bonds. For example, in a sample of helium gas subjected to a high-energy source, some He_2 molecules exist momentarily and then dissociate. Use MO theory (and diagrams) to explain why He_2 molecules can come to exist and why they dissociate.

120. Arrange the following from lowest to highest ionization energy: O, O_2, O_2^-, O_2^+. Explain your answer.

121. Use the MO model to determine which of the following has the smallest ionization energy: $N_2, O_2, N_2^{2-}, N_2^-, O_2^+$. Explain your answer.

122. Given that the ionization energy of F_2^- is 290 kJ/mol, do the following:
  a. Calculate the bond energy of F_2^-. You will need to look up the bond energy of F_2 and ionization energy of F^-.
  b. Explain the difference in bond energy between F_2^- and F_2 using MO theory.

123. Carbon monoxide (CO) forms bonds to a variety of metals and metal ions. Its ability to bond to iron in hemoglobin is the reason that CO is so toxic. The bond carbon monoxide forms to metals is through the carbon atom:

$$M—C\equiv O$$

  a. On the basis of electronegativities, would you expect the carbon atom or the oxygen atom to form bonds to metals?
  b. Assign formal charges to the atoms in CO. Which atom would you expect to bond to a metal on this basis?
  c. In the MO model, bonding MOs place more electron density near the more electronegative atom. (See the HF molecule in Figs. 4-54 and 4-55.) Antibonding MOs place more electron density near the less electronegative atom in the diatomic molecule. Use the MO model to predict which atom of carbon monoxide should form bonds to metals.

124. The space-filling model for benzoic acid, a food preservative, is shown below.

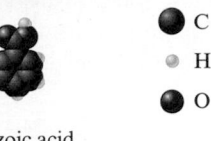

Benzoic acid
($C_6H_5CO_2H$)

Describe the bonding in benzoic acid using the localized electron model combined with the molecular orbital model.

Integrative Problems

These problems require the integration of multiple concepts to find the solutions.

125. As the head engineer of your starship in charge of the warp drive, you notice that the supply of dilithium is critically low. While searching for a replacement fuel, you discover some diboron, B_2.
  a. What is the bond order in Li_2 and B_2?
  b. How many electrons must be removed from B_2 to make it isoelectronic with Li_2 so that it might be used in the warp drive?
  c. The reaction to make B_2 isoelectronic with Li_2 is generalized (where n = number of electrons determined in part b) as follows:

$$B_2 \longrightarrow B_2^{n+} + ne^- \quad \Delta E = 6455 \text{ kJ/mol}$$

  How much energy is needed to ionize 1.5 kg B_2 to the desired isoelectronic species?

126. A flask containing gaseous N_2 is irradiated with 25-nm light.
  a. Using the following information, indicate what species can form in the flask during irradiation.

$$N_2(g) \longrightarrow 2N(g) \quad \Delta E = 941 \text{ kJ/mol}$$
$$N_2(g) \longrightarrow N_2^+(g) + e^- \quad \Delta E = 1501 \text{ kJ/mol}$$
$$N(g) \longrightarrow N^+(g) + e^- \quad \Delta E = 1402 \text{ kJ/mol}$$

  b. What range of wavelengths will produce atomic nitrogen in the flask but will not produce any ions?
  c. Explain why the first ionization energy of N_2 (1501 kJ/mol) is greater than the first ionization energy of atomic nitrogen (1402 kJ/mol).

127. Determine the molecular structure and hybridization of the central atom X in the polyatomic ion XY_3^+ given the following information: A neutral atom of X contains 36 electrons, and the element Y makes an anion with a 1– charge, which has the electron configuration $1s^2 2s^2 2p^6$.

C H A P T E R   F I V E

Stoichiometry

Fireworks provide a spectacular example of chemical reactions. (Alexander Nemenov/AFP/Getty Images)

Chemical reactions have a profound effect on our lives. There are many examples: Food is converted to energy in the human body; nitrogen and hydrogen are combined to form ammonia, which is used as a fertilizer; fuels and plastics are produced from petroleum; the starch in plants is synthesized from carbon dioxide and water using energy from sunlight; human insulin is produced in laboratories by bacteria; cancer is induced in humans by substances from our environment; and so on, in a seemingly endless list. The central activity of chemistry is to understand chemical changes such as these, and the study of reactions occupies a central place in this book. We will examine why reactions occur, how fast they occur, and which specific pathways they follow.

Because chemical reactions involve rearrangement of atoms, we need to know how many atoms are in given samples of starting materials when we run a reaction. However, because atoms are too small to count directly, we need an indirect way to count them that is convenient in factories and laboratories. In this chapter we will learn how atoms are counted by weighing samples of them. We will also show how to compute the number of moles of an element or compound from the mass of the sample.

We will introduce the topics in this chapter emphasizing molecular-level views in order to help you better understand these concepts. We will make explicit links between the micro-world (atomic-molecular level) and quantitative measurements in the macro-world (what we observe).

5-1 | Counting by Weighing

Suppose you work in a candy store that sells gourmet jelly beans by the bean. People come in and ask for 50 beans, 100 beans, 1000 beans, and so on, and you have to count them out—a tedious process at best. As a good problem solver, you try to come up with a better system. It occurs to you that it might be far more efficient to buy a scale and count the jelly beans by weighing them. How can you count jelly beans by weighing them? ◄ What information about the individual beans do you need to know?

Assume that all of the jelly beans are identical and that each has a mass of 5 g. If a customer asks for 1000 jelly beans, what mass of jelly beans would be required? Each bean has a mass of 5 g, so you would need 1000 beans × 5 g/bean, or 5000 g (5 kg). It takes just a few seconds to weigh out 5 kg of jelly beans. It would take much longer to count out 1000 of them.

In reality, jelly beans are not identical. For example, let's assume that you weigh 10 beans individually and get the following results:

Bean	Mass	Bean	Mass
1	5.1 g	6	5.0 g
2	5.2 g	7	5.0 g
3	5.0 g	8	5.1 g
4	4.8 g	9	4.9 g
5	4.9 g	10	5.0 g

▲ Jelly beans can be counted by weighing.

Can we count these nonidentical beans by weighing? Yes. The key piece of information we need is the *average mass* of the jelly beans. Let's compute the average mass for our 10-bean sample.

$$\text{Average mass} = \frac{\text{total mass of beans}}{\text{number of beans}}$$

$$= \frac{5.1\text{ g} + 5.2\text{ g} + 5.0\text{ g} + 4.8\text{ g} + 4.9\text{ g} + 5.0\text{ g} + 5.0\text{ g} + 5.1\text{ g} + 4.9\text{ g} + 5.0\text{ g}}{10}$$

$$= \frac{50.0}{10} = 5.0\text{ g}$$

The average mass of a jelly bean is 5.0 g. Thus, to count out 1000 beans, we need to weigh out 5000 g of beans. This sample of beans, in which the beans have an average mass of 5.0 g, can be treated exactly like a sample where all of the beans are identical. Objects do not need to have identical masses to be counted by weighing. We simply need to know the average mass of the objects. For purposes of counting, the objects *behave as though they were all identical*, as though they each actually had the average mass.

We count atoms in exactly the same way. Because atoms are so small, we deal with samples of matter that contain huge numbers of atoms. Even if we could see the atoms it would not be possible to count them directly. Thus we determine the number of atoms in a given sample by finding its mass. However, just as with jelly beans, to relate the mass to a number of atoms, we must know the average mass of the atoms.

5-2 | Atomic Masses

As we saw in Chapter 1, the first quantitative information about atomic masses came from the work of Dalton, Gay-Lussac, Lavoisier, Avogadro, and Berzelius. By observing the proportions in which elements combine to form various compounds, nineteenth-century chemists calculated relative atomic masses. The modern system of atomic masses, instituted in 1961, is based on ^{12}C ("carbon twelve") as the standard. In this system, *^{12}C is assigned a mass of exactly 12 atomic mass units* (u), and the masses of all other atoms are given relative to this standard.

The most accurate method currently available for comparing the masses of atoms involves the use of the **mass spectrometer**. In this instrument, diagrammed in Fig. 5-1, atoms or molecules are passed into a beam of high-speed electrons, which knock

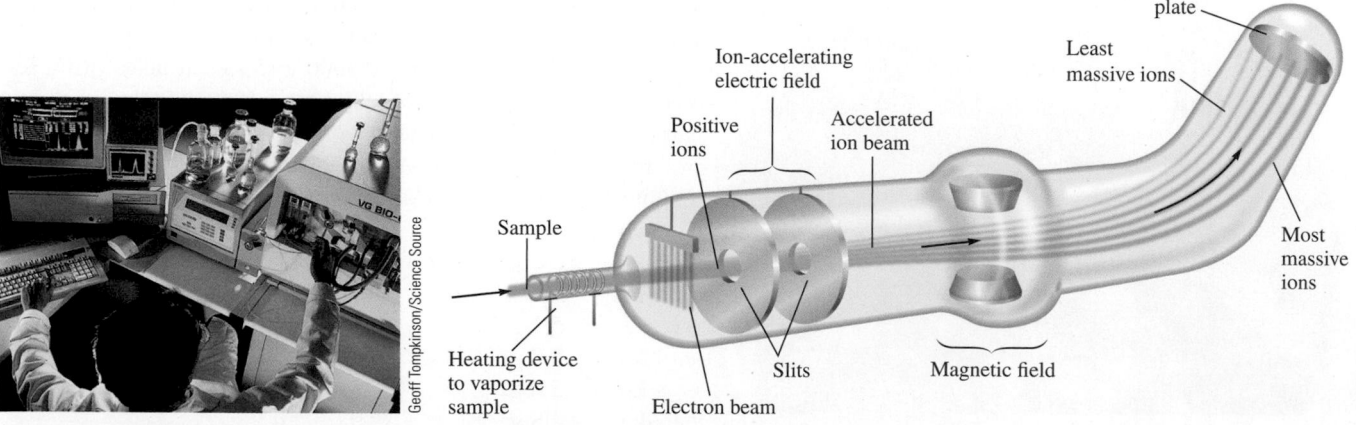

Figure 5-1 | (left) A scientist injecting a sample into a mass spectrometer. (right) Schematic diagram of a mass spectrometer.

electrons off the atoms or molecules being analyzed and change them into positive ions. An applied electric field then accelerates these ions into a magnetic field. Because an accelerating ion produces its own magnetic field, an interaction with the applied magnetic field occurs, which tends to change the path of the ion. The amount of path deflection for each ion depends on its mass—the most massive ions are deflected the smallest amount—which causes the ions to separate, as shown in Fig. 5-1. A comparison of the positions where the ions hit the detector plate gives very accurate values of their relative masses. For example, when ^{12}C and ^{13}C are analyzed in a mass spectrometer, the ratio of their masses is found to be

$$\frac{\text{Mass }^{13}C}{\text{Mass }^{12}C} = 1.0836129$$

Since the atomic mass unit is defined such that the mass of ^{12}C is *exactly* 12 atomic mass units, then on this same scale,

$$\text{Mass of }^{13}C = (1.0836129)(12 \text{ u}) = 13.003355 \text{ u}$$
$$\uparrow$$
Exact number
by definition

The masses of other atoms can be determined in a similar fashion.

The mass for each element is given in the table inside the front cover of this text. This value, even though it is actually a mass, is (for historical reasons) sometimes called the *atomic weight* for each element. ◄ *i*

Look at the value of the atomic mass of carbon given in this table. You might expect to see 12, since we said the system of atomic masses is based on ^{12}C. However, the number given for carbon is not 12 but 12.01. Why? The reason for this apparent discrepancy is that the carbon found on the earth (natural carbon) is a mixture of the isotopes ^{12}C, ^{13}C, and ^{14}C. All three isotopes have six protons, but they have six, seven, and eight neutrons, respectively. Because natural carbon is a mixture of isotopes, the atomic mass we use for carbon is an *average value* reflecting the average of the isotopes composing it. ◄ *i*

The average atomic mass for carbon is computed as follows. It is known that natural carbon is composed of 98.89% ^{12}C atoms and 1.11% ^{13}C atoms. The amount of ^{14}C is negligibly small at this level of precision. Using the masses of ^{12}C (exactly 12 u) and ^{13}C (13.003355 u), we can calculate the average atomic mass for natural carbon as follows:

98.89% of 12 u + 1.11% of 13.0034 u =
$$(0.9889)(12 \text{ u}) + (0.0111)(13.0034 \text{ u}) = 12.01 \text{ u}$$

In this text we will call the average mass for an element the **average atomic mass** or, simply, the *atomic mass* for that element.

Even though natural carbon does not contain a single atom with mass 12.01, for stoichiometric purposes, we can consider carbon to be composed of only one type of atom with a mass of 12.01. This enables us to count atoms of natural carbon by weighing a sample of carbon. ◄

Recall from Section 5-1 that counting by weighing works if you know the *average* mass of the units being counted. Counting by weighing works just the same for atoms as for jelly beans. For natural carbon with an average mass of 12.01 atomic mass units, to obtain 1000 atoms would require weighing out 12,010 atomic mass units of natural carbon (a mixture of ^{12}C and ^{13}C).

As in the case of carbon, the mass for each element listed in the table inside the front cover of the text is an average value based on the isotopic composition of the naturally occurring element. For instance, the mass listed for hydrogen (1.008) is the average mass for natural hydrogen, which is a mixture of ^{1}H and ^{2}H (deuterium). *No* atom of hydrogen actually has the mass 1.008.

i The International Union of Pure and Applied Chemistry (IUPAC) has declared that what we refer to as the "average atomic mass" should be called the "atomic weight" of an element, which is dimensionless, by custom. However, we will retain the term "average atomic mass" because this name accurately describes what the term represents.

i Most elements occur in nature as mixtures of isotopes; thus atomic masses are usually average values.

© Cengage Learning

▲ If we know the average mass of a hex nut, it is much easier to weigh out 600 hex nuts than count them one by one.

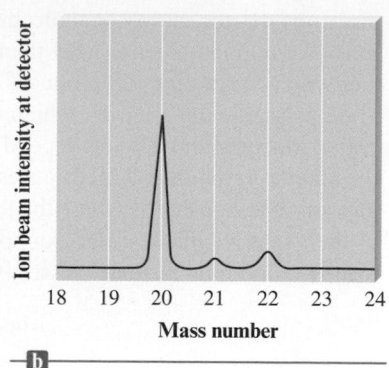

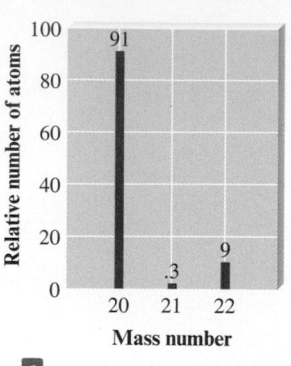

David Young-Wolff/Alamy

a   **b**   **c**

Figure 5-2 | (a) Neon gas glowing in a discharge tube. The relative intensities of the signals recorded when natural neon is injected into a mass spectrometer are represented in terms of (b) "peaks" and (c) a bar graph. The relative areas of the peaks are 0.9092 (^{20}Ne), 0.00257 (^{21}Ne), and 0.0882 (^{22}Ne); natural neon is therefore 90.92% ^{20}Ne, 0.257% ^{21}Ne, and 8.82% ^{22}Ne.

In addition to being useful for determining accurate mass values for individual atoms, the mass spectrometer is used to determine the isotopic composition of a natural element. For example, when a sample of natural neon is injected into a mass spectrometer, the mass spectrum shown in Fig. 5-2 is obtained. The areas of the "peaks" or the heights of the bars indicate the relative abundances of $^{20}_{10}$Ne, $^{21}_{10}$Ne, and $^{22}_{10}$Ne atoms.

EXAMPLE 5-1

The Average Mass of an Element

When a sample of natural copper is vaporized and injected into a mass spectrometer, the results shown in Fig. 5-3 are obtained. Use these data to compute the average mass of natural copper. (The mass values for ^{63}Cu and ^{65}Cu are 62.93 u and 64.93 u, respectively.) ◄

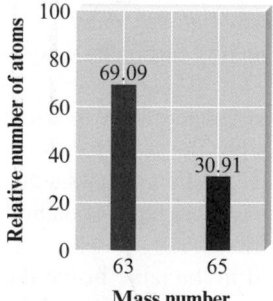

Arturo Limon/Shutterstock.com

▲ Copper nugget.

Solution

Where are we going?
To calculate the average mass of natural copper

What do we know?
- ^{63}Cu mass = 62.93 u
- ^{65}Cu mass = 64.93 u

How do we get there?
As shown by the graph, of every 100 atoms of natural copper, 69.09 are ^{63}Cu and 30.91 are ^{65}Cu. Thus the mass of 100 atoms of natural copper is

$$(69.09 \text{ atoms})\left(62.93 \ \frac{u}{\text{atom}}\right) + (30.91 \text{ atoms})\left(64.93 \ \frac{u}{\text{atom}}\right) = 6355 \text{ u}$$

The *average mass* of a copper atom is

$$\frac{6355 \text{ u}}{100 \text{ atoms}} = 63.55 \text{ u/atom}$$

Figure 5-3 | Mass spectrum of natural copper.

This mass value is used in doing calculations involving the reactions of copper and is the value given in the table inside the front cover of this book.

Always check to see if your answer is sensible (Reality Check). Paying careful attention to units and making sure the answer is reasonable can help you detect an inverted conversion factor or a number that was incorrectly entered in your calculator.

REALITY CHECK | When you finish a calculation, you should always check whether your answer makes sense. In this case our answer of 63.55 u is between the masses of the atoms that make up natural copper. This makes sense. The answer could not be smaller than 62.93 u or larger than 64.93 u. ◀ ⓘ

See Exercises 5-37 and 5-38

5-3 | Learning to Solve Problems

One of the great rewards of studying chemistry is to become a good problem-solver. Being able to solve complex problems is a talent that will serve you well in all walks of life. It is our purpose in this text to help you learn to solve problems in a flexible, creative way based on understanding the fundamental ideas of chemistry. We call this approach **conceptual problem solving**.

The ultimate goal is to be able to solve new problems (that is, problems you have not seen before) on your own. In this text we will provide problems and offer solutions by explaining how to think about the problems. While the answers to these problems are important, it is perhaps even more important to understand the process—the thinking necessary to get the answer. Although at first we will be solving the problem for you, do not take a passive role. While studying the solution, it is crucial that you interactively think through the problem with us. Do not skip the discussion and jump to the answer. Usually, the solution will involve asking a series of questions. Make sure that you understand each step in the process. This active approach should apply to problems outside of chemistry as well. For example, imagine riding with someone in a car to an unfamiliar destination. If your goal is simply to have the other person get you to that destination, you will probably not pay much attention to how to get there (passive), and if you have to find this same place in the future on your own, you will probably not be able to do it. If, however, your goal is to learn how to get there, you would pay attention to distances, signs, and turns (active). This is how you should read the solutions in the text (and the text in general).

While actively studying our solutions to problems is helpful, at some point you will need to know how to think through these problems on your own. If we help you too much as you solve a problem, you won't really learn effectively. If we always "drive," you won't interact as meaningfully with the material. Eventually you need to learn to drive yourself. We will provide more help at the beginning of the text and less as we proceed to later chapters.

There are two fundamentally different ways you might use to approach a problem. One way emphasizes memorization. We might call this the "pigeonholing method." In this approach, the first step is to label the problem—to decide in which pigeonhole it fits. ◀ The pigeonholing method requires that we provide you with a set of steps that you memorize and store in the appropriate slot for each different problem you encounter. The difficulty with this method is that it requires a new pigeonhole each time a problem is changed by even a small amount.

Consider the driving analogy again. Suppose you have memorized how to drive from your house to the grocery store. Do you know how to drive back from the grocery store to your house? Not necessarily. If you have only memorized the directions and do not understand fundamental principles such as "I traveled north to get to the store, so my house is south of the store," you may find yourself stranded. In a more complicated example, suppose you know how to get from your house to the store (and back) and from your house to the library (and back). Can you get from the library to the store without having to go back home? Probably not if you have only memorized directions and you do not have a "big picture" of where your house, the store, and the library are relative to one another.

▲ Pigeon holes can be used for sorting and classifying objects like mail.

The second approach is conceptual problem-solving, in which we help you get the "big picture"—a real understanding of the situation. This approach to problem-solving looks within the problem for a solution. In this method we assume that the problem is a new one, and we let the problem guide us as we solve it. In this approach we ask a series of questions as we proceed and use our knowledge of fundamental principles to answer these questions. Learning this approach requires some patience, but the reward for learning to solve problems this way is that we each become an effective solver of any new problem that confronts us in daily life or in our work in any field. In summary, instead of looking outside the problem for a memorized solution, we will look inside the problem and let the problem help us as we proceed to a solution.

To help us as we proceed to solve a problem, we will summarize the organizing principles which we introduced earlier:

1. We need to read the problem and decide on the final goal. Then we sort through the facts given, focusing on the key words and often drawing a diagram of the problem. We might summarize this process by the question, *What do we know?* In this part of the analysis we need to state the problem as simply and as visually as possible. We could summarize this process as "**Where are we going?**"
2. We need to work backwards from the final goal to decide where to start. We will ask a series of questions as we proceed. Our understanding of the fundamental principles of chemistry will enable us to answer each of these questions and will eventually lead us to the final solution. We might summarize this process as "**How do we get there?**"
3. Once we get the solution of the problem, then we ask ourselves, "Does it make sense?" That is, does our answer seem reasonable? We call this the **Reality Check**. It always pays to check your answer.

Using a conceptual approach to problem-solving will enable you to develop real confidence as a problem-solver. You will no longer panic when you see a problem that is different in some ways from those you have solved in the past. Although you might be frustrated at times as you learn this method, we guarantee that it will pay dividends later and should make your experience with chemistry a positive one that will prepare you for any career you choose.

To summarize, one of our major goals in this text is to help you become a creative problem-solver. We will do this by, at first, giving you lots of guidance in how to solve problems. We will "drive," but we hope you will be paying attention instead of just "riding along." As we move forward, we will gradually shift more of the responsibility to you. This process will be supported by interactive exercises, which will help you learn to think your way through a problem. As you gain confidence in letting the problem guide you, you will be amazed at how effective you can be at solving some really complex problems—just like the ones you will confront in "real life."

5-4 | The Mole

As we saw in Section R-10, because samples of matter typically contain so many atoms, a unit of measure called the *mole* has been established for use in counting atoms. For our purposes, it is most convenient to define the **mole** (abbreviated mol) as *the number equal to the number of carbon atoms in exactly 12 grams of pure ^{12}C*. ◄ ⓘ Techniques such as mass spectrometry, which count atoms very precisely, have been used to determine this number as 6.02214×10^{23} (6.022×10^{23} will be sufficient for our purposes). This number is called **Avogadro's number** to honor his contributions to chemistry. One mole of something consists of 6.022×10^{23} units of that substance. ◄ ⓘ Just as a dozen eggs is 12 eggs, a mole of eggs is 6.022×10^{23} eggs.

The magnitude of the number 6.022×10^{23} is very difficult to imagine. To give you some idea, 1 mole of seconds represents a span of time 4 million times as long as the earth has already existed, and 1 mole of marbles is enough to cover the entire earth to

ⓘ The SI definition of the mole is the amount of a substance that contains as many entities as there are in exactly 12 g of carbon-12.

ⓘ Avogadro's number is 6.022×10^{23}. One mole of anything is 6.022×10^{23} units of that substance.

Figure 5-4 | One-mole samples of several elements.

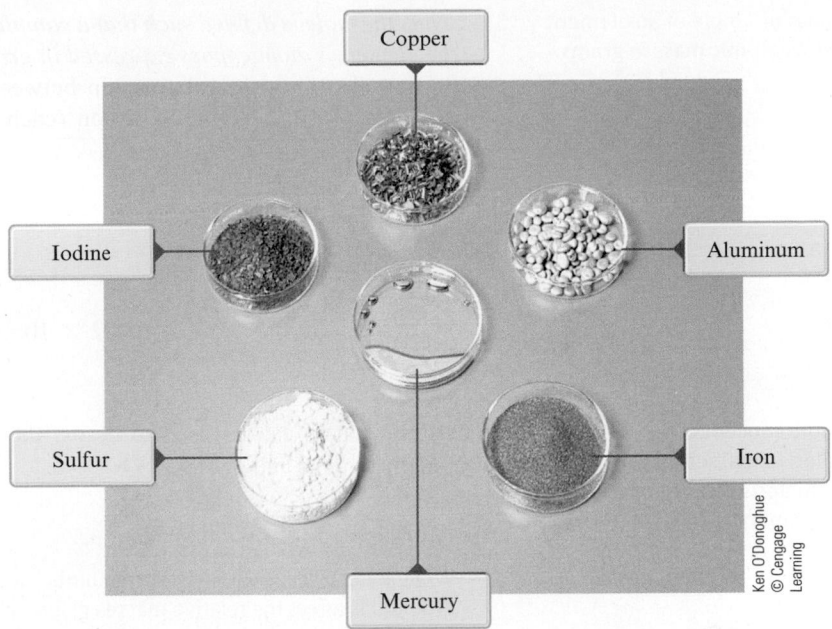

a depth of 50 miles! However, since atoms are so tiny, a mole of atoms or molecules is a perfectly manageable quantity to use in a reaction (see Fig. 5-4).

Critical Thinking

What if you were offered $1 million to count from 1 to 6×10^{23} at a rate of one number each second? Determine your hourly wage. Would you do it? Could you do it?

How do we use the mole in chemical calculations? Recall that Avogadro's number is defined as the number of atoms in exactly 12 grams of ^{12}C. This means that 12 grams of ^{12}C contains 6.022×10^{23} atoms. It also means that a 12.01-gram sample of natural carbon contains 6.022×10^{23} atoms (a mixture of ^{12}C, ^{13}C, and ^{14}C atoms, with an average atomic mass of 12.01). Since the ratio of the masses of the samples (12 g/12.01 g) is the same as the ratio of the masses of the individual components (12 u/12.01 u), the two samples contain the *same number* of atoms (6.022×10^{23}).

To be sure this point is clear, think of oranges with an average mass of 0.5 pound each and grapefruit with an average mass of 1.0 pound each. Any two sacks for which the sack of grapefruit weighs twice as much as the sack of oranges will contain the same number of pieces of fruit. The same idea extends to atoms. Compare natural carbon (average mass of 12.01) and natural helium (average mass of 4.003). A sample of 12.01 grams of natural carbon contains the same number of atoms as 4.003 grams of natural helium. Both samples contain 1 mole of atoms (6.022×10^{23}). Table 5-1 gives more examples that illustrate this basic idea.

Table 5-1 | Comparison of 1-Mole Samples of Various Elements

Element	Number of Atoms Present	Mass of Sample (g)
Aluminum	6.022×10^{23}	26.98
Copper	6.022×10^{23}	63.55
Iron	6.022×10^{23}	55.85
Sulfur	6.022×10^{23}	32.07
Iodine	6.022×10^{23}	126.9
Mercury	6.022×10^{23}	200.6

ⓘ The mass of 1 mole of an element is equal to its atomic mass in grams.

Thus the mole is defined such that a sample of a natural element with a mass equal to the element's atomic mass expressed in grams contains 1 mole of atoms. ◄ ⓘ This definition also fixes the relationship between the atomic mass unit and the gram. Since 6.022×10^{23} atoms of carbon (each with a mass of 12 u) have a mass of 12 g, then

$$(6.022 \times 10^{23} \text{ atoms})\left(\frac{12 \text{ u}}{\text{atom}}\right) = 12 \text{ g}$$

and

$$6.022 \times 10^{23} \text{ u} = 1 \text{ g}$$
$$\uparrow$$
$$\text{Exact number}$$

ⓘ The mole concept enables us to use the relative masses on the periodic table in units that we can measure (grams).

This relationship can be used to derive the unit factor needed to convert between atomic mass units and grams. ◄ ⓘ

Critical Thinking

What if you discovered Avogadro's number was not 6.02×10^{23} but 3.01×10^{23}? Would this affect the relative masses given on the periodic table? If so, how? If not, why not?

INTERACTIVE EXAMPLE 5-2

Determining the Mass of a Sample of Atoms

Americium is an element that does not occur naturally. It can be made in very small amounts in a device known as a *particle accelerator*. Compute the mass in grams of a sample of americium containing six atoms.

Solution

Where are we going?
To calculate the mass of six americium atoms

What do we know?

- Mass of 1 atom of Am = 243 u (from the periodic table inside the front cover)

How do we get there?
The mass of six atoms is

$$6 \text{ atoms} \times 243 \frac{\text{u}}{\text{atom}} = 1.46 \times 10^3 \text{ u}$$

Using the relationship

$$6.022 \times 10^{23} \text{ u} = 1 \text{ g}$$

we write the conversion factor for converting atomic mass units to grams:

$$\frac{1 \text{ g}}{6.022 \times 10^{23} \text{ u}}$$

The mass of six americium atoms in grams is

$$1.46 \times 10^3 \text{ u} \times \frac{1 \text{ g}}{6.022 \times 10^{23} \text{ u}} = 2.42 \times 10^{-21} \text{ g}$$

REALITY CHECK | Since this sample contains only six atoms, the mass should be very small as the amount 2.42×10^{-21} g indicates.

See Exercise 5-45

To do chemical calculations, you must understand what the mole means and how to determine the number of moles in a given mass of a substance. These procedures are illustrated in Examples 5-3 and 5-4.

INTERACTIVE EXAMPLE 5-3

Determining Moles of Atoms

Aluminum (Al) is a metal with a high strength-to-mass ratio and a high resistance to corrosion; thus it is often used for structural purposes. ◄ Compute both the number of moles of atoms and the number of atoms in a 10.0-g sample of aluminum. ◄

▲ (left) Pure aluminum. (right) Aluminum alloys are used for many products used in our kitchens.

Solution

Where are we going?
To calculate the moles and number of atoms in a sample of Al

What do we know?
- Sample contains 10.0 g of Al
- Mass of 1 mole (6.022×10^{23} atoms) of Al = 26.93 g

How do we get there?
We can calculate the number of moles of Al in a 10.0-g sample as follows:

$$10.0 \ \text{g Al} \times \frac{1 \ \text{mol Al}}{26.98 \ \text{g Al}} = 0.371 \ \text{mol Al atoms}$$

The number of atoms in 10.0 g (0.371 mole) of aluminum is

$$0.371 \ \text{mol Al} \times \frac{6.022 \times 10^{23} \ \text{atoms}}{1 \ \text{mol Al}} = 2.23 \times 10^{23} \ \text{atoms}$$

REALITY CHECK | One mole of Al has a mass of 26.98 g and contains 6.022×10^{23} atoms. Our sample is 10.0 g, which is roughly 1/3 of 26.98. Thus the calculated amount should be on the order of 1/3 of 6×10^{23}, which it is.

See Exercise 5-46

INTERACTIVE EXAMPLE 5-4

Calculating Numbers of Atoms

A silicon chip used in an integrated circuit of a microcomputer has a mass of 5.68 mg. How many silicon (Si) atoms are present in the chip?

Solution

Where are we going?
To calculate the atoms of Si in the chip

What do we know?

- The chip has 5.68 mg of Si
- Mass of 1 mole (6.022×10^{23} atoms) of Si = 28.09 g

How do we get there?
The strategy for doing this problem is to convert from milligrams of silicon to grams of silicon, then to moles of silicon, and finally to atoms of silicon:

$$5.68 \text{ mg Si} \times \frac{1 \text{ g Si}}{1000 \text{ mg Si}} = 5.68 \times 10^{-3} \text{ g Si}$$

$$5.68 \times 10^{-3} \text{ g Si} \times \frac{1 \text{ mol Si}}{28.09 \text{ g Si}} = 2.02 \times 10^{-4} \text{ mol Si}$$

$$2.02 \times 10^{-4} \text{ mol Si} \times \frac{6.022 \times 10^{23} \text{ atoms}}{1 \text{ mol Si}} = 1.22 \times 10^{20} \text{ atoms}$$

REALITY CHECK | Note that 5.68 mg of silicon is clearly much less than 1 mole of silicon (which has a mass of 28.09 g), so the final answer of 1.22×10^{20} atoms (compared with 6.022×10^{23} atoms) is in the right direction. ◄ ⓘ

See Exercise 5-47

ⓘ Always check to see if your answer is sensible. Paying careful attention to units and making sure the answer is reasonable can help you detect an inverted conversion factor or a number that was incorrectly entered in your calculator.

INTERACTIVE EXAMPLE 5-5

Calculating the Number of Moles and Mass

Cobalt (Co) is a metal that is added to steel to improve its resistance to corrosion. ◄ Calculate both the number of moles in a sample of cobalt containing 5.00×10^{20} atoms and the mass of the sample.

Solution

Where are we going?
To calculate the number of moles and the mass of a sample of Co

What do we know?

- Sample contains 5.00×10^{20} atoms of Co

How do we get there?
Note that the sample of 5.00×10^{20} atoms of cobalt is less than 1 mole (6.022×10^{23} atoms) of cobalt. What fraction of a mole it represents can be determined as follows:

$$5.00 \times 10^{20} \text{ atoms Co} \times \frac{1 \text{ mol Co}}{6.022 \times 10^{23} \text{ atoms Co}} = 8.30 \times 10^{-4} \text{ mol Co}$$

Since the mass of 1 mole of cobalt atoms is 58.93 g, the mass of 5.00×10^{20} atoms can be determined as follows:

$$8.30 \times 10^{-4} \text{ mol Co} \times \frac{58.93 \text{ g Co}}{1 \text{ mol Co}} = 4.89 \times 10^{-2} \text{ g Co}$$

REALITY CHECK | In this case the sample contains 5×10^{20} atoms, which is approximately 1/1000 of a mole. Thus the sample should have a mass of about $(1/1000)(58.93) \cong 0.06$. Our answer of $\sim$0.05 makes sense.

See Exercise 5-48

▲ Fragments of cobalt metal.

Russ Lappa/Science Source

5-5 | Molar Mass

A chemical compound is, ultimately, a collection of atoms. For example, methane (the major component of natural gas) consists of molecules that each contain one carbon and four hydrogen atoms (CH_4). How can we calculate the mass of 1 mole of methane? That is, what is the mass of 6.022×10^{23} CH_4 molecules? Since each CH_4 molecule contains one carbon atom and four hydrogen atoms, 1 mole of CH_4 molecules contains 1 mole of carbon atoms and 4 moles of hydrogen atoms. The mass of 1 mole of methane can be found by summing the masses of carbon and hydrogen present:

In this case, the term 12.01 limits the number of significant figures.

$$\text{Mass of 1 mol C} = 12.01 \text{ g}$$
$$\text{Mass of 4 mol H} = 4 \times 1.008 \text{ g}$$
$$\text{Mass of 1 mol } CH_4 = 16.04 \text{ g}$$

Because 16.04 g represents the mass of 1 mole of methane molecules, it makes sense to call it the *molar mass* for methane. Thus the **molar mass** of a substance is *the mass in grams of one mole of the compound.* Traditionally, the term *molecular weight* has been used for this quantity. However, we will use molar mass exclusively in this text. The molar mass of a known substance is obtained by summing the masses of the component atoms as we did for methane.

A substance's molar mass is the mass in grams of 1 mole of the substance.

Methane is a molecular compound—its components are molecules. Many substances are ionic—they contain simple ions or polyatomic ions. Examples are NaCl (which contains Na^+ and Cl^-) and $CaCO_3$ (which contains Ca^{2+} and CO_3^{2-}). Because ionic compounds do not contain molecules, we need a special name for the fundamental unit of these materials. Instead of molecule, we use the term *formula unit.* Thus $CaCO_3$ is the formula unit for calcium carbonate, and NaCl is the formula unit for sodium chloride.

INTERACTIVE EXAMPLE 5-6

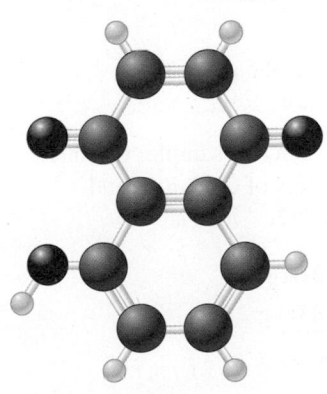

Juglone

Carbon

Oxygen

Hydrogen

Calculating Molar Mass I

Juglone, a dye known for centuries, is produced from the husks of black walnuts. It is also a natural herbicide (weed killer) that kills off competitive plants around the black walnut tree but does not affect grass and other noncompetitive plants. The formula for juglone is $C_{10}H_6O_3$. ◄

a. Calculate the molar mass of juglone.

b. A sample of 1.56×10^{-2} g of pure juglone was extracted from black walnut husks. How many moles of juglone does this sample represent?

Solution

a. The molar mass is obtained by summing the masses of the component atoms. In 1 mole of juglone there are 10 moles of carbon atoms, 6 moles of hydrogen atoms, and 3 moles of oxygen atoms:

$$
\begin{aligned}
10 \text{ C:} & \quad 10 \times 12.01 \text{ g} = 120.1 \text{ g} \\
6 \text{ H:} & \quad 6 \times 1.008 \text{ g} = 6.048 \text{ g} \\
3 \text{ O:} & \quad 3 \times 16.00 \text{ g} = \underline{48.00 \text{ g}} \\
\text{Mass of 1 mol } C_{10}H_6O_3 & = 174.1 \text{ g}
\end{aligned}
$$

The mass of 1 mole of juglone is 174.1 g, which is the molar mass.

b. The mass of 1 mole of this compound is 174.1 g; thus 1.56×10^{-2} g is much less than a mole. The exact fraction of a mole can be determined as follows:

$$1.56 \times 10^{-2} \text{ g juglone} \times \frac{1 \text{ mol juglone}}{174.1 \text{ g juglone}} = 8.96 \times 10^{-5} \text{ mol juglone}$$

See Exercises 5-51 through 5-54

INTERACTIVE EXAMPLE 5-7

▲ Calcite ($CaCO_3$) crystal.

Charles D. Winters

Calculating Molar Mass II

Calcium carbonate ($CaCO_3$), also called *calcite*, is the principal mineral found in limestone, marble, chalk, pearls, and the shells of marine animals such as clams. ◄

a. Calculate the molar mass of calcium carbonate.

b. A certain sample of calcium carbonate contains 4.86 moles. What is the mass in grams of this sample? What is the mass of the CO_3^{2-} ions present?

Solution

a. Calcium carbonate is an ionic compound composed of Ca^{2+} and CO_3^{2-} ions. In 1 mole of calcium carbonate there are 1 mole of Ca^{2+} ions and 1 mole of CO_3^{2-} ions. The molar mass is calculated by summing the masses of the components:

$$Ca^{2+}: \quad 1 \times 40.08 \text{ g} = \quad 40.08 \text{ g}$$

CO_3^{2-}:

$$1 \text{ C}: \quad 1 \times 12.01 \text{ g} = \quad 12.01 \text{ g}$$
$$3 \text{ O}: \quad 3 \times 16.00 \text{ g} = \quad \underline{48.00 \text{ g}}$$
$$\text{Mass of 1 mol } CaCO_3 = 100.09 \text{ g}$$

Thus the mass of 1 mole of $CaCO_3$ (1 mol Ca^{2+} plus 1 mol CO_3^{2-}) is 100.09 g. This is the molar mass.

b. The mass of 1 mole of $CaCO_3$ is 100.09 g. The sample contains nearly 5 moles, or close to 500 g. The exact amount is determined as follows:

$$4.86 \text{ mol } CaCO_3 \times \frac{100.09 \text{ g } CaCO_3}{1 \text{ mol } CaCO_3} = 486 \text{ g } CaCO_3$$

To find the mass of carbonate ions (CO_3^{2-}) present in this sample, we must realize that 4.86 moles of $CaCO_3$ contains 4.86 moles of Ca^{2+} ions and 4.86 moles of CO_3^{2-} ions. The mass of 1 mole of CO_3^{2-} ions is

$$1 \text{ C}: \quad 1 \times 12.01 = 12.01 \text{ g}$$
$$3 \text{ O}: \quad 3 \times 16.00 = \underline{48.00 \text{ g}}$$
$$\text{Mass of 1 mol } CO_3^{2-} = 60.01 \text{ g}$$

Thus the mass of 4.86 moles of CO_3^{2-} ions is

$$4.86 \text{ mol } CO_3^{2-} \times \frac{60.01 \text{ g } CO_3^{2-}}{1 \text{ mol } CO_3^{2-}} = 292 \text{ g } CO_3^{2-}$$

See Exercises 5-55 through 5-58

INTERACTIVE EXAMPLE 5-8

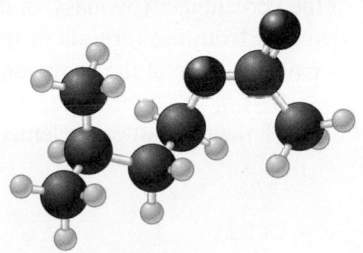

Isopentyl acetate

 Carbon

 Oxygen

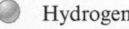 Hydrogen

Kenneth Lorenzen

▲ Isopentyl acetate is released when a bee stings.

Molar Mass and Numbers of Molecules

Isopentyl acetate ($C_7H_{14}O_2$) is the compound responsible for the scent of bananas. ◄ Interestingly, bees release about 1 μg (1×10^{-6} g) of this compound when they sting. The resulting scent attracts other bees to join the attack. How many molecules of isopentyl acetate are released in a typical bee sting? ◄ How many atoms of carbon are present?

Solution

Where are we going?
To calculate the number of molecules of isopentyl acetate and the number of carbon atoms in a bee sting

What do we know?
- Mass of isopentyl acetate in a typical bee sting is 1 microgram = 1×10^{-6} g

How do we get there?
Since we are given a mass of isopentyl acetate and want to find the number of molecules, we must first compute the molar mass of $C_7H_{14}O_2$:

$$7 \text{ mol C} \times 12.01\frac{\text{g}}{\text{mol}} = 84.07 \text{ g C}$$

$$14 \text{ mol H} \times 1.008\frac{\text{g}}{\text{mol}} = 14.11 \text{ g H}$$

$$2 \text{ mol O} \times 16.00\frac{\text{g}}{\text{mol}} = \frac{32.00 \text{ g O}}{130.18 \text{ g}}$$

This means that 1 mole of isopentyl acetate (6.022×10^{23} molecules) has a mass of 130.18 g.

To find the number of molecules released in a sting, we must first determine the number of moles of isopentyl acetate in 1×10^{-6} g:

$$1 \times 10^{-6} \text{ g } C_7H_{14}O_2 \times \frac{1 \text{ mol } C_7H_{14}O_2}{130.18 \text{ g } C_7H_{14}O_2} = 8 \times 10^{-9} \text{ mol } C_7H_{14}O_2$$

Since 1 mole is 6.022×10^{23} units, we can determine the number of molecules:

$$8 \times 10^{-9} \text{ mol } C_7H_{14}O_2 \times \frac{6.022 \times 10^{23} \text{ molecules}}{1 \text{ mol } C_7H_{14}O_2} = 5 \times 10^{15} \text{ molecules}$$

To determine the number of carbon atoms present, we must multiply the number of molecules by 7, since each molecule of isopentyl acetate contains seven carbon atoms:

$$5 \times 10^{15} \text{ molecules} \times \frac{7 \text{ carbon atoms}}{\text{molecule}} = 4 \times 10^{16} \text{ carbon atoms}$$

Note: In keeping with our practice of always showing the correct number of significant figures, we have rounded after each step. However, if extra digits are carried throughout this problem, the final answer rounds to 3×10^{16}. ◄ *ⓘ*

ⓘ To show the correct number of significant figures in each calculation, we round after each step. In your calculations, always carry extra significant figures through to the end, then round.

See Exercises 5-59 through 5-64

5-6 | Percent Composition of Compounds

There are two common ways of describing the composition of a compound: in terms of the numbers of its constituent atoms and in terms of the percentages (by mass) of its elements. We can obtain the mass percents of the elements from the formula of the compound by comparing the mass of each element present in 1 mole of the compound to the total mass of 1 mole of the compound.

For example, for ethanol, which has the formula C_2H_5OH, the mass of each element present and the molar mass are obtained as follows:

$$\text{Mass of C} = 2 \text{ mol} \times 12.01\frac{\text{g}}{\text{mol}} = 24.02 \text{ g}$$

$$\text{Mass of H} = 6 \text{ mol} \times 1.008\frac{\text{g}}{\text{mol}} = 6.048 \text{ g}$$

$$\text{Mass of O} = 1 \text{ mol} \times 16.00\frac{\text{g}}{\text{mol}} = \underline{16.00 \text{ g}}$$

$$\text{Mass of 1 mol } C_2H_5OH = 46.07 \text{ g}$$

The **mass percent** (often called the *weight percent*) of carbon in ethanol can be computed by comparing the mass of carbon in 1 mole of ethanol to the total mass of 1 mole of ethanol and multiplying the result by 100%:

$$\text{Mass percent of C} = \frac{\text{mass of C in 1 mol } C_2H_5OH}{\text{mass of 1 mol } C_2H_5OH} \times 100\%$$

$$= \frac{24.02 \text{ g}}{46.07 \text{ g}} \times 100\% = 52.14\%$$

The mass percents of hydrogen and oxygen in ethanol are obtained in a similar manner:

$$\text{Mass percent of H} = \frac{\text{mass of H in 1 mol } C_2H_5OH}{\text{mass of 1 mol } C_2H_5OH} \times 100\%$$

$$= \frac{6.048 \text{ g}}{46.07 \text{ g}} \times 100\% = 13.13\%$$

$$\text{Mass percent of O} = \frac{\text{mass of O in 1 mol } C_2H_5OH}{\text{mass of 1 mol } C_2H_5OH} \times 100\%$$

$$= \frac{16.00 \text{ g}}{46.07 \text{ g}} \times 100\% = 34.73\%$$

REALITY CHECK | Notice that the percentages add up to 100.00%; this provides a check that the calculations are correct.

INTERACTIVE EXAMPLE 5-9

Calculating Mass Percent

Carvone is a substance that occurs in two forms having different arrangements of atoms but the same molecular formula ($C_{10}H_{14}O$) and mass. ◄ One type of carvone gives caraway seeds their characteristic smell, and the other type is responsible for the smell of spearmint oil. Compute the mass percent of each element in carvone.

Solution

Where are we going?

To find the mass percent of each element in carvone

What do we know?

- Molecular formula, $C_{10}H_{14}O$

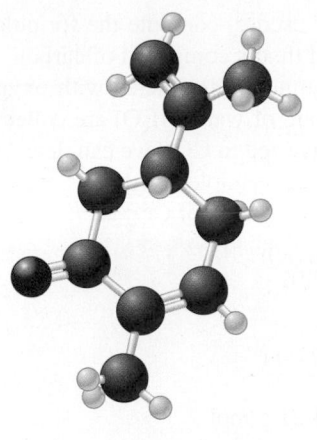

Carvone

What information do we need to find the mass percent?

- Mass of each element (we'll use 1 mole of carvone)
- Molar mass of carvone

How do we get there?
What is the mass of each element in 1 mole of $C_{10}H_{14}O$?

$$\text{Mass of C in 1 mol} = 10 \ \cancel{\text{mol}} \times 12.01 \frac{g}{\cancel{\text{mol}}} = 120.1 \ g$$

$$\text{Mass of H in 1 mol} = 14 \ \cancel{\text{mol}} \times 1.008 \frac{g}{\cancel{\text{mol}}} = 14.11 \ g$$

$$\text{Mass of O in 1 mol} = \ \ 1 \ \cancel{\text{mol}} \times 16.00 \frac{g}{\cancel{\text{mol}}} = 16.00 \ g$$

What is the molar mass of $C_{10}H_{14}O$?

$$120.1 \ g \ + 14.11 \ g \ + 16.00 \ g = 150.2 \ g$$
$$C_{10} \ \ + \ \ H_{14} \ \ + \ \ O \ \ = C_{10}H_{14}O$$

What is the mass percent of each element?
We find the fraction of the total mass contributed by each element and convert it to a percentage:

- Mass percent of C $= \dfrac{120.1 \ g \ C}{150.2 \ g \ C_{10}H_{14}O} \times 100\% = 79.96\%$

- Mass percent of H $= \dfrac{14.11 \ g \ H}{150.2 \ g \ C_{10}H_{14}O} \times 100\% = 9.394\%$

- Mass percent of O $= \dfrac{16.00 \ g \ O}{150.2 \ g \ C_{10}H_{14}O} \times 100\% = 10.65\%$

REALITY CHECK | Sum the individual mass percent values—they should total to 100% within round-off errors. In this case, the percentages add up to 100.00%.

See Exercises 5-73 and 5-74

5-7 | Determining the Formula of a Compound

When a new compound is prepared, one of the first items of interest is its formula. This is most often determined by taking a weighed sample of the compound and either decomposing it into its component elements or reacting it with oxygen to produce substances such as CO_2, H_2O, and N_2, which are then collected and weighed. A device for doing this type of analysis is shown in Fig. 5-5. The results of such analyses provide the mass of each type of element in the compound, which can be used to determine the mass percent of each element.

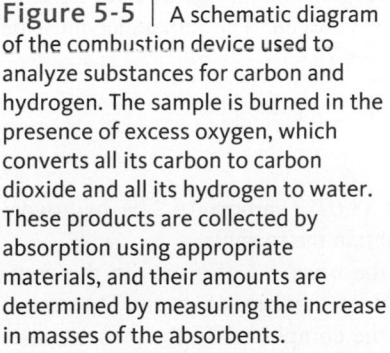

Figure 5-5 | A schematic diagram of the combustion device used to analyze substances for carbon and hydrogen. The sample is burned in the presence of excess oxygen, which converts all its carbon to carbon dioxide and all its hydrogen to water. These products are collected by absorption using appropriate materials, and their amounts are determined by measuring the increase in masses of the absorbents.

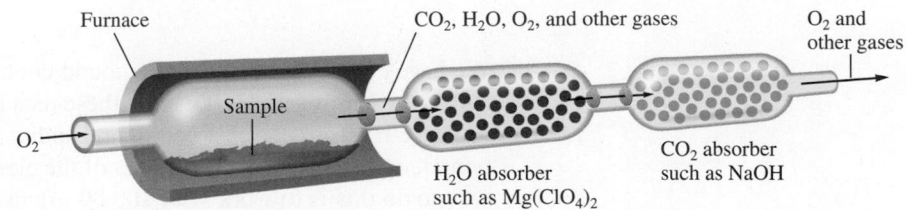

We will see how information of this type can be used to compute the formula of a compound. Suppose a substance has been prepared that is composed of carbon, hydrogen, and nitrogen. When 0.1156 gram of this compound is reacted with oxygen, 0.1638 gram of carbon dioxide (CO_2) and 0.1676 gram of water (H_2O) are collected. Assuming that all the carbon in the compound is converted to CO_2, we can determine the mass of carbon originally present in the 0.1156-gram sample. To do this, we must use the fraction (by mass) of carbon in CO_2. ◄ The molar mass of CO_2 is

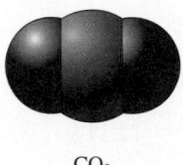

CO_2

$$\text{C:}\quad 1\ \cancel{\text{mol}} \times 12.01\frac{\text{g}}{\cancel{\text{mol}}} = 12.01\text{ g}$$

$$\text{O:}\quad 2\ \cancel{\text{mol}} \times 16.00\frac{\text{g}}{\cancel{\text{mol}}} = \underline{32.00\text{ g}}$$

$$\text{Molar mass of } CO_2 = 44.01\text{ g/mol}$$

The fraction of carbon present by mass is

$$\frac{\text{Mass of C}}{\text{Total mass of } CO_2} = \frac{12.01\text{ g C}}{44.01\text{ g } CO_2}$$

This factor can now be used to determine the mass of carbon in 0.1638 gram of CO_2:

$$0.1638\ \text{g } \cancel{CO_2} \times \frac{12.01\text{ g C}}{44.01\ \text{g } \cancel{CO_2}} = 0.04470\text{ g C}$$

Remember that this carbon originally came from the 0.1156-gram sample of unknown compound. Thus the mass percent of carbon in this compound is

$$\frac{0.04470\text{ g C}}{0.1156\text{ g compound}} \times 100\% = 38.67\%\text{ C}$$

The same procedure can be used to find the mass percent of hydrogen in the unknown compound. We assume that all the hydrogen present in the original 0.1156 gram of compound was converted to H_2O. ◄ The molar mass of H_2O is 18.02 grams, and the fraction of hydrogen by mass in H_2O is

H_2O

$$\frac{\text{Mass of H}}{\text{Mass of } H_2O} = \frac{2.016\text{ g H}}{18.02\text{ g } H_2O}$$

Therefore, the mass of hydrogen in 0.1676 gram of H_2O is

$$0.1676\ \text{g } \cancel{H_2O} \times \frac{2.016\text{ g H}}{18.02\ \text{g } \cancel{H_2O}} = 0.01875\text{ g H}$$

The mass percent of hydrogen in the compound is

$$\frac{0.01875\text{ g H}}{0.1156\text{ g compound}} \times 100\% = 16.22\%\text{ H}$$

The unknown compound contains only carbon, hydrogen, and nitrogen. So far we have determined that it is 38.67% carbon and 16.22% hydrogen. The remainder must be nitrogen:

$$100.00\% - (\underset{\substack{\uparrow \\ \%\ \text{C}}}{38.67\%} + \underset{\substack{\uparrow \\ \%\ \text{H}}}{16.22\%}) = 45.11\%\text{ N}$$

We have determined that the compound contains 38.67% carbon, 16.22% hydrogen, and 45.11% nitrogen. Next we use these data to obtain the formula.

Since the formula of a compound indicates the *numbers* of atoms in the compound, we must convert the masses of the elements to numbers of atoms. The easiest way to do this is to work with 100.00 grams of the compound. In the present case, 38.67% carbon by mass means 38.67 grams of carbon per 100.00 grams of compound;

16.22% hydrogen means 16.22 grams of hydrogen per 100.00 grams of compound; and so on. To determine the formula, we must calculate the number of carbon atoms in 38.67 grams of carbon, the number of hydrogen atoms in 16.22 grams of hydrogen, and the number of nitrogen atoms in 45.11 grams of nitrogen. We can do this as follows:

$$38.67 \text{ g C} \times \frac{1 \text{ mol C}}{12.01 \text{ g C}} = 3.220 \text{ mol C}$$

$$16.22 \text{ g H} \times \frac{1 \text{ mol H}}{1.008 \text{ g H}} = 16.09 \text{ mol H}$$

$$45.11 \text{ g N} \times \frac{1 \text{ mol N}}{14.01 \text{ g N}} = 3.220 \text{ mol N}$$

Thus 100.00 grams of this compound contains 3.220 moles of carbon atoms, 16.09 moles of hydrogen atoms, and 3.220 moles of nitrogen atoms.

We can find the smallest *whole-number ratio* of atoms in this compound by dividing each of the mole values above by the smallest of the three:

$$\text{C: } \frac{3.220}{3.220} = 1.000 = 1$$

$$\text{H: } \frac{16.09}{3.220} = 4.997 = 5$$

$$\text{N: } \frac{3.220}{3.220} = 1.000 = 1$$

Thus the formula might well be CH_5N. However, it also could be $C_2H_{10}N_2$ or $C_3H_{15}N_3$, and so on—that is, some multiple of the smallest whole-number ratio. Each of these alternatives also has the correct relative numbers of atoms. That is, any molecule that can be represented as $(CH_5N)_n$, where n is an integer, has the **empirical formula** CH_5N. To be able to specify the exact formula of the molecule involved, the **molecular formula**, we must know the molar mass.

Suppose we know that this compound with empirical formula CH_5N has a molar mass of 31.06 g/mol. How do we determine which of the possible choices represents the molecular formula? Since the molecular formula is always a whole-number multiple of the empirical formula, we must first find the empirical formula mass for CH_5N:

$$1 \text{ C: } 1 \times 12.01 \text{ g} = 12.01 \text{ g}$$
$$5 \text{ H: } 5 \times 1.008 \text{ g} = 5.040 \text{ g}$$
$$1 \text{ N: } 1 \times 14.01 \text{ g} = 14.01 \text{ g}$$
$$\text{Formula mass of } CH_5N = 31.06 \text{ g/mol}$$

This is the same as the known molar mass of the compound. Thus, in this case, the empirical formula and the molecular formula are the same; this substance consists of molecules with the formula CH_5N. It is quite common for the empirical and molecular formulas to be different; some examples are shown in Fig. 5-6. ◄ ⓘ

ⓘ Molecular formula = (empirical formula)$_n$, where n is an integer.

Figure 5-6 | Examples of substances whose empirical and molecular formulas differ. Notice that molecular formula = (empirical formula)$_n$, where n is an integer.

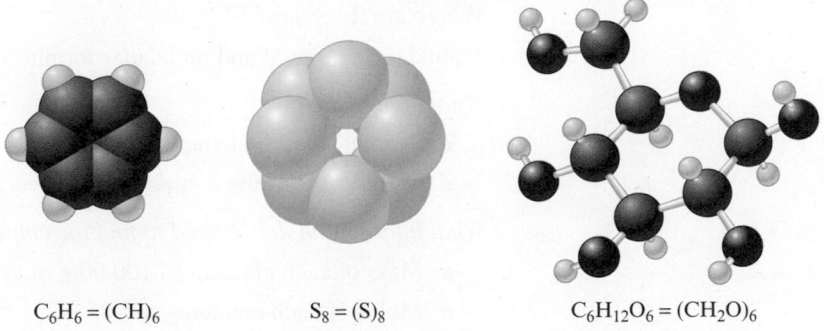

$C_6H_6 = (CH)_6$ $S_8 = (S)_8$ $C_6H_{12}O_6 = (CH_2O)_6$

PROBLEM-SOLVING STRATEGY

Empirical Formula Determination

- Since mass percentage gives the number of grams of a particular element per 100 grams of compound, base the calculation on 100 grams of compound. Each percent will then represent the mass in grams of that element.
- Determine the number of moles of each element present in 100 grams of compound using the atomic masses of the elements present.
- Divide each value of the number of moles by the smallest of the values. If each resulting number is a whole number (after appropriate rounding), these numbers represent the subscripts of the elements in the empirical formula.
- If the numbers obtained in the previous step are not whole numbers, multiply each number by an integer so that the results are all whole numbers. ◀ 𝑖

𝑖 Numbers very close to whole numbers, such as 9.92 and 1.08, should be rounded to whole numbers. Numbers such as 2.25, 4.33, and 2.72 should not be rounded to whole numbers.

PROBLEM-SOLVING STRATEGY

Determining Molecular Formula from Empirical Formula

- Obtain the empirical formula.
- Compute the mass corresponding to the empirical formula.
- Calculate the ratio

$$\frac{\text{Molar mass}}{\text{Empirical formula mass}}$$

- The integer from the previous step represents the number of empirical formula units in one molecule. When the empirical formula subscripts are multiplied by this integer, the molecular formula results. This procedure is summarized by the following equation:

$$\text{Molecular formula} = (\text{empirical formula}) \times \frac{\text{molar mass}}{\text{empirical formula mass}}$$

INTERACTIVE EXAMPLE 5-10

Determining Empirical and Molecular Formulas I

Determine the empirical and molecular formulas for a compound that gives the following percentages on analysis (in mass percents):

$$71.65\% \text{ Cl} \qquad 24.27\% \text{ C} \qquad 4.07\% \text{ H}$$

The molar mass is known to be 98.96 g/mol.

Solution

Where are we going?

To find the empirical and molecular formulas for the given compound

What do we know?

- Percent of each element
- Molar mass of the compound is 98.96 g/mol

What information do we need to find the empirical formula?

- Mass of each element in 100.00 g of compound
- Moles of each element

How do we get there?
What is the mass of each element in 100.00 g of compound?

Cl   71.65 g   C   24.27 g   H   4.07 g

What are the moles of each element in 100.00 g of compound?

$$71.65 \text{ g Cl} \times \frac{1 \text{ mol Cl}}{35.45 \text{ g Cl}} = 2.021 \text{ mol Cl}$$

$$24.27 \text{ g C} \times \frac{1 \text{ mol C}}{12.01 \text{ g C}} = 2.021 \text{ mol C}$$

$$4.07 \text{ g H} \times \frac{1 \text{ mol H}}{1.008 \text{ g H}} = 4.04 \text{ mol H}$$

What is the empirical formula for the compound?
Dividing each mole value by 2.021 (the smallest number of moles present), we find the empirical formula $ClCH_2$.

What is the molecular formula for the compound?
Compare the empirical formula mass to the molar mass.

Empirical formula mass = 49.48 g/mol (Confirm this!)

Molar mass is given = 98.96 g/mol

$$\frac{\text{Molar mass}}{\text{Empirical formula mass}} = \frac{98.96 \text{ g/mol}}{49.48 \text{ g/mol}} = 2$$

$$\text{Molecular formula} = (ClCH_2)_2 = Cl_2C_2H_4$$

This substance is composed of molecules with the formula $Cl_2C_2H_4$.

Note: The method we use here allows us to determine the molecular formula of a compound but not its structural formula. The compound $Cl_2C_2H_4$ is called *dichloroethane*. There are two forms of this compound, shown in Fig. 5-7. The form on the right was formerly used as an additive in leaded gasoline.

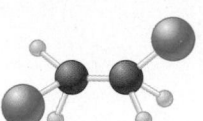
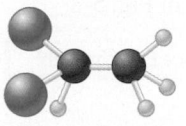

Figure 5-7 | The two forms of dichloroethane.

See Exercises 5-87 and 5-88

INTERACTIVE EXAMPLE 5-11

Determining Empirical and Molecular Formulas II

A white powder is analyzed and found to contain 43.64% phosphorus and 56.36% oxygen by mass. The compound has a molar mass of 283.88 g/mol. What are the compound's empirical and molecular formulas?

Solution

Where are we going?

To find the empirical and molecular formulas for the given compound

What do we know?
- Percent of each element
- Molar mass of the compound is 283.88 g/mol

What information do we need to find the empirical formula?
- Mass of each element in 100.00 g of compound
- Moles of each element

How do we get there?
What is the mass of each element in 100.00 g of compound?

P   43.64 g   O   56.36 g

What are the moles of each element in 100.00 g of compound?

$$43.64 \text{ g P} \times \frac{1 \text{ mol P}}{30.97 \text{ g P}} = 1.409 \text{ mol P}$$

$$56.36 \text{ g O} \times \frac{1 \text{ mol O}}{16.00 \text{ g O}} = 3.523 \text{ mol O}$$

What is the empirical formula for the compound?
Dividing each mole value by the smaller one gives

$$\frac{1.409}{1.409} = 1 \text{ P} \quad \text{and} \quad \frac{3.523}{1.409} = 2.5 \text{ O}$$

This yields the formula $PO_{2.5}$. Since compounds must contain whole numbers of atoms, the empirical formula should contain only whole numbers. To obtain the simplest set of whole numbers, we multiply both numbers by 2 to give the empirical formula P_2O_5.

What is the molecular formula for the compound?
Compare the empirical formula mass to the molar mass.

Empirical formula mass = 141.94 g/mol (Confirm this!)

Molar mass is given = 283.88 g/mol

$$\frac{\text{Molar mass}}{\text{Empirical formula mass}} = \frac{283.88}{141.94} = 2$$

The molecular formula is $(P_2O_5)_2$, or P_4O_{10}.

Note: The structural formula for this interesting compound is given in Fig. 5-8.

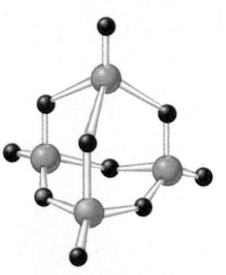

Figure 5-8 | The structure of P_4O_{10}. Note that some of the oxygen atoms act as "bridges" between the phosphorus atoms. This compound has a great affinity for water and is often used as a desiccant, or drying agent.

See Exercise 5-89

In Examples 5-10 and 5-11 we found the molecular formula by comparing the empirical formula mass with the molar mass. There is an alternate way to obtain the molecular formula. For instance, in Example 5-10 we know the molar mass of the compound is 98.96 g/mol. This means that 1 mole of the compound weighs 98.96 grams. Since we also know the mass percentages of each element, we can compute the mass of each element present in 1 mole of compound:

$$\text{Chlorine:} \quad \frac{71.65 \text{ g Cl}}{100.0 \text{ g compound}} \times \frac{98.96 \text{ g}}{\text{mol}} = \frac{70.90 \text{ g Cl}}{\text{mol compound}}$$

$$\text{Carbon:} \quad \frac{24.27 \text{ g C}}{100.0 \text{ g compound}} \times \frac{98.96 \text{ g}}{\text{mol}} = \frac{24.02 \text{ g C}}{\text{mol compound}}$$

$$\text{Hydrogen:} \quad \frac{4.07 \text{ g H}}{100.0 \text{ g compound}} \times \frac{98.96 \text{ g}}{\text{mol}} = \frac{4.03 \text{ g H}}{\text{mol compound}}$$

Now we can compute moles of atoms present per mole of compound:

$$\text{Chlorine:} \quad \frac{70.90 \text{ g Cl}}{\text{mol compound}} \times \frac{1 \text{ mol Cl}}{35.45 \text{ g Cl}} = \frac{2.000 \text{ mol Cl}}{\text{mol compound}}$$

$$\text{Carbon:} \quad \frac{24.02 \text{ g C}}{\text{mol compound}} \times \frac{1 \text{ mol C}}{12.01 \text{ g C}} = \frac{2.000 \text{ mol C}}{\text{mol compound}}$$

$$\text{Hydrogen:} \quad \frac{4.03 \text{ g H}}{\text{mol compound}} \times \frac{1 \text{ mol H}}{1.008 \text{ g H}} = \frac{4.00 \text{ mol H}}{\text{mol compound}}$$

Thus 1 mole of the compound contains 2 moles of Cl atoms, 2 moles of C atoms, and 4 moles of H atoms, and the molecular formula is $Cl_2C_2H_4$, as obtained in Example 5-10.

> **PROBLEM-SOLVING STRATEGY**
>
> **Determining Molecular Formula from Mass Percent and Molar Mass**
>
> - Using the mass percentages and the molar mass, determine the mass of each element present in one mole of compound.
> - Determine the number of moles of each element present in one mole of compound.
> - The integers from the previous step represent the subscripts in the molecular formula.

INTERACTIVE EXAMPLE 5-12

▲ Computer-generated molecule of caffeine.

Ken O'Donoghue © Cengage Learning

Determining a Molecular Formula

Caffeine, a stimulant found in coffee, tea, and chocolate, contains 49.48% carbon, 5.15% hydrogen, 28.87% nitrogen, and 16.49% oxygen by mass and has a molar mass of 194.2 g/mol. Determine the molecular formula of caffeine.

Solution

Where are we going?
To find the molecular formula for caffeine ◄

What do we know?
- Percent of each element

  - 49.48% C
  - 5.15% H
  - 28.87% N
  - 16.49% O

- Molar mass of caffeine is 194.2 g/mol

What information do we need to find the molecular formula?
- Mass of each element (in 1 mole of caffeine)
- Mole of each element (in 1 mole of caffeine)

How do we get there?
What is the mass of each element in 1 mole (194.2 g) of caffeine?

$$\frac{49.48 \text{ g C}}{100.0 \text{ g caffeine}} \times \frac{194.2 \text{ g}}{\text{mol}} = \frac{96.09 \text{ g C}}{\text{mol caffeine}}$$

$$\frac{5.15 \text{ g H}}{100.0 \text{ g caffeine}} \times \frac{194.2 \text{ g}}{\text{mol}} = \frac{10.0 \text{ g H}}{\text{mol caffeine}}$$

$$\frac{28.87 \text{ g N}}{100.0 \text{ g caffeine}} \times \frac{194.2 \text{ g}}{\text{mol}} = \frac{56.07 \text{ g N}}{\text{mol caffeine}}$$

$$\frac{16.49 \text{ g O}}{100.0 \text{ g caffeine}} \times \frac{194.2 \text{ g}}{\text{mol}} = \frac{32.02 \text{ g O}}{\text{mol caffeine}}$$

What are the moles of each element in 1 mole of caffeine?

$$\text{Carbon:} \quad \frac{96.09 \text{ g C}}{\text{mol caffeine}} \times \frac{1 \text{ mol C}}{12.01 \text{ g C}} = \frac{8.001 \text{ mol C}}{\text{mol caffeine}}$$

$$\text{Hydrogen:} \quad \frac{10.0 \text{ g H}}{\text{mol caffeine}} \times \frac{1 \text{ mol H}}{1.008 \text{ g H}} = \frac{9.92 \text{ mol H}}{\text{mol caffeine}}$$

Nitrogen: $\dfrac{56.07 \text{ g N}}{\text{mol caffeine}} \times \dfrac{1 \text{ mol N}}{14.01 \text{ g N}} = \dfrac{4.002 \text{ mol N}}{\text{mol caffeine}}$

Oxygen: $\dfrac{32.02 \text{ g O}}{\text{mol caffeine}} \times \dfrac{1 \text{ mol O}}{16.00 \text{ g O}} = \dfrac{2.001 \text{ mol O}}{\text{mol caffeine}}$

Rounding the numbers to integers gives the molecular formula for caffeine: $C_8H_{10}N_4O_2$.

See Exercise 5-90

5-8 | Chemical Equations

Chemical Reactions

A chemical change involves a reorganization of the atoms in one or more substances. For example, when the methane (CH_4) in natural gas combines with oxygen (O_2) in the air and burns, carbon dioxide (CO_2) and water (H_2O) are formed. This process is represented by a **chemical equation** with the **reactants** (here methane and oxygen) on the left side of an arrow and the **products** (carbon dioxide and water) on the right side:

$$CH_4 + O_2 \longrightarrow CO_2 + H_2O$$
$$\underset{\text{Reactants}}{} \qquad \underset{\text{Products}}{}$$

Notice that the atoms have been reorganized. *Bonds have been broken, and new ones have been formed.* It is important to recognize that *in a chemical reaction, atoms are neither created nor destroyed.* All atoms present in the reactants must be accounted for among the products. In other words, there must be the same number of each type of atom on the product side and on the reactant side of the arrow. Making sure that this rule is obeyed is called **balancing a chemical equation** for a reaction.

The equation (shown above) for the reaction between CH_4 and O_2 is not balanced. We can see this from the following representation of the reaction:

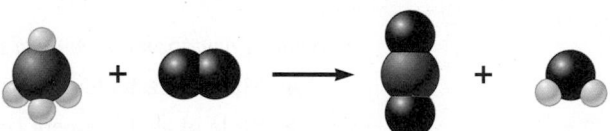

Notice that the number of oxygen atoms (in O_2) on the left of the arrow is two, while on the right there are three O atoms (in CO_2 and H_2O). Also, there are four hydrogen atoms (in CH_4) on the left and only two (in H_2O) on the right. Remember that a chemical reaction is simply a rearrangement of the atoms (a change in the way they are organized). Atoms are not created or destroyed in a chemical reaction. Thus the reactants and products must occur in numbers that give the same number of each type of atom among both the reactants and products. Simple trial and error will allow us to figure this out for the reaction of methane with oxygen. The needed numbers of molecules are

Notice that now we have the same number of each type of atom represented among the reactants and the products. ◀ ⓘ

We can represent the preceding situation in a shorthand manner by the following chemical equation:

$$CH_4 + 2O_2 \longrightarrow CO_2 + 2H_2O$$

ⓘ Thinking in terms of the molecular level helps you differentiate between the coefficients and subscripts in a balanced chemical equation. Remember that subscripts define a particular molecule and coefficients address the number of molecules in a reaction.

We can check that the equation is balanced by comparing the number of each type of atom on both sides:

$$CH_4 + 2O_2 \longrightarrow CO_2 + 2H_2O$$

1 C 4 H   4 O   1 C 4 H   2 O 2 O

To summarize, we have

Reactants	Products
1 C	1 C
4 H	4 H
4 O	4 O

The Meaning of a Chemical Equation

The chemical equation for a reaction gives two important types of information: the nature of the reactants and products and the relative numbers of each.

The reactants and products in a specific reaction must be identified by experiment. Besides specifying the compounds involved in the reaction, the equation often gives the *physical states* of the reactants and products:

State	Symbol
Solid	(*s*)
Liquid	(*l*)
Gas	(*g*)
Dissolved in water (in aqueous solution)	(*aq*)

For example, when hydrochloric acid in aqueous solution is added to solid sodium hydrogen carbonate, the products carbon dioxide gas, liquid water, and sodium chloride (which dissolves in the water) are formed:

$$HCl(aq) + NaHCO_3(s) \longrightarrow CO_2(g) + H_2O(l) + NaCl(aq)$$

The relative numbers of reactants and products in a reaction are indicated by the *coefficients* in the balanced equation. (The coefficients can be determined because we know that the same number of each type of atom must occur on both sides of the equation.) For example, the balanced equation

$$CH_4(g) + 2O_2(g) \longrightarrow CO_2(g) + 2H_2O(g)$$

can be interpreted in several equivalent ways, as shown in Table 5-2. Note that the total mass is 80 grams for both reactants and products. We expect the mass to remain constant, since chemical reactions involve only a rearrangement of atoms. Atoms, and therefore mass, are conserved in a chemical reaction.

From this discussion you can see that a balanced chemical equation gives you a great deal of information.

Table 5-2 | Information Conveyed by the Balanced Equation for the Combustion of Methane

Reactants		Products
$CH_4(g) + 2O_2(g)$	$\longrightarrow$	$CO_2(g) + 2H_2O(g)$
1 molecule + 2 molecules	$\longrightarrow$	1 molecule + 2 molecules
1 mol + 2 mol	$\longrightarrow$	1 mol + 2 mol
6.022×10^{23} molecules + 2 (6.022×10^{23} molecules)	$\longrightarrow$	6.022×10^{23} molecules + 2 (6.022×10^{23} molecules)
16 g + 2 (32 g)	$\longrightarrow$	44 g + 2 (18 g)
80 g reactants	$\longrightarrow$	80 g products

5-9 | Balancing Chemical Equations

Recall that a chemical equation is used to describe a chemical reaction. Remember that a chemical reaction involves a rearrangement of atoms. This means that atoms are neither created nor destroyed in a reaction. All atoms present in the reactants must be accounted for in the products.

An unbalanced chemical equation is of limited use. Whenever you see an equation, you should ask yourself whether it is balanced. The principle that lies at the heart of the balancing process is that atoms are conserved in a chemical reaction. The same number of each type of atom must be found among the reactants and products. It is also important to recognize that the identities of the reactants and products of a reaction are determined by experimental observation. For example, when liquid ethanol is burned in the presence of sufficient oxygen gas, the products are always carbon dioxide and water. When the equation for this reaction is balanced, the *identities* of the reactants and products must not be changed. The formulas of the compounds must never be changed in balancing a chemical equation. That is, the subscripts in a formula cannot be changed, nor can atoms be added or subtracted from a formula.

> **Critical Thinking**
>
> What if a friend was balancing chemical equations by changing the values of the subscripts instead of using the coefficients? How would you explain to your friend that this was the wrong thing to do?

ⓘ In balancing equations, start with the most complicated molecule.

Most chemical equations can be balanced by inspection, that is, by trial and error. It is always best to start with the most complicated molecules (those containing the greatest number of atoms). ◄ ⓘ For example, consider the reaction of ethanol with oxygen, given by the unbalanced equation

$$C_2H_5OH(l) + O_2(g) \longrightarrow CO_2(g) + H_2O(g)$$

which can be represented by the following molecular models:

Notice that the carbon and hydrogen atoms are not balanced. There are two carbon atoms on the left and one on the right, and there are six hydrogens on the left and two on the right. We need to find the correct numbers of reactants and products so that we have the same number of all types of atoms among the reactants and products. We will balance the equation "by inspection" (a systematic trial-and-error procedure).

The most complicated molecule here is C_2H_5OH. We will begin by balancing the products that contain the atoms in C_2H_5OH. Since C_2H_5OH contains two carbon atoms, we place the coefficient 2 before the CO_2 to balance the carbon atoms:

$$\underset{\text{2 C atoms}}{C_2H_5OH(l)} + O_2(g) \longrightarrow \underset{\text{2 C atoms}}{2CO_2(g)} + H_2O(g)$$

Since C_2H_5OH contains six hydrogen atoms, the hydrogen atoms can be balanced by placing a 3 before the H_2O:

$$\underset{(5 + 1)\, H}{C_2H_5OH(l)} + O_2(g) \longrightarrow 2CO_2(g) + \underset{(3 \times 2)\, H}{3H_2O(g)}$$

Last, we balance the oxygen atoms. Note that the right side of the preceding equation contains seven oxygen atoms, whereas the left side has only three. We can correct this by putting a 3 before the O_2 to produce the balanced equation:

$$\underbrace{C_2H_5OH(l) + \underset{6\,O}{3O_2(g)}}_{7\,O} \longrightarrow \underbrace{\underset{(2\times2)\,O}{2CO_2(g)} + \underset{3\,O}{3H_2O(g)}}_{7\,O}$$

Now we check:

$$C_2H_5OH(l) + 3O_2(g) \longrightarrow 2CO_2(g) + 3H_2O(g)$$

2 C atoms	2 C atoms
6 H atoms	6 H atoms
7 O atoms	7 O atoms

The equation is balanced.

The balanced equation can be represented as follows:

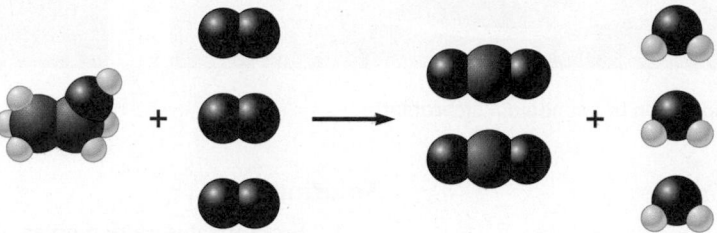

You can see that all the elements balance.

PROBLEM-SOLVING STRATEGY

Writing and Balancing the Equation for a Chemical Reaction

1. Determine what reaction is occurring. What are the reactants, the products, and the physical states involved?

2. Write the *unbalanced* equation that summarizes the reaction described in step 1.

3. Balance the equation by inspection, starting with the most complicated molecule(s). Determine what coefficients are necessary so that the same number of each type of atom appears on both reactant and product sides. Do not change the identities (formulas) of any of the reactants or products.

Critical Thinking

One part of the problem-solving strategy for balancing chemical equations is "starting with the most complicated molecule." What if you started with a different molecule? Could you still eventually balance the chemical equation? How would this approach be different from the suggested technique?

INTERACTIVE EXAMPLE 5-13

i Chromate and dichromate compounds are carcinogens (cancer-inducing agents) and should be handled very carefully.

Balancing a Chemical Equation I

Chromium compounds exhibit a variety of bright colors. When solid ammonium dichromate, $(NH_4)_2Cr_2O_7$, a vivid orange compound, is ignited, a spectacular reaction occurs. ◄ *i* ◄ Although the reaction is actually somewhat more complex, let's assume here that the products are solid chromium(III) oxide, nitrogen gas (consisting of N_2 molecules), and water vapor. Balance the equation for this reaction.

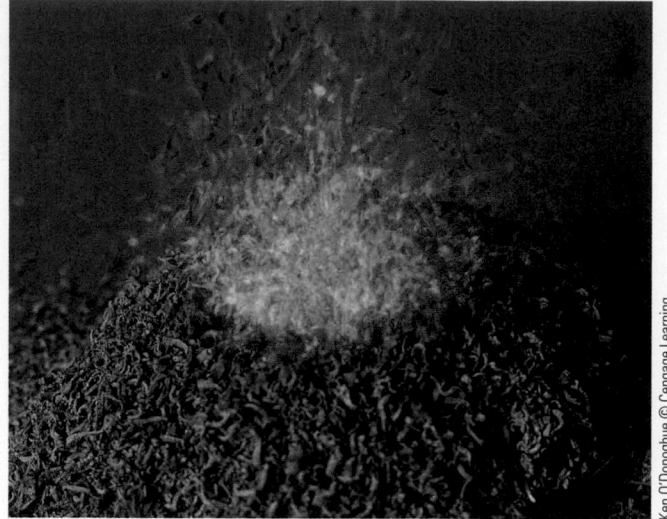

Ken O'Donoghue © Cengage Learning

Ken O'Donoghue © Cengage Learning

▲ Decomposition of ammonium dichromate.

Solution

1. From the description given, the reactant is solid ammonium dichromate, $(NH_4)_2Cr_2O_7(s)$, and the products are nitrogen gas, $N_2(g)$, water vapor, $H_2O(g)$, and solid chromium(III) oxide, $Cr_2O_3(s)$. The formula for chromium(III) oxide can be determined by recognizing that the Roman numeral III means that Cr^{3+} ions are present. For a neutral compound, the formula must then be Cr_2O_3, since each oxide ion is O^{2-}.

2. The unbalanced equation is

$$(NH_4)_2Cr_2O_7(s) \rightarrow Cr_2O_3(s) + N_2(g) + H_2O(g)$$

3. Note that nitrogen and chromium are balanced (two nitrogen atoms and two chromium atoms on each side), but hydrogen and oxygen are not. A coefficient of 4 for H_2O balances the hydrogen atoms:

$$\underset{(4 \times 2)\ H}{(NH_4)_2Cr_2O_7(s)} \rightarrow Cr_2O_3(s) + N_2(g) + \underset{(4 \times 2)\ H}{4H_2O(g)}$$

Note that in balancing the hydrogen we also have balanced the oxygen, since there are seven oxygen atoms in the reactants and in the products.

REALITY CHECK |

$$2\ N,\ 8\ H,\ 2\ Cr,\ 7\ O \rightarrow 2\ N,\ 8\ H,\ 2\ Cr,\ 7\ O$$

Reactant   Product
atoms   atoms

The equation is balanced.

See Exercises 5-95 through 5-98

INTERACTIVE EXAMPLE 5-14

Balancing a Chemical Equation II

At 1000°C, ammonia gas, $NH_3(g)$, reacts with oxygen gas to form gaseous nitric oxide, $NO(g)$, and water vapor. This reaction is the first step in the commercial production of nitric acid by the Ostwald process. Balance the equation for this reaction.

Solution

1, 2. The unbalanced equation for the reaction is

$$NH_3(g) + O_2(g) \rightarrow NO(g) + H_2O(g)$$

3. Because all the molecules in this equation are of about equal complexity, where we start in balancing it is rather arbitrary. Let's begin by balancing the hydrogen. A coefficient of 2 for NH_3 and a coefficient of 3 for H_2O give six atoms of hydrogen on both sides:

$$2NH_3(g) + O_2(g) \rightarrow NO(g) + 3H_2O(g)$$

The nitrogen can be balanced with a coefficient of 2 for NO:

$$2NH_3(g) + O_2(g) \rightarrow 2NO(g) + 3H_2O(g)$$

Finally, note that there are two atoms of oxygen on the left and five on the right. The oxygen can be balanced with a coefficient of $\frac{5}{2}$ for O_2:

$$2NH_3(g) + \frac{5}{2}O_2(g) \rightarrow 2NO(g) + 3H_2O(g)$$

However, the usual custom is to have whole-number coefficients. We simply multiply the entire equation by 2.

$$4NH_3(g) + 5O_2(g) \rightarrow 4NO(g) + 6H_2O(g)$$

REALITY CHECK | There are 4 N, 12 H, and 10 O on both sides, so the equation is balanced.

We can represent this balanced equation visually as

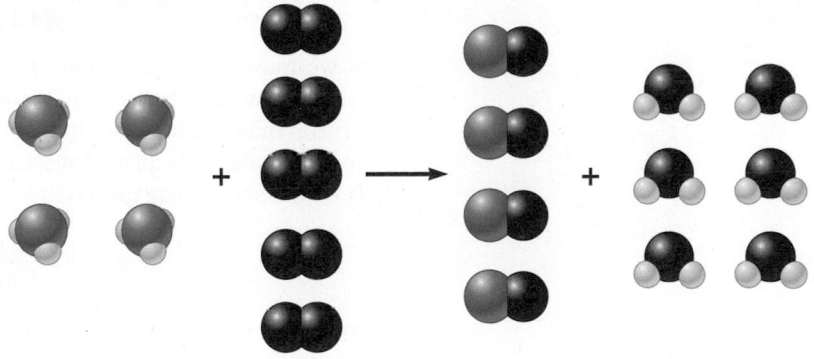

See Exercises 5-99 through 5-104

5-10 | Stoichiometric Calculations: Amounts of Reactants and Products

As we have seen in previous sections of this chapter, the coefficients in chemical equations represent *numbers* of molecules, not masses of molecules. However, when a reaction is to be run in a laboratory or chemical plant, the amounts of substances needed cannot be determined by counting molecules directly. Counting is always done by weighing. In this section we will see how chemical equations can be used to determine the *masses* of reacting chemicals.

To develop the principles for dealing with the stoichiometry of reactions, we will consider the reaction of propane with oxygen to produce carbon dioxide and water. We will consider the question: *"What mass of oxygen will react with 96.1 grams of*

i Before doing any calculations involving a chemical reaction, be sure the equation for the reaction is balanced.

propane?" In doing stoichiometry, the first thing we must do is *write the balanced chemical equation* for the reaction. ◄ *i* In this case the balanced equation is

$$C_3H_8(g) + 5O_2(g) \longrightarrow 3CO_2(g) + 4H_2O(g)$$

which can be visualized as

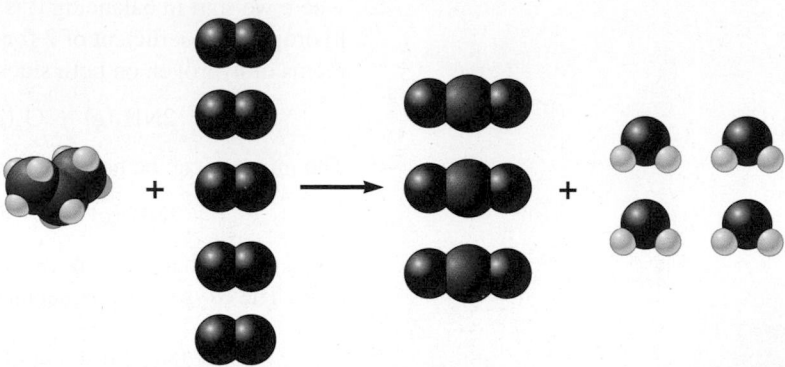

This equation means that 1 mole of C_3H_8 reacts with 5 moles of O_2 to produce 3 moles of CO_2 and 4 moles of H_2O. To use this equation to find the masses of reactants and products, we must be able to convert between masses and moles of substances. Thus we must first ask: *"How many moles of propane are present in 96.1 grams of propane?"* The molar mass of propane to three significant figures is 44.1 (that is, $3 \times 12.01 + 8 \times 1.008$). The moles of propane can be calculated as follows:

$$96.1 \text{ g } C_3H_8 \times \frac{1 \text{ mol } C_3H_8}{44.1 \text{ g } C_3H_8} = 2.18 \text{ mol } C_3H_8$$

Next we must take into account the fact that each mole of propane reacts with 5 moles of oxygen. The best way to do this is to use the balanced equation to construct a **mole ratio**. In this case we want to convert from moles of propane to moles of oxygen. From the balanced equation we see that 5 moles of O_2 is required for each mole of C_3H_8, so the appropriate ratio is

$$\frac{5 \text{ mol } O_2}{1 \text{ mol } C_3H_8}$$

Multiplying the number of moles of C_3H_8 by this factor gives the number of moles of O_2 required:

$$2.18 \text{ mol } C_3H_8 \times \frac{5 \text{ mol } O_2}{1 \text{ mol } C_3H_8} = 10.9 \text{ mol } O_2$$

Notice that the mole ratio is set up so that the moles of C_3H_8 cancel out, and the units that result are moles of O_2.

Since the original question asked for the mass of oxygen needed to react with 96.1 grams of propane, the 10.9 moles of O_2 must be converted to *grams.* Since the molar mass of O_2 is 32.0 g/mol,

$$10.9 \text{ mol } O_2 \times \frac{32.0 \text{ g } O_2}{1 \text{ mol } O_2} = 349 \text{ g } O_2$$

Therefore, 349 grams of oxygen is required to burn 96.1 grams of propane.

This example can be extended by asking: *"What mass of carbon dioxide is produced when 96.1 grams of propane is combusted with oxygen?"* In this case we must convert between moles of propane and moles of carbon dioxide. This can be accomplished by looking at the balanced equation, which shows that 3 moles of CO_2 is produced for each mole of C_3H_8 reacted. The mole ratio needed to convert from moles of propane to moles of carbon dioxide is

$$\frac{3 \text{ mol CO}_2}{1 \text{ mol C}_3\text{H}_8}$$

The conversion is

$$2.18 \text{ mol C}_3\text{H}_8 \times \frac{3 \text{ mol CO}_2}{1 \text{ mol C}_3\text{H}_8} = 6.54 \text{ mol CO}_2$$

Then, using the molar mass of CO_2 (44.0 g/mol), we calculate the mass of CO_2 produced:

$$6.54 \text{ mol CO}_2 \times \frac{44.0 \text{ g CO}_2}{1 \text{ mol CO}_2} = 288 \text{ g CO}_2$$

We will now summarize the sequence of steps needed to carry out stoichiometric calculations.

$$96.1 \text{ g C}_3\text{H}_8 \quad \boxed{\frac{1 \text{ mol C}_3\text{H}_8}{44.1 \text{ g C}_3\text{H}_8}} \quad 2.18 \text{ mol C}_3\text{H}_8 \quad \boxed{\frac{3 \text{ mol CO}_2}{1 \text{ mol C}_3\text{H}_8}} \quad 6.54 \text{ mol CO}_2 \quad \boxed{\frac{44.0 \text{ g CO}_2}{1 \text{ mol CO}_2}} \quad 288 \text{ g CO}_2$$

PROBLEM-SOLVING STRATEGY

Calculating Masses of Reactants and Products in Chemical Reactions

1. Balance the equation for the reaction.
2. Convert the known mass of the reactant or product to moles of that substance.
3. Use the balanced equation to set up the appropriate mole ratios.
4. Use the appropriate mole ratios to calculate the number of moles of the desired reactant or product.
5. Convert from moles back to grams if required by the problem.

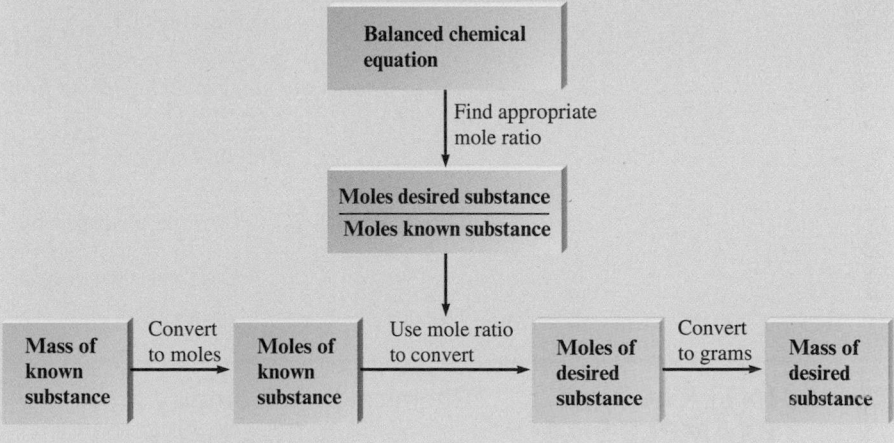

INTERACTIVE EXAMPLE 5-15

Chemical Stoichiometry I

Solid lithium hydroxide is used in space vehicles to remove exhaled carbon dioxide from the living environment by forming solid lithium carbonate and liquid water. What mass of gaseous carbon dioxide can be absorbed by 1.00 kg of lithium hydroxide?

Solution

Where are we going?

To find the mass of CO_2 absorbed by 1.00 kg LiOH

What do we know?

- Chemical reaction

$$LiOH(s) + CO_2(g) \longrightarrow Li_2CO_3(s) + H_2O(l)$$

- 1.00 kg LiOH

What information do we need to find the mass of CO_2?

- Balanced equation for the reaction

How do we get there?

1. *What is the balanced equation?*

$$2LiOH(s) + CO_2(g) \rightarrow Li_2CO_3(s) + H_2O(l)$$

2. *What are the moles of LiOH?*
   To find the moles of LiOH, we need to know the molar mass.

   What is the molar mass for LiOH?

   $$6.941 + 16.00 + 1.008 = 23.95 \text{ g/mol}$$

   Now we use the molar mass to find the moles of LiOH:

   $$1.00 \text{ kg LiOH} \times \frac{1000 \text{ g LiOH}}{1 \text{ kg LiOH}} \times \frac{1 \text{ mol LiOH}}{23.95 \text{ g LiOH}} = 41.8 \text{ mol LiOH}$$

3. *What is the mole ratio between CO_2 and LiOH in the balanced equation?*

   $$\frac{1 \text{ mol CO}_2}{2 \text{ mol LiOH}}$$

4. *What are the moles of CO_2?*

   $$41.8 \text{ mol LiOH} \times \frac{1 \text{ mol CO}_2}{2 \text{ mol LiOH}} = 20.9 \text{ mol CO}_2$$

5. *What is the mass of CO_2 formed from 1.00 kg LiOH?*

   $$20.9 \text{ mol CO}_2 \times \frac{44.0 \text{ g CO}_2}{1 \text{ mol CO}_2} = 9.20 \times 10^2 \text{ g CO}_2$$

Thus 920. g of $CO_2(g)$ will be absorbed by 1.00 kg of LiOH(s).

See Exercises 5-105 and 5-106

INTERACTIVE EXAMPLE 5-16

Chemical Stoichiometry II

Baking soda ($NaHCO_3$) is often used as an antacid. It neutralizes excess hydrochloric acid secreted by the stomach:

$$NaHCO_3(s) + HCl(aq) \longrightarrow NaCl(aq) + H_2O(l) + CO_2(aq)$$

Milk of magnesia, which is an aqueous suspension of magnesium hydroxide, is also used as an antacid:

$$Mg(OH)_2(s) + 2HCl(aq) \longrightarrow 2H_2O(l) + MgCl_2(aq)$$

Which is the more effective antacid per gram, $NaHCO_3$ or $Mg(OH)_2$?

Solution

Where are we going?

To compare the acid neutralizing power of $NaHCO_3$ and $Mg(OH)_2$ per gram

What do we know?

- Balanced equations for the reactions
- 1.00 g $NaHCO_3$
- 1.00 g $Mg(OH)_2$

How do we get there?

For $NaHCO_3$

1. *What is the balanced equation?*

$$NaHCO_3(s) + HCl(aq) \longrightarrow NaCl(aq) + H_2O(l) + CO_2(aq)$$

2. *What are the moles of $NaHCO_3$ in 1.00 g?*
   To find the moles of $NaHCO_3$, we need to know the molar mass (84.01 g/mol).

$$1.00 \text{ g NaHCO}_3 \times \frac{1 \text{ mol NaHCO}_3}{84.01 \text{ g NaHCO}_3} = 1.19 \times 10^{-2} \text{ mol NaHCO}_3$$

3. *What is the mole ratio between HCl and $NaHCO_3$ in the balanced equation?*

$$\frac{1 \text{ mol HCl}}{1 \text{ mol NaHCO}_3}$$

4. *What are the moles of HCl?*

$$1.19 \times 10^{-2} \text{ mol NaHCO}_3 \times \frac{1 \text{ mol HCl}}{1 \text{ mol NaHCO}_3} = 1.19 \times 10^{-2} \text{ moles of HCl}$$

Thus 1.00 g of $NaHCO_3$ will neutralize 1.19×10^{-2} mol HCl.

For $Mg(OH)_2$

1. *What is the balanced equation?*

$$Mg(OH)_2(s) + 2HCl(aq) \longrightarrow 2H_2O(l) + MgCl_2(aq)$$

2. *What are the moles of $Mg(OH)_2$ in 1.00 g?*
   To find the moles of $Mg(OH)_2$, we need to know the molar mass (58.32 g/mol).

$$1.00 \text{ g Mg(OH)}_2 \times \frac{1 \text{ mol Mg(OH)}_2}{58.32 \text{ g Mg(OH)}_2} = 1.71 \times 10^{-2} \text{ mol Mg(OH)}_2$$

3. *What is the mole ratio between HCl and $Mg(OH)_2$ in the balanced equation?*

$$\frac{2 \text{ mol HCl}}{1 \text{ mol Mg(OH)}_2}$$

4. *What are the moles of HCl?*

$$1.71 \times 10^{-2} \text{ mol Mg(OH)}_2 \times \frac{2 \text{ mol HCl}}{1 \text{ mol Mg(OH)}_2} = 3.42 \times 10^{-2} \text{ mol HCl}$$

Thus 1.00 g of $Mg(OH)_2$ will neutralize 3.42×10^{-2} moles of HCl.

Since 1.00 g $NaHCO_3$ neutralizes 1.19×10^{-2} moles of HCl and 1.00 g $Mg(OH)_2$ neutralizes 3.42×10^{-2} moles of HCl, $Mg(OH)_2$ is the more effective antacid.

Charles D. Winters

▲ Milk of magnesia contains a suspension of $Mg(OH)_2(s)$.

See Exercises 5-107 and 5-108

5-11 | The Concept of Limiting Reactant

Suppose you have a part-time job in a sandwich shop. One very popular sandwich is always made as follows:

$$2 \text{ slices bread} + 3 \text{ slices meat} + 1 \text{ slice cheese} \longrightarrow \text{sandwich}$$

Assume that you come to work one day and find the following quantities of ingredients:

$$8 \text{ slices bread} + 9 \text{ slices meat} + 5 \text{ slices cheese}$$

How many sandwiches can you make? What will be left over?

To solve this problem, let's see how many sandwiches we can make with each component:

Bread: $8 \text{ slices bread} \times \dfrac{1 \text{ sandwich}}{2 \text{ slices bread}} = 4 \text{ sandwiches}$

Meat: $9 \text{ slices meat} \times \dfrac{1 \text{ sandwich}}{3 \text{ slices meat}} = 3 \text{ sandwiches}$

Cheese: $5 \text{ slices cheese} \times \dfrac{1 \text{ sandwich}}{1 \text{ slice cheese}} = 5 \text{ sandwiches}$

How many sandwiches can you make? The answer is three. When you run out of meat, you must stop making sandwiches. The meat is the limiting ingredient (Fig. 5-9). ◄ ⓘ

What do you have left over? Making three sandwiches requires six pieces of bread. You started with eight slices, so you have two slices of bread left. You also

ⓘ The sandwich example shows how the concept of limiting reactants occurs in real life. We will use this same concept in chemical equations, in which the amounts of reactants will be given as masses. However, we will first have to convert these masses to numbers of molecules (moles) to apply this concept. Again, this illustrates why thinking about chemical equations at a molecular level is so important.

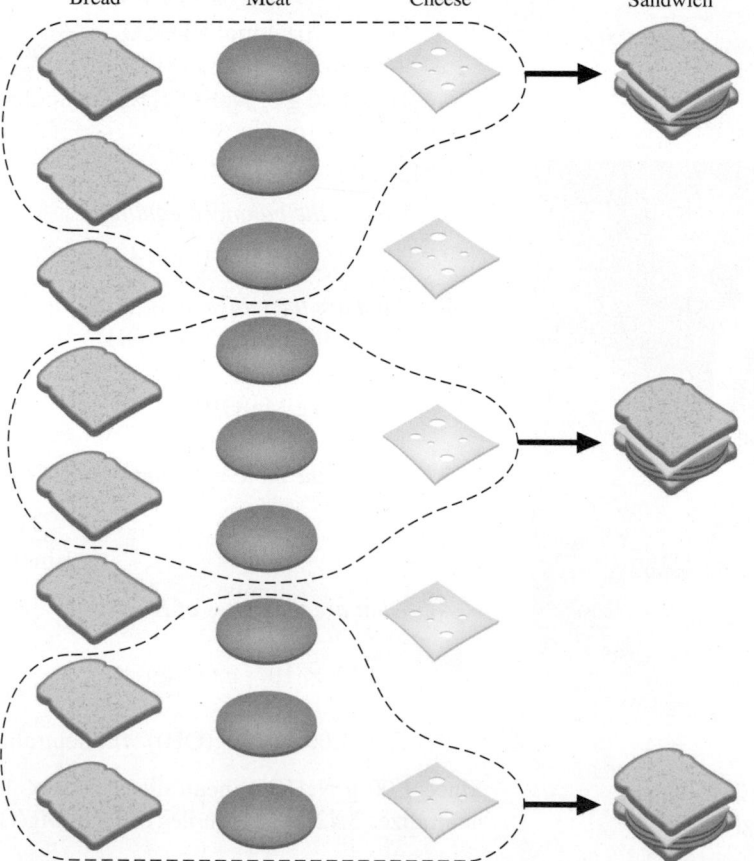

Figure 5-9 | Making sandwiches.

used three pieces of cheese for the three sandwiches, so you have two pieces of cheese left.

In this example, the ingredient present in the largest number (the meat) was actually the component that limited the number of sandwiches you could make. This situation arose because each sandwich required three slices of meat—more than the quantity required of any other ingredient.

When molecules react with each other to form products, considerations very similar to those involved in making sandwiches arise. We can illustrate these ideas with the reaction of $N_2(g)$ and $H_2(g)$ to form $NH_3(g)$:

$$N_2(g) + 3H_2(g) \longrightarrow 2NH_3(g)$$

Consider the following container of $N_2(g)$ and $H_2(g)$:

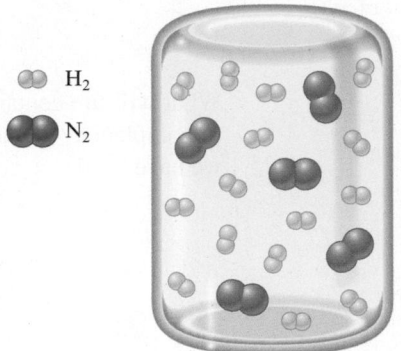

What will this container look like if the reaction between N_2 and H_2 proceeds to completion? To answer this question, you need to remember that each N_2 requires 3 H_2 molecules to form 2 NH_3. To make things clear, we will circle groups of reactants:

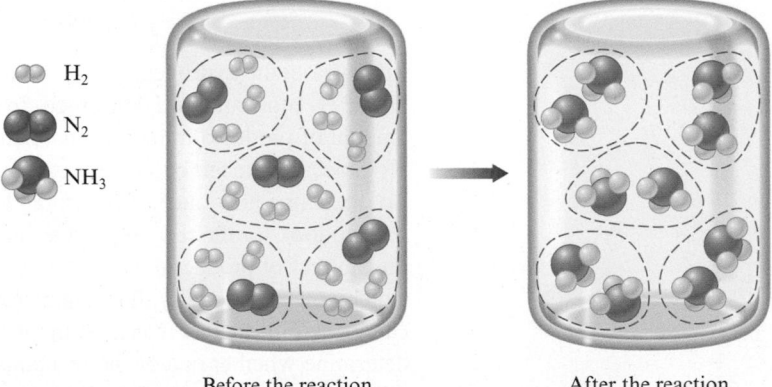

Before the reaction   After the reaction

In this case, the mixture of N_2 and H_2 contained just the number of molecules needed to form NH_3 with nothing left over. That is, the ratio of the number of H_2 molecules to N_2 molecules was

$$\frac{15H_2}{5N_2} = \frac{3H_2}{1N_2}$$

This ratio exactly matches the numbers in the balanced equation

$$3H_2(g) + N_2(g) \longrightarrow 2NH_3(g)$$

This type of mixture is called a **stoichiometric mixture**—one that contains the relative amounts of reactants that match the numbers in the balanced equation. In this case all reactants will be consumed to form products.

Now consider another container of $N_2(g)$ and $H_2(g)$:

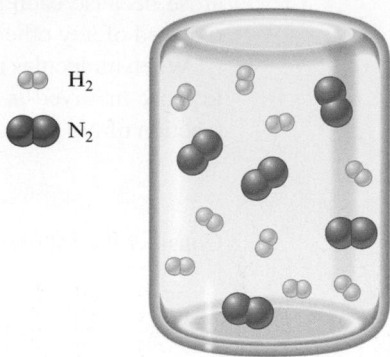

What will the container look like if the reaction between $N_2(g)$ and $H_2(g)$ proceeds to completion? Remember that each N_2 requires 3 H_2. Circling groups of reactants, we have

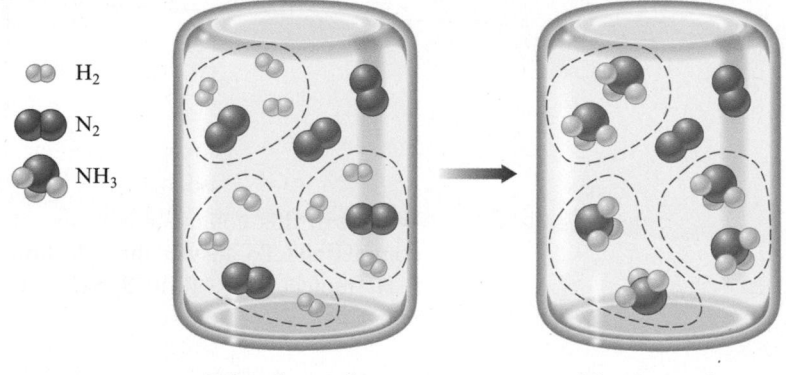

Before the reaction   After the reaction

In this case, the hydrogen (H_2) is limiting. That is, the H_2 molecules are used up before all the N_2 molecules are consumed. In this situation the amount of hydrogen limits the amount of product (ammonia) that can form—hydrogen is the limiting reactant. Some N_2 molecules are left over in this case because the reaction runs out of H_2 molecules first. To determine how much product can be formed from a given mixture of reactants, we have to look for the reactant that is limiting—the one that runs out first and thus limits the amount of product that can form. In some cases, the mixture of reactants might be stoichiometric—that is, all reactants run out at the same time. In general, however, you cannot assume that a given mixture of reactants is a stoichiometric mixture, so you must determine whether one of the reactants is limiting. The reactant that runs out first and thus limits the amounts of products that can form is called the **limiting reactant**.

To this point we have considered examples where the numbers of reactant molecules could be counted. In "real life" you can't count the molecules directly—you can't see them, and even if you could, there would be far too many to count. Instead, you must count by weighing. We must therefore explore how to find the limiting reactant, given the masses of the reactants.

A. Determination of Limiting Reactant Using Reactant Quantities

There are two ways to determine the limiting reactant in a chemical reaction. One involves comparing the moles of reactants to see which runs out first. We will consider this approach here.

In the laboratory or chemical plant, we work with much larger quantities than the few molecules of the preceding example. Therefore, we must learn to deal with limiting reactants using moles. The ideas are exactly the same, except that we are using moles of molecules instead of individual molecules. For example, suppose 25.0 kg of nitrogen and 5.00 kg of hydrogen are mixed and reacted to form ammonia. How do we calculate the mass of ammonia produced when this reaction is run to completion (until one of the reactants is completely consumed)?

As in the preceding example, we must use the balanced equation

$$N_2(g) + 3H_2(g) \longrightarrow 2NH_3(g)$$

to determine whether nitrogen or hydrogen is the limiting reactant and then to determine the amount of ammonia that is formed. We first calculate the moles of reactants present:

$$25.0 \text{ kg N}_2 \times \frac{1000 \text{ g N}_2}{1 \text{ kg N}_2} \times \frac{1 \text{ mol N}_2}{28.0 \text{ g N}_2} = 8.93 \times 10^2 \text{ mol N}_2$$

$$5.00 \text{ kg H}_2 \times \frac{1000 \text{ g H}_2}{1 \text{ kg H}_2} \times \frac{1 \text{ mol H}_2}{2.016 \text{ g H}_2} = 2.48 \times 10^3 \text{ mol H}_2$$

Since 1 mole of N_2 reacts with 3 moles of H_2, the number of moles of H_2 that will react exactly with 8.93×10^2 moles of N_2 is

$$8.93 \times 10^2 \text{ mol N}_2 \times \frac{3 \text{ mol H}_2}{1 \text{ mol N}_2} = 2.68 \times 10^3 \text{ mol H}_2$$

Thus 8.93×10^2 moles of N_2 requires 2.68×10^3 moles of H_2 to react completely. However, in this case, only 2.48×10^3 moles of H_2 are present. This means that the hydrogen will be consumed before the nitrogen. ◄ ⓘ Thus hydrogen is the *limiting reactant* in this particular situation, and we must use the amount of hydrogen to compute the quantity of ammonia formed:

ⓘ Always determine which reactant is limiting.

$$2.48 \times 10^3 \text{ mol H}_2 \times \frac{2 \text{ mol NH}_3}{3 \text{ mol H}_2} = 1.65 \times 10^3 \text{ mol NH}_3$$

Converting moles to kilograms gives

$$1.65 \times 10^3 \text{ mol NH}_3 \times \frac{17.0 \text{ g NH}_3}{1 \text{ mol NH}_3} = 2.80 \times 10^4 \text{ g NH}_3 = 28.0 \text{ kg NH}_3$$

Note that to determine the limiting reactant, we could have started instead with the given amount of hydrogen and calculated the moles of nitrogen required:

$$2.48 \times 10^3 \text{ mol H}_2 \times \frac{1 \text{ mol N}_2}{3 \text{ mol H}_2} = 8.27 \times 10^2 \text{ mol N}_2$$

Thus 2.48×10^3 moles of H_2 requires 8.27×10^2 moles of N_2. Since 8.93×10^2 moles of N_2 are actually present, the nitrogen is in excess. The hydrogen will run out first, and thus again we find that hydrogen limits the amount of ammonia formed.

A related but simpler way to determine which reactant is limiting is to compare the mole ratio of the substances required by the balanced equation with the mole ratio of reactants actually present. For example, in this case the mole ratio of H_2 to N_2 required by the balanced equation is

$$\frac{3 \text{ mol H}_2}{1 \text{ mol N}_2}$$

That is,

$$\frac{\text{mol H}_2}{\text{mol N}_2} \text{ (required)} = \frac{3}{1} = 3$$

In this experiment we have 2.48×10^3 moles of H_2 and 8.93×10^2 moles of N_2. Thus the ratio

$$\frac{\text{mol } H_2}{\text{mol } N_2} \text{ (actual)} = \frac{2.48 \times 10^3}{8.93 \times 10^2} = 2.78$$

Since 2.78 is less than 3, the actual mole ratio of H_2 to N_2 is too small, and H_2 must be limiting. If the actual H_2 to N_2 mole ratio had been greater than 3, then the H_2 would have been in excess and the N_2 would be limiting.

B. Determination of Limiting Reactant Using Quantities of Products Formed

A second method for determining which reactant in a chemical reaction is limiting is to consider the amounts of products that can be formed by completely consuming each reactant. The reactant that produces the smallest amount of product must run out first and thus be limiting. To see how this works, consider again the reaction of 25.0 kg (8.93×10^2 moles) of nitrogen with 5.00 kg (2.48×10^3 moles) of hydrogen.

We will now use these amounts of reactants to determine how much NH_3 would form. Since 1 mole of N_2 forms 2 moles of NH_3, the amount of NH_3 that would be formed if all of the N_2 was used up is calculated as follows:

$$8.93 \times 10^2 \text{ mol } N_2 \times \frac{2 \text{ mol } NH_3}{1 \text{ mol } N_2} = 1.79 \times 10^3 \text{ mol } NH_3$$

Next we will calculate how much NH_3 would be formed if the H_2 was completely used up:

$$2.48 \times 10^3 \text{ mol } H_2 \times \frac{2 \text{ mol } NH_3}{3 \text{ mol } H_2} = 1.65 \times 10^3 \text{ mol } NH_3$$

Because a smaller amount of NH_3 is produced from the H_2 than from the N_2, the amount of H_2 must be limiting.

Thus because the H_2 is the limiting reactant, the amount of NH_3 that can form is 1.65×10^3 moles. Converting moles to kilograms gives:

$$1.65 \times 10^3 \text{ mol } NH_3 \times \frac{17 \text{ g } NH_3}{1 \text{ mol } NH_3} = 2.80 \times 10^4 \text{ g } NH_3 = 28.0 \text{ kg } NH_3$$

INTERACTIVE EXAMPLE 5-17

Stoichiometry: Limiting Reactant

Nitrogen gas can be prepared by passing gaseous ammonia over solid copper(II) oxide at high temperatures. The other products of the reaction are solid copper and water vapor. If a sample containing 18.1 g of NH_3 is reacted with 90.4 g of CuO, which is the limiting reactant? How many grams of N_2 will be formed?

Solution

Where are we going?
To find the limiting reactant
To find the mass of N_2 produced

What do we know?

- The chemical reaction

$$NH_3(g) + CuO(s) \longrightarrow N_2(g) + Cu(s) + H_2O(g)$$

- 18.1 g NH_3
- 90.4 g CuO

What information do we need?

■ Balanced equation for the reaction

■ Moles of NH_3

■ Moles of CuO

How do we get there?
To find the limiting reactant

What is the balanced equation?

$$2NH_3(g) + 3CuO(s) \longrightarrow N_2(g) + 3Cu(s) + 3H_2O(g)$$

What are the moles of NH_3 and CuO?
To find the moles, we need to know the molar masses.

$$NH_3 \quad 17.03 \text{ g/mol}$$
$$CuO \quad 79.55 \text{ g/mol}$$

$$18.1 \text{ g NH}_3 \times \frac{1 \text{ mol NH}_3}{17.03 \text{ g NH}_3} = 1.06 \text{ mol NH}_3$$

$$90.4 \text{ g CuO} \times \frac{1 \text{ mol CuO}}{79.55 \text{ g CuO}} = 1.14 \text{ mol CuO}$$

A. First we will determine the limiting reactant by comparing the moles of reactants to see which one is consumed first.

What is the mole ratio between NH_3 and CuO in the balanced equation?

$$\frac{3 \text{ mol CuO}}{2 \text{ mol NH}_3}$$

How many moles of CuO are required to react with 1.06 moles of NH_3?

$$1.06 \text{ mol NH}_3 \times \frac{3 \text{ mol CuO}}{2 \text{ mol NH}_3} = 1.59 \text{ mol CuO}$$

■ Thus 1.59 moles of CuO are required to react with 1.06 moles of NH_3. Since only 1.14 moles of CuO are actually present, the amount of CuO is limiting; CuO will run out before NH_3 does. We can verify this conclusion by comparing the mole ratio of CuO and NH_3 required by the balanced equation:

$$\frac{\text{mol CuO}}{\text{mol NH}_3} \text{ (required)} = \frac{3}{2} = 1.5$$

with the mole ratio actually present:

$$\frac{\text{mol CuO}}{\text{mol NH}_3} \text{ (actual)} = \frac{1.14}{1.06} = 1.08$$

■ Since the actual ratio is too small (less than 1.5), CuO is the limiting reactant.

B. Alternatively we can determine the limiting reactant by computing the moles of N_2 that would be formed by complete consumption of NH_3 and CuO:

$$1.06 \text{ mol NH}_3 \times \frac{1 \text{ mol N}_2}{2 \text{ mol NH}_3} = 0.530 \text{ mol N}_2$$

$$1.14 \text{ mol CuO} \times \frac{1 \text{ mol N}_2}{3 \text{ mol CuO}} = 0.380 \text{ mol N}_2$$

As before, we see that the CuO is limiting since it produces the smaller amount of N_2.

To find the mass of N_2 produced

What are the moles of N_2 formed?
Because CuO is the limiting reactant, we must use the amount of CuO to calculate the amount of N_2 formed.

What is the mole ratio between N_2 and CuO in the balanced equation?

$$\frac{1 \text{ mol } N_2}{3 \text{ mol CuO}}$$

What are the moles of N_2?

$$1.14 \text{ mol CuO} \times \frac{1 \text{ mol } N_2}{3 \text{ mol CuO}} = 0.380 \text{ mol } N_2$$

What mass of N_2 is produced?
Using the molar mass of N_2 (28.02 g/mol), we can calculate the mass of N_2 produced:

$$\blacksquare \quad 0.380 \text{ mol } N_2 \times \frac{28.02 \text{ g } N_2}{1 \text{ mol } N_2} = 10.6 \text{ g } N_2$$

See Exercises 5-117 through 5-122

The amount of a product formed when the limiting reactant is completely consumed is called the **theoretical yield** of that product. In Example 5-17, 10.6 g of nitrogen represent the theoretical yield. This is the *maximum amount* of nitrogen that can be produced from the quantities of reactants used. Actually, the amount of product predicted by the theoretical yield is seldom obtained because of side reactions (other reactions that involve one or more of the reactants or products) and other complications. The *actual yield* of product is often given as a percentage of the theoretical yield. This is called the **percent yield**: ◄ ⓘ

$$\frac{\text{Actual yield}}{\text{Theoretical yield}} \times 100\% = \text{percent yield}$$

For example, if the reaction considered in Example 3.17 actually gave 6.63 g of nitrogen instead of the predicted 10.6 g, the percent yield of nitrogen would be

$$\frac{6.63 \text{ g } N_2}{10.6 \text{ g } N_2} \times 100\% = 62.5\%$$

INTERACTIVE EXAMPLE 5-18

Methanol

Calculating Percent Yield

Methanol (CH_3OH), also called *methyl alcohol*, is the simplest alcohol. ◄ It is used as a fuel in race cars and is a potential replacement for gasoline. Methanol can be manufactured by combining gaseous carbon monoxide and hydrogen. Suppose 68.5 kg CO(*g*) is reacted with 8.60 kg H_2(*g*). Calculate the theoretical yield of methanol. If 3.57×10^4 g CH_3OH is actually produced, what is the percent yield of methanol?

Solution

Where are we going?
To calculate the theoretical yield of methanol
To calculate the percent yield of methanol

What do we know?

■ The chemical reaction

$$H_2(g) + CO(g) \longrightarrow CH_3OH(l)$$

■ 68.5 kg CO(*g*)

- 8.60 kg H$_2$(g)
- 3.57 × 10^4 g CH$_3$OH is produced

What information do we need?

- Balanced equation for the reaction
- Moles of H$_2$
- Moles of CO
- Which reactant is limiting
- Amount of CH$_3$OH produced

How do we get there?
To find the limiting reactant

What is the balanced equation?

$$2H_2(g) + CO(g) \longrightarrow CH_3OH(l)$$

What are the moles of H$_2$ and CO?
To find the moles, we need to know the molar masses.

$$H_2 \quad 2.016 \text{ g/mol}$$
$$CO \quad 28.02 \text{ g/mol}$$

$$68.5 \text{ kg CO} \times \frac{1000 \text{ g CO}}{1 \text{ kg CO}} \times \frac{1 \text{ mol CO}}{28.02 \text{ g CO}} = 2.44 \times 10^3 \text{ mol CO}$$

$$8.60 \text{ kg H}_2 \times \frac{1000 \text{ g H}_2}{1 \text{ kg H}_2} \times \frac{1 \text{ mol H}_2}{2.016 \text{ g H}_2} = 4.27 \times 10^3 \text{ mol H}_2$$

A. Determination of Limiting Reactant Using Reactant Quantities

What is the mole ratio between H$_2$ and CO in the balanced equation?

$$\frac{2 \text{ mol H}_2}{1 \text{ mol CO}}$$

How does the actual mole ratio compare to the stoichiometric ratio?
To determine which reactant is limiting, we compare the mole ratio of H$_2$ and CO required by the balanced equation

$$\frac{\text{mol H}_2}{\text{mol CO}} \text{ (required)} = \frac{2}{1} = 2$$

with the actual mole ratio

$$\frac{\text{mol H}_2}{\text{mol CO}} \text{ (actual)} = \frac{4.27 \times 10^3}{2.44 \times 10^3} = 1.75$$

- Since the actual mole ratio of H$_2$ to CO is smaller than the required ratio, *H$_2$ is limiting.*

B. Determination of Limiting Reactant Using Quantities of Products Formed

We can also determine the limiting reactant by calculating the amounts of CH$_3$OH formed by complete consumption of CO(g) and H$_2$(g):

$$2.44 \times 10^3 \text{ mol CO} \times \frac{1 \text{ mol CH}_3\text{OH}}{1 \text{ mol CO}} = 2.44 \times 10^3 \text{ mol CH}_3\text{OH}$$

$$4.27 \times 10^3 \text{ mol H}_2 \times \frac{1 \text{ mol CH}_3\text{OH}}{2 \text{ mol H}_2} = 2.14 \times 10^3 \text{ mol CH}_3\text{OH}$$

Since complete consumption of the H$_2$ produces the smaller amount of CH$_3$OH, the H$_2$ is the limiting reactant as we determined above.

To calculate the theoretical yield of methanol

What are the moles of CH₃OH formed?

We must use the amount of H_2 and the mole ratio between H_2 and CH_3OH to determine the maximum amount of methanol that can be produced:

$$4.27 \times 10^3 \ \cancel{\text{mol } H_2} \times \frac{1 \text{ mol } CH_3OH}{2 \ \cancel{\text{mol } H_2}} = 2.14 \times 10^3 \text{ mol } CH_3OH$$

What is the theoretical yield of CH₃OH in grams?

$$2.14 \times 10^3 \ \cancel{\text{mol } CH_3OH} \times \frac{32.04 \text{ g } CH_3OH}{1 \ \cancel{\text{mol } CH_3OH}} = 6.86 \times 10^4 \text{ g } CH_3OH$$

■ Thus, from the amount of reactants given, the maximum amount of CH_3OH that can be formed is 6.86×10^4 g. This is the *theoretical yield*.

What is the percent yield of CH₃OH?

■ $\dfrac{\text{Actual yield (grams)}}{\text{Theoretical yield (grams)}} \times 100 = \dfrac{3.57 \times 10^4 \ \text{g } \cancel{CH_3OH}}{6.86 \times 10^4 \ \text{g } \cancel{CH_3OH}} \times 100\% = 52.0\%$

See Exercises 5-123 and 5-124

PROBLEM-SOLVING STRATEGY

Solving a Stoichiometry Problem Involving Masses of Reactants and Products

1. Write and balance the equation for the reaction.

2. Convert the known masses of substances to moles.

3. Determine which reactant is limiting.

4. Using the amount of the limiting reactant and the appropriate mole ratios, compute the number of moles of the desired product.

5. Convert from moles to grams, using the molar mass.

This process is summarized in the diagram below:

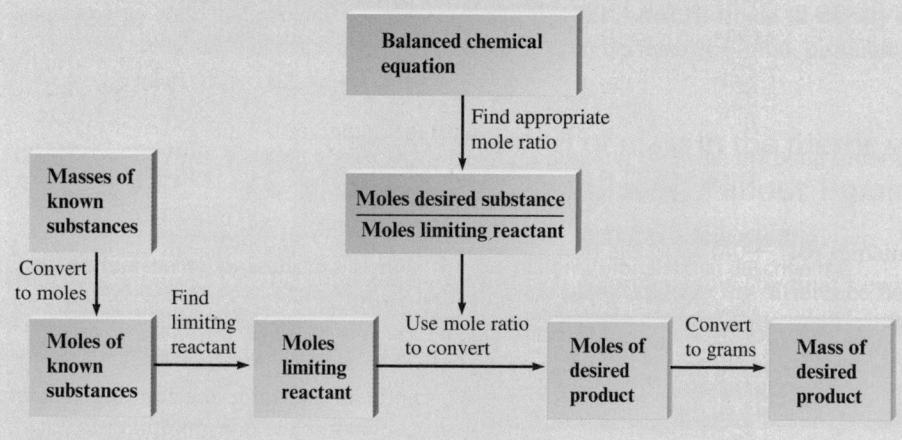

FOR REVIEW

Key Terms

Section 5-2
mass spectrometer
average atomic mass

Section 5-3
conceptual problem solving

Section 5-4
mole
Avogadro's number

Section 5-5
molar mass

Section 5-6
mass percent

Section 5-7
empirical formula
molecular formula

Section 5-8
chemical equation
reactants
products
balancing a chemical equation

Section 5-10
mole ratio

Section 5-11
stoichiometric mixture
limiting reactant
theoretical yield
percent yield

Stoichiometry

- Deals with the amounts of substances consumed and/or produced in a chemical reaction.
- We count atoms by measuring the mass of the sample.
- To relate mass and the number of atoms, the average atomic mass is required.

Mole

- The amount of carbon atoms in exactly 12 g of pure ^{12}C
- 6.022×10^{23} units of a substance
- The mass of 1 mole of an element = the atomic mass in grams

Molar Mass

- Mass (g) of 1 mole of a compound or element
- Obtained for a compound by finding the sum of the average masses of its constituent atoms

Percent Composition

- The mass percent of each element in a compound
- $\text{Mass percent} = \dfrac{\text{mass of element in 1 mole of substance}}{\text{mass of 1 mole of substance}} \times 100\%$

Empirical Formula

- The simplest whole-number ratio of the various types of atoms in a compound
- Can be obtained from the mass percent of elements in a compound

Molecular Formula

- For molecular substances:
  - The formula of the constituent molecules
  - Always an integer multiple of the empirical formula
- For ionic substances:
  - The same as the empirical formula

Chemical Reactions

- Reactants are turned into products.
- Atoms are neither created nor destroyed.
- All of the atoms present in the reactants must also be present in the products.

Characteristics of a Chemical Equation

- Represents a chemical reaction
- Reactants on the left side of the arrow, products on the right side
- When balanced, gives the relative numbers of reactant and product molecules or ions

Stoichiometry Calculations

- Amounts of reactants consumed and products formed can be determined from the balanced chemical equation.
- The limiting reactant is the one consumed first, thus limiting the amount of product that can form.

Yield

- The theoretical yield is the maximum amount that can be produced from a given amount of the limiting reactant.

- The actual yield, the amount of product actually obtained, is always less than the theoretical yield.

- Percent yield $= \dfrac{\text{actual yield (g)}}{\text{theoretical yield (g)}} \times 100\%$

Review Questions Answers to the Review Questions can be found on the Student Web site (accessible from www.cengagebrain.com).

1. Explain the concept of "counting by weighing" using marbles as your example.

2. Atomic masses are relative masses. What does this mean?

3. The atomic mass of boron (B) is given in the periodic table as 10.81, yet no single atom of boron has a mass of 10.81 u. Explain.

4. What three conversion factors and in what order would you use them to convert the mass of a compound into atoms of a particular element in that compound—for example, from 1.00 g aspirin ($C_9H_8O_4$) to number of hydrogen atoms in the 1.00-g sample?

5. Fig. 5-5 illustrates a schematic diagram of a combustion device used to analyze organic compounds. Given that a certain amount of a compound containing carbon, hydrogen, and oxygen is combusted in this device, explain how the data relating to the mass of CO_2 produced and the mass of H_2O produced can be manipulated to determine the empirical formula.

6. What is the difference between the empirical and molecular formulas of a compound? Can they ever be the same? Explain.

7. Consider the hypothetical reaction between A_2 and AB pictured below.

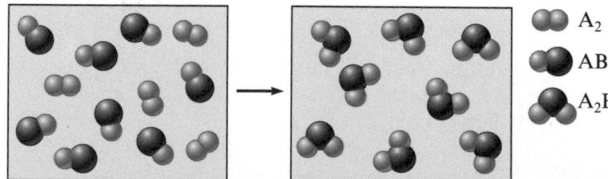

What is the balanced equation? If 2.50 moles of A_2 are reacted with excess AB, what amount (moles) of product will form? If the mass of AB is 30.0 u and the mass of A_2 are 40.0 u, what is the mass of the product? If 15.0 g of AB is reacted, what mass of A_2 is required to react with all of the AB, and what mass of product is formed?

8. What is a limiting reactant problem? Explain the method you are going to use to solve limiting reactant problems.

9. Consider the following mixture of $SO_2(g)$ and $O_2(g)$.

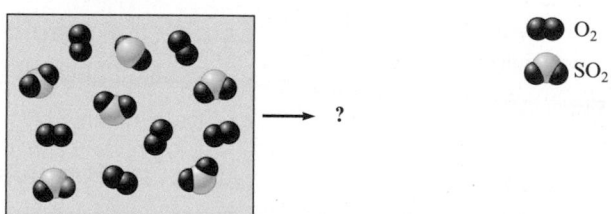

If $SO_2(g)$ and $O_2(g)$ react to form $SO_3(g)$, draw a representation of the product mixture assuming the reaction goes to completion. What is the limiting reactant in the reaction? If 96.0 g of SO_2 react with 32.0 g O_2, what mass of product will form?

10. Why is the actual yield of a reaction often less than the theoretical yield?

Active Learning Questions

These questions are designed to be used by groups of students in class.

1. The following are actual student responses to the question: Why is it necessary to balance chemical equations?
   a. The chemicals will not react until you have added the correct mole ratios.
   b. The correct products will not be formed unless the right amount of reactants have been added.
   c. A certain number of products cannot be formed without a certain number of reactants.
   d. The balanced equation tells you how much reactant you need and allows you to predict how much product you'll make.
   e. A mole-to-mole ratio must be established for the reaction to occur as written.

   Justify the best choice, and for choices you did not pick, explain what is wrong with them.

2. What information do we get from a chemical formula? From a chemical equation?

3. You are making cookies and are missing a key ingredient—eggs. You have most of the other ingredients needed to make the cookies, except you have only 1.33 cups of butter and no eggs. You note that the recipe calls for two cups of butter and three eggs (plus the other ingredients) to make six dozen cookies. You call a friend and have him bring you some eggs.
   a. What number of eggs do you need?
   b. If you use all the butter (and get enough eggs), what number of cookies will you make?

   Unfortunately, your friend hangs up before you tell him how many eggs you need. When he arrives, he has a surprise for you—to save time, he has broken them all in a bowl for you. You ask him how many he brought, and he replies, "I can't remember." You weigh the eggs and find that they weigh 62.1 g. Assuming that an average egg weighs 34.21 g,
   a. What quantity of butter is needed to react with all the eggs?
   b. What number of cookies can you make?
   c. Which will you have left over, eggs or butter?
   d. What quantity is left over?

4. Nitrogen gas (N_2) and hydrogen gas (H_2) react to form ammonia gas (NH_3).

   Consider the mixture of N_2 () and H_2 ( ) in a closed container as illustrated below:

   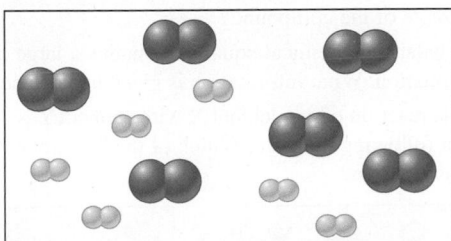

   Assuming the reaction goes to completion, draw a representation of the product mixture. Explain how you arrived at this representation.

5. For the preceding question, which of the following equations best represents the reaction?
   a. $6N_2 + 6H_2 \longrightarrow 4NH_3 + 4N_2$
   b. $N_2 + H_2 \longrightarrow NH_3$
   c. $N + 3H \longrightarrow NH_3$
   d. $N_2 + 3H_2 \longrightarrow 2NH_3$
   e. $2N_2 + 6H_2 \longrightarrow 4NH_3$

   Justify your choice, and for choices you did not pick, explain what is wrong with them.

6. You know that chemical A reacts with chemical B. You react 10.0 g A with 10.0 g B. What information do you need to determine the amount of product that will be produced? Explain.

7. A new grill has a mass of 30.0 kg. You put 3.0 kg of charcoal in the grill. You burn all the charcoal and the grill has a mass of 30.0 kg. What is the mass of the gases given off? Explain.

8. Consider an iron bar on a balance as shown.

   As the iron bar rusts, which of the following is true? Explain your answer.
   a. The balance will read less than 75.0 g.
   b. The balance will read 75.0 g.
   c. The balance will read greater than 75.0 g.
   d. The balance will read greater than 75.0 g, but if the bar is removed, the rust is scraped off, and the bar replaced, the balance will read 75.0 g.

9. You may have noticed that water sometimes drips from the exhaust of a car as it is running. Is this evidence that there is at least a small amount of water originally present in the gasoline? Explain.

Questions 10 and 11 deal with the following situation: You react chemical A with chemical B to make one product. It takes 100 g of A to react completely with 20 g of B.

10. What is the mass of the product?
    a. less than 10 g
    b. between 20 and 100 g
    c. between 100 and 120 g
    d. exactly 120 g
    e. more than 120 g

11. What is true about the chemical properties of the product?
    a. The properties are more like chemical A.
    b. The properties are more like chemical B.
    c. The properties are an average of those of chemical A and chemical B.
    d. The properties are not necessarily like either chemical A or chemical B.
    e. The properties are more like chemical A or more like chemical B, but more information is needed.

    Justify your choice, and for choices you did not pick, explain what is wrong with them.

12. Is there a difference between a homogeneous mixture of hydrogen and oxygen in a 2:1 mole ratio and a sample of water vapor? Explain.

13. Chlorine exists mainly as two isotopes, ^{37}Cl and ^{35}Cl. Which is more abundant? How do you know?

14. The average mass of a carbon atom is 12.011. Assuming you could pick up one carbon atom, estimate the chance that you would randomly get one with a mass of 12.011. Support your answer.

15. Can the subscripts in a chemical formula be fractions? Explain. Can the coefficients in a balanced chemical equation be fractions? Explain. Changing the subscripts of chemicals can balance the equations mathematically. Why is this unacceptable?

16. Consider the equation $2A + B \longrightarrow A_2B$. If you mix 1.0 mole of A with 1.0 mole of B, what amount (moles) of A_2B can be produced?

17. According to the law of conservation of mass, mass cannot be gained or destroyed in a chemical reaction. Why can't you simply add the masses of two reactants to determine the total mass of product?

18. Which of the following pairs of compounds have the same *empirical* formula?
    a. acetylene, C_2H_2, and benzene, C_6H_6
    b. ethane, C_2H_6, and butane, C_4H_{10}
    c. nitrogen dioxide, NO_2, and dinitrogen tetroxide, N_2O_4
    d. diphenyl ether, $C_{12}H_{10}O$, and phenol, C_6H_5OH

19. Atoms of three different elements are represented by O, □, and Δ. Which compound is left over when three molecules of OΔ and three molecules of □□Δ react to form O□Δ and OΔΔ?

20. In chemistry, what is meant by the term "mole"? What is the importance of the mole concept?

21. Which (if any) of the following is (are) *true* regarding the limiting reactant in a chemical reaction?
    a. The limiting reactant has the lowest coefficient in a balanced equation.
    b. The limiting reactant is the reactant for which you have the fewest number of moles.
    c. The limiting reactant has the lowest ratio of moles available/coefficient in the balanced equation.
    d. The limiting reactant has the lowest ratio of coefficient in the balanced equation/moles available.

    Justify your choice. For those you did not choose, explain why they are incorrect.

22. Consider the equation $3A + B \rightarrow C + D$. You react 4 moles of A with 2 moles of B. Which of the following is true?
    a. The limiting reactant is the one with the higher molar mass.
    b. A is the limiting reactant because you need 6 moles of A and have 4 moles.
    c. B is the limiting reactant because you have fewer moles of B than A.
    d. B is the limiting reactant because three A molecules react with each B molecule.
    e. Neither reactant is limiting.

    Justify your choice. For those you did not choose, explain why they are incorrect.

Questions

23. Reference Section 5-2 to find the atomic masses of ^{12}C and ^{13}C, the relative abundance of ^{12}C and ^{13}C in natural carbon, and the average mass (in u) of a carbon atom. If you had a sample of natural carbon containing exactly 10,000 atoms, determine the number of ^{12}C and ^{13}C atoms present. What would be the average mass (in u) and the total mass (in u) of the carbon atoms in this 10,000-atom sample? If you had a sample of natural carbon containing 6.0221×10^{23} atoms, determine the number of ^{12}C and ^{13}C atoms present. What would be the average mass (in u) and the total mass (in u) of this 6.0221×10^{23} atom sample? Given that $1 \text{ g} = 6.0221 \times 10^{23}$ u, what is the total mass of 1 mole of natural carbon in units of grams?

24. Avogadro's number, molar mass, and the chemical formula of a compound are three useful conversion factors. What unit conversions can be accomplished using these conversion factors?

25. If you had a mole of U.S. dollar bills and equally distributed the money to all of the people of the world, how rich would every person be? Assume a world population of 7 billion.

26. Describe 1 mole of CO_2 in as many ways as you can.

27. Which of the following compounds have the same empirical formulas?

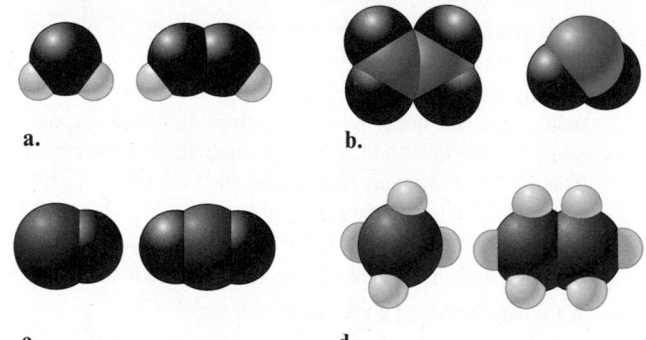

a.   b.

c.   d.

28. What is the difference between the molar mass and the empirical formula mass of a compound? When are these masses the same, and when are they different? When different, how is the molar mass related to the empirical formula mass?

29. How is the mass percent of elements in a compound different for a 1.0-g sample versus a 100.-g sample versus a 1-mole sample of the compound?

30. A balanced chemical equation contains a large amount of information. What information is given in a balanced equation?

31. The reaction of an element X with element Y is represented in the following diagram. Which of the equations best describes this reaction?

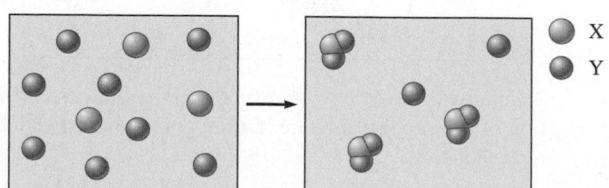

a. $3X + 8Y \rightarrow X_3Y_8$
b. $3X + 6Y \rightarrow X_3Y_6$
c. $X + 2Y \rightarrow XY_2$
d. $3X + 8Y \rightarrow 3XY_2 + 2Y$

32. Hydrogen gas and oxygen gas react to form water, and this reaction can be depicted as follows:

Explain why this equation is not balanced, and draw a picture of the balanced equation.

33. What is the theoretical yield for a reaction, and how does this quantity depend on the limiting reactant?

34. What does it mean to say a reactant is present "in excess" in a process? Can the limiting reactant be present in excess? Does the presence of an excess of a reactant affect the mass of products expected for a reaction?

35. Consider the following generic reaction:

$$A_2B_2 + 2C \longrightarrow 2CB + 2A$$

What steps and information are necessary to perform the following determinations assuming that 1.00×10^4 molecules of A_2B_2 are reacted with excess C?

a. mass of CB produced
b. atoms of A produced
c. moles of C reacted
d. percent yield of CB

36. Consider the following generic reaction:

$$Y_2 + 2XY \longrightarrow 2XY_2$$

In a limiting reactant problem, a certain quantity of each reactant is given and you are usually asked to calculate the mass of product formed. If 10.0 g of Y_2 is reacted with 10.0 g of XY, outline two methods you could use to determine which reactant is limiting (runs out first) and thus determines the mass of product formed.

Exercises

In this section similar exercises are paired.

Atomic Masses and the Mass Spectrometer

37. An element consists of 1.40% of an isotope with mass 203.973 u, 24.10% of an isotope with mass 205.9745 u, 22.10% of an isotope with mass 206.9759 u, and 52.40% of an isotope with mass 207.9766 u. Calculate the average atomic mass, and identify the element.

38. An element "X" has five major isotopes, which are listed below along with their abundances. What is the element?

Isotope	Percent Natural Abundance	Mass (u)
^{46}X	8.00%	45.95232
^{47}X	7.30%	46.951764
^{48}X	73.80%	47.947947
^{49}X	5.50%	48.947841
^{50}X	5.40%	49.944792

39. The element rhenium (Re) has two naturally occurring isotopes, ^{185}Re and ^{187}Re, with an average atomic mass of 186.207 u. Rhenium is 62.60% ^{187}Re, and the atomic mass of ^{187}Re is 186.956 u. Calculate the mass of ^{185}Re.

40. Assume silicon has three major isotopes in nature as shown in the table below. Fill in the missing information.

Isotope	Mass (u)	Abundance
^{28}Si	27.98	_____
^{29}Si	_____	4.70%
^{30}Si	29.97	3.09%

41. The element europium exists in nature as two isotopes: ^{151}Eu has a mass of 150.9196 u and ^{153}Eu has a mass of 152.9209 u. The average atomic mass of europium is 151.96 u. Calculate the relative abundance of the two europium isotopes.

42. The element silver (Ag) has two naturally occurring isotopes: ^{109}Ag and ^{107}Ag with a mass of 106.905 u. Silver consists of 51.82% ^{107}Ag and has an average atomic mass of 107.868 u. Calculate the mass of ^{109}Ag.

43. The mass spectrum of bromine (Br_2) consists of three peaks with the following characteristics:

Mass (u)	Relative Size
157.84	0.2534
159.84	0.5000
161.84	0.2466

How do you interpret these data?

44. The stable isotopes of iron are ^{54}Fe, ^{56}Fe, ^{57}Fe, and ^{58}Fe. The mass spectrum of iron looks like the following:

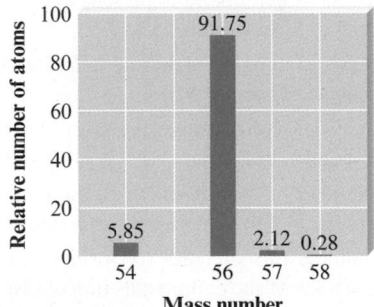

Use the data on the mass spectrum to estimate the average atomic mass of iron, and compare it to the value given in the table inside the front cover of this book.

Moles and Molar Masses

45. Calculate the mass of 500. atoms of iron (Fe).

46. What number of Fe atoms and what amount (moles) of Fe atoms are in 500.0 g of iron?

47. Diamond is a natural form of pure carbon. What number of atoms of carbon are in a 1.00-carat diamond (1.00 carat = 0.200 g)?

48. A diamond contains 5.0×10^{21} atoms of carbon. What amount (moles) of carbon and what mass (grams) of carbon are in this diamond?

49. Aluminum metal is produced by passing an electric current through a solution of aluminum oxide (Al_2O_3) dissolved in molten cryolite (Na_3AlF_6). Calculate the molar masses of Al_2O_3 and Na_3AlF_6.

50. The Freons are a class of compounds containing carbon, chlorine, and fluorine. While they have many valuable uses, they have been shown to be responsible for depletion of the ozone in the upper atmosphere. In 1991, two replacement compounds for Freons went into production: HFC-134a (CH_2FCF_3) and HCFC-124 ($CHClFCF_3$). Calculate the molar masses of these two compounds.

51. Calculate the molar mass of the following substances.

    a. 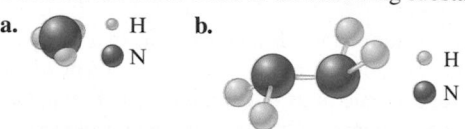 ○ H
    ● N

    b. ○ H
    ● N

    c. $(NH_4)_2Cr_2O_7$

52. Calculate the molar mass of the following substances.

    a. 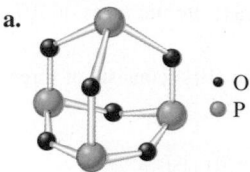 ● O
    ○ P

    b. $Ca_3(PO_4)_2$ c. Na_2HPO_4

53. What amount (moles) of compound is present in 1.00 g of each of the compounds in Exercise 51?

54. What amount (moles) of compound is present in 1.00 g of each of the compounds in Exercise 52?

55. What mass of compound is present in 5.00 moles of each of the compounds in Exercise 51?

56. What mass of compound is present in 5.00 moles of each of the compounds in Exercise 52?

57. What mass of nitrogen is present in 5.00 moles of each of the compounds in Exercise 51?

58. What mass of phosphorus is present in 5.00 moles of each of the compounds in Exercise 52?

59. What number of molecules (or formula units) are present in 1.00 g of each of the compounds in Exercise 51?

60. What number of molecules (or formula units) are present in 1.00 g of each of the compounds in Exercise 52?

61. What number of atoms of nitrogen are present in 1.00 g of each of the compounds in Exercise 51?

62. What number of atoms of phosphorus are present in 1.00 g of each of the compounds in Exercise 52?

63. Freon-12 (CCl_2F_2) is used as a refrigerant in air conditioners and as a propellant in aerosol cans. Calculate the number of molecules of Freon-12 in 5.56 mg of Freon-12. What is the mass of chlorine in 5.56 mg of Freon-12?

64. Bauxite, the principal ore used in the production of aluminum, has a molecular formula of $Al_2O_3 \cdot 2H_2O$. The $\cdot H_2O$ in the formula are called waters of hydration. Each formula unit of the compound contains two water molecules.

    a. What is the molar mass of bauxite?

    b. What is the mass of aluminum in 0.58 mole of bauxite?

    c. How many atoms of aluminum are in 0.58 mole of bauxite?

    d. What is the mass of 2.1×10^{24} formula units of bauxite?

65. What amount (moles) is represented by each of these samples?

    a. 150.0 g Fe_2O_3

    b. 10.0 mg NO_2

    c. 1.5×10^{16} molecules of BF_3

66. What amount (moles) is represented by each of these samples?

    a. 20.0 mg caffeine, $C_8H_{10}N_4O_2$

    b. 2.72×10^{21} molecules of ethanol, C_2H_5OH

    c. 1.50 g of dry ice, CO_2

67. What number of atoms of nitrogen are present in 5.00 g of each of the following?

    a. glycine, $C_2H_5O_2N$

    b. magnesium nitride

    c. calcium nitrate

    d. dinitrogen tetroxide

68. Complete the following table.

Mass of Sample	Moles of Sample	Molecules in Sample	Total Atoms in Sample
4.24 g C_6H_6	_____	_____	_____
_____	0.224 mol H_2O	_____	_____
_____	_____	2.71×10^{22} molecules CO_2	_____
_____	_____	_____	3.35×10^{22} total atoms in CH_3OH sample

69. Ascorbic acid, or vitamin C ($C_6H_8O_6$), is an essential vitamin. It cannot be stored by the body and must be present in the diet. What is the molar mass of ascorbic acid? Vitamin C tablets are taken as a dietary supplement. If a typical tablet contains 500.0 mg vitamin C, what amount (moles) and what number of molecules of vitamin C does it contain?

70. The molecular formula of acetylsalicylic acid (aspirin), one of the most commonly used pain relievers, is $C_9H_8O_4$.

    a. Calculate the molar mass of aspirin.

    b. A typical aspirin tablet contains 500. mg $C_9H_8O_4$. What amount (moles) of $C_9H_8O_4$ molecules and what number of molecules of acetylsalicylic acid are in a 500.-mg tablet?

71. Chloral hydrate ($C_2H_3Cl_3O_2$) is a drug formerly used as a sedative and hypnotic. It is the compound used to make "Mickey Finns" in detective stories.

    a. Calculate the molar mass of chloral hydrate.

    b. What amount (moles) of $C_2H_3Cl_3O_2$ molecules are in 500.0 g chloral hydrate?

    c. What is the mass in grams of 2.0×10^{-2} mole of chloral hydrate?

    d. What number of chlorine atoms are in 5.0 g chloral hydrate?

    e. What mass of chloral hydrate would contain 1.0 g Cl?

    f. What is the mass of exactly 500 molecules of chloral hydrate?

72. Dimethylnitrosamine, $(CH_3)_2N_2O$, is a carcinogenic (cancer-causing) substance that may be formed in foods, beverages, or gastric juices from the reaction of nitrite ion (used as a food preservative) with other substances.

    a. What is the molar mass of dimethylnitrosamine?

    b. How many moles of $(CH_3)_2N_2O$ molecules are present in 250 mg dimethylnitrosamine?

    c. What is the mass of 0.050 mole of dimethylnitrosamine?

    d. How many atoms of hydrogen are in 1.0 mole of dimethylnitrosamine?

    e. What is the mass of 1.0×10^6 molecules of dimethylnitrosamine?

    f. What is the mass in grams of one molecule of dimethylnitrosamine?

Percent Composition

73. Calculate the percent composition by mass of the following compounds that are important starting materials for synthetic polymers:

    a. $C_3H_4O_2$ (acrylic acid, from which acrylic plastics are made)

    b. $C_4H_6O_2$ (methyl acrylate, from which Plexiglas is made)

    c. C_3H_3N (acrylonitrile, from which Orlon is made)

74. In 1987 the first substance to act as a superconductor at a temperature above that of liquid nitrogen (77 K) was discovered. The approximate formula of this substance is $YBa_2Cu_3O_7$. Calculate the percent composition by mass of this material.

75. The percent by mass of nitrogen for a compound is found to be 46.7%. Which of the following could be this species?

                             ● N

                             ● O

76. Arrange the following substances in order of increasing mass percent of carbon.

    a. caffeine, $C_8H_{10}N_4O_2$

    b. sucrose, $C_{12}H_{22}O_{11}$

    c. ethanol, C_2H_5OH

77. Fungal laccase, a blue protein found in wood-rotting fungi, is 0.390% Cu by mass. If a fungal laccase molecule contains four copper atoms, what is the molar mass of fungal laccase?

78. Hemoglobin is the protein that transports oxygen in mammals. Hemoglobin is 0.347% Fe by mass, and each hemoglobin molecule contains four iron atoms. Calculate the molar mass of hemoglobin.

Empirical and Molecular Formulas

79. Express the composition of each of the following compounds as the mass percents of its elements.

    a. formaldehyde, CH_2O

    b. glucose, $C_6H_{12}O_6$

    c. acetic acid, $HC_2H_3O_2$

80. Considering your answer to Exercise 79, which type of formula, empirical or molecular, can be obtained from elemental analysis that gives percent composition?

81. Give the empirical formula for each of the compounds represented below.

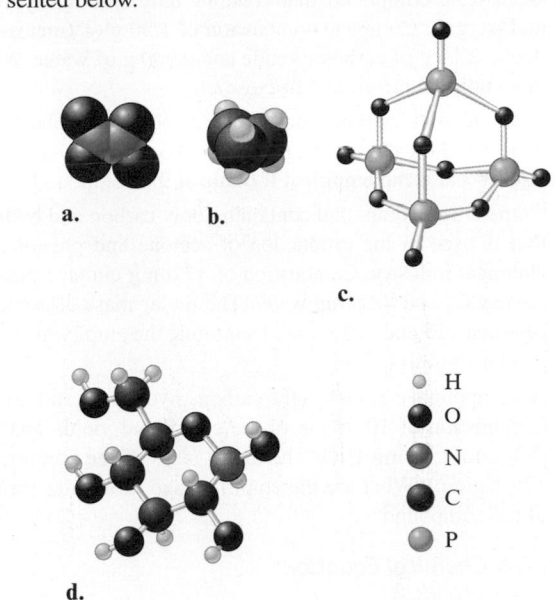

    ○ H

    ● O

    ● N

    ● C

    ○ P

82. Determine the molecular formulas to which the following empirical formulas and molar masses pertain.

    a. SNH (188.35 g/mol)   **c.** CoC_4O_4 (341.94 g/mol)

    b. $NPCl_2$ (347.64 g/mol)   **d.** SN (184.32 g/mol)

83. A compound that contains only carbon, hydrogen, and oxygen is 48.64% C and 8.16% H by mass. What is the empirical formula of this substance?

84. The most common form of nylon (nylon-6) is 63.68% carbon, 12.38% nitrogen, 9.80% hydrogen, and 14.14% oxygen. Calculate the empirical formula for nylon-6.

85. There are two binary compounds of mercury and oxygen. Heating either of them results in the decomposition of the compound, with oxygen gas escaping into the atmosphere while leaving a residue of pure mercury. Heating 0.6498 g of one of the compounds leaves a residue of 0.6018 g. Heating 0.4172 g of the other compound results in a mass loss of 0.016 g. Determine the empirical formula of each compound.

86. A sample of urea contains 1.121 g N, 0.161 g H, 0.480 g C, and 0.640 g O. What is the empirical formula of urea?

87. A compound containing only sulfur and nitrogen is 69.6% S by mass; the molar mass is 184 g/mol. What are the empirical and molecular formulas of the compound?

88. Determine the molecular formula of a compound that contains 26.7% P, 12.1% N, and 61.2% Cl, and has a molar mass of 580 g/mol.

89. A compound contains 47.08% carbon, 6.59% hydrogen, and 46.33% chlorine by mass; the molar mass of the compound is 153 g/mol. What are the empirical and molecular formulas of the compound?

90. Maleic acid is an organic compound composed of 41.39% C, 3.47% H, and the rest oxygen. If 0.129 mole of maleic acid has a mass of 15.0 g, what are the empirical and molecular formulas of maleic acid?

91. One of the components that make up common table sugar is fructose, a compound that contains only carbon, hydrogen, and oxygen. Complete combustion of 1.50 g of fructose produced 2.20 g of carbon dioxide and 0.900 g of water. What is the empirical formula of fructose?

92. A compound contains only C, H, and N. Combustion of 35.0 mg of the compound produces 33.5 mg CO_2 and 41.1 mg H_2O. What is the empirical formula of the compound?

93. Cumene is a compound containing only carbon and hydrogen that is used in the production of acetone and phenol in the chemical industry. Combustion of 47.6 mg cumene produces some CO_2 and 42.8 mg water. The molar mass of cumene is between 115 and 125 g/mol. Determine the empirical and molecular formulas.

94. A compound contains only carbon, hydrogen, and oxygen. Combustion of 10.68 mg of the compound yields 16.01 mg CO_2 and 4.37 mg H_2O. The molar mass of the compound is 176.1 g/mol. What are the empirical and molecular formulas of the compound?

Balancing Chemical Equations

95. Give the balanced equation for each of the following chemical reactions:

    a. Glucose ($C_6H_{12}O_6$) reacts with oxygen gas to produce gaseous carbon dioxide and water vapor.

    b. Solid iron(III) sulfide reacts with gaseous hydrogen chloride to form solid iron(III) chloride and hydrogen sulfide gas.

    c. Carbon disulfide liquid reacts with ammonia gas to produce hydrogen sulfide gas and solid ammonium thiocyanate (NH_4SCN).

96. Give the balanced equation for each of the following.

    a. The combustion of ethanol (C_2H_5OH) forms carbon dioxide and water vapor. A combustion reaction refers to a reaction of a substance with oxygen gas.

    b. Aqueous solutions of lead(II) nitrate and sodium phosphate are mixed, resulting in the precipitate formation of lead(II) phosphate with aqueous sodium nitrate as the other product.

    c. Solid zinc reacts with aqueous HCl to form aqueous zinc chloride and hydrogen gas.

    d. Aqueous strontium hydroxide reacts with aqueous hydrobromic acid to produce water and aqueous strontium bromide.

97. A common demonstration in chemistry courses involves adding a tiny speck of manganese(IV) oxide to a concentrated hydrogen peroxide (H_2O_2) solution. Hydrogen peroxide decomposes quite spectacularly under these conditions to produce oxygen gas and steam (water vapor). Manganese(IV) oxide is a catalyst for the decomposition of hydrogen peroxide and is not consumed in the reaction. Write the balanced equation for the decomposition reaction of hydrogen peroxide.

98. Iron oxide ores, commonly a mixture of FeO and Fe_2O_3, are given the general formula Fe_3O_4. They yield elemental iron when heated to a very high temperature with either carbon monoxide or elemental hydrogen. Balance the following equations for these processes:

$$Fe_3O_4(s) + H_2(g) \longrightarrow Fe(s) + H_2O(g)$$
$$Fe_3O_4(s) + CO(g) \longrightarrow Fe(s) + CO_2(g)$$

99. Balance the following equations:

    a. $Ca(OH)_2(aq) + H_3PO_4(aq) \rightarrow H_2O(l) + Ca_3(PO_4)_2(s)$

    b. $Al(OH)_3(s) + HCl(aq) \rightarrow AlCl_3(aq) + H_2O(l)$

    c. $AgNO_3(aq) + H_2SO_4(aq) \rightarrow Ag_2SO_4(s) + HNO_3(aq)$

100. Balance each of the following chemical equations.

    a. $KO_2(s) + H_2O(l) \rightarrow KOH(aq) + O_2(g) + H_2O_2(aq)$

    b. $Fe_2O_3(s) + HNO_3(aq) \rightarrow Fe(NO_3)_3(aq) + H_2O(l)$

    c. $NH_3(g) + O_2(g) \rightarrow NO(g) + H_2O(g)$

    d. $PCl_5(l) + H_2O(l) \rightarrow H_3PO_4(aq) + HCl(g)$

    e. $CaO(s) + C(s) \rightarrow CaC_2(s) + CO_2(g)$

    f. $MoS_2(s) + O_2(g) \rightarrow MoO_3(s) + SO_2(g)$

    g. $FeCO_3(s) + H_2CO_3(aq) \rightarrow Fe(HCO_3)_2(aq)$

101. Balance the following equations representing combustion reactions:

    a.
    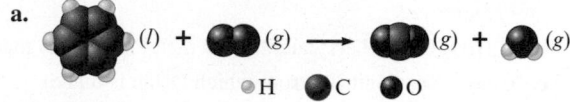

    b.
    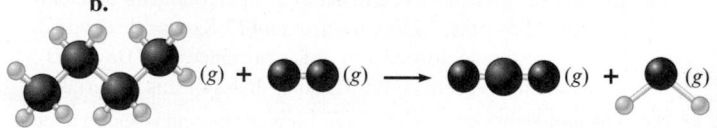

    c. $C_{12}H_{22}O_{11}(s) + O_2(g) \rightarrow CO_2(g) + H_2O(g)$

    d. $Fe(s) + O_2(g) \rightarrow Fe_2O_3(s)$

    e. $FeO(s) + O_2(g) \rightarrow Fe_2O_3(s)$

102. Balance the following equations:

    a. $Cr(s) + S_8(s) \rightarrow Cr_2S_3(s)$

    b. $NaHCO_3(s) \xrightarrow{\text{Heat}} Na_2CO_3(s) + CO_2(g) + H_2O(g)$

    c. $KClO_3(s) \xrightarrow{\text{Heat}} KCl(s) + O_2(g)$

    d. $Eu(s) + HF(g) \rightarrow EuF_3(s) + H_2(g)$

103. Silicon is produced for the chemical and electronics industries by the following reactions. Give the balanced equation for each reaction.

    a. $SiO_2(s) + C(s) \xrightarrow[\text{arc furnace}]{\text{Electric}} Si(s) + CO(g)$

    b. Liquid silicon tetrachloride is reacted with very pure solid magnesium, producing solid silicon and solid magnesium chloride.

    c. $Na_2SiF_6(s) + Na(s) \rightarrow Si(s) + NaF(s)$

104. Glass is a mixture of several compounds, but a major constituent of most glass is calcium silicate, $CaSiO_3$. Glass can be etched by treatment with hydrofluoric acid; HF attacks the calcium silicate of the glass, producing gaseous and water-soluble products (which can be removed by washing the glass). For example, the volumetric glassware in chemistry laboratories is often graduated by using this process. Balance the following equation for the reaction of hydrofluoric acid with calcium silicate.

$$CaSiO_3(s) + HF(aq) \longrightarrow CaF_2(aq) + SiF_4(g) + H_2O(l)$$

Reaction Stoichiometry

105. Over the years, the thermite reaction has been used for welding railroad rails, in incendiary bombs, and to ignite solid-fuel rocket motors. The reaction is

$$Fe_2O_3(s) + 2Al(s) \longrightarrow 2Fe(l) + Al_2O_3(s)$$

What masses of iron(III) oxide and aluminum must be used to produce 15.0 g iron? What is the maximum mass of aluminum oxide that could be produced?

106. The reaction between potassium chlorate and red phosphorus takes place when you strike a match on a matchbox. If you were to react 52.9 g of potassium chlorate ($KClO_3$) with excess red phosphorus, what mass of tetraphosphorus decaoxide (P_4O_{10}) could be produced?

$$KClO_3(s) + P_4(s) \longrightarrow P_4O_{10}(s) + KCl(s) \quad \text{(unbalanced)}$$

107. The reusable booster rockets of the U.S. space shuttle employ a mixture of aluminum and ammonium perchlorate for fuel. A possible equation for this reaction is

$$3Al(s) + 3NH_4ClO_4(s) \longrightarrow$$
$$Al_2O_3(s) + AlCl_3(s) + 3NO(g) + 6H_2O(g)$$

What mass of NH_4ClO_4 should be used in the fuel mixture for every kilogram of Al?

108. One of relatively few reactions that takes place directly between two solids at room temperature is

$$Ba(OH)_2 \cdot 8H_2O(s) + NH_4SCN(s) \longrightarrow$$
$$Ba(SCN)_2(s) + H_2O(l) + NH_3(g)$$

In this equation, the $\cdot 8H_2O$ in $Ba(OH)_2 \cdot 8H_2O$ indicates the presence of eight water molecules. This compound is called barium hydroxide octahydrate.

a. Balance the equation.

b. What mass of ammonium thiocyanate (NH_4SCN) must be used if it is to react completely with 6.5 g barium hydroxide octahydrate?

109. Elixirs such as Alka-Seltzer use the reaction of sodium bicarbonate with citric acid in aqueous solution to produce a fizz:

$$3NaHCO_3(aq) + C_6H_8O_7(aq) \longrightarrow$$
$$3CO_2(g) + 3H_2O(l) + Na_3C_6H_5O_7(aq)$$

a. What mass of $C_6H_8O_7$ should be used for every 1.0×10^2 mg $NaHCO_3$?

b. What mass of $CO_2(g)$ could be produced from such a mixture?

110. Aspirin ($C_9H_8O_4$) is synthesized by reacting salicylic acid ($C_7H_6O_3$) with acetic anhydride ($C_4H_6O_3$). The balanced equation is

$$C_7H_6O_3 + C_4H_6O_3 \longrightarrow C_9H_8O_4 + HC_2H_3O_2$$

a. What mass of acetic anhydride is needed to completely consume 1.00×10^2 g salicylic acid?

b. What is the maximum mass of aspirin (the theoretical yield) that could be produced in this reaction?

111. Bacterial digestion is an economical method of sewage treatment. The reaction

$$5CO_2(g) + 55NH_4^+(aq) + 76O_2(g) \xrightarrow{\text{bacteria}}$$
$$\underset{\text{bacterial tissue}}{C_5H_7O_2N(s)} + 54NO_2^-(aq) + 52H_2O(l) + 109H^+(aq)$$

is an intermediate step in the conversion of the nitrogen in organic compounds into nitrate ions. What mass of bacterial tissue is produced in a treatment plant for every 1.0×10^4 kg of wastewater containing 3.0% NH_4^+ ions by mass? Assume that 95% of the ammonium ions are consumed by the bacteria.

112. Phosphorus can be prepared from calcium phosphate by the following reaction:

$$2Ca_3(PO_4)_2(s) + 6SiO_2(s) + 10C(s) \longrightarrow$$
$$6CaSiO_3(s) + P_4(s) + 10CO(g)$$

Phosphorite is a mineral that contains $Ca_3(PO_4)_2$ plus other non-phosphorus-containing compounds. What is the maximum amount of P_4 that can be produced from 1.0 kg of phosphorite if the phorphorite sample is 75% $Ca_3(PO_4)_2$ by mass? Assume an excess of the other reactants.

113. Coke is an impure form of carbon that is often used in the industrial production of metals from their oxides. If a sample of coke is 95% carbon by mass, determine the mass of coke needed to react completely with 1.0 ton of copper(II) oxide.

$$2CuO(s) + C(s) \longrightarrow 2Cu(s) + CO_2(g)$$

114. The space shuttle environmental control system handles excess CO_2 (which the astronauts breathe out; it is 4.0% by mass of exhaled air) by reacting it with lithium hydroxide, LiOH, pellets to form lithium carbonate, Li_2CO_3, and water. If there are seven astronauts on board the shuttle, and each exhales 20. L of air per minute, how long could clean air be generated if there were 25,000 g of LiOH pellets available for each shuttle mission? Assume the density of air is 0.0010 g/mL.

Limiting Reactants and Percent Yield

115. Consider the reaction between $NO(g)$ and $O_2(g)$ represented below.

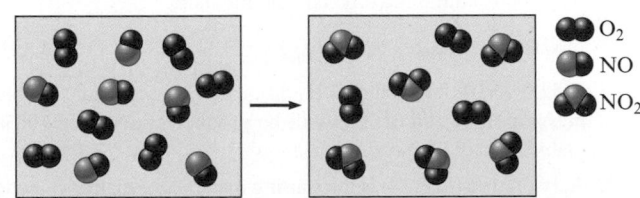

What is the balanced equation for this reaction, and what is the limiting reactant?

116. Consider the following reaction:

$$4NH_3(g) + 5O_2(g) \longrightarrow 4NO(g) + 6H_2O(g)$$

If a container were to have 10 molecules of O_2 and 10 molecules of NH_3 initially, how many total molecules (reactants plus products) would be present in the container after this reaction goes to completion?

117. Ammonia is produced from the reaction of nitrogen and hydrogen according to the following balanced equation:

$$N_2(g) + 3H_2(g) \longrightarrow 2NH_3(g)$$

    a. What is the maximum mass of ammonia that can be produced from a mixture of 1.00×10^3 g N_2 and 5.00×10^2 g H_2?

    b. What mass of which starting material would remain unreacted?

118. Consider the following unbalanced equation:

$$Ca_3(PO_4)_2(s) + H_2SO_4(aq) \longrightarrow CaSO_4(s) + H_3PO_4(aq)$$

What masses of calcium sulfate and phosphoric acid can be produced from the reaction of 1.0 kg calcium phosphate with 1.0 kg concentrated sulfuric acid (98% H_2SO_4 by mass)?

119. Hydrogen peroxide is used as a cleansing agent in the treatment of cuts and abrasions for several reasons. It is an oxidizing agent that can directly kill many microorganisms; it decomposes on contact with blood, releasing elemental oxygen gas (which inhibits the growth of anaerobic microorganisms); and it foams on contact with blood, which provides a cleansing action. In the laboratory, small quantities of hydrogen peroxide can be prepared by the action of an acid on an alkaline earth metal peroxide, such as barium peroxide:

$$BaO_2(s) + 2HCl(aq) \longrightarrow H_2O_2(aq) + BaCl_2(aq)$$

What mass of hydrogen peroxide should result when 1.50 g barium peroxide is treated with 25.0 mL hydrochloric acid solution containing 0.0272 g HCl per mL? What mass of which reagent is left unreacted?

120. Silver sulfadiazine burn-treating cream creates a barrier against bacterial invasion and releases antimicrobial agents directly into the wound. If 25.0 g Ag_2O is reacted with 50.0 g $C_{10}H_{10}N_4SO_2$, what mass of silver sulfadiazine, $AgC_{10}H_9N_4SO_2$, can be produced, assuming 100% yield?

$$Ag_2O(s) + 2C_{10}H_{10}N_4SO_2(s) \longrightarrow 2AgC_{10}H_9N_4SO_2(s) + H_2O(l)$$

121. Hydrogen cyanide is produced industrially from the reaction of gaseous ammonia, oxygen, and methane:

$$2NH_3(g) + 3O_2(g) + 2CH_4(g) \longrightarrow 2HCN(g) + 6H_2O(g)$$

If 5.00×10^3 kg each of NH_3, O_2, and CH_4 are reacted, what mass of HCN and of H_2O will be produced, assuming 100% yield?

122. Acrylonitrile (C_3H_3N) is the starting material for many synthetic carpets and fabrics. It is produced by the following reaction.

$$2C_3H_6(g) + 2NH_3(g) + 3O_2(g) \longrightarrow 2C_3H_3N(g) + 6H_2O(g)$$

If 15.0 g C_3H_6, 10.0 g O_2, and 5.00 g NH_3 are reacted, what mass of acrylonitrile can be produced, assuming 100% yield?

123. The reaction of ethane gas (C_2H_6) with chlorine gas produces C_2H_5Cl as its main product (along with HCl). In addition, the reaction invariably produces a variety of other minor products, including $C_2H_4Cl_2$, $C_2H_3Cl_3$, and others. Naturally, the production of these minor products reduces the yield of the main product. Calculate the percent yield of C_2H_5Cl if the reaction of 300. g of ethane with 650. g of chlorine produced 490. g of C_2H_5Cl.

124. DDT, an insecticide harmful to fish, birds, and humans, is produced by the following reaction:

$$2C_6H_5Cl + C_2HOCl_3 \longrightarrow C_{14}H_9Cl_5 + H_2O$$
$$\text{chlorobenzene} \quad \text{chloral} \qquad\qquad \text{DDT}$$

In a government lab, 1142 g of chlorobenzene is reacted with 485 g of chloral.

    a. What mass of DDT is formed, assuming 100% yield?

    b. Which reactant is limiting? Which is in excess?

    c. What mass of the excess reactant is left over?

    d. If the actual yield of DDT is 200.0 g, what is the percent yield?

125. Bornite (Cu_3FeS_3) is a copper ore used in the production of copper. When heated, the following reaction occurs:

$$2Cu_3FeS_3(s) + 7O_2(g) \longrightarrow 6Cu(s) + 2FeO(s) + 6SO_2(g)$$

If 2.50 metric tons of bornite is reacted with excess O_2 and the process has an 86.3% yield of copper, what mass of copper is produced?

126. Consider the following unbalanced reaction:

$$P_4(s) + F_2(g) \longrightarrow PF_3(g)$$

What mass of F_2 is needed to produce 120. g of PF_3 if the reaction has a 78.1% yield?

Additional Exercises

127. In using a mass spectrometer, a chemist sees a peak at a mass of 30.0106. Of the choices $^{12}C_2{}^1H_6$, $^{12}C^1H_2{}^{16}O$, and $^{14}N^{16}O$, which is responsible for this peak? Pertinent masses are ^1H, 1.007825; ^{16}O, 15.994915; and ^{14}N, 14.003074.

128. Boron consists of two isotopes, ^{10}B and ^{11}B. Chlorine also has two isotopes, ^{35}Cl and ^{37}Cl. Consider the mass spectrum of BCl_3. How many peaks would be present, and what approximate mass would each peak correspond to in the BCl_3 mass spectrum?

129. A given sample of a xenon fluoride compound contains molecules of the type XeF_n, where n is some whole number. Given that 9.03×10^{20} molecules of XeF_n weigh 0.368 g, determine the value for n in the formula.

130. Aspartame is an artificial sweetener that is 160 times sweeter than sucrose (table sugar) when dissolved in water. It is marketed as NutraSweet. The molecular formula of aspartame is $C_{14}H_{18}N_2O_5$.

    a. Calculate the molar mass of aspartame.

    b. What amount (moles) of molecules are present in 10.0 g aspartame?

    c. Calculate the mass in grams of 1.56 mole of aspartame.

    d. What number of molecules are in 5.0 mg aspartame?

    e. What number of atoms of nitrogen are in 1.2 g aspartame?

    f. What is the mass in grams of 1.0×10^9 molecules of aspartame?

    g. What is the mass in grams of one molecule of aspartame?

131. Anabolic steroids are performance enhancement drugs whose use has been banned from most major sporting activities. One anabolic steroid is fluoxymesterone ($C_{20}H_{29}FO_3$). Calculate the percent composition by mass of fluoxymesterone.

132. Many cereals are made with high moisture content so that the cereal can be formed into various shapes before it is dried. A cereal product containing 58% H_2O by mass is produced at the rate of 1000. kg/h. What mass of water must be evaporated per hour if the final product contains only 20.% water?

133. The compound adrenaline contains 56.79% C, 6.56% H, 28.37% O, and 8.28% N by mass. What is the empirical formula for adrenaline?

134. Adipic acid is an organic compound composed of 49.31% C, 43.79% O, and the rest hydrogen. If the molar mass of adipic acid is 146.1 g/mol, what are the empirical and molecular formulas for adipic acid?

135. Vitamin B_{12}, cyanocobalamin, is essential for human nutrition. It is concentrated in animal tissue but not in higher plants. Although nutritional requirements for the vitamin are quite low, people who abstain completely from animal products may develop a deficiency anemia. Cyanocobalamin is the form used in vitamin supplements. It contains 4.34% cobalt by mass. Calculate the molar mass of cyanocobalamin, assuming that there is one atom of cobalt in every molecule of cyanocobalamin.

136. Some bismuth tablets, a medication used to treat upset stomachs, contain 262 mg of bismuth subsalicylate, $C_7H_5BiO_4$, per tablet. Assuming two tablets are digested, calculate the mass of bismuth consumed.

137. The empirical formula of styrene is CH; the molar mass of styrene is 104.14 g/mol. What number of H atoms are present in a 2.00-g sample of styrene?

138. Terephthalic acid is an important chemical used in the manufacture of polyesters and plasticizers. It contains only C, H, and O. Combustion of 19.81 mg terephthalic acid produces 41.98 mg CO_2 and 6.45 mg H_2O. If 0.250 mole of terephthalic acid has a mass of 41.5 g, determine the molecular formula for terephthalic acid.

139. A sample of a hydrocarbon (a compound consisting of only carbon and hydrogen) contains 2.59×10^{23} atoms of hydrogen and is 17.3% hydrogen by mass. If the molar mass of the hydrocarbon is between 55 and 65 g/mol, what amount (moles) of compound is present, and what is the mass of the sample?

140. A binary compound between an unknown element E and hydrogen contains 91.27% E and 8.73% H by mass. If the formula of the compound is E_3H_8, calculate the atomic mass of E.

141. A 0.755-g sample of hydrated copper(II) sulfate

$$CuSO_4 \cdot xH_2O$$

was heated carefully until it had changed completely to anhydrous copper(II) sulfate ($CuSO_4$) with a mass of 0.483 g. Determine the value of x. [This number is called the *number of waters of hydration* of copper(II) sulfate. It specifies the number of water molecules per formula unit of $CuSO_4$ in the hydrated crystal.]

142. ABS plastic is a tough, hard plastic used in applications requiring shock resistance. The polymer consists of three monomer units: acrylonitrile (C_3H_3N), butadiene (C_4H_6), and styrene (C_8H_8).

   a. A sample of ABS plastic contains 8.80% N by mass. It took 0.605 g of Br_2 to react completely with a 1.20-g sample of ABS plastic. Bromine reacts 1:1 (by moles) with the butadiene molecules in the polymer and nothing

else. What is the percent by mass of acrylonitrile and butadiene in this polymer?

   b. What are the relative numbers of each of the monomer units in this polymer?

143. A sample of LSD (D-lysergic acid diethylamide, $C_{24}H_{30}N_3O$) is added to some table salt (sodium chloride) to form a mixture. Given that a 1.00-g sample of the mixture undergoes combustion to produce 1.20 g of CO_2, what is the mass percent of LSD in the mixture?

144. Methane (CH_4) is the main component of marsh gas. Heating methane in the presence of sulfur produces carbon disulfide and hydrogen sulfide as the only products.

   a. Write the balanced chemical equation for the reaction of methane and sulfur.

   b. Calculate the theoretical yield of carbon disulfide when 120. g of methane is reacted with an equal mass of sulfur.

145. A potential fuel for rockets is a combination of B_5H_9 and O_2. The two react according to the following balanced equation:

$$2B_5H_9(l) + 12O_2(g) \longrightarrow 5B_2O_3(s) + 9H_2O(g)$$

If one tank in a rocket holds 126 g B_5H_9 and another tank holds 192 g O_2, what mass of water can be produced when the entire contents of each tank react together?

146. A 0.4230-g sample of impure sodium nitrate was heated, converting all the sodium nitrate to 0.2864 g of sodium nitrite and oxygen gas. Determine the percent of sodium nitrate in the original sample.

147. An iron ore sample contains Fe_2O_3 plus other impurities. A 752-g sample of impure iron ore is heated with excess carbon, producing 453 g of pure iron by the following reaction:

$$Fe_2O_3(s) + 3C(s) \longrightarrow 2Fe(s) + 3CO(g)$$

What is the mass percent of Fe_2O_3 in the impure iron ore sample? Assume that Fe_2O_3 is the only source of iron and that the reaction is 100% efficient.

148. Commercial brass, an alloy of Zn and Cu, reacts with hydrochloric acid as follows:

$$Zn(s) + 2HCl(aq) \longrightarrow ZnCl_2(aq) + H_2(g)$$

(Cu does not react with HCl.) When 0.5065 g of a certain brass alloy is reacted with excess HCl, 0.0985 g $ZnCl_2$ is eventually isolated.

   a. What is the composition of the brass by mass?

   b. How could this result be checked without changing the above procedure?

149. Vitamin A has a molar mass of 286.4 g/mol and a general molecular formula of C_xH_yE, where E is an unknown element. If vitamin A is 83.86% C and 10.56% H by mass, what is the molecular formula of vitamin A?

150. You have seven closed containers, each with equal masses of chlorine gas (Cl_2). You add 10.0 g of sodium to the first sample, 20.0 g of sodium to the second sample, and so on (adding 70.0 g of sodium to the seventh sample). Sodium and chlorine react to form sodium chloride according to the equation

$$2Na(s) + Cl_2(g) \longrightarrow 2NaCl(s)$$

After each reaction is complete, you collect and measure the amount of sodium chloride formed. A graph of your results is shown below.

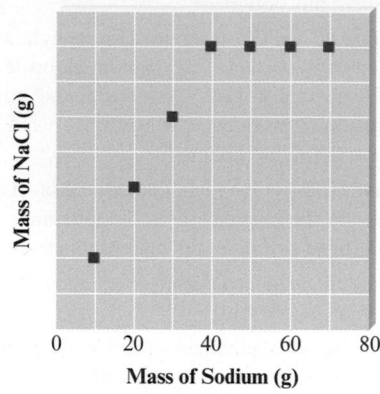

Answer the following questions:

a. Explain the shape of the graph.

b. Calculate the mass of NaCl formed when 20.0 g of sodium is used.

c. Calculate the mass of Cl_2 in each container.

d. Calculate the mass of NaCl formed when 50.0 g of sodium is used.

e. Identify the leftover reactant, and determine its mass for parts b and d above.

151. A substance X_2Z has the composition (by mass) of 40.0% X and 60.0% Z. What is the composition (by mass) of the compound XZ_2?

ChemWork Problems

These multiconcept problems (and additional ones) are found interactively online with the same type of assistance a student would get from an instructor.

152. Consider samples of phosphine (PH_3), water (H_2O), hydrogen sulfide (H_2S), and hydrogen fluoride (HF), each with a mass of 119 g. Rank the compounds from the least to the greatest number of hydrogen atoms contained in the samples.

153. Calculate the number of moles for each compound in the following table.

Compound	Mass	Moles
Magnesium phosphate	326.4 g	_____
Calcium nitrate	303.0 g	_____
Potassium chromate	141.6 g	_____
Dinitrogen pentoxide	406.3 g	_____

154. Arrange the following substances in order of increasing mass percent of nitrogen.

a. NO
b. N_2O
c. NH_3
d. SNH

155. Para-cresol, a substance used as a disinfectant and in the manufacture of several herbicides, is a molecule that contains the elements carbon, hydrogen, and oxygen. Complete combustion of a 0.345-g sample of p-cresol produced 0.983 g carbon dioxide and 0.230 g water. Determine the empirical formula for p-cresol.

156. A compound with molar mass 180.1 g/mol has the following composition by mass:

C	40.0%
H	6.70%
O	53.3%

Determine the empirical and molecular formulas of the compound.

157. Which of the following statements about chemical equations is(are) *true*?

a. When balancing a chemical equation, you can never change the coefficient in front of any chemical formula.

b. The coefficients in a balanced chemical equation refer to the number of grams of reactants and products.

c. In a chemical equation, the reactants are on the right and the products are on the left.

d. When balancing a chemical equation, you can never change the subscripts of any chemical formula.

e. In chemical reactions, matter is neither created nor destroyed so a chemical equation must have the same number of atoms on both sides of the equation.

158. Consider the following unbalanced chemical equation for the combustion of pentane (C_5H_{12}):

$$C_5H_{12}(l) + O_2(g) \longrightarrow CO_2(g) + H_2O(l)$$

If 20.4 g of pentane are burned in excess oxygen, what mass of water can be produced, assuming 100% yield?

159. Sulfur dioxide gas reacts with sodium hydroxide to form sodium sulfite and water. The *unbalanced* chemical equation for this reaction is given below:

$$SO_2(g) + NaOH(s) \longrightarrow Na_2SO_3(s) + H_2O(l)$$

Assuming you react 38.3 g sulfur dioxide with 32.8 g sodium hydroxide and assuming that the reaction goes to completion, calculate the mass of each product formed.

Challenge Problems

160. Gallium arsenide, GaAs, has gained widespread use in semiconductor devices that convert light and electrical signals in fiber-optic communications systems. Gallium consists of 60.% ^{69}Ga and 40.% ^{71}Ga. Arsenic has only one naturally occurring isotope, ^{75}As. Gallium arsenide is a polymeric material, but its mass spectrum shows fragments with the formulas GaAs and Ga_2As_2. What would the distribution of peaks look like for these two fragments?

161. Consider the following data for three binary compounds of hydrogen and nitrogen:

	% H (by Mass)	% N (by Mass)
I	17.75	82.25
II	12.58	87.42
III	2.34	97.66

When 1.00 L of each gaseous compound is decomposed to its elements, the following volumes of $H_2(g)$ and $N_2(g)$ are obtained:

	H_2 (L)	N_2 (L)
I	1.50	0.50
II	2.00	1.00
III	0.50	1.50

Use these data to determine the molecular formulas of compounds I, II, and III and to determine the relative values for the atomic masses of hydrogen and nitrogen.

162. Natural rubidium has the average mass of 85.4678 u and is composed of isotopes ^{85}Rb (mass = 84.9117 u) and ^{87}Rb. The ratio of atoms $^{85}Rb/^{87}Rb$ in natural rubidium is 2.591. Calculate the mass of ^{87}Rb.

163. A compound contains only carbon, hydrogen, nitrogen, and oxygen. Combustion of 0.157 g of the compound produced 0.213 g CO_2 and 0.0310 g H_2O. In another experiment, it is found that 0.103 g of the compound produces 0.0230 g NH_3. What is the empirical formula of the compound? *Hint:* Combustion involves reacting with excess O_2. Assume that all the carbon ends up in CO_2 and all the hydrogen ends up in H_2O. Also assume that all the nitrogen ends up in the NH_3 in the second experiment.

164. Nitric acid is produced commercially by the Ostwald process, represented by the following equations:

$$4NH_3(g) + 5O_2(g) \longrightarrow 4NO(g) + 6H_2O(g)$$
$$2NO(g) + O_2(g) \longrightarrow 2NO_2(g)$$
$$3NO_2(g) + H_2O(l) \longrightarrow 2HNO_3(aq) + NO(g)$$

What mass of NH_3 must be used to produce 1.0×10^6 kg HNO_3 by the Ostwald process? Assume 100% yield in each reaction, and assume that the NO produced in the third step is not recycled.

165. When the supply of oxygen is limited, iron metal reacts with oxygen to produce a mixture of FeO and Fe_2O_3. In a certain experiment, 20.00 g iron metal was reacted with 11.20 g oxygen gas. After the experiment, the iron was totally consumed, and 3.24 g oxygen gas remained. Calculate the amounts of FeO and Fe_2O_3 formed in this experiment.

166. A 9.780-g gaseous mixture contains ethane (C_2H_6) and propane (C_3H_8). Complete combustion to form carbon dioxide and water requires 1.120 mole of oxygen gas. Calculate the mass percent of ethane in the original mixture.

167. Zinc and magnesium metal each reacts with hydrochloric acid to make chloride salts of the respective metals, and hydrogen gas. A 10.00-g mixture of zinc and magnesium produces 0.5171 g of hydrogen gas upon being mixed with an excess of hydrochloric acid. Determine the percent magnesium by mass in the original mixture.

168. A gas contains a mixture of $NH_3(g)$ and $N_2H_4(g)$, both of which react with $O_2(g)$ to form $NO_2(g)$ and $H_2O(g)$. The gaseous mixture (with an initial mass of 61.00 g) is reacted with 10.00 moles O_2, and after the reaction is complete, 4.062 moles of O_2 remains. Calculate the mass percent of $N_2H_4(g)$ in the original gaseous mixture.

169. Consider a gaseous binary compound with a molar mass of 62.09 g/mol. When 1.39 g of this compound is completely burned in excess oxygen, 1.21 g of water is formed. Determine the formula of the compound. Assume water is the only product that contains hydrogen.

170. A 2.25-g sample of scandium metal is reacted with excess hydrochloric acid to produce 0.1502 g hydrogen gas. What is the formula of the scandium chloride produced in the reaction?

171. In the production of printed circuit boards for the electronics industry, a 0.60-mm layer of copper is laminated onto an insulating plastic board. Next, a circuit pattern made of a chemically resistant polymer is printed on the board. The unwanted copper is removed by chemical etching, and the protective polymer is finally removed by solvents. One etching reaction is

$$Cu(NH_3)_4Cl_2(aq) + 4NH_3(aq) + Cu(s) \longrightarrow 2Cu(NH_3)_4Cl(aq)$$

A plant needs to manufacture 10,000 printed circuit boards, each 8.0×16.0 cm in area. An average of 80.% of the copper is removed from each board (density of copper = 8.96 g/cm^3). What masses of $Cu(NH_3)_4Cl_2$ and NH_3 are needed to do this? Assume 100% yield.

172. The aspirin substitute, acetaminophen ($C_8H_9O_2N$), is produced by the following three-step synthesis:

I. $C_6H_5O_3N(s) + 3H_2(g) + HCl(aq) \longrightarrow$
$$C_6H_8ONCl(s) + 2H_2O(l)$$

II. $C_6H_8ONCl(s) + NaOH(aq) \longrightarrow$
$$C_6H_7ON(s) + H_2O(l) + NaCl(aq)$$

III. $C_6H_7ON(s) + C_4H_6O_3(l) \longrightarrow$
$$C_8H_9O_2N(s) + HC_2H_3O_2(l)$$

The first two reactions have percent yields of 87% and 98% by mass, respectively. The overall reaction yields 3 moles of acetaminophen product for every 4 moles of $C_6H_5O_3N$ reacted.

a. What is the percent yield by mass for the overall process?

b. What is the percent yield by mass of Step III?

173. An element X forms both a dichloride (XCl_2) and a tetrachloride (XCl_4). Treatment of 10.00 g XCl_2 with excess chlorine forms 12.55 g XCl_4. Calculate the atomic mass of X, and identify X.

174. When $M_2S_3(s)$ is heated in air, it is converted to $MO_2(s)$. A 4.000-g sample of $M_2S_3(s)$ shows a decrease in mass of 0.277 g when it is heated in air. What is the average atomic mass of M?

175. When aluminum metal is heated with an element from Group 6A of the periodic table, an ionic compound forms. When the experiment is performed with an unknown Group 6A element, the product is 18.56% Al by mass. What is the formula of the compound?

176. Consider a mixture of potassium chloride and potassium nitrate that is 43.2% potassium by mass. What is the percent KCl by mass of the original mixture?

177. Ammonia reacts with O_2 to form either $NO(g)$ or $NO_2(g)$ according to these unbalanced equations:

$$NH_3(g) + O_2(g) \longrightarrow NO(g) + H_2O(g)$$
$$NH_3(g) + O_2(g) \longrightarrow NO_2(g) + H_2O(g)$$

In a certain experiment 2.00 moles of $NH_3(g)$ and 10.00 moles of $O_2(g)$ are contained in a closed flask. After the reaction is complete, 6.75 moles of $O_2(g)$ remains. Calculate the number of moles of NO(*g*) in the product mixture: (*Hint:* You cannot do this problem by adding the balanced equations because you cannot assume that the two reactions will occur with equal probability.)

178. You take 1.00 g of an aspirin tablet (a compound consisting solely of carbon, hydrogen, and oxygen), burn it in air, and collect 2.20 g CO_2 and 0.400 g H_2O. You know that the molar mass of aspirin is between 170 and 190 g/mol. Reacting 1 mole of salicylic acid with 1 mole of acetic anhydride ($C_4H_6O_3$) gives you 1 mole of aspirin and 1 mole of acetic acid ($C_2H_4O_2$). Use this information to determine the molecular formula of salicylic acid.

Integrative Problems

These problems require the integration of multiple concepts to find the solutions.

179. With the advent of techniques such as scanning tunneling microscopy, it is now possible to "write" with individual atoms by manipulating and arranging atoms on an atomic surface.

   a. If an image is prepared by manipulating iron atoms and their total mass is 1.05×10^{-20} g, what number of iron atoms were used?

   b. If the image is prepared on a platinum surface that is exactly 20 platinum atoms high and 14 platinum atoms wide, what is the mass (grams) of the atomic surface?

   c. If the atomic surface were changed to ruthenium atoms and the same surface mass as determined in part b is used, what number of ruthenium atoms is needed to construct the surface?

180. Tetrodotoxin is a toxic chemical found in fugu pufferfish, a popular but rare delicacy in Japan. This compound has an LD_{50} (the amount of substance that is lethal to 50.% of a population sample) of 10. μg per kg of body mass. Tetrodotoxin is 41.38% carbon by mass, 13.16% nitrogen by mass, and 5.37% hydrogen by mass, with the remaining amount consisting of oxygen. What is the empirical formula of tetrodotoxin? If three molecules of tetrodotoxin have a mass of 1.59×10^{-21} g, what is the molecular formula of tetrodotoxin? What number of molecules of tetrodotoxin would be the LD_{50} dosage for a person weighing 165 lb?

181. An ionic compound MX_3 is prepared according to the following unbalanced chemical equation.

$$M + X_2 \longrightarrow MX_3$$

A 0.105-g sample of X_2 contains 8.92×10^{20} molecules. The compound MX_3 consists of 54.47% X by mass. What are the identities of M and X, and what is the correct name for MX_3? Starting with 1.00 g each of M and X_2, what mass of MX_3 can be prepared?

182. The compound As_2I_4 is synthesized by reaction of arsenic metal with arsenic triiodide. If a solid cubic block of arsenic ($d = 5.72$ g/cm^3) that is 3.00 cm on edge is allowed to react with 1.01×10^{24} molecules of arsenic triiodide, what mass of As_2I_4 can be prepared? If the percent yield of As_2I_4 was 75.6%, what mass of As_2I_4 was actually isolated?

183. A compound XF_5 is 42.81% fluorine by mass. Identify the element X and draw the Lewis structure for the compound. What is the molecular structure of XF_5?

Marathon Problems

These problems are designed to incorporate several concepts and techniques into one situation.

184. A 2.077-g sample of an element, which has an atomic mass between 40 and 55, reacts with oxygen to form 3.708 g of an oxide. Determine the formula of the oxide (and identify the element).

185. Consider the following balanced chemical equation:

$$A + 5B \longrightarrow 3C + 4D$$

   a. Equal masses of A and B are reacted. Complete each of the following with either "A is the limiting reactant because _____"; "B is the limiting reactant because _____"; or "we cannot determine the limiting reactant because _____."

     i. If the molar mass of A is greater than the molar mass of B, then

     ii. If the molar mass of B is greater than the molar mass of A, then

   b. The products of the reaction are carbon dioxide (C) and water (D). Compound A has a similar molar mass to carbon dioxide. Compound B is a diatomic molecule. Identify compound B, and support your answer.

   c. Compound A is a hydrocarbon that is 81.71% carbon by mass. Determine its empirical and molecular formulas.

Types of Chemical Reactions and Solution Stoichiometry

Sodium reacts violently when water is dripped on it. (Charles D. Winters)

Much of the chemistry that affects each of us occurs among substances dissolved in water. For example, virtually all the chemistry that makes life possible occurs in an aqueous environment. Also, various medical tests involve aqueous reactions, depending heavily on analyses of blood and other body fluids. In addition to the common tests for sugar, cholesterol, and iron, analyses for specific chemical markers allow detection of many diseases before obvious symptoms occur.

Aqueous chemistry is also important in our environment. In recent years, contamination of the groundwater by substances such as chloroform and nitrates has been widely publicized. Water is essential for life, and the maintenance of an ample supply of clean water is crucial to all civilization.

To understand the chemistry that occurs in such diverse places as the human body, the atmosphere, the groundwater, the oceans, the local water treatment plant, your hair as you shampoo it, and so on, we must understand how substances dissolved in water react with each other.

However, before we can understand solution reactions, we need to discuss the nature of solutions in which water is the dissolving medium, or *solvent*. These solutions are called **aqueous solutions**. In this chapter we will study the nature of materials after they are dissolved in water and various types of reactions that occur among these substances. You will see that the procedures developed in Chapter 5 to deal with chemical reactions work very well for reactions that take place in aqueous solutions. To understand the types of reactions that occur in aqueous solutions, we must first explore the types of species present. This requires an understanding of the polarity of water which depends on its structure as predicted by the VSEPR model. We will take an atomic-molecular level view of solutions when appropriate to better understand the nature of solutions and the calculations involved.

6-1 | Water, the Common Solvent

Water is one of the most important substances on the earth. It is essential for sustaining the reactions that keep us alive, but it also affects our lives in many indirect ways. Water helps moderate the earth's temperature; it cools automobile engines, nuclear power plants, and many industrial processes; it provides a means of transportation on the earth's surface and a medium for the growth of myriad creatures we use as food; and much more.

One of the most valuable properties of water is its ability to dissolve many different substances. For example, salt "disappears" when you sprinkle it into the water used to cook vegetables, as does sugar when you add it to your iced tea. In each case the "disappearing" substance is obviously still present—you can taste it. What happens when a solid dissolves? To understand this process, we need to consider the nature of water. Liquid water consists of a collection of H_2O molecules. Recall that an individual H_2O molecule is "bent" or V-shaped, with an H—O—H angle of approximately 105 degrees:

$$\underset{O}{H \overset{105°}{\frown} H}$$

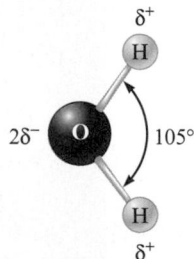

Figure 6-1 | (top) The water molecule is polar. (bottom) A space-filling model of the water molecule.

If needed you can review the discussion of VSEPR in Chapter 4.

The O—H bonds in the water molecule are covalent polar bonds in which the oxygen atom has a greater attraction for the shared electrons than the hydrogen atom. ◄ As we saw in Chapter 3, the result is a **polar molecule** as shown in Fig. 6-1.

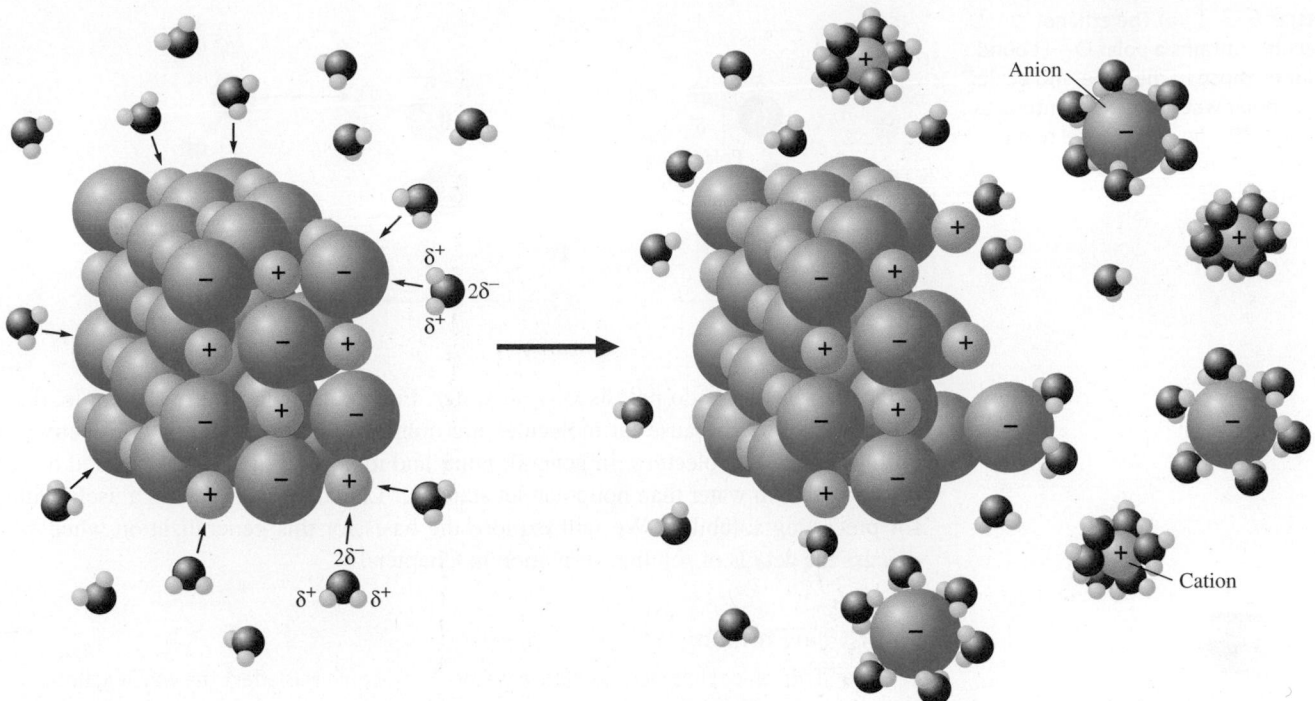

Figure 6-2 | Polar water molecules interact with the positive and negative ions of a salt, assisting in the dissolving process.

A schematic of an ionic solid dissolving in water is shown in Fig. 6-2. Note that the "positive ends" of the water molecules are attracted to the negatively charged anions and that the "negative ends" are attracted to the positively charged cations. This process is called **hydration**. The hydration of its ions tends to cause a salt to "fall apart" in the water, or to dissolve. The strong forces present among the positive and negative ions of the solid are replaced by strong water–ion interactions.

It is very important to recognize that when ionic substances (salts) dissolve in water, they break up into the *individual* cations and anions. For instance, when ammonium nitrate (NH_4NO_3) dissolves in water, the resulting solution contains NH_4^+ and NO_3^- ions moving around independently. This process can be represented as

$$NH_4NO_3(s) \xrightarrow{H_2O(l)} NH_4^+(aq) + NO_3^-(aq)$$

where (*aq*) designates that the ions are hydrated by unspecified numbers of water molecules.

The **solubility** of ionic substances in water varies greatly. For example, sodium chloride is quite soluble in water, whereas silver chloride (contains Ag^+ and Cl^- ions) is only very slightly soluble. The differences in the solubilities of ionic compounds in water typically depend on the relative attractions of the ions for each other (these forces hold the solid together) and the attractions of the ions for water molecules (which cause the solid to disperse [dissolve] in water). Solubility is a complex topic that we will explore in much more detail in Chapter 10. However, the most important thing to remember at this point is that when an ionic solid does dissolve in water, the ions become hydrated and are dispersed (move around independently).

Water also dissolves many nonionic substances. Ethanol (C_2H_5OH), for example, is very soluble in water. Wine, beer, and mixed drinks are aqueous solutions of ethanol and other substances. Why is ethanol so soluble in water? The answer lies in the structure of the alcohol molecules, which is shown in Fig. 6-3(a). The molecule contains a polar O—H bond like those in water, which makes it very compatible with water. The interaction of water with ethanol is represented in Fig. 6-3(b).

Figure 6-3 | (a) The ethanol molecule contains a polar O—H bond similar to those in the water molecule. (b) The polar water molecule interacts strongly with the polar O—H bond in ethanol. This is a case of "like dissolving like."

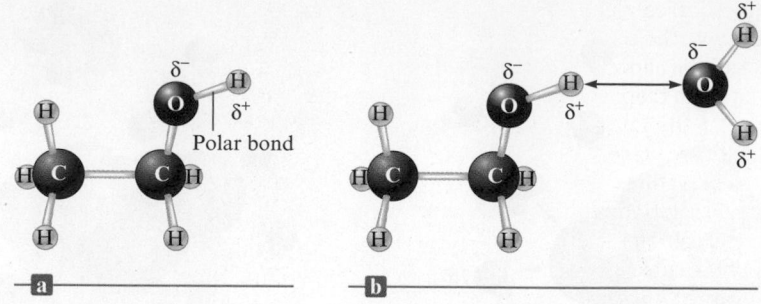

Many substances do not dissolve in water. Pure water will not, for example, dissolve animal fat, because fat molecules are nonpolar and do not interact effectively with polar water molecules. In general, polar and ionic substances are expected to be more soluble in water than nonpolar substances. "Like dissolves like" is a useful rule for predicting solubility. We will explore the basis for this generalization when we discuss the details of solution formation in Chapter 10.

Critical Thinking

What if no ionic solids were soluble in water? How would this affect the way reactions occur in aqueous solutions?

6-2 | The Nature of Aqueous Solutions: Strong and Weak Electrolytes

A solution is a homogeneous mixture. It is the same throughout (the first sip of a cup of coffee is the same as the last), but its composition can be varied by changing the amount of dissolved substances (one can make weak or strong coffee). In this section we will consider what happens when a substance, the **solute**, is dissolved in liquid water, the **solvent**.

One useful property for characterizing a solution is its **electrical conductivity**, its ability to conduct an electric current. ◄ 𝒊 This characteristic can be checked conveniently by using an apparatus like the ones shown in Fig. 6-4. If the solution in the container conducts electricity, the bulb lights. Pure water is not an electrical conductor. However, some aqueous solutions conduct current very efficiently, and the bulb shines very brightly; these solutions contain **strong electrolytes**. Other solutions conduct only a small current, and the bulb glows dimly; these solutions contain **weak electrolytes**. Some solutions permit no current to flow, and the bulb remains unlit; these solutions contain **nonelectrolytes**.

𝒊 An electrolyte is a substance that when dissolved in water produces a solution that can conduct electricity.

The basis for the conductivity properties of solutions was first correctly identified by Svante Arrhenius (1859–1927), then a Swedish graduate student in physics, who carried out research on the nature of solutions at the University of Uppsala in the early 1880s. Arrhenius came to believe that the conductivity of solutions arose from the presence of ions, an idea that was at first scorned by the majority of the scientific establishment. However, in the late 1890s when atoms were found to contain charged particles, the ionic theory suddenly made sense and became widely accepted.

As Arrhenius postulated, the extent to which a solution can conduct an electric current depends directly on the number of ions present. Some materials, such as sodium chloride, readily produce ions in aqueous solution and thus are strong electrolytes. Other substances, such as acetic acid, produce relatively few ions when dissolved in water and are weak electrolytes. A third class of materials, such as sugar, form virtually no ions when dissolved in water and are nonelectrolytes.

Figure 6-4 | Electrical conductivity of aqueous solutions. The circuit will be completed and will allow current to flow only when there are charge carriers (ions) in the solution. *Note*: Water molecules are present but not shown in these pictures. (a) A hydrochloric acid solution, which is a strong electrolyte, contains ions that readily conduct the current and give a brightly lit bulb. (b) An acetic acid solution, which is a weak electrolyte, contains only a few ions and does not conduct as much current as a strong electrolyte. The bulb is only dimly lit. (c) A sucrose solution, which is a nonelectrolyte, contains no ions and does not conduct a current. The bulb remains unlit.

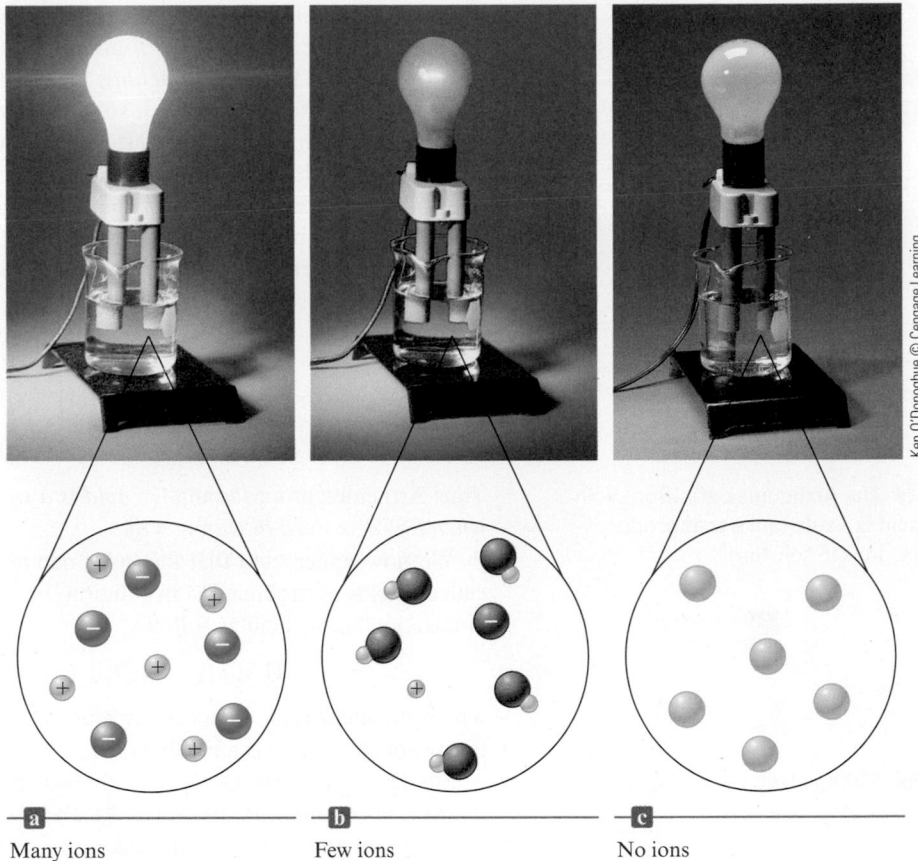

Ken O'Donoghue © Cengage Learning

a   Many ions   b   Few ions   c   No ions

ⓘ It is important to picture solutions at an atomic-molecular level.

Strong Electrolytes

Strong electrolytes are substances that are completely ionized when they are dissolved in water, as represented in Fig. 6-4(a). We will consider several classes of strong electrolytes: (1) soluble salts, (2) strong acids, and (3) strong bases.

As shown in Fig. 6-2, a salt consists of an array of cations and anions that separate and become hydrated when the salt dissolves. For example, when NaCl dissolves in water, it produces hydrated Na^+ and Cl^- ions in the solution (Fig. 6-5). Virtually no NaCl units are present. Thus NaCl is a strong electrolyte. It is important to recognize that these aqueous solutions contain millions of water molecules, which we will not include in our molecular-level drawings. ◂ *ⓘ*

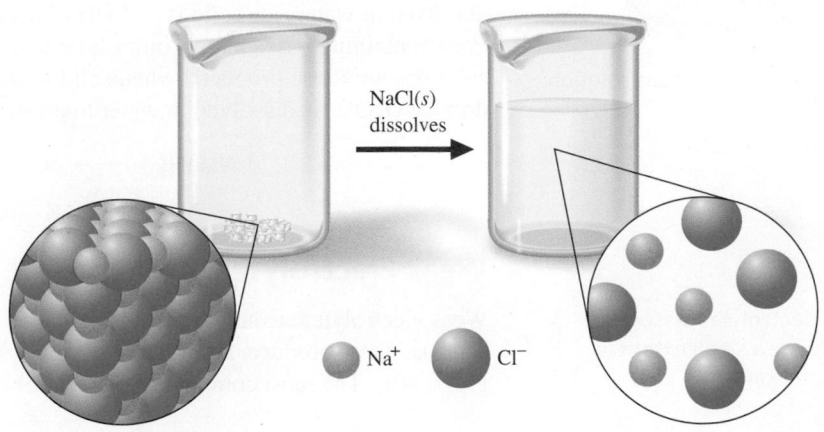

NaCl(*s*) dissolves

Na^+ Cl^-

Figure 6-5 | When solid NaCl dissolves, the Na^+ and Cl^- ions are randomly dispersed in the water.

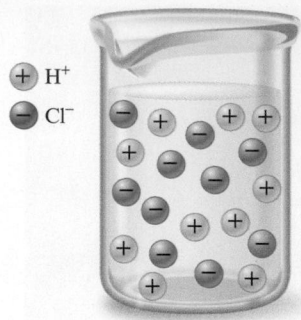

Figure 6-6 | HCl(aq) is completely ionized.

The Arrhenius definition of an acid is a substance that produces H^+ ions in solution.

Strong electrolytes dissociate (ionize) completely in aqueous solution.

Perchloric acid, $HClO_4(aq)$, is another strong acid.

One of Arrhenius's most important discoveries concerned the nature of acids. Acidity was first associated with the sour taste of citrus fruits. In fact, the word *acid* comes directly from the Latin word *acidus*, meaning "sour." The mineral acids sulfuric acid (H_2SO_4) and nitric acid (HNO_3), so named because they were originally obtained by the treatment of minerals, were discovered around 1300.

Although acids were known for hundreds of years before the time of Arrhenius, no one had recognized their essential nature. In his studies of solutions, Arrhenius found that when the substances HCl, HNO_3, and H_2SO_4 were dissolved in water, they behaved as strong electrolytes. He postulated that this was the result of ionization reactions in water, for example:

$$HCl \xrightarrow{H_2O} H^+(aq) + Cl^-(aq)$$
$$HNO_3 \xrightarrow{H_2O} H^+(aq) + NO_3^-(aq)$$
$$H_2SO_4 \xrightarrow{H_2O} H^+(aq) + HSO_4^-(aq)$$

Thus Arrhenius proposed that an **acid** *is a substance that produces H^+ ions (protons) when it is dissolved in water.*

We now understand that the polar nature of water plays a very important role in causing acids to produce H^+ in solution. In fact, it is most appropriate to represent the "ionization" of an acid as follows:

$$HA(aq) + H_2O(l) \longrightarrow H_3O^+(aq) + A^-(aq)$$

which emphasizes the important role of water in this process. We will have much more to say about this process in Chapter 13.

Studies of conductivity show that when HCl, HNO_3, and H_2SO_4 are placed in water, *virtually every molecule ionizes.* These substances are strong electrolytes and are thus called **strong acids**. All three are very important chemicals, and much more will be said about them as we proceed. However, at this point the following facts are important:

Sulfuric acid, nitric acid, and hydrochloric acid are aqueous solutions and should be written in chemical equations as $H_2SO_4(aq)$, $HNO_3(aq)$, and $HCl(aq)$, respectively, although they often appear without the (aq) symbol.

A strong acid is one that completely dissociates into its ions. Thus, if 100 molecules of HCl are dissolved in water, 100 H^+ ions and 100 Cl^- ions are produced. Virtually no HCl molecules exist in aqueous solutions (see Fig. 6-6).

Sulfuric acid is a special case. The formula H_2SO_4 indicates that this acid can produce two H^+ ions per molecule when dissolved in water. However, only the first H^+ ion is completely dissociated. The second H^+ ion can be pulled off under certain conditions, which we will discuss later. Thus an aqueous solution of H_2SO_4 contains mostly H^+ ions and HSO_4^- ions.

Another important class of strong electrolytes consists of the **strong bases**, soluble ionic compounds containing the hydroxide ion (OH^-). When these compounds are dissolved in water, the cations and OH^- ions separate and move independently. Solutions containing bases have a bitter taste and a slippery feel. The most common basic solutions are those produced when solid sodium hydroxide (NaOH) or potassium hydroxide (KOH) is dissolved in water to produce ions, as follows (see Fig. 6-7):

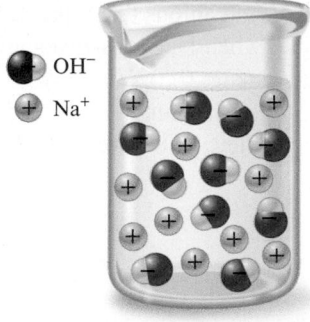

Figure 6-7 | An aqueous solution of sodium hydroxide.

$$NaOH(s) \xrightarrow{H_2O} Na^+(aq) + OH^-(aq)$$
$$KOH(s) \xrightarrow{H_2O} K^+(aq) + OH^-(aq)$$

Weak Electrolytes

Weak electrolytes dissociate (ionize) only to a small extent in aqueous solution.

Weak electrolytes are substances that exhibit a small degree of ionization in water. That is, they produce relatively few ions when dissolved in water, as shown in Fig. 6-4(b). The most common weak electrolytes are weak acids and weak bases.

Science is a human endeavor, subject to human frailties and governed by personalities, politics, and prejudices. One of the best illustrations of the often bumpy path of the advancement of scientific knowledge is the story of Swedish chemist Svante Arrhenius.

When Arrhenius began studies toward his doctorate at the University of Uppsala around 1880, he chose to investigate the passage of electricity through solutions, a mystery that had baffled scientists for a century. The first experiments had been done in the 1770s by Cavendish, who compared the conductivity of salt solution with that of rain water using his own physiologic reaction to the electric shocks he received! Arrhenius had an array of instruments to measure electric current, but the process of carefully weighing, measuring, and recording data from a multitude of experiments was a tedious one.

After his long series of experiments was performed, Arrhenius quit his laboratory bench and returned to his country home to try to formulate a model that could account for his data. He wrote, "I got the idea in the night of the 17th of May in the year 1883, and I could not sleep that night until I had worked through the whole problem." His idea was that ions were responsible for conducting electricity through a solution.

Back at Uppsala, Arrhenius took his doctoral dissertation containing the new theory to his advisor, Professor Cleve, an eminent chemist and the discoverer of the elements holmium and thulium. Cleve's uninterested response was what Arrhenius had expected. It was in keeping with Cleve's resistance to new ideas—he had not even accepted Mendeleev's periodic table, introduced 10 years earlier.

It is a long-standing custom that before a doctoral degree is granted, the dissertation must be defended before a panel of professors. Although this procedure is still followed at most universities today, the problems are usually worked out in private with the evaluating professors before the actual defense. However, when Arrhenius did it, the dissertation defense was an open debate, which could be rancorous and humiliating. Knowing that it would be unwise to antagonize his professors, Arrhenius downplayed his convictions about

Svante August Arrhenius.

his new theory as he defended his dissertation. His diplomacy paid off: He was awarded his degree, albeit reluctantly, because the professors still did not believe his model and considered him to be a marginal scientist, at best.

Such a setback could have ended his scientific career, but Arrhenius was a crusader; he was determined to see his theory triumph. He promptly embarked on a political campaign, enlisting the aid of several prominent scientists, to get his theory accepted.

Ultimately, the ionic theory triumphed. Arrhenius's fame spread, and honors were heaped on him, culminating in the Nobel Prize in chemistry in 1903. Not one to rest on his laurels, Arrhenius turned to new fields, including astronomy; he formulated a new theory that the solar system may have come into being through the collision of stars. His exceptional versatility led him to study the use of serums to fight disease, energy resources and conservation, and the origin of life.

Additional insight on Arrhenius and his scientific career can be obtained from his address on receiving the Willard Gibbs Award. See *Journal of the American Chemical Society* 36 (1912): 353.

The main acidic component of vinegar is acetic acid ($HC_2H_3O_2$). The formula is written to indicate that acetic acid has two chemically distinct types of hydrogen atoms. Formulas for acids are often written with the acidic hydrogen atom or atoms (any that will produce H^+ ions in solution) listed first. If any nonacidic hydrogens are present, they are written later in the formula. Thus the formula $HC_2H_3O_2$ indicates one

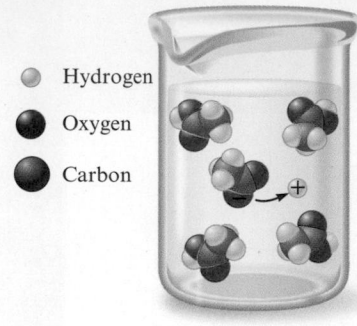

Figure 6-8 | Acetic acid ($HC_2H_3O_2$) exists in water mostly as undissociated molecules. Only a small percentage of the molecules are ionized.

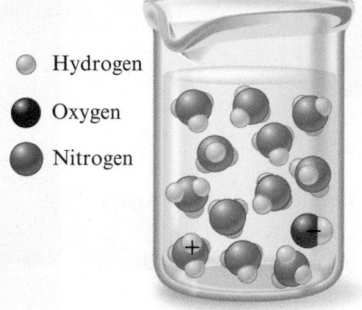

Figure 6-9 | The reaction of NH_3 in water.

acidic and three nonacidic hydrogen atoms. The dissociation reaction for acetic acid in water can be written as follows:

$$HC_2H_3O_2(aq) + H_2O(l) \rightleftharpoons H_3O^+(aq) + C_2H_3O_2^-(aq)$$

Acetic acid is very different from the strong acids because only about 1% of its molecules dissociate in aqueous solutions at typical concentrations. For example, in a solution containing 0.1 mole of $HC_2H_3O_2$ per liter, for every 100 molecules of $HC_2H_3O_2$ originally dissolved in water, approximately 99 molecules of $HC_2H_3O_2$ remain intact (Fig. 6-8). That is, only one molecule out of every 100 dissociates (to produce one H^+ ion and one $C_2H_3O_2^-$ ion). The double arrow indicates the reaction can occur in either direction.

Because acetic acid is a weak electrolyte, it is called a **weak acid**. Any acid, such as acetic acid, that *dissociates (ionizes) only to a slight extent in aqueous solutions is called a weak acid*. In Chapter 13 we will explore the subject of weak acids in detail.

The most common weak base is ammonia (NH_3). When ammonia is dissolved in water, it reacts as follows:

$$NH_3(aq) + H_2O(l) \longrightarrow NH_4^+(aq) + OH^-(aq)$$

The solution is *basic* because OH^- ions are produced. Ammonia is called a **weak base** because *the resulting solution is a weak electrolyte*; that is, very few ions are formed. In fact, in a solution containing 0.1 mole of NH_3 per liter, for every 100 molecules of NH_3 originally dissolved, only one NH_4^+ ion and one OH^- ion are produced; 99 molecules of NH_3 remain unreacted (Fig. 6-9).

Nonelectrolytes

Nonelectrolytes are substances that dissolve in water but do not produce any ions, as shown in Fig. 6-4(c). An example of a nonelectrolyte is ethanol (see Fig. 6-3 for the structural formula). When ethanol dissolves, entire C_2H_5OH molecules are dispersed in the water. Since the molecules do not break up into ions, the resulting solution does not conduct an electric current. Another common nonelectrolyte is table sugar (sucrose, $C_{12}H_{22}O_{11}$), which is very soluble in water but which produces no ions when it dissolves. The sucrose molecules remain intact.

6-3 | The Composition of Solutions

Chemical reactions often take place when two solutions are mixed. To perform stoichiometric calculations in such cases, we must know two things: (1) the *nature of the reaction*, which depends on the exact forms the chemicals take when dissolved, and (2) the *amounts of chemicals* present in the solutions, usually expressed as concentrations.

The concentration of a solution can be described in many different ways, as we will see in Chapter 10. At this point we will consider only the most commonly used expression of concentration, **molarity** (*M*), which is defined as *moles of solute per volume of solution in liters*:

$$M = \text{molarity} = \frac{\text{moles of solute}}{\text{liters of solution}}$$

A solution that is 1.0 molar (written as 1.0 *M*) contains 1.0 mole of solute per liter of solution.

INTERACTIVE EXAMPLE 6-1

Calculation of Molarity I

Calculate the molarity of a solution prepared by dissolving 11.5 g of solid NaOH in enough water to make 1.50 L of solution.

Solution

Where are we going?
To find the molarity of NaOH solution

What do we know?

- 11.5 g NaOH
- 1.50 L solution

What information do we need to find molarity?

- Moles solute

- Molarity = $\dfrac{\text{mol solute}}{\text{L solution}}$

How do we get there?
What are the moles of NaOH (40.00 g/mol)?

$$11.5 \text{ g } \cancel{\text{NaOH}} \times \frac{1 \text{ mol NaOH}}{40.00 \text{ g } \cancel{\text{NaOH}}} = 0.288 \text{ mol NaOH}$$

What is the molarity of the solution?

$$\text{Molarity} = \frac{\text{mol solute}}{\text{L solution}} = \frac{0.288 \text{ mol NaOH}}{1.50 \text{ L solution}} = 0.192 \, M \text{ NaOH}$$

REALITY CHECK | The units are correct for molarity.

See Exercises 6-29 and 6-30

INTERACTIVE EXAMPLE 6-2

Calculation of Molarity II

Calculate the molarity of a solution prepared by dissolving 1.56 g of gaseous HCl in enough water to make 26.8 mL of solution.

Solution

Where are we going?
To find the molarity of HCl solution

What do we know?

- 1.56 g HCl
- 26.8 mL solution

What information do we need to find molarity?

- Moles solute

- Molarity = $\dfrac{\text{mol solute}}{\text{L solution}}$

How do we get there?
What are the moles of HCl (36.46 g/mol)?

$$1.56 \text{ g } \cancel{\text{HCl}} \times \frac{1 \text{ mol HCl}}{36.46 \text{ g } \cancel{\text{HCl}}} = 4.28 \times 10^{-2} \text{ mol HCl}$$

What is the volume of solution (in liters)?

$$26.8 \ \cancel{mL} \times \frac{1 \ L}{1000 \ \cancel{mL}} = 2.68 \times 10^{-2} \ L$$

What is the molarity of the solution?

$$\text{Molarity} = \frac{4.28 \times 10^{-2} \ \text{mol HCl}}{2.68 \times 10^{-2} \ \text{L solution}} = 1.60 \ M \ \text{HCl}$$

REALITY CHECK | The units are correct for molarity.

See Exercises 6-29 and 6-30

INTERACTIVE EXAMPLE 6-3

Concentration of Ions I

Give the concentration of each type of ion in the following solutions:

a. 0.50 M Co(NO$_3$)$_2$

b. 1 M Fe(ClO$_4$)$_3$

Solution

Where are we going?
To find the molarity of each ion in the solution

What do we know?

- 0.50 M Co(NO$_3$)$_2$
- 1 M Fe(ClO$_4$)$_3$

What information do we need to find the molarity of each ion?

- Moles of each ion

How do we get there?

For Co(NO$_3$)$_2$ ◄
What is the balanced equation for dissolving the ions?

$$\text{Co(NO}_3)_2(s) \xrightarrow{\text{H}_2\text{O}} \text{Co}^{2+}(aq) + 2\text{NO}_3^-(aq)$$

What is the molarity for each ion?

$$\begin{array}{lll} \text{Co}^{2+} & 1 \times 0.50 \ M = 0.50 \ M \ \text{Co}^{2+} \\ \text{NO}_3^- & 2 \times 0.50 \ M = 1.0 \ M \ \text{NO}_3^- \end{array}$$

For Fe(ClO$_4$)$_3$
What is the balanced equation for dissolving the ions?

$$\text{Fe(ClO}_4)_3(s) \xrightarrow{\text{H}_2\text{O}} \text{Fe}^{3+}(aq) + 3\text{ClO}_4^-(aq)$$

What is the molarity for each ion?

$$\begin{array}{lll} \text{Fe}^{3+} & 1 \times 1 \ M = 1 \ M \ \text{Fe}^{3+} \\ \text{ClO}_4^- & 3 \times 1 \ M = 3 \ M \ \text{ClO}_4^- \end{array}$$

See Exercises 6-31 and 6-32

▲ An aqueous solution of Co(NO$_3$)$_2$.

Often chemists need to determine the number of moles of solute present in a given volume of a solution of known molarity. The procedure for doing this is easily derived from the definition of molarity. If we multiply the molarity of a solution by the volume (in liters) of a particular sample of the solution, we get the moles of solute present in that sample:

Liters of solution × molarity =

$$\text{liters of solution} \times \frac{\text{moles of solute}}{\text{liters of solution}} = \text{moles of solute} \blacktriangleleft \textit{i}$$

$\textit{i}$ $M = \dfrac{\text{moles of solute}}{\text{liters of solution}}$

This procedure is demonstrated in Examples 6-4 and 6-5.

INTERACTIVE EXAMPLE 6-4

Concentration of Ions II

Calculate the number of moles of Cl^- ions in 1.75 L of 1.0×10^{-3} M $ZnCl_2$.

Solution

Where are we going?
To find the moles of Cl^- ion in the solution

What do we know?
- 1.0×10^{-3} M $ZnCl_2$
- 1.75 L

What information do we need to find moles of Cl^-?
- Balanced equation for dissolving $ZnCl_2$

How do we get there?
What is the balanced equation for dissolving the ions?

$$ZnCl_2(s) \xrightarrow{H_2O} Zn^{2+}(aq) + 2Cl^-(aq)$$

What is the molarity of Cl^- ion in the solution?

$$2 \times (1.0 \times 10^{-3} M) = 2.0 \times 10^{-3} M \, Cl^-$$

How many moles of Cl^-?

$$1.75 \text{ L solution} \times 2.0 \times 10^{-3} M \, Cl^- = 1.75 \text{ L solution} \times \frac{2.0 \times 10^{-3} \text{ mol } Cl^-}{\text{L solution}}$$

$$= 3.5 \times 10^{-3} \text{ mol } Cl^-$$

See Exercise 6-33

INTERACTIVE EXAMPLE 6-5

Concentration and Volume

Typical blood serum is about 0.14 M NaCl. What volume of blood contains 1.0 mg NaCl?

Solution

Where are we going?
To find the volume of blood containing 1.0 mg NaCl

What do we know?
- 0.14 M NaCl
- 1.0 mg NaCl

What information do we need to find volume of blood containing 1.0 mg NaCl?
- Moles of NaCl (in 1.0 mg)

How do we get there?
What are the moles of NaCl (58.44 g/mol)?

$$1.0 \text{ mg NaCl} \times \frac{1 \text{ g NaCl}}{1000 \text{ mg NaCl}} \times \frac{1 \text{ mol NaCl}}{58.44 \text{ g NaCl}} = 1.7 \times 10^{-5} \text{ mol NaCl}$$

What volume of 0.14 M NaCl contains 1.0 mg (1.7 × 10⁻⁵ mol) NaCl?
There is some volume, call it V, that when multiplied by the molarity of this solution will yield 1.7×10^{-5} moles of NaCl. That is,

$$V \times \frac{0.14 \text{ mol NaCl}}{\text{L solution}} = 1.7 \times 10^{-5} \text{ mol NaCl}$$

We want to solve for the volume:

$$V = \frac{1.7 \times 10^{-5} \text{ mol NaCl}}{\dfrac{0.14 \text{ mol NaCl}}{\text{L solution}}} = 1.2 \times 10^{-4} \text{ L solution}$$

Thus 0.12 mL of blood contains 1.7×10^{-5} moles of NaCl or 1.0 mg NaCl.

See Exercises 6-35 and 6-36

A **standard solution** is a *solution whose concentration is accurately known.* Standard solutions, often used in chemical analysis, can be prepared as shown in Fig. 6-10 and in Example 6-6.

INTERACTIVE EXAMPLE 6-6

Solutions of Known Concentration

To analyze the alcohol content of a certain wine, a chemist needs 1.00 L of an aqueous 0.200 M $K_2Cr_2O_7$ (potassium dichromate) solution. How much solid $K_2Cr_2O_7$ must be weighed out to make this solution?

Solution

Where are we going?
To find the mass of $K_2Cr_2O_7$ required for the solution

What do we know?

- 1.00 L of 0.200 M $K_2Cr_2O_7$ is required

What information do we need to find the mass of $K_2Cr_2O_7$?

- Moles of $K_2Cr_2O_7$ in the required solution

Figure 6-10 | Steps involved in the preparation of a standard aqueous solution. (a) Put a weighed amount of a substance (the solute) into the volumetric flask, and add a small quantity of water. (b) Dissolve the solid in the water by gently swirling the flask (*with the stopper in place*). (c) Add more water (with gentle swirling) until the level of the solution just reaches the mark etched on the neck of the flask. Then mix the solution thoroughly by inverting the flask several times.

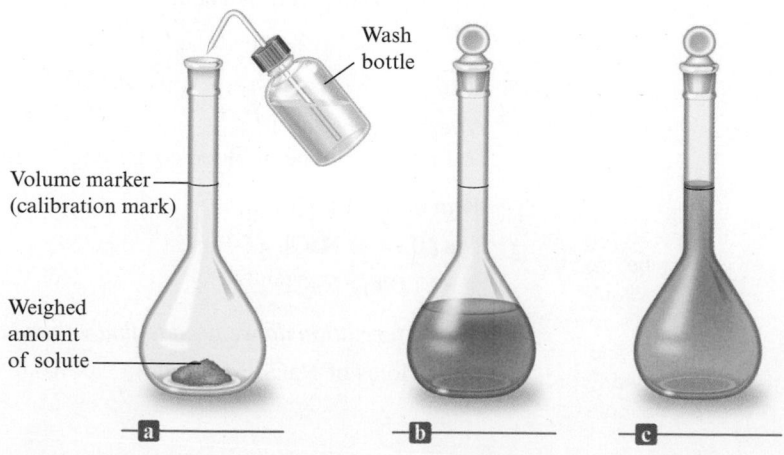

How do we get there?
What are the moles of $K_2Cr_2O_7$ required?

$$M \times V = mol$$

$$1.00 \text{ L solution} \times \frac{0.200 \text{ mol K}_2\text{Cr}_2\text{O}_7}{\text{L solution}} = 0.200 \text{ mol K}_2\text{Cr}_2\text{O}_7$$

What mass of $K_2Cr_2O_7$ is required for the solution?

$$0.200 \text{ mol K}_2\text{Cr}_2\text{O}_7 \times \frac{294.20 \text{ g K}_2\text{Cr}_2\text{O}_7}{\text{mol K}_2\text{Cr}_2\text{O}_7} = 58.8 \text{ g K}_2\text{Cr}_2\text{O}_7$$

To make 1.00 L of 0.200 M $K_2Cr_2O_7$, the chemist must weigh out 58.8 g $K_2Cr_2O_7$, transfer it to a 1.00-L volumetric flask, and add distilled water to the mark on the flask.

See Exercises 6-37a,c and 6-38c,e

Dilution

To save time and space in the laboratory, routinely used solutions are often purchased or prepared in concentrated form (called *stock solutions*). Water is then added to achieve the molarity desired for a particular solution. This process is called **dilution**. For example, the common acids are purchased as concentrated solutions and diluted as needed. A typical dilution calculation involves determining how much water must be added to an amount of stock solution to achieve a solution of the desired concentration. The key to doing these calculations is to remember that

> *i* Dilution with water does not alter the number of moles of solute present.

Moles of solute after dilution = moles of solute before dilution

because only water (no solute) is added to accomplish the dilution. ◄ *i*

For example, suppose we need to prepare 500 mL of 1.00 M acetic acid ($HC_2H_3O_2$) from a 17.4 M stock solution of acetic acid. What volume of the stock solution is required? The first step is to determine the number of moles of acetic acid in the final solution by multiplying the volume by the molarity (remembering that the volume must be changed to liters):

$$500. \text{ mL solution} \times \frac{1 \text{ L solution}}{1000 \text{ mL solution}} \times \frac{1.00 \text{ mol HC}_2\text{H}_3\text{O}_2}{\text{L solution}} = 0.500 \text{ mol HC}_2\text{H}_3\text{O}_2$$

Thus we need to use a volume of 17.4 M acetic acid that contains 0.500 mole of $HC_2H_3O_2$. That is,

$$V \times \frac{17.4 \text{ mol HC}_2\text{H}_3\text{O}_2}{\text{L solution}} = 0.500 \text{ mol HC}_2\text{H}_3\text{O}_2$$

Solving for V gives

$$V = \frac{0.500 \text{ mol HC}_2\text{H}_3\text{O}_2}{\frac{17.4 \text{ mol HC}_2\text{H}_3\text{O}_2}{\text{L solution}}} = 0.0287 \text{ L or } 28.7 \text{ mL solution}$$

Thus, to make 500 mL of a 1.00 M acetic acid solution, we can take 28.7 mL of 17.4 M acetic acid and dilute it to a total volume of 500 mL with distilled water.

A dilution procedure typically involves two types of glassware: a pipet and a volumetric flask. A *pipet* is a device for accurately measuring and transferring a given volume of solution. There are two common types of pipets: *volumetric* (or *transfer*) *pipets* and *measuring pipets*, as shown in Fig. 6-11. Volumetric pipets come in specific sizes, such as 5 mL, 10 mL, 25 mL, and so on. Measuring pipets are used to measure volumes for which a volumetric pipet is not available. For example, we would use a measuring

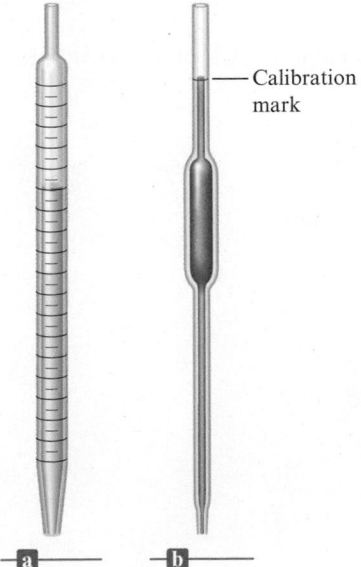

Calibration mark

a   b

Figure 6-11 | (a) A measuring pipet is graduated and can be used to measure various volumes of liquid accurately. (b) A volumetric (transfer) pipet is designed to measure one volume accurately. When filled to the mark, it delivers the volume indicated on the pipet.

Figure 6-12 | (a) A measuring pipet is used to transfer 28.7 mL of 17.4 M acetic acid solution to a volumetric flask. (b) Water is added to the flask to the calibration mark. (c) The resulting solution is 1.00 M acetic acid.

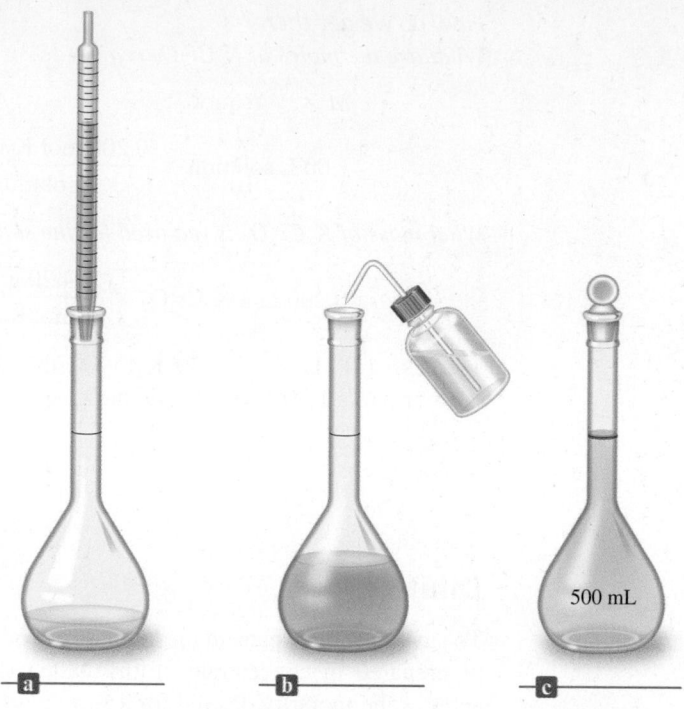

pipet as shown in Fig. 6-12 to deliver 28.7 mL of 17.4 M acetic acid into a 500-mL volumetric flask and then add water to the mark to perform the dilution described above.

INTERACTIVE EXAMPLE 6-7

Concentration and Volume

What volume of 16 M sulfuric acid must be used to prepare 1.5 L of a 0.10 M H_2SO_4 solution?

Solution

Where are we going?
To find the volume of H_2SO_4 required to prepare the solution

What do we know?

- 1.5 L of 0.10 M H_2SO_4 is required
- We have 16 M H_2SO_4

What information do we need to find the volume of H_2SO_4?

- Moles of H_2SO_4 in the required solution

How do we get there?
What are the moles of H_2SO_4 required?

$$M \times V = \text{mol}$$

$$1.5 \text{ L solution} \times \frac{0.10 \text{ mol } H_2SO_4}{\text{L solution}} = 0.15 \text{ mol } H_2SO_4$$

What volume of 16 M H_2SO_4 contains 0.15 mole of H_2SO_4?

$$V \times \frac{16 \text{ mol } H_2SO_4}{\text{L solution}} = 0.15 \text{ mol } H_2SO_4$$

Solving for V gives

$$V = \frac{0.15 \text{ mol H}_2\text{SO}_4}{\dfrac{16 \text{ mol H}_2\text{SO}_4}{1 \text{ L solution}}} = 9.4 \times 10^{-3} \text{ L or 9.4 mL solution}$$

To make 1.5 L of 0.10 M H_2SO_4 using 16 M H_2SO_4, we must take 9.4 mL of the concentrated acid and dilute it with water to 1.5 L. The correct way to do this is to add the 9.4 mL of acid to about 1 L of distilled water and then dilute to 1.5 L by adding more water. ◄ ⓘ

ⓘ In diluting an acid, "Do what you oughta: always add acid to water."

See Exercises 6-37b,d, and 6-38a,b,d

As noted earlier, the central idea in performing the calculations associated with dilutions is to recognize that the moles of solute are not changed by the dilution. Another way to express this condition is by the following equation:

$$M_1 V_1 = M_2 V_2$$

where M_1 and V_1 represent the molarity and volume of the original solution (before dilution) and M_2 and V_2 represent the molarity and volume of the diluted solution. This equation makes sense because

$$M_1 \times V_1 = \text{mol solute before dilution}$$
$$= \text{mol solute after dilution} = M_2 \times V_2$$

Repeat Example 6-7 using the equation $M_1 V_1 = M_2 V_2$. Note that in doing so,

$$M_1 = 16 \, M \quad M_2 = 0.10 \, M \quad V_2 = 1.5 \text{ L}$$

and V_1 is the unknown quantity sought. The equation $M_1 V_1 = M_2 V_2$ always holds for a dilution. This equation will be easy for you to remember if you understand where it comes from.

6-4 | Types of Chemical Reactions

Although we have considered many reactions so far in this text, we have examined only a tiny fraction of the millions of possible chemical reactions. To make sense of all these reactions, we need some system for grouping reactions into classes. Although there are many different ways to do this, we will use the system most commonly used by practicing chemists:

Types of Solution Reactions

- Precipitation reactions
- Acid–base reactions
- Oxidation–reduction reactions

ⓘ Sometimes when a precipitate forms, the particles are so small that they remain suspended in the liquid.

Virtually all reactions can be put into one of these classes. We will define and illustrate each type in the following sections.

6-5 | Precipitation Reactions

ⓘ A precipitation reaction also can be called a double-displacement reaction.

When two solutions are mixed, an insoluble substance sometimes forms; that is, a solid forms and separates from the solution. ◄ ⓘ Such a reaction is called a **precipitation reaction**, and the solid that forms is called a **precipitate**. ◄ ⓘ For example, a

Figure 6-13 | When yellow aqueous potassium chromate is added to a colorless barium nitrate solution, yellow barium chromate precipitates.

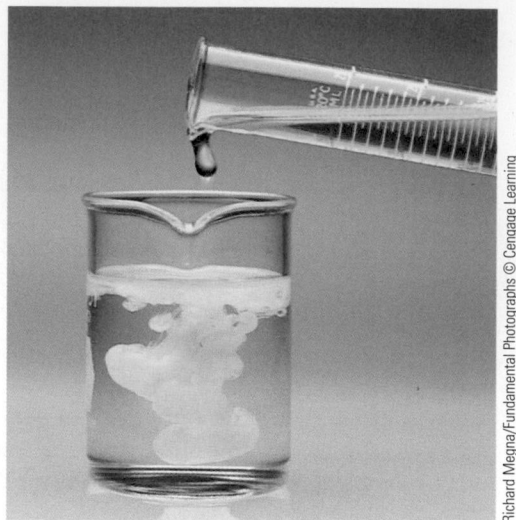

precipitation reaction occurs when an aqueous solution of potassium chromate, $K_2CrO_4(aq)$, which is yellow, is added to a colorless aqueous solution containing barium nitrate, $Ba(NO_3)_2(aq)$. As shown in Fig. 6-13, when these solutions are mixed, a yellow solid forms. What is the equation that describes this chemical change? To write the equation, we must know the identities of the reactants and products. The reactants have already been described: $K_2CrO_4(aq)$ and $Ba(NO_3)_2(aq)$. Is there some way we can predict the identities of the products? In particular, what is the yellow solid?

The best way to predict the identity of this solid is to think carefully about what products are possible. To do this, we need to know what species are present in the solution after the two reactant solutions are mixed. First, let's think about the nature of each reactant solution. The designation $Ba(NO_3)_2(aq)$ means that barium nitrate (a white solid) has been dissolved in water. Notice that barium nitrate contains the Ba^{2+} and NO_3^- ions. *Remember: In virtually every case, when a solid containing ions dissolves in water, the ions separate* and move around independently. That is, $Ba(NO_3)_2(aq)$ does not contain $Ba(NO_3)_2$ units; it contains separated Ba^{2+} and NO_3^- ions [Fig. 6-14(a)]. ◄ ⓘ

ⓘ When ionic compounds dissolve in water, the *resulting solution contains the separated ions.*

Figure 6-14 | Reactant solutions: (a) $Ba(NO_3)_2(aq)$ and (b) $K_2CrO_4(aq)$.

◉ K$^+$

◉ Ba^{2+}

◉ NO$_3^-$

◉ CrO$_4^{2-}$

Similarly, since solid potassium chromate contains the K^+ and CrO_4^{2-} ions, an aqueous solution of potassium chromate (which is prepared by dissolving solid K_2CrO_4 in water) contains these separated ions [Fig. 6-14(b)].

We can represent the mixing of $K_2CrO_4(aq)$ and $Ba(NO_3)_2(aq)$ in two ways. First, we can write

$$K_2CrO_4(aq) + Ba(NO_3)_2(aq) \longrightarrow \text{products}$$

However, a much more accurate representation is

$$\underbrace{2K^+(aq) + CrO_4^{2-}(aq)}_{\substack{\text{The ions in} \\ K_2CrO_4(aq)}} + \underbrace{Ba^{2+}(aq) + 2NO_3^-(aq)}_{\substack{\text{The ions in} \\ Ba(NO_3)_2(aq)}} \longrightarrow \text{products}$$

Thus the mixed solution contains the ions:

$$K^+ \qquad CrO_4^{2-} \qquad Ba^{2+} \qquad NO_3^-$$

as illustrated in Fig. 6-15(a).

How can some or all of these ions combine to form a yellow solid? This is not an easy question to answer. In fact, predicting the products of a chemical reaction is one of the hardest things a beginning chemistry student is asked to do. Even an experienced chemist, when confronted with a new reaction, is often not sure what will happen. The chemist tries to think of the various possibilities, considers the likelihood of each possibility, and then makes a prediction (an educated guess). Only after identifying each product *experimentally* is the chemist sure what reaction has taken place. However, an educated guess is very useful because it provides a place to start. It tells us what kinds of products we are most likely to find. We already know some things that will help us predict the products of the above reaction.

1. When ions form a solid compound, the compound must have a zero net charge. Thus the products of this reaction must contain *both anions and cations*. For example, K^+ and Ba^{2+} could not combine to form the solid, nor could CrO_4^{2-} and NO_3^-.

2. Most ionic materials contain only two types of ions: one type of cation and one type of anion (for example, NaCl, KOH, Na_2SO_4, K_2CrO_4, $Co(NO_3)_2$, NH_4Cl, Na_2CO_3).

The possible combinations of a given cation and a given anion from the list of ions K^+, CrO_4^{2-}, Ba^{2+}, and NO_3^- are

$$K_2CrO_4 \qquad KNO_3 \qquad BaCrO_4 \qquad Ba(NO_3)_2$$

Which of these possibilities is most likely to represent the yellow solid? We know it's not K_2CrO_4 or $Ba(NO_3)_2$. They are the reactants. They were present (dissolved) in the separate solutions that were mixed. The only real possibilities for the solid that formed are

$$KNO_3 \quad \text{and} \quad BaCrO_4$$

Figure 6-15 | The reaction of $K_2CrO_4(aq)$ and $Ba(NO_3)_2(aq)$. (a) The molecular-level "picture" of the mixed solution before any reaction has occurred. (b) The molecular-level "picture" of the solution after the reaction has occurred to form $BaCrO_4(s)$. *Note:* $BaCrO_4(s)$ is not molecular. It actually contains Ba^{2+} and CrO_4^{2-} ions packed together in a lattice. (c) A photo of the solution after the reaction has occurred, showing the solid $BaCrO_4$ on the bottom.

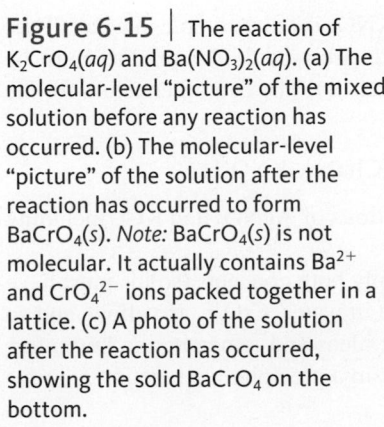

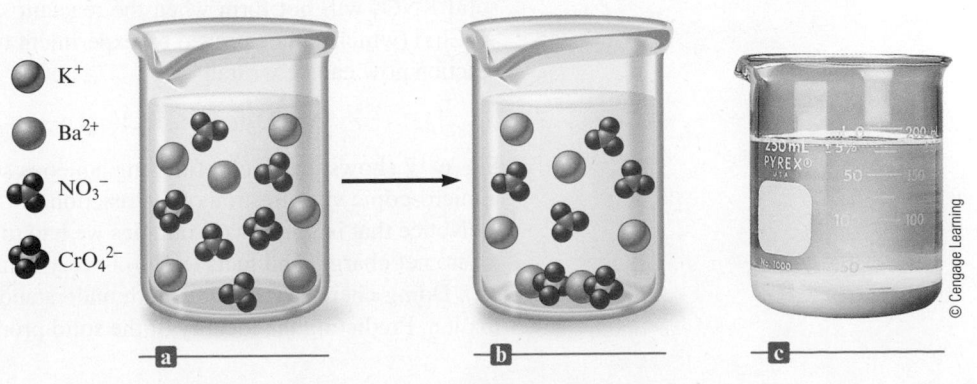

To decide which of these most likely represents the yellow solid, we need more facts. An experienced chemist knows that the K^+ ion and the NO_3^- ion are both colorless. Thus, if the solid is KNO_3, it should be white, not yellow. On the other hand, the CrO_4^{2-} ion is yellow (note in Fig. 6-14 that $K_2CrO_4(aq)$ is yellow). Thus the yellow solid is almost certainly $BaCrO_4$. Further tests show that this is the case.

So far we have determined that one product of the reaction between $K_2CrO_4(aq)$ and $Ba(NO_3)_2(aq)$ is $BaCrO_4(s)$, but what happened to the K^+ and NO_3^- ions? The answer is that these ions are left dissolved in the solution; KNO_3 does not form a solid when the K^+ and NO_3^- ions are present in this much water. In other words, if we took solid KNO_3 and put it in the same quantity of water as is present in the mixed solution, it would dissolve. Thus, when we mix $K_2CrO_4(aq)$ and $Ba(NO_3)_2(aq)$, $BaCrO_4(s)$ forms, but KNO_3 is left behind in solution [we write it as $KNO_3(aq)$]. Thus the overall equation for this precipitation reaction using the formulas of the reactants and products is

$$K_2CrO_4(aq) + Ba(NO_3)_2(aq) \longrightarrow BaCrO_4(s) + 2KNO_3(aq)$$

As long as water is present, the KNO_3 remains dissolved as separated ions. (See Fig. 6-15 to help visualize what is happening in this reaction. Note the solid $BaCrO_4$ on the bottom of the container, while the K^+ and NO_3^- ions remain dispersed in the solution.) If we removed the solid $BaCrO_4$ and then evaporated the water, white solid KNO_3 would be obtained; the K^+ and NO_3^- ions would assemble themselves into solid KNO_3 when the water is removed.

Now let's consider another example. When an aqueous solution of silver nitrate is added to an aqueous solution of potassium chloride, a white precipitate forms, as shown in Fig. 6-16. We can represent what we know so far as

$$AgNO_3(aq) + KCl(aq) \longrightarrow \text{unknown white solid}$$

Remembering that when ionic substances dissolve in water, the ions separate, we can write

$$\underbrace{Ag^+, NO_3^-}_{\substack{\text{In silver} \\ \text{nitrate} \\ \text{solution}}} + \underbrace{K^+, Cl^-}_{\substack{\text{In potassium} \\ \text{chloride} \\ \text{solution}}} \longrightarrow \underbrace{Ag^+, NO_3^-, K^+, Cl^-}_{\substack{\text{Combined solution,} \\ \text{before reaction}}} \longrightarrow \text{white solid}$$

Since we know the white solid must contain both positive and negative ions, the possible compounds that can be assembled from this collection of ions are

$$AgNO_3 \quad KCl \quad AgCl \quad KNO_3$$

Since $AgNO_3$ and KCl are the substances dissolved in the two reactant solutions, we know that they do not represent the white solid product. Therefore, the only real possibilities are

$$AgCl \quad \text{and} \quad KNO_3$$

From the first example considered, we know that KNO_3 is quite soluble in water. Thus solid KNO_3 will not form when the reactant solids are mixed. The product must be $AgCl(s)$ (which can be proved by experiment to be true). The overall equation for the reaction now can be written

$$AgNO_3(aq) + KCl(aq) \longrightarrow AgCl(s) + KNO_3(aq)$$

Fig. 6-17 shows the result of mixing aqueous solutions of $AgNO_3$ and KCl, including a microscopic visualization of the reaction.

Notice that in these two examples we had to apply both concepts (solids must have a zero net charge) and facts (KNO_3 is very soluble in water, CrO_4^{2-} is yellow, and so on). Doing chemistry requires both understanding ideas and remembering key information. Predicting the identity of the solid product in a precipitation reaction requires

Figure 6-16 | Precipitation of silver chloride by mixing solutions of silver nitrate and potassium chloride. The K^+ and NO_3^- ions remain in solution.

© Cengage Learning

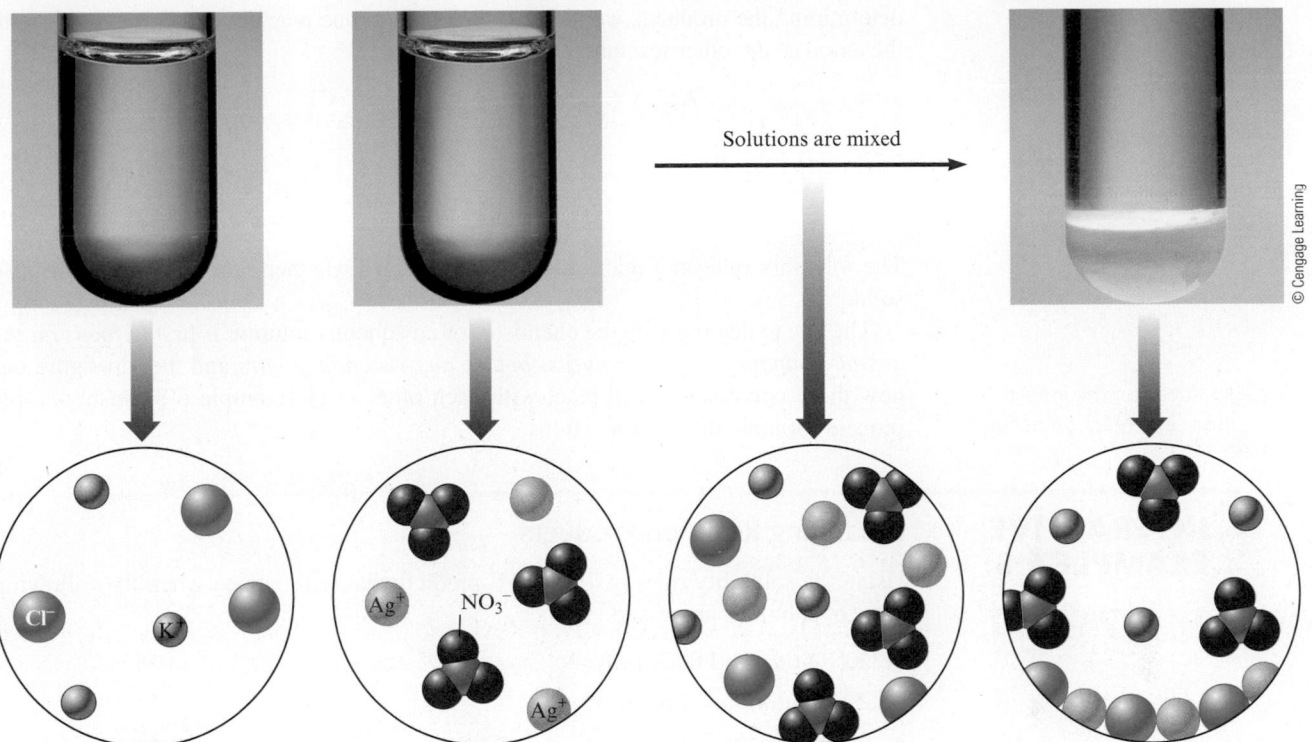

Figure 6-17 | Photos and accompanying molecular-level representations illustrating the reaction of KCl(aq) with AgNO$_3$(aq) to form AgCl(s). Note that it is not possible to have a photo of the mixed solution before the reaction occurs, because it is an imaginary step that we use to help visualize the reaction. Actually, the reaction occurs immediately when the two solutions are mixed.

knowledge of the solubilities of common ionic substances. As an aid in predicting the products of precipitation reactions, some simple solubility rules are given in Table 6-1. You should memorize these rules.

The phrase *slightly soluble* used in the solubility rules in Table 6-1 means that the tiny amount of solid that dissolves is not noticeable. The solid appears to be insoluble to the naked eye. Thus the terms *insoluble* and *slightly soluble* are often used interchangeably.

Note that the information in Table 6-1 allows us to predict that AgCl is the white solid formed when solutions of AgNO$_3$ and KCl are mixed. Rules 1 and 2 indicate that KNO$_3$ is soluble, and Rule 3 states that AgCl is insoluble.

When solutions containing ionic substances are mixed, it will be helpful in determining the products if you think in terms of *ion interchange*. For example, in the preceding discussion we considered the results of mixing AgNO$_3$(aq) and KCl(aq). In

Table 6-1 | Simple Rules for the Solubility of Salts in Water

1. Most nitrate (NO$_3^-$) salts are soluble.
2. Most salts containing the alkali metal ions (Li$^+$, Na$^+$, K$^+$, Cs$^+$, Rb$^+$) and the ammonium ion (NH$_4^+$) are soluble.
3. Most chloride, bromide, and iodide salts are soluble. Notable exceptions are salts containing the ions Ag$^+$, Pb^{2+}, and Hg$_2^{2+}$.
4. Most sulfate salts are soluble. Notable exceptions are BaSO$_4$, PbSO$_4$, Hg$_2$SO$_4$, and CaSO$_4$.
5. Most hydroxides are only slightly soluble. The important soluble hydroxides are NaOH and KOH. The compounds Ba(OH)$_2$, Sr(OH)$_2$, and Ca(OH)$_2$ are marginally soluble.
6. Most sulfide (S^{2-}), carbonate (CO$_3^{2-}$), chromate (CrO$_4^{2-}$), and phosphate (PO$_4^{3-}$) salts are only slightly soluble, except for those containing the cations in Rule 2.

determining the products, we took the cation from one reactant and combined it with the anion of the other reactant:

$$Ag^+ \;+\; NO_3^- \;+\; K^+ \;+\; Cl^- \;\longrightarrow$$

Possible
solid
products

The solubility rules in Table 6-1 allow us to predict whether either product forms as a solid.

The key to dealing with the chemistry of an aqueous solution is first to *focus on the actual components of the solution before any reaction occurs* and then to figure out how these components will react with each other. ◄ ⓘ Example 6-8 illustrates this process for three different reactions.

ⓘ To begin, focus on the ions in solution before any reaction occurs.

INTERACTIVE EXAMPLE 6-8

▲ Solid Fe(OH)₃ forms when aqueous KOH and Fe(NO₃)₃ are mixed.

© Cengage Learning

Predicting Reaction Products

Using the solubility rules in Table 6-1, predict what will happen when the following pairs of solutions are mixed.

a. $KNO_3(aq)$ and $BaCl_2(aq)$

b. $Na_2SO_4(aq)$ and $Pb(NO_3)_2(aq)$

c. $KOH(aq)$ and $Fe(NO_3)_3(aq)$ ◄

Solution

a. The formula $KNO_3(aq)$ represents an aqueous solution obtained by dissolving solid KNO_3 in water to form a solution containing the hydrated ions $K^+(aq)$ and $NO_3^-(aq)$. Likewise, $BaCl_2(aq)$ represents a solution formed by dissolving solid $BaCl_2$ in water to produce $Ba^{2+}(aq)$ and $Cl^-(aq)$. When these two solutions are mixed, the resulting solution contains the ions K^+, NO_3^-, Ba^{2+}, and Cl^-. All ions are hydrated, but the (aq) is omitted for simplicity. To look for possible solid products, combine the cation from one reactant with the anion from the other:

$$K^+ \;+\; NO_3^- \;+\; Ba^{2+} \;+\; Cl^- \;\longrightarrow$$

Possible
solid
products

Note from Table 6-1 that the rules predict that both KCl and Ba(NO₃)₂ are soluble in water. Thus no precipitate forms when $KNO_3(aq)$ and $BaCl_2(aq)$ are mixed. All the ions remain dissolved in solution. No chemical reaction occurs.

b. Using the same procedures as in part a, we find that the ions present in the combined solution before any reaction occurs are Na^+, SO_4^{2-}, Pb^{2+}, and NO_3^-. The possible salts that could form precipitates are

$$Na^+ \;+\; SO_4^{2-} \;+\; Pb^{2+} \;+\; NO_3^- \;\longrightarrow$$

The compound NaNO₃ is soluble, but PbSO₄ is insoluble (see Rule 4 in Table 6-1). When these solutions are mixed, PbSO₄ will precipitate from the solution. The balanced equation is

$$Na_2SO_4(aq) + Pb(NO_3)_2(aq) \longrightarrow PbSO_4(s) + 2NaNO_3(aq)$$

c. The combined solution (before any reaction occurs) contains the ions K^+, OH^-, Fe^{3+}, and NO_3^-. The salts that might precipitate are KNO_3 and $Fe(OH)_3$. The solubility rules in Table 6-1 indicate that both K^+ and NO_3^- salts are soluble.

However, $Fe(OH)_3$ is only slightly soluble (Rule 5) and hence will precipitate. The balanced equation is

$$3KOH(aq) + Fe(NO_3)_3(aq) \longrightarrow Fe(OH)_3(s) + 3KNO_3(aq)$$

See Exercises 6-47 and 6-48

6-6 | Describing Reactions in Solution

In this section we will consider the types of equations used to represent reactions in solution. For example, when we mix aqueous potassium chromate with aqueous barium nitrate, a reaction occurs to form a precipitate ($BaCrO_4$) and dissolved potassium nitrate. So far we have written the overall or **formula equation** for this reaction:

$$K_2CrO_4(aq) + Ba(NO_3)_2(aq) \longrightarrow BaCrO_4(s) + 2KNO_3(aq)$$

Although the formula equation shows the reactants and products of the reaction, it does not give a correct picture of what actually occurs in solution. As we have seen, aqueous solutions of potassium chromate, barium nitrate, and potassium nitrate contain individual ions, not collections of ions, as implied by the formula equation. Thus the **complete ionic equation**

$$2K^+(aq) + CrO_4^{2-}(aq) + Ba^{2+}(aq) + 2NO_3^-(aq) \longrightarrow$$
$$BaCrO_4(s) + 2K^+(aq) + 2NO_3^-(aq)$$

A strong electrolyte is a substance that completely breaks apart into ions when dissolved in water.

better represents the actual forms of the reactants and products in solution. *In a complete ionic equation, all substances that are strong electrolytes are represented as ions.* ◀ *i*

The complete ionic equation reveals that only some of the ions participate in the reaction. The K^+ and NO_3^- ions are present in solution both before and after the reaction. The ions that do not participate directly in the reaction are called **spectator ions**. The ions that participate in this reaction are the Ba^{2+} and CrO_4^{2-} ions, which combine to form solid $BaCrO_4$:

$$Ba^{2+}(aq) + CrO_4^{2-}(aq) \longrightarrow BaCrO_4(s)$$

Net ionic equations include only those components that undergo changes in the reaction.

This equation, called the **net ionic equation**, includes only those solution components directly involved in the reaction. ◀ *i* Chemists usually write the net ionic equation for a reaction in solution because it gives the actual forms of the reactants and products and includes only the species that undergo a change.

Three Types of Equations Are Used to Describe Reactions in Solution

- The **formula equation** gives the overall reaction stoichiometry but not necessarily the actual forms of the reactants and products in solution.
- The **complete ionic equation** represents as ions all reactants and products that are strong electrolytes.
- The **net ionic equation** includes only those solution components undergoing a change. Spectator ions are not included.

INTERACTIVE EXAMPLE 6-9

Writing Equations for Reactions

For each of the following reactions, write the formula equation, the complete ionic equation, and the net ionic equation.

a. Aqueous potassium chloride is added to aqueous silver nitrate to form a silver chloride precipitate plus aqueous potassium nitrate.

b. Aqueous potassium hydroxide is mixed with aqueous iron(III) nitrate to form a precipitate of iron(III) hydroxide and aqueous potassium nitrate.

Solution

a. Formula Equation

$$KCl(aq) + AgNO_3(aq) \longrightarrow AgCl(s) + KNO_3(aq)$$

Complete Ionic Equation
(*Remember:* Any ionic compound dissolved in water will be present as the separated ions.)

$$K^+(aq) + Cl^-(aq) + Ag^+(aq) + NO_3^-(aq) \longrightarrow AgCl(s) + K^+(aq) + NO_3^-(aq)$$

↑		↑	↑	↑	↑
Spectator ion		Spectator ion	Solid, not written as separate ions	Spectator ion	Spectator ion

Canceling the spectator ions

$$\cancel{K^+}(aq) + Cl^-(aq) + Ag^+(aq) + \cancel{NO_3^-}(aq) \longrightarrow AgCl(s) + \cancel{K^+}(aq) + \cancel{NO_3^-}(aq)$$

gives the following net ionic equation.

Net Ionic Equation

$$Cl^-(aq) + Ag^+(aq) \longrightarrow AgCl(s)$$

b. Formula Equation

$$3KOH(aq) + Fe(NO_3)_3(aq) \longrightarrow Fe(OH)_3(s) + 3KNO_3(aq)$$

Complete Ionic Equation

$$3K^+(aq) + 3OH^-(aq) + Fe^{3+}(aq) + 3NO_3^-(aq) \longrightarrow$$
$$Fe(OH)_3(s) + 3K^+(aq) + 3NO_3^-(aq)$$

Net Ionic Equation

$$3OH^-(aq) + Fe^{3+}(aq) \longrightarrow Fe(OH)_3(s)$$

See Exercises 6-49 through 6-54

6-7 | Stoichiometry of Precipitation Reactions

In Chapter 5 we covered the principles of chemical stoichiometry: the procedures for calculating quantities of reactants and products involved in a chemical reaction. Recall that in performing these calculations we first convert all quantities to moles and then use the coefficients of the balanced equation to assemble the appropriate mole ratios. In cases where reactants are mixed we must determine which reactant is limiting, since the reactant that is consumed first will limit the amounts of products formed. *These same principles apply to reactions that take place in solutions.* However, two points about solution reactions need special emphasis. The first is that it is sometimes difficult to tell immediately what reaction will occur when two solutions are mixed. Usually we must do some thinking about the various possibilities and then decide what probably will happen. The first step in this process *always* should be to write down the species that are actually present in the solution, as we did in Section 6-5. The second special point about solution reactions is that to obtain the moles of reactants we must use the volume of the solution and its molarity. This procedure was covered in Section 6-3.

We will introduce stoichiometric calculations for reactions in solution in Example 6-10.

> ### Critical Thinking
>
> What if all ionic solids were soluble in water? How would this affect stoichiometry calculations for reactions in aqueous solution?

INTERACTIVE EXAMPLE 6-10

Determining the Mass of Product Formed I

Calculate the mass of solid NaCl that must be added to 1.50 L of a 0.100 M $AgNO_3$ solution to precipitate all the Ag^+ ions in the form of AgCl.

Solution

Where are we going?
To find the mass of solid NaCl required to precipitate the Ag^+

What do we know?

■ 1.50 L of 0.100 M $AgNO_3$

What information do we need to find the mass of NaCl?

■ Moles of Ag^+ in the solution

How do we get there?
What are the ions present in the combined solution?

$$Ag^+ \quad NO_3^- \quad Na^+ \quad Cl^-$$

What is the balanced net ionic equation for the reaction?
Note from Table 6-1 that $NaNO_3$ is soluble and that AgCl is insoluble. Therefore, solid AgCl forms according to the following net ionic equation:

$$Ag^+(aq) + Cl^-(aq) \longrightarrow AgCl(s)$$

What are the moles of Ag^+ ions present in the solution?

$$1.50 \text{ L} \times \frac{0.100 \text{ mol } Ag^+}{\text{L}} = 0.150 \text{ mol } Ag^+$$

How many moles of Cl^- are required to react with all the Ag^+?
Because Ag^+ and Cl^- react in a 1:1 ratio, 0.150 mole of Cl^- and thus 0.150 mole of NaCl are required.

What mass of NaCl is required?

$$0.150 \text{ mol NaCl} \times \frac{58.44 \text{ g NaCl}}{\text{mol NaCl}} = 8.77 \text{ g NaCl}$$

See Exercise 6-57

Species present

↓ Write the reaction

Balanced net ionic equation

↓ Determine moles of reactants

Identify limiting reactant

↓ Determine moles of products

Check units of products

Notice from Example 6-10 that the procedures for doing stoichiometric calculations for solution reactions are very similar to those for other types of reactions. It is useful to think in terms of the following steps for reactions in solution.

PROBLEM-SOLVING STRATEGY

Solving Stoichiometry Problems for Reactions in Solution

1. Identify the species present in the combined solution, and determine what reaction occurs.
2. Write the balanced net ionic equation for the reaction.
3. Calculate the moles of reactants.
4. Determine which reactant is limiting.
5. Calculate the moles of product or products, as required.
6. Convert to grams or other units, as required.

INTERACTIVE EXAMPLE 6-11

Determining the Mass of Product Formed II

When aqueous solutions of Na_2SO_4 and $Pb(NO_3)_2$ are mixed, $PbSO_4$ precipitates. Calculate the mass of $PbSO_4$ formed when 1.25 L of 0.0500 M $Pb(NO_3)_2$ and 2.00 L of 0.0250 M Na_2SO_4 are mixed.

Solution

Where are we going?
To find the mass of solid $PbSO_4$ formed

What do we know?

- 1.25 L of 0.0500 M $Pb(NO_3)_2$
- 2.00 L of 0.0250 M Na_2SO_4
- Chemical reaction

$$Pb^{2+}(aq) + SO_4^{2-}(aq) \longrightarrow PbSO_4(s)$$

What information do we need?

- The limiting reactant

How do we get there?

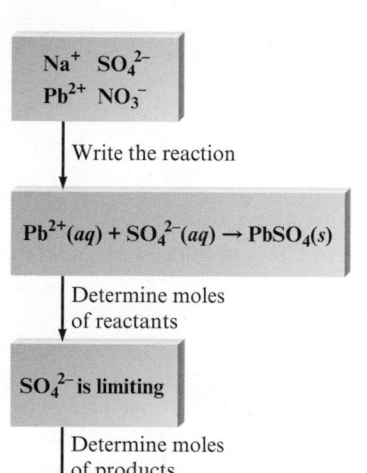

1. *What are the ions present in the combined solution?*

$$Na^+ \quad SO_4^{2-} \quad Pb^{2+} \quad NO_3^-$$

What is the reaction?
Since $NaNO_3$ is soluble and $PbSO_4$ is insoluble, solid $PbSO_4$ will form.

2. *What is the balanced net ionic equation for the reaction?*

$$Pb^{2+}(aq) + SO_4^{2-}(aq) \longrightarrow PbSO_4(s)$$

3. *What are the moles of reactants present in the solution?*

$$1.25 \, \cancel{L} \times \frac{0.0500 \text{ mol } Pb^{2+}}{\cancel{L}} = 0.0625 \text{ mol } Pb^{2+}$$

$$2.00 \, \cancel{L} \times \frac{0.0250 \text{ mol } SO_4^{2-}}{\cancel{L}} = 0.0500 \text{ mol } SO_4^{2-}$$

4. *Which reactant is limiting?*
Because Pb^{2+} and SO_4^{2-} react in a 1:1 ratio, the amount of SO_4^{2-} will be limiting (0.0500 mole of SO_4^{2-} is less than 0.0625 mole of Pb^{2+}).

5. *What number of moles of PbSO₄ will be formed?*
   Since SO_4^{2-} is limiting, only 0.0500 mole of solid $PbSO_4$ will be formed.

6. *What mass of PbSO₄ will be formed?*

$$0.0500 \ \text{mol} \ \cancel{\text{PbSO}_4} \times \frac{303.3 \ \text{g} \ PbSO_4}{1 \ \text{mol} \ \cancel{\text{PbSO}_4}} = 15.2 \ \text{g} \ PbSO_4$$

See Exercises 6-59 and 6-60

6-8 | Acid–Base Reactions

Earlier in this chapter we considered Arrhenius's concept of acids and bases: An acid is a substance that produces H^+ ions when dissolved in water, and a base is a substance that produces OH^- ions. Although these ideas are fundamentally correct, it is convenient to have a more general definition of a base, which includes substances that do not contain OH^- ions. Such a definition was provided by Johannes N. Brønsted (1879–1947) and Thomas M. Lowry (1874–1936), who defined acids and bases as follows: ◄ ⓘ

ⓘ The Brønsted–Lowry concept of acids and bases will be discussed in detail in Chapter 13.

An **acid** is a proton donor.

A **base** is a proton acceptor.

How do we know when to expect an acid–base reaction? One of the most difficult tasks for someone inexperienced in chemistry is to predict what reaction might occur when two solutions are mixed. With precipitation reactions, we found that the best way to deal with this problem is to focus on the species actually present in the mixed solution. This idea also applies to acid–base reactions. For example, when an aqueous solution of hydrogen chloride (HCl) is mixed with an aqueous solution of sodium hydroxide (NaOH), the combined solution contains the ions H^+, Cl^-, Na^+, and OH^-. The separated ions are present because HCl is a strong acid and NaOH is a strong base. How can we predict what reaction occurs, if any? First, will NaCl precipitate? From Table 6-1 we can see that NaCl is soluble in water and thus will not precipitate. Therefore, the Na^+ and Cl^- ions are spectator ions. On the other hand, because water is a nonelectrolyte, large quantities of H^+ and OH^- ions cannot coexist in solution. They react to form H_2O molecules:

$$H^+(aq) + OH^-(aq) \longrightarrow H_2O(l)$$

This is the net ionic equation for the reaction that occurs when aqueous solutions of HCl and NaOH are mixed.

Next, consider mixing an aqueous solution of acetic acid ($HC_2H_3O_2$) with an aqueous solution of potassium hydroxide (KOH). In our earlier discussion of conductivity we said that an aqueous solution of acetic acid is a weak electrolyte. This tells us that acetic acid does not dissociate into ions to any great extent. In fact, in 0.1 M $HC_2H_3O_2$ approximately 99% of the $HC_2H_3O_2$ molecules remain undissociated. However, when solid KOH is dissolved in water, it dissociates completely to produce K^+ and OH^- ions. Therefore, in the solution formed by mixing aqueous solutions of $HC_2H_3O_2$ and KOH, *before any reaction occurs*, the principal species are $HC_2H_3O_2$, K^+, and OH^-. What reaction will occur? A possible precipitation reaction could occur between K^+ and OH^-. However, we know that KOH is soluble, so precipitation does not occur. Another possibility is a reaction involving the hydroxide ion (a proton acceptor) and some proton donor. Is there a source of protons in the solution? The answer is yes—the $HC_2H_3O_2$ molecules. The OH^- ion has such a strong affinity for protons that it can strip them from the $HC_2H_3O_2$ molecules. The net ionic equation for this reaction is

$$OH^-(aq) + HC_2H_3O_2(aq) \longrightarrow H_2O(l) + C_2H_3O_2^-(aq)$$

This reaction illustrates a very important general principle: The hydroxide ion is such a strong base that for purposes of stoichiometric calculations it can be assumed to react completely with any weak acid that we will encounter. Of course, OH^- ions also react completely with the H^+ ions in solutions of strong acids.

We will now deal with the stoichiometry of acid–base reactions in aqueous solutions. The procedure is fundamentally the same as that used previously for precipitation reactions.

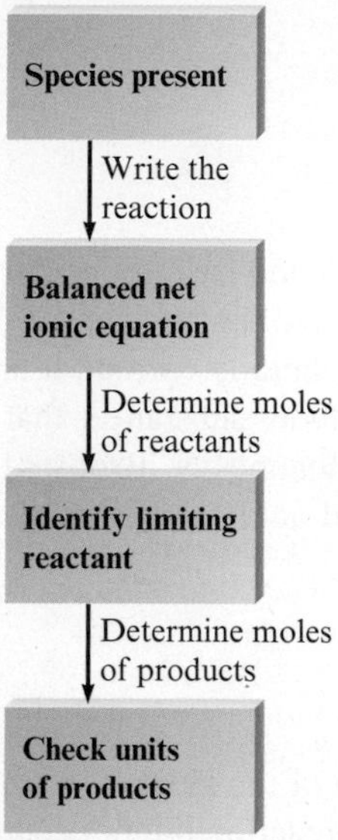

Species present

↓ Write the reaction

Balanced net ionic equation

↓ Determine moles of reactants

Identify limiting reactant

↓ Determine moles of products

Check units of products

PROBLEM-SOLVING STRATEGY

Performing Calculations for Acid–Base Reactions

1. List the species present in the combined solution *before any reaction occurs*, and decide what reaction will occur.

2. Write the balanced net ionic equation for this reaction.

3. Calculate the moles of reactants. For reactions in solution, use the volumes of the original solutions and their molarities.

4. Determine the limiting reactant where appropriate.

5. Calculate the moles of the required reactant or product.

6. Convert to grams or volume (of solution), as required.

An acid–base reaction is often called a **neutralization reaction**. When just enough base is added to react exactly with the acid in a solution, we say the acid has been *neutralized*.

INTERACTIVE EXAMPLE 6-12

Neutralization Reactions I

What volume of a 0.100 *M* HCl solution is needed to neutralize 25.0 mL of 0.350 *M* NaOH?

Solution

Where are we going?
To find the volume of 0.100 *M* HCl required for neutralization

What do we know?

- 25 mL of 0.350 *M* NaOH
- 0.100 *M* HCl
- The chemical reaction

$$H^+(aq) + OH^-(aq) \longrightarrow H_2O(l)$$

How do we get there?
Use the Problem-Solving Strategy for Performing Calculations for Acid–Base Reactions.

1. *What are the ions present in the combined solution?*

$$H^+ \quad Cl^- \quad Na^+ \quad OH^-$$

What is the reaction?

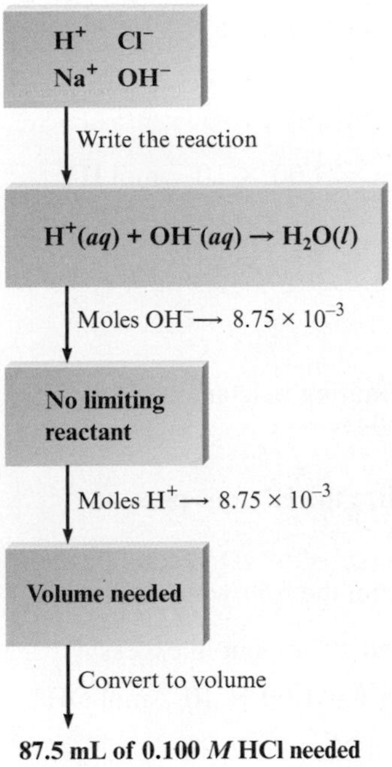

The two possibilities are

$$Na^+(aq) + Cl^-(aq) \longrightarrow NaCl(s)$$
$$H^+(aq) + OH^-(aq) \longrightarrow H_2O(l)$$

Since we know that NaCl is soluble, the first reaction does not take place (Na^+ and Cl^- are spectator ions). However, as we have seen before, the reaction of the H^+ and OH^- ions to form H_2O does occur.

2. *What is the balanced net ionic equation for the reaction?*

$$H^+(aq) + OH^-(aq) \longrightarrow H_2O(l)$$

3. *What are the moles of reactant present in the solution?*

$$25.0 \text{ mL NaOH} \times \frac{1 \text{ L}}{1000 \text{ mL}} \times \frac{0.350 \text{ mol OH}^-}{\text{L NaOH}} = 8.75 \times 10^{-3} \text{ mol OH}^-$$

4. *Which reactant is limiting?*
This problem requires the addition of just enough H^+ to react exactly with the OH^- ions present. We do not need to be concerned with limiting reactant here.

5. *What moles of H^+ are needed?*
Since H^+ and OH^- ions react in a 1:1 ratio, 8.75×10^{-3} moles of H^+ are required to neutralize the OH^- ions present.

6. *What volume of HCl is required?*

$$V \times \frac{0.100 \text{ mol H}^+}{\text{L}} = 8.75 \times 10^{-3} \text{ mol H}^+$$

Solving for V gives

$$V = \frac{8.75 \times 10^{-3} \text{ mol H}^+}{\dfrac{0.100 \text{ mol H}^+}{\text{L}}} = 8.75 \times 10^{-2} \text{ L}$$

See Exercises 6-71 and 6-72

H⁺ Cl⁻ Na⁺ OH⁻

Write the reaction

$H^+(aq) + OH^-(aq) \rightarrow H_2O(l)$

Moles OH⁻⟶ 8.75×10^{-3}

No limiting reactant

Moles H⁺⟶ 8.75×10^{-3}

Volume needed

Convert to volume

87.5 mL of 0.100 *M* HCl needed

INTERACTIVE EXAMPLE 6-13

Neutralization Reactions II

In a certain experiment, 28.0 mL of 0.250 *M* HNO_3 and 53.0 mL of 0.320 *M* KOH are mixed. Calculate the amount of water formed in the resulting reaction. What is the concentration of H^+ or OH^- ions in excess after the reaction goes to completion?

Solution

Where are we going?
To find the concentration of H^+ or OH^- in excess after the reaction is complete

What do we know?
- 28.0 mL of 0.250 *M* HNO_3
- 53.0 mL of 0.320 *M* KOH
- The chemical reaction

$$H^+(aq) + OH^-(aq) \longrightarrow H_2O(l)$$

How do we get there?
Use the Problem-Solving Strategy for Performing Calculations for Acid–Base Reactions.

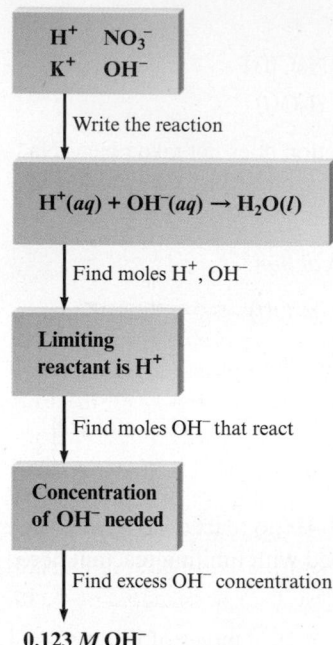

1. *What are the ions present in the combined solution?*

$$H^+ \quad NO_3^- \quad K^+ \quad OH^-$$

2. *What is the balanced net ionic equation for the reaction?*

$$H^+(aq) + OH^-(aq) \longrightarrow H_2O(l)$$

3. *What are the moles of reactant present in the solution?*

$$28.0 \text{ mL } HNO_3 \times \frac{1 \text{ L}}{1000 \text{ mL}} \times \frac{0.250 \text{ mol } H^+}{\text{L } HNO_3} = 7.00 \times 10^{-3} \text{ mol } H^+$$

$$53.0 \text{ mL } KOH \times \frac{1 \text{ L}}{1000 \text{ mL}} \times \frac{0.320 \text{ mol } OH^-}{\text{L } KOH} = 1.70 \times 10^{-2} \text{ mol } OH^-$$

4. *Which reactant is limiting?*
Since H^+ and OH^- ions react in a 1:1 ratio, the limiting reactant is H^+.

5. *What amount of OH^- will react?*
7.00×10^{-3} moles of OH^- is required to neutralize the H^+ ions present.

What amount of OH^- ions are in excess?
The amount of OH^- ions in excess is obtained from the following difference:

Original amount − amount consumed = amount in excess
$$1.70 \times 10^{-2} \text{ mol } OH^- - 7.00 \times 10^{-3} \text{ mol } OH^- = 1.00 \times 10^{-2} \text{ mol } OH^-$$

What is the volume of the combined solution?
The volume of the combined solution is the sum of the individual volumes:

Original volume of HNO_3 + original volume of KOH = total volume
$$28.0 \text{ mL} + 53.0 \text{ mL} = 81.0 \text{ mL} = 8.10 \times 10^{-2} \text{ L}$$

6. *What is the molarity of the OH^- ions in excess?*

$$\frac{\text{mol } OH^-}{\text{L solution}} = \frac{1.00 \times 10^{-2} \text{ mol } OH^-}{8.10 \times 10^{-2} \text{ L}} = 0.123 \text{ } M \text{ } OH^-$$

REALITY CHECK | This calculated molarity is less than the initial molarity, as it should be.

See Exercises 6-73 and 6-74

Acid–Base Titrations

Volumetric analysis is a technique for determining the amount of a certain substance by doing a titration. A **titration** involves delivery (from a buret) of a measured volume of a solution of known concentration (the *titrant*) into a solution containing the substance being analyzed (the *analyte*). The titrant contains a substance that reacts in a known manner with the analyte. The point in the titration where enough titrant has been added to react exactly with the analyte is called the **equivalence point** or the **stoichiometric point**. This point is often marked by an **indicator**, a substance added at the beginning of the titration that changes color at (or very near) the equivalence point. The point where the indicator *actually* changes color is called the **endpoint** of the titration. ◄ ⓘ The goal is to choose an indicator such that the endpoint (where the indicator changes color) occurs exactly at the equivalence point (where just enough titrant has been added to react with all the analyte).

ⓘ Ideally, the endpoint and stoichiometric point should coincide.

- The exact reaction between titrant and analyte must be known (and rapid).
- The stoichiometric (equivalence) point must be marked accurately.
- The volume of titrant required to reach the stoichiometric point must be known accurately.

When the analyte is a base or an acid, the required titrant is a strong acid or strong base, respectively. This procedure is called an *acid–base titration*. An indicator very commonly used for acid–base titrations is phenolphthalein, which is colorless in an acidic solution and pink in a basic solution. Thus, when an acid is titrated with a base, the phenolphthalein remains colorless until after the acid is consumed and the first drop of excess base is added. In this case, the endpoint (the solution changes from colorless to pink) occurs approximately one drop of base beyond the stoichiometric point. This type of titration is illustrated in the three photos in Fig. 6-18.

We will deal with the acid–base titrations only briefly here but will return to the topic of titrations and indicators in more detail in Chapter 14. The titration of an acid with a standard solution containing hydroxide ions is described in Example 6-15. In

a   b   c

© Cengage Learning

Figure 6-18 | The titration of an acid with a base. (a) The titrant (the base) is in the buret, and the flask contains the acid solution along with a small amount of indicator. (b) As base is added drop by drop to the acid solution in the flask during the titration, the indicator changes color, but the color disappears on mixing. (c) The stoichiometric (equivalence) point is marked by a permanent indicator color change. The volume of base added is the difference between the final and initial buret readings.

Example 6-14 we show how to determine accurately the concentration of a sodium hydroxide solution. This procedure is called *standardizing the solution*.

INTERACTIVE EXAMPLE 6-14

Neutralization Titration

A student carries out an experiment to standardize (determine the exact concentration of) a sodium hydroxide solution. To do this, the student weighs out a 1.3009-g sample of potassium hydrogen phthalate ($KHC_8H_4O_4$, often abbreviated KHP). KHP (molar mass 204.22 g/mol) has one acidic hydrogen. The student dissolves the KHP in distilled water, adds phenolphthalein as an indicator, and titrates the resulting solution with the sodium hydroxide solution to the phenolphthalein endpoint. The difference between the final and initial buret readings indicates that 41.20 mL of the sodium hydroxide solution is required to react exactly with the 1.3009 g KHP. Calculate the concentration of the sodium hydroxide solution.

Solution

Where are we going?
To find the concentration of NaOH solution

What do we know?

- 1.3009 g $KHC_8H_4O_4$ (KHP), molar mass (204.22 g/mol)
- 41.20 mL NaOH solution to neutralize KHP
- The chemical reaction

$$HC_8H_4O_4^-(aq) + OH^-(aq) \longrightarrow H_2O(l) + C_8H_4O_4^{2-}(aq)$$

How do we get there?
Use the Problem-Solving Strategy for Performing Calculations for Acid–Base Reactions.

1. *What are the ions present in the combined solution?*

$$K^+ \quad HC_8H_4O_4^- \quad Na^+ \quad OH^- \blacktriangleleft$$

2. *What is the balanced net ionic equation for the reaction?*

$$HC_8H_4O_4^-(aq) + OH^-(aq) \longrightarrow H_2O(l) + C_8H_4O_4^{2-}(aq)$$

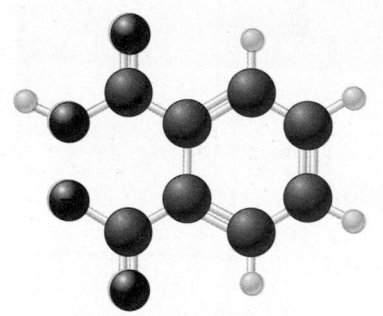

$HC_8H_4O_4^-$
Hydrogen phthalate ion

3. *What are the moles of KHP?*

$$1.3009 \text{ g } \cancel{KHC_8H_4O_4} \times \frac{1 \text{ mol } KHC_8H_4O_4}{204.22 \text{ g } \cancel{KHC_8H_4O_4}} = 6.3701 \times 10^{-3} \text{ mol } KHC_8H_4O_4$$

4. *Which reactant is limiting?*
    This problem requires the addition of just enough OH^- ions to react exactly with the KHP present. We do not need to be concerned with limiting reactant here.

5. *What moles of OH^- are required?*
    6.3701×10^{-3} moles of OH^- are required to neutralize the KHP present.

6. *What is the molarity of the NaOH solution?*

$$\text{Molarity of NaOH} = \frac{\text{mol NaOH}}{\text{L solution}} = \frac{6.3701 \times 10^{-3} \text{ mol NaOH}}{4.120 \times 10^{-2} \text{ L}}$$

$$= 0.1546 \, M$$

This standard sodium hydroxide solution can now be used in other experiments (see Example 6-15).

See Exercises 6-75, 6-76, and 6-80

INTERACTIVE EXAMPLE 6-15

Neutralization Analysis

An environmental chemist analyzed the effluent (the released waste material) from an industrial process known to produce the compounds carbon tetrachloride (CCl_4) and benzoic acid ($HC_7H_5O_2$), a weak acid that has one acidic hydrogen atom per molecule. A sample of this effluent weighing 0.3518 g was shaken with water, and the resulting aqueous solution required 10.59 mL of 0.1546 M NaOH for neutralization. Calculate the mass percent of $HC_7H_5O_2$ in the original sample.

Solution

Where are we going?
To find the mass percent of $HC_7H_5O_2$ in the original sample

What do we know?

- 0.3518 g effluent (original sample)
- 10.59 mL 0.1546 M NaOH for neutralization of $HC_7H_5O_2$
- The chemical reaction

$$HC_7H_5O_2(aq) + OH^-(aq) \longrightarrow H_2O(l) + C_7H_5O_2^-(aq)$$

How do we get there?
Use the Problem-Solving Strategy for Performing Calculations for Acid–Base Reactions.

1. *What are the species present in the combined solution?*

$$HC_7H_5O_2 \quad Na^+ \quad OH^-$$

2. *What is the balanced net ionic equation for the reaction?*

$$HC_7H_5O_2(aq) + OH^-(aq) \longrightarrow H_2O(l) + C_7H_5O_2^-(aq)$$

3. *What are the moles of OH^- required?*

$$10.59 \text{ mL NaOH} \times \frac{1 \text{ L}}{1000 \text{ mL}} \times \frac{0.1546 \text{ mol OH}^-}{\text{L NaOH}} = 1.637 \times 10^{-3} \text{ mol OH}^-$$

4. *Which reactant is limiting?*
   This problem requires the addition of just enough OH^- ions to react exactly with the $HC_7H_5O_2$ present. We do not need to be concerned with limiting reactant here.

5. *What mass of $HC_7H_5O_2$ is present?*

$$1.637 \times 10^{-3} \text{ mol HC}_7\text{H}_5\text{O}_2 \times \frac{122.12 \text{ g HC}_7\text{H}_5\text{O}_2}{1 \text{ mol HC}_7\text{H}_5\text{O}_2} = 0.1999 \text{ g HC}_7\text{H}_5\text{O}_2$$

6. *What is the mass percent of the $HC_7H_5O_2$ in the effluent?*

$$\frac{0.1999 \text{ g}}{0.3518 \text{ g}} \times 100\% = 56.82\%$$

REALITY CHECK | The calculated percent of $HC_7H_5O_2$ is less than 100%, as it should be.

See Exercise 6-79

ⓘ The first step in the analysis of a complex solution is to write down the components and focus on the chemistry of each one. When a strong electrolyte is present, write it as separated ions.

In doing problems involving titrations, you must first decide what reaction is occurring. Sometimes this seems difficult because the titration solution contains several components. The key to success in doing solution reactions is to first write down all the components in the solution and focus on the chemistry of each one. ◄ *ⓘ* We have been emphasizing this approach in dealing with the reactions between ions in solution. Make it a habit to write down the components of solutions before trying to decide what reaction(s) might take place as you attempt the end-of-chapter problems involving titrations.

6-9 | Oxidation–Reduction Reactions

We have seen that many important substances are ionic. Sodium chloride, for example, can be formed by the reaction of elemental sodium and chlorine:

$$2Na(s) + Cl_2(g) \longrightarrow 2NaCl(s)$$

In this reaction, solid sodium, which contains neutral sodium atoms, reacts with chlorine gas, which contains diatomic Cl_2 molecules, to form the ionic solid NaCl, which contains Na^+ and Cl^- ions. This process is represented in Fig. 6-19. *Reactions like this one, in which one or more electrons are transferred, are called* **oxidation–reduction reactions** *or* **redox reactions.**

Many important chemical reactions involve oxidation and reduction. Photosynthesis, which stores energy from the sun in plants by converting carbon dioxide and water to sugar, is a very important oxidation–reduction reaction. In fact, most reactions used for energy production are redox reactions. In humans, the oxidation of sugars, fats, and proteins provides the energy necessary for life. Combustion reactions, which provide

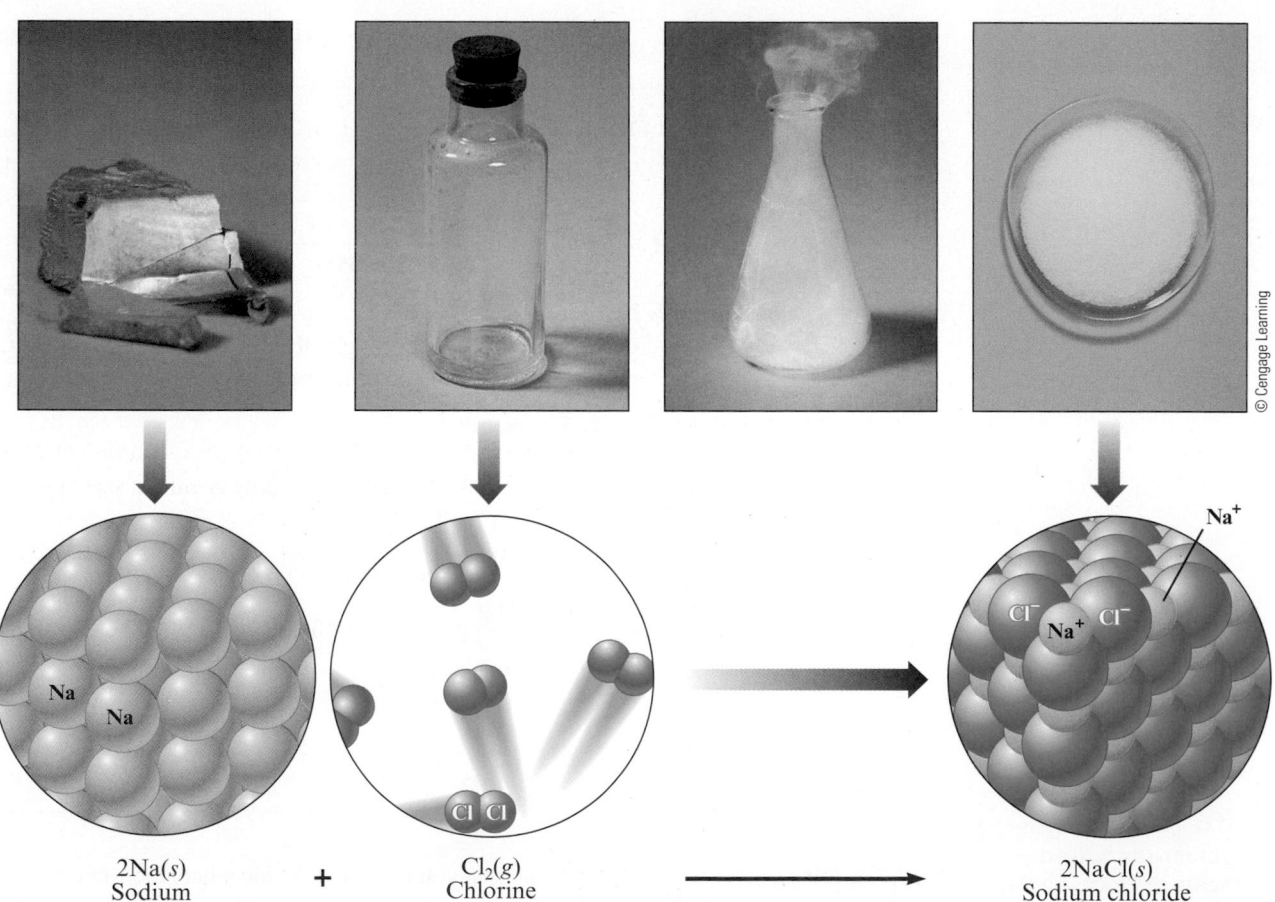

© Cengage Learning

$2Na(s)$
Sodium

$+$

$Cl_2(g)$
Chlorine

$2NaCl(s)$
Sodium chloride

Figure 6-19 | The reaction of solid sodium and gaseous chlorine to form solid sodium chloride.

most of the energy to power our civilization, also involve oxidation and reduction. An example is the reaction of methane with oxygen:

$$CH_4(g) + 2O_2(g) \longrightarrow CO_2(g) + 2H_2O(g) + energy$$

Even though none of the reactants or products in this reaction is ionic, the reaction is still assumed to involve a transfer of electrons from carbon to oxygen. To explain this, we must introduce the concept of oxidation states.

Oxidation States

The concept of **oxidation states** (also called *oxidation numbers*) provides a way to keep track of electrons in oxidation–reduction reactions, particularly redox reactions involving covalent substances. Recall that electrons are shared by atoms in covalent bonds. The oxidation states of atoms in covalent compounds are obtained by arbitrarily assigning the electrons (which are actually shared) to particular atoms. We do this as follows: For a covalent bond between two identical atoms, the electrons are split equally between the two. In cases where two different atoms are involved (and the electrons are thus shared unequally), the shared electrons are assigned completely to the atom that has the stronger attraction for electrons. For example, recall from the discussion of the water molecule in Section 6-1 that oxygen has a greater attraction for electrons than does hydrogen. Therefore, in assigning the oxidation state of oxygen and hydrogen in H_2O, we assume that the oxygen atom actually possesses all the electrons. Recall that a hydrogen atom has one electron. Thus, in water, oxygen has formally "taken" the electrons from two hydrogen atoms. This gives the oxygen an *excess* of two electrons (its oxidation state is -2) and leaves each hydrogen with no electrons (the oxidation state of each hydrogen is thus $+1$).

We define the *oxidation states* (or *oxidation numbers*) of the atoms in a covalent compound as the imaginary charges the atoms would have if the shared electrons were divided equally between identical atoms bonded to each other or, for different atoms, were all assigned to the atom in each bond that has the greater attraction for electrons. Of course, for ionic compounds containing monatomic ions, the oxidation states of the ions are equal to the ion charges.

These considerations lead to a series of rules for assigning oxidation states that are summarized in Table 6-2. Application of these simple rules allows the assignment of oxidation states in most compounds. To apply these rules recognize that *the sum of the oxidation states must be zero for an electrically neutral compound.* For an ion, the sum of the oxidation states must equal the charge of the ion. The principles are illustrated by Example 6-16.

It is worthwhile to note at this point that the convention is to write actual charges on ions as $n+$ or $n-$, the number being written *before* the plus or minus sign. On the other hand, oxidation states (not actual charges) are written $+n$ or $-n$, with the number being written *after* the plus or minus sign.

Table 6-2 | Rules for Assigning Oxidation States

The Oxidation State of . . .	Summary	Examples
• An atom in an element is zero	Element: 0	$Na(s)$, $O_2(g)$, $O_3(g)$, $Hg(l)$
• A monatomic ion is the same as its charge	Monatomic ion: charge of ion	Na^+, Cl^-
• Fluorine is -1 in its compounds	Fluorine: -1	HF, PF_3
• Oxygen is usually -2 in its compounds Exception: peroxides (containing O_2^{2-}), in which oxygen is -1	Oxygen: -2	H_2O, CO_2
• Hydrogen is $+1$ in its covalent compounds	Hydrogen: $+1$	H_2O, HCl, NH_3

INTERACTIVE EXAMPLE 6-16

Assigning Oxidation States

Assign oxidation states to all atoms in the following.

a. CO_2

b. SF_6

c. NO_3^-

Solution

a. Since we have a specific rule for the oxidation state of oxygen, we will assign its value first. The oxidation state of oxygen is -2. The oxidation state of the carbon atom can be determined by recognizing that since CO_2 has no charge, the sum of the oxidation states for oxygen and carbon must be zero. Since each oxygen is -2 and there are two oxygen atoms, the carbon atom must be assigned an oxidation state of $+4$:

$$CO_2$$
$$+4 \quad -2 \text{ for each oxygen}$$

REALITY CHECK | We can check the assigned oxidation states by noting that when the number of atoms is taken into account, the sum is zero as required:

$$1(+4) + 2(-2) = 0$$

No. of C atoms   No. of O atoms

b. Since we have no rule for sulfur, we first assign the oxidation state of each fluorine as -1. The sulfur must then be assigned an oxidation state of $+6$ to balance the total of -6 from the fluorine atoms:

$$SF_6$$
$$+6 \quad -1 \text{ for each fluorine}$$

REALITY CHECK | $+6 + 6(-1) = 0$

c. Oxygen has an oxidation state of -2. Because the sum of the oxidation states of the three oxygens is -6 and the net charge on the NO_3^- ion is $1-$, the nitrogen must have an oxidation state of $+5$:

$$NO_3^-$$
$$+5 \quad -2 \text{ for each oxygen}$$

REALITY CHECK | $+5 + 3(-2) = -1$

Note that in this case the sum must be -1 (the overall charge on the ion).

See Exercises 6-81 through 6-84

We need to make one more point about oxidation states, and this can be illustrated by the compound Fe_3O_4, which is the main component in magnetite, an iron ore that accounts for the reddish color of many types of rocks and soils. ◄ To determine the oxidation states in Fe_3O_4, we first assign each oxygen atom its usual oxidation state of -2. The three iron atoms must yield a total of $+8$ to balance the total of -8 from the four oxygens. This means that each iron atom has an oxidation state of $+\frac{8}{3}$. A noninteger value for the oxidation state may seem strange because charge is expressed

▲ Magnetite is a magnetic ore containing Fe_3O_4. Note that the compass needle points toward the ore.

© Cengage Learning

in whole numbers. However, although they are rare, noninteger oxidation states do occur because of the rather arbitrary way that electrons are divided up by the rules in Table 6-2. For Fe_3O_4, for example, the rules assume that all the iron atoms are equal, when in fact this compound can best be viewed as containing four O^{2-} ions, two Fe^{3+} ions, and one Fe^{2+} ion per formula unit. (Note that the "average" charge on iron works out to be $\frac{8}{3}+$, which is equal to the oxidation state we determined above.) Noninteger oxidation states should not intimidate you. They are used in the same way as integer oxidation states—for keeping track of electrons.

The Characteristics of Oxidation–Reduction Reactions

Oxidation–reduction reactions are characterized by a transfer of electrons. In some cases, the transfer occurs in a literal sense to form ions, such as in the reaction

$$2Na(s) + Cl_2(g) \longrightarrow 2NaCl(s)$$

However, sometimes the transfer is less obvious. For example, consider the combustion of methane (the oxidation state for each atom is given):

$$CH_4(g) + 2O_2(g) \longrightarrow CO_2(g) + 2H_2O(g)$$

Oxidation state: -4 $+1$ (each H) 0 $+4$ -2 (each O) $+1$ -2 (each H)

Note that the oxidation state for oxygen in O_2 is 0 because it is in elemental form. In this reaction there are no ionic compounds, but we can still describe the process in terms of a transfer of electrons. Note that carbon undergoes a change in oxidation state from -4 in CH_4 to $+4$ in CO_2. Such a change can be accounted for by a loss of eight electrons (the symbol e^- stands for an electron):

$$CH_4 \longrightarrow CO_2 + 8e^-$$
$$-4 \qquad +4$$

On the other hand, each oxygen changes from an oxidation state of 0 in O_2 to -2 in H_2O and CO_2, signifying a gain of two electrons per atom. Since four oxygen atoms are involved, this is a gain of eight electrons:

$$2O_2 + 8e^- \longrightarrow CO_2 + 2H_2O$$
$$0 \qquad 4(-2) = -8$$

No change occurs in the oxidation state of hydrogen, and it is not formally involved in the electron-transfer process.

With this background, we can now define some important terms. **Oxidation** is an *increase* in oxidation state (a loss of electrons). **Reduction** is a *decrease* in oxidation state (a gain of electrons). Thus in the reaction

$$2Na(s) + Cl_2(g) \longrightarrow 2NaCl(s)$$
$$0 \qquad 0 \qquad +1 \quad -1$$

sodium is oxidized and chlorine is reduced. In addition, Cl_2 is called the **oxidizing agent (electron acceptor)**, and Na is called the **reducing agent (electron donor)**. These terms are summarized in Fig. 6-20.

Concerning the reaction

$$CH_4(g) + 2O_2(g) \longrightarrow CO_2(g) + 2H_2O(g)$$
$$-4 \quad +1 \qquad 0 \qquad +4 \quad -2 \qquad +1 \quad -2$$

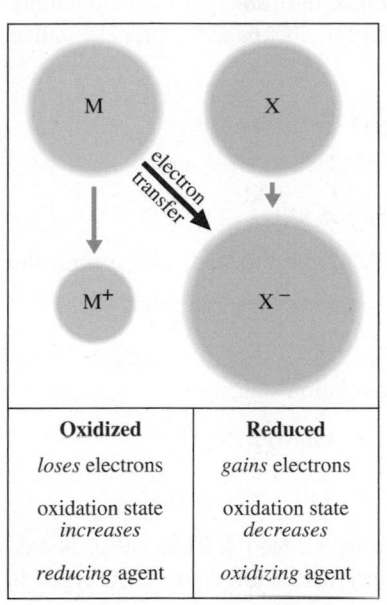

Oxidized	Reduced
loses electrons	*gains* electrons
oxidation state *increases*	oxidation state *decreases*
reducing agent	*oxidizing* agent

Figure 6-20 | A summary of an oxidation–reduction process, in which M is oxidized and X is reduced.

Oxidation is an increase in oxidation state. *Reduction* is a decrease in oxidation state.

An oxidizing agent is reduced and a reducing agent is oxidized in a redox reaction.

we can say the following:

> The carbon in methane is oxidized because there has been an increase in carbon's oxidation state (the carbon atom has formally lost electrons).
>
> Oxygen is reduced because there has been a decrease in its oxidation state (oxygen has formally gained electrons).
>
> CH_4 is the reducing agent.
>
> O_2 is the oxidizing agent.

i A helpful mnemonic device is OIL RIG (**O**xidation **I**nvolves **L**oss; **R**eduction **I**nvolves **G**ain). Another common mnemonic is LEO says GER. (**L**oss of **E**lectrons, **O**xidation; **G**ain of **E**lectrons, **R**eduction).

Note that when the oxidizing or reducing agent is named, the *whole compound* is specified, not just the element that undergoes the change in oxidation state. ◄ *i*

INTERACTIVE EXAMPLE 6-17

Oxidation–Reduction Reactions

Metallurgy, the process of producing a metal from its ore, always involves oxidation–reduction reactions. In the metallurgy of galena (PbS), the principal lead-containing ore, the first step is the conversion of lead sulfide to its oxide (a process called *roasting*):

$$2PbS(s) + 3O_2(g) \longrightarrow 2PbO(s) + 2SO_2(g)$$

The oxide is then treated with carbon monoxide to produce the free metal:

$$PbO(s) + CO(g) \longrightarrow Pb(s) + CO_2(g)$$

For each reaction, identify the atoms that are oxidized and reduced, and specify the oxidizing and reducing agents.

Solution

For the first reaction, we can assign the following oxidation states:

$$2PbS(s) + 3O_2(g) \longrightarrow 2PbO(s) + 2SO_2(g)$$

$$\begin{array}{ccccccc} {+2} & {-2} & {0} & {+2} & {-2} & {+4} & {-2 \text{ (each O)}} \end{array}$$

The oxidation state for the sulfur atom increases from -2 to $+4$. Thus sulfur is oxidized. The oxidation state for each oxygen atom decreases from 0 to -2. Oxygen is reduced. The oxidizing agent (that accepts the electrons) is O_2, and the reducing agent (that donates electrons) is PbS.

For the second reaction we have

$$PbO(s) + CO(g) \longrightarrow Pb(s) + CO_2(g)$$

$$\begin{array}{cccccc} {+2} & {-2} & {+2} & {-2} & {0} & {+4} & {-2 \text{ (each O)}} \end{array}$$

Lead is reduced (its oxidation state decreases from $+2$ to 0), and carbon is oxidized (its oxidation state increases from $+2$ to $+4$). PbO is the oxidizing agent, and CO is the reducing agent.

See Exercises 6-85 and 6-86

Critical Thinking

Dalton believed that atoms were indivisible. Thomson and Rutherford helped to show that this was not true. What if atoms were indivisible? How would this affect the types of reactions you have learned about in this chapter?

6-10 | Balancing Oxidation–Reduction Equations

It is important to be able to balance oxidation–reduction reactions. One method involves the use of oxidation states (discussed in this section), and the other method (normally used for more complex reactions) involves separating the reaction into two half-reactions. We'll discuss the second method for balancing oxidation–reduction reactions in Chapter 17.

Oxidation States Method of Balancing Oxidation–Reduction Reactions

Consider the reaction between solid copper and silver ions in aqueous solution:

$$Cu(s) + Ag^+(aq) \longrightarrow Ag(s) + Cu^{2+}(aq)$$

We can tell this is a redox reaction by assigning oxidation states as follows:

$$\underset{0}{Cu} + \underset{+1}{Ag^+} \longrightarrow \underset{0}{Ag} + \underset{+2}{Cu^{2+}}$$

1 e⁻ gained; 2 e⁻ lost

We know that in an oxidation–reduction reaction we must ultimately have equal numbers of electrons gained and lost, and we can use this principle to balance redox equations. For example, in this case, 2 Ag^+ ions must be reduced for every Cu atom oxidized:

$$Cu(s) + 2Ag^+(aq) \longrightarrow 2Ag(s) + Cu^{2+}(aq)$$

This gives us the balanced equation.

Now consider a more complex reaction:

$$H^+(aq) + Cl^-(aq) + Sn(s) + NO_3^-(aq) \longrightarrow SnCl_6^{2-}(aq) + NO_2(g) + H_2O(l)$$

To balance this equation by oxidation states, we first need to assign the oxidation states to all the atoms in the reactants and products.

$$\underset{+1}{H^+} + \underset{-1}{Cl^-} + \underset{0}{Sn} + \underset{+5,\,-2}{NO_3^-} \longrightarrow \underset{+4,\,-1}{SnCl_6^{2-}} + \underset{+4,\,-2}{NO_2} + \underset{+1,\,-2}{H_2O}$$

Note that hydrogen, chlorine, and oxygen do not change oxidation states and are not involved in electron exchange. Thus we focus our attention on Sn and N:

$$\underset{0}{Sn} + \underset{+5}{NO_3^-} \longrightarrow \underset{+4}{SnCl_6^{2-}} + \underset{+4}{NO_2}$$

4 e⁻ lost; 1 e⁻ gained

This means we need a coefficient of 4 for the N-containing species.

$$H^+ + Cl^- + Sn + 4NO_3^- \longrightarrow SnCl_6^{2-} + 4NO_2 + H_2O$$

Now we balance the rest of the equation by inspection.

Balance Cl^-:

$$H^+ + 6Cl^- + Sn + 4NO_3^- \longrightarrow SnCl_6^{2-} + 4NO_2 + H_2O$$

Balance O:

$$H^+ + 6Cl^- + Sn + 4NO_3^- \longrightarrow SnCl_6^{2-} + 4NO_2 + 4H_2O$$

Balance H:

$$8H^+ + 6Cl^- + Sn + 4NO_3^- \longrightarrow SnCl_6^{2-} + 4NO_2 + 4H_2O$$

This gives the final balanced equation. Now we will write the equation with the states included:

$$8H^+(aq) + 6Cl^-(aq) + Sn(s) + 4NO_3^-(aq) \longrightarrow SnCl_6^{2-}(aq) + 4NO_2(g) + 4H_2O(l)$$

PROBLEM-SOLVING STRATEGY

Balancing Oxidation–Reduction Reactions by Oxidation States

1. Write the unbalanced equation.
2. Determine the oxidation states of all atoms in the reactants and products.
3. Show electrons gained and lost using "tie lines."
4. Use coefficients to equalize the electrons gained and lost.
5. Balance the rest of the equation by inspection.
6. Add appropriate states.

EXAMPLE 6-18 Balancing Oxidation–Reduction Reactions

Balance the reaction between solid lead(II) oxide and ammonia gas to produce nitrogen gas, liquid water, and solid lead.

Solution

We'll use the Problem-Solving Strategy for Balancing Oxidation–Reduction Reactions by Oxidation States.

1. *What is the unbalanced equation?*

$$PbO(s) + NH_3(g) \longrightarrow N_2(g) + H_2O(l) + Pb(s)$$

2. *What are the oxidation states for each atom?*

$$\underset{+2}{Pb}\overset{-2}{O} + \overset{-3}{N}\underset{+1}{H_3} \longrightarrow \overset{0}{N_2} + \underset{+1}{H_2}\overset{-2}{O} + \underset{0}{Pb}$$

3. *How are electrons gained and lost?*

3 e^- lost (each atom)

$$\underset{+2}{Pb}O + \overset{-3}{N}H_3 \longrightarrow \overset{0}{N_2} + H_2O + \underset{0}{Pb}$$

2 e^- gained

The oxidation states of all other atoms are unchanged.

4. *What coefficents are needed to equalize the electrons gained and lost?*

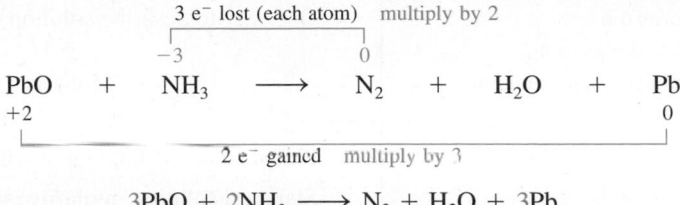

$$3PbO + 2NH_3 \longrightarrow N_2 + H_2O + 3Pb$$

5. *What coefficents are needed to balance the remaining elements?*
Balance O:

$$3PbO + 2NH_3 \longrightarrow N_2 + 3H_2O + 3Pb$$

All the elements are now balanced. The balanced equation with states is:

$$3PbO(s) + 2NH_3(g) \longrightarrow N_2(g) + 3H_2O(l) + 3Pb(s)$$

See Exercises 6-89 and 6-90

FOR REVIEW

Key Terms

aqueous solution

Section 6-1
polar molecule
hydration
solubility

Section 6-2
solute
solvent
electrical conductivity
strong electrolyte
weak electrolyte
nonelectrolyte
acid
strong acid
strong base
weak acid
weak base

Section 6-3
molarity
standard solution
dilution

Section 6-5
precipitation reaction
precipitate

Chemical reactions in solution are very important in everyday life.

Water is a polar solvent that dissolves many ionic and polar substances.

Electrolytes

- Strong electrolyte: 100% dissociated to produce separate ions; strongly conducts an electric current
- Weak electrolyte: Only a small percentage of dissolved molecules produce ions; weakly conducts an electric current
- Nonelectrolyte: Dissolved substance produces no ions; does not conduct an electric current

Acids and Bases

- Arrhenius model
  - Acid: produces H^+
  - Base: produces OH^-
- Brønsted–Lowry model
  - Acid: proton donor
  - Base: proton acceptor
- Strong acid: completely dissociates into separated H^+ and anions
- Weak acid: dissociates to a slight extent

Key Terms

Section 6-6
formula equation
complete ionic equation
spectator ions
net ionic equation

Section 6-8
acid
base
neutralization reaction
volumetric analysis
titration
stoichiometric (equivalence)
    point
indicator
endpoint

Section 6-9
oxidation–reduction (redox)
    reaction
oxidation state
oxidation
reduction
oxidizing agent (electron
    acceptor)
reducing agent (electron
    donor)

Molarity

- One way to describe solution composition

$$\text{Molarity } (M) = \frac{\text{moles of solute}}{\text{volume of solution (L)}}$$

- Moles solute = volume of solution (L) $\times$ molarity
- Standard solution: molarity is accurately known

Dilution

- Solvent is added to reduce the molarity
- Moles of solute after dilution = moles of solute before dilution

$$M_1V_1 = M_2V_2$$

Types of Equations That Describe Solution Reactions

- Formula equation: all reactants and products are written as complete formulas
- Complete ionic equation: all reactants and products that are strong electrolytes are written as separated ions
- Net ionic equation: only those compounds that undergo a change are written; spectator ions are not included

Solubility Rules

- Based on experiment observation
- Help predict the outcomes of precipitation reactions

Important Types of Solution Reactions

- Acid–base reactions: involve a transfer of H^+ ions
- Precipitation reactions: formation of a solid occurs
- Oxidation–reduction reactions: involve electron transfer

Titration

- Measures the volume of a standard solution (titrant) needed to react with a substance in solution
- Stoichiometric (equivalence) point: the point at which the required amount of titrant has been added to exactly react with the substance being analyzed
- Endpoint: the point at which a chemical indicator changes color

Oxidation–Reduction Reactions

- Oxidation states are assigned by using a set of rules to keep track of electron flow
- Oxidation: increase in oxidation state (a loss of electrons)
- Reduction: decrease in oxidation state (a gain of electrons)
- Oxidizing agent: gains electrons (is reduced)
- Reducing agent: loses electrons (is oxidized)
- Equations for oxidation–reduction reactions can be balanced by the oxidation states method

Review Questions Answers to the Review Questions can be found on the Student Web site (accessible from **www.cengagebrain.com**).

1. The (*aq*) designation listed after a solute indicates the process of hydration. Using KBr(*aq*) and C_2H_5OH(*aq*) as your examples, explain the process of hydration for soluble ionic compounds and for soluble covalent compounds.

2. Characterize strong electrolytes versus weak electrolytes versus nonelectrolytes. Give examples of each. How do you experimentally determine whether a soluble substance is a strong electrolyte, weak electrolyte, or nonelectrolyte?

3. Distinguish between the terms *slightly soluble* and *weak electrolyte*.

4. Molarity is a conversion factor relating moles of solute in solution to the volume of the solution. How does one use molarity as a conversion factor to convert from moles of solute to volume of solution, and from volume of solution to moles of solute present?

5. What is a dilution? What stays constant in a dilution? Explain why the equation $M_1V_1 = M_2V_2$ works for dilution problems.

6. When the following beakers are mixed, draw a molecular-level representation of the product mixture (see Fig. 6-17).

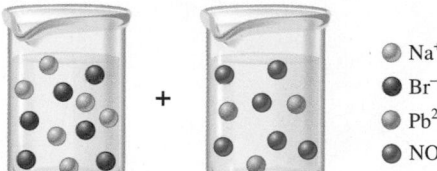

7. Differentiate between the formula equation, the complete ionic equation, and the net ionic equation. For each reaction in Question 6, write all three balanced equations.

8. What is an acid–base reaction? Strong bases are soluble ionic compounds that contain the hydroxide ion. List the strong bases. When a strong base reacts with an acid, what is always produced? Explain the terms *titration, stoichiometric point, neutralization,* and *standardization*.

9. Define the terms *oxidation, reduction, oxidizing agent,* and *reducing agent*. Given a chemical reaction, how can you tell if it is a redox reaction?

10. Consider the steps involved in balancing oxidation–reduction reactions by using oxidation states. The key to the oxidation states method is to equalize the electrons lost by the species oxidized with the electrons gained by the species reduced. First of all, how do you recognize what is oxidized and what is reduced? Second, how do you balance the electrons lost with the electrons gained? Once the electrons are balanced, what else is needed to balance the oxidation–reduction reaction?

Active Learning Questions

These questions are designed to be used by groups of students in class.

1. Assume you have a highly magnified view of a solution of HCl that allows you to "see" the HCl. Draw this magnified view. If you dropped in a piece of magnesium, the magnesium would disappear and hydrogen gas would be released. Represent this change using symbols for the elements, and write out the balanced equation.

2. You have a solution of table salt in water. What happens to the salt concentration (increases, decreases, or stays the same) as the solution boils? Draw pictures to explain your answer.

3. You have a sugar solution (solution *A*) with concentration *x*. You pour one-fourth of this solution into a beaker, and add an equivalent volume of water (solution *B*).

   a. What is the ratio of sugar in solutions *A* and *B*?

   b. Compare the volumes of solutions *A* and *B*.

   c. What is the ratio of the concentrations of sugar in solutions *A* and *B*?

4. You add an aqueous solution of lead nitrate to an aqueous solution of potassium iodide. Draw highly magnified views of each solution individually, and the mixed solution, including any product that forms. Write the balanced equation for the reaction.

5. Order the following molecules from lowest to highest oxidation state of the nitrogen atom: HNO_3, NH_4Cl, N_2O, NO_2, $NaNO_2$.

6. Why is it that when something gains electrons, it is said to be *reduced?* What is being reduced?

7. Consider separate aqueous solutions of HCl and H_2SO_4 with the same molar concentrations. You wish to neutralize an aqueous solution of NaOH. For which acid solution would you need to add more volume (in milliliters) to neutralize the base?

   a. the HCl solution

   b. the H_2SO_4 solution

c. You need to know the acid concentrations to answer this question.

d. You need to know the volume and concentration of the NaOH solution to answer this question.

e. c and d

Explain.

8. Draw molecular-level pictures to differentiate between concentrated and dilute solutions.

9. You need to make 150.0 mL of a 0.10-*M* NaCl solution. You have solid NaCl, and your lab partner has a 2.5-*M* NaCl solution. Explain how you each make the 0.10-*M* NaCl solution.

10. The exposed electrodes of a light bulb are placed in a solution of H_2SO_4 in an electrical circuit such that the light bulb is glowing. You add a dilute salt solution, and the bulb dims. Which of the following could be the salt in the solution?

a. $Ba(NO_3)_2$ **c.** K_2SO_4

b. $NaNO_3$ **d.** $Ca(NO_3)_2$

Justify your choices. For those you did not choose, explain why they are incorrect.

11. You have two solutions of chemical A. To determine which has the highest concentration of A (molarity), which of the following must you know (there may be more than one answer)?

a. the mass in grams of A in each solution

b. the molar mass of A

c. the volume of water added to each solution

d. the total volume of the solution

Explain.

12. Which of the following must be known to calculate the molarity of a salt solution (there may be more than one answer)?

a. the mass of salt added

b. the molar mass of the salt

c. the volume of water added

d. the total volume of the solution

Explain.

A blue question or exercise number indicates that the answer to that question or exercise appears at the back of this book and a solution appears in the *Solutions Guide*, as found on PowerLecture.

Questions

13. Differentiate between what happens when the following are added to water.

a. polar solute versus nonpolar solute

b. KF versus $C_6H_{12}O_6$

c. RbCl versus AgCl

d. HNO_3 versus CO

14. Consider the following electrostatic potential diagrams for some covalent compounds. Which of the represented compounds would not be soluble in water?

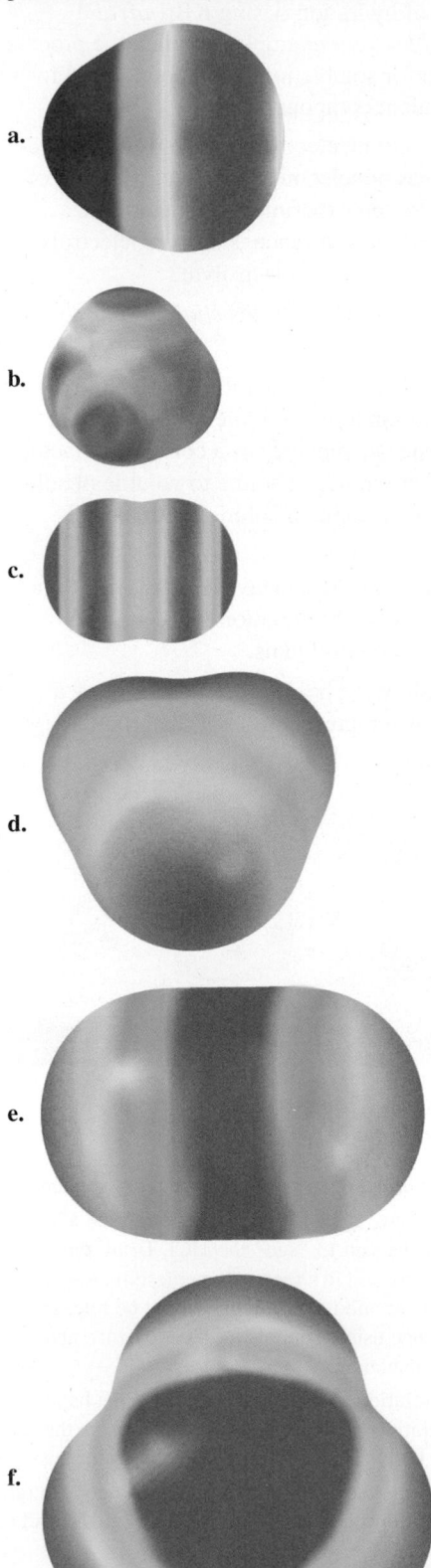

a.

b.

c.

d.

e.

f.

15. Of F_2, CF_4, and SF_2, which substance is most soluble in water? Explain.

16. A typical solution used in general chemistry laboratories is 3.0 M HCl. Describe, in detail, the composition of 2.0 L of a 3.0-M HCl solution. How would 2.0 L of a 3.0-M $HC_2H_3O_2$ solution differ from the same quantity of the HCl solution?

17. Which of the following statements is(are) *true*? For the false statements, correct them.

    a. A concentrated solution in water will always contain a strong or weak electrolyte.

    b. A strong electrolyte will break up into ions when dissolved in water.

    c. An acid is a strong electrolyte.

    d. All ionic compounds are strong electrolytes in water.

18. A student wants to prepare 1.00 L of a 1.00-M solution of NaOH (molar mass = 40.00 g/mol). If solid NaOH is available, how would the student prepare this solution? If 2.00 M NaOH is available, how would the student prepare the solution? To help ensure three significant figures in the NaOH molarity, to how many significant figures should the volumes and mass be determined?

19. List the formulas of three soluble bromide salts and three insoluble bromide salts. Do the same exercise for sulfate salts, hydroxide salts, and phosphate salts (list three soluble salts and three insoluble salts). List the formulas for six insoluble Pb^{2+} salts and one soluble Pb^{2+} salt.

20. When 1.0 mole of solid lead nitrate is added to 2.0 moles of aqueous potassium iodide, a yellow precipitate forms. After the precipitate settles to the bottom, does the solution above the precipitate conduct electricity? Explain. Write the complete ionic equation to help you answer this question.

21. What is an acid and what is a base? An acid–base reaction is sometimes called a proton-transfer reaction. Explain.

22. A student had 1.00 L of a 1.00-M acid solution. Much to the surprise of the student, it took 2.00 L of 1.00 M NaOH solution to react completely with the acid. Explain why it took twice as much NaOH to react with all of the acid.

    In a different experiment, a student had 10.0 mL of 0.020 M HCl. Again, much to the surprise of the student, it took only 5.00 mL of 0.020 M strong base to react completely with the HCl. Explain why it took only half as much strong base to react with all of the HCl.

23. Differentiate between the following terms.

    a. species reduced versus the reducing agent

    b. species oxidized versus the oxidizing agent

    c. oxidation state versus actual charge

24. How does one balance redox reactions by the oxidation states method?

Exercises

In this section similar exercises are paired.

Aqueous Solutions: Strong and Weak Electrolytes

25. Show how each of the following strong electrolytes "breaks up" into its component ions upon dissolving in water by drawing molecular-level pictures.

    a. NaBr   f. $FeSO_4$

    b. $MgCl_2$ g. $KMnO_4$

    c. $Al(NO_3)_3$ h. $HClO_4$

    d. $(NH_4)_2SO_4$ i. $NH_4C_2H_3O_2$ (ammonium acetate)

    e. NaOH

26. Match each name below with the following microscopic pictures of that compound in aqueous solution.

    i.   ii.   iii.   iv.

    a. barium nitrate   c. potassium carbonate

    b. sodium chloride   d. magnesium sulfate

    Which picture best represents $HNO_3(aq)$? Why aren't any of the pictures a good representation of $HC_2H_3O_2(aq)$?

27. Calcium chloride is a strong electrolyte and is used to "salt" streets in the winter to melt ice and snow. Write a reaction to show how this substance breaks apart when it dissolves in water.

28. Commercial cold packs and hot packs are available for treating athletic injuries. Both types contain a pouch of water and a dry chemical. When the pack is struck, the pouch of water breaks, dissolving the chemical, and the solution becomes either hot or cold. Many hot packs use magnesium sulfate, and many cold packs use ammonium nitrate. Write reactions to show how these strong electrolytes break apart when they dissolve in water.

Solution Concentration: Molarity

29. Calculate the molarity of each of these solutions.

    a. A 5.623-g sample of $NaHCO_3$ is dissolved in enough water to make 250.0 mL of solution.

    b. A 184.6-mg sample of $K_2Cr_2O_7$ is dissolved in enough water to make 500.0 mL of solution.

    c. A 0.1025-g sample of copper metal is dissolved in 35 mL of concentrated HNO_3 to form Cu^{2+} ions and then water is added to make a total volume of 200.0 mL. (Calculate the molarity of Cu^{2+}.)

30. A solution of ethanol (C_2H_5OH) in water is prepared by dissolving 75.0 mL of ethanol (density = 0.79 g/cm³) in enough water to make 250.0 mL of solution. What is the molarity of the ethanol in this solution?

31. Calculate the concentration of all ions present in each of the following solutions of strong electrolytes.
    a. 0.100 mole of $Ca(NO_3)_2$ in 100.0 mL of solution
    b. 2.5 moles of Na_2SO_4 in 1.25 L of solution
    c. 5.00 g of NH_4Cl in 500.0 mL of solution
    d. 1.00 g K_3PO_4 in 250.0 mL of solution

32. Calculate the concentration of all ions present in each of the following solutions of strong electrolytes.
    a. 0.0200 mole of sodium phosphate in 10.0 mL of solution
    b. 0.300 mole of barium nitrate in 600.0 mL of solution
    c. 1.00 g of potassium chloride in 0.500 L of solution
    d. 132 g of ammonium sulfate in 1.50 L of solution

33. Which of the following solutions of strong electrolytes contains the largest number of moles of chloride ions: 100.0 mL of 0.30 M $AlCl_3$, 50.0 mL of 0.60 M $MgCl_2$, or 200.0 mL of 0.40 M NaCl?

34. Which of the following solutions of strong electrolytes contains the largest number of ions: 100.0 mL of 0.100 M NaOH, 50.0 mL of 0.200 M $BaCl_2$, or 75.0 mL of 0.150 M Na_3PO_4?

35. What mass of NaOH is contained in 250.0 mL of a 0.400 M sodium hydroxide solution?

36. If 10. g of $AgNO_3$ is available, what volume of 0.25 M $AgNO_3$ solution can be prepared?

37. Describe how you would prepare 2.00 L of each of the following solutions.
    a. 0.250 M NaOH from solid NaOH
    b. 0.250 M NaOH from 1.00 M NaOH stock solution
    c. 0.100 M K_2CrO_4 from solid K_2CrO_4
    d. 0.100 M K_2CrO_4 from 1.75 M K_2CrO_4 stock solution

38. How would you prepare 1.00 L of a 0.50-M solution of each of the following?
    a. H_2SO_4 from "concentrated" (18 M) sulfuric acid
    b. HCl from "concentrated" (12 M) reagent
    c. $NiCl_2$ from the salt $NiCl_2 \cdot 6H_2O$
    d. HNO_3 from "concentrated" (16 M) reagent
    e. Sodium carbonate from the pure solid

39. A solution is prepared by dissolving 10.8 g ammonium sulfate in enough water to make 100.0 mL of stock solution. A 10.00-mL sample of this stock solution is added to 50.00 mL of water. Calculate the concentration of ammonium ions and sulfate ions in the final solution.

40. A solution was prepared by mixing 50.00 mL of 0.100 M HNO_3 and 100.00 mL of 0.200 M HNO_3. Calculate the molarity of the final solution of nitric acid.

41. Calculate the sodium ion concentration when 70.0 mL of 3.0 M sodium carbonate is added to 30.0 mL of 1.0 M sodium bicarbonate.

42. Suppose 50.0 mL of 0.250 M $CoCl_2$ solution is added to 25.0 mL of 0.350 M $NiCl_2$ solution. Calculate the concentration, in moles per liter, of each of the ions present after mixing. Assume that the volumes are additive.

43. A standard solution is prepared for the analysis of fluoxymesterone ($C_{20}H_{29}FO_3$), an anabolic steroid. A stock solution is first prepared by dissolving 10.0 mg of fluoxymesterone in enough water to give a total volume of 500.0 mL. A 100.0-μL aliquot (portion) of this solution is diluted to a final volume of 100.0 mL. Calculate the concentration of the final solution in terms of molarity.

44. A stock solution containing Mn^{2+} ions was prepared by dissolving 1.584 g pure manganese metal in nitric acid and diluting to a final volume of 1.000 L. The following solutions were then prepared by dilution:

    For solution A, 50.00 mL of stock solution was diluted to 1000.0 mL.

    For solution B, 10.00 mL of solution A was diluted to 250.0 mL.
    For solution C, 10.00 mL of solution B was diluted to 500.0 mL.

    Calculate the concentrations of the stock solution and solutions A, B, and C.

Precipitation Reactions

45. On the basis of the general solubility rules given in Table 6-1, predict which of the following substances are likely to be soluble in water.
    a. aluminum nitrate
    b. magnesium chloride
    c. rubidium sulfate
    d. nickel(II) hydroxide
    e. lead(II) sulfide
    f. magnesium hydroxide
    g. iron(III) phosphate

46. On the basis of the general solubility rules given in Table 6-1, predict which of the following substances are likely to be soluble in water.
    a. zinc chloride
    b. lead(II) nitrate
    c. lead(II) sulfate
    d. sodium iodide
    e. cobalt(III) sulfide
    f. chromium(III) hydroxide
    g. magnesium carbonate
    h. ammonium carbonate

47. When the following solutions are mixed together, what precipitate (if any) will form?
    a. $FeSO_4(aq) + KCl(aq)$
    b. $Al(NO_3)_3(aq) + Ba(OH)_2(aq)$
    c. $CaCl_2(aq) + Na_2SO_4(aq)$
    d. $K_2S(aq) + Ni(NO_3)_2(aq)$

48. When the following solutions are mixed together, what precipitate (if any) will form?
    a. $Hg_2(NO_3)_2(aq) + CuSO_4(aq)$
    b. $Ni(NO_3)_2(aq) + CaCl_2(aq)$
    c. $K_2CO_3(aq) + MgI_2(aq)$
    d. $Na_2CrO_4(aq) + AlBr_3(aq)$

49. For the reactions in Exercise 47, write the balanced formula equation, complete ionic equation, and net ionic equation. If no precipitate forms, write "No reaction."

50. For the reactions in Exercise 48, write the balanced formula equation, complete ionic equation, and net ionic equation. If no precipitate forms, write "No reaction."

51. Write the balanced formula and net ionic equation for the reaction that occurs when the contents of the two beakers are added together. What colors represent the spectator ions in each reaction?

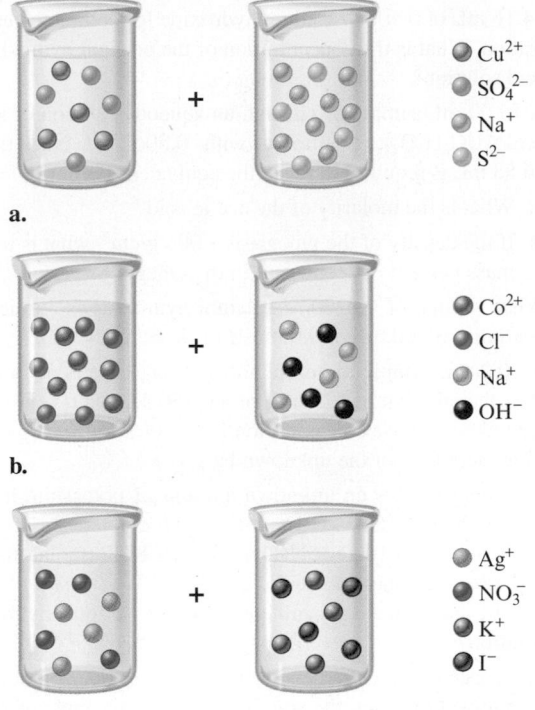

a.

b.

c.

52. Give an example how each of the following insoluble ionic compounds could be produced using a precipitation reaction. Write the balanced formula equation for each reaction.

   a. $Fe(OH)_3(s)$ c. $PbSO_4(s)$
   b. $Hg_2Cl_2(s)$ d. $BaCrO_4(s)$

53. Write net ionic equations for the reaction, if any, that occurs when aqueous solutions of the following are mixed.

   a. ammonium sulfate and barium nitrate
   b. lead(II) nitrate and sodium chloride
   c. sodium phosphate and potassium nitrate
   d. sodium bromide and rubidium chloride
   e. copper(II) chloride and sodium hydroxide

54. Write net ionic equations for the reaction, if any, that occurs when aqueous solutions of the following are mixed.

   a. chromium(III) chloride and sodium hydroxide
   b. silver nitrate and ammonium carbonate
   c. copper(II) sulfate and mercury(I) nitrate
   d. strontium nitrate and potassium iodide

55. Separate samples of a solution of an unknown soluble ionic compound are treated with KCl, Na_2SO_4, and NaOH. A precipitate forms only when Na_2SO_4 is added. Which cations could be present in the unknown soluble ionic compound?

56. A sample may contain any or all of the following ions: Hg_2^{2+}, Ba^{2+}, and Mn^{2+}.

   a. No precipitate formed when an aqueous solution of NaCl was added to the sample solution.
   b. No precipitate formed when an aqueous solution of Na_2SO_4 was added to the sample solution.
   c. A precipitate formed when the sample solution was made basic with NaOH.

   Which ion or ions are present in the sample solution?

57. What mass of Na_2CrO_4 is required to precipitate all of the silver ions from 75.0 mL of a 0.100-M solution of $AgNO_3$?

58. What volume of 0.100 M Na_3PO_4 is required to precipitate all the lead(II) ions from 150.0 mL of 0.250 M $Pb(NO_3)_2$?

59. What mass of solid aluminum hydroxide can be produced when 50.0 mL of 0.200 M $Al(NO_3)_3$ is added to 200.0 mL of 0.100 M KOH?

60. What mass of barium sulfate can be produced when 100.0 mL of a 0.100-M solution of barium chloride is mixed with 100.0 mL of a 0.100-M solution of iron(III) sulfate?

61. What mass of solid AgBr is produced when 100.0 mL of 0.150 M $AgNO_3$ is added to 20.0 mL of 1.00 M NaBr?

62. What mass of silver chloride can be prepared by the reaction of 100.0 mL of 0.20 M silver nitrate with 100.0 mL of 0.15 M calcium chloride? Calculate the concentrations of each ion remaining in solution after precipitation is complete.

63. A 100.0-mL aliquot of 0.200 M aqueous potassium hydroxide is mixed with 100.0 mL of 0.200 M aqueous magnesium nitrate.

   a. Write a balanced chemical equation for any reaction that occurs.
   b. What precipitate forms?
   c. What mass of precipitate is produced?
   d. Calculate the concentration of each ion remaining in solution after precipitation is complete.

64. The drawings below represent aqueous solutions. Solution A is 2.00 L of a 2.00-M aqueous solution of copper(II) nitrate. Solution B is 2.00 L of a 3.00-M aqueous solution of potassium hydroxide.

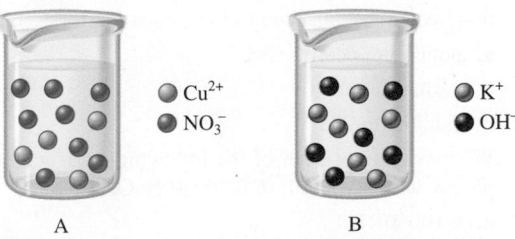

   A   B

   a. Draw a picture of the solution made by mixing solutions A and B together after the precipitation reaction takes place. Make sure this picture shows the correct relative volume compared to solutions A and B, and the correct relative number of ions, along with the correct relative amount of solid formed.

b. Determine the concentrations (in M) of all ions left in solution (from part a) and the mass of solid formed.

65. A 1.42-g sample of a pure compound, with formula M_2SO_4, was dissolved in water and treated with an excess of aqueous calcium chloride, resulting in the precipitation of all the sulfate ions as calcium sulfate. The precipitate was collected, dried, and found to weigh 1.36 g. Determine the atomic mass of M, and identify M.

66. You are given a 1.50-g mixture of sodium nitrate and sodium chloride. You dissolve this mixture into 100 mL of water and then add an excess of 0.500 M silver nitrate solution. You produce a white solid, which you then collect, dry, and measure. The white solid has a mass of 0.641 g.

   a. If you had an extremely magnified view of the solution (to the atomic-molecular level), list the species you would see (include charges, if any).

   b. Write the balanced net ionic equation for the reaction that produces the solid. Include phases and charges.

   c. Calculate the percent sodium chloride in the original unknown mixture.

Acid–Base Reactions

67. Write the balanced formula, complete ionic, and net ionic equations for each of the following acid–base reactions.

   a. $HClO_4(aq) + Mg(OH)_2(s) \rightarrow$

   b. $HCN(aq) + NaOH(aq) \rightarrow$

   c. $HCl(aq) + NaOH(aq) \rightarrow$

68. Write the balanced formula, complete ionic, and net ionic equations for each of the following acid–base reactions.

   a. $HNO_3(aq) + Al(OH)_3(s) \rightarrow$

   b. $HC_2H_3O_2(aq) + KOH(aq) \rightarrow$

   c. $Ca(OH)_2(aq) + HCl(aq) \rightarrow$

69. Write the balanced formula equation for the acid–base reactions that occur when the following are mixed.

   a. potassium hydroxide (aqueous) and nitric acid

   b. barium hydroxide (aqueous) and hydrochloric acid

   c. perchloric acid [$HClO_4(aq)$] and solid iron(III) hydroxide

   d. solid silver hydroxide and hydrobromic acid

   e. aqueous strontium hydroxide and hydroiodic acid

70. What acid and what base would react in aqueous solution so that the following salts appear as products in the formula equation? Write the balanced formula equation for each reaction.

   a. potassium perchlorate

   b. cesium nitrate

   c. calcium iodide

71. What volume of each of the following acids will react completely with 50.00 mL of 0.200 M NaOH?

   a. 0.100 M HCl

   b. 0.150 M HNO_3

   c. 0.200 M $HC_2H_3O_2$ (1 acidic hydrogen)

72. What volume of each of the following bases will react completely with 25.00 mL of 0.200 M HCl?

   a. 0.100 M NaOH   **c.** 0.250 M KOH

   b. 0.0500 M $Sr(OH)_2$

73. Hydrochloric acid (75.0 mL of 0.250 M) is added to 225.0 mL of 0.0550 M $Ba(OH)_2$ solution. What is the concentration of the excess H^+ or OH^- ions left in this solution?

74. A student mixes four reagents together, thinking that the solutions will neutralize each other. The solutions mixed together are 50.0 mL of 0.100 M hydrochloric acid, 100.0 mL of 0.200 M of nitric acid, 500.0 mL of 0.0100 M calcium hydroxide, and 200.0 mL of 0.100 M rubidium hydroxide. Did the acids and bases exactly neutralize each other? If not, calculate the concentration of excess H^+ or OH^- ions left in solution.

75. A 25.00-mL sample of hydrochloric acid solution requires 24.16 mL of 0.106 M sodium hydroxide for complete neutralization. What is the concentration of the original hydrochloric acid solution?

76. A 10.00-mL sample of vinegar, an aqueous solution of acetic acid ($HC_2H_3O_2$), is titrated with 0.5062 M NaOH, and 16.58 mL is required to reach the equivalence point.

   a. What is the molarity of the acetic acid?

   b. If the density of the vinegar is 1.006 g/cm³, what is the mass percent of acetic acid in the vinegar?

77. What volume of 0.0200 M calcium hydroxide is required to neutralize 35.00 mL of 0.0500 M nitric acid?

78. A 30.0-mL sample of an unknown strong base is neutralized after the addition of 12.0 mL of a 0.150 M HNO_3 solution. If the unknown base concentration is 0.0300 M, give some possible identities for the unknown base.

79. A student titrates an unknown amount of potassium hydrogen phthalate ($KHC_8H_4O_4$, often abbreviated KHP) with 20.46 mL of a 0.1000-M NaOH solution. KHP (molar mass = 204.22 g/mol) has one acidic hydrogen. What mass of KHP was titrated (reacted completely) by the sodium hydroxide solution?

80. The concentration of a certain sodium hydroxide solution was determined by using the solution to titrate a sample of potassium hydrogen phthalate (abbreviated as KHP). KHP is an acid with one acidic hydrogen and a molar mass of 204.22 g/mol. In the titration, 34.67 mL of the sodium hydroxide solution was required to react with 0.1082 g KHP. Calculate the molarity of the sodium hydroxide.

Oxidation–Reduction Reactions

81. Assign oxidation states for all atoms in each of the following compounds.

   a. $KMnO_4$ **f.** Fe_3O_4

   b. NiO_2 **g.** $XeOF_4$

   c. $Na_4Fe(OH)_6$ **h.** SF_4

   d. $(NH_4)_2HPO_4$ **i.** CO

   e. P_4O_6 **j.** $C_6H_{12}O_6$

82. Assign oxidation states for all atoms in each of the following compounds.

   a. UO_2^{2+} **f.** $Mg_2P_2O_7$

   b. As_2O_3 **g.** $Na_2S_2O_3$

   c. $NaBiO_3$ **h.** Hg_2Cl_2

   d. As_4 **i.** $Ca(NO_3)_2$

   e. $HAsO_2$

83. Assign the oxidation state for nitrogen in each of the following.
    a. Li_3N f. NO_2
    b. NH_3 g. NO_2^-
    c. N_2H_4 h. NO_3^-
    d. NO i. N_2
    e. N_2O

84. Assign oxidation numbers to all the atoms in each of the following.
    a. $SrCr_2O_7$ g. $PbSO_3$
    b. $CuCl_2$ h. PbO_2
    c. O_2 i. $Na_2C_2O_4$
    d. H_2O_2 j. CO_2
    e. $MgCO_3$ k. $(NH_4)_2Ce(SO_4)_3$
    f. Ag l. Cr_2O_3

85. Specify which of the following are oxidation–reduction reactions, and identify the oxidizing agent, the reducing agent, the substance being oxidized, and the substance being reduced.
    a. $Cu(s) + 2Ag^+(aq) \rightarrow 2Ag(s) + Cu^{2+}(aq)$
    b. $HCl(g) + NH_3(g) \rightarrow NH_4Cl(s)$
    c. $SiCl_4(l) + 2H_2O(l) \rightarrow 4HCl(aq) + SiO_2(s)$
    d. $SiCl_4(l) + 2Mg(s) \rightarrow 2MgCl_2(s) + Si(s)$
    e. $Al(OH)_4^-(aq) \rightarrow AlO_2^-(aq) + 2H_2O(l)$

86. Specify which of the following equations represent oxidation–reduction reactions, and indicate the oxidizing agent, the reducing agent, the species being oxidized, and the species being reduced.
    a. $CH_4(g) + H_2O(g) \rightarrow CO(g) + 3H_2(g)$
    b. $2AgNO_3(aq) + Cu(s) \rightarrow Cu(NO_3)_2(aq) + 2Ag(s)$
    c. $Zn(s) + 2HCl(aq) \rightarrow ZnCl_2(aq) + H_2(g)$
    d. $2H^+(aq) + 2CrO_4^{2-}(aq) \rightarrow Cr_2O_7^{2-}(aq) + H_2O(l)$

87. Consider the reaction between sodium metal and fluorine (F_2) gas to form sodium fluoride. Using oxidation states, how many electrons would each sodium atom lose, and how many electrons would each fluorine atom gain? How many sodium atoms are needed to react with one fluorine molecule? Write a balanced equation for this reaction.

88. Consider the reaction between oxygen (O_2) gas and magnesium metal to form magnesium oxide. Using oxidation states, how many electrons would each oxygen atom gain, and how many electrons would each magnesium atom lose? How many magnesium atoms are needed to react with one oxygen molecule? Write a balanced equation for this reaction.

89. Balance each of the following oxidation–reduction reactions by using the oxidation states method.
    a. $C_2H_6(g) + O_2(g) \rightarrow CO_2(g) + H_2O(g)$
    b. $Mg(s) + HCl(aq) \rightarrow Mg^{2+}(aq) + Cl^-(aq) + H_2(g)$
    c. $Co^{3+}(aq) + Ni(s) \rightarrow Co^{2+}(aq) + Ni^{2+}(aq)$
    d. $Zn(s) + H_2SO_4(aq) \rightarrow ZnSO_4(aq) + H_2(g)$

90. Balance each of the following oxidation–reduction reactions by using the oxidation states method.
    a. $Cl_2(g) + Al(s) \rightarrow Al^{3+}(aq) + Cl^-(aq)$
    b. $O_2(g) + H_2O(l) + Pb(s) \rightarrow Pb(OH)_2(s)$
    c. $H^+(aq) + MnO_4^-(aq) + Fe^{2+}(aq) \rightarrow$
    $Mn^{2+}(aq) + Fe^{3+}(aq) + H_2O(l)$

Additional Exercises

91. You wish to prepare 1 L of a 0.02-M potassium iodate solution. You require that the final concentration be within 1% of 0.02 M and that the concentration must be known accurately to the fourth decimal place. How would you prepare this solution? Specify the glassware you would use, the accuracy needed for the balance, and the ranges of acceptable masses of KIO_3 that can be used.

92. The figures below are molecular-level representations of four aqueous solutions of the same solute. Arrange the solutions from most to least concentrated.

Solution A (1.0 L)   Solution B (4.0 L)

Solution C (2.0 L)   Solution D (2.0 L)

93. An average human being has about 5.0 L of blood in his or her body. If an average person were to eat 32.0 g of sugar (sucrose, $C_{12}H_{22}O_{11}$, 342.30 g/mol), and all that sugar were dissolved into the bloodstream, how would the molarity of the blood sugar change?

94. A 230.-mL sample of a 0.275-M $CaCl_2$ solution is left on a hot plate overnight; the following morning, the solution is 1.10 M. What volume of water evaporated from the 0.275 M $CaCl_2$ solution?

95. Using the general solubility rules given in Table 6-1, name three reagents that would form precipitates with each of the following ions in aqueous solution. Write the net ionic equation for each of your suggestions.
    a. chloride ion   d. sulfate ion
    b. calcium ion   e. mercury(I) ion, Hg_2^{2+}
    c. iron(III) ion   f. silver ion

96. Consider a 1.50-g mixture of magnesium nitrate and magnesium chloride. After dissolving this mixture in water, 0.500 M silver nitrate is added dropwise until precipitate formation is complete. The mass of the white precipitate formed is 0.641 g.
    a. Calculate the mass percent of magnesium chloride in the mixture.
    b. Determine the minimum volume of silver nitrate that must have been added to ensure complete formation of the precipitate.

97. A 1.00-g sample of an alkaline earth metal chloride is treated with excess silver nitrate. All of the chloride is recovered as 1.38 g of silver chloride. Identify the metal.

98. A mixture contains only NaCl and $Al_2(SO_4)_3$. A 1.45-g sample of the mixture is dissolved in water and an excess of NaOH is added, producing a precipitate of $Al(OH)_3$. The precipitate is filtered, dried, and weighed. The mass of the precipitate is 0.107 g. What is the mass percent of $Al_2(SO_4)_3$ in the sample?

99. The thallium (present as Tl_2SO_4) in a 9.486-g pesticide sample was precipitated as thallium(I) iodide. Calculate the mass percent of Tl_2SO_4 in the sample if 0.1824 g of TlI was recovered.

100. A mixture contains only NaCl and $Fe(NO_3)_3$. A 0.456-g sample of the mixture is dissolved in water, and an excess of NaOH is added, producing a precipitate of $Fe(OH)_3$. The precipitate is filtered, dried, and weighed. Its mass is 0.107 g. Calculate the following.

   a. the mass of iron in the sample

   b. the mass of $Fe(NO_3)_3$ in the sample

   c. the mass percent of $Fe(NO_3)_3$ in the sample

101. A student added 50.0 mL of an NaOH solution to 100.0 mL of 0.400 M HCl. The solution was then treated with an excess of aqueous chromium(III) nitrate, resulting in formation of 2.06 g of precipitate. Determine the concentration of the NaOH solution.

102. Some of the substances commonly used in stomach antacids are MgO, $Mg(OH)_2$, and $Al(OH)_3$.

   a. Write a balanced equation for the neutralization of hydrochloric acid by each of these substances.

   b. Which of these substances will neutralize the greatest amount of 0.10 M HCl per gram?

103. Acetylsalicylic acid is the active ingredient in aspirin. It took 35.17 mL of 0.5065 M sodium hydroxide to react completely with 3.210 g of acetylsalicylic acid. Acetylsalicylic acid has one acidic hydrogen. What is the molar mass of acetylsalicylic acid?

104. When hydrochloric acid reacts with magnesium metal, hydrogen gas and aqueous magnesium chloride are produced. What volume of 5.0 M HCl is required to react completely with 3.00 g of magnesium?

105. A 2.20-g sample of an unknown acid (empirical formula = $C_3H_4O_3$) is dissolved in 1.0 L of water. A titration required 25.0 mL of 0.500 M NaOH to react completely with all the acid present. Assuming the unknown acid has one acidic proton per molecule, what is the molecular formula of the unknown acid?

106. Carminic acid, a naturally occurring red pigment extracted from the cochineal insect, contains only carbon, hydrogen, and oxygen. It was commonly used as a dye in the first half of the nineteenth century. It is 53.66% C and 4.09% H by mass. A titration required 18.02 mL of 0.0406 M NaOH to neutralize 0.3602 g carminic acid. Assuming that there is only one acidic hydrogen per molecule, what is the molecular formula of carminic acid?

107. Chlorisondamine chloride ($C_{14}H_{20}Cl_6N_2$) is a drug used in the treatment of hypertension. A 1.28-g sample of a medication containing the drug was treated to destroy the organic material and to release all the chlorine as chloride ion. When the filtered solution containing chloride ion was treated with an excess of silver nitrate, 0.104 g silver chloride was recovered.

Calculate the mass percent of chlorisondamine chloride in the medication, assuming the drug is the only source of chloride.

108. Saccharin ($C_7H_5NO_3S$) is sometimes dispensed in tablet form. Ten tablets with a total mass of 0.5894 g were dissolved in water. The saccharin was oxidized to convert all the sulfur to sulfate ion, which was precipitated by adding an excess of barium chloride solution. The mass of $BaSO_4$ obtained was 0.5032 g. What is the average mass of saccharin per tablet? What is the average mass percent of saccharin in the tablets?

109. Douglasite is a mineral with the formula $2KCl \cdot FeCl_2 \cdot 2H_2O$. Calculate the mass percent of douglasite in a 455.0-mg sample if it took 37.20 mL of a 0.1000-M $AgNO_3$ solution to precipitate all the Cl^- as AgCl. Assume the douglasite is the only source of chloride ion.

110. Many oxidation–reduction reactions can be balanced by inspection. Try to balance the following reactions by inspection. In each reaction, identify the substance reduced and the substance oxidized.

   a. $Al(s) + HCl(aq) \rightarrow AlCl_3(aq) + H_2(g)$

   b. $CH_4(g) + S(s) \rightarrow CS_2(l) + H_2S(g)$

   c. $C_3H_8(g) + O_2(g) \rightarrow CO_2(g) + H_2O(l)$

   d. $Cu(s) + Ag^+(aq) \rightarrow Ag(s) + Cu^{2+}(aq)$

111. The blood alcohol (C_2H_5OH) level can be determined by titrating a sample of blood plasma with an acidic potassium dichromate solution, resulting in the production of $Cr^{3+}(aq)$ and carbon dioxide. The reaction can be monitored because the dichromate ion ($Cr_2O_7^{2-}$) is orange in solution, and the Cr^{3+} ion is green. The balanced equation is

$$16H^+(aq) + 2Cr_2O_7^{2-}(aq) + C_2H_5OH(aq) \longrightarrow$$
$$4Cr^{3+}(aq) + 2CO_2(g) + 11H_2O(l)$$

This reaction is an oxidation–reduction reaction. What species is reduced, and what species is oxidized? How many electrons are transferred in the balanced equation above?

ChemWork Problems

These multiconcept problems (and additional ones) are found interactively online with the same type of assistance a student would get from an instructor.

112. Calculate the concentration of all ions present when 0.160 g of $MgCl_2$ is dissolved in 100.0 mL of solution.

113. A solution is prepared by dissolving 0.6706 g oxalic acid ($H_2C_2O_4$) in enough water to make 100.0 mL of solution. A 10.00-mL aliquot (portion) of this solution is then diluted to a final volume of 250.0 mL. What is the final molarity of the oxalic acid solution?

114. For the following chemical reactions, determine the precipitate produced when the two reactants listed below are mixed together. Indicate "none" if no precipitate will form.

	Formula of Precipitate
$Sr(NO_3)_2(aq) + K_3PO_4(aq) \longrightarrow$	_____ (s)
$K_2CO_3(aq) + AgNO_3(aq) \longrightarrow$	_____ (s)
$NaCl(aq) + KNO_3(aq) \longrightarrow$	_____ (s)
$KCl(aq) + AgNO_3(aq) \longrightarrow$	_____ (s)
$FeCl_3(aq) + Pb(NO_3)_2(aq) \longrightarrow$	_____ (s)

115. What volume of 0.100 *M* NaOH is required to precipitate all of the nickel(II) ions from 150.0 mL of a 0.249-*M* solution of $Ni(NO_3)_2$?

116. A 500.0-mL sample of 0.200 *M* sodium phosphate is mixed with 400.0 mL of 0.289 *M* barium chloride. What is the mass of the solid produced?

117. A 450.0-mL sample of a 0.257-*M* solution of silver nitrate is mixed with 400.0 mL of 0.200 *M* calcium chloride. What is the concentration of Cl^- in solution after the reaction is complete?

118. The zinc in a 1.343-g sample of a foot powder was precipitated as $ZnNH_4PO_4$. Strong heating of the precipitate yielded 0.4089 g $Zn_2P_2O_7$. Calculate the mass percent of zinc in the sample of foot powder.

119. A 50.00-mL sample of aqueous $Ca(OH)_2$ requires 34.66 mL of a 0.944-*M* nitric acid for neutralization. Calculate the concentration (molarity) of the original solution of calcium hydroxide.

120. When organic compounds containing sulfur are burned, sulfur dioxide is produced. The amount of SO_2 formed can be determined by the reaction with hydrogen peroxide:

$$H_2O_2(aq) + SO_2(g) \longrightarrow H_2SO_4(aq)$$

The resulting sulfuric acid is then titrated with a standard NaOH solution. A 1.302-g sample of coal is burned and the SO_2 is collected in a solution of hydrogen peroxide. It took 28.44 mL of a 0.1000-*M* NaOH solution to titrate the resulting sulfuric acid. Calculate the mass percent of sulfur in the coal sample. Sulfuric acid has two acidic hydrogens.

121. Assign the oxidation state for the element listed in each of the following compounds:

	Oxidation State
S in $MgSO_4$	_____
Pb in $PbSO_4$	_____
O in O_2	_____
Ag in Ag	_____
Cu in $CuCl_2$	_____

Challenge Problems

122. A 10.00-g sample consisting of a mixture of sodium chloride and potassium sulfate is dissolved in water. This aqueous mixture then reacts with excess aqueous lead(II) nitrate to form 21.75 g of solid. Determine the mass percent of sodium chloride in the original mixture.

123. The units of parts per million (ppm) and parts per billion (ppb) are commonly used by environmental chemists. In general, 1 ppm means 1 part of solute for every 10^6 parts of solution. Mathematically, by mass:

$$ppm = \frac{\mu g\ solute}{g\ solution} = \frac{mg\ solute}{kg\ solution}$$

In the case of very dilute aqueous solutions, a concentration of 1.0 ppm is equal to 1.0 μg of solute per 1.0 mL, which equals 1.0 g solution. Parts per billion is defined in a similar fashion. Calculate the molarity of each of the following aqueous solutions.

a. 5.0 ppb Hg in H_2O

b. 1.0 ppb $CHCl_3$ in H_2O

c. 10.0 ppm As in H_2O

d. 0.10 ppm DDT ($C_{14}H_9Cl_5$) in H_2O

124. In the spectroscopic analysis of many substances, a series of standard solutions of known concentration are measured to generate a calibration curve. How would you prepare standard solutions containing 10.0, 25.0, 50.0, 75.0, and 100. ppm of copper from a commercially produced 1000.0-ppm solution? Assume each solution has a final volume of 100.0 mL. (See Exercise 123 for definitions.)

125. In most of its ionic compounds, cobalt is either Co(II) or Co(III). One such compound, containing chloride ion and waters of hydration, was analyzed, and the following results were obtained. A 0.256-g sample of the compound was dissolved in water, and excess silver nitrate was added. The silver chloride was filtered, dried, and weighed, and it had a mass of 0.308 g. A second sample of 0.416 g of the compound was dissolved in water, and an excess of sodium hydroxide was added. The hydroxide salt was filtered and heated in a flame, forming cobalt(III) oxide. The mass of cobalt(III) oxide formed was 0.145 g.

a. What is the percent composition, by mass, of the compound?

b. Assuming the compound contains one cobalt ion per formula unit, what is the formula?

c. Write balanced equations for the three reactions described.

126. Polychlorinated biphenyls (PCBs) have been used extensively as dielectric materials in electrical transformers. Because PCBs have been shown to be potentially harmful, analysis for their presence in the environment has become very important. PCBs are manufactured according to the following generic reaction:

$$C_{12}H_{10} + nCl_2 \rightarrow C_{12}H_{10-n}Cl_n + nHCl$$

This reaction results in a mixture of PCB products. The mixture is analyzed by decomposing the PCBs and then precipitating the resulting Cl^- as AgCl.

a. Develop a general equation that relates the average value of *n* to the mass of a given mixture of PCBs and the mass of AgCl produced.

b. A 0.1947-g sample of a commercial PCB yielded 0.4791 g of AgCl. What is the average value of *n* for this sample?

127. Consider the reaction of 19.0 g of zinc with excess silver nitrite to produce silver metal and zinc nitrite. The reaction is stopped before all the zinc metal has reacted and 29.0 g of solid metal is present. Calculate the mass of each metal in the 29.0-g mixture.

128. A mixture contains only sodium chloride and potassium chloride. A 0.1586-g sample of the mixture was dissolved in water. It took 22.90 mL of 0.1000 *M* $AgNO_3$ to completely precipitate all the chloride present. What is the composition (by mass percent) of the mixture?

129. You are given a solid that is a mixture of Na_2SO_4 and K_2SO_4. A 0.205-g sample of the mixture is dissolved in water. An excess of an aqueous solution of $BaCl_2$ is added. The $BaSO_4$ that is formed is filtered, dried, and weighed. Its mass is 0.298 g. What mass of SO_4^{2-} ion is in the sample? What is the mass percent of SO_4^{2-} ion in the sample? What are the percent compositions by mass of Na_2SO_4 and K_2SO_4 in the sample?

130. Zinc and magnesium metal each react with hydrochloric acid according to the following equations:

$$Zn(s) + 2HCl(aq) \longrightarrow ZnCl_2(aq) + H_2(g)$$
$$Mg(s) + 2HCl(aq) \longrightarrow MgCl_2(aq) + H_2(g)$$

A 10.00-g mixture of zinc and magnesium is reacted with the stoichiometric amount of hydrochloric acid. The reaction mixture is then reacted with 156 mL of 3.00 M silver nitrate to produce the maximum possible amount of silver chloride.

a. Determine the percent magnesium by mass in the original mixture.

b. If 78.0 mL of HCl was added, what was the concentration of the HCl?

131. You made 100.0 mL of a lead(II) nitrate solution for lab but forgot to cap it. The next lab session you noticed that there was only 80.0 mL left (the rest had evaporated). In addition, you forgot the initial concentration of the solution. You decide to take 2.00 mL of the solution and add an excess of a concentrated sodium chloride solution. You obtain a solid with a mass of 3.407 g. What was the concentration of the original lead(II) nitrate solution?

132. Consider reacting copper(II) sulfate with iron. Two possible reactions can occur, as represented by the following equations.

copper(II) sulfate(aq) + iron(s) $\longrightarrow$
$\qquad\qquad\qquad$ copper(s) + iron(II) sulfate(aq)

copper(II) sulfate(aq) + iron(s) $\longrightarrow$
$\qquad\qquad\qquad$ copper(s) + iron(III) sulfate(aq)

You place 87.7 mL of a 0.500-M solution of copper(II) sulfate in a beaker. You then add 2.00 g of iron filings to the copper(II) sulfate solution. After one of the above reactions occurs, you isolate 2.27 g of copper. Which equation above describes the reaction that occurred? Support your answer.

133. Consider an experiment in which two burets, Y and Z, are simultaneously draining into a beaker that initially contained 275.0 mL of 0.300 M HCl. Buret Y contains 0.150 M NaOH and buret Z contains 0.250 M KOH. The stoichiometric point in the titration is reached 60.65 minutes after Y and Z were started simultaneously. The total volume in the beaker at the stoichiometric point is 655 mL. Calculate the flow rates of burets Y and Z. Assume the flow rates remain constant during the experiment.

134. Complete and balance each acid–base reaction.

a. $H_3PO_4(aq) + NaOH(aq) \rightarrow$
Contains three acidic hydrogens

b. $H_2SO_4(aq) + Al(OH)_3(s) \rightarrow$
Contains two acidic hydrogens

c. $H_2Se(aq) + Ba(OH)_2(aq) \rightarrow$
Contains two acidic hydrogens

d. $H_2C_2O_4(aq) + NaOH(aq) \rightarrow$
Contains two acidic hydrogens

135. What volume of 0.0521 M $Ba(OH)_2$ is required to neutralize exactly 14.20 mL of 0.141 M H_3PO_4? Phosphoric acid contains three acidic hydrogens.

136. A 10.00-mL sample of sulfuric acid from an automobile battery requires 35.08 mL of 2.12 M sodium hydroxide solution for complete neutralization. What is the molarity of the sulfuric acid? Sulfuric acid contains two acidic hydrogens.

137. A 0.500-L sample of H_2SO_4 solution was analyzed by taking a 100.0-mL aliquot and adding 50.0 mL of 0.213 M NaOH. After the reaction occurred, an excess of OH^- ions remained in the solution. The excess base required 13.21 mL of 0.103 M HCl for neutralization. Calculate the molarity of the original sample of H_2SO_4. Sulfuric acid has two acidic hydrogens.

138. A 6.50-g sample of a diprotic acid requires 137.5 mL of a 0.750 M NaOH solution for complete neutralization. Determine the molar mass of the acid.

139. Citric acid, which can be obtained from lemon juice, has the molecular formula $C_6H_8O_7$. A 0.250-g sample of citric acid dissolved in 25.0 mL of water requires 37.2 mL of 0.105 M NaOH for complete neutralization. What number of acidic hydrogens per molecule does citric acid have?

140. A stream flows at a rate of 5.00×10^4 liters per second (L/s) upstream of a manufacturing plant. The plant discharges 3.50×10^3 L/s of water that contains 65.0 ppm HCl into the stream. (See Exercise 123 for definitions.)

a. Calculate the stream's total flow rate downstream from this plant.

b. Calculate the concentration of HCl in ppm downstream from this plant.

c. Further downstream, another manufacturing plant diverts 1.80×10^4 L/s of water from the stream for its own use. This plant must first neutralize the acid and does so by adding lime:

$$CaO(s) + 2H^+(aq) \longrightarrow Ca^{2+}(aq) + H_2O(l)$$

What mass of CaO is consumed in an 8.00-h work day by this plant?

d. The original stream water contained 10.2 ppm Ca^{2+}. Although no calcium was in the waste water from the first plant, the waste water of the second plant contains Ca^{2+} from the neutralization process. If 90.0% of the water used by the second plant is returned to the stream, calculate the concentration of Ca^{2+} in ppm downstream of the second plant.

141. It took 25.06 ± 0.05 mL of a sodium hydroxide solution to titrate a 0.4016-g sample of KHP (see Exercise 79). Calculate the concentration and uncertainty in the concentration of the sodium hydroxide solution. (See Appendix 1.5.) Neglect any uncertainty in the mass.

Integrative Problems

These problems require the integration of multiple concepts to find the solutions.

142. Tris(pentafluorophenyl)borane, commonly known by its acronym BARF, is frequently used to initiate polymerization of ethylene or propylene in the presence of a catalytic transition metal compound. It is composed solely of C, F, and B; it is 42.23% C and 55.66% F by mass.

   a. What is the empirical formula of BARF?

   b. A 2.251-g sample of BARF dissolved in 347.0 mL of solution produces a 0.01267-M solution. What is the molecular formula of BARF?

143. In a 1-L beaker, 203 mL of 0.307 M ammonium chromate was mixed with 137 mL of 0.269 M chromium(III) nitrite to produce ammonium nitrite and chromium(III) chromate. Write the balanced chemical equation for the reaction occurring here. If the percent yield of the reaction was 88.0%, what mass of chromium(III) chromate was isolated?

144. The vanadium in a sample of ore is converted to VO^{2+}. The VO^{2+} ion is subsequently titrated with MnO_4^- in acidic solution to form $V(OH)_4^+$ and manganese(II) ion. The unbalanced titration reaction is

$$MnO_4^-(aq) + VO^{2+}(aq) + H_2O(l) \longrightarrow$$
$$V(OH)_4^+(aq) + Mn^{2+}(aq) + H^+(aq)$$

To titrate the solution, 26.45 mL of 0.02250 M MnO_4^- was required. If the mass percent of vanadium in the ore was 58.1%, what was the mass of the ore sample? *Hint:* Balance the titration reaction by the oxidation states method.

145. The unknown acid H_2X can be neutralized completely by OH^- according to the following (unbalanced) equation:

$$H_2X(aq) + OH^-(aq) \longrightarrow X^{2-}(aq) + H_2O(l)$$

The ion formed as a product, X^{2-}, was shown to have 36 total electrons. What is element X? Propose a name for H_2X. To completely neutralize a sample of H_2X, 35.6 mL of 0.175 M OH^- solution was required. What was the mass of the H_2X sample used?

Marathon Problems

These problems are designed to incorporate several concepts and techniques into one situation.

146. Three students were asked to find the identity of the metal in a particular sulfate salt. They dissolved a 0.1472-g sample of the salt in water and treated it with excess barium chloride, resulting in the precipitation of barium sulfate. After the precipitate had been filtered and dried, it weighed 0.2327 g.

   Each student analyzed the data independently and came to different conclusions. Pat decided that the metal was titanium. Chris thought it was sodium. Randy reported that it was gallium. What formula did each student assign to the sulfate salt?

   Look for information on the sulfates of gallium, sodium, and titanium in this text and reference books such as the *CRC Handbook of Chemistry and Physics*. What further tests would you suggest to determine which student is most likely correct?

147. You have two 500.0-mL aqueous solutions. Solution A is a solution of a metal nitrate that is 8.246% nitrogen by mass. The ionic compound in solution B consists of potassium, chromium, and oxygen; chromium has an oxidation state of +6 and there are 2 potassiums and 1 chromium in the formula. The masses of the solutes in each of the solutions are the same. When the solutions are added together, a blood-red precipitate forms. After the reaction has gone to completion, you dry the solid and find that it has a mass of 331.8 g.

   a. Identify the ionic compounds in solution A and solution B.

   b. Identify the blood-red precipitate.

   c. Calculate the concentration (molarity) of all ions in the original solutions.

   d. Calculate the concentration (molarity) of all ions in the final solution.

CHAPTER SEVEN

Chemical Energy

A burning match is an example of exothermic reaction. This double exposure shows the match lit and the match blown out. (Milos Jokic/Getty Images)

Energy is the essence of our very existence as individuals and as a society. The food that we eat furnishes the energy to live, work, and play, just as the coal and oil consumed by manufacturing and transportation systems power our modern industrialized civilization.

In the past, huge quantities of carbon-based fossil fuels have been available for the taking. This abundance of fuels has led to a world society with a voracious appetite for energy, consuming millions of barrels of petroleum every day. We are now dangerously dependent on the dwindling supplies of oil, and this dependence is an important source of tension among nations in today's world. In an incredibly short time we have moved from a period of ample, cheap supplies of petroleum to one of high-priced, uncertain supplies. If our present standard of living is to be maintained, we must find alternatives to petroleum. To do this, we need to know the relationship between chemistry and energy, which we explore in this chapter.

There are additional problems with fossil fuels. The waste products from burning fossil fuels significantly affect our environment. For example, when a carbon-based fuel is burned, the carbon reacts with oxygen to form carbon dioxide, which is released into the atmosphere. Although much of this carbon dioxide is consumed in various natural processes such as photosynthesis and the formation of carbonate materials, the amount of carbon dioxide in the atmosphere is steadily increasing. This increase is significant because atmospheric carbon dioxide absorbs heat radiated from the earth's surface and radiates it back toward the earth. Since this is an important mechanism for controlling the earth's temperature, many scientists fear that an increase in the concentration of carbon dioxide will warm the earth, causing significant changes in climate. In addition, impurities in the fossil fuels react with components of the air to produce air pollution.

Just as energy is important to our society on a macroscopic scale, it is critically important to each living organism on a microscopic scale. The living cell is a miniature chemical factory powered by energy from chemical reactions. The process of cellular respiration extracts the energy stored in sugars and other nutrients to drive the various tasks of the cell. Although the extraction process is more complex and more subtle, the energy obtained from "fuel" molecules by the cell is the same as would be obtained from burning the fuel to power an internal combustion engine.

The potential energy for a chemical reaction is the difference between the energy stored in the chemical bonds of the products and the chemical bonds of the reactants. An understanding of the nature of these chemical bonds (see Chapter 3) is important in understanding energy changes in chemical reactions.

Whether it is an engine or a cell that is converting energy from one form to another, the processes are all governed by the same principles, which we will begin to explore in this chapter. Additional aspects of energy transformation will be covered in Chapter 16.

7-1 | The Nature of Chemical Energy

Although the concept of energy is quite familiar, energy itself is rather difficult to define precisely. Recall from Section R-9 that we define **energy** as the *capacity to do work or to produce heat.* In this chapter we will concentrate specifically on the heat transfer that accompanies chemical processes.

One of the most important characteristics of energy is that it is conserved. The **law of conservation of energy** states that *energy can be converted from one form to*

ⓘ The total energy content of the universe is constant.

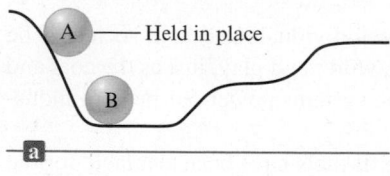

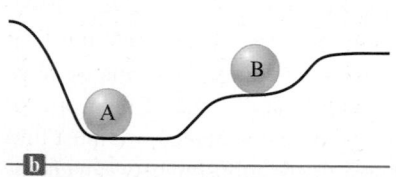

Figure 7-1 | (a) In the initial positions, ball A has a higher potential energy than ball B. (b) After A has rolled down the hill, the potential energy lost by A has been converted to random motions of the components of the hill (frictional heating) and to the increase in the potential energy of B.

ⓘ Heat involves a transfer of energy.

▲ This infrared photo of a house shows where energy leaks occur. The more red the color, the more energy (heat) is leaving the house.

another but can be neither created nor destroyed. That is, the energy of the universe is constant. ◀ ⓘ As we discussed in Section R-9, energy can be classified as either potential or kinetic energy. **Potential energy** is energy due to position or composition. For example, water behind a dam has potential energy that can be converted to work when the water flows down through turbines, thereby creating electricity. Attractive and repulsive forces also lead to potential energy. The energy released when gasoline is burned results from differences in attractive forces (bond energies) between the nuclei and electrons in the reactants and products. The **kinetic energy** of an object is energy due to the motion of the object and depends on the mass of the object m and its velocity v: $KE = \frac{1}{2}mv^2$.

Energy can be converted from one form to another. For example, consider the two balls in Fig. 7-1(a). Ball A, because of its higher position initially, has more potential energy than ball B. When A is released, it moves down the hill and strikes B. Eventually, the arrangement shown in Fig. 7-1(b) is achieved. What has happened in going from the initial to the final arrangement? The potential energy of A has decreased, but since energy is conserved, all the energy lost by A must be accounted for. How is this energy distributed?

Initially, the potential energy of A is changed to kinetic energy as the ball rolls down the hill. Part of this kinetic energy is then transferred to B, causing it to be raised to a higher final position. Thus the potential energy of B has been increased. However, since the final position of B is lower than the original position of A, some of the energy is still unaccounted for. Both balls in their final positions are at rest, so the missing energy cannot be due to their motions. What has happened to the remaining energy?

The answer lies in the interaction between the hill's surface and the ball. As ball A rolls down the hill, some of its kinetic energy is transferred to the surface of the hill as heat. This transfer of energy is called *frictional heating*. The temperature of the hill increases very slightly as the ball rolls down.

Before we proceed further, it is important to recognize that heat and temperature are decidedly different. As we will see in detail in Chapter 8, *temperature* is a property that reflects the random motions of the particles in a particular substance. **Heat**, on the other hand, involves the *transfer* of energy between two objects due to a temperature difference. ◀ ⓘ Heat is not a substance contained by an object, although we often talk of heat as if this were true. ◀

Note that in going from the initial to the final arrangements in Fig. 7-1, ball B gains potential energy because work was done by ball A on B. **Work** is defined as force acting over a distance. Work is required to raise B from its original position to its final one. Part of the original energy stored as potential energy in A has been transferred through work to B, thereby increasing B's potential energy. Thus there are two ways to transfer energy: through work and through heat.

In rolling to the bottom of the hill as shown as in Fig. 7-1, ball A will always lose the same amount of potential energy. However, the way that this energy transfer is divided between work and heat depends on the specific conditions—the **pathway**. For example, the surface of the hill might be so rough that the energy of A is expended completely through frictional heating; A is moving so slowly when it hits B that it cannot move B to the next level. In this case, no work is done. Regardless of the condition of the hill's surface, the *total energy* transferred will be constant. However, the amounts of heat and work will differ. Energy change is independent of the pathway; however, work and heat are both dependent on the pathway.

This brings us to a very important concept: the **state function** or **state property**. A state function refers to a property of the system that depends only on its *present state*. A state function (property) does not depend in any way on the system's past (or future). In other words, the value of a state function does not depend on how the system arrived at the present state; it depends only on the characteristics of the present state.

This leads to a very important characteristic of a state function: A change in this function (property) in going from one state to another state is independent of the particular pathway taken between the two states.

A nonscientific analogy that illustrates the difference between a state function and a nonstate function is elevation on the earth's surface and distance between two points. In traveling from Chicago (elevation 674 ft) to Denver (elevation 5280 ft), the change in elevation is always $5280 - 674 = 4606$ ft, regardless of the route taken between the two cities. The distance traveled, however, depends on how you make the trip. Thus elevation is a function that does not depend on the route (pathway), but distance is pathway dependent. Elevation is a state function and distance is not.

Of the functions considered in our present example, energy is a state function, but work and heat are not state functions. ◄ ⓘ

Chemical Energy

The ideas we have just illustrated using mechanical examples also apply to chemical systems. The combustion of methane, for example, is used to heat many homes in the United States:

$$CH_4(g) + 2O_2(g) \longrightarrow CO_2(g) + 2H_2O(g) + \text{energy (heat)}$$

To discuss this reaction, we divide the universe into two parts: the system and the surroundings. The **system** is the part of the universe on which we wish to focus attention; the **surroundings** include everything else in the universe. In this case we define the system as the reactants and products of the reaction. The surroundings consist of the reaction container (a furnace, for example), the room, and anything else other than the reactants and products.

When a reaction results in the evolution of heat, it is said to be **exothermic** (*exo-* is a prefix meaning "out of"); that is, energy flows *out of the system*. For example, in the combustion of methane, energy flows out of the system as heat. Reactions that absorb energy from the surroundings are said to be **endothermic**. When the heat flow is *into a system*, the process is endothermic. For example, the formation of nitric oxide from nitrogen and oxygen is endothermic:

$$N_2(g) + O_2(g) + \text{energy (heat)} \longrightarrow 2NO(g)$$

Where does the energy, released as heat, come from in an exothermic reaction? The answer lies in the difference in potential energies between the products and the reactants. Which has lower potential energy, the reactants or the products? We know that total energy is conserved and that energy flows from the system into the surroundings in an exothermic reaction. This means that *the energy gained by the surroundings must be equal to the energy lost by the system*. In the combustion of methane, the energy content of the system decreases, which means that 1 mole of CO_2 and 2 moles of H_2O molecules (the products) possess less potential energy than do 1 mole of CH_4 and 2 moles of O_2 molecules (the reactants). The heat flow into the surroundings results from a lowering of the potential energy of the reaction system. This always holds true. *In any exothermic reaction, some of the potential energy is being converted to thermal energy (random kinetic energy) via heat.*

The energy diagram for the combustion of methane is shown in Fig. 7-2, where $\Delta(PE)$ represents the *change* in potential energy that occurs in the chemical reaction with the bonds of the reactants. In other words, this quantity represents the difference between the energy required to break the bonds in the reactants and the energy released when the bonds in the products are formed. In an exothermic process, the bonds in the products are stronger (on average) than those of the reactants. That is, more energy is released by forming the new bonds in the products than is consumed to break

Figure 7-2 | The combustion of methane releases the quantity of energy $\Delta(PE)$ to the surroundings via heat flow. This is an exothermic process.

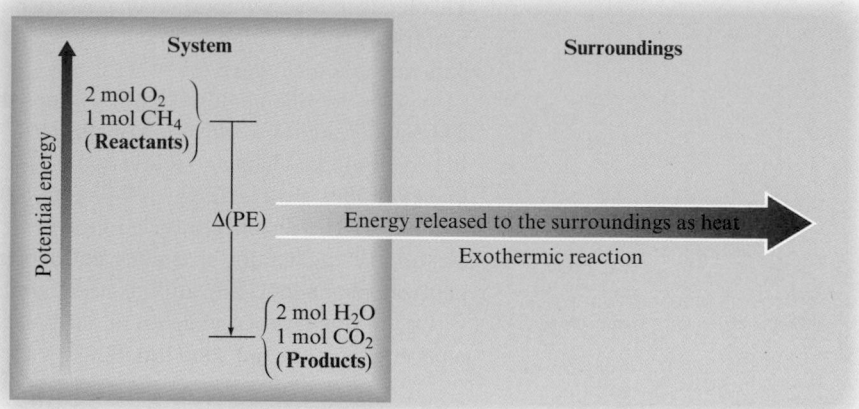

the bonds in the reactants. The net result is that the quantity of energy $\Delta(PE)$ is transferred to the surroundings through heat.

For an endothermic reaction, the situation is reversed, as shown in Fig. 7-3. Energy that flows into the system as heat is used to increase the potential energy of the system. In this case the products have higher potential energy (weaker bonds on average) than the reactants.

The study of energy and its interconversions is called **thermodynamics**. The law of conservation of energy is often called the **first law of thermodynamics** and is stated as follows: *The energy of the universe is constant.*

The **internal energy** E of a system can be defined most precisely as the sum of the kinetic and potential energies of all the "particles" in the system. The internal energy of a system can be changed by a flow of work, heat, or both. That is,

$$\Delta E = q + w$$

where ΔE represents the change in the system's internal energy, q represents heat, and w represents work.

Thermodynamic quantities always consist of two parts: a *number*, giving the magnitude of the change, and a *sign*, indicating the direction of the flow. *The sign reflects the system's point of view.* For example, if a quantity of energy flows *into* the system via heat (an endothermic process), q is equal to $+x$, where the *positive* sign indicates that the *system's energy is increasing*. On the other hand, when energy flows *out of* the

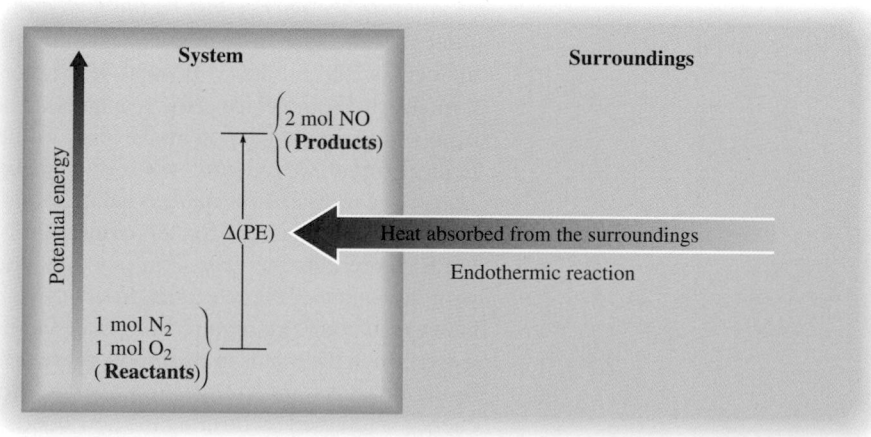

Figure 7-3 | The energy diagram for the reaction of nitrogen and oxygen to form nitric oxide. This is an endothermic process: Heat [equal in magnitude to $\Delta(PE)$] flows into the system from the surroundings.

system via heat (an exothermic process), q is equal to $-x$, where the *negative* sign indicates that the *system's energy is decreasing.*

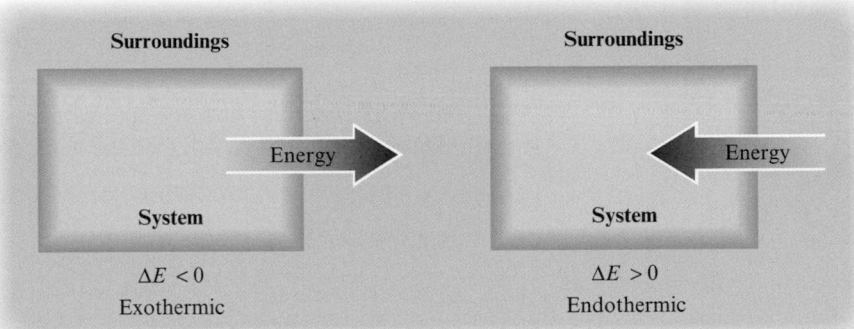

In this text the same conventions also apply to the flow of work. If the system does work on the surroundings (energy flows out of the system), w is negative. If the surroundings do work on the system (energy flows into the system), w is positive. We define work from the system's point of view to be consistent for all thermodynamic quantities. That is, in this convention the signs of both q and w reflect what happens to the system; thus we use $\Delta E = q + w$.

In this text we *always* take the system's point of view. ◄ ⓘ This convention is not followed in every area of science. For example, engineers are in the business of designing machines to do work, that is, to make the system (the machine) transfer energy to its surroundings through work. Consequently, engineers define work from the surroundings' point of view. In their convention, work that flows out of the system is treated as positive because the energy of the surroundings has increased. The first law of thermodynamics is then written $\Delta E = q - w'$, where w' signifies work from the surroundings' point of view.

ⓘ The convention in this text is to take the system's point of view; $q = -x$ denotes an exothermic process, and $q = +x$ denotes an endothermic one.

INTERACTIVE EXAMPLE 7-1

ⓘ The joule (J) is the fundamental SI unit for energy:

$$J = \frac{kg \cdot m^2}{s^2}$$

One kilojoule (kJ) = 10^3 J.

Internal Energy

Calculate ΔE for a system undergoing an endothermic process in which 15.6 kJ of heat flows and where 1.4 kJ of work is done on the system. ◄ ⓘ

Solution

We use the equation

$$\Delta E = q + w$$

where $q = +15.6$ kJ, since the process is endothermic, and $w = +1.4$ kJ, since work is done on the system. Thus

$$\Delta E = 15.6 \text{ kJ} + 1.4 \text{ kJ} = 17.0 \text{ kJ}$$

The system has gained 17.0 kJ of energy.

See Exercises 7-29 through 7-32

ⓘ We will discuss the properties of gases in detail in Chapter 8.

A common type of work associated with chemical processes is work done by a gas (through *expansion*) or work done to a gas (through *compression*). ◄ ⓘ For example, in an automobile engine, the heat from the combustion of the gasoline expands the gases in the cylinder to push back the piston, and this motion is then translated into the motion of the car.

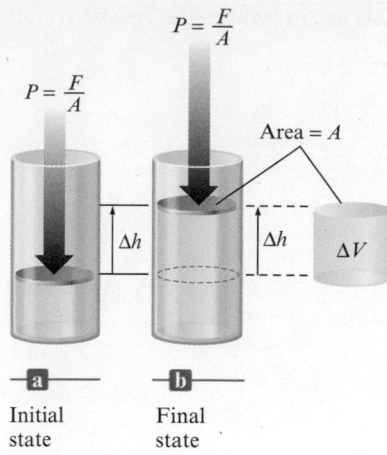

$$P = \frac{F}{A}$$

$$P = \frac{F}{A}$$

Area = A

Δh Δh ΔV

a   **b**

Initial   Final
state   state

Figure 7-4 | (a) The piston, moving a distance Δh against a pressure P, does work on the surroundings. (b) Since the volume of a cylinder is the area of the base times its height, the change in volume of the gas is given by $\Delta h \times A = \Delta V$.

Suppose we have a gas confined to a cylindrical container with a movable piston as shown in Fig. 7-4, where F is the force acting on a piston of area A. Pressure is defined as force per unit area; thus the pressure of the gas is

$$P = \frac{F}{A}$$

Work is defined as force applied over a distance, so if the piston moves a distance Δh, as shown in Fig. 7-4, then the work done is

$$\text{Work} = \text{force} \times \text{distance} = F \times \Delta h$$

Since $P = F/A$ or $F = P \times A$, then

$$\text{Work} = F \times \Delta h = P \times A \times \Delta h$$

Since the volume of a cylinder equals the area of the piston times the height of the cylinder (see Fig. 7-4), the change in volume ΔV resulting from the piston moving a distance Δh is

$$\Delta V = \text{final volume} - \text{initial volume} = A \times \Delta h$$

Substituting $\Delta V = A \times \Delta h$ into the expression for work gives

$$\text{Work} = P \times A \times \Delta h = P\Delta V$$

This gives us the *magnitude* (size) of the work required to expand a gas ΔV against a pressure P.

What about the sign of the work? The gas (the system) is expanding, moving the piston against the pressure. Thus the system is doing work on the surroundings, so from the system's point of view the sign of the work should be negative.

For an *expanding* gas, ΔV is a positive quantity because the volume is increasing. Thus ΔV and w must have opposite signs, which leads to the equation ◄ ⓘ

$$w = -P\Delta V$$

ⓘ *w and $P\Delta V$ have opposite signs because when the gas expands (ΔV is positive), work flows into the surroundings (w is negative).*

Note that for a gas expanding against an external pressure P, w is a negative quantity as required, since work flows out of the system. When a gas is *compressed*, ΔV is a negative quantity (the volume decreases), which makes w a positive quantity (work flows into the system).

Historically, instruments for measuring pressure are based on the height of a mercury column (in mm) that the gas pressure can support. The most commonly used unit for pressure in chemistry is the atmosphere, which is defined as

$$1 \text{ standard atmosphere} = 1 \text{ atm} = 760 \text{ mm Hg}$$

Critical Thinking

You are calculating ΔE in a chemistry problem. What if you confuse the system and the surroundings? How would this affect the magnitude of the answer you calculate? The sign?

INTERACTIVE EXAMPLE 7-2

PV Work

Calculate the work associated with the expansion of a gas from 46 L to 64 L at a constant external pressure of 15 atm.

Solution

For a gas at constant pressure,

$$w = -P\Delta V$$

In this case $P = 15$ atm and $\Delta V = 64 - 46 = 18$ L. Hence

$$w = -15 \text{ atm} \times 18 \text{ L} = -270 \text{ L} \cdot \text{atm}$$

Note that since the gas expands, it does work on its surroundings.

REALITY CHECK | Energy flows out of the gas, so w is a negative quantity.

See Exercises 7-35 through 7-37

In dealing with "PV work," keep in mind that the P in $P\Delta V$ always refers to the external pressure—the pressure that causes a compression or that resists an expansion.

INTERACTIVE EXAMPLE 7-3

▲ A propane burner is used to heat the air in a hot-air balloon.

Carlos Caetano/Shutterstock

Internal Energy, Heat, and Work

A balloon is being inflated to its full extent by heating the air inside it. ◀ In the final stages of this process, the volume of the balloon changes from 4.00×10^6 L to 4.50×10^6 L by the addition of 1.3×10^8 J of energy as heat. Assuming that the balloon expands against a constant pressure of 1.0 atm, calculate ΔE for the process. (To convert between L · atm and J, use 1 L · atm $= 101.3$ J.)

Solution

To calculate ΔE

We know the following:

$$V_1 = 4.00 \times 10^6 \text{ L}$$
$$q = +1.3 \times 10^8 \text{ J}$$
$$P = 1.0 \text{ atm}$$
$$1 \text{ L} \cdot \text{atm} = 101.3 \text{ J}$$
$$V_2 = 4.50 \times 10^6 \text{ L}$$

First we need to calculate the work done on the gas.

$$w = -P\Delta V$$

What is ΔV?

$$\Delta V = V_2 - V_1 = 4.50 \times 10^6 \text{ L} - 4.00 \times 10^6 \text{ L} = 5.0 \times 10^5 \text{ L}$$

What is the work?

$$w = -P\Delta V = -1.0 \text{ atm} \times 5.0 \times 10^5 \text{ L} = -5.0 \times 10^5 \text{ L} \cdot \text{atm}$$

The negative sign makes sense because the gas is expanding and doing work on the surroundings. To calculate ΔE, we use the following equation:

$$\Delta E = q + w$$

However, since q is given in units of J and w is given in units of L · atm, we must change the work to units of joules:

$$w = -5.0 \times 10^5 \text{ L} \cdot \text{atm} \times \frac{101.3 \text{ J}}{\text{L} \cdot \text{atm}} = -5.1 \times 10^7 \text{ J}$$

Then

$$\Delta E = q + w = (+1.3 \times 10^8 \text{ J}) + (-5.1 \times 10^7 \text{ J}) = 8 \times 10^7 \text{ J}$$

Since more energy is added through heating than the gas expends doing work, there is a net increase in the internal energy of the gas in the balloon. Hence ΔE is positive.

See Exercises 7-38 through 7-40

7-2 | Enthalpy

So far we have discussed the internal energy of a system. A less familiar property of a system is its **enthalpy**, H, which is defined as

$$H = E + PV$$

where E is the internal energy of the system, P is the pressure of the system, and V is the volume of the system.

Since internal energy, pressure, and volume are all state functions, *enthalpy is also a state function.* ◄ ⓘ But what exactly is enthalpy? To help answer this question, consider a process carried out at constant pressure and where the only work allowed is pressure–volume work ($w = -P\Delta V$). ◄ ⓘ Under these conditions, the expression

$$\Delta E = q_P + w$$

becomes

$$\Delta E = q_P - P\Delta V$$

or

$$q_P = \Delta E + P\Delta V$$

where q_P is the heat at constant pressure.

We will now relate q_P to a change in enthalpy. The definition of enthalpy is $H = E + PV$. Therefore, we can say

$$\text{Change in } H = (\text{change in } E) + (\text{change in } PV)$$

or

$$\Delta H = \Delta E + \Delta(PV)$$

Since P is constant, the change in PV is due only to a change in volume. Thus

$$\Delta(PV) = P\Delta V$$

and

$$\Delta H = \Delta E + P\Delta V$$

This expression is identical to the one we obtained for q_P:

$$q_P = \Delta E + P\Delta V$$

Thus, for a process carried out at constant pressure and where the only work allowed is that from a volume change, we have

$$\Delta H = q_P$$

At constant pressure (where only PV work is allowed), the change in enthalpy, ΔH, of the system is equal to the energy flow as heat. ◄ ⓘ This means that for a reaction studied at constant pressure, the flow of heat is a measure of the change in enthalpy for the system. For this reason, the terms *heat of reaction* and *change in enthalpy* are used interchangeably for reactions studied at constant pressure. ◄ ⓘ

For a chemical reaction, the enthalpy change is given by the equation

$$\Delta H = H_{\text{products}} - H_{\text{reactants}}$$

In a case in which the products of a reaction have a greater enthalpy than the reactants, ΔH will be positive. Thus heat will be absorbed by the system, and the reaction is endothermic. On the other hand, if the enthalpy of the products is less than that of the reactants, ΔH will be negative. ◄ ⓘ In this case the overall decrease in enthalpy is achieved by the generation of heat, and the reaction is exothermic.

ⓘ Enthalpy is a state function. A change in enthalpy does not depend on the pathway between two states.

ⓘ Recall from Section 7-1 that w and $P\Delta V$ have opposite signs:

$$w = -P\Delta V$$

ⓘ $\Delta H = q$ only at constant pressure.

ⓘ The change in enthalpy of a system has no easily interpreted meaning except at constant pressure, where $\Delta H =$ heat.

ⓘ At constant pressure, exothermic means ΔH is negative; endothermic means ΔH is positive.

INTERACTIVE EXAMPLE 7-4

Enthalpy

When 1 mole of methane (CH_4) is burned at constant pressure, 890 kJ of energy is released as heat. Calculate ΔH for a process in which a 5.8-g sample of methane is burned at constant pressure.

Solution

Where are we going?
To calculate ΔH

What do we know?

- $q_P = \Delta H = -890$ kJ/mol CH_4
- Mass = 5.8 g CH_4
- Molar mass CH_4 = 16.04 g ◄ ⓘ

i Molar mass was introduced in Chapter 5.

How do we get there?
What are the moles of CH_4 burned?

$$5.8 \text{ g } CH_4 \times \frac{1 \text{ mol } CH_4}{16.04 \text{ g } CH_4} = 0.36 \text{ mol } CH_4$$

What is ΔH?

$$\Delta H = 0.36 \text{ mol } CH_4 \times \frac{-890 \text{ kJ}}{\text{mol } CH_4} = -320 \text{ kJ}$$

Thus, when a 5.8-g sample of CH_4 is burned at constant pressure,

$$\Delta H = \text{heat flow} = -320 \text{ kJ}$$

REALITY CHECK | In this case, a 5.8-g sample of CH4 is burned. Since this amount is smaller than 1 mole, less than 890 kJ will be released as heat.

See Exercises 7-45 through 7-48

7-3 | Calorimetry

In Chapters 5 and 6 we have considered the stoichiometry of various types of reactions. Another important property of reactions is the heat generated when a reaction occurs. We already considered this topic in Chapter 6, but at that point we did not show how the heat of a reaction is actually measured. In this section we will consider how to measure the heat produced by a reaction that occurs in solution.

The device used experimentally to determine the heat associated with a chemical reaction is called a **calorimeter**. **Calorimetry**, the science of measuring heat, is based on observing the temperature change when a body absorbs or discharges energy as heat. Substances respond differently to being heated. One substance might require a great deal of heat energy to raise its temperature by one degree, whereas another will exhibit the same temperature change after absorbing relatively little heat. The **heat capacity** C of a substance, which is a measure of this property, is defined as

$$C = \frac{\text{heat absorbed}}{\text{increase in temperature}}$$

When an element or a compound is heated, the energy required will depend on the amount of the substance present (for example, it takes twice as much energy to raise the temperature of two grams of water by one degree than it takes to raise the

Specific heat capacity: the energy required to raise the temperature of one gram of a substance by one degree Celsius.

Molar heat capacity: the energy required to raise the temperature of one mole of a substance by one degree Celsius.

Table 7-1 | The Specific Heat Capacities of Some Common Substances

Substance	Specific Heat Capacity (J/°C · g)
$H_2O(l)$	4.18
$H_2O(s)$	2.03
$Al(s)$	0.89
$Fe(s)$	0.45
$Hg(l)$	0.14
$C(s)$	0.71

If two reactants at the same temperature are mixed and the resulting solution gets warmer, this means the reaction taking place is exothermic. An endothermic reaction cools the solution.

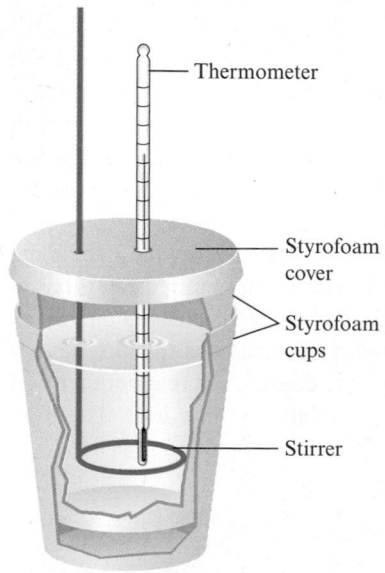

Figure 7-5 | A coffee-cup calorimeter made of two Styrofoam cups.

— Thermometer

— Styrofoam cover

— Styrofoam cups

— Stirrer

temperature of one gram of water by one degree). Thus, in defining the heat capacity of a substance, the amount of substance must be specified. If the heat capacity is given *per gram* of substance, it is called the **specific heat capacity**, and its units are J/°C · g or J/K · g. If the heat capacity is given *per mole* of the substance, it is called the **molar heat capacity**, and it has the units J/°C · mol or J/K · mol. The specific heat capacities of some common substances are given in Table 7-1. Note from this table that the heat capacities of metals are very different from that of water. It takes much less energy to change the temperature of a gram of a metal by 1°C than for a gram of water.

Although the calorimeters used for highly accurate work are precision instruments, a very simple calorimeter can be used to examine the fundamentals of calorimetry. All we need are two nested Styrofoam cups with a cover through which a stirrer and thermometer can be inserted, as shown in Fig. 7-5. This device is called a "coffee-cup calorimeter." The outer cup is used to provide extra insulation. The inner cup holds the solution in which the reaction occurs.

The measurement of heat using a simple calorimeter such as that shown in Fig. 7-5 is an example of **constant-pressure calorimetry**, since the pressure (atmospheric pressure) remains constant during the process. Constant-pressure calorimetry is used in determining the changes in enthalpy (heats of reactions) for reactions occurring in solution. Under these conditions, the change in enthalpy equals the heat; that is, $\Delta H = q_P$.

For example, suppose we mix 50.0 mL of 1.0 *M* HCl at 25.0°C with 50.0 mL of 1.0 *M* NaOH, also at 25°C, in a calorimeter. After the reactants are mixed by stirring, the temperature is observed to increase to 31.9°C. As we saw in Section 6-6, the net ionic equation for this reaction is

$$H^+(aq) + OH^-(aq) \longrightarrow H_2O(l)$$

When these reactants (each originally at the same temperature) are mixed, the temperature of the mixed solution is observed to increase. Therefore, the chemical reaction must be releasing energy as heat. This released energy increases the random motions of the solution components, which in turn increases the temperature. The quantity of energy released can be determined from the temperature increase, the mass of solution, and the specific heat capacity of the solution. For an approximate result, we will assume that the calorimeter does not absorb or leak any heat and that the solution can be treated as if it were pure water with a density of 1.0 g/mL.

We also need to know the heat required to raise the temperature of a given amount of water by 1°C. Table 7-1 lists the specific heat capacity of water as 4.18 J/°C · g. This means that 4.18 J of energy is required to raise the temperature of 1 g of water by 1°C.

From these assumptions and definitions, we can calculate the heat (change in enthalpy) for the neutralization reaction:

Energy (as heat) released by the reaction

= energy (as heat) absorbed by the solution

= specific heat capacity × mass of solution × increase in temperature

= $s \times m \times \Delta T$

In this case the increase in temperature $(\Delta T) = 31.9°C - 25.0°C = 6.9°C$, and the mass of solution $(m) = 100.0 \text{ mL} \times 1.0 \text{ g/mL} = 1.0 \times 10^2 \text{ g}$. Thus

Energy released (as heat) = $s \times m \times \Delta T$

$$= \left(4.18 \frac{J}{°C \cdot g}\right)(1.0 \times 10^2 \, g)(6.9°C)$$

$$= 2.9 \times 10^3 \text{ J}$$

The voodoo lily is a beautiful, seductive—and foul-smelling—plant. The exotic-looking lily features an elaborate reproductive mechanism—a purple spike that can reach nearly 3 feet in length and is cloaked by a hoodlike leaf. But approaching the plant reveals bad news—it smells terrible!

Despite its antisocial odor, this putrid plant has fascinated biologists for many years because of its ability to generate heat. At the peak of its metabolic activity, the plant's blossom can be as much as 15°C above its ambient temperature. To generate this much heat, the metabolic rate of the plant must be close to that of a flying hummingbird!

What's the purpose of this intense heat production? For a plant faced with limited food supplies in the very competitive tropical climate where it grows, heat production seems like a great waste of energy. The answer to this mystery is that the voodoo lily is pollinated mainly by carrion-loving insects. Thus the lily prepares a malodorous mixture of chemicals characteristic of rotting meat, which it then "cooks" off into the surrounding air to attract flesh-feeding beetles and flies. Then, once the insects enter the pollination chamber, the high temperatures there (as high as 110°F) cause the insects to remain very active to better carry out their pollination duties.

The voodoo lily is only one of many such thermogenic (heat-producing) plants. Another interesting example is the eastern skunk cabbage, which produces enough heat to bloom inside of a snow bank by creating its own ice caves. These plants are of special interest to biologists because they provide opportunities to study metabolic reactions that are quite subtle in "normal" plants. For example, recent studies have shown that salicylic acid, the active form of aspirin, is probably very important in producing the metabolic bursts in thermogenic plants.

Besides studying the dramatic heat effects in thermogenic plants, biologists are also interested in calorimetric studies of regular plants. For example, very precise calorimeters have been designed that can be used to study the heat produced, and thus the metabolic activities, of clumps of cells no larger than a bread crumb. Several scientists have suggested that a single calorimetric measurement taking just a few minutes on a tiny plant might be useful in predicting the growth rate of the mature plant throughout its lifetime. If true, this would provide a very efficient method for selecting the plants most likely to thrive as adults.

Because the study of the heat production by plants is an excellent way to learn about plant metabolism, this continues to be a "hot" area of research.

The voodoo lily attracts pollinating insects with its foul odor.

How much energy would have been released if twice these amounts of solutions had been mixed? The answer is that twice as much heat would have been produced. The heat of a reaction is an *extensive property*; it depends directly on the amount of substance, in this case on the amounts of reactants. In contrast, an *intensive property* is not related to the amount of a substance. For example, temperature is an intensive property.

Enthalpies of reaction are often expressed in terms of moles of reacting substances. The number of moles of H^+ ions consumed in the preceding experiment is

$$50.0 \text{ mL} \times \frac{1 \text{ L}}{1000 \text{ mL}} \times \frac{1.0 \text{ mol } H^+}{L} = 5.0 \times 10^{-2} \text{ mol } H^+$$

Thus 2.9×10^3 J heat was released when 5.0×10^{-2} moles of H^+ ions reacted, or

$$\frac{2.9 \times 10^3 \text{ J}}{5.0 \times 10^{-2} \text{ mol } H^+} = 5.8 \times 10^4 \text{ J/mol}$$

of heat released per 1.0 mole of H^+ ions neutralized. Thus the *magnitude* of the enthalpy change per mole for the reaction

$$H^+(aq) + OH^-(aq) \longrightarrow H_2O(l)$$

ⓘ Notice that in this example we mentally keep track of the direction of the energy flow and assign the correct sign at the end of the calculation.

is 58 kJ/mol. Since heat is *evolved*, $\Delta H = -58$ kJ/mol. ◄ ⓘ

To summarize, we have found that when an object changes temperature, the heat can be calculated from the equation

$$q = s \times m \times \Delta T$$

where s is the specific heat capacity, m is the mass, and ΔT is the change in temperature. Note that when ΔT is positive, q also has a positive sign. The object has absorbed heat, so its temperature increases.

INTERACTIVE EXAMPLE 7-5

Constant-Pressure Calorimetry

When 1.00 L of 1.00 M $Ba(NO_3)_2$ solution at 25.0°C is mixed with 1.00 L of 1.00 M Na_2SO_4 solution at 25.0°C in a calorimeter, the white solid $BaSO_4$ forms, and the temperature of the mixture increases to 28.1°C. Assuming that the calorimeter absorbs only a negligible quantity of heat, that the specific heat capacity of the solution is 4.18 J/°C · g, and that the density of the final solution is 1.0 g/mL, calculate the enthalpy change per mole of $BaSO_4$ formed.

Solution

Where are we going?
To calculate ΔH per mole of $BaSO_4$ formed

What do we know?

- 1.00 L of 1.00 M $Ba(NO_3)_2$
- 1.00 L of 1.00 M Na_2SO_4
- $T_{initial} = 25.0°C$
- $T_{final} = 28.1°C$
- Heat capacity of solution = 4.18 J/°C · g
- Density of final solution = 1.0 g/mL

What do we need?

- Net ionic equation for the reaction

The ions present before any reaction occurs are Ba^{2+}, NO_3^-, Na^+, and SO_4^{2-}. The Na^+ and NO_3^- ions are spectator ions, since $NaNO_3$ is very soluble in water and will not precipitate under these conditions. The net ionic equation for the reaction is therefore

$$Ba^{2+}(aq) + SO_4^{2-}(aq) \longrightarrow BaSO_4(s)$$

How do we get there?
What is ΔH?
Since the temperature increases, formation of solid $BaSO_4$ must be exothermic; ΔH is negative.

Heat evolved by the reaction

= heat absorbed by the solution

= specific heat capacity × mass of solution × increase in temperature

For millennia people have been amazed at the ability of Eastern mystics to walk across beds of glowing coals without any apparent discomfort. Even in the United States, thousands of people have performed feats of firewalking as part of motivational seminars. How is this possible? Do firewalkers have supernatural powers?

Actually, there are good scientific explanations, based on the concepts covered in this chapter, of why firewalking is possible. The first important factor concerns the heat capacity of feet. Because human tissue is mainly composed of water, it has a relatively large specific heat capacity. This means that a large amount of energy must be transferred from the coals to significantly change the temperature of the feet. During the brief contact between feet and coals, there is relatively little time for energy flow so the feet do not reach a high enough temperature to cause damage.

Second, although the surface of the coals has a very high temperature, the red hot layer is very thin. Therefore, the quantity of energy available to heat the feet is smaller than might be expected. This factor points to the difference between temperature and heat. Temperature reflects the *intensity* of the random kinetic energy in a given sample of matter. The amount of energy available for heat flow, on the other hand, depends on the quantity of matter at a given temperature—10 grams of matter at a given temperature contains 10 times as much thermal energy as 1 gram of the same matter. This is why the tiny spark from a sparkler does not hurt when it hits your hand. The spark has a very high temperature but has so little mass that no significant energy transfer occurs to your hand. This same argument applies to the very thin hot layer on the coals.

Thus, although firewalking is an impressive feat, there are several sound scientific reasons why it is possible (with the proper training and a properly prepared bed of coals).

Firewalkers in Thailand

What is the mass of the final solution?

$$\text{mass of solution} = 2.00\ \text{L} \times \frac{1000\ \text{mL}}{1\ \text{L}} \times \frac{1.0\ \text{g}}{\text{mL}} = 2.0 \times 10^3\ \text{g}$$

What is the temperature increase?

$$\Delta T = T_{\text{final}} - T_{\text{initial}} = 28.1°\text{C} - 25.0°\text{C} = 3.1°\text{C}$$

How much heat is evolved by the reaction?

$$\text{Heat evolved} = (4.18\ \text{J/}°\text{C} \cdot \text{g})(2.0 \times 10^3\ \text{g})(3.1°\text{C}) = 2.6 \times 10^4\ \text{J}$$

Thus

$$q = q_P = \Delta H = -2.6 \times 10^4\ \text{J}$$

What is ΔH per mole of $BaSO_4$ formed?

Since 1.0 L of 1.0 M $Ba(NO_3)_2$ contains 1 mole of Ba^{2+} ions and 1.0 L of 1.0 M Na_2SO_4 contains 1.0 mole of SO_4^{2-} ions, 1.0 mole of solid $BaSO_4$ is formed in this experiment. Thus the enthalpy change per mole of $BaSO_4$ formed is

$$\Delta H = -2.6 \times 10^4\ \text{J/mol} = -26\ \text{kJ/mol}$$

See Exercises 7-59 through 7-62

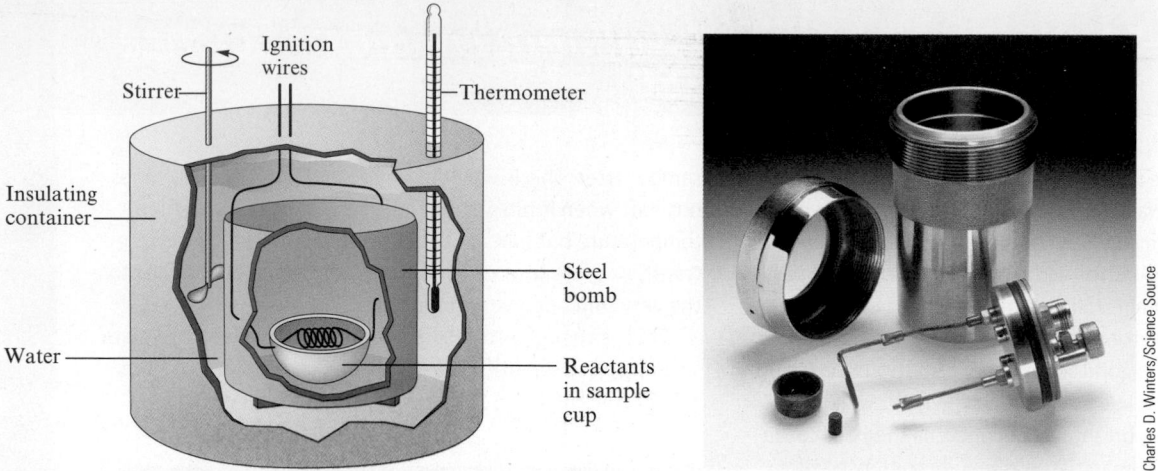

Charles D. Winters/Science Source

Figure 7-6 │ A bomb calorimeter. The reaction is carried out inside a rigid steel "bomb" (photo of actual disassembled "bomb" shown on right), and the heat evolved is absorbed by the surrounding water and other calorimeter parts. The quantity of energy produced by the reaction can be calculated from the temperature increase.

Constant-volume calorimetry experiments can also be performed. For example, when a photographic flashbulb flashes, the bulb becomes very hot, because the reaction of the zirconium or magnesium wire with the oxygen inside the bulb is exothermic. The reaction occurs inside the flashbulb, which is rigid and does not change volume. Under these conditions, no work is done (because the volume must change for pressure–volume work to be performed). To study the energy changes in reactions under conditions of constant volume, a "bomb calorimeter" (Fig. 7-6) is used. Weighed reactants are placed inside a rigid steel container (the "bomb") and ignited. The energy change is determined by measuring the increase in the temperature of the water and other calorimeter parts. For a constant-volume process, the change in volume ΔV is equal to zero, so work (which is $-P\Delta V$) is also equal to zero. Therefore,

$$\Delta E = q + w = q = q_V \qquad \text{(constant volume)}$$

Suppose we wish to measure the energy of combustion of octane (C_8H_{18}), a component of gasoline. A 0.5269-g sample of octane is placed in a bomb calorimeter known to have a heat capacity of 11.3 kJ/°C. This means that 11.3 kJ of energy is required to raise the temperature of the water and other parts of the calorimeter by 1°C. The octane is ignited in the presence of excess oxygen, and the temperature increase of the calorimeter is 2.25°C. The amount of energy released is calculated as follows:

Energy released by the reaction

= temperature increase × energy required to change the temperature by 1°C

= ΔT × heat capacity of calorimeter

= 2.25°$\cancel{C}$ × 11.3 kJ/°$\cancel{C}$ = 25.4 kJ

This means that 25.4 kJ of energy was released by the combustion of 0.5269 g octane. The number of moles of octane is

$$0.5269 \ \cancel{\text{g octane}} \times \frac{1 \ \text{mol octane}}{114.2 \ \cancel{\text{g octane}}} = 4.614 \times 10^{-3} \ \text{mol octane}$$

Since 25.4 kJ of energy was released for 4.614×10^{-3} mole of octane, the energy released per mole is

$$\frac{25.4 \ \text{kJ}}{4.614 \times 10^{-3} \ \text{mol}} = 5.50 \times 10^3 \ \text{kJ/mol}$$

Since the reaction is exothermic, ΔE is negative:

$$\Delta E_{combustion} = -5.50 \times 10^3 \text{ kJ/mol}$$

Note that since no work is done in this case, ΔE is equal to the heat.

$$\Delta E = q + w = q \qquad \text{since } w = 0$$

Thus $q = -5.50 \times 10^3$ kJ/mol.

EXAMPLE 7-6

Constant-Volume Calorimetry

It has been suggested that hydrogen gas obtained by the decomposition of water might be a substitute for natural gas (principally methane). To compare the energies of combustion of these fuels, the following experiment was carried out using a bomb calorimeter with a heat capacity of 11.3 kJ/°C. When a 1.50-g sample of methane gas was burned with excess oxygen in the calorimeter, the temperature increased by 7.3°C. When a 1.15-g sample of hydrogen gas was burned with excess oxygen, the temperature increase was 14.3°C. Compare the energies of combustion (per gram) for hydrogen and methane.

Solution

Where are we going?
To calculate ΔH of combustion per gram for H_2 and CH_4

What do we know?

- 1.50 g $CH_4 \Rightarrow \Delta T = 7.3$°C
- 1.15 g $H_2 \Rightarrow \Delta T = 14.3$°C
- Heat capacity of calorimeter = 11.3 kJ/°C

What do we need?

- $\Delta E = \Delta T \times$ heat capacity of calorimeter

How do we get there?
What is the energy released for each combustion?
For CH_4, we calculate the energy of combustion for methane using the heat capacity of the calorimeter (11.3 kJ/°C) and the observed temperature increase of 7.3°C:

$$\text{Energy } \textit{released} \text{ in the combustion of 1.5 g } CH_4 = (11.3 \text{ kJ/°C})(7.3°C)$$
$$= 83 \text{ kJ}$$

$$\text{Energy } \textit{released} \text{ in the combustion of 1 g } CH_4 = \frac{83 \text{ kJ}}{1.5 \text{ g}} = 55 \text{ kJ/g}$$

For H_2,

$$\text{Energy } \textit{released} \text{ in the combustion of 1.15 g } H_2 = (11.3 \text{ kJ/°C})(14.3°C)$$
$$= 162 \text{ kJ}$$

$$\text{Energy } \textit{released} \text{ in the combustion of 1 g } H_2 = \frac{162 \text{ kJ}}{1.15 \text{ g}} = 141 \text{ kJ/g} \blacktriangleleft \text{ⓘ}$$

How do the energies of combustion compare?

- The energy released in the combustion of 1 g hydrogen is approximately 2.5 times that for 1 g methane, indicating that hydrogen gas is a potentially useful fuel.

ⓘ The direction of energy flow is indicated by words in this example. Using signs to designate the direction of energy flow:

$$\Delta E_{combustion} = -55 \text{ kJ/g}$$

for methane and

$$\Delta E_{combustion} = -141 \text{ kJ/g}$$

for hydrogen.

See Exercises 7-65 and 7-66

7-4 | Hess's Law

ⓘ ΔH is not dependent on the reaction pathway.

Since enthalpy is a state function, the change in enthalpy in going from some initial state to some final state is independent of the pathway. This means that *in going from a particular set of reactants to a particular set of products, the change in enthalpy is the same whether the reaction takes place in one step or in a series of steps.* ◀ ⓘ This principle is known as **Hess's law** and can be illustrated by examining the oxidation of nitrogen to produce nitrogen dioxide. The overall balanced equation for this reaction can be written in one step, where the enthalpy change is represented by ΔH_1.

$$N_2(g) + 2O_2(g) \longrightarrow 2NO_2(g) \quad \Delta H_1 = 68 \text{ kJ}$$

This reaction also can be carried out in two distinct steps, with enthalpy changes designated by ΔH_2 and ΔH_3:

$$
\begin{aligned}
N_2(g) + O_2(g) &\longrightarrow 2NO(g) & \Delta H_2 &= 180 \text{ kJ} \\
2NO(g) + O_2(g) &\longrightarrow 2NO_2(g) & \Delta H_3 &= -112 \text{ kJ}
\end{aligned}
$$

Net reaction: $N_2(g) + 2O_2(g) \longrightarrow 2NO_2(g)$ $\Delta H_2 + \Delta H_3 = 68 \text{ kJ}$

Note that the sum of the two steps gives the net, or overall, reaction and that

$$\Delta H_1 = \Delta H_2 + \Delta H_3 = 68 \text{ kJ}$$

The principle of Hess's law is shown schematically in Fig. 7-7.

Characteristics of Enthalpy Changes

To use Hess's law to compute enthalpy changes for reactions, it is important to understand two characteristics of ΔH for a reaction:

Characteristics of ΔH for a Reaction

- If a reaction is reversed, the sign of ΔH is also reversed. ◀ ⓘ
- The magnitude of ΔH is directly proportional to the quantities of reactants and products in a reaction. If the coefficients in a balanced reaction are multiplied by an integer, the value of ΔH is multiplied by the same integer.

ⓘ Reversing the direction of a reaction changes the sign of ΔH.

Both these rules follow in a straightforward way from the properties of enthalpy changes. The first rule can be explained by recalling that the *sign* of ΔH indicates the *direction* of the heat flow at constant pressure. If the direction of the reaction is reversed, the direction of the heat flow also will be reversed. To see this, consider the

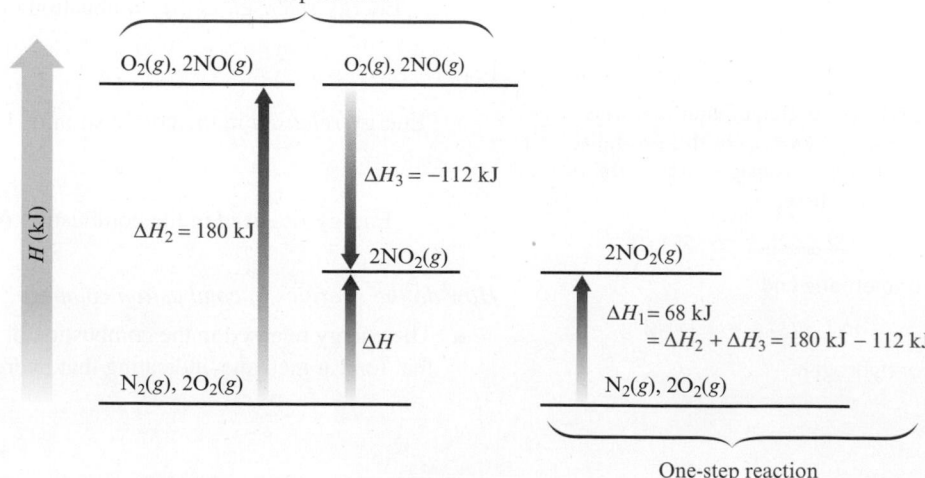

Figure 7-7 | The principle of Hess's law. The same change in enthalpy occurs when nitrogen and oxygen react to form nitrogen dioxide, regardless of whether the reaction occurs in one (red) or two (blue) steps.

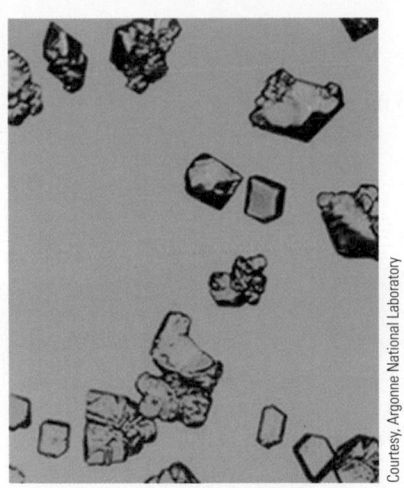

▲ Crystals of xenon tetrafluoride, the first reported binary compound containing a noble gas element.

preparation of xenon tetrafluoride, which was the first binary compound made from a noble gas:

$$Xe(g) + 2F_2(g) \longrightarrow XeF_4(s) \quad \Delta H = -251 \text{ kJ}$$

This reaction is exothermic, and 251 kJ of energy flows into the surroundings as heat. On the other hand, if the colorless XeF_4 crystals are decomposed into the elements, according to the equation ◄

$$XeF_4(s) \longrightarrow Xe(g) + 2F_2(g)$$

the opposite energy flow occurs because 251 kJ of energy must be added to the system to produce this endothermic reaction. Thus, for this reaction, $\Delta H = +251$ kJ.

The second rule comes from the fact that ΔH is an extensive property, depending on the amount of substances reacting. For example, since 251 kJ of energy is evolved for the reaction

$$Xe(g) + 2F_2(g) \longrightarrow XeF_4(s)$$

then for a preparation involving twice the quantities of reactants and products, or

$$2Xe(g) + 4F_2(g) \longrightarrow 2XeF_4(s)$$

twice as much heat would be evolved:

$$\Delta H = 2(-251 \text{ kJ}) = -502 \text{ kJ}$$

Critical Thinking

What if Hess's law were not true? What are some possible repercussions this would have?

INTERACTIVE EXAMPLE 7-7

Hess's Law I

Two forms of carbon are *graphite*, the soft, black, slippery material used in "lead" pencils and as a lubricant for locks, and *diamond*, the brilliant, hard gemstone. Using the enthalpies of combustion for graphite (-394 kJ/mol) and diamond (-396 kJ/mol), calculate ΔH for the conversion of graphite to diamond:

$$C_{graphite}(s) \longrightarrow C_{diamond}(s)$$

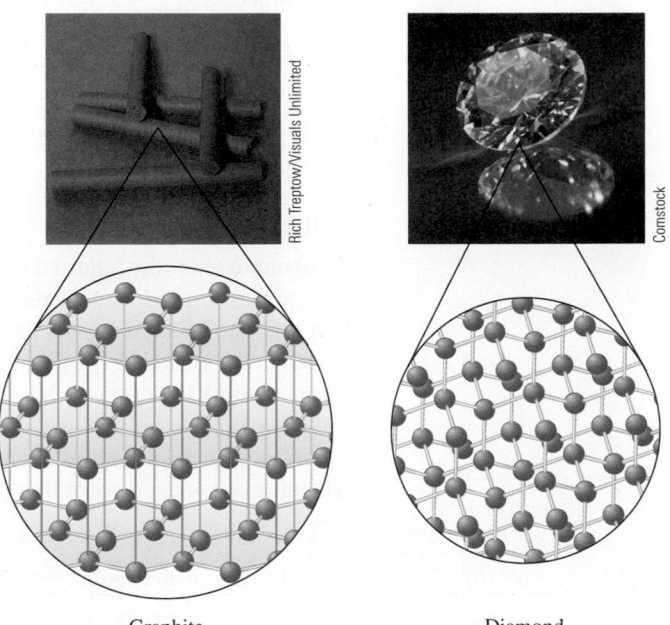

Graphite   Diamond

Solution

Where are we going?
To calculate ΔH for the conversion of graphite to diamond

What do we know?
The combustion reactions are

- $C_{graphite}(s) + O_2(g) \longrightarrow CO_2(g) \quad \Delta H = -394 \text{ kJ}$
- $C_{diamond}(s) + O_2(g) \longrightarrow CO_2(g) \quad \Delta H = -396 \text{ kJ}$

How do we get there?
How do we combine the combustion equations to obtain the equation for the conversion of graphite to diamond?
Note that if we reverse the second reaction (which means we must change the sign of ΔH) and sum the two reactions, we obtain the desired reaction:

$$C_{graphite}(s) + O_2(g) \longrightarrow CO_2(g) \qquad\qquad \Delta H = -394 \text{ kJ}$$
$$\underline{CO_2(g) \longrightarrow C_{diamond}(s) + O_2(g) \quad \Delta H = -(-396 \text{ kJ})}$$
$$C_{graphite}(s) \longrightarrow C_{diamond}(s) \qquad\qquad \Delta H = 2 \text{ kJ}$$

The ΔH for the conversion of graphite to diamond is

$$\Delta H = 2 \text{ kJ/mol graphite}$$

We obtain this value by summing the ΔH values for the equations as shown above. This process is endothermic since the sign of ΔH is positive.

See Exercises 7-67 and 7-68

INTERACTIVE EXAMPLE 7-8

Hess's Law II

Diborane (B_2H_6) is a highly reactive boron hydride that was once considered as a possible rocket fuel for the U.S. space program. Calculate ΔH for the synthesis of diborane from its elements, according to the equation

$$2B(s) + 3H_2(g) \longrightarrow B_2H_6(g)$$

using the following data:

Reaction	ΔH
(a) $2B(s) + \frac{3}{2}O_2(g) \longrightarrow B_2O_3(s)$	-1273 kJ
(b) $B_2H_6(g) + 3O_2(g) \longrightarrow B_2O_3(s) + 3H_2O(g)$	-2035 kJ
(c) $H_2(g) + \frac{1}{2}O_2(g) \longrightarrow H_2O(l)$	-286 kJ
(d) $H_2O(l) \longrightarrow H_2O(g)$	44 kJ

Solution

To obtain ΔH for the required reaction, we must somehow combine equations (a), (b), (c), and (d) to produce that reaction and add the corresponding ΔH values. This can best be done by focusing on the reactants and products of the required reaction. The reactants are B(s) and $H_2(g)$, and the product is $B_2H_6(g)$. How can we obtain the correct equation? Reaction (a) has B(s) as a reactant, as needed in the required equation. Thus reaction (a) will be used as it is. Reaction (b) has $B_2H_6(g)$ as a reactant, but this substance is needed as a product. Thus reaction (b) must be reversed, and the sign of ΔH must be changed accordingly. Up to this point we have

(a)	$2B(s) + \frac{3}{2}O_2(g) \longrightarrow B_2O_3(s)$	$\Delta H = -1273$ kJ
$-$(b)	$B_2O_3(s) + 3H_2O(g) \longrightarrow B_2H_6(g) + 3O_2(g)$	$\Delta H = -(-2035 \text{ kJ})$

Sum: $B_2O_3(s) + 2B(s) + \frac{3}{2}O_2(g) + 3H_2O(g) \longrightarrow B_2O_3(s) + B_2H_6(g) + 3O_2(g) \quad \Delta H = 762$ kJ

Deleting the species that occur on both sides gives

$$2B(s) + 3H_2O(g) \longrightarrow B_2H_6(g) + \tfrac{3}{2}O_2(g) \quad \Delta H = 762 \text{ kJ}$$

We are closer to the required reaction, but we still need to remove $H_2O(g)$ and $O_2(g)$ and introduce $H_2(g)$ as a reactant. We can do this using reactions (c) and (d). If we multiply reaction (c) and its ΔH value by 3 and add the result to the preceding equation, we have

	$2B(s) + 3H_2O(g) \longrightarrow B_2H_6(g) + \tfrac{3}{2}O_2(g)$	$\Delta H = 762$ kJ
$3 \times$ (c)	$3[H_2(g) + \tfrac{1}{2}O_2(g) \longrightarrow H_2O(l)]$	$\Delta H = 3(-286$ kJ$)$

Sum: $2B(s) + 3H_2(g) + \tfrac{3}{2}O_2(g) + 3H_2O(g) \longrightarrow B_2H_6(g) + \tfrac{3}{2}O_2(g) + 3H_2O(l) \quad \Delta H = -96$ kJ

We can cancel the $\tfrac{3}{2}O_2(g)$ on both sides, but we cannot cancel the H_2O because it is gaseous on one side and liquid on the other. This can be solved by adding reaction (d), multiplied by 3:

	$2B(s) + 3H_2(g) + 3H_2O(g) \longrightarrow B_2H_6(g) + 3H_2O(l)$	$\Delta H = -96$ kJ
$3 \times$ (d)	$3[H_2O(l) \longrightarrow H_2O(g)]$	$\Delta H = 3(44$ kJ$)$

Sum: $2B(s) + 3H_2(g) + 3H_2O(g) + 3H_2O(l) \longrightarrow B_2H_6(g) + 3H_2O(l) + 3H_2O(g) \quad \Delta H = +36$ kJ

This gives the reaction required by the problem:

$$2B(s) + 3H_2(g) \longrightarrow B_2H_6(g) \quad \Delta H = +36 \text{ kJ}$$

Thus ΔH for the synthesis of 1 mole of diborane from the elements is $+36$ kJ.

See Exercises 7-69 through 7-74

PROBLEM-SOLVING STRATEGY

Hess's Law

Calculations involving Hess's law typically require that several reactions be manipulated and combined to finally give the reaction of interest. In doing this procedure you should.

■ Work *backward* from the required reaction, using the reactants and products to decide how to manipulate the other given reactions at your disposal.

■ Reverse any reactions as needed to give the required reactants and products.

■ Multiply reactions to give the correct numbers of reactants and products.

This process involves some trial and error, but it can be very systematic if you always allow the final reaction to guide you.

7-5 | Standard Enthalpies of Formation

For a reaction studied under conditions of constant pressure, we can obtain the enthalpy change using a device called a calorimeter, as was shown in Section 7-3. However, this process can be very difficult. In fact, in some cases it is impossible, since certain reactions do not lend themselves to such study. An example is the conversion of solid carbon from its graphite form to its diamond form:

$$C_{graphite}(s) \longrightarrow C_{diamond}(s)$$

The value of ΔH for this process cannot be obtained by direct measurement because the process is much too slow under normal conditions. However, as we saw in Example 7-7, ΔH for this process can be calculated from heats of combustion. This is only one example of how useful it is to be able to *calculate* ΔH values for chemical reactions. We will next show how to do this by using standard enthalpies of formation.

The **standard enthalpy of formation** (ΔH_f°) of a compound is defined as the *change in enthalpy that accompanies the formation of one mole of a compound from its elements with all substances in their standard states.*

A *degree symbol* on a thermodynamic function, for example, ΔH°, indicates that the corresponding process has been carried out under standard conditions. The **standard state** for a substance is a precisely defined reference state. Because thermodynamic functions often depend on the concentrations (or pressures) of the substances involved, we must use a common reference state to properly compare the thermodynamic properties of two substances. This is especially important because, for most thermodynamic properties, we can measure only *changes* in the property. For example, we have no method for determining absolute values of enthalpy. We can measure enthalpy changes (ΔH values) only by performing heat-flow experiments.

Conventional Definitions of Standard States

For a Compound

- The standard state of a gaseous substance is a pressure of exactly 1 atmosphere.
- For a pure substance in a condensed state (liquid or solid), the standard state is the pure liquid or solid.
- For a substance present in a solution, the standard state is a concentration of exactly 1 mole/L.

For an Element

- The standard state of an element is the form in which the element exists under conditions of 1 atmosphere and 25°C. (The standard state for oxygen is $O_2(g)$ at a pressure of 1 atmosphere; the standard state for sodium is $Na(s)$; the standard state for mercury is $Hg(l)$; and so on.)

Several important characteristics of the definition of the enthalpy of formation will become clearer if we again consider the formation of nitrogen dioxide from the elements in their standard states: ◄

$$\tfrac{1}{2}N_2(g) + O_2(g) \longrightarrow NO_2(g) \quad \Delta H_f^\circ = 34 \text{ kJ/mol}$$

Note that the reaction is written so that both elements are in their standard states, and 1 mole of product is formed. Enthalpies of formation are always given per mole of product with the product in its standard state.

The formation reaction for methanol is written as

$$C(s) + 2H_2(g) + \tfrac{1}{2}O_2(g) \longrightarrow CH_3OH(l) \quad \Delta H_f^\circ = -239 \text{ kJ/mol}$$

The standard state of carbon is graphite, the standard states for oxygen and hydrogen are the diatomic gases, and the standard state for methanol is the liquid.

The ΔH_f° values for some common substances are shown in Table 7-2. More values are found in Appendix 6. The importance of the tabulated ΔH_f° values is that enthalpies for many reactions can be calculated using these numbers. To see how this is done, we will calculate the standard enthalpy change for the combustion of methane:

$$CH_4(g) + 2O_2(g) \longrightarrow CO_2(g) + 2H_2O(l)$$

Enthalpy is a state function, so we can invoke Hess's law and choose *any* convenient pathway from reactants to products and then sum the enthalpy changes along the chosen pathway. A convenient pathway, shown in Fig. 7-8, involves taking the reactants apart to the respective elements in their standard states in reactions (a) and (b) and then forming the products from these elements in reactions (c) and (d). This general pathway will work for any reaction, since atoms are conserved in a chemical reaction.

▲ Brown nitrogen dioxide gas.

Table 7-2 | Standard Enthalpies of Formation for Several Compounds at 25°C

Compound	ΔH_f° (kJ/mol)
$NH_3(g)$	−46
$NO_2(g)$	34
$H_2O(l)$	−286
$Al_2O_3(s)$	−1676
$Fe_2O_3(s)$	−826
$CO_2(g)$	−394
$CH_3OH(l)$	−239
$C_8H_{18}(l)$	−269

Figure 7-8 | In this pathway for the combustion of methane, the reactants are first taken apart in reactions (a) and (b) to form the constituent elements in their standard states, which are then used to assemble the products in reactions (c) and (d).

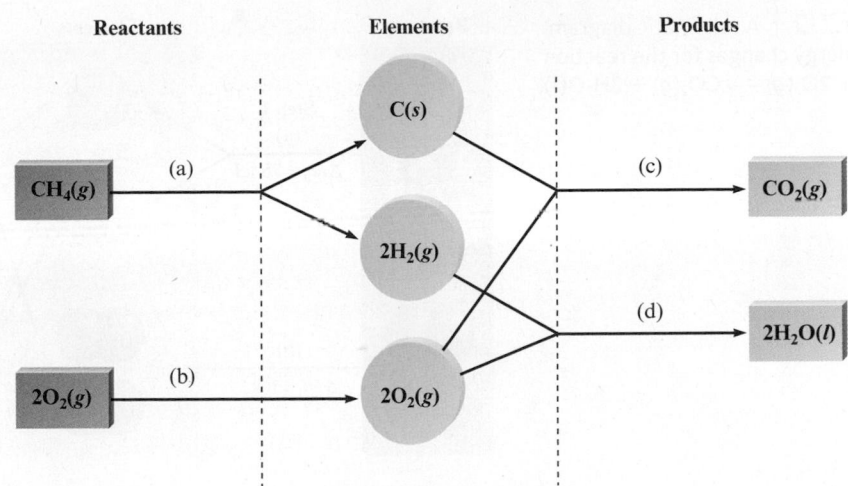

Note from Fig. 7-8 that reaction (a), where methane is taken apart into its elements,

$$CH_4(g) \longrightarrow C(s) + 2H_2(g)$$

is just the reverse of the formation reaction for methane:

$$C(s) + 2H_2(g) \longrightarrow CH_4(g) \quad \Delta H_f^\circ = -75 \text{ kJ/mol}$$

Since reversing a reaction means changing the sign of ΔH but keeping the magnitude the same, ΔH for reaction (a) is $-\Delta H_f^\circ$, or 75 kJ. Thus $\Delta H^\circ_{(a)} = 75$ kJ.

Next we consider reaction (b). Here oxygen is already an element in its standard state, so no change is needed. Thus $\Delta H^\circ_{(b)} = 0$.

The next steps, reactions (c) and (d), use the elements formed in reactions (a) and (b) to form the products. Note that reaction (c) is simply the formation reaction for carbon dioxide:

$$C(s) + O_2(g) \longrightarrow CO_2(g) \quad \Delta H_f^\circ = -394 \text{ kJ/mol}$$

and

$$\Delta H^\circ_{(c)} = \Delta H_f^\circ \text{ for } CO_2(g) = -394 \text{ kJ}$$

Reaction (d) is the formation reaction for water:

$$H_2(g) + \tfrac{1}{2} O_2(g) \longrightarrow H_2O(l) \quad \Delta H_f^\circ = -286 \text{ kJ/mol}$$

However, since 2 moles of water are required in the balanced equation, we must form 2 moles of water from the elements:

$$2H_2(g) + O_2(g) \longrightarrow 2H_2O(l)$$

Thus

$$\Delta H^\circ_{(d)} = 2 \times \Delta H_f^\circ \text{ for } H_2O(l) = 2(-286 \text{ kJ}) = -572 \text{ kJ}$$

We have now completed the pathway from the reactants to the products. The change in enthalpy for the reaction is the sum of the ΔH values (including their signs) for the steps:

$\Delta H^\circ_{\text{reaction}}$

$$= \Delta H^\circ_{(a)} + \Delta H^\circ_{(b)} + \Delta H^\circ_{(c)} + \Delta H^\circ_{(d)}$$

$$= [-\Delta H_f^\circ \text{ for } CH_4(g)] + 0 + [\Delta H_f^\circ \text{ for } CO_2(g)] + [2 \times \Delta H_f^\circ \text{ for } H_2O(l)]$$

$$= -(-75 \text{ kJ}) + 0 + (-394 \text{ kJ}) + (-572 \text{ kJ})$$

$$= -891 \text{ kJ}$$

Figure 7-9 | A schematic diagram of the energy changes for the reaction $CH_4(g) + 2O_2(g) \longrightarrow CO_2(g) + 2H_2O(l)$.

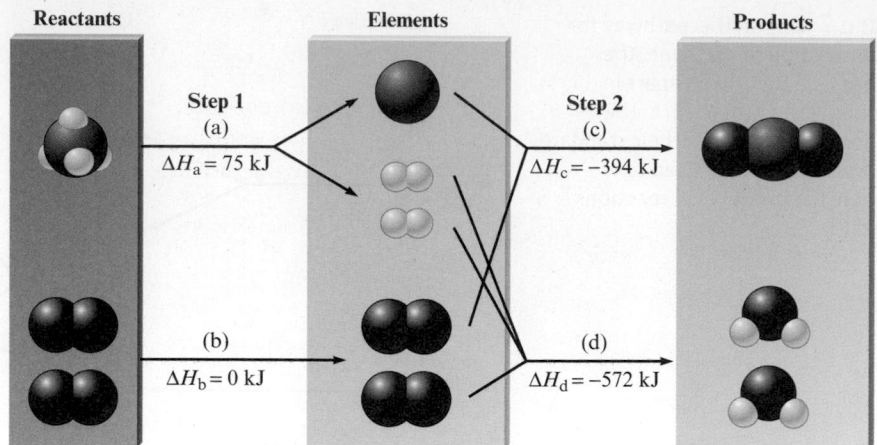

This process is diagramed in Fig. 7-9. Notice that the reactants are taken apart and converted to elements [not necessary for $O_2(g)$] that are then used to form products. You can see that this is a very exothermic reaction because very little energy is required to convert the reactants to the respective elements but a great deal of energy is released when these elements form the products. This is why this reaction is so useful for producing heat to warm homes and offices.

Let's examine carefully the pathway we used in this example. First, the reactants were broken down into the elements in their standard states. This process involved reversing the formation reactions and thus switching the signs of the enthalpies of formation. ◄ *i* The products were then constructed from these elements. This involved formation reactions and thus enthalpies of formation. We can summarize this entire process as follows: The enthalpy change for a given reaction can be calculated by subtracting the enthalpies of formation of the reactants from the enthalpies of formation of the products. Remember to multiply the enthalpies of formation by integers as required by the balanced equation. This statement can be represented symbolically as follows:

i Subtraction means to reverse the sign and add.

$$\Delta H^{\circ}_{\text{reaction}} = \Sigma n_p \Delta H^{\circ}_f \text{ (products)} - \Sigma n_r \Delta H^{\circ}_f \text{ (reactants)} \qquad (7\text{-}1)$$

where the symbol Σ (sigma) means "to take the sum of the terms," and n_p and n_r represent the moles of each product or reactant, respectively. ◄ *i*

i Energy calculations for chemical reactions can also be done using bond energies as shown in Section 3-8.

i Elements in their standard states are not included in enthalpy calculations using ΔH°_f values.

Elements are not included in the calculation because elements require no change in form. ◄ *i* We have in effect *defined* the enthalpy of formation of an element in its standard state as zero, since we have chosen this as our reference point for calculating enthalpy changes in reactions.

PROBLEM-SOLVING STRATEGY

Enthalpy Calculations

- When a reaction is reversed, the magnitude of ΔH remains the same, but its sign changes.
- When the balanced equation for a reaction is multiplied by an integer, the value of ΔH for that reaction must be multiplied by the same integer.
- The change in enthalpy for a given reaction can be calculated from the enthalpies of formation of the reactants and products:

$$\Delta H^{\circ}_{\text{reaction}} = \Sigma n_p \Delta H^{\circ}_f \text{(products)} - \Sigma n_r \Delta H^{\circ}_f \text{(reactants)}$$

- Elements in their standard states are not included in the $\Delta H_{\text{reaction}}$ calculations. That is, ΔH°_f for an element in its standard state is zero.

INTERACTIVE EXAMPLE 7-9

Enthalpies from Standard Enthalpies of Formation I

Using the standard enthalpies of formation listed in Table 7-2, calculate the standard enthalpy change for the overall reaction that occurs when ammonia is burned in air to form nitrogen dioxide and water. This is the first step in the manufacture of nitric acid.

$$4NH_3(g) + 7O_2(g) \longrightarrow 4NO_2(g) + 6H_2O(l)$$

Solution

We will use the pathway in which the reactants are broken down into elements in their standard states, which are then used to form the products (see Fig. 7-10).

Decomposition of $NH_3(g)$ into elements [reaction (a) in Fig. 7-10]. The first step is to decompose 4 moles of NH_3 into N_2 and H_2:

$$4NH_3(g) \longrightarrow 2N_2(g) + 6H_2(g)$$

The preceding reaction is 4 times the *reverse* of the formation reaction for NH_3: ◀ ⓘ

$$\tfrac{1}{2}N_2(g) + \tfrac{3}{2}H_2(g) \longrightarrow NH_3(g) \quad \Delta H_f^\circ = -46 \text{ kJ/mol}$$

Thus

$$\Delta H^\circ_{(a)} = 4 \text{ mol}[-(-46 \text{ kJ/mol})] = 184 \text{ kJ}$$

Elemental oxygen [reaction (b) in Fig. 7-10]. Since $O_2(g)$ is an element in its standard state, $\Delta H^\circ_{(b)} = 0$.

We now have the elements $N_2(g)$, $H_2(g)$, and $O_2(g)$, which can be combined to form the products of the overall reaction.

Synthesis of $NO_2(g)$ from elements [reaction (c) in Fig. 7-10]. The overall reaction equation has 4 moles of NO_2. Thus the required reaction is 4 times the formation reaction for NO_2:

$$4 \times [\tfrac{1}{2}N_2(g) + O_2(g) \longrightarrow NO_2(g)]$$

and

$$\Delta H^\circ_{(c)} = 4 \times \Delta H_f^\circ \text{ for } NO_2(g)$$

From Table 7-2, ΔH_f° for $NO_2(g) = 34$ kJ/mol and

$$\Delta H^\circ_{(c)} = 4 \text{ mol} \times 34 \text{ kJ/mol} = 136 \text{ kJ}$$

ⓘ If you use the bond energies from Table 3-4 to determine the enthalpy of formation of NH_3 from N_2 and H_2, you get an answer very close to -46 kJ/mol (about -54 kJ/mol). This shows us that it is acceptable to consider potential energy as energy stored in the bonds and that our simple model of bond energies discussed in Chapter 3 works reasonably well.

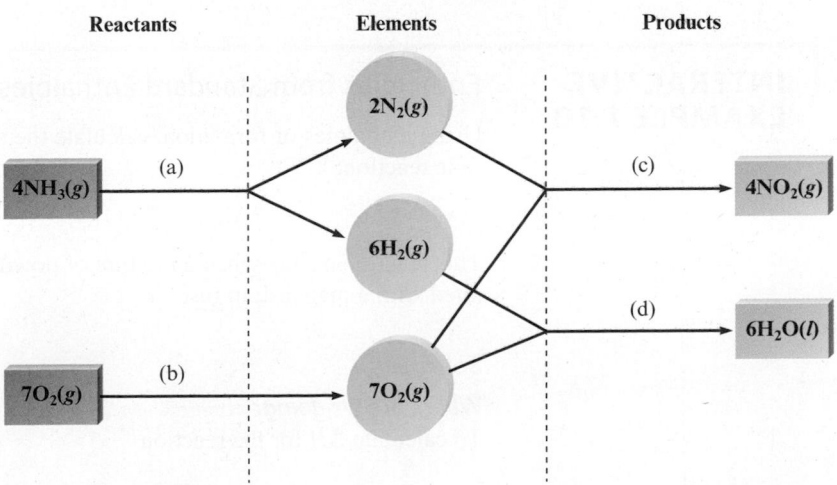

Figure 7-10 | A pathway for the combustion of ammonia.

Synthesis of $H_2O(l)$ from elements [reaction (d) in Fig. 7-10]. Since the overall equation for the reaction has 6 moles of $H_2O(l)$, the required reaction is 6 times the formation reaction for $H_2O(l)$:

$$6 \times [H_2(g) + \tfrac{1}{2}O_2(g) \longrightarrow H_2O(l)]$$

and

$$\Delta H°_{(d)} = 6 \times \Delta H°_f \text{ for } H_2O(l)$$

From Table 7-2, $\Delta H°_f$ for $H_2O(l) = -286$ kJ/mol and

$$\Delta H°_{(d)} = 6 \text{ mol} (-286 \text{ kJ/mol}) = -1716 \text{ kJ}$$

To summarize, we have done the following:

$$4NH_3(g) \xrightarrow{\Delta H°_{(a)}} \left\{ \begin{array}{l} 2N_2(g) + 6H_2(g) \\ 7O_2(g) \end{array} \right. \xrightarrow{\Delta H°_{(c)}} 4NO_2(g)$$

$$7O_2(g) \xrightarrow{\Delta H°_{(b)}} \left. \right\} \xrightarrow{\Delta H°_{(d)}} 6H_2O(l)$$

Elements in their
standard states

We add the $\Delta H°$ values for the steps to get $\Delta H°$ for the overall reaction:

$$\begin{aligned}
\Delta H°_{reaction} &= \Delta H°_{(a)} + \Delta H°_{(b)} + \Delta H°_{(c)} + \Delta H°_{(d)} \\
&= [4 \times -\Delta H°_f \text{ for } NH_3(g)] + 0 + [4 \times \Delta H°_f \text{ for } NO_2(g)] \\
&\quad + [6 \times \Delta H°_f \text{ for } H_2O(l)] \\
&= [4 \times \Delta H°_f \text{ for } NO_2(g)] + [6 \times \Delta H°_f \text{ for } H_2O(l)] \\
&\quad - [4 \times \Delta H°_f \text{ for } NH_3(g)] \\
&= \Sigma n_p \Delta H°_f \text{ (products)} - \Sigma n_r \Delta H°_f \text{ (reactants)}
\end{aligned}$$

Remember that elemental reactants and products do not need to be included, since $\Delta H°_f$ for an element in its standard state is zero. Note that we have again obtained Equation (7-1). The final solution is

$$\begin{aligned}
\Delta H°_{reaction} &= [4 \times (34 \text{ kJ})] + [6 \times (-286 \text{ kJ})] - [4 \times (-46 \text{ kJ})] \\
&= -1396 \text{ kJ}
\end{aligned}$$

See Exercises 7-77 and 7-78

Now that we have shown the basis for Equation (7-1), we will make direct use of it to calculate ΔH for reactions in succeeding exercises.

INTERACTIVE EXAMPLE 7-10

Enthalpies from Standard Enthalpies of Formation II

Using enthalpies of formation, calculate the standard change in enthalpy for the thermite reaction:

$$2Al(s) + Fe_2O_3(s) \longrightarrow Al_2O_3(s) + 2Fe(s)$$

This reaction occurs when a mixture of powdered aluminum and iron(III) oxide is ignited with a magnesium fuse. ◄

Solution

Where are we going?
To calculate ΔH for the reaction

▲ The thermite reaction is one of the most energetic chemical reactions known.

What do we know?

- ΔH_f° for $Fe_2O_3(s) = -826$ kJ/mol
- ΔH_f° for $Al_2O_3(s) = -1676$ kJ/mol
- ΔH_f° for $Al(s) = \Delta H_f^\circ$ for $Fe(s) = 0$

What do we need?

- We use Equation (7-1):

$$\Delta H^\circ = \Sigma n_p \Delta H_f^\circ(\text{products}) - \Sigma n_r \Delta H_f^\circ(\text{reactants})$$

How do we get there?

$$\Delta H^\circ_{\text{reaction}} = \Delta H_f^\circ \text{ for } Al_2O_3(s) - \Delta H_f^\circ \text{ for } Fe_2O_3(s)$$
$$= -1676 \text{ kJ} - (-826 \text{ kJ}) = -850. \text{ kJ}$$

This reaction is so highly exothermic that the iron produced is initially molten. This process is often used as a lecture demonstration and also has been used in welding massive steel objects such as ships' propellers.

See Exercises 7-81 and 7-82

Critical Thinking

For $\Delta H_{\text{reaction}}$ calculations, we define ΔH_f° for an element in its standard state as zero. What if we define ΔH_f° for an element in its standard state as 10 kJ/mol? How would this affect your determination of $\Delta H_{\text{reaction}}$? Provide support for your answer with a sample calculation.

EXAMPLE 7-11

Enthalpies from Standard Enthalpies of Formation III

Until recently, methanol (CH_3OH) was used as a fuel in high-performance engines in race cars. Using the data in Table 7-2, compare the standard enthalpy of combustion per gram of methanol with that per gram of gasoline. Gasoline is actually a mixture of compounds, but assume for this problem that gasoline is pure liquid octane (C_8H_{18}).

Solution

Where are we going?

To compare ΔH of combustion for methanol and octane

What do we know?

- Standard enthalpies of formation from Table 7-2

How do we get there?

For methanol:

What is the combustion reaction?

$$2CH_3OH(l) + 3O_2(g) \longrightarrow 2CO_2(g) + 4H_2O(l)$$

What is the $\Delta H^\circ_{reaction}$?

Using the standard enthalpies of formation from Table 7-2 and Equation (7-1), we have

$$\Delta H^\circ_{\text{reaction}} = 2 \times \Delta H_f^\circ \text{ for } CO_2(g) + 4 \times \Delta H_f^\circ \text{ for } H_2O(l)$$
$$- 2 \times \Delta H_f^\circ \text{ for } CH_3OH(l)$$
$$= 2 \times (-394 \text{ kJ}) + 4 \times (-286 \text{ kJ}) - 2 \times (-239 \text{ kJ})$$
$$= -1.45 \times 10^3 \text{ kJ}$$

© Cengage Learning

What is the enthalpy of combustion per gram?

Thus 1.45×10^3 kJ of heat is evolved when 2 moles of methanol burn. The molar mass of methanol is 32.04 g/mol. This means that 1.45×10^3 kJ of energy is produced when 64.08 g methanol burns. The enthalpy of combustion per gram of methanol is

$$\frac{-1.45 \times 10^3 \text{ kJ}}{64.08 \text{ g}} = -22.6 \text{ kJ/g}$$

For octane:

What is the combustion reaction?

$$2C_8H_{18}(l) + 25O_2(g) \longrightarrow 16CO_2(g) + 18H_2O(l)$$

What is the $\Delta H°_{reaction}$?

Using the standard enthalpies of formation from Table 7-2 and Equation (7-1), we have

$$
\begin{aligned}
\Delta H°_{reaction} &= 16 \times \Delta H°_f \text{ for } CO_2(g) + 18 \times \Delta H°_f \text{ for } H_2O(l) \\
&\quad - 2 \times \Delta H°_f \text{ for } C_8H_{18}(l) \\
&= 16 \times (-394 \text{ kJ}) + 18 \times (-286 \text{ kJ}) - 2 \times (-269 \text{ kJ}) \\
&= -1.09 \times 10^4 \text{ kJ}
\end{aligned}
$$

What is the enthalpy of combustion per gram?

This is the amount of heat evolved when 2 moles of octane burn. Since the molar mass of octane is 114.22 g/mol, the enthalpy of combustion per gram of octane is

$$\frac{-1.09 \times 10^4 \text{ kJ}}{2(114.22 \text{ g})} = -47.7 \text{ kJ/g}$$

The enthalpy of combustion per gram of octane is approximately twice that per gram of methanol. On this basis, gasoline appears to be superior to methanol for use in a racing car, where weight considerations are usually very important. Why, then, was methanol used in racing cars? ◄ 🅘 The answer was that methanol burns much more smoothly than gasoline in high-performance engines, and this advantage more than compensates for its weight disadvantage.

🅘 In the cars used in the Indianapolis 500, ethanol is now used instead of methanol.

See Exercise 7-89

7-6 | Present Sources of Energy

Woody plants, coal, petroleum, and natural gas hold a vast amount of energy that originally came from the sun. By the process of photosynthesis, plants store energy that can be claimed by burning the plants themselves or the decay products that have been converted over millions of years to **fossil fuels**. Although the United States currently depends heavily on petroleum for energy, this dependency is a relatively recent phenomenon, as shown in Fig. 7-11. In this section we discuss some sources of energy and their effects on the environment.

Petroleum and Natural Gas

Although how they were produced is not completely understood, petroleum and natural gas were most likely formed from the remains of marine organisms that lived approximately 500 million years ago. ◄ **Petroleum** is a thick, dark liquid composed mostly of compounds called *hydrocarbons* that contain carbon and hydrogen. (Carbon is unique among elements in the extent to which it can bond to itself to form chains of various lengths.) Table 7-3 gives the formulas and names for several common hydrocarbons. **Natural gas**, usually associated with petroleum deposits, consists mostly of methane, but it also contains significant amounts of ethane, propane, and butane. Over

Keith Wood/Stone/Getty Images

▲ Oil rig in Gulf of Mexico.

Figure 7-11 | Energy sources used in the United States.

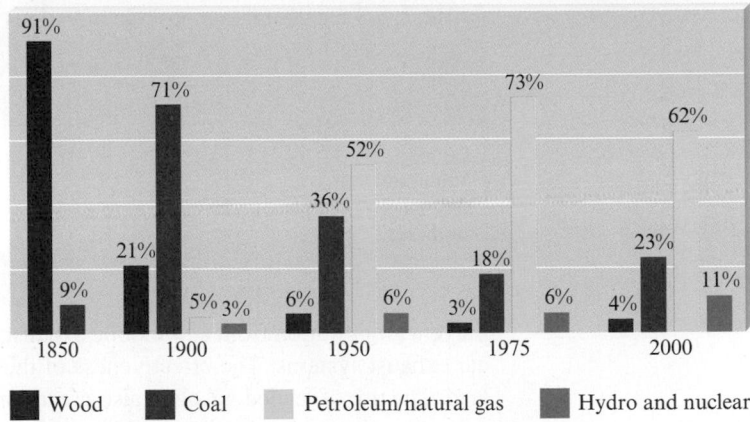

| Wood | Coal | Petroleum/natural gas | Hydro and nuclear |

Table 7-3 | Names and Formulas for Some Common Hydrocarbons

Formula	Name
CH_4	Methane
C_2H_6	Ethane
C_3H_8	Propane
C_4H_{10}	Butane
C_5H_{12}	Pentane
C_6H_{14}	Hexane
C_7H_{16}	Heptane
C_8H_{18}	Octane

Table 7-4 | Uses of the Various Petroleum Fractions

Petroleum Fraction in Terms of Numbers of Carbon Atoms	Major Uses
$C_5–C_{10}$	Gasoline
$C_{10}–C_{18}$	Kerosene
	Jet fuel
$C_{15}–C_{25}$	Diesel fuel
	Heating oil
	Lubricating oil
$>C_{25}$	Asphalt

the last several years it has become clear that there are tremendous reserves of natural gas deep in shale deposits. Estimates indicate that there may be as much as 200 trillion cubic meters of recoverable natural gas in these deposits around the globe. In the eastern United States, the 1-mile-deep Marsellus Shale deposit may contain as much as two trillion cubic meters of recoverable gas. The problem with these gas deposits is that the shale is very impermeable, and the gas does not flow out into wells on its own. A technique called hydraulic fracturing, or "fracking," is now being used to access these gas deposits. Fracking involves injecting a slurry of water, sand, and chemical additives under pressure through a well bore drilled into the shale. This produces fractures in the rock that allow the gas to flow out into wells. With the increased use of fracking has come environmental concerns, including potential contamination of ground water, risks to air quality, and possible mishandling of wastes associated with the process. However, even in view of these potential hazards, it is expected that natural gas obtained from fracking may supply as much as half of the U.S. natural gas supply by the year 2020.

The composition of petroleum varies somewhat, but it consists mostly of hydrocarbons having chains that contain from 5 to more than 25 carbons. To be used efficiently, the petroleum must be separated into fractions by boiling. The lighter molecules (having the lowest boiling points) can be boiled off, leaving the heavier ones behind. The commercial uses of various petroleum fractions are shown in Table 7-4.

The petroleum era began when the demand for lamp oil during the Industrial Revolution outstripped the traditional sources: animal fats and whale oil. In response to this increased demand, Edwin Drake drilled the first oil well in 1859 at Titusville, Pennsylvania. The petroleum from this well was refined to produce *kerosene* (fraction $C_{10}–C_{18}$), which served as an excellent lamp oil. *Gasoline* (fraction $C_5–C_{10}$) had limited use and was often discarded. However, this situation soon changed. The development of the electric light decreased the need for kerosene, and the advent of the "horseless carriage" with its gasoline-powered engine signaled the birth of the gasoline age.

As gasoline became more important, new ways were sought to increase the yield of gasoline obtained from each barrel of petroleum. William Burton invented a process at Standard Oil of Indiana called *pyrolytic (high-temperature) cracking*. In this process, the heavier molecules of the kerosene fraction are heated to about 700°C, causing them to break (crack) into the smaller molecules of hydrocarbons in the gasoline fraction. As cars became larger, more efficient internal combustion engines were designed. Because of the uneven burning of the gasoline then available, these engines "knocked," producing unwanted noise and even engine damage. Intensive research to find additives that would promote smoother burning produced tetraethyl lead, $(C_2H_5)_4Pb$, a very effective "antiknock" agent.

The addition of tetraethyl lead to gasoline became a common practice, and by 1960, gasoline contained as much as 3 grams of lead per gallon. As we have discovered so often in recent years, technological advances can produce environmental problems. To

Table 7-5 | Elemental Composition of Various Types of Coal

Type of Coal	Mass Percent of Each Element				
	C	H	O	N	S
Lignite	71	4	23	1	1
Subbituminous	77	5	16	1	1
Bituminous	80	6	8	1	5
Anthracite	92	3	3	1	1

prevent air pollution from automobile exhaust, catalytic converters have been added to car exhaust systems. The effectiveness of these converters, however, is destroyed by lead. The use of leaded gasoline also greatly increased the amount of lead in the environment, where it can be ingested by animals and humans. For these reasons, the use of lead in gasoline has been phased out, requiring extensive (and expensive) modifications of engines and of the gasoline refining process.

Coal

Coal was formed from the remains of plants that were buried and subjected to high pressure and heat over long periods of time. Plant materials have a high content of cellulose, a complex molecule whose empirical formula is CH_2O but whose molar mass is around 500,000 g/mol. After the plants and trees that flourished on the earth at various times and places died and were buried, chemical changes gradually lowered the oxygen and hydrogen content of the cellulose molecules. Coal "matures" through four stages: lignite, subbituminous, bituminous, and anthracite. Each stage has a higher carbon-to-oxygen and carbon-to-hydrogen ratio; that is, the relative carbon content gradually increases. Typical elemental compositions of the various coals are given in Table 7-5. The energy available from the combustion of a given mass of coal increases as the carbon content increases. Therefore, anthracite is the most valuable coal, and lignite the least valuable.

Coal is an important and plentiful fuel in the United States, currently furnishing approximately 23% of our energy. However, coal is expensive and dangerous to mine underground, and the strip mining of fertile farmland in the Midwest or of scenic land in the West causes obvious problems. In addition, the burning of coal, especially high-sulfur coal, yields air pollutants such as sulfur dioxide, which, in turn, can lead to acid rain.

Effects of Carbon Dioxide on Climate

The earth receives a tremendous quantity of radiant energy from the sun, about 30% of which is reflected back into space by the earth's atmosphere. The remaining energy passes through the atmosphere to the earth's surface. Some of this energy is absorbed by plants for photosynthesis and some by the oceans to evaporate water, but most of it is absorbed by soil, rocks, and water, increasing the temperature of the earth's surface. This energy is in turn radiated from the heated surface mainly as *infrared radiation*, often called *heat radiation*.

The atmosphere, like window glass, is transparent to visible light but does not allow all the infrared radiation to pass back into space. Molecules in the atmosphere, principally H_2O and CO_2, strongly absorb infrared radiation and radiate it back toward the earth, as shown in Fig. 7-12, so a net amount of thermal energy is retained by the earth's atmosphere, causing the earth to be much warmer than it would be without its atmosphere. In a way, the atmosphere acts like the glass of a greenhouse, which is transparent to visible light but absorbs infrared radiation, thus raising the temperature inside the building. This **greenhouse effect** is seen even more spectacularly on Venus,

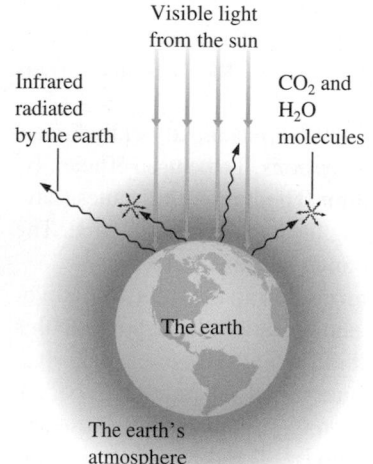

Figure 7-12 | The earth's atmosphere is transparent to visible light from the sun. This visible light strikes the earth, and part of it is changed to infrared radiation. The infrared radiation from the earth's surface is strongly absorbed by CO_2, H_2O, and other molecules present in smaller amounts (for example, CH_4 and N_2O) in the atmosphere. In effect, the atmosphere traps some of the energy, acting like the glass in a greenhouse and keeping the earth warmer than it would otherwise be.

Visible light from the sun

Infrared radiated by the earth

CO_2 and H_2O molecules

The earth

The earth's atmosphere

ⓘ The average temperature of the earth's surface is 25°C. It would be −18°C without the "greenhouse gases."

ⓘ The electromagnetic spectrum, including visible and infrared radiation, is discussed in Chapter 2.

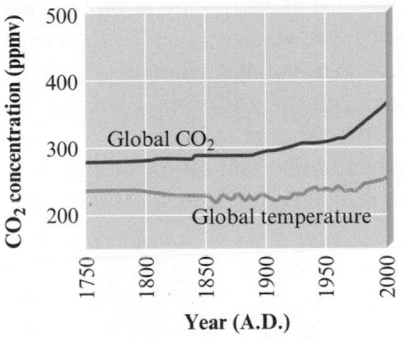

Figure 7-13 | The atmospheric CO_2 concentration and the average global temperature over the last 250 years. Note the significant increase in CO_2 concentration in the last 50 years.

(Source: National Assessment Synthesis Team, *Climate Change Impacts on the United States: The Potential Consequences of Climate, Variability and Change, Overview,* Report for the U.S. Global Change Research Program, Cambridge University Press, Cambridge, UK, p. 13, 2000.)

where the dense atmosphere is thought to be responsible for the high surface temperature of that planet. ◀ ⓘ

Thus the temperature of the earth's surface is controlled to a significant extent by the carbon dioxide and water content of the atmosphere. The effect of atmospheric moisture (humidity) is apparent in the Midwest. In summer, when the humidity is high, the heat of the sun is retained well into the night, giving very high nighttime temperatures. On the other hand, in winter, the coldest temperatures always occur on clear nights, when the low humidity allows efficient radiation of energy back into space.

The atmosphere's water content is controlled by the water cycle (evaporation and precipitation), and the average remains constant over the years. However, as fossil fuels have been used more extensively, the carbon dioxide concentration has increased by about 16% from 1880 to 1980. Comparisons of satellite data have now produced evidence that the greenhouse effect has significantly warmed the earth's atmosphere. The data compare the same areas in both 1979 and 1997. The analysis shows that more infrared radiation was blocked by CO_2, methane, and other greenhouse gases. ◀ ⓘ This *could* increase the earth's average temperature by as much as 3°C, causing dramatic changes in climate and greatly affecting the growth of food crops.

How well can we predict long-term effects? Because weather has been studied for a period of time that is minuscule compared with the age of the earth, the factors that control the earth's climate in the long range are not clearly understood. For example, we do not understand what causes the earth's periodic ice ages. So it is difficult to estimate the impact of the increasing carbon dioxide levels.

In fact, the variation in the earth's average temperature over the past century is somewhat confusing. In the northern latitudes during the past century, the average temperature rose by 0.8°C over a period of 60 years, then cooled by 0.5°C during the next 25 years, and finally warmed by 0.2°C in the succeeding 15 years. Such fluctuations do not match the steady increase in carbon dioxide. However, in southern latitudes and near the equator during the past century, the average temperature showed a steady rise totaling 0.4°C. This figure is in reasonable agreement with the predicted effect of the increasing carbon dioxide concentration over that period. Another significant fact is that the past 10 years constitute the warmest decade on record, although the warming has not been as great as the models predicted for the increase in carbon dioxide concentration.

The relationship between the carbon dioxide concentration in the atmosphere and the earth's temperature is not well-understood at present. The increase in the atmospheric concentration of carbon dioxide is shown in Fig. 7-13, and we must consider the implications of this increase as we consider our future energy needs.

Methane is another greenhouse gas that is 21 times more potent than carbon dioxide. This fact is particularly significant for countries with lots of animals, because methane is produced by methanogenic archaea that live in the animals' rumen. For example, sheep and cattle produce about 14% of Australia's total greenhouse emissions. To reduce this level, Australia has initiated a program to vaccinate sheep and cattle to lower the number of archaea present in their digestive systems. It is hoped that this effort will reduce by 20% the amount of methane emitted by these animals.

7-7 | New Energy Sources

As we search for the energy sources of the future, we need to consider economic, climatic, and supply factors. There are several potential energy sources: the sun (solar), nuclear processes (fission and fusion), biomass (plants), and synthetic fuels. Direct use of the sun's radiant energy to heat our homes and run our factories and transportation systems seems a sensible long-term goal. But what do we do now? Conservation of fossil fuels is one obvious step, but substitutes for fossil fuels also must be found. We will discuss some alternative sources of energy here. Nuclear power will be considered in Chapter 18.

Farming the Wind

▲ This State Line Wind Project along the Oregon–Washington border uses approximately 399 wind turbines to create enough electricity to power some 70,000 households.

There is plenty of untapped wind power in the United States. Wind mappers rate regions on a scale from 1 to 6 (with 6 being the best) to indicate the quality of the wind resource. Wind farms are now being developed in areas rated from 4 to 6. The farmers who own the land welcome the increased income derived from the wind blowing across their land. Economists estimate that each acre devoted to wind turbines can pay royalties to the farmers that amount to many times the revenue from growing corn on that same land. As a result, hundreds of "wind farms" are springing up in the Midwest. The largest of these installations can produce more than 1000 MW of energy. The total wind capacity in the United States is now more than 60,000 MW. In addition to the many land-based wind installations, there are many off-shore wind farms in places such as the United Kingdom, Germany, and Denmark. Worldwide there are now more than 200,000 wind turbines in operation producing about 300,000 MW. ◄

Although wind power has many environmental advantages, there are some major disadvantages. One problem is the variability of the wind on a short-term basis. Because instantaneous electrical generation and consumption must remain in balance to have a stable electrical grid, the variability of wind requires backup systems such as natural gas-powered generators. Another problem is that wind farms are often constructed where electrical transmission is limited. Thus, the infrastructure must be enhanced in these areas. A very important issue is the cost of energy produced by wind farms. Currently in the United States, electricity from wind sources is somewhat higher in cost than from more traditional sources. Because of this, wind power energy suppliers have been subsidized by the federal government. However, costs are decreasing and are approaching the costs of electricity from other sources.

Everything considered, wind power has become an important source of energy in the world. Predictions are that wind installation will produce almost 10% of total worldwide energy use by 2020.

Hydrogen as a Fuel ◄

▲ The main engines in the space shuttle *Endeavor* used hydrogen and oxygen as fuel in the shuttle's many trips into space.

If you have ever seen a lecture demonstration where hydrogen–oxygen mixtures were ignited, you have witnessed a demonstration of hydrogen's potential as a fuel. The combustion reaction is

$$H_2(g) + \tfrac{1}{2}O_2(g) \longrightarrow H_2O(l) \quad \Delta H° = -286 \text{ kJ}$$

As we saw in Example 7-11, the heat of combustion of $H_2(g)$ per gram is approximately 2.5 times that of natural gas. In addition, hydrogen has a real advantage over fossil fuels in that the only product of hydrogen combustion is water; fossil fuels also produce carbon dioxide. However, even though it appears that hydrogen is a very logical choice as a major fuel for the future, there are three main problems: the cost of production, storage, and transport.

First let's look at the production problem. Although hydrogen is very abundant on the earth, virtually none of it exists as the free gas. Currently, the main source of hydrogen gas is from the treatment of natural gas with steam:

$$CH_4(g) + H_2O(g) \longrightarrow 3H_2(g) + CO(g)$$

We can calculate ΔH for this reaction using Equation (7-1):

$$\Delta H° = \Sigma n_p \Delta H_f°(\text{products}) - \Sigma n_r \Delta H_f°(\text{reactants})$$
$$= \Delta H_f° \text{ for } CO(g) - \Delta H_f° \text{ for } CH_4(g) - \Delta H_f° \text{ for } H_2O(g)$$
$$= -111 \text{ kJ} - (-75 \text{ kJ}) - (-242 \text{ kJ}) = 206 \text{ kJ}$$

Note that this reaction is highly endothermic; treating methane with steam is not an efficient way to obtain hydrogen for fuel. It would be much more economical to burn the methane directly.

A virtually inexhaustible supply of hydrogen exists in the waters of the world's oceans. However, the reaction

$$H_2O(l) \longrightarrow H_2(g) + \tfrac{1}{2}O_2(g)$$

requires 286 kJ of energy per mole of liquid water, and under current circumstances, large-scale production of hydrogen from water is not economically feasible. However, several methods for such production are currently being studied: electrolysis of water, thermal decomposition of water, thermochemical decomposition of water, and biological decomposition of water.

Electrolysis of water involves passing an electric current through it. The present cost of electricity makes the hydrogen produced by electrolysis too expensive to be competitive as a fuel. However, if in the future we develop more efficient sources of electricity, this situation could change.

Recent research at the University of Minnesota by Lanny Schmidt and his coworkers suggests that corn could be a feasible source of hydrogen. In this process the starch from the corn is fermented to produce alcohol, which is then decomposed in a special reactor at 140°C with a rhodium and cerium oxide catalyst to give hydrogen. These scientists indicate that enough hydrogen gas can be obtained from a few ounces of ethanol to generate electricity to run six 60-watt bulbs for an hour.

Thermal decomposition is another method for producing hydrogen from water. This involves heating the water to several thousand degrees, where it spontaneously decomposes into hydrogen and oxygen. However, attaining temperatures in this range would be very expensive even if a practical heat source and a suitable reaction container were available.

In the thermochemical decomposition of water, chemical reactions, as well as heat, are used to "split" water into its components. One such system involves the following reactions (the temperature required for each is given in parentheses):

$$2HI \longrightarrow I_2 + H_2 \qquad (425°C)$$
$$2H_2O + SO_2 + I_2 \longrightarrow H_2SO_4 + 2HI \qquad (90°C)$$
$$\underline{H_2SO_4 \longrightarrow SO_2 + H_2O + \tfrac{1}{2}O_2 \quad (825°C)}$$
$$\text{Net reaction: } H_2O \longrightarrow H_2 + \tfrac{1}{2}O_2$$

Note that the HI is not consumed in the net reaction. Note also that the maximum temperature required is 825°C, a temperature that is feasible if a nuclear reactor is used as a heat source. A current research goal is to find a system for which the required temperatures are low enough that sunlight can be used as the energy source.

But what about the organisms that decompose water without the aid of electricity or high temperatures? In the process of photosynthesis, green plants absorb carbon dioxide and water and use them along with energy from the sun to produce the substances needed for growth. Scientists have studied photosynthesis for years, hoping to get answers to humanity's food and energy shortages. At present, much of this research involves attempts to modify the photosynthetic process so that plants will release hydrogen gas from water instead of using the hydrogen to produce complex compounds. Small-scale experiments have shown that under certain conditions plants do produce hydrogen gas, but the yields are far from being commercially useful. At this point the economical production of hydrogen gas remains unrealized.

The storage and transportation of hydrogen also present problems. First, on metal surfaces the H_2 molecule decomposes to atoms. Since the atoms are so small, they can migrate into the metal, causing structural changes that make it brittle. This might lead to a pipeline failure if hydrogen were pumped under high pressure.

An additional problem is the relatively small amount of energy that is available *per unit volume* of hydrogen gas. Although the energy available per gram of hydrogen is significantly greater than that per gram of methane, the energy available per given volume of hydrogen is about one-third that available from the same volume of methane.

Although the use of hydrogen as a fuel solves some of the problems associated with fossil fuels, it does present some potential environmental problems of its own. Studies by John M. Eiler and his colleagues at California Institute of Technology indicate that, if hydrogen becomes a major source of energy, accidental leakage of the gas into the atmosphere could pose a threat. The Cal Tech scientists calculate that leakage could raise the concentration of H_2 in the atmosphere from its natural level of 0.5 part per million to more than 2 parts per million. As some of the H_2 eventually finds its way into the upper atmosphere, it would react with O_2 to form water, which would increase the number of ice crystals. This could lead to the destruction of some of the protective ozone because many of the chemical reactions that destroy ozone occur on the surfaces of ice crystals. However, as is the usual case with environmental issues, the situation is complicated. The scenario suggested by Eiler's team may not happen because the leaked H_2 could be consumed by soil microbes that use hydrogen as a nutrient. In fact, Eiler's studies show that 90% of the H_2 emitted into the atmosphere today from sources such as motor vehicles and forest fires is eventually absorbed by soil organisms.

The evaluation of hydrogen as a fuel illustrates how complex and interconnected the economic and environmental issues are.

Could hydrogen be considered as a potential fuel for automobiles? This is an intriguing question. The internal combustion engines in automobiles can be easily adapted to burn hydrogen. In fact, BMW has experimented with a fleet of cars powered by hydrogen-burning internal combustion engines. However, the primary difficulty is the storage of enough hydrogen to give an automobile a reasonable range. This is illustrated by Example 7-12.

INTERACTIVE EXAMPLE 7-12

Comparing Enthalpies of Combustion

Assuming that the combustion of hydrogen gas provides three times as much energy per gram as gasoline, calculate the volume of liquid H_2 (density = 0.0710 g/mL) required to furnish the energy contained in 80.0 L (about 20 gal) of gasoline (density = 0.740 g/mL).

Solution

Where are we going?
To calculate the volume of $H_2(l)$ required

What do we know?

- Density for $H_2(l)$ = 0.0710 g/mL
- 80.0 L gasoline
- Density for gasoline = 0.740 g/mL

How do we get there?
What is the mass of gasoline?

$$80.0 \text{ L} \times \frac{1000 \text{ mL}}{1 \text{ L}} \times \frac{0.740 \text{ g}}{\text{mL}} = 59{,}200 \text{ g}$$

How much $H_2(l)$ is needed?
Since H_2 furnishes three times as much energy per gram as gasoline, only a third as much liquid hydrogen is needed to furnish the same energy:

$$\text{Mass of } H_2(l) \text{ needed} = \frac{59{,}200 \text{ g}}{3} = 19{,}700 \text{ g}$$

Since density = mass/volume, then volume = mass/density, and the volume of $H_2(l)$ needed is

$$V = \frac{19,700 \text{ g}}{0.0710 \text{ g/mL}}$$

$$= 2.77 \times 10^5 \text{ mL} = 277 \text{ L}$$

Thus 277 L of liquid H_2 is needed to furnish the same energy of combustion as 80.0 L of gasoline.

See Exercises 7-87 through 7-90

You can see from Example 7-12 that an automobile would need a significantly larger tank to hold enough liquid hydrogen to have a typical mileage range. In addition, storing liquid hydrogen in an automobile is totally impractical. Because of its very low boiling point (20 K), storage of liquid hydrogen requires a superinsulated container that can withstand high pressures. Storage in this manner would be both expensive and hazardous because of the potential for explosion. Thus storage of hydrogen in the individual automobile as a liquid does not seem practical.

What about using hydrogen as a gas to fuel automobiles? One obvious problem with this is that when a liquid changes to a gas, its volume greatly increases. Storing this much hydrogen as a gas would require a huge tank. A much better alternative seems to be the use of metals that absorb hydrogen to form solid metal hydrides:

$$H_2(g) + M(s) \longrightarrow MH_2(s)$$

To use this method of storage, hydrogen gas would be pumped into a tank containing the solid metal in powdered form, where it would be absorbed to form the hydride, whose volume would be little more than that of the metal alone. This hydrogen would then be available for combustion in the engine by release of $H_2(g)$ from the hydride as needed:

$$MH_2(s) \longrightarrow M(s) + H_2(g)$$

Several types of solids that absorb hydrogen to form hydrides are being studied for use in hydrogen-powered vehicles. The most likely use of hydrogen in automobiles will be to power fuel cells. Several automobile manufacturers are experimenting with cars powered by hydrogen fuel cells.

Other Energy Alternatives

Many other energy sources are being considered for future use. The western states, especially Colorado, contain huge deposits of *oil shale*, which consists of a complex carbon-based material called *kerogen* contained in porous rock formations. These deposits have the potential of being a larger energy source than the vast petroleum deposits of the Middle East. The main problem with oil shale is that the trapped fuel is not fluid and cannot be pumped. To recover the fuel, the rock must be heated to a temperature of 250°C or higher to decompose the kerogen to smaller molecules that produce gaseous and liquid products. This process is expensive and yields large quantities of waste rock, which have a negative environmental impact.

Ethanol (C_2H_5OH) is another fuel with the potential to supplement, if not replace, gasoline. The most common method of producing ethanol is fermentation, a process in which sugar is changed to alcohol by the action of yeast. The sugar can come from virtually any source, including fruits and grains, although fuel-grade ethanol would probably come mostly from corn. Car engines can burn pure alcohol or *gasohol*, an alcohol–gasoline mixture (10% ethanol in gasoline), with little modification. Gasohol is now widely available in the United States. A fuel called E85, which is 85% ethanol and 15% gasoline, is also widely available for cars with "flex-fuel" engines. The use of pure alcohol as a motor fuel is not feasible in most of the United States because it does

Gasoline usage is as high as ever, and world petroleum supplies will eventually dwindle. One possible alternative to petroleum as a source of fuels and lubricants is vegetable oil—the same vegetable oil we now use to cook french fries. Researchers believe that the oils from soybeans, corn, canola, and sunflowers all have the potential to be used in cars as well as on salads.

The use of vegetable oil for fuel is not a new idea. Rudolf Diesel reportedly used peanut oil to run one of his engines at the Paris Exposition in 1900. In addition, ethyl alcohol has been used widely as a fuel in South America and as a fuel additive in the United States.

Biodiesel, a fuel made by esterifying the fatty acids found in vegetable oil, has some real advantages over regular diesel fuel. Biodiesel produces fewer pollutants such as particulates, carbon monoxide, and complex organic molecules, and since vegetable oils have no sulfur, there is no noxious sulfur dioxide in the exhaust gases. Also, biodiesel can run in existing engines with little modification. In addition, biodiesel is much more biodegradable than petroleum-based fuels, so spills cause less environmental damage.

Of course, biodiesel also has some serious drawbacks. The main one is that it costs about three times as much as regular diesel fuel. Biodiesel also produces more nitrogen oxides in the exhaust than conventional diesel fuel and is less stable in storage. Biodiesel also can leave more gummy deposits in engines and must be "winterized" by removing components that tend to solidify at low temperatures.

The best solution may be to use biodiesel as an additive to regular diesel fuel. One such fuel is known as B20 because it is 20% biodiesel and 80% conventional diesel fuel. B20 is especially attractive because of the higher lubricating ability of vegetable oils, thus reducing diesel engine wear.

Vegetable oils are also being looked at as replacements for motor oils and hydraulic fluids. Tests of a sunflower seed–based engine lubricant manufactured by Renewable Lubricants of Hartville, Ohio, have shown satisfactory lubricating ability while lowering particle emissions. In addition, Lou Honary and his colleagues at the University of Northern Iowa have developed BioSOY, a vegetable oil–based hydraulic fluid for use in heavy machinery.

Veggie oil fuels and lubricants seem to have a growing market as petroleum supplies wane and as environmental laws become more stringent. In Germany's Black Forest region, for example, environmental protection laws require that farm equipment use only vegetable oil fuels and lubricants. In the near future there may be veggie oil in your garage as well as in your kitchen.

Adapted from "Fill 'Er Up . . . with Veggie Oil," by Corinna Wu, as appeared in *Science News*, Vol. 154, December 5, 1998, p. 364.

Courtesy, National Biodiesel Board, Inc.

This promotion bus both advertises biodiesel and demonstrates its usefulness.

Xico Putini/Shutterstock.com

▲ Castor seed oil is one of the raw materials used in the production of biofuels.

not vaporize easily when temperatures are low. However, pure ethanol could be a very practical fuel in warm climates. For example, in Brazil, large quantities of ethanol fuel are being produced for cars.

Another potential source of liquid fuels is oil squeezed from seeds (*seed oil*). ◄ For example, some farmers in North Dakota, South Africa, and Australia are now using sunflower oil to replace diesel fuel. Oil seeds, found in a wide variety of plants, can be processed to produce an oil composed mainly of carbon and hydrogen, which of course reacts with oxygen to produce carbon dioxide, water, and heat. It is hoped that oil-seed plants can be developed that will thrive under soil and climatic conditions unsuitable for corn and wheat. The main advantage of seed oil as a fuel is that it is renewable. Ideally, fuel would be grown just like food crops.

FOR REVIEW

Key Terms

Section 7-1
energy
law of conservation of energy
potential energy
kinetic energy
heat
work
pathway
state function (property)
system
surroundings
exothermic
endothermic
thermodynamics
first law of thermodynamics
internal energy

Section 7-2
enthalpy

Section 7-3
calorimeter
calorimetry
heat capacity
specific heat capacity
molar heat capacity
constant-pressure calorimetry
constant-volume calorimetry

Section 7-4
Hess's law

Section 7-5
standard enthalpy of formation
standard state

Section 7-6
fossil fuels
petroleum
natural gas
coal
greenhouse effect

Energy

- The capacity to do work or produce heat
- Is conserved (first law of thermodynamics)
- Can be converted from one form to another
- Is a state function
- Potential energy: stored energy
- Kinetic energy: energy due to motion
- The internal energy for a system is the sum of its potential and kinetic energies
- The internal energy of a system can be changed by work and heat:

$$\Delta E = q + w$$

Work

- Force applied over a distance
- For an expanding/contracting gas
- Not a state function

$$w = -P\Delta V$$

Heat

- Energy flow due to a temperature difference
- Exothermic: energy as heat flows out of a system
- Endothermic: energy as heat flows into a system
- Not a state function
- Measured for chemical reactions by calorimetry

Enthalpy

- $H = E + PV$
- Is a state function
- Hess's law: the change in enthalpy in going from a given set of reactants to a given set of products is the same whether the process takes place in one step or a series of steps
- Standard enthalpies of formation ($\Delta H°_f$) can be used to calculate ΔH for a chemical reaction

$$\Delta H°_{\text{reaction}} = \Sigma\, n_p \Delta H_f^\circ(\text{products}) - \Sigma\, n_r \Delta H^\circ(\text{reactants})$$

Energy use

- Energy sources from fossil fuels are associated with difficult supply and environmental impact issues
- The greenhouse effect results from release into the atmosphere of gases, including carbon dioxide, that strongly absorb infrared radiation, thus warming the earth
- Alternative fuels are being sought to replace fossil fuels:
  - Hydrogen
  - Biofuels from plants such as corn and certain seed-producing plants

Review Questions
Answers to the Review Questions can be found on the Student Web site (accessible from **www.cengagebrain.com**).

1. Define the following terms: potential energy, kinetic energy, path-dependent function, state function, system, surroundings.

2. Consider the following potential energy diagrams for two different reactions.

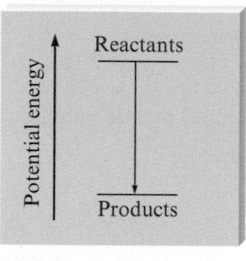
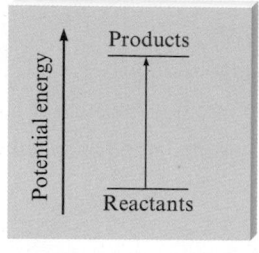

Which plot represents an exothermic reaction? In plot a, do the reactants on average have stronger or weaker bonds than the products? In plot b, reactants must gain potential energy to convert to products. How does this occur?

3. What is the first law of thermodynamics? How can a system change its internal energy, E? What are the sign conventions for thermodynamic quantities used in this text?

4. When a gas expands, what is the sign of w? Why? When a gas contracts, what is the sign of w? Why? What are the signs of q and w for the process of boiling water?

5. What is the heat gained/released at constant pressure equal to ($q_P = ?$)? What is the heat gained/released at constant volume equal to ($q_V = ?$)? Explain why ΔH is obtained directly from a coffee-cup calorimeter, whereas ΔE is obtained directly from a bomb calorimeter.

6. High-quality audio amplifiers generate large amounts of heat. To dissipate the heat and prevent damage to the electronic components, heat-radiating metal fins are used. Would it be better to make these fins out of iron or aluminum? Why? (See Table 7-1 for specific heat capacities.)

7. Explain how calorimetry works to calculate ΔH or ΔE for a reaction. Does the temperature of the calorimeter increase or decrease for an endothermic reaction? For an exothermic reaction? Explain.

8. What is Hess's law? When a reaction is reversed, what happens to the sign and magnitude of ΔH for that reversed reaction? When the coefficients in a balanced reaction are multiplied by a factor n, what happens to the sign and magnitude of ΔH for that multiplied reaction?

9. Define the standard enthalpy of formation. What are standard states for elements and for compounds? Using Hess's law, illustrate why the formula $\Delta H°_{reaction} = \Sigma n_p \Delta H°_f (\text{products}) - \Sigma n_r \Delta H°_f (\text{reactants})$ works to calculate $\Delta H°$ for a reaction.

10. What are some of the problems associated with the world's dependence on fossil fuels? What are some alternative fuels for petroleum products?

Active Learning Questions

These questions are designed to be used by groups of students in class.

1. Objects placed together eventually reach the same temperature. When you go into a room and touch a piece of metal in that room, it feels colder than a piece of plastic. Explain.

2. What is meant by the term *lower in energy*? Which is lower in energy, a mixture of hydrogen and oxygen gases or liquid water? How do you know? Which of the two is more stable? How do you know?

3. A fire is started in a fireplace by striking a match and lighting crumpled paper under some logs. Explain all the energy transfers in this scenario using the terms *exothermic, endothermic, system, surroundings, potential energy,* and *kinetic energy* in the discussion.

4. Liquid water turns to ice. Is this process endothermic or exothermic? Explain what is occurring using the terms *system, surroundings, heat, potential energy,* and *kinetic energy* in the discussion.

5. Consider the following statements: "Heat is a form of energy, and energy is conserved. The heat lost by a system must be equal to the amount of heat gained by the surroundings. Therefore, heat is conserved." Indicate everything you think is correct in these statements. Indicate everything you think is incorrect. Correct the incorrect statements and explain.

6. Consider 5.5 L of a gas at a pressure of 3.0 atm in a cylinder with a movable piston. The external pressure is changed so that the volume changes to 10.5 L.

a. Calculate the work done, and indicate the correct sign.

b. Use the preceding data but consider the process to occur in two steps. At the end of the first step, the volume is 7.0 L. The second step results in a final volume of 10.5 L. Calculate the work done, and indicate the correct sign.

c. Calculate the work done if after the first step the volume is 8.0 L and the second step leads to a volume of 10.5 L. Does the work differ from that in part b? Explain.

7. In Question 6 the work calculated for the different conditions in the various parts of the question was different even though the system had the same initial and final conditions. Based on this information, is work a state function?

a. Explain how you know that work is not a state function.

b. Why does the work increase with an increase in the number of steps?

c. Which two-step process resulted in more work, when the first step had the bigger change in volume or when the second step had the bigger change in volume? Explain.

8. Explain why oceanfront areas generally have smaller temperature fluctuations than inland areas.

9. Hess's law is really just another statement of the first law of thermodynamics. Explain.

10. In the equation $w = -P\Delta V$, why is there a negative sign?

A blue question or exercise number indicates that the answer to that question or exercise appears at the back of this book and a solution appears in the *Solutions Guide*, as found on PowerLecture.

Questions

11. Consider an airplane trip from Chicago, Illinois, to Denver, Colorado. List some path-dependent functions and some state functions for the plane trip.

12. How is average bond strength related to relative potential energies of the reactants and the products?

13. Assuming gasoline is pure $C_8H_{18}(l)$, predict the signs of q and w for the process of combusting gasoline into $CO_2(g)$ and $H_2O(g)$.

14. What is the difference between ΔH and ΔE?

15. The enthalpy change for the reaction

$$CH_4(g) + 2O_2(g) \longrightarrow CO_2(g) + 2H_2O(l)$$

is -891 kJ for the reaction *as written*.

a. What quantity of heat is released for each mole of water formed?

b. What quantity of heat is released for each mole of oxygen reacted?

16. For the reaction $HgO(s) \rightarrow Hg(l) + \frac{1}{2}O_2(g)$, $\Delta H = +90.7$ kJ:

a. What quantity of heat is required to produce 1 mole of mercury by this reaction?

b. What quantity of heat is required to produce 1 mole of oxygen gas by this reaction?

c. What quantity of heat would be released in the following reaction as written?

$$2Hg(l) + O_2(g) \longrightarrow 2HgO(s)$$

17. The enthalpy of combustion of $CH_4(g)$ when $H_2O(l)$ is formed is -891 kJ/mol and the enthalpy of combustion of $CH_4(g)$ when $H_2O(g)$ is formed is -803 kJ/mol. Use these data and Hess's law to determine the enthalpy of vaporization for water.

18. The enthalpy change for a reaction is a state function and it is an extensive property. Explain.

19. Standard enthalpies of formation are relative values. What are ΔH_f° values relative to?

20. The combustion of methane can be represented as follows:

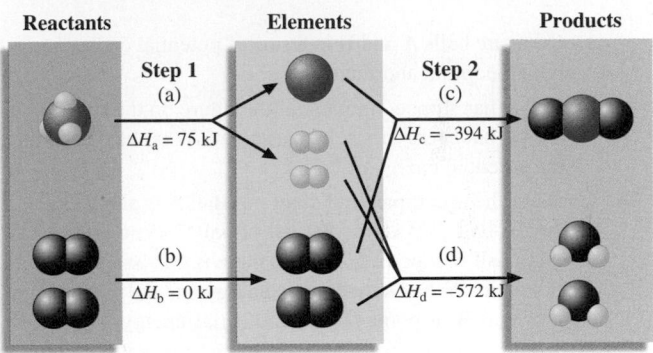

a. Use the information given above to determine the value of ΔH for the combustion of methane to form $CO_2(g)$ and $2H_2O(l)$.

b. What is ΔH_f° for an element in its standard state? Why is this? Use the figure above to support your answer.

c. How does ΔH for the reaction $CO_2(g) + 2H_2O(l) \rightarrow CH_4(g) + O_2(g)$ compare to that of the combustion of methane? Why is this?

21. Why is it a good idea to rinse your thermos bottle with hot water before filling it with hot coffee?

22. Photosynthetic plants use the following reaction to produce glucose, cellulose, and so forth:

$$6CO_2(g) + 6H_2O(l) \xrightarrow{\text{Sunlight}} C_6H_{12}O_6(s) + 6O_2(g)$$

How might extensive destruction of forests exacerbate the greenhouse effect?

23. What is incomplete combustion of fossil fuels? Why can this be a problem?

24. Explain the advantages and disadvantages of hydrogen as an alternative fuel.

Exercises

In this section similar exercises are paired.

Potential and Kinetic Energy

25. Calculate the kinetic energy of a baseball (mass = 5.25 oz) with a velocity of 1.0×10^2 mi/h.

26. Which has the greater kinetic energy, an object with a mass of 2.0 kg and a velocity of 1.0 m/s or an object with a mass of 1.0 kg and a velocity of 2.0 m/s?

27. Consider the following diagram when answering the questions below.

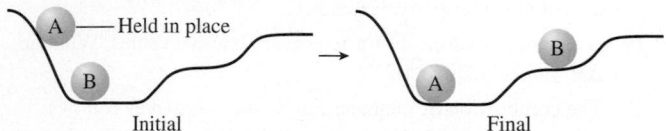

    Initial   Final

  a. Compare balls A and B in terms of potential energy in both the initial and final setups.

  b. Ball A has stopped moving in the figure on the right above, but energy must be conserved. What happened to the potential energy of ball A?

28. Consider the accompanying diagram. Ball A is allowed to fall and strike ball B. Assume that all of ball A's energy is transferred to ball B at point I, and that there is no loss of energy to other sources. What is the kinetic energy and the potential energy of ball B at point II? The potential energy is given by $PE = mgz$, where m is the mass in kilograms, g is the gravitational constant (9.81 m/s²), and z is the distance in meters.

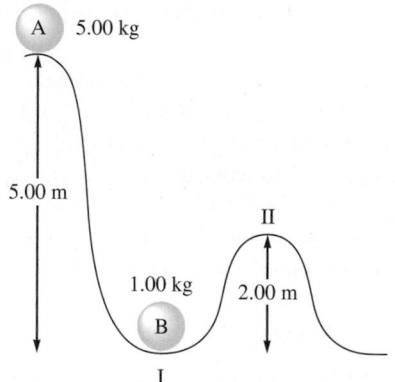

Heat and Work

29. A gas absorbs 45 kJ of heat and does 29 kJ of work. Calculate ΔE.

30. A system releases 125 kJ of heat while 104 kJ of work is done on it. Calculate ΔE.

31. Calculate ΔE for each of the following.

  a. $q = -47$ kJ, $w = +88$ kJ

  b. $q = +82$ kJ, $w = -47$ kJ

  c. $q = +47$ kJ, $w = 0$

  d. In which of these cases do the surroundings do work on the system?

32. A system undergoes a process consisting of the following two steps:

Step 1: The system absorbs 72 J of heat while 35 J of work is done on it.

Step 2: The system absorbs 35 J of heat while performing 72 J of work.

Calculate ΔE for the overall process.

33. If the internal energy of a thermodynamic system is increased by 300. J while 75 J of expansion work is done, how much heat was transferred and in which direction, to or from the system?

34. Calculate the internal energy change for each of the following.

  a. One hundred (100.) joules of work is required to compress a gas. At the same time, the gas releases 23 J of heat.

  b. A piston is compressed from a volume of 8.30 L to 2.80 L against a constant pressure of 1.90 atm. In the process, there is a heat gain by the system of 350. J.

  c. A piston expands against 1.00 atm of pressure from 11.2 L to 29.1 L. In the process, 1037 J of heat is absorbed.

35. A sample of an ideal gas at 15.0 atm and 10.0 L is allowed to expand against a constant external pressure of 2.00 atm to a volume of 75.0 L. Calculate the work in units of kJ for the gas expansion.

36. A piston performs work of 210. L · atm on the surroundings, while the cylinder in which it is placed expands from 10. L to 25 L. At the same time, 45 J of heat is transferred from the surroundings to the system. Against what pressure was the piston working?

37. Consider a mixture of air and gasoline vapor in a cylinder with a piston. The original volume is 40. cm³. If the combustion of this mixture releases 950. J of energy, to what volume will the gases expand against a constant pressure of 650. torr if all the energy of combustion is converted into work to push back the piston?

38. As a system increases in volume, it absorbs 52.5 J of energy in the form of heat from the surroundings. The piston is working against a pressure of 0.500 atm. The final volume of the system is 58.0 L. What was the initial volume of the system if the internal energy of the system decreased by 102.5 J?

39. A balloon filled with 39.1 moles of helium has a volume of 876 L at 0.0°C and 1.00 atm pressure. The temperature of the balloon is increased to 38.0°C as it expands to a volume of 998 L, the pressure remaining constant. Calculate q, w, and ΔE for the helium in the balloon. (The molar heat capacity for helium gas is 20.8 J/°C · mol.)

40. One mole of $H_2O(g)$ at 1.00 atm and 100.°C occupies a volume of 30.6 L. When 1 mole of $H_2O(g)$ is condensed to 1 mole of $H_2O(l)$ at 1.00 atm and 100.°C, 40.66 kJ of heat is released. If the density of $H_2O(l)$ at this temperature and pressure is 0.996 g/cm³, calculate ΔE for the condensation of 1 mole of water at 1.00 atm and 100.°C.

Properties of Enthalpy

41. One of the components of polluted air is NO. It is formed in the high-temperature environment of internal combustion engines by the following reaction:

$$N_2(g) + O_2(g) \longrightarrow 2NO(g) \qquad \Delta H = 180 \text{ kJ}$$

Why are high temperatures needed to convert N_2 and O_2 to NO?

42. The reaction

$$SO_3(g) + H_2O(l) \longrightarrow H_2SO_4(aq)$$

is the last step in the commercial production of sulfuric acid. The enthalpy change for this reaction is -227 kJ. In designing a sulfuric acid plant, is it necessary to provide for heating or cooling of the reaction mixture? Explain.

43. Are the following processes exothermic or endothermic?
    a. When solid KBr is dissolved in water, the solution gets colder.
    b. Natural gas (CH_4) is burned in a furnace.
    c. When concentrated H_2SO_4 is added to water, the solution gets very hot.
    d. Water is boiled in a teakettle.

44. Are the following processes exothermic or endothermic?
    a. the combustion of gasoline in a car engine
    b. water condensing on a cold pipe
    c. $CO_2(s) \longrightarrow CO_2(g)$
    d. $F_2(g) \longrightarrow 2F(g)$

45. The overall reaction in a commercial heat pack can be represented as

$$4Fe(s) + 3O_2(g) \longrightarrow 2Fe_2O_3(s) \qquad \Delta H = -1652 \text{ kJ}$$

    a. How much heat is released when 4.00 moles of iron are reacted with excess O_2?
    b. How much heat is released when 1.00 mole of Fe_2O_3 is produced?
    c. How much heat is released when 1.00 g iron is reacted with excess O_2?
    d. How much heat is released when 10.0 g Fe and 2.00 g O_2 are reacted?

46. Consider the following reaction:

$$2H_2(g) + O_2(g) \longrightarrow 2H_2O(l) \qquad \Delta H = -572 \text{ kJ}$$

    a. How much heat is evolved for the production of 1.00 mole of $H_2O(l)$?
    b. How much heat is evolved when 4.03 g hydrogen are reacted with excess oxygen?
    c. How much heat is evolved when 186 g oxygen are reacted with excess hydrogen?

47. Consider the combustion of propane:

$$C_3H_8(g) + 5O_2(g) \longrightarrow 3CO_2(g) + 4H_2O(l) \qquad \Delta H = -2221 \text{ kJ}$$

Assume that all the heat in Example 7-3 comes from the combustion of propane. What mass of propane must be burned to furnish this amount of energy assuming the heat transfer process is 60.% efficient?

48. Consider the following reaction:

$$CH_4(g) + 2O_2(g) \longrightarrow CO_2(g) + 2H_2O(l) \qquad \Delta H = -891 \text{ kJ}$$

Calculate the enthalpy change for each of the following cases:
    a. 1.00 g methane is burned in excess oxygen.
    b. 1.00×10^3 L methane gas at 740. torr and 25°C are burned in excess oxygen. The density of $CH_4(g)$ at these conditions is 0.639 g/L.

Calorimetry and Heat Capacity

49. Consider the substances in Table 7-1. Which substance requires the largest amount of energy to raise the temperature of 25.0 g of the substance from 15.0°C to 37.0°C? Calculate the energy. Which substance in Table 7-1 has the largest temperature change when 550. g of the substance absorbs 10.7 kJ of energy? Calculate the temperature change.

50. The specific heat capacity of silver is 0.24 J/°C · g.
    a. Calculate the energy required to raise the temperature of 150.0 g Ag from 273 K to 298 K.
    b. Calculate the energy required to raise the temperature of 1.0 mole of Ag by 1.0°C (called the *molar heat capacity* of silver).
    c. It takes 1.25 kJ of energy to heat a sample of pure silver from 12.0°C to 15.2°C. Calculate the mass of the sample of silver.

51. A 5.00-g sample of one of the substances listed in Table 7-1 was heated from 25.2°C to 55.1°C, requiring 133 J to do so. Which substance was it?

52. It takes 585 J of energy to raise the temperature of 125.6 g mercury from 20.0°C to 53.5°C. Calculate the specific heat capacity and the molar heat capacity of mercury.

53. A 30.0-g sample of water at 280. K is mixed with 50.0 g water at 330. K. Calculate the final temperature of the mixture assuming no heat loss to the surroundings.

54. A biology experiment requires the preparation of a water bath at 37.0°C (body temperature). The temperature of the cold tap water is 22.0°C, and the temperature of the hot tap water is 55.0°C. If a student starts with 90.0 g cold water, what mass of hot water must be added to reach 37.0°C?

55. A 5.00-g sample of aluminum pellets (specific heat capacity = 0.89 J/°C · g) and a 10.00-g sample of iron pellets (specific heat capacity = 0.45 J/°C · g) are heated to 100.0°C. The mixture of hot iron and aluminum is then dropped into 97.3 g water at 22.0°C. Calculate the final temperature of the metal and water mixture, assuming no heat loss to the surroundings.

56. Hydrogen gives off 120. J/g of energy when burned in oxygen, and methane gives off 50. J/g under the same circumstances. If a mixture of 5.0 g hydrogen and 10. g methane is burned, and the heat released is transferred to 50.0 g water at 25.0°C, what final temperature will be reached by the water?

57. A 150.0-g sample of a metal at 75.0°C is added to 150.0 g H_2O at 15.0°C. The temperature of the water rises to 18.3°C. Calculate the specific heat capacity of the metal, assuming that all the heat lost by the metal is gained by the water.

58. A 110.-g sample of copper (specific heat capacity = 0.20 J/°C · g) is heated to 82.4°C and then placed in a container of water at 22.3°C. The final temperature of the water and copper is 24.9°C. What is the mass of the water in the container, assuming that all the heat lost by the copper is gained by the water?

59. In a coffee-cup calorimeter, 50.0 mL of 0.100 *M* $AgNO_3$ and 50.0 mL of 0.100 *M* HCl are mixed to yield the following reaction:

$$Ag^+(aq) + Cl^-(aq) \longrightarrow AgCl(s)$$

The two solutions were initially at 22.60°C, and the final temperature is 23.40°C. Calculate the heat that accompanies this reaction in kJ/mol of AgCl formed. Assume that the combined solution has a mass of 100.0 g and a specific heat capacity of 4.18 J/°C · g.

60. In a coffee-cup calorimeter, 100.0 mL of 1.0 *M* NaOH and 100.0 mL of 1.0 *M* HCl are mixed. Both solutions were originally at 24.6°C. After the reaction, the final temperature is 31.3°C. Assuming that all the solutions have a density of 1.0 g/cm³ and a specific heat capacity of 4.18 J/°C · g, calculate the enthalpy change for the neutralization of HCl by NaOH. Assume that no heat is lost to the surroundings or to the calorimeter.

61. A coffee-cup calorimeter initially contains 125 g water at 24.2°C. Potassium bromide (10.5 g), also at 24.2°C, is added to the water, and after the KBr dissolves, the final temperature is 21.1°C. Calculate the enthalpy change for dissolving the salt in J/g and kJ/mol. Assume that the specific heat capacity of the solution is 4.18 J/°C · g and that no heat is transferred to the surroundings or to the calorimeter.

62. In a coffee-cup calorimeter, 1.60 g NH_4NO_3 is mixed with 75.0 g water at an initial temperature of 25.00°C. After dissolution of the salt, the final temperature of the calorimeter contents is 23.34°C. Assuming the solution has a heat capacity of 4.18 J/°C · g and assuming no heat loss to the calorimeter, calculate the enthalpy change for the dissolution of NH_4NO_3 in units of kJ/mol.

63. Consider the dissolution of $CaCl_2$:

$$CaCl_2(s) \longrightarrow Ca^{2+}(aq) + 2Cl^-(aq) \quad \Delta H = -81.5 \text{ kJ}$$

An 11.0-g sample of $CaCl_2$ is dissolved in 125 g water, with both substances at 25.0°C. Calculate the final temperature of the solution assuming no heat loss to the surroundings

and assuming the solution has a specific heat capacity of 4.18 J/°C · g.

64. Consider the reaction

$$2HCl(aq) + Ba(OH)_2(aq) \longrightarrow BaCl_2(aq) + 2H_2O(l)$$
$$\Delta H = -118 \text{ kJ}$$

Calculate the heat when 100.0 mL of 0.500 *M* HCl is mixed with 300.0 mL of 0.100 *M* $Ba(OH)_2$. Assuming that the temperature of both solutions was initially 25.0°C and that the final mixture has a mass of 400.0 g and a specific heat capacity of 4.18 J/°C · g, calculate the final temperature of the mixture.

65. The heat capacity of a bomb calorimeter was determined by burning 6.79 g methane (energy of combustion = −802 kJ/mol CH_4) in the bomb. The temperature changed by 10.8°C.

a. What is the heat capacity of the bomb?

b. A 12.6-g sample of acetylene, C_2H_2, produced a temperature increase of 16.9°C in the same calorimeter. What is the energy of combustion of acetylene (in kJ/mol)?

66. The combustion of 0.1584 g benzoic acid increases the temperature of a bomb calorimeter by 2.54°C. Calculate the heat capacity of this calorimeter. (The energy released by combustion of benzoic acid is 26.42 kJ/g.) A 0.2130-g sample of vanillin ($C_8H_8O_3$) is then burned in the same calorimeter, and the temperature increases by 3.25°C. What is the energy of combustion per gram of vanillin? Per mole of vanillin?

Hess's Law

67. The enthalpy of combustion of solid carbon to form carbon dioxide is −393.7 kJ/mol carbon, and the enthalpy of combustion of carbon monoxide to form carbon dioxide is −283.3 kJ/mol CO. Use these data to calculate ΔH for the reaction

$$2C(s) + O_2(g) \longrightarrow 2CO(g)$$

68. Combustion reactions involve reacting a substance with oxygen. When compounds containing carbon and hydrogen are combusted, carbon dioxide and water are the products. Using the enthalpies of combustion for C_4H_4 (−2341 kJ/mol), C_4H_8 (−2755 kJ/mol), and H_2 (−286 kJ/mol), calculate ΔH for the reaction

$$C_4H_4(g) + 2H_2(g) \longrightarrow C_4H_8(g)$$

69. Given the following data

● N   ○ H   ● O

calculate ΔH for the reaction

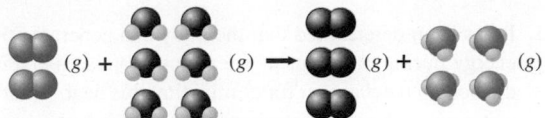

On the basis of the enthalpy change, is this a useful reaction for the synthesis of ammonia?

70. Given the following data

$$2ClF(g) + O_2(g) \longrightarrow Cl_2O(g) + F_2O(g) \qquad \Delta H = 167.4 \text{ kJ}$$
$$2ClF_3(g) + 2O_2(g) \longrightarrow Cl_2O(g) + 3F_2O(g) \qquad \Delta H = 341.4 \text{ kJ}$$
$$2F_2(g) + O_2(g) \longrightarrow 2F_2O(g) \qquad \Delta H = -43.4 \text{ kJ}$$

calculate ΔH for the reaction

$$ClF(g) + F_2(g) \longrightarrow ClF_3(g)$$

71. Given the following data

$$2O_3(g) \longrightarrow 3O_2(g) \qquad \Delta H = -427 \text{ kJ}$$
$$O_2(g) \longrightarrow 2O(g) \qquad \Delta H = 495 \text{ kJ}$$
$$NO(g) + O_3(g) \longrightarrow NO_2(g) + O_2(g) \qquad \Delta H = -199 \text{ kJ}$$

calculate ΔH for the reaction

$$NO(g) + O(g) \longrightarrow NO_2(g)$$

72. Calculate ΔH for the reaction

$$2NH_3(g) + \tfrac{1}{2}O_2(g) \longrightarrow N_2H_4(l) + H_2O(l)$$

given the following data:

$$2NH_3(g) + 3N_2O(g) \longrightarrow 4N_2(g) + 3H_2O(l) \qquad \Delta H = -1010.\text{ kJ}$$
$$N_2O(g) + 3H_2(g) \longrightarrow N_2H_4(l) + H_2O(l) \qquad \Delta H = -317 \text{ kJ}$$
$$N_2H_4(l) + O_2(g) \longrightarrow N_2(g) + 2H_2O(l) \qquad \Delta H = -623 \text{ kJ}$$
$$H_2(g) + \tfrac{1}{2}O_2(g) \longrightarrow H_2O(l) \qquad \Delta H = -286 \text{ kJ}$$

73. Given the following data

$$Ca(s) + 2C(graphite) \longrightarrow CaC_2(s) \qquad \Delta H = -62.8 \text{ kJ}$$
$$Ca(s) + \tfrac{1}{2}O_2(g) \longrightarrow CaO(s) \qquad \Delta H = -635.5 \text{ kJ}$$
$$CaO(s) + H_2O(l) \longrightarrow Ca(OH)_2(aq) \qquad \Delta H = -653.1 \text{ kJ}$$
$$C_2H_2(g) + \tfrac{5}{2}O_2(g) \longrightarrow 2CO_2(g) + H_2O(l) \qquad \Delta H = -1300.\text{ kJ}$$
$$C(graphite) + O_2(g) \longrightarrow CO_2(g) \qquad \Delta H = -393.5 \text{ kJ}$$

calculate ΔH for the reaction

$$CaC_2(s) + 2H_2O(l) \longrightarrow Ca(OH)_2(aq) + C_2H_2(g)$$

74. Given the following data

$$P_4(s) + 6Cl_2(g) \longrightarrow 4PCl_3(g) \qquad \Delta H = -1225.6 \text{ kJ}$$
$$P_4(s) + 5O_2(g) \longrightarrow P_4O_{10}(s) \qquad \Delta H = -2967.3 \text{ kJ}$$
$$PCl_3(g) + Cl_2(g) \longrightarrow PCl_5(g) \qquad \Delta H = -84.2 \text{ kJ}$$
$$PCl_3(g) + \tfrac{1}{2}O_2(g) \longrightarrow Cl_3PO(g) \qquad \Delta H = -285.7 \text{ kJ}$$

calculate ΔH for the reaction

$$P_4O_{10}(s) + 6PCl_5(g) \longrightarrow 10Cl_3PO(g)$$

Standard Enthalpies of Formation

75. Give the definition of the standard enthalpy of formation for a substance. Write separate reactions for the formation of NaCl, H_2O, $C_6H_{12}O_6$, and $PbSO_4$ that have $\Delta H°$ values equal to $\Delta H_f°$ for each compound.

76. Write reactions for which the enthalpy change will be

  a. $\Delta H_f°$ for solid aluminum oxide.

  b. the standard enthalpy of combustion of liquid ethanol, $C_2H_5OH(l)$.

  c. the standard enthalpy of neutralization of sodium hydroxide solution by hydrochloric acid.

  d. $\Delta H_f°$ for gaseous vinyl chloride, $C_2H_3Cl(g)$.

  e. the enthalpy of combustion of liquid benzene, $C_6H_6(l)$.

  f. the enthalpy of solution of solid ammonium bromide.

77. Use the values of $\Delta H_f°$ in Appendix 4 to calculate $\Delta H°$ for the following reactions.

  a.

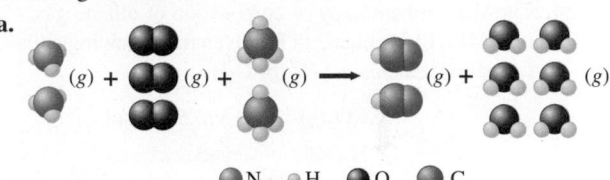

$\bullet$ N $\quad \circ$ H $\quad \bullet$ O $\quad \bullet$ C

  b. $Ca_3(PO_4)_2(s) + 3H_2SO_4(l) \longrightarrow 3CaSO_4(s) + 2H_3PO_4(l)$

  c. $NH_3(g) + HCl(g) \longrightarrow NH_4Cl(s)$

78. Use the values of $\Delta H_f°$ in Appendix 4 to calculate $\Delta H°$ for the following reactions. (See Exercise 77.)

  a.

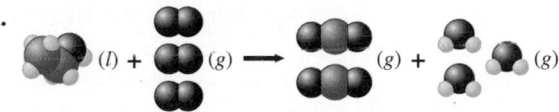

  b. $SiCl_4(l) + 2H_2O(l) \longrightarrow SiO_2(s) + 4HCl(aq)$

  c. $MgO(s) + H_2O(l) \longrightarrow Mg(OH)_2(s)$

79. The Ostwald process for the commercial production of nitric acid from ammonia and oxygen involves the following steps:

$$4NH_3(g) + 5O_2(g) \longrightarrow 4NO(g) + 6H_2O(g)$$
$$2NO(g) + O_2(g) \longrightarrow 2NO_2(g)$$
$$3NO_2(g) + H_2O(l) \longrightarrow 2HNO_3(aq) + NO(g)$$

  a. Use the values of $\Delta H_f°$ in Appendix 4 to calculate the value of $\Delta H°$ for each of the preceding reactions.

  b. Write the overall equation for the production of nitric acid by the Ostwald process by combining the preceding equations. (Water is also a product.) Is the overall reaction exothermic or endothermic?

80. Calculate $\Delta H°$ for each of the following reactions using the data in Appendix 4:

$$4Na(s) + O_2(g) \longrightarrow 2Na_2O(s)$$
$$2Na(s) + 2H_2O(l) \longrightarrow 2NaOH(aq) + H_2(g)$$
$$2Na(s) + CO_2(g) \longrightarrow Na_2O(s) + CO(g)$$

Explain why a water or carbon dioxide fire extinguisher might not be effective in putting out a sodium fire.

81. The reusable booster rockets of the space shuttle use a mixture of aluminum and ammonium perchlorate as fuel. A possible reaction is

$$3Al(s) + 3NH_4ClO_4(s) \longrightarrow$$
$$Al_2O_3(s) + AlCl_3(s) + 3NO(g) + 6H_2O(g)$$

Calculate $\Delta H°$ for this reaction.

82. The space shuttle *Orbiter* utilizes the oxidation of methylhydrazine by dinitrogen tetroxide for propulsion:

$$4N_2H_3CH_3(l) + 5N_2O_4(l) \longrightarrow 12H_2O(g) + 9N_2(g) + 4CO_2(g)$$

Calculate $\Delta H°$ for this reaction.

83. Consider the reaction

$$2ClF_3(g) + 2NH_3(g) \longrightarrow N_2(g) + 6HF(g) + Cl_2(g)$$
$$\Delta H° = -1196 \text{ kJ}$$

Calculate $\Delta H_f°$ for $ClF_3(g)$.

84. The standard enthalpy of combustion of ethene gas, $C_2H_4(g)$, is -1411.1 kJ/mol at 298 K. Given the following enthalpies of formation, calculate $\Delta H_f°$ for $C_2H_4(g)$.

$$CO_2(g) \quad -393.5 \text{ kJ/mol}$$
$$H_2O(l) \quad -285.8 \text{ kJ/mol}$$

Energy Consumption and Sources

85. Water gas is produced from the reaction of steam with coal:

$$C(s) + H_2O(g) \longrightarrow H_2(g) + CO(g)$$

Assuming that coal is pure graphite, calculate $\Delta H°$ for this reaction.

86. Syngas can be burned directly or converted to methanol. Calculate $\Delta H°$ for the reaction

$$CO(g) + 2H_2(g) \longrightarrow CH_3OH(l)$$

87. Ethanol (C_2H_5OH) has been proposed as an alternative fuel. Calculate the standard enthalpy of combustion per gram of liquid ethanol.

88. Methanol (CH_3OH) has also been proposed as an alternative fuel. Calculate the standard enthalpy of combustion per gram of liquid methanol, and compare this answer to that for ethanol in Exercise 87.

89. Some automobiles and buses have been equipped to burn propane (C_3H_8). Compare the amounts of energy that can be obtained per gram of $C_3H_8(g)$ and per gram of gasoline, assuming that gasoline is pure octane, $C_8H_{18}(l)$. (See Example 7-11.) Look up the boiling point of propane. What disadvantages are there to using propane instead of gasoline as a fuel?

90. The complete combustion of acetylene, $C_2H_2(g)$, produces 1300. kJ of energy per mole of acetylene consumed. How many grams of acetylene must be burned to produce enough heat to raise the temperature of 1.00 gal water by 10.0°C if the process is 80.0% efficient? Assume the density of water is 1.00 g/cm³.

Additional Exercises

91. It has been determined that the body can generate 5500 kJ of energy during one hour of strenuous exercise. Perspiration is the body's mechanism for eliminating this heat. What mass of water would have to be evaporated through perspiration to rid the body of the heat generated during 2 hours of exercise? (The heat of vaporization of water is 40.6 kJ/mol.)

92. One way to lose weight is to exercise! Walking briskly at 4.0 miles per hour for an hour consumes about 400 kcal of energy. How many hours would you have to walk at 4.0 miles per hour to lose one pound of body fat? One gram of body fat is equivalent to 7.7 kcal of energy. There are 454 g in 1 lb.

93. Three gas-phase reactions were run in a constant-pressure piston apparatus as shown in the following illustration. For each reaction, give the balanced reaction and predict the sign of w (the work done) for the reaction.

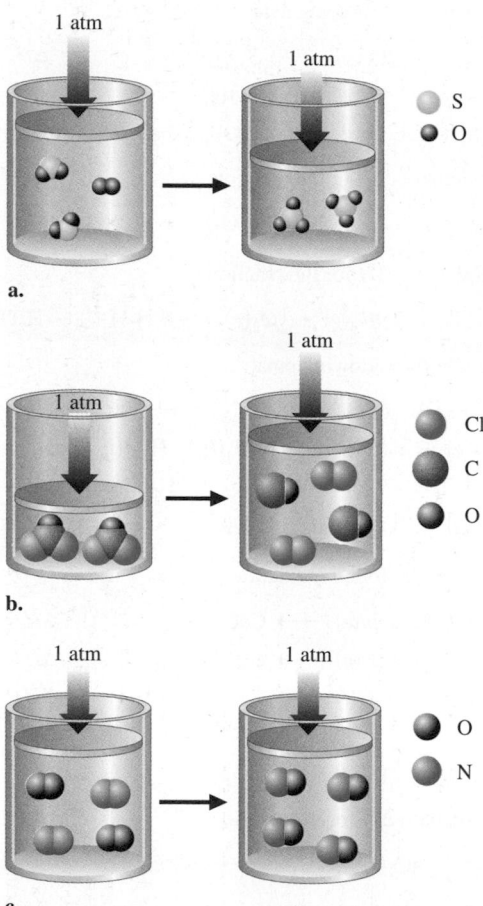

a.

b.

c.

If just the balanced reactions were given, how could you predict the sign of w for a reaction?

94. Nitrogen gas ⬤ reacts with hydrogen gas ◯ to form ammonia gas ◑. Consider the reaction between nitrogen and hydrogen as depicted below:

1 atm

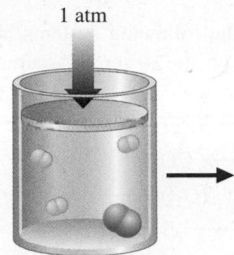

a. Draw what the container will look like after the reaction has gone to completion. Assume a constant pressure of 1 atm.

b. Is the sign of work positive or negative, or is the value of work equal to zero for the reaction? Explain your answer.

95. Combustion of table sugar produces $CO_2(g)$ and $H_2O(l)$. When 1.46 g table sugar is combusted in a constant-volume (bomb) calorimeter, 24.00 kJ of heat is liberated.

a. Assuming that table sugar is pure sucrose, $C_{12}H_{22}O_{11}(s)$, write the balanced equation for the combustion reaction.

b. Calculate ΔE in kJ/mol $C_{12}H_{22}O_{11}$ for the combustion reaction of sucrose.

96. Consider the following changes:

a. $N_2(g) \longrightarrow N_2(l)$
b. $CO(g) + H_2O(g) \longrightarrow H_2(g) + CO_2(g)$
c. $Ca_3P_2(s) + 6H_2O(l) \longrightarrow 3Ca(OH)_2(s) + 2PH_3(g)$
d. $2CH_3OH(l) + 3O_2(g) \longrightarrow 2CO_2(g) + 4H_2O(l)$
e. $I_2(s) \longrightarrow I_2(g)$

At constant temperature and pressure, in which of these changes is work done by the system on the surroundings? By the surroundings on the system? In which of them is no work done?

97. Consider the following cyclic process carried out in two steps on a gas:

Step 1: 45 J of heat is added to the gas, and 10. J of expansion work is performed.

Step 2: 60. J of heat is removed from the gas as the gas is compressed back to the initial state.

Calculate the work for the gas compression in Step 2.

98. Calculate $\Delta H°$ for the reaction

$$2K(s) + 2H_2O(l) \longrightarrow 2KOH(aq) + H_2(g)$$

A 5.00-g chunk of potassium is dropped into 1.00 kg water at 24.0°C. What is the final temperature of the water after the preceding reaction occurs? Assume that all the heat is used to raise the temperature of the water. (Never run this reaction. It is very dangerous; it bursts into flame!)

99. The enthalpy of neutralization for the reaction of a strong acid with a strong base is −56 kJ/mol water produced. How much energy will be released when 200.0 mL of 0.400 M HNO_3 is mixed with 150.0 mL of 0.500 M KOH?

100. When 1.00 L of 2.00 M Na_2SO_4 solution at 30.0°C is added to 2.00 L of 0.750 M $Ba(NO_3)_2$ solution at 30.0°C in a calorimeter, a white solid ($BaSO_4$) forms. The temperature of the mixture increases to 42.0°C. Assuming that the specific heat capacity of the solution is 6.37 J/°C · g and that the density of the final solution is 2.00 g/mL, calculate the enthalpy change per mole of $BaSO_4$ formed.

101. If a student performs an endothermic reaction in a calorimeter, how does the calculated value of ΔH differ from the actual value if the heat exchanged with the calorimeter is not taken into account?

102. In a bomb calorimeter, the reaction vessel is surrounded by water that must be added for each experiment. Since the amount of water is not constant from experiment to experiment, the mass of water must be measured in each case. The heat capacity of the calorimeter is broken down into two parts: the water and the calorimeter components. If a calorimeter contains 1.00 kg water and has a total heat capacity of 10.84 kJ/°C, what is the heat capacity of the calorimeter components?

103. The bomb calorimeter in Exercise 102 is filled with 987 g water. The initial temperature of the calorimeter contents is 23.32°C. A 1.056-g sample of benzoic acid ($\Delta E_{comb} = -26.42$ kJ/g) is combusted in the calorimeter. What is the final temperature of the calorimeter contents?

104. Consider the two space shuttle fuel reactions in Exercises 81 and 82. Which reaction produces more energy per kilogram of reactant mixture (stoichiometric amounts)?

105. Consider the following equations:

$$3A + 6B \longrightarrow 3D \qquad \Delta H = -403 \text{ kJ/mol}$$
$$E + 2F \longrightarrow A \qquad \Delta H = -105.2 \text{ kJ/mol}$$
$$C \longrightarrow E + 3D \qquad \Delta H = 64.8 \text{ kJ/mol}$$

Suppose the first equation is reversed and multiplied by $\frac{1}{6}$, the second and third equations are divided by 2, and the three adjusted equations are added. What is the net reaction and what is the overall heat of this reaction?

106. Given the following data

$$Fe_2O_3(s) + 3CO(g) \longrightarrow 2Fe(s) + 3CO_2(g) \qquad \Delta H° = -23 \text{ kJ}$$
$$3Fe_2O_3(s) + CO(g) \longrightarrow 2Fe_3O_4(s) + CO_2(g) \qquad \Delta H° = -39 \text{ kJ}$$
$$Fe_3O_4(s) + CO(g) \longrightarrow 3FeO(s) + CO_2(g) \qquad \Delta H° = 18 \text{ kJ}$$

calculate $\Delta H°$ for the reaction

$$FeO(s) + CO(g) \longrightarrow Fe(s) + CO_2(g)$$

107. At 298 K, the standard enthalpies of formation for $C_2H_2(g)$ and $C_6H_6(l)$ are 227 kJ/mol and 49 kJ/mol, respectively.

    a. Calculate $\Delta H°$ for

    $$C_6H_6(l) \longrightarrow 3C_2H_2(g)$$

    b. Both acetylene (C_2H_2) and benzene (C_6H_6) can be used as fuels. Which compound would liberate more energy per gram when combusted in air?

108. Using the following data, calculate the standard heat of formation of $ICl(g)$ in kJ/mol:

$Cl_2(g) \longrightarrow 2Cl(g)$	$\Delta H° = 242.3$ kJ
$I_2(g) \longrightarrow 2I(g)$	$\Delta H° = 151.0$ kJ
$ICl(g) \longrightarrow I(g) + Cl(g)$	$\Delta H° = 211.3$ kJ
$I_2(s) \longrightarrow I_2(g)$	$\Delta H° = 62.8$ kJ

109. A sample of nickel is heated to 99.8°C and placed in a coffee-cup calorimeter containing 150.0 g water at 23.5°C. After the metal cools, the final temperature of metal and water mixture is 25.0°C. If the specific heat capacity of nickel is 0.444 J/°C · g, what mass of nickel was originally heated? Assume no heat loss to the surroundings.

110. Quinone is an important type of molecule that is involved in photosynthesis. The transport of electrons mediated by quinone in certain enzymes allows plants to take water, carbon dioxide, and the energy of sunlight to create glucose. A 0.1964-g sample of quinone ($C_6H_4O_2$) is burned in a bomb calorimeter with a heat capacity of 1.56 kJ/°C. The temperature of the calorimeter increases by 3.2°C. Calculate the energy of combustion of quinone per gram and per mole.

111. Calculate $\Delta H°$ for each of the following reactions, which occur in the atmosphere.

    a. $C_2H_4(g) + O_3(g) \longrightarrow CH_3CHO(g) + O_2(g)$

    b. $O_3(g) + NO(g) \longrightarrow NO_2(g) + O_2(g)$

    c. $SO_3(g) + H_2O(l) \longrightarrow H_2SO_4(aq)$

    d. $2NO(g) + O_2(g) \longrightarrow 2NO_2(g)$

112. Compare your answers from parts a and b of Exercise 69 of Chapter 3 with ΔH values calculated for each reaction using standard enthalpies of formation in Appendix 4. Do enthalpy changes calculated from bond energies give a reasonable estimate of the actual values?

113. Compare your answer from Exercise 72 of Chapter 3 to the ΔH value calculated from standard enthalpies of formation in Appendix 4. Explain any discrepancies.

ChemWork Problems

These multiconcept problems (and additional ones) are found interactively online with the same type of assistance a student would get from an instructor.

114. Consider a balloon filled with helium at the following conditions.

    > 313 g He
    > 1.00 atm
    > 1910. L
    > Molar Heat Capacity = 20.8 J/°C · mol

The temperature of this balloon is decreased by 41.6°C as the volume decreases to 1643 L, with the pressure remaining constant. Determine q, w, and ΔE (in kJ) for the compression of the balloon.

115. In which of the following systems is(are) work done by the surroundings on the system? Assume pressure and temperature are constant.

    a. $2SO_2(g) + O_2(g) \longrightarrow 2SO_3(g)$

    b. $CO_2(s) \longrightarrow CO_2(g)$

    c. $4NH_3(g) + 7O_2(g) \longrightarrow 4NO_2(g) + 6H_2O(g)$

    d. $N_2O_4(g) \longrightarrow 2NO_2(g)$

    e. $CaCO_3(s) \longrightarrow CaCO(s) + CO_2(g)$

116. Which of the following processes are exothermic?

    a. $N_2(g) \longrightarrow 2N(g)$

    b. $H_2O(l) \longrightarrow H_2O(s)$

    c. $Cl_2(g) \longrightarrow 2Cl(g)$

    d. $2H_2(g) + O_2(g) \longrightarrow 2H_2O(g)$

    e. $O_2(g) \longrightarrow 2O(g)$

117. Consider the reaction

    $$B_2H_6(g) + 3O_2(g) \longrightarrow B_2O_3(s) + 3H_2O(g) \qquad \Delta H = -2035 \text{ kJ}$$

    Calculate the amount of heat released when 54.0 g of diborane is combusted.

118. A swimming pool, 10.0 m by 4.0 m, is filled with water to a depth of 3.0 m at a temperature of 20.2°C. How much energy is required to raise the temperature of the water to 24.6°C?

119. In a coffee-cup calorimeter, 150.0 mL of 0.50 M HCl is added to 50.0 mL of 1.00 M NaOH to make 200.0 g solution at an initial temperature of 48.2°C. If the enthalpy of neutralization for the reaction between a strong acid and a strong base is -56 kJ/mol, calculate the final temperature of the calorimeter contents. Assume the specific heat capacity of the solution is 4.184 J/g · °C and assume no heat loss to the surroundings.

120. Calculate ΔH for the reaction

    $$N_2H_4(l) + O_2(g) \longrightarrow N_2(g) + 2H_2O(l)$$

    given the following data:

Equation	ΔH (kJ)
$2NH_3(g) + 3N_2O(g) \longrightarrow 4N_2(g) + 3H_2O(l)$	-1010
$N_2O(g) + 3H_2(g) \longrightarrow N_2H_4(l) + H_2O(l)$	-317
$2NH_3(g) + \frac{1}{2}O_2(g) \longrightarrow N_2H_4(l) + H_2O(l)$	-143
$H_2(g) + \frac{1}{2}O_2(g) \longrightarrow H_2O(l)$	-286

121. Which of the following substances have an enthalpy of formation equal to zero?

    a. $Cl_2(g)$

    b. $H_2(g)$

    c. $N_2(l)$

    d. $Cl(g)$

Challenge Problems

122. Consider 2.00 moles of an ideal gas that are taken from state A ($P_A = 2.00$ atm, $V_A = 10.0$ L) to state B ($P_B = 1.00$ atm, $V_B = 30.0$ L) by two different pathways:

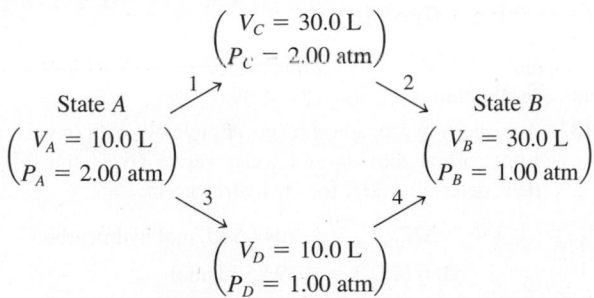

These pathways are summarized on the following graph of P versus V:

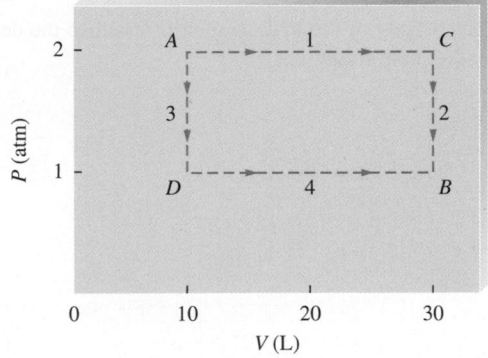

Calculate the work (in units of J) associated with the two pathways. Is work a state function? Explain.

123. For the process $H_2O(l) \longrightarrow H_2O(g)$ at 298 K and 1.0 atm, ΔH is more positive than ΔE by 2.5 kJ/mol. What does the 2.5 kJ/mol quantity represent?

124. The sun supplies energy at a rate of about 1.0 kilowatt per square meter of surface area (1 watt = 1 J/s). The plants in an agricultural field produce the equivalent of 20. kg sucrose ($C_{12}H_{22}O_{11}$) per hour per hectare (1 ha = 10,000 m²). Assuming that sucrose is produced by the reaction

$$12CO_2(g) + 11H_2O(l) \longrightarrow C_{12}H_{22}O_{11}(s) + 12O_2(g)$$
$$\Delta H = 5640 \text{ kJ}$$

calculate the percentage of sunlight used to produce the sucrose—that is, determine the efficiency of photosynthesis.

125. The best solar panels currently available are about 15% efficient in converting sunlight to electricity. A typical home will use about 40. kWh of electricity per day (1 kWh = 1 kilowatt hour; 1 kW = 1000 J/s). Assuming 8.0 hours of useful sunlight per day, calculate the minimum solar panel surface area necessary to provide all of a typical home's electricity. (See Exercise 124 for the energy rate supplied by the sun.)

126. The standard enthalpies of formation for $S(g)$, $F(g)$, $SF_4(g)$, and $SF_6(g)$ are $+278.8$, $+79.0$, -775, and $+1209$ kJ/mol, respectively.

    a. Use these data to estimate the energy of an S—F bond.

    b. Compare your calculated value to the value given in Table 3-3. What conclusions can you draw?

    c. Why are the ΔH_f° values for $S(g)$ and $F(g)$ not equal to zero, since sulfur and fluorine are elements?

127. Use the following standard enthalpies of formation to estimate the N—H bond energy in ammonia: $N(g)$, 472.7 kJ/mol; $H(g)$, 216.0 kJ/mol; $NH_3(g)$, -46.1 kJ/mol. Compare your value to the one in Table 3-3.

128. The standard enthalpy of formation for $N_2H_4(g)$ is 95.4 kJ/mol. Use this and the data in Exercise 127 to estimate the N—N single bond energy. Compare this with the value in Table 3-3.

129. The standard enthalpy of formation for $NO(g)$ is 90. kJ/mol. Use this and the values for the O=O and N≡N bond energies to estimate the bond strength in NO.

130. A piece of chocolate cake contains about 400 calories. A nutritional calorie is equal to 1000 calories (thermochemical calories), which is equal to 4.184 kJ. How many 8-in-high steps must a 180-lb man climb to expend the 400 Cal from the piece of cake? See Exercise 28 for the formula for potential energy.

131. You have a 1.00-mole sample of water at $-30.°C$ and you heat it until you have gaseous water at 140.°C. Calculate q for the entire process. Use the following data.

Specific heat capacity of ice = 2.03 J/°C · g

Specific heat capacity of water = 4.18 J/°C · g

Specific heat capacity of steam = 2.02 J/°C · g

$H_2O(s) \longrightarrow H_2O(l)$ $\Delta H_{fusion} = 6.02$ kJ/mol (at 0°C)

$H_2O(l) \longrightarrow H_2O(g)$ $\Delta H_{vaporization} = 40.7$ kJ/mol (at 100.°C)

132. A 500.0-g sample of an element at 195°C is dropped into an ice–water mixture; 109.5 g ice melts and an ice–water mixture remains. Calculate the specific heat of the element. See Exercise 131 for pertinent information.

Integrative Problems

These problems require the integration of multiple concepts to find the solutions.

133. A cubic piece of uranium metal (specific heat capacity = 0.117 J/°C · g) at 200.0°C is dropped into 1.00 L deuterium oxide ("heavy water," specific heat capacity = 4.211 J/°C · g) at 25.5°C. The final temperature of the uranium and deuterium oxide mixture is 28.5°C. Given the densities of uranium (19.05 g/cm³) and deuterium oxide (1.11 g/mL), what is the edge length of the cube of uranium?

134. On Easter Sunday, April 3, 1983, nitric acid spilled from a tank car near downtown Denver, Colorado. The spill was neutralized with sodium carbonate:

$$2HNO_3(aq) + Na_2CO_3(s) \longrightarrow 2NaNO_3(aq) + H_2O(l) + CO_2(g)$$

a. Calculate $\Delta H°$ for this reaction. Approximately 2.0×10^4 gal nitric acid was spilled. Assume that the acid was an aqueous solution containing 70.0% HNO_3 by mass with a density of 1.42 g/cm³. What mass of sodium carbonate was required for complete neutralization of the spill, and what quantity of heat was evolved? ($\Delta H_f°$ for $NaNO_3(aq) = -467$ kJ/mol)

b. According to *The Denver Post* for April 4, 1983, authorities feared that dangerous air pollution might occur during the neutralization. Considering the magnitude of $\Delta H°$, what was their major concern?

135. Using data from Chapter 2, calculate the change in energy expected for each of the following processes.

a. $Na(g) + Cl(g) \longrightarrow Na^+(g) + Cl^-(g)$

b. $Mg(g) + F(g) \longrightarrow Mg^+(g) + F^-(g)$

c. $Mg^+(g) + F(g) \longrightarrow Mg^{2+}(g) + F^-(g)$

d. $Mg(g) + 2F(g) \longrightarrow Mg^{2+}(g) + 2F^-(g)$

136. In Exercise 89 in Chapter 3, the Lewis structures for benzene (C_6H_6) were drawn. Using one of the Lewis Structures, estimate $\Delta H_f°$ for $C_6H_6(g)$ using bond energies and given that the standard enthalpy of formation of $C(g)$ is 717 kJ/mol. The experimental $\Delta H_f°$ value for $C_6H_6(g)$ is 83 kJ/mol. Explain the discrepancy between the experimental value and the calculated $\Delta H_f°$ value for $C_6H_6(g)$.

Marathon Problem

This problem is designed to incorporate several concepts and techniques into one situation.

137. A gaseous hydrocarbon reacts completely with oxygen gas to form carbon dioxide and water vapor. Given the following data, determine $\Delta H_f°$ for the hydrocarbon:

$$\Delta H_{reaction}° = -2044.5 \text{ kJ/mol hydrocarbon}$$
$$\Delta H_f°(CO_2) = -393.5 \text{ kJ/mol}$$
$$\Delta H_f°(H_2O) = -242 \text{ kJ/mol}$$

Density of CO_2 and H_2O product mixture at 1.00 atm, 200.°C = 0.751 g/L.

The density of the hydrocarbon is less than the density of Kr at the same conditions.

CHAPTER EIGHT

Gases

Underwater volcanic hotspot near Galapagos Islands in the Pacific. Gas bubbles rising from underwater volcanic vents among rocks. (© imageBROKER/Superstock)

Matter exists in three distinct physical states: gas, liquid, and solid. Although relatively few substances exist in the gaseous state under typical conditions, gases are very important. For example, we live immersed in a gaseous solution. The earth's atmosphere is a mixture of gases that consists mainly of elemental nitrogen (N_2) and oxygen (O_2). The atmosphere both supports life and acts as a waste receptacle of the exhaust gases that accompany many industrial processes. The chemical reactions of these waste gases in the atmosphere lead to various types of pollution, including smog and acid rain. The gases in the atmosphere also shield us from harmful radiation from the sun and keep the earth warm by reflecting heat radiation back toward the earth. In fact, there is now great concern that an increase in atmospheric carbon dioxide, a product of the combustion of fossil fuels, is causing a significant warming of the earth.

In this chapter we will look carefully at the properties of gases. We are particularly interested in how the behavior of the atoms and molecules that are the basic components of gases determine the properties of gases. However, we will not start by considering the microscopic view of gases. Instead we will proceed in the way that science usually works, by going from the macroscopic to the microscopic. That is, we will examine the observable properties of gases, such as pressure and volume, and work toward an understanding of the behavior of gases in terms of the motions and interactions of the individual articles (atoms and/or molecules) that the gases comprise.

The study of gases provides an excellent example of the scientific method in action. It illustrates how observations lead to natural laws, which can in turn be accounted for by models.

As we will see, if we think of a gas as a collection of particles (atoms/molecules) that do not interact with one another we can explain the simple gas laws that we observe. Although the simple theory explains the ideal case, when we consider the polarity of the molecules (as we discussed in Chapter 4), we will need to make the model more complex.

8-1 | Pressure

i As a gas, water occupies 1200 times as much space as it does as a liquid at 25°C and atmospheric pressure.

A gas uniformly fills any container, is easily compressed, and mixes completely with any other gas. ◄ *i* One of the most obvious properties of a gas is that it exerts pressure on its surroundings. For example, when you blow up a balloon, the air inside pushes against the elastic sides of the balloon and keeps it firm.

As mentioned earlier, the gases most familiar to us form the earth's atmosphere. The pressure exerted by this gaseous mixture that we call air can be dramatically demonstrated by the experiment shown in Fig. 8-1. A small volume of water is placed in a metal can, and the water is boiled, which fills the can with steam. The can is then sealed and allowed to cool. Why does the can collapse as it cools? It is the atmospheric pressure that crumples the can. When the can is cooled after being sealed so that no air can flow in, the water vapor (steam) condenses to a very small volume of liquid water. As a gas, the water filled the can, but when it is condensed to a liquid, the liquid does not come close to filling the can. The H_2O molecules formerly present as a gas are now collected in a very small volume of liquid, and there are very few molecules of gas left to exert pressure outward and counteract the air pressure. As a result, the pressure exerted by the gas molecules in the atmosphere smashes the can.

A device to measure atmospheric pressure, the **barometer**, was invented in 1643 by an Italian scientist named Evangelista Torricelli (1608–1647), who had been a

Figure 8-1 | The pressure exerted by the gases in the atmosphere can be demonstrated by boiling water in a large metal can (a) and then turning off the heat and sealing the can. As the can cools, the water vapor condenses, lowering the gas pressure inside the can. This causes the can to crumple (b).

a

b

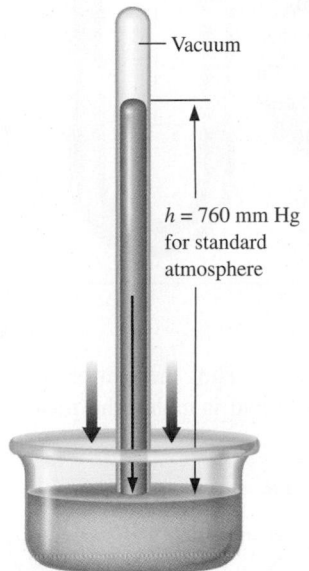

Figure 8-2 | A torricellian barometer. The tube, completely filled with mercury, is inverted in a dish of mercury. Mercury flows out of the tube until the pressure of the column of mercury (shown by the black arrow) "standing on the surface" of the mercury in the dish is equal to the pressure of the air (shown by the purple arrows) on the rest of the surface of the mercury in the dish.

ⓘ Soon after Torricelli died, a German physicist named Otto von Guericke invented an air pump. In a famous demonstration for the King of Prussia in 1663, Guericke placed two hemispheres together, pumped the air out of the resulting sphere through a valve, and showed that teams of horses could not pull the hemispheres apart. Then, after secretly opening the air valve, Guericke easily separated the hemispheres by hand. The King of Prussia was so impressed that he awarded Guericke a lifetime pension!

student of Galileo. Torricelli's barometer is constructed by filling a glass tube with liquid mercury and inverting it in a dish of mercury, as shown in Fig. 8-2. Notice that a large quantity of mercury stays in the tube. In fact, at sea level the height of this column of mercury averages 760 mm. Why does this mercury stay in the tube, seemingly in defiance of gravity? Fig. 8-2 illustrates how the pressure exerted by the atmospheric gases on the surface of mercury in the dish keeps the mercury in the tube.

Atmospheric pressure results from the mass of the air being pulled toward the center of the earth by gravity—in other words, it results from the weight of the air. Changing weather conditions cause the atmospheric pressure to vary, so the height of the column of Hg supported by the atmosphere at sea level varies; it is not always 760 mm. The meteorologist who says a "low" is approaching means that the atmospheric pressure is going to decrease. This condition often occurs in conjunction with a storm.

Atmospheric pressure also varies with altitude. For example, when Torricelli's experiment is done in Breckenridge, Colorado (elevation 9600 feet), the atmosphere supports a column of mercury only about 520 mm high because the air is "thinner." That is, there is less air pushing down on the earth's surface at Breckenridge than at sea level.

Units of Pressure

Because instruments used for measuring pressure, such as the **manometer** (Fig. 8-3), often contain mercury, the most commonly used units for pressure are based on the height of the mercury column (in millimeters) that the gas pressure can support. The unit **mm Hg** (millimeter of mercury) is often called the **torr** in honor of Torricelli. ◄ ⓘ The terms *torr* and *mm Hg* are used interchangeably by chemists. A related unit for pressure is the **standard atmosphere** (abbreviated atm):

$$1 \text{ standard atmosphere} = 1 \text{ atm} = 760 \text{ mm Hg} = 760 \text{ torr}$$

However, since pressure is defined as force per unit area,

$$\text{Pressure} = \frac{\text{force}}{\text{area}}$$

the fundamental units of pressure involve units of force divided by units of area. In the SI system, the unit of force is the newton (N) and the unit of area is meters squared

Figure 8-3 | A simple manometer, a device for measuring the pressure of a gas in a container. The pressure of the gas is given by h (the difference in mercury levels) in units of torr (equivalent to mm Hg). (a) Gas pressure = atmospheric pressure − h. (b) Gas pressure = atmospheric pressure + h.

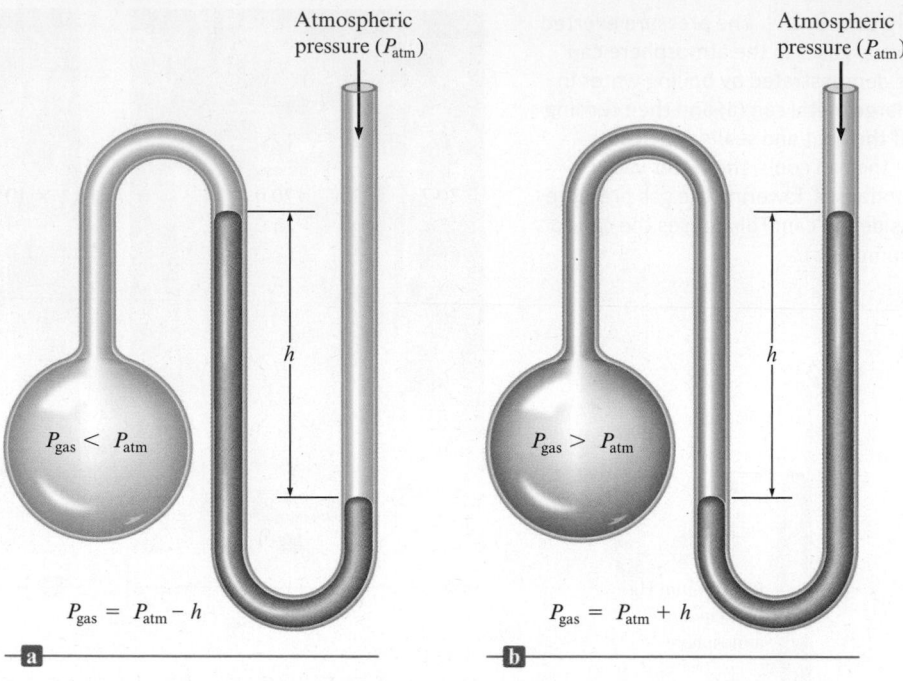

$$P_{gas} = P_{atm} - h$$

$$P_{gas} = P_{atm} + h$$

(m^2). (For a review of the SI system, see the Review Chapter.) Thus the unit of pressure in the SI system is newtons per meter squared (N/m^2) and is called the **pascal** (Pa). In terms of pascals, the standard atmosphere is

$$1 \text{ standard atmosphere} = 101{,}325 \text{ Pa}$$

Thus 1 atmosphere is about 10^5 pascals. Since the pascal is so small, and since it is not commonly used in the United States, we will use it sparingly in this book. However, converting from torrs or atmospheres to pascals is straightforward, as shown in Example 8-1. ◄ *i*

i 1 atm = 760 mm Hg
       = 760 torr
       = 101,325 Pa
       = 29.92 in Hg
       = 14.7 lb/in²

INTERACTIVE EXAMPLE 8-1

Pressure Conversions

The pressure of a gas is measured as 49 torr. Represent this pressure in both atmospheres and pascals.

Solution

$$49 \text{ torr} \times \frac{1 \text{ atm}}{760 \text{ torr}} = 6.4 \times 10^{-2} \text{ atm}$$

$$6.4 \times 10^{-2} \text{ atm} \times \frac{101{,}325 \text{ Pa}}{1 \text{ atm}} = 6.5 \times 10^3 \text{ Pa}$$

See Exercises 8-37 and 8-38

8-2 | The Gas Laws of Boyle, Charles, and Avogadro

In this section we will consider several mathematical laws that relate the properties of gases. These laws derive from experiments involving careful measurements of the relevant gas properties. From these experimental results, the mathematical relation-

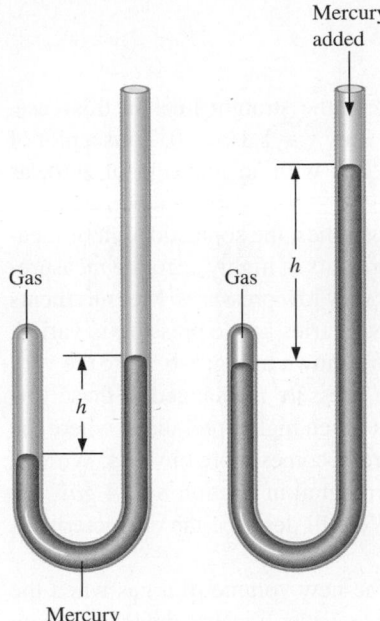

Figure 8-4 | A J-tube similar to the one used by Boyle. When mercury is added to the tube, pressure on the trapped gas is increased, resulting in a decreased volume.

Table 8-1 | Actual Data from Boyle's Experiment

Volume (in³)	Pressure (in Hg)	Pressure × Volume (in Hg × in³)
117.5	12.0	14.1×10^2
87.2	16.0	14.0×10^2
70.7	20.0	14.1×10^2
58.8	24.0	14.1×10^2
44.2	32.0	14.1×10^2
35.3	40.0	14.1×10^2
29.1	48.0	14.0×10^2

ships among the properties can be discovered. These relationships are often represented pictorially by means of graphs (plots).

We will take a historical approach to these laws to give you some perspective on the scientific method in action.

Boyle's Law

The first quantitative experiments on gases were performed by an Irish chemist, Robert Boyle (1627–1691). Using a J-shaped tube closed at one end (Fig. 8-4), which he reportedly set up in the multistory entryway of his house, Boyle studied the relationship between the pressure of the trapped gas and its volume. Representative values from Boyle's experiments are given in Table 8-1. These data show that the product of the pressure and volume for the trapped air sample is constant within the accuracies of Boyle's measurements (note the third column in Table 8-1). This behavior can be represented by the equation

$$PV = k$$

ⓘ Boyle's law: $V \propto 1/P$ at constant temperature

ⓘ Graphing is reviewed in Appendix 1-3.

which is called **Boyle's law** and where k is a constant for a given sample of air at a specific temperature. ◄ ⓘ

It is convenient to represent the data in Table 8-1 by using two different plots. ◄ ⓘ The first type of plot, P versus V, forms a curve called a *hyperbola*, shown in Fig. 8-5(a). Looking at this plot, note that as the volume drops by about half (from 58.8 to 29.1), the pressure doubles (from 24.0 to 48.0). In other words, there is an *inverse relationship* between pressure and volume. The second type of plot can be obtained by rearranging Boyle's law to give

$$V = \frac{k}{P} = k\frac{1}{P}$$

Figure 8-5 | Plotting Boyle's data from Table 8-1. (a) A plot of P versus V shows that the volume doubles as the pressure is halved. (b) A plot of V versus $1/P$ gives a straight line. The slope of this line equals the value of the constant k.

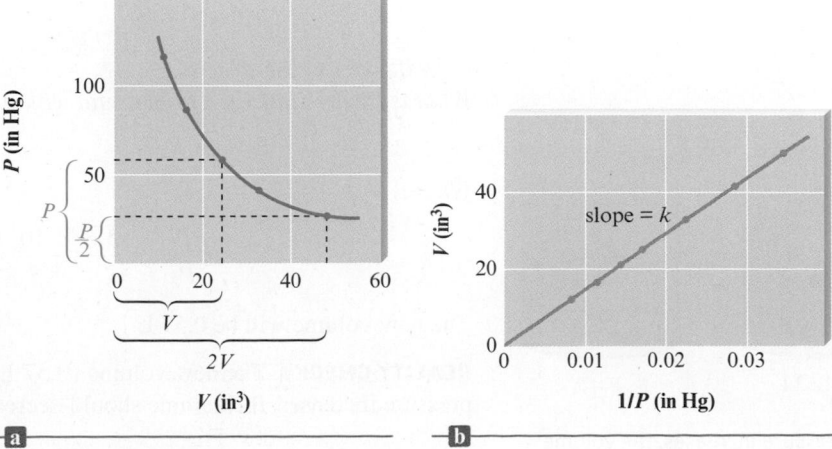

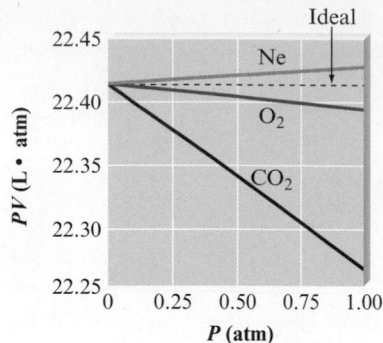

Figure 8-6 | A plot of PV versus P for several gases at pressures below 1 atm. An ideal gas is expected to have a constant value of PV, as shown by the dotted line. Carbon dioxide shows the largest change in PV, and this change is actually quite small: PV changes from about 22.39 L · atm at 0.25 atm to 22.26 L · atm at 1.00 atm. Thus Boyle's law is a good approximation at these relatively low pressures.

which is the equation for a straight line of the type

$$y = mx + b$$

where m represents the slope and b the intercept of the straight line. In this case, $y = V$, $x = 1/P$, $m = k$, and $b = 0$. See for Boyle's Law $v = k\,1/p + 0$. Thus a plot of V versus $1/P$ using Boyle's data gives a straight line with an intercept of zero, as shown in Fig. 8-5(b).

In the three centuries since Boyle carried out his studies, the sophistication of measuring techniques has increased tremendously. The results of highly accurate measurements show that Boyle's law holds precisely only at very low pressures. Measurements at higher pressures reveal that PV is not constant but varies as the pressure is varied. Results for several gases at pressures below 1 atm are shown in Fig. 8-6. Note the very small changes that occur in the product PV as the pressure is changed at these low pressures. Such changes become more significant at much higher pressures, where the complex nature of the dependence of PV on pressure becomes more obvious. We will discuss these deviations and the reasons for them in detail in Section 8-7. *A gas that strictly obeys Boyle's law is called an **ideal gas**.* We will describe the characteristics of an ideal gas more completely in Section 8-3.

One common use of Boyle's law is to predict the new volume of a gas when the pressure is changed (at constant temperature), or vice versa. Because deviations from Boyle's law are so slight at pressures close to 1 atm, in our calculations we will assume that gases obey Boyle's law (unless stated otherwise).

INTERACTIVE EXAMPLE 8-2

Boyle's Law I

Sulfur dioxide (SO_2), a gas that plays a central role in the formation of acid rain, is found in the exhaust of automobiles and power plants. Consider a 1.53-L sample of gaseous SO_2 at a pressure of 5.6×10^3 Pa. If the pressure is changed to 1.5×10^4 Pa at a constant temperature, what will be the new volume of the gas?

Solution

Where are we going?
To calculate the new volume of gas

What do we know?

- $P_1 = 5.6 \times 10^3$ Pa   - $P_2 = 1.5 \times 10^4$ Pa
- $V_1 = 1.53$ L   - $V_2 = ?$

What information do we need?

- Boyle's law

$$PV = k \blacktriangleleft \textit{i}$$

(ⓘ) Boyle's law also can be written as

$$P_1V_1 = P_2V_2$$

How do we get there?
What is Boyle's law (in a form useful with our knowns)?

$$P_1V_1 = P_2V_2$$

What is V_2?

$$V_2 = \frac{P_1V_1}{P_2} = \frac{5.6 \times 10^3 \text{ Pa} \times 1.53 \text{ L}}{1.5 \times 10^4 \text{ Pa}} = 0.57 \text{ L}$$

The new volume will be 0.57 L.

REALITY CHECK | The new volume (0.57 L) is smaller than the original volume. As pressure increases, the volume should decrease, so our answer is reasonable. ◄

5.6×10^3 Pa 1.5×10^4 Pa

$V = 1.53$ L $V = ?$

▲ As pressure increases, the volume of SO_2 decreases.

See Exercise 8-43

The fact that the volume decreases in Example 8-2 makes sense because the pressure was increased. To help eliminate errors, make it a habit to check whether an answer to a problem makes physical (common!) sense.

We mentioned before that Boyle's law is only approximately true for real gases. To determine the significance of the deviations, studies of the effect of changing pressure on the volume of a gas are often done, as shown in Example 8-3.

EXAMPLE 8-3

Boyle's Law II

In a study to see how closely gaseous ammonia obeys Boyle's law, several volume measurements were made at various pressures, using 1.0 mole of NH_3 gas at a temperature of 0°C. Using the results listed below, calculate the Boyle's law constant for NH_3 at the various pressures.

Experiment	Pressure (atm)	Volume (L)
1	0.1300	172.1
2	0.2500	89.28
3	0.3000	74.35
4	0.5000	44.49
5	0.7500	29.55
6	1.000	22.08

Solution

To determine how closely NH_3 gas follows Boyle's law under these conditions, we calculate the value of k (in L · atm) for each set of values:

Experiment	1	2	3	4	5	6
$k = PV$	22.37	22.32	22.31	22.25	22.16	22.08

Although the deviations from true Boyle's law behavior are quite small at these low pressures, note that the value of k changes regularly in one direction as the pressure is increased. Thus, to calculate the "ideal" value of k for NH_3, we can plot PV versus P, as shown in Fig. 8-7, and extrapolate (extend the line beyond the experimental points) back to zero pressure, where, for reasons we will discuss later, a gas behaves most ideally. The value of k obtained by this extrapolation is 22.41 L · atm. Notice that this is the same value obtained from similar plots for the gases CO_2, O_2, and Ne at 0°C, as shown in Fig. 8-6.

See Exercise 8-125

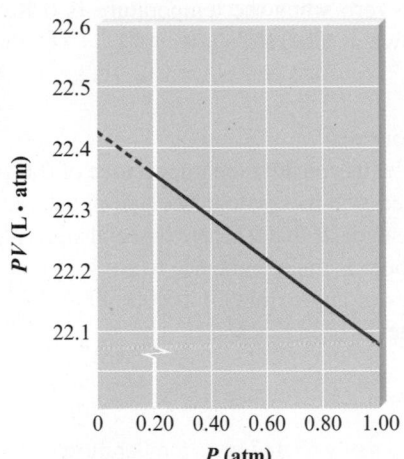

Figure 8-7 | A plot of *PV* versus *P* for 1 mole of ammonia. The dashed line shows the extrapolation of the data to zero pressure to give the "ideal" value of *PV* of 22.41 L · atm.

Charles's Law

In the century following Boyle's findings, scientists continued to study the properties of gases. One of these scientists was a French physicist, Jacques Charles (1746–1823), who was the first person to fill a balloon with hydrogen gas and who made the first solo balloon flight. Charles found in 1787 that the volume of a gas at constant pressure increases *linearly* with the temperature of the gas. That is, a plot of the volume of a gas (at constant pressure) versus its temperature (°C) gives a straight line. This behavior is shown for samples of several gases in Fig. 8-8. The slopes of the lines in this graph are different because the samples contain different numbers of moles of gas. A very interesting feature of these plots is that the volumes of all the gases extrapolate to zero at the same temperature, −273°C. On the Kelvin temperature scale this point is defined as 0 K, which leads to the following relationship between the Kelvin and Celsius scales:

$$K = °C + 273$$

Figure 8-8 | Plots of V versus T (°C) for several gases. The solid lines represent experimental measurements of gases. The dashed lines represent extrapolation of the data into regions where these gases would become liquids or solids. Note that the samples of the various gases contain different numbers of moles.

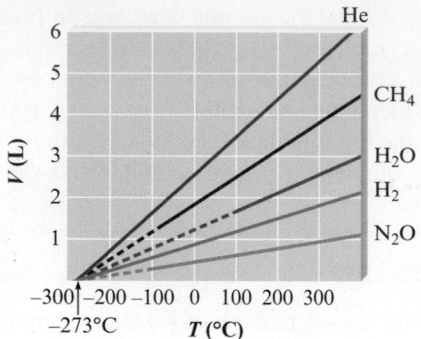

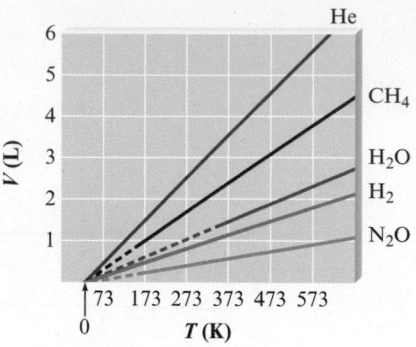

Figure 8-9 | Plots of V versus T as in Fig. 8-8, except here the Kelvin scale is used for temperature.

When the volumes of the gases shown in Fig. 8-8 are plotted versus temperature on the Kelvin scale, the plots in Fig. 8-9 result. In this case, the volume of each gas is *directly proportional to temperature* and extrapolates to zero when the temperature is 0 K. This behavior is represented by the equation known as **Charles's law**, ◄ ⓘ

ⓘ Charles's law: $V \propto T$ (expressed in K) of constant pressure.

$$V = bT$$

where T is in kelvins and b is a proportionality constant.

Before we illustrate the uses of Charles's law, let us consider the importance of 0 K. At temperatures below this point, the extrapolated volumes would become negative. The fact that a gas cannot have a negative volume suggests that 0 K has a special significance. In fact, 0 K is called **absolute zero**, and there is much evidence to suggest that this temperature cannot be attained. Temperatures of approximately 0.0000000001 K have been produced in laboratories, but 0 K has never been reached.

> **Critical Thinking**
>
> According to Charles's law, the volume of a gas is directly related to its temperature in kelvins at constant pressure and number of moles. What if the volume of a gas was directly related to its temperature in degrees Celsius at constant pressure and number of moles? What differences would you notice?

INTERACTIVE EXAMPLE 8-4

Charles's Law

A sample of gas at 15°C and 1 atm has a volume of 2.58 L. What volume will this gas occupy at 38°C and 1 atm?

Solution

Where are we going?
To calculate the new volume of gas

What do we know?

- $T_1 = 15°C + 273 = 288$ K
- $T_2 = 38°C + 273 = 311$ K
- $V_1 = 2.58$ L
- $V_2 = ?$

What information do we need?

- Charles's law

ⓘ Charles's law also can be written as

$$\frac{V_1}{T_1} = \frac{V_2}{T_2}$$

$$\frac{V}{T} = b \blacktriangleleft ⓘ$$

How do we get there?
What is Charles's law (in a form useful with our knowns)?

$$\frac{V_1}{T_1} = \frac{V_2}{T_2}$$

What is V_2?

$$V_2 = \left(\frac{T_2}{T_1}\right) V_1 = \left(\frac{311\ \text{K}}{288\ \text{K}}\right) 2.58\ \text{L} = 2.79\ \text{L}$$

REALITY CHECK | The new volume is greater than the original volume, which makes physical sense because the gas will expand as it is heated.

See Exercise 8-44

Avogadro's Law

In Chapter 1 we noted that in 1811 the Italian chemist Avogadro postulated that equal volumes of gases at the same temperature and pressure contain the same number of "particles." This observation is called **Avogadro's law**, which is illustrated by Fig. 8-10. Stated mathematically, Avogadro's law is

$$V = an$$

where V is the volume of the gas, n is the number of moles of gas particles, and a is a proportionality constant. This equation states that *for a gas at constant temperature and pressure, the volume is directly proportional to the number of moles of gas*. This relationship is obeyed closely by gases at low pressures.

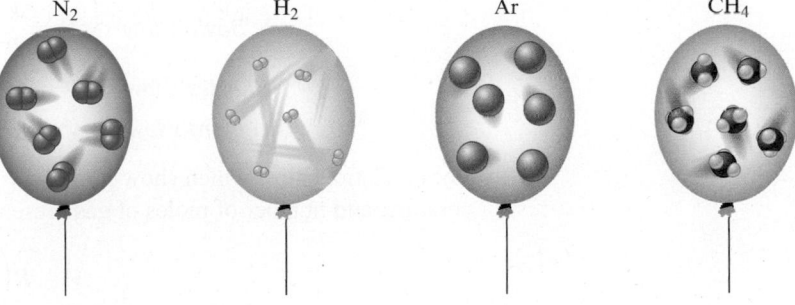

N₂   H₂   Ar   CH₄

Figure 8-10 | These balloons each hold 1.0 L gas at 25°C and 1 atm. Each balloon contains 0.041 mole of gas, or 2.5×10^{22} molecules.

INTERACTIVE EXAMPLE 8-5

Avogadro's Law

Suppose we have a 12.2-L sample containing 0.50 mole of oxygen gas (O_2) at a pressure of 1 atm and a temperature of 25°C. If all this O_2 were converted to 0.33 mole of ozone (O_3) at the same temperature and pressure, what would be the volume of the ozone?

Solution

Where are we going?
To calculate the volume of the ozone produced by 0.50 mole of oxygen

What do we know?

- $n_1 = 0.50$ mol O_2
- $V_1 = 12.2$ L O_2
- $n_2 = 0.33$ mol O_3
- $V_2 = ?$ L O_3

What information do we need?

- Avogadro's law

$$V = an \quad \blacktriangleleft \ \textit{i}$$

ⓘ Avogadro's law also can be written as

$$\frac{V_1}{n_1} = \frac{V_2}{n_2}$$

How do we get there?
What is the volume of O_3 produced?
Avogadro's law states that $V = an$, which can be rearranged to give

$$\frac{V}{n} = a$$

Since a is a constant, an alternative representation is

$$\frac{V_1}{n_1} = a = \frac{V_2}{n_2}$$

where V_1 is the volume of n_1 moles of O_2 gas and V_2 is the volume of n_2 moles of O_3 gas.

Solving for V_2 gives

$$V_2 = \left(\frac{n_2}{n_1}\right)V_1 = \left(\frac{0.33 \text{ mol}}{0.50 \text{ mol}}\right)12.2 \text{ L} = 8.1 \text{ L}$$

REALITY CHECK | Note that the volume decreases, as it should, since fewer moles of gas molecules will be present after O_2 is converted to O_3.

See Exercises 8-45 and 8-46

8-3 | The Ideal Gas Law

We have considered three laws that describe the behavior of gases as revealed by experimental observations:

$$\text{Boyle's law:} \quad V = \frac{k}{P} \quad \text{(at constant } T \text{ and } n\text{)}$$

$$\text{Charles's law:} \quad V = bT \quad \text{(at constant } P \text{ and } n\text{)}$$

$$\text{Avogadro's law:} \quad V = an \quad \text{(at constant } T \text{ and } P\text{)}$$

These relationships, which show how the volume of a gas depends on pressure, temperature, and number of moles of gas present, can be combined as follows:

$$V = R\left(\frac{Tn}{P}\right)$$

where R is the combined proportionality constant called the **universal gas constant**. When the pressure is expressed in atmospheres and the volume in liters, R has the value 0.08206 L · atm/K · mol. ◄ ⓘ The preceding equation can be rearranged to the more familiar form of the **ideal gas law**:

$$PV = nRT$$

ⓘ $R = 0.08206 \dfrac{\text{L} \cdot \text{atm}}{\text{K} \cdot \text{mol}}$

The ideal gas law is an *equation of state* for a gas, where the state of the gas is its condition at a given time. A particular *state* of a gas is described by its pressure, volume, temperature, and number of moles. Knowledge of any three of these properties is enough to completely define the state of a gas, since the fourth property can then be determined from the equation for the ideal gas law.

It is important to recognize that the ideal gas law is an empirical equation—it is based on experimental measurements of the properties of gases. A gas that obeys this equation is said to behave *ideally*. The ideal gas equation is best regarded as a limiting law—it expresses behavior that real gases *approach* at low pressures and high temperatures. Therefore, an ideal gas is a hypothetical substance. However, most gases obey the ideal gas equation closely enough at pressures below 1 atm that only minimal errors result from assuming ideal behavior. ◄ ⓘ Unless you are given information to

ⓘ The ideal gas law applies best at pressures smaller than 1 atm.

the contrary, you should assume ideal gas behavior when solving problems involving gases in this text.

The ideal gas law can be used to solve a variety of problems. Example 8-6 demonstrates one type, where you are asked to find one property characterizing the state of a gas, given the other three.

INTERACTIVE EXAMPLE 8-6

▲ The reaction of zinc with hydrochloric acid to produce bubbles of hydrogen gas.

ⓘ Always convert the temperature to the Kelvin scale when applying the ideal gas law.

Ideal Gas Law I

A sample of hydrogen gas (H_2) has a volume of 8.56 L at a temperature of 0°C and a pressure of 1.5 atm. ◄ Calculate the moles of H_2 molecules present in this gas sample.

Solution

Where are we going?
To calculate the moles of H_2

What do we know?

- $n = ?$ mol H_2
- $V = 8.56$ L
- $P = 1.5$ atm
- $T = 0°C + 273 = 273$ K ◄ ⓘ

What information do we need?

- Ideal gas law

$$PV = nRT$$

- $R = 0.08206$ L · atm/K · mol

How do we get there?
How many moles of H_2 are present in the sample?

Solve the ideal gas equation for n:

$$n = \frac{(1.5\ \cancel{atm})(8.56\ \cancel{L})}{\left(0.08206\dfrac{\cancel{L} \cdot \cancel{atm}}{\cancel{K} \cdot mol}\right)(273\ \cancel{K})} = 0.57\ \text{mol}$$

See Exercises 8-47 through 8-54

Gas law problems can be solved in a variety of ways. They can be classified as a Boyle's law, Charles's law, or Avogadro's law problem and solved, but now we need to remember the specific law and when it applies. The real advantage of using the ideal gas law is that it applies to virtually any problem dealing with gases and is easy to remember.

The basic assumption we make when using the ideal gas law to describe a change in state for a gas is that the equation applies equally well to both the initial and final states. In dealing with changes in state, we always *place the variables that change on one side of the equal sign and the constants on the other*. Let's see how this might work in several examples.

INTERACTIVE EXAMPLE 8-7

Ideal Gas Law II

Suppose we have a sample of ammonia gas with a volume of 7.0 mL at a pressure of 1.68 atm. The gas is compressed to a volume of 2.7 mL at a constant temperature. Use the ideal gas law to calculate the final pressure.

Solution

Where are we going?
To use the ideal gas equation to determine the final pressure

What do we know?

- $P_1 = 1.68$ atm   - $P_2 = ?$
- $V_1 = 7.0$ mL   - $V_2 = 2.7$ mL

What information do we need?

- Ideal gas law

$$PV = nRT$$

- $R = 0.08206$ L · atm/K · mol

How do we get there?
What are the variables that change?

$$P, V$$

What are the variables that remain constant?

$$n, R, T$$

Write the ideal gas law, collecting the change variables on one side of the equal sign and the variables that do not change on the other.

$$PV = nRT$$

Change   Remain constant

Since n and T remain the same in this case, we can write $P_1V_1 = nRT$ and $P_2V_2 = nRT$. Combining these gives

$$P_1V_1 = nRT = P_2V_2 \quad \text{or} \quad P_1V_1 = P_2V_2$$

We are given $P_1 = 1.68$ atm, $V_1 = 7.0$ mL, and $V_2 = 2.7$ mL. Solving for P_2 thus gives

$$P_2 = \left(\frac{V_1}{V_2}\right)P_1 = \left(\frac{7.0 \text{ mL}}{2.7 \text{ mL}}\right) 1.68 \text{ atm} = 4.4 \text{ atm}$$

REALITY CHECK | Does this answer make sense? The volume decreased (at constant temperature), so the pressure should increase, as the result of the calculation indicates. Note that the calculated final pressure is 4.4 atm. Most gases do not behave ideally above 1 atm. Therefore, we might find that if we *measured* the pressure of this gas sample, the observed pressure would differ slightly from 4.4 atm. ◄

See Exercises 8-55 and 8-56

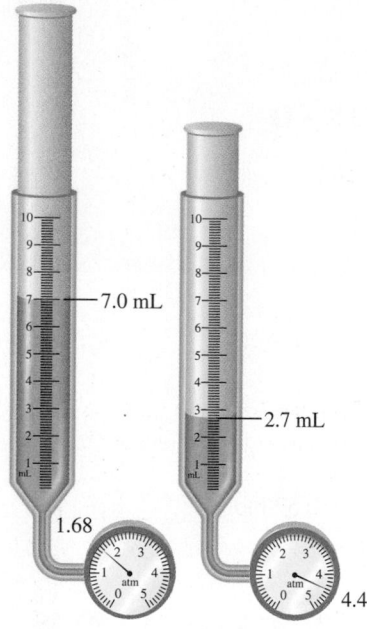

▲ As pressure increases, the volume decreases.

| **INTERACTIVE EXAMPLE 8-8** | **Ideal Gas Law III** |

A sample of methane gas that has a volume of 3.8 L at 5°C is heated to 86°C at constant pressure. Calculate its new volume.

Solution

Where are we going?
To use the ideal gas equation to determine the final volume

What do we know?

- $T_1 = 5°C + 273 = 278$ K   - $T_2 = 86°C + 273 = 359$ K
- $V_1 = 3.8$ L   - $V_2 = ?$

What information do we need?

- Ideal gas law

$$PV = nRT$$

- $R = 0.08206 \text{ L} \cdot \text{atm/K} \cdot \text{mol}$

How do we get there?
What are the variables that change?

$$V, T$$

What are the variables that remain constant?

$$n, R, P$$

Write the ideal gas law, collecting the change variables on one side of the equal sign and the variables that do not change on the other.

$$\frac{V}{T} = \frac{nR}{P}$$

which leads to

$$\frac{V_1}{T_1} = \frac{nR}{P} \quad \text{and} \quad \frac{V_2}{T_2} = \frac{nR}{P}$$

Combining these gives

$$\frac{V_1}{T_1} = \frac{nR}{P} = \frac{V_2}{T_2} \quad \text{or} \quad \frac{V_1}{T_1} = \frac{V_2}{T_2}$$

Solving for V_2:

$$V_2 = \frac{T_2 V_1}{T_1} = \frac{(359 \text{ K})(3.8 \text{ L})}{278 \text{ K}} = 4.9 \text{ L}$$

REALITY CHECK | Is the answer sensible? In this case the temperature increased (at constant pressure), so the volume should increase. Thus the answer makes sense.

See Exercises 8-57 and 8-59

The problem in Example 8-8 could be described as a Charles's law problem, whereas the problem in Example 8-7 might be called a Boyle's law problem. In both cases, however, we started with the ideal gas law. The real advantage of using the ideal gas law is that it applies to virtually any problem dealing with gases and is easy to remember.

INTERACTIVE EXAMPLE 8-9

Ideal Gas Law IV

A sample of diborane gas (B_2H_6), a substance that bursts into flame when exposed to air, has a pressure of 345 torr at a temperature of $-15°C$ and a volume of 3.48 L. If conditions are changed so that the temperature is $36°C$ and the pressure is 468 torr, what will be the volume of the sample?

Solution

Where are we going?
To use the ideal gas equation to determine the final volume

What do we know?

- $T_1 = -15°C + 273 = 258 \text{ K}$
- $V_1 = 3.48 \text{ L}$
- $P_1 = 345 \text{ torr}$

- $T_2 = 36°C + 273 = 309 \text{ K}$
- $V_2 = ?$
- $P_2 = 468 \text{ torr}$

What information do we need?

- Ideal gas law

$$PV = nRT$$

- $R = 0.08206 \ \text{L} \cdot \text{atm/K} \cdot \text{mol}$

How do we get there?
What are the variables that change?

$$P, V, T$$

What are the variables that remain constant?

$$n, R$$

Write the ideal gas law, collecting the change variables on one side of the equal sign and the variables that do not change on the other.

$$\frac{PV}{T} = nR$$

which leads to

$$\frac{P_1 V_1}{T_1} = nR = \frac{P_2 V_2}{T_2} \quad \text{or} \quad \frac{P_1 V_1}{T_1} = \frac{P_2 V_2}{T_2}$$

Solving for V_2:

$$V_2 = \frac{T_2 P_1 V_1}{T_1 P_2} = \frac{(309 \ \cancel{K})(345 \ \cancel{\text{torr}})(3.48 \ \text{L})}{(258 \ \cancel{K})(468 \ \cancel{\text{torr}})} = 3.07 \ \text{L}$$

See Exercises 8-61 and 8-62

Since the equation used in Example 8-9 involves a *ratio* of pressures, it was unnecessary to convert pressures to units of atmospheres. The units of torrs cancel. (You will obtain the same answer by inserting $P_1 = \dfrac{345}{760}$ and $P_2 = \dfrac{468}{760}$ into the equation.) However, temperature *must always* be converted to the Kelvin scale; since this conversion involves *addition* of 273, the conversion factor does not cancel. Be careful.

One of the many other types of problems dealing with gases that can be solved using the ideal gas law is illustrated in Example 8-10.

INTERACTIVE EXAMPLE 8-10

▲ A Twyman–Green interferometer emits green argon laser light. Interferometers can measure extremely small distances, useful in configuring telescope mirrors.

Georgy Shafeev/Shutterstock.com

Ideal Gas Law V

A sample containing 0.35 mole of argon gas at a temperature of 13°C and a pressure of 568 torr is heated to 56°C and a pressure of 897 torr. ◄ Calculate the change in volume that occurs.

Solution

Where are we going?
To use the ideal gas equation to determine the final volume

What do we know?

State 1	State 2
$n_1 = 0.35 \ \text{mol}$	$n_2 = 0.35 \ \text{mol}$
$P_1 = 568 \ \cancel{\text{torr}} \times \dfrac{1 \ \text{atm}}{760 \ \cancel{\text{torr}}} = 0.747 \ \text{atm}$	$P_2 = 897 \ \cancel{\text{torr}} \times \dfrac{1 \ \text{atm}}{760 \ \cancel{\text{torr}}} = 1.18 \ \text{atm}$
$T_1 = 13°\text{C} + 273 = 286 \ \text{K}$	$T_2 = 56°\text{C} + 273 = 329 \ \text{K}$

What information do we need?

- Ideal gas law

$$PV = nRT$$

- $R = 0.08206 \text{ L} \cdot \text{atm/K} \cdot \text{mol}$
- V_1 and V_2

How do we get there?
What is V_1?

$$V_1 = \frac{n_1RT_1}{P_1} = \frac{(0.35 \text{ mol})(0.08206 \text{ L} \cdot \text{atm/K} \cdot \text{mol})(286 \text{ K})}{(0.747 \text{ atm})} = 11 \text{ L}$$

What is V_2?

$$V_2 = \frac{n_2RT_2}{P_2} = \frac{(0.35 \text{ mol})(0.08206 \text{ L} \cdot \text{atm/K} \cdot \text{mol})(329 \text{ K})}{(1.18 \text{ atm})} = 8.0 \text{ L}$$

What is the change in volume ΔV?

$$\Delta V = V_2 - V_1 = 8.0 \text{ L} - 11 \text{ L} = -3 \text{ L}$$

The *change* in volume is negative because the volume decreases.

Note: For this problem (unlike Example 8-9), the pressures must be converted from torrs to atmospheres, as required by the atmospheres part of the units for R since each volume was found separately, and the conversion factor does not cancel.

See Exercise 8-63

Figure 8-11 | 22.42 L of a gas would just fit into this beach ball.

ⓘ STP: 0°C and 1 atm

8-4 | Gas Stoichiometry

So far in this chapter we have dealt with masses of reactants and products. In this section we will discuss how volume enters into the stoichiometry of gases. Suppose we have 1 mole of an ideal gas at 0°C (273.2 K) and 1 atm. From the ideal gas law, the volume of the gas is given by

$$V = \frac{nRT}{P} = \frac{(1.000 \text{ mol})(0.08206 \text{ L} \cdot \text{atm/K} \cdot \text{mol})(273.2 \text{ K})}{1.000 \text{ atm}} = 22.42 \text{ L}$$

This volume of 22.42 liters is the **molar volume** of an ideal gas (at 0°C and 1 atm). The measured molar volumes of several gases are listed in Table 8-2. Note that the molar volumes of some of the gases are very close to the ideal value, while others deviate significantly. Later in this chapter we will discuss some of the reasons for the deviations.

The conditions 0°C and 1 atm, called **standard temperature and pressure** (abbreviated **STP**), are common reference conditions for the properties of gases. For example, the molar volume of an ideal gas is 22.42 liters at STP (Fig. 8-11). ◂ ⓘ

Table 8-2 | Molar Volumes for Various Gases at 0°C and 1 atm

Gas	Molar Volume (L)
Oxygen (O_2)	22.397
Nitrogen (N_2)	22.402
Hydrogen (H_2)	22.433
Helium (He)	22.434
Argon (Ar)	22.397
Carbon dioxide (CO_2)	22.260
Ammonia (NH_3)	22.079

The next time that you visit a gas station, take a moment to note the octane rating that accompanies the grade of gasoline that you are purchasing. The gasoline is priced according to its octane rating—a measure of the fuel's antiknock properties. In a conventional internal combustion engine, gasoline vapors and air are drawn into the combustion cylinder on the downward stroke of the piston. This air–fuel mixture is compressed on the upward piston stroke (compression stroke), and a spark from the sparkplug ignites the mix. The rhythmic combustion of the air–fuel mix occurring sequentially in several cylinders furnishes the power to propel the vehicle down the road. Excessive heat and pressure (or poor-quality fuel) within the cylinder may cause the premature combustion of the mixture—commonly known as engine "knock" or "ping." Over time, this engine knock can damage the engine, resulting in inefficient performance and costly repairs.

A consumer typically is faced with three choices of gasoline, with octane ratings of 87 (regular), 89 (midgrade), and 93 (premium). But if you happen to travel or live in the higher elevations of the Rocky Mountain states, you might be surprised to find different octane ratings at the gasoline pumps. The reason for this provides a lesson in stoichiometry. At higher elevations the air is less dense—the volume of oxygen per unit volume of air is smaller. Most engines are designed to achieve a 14:1 oxygen-to-fuel ratio in the cylinder prior to combustion. If less oxygen is available, then less fuel is required to achieve this optimal ratio. In turn, the lower volumes of oxygen and fuel result in a lower pressure in the cylinder. Because high pressure tends to promote knocking, the lower pressure within engine cylinders at higher elevations promotes a more controlled combustion of the air–fuel mixture, and therefore octane requirements are lower. While consumers in the Rocky Mountain states can purchase three grades of gasoline, the octane ratings of these fuel blends are different from those in the rest of the United States. In Denver, Colorado, regular gasoline is 85 octane, midgrade is 87 octane, and premium is 91 octane—2 points lower than gasoline sold in most of the rest of the country.

INTERACTIVE EXAMPLE 8-11

Gas Stoichiometry I

A sample of nitrogen gas has a volume of 1.75 L at STP. How many moles of N_2 are present?

Solution

We could solve this problem by using the ideal gas equation, but we can take a shortcut by using the molar volume of an ideal gas at STP. Since 1 mole of an ideal gas at STP has a volume of 22.42 L, 1.75 L N_2 at STP will contain less than 1 mole. We can find how many moles using the ratio of 1.75 L to 22.42 L:

$$1.75 \ \cancel{L \ N_2} \times \frac{1 \ \text{mol} \ N_2}{22.42 \ \cancel{L \ N_2}} = 7.81 \times 10^{-2} \ \text{mol} \ N_2$$

See Exercises 8-65 and 8-66

Many chemical reactions involve gases. By assuming ideal behavior for these gases, we can carry out stoichiometric calculations if the pressure, volume, and temperature of the gases are known.

INTERACTIVE EXAMPLE 8-12

Gas Stoichiometry II

Quicklime (CaO) is produced by the thermal decomposition of calcium carbonate ($CaCO_3$). Calculate the volume of CO_2 at STP produced from the decomposition of 152 g $CaCO_3$ by the reaction

$$CaCO_3(s) \longrightarrow CaO(s) + CO_2(g)$$

Solution

Where are we going?
To use stoichiometry to determine the volume of CO_2 produced

What do we know?

$$\blacksquare \; CaCO_3(s) \longrightarrow CaO(s) + CO_2(g)$$

What information do we need?

- Molar volume of a gas at STP is 22.42 L

How do we get there?
We need to use the strategy for solving stoichiometry problems that we learned earlier in this chapter.

1. *What is the balanced equation?*

$$CaCO_3(s) \longrightarrow CaO(s) + CO_2(g)$$

2. *What are the moles of $CaCO_3$ (100.9 g/mol)?*

$$152 \text{ g } \cancel{CaCO_3} \times \frac{1 \text{ mol } CaCO_3}{100.09 \text{ g } \cancel{CaCO_3}} = 1.52 \text{ mol } CaCO_3$$

3. *What is the mole ratio between CO_2 and $CaCO_3$ in the balanced equation?*

$$\frac{1 \text{ mol } CO_2}{1 \text{ mol } CaCO_3}$$

4. *What are the moles of CO_2?*
1.52 moles of CO_2, which is the same as the moles of $CaCO_3$ because the mole ratio is 1.

5. *What is the volume of CO_2 produced?*
We can compute this by using the molar volume, since the sample is at STP: ◄ ⓘ

$$1.52 \text{ mol } \cancel{CO_2} \times \frac{22.42 \text{ L } CO_2}{1 \text{ mol } \cancel{CO_2}} = 34.1 \text{ L } CO_2$$

ⓘ Remember that the molar volume of an ideal gas is 22.42 L when measured at STP.

Thus the decomposition of 152 g $CaCO_3$ produces 34.1 L CO_2 at STP.

See Exercises 8-67 through 8-70

Note that in Example 8-12 the final step involved calculation of the volume of gas from the number of moles. Since the conditions were specified as STP, we were able to use the molar volume of a gas at STP. If the conditions of a problem are different from STP, the ideal gas law must be used to compute the volume.

INTERACTIVE EXAMPLE 8-13

Gas Stoichiometry III

A sample of methane gas having a volume of 2.80 L at 25°C and 1.65 atm was mixed with a sample of oxygen gas having a volume of 35.0 L at 31°C and 1.25 atm. The mixture was then ignited to form carbon dioxide and water. Calculate the volume of CO_2 formed at a pressure of 2.50 atm and a temperature of 125°C.

Solution

Where are we going?
To determine the volume of CO_2 produced

What do we know?

	CH$_4$	O$_2$	CO$_2$
P	1.65 atm	1.25 atm	2.50 atm
V	2.80 L	35.0 L	?
T	25°C + 273 = 298 K	31°C + 273 = 304 K	125°C + 273 = 398 K

What information do we need?

- Balanced chemical equation for the reaction
- Ideal gas law

$$PV = nRT$$

- $R = 0.08206 \text{ L} \cdot \text{atm/K} \cdot \text{mol}$

How do we get there?

We need to use the strategy for solving stoichiometry problems that we learned earlier in this chapter.

1. *What is the balanced equation?*
   From the description of the reaction, the unbalanced equation is

$$CH_4(g) + O_2(g) \longrightarrow CO_2(g) + H_2O(g)$$

   which can be balanced to give

$$CH_4(g) + 2O_2(g) \longrightarrow CO_2(g) + 2H_2O(g)$$

2. *What is the limiting reactant?*
   We can determine this by using the ideal gas law to determine the moles for each reactant:

$$n_{CH_4} = \frac{PV}{RT} = \frac{(1.65 \text{ atm})(2.80 \text{ L})}{(0.08206 \text{ L} \cdot \text{atm/K} \cdot \text{mol})(298 \text{ K})} = 0.189 \text{ mol}$$

$$n_{O_2} = \frac{PV}{RT} = \frac{(1.25 \text{ atm})(35.0 \text{ L})}{(0.08206 \text{ L} \cdot \text{atm/K} \cdot \text{mol})(304 \text{ K})} = 1.75 \text{ mol}$$

   In the balanced equation for the combustion reaction, 1 mole of CH$_4$ requires 2 moles of O$_2$. Thus the moles of O$_2$ required by 0.189 mole of CH$_4$ can be calculated as follows:

$$0.189 \text{ mol CH}_4 \times \frac{2 \text{ mol O}_2}{1 \text{ mol CH}_4} = 0.378 \text{ mol O}_2$$

   The limiting reactant is CH$_4$.

3. *What are the moles of CO$_2$?*
   Since CH$_4$ is limiting, we use the moles of CH$_4$ to determine the moles of CO$_2$ produced:

$$0.189 \text{ mol CH}_4 \times \frac{1 \text{ mol CO}_2}{1 \text{ mol CH}_4} = 0.189 \text{ mol CO}_2$$

4. *What is the volume of CO$_2$ produced?*
   Since the conditions stated are not STP, we must use the ideal gas law to calculate the volume:

$$V = \frac{nRT}{P}$$

In this case $n = 0.189$ mol, $T = 125°C + 273 = 398$ K, $P = 2.50$ atm, and $R = 0.08206$ L · atm/K · mol. Thus

$$V = \frac{(0.189 \text{ mol})(0.08206 \text{ L} \cdot \text{atm/K} \cdot \text{mol})(398 \text{ K})}{2.50 \text{ atm}} = 2.47 \text{ L}$$

This represents the volume of CO_2 produced under these conditions.

See Exercises 8-71 through 8-74

Molar Mass of a Gas

One very important use of the ideal gas law is in the calculation of the molar mass (molecular weight) of a gas from its measured density. To see the relationship between gas density and molar mass, consider that the number of moles of gas n can be expressed as

$$n = \frac{\text{grams of gas}}{\text{molar mass}} = \frac{\text{mass}}{\text{molar mass}} = \frac{m}{\text{molar mass}}$$

Substitution into the ideal gas equation gives

$$P = \frac{nRT}{V} = \frac{(m/\text{molar mass}) RT}{V} = \frac{m(RT)}{V(\text{molar mass})}$$

ⓘ Density = $\frac{\text{mass}}{\text{volume}}$

However, m/V is the gas density d in units of grams per liter. ◄ **ⓘ** Thus

$$P = \frac{dRT}{\text{molar mass}}$$

or

$$\text{Molar mass} = \frac{dRT}{P}$$

Thus, if the density of a gas at a given temperature and pressure is known, its molar mass can be calculated.

INTERACTIVE EXAMPLE 8-14

Gas Density/Molar Mass

The density of a gas was measured at 1.50 atm and 27°C and found to be 1.95 g/L. Calculate the molar mass of the gas.

Solution

Where are we going?
To determine the molar mass of the gas

What do we know?
- $P = 1.50$ atm
- $T = 27°C + 273 = 300.$ K
- $d = 1.95$ g/L

What information do we need?
- Molar mass = $\frac{dRT}{P}$
- $R = 0.08206$ L · atm/K · mol

How do we get there?

$$\text{Molar mass} = \frac{dRT}{P} = \frac{\left(1.95\frac{\text{g}}{\text{L}}\right)\left(0.08206\frac{\text{L} \cdot \text{atm}}{\text{K} \cdot \text{mol}}\right)(300.\ \text{K})}{1.50\ \text{atm}} = 32.0\ \text{g/mol}$$

REALITY CHECK | These are the units expected for molar mass.

See Exercises 8-77 through 8-80

You could memorize the equation involving gas density and molar mass, but it is better simply to remember the ideal gas equation, the definition of density, and the relationship between number of moles and molar mass. You can then derive the appropriate equation when you need it. This approach ensures that you understand the concepts, and it means you have one less equation to memorize.

8-5 | Dalton's Law of Partial Pressures

Among the experiments that led John Dalton to propose the atomic theory were his studies of mixtures of gases. In 1803 Dalton summarized his observations as follows: *For a mixture of gases in a container, the total pressure exerted is the sum of the pressures that each gas would exert if it were alone.* This statement, known as **Dalton's law of partial pressures**, can be expressed as follows:

$$P_{\text{TOTAL}} = P_1 + P_2 + P_3 + \cdots$$

where the subscripts refer to the individual gases (gas 1, gas 2, and so on). The symbols P_1, P_2, P_3, and so on represent each **partial pressure**, the pressure that a particular gas would exert if it were alone in the container.

Assuming that each gas behaves ideally, the partial pressure of each gas can be calculated from the ideal gas law:

$$P_1 = \frac{n_1RT}{V}, \quad P_2 = \frac{n_2RT}{V}, \quad P_3 = \frac{n_3RT}{V}, \quad \cdots$$

The total pressure of the mixture P_{TOTAL} can be represented as

$$P_{\text{TOTAL}} = P_1 + P_2 + P_3 + \cdots = \frac{n_1RT}{V} + \frac{n_2RT}{V} + \frac{n_3RT}{V} + \cdots$$

$$= (n_1 + n_2 + n_3 + \cdots)\left(\frac{RT}{V}\right)$$

$$= n_{\text{TOTAL}}\left(\frac{RT}{V}\right)$$

We can use a simple model of the atom without regard to sub-atomic particles for ideal gases. As we will see in Sections 8-8 and 8-9, this model does not explain the properties of real gases under certain conditions.

where n_{TOTAL} is the sum of the numbers of moles of the various gases. Thus, for a mixture of ideal gases, it is the *total number of moles of particles* that is important, not the identity or composition of the involved gas particles. ◄ ⓘ This idea is illustrated in Fig. 8-12.

This important observation indicates some fundamental characteristics of an ideal gas. The fact that the pressure exerted by an ideal gas is not affected by the identity (composition) of the gas particles reveals two things about ideal gases: (1) the volume of the individual gas particle must not be important, and (2) the forces among the particles must not be important. If these factors were important, the pressure exerted by the gas would depend on the nature of the individual particles. These observations will strongly influence the model that we will eventually construct to explain ideal gas behavior.

Figure 8-12 | The partial pressure of each gas in a mixture of gases in a container depends on the number of moles of that gas. The total pressure is the sum of the partial pressures and depends on the total moles of gas particles present, no matter what they are. Note that the volume remains constant.

 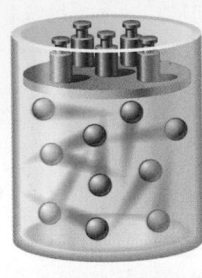 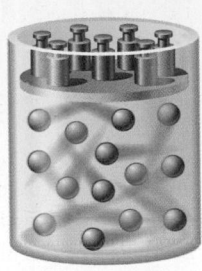

INTERACTIVE EXAMPLE 8-15

Dalton's Law I

Mixtures of helium and oxygen can be used in scuba diving tanks to help prevent "the bends." For a particular dive, 46 L He at 25°C and 1.0 atm and 12 L O_2 at 25°C and 1.0 atm were pumped into a tank with a volume of 5.0 L. Calculate the partial pressure of each gas and the total pressure in the tank at 25°C.

Solution

Where are we going?
To determine the partial pressure of each gas
To determine the total pressure in the tank at 25°C

What do we know?

	He	O_2	Tank
P	1.00 atm	1.00 atm	? atm
V	46 L	12 L	5.0 L
T	25°C + 273 = 298 K	25°C + 273 = 298 K	25°C + 273 = 298 K

What information do we need?

■ Ideal gas law

$$PV = nRT$$

■ $R = 0.08206$ L · atm/K · mol

How do we get there?
How many moles are present for each gas?

$$n = \frac{PV}{RT}$$

$$n_{He} = \frac{(1.0\ \text{atm})(46\ \text{L})}{(0.08206\ \text{L} \cdot \text{atm/K} \cdot \text{mol})(298\ \text{K})} = 1.9\ \text{mol}$$

$$n_{O_2} = \frac{(1.0\ \text{atm})(12\ \text{L})}{(0.08206\ \text{L} \cdot \text{atm/K} \cdot \text{mol})(298\ \text{K})} = 0.49\ \text{mol}$$

What is the partial pressure for each gas in the tank?
The tank containing the mixture has a volume of 5.0 L, and the temperature is 25°C. We can use these data and the ideal gas law to calculate the partial pressure of each gas:

$$P = \frac{nRT}{V}$$

$$P_{He} = \frac{(1.9\ \text{mol})(0.08206\ \text{L} \cdot \text{atm/K} \cdot \text{mol})(298\ \text{K})}{5.0\ \text{L}} = 9.3\ \text{atm}$$

$$P_{O_2} = \frac{(0.49\ \text{mol})(0.08206\ \text{L} \cdot \text{atm/K} \cdot \text{mol})(298\ \text{K})}{5.0\ \text{L}} = 2.4\ \text{atm}$$

Assume you work for an oil company that owns a huge natural gas reservoir containing a mixture of methane and nitrogen gases. In fact, the gas mixture contains so much nitrogen that it is unusable as a fuel. Your job is to separate the nitrogen (N_2) from the methane (CH_4). How might you accomplish this task? You clearly need some sort of "molecular filter" that will stop the slightly larger methane molecules (size ≈ 430 pm) and allow the nitrogen molecules (size ≈ 410 pm) to pass through. To accomplish the separation of molecules so similar in size will require a very precise "filter."

The good news is that such a filter exists. Recent work by Steven Kuznicki and Valerie Bell at Engelhard Corporation in New Jersey and Michael Tsapatsis at the University of Massachusetts has produced a "molecular sieve" in which the pore (passage) sizes can be adjusted precisely enough to separate N_2 molecules from CH_4 molecules. The material involved is a special hydrated titanosilicate (contains H_2O, Ti, Si, O, and Sr) compound patented by Engelhard known as ETS-4

(Engelhard TitanoSilicate-4). When sodium ions are substituted for the strontium ions in ETS-4 and the new material is carefully dehydrated, a uniform and controllable pore-size reduction occurs (see figure). The researchers have shown that the material can be used to separate N_2 (≈ 410 pm) from O_2 (≈ 390 pm). They have also shown that it is possible to reduce the nitrogen content of natural gas from 18% to less than 5% with a 90% recovery of methane.

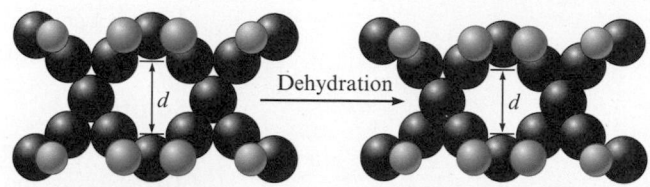

Molecular sieve framework of titanium (blue), silicon (green), and oxygen (red) atoms contracts on heating—at room temperature (left), d = 4.27 Å; at 250°C (right), d = 3.94 Å.

What is the total pressure of the mixture of gases in the tank?
The total pressure is the sum of the partial pressures:

$$P_{TOTAL} = P_{He} + P_{O_2} = 9.3 \text{ atm} + 2.4 \text{ atm} = 11.7 \text{ atm}$$

See Exercises 8-83 and 8-84

At this point we need to define the **mole fraction**: *the ratio of the number of moles of a given component in a mixture to the total number of moles in the mixture.* The Greek lowercase letter chi (χ) is used to symbolize the mole fraction. For example, for a given component in a mixture, the mole fraction χ_1 is

$$\chi_1 = \frac{n_1}{n_{TOTAL}} = \frac{n_1}{n_1 + n_2 + n_3 + \cdots}$$

From the ideal gas equation we know that the number of moles of a gas is directly proportional to the pressure of the gas, since

$$n = P\left(\frac{V}{RT}\right)$$

That is, for each component in the mixture,

$$n_1 = P_1\left(\frac{V}{RT}\right), \quad n_2 = P_2\left(\frac{V}{RT}\right), \quad \cdots$$

Therefore, we can represent the mole fraction in terms of pressures:

$$\chi_1 = \frac{n_1}{n_{\text{TOTAL}}} = \frac{\overset{n_1}{\overbrace{P_1(V/RT)}}}{\underset{n_1}{\underbrace{P_1(V/RT)}} + \underset{n_2}{\underbrace{P_2(V/RT)}} + \underset{n_3}{\underbrace{P_3(V/RT)}} + \cdots}$$

$$= \frac{(V/RT)P_1}{(V/RT)(P_1 + P_2 + P_3 + \cdots)}$$

$$= \frac{P_1}{P_1 + P_2 + P_3 + \cdots} = \frac{P_1}{P_{\text{TOTAL}}}$$

In fact, the mole fraction of each component in a mixture of ideal gases is directly related to its partial pressure:

$$\chi_2 = \frac{n_2}{n_{\text{TOTAL}}} = \frac{P_2}{P_{\text{TOTAL}}}$$

INTERACTIVE EXAMPLE 8-16

Dalton's Law II

The partial pressure of oxygen was observed to be 156 torr in air with a total atmospheric pressure of 743 torr. Calculate the mole fraction of O_2 present.

Solution

Where are we going?
To determine the mole fraction of O_2

What do we know?

- P_{O_2} = 156 torr
- P_{TOTAL} = 743 torr

How do we get there?
The mole fraction of O_2 can be calculated from the equation

$$\chi_{O_2} = \frac{P_{O_2}}{P_{\text{TOTAL}}} = \frac{156 \text{ torr}}{743 \text{ torr}} = 0.210$$

Note that the mole fraction has no units.

See Exercise 8-89

The expression for the mole fraction,

$$\chi_1 = \frac{P_1}{P_{\text{TOTAL}}}$$

can be rearranged to give

$$P_1 = \chi_1 \times P_{\text{TOTAL}}$$

That is, *the partial pressure of a particular component of a gaseous mixture is the mole fraction of that component times the total pressure.*

INTERACTIVE EXAMPLE 8-17

Dalton's Law III

The mole fraction of nitrogen in the air is 0.7808. Calculate the partial pressure of N_2 in air when the atmospheric pressure is 760. torr.

Solution

The partial pressure of N_2 can be calculated as follows:

$$P_{N_2} = \chi_{N_2} \times P_{TOTAL} = 0.7808 \times 760. \text{ torr} = 593 \text{ torr}$$

See Exercise 8-90

Collecting a Gas over Water

A mixture of gases results whenever a gas is collected by displacement of water. For example, Fig. 8-13 shows the collection of oxygen gas produced by the decomposition of solid potassium chlorate. In this situation, the gas in the bottle is a mixture of water vapor and the oxygen being collected. Water vapor is present because molecules of water escape from the surface of the liquid and collect in the space above the liquid. Molecules of water also return to the liquid. When the rate of escape equals the rate of return, the number of water molecules in the vapor state remains constant, and thus the pressure of water vapor remains constant. This pressure, which depends on temperature, is called the *vapor pressure of water.*

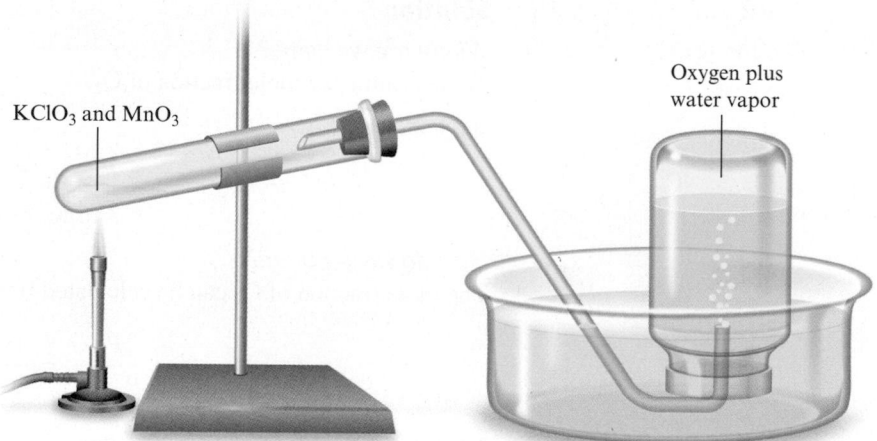

KClO$_3$ and MnO$_3$

Oxygen plus water vapor

Figure 8-13 | The production of oxygen by thermal decomposition of KClO$_3$. The MnO$_2$ is mixed with the KClO$_3$ to make the reaction faster.

INTERACTIVE EXAMPLE 8-18

Gas Collection over Water

A sample of solid potassium chlorate (KClO$_3$) was heated in a test tube (see Fig. 8-13) and decomposed by the following reaction:

$$2KClO_3(s) \longrightarrow 2KCl(s) + 3O_2(g)$$

The oxygen produced was collected by displacement of water at 22°C at a total pressure of 754 torr. The volume of the gas collected was 0.650 L, and the vapor pressure of water at 22°C is 21 torr. Calculate the partial pressure of O_2 in the gas collected and the mass of KClO$_3$ in the sample that was decomposed.

Solution

Where are we going?
To determine the partial pressure of O_2 in the gas collected
To calculate the mass of KClO$_3$ in the original sample

What do we know?

	Gas Collected	Water Vapor
P	754 torr	21 torr
V	0.650 L	
T	22°C + 273 = 295 K	22°C + 273 = 295 K

How do we get there?

What is the partial pressure of O_2?

$$P_{\text{TOTAL}} = P_{O_2} + P_{H_2O} = P_{O_2} + 21 \text{ torr} = 754 \text{ torr}$$

$$P_{O_2} = 754 \text{ torr} - 21 \text{ torr} = 733 \text{ torr}$$

What is the number of moles of O_2?

Now we use the ideal gas law to find the number of moles of O_2:

$$n_{O_2} = \frac{P_{O_2}V}{RT}$$

In this case, the partial pressure of the O_2 is

$$P_{O_2} = 733 \text{ torr} = \frac{733 \text{ torr}}{760 \text{ torr/atm}} = 0.964 \text{ atm}$$

To find the moles of O_2 produced, we use

$$V = 0.650 \text{ L}$$

$$T = 22°C + 273 = 295 \text{ K}$$

$$R = 0.08206 \text{ L} \cdot \text{atm/K} \cdot \text{mol}$$

$$n_{O_2} = \frac{(0.964 \text{ atm})(0.650 \text{ L})}{(0.08206 \text{ L} \cdot \text{atm/K} \cdot \text{mol})(295 \text{ K})} = 2.59 \times 10^{-2} \text{ mol}$$

How many moles of $KClO_3$ are required to produce this amount of O_2?

Use the stoichiometry problem-solving strategy:

1. *What is the balanced equation?*

$$2KClO_3(s) \longrightarrow 2KCl(s) + 3O_2(g)$$

2. *What is the mole ratio between $KClO_3$ and O_2 in the balanced equation?*

$$\frac{2 \text{ mol KClO}_3}{3 \text{ mol O}_2}$$

3. *What are the moles of $KClO_3$?*

$$2.59 \times 10^{-2} \text{ mol O}_2 \times \frac{2 \text{ mol KClO}_3}{3 \text{ mol O}_2} = 1.73 \times 10^{-2} \text{ mol KClO}_3$$

4. *What is the mass of $KClO_3$ (molar mass 122.6 g/mol) in the original sample?*

$$1.73 \times 10^{-2} \text{ mol KClO}_3 \times \frac{122.6 \text{ g KClO}_3}{1 \text{ mol KClO}_3} = 2.12 \text{ g KClO}_3$$

Thus the original sample contained 2.12 g $KClO_3$.

See Exercises 8-91 through 8-93

Most experts agree that air bags represent a very important advance in automobile safety. First patented by American inventor John W. Hetrick in 1953, air bags are now required in all cars and trucks in the United States. These bags, which are stored in the auto's steering wheel or dash, are designed to inflate rapidly (within about 40 ms) in the event of a crash, cushioning the front-seat occupants against impact. The bags then deflate immediately to allow vision and movement after the crash. Air bags are activated when a severe deceleration (an impact) causes a steel ball to compress a spring and electrically ignite a detonator cap, which, in turn, causes sodium azide (NaN_3) to decompose explosively, forming sodium and nitrogen gas:

$$2NaN_3(s) \longrightarrow 2Na(s) + 3N_2(g)$$

This system works very well and requires only a relatively small amount of sodium azide [100 g yields 56 L $N_2(g)$ at 25°C and 1 atm].

In addition to being located in the steering wheel and dash, air bags are now found in many other sites in motor vehicles. Air bags to protect against side impacts are located above the windows (called *curtain air bags*) to protect against head injuries and in the doors (called *torso air bags*). Some vehicles also have bags to prevent knee injuries. Because the explosive deployment of air bags can cause serious injuries, especially to children, variable force air bags have been developed that depend on the weight of the person occupying the front seat.

When a vehicle containing air bags reaches the end of its useful life, the sodium azide present in the activators must be given proper disposal. Besides being explosive, sodium azide has a toxicity roughly equal to that of sodium cyanide. It also forms hydrazoic acid (HN_3), a toxic and explosive liquid, when treated with acid.

The air bag represents an application of chemistry that has already saved thousands of lives.

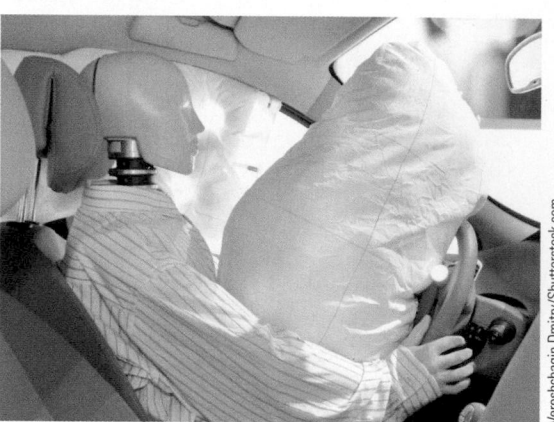

Inflated dual air bags.

8-6 | The Kinetic Molecular Theory of Gases

We have so far considered the behavior of gases from an experimental point of view. Based on observations from different types of experiments, we know that at pressures of less than 1 atm most gases closely approach the behavior described by the ideal gas law. Now we want to construct a model to explain this behavior.

Before we do this, let's briefly review the scientific method. Recall that a law is a way of generalizing behavior that has been observed in many experiments. Laws are very useful, since they allow us to predict the behavior of similar systems. For example, if a chemist prepares a new gaseous compound, a measurement of the gas density at known pressure and temperature can provide a reliable value for the compound's molar mass.

However, although laws summarize observed behavior, they do not tell us *why* nature behaves in the observed fashion. This is the central question for scientists. To try to answer this question, we construct theories (build models). The models in chemistry consist of speculations about what the individual atoms or molecules (microscopic particles) might be doing to cause the observed behavior of the macroscopic systems (collections of very large numbers of atoms and molecules).

A model is considered successful if it explains the observed behavior in question and predicts correctly the results of future experiments. It is important to understand that a model can never be proved absolutely true. In fact, *any model is an approximation* by its very nature and is bound to fail at some point. Models range from the simple to the extraordinarily complex. We use simple models to predict approximate behavior and more complicated models to account very precisely for observed quantitative behavior. In this text we will stress simple models that provide an approximate picture of what might be happening and that fit the most important experimental results.

An example of this type of model is the **kinetic molecular theory (KMT)**, a simple model that attempts to explain the properties of an ideal gas. This model is based on speculations about the behavior of the individual gas particles (atoms or molecules). The postulates of the kinetic molecular theory as they relate to the particles of an ideal gas can be stated as follows:

Postulates of the Kinetic Molecular Theory

1. The particles are so small compared with the distances between them that *the volume of the individual particles can be assumed to be negligible* (zero) (Fig. 8-14).

2. *The particles are in constant motion. The collisions of the particles with the walls of the container are the cause of the pressure exerted by the gas.*

3. *The particles are assumed to exert no forces on each other; they are assumed neither to attract nor to repel each other.*

4. *The average kinetic energy of a collection of gas particles is assumed to be directly proportional to the Kelvin temperature of the gas.*

Of course, the molecules in a real gas have finite volumes and do exert forces on each other. Thus *real gases* do not conform to these assumptions. However, we will see that these postulates do indeed explain *ideal gas* behavior.

The true test of a model is how well its predictions fit the experimental observations. The postulates of the kinetic molecular model picture an ideal gas as consisting of particles having no volume and no attractions for each other, and the model assumes that the gas produces pressure on its container by collisions with the walls.

Let's consider how this model accounts for the properties of gases as summarized by the ideal gas law: $PV = nRT$.

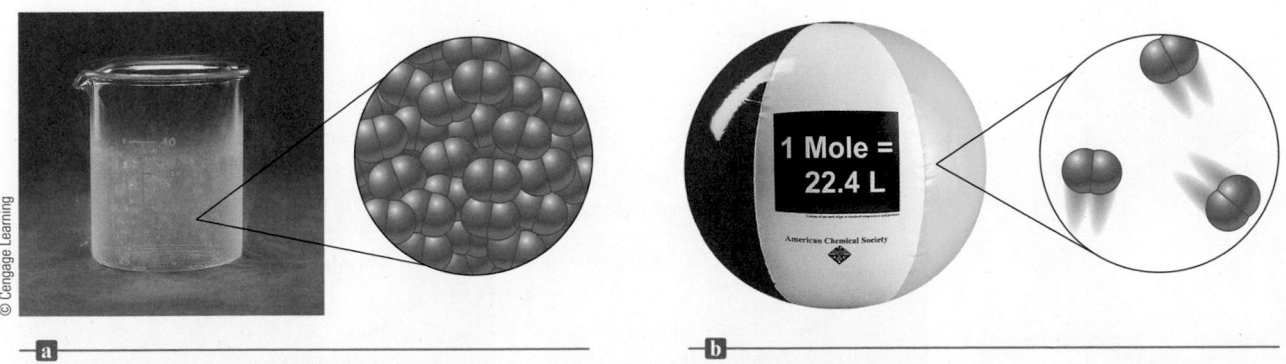

Figure 8-14 | (a) One mole of $N_2(l)$ has a volume of approximately 35 mL and a density of 0.81 g/mL. (b) One mole of $N_2(g)$ has a volume of 22.42 L (STP) and a density of 1.2×10^{-3} g/mL. Thus the ratio of the volumes of gaseous N_2 and liquid N_2 is 22.42/0.035 = 640, and the spacing of the molecules is 9 times farther apart in $N_2(g)$.

Figure 8-15 | The effects of decreasing the volume of a sample of gas at constant temperature.

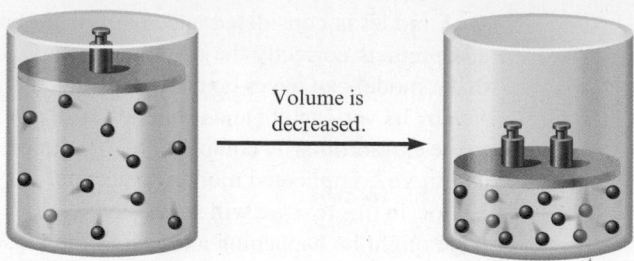

Volume is decreased.

Pressure and Volume (Boyle's Law)

We have seen that for a given sample of gas at a given temperature (n and T are constant), if the volume of a gas is decreased, then the pressure increases:

$$P = \underbrace{(nRT)}_{\text{Constant}}\frac{1}{V}$$

This makes sense based on the kinetic molecular theory because a decrease in volume means that the gas particles will hit the wall more often, thus increasing pressure, as illustrated in Fig. 8-15.

Pressure and Temperature ◄

From the ideal gas law we can predict that for a given sample of an ideal gas at a constant volume, the pressure will be directly proportional to the temperature:

$$P = \left(\frac{nR}{V}\right)T$$

$$\underset{\text{Constant}}{\uparrow}$$

Ken O'Donoghue

a   b   c

▲ (a) A balloon filled with air at room temperature. (b) The balloon is dipped into liquid nitrogen at 77 K. (c) The balloon collapses as the molecules inside slow down due to the decreased temperature. Slower molecules produce a lower pressure.

Figure 8-16 | The effects of increasing the temperature of a sample of gas at constant volume.

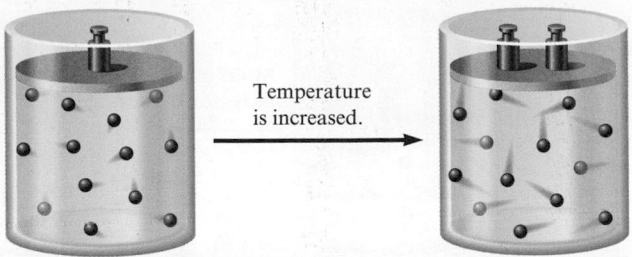

The KMT accounts for this behavior because when the temperature of a gas increases, the speeds of its particles increase; the particles hit the wall with greater force and greater frequency. Since the volume remains the same, this would result in increased gas pressure, as illustrated in Fig. 8-16.

Critical Thinking

You have learned the postulates of the KMT. What if we could not assume the third postulate to be true? How would this affect the measured pressure of a gas?

Volume and Temperature (Charles's Law)

The ideal gas law indicates that for a given sample of gas at a constant pressure, the volume of the gas is directly proportional to the temperature in kelvins:

$$V = \left(\frac{nR}{P}\right)T$$

$$\uparrow$$
Constant

This can be visualized from the KMT, as shown in Fig. 8-17. When the gas is heated to a higher temperature, the speeds of its molecules increase and thus they hit the walls more often and with more force. The only way to keep the pressure constant in this situation is to increase the volume of the container. This compensates for the increased particle speeds.

Volume and Number of Moles (Avogadro's Law)

The ideal gas law predicts that the volume of a gas at a constant temperature and pressure depends directly on the number of gas particles present:

$$V = \left(\frac{RT}{P}\right)n$$

$$\uparrow$$
Constant

Figure 8-17 | The effects of increasing the temperature of a sample of gas at constant pressure.

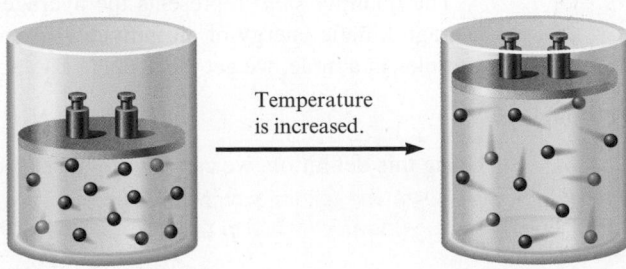

Figure 8-18 | The effects of increasing the number of moles of gas particles at constant temperature and pressure.

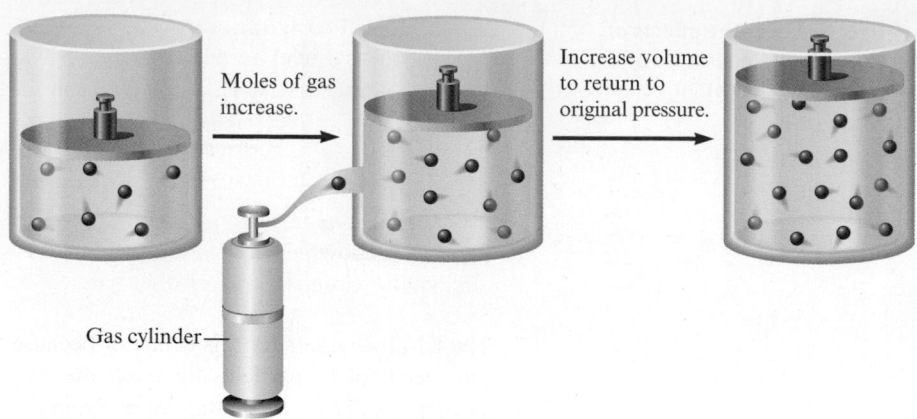

Moles of gas increase.

Increase volume to return to original pressure.

Gas cylinder

This makes sense in terms of the KMT because an increase in the number of gas particles at the same temperature would cause the pressure to increase if the volume were held constant (see Fig. 8-18). The only way to return the pressure to its original value is to increase the volume.

It is important to recognize that the volume of a gas (at constant P and T) depends only on the *number* of gas particles present. The individual volumes of the particles are not a factor because the particle volumes are so small compared with the distances between the particles (for a gas behaving ideally).

Mixture of Gases (Dalton's Law)

The observation that the total pressure exerted by a mixture of gases is the sum of the pressures of the individual gases is expected because the KMT assumes that all gas particles are independent of each other and that the volumes of the individual particles are unimportant. Thus the identities of the gas particles do not matter.

Deriving the Ideal Gas Law

We have shown qualitatively that the assumptions of the KMT successfully account for the observed behavior of an ideal gas. We can go further. By applying the principles of physics to the assumptions of the KMT, we can in effect derive the ideal gas law.

As shown in detail in Appendix 2, we can apply the definitions of velocity, momentum, force, and pressure to the collection of particles in an ideal gas and *derive* the following expression for pressure:

$$P = \frac{2}{3}\left[\frac{nN_A\left(\frac{1}{2}m\overline{u^2}\right)}{V}\right]$$

where P is the pressure of the gas, n is the number of moles of gas, N_A is Avogadro's number, m is the mass of each particle, $\overline{u^2}$ is the average of the square of the velocities of the particles, and V is the volume of the container.

The quantity $\frac{1}{2}m\overline{u^2}$ represents the average kinetic energy of a gas particle. If the average kinetic energy of an individual particle is multiplied by N_A, the number of particles in a mole, we get the average kinetic energy for a mole of gas particles:

$$(KE)_{avg} = N_A\left(\tfrac{1}{2}m\overline{u^2}\right) \blacktriangleleft \; ⓘ$$

ⓘ Kinetic energy (KE) given by the equation $KE = \frac{1}{2}mu^2$ is the energy due to the motion of a particle.

Using this definition, we can rewrite the expression for pressure as

$$P = \frac{2}{3}\left[\frac{n(KE)_{avg}}{V}\right] \quad \text{or} \quad \frac{PV}{n} = \frac{2}{3}(KE)_{avg}$$

The fourth postulate of the kinetic molecular theory is that the average kinetic energy of the particles in the gas sample is directly proportional to the temperature in kelvins. Thus, since $(KE)_{avg} \propto T$, we can write

$$\frac{PV}{n} = \frac{2}{3}(KE)_{avg} \propto T \quad \text{or} \quad \frac{PV}{n} \propto T$$

Note that this expression has been *derived* from the assumptions of the kinetic molecular theory. How does it compare to the ideal gas law—the equation obtained from experiments? Compare the ideal gas law,

$$\frac{PV}{n} = RT \quad \text{From experiment}$$

with the result from the kinetic molecular theory,

$$\frac{PV}{n} \propto T \quad \text{From theory}$$

These expressions have exactly the same form if R, the universal gas constant, is considered the proportionality constant in the second case.

The agreement between the ideal gas law and the predictions of the kinetic molecular theory gives us confidence in the validity of the model. The characteristics we have assumed for ideal gas particles must agree, at least under certain conditions, with their actual behavior.

The Meaning of Temperature

We have seen from the kinetic molecular theory that the Kelvin temperature indicates the average kinetic energy of the gas particles. The exact relationship between temperature and average kinetic energy can be obtained by combining the equations:

$$\frac{PV}{n} = RT = \frac{2}{3}(KE)_{avg}$$

which yields the expression

$$(KE)_{avg} = \frac{3}{2}RT$$

This is a very important relationship. It summarizes the meaning of the Kelvin temperature of a gas: The Kelvin temperature is an index of the random motions of the particles of a gas, with higher temperature meaning greater motion. (As we will see in Chapter 9, temperature is an index of the random motions in solids and liquids as well as in gases.)

Root Mean Square Velocity

In the equation from the kinetic molecular theory, the average velocity of the gas particles is a special kind of average. The symbol $\overline{u^2}$ means the average of the *squares* of the particle velocities. The square root of $\overline{u^2}$ is called the **root mean square velocity** and is symbolized by u_{rms}:

$$u_{rms} = \sqrt{\overline{u^2}}$$

We can obtain an expression for u_{rms} from the equations

$$(KE)_{avg} = N_A(\tfrac{1}{2}m\overline{u^2}) \quad \text{and} \quad (KE)_{avg} = \frac{3}{2}RT$$

Combination of these equations gives

$$N_A(\tfrac{1}{2}m\overline{u^2}) = \frac{3}{2}RT \quad \text{or} \quad \overline{u^2} = \frac{3RT}{N_A m}$$

Taking the square root of both sides of the last equation produces

$$\sqrt{\overline{u^2}} = u_{\text{rms}} = \sqrt{\frac{3RT}{N_A m}}$$

In this expression m represents the mass in kilograms of a single gas particle. When N_A, the number of particles in a mole, is multiplied by m, the product is the mass of a *mole* of gas particles in *kilograms*. We will call this quantity M. Substituting M for $N_A m$ in the equation for u_{rms}, we obtain

$$u_{\text{rms}} = \sqrt{\frac{3RT}{M}}$$

Before we can use this equation, we need to consider the units for R. So far we have used $0.08206 \, \text{L} \cdot \text{atm/K} \cdot \text{mol}$ as the value of R. But to obtain the desired units (meters per second) for u_{rms}, R must be expressed in different units. As we saw in Chapter 7, the energy unit most often used in the SI system is the joule (J). A **joule** is defined as a kilogram meter squared per second squared ($\text{kg} \cdot \text{m}^2/\text{s}^2$). When R is converted to include the unit of joules, it has the value $8.3145 \, \text{J/K} \cdot \text{mol}$. ◄ 🛈 When R in these units is used in the expression $\sqrt{3RT/M}$, u_{rms} is obtained in the units of meters per second as desired.

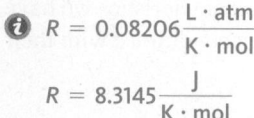

🛈 $R = 0.08206 \dfrac{\text{L} \cdot \text{atm}}{\text{K} \cdot \text{mol}}$

$R = 8.3145 \dfrac{\text{J}}{\text{K} \cdot \text{mol}}$

INTERACTIVE EXAMPLE 8-19

Root Mean Square Velocity

Calculate the root mean square velocity for the atoms in a sample of helium gas at 25°C.

Solution

Where are we going?
To determine the root mean square velocity for the atoms of He

What do we know?

- $T = 25°C + 273 = 298 \text{ K}$
- $R = 8.3145 \, \text{J/K} \cdot \text{mol}$

What information do we need?

- Root mean square velocity is $u_{\text{rms}} = \sqrt{\dfrac{3RT}{M}}$

How do we get there?
What is the mass of a mole of He in kilograms?

$$M = 4.00\frac{g}{\text{mol}} \times \frac{1 \text{ kg}}{1000 \text{ g}} = 4.00 \times 10^{-3} \text{ kg/mol}$$

What is the root mean square velocity for the atoms of He?

$$u_{\text{rms}} = \sqrt{\frac{3\left(8.3145\dfrac{\text{J}}{\text{K} \cdot \text{mol}}\right)(298 \text{ K})}{4.00 \times 10^{-3}\dfrac{\text{kg}}{\text{mol}}}} = \sqrt{1.86 \times 10^6\frac{\text{J}}{\text{kg}}}$$

Since the units of J are kg · m²/s², this expression gives

$$u_{rms} = \sqrt{1.86 \times 10^6 \frac{kg \cdot m^2}{kg \cdot s^2}} = 1.36 \times 10^3 \text{ m/s}$$

REALITY CHECK | The resulting units are appropriate for velocity.

See Exercises 8-103 and 8-104

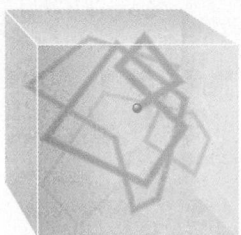

Figure 8-19 | Path of one particle in a gas. Any given particle will continuously change its course as a result of collisions with other particles, as well as with the walls of the container.

So far we have said nothing about the range of velocities actually found in a gas sample. In a real gas there are large numbers of collisions between particles. For example, as we will see in the next section, when an odorous gas such as ammonia is released in a room, it takes some time for the odor to permeate the air. This delay results from collisions between the NH_3 molecules and the O_2 and N_2 molecules in the air, which greatly slow the mixing process.

If the path of a particular gas particle could be monitored, it would look very erratic, something like that shown in Fig. 8-19. The average distance a particle travels between collisions in a particular gas sample is called the *mean free path*. It is typically a very small distance (1×10^{-7} m for O_2 at STP). One effect of the many collisions among gas particles is to produce a large range of velocities as the particles collide and exchange kinetic energy. Although u_{rms} for oxygen gas at STP is approximately 500 meters per second, the majority of O_2 molecules do not have this velocity. The actual distribution of molecular velocities for oxygen gas at STP is shown in Fig. 8-20. This figure shows the relative number of gas molecules having each particular velocity.

We are also interested in the effect of *temperature* on the velocity distribution in a gas. Fig. 8-21 shows the velocity distribution for nitrogen gas at three temperatures. Note that as the temperature is increased, the curve peak moves toward higher values and the range of velocities becomes much larger. The peak of the curve reflects the most probable velocity (the velocity found most often as we sample the movement of the various particles in the gas). Because the kinetic energy increases with temperature, it makes sense that the peak of the curve should move to higher values as the temperature of the gas is increased.

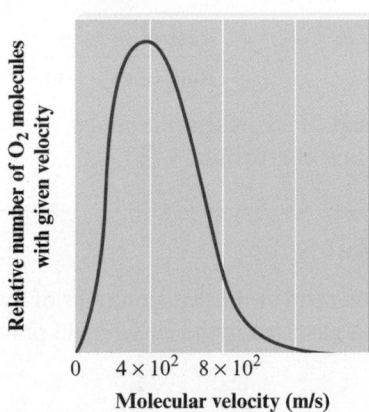

Figure 8-20 | A plot of the relative number of O_2 molecules that have a given velocity at STP.

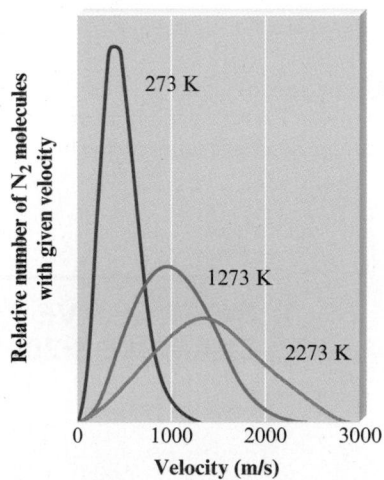

Figure 8-21 | A plot of the relative number of N_2 molecules that have a given velocity at three temperatures. Note that as the temperature increases, both the average velocity and the spread of velocities increase.

Figure 8-22 | The effusion of a gas into an evacuated chamber. The rate of effusion (the rate at which the gas is transferred across the barrier through the pin hole) is inversely proportional to the square root of the mass of the gas molecules.

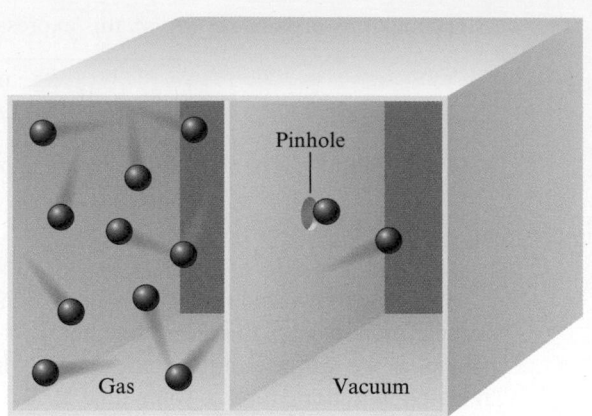

8-7 | Effusion and Diffusion

We have seen that the postulates of the kinetic molecular theory, when combined with the appropriate physical principles, produce an equation that successfully fits the experimentally observed behavior of gases as they approach ideal behavior. Two phenomena involving gases provide further tests of this model.

Diffusion is the term used to describe the mixing of gases. When a small amount of pungent-smelling ammonia is released at the front of a classroom, it takes some time before everyone in the room can smell it, because time is required for the ammonia to mix with the air. The rate of diffusion is the rate of the mixing of gases. **Effusion** is the term used to describe the passage of a gas through a tiny orifice into an evacuated chamber, as shown in Fig. 8-22. The rate of effusion measures the speed at which the gas is transferred into the chamber.

Effusion

Thomas Graham (1805–1869), a Scottish chemist, found experimentally that the rate of effusion of a gas is inversely proportional to the square root of the mass of its particles. Stated in another way, the relative rates of effusion of two gases at the same temperature and pressure are given by the inverse ratio of the square roots of the masses of the gas particles:

ⓘ In Graham's law the units for molar mass can be g/mol or kg/mol, since the units cancel in the ratio $\dfrac{\sqrt{M_2}}{\sqrt{M_1}}$

$$\frac{\text{Rate of effusion for gas 1}}{\text{Rate of effusion for gas 2}} = \frac{\sqrt{M_2}}{\sqrt{M_1}}$$

where M_1 and M_2 represent the molar masses of the gases. This equation is called **Graham's law of effusion.** ◀ *ⓘ*

INTERACTIVE EXAMPLE 8-20

Effusion Rates

Calculate the ratio of the effusion rates of hydrogen gas (H_2) and uranium hexafluoride (UF_6), a gas used in the enrichment process to produce fuel for nuclear reactors (Fig. 8-23).

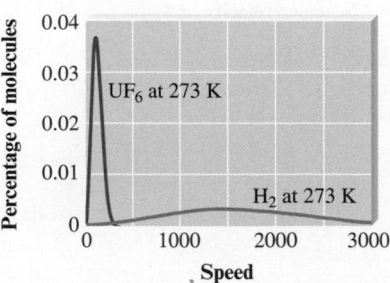

Figure 8-23 | Relative molecular speed distribution of H_2 and UF_6.

Solution

First we need to compute the molar masses: Molar mass of H_2 = 2.016 g/mol, and molar mass of UF_6 = 352.02 g/mol. Using Graham's law, we find

$$\frac{\text{Rate of effusion for } H_2}{\text{Rate of effusion for } UF_6} = \frac{\sqrt{M_{UF_6}}}{\sqrt{M_{H_2}}} = \sqrt{\frac{352.02}{2.016}} = 13.2$$

The effusion rate of the very light H_2 molecules is about 13 times that of the massive UF_6 molecules.

See Exercises 8-111 through 8-114

Does the kinetic molecular model for gases correctly predict the relative effusion rates of gases summarized by Graham's law? To answer this question, we must recognize that the effusion rate for a gas depends directly on the average velocity of its particles. The faster the gas particles are moving, the more likely they are to pass through the effusion orifice. This reasoning leads to the following *prediction* for two gases at the same pressure and temperature (T):

$$\frac{\text{Effusion rate for gas 1}}{\text{Effusion rate for gas 2}} = \frac{u_{rms} \text{ for gas 1}}{u_{rms} \text{ for gas 2}} = \frac{\sqrt{\dfrac{3RT}{M_1}}}{\sqrt{\dfrac{3RT}{M_2}}} = \frac{\sqrt{M_2}}{\sqrt{M_1}}$$

This equation is identical to Graham's law. Thus the kinetic molecular model does fit the experimental results for the effusion of gases.

Diffusion

Diffusion is frequently illustrated by the lecture demonstration represented in Fig. 8-24, in which two cotton plugs soaked in ammonia and hydrochloric acid are simultaneously placed at the ends of a long tube. A white ring of ammonium chloride (NH_4Cl) forms where the NH_3 and HCl molecules meet several minutes later:

$$NH_3(g) + HCl(g) \longrightarrow NH_4Cl(s)$$
$$\text{White solid}$$

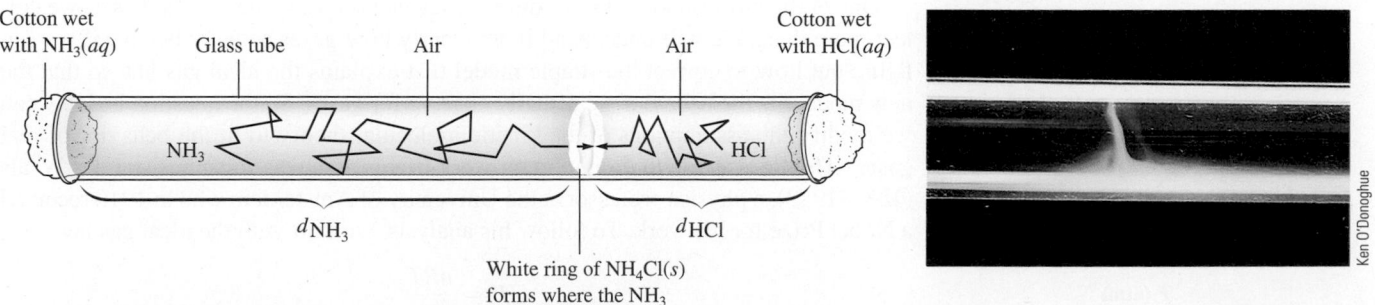

Figure 8-24 | (above right) When HCl(g) and NH_3(g) meet in the tube, a white ring of NH_4Cl(s) forms. (above left) A demonstration of the relative diffusion rates of NH_3 and HCl molecules through air. Two cotton plugs, one dipped in HCl(aq) and one dipped in NH_3(aq), are simultaneously inserted into the ends of the tube. Gaseous NH_3 and HCl vaporizing from the cotton plugs diffuse toward each other and, where they meet, react to form NH_4Cl(s).

As a first approximation we might expect that the distances traveled by the two gases are related to the relative velocities of the gas molecules:

$$\frac{\text{Distance traveled by NH}_3}{\text{Distance traveled by HCl}} = \frac{u_{\text{rms}} \text{ for NH}_3}{u_{\text{rms}} \text{ for HCl}} = \sqrt{\frac{M_{\text{HCl}}}{M_{\text{NH}_3}}} = \sqrt{\frac{36.5}{17}} = 1.5$$

However, careful experiments produce an observed ratio of less than 1.5, indicating that a quantitative analysis of diffusion requires a more complex analysis.

The diffusion of the gases through the tube is surprisingly slow in light of the fact that the velocities of HCl and NH$_3$ molecules at 25°C are about 450 and 660 meters per second, respectively. Why does it take several minutes for the NH$_3$ and HCl molecules to meet? The answer is that the tube contains air and thus the NH$_3$ and HCl molecules undergo many collisions with O$_2$ and N$_2$ molecules as they travel through the tube. Because so many collisions occur when gases mix, diffusion is quite complicated to describe theoretically.

8-8 | Real Gases

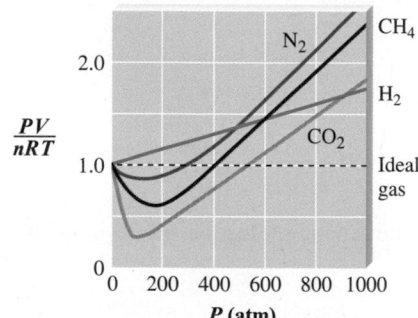

Figure 8-25 | Plots of PV/nRT versus P for several gases (200 K). Note the significant deviations from ideal behavior ($PV/nRT = 1$). The behavior is close to ideal only at low pressures (less than 1 atm).

An ideal gas is a hypothetical concept. No gas *exactly* follows the ideal gas law, although many gases come very close at low pressures and/or high temperatures. Thus ideal gas behavior can best be thought of as the behavior *approached by real gases* under certain conditions.

We have seen that a very simple model, the kinetic molecular theory, by making some rather drastic assumptions (no interparticle interactions and zero volume for the gas particles), successfully explains ideal behavior. However, it is important that we examine real gas behavior to see how it differs from that predicted by the ideal gas law and to determine what modifications are needed in the kinetic molecular theory to explain the observed behavior. Since a model is an approximation and will inevitably fail, we must be ready to learn from such failures. In fact, we often learn more about nature from the failures of our models than from their successes.

We will examine the experimentally observed behavior of real gases by measuring the pressure, volume, temperature, and number of moles for a gas and noting how the quantity PV/nRT depends on pressure. Plots of PV/nRT versus P are shown for several gases in Fig. 8-25. For an ideal gas, PV/nRT equals 1 under all conditions, but notice that for real gases, PV/nRT approaches 1 only at very low pressures (typically below 1 atm). To illustrate the effect of temperature, PV/nRT is plotted versus P for nitrogen gas at several temperatures in Fig. 8-26. Note that the behavior of the gas appears to become more nearly ideal as the temperature is increased. The most important conclusion to be drawn from these figures is that a **real gas** typically exhibits behavior that is closest to ideal behavior at *low pressures* and *high temperatures*.

One of the most important procedures in science is correcting our models as we collect more data. We will understand more clearly how gases actually behave if we can figure out how to correct the simple model that explains the ideal gas law so that the new model fits the behavior we actually observe for gases. So the question is: How can we modify the assumptions of the kinetic molecular theory to fit the behavior of real gases? The first person to do important work in this area was Johannes van der Waals (1837–1923), a physics professor at the University of Amsterdam who in 1910 received a Nobel Prize for his work. To follow his analysis, we start with the ideal gas law,

$$P = \frac{nRT}{V}$$

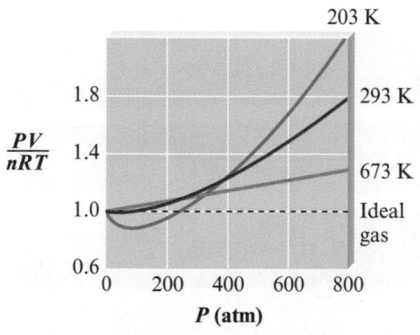

Figure 8-26 | Plots of PV/nRT versus P for nitrogen gas at three temperatures. Note that although nonideal behavior is evident in each case, the deviations are smaller at the higher temperatures.

Remember that this equation describes the behavior of a hypothetical gas consisting of volumeless entities that do not interact with each other. In contrast, a real gas consists of atoms or molecules that have finite volumes. Therefore, the volume available to a given particle in a real gas is less than the volume of the container because the gas particles themselves take up some of the space. To account for this discrepancy, van der Waals represented the actual volume as the volume of the container V minus a correction factor for the volume of the molecules nb, where n is the number of moles of gas and b is an empirical constant (one determined by fitting the equation to the experimental results). Thus the volume *actually available* to a given gas molecule is given by the difference $V - nb$.

This modification of the ideal gas equation leads to the equation ◄ ⓘ

$$P' = \frac{nRT}{V - nb}$$

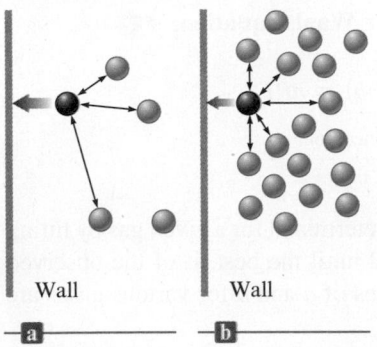

ⓘ P' is corrected for the finite volume of the particles. The attractive forces have not yet been taken into account.

The volume of the gas particles has now been taken into account.

The next step is to allow for the attractions that occur among the particles in a real gas. The effect of these attractions is to make the observed pressure P_{obs} smaller than it would be if the gas particles did not interact:

$$P_{obs} = (P' - \text{correction factor}) = \left(\frac{nRT}{V - nb} - \text{correction factor} \right)$$

This effect can be understood using the following model. When gas particles come close together, attractive forces occur, which cause the particles to hit the wall very slightly less often than they would in the absence of these interactions (Fig. 8-27).

The size of the correction factor depends on the concentration of gas molecules defined in terms of moles of gas particles per liter (n/V). The higher the concentration, the more likely a pair of gas particles will be close enough to attract each other. For large numbers of particles, the number of interacting *pairs* of particles depends on the square of the number of particles and thus on the square of the concentration, or $(n/V)^2$. This can be justified as follows: In a gas sample containing N particles, there are $N - 1$ partners available for each particle, as shown in Fig. 8-28. Since the $1 \cdots 2$ pair is the same as the $2 \cdots 1$ pair, this analysis counts each pair twice. Thus, for N particles, there are $N(N - 1)/2$ pairs. If N is a very large number, $N - 1$ approximately equals N, giving $N^2/2$ possible pairs. Thus the pressure, corrected for the attractions of the particles, has the form

Figure 8-27 | (a) Gas at low concentration—relatively few interactions between particles. The indicated gas particle exerts a pressure on the wall close to that predicted for an ideal gas. (b) Gas at high concentration—many more interactions between particles. The indicated gas particle exerts a much lower pressure on the wall than would be expected in the absence of interactions.

$$P_{obs} = P' - a\left(\frac{n}{V} \right)^2$$

where a is a proportionality constant (which includes the factor of $\frac{1}{2}$ from $N^2/2$). The value of a for a given real gas can be determined from observing the actual behavior of that gas. Inserting the corrections for both the volume of the particles and the attractions of the particles gives the equation ◄ ⓘ

ⓘ We have now corrected for both the finite volume and the attractive forces of the particles.

$$P_{obs} = \frac{nRT}{V - nb} - a\left(\frac{n}{V} \right)^2$$

Observed pressure   Volume of the container   Volume correction   Pressure correction

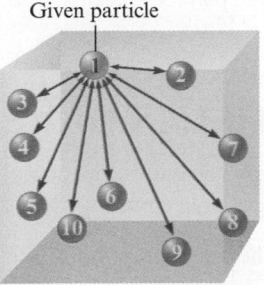

Given particle

Gas sample with 10 particles

Figure 8-28 | Illustration of pairwise interactions among gas particles. In a sample with 10 particles, each particle has 9 possible partners, to give $10(9)/2 = 45$ distinct pairs. The factor of $\frac{1}{2}$ arises because when particle ① is the particle of interest, we count the ① $\cdots$ ② pair, and when particle ② is the particle of interest, we count the ② $\cdots$ ① pair. However, ① $\cdots$ ② and ② $\cdots$ ① are the same pair, which we thus have counted twice. Therefore, we must divide by 2 to get the correct number of pairs.

Figure 8-29 | The volume taken up by the gas particles themselves is less important at (a) large container volume (low pressure) than at (b) small container volume (high pressure).

P_{obs} is usually called just P.

This equation can be rearranged to give the **van der Waals equation**:

$$\left[P_{obs} + a\left(\frac{n}{V}\right)^2\right] \times \underbrace{(V - nb)}_{} = nRT$$

$$\underbrace{\qquad\qquad\qquad}_{\text{Corrected pressure}}\qquad\underbrace{\qquad\qquad}_{\text{Corrected volume}}$$

$$P_{\text{ideal}}\qquad\qquad\qquad V_{\text{ideal}}$$

Table 8-3 | Values of the van der Waals Constants for Some Common Gases

Gas	$a\left(\dfrac{atm \cdot L^2}{mol^2}\right)$	$b\left(\dfrac{L}{mol}\right)$
He	0.0341	0.0237
Ne	0.211	0.0171
Ar	1.35	0.0322
Kr	2.32	0.0398
Xe	4.19	0.0511
H_2	0.244	0.0266
N_2	1.39	0.0391
O_2	1.36	0.0318
Cl_2	6.49	0.0562
CO_2	3.59	0.0427
CH_4	2.25	0.0428
NH_3	4.17	0.0371
H_2O	5.46	0.030

The values of the weighting factors a and b are determined for a given gas by fitting experimental behavior. That is, a and b are varied until the best fit of the observed pressure is obtained under all conditions. The values of a and b for various gases are given in Table 8-3.

Experimental studies indicate that the changes van der Waals made in the basic assumptions of the kinetic molecular theory correct the major flaws in the model. First, consider the effects of volume. For a gas at low pressure (large volume), the volume of the container is very large compared with the volumes of the gas particles. That is, in this case the volume available to the gas is essentially equal to the volume of the container, and the gas behaves ideally. On the other hand, for a gas at high pressure (small container volume), the volume of the particles becomes significant so that the volume available to the gas is significantly less than the container volume. These cases are illustrated in Fig. 8-29. Note from Table 8-3 that the volume correction constant b generally increases with the size of the gas molecule, which gives further support to these arguments.

The fact that a real gas tends to behave more ideally at high temperatures also can be explained in terms of the van der Waals model. At high temperatures the particles are moving so rapidly that the effects of interparticle interactions are not very important.

The corrections to the kinetic molecular theory that van der Waals found necessary to explain real gas behavior make physical sense, which makes us confident that we understand the fundamentals of gas behavior at the particle level. This is significant because so much important chemistry takes place in the gas phase. In fact, the mixture of gases called the atmosphere is vital to our existence. In Section 8-10 we consider some of the important reactions that occur in the atmosphere.

Critical Thinking

You have learned that no gases behave perfectly ideally, but under conditions of high temperature and low pressure (high volume), gases behave more ideally. What if all gases always behaved perfectly ideally? How would the world be different?

8-9 | Characteristics of Several Real Gases

We can understand gas behavior more completely if we examine the characteristics of several common gases. Note from Fig. 8-25 that the gases H_2, N_2, CH_4, and CO_2 show different behavior when the compressibility $\left(\dfrac{PV}{nRT}\right)$ is plotted versus P. For example, notice that the plot for $H_2(g)$ never drops below the ideal value (1.0) in contrast to all the other gases. What is special about H_2 compared to these other gases? Recall from Section 8-8 that the reason that the compressibility of a real gas falls below 1.0 is that the actual (observed) pressure is lower than the pressure expected for an ideal gas due to the intermolecular attractions that occur in real gases. This must mean that H_2 molecules have very low attractive forces for each other. This idea is borne out by looking at the van der Waals a value for H_2 in Table 8-3. Note that H_2 has the lowest value among the gases H_2, N_2, CH_4, and CO_2. Remember that the value of a reflects how much of a correction must be made to adjust the observed pressure up to the expected ideal pressure:

$$P_{ideal} = P_{observed} + a\left(\frac{n}{V}\right)^2$$

❿ We will be able to better explain relative *a* values for real gases after discussing intermolecular forces in Section 9-1.

A low value for a reflects weak intermolecular forces among the gas molecules. ◀ ❿

Also notice that although the compressibility for N_2 dips below 1.0, it does not show as much deviation as that for CH_4, which in turn does not show as much deviation as the compressibility for CO_2. Based on this behavior, we can surmise that the importance of intermolecular interactions increases in this order:

$$H_2 < N_2 < CH_4 < CO_2$$

This order is reflected by the relative a values for these gases in Table 8-3. In Section 9-1, we will see how these variations in intermolecular interactions can be explained. The main point to be made here is that real gas behavior can tell us about the relative importance of intermolecular attractions among gas molecules.

8-10 | Chemistry in the Atmosphere

The most important gases to us are those in the **atmosphere** that surrounds the earth's surface. The principal components are N_2 and O_2, but many other important gases, such as H_2O and CO_2, are also present. The average composition of the earth's atmosphere near sea level, with the water vapor removed, is shown in Table 8-4. Because of gravitational effects, the composition of the earth's atmosphere is not constant; heavier molecules tend to be near the earth's surface, and light molecules tend to migrate to higher altitudes, with some eventually escaping into space. The atmosphere is a highly complex and dynamic system, but for convenience we divide it into several layers based on the way the temperature changes with altitude. (The lowest layer, called the *troposphere*, is shown in Fig. 8-30.) Note that in contrast to the complex temperature profile of the atmosphere, the pressure decreases in a regular way with increasing altitude.

Table 8-4 | Atmospheric Composition Near Sea Level (Dry Air)*

Component	Mole Fraction
N_2	0.78084
O_2	0.20948
Ar	0.00934
CO_2	0.000345
Ne	0.00001818
He	0.00000524
CH_4	0.00000168
Kr	0.00000114
H_2	0.0000005
NO	0.0000005
Xe	0.000000087

*The atmosphere contains various amounts of water vapor depending on conditions.

The chemistry occurring in the higher levels of the atmosphere is mostly determined by the effects of high-energy radiation and particles from the sun and other sources in space. In fact, the upper atmosphere serves as an important shield to prevent this high-energy radiation from reaching the earth, where it would damage the relatively fragile molecules sustaining life. In particular, the ozone in the upper atmosphere helps prevent high-energy ultraviolet radiation from penetrating to the earth. Intensive research is in progress to determine the natural factors that control the ozone concentration and how it is affected by chemicals released into the atmosphere.

The chemistry occurring in the troposphere, the layer of atmosphere closest to the earth's surface, is strongly influenced by human activities. Millions of tons of gases and particulates are released into the troposphere by our highly industrial civilization. Actually, it is amazing that the atmosphere can absorb so much material with relatively small permanent changes (so far).

Significant changes, however, are occurring. Severe **air pollution** is found around many large cities, and it is probable that long-range changes in our planet's weather

Figure 8-30 | The variation of temperature with altitude.

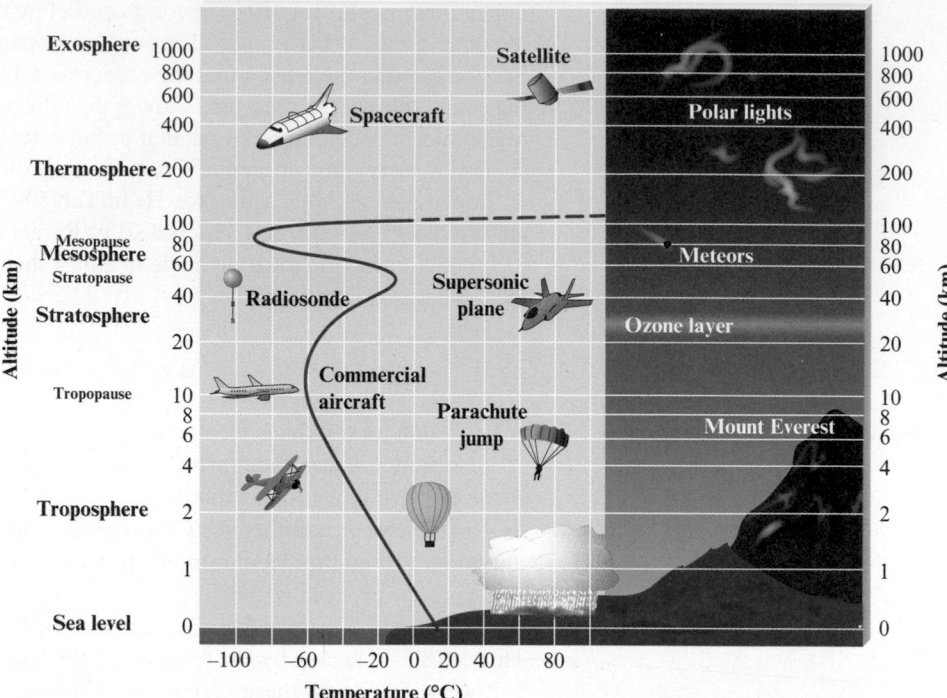

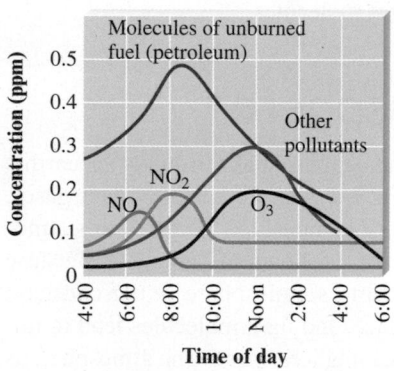

Molecules of unburned fuel (petroleum)

Figure 8-31 | Concentration (in molecules per million molecules of "air") for some smog components versus time of day.

(Source: After P.A. Leighton, "Photochemistry of Air Pollution," in *Physical Chemistry: A Series of Monographs,* edited by Eric Hutchinson and P. Van Rysselberghe, Vol. IX. New York: Academic Press, 1961.)

ⓘ The OH radical has no charge [it has one fewer electron than the hydroxide ion (OH⁻)].

are taking place. We discussed some of the long-range effects of pollution in Chapter 7. In this section we will deal with short-term, localized effects of pollution.

The two main sources of pollution are transportation and the production of electricity. The combustion of petroleum in vehicles produces CO, CO_2, NO, and NO_2, along with unburned molecules from petroleum. When this mixture is trapped close to the ground in stagnant air, reactions occur, producing chemicals that are potentially irritating and harmful to living systems.

The complex chemistry of polluted air appears to center around the nitrogen oxides (NO_x). At the high temperatures found in the gasoline and diesel engines of cars and trucks, N_2 and O_2 react to form a small quantity of NO that is emitted into the air with the exhaust gases (Fig. 8-31). This NO is immediately oxidized in air to NO_2, which, in turn, absorbs energy from sunlight and breaks up into nitric oxide and free oxygen atoms:

$$NO_2(g) \xrightarrow{\text{Radiant energy}} NO(g) + O(g)$$

Oxygen atoms are very reactive and can combine with O_2 to form *ozone:*

$$O(g) + O_2(g) \longrightarrow O_3(g)$$

Ozone is also very reactive and can react directly with other pollutants, or the ozone can absorb light and break up to form an energetically excited O_2 molecule (O_2^*) and an energetically excited oxygen atom (O^*). The latter species readily reacts with a water molecule to form two hydroxyl radicals (OH): ◄ ⓘ

$$O^* + H_2O \longrightarrow 2OH$$

The hydroxyl radical is a very reactive oxidizing agent. For example, OH can react with NO_2 to form nitric acid:

$$OH + NO_2 \longrightarrow HNO_3$$

The OH radical also can react with the unburned hydrocarbons in the polluted air to produce chemicals that cause the eyes to water and burn and are harmful to the respiratory system.

The end product of this whole process is often referred to as **photochemical smog**, so called because light is required to initiate some of the reactions. The production of photochemical smog can be understood more clearly by examining as a group the reactions discussed above:

$$NO_2(g) \longrightarrow NO(g) + O(g)$$
$$O(g) + O_2(g) \longrightarrow O_3(g) \blacktriangleleft \textit{i}$$
$$\underline{NO(g) + \tfrac{1}{2}O_2(g) \longrightarrow NO_2(g)}$$

Net reaction: $\qquad \tfrac{3}{2}O_2(g) \longrightarrow O_3(g)$

Note that the NO_2 molecules assist in the formation of ozone without being themselves used up. The ozone formed then leads to the formation of OH and other pollutants.

We can observe this process by analyzing polluted air at various times during a day (see Fig. 8-31). As people drive to work between 6 and 8 a.m., the amounts of NO, NO_2, and unburned molecules from petroleum increase. Later, as the decomposition of NO_2 occurs, the concentration of ozone and other pollutants builds up. Current efforts to combat the formation of photochemical smog are focused on cutting down the amounts of molecules from unburned fuel in automobile exhaust and designing engines that produce less nitric oxide.

The other major source of pollution results from burning coal to produce electricity. Much of the coal found in the Midwest contains significant quantities of sulfur, which, when burned, produces sulfur dioxide:

$$S \text{ (in coal)} + O_2(g) \longrightarrow SO_2(g)$$

A further oxidation reaction occurs when sulfur dioxide is changed to sulfur trioxide in the air:*

$$2SO_2(g) + O_2(g) \longrightarrow 2SO_3(g)$$

The production of sulfur trioxide is significant because it can combine with droplets of water in the air to form sulfuric acid:

$$SO_3(g) + H_2O(l) \longrightarrow H_2SO_4(aq)$$

Sulfuric acid is very corrosive to both living things and building materials. Another result of this type of pollution is **acid rain**. In many parts of the northeastern United States and southeastern Canada, acid rain has caused some freshwater lakes to become too acidic to support any life (Fig. 8-32).

The problem of sulfur dioxide pollution is made more complicated by the energy crisis. As petroleum supplies dwindle and the price increases, our dependence on coal will probably grow. As supplies of low-sulfur coal are used up, high-sulfur coal will be utilized. One way to use high-sulfur coal without further harming the air quality is to remove the sulfur dioxide from the exhaust gas by means of a system called a *scrubber* before it is emitted from the power plant stack. A common method of scrubbing is to blow powdered limestone ($CaCO_3$) into the combustion chamber, where it is decomposed to lime and carbon dioxide:

$$CaCO_3(s) \longrightarrow CaO(s) + CO_2(g)$$

The lime then combines with the sulfur dioxide to form calcium sulfite:

$$CaO(s) + SO_2(g) \longrightarrow CaSO_3(s)$$

To remove the calcium sulfite and any remaining unreacted sulfur dioxide, an aqueous suspension of lime is injected into the exhaust gases to produce a *slurry* (a thick suspension), as shown in Fig. 8-33.

i Although represented here as O_2, the actual oxidant for NO is OH or an organic peroxide such as CH_3COO, formed by oxidation of organic pollutants.

Robert Krueger/US Forest Service/Science Source

Figure 8-32 | Testing for acid rain in Wilderness Lake, Colorado.

---

*This reaction is very slow unless solid particles are present. See Chapter 11 for a discussion.

Figure 8-33 | A schematic diagram of the process for scrubbing sulfur dioxide from stack gases in power plants.

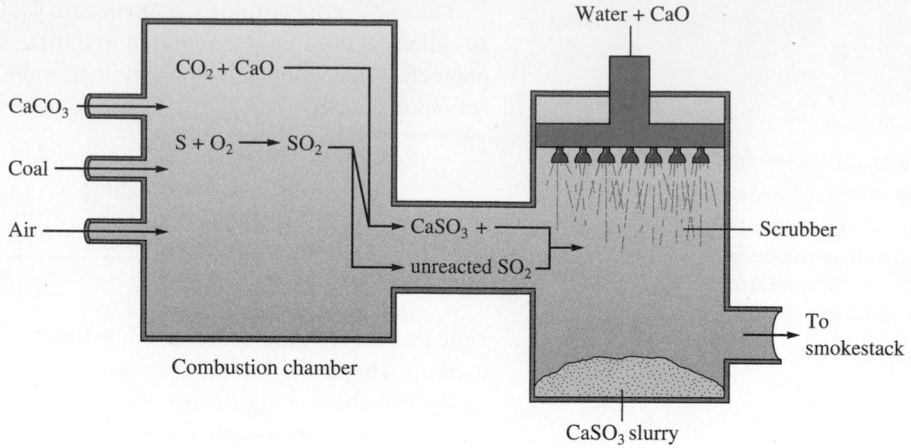

Unfortunately, there are many problems associated with scrubbing. The systems are complicated and expensive and consume a great deal of energy. The large quantities of calcium sulfite produced in the process present a disposal problem. With a typical scrubber, approximately 1 ton of calcium sulfite per year is produced per person served by the power plant. Since no use has yet been found for this calcium sulfite, it is usually buried in a landfill. As a result of these difficulties, air pollution by sulfur dioxide continues to be a major problem, one that is expensive in terms of damage to the environment and human health as well as in monetary terms.

FOR REVIEW

Key Terms

Section 8-1
barometer
manometer
mm Hg
torr
standard atmosphere
pascal

Section 8-2
Boyle's law
ideal gas
Charles's law
absolute zero
Avogadro's law

Section 8-3
universal gas constant
ideal gas law

Section 8-4
molar volume
standard temperature and pressure (STP)

State of a Gas

- The state of a gas can be described completely by specifying its pressure (P), volume (V), temperature (T), and the amount (moles) of gas present (n)
- Pressure
  - Common units

$$1 \text{ torr} = 1 \text{ mm Hg}$$
$$1 \text{ atm} = 760 \text{ torr}$$

  - SI unit: pascal

$$1 \text{ atm} = 101{,}325 \text{ Pa}$$

Gas Laws

- Discovered by observing the properties of gases
- Boyle's law: $PV = k$
- Charles's law: $V = bT$
- Avogadro's law: $V = an$
- Ideal gas law: $PV = nRT$
- Dalton's law of partial pressures: $P_{TOTAL} = P_1 + P_2 + P_3 + \cdots$, where P_n represents the partial pressure of component n in a mixture of gases

Key Terms

Kinetic Molecular Theory (KMT)

- Model that accounts for ideal gas behavior
- Postulates of the KMT:
  - Volume of gas particles is zero
  - No particle interactions
  - Particles are in constant motion, colliding with the container walls to produce pressure
  - The average kinetic energy of the gas particles is directly proportional to the temperature of the gas in kelvins

Gas Properties

- The particles in any gas sample have a range of velocities
- The root mean square (rms) velocity for a gas represents the average of the squares of the particle velocities

$$u_{\text{rms}} = \sqrt{\frac{3RT}{M}}$$

- Diffusion: the mixing of two or more gases
- Effusion: the process in which a gas passes through a small hole into an empty chamber

Real Gas Behavior

- Real gases behave ideally only at high temperatures and low pressures
- Understanding how the ideal gas equation must be modified to account for real gas behavior helps us understand how gases behave on a molecular level
- Van der Waals found that to describe real gas behavior we must consider particle interactions and particle volumes

---

Review Questions Answers to the Review Questions can be found on the Student Web site (accessible from www.cengagebrain.com).

1. Explain how a barometer and a manometer work to measure the pressure of the atmosphere or the pressure of a gas in a container.

2. What are Boyle's law, Charles's law, and Avogadro's law? What plots do you make to show a linear relationship for each law?

3. Show how Boyle's law, Charles's law, and Avogadro's law are special cases of the ideal gas law. Using the ideal gas law, determine the relationship between P and n (at constant V and T) and between P and T (at constant V and n).

4. Rationalize the following observations.
   a. Aerosol cans will explode if heated.
   b. You can drink through a soda straw.
   c. A thin-walled can will collapse when the air inside is removed by a vacuum pump.
   d. Manufacturers produce different types of tennis balls for high and low elevations.

5. Consider the following balanced equation in which gas X forms gas X_2:

$$2X(g) \longrightarrow X_2(g)$$

Equal moles of X are placed in two separate containers. One container is rigid so the volume cannot change; the other container is flexible so the volume changes to keep the internal pressure equal to the external pressure. The above reaction is run in each container. What happens to the pressure and density of the gas inside each container as reactants are converted to products?

6. Use the postulates of the kinetic molecular theory (KMT) to explain why Boyle's law, Charles's law, Avogadro's law, and Dalton's law of partial pressures hold true for ideal gases. Use the KMT to explain the P versus n (at constant V and T) relationship and the P versus T (at constant V and n) relationship.

7. Consider the following velocity distribution curves *A* and *B*.

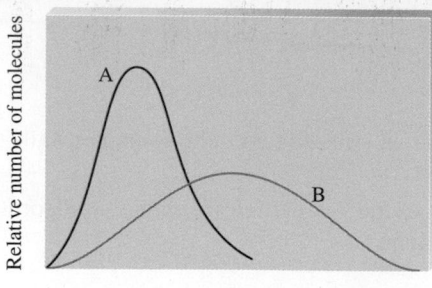

Velocity (m/s)

a. If the plots represent the velocity distribution of 1.0 L of He(g) at STP versus 1.0 L of Cl$_2$(g) at STP, which plot corresponds to each gas? Explain your reasoning.

b. If the plots represent the velocity distribution of 1.0 L of O$_2$(g) at temperatures of 273 K versus 1273 K, which plot corresponds to each temperature? Explain your reasoning. Under which temperature condition would the O$_2$(g) sample behave most ideally? Explain.

8. Briefly describe two methods one might use to find the molar mass of a newly synthesized gas for which a molecular formula was not known.

9. In the van der Waals equation, why is a term added to the observed pressure and why is a term subtracted from the container volume to correct for nonideal gas behavior?

10. Why do real gases not always behave ideally? Under what conditions does a real gas behave most ideally? Why?

Active Learning Questions

These questions are designed to be used by groups of students in class.

1. Consider the following apparatus: a test tube covered with a nonpermeable elastic membrane inside a container that is closed with a cork. A syringe goes through the cork.

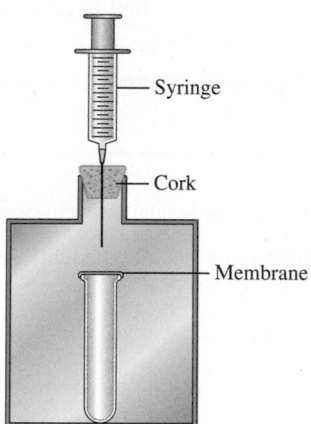

— Syringe

— Cork

— Membrane

a. As you push down on the syringe, how does the membrane covering the test tube change?

b. You stop pushing the syringe but continue to hold it down. In a few seconds, what happens to the membrane?

2. Figure 8-2 shows a picture of a barometer. Which of the following statements is the best explanation of how this barometer works?

a. Air pressure outside the tube causes the mercury to move in the tube until the air pressure inside and outside the tube is equal.

b. Air pressure inside the tube causes the mercury to move in the tube until the air pressure inside and outside the tube is equal.

c. Air pressure outside the tube counterbalances the weight of the mercury in the tube.

d. Capillary action of the mercury causes the mercury to go up the tube.

e. The vacuum that is formed at the top of the tube holds up the mercury.

Justify your choice, and for the choices you did not pick, explain what is wrong with them. Pictures help!

3. The barometer below shows the level of mercury at a given atmospheric pressure. Fill all the other barometers with mercury for that same atmospheric pressure. Explain your answer.

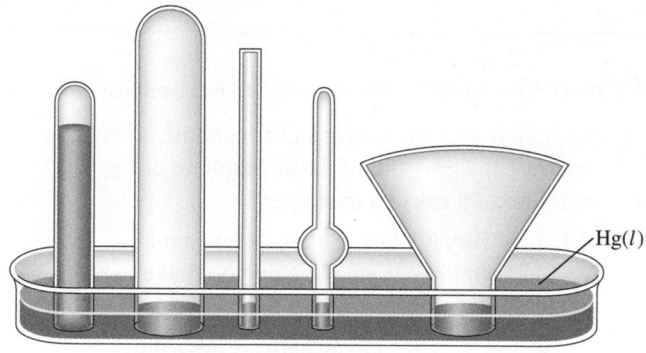

Hg(l)

4. As you increase the temperature of a gas in a sealed, rigid container, what happens to the density of the gas? Would the results be the same if you did the same experiment in a container with a piston at constant pressure? (See Fig. 8-17.)

5. A diagram in a chemistry book shows a magnified view of a flask of air as follows:

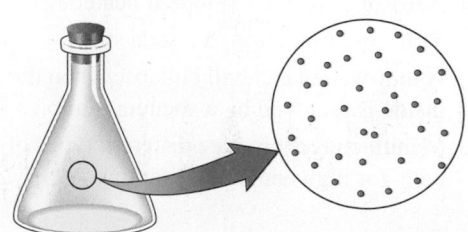

What do you suppose is between the dots (the dots represent air molecules)?

a. air

b. dust

c. pollutants

d. oxygen

e. nothing

6. If you put a drinking straw in water, place your finger over the opening, and lift the straw out of the water, some water stays in the straw. Explain.

7. A chemistry student relates the following story: I noticed my tires were a bit low and went to the gas station. As I was filling the tires, I thought about the kinetic molecular theory (KMT). I noticed the tires because the volume was low, and I realized that I was increasing both the pressure and volume of the tires. "Hmmm," I thought, "that goes against what I learned in chemistry, where I was told pressure and volume are inversely proportional." What is the fault in the logic of the chemistry student in this situation? Explain *why* we think pressure and volume to be inversely related (draw pictures and use the KMT).

8. Chemicals *X* and *Y* (both gases) react to form the gas *XY*, but it takes a bit of time for the reaction to occur. Both *X* and *Y* are placed in a container with a piston (free to move), and you note the volume. As the reaction occurs, what happens to the volume of the container? (See Fig. 8-18.)

9. Which statement best explains why a hot-air balloon rises when the air in the balloon is heated?

a. According to Charles's law, the temperature of a gas is directly related to its volume. Thus the volume of the balloon increases, making the density smaller. This lifts the balloon.

b. Hot air rises inside the balloon, and this lifts the balloon.

c. The temperature of a gas is directly related to its pressure. The pressure therefore increases, and this lifts the balloon.

d. Some of the gas escapes from the bottom of the balloon, thus decreasing the mass of gas in the balloon. This decreases the density of the gas in the balloon, which lifts the balloon.

e. Temperature is related to the root mean square velocity of the gas molecules. Thus the molecules are moving faster, hitting the balloon more, and thus lifting the balloon.

Justify your choice, and for the choices you did not pick, explain what is wrong with them.

10. Draw a highly magnified view of a sealed, rigid container filled with a gas. Then draw what it would look like if you cooled the gas significantly but kept the temperature above the boiling point of the substance in the container. Also draw what it would look like if you heated the gas significantly. Finally, draw what each situation would look like if you evacuated enough of the gas to decrease the pressure by a factor of 2.

11. If you release a helium balloon, it soars upward and eventually pops. Explain this behavior.

12. If you have any two gases in different containers that are the same size at the same pressure and same temperature, what is true about the moles of each gas? Why is this true?

13. Explain the following seeming contradiction: You have two gases, *A* and *B*, in two separate containers of equal volume and at equal pressure and temperature. Therefore, you must have the same number of moles of each gas. Because the two temperatures are equal, the average kinetic energies of the two samples are equal. Therefore, since the energy given such a system will be converted to translational motion (that is, move the molecules), the root mean square velocities of the two are equal, and thus the particles in each sample move, on average, with the same relative speed. Since *A* and *B* are different gases, they each must have a different molar mass. If *A* has a higher molar mass than *B*, the particles of *A* must be hitting the sides of the container with more force. Thus the pressure in the container of gas *A* must be higher than that in the container with gas *B*. However, one of our initial assumptions was that the pressures were equal.

14. You have a balloon covering the mouth of a flask filled with air at 1 atm. You apply heat to the bottom of the flask until the volume of the balloon is equal to that of the flask.

a. Which has more air in it, the balloon or the flask? Or do both have the same amount? Explain.

b. In which is the pressure greater, the balloon or the flask? Or is the pressure the same? Explain.

15. How does Dalton's law of partial pressures help us with our model of ideal gases? That is, what postulates of the kinetic molecular theory does it support?

16. At the same conditions of pressure and temperature, ammonia gas is less dense than air. Why is this true?

17. For each of the quantities listed below, explain which of the following properties (mass of the molecule, density of the gas sample, temperature of the gas sample, size of the molecule, and number of moles of gas) must be known to calculate the quantity.

a. average kinetic energy

b. average number of collisions per second with other gas molecules

c. average force of each impact with the wall of the container

d. root mean square velocity

e. average number of collisions with a given area of the container

f. distance between collisions

18. You have two containers each with 1 mole of xenon gas at 15°C. Container A has a volume of 3.0 L, and container B has a volume of 1.0 L. Explain how the following quantities compare between the two containers.

a. the average kinetic energy of the Xe atoms

b. the force with which the Xe atoms collide with the container walls

c. the root mean square velocity of the Xe atoms

d. the collision frequency of the Xe atoms (with other atoms)

e. the pressure of the Xe sample

19. Draw molecular-level views that show the differences among solids, liquids, and gases.

A blue question or exercise number indicates that the answer to that question or exercise appears at the back of the book and a solution appears in the *Solutions Guide*, as found on PowerLecture.

Questions

20. At room temperature, water is a liquid with a molar volume of 18 mL. At 105°C and 1 atm pressure, water is a gas and has a molar volume of over 30 L. Explain the large difference in molar volumes.

21. If a barometer were built using water ($d = 1.0$ g/cm^3) instead of mercury ($d = 13.6$ g/cm^3), would the column of water be higher than, lower than, or the same as the column of mercury at 1.00 atm? If the level is different, by what factor? Explain.

22. A bag of potato chips is packed and sealed in Los Angeles, California, and then shipped to Lake Tahoe, Nevada, during ski season. It is noticed that the volume of the bag of potato chips has increased upon its arrival in Lake Tahoe. What external conditions would most likely cause the volume increase?

23. Boyle's law can be represented graphically in several ways. Which of the following plots does *not* correctly represent Boyle's law (assuming constant T and n)? Explain.

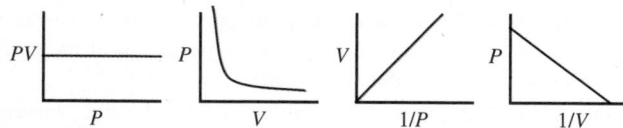

24. As weather balloons rise from the earth's surface, the pressure of the atmosphere becomes less, tending to cause the volume of the balloons to expand. However, the temperature is much lower in the upper atmosphere than at sea level. Would this temperature effect tend to make such a balloon expand or contract? Weather balloons do, in fact, expand as they rise. What does this tell you?

25. Which noble gas has the smallest density at STP? Explain.

26. Consider two different containers, each filled with 2 moles of Ne(g). One of the containers is rigid and has constant volume. The other container is flexible (like a balloon) and is capable of changing its volume to keep the external pressure and internal pressure equal to each other. If you raise the temperature in both containers, what happens to the pressure and density of the gas inside each container? Assume a constant external pressure.

27. In Example 8-11 of the text, the molar volume of N$_2$(g) at STP is given as 22.42 L/mol N$_2$. How is this number calculated? How does the molar volume of He(g) at STP compare to the molar volume of N$_2$(g) at STP (assuming ideal gas behavior)? Is the molar volume of N$_2$(g) at 1.000 atm and 25.0°C equal to, less than, or greater than 22.42 L/mol? Explain. Is the molar volume of N$_2$(g) collected over water at a total pressure of 1.000 atm and 0.0°C equal to, less than, or greater than 22.42 L/mol? Explain.

28. Consider the flasks in the following diagrams.

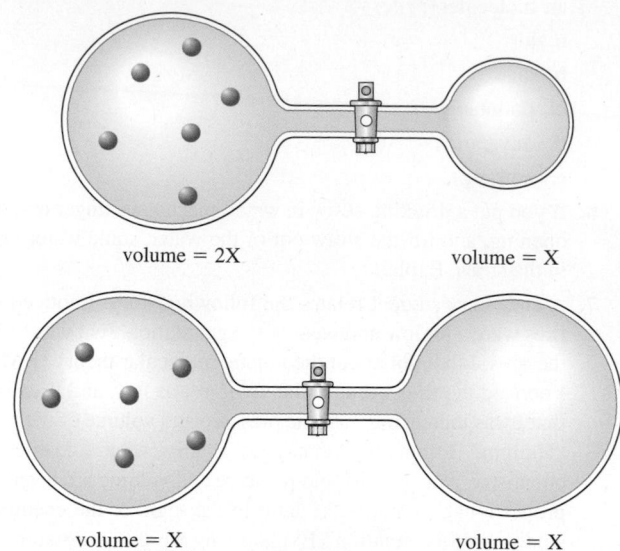

Assuming the connecting tube has negligible volume, draw what each diagram will look like after the stopcock between the two flasks is opened. Also, solve for the final pressure in each case, in terms of the original pressure. Assume temperature is constant.

29. Do all the molecules in a 1-mole sample of CH$_4$(g) have the same kinetic energy at 273 K? Do all molecules in a 1-mole sample of N$_2$(g) have the same velocity at 546 K? Explain.

30. Consider the following samples of gases at the same temperature.

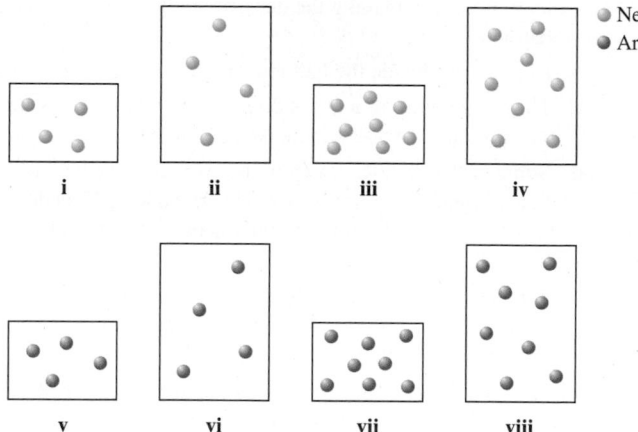

Arrange each of these samples in order from lowest to highest:

a. pressure

b. average kinetic energy

c. density

d. root mean square velocity

Note: Some samples of gases may have equal values for these attributes. Assume the larger containers have a volume twice the volume of the smaller containers, and assume the mass of an argon atom is twice the mass of a neon atom.

31. As $NH_3(g)$ is decomposed into nitrogen gas and hydrogen gas at constant pressure and temperature, the volume of the product gases collected is twice the volume of NH_3 reacted. Explain. As $NH_3(g)$ is decomposed into nitrogen gas and hydrogen gas at constant volume and temperature, the total pressure increases by some factor. Why the increase in pressure and by what factor does the total pressure increase when reactants are completely converted into products? How do the partial pressures of the product gases compare to each other and to the initial pressure of NH_3?

32. Which of the following statements is(are) *true?* For the false statements, correct them.

    a. At constant temperature, the lighter the gas molecules, the faster the average velocity of the gas molecules.

    b. At constant temperature, the heavier the gas molecules, the larger the average kinetic energy of the gas molecules.

    c. A real gas behaves most ideally when the container volume is relatively large and the gas molecules are moving relatively quickly.

    d. As temperature increases, the effect of interparticle interactions on gas behavior is increased.

    e. At constant V and T, as gas molecules are added into a container, the number of collisions per unit area increases resulting in a higher pressure.

    f. The kinetic molecular theory predicts that pressure is inversely proportional to temperature at constant volume and moles of gas.

33. From the values in Table 8-3 for the van der Waals constant a for the gases H_2, CO_2, N_2, and CH_4, predict which of these gas molecules show the strongest intermolecular attractions.

34. Without looking at a table of values, which of the following gases would you expect to have the largest value of the van der Waals constant b: H_2, N_2, CH_4, C_2H_6, or C_3H_8?

35. Figure 8-6 shows the PV versus P plot for three different gases. Which gas behaves most ideally? Explain.

36. Ideal gas particles are assumed to be volumeless and to neither attract nor repel each other. Why are these assumptions crucial to the validity of Dalton's law of partial pressures?

Exercises

In this section similar exercises are paired.

Pressure

37. Freon-12 (CF_2Cl_2) is commonly used as the refrigerant in central home air conditioners. The system is initially charged to a pressure of 4.8 atm. Express this pressure in each of the following units (1 atm = 14.7 psi).

    a. mm Hg   c. Pa

    b. torr   d. psi

38. A gauge on a compressed gas cylinder reads 2200 psi (pounds per square inch; 1 atm = 14.7 psi). Express this pressure in each of the following units.

    a. standard atmospheres

    b. megapascals (MPa)

    c. torr

39. A sealed-tube manometer (as shown below) can be used to measure pressures below atmospheric pressure. The tube above the mercury is evacuated. When there is a vacuum in the flask, the mercury levels in both arms of the U-tube are equal. If a gaseous sample is introduced into the flask, the mercury levels are different. The difference h is a measure of the pressure of the gas inside the flask. If h is equal to 6.5 cm, calculate the pressure in the flask in torr, pascals, and atmospheres.

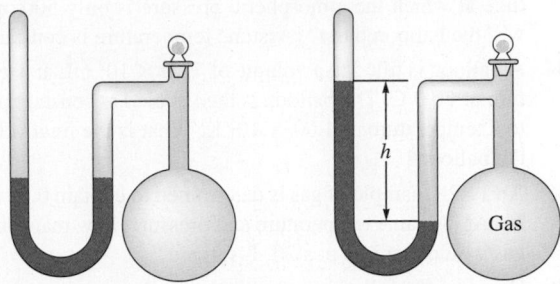

40. If the sealed-tube manometer in Exercise 39 had a height difference of 20.0 inches between the mercury levels, what is the pressure in the flask in torr and atmospheres?

41. A diagram for an open-tube manometer is shown below.

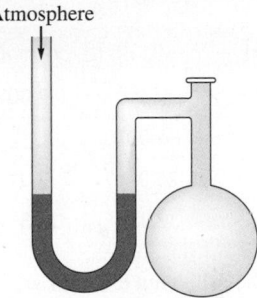

If the flask is open to the atmosphere, the mercury levels are equal. For each of the following situations where a gas is contained in the flask, calculate the pressure in the flask in torr, atmospheres, and pascals.

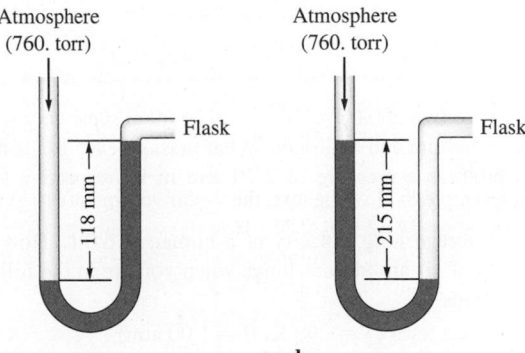

a.   b.

c. Calculate the pressures in the flask in parts a and b (in torr) if the atmospheric pressure is 635 torr.

42. a. If the open-tube manometer in Exercise 41 contains a non-volatile silicone oil (density = 1.30 g/cm^3) instead of mercury (density = 13.6 g/cm^3), what are the pressures in the flask as shown in parts a and b in torr, atmospheres, and pascals?

b. What advantage would there be in using a less dense fluid than mercury in a manometer used to measure relatively small differences in pressure?

Gas Laws

43. A particular balloon is designed by its manufacturer to be inflated to a volume of no more than 2.5 L. If the balloon is filled with 2.0 L helium at sea level, is released, and rises to an altitude at which the atmospheric pressure is only 500. mm Hg, will the balloon burst? (Assume temperature is constant.)

44. A balloon is filled to a volume of 7.00×10^2 mL at a temperature of 20.0°C. The balloon is then cooled at constant pressure to a temperature of 1.00×10^2 K. What is the final volume of the balloon?

45. An 11.2-L sample of gas is determined to contain 0.50 mole of N_2. At the same temperature and pressure, how many moles of gas would there be in a 20.-L sample?

46. Consider the following chemical equation.

$$2NO_2(g) \longrightarrow N_2O_4(g)$$

If 25.0 mL NO_2 gas is completely converted to N_2O_4 gas under the same conditions, what volume will the N_2O_4 occupy?

47. Complete the following table for an ideal gas.

	P (atm)	V (L)	n (mol)	T
a.	5.00		2.00	155°C
b.	0.300	2.00		155 K
c.	4.47	25.0	2.01	
d.		2.25	10.5	75°C

48. Complete the following table for an ideal gas.

	P	V	n	T
a.	7.74×10^3 Pa	12.2 mL		25°C
b.		43.0 mL	0.421 mol	223 K
c.	455 torr		4.4×10^{-2} mol	331°C
d.	745 mm Hg	11.2 L	0.401 mol	

49. Suppose two 200.0-L tanks are to be filled separately with the gases helium and hydrogen. What mass of each gas is needed to produce a pressure of 2.70 atm in its respective tank at 24°C?

50. The average lung capacity of a human is 6.0 L. How many moles of air are in your lungs when you are in the following situations?

a. At sea level ($T = 298$ K, $P = 1.00$ atm).

b. 10. m below water ($T = 298$ K, $P = 1.97$ atm).

c. At the top of Mount Everest ($T = 200.$ K, $P = 0.296$ atm).

51. The steel reaction vessel of a bomb calorimeter, which has a volume of 75.0 mL, is charged with oxygen gas to a pressure of 14.5 atm at 22°C. Calculate the moles of oxygen in the reaction vessel.

52. A 5.0-L flask contains 0.60 g O_2 at a temperature of 22°C. What is the pressure (in atm) inside the flask?

53. A 2.50-L container is filled with 175 g argon.

a. If the pressure is 10.0 atm, what is the temperature?

b. If the temperature is 225 K, what is the pressure?

54. A person accidentally swallows a drop of liquid oxygen, $O_2(l)$, which has a density of 1.149 g/mL. Assuming the drop has a volume of 0.050 mL, what volume of gas will be produced in the person's stomach at body temperature (37°C) and a pressure of 1.0 atm?

55. A gas sample containing 1.50 moles at 25°C exerts a pressure of 400. torr. Some gas is added to the same container and the temperature is increased to 50.°C. If the pressure increases to 800. torr, how many moles of gas were added to the container? Assume a constant-volume container.

56. A bicycle tire is filled with air to a pressure of 75 psi at a temperature of 19°C. Riding the bike on asphalt on a hot day increases the temperature of the tire to 58°C. The volume of the tire increases by 4.0%. What is the new pressure in the bicycle tire?

57. Consider two separate gas containers at the following conditions:

Container A	Container B
Contents: $SO_2(g)$	Contents: unknown gas
Pressure = P_A	Pressure = P_B
Moles of gas = 1.0 mol	Moles of gas = 2.0 mol
Volume = 1.0 L	Volume = 2.0 L
Temperature = 7°C	Temperature = 287°C

How is the pressure in container B related to the pressure in container A?

58. What will be the effect on the volume of an ideal gas if the pressure is doubled and the absolute temperature is halved?

59. A container is filled with an ideal gas to a pressure of 11.0 atm at 0°C.

a. What will be the pressure in the container if it is heated to 45°C?

b. At what temperature would the pressure be 6.50 atm?

c. At what temperature would the pressure be 25.0 atm?

60. An ideal gas at 7°C is in a spherical flexible container having a radius of 1.00 cm. The gas is heated at constant pressure to 88°C. Determine the radius of the spherical container after the gas is heated. [Volume of a sphere = $(4/3)\pi r^3$.]

61. An ideal gas is contained in a cylinder with a volume of 5.0×10^2 mL at a temperature of 30.°C and a pressure of 710. torr. The gas is then compressed to a volume of 25 mL, and the temperature is raised to 820.°C. What is the new pressure of the gas?

62. A compressed gas cylinder contains 1.00×10^3 g argon gas. The pressure inside the cylinder is 2050. psi (pounds per square inch) at a temperature of 18°C. How much gas remains in the cylinder if the pressure is decreased to 650. psi at a temperature of 26°C?

63. A sealed balloon is filled with 1.00 L helium at 23°C and 1.00 atm. The balloon rises to a point in the atmosphere where the pressure is 220. torr and the temperature is −31°C. What is the change in volume of the balloon as it ascends from 1.00 atm to a pressure of 220. torr?

64. A hot-air balloon is filled with air to a volume of 4.00×10^3 m³ at 745 torr and 21°C. The air in the balloon is then heated to 62°C, causing the balloon to expand to a volume of 4.20×10^3 m³. What is the ratio of the number of moles of air in the heated balloon to the original number of moles of air in the balloon? (*Hint:* Openings in the balloon allow air to flow in and out. Thus the pressure in the balloon is always the same as that of the atmosphere.)

Gas Density, Molar Mass, and Reaction Stoichiometry

65. Consider the following reaction:

$$4Al(s) + 3O_2(g) \longrightarrow 2Al_2O_3(s)$$

It takes 2.00 L of pure oxygen gas at STP to react completely with a certain sample of aluminum. What is the mass of aluminum reacted?

66. A student adds 4.00 g of dry ice (solid CO_2) to an empty balloon. What will be the volume of the balloon at STP after all the dry ice sublimes (converts to gaseous CO_2)?

67. Air bags are activated when a severe impact causes a steel ball to compress a spring and electrically ignite a detonator cap. This causes sodium azide (NaN_3) to decompose explosively according to the following reaction:

$$2NaN_3(s) \longrightarrow 2Na(s) + 3N_2(g)$$

What mass of $NaN_3(s)$ must be reacted to inflate an air bag to 70.0 L at STP?

68. Concentrated hydrogen peroxide solutions are explosively decomposed by traces of transition metal ions (such as Mn or Fe):

$$2H_2O_2(aq) \longrightarrow 2H_2O(l) + O_2(g)$$

What volume of pure $O_2(g)$, collected at 27°C and 746 torr, would be generated by decomposition of 125 g of a 50.0% by mass hydrogen peroxide solution? Ignore any water vapor that may be present.

69. In 1897 the Swedish explorer Andreé tried to reach the North Pole in a balloon. The balloon was filled with hydrogen gas. The hydrogen gas was prepared from iron splints and diluted sulfuric acid. The reaction is

$$Fe(s) + H_2SO_4(aq) \longrightarrow FeSO_4(aq) + H_2(g)$$

The volume of the balloon was 4800 m³ and the loss of hydrogen gas during filling was estimated at 20.%. What mass of iron splints and 98% (by mass) H_2SO_4 were needed to ensure the complete filling of the balloon? Assume a temperature of 0°C, a pressure of 1.0 atm during filling, and 100% yield.

70. Sulfur trioxide, SO_3, is produced in enormous quantities each year for use in the synthesis of sulfuric acid.

$$S(s) + O_2(g) \longrightarrow SO_2(g)$$
$$2SO_2(g) + O_2(g) \longrightarrow 2SO_3(g)$$

What volume of $O_2(g)$ at 350.°C and a pressure of 5.25 atm is needed to completely convert 5.00 g sulfur to sulfur trioxide?

71. A 15.0-L rigid container was charged with 0.500 atm of krypton gas and 1.50 atm of chlorine gas at 350.°C. The krypton and chlorine react to form krypton tetrachloride. What mass of krypton tetrachloride can be produced assuming 100% yield?

72. An important process for the production of acrylonitrile (C_3H_3N) is given by the following equation:

$$2C_3H_6(g) + 2NH_3(g) + 3O_2(g) \longrightarrow 2C_3H_3N(g) + 6H_2O(g)$$

A 150.-L reactor is charged to the following partial pressures at 25°C:

$$P_{C_3H_6} = 0.500 \text{ MPa}$$
$$P_{NH_3} = 0.800 \text{ MPa}$$
$$P_{O_2} = 1.500 \text{ MPa}$$

What mass of acrylonitrile can be produced from this mixture (MPa = 10^6 Pa)?

73. Consider the reaction between 50.0 mL liquid methanol, CH_3OH (density = 0.850 g/mL), and 22.8 L O_2 at 27°C and a pressure of 2.00 atm. The products of the reaction are $CO_2(g)$ and $H_2O(g)$. Calculate the number of moles of H_2O formed if the reaction goes to completion.

74. Urea (H_2NCONH_2) is used extensively as a nitrogen source in fertilizers. It is produced commercially from the reaction of ammonia and carbon dioxide:

$$2NH_3(g) + CO_2(g) \xrightarrow[\text{Pressure}]{\text{Heat}} H_2NCONH_2(s) + H_2O(g)$$

Ammonia gas at 223°C and 90. atm flows into a reactor at a rate of 500. L/min. Carbon dioxide at 223°C and 45 atm flows into the reactor at a rate of 600. L/min. What mass of urea is produced per minute by this reaction assuming 100% yield?

75. Hydrogen cyanide is prepared commercially by the reaction of methane, $CH_4(g)$, ammonia, $NH_3(g)$, and oxygen, $O_2(g)$, at high temperature. The other product is gaseous water.

a. Write a chemical equation for the reaction.

b. What volume of $HCN(g)$ can be obtained from the reaction of 20.0 L $CH_4(g)$, 20.0 L $NH_3(g)$, and 20.0 L $O_2(g)$? The volumes of all gases are measured at the same temperature and pressure.

76. Ethene is converted to ethane by the reaction

$$C_2H_4(g) + H_2(g) \xrightarrow{\text{Catalyst}} C_2H_6(g)$$

C_2H_4 flows into a catalytic reactor at 25.0 atm and 300.°C with a flow rate of 1000. L/min. Hydrogen at 25.0 atm and 300.°C flows into the reactor at a flow rate of 1500. L/min. If 15.0 kg C_2H_6 is collected per minute, what is the percent yield of the reaction?

77. An unknown diatomic gas has a density of 3.164 g/L at STP. What is the identity of the gas?

78. A compound has the empirical formula CHCl. A 256-mL flask, at 373 K and 750. torr, contains 0.800 g of the gaseous compound. Give the molecular formula.

79. Uranium hexafluoride is a solid at room temperature, but it boils at 56°C. Determine the density of uranium hexafluoride at 60.°C and 745 torr.

80. Given that a sample of air is made up of nitrogen, oxygen, and argon in the mole fractions 0.78 N_2, 0.21 O_2, and 0.010 Ar, what is the density of air at standard temperature and pressure?

Partial Pressure

81. Determine the partial pressure of each gas as shown in the figure below. *Note:* The relative numbers of each type of gas are depicted in the figure.

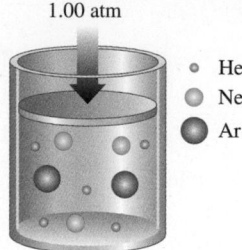

1.00 atm

- He
- Ne
- Ar

82. Consider the flasks in the following diagrams.

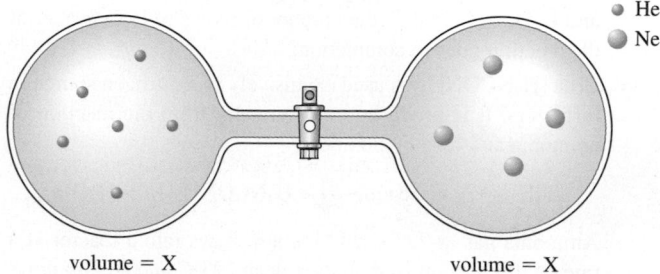

- He
- Ne

volume = X   volume = X

   a. Which is greater, the initial pressure of helium or the initial pressure of neon? How much greater?

   b. Assuming the connecting tube has negligible volume, draw what each diagram will look like after the stopcock between the two flasks is opened.

   c. Solve for the final pressure in terms of the original pressures of helium and neon. Assume temperature is constant.

   d. Solve for the final partial pressures of helium and neon in terms of their original pressures. Assume the temperature is constant.

83. A piece of solid carbon dioxide, with a mass of 7.8 g, is placed in a 4.0-L otherwise empty container at 27°C. What is the pressure in the container after all the carbon dioxide vaporizes? If 7.8 g solid carbon dioxide were placed in the same container but it already contained air at 740 torr, what would be the partial pressure of carbon dioxide and the total pressure in the container after the carbon dioxide vaporizes?

84. A mixture of 1.00 g H_2 and 1.00 g He is placed in a 1.00-L container at 27°C. Calculate the partial pressure of each gas and the total pressure.

85. Consider the flasks in the following diagram. What are the final partial pressures of H_2 and N_2 after the stopcock between the two flasks is opened? (Assume the final volume is 3.00 L.) What is the total pressure (in torr)?

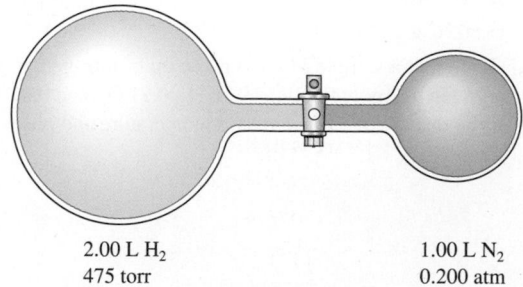

2.00 L H_2 1.00 L N_2
475 torr   0.200 atm

86. Consider the flask apparatus in Exercise 85, which now contains 2.00 L H_2 at a pressure of 360. torr and 1.00 L N_2 at an unknown pressure. If the total pressure in the flasks is 320. torr after the stopcock is opened, determine the initial pressure of N_2 in the 1.00-L flask.

87. Consider the three flasks in the diagram below. Assuming the connecting tubes have negligible volume, what is the partial pressure of each gas and the total pressure after all the stopcocks are opened?

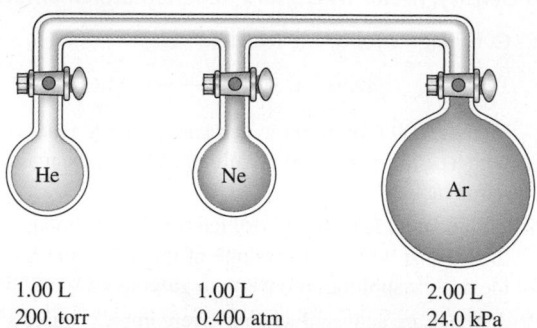

He   Ne   Ar

1.00 L   1.00 L   2.00 L
200. torr   0.400 atm   24.0 kPa

88. At 0°C a 1.0-L flask contains 5.0×10^{-2} mole of N_2, 1.5×10^2 mg O_2, and 5.0×10^{21} molecules of NH_3. What is the partial pressure of each gas, and what is the total pressure in the flask?

89. The partial pressure of $CH_4(g)$ is 0.175 atm and that of $O_2(g)$ is 0.250 atm in a mixture of the two gases.

   a. What is the mole fraction of each gas in the mixture?

   b. If the mixture occupies a volume of 10.5 L at 65°C, calculate the total number of moles of gas in the mixture.

   c. Calculate the number of grams of each gas in the mixture.

90. A tank contains a mixture of 52.5 g oxygen gas and 65.1 g carbon dioxide gas at 27°C. The total pressure in the tank is 9.21 atm. Calculate the partial pressures of each gas in the container.

91. Small quantities of hydrogen gas can be prepared in the laboratory by the addition of aqueous hydrochloric acid to metallic zinc.

$$Zn(s) + 2HCl(aq) \longrightarrow ZnCl_2(aq) + H_2(g)$$

Typically, the hydrogen gas is bubbled through water for collection and becomes saturated with water vapor. Suppose 240. mL of hydrogen gas is collected at 30.°C and has a total pressure of 1.032 atm by this process. What is the partial pressure of hydrogen gas in the sample? How many grams of zinc must have reacted to produce this quantity of hydrogen? (The vapor pressure of water is 32 torr at 30.°C.)

92. Helium is collected over water at 25°C and 1.00 atm total pressure. What total volume of gas must be collected to obtain 0.586 g helium? (At 25°C the vapor pressure of water is 23.8 torr.)

93. At elevated temperatures, sodium chlorate decomposes to produce sodium chloride and oxygen gas. A 0.8765-g sample of impure sodium chlorate was heated until the production of oxygen gas ceased. The oxygen gas collected over water occupied 57.2 mL at a temperature of 22°C and a pressure of 734 torr. Calculate the mass percent of $NaClO_3$ in the original sample. (At 22°C the vapor pressure of water is 19.8 torr.)

94. Xenon and fluorine will react to form binary compounds when a mixture of these two gases is heated to 400°C in a nickel reaction vessel. A 100.0-mL nickel container is filled with xenon and fluorine, giving partial pressures of 1.24 atm and 10.10 atm, respectively, at a temperature of 25°C. The reaction vessel is heated to 400°C to cause a reaction to occur and then cooled to a temperature at which F_2 is a gas and the xenon fluoride compound produced is a nonvolatile solid. The remaining F_2 gas is transferred to another 100.0-mL nickel container, where the pressure of F_2 at 25°C is 7.62 atm. Assuming all of the xenon has reacted, what is the formula of the product?

95. Methanol (CH_3OH) can be produced by the following reaction:

$$CO(g) + 2H_2(g) \longrightarrow CH_3OH(g)$$

Hydrogen at STP flows into a reactor at a rate of 16.0 L/min. Carbon monoxide at STP flows into the reactor at a rate of 25.0 L/min. If 5.30 g methanol is produced per minute, what is the percent yield of the reaction?

96. In the "Méthode Champenoise," grape juice is fermented in a wine bottle to produce sparkling wine. The reaction is

$$C_6H_{12}O_6(aq) \longrightarrow 2C_2H_5OH(aq) + 2CO_2(g)$$

Fermentation of 750. mL grape juice (density = 1.0 g/cm³) is allowed to take place in a bottle with a total volume of 825 mL until 12% by volume is ethanol (C_2H_5OH). Assuming that the CO_2 is insoluble in H_2O (actually, a wrong assumption), what would be the pressure of CO_2 inside the wine bottle at 25°C? (The density of ethanol is 0.79 g/cm³.)

97. Hydrogen azide, HN_3, decomposes on heating by the following *unbalanced* equation:

$$HN_3(g) \longrightarrow N_2(g) + H_2(g)$$

If 3.0 atm of pure $HN_3(g)$ is decomposed initially, what is the final total pressure in the reaction container? What are the partial pressures of nitrogen and hydrogen gas? Assume the volume and temperature of the reaction container are constant.

98. Equal moles of sulfur dioxide gas and oxygen gas are mixed in a flexible reaction vessel and then sparked to initiate the formation of gaseous sulfur trioxide. Assuming that the reaction goes to completion, what is the ratio of the final volume of the gas mixture to the initial volume of the gas mixture if both volumes are measured at the same temperature and pressure?

99. Some very effective rocket fuels are composed of lightweight liquids. The fuel composed of dimethylhydrazine [$(CH_3)_2N_2H_2$] mixed with dinitrogen tetroxide was used to power the Lunar Lander in its missions to the moon. The two components react according to the following equation:

$$(CH_3)_2N_2H_2(l) + 2N_2O_4(l) \longrightarrow 3N_2(g) + 4H_2O(g) + 2CO_2(g)$$

If 150 g dimethylhydrazine reacts with excess dinitrogen tetroxide and the product gases are collected at 127°C in an evacuated 250-L tank, what is the partial pressure of nitrogen gas produced and what is the total pressure in the tank assuming the reaction has 100% yield?

100. The oxides of Group 2A metals (symbolized by M here) react with carbon dioxide according to the following reaction:

$$MO(s) + CO_2(g) \longrightarrow MCO_3(s)$$

A 2.85-g sample containing only MgO and CuO is placed in a 3.00-L container. The container is filled with CO_2 to a pressure of 740. torr at 20.°C. After the reaction has gone to completion, the pressure inside the flask is 390. torr at 20.°C. What is the mass percent of MgO in the mixture? Assume that only the MgO reacts with CO_2.

Kinetic Molecular Theory and Real Gases

101. Calculate the average kinetic energies of $CH_4(g)$ and $N_2(g)$ molecules at 273 K and 546 K.

102. A 100.-L flask contains a mixture of methane (CH_4) and argon gases at 25°C. The mass of argon present is 228 g and the mole fraction of methane in the mixture is 0.650. Calculate the total kinetic energy of the gaseous mixture.

103. Calculate the root mean square velocities of $CH_4(g)$ and $N_2(g)$ molecules at 273 K and 546 K.

104. Consider separate 1.0-L samples of He(g) and $UF_6(g)$, both at 1.00 atm and containing the same number of moles. What ratio of temperatures for the two samples would produce the same root mean square velocity?

105. You have a gas in a container fitted with a piston, and you change one of the conditions of the gas such that a change takes place, as shown below:

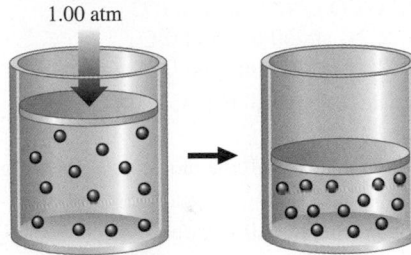

State two distinct changes you can make to accomplish this, and explain why each would work.

106. You have a gas in a container fitted with a piston, and you change one of the conditions of the gas such that a change takes place, as shown below:

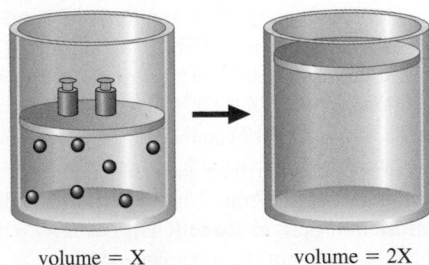

volume = X   volume = 2X

State three distinct changes you can make to accomplish this, and explain why each would work.

107. Consider a 1.0-L container of neon gas at STP. Will the average kinetic energy, average velocity, and frequency of collisions of gas molecules with the walls of the container increase, decrease, or remain the same under each of the following conditions?

   a. The temperature is increased to 100°C.
   b. The temperature is decreased to -50°C.
   c. The volume is decreased to 0.5 L.
   d. The number of moles of neon is doubled.

108. Consider two gases, A and B, each in a 1.0-L container with both gases at the same temperature and pressure. The mass of gas A in the container is 0.34 g and the mass of gas B in the container is 0.48 g.

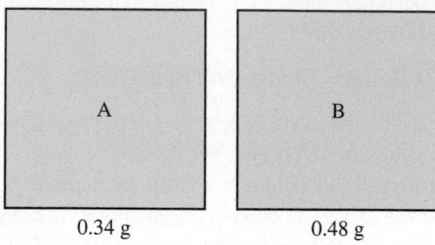

a. Which gas sample has the most molecules present? Explain.

b. Which gas sample has the largest average kinetic energy? Explain.

c. Which gas sample has the fastest average velocity? Explain.

d. How can the pressure in the two containers be equal to each other since the larger gas B molecules collide with the container walls more forcefully?

109. Consider three identical flasks filled with different gases.

Flask A: CO at 760 torr and 0°C
Flask B: N_2 at 250 torr and 0°C
Flask C: H_2 at 100 torr and 0°C

a. In which flask will the molecules have the greatest average kinetic energy?

b. In which flask will the molecules have the greatest average velocity?

110. Consider separate 1.0-L gaseous samples of H_2, Xe, Cl_2, and O_2 all at STP.

a. Rank the gases in order of increasing average kinetic energy.

b. Rank the gases in order of increasing average velocity.

c. How can separate 1.0-L samples of O_2 and H_2 each have the same average velocity?

111. Freon-12 is used as a refrigerant in central home air conditioners. The rate of effusion of Freon-12 to Freon-11 (molar mass = 137.4 g/mol) is 1.07:1. The formula of Freon-12 is one of the following: CF_4, CF_3Cl, CF_2Cl_2, $CFCl_3$, or CCl_4. Which formula is correct for Freon-12?

112. The rate of effusion of a particular gas was measured and found to be 24.0 mL/min. Under the same conditions, the rate of effusion of pure methane (CH_4) gas is 47.8 mL/min. What is the molar mass of the unknown gas?

113. One way of separating oxygen isotopes is by gaseous diffusion of carbon monoxide. The gaseous diffusion process behaves like an effusion process. Calculate the relative rates of effusion of $^{12}C^{16}O$, $^{12}C^{17}O$, and $^{12}C^{18}O$. Name some advantages and disadvantages of separating oxygen isotopes by gaseous diffusion of carbon dioxide instead of carbon monoxide.

114. It took 4.5 minutes for 1.0 L helium to effuse through a porous barrier. How long will it take for 1.0 L Cl_2 gas to effuse under identical conditions?

115. Calculate the pressure exerted by 0.5000 mole of N_2 in a 1.0000-L container at 25.0°C

a. using the ideal gas law.

b. using the van der Waals equation.

c. Compare the results.

116. Calculate the pressure exerted by 0.5000 mole of N_2 in a 10.000-L container at 25.0°C

a. using the ideal gas law.

b. using the van der Waals equation.

c. Compare the results.

d. Compare the results with those in Exercise 115.

Atmosphere Chemistry

117. Use the data in Table 8-4 to calculate the partial pressure of He in dry air assuming that the total pressure is 1.0 atm. Assuming a temperature of 25°C, calculate the number of He atoms per cubic centimeter.

118. A 1.0-L sample of air is collected at 25°C at sea level (1.00 atm). Estimate the volume this sample of air would have at an altitude of 15 km (see Fig. 8-30). At 15 km, the pressure is about 0.1 atm.

119. Write an equation to show how sulfuric acid is produced in the atmosphere.

120. Write an equation to show how sulfuric acids in acid rain reacts with marble and limestone. (Both marble and limestone are primarily calcium carbonate.)

121. Atmospheric scientists often use mixing ratios to express the concentrations of trace compounds in air. Mixing ratios are often expressed as ppmv (parts per million volume):

$$\text{ppmv of } X = \frac{\text{vol of } X \text{ at STP}}{\text{total vol of air at STP}} \times 10^6$$

On a certain November day, the concentration of carbon monoxide in the air in downtown Denver, Colorado, reached 3.0×10^2 ppmv. The atmospheric pressure at that time was 628 torr and the temperature was 0°C.

a. What was the partial pressure of CO?

b. What was the concentration of CO in molecules per cubic meter?

c. What was the concentration of CO in molecules per cubic centimeter?

122. Trace organic compounds in the atmosphere are first concentrated and then measured by gas chromatography. In the concentration step, several liters of air are pumped through a tube containing a porous substance that traps organic compounds. The tube is then connected to a gas chromatograph and heated to release the trapped compounds. The organic compounds are separated in the column and the amounts are measured. In an analysis for benzene and toluene in air, a 3.00-L sample of air at 748 torr and 23°C was passed through the trap. The gas chromatography analysis showed that this air sample contained 89.6 ng benzene (C_6H_6) and 153 ng toluene (C_7H_8). Calculate the mixing ratio (see Exercise 121) and number of molecules per cubic centimeter for both benzene and toluene.

Additional Exercises

123. Draw a qualitative graph to show how the first property varies with the second in each of the following (assume 1 mole of an ideal gas and T in kelvin).

  a. PV versus V with constant T

  b. P versus T with constant V

  c. T versus V with constant P

  d. P versus V with constant T

  e. P versus $1/V$ with constant T

  f. PV/T versus P

124. At STP, 1.0 L Br_2 reacts completely with 3.0 L F_2, producing 2.0 L of a product. What is the formula of the product? (All substances are gases.)

125. A form of Boyle's law is $PV = k$ (at constant T and n). Table 8-1 contains actual data from pressure–volume experiments conducted by Robert Boyle. The value of k in most experiments is 14.1×10^2 in Hg $\cdot$ in^3. Express k in units of atm $\cdot$ L. In Example 8-3, k was determined for NH_3 at various pressures and volumes. Give some reasons why the k values differ so dramatically between Example 8-3 and Table 8-1.

126. A 2.747-g sample of manganese metal is reacted with excess HCl gas to produce 3.22 L $H_2(g)$ at 373 K and 0.951 atm and a manganese chloride compound ($MnCl_x$). What is the formula of the manganese chloride compound produced in the reaction?

127. A 1.00-L gas sample at 100.°C and 600. torr contains 50.0% helium and 50.0% xenon by mass. What are the partial pressures of the individual gases?

128. Cyclopropane, a gas that when mixed with oxygen is used as a general anesthetic, is composed of 85.7% C and 14.3% H by mass. If the density of cyclopropane is 1.88 g/L at STP, what is the molecular formula of cyclopropane?

129. The nitrogen content of organic compounds can be determined by the Dumas method. The compound in question is first reacted by passage over hot CuO(s):

$$\text{Compound} \xrightarrow[\text{CuO}(s)]{\text{Hot}} N_2(g) + CO_2(g) + H_2O(g)$$

The product gas is then passed through a concentrated solution of KOH to remove the CO_2. After passage through the KOH solution, the gas contains N_2 and is saturated with water vapor. In a given experiment a 0.253-g sample of a compound produced 31.8 mL N_2 saturated with water vapor at 25°C and 726 torr. What is the mass percent of nitrogen in the compound? (The vapor pressure of water at 25°C is 23.8 torr.)

130. An organic compound containing only C, H, and N yields the following data.

  i. Complete combustion of 35.0 mg of the compound produced 33.5 mg CO_2 and 41.1 mg H_2O.

  ii. A 65.2-mg sample of the compound was analyzed for nitrogen by the Dumas method (see Exercise 129), giving 35.6 mL of dry N_2 at 740. torr and 25°C.

  iii. The effusion rate of the compound as a gas was measured and found to be 24.6 mL/min. The effusion rate of argon gas, under identical conditions, is 26.4 mL/min.

What is the molecular formula of the compound?

131. A 15.0-L tank is filled with H_2 to a pressure of 2.00×10^2 atm. How many balloons (each 2.00 L) can be inflated to a pressure of 1.00 atm from the tank? Assume that there is no temperature change and that the tank cannot be emptied below 1.00 atm pressure.

132. A spherical glass container of unknown volume contains helium gas at 25°C and 1.960 atm. When a portion of the helium is withdrawn and adjusted to 1.00 atm at 25°C, it is found to have a volume of 1.75 cm^3. The gas remaining in the first container shows a pressure of 1.710 atm. Calculate the volume of the spherical container.

133. A 2.00-L sample of $O_2(g)$ was collected over water at a total pressure of 785 torr and 25°C. When the $O_2(g)$ was dried (water vapor removed), the gas had a volume of 1.94 L at 25°C and 785 torr. Calculate the vapor pressure of water at 25°C.

134. A 20.0-L stainless steel container at 25°C was charged with 2.00 atm of hydrogen gas and 3.00 atm of oxygen gas. A spark ignited the mixture, producing water. What is the pressure in the tank at 25°C? If the exact same experiment were performed, but the temperature was 125°C instead of 25°C, what would be the pressure in the tank?

135. Metallic molybdenum can be produced from the mineral molybdenite, MoS_2. The mineral is first oxidized in air to molybdenum trioxide and sulfur dioxide. Molybdenum trioxide is then reduced to metallic molybdenum using hydrogen gas. The balanced equations are

$$MoS_2(s) + \tfrac{7}{2}O_2(g) \longrightarrow MoO_3(s) + 2SO_2(g)$$
$$MoO_3(s) + 3H_2(g) \longrightarrow Mo(s) + 3H_2O(l)$$

Calculate the volumes of air and hydrogen gas at 17°C and 1.00 atm that are necessary to produce 1.00×10^3 kg pure molybdenum from MoS_2. Assume air contains 21% oxygen by volume, and assume 100% yield for each reaction.

136. Nitric acid is produced commercially by the Ostwald process. In the first step ammonia is oxidized to nitric oxide:

$$4NH_3(g) + 5O_2(g) \longrightarrow 4NO(g) + 6H_2O(g)$$

Assume this reaction is carried out in the apparatus diagramed below.

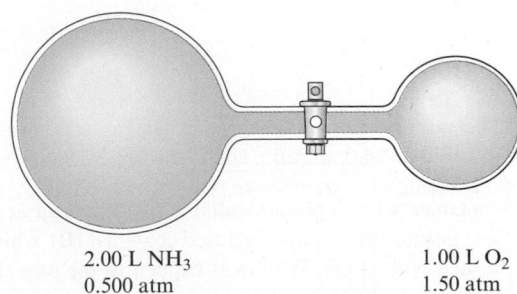

2.00 L NH_3 1.00 L O_2
0.500 atm   1.50 atm

The stopcock between the two reaction containers is opened, and the reaction proceeds using proper catalysts. Calculate the partial pressure of NO after the reaction is complete. Assume 100% yield for the reaction, assume the final container volume is 3.00 L, and assume the temperature is constant.

137. A compound contains only C, H, and N. It is 58.51% C and 7.37% H by mass. Helium effuses through a porous frit 3.20 times as fast as the compound does. Determine the empirical and molecular formulas of this compound.

138. One of the chemical controversies of the nineteenth century concerned the element beryllium (Be). Berzelius originally claimed that beryllium was a trivalent element (forming Be^{3+} ions) and that it gave an oxide with the formula Be_2O_3. This resulted in a calculated atomic mass of 13.5 for beryllium. In formulating his periodic table, Mendeleev proposed that beryllium was divalent (forming Be^{2+} ions) and that it gave an oxide with the formula BeO. This assumption gives an atomic mass of 9.0. In 1894, A. Combes (*Comptes Rendus* 1894, p. 1221) reacted beryllium with the anion $C_5H_7O_2^-$ and measured the density of the gaseous product. Combes's data for two different experiments are as follows:

	I	II
Mass	0.2022 g	0.2224 g
Volume	22.6 cm^3	26.0 cm^3
Temperature	13°C	17°C
Pressure	765.2 mm Hg	764.6 mm

If beryllium is a divalent metal, the molecular formula of the product will be $Be(C_5H_7O_2)_2$; if it is trivalent, the formula will be $Be(C_5H_7O_2)_3$. Show how Combes's data help to confirm that beryllium is a divalent metal.

139. An organic compound contains C, H, N, and O. Combustion of 0.1023 g of the compound in excess oxygen yielded 0.2766 g CO_2 and 0.0991 g H_2O. A sample of 0.4831 g of the compound was analyzed for nitrogen by the Dumas method (see Exercise 129). At STP, 27.6 mL of dry N_2 was obtained. In a third experiment, the density of the compound as a gas was found to be 4.02 g/L at 127°C and 256 torr. What are the empirical and molecular formulas of the compound?

140. Consider the following diagram:

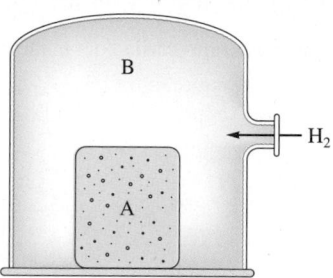

Container A (with porous walls) is filled with air at STP. It is then inserted into a large enclosed container (B), which is then flushed with $H_2(g)$. What will happen to the pressure inside container A? Explain your answer.

141. The total volume of hydrogen gas needed to fill the *Hindenburg* was 2.0×10^8 L at 1.0 atm and 25°C. Given that ΔH_f° for $H_2O(l)$ is -286 kJ/mol, how much heat was evolved when the *Hindenburg* exploded, assuming all of the hydrogen reacted to form water?

142. Assume that 4.19×10^6 kJ of energy is needed to heat a home. If this energy is derived from the combustion of methane (CH_4), what volume of methane, measured at 1.00 atm and 0°C, must be burned? ($\Delta H_{combustion}^\circ$ for $CH_4 = -891$ kJ/mol)

ChemWork Problems

These multiconcept problems (and additional ones) are found interactively online with the same type of assistance a student would get from an instructor.

143. A glass vessel contains 28 g of nitrogen gas. Assuming ideal behavior, which of the processes listed below would double the pressure exerted on the walls of the vessel?

  a. Adding 28 g of oxygen gas

  b. Raising the temperature of the container from -73°C to 127°C

  c. Adding enough mercury to fill one-half the container

  d. Adding 32 g of oxygen gas

  e. Raising the temperature of the container from 30.°C to 60.°C

144. A steel cylinder contains 150.0 moles of argon gas at a temperature of 25°C and a pressure of 8.93 MPa. After some argon has been used, the pressure is 2.00 MPa at a temperature of 19°C. What mass of argon remains in the cylinder?

145. A certain flexible weather balloon contains helium gas at a volume of 855 L. Initially, the balloon is at sea level where the temperature is 25°C and the barometric pressure is 730 torr. The balloon then rises to an altitude of 6000 ft, where the pressure is 605 torr and the temperature is 15°C. What is the change in volume of the balloon as it ascends from sea level to 6000 ft?

146. A large flask with a volume of 936 mL is evacuated and found to have a mass of 134.66 g. It is then filled to a pressure of 0.967 atm at 31°C with a gas of unknown molar mass and then reweighed to give a new mass of 135.87 g. What is the molar mass of this gas?

147. A 20.0-L nickel container was charged with 0.859 atm of xenon gas and 1.37 atm of fluorine gas at 400°C. The xenon and fluorine react to form xenon tetrafluoride. What mass of xenon tetrafluoride can be produced assuming 100% yield?

148. Consider the *unbalanced* chemical equation below:

$$CaSiO_3(s) + HF(g) \longrightarrow CaF_2(aq) + SiF_4(g) + H_2O(l)$$

Suppose a 32.9-g sample of $CaSiO_3$ is reacted with 31.8 L of HF at 27.0°C and 1.00 atm. Assuming the reaction goes to completion, calculate the mass of the SiF_4 and H_2O produced in the reaction.

149. Consider separate 1.0-L gaseous samples of He, N_2, and O_2, all at STP and all acting ideally. Rank the gases in order of increasing average kinetic energy and in order of increasing average velocity.

150. Which of the following statements is(are) *true*?

  a. If the number of moles of a gas is doubled, the volume will double, assuming the pressure and temperature of the gas remain constant.

b. If the temperature of a gas increases from 25°C to 50°C, the volume of the gas would double, assuming that the pressure and the number of moles of gas remain constant.

c. The device that measures atmospheric pressure is called a barometer.

d. If the volume of a gas decreases by one half, then the pressure would double, assuming that the number of moles and the temperature of the gas remain constant.

Challenge Problems

151. A chemist weighed out 5.14 g of a mixture containing unknown amounts of $BaO(s)$ and $CaO(s)$ and placed the sample in a 1.50-L flask containing $CO_2(g)$ at 30.0°C and 750. torr. After the reaction to form $BaCO_3(s)$ and $CaCO_3(s)$ was completed, the pressure of $CO_2(g)$ remaining was 230. torr. Calculate the mass percentages of $CaO(s)$ and $BaO(s)$ in the mixture.

152. A mixture of chromium and zinc weighing 0.362 g was reacted with an excess of hydrochloric acid. After all the metals in the mixture reacted, 225 mL dry of hydrogen gas was collected at 27°C and 750. torr. Determine the mass percent of Zn in the metal sample. [Zinc reacts with hydrochloric acid to produce zinc chloride and hydrogen gas; chromium reacts with hydrochloric acid to produce chromium(III) chloride and hydrogen gas.]

153. Consider a sample of a hydrocarbon (a compound consisting of only carbon and hydrogen) at 0.959 atm and 298 K. Upon combusting the entire sample in oxygen, you collect a mixture of gaseous carbon dioxide and water vapor at 1.51 atm and 375 K. This mixture has a density of 1.391 g/L and occupies a volume four times as large as that of the pure hydrocarbon. Determine the molecular formula of the hydrocarbon.

154. You have an equimolar mixture of the gases SO_2 and O_2, along with some He, in a container fitted with a piston. The density of this mixture at STP is 1.924 g/L. Assume ideal behavior and constant temperature and pressure.

a. What is the mole fraction of He in the original mixture?

b. The SO_2 and O_2 react to completion to form SO_3. What is the density of the gas mixture after the reaction is complete?

155. Methane (CH_4) gas flows into a combustion chamber at a rate of 200. L/min at 1.50 atm and ambient temperature. Air is added to the chamber at 1.00 atm and the same temperature, and the gases are ignited.

a. To ensure complete combustion of CH_4 to $CO_2(g)$ and $H_2O(g)$, three times as much oxygen as is necessary is reacted. Assuming air is 21 mole percent O_2 and 79 mole percent N_2, calculate the flow rate of air necessary to deliver the required amount of oxygen.

b. Under the conditions in part a, combustion of methane was not complete as a mixture of $CO_2(g)$ and $CO(g)$ was produced. It was determined that 95.0% of the carbon in the exhaust gas was present in CO_2. The remainder was present as carbon in CO. Calculate the composition of the exhaust gas in terms of mole fraction of CO, CO_2, O_2, N_2, and H_2O. Assume CH_4 is completely reacted and N_2 is unreacted.

156. A steel cylinder contains 5.00 moles of graphite (pure carbon) and 5.00 moles of O_2. The mixture is ignited and all the graphite reacts. Combustion produces a mixture of CO gas and CO_2 gas. After the cylinder has cooled to its original temperature, it is found that the pressure of the cylinder has increased by 17.0%. Calculate the mole fractions of CO, CO_2, and O_2 in the final gaseous mixture.

157. The total mass that can be lifted by a balloon is given by the difference between the mass of air displaced by the balloon and the mass of the gas inside the balloon. Consider a hot-air balloon that approximates a sphere 5.00 m in diameter and contains air heated to 65°C. The surrounding air temperature is 21°C. The pressure in the balloon is equal to the atmospheric pressure, which is 745 torr.

a. What total mass can the balloon lift? Assume that the average molar mass of air is 29.0 g/mol. (*Hint:* Heated air is less dense than cool air.)

b. If the balloon is filled with enough helium at 21°C and 745 torr to achieve the same volume as in part a, what total mass can the balloon lift?

c. What mass could the hot-air balloon in part a lift if it were on the ground in Denver, Colorado, where a typical atmospheric pressure is 630. torr?

158. You have a sealed, flexible balloon filled with argon gas. The atmospheric pressure is 1.00 atm and the temperature is 25°C. Assume that air has a mole fraction of nitrogen of 0.790, the rest being oxygen.

a. Explain why the balloon would float when heated. Make sure to discuss which factors change and which remain constant, and why this matters.

b. Above what temperature would you heat the balloon so that it would float?

159. You have a helium balloon at 1.00 atm and 25°C. You want to make a hot-air balloon with the same volume and same lift as the helium balloon. Assume air is 79.0% nitrogen and 21.0% oxygen by volume. The "lift" of a balloon is given by the difference between the mass of air displaced by the balloon and the mass of gas inside the balloon.

a. Will the temperature in the hot-air balloon have to be higher or lower than 25°C? Explain.

b. Calculate the temperature of the air required for the hot-air balloon to provide the same lift as the helium balloon at 1.00 atm and 25°C. Assume atmospheric conditions are 1.00 atm and 25°C.

160. We state that the ideal gas law tends to hold best at low pressures and high temperatures. Show how the van der Waals equation simplifies to the ideal gas law under these conditions.

161. You are given an unknown gaseous binary compound (that is, a compound consisting of two different elements). When 10.0 g of the compound is burned in excess oxygen, 16.3 g of water is produced. The compound has a density 1.38 times that of oxygen gas at the same conditions of temperature and pressure. Give a possible identity for the unknown compound.

162. Nitrogen gas (N_2) reacts with hydrogen gas (H_2) to form ammonia gas (NH_3). You have nitrogen and hydrogen gases in a 15.0-L container fitted with a movable piston (the piston allows the container volume to change so as to keep the pressure constant inside the container). Initially the partial pressure of each reactant gas is 1.00 atm. Assume the temperature is constant and that the reaction goes to completion.

  a. Calculate the partial pressure of ammonia in the container after the reaction has reached completion.

  b. Calculate the volume of the container after the reaction has reached completion.

163. Calculate w and ΔE when 1 mole of a liquid is vaporized at its boiling point (80.°C) and 1.00 atm pressure. ΔH for the vaporization of the liquid is 30.7 kJ/mol at 80.°C. Assume the volume of 1 mole of liquid is negligible as compared to the volume of 1 mole of gas at 80.°C and 1.00 atm.

Integrative Problems

These problems require the integration of multiple concepts to find the solutions.

164. The preparation of $NO_2(g)$ from $N_2(g)$ and $O_2(g)$ is an endothermic reaction:

$$N_2(g) + O_2(g) \longrightarrow NO_2(g) \quad \text{(unbalanced)}$$

The enthalpy change of reaction for the balanced equation (with lowest whole-number coefficients) is $\Delta H = 67.7$ kJ. If 2.50×10^2 mL $N_2(g)$ at 100.°C and 3.50 atm and 4.50×10^2 mL $O_2(g)$ at 100.°C and 3.50 atm are mixed, what amount of heat is necessary to synthesize the maximum yield of $NO_2(g)$?

165. In the presence of nitric acid, UO^{2+} undergoes a redox process. It is converted to UO_2^{2+} and nitric oxide (NO) gas is produced according to the following unbalanced equation:

$$H^+(aq) + NO_3^-(aq) + UO^{2+}(aq) \longrightarrow$$
$$NO(g) + UO_2^{2+}(aq) + H_2O(l)$$

If 2.55×10^2 mL $NO(g)$ is isolated at 29°C and 1.5 atm, what amount (moles) of UO^{2+} was used in the reaction? (*Hint:* Balance the reaction by the oxidation states method.)

166. Silane, SiH_4, is the silicon analogue of methane, CH_4. It is prepared industrially according to the following equations:

$$Si(s) + 3HCl(g) \longrightarrow HSiCl_3(l) + H_2(g)$$
$$4HSiCl_3(l) \longrightarrow SiH_4(g) + 3SiCl_4(l)$$

  a. If 156 mL $HSiCl_3$ ($d = 1.34$ g/mL) is isolated when 15.0 L HCl at 10.0 atm and 35°C is used, what is the percent yield of $HSiCl_3$?

  b. When 156 mL $HSiCl_3$ is heated, what volume of SiH_4 at 10.0 atm and 35°C will be obtained if the percent yield of the reaction is 93.1%?

167. Solid thorium(IV) fluoride has a boiling point of 1680°C. What is the density of a sample of gaseous thorium(IV) fluoride at its boiling point under a pressure of 2.5 atm in a 1.7-L container? Which gas will effuse faster at 1680°C, thorium(IV) fluoride or uranium(III) fluoride? How much faster?

168. Natural gas is a mixture of hydrocarbons, primarily methane (CH_4) and ethane (C_2H_6). A typical mixture might have $\chi_{\text{methane}} = 0.915$ and $\chi_{\text{ethane}} = 0.085$. What are the partial pressures of the two gases in a 15.00-L container of natural gas at 20.°C and 1.44 atm? Assuming complete combustion of both gases in the natural gas sample, what is the total mass of water formed?

Marathon Problem

This problem is designed to incorporate several concepts and techniques into one situation.

169. Consider an equimolar mixture (equal number of moles) of two diatomic gases (A_2 and B_2) in a container fitted with a piston. The gases react to form one product (which is also a gas) with the formula A_xB_y. The density of the sample after the reaction is complete (and the temperature returns to its original state) is 1.50 times greater than the density of the reactant mixture.

  a. Specify the formula of the product, and explain if more than one answer is possible based on the given data.

  b. Can you determine the molecular formula of the product with the information given or only the empirical formula?

C H A P T E R   N I N E

Liquids and Solids

Frozen soap bubbles at sunrise. (© Angela Kelly)

W ater is interesting to consider because we have experience with water in three states of matter. Flying, swimming, and ice skating are all done in contact with water in its various forms. Clearly, the arrangements of the water molecules must be significantly different in its gas, liquid, and solid forms.

Why is water a liquid at room temperature and pressure? To answer this question we need to understand the structure of the water molecule (Chapter 4), which requires us to understand the nature of chemical bonds (Chapter 3), which then requires an understanding of atomic structure (Chapter 2). As you can see, answering this question required us to take an "atoms first" approach to learning chemistry.

In Chapter 8 we saw that a gas can be pictured as a substance whose component particles are far apart, are in rapid random motion, and exert relatively small forces on each other. The kinetic molecular model was constructed to account for the ideal behavior that most gases approach at high temperatures and low pressures.

Solids are obviously very different from gases. A gas has low density and high compressibility and completely fills its container. Solids have much greater densities, are compressible only to a very slight extent, and are rigid—a solid maintains its shape irrespective of its container. These properties indicate that the components of a solid are close together and exert large attractive forces on each other.

The properties of liquids lie somewhere between those of solids and gases but not midway between, as can be seen from some of the properties of the three states of water. For example, compare the enthalpy change for the melting of ice at 0°C (the heat of fusion) with that for vaporizing liquid water at 100°C (the heat of vaporization):

$$H_2O(s) \longrightarrow H_2O(l) \qquad \Delta H°_{fus} = 6.02 \text{ kJ/mol}$$
$$H_2O(l) \longrightarrow H_2O(g) \qquad \Delta H°_{vap} = 40.7 \text{ kJ/mol}$$

These values show a much greater change in structure in going from the liquid to the gaseous state than in going from the solid to the liquid state. This suggests that there are extensive attractive forces among the molecules in liquid water, similar to but not as strong as those in the solid state.

The relative similarity of the liquid and solid states also can be seen in the densities of the three states of water. As shown in Table 9-1, the densities for liquid and solid water are quite close.* Compressibilities also can be used to explore the relationship among water's states. At 25°C, the density of liquid water changes from 0.99707 g/cm^3 at a pressure of 1 atm to 1.046 g/cm^3 at 1065 atm. Given the large change in pressure, this is a very small variation in the density. Ice also shows little variation in density with increased pressure. On the other hand, at 400°C, the density of gaseous water changes from 3.26 × 10^{-4} g/cm^3 at 1 atm pressure to 0.157 g/cm^3 at 242 atm—a huge variation.

The conclusion is clear. The liquid and solid states show many similarities and are strikingly different from the gaseous state, as shown schematically in Fig. 9-1. We must bear this in mind as we develop models for the structures of solids and liquids.

We will proceed in our study of liquids and solids by first considering the properties and structures of liquids and solids. Then we will consider the changes in state that occur between solid and liquid, liquid and gas, and solid and gas.

Table 9-1 | Densities of the Three States of Water

State	Density (g/cm³)
Solid (0°C, 1 atm)	0.9168
Liquid (25°C, 1 atm)	0.9971
Gas (400°C, 1 atm)	3.26 × 10^{-4}

*Although the densities of solid and liquid water are quite similar, as is typical for most substances, water is quite unusual in that the density of its solid state is slightly less than that of its liquid state. For most substances, the density of the solid state is slightly greater than that of the liquid state.

Figure 9-1 | Schematic representations of the three states of matter.

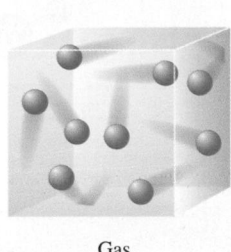

Gas

Liquid

Solid

9-1 | Intermolecular Forces

In Chapters 3 and 4 we saw that atoms can form stable units called *molecules* by sharing electrons. This is called *intramolecular* (within the molecule) *bonding*. In this chapter we consider the properties of the **condensed states** of matter (liquids and solids) and the forces that cause the aggregation of the components of a substance to form a liquid or a solid. These forces may involve covalent or ionic bonding, or they may involve weaker interactions usually called **intermolecular forces** (because they occur between, rather than within, molecules). ◄ ⓘ We will also talk about forces in ionic and metallic substances that might better be called interparticle forces.

It is important to recognize that when a substance such as water changes from solid to liquid to gas, *the molecules remain intact*. The changes in states are due to changes in the forces *among* the molecules rather than in those *within* the molecules. In ice, as we will see later in this chapter, the molecules are virtually locked in place, although they can vibrate about their positions. If energy is added, the motions of the molecules increase, and they eventually achieve the greater movement and disorder characteristic of liquid water. ◄ ⓘ The ice has melted. As more energy is added, the gaseous state is eventually reached, with the individual molecules far apart and interacting relatively little. However, the gas still consists of water molecules. It would take much energy to overcome the covalent bonds and decompose the water molecules into their component atoms. This can be seen by comparing the energy needed to vaporize 1 mole of liquid water (40.7 kJ) with that needed to break the O—H bonds in 1 mole of water molecules (934 kJ).

Dipole–Dipole Forces

As we saw in Section 4-2, molecules with polar bonds often behave in an electric field as if they had a center of positive charge and a center of negative charge. That is, they exhibit a dipole moment. Molecules with dipole moments can attract each other electrostatically by lining up so that the positive and negative ends are close to each other, as shown in Fig. 9-2(a). This is called a **dipole–dipole attraction**. In a condensed state such as a liquid, where many molecules are in close proximity, the dipoles find the best compromise between attraction and repulsion. That is, the molecules orient themselves to maximize the ⊕---⊖ interactions and to minimize ⊕---⊕ and ⊖---⊖ interactions, as represented in Fig. 9-2(b).

Dipole–dipole forces are typically only about 1% as strong as covalent or ionic bonds, and they rapidly become weaker as the distance between the dipoles increases. At low pressures in the gas phase, where the molecules are far apart, these forces are relatively unimportant.

Particularly strong dipole–dipole forces, however, are seen among molecules in which hydrogen is bound to a highly electronegative atom, such as nitrogen, oxygen, or fluorine. ◄ ⓘ Two factors account for the strengths of these interactions: the great

ⓘ Intermolecular forces were introduced in Chapter 8 to explain non-ideal gas behavior.

ⓘ Remember that temperature is a measure of the random motions of the particles in a substance.

ⓘ Electronegativity was discussed in Chapter 4. Notice how atomic properties affect molecular properties.

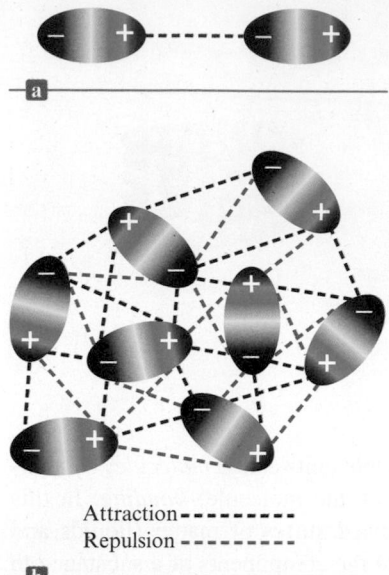

Attraction --------
Repulsion --------

Figure 9-2 | (a) The electrostatic interaction of two polar molecules. (b) The interaction of many dipoles in a condensed state.

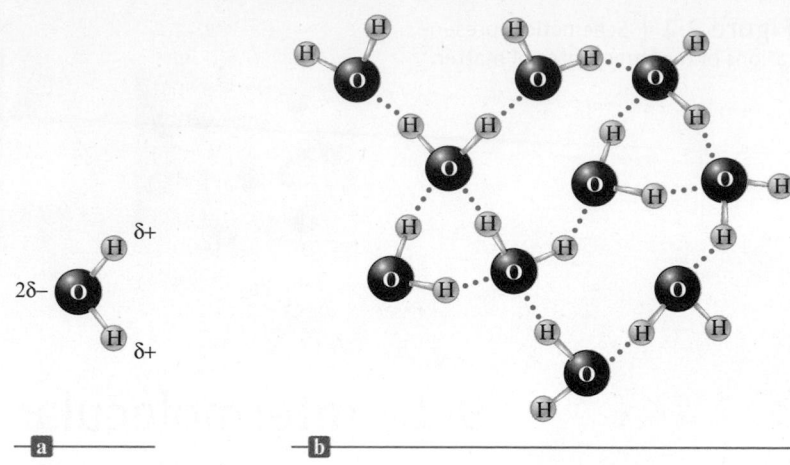

Figure 9-3 | (a) The polar water molecule. (b) Hydrogen bonding (••) among water molecules. Note that the small size of the hydrogen atom allows for close interactions.

polarity of the bond and the close approach of the dipoles, allowed by the very small size of the hydrogen atom. Because dipole–dipole attractions of this type are so unusually strong, they are given a special name—**hydrogen bonding**. In this type of bonding, the hydrogen atom is bound to a lone pair on an adjacent molecule. Fig. 9-3 shows hydrogen bonding among water molecules, which occurs between the partially positive H atoms and the lone pairs on adjacent water molecules.

Hydrogen bonding has a very important effect on physical properties. For example, the boiling points for the covalent hydrides of the elements in Groups 4A, 5A, 6A, and 7A are given in Fig. 9-4. Note that the nonpolar tetrahedral hydrides of Group 4A show a steady increase in boiling point with molar mass (that is, in going down the group), whereas, for the other groups, the lightest member has an unexpectedly high boiling point. Why? The answer lies in the especially large hydrogen bonding interactions that exist among the smallest molecules with the most polar X—H bonds. These

Figure 9-4 | The boiling points of the covalent hydrides of the elements in Groups 4A, 5A, 6A, and 7A. The dashed line shows the expected boiling point of water if it had no hydrogen bonding.

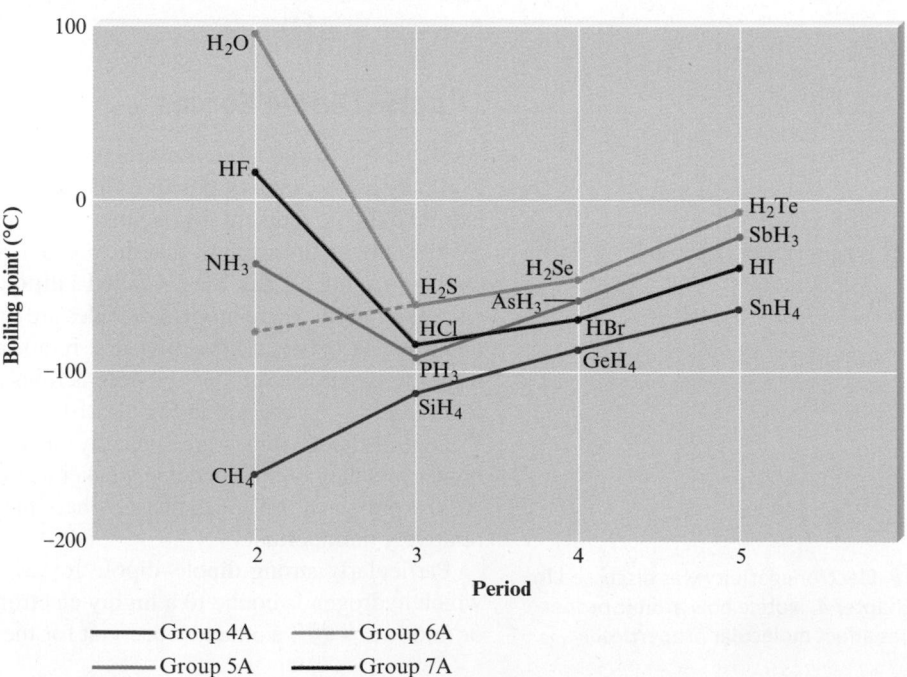

unusually strong hydrogen bonding forces are due primarily to two factors. One factor is the relatively large electronegativity values of the lightest elements in each group, which leads to especially polar X—H bonds. The second factor is the small size of the first element of each group, which allows for the close approach of the dipoles, further strengthening the intermolecular forces. Because the interactions among the molecules containing the lightest elements in Groups 5A and 6A are so strong, an unusually large quantity of energy must be supplied to overcome these interactions and separate the molecules to produce the gaseous state. These molecules will remain together in the liquid state even at high temperatures—hence the very high boiling points. ◄ *i*

Hydrogen bonding is also important in organic molecules (molecules with a carbon chain backbone). For example, the alcohols methanol (CH_3OH) and ethanol (CH_3CH_2OH) have much higher boiling points than would be expected from their molar masses because of the polar O—H bonds in these molecules, which produce hydrogen bonding.

London Dispersion Forces

Even molecules without dipole moments must exert forces on each other. We know this because all substances—even the noble gases—exist in the liquid and solid states under certain conditions. The forces that exist among noble gas atoms and nonpolar molecules are called **London dispersion forces**. ◄ *i* To understand the origin of these forces, let's consider a pair of noble gas atoms. Although we usually assume that the electrons of an atom are uniformly distributed about the nucleus, this is apparently not true at every instant. As the electrons move about the nucleus, a momentary nonsymmetrical electron distribution can develop that produces a temporary dipolar arrangement of charge. The formation of this temporary dipole can, in turn, affect the electron distribution of a neighboring atom. That is, this *instantaneous dipole* that occurs accidentally in a given atom can then *induce* a similar dipole in a neighboring atom, as represented in Fig. 9-5(a). This phenomenon leads to

i Boiling point will be defined precisely in Section 9-8.

i All molecules exhibit London dispersion forces. For polar molecules, however, dipole–dipole interactions are much stronger.

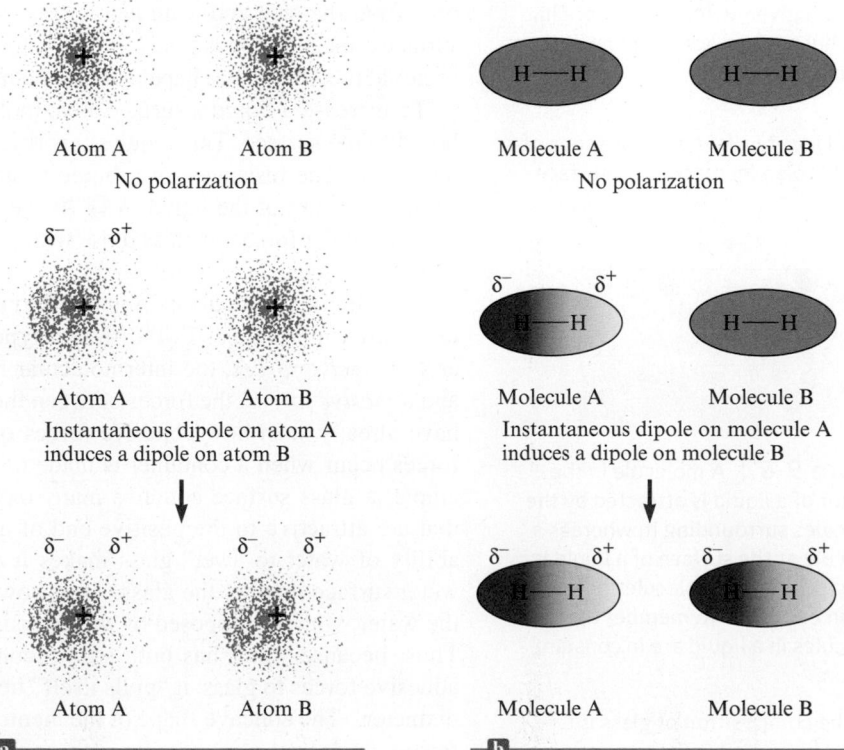

Figure 9-5 | (a) An instantaneous polarization can occur on atom A, creating an instantaneous dipole. This dipole creates an induced dipole on neighboring atom B. (b) Nonpolar molecules such as H_2 also can develop instantaneous and induced dipoles.

Table 9-2 | The Freezing Points of the Group 8A Elements

Element	Freezing Point (°C)
Helium*	−269.7
Neon	−248.6
Argon	−189.4
Krypton	−157.3
Xenon	−111.9

*Helium is the only element that will not freeze with the lowering of its temperature at 1 atm. Pressure must be applied to freeze helium.

🛈 The dispersion forces in molecules with large atoms are quite significant and are often actually more important than dipole–dipole forces.

an interatomic attraction that is relatively weak and short-lived but that can be very significant, especially for large atoms (see below). For these interactions to become strong enough to produce a solid, the motions of the atoms must be greatly slowed down. This explains, for instance, why the noble gas elements have such low freezing points (Table 9-2).

Note from Table 9-2 that the freezing point rises going down the group. The principal cause for this trend is that as the atomic number increases, the number of electrons increases, and there is an increased chance of the occurrence of momentary dipole interactions. We describe this phenomenon using the term *polarizability*, which indicates the ease with which the electron "cloud" of an atom can be distorted to give a dipolar charge distribution. Thus we say that large atoms with many electrons exhibit a higher polarizability than small atoms. This means that the importance of London dispersion forces increases greatly as the size of the atom increases.

These same ideas also apply to nonpolar molecules such as H_2, CH_4, CCl_4, and CO_2 [see Fig. 9-5(b)]. Since none of these molecules has a permanent dipole moment, their principal means of attracting each other is through London dispersion forces. ◄ 🛈

9-2 | The Liquid State

Liquids and liquid solutions are vital to our lives. Of course, water is the most important liquid. Besides being essential to life, water provides a medium for food preparation, for transportation, for cooling in many types of machines and industrial processes, for recreation, for cleaning, and for myriad other uses.

Liquids exhibit many characteristics that help us understand their nature. We have already mentioned their low compressibility, lack of rigidity, and high density compared with gases. Many of the properties of liquids give us direct information about the forces that exist among the particles. ◄ 🛈 For example, when most liquids are poured onto a solid surface, they tend to bead as droplets, a phenomenon that depends on the intermolecular forces. Although molecules in the interior of the liquid are completely surrounded by other molecules, those at the liquid surface are subject to attractions only from the side and from below (Fig. 9-6). The effect of this uneven pull on the surface molecules tends to draw them into the body of the liquid and causes a droplet of liquid to assume the shape that has the minimum surface area—a sphere. ◄ 🛈

🛈 We need to consider molecular structure in order to explain the observable properties for liquids.

🛈 For a given volume, a sphere has a smaller surface area than any other shape.

🛈 *Surface tension:* the resistance of a liquid to an increase in its surface area.

Surface

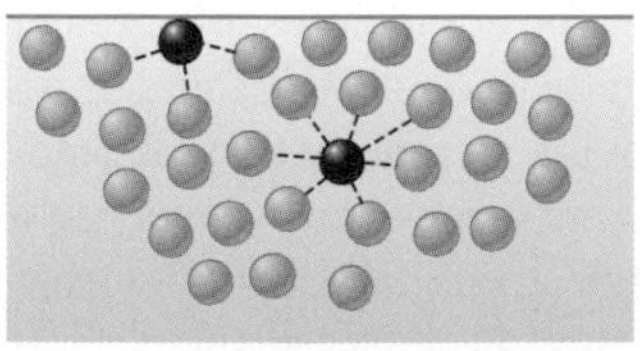

Figure 9-6 | A molecule in the interior of a liquid is attracted by the molecules surrounding it, whereas a molecule at the surface of a liquid is attracted only by molecules below it and on each side. Remember that the molecules in a liquid are in constant motion.

🛈 The composition of glass is discussed in Section 9-5.

To increase a liquid's surface area, molecules must move from the interior of the liquid to the surface. This requires energy, since some intermolecular forces must be overcome. The resistance of a liquid to an increase in its surface area is called the **surface tension** of the liquid. ◄ 🛈 As we would expect, liquids with relatively large intermolecular forces, such as those with polar molecules, tend to have relatively high surface tensions.

Polar liquids typically exhibit **capillary action**, the spontaneous rising of a liquid in a narrow glass tube. Two different types of forces are responsible for this property: *cohesive forces*, the intermolecular forces among the molecules of the liquid, and *adhesive forces*, the forces between the liquid molecules and their container. We have already seen how cohesive forces operate among polar molecules. Adhesive forces occur when a container is made of a substance that has polar bonds. For example, a glass surface contains many oxygen atoms with partial negative charges that are attractive to the positive end of a polar molecule such as water. ◄ 🛈 This ability of water to "wet" glass makes it creep up the walls of the tube where the water surface touches the glass. This, however, tends to increase the surface area of the water, which is opposed by the cohesive forces that try to minimize the surface. Thus, because water has both strong cohesive (intermolecular) forces and strong adhesive forces to glass, it "pulls itself" up a glass capillary tube (a tube with a small diameter). The concave shape of the meniscus (Fig. 9-7) shows that water's adhesive forces toward the glass are stronger than its cohesive forces. A nonpolar liquid such

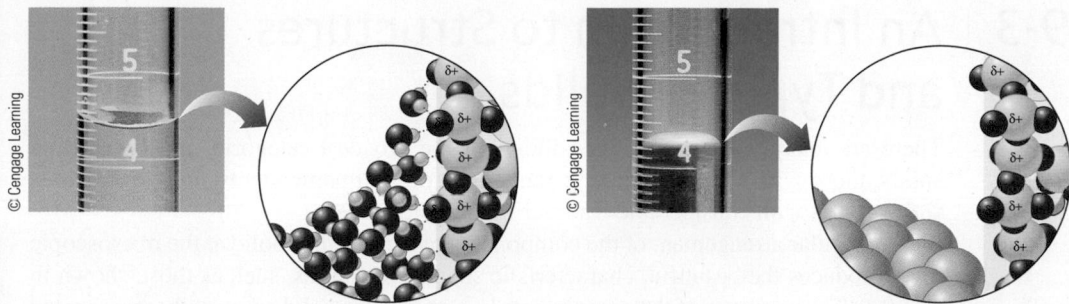

Figure 9-7 | Nonpolar liquid mercury forms a convex meniscus in a glass tube, whereas polar water forms a concave meniscus.

as mercury (see Fig. 9-7) shows a convex meniscus. This behavior is characteristic of a liquid in which the cohesive forces are stronger than the adhesive forces toward glass.

Another property of liquids strongly dependent on intermolecular forces is **viscosity**, a measure of a liquid's resistance to flow. As might be expected, liquids with large intermolecular forces tend to be highly viscous. For example, glycerol, whose structure is

$$
\begin{array}{c}
H \\
| \\
H-C-O-H \\
| \\
H-C-O-H \\
| \\
H-C-O-H \\
| \\
H
\end{array}
$$

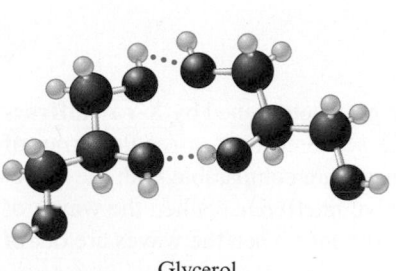

Glycerol

has an unusually high viscosity due mainly to its high capacity to form hydrogen bonds using its O—H groups (see margin). ◄

Molecular complexity also leads to higher viscosity because very large molecules can become entangled with each other. For example, gasoline, a nonviscous liquid, contains hydrocarbon molecules of the type $CH_3-(CH_2)_n-CH_3$, where n varies from about 3 to 8. However, grease, which is very viscous, contains much larger hydrocarbon molecules in which n varies from 20 to 25.

Structural Model for Liquids

In many respects, the development of a structural model for liquids presents greater challenges than the development of such a model for the other two states of matter. In the gaseous state the particles are so far apart and are moving so rapidly that intermolecular forces are negligible under most circumstances. This means that we can use a relatively simple model for gases. In the solid state, although the intermolecular forces are large, the molecular motions are minimal, and fairly simple models are again possible. The liquid state, however, has both strong intermolecular forces *and* significant molecular motions. Such a situation precludes the use of really simple models for liquids. Recent advances in *spectroscopy*, the study of the manner in which substances interact with electromagnetic radiation, make it possible to follow the very rapid changes that occur in liquids. As a result, our models of liquids are becoming more accurate. As a starting point, a typical liquid might best be viewed as containing a large number of regions where the arrangements of the components are similar to those found in the solid, but with more disorder, and a smaller number of regions where holes are present. The situation is highly dynamic, with rapid fluctuations occurring in both types of regions.

9-3 | An Introduction to Structures and Types of Solids

There are many ways to classify solids, but the broadest categories are **crystalline solids**, those with a highly regular arrangement of their components, and **amorphous solids**, those with considerable disorder in their structures.

The regular arrangement of the components of a crystalline solid at the microscopic level produces the beautiful, characteristic shapes of crystals, such as those shown in Fig. 9-8. The positions of the components in a crystalline solid are usually represented by a **lattice**, a three-dimensional system of points designating the positions of the components (atoms, ions, or molecules) that make up the substance. The *smallest repeating unit* of the lattice is called the **unit cell**. Thus a particular lattice can be generated by repeating the unit cell in all three dimensions to form the extended structure. Three common unit cells and their lattices are shown in Fig. 9-9. Note from Fig. 9-9 that the extended structure in each case can be viewed as a series of repeating unit cells that share common faces in the interior of the solid.

Although we will concentrate on crystalline solids in this book, there are many important noncrystalline (amorphous) materials. An example is common glass, which is best pictured as a solution in which the components are "frozen in place" before they can achieve an ordered arrangement. Although glass is a solid (it has a rigid shape), a great deal of disorder exists in its structure.

X-Ray Analysis of Solids

The structures of crystalline solids are most commonly determined by **X-ray diffraction**. Diffraction occurs when beams of light are scattered from a regular array of points in which the spacings between the components are comparable with the wavelength of the light. Diffraction is due to constructive interference when the waves of parallel beams are in phase and to destructive interference when the waves are out of phase.

When X rays of a single wavelength are directed at a crystal, a diffraction pattern is obtained, as we saw in Fig. 2-6. The light and dark areas on the photographic plate occur because the waves scattered from various atoms may reinforce or cancel each other (see Fig. 9-10). The key to whether the waves reinforce or cancel is the difference in distance traveled by the waves after they strike the atoms. The waves are in

Brian Parker/Tom Stock & Associates

Darlldo/iStockPhoto.com

Figure 9-8 | Two crystalline solids: pyrite (left), amethyst (right).

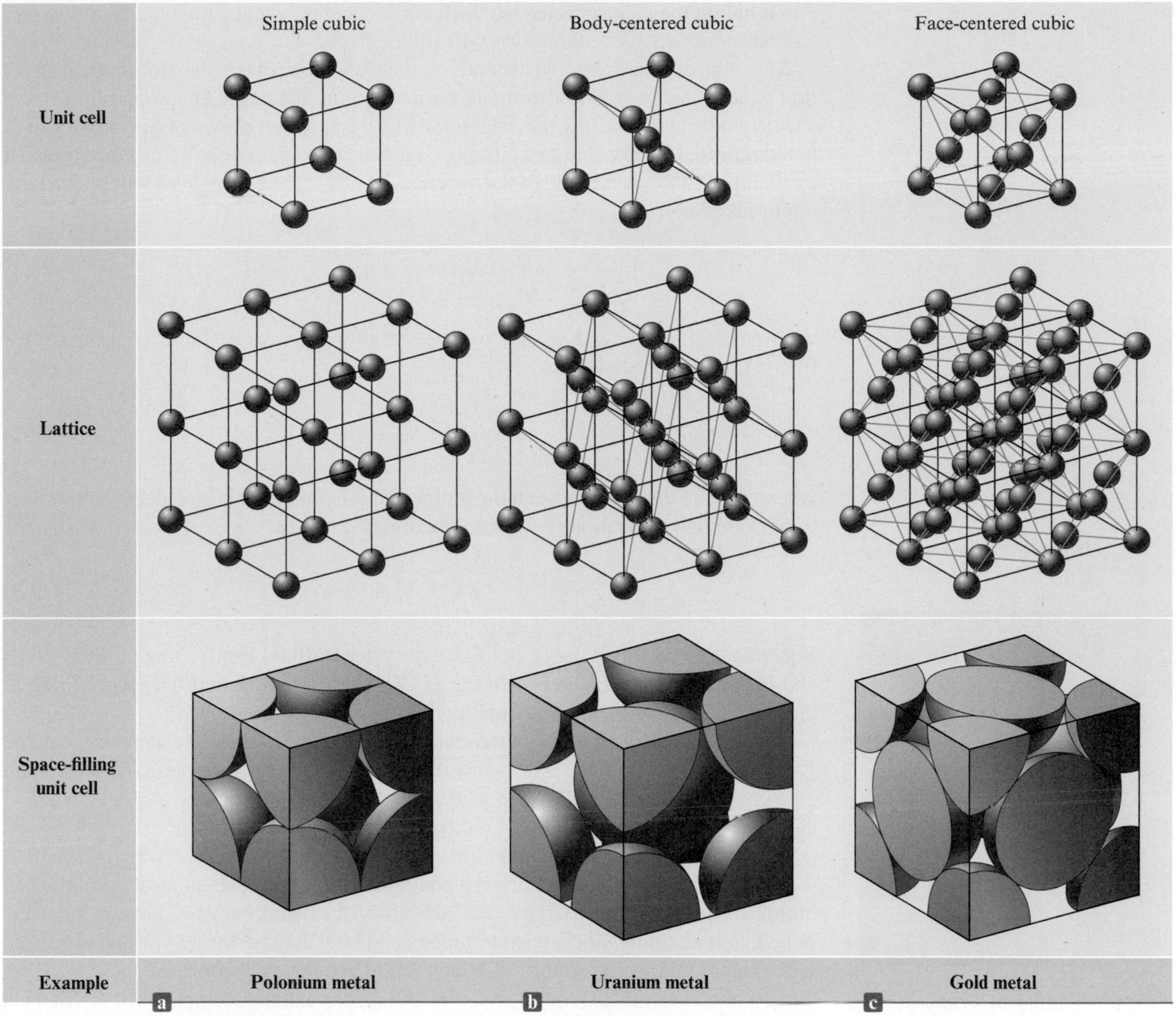

Simple cubic
Body-centered cubic
Face-centered cubic
Unit cell
Lattice
Space-filling unit cell
Example
Polonium metal
Uranium metal
Gold metal
a
b
c

Figure 9-9 | Three cubic unit cells and the corresponding lattices. Note that only parts of spheres on the corners and faces of the unit cells reside inside the unit cell, as shown by the "cutoff" versions.

In-phase
In-phase
d_1
a
In-phase
Out of phase
d_2
b

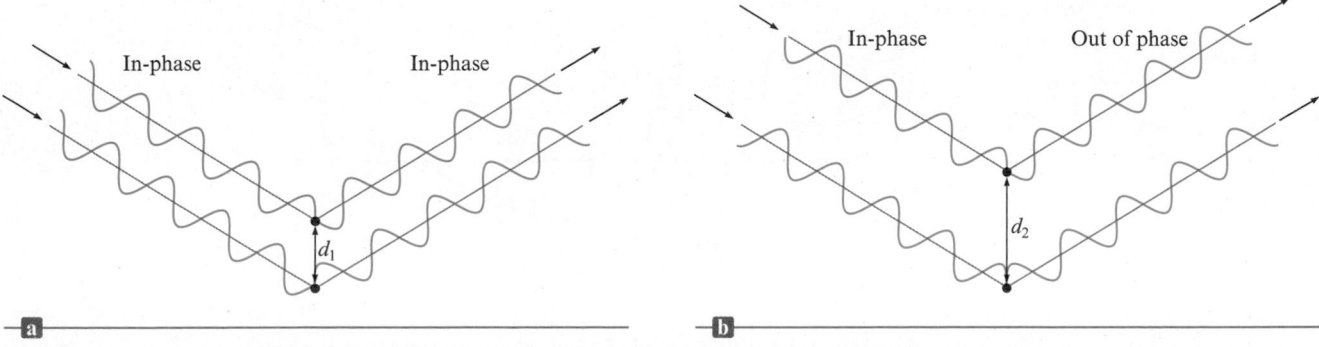

Figure 9-10 | X rays scattered from two different atoms may reinforce (constructive interference) or cancel (destructive interference) one another. (a) Both the incident rays and the reflected rays are also in phase. In this case, d_1 is such that the difference in the distances traveled by the two rays is a whole number of wavelengths. (b) The incident rays are in phase but the reflected rays are exactly out of phase. In this case d_2 is such that the difference in distances traveled by the two rays is an odd number of half wavelengths.

phase before they are reflected, so if the difference in distance traveled is an *integral number of wavelengths*, the waves will still be in phase.

Since the distance traveled depends on the distance between the atoms, the diffraction pattern can be used to determine the interatomic spacings. The exact relationship can be worked out by using the diagram in Fig. 9-11, which shows two in-phase waves being reflected by atoms in two different layers in a crystal. The extra distance traveled by the lower wave is the sum of the distances *xy* and *yz*, and the waves will be in phase after reflection if

$$xy + yz = n\lambda \tag{9-1}$$

where *n* is an integer and λ is the wavelength of the X rays. Using trigonometry (see Fig. 9-11), we can show that

$$xy + yz = 2d \sin \theta \tag{9-2}$$

where *d* is the distance between the atoms and θ is the angle of incidence and reflection. Combining Equation (9-1) and Equation (9-2) gives

$$n\lambda = 2d \sin \theta \tag{9-3}$$

Equation (9-3) is called the *Bragg equation*, after William Henry Bragg (1862–1942) and his son William Lawrence Bragg (1890–1972), who shared the Nobel Prize in physics in 1915 for their pioneering work in X-ray crystallography.

A diffractometer is a computer-controlled instrument used for carrying out the X-ray analysis of crystals. ◄ It rotates the crystal with respect to the X-ray beam and collects the data produced by the scattering of the X rays from the various planes of atoms in the crystal. The results are then analyzed by computer.

The techniques for crystal structure analysis have reached a level of sophistication that allows the determination of very complex structures, such as those important in biological systems. For example, the structures of several enzymes have been determined, thus enabling biochemists to understand how they perform their functions. We will explore this topic further in Chapter 11. Using X-ray diffraction, we can gather data on bond lengths and angles and, in so doing, can test the predictions of our models of molecular geometry.

▲ Student operating an X-ray diffractometer at Colgate University.

Colgate University Geology Department. Photo by Di Keller.

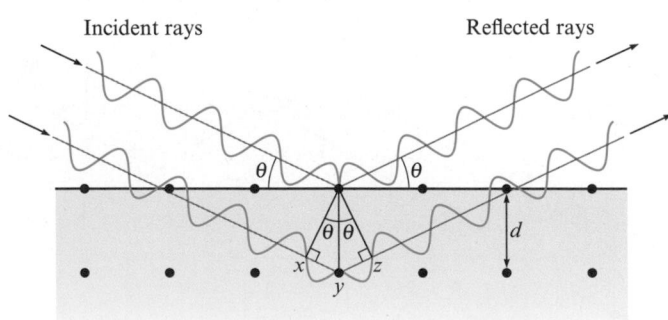

Figure 9-11 | Reflection of X rays of wavelength λ from a pair of atoms in two different layers of a crystal. The lower wave travels an extra distance equal to the sum of *xy* and *yz*. If this distance is an integral number of wavelengths (n = 1, 2, 3, . . .), the waves will reinforce each other when they exit the crystal.

Matter seems to be getting smarter these days. Increasingly, we have discovered materials that can remember their initial shape after being deformed or can sense and respond to their environment. In particular, valuable new materials have been formulated whose properties can be changed instantly by applying a magnetic or electric field.

One example of such a substance is a fluid whose flow characteristics (rheology) can be changed from free flowing to almost solid in about 0.01 second by the application of an electromagnetic field. This "magnetorheological" (MR) fluid was developed by Lord Corporation. Working in collaboration with Delphi Corporation, the company is applying the fluid in

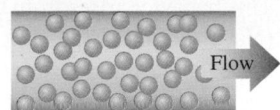

Magnetic field off
Magnetic particles
flow randomly

Magnetic field on
Applied field (*H*)
creates structure that
increases viscosity

suspension control of General Motors automobiles such as Cadillacs and Corvettes. The so-called Magneride system has sensors that monitor the road surface and provide information about what suspension damping is needed. In response, a message is instantly sent to an electromagnetic coil in the shock absorbers, which adjusts the viscosity of the MR fluid to provide continuously variable damping. The result: an amazingly smooth ride and unerring road-holding ability.

The MR fluid is composed of a synthetic oil in which particles of an iron-containing compound are suspended. When the magnetic field is turned off, these particles flow freely in all directions (see the figure above). When the field is turned on, the particles aggregate into chains that line up perpendicular to the flow of the fluid, thereby increasing its viscosity in proportion to the strength of the applied field.

Many other applications of MR fluids besides auto suspensions are under development. One very large-scale application is in Japan's National Museum of Emerging Science and Innovation, where an MR fluid is being used in dampers to protect the building against earthquake damage. Large MR-fluid dampers are also being used for stabilizing bridges such as the Dong Ting Lake Bridge in China's Hunan province to steady it in high winds.

This Corvette ZR1 uses magnetorheological fluid in its suspension system.

Frederic J. Brown/AFP/Getty Images/Newscom.com

INTERACTIVE EXAMPLE 9-1

Using the Bragg Equation

X rays of wavelength 1.54 Å were used to analyze an aluminum crystal. A reflection was produced at $\theta = 19.3$ degrees. Assuming $n = 1$, calculate the distance d between the planes of atoms producing this reflection.

Solution

To determine the distance between the planes, we use Equation (9-3) with $n = 1$, $\lambda = 1.54$ Å, and $\theta = 19.3$ degrees. Since $2d \sin \theta = n\lambda$,

$$d = \frac{n\lambda}{2 \sin \theta} = \frac{(1)(1.54 \text{ Å})}{(2)(0.3305)} = 2.33 \text{ Å} = 233 \text{ pm}$$

See Exercises 9-47 through 9-50

Types of Crystalline Solids

There are many different types of crystalline solids. For example, although both sugar and salt dissolve readily in water, the properties of the resulting solutions are quite different. The salt solution readily conducts an electric current, whereas the sugar solution does not. This behavior arises from the nature of the components in these two solids. Common salt (NaCl) is an ionic solid; it contains Na^+ and Cl^- ions. When solid sodium chloride dissolves in the polar water, sodium and chloride ions are distributed throughout the resulting solution and are free to conduct electric current. Table sugar (sucrose), on the other hand, is composed of neutral molecules that are dispersed throughout the water when the solid dissolves. No ions are present, and the resulting solution does not conduct electricity. These examples illustrate two important types of solids: **ionic solids**, represented by sodium chloride, and **molecular solids**, represented by sucrose. Ionic solids have ions at the points of the lattice that describes the structure of the solid. A molecular solid, on the other hand, has discrete covalently bonded molecules at each of its lattice points. Ice is a molecular solid that has an H_2O molecule at each point (Fig. 9-12).

A third type of solid is represented by elements such as carbon (which exists in the forms graphite, diamond, and the fullerenes), boron, silicon, and all metals. ◀ ⓘ These substances all have atoms at the lattice points that describe the structure of the solid. Therefore, we call solids of this type **atomic solids**. Examples of these three types of solids are shown in Fig. 9-12.

ⓘ Buckminsterfullerene, C_{60}, is a particular member of the fullerene family.

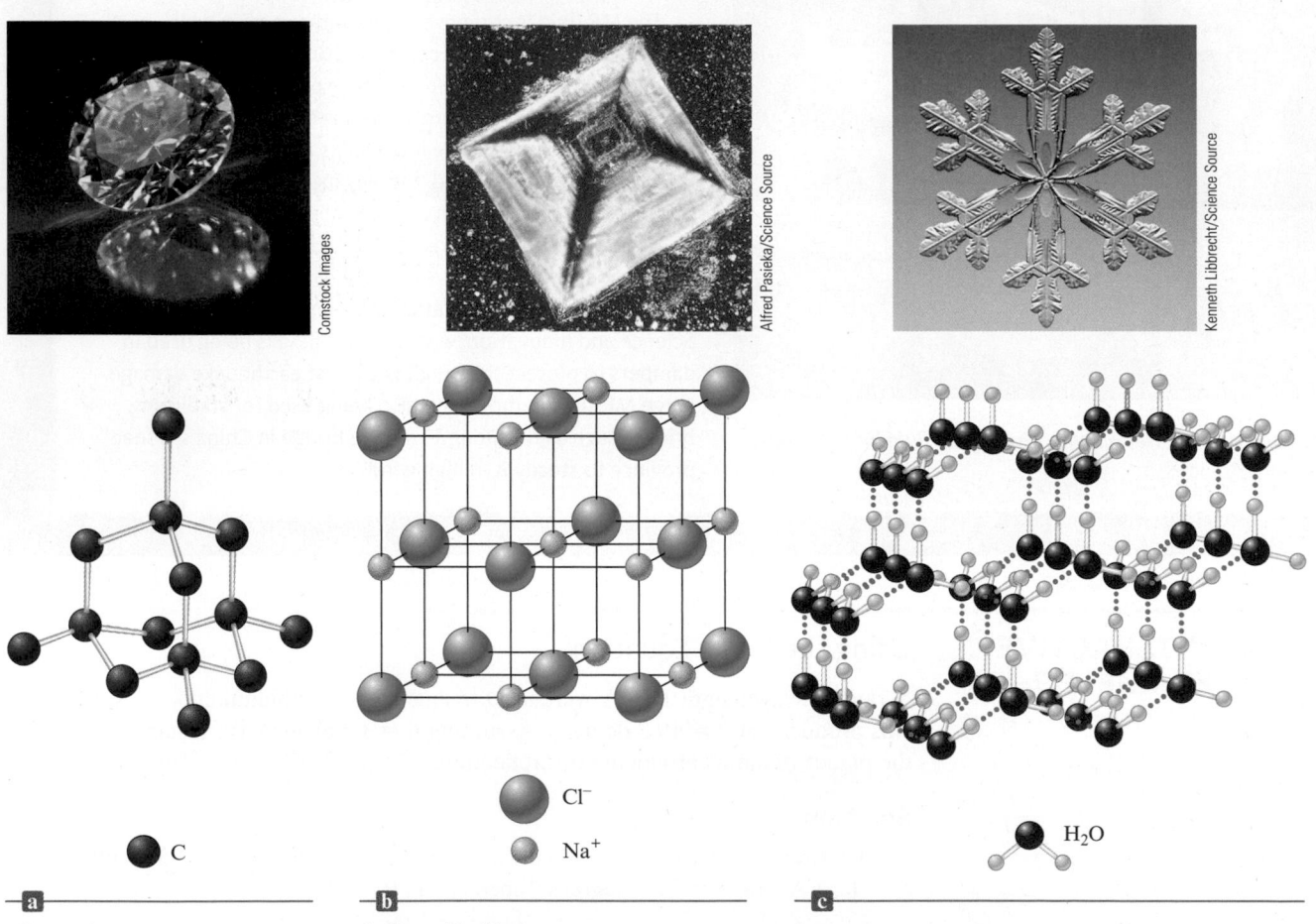

Figure 9-12 | Examples of three types of crystalline solids. Only part of the structure is shown in each case. (a) An atomic solid. (b) An ionic solid. (c) A molecular solid. The dotted lines show the hydrogen bonding interactions among the polar water molecules.

Table 9-3 | Classification of Solids

	Atomic Solids			Molecular Solids	Ionic Solids
	Metallic	Network	Group 8A (18)		
Components That Occupy the Lattice Points:	Metal atoms	Nonmetal atoms	Group 8A (18) atoms	Discrete molecules	Ions
Bonding:	Delocalized covalent	Directional covalent (leading to giant molecules)	London dispersion forces	Dipole–dipole and/or London dispersion forces	Ionic

To summarize, we find it convenient to classify solids according to what type of component occupies the lattice points. This leads to the classifications *atomic solids* (atoms at the lattice points), *molecular solids* (discrete, relatively small molecules at the lattice points), and *ionic solids* (ions at the lattice points). In addition, atomic solids are placed into the following subgroups based on the bonding that exists among the atoms in the solid: *metallic solids, network solids*, and *Group 8A (18) solids*. In metallic solids, a special type of delocalized nondirectional covalent bonding occurs. In network solids, the atoms bond to each other with strong directional covalent bonds that lead to giant molecules, or networks, of atoms. In the Group 8A (18) solids, the noble gas elements are attracted to each other with London dispersion forces. The classification of solids is summarized in Table 9-3.

ⓘ The internal forces in a solid determine the properties of the solid.

The markedly different bonding present in the various atomic solids leads to dramatically different properties for the resulting solids. ◄ ⓘ For example, although argon, copper, and diamond all are atomic solids, they have strikingly different properties. Argon (a Group 8A (18) solid) has a very low melting point ($-189°C$), whereas diamond (a network solid) and copper (a metallic solid) melt at high temperatures (about 3500 and 1083°C, respectively). Copper is an excellent conductor of electricity, whereas argon and diamond are both insulators. Copper can be easily changed in shape; it is both malleable (can be formed into thin sheets) and ductile (can be pulled into a wire). Diamond, on the other hand, is the hardest natural substance known. We will explore the structure and bonding of atomic solids in the next two sections.

9-4 | Structure and Bonding in Metals

Metals are characterized by high thermal and electrical conductivity, malleability, and ductility. As we will see, these properties can be traced to the nondirectional covalent bonding found in metallic crystals.

ⓘ The closest packing model for metallic crystals assumes that metal atoms are uniform, hard spheres.

A metallic crystal can be pictured as containing spherical atoms packed together and bonded to each other equally in all directions. We can model such a structure by packing uniform, hard spheres in a manner that most efficiently uses the available space. Such an arrangement is called **closest packing**. ◄ ⓘ The spheres are packed in layers, as shown in Fig. 9-13, in which each sphere is surrounded by six others. In the second layer the spheres do not lie directly over those in the first layer. Instead, each one occupies an indentation (or dimple) formed by three spheres in the first layer. In the third layer the spheres can occupy the dimples of the second layer in two possible ways: They can occupy positions so that each sphere in the third layer lies directly over a sphere in the first layer [the *aba* arrangement; Fig. 9-13(a)], or they can occupy positions so that no sphere in the third layer lies over one in the first layer [the *abc* arrangement; Fig. 9-13(b)].

The *aba* arrangement has the *hexagonal* unit cell shown in Fig. 9-14, and the resulting structure is called the **hexagonal closest packed (hcp) structure**. The *abc* arrangement has a *face-centered cubic* unit cell, as shown in Fig. 9-15, and the resulting structure is called the **cubic closest packed (ccp) structure**. Note that in the hcp structure the spheres in every other layer occupy the same vertical position ($ababab \cdots$), whereas in

Figure 9-13 | The closest packing arrangement of uniform spheres. In each layer a given sphere is surrounded by six others, creating six dimples, only three of which can be occupied in the next layer. (a) *aba* packing: The second layer is like the first, but it is displaced so that each sphere in the second layer occupies a dimple in the first layer. The spheres in the third layer occupy dimples in the second layer so that the spheres in the third layer lie directly over those in the first layer (*aba*). (b) *abc* packing: The spheres in the third layer occupy dimples in the second layer so that no spheres in the third layer lie above any in the first layer (*abc*). The fourth layer is like the first.

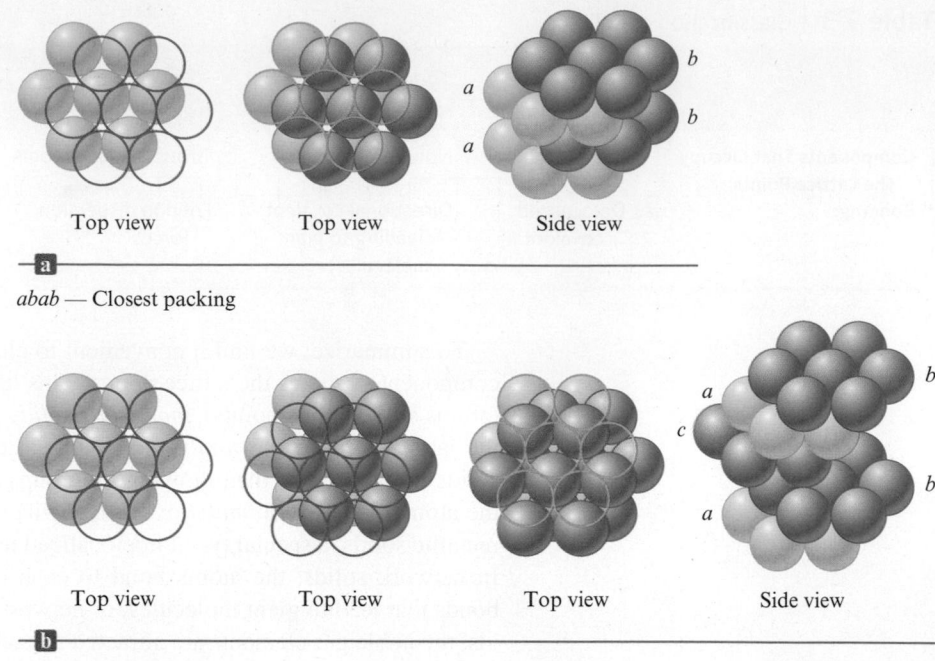

abab — Closest packing

abca — Closest packing

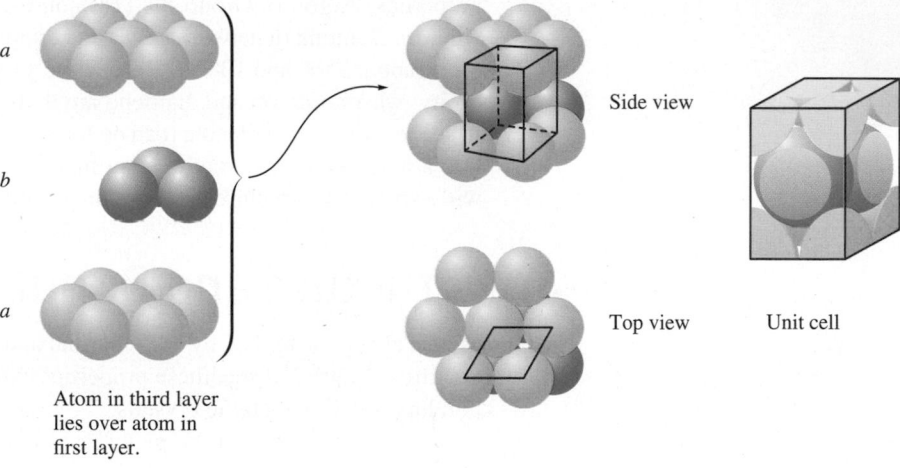

Figure 9-14 | When spheres are closest packed so that the spheres in the third layer are directly over those in the first layer (*aba*), the unit cell is the hexagonal prism illustrated here in red.

Atom in third layer lies over atom in first layer.

Side view

Top view

Unit cell

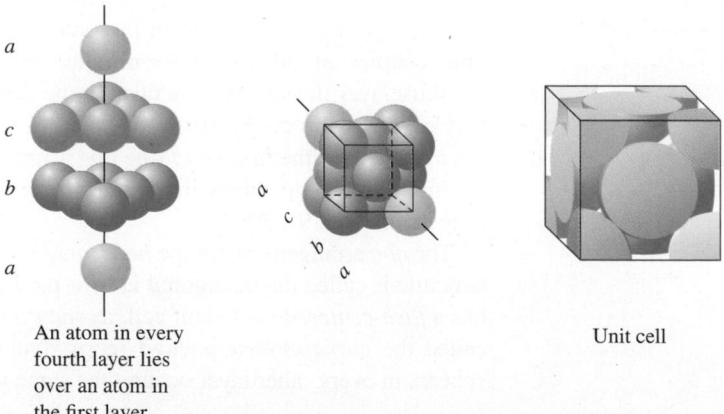

Figure 9-15 | When spheres are packed in the *abc* arrangement, the unit cell is face-centered cubic. To make the cubic arrangement easier to see, the vertical axis has been tilted as shown.

An atom in every fourth layer lies over an atom in the first layer.

Unit cell

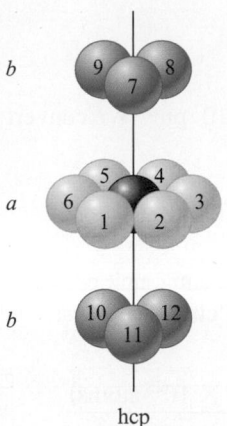

Figure 9-16 | The indicated sphere has 12 nearest neighbors.

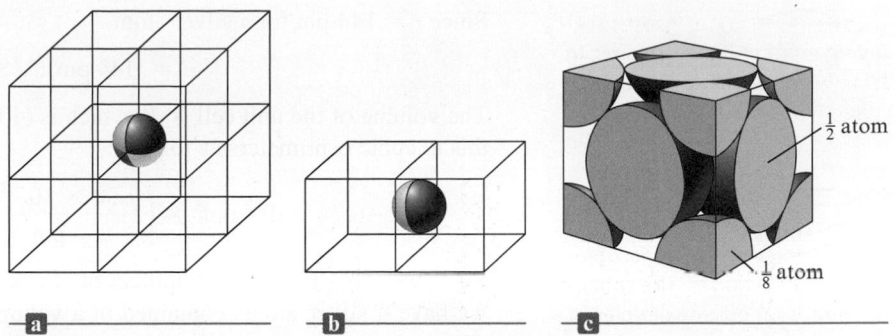

$\frac{1}{2}$ atom

$\frac{1}{8}$ atom

Figure 9-17 | The net number of spheres in a face-centered cubic unit cell. (a) Note that the sphere on a corner of the colored cell is shared with 7 other unit cells (a total of 8). Thus $\frac{1}{8}$ of such a sphere lies within a given unit cell. Since there are 8 corners in a cube, there are 8 of these $\frac{1}{8}$ pieces, or 1 net sphere. (b) The sphere on the center of each face is shared by 2 unit cells, and thus each unit cell has $\frac{1}{2}$ of each of these types of spheres. There are 6 of these $\frac{1}{2}$ spheres to give 3 net spheres. (c) Thus the face-centered cubic unit cell contains 4 net spheres (all of the pieces can be assembled to give 4 spheres).

the ccp structure the spheres in every fourth layer occupy the same vertical position ($abcabca \cdots$). A characteristic of both structures is that each sphere has 12 equivalent nearest neighbors: 6 in the same layer, 3 in the layer above, and 3 in the layer below (that form the dimples). This is illustrated for the hcp structure in Fig. 9-16.

Knowing the *net* number of spheres (atoms) in a particular unit cell is important for many applications involving solids. To illustrate how to find the net number of spheres in a unit cell, we will consider a face-centered cubic unit cell (Fig. 9-17). Note that this unit cell is defined by the *centers* of the spheres on the cube's corners. Thus 8 cubes share a given sphere, so $\frac{1}{8}$ of this sphere lies inside each unit cell. Since a cube has 8 corners, there are $8 \times \frac{1}{8}$ pieces, or enough to put together 1 whole sphere. The spheres at the center of each face are shared by 2 unit cells, so $\frac{1}{2}$ of each lies inside a particular unit cell. Since the cube has 6 faces, we have $6 \times \frac{1}{2}$ pieces, or enough to construct 3 whole spheres. Thus the net number of spheres in a face-centered cubic unit cell is

$$\left(8 \times \frac{1}{8}\right) + \left(6 \times \frac{1}{2}\right) = 4$$

INTERACTIVE EXAMPLE 9-2

▲ Crystalline silver contains cubic closest packed silver atoms.

Calculating the Density of a Closest Packed Solid

Silver crystallizes in a cubic closest packed structure. ◄ The radius of a silver atom is 144 pm. Calculate the density of solid silver.

Solution

Density is mass per unit volume. Thus we need to know how many silver atoms occupy a given volume in the crystal. The structure is cubic closest packed, which means the unit cell is face-centered cubic, as shown in the accompanying figure.

We must find the volume of this unit cell for silver and the net number of atoms it contains. Note that in this structure the atoms touch along the diagonals for each face and not along the edges of the cube. Thus the length of the diagonal is $r + 2r + r$, or $4r$. We use this fact to find the length of the edge of the cube by the Pythagorean theorem: ◄

$$d^2 + d^2 = (4r)^2$$
$$2d^2 = 16r^2$$
$$d^2 = 8r^2$$
$$d = \sqrt{8r^2} = r\sqrt{8}$$

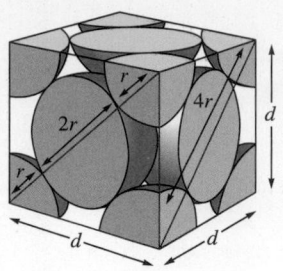

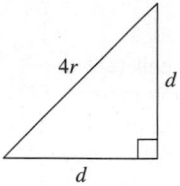

Since $r = 144$ pm for a silver atom,

$$d = (144 \text{ pm})(\sqrt{8}) = 407 \text{ pm}$$

The volume of the unit cell is d^3, which is $(407 \text{ pm})^3$, or $6.74 \times 10^7 \text{ pm}^3$. We convert this to cubic centimeters as follows:

$$6.74 \times 10^7 \text{ pm}^3 \times \left(\frac{1.00 \times 10^{-10} \text{ cm}}{\text{pm}}\right)^3 = 6.74 \times 10^{-23} \text{ cm}^3$$

Since we know that the net number of atoms in the face-centered cubic unit cell is 4, we have 4 silver atoms contained in a volume of $6.74 \times 10^{-23} \text{ cm}^3$. The density is therefore

$$\text{Density} = \frac{\text{mass}}{\text{volume}} = \frac{(4 \text{ atoms})(107.9 \text{ g/mol})(1 \text{ mol}/6.022 \times 10^{23} \text{ atoms})}{6.74 \times 10^{-23} \text{ cm}^3}$$

$$= 10.6 \text{ g/cm}^3$$

See Exercises 9-51 through 9-54

Examples of metals that form cubic closest packed solids are aluminum, iron, copper, cobalt, and nickel. Magnesium and zinc are hexagonal closest packed. Calcium and certain other metals can crystallize in either of these structures. Some metals, however, assume structures that are not closest packed. For example, the alkali metals have structures characterized by a *body-centered cubic (bcc) unit cell* (see Fig. 9-9), where the spheres touch along the body diagonal of the cube. In this structure, each sphere has 8 nearest neighbors (count the number of atoms around the atom at the center of the unit cell), as compared with 12 in the closest packed structures. Why a particular metal adopts the structure it does is not well understood.

Bonding Models for Metals

Any successful bonding model for metals must account for the typical physical properties of metals: malleability, ductility, and the efficient and uniform conduction of heat and electricity in all directions. ◀ ⓘ Although the shapes of most pure metals can be changed relatively easily, most metals are durable and have high melting points. These facts indicate that the bonding in most metals is both *strong* and *nondirectional*. That is, although it is difficult to separate metal atoms, it is relatively easy to move them, provided the atoms stay in contact with each other.

The simplest picture that explains these observations is the *electron sea model*, which envisions a regular array of metal cations in a "sea" of valence electrons (Fig. 9-18). The mobile electrons can conduct heat and electricity, and the metal ions can be easily moved around as the metal is hammered into a sheet or pulled into a wire.

ⓘ *Malleable:* can be pounded into thin sheets.

Ductile: can be drawn to form a wire.

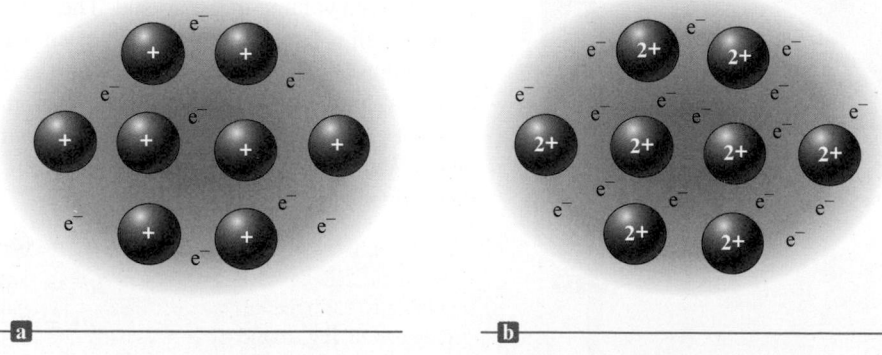

Figure 9-18 | The electron sea model for metals postulates a regular array of cations in a "sea" of valence electrons. (a) Representation of an alkali metal (Group 1A) with one valence electron. (b) Representation of an alkaline earth metal (Group 2A) with two valence electrons.

Although we usually think of scientists as dealing with eso-teric and often toxic materials, sometimes they surprise us. For example, scientists at several prestigious universities have lately shown a lot of interest in M & M candies.

To appreciate the scientists' interest in M & Ms, we must consider the importance of packing atoms, molecules, or microcrystals in understanding the structures of solids. The most efficient use of space is the closest packing of uniform spheres, where 74% of the space is occupied by the spheres and 26% of space is left unoccupied. Although the structures of most pure metals can be explained in terms of closest packing, most other substances—such as many alloys and ceramics—consist of random arrays of microscopic particles. For this reason, it is of interest to study how such objects pack in a random way.

When uniform spheres, such as marbles, are poured into a large container, the resulting random packing of the spheres results in only 64% of the space being occupied by the spheres. Thus it was very surprising when Princeton Univer-sity chemist Salvatore Torquato and his colleagues at Cornell and North Carolina Central Universities discovered that, when the ellipsoidal-shaped M & Ms are poured into a large container, the candies occupy 73.5% of the available space. In other words, the randomly packed M & Ms occupy space with almost the same efficiency as closest packed spheres do.

Why do randomly packed ellipsoids occupy space so much more efficiently than randomly packed spheres? The scien-tists speculate that because the ellipsoids can tip and rotate

in ways that spheres cannot, they can pack more closely to their neighbors.

According to Torquato, these results are important because they will help us better understand the properties of disordered materials ranging from powders to glassy solids. He also says that M & Ms make ideal test objects because they are inexpensive and uniform and "you can eat the exper-iment afterward."

Princeton University. Photo by Denise Applewhite.

Number of interacting atomic orbitals

2   4   16 6.02×10^{23}

E

A related model that gives a more detailed view of the electron energies and mo-tions is the **band model**, or **molecular orbital (MO) model**, for metals. In this model, the electrons are assumed to travel around the metal crystal in molecular orbitals formed from the valence atomic orbitals of the metal atoms (Fig. 9-19).

Recall that in the MO model for the gaseous Li_2 molecule (Section 4-5), two widely spaced molecular orbital energy levels (bonding and antibonding) result when two identical atomic orbitals interact. However, when many metal atoms interact, as in a metal crystal, the large number of resulting molecular orbitals become more closely spaced and finally form a virtual continuum of levels, called *bands*, as shown in Fig. 9-19.

Figure 9-19 | The molecular orbital energy levels produced when various numbers of atomic orbitals interact. Note that for two atomic orbitals two rather widely spaced energy levels result. (Recall the description of H_2 in Section 4-4.) As more atomic orbitals are available to form molecular orbitals, the resulting energy levels are more closely spaced, finally producing a band of very closely spaced orbitals.

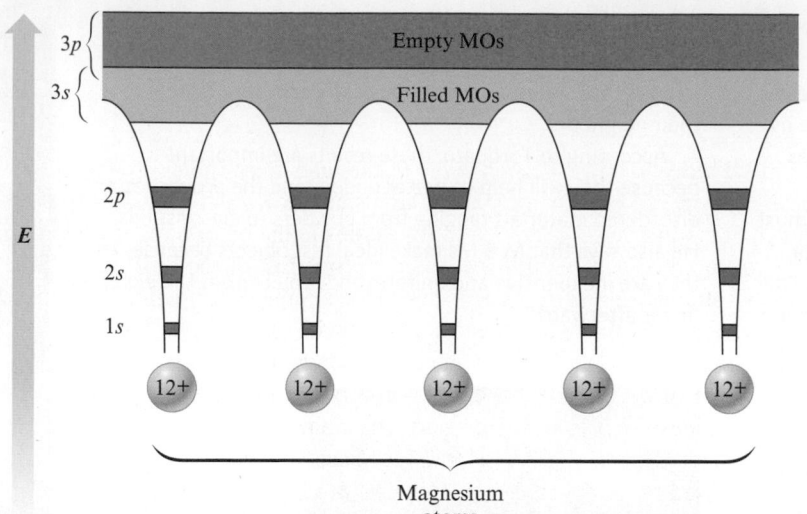

Figure 9-20 | (left) A representation of the energy levels (bands) in a magnesium crystal. The electrons in the 1s, 2s, and 2p orbitals are close to the nuclei and thus are localized on each magnesium atom as shown. However, the 3s and 3p valence orbitals overlap and mix to form molecular orbitals. Electrons in these energy levels can travel throughout the crystal. (right) Crystals of magnesium grown from a vapor.

As an illustration, picture a magnesium metal crystal, which has an hcp structure. Since each magnesium atom has one 3s and three 3p valence atomic orbitals, a crystal with n magnesium atoms has available $n(3s)$ and $3n(3p)$ orbitals to form the molecular orbitals, as illustrated in Fig. 9-20. Note that the core electrons are localized, as shown by their presence in the energy "well" around each magnesium atom. However, the valence electrons occupy closely spaced molecular orbitals, which are only partially filled.

The existence of empty molecular orbitals close in energy to filled molecular orbitals explains the thermal and electrical conductivity of metal crystals. Metals conduct electricity and heat very efficiently because of the availability of highly mobile electrons. For example, when an electric potential is placed across a strip of metal, for current to flow, electrons must be free to move. In the band model for metals, the electrons in partially filled bonds are mobile. These conduction electrons are free to travel throughout the metal crystal as dictated by the potential imposed on the metal. The molecular orbitals occupied by these conducting electrons are called *conduction bands*. These mobile electrons also account for the efficiency of the conduction of heat through metals. When one end of a metal rod is heated, the mobile electrons can rapidly transmit the thermal energy to the other end.

Metal Alloys

Because of the nature of the structure and bonding of metals, other elements can be introduced into a metallic crystal relatively easily to produce substances called alloys. An **alloy** is best defined as *a substance that contains a mixture of elements and has metallic properties*. Alloys can be conveniently classified into two types.

In a **substitutional alloy** some of the host metal atoms are *replaced* by other metal atoms of similar size. For example, in brass, approximately one-third of the atoms in the host copper metal have been replaced by zinc atoms, as shown in Fig. 9-21(a). Sterling silver (93% silver and 7% copper), pewter (85% tin, 7% copper, 6% bismuth, and 2% antimony), and plumber's solder (95% tin and 5% antimony) are other examples of substitutional alloys.

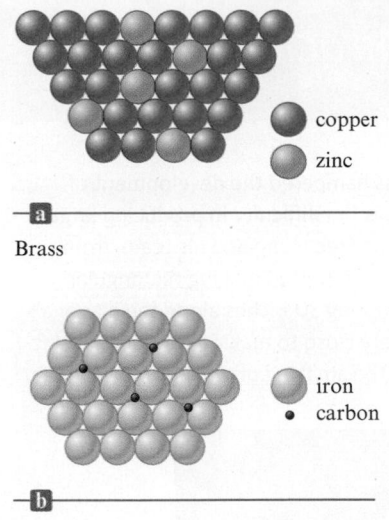

Brass

iron
• carbon

Steel

Figure 9-21 | Two types of alloys.

Table 9-4 | The Composition of the Two Brands of Steel Tubing Commonly Used to Make Lightweight Racing Bicycles

Brand of Tubing	% C	% Si	% Mn	% Mo	% Cr
Reynolds	0.25	0.25	1.3	0.20	—
Columbus	0.25	0.30	0.65	0.20	1.0

An **interstitial alloy** is formed when some of the interstices (holes) in the closest packed metal structure are occupied by small atoms, as shown in Fig. 9-21(b). Steel, the best-known interstitial alloy, contains carbon atoms in the holes of an iron crystal. The presence of the interstitial atoms changes the properties of the host metal. Pure iron is relatively soft, ductile, and malleable due to the absence of directional bonding. The spherical metal atoms can be rather easily moved with respect to each other. However, when carbon, which forms strong directional bonds, is introduced into an iron crystal, the presence of the directional carbon–iron bonds makes the resulting alloy harder, stronger, and less ductile than pure iron. The amount of carbon directly affects the properties of steel. *Mild steels*, containing less than 0.2% carbon, are ductile and malleable and are used for nails, cables, and chains. *Medium steels*, containing 0.2 to 0.6% carbon, are harder than mild steels and are used in rails and structural steel beams. *High-carbon steels*, containing 0.6 to 1.5% carbon, are tough and hard and are used for springs, tools, and cutlery.

Many types of steel also contain elements in addition to iron and carbon. Such steels are often called *alloy steels*, and they can be viewed as being mixed interstitial (carbon) and substitutional (other metals) alloys. Bicycle frames, for example, are constructed from a wide variety of alloy steels. The compositions of the two brands of steel tubing most commonly used in expensive racing bicycles are given in Table 9-4.

9-5 | Carbon and Silicon: Network Atomic Solids

Many atomic solids contain strong directional covalent bonds to form a solid that might best be viewed as a "giant molecule." We call these substances **network solids**. In contrast to metals, these materials are typically brittle and do not efficiently conduct heat or electricity. To illustrate network solids, in this section we will discuss two very important elements, carbon and silicon, and some of their compounds.

The two most common forms of carbon, diamond and graphite, are typical network solids. In diamond, the hardest naturally occurring substance, each carbon atom is surrounded by a tetrahedral arrangement of other carbon atoms to form a huge molecule [Fig. 9-22(a)]. This structure is stabilized by covalent bonds, which, in terms of the

Figure 9-22 | The structures of diamond and graphite. In each case only a small part of the entire structure is shown.

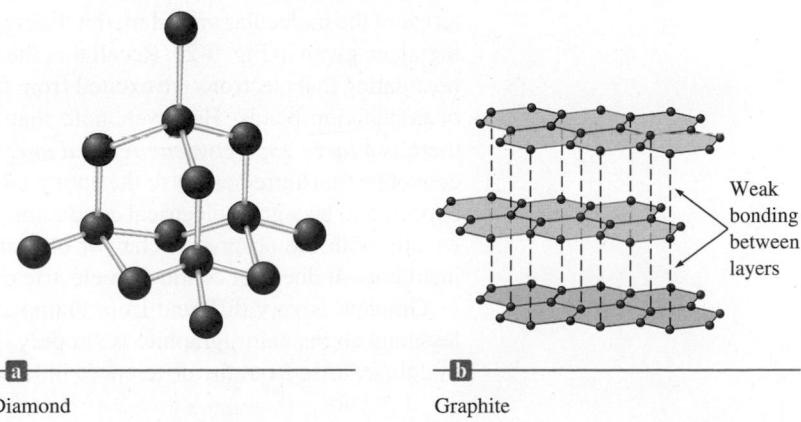

Weak bonding between layers

Diamond

Graphite

Graphite is a fascinating substance consisting of "chicken-wire" layers of carbon atoms. Due to the strong carbon–carbon bonds in each of its layers, graphite is even more thermodynamically stable than diamond. Because of the unusual stability of graphite, scientists have long wondered whether graphene (the individual chicken-wire layers) could exist independently. This question was answered only recently by Andre Gein and Konstantin Novoselov, two scientists working at the University of Manchester in the United Kingdom, when they were able to pull layers a single atom thick from graphite using Scotch tape. This work was deemed so important that Gein and Novoselov were awarded the Nobel Prize in physics in 2010.

Since the isolation of sheets of graphene by Gein and Novoselov, research in the properties and possible uses of this material has virtually exploded. Graphene has truly amazing properties. For example, it is believed to be the strongest material known, more than 100 times stronger than steel. Also, it is stiffer than diamond, yet it can be stretched like rubber.

Much of the interest in graphene centers on the fact that it is the best conductor of heat and electricity known. Electrons travel through graphene at ultrafast speeds. In fact, electrons in graphene exhibit the fractional quantum Hall effect: The electrons act collectively as if they are particles with only a fraction of the charge of an electron. Because of its exceptional thermal and electrical conductivities, graphene appears to be an ideal candidate for future electronic devices. Scientists at IBM have already built graphene-based transistors that can switch on and off 26 billion times per second, far faster than conventional silicon-based devices.

One problem that has hampered the development of graphene-based devices is the difficulty in producing large sheets of graphene. Byung Hee Hong and his team, from South Korea, have recently reported making rectangular sheets of graphene measuring 30 inches along the diagonal. It appears that we are very close to making commercial electronic devices based on the amazing properties of graphene.

Byung Hee Hong. Photo by Ji Hye Hong

A flexible electrode made of graphene.

localized electron model, are formed by the overlap of sp^3 hybridized carbon atomic orbitals.

It is also useful to consider the bonding among the carbon atoms in diamond in terms of the molecular orbital model. Energy-level diagrams for diamond and a typical metal are given in Fig. 9-23. Recall that the conductivity of metals can be explained by postulating that electrons are excited from filled levels into the very near empty levels, or conduction bands. However, note that in the energy-level diagram for diamond there is *a large gap between the filled and the empty levels*. This means that electrons cannot be transferred easily to the empty conduction bands. As a result, diamond is not expected to be a good electrical conductor. In fact, this prediction of the model agrees exactly with the observed behavior of diamond, which is known to be an electrical insulator—it does not conduct an electric current.

Graphite is very different from diamond. While diamond is hard, basically colorless, and an insulator, graphite is slippery, black, and a conductor. These differences, of course, arise from the differences in bonding in the two types of solids. In contrast

Figure 9-23 | Partial representation of the molecular orbital energies in (a) diamond and (b) a typical metal.

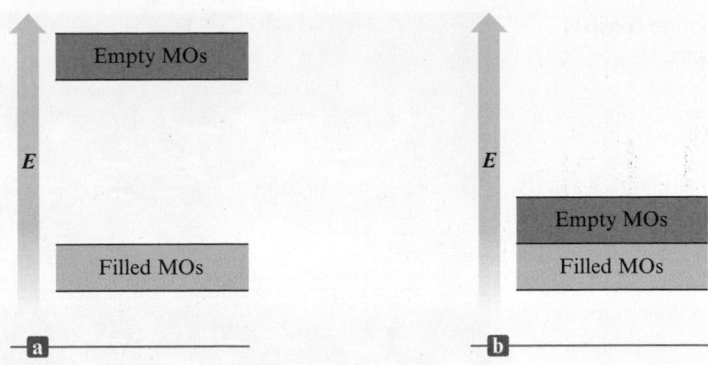

to the tetrahedral arrangement of carbon atoms in diamond, the structure of graphite is based on layers of carbon atoms arranged in fused six-membered rings, as shown in Fig. 9-22(b). Each carbon atom in a particular layer of graphite is surrounded by the three other carbon atoms in a trigonal planar arrangement with 120-degree bond angles. The localized electron model predicts sp^2 hybridization in this case. The three sp^2 orbitals on each carbon are used to form σ bonds with three other carbon atoms. One $2p$ orbital remains unhybridized on each carbon and is perpendicular to the plane of carbon atoms, as shown in Fig. 9-24. These orbitals combine to form a group of closely spaced π molecular orbitals that are important in two ways. First, they contribute significantly to the stability of the graphite layers because of the π bond formation. Second, the π molecular orbitals with their delocalized electrons account for the electrical conductivity of graphite. These closely spaced orbitals are exactly analogous to the conduction bands found in metal crystals.

Graphite is often used as a lubricant in locks (where oil is undesirable because it collects dirt). The slipperiness that is characteristic of graphite can be explained by noting that graphite has very strong bonding *within* the layers of carbon atoms but little bonding *between* the layers (the valence electrons are all used to form σ and π bonds among carbons within the layers). This arrangement allows the layers to slide past one another quite readily. Graphite's layered structure is shown in Fig. 9-25. This is in contrast to diamond, which has uniform bonding in all directions in the crystal.

Because of their extreme hardness, diamonds are used extensively in industrial cutting implements. Thus it is desirable to convert cheaper graphite to diamond. As we might expect from the higher density of diamond (3.5 g/cm^3) compared with that of

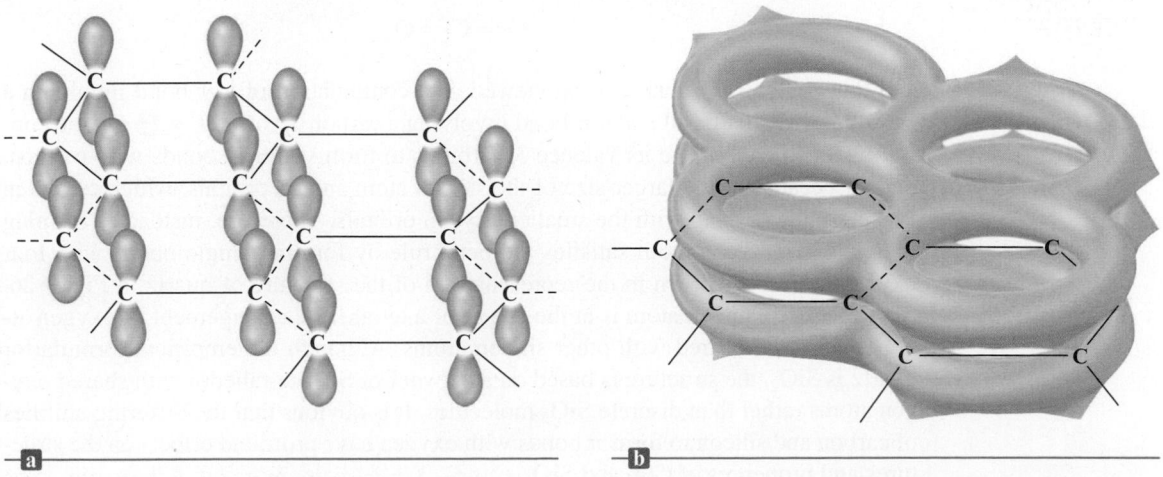

Figure 9-24 | The *p* orbitals (a) perpendicular to the plane of the carbon ring system in graphite can combine to form (b) an extensive π-bonding network.

Figure 9-25 | Graphite consists of layers of carbon atoms.

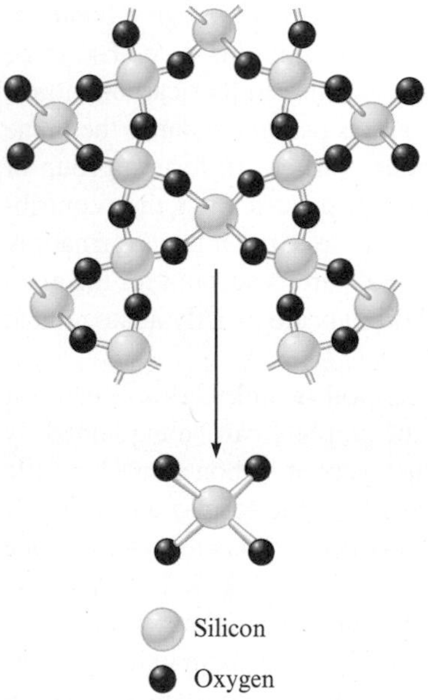

Ken Eward/Science Source

Figure 9-26 | (top) The structure of quartz (empirical formula SiO_2). Quartz contains chains of SiO_4 tetrahedra (bottom) that share oxygen atoms.

○ Silicon

● Oxygen

🛈 The bonding in the CO_2 molecule was described in Section 4-3.

graphite (2.2 g/cm^3), this transformation can be accomplished by applying very high pressures to graphite. The application of 150,000 atm of pressure at 2800°C converts graphite virtually completely to diamond. The high temperature is required to break the strong bonds in graphite so the rearrangement can occur.

Silicon is an important constituent of the compounds that make up the earth's crust. In fact, silicon is to geology as carbon is to biology. Just as carbon compounds are the basis for most biologically significant systems, silicon compounds are fundamental to most of the rocks, sands, and soils found in the earth's crust. However, although carbon and silicon are next to each other in Group 4A (14) of the periodic table, the carbon-based compounds of biology and the silicon-based compounds of geology have markedly different structures. Carbon compounds typically contain long strings of carbon–carbon bonds, whereas the most stable silicon compounds involve chains with silicon–oxygen bonds.

The fundamental silicon–oxygen compound is **silica**, which has the empirical formula SiO_2. Knowing the properties of the similar compound carbon dioxide, one might expect silica to be a gas that contains discrete SiO_2 molecules. In fact, nothing could be further from the truth—quartz and some types of sand are typical of the materials composed of silica. What accounts for this difference? The answer lies in the bonding.

Recall that the Lewis structure of CO_2 is

$$\ddot{\text{O}}\!=\!\text{C}\!=\!\ddot{\text{O}}$$

and that each C=O bond can be viewed as a combination of a σ bond involving a carbon sp hybrid orbital and a π bond involving a carbon $2p$ orbital. ◀ 🛈 On the contrary, silicon cannot use its valence $3p$ orbitals to form strong π bonds with oxygen, mainly because of the larger size of the silicon atom and its orbitals, which results in less effective overlap with the smaller oxygen orbitals. Therefore, instead of forming π bonds, the silicon atom satisfies the octet rule by forming single bonds with four oxygen atoms, as shown in the representation of the structure of quartz in Fig. 9-26. Note that each silicon atom is at the center of a tetrahedral arrangement of oxygen atoms, which are shared with other silicon atoms. Although the empirical formula for quartz is SiO_2, the structure is based on a *network* of SiO_4 tetrahedra with shared oxygen atoms rather than discrete SiO_2 molecules. It is obvious that the differing abilities of carbon and silicon to form π bonds with oxygen have profound effects on the structures and properties of CO_2 and SiO_2.

Compounds closely related to silica and found in most rocks, soils, and clays are the **silicates**. Like silica, the silicates are based on interconnected SiO_4 tetrahedra.

Figure 9-27 | Examples of silicate anions, all of which are based on SiO_4^{4-} tetrahedra.

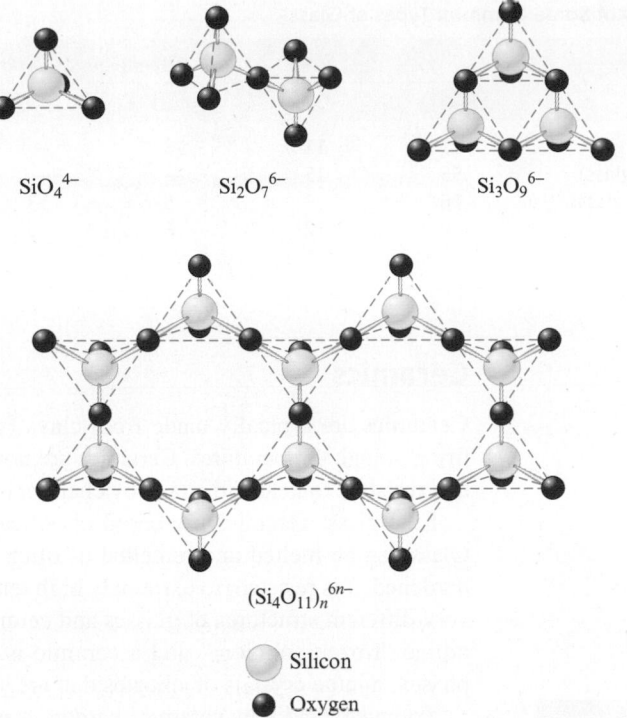

SiO_4^{4-} $Si_2O_7^{6-}$ $Si_3O_9^{6-}$

$(Si_4O_{11})_n^{6n-}$

○ Silicon
● Oxygen

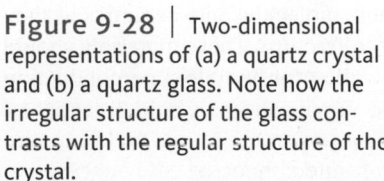

▲ A glassblower in Venice, Italy.

Plcavet/Science Source

However, in contrast to silica, where the O/Si ratio is 2:1, silicates have O/Si ratios greater than 2:1 and contain silicon–oxygen *anions*. This means that to form the neutral solid silicates, cations are needed to balance the excess negative charge. In other words, silicates are salts containing metal cations and polyatomic silicon–oxygen anions. Examples of important silicate anions are shown in Fig. 9-27.

When silica is heated above its melting point (about 1600°C) and cooled rapidly, an amorphous solid called a **glass** results (Fig. 9-28). Note that a glass contains a good deal of disorder, in contrast to the crystalline nature of quartz. Glass more closely resembles a very viscous solution than it does a crystalline solid. Common glass results when substances such as Na_2CO_3 are added to the silica melt, which is then cooled. The properties of glass can be varied greatly by varying the additives. For example, addition of B_2O_3 produces a glass (called *borosilicate glass*) that expands and contracts little under large temperature changes. ◀ Thus it is useful for labware and cooking utensils. The most common brand name for this glass is Pyrex. The addition of K_2O produces an especially hard glass that can be ground to the precise shapes needed for eyeglass and contact lenses. The compositions of several types of glass are shown in Table 9-5.

Figure 9-28 | Two-dimensional representations of (a) a quartz crystal and (b) a quartz glass. Note how the irregular structure of the glass contrasts with the regular structure of the crystal.

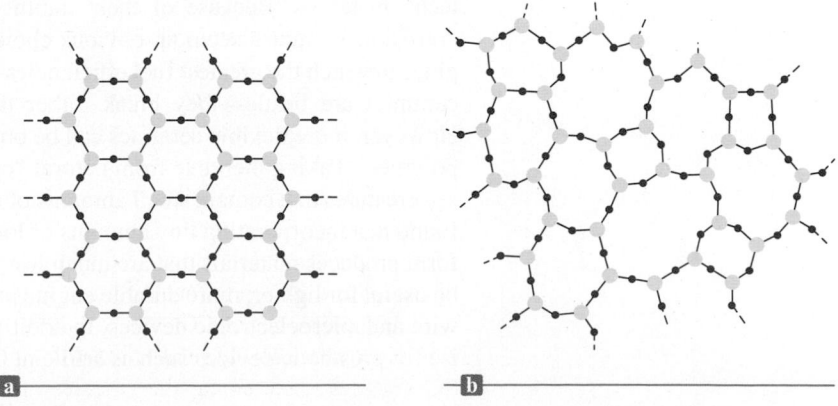

a   b

Table 9-5 | Compositions of Some Common Types of Glass

Type of Glass	Percentages of Various Components						
	SiO$_2$	CaO	Na$_2$O	B$_2$O$_3$	Al$_2$O$_3$	K$_2$O	MgO
Window (soda-lime glass)	72	11	13	—	0.3	3.8	—
Cookware (aluminosilicate glass)	55	15	—	—	20	—	10
Heat-resistant (borosilicate glass)	76	3	5	13	2	0.5	—
Optical	69	12	6	0.3	—	12	—

Ceramics

Ceramics are typically made from clays (which contain silicates) and hardened by firing at high temperatures. Ceramics are nonmetallic materials that are strong, brittle, and resistant to heat and attack by chemicals.

Like glass, ceramics are based on silicates, but with that the resemblance ends. Glass can be melted and remelted as often as desired, but once a ceramic has been hardened, it is resistant to extremely high temperatures. This behavior results from the very different structures of glasses and ceramics. A glass is a *homogeneous*, noncrystalline "frozen solution," and a ceramic is *heterogeneous*. A ceramic contains two phases: minute crystals of silicates that are suspended in a glassy cement.

To understand how ceramics harden, it is necessary to know something about the structure of clays. Clays are formed by the weathering action of water and carbon dioxide on the mineral feldspar, which is a mixture of silicates with empirical formulas such as K$_2$O · Al$_2$O$_3$ · 6SiO$_2$ and Na$_2$O · Al$_2$O$_3$ · 6SiO$_2$. Feldspar is really an *aluminosilicate* in which aluminum as well as silicon atoms are part of the oxygen-bridged polyanion. The weathering of feldspar produces kaolinite, consisting of tiny, thin platelets with the empirical formula Al$_2$Si$_2$O$_5$(OH)$_4$. When dry, the platelets cling together; when water is present, they can slide over one another, giving clay its plasticity. As clay dries, the platelets begin to interlock again. When the remaining water is driven off during firing, the silicates and cations form a glass that binds the tiny crystals of kaolinite.

▲ An artist paints a ceramic vase before glazing.

Ceramics have a very long history. Rocks, which are natural ceramic materials, served as the earliest tools. Later, clay vessels dried in the sun or baked in fires served as containers for food and water. These early vessels were no doubt crude and quite porous. With the discovery of glazing, which probably occurred about 3000 B.C. in Egypt, pottery became more serviceable as well as more beautiful. Prized porcelain is essentially the same material as crude earthenware, but specially selected clays and glazings are used for porcelain and the clay object is fired at a very high temperature. ◄

Although ceramics have been known since antiquity, they are not obsolete materials. On the contrary, ceramics constitute one of the most important classes of "high-tech" materials. Because of their stability at high temperatures and resistance to corrosion, ceramics seem an obvious choice for constructing jet and automobile engines in which the greatest fuel efficiencies are possible at very high temperatures. But ceramics are brittle—they break rather than bend—which limits their usefulness. However, more flexible ceramics can be obtained by adding small amounts of organic polymers. Taking their cue from natural "organoceramics" such as teeth and shells of sea creatures that contain small amounts of organic polymers, materials scientists have found that incorporating tiny amounts of long organic molecules into ceramics as they form produces materials that are much less subject to fracture. These materials should be useful for lighter, more durable engine parts, as well as for flexible superconducting wire and microelectronic devices. In addition, these organoceramics hold great promise for prosthetic devices such as artificial bones.

Semiconductors

Elemental silicon has the same structure as diamond, as might be expected from its position in the periodic table (in Group 4A directly under carbon). Recall that in diamond there is a large energy gap between the filled and empty molecular orbitals (see Fig. 9-23). This gap prevents excitation of electrons to the empty molecular orbitals (conduction bands) and makes diamond an insulator. In silicon the situation is similar, but the energy gap is smaller. A few electrons can cross the gap at 25°C, making silicon a **semiconducting element**, or **semiconductor**. In addition, at higher temperatures, where more energy is available to excite electrons into the conduction bands, the conductivity of silicon increases. This is typical behavior for a semiconducting element and is in contrast to that of metals, whose conductivity decreases with increasing temperature.

The small conductivity of silicon can be enhanced at normal temperatures if the silicon crystal is *doped* with certain other elements. For example, when a small fraction of silicon atoms is replaced by arsenic atoms, each having *one more* valence electron than silicon, extra electrons become available for conduction. This produces an **n-type semiconductor**, a substance whose conductivity is increased by doping it with atoms having more valence electrons than the atoms in the host crystal. These extra electrons lie close in energy to the conduction bands and can be easily excited into these levels, where they can conduct an electric current [Fig. 9-29(a)].

We also can enhance the conductivity of silicon by doping the crystal with an element such as boron, which has only three valence electrons, *one less* than silicon. Because boron has one less electron than is required to form the bonds with the surrounding silicon atoms, an electron vacancy, or *hole*, is created. As an electron fills this hole, it leaves a new hole, and this process can be repeated. Thus the hole advances through the crystal in a direction opposite to movement of the electrons jumping to fill the hole. Another way of thinking about this phenomenon is that in pure silicon each atom has four valence electrons and the low-energy molecular orbitals are exactly filled. Replacing silicon atoms with boron atoms leaves vacancies in these molecular orbitals, as shown in Fig. 9-29(b). This means that there is only one electron in some of the molecular orbitals, and these unpaired electrons can function as conducting electrons. ◄ ⓘ Thus the substance becomes a better conductor. When semiconductors are doped with atoms having fewer valence electrons than the atoms of the host crystal, they are called **p-type semiconductors**, so named because the positive holes can be viewed as the charge carriers.

> ⓘ Electrons must be in singly occupied molecular orbitals to conduct a current.

Most important applications of semiconductors involve connection of a p-type and an n-type to form a **p–n junction**. Figure 9-30(a) shows a typical junction; the red dots represent excess electrons in the n-type semiconductor, and the white circles represent holes (electron vacancies) in the p-type semiconductor. At the junction, a small

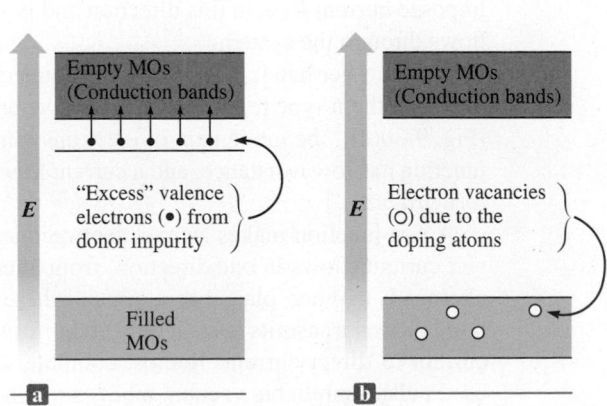

Figure 9-29 | Energy-level diagrams for (a) an n-type semiconductor and (b) a p-type semiconductor.

Figure 9-30 | The p–n junction involves the contact of a p-type and an n-type semiconductor. (a) The charge carriers of the p-type region are holes (ART). In the n-type region the charge carriers are electrons (ART). (b) No current flows (reverse bias). (c) Current readily flows (forward bias). Note that each electron that crosses the boundary leaves a hole behind. Thus the electrons and the holes move in opposite directions.

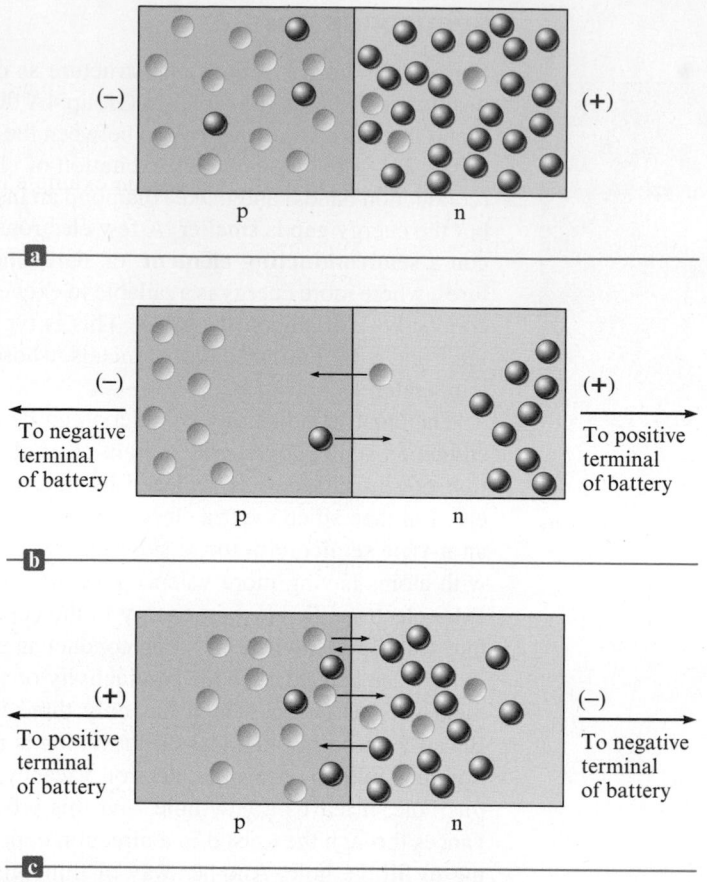

number of electrons migrate from the n-type region into the p-type region, where there are vacancies in the low-energy molecular orbitals. The effect of these migrations is to place a negative charge on the p-type region (since it now has a surplus of electrons) and a positive charge on the n-type region (since it has lost electrons, leaving holes in its low-energy molecular orbitals). This charge buildup, called the *contact potential*, or *junction potential*, prevents further migration of electrons.

Now suppose an external electric potential is applied by connecting the negative terminal of a battery to the p-type region and the positive terminal to the n-type region. The situation represented in Fig. 9-30(b) results. Electrons are drawn toward the positive terminal, and the resulting holes move toward the negative terminal—exactly opposite to the natural flow of electrons at the p–n junction. The junction resists the imposed current flow in this direction and is said to be under *reverse bias*. No current flows through the system.

On the other hand, if the battery is connected so that the negative terminal is connected to the n-type region and the positive terminal is connected to the p-type region [Fig. 9-30(c)], the movement of electrons (and holes) is in the favored direction. The junction has low resistance, and a current flows easily. The junction is said to be under *forward bias*.

A p–n junction makes an excellent *rectifier*, a device that produces a pulsating direct current (flows in one direction) from alternating current (flows in both directions alternately). When placed in a circuit where the potential is constantly reversing, a p–n junction transmits current only under forward bias, thus converting the alternating current to direct current. Radios, computers, and other electronic devices formerly used bulky, unreliable vacuum tubes as rectifiers. The p–n junction has revolutionized electronics; modern solid-state components contain p–n junctions in printed circuits.

9-6 | Molecular Solids

So far we have considered solids in which atoms occupy the lattice positions. In some of these substances (network solids), the solid can be considered to be one giant molecule. In addition, there are many types of solids that contain discrete molecular units at each lattice position. A common example is ice, where the lattice positions are occupied by water molecules [see Fig. 9-12(c)]. Other examples are dry ice (solid carbon dioxide), ◀ some forms of sulfur that contain S_8 molecules [Fig. 9-31(a)], and certain forms of phosphorus that contain P_4 molecules [Fig. 9-31(b)]. These substances are characterized by strong covalent bonding *within* the molecules but relatively weak forces *between* the molecules. For example, it takes only 6 kJ of energy to melt 1 mole of solid water (ice) because only intermolecular (H_2O—H_2O) interactions must be overcome. However, 470 kJ of energy is required to break 1 mole of covalent O—H bonds. The differences between the covalent bonds within the molecules and the forces between the molecules are apparent from the comparison of the interatomic and intermolecular distances in solids shown in Table 9-6.

The forces that exist among the molecules in a molecular solid depend on the nature of the molecules. Many molecules such as CO_2, I_2, P_4, and S_8 have no dipole moment, and the intermolecular forces are London dispersion forces. Because these forces are often relatively small, we might expect all these substances to be gaseous at 25°C, as is the case for carbon dioxide. However, as the size of the molecules increases, the London forces become quite large, causing many of these substances to be solids at 25°C.

▲ A "steaming" piece of dry ice.

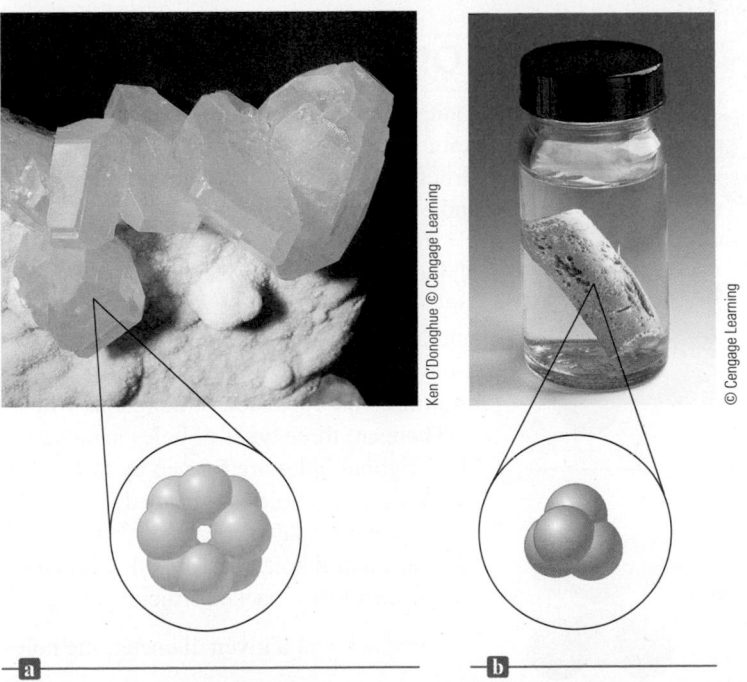

Figure 9-31 | (a) Sulfur crystals (yellow) contain S_8 molecules. (b) White phosphorus (containing P_4 molecules) is so reactive with the oxygen in air that it must be stored under water.

Table 9-6 | Comparison of Atomic Separations Within Molecules (Covalent Bonds) and Between Molecules (Intermolecular Interactions)

Solid	Distance Between Atoms in Molecule*	Closest Distance Between Molecules in the Solid
P_4	220 pm	380 pm
S_8	206 pm	370 pm
Cl_2	199 pm	360 pm

*The shorter distances within the molecules indicate stronger bonding.

When molecules do have dipole moments, their intermolecular forces are significantly greater, especially when hydrogen bonding is possible. Water molecules are particularly well suited to interact with each other because each molecule has two polar O—H bonds and two lone pairs on the oxygen atom. This can lead to the association of four hydrogen atoms with each oxygen, two by covalent bonds and two by dipole forces:

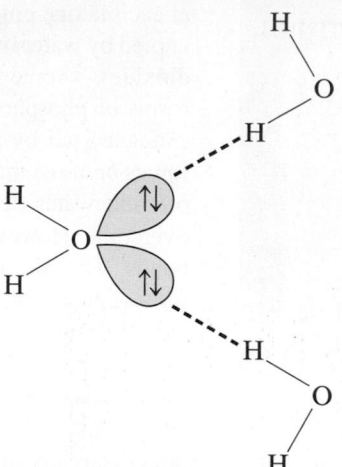

Note the two relatively short covalent oxygen–hydrogen bonds and the two longer oxygen–hydrogen dipole interactions that can be seen in the ice structure in Fig. 9-12(c).

9-7 | Ionic Solids

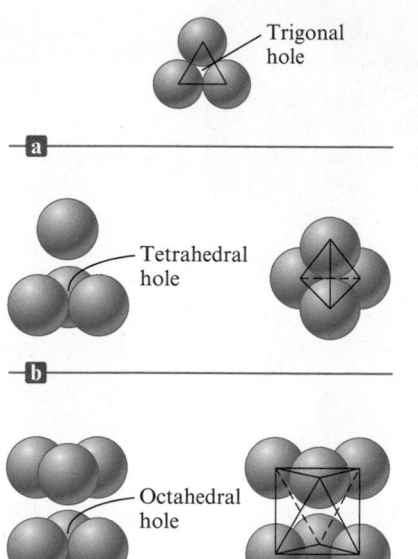

Figure 9-32 | The holes that exist among closest packed uniform spheres. (a) The trigonal hole formed by three spheres in a given plane. (b) The tetrahedral hole formed when a sphere occupies a dimple formed by three spheres in an adjacent layer. (c) The octahedral hole formed by six spheres in two adjacent layers.

Ionic solids are stable, high-melting substances held together by the strong electrostatic forces that exist between oppositely charged ions. The principles governing the structures of ionic solids were introduced in Section 3-4. In this section we will review and extend these principles.

The structures of most binary ionic solids, such as sodium chloride, can be explained by the closest packing of spheres. Typically, the larger ions, usually the anions, are packed in one of the closest packing arrangements (hcp or ccp), and the smaller cations fit into holes among the closest packed anions. The packing is done in a way that maximizes the electrostatic attractions among oppositely charged ions and minimizes the repulsions among ions with like charges.

There are three types of holes in closest packed structures:

1. Trigonal holes are formed by three spheres in the same layer [Fig. 9-32(a)].
2. Tetrahedral holes are formed when a sphere sits in the dimple of three spheres in an adjacent layer [Fig. 9-32(b)].
3. Octahedral holes are formed between two sets of three spheres in adjoining layers of the closest packed structures [Fig. 9-32(c)].

For spheres of a given diameter, the holes increase in size in the order

$$\text{trigonal} < \text{tetrahedral} < \text{octahedral}$$

In fact, trigonal holes are so small that they are never occupied in binary ionic compounds. Whether the tetrahedral or octahedral holes in a given binary ionic solid are occupied depends mainly on the *relative* sizes of the anion and cation. For example, in zinc sulfide the S^{2-} ions (ionic radius = 180 pm) are arranged in a cubic closest packed structure with the smaller Zn^{2+} ions (ionic radius = 70 pm) in the tetrahedral holes. The locations of the tetrahedral holes in the face-centered cubic unit cell of the ccp structure are shown in Fig. 9-33(a). Note from this figure that there are eight tetrahedral holes in the unit cell. Also recall from the discussion in Section 9-4 that there are four net spheres in the face-centered cubic unit cell. Thus there are *twice as many*

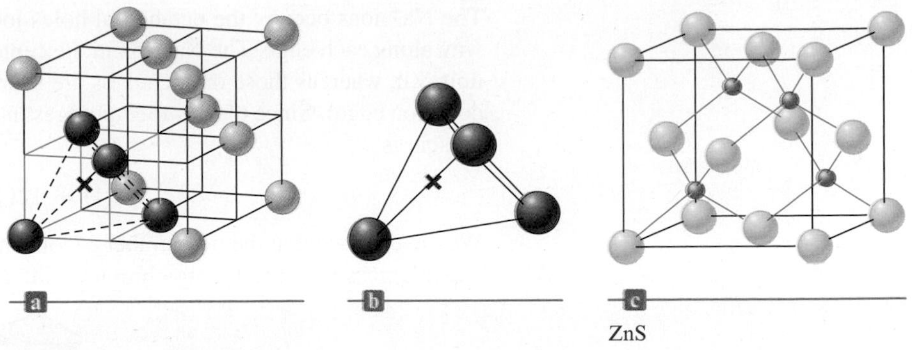

Figure 9-33 | (a) The location (red X) of a tetrahedral hole in the face-centered cubic unit cell. (b) One of the tetrahedral holes. (c) The unit cell for ZnS where the S^{2-} ions (yellow) are closest packed with the Zn^{2+} ions (red) in alternating tetrahedral holes.

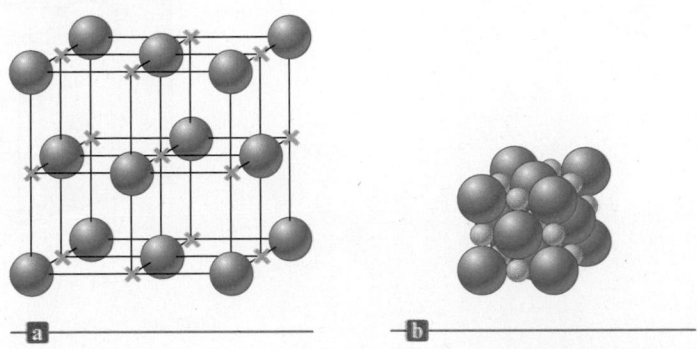

ZnS

Figure 9-34 | (a) The locations (gray X) of the octahedral holes in the face-centered cubic unit cell. (b) Representation of the unit cell for solid NaCl. The Cl^- ions (green spheres) have a ccp arrangement with Na^+ ions (gray spheres) in all the octahedral holes. Note that this representation shows the idealized closest packed structure of NaCl. In the actual structure, the Cl^- ions do not quite touch.

ⓘ Closest packed structures contain twice as many tetrahedral holes as packed spheres. Closest packed structures contain the same number of octahedral holes as packed spheres.

tetrahedral holes as packed anions in the closest packed structure. ◄ *ⓘ* Zinc sulfide must have the same number of S^{2-} ions and Zn^{2+} ions to achieve electrical neutrality. Thus in the zinc sulfide structure only *half* the tetrahedral holes contain Zn^{2+} ions, as shown in Fig. 9-33(c).

The structure of sodium chloride can be described in terms of a cubic closest packed array of Cl^- ions with Na^+ ions in all the octahedral holes. The locations of the octahedral holes in the face-centered cubic unit cell are shown in Fig. 9-34(a). The easiest octahedral hole to find in this structure is the one at the center of the cube. Note that this hole is surrounded by six spheres, as is required to form an octahedron. The remaining octahedral holes are shared with other unit cells and are more difficult to visualize. However, it can be shown that the number of octahedral holes in the ccp structure is the *same* as the number of packed anions. Figure 9-34(b) shows the structure for sodium chloride that results from Na^+ ions filling all the octahedral holes in a ccp array of Cl^- ions.

A great variety of ionic solids exists. Our purpose in this section is not to give an exhaustive treatment of ionic solids, but to emphasize the fundamental principles governing their structures. As we have seen, the most useful model for explaining the structures of these solids regards the ions as hard spheres that are packed to maximize attractions and minimize repulsions.

EXAMPLE 9-3   Determining the Number of Ions in a Unit Cell

Determine the net number of Na^+ and Cl^- ions in the sodium chloride unit cell.

Solution

Note from Fig. 9-34(b) that the Cl^- ions are cubic closest packed and thus form a face-centered cubic unit cell. There is a Cl^- ion on each corner and one at the center of each face of the cube. Thus the net number of Cl^- ions present in a unit cell is

$$8\left(\tfrac{1}{8}\right) + 6\left(\tfrac{1}{2}\right) = 4$$

The Na^+ ions occupy the octahedral holes located in the center of the cube and midway along each edge. The Na^+ ion in the center of the cube is contained entirely in the unit cell, whereas those on the edges are shared by four unit cells (four cubes share a common edge). Since the number of edges in a cube is 12, the net number of Na^+ ions present is

$$1(1) + 12(\tfrac{1}{4}) = 4$$

We have shown that the net number of ions in a unit cell is 4 Na^+ ions and 4 Cl^- ions, which agrees with the 1:1 stoichiometry of sodium chloride.

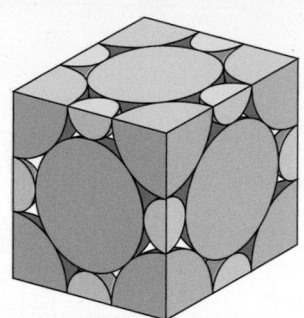

See Exercises 9-69 through 9-76

In this chapter we have considered various types of solids. Table 9-7 summarizes these types of solids and some of their properties.

Table 9-7 | Types and Properties of Solids

Type of Solid:	Atomic			Molecular	Ionic
	Network	**Metallic**	**Group 8A**	Molecule	Ion
Structural Unit:	Atom	Atom	Atom	Molecule	Ion
Type of Bonding:	Directional covalent bonds	Nondirectional covalent bonds involving electrons that are delocalized throughout the crystal	London dispersion forces	Polar molecules: dipole–dipole interactions Nonpolar molecules: London dispersion forces	Ionic
Typical Properties:	Hard High melting point	Wide range of hardness Wide range of melting points	Very low melting point	Soft Low melting point	Hard High melting point
Examples:	Insulator Diamond	Conductor Silver Iron Brass	Argon(s)	Insulator Ice (solid H_2O) Dry ice (solid CO_2)	Insulator Sodium chloride Calcium fluoride

INTERACTIVE EXAMPLE 9-4

Types of Solids

Using Table 9-7, classify each of the following substances according to the type of solid it forms.

a. Gold

b. Carbon dioxide

c. Lithium fluoride

d. Krypton

Solution

a. Solid gold is an atomic solid with metallic properties.

b. Solid carbon dioxide contains nonpolar carbon dioxide molecules and is a molecular solid.

c. Solid lithium fluoride contains Li^+ and F^- ions and is a binary ionic solid.

d. Solid krypton contains krypton atoms that can interact only through London dispersion forces. It is an atomic solid but has properties characteristic of a molecular solid with nonpolar molecules.

See Exercises 9-81 and 9-82

9-8 | The Various Forces in Substances: Summary and Comparison

In this chapter we have considered the forces that are most important in determining the properties of all substances. At this point it is useful to summarize these forces and compare the energies involved in the various types of interactions. In Table 9-8 we show the intramolecular forces (bonding forces) that hold substances together. Typically these are the strongest of the forces. In substances such as H_2O, CH_4, silica, and diamond, these forces are covalent bonds. Metals are best pictured as being held together by nondirectional covalent bonds best described by the molecular orbital model.

Table 9-8 | Intramolecular (Bonding) Forces

Type of Bonding	Nature of Attraction	Model	Range of Energies (kJ/mol)	Examples
Ionic	Interactions among ions (cations and anions)	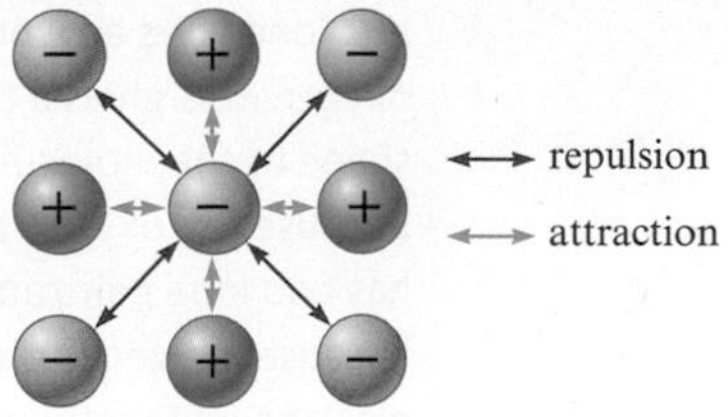 repulsion / attraction	400–4,000	LiF 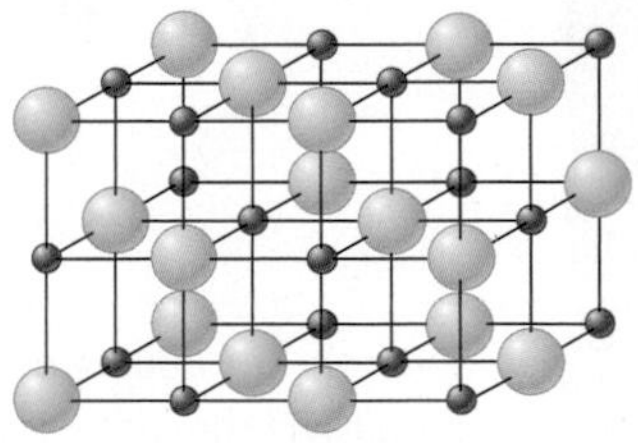
Covalent	Atoms sharing electrons	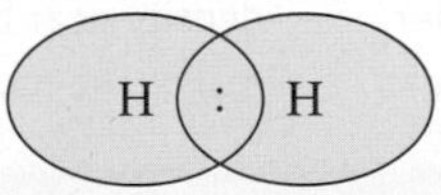 H : H	150–1,100	H_2  H H
Metallic	Atoms sharing electrons in a nondirectional way (ions in a "sea of electrons")	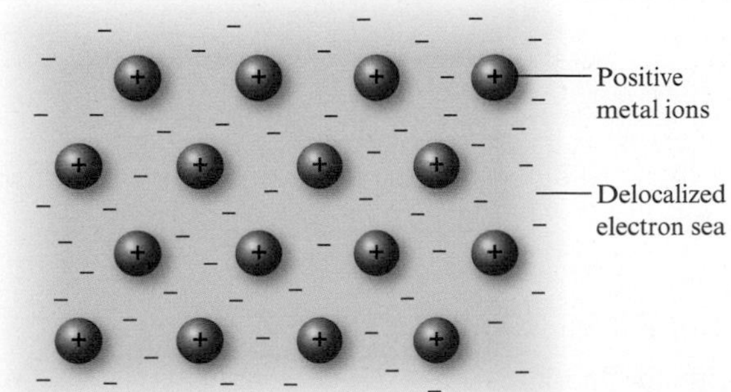 Positive metal ions / Delocalized electron sea	75–100	Au 

CONNECTING TO ATOMS 9-1 | *Intermolecular Forces, Physical Properties, and Real Gases*

Why are some substances liquids, some gases, and some solids at the same conditions of temperature and pressure? It has to do with the strength and number of intermolecular forces between the molecules (or atoms) of the substance.

In this chapter you have learned about the various intermolecular forces. In order to decide which forces apply for a given molecule, we need to decide if the molecule is polar or nonpolar. To do this we use the VSEPR theory, which means we need to start with a Lewis structure. Drawing Lewis structures requires us to be able to determine the number of valence electrons from the atoms in the molecule. So, answering our original question requires us to take an atoms first approach and to bring together a great deal of what we have been learning throughout the text.

Let's consider three molecules of similar size: propane, dimethyl ether, and ethanol, with the boiling points given beneath each substance:

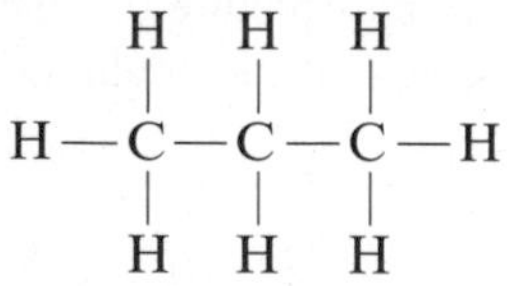

Propane
bp –42°C

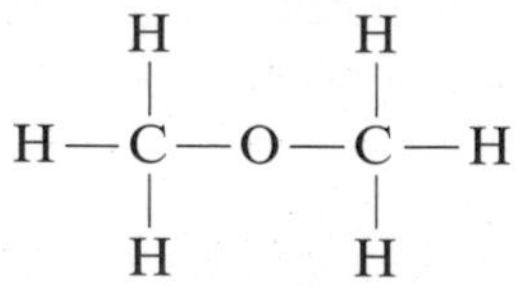

Dimethyl ether
bp –24°C

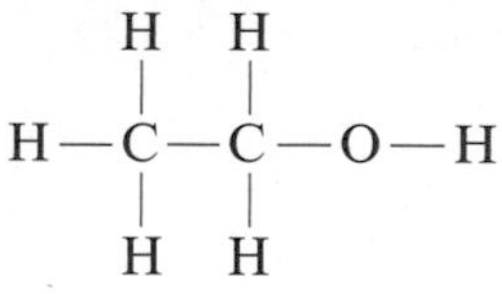

Ethanol
bp 79°C

From the boiling points we can see that propane and dimethyl ether are gases, and that propane has a lower boiling point than dimethyl ether. Ethanol is a liquid under normal conditions. Let's look at the nature of the strongest intermolecular force for each of these substances to explain these data.

First, consider propane. Propane is a hydrocarbon with a tetrahedral shape around each carbon atom. While carbon has a slightly higher electronegativity value than hydrogen, the hydrogen atoms are symmetrical around the molecule. As we discussed in Chapter 4, this symmetry makes this molecule nonpolar.

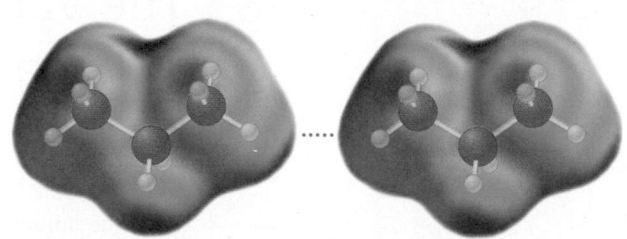

London dispersion force

[Source: Propane: http://academic .reed.edu/chemistry/roco/Potential/ standard_color_scale.html]

Because propane is nonpolar the strongest intermolecular forces are London dispersion forces, which are relatively weak in this case.

Now consider dimethyl ether. There is tetrahedral geometry around each carbon atom and the oxygen atom, but since there are two lone pairs around the oxygen, this gives a bent shape to the molecule. Because the oxygen atom has two lone pairs and because oxygen is more electronegative than carbon, the relatively high electron density near the oxygen makes the molecule polar.

dipole-dipole interaction

[Source: Dimethyl ether: http://www.mhhe.com/physsci/chemistry/ carey/student/olc/graphics/carey04oc/ref/ch16ethers.html]

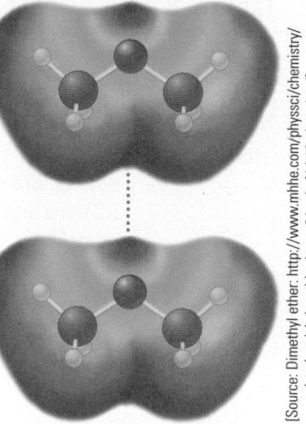

The strongest intermolecular forces for dimethyl ether are dipole–dipole interactions, which are stronger than the London dispersion forces of propane. It makes sense, then, that dimethyl ether has a higher boiling point than propane.

Finally, let's consider ethanol. Since, in this case, the oxygen is bonded to a hydrogen, a very polar bond results and this leads to hydrogen bonding.

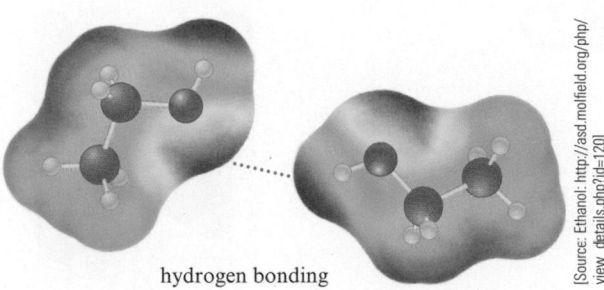

hydrogen bonding

Hydrogen bonding is the strongest of the intermolecular forces for these three molecules. It makes sense that ethanol has a higher boiling point than the other two. This explains why ethanol is a liquid under normal conditions.

We can also see how the nature of intermolecular forces affects the behavior of gases, specifically how stronger forces lead to nonideal behavior. Recall from Chapter 8 that the van der Waals equation corrects for the lowering of pressure due to the attractions among gas particles:

$$P_{ideal} = P_{observed} + a\left(\frac{n}{V}\right)^2$$

When comparing two gases, the one with the smallest value for a will behave more ideally at a given value for n and

V. If we compare the structures of methane and water, we see that methane is nonpolar but water is a polar molecule.

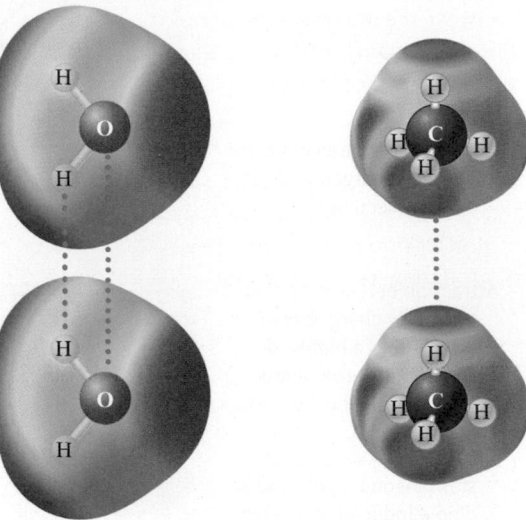

Hydrogen bonding   London dispersion force

Because methane is nonpolar, it exhibits only London dispersion forces. Water is a polar molecule with strong dipole–dipole interactions and exhibits hydrogen bonding. Therefore we would expect methane gas to be more ideal than gaseous water. Indeed, from Table 8.3 we see the value for a for methane is 2.25, while the a value for water is 5.46.

Therefore, thinking about how molecules interact helps us explain their properties.

Table 9-9 | Intermolecular Forces

Type of Interaction	Nature of Attraction	Model	Range of Energies (kJ/mol)	Examples
Ion–dipole	Attraction between charge on anion and end of dipole with opposite partial charge	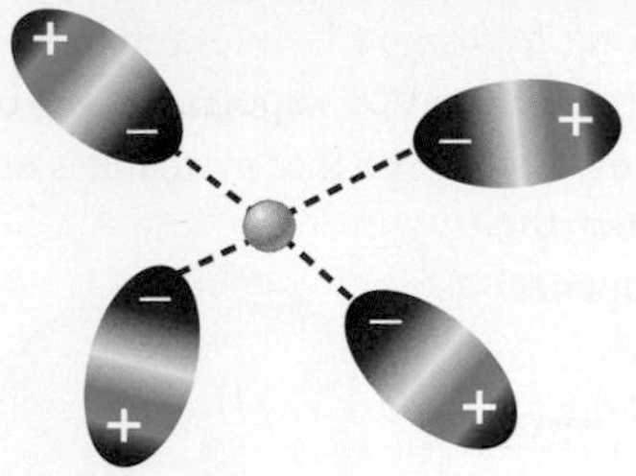	35–400	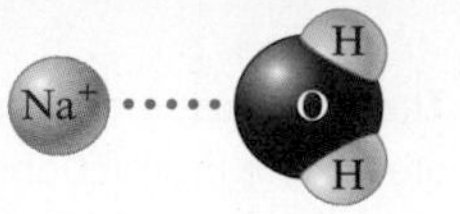
Dipole–dipole	Attraction between opposite partially charged ends of polar molecules		5–30	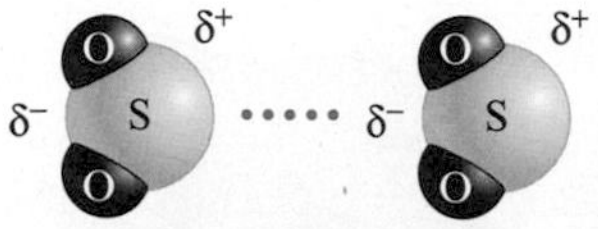
Hydrogen bond	Attraction between partially positively charged H atom attached to a highly electronegative atom and a lone pair on another atom	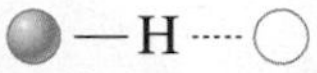 where ● is N, O, or F and ○ is an atom with a lone pair of electrons	10–40	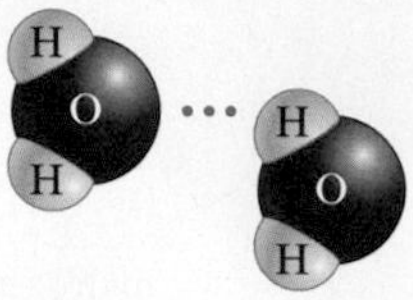
London dispersion	Attraction between an instantaneous dipole and a dipole induced in a neighboring atom or molecule	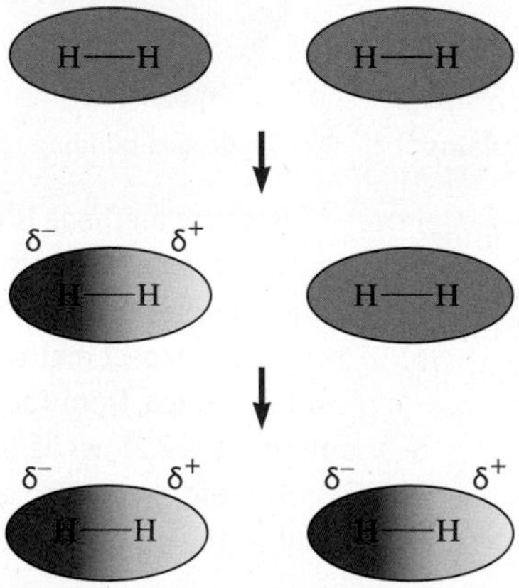	<1–40	

In ionic substances these interactions might better be called *interparticle forces*, where the interactions involve oppositely charged ions.

In Table 9-9 we summarize and compare forces that are typically called *intermolecular interactions*. These usually are weaker than bonding forces but are very important in determining the properties of substances.

9-9 | Vapor Pressure and Changes of State

Now that we have considered the general properties of the three states of matter, we can explore the processes by which matter changes state. One very familiar example of a change in state occurs when a liquid evaporates from an open container. This is clear evidence that the molecules of a liquid can escape the liquid's surface and form a gas, a process called **vaporization**, or **evaporation**. ◄ ⓘ Vaporization is endothermic because energy is required to overcome the relatively strong intermolecular forces in the liquid. The energy required to vaporize 1 mole of a liquid at a pressure of 1 atm

ⓘ *Vapor* is the usual term for the gas phase of a substance that exists as a solid or liquid at 25°C and 1 atm.

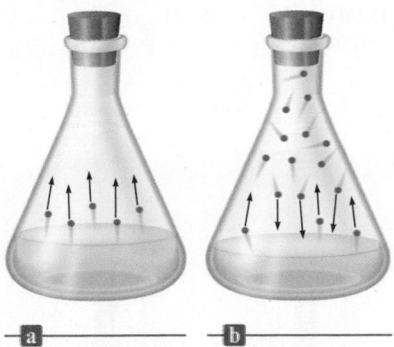

$\boldsymbol{i}$ ΔH_{vap} for water at 100°C is 40.7 kJ/mol.

Figure 9-35 | Behavior of a liquid in a closed container. (a) Initially, net evaporation occurs as molecules are transferred from the liquid to the vapor phase, so the amount of liquid decreases. (b) As the number of vapor molecules increases, the rate of return to the liquid (condensation) increases, until finally the rate of condensation equals the rate of evaporation. The system is at equilibrium, and no further changes occur in the amounts of vapor or liquid.

$\boldsymbol{i}$ A system at equilibrium is dynamic on the molecular level but shows no macroscopic changes.

Figure 9-36 | The rates of condensation and evaporation over time for a liquid sealed in a closed container. The rate of evaporation remains constant and the rate of condensation increases as the number of molecules in the vapor phase increases, until the two rates become equal. At this point, the equilibrium vapor pressure is attained.

is called the **heat of vaporization**, or the **enthalpy of vaporization**, and is usually symbolized as ΔH_{vap}. ◄ $\boldsymbol{i}$

The endothermic nature of vaporization has great practical significance; in fact, one of the most important roles that water plays in our world is to act as a coolant. Because of the strong hydrogen bonding among its molecules in the liquid state, water has an unusually large heat of vaporization (40.7 kJ/mol). A significant portion of the sun's energy that reaches the earth is spent evaporating water from the oceans, lakes, and rivers rather than warming the earth. The vaporization of water is also crucial to the body's temperature-control system through evaporation of perspiration.

Vapor Pressure

When a liquid is placed in a closed container, the amount of liquid at first decreases but eventually becomes constant. The decrease occurs because there is an initial net transfer of molecules from the liquid to the vapor phase (Fig. 9-35). This evaporation process occurs at a constant rate at a given temperature (Fig. 9-36). However, the reverse process is different. Initially, as the number of vapor molecules increases, so does the rate of return of these molecules to the liquid. The process by which vapor molecules re-form a liquid is called **condensation**. Eventually, enough vapor molecules are present above the liquid so that the rate of condensation equals the rate of evaporation (see Fig. 9-36). *At this point no further net change occurs in the amount of liquid or vapor because the two opposite processes exactly balance each other*; the system is at **equilibrium**. Note that this system is highly *dynamic* on the molecular level—molecules are constantly escaping from and entering the liquid at a high rate. However, there is no *net* change because the two opposite processes just *balance* each other. ◄ $\boldsymbol{i}$

The pressure of the vapor present at equilibrium is called the **equilibrium vapor pressure**, or more commonly, the **vapor pressure** of the liquid. A simple barometer can measure the vapor pressure of a liquid, as shown in Fig. 9-37(a). The liquid is injected at the bottom of the tube of mercury and floats to the surface because the mercury is so dense. A portion of the liquid evaporates at the top of the column, producing a vapor whose pressure pushes some mercury out of the tube. When the system reaches equilibrium, the vapor pressure can be determined from the change in the height of the mercury column since

$$P_{atmosphere} = P_{vapor} + P_{Hg\ column}$$

Thus
$$P_{vapor} = P_{atmosphere} - P_{Hg\ column}$$

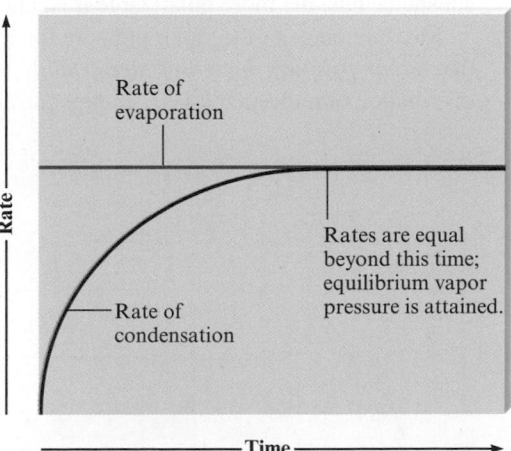

Figure 9-37 | (a) The approximate vapor pressure of a liquid can be measured easily using a simple barometer of the type shown here. (b) The three liquids, water, ethanol (C₂H₅OH), and diethyl ether [(C₂H₅)₂O], have quite different vapor pressures. Ether is by far the most volatile of the three. Note that in each case a little liquid remains (floating on the mercury).

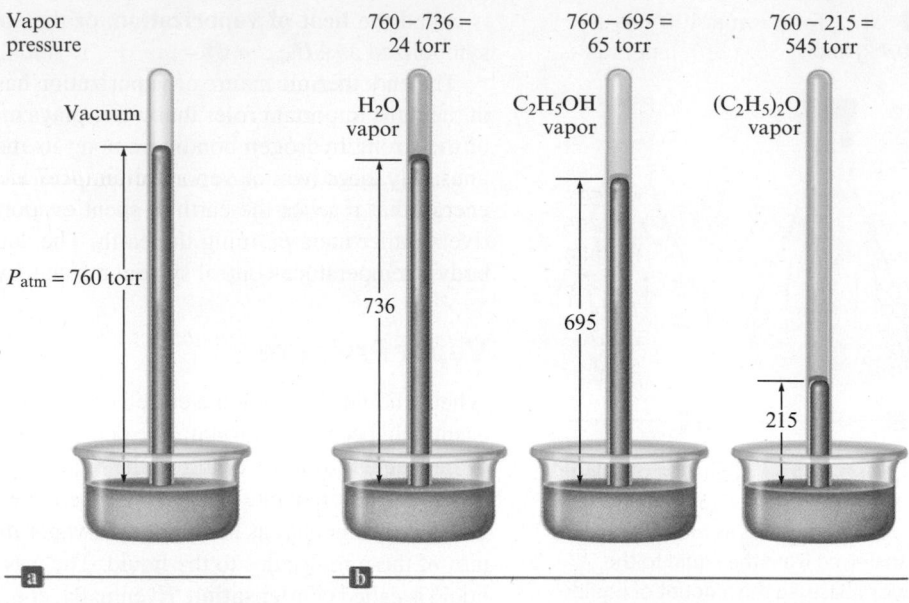

Critical Thinking

You have seen that the water molecule has a bent shape and therefore is a polar molecule. This accounts for many of water's interesting properties. What if the water molecule was linear? How would this affect the properties of water, such as its surface tension, heat of vaporization, and vapor pressure? How would life be different?

The vapor pressures of liquids vary widely [see Fig. 9-37(b)]. Liquids with high vapor pressures are said to be *volatile*—they evaporate rapidly from an open dish. The vapor pressure of a liquid is principally determined by the size of the *intermolecular forces* in the liquid. Liquids in which the intermolecular forces are large have relatively low vapor pressures because the molecules need high energies to escape to the vapor phase. For example, although water has a much lower molar mass than diethyl ether, the strong hydrogen-bonding forces that exist among water molecules in the liquid cause water's vapor pressure to be much lower than that of diethyl ether [see Fig. 9-37(b)]. In general, substances with large molar masses have relatively low vapor pressures, mainly because of the large dispersion forces. The more electrons a substance has, the more polarizable it is, and the greater the dispersion forces are.

Measurements of the vapor pressure for a given liquid at several temperatures show that *vapor pressure increases significantly with temperature.* Fig. 9-38 illustrates the distribution of molecular kinetic energy present in a liquid at two different temperatures.

Figure 9-38 | The number of molecules in a liquid with a given energy versus kinetic energy at two temperatures. Part (a) shows a lower temperature than that in part (b). Note that the proportion of molecules with enough energy to escape the liquid to the vapor phase (indicated by shaded areas) increases dramatically with temperature. This causes vapor pressure to increase markedly with temperature.

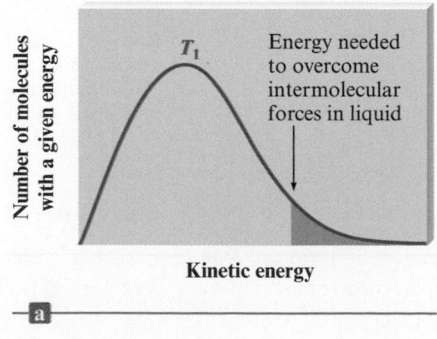

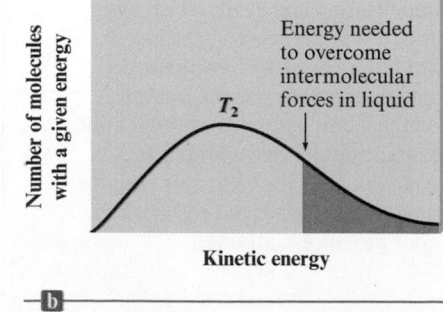

Table 9-10 | The Vapor Pressure of Water as a Function of Temperature

T (°C)	P (torr)
0.0	4.579
10.0	9.209
20.0	17.535
25.0	23.756
30.0	31.824
40.0	55.324
60.0	149.4
70.0	233.7
90.0	525.8

To overcome the intermolecular forces in a liquid, a molecule must have sufficient kinetic energy. As the temperature of the liquid is increased, the fraction of molecules having the minimum energy needed to overcome these forces and escape to the vapor phase increases markedly. Thus the vapor pressure of a liquid increases dramatically with temperature. Values for water at several temperatures are given in Table 9-10.

The quantitative nature of the temperature dependence of vapor pressure can be represented graphically. Plots of vapor pressure versus temperature for water, ethanol, and diethyl ether are shown in Fig. 9-39(a). Note the nonlinear increase in vapor pressure for all the liquids as the temperature is increased. We find that a straight line can be obtained by plotting $\ln(P_{vap})$ versus $1/T$, where T is the Kelvin temperature, as shown in Fig. 9-39(b). We can represent this behavior by the equation

$$\ln(P_{vap}) = -\frac{\Delta H_{vap}}{R}\left(\frac{1}{T}\right) + C \tag{9-4}$$

where ΔH_{vap} is the enthalpy of vaporization, R is the universal gas constant, and C is a constant characteristic of a given liquid. The symbol *ln* means that the natural logarithm of the vapor pressure is taken.

Equation (9-4) is the equation for a straight line of the form $y = mx + b$, where

$$y = \ln(P_{vap})$$

$$x = \frac{1}{T}$$

$$m = \text{slope} = -\frac{\Delta H_{vap}}{R}$$

$$b = \text{intercept} = C$$

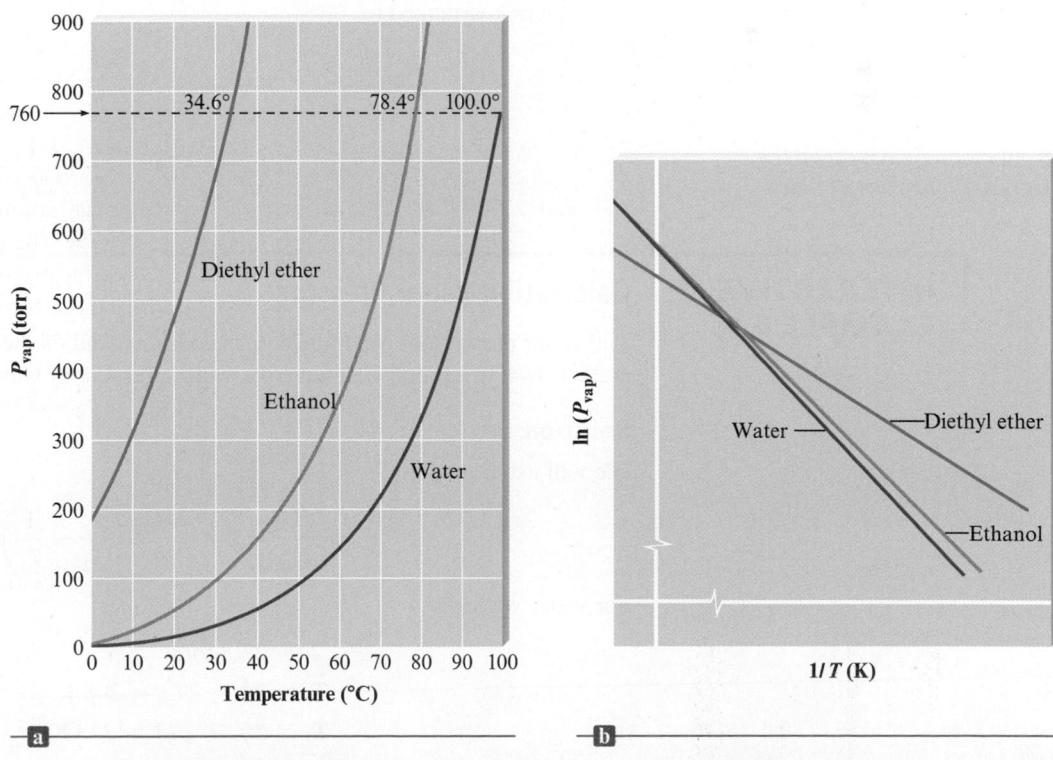

Figure 9-39 | (a) The vapor pressure of water, ethanol, and diethyl ether as a function of temperature. (b) Plots of $\ln(P_{vap})$ versus $1/T$ (Kelvin temperature) for water, ethanol, and diethyl ether.

EXAMPLE 9-5

Determining Enthalpies of Vaporization

Using the plots in Fig. 9-39(b), determine whether water or diethyl ether has the larger enthalpy of vaporization.

Solution

When $\ln(P_{vap})$ is plotted versus $1/T$, the slope of the resulting straight line is

$$-\frac{\Delta H_{vap}}{R}$$

Note from Fig. 9-39(b) that the slopes of the lines for water and diethyl ether are both negative, as expected, and that the line for ether has the smaller slope. Thus ether has the smaller value of ΔH_{vap}. This makes sense because the hydrogen bonding in water causes it to have a relatively large enthalpy of vaporization.

See Exercise 9-89

Equation (9-4) is important for several reasons. For example, we can determine the heat of vaporization for a liquid by measuring P_{vap} at several temperatures and then evaluating the slope of a plot of $\ln(P_{vap})$ versus $1/T$. On the other hand, if we know the values of ΔH_{vap} and P_{vap} at one temperature, we can use Equation (9-4) to calculate P_{vap} at another temperature. This can be done by recognizing that the constant C does not depend on temperature. Thus at two temperatures T_1 and T_2 we can solve Equation (9-4) for C and then write the equality

$$\ln(P_{vap,T_1}) + \frac{\Delta H_{vap}}{RT_1} = C = \ln(P_{vap,T_2}) + \frac{\Delta H_{vap}}{RT_2}$$

This can be rearranged to

$$\ln(P_{vap,T_1}) - \ln(P_{vap,T_2}) = \frac{\Delta H_{vap}}{R}\left(\frac{1}{T_2} - \frac{1}{T_1}\right)$$

ⓘ Equation (9-5) is called the *Clausius–Clapeyron equation.*

or

$$\ln\left(\frac{P_{vap,T_1}}{P_{vap,T_2}}\right) = \frac{\Delta H_{vap}}{R}\left(\frac{1}{T_2} - \frac{1}{T_1}\right) \qquad (9\text{-}5) \blacktriangleleft \text{ⓘ}$$

INTERACTIVE EXAMPLE 9-6

Calculating Vapor Pressure

The vapor pressure of water at 25°C is 23.8 torr, and the heat of vaporization of water at 25°C is 43.9 kJ/mol. Calculate the vapor pressure of water at 50.°C.

Solution

We will use Equation (9-5):

$$\ln\left(\frac{P_{vap,T_1}}{P_{vap,T_2}}\right) = \frac{\Delta H_{vap}}{R}\left(\frac{1}{T_2} - \frac{1}{T_1}\right)$$

For water we have

$$P_{vap,T_1} = 23.8 \text{ torr}$$
$$T_1 = 25 + 273 = 298 \text{ K}$$
$$T_2 = 50. + 273 = 323 \text{ K}$$
$$\Delta H_{vap} = 43.9 \text{ kJ/mol} = 43,900 \text{ J/mol}$$
$$R = 8.3145 \text{ J/K} \cdot \text{mol}$$

ⓘ In solving this problem, we ignore the fact that ΔH_{vap} is slightly temperature dependent.

Thus

$$\ln\left(\frac{23.8 \text{ torr}}{P_{vap,T_2}(\text{torr})}\right) = \frac{43{,}900 \text{ J/mol}}{8.3145 \text{ J/K} \cdot \text{mol}}\left(\frac{1}{323 \text{ K}} - \frac{1}{298 \text{ K}}\right)$$ ◂ ⓘ

$$\ln\left(\frac{23.8}{P_{vap,T_2}}\right) = -1.37$$

Taking the antilog (see Appendix 1-2) of both sides gives

$$\frac{23.8}{P_{vap,T_2}} = 0.254$$

$$P_{vap,T_2} = 93.7 \text{ torr}$$

See Exercises 9-91 through 9-94

Figure 9-40 | Iodine being heated, causing it to sublime, forming crystals of $I_2(s)$ on the bottom of a test tube cooled by ice.

Like liquids, solids have vapor pressures. Fig. 9-40 shows iodine vapor forming solid iodine on the bottom of a cooled dish. Under normal conditions iodine **sublimes**; that is, it goes directly from the solid to the gaseous state without passing through the liquid state. **Sublimation** also occurs with dry ice (solid carbon dioxide).

Changes of State

What happens when a solid is heated? Typically, it will melt to form a liquid. If the heating continues, the liquid will at some point boil and form the vapor phase. This process can be represented by a **heating curve**: a plot of temperature versus time for a process where energy is added at a constant rate.

The heating curve for water is given in Fig. 9-41. As energy flows into the ice, the random vibrations of the water molecules increase as the temperature rises. Eventually, the molecules become so energetic that they break loose from their lattice positions, and the change from solid to liquid occurs. This is indicated by a plateau at 0°C on the heating curve. At this temperature, called the *melting point*, all the added energy is used to disrupt the ice structure by breaking the hydrogen bonds, thus increasing the potential energy of the water molecules. The enthalpy change that occurs at the melting point when a solid melts is called the **heat of fusion**, or more accurately, the **enthalpy of fusion**, ΔH_{fus}. The melting points and enthalpies of fusion for several representative solids are listed in Table 9-11. ◂ ⓘ

The temperature remains constant until the solid has completely changed to liquid; then it begins to increase again. At 100°C the liquid water reaches its *boiling point*, and the temperature then remains constant as the added energy is used to vaporize the liquid. When the liquid is completely changed to vapor, the temperature again begins to rise. Note that changes of state are physical changes; although intermolecular forces

ⓘ Ionic solids such as NaCl and NaF have very high melting points and enthalpies of fusion because of the strong ionic forces in these solids. At the other extreme is $O_2(s)$, a molecular solid containing nonpolar molecules with weak intermolecular forces. (See Table 9-11.)

Figure 9-41 | The heating curve (not drawn to scale) for a given quantity of water where energy is added at a constant rate. The plateau at the boiling point is longer than the plateau at the melting point because it takes almost seven times more energy (and thus seven times the heating time) to vaporize liquid water than to melt ice. The slopes of the other lines are different because the different states of water have different molar heat capacities (the energy required to raise the temperature of 1 mole of a substance by 1°C).

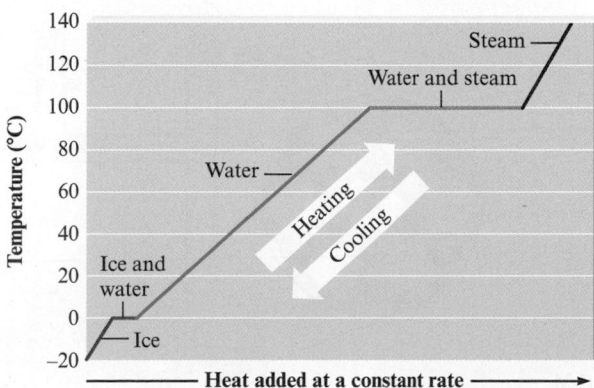

Table 9-11 | Melting Points and Enthalpies of Fusion for Several Representative Solids

Compound	Melting Point (°C)	Enthalpy of Fusion (kJ/mol)
O_2	−218	0.45
HCl	−114	1.99
HI	−51	2.87
CCl_4	−23	2.51
$CHCl_3$	−64	9.20
H_2O	0	6.02
NaF	992	29.3
NaCl	801	30.2

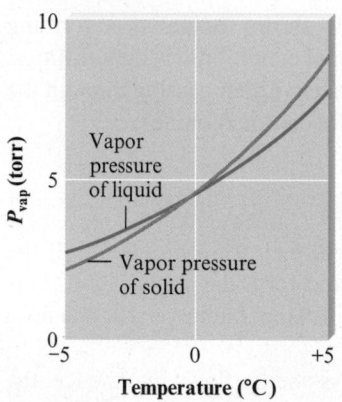

Figure 9-42 | The vapor pressures of solid and liquid water as a function of temperature. The data for liquid water below 0°C are obtained from supercooled water. The data for solid water above 0°C are estimated by extrapolation of vapor pressure from below 0°C.

have been overcome, no chemical bonds have been broken. If the water vapor were heated to much higher temperatures, the water molecules would break down into the individual atoms. This would be a chemical change, since covalent bonds are broken. We no longer have water after this occurs.

The melting and boiling points for a substance are determined by the vapor pressures of the solid and liquid states. Fig. 9-42 shows the vapor pressures of solid and liquid water as functions of temperature near 0°C. Note that below 0°C the vapor pressure of ice is less than the vapor pressure of liquid water. Also note that the vapor pressure of ice has a larger temperature dependence than that of the liquid. That is, the vapor pressure of ice increases more rapidly for a given rise in temperature than does the vapor pressure of water. Thus, as the temperature of the solid is increased, a point is eventually reached where the *liquid and solid have identical vapor pressures*. This is the melting point.

These concepts can be demonstrated experimentally using the apparatus illustrated in Fig. 9-43, where ice occupies one compartment and liquid water the other. Consider the following cases.

Case 1

A temperature at which the vapor pressure of the solid is greater than that of the liquid. At this temperature the solid requires a higher pressure than the liquid does to be in equilibrium with the vapor. Thus, as vapor is released from the solid to try to achieve equilibrium, the liquid will absorb vapor in an attempt to reduce the vapor pressure to its equilibrium value. The net effect is a conversion from solid to liquid through the vapor phase. In fact, no solid can exist under these conditions. The amount of solid

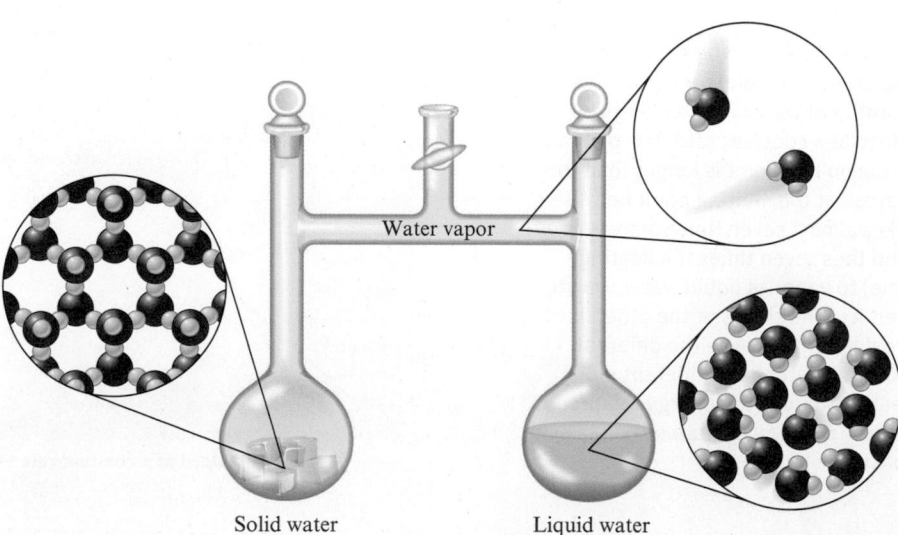

Figure 9-43 | An apparatus that allows solid and liquid water to interact only through the vapor state.

Solid water   Liquid water

will steadily decrease and the volume of liquid will increase. Finally, there will be only liquid in the right compartment, which will come to equilibrium with the water vapor, and no further changes will occur in the system. This temperature must be above the melting point of ice, since only the liquid state can exist.

Case 2

A temperature at which the vapor pressure of the solid is less than that of the liquid. This is the opposite of the situation in case 1. In this case, the liquid requires a higher pressure than the solid does to be in equilibrium with the vapor, so the liquid will gradually disappear, and the amount of ice will increase. Finally, only the solid will remain, which will achieve equilibrium with the vapor. This temperature must be *below the melting point* of ice, since only the solid state can exist.

Case 3

A temperature at which the vapor pressures of the solid and liquid are identical. In this case, the solid and liquid states have the same vapor pressure, so they can coexist in the apparatus at equilibrium simultaneously with the vapor. This temperature represents the *freezing point* where both the solid and liquid states can exist.

We can now describe the melting point of a substance more precisely. The **normal melting point** is defined as *the temperature at which the solid and liquid states have the same vapor pressure under conditions where the total pressure is 1 atmosphere.*

Boiling occurs when the vapor pressure of a liquid becomes equal to the pressure of its environment. The **normal boiling point** of a liquid is *the temperature at which the vapor pressure of the liquid is exactly 1 atmosphere.* This concept is illustrated in Fig. 9-44. At temperatures where the vapor pressure of the liquid is less than 1 atmosphere, no bubbles of vapor can form because the pressure on the surface of the liquid is greater than the pressure in any spaces in the liquid where the bubbles are trying to form. Only when the liquid reaches a temperature at which the pressure of vapor in the spaces in the liquid is 1 atmosphere can bubbles form and boiling occur.

However, changes of state do not always occur exactly at the boiling point or melting point. For example, water can be readily **supercooled**; that is, it can be cooled below 0°C at 1 atm pressure and remain in the liquid state. Supercooling occurs because, as it is cooled, the water may not achieve the degree of organization necessary to form ice at 0°C, and thus it continues to exist as the liquid. At some point the correct ordering occurs and ice rapidly forms, releasing energy in the exothermic process and bringing the temperature back up to the melting point, where the remainder of the water freezes (Fig. 9-45).

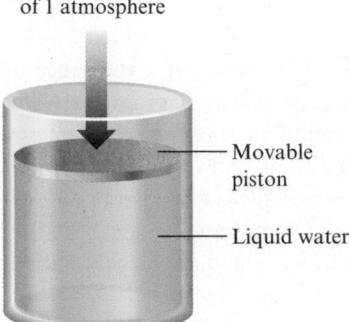

Constant pressure of 1 atmosphere

— Movable piston

— Liquid water

Figure 9-44 | Water in a closed system with a pressure of 1 atm exerted on the piston. No bubbles can form within the liquid as long as the vapor pressure is less than 1 atm.

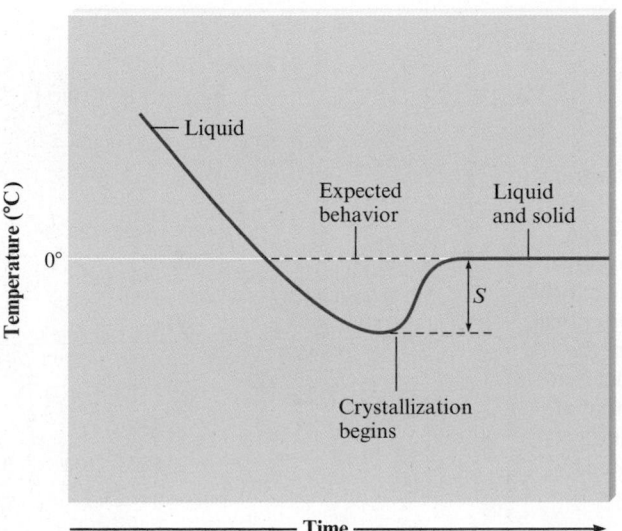

Figure 9-45 | The supercooling of water. The extent of supercooling is given by *S*.

A liquid also can be **superheated**, or raised to temperatures above its boiling point, especially if it is heated rapidly. Superheating can occur because bubble formation in the interior of the liquid requires that many high-energy molecules gather in the same vicinity, and this may not happen at the boiling point, especially if the liquid is heated rapidly. If the liquid becomes superheated, the vapor pressure in the liquid is greater than the atmospheric pressure. Once a bubble does form, since its internal pressure is greater than that of the atmosphere, it can burst before rising to the surface, blowing the surrounding liquid out of the container. This is called *bumping* and has ruined many experiments. It can be avoided by adding boiling chips to the flask containing the liquid. Boiling chips are bits of porous ceramic material containing trapped air that escapes on heating, forming tiny bubbles that act as "starters" for vapor bubble formation. This allows a smooth onset of boiling as the boiling point is reached.

9-10 | Phase Diagrams

A **phase diagram** is a convenient way of representing the phases of a substance as a function of temperature and pressure. For example, the phase diagram for water (Fig. 9-46) shows which state exists at a given temperature and pressure. It is important to recognize that a phase diagram describes conditions and events in a *closed* system of the type represented in Fig. 9-44, where no material can escape into the surroundings and no air is present. Notice that the diagram is not drawn to scale (neither axis is linear). This is done to emphasize certain features of the diagram that will be discussed below.

To show how to interpret the phase diagram for water, we will consider heating experiments at several pressures, shown by the dashed lines in Fig. 9-47.

Experiment 1

Pressure is 1 atm. This experiment begins with the cylinder shown in Fig. 9-44 completely filled with ice at a temperature of $-20°C$ and the piston exerting a pressure of 1 atm directly on the ice (there is no air space). Since at temperatures below $0°C$ the vapor pressure of ice is less than 1 atm—which is the constant external pressure on the piston—no vapor is present in the cylinder. As the cylinder is heated, ice is the only component until the temperature reaches $0°C$, where the ice changes to liquid water as energy is added. This is the normal melting point of water. Note that under these conditions no vapor exists in the system. The vapor pressures of the solid and liquid are equal, but this vapor pressure is less than 1 atm, so no water vapor can exist. This is true on the solid/liquid line everywhere except at the triple point (see Experiment 3). When the solid has completely changed to liquid, the temperature again rises. At this point,

Figure 9-46 | The phase diagram for water. T_m represents the normal melting point; T_3 and P_3 denote the triple point; T_b represents the normal boiling point; T_c represents the critical temperature; P_c represents the critical pressure. The negative slope of the solid/liquid line reflects the fact that the density of ice is less than that of liquid water. (Note that this line extends indefinitely, as indicated by the arrow.)

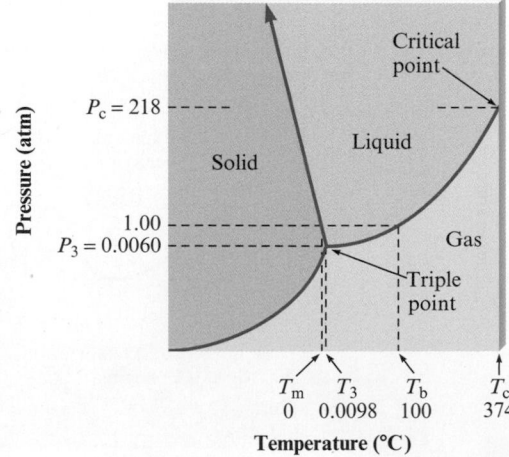

Figure 9-47 | Diagrams of various heating experiments on samples of water in a closed system.

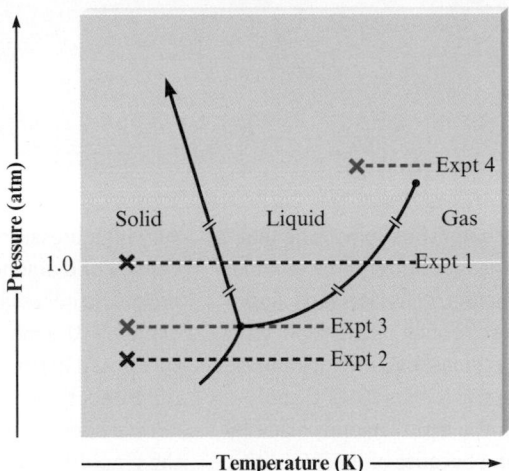

the cylinder contains only liquid water. *No vapor is present* because the vapor pressure of liquid water under these conditions is less than 1 atm, the constant external pressure on the piston. Heating continues until the temperature of the liquid water reaches 100°C. At this point, the vapor pressure of liquid water is 1 atm, and boiling occurs, with the liquid changing to vapor. This is the normal boiling point of water. After the liquid has been completely converted to steam, the temperature again rises as the heating continues. The cylinder now contains only water vapor.

Experiment 2

Pressure is 2.0 torr. Again, we start with ice as the only component in the cylinder at −20°C. The pressure exerted by the piston in this case is only 2.0 torr. As heating proceeds, the temperature rises to −10°C, where the ice changes directly to vapor, a process known as *sublimation*. Sublimation occurs when the vapor pressure of ice is equal to the external pressure, which in this case is only 2.0 torr. No liquid water appears under these conditions because the vapor pressure of liquid water is always greater than 2.0 torr, and thus it cannot exist at this pressure. If liquid water were placed in a cylinder under such a low pressure, it would vaporize immediately at temperatures above −10°C or freeze at temperatures below −10°C.

Experiment 3

Pressure is 4.58 torr. Again, we start with ice as the only component in the cylinder at −20°C. In this case the pressure exerted on the ice by the piston is 4.58 torr. As the cylinder is heated, no new phase appears until the temperature reaches 0.01°C (273.16 K). At this point, called the **triple point**, solid and liquid water have identical vapor pressures of 4.58 torr. Thus *at 0.01°C (273.16 K) and 4.58 torr all three states of water are present.* In fact, *only* under these conditions can all three states of water coexist in a closed system.

Experiment 4

Pressure is 225 atm. In this experiment we start with liquid water in the cylinder at 300°C; the pressure exerted by the piston on the water is 225 atm. Liquid water can be present at this temperature because of the high external pressure. As the temperature increases, something happens that we did not see in the first three experiments: The liquid gradually changes into a vapor but goes through an intermediate "fluid" region, which is neither true liquid nor vapor. This is quite unlike the behavior at lower temperatures and pressures, say at 100°C and 1 atm, where the temperature remains constant while a definite phase change from liquid to vapor occurs. This unusual behavior

In 1955 Robert H. Wentorf, Jr. accomplished something that borders on alchemy—he turned peanut butter into diamonds. He and his coworkers at the General Electric Research and Development Center also changed roofing pitch, wood, coal, and many other carbon-containing materials into diamonds using a process involving temperatures of $\approx 2000°C$ and pressures of $\approx 10^5$ atm. Although the first diamonds made by this process looked like black sand because of the impurities present, the process has now been developed to a point such that beautiful, clear, gem-quality diamonds can be produced. General Electric now has the capacity to produce 150 million carats (30,000 kg) of diamonds annually (virtually all of which is "diamond grit" used for industrial purposes such as abrasive coatings on cutting tools). The production of large, gem-quality diamonds by this process is still too expensive to compete with the natural sources of these stones. However, this may change as methods are developed for making diamonds at low pressures.

The high temperatures and pressures used in the GE process for making diamonds make sense if one looks at the accompanying phase diagram for carbon. Note that graphite—not diamond—is the most stable form of carbon under ordinary conditions of temperature and pressure. However, diamond becomes more stable than graphite at very high pressures (as one would expect from the greater density of diamond). The high temperature used in the GE process is necessary to disrupt the bonds in graphite so that diamond (the most stable form of carbon at the high pressures used in the process) can form. Once the diamond is produced, the elemental carbon is "trapped" in this form at normal conditions (25°C, 1 atm) because the reaction back to the graphite form is so slow. That is, even though graphite is more stable than diamond at 25°C and 1 atm, diamond can exist almost indefinitely because the conversion to graphite is a very *slow* reaction. As a result, diamonds formed at the high pressures found deep in the earth's crust can be brought to the earth's surface by natural geologic processes and continue to exist for millions of years.*

We have seen that diamond formed in the laboratory at high pressures is "trapped" in this form, but this process is

---

*In Morocco, a 50-km-long slab called Beni Bousera contains chunks of graphite that were probably once diamonds formed in the deposit when it was buried 150 km underground. As this slab slowly rose to the surface over millions of years, the very slow reaction changing diamond to graphite had time to occur. On the other hand, in the diamond-rich kimberlite deposits in South Africa, which rise to the surface much faster, the diamonds have not had sufficient time to revert to graphite.

---

occurs because the conditions are beyond the critical point for water. The **critical temperature** can be defined as the temperature above which the vapor cannot be liquefied no matter what pressure is applied. The **critical pressure** is the pressure required to produce liquefaction *at* the critical temperature. Together, the critical temperature and the critical pressure define the **critical point**. For water the critical point is 374°C and 218 atm. Note that the liquid/vapor line on the phase diagram for water ends at the critical point. Beyond this point the transition from one state to another involves the intermediate "fluid" region just described.

Applications of the Phase Diagram for Water

There are several additional interesting features of the phase diagram for water. Note that the solid/liquid boundary line has a negative slope. This means that the melting point of ice *decreases* as the external pressure *increases*. This behavior, which is opposite to that observed for most substances, occurs because the density of ice is *less* than that of liquid water at the melting point. The maximum density of water occurs at 4°C; when liquid water freezes, its volume increases.

We can account for the effect of pressure on the melting point of water using the following reasoning. At the melting point, liquid and solid water coexist—they are in dynamic equilibrium, since the rate at which ice is melting is just balanced by the rate

very expensive. Can diamond be formed at low pressures? The phase diagram for carbon says no. However, researchers have found that under the right conditions diamonds can be "grown" at low pressures. The process used is called *chemical vapor deposition* (*CVD*). CVD uses an energy source to release carbon atoms from a compound such as methane into a steady flow of hydrogen gas (some of which is dissociated to produce hydrogen atoms). The carbon atoms then deposit as a diamond film on a surface maintained at a temperature between 600 and 900°C. Why does diamond form on this surface rather than the favored graphite? Nobody is sure, but it has been suggested that at these relatively high temperatures the diamond structure grows faster than the graphite structure, so diamond is favored under these conditions. It also has been suggested that the hydrogen atoms present react much faster with graphite fragments than with diamond fragments, effectively removing any graphite from the growing film. Once it forms, of course, diamond is trapped. The major advantage of CVD is that there is no need for the extraordinarily high pressures used in the traditional process for synthesizing diamonds.

The first products with diamond films are already on the market. Audiophiles can buy tweeters that have diaphragms coated with a thin diamond film that limits sound distortion.

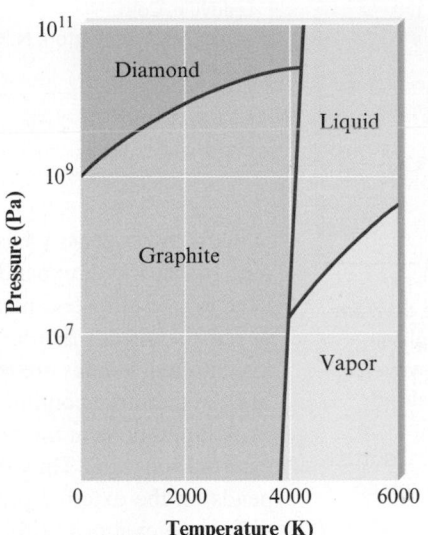

The phase diagram for carbon.

Watches with diamond-coated crystals are planned, as are diamond-coated windows in infrared scanning devices used in analytical instruments and missile-guidance systems. These applications represent only the beginning for diamond-coated products.

at which the water is freezing. What happens if we apply pressure to this system? When subjected to increased pressure, matter reduces its volume. This behavior is most dramatic for gases but also occurs for condensed states. Since a given mass of ice at 0°C has a larger volume than the same mass of liquid water, the system can reduce its volume in response to the increased pressure by changing to liquid. Thus at 0°C and an external pressure greater than 1 atm, water is liquid. In other words, the freezing point of water is less than 0°C when the pressure is greater than 1 atm.

Fig. 9-48 illustrates the effect of pressure on ice. At the point *X* on the phase diagram, ice is subjected to increased pressure at constant temperature. Note that as the pressure is increased, the solid/liquid line is crossed, indicating that the ice melts. This phenomenon may be important in ice skating. The narrow blade of the skate exerts a large pressure, since the skater's weight is supported by the small area of the blade. Also, the frictional heating due to the moving skate contributes to the melting of the ice.* After the blade passes, the liquid refreezes as normal pressure and temperature return. Without this lubrication effect due to the thawing ice, ice skating would not be the smooth, graceful activity that many people enjoy.

---

*The physics of ice skating is quite complex, and there is disagreement about whether the pressure or the frictional heating of the ice skate is most important. See "Letter to the Editor," by R. Silberman, *J. Chem. Ed.* **65** (1988): 186.

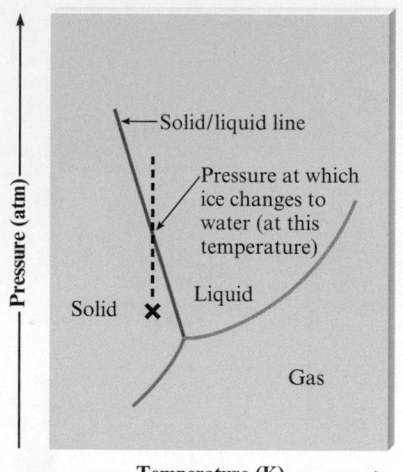

Figure 9-48 | The phase diagram for water. At point *X* on the phase diagram, water is a solid. However, as the external pressure is increased while the temperature remains constant (indicated by the vertical dotted line), the solid/liquid line is crossed and the ice melts.

ⓘ Water boils at 89°C in Leadville, Colorado.

Table 9-12 | Boiling Point of Water at Various Locations

Location	Feet Above Sea Level	P_{atm} (torr)	Boiling Point (°C)
Top of Mt. Everest, Tibet	29,028	240	70
Top of Mt. McKinley, Alaska	20,320	340	79
Top of Mt. Whitney, Calif.	14,494	430	85
Leadville, Colo.	10,150	510	89
Top of Mt. Washington, N.H.	6293	590	93
Boulder, Colo.	5430	610	94
Madison, Wis.	900	730	99
New York City, N.Y.	10	760	100
Death Valley, Calif.	−282	770	100.3

Ice's lower density has other implications. When water freezes in a pipe or an engine block, it will expand and break the container. This is why water pipes are insulated in cold climates and antifreeze is used in water-cooled engines. The lower density of ice also means that ice formed on rivers and lakes will float, providing a layer of insulation that helps prevent bodies of water from freezing solid in the winter. Aquatic life can therefore continue to live through periods of freezing temperatures.

A liquid boils at the temperature where the vapor pressure of the liquid equals the external pressure. Thus the boiling point of a substance, like the melting point, depends on the external pressure. This is why water boils at different temperatures at different elevations (Table 9-12), and any cooking carried out in boiling water will be affected by this variation. For example, it takes longer to hard-boil an egg in Leadville, Colorado (elevation: 10,150 ft), than in San Diego, California (sea level), since water boils at a lower temperature in Leadville. ◀ ⓘ

Critical Thinking

Ice is less dense than liquid water, as evidenced by the fact that ice floats in a glass of water. What if ice was more dense than liquid water? How would this affect the phase diagram for water?

As we mentioned earlier, the phase diagram for water describes a closed system. Therefore, we must be very cautious in using the phase diagram to explain the behavior of water in a natural setting, such as on the earth's surface. For example, in dry climates (low humidity), snow and ice seem to sublime—a minimum amount of slush is produced. Wet clothes put on an outside line at temperatures below 0°C freeze and then dry while frozen. However, the phase diagram (Fig. 9-44) shows that ice should *not* be able to sublime at normal atmospheric pressures. What is happening in these cases? Ice in the natural environment is not in a closed system. The pressure is provided by the atmosphere rather than by a solid piston. This means that the vapor produced over the ice can escape from the immediate region as soon as it is formed. The vapor does not come to equilibrium with the solid, and the ice slowly disappears. Sublimation, which seems forbidden by the phase diagram, does in fact occur under these conditions, although it is not the sublimation under equilibrium conditions described by the phase diagram.

The Phase Diagram for Carbon Dioxide

The phase diagram for carbon dioxide (Fig. 9-49) differs from that for water. The solid/liquid line has a positive slope, since solid carbon dioxide is more dense than liquid carbon dioxide. The triple point for carbon dioxide occurs at 5.1 atm and −56.6°C, and the critical point occurs at 72.8 atm and 31°C. At a pressure of 1 atm, solid carbon dioxide sublimes at −78°C, a property that leads to its common name,

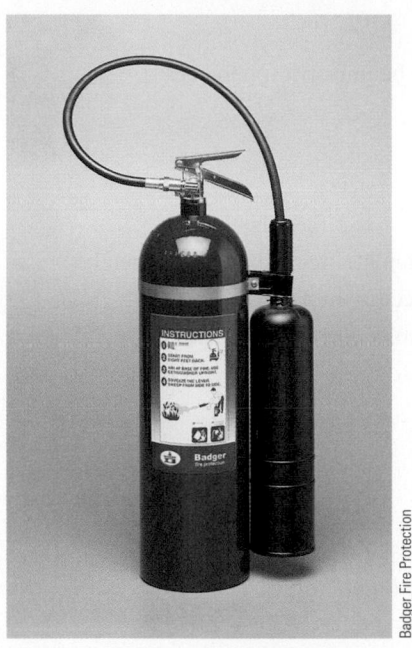

▲ A carbon dioxide fire extinguisher.

Figure 9-49 | The phase diagram for carbon dioxide. The liquid state does not exist at a pressure of 1 atm. The solid/liquid line has a positive slope, since the density of solid carbon dioxide is greater than that of liquid carbon dioxide.

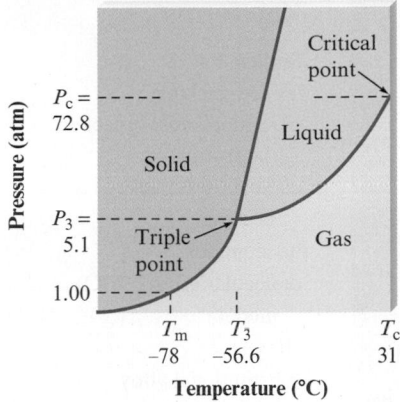

dry ice. No liquid phase occurs under normal atmospheric conditions, making dry ice a convenient refrigerant.

Carbon dioxide is often used in fire extinguishers, where it exists as a liquid at 25°C under high pressures. ◄ Liquid carbon dioxide released from the extinguisher into the environment at 1 atm immediately changes to a vapor. Being heavier than air, this vapor smothers the fire by keeping oxygen away from the flame. The liquid/vapor transition is highly endothermic, so cooling also results, which helps to put out the fire.

FOR REVIEW

Key Terms

Section 9-1
condensed states
intermolecular forces
dipole–dipole attraction
hydrogen bonding
London dispersion forces

Section 9-2
surface tension
capillary action
viscosity

Section 9-3
crystalline solid
amorphous solid
lattice
unit cell
X-ray diffraction
ionic solid
molecular solid
atomic solid

Condensed States of Matter: Liquids and Solids

- Held together by forces among the component molecules, atoms, or ions
- Liquids exhibit properties such as surface tension, capillary action, and viscosity that depend on the forces among the components

Dipole–Dipole Forces

- Attractions among molecules with dipole moments
- Hydrogen bonding is a particularly strong form of dipole–dipole attraction
  - Occurs in molecules containing hydrogen bonded to a highly electronegative element such as nitrogen, oxygen, or fluorine
  - Produces unusually high boiling points

London Dispersion Forces

- Caused by instantaneous dipoles that form in atoms or nonpolar molecules

Crystalline Solids

- Have a regular arrangement of components often represented as a lattice; the smallest repeating unit of the lattice is called the unit cell
- Classified by the types of components:
  - Atomic solids (atoms)
  - Ionic solids (ions)
  - Molecular solids (molecules)
- Arrangement of the components can be determined by X-ray analysis

Key Terms

Section 9-4

closest packing
hexagonal closest packed (hcp) structure
cubic closest packed (ccp) structure
band model
molecular orbital (MO) model
alloy
substitutional alloy
interstitial alloy

Section 9-5

network solid
silica
silicate
glass
ceramic
semiconductor
n-type semiconductor
p-type semiconductor
p–n junction

Section 9-9

vaporization (evaporation)
heat of vaporization
enthalpy of vaporization (ΔH_{vap})
condensation
equilibrium
equilibrium vapor pressure
sublimation
heating curve
enthalpy (heat) of fusion (ΔH_{fus})
normal melting point
normal boiling point
supercooled
superheated

Section 9-10

phase diagram
triple point
critical temperature
critical pressure
critical point

Metals

- Structure is modeled by assuming atoms to be uniform spheres
  - Closest packing
    - Hexagonal
    - Cubic
- Metallic bonding can be described in terms of two models
  - Electron sea model: valence electrons circulate freely among the metal cations
  - Band model: electrons are assumed to occupy molecular orbitals
    - Conduction bands: closely spaced molecular orbitals with empty electron spaces
- Alloys: mixtures with metallic properties
  - Substitutional
  - Interstitial

Network Solids

- Contain giant networks of atoms covalently bound together
- Examples are diamond and graphite
- Silicates are network solids containing Si—O—Si bridges that form the basis for many rocks, clays, and ceramics

Semiconductors

- Very pure silicon is "doped" with other elements
  - n-type: doping atoms typically contain five valence electrons (one more than silicon)
  - p-type: doping elements typically contain three valence electrons
- Modern electronics are based on devices with p–n junctions

Molecular Solids

- Components are discrete molecules
- Intermolecular forces are typically weak, leading to relatively low boiling and melting points

Ionic Solids

- Components are ions
- Interionic forces are relatively strong, leading to solids with high melting and boiling points
- Many structures consist of closest packing of the larger ions with the smaller ions in tetrahedral or octahedral holes

Phase Changes

- The change from liquid to gas (vapor) is called vaporization or evaporation
- Condensation is the reverse of vaporization
- Equilibrium vapor pressure: the pressure that occurs over a liquid or solid in a closed system when the rate of evaporation equals the rate of condensation
  - Liquids whose components have high intermolecular forces have relatively low vapor pressures
  - Normal boiling point: the temperature at which the vapor pressure of a liquid equals one atmosphere

- Normal melting point: the temperature at which a solid and its liquid have the same vapor pressure (at 1 atm external pressure)
- Phase diagram
  - Shows what state exists at a given temperature and pressure in a closed system
  - Triple point: temperature at which all three phases exist simultaneously
  - Critical point: defined by the critical temperature and pressure
    - Critical temperature: the temperature above which the vapor cannot be liquefied no matter the applied pressure
    - Critical pressure: the pressure required to produce liquefaction at the critical temperature

Review Questions
Answers to the Review Questions can be found on the Student Web site (accessible from **www.cengagebrain.com**).

1. What are intermolecular forces? How do they differ from intramolecular forces? What are dipole–dipole forces? How do typical dipole–dipole forces differ from hydrogen bonding interactions? In what ways are they similar? What are London dispersion forces? How do typical London dispersion forces differ from dipole–dipole forces? In what ways are they similar? Describe the relationship between molecular size and strength of London dispersion forces. Place the major types of intermolecular forces in order of increasing strength. Is there some overlap? That is, can the strongest London dispersion forces be greater than some dipole–dipole forces? Give an example of such an instance.

2. Define the following terms, and describe how each depends on the strength of the intermolecular forces.
   a. surface tension
   b. viscosity
   c. melting point
   d. boiling point
   e. vapor pressure

3. Compare and contrast solids, liquids, and gases.

4. Distinguish between the items in the following pairs.
   a. crystalline solid; amorphous solid
   b. ionic solid; molecular solid
   c. molecular solid; network solid
   d. metallic solid; network solid

5. What is a lattice? What is a unit cell? Describe a simple cubic unit cell. How many net atoms are contained in a simple cubic unit cell? How is the radius of the atom related to the cube edge length for a simple cubic unit cell? Answer the same questions for the body-centered cubic unit cell and for the face-centered unit cell.

6. What is closest packing? What is the difference between hexagonal closest packing and cubic closest packing? What is the unit cell for each closest packing?

7. Use the band model to describe differences among insulators, conductors, and semiconductors. Also use the band model to explain why each of the following increases the conductivity of a semiconductor.
   a. increasing the temperature
   b. irradiating with light
   c. adding an impurity

   How do conductors and semiconductors differ as to the effect of temperature on electrical conductivity? How can an n-type semiconductor be produced from pure germanium? How can a p-type semiconductor be produced from pure germanium?

8. Describe, in general, the structures of ionic solids. Compare and contrast the structure of sodium chloride and zinc sulfide. How many tetrahedral holes and octahedral holes are there per closest packed anion? In zinc sulfide, why are only one-half of the tetrahedral holes filled with cations?

9. Define each of the following.
   a. evaporation
   b. condensation
   c. sublimation
   d. boiling
   e. melting
   f. enthalpy of vaporization
   g. enthalpy of fusion
   h. heating curve

10. Why is the enthalpy of vaporization for water much greater than its enthalpy of fusion? What does this say about the changes in intermolecular forces in going from solid to liquid to vapor? What do we mean when we say that a liquid is *volatile?* Do

volatile liquids have large or small vapor pressures at room temperature? What strengths of intermolecular forces occur in highly volatile liquids?

11. Compare and contrast the phase diagrams of water and carbon dioxide. Why doesn't CO_2 have a normal melting point and a normal boiling point, whereas water does? The slopes of the solid–liquid lines in the phase diagrams of H_2O and CO_2 are different. What do the slopes of the solid–liquid lines indicate in terms of the relative densities of the solid and liquid states for each substance? How do the melting points of H_2O and CO_2 depend on pressure? How do the boiling points of H_2O and CO_2 depend on pressure? Rationalize why the critical temperature for H_2O is greater than that for CO_2.

Active Learning Questions

These questions are designed to be used by groups of students in class.

1. It is possible to balance a paper clip on the surface of water in a beaker. If you add a bit of soap to the water, however, the paper clip sinks. Explain how the paper clip can float and why it sinks when soap is added.

2. Consider a sealed container half-filled with water. Which statement best describes what occurs in the container?

   a. Water evaporates until the air is saturated with water vapor; at this point, no more water evaporates.

   b. Water evaporates until the air is overly saturated (supersaturated) with water, and most of this water recondenses; this cycle continues until a certain amount of water vapor is present, and then the cycle ceases.

   c. Water does not evaporate because the container is sealed.

   d. Water evaporates, and then water evaporates and recondenses simultaneously and continuously.

   e. Water evaporates until it is eventually all in vapor form.

   Explain each choice. Justify your choice, and for choices you did not pick, explain what is wrong with them.

3. Explain the following: You add 100 mL water to a 500-mL round-bottom flask and heat the water until it is boiling. You remove the heat and stopper the flask, and the boiling stops. You then run cool water over the neck of the flask, and the boiling begins again. It seems as though you are boiling water by cooling it.

4. Is it possible for the dispersion forces in a particular substance to be stronger than the hydrogen bonding forces in another substance? Explain your answer.

5. Does the nature of intermolecular forces change when a substance goes from a solid to a liquid, or from a liquid to a gas? What causes a substance to undergo a phase change?

6. Why do liquids have a vapor pressure? Do all liquids have vapor pressures? Explain. Do solids exhibit vapor pressure? Explain. How does vapor pressure change with changing temperature? Explain.

7. Water in an open beaker evaporates over time. As the water is evaporating, is the vapor pressure increasing, decreasing, or staying the same? Why?

8. What is the vapor pressure of water at 100°C? How do you know?

9. Refer to Fig. 9-41. Why doesn't temperature increase continuously over time? That is, why does the temperature stay constant for periods of time?

10. Which are stronger, intermolecular or intramolecular forces for a given molecule? What observation(s) have you made that support this? Explain.

11. Why does water evaporate?

A blue question or exercise number indicates that the answer to that question or exercise appears at the back of this book and a solution appears in the *Solutions Guide*, as found on PowerLecture.

Questions

12. Rationalize why chalk (calcium carbonate) has a higher melting point than motor oil (large compound made from carbon and hydrogen), which has a higher melting point than water and engages in relatively strong hydrogen bonding interactions.

13. In the diagram below, which lines represent the hydrogen bonding?

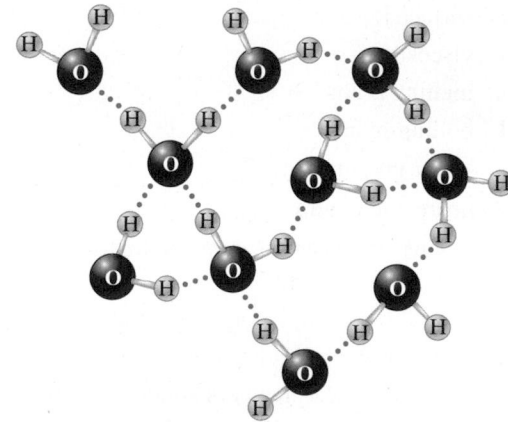

   a. the dotted lines between the hydrogen atoms of one water molecule and the oxygen atoms of a different water molecule

   b. the solid lines between a hydrogen atom and oxygen atom in the same water molecule

   c. Both the solid lines and dotted lines represent hydrogen bonding.

   d. There are no hydrogen bonds represented in the diagram.

14. Hydrogen bonding is a special case of very strong dipole–dipole interactions possible among only certain atoms. What atoms in addition to hydrogen are necessary for hydrogen bonding? How does the small size of the hydrogen atom contribute to the unusual strength of the dipole–dipole forces involved in hydrogen bonding?

15. Atoms are assumed to touch in closest packed structures, yet every closest packed unit cell contains a significant amount of empty space. Why?

16. Define *critical temperature* and *critical pressure*. In terms of the kinetic molecular theory, why is it impossible for a substance to exist as a liquid above its critical temperature?

17. Use the kinetic molecular theory to explain why a liquid gets cooler as it evaporates from an insulated container.

18. Will a crystalline solid or an amorphous solid give a simpler X-ray diffraction pattern? Why?

19. What is an alloy? Explain the differences in structure between substitutional and interstitial alloys. Give an example of each type.

20. Describe what is meant by a dynamic equilibrium in terms of the vapor pressure of a liquid.

21. How does each of the following affect the rate of evaporation of a liquid in an open dish?
   a. intermolecular forces
   b. temperature
   c. surface area

22. A common response to hearing that the temperature in New Mexico is 105°F is, "It's not that bad; it's a dry heat," whereas at the same time the summers in Atlanta, Georgia, are characterized as "dreadful," even though the air temperature is typically lower. What role does humidity play in how our bodies regulate temperature?

23. When a person has a severe fever, one therapy used to reduce the fever is an "alcohol rub." Explain how the evaporation of alcohol from a person's skin removes heat energy from the body.

24. Why is a burn from steam typically much more severe than a burn from boiling water?

25. When wet laundry is hung on a clothesline on a cold winter day, it will freeze but eventually dry. Explain.

26. Cake mixes and other packaged foods that require cooking often contain special directions for use at high elevations. Typically these directions indicate that the food should be cooked longer above 5000 ft. Explain why it takes longer to cook something at higher elevations.

27. You have three covalent compounds with three very different boiling points. All of the compounds have similar molar mass and relative shape. Explain how these three compounds could have very different boiling points.

28. Compare and contrast the structures of the following solids.
   a. diamond versus graphite
   b. silica versus silicates versus glass

29. Compare and contrast the structures of the following solids.
   a. $CO_2(s)$ versus $H_2O(s)$
   b. $NaCl(s)$ versus $CsCl(s)$; see Exercise 69 for the structures.

30. Silicon carbide (SiC) is an extremely hard substance that acts as an electrical insulator. Propose a structure for SiC.

31. How could you tell experimentally if TiO_2 is an ionic solid or a network solid?

32. A common prank on college campuses is to switch the salt and sugar on dining hall tables, which is usually easy because the substances look so much alike. Yet, despite the similarity in their appearance, these two substances differ greatly in their properties, since one is a molecular solid and the other is an ionic solid. How do the properties differ and why?

33. A plot of ln (P_{vap}) versus $1/T$ (K) is linear with a negative slope. Why is this the case?

34. Iodine, like most substances, exhibits only three phases: solid, liquid, and vapor. The triple point of iodine is at 90 torr and 115°C. Which of the following statements concerning liquid I_2 must be true? Explain your answer.
   a. $I_2(l)$ is more dense than $I_2(g)$.
   b. $I_2(l)$ cannot exist above 115°C.
   c. $I_2(l)$ cannot exist at 1 atmosphere pressure.
   d. $I_2(l)$ cannot have a vapor pressure greater than 90 torr.
   e. $I_2(l)$ cannot exist at a pressure of 10 torr.

Exercises

In this section similar exercises are paired.

Intermolecular Forces and Physical Properties

35. Identify the most important types of interparticle forces present in the solids of each of the following substances.
   a. Ar   e. CH_4
   b. HCl   f. CO
   c. HF   g. $NaNO_3$
   d. $CaCl_2$

36. Identify the most important types of interparticle forces present in the solids of each of the following substances.
   a. $BaSO_4$ e. CsI
   b. H_2S f. P_4
   c. Xe   g. NH_3
   d. C_2H_6

37. Predict which substance in each of the following pairs would have the greater intermolecular forces.
   a. CO_2 or OCS
   b. SeO_2 or SO_2
   c. $CH_3CH_2CH_2NH_2$ or $H_2NCH_2CH_2NH_2$
   d. CH_3CH_3 or H_2CO
   e. CH_3OH or H_2CO

38. Consider the compounds Cl_2, HCl, F_2, NaF, and HF. Which compound has a boiling point closest to that of argon? Explain.

39. Rationalize the difference in boiling points for each of the following pairs of substances:
   a. *n*-pentane $CH_3CH_2CH_2CH_2CH_3$ 36.2°C

   neopentane $H_3C-\overset{\displaystyle CH_3}{\underset{\displaystyle CH_3}{\overset{|}{\underset{|}{C}}}}-CH_3$ 9.5°C

b. HF   20°C
     HCl   −85°C

c. HCl   −85°C
     LiCl   1360°C

d. *n*-pentane $CH_3CH_2CH_2CH_2CH_3$ 36.2°C
     n-hexane $CH_3CH_2CH_2CH_2CH_2CH_3$ 69°C

40. Consider the following electrostatic potential diagrams:

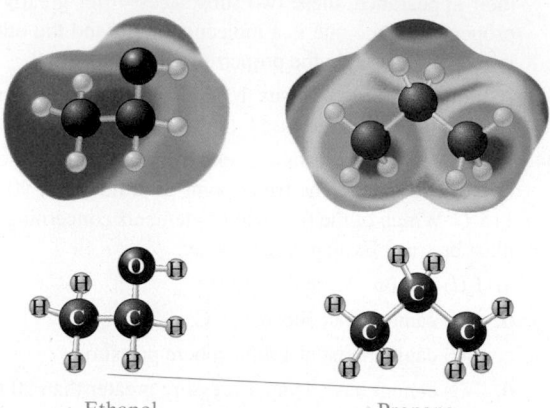

Ethanol   Propane

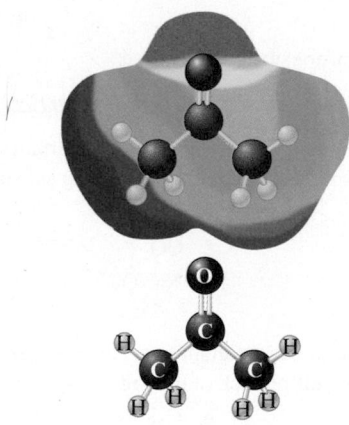

Acetone

Rank the compounds from lowest to highest boiling point and explain your answer.

41. In each of the following groups of substances, pick the one that has the given property. Justify your answer.

a. highest boiling point: HBr, Kr, or Cl_2

b. highest freezing point: H_2O, NaCl, or HF

c. lowest vapor pressure at 25°C: Cl_2, Br_2, or I_2

d. lowest freezing point: N_2, CO, or CO_2

e. lowest boiling point: CH_4, CH_3CH_3, or $CH_3CH_2CH_3$

f. highest boiling point: HF, HCl, or HBr

g. lowest vapor pressure at 25°C: $CH_3CH_2CH_3$, $CH_3\overset{\displaystyle O}{\overset{\displaystyle \|}{C}}CH_3$, or $CH_3CH_2CH_2OH$

42. In each of the following groups of substances, pick the one that has the given property. Justify each answer.

a. highest boiling point: CCl_4, CF_4, CBr_4

b. lowest freezing point: LiF, F_2, HCl

c. smallest vapor pressure at 25°C: CH_3OCH_3, CH_3CH_2OH, $CH_3CH_2CH_3$

d. greatest viscosity: H_2S, HF, H_2O_2

e. greatest heat of vaporization: H_2CO, CH_3CH_3, CH_4

f. smallest enthalpy of fusion: I_2, CsBr, CaO

Properties of Liquids

43. The shape of the meniscus of water in a glass tube is different from that of mercury in a glass tube. Why?

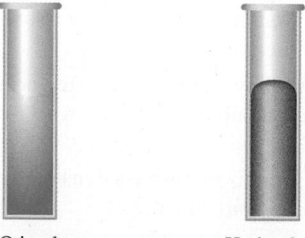

H_2O in glass   Hg in glass

44. Explain why water forms into beads on a waxed car finish.

45. Hydrogen peroxide (H_2O_2) is a syrupy liquid with a relatively low vapor pressure and a normal boiling point of 152.2°C. Rationalize the differences of these physical properties from those of water.

46. Carbon diselenide (CSe_2) is a liquid at room temperature. The normal boiling point is 125°C, and the melting point is −45.5°C. Carbon disulfide (CS_2) is also a liquid at room temperature with normal boiling and melting points of 46.5°C and −111.6°C, respectively. How do the strengths of the intermolecular forces vary from CO_2 to CS_2 to CSe_2? Explain.

Structures and Properties of Solids

47. X rays from a copper X-ray tube ($\lambda = 154$ pm) were diffracted at an angle of 14.22 degrees by a crystal of silicon. Assuming first-order diffraction ($n = 1$ in the Bragg equation), what is the interplanar spacing in silicon?

48. The second-order diffraction ($n = 2$) for a gold crystal is at an angle of 22.20° for X rays of 154 pm. What is the spacing between these crystal planes?

49. A topaz crystal has an interplanar spacing (d) of 1.36 Å (1 Å $= 1 \times 10^{-10}$ m). Calculate the wavelength of the X ray that should be used if $\theta = 15.0°$ (assume $n = 1$).

50. X rays of wavelength 2.63 Å were used to analyze a crystal. The angle of first-order diffraction ($n = 1$ in the Bragg equation) was 15.55 degrees. What is the spacing between crystal planes, and what would be the angle for second-order diffraction ($n = 2$)?

51. Calcium has a cubic closest packed structure as a solid. Assuming that calcium has an atomic radius of 197 pm, calculate the density of solid calcium.

52. Nickel has a face-centered cubic unit cell. The density of nickel is 6.84 g/cm^3. Calculate a value for the atomic radius of nickel.

53. A certain form of lead has a cubic closest packed structure with an edge length of 492 pm. Calculate the value of the atomic radius and the density of lead.

54. Iridium (Ir) has a face-centered cubic unit cell with an edge length of 383.3 pm. Calculate the density of solid iridium.

55. You are given a small bar of an unknown metal X. You find the density of the metal to be 10.5 g/cm^3. An X-ray diffraction experiment measures the edge of the face-centered cubic unit cell as 4.09 Å (1 Å = 10^{-10} m). Identify X.

56. A metallic solid with atoms in a face-centered cubic unit cell with an edge length of 392 pm has a density of 21.45 g/cm^3. Calculate the atomic mass and the atomic radius of the metal. Identify the metal.

57. Titanium metal has a body-centered cubic unit cell. The density of titanium is 4.50 g/cm^3. Calculate the edge length of the unit cell and a value for the atomic radius of titanium. (*Hint:* In a body-centered arrangement of spheres, the spheres touch across the body diagonal.)

58. Barium has a body-centered cubic structure. If the atomic radius of barium is 222 pm, calculate the density of solid barium.

59. The radius of gold is 144 pm, and the density is 19.32 g/cm^3. Does elemental gold have a face-centered cubic structure or a body-centered cubic structure?

60. The radius of tungsten is 137 pm and the density is 19.3 g/cm^3. Does elemental tungsten have a face-centered cubic structure or a body-centered cubic structure?

61. What fraction of the total volume of a cubic closest packed structure is occupied by atoms? (*Hint:* $V_{sphere} = \frac{4}{3}\pi r^3$.) What fraction of the total volume of a simple cubic structure is occupied by atoms? Compare the answers.

62. Iron has a density of 7.86 g/cm^3 and crystallizes in a body-centered cubic lattice. Show that only 68% of a body-centered lattice is actually occupied by atoms, and determine the atomic radius of iron.

63. Explain how doping silicon with either phosphorus or gallium increases the electrical conductivity over that of pure silicon.

64. Explain how a p–n junction makes an excellent rectifier.

65. Selenium is a semiconductor used in photocopying machines. What type of semiconductor would be formed if a small amount of indium impurity is added to pure selenium?

66. The Group 3A/Group 5A semiconductors are composed of equal amounts of atoms from Group 3A and Group 5A—for example, InP and GaAs. These types of semiconductors are used in light-emitting diodes and solid-state lasers. What would you add to make a p-type semiconductor from pure GaAs? How would you dope pure GaAs to make an n-type semiconductor?

67. The band gap in aluminum phosphide (AlP) is 2.5 electron-volts (1 eV = 1.6 × 10^{-19} J). What wavelength of light is emitted by an AlP diode?

68. An aluminum antimonide solid-state laser emits light with a wavelength of 730. nm. Calculate the band gap in joules.

69. The structures of some common crystalline substances are shown below. Show that the net composition of each unit cell corresponds to the correct formula of each substance.

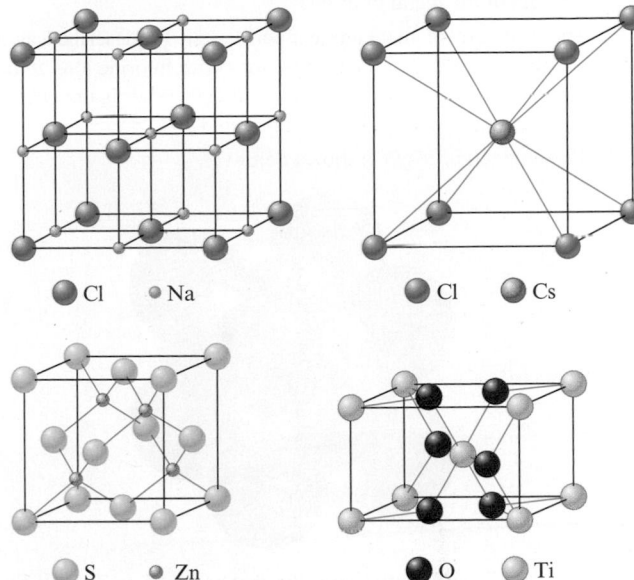

⬤ Cl   • Na   ⬤ Cl   ⬤ Cs

◯ S   • Zn   ⬤ O   ◯ Ti

70. The unit cell for nickel arsenide is shown below. What is the formula of this compound?

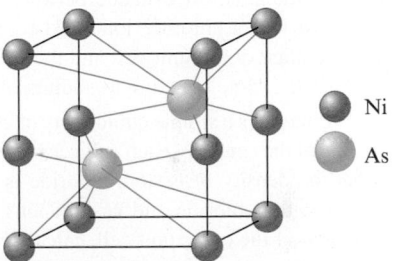

⬤ Ni
◯ As

71. Cobalt fluoride crystallizes in a closest packed array of fluoride ions with the cobalt ions filling one-half of the octahedral holes. What is the formula of this compound?

72. The compounds Na$_2$O, CdS, and ZrI$_4$ all can be described as cubic closest packed anions with the cations in tetrahedral holes. What fraction of the tetrahedral holes is occupied for each case?

73. What is the formula for the compound that crystallizes with a cubic closest packed array of sulfur ions, and that contains zinc ions in $\frac{1}{8}$ of the tetrahedral holes and aluminum ions in $\frac{1}{2}$ of the octahedral holes?

74. Assume the two-dimensional structure of an ionic compound, M$_x$A$_y$, is

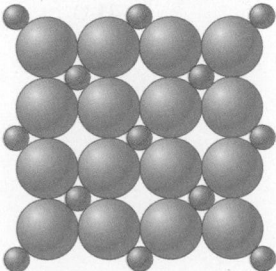

What is the empirical formula of this ionic compound?

75. A certain metal fluoride crystallizes in such a way that the fluoride ions occupy simple cubic lattice sites, while the metal ions occupy the body centers of *half* the cubes. What is the formula of the metal fluoride?

76. The structure of manganese fluoride can be described as a simple cubic array of manganese ions with fluoride ions at the center of each edge of the cubic unit cell. What is the charge of the manganese ions in this compound?

77. The unit cell of MgO is shown below.

Does MgO have a structure like that of NaCl or ZnS? If the density of MgO is 3.58 g/cm^3, estimate the radius (in centimeters) of the O^{2-} anions and the Mg^{2+} cations.

78. In solid KCl the smallest distance between the centers of a potassium ion and a chloride ion is 314 pm. Calculate the length of the edge of the unit cell and the density of KCl, assuming it has the same structure as sodium chloride.

79. The CsCl structure is a simple cubic array of chloride ions with a cesium ion at the center of each cubic array (see Exercise 69). Given that the density of cesium chloride is 3.97 g/cm^3, and assuming that the chloride and cesium ions touch along the body diagonal of the cubic unit cell, calculate the distance between the centers of adjacent Cs$^+$ and Cl$^-$ ions in the solid. Compare this value with the expected distance based on the sizes of the ions. The ionic radius of Cs$^+$ is 169 pm, and the ionic radius of Cl$^-$ is 181 pm.

80. MnO has either the NaCl type structure or the CsCl type structure (see Exercise 69). The edge length of the MnO unit cell is 4.47×10^{-8} cm and the density of MnO is 5.28 g/cm^3.
    a. Does MnO crystallize in the NaCl or the CsCl type structure?
    b. Assuming that the ionic radius of oxygen is 140. pm, estimate the ionic radius of manganese.

81. What type of solid will each of the following substances form?
    a. CO$_2$ g. KBr
    b. SiO$_2$ h. H$_2$O
    c. Si i. NaOH
    d. CH$_4$ j. U
    e. Ru k. CaCO$_3$
    f. I$_2$ l. PH$_3$

82. What type of solid will each of the following substances form?
    a. diamond g. NH$_4$NO$_3$
    b. PH$_3$ h. SF$_2$
    c. H$_2$ i. Ar
    d. Mg j. Cu
    e. KCl k. C$_6$H$_{12}$O$_6$
    f. quartz

83. The memory metal, nitinol, is an alloy of nickel and titanium. It is called a *memory metal* because after being deformed, a piece of nitinol wire will return to its original shape. The structure of nitinol consists of a simple cubic array of Ni atoms and an inner penetrating simple cubic array of Ti atoms. In the extended lattice, a Ti atom is found at the center of a cube of Ni atoms; the reverse is also true.
    a. Describe the unit cell for nitinol.
    b. What is the empirical formula of nitinol?
    c. What are the coordination numbers (number of nearest neighbors) of Ni and Ti in nitinol?

84. Superalloys have been made of nickel and aluminum. The alloy owes its strength to the formation of an ordered phase, called the *gamma-prime phase,* in which Al atoms are at the corners of a cubic unit cell and Ni atoms are at the face centers. What is the composition (relative numbers of atoms) for this phase of the nickel–aluminum superalloy?

85. Perovskite is a mineral containing calcium, titanium, and oxygen. Two different representations of the unit cell are shown below. Show that both these representations give the same formula and the same number of oxygen atoms around each titanium atom.

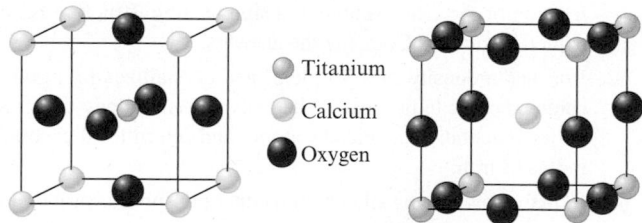

86. A mineral crystallizes in a cubic closest packed array of oxygen ions with aluminum ions in some of the octahedral holes and magnesium ions in some of the tetrahedral holes. Deduce the formula of this mineral and predict the fraction of octahedral holes and tetrahedral holes that are filled by the various cations.

87. Materials containing the elements Y, Ba, Cu, and O that are superconductors (electrical resistance equals zero) at temperatures above that of liquid nitrogen were recently discovered. The structures of these materials are based on the perovskite structure. Were they to have the ideal perovskite structure, the

superconductor would have the structure shown in part (a) of the following figure.

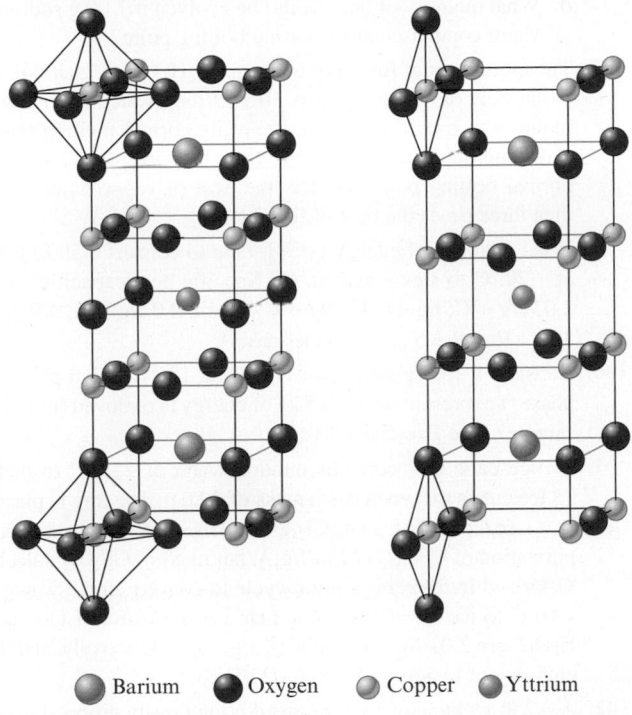

(a) Ideal perovskite structure

(b) Actual structure of superconductor

Legend: Barium, Oxygen, Copper, Yttrium

a. What is the formula of this ideal perovskite material?

b. How is this structure related to the perovskite structure shown in Exercise 85?

These materials, however, do not act as superconductors unless they are deficient in oxygen. The structure of the actual superconducting phase appears to be that shown in part (b) of the figure.

c. What is the formula of this material?

88. The structures of another class of ceramic, high-temperature superconductors are shown in figures a-d.

a. Determine the formula of each of these four superconductors.

b. One of the structural features that appears to be essential for high-temperature superconductivity is the presence of planar sheets of copper and oxygen atoms. As the number of sheets in each unit cell increases, the temperature for the onset of superconductivity increases. Order the four structures from lowest to the highest superconducting temperature.

c. Assign oxidation states to Cu in each structure assuming Tl exists as Tl^{3+}. The oxidation states of Ca, Ba, and O are assumed to be +2, +2, and −2, respectively.

d. It also appears that copper must display a mixture of oxidation states for a material to exhibit superconductivity. Explain how this occurs in these materials as well as in the superconductor in Exercise 87.

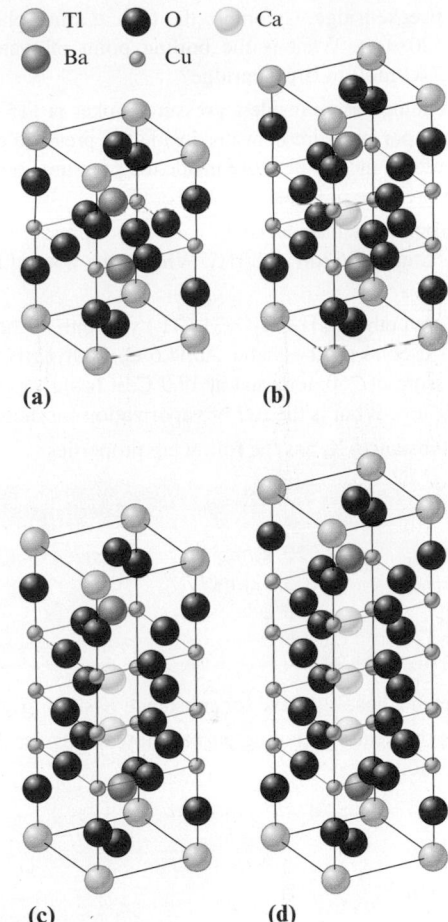

Legend: Tl, O, Ca, Ba, Cu

(a)   (b)

(c)   (d)

Phase Changes and Phase Diagrams

89. Plot the following data and determine ΔH_{vap} for magnesium and lithium. In which metal is the bonding stronger?

Vapor Pressure (mm Hg)	Temperature (°C)	
	Li	Mg
1.	750.	620.
10.	890.	740.
100.	1080.	900.
400.	1240.	1040.
760.	1310.	1110.

90. From the following data for liquid nitric acid, determine its heat of vaporization and normal boiling point.

Temperature (°C)	Vapor Pressure (mm Hg)
0.	14.4
10.	26.6
20.	47.9
30.	81.3
40.	133
50.	208
80.	670.

91. In Breckenridge, Colorado, the typical atmospheric pressure is 520. torr. What is the boiling point of water (ΔH_{vap} = 40.7 kJ/mol) in Breckenridge?

92. The temperature inside a pressure cooker is 115°C. Calculate the vapor pressure of water inside the pressure cooker. What would be the temperature inside the pressure cooker if the vapor pressure of water was 3.50 atm?

93. Carbon tetrachloride, CCl_4, has a vapor pressure of 213 torr at 40.°C and 836 torr at 80.°C. What is the normal boiling point of CCl_4?

94. Diethyl ether ($CH_3CH_2OCH_2CH_3$) was one of the first chemicals used as an anesthetic. At 34.6°C, diethyl ether has a vapor pressure of 760. torr, and at 17.9°C, it has a vapor pressure of 400. torr. What is the ΔH of vaporization for diethyl ether?

95. A substance, X, has the following properties:

		Specific Heat Capacities	
ΔH_{vap}	20. kJ/mol	C(s)	3.0 J/g · °C
ΔH_{fus}	5.0 kJ/mol	C(l)	2.5 J/g · °C
bp	75°C	C(g)	1.0 J/g · °C
mp	−15°C		

Sketch a heating curve for substance X starting at −50.°C.

96. Use the heating–cooling curve below to answer the following questions.

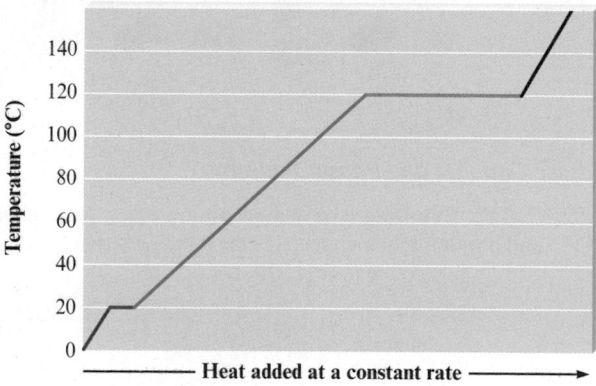

a. What is the freezing point of the liquid?

b. What is the boiling point of the liquid?

c. Which is greater, the heat of fusion or the heat of vaporization? Explain each term and explain how the heating–cooling curve above helps you to answer the question.

97. The molar heat of fusion of sodium metal is 2.60 kJ/mol, whereas its heat of vaporization is 97.0 kJ/mol.

a. Why is the heat of vaporization so much larger than the heat of fusion?

b. What quantity of heat would be needed to *melt* 1.00 g sodium at its normal melting point?

c. What quantity of heat would be needed to *vaporize* 1.00 g sodium at its normal boiling point?

d. What quantity of heat would be evolved if 1.00 g sodium vapor condensed at its normal boiling point?

98. The molar heat of fusion of benzene (C_6H_6) is 9.92 kJ/mol. Its molar heat of vaporization is 30.7 kJ/mol. Calculate the heat required to melt 8.25 g benzene at its normal melting point. Calculate the heat required to vaporize 8.25 g benzene at its normal boiling point. Why is the heat of vaporization more than three times the heat of fusion?

99. What quantity of energy does it take to convert 0.500 kg ice at −20.°C to steam at 250.°C? Specific heat capacities: ice, 2.03 J/g · °C; liquid, 4.2 J/g · °C; steam, 2.0 J/g · °C; ΔH_{vap} = 40.7 kJ/mol; ΔH_{fus} = 6.02 kJ/mol.

100. Consider a 75.0-g sample of $H_2O(g)$ at 125°C. What phase or phases are present when 215 kJ of energy is removed from this sample? (See Exercise 99.)

101. An ice cube tray contains enough water at 22.0°C to make 18 ice cubes that each has a mass of 30.0 g. The tray is placed in a freezer that uses CF_2Cl_2 as a refrigerant. The heat of vaporization of CF_2Cl_2 is 158 J/g. What mass of CF_2Cl_2 must be vaporized in the refrigeration cycle to convert all the water at 22.0°C to ice at −5.0°C? The heat capacities for $H_2O(s)$ and $H_2O(l)$ are 2.03 J/g · °C and 4.18 J/g · °C, respectively, and the enthalpy of fusion for ice is 6.02 kJ/mol.

102. A 0.250-g chunk of sodium metal is cautiously dropped into a mixture of 50.0 g water and 50.0 g ice, both at 0°C. The reaction is

$$2Na(s) + 2H_2O(l) \longrightarrow 2NaOH(aq) + H_2(g) \quad \Delta H = -368 \text{ kJ}$$

Assuming no heat loss to the surroundings, will the ice melt? Assuming the final mixture has a specific heat capacity of 4.18 J/g · °C, calculate the final temperature. The enthalpy of fusion for ice is 6.02 kJ/mol.

103. Consider the phase diagram given below. What phases are present at points A through H? Identify the triple point, normal boiling point, normal freezing point, and critical point. Which phase is denser, solid or liquid?

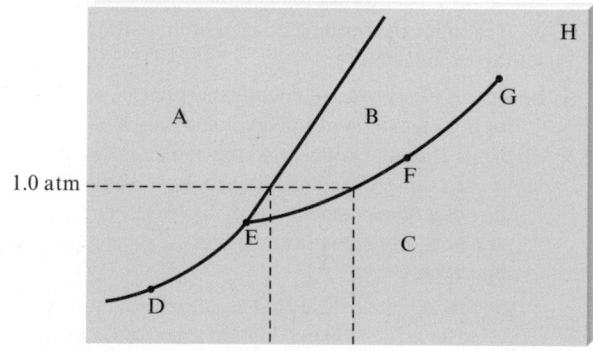

104. Sulfur exhibits two solid phases, rhombic and monoclinic. Use the accompanying phase diagram for sulfur to answer the following questions. (The phase diagram is not to scale.)

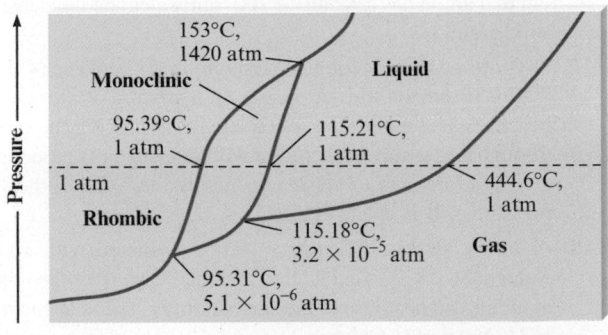

a. How many triple points are in the phase diagram?

b. What phases are in equilibrium at each of the triple points?

c. What is the stable phase at 1 atm and 100.°C?

d. What are the normal melting point and the normal boiling point of sulfur?

e. Which is the densest phase?

f. At a pressure of 1.0×10^{-5} atm, can rhombic sulfur sublime?

g. What phase changes occur when the pressure on a sample of sulfur at 100.°C is increased from 1.0×10^{-8} atm to 1500 atm?

105. Use the accompanying phase diagram for carbon to answer the following questions.

a. How many triple points are in the phase diagram?

b. What phases can coexist at each triple point?

c. What happens if graphite is subjected to very high pressures at room temperature?

d. If we assume that the density increases with an increase in pressure, which is more dense, graphite or diamond?

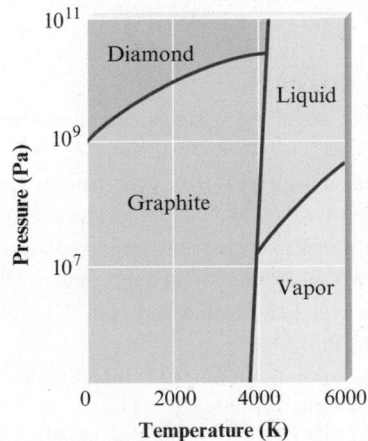

106. Like most substances, bromine exists in one of the three typical phases. Br_2 has a normal melting point of $-7.2°C$ and a normal boiling point of 59°C. The triple point for Br_2 is $-7.3°C$ and 40 torr, and the critical point is 320°C and 100 atm. Using this information, sketch a phase diagram for bromine indicating the points described above. Based on your phase diagram, order the three phases from least dense to most dense. What is the stable phase of Br_2 at room temperature and 1 atm? Under what temperature conditions can liquid bromine never exist? What phase changes occur as the temperature of a sample of bromine at 0.10 atm is increased from $-50°C$ to 200°C?

107. The melting point of a fictional substance X is 225°C at 10.0 atm. If the density of the solid phase of X is 2.67 g/cm³ and the density of the liquid phase is 2.78 g/cm³ at 10.0 atm, predict whether the normal melting point of X will be less than, equal to, or greater than 225°C. Explain.

108. Consider the following data for xenon:

Triple point:	$-121°C$, 280 torr
Normal melting point:	$-112°C$
Normal boiling point:	$-107°C$

Which is more dense, Xe(s) or Xe(l)? How do the melting point and boiling point of xenon depend on pressure?

Additional Exercises

109. Some of the physical properties of H_2O and D_2O are as follows:

Property	H_2O	D_2O
Density at 20°C (g/mL)	0.997	1.108
Boiling point (°C)	100.00	101.41
Melting point (°C)	0.00	3.79
$\Delta H°_{vap}$ (kJ/mol)	40.7	41.61
$\Delta H°_{fus}$ (kJ/mol)	6.02	6.3

Account for the differences. (*Note:* D is a symbol often used for ^2H, the deuterium isotope of hydrogen.)

110. Rationalize the following boiling points:

$$CH_3C \overset{O}{\underset{OH}{\Big\langle}} \quad 118°C$$

$$ClCH_2C \overset{O}{\underset{OH}{\Big\langle}} \quad 189°C$$

$$CH_3C \overset{O}{\underset{OCH_3}{\Big\langle}} \quad 57°C$$

111. Consider the following vapor pressure versus temperature plot for three different substances: A, B, and C.

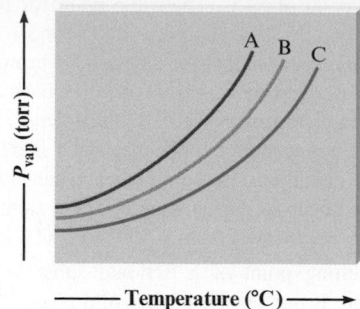

If the three substances are CH_4, SiH_4, and NH_3, match each curve to the correct substance.

112. Consider the following enthalpy changes:

$$F^- + HF \longrightarrow FHF^- \qquad \Delta H = -155 \text{ kJ/mol}$$

$$(CH_3)_2C\!=\!O + HF \longrightarrow (CH_3)_2C\!=\!O\text{---}HF$$
$$\Delta H = -46 \text{ kJ/mol}$$

$$H_2O(g) + HOH(g) \longrightarrow H_2O\text{---}HOH \text{ (in ice)}$$
$$\Delta H = -21 \text{ kJ/mol}$$

How do the strengths of hydrogen bonds vary with the electronegativity of the element to which hydrogen is bonded? Where in the preceding series would you expect hydrogen bonds of the following type to fall?

$$\overset{|}{\underset{|}{N}}\text{---HO---}\quad \text{and} \quad \overset{|}{\underset{|}{N}}\text{---H---}N\underset{|}{\overset{|}{}}$$

113. The unit cell for a pure xenon fluoride compound is shown below. What is the formula of the compound?

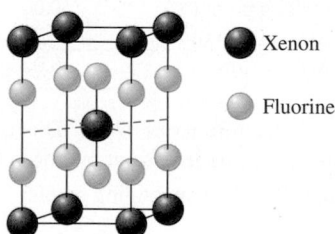

Xenon

Fluorine

114. Boron nitride (BN) exists in two forms. The first is a slippery solid formed from the reaction of BCl_3 with NH_3, followed by heating in an ammonia atmosphere at 750°C. Subjecting the first form of BN to a pressure of 85,000 atm at 1800°C produces a second form that is the second hardest substance known. Both forms of BN remain solids to 3000°C. Suggest structures for the two forms of BN.

115. Consider the following data concerning four different substances.

Compound	Conducts Electricity as a Solid	Other Properties
B_2H_6	No	Gas at 25°C
SiO_2	No	High mp
CsI	No	Aqueous solution Conducts electricity
W	Yes	High mp

Label the four substances as either ionic, network, metallic, or molecular solids.

116. Argon has a cubic closest packed structure as a solid. Assuming that argon has a radius of 190. pm, calculate the density of solid argon.

117. Dry nitrogen gas is bubbled through liquid benzene (C_6H_6) at 20.0°C. From 100.0 L of the gaseous mixture of nitrogen and benzene, 24.7 g benzene is condensed by passing the mixture through a trap at a temperature where nitrogen is gaseous and the vapor pressure of benzene is negligible. What is the vapor pressure of benzene at 20.0°C?

118. A 20.0-g sample of ice at -10.0°C is mixed with 100.0 g water at 80.0°C. Calculate the final temperature of the mixture assuming no heat loss to the surroundings. The heat capacities of $H_2O(s)$ and $H_2O(l)$ are 2.03 and 4.18 J/g · °C, respectively, and the enthalpy of fusion for ice is 6.02 kJ/mol.

119. In regions with dry climates, evaporative coolers are used to cool air. A typical electric air conditioner is rated at 1.00×10^4 Btu/h (1 Btu, or British thermal unit = amount of energy to raise the temperature of 1 lb water by 1°F). What quantity of water must be evaporated each hour to dissipate as much heat as a typical electric air conditioner?

120. The critical point of NH_3 is 132°C and 111 atm, and the critical point of N_2 is -147°C and 34 atm. Which of these substances cannot be liquefied at room temperature no matter how much pressure is applied? Explain.

ChemWork Problems

These multiconcept problems (and additional ones) are found interactively online with the same type of assistance a student would get from an instructor.

121. Which of the following compound(s) exhibit only London dispersion intermolecular forces? Which compound(s) exhibit hydrogen-bonding forces? Considering only the compounds without hydrogen-bonding interactions, which compounds have dipole–dipole intermolecular forces?

a. SF_4 d. HF

b. CO_2 e. ICl_5

c. CH_3CH_2OH f. XeF_4

122. Which of the following statements about intermolecular forces is(are) *true*?

a. London dispersion forces are the only type of intermolecular force that nonpolar molecules exhibit.

b. Molecules that have only London dispersion forces will always be gases at room temperature (25°C).

c. The hydrogen-bonding forces in NH_3 are stronger than those in H_2O.

d. The molecules in $SO_2(g)$ exhibit dipole–dipole intermolecular interactions.

e. $CH_3CH_2CH_3$ has stronger London dispersion forces than does CH_4.

123. Which of the following statements is(are) *true*?

a. LiF will have a higher vapor pressure at 25°C than H_2S.

b. HF will have a lower vapor pressure at -50°C than HBr.

c. Cl_2 will have a higher boiling point than Ar.

d. HCl is more soluble in water than in CCl₄.

e. MgO will have a higher vapor pressure at 25°C than CH₃CH₂OH.

124. Aluminum has an atomic radius of 143 pm and forms a solid with a cubic closest packed structure. Calculate the density of solid aluminum in g/cm³.

125. Pyrolusite is a mineral containing manganese ions and oxide ions. Its structure can best be described as a body-centered cubic array of manganese ions with two oxide ions inside the unit cell and two oxide ions each on two faces of the cubic unit cell. What is the charge on the manganese ions in pyrolusite?

126. The structure of the compound K₂O is best described as a cubic closest packed array of oxide ions with the potassium ions in tetrahedral holes. What percent of the tetrahedral holes are occupied in this solid?

127. What type of solid (network, metallic, Group 8A, ionic, or molecular) will each of the following substances form?

a. Kr   **d.** SiO₂

b. SO₂   **e.** NH₃

c. Ni   **f.** Pt

128. Some ice cubes at 0°C with a total mass of 403 g are placed in a microwave oven and subjected to 750. W (750. J/s) of energy for 5.00 minutes. What is the final temperature of the water? Assume all the energy of the microwave is absorbed by the water, and assume no heat loss by the water.

129. The enthalpy of vaporization for acetone is 32.0 kJ/mol. The normal boiling point for acetone is 56.5°C. What is the vapor pressure of acetone at 23.5°C?

130. Choose the statements that correctly describe the following phase diagram.

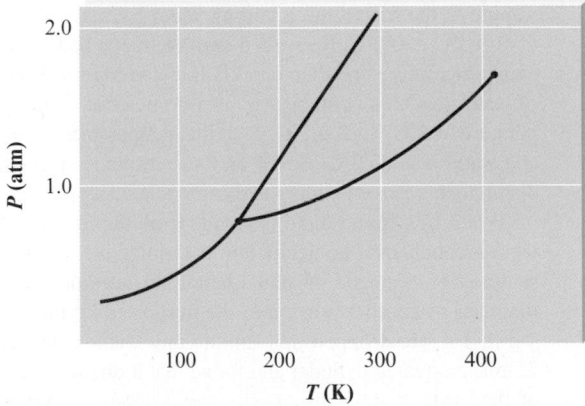

a. If the temperature is raised from 50 K to 400 K at a pressure of 1 atm, the substance boils at approximately 185 K.

b. The liquid phase of this substance cannot exist under conditions of 2 atm at any temperature.

c. The triple point occurs at approximately 165 K.

d. At a pressure of 1.5 atm, the melting point of the substance is approximately 370 K.

e. The critical point occurs at approximately 1.7 atm and 410 K.

Challenge Problems

131. When 1 mole of benzene is vaporized at a constant pressure of 1.00 atm and at its boiling point of 353.0 K, 30.79 kJ of energy (heat) is absorbed and the volume change is +28.90 L. What are ΔE and ΔH for this process?

132. You and a friend each synthesize a compound with the formula XeCl₂F₂. Your compound is a liquid and your friend's compound is a gas (at the same conditions of temperature and pressure). Explain how the two compounds with the same formulas can exist in different phases at the same conditions of pressure and temperature.

133. Using the heats of fusion and vaporization for water given in Exercise 99, calculate the change in enthalpy for the sublimation of water:

$$H_2O(s) \longrightarrow H_2O(g)$$

Using the ΔH value given in Exercise 112 and the number of hydrogen bonds formed with each water molecule, estimate what portion of the intermolecular forces in ice can be accounted for by hydrogen bonding.

134. Consider a perfectly insulated and sealed container. Determine the minimum volume of a container such that a gallon of water at 25°C will evaporate completely. If the container is a cube, determine the dimensions in feet. Assume the density of water is 0.998 g/cm³.

135. Consider two different organic compounds, each with the formula C₂H₆O. One of these compounds is a liquid at room conditions and the other is a gas. Write Lewis structures consistent with this observation, and explain your answer. (*Hint:* The oxygen atom in both structures satisfies the octet rule with two bonds and two lone pairs.)

136. Rationalize the differences in physical properties in terms of intermolecular forces for the following organic compounds. Compare the first three substances with each other, compare the last three with each other, and then compare all six. Can you account for any anomalies?

	bp (°C)	mp (°C)	ΔH$_{vap}$ (kJ/mol)
Benzene, C₆H₆	80	6	33.9
Naphthalene, C₁₀H₈	218	80	51.5
Carbon tetrachloride	76	−23	31.8
Acetone, CH₃COCH₃	56	−95	31.8
Acetic acid, CH₃CO₂H	118	17	39.7
Benzoic acid, C₆H₅CO₂H	249	122	68.2

137. Consider the following melting point data:

Compound	NaCl	MgCl₂	AlCl₃	SiCl₄	PCl₃	SCl₂	Cl₂
mp (°C)	801	708	190	−70	−91	−78	−101

Compound	NaF	MgF₂	AlF₃	SiF₄	PF₅	SF₆	F₂
mp (°C)	997	1396	1040	−90	−94	−56	−220

Account for the trends in melting points in terms of interparticle forces.

138. Some ionic compounds contain a mixture of different charged cations. For example, wüstite is an oxide that contains both Fe^{2+} and Fe^{3+} cations and has a formula of $Fe_{0.950}O_{1.00}$. Calculate the fraction of iron ions present as Fe^{3+}. What fraction of the sites normally occupied by Fe^{2+} must be vacant in this solid?

139. Some ionic compounds contain a mixture of different charged cations. For example, some titanium oxides contain a mixture of Ti^{2+} and Ti^{3+} ions. Consider a certain oxide of titanium that is 28.31% oxygen by mass and contains a mixture of Ti^{2+} and Ti^{3+} ions. Determine the formula of the compound and the relative numbers of Ti^{2+} and Ti^{3+} ions.

140. Spinel is a mineral that contains 37.9% aluminum, 17.1% magnesium, and 45.0% oxygen, by mass, and has a density of 3.57 g/cm^3. The edge of the cubic unit cell measures 809 pm. How many of each type of ion are present in the unit cell?

141. Mn crystallizes in the same type of cubic unit cell as Cu. Assuming that the radius of Mn is 5.6% larger than the radius of Cu and the density of copper is 8.96 g/cm^3, calculate the density of Mn.

142. You are asked to help set up a historical display in the park by stacking some cannonballs next to a Revolutionary War cannon. You are told to stack them by starting with a triangle in which each side is composed of four touching cannonballs. You are to continue stacking them until you have a single ball on the top centered over the middle of the triangular base.

    a. How many cannonballs do you need?

    b. What type of closest packing is displayed by the cannonballs?

    c. The four corners of the pyramid of cannonballs form the corners of what type of regular geometric solid?

143. Some water is placed in a sealed glass container connected to a vacuum pump (a device used to pump gases from a container), and the pump is turned on. The water appears to boil and then freezes. Explain these changes using the phase diagram for water. What would happen to the ice if the vacuum pump was left on indefinitely?

144. The molar enthalpy of vaporization of water at 373 K and 1.00 atm is 40.7 kJ/mol. What fraction of this energy is used to change the internal energy of the water, and what fraction is used to do work against the atmosphere? (*Hint:* Assume that water vapor is an ideal gas.)

145. For a simple cubic array, solve for the volume of an interior sphere (cubic hole) in terms of the radius of a sphere in the array.

146. Rubidium chloride has the sodium chloride structure at normal pressures but assumes the cesium chloride structure at high pressures. (See Exercise 69.) What ratio of densities is expected for these two forms? Does this change in structure make sense on the basis of simple models? The ionic radius is 148 pm for Rb^+ and 181 pm for Cl^-.

Integrative Problems

These problems require the integration of multiple concepts to find the solutions.

147. A 0.132-mole sample of an unknown semiconducting material with the formula XY has a mass of 19.0 g. The element X has an electron configuration of $[Kr]5s^24d^{10}$. What is this semiconducting material? A small amount of the Y atoms in the semiconductor is replaced with an equivalent amount of atoms with an electron configuration of $[Ar]4s^23d^{10}4p^5$. Does this correspond to n-type or p-type doping?

148. A metal burns in air at 600°C under high pressure to form an oxide with formula MO_2. This compound is 23.72% oxygen by mass. The distance between the centers of touching atoms in a cubic closest packed crystal of this metal is 269.0 pm. What is this metal? What is its density?

149. One method of preparing elemental mercury involves roasting cinnabar (HgS) in quicklime (CaO) at 600.°C followed by condensation of the mercury vapor. Given the heat of vaporization of mercury (296 J/g) and the vapor pressure of mercury at 25.0°C (2.56×10^{-3} torr), what is the vapor pressure of the condensed mercury at 300.°C? How many atoms of mercury are present in the mercury vapor at 300.°C if the reaction is conducted in a closed 15.0-L container?

Marathon Problem

This problem is designed to incorporate several concepts and techniques into one situation.

150. General Zod has sold Lex Luthor what Zod claims to be a new copper-colored form of kryptonite, the only substance that can harm Superman. Lex, not believing in honor among thieves, decided to carry out some tests on the supposed kryptonite. From previous tests, Lex knew that kryptonite is a metal having a specific heat capacity of 0.082 J/g · °C and a density of 9.2 g/cm^3.

    Lex Luthor's first experiment was an attempt to find the specific heat capacity of kryptonite. He dropped a 10 g ± 3 g sample of the metal into a boiling water bath at a temperature of 100.0°C ± 0.2°C. He waited until the metal had reached the bath temperature and then quickly transferred it to 100 g ± 3 g of water that was contained in a calorimeter at an initial temperature of 25.0°C ± 0.2°C. The final temperature of the metal and water was 25.2°C. Based on these results, is it possible to distinguish between copper and kryptonite? Explain.

    When Lex found that his results from the first experiment were inconclusive, he decided to determine the density of the sample. He managed to steal a better balance and determined the mass of another portion of the purported kryptonite to be 4 g ± 1 g. He dropped this sample into water contained in a 25-mL graduated cylinder and found that it displaced a volume of 0.42 mL ± 0.02 mL. Is the metal copper or kryptonite? Explain.

    Lex was finally forced to determine the crystal structure of the metal General Zod had given him. He found that the cubic unit cell contained four atoms and had an edge length of 600. pm. Explain how this information enabled Lex to identify the metal as copper or kryptonite.

    Will Lex be going after Superman with the kryptonite or seeking revenge on General Zod? What improvements could he have made in his experimental techniques to avoid performing the crystal structure determination?

CHAPTER TEN

Properties of Solutions

Underwater photo of paint mixing with water (Mark Mawson/Stone/Getty Images)

413

Most of the substances we encounter in daily life are mixtures: Wood, milk, gasoline, champagne, seawater, shampoo, steel, and air are common examples. When the components of a mixture are uniformly intermingled—that is, when a mixture is homogeneous—it is called a *solution.* Solutions can be gases, liquids, or solids, as shown in Table 10-1. However, we will be concerned in this chapter with the properties of liquid solutions, particularly those containing water. As we have seen earlier in the text, many essential chemical reactions occur in aqueous solutions because water is capable of dissolving so many substances.

As we will see, the properties of solutions depend on the structure of the molecules making up the solution. That is, we need to know the nature of the solution components and to determine the intermolecular forces involved in order to better understand solubility, vapor pressure of solutions, and colligative properties such as freezing-point depression, boiling-point elevation, and osmotic pressure.

10-1 | Solution Composition

(i) A solute is the substance being dissolved. The solvent is the dissolving medium.

(i) $\text{Molarity} = \dfrac{\text{moles of solute}}{\text{liters of solution}}$

Because a mixture, unlike a chemical compound, has a variable composition, the relative amounts of substances in a solution must be specified. The qualitative terms *dilute* (relatively little solute present) and *concentrated* (relatively large amount of solute) are often used to describe solution content, but we need to define solution composition more precisely to perform calculations. ◄ *(i)* For example, in dealing with the stoichiometry of solution reactions in Chapter 3, we found it useful to describe solution composition in terms of **molarity**, or the number of moles of solute per liter of solution (symbolized by *M*). ◄ *(i)*

Other ways of describing solution composition are also useful. **Mass percent** (sometimes called *weight percent*) is the percent by mass of the solute in the solution:

$$\text{Mass percent} = \left(\frac{\text{mass of solute}}{\text{mass of solution}} \right) \times 100\%$$

Another way of describing solution composition is the **mole fraction** (symbolized by the lowercase Greek letter chi, χ), the ratio of the number of moles of a given component to the total number of moles of solution. For a two-component solution, where n_A and n_B represent the number of moles of the two components,

$$\text{Mole fraction of component A} = \chi_A = \frac{n_A}{n_A + n_B}$$

Table 10-1 | Various Types of Solutions

Example	State of Solution	State of Solute	State of Solvent
Air, natural gas	Gas	Gas	Gas
Vodka, antifreeze	Liquid	Liquid	Liquid
Brass	Solid	Solid	Solid
Carbonated water	Liquid	Gas	Liquid
Seawater, sugar solution	Liquid	Solid	Liquid
Hydrogen in platinum	Solid	Gas	Solid

Still another way of describing solution composition is **molality** (symbolized by m), the number of moles of solute per *kilogram of solvent:*

$$\text{Molality} = \frac{\text{moles of solute}}{\text{kilogram of solvent}}$$

INTERACTIVE EXAMPLE 10-1

Various Methods for Describing Solution Composition

A solution is prepared by mixing 1.00 g ethanol (C_2H_5OH) with 100.0 g water to give a final volume of 101 mL. Calculate the molarity, mass percent, mole fraction, and molality of ethanol in this solution.

Solution

Molarity

The moles of ethanol can be obtained from its molar mass (46.07 g/mol):

$$1.00 \text{ g } C_2H_5OH \times \frac{1 \text{ mol } C_2H_5OH}{46.07 \text{ g } C_2H_5OH} = 2.17 \times 10^{-2} \text{ mol } C_2H_5OH$$

$$\text{Volume} = 101 \text{ mL} \times \frac{1 \text{ L}}{1000 \text{ mL}} = 0.101 \text{ L}$$

$$\text{Molarity of } C_2H_5OH = \frac{\text{moles of } C_2H_5OH}{\text{liters of solution}} = \frac{2.17 \times 10^{-2} \text{ mol}}{0.101 \text{ L}}$$

$$= 0.215 \ M$$

Mass Percent

$$\text{Mass percent } C_2H_5OH = \left(\frac{\text{mass of } C_2H_5OH}{\text{mass of solution}}\right) \times 100\%$$

$$= \left(\frac{1.00 \text{ g } C_2H_5OH}{100.0 \text{ g } H_2O + 1.00 \text{ g } C_2H_5OH}\right) \times 100\%$$

$$= 0.990\% \ C_2H_5OH$$

Mole Fraction

$$\text{Mole fraction of } C_2H_5OH = \frac{n_{C_2H_5OH}}{n_{C_2H_5OH} + n_{H_2O}}$$

$$n_{H_2O} = 100.0 \text{ g } H_2O \times \frac{1 \text{ mol } H_2O}{18.0 \text{ g } H_2O} = 5.56 \text{ mol}$$

$$\chi_{C_2H_5OH} = \frac{2.17 \times 10^{-2} \text{ mol}}{2.17 \times 10^{-2} \text{ mol} + 5.56 \text{ mol}}$$

$$= \frac{2.17 \times 10^{-2}}{5.58} = 0.00389$$

Molality

$$\text{Molality of } C_2H_5OH = \frac{\text{moles of } C_2H_5OH}{\text{kilogram of } H_2O} = \frac{2.17 \times 10^{-2} \text{ mol}}{100.0 \text{ g} \times \dfrac{1 \text{ kg}}{1000 \text{ g}}}$$

$$= \frac{2.17 \times 10^{-2} \text{ mol}}{0.1000 \text{ kg}}$$

$$= 0.217 \ m$$

See Exercises 10-29 through 10-31

Table 10-2 | The Molar Mass, Equivalent Mass, and Relationship of Molarity and Normality for Several Acids and Bases

Acid or Base	Molar Mass	Equivalent Mass	Relationship of Molarity and Normality
HCl	36.5	36.5	$1\,M = 1\,N$
H_2SO_4	98	$\dfrac{98}{2} = 49$	$1\,M = 2\,N$
NaOH	40	40	$1\,M = 1\,N$
$Ca(OH)_2$	74	$\dfrac{74}{2} = 37$	$1\,M = 2\,N$

Critical Thinking

You are given two aqueous solutions with different ionic solutes (Solution A and Solution B). What if you are told that Solution A has a greater concentration than Solution B by mass percent, but Solution B has a greater concentration than Solution A in terms of molality? Is this possible? If not, explain why not. If it is possible, provide example solutes for A and B and justify your answer with calculations.

i The definition of an equivalent depends on the reaction taking place in the solution.

i The quantity we call *equivalent mass* here traditionally has been called *equivalent weight*.

Another concentration measure sometimes encountered is **normality** (symbolized by N). Normality is defined as the number of *equivalents* per liter of solution, where the definition of an equivalent depends on the reaction taking place in the solution. ◄ *i* *For an acid–base reaction,* the equivalent is the mass of acid or base that can furnish or accept exactly 1 mole of protons (H^+ ions). ◄ *i* In Table 10-2, note, for example, that the equivalent mass of sulfuric acid is the molar mass divided by 2, since each mole of H_2SO_4 can furnish 2 moles of protons. The equivalent mass of calcium hydroxide is also half the molar mass, since each mole of $Ca(OH)_2$ contains 2 moles of OH^- ions that can react with 2 moles of protons. The equivalent is defined so that 1 equivalent of acid will react with exactly 1 equivalent of base.

For oxidation–reduction reactions, the equivalent is defined as the quantity of oxidizing or reducing agent that can accept or furnish 1 mole of electrons. Thus 1 equivalent of reducing agent will react with exactly 1 equivalent of oxidizing agent. The equivalent mass of an oxidizing or reducing agent can be calculated from the number of electrons in its half-reaction. For example, MnO_4^- reacting in acidic solution absorbs five electrons to produce Mn^{2+}:

$$MnO_4^- + 5e^- + 8H^+ \longrightarrow Mn^{2+} + 4H_2O$$

Since the MnO_4^- ion present in 1 mole of $KMnO_4$ consumes 5 moles of electrons, the equivalent mass is the molar mass divided by 5:

$$\text{Equivalent mass of } KMnO_4 = \frac{\text{molar mass}}{5} = \frac{158\ \text{g}}{5} = 31.6\ \text{g}$$

INTERACTIVE EXAMPLE 10-2

Calculating Various Methods of Solution Composition from the Molarity

The electrolyte in automobile lead storage batteries is a 3.75 M sulfuric acid solution that has a density of 1.230 g/mL. Calculate the mass percent, molality, and normality of the sulfuric acid. ◄

▲ A modern 12-volt lead storage battery of the type used in automobiles.

© Cengage Learning

Solution

What is the density of the solution in grams per liter?

$$1.230\frac{g}{mL} \times \frac{1000 \text{ mL}}{1 \text{ L}} = 1.230 \times 10^3 \text{ g/L}$$

What mass of H_2SO_4 is present in 1.00 L of solution?

We know 1 liter of this solution contains 1230. g of the mixture of sulfuric acid and water. Since the solution is 3.75 *M*, we know that 3.75 moles of H_2SO_4 is present per liter of solution. The number of grams of H_2SO_4 present is

$$3.75 \text{ mol} \times \frac{98.0 \text{ g } H_2SO_4}{1 \text{ mol}} = 368 \text{ g } H_2SO_4$$

How much water is present in 1.00 L of solution?

The amount of water present in 1 liter of solution is obtained from the difference

$$1230. \text{ g solution} - 368 \text{ g } H_2SO_4 = 862 \text{ g } H_2O$$

What is the mass percent?

Since we now know the masses of the solute and solvent, we can calculate the mass percent.

- Mass percent $H_2SO_4 = \dfrac{\text{mass of } H_2SO_4}{\text{mass of solution}} \times 100\% = \dfrac{368 \text{ g}}{1230. \text{ g}} \times 100\%$

$$= 29.9\% \ H_2SO_4$$

What is the molality?

From the moles of solute and the mass of solvent, we can calculate the molality.

- Molality of $H_2SO_4 = \dfrac{\text{moles } H_2SO_4}{\text{kilogram of } H_2O}$

$$= \frac{3.75 \text{ mol } H_2SO_4}{862 \text{ g } H_2O \times \dfrac{1 \text{ kg } H_2O}{1000 \text{ g } H_2O}} = 4.35 \ m$$

What is the normality?

Since each sulfuric acid molecule can furnish two protons, 1 mole of H_2SO_4 represents 2 equivalents. Thus a solution with 3.75 moles of H_2SO_4 per liter contains $2 \times 3.75 = 7.50$ equivalents per liter.

- The normality is 7.50 *N*.

See Exercise 10-37

10-2 | The Energies of Solution Formation

Dissolving solutes in liquids is very common. We dissolve salt in the water used to cook vegetables, sugar in iced tea, stains in cleaning fluid, gaseous carbon dioxide in water to make carbonated water, ethanol in gasoline to make gasohol, and so on.

Solubility is important in other ways. For example, because the pesticide DDT is fat-soluble, it is retained and concentrated in animal tissues, where it causes detrimental effects. This is why DDT, even though it is effective for killing mosquitos, has been banned in the United States. ◄ Also, the solubility of various vitamins is important in determining correct dosages. The insolubility of barium sulfate means it can be used safely to improve X rays of the gastrointestinal tract, even though Ba^{2+} ions are quite toxic.

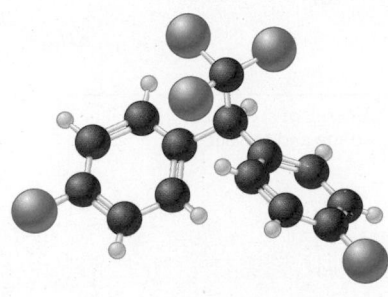

DDT

Figure 10-1 | The formation of a liquid solution can be divided into three steps: (1) expanding the solute, (2) expanding the solvent, and (3) combining the expanded solute and solvent to form the solution. ◄ *i*

i The magnitudes of ΔH_1 and ΔH_2 will depend on the intermolecular forces between the molecules of the solute and solvent, respectively.

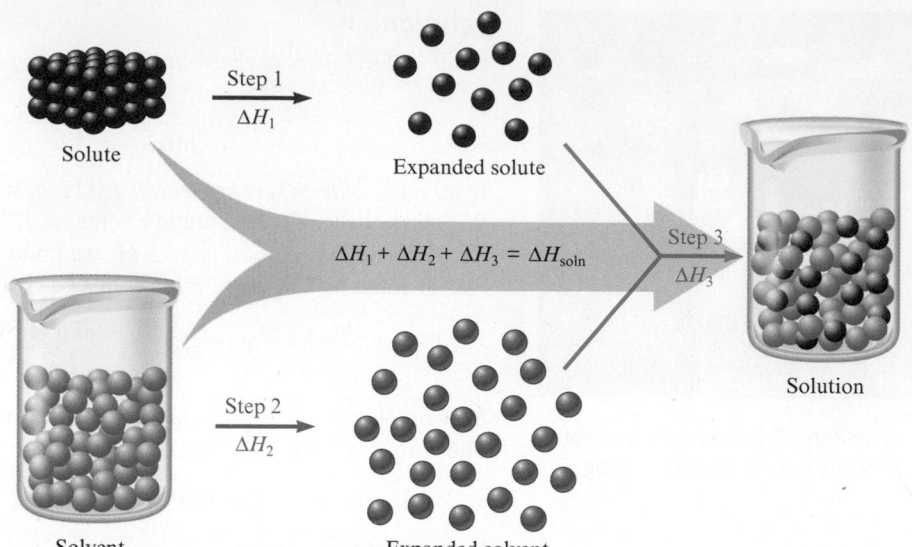

i Polar solvents dissolve polar solutes; nonpolar solvents dissolve nonpolar solutes.

i The enthalpy of solution is the sum of the energies used in expanding both solvent and solute and the energy of solvent–solute interaction.

What factors affect solubility? The cardinal rule of solubility is *like dissolves like*. We find that we must use a polar solvent to dissolve a polar or ionic solute and a nonpolar solvent to dissolve a nonpolar solute. ◄ *i* Now we will try to understand why this behavior occurs. To simplify the discussion, we will assume that the formation of a liquid solution takes place in three distinct steps:

1. Separating the solute into its individual components (expanding the solute)
2. Overcoming intermolecular forces in the solvent to make room for the solute (expanding the solvent)
3. Allowing the solute and solvent to interact to form the solution

These steps are illustrated in Fig. 10-1. Steps 1 and 2 require energy, since forces must be overcome to expand the solute and solvent. Step 3 usually releases energy. In other words, Steps 1 and 2 are endothermic, and Step 3 is often exothermic. The enthalpy change associated with the formation of the solution, called the **enthalpy (heat) of solution** (ΔH_{soln}), is the sum of the ΔH values for the steps:

$$\Delta H_{\text{soln}} = \Delta H_1 + \Delta H_2 + \Delta H_3$$

where ΔH_{soln} may have a positive sign (energy absorbed) or a negative sign (energy released) (Fig. 10-2). ◄ *i*

Figure 10-2 | The heat of solution (a) ΔH_{soln} has a negative sign (the process is exothermic) if Step 3 releases more energy than that required by Steps 1 and 2. (b) ΔH_{soln} has a positive sign (the process is endothermic) if Steps 1 and 2 require more energy than is released in Step 3. (If the energy changes for Steps 1 and 2 equal that for Step 3, then ΔH_{soln} is zero.)

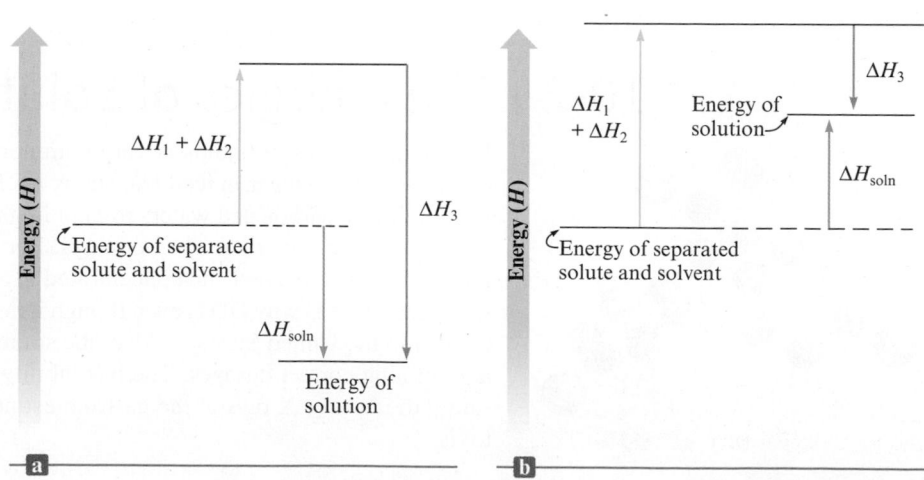

To illustrate the importance of the various energy terms in the equation for ΔH_{soln}, we will consider two specific cases. First, we know that oil is not soluble in water. When oil tankers leak, the petroleum forms an oil slick that floats on the water and is eventually carried onto the beaches. We can explain the immiscibility of oil and water by considering the energy terms involved. Oil is a mixture of nonpolar molecules that interact through London dispersion forces, which depend on molecule size. We expect ΔH_1 to be small for a typical nonpolar solute, but it will be relatively large for the large oil molecules. ◄ ⓘ The term ΔH_3 will be small, since interactions between the nonpolar solute molecules and the polar water molecules will be negligible. However, ΔH_2 will be large and positive because it takes considerable energy to overcome the hydrogen-bonding forces among the water molecules to expand the solvent. Thus ΔH_{soln} will be large and positive because of the ΔH_1 and ΔH_2 terms. Since a large amount of energy would have to be expended to form an oil–water solution, this process does not occur to any appreciable extent. These same arguments hold true for any nonpolar solute and polar solvent—the combination of a nonpolar solute and a highly polar solvent is not expected to produce a solution. ◄

As a second case, let's consider the solubility of an ionic solute, such as sodium chloride, in water. Here the term ΔH_1 is large and positive because the strong ionic forces in the crystal must be overcome, and ΔH_2 is large and positive because hydrogen bonds must be broken in the water. Finally, ΔH_3 is large and negative because of the strong interactions between the ions and the water molecules. In fact, the exothermic and endothermic terms essentially cancel, as shown from the known values:

$$NaCl(s) \longrightarrow Na^+(g) + Cl^-(g) \qquad \Delta H_1 = 786 \text{ kJ/mol}$$

$$H_2O(l) + Na^+(g) + Cl^-(g) \longrightarrow Na^+(aq) + Cl^-(aq) \qquad \Delta H_{hyd} = \Delta H_2 + \Delta H_3$$
$$= -783 \text{ kJ/mol}$$

Here the **enthalpy (heat) of hydration** (ΔH_{hyd}) combines the terms ΔH_2 (for expanding the solvent) and ΔH_3 (for solvent–solute interactions). The heat of hydration represents the enthalpy change associated with the dispersal of a gaseous solute in water. Thus the heat of solution for dissolving sodium chloride is the sum of ΔH_1 and ΔH_{hyd}:

$$\Delta H_{soln} = 786 \text{ kJ/mol} - 783 \text{ kJ/mol} = 3 \text{ kJ/mol}$$

Note that ΔH_{soln} is small but positive; the dissolving process requires a small amount of energy. Then why is NaCl so soluble in water? The answer lies in nature's tendency toward higher probability of the mixed state. That is, processes naturally run in the direction that leads to the most probable state. For example, imagine equal numbers of orange and yellow spheres separated by a partition [Fig. 10-3(a)]. If we remove the partition and shake the container, the spheres will mix [Fig. 10-3(b)], and no amount of shaking will cause them to return to the state of separated orange and yellow. Why? The mixed state is simply much more likely to occur (more probable) than the original separate state because there are many more ways of placing the spheres to give a mixed state than a separated state. This is a general principle. *One factor that favors a process is an increase in probability.*

ⓘ ΔH_1 is expected to be small for nonpolar solutes but can be large for large molecules.

Anglianart/Dreamstime.com

▲ Gasoline floating on water. Since gasoline is nonpolar, it is immiscible with water, because water contains polar molecules.

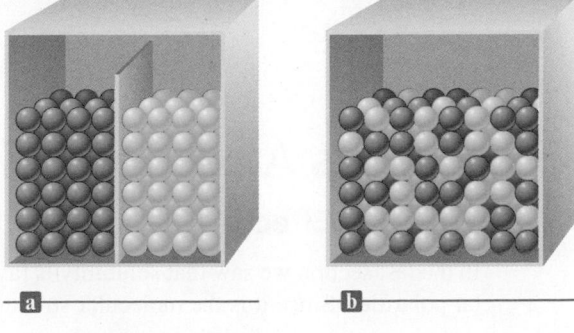

Figure 10-3 | (a) Orange and yellow spheres separated by a partition in a closed container. (b) The spheres after the partition is removed and the container has been shaken for some time.

a   b

Table 10-3 | The Energy Terms for Various Types of Solutes and Solvents

	ΔH_1	ΔH_2	ΔH_3	ΔH_{soln}	Outcome
Polar solute, polar solvent	Large	Large	Large, negative	Small	Solution forms
Nonpolar solute, polar solvent	Small	Large	Small	Large, positive	No solution forms
Nonpolar solute, nonpolar solvent	Small	Small	Small	Small	Solution forms
Polar solute, nonpolar solvent	Large	Small	Small	Large, positive	No solution forms

ℹ️ The factors that act as driving forces for a process are discussed more fully in Chapter 16.

But energy considerations are also important. *Processes that require large amounts of energy tend not to occur.* ◄ ℹ️ Since dissolving 1 mole of solid NaCl requires only a small amount of energy, the solution forms, presumably because of the large increase in the probability of the state when the solute and solvent are mixed.

The various possible cases for solution formation are summarized in Table 10-3. Note that in two cases, polar–polar and nonpolar–nonpolar, the heat of solution is expected to be small. In these cases, the solution forms because of the increase in the probability of the mixed state. In the other cases (polar–nonpolar and nonpolar–polar), the heat of solution is expected to be large and positive, and the large quantity of energy required acts to prevent the solution from forming. Although this discussion has greatly oversimplified the complex driving forces for solubility, these ideas are a useful starting point for understanding the observation that *like dissolves like.*

Critical Thinking

You and a friend are studying for a chemistry exam. What if your friend tells you, "Since exothermic processes are favored and the sign of the enthalpy change tells us whether or not a process is endothermic or exothermic, the sign of ΔH_{soln} tells us whether or not a solution will form"? How would you explain to your friend that this conclusion is not correct? What part, if any, of what your friend says is correct?

INTERACTIVE EXAMPLE 10-3

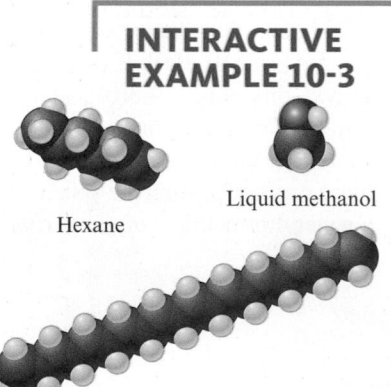

Hexane

Liquid methanol

Grease

Differentiating Solvent Properties

Decide whether liquid hexane (C_6H_{14}) or liquid methanol (CH_3OH) is the more appropriate solvent for the substances grease ($C_{20}H_{42}$) and potassium iodide (KI). ◄

Solution

Hexane is a nonpolar solvent because it contains C—H bonds. Thus hexane will work best for the nonpolar solute grease. Methanol has an O—H group that makes it significantly polar. Thus it will serve as the better solvent for the ionic solid KI.

See Exercises 10-43 through 10-45

10-3 | Factors Affecting Solubility

Structure Effects

In the last section we saw that solubility is favored if the solute and solvent have similar polarities. Since it is the molecular structure that determines polarity, there should

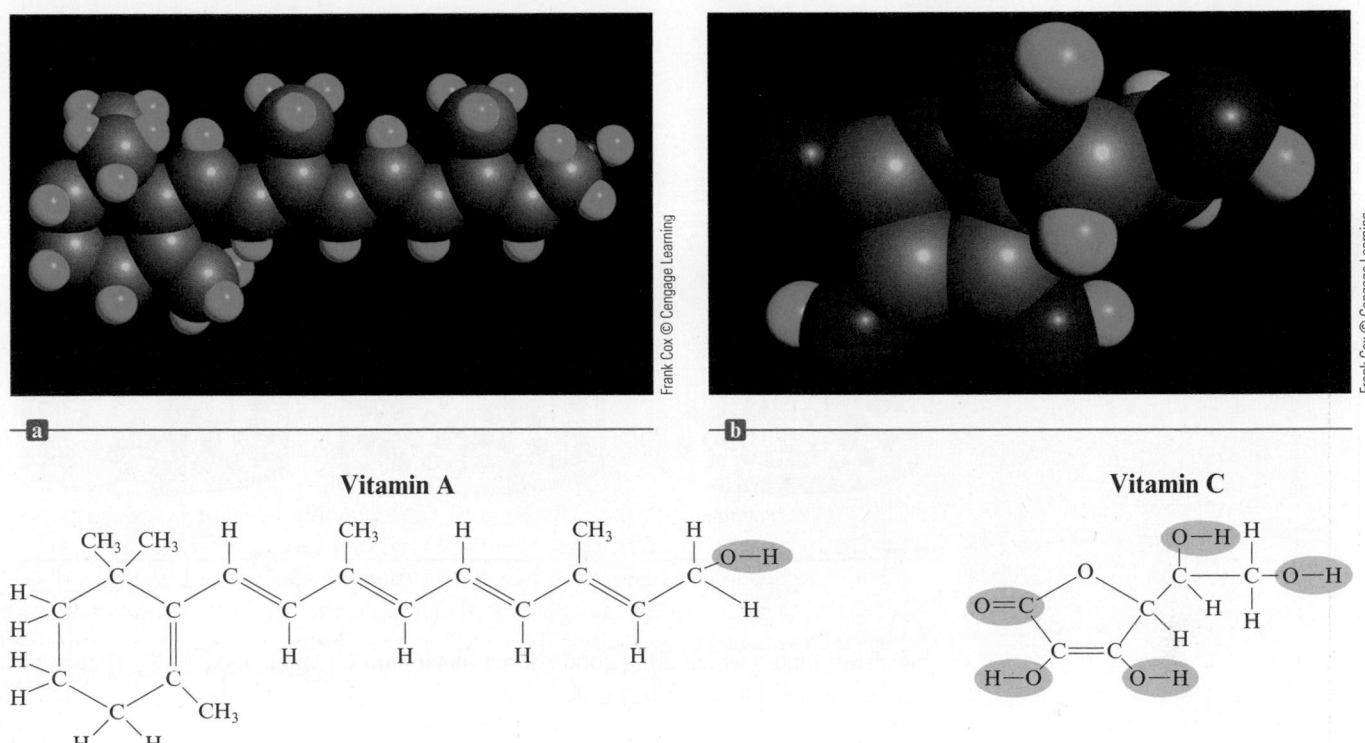

Figure 10-4 | The molecular structures of (a) vitamin A (nonpolar, fat-soluble) and (b) vitamin C (polar, water-soluble). The circles in the structural formulas indicate polar bonds. Note that vitamin C contains far more polar bonds than vitamin A. Frank Cox © Cengage Learning.

be a definite connection between structure and solubility. Vitamins provide an excellent example of the relationship among molecular structure, polarity, and solubility.

Recently, there has been considerable publicity about the pros and cons of consuming large quantities of vitamins. For example, large doses of vitamin C have been advocated to combat various illnesses, including the common cold. Vitamin E has been extolled as a youth-preserving elixir and a protector against the carcinogenic (cancer-causing) effects of certain chemicals. However, there are possible detrimental effects from taking large amounts of some vitamins, depending on their solubilities.

Vitamins can be divided into two classes: *fat-soluble* (vitamins A, D, E, and K) and *water-soluble* (vitamins B and C). The reason for the differing solubility characteristics can be seen by comparing the structures of vitamins A and C (Fig. 10-4). Vitamin A, composed mostly of carbon and hydrogen atoms that have similar electronegativities, is virtually nonpolar. This causes it to be soluble in nonpolar materials such as body fat, which is also largely composed of carbon and hydrogen, but not soluble in polar solvents such as water. On the other hand, vitamin C has many polar O—H and C—O bonds, making the molecule polar and thus water-soluble. We often describe nonpolar materials such as vitamin A as *hydrophobic* (water-fearing) and polar substances such as vitamin C as *hydrophilic* (water-loving).

Because of their solubility characteristics, the fat-soluble vitamins can build up in the fatty tissues of the body. This has both positive and negative effects. Since these vitamins can be stored, the body can tolerate for a time a diet deficient in vitamin A, D, E, or K. Conversely, if excessive amounts of these vitamins are consumed, their buildup can lead to the illness *hypervitaminosis.*

In contrast, the water-soluble vitamins are excreted by the body and must be consumed regularly. This fact was first recognized when the British navy discovered that scurvy, a disease often suffered by sailors, could be prevented if the sailors regularly

Figure 10-5 | (a) A gaseous solute in equilibrium with a solution. (b) The piston is pushed in, which increases the pressure of the gas and the number of gas molecules per unit volume. This causes an increase in the rate at which the gas enters the solution, so the concentration of dissolved gas increases. (c) The greater gas concentration in the solution causes an increase in the rate of escape. A new equilibrium is reached.

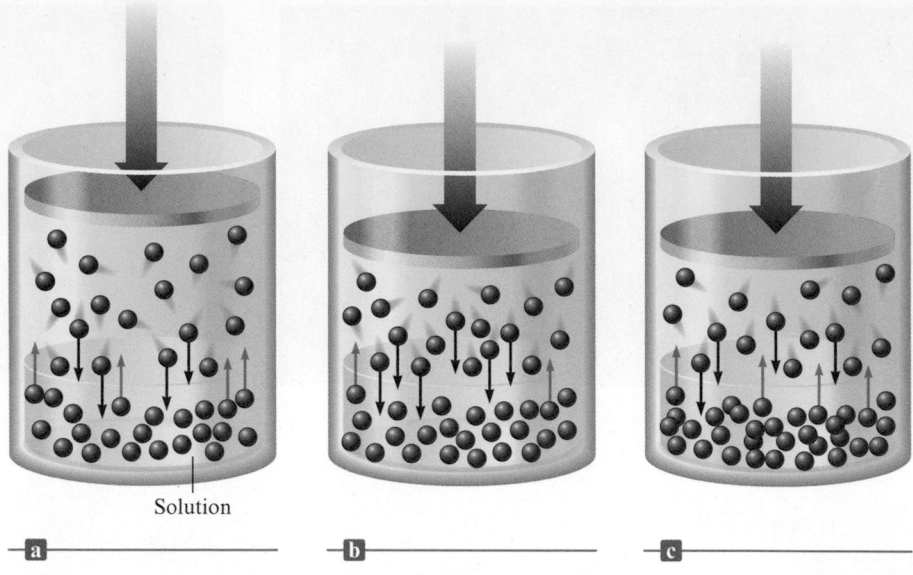

Solution

ate fresh limes (which are a good source of vitamin C) when aboard ship (hence the name "limey" for the British sailor).

Pressure Effects

While pressure has little effect on the solubilities of solids or liquids, it does significantly increase the solubility of a gas. Carbonated beverages, for example, are always bottled at high pressures of carbon dioxide to ensure a high concentration of carbon dioxide in the liquid. The fizzing that occurs when you open a can of soda results from the escape of gaseous carbon dioxide because under these conditions the pressure of CO_2 above the solution is now much lower than that used in the bottling process.

The increase in gas solubility with pressure can be understood from Fig. 10-5. Figure 10-5(a) shows a gas in equilibrium with a solution; that is, the gas molecules are entering and leaving the solution at the same rate. If the pressure is suddenly increased [Fig. 10-5(b)], the number of gas molecules per unit volume increases, and the gas enters the solution at a higher rate than it leaves. As the concentration of dissolved gas increases, the rate of the escape of the gas also increases until a new equilibrium is reached [Fig. 10-5(c)], where the solution contains more dissolved gas than before.

The relationship between gas pressure and the concentration of dissolved gas is given by **Henry's law**:

$$C = kP$$

where C represents the concentration of the dissolved gas, k is a constant characteristic of a particular solution, and P represents the partial pressure of the gaseous solute above the solution. In words, Henry's law states that *the amount of a gas dissolved in a solution is directly proportional to the pressure of the gas above the solution.* ◄ ⓘ

Henry's law is obeyed most accurately for dilute solutions of gases that do not dissociate in or react with the solvent. ◄ ⓘ For example, Henry's law is obeyed by oxygen gas in water, but it does *not* correctly represent the behavior of gaseous hydrogen chloride in water because of the dissociation reaction

$$HCl(g) \xrightarrow{\text{H}_2\text{O}} H^+(aq) + Cl^-(aq)$$

ⓘ William Henry (1774–1836), a close friend of John Dalton, formulated his law in 1801.

ⓘ Henry's law holds only when there is no chemical reaction between the solute and solvent.

INTERACTIVE EXAMPLE 10-4

▲ Carbonation in a glass of soda.

doug3437/iStockphoto.com

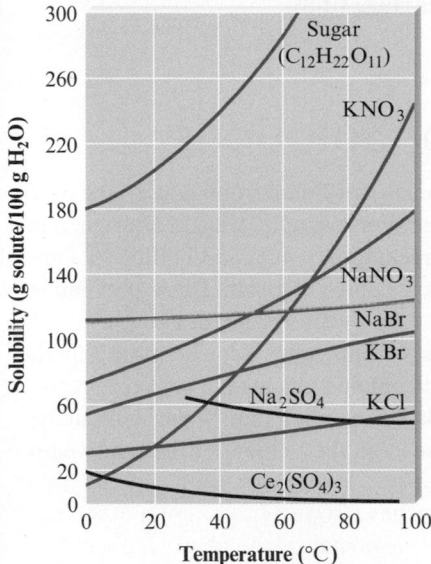

Figure 10-6 | The solubilities of several solids as a function of temperature. Note that while most substances become more soluble in water with increasing temperature, sodium sulfate and cerium sulfate become less soluble.

ⓘ ΔH°_{soln} refers to the formation of a 1.0-M ideal solution and is not necessarily relevant to the process of dissolving a solid in a saturated solution. Thus ΔH°_{soln} is of limited use in predicting the variation of solubility with temperature.

Calculations Using Henry's Law

A certain soft drink is bottled so that a bottle at 25°C contains CO_2 gas at a pressure of 5.0 atm over the liquid. ◄ Assuming that the partial pressure of CO_2 in the atmosphere is 4.0×10^{-4} atm, calculate the equilibrium concentrations of CO_2 in the soda both before and after the bottle is opened. The Henry's law constant for CO_2 in aqueous solution is 3.1×10^{-2} mol/L · atm at 25°C.

Solution

What is Henry's law for CO_2?

$$C_{CO_2} = k_{CO_2}P_{CO_2}$$

where $k_{CO_2} = 3.1 \times 10^{-2}$ mol/L · atm.

What is the C_{CO_2} in the unopened bottle?
In the *unopened* bottle, $P_{CO_2} = 5.0$ atm and

■ $C_{CO_2} = k_{CO_2}P_{CO_2} = (3.1 \times 10^{-2} \text{ mol/L} \cdot \text{atm})(5.0 \text{ atm}) = 0.16$ mol/L

What is the C_{CO_2} in the opened bottle?
In the *opened* bottle, the CO_2 in the soda eventually reaches equilibrium with the atmospheric CO_2, so $P_{CO_2} = 4.0 \times 10^{-4}$ atm and

■ $C_{CO_2} = k_{CO_2}P_{CO_2} = \left(3.1 \times 10^{-2} \dfrac{\text{mol}}{\text{L} \cdot \text{atm}}\right)(4.0 \times 10^{-4} \text{ atm}) = 1.2 \times 10^{-5}$ mol/L

Note the large change in concentration of CO_2. This is why soda goes "flat" after being open for a while.

See Exercises 10-49 and 10-50

Temperature Effects (for Aqueous Solutions)

Everyday experiences of dissolving substances such as sugar may lead you to think that solubility always increases with temperature. This is not the case. The dissolving of a solid occurs *more rapidly* at higher temperatures, but the amount of solid that can be dissolved may increase or decrease with increasing temperature. The effect of temperature on the solubility in water of several solids is shown in Fig. 10-6. Note that although the solubility of most solids in water increases with temperature, the solubilities of some substances (such as sodium sulfate and cerium sulfate) decrease with increasing temperature.

Predicting the temperature dependence of solubility is very difficult. For example, although there is some correlation between the sign of ΔH°_{soln} and the variation of solubility with temperature, important exceptions exist.* The only sure way to determine the temperature dependence of a solid's solubility is by experiment. ◄ ⓘ

The behavior of gases dissolving in water appears less complex. The solubility of a gas in water typically decreases with increasing temperature,[†] as is shown for several cases in Fig. 10-7. This temperature effect has important environmental implications because of the widespread use of water from lakes and rivers for industrial cooling. After being used, the water is returned to its natural source at a higher than ambient temperature (**thermal pollution** has occurred). Because it is warmer, this water contains less than the normal concentration of oxygen and is also less dense; it tends to

*For more information see R. S. Treptow, "Le Châtelier's Principle Applied to the Temperature Dependence of Solubility," *J. Chem. Ed.* **61** (1984): 499.
[†]The opposite behavior is observed for most nonaqueous solvents.

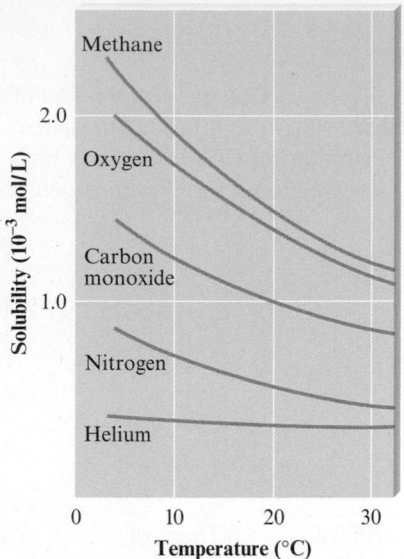

Figure 10-7 | The solubilities of several gases in water as a function of temperature at a constant pressure of 1 atm of gas above the solution.

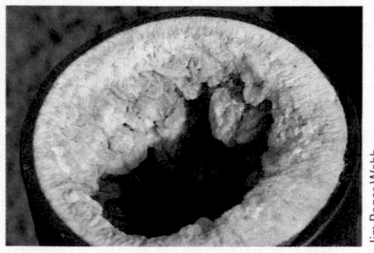

Figure 10-8 | A pipe with accumulated mineral deposits. The cross section clearly indicates the reduction in pipe capacity.

"float" on the colder water below, thus blocking normal oxygen absorption. This effect can be especially important in deep lakes. The warm upper layer can seriously decrease the amount of oxygen available to aquatic life in the deeper layers of the lake.

The decreasing solubility of gases with increasing temperature is also responsible for the formation of *boiler scale*. As we will see in more detail in Chapter 13, the bicarbonate ion is formed when carbon dioxide is dissolved in water containing the carbonate ion:

$$CO_3^{2-}(aq) + CO_2(aq) + H_2O(l) \longrightarrow 2HCO_3^-(aq)$$

When the water also contains Ca^{2+} ions, this reaction is especially important—calcium bicarbonate is soluble in water, but calcium carbonate is insoluble. When the water is heated, the carbon dioxide is driven off. For the system to replace the lost carbon dioxide, the reverse reaction must occur:

$$2HCO_3^-(aq) \longrightarrow H_2O(l) + CO_2(aq) + CO_3^{2-}(aq)$$

This reaction, however, also increases the concentration of carbonate ions, causing solid calcium carbonate to form. This solid is the boiler scale that coats the walls of containers such as industrial boilers and tea kettles. Boiler scale reduces the efficiency of heat transfer and can lead to blockage of pipes (Fig. 10-8).

10-4 | The Vapor Pressures of Solutions

Liquid solutions have physical properties significantly different from those of the pure solvent, a fact that has great practical importance. For example, we add antifreeze to the water in a car's cooling system to prevent freezing in winter and boiling in summer. We also melt ice on sidewalks and streets by spreading salt. These preventive measures work because of the solute's effect on the solvent's properties.

To explore how a nonvolatile solute affects a solvent, we will consider the experiment represented in Fig. 10-9, in which a sealed container encloses a beaker containing an aqueous sulfuric acid solution and a beaker containing pure water. Gradually, the volume of the sulfuric acid solution increases and the volume of the pure water

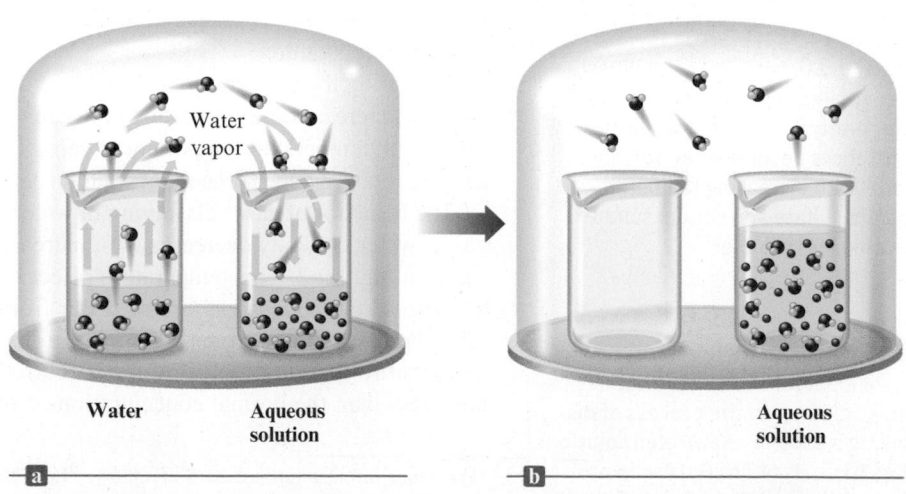

Figure 10-9 | An aqueous solution and pure water in a closed environment. (a) Initial stage. (b) After a period of time, the water is transferred to the solution.

On August 21, 1986, a cloud of gas suddenly boiled from Lake Nyos in Cameroon, killing nearly 2000 people. Although at first it was speculated that the gas was hydrogen sulfide, it now seems clear it was carbon dioxide. What would cause Lake Nyos to emit this huge, suffocating cloud of CO_2? Although the answer may never be known for certain, many scientists believe that the lake suddenly "turned over," bringing to the surface water that contained huge quantities of dissolved carbon dioxide. Lake Nyos is a deep lake that is thermally stratified: Layers of warm, less dense water near the surface float on the colder, denser water layers near the lake's bottom. Under normal conditions the lake stays this way; there is little mixing among the different layers. Scientists believe that over hundreds or thousands of years, carbon dioxide gas had seeped into the cold water at the lake's bottom and dissolved in great amounts because of the large pressure of CO_2 present (in accordance with Henry's law). For some reason on August 21, 1986, the lake apparently suffered an overturn, possibly due to wind or to unusual cooling of the lake's surface by monsoon clouds. This caused water that was greatly supersaturated with CO_2 to reach the surface and release tremendous quantities of gaseous CO_2 that suffocated thousands of humans and animals before they knew what hit them—a tragic, monumental illustration of Henry's law.

Since 1986 the scientists studying Lake Nyos and nearby Lake Monoun have observed a rapid recharging of the CO_2 levels in the deep waters of these lakes, causing concern that another deadly gas release could occur at any time. Apparently the only way to prevent such a disaster is to pump away the CO_2-charged deep water in the two lakes. Scientists at a conference to study this problem in 1994 recommended such a solution, but it has not yet been funded by Cameroon.

Lake Nyos in Cameroon one month after a carbon dioxide eruption

decreases. Why? We can explain this observation if the vapor pressure of the pure solvent is greater than that of the solution. Under these conditions, the pressure of vapor necessary to achieve equilibrium with the pure solvent is greater than that required to reach equilibrium with the aqueous acid solution. Thus as the pure solvent emits vapor to attempt to reach equilibrium, the aqueous sulfuric acid solution absorbs vapor to try to lower the vapor pressure toward its equilibrium value. This process results in a net transfer of water from the pure water through the vapor phase to the sulfuric acid solution. The system can reach an equilibrium vapor pressure only when all the water is transferred to the solution. This experiment is just one of many observations indicating that the presence of a *nonvolatile solute lowers the vapor pressure of a solvent.* ◄ ⓘ

ⓘ A nonvolatile solute has no tendency to escape from solution into the vapor phase.

We can account for this behavior in terms of the simple model shown in Fig. 10-10. The dissolved nonvolatile solute decreases the number of solvent molecules per unit volume, and it should proportionately lower the escaping tendency of the solvent molecules. For example, in a solution consisting of half nonvolatile solute molecules and half solvent molecules, we might expect the observed vapor pressure to be half that of the pure solvent, since only half as many molecules can escape. In fact, this is what is observed.

Detailed studies of the vapor pressures of solutions containing nonvolatile solutes were carried out by François M. Raoult (1830–1901). His results are described by the equation known as **Raoult's law**:

$$P_{\text{soln}} = \chi_{\text{solvent}} P^0_{\text{solvent}}$$

425

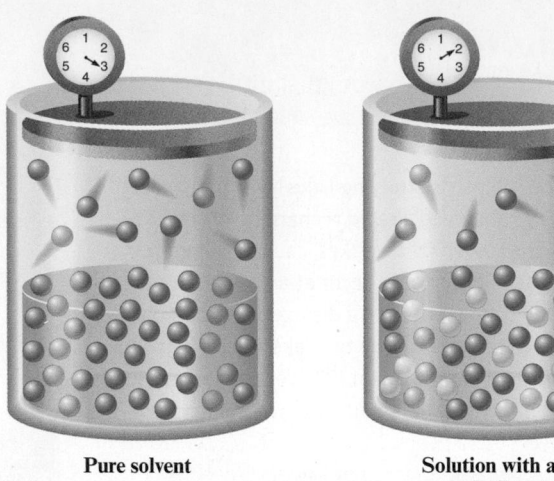

Pure solvent   **Solution with a nonvolatile solute**

Figure 10-10 | The presence of a nonvolatile solute inhibits the escape of solvent molecules from the liquid and so lowers the vapor pressure of the solvent.

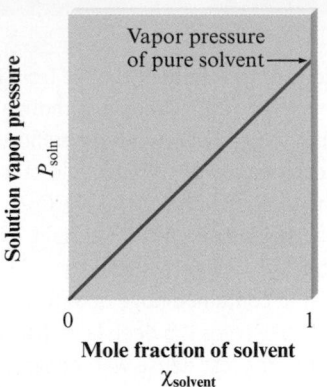

Figure 10-11 | For a solution that obeys Raoult's law, a plot of P_{soln} versus $\chi_{solvent}$ gives a straight line.

where P_{soln} is the observed vapor pressure of the solution, $\chi_{solvent}$ is the mole fraction of solvent, and $P^0_{solvent}$ is the vapor pressure of the pure solvent. Note that for a solution of half solute and half solvent molecules, $\chi_{solvent}$ is 0.5, so the vapor pressure of the solution is half that of the pure solvent. On the other hand, for a solution in which three-fourths of the solution molecules are solvent, $\chi_{solvent} = \frac{3}{4} = 0.75$, and $P_{soln} = 0.75P^0_{solvent}$. The idea is that the nonvolatile solute simply dilutes the solvent.

Raoult's law is a linear equation of the form $y = mx + b$, where $y = P_{soln}$, $x = \chi_{solvent}$, $m = P^0_{solvent}$, and $b = 0$. Thus a plot of P_{soln} versus $\chi_{solvent}$ gives a straight line with a slope equal to $P^0_{solvent}$ (Fig. 10-11).

INTERACTIVE EXAMPLE 10-5

Calculating the Vapor Pressure of a Solution

Calculate the expected vapor pressure at 25°C for a solution prepared by dissolving 158.0 g common table sugar (sucrose, molar mass = 342.3 g/mol) in 643.5 cm³ of water. At 25°C, the density of water is 0.9971 g/cm³ and the vapor pressure is 23.76 torr.

Solution

What is Raoult's law for this case?

$$P_{soln} = \chi_{H_2O}P^0_{H_2O}$$

To calculate the mole fraction of water in the solution, we must first determine the number of moles of sucrose and the moles of water present.

What are the moles of sucrose?

$$\text{Moles of sucrose} = 158.0 \text{ g sucrose} \times \frac{1 \text{ mol sucrose}}{342.3 \text{ g sucrose}}$$

$$= 0.4616 \text{ mol sucrose}$$

What are the moles of water?

To determine the moles of water present, we first convert volume to mass using the density:

$$643.5 \text{ cm}^3 \text{ H}_2\text{O} \times \frac{0.9971 \text{ g H}_2\text{O}}{\text{cm}^3 \text{ H}_2\text{O}} = 641.6 \text{ g H}_2\text{O}$$

The number of moles of water is therefore

$$641.6 \text{ g H}_2\text{O} \times \frac{1 \text{ mol H}_2\text{O}}{18.02 \text{ g H}_2\text{O}} = 35.60 \text{ mol H}_2\text{O}$$

What is the mole fraction of water in the solution?

$$\chi_{\text{H}_2\text{O}} = \frac{\text{mol H}_2\text{O}}{\text{mol H}_2\text{O} + \text{mol sucrose}} = \frac{35.60 \text{ mol}}{35.60 \text{ mol} + 0.4616 \text{ mol}}$$

$$= \frac{35.60 \text{ mol}}{36.06 \text{ mol}} = 0.9873$$

■ The vapor pressure of the solution is: $P_{\text{soln}} = \chi_{\text{H}_2\text{O}} P^0_{\text{H}_2\text{O}} = (0.9872)(23.76 \text{ torr})$
    $= 23.46 \text{ torr}$

Thus the vapor pressure of water has been lowered from 23.76 torr in the pure state to 23.46 torr in the solution. The vapor pressure has been lowered by 0.30 torr.

See Exercises 10-51 and 10-52

The phenomenon of the lowering of the vapor pressure gives us a convenient way to "count" molecules and thus provides a means for experimentally determining molar masses. Suppose a certain mass of a compound is dissolved in a solvent and the vapor pressure of the resulting solution is measured. Using Raoult's law, we can determine the number of moles of solute present. Since the mass of this number of moles is known, we can calculate the molar mass.

ⓘ The lowering of vapor pressure depends on the number of solute particles present in the solution.

We also can use vapor pressure measurements to characterize solutions. For example, 1 mole of sodium chloride dissolved in water lowers the vapor pressure approximately twice as much as expected because the solid has two ions per formula unit, which separate when it dissolves. Thus vapor pressure measurements can give valuable information about the nature of the solute after it dissolves. ◄ *ⓘ*

INTERACTIVE EXAMPLE 10-6

Calculating the Vapor Pressure of a Solution Containing Ionic Solute

Predict the vapor pressure of a solution prepared by mixing 35.0 g solid Na_2SO_4 (molar mass = 142.05 g/mol) with 175 g water at 25°C. The vapor pressure of pure water at 25°C is 23.76 torr.

Solution

First, we need to know the mole fraction of H_2O.

$$n_{\text{H}_2\text{O}} = 175 \text{ g H}_2\text{O} \times \frac{1 \text{ mol H}_2\text{O}}{18.02 \text{ g H}_2\text{O}} = 9.71 \text{ mol H}_2\text{O}$$

$$n_{\text{Na}_2\text{SO}_4} = 35.0 \text{ g Na}_2\text{SO}_4 \times \frac{1 \text{ mol Na}_2\text{SO}_4}{142.05 \text{ g Na}_2\text{SO}_4} = 0.246 \text{ mol Na}_2\text{SO}_4$$

It is essential to recognize that when 1 mole of solid Na_2SO_4 dissolves, it produces 2 moles of Na^+ ions and 1 mole of SO_4^{2-} ions. Thus the number of solute particles present in this solution is three times the number of moles of solute dissolved:

$$n_{\text{solute}} = 3(0.246) = 0.738 \text{ mol}$$

$$\chi_{\text{H}_2\text{O}} = \frac{n_{\text{H}_2\text{O}}}{n_{\text{solute}} + n_{\text{H}_2\text{O}}} = \frac{9.71 \text{ mol}}{0.738 \text{ mol} + 9.72 \text{ mol}} = \frac{9.71}{10.458} = 0.929$$

Now we can use Raoult's law to predict the vapor pressure:

$$P_{\text{soln}} = \chi_{\text{H}_2\text{O}} P^0_{\text{H}_2\text{O}} = (0.929)(23.76 \text{ torr}) = 22.1 \text{ torr}$$

See Exercise 10-56

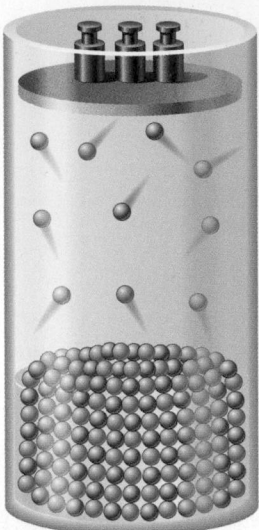

Figure 10-12 | When a solution contains two volatile components, both contribute to the total vapor pressure. Note that in this case the solution contains equal numbers of the components ● and ●, but the vapor contains more ● than ●. This means that component ● is more volatile (has a higher vapor pressure as a pure liquid) than component ●.

ⓘ Strong solute–solvent interaction gives a vapor pressure lower than that predicted by Raoult's law.

ⓘ The strength of the interactions is determined by the nature of the intermolecular forces. This knowledge comes from a consideration of the structure of the molecules in question.

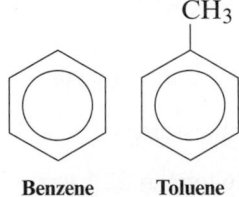

Benzene   Toluene

Nonideal Solutions

So far we have assumed that the solute is nonvolatile and so does not contribute to the vapor pressure over the solution. However, for liquid–liquid solutions where both components are volatile, a modified form of Raoult's law applies:

$$P_{TOTAL} = P_A + P_B = \chi_A P_A^0 + \chi_B P_B^0$$

where P_{TOTAL} represents the total vapor pressure of a solution containing A and B, χ_A and χ_B are the mole fractions of A and B, P_A^0 and P_B^0 are the vapor pressures of pure A and pure B, and P_A and P_B are the partial pressures resulting from molecules of A and of B in the vapor above the solution (Fig. 10-12).

A liquid–liquid solution that obeys Raoult's law is called an **ideal solution**. Raoult's law is to solutions what the ideal gas law is to gases. As with gases, ideal behavior for solutions is never perfectly achieved but is sometimes closely approached. Nearly ideal behavior is often observed when the solute–solute, solvent–solvent, and solute–solvent interactions are very similar. That is, in solutions where the solute and solvent are very much alike, the solute simply acts to dilute the solvent. However, if the solvent has a special affinity for the solute, such as if hydrogen bonding occurs, the tendency of the solvent molecules to escape will be lowered more than expected. The observed vapor pressure will be *lower* than the value predicted by Raoult's law; there will be a *negative deviation from Raoult's law*. ◄ ⓘ

When a solute and solvent release large quantities of energy in the formation of a solution, that is, when ΔH_{soln} is large and negative, we can assume that strong interactions exist between the solute and solvent. In this case we expect a negative deviation from Raoult's law, because both components will have a lower escaping tendency in the solution than in the pure liquids. This behavior is illustrated by an acetone–water solution, where the molecules can hydrogen-bond effectively:

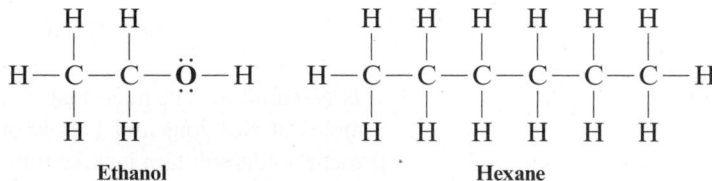

In contrast, if two liquids mix endothermically, it indicates that the solute–solvent interactions are weaker than the interactions among the molecules in the pure liquids. ◄ ⓘ More energy is required to expand the liquids than is released when the liquids are mixed. In this case the molecules in the solution have a higher tendency to escape than expected, and *positive* deviations from Raoult's law are observed (Fig. 10-13). An example of this case is provided by a solution of ethanol and hexane, whose Lewis structures are as follows:

Ethanol   Hexane

The polar ethanol and the nonpolar hexane molecules are not able to interact effectively. Thus the enthalpy of solution is positive, as is the deviation from Raoult's law.

Finally, for a solution of very similar liquids, such as benzene and toluene (shown in margin), the enthalpy of solution is very close to zero, and thus the solution closely obeys Raoult's law (ideal behavior). ◄

A summary of the behavior of various types of solutions is given in Table 10-4.

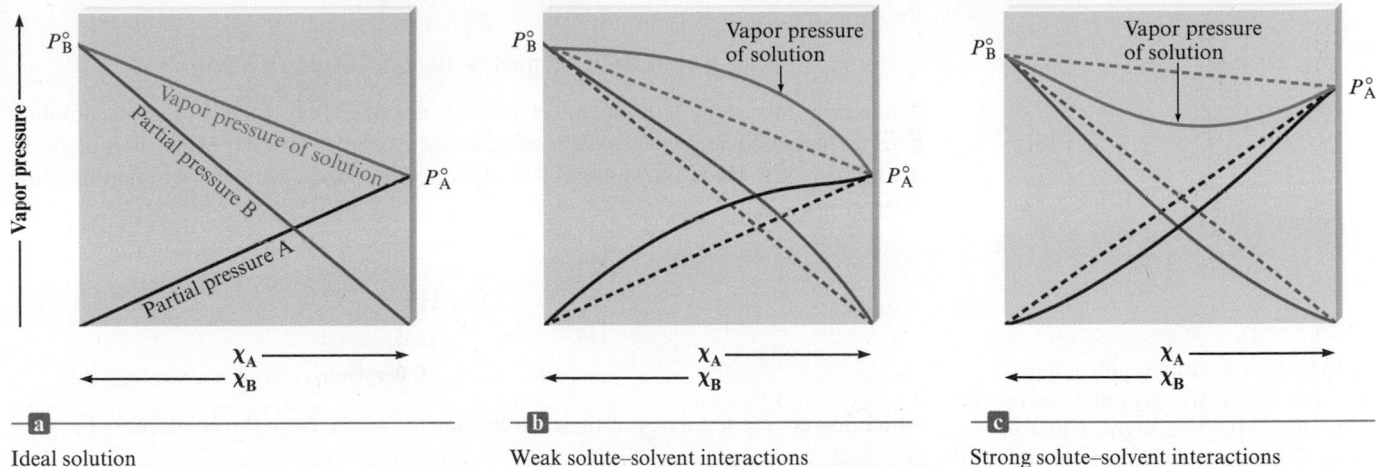

Figure 10-13 | Vapor pressure for a solution of two volatile liquids. (a) The behavior predicted for an ideal liquid–liquid solution by Raoult's law. (b) A solution for which P_{TOTAL} is larger than the value calculated from Raoult's law. This solution shows a positive deviation from Raoult's law. (c) A solution for which P_{TOTAL} is smaller than the value calculated from Raoult's law. This solution shows a negative deviation from Raoult's law.

Table 10-4 | Summary of the Behavior of Various Types of Solutions

Interactive Forces Between Solute (A) and Solvent (B) Particles	ΔH_{soln}	ΔT for Solution Formation	Deviation from Raoult's Law	Example
$A \leftrightarrow A, B \leftrightarrow B \equiv A \leftrightarrow B$	Zero	Zero	None (ideal solution)	Benzene–toluene
$A \leftrightarrow A, B \leftrightarrow B < A \leftrightarrow B$	Negative (exothermic)	Positive	Negative	Acetone–water
$A \leftrightarrow A, B \leftrightarrow B > A \leftrightarrow B$	Positive (endothermic)	Negative	Positive	Ethanol–hexane

INTERACTIVE EXAMPLE 10-7

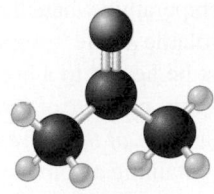

Acetone

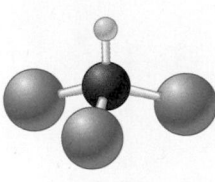

Chloroform

Calculating the Vapor Pressure of a Solution Containing Two Liquids

A solution is prepared by mixing 5.81 g acetone (C_3H_6O, molar mass = 58.1 g/mol) and 11.9 g chloroform ($HCCl_3$, molar mass = 119.4 g/mol). ◄ At 35°C, this solution has a total vapor pressure of 260. torr. Is this an ideal solution? The vapor pressures of pure acetone and pure chloroform at 35°C are 345 and 293 torr, respectively.

Solution

To decide whether this solution behaves ideally, we first calculate the expected vapor pressure using Raoult's law:

$$P_{TOTAL} = \chi_A P_A^0 + \chi_C P_C^0$$

where A stands for acetone and C stands for chloroform. The calculated value can then be compared with the observed vapor pressure.

First, we must calculate the number of moles of acetone and chloroform:

$$5.81 \text{ g acetone} \times \frac{1 \text{ mol acetone}}{58.1 \text{ g acetone}} = 0.100 \text{ mol acetone}$$

$$11.9 \text{ g chloroform} \times \frac{1 \text{ mol chloroform}}{119 \text{ g chloroform}} = 0.100 \text{ mol chloroform}$$

Since the solution contains equal numbers of moles of acetone and chloroform, that is,

$$\chi_A = 0.500 \quad \text{and} \quad \chi_C = 0.500$$

the expected vapor pressure is

$$P_{\text{TOTAL}} = (0.500)(345 \text{ torr}) + (0.500)(293 \text{ torr}) = 319 \text{ torr}$$

Comparing this value with the observed pressure of 260. torr shows that the solution does not behave ideally. The observed value is lower than that expected. This negative deviation from Raoult's law can be explained in terms of the hydrogen-bonding interaction

In this case the usually nonpolar C—H bond is strongly polarized by the three attached, highly electronegative chlorine atoms, thus producing hydrogen bonding.

$$
\begin{array}{ccc}
\text{CH}_3 & & \text{Cl} \\
& \diagdown & | \\
& \text{C}=\text{O}\cdots\text{H}-\text{C}-\text{Cl} \\
& \diagup & | \\
\text{CH}_3 & \delta- \quad \delta+ & \text{Cl}
\end{array}
$$

Acetone   **Chloroform**

which lowers the tendency of these molecules to escape from the solution. ◄ *ℹ*

See Exercises 10-57, 10-58, 10-64, and 10-99

10-5 | Boiling-Point Elevation and Freezing-Point Depression

In the preceding section we saw how a solute affects the vapor pressure of a liquid solvent. Because changes of state depend on vapor pressure, the presence of a solute also affects the freezing point and boiling point of a solvent. Freezing-point depression, boiling-point elevation, and osmotic pressure (discussed in Section 10-6) are called **colligative properties**. As we will see, they are grouped together because they depend only on the number, and not on the identity, of the solute particles in an ideal solution. Because of their direct relationship to the number of solute particles, the colligative properties are very useful for characterizing the nature of a solute after it is dissolved in a solvent and for determining molar masses of substances.

Boiling-Point Elevation

The normal boiling point of a liquid occurs at the temperature where the vapor pressure is equal to 1 atmosphere. We have seen that a nonvolatile solute lowers the vapor pressure of the solvent. Therefore, such a solution must be heated to a higher temperature than the boiling point of the pure solvent to reach a vapor pressure of 1 atmosphere. This means that *a nonvolatile solute elevates the boiling point of the solvent.* Figure 10-14 shows the phase diagram for an aqueous solution containing a nonvolatile solute. Note that the liquid/vapor line is shifted to higher temperatures than those for pure water.

Figure 10-14 | Phase diagrams for pure water (purple lines) and for an aqueous solution containing a nonvolatile solute (blue lines). Note that the boiling point of the solution is higher than that of pure water. Conversely, the freezing point of the solution is lower than that of pure water. The effect of a nonvolatile solute is to extend the liquid range of a solvent. These lines are not drawn to scale.

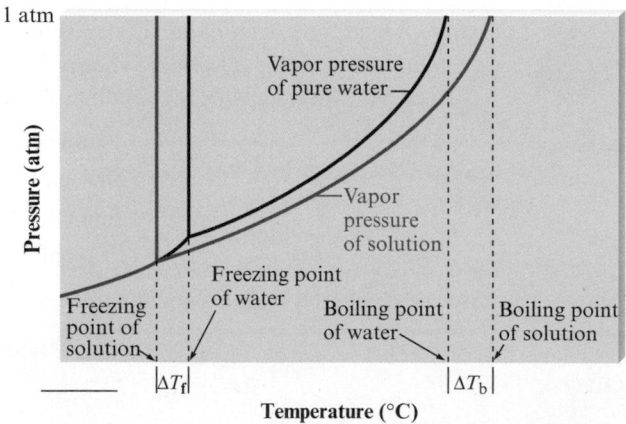

Table 10-5 | Molal Boiling-Point Elevation Constants (K_b) and Freezing-Point Depression Constants (K_f) for Several Solvents

Solvent	Boiling Point (°C)	K_b (°C · kg/mol)	Freezing Point (°C)	K_f (°C · kg/mol)
Water (H_2O)	100.0	0.51	0	1.86
Carbon tetrachloride (CCl_4)	76.5	5.03	−22.99	30.
Chloroform ($CHCl_3$)	61.2	3.63	−63.5	4.70
Benzene (C_6H_6)	80.1	2.53	5.5	5.12
Carbon disulfide (CS_2)	46.2	2.34	−111.5	3.83
Ethyl ether ($C_4H_{10}O$)	34.5	2.02	−116.2	1.79
Camphor ($C_{10}H_{16}O$)	208.0	5.95	179.8	40.

As you might expect, the magnitude of the boiling-point elevation depends on the concentration of the solute. The change in boiling point can be represented by the equation

$$\Delta T = K_b m_{solute}$$

where ΔT is the boiling-point elevation, or the difference between the boiling point of the solution and that of the pure solvent, K_b is a constant that is characteristic of the solvent and is called the **molal boiling-point elevation constant**, and m_{solute} is the *molality* of the solute in the solution.

Values of K_b for some common solvents are given in Table 10-5. The molar mass of a solute can be determined from the observed boiling-point elevation, as shown in Example 10-8.

INTERACTIVE EXAMPLE 10-8

Calculating the Molar Mass by Boiling-Point Elevation

A solution was prepared by dissolving 18.00 g glucose in 150.0 g water. The resulting solution was found to have a boiling point of 100.34°C. ◄ Calculate the molar mass of glucose. Glucose is a molecular solid that is present as individual molecules in solution.

Solution

We make use of the equation

$$\Delta T = K_b m_{solute}$$

where

$$\Delta T = 100.34°C - 100.00°C = 0.34°C$$

From Table 10-5, for water $K_b = 0.51$. The molality of this solution then can be calculated by rearranging the boiling-point elevation equation to give

$$m_{solute} = \frac{\Delta T}{K_b} = \frac{0.34°C}{0.51°C \cdot kg/mol} = 0.67 \text{ mol/kg}$$

The solution was prepared using 0.1500 kg water. Using the definition of molality, we can find the number of moles of glucose in the solution.

$$m_{solute} = 0.67 \text{ mol/kg} = \frac{\text{mol solute}}{\text{kg solvent}} = \frac{n_{glucose}}{0.1500 \text{ kg}}$$

$$n_{glucose} = (0.67 \text{ mol/kg})(0.1500 \text{ kg}) = 0.10 \text{ mol}$$

Thus 0.10 mole of glucose has a mass of 18.00 g, and 1.0 mole of glucose has a mass of 180 g (10×18.00 g). The molar mass of glucose is 180 g/mol.

▲ Sugar dissolved in water to make candy causes the boiling point to be elevated above 100°C.

© Cengage Learning

See Exercise 10-66

ℹ️ *Melting point* and *freezing point* both refer to the temperature where the solid and liquid coexist.

Figure 10-15 | (a) Ice in equilibrium with liquid water. (b) Ice in equilibrium with liquid water containing a dissolved solute (shown in red). ◀ ℹ️

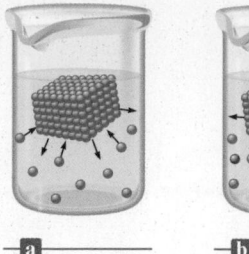
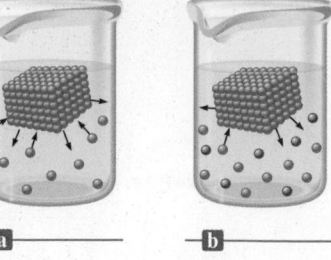

Freezing-Point Depression

When a solute is dissolved in a solvent, the freezing point of the solution is lower than that of the pure solvent. Why? Recall that the vapor pressures of ice and liquid water are the same at 0°C. Suppose a solute is dissolved in water. The resulting solution will not freeze at 0°C because *the water in the solution has a lower vapor pressure than that of pure ice.* No ice will form under these conditions. However, the vapor pressure of ice decreases more rapidly than that of liquid water as the temperature decreases. Therefore, as the solution is cooled, the vapor pressure of the ice and that of the liquid water in the solution will eventually become equal. The temperature at which this occurs is the new freezing point of the solution and is below 0°C. The freezing point has been *depressed.*

We can account for this behavior in terms of the simple model shown in Fig. 10-15. The presence of the solute lowers the rate at which molecules in the liquid return to the solid state. Thus for an aqueous solution, only the liquid state is found at 0°C. As the solution is cooled, the rate at which water molecules leave the solid ice decreases until this rate and the rate of formation of ice become equal and equilibrium is reached. This is the freezing point of the water in the solution.

Because a solute lowers the freezing point of water, compounds such as sodium chloride and calcium chloride are often spread on streets and sidewalks to prevent ice from forming in freezing weather. Of course, if the outside temperature is lower than the freezing point of the resulting salt solution, ice forms anyway. So this procedure is not effective at extremely cold temperatures. ◀

The solid/liquid line for an aqueous solution is shown on the phase diagram for water in Fig. 10-14. Since the presence of a solute elevates the boiling point and depresses the freezing point of the solvent, adding a solute has the effect of extending the liquid range.

The equation for freezing-point depression is analogous to that for boiling-point elevation:

$$\Delta T = K_f m_{solute}$$

where ΔT is the freezing-point depression, or the difference between the freezing point of the pure solvent and that of the solution, and K_f is a constant that is characteristic of a particular solvent and is called the **molal freezing-point depression constant**. Values of K_f for common solvents are listed in Table 10-5.

Like the boiling-point elevation, the observed freezing-point depression can be used to determine molar masses and to characterize solutions.

▲ Spreading salt on a highway.

INTERACTIVE EXAMPLE 10-9

Ethylene glycol

Freezing-Point Depression

What mass of ethylene glycol ($C_2H_6O_2$, molar mass = 62.1 g/mol), the main component of antifreeze, must be added to 10.0 L water to produce a solution for use in a car's radiator that freezes at −10.0°F (−23.3°C)? Assume the density of water is exactly 1 g/mL.

Solution

The freezing point must be lowered from 0°C to −23.3°C. ◀ To determine the molality of ethylene glycol needed to accomplish this, we can use the equation

$$\Delta T = K_f m_{solute}$$

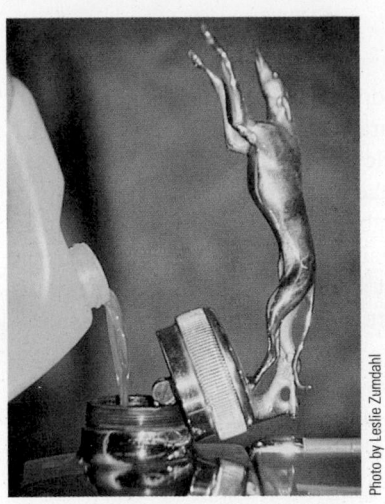

▲ The addition of antifreeze lowers the freezing point of water in a car's radiator.

where $\Delta T = 23.3°$ and $K_f = 1.86$ (from Table 10-5). Solving for the molality gives

$$m_{solute} = \frac{\Delta T}{K_f} = \frac{23.3°C}{1.86°C \cdot kg/mol} = 12.5 \text{ mol/kg}$$

This means that 12.5 moles of ethylene glycol must be added per kilogram of water. We have 10.0 L, or 10.0 kg, of water. Therefore, the total number of moles of ethylene glycol needed is

$$\frac{12.5 \text{ mol}}{kg} \times 10.0 \text{ kg} = 1.25 \times 10^2 \text{ mol}$$

The mass of ethylene glycol needed is

$$1.25 \times 10^2 \text{ mol} \times \frac{62.1 \text{ g}}{mol} = 7.76 \times 10^3 \text{ g (or } 7.76 \text{ kg)}$$

See Exercises 10-69 and 10-70

INTERACTIVE EXAMPLE 10-10

Determining Molar Mass by Freezing-Point Depression

A chemist is trying to identify a human hormone that controls metabolism by determining its molar mass. A sample weighing 0.546 g was dissolved in 15.0 g benzene, and the freezing-point depression was determined to be 0.240°C. Calculate the molar mass of the hormone.

Solution

From Table 10-5, K_f for benzene is $5.12°C \cdot kg/mol$, so the molality of the hormone is

$$m_{hormone} = \frac{\Delta T}{K_f} = \frac{0.240°C}{5.12°C \cdot kg/mol} = 4.69 \times 10^{-2} \text{ mol/kg}$$

The moles of hormone can be obtained from the definition of molality:

$$4.69 \times 10^{-2} \text{ mol/kg} = m_{solute} = \frac{\text{mol hormone}}{0.0150 \text{ kg benzene}}$$

or

$$\text{mol hormone} = \left(4.69 \times 10^{-2} \frac{\text{mol}}{kg} \right) (0.0150 \text{ kg}) = 7.04 \times 10^{-4} \text{ mol}$$

Since 0.546 g hormone was dissolved, 7.04×10^{-4} mole of hormone has a mass of 0.546 g, and

$$\frac{0.546 \text{ g}}{7.04 \times 10^{-4} \text{ mol}} = \frac{x}{1.00 \text{ mol}}$$

$$x = 776$$

Thus the molar mass of the hormone is 776 g/mol.

See Exercise 10-71

10-6 | Osmotic Pressure

Osmotic pressure, another of the colligative properties, can be understood from Fig. 10-16. A solution and pure solvent are separated by a **semipermeable membrane**, which allows *solvent but not solute* molecules to pass through. As time passes, the volume of the solution increases and that of the solvent decreases. This flow of solvent into the solution through the semipermeable membrane is called **osmosis**. Eventually the liquid levels stop changing, indicating that the system has reached equilibrium. Because the liquid levels are different at this point, there is a greater hydrostatic pressure on the solution than on the pure solvent. This excess pressure is called the **osmotic pressure**.

We can take another view of this phenomenon, as illustrated in Fig. 10-17. Osmosis can be prevented by applying a pressure to the solution. *The minimum pressure that stops the osmosis is equal to the osmotic pressure of the solution.* A simple model to explain osmotic pressure can be constructed as shown in Fig. 10-18. The membrane allows only solvent molecules to pass through. However, the initial rates of solvent transfer to and from the solution are not the same. The solute particles interfere with the passage of solvent, so the rate of transfer is slower from the solution to the solvent than in the reverse direction. Thus there is a net transfer of solvent molecules into the solution, which causes the solution volume to increase. As the solution level rises in the tube, the resulting pressure exerts an extra "push" on the solvent molecules in the solution, forcing them back through the membrane. Eventually, enough pressure develops so that the solvent transfer becomes equal in both directions. At this point, equilibrium is achieved and the levels stop changing.

Osmotic pressure can be used to characterize solutions and determine molar masses, as can the other colligative properties, but osmotic pressure is particularly useful because a small concentration of solute produces a relatively large osmotic pressure.

Experiments show that the dependence of the osmotic pressure on solution concentration is represented by the equation

$$\Pi = MRT$$

where Π is the osmotic pressure in atmospheres, M is the molarity of the solution, R is the gas law constant, and T is the Kelvin temperature.

A molar mass determination using osmotic pressure is illustrated in Example 10-11.

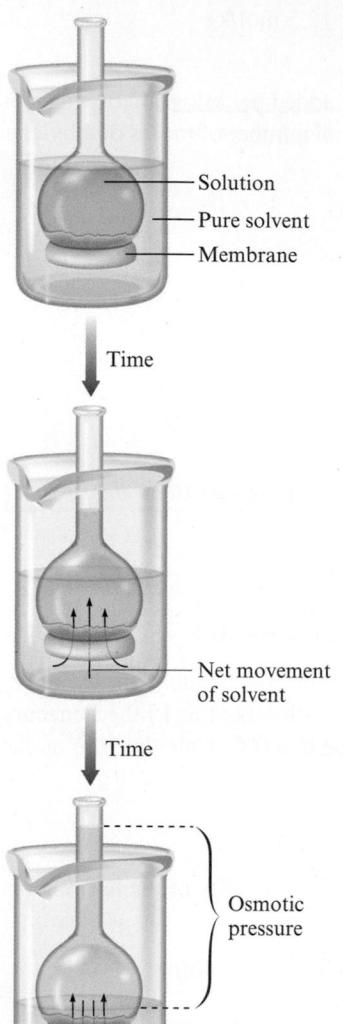

(at equilibrium)

Figure 10-16 | A tube with a bulb on the end that is covered by a semipermeable membrane. The solution is inside the tube and is bathed in the pure solvent. There is a net transfer of solvent molecules into the solution until the hydrostatic pressure equalizes the solvent flow in both directions.

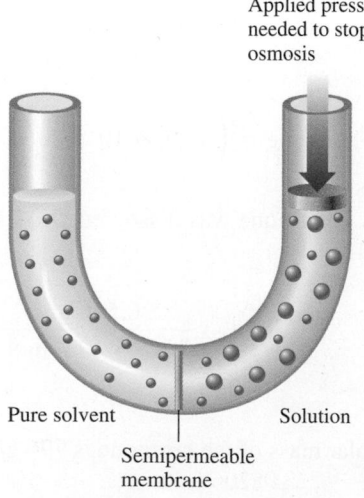

Figure 10-17 | The normal flow of solvent into the solution (osmosis) can be prevented by applying an external pressure to the solution. The minimum pressure required to stop the osmosis is equal to the osmotic pressure of the solution.

Figure 10-18 | (a) A pure solvent and its solution (containing a nonvolatile solute) are separated by a semipermeable membrane through which solvent molecules (blue) can pass but solute molecules (green) cannot. The rate of solvent transfer is greater from solvent to solution than from solution to solvent. (b) The system at equilibrium, where the rate of solvent transfer is the same in both directions.

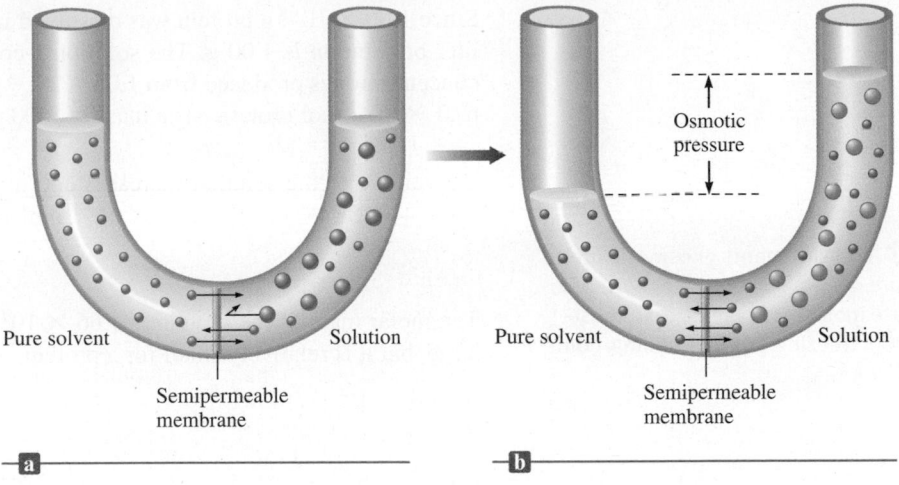

Critical Thinking

Consider the model of osmotic pressure as shown in Fig. 10-18. What if each side contained a different pure solvent, each with a different vapor pressure? What would the system look like at equilibrium? Assume the different solvent molecules are able to pass through the membrane.

INTERACTIVE EXAMPLE 10-11

Determining Molar Mass from Osmotic Pressure

To determine the molar mass of a certain protein, 1.00×10^{-3} g of it was dissolved in enough water to make 1.00 mL of solution. The osmotic pressure of this solution was found to be 1.12 torr at 25.0°C. Calculate the molar mass of the protein.

Solution

We use the equation

$$\Pi = MRT$$

In this case we have

$$\Pi = 1.12 \text{ torr} \times \frac{1 \text{ atm}}{760 \text{ torr}} = 1.47 \times 10^{-3} \text{ atm}$$

$$R = 0.08206 \text{ L} \cdot \text{atm/K} \cdot \text{mol}$$

$$T = 25.0 + 273 = 298 \text{ K}$$

Note that the osmotic pressure must be converted to atmospheres because of the units of R.

Solving for M gives

$$M = \frac{1.47 \times 10^{-3} \text{ atm}}{(0.08206 \text{ L} \cdot \text{atm/K} \cdot \text{mol})(298 \text{ K})} = 6.01 \times 10^{-5} \text{ mol/L}$$

Since 1.00×10^{-3} g protein was dissolved in 1 mL solution, the mass of protein per liter of solution is 1.00 g. The solution's concentration is 6.01×10^{-5} mol/L. This concentration is produced from 1.00×10^{-3} g protein per milliliter, or 1.00 g/L. Thus 6.01×10^{-5} mol protein has a mass of 1.00 g and

$$\frac{1.00 \text{ g}}{6.01 \times 10^{-5} \text{ mol}} = \frac{x}{1.00 \text{ mol}}$$

$$x = 1.66 \times 10^4 \text{ g}$$

The molar mass of the protein is 1.66×10^4 g/mol. This molar mass may seem very large, but it is relatively small for a protein. ◄ ⓘ

See Exercises 10-75 and 10-76

ⓘ Measurements of osmotic pressure generally give much more accurate molar mass values than those from freezing-point or boiling-point changes.

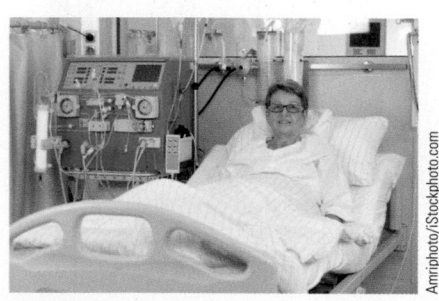

▲ Patient undergoing dialysis.

ⓘ The brine used in pickling causes the cucumbers to shrivel.

In osmosis, a semipermeable membrane prevents transfer of *all* solute particles. A similar phenomenon, called **dialysis**, occurs at the walls of most plant and animal cells. However, in this case the membrane allows transfer of both solvent molecules and *small* solute molecules and ions. One of the most important applications of dialysis is the use of artificial kidney machines to purify the blood. ◄ The blood is passed through a cellophane tube, which acts as the semipermeable membrane. The tube is immersed in a dialyzing solution (Fig. 10-19). This "washing" solution contains the same concentrations of ions and small molecules as blood but has none of the waste products normally removed by the kidneys. The resulting dialysis (movement of waste molecules into the washing solution) cleanses the blood.

Solutions that have identical osmotic pressures are said to be **isotonic solutions**. Fluids administered intravenously must be isotonic with body fluids. For example, if red blood cells are bathed in a hypertonic solution, which is a solution having an osmotic pressure higher than that of the cell fluids, the cells will shrivel because of a net transfer of water out of the cells. This phenomenon is called *crenation*. The opposite phenomenon, called *hemolysis,* occurs when cells are bathed in a hypotonic solution, a solution with an osmotic pressure lower than that of the cell fluids. In this case, the cells rupture because of the flow of water into the cells. ◄

We can use the phenomenon of crenation to our advantage. Food can be preserved by treating its surface with a solute that gives a solution that is hypertonic to bacteria cells. Bacteria on the food then tend to shrivel and die. This is why salt can be used to protect meat and sugar can be used to protect fruit. ◄ ⓘ

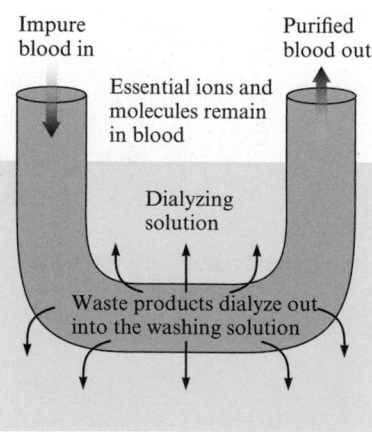

Figure 10-19 | Representation of the functioning of an artificial kidney.

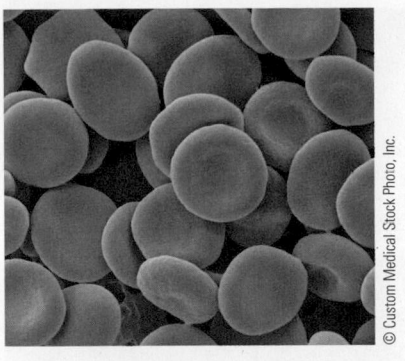
Normal

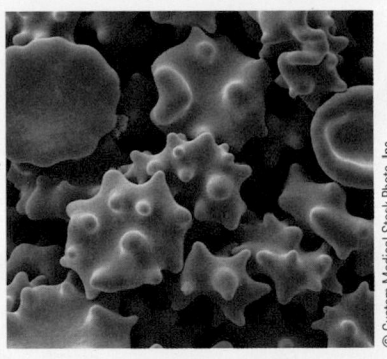
Shriveled

Swollen

▲ Normal red blood cells have a concave shape in an isotonic solution. (left) When red blood cells are placed in a hypertonic solution they crenate (shrink or shrivel) and are less concave. (center) When red blood cells are placed in a hypotonic solution they swell and are less concave. (right)

INTERACTIVE EXAMPLE 10-12

Isotonic Solutions

What concentration of sodium chloride in water is needed to produce an aqueous solution isotonic with blood ($\Pi = 7.70$ atm at 25°C)?

Solution

We can calculate the molarity of the solute from the equation

$$\Pi = MRT \quad \text{or} \quad M = \frac{\Pi}{RT}$$

$$M = \frac{7.70 \text{ atm}}{(0.08206 \text{ L} \cdot \text{atm/K} \cdot \text{mol})(298 \text{ K})} = 0.315 \text{ mol/L}$$

This represents the total molarity of solute particles. But NaCl gives two ions per formula unit. Therefore, the concentration of NaCl needed is $\frac{0.315 \, M}{2} = 0.1575 \, M = 0.158 \, M$. That is,

$$NaCl \longrightarrow Na^+ + Cl^-$$
$$0.1575 \, M \quad \underbrace{0.1575 \, M \quad 0.1575 \, M}_{0.315 \, M}$$

See Exercise 10-78

Figure 10-20 | Reverse osmosis. A pressure greater than the osmotic pressure of the solution is applied, which causes a net flow of solvent molecules (blue) from the solution to the pure solvent. The solute molecules (green) remain behind.

Pressure greater than Π_{soln}

Pure solvent　Solution

Semipermeable membrane

Reverse Osmosis

If a solution in contact with pure solvent across a semipermeable membrane is subjected to an external pressure larger than its osmotic pressure, **reverse osmosis** occurs. The pressure will cause a net flow of solvent from the solution to the solvent (Fig. 10-20). In reverse osmosis, the semipermeable membrane acts as a "molecular filter" to remove solute particles. This fact is applicable to the **desalination** (removal of dissolved salts) of seawater, which is highly hypertonic to body fluids and thus is not drinkable.

As the population of the Sun Belt areas of the United States increases, more demand will be placed on the limited supplies of fresh water there. One obvious source of fresh water is from the desalination of seawater. Various schemes have been suggested, including solar evaporation, reverse osmosis, and even a plan for towing ice-

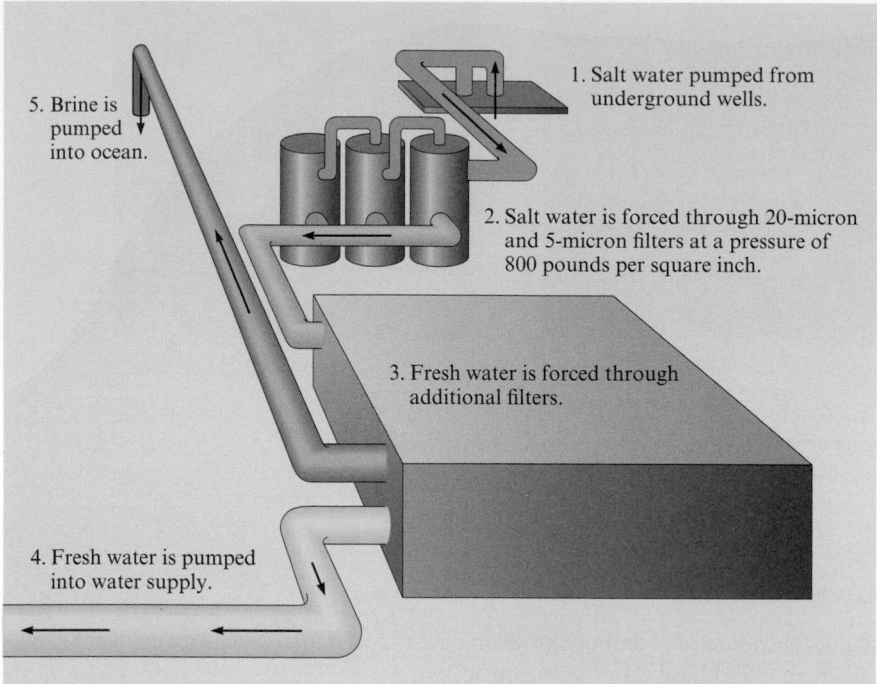

5. Brine is pumped into ocean.

1. Salt water pumped from underground wells.

2. Salt water is forced through 20-micron and 5-micron filters at a pressure of 800 pounds per square inch.

3. Fresh water is forced through additional filters.

4. Fresh water is pumped into water supply.

Courtesy Southern California Electric

Figure 10-21 │ (a) Residents of Catalina Island off the coast of southern California are benefiting from a desalination plant that can supply 132,000 gallons a day, or one-third of the island's daily needs. (b) Machinery in the desalination plant for Catalina Island.

bergs from Antarctica. The problem, of course, is that all the available processes are expensive. However, as water shortages increase, desalination is becoming necessary. For example, the first full-time public desalination plant in the United States started operations on Catalina Island, just off the coast of California (Fig. 10-21). This plant, which can produce 132,000 gallons of drinkable water from the Pacific Ocean every day, operates by reverse osmosis. Powerful pumps, developing over 800 lb/in² of pressure, are used to force seawater through synthetic semipermeable membranes.

Catalina Island's plant may be just the beginning. The city of Santa Barbara opened a $40 million desalination plant in 1992 that can produce 8 million gallons of drinking water per day. The southern California city of Carlsbad opened a reverse osmosis desalination plant in 2012 that can produce 50 million gallons of drinking water daily from seawater. Desalination plants are also in the works for Huntington Beach, California, and Camp Pendleton, a military base just north of Carlsbad.

A small-scale, manually operated reverse osmosis desalinator has been developed by the U.S. Navy to provide fresh water on life rafts. Potable water can be supplied by this desalinator at the rate of 1.25 gallons of water per hour—enough to keep 25 people alive. This compact desalinator, which weighs only 10 pounds, can now replace the bulky cases of fresh water formerly stored in Navy life rafts.

10-7 │ Colligative Properties of Electrolyte Solutions

As we have seen previously, the colligative properties of solutions depend on the total concentration of solute particles. For example, a 0.10-m glucose solution shows a freezing-point depression of 0.186°C:

$$\Delta T = K_\mathrm{f} m = (1.86°\mathrm{C} \cdot \mathrm{kg/mol})(0.100 \ \mathrm{mol/kg}) = 0.186°\mathrm{C}$$

On the other hand, a 0.10-*m* sodium chloride solution should show a freezing-point depression of 0.37°C, since the solution is 0.10 *m* Na^+ ions and 0.10 *m* Cl^- ions. Therefore, the solution contains a total of 0.20 *m* solute particles, and $\Delta T = (1.86°C \cdot \text{kg/mol})(0.20 \text{ mol/kg}) = 0.37°C$.

The relationship between the moles of solute dissolved and the moles of particles in solution is usually expressed using the **van't Hoff factor**, *i:* ◄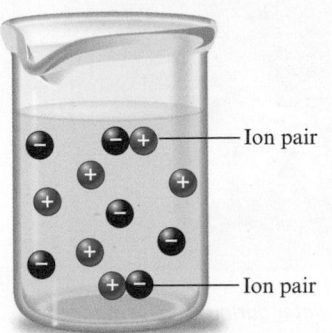

$$i = \frac{\text{moles of particles in solution}}{\text{moles of solute dissolved}}$$

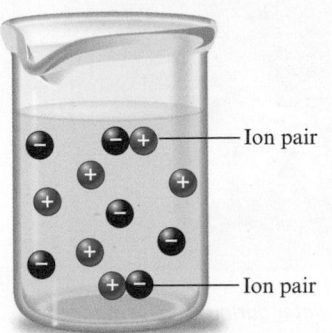

The *expected* value for *i* can be calculated for a salt by noting the number of ions per formula unit. For example, for NaCl, *i* is 2; for K_2SO_4, *i* is 3; and for $Fe_3(PO_4)_2$, *i* is 5. These calculated values assume that when a salt dissolves, it completely dissociates into its component ions, which then move around independently. This assumption is not always true. For example, the freezing-point depression observed for 0.10 *m* NaCl is 1.87 times that for 0.10 *m* glucose rather than twice as great. That is, for a 0.10-*m* NaCl solution, the observed value for *i* is 1.87 rather than 2. Why? The best explanation is that **ion pairing** occurs in solution (Fig. 10-22). At a given instant a small percentage of the sodium and chloride ions are paired and thus count as a single particle. In general, ion pairing is most important in concentrated solutions. As the solution becomes more dilute, the ions are farther apart and less ion pairing occurs. For example, in a 0.0010-*m* NaCl solution, the observed value of *i* is 1.97, which is very close to the expected value. ◄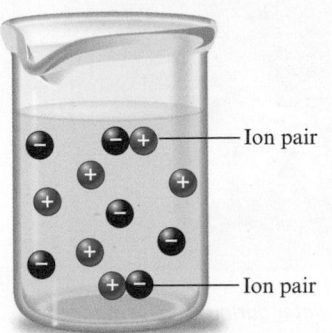

Ion pairing occurs to some extent in all electrolyte solutions. Table 10-6 shows expected and observed values of *i* for a given concentration of various electrolytes. Note that the deviation of *i* from the expected value tends to be greatest where the ions have multiple charges. This is expected because ion pairing ought to be most important for highly charged ions.

The colligative properties of electrolyte solutions are described by including the van't Hoff factor in the appropriate equation. For example, for changes in freezing and boiling points, the modified equation is

$$\Delta T = imK$$

where *K* represents the freezing-point depression or boiling-point elevation constant for the solvent.

For the osmotic pressure of electrolyte solutions, the equation is

$$\Pi = iMRT$$

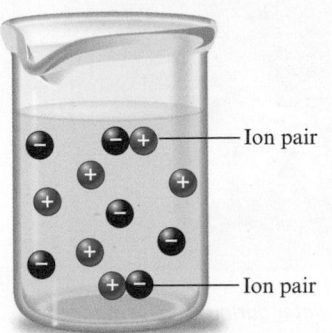

Dutch chemist J. H. van't Hoff (1852–1911) received the first Nobel Prize in chemistry in 1901.

Figure 10-22 | In an aqueous solution a few ions aggregate, forming ion pairs that behave as a unit.

Note that we need to understand the nature of the solute at a "molecular level" to explain the value of the van't Hoff factor.

Table 10-6 | Expected and Observed Values of the van't Hoff Factor for 0.05 *m* Solutions of Several Electrolytes

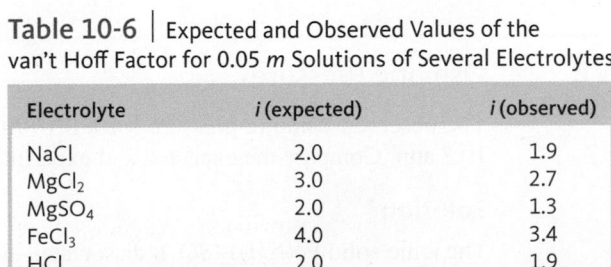

Electrolyte	*i* (expected)	*i* (observed)
NaCl	2.0	1.9
$MgCl_2$	3.0	2.7
$MgSO_4$	2.0	1.3
$FeCl_3$	4.0	3.4
HCl	2.0	1.9
Glucose*	1.0	1.0

*A nonelectrolyte shown for comparison.

In 1965, the University of Florida football team, the Gators, participated in a research program to test a sports drink formula containing a mixture of carbohydrates and electrolytes. The drink was used to help prevent dehydration caused by extreme workouts in the hot Florida climate. The Gators' success that season was in part attributed to their use of the sports drink formula. In 1967, a modified form of this formula was marketed with the name Gatorade. Today, Gatorade leads sales in sports drinks, but many other brands have entered a market where annual sales exceed $700 million!

During moderate- to high-intensity exercise, glycogen (a fuel reserve that helps maintain normal body processes) can be depleted within 60 to 90 minutes. Blood-sugar levels drop as the glycogen reserves are used up, and lactic acid (a by-product of glucose metabolism) builds up in muscle tissue causing fatigue and muscle cramps. Muscles also generate a large amount of heat that must be dissipated. Water, which has a large specific heat capacity, is used to take heat away from these muscles. Sweating and evaporative cooling help the body maintain a constant temperature, but at a huge cost. During a high-intensity workout in hot weather, anywhere from 1 to 3 quarts of water can be lost from sweating per hour. Sweating away more than 2% of your body weight—a quart for every 100 pounds—can put a large stress on the heart, increasing body temperature and decreasing performance. Excessive sweating also results in the loss of sodium and potassium ions—two very important electrolytes that are present in the fluids inside and outside cells.

All the major sports drinks contain three main ingredients—carbohydrates in the form of simple sugars such as sucrose, glucose, and fructose; electrolytes, including sodium and potassium ions; and water. Because these are the three major substances lost through sweating, good scientific reasoning suggests that drinking sports drinks should improve perfor-

For healthy athletes, drinking water during exercise may be as effective as drinking sports drinks.

mance. But just how effectively do sports drinks deliver on their promises?

Recent studies have confirmed that athletes who eat a balanced diet and drink plenty of water are just as well off as those who consume sports drinks. A sports drink may have only one advantage over drinking water—it tastes better than water to most athletes. And if a drink tastes better, it will encourage more consumption, thus keeping cells hydrated.

Since most of the leading sports drinks contain the same ingredients in similar concentrations, taste may be the single most important factor in choosing your drink. If you are not interested in any particular sports drink, drink plenty of water. The key to quality performance is to keep your cells hydrated.

From "Sports Drinks: Don't Sweat the Small Stuff," by Tim Graham, *ChemMatters*, February 1999, p. 11.

INTERACTIVE EXAMPLE 10-13

Osmotic Pressure

The observed osmotic pressure for a 0.10-M solution of $Fe(NH_4)_2(SO_4)_2$ at 25°C is 10.8 atm. Compare the expected and experimental values for i.

Solution

The ionic solid $Fe(NH_4)_2(SO_4)_2$ dissociates in water to produce 5 ions:

$$Fe(NH_4)_2(SO_4)_2 \xrightarrow{H_2O} Fe^{2+} + 2NH_4^+ + 2SO_4^{2-}$$

Thus the expected value for i is 5. We can obtain the experimental value for i by using the equation for osmotic pressure:

$$\Pi = iMRT \quad \text{or} \quad i = \frac{\Pi}{MRT}$$

where $\Pi = 10.8$ atm, $M = 0.10$ mol/L, $R = 0.08206$ L $\cdot$ atm/K $\cdot$ mol, and $T = 25 + 273 = 298$ K. Substituting these values into the equation gives

$$i = \frac{\Pi}{MRT} = \frac{10.8 \text{ atm}}{(0.10 \text{ mol/L})(0.08206 \text{ L} \cdot \text{atm/K} \cdot \text{mol})(298 \text{ K})} = 4.4$$

The experimental value for i is less than the expected value, presumably because of ion pairing.

See Exercises 10-87 and 10-88

10-8 | Colloids

Mud can be suspended in water by vigorous stirring. When the stirring stops, most of the particles rapidly settle out, but even after several days some of the smallest particles remain suspended. Although undetected in normal lighting, their presence can be demonstrated by shining a beam of intense light through the suspension. The beam is visible from the side because the light is scattered by the suspended particles (Fig. 10-23). In a true solution, on the other hand, the beam is invisible from the side because the individual ions and molecules dispersed in the solution are too small to scatter visible light.

The scattering of light by particles is called the **Tyndall effect** and is often used to distinguish between a suspension and a true solution.

A suspension of tiny particles in some medium is called a **colloidal dispersion**, or a **colloid**. The suspended particles are single large molecules or aggregates of molecules or ions ranging in size from 1 to 1000 nm. Colloids are classified according to the states of the dispersed phase and the dispersing medium. Table 10-7 summarizes various types of colloids.

What stabilizes a colloid? Why do the particles remain suspended rather than forming larger aggregates and precipitating out? The answer is complicated, but the main factor seems to be *electrostatic repulsion*. A colloid, like all other macroscopic substances, is electrically neutral. However, when a colloid is placed in an electric field, the dispersed particles all migrate to the same electrode and thus must all have the same charge. How is this possible? The center of a colloidal particle (a tiny ionic crystal, a group of molecules, or a single large molecule) attracts from the medium a layer of ions, all of the same charge. This group of ions, in turn, attracts another layer of oppositely charged ions (Fig. 10-24). Because the colloidal particles all have an outer

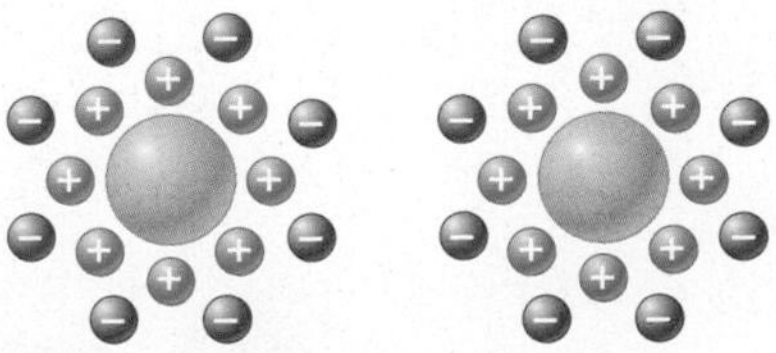

© Cengage Learning

Figure 10-23 | The Tyndall effect.

Figure 10-24 | A representation of two colloidal particles. In each the center particle is surrounded by a layer of positive ions, with negative ions in the outer layer. Thus, although the particles are electrically neutral, they still repel each other because of their outer negative layer of ions.

Table 10-7 | Types of Colloids

Examples	Dispersing Medium	Dispersed Substance	Colloid Type
Fog, aerosol sprays	Gas	Liquid	Aerosol
Smoke, airborne bacteria	Gas	Solid	Aerosol
Whipped cream, soap suds	Liquid	Gas	Foam
Milk, mayonnaise	Liquid	Liquid	Emulsion
Paint, clays, gelatin	Liquid	Solid	Sol
Marshmallow, polystyrene foam	Solid	Gas	Solid foam
Butter, cheese	Solid	Liquid	Solid emulsion
Ruby glass	Solid	Solid	Solid sol

layer of ions with the same charge, they repel each other and do not easily aggregate to form particles that are large enough to precipitate.

The destruction of a colloid, called **coagulation**, usually can be accomplished either by heating or by adding an electrolyte. Heating increases the velocities of the colloidal particles, causing them to collide with enough energy that the ion barriers are penetrated and the particles can aggregate. Because this process is repeated many times, the particle grows to a point where it settles out. Adding an electrolyte neutralizes the adsorbed ion layers. This is why clay suspended in rivers is deposited where the river reaches the ocean, forming the deltas characteristic of large rivers like the Mississippi. The high salt content of the seawater causes the colloidal clay particles to coagulate.

The removal of soot from smoke is another example of the coagulation of a colloid. When smoke is passed through an electrostatic precipitator (Fig. 10-25), the suspended solids are removed. The use of precipitators has produced an immense improvement in the air quality of heavily industrialized cities.

Figure 10-25 | The Cottrell precipitator installed in a smokestack. The charged plates attract the colloidal particles because of their ion layers and thus remove them from the smoke.

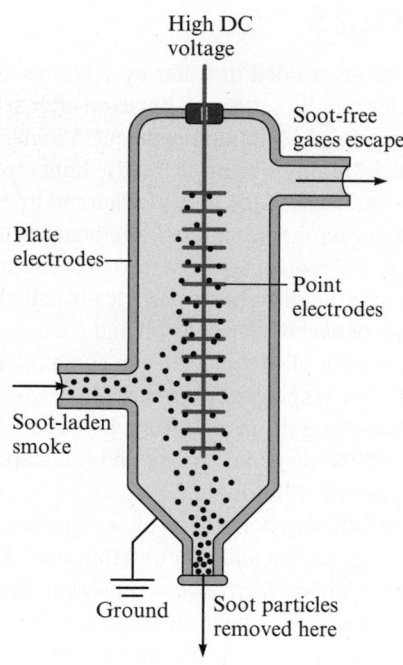

High DC voltage

Soot-free gases escape

Plate electrodes

Point electrodes

Soot-laden smoke

Ground

Soot particles removed here

The ice-cold waters of the polar oceans are teeming with fish that seem immune to freezing. One might think that these fish have some kind of antifreeze in their blood. However, studies show that they are protected from freezing in a very different way from the way antifreeze protects our cars. As we have seen in this chapter, solutes such as sugar, salt, and ethylene glycol lower the temperature at which the solid and liquid phases of water can coexist. However, the fish could not tolerate high concentrations of solutes in their blood because of the osmotic pressure effects. Instead, they are protected by proteins in their blood. These proteins allow the water in the bloodstream to be supercooled—exist below 0°C—without forming ice. They apparently coat the surface of each tiny ice crystal as soon as it begins to form, preventing it from growing to a size that would cause biological damage.

Although it might at first seem surprising, this research on polar fish has attracted the attention of ice cream manufacturers. Premium quality ice cream is smooth; it does not have large ice crystals in it. The makers of ice cream would like to incorporate these polar fish proteins, or molecules that behave similarly, into ice cream to prevent the growth of ice crystals during storage.

Fruit and vegetable growers have a similar interest: They also want to prevent ice formation that damages their crops during an unusual cold wave. However, this is a very different kind of problem than keeping polar fish from freezing. Many types of fruits and vegetables are colonized by bacteria that

Blackfin icefish, Chaenocephalus aceratus.

manufacture a protein that *encourages* freezing by acting as a nucleating agent to start an ice crystal. Chemists have identified the offending protein in the bacteria and the gene that is responsible for making it. They have learned to modify the genetic material of these bacteria in a way that removes their ability to make the protein that encourages ice crystal formation. If testing shows that these modified bacteria have no harmful effects on the crop or the environment, the original bacteria strain will be replaced with the new form so that ice crystals will not form so readily when a cold snap occurs.

FOR REVIEW

Key Terms

Section 10-1
molarity (M)
mass percent
mole fraction (X)
molality (m)
normality (N)

Section 10-2
enthalpy (heat) of solution
enthalpy (heat) of hydration

Solution Composition

- Molarity (M): moles solute per liter of solution
- Mass percent: ratio of mass of solute to mass of solution times 100%
- Mole fraction (X): ratio of moles of a given component to total moles of all components
- Molality (m): moles solute per mass of solvent (in kg)
- Normality (N): number of equivalents per liter of solution

Key Terms

Section 10-3
Henry's law
thermal pollution

Section 10-4
Raoult's law
ideal solution

Section 10-5
colligative properties
molal boiling-point elevation constant
molal freezing-point depression constant

Section 10-6
semipermeable membrane
osmosis
osmotic pressure
dialysis
isotonic solution
reverse osmosis
desalination

Section 10-7
van't Hoff factor
ion pairing

Section 10-8
Tyndall effect
colloid (colloidal dispersion)
coagulation

Enthalpy of Solution (ΔH_{soln})

- The enthalpy change accompanying solution formation
- Can be partitioned into
  - The energy required to overcome the solute–solute interactions
  - The energy required to "make holes" in the solvent
  - The energy associated with solute–solvent interactions

Factors That Affect Solubility

- Polarity of solute and solvent
  - "Like dissolves like" is a useful generalization
- Pressure increases the solubility of gases in a solvent
  - Henry's law: $C = kP$
- Temperature effects
  - Increased temperature decreases the solubility of a gas in water
  - Most solids are more soluble at higher temperatures, but important exceptions exist

Vapor Pressure of Solutions

- A solution containing a nonvolatile solute has a lower vapor pressure than a solution of the pure solvent
- Raoult's law defines an ideal solution:

$$P_{vapor}^{soln} = \chi_{solvent} P_{vapor}^{solvent}$$

  - Solutions in which the solute–solvent attractions differ from the solute–solute and solvent–solvent attractions violate Raoult's law

Colligative Properties

- Depend on the number of solute particles present
- Boiling-point elevation: $\Delta T = K_b m_{solute}$
- Freezing-point lowering: $\Delta T = K_f m_{solute}$
- Osmotic pressure: $\Pi = MRT$
  - Osmosis occurs when a solution and pure solvent are separated by a semipermeable membrane that allows solvent molecules to pass but not solute particles
  - Reverse osmosis occurs when the applied pressure is greater than the osmotic pressure of the solution
- Because colligative properties depend on the number of particles, solutes that break into several ions when they dissolve have an effect proportional to the number of ions produced
  - The van't Hoff factor i represents the number of ions produced by each formula unit of solute

Colloids

- A suspension of tiny particles stabilized by electrostatic repulsion among the ion layers surrounding the individual particles
- Can be coagulated (destroyed) by heating or adding an electrolyte

1. The four most common ways to describe solution composition are mass percent, mole fraction, molarity, and molality. Define each of these solution composition terms. Why is molarity temperature-dependent, whereas the other three solution composition terms are temperature-independent?

2. Using KF as an example, write equations that refer to ΔH_{soln} and ΔH_{hyd}. Lattice energy was defined in Chapter 3 as ΔH for the reaction $K^+(g) + F^-(g) \longrightarrow KF(s)$. Show how you would utilize Hess's law to calculate ΔH_{soln} from ΔH_{hyd} and ΔH_{LE} for KF, where ΔH_{LE} = lattice energy. ΔH_{soln} for KF, as for other soluble ionic compounds, is a relatively small number. How can this be since ΔH_{hyd} and ΔH_{LE} are relatively large negative numbers?

3. What does the axiom "like dissolves like" mean? There are four types of solute/solvent combinations: polar solutes in polar solvents, nonpolar solutes in polar solvents, and so on. For each type of solution, discuss the magnitude of ΔH_{soln}.

4. Structure, pressure, and temperature all have an effect on solubility. Discuss each of their effects. What is Henry's law? Why does Henry's law not work for HCl(g)? What do the terms *hydrophobic* and *hydrophilic* mean?

5. Define the terms in Raoult's law. Figure 10-9 illustrates the net transfer of water molecules from pure water to an aqueous solution of a nonvolatile solute. Explain why eventually all of the water from the beaker of pure water will transfer to the aqueous solution. If the experiment illustrated in Fig. 10-9 was performed using a volatile solute, what would happen? How do you calculate the total vapor pressure when both the solute and solvent are volatile?

6. In terms of Raoult's law, distinguish between an ideal liquid–liquid solution and a nonideal liquid–liquid solution. If a solution is ideal, what is true about ΔH_{soln}, ΔT for the solution formation, and the interactive forces within the pure solute and pure solvent as compared to the interactive forces within the solution? Give an example of an ideal solution. Answer the previous two questions for solutions that exhibit either negative or positive deviations from Raoult's law.

7. Vapor-pressure lowering is a colligative property, as are freezing-point depression and boiling-point elevation. What is a colligative property? Why is the freezing point depressed for a solution as compared to the pure solvent? Why is the boiling point elevated for a solution as compared to the pure solvent? Explain how to calculate ΔT for a freezing-point depression problem or a boiling-point elevation problem. Of the solvents listed in Table 10-5, which would have the largest freezing-point depression for a 0.50 molal solution? Which would have the smallest boiling-point elevation for a 0.50 molal solution?

A common application of freezing-point depression and boiling-point elevation experiments is to provide a means to calculate the molar mass of a nonvolatile solute. What data are needed to calculate the molar mass of a nonvolatile solute? Explain how you would manipulate these data to calculate the molar mass of the nonvolatile solute.

8. What is osmotic pressure? How is osmotic pressure calculated? Molarity units are used in the osmotic pressure equation. When does the molarity of a solution approximately equal the molality of the solution? Before refrigeration was common, many foods were preserved by salting them heavily, and many fruits were preserved by mixing them with a large amount of sugar (fruit preserves). How do salt and sugar act as preservatives? Two applications of osmotic pressure are dialysis and desalination. Explain these two processes.

9. Distinguish between a strong electrolyte, a weak electrolyte, and a nonelectrolyte. How can colligative properties be used to distinguish between them? What is the van't Hoff factor? Why is the observed freezing-point depression for electrolyte solutions sometimes less than the calculated value? Is the discrepancy greater for concentrated or dilute solutions?

10. What is a colloidal dispersion? Give some examples of colloids. The Tyndall effect is often used to distinguish between a colloidal suspension and a true solution. Explain. The destruction of a colloid is done through a process called coagulation. What is coagulation?

Active Learning Questions

These questions are designed to be used by groups of students in class.

1. Consider Fig. 10-9. According to the caption and picture, water seems to go from one beaker to another.
   a. Explain why this occurs.
   b. The explanation in the text uses terms such as *vapor pressure* and *equilibrium*. Explain what these have to do with the phenomenon. For example, what is coming to equilibrium?
   c. Does all the water end up in the second beaker?
   d. Is water evaporating from the beaker containing the solution? If so, is the rate of evaporation increasing, decreasing, or staying constant?

   Draw pictures to illustrate your explanations.

2. Once again, consider Fig. 10-9. Suppose instead of having a nonvolatile solute in the solvent in one beaker, the two beakers contain different volatile liquids. That is, suppose one beaker contains liquid A (P_{vap} = 50 torr) and the other beaker contains liquid B (P_{vap} = 100 torr). Explain what happens as time passes. How is this similar to the first case (shown in the figure)? How is it different?

3. Assume that you place a freshwater plant into a saltwater solution and examine it under a microscope. What happens to the plant cells? What if you placed a saltwater plant in pure water? Explain. Draw pictures to illustrate your explanations.

4. How does ΔH_{soln} relate to deviations from Raoult's law? Explain.

5. You have read that adding a solute to a solvent can both increase the boiling point and decrease the freezing point. A friend of yours explains it to you like this: "The solute and solvent can be like salt in water. The salt gets in the way of freezing in that it blocks the water molecules from joining together. The salt acts like a strong bond holding the water molecules together so that it is harder to boil." What do you say to your friend?

6. You drop an ice cube (made from pure water) into a saltwater solution at 0°C. Explain what happens and why.

7. Using the phase diagram for water and Raoult's law, explain why salt is spread on the roads in winter (even when it is below freezing).

8. You and your friend are each drinking cola from separate 2-L bottles. Both colas are equally carbonated. You are able to drink 1 L of cola, but your friend can drink only about half a liter. You each close the bottles and place them in the refrigerator. The next day when you each go to get the colas, whose will be more carbonated and why?

9. Is molality or molarity dependent on temperature? Explain your answer. Why is molality, and not molarity, used in the equations describing freezing-point depression and boiling-point elevation?

10. Consider a beaker of salt water sitting open in a room. Over time, does the vapor pressure increase, decrease, or stay the same? Explain.

A blue question or exercise number indicates that the answer to that question or exercise appears at the back of this book and a solution appears in the *Solutions Guide*, as found on PowerLecture.

Solution Review

If you have trouble with these exercises, review Sections 6-1 to 6-3 in Chapter 6.

11. Rubbing alcohol contains 585 g isopropanol (C_3H_7OH) per liter (aqueous solution). Calculate the molarity.

12. What mass of sodium oxalate ($Na_2C_2O_4$) is needed to prepare 0.250 L of a 0.100-*M* solution?

13. What volume of 0.25 *M* HCl solution must be diluted to prepare 1.00 L of 0.040 *M* HCl?

14. What volume of a 0.580-*M* solution of $CaCl_2$ contains 1.28 g solute?

15. Calculate the sodium ion concentration when 70.0 mL of 3.0 *M* sodium carbonate is added to 30.0 mL of 1.0 *M* sodium bicarbonate.

16. Write equations showing the ions present after the following strong electrolytes are dissolved in water.
    a. HNO_3 d. $SrBr_2$ g. NH_4NO_3
    b. Na_2SO_4 e. $KClO_4$ h. $CuSO_4$
    c. $Al(NO_3)_3$ f. NH_4Br i. $NaOH$

Questions

17. Rationalize the temperature dependence of the solubility of a gas in water in terms of the kinetic molecular theory.

18. The weak electrolyte $NH_3(g)$ does not obey Henry's law. Why? $O_2(g)$ obeys Henry's law in water but not in blood (an aqueous solution). Why?

19. The two beakers in the sealed container illustrated below contain pure water and an aqueous solution of a volatile solute.

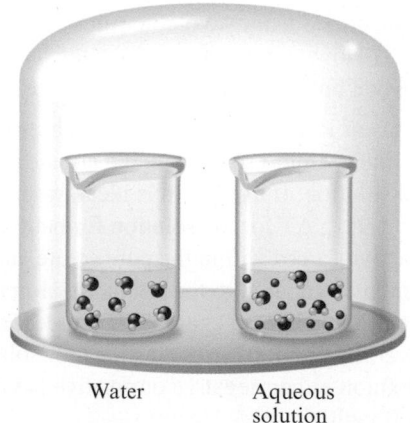

Water   Aqueous solution

If the solute is less volatile than water, explain what will happen to the volumes in the two containers as time passes.

20. The following plot shows the vapor pressure of various solutions of components A and B at some temperature.

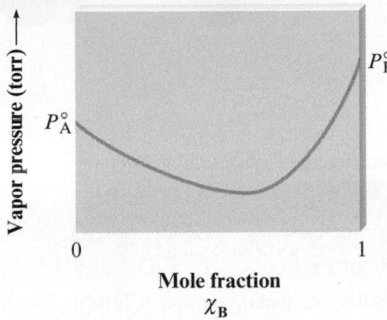

Mole fraction
χ_B

Which of the following statements is false concerning solutions of A and B?

a. The solutions exhibit negative deviations from Raoult's law.

b. ΔH_{soln} for the solutions should be exothermic.

c. The intermolecular forces are stronger in solution than in either pure A or pure B.

d. Pure liquid B is more volatile than pure liquid A.

e. The solution with $\chi_B = 0.6$ will have a lower boiling point than either pure A or pure B.

21. When pure methanol is mixed with water, the resulting solution feels warm. Would you expect this solution to be ideal? Explain.

22. Detergent molecules can stabilize the emulsion of oil in water as well as remove dirt from soiled clothes. A typical detergent is sodium dodecylsulfate, or SDS, and it has a formula of $CH_3(CH_2)_{10}CH_2SO_4^-Na^+$. In aqueous solution, SDS suspends oil or dirt by forming small aggregates of detergent anions called *micelles*. Propose a structure for micelles.

23. For an acid or a base, when is the normality of a solution equal to the molarity of the solution and when are the two concentration units different?

24. In order for sodium chloride to dissolve in water, a small amount of energy must be added during solution formation. This is not energetically favorable. Why is NaCl so soluble in water?

25. Which of the following statements is(are) *true*? Correct the false statements.

a. The vapor pressure of a solution is directly related to the mole fraction of solute.

b. When a solute is added to water, the water in solution has a lower vapor pressure than that of pure ice at 0°C.

c. Colligative properties depend only on the identity of the solute and not on the number of solute particles present.

d. When sugar is added to water, the boiling point of the solution increases above 100°C because sugar has a higher boiling point than water.

26. Is the following statement true or false? Explain your answer. When determining the molar mass of a solute using boiling-point or freezing-point data, camphor would be the best solvent choice of all of the solvents listed in Table 10-5.

27. Explain the terms *isotonic solution, crenation,* and *hemolysis.*

28. What is ion pairing?

Exercises

In this section similar exercises are paired.

Solution Composition

29. A solution of phosphoric acid was made by dissolving 10.0 g H_3PO_4 in 100.0 mL water. The resulting volume was 104 mL. Calculate the density, mole fraction, molarity, and molality of the solution. Assume water has a density of 1.00 g/cm³.

30. An aqueous antifreeze solution is 40.0% ethylene glycol ($C_2H_6O_2$) by mass. The density of the solution is 1.05 g/cm³. Calculate the molality, molarity, and mole fraction of the ethylene glycol.

31. Common commercial acids and bases are aqueous solutions with the following properties:

	Density (g/cm³)	Mass Percent of Solute
Hydrochloric acid	1.19	38
Nitric acid	1.42	70.
Sulfuric acid	1.84	95
Acetic acid	1.05	99
Ammonia	0.90	28

Calculate the molarity, molality, and mole fraction of each of the preceding reagents.

32. In lab you need to prepare at least 100 mL of each of the following solutions. Explain how you would proceed using the given information.

a. 2.0 *m* KCl in water (density of H_2O = 1.00 g/cm³)

b. 15% NaOH by mass in water (*d* = 1.00 g/cm³)

c. 25% NaOH by mass in CH_3OH (*d* = 0.79 g/cm³)

d. 0.10 mole fraction of $C_6H_{12}O_6$ in water (*d* = 1.00 g/cm³)

33. A solution is prepared by mixing 25 mL pentane (C_5H_{12}, *d* = 0.63 g/cm³) with 45 mL hexane (C_6H_{14}, *d* = 0.66 g/cm³). Assuming that the volumes add on mixing, calculate the mass percent, mole fraction, molality, and molarity of the pentane.

34. A solution is prepared by mixing 50.0 mL toluene ($C_6H_5CH_3$, *d* = 0.867 g/cm³) with 125 mL benzene (C_6H_6, *d* = 0.874 g/cm³). Assuming that the volumes add on mixing, calculate the mass percent, mole fraction, molality, and molarity of the toluene.

35. A bottle of wine contains 12.5% ethanol by volume. The density of ethanol (C_2H_5OH) is 0.789 g/cm³. Calculate the concentration of ethanol in wine in terms of mass percent and molality.

36. Calculate the molarity and mole fraction of acetone in a 1.00-*m* solution of acetone (CH_3COCH_3) in ethanol (C_2H_5OH). (Density of acetone = 0.788 g/cm³; density of ethanol = 0.789 g/cm³.) Assume that the volumes of acetone and ethanol add.

37. A 1.37-*M* solution of citric acid ($H_3C_6H_5O_7$) in water has a density of 1.10 g/cm³. Calculate the mass percent, molality, mole fraction, and normality of the citric acid. Citric acid has three acidic protons.

38. Calculate the normality of each of the following solutions.

    a. 0.250 M HCl

    b. 0.105 M H_2SO_4

    c. 5.3×10^{-2} M H_3PO_4

    d. 0.134 M NaOH

    e. 0.00521 M $Ca(OH)_2$

What is the equivalent mass for each of the acids or bases listed above?

Energetics of Solutions and Solubility

39. The lattice energy* of NaI is −686 kJ/mol, and the enthalpy of hydration is −694 kJ/mol. Calculate the enthalpy of solution per mole of solid NaI. Describe the process to which this enthalpy change applies.

40. a. Use the following data to calculate the enthalpy of hydration for calcium chloride and calcium iodide.

	Lattice Energy	ΔH_{soln}
$CaCl_2(s)$	−2247 kJ/mol	−46 kJ/mol
$CaI_2(s)$	−2059 kJ/mol	−104 kJ/mol

    b. Based on your answers to part a, which ion, Cl^- or I^-, is more strongly attracted to water?

41. Although $Al(OH)_3$ is insoluble in water, NaOH is very soluble. Explain in terms of lattice energies.

42. The high melting points of ionic solids indicate that a lot of energy must be supplied to separate the ions from one another. How is it possible that the ions can separate from one another when soluble ionic compounds are dissolved in water, often with essentially no temperature change?

43. Which solvent, water or carbon tetrachloride, would you choose to dissolve each of the following?

    a. KrF_2 **e.** MgF_2

    b. SF_2 **f.** CH_2O

    c. SO_2 **g.** $CH_2{=}CH_2$

    d. CO_2

44. Which solvent, water or hexane (C_6H_{14}), would you choose to dissolve each of the following?

    a. $Cu(NO_3)_2$ **d.** $CH_3(CH_2)_{16}CH_2OH$

    b. CS_2 **e.** HCl

    c. CH_3OH **f.** C_6H_6

45. For each of the following pairs, predict which substance would be more soluble in water.

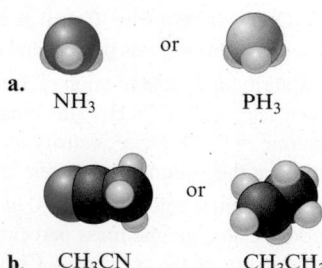

    a. NH_3 or PH_3

    b. CH_3CN or CH_3CH_3

*Lattice energy was defined in Chapter 3 as the energy change for the process $M^+(g) + X^-(g) \rightarrow MX(s)$.

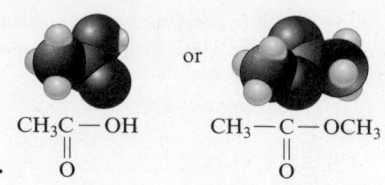

                            or

          $CH_3C{-}OH$ $CH_3{-}C{-}OCH_3$

c.   O   O

46. Which ion in each of the following pairs would you expect to be more strongly hydrated? Why?

    a. Na^+ or Mg^{2+} **d.** F^- or Br^-

    b. Mg^{2+} or Be^{2+} **e.** Cl^- or ClO_4^-

    c. Fe^{2+} or Fe^{3+} **f.** ClO_4^- or SO_4^{2-}

47. Rationalize the trend in water solubility for the following simple alcohols:

Alcohol	Solubility (g/100 g H_2O at 20°C)
Methanol, CH_3OH	Soluble in all proportions
Ethanol, CH_3CH_2OH	Soluble in all proportions
Propanol, $CH_3CH_2CH_2OH$	Soluble in all proportions
Butanol, $CH_3(CH_2)_2CH_2OH$	8.14
Pentanol, $CH_3(CH_2)_3CH_2OH$	2.64
Hexanol, $CH_3(CH_2)_4CH_2OH$	0.59
Heptanol, $CH_3(CH_2)_5CH_2OH$	0.09

48. In flushing and cleaning columns used in liquid chromatography to remove adsorbed contaminants, a series of solvents is used. Hexane (C_6H_{14}), chloroform ($CHCl_3$), methanol (CH_3OH), and water are passed through the column in that order. Rationalize the order in terms of intermolecular forces and the mutual solubility (miscibility) of the solvents.

49. The solubility of nitrogen in water is 8.21×10^{-4} mol/L at 0°C when the N_2 pressure above water is 0.790 atm. Calculate the Henry's law constant for N_2 in units of mol/L · atm for Henry's law in the form $C = kP$, where C is the gas concentration in mol/L. Calculate the solubility of N_2 in water when the partial pressure of nitrogen above water is 1.10 atm at 0°C.

50. Calculate the solubility of O_2 in water at a partial pressure of O_2 of 120 torr at 25°C. The Henry's law constant for O_2 is 1.3×10^{-3} mol/L · atm for Henry's law in the form $C = kP$, where C is the gas concentration (mol/L).

Vapor Pressures of Solutions

51. Glycerin, $C_3H_8O_3$, is a nonvolatile liquid. What is the vapor pressure of a solution made by adding 164 g glycerin to 338 mL H_2O at 39.8°C? The vapor pressure of pure water at 39.8°C is 54.74 torr and its density is 0.992 g/cm³.

52. The vapor pressure of a solution containing 53.6 g glycerin ($C_3H_8O_3$) in 133.7 g ethanol (C_2H_5OH) is 113 torr at 40°C. Calculate the vapor pressure of pure ethanol at 40°C assuming that glycerin is a nonvolatile, nonelectrolyte solute in ethanol.

53. The normal boiling point of diethyl ether is 34.5°C. A solution containing a nonvolatile solute dissolved in diethyl ether has a vapor pressure of 698 torr at 34.5°C. What is the mole fraction of diethyl ether in this solution?

54. At a certain temperature, the vapor pressure of pure benzene (C_6H_6) is 0.930 atm. A solution was prepared by dissolving 10.0 g of a nondissociating, nonvolatile solute in 78.11 g of benzene at that temperature. The vapor pressure of the solution was found to be 0.900 atm. Assuming the solution behaves ideally, determine the molar mass of the solute.

55. A solution is made by dissolving 25.8 g urea (CH_4N_2O), a nonelectrolyte, in 275 g water. Calculate the vapor pressures of this solution at 25°C and 45°C. (The vapor pressure of pure water is 23.8 torr at 25°C and 71.9 torr at 45°C.)

56. A solution of sodium chloride in water has a vapor pressure of 19.6 torr at 25°C. What is the mole fraction of solute particles in this solution? What would be the vapor pressure of this solution at 45°C? The vapor pressure of pure water is 23.8 torr at 25°C and 71.9 torr at 45°C, and assume sodium chloride exists as Na^+ and Cl^- ions in solution.

57. Pentane (C_5H_{12}) and hexane (C_6H_{14}) form an ideal solution. At 25°C the vapor pressures of pentane and hexane are 511 and 150. torr, respectively. A solution is prepared by mixing 25 mL pentane (density, 0.63 g/mL) with 45 mL hexane (density, 0.66 g/mL).

 a. What is the vapor pressure of the resulting solution?

 b. What is the composition by mole fraction of pentane in the vapor that is in equilibrium with this solution?

58. A solution is prepared by mixing 0.0300 mole of CH_2Cl_2 and 0.0500 mole of CH_2Br_2 at 25°C. Assuming the solution is ideal, calculate the composition of the vapor (in terms of mole fractions) at 25°C. At 25°C, the vapor pressures of pure CH_2Cl_2 and pure CH_2Br_2 are 133 and 11.4 torr, respectively.

59. What is the composition of a methanol (CH_3OH)–propanol ($CH_3CH_2CH_2OH$) solution that has a vapor pressure of 174 torr at 40°C? At 40°C, the vapor pressures of pure methanol and pure propanol are 303 and 44.6 torr, respectively. Assume the solution is ideal.

60. Benzene and toluene form an ideal solution. Consider a solution of benzene and toluene prepared at 25°C. Assuming the mole fractions of benzene and toluene in the vapor phase are equal, calculate the composition of the solution. At 25°C the vapor pressures of benzene and toluene are 95 and 28 torr, respectively.

61. Which of the following will have the lowest total vapor pressure at 25°C?

 a. pure water (vapor pressure = 23.8 torr at 25°C)

 b. a solution of glucose in water with $X_{C_6H_{12}O_6} = 0.01$

 c. a solution of sodium chloride in water with $X_{NaCl} = 0.01$

 d. a solution of methanol in water with $X_{CH_3OH} = 0.2$ (Consider the vapor pressure of both methanol [143 torr at 25°C] and water.)

62. Which of the choices in Exercise 61 has the highest vapor pressure?

63. Match the vapor pressure diagrams with the solute–solvent combinations and explain your answers.

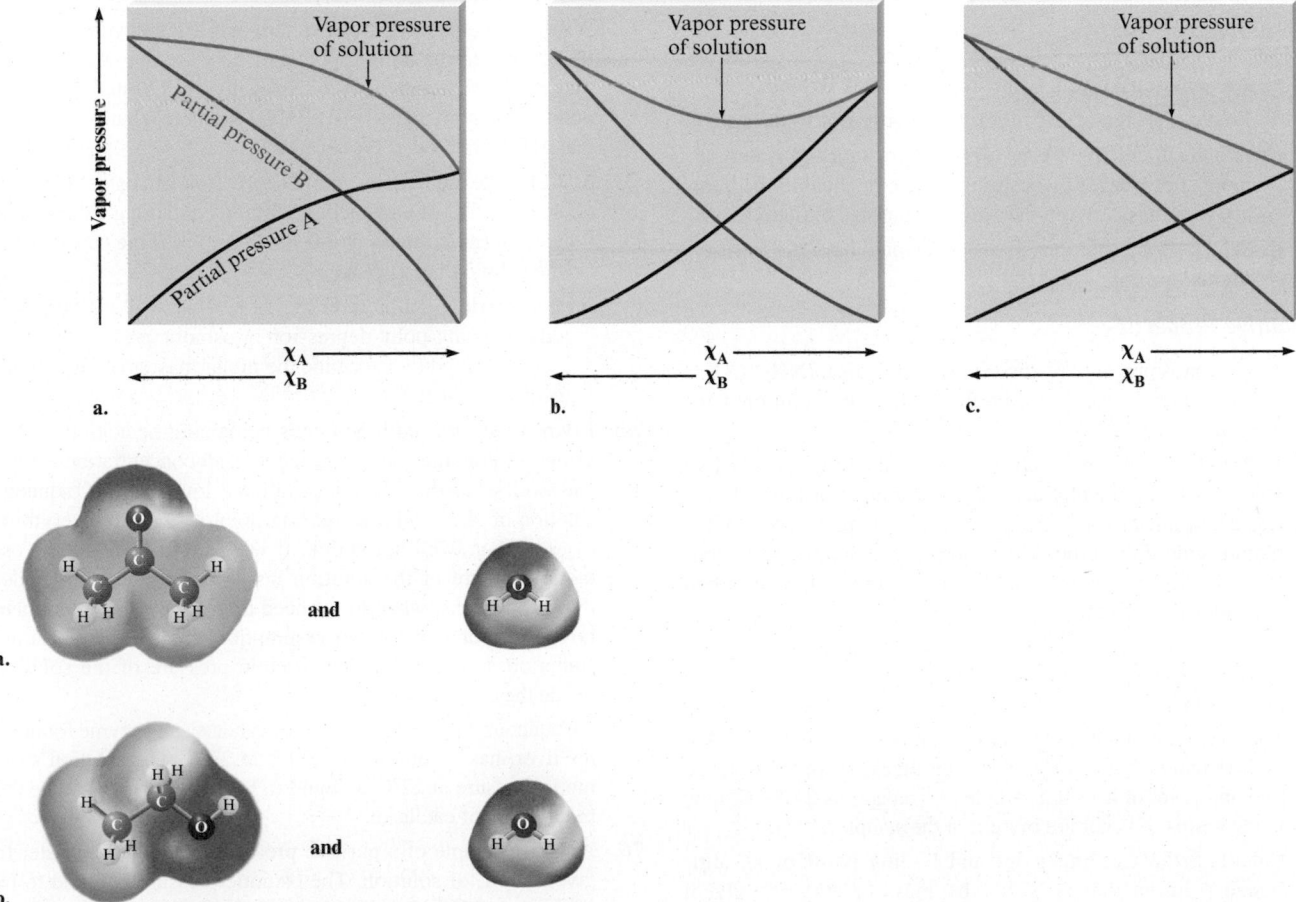

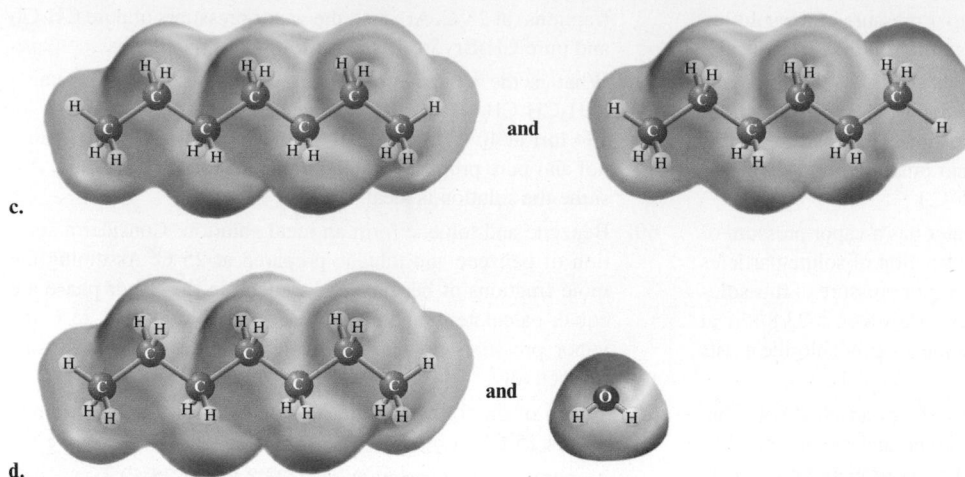

c.

d.

64. The vapor pressures of several solutions of water–propanol ($CH_3CH_2CH_2OH$) were determined at various compositions, with the following data collected at 45°C:

X_{H_2O}	Vapor Pressure (torr)
0	74.0
0.15	77.3
0.37	80.2
0.54	81.6
0.69	80.6
0.83	78.2
1.00	71.9

a. Are solutions of water and propanol ideal? Explain.

b. Predict the sign of ΔH_{soln} for water–propanol solutions.

c. Are the interactive forces between propanol and water molecules weaker than, stronger than, or equal to the interactive forces between the pure substances? Explain.

d. Which of the solutions in the data would have the lowest normal boiling point?

Colligative Properties

65. A solution is prepared by dissolving 27.0 g urea, $(NH_2)_2CO$, in 150.0 g water. Calculate the boiling point of the solution. Urea is a nonelectrolyte.

66. A 2.00-g sample of a large biomolecule was dissolved in 15.0 g carbon tetrachloride. The boiling point of this solution was determined to be 77.85°C. Calculate the molar mass of the biomolecule. For carbon tetrachloride, the boiling-point constant is 5.03°C · kg/mol, and the boiling point of pure carbon tetrachloride is 76.50°C.

67. What mass of glycerin ($C_3H_8O_3$), a nonelectrolyte, must be dissolved in 200.0 g water to give a solution with a freezing point of −1.50°C?

68. The freezing point of *t*-butanol is 25.50°C and K_f is 9.1°C · kg/mol. Usually *t*-butanol absorbs water on exposure to air. If the freezing point of a 10.0-g sample of *t*-butanol is 24.59°C, how many grams of water are present in the sample?

69. Calculate the freezing point and boiling point of an antifreeze solution that is 50.0% by mass of ethylene glycol ($HOCH_2CH_2OH$) in water. Ethylene glycol is a nonelectrolyte.

70. What volume of ethylene glycol ($C_2H_6O_2$), a nonelectrolyte, must be added to 15.0 L water to produce an antifreeze solution with a freezing point of −25.0°C? What is the boiling point of this solution? (The density of ethylene glycol is 1.11 g/cm³, and the density of water is 1.00 g/cm³.)

71. Reserpine is a natural product isolated from the roots of the shrub *Rauwolfia serpentina*. It was first synthesized in 1956 by Nobel Prize winner R. B. Woodward. It is used as a tranquilizer and sedative. When 1.00 g reserpine is dissolved in 25.0 g camphor, the freezing-point depression is 2.63°C (K_f for camphor is 40.°C · kg/mol). Calculate the molality of the solution and the molar mass of reserpine.

72. A solution contains 3.75 g of a nonvolatile pure hydrocarbon in 95 g acetone. The boiling points of pure acetone and the solution are 55.95°C and 56.50°C, respectively. The molal boiling-point constant of acetone is 1.71°C · kg/mol. What is the molar mass of the hydrocarbon?

73. a. Calculate the freezing-point depression and osmotic pressure at 25°C of an aqueous solution containing 1.0 g/L of a protein (molar mass = 9.0×10^4 g/mol) if the density of the solution is 1.0 g/cm³.

b. Considering your answer to part a, which colligative property, freezing-point depression or osmotic pressure, would be better used to determine the molar masses of large molecules? Explain.

74. Erythrocytes are red blood cells containing hemoglobin. In a saline solution they shrivel when the salt concentration is high and swell when the salt concentration is low. In a 25°C aqueous solution of NaCl, whose freezing point is −0.406°C, erythrocytes neither swell nor shrink. If we want to calculate the osmotic pressure of the solution inside the erythrocytes under these conditions, what do we need to assume? Why? Estimate how good (or poor) of an assumption this is. Make this assumption and calculate the osmotic pressure of the solution inside the erythrocytes.

75. An aqueous solution of 10.00 g of catalase, an enzyme found in the liver, has a volume of 1.00 L at 27°C. The solution's osmotic pressure at 27°C is found to be 0.745 torr. Calculate the molar mass of catalase.

76. A 0.15-g sample of a purified protein is dissolved in water to give 2.0 mL of solution. The osmotic pressure is found to be 18.6 torr at 25°C. Calculate the protein's molar mass.

77. How would you prepare 1.0 L of an aqueous solution of sucrose ($C_{12}H_{22}O_{11}$) having an osmotic pressure of 15 atm at a temperature of 22°C? Sucrose is a nonelectrolyte.

78. How would you prepare 1.0 L of an aqueous solution of sodium chloride having an osmotic pressure of 15 atm at 22°C? Assume sodium chloride exists as Na^+ and Cl^- ions in solution.

Properties of Electrolyte Solutions

79. Consider the following solutions:

    0.010 m Na_3PO_4 in water

    0.020 m $CaBr_2$ in water

    0.020 m KCl in water

    0.020 m HF in water (HF is a weak acid.)

    a. Assuming complete dissociation of the soluble salts, which solution(s) would have the same boiling point as 0.040 m $C_6H_{12}O_6$ in water? $C_6H_{12}O_6$ is a nonelectrolyte.

    b. Which solution would have the highest vapor pressure at 28°C?

    c. Which solution would have the largest freezing-point depression?

80. From the following:

    pure water

    solution of $C_{12}H_{22}O_{11}$ ($m = 0.01$) in water

    solution of NaCl ($m = 0.01$) in water

    solution of $CaCl_2$ ($m = 0.01$) in water

    Choose the one with the

    a. highest freezing point.

    b. lowest freezing point.

    c. highest boiling point.

    d. lowest boiling point.

    e. highest osmotic pressure.

81. Calculate the freezing point and the boiling point of each of the following solutions. (Assume complete dissociation.)

    a. 5.0 g NaCl in 25 g H_2O

    b. 2.0 g $Al(NO_3)_3$ in 15 g H_2O

82. A water desalination plant is set up near a salt marsh containing water that is 0.10 M NaCl. Calculate the minimum pressure that must be applied at 20.°C to purify the water by reverse osmosis. Assume NaCl is completely dissociated.

83. Determine the van't Hoff factor for the following ionic solute dissolved in water.

84. Consider the following representations of an ionic solute in water. Which flask contains $MgSO_4$, and which flask contains NaCl? How can you tell?

85. Calculate the freezing point and the boiling point of each of the following aqueous solutions. (Assume complete dissociation.)

    a. 0.050 m $MgCl_2$

    b. 0.050 m $FeCl_3$

86. Calculate the freezing point and the boiling point of each of the following solutions using the observed van't Hoff factors in Table 10-6.

    a. 0.050 m $MgCl_2$

    b. 0.050 m $FeCl_3$

87. Use the following data for three aqueous solutions of $CaCl_2$ to calculate the apparent value of the van't Hoff factor.

Molality	Freezing-Point Depression (°C)
0.0225	0.110
0.0910	0.440
0.278	1.330

88. The freezing-point depression of a 0.091-m solution of CsCl is 0.320°C. The freezing-point depression of a 0.091-m solution of $CaCl_2$ is 0.440°C. In which solution does ion association appear to be greater? Explain.

89. In the winter of 1994, record low temperatures were registered throughout the United States. For example, in Champaign, Illinois, a record low of -29°F was registered. At this temperature can salting icy roads with $CaCl_2$ be effective in melting the ice?

    a. Assume $i = 3.00$ for $CaCl_2$.

    b. Assume the average value of i from Exercise 87.

    (The solubility of $CaCl_2$ in cold water is 74.5 g per 100.0 g water.)

90. A 0.500-g sample of a compound is dissolved in enough water to form 100.0 mL of solution. This solution has an osmotic pressure of 2.50 atm at 25°C. If each molecule of the solute dissociates into two particles (in this solvent), what is the molar mass of this solute?

Additional Exercises

91. The solubility of benzoic acid ($HC_7H_5O_2$),

![benzoic acid structure: benzene ring—C(=O)—O—H]

is 0.34 g/100 mL in water at 25°C and is 10.0 g/100 mL in benzene (C_6H_6) at 25°C. Rationalize this solubility behavior. (*Hint:* Benzoic acid forms a dimer in benzene.) Would benzoic acid be more or less soluble in a 0.1-*M* NaOH solution than it is in water? Explain.

92. Given the following electrostatic potential diagrams, comment on the expected solubility of CH_4 in water and NH_3 in water.

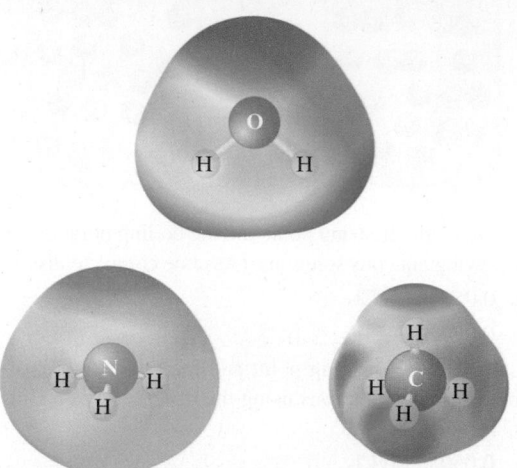

93. In a coffee-cup calorimeter, 1.60 g NH_4NO_3 was mixed with 75.0 g water at an initial temperature 25.00°C. After dissolution of the salt, the final temperature of the calorimeter contents was 23.34°C.

    a. Assuming the solution has a heat capacity of 4.18 J/g · °C, and assuming no heat loss to the calorimeter, calculate the enthalpy of solution (ΔH_{soln}) for the dissolution of NH_4NO_3 in units of kJ/mol.

    b. If the enthalpy of hydration for NH_4NO_3 is −630. kJ/mol, calculate the lattice energy of NH_4NO_3.

94. In Exercise 96 in Chapter 8, the pressure of CO_2 in a bottle of sparkling wine was calculated assuming that the CO_2 was insoluble in water. This was a bad assumption. Redo this problem by assuming that CO_2 obeys Henry's law. Use the data given in that problem to calculate the partial pressure of CO_2 in the gas phase and the solubility of CO_2 in the wine at 25°C. The Henry's law constant for CO_2 is 3.1×10^{-2} mol/L · atm at 25°C with Henry's law in the form $C = kP$, where C is the concentration of the gas in mol/L.

95. Explain the following on the basis of the behavior of atoms and/or ions.

    a. Cooking with water is faster in a pressure cooker than in an open pan.

    b. Salt is used on icy roads.

    c. Melted sea ice from the Arctic Ocean produces fresh water.

    d. $CO_2(s)$ (dry ice) does not have a normal boiling point under normal atmospheric conditions, even though CO_2 is a liquid in fire extinguishers.

    e. Adding a solute to a solvent extends the liquid phase over a larger temperature range.

96. The term *proof* is defined as twice the percent by volume of pure ethanol in solution. Thus, a solution that is 95% (by volume) ethanol is 190 proof. What is the molarity of ethanol in a 92 proof ethanol–water solution? Assume the density of ethanol, C_2H_5OH, is 0.79 g/cm^3 and the density of water is 1.0 g/cm^3.

97. At 25°C, the vapor in equilibrium with a solution containing carbon disulfide and acetonitrile has a total pressure of 263 torr and is 85.5 mole percent carbon disulfide. What is the mole fraction of carbon disulfide in the solution? At 25°C, the vapor pressure of carbon disulfide is 375 torr. Assume the solution and vapor exhibit ideal behavior.

98. For each of the following solute–solvent combinations, state the sign and relative magnitudes for ΔH_1, ΔH_2, ΔH_3, and ΔH_{soln} (as defined in Fig. 10-1 of the text). Explain your answers.

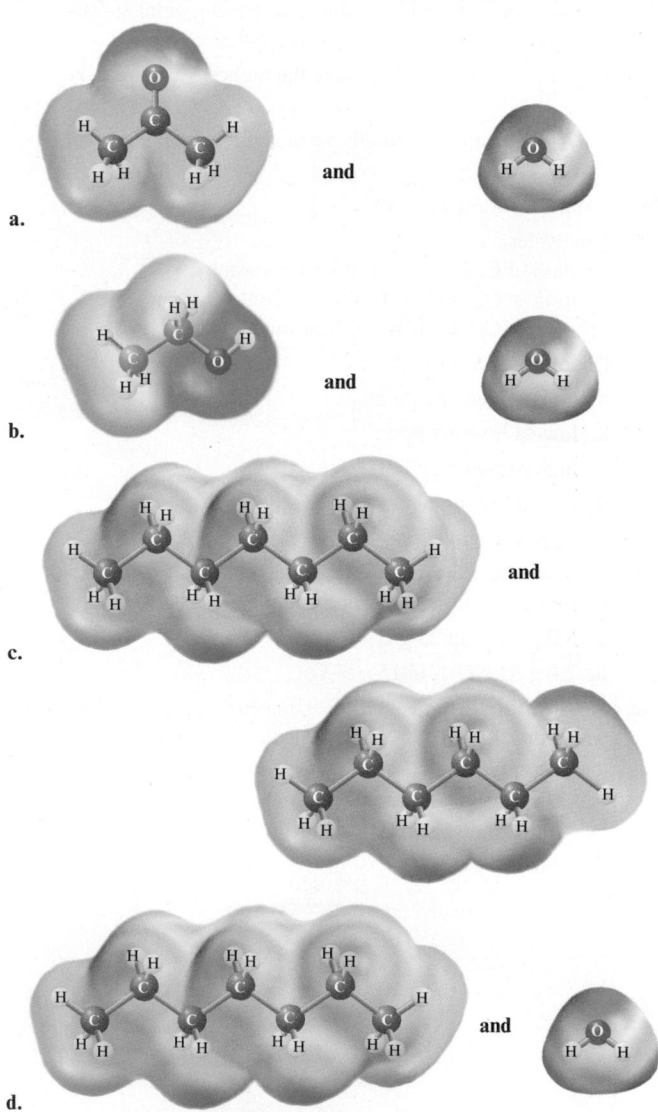

a.

b.

c.

d.

99. A solution is made by mixing 50.0 g acetone (CH_3COCH_3) and 50.0 g methanol (CH_3OH). What is the vapor pressure of this solution at 25°C? What is the composition of the vapor expressed as a mole fraction? Assume ideal solution and gas behavior. (At 25°C the vapor pressures of pure acetone and pure methanol are 271 and 143 torr, respectively.) The actual vapor pressure of this solution is 161 torr. Explain any discrepancies.

100. If the fluid inside a tree is about 0.1 M more concentrated in solute than the groundwater that bathes the roots, how high will a column of fluid rise in the tree at 25°C? Assume that the density of the fluid is 1.0 g/cm³. (The density of mercury is 13.6 g/cm³.)

101. Thyroxine, an important hormone that controls the rate of metabolism in the body, can be isolated from the thyroid gland. When 0.455 g thyroxine is dissolved in 10.0 g benzene, the freezing point of the solution is depressed by 0.300°C. What is the molar mass of thyroxine? See Table 10-5.

102. If the human eye has an osmotic pressure of 8.00 atm at 25°C, what concentration of solute particles in water will provide an isotonic eyedrop solution (a solution with equal osmotic pressure)?

103. An unknown compound contains only carbon, hydrogen, and oxygen. Combustion analysis of the compound gives mass percents of 31.57% C and 5.30% H. The molar mass is determined by measuring the freezing-point depression of an aqueous solution. A freezing point of −5.20°C is recorded for a solution made by dissolving 10.56 g of the compound in 25.0 g water. Determine the empirical formula, molar mass, and molecular formula of the compound. Assume that the compound is a nonelectrolyte.

104. Consider the following:

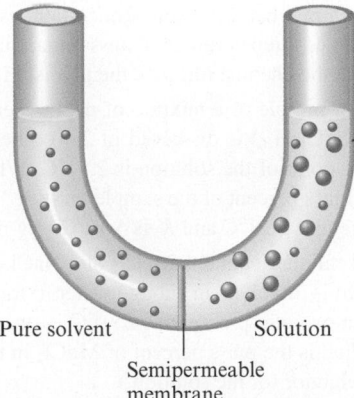

Pure solvent   Solution

Semipermeable membrane

What would happen to the level of liquid in the two arms if the semipermeable membrane separating the two liquids were permeable to

a. H_2O (the solvent) only?

b. H_2O and solute?

105. Consider an aqueous solution containing sodium chloride that has a density of 1.01 g/mL. Assume the solution behaves ideally. The freezing point of this solution at 1.0 atm is −1.28°C. Calculate the percent composition of this solution (by mass).

106. What stabilizes a colloidal suspension? Explain why adding heat or adding an electrolyte can cause the suspended particles to settle out.

107. The freezing point of an aqueous solution is −2.79°C.

a. Determine the boiling point of this solution.

b. Determine the vapor pressure (in mm Hg) of this solution at 25°C (the vapor pressure of pure water at 25°C is 23.76 mm Hg).

c. Explain any assumptions you make in solving parts a and b.

108. Specifications for lactated Ringer's solution, which is used for intravenous (IV) injections, are as follows to reach 100. mL of solution:

285–315 mg Na^+
14.1–17.3 mg K^+
4.9–6.0 mg Ca^{2+}
368–408 mg Cl^-
231–261 mg lactate, $C_3H_5O_3^-$

a. Specify the amount of NaCl, KCl, $CaCl_2 \cdot 2H_2O$, and $NaC_3H_5O_3$ needed to prepare 100. mL lactated Ringer's solution.

b. What is the range of the osmotic pressure of the solution at 37°C, given the preceding specifications?

109. Patients undergoing an upper gastrointestinal tract laboratory test are typically given an X-ray contrast agent that aids with the radiologic imaging of the anatomy. One such contrast agent is sodium diatrizoate, a nonvolatile water-soluble compound. A 0.378-m solution is prepared by dissolving 38.4 g sodium diatrizoate (NaDTZ) in 1.60×10^2 mL water at 31.2°C (the density of water at 31.2°C is 0.995 g/cm³). What is the molar mass of sodium diatrizoate? What is the vapor pressure of this solution if the vapor pressure of pure water at 31.2°C is 34.1 torr?

ChemWork Problems

These multiconcept problems (and additional ones) are found interactively online with the same type of assistance a student would get from an instructor.

110. A solution is prepared by dissolving 52.3 g cesium chloride in 60.0 g water. The volume of the solution is 63.3 mL. Calculate the mass percent, molarity, molality, and mole fraction of the CsCl solution.

111. The lattice energy of NaCl is −786 kJ/mol, and the enthalpy of hydration of 1 mole of gaseous Na^+ and 1 mole of gaseous Cl^- ions is −783 kJ/mol. Calculate the enthalpy of solution per mole of solid NaCl.

112. For each of the following pairs, predict which substance is more soluble in water.

a. CH_3NH_2 or NH_3

b. CH_3CN or CH_3OCH_3

c. CH_3CH_2OH or $CH_3CH_2CH_3$

d. CH_3OH or CH_3CH_2OH

e. $(CH_3)_3CCH_2OH$ or $CH_3(CH_2)_6OH$

f. CH_3OCH_3 or CH_3CO_2H

113. The normal boiling point of methanol is 64.7°C. A solution containing a nonvolatile solute dissolved in methanol has a vapor pressure of 556.0 torr at 64.7°C. What is the mole fraction of methanol in this solution?

114. A solution is prepared by mixing 1.000 mole of methanol (CH_3OH) and 3.18 moles of propanol ($CH_3CH_2CH_2OH$). What is the composition of the vapor (in mole fractions) at 40°C? At 40°C, the vapor pressure of pure methanol is 303 torr, and the vapor pressure of pure propanol is 44.6 torr.

115. The molar mass of a nonelectrolyte is 58.0 g/mol. Determine the boiling point of a solution containing 35.0 g of this compound and 600.0 g of water. The barometric pressure during the experiment was such that the boiling point of pure water was 99.725°C.

116. A 4.7×10^{-2} mg sample of a protein is dissolved in water to make 0.25 mL of solution. The osmotic pressure of the solution is 0.56 torr at 25°C. What is the molar mass of the protein?

117. A solid consists of a mixture of $NaNO_3$ and $Mg(NO_3)_2$. When 6.50 g of the solid is dissolved in 50.0 g of water, the freezing point of the solution is lowered by 5.23°C. What is the composition by mass of the solid mixture?

Challenge Problems

118. The vapor pressure of pure benzene is 750.0 torr and the vapor pressure of toluene is 300.0 torr at a certain temperature. You make a solution by pouring "some" benzene with "some" toluene. You then place this solution in a closed container and wait for the vapor to come into equilibrium with the solution. Next, you condense the vapor. You put this liquid (the condensed vapor) in a closed container and wait for the vapor to come into equilibrium with the solution. You then condense this vapor and find the mole fraction of benzene in this vapor to be 0.714. Determine the mole fraction of benzene in the original solution assuming the solution behaves ideally.

119. Liquid A has vapor pressure x, and liquid B has vapor pressure y. What is the mole fraction of the liquid mixture if the vapor above the solution is 30.% A by moles? 50.% A? 80.% A? (Calculate in terms of x and y.)

    Liquid A has vapor pressure x, liquid B has vapor pressure y. What is the mole fraction of the vapor above the solution if the liquid mixture is 30.% A by moles? 50.% A? 80.% A? (Calculate in terms of x and y.)

120. Plants that thrive in salt water must have internal solutions (inside the plant cells) that are isotonic with (have the same osmotic pressure as) the surrounding solution. A leaf of a saltwater plant is able to thrive in an aqueous salt solution (at 25°C) that has a freezing point equal to −0.621°C. You would like to use this information to calculate the osmotic pressure of the solution in the cell.

    a. In order to use the freezing-point depression to calculate osmotic pressure, what assumption must you make (in addition to ideal behavior of the solutions, which we will assume)?

    b. Under what conditions is the assumption (in part a) reasonable?

    c. Solve for the osmotic pressure (at 25°C) of the solution in the plant cell.

    d. The plant leaf is placed in an aqueous salt solution (at 25°C) that has a boiling point of 102.0°C. What will happen to the plant cells in the leaf?

121. You make 20.0 g of a sucrose ($C_{12}H_{22}O_{11}$) and NaCl mixture and dissolve it in 1.00 kg water. The freezing point of this solution is found to be −0.426°C. Assuming ideal behavior, calculate the mass percent composition of the original mixture, and the mole fraction of sucrose in the original mixture.

122. An aqueous solution is 1.00% NaCl by mass and has a density of 1.071 g/cm^3 at 25°C. The observed osmotic pressure of this solution is 7.83 atm at 25°C.

    a. What fraction of the moles of NaCl in this solution exist as ion pairs?

    b. Calculate the freezing point that would be observed for this solution.

123. The vapor in equilibrium with a pentane–hexane solution at 25°C has a mole fraction of pentane equal to 0.15 at 25°C. What is the mole fraction of pentane in the solution? (See Exercise 57 for the vapor pressures of the pure liquids.)

124. A forensic chemist is given a white solid that is suspected of being pure cocaine ($C_{17}H_{21}NO_4$, molar mass = 303.35 g/mol). She dissolves 1.22 ± 0.01 g of the solid in 15.60 ± 0.01 g benzene. The freezing point is lowered by 1.32 ± 0.04°C.

    a. What is the molar mass of the substance? Assuming that the percent uncertainty in the calculated molar mass is the same as the percent uncertainty in the temperature change, calculate the uncertainty in the molar mass.

    b. Could the chemist unequivocally state that the substance is cocaine? For example, is the uncertainty small enough to distinguish cocaine from codeine ($C_{18}H_{21}NO_3$, molar mass = 299.36 g/mol)?

    c. Assuming that the absolute uncertainties in the measurements of temperature and mass remain unchanged, how could the chemist improve the precision of her results?

125. A 1.60-g sample of a mixture of naphthalene ($C_{10}H_8$) and anthracene ($C_{14}H_{10}$) is dissolved in 20.0 g benzene (C_6H_6). The freezing point of the solution is 2.81°C. What is the composition as mass percent of the sample mixture? The freezing point of benzene is 5.51°C and K_f is 5.12°C · kg/mol.

126. A solid mixture contains $MgCl_2$ and NaCl. When 0.5000 g of this solid is dissolved in enough water to form 1.000 L of solution, the osmotic pressure at 25.0°C is observed to be 0.3950 atm. What is the mass percent of $MgCl_2$ in the solid? (Assume ideal behavior for the solution.)

127. Formic acid (HCO_2H) is a monoprotic acid that ionizes only partially in aqueous solutions. A 0.10-M formic acid solution is 4.2% ionized. Assuming that the molarity and molality of the solution are the same, calculate the freezing point and the boiling point of 0.10 M formic acid.

128. You have a solution of two volatile liquids, A and B (assume ideal behavior). Pure liquid A has a vapor pressure of 350.0 torr and pure liquid B has a vapor pressure of 100.0 torr at the temperature of the solution. The vapor at equilibrium above the solution has double the mole fraction of substance A that the solution does. What is the mole fraction of liquid A in the solution?

129. In some regions of the southwest United States, the water is very hard. For example, in Las Cruces, New Mexico, the tap water contains about 560 μg of dissolved solids per milliliter. Reverse osmosis units are marketed in this area to soften water. A typical unit exerts a pressure of 8.0 atm and can produce 45 L water per day.

    a. Assuming all of the dissolved solids are $MgCO_3$ and assuming a temperature of 27°C, what total volume of water must be processed to produce 45 L pure water?

    b. Would the same system work for purifying seawater? (Assume seawater is 0.60 M NaCl.)

Integrative Problems

These problems require the integration of multiple concepts to find the solutions.

130. Creatinine, $C_4H_7N_3O$, is a by-product of muscle metabolism, and creatinine levels in the body are known to be a fairly reliable indicator of kidney function. The normal level of creatinine in the blood for adults is approximately 1.0 mg per deciliter (dL) of blood. If the density of blood is 1.025 g/mL, calculate the molality of a normal creatinine level in a 10.0-mL blood sample. What is the osmotic pressure of this solution at 25.0°C?

131. An aqueous solution containing 0.250 mole of Q, a strong electrolyte, in 5.00×10^2 g water freezes at −2.79°C. What is the van't Hoff factor for Q? The molal freezing-point depression constant for water is 1.86°C · kg/mol. What is the formula of Q if it is 38.68% chlorine by mass and there are twice as many anions as cations in one formula unit of Q?

132. Anthraquinone contains only carbon, hydrogen, and oxygen. When 4.80 mg anthraquinone is burned, 14.2 mg CO_2 and 1.65 mg H_2O are produced. The freezing point of camphor is lowered by 22.3°C when 1.32 g anthraquinone is dissolved in 11.4 g camphor. Determine the empirical and molecular formulas of anthraquinone.

CHAPTER ELEVEN

Chemical Kinetics

These wheelchair athletes generate a great amount of kinetic energy as they race. (Simon Balson/Alamy)

446

The applications of chemistry focus largely on chemical reactions, and the commercial use of a reaction requires knowledge of several of its characteristics, including its stoichiometry, energetics, and rate. ◄ A reaction is defined by its reactants and products, whose identity must be learned by experiment. Once the reactants and products are known, the equation for the reaction can be written and balanced, and stoichiometric calculations can be carried out. Another very important characteristic of a reaction is its spontaneity. Spontaneity refers to the *inherent tendency* for the process to occur; however, it implies nothing about speed. *Spontaneous does not mean fast.* There are many spontaneous reactions that are so slow that no apparent reaction occurs over a period of weeks or years at normal temperatures. For example, there is a strong inherent tendency for gaseous hydrogen and oxygen to combine, that is,

$$2H_2(g) + O_2(g) \longrightarrow 2H_2O(l)$$

but in fact the two gases can coexist indefinitely at 25°C. Similarly, the gaseous reactions

$$H_2(g) + Cl_2(g) \longrightarrow 2HCl(g)$$
$$N_2(g) + 3H_2(g) \longrightarrow 2NH_3(g)$$

are both highly likely to occur from a thermodynamic standpoint, but we observe no reactions under normal conditions. In addition, the process of changing diamond to graphite is spontaneous but is so slow that it is not detectable.

To be useful, reactions must occur at a reasonable rate. To produce the 20 million tons of ammonia needed each year for fertilizer, we cannot simply mix nitrogen and hydrogen gases at 25°C and wait for them to react. It is not enough to understand the stoichiometry and thermodynamics of a reaction; we also must understand the factors that govern the rate of the reaction. The area of chemistry that concerns reaction rates is called **chemical kinetics**.

One of the main goals of chemical kinetics is to understand the steps by which a reaction takes place. This series of steps is called the *reaction mechanism*. Notice that the balanced chemical equation does not give us information on *how* the reaction proceeds. For example, when hydrogen and oxygen gases combine to form water, is this reaction initiated by the breaking of the chemical bond that holds the hydrogen atoms

▲ The energy required for athletic exertion and the combustion of fuel in a race car both result from chemical reactions.

447

Table 11-1 | Concentrations of Reactant and Products as a Function of Time for the Reaction $2NO_2(g) \rightarrow 2NO(g) + O_2(g)$ (at 300°C)

Time (±1 s)	Concentration (mol/L)		
	NO_2	NO	O_2
0	0.0100	0	0
50	0.0079	0.0021	0.0011
100	0.0065	0.0035	0.0018
150	0.0055	0.0045	0.0023
200	0.0048	0.0052	0.0026
250	0.0043	0.0057	0.0029
300	0.0038	0.0062	0.0031
350	0.0034	0.0066	0.0033
400	0.0031	0.0069	0.0035

together? Or is the O_2 bond broken? Or is it something else entirely? It is possible to use data about reaction rates in order to support a reaction mechanism. Understanding the mechanism allows us to find ways to facilitate the reaction. For example, the Haber process for the production of ammonia requires high temperatures to achieve

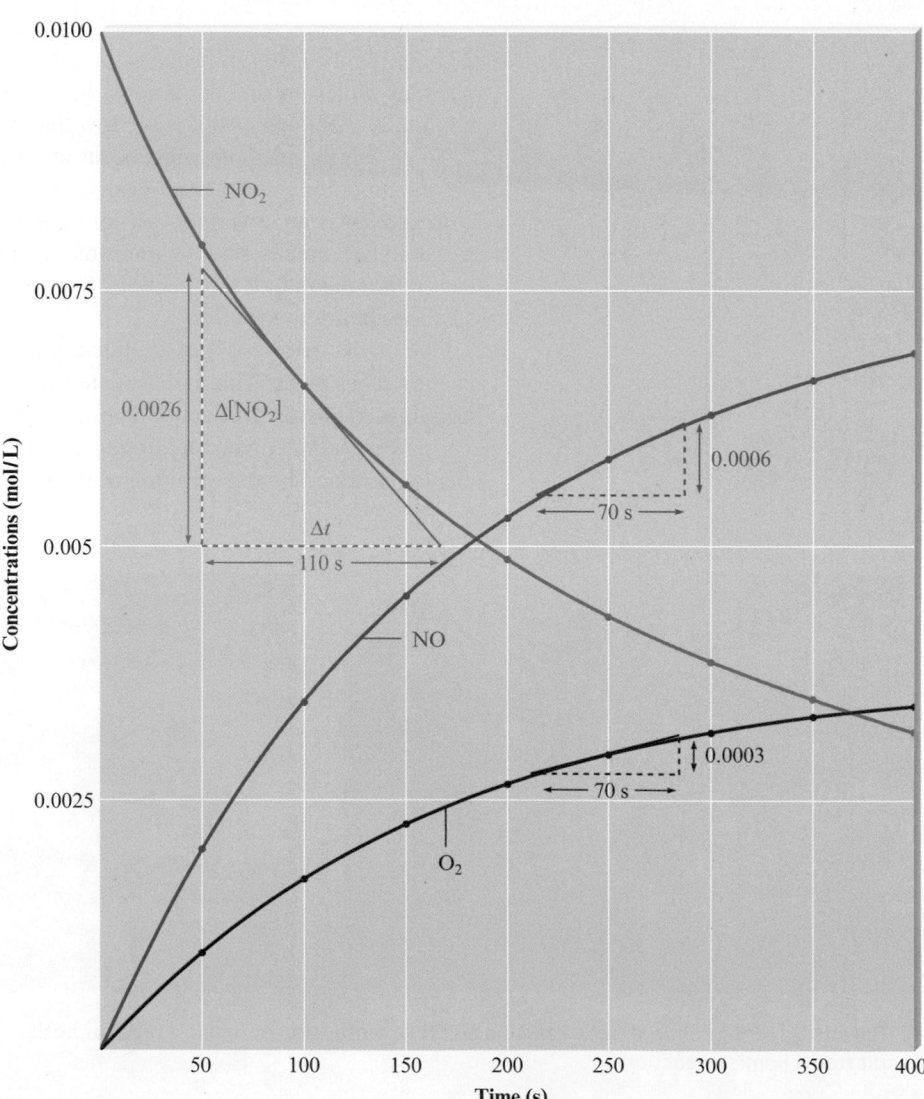

Figure 11-1 | Starting with a flask of nitrogen dioxide at 300°C, the concentrations of nitrogen dioxide, nitric oxide, and oxygen are plotted versus time.

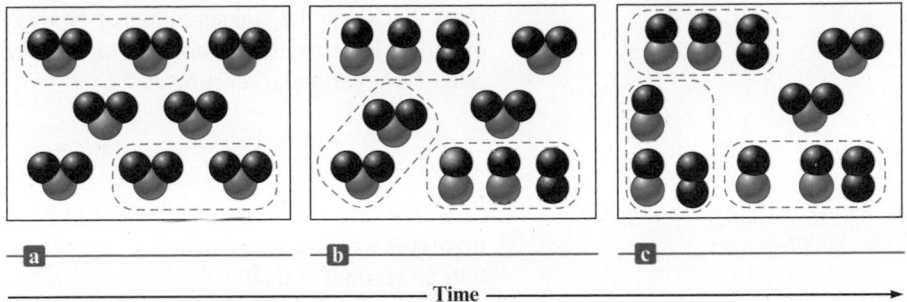

Figure 11-2 | Representation of the reaction $2NO_2(g) \rightarrow 2NO(g) + O_2(g)$. (a) The reaction at the very beginning ($t = 0$). (b) and (c) As time passes, NO_2 is converted to NO and O_2.

Time

commercially feasible reaction rates. However, even higher temperatures (and more cost) would be required without the use of iron oxide, which speeds up the reaction.

In this chapter we will consider the main ideas of chemical kinetics. We will explore rate laws, reaction mechanisms, and simple models for chemical reactions.

11-1 | Reaction Rates

𝒾 The kinetics of air pollution is discussed in Section 11-7.

To introduce the concept of the rate of a reaction, we will consider the decomposition of nitrogen dioxide, a gas that causes air pollution. ◄ 𝒾 ◄ Nitrogen dioxide decomposes to nitric oxide and oxygen as follows:

$$2NO_2(g) \longrightarrow 2NO(g) + O_2(g)$$

Suppose in a particular experiment we start with a flask of nitrogen dioxide at 300°C and measure the concentrations of nitrogen dioxide, nitric oxide, and oxygen as the nitrogen dioxide decomposes. The results of this experiment are summarized in Table 11-1, and the data are plotted in Fig. 11-1.

Note from these results that the concentration of the reactant (NO_2) decreases with time and the concentrations of the products (NO and O_2) increase with time (Fig. 11-2). Chemical kinetics deals with the speed at which these changes occur. The speed, or *rate,* of a process is defined as the change in a given quantity over a specific period of time. For chemical reactions, the quantity that changes is the amount or concentration of a reactant or product. So the **reaction rate** of a chemical reaction is defined as the *change in concentration of a reactant or product per unit time:*

$$\text{Rate} = \frac{\text{concentration of A at time } t_2 - \text{concentration of A at time } t_1}{t_2 - t_1}$$

𝒾 [A] means concentration of A in mol/L.

$$= \frac{\Delta[A]}{\Delta t} \blacktriangleleft 𝒾$$

where A is the reactant or product being considered, and the square brackets indicate concentration in mol/L. As usual, the symbol Δ indicates a *change* in a given quantity. Note that a change can be positive (increase) or negative (decrease), thus leading to a positive or negative reaction rate by this definition. However, for convenience, we will always define the rate as a positive quantity, as we will see.

Now let us calculate the average rate at which the concentration of NO_2 changes over the first 50 seconds of the reaction using the data given in Table 11-1.

$$\frac{\text{Change in } [NO_2]}{\text{Time elapsed}} = \frac{\Delta[NO_2]}{\Delta t}$$

$$= \frac{[NO_2]_{t=50} - [NO_2]_{t=0}}{50.\ \text{s} - 0\ \text{s}}$$

$$= \frac{0.0079\ \text{mol/L} - 0.0100\ \text{mol/L}}{50.\ \text{s}}$$

$$= -4.2 \times 10^{-5}\ \text{mol/L} \cdot \text{s}$$

Note that since the concentration of NO_2 decreases with time, $\Delta[NO_2]$ is a negative quantity. Because it is customary to work with *positive* reaction rates, we define the rate of this particular reaction as

$$\text{Rate} = -\frac{\Delta[NO_2]}{\Delta t}$$

Since the concentrations of reactants always decrease with time, any rate expression involving a reactant will include a negative sign. The average rate of this reaction from 0 to 50 seconds is then

$$
\begin{aligned}
\text{Rate} &= -\frac{\Delta[NO_2]}{\Delta t} \\
&= -(-4.2 \times 10^{-5}\,\text{mol/L} \cdot \text{s}) \\
&= 4.2 \times 10^{-5}\,\text{mol/L} \cdot \text{s}
\end{aligned}
$$

Appendix 1.3 reviews slopes of straight lines.

The average rates for this reaction during several other time intervals are given in Table 11-2. *Note that the rate is not constant but decreases with time.* The rates given in Table 11-2 are *average rates* over 50-second time intervals. The value of the rate at a particular time (the **instantaneous rate**) can be obtained by computing the slope of a line tangent to the curve at that point. ◄ *i* Figure 11-1 shows a tangent drawn at $t = 100$ seconds. The *slope* of this line gives the rate at $t = 100$ seconds as follows:

$$
\begin{aligned}
\text{Slope of the tangent line} &= \frac{\text{change in } y}{\text{change in } x} \\
&= \frac{\Delta[NO_2]}{\Delta t}
\end{aligned}
$$

Table 11-2 │ Average Rate (in mol/L · s) of Decomposition of Nitrogen Dioxide as a Function of Time*

$\dfrac{\Delta[NO_2]}{\Delta t}$	Time Period (s)
4.2×10^{-5}	$0 \rightarrow 50$
2.8×10^{-5}	$50 \rightarrow 100$
2.0×10^{-5}	$100 \rightarrow 150$
1.4×10^{-5}	$150 \rightarrow 200$
1.0×10^{-5}	$200 \rightarrow 250$

*Note that the *rate* decreases with time.

But

$$\text{Rate} = -\frac{\Delta[NO_2]}{\Delta t}$$

Therefore,

$$
\begin{aligned}
\text{Rate} &= -(\text{slope of the tangent line}) \\
&= -\left(\frac{-0.0026\,\text{mol/L}}{110\,\text{s}}\right) \\
&= 2.4 \times 10^{-5}\,\text{mol/L} \cdot \text{s}
\end{aligned}
$$

So far we have discussed the rate of this reaction only in terms of the reactant. The rate also can be defined in terms of the products. However, in doing so we must take

▲ Los Angeles on a clear day, and on a day when air pollution is significant.

into account the coefficients in the balanced equation for the reaction, because the stoichiometry determines the relative rates of consumption of reactants and generation of products. For example, in the reaction we are considering,

$$2NO_2(g) \longrightarrow 2NO(g) + O_2(g)$$

both the reactant NO_2 and the product NO have a coefficient of 2, so NO is produced at the same rate as NO_2 is consumed. We can verify this from Fig. 11-1. Note that the curve for NO is the same shape as the curve for NO_2, except that it is inverted, or flipped over. This means that, at any point in time, the slope of the tangent to the curve for NO will be the negative of the slope to the curve for NO_2. (Verify this at the point $t = 100$ seconds on both curves.) In the balanced equation, the product O_2 has a coefficient of 1, which means it is produced half as fast as NO, since NO has a coefficient of 2. That is, the rate of NO production is twice the rate of O_2 production.

We also can verify this fact from Fig. 11-1. For example, at $t = 250$ seconds,

$$\text{Slope of the tangent to the NO curve} = \frac{6.0 \times 10^{-4} \text{ mol/L}}{70. \text{ s}}$$

$$= 8.6 \times 10^{-6} \text{ mol/L} \cdot \text{s}$$

$$\text{Slope of the tangent to the O}_2 \text{ curve} = \frac{3.0 \times 10^{-4} \text{ mol/L}}{70. \text{ s}}$$

$$= 4.3 \times 10^{-6} \text{ mol/L} \cdot \text{s}$$

The slope at $t = 250$ seconds on the NO curve is twice the slope of that point on the O_2 curve, showing that the rate of production of NO is twice that of O_2.

The rate information can be summarized as follows:

Rate of consumption of NO_2	=	rate of production of NO	=	2(rate of production of O_2)
$-\dfrac{\Delta[NO_2]}{\Delta t}$	=	$\dfrac{\Delta[NO]}{\Delta t}$	=	$2\left(\dfrac{\Delta[O_2]}{\Delta t}\right)$

i As we will see later, there is a certain type of rate that remains constant over time.

We have seen that the rate of a reaction is typically not constant. ◄ *i* Most reaction rates change with time. This is so because the concentrations change with time (see Fig. 11-1).

Because the reaction rate changes with time, and because the rate is different (by factors that depend on the coefficients in the balanced equation) depending on which reactant or product is being studied, we must be very specific when we describe a rate for a chemical reaction.

11-2 | Rate Laws: An Introduction

Chemical reactions are *reversible*. In our discussion of the decomposition of nitrogen dioxide, we have so far considered only the *forward reaction,* as shown here:

$$2NO_2(g) \longrightarrow 2NO(g) + O_2(g)$$

However, the *reverse reaction* also can occur. As NO and O_2 accumulate, they can react to re-form NO_2:

$$O_2(g) + 2NO(g) \longrightarrow 2NO_2(g)$$

When gaseous NO_2 is placed in an otherwise empty container, initially the dominant reaction is

$$2NO_2(g) \longrightarrow 2NO(g) + O_2(g)$$

ⓘ When forward and reverse reaction rates are equal, there will be no changes in the concentrations of reactants or products. This is called *chemical equilibrium* and is discussed fully in Chapter 12.

and the change in the concentration of NO_2 ($\Delta[NO_2]$) depends only on the forward reaction. However, after a period of time, enough products accumulate so that the reverse reaction becomes important. Now $\Delta[NO_2]$ depends on the *difference in the rates of the forward and reverse reactions.* ◄ ⓘ This complication can be avoided if we study the rate of a reaction under conditions where the reverse reaction makes only a negligible contribution. Typically, this means that we must study a reaction at a point soon after the reactants are mixed, before the products have had time to build up to significant levels.

If we choose conditions where the reverse reaction can be neglected, the *reaction rate will depend only on the concentrations of the reactants.* For the decomposition of nitrogen dioxide, we can write

$$Rate = k[NO_2]^n \tag{11-1}$$

Such an expression, which shows how the rate depends on the concentrations of reactants, is called a **rate law**. The proportionality constant k, called the **rate constant**, and n, called the **order** of the reactant, must both be determined by experiment. The order of a reactant can be an integer (including zero) or a fraction. For the relatively simple reactions we will consider in this book, the orders will often be positive integers.

Note two important points about Equation (11-1):

1. The concentrations of the products do not appear in the rate law because the reaction rate is being studied under conditions where the reverse reaction does not contribute to the overall rate.
2. The value of the exponent n must be determined by experiment; it cannot be written from the balanced equation.

Before we go further we must define exactly what we mean by the term *rate* in Equation (11-1). In Section 11-1 we saw that reaction rate means a change in concentration per unit time. However, which reactant or product concentration do we choose in defining the rate? For example, for the decomposition of NO_2 to produce O_2 and NO considered in Section 11-1, we could define the rate in terms of any of these three species. However, since O_2 is produced only half as fast as NO, we must be careful to specify which species we are talking about in a given case. For instance, we might choose to define the reaction rate in terms of the consumption of NO_2:

$$Rate = -\frac{\Delta[NO_2]}{\Delta t} = k[NO_2]^n$$

On the other hand, we could define the rate in terms of the production of O_2:

$$Rate' = \frac{\Delta[O_2]}{\Delta t} = k'[NO_2]^n$$

Note that because $2NO_2$ molecules are consumed for every O_2 molecule produced,

$$Rate = 2 \times rate'$$
or
$$k[NO_2]^n = 2k'[NO_2]^n$$
and
$$k = 2 \times k'$$

Thus the value of the rate constant depends on how the rate is defined.

In this text we will always be careful to define exactly what is meant by the rate for a given reaction so that there will be no confusion about which specific rate constant is being used.

Types of Rate Laws

Notice that the rate law we have used to this point expresses rate as a function of concentration. For example, for the decomposition of NO_2 we have defined

$$Rate = -\frac{\Delta[NO_2]}{\Delta t} = k[NO_2]^n$$

which tells us (once we have determined the value of n) exactly how the rate depends on the concentration of the reactant, NO_2. A rate law that expresses how the *rate depends on concentration* is technically called the **differential rate law**, but it is often simply called the **rate law**. ◄ ❸ Thus when we use the term *rate law* in this text, we mean the expression that gives the rate as a function of concentration.

A second kind of rate law, the **integrated rate law**, also will be important in our study of kinetics. The integrated rate law expresses how the *concentrations depend on time*. Although we will not consider the details here, a given differential rate law is always related to a certain type of integrated rate law, and vice versa. That is, if we determine the differential rate law for a given reaction, we automatically know the form of the integrated rate law for the reaction. This means that once we determine experimentally either type of rate law for a reaction, we also know the other one.

Which rate law we choose to determine by experiment often depends on what types of data are easiest to collect. If we can conveniently measure how the rate changes as the concentrations are changed, we can readily determine the differential (rate/concentration) rate law. On the other hand, if it is more convenient to measure the concentration as a function of time, we can determine the form of the integrated (concentration/time) rate law. We will discuss how rate laws are actually determined in the next several sections.

Why are we interested in determining the rate law for a reaction? How does it help us? It helps us because we can work backward from the rate law to infer the steps by which the reaction occurs. Most chemical reactions do not take place in a single step but result from a series of sequential steps. To understand a chemical reaction, we must learn what these steps are. For example, a chemist who is designing an insecticide may study the reactions involved in the process of insect growth to see what type of molecule might interrupt this series of reactions. Or an industrial chemist may be trying to make a given reaction occur faster. To accomplish this, he or she must know which step is slowest, because it is that step that must be speeded up. Thus a chemist is usually not interested in a rate law for its own sake but because of what it reveals about the steps by which a reaction occurs. We will develop a process for finding the reaction steps in this chapter.

❸ The name *differential rate law* comes from a mathematical term. We will regard it simply as a label. The terms *differential rate law* and *rate law* will be used interchangeably in this text.

LET'S REVIEW | *Rate Laws: A Summary*

- There are two types of rate laws.
  1. The *differential rate law* (often called simply the *rate law*) shows how the rate of a reaction depends on concentrations.
  2. The *integrated rate law* shows how the concentrations of species in the reaction depend on time.
- Because we typically consider reactions only under conditions where the reverse reaction is unimportant, our rate laws will involve only concentrations of reactants.
- Because the differential and integrated rate laws for a given reaction are related in a well-defined way, the experimental determination of either of the rate laws is sufficient.
- Experimental convenience usually dictates which type of rate law is determined experimentally.
- Knowing the rate law for a reaction is important mainly because we can usually infer the individual steps involved in the reaction from the specific form of the rate law.

11-3 | Determining the Form of the Rate Law

The first step in understanding how a given chemical reaction occurs is to determine the *form* of the rate law. That is, we need to determine experimentally the power to which each reactant concentration must be raised in the rate law. In this section we will

Figure 11-3 | A plot of the concentration of N_2O_5 as a function of time for the reaction $2N_2O_5(soln) \rightarrow 4NO_2(soln) + O_2(g)$ (at 45°C). Note that the reaction rate at $[N_2O_5] = 0.90\ M$ is twice that at $[N_2O_5] = 0.45\ M$.

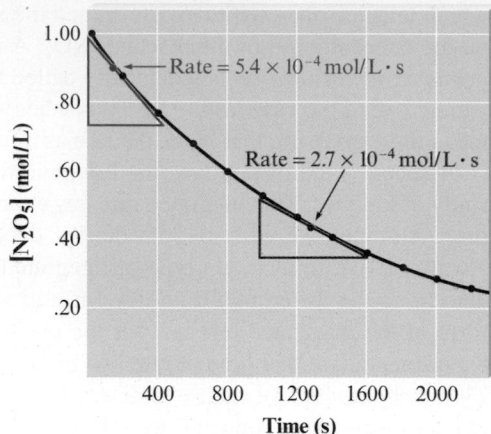

Table 11-3 | Concentration/Time Data for the Reaction $2N_2O_5(soln) \rightarrow 4NO_2(soln) + O_2(g)$ (at 45°C)

$[N_2O_5]$ (mol/L)	Time (s)
1.00	0
0.88	200
0.78	400
0.69	600
0.61	800
0.54	1000
0.48	1200
0.43	1400
0.38	1600
0.34	1800
0.30	2000

ⓘ First order: rate = $k[A]$. Doubling the concentration of A doubles the reaction rate.

ⓘ The value of the initial rate is determined for each experiment at the same value of t as close to $t = 0$ as possible.

explore ways to obtain the differential rate law for a reaction. First, we will consider the decomposition of dinitrogen pentoxide in carbon tetrachloride solution:

$$2N_2O_5(soln) \longrightarrow 4NO_2(soln) + O_2(g)$$

Data for this reaction at 45°C are listed in Table 11-3 and plotted in Fig. 11-3. In this reaction the oxygen gas escapes from the solution and thus does not react with the nitrogen dioxide, so we do not have to be concerned about the effects of the reverse reaction at any time over the life of the reaction. That is, the reverse reaction is negligible at all times over the course of this reaction.

Evaluation of the reaction rates at concentrations of N_2O_5 of 0.90 M and 0.45 M, by taking the slopes of the tangents to the curve at these points (see Fig. 11-3), yields the following data:

$[N_2O_5]$	Rate (mol/L · s)
0.90 M	5.4×10^{-4}
0.45 M	2.7×10^{-4}

Note that when $[N_2O_5]$ is halved, the rate is also halved. This means that the rate of this reaction depends on the concentration of N_2O_5 to the *first power*. In other words, the (differential) rate law for this reaction is

$$\text{Rate} = -\frac{\Delta[N_2O_5]}{\Delta t} = k[N_2O_5]^1 = k[N_2O_5]$$

Thus the reaction is *first order* in N_2O_5. ◄ *ⓘ* Note that for this reaction the order is *not* the same as the coefficient of N_2O_5 in the balanced equation for the reaction. This reemphasizes the fact that the order of a particular reactant must be obtained by *observing* how the reaction rate depends on the concentration of that reactant.

We have seen that by determining the instantaneous rate at two different reactant concentrations, the rate law for the decomposition of N_2O_5 is shown to have the form

$$\text{Rate} = -\frac{\Delta[A]}{\Delta t} = k[A]$$

where A represents N_2O_5.

Method of Initial Rates

One common method for experimentally determining the form of the rate law for a reaction is the **method of initial rates**. ◄ *ⓘ* The **initial rate** of a reaction is the instantaneous rate determined just after the reaction begins (just after $t = 0$). The idea is to

Table 11-4 | Initial Rates from Three Experiments for the Reaction $NH_4^+(aq) + NO_2^-(aq) \rightarrow N_2(g) + 2H_2O(l)$

Experiment	Initial Concentration of NH_4^+	Initial Concentration of NO_2^-	Initial Rate (mol/L · s)
1	0.100 M	0.0050 M	1.35×10^{-7}
2	0.100 M	0.010 M	2.70×10^{-7}
3	0.200 M	0.010 M	5.40×10^{-7}

determine the instantaneous rate before the initial concentrations of reactants have changed significantly. Several experiments are carried out using different initial concentrations, and the initial rate is determined for each run. The results are then compared to see how the initial rate depends on the initial concentrations. This allows the form of the rate law to be determined. We will illustrate the method of initial rates using the following equation:

$$NH_4^+(aq) + NO_2^-(aq) \longrightarrow N_2(g) + 2H_2O(l)$$

Table 11-4 gives initial rates obtained from three experiments involving different initial concentrations of reactants. The general form of the rate law for this reaction is

$$\text{Rate} = -\frac{\Delta[NH_4^+]}{\Delta t} = k[NH_4^+]^n[NO_2^-]^m$$

We can determine the values of n and m by observing how the initial rate depends on the initial concentrations of NH_4^+ and NO_2^-. In Experiments 1 and 2, where the initial concentration of NH_4^+ remains the same but the initial concentration of NO_2^- doubles, the observed initial rate also doubles. Since

$$\text{Rate} = k[NH_4^+]^n[NO_2^-]^m$$

we have for Experiment 1

$$\text{Rate} = 1.35 \times 10^{-7} \text{ mol/L} \cdot \text{s} = k(0.100 \text{ mol/L})^n(0.0050 \text{ mol/L})^m$$

and for Experiment 2

$$\text{Rate} = 2.70 \times 10^{-7} \text{ mol/L} \cdot \text{s} = k(0.100 \text{ mol/L})^n(0.010 \text{ mol/L})^m$$

The ratio of these rates is

$$\frac{\text{Rate 2}}{\text{Rate 1}} = \underbrace{\frac{2.70 \times 10^{-7} \text{ mol/L} \cdot \text{s}}{1.35 \times 10^{-7} \text{ mol/L} \cdot \text{s}}}_{2.00} = \frac{k(0.100 \text{ mol/L})^n(0.010 \text{ mol/L})^m}{k(0.100 \text{ mol/L})^n(0.0050 \text{ mol/L})^m}$$

$$= \frac{(0.010 \text{ mol/L})^m}{(0.0050 \text{ mol/L})^m} = (2.0)^m$$

Rates 1, 2, and 3 were determined at the same value of t (very close to $t = 0$).

Thus

$$\frac{\text{Rate 2}}{\text{Rate 1}} = 2.00 = (2.0)^m$$

which means the value of m is 1. ◄ ⓘ The rate law for this reaction is first order in the reactant NO_2^-.

A similar analysis of the results for Experiments 2 and 3 yields the ratio

$$\frac{\text{Rate 3}}{\text{Rate 2}} = \frac{5.40 \times 10^{-7} \text{ mol/L} \cdot \text{s}}{2.70 \times 10^{-7} \text{ mol/L} \cdot \text{s}} = \frac{(0.200 \text{ mol/L})^n}{(0.100 \text{ mol/L})^n}$$

$$= 2.00 = \left(\frac{0.200}{0.100}\right)^n = (2.00)^n$$

The value of n is also 1.

We have shown that the values of n and m are both 1 and the rate law is

$$\text{Rate} = k[\text{NH}_4^+][\text{NO}_2^-]$$

This rate law is first order in both NO_2^- and NH_4^+. Note that it is merely a coincidence that n and m have the same values as the coefficients of NH_4^+ and NO_2^- in the balanced equation for the reaction.

ⓘ Overall reaction order is the sum of the orders for the various reactants.

The **overall reaction order** is the sum of n and m. For this reaction, $n + m = 2$. ◄ ⓘ The reaction is second order overall.

The value of the rate constant k can now be calculated using the results of *any* of the three experiments shown in Table 11-4. From the data for Experiment 1, we know that

$$\text{Rate} = k[\text{NH}_4^+][\text{NO}_2^-]$$
$$1.35 \times 10^{-7}\,\text{mol/L} \cdot \text{s} = k(0.100\,\text{mol/L})(0.0050\,\text{mol/L})$$

Then

$$k = \frac{1.35 \times 10^{-7}\,\text{mol/L} \cdot \text{s}}{(0.100\,\text{mol/L})(0.0050\,\text{mol/L})} = 2.7 \times 10^{-4}\,\text{L/mol} \cdot \text{s}$$

EXAMPLE 11-1   Determining a Rate Law

The reaction between bromate ions and bromide ions in acidic aqueous solution is given by the equation

$$\text{BrO}_3^-(aq) + 5\text{Br}^-(aq) + 6\text{H}^+(aq) \longrightarrow 3\text{Br}_2(l) + 3\text{H}_2\text{O}(l)$$

Table 11-5 gives the results from four experiments. Using these data, determine the orders for all three reactants, the overall reaction order, and the value of the rate constant.

Solution

The general form of the rate law for this reaction is

$$\text{Rate} = k[\text{BrO}_3^-]^n[\text{Br}^-]^m[\text{H}^+]^p$$

We can determine the values of n, m, and p by comparing the rates from the various experiments. To determine the value of n, we use the results from Experiments 1 and 2, in which only $[\text{BrO}_3^-]$ changes:

$$\frac{\text{Rate 2}}{\text{Rate 1}} = \frac{1.6 \times 10^{-3}\,\text{mol/L} \cdot \text{s}}{8.0 \times 10^{-4}\,\text{mol/L} \cdot \text{s}} = \frac{k(0.20\,\text{mol/L})^n(0.10\,\text{mol/L})^m(0.10\,\text{mol/L})^p}{k(0.10\,\text{mol/L})^n(0.10\,\text{mol/L})^m(0.10\,\text{mol/L})^p}$$

$$2.0 = \left(\frac{0.20\,\text{mol/L}}{0.10\,\text{mol/L}}\right)^n = (2.0)^n$$

Thus n is equal to 1.

Table 11-5 | The Results from Four Experiments to Study the Reaction $\text{BrO}_3^-(aq) + 5\text{Br}^-(aq) + 6\text{H}^+(aq) \rightarrow 3\text{Br}_2(l) + 3\text{H}_2\text{O}(l)$

Experiment	Initial Concentration of BrO_3^- (mol/L)	Initial Concentration of Br^- (mol/L)	Initial Concentration of H^+ (mol/L)	Measured Initial Rate (mol/L · s)
1	0.10	0.10	0.10	8.0×10^{-4}
2	0.20	0.10	0.10	1.6×10^{-3}
3	0.20	0.20	0.10	3.2×10^{-3}
4	0.10	0.10	0.20	3.2×10^{-3}

To determine the value of m, we use the results from Experiments 2 and 3, in which only $[Br^-]$ changes:

$$\frac{\text{Rate } 3}{\text{Rate } 2} = \frac{3.2 \times 10^{-3} \text{ mol/L} \cdot \text{s}}{1.6 \times 10^{-3} \text{ mol/L} \cdot \text{s}} = \frac{k(0.20 \text{ mol/L})^n(0.20 \text{ mol/L})^m(0.10 \text{ mol/L})^p}{k(0.20 \text{ mol/L})^n(0.10 \text{ mol/L})^m(0.10 \text{ mol/L})^p}$$

$$2.0 = \left(\frac{0.20 \text{ mol/L}}{0.10 \text{ mol/L}}\right)^m = (2.0)^m$$

Thus m is equal to 1.

To determine the value of p, we use the results from Experiments 1 and 4, in which $[BrO_3^-]$ and $[Br^-]$ are constant but $[H^+]$ differs:

$$\frac{\text{Rate } 4}{\text{Rate } 1} = \frac{3.2 \times 10^{-3} \text{ mol/L} \cdot \text{s}}{8.0 \times 10^{-4} \text{ mol/L} \cdot \text{s}} = \frac{k(0.10 \text{ mol/L})^n(0.10 \text{ mol/L})^m(0.20 \text{ mol/L})^p}{k(0.10 \text{ mol/L})^n(0.10 \text{ mol/L})^m(0.10 \text{ mol/L})^p}$$

$$4.0 = \left(\frac{0.20 \text{ mol/L}}{0.10 \text{ mol/L}}\right)^p$$

$$4.0 = (2.0)^p = (2.0)^2$$

Thus p is equal to 2.

The rate of this reaction is first order in BrO_3^- and Br^- and second order in H^+. The overall reaction order is $n + m + p = 4$.

The rate law can now be written

$$\text{Rate} = k[BrO_3^-][Br^-][H^+]^2$$

The value of the rate constant k can be calculated from the results of any of the four experiments. For Experiment 1, the initial rate is 8.0×10^{-4} mol/L $\cdot$ s and $[BrO_3^-] = 0.100\ M$, $[Br^-] = 0.10\ M$, and $[H^+] = 0.10\ M$. Using these values in the rate law gives

$$8.0 \times 10^{-4} \text{ mol/L} \cdot \text{s} = k(0.10 \text{ mol/L})(0.10 \text{ mol/L})(0.10 \text{ mol/L})^2$$

$$8.0 \times 10^{-4} \text{ mol/L} \cdot \text{s} = k(1.0 \times 10^{-4} \text{ mol}^4/\text{L}^4)$$

$$k = \frac{8.0 \times 10^{-4} \text{ mol/L} \cdot \text{s}}{1.0 \times 10^{-4} \text{ mol}^4/\text{L}^4} = 8.0 \text{ L}^3/\text{mol}^3 \cdot \text{s}$$

REALITY CHECK | Verify that the same value of k can be obtained from the results of the other experiments.

See Exercises 11-29 through 11-32

11-4 | The Integrated Rate Law

The rate laws we have considered so far express the rate as a function of the reactant concentrations. It is also useful to be able to express the reactant concentrations as a function of time, given the (differential) rate law for the reaction. In this section we show how this is done.

We will proceed by first looking at reactions involving a single reactant:

$$aA \longrightarrow \text{products}$$

all of which have a rate law of the form

$$\text{Rate} = -\frac{\Delta[A]}{\Delta t} = k[A]^n$$

We will develop the integrated rate laws individually for the cases $n = 1$ (first order), $n = 2$ (second order), and $n = 0$ (zero order).

First-Order Rate Laws

For the reaction

$$2N_2O_5(soln) \longrightarrow 4NO_2(soln) + O_2(g)$$

we have found that the rate law is

$$\text{Rate} = -\frac{\Delta[N_2O_5]}{\Delta t} = k[N_2O_5]$$

Since the rate of this reaction depends on the concentration of N_2O_5 to the first power, it is a **first-order reaction**. This means that if the concentration of N_2O_5 in a flask were suddenly doubled, the rate of production of NO_2 and O_2 also would double. This rate law can be put into a different form using a calculus operation known as integration, which yields the expression

$$\ln[N_2O_5] = -kt + \ln[N_2O_5]_0$$

> *i* Appendix 1.2 contains a review of logarithms.

where ln indicates the natural logarithm, t is the time, $[N_2O_5]$ is the concentration of N_2O_5 at time t, and $[N_2O_5]_0$ is the initial concentration of N_2O_5 (at $t = 0$, the start of the experiment). ◄ *i* Note that such an equation, called the *integrated rate law,* expresses the *concentration of the reactant as a function of time.*

For a chemical reaction of the form

$$aA \longrightarrow \text{products}$$

where the kinetics are first order in [A], the rate law is

$$\text{Rate} = -\frac{\Delta[A]}{\Delta t} = k[A]$$

and the **integrated first-order rate law** is

$$\ln[A] = -kt + \ln[A]_0 \tag{11-2}$$

There are several important things to note about Equation (11-2):

> *i* An integrated rate law relates concentration to reaction time.

1. The equation shows how the concentration of A depends on time. If the initial concentration of A and the rate constant k are known, the concentration of A at any time can be calculated. ◄ *i*
2. Equation (11-2) is of the form $y = mx + b$, where a plot of y versus x is a straight line with slope m and intercept b. In Equation (11-2),

$$y = \ln[A] \qquad x = t \qquad m = -k \qquad b = \ln[A]_0$$

> *i* For a first-order reaction, a plot of ln[A] versus *t* is always a straight line.

Thus for a first-order reaction, plotting the natural logarithm of concentration versus time always gives a straight line. ◄ *i* This fact is often used to test whether a reaction is first order. For the reaction

$$aA \longrightarrow \text{products}$$

the *reaction is first order in A if a plot of ln[A] versus t is a straight line.* Conversely, if this plot is not a straight line, the reaction is not first order in A.
3. This integrated rate law for a first-order reaction also can be expressed in terms of a *ratio* of [A] and [A]$_0$ as follows:

$$\ln\left(\frac{[A]_0}{[A]}\right) = kt$$

EXAMPLE 11-2

First-Order Rate Laws I

The decomposition of N_2O_5 in the gas phase was studied at constant temperature.

$$2N_2O_5(g) \longrightarrow 4NO_2(g) + O_2(g)$$

The following results were collected:

$[N_2O_5]$ (mol/L)	Time (s)
0.1000	0
0.0707	50
0.0500	100
0.0250	200
0.0125	300
0.00625	400

Using these data, verify that the rate law is first order in $[N_2O_5]$, and calculate the value of the rate constant, where the rate $= -\Delta[N_2O_5]/\Delta t$.

Solution

We can verify that the rate law is first order in $[N_2O_5]$ by constructing a plot of $\ln[N_2O_5]$ versus time.

$\ln[N_2O_5]$	Time (s)
-2.303	0
-2.649	50
-2.996	100
-3.689	200
-4.382	300
-5.075	400

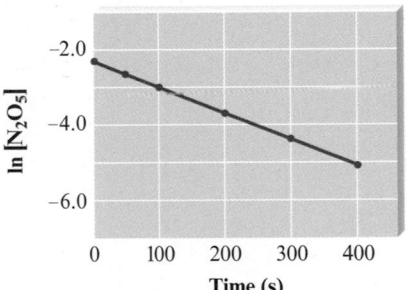

The values of $\ln[N_2O_5]$ at various times are given in the table above and shown in the plot of $\ln[N_2O_5]$. The fact that the plot is a straight line confirms that the reaction is first order in N_2O_5, since it follows the equation $\ln[N_2O_5] = -kt + \ln[N_2O_5]_0$.

Since the reaction is first order, the slope of the line equals $-k$, where

$$\text{Slope} = \frac{\text{change in } y}{\text{change in } x} = \frac{\Delta y}{\Delta x} = \frac{\Delta(\ln[N_2O_5])}{\Delta t}$$

Since the first and last points are exactly on the line, we will use these points to calculate the slope:

$$\text{Slope} = \frac{-5.075 - (-2.303)}{400.\text{ s} - 0\text{ s}} = \frac{-2.772}{400.\text{ s}} = -6.93 \times 10^{-3}\text{ s}^{-1}$$

$$k = -(\text{slope}) = 6.93 \times 10^{-3}\text{ s}^{-1}$$

See Exercise 11-37

| **EXAMPLE 11-3** | **First-Order Rate Laws II** |

Using the data given in Example 11-2, calculate $[N_2O_5]$ at 150 s after the start of the reaction.

Solution

We know from Example 11-2 that $[N_2O_5] = 0.0500$ mol/L at 100 s and $[N_2O_5] = 0.0250$ mol/L at 200 s. Since 150 s is halfway between 100 and 200 s, it is tempting to assume that we can simply use an arithmetic average to obtain $[N_2O_5]$ at that time. This is incorrect because it is $\ln[N_2O_5]$, not $[N_2O_5]$, that is directly proportional to t. To calculate $[N_2O_5]$ after 150 s, we use Equation (11-2):

$$\ln[N_2O_5] = -kt + \ln[N_2O_5]_0$$

ⓘ The antilog operation means to exponentiate (see Appendix 1.2).

where $t = 150.$ s, $k = 6.93 \times 10^{-3}$ s^{-1} (as determined in Example 11-2), and $[N_2O_5]_0 = 0.1000$ mol/L. ◀ ⓘ

$$\ln([N_2O_5]_{t=150}) = -(6.93 \times 10^{-3}\,\text{s}^{-1})(150.\,\text{s}) + \ln(0.100)$$
$$= -1.040 - 2.303 = -3.343$$
$$[N_2O_5]_{t=150} = \text{antilog}(-3.343) = 0.0353\,\text{mol/L}$$

Note that this value of $[N_2O_5]$ is *not* halfway between 0.0500 and 0.0250 mol/L.

See Exercise 11-37

Half-Life of a First-Order Reaction

The time required for a reactant to reach half its original concentration is called the **half-life of a reactant** and is designated by the symbol $t_{1/2}$. For example, we can calculate the half-life of the decomposition reaction discussed in Example 11-2. The data plotted in Fig. 11-4 show that the half-life for this reaction is 100 seconds. We can see this by considering the following numbers:

$[N_2O_5]$ (mol/L)	Δt (s)	Ratio of Concentrations
0.100	0	
		$\Delta t = 100$ s; $\dfrac{[N_2O_5]_{t=100}}{[N_2O_5]_{t=0}} = \dfrac{0.050}{0.100} = \dfrac{1}{2}$
0.0500	100	
		$\Delta t = 100$ s; $\dfrac{[N_2O_5]_{t=200}}{[N_2O_5]_{t=100}} = \dfrac{0.025}{0.050} = \dfrac{1}{2}$
0.0250	200	
		$\Delta t = 100$ s; $\dfrac{[N_2O_5]_{t=300}}{[N_2O_5]_{t=200}} = \dfrac{0.0125}{0.0250} = \dfrac{1}{2}$
0.0125	300	

Note that it *always* takes 100 seconds for $[N_2O_5]$ to be halved in this reaction.

A general formula for the half-life of a first-order reaction can be derived from the integrated rate law for the general reaction

$$aA \longrightarrow \text{products}$$

If the reaction is first order in $[A]$,

$$\ln\left(\frac{[A]_0}{[A]}\right) = kt$$

Figure 11-4 | A plot of $[N_2O_5]$ versus time for the decomposition reaction of N_2O_5.

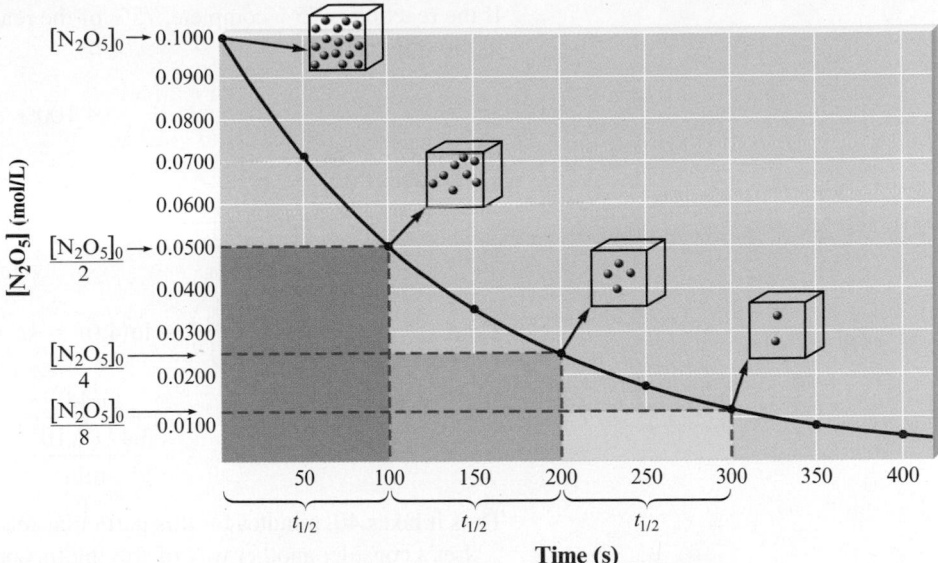

By definition, when $t = t_{1/2}$,

$$[A] = \frac{[A]_0}{2}$$

Then, for $t = t_{1/2}$, the integrated rate law becomes

$$\ln\left(\frac{[A]_0}{[A]_0/2}\right) = kt_{1/2}$$

or

$$\ln(2) = kt_{1/2}$$

Substituting the value of $\ln(2)$ and solving for $t_{1/2}$ gives

$$t_{1/2} = \frac{0.693}{k} \tag{11-3}$$

ⓘ For a first-order reaction, $t_{1/2}$ is independent of the initial concentration.

This is the *general equation for the half-life of a first-order reaction.* Equation (11-3) can be used to calculate $t_{1/2}$ if k is known or k if $t_{1/2}$ is known. Note that for a first-order reaction, *the half-life does not depend on concentration.* ◄ ⓘ

INTERACTIVE EXAMPLE 11-4

Half-Life for a First-Order Reaction

A certain first-order reaction has a half-life of 20.0 minutes.

a. Calculate the rate constant for this reaction.

b. How much time is required for this reaction to be 75% complete?

Solution

a. Solving Equation (11-3) for k gives

$$k = \frac{0.693}{t_{1/2}} = \frac{0.693}{20.0 \text{ min}} = 3.47 \times 10^{-2} \text{ min}^{-1}$$

b. We use the integrated rate law in the form

$$\ln\left(\frac{[A]_0}{[A]}\right) = kt$$

If the reaction is 75% complete, 75% of the reactant has been consumed, leaving 25% in the original form:

$$\frac{[A]}{[A]_0} \times 100\% = 25\%$$

This means that

$$\frac{[A]}{[A]_0} = 0.25 \quad \text{or} \quad \frac{[A]_0}{[A]} = \frac{1}{0.25} = 4.0$$

Then

$$\ln\left(\frac{[A]_0}{[A]}\right) = \ln(4.0) = kt = \left(\frac{3.47 \times 10^{-2}}{\min}\right)t$$

and

$$t = \frac{\ln(4.0)}{\dfrac{3.47 \times 10^{-2}}{\min}} = 40. \ \min$$

Thus it takes 40. minutes for this particular reaction to reach 75% completion.

Let's consider another way of solving this problem using the definition of half-life. After one half-life the reaction has gone 50% to completion. If the initial concentration were 1.0 mol/L, after one half-life the concentration would be 0.50 mol/L. One more half-life would produce a concentration of 0.25 mol/L. Comparing 0.25 mol/L with the original 1.0 mol/L shows that 25% of the reactant is left after two half-lives. This is a general result. (What percentage of reactant remains after three half-lives?) Two half-lives for this reaction is 2(20.0 min), or 40.0 min, which agrees with the preceding answer.

See Exercises 11-38 and 11-49 through 11-52

Second-Order Rate Laws

For a general reaction involving a single reactant, that is,

$$aA \longrightarrow \text{products}$$

that is second order in A, the rate law is

> ⓘ Second order: rate = $k[A]^2$. Doubling the concentration of A quadruples the reaction rate; tripling the concentration of A increases the rate by nine times.

$$\text{Rate} = -\frac{\Delta[A]}{\Delta t} = k[A]^2 \quad \blacktriangleleft \ \textit{ⓘ} \tag{11-4}$$

The **integrated second-order rate law** has the form

$$\frac{1}{[A]} = kt + \frac{1}{[A]_0} \tag{11-5}$$

Note the following characteristics of Equation (11-5):

> ⓘ For second-order reactions, a plot of 1/[A] versus t will be linear.

1. A plot of $1/[A]$ versus t will produce a straight line with a slope equal to k. ◀ ⓘ
2. Equation (11-5) shows how [A] depends on time and can be used to calculate [A] at any time t, provided k and $[A]_0$ are known.

When one half-life of the second-order reaction has elapsed ($t = t_{1/2}$), by definition,

$$[A] = \frac{[A]_0}{2}$$

Equation (11-5) then becomes

$$\frac{1}{\frac{[A]_0}{2}} = kt_{1/2} + \frac{1}{[A]_0}$$

$$\frac{2}{[A]_0} - \frac{1}{[A]_0} = kt_{1/2}$$

$$\frac{1}{[A]_0} = kt_{1/2}$$

Solving for $t_{1/2}$ gives *the expression for the half-life of a second-order reaction:*

$$t_{1/2} = \frac{1}{k[A]_0} \tag{11-6}$$

EXAMPLE 11-5

ℹ️ When two identical molecules combine, the resulting molecule is called a *dimer*.

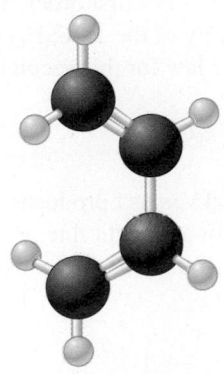

Butadiene (C_4H_6)

Determining Rate Laws

Butadiene reacts to form its dimer according to the equation ◄

$$2C_4H_6(g) \longrightarrow C_8H_{12}(g) \ ◄ \ ℹ️$$

The following data were collected for this reaction at a given temperature:

$[C_4H_6]$ (mol/L)	Time (± 1 s)
0.01000	0
0.00625	1000
0.00476	1800
0.00370	2800
0.00313	3600
0.00270	4400
0.00241	5200
0.00208	6200

a. Is this reaction first order or second order?

b. What is the value of the rate constant for the reaction?

c. What is the half-life for the reaction under the initial conditions of this experiment?

Solution

a. To decide whether the rate law for this reaction is first order or second order, we must see whether the plot of $\ln[C_4H_6]$ versus time is a straight line (first order) or the plot of $1/[C_4H_6]$ versus time is a straight line (second order). The data necessary to make these plots are as follows:

t (s)	$\frac{1}{[C_4H_6]}$	$\ln[C_4H_6]$
0	100	−4.605
1000	160	−5.075
1800	210	−5.348
2800	270	−5.599
3600	320	−5.767
4400	370	−5.915
5200	415	−6.028
6200	481	−6.175

Figure 11-5 | (a) A plot of $\ln[C_4H_6]$ versus t. (b) A plot of $1/[C_4H_6]$ versus t.

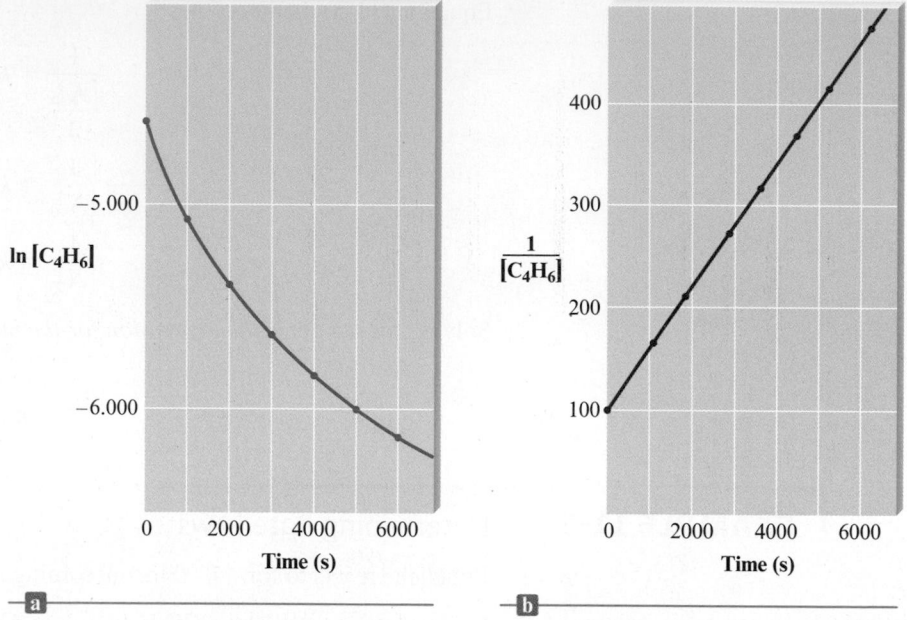

The resulting plots are shown in Fig. 11-5. Since the $\ln[C_4H_6]$ versus t plot [Fig. 11-5(a)] is not a straight line, the reaction is *not* first order. The reaction is, however, second order, as shown by the linearity of the $1/[C_4H_6]$ versus t plot [Fig. 11-5(b)]. Thus we can now write the rate law for this second-order reaction:

$$\text{Rate} = -\frac{\Delta[C_4H_6]}{\Delta t} = k[C_4H_6]^2$$

b. For a second-order reaction, a plot of $1/[C_4H_6]$ versus t produces a straight line of slope k. In terms of the standard equation for a straight line, $y = mx + b$, we have $y = 1/[C_4H_6]$ and $x = t$. Thus the slope of the line can be expressed as follows:

$$\text{Slope} = \frac{\Delta y}{\Delta x} = \frac{\Delta\left(\frac{1}{[C_4H_6]}\right)}{\Delta t}$$

Using the points at $t = 0$ and $t = 6200$, we can find the rate constant for the reaction:

$$k = \text{slope} = \frac{(481 - 100)\ \text{L/mol}}{(6200. - 0)\ \text{s}} = \frac{381}{6200.}\text{L/mol} \cdot \text{s} = 6.14 \times 10^{-2}\ \text{L/mol} \cdot \text{s}$$

c. The expression for the half-life of a second-order reaction is

$$t_{1/2} = \frac{1}{k[A]_0}$$

In this case $k = 6.14 \times 10^{-2}\ \text{L/mol} \cdot \text{s}$ (from part b) and $[A]_0 = [C_4H_6]_0 = 0.01000\ M$ (the concentration at $t = 0$). Thus

$$t_{1/2} = \frac{1}{(6.14 \times 10^{-2}\ \text{L/mol} \cdot \text{s})(1.000 \times 10^{-2}\ \text{mol/L})} = 1.63 \times 10^3\ \text{s}$$

The initial concentration of C_4H_6 is halved in 1630 s.

See Exercises 11-39, 11-40, 11-53, and 11-54

ⓘ For a second-order reaction, $t_{1/2}$ is dependent on $[A]_0$. For a first-order reaction, $t_{1/2}$ is independent of $[A]_0$.

ⓘ For each successive half-life, $[A]_0$ is halved. Since $t_{1/2} = 1/k[A]_0$, $t_{1/2}$ doubles.

ⓘ A zero-order reaction has a constant rate.

It is important to recognize the difference between the half-life for a first-order reaction and the half-life for a second-order reaction. For a second-order reaction, $t_{1/2}$ depends on both k and $[A]_0$; for a first-order reaction, $t_{1/2}$ depends only on k. ◄ **ⓘ** For a first-order reaction, a constant time is required to reduce the concentration of the reactant by half, and then by half again, and so on, as the reaction proceeds. From Example 11-5 we can see that this is *not* true for a second-order reaction. For that second-order reaction, we found that the first half-life (the time required to go from $[C_4H_6] = 0.010$ M to $[C_4H_6] = 0.0050$ M) is 1630 seconds. We can estimate the second half-life from the concentration data as a function of time. Note that to reach 0.0024 M C_4H_6 (approximately 0.0050/2) requires 5200 seconds of reaction time. Thus to get from 0.0050 M C_4H_6 to 0.0024 M C_4H_6 takes 3570 seconds (5200 − 1630). The second half-life is much longer than the first. This pattern is characteristic of second-order reactions. In fact, *for a second-order reaction, each successive half-life is double the preceding one* (provided the effects of the reverse reaction can be ignored, as we are assuming here). ◄ **ⓘ** Prove this to yourself by examining the equation $t_{1/2} = 1/(k[A]_0)$.

Zero-Order Rate Laws

Most reactions involving a single reactant show either first-order or second-order kinetics. However, sometimes such a reaction can be a **zero-order reaction**. The rate law for a zero-order reaction is

$$\text{Rate} = k[A]^0 = k(1) = k$$

For a zero-order reaction, the rate is constant. ◄ **ⓘ** It does not change with concentration as it does for first-order or second-order reactions.

The **integrated rate law for a zero-order reaction** is

$$[A] = -kt + [A]_0 \tag{11-7}$$

In this case a plot of $[A]$ versus t gives a straight line of slope $-k$ (Fig. 11-6).

The expression for the half-life of a zero-order reaction can be obtained from the integrated rate law. By definition, $[A] = [A]_0/2$ when $t = t_{1/2}$, so

$$\frac{[A]_0}{2} = -kt_{1/2} + [A]_0$$

or

$$kt_{1/2} = \frac{[A]_0}{2k}$$

Solving for $t_{1/2}$ gives

$$t_{1/2} = \frac{[A]_0}{2k} \tag{11-8}$$

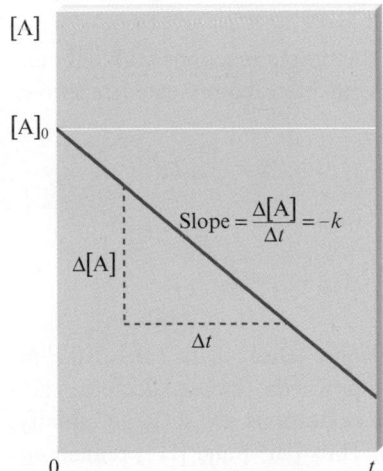

Figure 11-6 | A plot of [A] versus t for a zero-order reaction.

Zero-order reactions are most often encountered when a substance such as a metal surface or an enzyme is required for the reaction to occur. For example, the decomposition reaction

$$2N_2O(g) \longrightarrow 2N_2(g) + O_2(g)$$

occurs on a hot platinum surface. When the platinum surface is completely covered with N_2O molecules, an increase in the concentration of N_2O has no effect on the rate, since only those N_2O molecules on the surface can react. Under these conditions, *the rate is a constant* because it is controlled by what happens on the platinum surface rather than by the total concentration of N_2O (Fig. 11-7). This reaction also can occur at high temperatures with no platinum surface present, but under these conditions, it is not zero order.

Figure 11-7 | The decomposition reaction $2N_2O(g) \rightarrow 2N_2(g) + O_2(g)$ takes place on a platinum surface. Although $[N_2O]$ is three times as great in (b) as in (a), the rate of decomposition of N_2O is the same in both cases because the platinum surface can accommodate only a certain number of molecules. As a result, this reaction is zero order.

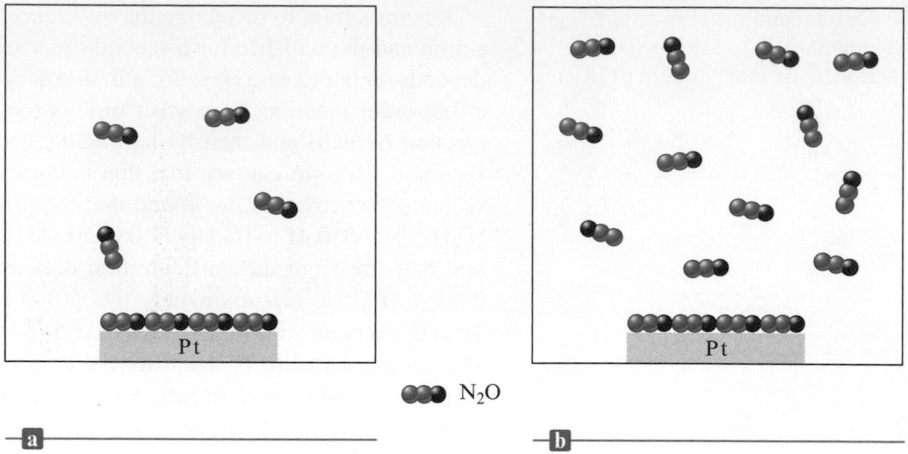

$\bullet\bullet\bullet$ N_2O

a ──────────────────────── b ────────────────────────

Critical Thinking

Consider the simple reaction aA $\rightarrow$ products. You run this reaction and wish to determine its order. What if you made a graph of reaction rate versus time? Could you use this to determine the order? Sketch three plots of rate versus time for the reaction if it is zero, first, or second order. Sketch these plots on the same graph and compare them. Defend your answer.

Integrated Rate Laws for Reactions with More Than One Reactant

So far we have considered the integrated rate laws for simple reactions with only one reactant. Special techniques are required to deal with more complicated reactions. Let's consider the reaction

$$BrO_3^-(aq) + 5Br^-(aq) + 6H^+(aq) \longrightarrow 3Br_2(l) + 3H_2O(l)$$

From experimental evidence we know that the rate law is

$$\text{Rate} = -\frac{\Delta[BrO_3^-]}{\Delta t} = k[BrO_3^-][Br^-][H^+]^2$$

Suppose we run this reaction under conditions where $[BrO_3^-]_0 = 1.0 \times 10^{-3}\ M$, $[Br^-]_0 = 1.0\ M$, and $[H^+]_0 = 1.0\ M$. As the reaction proceeds, $[BrO_3^-]$ decreases significantly, but because the Br^- ion and H^+ ion concentrations are so large initially, relatively little of these two reactants is consumed. Thus $[Br^-]$ and $[H^+]$ remain *approximately constant*. In other words, under the conditions where the Br^- ion and H^+ ion concentrations are much larger than the BrO_3^- ion concentration, we can assume that throughout the reaction

$$[Br^-] = [Br^-]_0 \quad \text{and} \quad [H^+] = [H^+]_0$$

This means that the rate law can be written

$$\text{Rate} = k[Br^-]_0[H^+]_0^2[BrO_3^-] = k'[BrO_3^-]$$

where, since $[Br^-]_0$ and $[H^+]_0$ are constant,

$$k' = k[Br^-]_0[H^+]_0^2$$

The rate law

$$\text{Rate} = k'[\text{BrO}_3^-]$$

is first order. However, since this law was obtained by simplifying a more complicated one, it is called a **pseudo-first-order rate law**. Under the conditions of this experiment, a plot of $\ln[\text{BrO}_3^-]$ versus t will give a straight line where the slope is equal to $-k'$. Since $[\text{Br}^-]_0$ and $[\text{H}^+]_0$ are known, the value of k can be calculated from the equation

$$k' = k[\text{Br}^-]_0[\text{H}^+]_0^2$$

which can be rearranged to give

$$k = \frac{k'}{[\text{Br}^-]_0[\text{H}^+]_0^2}$$

Note that the kinetics of complicated reactions can be studied by observing the behavior of one reactant at a time. If the concentration of one reactant is much smaller than the concentrations of the others, then the amounts of those reactants present in large concentrations will not change significantly and can be regarded as constant. The change in concentration with time of the reactant present in a relatively small amount can then be used to determine the order of the reaction in that component. This technique allows us to determine rate laws for complex reactions.

LET'S REVIEW | *Rate Laws: A Summary*

1. To simplify the rate laws for reactions, we have always assumed that the rate is being studied under conditions where only the forward reaction is important. This produces rate laws that contain only reactant concentrations.

2. There are two types of rate laws.
   a. The *differential rate law* (often called the *rate law*) shows how the rate depends on the concentrations. The forms of the rate laws for zero-order, first-order, and second-order kinetics of reactions with single reactants are shown in Table 11-6.
   b. The *integrated rate law* shows how concentration depends on time. The integrated rate laws corresponding to zero-order, first-order, and second-order kinetics of one-reactant reactions are given in Table 11-6.

3. Whether we determine the differential rate law or the integrated rate law depends on the type of data that can be collected conveniently and accurately. Once we have experimentally determined either type of rate law, we can write the other for a given reaction.

4. The most common method for experimentally determining the differential rate law is the method of initial rates. In this method several experiments are run at different initial concentrations and the instantaneous rates are determined for each at the same value of t (as close to $t = 0$ as possible). The point is to evaluate the rate before the concentrations change significantly from the initial values. From a comparison of the initial rates and the initial concentrations, the dependence of the rate on the concentrations of various reactants can be obtained—that is, the order in each reactant can be determined.

5. To experimentally determine the integrated rate law for a reaction, concentrations are measured at various values of t as the reaction proceeds. Then the job is to see which integrated rate law correctly fits the data. Typically this is done visually by ascertaining which type of plot gives a straight line. A summary for one-reactant reactions is given in Table 11-6. Once the correct straight-line plot is found, the correct integrated rate law can be chosen and the value of k obtained from the slope. Also, the (differential) rate law for the reaction can then be written.

(continued)

LET'S REVIEW | *Rate Laws: A Summary (continued)*

6. The integrated rate law for a reaction that involves several reactants can be treated by choosing conditions such that the concentration of only one reactant varies in a given experiment. This is done by having the concentration of one reactant remain small compared with the concentrations of all the others, causing a rate law such as

$$\text{Rate} = k[A]^n[B]^m[C]^p$$

to reduce to

$$\text{Rate} = k'[A]^n$$

where $k' = k[B]_0^m[C]_0^p$ and $[B]_0 \gg [A]_0$ and $[C]_0 \gg [A]_0$. The value of n is obtained by determining whether a plot of $[A]$ versus t is linear ($n = 0$), a plot of $\ln[A]$ versus t is linear ($n = 1$), or a plot of $1/[A]$ versus t is linear ($n = 2$). The value of k' is determined from the slope of the appropriate plot. The values of m, p, and k can be found by determining the value of k' at several different concentrations of B and C.

Table 11-6 | Summary of the Kinetics for Reactions of the Type aA → Products That Are Zero, First, or Second Order in [A]

	Order		
	Zero	First	Second
Rate law	$\text{Rate} = k$	$\text{Rate} = k[A]$	$\text{Rate} = k[A]^2$
Integrated rate law	$[A] = -kt + [A]_0$	$\ln[A] = -kt + \ln[A]_0$	$\dfrac{1}{[A]} = kt + \dfrac{1}{[A]_0}$
Plot needed to give a straight line	$[A]$ versus t	$\ln[A]$ versus t	$\dfrac{1}{[A]}$ versus t
Relationship of rate constant to the slope of straight line	Slope $= -k$	Slope $= -k$	Slope $= k$
Half-life	$t_{1/2} = \dfrac{[A]_0}{2k}$	$t_{1/2} = \dfrac{0.693}{k}$	$t_{1/2} = \dfrac{1}{k[A]_0}$

11-5 | Reaction Mechanisms

Most chemical reactions occur by a *series of steps* called the **reaction mechanism**. To understand a reaction, we must know its mechanism, and one of the main purposes for studying kinetics is to learn as much as possible about the steps involved in a reaction. In this section we explore some of the fundamental characteristics of reaction mechanisms.

Consider the reaction between nitrogen dioxide and carbon monoxide:

$$NO_2(g) + CO(g) \longrightarrow NO(g) + CO_2(g)$$

The rate law for this reaction is known from experiment to be

$$\text{Rate} = k[NO_2]^2$$

A balanced equation does not tell us *how* the reactants become products.

As we will see, this reaction is more complicated than it appears from the balanced equation. This is quite typical; the balanced equation for a reaction tells us the reactants, the products, and the stoichiometry but gives no direct information about the reaction mechanism.

Figure 11-8 | A molecular representation of the elementary steps in the reaction of NO_2 and CO.

Step 1

Step 2

For the reaction between nitrogen dioxide and carbon monoxide, the mechanism is thought to involve the following steps:

$$NO_2(g) + NO_2(g) \xrightarrow{k_1} NO_3(g) + NO(g)$$

$$NO_3(g) + CO(g) \xrightarrow{k_2} NO_2(g) + CO_2(g)$$

ⓘ An intermediate is formed in one step and used up in a subsequent step and so is never seen as a product.

where k_1 and k_2 are the rate constants of the individual reactions. In this mechanism, gaseous NO_3 is an **intermediate**, a species that is neither a reactant nor a product but that is formed and consumed during the reaction sequence. ◄ *ⓘ* This reaction is illustrated in Fig. 11-8.

Each of these two reactions is called an **elementary step**, *a reaction whose rate law can be written from its molecularity.* **Molecularity** is defined as the number of species that must collide to produce the reaction indicated by that step. A reaction involving one molecule is called a **unimolecular step**. Reactions involving the collision of two and three species are termed **bimolecular** and **termolecular**, respectively. ◄ *ⓘ* Termolecular steps are quite rare, because the probability of three molecules colliding simultaneously is very small. Examples of these three types of elementary steps and the corresponding rate laws are shown in Table 11-7. Note from Table 11-7 that the rate law for an elementary step follows *directly* from the molecularity of that step. ◄ *ⓘ* For example, for a bimolecular step the rate law is always second order, either of the form $k[A]^2$ for a step with a single reactant or of the form $k[A][B]$ for a step involving two reactants.

ⓘ The prefix *uni-* means one, *bi-* means two, and *ter-* means three.

ⓘ A unimolecular elementary step is always first order, a bimolecular step is always second order, and so on.

We can now define a reaction mechanism more precisely. It is a *series of elementary steps that must satisfy two requirements:*

1. The sum of the elementary steps must give the overall balanced equation for the reaction.
2. The mechanism must agree with the experimentally determined rate law.

Table 11-7 | Examples of Elementary Steps

Elementary Step	Molecularity	Rate Law
A → products	*Uni*molecular	Rate = $k[A]$
A + A → products	*Bi*molecular	Rate = $k[A]^2$
(2A → products)		
A + B → products	*Bi*molecular	Rate = $k[A][B]$
A + A + B → products	*Ter*molecular	Rate = $k[A]^2[B]$
(2A + B → products)		
A + B + C → products	*Ter*molecular	Rate = $k[A][B][C]$

To see how these requirements are applied, we will consider the mechanism given above for the reaction of nitrogen dioxide and carbon monoxide. First, note that the sum of the two steps gives the overall balanced equation:

$$NO_2(g) + NO_2(g) \longrightarrow NO_3(g) + NO(g)$$
$$NO_3(g) + CO(g) \longrightarrow NO_2(g) + CO_2(g)$$
$$\overline{NO_2(g) + NO_2(g) + NO_3(g) + CO(g) \longrightarrow NO_3(g) + NO(g) + NO_2(g) + CO_2(g)}$$

Overall reaction: $NO_2(g) + CO(g) \longrightarrow NO(g) + CO_2(g)$

The first requirement for a correct mechanism is met. To see whether the mechanism meets the second requirement, we need to introduce a new idea: the **rate-determining step**. Multistep reactions often have one step that is much slower than all the others. Reactants can become products only as fast as they can get through this slowest step. ◄ ⓘ That is, the overall reaction can be no faster than the slowest, or rate-determining, step in the sequence. An analogy for this situation is the pouring of water rapidly into a container through a funnel. The water collects in the container at a rate that is essentially determined by the size of the funnel opening and not by the rate of pouring. ◄

Which is the rate-determining step in the reaction of nitrogen dioxide and carbon monoxide? Let's *assume* that the first step is rate-determining and the second step is relatively fast:

$$NO_2(g) + NO_2(g) \longrightarrow NO_3(g) + NO(g) \quad \text{Slow (rate-determining)}$$
$$NO_3(g) + CO(g) \longrightarrow NO_2(g) + CO_2(g) \quad \text{Fast}$$

What we have really assumed here is that the formation of NO_3 occurs much more slowly than its reaction with CO. The rate of CO_2 production is then controlled by the rate of formation of NO_3 in the first step. Since this is an elementary step, we can write the rate law from the molecularity. The bimolecular first step has the rate law

$$\text{Rate of formation of } NO_3 = \frac{\Delta[NO_3]}{\Delta t} = k_1[NO_2]^2$$

Since the overall reaction rate can be no faster than the slowest step,

$$\text{Overall rate} = k_1[NO_2]^2$$

Note that this rate law agrees with the experimentally determined rate law given earlier. The mechanism we assumed above satisfies the two requirements stated earlier and *may* be the correct mechanism for the reaction.

How does a chemist deduce the mechanism for a given reaction? The rate law is always determined first. Then, using chemical intuition and following the two rules given on the previous page, the chemist constructs possible mechanisms and tries, with further experiments, to eliminate those that are least likely. *A mechanism can never be proved absolutely.* We can say only that a mechanism that satisfies the two requirements is *possibly* correct. Deducing mechanisms for chemical reactions can be difficult and requires skill and experience. We will only touch on this process in this text.

ⓘ A reaction is only as fast as its slowest step.

▲ The rate at which this colored solution enters the flask is determined by the size of the funnel stem, not how fast the solution is poured.

© Cengage Learning

EXAMPLE 11-6   Reaction Mechanisms

The balanced equation for the reaction of the gases nitrogen dioxide and fluorine is

$$2NO_2(g) + F_2(g) \longrightarrow 2NO_2F(g)$$

The experimentally determined rate law is

$$\text{Rate} = k[NO_2][F_2]$$

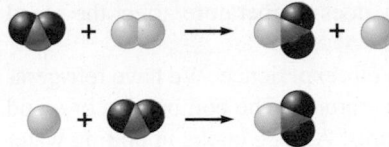

A suggested mechanism for this reaction is

$$NO_2 + F_2 \xrightarrow{k_1} NO_2F + F \qquad \text{Slow} \blacktriangleleft$$

$$F + NO_2 \xrightarrow{k_2} NO_2F \qquad \text{Fast} \blacktriangleleft$$

Is this an acceptable mechanism? That is, does it satisfy the two requirements?

Solution

The first requirement for an acceptable mechanism is that the sum of the steps should give the balanced equation:

$$NO_2 + F_2 \longrightarrow NO_2F + F$$
$$\underline{F + NO_2 \longrightarrow NO_2F}$$
$$2NO_2 + F_2 + \cancel{F} \longrightarrow 2NO_2F + \cancel{F}$$

Overall reaction: $2NO_2 + F_2 \longrightarrow 2NO_2F$

The first requirement is met.

The second requirement is that the mechanism must agree with the experimentally determined rate law. Since the proposed mechanism states that the first step is rate-determining, the overall reaction rate must be that of the first step. The first step is bimolecular, so the rate law is

$$\text{Rate} = k_1[NO_2][F_2]$$

This has the same form as the experimentally determined rate law. The proposed mechanism is acceptable because it satisfies both requirements. (Note that we have not proved that it is *the correct* mechanism.)

See Exercises 11-61 and 11-62

Although the mechanism given in Example 11-6 has the correct stoichiometry and fits the observed rate law, other mechanisms may also satisfy these requirements. For example, the mechanism might be

$$NO_2 + F_2 \longrightarrow NOF_2 + O \qquad \text{Slow}$$
$$NO_2 + O \longrightarrow NO_3 \qquad \text{Fast}$$
$$NOF_2 + NO_2 \longrightarrow NO_2F + NOF \qquad \text{Fast}$$
$$NO_3 + NOF \longrightarrow NO_2F + NO_2 \qquad \text{Fast}$$

To decide on the most probable mechanism for the reaction, the chemist doing the study would have to perform additional experiments.

11-6 | A Model for Chemical Kinetics

How do chemical reactions occur? We already have given some indications. For example, we have seen that the rates of chemical reactions depend on the concentrations of the reacting species. The initial rate for the reaction

$$aA + bB \longrightarrow \text{products}$$

can be described by the rate law

$$\text{Rate} = k[A]^n[B]^m$$

where the order of each reactant depends on the detailed reaction mechanism. This explains why reaction rates depend on concentration. But what about some of the other

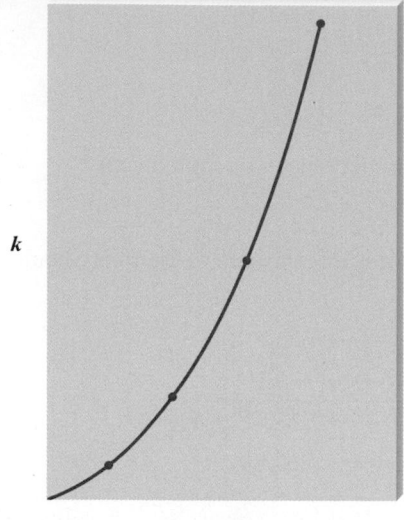

Figure 11-9 | A plot showing the exponential dependence of the rate constant on absolute temperature. The exact temperature dependence of k is different for each reaction. This plot represents the behavior of a rate constant that doubles for every increase in temperature of 10 K.

i The activation energy is often related to bond energy and can sometimes be used in supporting one mechanism over another. This is another case where knowing something about the nature of the molecules in question helps us better understand a phenomenon.

i The higher the activation energy, the slower the reaction at a given temperature.

factors affecting reaction rates? For example, how does temperature affect the speed of a reaction?

We can answer this question qualitatively from our experience. We have refrigerators because food spoilage is retarded at low temperatures. The combustion of wood occurs at a measurable rate only at high temperatures. An egg cooks in boiling water much faster at sea level than in Leadville, Colorado (elevation 10,000 ft), where the boiling point of water is approximately 90°C. These observations and others lead us to conclude that *chemical reactions speed up when the temperature is increased.* Experiments have shown that virtually all rate constants show an exponential increase with absolute temperature, as represented in Fig. 11-9.

In this section we discuss a model used to account for the observed characteristics of reaction rates. This model, called the **collision model**, is built around the central idea that *molecules must collide to react.* We have already seen how this assumption explains the concentration dependence of reaction rates. Now we need to consider whether this model can account for the observed temperature dependence of reaction rates.

The kinetic molecular theory of gases predicts that an increase in temperature raises molecular velocities and so increases the frequency of collisions between molecules. This idea agrees with the observation that reaction rates are greater at higher temperatures. Thus there is qualitative agreement between the collision model and experimental observations. However, it is found that the rate of reaction is much smaller than the calculated collision frequency in a collection of gas particles. This must mean that *only a small fraction of the collisions produces a reaction.* Why?

This question was first addressed in the 1880s by Svante Arrhenius. He proposed the existence of a *threshold energy,* called the **activation energy**, that must be overcome to produce a chemical reaction. ◄ *i* Such a proposal makes sense, as we can see by considering the decomposition of BrNO in the gas phase:

$$2BrNO(g) \longrightarrow 2NO(g) + Br_2(g)$$

In this reaction two Br—N bonds must be broken and one Br—Br bond must be formed. Breaking a Br—N bond requires considerable energy (243 kJ/mol), which must come from somewhere. The collision model postulates that the energy comes from the kinetic energies possessed by the reacting molecules before the collision. This kinetic energy is changed into potential energy as the molecules are distorted during a collision to break bonds and rearrange the atoms into the product molecules.

We can envision the reaction progress as shown in Fig. 11-10. The arrangement of atoms found at the top of the potential energy "hill," or barrier, is called the **activated complex**, or **transition state**. The conversion of BrNO to NO and Br_2 is exothermic, as indicated by the fact that the products have lower potential energy than the reactant. However, ΔE has no effect on the rate of the reaction. Rather, the rate depends on the size of the activation energy E_a. ◄ *i*

The main point here is that a certain minimum energy is required for two BrNO molecules to "get over the hill" so that products can form. This energy is furnished by the energy of the collision. A collision between two BrNO molecules with small kinetic energies will not have enough energy to get over the barrier. At a given temperature only a certain fraction of the collisions possesses enough energy to be effective (to result in product formation).

We can be more precise by recalling from Chapter 8 that a distribution of velocities exists in a sample of gas molecules. Therefore, a distribution of collision energies also exists, as shown in Fig. 11-11 for two different temperatures. Figure 11-11 also shows the activation energy for the reaction in question. Only collisions with energy greater than the activation energy are able to react (get over the barrier). At the lower temperature, T_1, the fraction of effective collisions is quite small. However, as the temperature is increased to T_2, the fraction of collisions with the required activation energy increases dramatically. When the temperature is doubled, the fraction of effective

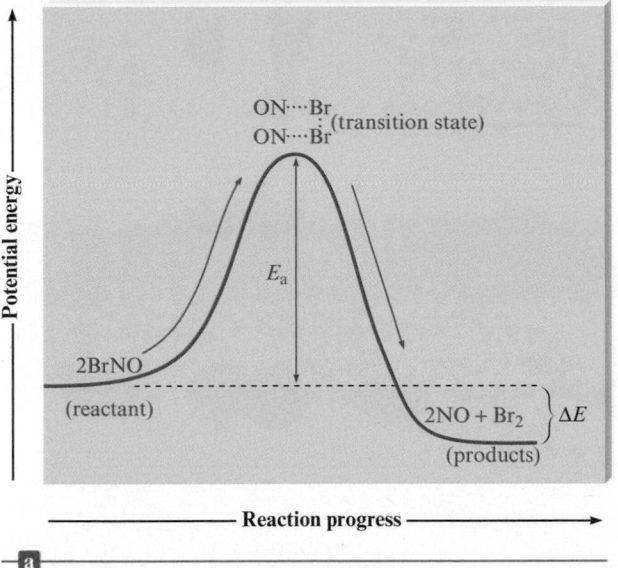

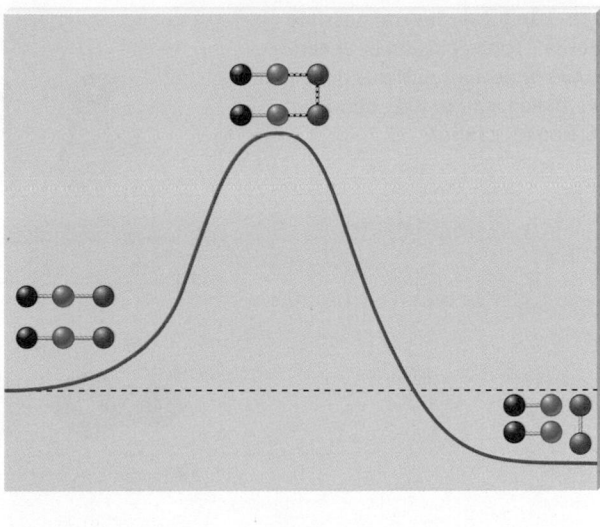

Figure 11-10 | (a) The change in potential energy as a function of reaction progress for the reaction 2BrNO → 2NO + Br$_2$. The activation energy E_a represents the energy needed to disrupt the BrNO molecules so that they can form products. The quantity ΔE represents the net change in energy in going from reactant to products. (b) A molecular representation of the reaction.

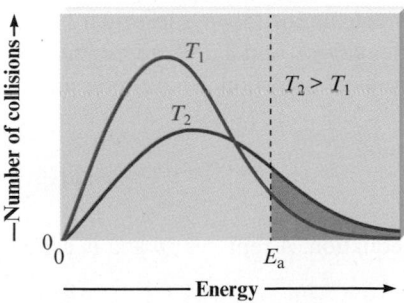

Figure 11-11 | Plot showing the number of collisions with a particular energy at T_1 and T_2, where $T_2 > T_1$.

collisions much more than doubles. In fact, the fraction of effective collisions increases *exponentially* with temperature. This is encouraging for our theory; remember that rates of reactions are observed to increase exponentially with temperature. Arrhenius postulated that the number of collisions having an energy greater than or equal to the activation energy is given by the expression:

Number of collisions with the activation energy = (total number of collisions)$e^{-E_a/RT}$

where E_a is the activation energy, R is the universal gas constant, and T is the Kelvin temperature. The factor $e^{-E_a/RT}$ represents the fraction of collisions with energy E_a or greater at temperature T. ◄

We have seen that not all molecular collisions are effective in producing chemical reactions because a minimum energy is required for the reaction to occur. There is, however, another complication. Experiments show that the *observed reaction rate is considerably smaller than the rate of collisions with enough energy to surmount the barrier.* This means that many collisions, even though they have the required energy, still do not produce a reaction. Why not?

The answer lies in the **molecular orientations** during collisions. We can illustrate this using the reaction between two BrNO molecules (Fig. 11-12). Some collision orientations can lead to reaction, and others cannot. Therefore, we must include a correction factor to allow for collisions with nonproductive molecular orientations.

To summarize, two requirements must be satisfied for reactants to collide successfully (to rearrange to form products):

1. The collision must involve enough energy to produce the reaction; that is, the collision energy must equal or exceed the activation energy.
2. The relative orientation of the reactants must allow formation of any new bonds necessary to produce products.

Taking these factors into account, we can represent the rate constant as

$$k = zpe^{-E_a/RT}$$

Figure 11-12 | Several possible orientations for a collision between two BrNO molecules. Orientations (a) and (b) can lead to a reaction, but orientation (c) cannot.

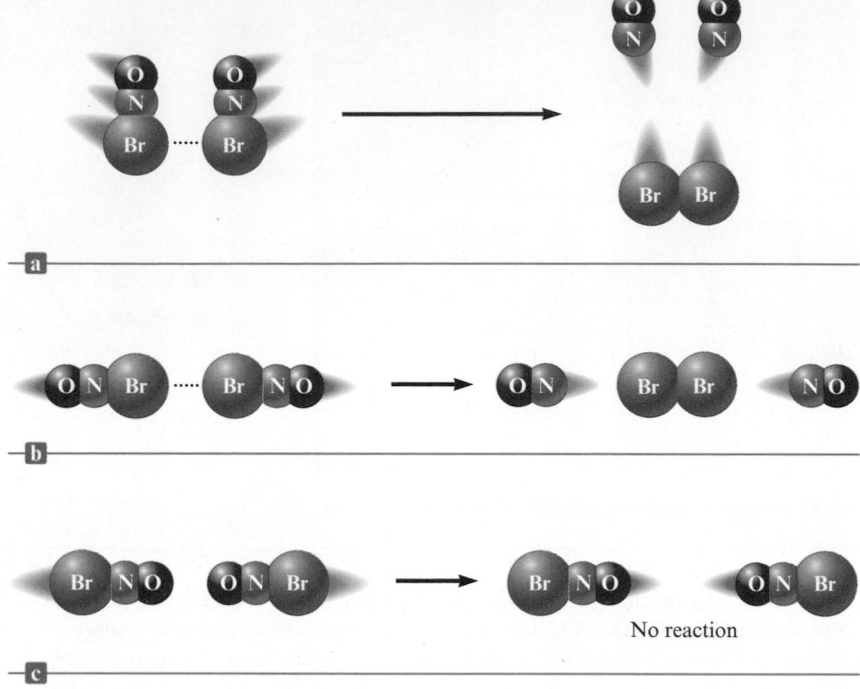

▲ A snowy tree cricket. The frequency of a cricket's chirps depends on the temperature of the cricket.

Townsend P. Dickinson/The Image Works

where z is the collision frequency, p is called the **steric factor** (always less than 1) and reflects the fraction of collisions with effective orientations, and $e^{-E_a/RT}$ represents the fraction of collisions with sufficient energy to produce a reaction. This expression is most often written in form

$$k = Ae^{-E_a/RT} \qquad (11\text{-}9)$$

which is called the **Arrhenius equation**. In this equation, A replaces zp and is called the **frequency factor** for the reaction.

Taking the natural logarithm of each side of the Arrhenius equation gives

$$\ln(k) = -\frac{E_a}{R}\left(\frac{1}{T}\right) + \ln(A) \qquad (11\text{-}10)$$

Equation (11-10) is a linear equation of the type $y = mx + b$, where $y = \ln(k)$, $m = -E_a/R =$ slope, $x = 1/T$, and $b = \ln(A) =$ intercept. Thus for a reaction where the rate constant obeys the Arrhenius equation, a plot of $\ln(k)$ versus $1/T$ gives a straight line. The slope and intercept can be used to determine, respectively, the values of E_a and A characteristic of that reaction. The fact that most rate constants obey the Arrhenius equation to a good approximation indicates that the collision model for chemical reactions is physically reasonable.

Critical Thinking

There are many conditions that need to be met to result in a chemical reaction between molecules. What if all collisions between molecules resulted in a chemical reaction? How would life be different?

EXAMPLE 11-7   Determining Activation Energy I

The reaction

$$2N_2O_5(g) \longrightarrow 4NO_2(g) + O_2(g)$$

was studied at several temperatures, and the following values of k were obtained:

k (s^{-1})	T (°C)
2.0×10^{-5}	20
7.3×10^{-5}	30
2.7×10^{-4}	40
9.1×10^{-4}	50
2.9×10^{-3}	60

Calculate the value of E_a for this reaction.

Solution

To obtain the value of E_a, we need to construct a plot of $\ln(k)$ versus $1/T$. First, we must calculate values of $\ln(k)$ and $1/T$, as shown below:

T (°C)	T (K)	$1/T$ (K)	k (s^{-1})	$\ln(k)$
20	293	3.41×10^{-3}	2.0×10^{-5}	-10.82
30	303	3.30×10^{-3}	7.3×10^{-5}	-9.53
40	313	3.19×10^{-3}	2.7×10^{-4}	-8.22
50	323	3.10×10^{-3}	9.1×10^{-4}	-7.00
60	333	3.00×10^{-3}	2.9×10^{-3}	-5.84

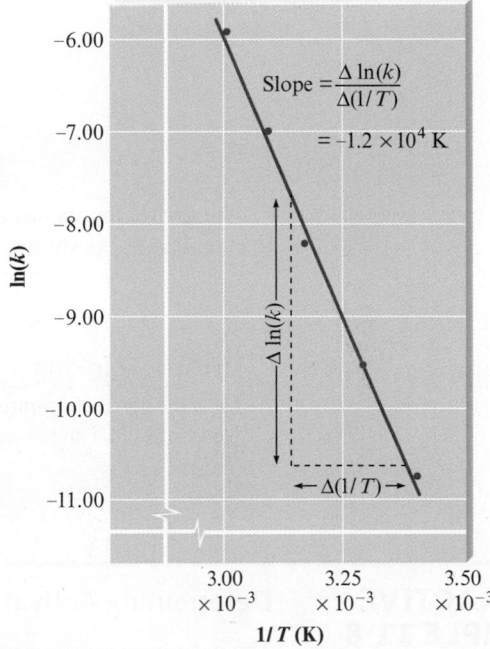

The plot of $\ln(k)$ versus $1/T$ is shown above, where the slope

$$\frac{\Delta \ln(k)}{\Delta \left(\dfrac{1}{T} \right)}$$

is found to be -1.2×10^4 K. The value of E_a can be determined by solving the following equation:

$$\text{Slope} = -\frac{E_a}{R}$$

$$E_a = -R(\text{slope}) = -(8.3145 \text{ J/K} \cdot \text{mol})(-1.2 \times 10^4 \text{ K})$$
$$= 1.0 \times 10^5 \text{ J/mol}$$

Thus the value of the activation energy for this reaction is 1.0×10^5 J/mol.

See Exercises 11-67 and 11-68

The most common procedure for finding E_a for a reaction involves measuring the rate constant k at several temperatures and then plotting $\ln(k)$ versus $1/T$, as shown in Example 11-7. However, E_a also can be calculated from the values of k at only two temperatures by using a formula that can be derived as follows from Equation (11-10).

At temperature T_1, where the rate constant is k_1,

$$\ln(k_1) = -\frac{E_a}{RT_1} + \ln(A)$$

At temperature T_2, where the rate constant is k_2,

$$\ln(k_2) = -\frac{E_a}{RT_2} + \ln(A)$$

Subtracting the first equation from the second gives

$$\ln(k_2) - \ln(k_1) = \left[-\frac{E_a}{RT_2} + \ln(A) \right] - \left[-\frac{E_a}{RT_1} + \ln(A) \right]$$

$$= -\frac{E_a}{RT_2} + \frac{E_a}{RT_1}$$

And

$$\ln\left(\frac{k_2}{k_1}\right) = \frac{E_a}{R}\left(\frac{1}{T_1} - \frac{1}{T_2}\right) \tag{11-11}$$

Therefore, the values of k_1 and k_2 measured at temperatures T_1 and T_2 can be used to calculate E_a, as shown in Example 11-8.

Critical Thinking

Most modern refrigerators have an internal temperature of 45°F. What if refrigerators were set at 55°F in the factory? How would this affect our lives?

INTERACTIVE EXAMPLE 11-8

Determining Activation Energy II

The gas-phase reaction between methane and diatomic sulfur is given by the equation

$$CH_4(g) + 2S_2(g) \longrightarrow CS_2(g) + 2H_2S(g)$$

At 550°C the rate constant for this reaction is 1.1 L/mol · s, and at 625°C the rate constant is 6.4 L/mol · s. Using these values, calculate E_a for this reaction.

Solution

The relevant data are shown in the following table:

k (L/mol · s)	T (°C)	T (K)
$1.1 = k_1$	550	$823 = T_1$
$6.4 = k_2$	625	$898 = T_2$

Substituting these values into Equation (11-11) gives

$$\ln\left(\frac{6.4}{1.1}\right) = \frac{E_a}{8.3145 \text{ J/K} \cdot \text{mol}}\left(\frac{1}{823 \text{ K}} - \frac{1}{898 \text{ K}}\right)$$

Solving for E_a gives

$$E_a = \frac{(8.3145 \text{ J/K} \cdot \text{mol})\ln\left(\dfrac{6.4}{1.1}\right)}{\left(\dfrac{1}{823 \text{ K}} - \dfrac{1}{898 \text{ K}}\right)}$$

$$= 1.4 \times 10^5 \text{ J/mol}$$

See Exercises 11-69 through 11-72

11-7 | Catalysis

We have seen that the rate of a reaction increases dramatically with temperature. If a particular reaction does not occur fast enough at normal temperatures, we can speed it up by raising the temperature. However, sometimes this is not feasible. For example, living cells can survive only in a rather narrow temperature range, and the human body is designed to operate at an almost constant temperature of 98.6°F. But many of the complicated biochemical reactions keeping us alive would be much too slow at this temperature without intervention. We exist only because the body contains many substances called **enzymes**, which increase the rates of these reactions. In fact, almost every biologically important reaction is assisted by a specific enzyme.

Although it is possible to use higher temperatures to speed up commercially important reactions, such as the Haber process for synthesizing ammonia, this is very expensive. In a chemical plant an increase in temperature means significantly increased costs for energy. The use of an appropriate catalyst allows a reaction to proceed rapidly at a relatively low temperature and can therefore hold down production costs.

A **catalyst** is *a substance that speeds up a reaction without being consumed itself.* Just as virtually all vital biological reactions are assisted by enzymes (biological catalysts), almost all industrial processes also involve the use of catalysts. For example, the production of sulfuric acid uses vanadium(V) oxide, and the Haber process uses a mixture of iron and iron oxide.

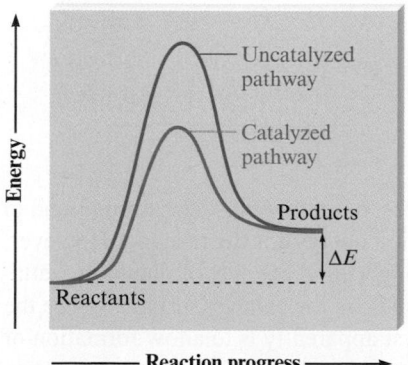

Figure 11-13 | Energy plots for a catalyzed and an uncatalyzed pathway for a given reaction.

How does a catalyst work? Remember that for each reaction a certain energy barrier must be surmounted. How can we make a reaction occur faster without raising the temperature to increase the molecular energies? The solution is to provide a new pathway for the reaction, one with a *lower activation energy.* This is what a catalyst does, as is shown in Fig. 11-13. Because the catalyst allows the reaction to occur with a lower activation energy, a much larger fraction of collisions is effective at a given temperature, and the reaction rate is increased. This effect is illustrated in Fig. 11-14. Note from this diagram that although a catalyst lowers the activation energy E_a for a reaction, it does not affect the energy difference ΔE between products and reactants.

Figure 11-14 | Effect of a catalyst on the number of reaction-producing collisions. Because a catalyst provides a reaction pathway with a lower activation energy, a much greater fraction of the collisions is effective for the catalyzed pathway (b) than for the uncatalyzed pathway (a) (at a given temperature). This allows reactants to become products at a much higher rate, even though there is no temperature increase.

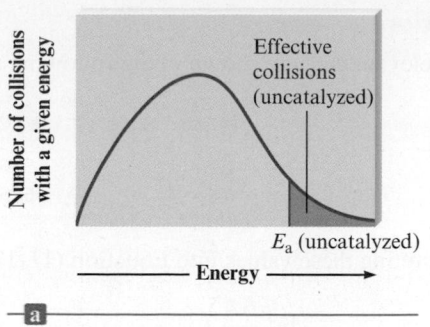

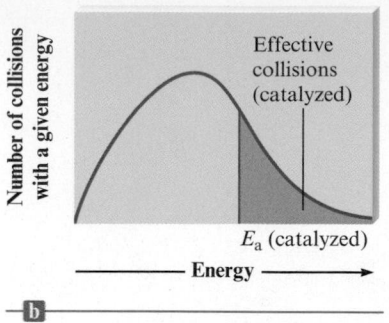

Catalysts are classified as homogeneous or heterogeneous. A **homogeneous catalyst** is one that is *present in the same phase as the reacting molecules.* A **heterogeneous catalyst** exists *in a different phase,* usually as a solid.

Heterogeneous Catalysis

Heterogeneous catalysis most often involves gaseous reactants being adsorbed on the surface of a solid catalyst. **Adsorption** refers to the collection of one substance on the surface of another substance; *absorption* refers to the penetration of one substance into another. Water is *absorbed* by a sponge.

An important example of heterogeneous catalysis occurs in the hydrogenation of unsaturated hydrocarbons, compounds composed mainly of carbon and hydrogen with some carbon–carbon double bonds. Hydrogenation is an important industrial process used to change unsaturated fats, occurring as oils, to saturated fats (solid shortenings such as Crisco) in which the C=C bonds have been converted to C—C bonds through addition of hydrogen.

A simple example of hydrogenation involves ethylene:

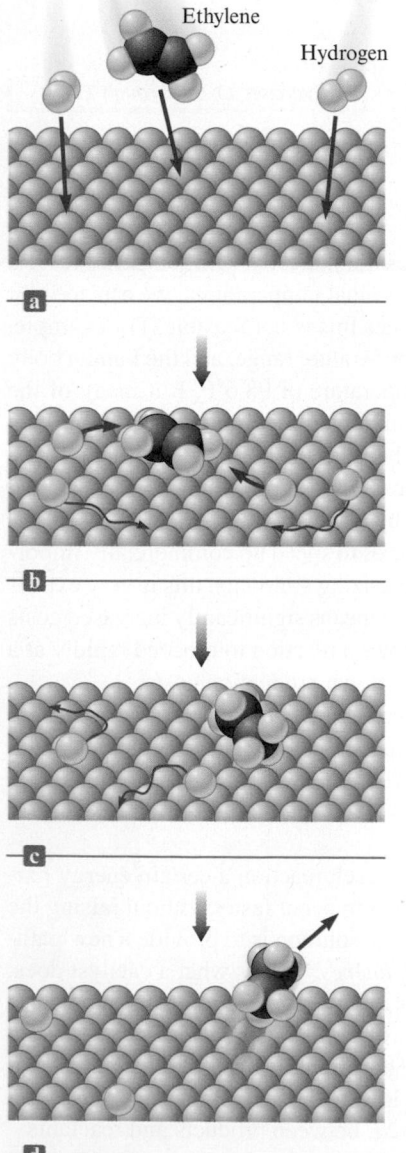

$$
\begin{array}{c}
H \\
 \\
H
\end{array}
C{=}C
\begin{array}{c}
H \\
 \\
H
\end{array}
(g) + H_2(g) \longrightarrow
H{-}
\begin{array}{c}
H \\
 \\
H
\end{array}
C{-}
\begin{array}{c}
H \\
 \\
H
\end{array}
C{-}H(g)
$$

Ethylene   Ethane

This reaction is quite slow at normal temperatures, mainly because the strong bond in the hydrogen molecule results in a large activation energy for the reaction. However, the reaction rate can be greatly increased by using a solid catalyst of platinum, palladium, or nickel. The hydrogen and ethylene adsorb on the catalyst surface, where the reaction occurs. The main function of the catalyst apparently is to allow formation of metal–hydrogen interactions that weaken the H—H bonds and facilitate the reaction. The mechanism is illustrated in Fig. 11-15.

Figure 11-15 | Heterogeneous catalysis of the hydrogenation of ethylene. (a) The reactants above the metal surface. (b) Hydrogen is adsorbed onto the metal surface, forming metal–hydrogen bonds and breaking the H—H bonds. The π bond in ethylene is broken and metal–hydrogen bonds are formed during adsorption. (c) The adsorbed molecules and atoms migrate toward each other on the metal surface, forming new C—H bonds. (d) The C atoms in ethane (C_2H_6) have completely saturated bonding capacities and so cannot bind strongly to the metal surfaces. The C_2H_6 molecule thus escapes.

Heterogeneous catalysis also occurs in the oxidation of gaseous sulfur dioxide to gaseous sulfur trioxide. This process is especially interesting because it illustrates both positive and negative consequences of chemical catalysis.

The negative side is the formation of damaging air pollutants. Recall that sulfur dioxide, a toxic gas with a choking odor, is formed whenever sulfur-containing fuels are burned. However, it is sulfur trioxide that causes most of the environmental damage, mainly through the production of acid rain. When sulfur trioxide combines with a droplet of water, sulfuric acid is formed:

$$H_2O(l) + SO_3(g) \longrightarrow H_2SO_4(aq)$$

This sulfuric acid can cause considerable damage to vegetation, buildings and statues, and fish populations.

Sulfur dioxide is *not* rapidly oxidized to sulfur trioxide in clean, dry air. Why, then, is there a problem? The answer is catalysis. Dust particles and water droplets catalyze the reaction between SO_2 and O_2 in the air.

On the positive side, the heterogeneous catalysis of the oxidation of SO_2 is used to advantage in the manufacture of sulfuric acid, where the reaction of O_2 and SO_2 to form SO_3 is catalyzed by a solid mixture of platinum and vanadium(V) oxide.

Heterogeneous catalysis is also utilized in the catalytic converters in automobile exhaust systems. The exhaust gases, containing compounds such as nitric oxide, carbon monoxide, and unburned hydrocarbons, are passed through a converter containing beads of solid catalyst (Fig. 11-16). The catalyst promotes the conversion of carbon monoxide to carbon dioxide, hydrocarbons to carbon dioxide and water, and nitric oxide to nitrogen gas to lessen the environmental impact of the exhaust gases. However, this beneficial catalysis can, unfortunately, be accompanied by the unwanted catalysis of the oxidation of SO_2 to SO_3, which reacts with the moisture present to form sulfuric acid.

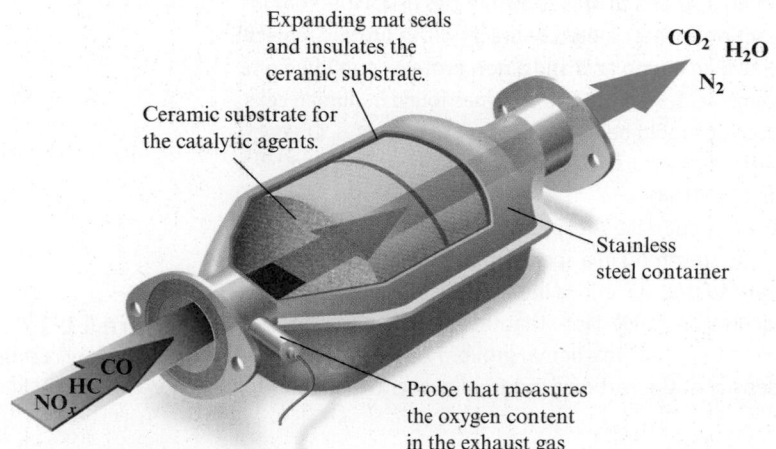

Figure 11-16 │ The exhaust gases (HC, hydrocarbons; NO_x, nitrous oxides; and CO) from an automobile engine are passed through a catalytic converter to minimize environmental damage.

The most impressive examples of homogeneous catalysis occur in nature, where the complex reactions necessary for plant and animal life are made possible by enzymes. Enzymes are large molecules specifically tailored to facilitate a given type of reaction. Usually enzymes are proteins, an important class of biomolecules constructed from α-amino acids that have the general structure

where R represents any one of 20 different substituents. These amino acid molecules can be "hooked together" to form a *polymer* (a word meaning "many parts") called a *protein*. The general structure of a protein can be represented as follows:

Many amino acid fragments

Fragment from an amino acid with substituent R

Fragment from an amino acid with substituent R'

Fragment from an amino acid with substituent R"

Since specific proteins are needed by the human body, the proteins in food must be broken into their constituent amino acids, which are then used to construct new proteins in the body's cells. The reaction in which a protein is broken down one amino acid at a time is shown in Fig. 11-17. Note that in this reaction a water molecule reacts with a protein molecule to produce an amino acid and a new protein containing one less amino acid. Without the enzymes found in human cells, this reaction would be much too slow to be useful. One of these enzymes is *carboxypeptidase-A*, a zinc-containing protein (Fig. 11-18).

Carboxypeptidase-A captures the protein to be acted on (called the *substrate*) in a special groove and positions the substrate so that the end is in the active site, where the catalysis occurs (Fig. 11-19). Note that the Zn^{2+} ion bonds to the oxygen of the C=O (carbonyl) group. This polarizes the electron density in the carbonyl group, allowing the neighboring

C—N bond to be broken much more easily. When the reaction is completed, the remaining portion of the substrate protein and the newly formed amino acid are released by the enzyme.

The process just described for carboxypeptidase-A is characteristic of the behavior of other enzymes. Enzyme catalysis can be represented by the series of reactions shown below:

$$E + S \longrightarrow E \cdot S$$
$$E \cdot S \longrightarrow E + P$$

where E represents the enzyme, S represents the substrate, E · S represents the enzyme–substrate complex, and P represents the products. The enzyme and substrate form a complex, where the reaction occurs. The enzyme then releases the product and is ready to repeat the process. The most amazing thing about enzymes is their efficiency. Because an enzyme plays its catalytic role over and over and very rapidly, only a tiny amount of enzyme is required. This makes the isolation of enzymes for study quite difficult.

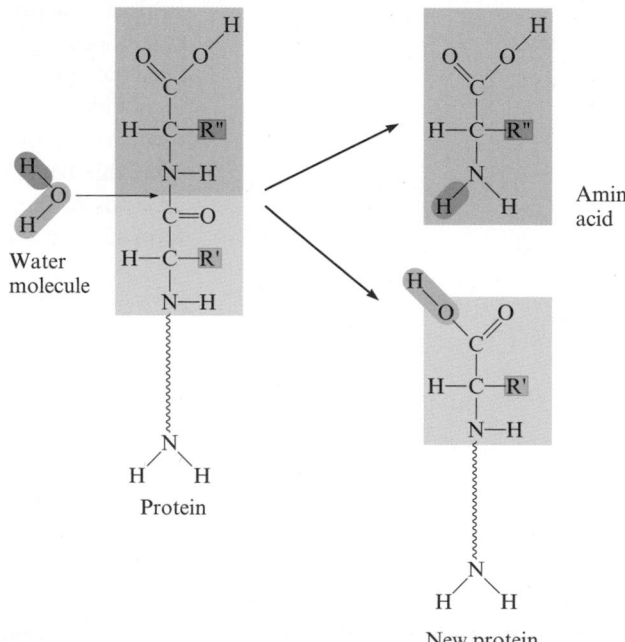

Figure 11-17 | The removal of the end amino acid from a protein by reaction with a molecule of water. The products are an amino acid and a new, smaller protein.

Kenneth Eward / BioGrafx / Science Source

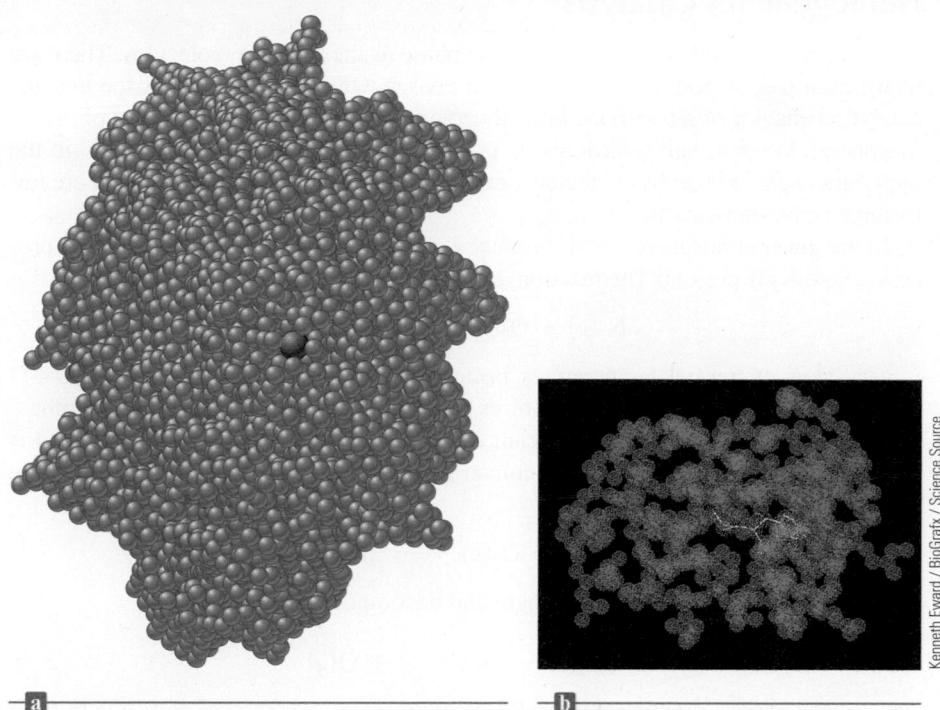

Figure 11-18 | (a) The structure of the enzyme carboxypeptidase-A, which contains 307 amino acids. The zinc ion is shown above as a black sphere in the center. (b) Carboxypeptidase A with a substrate (pink) in place.

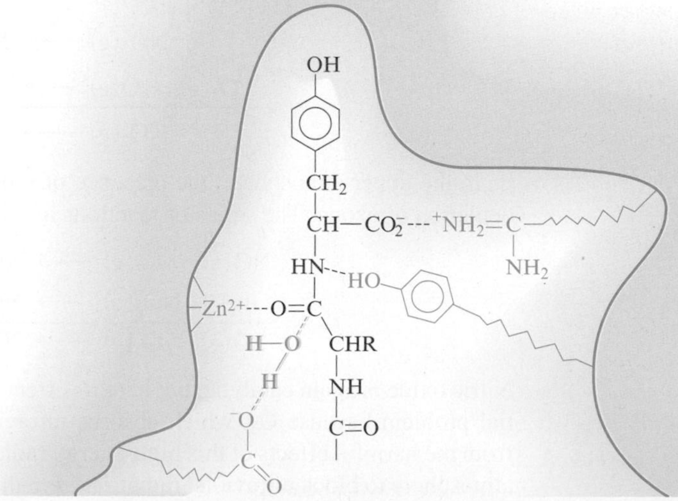

Figure 11-19 | Protein–substrate interaction. The substrate is shown in black and red, with the red representing the terminal amino acid. Blue indicates side chains from the enzyme that help bind the substrate.

Because of the complex nature of the reactions that take place in the converter, a mixture of catalysts is used. The most effective catalytic materials are transition metal oxides and noble metals such as palladium and platinum.

Homogeneous Catalysis

A homogeneous catalyst exists in the same phase as the reacting molecules. There are many examples in both the gas and liquid phases. One such example is the unusual catalytic behavior of nitric oxide toward ozone. In the troposphere, that part of the atmosphere closest to earth, nitric oxide catalyzes ozone production. However, in the upper atmosphere it catalyzes the decomposition of ozone. Both these effects are unfortunate environmentally.

In the lower atmosphere, NO is produced in any high-temperature combustion process where N_2 is present. The reaction

$$N_2(g) + O_2(g) \longrightarrow 2NO(g)$$

is very slow at normal temperatures because of the very strong $N\equiv N$ and $O\equiv O$ bonds. However, at elevated temperatures, such as those found in the internal combustion engines of automobiles, significant quantities of NO form. Some of this NO is converted back to N_2 in the catalytic converter, but significant amounts escape into the atmosphere to react with oxygen:

$$2NO(g) + O_2(g) \longrightarrow 2NO_2(g)$$

In the atmosphere, NO_2 can absorb light and decompose as follows:

$$NO_2(g) \xrightarrow{\text{Light}} NO(g) + O(g)$$

The oxygen atom is very reactive and can combine with an oxygen molecule to form ozone: ◄ 𝒊

$$O_2(g) + O(g) \longrightarrow O_3(g)$$

Ozone is a powerful oxidizing agent that can react with other air pollutants to form substances irritating to the eyes and lungs, and is itself very toxic.

In this series of reactions, nitric oxide is acting as a true catalyst because it assists the production of ozone without being consumed itself. This can be seen by summing the reactions:

$$NO(g) + \tfrac{1}{2}O_2(g) \longrightarrow NO_2(g)$$
$$NO_2(g) \xrightarrow{\text{Light}} NO(g) + O(g)$$
$$\underline{O_2(g) + O(g) \longrightarrow O_3(g)}$$
$$\tfrac{3}{2}O_2(g) \longrightarrow O_3(g)$$

In the upper atmosphere, the presence of nitric oxide has the opposite effect—the depletion of ozone. The series of reactions involved is

$$NO(g) + O_3(g) \longrightarrow NO_2(g) + O_2(g)$$
$$\underline{O(g) + NO_2(g) \longrightarrow NO(g) + O_2(g)}$$
$$O(g) + O_3(g) \longrightarrow 2O_2(g)$$

Nitric oxide is again catalytic, but here its effect is to change O_3 to O_2. This is a potential problem because O_3, which absorbs ultraviolet light, is necessary to protect us from the harmful effects of this high-energy radiation. That is, we want O_3 in the upper atmosphere to block ultraviolet radiation from the sun but not in the lower atmosphere, where we would have to breathe it and its oxidation products.

𝒊 Although O_2 is represented here as the oxidizing agent for NO, the actual oxidizing agent is probably some type of peroxide compound produced by reaction of oxygen with pollutants. The direct reaction of NO and O_2 is very slow.

▲ This graphic shows data from the Total Ozone Mapping Spectrometer (TOMS) Earth Probe.

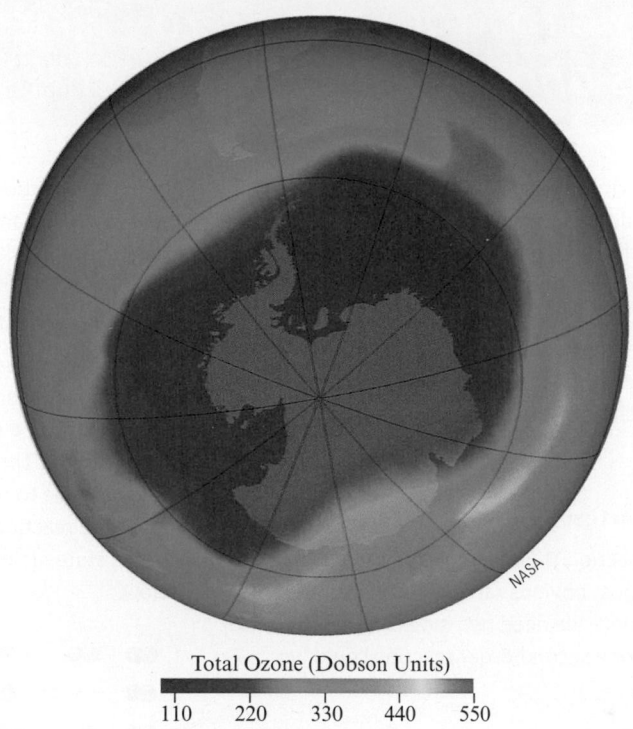

Total Ozone (Dobson Units)

110   220   330   440   550

The ozone layer is also threatened by *Freons,* a group of stable, noncorrosive compounds, formerly used as refrigerants and as propellants in aerosol cans. The most commonly used substance of this type was Freon-12 (CCl_2F_2). ◄ The chemical inertness of Freons makes them valuable but also creates a problem, since they remain in the environment a long time. Eventually, they migrate into the upper atmosphere to be decomposed by high-energy light. Among the decomposition products are chlorine atoms:

Freon-12

$$CCl_2F_2(g) \xrightarrow{\text{Light}} CClF_2(g) + Cl(g)$$

These chlorine atoms can catalyze the decomposition of ozone: ◄

Ozone

$$Cl(g) + O_3(g) \longrightarrow ClO(g) + O_2(g)$$
$$\underline{O(g) + ClO(g) \longrightarrow Cl(g) + O_2(g)}$$
$$O(g) + O_3(g) \longrightarrow 2O_2(g)$$

The problem of Freons has been brought into strong focus by the discovery of a mysterious "hole" in the ozone layer in the stratosphere over Antarctica. ◄ Studies performed there to find the reason for the hole have found unusually high levels of chlorine monoxide (ClO). This strongly implicates the Freons in the atmosphere as being responsible for the ozone destruction.

Because they pose environmental problems, Freons have been banned by international agreement. Substitute compounds are now being used.

In Section 11-4 we derived integrated rate laws for zero-, first-, and second-order reactions, and in Section 11-5 we discussed mechanisms for such reactions. Let's consider what each of these types of reactions would "look" like at the molecular level.

Let's consider the generic reaction $2AB \rightarrow AB_2 + A$, which we can represent as ⬤⬤ + ⬤⬤ → ⬤⬤⬤ + ⬤. How could various possible mechanisms lead to different rate laws for this reaction?

Possible mechanism that yields a second order rate law

As we discussed in Section 11-6, the molecules must collide in order to react. The most obvious rate law, then, would seem to be second order since we need two molecules to collide. A simple mechanism for a second-order reaction could be:

1. ⬤⬤ + ⬤⬤ ⟶ ⬤⬤⬤⬤ (slow)
2. ⬤⬤⬤⬤ ⟶ ⬤⬤⬤ + ⬤ (fast)

The first step is the rate-determining step, making the rate = $k[AB]^2$, or second order.

Possible mechanism that yields a zero order rate law

Zero-order reactions are often the result of catalysis. For example, suppose the reaction occurs on the surface of a metal catalyst, such as platinum (Pt). Once the surface is completely covered with AB molecules, increasing the concentration of AB would have no effect on the rate since only the AB molecules on the surface can react.

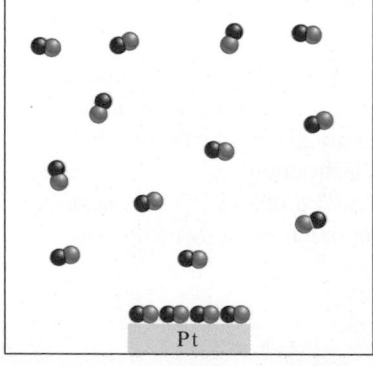

Note the molecules still must collide to react, but since the rate does not change with concentration of AB, then rate = $k[AB]^0$, or rate = k.

Possible mechanism that yields a first order rate law

Because molecules must collide in order to react, first-order reactions may seem impossible at first. For example, how can the slow step of a mechanism be ⬤⬤ ⟶ ⬤ + ⬤, if molecules do not collide? The key to understanding this is to realize that for a molecule to react, it must have enough energy to surmount the reaction energy barrier. We will represent such an excited state molecule with an asterisk in the mechanism below.

1. ⬤⬤ + ⬤⬤ ⟶ ⬤⬤* + ⬤⬤ (fast)
2. ⬤⬤* ⟶ ⬤ + ⬤ (slow)
3. ⬤⬤ + ⬤ ⟶ ⬤⬤⬤ (fast)

Note that the AB molecules must collide in order to react. However, the collisions between AB molecules only transfer energy from one AB to another. If the first and third steps in the above mechanism occur at a much faster rate than the second step, the decomposition of the excited state is the rate-determining step. Thus the second step dictates the rate and the rate law = $k[AB]$, which is first order.

In all three mechanisms, then, the molecules must collide for a reaction to occur.

Collision Model

In Section 11-6, you learned about the collision model, which tells us that in order for a reaction to occur, molecules must collide. We can see that even with this simple model, we can develop different mechanisms that give rise to different rate laws. As we have discussed, however, while a collision is necessary, it is not the only factor required to result in a reaction. The molecules must be oriented correctly (which is one way in which a catalyst works to speed up a reaction), and activation energy must be overcome (which is why reactions proceed at a higher rate with increased temperature). To better understand how reactions proceed and what factors are important in the rate of a reaction, we need to take a molecular-level approach to consider how the molecules are interacting.

FOR REVIEW

Key Terms

chemical kinetics

Section 11-1
reaction rate
instantaneous rate

Section 11-2
rate law
rate constant
order
(differential) rate law
integrated rate law

Section 11-3
method of initial rates
initial rate
overall reaction order

Section 11-4
first-order reaction
integrated first-order rate law
half-life of a reactant
integrated second-order rate law
zero-order reaction
integrated zero-order rate law
pseudo-first-order rate law

Section 11-5
reaction mechanism
intermediate
elementary step
molecularity
unimolecular step
bimolecular step
termolecular step
rate-determining step

Section 11-6
collision model
activation energy
activated complex (transition state)
molecular orientations
steric factor
Arrhenius equation
frequency factor

Section 11-7
enzyme
catalyst
homogeneous catalyst
heterogeneous catalyst
adsorption

Chemical Kinetics

- The study of the factors that control the rate (speed) of a chemical reaction
  - Rate is defined in terms of the change in concentration of a given reaction component per unit time
  - Kinetic measurements are often made under conditions where the reverse reaction is insignificant
- The kinetic and thermodynamic properties of a reaction are not fundamentally related

Rate Laws

- Differential rate law: describes the rate as a function of concentration

$$\text{Rate} = -\frac{\Delta[\text{A}]}{\Delta t} = k[\text{A}]^n$$

  - k is the rate constant
  - n is the order; not related to the coefficients in the balanced equation
- Integrated rate law: describes the concentration as a function of time
  - For a reaction of the type

$$a\text{A} \longrightarrow \text{products}$$

  for which

$$\text{Rate} = k[\text{A}]^n$$

$n = 0$:

$$[\text{A}] = -kt + [\text{A}]_0$$

$$t_{1/2} = \frac{[\text{A}]_0}{2k}$$

$n = 1$:

$$\ln[\text{A}] = -kt + \ln[\text{A}]_0$$

$$t_{1/2} = \frac{0.693}{k}$$

$n = 2$:

$$\frac{1}{[\text{A}]} = kt + \frac{1}{[\text{A}]_0}$$

$$t_{1/2} = \frac{1}{k[\text{A}]_0}$$

  - The value of k can be determined from the plot of the appropriate function of [A] versus t

Reaction Mechanism

- Series of elementary steps by which an overall reaction occurs
  - Elementary step: rate law for the step can be written from the molecularity of the reaction
- Two requirements for an acceptable mechanism:
  - The elementary steps sum to give the correct overall balanced equation
  - The mechanism agrees with the experimentally determined rate law
- Simple reactions can have an elementary step that is slower than all of the other steps; this is called the rate-determining step.

Kinetic Models

- The simplest model to account for reaction kinetics is the collision model
  - Molecules must collide to react
  - The collision kinetic energy furnishes the potential energy needed to enable the reactants to rearrange to form products
  - A certain threshold energy called the activation energy (E_a) is necessary for a reaction to occur
  - The relative orientations of the colliding reactants are also a determining factor in the reaction rate
  - This model leads to the Arrhenius equation:

$$k = Ae^{-E_a/RT}$$

  - A depends on the collision frequency and relative orientation of the molecules
  - The value of E_a can be found by obtaining the values of k at several temperatures

Catalyst

- Speeds up a reaction without being consumed
- Works by providing a lower-energy pathway for the reaction
- Enzymes are biological catalysts
- Catalysts can be classified as homogeneous or heterogeneous
  - Homogeneous: exist in the same phase as the reactants
  - Heterogeneous: exist in a different phase than the reactants

Review Questions Answers to the Review Questions can be found on the Student Web site (accessible from www.cengagebrain.com).

1. Define *reaction rate*. Distinguish between the initial rate, average rate, and instantaneous rate of a chemical reaction. Which of these rates is usually fastest? The initial rate is the rate used by convention. Give a possible explanation as to why.

2. Distinguish between the differential rate law and the integrated rate law. Which of these is often called just the "rate law"? What is k in a rate law, and what are orders in a rate law? Explain.

3. One experimental procedure that can be used to determine the rate law of a reaction is the method of initial rates. What data are gathered in the method of initial rates, and how are these data manipulated to determine k and the orders of the species in the rate law? Are the units for k, the rate constant, the same for all rate laws? Explain. If a reaction is first order in A, what happens to the rate if [A] is tripled? If the initial rate for a reaction increases by a factor of 16 when [A] is quadrupled, what is the order of n? If a reaction is third order in A and [A] is doubled, what happens to the initial rate? If a reaction is zero order, what effect does [A] have on the initial rate of a reaction?

4. The initial rate for a reaction is equal to the slope of the tangent line at $t \approx 0$ in a plot of [A] versus time. From calculus, initial rate $= \dfrac{-d[A]}{dt}$. Therefore, the differential rate law for a reaction is

   Rate $= \dfrac{-d[A]}{dt} = k[A]^n$.

   Assuming you have some calculus in your background, derive the zero-, first-, and second-order integrated rate laws using the differential rate law.

5. Consider the zero-, first-, and second-order integrated rate laws. If you have concentration versus time data for some species in a reaction, what plots would you make to "prove" a reaction is either zero, first, or second order? How would the rate constant, k, be determined from such a plot? What does the y-intercept equal in each plot? When a rate law contains the concentration of two or more species, how can plots be used to determine k and the orders of the species in the rate law?

6. Derive expressions for the half-life of zero-, first-, and second-order reactions using the integrated rate

law for each order. How does each half-life depend on concentration? If the half-life for a reaction is 20. seconds, what would be the second half-life assuming the reaction is either zero, first, or second order?

7. Define each of the following.
   a. elementary step
   b. molecularity
   c. reaction mechanism
   d. intermediate
   e. rate-determining step

8. What two requirements must be met to call a mechanism plausible? Why say a "plausible" mechanism instead of the "correct" mechanism? Is it true that most reactions occur by a one-step mechanism? Explain.

9. What is the premise underlying the collision model? How is the rate affected by each of the following?
   a. activation energy
   b. temperature
   c. frequency of collisions
   d. orientation of collisions

Sketch a potential energy versus reaction progress plot for an endothermic reaction and for an exothermic reaction. Show ΔE and E_a in both plots. When concentrations and temperatures are equal, would you expect the rate of the forward reaction to be equal to, greater than, or less than the rate of the reverse reaction if the reaction is exothermic? Endothermic?

10. Give the Arrhenius equation. Take the natural log of both sides and place this equation in the form of a straight-line equation ($y = mx + b$). What data would you need and how would you graph those data to get a linear relationship using the Arrhenius equation? What does the slope of the straight line equal? What does the y-intercept equal? What are the units of R in the Arrhenius equation? Explain how if you know the rate constant value at two different temperatures, you can determine the activation energy for the reaction.

11. Why does a catalyst increase the rate of a reaction? What is the difference between a homogeneous catalyst and a heterogeneous catalyst? Would a given reaction necessarily have the same rate law for both a catalyzed and an uncatalyzed pathway? Explain.

Active Learning Questions*

These questions are designed to be used by groups of students in class.

1. Define *stability* from both a kinetic and thermodynamic perspective. Give examples to show the differences in these concepts.

2. Describe at least two experiments you could perform to determine a rate law.

3. Make a graph of [A] versus time for zero-, first-, and second-order reactions. From these graphs, compare successive half-lives.

4. How does temperature affect k, the rate constant? Explain.

5. Consider the following statements: "In general, the rate of a chemical reaction increases a bit at first because it takes a while for the reaction to get 'warmed up.' After that, however, the rate of the reaction decreases because its rate is dependent on the concentrations of the reactants, and these are decreasing." Indicate everything that is correct in these statements, and indicate everything that is incorrect. Correct the incorrect statements and explain.

6. For the reaction A + B → C, explain at least two ways in which the rate law could be zero order in chemical A.

7. A friend of yours states, "A balanced equation tells us how chemicals interact. Therefore, we can determine the rate law directly from the balanced equation." What do you tell your friend?

8. Provide a conceptual rationale for the differences in the half-lives of zero-, first-, and second-order reactions.

9. The rate constant (k) depends on which of the following (there may be more than one answer)?
   a. the concentration of the reactants
   b. the nature of the reactants
   c. the temperature
   d. the order of the reaction
   Explain.

A blue question or exercise number indicates that the answer to that question or exercise appears at the back of this book and a solution appears in the *Solutions Guide*, as found on PowerLecture.

Questions

10. Each of the statements given below is false. Explain why.
   a. The activation energy of a reaction depends on the overall energy change (ΔE) for the reaction.

---

*In the Questions and the Exercises, the term *rate law* always refers to the differential rate law.

b. The rate law for a reaction can be deduced from examination of the overall balanced equation for the reaction.

c. Most reactions occur by one-step mechanisms.

11. Define what is meant by unimolecular and bimolecular steps. Why are termolecular steps infrequently seen in chemical reactions?

12. The plot below shows the number of collisions with a particular energy for two different temperatures.

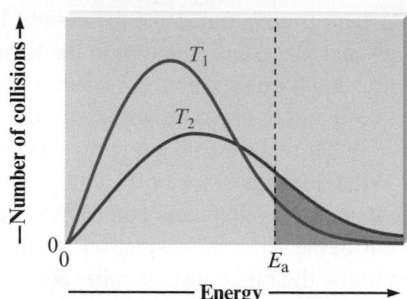

a. Which is greater, T_2 or T_1? How can you tell?

b. What does this plot tell us about the temperature of the rate of a chemical reaction? Explain your answer.

13. For the reaction

$$O_2(g) + 2NO(g) \longrightarrow 2NO_2(g)$$

the observed rate law is

$$\text{Rate} = k[NO]^2[O_2]$$

Which of the changes listed below would affect the value of the rate constant k?

a. increasing the partial pressure of oxygen gas

b. changing the temperature

c. using an appropriate catalyst

14. The rate law for a reaction can be determined only from experiment and not from the balanced equation. Two experimental procedures were outlined in Chapter 11. What are these two procedures? Explain how each method is used to determine rate laws.

15. Table 11-2 illustrates how the average rate of a reaction decreases with time. Why does the average rate decrease with time? How does the instantaneous rate of a reaction depend on time? Why are initial rates used by convention?

16. The type of rate law for a reaction, either the differential rate law or the integrated rate law, is usually determined by which data is easiest to collect. Explain.

17. The initial rate of a reaction doubles as the concentration of one of the reactants is quadrupled. What is the order of this reactant? If a reactant has a -1 order, what happens to the initial rate when the concentration of that reactant increases by a factor of two?

18. Hydrogen reacts explosively with oxygen. However, a mixture of H_2 and O_2 can exist indefinitely at room temperature. Explain why H_2 and O_2 do not react under these conditions.

19. The central idea of the collision model is that molecules must collide in order to react. Give two reasons why not all collisions of reactant molecules result in product formation.

20. Consider the following energy plots for a chemical reaction when answering the questions below.

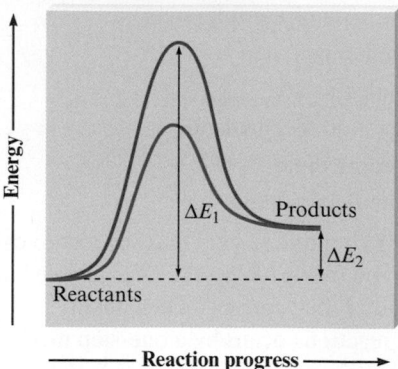

a. Which plot (purple or blue) is the catalyzed pathway? How do you know?

b. What does ΔE_1 represent?

c. What does ΔE_2 represent?

d. Is the reaction endothermic or exothermic?

21. Enzymes are kinetically important for many of the complex reactions necessary for plant and animal life to exist. However, only a tiny amount of any particular enzyme is required for these complex reactions to occur. Explain.

22. Would the slope of a $\ln(k)$ versus $1/T$ plot (with temperature in kelvin) for a catalyzed reaction be more or less negative than the slope of the $\ln(k)$ versus $1/T$ plot for the uncatalyzed reaction? Explain. Assume both rate laws are first-order overall.

Exercises

In this section similar exercises are paired.

Reaction Rates

23. Consider the reaction

$$4PH_3(g) \longrightarrow P_4(g) + 6H_2(g)$$

If, in a certain experiment, over a specific time period, 0.0048 mole of PH_3 is consumed in a 2.0-L container each second of reaction, what are the rates of production of P_4 and H_2 in this experiment?

24. In the Haber process for the production of ammonia,

$$N_2(g) + 3H_2(g) \longrightarrow 2NH_3(g)$$

what is the relationship between the rate of production of ammonia and the rate of consumption of hydrogen?

25. At 40°C, $H_2O_2(aq)$ will decompose according to the following reaction:

$$2H_2O_2(aq) \longrightarrow 2H_2O(l) + O_2(g)$$

The following data were collected for the concentration of H_2O_2 at various times.

Time (s)	$[H_2O_2]$ (mol/L)
0	1.000
2.16×10^4	0.500
4.32×10^4	0.250

a. Calculate the average rate of decomposition of H_2O_2 between 0 and 2.16×10^4 s. Use this rate to calculate the average rate of production of $O_2(g)$ over the same time period.

b. What are these rates for the time period 2.16×10^4 s to 4.32×10^4 s?

26. Consider the general reaction

$$aA + bB \longrightarrow cC$$

and the following average rate data over some time period Δt:

$$-\frac{\Delta A}{\Delta t} = 0.0080 \text{ mol/L} \cdot \text{s}$$

$$-\frac{\Delta B}{\Delta t} = 0.0120 \text{ mol/L} \cdot \text{s}$$

$$\frac{\Delta C}{\Delta t} = 0.0160 \text{ mol/L} \cdot \text{s}$$

Determine a set of possible coefficients to balance this general reaction.

27. What are the units for each of the following if the concentrations are expressed in moles per liter and the time in seconds?

a. rate of a chemical reaction

b. rate constant for a zero-order rate law

c. rate constant for a first-order rate law

d. rate constant for a second-order rate law

e. rate constant for a third-order rate law

28. The rate law for the reaction

$$Cl_2(g) + CHCl_3(g) \longrightarrow HCl(g) + CCl_4(g)$$

is   Rate $= k[Cl_2]^{1/2}[CHCl_3]$

What are the units for k, assuming time in seconds and concentration in mol/L?

Rate Laws from Experimental Data: Initial Rates Method

29. The reaction

$$2NO(g) + Cl_2(g) \longrightarrow 2NOCl(g)$$

was studied at $-10°C$. The following results were obtained where

$$\text{Rate} = -\frac{\Delta[Cl_2]}{\Delta t}$$

$[NO]_0$ (mol/L)	$[Cl_2]_0$ (mol/L)	Initial Rate (mol/L · min)
0.10	0.10	0.18
0.10	0.20	0.36
0.20	0.20	1.45

a. What is the rate law?

b. What is the value of the rate constant?

30. The reaction

$$2I^-(aq) + S_2O_8{}^{2-}(aq) \longrightarrow I_2(aq) + 2SO_4{}^{2-}(aq)$$

was studied at $25°C$. The following results were obtained where

$$\text{Rate} = -\frac{\Delta[S_2O_8{}^{2-}]}{\Delta t}$$

$[I^-]_0$ (mol/L)	$[S_2O_8{}^{2-}]_0$ (mol/L)	Initial Rate (mol/L · s)
0.080	0.040	12.5×10^{-6}
0.040	0.040	6.25×10^{-6}
0.080	0.020	6.25×10^{-6}
0.032	0.040	5.00×10^{-6}
0.060	0.030	7.00×10^{-6}

a. Determine the rate law.

b. Calculate a value for the rate constant for each experiment and an average value for the rate constant.

31. The decomposition of nitrosyl chloride was studied:

$$2NOCl(g) \rightleftharpoons 2NO(g) + Cl_2(g)$$

The following data were obtained where

$$\text{Rate} = -\frac{\Delta[NOCl]}{\Delta t}$$

$[NOCl]_0$ (molecules/cm³)	Initial Rate (molecules/cm³ · s)
3.0×10^{16}	5.98×10^4
2.0×10^{16}	2.66×10^4
1.0×10^{16}	6.64×10^3
4.0×10^{16}	1.06×10^5

a. What is the rate law?

b. Calculate the value of the rate constant.

c. Calculate the value of the rate constant when concentrations are given in moles per liter.

32. The following data were obtained for the gas-phase decomposition of dinitrogen pentoxide,

$$2N_2O_5(g) \longrightarrow 4NO_2(g) + O_2(g)$$

$[N_2O_5]_0$ (mol/L)	Initial Rate (mol/L · s)
0.0750	8.90×10^{-4}
0.190	2.26×10^{-3}
0.275	3.26×10^{-3}
0.410	4.85×10^{-3}

Defining the rate as $-\Delta[N_2O_5]/\Delta t$, write the rate law and calculate the value of the rate constant.

33. The reaction

$$I^-(aq) + OCl^-(aq) \longrightarrow IO^-(aq) + Cl^-(aq)$$

was studied, and the following data were obtained:

$[I^-]_0$ (mol/L)	$[OCl^-]_0$ (mol/L)	Initial Rate (mol/L · s)
0.12	0.18	7.91×10^{-2}
0.060	0.18	3.95×10^{-2}
0.030	0.090	9.88×10^{-3}
0.24	0.090	7.91×10^{-2}

a. What is the rate law?

b. Calculate the value of the rate constant.

c. Calculate the initial rate for an experiment where both I^- and OCl^- are initially present at 0.15 mol/L.

34. The reaction

$$2NO(g) + O_2(g) \longrightarrow 2NO_2(g)$$

was studied, and the following data were obtained where

$$\text{Rate} = -\frac{\Delta[O_2]}{\Delta t}$$

$[NO]_0$ (molecules/cm³)	$[O_2]_0$ (molecules/cm³)	Initial Rate (molecules/cm³ · s)
1.00×10^{18}	1.00×10^{18}	2.00×10^{16}
3.00×10^{18}	1.00×10^{18}	1.80×10^{17}
2.50×10^{18}	2.50×10^{18}	3.13×10^{17}

What would be the initial rate for an experiment where $[NO]_0 = 6.21 \times 10^{18}$ molecules/cm³ and $[O_2]_0 = 7.36 \times 10^{18}$ molecules/cm³?

35. The rate of the reaction between hemoglobin (Hb) and carbon monoxide (CO) was studied at 20°C. The following data were collected with all concentration units in μmol/L. (A hemoglobin concentration of 2.21 μmol/L is equal to 2.21×10^{-6} mol/L.)

$[Hb]_0$ (μmol/L)	$[CO]_0$ (μmol/L)	Initial Rate (μmol/L · s)
2.21	1.00	0.619
4.42	1.00	1.24
4.42	3.00	3.71

a. Determine the orders of this reaction with respect to Hb and CO.

b. Determine the rate law.

c. Calculate the value of the rate constant.

d. What would be the initial rate for an experiment with $[Hb]_0 = 3.36 \ \mu$mol/L and $[CO]_0 = 2.40 \ \mu$mol/L?

36. The following data were obtained for the reaction

$$2ClO_2(aq) + 2OH^-(aq) \longrightarrow ClO_3^-(aq) + ClO_2^-(aq) + H_2O(l)$$

where

$$\text{Rate} = -\frac{\Delta[ClO_2]}{\Delta t}$$

$[ClO_2]_0$ (mol/L)	$[OH^-]_0$ (mol/L)	Initial Rate (mol/L · s)
0.0500	0.100	5.75×10^{-2}
0.100	0.100	2.30×10^{-1}
0.100	0.0500	1.15×10^{-1}

a. Determine the rate law and the value of the rate constant.

b. What would be the initial rate for an experiment with $[ClO_2]_0 = 0.175$ mol/L and $[OH^-]_0 = 0.0844$ mol/L?

Integrated Rate Laws

37. The decomposition of hydrogen peroxide was studied, and the following data were obtained at a particular temperature:

Time (s)	$[H_2O_2]$ (mol/L)
0	1.00
120 ± 1	0.91
300 ± 1	0.78
600 ± 1	0.59
1200 ± 1	0.37
1800 ± 1	0.22
2400 ± 1	0.13
3000 ± 1	0.082
3600 ± 1	0.050

Assuming that

$$\text{Rate} = -\frac{\Delta[H_2O_2]}{\Delta t}$$

determine the rate law, the integrated rate law, and the value of the rate constant. Calculate $[H_2O_2]$ at 4000. s after the start of the reaction.

38. A certain reaction has the following general form:

$$aA \longrightarrow bB$$

At a particular temperature and $[A]_0 = 2.00 \times 10^{-2} \ M$, concentration versus time data were collected for this reaction, and a plot of $\ln[A]$ versus time resulted in a straight line with a slope value of $-2.97 \times 10^{-2} \ \text{min}^{-1}$.

a. Determine the rate law, the integrated rate law, and the value of the rate constant for this reaction.

b. Calculate the half-life for this reaction.

c. How much time is required for the concentration of A to decrease to $2.50 \times 10^{-3}\ M$?

39. The rate of the reaction

$$NO_2(g) + CO(g) \longrightarrow NO(g) + CO_2(g)$$

depends only on the concentration of nitrogen dioxide below 225°C. At a temperature below 225°C, the following data were collected:

Time (s)	[NO$_2$] (mol/L)
0	0.500
1.20×10^3	0.444
3.00×10^3	0.381
4.50×10^3	0.340
9.00×10^3	0.250
1.80×10^4	0.174

Determine the rate law, the integrated rate law, and the value of the rate constant. Calculate [NO$_2$] at 2.70×10^4 s after the start of the reaction.

40. A certain reaction has the following general form:

$$aA \longrightarrow bB$$

At a particular temperature and $[A]_0 = 2.80 \times 10^{-3}\ M$, concentration versus time data were collected for this reaction, and a plot of $1/[A]$ versus time resulted in a straight line with a slope value of $+3.60 \times 10^{-2}$ L/mol · s.

a. Determine the rate law, the integrated rate law, and the value of the rate constant for this reaction.

b. Calculate the half-life for this reaction.

c. How much time is required for the concentration of A to decrease to $7.00 \times 10^{-4}\ M$?

41. The decomposition of ethanol (C$_2$H$_5$OH) on an alumina (Al$_2$O$_3$) surface

$$C_2H_5OH(g) \longrightarrow C_2H_4(g) + H_2O(g)$$

was studied at 600 K. Concentration versus time data were collected for this reaction, and a plot of [A] versus time resulted in a straight line with a slope of -4.00×10^{-5} mol/L · s.

a. Determine the rate law, the integrated rate law, and the value of the rate constant for this reaction.

b. If the initial concentration of C$_2$H$_5$OH was 1.25×10^{-2} M, calculate the half-life for this reaction.

c. How much time is required for all the $1.25 \times 10^{-2}\ M$ C$_2$H$_5$OH to decompose?

42. At 500 K in the presence of a copper surface, ethanol decomposes according to the equation

$$C_2H_5OH(g) \longrightarrow CH_3CHO(g) + H_2(g)$$

The pressure of C$_2$H$_5$OH was measured as a function of time and the following data were obtained:

Time (s)	$P_{C_2H_5OH}$ (torr)
0	250.
100.	237
200.	224
300.	211
400.	198
500.	185

Since the pressure of a gas is directly proportional to the concentration of gas, we can express the rate law for a gaseous reaction in terms of partial pressures. Using the above data, deduce the rate law, the integrated rate law, and the value of the rate constant, all in terms of pressure units in atm and time in seconds. Predict the pressure of C$_2$H$_5$OH after 900. s from the start of the reaction. (*Hint:* To determine the order of the reaction with respect to C$_2$H$_5$OH, compare how the pressure of C$_2$H$_5$OH decreases with each time listing.)

43. The dimerization of butadiene

$$2C_4H_6(g) \longrightarrow C_8H_{12}(g)$$

was studied at 500. K, and the following data were obtained:

Time (s)	[C$_4$H$_6$] (mol/L)
195	1.6×10^{-2}
604	1.5×10^{-2}
1246	1.3×10^{-2}
2180	1.1×10^{-2}
6210	0.68×10^{-2}

Assuming that

$$Rate = -\frac{\Delta[C_4H_6]}{\Delta t}$$

determine the form of the rate law, the integrated rate law, and the value of the rate constant for this reaction.

44. The rate of the reaction

$$O(g) + NO_2(g) \longrightarrow NO(g) + O_2(g)$$

was studied at a certain temperature.

a. In one experiment, NO_2 was in large excess, at a concentration of 1.0×10^{13} molecules/cm³ with the following data collected:

Time (s)	[O] (atoms/cm³)
0	5.0×10^9
1.0×10^{-2}	1.9×10^9
2.0×10^{-2}	6.8×10^8
3.0×10^{-2}	2.5×10^8

What is the order of the reaction with respect to oxygen atoms?

b. The reaction is known to be first order with respect to NO_2. Determine the overall rate law and the value of the rate constant.

45. Experimental data for the reaction

$$A \longrightarrow 2B + C$$

have been plotted in the following three different ways (with concentration units in mol/L):

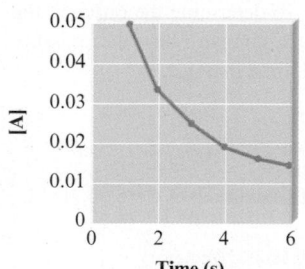
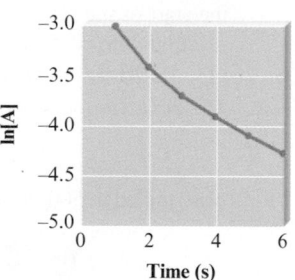

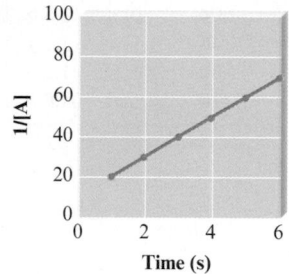

What is the order of the reaction with respect to A, and what is the initial concentration of A?

46. Consider the data plotted in Exercise 45 when answering the following questions.

a. What is the concentration of A after 9 s?

b. What are the first three half-lives for this experiment?

47. The reaction

$$A \longrightarrow B + C$$

is known to be zero order in A and to have a rate constant of 5.0×10^{-2} mol/L · s at 25°C. An experiment was run at 25°C where $[A]_0 = 1.0 \times 10^{-3}$ M.

a. Write the integrated rate law for this reaction.

b. Calculate the half-life for the reaction.

c. Calculate the concentration of B after 5.0×10^{-3} s has elapsed assuming $[B]_0 = 0$.

48. The decomposition of hydrogen iodide on finely divided gold at 150°C is zero order with respect to HI. The rate defined below is constant at 1.20×10^{-4} mol/L · s.

$$2HI(g) \xrightarrow{Au} H_2(g) + I_2(g)$$

$$Rate = -\frac{\Delta[HI]}{\Delta t} = k = 1.20 \times 10^{-4} \text{ mol/L} \cdot s$$

a. If the initial HI concentration was 0.250 mol/L, calculate the concentration of HI at 25 minutes after the start of the reaction.

b. How long will it take for all of the 0.250 M HI to decompose?

49. A certain first-order reaction is 45.0% complete in 65 s. What are the values of the rate constant and the half-life for this process?

50. A first-order reaction is 75.0% complete in 320. s.

a. What are the first and second half-lives for this reaction?

b. How long does it take for 90.0% completion?

51. The rate law for the decomposition of phosphine (PH_3) is

$$Rate = -\frac{\Delta[PH_3]}{\Delta t} = k[PH_3]$$

It takes 120. s for 1.00 M PH_3 to decrease to 0.250 M. How much time is required for 2.00 M PH_3 to decrease to a concentration of 0.350 M?

52. DDT (molar mass = 354.49 g/mol) was a widely used insecticide that was banned from use in the United States in 1973. This ban was brought about due to the persistence of DDT in many different ecosystems, leading to high accumulations of the substance in many birds of prey. The insecticide was shown to cause a thinning of egg shells, pushing many birds toward extinction. If a 20-L drum of DDT was spilled into a pond, resulting in a DDT concentration of 8.75×10^{-5} M, how long would it take for the levels of DDT to reach a concentration of 1.41×10^{-7} M (a level that is generally assumed safe in mammals)? Assume the decomposition of DDT is a first-order process with a half-life of 56.0 days.

53. Consider the following initial rate data for the decomposition of compound AB to give A and B:

$[AB]_0$ (mol/L)	Initial Rate (mol/L · s)
0.200	3.20×10^{-3}
0.400	1.28×10^{-2}
0.600	2.88×10^{-2}

Determine the half-life for the decomposition reaction initially having 1.00 M AB present.

54. The rate law for the reaction

$$2NOBr(g) \longrightarrow 2NO(g) + Br_2(g)$$

at some temperature is

$$Rate = -\frac{\Delta[NOBr]}{\Delta t} = k[NOBr]^2$$

a. If the half-life for this reaction is 2.00 s when $[NOBr]_0 = 0.900\ M$, calculate the value of k for this reaction.

b. How much time is required for the concentration of NOBr to decrease to $0.100\ M$?

55. For the reaction A $\rightarrow$ products, successive half-lives are observed to be 10.0, 20.0, and 40.0 min for an experiment in which $[A]_0 = 0.10\ M$. Calculate the concentration of A at the following times.

a. 80.0 min

b. 30.0 min

56. Theophylline is a pharmaceutical drug that is sometimes used to help with lung function. You observe a case where the initial lab results indicate that the concentration of theophylline in a patient's body decreased from $2.0 \times 10^{-3}\ M$ to $1.0 \times 10^{-3}\ M$ in 24 hours. In another 12 hours the drug concentration was found to be $5.0 \times 10^{-4}\ M$. What is the value of the rate constant for the metabolism of this drug in the body?

57. You and a coworker have developed a molecule that has shown potential as cobra antivenin (AV). This antivenin works by binding to the venom (V), thereby rendering it nontoxic. This reaction can be described by the rate law

$$Rate = k[AV]^1[V]^1$$

You have been given the following data from your coworker:

$$[V]_0 = 0.20\ M$$
$$[AV]_0 = 1.0 \times 10^{-4}\ M$$

A plot of $\ln[AV]$ versus t (s) gives a straight line with a slope of $-0.32\ s^{-1}$. What is the value of the rate constant (k) for this reaction?

58. Consider the hypothetical reaction

$$A + B + 2C \longrightarrow 2D + 3E$$

where the rate law is

$$Rate = -\frac{\Delta[A]}{\Delta t} = k[A][B]^2$$

An experiment is carried out where $[A]_0 = 1.0 \times 10^{-2}\ M$, $[B]_0 = 3.0\ M$, and $[C]_0 = 2.0\ M$. The reaction is started, and after 8.0 seconds, the concentration of A is $3.8 \times 10^{-3}\ M$.

a. Calculate the value of k for this reaction.

b. Calculate the half-life for this experiment.

c. Calculate the concentration of A after 13.0 seconds.

d. Calculate the concentration of C after 13.0 seconds.

Reaction Mechanisms

59. Write the rate laws for the following elementary reactions.

a. $CH_3NC(g) \rightarrow CH_3CN(g)$

b. $O_3(g) + NO(g) \rightarrow O_2(g) + NO_2(g)$

c. $O_3(g) \rightarrow O_2(g) + O(g)$

d. $O_3(g) + O(g) \rightarrow 2O_2(g)$

60. A possible mechanism for the decomposition of hydrogen peroxide is

$$H_2O_2 \longrightarrow 2OH$$
$$H_2O_2 + OH \longrightarrow H_2O + HO_2$$
$$HO_2 + OH \longrightarrow H_2O + O_2$$

Using your results from Exercise 37, specify which step is the rate-determining step. What is the overall balanced equation for the reaction?

61. A proposed mechanism for a reaction is

$$C_4H_9Br \longrightarrow C_4H_9^+ + Br^- \quad\quad \text{Slow}$$
$$C_4H_9^+ + H_2O \longrightarrow C_4H_9OH_2^+ \quad\quad \text{Fast}$$
$$C_4H_9OH_2^+ + H_2O \longrightarrow C_4H_9OH + H_3O^+ \quad \text{Fast}$$

Write the rate law expected for this mechanism. What is the overall balanced equation for the reaction? What are the intermediates in the proposed mechanism?

62. The mechanism for the gas-phase reaction of nitrogen dioxide with carbon monoxide to form nitric oxide and carbon dioxide is thought to be

$$NO_2 + NO_2 \longrightarrow NO_3 + NO \quad\quad \text{Slow}$$
$$NO_3 + CO \longrightarrow NO_2 + CO_2 \quad\quad \text{Fast}$$

Write the rate law expected for this mechanism. What is the overall balanced equation for the reaction?

Temperature Dependence of Rate Constants and the Collision Model

63. For the following reaction profile, indicate

a. the positions of reactants and products.

b. the activation energy.

c. ΔE for the reaction.

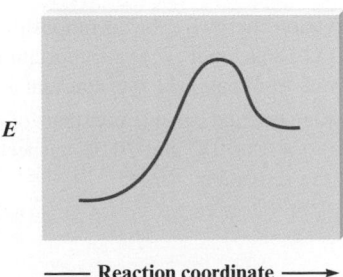

E

Reaction coordinate $\longrightarrow$

64. Draw a rough sketch of the energy profile for each of the following cases:

    a. $\Delta E = +10$ kJ/mol, $E_a = 25$ kJ/mol

    b. $\Delta E = -10$ kJ/mol, $E_a = 50$ kJ/mol

    c. $\Delta E = -50$ kJ/mol, $E_a = 50$ kJ/mol

65. The activation energy for the reaction

    $$NO_2(g) + CO(g) \longrightarrow NO(g) + CO_2(g)$$

    is 125 kJ/mol, and ΔE for the reaction is -216 kJ/mol. What is the activation energy for the reverse reaction $[NO(g) + CO_2(g) \longrightarrow NO_2(g) + CO(g)]$?

66. The activation energy for some reaction

    $$X_2(g) + Y_2(g) \longrightarrow 2XY(g)$$

    is 167 kJ/mol, and ΔE for the reaction is $+28$ kJ/mol. What is the activation energy for the decomposition of XY?

67. The rate constant for the gas-phase decomposition of N_2O_5,

    $$N_2O_5 \longrightarrow 2NO_2 + \tfrac{1}{2}O_2$$

    has the following temperature dependence:

T (K)	k (s^{-1})
338	4.9×10^{-3}
318	5.0×10^{-4}
298	3.5×10^{-5}

    Make the appropriate graph using these data, and determine the activation energy for this reaction.

68. The reaction

    $$(CH_3)_3CBr + OH^- \longrightarrow (CH_3)_3COH + Br^-$$

    in a certain solvent is first order with respect to $(CH_3)_3CBr$ and zero order with respect to OH^-. In several experiments, the rate constant k was determined at different temperatures. A plot of $\ln(k)$ versus $1/T$ was constructed resulting in a straight line with a slope value of -1.10×10^4 K and y-intercept of 33.5. Assume k has units of s^{-1}.

    a. Determine the activation energy for this reaction.

    b. Determine the value of the frequency factor A.

    c. Calculate the value of k at 25°C.

69. The activation energy for the decomposition of $HI(g)$ to $H_2(g)$ and $I_2(g)$ is 186 kJ/mol. The rate constant at 555 K is 3.52×10^{-7} L/mol · s. What is the rate constant at 645 K?

70. A first-order reaction has rate constants of 4.6×10^{-2} s^{-1} and 8.1×10^{-2} s^{-1} at 0°C and 20.°C, respectively. What is the value of the activation energy?

71. A certain reaction has an activation energy of 54.0 kJ/mol. As the temperature is increased from 22°C to a higher temperature, the rate constant increases by a factor of 7.00. Calculate the higher temperature.

72. Chemists commonly use a rule of thumb that an increase of 10 K in temperature doubles the rate of a reaction. What must the activation energy be for this statement to be true for a temperature increase from 25 to 35°C?

73. Which of the following reactions would you expect to proceed at a faster rate at room temperature? Why? (*Hint:* Think about which reaction would have the lower activation energy.)

    $$2Ce^{4+}(aq) + Hg_2^{2+}(aq) \longrightarrow 2Ce^{3+}(aq) + 2Hg^{2+}(aq)$$
    $$H_3O^+(aq) + OH^-(aq) \longrightarrow 2H_2O(l)$$

74. One reason suggested for the instability of long chains of silicon atoms is that the decomposition involves the transition state shown below:

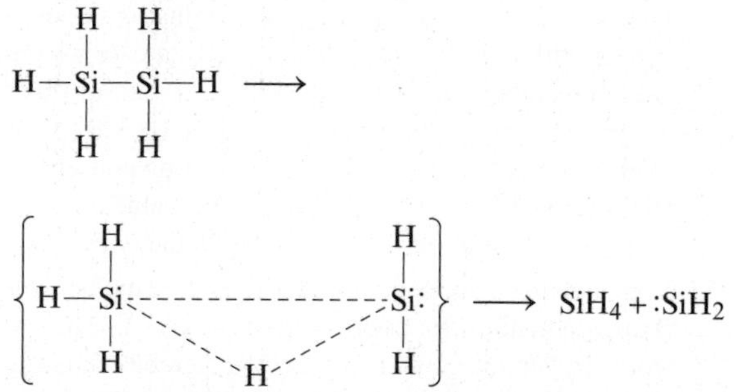

    The activation energy for such a process is 210 kJ/mol, which is less than either the Si—Si or the Si—H bond energy. Why would a similar mechanism not be expected to play a very important role in the decomposition of long chains of carbon atoms as seen in organic compounds?

Catalysts

75. One mechanism for the destruction of ozone in the upper atmosphere is

$O_3(g) + NO(g) \longrightarrow NO_2(g) + O_2(g)$	Slow
$NO_2(g) + O(g) \longrightarrow NO(g) + O_2(g)$	Fast

    Overall reaction $\quad O_3(g) + O(g) \longrightarrow 2O_2(g)$

    a. Which species is a catalyst?

    b. Which species is an intermediate?

    c. E_a for the uncatalyzed reaction

    $$O_3(g) + O(g) \longrightarrow 2O_2(g)$$

    is 14.0 kJ. E_a for the same reaction when catalyzed is 11.9 kJ. What is the ratio of the rate constant for the catalyzed reaction to that for the uncatalyzed reaction at 25°C? Assume that the frequency factor A is the same for each reaction.

76. One of the concerns about the use of Freons is that they will migrate to the upper atmosphere, where chlorine atoms can be generated by the following reaction:

$$CCl_2F_2(g) \xrightarrow{h\nu} CF_2Cl(g) + Cl(g)$$
Freon-12

Chlorine atoms can act as a catalyst for the destruction of ozone. The activation energy for the reaction

$$Cl(g) + O_3(g) \longrightarrow ClO(g) + O_2(g)$$

is 2.1 kJ/mol. Which is the more effective catalyst for the destruction of ozone, Cl or NO? (See Exercise 75.)

77. Assuming that the mechanism for the hydrogenation of C_2H_4 given in Section 11-7 is correct, would you predict that the product of the reaction of C_2H_4 with D_2 would be CH_2D—CH_2D or CHD_2—CH_3? How could the reaction of C_2H_4 with D_2 be used to confirm the mechanism for the hydrogenation of C_2H_4 given in Section 11-7?

78. The decomposition of NH_3 to N_2 and H_2 was studied on two surfaces:

Surface	E_a (kJ/mol)
W	163
Os	197

Without a catalyst, the activation energy is 335 kJ/mol.

a. Which surface is the better heterogeneous catalyst for the decomposition of NH_3? Why?

b. How many times faster is the reaction at 298 K on the W surface compared with the reaction with no catalyst present? Assume that the frequency factor A is the same for each reaction.

c. The decomposition reaction on the two surfaces obeys a rate law of the form

$$\text{Rate} = k \frac{[NH_3]}{[H_2]}$$

How can you explain the inverse dependence of the rate on the H_2 concentration?

79. The decomposition of many substances on the surface of a heterogeneous catalyst shows the following behavior:

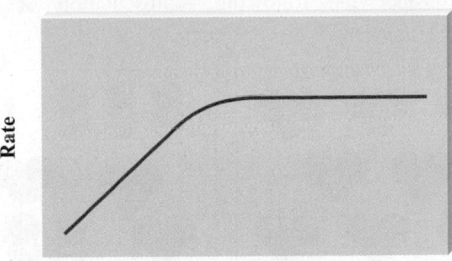

Concentration of reactant

How do you account for the rate law changing from first order to zero order in the concentration of reactant?

80. For enzyme-catalyzed reactions that follow the mechanism

$$E + S \rightleftharpoons E \cdot S$$
$$E \cdot S \rightleftharpoons E + P$$

a graph of the rate as a function of [S], the concentration of the substrate, has the following appearance:

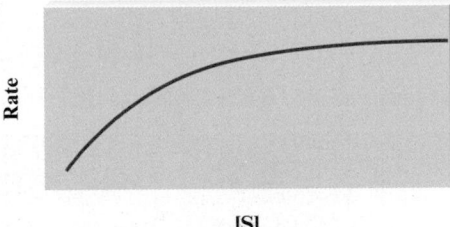

[S]

Note that at higher substrate concentrations the rate no longer changes with [S]. Suggest a reason for this.

81. A popular chemical demonstration is the "magic genie" procedure, in which hydrogen peroxide decomposes to water and oxygen gas with the aid of a catalyst. The activation energy of this (uncatalyzed) reaction is 70.0 kJ/mol. When the catalyst is added, the activation energy (at 20.°C) is 42.0 kJ/mol. Theoretically, to what temperature (°C) would one have to heat the hydrogen peroxide solution so that the rate of the uncatalyzed reaction is equal to the rate of the catalyzed reaction at 20.°C? Assume the frequency factor A is constant, and assume the initial concentrations are the same.

82. The activation energy for a reaction is changed from 184 kJ/mol to 59.0 kJ/mol at 600. K by the introduction of a catalyst. If the uncatalyzed reaction takes about 2400 years to occur, about how long will the catalyzed reaction take? Assume the frequency factor A is constant, and assume the initial concentrations are the same.

Additional Exercises

83. Consider the following representation of the reaction
$2NO_2(g) \rightarrow 2NO(g) + O_2(g)$.

  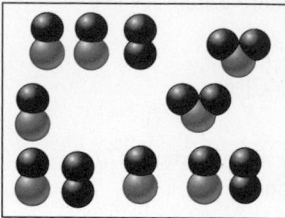

(a) time = 0 minutes   **(b)** time = 10 minutes   **(c)** time = ? minutes

⸻ Time ⸻→

Determine the time for the final representation above if the reaction is

a. first order

b. second order

c. zero order

84. The reaction

$H_2SeO_3(aq) + 6I^-(aq) + 4H^+(aq)$
$$\longrightarrow Se(s) + 2I_3^-(aq) + 3H_2O(l)$$

was studied at 0°C, and the following data were obtained:

$[H_2SeO_3]_0$ (mol/L)	$[H^+]_0$ (mol/L)	$[I^-]_0$ (mol/L)	Initial Rate (mol/L · s)
1.0×10^{-4}	2.0×10^{-2}	2.0×10^{-2}	1.66×10^{-7}
2.0×10^{-4}	2.0×10^{-2}	2.0×10^{-2}	3.33×10^{-7}
3.0×10^{-4}	2.0×10^{-2}	2.0×10^{-2}	4.99×10^{-7}
1.0×10^{-4}	4.0×10^{-2}	2.0×10^{-2}	6.66×10^{-7}
1.0×10^{-4}	1.0×10^{-2}	2.0×10^{-2}	0.42×10^{-7}
1.0×10^{-4}	2.0×10^{-2}	4.0×10^{-2}	13.2×10^{-7}
1.0×10^{-4}	1.0×10^{-2}	4.0×10^{-2}	3.36×10^{-7}

These relationships hold only if there is a very small amount of I_3^- present. What is the rate law and the value of the rate constant? $\left(\text{Assume that rate} = -\dfrac{\Delta[H_2SeO_3]}{\Delta t}. \right)$

85. Consider two reaction vessels, one containing A and the other containing B, with equal concentrations at $t = 0$. If both substances decompose by first-order kinetics, where

$$k_A = 4.50 \times 10^{-4} \text{ s}^{-1}$$
$$k_B = 3.70 \times 10^{-3} \text{ s}^{-1}$$

how much time must pass to reach a condition such that $[A] = 4.00[B]$?

86. Sulfuryl chloride (SO_2Cl_2) decomposes to sulfur dioxide (SO_2) and chlorine (Cl_2) by reaction in the gas phase. The following pressure data were obtained when a sample containing 5.00×10^{-2} mol sulfuryl chloride was heated to 600. K in a 5.00×10^{-1}-L container.

Time (hours):	0.00	1.00	2.00	4.00	8.00	16.00
$P_{SO_2Cl_2}$ (atm):	4.93	4.26	3.52	2.53	1.30	0.34

Defining the rate as $-\dfrac{\Delta[SO_2Cl_2]}{\Delta t}$,

a. determine the value of the rate constant for the decomposition of sulfuryl chloride at 600. K.

b. what is the half-life of the reaction?

c. what fraction of the sulfuryl chloride remains after 20.0 h?

87. For the reaction

$$2N_2O_5(g) \longrightarrow 4NO_2(g) + O_2(g)$$

the following data were collected, where

$$\text{Rate} = -\dfrac{\Delta[N_2O_5]}{\Delta t}$$

Time (s)	$T = 338$ K $[N_2O_5]$	$T = 318$ K $[N_2O_5]$
0	1.00×10^{-1} M	1.00×10^{-1} M
100.	6.14×10^{-2} M	9.54×10^{-2} M
300.	2.33×10^{-2} M	8.63×10^{-2} M
600.	5.41×10^{-3} M	7.43×10^{-2} M
900.	1.26×10^{-3} M	6.39×10^{-2} M

Calculate E_a for this reaction.

88. Experimental values for the temperature dependence of the rate constant for the gas-phase reaction

$$NO + O_3 \longrightarrow NO_2 + O_2$$

are as follows:

T (K)	k (L/mol · s)
195	1.08×10^9
230.	2.95×10^9
260.	5.42×10^9
298	12.0×10^9
369	35.5×10^9

Make the appropriate graph using these data, and determine the activation energy for this reaction.

89. Cobra venom helps the snake secure food by binding to acetylcholine receptors on the diaphragm of a bite victim, leading to the loss of function of the diaphragm muscle tissue and eventually death. In order to develop more potent antivenins, scientists have studied what happens to the toxin once it has bound the acetylcholine receptors. They have found that the toxin is released from the receptor in a process that can be described by the rate law

Rate = k[acetylcholine receptor–toxin complex]

If the activation energy of this reaction at 37.0°C is 26.2 kJ/mol and $A = 0.850\ s^{-1}$, what is the rate of reaction if you have a 0.200-M solution of receptor–toxin complex at 37.0°C?

90. Iodomethane (CH_3I) is a commonly used reagent in organic chemistry. When used properly, this reagent allows chemists to introduce methyl groups in many different useful applications. The chemical does pose a risk as a carcinogen, possibly owing to iodomethane's ability to react with portions of the DNA strand (if they were to come in contact). Consider the following hypothetical initial rates data:

$[DNA]_0$ (μmol/L)	$[CH_3I]_0$ (μmol/L)	Initial Rate (μmol/L · s)
0.100	0.100	3.20×10^{-4}
0.100	0.200	6.40×10^{-4}
0.200	0.200	1.28×10^{-3}

Which of the following could be a possible mechanism to explain the initial rate data?

Mechanism I $DNA + CH_3I \longrightarrow DNA—CH_3^+ + I^-$

Mechanism II $CH_3I \longrightarrow CH_3^+ + I^-$ Slow

$DNA + CH_3^+ \longrightarrow DNA—CH_3^+$ Fast

91. Experiments during a recent summer on a number of fireflies (small beetles, *Lampyridaes photinus*) showed that the average interval between flashes of individual insects was 16.3 s at 21.0°C and 13.0 s at 27.8°C.

 a. What is the apparent activation energy of the reaction that controls the flashing?

 b. What would be the average interval between flashes of an individual firefly at 30.0°C?

 c. Compare the observed intervals and the one you calculated in part b to the rule of thumb that the Celsius temperature is 54 minus twice the interval between flashes.

92. The activation energy of a certain uncatalyzed biochemical reaction is 50.0 kJ/mol. In the presence of a catalyst at 37°C, the rate constant for the reaction increases by a factor of 2.50×10^3 as compared with the uncatalyzed reaction. Assuming the frequency factor A is the same for both the catalyzed and uncatalyzed reactions, calculate the activation energy for the catalyzed reaction.

93. Consider the reaction

$$3A + B + C \longrightarrow D + E$$

where the rate law is defined as

$$-\frac{\Delta[A]}{\Delta t} = k[A]^2[B][C]$$

An experiment is carried out where $[B]_0 = [C]_0 = 1.00\ M$ and $[A]_0 = 1.00 \times 10^{-4}\ M$.

 a. If after 3.00 min, $[A] = 3.26 \times 10^{-5}\ M$, calculate the value of k.

 b. Calculate the half-life for this experiment.

 c. Calculate the concentration of B and the concentration of A after 10.0 min.

ChemWork Problems

These multiconcept problems (and additional ones) are found interactively online with the same type of assistance a student would get from an instructor.

94. The thiosulfate ion ($S_2O_3^{2-}$) is oxidized by iodine as follows:

$$2S_2O_3^{2-}(aq) + I_2(aq) \longrightarrow S_4O_6^{2-}(aq) + 2I^-(aq)$$

In a certain experiment, 7.05×10^{-3} mol/L of $S_2O_3^{2-}$ is consumed in the first 11.0 seconds of the reaction. Calculate the rate of consumption of $S_2O_3^{2-}$. Calculate the rate of production of iodide ion.

95. The reaction $A(aq) + B(aq) \longrightarrow$ products(aq) was studied, and the following data were obtained:

$[A]_0$ (mol/L)	$[B]_0$ (mol/L)	Initial Rate (mol/L · s)
0.12	0.18	3.46×10^{-2}
0.060	0.12	1.15×10^{-2}
0.030	0.090	4.32×10^{-3}
0.24	0.090	3.46×10^{-2}

What is the order of the reaction with respect to A? What is the order of the reaction with respect to B? What is the value of the rate constant for the reaction?

96. A certain substance, initially present at 0.0800 M, decomposes by zero-order kinetics with a rate constant of 2.50×10^{-2} mol/L · s. Calculate the time (in seconds) required for the system to reach a concentration of 0.0210 M.

97. A reaction of the form

$$aA \longrightarrow Products$$

gives a plot of ln[A] versus time (in seconds), which is a straight line with a slope of -7.35×10^{-3}. Assuming $[A]_0 = 0.0100\ M$, calculate the time (in seconds) required for the reaction to reach 22.9% completion.

98. A certain reaction has the form

$$aA \longrightarrow Products$$

At a particular temperature, concentration versus time data were collected. A plot of 1/[A] versus time (in seconds) gave a straight line with a slope of 6.90×10^{-2}. What is the differential rate law for this reaction? What is the integrated rate law for this reaction? What is the value of the rate constant for this reaction? If $[A]_0$ for this reaction is 0.100 M, what is the first half-life (in seconds)? If the original concentration (at $t = 0$) is 0.100 M, what is the second half-life (in seconds)?

99. Which of the following statement(s) is(are) *true*?

  a. The half-life for a zero-order reaction increases as the reaction proceeds.

  b. A catalyst does not change the value of the rate constant.

  c. The half-life for a reaction, aA ⟶ products, that is first order in A increases with increasing $[A]_0$.

  d. The half-life for a second-order reaction increases as the reaction proceeds.

100. Consider the hypothetical reaction $A_2(g) + B_2(g) \longrightarrow 2AB(g)$, where the rate law is:

$$-\frac{\Delta[A_2]}{\Delta t} = k[A_2][B_2]$$

The value of the rate constant at 302°C is 2.45×10^{-4} L/mol · s, and at 508°C the rate constant is 0.891 L/mol · s. What is the activation energy for this reaction? What is the value of the rate constant for this reaction at 375°C?

101. Experiments have shown that the average frequency of chirping by a snowy tree cricket (*Oecanthus fultoni*) depends on temperature as shown in the table.

Chirping Rate (per min)	Temperature (°C)
178	25.0
126	20.3
100.	17.3

What is the apparent activation energy of the process that controls the chirping? What is the rate of chirping expected at a temperature of 7.5°C?

Challenge Problems

102. Consider a reaction of the type aA ⟶ products, in which the rate law is found to be rate = $k[A]^3$ (termolecular reactions are improbable but possible). If the first half-life of the reaction is found to be 40. s, what is the time for the second half-life? *Hint:* Using your calculus knowledge, derive the integrated rate law from the differential rate law for a termolecular reaction:

$$\text{Rate} = \frac{-d[A]}{dt} = k[A]^3$$

103. A study was made of the effect of the hydroxide concentration on the rate of the reaction

$$I^-(aq) + OCl^-(aq) \longrightarrow IO^-(aq) + Cl^-(aq)$$

The following data were obtained:

$[I^-]_0$ (mol/L)	$[OCl^-]_0$ (mol/L)	$[OH^-]_0$ (mol/L)	Initial Rate (mol/L · s)
0.0013	0.012	0.10	9.4×10^{-3}
0.0026	0.012	0.10	18.7×10^{-3}
0.0013	0.0060	0.10	4.7×10^{-3}
0.0013	0.018	0.10	14.0×10^{-3}
0.0013	0.012	0.050	18.7×10^{-3}
0.0013	0.012	0.20	4.7×10^{-3}
0.0013	0.018	0.20	7.0×10^{-3}

Determine the rate law and the value of the rate constant for this reaction.

104. Two isomers (A and B) of a given compound dimerize as follows:

$$2A \xrightarrow{k_1} A_2$$
$$2B \xrightarrow{k_2} B_2$$

Both processes are known to be second order in reactant, and k_1 is known to be 0.250 L/mol · s at 25°C. In a particular experiment A and B were placed in separate containers at 25°C, where $[A]_0 = 1.00 \times 10^{-2} M$ and $[B]_0 = 2.50 \times 10^{-2} M$. It was found that after each reaction had progressed for 3.00 min, $[A] = 3.00[B]$. In this case the rate laws are defined as

$$\text{Rate} = -\frac{\Delta[A]}{\Delta t} = k_1[A]^2$$

$$\text{Rate} = -\frac{\Delta[B]}{\Delta t} = k_2[B]^2$$

  a. Calculate the concentration of A_2 after 3.00 min.

  b. Calculate the value of k_2.

  c. Calculate the half-life for the experiment involving A.

105. The reaction

$$NO(g) + O_3(g) \longrightarrow NO_2(g) + O_2(g)$$

was studied by performing two experiments. In the first experiment the rate of disappearance of NO was followed in the presence of a large excess of O_3. The results were as follows ($[O_3]$ remains effectively constant at 1.0×10^{14} molecules/cm³):

Time (ms)	[NO] (molecules/cm³)
0	6.0×10^8
100 ± 1	5.0×10^8
500 ± 1	2.4×10^8
700 ± 1	1.7×10^8
1000 ± 1	9.9×10^7

In the second experiment [NO] was held constant at 2.0×10^{14} molecules/cm³. The data for the disappearance of O_3 are as follows:

Time (ms)	[O₃] (molecules/cm³)
0	1.0×10^{10}
50 ± 1	8.4×10^9
100 ± 1	7.0×10^9
200 ± 1	4.9×10^9
300 ± 1	3.4×10^9

  a. What is the order with respect to each reactant?

  b. What is the overall rate law?

  c. What is the value of the rate constant from each set of experiments?

$$\text{Rate} = k'[NO]^x \qquad \text{Rate} = k''[O_3]^y$$

d. What is the value of the rate constant for the overall rate law?

$$\text{Rate} = k[\text{NO}]^x[\text{O}_3]^y$$

106. Most reactions occur by a series of steps. The energy profile for a certain reaction that proceeds by a two-step mechanism is

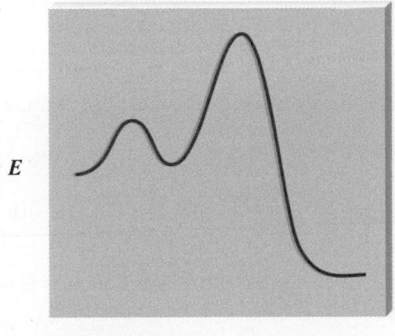

On the energy profile, indicate

a. the positions of reactants and products.

b. the activation energy for the overall reaction.

c. ΔE for the reaction.

d. Which point on the plot represents the energy of the intermediate in the two-step reaction?

e. Which step in the mechanism for this reaction is rate determining, the first or the second step? Explain.

107. You are studying the kinetics of the reaction $H_2(g) + F_2(g) \rightarrow 2HF(g)$ and you wish to determine a mechanism for the reaction. You run the reaction twice by keeping one reactant at a much higher pressure than the other reactant (this lower-pressure reactant begins at 1.000 atm). Unfortunately, you neglect to record which reactant was at the higher pressure, and you forget which it was later. Your data for the first experiment are:

Pressure of HF (atm)	Time (min)
0	0
0.300	30.0
0.600	65.8
0.900	110.4
1.200	169.1
1.500	255.9

When you ran the second experiment (in which the higher-pressure reactant was run at a much higher pressure), you determine the values of the apparent rate constants to be the same. It also turns out that you find data taken from another person in the lab. This individual found that the reaction proceeds 40.0 times faster at 55°C than at 35°C. You also know, from the energy-level diagram, that there are three steps to the mechanism, and the first step has the highest activation energy. You look up the bond energies of the species involved and they are (in kJ/mol): H—H (432), F—F (154), and H—F (565).

a. Sketch an energy-level diagram (qualitative) that is consistent with the one described previously. *Hint:* See Exercise 106.

b. Develop a reasonable mechanism for the reaction.

c. Which reactant was limiting in the experiments?

108. The decomposition of $NO_2(g)$ occurs by the following bimolecular elementary reaction:

$$2NO_2(g) \longrightarrow 2NO(g) + O_2(g)$$

The rate constant at 273 K is 2.3×10^{-12} L/mol · s, and the activation energy is 111 kJ/mol. How long will it take for the concentration of $NO_2(g)$ to decrease from an initial partial pressure of 2.5 atm to 1.5 atm at 500. K? Assume ideal gas behavior.

109. The following data were collected in two studies of the reaction

$$2A + B \longrightarrow C + D$$

Time (s)	Experiment 1 [A] (mol/L) × 10⁻²	Experiment 2 [A] (mol/L) × 10⁻²
0	10.0	10.0
20.	6.67	5.00
40.	5.00	3.33
60.	4.00	2.50
80.	3.33	2.00
100.	2.86	1.67
120.	2.50	1.43

In Experiment 1, $[B]_0 = 5.0 \, M$.
In Experiment 2, $[B]_0 = 10.0 \, M$.

$$\text{Rate} = \frac{-\Delta[A]}{\Delta t}$$

a. Why is [B] much greater than [A]?

b. Give the rate law and value for k for this reaction.

110. Consider the following hypothetical data collected in two studies of the reaction

$$2A + 2B \longrightarrow C + 2D$$

Time (s)	Experiment 1 [A] (mol/L)	Experiment 2 [A] (mol/L)
0	1.0×10^{-2}	1.0×10^{-2}
10.	8.4×10^{-3}	5.0×10^{-3}
20.	7.1×10^{-3}	2.5×10^{-3}
30.	?	1.3×10^{-3}
40.	5.0×10^{-3}	6.3×10^{-4}

In Experiment 1, $[B]_0 = 10.0 \, M$.
In Experiment 2, $[B]_0 = 20.0 \, M$.

$$\text{Rate} = \frac{-\Delta[A]}{\Delta t}$$

a. Use the concentration versus time data to determine the rate law for the reaction.

b. Solve for the value of the rate constant (k) for the reaction. Include units.

c. Calculate the concentration of A in Experiment 1 at $t = 30.$ s.

111. Consider the hypothetical reaction

$$A + B + 2C \longrightarrow 2D + 3E$$

In a study of this reaction three experiments were run at the same temperature. The rate is defined as $-\Delta[B]/\Delta t$.

Experiment 1:

$$[A]_0 = 2.0\,M \quad [B]_0 = 1.0 \times 10^{-3}\,M \quad [C]_0 = 1.0\,M$$

[B] (mol/L)	Time (s)
2.7×10^{-4}	1.0×10^5
1.6×10^{-4}	2.0×10^5
1.1×10^{-4}	3.0×10^5
8.5×10^{-5}	4.0×10^5
6.9×10^{-5}	5.0×10^5
5.8×10^{-5}	6.0×10^5

Experiment 2:

$$[A]_0 = 1.0 \times 10^{-2}\,M \quad [B]_0 = 3.0\,M \quad [C]_0 = 1.0\,M$$

[A] (mol/L)	Time (s)
8.9×10^{-3}	1.0
7.1×10^{-3}	3.0
5.5×10^{-3}	5.0
3.8×10^{-3}	8.0
2.9×10^{-3}	10.0
2.0×10^{-3}	13.0

Experiment 3:

$$[A]_0 = 10.0\,M \quad [B]_0 = 5.0\,M \quad [C]_0 = 5.0 \times 10^{-1}\,M$$

[C] (mol/L)	Time (s)
0.43	1.0×10^{-2}
0.36	2.0×10^{-2}
0.29	3.0×10^{-2}
0.22	4.0×10^{-2}
0.15	5.0×10^{-2}
0.08	6.0×10^{-2}

Write the rate law for this reaction, and calculate the value of the rate constant.

112. Hydrogen peroxide and the iodide ion react in acidic solution as follows:

$$H_2O_2(aq) + 3I^-(aq) + 2H^+(aq) \longrightarrow I_3^-(aq) + 2H_2O(l)$$

The kinetics of this reaction were studied by following the decay of the concentration of H_2O_2 and constructing plots of $\ln[H_2O_2]$ versus time. All the plots were linear and all solutions had $[H_2O_2]_0 = 8.0 \times 10^{-4}$ mol/L. The slopes of these straight lines depended on the initial concentrations of I^- and H^+. The results follow:

$[I^-]_0$ (mol/L)	$[H^+]_0$ (mol/L)	Slope (min^{-1})
0.1000	0.0400	-0.120
0.3000	0.0400	-0.360
0.4000	0.0400	-0.480
0.0750	0.0200	-0.0760
0.0750	0.0800	-0.118
0.0750	0.1600	-0.174

The rate law for this reaction has the form

$$\text{Rate} = \frac{-\Delta[H_2O_2]}{\Delta t} = (k_1 + k_2[H^+])[I^-]^m[H_2O_2]^n$$

a. Specify the order of this reaction with respect to $[H_2O_2]$ and $[I^-]$.

b. Calculate the values of the rate constants, k_1 and k_2.

c. What reason could there be for the two-term dependence of the rate on $[H^+]$?

Integrative Problems

These problems require the integration of multiple concepts to find the solutions.

113. Sulfuryl chloride undergoes first-order decomposition at 320.°C with a half-life of 8.75 h.

$$SO_2Cl_2(g) \longrightarrow SO_2(g) + Cl_2(g)$$

What is the value of the rate constant, k, in s^{-1}? If the initial pressure of SO_2Cl_2 is 791 torr and the decomposition occurs in a 1.25-L container, how many molecules of SO_2Cl_2 remain after 12.5 h?

114. Upon dissolving InCl(s) in HCl, In$^+$(aq) undergoes a disproportionation reaction according to the following unbalanced equation:

$$In^+(aq) \longrightarrow In(s) + In^{3+}(aq)$$

This disproportionation follows first-order kinetics with a half-life of 667 s. What is the concentration of In$^+$(aq) after 1.25 h if the initial solution of In$^+$(aq) was prepared by dissolving 2.38 g InCl(s) in dilute HCl to make 5.00×10^2 mL of solution? What mass of In(s) is formed after 1.25 h?

115. The decomposition of iodoethane in the gas phase proceeds according to the following equation:

$$C_2H_5I(g) \longrightarrow C_2H_4(g) + HI(g)$$

At 660. K, $k = 7.2 \times 10^{-4}$ s^{-1}; at 720. K, $k = 1.7 \times 10^{-2}$ s^{-1}. What is the value of the rate constant for this first-order decomposition at 325°C? If the initial pressure of iodoethane is 894 torr at 245°C, what is the pressure of iodoethane after three half-lives?

Marathon Problem

This problem is designed to incorporate several concepts and techniques into one situation.

116. Consider the following reaction:

$$CH_3X + Y \longrightarrow CH_3Y + X$$

At 25°C, the following two experiments were run, yielding the following data:

Experiment 1: $[Y]_0 = 3.0\ M$

$[CH_3X]$ (mol/L)	Time (h)
7.08×10^{-3}	1.0
4.52×10^{-3}	1.5
2.23×10^{-3}	2.3
4.76×10^{-4}	4.0
8.44×10^{-5}	5.7
2.75×10^{-5}	7.0

Experiment 2: $[Y]_0 = 4.5\ M$

$[CH_3X]$ (mol/L)	Time (h)
4.50×10^{-3}	0
1.70×10^{-3}	1.0
4.19×10^{-4}	2.5
1.11×10^{-4}	4.0
2.81×10^{-5}	5.5

Experiments also were run at 85°C. The value of the rate constant at 85°C was found to be 7.88×10^8 (with the time in units of hours), where $[CH_3X]_0 = 1.0 \times 10^{-2}\ M$ and $[Y]_0 = 3.0\ M$.

a. Determine the rate law and the value of k for this reaction at 25°C.

b. Determine the half-life at 85°C.

c. Determine E_a for the reaction.

d. Given that the C—X bond energy is known to be about 325 kJ/mol, suggest a mechanism that explains the results in parts a and c.

CHAPTER TWELVE

Chemical Equilibrium

The equilibrium in a salt water aquarium must be carefully maintained to keep the sea life healthy. (Borissos/Dreamstime.com)

488

n doing stoichiometry calculations we assumed that reactions proceed to completion, that is, until one of the reactants runs out. Many reactions do proceed essentially to completion. For such reactions it can be assumed that the reactants are quantitatively converted to products and that the amount of limiting reactant that remains is negligible. On the other hand, there are many chemical reactions that stop far short of completion. An example is the dimerization of nitrogen dioxide:

$$NO_2(g) + NO_2(g) \longrightarrow N_2O_4(g)$$

The reactant, NO_2, is a dark brown gas, and the product, N_2O_4, is a colorless gas. When NO_2 is placed in an evacuated, sealed glass vessel at 25°C, the initial dark brown color decreases in intensity as it is converted to colorless N_2O_4. However, even over a long period of time, the contents of the reaction vessel do not become colorless. Instead, the intensity of the brown color eventually becomes constant, which means that the concentration of NO_2 is no longer changing. This is illustrated on the molecular level in Fig. 12-1. This observation is a clear indication that the reaction has stopped short of completion. In fact, the system has reached **chemical equilibrium**, *the state where the concentrations of all reactants and products remain constant with time.*

Any chemical reactions carried out in a closed vessel will reach equilibrium. For some reactions the equilibrium position so favors the products that the reaction appears to have gone to completion. We say that the equilibrium position for such reactions lies *far to the right* (in the direction of the products). For example, when gaseous hydrogen and oxygen are mixed in stoichiometric quantities and react to form water vapor, the reaction proceeds essentially to completion. The amounts of the reactants that remain when the system reaches equilibrium are so tiny as to be negligible. By contrast, some reactions occur only to a slight extent. For example, when solid CaO is placed in a closed vessel at 25°C, the decomposition to solid Ca and gaseous O_2 is virtually undetectable. In cases like this, the equilibrium position is said to lie *far to the left* (in the direction of the reactants).

In this chapter we will discuss how and why a chemical system comes to equilibrium and the characteristics of equilibrium. In particular, we will discuss how to calculate the concentrations of the reactants and products present for a given system at equilibrium. It is important not to get "lost in the math" when we discuss these calculations. Always think about what is happening to the system at the molecular level, and you will better understand the calculations. We will consider a molecular-level perspective when we first discuss equilibrium constants and equilibrium conditions.

12-1 | The Equilibrium Condition

Since no changes occur in the concentrations of reactants or products in a reaction system at equilibrium, it may appear that everything has stopped. However, this is not the case. On the molecular level, there is frantic activity. Equilibrium is not static but is a highly *dynamic* situation. ◄ ⓘ The concept of chemical equilibrium is analogous to the flow of cars across a bridge connecting two island cities. Suppose the traffic flow on the bridge is the same in both directions. It is obvious that there is motion, since one can see the cars traveling back and forth across the bridge, but the number of cars in each city is not changing because equal numbers of cars are entering and leaving. The result is no *net* change in the car population.

ⓘ Equilibrium is a dynamic situation.

489

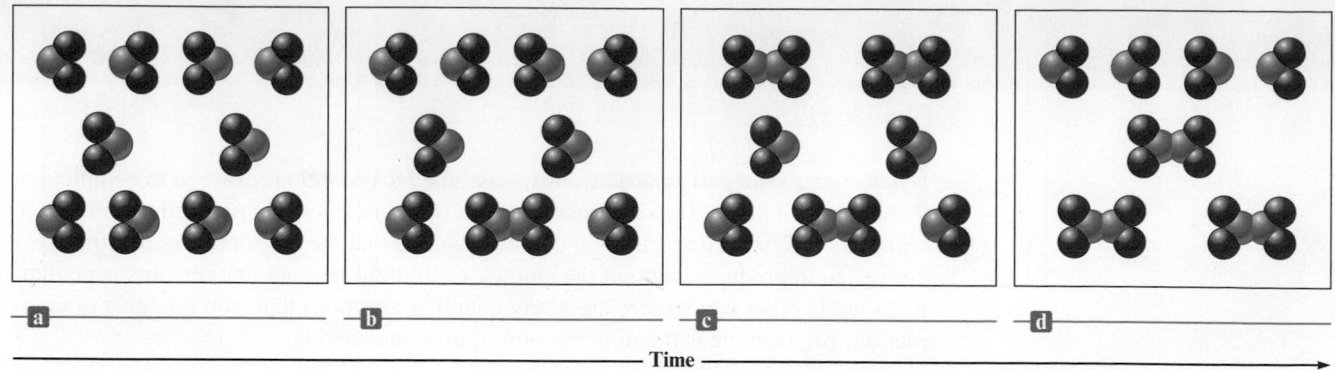

Figure 12-1 | A molecular representation of the reaction $2NO_2(g) \rightarrow N_2O_4(g)$ over time in a closed vessel. Note that the numbers of NO_2 and N_2O_4 in the container become constant (c and d) after sufficient time has passed.

To see how this concept applies to chemical reactions, consider the reaction between steam and carbon monoxide in a closed vessel at a high temperature where the reaction takes place rapidly:

$$H_2O(g) + CO(g) \rightleftharpoons H_2(g) + CO_2(g)$$

Assume that the same number of moles of gaseous CO and gaseous H_2O are placed in a closed vessel and allowed to react. The plots of the concentrations of reactants and products versus time are shown in Fig. 12-2. Note that since CO and H_2O were originally present in equal molar quantities, and since they react in a 1:1 ratio, the concentrations of the two gases are always equal. Also, since H_2 and CO_2 are formed in equal amounts, they are always present in the same concentrations.

Figure 12-2 is a profile of the progress of the reaction. When CO and H_2O are mixed, they immediately begin to react to form H_2 and CO_2. This leads to a decrease in the concentrations of the reactants, but the concentrations of the products, which were initially at zero, are increasing. Beyond a certain time, indicated by the dashed line in Fig. 12-2, the concentrations of reactants and products no longer change—equilibrium has been reached. Unless the system is somehow disturbed, no further changes in concentrations will occur. ◀ ⓘ Note that although the equilibrium position lies far to the right, the concentrations of reactants never go to zero; the reactants will always be present in small but constant concentrations. This is shown on the microscopic level in Fig. 12-3.

What would happen to the gaseous equilibrium mixture of reactants and products represented in Fig. 12-3, parts (c) and (d), if we injected some $H_2O(g)$ into the box? To answer this question, we need to be sure we understand the equilibrium condition: The concentrations of reactants and products remain constant at equilibrium because the forward and reverse reaction rates are equal. If we inject some H_2O molecules, what will happen to the forward reaction: $H_2O + CO \rightarrow H_2 + CO_2$? It will speed

ⓘ In order to appreciate that chemical equilibrium is a dynamic situation, we need to consider what is happening at the molecular level. Figure 12-2 shows us that once equilibrium is reached, the concentrations are constant over time. But if we look at Figure 12-1, we can see that both the forward and reverse reactions occur even after equilibrium is reached.

Figure 12-2 | The changes in concentrations with time for the reaction $H_2O(g) + CO(g) \rightleftharpoons$ $H_2(g) + CO_2(g)$ when equimolar quantities of $H_2O(g)$ and $CO(g)$ are mixed.

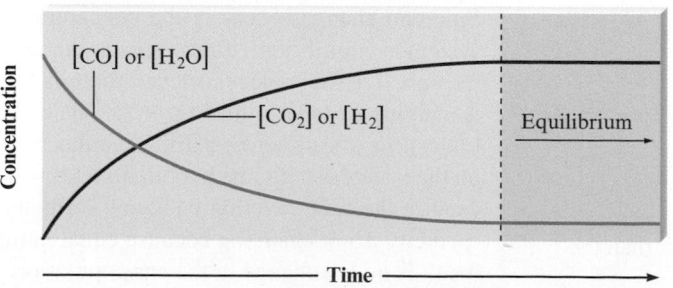

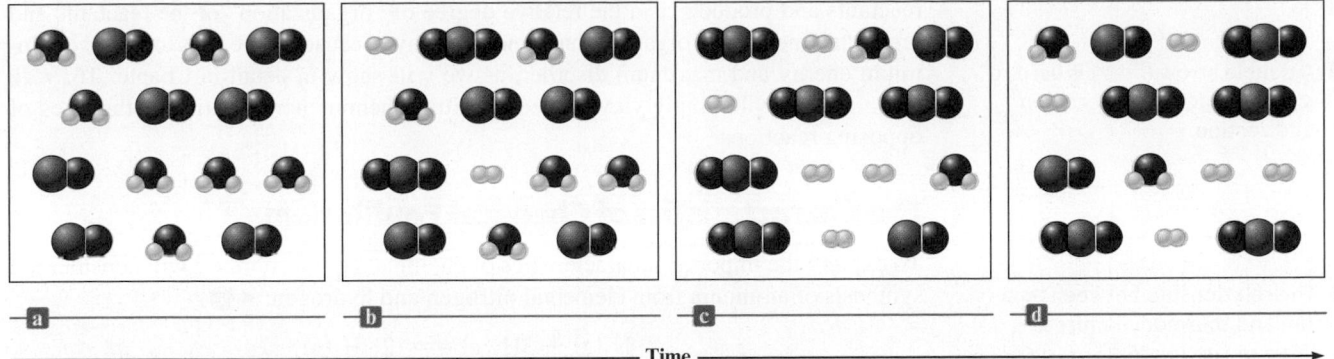

Figure 12-3 | (a) H_2O and CO are mixed in equal numbers and begin to react (b) to form CO_2 and H_2. After time has passed, equilibrium is reached (c) and the numbers of reactant and product molecules then remain constant over time (d).

up because more H_2O molecules means more collisions between H_2O and CO molecules. This in turn will form more products and will cause the reverse reaction $H_2O + CO \leftarrow H_2 + CO_2$ to speed up. Thus the system will change until the forward and reverse reaction rates again become equal. Will this new equilibrium position contain more or fewer product molecules than are shown in Fig. 12-3(c) and (d)? Think about this carefully. If you are not sure of the answer now, keep reading. We will consider this type of situation in more detail later in this chapter.

Why does equilibrium occur? We saw in Chapter 11 that molecules react by colliding with one another, and the more collisions, the faster the reaction. This is why reaction rates depend on concentrations. In this case the concentrations of H_2O and CO are lowered by the forward reaction:

$$H_2O + CO \longrightarrow H_2 + CO_2$$

As the concentrations of the reactants decrease, the forward reaction slows down (Fig. 12-4). As in the bridge traffic analogy, there is also a reverse direction:

$$H_2O + CO \longleftarrow H_2 + CO_2$$

Initially in this experiment no H_2 and CO_2 were present, and this reverse reaction could not occur. However, as the forward reaction proceeds, the concentrations of H_2 and CO_2 build up, and the rate of the reverse reaction increases (Fig. 12-4) as the forward reaction slows down. Eventually, the concentrations reach levels where the rate of the forward reaction equals the rate of the reverse reaction. The system has reached equilibrium.

The equilibrium position of a reaction—left, right, or somewhere in between—is determined by many factors: the initial concentrations, the relative energies of the

Figure 12-4 | The changes with time in the rates of forward and reverse reactions for $H_2O(g) + CO(g) \rightleftharpoons H_2(g) + CO_2(g)$ when equimolar quantities of $H_2O(g)$ and $CO(g)$ are mixed. The rates do not change in the same way with time because the forward reaction has a much larger rate constant than the reverse reaction.

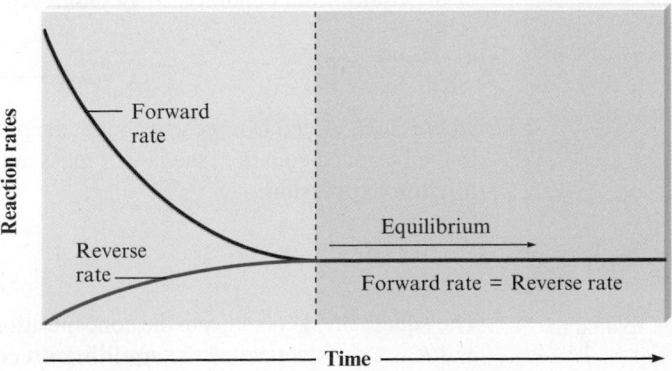

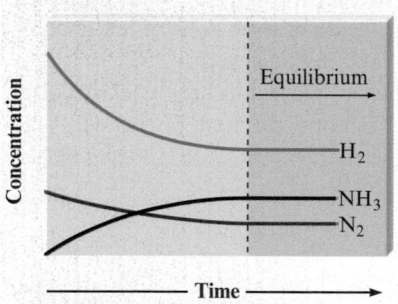

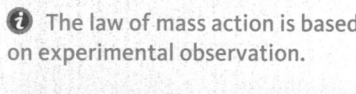

ⓘ A double arrow (⇌) is used to show that a reaction can occur in either direction.

ⓘ The relationship between equilibrium and thermodynamics is explored in Section 16-8.

ⓘ The United States produces about 20 million tons of ammonia annually.

ⓘ Molecules with strong bonds produce large activation energies and tend to react slowly at 25°C.

Figure 12-5 | A concentration profile for the reaction $N_2(g)$ + $3H_2(g) \rightleftharpoons 2NH_3(g)$ when only $N_2(g)$ and $H_2(g)$ are mixed initially.

ⓘ The law of mass action is based on experimental observation.

▲ Anhydrous ammonia is applied to soil to act as a fertilizer.

reactants and products, and the relative degree of "organization" of the reactants and products. Energy and organization come into play because nature tries to achieve minimum energy and maximum disorder, as we will show in detail in Chapter 16. ◄ ⓘ For now, we will simply view the equilibrium phenomenon in terms of the rates of opposing reactions.

The Characteristics of Chemical Equilibrium

To explore the important characteristics of chemical equilibrium, we will consider the synthesis of ammonia from elemental nitrogen and hydrogen: ◄ ⓘ

$$N_2(g) + 3H_2(g) \rightleftharpoons 2NH_3(g)$$

This process is of great commercial value because ammonia is an important fertilizer for the growth of corn and other crops. ◄ ⓘ Ironically, this beneficial process was discovered in Germany just before World War I in a search for ways to produce nitrogen-based explosives. In the course of this work, German chemist Fritz Haber (1868–1934) pioneered the large-scale production of ammonia. ◄

When gaseous nitrogen, hydrogen, and ammonia are mixed in a closed vessel at 25°C, no apparent change in the concentrations occurs over time, regardless of the original amounts of the gases. Why? There are two possible reasons why the concentrations of the reactants and products of a given chemical reaction remain unchanged when mixed.

1. The system is at chemical equilibrium.
2. The forward and reverse reactions are so slow that the system moves toward equilibrium at a rate that cannot be detected.

The second reason applies to the nitrogen, hydrogen, and ammonia mixture at 25°C. As we saw in Chapters 3 and 4, the N_2 molecule has a very strong triple bond (941 kJ/mol) and thus is very unreactive. Also, the H_2 molecule has an unusually strong single bond (432 kJ/mol). Therefore, mixtures of N_2, H_2, and NH_3 at 25°C can exist with no apparent change over long periods of time, unless a catalyst is introduced to speed up the forward and reverse reactions. ◄ ⓘ Under appropriate conditions, the system does reach equilibrium, as shown in Fig. 12-5. Note that because of the reaction stoichiometry, H_2 disappears three times as fast as N_2 does and NH_3 forms twice as fast as N_2 disappears.

12-2 | The Equilibrium Constant

Science is fundamentally empirical—it is based on experiment. The development of the equilibrium concept is typical. From their observations of many chemical reactions, two Norwegian chemists, Cato Maximilian Guldberg (1836–1902) and Peter Waage (1833–1900), proposed in 1864 the **law of mass action** as a general description of the equilibrium condition. ◄ ⓘ Guldberg and Waage postulated that for a reaction of the type

$$jA + kB \rightleftharpoons lC + mD$$

where A, B, C, and D represent chemical species and j, k, l, and m are their coefficients in the balanced equation, the law of mass action is represented by the following **equilibrium expression**:

$$K = \frac{[C]^l[D]^m}{[A]^j[B]^k}$$

The square brackets indicate the concentrations of the chemical species *at equilibrium*, and K is a constant called the **equilibrium constant**.

INTERACTIVE EXAMPLE 12-1

Writing Equilibrium Expressions

Write the equilibrium expression for the following reaction:

$$4NH_3(g) + 7O_2(g) \rightleftharpoons 4NO_2(g) + 6H_2O(g)$$

Solution

Applying the law of mass action gives

$$K = \frac{[NO_2]^4[H_2O]^6}{[NH_3]^4[O_2]^7}$$

Coefficient of NO_2

Coefficient of H_2O

Coefficient of NH_3

Coefficient of O_2

See Exercise 12-21

The value of the equilibrium constant at a given temperature can be calculated if we know the equilibrium concentrations of the reaction components, as illustrated in Example 12-2.

It is very important to note at this point that the equilibrium constants are customarily given without units. The reason for this is beyond the scope of this text, but it involves corrections for the nonideal behavior of the substances taking part in the reaction. When these corrections are made, the units cancel out and the corrected K has no units. Thus we will not use units for K in this text.

INTERACTIVE EXAMPLE 12-2

Calculating the Values of K

The following equilibrium concentrations were observed for the Haber process for synthesis of ammonia at 127°C:

$$[NH_3] = 3.1 \times 10^{-2} \text{ mol/L}$$
$$[N_2] = 8.5 \times 10^{-1} \text{ mol/L}$$
$$[H_2] = 3.1 \times 10^{-3} \text{ mol/L}$$

a. Calculate the value of K at 127°C for this reaction.

b. Calculate the value of the equilibrium constant at 127°C for the reaction

$$2NH_3(g) \rightleftharpoons N_2(g) + 3H_2(g)$$

c. Calculate the value of the equilibrium constant at 127°C for the reaction given by the equation

$$\tfrac{1}{2}N_2(g) + \tfrac{3}{2}H_2(g) \rightleftharpoons NH_3(g)$$

Solution

a. What is the balanced equation for the Haber process?

$$N_2(g) + 3H_2(g) \rightleftharpoons 2NH_3(g)$$

Thus

$$K = \frac{[NH_3]^2}{[N_2][H_2]^3} = \frac{(3.1 \times 10^{-2})^2}{(8.5 \times 10^{-1})(3.1 \times 10^{-3})^3}$$
$$= 3.8 \times 10^4$$

Note that K is written without units.

b. What is the equilibrium expression? This reaction is written in the reverse order from the equation given in part a. This leads to the equilibrium expression

$$K' = \frac{[N_2][H_2]^3}{[NH_3]^2}$$

which is the reciprocal of the expression used in part a. Therefore,

$$K' = \frac{[N_2][H_2]^3}{[NH_3]^2} = \frac{1}{K} = \frac{1}{3.8 \times 10^4} = 2.6 \times 10^{-5}$$

c. What is the equilibrium constant? We use the law of mass action:

$$K'' = \frac{[NH_3]}{[N_2]^{1/2}[H_2]^{3/2}}$$

If we compare this expression to that obtained in part a, we see that since

$$\frac{[NH_3]}{[N_2]^{1/2}[H_2]^{3/2}} = \left(\frac{[NH_3]^2}{[N_2][H_2]^3}\right)^{1/2}$$
$$K'' = K^{1/2}$$

Thus

$$K'' = K^{1/2} = (3.8 \times 10^4)^{1/2} = 1.9 \times 10^2$$

See Exercises 12-23 and 12-25 through 12-27

We can draw some important conclusions from the results of Example 12-2. For a reaction of the form

$$jA + kB \rightleftharpoons lC + mD$$

the equilibrium expression is

$$K = \frac{[C]^l[D]^m}{[A]^j[B]^k}$$

If this reaction is reversed, then the new equilibrium expression is

$$K' = \frac{[A]^j[B]^k}{[C]^l[D]^m} = \frac{1}{K}$$

If the original reaction is multiplied by some factor n to give

$$njA + nkB \rightleftharpoons nlC + nmD$$

the equilibrium expression becomes

$$K'' = \frac{[C]^{nl}[D]^{nm}}{[A]^{nj}[B]^{nk}} = K^n$$

> **LET'S REVIEW** | *Conclusions About the Equilibrium Expression*
>
> - The equilibrium expression for a reaction is the reciprocal of that for the reaction written in reverse.
> - When the balanced equation for a reaction is multiplied by a factor n, the equilibrium expression for the new reaction is the original expression raised to the nth power. Thus $K_{new} = (K_{original})^n$.
> - K values are customarily written without units.

The law of mass action applies to solution and gaseous equilibria.

The law of mass action is widely applicable. ◄ *It* correctly describes the equilibrium behavior of an amazing variety of chemical systems in solution and in the gas phase. Although, as we will see later, corrections must be applied in certain cases, such as for concentrated aqueous solutions and for gases at high pressures, the law of mass action provides a remarkably accurate description of all types of chemical equilibria.

Consider again the ammonia synthesis reaction. The equilibrium constant K always has the same value at a given temperature. At 500°C the value of K is 6.0×10^{-2}. Whenever N_2, H_2, and NH_3 are mixed together at this temperature, the system will always come to an equilibrium position such that

$$\frac{[NH_3]^2}{[N_2][H_2]^3} = 6.0 \times 10^{-2}$$

This expression has the same value at 500°C, *regardless of the amounts of the gases that are mixed together initially.*

Although the special ratio of products to reactants defined by the equilibrium expression is constant for a given reaction system at a given temperature, the *equilibrium concentrations will not always be the same.* Table 12-1 gives three sets of data for the synthesis of ammonia, showing that even though the individual sets of equilibrium concentrations are quite different for the different situations, the *equilibrium constant, which depends on the ratio of the concentrations, remains the same* (within experimental error). Note that subscript zeros indicate initial concentrations.

Each *set of equilibrium concentrations* is called an **equilibrium position**. It is essential to distinguish between the equilibrium constant and the equilibrium positions for a given reaction system. There is only *one* equilibrium constant for a particular system at a particular temperature, but there are an *infinite* number of equilibrium positions. The specific equilibrium position adopted by a system depends on the initial concentrations, but the equilibrium constant does not. ◄

For a reaction at a given temperature, there are many equilibrium positions but only one value for K.

Table 12-1 | Results of Three Experiments for the Reaction $N_2(g) + 3H_2(g) \rightleftharpoons 2NH_3(g)$

Experiment	Initial Concentrations	Equilibrium Concentrations	$K = \dfrac{[NH_3]^2}{[N_2][H_2]^3}$
I	$[N_2]_0 = 1.000\ M$ $[H_2]_0 = 1.000\ M$ $[NH_3]_0 = 0$	$[N_2] = 0.921\ M$ $[H_2] = 0.763\ M$ $[NH_3] = 0.157\ M$	$K = 6.02 \times 10^{-2}$
II	$[N_2]_0 = 0$ $[H_2]_0 = 0$ $[NH_3]_0 = 1.000\ M$	$[N_2] = 0.399\ M$ $[H_2] = 1.197\ M$ $[NH_3] = 0.203\ M$	$K = 6.02 \times 10^{-2}$
III	$[N_2]_0 = 2.00\ M$ $[H_2]_0 = 1.00\ M$ $[NH_3]_0 = 3.00\ M$	$[N_2] = 2.59\ M$ $[H_2] = 2.77\ M$ $[NH_3] = 1.82\ M$	$K = 6.02 \times 10^{-2}$

EXAMPLE 12-3   Equilibrium Positions

The following results were collected for two experiments involving the reaction at 600°C between gaseous sulfur dioxide and oxygen to form gaseous sulfur trioxide:

Experiment 1		Experiment 2	
Initial	Equilibrium	Initial	Equilibrium
$[SO_2]_0 = 2.00\ M$	$[SO_2] = 1.50\ M$	$[SO_2]_0 = 0.500\ M$	$[SO_2] = 0.590\ M$
$[O_2]_0 = 1.50\ M$	$[O_2] = 1.25\ M$	$[O_2]_0 = 0$	$[O_2] = 0.0450\ M$
$[SO_3]_0 = 3.00\ M$	$[SO_3] = 3.50\ M$	$[SO_3]_0 = 0.350\ M$	$[SO_3] = 0.260\ M$

Show that the equilibrium constant is the same in both cases.

Solution

The balanced equation for the reaction is

$$2SO_2(g) + O_2(g) \rightleftharpoons 2SO_3(g)$$

From the law of mass action,

$$K = \frac{[SO_3]^2}{[SO_2]^2[O_2]}$$

For Experiment 1,

$$K_1 = \frac{(3.50)^2}{(1.50)^2(1.25)} = 4.36$$

For Experiment 2,

$$K_2 = \frac{(0.260)^2}{(0.590)^2(0.0450)} = 4.32$$

The value of K is constant, within experimental error.

See Exercise 12-28

12-3 | Equilibrium Expressions Involving Pressures

So far we have been describing equilibria involving gases in terms of concentrations. Equilibria involving gases also can be described in terms of pressures. The relationship between the pressure and the concentration of a gas can be seen from the ideal gas equation:

$$PV = nRT \qquad \text{or} \qquad P = \left(\frac{n}{V}\right)RT = CRT$$

where C equals n/V, or the number of moles n of gas per unit volume V. Thus C represents the *molar concentration of the gas*.

For the ammonia synthesis reaction, the equilibrium expression can be written in terms of concentrations, that is,

$$K = \frac{[NH_3]^2}{[N_2][H_2]^3} = \frac{C_{NH_3}^{\ 2}}{(C_{N_2})(C_{H_2}^{\ 3})} = K_c$$

or in terms of the *equilibrium partial pressures of the gases,* that is,

$$K_p = \frac{P_{NH_3}^2}{(P_{N_2})(P_{H_2}^3)}$$

ⓘ *K involves concentrations; K_p involves pressures. In some books, the symbol K_c is used instead of K.*

Both the symbols K and K_c are used commonly for an equilibrium constant in terms of concentrations. We will always use K in this book. The symbol K_p represents an equilibrium constant in terms of partial pressures. ◄ **ⓘ**

INTERACTIVE EXAMPLE 12-4

Calculating Values of K_p

The reaction for the formation of nitrosyl chloride

$$2NO(g) + Cl_2(g) \rightleftharpoons 2NOCl(g)$$

was studied at 25°C. The pressures at equilibrium were found to be

$$P_{NOCl} = 1.2 \text{ atm}$$
$$P_{NO} = 5.0 \times 10^{-2} \text{ atm}$$
$$P_{Cl_2} = 3.0 \times 10^{-1} \text{ atm}$$

Calculate the value of K_p for this reaction at 25°C.

Solution

For this reaction,

$$K_p = \frac{P_{NOCl}^2}{(P_{NO_2})^2(P_{Cl_2})} = \frac{(1.2)^2}{(5.0 \times 10^{-2})^2(3.0 \times 10^{-1})}$$
$$= 1.9 \times 10^3$$

See Exercises 12-29 and 12-30

The relationship between K and K_p for a particular reaction follows from the fact that for an ideal gas, $C = P/RT$. For example, for the ammonia synthesis reaction,

$$K = \frac{[NH_3]^2}{[N_2][H_2]^3} = \frac{C_{NH_3}^2}{(C_{N_2})(C_{H_2}^3)}$$

$$= \frac{\left(\dfrac{P_{NH_3}}{RT}\right)^2}{\left(\dfrac{P_{N_2}}{RT}\right)\left(\dfrac{P_{H_2}}{RT}\right)^3} = \frac{P_{NH_3}^2}{(P_{N_2})(P_{H_2}^3)} \times \frac{\left(\dfrac{1}{RT}\right)^2}{\left(\dfrac{1}{RT}\right)^4}$$

$$= \frac{P_{NH_3}^2}{(P_{N_2})(P_{H_2}^3)}(RT)^2$$

$$= K_p(RT)^2$$

However, for the synthesis of hydrogen fluoride from its elements,

$$H_2(g) + F_2(g) \rightleftharpoons 2HF(g)$$

the relationship between K and K_p is given by

$$K = \frac{[HF]^2}{[H_2][F_2]} = \frac{C_{HF}^{2}}{(C_{H_2})(C_{F_2})}$$

$$= \frac{\left(\dfrac{P_{HF}}{RT}\right)^2}{\left(\dfrac{P_{H_2}}{RT}\right)\left(\dfrac{P_{F_2}}{RT}\right)} = \frac{P_{HF}^{2}}{(P_{H_2})(P_{F_2})}$$

$$= K_p$$

Thus for this reaction, K is equal to K_p. This equality occurs because the sum of the coefficients on either side of the balanced equation is identical, so the terms in RT cancel out. In the equilibrium expression for the ammonia synthesis reaction, the sum of the powers in the numerator is different from that in the denominator, and K does not equal K_p.

For the general reaction

$$jA + kB \rightleftharpoons lC + mD$$

the relationship between K and K_p is

$$K_p = K(RT)^{\Delta n}$$

where Δn is the sum of the coefficients of the *gaseous* products minus the sum of the coefficients of the *gaseous* reactants. This equation is quite easy to derive from the definitions of K and K_p and the relationship between pressure and concentration. For the preceding general reaction,

$$K_p = \frac{(P_C^{l})(P_D^{m})}{(P_A^{j})(P_B^{k})} = \frac{(C_C \times RT)^{l}(C_D \times RT)^{m}}{(C_A \times RT)^{j}(C_B \times RT)^{k}}$$

$$= \frac{(C_C^{l})(C_D^{m})}{(C_A^{j})(C_B^{k})} \times \frac{(RT)^{l+m}}{(RT)^{j+k}} = K(RT)^{(l+m)-(j+k)}$$

$$= K(RT)^{\Delta n}$$

ⓘ Δn always involves products minus reactants.

where $\Delta n = (l + m) - (j + k)$, the difference in the sums of the coefficients for the gaseous products and reactants. ◂ *ⓘ*

> ## Critical Thinking
>
> The text gives an example reaction for which $K = K_p$. The text states this is true "because the sum of the coefficients on either side of the balanced equation is identical. . . ." What if you are told that for a reaction, $K = K_p$, and the sum of the coefficients on either side of the balanced equation is not equal? How is this possible?

INTERACTIVE EXAMPLE 12-5

Calculating K from K_p

Using the value of K_p obtained in Example 12-4, calculate the value of K at 25°C for the reaction

$$2NO(g) + Cl_2(g) \rightleftharpoons 2NOCl(g)$$

Solution

From the value of K_p, we can calculate K using

$$K_p = K(RT)^{\Delta n}$$

where $T = 25 + 273 = 298$ K and

$$\Delta n = 2 - (2 + 1) = -1$$

↗ Sum of product coefficients   ↖ Sum of reactant coefficients

Δn
↓

Thus

$$K_p = K(RT)^{-1} = \frac{K}{RT}$$

and

$$K = K_p(RT)$$
$$= (1.9 \times 10^3)(0.08206)(298)$$
$$= 4.6 \times 10^4$$

See Exercises 12-31 and 12-32

12-4 | Heterogeneous Equilibria

So far we have discussed equilibria only for systems in the gas phase, where all reactants and products are gases. These are **homogeneous equilibria**. However, many equilibria involve more than one phase and are called **heterogeneous equilibria**. For example, the thermal decomposition of calcium carbonate in the commercial preparation of lime occurs by a reaction involving both solid and gas phases: ◀ *i*

ⓘ Lime is among the top five chemicals manufactured in the United States in terms of the amount produced.

$$CaCO_3(s) \rightleftharpoons CaO(s) + CO_2(g)$$

↑
Lime

Straightforward application of the law of mass action leads to the equilibrium expression

$$K' = \frac{[CO_2][CaO]}{[CaCO_3]}$$

However, experimental results show that the *position of a heterogeneous equilibrium does not depend on the amounts of pure solids or liquids present* (Fig. 12-6).

Figure 12-6 | The position of the equilibrium $CaCO_3(s) \rightleftharpoons CaO(s) + CO_2(g)$ does not depend on the amounts of $CaCO_3(s)$ and $CaO(s)$ present.

CaCO₃ ——   —— CO₂
—— CaO

a   b

The concentrations of pure liquids and solids are constant.

▲ The Seven Sisters chalk cliffs in East Sussex, England. The chalk is made up of compressed calcium carbonate skeletons of microscopic algae from the late Cretaceous Period.

The fundamental reason for this behavior is that the concentrations of pure solids and liquids cannot change. ◄ *i* Thus the equilibrium expression for the decomposition of solid calcium carbonate might be represented as ◄

$$K' = \frac{[CO_2]C_1}{C_2}$$

where C_1 and C_2 are constants representing the concentrations of the solids CaO and $CaCO_3$, respectively. This expression can be rearranged to give

$$\frac{C_2 K'}{C_1} = K = [CO_2]$$

We can generalize from this result as follows: If pure solids or pure liquids are involved in a chemical reaction, their concentrations *are not included in the equilibrium expression* for the reaction. This simplification occurs *only* with pure solids or liquids, not with solutions or gases, since in these last two cases the concentrations can vary.

For example, in the decomposition of liquid water to gaseous hydrogen and oxygen,

$$2H_2O(l) \rightleftharpoons 2H_2(g) + O_2(g)$$

where

$$K = [H_2]^2[O_2] \quad \text{and} \quad K_p = (P_{H_2}{}^2)(P_{O_2})$$

water is not included in either equilibrium expression because it is a pure liquid. However, if the reaction were carried out under conditions where the water is a gas rather than a liquid, that is,

$$2H_2O(g) \rightleftharpoons 2H_2(g) + O_2(g)$$

then

$$K = \frac{[H_2]^2[O_2]}{[H_2O]^2} \quad \text{and} \quad K_p = \frac{(P_{H_2}{}^2)(P_{O_2})}{P_{H_2O}{}^2}$$

because the concentration or pressure of water vapor can change.

INTERACTIVE EXAMPLE 12-6

▲ Water applied to anhydrous copper(II) sulfate forms the dark blue hydrated compound.

Equilibrium Expressions for Heterogeneous Equilibria

Write the expressions for K and K_p for the following processes:

a. Solid phosphorus pentachloride decomposes to liquid phosphorus trichloride and chlorine gas.

b. Deep blue solid copper(II) sulfate pentahydrate is heated to drive off water vapor to form white solid copper(II) sulfate. ◄

Solution

a. What is the balanced equation for the reaction?

$$PCl_5(s) \rightleftharpoons PCl_3(l) + Cl_2(g)$$

What are the equilibrium expressions?

$$K = [Cl_2] \quad \text{and} \quad K_p = P_{Cl_2}$$

In this case neither the pure solid PCl_5 nor the pure liquid PCl_3 is included in the equilibrium expressions.

b. What is the balanced equation for the reaction?

$$CuSO_4 \cdot 5H_2O(s) \rightleftharpoons CuSO_4(s) + 5H_2O(g)$$

What are the equilibrium expressions?

$$K = [H_2O]^5 \quad \text{and} \quad K_p = (P_{H_2O})^5$$

The solids are not included.

See Exercises 12-33 and 12-34

12-5 | Applications of the Equilibrium Constant

Knowing the equilibrium constant for a reaction allows us to predict several important features of the reaction: the tendency of the reaction to occur (but not the speed of the reaction), whether a given set of concentrations represents an equilibrium condition, and the equilibrium position that will be achieved from a given set of initial concentrations.

To introduce some of these ideas, we will first consider the reaction

where ○ and ● represent two different types of atoms. Assume that this reaction has an equilibrium constant equal to 16.

In a given experiment, the two types of molecules are mixed together in the following amounts:

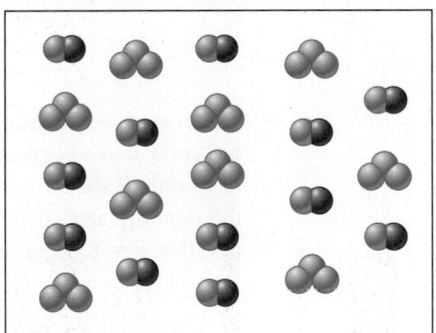

After the system reacts and comes to equilibrium, what will the system look like? We know that at equilibrium the ratio

$$\frac{(N_{\text{○●}})(N_{\text{○○○}})}{(N_{\text{○○○}})(N_{\text{●●}})} = 16$$

must be satisfied, where each N represents the number of molecules of each type. We originally have 9 ○○○ molecules and 12 ●● molecules. As a place to start, let's just assume that 5 ○○○ molecules disappear for the system to reach equilibrium. Since equal numbers of the ○○○ and ●● molecules react, this means that 5 ●● molecules also will disappear. This also means that 5 ●● molecules and 5 ○○○ molecules will be formed. We can summarize as follows:

Initial Conditions	New Conditions
9 ○○○ molecules	9 − 5 = 4 ○○○ molecules
12 ●● molecules	12 − 5 = 7 ●● molecules
0 ○○○ molecules	0 + 5 = 5 ○○○ molecules
0 ●● molecules	0 + 5 = 5 ●● molecules

Do the new conditions represent equilibrium for this reaction system? We can find out by taking the ratio of the numbers of molecules:

$$\frac{(N_{\text{⬤⬤}})(N_{\text{⬤⬤⬤}})}{(N_{\text{⬤⬤⬤}})(N_{\text{⬤⬤}})} = \frac{(5)(5)}{(4)(7)} = 0.9$$

Thus this is not an equilibrium position because the ratio is not 16, as required for equilibrium. In which direction must the system move to achieve equilibrium? Since the observed ratio is smaller than 16, we must increase the numerator and decrease the denominator: The system needs to move to the right (toward more products) to achieve equilibrium. That is, more than 5 of the original reactant molecules must disappear to reach equilibrium for this system. How can we find the correct number? Since we do not know the number of molecules that need to disappear to reach equilibrium, let's call this number x. Now we can set up a table similar to the one we used earlier:

Initial Conditions		Equilibrium Conditions
9 ⬤⬤ molecules	x ⬤⬤ disappear	$9 - x$ ⬤⬤ molecules
12 ⬤⬤ molecules	x ⬤⬤ disappear	$12 - x$ ⬤⬤ molecules
0 ⬤⬤ molecules	x ⬤⬤ form	x ⬤⬤ molecules
0 ⬤⬤ molecules	x ⬤⬤ form	x ⬤⬤ molecules

For the system to be at equilibrium, we know that the following ratio must be satisfied:

$$\frac{(N_{\text{⬤⬤}})(N_{\text{⬤⬤⬤}})}{(N_{\text{⬤⬤⬤}})(N_{\text{⬤⬤}})} = 16 = \frac{(x)(x)}{(9 - x)(12 - x)}$$

The easiest way to solve for x here is by trial and error. From our previous discussion, we know that x is greater than 5. Also, we know that it must be less than 9 because we have only 9 ⬤⬤ molecules to start. We can't use all of them or we will have a zero in the denominator, which causes the ratio to be infinitely large. By trial and error, we find that $x = 8$ because

$$\frac{(x)(x)}{(9 - x)(12 - x)} = \frac{(8)(8)}{(9 - 8)(12 - 8)} = \frac{64}{4} = 16$$

The equilibrium mixture can be pictured as follows:

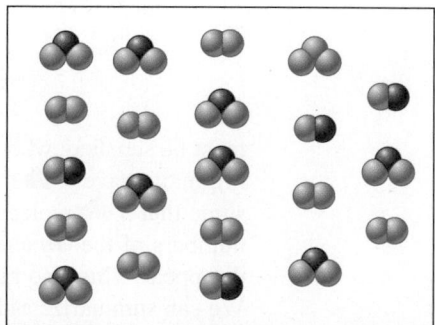

Note that it contains 8 ⬤⬤ molecules, 8 ⬤⬤ molecules, 1 ⬤⬤ molecule, and 4 ⬤⬤ molecules as required. ◀ ⓘ

This pictorial example should help you understand the fundamental ideas of equilibrium. Now we will proceed to a more systematic, quantitative treatment of chemical equilibrium.

ⓘ Note that if we started with 10 molecules of each reactant, the equilibrium mixture would include 8 molecules of each product and 2 molecules of each reactant. The value for the equilibrium constant would be the same (16), but the equilibrium condition would be different.

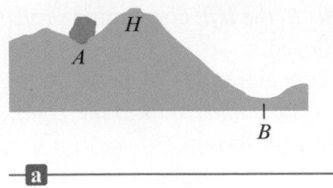

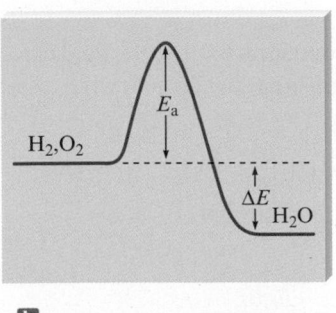

Figure 12-7 | (a) A physical analogy illustrating the difference between thermodynamic and kinetic stabilities. The boulder is thermo-dynamically more stable (lower potential energy) in position *B* than in position *A* but cannot get over the hump *H*. (b) The reactants H_2 and O_2 have a strong tendency to form H_2O. That is, H_2O has lower energy than H_2 and O_2. However, the large activation energy E_a prevents the reaction at 25°C. In other words, the magnitude of *K* for the reaction depends on ΔE, but the reaction rate depends on E_a.

The Extent of a Reaction

The inherent tendency for a reaction to occur is indicated by the magnitude of the equilibrium constant. A value of *K* much larger than 1 means that at equilibrium the reaction system will consist of mostly products—the equilibrium lies to the right. Another way of saying this is that reactions with very large equilibrium constants go essentially to completion. On the other hand, a very small value of *K* means that the system at equilibrium will consist of mostly reactants—the equilibrium position is far to the left. The given reaction does not occur to any significant extent.

It is important to understand that *the size of K and the time required to reach equilibrium are not directly related.* The time required to achieve equilibrium depends on the reaction rate, which is determined by the size of the activation energy. The size of *K* is determined by thermodynamic factors such as the difference in energy between products and reactants. This difference is represented in Fig. 12-7 and will be discussed in detail in Chapter 15.

Reaction Quotient

When the reactants and products of a given chemical reaction are mixed, it is useful to know whether the mixture is at equilibrium or, if not, the direction in which the system must shift to reach equilibrium. If the concentration of one of the reactants or products is zero, the system will shift in the direction that produces the missing component. However, if all the initial concentrations are nonzero, it is more difficult to determine the direction of the move toward equilibrium. To determine the shift in such cases, we use the **reaction quotient**, *Q*. The reaction quotient is obtained by applying the law of mass action using *initial concentrations* instead of equilibrium concentrations. For example, for the synthesis of ammonia

$$N_2(g) + 3H_2(g) \rightleftharpoons 2NH_3(g)$$

the expression for the reaction quotient is

$$Q = \frac{[NH_3]_0^2}{[N_2]_0[H_2]_0^3}$$

where the subscript zeros indicate initial concentrations.

To determine in which direction a system will shift to reach equilibrium, we compare the values of *Q* and *K*. There are three possible cases (Fig. 12-8):

1. *Q is equal to K.* The system is at equilibrium; no shift will occur.
2. *Q is greater than K.* In this case, the ratio of initial concentrations of products to initial concentrations of reactants is too large. To reach equilibrium, a net change

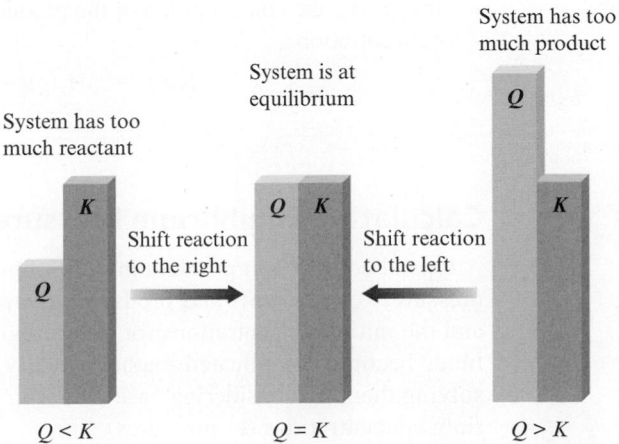

Figure 12-8 | The relationship between reaction quotient *Q* and the equilibrium constant *K*.

of products to reactants must occur. The system *shifts to the left*, consuming products and forming reactants, until equilibrium is achieved.

3. *Q is less than K.* In this case, the ratio of initial concentrations of products to initial concentrations of reactants is too small. The *system must shift to the right*, consuming reactants and forming products, to attain equilibrium.

INTERACTIVE EXAMPLE 12-7

Using the Reaction Quotient

For the synthesis of ammonia at 500°C, the equilibrium constant is 6.0×10^{-2}. Predict the direction in which the system will shift to reach equilibrium in each of the following cases:

a. $[NH_3]_0 = 1.0 \times 10^{-3} M$; $[N_2]_0 = 1.0 \times 10^{-5} M$; $[H_2]_0 = 2.0 \times 10^{-3} M$
b. $[NH_3]_0 = 2.00 \times 10^{-4} M$; $[N_2]_0 = 1.50 \times 10^{-5} M$; $[H_2]_0 = 3.54 \times 10^{-1} M$
c. $[NH_3]_0 = 1.0 \times 10^{-4} M$; $[N_2]_0 = 5.0 M$; $[H_2]_0 = 1.0 \times 10^{-2} M$

Solution

a. What is the value of Q?

$$Q = \frac{[NH_3]_0^2}{[N_2]_0[H_2]_0^3} = \frac{(1.0 \times 10^{-3})^2}{(1.0 \times 10^{-5})(2.0 \times 10^{-3})^3}$$
$$= 1.3 \times 10^7$$

Since $K = 6.0 \times 10^{-2}$, Q is much greater than K. To attain equilibrium, the concentrations of the products must be decreased and the concentrations of the reactants increased. The system will shift to the left:

$$N_2(g) + 3H_2(g) \longleftarrow 2NH_3(g)$$

b. What is the value of Q?

$$Q = \frac{[NH_3]_0^2}{[N_2]_0[H_2]_0^3} = \frac{(2.00 \times 10^{-4})^2}{(1.50 \times 10^{-5})(3.54 \times 10^{-1})^3}$$
$$= 6.01 \times 10^{-2}$$

In this case $Q = K$, so the system is at equilibrium. No shift will occur.

c. What is the value of Q?

$$Q = \frac{[NH_3]_0^2}{[N_2]_0[H_2]_0^3} = \frac{(1.0 \times 10^{-4})^2}{(5.0)(1.0 \times 10^{-2})^3}$$
$$= 2.0 \times 10^{-3}$$

Here Q is less than K, so the system will shift to the right to attain equilibrium by increasing the concentration of the product and decreasing the reactant concentrations:

$$N_2(g) + 3H_2(g) \longrightarrow 2NH_3(g)$$

See Exercises 12-39 through 12-42

Calculating Equilibrium Pressures and Concentrations

A typical equilibrium problem involves finding the equilibrium concentrations (or pressures) of reactants and products, given the value of the equilibrium constant and the initial concentrations (or pressures). However, since such problems sometimes become complicated mathematically, we will develop useful strategies for solving them by considering cases for which we know one or more of the equilibrium concentrations (or pressures).

INTERACTIVE EXAMPLE 12-8

NASA

▲ *Apollo II* lunar landing module at Tranquility Base, 1969.

Calculating Equilibrium Pressures I

Dinitrogen tetroxide in its liquid state was used as one of the fuels on the lunar lander for the NASA Apollo missions. ◄ In the gas phase, it decomposes to gaseous nitrogen dioxide:

$$N_2O_4(g) \rightleftharpoons 2NO_2(g)$$

Consider an experiment in which gaseous N_2O_4 was placed in a flask and allowed to reach equilibrium at a temperature where $K_p = 0.133$. At equilibrium, the pressure of N_2O_4 was found to be 2.71 atm. Calculate the equilibrium pressure of $NO_2(g)$.

Solution

We know that the equilibrium pressures of the gases NO_2 and N_2O_4 must satisfy the relationship

$$K_p = \frac{P_{NO_2}^{\,2}}{P_{N_2O_4}} = 0.133$$

Since we know $P_{N_2O_4}$, we can simply solve for P_{NO_2}:

$$P_{NO_2}^{\,2} = K_p(P_{N_2O_4}) = (0.133)(2.71) = 0.360$$

Therefore,

$$P_{NO_2} = \sqrt{0.360} = 0.600$$

See Exercises 12-43 and 12-44

INTERACTIVE EXAMPLE 12-9

Calculating Equilibrium Pressures II

At a certain temperature, a 1.00-L flask initially contained 0.298 mole of $PCl_3(g)$ and 8.70×10^{-3} mole of $PCl_5(g)$. After the system had reached equilibrium, 2.00×10^{-3} mole of $Cl_2(g)$ was found in the flask. Gaseous PCl_5 decomposes according to the reaction

$$PCl_5(g) \rightleftharpoons PCl_3(g) + Cl_2(g)$$

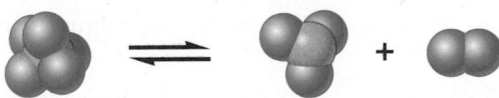

Calculate the equilibrium concentrations of all species and the value of K.

Solution

What is the equilibrium expression for this reaction?

$$K = \frac{[Cl_2][PCl_3]}{[PCl_5]}$$

To find the value of K, we must calculate the equilibrium concentrations of all species and then substitute these quantities into the equilibrium expression. The best method for finding the equilibrium concentrations is to begin with the initial concentrations, which we will define as the concentrations before any shift toward equilibrium has occurred. We will then modify these initial concentrations appropriately to find the equilibrium concentrations.

What are the initial concentrations?

$$[Cl_2]_0 = 0$$

$$[PCl_3]_0 = \frac{0.298 \text{ mol}}{1.00 \text{ L}} = 0.298 \ M$$

$$[PCl_5]_0 = \frac{8.70 \times 10^{-3} \text{ mol}}{1.00 \text{ L}} = 8.70 \times 10^{-3} \ M$$

What change is required to reach equilibrium?
Since no Cl_2 was initially present but $2.00 \times 10^{-3} \ M$ Cl_2 is present at equilibrium, 2.00×10^{-3} mole of PCl_5 must have decomposed to form 2.00×10^{-3} mole of Cl_2 and 2.00×10^{-3} mole of PCl_3. In other words, to reach equilibrium, the reaction shifted to the right:

$$PCl_5(g) \longrightarrow PCl_3(g) + Cl_2(g)$$

$$2.00 \times 10^{-3} \text{ mol} \longrightarrow 2.00 \times 10^{-3} \text{ mol} + 2.00 \times 10^{-3} \text{ mol}$$

Net amount of PCl₅   Net amounts of
decomposed   products formed

Now we apply this change to the initial concentrations.

What are the equilibrium concentrations?

- $$[Cl_2] = 0 + \frac{2.00 \times 10^{-3} \text{ mol}}{1.00 \text{ L}} = 2.00 \times 10^{-3} \ M$$
  $[Cl_2]_0$

- $$[PCl_3] = 0.298 \ M + \frac{2.00 \times 10^{-3} \text{ mol}}{1.00 \text{ L}} = 0.300 \ M$$
  $[PCl_3]_0$

- $$[PCl_5] = 8.70 \times 10^{-3} \ M - \frac{2.00 \times 10^{-3} \text{ mol}}{1.00 \text{ L}} = 6.70 \times 10^{-3} \ M$$
  $[PCl_5]_0$

What is the value of K?

- The equilibrium concentrations are substituted into the equilibrium expression:

$$K = \frac{[Cl_2][PCl_3]}{[PCl_5]} = \frac{(2.00 \times 10^{-3})(0.300)}{6.70 \times 10^{-3}}$$

$$= 8.96 \times 10^{-2}$$

See Exercises 12-45 through 12-48

Sometimes we are not given any of the equilibrium concentrations (or pressures), only the initial values. Then we must use the stoichiometry of the reaction to express concentrations (or pressures) at equilibrium in terms of the initial values. This is illustrated in Example 12-10.

INTERACTIVE EXAMPLE 12-10

Calculating Equilibrium Concentrations I

Carbon monoxide reacts with steam to produce carbon dioxide and hydrogen. At 700 K the equilibrium constant is 5.10. Calculate the equilibrium concentrations of all species if 1.000 mol of each component is mixed in a 1.000-L flask.

Solution

What is the balanced equation for the reaction?

$$CO(g) + H_2O(g) \rightleftharpoons CO_2(g) + H_2(g) \blacktriangleleft$$

What is the equilibrium expression?

$$K = \frac{[CO_2][H_2]}{[CO][H_2O]} = 5.10$$

What are the initial concentrations?

$$[CO]_0 = [H_2O]_0 = [CO_2]_0 = [H_2]_0 = \frac{1.000 \text{ mol}}{1.000 \text{ L}} = 1.000 \ M$$

Is the system at equilibrium, and if not, which way will it shift to reach the equilibrium position? These questions can be answered by calculating Q:

$$Q = \frac{[CO_2]_0[H_2]_0}{[CO]_0[H_2O]_0} = \frac{(1.000 \text{ mol/L})(1.000 \text{ mol/L})}{(1.000 \text{ mol/L})(1.000 \text{ mol/L})} = 1.000$$

Since Q is less than K, the system is not at equilibrium initially but must shift to the right.

What are the equilibrium concentrations?
As before, we start with the initial concentrations and modify them to obtain the equilibrium concentrations. We must ask this question: How much will the system shift to the right to attain the equilibrium condition? In Example 12-9 the change needed for the system to reach equilibrium was given. However, in this case we do not have this information.

Since the required change in concentrations is unknown at this point, we will define it in terms of x. Let's assume that x mol/L CO must react for the system to reach equilibrium. This means that the initial concentration of CO will decrease by x mol/L:

$$[CO] = [CO]_0 - x$$

    ↑   ↑   ↑
Equilibrium Initial Change

Since each CO molecule reacts with one H_2O molecule, the concentration of water vapor also must decrease by x mol/L:

$$[H_2O] = [H_2O]_0 - x$$

As the reactant concentrations decrease, the product concentrations increase. Since all the coefficients are 1 in the balanced reaction, 1 mole of CO_2 reacting with 1 mole of H_2O will produce 1 mole of CO_2 and 1 mole of H_2. Or in the present case, to reach equilibrium, x mol/L CO will react with x mol/L H_2O to give an additional x mol/L CO_2 and x mol/L H_2:

$$xCO + xH_2O \longrightarrow xCO_2 + xH_2$$

Thus the initial concentrations of CO_2 and H_2 will increase by x mol/L:

$$[CO_2] = [CO_2]_0 + x$$
$$[H_2] = [H_2]_0 + x$$

Now we have all the equilibrium concentrations defined in terms of the initial concentrations and the change x:

Initial Concentration (mol/L)	Change (mol/L)	Equilibrium Concentration (mol/L)
$[CO]_0 = 1.000$	$-x$	$1.000 - x$
$[H_2O]_0 = 1.000$	$-x$	$1.000 - x$
$[CO_2]_0 = 1.000$	$+x$	$1.000 + x$
$[H_2]_0 = 1.000$	$+x$	$1.000 + x$

Note that the sign of x is determined by the direction of the shift. In this example, the system shifts to the right, so the product concentrations increase and the reactant concentrations decrease. Also note that because the coefficients in the balanced equation are all 1, the magnitude of the change is the same for all species.

Now since we know that the equilibrium concentrations must satisfy the equilibrium expression, we can find the value of x by substituting these concentrations into the expression

$$K = 5.10 = \frac{[CO_2][H_2]}{[CO][H_2O]} = \frac{(1.000 + x)(1.000 + x)}{(1.000 - x)(1.000 - x)} = \frac{(1.000 + x)^2}{(1.000 - x)^2}$$

Since the right side of the equation is a perfect square, the solution of the problem can be simplified by taking the square root of both sides:

$$\sqrt{5.10} = 2.26 = \frac{1.000 + x}{1.000 - x}$$

Multiplying and collecting terms gives

$$x = 0.387 \text{ mol/L}$$

Thus the system shifts to the right, consuming 0.387 mol/L CO and 0.387 mol/L H_2O and forming 0.387 mol/L CO_2 and 0.387 mol/L H_2.

Now the equilibrium concentrations can be calculated:

- $[CO] = [H_2O] = 1.000 - x = 1.000 - 0.387 = 0.613 \, M$
- $[CO_2] = [H_2] = 1.000 + x = 1.000 + 0.387 = 1.387 \, M$

REALITY CHECK | These values can be checked by substituting them back into the equilibrium expression to make sure they give the correct value for K:

$$K = \frac{[CO_2][H_2]}{[CO][H_2O]} = \frac{(1.387)^2}{(0.613)^2} = 5.12$$

This result is the same as the given value of K (5.10) within round-off error, so the answer must be correct.

See Exercise 12-51

INTERACTIVE EXAMPLE 12-11

Calculating Equilibrium Concentrations II

Assume that the reaction for the formation of gaseous hydrogen fluoride from hydrogen and fluorine has an equilibrium constant of 1.15×10^2 at a certain temperature. In a particular experiment, 3.000 moles of each component were added to a 1.500-L flask. Calculate the equilibrium concentrations of all species.

Solution

What is the balanced equation for the reaction?

$$H_2(g) + F_2(g) \rightleftharpoons 2HF(g)$$

What is the equilibrium expression?

$$K = 1.15 \times 10^2 - \frac{[HF]^2}{[H_2][F_2]}$$

What are the initial concentrations?

$$[HF]_0 = [H_2]_0 = [F_2]_0 = \frac{3.000 \text{ mol}}{1.500 \text{ L}} = 2.000 \, M$$

What is the value of Q?

$$Q = \frac{[HF]_0^2}{[H_2]_0[F_2]_0} = \frac{(2.000)^2}{(2.000)(2.000)} = 1.000$$

Since Q is much less than K, the system must shift to the right to reach equilibrium.

What change in the concentrations is necessary?

Since this is presently unknown, we will define the change needed in terms of x. Let x equal the number of moles per liter of H_2 consumed to reach equilibrium. The stoichiometry of the reaction shows that x mol/L F_2 also will be consumed and $2x$ mol/L HF will be formed:

$$H_2(g) + F_2(g) \longrightarrow 2HF(g)$$
$$x \text{ mol/L} + x \text{ mol/L} \longrightarrow 2x \text{ mol/L}$$

Now the equilibrium concentrations can be expressed in terms of x:

Initial Concentration (mol/L)	Change (mol/L)	Equilibrium Concentration (mol/L)
$[H_2]_0 = 2.000$	$-x$	$[H_2] = 2.000 - x$
$[F_2]_0 = 2.000$	$-x$	$[F_2] = 2.000 - x$
$[HF]_0 = 2.000$	$+2x$	$[HF] = 2.000 + 2x$

These concentrations can be represented in a shorthand table as follows: ◀

	$H_2(g)$	$+$	$F_2(g)$	$\rightleftharpoons$	$2HF(g)$
Initial	2.000		2.000		2.000
Change	$-x$		$-x$		$+2x$
Equilibrium	$2.000 - x$		$2.000 - x$		$2.000 + 2x$

What is the value of x?

To solve for x, we substitute the equilibrium concentrations into the equilibrium expression:

$$K = 1.15 \times 10^2 = \frac{[HF]^2}{[H_2][F_2]} = \frac{(2.000 + 2x)^2}{(2.000 - x)^2}$$

The right side of this equation is a perfect square, so taking the square root of both sides gives

$$\sqrt{1.15 \times 10^2} = \frac{2.000 + 2x}{2.000 - x}$$

which yields $x = 1.528$.

We often refer to this form as an **ICE** table (indicated by the first letters of Initial, Change, and Equilibrium).

What are the equilibrium concentrations?

- $[H_2] = [F_2] = 2.000\ M - x = 0.472\ M$
- $[HF] = 2.000\ M + 2x = 5.056\ M$

REALITY CHECK | Checking these values by substituting them into the equilibrium expression gives

$$\frac{[HF]^2}{[H_2][F_2]} = \frac{(5.056)^2}{(0.472)^2} = 1.15 \times 10^2$$

which agrees with the given value of K.

See Exercise 12-52

12-6 | Solving Equilibrium Problems

We have already considered most of the strategies needed to solve equilibrium problems. The typical procedure for analyzing a chemical equilibrium problem can be summarized as follows:

PROBLEM-SOLVING STRATEGY

Solving Equilibrium Problems

1. Write the balanced equation for the reaction.
2. Write the equilibrium expression using the law of mass action.
3. List the initial concentrations.
4. Calculate Q, and determine the direction of the shift to equilibrium.
5. Define the change needed to reach equilibrium, and define the equilibrium concentrations by applying the change to the initial concentrations.
6. Substitute the equilibrium concentrations into the equilibrium expression, and solve for the unknown.
7. Check your calculated equilibrium concentrations by making sure they give the correct value of K.

So far we have been careful to choose systems in which we can solve for the unknown by taking the square root of both sides of the equation. However, this type of system is not really very common, and we must now consider a more typical problem. Suppose for a synthesis of hydrogen fluoride from hydrogen and fluorine, 3.000 moles of H_2 and 6.000 moles of F_2 are mixed in a 3.000-L flask. Assume that the equilibrium constant for the synthesis reaction at this temperature is 1.15×10^2. We calculate the equilibrium concentration of each component as follows:

1. *What is the balanced equation for the reaction?*

$$H_2(g) + F_2(g) \rightleftharpoons 2HF(g)$$

2. *What is the equilibrium expression?*

$$K = 1.15 \times 10^2 = \frac{[HF]^2}{[H_2][F_2]}$$

3. *What are the initial concentrations?*

$$[H_2]_0 = \frac{3.000 \text{ mol}}{3.000 \text{ L}} = 1.000 \ M$$

$$[F_2]_0 = \frac{6.000 \text{ mol}}{3.000 \text{ L}} = 2.000 \ M$$

$$[HF]_0 = 0$$

4. *What is Q?*
    There is no need to calculate Q because no HF is present initially, and we know that the system must shift to the right to reach equilibrium.

5. *What change is required to reach equilibrium?*
    If we let x represent the number of moles per liter of H_2 consumed to reach equilibrium, we can represent the equilibrium concentrations as follows:

	$H_2(g)$	+	$F_2(g)$	⇌	$2HF(g)$
Initial	1.000		2.000		0
Change	$-x$		$-x$		$+2x$
Equilibrium	$1.000 - x$		$2.000 - x$		$2x$

6. *What is the value of K?*
    Substituting the equilibrium concentrations into the equilibrium expression gives

    $$K = 1.15 \times 10^2 = \frac{[HF]^2}{[H_2][F_2]} = \frac{(2x)^2}{(1.000 - x)(2.000 - x)}$$

    Since the right side of this equation is not a perfect square, we cannot take the square root of both sides, but must use some other procedure.
    First, do the indicated multiplication:

    $$(1.000 - x)(2.000 - x)(1.15 \times 10^2) = (2x)^2$$

    or $(1.15 \times 10^2)x^2 - 3.000(1.15 \times 10^2)x + 2.000(1.15 \times 10^2) = 4x^2$
    and collect terms

    $$(1.11 \times 10^2)x^2 - (3.45 \times 10^2)x + 2.30 \times 10^2 = 0$$

    This is a quadratic equation of the general form

    $$ax^2 + bx + c = 0$$

Use of the quadratic formula is explained in Appendix 1.4.

    where the roots can be obtained from the quadratic formula: ◀ ⓘ

    $$x = \frac{-b \pm \sqrt{b^2 - 4ac}}{2a}$$

    In this example, $a = 1.11 \times 10^2$, $b = -3.45 \times 10^2$, and $c = 2.30 \times 10^2$. Substituting these values into the quadratic formula gives two values for x:

    $$x = 2.14 \text{ mol/L} \quad \text{and} \quad x = 0.968 \text{ mol/L}$$

    Both of these results cannot be valid (since a *given* set of initial concentrations leads to only *one* equilibrium position). How can we choose between them? Since the expression for the equilibrium concentration of H_2 is

    $$[H_2] = 1.000 \ M - x$$

    the value of x cannot be 2.14 mol/L (because subtracting 2.14 M from 1.000 M gives a negative concentration of H_2, which is physically impossible). Thus the correct value for x is 0.968 mol/L, and the equilibrium concentrations are as follows:

- $[H_2] = 1.000\ M - 0.968\ M = 3.2 \times 10^{-2}\ M$
- $[F_2] = 2.000\ M - 0.968\ M = 1.032\ M$
- $[HF] = 2(0.968\ M) = 1.936\ M$

REALITY CHECK

7. *We can check these concentrations by substituting them into the equilibrium expression:*

$$\frac{[HF]^2}{[H_2][F_2]} = \frac{(1.936)^2}{(3.2 \times 10^{-2})(1.032)} = 1.13 \times 10^2$$

This value is in close agreement with the given value for K (1.15×10^2), so the calculated equilibrium concentrations are correct.

This procedure is further illustrated for a problem involving pressures in Example 12-12.

INTERACTIVE EXAMPLE 12-12

Calculating Equilibrium Pressures

Assume that gaseous hydrogen iodide is synthesized from hydrogen gas and iodine vapor at a temperature where the equilibrium constant is 1.00×10^2. Suppose HI at 5.000×10^{-1} atm, H_2 at 1.000×10^{-2} atm, and I_2 at 5.000×10^{-3} atm are mixed in a 5.000-L flask. Calculate the equilibrium pressures of all species.

Solution

1. *What is the balanced equation for this process?*

$$H_2(g) + I_2(g) \rightleftharpoons 2HI(g)$$

2. *What is the equilibrium expression in terms of pressure?*

$$K_p = \frac{P_{HI}^2}{(P_{H_2})(P_{I_2})} = 1.00 \times 10^2$$

3. *What are the given initial pressures?*

$$P_{HI}^0 = 5.000 \times 10^{-1}\ \text{atm}$$
$$P_{H_2}^0 = 1.000 \times 10^{-2}\ \text{atm}$$
$$P_{I_2}^0 = 5.000 \times 10^{-3}\ \text{atm}$$

4. *What is the value of Q for this system?*

$$Q = \frac{(P_{HI}^0)^2}{(P_{H_2}^0)(P_{I_2}^0)} = \frac{(5.000 \times 10^{-1}\ \text{atm})^2}{(1.000 \times 10^{-2}\ \text{atm})(5.000 \times 10^{-3}\ \text{atm})} = 5.000 \times 10^3$$

Since Q is greater than K, the system will shift to the left to reach equilibrium.

So far we have used moles or concentrations in stoichiometric calculations. However, it is equally valid to use pressures for a gas-phase system at constant temperature and volume because in this case pressure is directly proportional to the number of moles:

$$P = n\left(\frac{RT}{V}\right) \quad \longleftarrow \text{Constant if constant } T \text{ and } V$$

Thus we can represent the change needed to achieve equilibrium in terms of pressures.

5. *What change is required to reach equilibrium?*
   Let x be the change in pressure (in atm) of H_2 as the system shifts left toward equilibrium. This leads to the following equilibrium pressures:

	$H_2(g)$	$+$	$I_2(g)$	$\rightleftharpoons$	$2HI(g)$
Initial	1.000×10^{-2}		5.000×10^{-3}		5.000×10^{-1}
Change	$\mid x$		$+x$		$-2x$
Equilibrium	$1.000 \times 10^{-2} + x$		$5.000 \times 10^{-3} + x$		$5.000 \times 10^{-1} - 2x$

6. *What is the value of K_p?*
   Substitution into the equilibrium expression gives

$$K_p = \frac{(P_{HI})^2}{(P_{H_2})(P_{I_2})} = \frac{(5.000 \times 10^{-1} - 2x)^2}{(1.000 \times 10^{-2} + x)(5.000 \times 10^{-3} + x)}$$

   Multiplying and collecting terms yield the quadratic equation where $a = 9.60 \times 10^1$, $b = 3.5$, and $c = -2.45 \times 10^{-1}$:

$$(9.60 \times 10^1)x^2 + 3.5x - (2.45 \times 10^{-1}) = 0$$

   From the quadratic formula, the correct value for x is $x = 3.55 \times 10^{-2}$ atm.

7. *What are the equilibrium pressures?*
   The equilibrium pressures can now be calculated from the expressions involving x:

   - $P_{HI} = 5.000 \times 10^{-1}$ atm $- 2(3.55 \times 10^{-2})$ atm $= 4.29 \times 10^{-1}$ atm
   - $P_{H_2} = 1.000 \times 10^{-2}$ atm $+ 3.55 \times 10^{-2}$ atm $= 4.55 \times 10^{-2}$ atm
   - $P_{I_2} = 5.000 \times 10^{-3}$ atm $+ 3.55 \times 10^{-2}$ atm $= 4.05 \times 10^{-2}$ atm

REALITY CHECK

$$\frac{P_{HI}^2}{P_{H_2} \cdot P_{I_2}} = \frac{(4.29 \times 10^{-1})^2}{(4.55 \times 10^{-2})(4.05 \times 10^{-2})} = 99.9$$

This agrees with the given value of K (1.00×10^2), so the calculated equilibrium concentrations are correct.

See Exercises 12-53 through 12-56

Treating Systems That Have Small Equilibrium Constants

We have seen that fairly complicated calculations are often necessary to solve equilibrium problems. However, under certain conditions, simplifications are possible that greatly reduce the mathematical difficulties. For example, gaseous NOCl decomposes to form the gases NO and Cl_2. At 35°C the equilibrium constant is 1.6×10^{-5}. In an experiment in which 1.0 mole of NOCl is placed in a 2.0-L flask, what are the equilibrium concentrations?
   The balanced equation is

$$2NOCl(g) \rightleftharpoons 2NO(g) + Cl_2(g)$$

and
$$K = \frac{[NO]^2[Cl_2]}{[NOCl]^2} = 1.6 \times 10^{-5}$$

The initial concentrations are

$$[NOCl]_0 = \frac{1.0 \text{ mol}}{2.0 \text{ L}} = 0.50 \ M \qquad [NO]_0 = 0 \qquad [Cl_2]_0 = 0$$

Since there are no products initially, the system will move to the right to reach equilibrium. We will define x as the change in concentration of Cl_2 needed to reach equilibrium. The changes in the concentrations of NOCl and NO can then be obtained from the balanced equation:

$$2NOCl(g) \longrightarrow 2NO(g) + Cl_2(g)$$
$$2x \longrightarrow 2x + x$$

The concentrations can be summarized as follows:

	2NOCl(g)	$\rightleftharpoons$	2NO(g)	+	Cl$_2$(g)
Initial	0.50		0		0
Change	−2x		+2x		+x
Equilibrium	0.50 − 2x		2x		x

The equilibrium concentrations must satisfy the equilibrium expression

$$K = 1.6 \times 10^{-5} = \frac{[NO]^2[Cl_2]}{[NOCl]^2} = \frac{(2x)^2(x)}{(0.50 - 2x)^2}$$

Multiplying and collecting terms will give an equation with terms containing x^3, x^2, and x, which requires complicated methods to solve directly. However, we can avoid this situation by recognizing that since K is so small (1.6×10^{-5}), the system will not proceed far to the right to reach equilibrium. That is, x *represents a relatively small number*. The consequence of this fact is that the term $(0.50 - 2x)$ can be approximated by 0.50. That is, when x is small,

$$0.50 - 2x \approx 0.50$$

Making this approximation allows us to simplify the equilibrium expression:

$$1.6 \times 10^{-5} = \frac{(2x)^2(x)}{(0.50 - 2x)^2} \approx \frac{(2x)^2(x)}{(0.50)^2} = \frac{4x^3}{(0.50)^2}$$

Solving for x^3 gives

$$x^3 = \frac{(1.6 \times 10^{-5})(0.50)^2}{4} = 1.0 \times 10^{-6}$$

ⓘ Approximations can simplify complicated math, but their validity should be checked carefully.

and $x = 1.0 \times 10^{-2}$.

How valid is this approximation? ◀ ⓘ If $x = 1.0 \times 10^{-2}$, then

$$0.50 - 2x = 0.50 - 2(1.0 \times 10^{-2}) = 0.48$$

The difference between 0.50 and 0.48 is 0.02, or 4% of the initial concentration of NOCl, a relatively small discrepancy that will have little effect on the outcome. That is, since $2x$ is very small compared with 0.50, the value of x obtained in the approximate solution should be very close to the exact value. We use this approximate value of x to calculate the equilibrium concentrations:

- $[NOCl] = 0.50 - 2x \approx 0.50 \, M$
- $[NO] = 2x = 2(1.0 \times 10^{-2} \, M) = 2.0 \times 10^{-2} \, M$
- $[Cl_2] = x = 1.0 \times 10^{-2} \, M$

REALITY CHECK

$$\frac{[NO]^2[Cl_2]}{[NOCl]^2} = \frac{(2.0 \times 10^{-2})^2(1.0 \times 10^{-2})}{(0.50)^2} = 1.6 \times 10^{-5}$$

Since the given value of K is 1.6×10^{-5}, these calculated concentrations are correct.

This problem was much easier to solve than it appeared at first because the *small value of K and the resulting small shift to the right to reach equilibrium allowed simplification*.

Critical Thinking

You have learned how to treat systems that have small equilibrium constants by making approximations to simplify the math. What if the system has a very large equilibrium constant? What can you do to simplify the math for this case? Use the example from the text, but change the value of the equilibrium constant to 1.6×10^5 and rework the problem. Why can you not use approximations for the case in which $K = 1.6$?

12-7 | Le Châtelier's Principle

It is important to understand the factors that control the *position* of a chemical equilibrium. For example, when a chemical is manufactured, the chemists and chemical engineers in charge of production want to choose conditions that favor the desired product as much as possible. That is, they want the equilibrium to lie far to the right. When Fritz Haber was developing the process for the synthesis of ammonia, he did extensive studies on how temperature and pressure affect the equilibrium concentration of ammonia. Some of his results are given in Table 12-2. Note that the equilibrium amount of NH_3 increases with an increase in pressure but decreases as the temperature is increased. Thus the amount of NH_3 present at equilibrium is favored by conditions of low temperature and high pressure.

However, this is not the whole story. Carrying out the process at low temperatures is not feasible because then the reaction is too slow. Even though the equilibrium tends to shift to the right as the temperature is lowered, the attainment of equilibrium would be much too slow at low temperatures to be practical. This emphasizes once again that we must study both the thermodynamics and the kinetics of a reaction before we really understand the factors that control it.

We can qualitatively predict the effects of changes in concentration, pressure, and temperature on a system at equilibrium by using **Le Châtelier's principle**, which states that if a change is imposed on a system at equilibrium, the position of the equilibrium will shift in a direction that tends to reduce that change. Although this rule sometimes oversimplifies the situation, it works remarkably well. ◄

The Effect of a Change in Concentration

To see how we can predict the effect of change in concentration on a system at equilibrium, we will consider the ammonia synthesis reaction. Suppose there is an equilibrium position described by these concentrations:

$$[N_2] = 0.399\ M \qquad [H_2] = 1.197\ M \qquad [NH_3] = 0.202\ M$$

▲ Henri Louis Le Châtelier (1850–1936), the French physical chemist and metallurgist, seen here while a student at the École Polytechnique.

SPL/Science Source

Table 12-2 | The Percent by Mass of NH_3 at Equilibrium in a Mixture of N_2, H_2, and NH_3 as a Function of Temperature and Total Pressure*

Temperature (°C)	Total Pressure		
	300 atm	400 atm	500 atm
400	48% NH_3	55% NH_3	61% NH_3
500	26% NH_3	32% NH_3	38% NH_3
600	13% NH_3	17% NH_3	21% NH_3

*Each experiment was begun with a $3:1$ mixture of H_2 and N_2.

What will happen if 1.000 mol/L N_2 is suddenly injected into the system? We can answer this question by calculating the value of Q. The concentrations before the system adjusts are

$$[N_2]_0 = 0.399 \, M + 1.000 \, M = 1.399 \, M$$
$$\uparrow$$
$$\text{Added } N_2$$

$$[H_2]_0 = 1.197 \, M$$
$$[NH_3]_0 = 0.202 \, M$$

Note we are labeling these as "initial concentrations" because the system is no longer at equilibrium. Then

$$Q = \frac{[NH_3]_0^2}{[N_2]_0[H_2]_0^3} = \frac{(0.202)^2}{(1.399)(1.197)^3} = 1.70 \times 10^{-2}$$

Since we are not given the value of K, we must calculate it from the first set of equilibrium concentrations:

$$K = \frac{[NH_3]^2}{[N_2][H_2]^3} = \frac{(0.202)^2}{(0.399)(1.197)^3} = 5.96 \times 10^{-2}$$

As expected, Q is less than K because the concentration of N_2 was increased.

The system will shift to the right to come to the new equilibrium position. Rather than do the calculations, we simply summarize the results:

Equilibrium Position I		Equilibrium Position II
$[N_2] = 0.399 \, M$	$\xrightarrow[\text{of } N_2 \text{ added}]{1.000 \text{ mol/L}}$	$[N_2] = 1.348 \, M$
$[H_2] = 1.197 \, M$		$[H_2] = 1.044 \, M$
$[NH_3] = 0.202 \, M$		$[NH_3] = 0.304 \, M$

Note from these data that the equilibrium position does in fact shift to the right: The concentration of H_2 decreases, the concentration of NH_3 increases, and, of course, since nitrogen is added, the concentration of N_2 shows an increase relative to the amount present in the original equilibrium position. (However, notice that the nitrogen showed a decrease relative to the amount present immediately after addition of the 1.000 mole of N_2.)

We can understand this shift by thinking about reaction rates. When we add N_2 molecules to the system, the number of collisions between N_2 and H_2 will increase, thus increasing the rate of the forward reaction and in turn increasing the rate of formation of NH_3 molecules. More NH_3 molecules will in turn lead to a higher rate for the reverse reaction. Eventually, the forward and reverse reaction rates will again become equal, and the system will reach its new equilibrium position.

We can predict this shift qualitatively by using Le Châtelier's principle. Since the change imposed is the addition of nitrogen, Le Châtelier's principle predicts that the system will shift in a direction that consumes nitrogen. This reduces the effect of the addition. Thus Le Châtelier's principle correctly predicts that adding nitrogen will cause the equilibrium to shift to the right (Fig. 12-9).

If ammonia had been added instead of nitrogen, the system would have shifted to the left to consume ammonia. So another way of stating Le Châtelier's principle is to say that if a component (reactant or product) is added to a reaction system at equilibrium (at constant T and P or constant T and V), the equilibrium position will shift in the direction that lowers the concentration of that component. If a component is removed, the opposite effect occurs. ◀ ⓘ

ⓘ The system shifts in the direction that compensates for the imposed change.

Figure 12-9 | (a) The initial equilibrium mixture of N_2, H_2, and NH_3. (b) Addition of N_2. (c) The new equilibrium position for the system containing more N_2 (due to addition of N_2), less H_2, and more NH_3 than the mixture in (a).

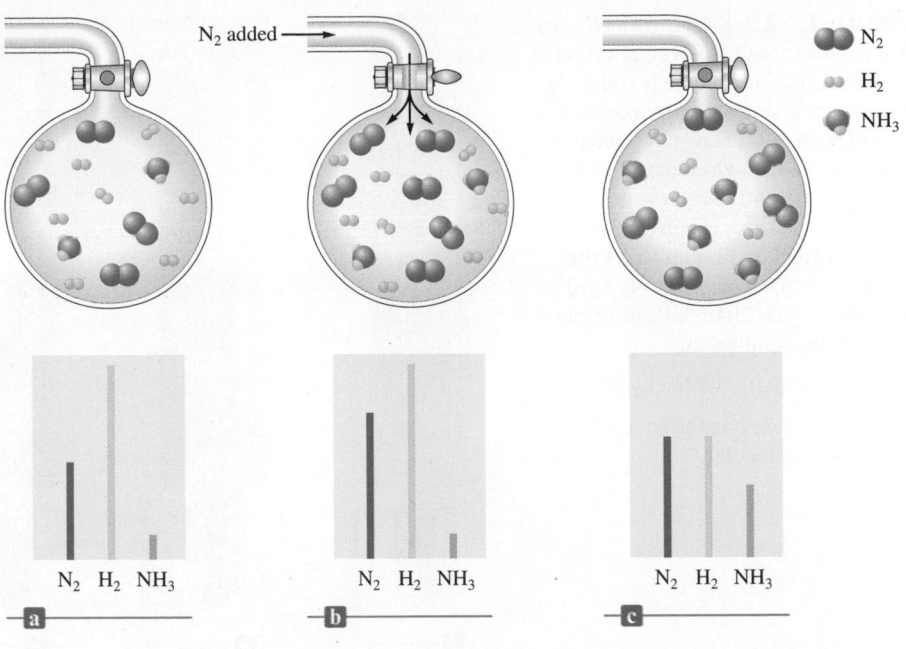

INTERACTIVE EXAMPLE 12-13

Using Le Châtelier's Principle I

Arsenic can be extracted from its ores by first reacting the ore with oxygen (called *roasting*) to form solid As_4O_6, which is then reduced using carbon:

$$As_4O_6(s) + 6C(s) \rightleftharpoons As_4(g) + 6CO(g)$$

Predict the direction of the shift of the equilibrium position in response to each of the following changes in conditions.

a. Addition of carbon monoxide

b. Addition or removal of carbon or tetraarsenic hexoxide (As_4O_6)

c. Removal of gaseous arsenic (As_4)

Solution

a. Le Châtelier's principle predicts that the shift will be away from the substance whose concentration is increased. The equilibrium position will shift to the left when carbon monoxide is added.

b. Since the amount of a pure solid has no effect on the equilibrium position, changing the amount of carbon or tetraarsenic hexoxide will have no effect.

c. If gaseous arsenic is removed, the equilibrium position will shift to the right to form more products. In industrial processes, the desired product is often continuously removed from the reaction system to increase the yield.

See Exercise 12-63

The Effect of a Change in Pressure

Basically, there are three ways to change the pressure of a reaction system involving gaseous components:

1. Add or remove a gaseous reactant or product.
2. Add an inert gas (one not involved in the reaction).
3. Change the volume of the container.

Figure 12-10 | (a) Brown $NO_2(g)$ and colorless $N_2O_4(g)$ in equilibrium in a syringe. (b) The volume is suddenly decreased, giving a greater concentration of both N_2O_4 and NO_2 (indicated by the darker brown color). (c) A few seconds after the sudden volume decrease, the color is much lighter brown as the equilibrium shifts the brown $NO_2(g)$ to colorless $N_2O_4(g)$ as predicted by Le Châtelier's principle, since in the equilibrium

$$2NO_2(g) \rightleftharpoons N_2O_4(g)$$

the product side has the smaller number of molecules.

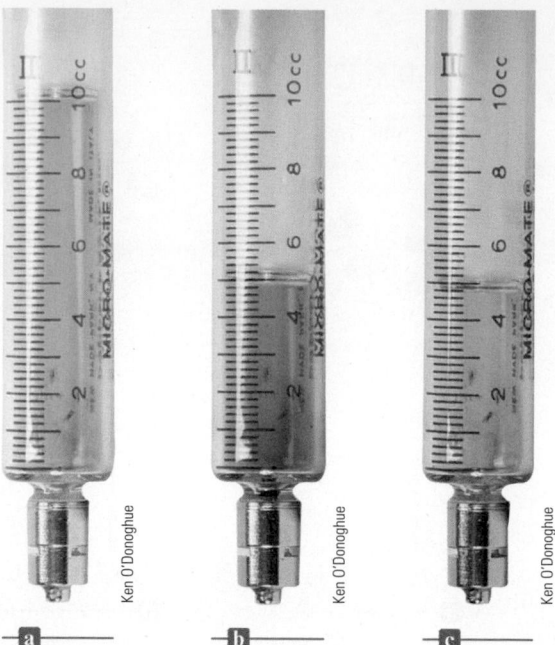

Ken O'Donoghue

We have already considered the addition or removal of a reactant or product. When an inert gas is added, there is no effect on the equilibrium position. The addition of an inert gas increases the total pressure but has no effect on the concentrations or partial pressures of the reactants or products. That is, in this case the added molecules do not participate in the reaction in any way and thus cannot affect the equilibrium in any way. Thus the system remains at the original equilibrium position.

When the volume of the container is changed, the concentrations (and thus the partial pressures) of both reactants and products are changed. We could calculate Q and predict the direction of the shift. However, for systems involving gaseous components, there is an easier way: We focus on the volume. The central idea is that when the volume of the container holding a gaseous system is reduced, the system responds by reducing its own volume. This is done by decreasing the total number of gaseous molecules in the system. This is illustrated by the NO_2/N_2O_4 system shown in Fig. 12-10.

To see that this is true, we can rearrange the ideal gas law to give

$$V = \left(\frac{RT}{P}\right) n$$

or at constant T and P,

$$V \propto n$$

That is, at constant temperature and pressure, the volume of a gas is directly proportional to the number of moles of gas present.

Suppose we have a mixture of the gases nitrogen, hydrogen, and ammonia at equilibrium (Fig. 12-11). If we suddenly reduce the volume, what will happen to the equilibrium position? The reaction system can reduce its volume by reducing the number of molecules present. This means that the reaction

$$N_2(g) + 3H_2(g) \rightleftharpoons 2NH_3(g)$$

will shift to the right, since in this direction four molecules (one of nitrogen and three of hydrogen) react to produce two molecules (of ammonia), thus *reducing the total number of gaseous molecules present*. The new equilibrium position will be farther to the right than the original one. That is, the equilibrium position will shift toward the

Figure 12-11 | (a) A mixture of $NH_3(g)$, $N_2(g)$, and $H_2(g)$ at equilibrium. (b) The volume is suddenly decreased. (c) The new equilibrium position for the system containing more NH_3 and less N_2 and H_2. The reaction $N_2(g) + 3H_2(g) \rightleftharpoons 2NH_3(g)$ shifts to the right (toward the side with fewer molecules) when the container volume is decreased.

side of the reaction involving the smaller number of gaseous molecules in the balanced equation.

The opposite is also true. When the container volume is increased, the system will shift so as to increase its volume. An increase in volume in the ammonia synthesis system will produce a shift to the left to increase the total number of gaseous molecules present.

Critical Thinking

You and a friend are studying for a chemistry exam. What if your friend says, "Adding an inert gas to a system of gaseous components at equilibrium never changes the equilibrium position"? How do you explain to your friend that this holds true for a system at constant volume but is not necessarily true for a system at constant pressure? When would it hold true for a system at constant pressure?

INTERACTIVE EXAMPLE 12-14

Using Le Châtelier's Principle II

Predict the shift in equilibrium position that will occur for each of the following processes when the volume is reduced.

a. The preparation of liquid phosphorus trichloride by the reaction

$$P_4(s) + 6Cl_2(g) \rightleftharpoons 4PCl_3(l)$$

b. The preparation of gaseous phosphorus pentachloride according to the equation

$$PCl_3(g) + Cl_2(g) \rightleftharpoons PCl_5(g)$$

c. The reaction of phosphorus trichloride with ammonia:

$$PCl_3(g) + 3NH_3(g) \rightleftharpoons P(NH_2)_3(g) + 3HCl(g)$$

Solution

a. Since P_4 and PCl_3 are a pure solid and a pure liquid, respectively, we need to consider only the effect of the change in volume on Cl_2. The volume is decreased, so the position of the equilibrium will shift to the right, since the reactant side contains six gaseous molecules and the product side has none.

b. Decreasing the volume will shift the given reaction to the right, since the product side contains only one gaseous molecule while the reactant side has two.

c.   Both sides of the balanced reaction equation have four gaseous molecules. A change in volume will have no effect on the equilibrium position. There is no shift in this case.

See Exercise 12-64

The Effect of a Change in Temperature

It is important to realize that although the changes we have just discussed may alter the equilibrium *position,* they do not alter the equilibrium *constant.* For example, the addition of a reactant shifts the equilibrium position to the right but has no effect on the value of the equilibrium constant; the new equilibrium concentrations satisfy the original equilibrium constant.

The effect of temperature on equilibrium is different, however, because *the value of K changes with temperature.* We can use Le Châtelier's principle to predict the direction of the change.

The synthesis of ammonia from nitrogen and hydrogen is exothermic. We can represent this by treating energy as a product:

$$N_2(g) + 3H_2(g) \rightleftharpoons 2NH_3(g) + 92 \text{ kJ}$$

ⓘ Of course, energy is not a chemical product of this reaction, but thinking of it in this way makes it easy to apply Le Châtelier's principle.

If energy is added to this system at equilibrium by heating it, Le Châtelier's principle predicts that the shift will be in the direction that consumes energy, that is, to the left. ◄ ⓘ Note that this shift decreases the concentration of NH_3 and increases the concentrations of N_2 and H_2, thus *decreasing the value of K.* The experimentally observed change in K with temperature for this reaction is indicated in Table 12-3. The value of K decreases with increased temperature, as predicted.

On the other hand, for an endothermic reaction, such as the decomposition of calcium carbonate,

$$556 \text{ kJ} + CaCO_3(s) \rightleftharpoons CaO(s) + CO_2(g)$$

Table 12-3 | Observed Value of K for the Ammonia Synthesis Reaction as a Function of Temperature*

Temperature (K)	K
500	90
600	3
700	0.3
800	0.04

*For this exothermic reaction, the value of K decreases as the temperature increases, as predicted by Le Châtelier's principle.

an increase in temperature will cause the equilibrium to shift to the right and the value of K to increase.

In summary, to use Le Châtelier's principle to describe the effect of a temperature change on a system at equilibrium, treat energy as a reactant (in an endothermic process) or as a product (in an exothermic process), and predict the direction of the shift in the same way as when an actual reactant or product is added or removed. Although Le Châtelier's principle cannot predict the size of the change in K, it does correctly predict the direction of the change.

INTERACTIVE EXAMPLE 12-15

Using Le Châtelier's Principle III

For each of the following reactions, predict how the value of K changes as the temperature is increased.

a. $N_2(g) + O_2(g) \rightleftharpoons 2NO(g)$ $\Delta H° = 181 \text{ kJ}$

b. $2SO_2(g) + O_2(g) \rightleftharpoons 2SO_3(g)$ $\Delta H° = -198 \text{ kJ}$

Solution

a.   This is an endothermic reaction, as indicated by the positive value for $\Delta H°$. Energy can be viewed as a reactant, and K increases (the equilibrium shifts to the right) as the temperature is increased.

Figure 12-12 | Shifting the $N_2O_4(g) \rightarrow 2NO_2(g)$ equilibrium by changing the temperature.
(a) At 100°C the flask is definitely reddish brown due to a large amount of NO_2 present. (b) At 0°C the
equilibrium is shifted toward colorless $N_2O_4(g)$.

b. This is an exothermic reaction (energy can be regarded as a product). As the temperature is increased, the value of K decreases (the equilibrium shifts to the left).

See Exercises 12-69 and 12-70

We have seen how Le Châtelier's principle can be used to predict the effect of several
types of changes on a system at equilibrium. To summarize these ideas, Table 12-4
shows how various changes affect the equilibrium position of the endothermic reaction

$$N_2O_4(g) \rightleftharpoons 2NO_2(g) \qquad \Delta H° = 58 \text{ kJ}$$

For an illustration of the effect of temperature on this reaction, see Fig. 12-12.

Table 12-4 | Shifts in the Equilibrium Position
for the Reaction 58 kJ + $N_2O_4(g) \rightleftharpoons 2NO_2(g)$

Change	Shift
Addition of $N_2O_4(g)$	Right
Addition of $NO_2(g)$	Left
Removal of $N_2O_4(g)$	Left
Removal of $NO_2(g)$	Right
Addition of He(g)	None
Decrease container volume	Left
Increase container volume	Right
Increase temperature	Right
Decrease temperature	Left

Now that you have completed the chapter, let's review the concepts of *equilibrium condition, reaction quotient,* and *equilibrium constant* in the context of understanding Le Châtelier's principle. Consider the reaction

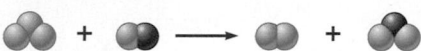

where ⚪ and ⚫ represent two different types of atoms. Suppose we mix the two types of molecules together as seen in the left box, and the equilibrium mixture, as seen in the right box, results.

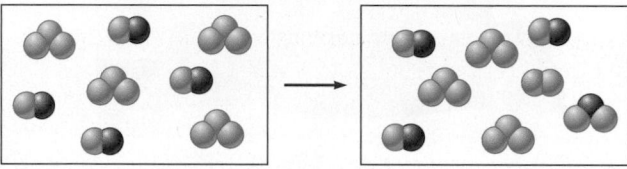

Since there is a 1:1:1:1 mole ratio in the balanced equation, we can calculate the equilibrium constant by finding the ratio of products to reactants. Doing so, we get

$$\frac{(N_{⚪⚫})(N_{⚫⚪})}{(N_{⚪⚪})(N_{⚫⚪})} = \frac{(1)(1)}{(3)(3)} = 0.111$$

Suppose we then add 16 more molecules of ⚪⚪ to the mixture at equilibrium. Initially, the mixture will look like this:

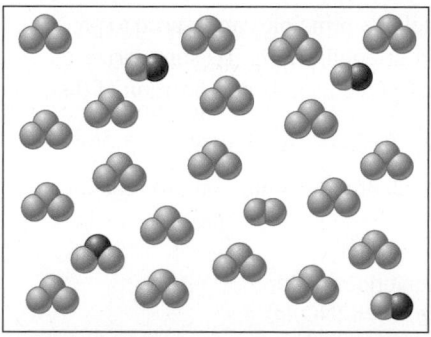

We can see this reaction is not at equilibrium because the reaction quotient is now:

$$\frac{(N_{⚪⚫})(N_{⚫⚪})}{(N_{⚪⚪})(N_{⚫⚪})} = \frac{(1)(1)}{(19)(3)} = 0.0175$$

Note that the value of the reaction quotient is 0.0175, which is less than the equilibrium constant. This means that the reaction should proceed to the right to reach equilibrium. This matches our prediction using Le Châtelier's principle since we have disturbed the equilibrium system by increasing the concentration of one of the reactants. That is, the system will respond by shifting equilibrium "to the right" (product side). We can see the new equilibrium system here:

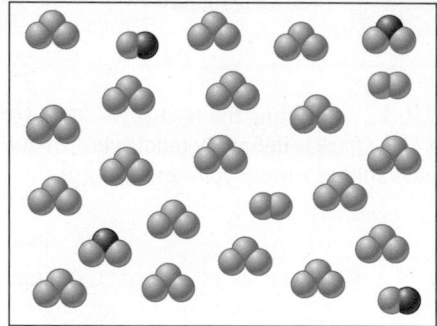

In this case, the ratio of products to reactants is

$$\frac{(N_{⚪⚫})(N_{⚫⚪})}{(N_{⚪⚪})(N_{⚫⚪})} = \frac{(2)(2)}{(18)(2)} = 0.111$$

And thus, the system is at equilibrium. The equilibrium condition is different from the first case, but the equilibrium constant is the same.

FOR REVIEW

Key Terms

chemical equilibrium

Section 12-2

law of mass action
equilibrium expression
equilibrium constant
equilibrium position

Section 12-4

homogeneous equilibria
heterogeneous equilibria

Section 12-5

reaction quotient (Q)

Section 12-7

Le Châtelier's principle

Chemical Equilibrium

- When a reaction takes place in a closed system, it reaches a condition where the concentrations of the reactants and products remain constant over time
- Dynamic state: reactants and products are interconverted continually
  - Forward rate = reverse rate
- The law of mass action: for the reaction

$$jA + kB \rightleftharpoons mC + nD$$

$$K = \frac{[C]^m[D]^n}{[A]^j[B]^k} = \text{equilibrium constant}$$

- A pure liquid or solid is never included in the equilibrium expression
- For a gas-phase reaction, the reactants and products can be described in terms of their partial pressures and the equilibrium constant is called K_p:

$$K_p = K(RT)^{\Delta n}$$

where Δn is the sum of the coefficients of the gaseous products minus the sum of the coefficients of the gaseous reactants

Equilibrium Position

- A set of reactant and product concentrations that satisfies the equilibrium constant expression
  - There is one value of K for a given system at a given temperature
  - There are an infinite number of equilibrium positions at a given temperature depending on the initial concentrations
- A small value of K means the equilibrium lies to the left; a large value of K means the equilibrium lies to the right
  - The size of K has no relationship to the speed at which equilibrium is achieved
- Q, the reaction quotient, applies the law of mass action to initial concentrations rather than equilibrium concentrations
  - If Q > K, the system will shift to the left to achieve equilibrium
  - If Q < K, the system will shift to the right to achieve equilibrium
- Finding the concentrations that characterize a given equilibrium position:
  - Start with the given initial concentrations (pressures)
  - Define the change needed to reach equilibrium
  - Apply the change to the initial concentrations (pressures) and solve for the equilibrium concentrations (pressures)

Le Châtelier's Principle

- Enables qualitative prediction of the effects of changes in concentration, pressure, and temperature on a system at equilibrium
- If a change in conditions is imposed on a system at equilibrium, the system will shift in a direction that compensates for the imposed change
  - In other words, when a stress is placed on a system at equilibrium, the system shifts in the direction that relieves the stress

Review Questions

1. Characterize a system at chemical equilibrium with respect to each of the following:

   a. the rates of the forward and reverse reactions

   b. the overall composition of the reaction mixture

   For a general reaction $3A(g) + B(g) \longrightarrow 2C(g)$, if one starts an experiment with only reactants present, show what the plot of concentrations of A, B, and C versus time would look like. Also sketch the plot illustrating the rate of the forward reaction and rate of the reverse reaction versus time.

2. What is the law of mass action? Is it true that the value of K depends on the amounts of reactants and products mixed together initially? Explain. Is it true that reactions with large equilibrium constant values are very fast? Explain. There is only one value of the equilibrium constant for a particular system at a particular temperature, but there is an infinite number of equilibrium positions. Explain.

3. Consider the following reactions at some temperature:

   $$2NOCl(g) \rightleftharpoons 2NO(g) + Cl_2(g) \qquad K = 1.6 \times 10^{-5}$$
   $$2NO(g) \rightleftharpoons N_2(g) + O_2(g) \qquad K = 1 \times 10^{31}$$

   For each reaction, assume some quantities of the reactants were placed in separate containers and allowed to come to equilibrium. Describe the relative amounts of reactants and products that would be present at equilibrium. At equilibrium, which is faster, the forward or reverse reaction in each case?

4. What is the difference between K and K_p? When does $K = K_p$ for a reaction? When does $K \neq K_p$ for a reaction? If the coefficients in a reaction equation are tripled, how is the new value of K related to the initial value of K? If an equation for a reaction is reversed, how is the value of K_p for the reversed equation related to the value of K_p for the initial equation?

5. What are homogeneous equilibria? Heterogeneous equilibria? What is the difference in writing K expressions for homogeneous versus heterogeneous reactions? Summarize which species are included in the K expression and which species are not included.

6. Distinguish between the terms *equilibrium constant* and *reaction quotient*. When $Q = K$, what does this say about a reaction? When $Q < K$, what does this say about a reaction? When $Q > K$, what does this say about a reaction?

7. Summarize the steps for solving equilibrium problems (see the beginning of Section 12-6). In general, when solving an equilibrium problem, you should always set up an ICE table. What is an ICE table?

8. A common type of reaction we will study is that having a very small K value ($K \ll 1$). Solving for equilibrium concentrations in an equilibrium problem usually requires many mathematical operations to be performed. However, the math involved when solving equilibrium problems for reactions having small K values ($K \ll 1$) is simplified. What assumption is made when solving the equilibrium concentrations for reactions with small K values? Whenever assumptions are made, they must be checked for validity. In general, the "5% rule" is used to check the validity of assuming x (or $2x$, $3x$, and so on) is very small compared to some number. When x (or $2x$, $3x$, and so on) is less than 5% of the number the assumption was made against, then the assumption is said to be valid. If the 5% rule fails, what do you do to solve for the equilibrium concentrations?

9. What is Le Châtelier's principle? Consider the reaction $2NOCl(g) \rightleftharpoons 2NO(g) + Cl_2(g)$. If this reaction is at equilibrium, what happens when the following changes occur?

   a. $NOCl(g)$ is added.

   b. $NO(g)$ is added.

   c. $NOCl(g)$ is removed.

   d. $Cl_2(g)$ is removed.

   e. The container volume is decreased.

   For each of these changes, what happens to the value of K for the reaction as equilibrium is reached again? Give an example of a reaction for which the addition or removal of one of the reactants or products has no effect on the equilibrium position.

   In general, how will the equilibrium position of a gas-phase reaction be affected if the volume of the reaction vessel changes? Are there reactions that will not have their equilibria shifted by a change in volume? Explain. Why does changing the pressure in a rigid container by adding an inert gas not shift the equilibrium position for a gas-phase reaction?

10. The only "stress" (change) that also changes the value of K is a change in temperature. For an exothermic reaction, how does the equilibrium position change as temperature increases, and what happens to the value of K? Answer the same questions for an endothermic reaction. If the value of K increases with a decrease in temperature, is the reaction exothermic or endothermic? Explain.

Active Learning Questions

These questions are designed to be used by groups of students in class.

1. Consider an equilibrium mixture of four chemicals (A, B, C, and D, all gases) reacting in a closed flask according to the equation:

$$A(g) + B(g) \rightleftharpoons C(g) + D(g)$$

   a. You add more A to the flask. How does the concentration of each chemical compare to its original concentration after equilibrium is reestablished? Justify your answer.

   b. You have the original setup at equilibrium, and you add more D to the flask. How does the concentration of each chemical compare to its original concentration after equilibrium is reestablished? Justify your answer.

2. The boxes shown below represent a set of initial conditions for the reaction:

   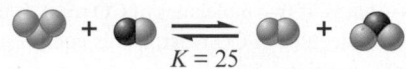

   Draw a quantitative molecular picture that shows what this system looks like after the reactants are mixed in one of the boxes and the system reaches equilibrium. Support your answer with calculations.

   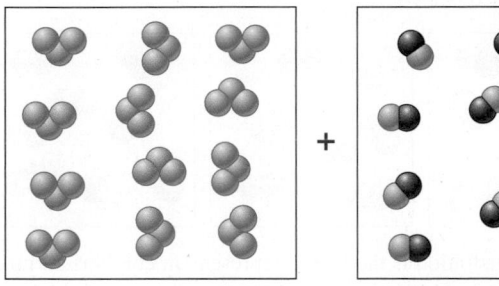 +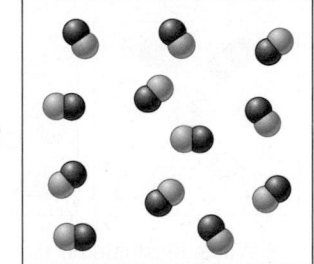

3. For the reaction $H_2(g) + I_2(g) \rightleftharpoons 2HI(g)$, consider two possibilities: (a) you mix 0.5 mole of each reactant, allow the system to come to equilibrium, and then add another mole of H_2 and allow the system to reach equilibrium again, or (b) you mix 1.5 moles of H_2 and 0.5 mole of I_2 and allow the system to reach equilibrium. Will the final equilibrium mixture be different for the two procedures? Explain.

4. Given the reaction $A(g) + B(g) \rightleftharpoons C(g) + D(g)$, consider the following situations:

   i. You have 1.3 M A and 0.8 M B initially.

   ii. You have 1.3 M A, 0.8 M B, and 0.2 M C initially.

   iii. You have 2.0 M A and 0.8 M B initially.

   Order the preceding situations in terms of increasing equilibrium concentration of D. Explain your order. Then give the order in terms of increasing equilibrium concentration of B and explain.

5. Consider the reaction $A(g) + 2B(g) \rightleftharpoons C(g) + D(g)$ in a 1.0-L rigid flask. Answer the following questions for each situation (a–d):

   i. Estimate a range (as small as possible) for the requested substance. For example, [A] could be between 95 M and 100 M.

   ii. Explain how you decided on the limits for the estimated range.

   iii. Indicate what other information would enable you to narrow your estimated range.

   iv. Compare the estimated concentrations for a through d, and explain any differences.

   a. If at equilibrium [A] = 1 M, and then 1 mole of C is added, estimate the value for [A] once equilibrium is reestablished.

   b. If at equilibrium [B] = 1 M, and then 1 mole of C is added, estimate the value for [B] once equilibrium is reestablished.

   c. If at equilibrium [C] = 1 M, and then 1 mole of C is added, estimate the value for [C] once equilibrium is reestablished.

   d. If at equilibrium [D] = 1 M, and then 1 mole of C is added, estimate the value for [D] once equilibrium is reestablished.

6. Consider the reaction $A(g) + B(g) \rightleftharpoons C(g) + D(g)$. A friend asks the following: "I know we have been told that if a mixture of A, B, C, and D is at equilibrium and more of A is added, more C and D will form. But how can more C and D form if we do not add more B?" What do you tell your friend?

7. Consider the following statements: "Consider the reaction $A(g) + B(g) \rightleftharpoons C(g)$, for which at equilibrium [A] = 2 M, [B] = 1 M, and [C] = 4 M. To a 1-L container of the system at equilibrium, you add 3 moles of B. A possible equilibrium condition is [A] = 1 M, [B] = 3 M, and [C] = 6 M because in both cases K = 2." Indicate everything that is correct in these statements and everything that is incorrect. Correct the incorrect statements, and explain.

8. Le Châtelier's principle is stated (Section 12-7) as follows: "If a change is imposed on a system at equilibrium, the position of the equilibrium will shift in a direction that tends to reduce that change." The system $N_2(g) + 3H_2(g) \rightleftharpoons 2NH_3(g)$ is used as an example in which the addition of nitrogen gas at equilibrium results in a decrease in H_2 concentration and an increase in NH_3 concentration. In the experiment the volume is assumed to be constant. On the other hand, if N_2 is added to the reaction system in a container with a piston so that the pressure can be held constant, the amount of NH_3 actually could decrease, and the concentration of H_2 would increase as equilibrium is reestablished. Explain how this can happen. Also, if you consider this same system at equilibrium, the addition of an inert gas, holding the pressure constant, *does* affect the equilibrium position. Explain why the addition of an inert gas to this system in a rigid container does not affect the equilibrium position.

9. The value of the equilibrium constant K depends on which of the following (more than one answer may be correct)?

   a. the initial concentrations of the reactants

   b. the initial concentrations of the products

   c. the temperature of the system

   d. the nature of the reactants and products

   Explain.

A blue question or exercise number indicates that the answer to that question or exercise appears at the back of this book and a solution appears in the *Solutions Guide*, as found on PowerLecture.

Questions

10. Consider an initial mixture of N_2 and H_2 gases that can be represented as follows:

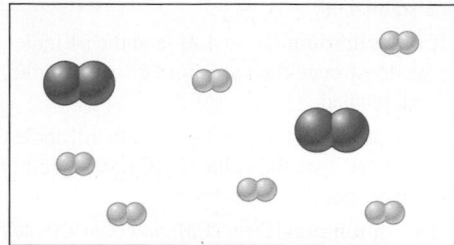

The gases react to form ammonia gas (NH_3) as represented by the following concentration profile:

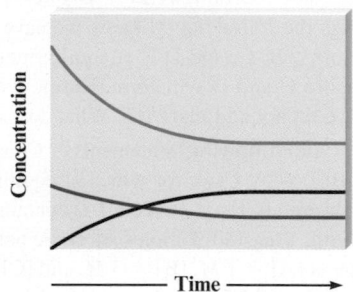

a. Label each plot on the graph as N_2, H_2, or NH_3, and explain your answers.

b. Explain the relative shapes of the plots.

c. When is equilibrium reached? How do you know?

11. Consider the following reaction:

$$H_2O(g) + CO(g) \rightleftharpoons H_2(g) + CO_2(g)$$

Amounts of H_2O, CO, H_2, and CO_2 are put into a flask so that the composition corresponds to an equilibrium position. If the CO placed in the flask is labeled with radioactive ^{14}C, will ^{14}C be found only in CO molecules for an indefinite period of time? Explain.

12. Consider the same reaction as in Question 11. In one experiment 1.0 mole of $H_2O(g)$ and 1.0 mole of $CO(g)$ are put into a flask and heated to 350°C. In a second experiment 1.0 mole of $H_2(g)$ and 1.0 mole of $CO_2(g)$ are put into another flask with the same volume as the first. This mixture is also heated to 350°C. After equilibrium is reached, will there be any difference in the composition of the mixtures in the two flasks?

13. Suppose a reaction has the equilibrium constant $K = 1.3 \times 10^8$. What does the magnitude of this constant tell you about the relative concentrations of products and reactants that will be present once equilibrium is reached? Is this reaction likely to be a good source of the products?

14. Suppose a reaction has the equilibrium constant $K = 1.7 \times 10^{-8}$ at a particular temperature. Will there be a large or small amount of unreacted starting material present when this reaction reaches equilibrium? Is this reaction likely to be a good source of products at this temperature?

15. Consider the following reaction at some temperature:

$$H_2O(g) + CO(g) \rightleftharpoons H_2(g) + CO_2(g) \qquad K = 2.0$$

Some molecules of H_2O and CO are placed in a 1.0-L container as shown below.

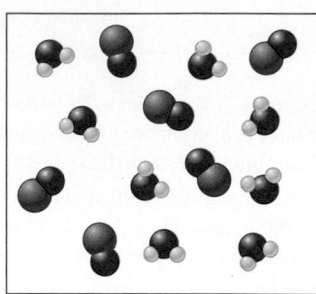

When equilibrium is reached, how many molecules of H_2O, CO, H_2, and CO_2 are present? Do this problem by trial and error—that is, if two molecules of CO react, is this equilibrium; if three molecules of CO react, is this equilibrium; and so on.

16. Consider the following generic reaction:

$$2A_2B(g) \rightleftharpoons 2A_2(g) + B_2(g)$$

Some molecules of A_2B are placed in a 1.0-L container. As time passes, several snapshots of the reaction mixture are taken as illustrated below.

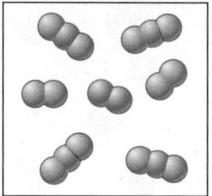 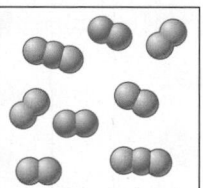 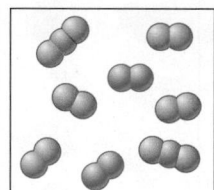

Which illustration is the first to represent an equilibrium mixture? Explain. How many molecules of A_2B reacted initially?

17. Explain the difference between K, K_p, and Q.

18. Consider the following reactions:

$$H_2(g) + I_2(g) \longrightarrow 2HI(g) \quad \text{and} \quad H_2(g) + I_2(s) \longrightarrow 2HI(g)$$

List two property differences between these two reactions that relate to equilibrium.

19. For a typical equilibrium problem, the value of K and the initial reaction conditions are given for a specific reaction, and you are asked to calculate the equilibrium concentrations. Many of these calculations involve solving a quadratic or cubic equation. What can you do to avoid solving a quadratic or cubic equation and still come up with reasonable equilibrium concentrations?

20. Which of the following statements is(are) *true*? Correct the false statement(s).

a. When a reactant is added to a system at equilibrium at a given temperature, the reaction will shift right to reestablish equilibrium.

b. When a product is added to a system at equilibrium at a given temperature, the value of K for the reaction will increase when equilibrium is reestablished.

c. When temperature is increased for a reaction at equilibrium, the value of K for the reaction will increase.

d. When the volume of a reaction container is increased for a system at equilibrium at a given temperature, the reaction will shift left to reestablish equilibrium.

e. Addition of a catalyst (a substance that increases the speed of the reaction) has no effect on the equilibrium position.

Exercises

In this section similar exercises are paired.

The Equilibrium Constant

21. Write the equilibrium expression (K) for each of the following gas-phase reactions.

a. $N_2(g) + O_2(g) \rightleftharpoons 2NO(g)$

b. $N_2O_4(g) \rightleftharpoons 2NO_2(g)$

c. $SiH_4(g) + 2Cl_2(g) \rightleftharpoons SiCl_4(g) + 2H_2(g)$

d. $2PBr_3(g) + 3Cl_2(g) \rightleftharpoons 2PCl_3(g) + 3Br_2(g)$

22. Write the equilibrium expression (K_p) for each reaction in Exercise 21.

23. At a given temperature, $K = 1.3 \times 10^{-2}$ for the reaction

$$N_2(g) + 3H_2(g) \rightleftharpoons 2NH_3(g)$$

Calculate values of K for the following reactions at this temperature.

a. $\frac{1}{2}N_2(g) + \frac{3}{2}H_2(g) \rightleftharpoons NH_3(g)$

b. $2NH_3(g) \rightleftharpoons N_2(g) + 3H_2(g)$

c. $NH_3(g) \rightleftharpoons \frac{1}{2}N_2(g) + \frac{3}{2}H_2(g)$

d. $2N_2(g) + 6H_2(g) \rightleftharpoons 4NH_3(g)$

24. For the reaction

$$H_2(g) + Br_2(g) \rightleftharpoons 2HBr(g)$$

$K_p = 3.5 \times 10^4$ at 1495 K. What is the value of K_p for the following reactions at 1495 K?

a. $HBr(g) \rightleftharpoons \frac{1}{2}H_2(g) + \frac{1}{2}Br_2(g)$

b. $2HBr(g) \rightleftharpoons H_2(g) + Br_2(g)$

c. $\frac{1}{2}H_2(g) + \frac{1}{2}Br_2(g) \rightleftharpoons HBr(g)$

25. For the reaction

$$2NO(g) + 2H_2(g) \rightleftharpoons N_2(g) + 2H_2O(g)$$

it is determined that, at equilibrium at a particular temperature, the concentrations are as follows: $[NO(g)] = 8.1 \times 10^{-3}\ M$, $[H_2(g)] = 4.1 \times 10^{-5}\ M$, $[N_2(g)] = 5.3 \times 10^{-2}\ M$, and $[H_2O(g)] = 2.9 \times 10^{-3}\ M$. Calculate the value of K for the reaction at this temperature.

26. At high temperatures, elemental nitrogen and oxygen react with each other to form nitrogen monoxide:

$$N_2(g) + O_2(g) \rightleftharpoons 2NO(g)$$

Suppose the system is analyzed at a particular temperature, and the equilibrium concentrations are found to be $[N_2] = 0.041\ M$, $[O_2] = 0.0078\ M$, and $[NO] = 4.7 \times 10^{-4}\ M$. Calculate the value of K for the reaction.

27. At a particular temperature, a 3.0-L flask contains 2.4 moles of Cl_2, 1.0 mole of $NOCl$, and 4.5×10^{-3} mole of NO. Calculate K at this temperature for the following reaction:

$$2NOCl(g) \rightleftharpoons 2NO(g) + Cl_2(g)$$

28. At a particular temperature a 2.00-L flask at equilibrium contains 2.80×10^{-4} mole of N_2, 2.50×10^{-5} mole of O_2, and 2.00×10^{-2} mole of N_2O. Calculate K at this temperature for the reaction

$$2N_2(g) + O_2(g) \rightleftharpoons 2N_2O(g)$$

If $[N_2] - 2.00 \times 10^{-4}\ M$, $[N_2O] = 0.200\ M$, and $[O_2] = 0.00245\ M$, does this represent a system at equilibrium?

29. The following equilibrium pressures at a certain temperature were observed for the reaction

$$2NO_2(g) \rightleftharpoons 2NO(g) + O_2(g)$$

$$P_{NO_2} = 0.55\ \text{atm}$$
$$P_{NO} = 6.5 \times 10^{-5}\ \text{atm}$$
$$P_{O_2} = 4.5 \times 10^{-5}\ \text{atm}$$

Calculate the value for the equilibrium constant K_p at this temperature.

30. The following equilibrium pressures were observed at a certain temperature for the reaction

$$N_2(g) + 3H_2(g) \rightleftharpoons 2NH_3(g)$$

$$P_{NH_3} = 3.1 \times 10^{-2}\ \text{atm}$$
$$P_{N_2} = 8.5 \times 10^{-1}\ \text{atm}$$
$$P_{H_2} = 3.1 \times 10^{-3}\ \text{atm}$$

Calculate the value for the equilibrium constant K_p at this temperature.

If $P_{N_2} = 0.525$ atm, $P_{NH_3} = 0.0167$ atm, and $P_{H_2} = 0.00761$ atm, does this represent a system at equilibrium?

31. At 327°C, the equilibrium concentrations are $[CH_3OH] = 0.15\ M$, $[CO] = 0.24\ M$, and $[H_2] = 1.1\ M$ for the reaction

$$CH_3OH(g) \rightleftharpoons CO(g) + 2H_2(g)$$

Calculate K_p at this temperature.

32. At 1100 K, $K_p = 0.25$ for the reaction

$$2SO_2(g) + O_2(g) \rightleftharpoons 2SO_3(g)$$

What is the value of K at this temperature?

33. Write expressions for K and K_p for the following reactions.

a. $2NH_3(g) + CO_2(g) \rightleftharpoons N_2CH_4O(s) + H_2O(g)$

b. $2NBr_3(s) \rightleftharpoons N_2(g) + 3Br_2(g)$

c. $2KClO_3(s) \rightleftharpoons 2KCl(s) + 3O_2(g)$

d. $CuO(s) + H_2(g) \rightleftharpoons Cu(l) + H_2O(g)$

34. Write expressions for K_p for the following reactions.

a. $2Fe(s) + \frac{3}{2}O_2(g) \rightleftharpoons Fe_2O_3(s)$

b. $CO_2(g) + MgO(s) \rightleftharpoons MgCO_3(s)$

c. $C(s) + H_2O(g) \rightleftharpoons CO(g) + H_2(g)$

d. $4KO_2(s) + 2H_2O(g) \rightleftharpoons 4KOH(s) + 3O_2(g)$

35. For which reactions in Exercise 33 is K_p equal to K?

36. For which reactions in Exercise 34 is K_p equal to K?

37. Consider the following reaction at a certain temperature:

$$4Fe(s) + 3O_2(g) \rightleftharpoons 2Fe_2O_3(s)$$

An equilibrium mixture contains 1.0 mole of Fe, 1.0×10^{-3} mole of O_2, and 2.0 moles of Fe_2O_3 all in a 2.0-L container. Calculate the value of K for this reaction.

38. In a study of the reaction

$$3Fe(s) + 4H_2O(g) \rightleftharpoons Fe_3O_4(s) + 4H_2(g)$$

at 1200 K it was observed that when the equilibrium partial pressure of water vapor is 15.0 torr, the total pressure at equilibrium is 36.3 torr. Calculate the value of K_p for this reaction at 1200 K. (*Hint:* Apply Dalton's law of partial pressures.)

Equilibrium Calculations

39. The equilibrium constant is 0.0900 at 25°C for the reaction

$$H_2O(g) + Cl_2O(g) \rightleftharpoons 2HOCl(g)$$

For which of the following sets of conditions is the system at equilibrium? For those that are not at equilibrium, in which direction will the system shift?

a. A 1.0-L flask contains 1.0 mole of HOCl, 0.10 mole of Cl_2O, and 0.10 mole of H_2O.

b. A 2.0-L flask contains 0.084 mole of HOCl, 0.080 mole of Cl_2O, and 0.98 mole of H_2O.

c. A 3.0-L flask contains 0.25 mole of HOCl, 0.0010 mole of Cl_2O, and 0.56 mole of H_2O.

40. The equilibrium constant is 0.0900 at 25°C for the reaction

$$H_2O(g) + Cl_2O(g) \rightleftharpoons 2HOCl(g)$$

For which of the following sets of conditions is the system at equilibrium? For those that are not at equilibrium, in which direction will the system shift?

a. $P_{H_2O} = 1.00$ atm, $P_{Cl_2O} = 1.00$ atm, $P_{HOCl} = 1.00$ atm

b. $P_{H_2O} = 200.$ torr, $P_{Cl_2O} = 49.8$ torr, $P_{HOCl} = 21.0$ torr

c. $P_{H_2O} = 296$ torr, $P_{Cl_2O} = 15.0$ torr, $P_{HOCl} = 20.0$ torr

41. At 900°C, $K_p = 1.04$ for the reaction

$$CaCO_3(s) \rightleftharpoons CaO(s) + CO_2(g)$$

At a low temperature, dry ice (solid CO_2), calcium oxide, and calcium carbonate are introduced into a 50.0-L reaction chamber. The temperature is raised to 900°C, resulting in the dry ice converting to gaseous CO_2. For the following mixtures, will the initial amount of calcium oxide increase, decrease, or remain the same as the system moves toward equilibrium at 900°C?

a. 655 g $CaCO_3$, 95.0 g CaO, $P_{CO_2} = 2.55$ atm

b. 780 g $CaCO_3$, 1.00 g CaO, $P_{CO_2} = 1.04$ atm

c. 0.14 g $CaCO_3$, 5000 g CaO, $P_{CO_2} = 1.04$ atm

d. 715 g $CaCO_3$, 813 g CaO, $P_{CO_2} = 0.211$ atm

42. Ethyl acetate is synthesized in a nonreacting solvent (not water) according to the following reaction:

$$CH_3CO_2H + C_2H_5OH \rightleftharpoons CH_3CO_2C_2H_5 + H_2O \quad K = 2.2$$
Acetic acid   Ethanol   Ethyl acetate

For the following mixtures (a–d), will the concentration of H_2O increase, decrease, or remain the same as equilibrium is established?

a. $[CH_3CO_2C_2H_5] = 0.22\ M$, $[H_2O] = 0.10\ M$,
$[CH_3CO_2H] = 0.010\ M$, $[C_2H_5OH] = 0.010\ M$

b. $[CH_3CO_2C_2H_5] = 0.22\ M$, $[H_2O] = 0.0020\ M$,
$[CH_3CO_2H] = 0.0020\ M$, $[C_2H_5OH] = 0.10\ M$

c. $[CH_3CO_2C_2H_5] = 0.88\ M$, $[H_2O] = 0.12\ M$,
$[CH_3CO_2H] = 0.044\ M$, $[C_2H_5OH] = 6.0\ M$

d. $[CH_3CO_2C_2H_5] = 4.4\ M$, $[H_2O] = 4.4\ M$,
$[CH_3CO_2H] = 0.88\ M$, $[C_2H_5OH] = 10.0\ M$

e. What must the concentration of water be for a mixture with $[CH_3CO_2C_2H_5] = 2.0\ M$, $[CH_3CO_2H] = 0.10\ M$, and $[C_2H_5OH] = 5.0\ M$ to be at equilibrium?

f. Why is water included in the equilibrium expression for this reaction?

43. For the reaction

$$2H_2O(g) \rightleftharpoons 2H_2(g) + O_2(g)$$

$K = 2.4 \times 10^{-3}$ at a given temperature. At equilibrium in a 2.0-L container it is found that $[H_2O(g)] = 1.1 \times 10^{-1}\ M$ and $[H_2(g)] = 1.9 \times 10^{-2}\ M$. Calculate the moles of $O_2(g)$ present under these conditions.

44. The reaction

$$2NO(g) + Br_2(g) \rightleftharpoons 2NOBr(g)$$

has $K_p = 109$ at 25°C. If the equilibrium partial pressure of Br_2 is 0.0159 atm and the equilibrium partial pressure of NOBr is 0.0768 atm, calculate the partial pressure of NO at equilibrium.

45. A 1.00-L flask was filled with 2.00 moles of gaseous SO_2 and 2.00 moles of gaseous NO_2 and heated. After equilibrium was reached, it was found that 1.30 moles of gaseous NO was present. Assume that the reaction

$$SO_2(g) + NO_2(g) \rightleftharpoons SO_3(g) + NO(g)$$

occurs under these conditions. Calculate the value of the equilibrium constant, K, for this reaction.

46. A sample of $S_8(g)$ is placed in an otherwise empty rigid container at 1325 K at an initial pressure of 1.00 atm, where it decomposes to $S_2(g)$ by the reaction

$$S_8(g) \rightleftharpoons 4S_2(g)$$

At equilibrium, the partial pressure of S_8 is 0.25 atm. Calculate K_p for this reaction at 1325 K.

47. At a particular temperature, 12.0 moles of SO_3 is placed into a 3.0-L rigid container, and the SO_3 dissociates by the reaction

$$2SO_3(g) \rightleftharpoons 2SO_2(g) + O_2(g)$$

At equilibrium, 3.0 moles of SO_2 is present. Calculate K for this reaction.

48. At a particular temperature, 8.0 moles of NO_2 is placed into a 1.0-L container and the NO_2 dissociates by the reaction

$$2NO_2(g) \rightleftharpoons 2NO(g) + O_2(g)$$

At equilibrium the concentration of NO(g) is 2.0 M. Calculate K for this reaction.

49. An initial mixture of nitrogen gas and hydrogen gas is reacted in a rigid container at a certain temperature by the reaction

$$3H_2(g) + N_2(g) \rightleftharpoons 2NH_3(g)$$

At equilibrium, the concentrations are $[H_2] = 5.0\ M$, $[N_2] = 8.0\ M$, and $[NH_3] = 4.0\ M$. What were the concentrations of nitrogen gas and hydrogen gas that were reacted initially?

50. Nitrogen gas (N_2) reacts with hydrogen gas (H_2) to form ammonia (NH_3). At 200°C in a closed container, 1.00 atm of nitrogen gas is mixed with 2.00 atm of hydrogen gas. At equilibrium, the total pressure is 2.00 atm. Calculate the partial pressure of hydrogen gas at equilibrium, and calculate the K_p value for this reaction.

51. At a particular temperature, $K = 3.75$ for the reaction

$$SO_2(g) + NO_2(g) \rightleftharpoons SO_3(g) + NO(g)$$

If all four gases had initial concentrations of 0.800 M, calculate the equilibrium concentrations of the gases.

52. At a particular temperature, $K = 1.00 \times 10^2$ for the reaction

$$H_2(g) + I_2(g) \rightleftharpoons 2HI(g)$$

In an experiment, 1.00 mole of H_2, 1.00 mole of I_2, and 1.00 mole of HI are introduced into a 1.00-L container. Calculate the concentrations of all species when equilibrium is reached.

53. At 2200°C, $K_p = 0.050$ for the reaction

$$N_2(g) + O_2(g) \rightleftharpoons 2NO(g)$$

What is the partial pressure of NO in equilibrium with N_2 and O_2 that were placed in a flask at initial pressures of 0.80 and 0.20 atm, respectively?

54. At 25°C, $K = 0.090$ for the reaction

$$H_2O(g) + Cl_2O(g) \rightleftharpoons 2HOCl(g)$$

Calculate the concentrations of all species at equilibrium for each of the following cases.

a. 1.0 g H_2O and 2.0 g Cl_2O are mixed in a 1.0-L flask.

b. 1.0 mole of pure HOCl is placed in a 2.0-L flask.

55. At 1100 K, $K_p = 0.25$ for the reaction

$$2SO_2(g) + O_2(g) \rightleftharpoons 2SO_3(g)$$

Calculate the equilibrium partial pressures of SO_2, O_2, and SO_3 produced from an initial mixture in which $P_{SO_2} = P_{O_2} = 0.50$ atm and $P_{SO_3} = 0$. (*Hint:* If you don't have a graphing calculator, then use the method of successive approximations to solve, as discussed in Appendix 1.4.)

56. At a particular temperature, $K_p = 0.25$ for the reaction

$$N_2O_4(g) \rightleftharpoons 2NO_2(g)$$

a. A flask containing only N_2O_4 at an initial pressure of 4.5 atm is allowed to reach equilibrium. Calculate the equilibrium partial pressures of the gases.

b. A flask containing only NO_2 at an initial pressure of 9.0 atm is allowed to reach equilibrium. Calculate the equilibrium partial pressures of the gases.

c. From your answers to parts a and b, does it matter from which direction an equilibrium position is reached?

57. At 35°C, $K = 1.6 \times 10^{-5}$ for the reaction

$$2NOCl(g) \rightleftharpoons 2NO(g) + Cl_2(g)$$

Calculate the concentrations of all species at equilibrium for each of the following original mixtures.

a. 2.0 moles of pure NOCl in a 2.0-L flask

b. 1.0 mole of NOCl and 1.0 mole of NO in a 1.0-L flask

c. 2.0 moles of NOCl and 1.0 mole of Cl_2 in a 1.0-L flask

58. At a particular temperature, $K = 4.0 \times 10^{-7}$ for the reaction

$$N_2O_4(g) \rightleftharpoons 2NO_2(g)$$

In an experiment, 1.0 mole of N_2O_4 is placed in a 10.0-L vessel. Calculate the concentrations of N_2O_4 and NO_2 when this reaction reaches equilibrium.

59. At a particular temperature, $K = 2.0 \times 10^{-6}$ for the reaction

$$2CO_2(g) \rightleftharpoons 2CO(g) + O_2(g)$$

If 2.0 moles of CO_2 is initially placed into a 5.0-L vessel, calculate the equilibrium concentrations of all species.

60. Lexan is a plastic used to make compact discs, eyeglass lenses, and bulletproof glass. One of the compounds used to make Lexan is phosgene ($COCl_2$), an extremely poisonous gas. Phosgene decomposes by the reaction

$$COCl_2(g) \rightleftharpoons CO(g) + Cl_2(g)$$

for which $K_p = 6.8 \times 10^{-9}$ at 100°C. If pure phosgene at an initial pressure of 1.0 atm decomposes, calculate the equilibrium pressures of all species.

61. At 25°C, $K_p = 2.9 \times 10^{-3}$ for the reaction

$$NH_4OCONH_2(s) \rightleftharpoons 2NH_3(g) + CO_2(g)$$

In an experiment carried out at 25°C, a certain amount of NH_4OCONH_2 is placed in an evacuated rigid container and allowed to come to equilibrium. Calculate the total pressure in the container at equilibrium.

62. A sample of solid ammonium chloride was placed in an evacuated container and then heated so that it decomposed to ammonia gas and hydrogen chloride gas. After heating, the total pressure in the container was found to be 4.4 atm. Calculate K_p at this temperature for the decomposition reaction

$$NH_4Cl(s) \rightleftharpoons NH_3(g) + HCl(g)$$

Le Châtelier's Principle

63. Suppose the reaction system

$$UO_2(s) + 4HF(g) \rightleftharpoons UF_4(g) + 2H_2O(g)$$

has already reached equilibrium. Predict the effect that each of the following changes will have on the equilibrium position. Tell whether the equilibrium will shift to the right, will shift to the left, or will not be affected.

a. Additional $UO_2(s)$ is added to the system.

b. The reaction is performed in a glass reaction vessel; HF(g) attacks and reacts with glass.

c. Water vapor is removed.

64. Predict the shift in the equilibrium position that will occur for each of the following reactions when the volume of the reaction container is increased.

a. $N_2(g) + 3H_2(g) \rightleftharpoons 2NH_3(g)$

b. $PCl_5(g) \rightleftharpoons PCl_3(g) + Cl_2(g)$

c. $H_2(g) + F_2(g) \rightleftharpoons 2HF(g)$

d. $COCl_2(g) \rightleftharpoons CO(g) + Cl_2(g)$

e. $CaCO_3(s) \rightleftharpoons CaO(s) + CO_2(g)$

65. An important reaction in the commercial production of hydrogen is

$$CO(g) + H_2O(g) \rightleftharpoons H_2(g) + CO_2(g)$$

How will this system at equilibrium shift in each of the five following cases?

a. Gaseous carbon dioxide is removed.

b. Water vapor is added.

c. In a rigid reaction container, the pressure is increased by adding helium gas.

d. The temperature is increased (the reaction is exothermic).

e. The pressure is increased by decreasing the volume of the reaction container.

66. What will happen to the number of moles of SO_3 in equilibrium with SO_2 and O_2 in the reaction

$$2SO_3(g) \rightleftharpoons 2SO_2(g) + O_2(g)$$

in each of the following cases?

a. Oxygen gas is added.

b. The pressure is increased by decreasing the volume of the reaction container.

c. In a rigid reaction container, the pressure is increased by adding argon gas.

d. The temperature is decreased (the reaction is endothermic).

e. Gaseous sulfur dioxide is removed.

67. In which direction will the position of the equilibrium

$$2HI(g) \rightleftharpoons H_2(g) + I_2(g)$$

be shifted for each of the following changes?

a. $H_2(g)$ is added.

b. $I_2(g)$ is removed.

c. $HI(g)$ is removed.

d. In a rigid reaction container, some $Ar(g)$ is added.

e. The volume of the container is doubled.

f. The temperature is decreased (the reaction is exothermic).

68. Hydrogen for use in ammonia production is produced by the reaction

$$CH_4(g) + H_2O(g) \xrightarrow[750°C]{\text{Ni catalyst}} CO(g) + 3H_2(g)$$

What will happen to a reaction mixture at equilibrium if

a. $H_2O(g)$ is removed?

b. the temperature is increased (the reaction is endothermic)?

c. an inert gas is added to a rigid reaction container?

d. $CO(g)$ is removed?

e. the volume of the container is tripled?

69. Old-fashioned "smelling salts" consist of ammonium carbonate, $(NH_4)_2CO_3$. The reaction for the decomposition of ammonium carbonate

$$(NH_4)_2CO_3(s) \rightleftharpoons 2NH_3(g) + CO_2(g) + H_2O(g)$$

is endothermic. Would the smell of ammonia increase or decrease as the temperature is increased?

70. Ammonia is produced by the Haber process, in which nitrogen and hydrogen are reacted directly using an iron mesh impregnated with oxides as a catalyst. For the reaction

$$N_2(g) + 3H_2(g) \rightleftharpoons 2NH_3(g)$$

equilibrium constants (K_p values) as a function of temperature are

300°C, 4.34×10^{-3}
500°C, 1.45×10^{-5}
600°C, 2.25×10^{-6}

Is the reaction exothermic or endothermic?

Additional Exercises

71. Calculate a value for the equilibrium constant for the reaction

$$O_2(g) + O(g) \rightleftharpoons O_3(g)$$

given

$$NO_2(g) \xrightarrow{hv} NO(g) + O(g) \qquad K = 6.8 \times 10^{-49}$$
$$O_3(g) + NO(g) \rightleftharpoons NO_2(g) + O_2(g) \qquad K = 5.8 \times 10^{-34}$$

(*Hint:* When reactions are added together, the equilibrium expressions are multiplied.)

72. Given the following equilibrium constants at 427°C,

$$Na_2O(s) \rightleftharpoons 2Na(l) + \tfrac{1}{2}O_2(g) \qquad K_1 = 2 \times 10^{-25}$$
$$NaO(g) \rightleftharpoons Na(l) + \tfrac{1}{2}O_2(g) \qquad K_2 = 2 \times 10^{-5}$$
$$Na_2O_2(s) \rightleftharpoons 2Na(l) + O_2(g) \qquad K_3 = 5 \times 10^{-29}$$
$$NaO_2(s) \rightleftharpoons Na(l) + O_2(g) \qquad K_4 = 3 \times 10^{-14}$$

determine the values for the equilibrium constants for the following reactions:

a. $Na_2O(s) + \tfrac{1}{2}O_2(g) \rightleftharpoons Na_2O_2(s)$

b. $NaO(g) + Na_2O(s) \rightleftharpoons Na_2O_2(s) + Na(l)$

c. $2NaO(g) \rightleftharpoons Na_2O_2(s)$

(*Hint:* When reaction equations are added, the equilibrium expressions are multiplied.)

73. Consider the decomposition of the compound $C_5H_6O_3$ as follows:

$$C_5H_6O_3(g) \longrightarrow C_2H_6(g) + 3CO(g)$$

When a 5.63-g sample of pure $C_5H_6O_3(g)$ was sealed into an otherwise empty 2.50-L flask and heated to 200.°C, the pressure in the flask gradually rose to 1.63 atm and remained at that value. Calculate K for this reaction.

74. At 25°C, $K_p \approx 1 \times 10^{-31}$ for the reaction

$$N_2(g) + O_2(g) \rightleftharpoons 2NO(g)$$

a. Calculate the concentration of NO, in molecules/cm³, that can exist in equilibrium in air at 25°C. In air, $P_{N_2} = 0.8$ atm and $P_{O_2} = 0.2$ atm.

b. Typical concentrations of NO in relatively pristine environments range from 10^8 to 10^{10} molecules/cm³. Why is there a discrepancy between these values and your answer to part a?

75. The gas arsine, AsH_3, decomposes as follows:

$$2AsH_3(g) \rightleftharpoons 2As(s) + 3H_2(g)$$

In an experiment at a certain temperature, pure $AsH_3(g)$ was placed in an empty, rigid, sealed flask at a pressure of 392.0 torr.

After 48 hours the pressure in the flask was observed to be constant at 488.0 torr.

a. Calculate the equilibrium pressure of $H_2(g)$.

b. Calculate K_p for this reaction.

76. At a certain temperature, $K = 9.1 \times 10^{-4}$ for the reaction

$$FeSCN^{2+}(aq) \rightleftharpoons Fe^{3+}(aq) + SCN^-(aq)$$

Calculate the concentrations of Fe^{3+}, SCN^-, and $FeSCN^{2+}$ in a solution that is initially 2.0 M $FeSCN^{2+}$.

77. At a certain temperature, $K = 1.1 \times 10^3$ for the reaction

$$Fe^{3+}(aq) + SCN^-(aq) \rightleftharpoons FeSCN^{2+}(aq)$$

Calculate the concentrations of Fe^{3+}, SCN^-, and $FeSCN^{2+}$ at equilibrium if 0.020 mole of $Fe(NO_3)_3$ is added to 1.0 L of 0.10 M KSCN. (Neglect any volume change.)

78. For the reaction

$$PCl_5(g) \rightleftharpoons PCl_3(g) + Cl_2(g)$$

at 600. K, the equilibrium constant, K_p, is 11.5. Suppose that 2.450 g PCl_5 is placed in an evacuated 500.-mL bulb, which is then heated to 600. K.

a. What would be the pressure of PCl_5 if it did not dissociate?

b. What is the partial pressure of PCl_5 at equilibrium?

c. What is the total pressure in the bulb at equilibrium?

d. What is the percent dissociation of PCl_5 at equilibrium?

79. At 25°C, gaseous SO_2Cl_2 decomposes to $SO_2(g)$ and $Cl_2(g)$ to the extent that 12.5% of the original SO_2Cl_2 (by moles) has decomposed to reach equilibrium. The total pressure (at equilibrium) is 0.900 atm. Calculate the value of K_p for this system.

80. For the following reaction at a certain temperature

$$H_2(g) + F_2(g) \rightleftharpoons 2HF(g)$$

it is found that the equilibrium concentrations in a 5.00-L rigid container are $[H_2] = 0.0500\ M$, $[F_2] = 0.0100\ M$, and $[HF] = 0.400\ M$. If 0.200 mole of F_2 is added to this equilibrium mixture, calculate the concentrations of all gases once equilibrium is reestablished.

81. Novelty devices for predicting rain contain cobalt(II) chloride and are based on the following equilibrium:

$$\underset{\text{Purple}}{CoCl_2(s)} + 6H_2O(g) \rightleftharpoons \underset{\text{Pink}}{CoCl_2 \cdot 6H_2O(s)}$$

What color will such an indicator be if rain is imminent?

82. Consider the reaction

$$Fe^{3+}(aq) + SCN^-(aq) \rightleftharpoons FeSCN^{2+}(aq)$$

How will the equilibrium position shift if

a. water is added, doubling the volume?

b. $AgNO_3(aq)$ is added? (AgSCN is insoluble.)

c. $NaOH(aq)$ is added? [$Fe(OH)_3$ is insoluble.]

d. $Fe(NO_3)_3(aq)$ is added?

83. Chromium(VI) forms two different oxyanions, the orange dichromate ion, $Cr_2O_7^{2-}$, and the yellow chromate ion, CrO_4^{2-}.

(See the following photos.) The equilibrium reaction between the two ions is

$$Cr_2O_7^{2-}(aq) + H_2O(l) \rightleftharpoons 2CrO_4^{2-}(aq) + 2H^+(aq)$$

Explain why orange dichromate solutions turn yellow when sodium hydroxide is added.

© Cengage Learning

84. The synthesis of ammonia gas from nitrogen gas and hydrogen gas represents a classic case in which a knowledge of kinetics and equilibrium was used to make a desired chemical reaction economically feasible. Explain how each of the following conditions helps to maximize the yield of ammonia.

a. running the reaction at an elevated temperature

b. removing the ammonia from the reaction mixture as it forms

c. using a catalyst

d. running the reaction at high pressure

85. Suppose $K = 4.5 \times 10^{-3}$ at a certain temperature for the reaction

$$PCl_5(g) \rightleftharpoons PCl_3(g) + Cl_2(g)$$

If it is found that the concentration of PCl_5 is twice the concentration of PCl_3, what must be the concentration of Cl_2 under these conditions?

86. For the reaction below, $K_p = 1.16$ at 800.°C.

$$CaCO_3(s) \rightleftharpoons CaO(s) + CO_2(g)$$

If a 20.0-g sample of $CaCO_3$ is put into a 10.0-L container and heated to 800.°C, what percentage by mass of the $CaCO_3$ will react to reach equilibrium?

87. Many sugars undergo a process called *mutarotation,* in which the sugar molecules interconvert between two isomeric forms, finally reaching an equilibrium between them. This is true for the simple sugar glucose, $C_6H_{12}O_6$, which exists in solution in isomeric forms called alpha-glucose and beta-glucose. If a solution of glucose at a certain temperature is analyzed, and it is found that the concentration of alpha-glucose is twice the concentration of beta-glucose, what is the value of K for the interconversion reaction?

88. Peptide decomposition is one of the key processes of digestion, where a peptide bond is broken into an acid group and an amine group. We can describe this reaction as follows:

$$\text{Peptide}(aq) + H_2O(l) \rightleftharpoons \text{acid group}(aq) + \text{amine group}(aq)$$

If we place 1.0 mole of peptide into 1.0 L water, what will be the equilibrium concentrations of all species in this reaction? Assume the K value for this reaction is 3.1×10^{-5}.

89. The creation of shells by mollusk species is a fascinating process. By utilizing the Ca^{2+} in their food and aqueous environment, as well as some complex equilibrium processes, a hard calcium carbonate shell can be produced. One important equilibrium reaction in this complex process is

$$HCO_3^-(aq) \rightleftharpoons H^+(aq) + CO_3^{2-}(aq) \qquad K = 5.6 \times 10^{-11}$$

If 0.16 mole of HCO_3^- is placed into 1.00 L of solution, what will be the equilibrium concentration of CO_3^{2-}?

90. Methanol, a common laboratory solvent, poses a threat of blindness or death if consumed in sufficient amounts. Once in the body, the substance is oxidized to produce formaldehyde (embalming fluid) and eventually formic acid. Both of these substances are also toxic in varying levels. The equilibrium between methanol and formaldehyde can be described as follows:

$$CH_3OH(aq) \rightleftharpoons H_2CO(aq) + H_2(aq)$$

Assuming the value of K for this reaction is 3.7×10^{-10}, what are the equilibrium concentrations of each species if you start with a 1.24 M solution of methanol? What will happen to the concentration of methanol as the formaldehyde is further converted to formic acid?

ChemWork Problems

These multiconcept problems (and additional ones) are found interactively online with the same type of assistance a student would get from an instructor.

91. For the reaction:

$$3O_2(g) \rightleftharpoons 2O_3(g)$$

$K = 1.8 \times 10^{-7}$ at a certain temperature. If at equilibrium $[O_2] = 0.062\ M$, calculate the equilibrium O_3 concentration.

92. An equilibrium mixture contains 0.60 g solid carbon and the gases carbon dioxide and carbon monoxide at partial pressures of 2.60 atm and 2.89 atm, respectively. Calculate the value of K_p for the reaction $C(s) + CO_2(g) \rightleftharpoons 2CO(g)$.

93. At a particular temperature, 8.1 moles of NO_2 gas is placed in a 3.0-L container. Over time the NO_2 decomposes to NO and O_2:

$$2NO_2(g) \rightleftharpoons 2NO(g) + O_2(g)$$

At equilibrium the concentration of $NO(g)$ was found to be 1.4 mol/L. Calculate the value of K for this reaction.

94. A sample of solid ammonium chloride was placed in an evacuated chamber and then heated, causing it to decompose according to the following reaction:

$$NH_4Cl(s) \rightleftharpoons NH_3(g) + HCl(g)$$

In a particular experiment, the equilibrium partial pressure of $NH_3(g)$ in the container was 2.9 atm. Calculate the value of K_p for the decomposition of $NH_4Cl(s)$ at this temperature.

95. In a given experiment, 5.2 moles of pure NOCl was placed in an otherwise empty 2.0-L container. Equilibrium was established by the following reaction:

$$2NOCl(g) \rightleftharpoons 2NO(g) + Cl_2(g) \qquad K = 1.6 \times 10^{-5}$$

a. Using numerical values for the concentrations in the Initial row and expressions containing the variable x in both the Change and Equilibrium rows, complete the following table summarizing what happens as this reaction reaches equilibrium. Let x = the concentration of Cl_2 that is present at equilibrium.

	NOCl	NO	Cl₂
Initial			
Change			
Equilibrium	$2.6 - 2x$		

b. Calculate the equilibrium concentrations for all species.

96. For the reaction $N_2O_4(g) \rightleftharpoons 2NO_2(g)$, $K_p = 0.25$ at a certain temperature. If 0.040 atm of N_2O_4 is reacted initially, calculate the equilibrium partial pressures of $NO_2(g)$ and $N_2O_4(g)$.

97. Consider the following exothermic reaction at equilibrium:

$$N_2(g) + 3H_2(g) \rightleftharpoons 2NH_3(g)$$

Predict how the following changes affect the number of moles of each component of the system after equilibrium is reestablished by completing the table below. Complete the table with the terms *increase*, *decrease*, or *no change*.

	N₂	H₂	NH₃
Add N₂(g)			
Remove H₂(g)			
Add NH₃(g)			
Add Ne(g) (constant V)			
Increase the temperature			
Decrease the volume (constant T)			
Add a catalyst			

98. For the following endothermic reaction at equilibrium:

$$2SO_3(g) \rightleftharpoons 2SO_2(g) + O_2(g)$$

which of the following changes will increase the value of K?

a. increasing the temperature
b. decreasing the temperature
c. removing $SO_3(g)$ (constant T)
d. decreasing the volume (constant T)
e. adding $Ne(g)$ (constant T)
f. adding $SO_2(g)$ (constant T)
g. adding a catalyst (constant T)

Challenge Problems

99. A 1.604-g sample of methane (CH_4) gas and 6.400 g oxygen gas are sealed into a 2.50-L vessel at 411°C and are allowed to reach equilibrium. Methane can react with oxygen to form gaseous carbon dioxide and water vapor, or methane can react with oxygen to form gaseous carbon monoxide and water vapor. At equilibrium, the pressure of oxygen is 0.326 atm, and the pressure of water vapor is 4.45 atm. Calculate the pressures of carbon monoxide and carbon dioxide present at equilibrium.

100. A 4.72-g sample of methanol (CH_3OH) was placed in an otherwise empty 1.00-L flask and heated to 250.°C to vaporize the methanol. Over time, the methanol vapor decomposed by the following reaction:

$$CH_3OH(g) \rightleftharpoons CO(g) + 2H_2(g)$$

After the system has reached equilibrium, a tiny hole is drilled in the side of the flask allowing gaseous compounds to effuse out of the flask. Measurements of the effusing gas show that it contains 33.0 times as much $H_2(g)$ as $CH_3OH(g)$. Calculate K for this reaction at 250.°C.

101. At 35°C, $K = 1.6 \times 10^{-5}$ for the reaction

$$2NOCl(g) \rightleftharpoons 2NO(g) + Cl_2(g)$$

If 2.0 moles of NO and 1.0 mole of Cl_2 are placed into a 1.0-L flask, calculate the equilibrium concentrations of all species.

102. Nitric oxide and bromine at initial partial pressures of 98.4 and 41.3 torr, respectively, were allowed to react at 300. K. At equilibrium the total pressure was 110.5 torr. The reaction is

$$2NO(g) + Br_2(g) \rightleftharpoons 2NOBr(g)$$

a. Calculate the value of K_p.

b. What would be the partial pressures of all species if NO and Br_2, both at an initial partial pressure of 0.30 atm, were allowed to come to equilibrium at this temperature?

103. At 25°C, $K_p = 5.3 \times 10^5$ for the reaction

$$N_2(g) + 3H_2(g) \rightleftharpoons 2NH_3(g)$$

When a certain partial pressure of $NH_3(g)$ is put into an otherwise empty rigid vessel at 25°C, equilibrium is reached when 50.0% of the original ammonia has decomposed. What was the original partial pressure of ammonia before any decomposition occurred?

104. Consider the reaction

$$P_4(g) \longrightarrow 2P_2(g)$$

where $K_p = 1.00 \times 10^{-1}$ at 1325 K. In an experiment where $P_4(g)$ is placed into a container at 1325 K, the equilibrium mixture of $P_4(g)$ and $P_2(g)$ has a total pressure of 1.00 atm. Calculate the equilibrium pressures of $P_4(g)$ and $P_2(g)$. Calculate the fraction (by moles) of $P_4(g)$ that has dissociated to reach equilibrium.

105. The partial pressures of an equilibrium mixture of $N_2O_4(g)$ and $NO_2(g)$ are $P_{N_2O_4} = 0.34$ atm and $P_{NO_2} = 1.20$ atm at a certain temperature. The volume of the container is doubled. Calculate the partial pressures of the two gases when a new equilibrium is established.

106. At 125°C, $K_p = 0.25$ for the reaction

$$2NaHCO_3(s) \rightleftharpoons Na_2CO_3(s) + CO_2(g) + H_2O(g)$$

A 1.00-L flask containing 10.0 g $NaHCO_3$ is evacuated and heated to 125°C.

a. Calculate the partial pressures of CO_2 and H_2O after equilibrium is established.

b. Calculate the masses of $NaHCO_3$ and Na_2CO_3 present at equilibrium.

c. Calculate the minimum container volume necessary for all of the $NaHCO_3$ to decompose.

107. A mixture of N_2, H_2, and NH_3 is at equilibrium [according to the equation $N_2(g) + 3H_2(g) \rightleftharpoons 2NH_3(g)$] as depicted below:

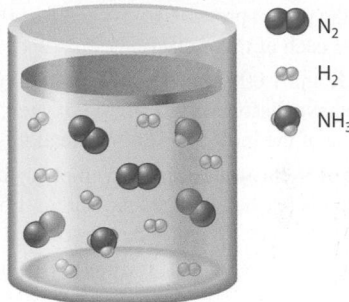

The volume is suddenly decreased (by increasing the external pressure) and a new equilibrium is established as depicted below:

a. If the volume of the final equilibrium mixture is 1.00 L, determine the value of the equilibrium constant, K, for the reaction. Assume temperature is constant.

b. Determine the volume of the initial equilibrium mixture assuming a final equilibrium volume of 1.00 L and assuming a constant temperature.

108. Consider the decomposition equilibrium for dinitrogen pentoxide:

$$2N_2O_5(g) \rightleftharpoons 4NO_2(g) + O_2(g)$$

At a certain temperature and a total pressure of 1.00 atm, the N_2O_5 is 0.50% decomposed (by moles) at equilibrium.

a. If the volume is increased by a factor of 10.0, will the mole percent of N_2O_5 decomposed at equilibrium be greater than, less than, or equal to 0.50%? Explain your answer.

b. Calculate the mole percent of N_2O_5 that will be decomposed at equilibrium if the volume is increased by a factor of 10.0.

109. An 8.00-g sample of SO_3 was placed in an evacuated container, where it decomposed at 600°C according to the following reaction:

$$SO_3(g) \rightleftharpoons SO_2(g) + \tfrac{1}{2}O_2(g)$$

At equilibrium the total pressure and the density of the gaseous mixtures were 1.80 atm and 1.60 g/L, respectively. Calculate K_p for this reaction.

110. A sample of iron(II) sulfate was heated in an evacuated container to 920 K, where the following reactions occurred:

$$2FeSO_4(s) \rightleftharpoons Fe_2O_3(s) + SO_3(g) + SO_2(g)$$
$$SO_3(g) \rightleftharpoons SO_2(g) + \tfrac{1}{2}O_2(g)$$

After equilibrium was reached, the total pressure was 0.836 atm and the partial pressure of oxygen was 0.0275 atm. Calculate K_p for each of these reactions.

111. At 5000 K and 1.000 atm, 83.00% of the oxygen molecules in a sample have dissociated to atomic oxygen. At what pressure will 95.0% of the molecules dissociate at this temperature?

112. A sample of $N_2O_4(g)$ is placed in an empty cylinder at 25°C. After equilibrium is reached the total pressure is 1.5 atm and 16% (by moles) of the original $N_2O_4(g)$ has dissociated to $NO_2(g)$.

  a. Calculate the value of K_p for this dissociation reaction at 25°C.

  b. If the volume of the cylinder is increased until the total pressure is 1.0 atm (the temperature of the system remains constant), calculate the equilibrium pressure of $N_2O_4(g)$ and $NO_2(g)$.

  c. What percentage (by moles) of the original $N_2O_4(g)$ is dissociated at the new equilibrium position (total pressure = 1.00 atm)?

113. A sample of gaseous nitrosyl bromide (NOBr) was placed in a container fitted with a frictionless, massless piston, where it decomposed at 25°C according to the following equation:

$$2NOBr(g) \rightleftharpoons 2NO(g) + Br_2(g)$$

The initial density of the system was recorded as 4.495 g/L. After equilibrium was reached, the density was noted to be 4.086 g/L.

  a. Determine the value of the equilibrium constant K for the reaction.

  b. If $Ar(g)$ is added to the system at equilibrium at constant temperature, what will happen to the equilibrium position? What happens to the value of K? Explain each answer.

114. The equilibrium constant K_p for the reaction

$$CCl_4(g) \rightleftharpoons C(s) + 2Cl_2(g)$$

at 700°C is 0.76. Determine the initial pressure of carbon tetrachloride that will produce a total equilibrium pressure of 1.20 atm at 700°C.

Integrative Problems

These problems require the integration of multiple concepts to find the solutions.

115. For the reaction

$$NH_3(g) + H_2S(g) \rightleftharpoons NH_4HS(s)$$

$K = 400.$ at 35.0°C. If 2.00 moles each of NH_3, H_2S, and NH_4HS are placed in a 5.00-L vessel, what mass of NH_4HS will be present at equilibrium? What is the pressure of H_2S at equilibrium?

116. Given $K = 3.50$ at 45°C for the reaction

$$A(g) + B(g) \rightleftharpoons C(g)$$

and $K = 7.10$ at 45°C for the reaction

$$2A(g) + D(g) \rightleftharpoons C(g)$$

what is the value of K at the same temperature for the reaction

$$C(g) + D(g) \rightleftharpoons 2B(g)$$

What is the value of K_p at 45°C for the reaction? Starting with 1.50 atm partial pressures of both C and D, what is the mole fraction of B once equilibrium is reached?

117. In a solution with carbon tetrachloride as the solvent, the compound VCl_4 undergoes dimerization:

$$2VCl_4 \rightleftharpoons V_2Cl_8$$

When 6.6834 g VCl_4 is dissolved in 100.0 g carbon tetrachloride, the freezing point is lowered by 5.97°C. Calculate the value of the equilibrium constant for the dimerization of VCl_4 at this temperature. (The density of the equilibrium mixture is 1.696 g/cm³, and $K_f = 29.8$°C kg/mol for CCl_4.)

118. The hydrocarbon naphthalene was frequently used in mothballs until recently, when it was discovered that human inhalation of naphthalene vapors can lead to hemolytic anemia. Naphthalene is 93.71% carbon by mass, and a 0.256-mole sample of naphthalene has a mass of 32.8 g. What is the molecular formula of naphthalene? This compound works as a pesticide in mothballs by sublimation of the solid so that it fumigates enclosed spaces with its vapors according to the equation

$$Naphthalene(s) \rightleftharpoons naphthalene(g)$$
$$K = 4.29 \times 10^{-6} \text{ (at 298 K)}$$

If 3.00 g solid naphthalene is placed into an enclosed space with a volume of 5.00 L at 25°C, what percentage of the naphthalene will have sublimed once equilibrium has been established?

Marathon Problem

This problem is designed to incorporate several concepts and techniques into one situation.

119. A gaseous material $XY(g)$ dissociates to some extent to produce $X(g)$ and $Y(g)$:

$$XY(g) \rightleftharpoons X(g) + Y(g)$$

A 2.00-g sample of XY (molar mass = 165 g/mol) is placed in a container with a movable piston at 25°C. The pressure is held constant at 0.967 atm. As XY begins to dissociate, the piston moves until 35.0 mole percent of the original XY has dissociated and then remains at a constant position. Assuming ideal behavior, calculate the density of the gas in the container after the piston has stopped moving, and determine the value of K for this reaction of 25°C.

CHAPTER THIRTEEN

Acids and Bases

The colors of hydrangea blossoms depend on the acidity of the soil. (Basie B/iStockphoto.com)

525

n this chapter we reencounter two very important classes of compounds, acids and bases. We will explore their interactions and apply the fundamentals of chemical equilibria discussed in Chapter 12 to systems involving proton-transfer reactions.

Acid–base chemistry is important in a wide variety of everyday applications. There are complex systems in our bodies that carefully control the acidity of our blood, since even small deviations may lead to serious illness and death. The same sensitivity exists in other life forms. If you have ever had tropical fish or goldfish, you know how important it is to monitor and control the acidity of the water in the aquarium.

Acids and bases are also important in industry. For example, the vast quantity of sulfuric acid manufactured in the United States each year is needed to produce fertilizers, polymers, steel, and many other materials.

The influence of acids on living things has assumed special importance in the United States, Canada, and Europe in recent years as a result of the phenomenon of acid rain. This problem is complex and has diplomatic and economic overtones that make it all the more difficult to solve.

We will discuss the nature of acids and bases, including how the molecular structure affects the acid–base properties of a substance. As you will see, even when solving quantitative problems, it is best to consider which molecules are present in the solution and how they behave. That is, we will ask you to consider the major species in the solutions.

13-1 | The Nature of Acids and Bases

ⓘ *Don't* taste chemicals!

ⓘ Acids and bases were first discussed in Section 6-2.

Acids were first recognized as a class of substances that taste sour. Vinegar tastes sour because it is a dilute solution of acetic acid; citric acid is responsible for the sour taste of a lemon. ◄ ⓘ ◄ Bases, sometimes called *alkalis,* are characterized by their bitter taste and slippery feel. ◄ Commercial preparations for unclogging drains are highly basic. ◄ ◄ ⓘ

▲ Acidic fruits.

▲ Liquids that contain acids and bases.

▲ Household products that contain acids and bases.

526

The first person to recognize the essential nature of acids and bases was Svante Arrhenius. Based on his experiments with electrolytes, Arrhenius postulated that *acids produce hydrogen ions in aqueous solution, while bases produce hydroxide ions.* At the time, the **Arrhenius concept** of acids and bases was a major step forward in quantifying acid–base chemistry, but this concept is limited because it applies only to aqueous solutions and allows for only one kind of base—the hydroxide ion. A more general definition of acids and bases was suggested by the Danish chemist Johannes Brønsted (1879–1947) and the English chemist Thomas Lowry (1874–1936). In terms of the **Brønsted–Lowry model**, *an acid is a proton (H^+) donor, and a base is a proton acceptor.* For example, when gaseous HCl dissolves in water, each HCl molecule donates a proton to a water molecule and so qualifies as a Brønsted–Lowry acid. The molecule that accepts the proton, in this case water, is a Brønsted–Lowry base.

i As we will see throughout the chapter, drawing the Lewis structures of the molecules helps us understand whether a molecule acts as an acid, base, both, or neither.

To understand how water can act as a base, we need to remember that the oxygen of the water molecule has two unshared electron pairs, either of which can form a covalent bond with an H^+ ion. When gaseous HCl dissolves, the following reaction occurs: ◄ *i*

$$H-\overset{\cdot\cdot}{\underset{|}{O}}: \; + \; H-\overset{\cdot\cdot}{\underset{\cdot\cdot}{C}l}: \longrightarrow \left[H-\overset{\cdot\cdot}{\underset{|}{O}}-H\right]^+ \; + \; \left[:\overset{\cdot\cdot}{\underset{\cdot\cdot}{C}l}:\right]^-$$
$$H H$$

Note that the proton is transferred from the HCl molecule to the water molecule to form H_3O^+, which is called the **hydronium ion***. This reaction is represented in Fig. 13-1 using molecular models.

The general reaction that occurs when an acid is dissolved in water can best be represented as ◄ *i*

i Recall that (*aq*) means the substance is hydrated.

$$HA(aq) + H_2O(l) \rightleftharpoons H_3O^+(aq) + A^-(aq) \qquad (13\text{-}1)$$

| Acid | Base | Conjugate acid | Conjugate base |

This representation emphasizes the significant role of the polar water molecule in pulling the proton from the acid. Note that the **conjugate base** is everything that remains of the acid molecule after a proton is lost. The **conjugate acid** is formed when the proton is transferred to the base. A **conjugate acid–base pair** consists of two substances related to each other by the donating and accepting of a single proton. In Equation (13-1) there are two conjugate acid–base pairs: HA and A^- as well as H_2O and H_3O^+. This reaction is represented by molecular models in Fig. 13-2.

It is important to note that Equation (13-1) really represents *a competition for the proton between the two bases H_2O and A^-*. If H_2O is a much stronger base than A^-, that is, if H_2O has a much greater affinity for H^+ than does A^-, the equilibrium position will be far to the right; most of the acid dissolved will be in the ionized form. Conversely, if A^- is a much stronger base than H_2O, the equilibrium position will lie far to the left. In this case most of the acid dissolved will be present at equilibrium as HA.

The equilibrium expression for the reaction given in Equation (13-1) is

$$K_a = \frac{[H_3O^+][A^-]}{[HA]} = \frac{[H^+][A^-]}{[HA]} \qquad (13\text{-}2)$$

Figure 13-1 | The reaction of HCl and H_2O.

---

*The hydronium ion is actually a more complicated species than H_3O^+. See "The Solvated Proton Is NOT H_3O^+!" (*J. Chem. Ed.*, 2011, 88(7), p 875).

Figure 13-2 | The reaction of an acid HA with water to form H_3O^+ and a conjugate base A^-.

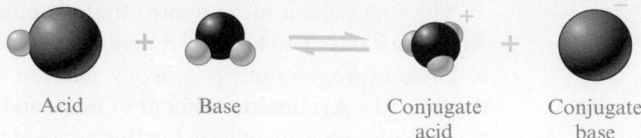

Acid   Base   Conjugate   Conjugate
                         acid   base

where K_a is called the **acid dissociation constant**. Both $H_3O^+(aq)$ and $H^+(aq)$ are commonly used to represent the hydrated proton. In this book we will often use simply H^+, but you should remember that it is hydrated in aqueous solutions.

In Chapter 12 we saw that the concentration of a pure solid or a pure liquid is always omitted from the equilibrium expression. In a dilute solution we can assume that the concentration of liquid water remains essentially constant when an acid is dissolved. Thus the term $[H_2O]$ is not included in Equation (13-2), and the equilibrium expression for K_a has the same form as that for the simple dissociation into ions:

$$HA(aq) \rightleftharpoons H^+(aq) + A^-(aq)$$

> *i* In this chapter we will always represent an acid as simply dissociating. This does not mean we are using the Arrhenius model for acids. Since water does not affect the equilibrium position, it is simply easier to leave it out of the acid dissociation reaction.

You should not forget, however, that water plays an important role in causing the acid to ionize. ◄ *i*

Note that K_a is the equilibrium constant for the reaction in which a proton is removed from HA to form the conjugate base A^-. We use K_a to represent *only* this type of reaction. Knowing this, you can write the K_a expression for any acid, even one that is totally unfamiliar to you. As you do Example 13-1, focus on the definition of the reaction corresponding to K_a.

INTERACTIVE EXAMPLE 13-1

Acid Dissociation (Ionization) Reactions

Write the simple dissociation (ionization) reaction (omitting water) for each of the following acids.

a. hydrochloric acid (HCl)

b. acetic acid ($HC_2H_3O_2$)

c. the ammonium ion (NH_4^+)

d. the anilinium ion ($C_6H_5NH_3^+$)

e. the hydrated aluminum(III) ion $[Al(H_2O)_6]^{3+}$

Solution

a. $HCl(aq) \rightleftharpoons H^+(aq) + Cl^-(aq)$

b. $HC_2H_3O_2(aq) \rightleftharpoons H^+(aq) + C_2H_3O_2^-(aq)$

c. $NH_4^+(aq) \rightleftharpoons H^+(aq) + NH_3(aq)$

d. $C_6H_5NH_3^+(aq) \rightleftharpoons H^+(aq) + C_6H_5NH_2(aq)$

e. Although this formula looks complicated, writing the reaction is simple if you concentrate on the meaning of K_a. Removing a proton, which can come only from one of the water molecules, leaves one OH^- and five H_2O molecules attached to the Al^{3+} ion. So the reaction is

$$Al(H_2O)_6^{3+}(aq) \rightleftharpoons H^+(aq) + Al(H_2O)_5OH^{2+}(aq)$$

See Exercises 13-35 and 13-36

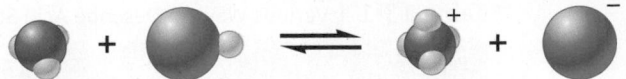

Figure 13-3 | The reaction of NH₃ with HCl to form NH₄⁺ and Cl⁻.

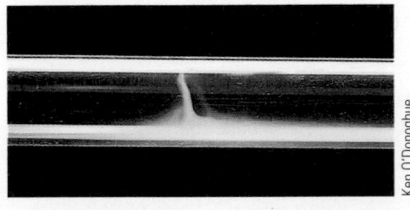

▲ When HCl(g) and NH₃(g) meet in a tube, a white ring of NH₄Cl(s) forms.

The Brønsted–Lowry model is not limited to aqueous solutions; it can be extended to reactions in the gas phase. For example, we discussed the reaction between gaseous hydrogen chloride and ammonia when we studied diffusion (see Section 8-7): ◄

$$NH_3(g) + HCl(g) \rightleftharpoons NH_4Cl(s)$$

In this reaction, a proton is donated by the hydrogen chloride to the ammonia, as shown by these Lewis structures:

$$
\begin{array}{c} H \\ | \\ H-\!N\!:\ +\ H-\!\ddot{\underset{\cdot\cdot}{Cl}}\!: \\ | \\ H \end{array}
\rightleftharpoons
\left[\begin{array}{c} H \\ | \\ H-\!N\!-\!H \\ | \\ H \end{array}\right]^{+}
\left[:\!\ddot{\underset{\cdot\cdot}{Cl}}\!:\right]^{-}
$$

Note that this is not considered an acid–base reaction according to the Arrhenius concept. Figure 13-3 shows a molecular representation of this reaction.

13-2 | Acid Strength

The strength of an acid is defined by the equilibrium position of its dissociation (ionization) reaction:

$$HA(aq) + H_2O(l) \rightleftharpoons H_3O^+(aq) + A^-(aq)$$

A **strong acid** is one for which *this equilibrium lies far to the right.* This means that almost all the original HA is dissociated (ionized) at equilibrium [Fig. 13-4(a)]. There is an important connection between the strength of an acid and that of its conjugate base. *A strong acid yields a weak conjugate base*—one that has a low affinity for a proton. ◄ *ⓘ* A strong acid also can be described as an acid whose conjugate base is a

ⓘ A strong acid has a weak conjugate base.

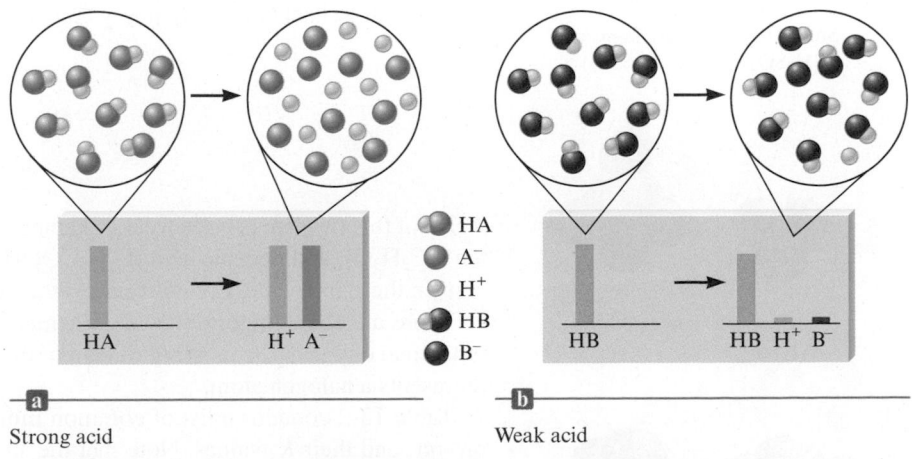

	HA
	A⁻
	H⁺
	HB
	B⁻

HA → H⁺ A⁻

HB → HB H⁺ B⁻

ⓐ Strong acid   ⓑ Weak acid

Figure 13-4 | (a) A strong acid HA is completely ionized in water. This is represented in terms of both molecules and a bar graph. (b) A weak acid HB consists of mostly undissociated HB molecules in water.

ℹ ≪ means much less than.
≫ means much greater than.

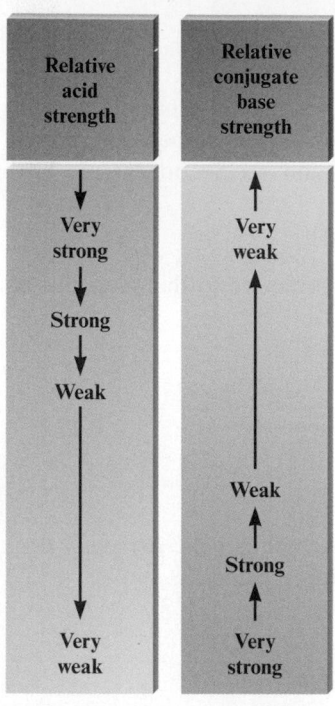

Figure 13-5 │ The relationship of acid strength and conjugate base strength for the reaction

HA(aq) + H$_2$O(l) ⇌

Acid

H$_3$O$^+$(aq) + A$^-$(aq)

Conjugate base

Sulfuric acid   Perchloric acid
(H$_2$SO$_4$)   (HClO$_4$)

Nitric acid   Phosphoric acid
(HNO$_3$)   (H$_3$PO$_4$)

Nitrous acid   Hypochlorous acid
(HNO$_2$)   (HOCl)

Table 13-1 │ Various Ways to Describe Acid Strength

Property	Strong Acid	Weak Acid
K_a value	K_a is large	K_a is small
Position of the dissociation (ionization) equilibrium	Far to the right	Far to the left
Equilibrium concentration of [H$^+$] compared with original concentration of HA	[H$^+$] ≈ [HA]$_0$	[H$^+$] ≪ [HA]$_0$ ◄ ℹ
Strength of conjugate base compared with that of water	A$^-$ much weaker base than H$_2$O	A$^-$ much stronger base than H$_2$O

much weaker base than water (Fig. 13-5). In this case the water molecules win the competition for the H$^+$ ions.

Conversely, a **weak acid** is one for which *the equilibrium lies far to the left.* Most of the acid originally placed in the solution is still present as HA at equilibrium. That is, a weak acid dissociates only to a very small extent in aqueous solution [see Fig. 13-4(b)]. In contrast to a strong acid, a weak acid has a conjugate base that is a much stronger base than water. In this case a water molecule is not very successful in pulling an H$^+$ ion from the conjugate base. The weaker the acid, the stronger its conjugate base. The various ways of describing the strength of an acid are summarized in Table 13-1.

The common strong acids are sulfuric acid [H$_2$SO$_4$(aq)], hydrochloric acid [HCl(aq)], nitric acid [HNO$_3$(aq)], and perchloric acid [HClO$_4$(aq)]. Sulfuric acid is actually a **diprotic acid**, an acid having two acidic protons. The acid H$_2$SO$_4$ is a strong acid, virtually 100% dissociated (ionized) in water:

$$H_2SO_4(aq) \longrightarrow H^+(aq) + HSO_4^-(aq)$$

The HSO$_4^-$ ion, however, is a weak acid:

$$HSO_4^-(aq) \rightleftharpoons H^+(aq) + SO_4^{2-}(aq)$$

Most acids are **oxyacids**, in which the acidic proton is attached to an oxygen atom. ◄ The strong acids mentioned above, except hydrochloric acid, are typical examples. Many common weak acids, such as phosphoric acid (H$_3$PO$_4$), nitrous acid (HNO$_2$), and hypochlorous acid (HOCl), are also oxyacids. **Organic acids**, those with a carbon atom backbone, commonly contain the **carboxyl group**:

$$-C\overset{\displaystyle O}{\underset{\displaystyle O-H}{\|}}$$

Acids of this type are usually weak. Examples are acetic acid (CH$_3$COOH), often written HC$_2$H$_3$O$_2$, and benzoic acid (C$_6$H$_5$COOH). Note that the remainder of the hydrogens in these molecules are not acidic—they do not form H$^+$ in water.

There are some important acids in which the acidic proton is attached to an atom other than oxygen. The most significant of these are the hydrohalic acids HX, where X represents a halogen atom.

Table 13-2 contains a list of common **monoprotic acids** (those having *one* acidic proton) and their K_a values. Note that the strong acids are not listed. When a strong acid molecule such as HCl, for example, is placed in water, the position of the dissociation equilibrium

$$HCl(aq) \rightleftharpoons H^+(aq) + Cl^-(aq)$$

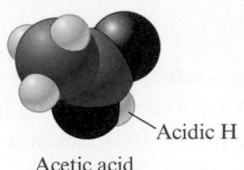

Acetic acid
(CH_3CO_2H)

Acidic H

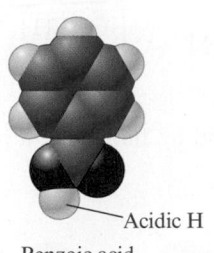

Benzoic acid
($C_6H_5CO_2H$)

Acidic H

Table 13-2 | Values of K_a for Some Common Monoprotic Acids

Formula	Name	Value of K_a*
HSO_4^-	Hydrogen sulfate ion	1.2×10^{-2}
$HClO_2$	Chlorous acid	1.2×10^{-2}
$HC_2H_2ClO_2$	Monochloracetic acid	1.35×10^{-3}
HF	Hydrofluoric acid	7.2×10^{-4}
HNO_2	Nitrous acid	4.0×10^{-4}
$HC_2H_3O_2$	Acetic acid	1.8×10^{-5}
$[Al(H_2O)_6]^{3+}$	Hydrated aluminum(III) ion	1.4×10^{-5}
HOCl	Hypochlorous acid	3.5×10^{-8}
HCN	Hydrocyanic acid	6.2×10^{-10}
NH_4^+	Ammonium ion	5.6×10^{-10}
HOC_6H_5	Phenol	1.6×10^{-10}

Increasing acid strength →

*The units of K_a are customarily omitted.

lies so far to the right that [HCl] cannot be measured accurately. This prevents an accurate calculation of K_a:

$$K_a = \frac{[H^+][Cl^-]}{[HCl]}$$

Very small and
highly uncertain

Critical Thinking

Vinegar contains acetic acid and is used in salad dressings. What if acetic acid was a strong acid instead of a weak acid? Would it be safe to use vinegar as a salad dressing?

INTERACTIVE EXAMPLE 13-2

Relative Base Strength

Using Table 13-2, arrange the following species according to their strengths as bases: H_2O, F^-, Cl^-, NO_2^-, and CN^-.

Solution

Remember that water is a stronger base than the conjugate base of a strong acid but a weaker base than the conjugate base of a weak acid. This leads to the following order:

$$Cl^- < H_2O < \text{conjugate bases of weak acids}$$

Weakest bases ⟶ Strongest bases

We can order the remaining conjugate bases by recognizing that the strength of an acid is *inversely related* to the strength of its conjugate base. Since from Table 13-2 we have

$$K_a \text{ for HF} > K_a \text{ for } HNO_2 > K_a \text{ for HCN} \; ◄ \; ⓘ$$

the base strengths increase as follows:

$$F^- < NO_2^- < CN^-$$

The combined order of increasing base strength is

$$Cl^- < H_2O < F^- < NO_2^- < CN^-$$

ⓘ Appendix 5-1 contains a table of K_a values.

See Exercises 13-41 through 13-44

Water as an Acid and a Base

A substance is said to be **amphoteric** if it can behave either as an acid or as a base. Water is the most common **amphoteric substance**. We can see this clearly in the **autoionization** of water, which involves the transfer of a proton from one water molecule to another to produce a hydroxide ion and a hydronium ion: ◄

$$\underset{\text{acid(1)}}{H_2O} \; + \; \underset{\text{base(1)}}{H_2O} \rightleftharpoons \underset{\text{acid(2)}}{H_3O^+} + \; \underset{\text{base(2)}}{OH^-}$$

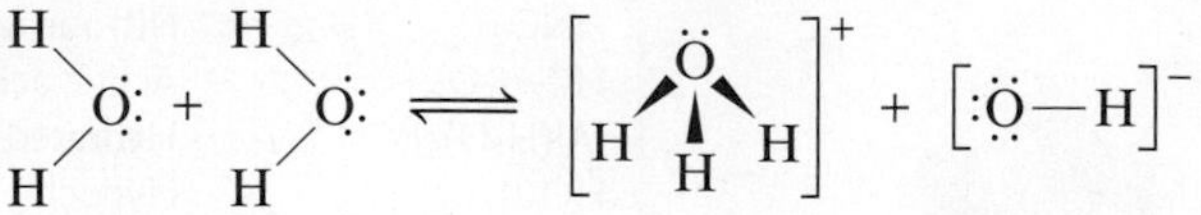

In this reaction, one water molecule acts as an acid by furnishing a proton, and the other acts as a base by accepting the proton.

Autoionization can occur in other liquids besides water. For example, in liquid ammonia the autoionization reaction is

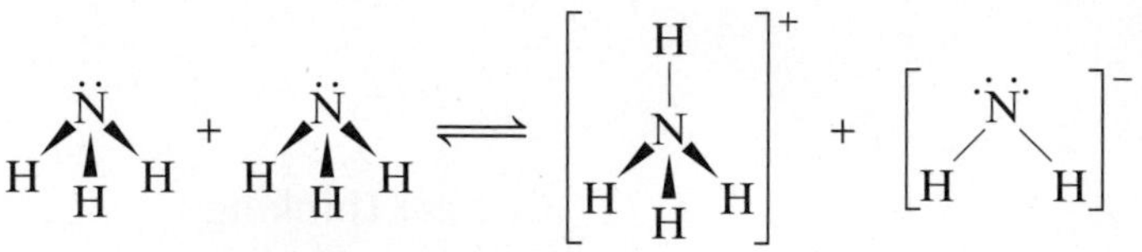

The autoionization reaction for water

$$2H_2O(l) \rightleftharpoons H_3O^+(aq) + OH^-(aq)$$

leads to the equilibrium expression

$$K_w = [H_3O^+][OH^-] = [H^+][OH^-]$$

where K_w, called the **ion-product constant** (or the **dissociation constant** for water), always refers to the autoionization of water.

Experiment shows that at 25°C in pure water,

$$[H^+] = [OH^-] = 1.0 \times 10^{-7}\,M$$

which means that at 25°C

$$\begin{aligned}
K_w = [H^+][OH^-] &= (1.0 \times 10^{-7})(1.0 \times 10^{-7}) \\
&= 1.0 \times 10^{-14}
\end{aligned}$$

LET'S REVIEW | Using K_w

It is important to recognize the meaning of K_w. In any aqueous solution at 25°C, *no matter what it contains*, the product of $[H^+]$ and $[OH^-]$ must always equal 1.0×10^{-14}. There are three possible situations:

- a neutral solution, where $[H^+] = [OH^-]$
- an acidic solution, where $[H^+] > [OH^-]$
- a basic solution, where $[OH^-] > [H^+]$

In each case, however, at 25°C,

$$K_w = [H^+][OH^-] = 1.0 \times 10^{-14}$$

We normally consider the pH of solutions at 25°C. However, it is important to realize that K_w is temperature dependent. For example, at 37°C (normal body temperature) the value of $K_w = 2.42 \times 10^{-14}$. This means for a neutral solution at 37°C

$$[H^+][OH^-] = 2.42 \times 10^{-14}$$

and

$$[H^+] = [OH^-] = 1.55 \times 10^{-7}$$

So the pH for a neutral solution at 37°C is 6.81.

INTERACTIVE EXAMPLE 13-3

Calculating [H⁺] and [OH⁻]

Calculate $[H^+]$ or $[OH^-]$ as required for each of the following solutions at 25°C, and state whether the solution is neutral, acidic, or basic.

a. $1.0 \times 10^{-5}\ M\ OH^-$

b. $1.0 \times 10^{-7}\ M\ OH^-$

c. $10.0\ M\ H^+$

Solution

a. $K_w = [H^+][OH^-] = 1.0 \times 10^{-14}$. Since $[OH^-]$ is $1.0 \times 10^{-5}\ M$, solving for $[H^+]$ gives

$$[H^+] = \frac{1.0 \times 10^{-14}}{[OH^-]} = \frac{1.0 \times 10^{-14}}{1.0 \times 10^{-5}} = 1.0 \times 10^{-9}\ M$$

Since $[OH^-] > [H^+]$, the solution is basic.

b. As in part a, solving for $[H^+]$ gives

$$[H^+] = \frac{1.0 \times 10^{-14}}{[OH^-]} = \frac{1.0 \times 10^{-14}}{1.0 \times 10^{-7}} = 1.0 \times 10^{-7}\ M$$

Since $[H^+] = [OH^-]$, the solution is neutral.

c. Solving for $[OH^-]$ gives

$$[OH^-] = \frac{1.0 \times 10^{-14}}{[H^+]} = \frac{1.0 \times 10^{-14}}{10.0} = 1.0 \times 10^{-15}\ M$$

Since $[H^+] > [OH^-]$, the solution is acidic.

See Exercises 13-45 and 13-46

Since K_w is an equilibrium constant, it varies with temperature. The effect of temperature is considered in Example 13-4.

EXAMPLE 13-4

Autoionization of Water

At 60°C, the value of K_w is 1×10^{-13}.

a. Using Le Châtelier's principle, predict whether the reaction

$$2H_2O(l) \rightleftharpoons H_3O^+(aq) + OH^-(aq)$$

is exothermic or endothermic.

b. Calculate $[H^+]$ and $[OH^-]$ in a neutral solution at 60°C.

Solution

a. K_w *increases* from 1×10^{-14} at 25°C to 1×10^{-13} at 60°C. Le Châtelier's principle states that if a system at equilibrium is heated, it will adjust to consume energy. Since the value of K_w increases with temperature, we must think of energy as a reactant, and so the process must be endothermic.

b. At 60°C,

$$[H^+][OH^-] = 1 \times 10^{-13}$$

For a neutral solution,

$$[H^+] = [OH^-] = \sqrt{1 \times 10^{-13}} = 3 \times 10^{-7}\,M$$

See Exercise 13-47

13-3 | The pH Scale ◄ ⓘ

ⓘ The pH scale is a compact way to represent solution acidity.

ⓘ Appendix 1-2 has a review of logs.

Because $[H^+]$ in an aqueous solution is typically quite small, the **pH scale** provides a convenient way to represent solution acidity. The pH is a log scale based on 10, where

$$pH = -\log[H^+]$$

Thus for a solution where

$$[H^+] = 1.0 \times 10^{-7}\,M$$
$$pH = -(-7.00) = 7.00$$

At this point we need to discuss significant figures for logarithms. ◄ ⓘ The number of decimal places in the log is equal to the number of significant figures in the original number. Thus

$$\overset{\frown\ \text{2 significant figures}}{[H^+] = 1.0 \times 10^{-9}\,M}$$
$$pH = 9.00$$
$$\underset{\text{2 decimal places}}{}$$

Similar log scales are used for representing other quantities; for example,

$$pOH = -\log[OH^-]$$
$$pK = -\log K$$

Since pH is a log scale based on 10, *the pH changes by 1 for every power of 10 change in* $[H^+]$. For example, a solution of pH 3 has an H^+ concentration 10 times that of a solution of pH 4 and 100 times that of a solution of pH 5. Also note that because pH is defined as $-\log[H^+]$, *the pH decreases as* $[H^+]$ *increases.* The pH scale and the pH values for several common substances are shown in Fig. 13-6.

The pH of a solution is usually measured using a pH meter, an electronic device with a probe that can be inserted into a solution of unknown pH. The probe contains an acidic aqueous solution enclosed by a special glass membrane that allows migration of H^+ ions. If the unknown solution has a different pH from the solution in the probe, an electric potential results, which is registered on the meter (Fig. 13-7).

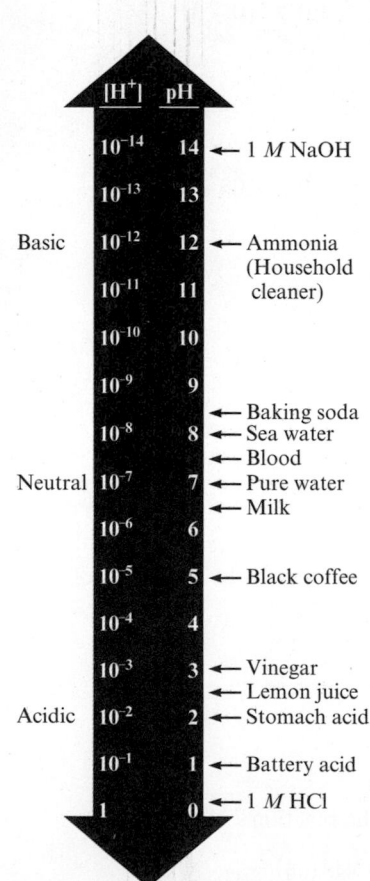

Figure 13-6 | The pH scale and pH values of some common substances.

> **Critical Thinking**
>
> What if you lived on a planet identical to the earth but for which room temperature was 50°C? How would the pH scale be different?

Figure 13-7 | pH meters are used to measure acidity. ◀ ⓘ

ⓘ The pH meter is discussed more fully in Section 17-5.

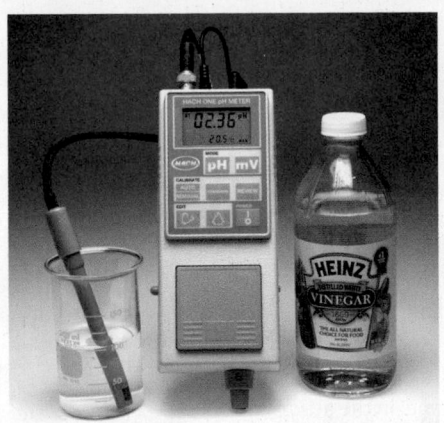
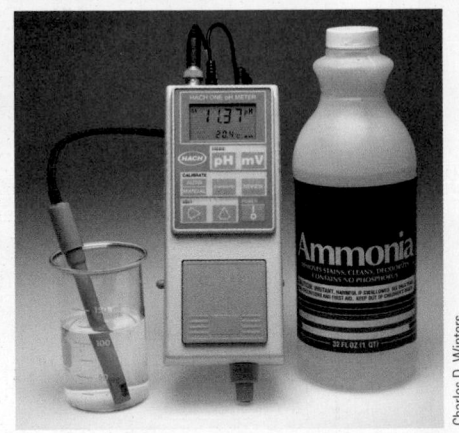

Charles D. Winters

a

b

INTERACTIVE EXAMPLE 13-5

Calculating pH and pOH

Calculate pH and pOH for each of the following solutions at 25°C.

a. $1.0 \times 10^{-3}\ M\ OH^-$

b. $1.0\ M\ OH^-$

Solution

a.
$$[H^+] = \frac{K_w}{[OH^-]} = \frac{1.0 \times 10^{-14}}{1.0 \times 10^{-3}} = 1.0 \times 10^{-11}\ M$$

$$pH = -\log[H^+] = -\log(1.0 \times 10^{-11}) = 11.00$$

$$pOH = -\log[OH^-] = -\log(1.0 \times 10^{-3}) = 3.00$$

b.
$$[OH^-] = \frac{K_w}{[H^+]} = \frac{1.0 \times 10^{-14}}{1.0} = 1.0 \times 10^{-14}\ M$$

$$pH = -\log[H^+] = -\log(1.0) = 0.00$$

$$pOH = -\log[OH^-] = -\log(1.0 \times 10^{-14}) = 14.00$$

See Exercise 13-49

It is useful to consider the log form of the expression

$$K_w = [H^+][OH^-]$$

That is,

$$\log K_w = \log[H^+] + \log[OH^-]$$

or

$$-\log K_w = -\log[H^+] - \log[OH^-]$$

Thus
$$pK_w = pH + pOH \qquad (13\text{-}3)$$

Since $K_w = 1.0 \times 10^{-14}$,

$$pK_w = -\log(1.0 \times 10^{-14}) = 14.00$$

Thus for *any* aqueous solution at 25°C, pH and pOH add up to 14.00:

$$pH + pOH = 14.00 \qquad (13\text{-}4)$$

Arnold Beckman died at age 104 in May 2004. Beckman's leadership in science and business spans virtually the entire twentieth century. He was born in 1900 in Cullom, Illinois, a town of 500 people that had no electricity or telephones. Beckman says, "In Cullom we were forced to improvise. I think it was a good thing."

The son of a blacksmith, Beckman had his interest in science awakened at age nine. At that time, in the attic of his house he discovered J. Dorman Steele's *Fourteen Weeks in Chemistry,* a book containing instructions for doing chemistry experiments. Beckman became so fascinated with chemistry that his father built him a small "chemistry shed" in the backyard for his tenth birthday.

Beckman's interest in chemistry was fostered by his high school teachers, and he eventually attended the University of Illinois, Urbana–Champaign. He graduated with a bachelor's degree in chemical engineering in 1922 and stayed one more year to get a master's degree. He then went to Caltech, where he earned a Ph.D. and became a faculty member.

Beckman was always known for his inventiveness. As a youth he designed a pressurized fuel system for his Model T Ford to overcome problems with its normal gravity feed fuel system—you had to *back* it up steep hills to keep it from starving for fuel. In 1927 he applied for his first patent: a buzzer to alert drivers that they were speeding.

In 1935 Beckman invented something that would cause a revolution in scientific instrumentation. A college friend who worked in a laboratory in the California citrus industry needed an accurate, convenient way to measure the acidity of orange juice. In response, Beckman invented the pH meter, which he initially called the acidimeter. This compact, sturdy device was an immediate hit. It signaled a new era in scientific instrumentation. In fact, business was so good that Beckman left Caltech to head his own company.

Over the years Beckman invented many other devices, including an improved potentiometer and an instrument for

Arnold Beckman.

Courtesy of the Archives, California Institute of Technology

measuring the light absorbed by molecules. At age 65 he retired as president of Beckman Instruments (headquartered in Fullerton, California). After a merger the company became Beckman Coulter; it had sales of more than $2 billion in 2003.

After stepping down as president of Beckman Instruments, Beckman began a new career—donating his wealth for the improvement of science. In 1984 he and Mabel, his wife of 58 years, donated $40 million to his alma mater—the University of Illinois—to fund the Beckman Institute. The Beckmans have also funded many other research institutes, including one at Caltech, and formed a foundation that currently gives $20 million each year to various scientific endeavors.

Arnold Beckman was a man known for his incredible creativity, but even more he was recognized as a man of absolute integrity. Beckman has important words for us: "Whatever you do, be enthusiastic about it."

Note: You can see Arnold Beckman's biography at the Chemical Heritage Foundation Web site (http://www.chemheritage.org).

INTERACTIVE EXAMPLE 13-6

Calculations Using pH

The pH of a sample of human blood was measured to be 7.41 at 25°C. Calculate pOH, [H^+], and [OH^-] for the sample.

Solution

Since pH + pOH = 14.00,

$$pOH = 14.00 - pH = 14.00 - 7.41 = 6.59$$

To find $[H^+]$ we must go back to the definition of pH:

$$pH = -\log[H^+]$$

Thus

$$7.41 = -\log[H^+] \quad \text{or} \quad \log[H^+] = -7.41$$

We need to know the *antilog* of -7.41. As shown in Appendix 1.2, taking the antilog is the same as exponentiation; that is,

$$\text{antilog}(n) = 10^n$$

Since $pH = -\log[H^+]$,

$$-pH = \log[H^+]$$

and $[H^+]$ can be calculated by taking the antilog of $-pH$:

$$[H^+] = \text{antilog}(-pH)$$

In the present case,

$$[H^+] = \text{antilog}(-pH) = \text{antilog}(-7.41) = 10^{-7.41} = 3.9 \times 10^{-8} \, M$$

Similarly, $[OH^-] = \text{antilog}(-pOH)$, and

$$[OH^-] = \text{antilog}(-6.59) = 10^{-6.59} = 2.6 \times 10^{-7} \, M$$

See Exercises 13-50 through 13-54

Now that we have considered all the fundamental definitions relevant to acid–base solutions, we can proceed to a quantitative description of the equilibria present in these solutions. The main reason that acid–base problems sometimes seem difficult is that a typical aqueous solution contains many components, so the problems tend to be complicated. However, you can deal with these problems successfully if you use the following general strategies:

PROBLEM-SOLVING STRATEGY

Solving Acid–Base Problems

- *Think chemistry.* Focus on the solution components and their reactions. It will almost always be possible to choose one reaction that is the most important.

- *Be systematic.* Acid–base problems require a step-by-step approach.

- *Be flexible.* Although all acid–base problems are similar in many ways, important differences do occur. Treat each problem as a separate entity. Do not try to force a given problem into matching any you have solved before. Look for both the similarities and the differences.

- *Be patient.* The complete solution to a complicated problem cannot be seen immediately in all its detail. Pick the problem apart into its workable steps.

- *Be confident.* Look within the problem for the solution, and let the problem guide you. Assume that you can think it out. Do not rely on memorizing solutions to problems. In fact, memorizing solutions is usually detrimental because you tend to try to force a new problem to be the same as one you have seen before. *Understand and think; don't just memorize.*

13-4 | Calculating the pH of Strong Acid Solutions

When we deal with acid–base equilibria, *we must focus on the solution components and their chemistry.* ◄ ⓘ For example, what species are present in a 1.0-*M* solution of HCl? Since hydrochloric acid is a strong acid, we assume that it is completely dissociated. Thus although the label on the bottle says 1.0 *M* HCl, the solution contains virtually no HCl molecules. Typically, container labels indicate the substance(s) used to make up the solution but do not necessarily describe the solution components after dissolution. Thus a 1.0-*M* HCl solution contains H^+ and Cl^- ions rather than HCl molecules.

The next step in dealing with aqueous solutions is to determine which components are significant and which can be ignored. We need to focus on the **major species**, those solution components present in relatively large amounts. In 1.0 *M* HCl, for example, the major species are H^+, Cl^-, and H_2O. Since this is a very acidic solution, OH^- is present only in tiny amounts and is classified as a minor species. In attacking acid–base problems, the importance of writing the major species in the solution as the first step cannot be overemphasized. This single step is the key to solving these problems successfully. ◄ ⓘ

To illustrate the main ideas involved, let us calculate the pH of 1.0 *M* HCl. We first list the major species: H^+, Cl^-, and H_2O. ◄ ⓘ Since we want to calculate the pH, we will focus on those major species that can furnish H^+. Obviously, we must consider H^+ from the dissociation of HCl. However, H_2O also furnishes H^+ by autoionization, which is often represented by the simple dissociation reaction

$$H_2O(l) \rightleftharpoons H^+(aq) + OH^-(aq)$$

However, is autoionization an important source of H^+ ions? In pure water at 25°C, $[H^+]$ is 10^{-7} *M*. In 1.0 *M* HCl solution, the water will produce even less than 10^{-7} *M* H^+, since by Le Châtelier's principle the H^+ from the dissociated HCl will drive the position of the water equilibrium to the left. ◄ ⓘ Thus the amount of H^+ contributed by water is negligible compared with the 1.0 *M* H^+ from the dissociation of HCl. Therefore, we can say that $[H^+]$ in the solution is 1.0 *M*. The pH is then

$$pH = -\log[H^+] = -\log(1.0) = 0$$

ⓘ **Common Strong Acids**

　HCl(*aq*)
　HNO_3(*aq*)
　H_2SO_4(*aq*)
　$HClO_4$(*aq*)

ⓘ Thinking about the molecules present in solution and how they behave (as an acid, base, both, or neither) is a crucial first step in solving acid–base problems.

ⓘ *Always* write the major species present in the solution.

ⓘ The H^+ from the strong acid drives the equilibrium $H_2O \rightleftharpoons H^+ + OH^-$ to the left.

INTERACTIVE EXAMPLE 13-7

pH of Strong Acids

a. Calculate the pH of 0.10 *M* HNO_3.

b. Calculate the pH of 1.0×10^{-10} *M* HCl.

Solution

a. Since HNO_3 is a strong acid, the major species in solution are ◄

$$H^+, \quad NO_3^-, \quad \text{and} \quad H_2O$$

The concentration of HNO_3 is virtually zero, since the acid completely dissociates in water. Also, $[OH^-]$ will be very small because the H^+ ions from the acid will drive the equilibrium

$$H_2O(l) \rightleftharpoons H^+(aq) + OH^-(aq)$$

Major Species

● H^+

●●● NO_3^-

●● H_2O

to the left. That is, this is an acidic solution where $[H^+] \gg [OH^-]$, so $[OH^-] \ll 10^{-7}\ M$. The sources of H^+ are

1. H^+ from HNO_3 (0.10 M)

2. H^+ from H_2O ◄ ⓘ

The number of H^+ ions contributed by the autoionization of water will be very small compared with the 0.10 M contributed by the HNO_3 and can be neglected. Since the dissolved HNO_3 is the only important source of H^+ ions in this solution,

$$[H^+] = 0.10\ M \quad \text{and} \quad pH = -\log(0.10) = 1.00$$

b. Normally, in an aqueous solution of HCl the major species are H^+, Cl^-, and H_2O. However, in this case the amount of HCl in solution is so small that it has no effect; the only major species is H_2O. Thus the pH will be that of pure water, or pH = 7.00.

See Exercises 13-55, 13-57, and 13-58

ⓘ In pure water, only $10^{-7}\ M\ H^+$ is produced.

13-5 | Calculating the pH of Weak Acid Solutions

Since a weak acid dissolved in water can be viewed as a prototype of almost any equilibrium occurring in aqueous solution, we will proceed carefully and systematically. Although some of the procedures we develop here may seem unnecessary, they will become essential as the problems become more complicated. We will develop the necessary strategies by calculating the pH of a 1.00-M solution of HF ($K_a = 7.2 \times 10^{-4}$).

The first step, as always, is to *write the major species in the solution.* ◄ ⓘ From its small K_a value, we know that hydrofluoric acid is a weak acid and will be dissociated only to a slight extent. Thus when we write the major species, the hydrofluoric acid will be represented in its dominant form, as HF. The major species in solution are HF and H_2O.

The next step (since this is a pH problem) is to decide which of the major species can furnish H^+ ions. Actually, both major species can do so: ◄

$$HF(aq) \rightleftharpoons H^+(aq) + F^-(aq) \qquad K_a = 7.2 \times 10^{-4}$$
$$H_2O(l) \rightleftharpoons H^+(aq) + OH^-(aq) \qquad K_w = 1.0 \times 10^{-14}$$

In aqueous solutions, however, typically one source of H^+ can be singled out as dominant. By comparing K_a for HF with K_w for H_2O, we see that hydrofluoric acid, although weak, is still a much stronger acid than water. Thus we will assume that hydrofluoric acid will be the dominant source of H^+. We will ignore the tiny contribution by water.

Therefore, it is the dissociation of HF that will determine the equilibrium concentration of H^+ and hence the pH:

$$HF(aq) \rightleftharpoons H^+(aq) + F^-(aq)$$

The equilibrium expression is

$$K_a = 7.2 \times 10^{-4} = \frac{[H^+][F^-]}{[HF]}$$

To solve the equilibrium problem, we follow the procedures developed in Chapter 12 for gas-phase equilibria. First, we list the initial concentrations, the *concentrations*

ⓘ First, *always* write the major species present in the solution.

Major Species

 HF

 H_2O

before the reaction of interest has proceeded to equilibrium. Before any HF dissociates, the concentrations of the species in the equilibrium are

$$[HF]_0 = 1.00 \, M \quad [F^-]_0 = 0 \quad [H^+]_0 = 10^{-7} \, M \approx 0$$

(Note that the zero value for $[H^+]_0$ is an approximation, since we are neglecting the H^+ ions from the autoionization of water.)

The next step is to determine the change required to reach equilibrium. Since some HF will dissociate to come to equilibrium (but this amount is presently unknown), we let x be the change in the concentration of HF that is required to achieve equilibrium. That is, we assume that x mol/L HF will dissociate to produce x mol/L H^+ and x mol/L F^- as the system adjusts to its equilibrium position. Now the equilibrium concentrations can be defined in terms of x:

$$[HF] = [HF]_0 - x = 1.00 - x$$
$$[F^-] = [F^-]_0 + x = 0 + x = x$$
$$[H^+] = [H^+]_0 + x \approx 0 + x = x$$

Substituting these equilibrium concentrations into the equilibrium expression gives

$$K_a = 7.2 \times 10^{-4} = \frac{[H^+][F^-]}{[HF]} = \frac{(x)(x)}{1.00 - x}$$

This expression produces a quadratic equation that can be solved using the quadratic formula, as for the gas-phase systems in Chapter 12. However, since K_a for HF is so small, HF will dissociate only slightly, and x is expected to be small. This will allow us to simplify the calculation. If x is very small compared to 1.00, the term in the denominator can be approximated as follows:

$$1.00 - x \approx 1.00$$

The equilibrium expression then becomes

$$7.2 \times 10^{-4} = \frac{(x)(x)}{1.00 - x} \approx \frac{(x)(x)}{1.00}$$

which yields

$$x^2 \approx (7.2 \times 10^{-4})(1.00) = 7.2 \times 10^{-4}$$
$$x \approx \sqrt{7.2 \times 10^{-4}} = 2.7 \times 10^{-2}$$

ⓘ The validity of an approximation should always be checked.

How valid is the approximation that $[HF] = 1.00 \, M$? ◄ ⓘ Because this question will arise often in connection with acid–base equilibrium calculations, we will consider it carefully. *The validity of the approximation depends on how much accuracy we demand for the calculated value of $[H^+]$.* Typically, the K_a values for acids are known to an accuracy of only about $\pm 5\%$. It is reasonable therefore to apply this figure when determining the validity of the approximation

$$[HA]_0 - x \approx [HA]_0$$

We will use the following test. First, we calculate the value of x by making the approximation

$$K_a = \frac{x^2}{[HA]_0 - x} \approx \frac{x^2}{[HA]_0}$$

where

$$x^2 \approx K_a[HA]_0 \quad \text{and} \quad x \approx \sqrt{K_a[HA]_0}$$

We then compare the sizes of x and $[HA]_0$. If the expression

$$\frac{x}{[HA]_0} \times 100\%$$

is less than or equal to 5%, the value of x is small enough that the approximation

$$[HA]_0 - x \approx [HA]_0$$

will be considered valid.

In our example,

$$x = 2.7 \times 10^{-2} \text{ mol/L}$$
$$[HA]_0 = [HF]_0 = 1.00 \text{ mol/L}$$

and

$$\frac{x}{[HA]_0} \times 100 = \frac{2.7 \times 10^{-2}}{1.00} \times 100\% = 2.7\%$$

The approximation we made is considered valid, and the value of x calculated using that approximation is acceptable. Thus

$$x = [H^+] = 2.7 \times 10^{-2} \, M \quad \text{and} \quad pH = -\log\,(2.7 \times 10^{-2}) = 1.57$$

This problem illustrates all the important steps for solving a typical equilibrium problem involving a weak acid. These steps are summarized as follows:

PROBLEM-SOLVING STRATEGY

Solving Weak Acid Equilibrium Problems

1. List the major species in the solution.
2. Choose the species that can produce H^+, and write balanced equations for the reactions producing H^+.
3. Using the values of the equilibrium constants for the reactions you have written, decide which equilibrium will dominate in producing H^+. ◀ ⓘ
4. Write the equilibrium expression for the dominant equilibrium.
5. List the initial concentrations of the species participating in the dominant equilibrium.
6. Define the change needed to achieve equilibrium; that is, define x.
7. Write the equilibrium concentrations in terms of x.
8. Substitute the equilibrium concentrations into the equilibrium expression.
9. Solve for x the "easy" way, that is, by assuming that $[HA]_0 - x \approx [HA]_0$.
10. Use the 5% rule to verify whether the approximation is valid.
11. Calculate $[H^+]$ and pH.

ⓘ The K_a values for various weak acids are given in Table 13-2 and in Appendix 5-1.

Critical Thinking

Consider two aqueous solutions of different weak acids, HA and HB. What if all you know about the two acids is that the K_a value for HA is greater than that for HB? Can you tell which of the acids is stronger than the other? Can you tell which of the acid solutions has the lower pH? Defend your answers.

INTERACTIVE EXAMPLE 13-8

The pH of Weak Acids

The hypochlorite ion (OCl^-) is a strong oxidizing agent often found in household bleaches and disinfectants. It is also the active ingredient that forms when swimming-pool water is treated with chlorine. ◀ In addition to its oxidizing abilities, the hypochlorite ion has a relatively high affinity for protons (it is a much stronger base than Cl^-, for example) and forms the weakly acidic hypochlorous acid (HOCl, $K_a = 3.5 \times 10^{-8}$). Calculate the pH of a 0.100-M aqueous solution of hypochlorous acid.

Solution

Major Species

 HOCl

  H_2O

▲ Swimming-pool water must be frequently tested for pH and chlorine content.

1. We list the major species. Since HOCl is a weak acid and remains mostly undissociated, the major species in a 0.100-M HOCl solution are ◀

$$HOCl \quad \text{and} \quad H_2O$$

2. Both species can produce H^+:

$$HOCl(aq) \rightleftharpoons H^+(aq) + OCl^-(aq) \qquad K_a = 3.5 \times 10^{-8}$$
$$H_2O(l) \rightleftharpoons H^+(aq) + OH^-(aq) \qquad K_w = 1.0 \times 10^{-14}$$

3. Since HOCl is a significantly stronger acid than H_2O, it will dominate in the production of H^+.

4. We therefore use the following equilibrium expression:

$$K_a = 3.5 \times 10^{-8} = \frac{[H^+][OCl^-]}{[HOCl]}$$

5. The initial concentrations appropriate for this equilibrium are

$$[HOCl]_0 = 0.100 \ M$$
$$[OCl^-]_0 = 0$$
$$[H^+]_0 \approx 0 \qquad \text{(We neglect the contribution from } H_2O.)$$

6. Since the system will reach equilibrium by the dissociation of HOCl, let x be the amount of HOCl (in mol/L) that dissociates in reaching equilibrium.

7. The equilibrium concentrations in terms of x are

$$[HOCl] = [HOCl]_0 - x = 0.100 - x$$
$$[OCl^-] = [OCl^-]_0 + x = 0 + x = x$$
$$[H^+] = [H^+]_0 + x \approx 0 + x = x$$

8. Substituting these concentrations into the equilibrium expression gives

$$K_a = 3.5 \times 10^{-8} = \frac{(x)(x)}{0.100 - x}$$

9. Since K_a is so small, we can expect a small value for x. Thus we make the approximation $[HA]_0 - x \approx [HA]_0$, or $0.100 - x \approx 0.100$, which leads to the expression

$$K_a = 3.5 \times 10^{-8} = \frac{x^2}{0.100 - x} \approx \frac{x^2}{0.100}$$

Solving for x gives

$$x = 5.9 \times 10^{-5}$$

10. The approximation $0.100 - x \approx 0.100$ must be validated. To do this, we compare x to $[\text{HOCl}]_0$:

$$\frac{x}{[\text{HA}]_0} \times 100 = \frac{x}{[\text{HOCl}]_0} \times 100 = \frac{5.9 \times 10^{-5}}{0.100} \times 100 = 0.059\%$$

Since this value is much less than 5%, the approximation is considered valid.

11. We calculate $[\text{H}^+]$ and pH:

$$[\text{H}^+] = x = 5.9 \times 10^{-5}\,M \qquad \text{and} \qquad \text{pH} = 4.23$$

See Exercises 13-63 and 13-64

The pH of a Mixture of Weak Acids

The same systematic approach applies to all solution equilibria.

Sometimes a solution contains two weak acids of very different strengths. This case is considered in Example 13-9. Note that the steps are again followed (though not labeled). ◀ ⓘ

INTERACTIVE EXAMPLE 13-9

The pH of Weak Acid Mixtures

Calculate the pH of a solution that contains 1.00 M HCN ($K_a = 6.2 \times 10^{-10}$) and 5.00 M HNO$_2$ ($K_a = 4.0 \times 10^{-4}$). Also calculate the concentration of cyanide ion (CN$^-$) in this solution at equilibrium.

Solution

Major Species

 HCN

 HNO$_2$

 H$_2$O

Since HCN and HNO$_2$ are both weak acids and are largely undissociated, the major species in the solution are ◀

$$\text{HCN}, \qquad \text{HNO}_2, \qquad \text{and} \qquad \text{H}_2\text{O}$$

All three of these components produce H$^+$:

$$\text{HCN}(aq) \rightleftharpoons \text{H}^+(aq) + \text{CN}^-(aq) \qquad K_a = 6.2 \times 10^{-10}$$
$$\text{HNO}_2(aq) \rightleftharpoons \text{H}^+(aq) + \text{NO}_2^-(aq) \qquad K_a = 4.0 \times 10^{-4}$$
$$\text{H}_2\text{O}(l) \rightleftharpoons \text{H}^+(aq) + \text{OH}^-(aq) \qquad K_w = 1.0 \times 10^{-14}$$

A mixture of three acids might lead to a very complicated problem. However, the situation is greatly simplified by the fact that even though HNO$_2$ is a weak acid, it is much stronger than the other two acids present (as revealed by the K values). Thus HNO$_2$ can be assumed to be the dominant producer of H$^+$, and we will focus on the equilibrium expression

$$K_a = 4.0 \times 10^{-4} = \frac{[\text{H}^+][\text{NO}_2^-]}{[\text{HNO}_2]}$$

The initial concentrations, the definition of x, and the equilibrium concentrations are as follows:

Initial Concentration (mol/L)		Equilibrium Concentration (mol/L)
$[\text{HNO}_2]_0 = 5.00$		$[\text{HNO}_2] = 5.00 - x$
$[\text{NO}_2^-]_0 = 0$	$\xrightarrow[\text{dissociates}]{x \text{ mol/L HNO}_2}$	$[\text{NO}_2^-] = x$
$[\text{H}^+]_0 \approx 0$		$[\text{H}^+] = x$

It is convenient to represent these concentrations in the following shorthand form (called an ICE table): ◄ ⓘ

	$HNO_2(aq)$	$\rightleftharpoons$	$H^+(aq)$	$+$	$NO_2^-(aq)$
Initial	5.00		0		0
Change	$-x$		$+x$		$+x$
Equilibrium	$5.00 - x$		x		x

Substituting the equilibrium concentrations in the equilibrium expression and making the approximation that $5.00 - x = 5.00$ give

$$K_a = 4.0 \times 10^{-4} = \frac{(x)(x)}{5.00 - x} \approx \frac{x^2}{5.00}$$

We solve for x:

$$x = 4.5 \times 10^{-2}$$

Using the 5% rule, we show that the approximation is valid:

$$\frac{x}{[HNO_2]_0} \times 100 = \frac{4.5 \times 10^{-2}}{5.00} \times 100 = 0.90\%$$

Therefore,

$$[H^+] = x = 4.5 \times 10^{-2}\,M \quad \text{and} \quad pH = 1.35$$

We also want to calculate the equilibrium concentration of cyanide ion in this solution. The CN^- ions in this solution come from the dissociation of HCN:

$$HCN(aq) \rightleftharpoons H^+(aq) + CN^-(aq)$$

Although the position of this equilibrium lies far to the left and does not contribute *significantly* to $[H^+]$, HCN is the *only source* of CN^-. Thus we must consider the extent of the dissociation of HCN to calculate $[CN^-]$. The equilibrium expression for the preceding reaction is

$$K_a = 6.2 \times 10^{-10} = \frac{[H^+][CN^-]}{[HCN]}$$

We know $[H^+]$ for this solution from the results of the first part of the problem. It is important to understand that *there is only one kind of H^+ in this solution*. It does not matter from which acid the H^+ ions originate. The equilibrium $[H^+]$ we need to insert into the HCN equilibrium expression is $4.5 \times 10^{-2}\,M$, even though the H^+ was contributed almost entirely from the dissociation of HNO_2. What is [HCN] at equilibrium? We know $[HCN]_0 = 1.00\,M$, and since K_a for HCN is so small, a negligible amount of HCN will dissociate. Thus

$$[HCN] = [HCN]_0 - \text{amount of HCN dissociated} \approx [HCN]_0 = 1.00\,M$$

Since $[H^+]$ and [HCN] are known, we can find $[CN^-]$ from the equilibrium expression:

$$K_a = 6.2 \times 10^{-10} = \frac{[H^+][CN^-]}{[HCN]} = \frac{(4.5 \times 10^{-2})[CN^-]}{1.00}$$

$$[CN^-] = \frac{(6.2 \times 10^{-10})(1.00)}{4.5 \times 10^{-2}} = 1.4 \times 10^{-8}\,M$$

Note the significance of this result. Since $[CN^-] = 1.4 \times 10^{-8}\,M$ and HCN is the only source of CN^-, this means that only 1.4×10^{-8} mol/L of HCN dissociated. This is a very small amount compared with the initial concentration of HCN, which is exactly what we would expect from its very small K_a value, and $[HCN] = 1.00\,M$ as assumed.

See Exercise 13-71

Percent Dissociation ◂ ⓘ

ⓘ Percent dissociation is also known as *percent ionization*.

It is often useful to specify the amount of weak acid that has dissociated in achieving equilibrium in an aqueous solution. The **percent dissociation** is defined as follows:

$$\text{Percent dissociation} = \frac{\text{amount dissociated (mol/L)}}{\text{initial concentration (mol/L)}} \times 100\% \qquad (13\text{-}5)$$

For example, we found earlier that in a 1.00-M solution of HF, $[H^+] = 2.7 \times 10^{-2}\,M$. To reach equilibrium, 2.7×10^{-2} mol/L of the original 1.00 M HF dissociates, so

$$\text{Percent dissociation} = \frac{2.7 \times 10^{-2}\ \text{mol/L}}{1.00\ \text{mol/L}} \times 100\% = 2.7\%$$

For a given weak acid, the percent dissociation increases as the acid becomes more dilute. For example, the percent dissociation of acetic acid ($HC_2H_3O_2$, $K_a = 1.8 \times 10^{-5}$) is significantly greater in a 0.10-M solution than in a 1.0-M solution, as demonstrated in Example 13-10.

INTERACTIVE EXAMPLE 13-10

Calculating Percent Dissociation

Calculate the percent dissociation of acetic acid ($K_a = 1.8 \times 10^{-5}$) in each of the following solutions: ◂

a. 1.00 M $HC_2H_3O_2$

b. 0.100 M $HC_2H_3O_2$

Major Species

 $HC_2H_3O_2$

 H_2O

Solution

a. Since acetic acid is a weak acid, the major species in this solution are $HC_2H_3O_2$ and H_2O. ◂ Both species are weak acids, but acetic acid is a much stronger acid than water. Thus the dominant equilibrium will be

$$HC_2H_3O_2(aq) \rightleftharpoons H^+(aq) + C_2H_3O_2^-(aq)$$

and the equilibrium expression is

$$K_a = 1.8 \times 10^{-5} = \frac{[H^+][C_2H_3O_2^-]}{[HC_2H_3O_2]}$$

The initial concentrations, definition of x, and equilibrium concentrations are:

	$HC_2H_3O_2(aq)$	$\rightleftharpoons$	$H^+(aq)$	$+$	$C_2H_3O_2^-(aq)$
Initial	1.00		0		0
Change	$-x$		x		x
Equilibrium	$1.00 - x$		x		x

Inserting the equilibrium concentrations into the equilibrium expression and making the usual approximation that x is small compared with $[HA]_0$ give

$$K_a = 1.8 \times 10^{-5} = \frac{[H^+][C_2H_3O_2^-]}{[HC_2H_3O_2]} = \frac{(x)(x)}{1.00 - x} \approx \frac{x^2}{1.00}$$

Thus

$$x^2 \approx 1.8 \times 10^{-5} \quad \text{and} \quad x \approx 4.2 \times 10^{-3}$$

The approximation $1.00 - x \approx 1.00$ is valid by the 5% rule (check this yourself), so

$$[H^+] = x = 4.2 \times 10^{-3}\,M$$

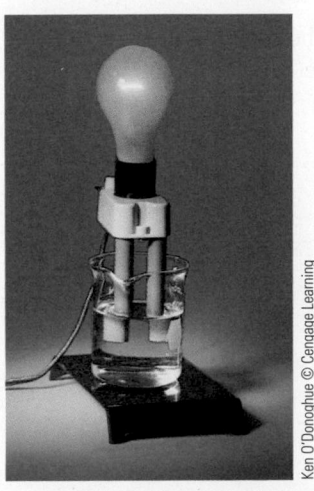

▲ An acetic acid solution, which is a weak electrolyte, contains only a few ions and does not conduct as much current as a strong electrolyte. The bulb is only dimly lit.

Ken O'Donoghue © Cengage Learning

The percent dissociation is

$$\frac{[H^+]}{[HC_2H_3O_2]_0} \times 100 = \frac{4.2 \times 10^{-3}}{1.00} \times 100\% = 0.42\%$$

b. This is a similar problem, except that in this case $[HC_2H_3O_2] = 0.100\ M$. Analysis of the problem leads to the expression

$$K_a = 1.8 \times 10^{-5} = \frac{[H^+][C_2H_3O_2^-]}{[HC_2H_3O_2]} = \frac{(x)(x)}{0.100 - x} \approx \frac{x^2}{0.100}$$

Thus

$$x = [H^+] = 1.3 \times 10^{-3}\ M$$

and

$$\text{Percent dissociation} = \frac{1.3 \times 10^{-3}}{0.10} \times 100\% = 1.3\%$$

See Exercises 13-73 and 13-74

The results in Example 13-10 show two important facts. The concentration of H^+ ion at equilibrium is smaller in the 0.10-M acetic acid solution than in the 1.0-M acetic acid solution, as we would expect. However, the percent dissociation is significantly greater in the 0.10-M solution than in the 1.0-M solution. This is a general result. *For solutions of any weak acid HA, $[H^+]$ decreases as $[HA]_0$ decreases, but the percent dissociation increases as $[HA]_0$ decreases.* ◄ ⓘ This phenomenon can be explained as follows.

Consider the weak acid HA with the initial concentration $[HA]_0$, where at equilibrium

$$[HA] = [HA]_0 - x \approx [HA]_0$$
$$[H^+] = [A^-] = x$$

Thus

$$K_a = \frac{[H^+][A^-]}{[HA]} \approx \frac{(x)(x)}{[HA]_0}$$

> ⓘ The more dilute the weak acid solution, the greater is the percent dissociation.

Now suppose enough water is added suddenly to dilute the solution by a factor of 10. The new concentrations before any adjustment occurs are

$$[A^-]_{new} = [H^+]_{new} = \frac{x}{10}$$

$$[HA]_{new} = \frac{[HA]_0}{10}$$

and Q, the reaction quotient, is

$$Q = \frac{\left(\dfrac{x}{10}\right)\left(\dfrac{x}{10}\right)}{\dfrac{[HA]_0}{10}} = \frac{1(x)(x)}{10[HA]_0} = \frac{1}{10}K_a$$

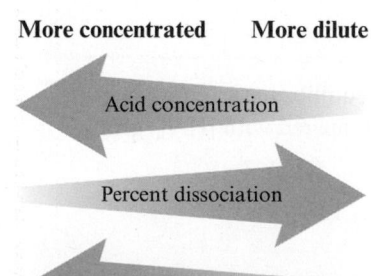

More concentrated   **More dilute**

Acid concentration

Percent dissociation

H^+ concentration

Figure 13-8 | The effect of dilution on the percent dissociation and $[H^+]$ of a weak acid solution.

Since Q is less than K_a, the system must adjust to the right to reach the new equilibrium position. Thus the percent dissociation increases when the acid is diluted. This behavior is summarized in Fig. 13-8. In Example 13-11 we see how the percent dissociation can be used to calculate the K_a value for a weak acid.

INTERACTIVE EXAMPLE 13-11

Calculating K_a from Percent Dissociation

Lactic acid ($HC_3H_5O_3$) is a chemical that accumulates in muscle tissue during exertion. ◄ In a 0.100-M aqueous solution, lactic acid is 3.7% dissociated. Calculate the value of K_a for this acid.

Solution

Major Species

 $HC_3H_5O_3$

 H_2O

From the small value for the percent dissociation, it is clear that $HC_3H_5O_3$ is a weak acid. Thus the major species in the solution are the undissociated acid and water: ◄

$$HC_3H_5O_3 \quad \text{and} \quad H_2O$$

However, even though $HC_3H_5O_3$ is a weak acid, it is a much stronger acid than water and will be the dominant source of H^+ in the solution. The dissociation reaction is

$$HC_3H_5O_3(aq) \rightleftharpoons H^+(aq) + C_3H_5O_3^-(aq)$$

and the equilibrium expression is

$$K_a = \frac{[H^+][C_3H_5O_3^-]}{[HC_3H_5O_3]}$$

The initial and equilibrium concentrations are as follows:

Initial Concentration (mol/L)		Equilibrium Concentration (mol/L)
$[HC_3H_5O_3]_0 = 0.10$	$\xrightarrow[\substack{HC_3H_5O_3 \\ \text{dissociates}}]{x \text{ mol/L}}$	$[HC_3H_5O_3] = 0.10 - x$
$[C_3H_5O_3^-]_0 = 0$		$[C_3H_5O_3^-] = x$
$[H^+]_0 \approx 0$		$[H^+] = x$

The change needed to reach equilibrium can be obtained from the percent dissociation and Equation (13-5). For this acid,

$$\text{Percent dissociation} = 3.7\% = \frac{x}{[HC_3H_5O_3]_0} \times 100\% = \frac{x}{0.10} \times 100\%$$

and

$$x = \frac{3.7}{100}(0.10) = 3.7 \times 10^{-3} \text{ mol/L}$$

Now we can calculate the equilibrium concentrations:

$$[HC_3H_5O_3] = 0.10 - x = 0.10 \, M \quad \text{(to the correct number of significant figures)}$$
$$[C_3H_5O_3^-] = [H^+] = x = 3.7 \times 10^{-3} \, M$$

These concentrations can now be used to calculate the value of K_a for lactic acid:

$$K_a = \frac{[H^+][C_3H_5O_3^-]}{[HC_3H_5O_3]} = \frac{(3.7 \times 10^{-3})(3.7 \times 10^{-3})}{0.10} = 1.4 \times 10^{-4}$$

See Exercises 13-75 and 13-76

Kevin Steele/Aurora Creative/Getty Images

▲ Strenuous exercise causes a buildup of lactic acid in muscle tissues.

PROBLEM-SOLVING STRATEGY

Solving Acid–Base Problems

1. List the major species in solution.

2. Look for reactions that can be assumed to go to completion—for example, a strong acid dissociating or H^+ reacting with OH^-.

(continued)

> **PROBLEM-SOLVING STRATEGY**
>
> **Solving Acid–Base Problems** *(continued)*
>
> 3. For a reaction that can be assumed to go to completion:
>
>   a. Determine the concentration of the products.
>
>   b. Write down the major species in solution after the reaction.
>
> 4. Look at each major component of the solution and decide if it is an acid or a base.
>
> 5. Pick the equilibrium that will control the pH. Use known values of the dissociation constants for the various species to help decide on the dominant equilibrium.
>
>   a. Write the equation for the reaction and the equilibrium expression.
>
>   b. Compute the initial concentrations (assuming the dominant equilibrium has not yet occurred, that is, no acid dissociation, and so on).
>
>   c. Define x.
>
>   d. Compute the equilibrium concentrations in terms of x.
>
>   e. Substitute the concentrations into the equilibrium expression, and solve for x.
>
>   f. Check the validity of the approximation.
>
>   g. Calculate the pH and other concentrations as required.
>
> Although these steps may seem somewhat cumbersome, especially for simpler problems, they will become increasingly helpful as the aqueous solutions become more complicated. If you develop the habit of approaching acid–base problems systematically, the more complex cases will be much easier to manage.

13-6 | Bases

According to the Arrhenius concept, a base is a substance that produces OH^- ions in aqueous solution. According to the Brønsted–Lowry model, a base is a proton acceptor. The bases sodium hydroxide (NaOH) and potassium hydroxide (KOH) fulfill both criteria. They contain OH^- ions in the solid lattice and, behaving as strong electrolytes, dissociate completely when dissolved in aqueous solution:

$$NaOH(s) \longrightarrow Na^+(aq) + OH^-(aq)$$

ⓘ In a basic solution at 25°C, pH > 7.

leaving virtually no undissociated NaOH. Thus a 1.0-M NaOH solution really contains 1.0 M Na^+ and 1.0 M OH^-. Because of their complete dissociation, NaOH and KOH are called **strong bases** in the same sense as we defined strong acids. ◄ ⓘ

All the hydroxides of the Group 1A elements (LiOH, NaOH, KOH, RbOH, and CsOH) are strong bases, but only NaOH and KOH are common laboratory reagents, because the lithium, rubidium, and cesium compounds are expensive. The alkaline earth (Group 2A) hydroxides—$Ca(OH)_2$, $Ba(OH)_2$, and $Sr(OH)_2$—are also strong bases. For these compounds, 2 moles of hydroxide ion are produced for every mole of metal hydroxide dissolved in aqueous solution.

The alkaline earth hydroxides are not very soluble and are used only when the solubility factor is not important. In fact, the low solubility of these bases can sometimes be an advantage. For example, many antacids are suspensions of metal hydroxides, such as aluminum hydroxide and magnesium hydroxide. ◄ The low solubility of these compounds prevents a large hydroxide ion concentration that would harm the tissues of the mouth, esophagus, and stomach. Yet these suspensions furnish plenty of hydroxide ion to react with the stomach acid, since the salts dissolve as this reaction proceeds.

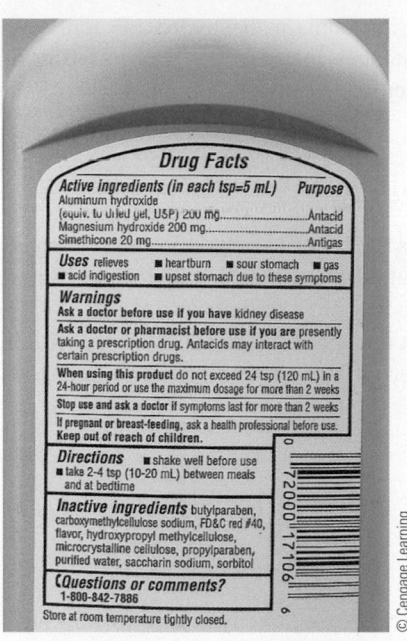

An antacid containing aluminum and magnesium hydroxides.

Calcium hydroxide, $Ca(OH)_2$, often called **slaked lime**, is widely used in industry because it is inexpensive and plentiful. For example, slaked lime is used in scrubbing stack gases to remove sulfur dioxide from the exhaust of power plants and factories. In the scrubbing process a suspension of slaked lime is sprayed into the stack gases to react with sulfur dioxide gas according to the following steps:

$$SO_2(g) + H_2O(l) \rightleftharpoons H_2SO_3(aq)$$
$$Ca(OH)_2(aq) + H_2SO_3(aq) \rightleftharpoons CaSO_3(s) + 2H_2O(l)$$

Slaked lime is also widely used in water treatment plants for softening hard water, which involves the removal of ions, such as Ca^{2+} and Mg^{2+}, that hamper the action of detergents. The softening method most often employed in water treatment plants is the **lime–soda process**, in which *lime* (CaO) and *soda ash* (Na_2CO_3) are added to the water. As we will see in more detail later in this chapter, the CO_3^{2-} ion reacts with water to produce the HCO_3^- ion. When the lime is added to the water, it forms slaked lime, that is,

$$CaO(s) + H_2O(l) \longrightarrow Ca(OH)_2(aq)$$

which then reacts with the HCO_3^- ion from the added soda ash and the Ca^{2+} ion in the hard water to produce calcium carbonate:

$$Ca(OH)_2(aq) + \underset{\substack{\nearrow \\ \text{From hard water}}}{Ca^{2+}(aq)} + 2HCO_3^-(aq) \longrightarrow 2CaCO_3(s) + 2H_2O(l)$$

Thus, for every mole of $Ca(OH)_2$ consumed, 1 mole of Ca^{2+} is removed from the hard water, thereby softening it. Some hard water naturally contains bicarbonate ions. In this case, no soda ash is needed—simply adding the lime produces the softening.

Calculating the pH of a strong base solution is relatively simple, as illustrated in Example 13-12.

INTERACTIVE EXAMPLE 13-12

The pH of Strong Bases

Calculate the pH of a 5.0×10^{-2}-M NaOH solution.

Solution

The major species in this solution are ◄

$$\underset{\text{From NaOH}}{\underbrace{Na^+, \quad OH^-,}} \quad \text{and} \quad H_2O$$

Major Species

🔵 Na^+

⚫ OH^-

🔵 H_2O

Although autoionization of water also produces OH^- ions, the pH will be dominated by the OH^- ions from the dissolved NaOH. Thus in the solution,

$$[OH^-] = 5.0 \times 10^{-2} M$$

and the concentration of H^+ can be calculated from K_w:

$$[H^+] = \frac{K_w}{[OH^-]} = \frac{1.0 \times 10^{-14}}{5.0 \times 10^{-2}} = 2.0 \times 10^{-13} M$$

$$pH = 12.70$$

Note that this is a basic solution for which

$$[OH^-] > [H^+] \quad \text{and} \quad pH > 7$$

The added OH^- from the salt has shifted the water autoionization equilibrium

$$H_2O(l) \rightleftharpoons H^+(aq) + OH^-(aq)$$

to the left, significantly lowering $[H^+]$ compared with that in pure water.

See Exercises 13-89 through 13-92

*A base does not have to contain hydroxide ion.

Many types of proton acceptors (bases) do not contain the hydroxide ion. ◄ *
However, when dissolved in water, these substances increase the concentration of hydroxide ion because of their reaction with water. For example, ammonia reacts with water as follows:

$$NH_3(aq) + H_2O(l) \rightleftharpoons NH_4^+(aq) + OH^-(aq)$$

The ammonia molecule accepts a proton and thus functions as a base. Water is the acid in this reaction. Note that even though the base ammonia contains no hydroxide ion, it still increases the concentration of hydroxide ion to yield a basic solution.

Bases such as ammonia typically have at least one unshared pair of electrons that is capable of forming a bond with a proton. The reaction of an ammonia molecule with a water molecule can be represented as follows:

$$
\begin{array}{c}
\text{H} \\
\text{H}-\text{N:}\quad \text{H}-\ddot{\text{O}}: \\
\text{H}\qquad\qquad \text{H}
\end{array}
\rightleftharpoons
\left[
\begin{array}{c}
\text{H} \\
\text{H}-\text{N}-\text{H} \\
\text{H}
\end{array}
\right]
+ \left[:\ddot{\text{O}}-\text{H}\right]^-
$$

There are many bases like ammonia that produce hydroxide ion by reaction with water. In most of these bases, the lone pair is located on a nitrogen atom. Some examples are

$$
\begin{array}{ccccc}
\overset{\ddot{\text{N}}}{\underset{\text{H}}{\text{H}_3\text{C}\diagup\big|\diagdown\text{H}}} &
\overset{\ddot{\text{N}}}{\underset{\text{H}}{\text{H}_3\text{C}\diagup\big|\diagdown\text{CH}_3}} &
\overset{\ddot{\text{N}}}{\underset{\text{CH}_3}{\text{H}_3\text{C}\diagup\big|\diagdown\text{CH}_3}} &
\overset{\ddot{\text{N}}}{\underset{\text{H}}{\text{H}\diagup\big|\diagdown\text{C}_2\text{H}_5}} &
\\
\text{Methylamine} & \text{Dimethylamine} & \text{Trimethylamine} & \text{Ethylamine} & \text{Pyridine}
\end{array}
$$

Note that the first four bases can be thought of as substituted ammonia molecules with hydrogen atoms replaced by methyl (CH_3) or ethyl (C_2H_5) groups. The pyridine molecule is like benzene

except that a nitrogen atom replaces one of the carbon atoms in the ring. The general reaction between a base B and water is given by

$$B(aq) + H_2O(l) \rightleftharpoons BH^+(aq) + OH^-(aq) \qquad (13\text{-}6)$$

$$\underset{\text{Base}}{}\quad \underset{\text{Acid}}{}\quad \underset{\substack{\text{Conjugate}\\\text{acid}}}{}\quad \underset{\substack{\text{Conjugate}\\\text{base}}}{}$$

The equilibrium constant for this general reaction is

*Appendix 5-3 contains a table of K_b values.

$$K_b = \frac{[BH^+][OH^-]}{[B]}$$

where K_b always refers to the reaction of a base with water to form the conjugate acid and the hydroxide ion. ◄ *

Bases of the type represented by B in Equation (13-6) compete with OH^-, a very strong base, for the H^+ ion. Thus their K_b values tend to be small (for example, for ammonia, $K_b = 1.8 \times 10^{-5}$), and they are called **weak bases**. The values of K_b for some common weak bases are listed in Table 13-3.

*Refer to the steps for solving weak acid equilibrium problems. Use the same systematic approach for weak base equilibrium problems.

Typically, pH calculations for solutions of weak bases are very similar to those for weak acids, as illustrated by Examples 13-13 and 13-14. ◄ *

Table 13-3 | Values of K_b for Some Common Weak Bases

Name	Formula	Conjugate Acid	K_b
Ammonia	NH_3	NH_4^+	1.8×10^{-5}
Methylamine	CH_3NH_2	$CH_3NH_3^+$	4.38×10^{-4}
Ethylamine	$C_2H_5NH_2$	$C_2H_5NH_3^+$	5.6×10^{-4}
Aniline	$C_6H_5NH_2$	$C_6H_5NH_3^+$	3.8×10^{-10}
Pyridine	C_5H_5N	$C_5H_5NH^+$	1.7×10^{-9}

INTERACTIVE EXAMPLE 13-13

The pH of Weak Bases I

Calculate the pH for a 15.0-M solution of NH_3 ($K_b = 1.8 \times 10^{-5}$).

Solution

Since ammonia is a weak base, as can be seen from its small K_b value, most of the dissolved NH_3 will remain as NH_3. Thus the major species in solution are ◄

Major Species

 NH_3

 H_2O

$$NH_3 \quad \text{and} \quad H_2O$$

Both these substances can produce OH^- according to the reactions

$$NH_3(aq) + H_2O(l) \rightleftharpoons NH_4^+(aq) + OH^-(aq) \qquad K_b = 1.8 \times 10^{-5}$$
$$H_2O(l) \rightleftharpoons H^+(aq) + OH^-(aq) \qquad K_w = 1.0 \times 10^{-14}$$

However, the contribution from water can be neglected, since $K_b \gg K_w$. The equilibrium for NH_3 will dominate, and the equilibrium expression to be used is

$$K_b = 1.8 \times 10^{-5} = \frac{[NH_4^+][OH^-]}{[NH_3]}$$

The appropriate concentrations are

Initial Concentration (mol/L)		Equilibrium Concentration (mol/L)
$[NH_3]_0 = 15.0$ $[NH_4^+]_0 = 0$ $[OH^-]_0 \approx 0$	$\xrightarrow[\text{H}_2\text{O to reach}]{\substack{x \text{ mol/L} \\ \text{NH}_3 \text{ reacts with} \\ \text{equilibrium}}}$	$[NH_3] = 15.0 - x$ $[NH_4^+] = x$ $[OH^-] = x$

In terms of an ICE table, these concentrations are

	$NH_3(aq)$	$+$	$H_2O(l)$	$\rightleftharpoons$	$NH_4^+(aq)$	$+$	$OH^-(aq)$
Initial	15.0		—		0		0
Change	$-x$		—		$+x$		$+x$
Equilibrium	$15.0 - x$		—		x		x

Substituting the equilibrium concentrations into the equilibrium expression and making the usual approximation gives

$$K_b = 1.8 \times 10^{-5} = \frac{[NH_4^+][OH^-]}{[NH_3]} = \frac{(x)(x)}{15.0 - x} \approx \frac{x^2}{15.0}$$

Thus

$$x \approx 1.6 \times 10^{-2}$$

The 5% rule validates the approximation (check it yourself), so

$$[OH^-] = 1.6 \times 10^{-2} \, M$$

We have seen that many bases have nitrogen atoms with one lone pair and can be viewed as substituted ammonia molecules, with the general formula $R_xNH_{(3-x)}$. Compounds of this type are called **amines**. Amines are widely distributed in animals and plants, and complex amines often serve as messengers or regulators. For example, in the human nervous system, there are two amine stimulants, *norepinephrine* and *adrenaline*.

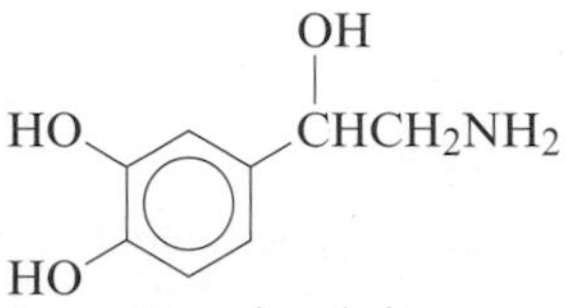

OH
HO—⬡—CHCH₂NH₂
HO
Norepinephrine

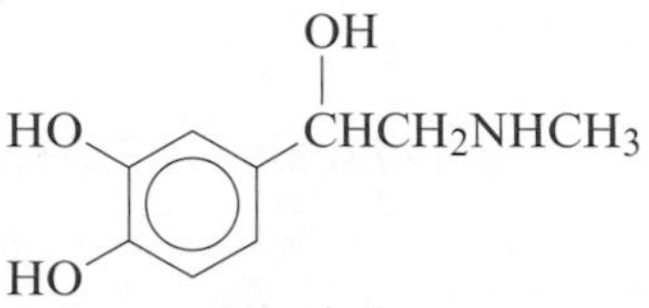

OH
HO—⬡—CHCH₂NHCH₃
HO
Adrenaline

Ephedrine, widely used as a decongestant, was a known drug in China over 2000 years ago. Indians in Mexico and the Southwest have used the hallucinogen *mescaline,* extracted from the peyote cactus, for centuries.

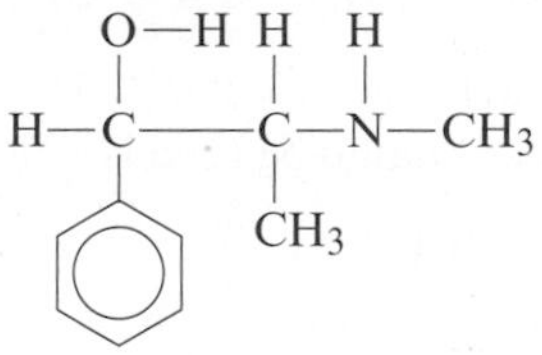

O—H H H
H—C—C—N—CH₃
⬡ CH₃
Ephedrine

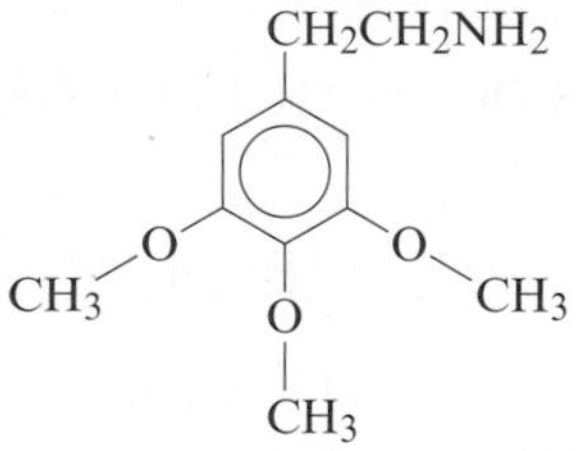

CH₂CH₂NH₂
CH₃O—⬡—OCH₃
O
CH₃
Mescaline

Peyote cactus growing on a rock.

Many other drugs, such as codeine and quinine, are amines, but they are usually not used in their pure amine forms. Instead, they are treated with an acid to become acid salts. An example of an acid salt is ammonium chloride, obtained by the reaction

$$NH_3 + HCl \longrightarrow NH_4Cl$$

Amines also can be protonated in this way. The resulting acid salt, written as AHCl (where A represents the amine), contains AH^+ and Cl^-. In general, the acid salts are more stable and more soluble in water than the parent amines. For instance, the parent amine of the well-known local anesthetic *novocaine* is insoluble in water, whereas the acid salt is much more soluble.

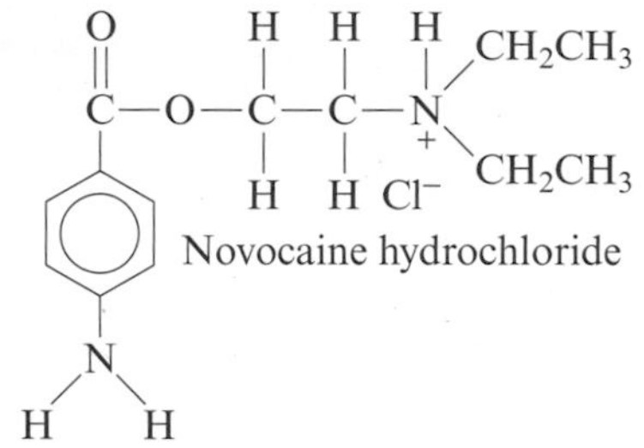

O H H H CH₂CH₃
C—O—C—C—N⁺
H H Cl⁻ CH₂CH₃
Novocaine hydrochloride
N
H H

Since we know that K_w must be satisfied for this solution, we can calculate $[H^+]$ as follows:

$$[H^+] = \frac{K_w}{[OH^-]} = \frac{1.0 \times 10^{-14}}{1.6 \times 10^{-2}} = 6.3 \times 10^{-13} \, M$$

Therefore, $pH = -\log(6.3 \times 10^{-13}) = 12.20$

See Exercises 13-95 and 13-96

A table of K_b values for bases is also given in Appendix 5-3.

Example 13-13 illustrates how a typical weak base equilibrium problem should be solved. ◄ ⓘ Note two additional important points:

1. We calculated $[H^+]$ from K_w and then calculated the pH, but another method is available. The pOH could have been calculated from $[OH^-]$ and then used in Equation (13-3):

$$pK_w = 14.00 = pH + pOH$$
$$pH = 14.00 - pOH$$

2. In a 15.0-M NH_3 solution, the equilibrium concentrations of NH_4^+ and OH^- are each 1.6×10^{-2} M. Only a small percentage,

$$\frac{1.6 \times 10^{-2}}{15.0} \times 100\% = 0.11\%$$

of the ammonia reacts with water. Bottles containing 15.0 M NH_3 solution are often labeled 15.0 M NH_4OH, but as you can see from these results, 15.0 M NH_3 is actually a much more accurate description of the solution contents.

INTERACTIVE EXAMPLE 13-14

The pH of Weak Bases II

Calculate the pH of a 1.0-M solution of methylamine ($K_b = 4.38 \times 10^{-4}$).

Solution

Since methylamine (CH_3NH_2) is a weak base, the major species in solution are ◄

$$CH_3NH_2 \quad \text{and} \quad H_2O$$

Major Species

 CH_3NH_2

 H_2O

Both are bases; however, water can be neglected as a source of OH^-, so the dominant equilibrium is

$$CH_3NH_2(aq) + H_2O(l) \rightleftharpoons CH_3NH_3^+(aq) + OH^-(aq)$$

and

$$K_b = 4.38 \times 10^{-4} = \frac{[CH_3NH_3^+][OH^-]}{[CH_3NH_2]}$$

The ICE table is:

	$CH_3NH_2(aq)$	+	$H_2O(l)$	$\rightleftharpoons$	$CH_3NH_3^+(aq)$	+	$OH^-(aq)$
Initial	1.0		—		0		0
Change	$-x$		—		$+x$		$+x$
Equilibrium	$1.0 - x$		—		x		x

Substituting the equilibrium concentrations in the equilibrium expression and making the usual approximation give

$$K_b = 4.38 \times 10^{-4} = \frac{[CH_3NH_3^+][OH^-]}{[CH_3NH_2]} = \frac{(x)(x)}{1.0 - x} \approx \frac{x^2}{1.0}$$

and

$$x \approx 2.1 \times 10^{-2}$$

The approximation is valid by the 5% rule, so

$$[OH^-] = x = 2.1 \times 10^{-2} \ M$$
$$pOH = 1.68$$
$$pH = 14.00 - 1.68 = 12.32$$

See Exercises 13-97 and 13-98

13-7 | Polyprotic Acids

Some important acids, such as sulfuric acid (H_2SO_4) and phosphoric acid (H_3PO_4), can furnish more than one proton and are called **polyprotic acids**. A polyprotic acid always dissociates in a *stepwise* manner, one proton at a time. For example, the diprotic (two-proton) acid carbonic acid (H_2CO_3), which is so important in maintaining a constant pH in human blood, dissociates in the following steps:

$$H_2CO_3(aq) \rightleftharpoons H^+(aq) + HCO_3^-(aq) \quad K_{a_1} = \frac{[H^+][HCO_3^-]}{[H_2CO_3]} = 4.3 \times 10^{-7}$$

$$HCO_3^-(aq) \rightleftharpoons H^+(aq) + CO_3^{2-}(aq) \quad K_{a_2} = \frac{[H^+][CO_3^{2-}]}{[HCO_3^-]} = 5.6 \times 10^{-11}$$

The successive K_a values for the dissociation equilibria are designated K_{a_1} and K_{a_2}. Note that the conjugate base HCO_3^- of the first dissociation equilibrium becomes the acid in the second step.

Carbonic acid is formed when carbon dioxide gas is dissolved in water. In fact, the first dissociation step for carbonic acid is best represented by the reaction

$$CO_2(aq) + H_2O(l) \rightleftharpoons H^+(aq) + HCO_3^-(aq)$$

since relatively little H_2CO_3 actually exists in solution. However, it is convenient to consider CO_2 in water as H_2CO_3 so that we can treat such solutions using the familiar dissociation reactions for weak acids.

Phosphoric acid is a **triprotic acid** (three protons) that dissociates in the following steps:

$$H_3PO_4(aq) \rightleftharpoons H^+(aq) + H_2PO_4^-(aq) \quad K_{a_1} = \frac{[H^+][H_2PO_4^-]}{[H_3PO_4]} = 7.5 \times 10^{-3}$$

$$H_2PO_4^-(aq) \rightleftharpoons H^+(aq) + HPO_4^{2-}(aq) \quad K_{a_2} = \frac{[H^+][HPO_4^{2-}]}{[H_2PO_4^-]} = 6.2 \times 10^{-8}$$

$$HPO_4^{2-}(aq) \rightleftharpoons H^+(aq) + PO_4^{3-}(aq) \quad K_{a_3} = \frac{[H^+][PO_4^{3-}]}{[HPO_4^{2-}]} = 4.8 \times 10^{-13}$$

For a typical weak polyprotic acid,

$$K_{a_1} > K_{a_2} > K_{a_3}$$

That is, the acid involved in each step of the dissociation is successively weaker, as shown by the stepwise dissociation constants given in Table 13-4. These values indicate that the loss of a second or third proton occurs less readily than loss of the first proton. This is not surprising; as the negative charge on the acid increases, it becomes more difficult to remove the positively charged proton. ◄ ⓘ

ⓘ A table of K_a values for polyprotic acids is also given in Appendix 5-2.

Table 13-4 | Stepwise Dissociation Constants for Several Common Polyprotic Acids

Name	Formula	K_{a_1}	K_{a_2}	K_{a_3}
Phosphoric acid	H_3PO_4	7.5×10^{-3}	6.2×10^{-8}	4.8×10^{-13}
Arsenic acid	H_3AsO_4	5.5×10^{-3}	1.7×10^{-7}	5.1×10^{-12}
Carbonic acid	H_2CO_3	4.3×10^{-7}	5.6×10^{-11}	
Sulfuric acid	H_2SO_4	Large	1.2×10^{-2}	
Sulfurous acid	H_2SO_3	1.5×10^{-2}	1.0×10^{-7}	
Hydrosulfuric acid*	H_2S	1.0×10^{-7}	$\sim 10^{-19}$	
Oxalic acid	$H_2C_2O_4$	6.5×10^{-2}	6.1×10^{-5}	
Ascorbic acid (vitamin C)	$H_2C_6H_6O_6$	7.9×10^{-5}	1.6×10^{-12}	

*The K_{a_2} value for H_2S is very uncertain. Because it is so small, the K_{a_2} value is very difficult to measure accurately.

Although we might expect the pH calculations for solutions of polyprotic acids to be complicated, the most common cases are surprisingly straightforward. To illustrate, we will consider a typical case, phosphoric acid, and a unique case, sulfuric acid.

Phosphoric Acid

Phosphoric acid is typical of most polyprotic acids in that the successive K_a values are very different. For example, the ratios of successive K_a values (from Table 13-4) are

$$\frac{K_{a_1}}{K_{a_2}} = \frac{7.5 \times 10^{-3}}{6.2 \times 10^{-8}} = 1.2 \times 10^5$$

$$\frac{K_{a_2}}{K_{a_3}} = \frac{6.2 \times 10^{-8}}{4.8 \times 10^{-13}} = 1.3 \times 10^5$$

Thus the relative acid strengths are

$$H_3PO_4 \gg H_2PO_4^- \gg HPO_4^{2-}$$

i For a typical polyprotic acid in water, only the first dissociation step is important in determining the pH.

This means that in a solution prepared by dissolving H_3PO_4 in water, *only the first dissociation step makes an important contribution to [H$^+$]*. This greatly simplifies the pH calculations for phosphoric acid solutions, as is illustrated in Example 13-15. ◄ i

---

INTERACTIVE EXAMPLE 13-15

The pH of a Polyprotic Acid

Calculate the pH of a 5.0-M H_3PO_4 solution and the equilibrium concentrations of the species H_3PO_4, $H_2PO_4^-$, HPO_4^{2-}, and PO_4^{3-}.

Solution

The major species in solution are ◄

$$H_3PO_4 \quad \text{and} \quad H_2O$$

Major Species

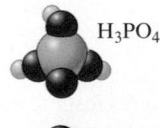

H_3PO_4

H_2O

None of the dissociation products of H_3PO_4 is written, since the K_a values are all so small that they will be minor species. The dominant equilibrium is the dissociation of H_3PO_4:

$$H_3PO_4(aq) \rightleftharpoons H^+(aq) + H_2PO_4^-(aq)$$

where

$$K_{a_1} = 7.5 \times 10^{-3} = \frac{[H^+][H_2PO_4^-]}{[H_3PO_4]}$$

The ICE table is:

	$H_3PO_4(aq)$	$\rightleftharpoons$	$H^+(aq)$	$+$	$H_2PO_4^-(aq)$
Initial	5.0		0		0
Change	$-x$		$+x$		$+x$
Equilibrium	$5.0 - x$		x		x

Substituting the equilibrium concentrations into the expression for K_{a_1} and making the usual approximation give

$$K_{a_1} = 7.5 \times 10^{-3} = \frac{[H^+][H_2PO_4^-]}{[H_3PO_4]} = \frac{(x)(x)}{5.0 - x} \approx \frac{x^2}{5.0}$$

Thus

$$x \approx 1.9 \times 10^{-1}$$

Since 1.9×10^{-1} is less than 5% of 5.0, the approximation is acceptable, and

$$[H^+] = x = 0.19 \, M$$

$$pH = 0.72$$

So far we have determined that

$$[H^+] = [H_2PO_4^-] = 0.19 \; M$$

and

$$[H_3PO_4] = 5.0 - x = 4.8 \; M$$

The concentration of HPO_4^{2-} can be obtained by using the expression for K_{a_2}:

$$K_{a_2} = 6.2 \times 10^{-8} = \frac{[H^+][HPO_4^{2-}]}{[H_2PO_4^-]}$$

where

$$[H^+] = [H_2PO_4^-] = 0.19 \; M$$

Thus

$$[HPO_4^{2-}] = K_{a_2} = 6.2 \times 10^{-8} \; M$$

To calculate $[PO_4^{3-}]$, we use the expression for K_{a_3} and the values of $[H^+]$ and $[HPO_4^{2-}]$ calculated previously:

$$K_{a_3} = \frac{[H^+][PO_4^{3-}]}{[HPO_4^{2-}]} = 4.8 \times 10^{-13} = \frac{0.19[PO_4^{3-}]}{(6.2 \times 10^{-8})}$$

$$[PO_4^{3-}] = \frac{(4.8 \times 10^{-13})(6.2 \times 10^{-8})}{0.19} = 1.6 \times 10^{-19} \; M$$

These results show that the second and third dissociation steps do not make an important contribution to $[H^+]$. This is apparent from the fact that $[HPO_4^{2-}]$ is $6.2 \times 10^{-8} \; M$, which means that only 6.2×10^{-8} mol/L $H_2PO_4^-$ has dissociated. The value of $[PO_4^{3-}]$ shows that the dissociation of HPO_4^{2-} is even smaller. We must, however, use the second and third dissociation steps to calculate $[HPO_4^{2-}]$ and $[PO_4^{3-}]$, since these steps are the only sources of these ions.

See Exercises 13-107 and 13-108

Critical Thinking

What if the three values of K_a for phosphoric acid were closer to each other in value? Why would this complicate the calculation of the pH for an aqueous solution of phosphoric acid?

Sulfuric Acid

Sulfuric acid is unique among the common acids in that it is *a strong acid in its first dissociation step and a weak acid in its second step:*

$$H_2SO_4(aq) \longrightarrow H^+(aq) + HSO_4^-(aq) \quad K_{a_1} \text{ is very large}$$
$$HSO_4^-(aq) \rightleftharpoons H^+(aq) + SO_4^{2-}(aq) \quad K_{a_2} = 1.2 \times 10^{-2}$$

Example 13-16 illustrates how to calculate the pH for sulfuric acid solutions.

INTERACTIVE EXAMPLE 13-16

The pH of Sulfuric Acid

Calculate the pH of a 1.0-M H_2SO_4 solution.

Solution

The major species in the solution are ◄

$$H^+, \quad HSO_4^-, \quad \text{and} \quad H_2O$$

Major Species

H⁺

HSO₄⁻

H₂O

where the first two ions are produced by the complete first dissociation step of H_2SO_4. The concentration of H^+ in this solution will be at least $1.0\ M$, since this amount is produced by the first dissociation step of H_2SO_4. We must now answer this question: Does the HSO_4^- ion dissociate enough to produce a significant contribution to the concentration of H^+? This question can be answered by calculating the equilibrium concentrations for the dissociation reactions of HSO_4^-:

$$HSO_4^-(aq) \rightleftharpoons H^+(aq) + SO_4^{2-}(aq)$$

where
$$K_{a_2} = 1.2 \times 10^{-2} = \frac{[H^+][SO_4^{2-}]}{[HSO_4^-]}$$

The ICE table is:

	$HSO_4^-(aq)$	$\rightleftharpoons$	$H^+(aq)$	+	$SO_4^{2-}(aq)$
Initial	1.0		1.0		0
Change	$-x$		$+x$		$+x$
Equilibrium	$1.0 - x$		$1.0 + x$		x

Note that $[H^+]_0$ is not equal to zero, as it usually is for a weak acid, because the first dissociation step has already occurred. Substituting the equilibrium concentrations into the expression for K_{a_2} and making the usual approximation give

$$K_{a_2} = 1.2 \times 10^{-2} = \frac{[H^+][SO_4^{2-}]}{[HSO_4^-]} = \frac{(1.0 + x)(x)}{1.0 - x} \approx \frac{(1.0)(x)}{(1.0)}$$

Thus
$$x \approx 1.2 \times 10^{-2}$$

Since 1.2×10^{-2} is 1.2% of 1.0, the approximation is valid according to the 5% rule. Note that x is not equal to $[H^+]$ in this case. Instead,

$$[H^+] = 1.0\ M + x = 1.0\ M + (1.2 \times 10^{-2})\ M$$
$$= 1.0\ M \quad \text{(to the correct number of significant figures)}$$

Thus the dissociation of HSO_4^- does not make a significant contribution to the concentration of H^+, and

$$[H^+] = 1.0\ M \quad \text{and} \quad pH = 0.00$$

See Exercise 13-111

ⓘ Only in dilute H_2SO_4 solutions does the second dissociation step contribute significantly to $[H^+]$.

Example 13-16 illustrates the most common case for sulfuric acid in which only the first dissociation makes an important contribution to the concentration of H^+. In solutions more dilute than $1.0\ M$ (for example, $0.10\ M\ H_2SO_4$), the dissociation of HSO_4^- is important, and solving the problem requires use of the quadratic formula, as shown in Example 13-17. ◀ ⓘ

EXAMPLE 13-17   The pH of Sulfuric Acid

Calculate the pH of a $1.00 \times 10^{-2}\ M\ H_2SO_4$ solution.

Solution

The major species in solution are ◀

$$H^+, \quad HSO_4^-, \quad \text{and} \quad H_2O$$

Major Species

H⁺

HSO₄⁻

H₂O

Proceeding as in Example 13-16, we consider the dissociation of HSO_4^-, which leads to the following ICE table:

	$HSO_4^-(aq)$	$\rightleftharpoons$	$H^+(aq)$	+	$SO_4^{2-}(aq)$
Initial	0.0100		0.0100		0
			From dissociation of H_2SO_4		
Change	$-x$		$+x$		$+x$
Equilibrium	$0.0100 - x$		$0.0100 + x$		x

Substituting the equilibrium concentrations into the expression for K_{a_2} gives

$$1.2 \times 10^{-2} = K_{a_2} = \frac{[H^+][SO_4^{2-}]}{[HSO_4^-]} = \frac{(0.0100 + x)(x)}{(0.0100 - x)}$$

If we make the usual approximation, then $0.0100 + x \approx 0.0100$ and $0.0100 - x \approx 0.0100$, and we have

$$1.2 \times 10^{-2} = \frac{(0.0100 + x)(x)}{(0.0100 - x)} \approx \frac{(0.0100)x}{(0.0100)}$$

The calculated value of x is

$$x = 1.2 \times 10^{-2} = 0.012$$

This value is larger than 0.010, clearly a ridiculous result. Thus we cannot make the usual approximation and must instead solve the quadratic equation. The expression

$$1.2 \times 10^{-2} = \frac{(0.0100 + x)(x)}{(0.0100 - x)}$$

leads to
$$(1.2 \times 10^{-2})(0.0100 - x) = (0.0100 + x)(x)$$
$$(1.2 \times 10^{-4}) - (1.2 \times 10^{-2})x = (1.0 \times 10^{-2})x + x^2$$
$$x^2 + (2.2 \times 10^{-2})x - (1.2 \times 10^{-4}) = 0$$

This equation can be solved using the quadratic formula

$$x = \frac{-b \pm \sqrt{b^2 - 4ac}}{2a}$$

where $a = 1$, $b = 2.2 \times 10^{-2}$, and $c = -1.2 \times 10^{-4}$. Use of the quadratic formula gives one negative root (which cannot be correct) and one positive root,

$$x = 4.5 \times 10^{-3}$$

Thus
$$[H^+] = 0.0100 + x = 0.0100 + 0.0045 = 0.0145$$

and
$$pH = 1.84$$

Note that in this case the second dissociation step produces about half as many H^+ ions as the initial step does.

This problem also can be solved by successive approximations, a method illustrated in Appendix 1.4.

See Exercise 13-112

13-8 | Acid–Base Properties of Salts

Salt is simply another name for *ionic compound*. When a salt dissolves in water, we assume that it breaks up into its ions, which move about independently, at least in dilute solutions. Under certain conditions, these ions can behave as acids or bases. In this section we explore such reactions.

Salts That Produce Neutral Solutions

Recall that the conjugate base of a strong acid has virtually no affinity for protons in water. This is why strong acids completely dissociate in aqueous solution. Thus, when anions such as Cl^- and NO_3^- are placed in water, they do not combine with H^+ and have no effect on the pH. Cations such as K^+ and Na^+ from strong bases have no affinity for H^+, nor can they produce H^+, so they too have no effect on the pH of an aqueous solution. Salts that consist of the cations of strong bases and the anions of strong acids have no effect on $[H^+]$ when dissolved in water. This means that aqueous solutions of salts such as KCl, NaCl, NaNO$_3$, and KNO$_3$ are neutral (have a pH of 7). ◄ ⓘ

ⓘ The salt of a strong acid and a strong base gives a neutral solution.

Salts That Produce Basic Solutions

In an aqueous solution of sodium acetate (NaC$_2$H$_3$O$_2$), the major species are ◄

$$Na^+, \quad C_2H_3O_2^-, \quad \text{and} \quad H_2O$$

Major Species

 Na$^+$

 C$_2$H$_3$O$_2^-$

 H$_2$O

What are the acid–base properties of each component? The Na$^+$ ion has neither acid nor base properties. The C$_2$H$_3$O$_2^-$ ion is the conjugate base of acetic acid, a weak acid. This means that C$_2$H$_3$O$_2^-$ has a significant affinity for a proton and is a base. Finally, water is a weakly amphoteric substance.

The pH of this solution will be determined by the C$_2$H$_3$O$_2^-$ ion. Since C$_2$H$_3$O$_2^-$ is a base, it will react with the best proton donor available. In this case, water is the *only* source of protons, and the reaction between the acetate ion and water is

$$C_2H_3O_2^-(aq) + H_2O(l) \rightleftharpoons HC_2H_3O_2(aq) + OH^-(aq) \tag{13-7}$$

Note that this reaction, which yields a base solution, involves a *base reacting with water to produce hydroxide ion and a conjugate acid*. We have defined K_b as the equilibrium constant for such a reaction. In this case,

$$K_b = \frac{[HC_2H_3O_2][OH^-]}{[C_2H_3O_2^-]}$$

The value of K_a for acetic acid is well known (1.8×10^{-5}). But how can we obtain the K_b value for the acetate ion? The answer lies in the relationships among K_a, K_b, and

K_w. Note that when the expression for K_a for acetic acid is multiplied by the expression for K_b for the acetate ion, the result is K_w:

$$K_a \times K_b = \frac{[\text{H}^+][\text{C}_2\text{H}_3\text{O}_2^-]}{[\text{HC}_2\text{H}_3\text{O}_2]} \times \frac{[\text{HC}_2\text{H}_3\text{O}_2][\text{OH}^-]}{[\text{C}_2\text{H}_3\text{O}_2^-]} = [\text{H}^+][\text{OH}^-] = K_w$$

This is a very important result. For any weak acid and its conjugate base,

$$K_a \times K_b = K_w$$

We can obtain a different form of this equation by taking $-\log$ of both sides of the above equation. This gives the equation

$$\text{p}K_a + \text{p}K_b = \text{p}K_w$$

At 25°C, $K_w = 1.0 \times 10^{-14}$, so $\text{p}K_w = 14.00$. In solutions at 25°C, we can use the relationship

$$\text{p}K_a + \text{p}K_b = 14.00$$

Thus when either K_a or K_b is known, the other can be calculated. For the acetate ion, since $K_w = K_b \times K_a$ (acetic acid), the K_b for acetate is:

$$K_b = \frac{K_w}{K_a \text{ (for HC}_2\text{H}_3\text{O}_2)} = \frac{1.0 \times 10^{-14}}{1.8 \times 10^{-5}} = 5.6 \times 10^{-10}$$

This is the K_b value for the reaction described by Equation (13-7). Note that it is obtained from the K_a value of the parent weak acid, in this case acetic acid. The sodium acetate solution is an example of an important general case. For any salt whose cation has neutral properties (such as Na^+ or K^+) and whose anion is the conjugate base of a weak acid, the aqueous solution will be basic. ◄ ⓘ The K_b value for the anion can be obtained from the relationship $K_b = K_w/K_a$. Equilibrium calculations of this type are illustrated in Example 13-18.

ⓘ A basic solution is formed if the anion of the salt is the conjugate base of a weak acid.

INTERACTIVE EXAMPLE 13-18

Salts as Weak Bases

Calculate the pH of a 0.30-M NaF solution. The K_a value for HF is 7.2×10^{-4}.

Solution

The major species in solution are ◄

$$\text{Na}^+, \quad \text{F}^-, \quad \text{and} \quad \text{H}_2\text{O}$$

Major Species

 Na^+

 F^-

⬤ H_2O

Since HF is a weak acid, the F^- ion must have a significant affinity for protons, and the dominant reaction will be

$$\text{F}^-(aq) + \text{H}_2\text{O}(l) \rightleftharpoons \text{HF}(aq) + \text{OH}^-(aq)$$

which yields the K_b expression

$$K_b = \frac{[\text{HF}][\text{OH}^-]}{[\text{F}^-]}$$

The value of K_b can be calculated from K_w and the K_a value for HF:

$$K_b = \frac{K_w}{K_a \text{ (for HF)}} = \frac{1.0 \times 10^{-14}}{7.2 \times 10^{-4}} = 1.4 \times 10^{-11}$$

The corresponding ICE table is:

	$\text{F}^-(aq)$	$+$	$\text{H}_2\text{O}(l)$	$\rightleftharpoons$	$\text{HF}(aq)$	$+$	$\text{OH}^-(aq)$
Initial	0.30		—		0		≈0
Change	$-x$		—		$+x$		$+x$
Equilibrium	$0.30 - x$		—		x		x

Thus $\qquad K_b = 1.4 \times 10^{-11} = \dfrac{[HF][OH^-]}{[F^-]} = \dfrac{(x)(x)}{0.30 - x} \approx \dfrac{x^2}{0.30}$

and

$$x \approx 2.0 \times 10^{-6}$$

The approximation is valid by the 5% rule, so

$$[OH^-] = x = 2.0 \times 10^{-6}\,M$$
$$\text{pOH} = 5.69$$
$$\text{pH} = 14.00 - 5.69 = 8.31$$

As expected, the solution is basic.

See Exercise 13-117

Base Strength in Aqueous Solutions

To emphasize the concept of base strength, let us consider the basic properties of the cyanide ion. One relevant reaction is the dissociation of hydrocyanic acid in water:

$$HCN(aq) + H_2O(l) \rightleftharpoons H_3O^+(aq) + CN^-(aq) \qquad K_a = 6.2 \times 10^{-10}$$

Since HCN is such a weak acid, CN^- appears to be a *strong* base, showing a very high affinity for H^+ *compared to H_2O*, with which it is competing. However, we also need to look at the reaction in which cyanide ion reacts with water:

$$CN^-(aq) + H_2O(l) \rightleftharpoons HCN(aq) + OH^-(aq)$$

where $\qquad K_b = \dfrac{K_w}{K_a} = \dfrac{1.0 \times 10^{-14}}{6.2 \times 10^{-10}} = 1.6 \times 10^{-5}$

In this reaction CN^- appears to be a weak base; the K_b value is only 1.6×10^{-5}. What accounts for this apparent difference in base strength? The key idea is that in the reaction of CN^- with H_2O, *CN^- is competing with OH^- for H^+, instead of competing with H_2O*, as it does in the HCN dissociation reaction. These equilibria show the following relative base strengths:

$$OH^- > CN^- > H_2O$$

Similar arguments can be made for other "weak" bases, such as ammonia, the acetate ion, the fluoride ion, and so on.

Salts That Produce Acidic Solutions

Some salts produce acidic solutions when dissolved in water. For example, when solid NH_4Cl is dissolved in water, NH_4^+ and Cl^- ions are present, with NH_4^+ behaving as a weak acid:

$$NH_4^+(aq) \rightleftharpoons NH_3(aq) + H^+(aq)$$

The Cl^- ion, having virtually no affinity for H^+ in water, does not affect the pH of the solution.

In general, salts in which the anion is not a base and the cation is the conjugate acid of a weak base produce acidic solutions.

INTERACTIVE EXAMPLE 13-19

Salts as Weak Acids I

Calculate the pH of a 0.10-M NH_4Cl solution. The K_b value for NH_3 is 1.8×10^{-5}.

Solution

The major species in solution are ◄

Major Species

Cl^-

NH_4^+

H_2O

$$NH_4^+, \quad Cl^-, \quad \text{and} \quad H_2O$$

Note that both NH_4^+ and H_2O can produce H^+. The dissociation reaction for the NH_4^+ ion is

$$NH_4^+(aq) \rightleftharpoons NH_3(aq) + H^+(aq)$$

for which

$$K_a = \frac{[NH_3][H^+]}{[NH_4^+]}$$

Note that although the K_b value for NH_3 is given, the reaction corresponding to K_b is not appropriate here, since NH_3 is not a major species in the solution. Instead, the given value of K_b is used to calculate K_a for NH_4^+ from the relationship

$$K_a \times K_b = K_w$$

Thus

$$K_a \text{ (for } NH_4^+) = \frac{K_w}{K_b \text{ (for } NH_3)} = \frac{1.0 \times 10^{-14}}{1.8 \times 10^{-5}} = 5.6 \times 10^{-10}$$

Although NH_4^+ is a very weak acid, as indicated by its K_a value, it is stronger than H_2O and will dominate in the production of H^+. Thus we will focus on the dissociation reaction of NH_4^+ to calculate the pH in this solution.

We solve the weak acid problem in the usual way:

	$NH_4^+(aq)$	$\rightleftharpoons$	$H^+(aq)$	$+$	$NH_3(aq)$
Initial	0.10		≈ 0		0
Change	$-x$		$+x$		$+x$
Equilibrium	$0.10 - x$		x		x

Thus

$$5.6 \times 10^{-10} = K_a = \frac{[H^+][NH_3]}{[NH_4^+]} = \frac{(x)(x)}{0.10 - x} \approx \frac{x^2}{0.10}$$

$$x \approx 7.5 \times 10^{-6}$$

The approximation is valid by the 5% rule, so

$$[H^+] = x = 7.5 \times 10^{-6} \, M \quad \text{and} \quad \text{pH} = 5.13$$

See Exercise 13-118

A second type of salt that produces an acidic solution is one that contains a *highly charged metal ion*. For example, when solid aluminum chloride ($AlCl_3$) is dissolved in water, the resulting solution is significantly acidic. Although the Al^{3+} ion is not itself a Brønsted–Lowry acid, the hydrated ion $Al(H_2O)_6^{3+}$ formed in water is a weak acid:

$$Al(H_2O)_6^{3+}(aq) \rightleftharpoons Al(OH)(H_2O)_5^{2+}(aq) + H^+(aq)$$

The high charge on the metal ion polarizes the O—H bonds in the attached water molecules, making the hydrogens in these water molecules more acidic than those in free water molecules. Typically, the higher the charge on the metal ion, the stronger the acidity of the hydrated ion.

INTERACTIVE EXAMPLE 13-20

Salts as Weak Acids II

Calculate the pH of a 0.010-M $AlCl_3$ solution. The K_a value for $Al(H_2O)_6^{3+}$ is 1.4×10^{-5}.

Solution

The major species in solution are ◄

Major Species

Cl^-

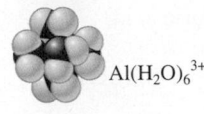

$Al(H_2O)_6^{3+}$

H_2O

$$Al(H_2O)_6^{3+}, \quad Cl^-, \quad and \quad H_2O$$

Since the $Al(H_2O)_6^{3+}$ ion is a stronger acid than water, the dominant equilibrium is

$$Al(H_2O)_6^{3+}(aq) \rightleftharpoons Al(OH)(H_2O)_5^{2+}(aq) + H^+(aq)$$

and

$$1.4 \times 10^{-5} = K_a = \frac{[Al(OH)(H_2O)_5^{2+}][H^+]}{[Al(H_2O)_6^{3+}]}$$

This is a typical weak acid problem, which we can solve with the usual procedure:

	$Al(H_2O)_6^{3+}(aq)$	$\rightleftharpoons$	$Al(OH)(H_2O)_5^{2+}(aq)$	+	$H^+(aq)$
Initial	0.010		0		≈ 0
Change	$-x$		$+x$		$+x$
Equilibrium	$0.010 - x$		x		x

Thus

$$1.4 \times 10^{-5} = K_a = \frac{[Al(OH)(H_2O)_5^{2+}][H^+]}{[Al(H_2O)_6^{3+}]} = \frac{(x)(x)}{0.010 - x} \approx \frac{x^2}{0.010}$$

$$x \approx 3.7 \times 10^{-4}$$

Since the approximation is valid by the 5% rule,

$$[H^+] = x = 3.7 \times 10^{-4} \, M \quad and \quad pH = 3.43$$

See Exercises 13-127 and 13-128

So far we have considered salts in which only one of the ions has acidic or basic properties. For many salts, such as ammonium acetate ($NH_4C_2H_3O_2$), both ions can affect the pH of the aqueous solution. Because the equilibrium calculations for these cases can be quite complicated, we will consider only the qualitative aspects of such problems. We can predict whether the solution will be basic, acidic, or neutral by comparing the K_a value for the acidic ion with the K_b value for the basic ion. If the K_a value for the acidic ion is larger than the K_b value for the basic ion, the solution will be acidic. If the K_b value is larger than the K_a value, the solution will be basic. Equal K_a and K_b values mean a neutral solution. These facts are summarized in Table 13-5.

Table 13-5 | Qualitative Prediction of pH for Solutions of Salts for Which Both Cation and Anion Have Acidic or Basic Properties

$K_a > K_b$	pH < 7 (acidic)
$K_b > K_a$	pH > 7 (basic)
$K_a = K_b$	pH = 7 (neutral)

INTERACTIVE EXAMPLE 13-21

The Acid–Base Properties of Salts

Predict whether an aqueous solution of each of the following salts will be acidic, basic, or neutral.

a. $NH_4C_2H_3O_2$ **b.** NH_4CN **c.** $Al_2(SO_4)_3$

Solution

a. The ions in solution are NH_4^+ and $C_2H_3O_2^-$. As we mentioned previously, K_a for NH_4^+ is 5.6×10^{-10} and K_b for $C_2H_3O_2^-$ is 5.6×10^{-10}. Thus K_a for NH_4^+ is equal to K_b for $C_2H_3O_2^-$, and the solution will be neutral (pH = 7).

b. The solution will contain NH_4^+ and CN^- ions. The K_a value for NH_4^+ is 5.6×10^{-10} and

$$K_b \text{ (for } CN^-) = \frac{K_w}{K_a \text{ (for HCN)}} = 1.6 \times 10^{-5}$$

Since K_b for CN^- is much larger than K_a for NH_4^+, CN^- is a much stronger base than NH_4^+ is an acid. This solution will be basic.

c. The solution will contain $Al(H_2O)_6^{3+}$ and SO_4^{2-} ions. The K_a value for $Al(H_2O)_6^{3+}$ is 1.4×10^{-5}, as given in Example 13-20. We must calculate K_b for SO_4^{2-}. The HSO_4^- ion is the conjugate acid of SO_4^{2-}, and its K_a value is K_{a_2} for sulfuric acid, or 1.2×10^{-2}. Therefore,

$$K_b \text{ (for } SO_4^{2-}) = \frac{K_w}{K_{a_2} \text{ (for sulfuric acid)}}$$

$$= \frac{1.0 \times 10^{-14}}{1.2 \times 10^{-2}} = 8.3 \times 10^{-13}$$

This solution will be acidic, since K_a for $Al(H_2O)_6^{3+}$ is much greater than K_b for SO_4^{2-}.

See Exercises 13-129 and 13-130

The acid–base properties of aqueous solutions of various salts are summarized in Table 13-6.

13-9 | The Effect of Structure on Acid–Base Properties

We have seen that when a substance is dissolved in water, it produces an acidic solution if it can donate protons and produces a basic solution if it can accept protons. What structural properties of a molecule cause it to behave as an acid or as a base?

Any molecule containing a hydrogen atom is potentially an acid. However, many such molecules show no acidic properties. For example, molecules containing C—H bonds, such as chloroform ($CHCl_3$) and nitromethane (CH_3NO_2), do not produce acidic aqueous solutions because a C—H bond is both strong and nonpolar and thus there is no tendency to donate protons. On the other hand, although the H—Cl bond in

Table 13-6 | Acid–Base Properties of Various Types of Salts

Type of Salt	Examples	Comment	pH of Solution
Cation is from strong base; anion is from strong acid	KCl, KNO₃, NaCl, NaNO₃	Neither acts as an acid or a base	Neutral
Cation is from strong base; anion is from weak acid	NaC₂H₃O₂, KCN, NaF	Anion acts as a base; cation has no effect on pH	Basic
Cation is conjugate acid of weak base; anion is from strong acid	NH₄Cl, NH₄NO₃	Cation acts as an acid; anion has no effect on pH	Acidic
Cation is conjugate acid of weak base; anion is conjugate base of weak acid	NH₄C₂H₃O₂, NH₄CN	Cation acts as an acid; anion acts as a base	Acidic if $K_a > K_b$, basic if $K_b > K_a$, neutral if $K_a = K_b$
Cation is highly charged metal ion; anion is from strong acid	Al(NO₃)₃, FeCl₃	Hydrated cation acts as an acid; anion has no effect on pH	Acidic

Table 13-7 | Bond Strengths and Acid Strengths for Hydrogen Halides

H—X Bond	Bond Strength (kJ/mol)	Acid Strength in Water
H—F	565	Weak
H—Cl	427	Strong
H—Br	363	Strong
H—I	295	Strong

ℹ Notice that knowledge of bond strengths helps us make sense of the strengths of the hydrogen halide acids.

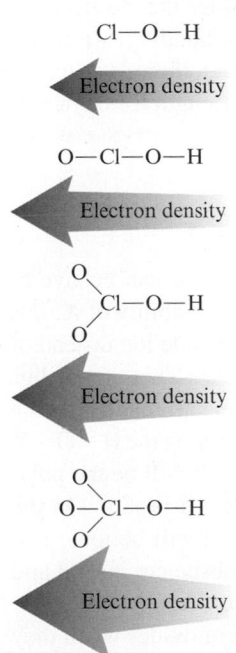

Figure 13-9 | The effect of the number of attached oxygens on the O—H bond in a series of chlorine oxyacids. As the number of oxygen atoms attached to the chlorine atom increases, they become more effective at withdrawing electron density from the O—H bond, thereby weakening and polarizing it. This increases the tendency for the molecule to produce a proton, and so its acid strength increases.

gaseous hydrogen chloride is slightly stronger than a C—H bond, it is much more polar, and this molecule readily dissociates when dissolved in water.

Thus there are two main factors that determine whether a molecule containing an X—H bond will behave as a Brønsted–Lowry acid: the strength of the bond and the polarity of the bond.

To explore these factors let's consider the relative acid strengths of the hydrogen halides. The bond polarities vary as shown

$$H—F > H—Cl > H—Br > H—I$$

Most polar   Least polar

because electronegativity decreases going down the group. Based on the high polarity of the H—F bond, we might expect hydrogen fluoride to be a very strong acid. In fact, among HX molecules, HF is the only weak acid ($K_a = 7.2 \times 10^{-4}$) when dissolved in water. The H—F bond is unusually strong, as shown in Table 13-7, and thus is difficult to break. This contributes significantly to the reluctance of the HF molecules to dissociate in water. ◄ ℹ

Another important class of acids are the oxyacids, which as we saw in Section 13-2 characteristically contain the grouping H—O—X. Several series of oxyacids are listed with their K_a values in Table 13-8. Note from these data that for a given series the acid strength increases with an increase in the number of oxygen atoms attached to the central atom. For example, in the series containing chlorine and a varying number of oxygen atoms, HOCl is a weak acid, but the acid strength is successively greater as the number of oxygen atoms increases. This happens because the very electronegative oxygen atoms are able to draw electrons away from the chlorine atom and the O—H bond (Fig. 13-9). The net effect is to both polarize and weaken the O—H bond; this effect becomes more important as the number of attached oxygen atoms increases.

Table 13-8 | Several Series of Oxyacids and Their K_a Values

Oxyacid	Structure	K_a Value
$HClO_4$	H—O—Cl with three O (two double-bond O and one O)	Large (~10^7)
$HClO_3$	H—O—Cl with two O	~1
$HClO_2$	H—O—Cl—O	1.2×10^{-2}
$HClO$	H—O—Cl	3.5×10^{-8}
H_2SO_4	H—O—S with O—H and two O	Large
H_2SO_3	H—O—S with O—H and one O	1.5×10^{-2}
HNO_3	H—O—N with two O	Large
HNO_2	H—O—N—O	4.0×10^{-4}

Table 13-9 | Comparison of Electronegativity of X and K_a Value for a Series of Oxyacids

Acid	X	Electronegativity of X	K_a for Acid
HOCl	Cl	3.0	4×10^{-8}
HOBr	Br	2.8	2×10^{-9}
HOI	I	2.5	2×10^{-11}
HOCH$_3$	CH$_3$	2.3 (for carbon in CH$_3$)	$\sim 10^{-15}$

This means that a proton is most readily produced by the molecule with the largest number of attached oxygen atoms ($HClO_4$).

This type of behavior is also observed for hydrated metal ions. Earlier in this chapter we saw that highly charged metal ions such as Al^{3+} produce acidic solutions. The acidity of the water molecules attached to the metal ion is increased by the attraction of electrons to the positive metal ion:

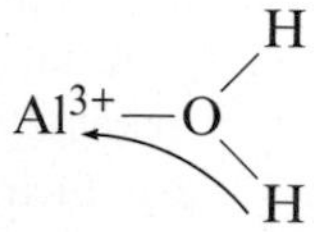

$$Al^{3+}\!\!\leftarrow\!\!O\diagdown^{\textstyle H}_{\textstyle H}$$

The greater the charge on the metal ion, the more acidic the hydrated ion becomes.

For acids containing the H—O—X grouping, the greater the ability of X to draw electrons toward itself, the greater the acidity of the molecule. Since the electronegativity of X reflects its ability to attract the electrons involved in bonding, we might expect acid strength to depend on the electronegativity of X. In fact, there is an excellent correlation between the electronegativity of X and the acid strength for oxyacids, as shown in Table 13-9.

13-10 | Acid–Base Properties of Oxides

We have just seen that molecules containing the grouping H—O—X can behave as acids and that the acid strength depends on the electron-withdrawing ability of X. But substances with this grouping also can behave as bases if the hydroxide ion instead of a proton is produced. What determines which behavior will occur? The answer lies mainly in the nature of the O—X bond. If X has a relatively high electronegativity, the O—X bond will be covalent and strong. When the compound containing the H—O—X grouping is dissolved in water, the O—X bond will remain intact. It will be the polar and relatively weak H—O bond that will tend to break, releasing a proton. On the other hand, if X has a very low electronegativity, the O—X bond will be ionic and subject to being broken in polar water. Examples are the ionic substances NaOH and KOH that dissolve in water to give the metal cation and the hydroxide ion. ◄ ⓘ

We can use these principles to explain the acid–base behavior of oxides when they are dissolved in water. For example, when a covalent oxide such as sulfur trioxide is dissolved in water, an acidic solution results because sulfuric acid is formed:

$$SO_3(g) + H_2O(l) \longrightarrow H_2SO_4(aq)$$

The structure of H_2SO_4 is shown in the margin. ◄ ⓘ In this case, the strong, covalent O—S bonds remain intact and the H—O bonds break to produce protons. Other common covalent oxides that react with water to form acidic solutions are sulfur dioxide, carbon dioxide, and nitrogen dioxide, as shown by the following reactions:

$$SO_2(g) + H_2O(l) \longrightarrow H_2SO_3(aq)$$
$$CO_2(g) + H_2O(l) \longrightarrow H_2CO_3(aq)$$
$$2NO_2(g) + H_2O(l) \longrightarrow HNO_3(aq) + HNO_2(aq)$$

ⓘ A compound containing the H—O—X group will produce an acidic solution in water if the O—X bond is strong and covalent. If the O—X bond is ionic, the compound will produce a basic solution in water.

Thus when a covalent oxide dissolves in water, an acidic solution forms. These oxides are called **acidic oxides**.

On the other hand, when an ionic oxide dissolves in water, a basic solution results, as shown by the following reactions:

$$CaO(s) + H_2O(l) \longrightarrow Ca(OH)_2(aq)$$
$$K_2O(s) + H_2O(l) \longrightarrow 2KOH(aq)$$

These reactions can be explained by recognizing that the oxide ion has a high affinity for protons and reacts with water to produce hydroxide ions:

$$O^{2-}(aq) + H_2O(l) \longrightarrow 2OH^-(aq)$$

Thus the most ionic oxides, such as those of the Group 1A and 2A metals, produce basic solutions when they are dissolved in water. As a result, these oxides are called **basic oxides**.

13-11 | The Lewis Acid–Base Model

We have seen that the first successful conceptualization of acid–base behavior was proposed by Arrhenius. This useful but limited model was replaced by the more general Brønsted–Lowry model. An even more general model for acid–base behavior was suggested by G. N. Lewis in the early 1920s. A **Lewis acid** is an *electron-pair acceptor,* and a **Lewis base** is an *electron-pair donor.* Another way of saying this is that a Lewis acid has an empty atomic orbital that it can use to accept (share) an electron pair from a molecule that has a lone pair of electrons (Lewis base). The three models for acids and bases are summarized in Table 13-10.

Note that Brønsted–Lowry acid–base reactions (proton donor–proton acceptor reactions) are encompassed by the Lewis model. For example, the reaction between a proton and an ammonia molecule, that is,

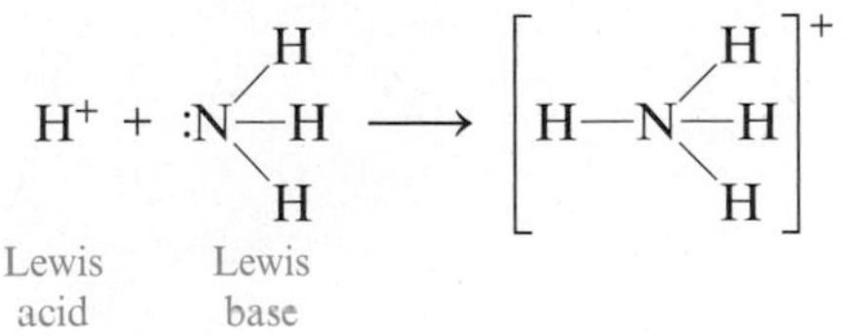

can be represented as a reaction between an electron-pair acceptor (H^+) and an electron-pair donor (NH_3). The same holds true for a reaction between a proton and a hydroxide ion:

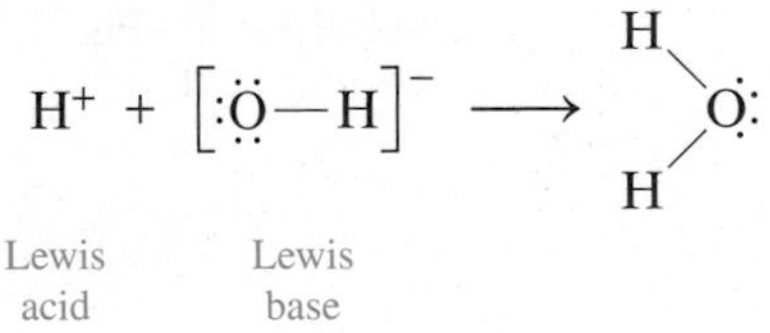

Table 13-10 | Three Models for Acids and Bases

Model	Definition of Acid	Definition of Base
Arrhenius	H^+ producer	OH^- producer
Brønsted–Lowry	H^+ donor	H^+ acceptor
Lewis	Electron-pair acceptor	Electron-pair donor

Figure 13-10 | Reaction of BF₃ with NH₃.

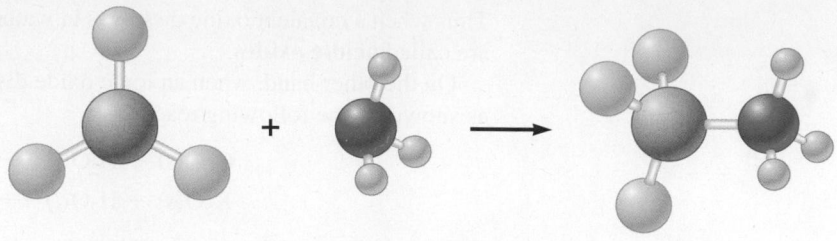

ⓘ The Lewis model encompasses the Brønsted–Lowry model, but the reverse is not true.

The real value of the Lewis model for acids and bases is that it covers many reactions that do not involve Brønsted–Lowry acids. ◄ ⓘ For example, consider the gas-phase reaction between boron trifluoride and ammonia:

$$
\begin{array}{ccc}
\ddot{\textrm{F}}: & & \text{H} \\
| & & | \\
\text{B}-\ddot{\textrm{F}}: + :\text{N}-\text{H} & \longrightarrow & :\ddot{\textrm{F}}-\text{B}-\text{N}-\text{H} \\
| & & | \\
:\ddot{\textrm{F}}: & & \text{H}
\end{array}
$$

Lewis acid   Lewis base

Here the electron-deficient BF₃ molecule (there are only six electrons around the boron) completes its octet by reacting with NH₃, which has a lone pair of electrons (Fig. 13-10). In fact, as mentioned in Chapter 3, the electron deficiency of boron trifluoride makes it very reactive toward any electron-pair donor. That is, it is a strong Lewis acid.

The hydration of a metal ion, such as Al^{3+}, also can be viewed as a Lewis acid–base reaction:

$$
Al^{3+} + 6\left(:\ddot{\textrm{O}}\begin{array}{c} \nearrow\text{H} \\ \searrow\text{H} \end{array}\right) \longrightarrow \left[Al-\left(\ddot{\textrm{O}}\begin{array}{c} \nearrow\text{H} \\ \searrow\text{H} \end{array}\right)_6\right]^{3+}
$$

Lewis acid   Lewis base

Here the Al^{3+} ion accepts one electron pair from each of six water molecules to form $Al(H_2O)_6^{3+}$ (Fig. 13-11).

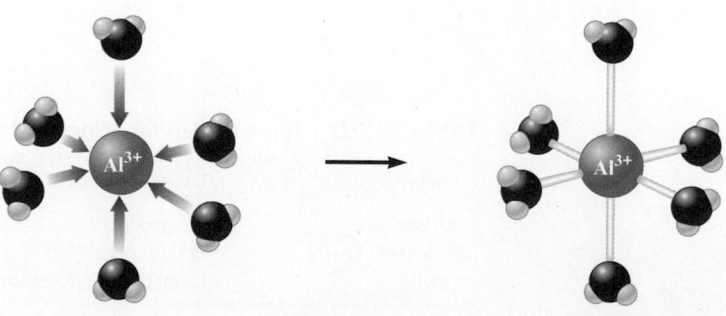

Figure 13-11 | The $Al(H_2O)_6^{3+}$ ion.

In addition, the reaction between a covalent oxide and water to form a Brønsted–Lowry acid can be defined as a Lewis acid–base reaction. An example is the reaction between sulfur trioxide and water:

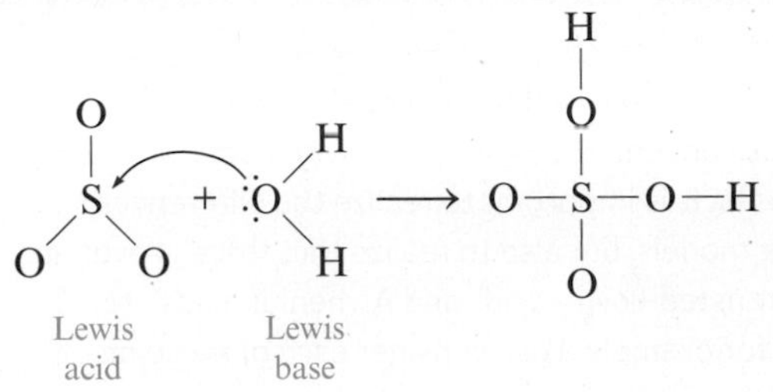

Note that as the water molecule attaches to sulfur trioxide, a proton shift occurs to form sulfuric acid.

EXAMPLE 13-22   Lewis Acids and Bases

For each reaction, identify the Lewis acid and base.

a. $Ni^{2+}(aq) + 6NH_3(aq) \longrightarrow Ni(NH_3)_6{}^{2+}(aq)$

b. $H^+(aq) + H_2O(aq) \rightleftharpoons H_3O^+(aq)$

Solution

a. Each NH_3 molecule donates an electron pair to the Ni^{2+} ion:

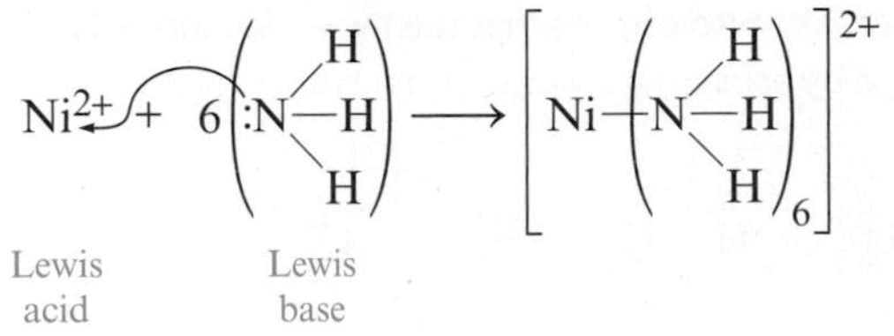

The nickel(II) ion is the Lewis acid, and ammonia is the Lewis base.

b. The proton is the Lewis acid and the water molecule is the Lewis base:

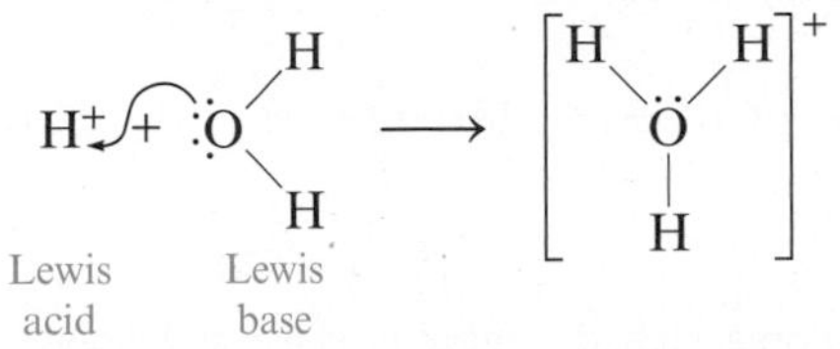

See Exercises 13-137 and 13-138

13-12 | Strategy for Solving Acid–Base Problems: A Summary

In this chapter we have encountered many different situations involving aqueous solutions of acids and bases, and in the next chapter we will encounter still more. In solving for the equilibrium concentrations in these aqueous solutions, it is tempting to create a pigeonhole for each possible situation and to memorize the procedures necessary to deal with that particular case. This approach is just not practical and usually leads to frustration: Too many pigeonholes are required—there seems

In this chapter we discussed three models of acids and bases: the Arrhenius concept, the Brønsted-Lowry model, and Lewis acids and bases. It is important to realize the differences among these models, but also to realize that there is overlap; that is, all Brønsted-Lowry acids and Arrhenius acids are Lewis acids, for example. Let's consider each of these models.

The Arrhenius Concept

In this model, acids produce hydrogen ions in aqueous solution, and bases produce hydroxide ions. A typical acid is hydrochloric acid, which dissociates in water as follows:

$$HCl \rightarrow H^+ + Cl^-$$

A typical Arrhenius base is sodium hydroxide, which dissociates in water as follows:

$$NaOH \rightarrow Na^+ + OH^-$$

The Brønsted-Lowry Model

In this model, an acid is a proton (H^+) donor, and a base is a proton acceptor. We can see that HCl is an acid according to this definition. We also can see that the hydroxide ion (OH^-) acts as a base by accepting a proton from HCl as follows:

$$\left[:\ddot{O} - H \right]^- + H - \ddot{\underset{\cdot\cdot}{C}}l: \longrightarrow \underset{H}{\overset{H}{\diagup}} \ddot{O}: + \left[:\ddot{\underset{\cdot\cdot}{C}}l: \right]^-$$

Thus, this model does not contradict the Arrhenius concept. However, this model is more general by allowing us to see that water can act as a base:

$$H - \underset{\underset{H}{\mid}}{\ddot{O}}: + H - \ddot{\underset{\cdot\cdot}{C}}l: \longrightarrow \left[H - \underset{\underset{H}{\mid}}{\ddot{O}} - H \right]^+ + \left[:\ddot{\underset{\cdot\cdot}{C}}l: \right]^-$$

In addition, ammonia, NH_3, can react as a base as follows:

$$H - \underset{\underset{H}{\mid}}{N}: + H - \ddot{\underset{\cdot\cdot}{C}}l: \rightleftharpoons \left[H - \underset{\underset{H}{\mid}}{N} - H \right]^+ \left[:\ddot{\underset{\cdot\cdot}{C}}l: \right]^-$$

Also, this model does not require an aqueous solution. Thus, the Brønsted-Lowry model is a more general model than the Arrhenius model.

Finally this model shows that water can act not only as a base but also as an acid; that is, it is amphoteric. In fact, it undergoes autoionization:

$$\underset{H}{\overset{H}{\diagup}} \ddot{O}: + \underset{H}{\overset{H}{\diagup}} \ddot{O}: \rightleftharpoons \left[H \underset{\underset{H}{\mid}}{\overset{\ddot{O}}{\diagdown}} H \right]^+ + \left[:\ddot{O} - H \right]^-$$

Lewis Model

A Lewis acid is an electron-pair acceptor, and a Lewis base is an electron-pair donor. This model incorporates both Arrhenius and Brønsted-Lowry acids and bases. We can view the simplest Arrhenius acids and bases such as H^+ and OH^- as Lewis acids and bases by considering the electron pairs as follows:

$$H^+ + \left[:\ddot{O} - H \right]^- \longrightarrow \underset{H}{\overset{H}{\diagup}} \ddot{O}:$$

Lewis	Lewis
acid	base

Note how H^+ accepts an electron pair and OH^- donates an electron pair. We also could say that OH^- accepts a proton, making it a Brønsted-Lowry base.

This model also allows us to see other reactants as acids that the Brønsted-Lowry model does not include. For example BF_3 is not a Brønsted-Lowry acid. However, we can view the following as a Lewis acid–base reaction:

$$\underset{:\ddot{F}:}{\overset{:\ddot{F}:}{\diagup}} B - \ddot{F}: + :N\underset{\underset{H}{\mid}}{\overset{H}{\mid}}H \longrightarrow :\ddot{F} - B\underset{\underset{:\ddot{F}:}{\mid}}{\overset{:\ddot{F}}{\mid}} - N\underset{\underset{H}{\mid}}{\overset{H}{\mid}}H$$

Lewis	Lewis
acid	base

Summary

There are various models for acids and bases. These models do not contradict one another, but some are more complicated than others. As we have discussed before, we always wish to keep our models as simple as possible. In aqueous solutions with simple acids and hydroxides, the Arrhenius definitions will suffice. However, to explain the entire acid–base behavior of water, ammonia, and other substances, more complicated models are required.

to be an infinite number of cases. But you can handle any case successfully by taking a systematic, patient, and thoughtful approach. When analyzing an acid–base equilibrium problem, do *not* ask yourself how a memorized solution can be used to solve the problem. Instead, ask this question: *What are the major species in the solution and what is their chemical behavior?*

The most important part of doing a complicated acid–base equilibrium problem is the analysis you do at the beginning of a problem.

What major species are present?

Does a reaction occur that can be assumed to go to completion?

What equilibrium dominates the solution?

Let the problem guide you. Be patient.

FOR REVIEW

Key Terms

Section 13-1
Arrhenius concept
Brønsted–Lowry model
hydronium ion
conjugate base
conjugate acid
conjugate acid–base pair
acid dissociation constant

Section 13-2
strong acid
weak acid
diprotic acid
oxyacids
organic acids
carboxyl group
monoprotic acids
amphoteric
amphoteric substance
autoionization
ion-product (dissociation)
    constant (K_w)

Section 13-3
pH scale

Section 13-4
major species

Section 13-5
percent dissociation

Models for Acids and Bases

- Arrhenius model
  - Acids produce H^+ in solution
  - Bases produce OH^- in solution
- Brønsted–Lowry model
  - An acid is a proton donor
  - A base is a proton acceptor
  - In this model an acid molecule reacts with a water molecule, which behaves as a base:

$$\underset{\text{Acid}}{HA(aq)} + \underset{\text{Base}}{H_2O(l)} \rightleftharpoons \underset{\substack{\text{Conjugate} \\ \text{acid}}}{H_3O^+(aq)} + \underset{\substack{\text{Conjugate} \\ \text{base}}}{A^-(aq)}$$

  to form a new acid (conjugate acid) and a new base (conjugate base).
- Lewis model
  - A Lewis acid is an electron-pair acceptor
  - A Lewis base is an electron-pair donor

Acid–Base Equilibrium

- The equilibrium constant for an acid dissociating (ionizing) in water is called K_a
- The K_a expression is

$$K_a = \frac{[H_3O^+][A^-]}{[HA]}$$

which is often simplified as

$$K_a = \frac{[H^+][A^-]}{[HA]}$$

- $[H_2O]$ is never included because it is assumed to be constant

Acid Strength

- A strong acid has a very large K_a value
  - The acid completely dissociates (ionizes) in water
  - The dissociation (ionization) equilibrium position lies all the way to the right
  - Strong acids have very weak conjugate bases
  - The common strong acids are nitric acid [$HNO_3(aq)$], hydrochloric acid [$HCl(aq)$], sulfuric acid [$H_2SO(aq)$], and perchloric acid [$HClO_4(aq)$]
- A weak acid has a small K_a value
  - The acid dissociates (ionizes) to only a slight extent
  - The dissociation (ionization) equilibrium position lies far to the left
  - Weak acids have relatively strong conjugate bases
  - Percent dissociation of a weak acid

$$\% \text{ dissociation} = \frac{\text{amount dissociated (mol/L)}}{\text{initial concentration (mol/L)}} \times 100\%$$

  - The smaller the percent dissociation, the weaker the acid
  - Dilution of a weak acid increases its percent dissociation

Autoionization of Water

- Water is an amphoteric substance: It behaves as both an acid and a base
- Water reacts with itself in an acid–base reaction

$$H_2O(l) + H_2O(l) \rightleftharpoons H_3O^+(aq) + OH^-(aq)$$

which leads to the equilibrium expression

$$K_w = [H_3O^+][OH^-] \quad \text{or} \quad [H^+][OH^-] = K_w$$

- K_w is the ion-product constant for water
- At 25°C in pure water $[H^+] = [OH^-] = 1.0 \times 10^{-7}$, so $K_w = 1.0 \times 10^{-14}$
- Acidic solution: $[H^+] > [OH^-]$
- Basic solution: $[OH^-] > [H^+]$
- Neutral solution: $[H^+] = [OH^-]$

The pH Scale

- $pH = -\log [H^+]$
- Since pH is a log scale, the pH changes by 1 for every 10-fold change in $[H^+]$
- The log scale is also used for $[OH^-]$ and for K_a values

$$pOH = -\log[OH^-]$$
$$pK_a = -\log K_a$$

Bases

- Strong bases are hydroxide salts, such as NaOH and KOH
- Weak bases react with water to produce OH^-

$$B(aq) + H_2O(l) \rightleftharpoons BH^+(aq) + OH^-(aq)$$

  - The equilibrium constant for this reaction is called K_b where

$$K_b = \frac{[BH^+][OH^-]}{[B]}$$

  - In water a base B is always competing with OH^- for a proton (H^+), so K_b values tend to be very small, thus making B a weak base (compared to OH^-)

Polyprotic Acids

- A polyprotic acid has more than one acidic proton
- Polyprotic acids dissociate one proton at a time
    - Each step has a characteristic K_a value
    - Typically for a weak polyprotic acid, $K_{a_1} > K_{a_2} > K_{a_3}$
- Sulfuric acid is unique:
    - It is a strong acid in the first dissociation step (K_{a_1} is very large)
    - It is a weak acid in the second step

Acid–Base Properties of Salts

- Can produce acidic, basic, or neutral solutions
- Salts that contain
    - cations of strong bases and anions of strong acids produce neutral solutions
    - cations of strong bases and anions of weak acids produce basic solutions
    - cations of weak bases and anions of strong acids produce acidic solutions
- Acidic solutions are produced by salts containing a highly charged metal cation— for example, Al^{3+} and Fe^{3+}

Effect of Structure on Acid–Base Properties

- Many substances that function as acids or bases contain the H—O—X grouping
    - Molecules in which the O—X bond is strong and covalent tend to behave as acids
    - As X becomes more electronegative, the acid becomes stronger
    - When the O—X bond is ionic, the substance behaves as a base, releasing OH^- ions in water

---

Review Questions   Answers to the Review Questions can be found on the Student Web site (accessible from www.cengagebrain.com).

1. Define each of the following:
   a. Arrhenius acid
   b. Brønsted–Lowry acid
   c. Lewis acid

   Which of the definitions is most general? Write reactions to justify your answer.

2. Define or illustrate the meaning of the following terms:
   a. K_a reaction
   b. K_a equilibrium constant
   c. K_b reaction
   d. K_b equilibrium constant
   e. conjugate acid–base pair

3. Define or illustrate the meaning of the following terms:
   a. amphoteric
   b. K_w reaction
   c. K_w equilibrium constant
   d. pH

   e. pOH
   f. pK_w

   Give the conditions for a neutral aqueous solution at 25°C, in terms of $[H^+]$, pH, and the relationship between $[H^+]$ and $[OH^-]$. Do the same for an acidic solution and for a basic solution. As a solution becomes more acidic, what happens to pH, pOH, $[H^+]$, and $[OH^-]$? As a solution becomes more basic, what happens to pH, pOH, $[H^+]$, and $[OH^-]$?

4. How is acid strength related to the value of K_a? What is the difference between strong acids and weak acids (see Table 13-1)? As the strength of an acid increases, what happens to the strength of the conjugate base? How is base strength related to the value of K_b? As the strength of a base increases, what happens to the strength of the conjugate acid?

5. Two strategies are followed when solving for the pH of an acid in water. What is the strategy for calculating the pH of a strong acid in water? What major assumptions are made when solving strong acid problems? The best way to recognize strong

acids is to memorize them. List the six common strong acids (the two not listed in the text are HBr and HI).

Most acids, by contrast, are weak acids. When solving for the pH of a weak acid in water, you must have the K_a value. List two places in this text that provide K_a values for weak acids. You can utilize these tables to help you recognize weak acids. What is the strategy for calculating the pH of a weak acid in water? What assumptions are generally made? What is the 5% rule? If the 5% rule fails, how do you calculate the pH of a weak acid in water?

6. Two strategies are also followed when solving for the pH of a base in water. What is the strategy for calculating the pH of a strong base in water? List the strong bases mentioned in the text that should be committed to memory. Why is calculating the pH of $Ca(OH)_2$ solutions a little more difficult than calculating the pH of NaOH solutions?

Most bases are weak bases. The presence of what element most commonly results in basic properties for an organic compound? What is present on this element in compounds that allows it to accept a proton?

Table 13-3 and Appendix 5 of the text list K_b values for some weak bases. What strategy is used to solve for the pH of a weak base in water? What assumptions are made when solving for the pH of weak base solutions? If the 5% rule fails, how do you calculate the pH of a weak base in water?

7. Table 13-4 lists the stepwise K_a values for some polyprotic acids. What is the difference between a monoprotic acid, a diprotic acid, and a triprotic acid? Most polyprotic acids are weak acids; the major exception is H_2SO_4. To solve for the pH of a solution of H_2SO_4, you must generally solve a strong acid problem as well as a weak acid problem. Explain. Write out the reactions that refer to K_{a_1} and K_{a_2} for H_2SO_4.

For H_3PO_4, $K_{a_1} = 7.5 \times 10^{-3}$, $K_{a_2} = 6.2 \times 10^{-8}$, and $K_{a_3} = 4.8 \times 10^{-13}$. Write out the reactions that refer to the K_{a_1}, K_{a_2}, and K_{a_3} equilibrium constants. What are the three acids in a solution of H_3PO_4? Which acid is strongest? What are the three conjugate bases in a solution of H_3PO_4? Which conjugate base is strongest? Summarize the strategy for calculating the pH of a polyprotic acid in water.

8. For conjugate acid–base pairs, how are K_a and K_b related? Consider the reaction of acetic acid in water

$$CH_3CO_2H(aq) + H_2O(l) \rightleftharpoons$$
$$CH_3CO_2^-(aq) + H_3O^+(aq)$$

where $K_a = 1.8 \times 10^{-5}$.

a. Which two bases are competing for the proton?

b. Which is the stronger base?

c. In light of your answer to part b, why do we classify the acetate ion ($CH_3CO_2^-$) as a weak base? Use an appropriate reaction to justify your answer.

In general, as base strength increases, conjugate acid strength decreases. Explain why the conjugate acid of the weak base NH_3 is a weak acid.

To summarize, the conjugate base of a weak acid is a weak base and the conjugate acid of a weak base is a weak acid (weak gives you weak). Assuming K_a for a monoprotic strong acid is 1×10^6, calculate K_b for the conjugate base of this strong acid. Why do conjugate bases of strong acids have no basic properties in water? List the conjugate bases of the six common strong acids. To tie it all together, some instructors have students think of Li^+, K^+, Rb^+, Cs^+, Ca^{2+}, Sr^{2+}, and Ba^{2+} as the conjugate acids of the strong bases LiOH, KOH, RbOH, CsOH, $Ca(OH)_2$, $Sr(OH)_2$, and $Ba(OH)_2$. Although not technically correct, the conjugate acid strength of these cations is similar to the conjugate base strength of the strong acids. That is, these cations have no acidic properties in water; similarly, the conjugate bases of strong acids have no basic properties (strong gives you worthless). Fill in the blanks with the correct response. The conjugate base of a weak acid is a _____ base. The conjugate acid of a weak base is a _____ acid. The conjugate base of a strong acid is a _____ base. The conjugate acid of a strong base is a _____ acid. (*Hint:* Weak gives you weak and strong gives you worthless.)

9. What is a salt? List some anions that behave as weak bases in water. List some anions that have no basic properties in water. List some cations that behave as weak acids in water. List some cations that have no acidic properties in water. Using these lists, give some formulas for salts that have only weak base properties in water. What strategy would you use to solve for the pH of these basic salt solutions? Identify some salts that have only weak acid properties in water. What strategy would you use to solve for the pH of these acidic salt solutions? Identify some salts that have no acidic or basic properties in water (produce neutral solutions). When a salt contains both a weak acid ion and a weak base ion, how do you predict whether the solution pH is acidic, basic, or neutral?

10. For oxyacids, how does acid strength depend on

  a. the strength of the bond to the acidic hydrogen atom?

  b. the electronegativity of the element bonded to the oxygen atom that bears the acidic hydrogen?

  c. the number of oxygen atoms?

How does the strength of a conjugate base depend on these factors?

  What type of solution forms when a nonmetal oxide dissolves in water? Give an example of such an oxide. What type of solution forms when a metal oxide dissolves in water? Give an example of such an oxide.

Active Learning Questions

These questions are designed to be used by groups of students in class.

1. Consider two beakers of pure water at different temperatures. How do their pH values compare? Which is more acidic? More basic? Explain.

2. Differentiate between the terms *strength* and *concentration* as they apply to acids and bases. When is HCl strong? Weak? Concentrated? Dilute? Answer the same questions for ammonia. Is the conjugate base of a weak acid a strong base?

3. Sketch two graphs: (a) percent dissociation for weak acid HA versus the initial concentration of HA ($[HA]_0$) and (b) H^+ concentration versus $[HA]_0$. Explain both.

4. Consider a solution prepared by mixing a weak acid HA and HCl. What are the major species? Explain what is occurring in solution. How would you calculate the pH? What if you added NaA to this solution? Then added NaOH?

5. Explain why salts can be acidic, basic, or neutral, and show examples. Do this without specific numbers.

6. Consider two separate aqueous solutions: one of a weak acid HA and one of HCl. Assuming you started with 10 molecules of each:

  a. Draw a picture of what each solution looks like at equilibrium.

  b. What are the major species in each beaker?

  c. From your pictures, calculate the K_a values of each acid.

  d. Order the following from the strongest to the weakest base: H_2O, A^-, Cl^-. Explain your order.

7. You are asked to calculate the H^+ concentration in a solution of NaOH(*aq*). Because sodium hydroxide is a base, can we say there is no H^+, since having H^+ would imply that the solution is acidic?

8. Consider a solution prepared by mixing a weak acid HA, HCl, and NaA. Which of the following statements best describes what happens?

  a. The H^+ from the HCl reacts completely with the A^- from the NaA. Then the HA dissociates somewhat.

  b. The H^+ from the HCl reacts somewhat with the A^- from the NaA to make HA, while the HA is dissociating. Eventually you have equal amounts of everything.

  c. The H^+ from the HCl reacts somewhat with the A^- from the NaA to make HA while the HA is dissociating. Eventually all the reactions have equal rates.

  d. The H^+ from the HCl reacts completely with the A^- from the NaA. Then the HA dissociates somewhat until "too much" H^+ and A^- are formed, so the H^+ and A^- react to form HA, and so on. Eventually equilibrium is reached.

Justify your choice, and for choices you did not pick, explain what is wrong with them.

9. Consider a solution formed by mixing 100.0 mL of 0.10 *M* HA ($K_a = 1.0 \times 10^{-6}$), 100.00 mL of 0.10 *M* NaA, and 100.0 mL of 0.10 *M* HCl. In calculating the pH for the final solution, you would make some assumptions about the order in which various reactions occur to simplify the calculations. State these assumptions. Does it matter whether the reactions actually occur in the assumed order? Explain.

10. A certain sodium compound is dissolved in water to liberate Na^+ ions and a certain negative ion. What evidence would you look for to determine whether the anion is behaving as an acid or a base? How could you tell whether the anion is a strong base? Explain how the anion could behave simultaneously as an acid and a base.

11. Acids and bases can be thought of as chemical opposites (acids are proton donors, and bases are proton acceptors). Therefore, one might think that $K_a = 1/K_b$. Why isn't this the case? What is the relationship between K_a and K_b? Prove it with a derivation.

12. Consider two solutions of the salts NaX(*aq*) and NaY(*aq*) at equal concentrations. What would you need to know to determine which solution has the higher pH? Explain how you would decide (perhaps even provide a sample calculation).

13. What is meant by pH? True or false: A strong acid solution always has a lower pH than a weak acid solution. Explain.

14. Why is the pH of water at 25°C equal to 7.00?

15. Can the pH of a solution be negative? Explain.

16. Is the conjugate base of a weak acid a strong base? Explain. Explain why Cl^- does not affect the pH of an aqueous solution.

17. Match the following pH values: 1, 2, 5, 6, 6.5, 8, 11, 11, and 13 with the following chemicals (of equal concentration): HBr, NaOH, NaF, NaCN, NH_4F, CH_3NH_3F, HF, HCN, and NH_3. Answer this question without performing calculations.

18. The salt BX, when dissolved in water, produces an acidic solution. Which of the following could be true? (There may be more than one correct answer.)

  a. The acid HX is a weak acid.

  b. The acid HX is a strong acid.

  c. The cation B^+ is a weak acid.

  Explain.

A blue question or exercise number indicates that the answer to that question or exercise appears at the back of this book and a solution appears in the *Solutions Guide*, as found on PowerLecture.

Questions

19. Anions containing hydrogen (for example, HCO_3^- and $H_2PO_4^-$) usually show amphoteric behavior. Write equations illustrating the amphoterism of these two anions.

20. Which of the following conditions indicate an *acidic* solution at 25°C?

    a. pH = 3.04

    b. $[H^+] > 1.0 \times 10^{-7} \, M$

    c. pOH = 4.51

    d. $[OH^-] = 3.21 \times 10^{-12} \, M$

21. Which of the following conditions indicate a *basic* solution at 25°C?

    a. pOH = 11.21

    b. pH = 9.42

    c. $[OH^-] > [H^+]$

    d. $[OH^-] > 1.0 \times 10^{-7} \, M$

22. Why is H_3O^+ the strongest acid and OH^- the strongest base that can exist in significant amounts in aqueous solutions?

23. How many significant figures are there in the following numbers: 10.78, 6.78, 0.78? If these were pH values, to how many significant figures can you express the $[H^+]$? Explain any discrepancies between your answers to the two questions.

24. In terms of orbitals and electron arrangements, what must be present for a molecule or an ion to act as a Lewis acid? What must be present for a molecule or an ion to act as a Lewis base?

25. Consider the autoionization of liquid ammonia:

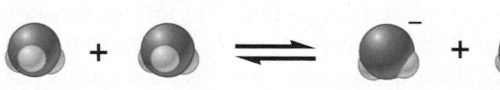

    Label each of the species in the equation as an acid or a base and explain your answer.

26. The following are representations of acid–base reactions:

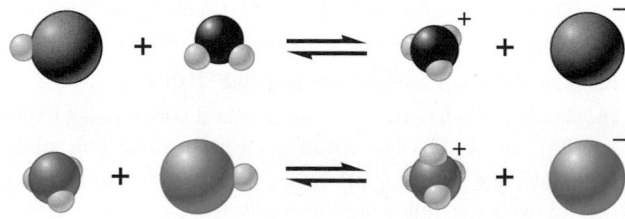

    a. Label each of the species in both equations as an acid or a base and explain your answers.

    b. For those species that are acids, which labels apply: Arrhenius acid, Brønsted–Lowry acid, Lewis acid? What about the bases?

27. Give three example solutions that fit each of the following descriptions.

    a. a strong electrolyte solution that is very acidic

    b. a strong electrolyte solution that is slightly acidic

    c. a strong electrolyte solution that is very basic

    d. a strong electrolyte solution that is slightly basic

    e. a strong electrolyte solution that is neutral

28. Derive an expression for the relationship between pK_a and pK_b for a conjugate acid–base pair. ($pK = -\log K$.)

29. Consider the following statements. Write out an example reaction and K expression that is associated with each statement.

    a. The autoionization of water.

    b. An acid reacts with water to produce the conjugate base of the acid and the hydronium ion.

    c. A base reacts with water to produce the conjugate acid of the base and the hydroxide ion.

30. Which of the following statements is(are) *true*? Correct the false statements.

    a. When a base is dissolved in water, the lowest possible pH of the solution is 7.0.

    b. When an acid is dissolved in water, the lowest possible pH is 0.

    c. A strong acid solution will have a lower pH than a weak acid solution.

    d. A 0.0010-M $Ba(OH)_2$ solution has a pOH that is twice the pOH value of a 0.0010-M KOH solution.

31. Consider the following mathematical expressions.

    a. $[H^+] = [HA]_0$

    b. $[H^+] = (K_a \times [HA]_0)^{1/2}$

    c. $[OH^-] = 2[B]_0$

    d. $[OH^-] = (K_b \times [B]_0)^{1/2}$

    For each expression, give three solutions where the mathematical expression would give a good approximation for the $[H^+]$ or $[OH^-]$. $[HA]_0$ and $[B]_0$ represent initial concentrations of an acid or a base.

32. Consider a 0.10-M H_2CO_3 solution and a 0.10-M H_2SO_4 solution. Without doing any detailed calculations, choose one of the following statements that best describes the $[H^+]$ of each solution and explain your answer.

    a. The $[H^+]$ is less than 0.10 M.

    b. The $[H^+]$ is 0.10 M.

    c. The $[H^+]$ is between 0.10 M and 0.20 M.

    d. The $[H^+]$ is 0.20 M.

33. Of the hydrogen halides, only HF is a weak acid. Give a possible explanation.

34. Explain why the following are done, both of which are related to acid–base chemistry.

  a. Power plants burning coal with high sulfur content use scrubbers to help eliminate sulfur emissions.

  b. A gardener mixes lime (CaO) into the soil of his garden.

Exercises

In this section similar exercises are paired.

Nature of Acids and Bases

35. Write balanced equations that describe the following reactions.

  a. the dissociation of perchloric acid in water

  b. the dissociation of propanoic acid ($CH_3CH_2CO_2H$) in water

  c. the dissociation of ammonium ion in water

36. Write the dissociation reaction and the corresponding K_a equilibrium expression for each of the following acids in water.

  a. HCN

  b. HOC_6H_5

  c. $C_6H_5NH_3^+$

37. For each of the following aqueous reactions, identify the acid, the base, the conjugate base, and the conjugate acid.

  a. $H_2O + H_2CO_3 \rightleftharpoons H_3O^+ + HCO_3^-$

  b. $C_5H_5NH^+ + H_2O \rightleftharpoons C_5H_5N + H_3O^+$

  c. $HCO_3^- + C_5H_5NH^+ \rightleftharpoons H_2CO_3 + C_5H_5N$

38. For each of the following aqueous reactions, identify the acid, the base, the conjugate base, and the conjugate acid.

  a. $Al(H_2O)_6^{3+} + H_2O \rightleftharpoons H_3O^+ + Al(H_2O)_5(OH)^{2+}$

  b. $H_2O + HONH_3^+ \rightleftharpoons HONH_2 + H_3O^+$

  c. $HOCl + C_6H_5NH_2 \rightleftharpoons OCl^- + C_6H_5NH_3^+$

39. Classify each of the following as a strong acid or a weak acid.

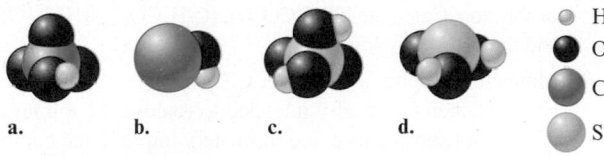

  a.   **b.**   **c.**   **d.**

H
O
Cl
S

40. Consider the following illustrations:

H^+
A^-
B^-

Which beaker best illustrates what happens when the following acids are dissolved in water?

  a. HNO_2 **d.** HF

  b. HNO_3 **e.** $HC_2H_3O_2$

  c. HCl

41. Use Table 13-2 to order the following from the strongest to the weakest acid.

$$HClO_2, \quad H_2O, \quad NH_4^+, \quad HClO_4$$

42. Use Table 13-2 to order the following from the strongest to the weakest base.

$$ClO_2^-, \quad H_2O, \quad NH_3, \quad ClO_4^-$$

43. You may need Table 13-2 to answer the following questions.

  a. Which is the stronger acid, HCl or H_2O?

  b. Which is the stronger acid, H_2O or HNO_2?

  c. Which is the stronger acid, HCN or HOC_6H_5?

44. You may need Table 13-2 to answer the following questions.

  a. Which is the stronger base, Cl^- or H_2O?

  b. Which is the stronger base, H_2O or NO_2^-?

  c. Which is the stronger base, CN^- or $OC_6H_5^-$?

Autoionization of Water and the pH Scale

45. Calculate the $[OH^-]$ of each of the following solutions at 25°C. Identify each solution as neutral, acidic, or basic.

  a. $[H^+] = 1.0 \times 10^{-7}\,M$ **c.** $[H^+] = 12\,M$

  b. $[H^+] = 8.3 \times 10^{-16}\,M$ **d.** $[H^+] = 5.4 \times 10^{-5}\,M$

46. Calculate the $[H^+]$ of each of the following solutions at 25°C. Identify each solution as neutral, acidic, or basic.

  a. $[OH^-] = 1.5\,M$

  b. $[OH^-] = 3.6 \times 10^{-15}\,M$

  c. $[OH^-] = 1.0 \times 10^{-7}\,M$

  d. $[OH^-] = 7.3 \times 10^{-4}\,M$

47. Values of K_w as a function of temperature are as follows:

Temperature (°C)	K_w
0	1.14×10^{-15}
25	1.00×10^{-14}
35	2.09×10^{-14}
40.	2.92×10^{-14}
50.	5.47×10^{-14}

  a. Is the autoionization of water exothermic or endothermic?

  b. Calculate $[H^+]$ and $[OH^-]$ in a neutral solution at 50.°C.

48. At 40.°C the value of K_w is 2.92×10^{-14}.

  a. Calculate the $[H^+]$ and $[OH^-]$ in pure water at 40.°C.

  b. What is the pH of pure water at 40.°C?

  c. If the hydroxide ion concentration in a solution is 0.10 M, what is the pH at 40.°C?

49. Calculate the pH and pOH of the solutions in Exercises 45 and 46.

50. Calculate $[H^+]$ and $[OH^-]$ for each solution at 25°C. Identify each solution as neutral, acidic, or basic.
    a. pH = 7.40 (the normal pH of blood)
    b. pH = 15.3
    c. pH = −1.0
    d. pH = 3.20
    e. pOH = 5.0
    f. pOH = 9.60

51. Fill in the missing information in the following table.

	pH	pOH	$[H^+]$	$[OH^-]$	Acidic, Basic, or Neutral?
Solution a	6.88	____	____	____	____
Solution b	____	____	____	$8.4 \times 10^{-14}\ M$	____
Solution c	____	3.11	____	____	____
Solution d	____	____	$1.0 \times 10^{-7}\ M$	____	____

52. Fill in the missing information in the following table.

	pH	pOH	$[H^+]$	$[OH^-]$	Acidic, Basic, or Neutral?
Solution a	9.63	____	____	____	____
Solution b	____	____	____	$3.9 \times 10^{-6}\ M$	____
Solution c	____	____	0.027 M	____	____
Solution d	____	1.22	____	____	____

53. The pH of a sample of gastric juice in a person's stomach is 2.1. Calculate the pOH, $[H^+]$, and $[OH^-]$ for this sample. Is gastric juice acidic or basic?

54. The pOH of a sample of baking soda dissolved in water is 5.74 at 25°C. Calculate the pH, $[H^+]$, and $[OH^-]$ for this sample. Is the solution acidic or basic?

Solutions of Acids

55. What are the major species present in 0.250 M solutions of each of the following acids? Calculate the pH of each of these solutions.
    a. $HClO_4$ b. HNO_3

56. A solution is prepared by adding 50.0 mL of 0.050 M HBr to 150.0 mL of 0.10 M HI. Calculate $[H^+]$ and the pH of this solution. HBr and HI are both considered strong acids.

57. Calculate the pH of each of the following solutions of a strong acid in water.
    a. 0.10 M HCl   c. $1.0 \times 10^{-11}\ M$ HCl
    b. 5.0 M HCl

58. Calculate the pH of each of the following solutions containing a strong acid in water.
    a. $2.0 \times 10^{-2}\ M$ HNO_3 c. $6.2 \times 10^{-12}\ M$ HNO_3
    b. 4.0 M HNO_3

59. Calculate the concentration of an aqueous HI solution that has pH = 2.50. HI is a strong acid.

60. Calculate the concentration of an aqueous HBr solution that has pH = 4.25. HBr is a strong acid.

61. How would you prepare 1600 mL of a pH = 1.50 solution using concentrated (12 M) HCl?

62. A solution is prepared by adding 50.0 mL concentrated hydrochloric acid and 20.0 mL concentrated nitric acid to 300 mL water. More water is added until the final volume is 1.00 L. Calculate $[H^+]$, $[OH^-]$, and the pH for this solution. [*Hint:* Concentrated HCl is 38% HCl (by mass) and has a density of 1.19 g/mL; concentrated HNO_3 is 70.% HNO_3 (by mass) and has a density of 1.42 g/mL.]

63. What are the major species present in 0.250 M solutions of each of the following acids? Calculate the pH of each of these solutions.
    a. HNO_2 b. CH_3CO_2H ($HC_2H_3O_2$)

64. What are the major species present in 0.250 M solutions of each of the following acids? Calculate the pH of each of these solutions.
    a. HOC_6H_5 b. HCN

65. Calculate the concentration of all species present and the pH of a 0.020-M HF solution.

66. Calculate the percent dissociation for a 0.22-M solution of chlorous acid ($HClO_2$, $K_a = 1.2 \times 10^{-2}$).

67. For propanoic acid ($HC_3H_5O_2$, $K_a = 1.3 \times 10^{-5}$), determine the concentration of all species present, the pH, and the percent dissociation of a 0.100-M solution.

68. A solution is prepared by dissolving 0.56 g benzoic acid ($C_6H_5CO_2H$, $K_a = 6.4 \times 10^{-5}$) in enough water to make 1.0 L of solution. Calculate $[C_6H_5CO_2H]$, $[C_6H_5CO_2^-]$, $[H^+]$, $[OH^-]$, and the pH of this solution.

69. Monochloroacetic acid, $HC_2H_2ClO_2$, is a skin irritant that is used in "chemical peels" intended to remove the top layer of dead skin from the face and ultimately improve the complexion. The value of K_a for monochloroacetic acid is 1.35×10^{-3}. Calculate the pH of a 0.10-M solution of monochloroacetic acid.

70. A typical aspirin tablet contains 325 mg acetylsalicylic acid ($HC_9H_7O_4$). Calculate the pH of a solution that is prepared by dissolving two aspirin tablets in enough water to make one cup (237 mL) of solution. Assume the aspirin tablets are pure acetylsalicylic acid, $K_a = 3.3 \times 10^{-4}$.

71. Calculate the pH of a solution that contains 1.0 M HF and 1.0 M HOC_6H_5. Also calculate the concentration of $OC_6H_5^-$ in this solution at equilibrium.

72. A solution is made by adding 50.0 mL of 0.200 M acetic acid ($K_a = 1.8 \times 10^{-5}$) to 50.0 mL of 1.00×10^{-3} M HCl.

   a. Calculate the pH of the solution.

   b. Calculate the acetate ion concentration.

73. Calculate the percent dissociation of the acid in each of the following solutions.

   a. 0.50 M acetic acid

   b. 0.050 M acetic acid

   c. 0.0050 M acetic acid

   d. Use Le Châtelier's principle to explain why percent dissociation increases as the concentration of a weak acid decreases.

   e. Even though the percent dissociation increases from solutions a to c, the [H$^+$] decreases. Explain.

74. Using the K_a values in Table 13-2, calculate the percent dissociation in a 0.20-M solution of each of the following acids.

   a. nitric acid (HNO$_3$)

   b. nitrous acid (HNO$_2$)

   c. phenol (HOC$_6$H$_5$)

   d. How is percent dissociation of an acid related to the K_a value for the acid (assuming equal initial concentrations of acids)?

75. A 0.15-M solution of a weak acid is 3.0% dissociated. Calculate K_a.

76. An acid HX is 25% dissociated in water. If the equilibrium concentration of HX is 0.30 M, calculate the K_a value for HX.

77. Trichloroacetic acid (CCl$_3$CO$_2$H) is a corrosive acid that is used to precipitate proteins. The pH of a 0.050-M solution of trichloroacetic acid is the same as the pH of a 0.040-M HClO$_4$ solution. Calculate K_a for trichloroacetic acid.

78. The pH of a 0.063-M solution of hypobromous acid (HOBr but usually written HBrO) is 4.95. Calculate K_a.

79. A solution of formic acid (HCOOH, $K_a = 1.8 \times 10^{-4}$) has a pH of 2.70. Calculate the initial concentration of formic acid in this solution.

80. A typical sample of vinegar has a pH of 3.0. Assuming that vinegar is only an aqueous solution of acetic acid ($K_a = 1.8 \times 10^{-5}$), calculate the concentration of acetic acid in vinegar.

81. One mole of a weak acid HA was dissolved in 2.0 L of solution. After the system had come to equilibrium, the concentration of HA was found to be 0.45 M. Calculate K_a for HA.

82. You have 100.0 g saccharin, a sugar substitute, and you want to prepare a pH = 5.75 solution. What volume of solution can be prepared? For saccharin, HC$_7$H$_4$NSO$_3$, pK_a = 11.70 (pK_a = $-\log K_a$).

Solutions of Bases

83. Write the reaction and the corresponding K_b equilibrium expression for each of the following substances acting as bases in water.

   a. NH$_3$ **b.** C$_5$H$_5$N

84. Write the reaction and the corresponding K_b equilibrium expression for each of the following substances acting as bases in water.

   a. aniline, C$_6$H$_5$NH$_2$ **b.** dimethylamine, (CH$_3$)$_2$NH

85. Use Table 13-3 to help order the following bases from strongest to weakest.

$$NO_3^-, \quad H_2O, \quad NH_3, \quad C_5H_5N$$

86. Use Table 13-3 to help order the following acids from strongest to weakest.

$$HNO_3, \quad H_2O, \quad NH_4^+, \quad C_5H_5NH^+$$

87. Use Table 13-3 to help answer the following questions.

   a. Which is the stronger base, ClO$_4^-$ or C$_6$H$_5$NH$_2$?

   b. Which is the stronger base, H$_2$O or C$_6$H$_5$NH$_2$?

   c. Which is the stronger base, OH$^-$ or C$_6$H$_5$NH$_2$?

   d. Which is the stronger base, C$_6$H$_5$NH$_2$ or CH$_3$NH$_2$?

88. Use Table 13-3 to help answer the following questions.

   a. Which is the stronger acid, HClO$_4$ or C$_6$H$_5$NH$_3^+$?

   b. Which is the stronger acid, H$_2$O or C$_6$H$_5$NH$_3^+$?

   c. Which is the stronger acid, C$_6$H$_5$NH$_3^+$ or CH$_3$NH$_3^+$?

89. Calculate the pH of the following solutions.

   a. 0.10 M NaOH

   b. 1.0×10^{-10} M NaOH

   c. 2.0 M NaOH

90. Calculate [OH$^-$], pOH, and pH for each of the following.

   a. 0.00040 M Ca(OH)$_2$

   b. a solution containing 25 g KOH per liter

   c. a solution containing 150.0 g NaOH per liter

91. What are the major species present in 0.015 M solutions of each of the following bases?

   a. KOH   **b.** Ba(OH)$_2$

   What is [OH$^-$] and the pH of each of these solutions?

92. What are the major species present in the following mixtures of bases?

   a. 0.050 M NaOH and 0.050 M LiOH

   b. 0.0010 M Ca(OH)$_2$ and 0.020 M RbOH

   What is [OH$^-$] and the pH of each of these solutions?

93. What mass of KOH is necessary to prepare 800.0 mL of a solution having a pH = 11.56?

94. Calculate the concentration of an aqueous Sr(OH)$_2$ that has pH = 10.50.

95. What are the major species present in a 0.150-M NH$_3$ solution? Calculate the [OH$^-$] and the pH of this solution.

96. For the reaction of hydrazine (N$_2$H$_4$) in water,

$$H_2NNH_2(aq) + H_2O(l) \rightleftharpoons H_2NNH_3^+(aq) + OH^-(aq)$$

K_b is 3.0×10^{-6}. Calculate the concentrations of all species and the pH of a 2.0-M solution of hydrazine in water.

97. Calculate [OH$^-$], [H$^+$], and the pH of 0.20 M solutions of each of the following amines.
    a. triethylamine [$(C_2H_5)_3N$, $K_b = 4.0 \times 10^{-4}$]
    b. hydroxylamine ($HONH_2$, $K_b = 1.1 \times 10^{-8}$)

98. Calculate [OH$^-$], [H$^+$], and the pH of 0.40 M solutions of each of the following amines (the K_b values are found in Table 13-3).
    a. aniline   b. methylamine

99. Calculate the pH of a 0.20-M $C_2H_5NH_2$ solution ($K_b = 5.6 \times 10^{-4}$).

100. Calculate the pH of a 0.050-M $(C_2H_5)_2NH$ solution ($K_b = 1.3 \times 10^{-3}$).

101. What is the percent ionization in each of the following solutions?
    a. 0.10 M NH_3 c. 0.10 M CH_3NH_2
    b. 0.010 M NH_3

102. Calculate the percentage of pyridine (C_5H_5N) that forms pyridinium ion, $C_5H_5NH^+$, in a 0.10-M aqueous solution of pyridine ($K_b = 1.7 \times 10^{-9}$).

103. The pH of a 0.016-M aqueous solution of p-toluidine ($CH_3C_6H_4NH_2$) is 8.60. Calculate K_b.

104. Calculate the mass of $HONH_2$ required to dissolve in enough water to make 250.0 mL of solution having a pH of 10.00 ($K_b = 1.1 \times 10^{-8}$).

Polyprotic Acids

105. Write out the stepwise K_a reactions for the diprotic acid H_2SO_3.

106. Write out the stepwise K_a reactions for citric acid ($H_3C_6H_5O_7$), a triprotic acid.

107. A typical vitamin C tablet (containing pure ascorbic acid, $H_2C_6H_6O_6$) weighs 500. mg. One vitamin C tablet is dissolved in enough water to make 200.0 mL of solution. Calculate the pH of this solution. Ascorbic acid is a diprotic acid.

108. Arsenic acid (H_3AsO_4) is a triprotic acid with $K_{a_1} = 5.5 \times 10^{-3}$, $K_{a_2} = 1.7 \times 10^{-7}$, and $K_{a_3} = 5.1 \times 10^{-12}$. Calculate [H$^+$], [OH$^-$], [$H_3AsO_4$], [$H_2AsO_4^-$], [$HAsO_4^{2-}$], and [$AsO_4^{3-}$] in a 0.20-$M$ arsenic acid solution.

109. Calculate the pH and [S^{2-}] in a 0.10-M H_2S solution. Assume $K_{a_1} = 1.0 \times 10^{-7}$; $K_{a_2} = 1.0 \times 10^{-19}$.

110. Calculate [CO_3^{2-}] in a 0.010-M solution of CO_2 in water (usually written as H_2CO_3). If all the CO_3^{2-} in this solution comes from the reaction

$$HCO_3^-(aq) \rightleftharpoons H^+(aq) + CO_3^{2-}(aq)$$

what percentage of the H$^+$ ions in the solution is a result of the dissociation of HCO_3^-? When acid is added to a solution of sodium hydrogen carbonate ($NaHCO_3$), vigorous bubbling occurs. How is this reaction related to the existence of carbonic acid (H_2CO_3) molecules in aqueous solution?

111. Calculate the pH of a 2.0-M H_2SO_4 solution.

112. Calculate the pH of a 5.0×10^{-3}-M solution of H_2SO_4.

Acid–Base Properties of Salts

113. Arrange the following 0.10 M solutions in order of most acidic to most basic.

$$KOH, \quad KNO_3, \quad KCN, \quad NH_4Cl, \quad HCl$$

114. Arrange the following 0.10 M solutions in order from most acidic to most basic. See Appendix 5 for K_a and K_b values.

$$CaBr_2, \quad KNO_2, \quad HClO_4, \quad HNO_2, \quad HONH_3ClO_4$$

115. Given that the K_a value for acetic acid is 1.8×10^{-5} and the K_a value for hypochlorous acid is 3.5×10^{-8}, which is the stronger base, OCl$^-$ or $C_2H_3O_2^-$?

116. The K_b values for ammonia and methylamine are 1.8×10^{-5} and 4.4×10^{-4}, respectively. Which is the stronger acid, NH_4^+ or $CH_3NH_3^+$?

117. Determine [OH$^-$], [H$^+$], and the pH of each of the following solutions.
    a. 1.0 M KCl   b. 1.0 M $KC_2H_3O_2$

118. Calculate the concentrations of all species present in a 0.25-M solution of ethylammonium chloride ($C_2H_5NH_3Cl$).

119. Calculate the pH of each of the following solutions.
    a. 0.10 M CH_3NH_3Cl b. 0.050 M NaCN

120. Calculate the pH of each of the following solutions.
    a. 0.12 M KNO_2 c. 0.40 M NH_4ClO_4
    b. 0.45 M NaOCl

121. Sodium azide (NaN_3) is sometimes added to water to kill bacteria. Calculate the concentration of all species in a 0.010-M solution of NaN_3. The K_a value for hydrazoic acid (HN_3) is 1.9×10^{-5}.

122. Papaverine hydrochloride (abbreviated papH$^+$Cl$^-$; molar mass = 378.85 g/mol) is a drug that belongs to a group of medicines called vasodilators, which cause blood vessels to expand, thereby increasing blood flow. This drug is the conjugate acid of the weak base papaverine (abbreviated pap; $K_b = 8.33 \times 10^{-9}$ at 35.0°C). Calculate the pH of a 30.0-mg/mL aqueous dose of papH$^+$Cl$^-$ prepared at 35.0°C. K_w at 35.0°C is 2.1×10^{-14}.

123. An unknown salt is either NaCN, $NaC_2H_3O_2$, NaF, NaCl, or NaOCl. When 0.100 mole of the salt is dissolved in 1.00 L of solution, the pH of the solution is 8.07. What is the identity of the salt?

124. Consider a solution of an unknown salt having the general formula BHCl, where B is one of the weak bases in Table 13-3. A 0.10-M solution of the unknown salt has a pH of 5.82. What is the actual formula of the salt?

125. A 0.050-M solution of the salt NaB has a pH of 9.00. Calculate the pH of a 0.010-M solution of HB.

126. A 0.20-*M* sodium chlorobenzoate ($NaC_7H_4ClO_2$) solution has a pH of 8.65. Calculate the pH of a 0.20-*M* chlorobenzoic acid ($HC_7H_4ClO_2$) solution.

127. Calculate the pH of a 0.050-*M* $Al(NO_3)_3$ solution. The K_a value for $Al(H_2O)_6^{3+}$ is 1.4×10^{-5}.

128. Calculate the pH of a 0.10-*M* $CoCl_3$ solution. The K_a value for $Co(H_2O)_6^{3+}$ is 1.0×10^{-5}.

129. Are solutions of the following salts acidic, basic, or neutral? For those that are not neutral, write balanced chemical equations for the reactions causing the solution to be acidic or basic. The relevant K_a and K_b values are found in Tables 13-2 and 13-3.
    a. $NaNO_3$ c. $C_5H_5NHClO_4$ e. KOCl
    b. $NaNO_2$ d. NH_4NO_2 f. NH_4OCl

130. Are solutions of the following salts acidic, basic, or neutral? For those that are not neutral, write balanced equations for the reactions causing the solution to be acidic or basic. The relevant K_a and K_b values are found in Tables 13-2 and 13-3.
    a. $Sr(NO_3)_2$ c. CH_3NH_3Cl e. NH_4F
    b. $NH_4C_2H_3O_2$ d. $C_6H_5NH_3ClO_2$ f. CH_3NH_3CN

Relationships Between Structure and Strengths of Acids and Bases

131. Place the species in each of the following groups in order of increasing acid strength. Explain the order you chose for each group.
    a. HIO_3, $HBrO_3$ c. HOCl, HOI
    b. HNO_2, HNO_3 d. H_3PO_4, H_3PO_3

132. Place the species in each of the following groups in order of increasing base strength. Give your reasoning in each case.
    a. IO_3^-, BrO_3^- b. NO_2^-, NO_3^- c. OCl^-, OI^-

133. Place the species in each of the following groups in order of increasing acid strength.
    a. H_2O, H_2S, H_2Se (bond energies: H—O, 467 kJ/mol; H—S, 363 kJ/mol; H—Se, 276 kJ/mol)
    b. CH_3CO_2H, FCH_2CO_2H, F_2CHCO_2H, F_3CCO_2H
    c. NH_4^+, $HONH_3^+$
    d. NH_4^+, PH_4^+ (bond energies: N—H, 391 kJ/mol; P—H, 322 kJ/mol)
    Give reasons for the orders you chose.

134. Using your results from Exercise 133, place the species in each of the following groups in order of increasing base strength.
    a. OH^-, SH^-, SeH^- b. NH_3, PH_3 c. NH_3, $HONH_2$

135. Will the following oxides give acidic, basic, or neutral solutions when dissolved in water? Write reactions to justify your answers.
    a. CaO   b. SO_2 c. Cl_2O

136. Will the following oxides give acidic, basic, or neutral solutions when dissolved in water? Write reactions to justify your answers.
    a. Li_2O b. CO_2 c. SrO

Lewis Acids and Bases

137. Identify the Lewis acid and the Lewis base in each of the following reactions.
    a. $B(OH)_3(aq) + H_2O(l) \rightleftharpoons B(OH)_4^-(aq) + H^+(aq)$
    b. $Ag^+(aq) + 2NH_3(aq) \rightleftharpoons Ag(NH_3)_2^+(aq)$
    c. $BF_3(g) + F^-(aq) \rightleftharpoons BF_4^-(aq)$

138. Identify the Lewis acid and the Lewis base in each of the following reactions.
    a. $Fe^{3+}(aq) + 6H_2O(l) \rightleftharpoons Fe(H_2O)_6^{3+}(aq)$
    b. $H_2O(l) + CN^-(aq) \rightleftharpoons HCN(aq) + OH^-(aq)$
    c. $HgI_2(s) + 2I^-(aq) \rightleftharpoons HgI_4^{2-}(aq)$

139. Aluminum hydroxide is an amphoteric substance. It can act as either a Brønsted–Lowry base or a Lewis acid. Write a reaction showing $Al(OH)_3$ acting as a base toward H^+ and as an acid toward OH^-.

140. Zinc hydroxide is an amphoteric substance. Write equations that describe $Zn(OH)_2$ acting as a Brønsted–Lowry base toward H^+ and as a Lewis acid toward OH^-.

141. Would you expect Fe^{3+} or Fe^{2+} to be the stronger Lewis acid? Explain.

142. Use the Lewis acid–base model to explain the following reaction.
$$CO_2(g) + H_2O(l) \longrightarrow H_2CO_3(aq)$$

Additional Exercises

143. A 10.0-mL sample of an HCl solution has a pH of 2.000. What volume of water must be added to change the pH to 4.000?

144. Which of the following represent conjugate acid–base pairs? For those pairs that are not conjugates, write the correct conjugate acid or base for each species in the pair.
    a. H_2O, OH^- c. H_3PO_4, $H_2PO_4^-$
    b. H_2SO_4, SO_4^{2-} d. $HC_2H_3O_2$, $C_2H_3O_2^-$

145. A solution is tested for pH and conductivity as pictured below:

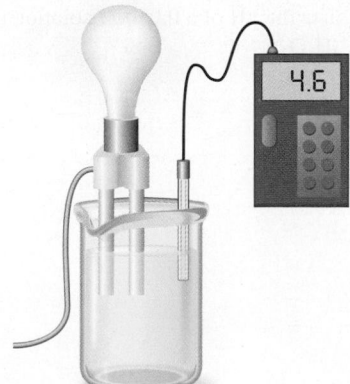

The solution contains one of the following substances: HCl, NaOH, NH_4Cl, HCN, NH_3, HF, or NaCN. If the solute concentration is about 1.0 M, what is the identity of the solute?

146. The pH of human blood is steady at a value of approximately 7.4 owing to the following equilibrium reactions:

$$CO_2(aq) + H_2O(l) \rightleftharpoons H_2CO_3(aq) \rightleftharpoons HCO_3^-(aq) + H^+(aq)$$

Acids formed during normal cellular respiration react with the HCO_3^- to form carbonic acid, which is in equilibrium with $CO_2(aq)$ and $H_2O(l)$. During vigorous exercise, a person's H_2CO_3 blood levels were 26.3 mM, whereas his CO_2 levels were 1.63 mM. On resting, the H_2CO_3 levels declined to 24.9 mM. What was the CO_2 blood level at rest?

147. Hemoglobin (abbreviated Hb) is a protein that is responsible for the transport of oxygen in the blood of mammals. Each hemoglobin molecule contains four iron atoms that are the binding sites for O_2 molecules. The oxygen binding is pH-dependent. The relevant equilibrium reaction is

$$HbH_4^{4+}(aq) + 4O_2(g) \rightleftharpoons Hb(O_2)_4(aq) + 4H^+(aq)$$

Use Le Châtelier's principle to answer the following.

a. What form of hemoglobin, HbH_4^{4+} or $Hb(O_2)_4$, is favored in the lungs? What form is favored in the cells?

b. When a person hyperventilates, the concentration of CO_2 in the blood is decreased. How does this affect the oxygen-binding equilibrium? How does breathing into a paper bag help to counteract this effect? (See Exercise 146.)

c. When a person has suffered a cardiac arrest, injection of a sodium bicarbonate solution is given. Why is this necessary? (*Hint:* CO_2 blood levels increase during cardiac arrest.)

148. A 0.25-g sample of lime (CaO) is dissolved in enough water to make 1500 mL of solution. Calculate the pH of the solution.

149. At 25°C, a saturated solution of benzoic acid ($K_a = 6.4 \times 10^{-5}$) has a pH of 2.80. Calculate the water solubility of benzoic acid in moles per liter.

150. Calculate the pH of an aqueous solution containing 1.0×10^{-2} M HCl, 1.0×10^{-2} M H_2SO_4, and 1.0×10^{-2} M HCN.

151. Acrylic acid ($CH_2{=}CHCO_2H$) is a precursor for many important plastics. K_a for acrylic acid is 5.6×10^{-5}.

a. Calculate the pH of a 0.10-M solution of acrylic acid.

b. Calculate the percent dissociation of a 0.10-M solution of acrylic acid.

c. Calculate the pH of a 0.050-M solution of sodium acrylate ($NaC_3H_3O_2$).

152. Classify each of the following as a strong acid, weak acid, strong base, or weak base in aqueous solution.

a. HNO_2

b. HNO_3

c. CH_3NH_2

d. NaOH

e. NH_3

f. HF

g. $HC{-}OH$ (with $\overset{O}{\overset{\|}{}}$ above HC)

h. $Ca(OH)_2$

i. H_2SO_4

153. The following illustration displays the relative number of species when an acid, HA, is added to water.

a. Is HA a weak or strong acid? How can you tell?

b. Using the relative numbers given in the illustration, determine the value for K_a and the percent dissociation of the acid. Assume the initial acid concentration is 0.20 M.

154. Quinine ($C_{20}H_{24}N_2O_2$) is the most important alkaloid derived from cinchona bark. It is used as an antimalarial drug. For quinine, $pK_{b_1} = 5.1$ and $pK_{b_2} = 9.7$ ($pK_b = -\log K_b$). Only 1 g quinine will dissolve in 1900.0 mL of solution. Calculate the pH of a saturated aqueous solution of quinine. Consider only the reaction $Q + H_2O \rightleftharpoons QH^+ + OH^-$ described by pK_{b_1}, where Q = quinine.

155. Codeine ($C_{18}H_{21}NO_3$) is a derivative of morphine that is used as an analgesic, narcotic, or antitussive. It was once commonly used in cough syrups but is now available only by prescription because of its addictive properties. If the pH of a 1.7×10^{-3}-M solution of codeine is 9.59, calculate K_b.

156. A codeine-containing cough syrup lists codeine sulfate as a major ingredient instead of codeine. *The Merck Index* gives $C_{36}H_{44}N_2O_{10}S$ as the formula for codeine sulfate. Describe the composition of codeine sulfate. (See Exercise 155.) Why is codeine sulfate used instead of codeine?

157. The equilibrium constant K_a for the reaction

$$Fe(H_2O)_6^{3+}(aq) + H_2O(l) \rightleftharpoons$$
$$Fe(H_2O)_5(OH)^{2+}(aq) + H_3O^+(aq)$$

is 6.0×10^{-3}.

a. Calculate the pH of a 0.10-M solution of $Fe(H_2O)_6^{3+}$.

b. Will a 1.0-M solution of iron(II) nitrate have a higher or lower pH than a 1.0-M solution of iron(III) nitrate? Explain.

158. Rank the following 0.10 M solutions in order of increasing pH.

a. HI, HF, NaF, NaI

b. NH_4Br, HBr, KBr, NH_3

c. $C_6H_5NH_3NO_3$, $NaNO_3$, NaOH, HOC_6H_5, KOC_6H_5, $C_6H_5NH_2$, HNO_3

159. Is an aqueous solution of $NaHSO_4$ acidic, basic, or neutral? What reaction occurs with water? Calculate the pH of a 0.10-M solution of $NaHSO_4$.

160. Calculate the value for the equilibrium constant for each of the following aqueous reactions.

a. $NH_3 + H_3O^+ \rightleftharpoons NH_4^+ + H_2O$

b. $NO_2^- + H_3O^+ \rightleftharpoons HNO_2 + H_2O$

c. $NH_4^+ + OH^- \rightleftharpoons NH_3 + H_2O$

d. $HNO_2 + OH^- \rightleftharpoons H_2O + NO_2^-$

161. Students are often surprised to learn that organic acids, such as acetic acid, contain —OH groups. Actually, all oxyacids contain hydroxyl groups. Sulfuric acid, usually written as H_2SO_4, has the structural formula $SO_2(OH)_2$, where S is the central atom. Identify the acids whose structural formulas are shown below. Why do they behave as acids, while NaOH and KOH are bases?

a. $SO(OH)_2$ **b.** $ClO_2(OH)$ **c.** $HPO(OH)_2$

ChemWork Problems

These multiconcept problems (and additional ones) are found interactively online with the same type of assistance a student would get from an instructor.

162. For solutions of the same concentration, as acid strength increases, indicate what happens to each of the following (increases, decreases, or doesn't change).

a. $[H^+]$ **d.** pOH

b. pH   **e.** K_a

c. $[OH^-]$

163. Complete the table for each of the following solutions:

	$[H^+]$	pH	pOH	$[OH^-]$
0.0070 M HNO_3	____	____	____	____
3.0 M KOH	____	____	____	____

164. Consider a 0.60-M solution of $HC_3H_5O_3$, lactic acid ($K_a = 1.4 \times 10^{-4}$).

a. Which of the following are major species in the solution?

    i. $HC_3H_5O_3$

    ii. $C_3H_5O_3^-$

    iii. H^+

    iv. H_2O

    v. OH^-

b. Complete the following ICE table in terms of x, the amount (mol/L) of lactic acid that dissociates to reach equilibrium.

	$[HC_3H_5O_3]$	$[H^+]$	$[C_3H_5O_3^-]$
Initial	____	____	____
Change	____	____	____
Equilibrium	$0.60 - x$	____	____

c. What is the equilibrium concentration for $C_3H_5O_3^-$?

d. Calculate the pH of the solution.

165. Consider a 0.67-M solution of $C_2H_5NH_2$ ($K_b = 5.6 \times 10^{-4}$).

a. Which of the following are major species in the solution?

    i. $C_2H_5NH_2$

    ii. H^+

    iii. OH^-

    iv. H_2O

    v. $C_2H_5NH_3^+$

b. Calculate the pH of this solution.

166. Rank the following 0.10 M solutions in order of increasing pH.

a. NH_3 **d.** KCl

b. KOH   **e.** HCl

c. $HC_2H_3O_2$

167. Consider 0.25 M solutions of the following salts: NaCl, RbOCl, KI, $Ba(ClO_4)_2$, and NH_4NO_3. For each salt, indicate whether the solution is acidic, basic, or neutral.

168. Calculate the pH of the following solutions:

a. 1.2 M $CaBr_2$

b. 0.84 M $C_6H_5NH_3NO_3$ (K_b for $C_6H_5NH_2 = 3.8 \times 10^{-10}$)

c. 0.57 M $KC_7H_5O_2$ (K_a for $HC_7H_5O_2 = 6.4 \times 10^{-5}$)

169. Consider 0.10 M solutions of the following compounds: $AlCl_3$, NaCN, KOH, $CsClO_4$, and NaF. Place these solutions in order of increasing pH.

Challenge Problems

170. The pH of 1.0×10^{-8} M hydrochloric acid is not 8.00. The correct pH can be calculated by considering the relationship between the molarities of the three principal ions in the solution (H^+, Cl^-, and OH^-). These molarities can be calculated from algebraic equations that can be derived from the considerations given below.

   a. The solution is electrically neutral.

   b. The hydrochloric acid can be assumed to be 100% ionized.

   c. The product of the molarities of the hydronium ions and the hydroxide ions must equal K_w.

   Calculate the pH of a 1.0×10^{-8}-M HCl solution.

171. Calculate the pH of a 1.0×10^{-7}-M solution of NaOH in water.

172. Calculate $[OH^-]$ in a 3.0×10^{-7}-M solution of $Ca(OH)_2$.

173. Consider 50.0 mL of a solution of weak acid HA ($K_a = 1.00 \times 10^{-6}$), which has a pH of 4.000. What volume of water must be added to make the pH = 5.000?

174. Making use of the assumptions we ordinarily make in calculating the pH of an aqueous solution of a weak acid, calculate the pH of a 1.0×10^{-6}-M solution of hypobromous acid (HBrO, $K_a = 2 \times 10^{-9}$). What is wrong with your answer? Why is it wrong? Without trying to solve the problem, explain what has to be included to solve the problem correctly.

175. Calculate the pH of a 0.200-M solution of C_5H_5NHF. *Hint:* C_5H_5NHF is a salt composed of $C_5H_5NH^+$ and F^- ions. The principal equilibrium in this solution is the best acid reacting with the best base; the reaction for the principal equilibrium is

$$C_5H_5NH^+(aq) + F^-(aq) \rightleftharpoons$$
$$C_5H_5N(aq) + HF(aq) \quad K = 8.2 \times 10^{-3}$$

176. Determine the pH of a 0.50-M solution of NH_4OCl. (See Exercise 175.)

177. Calculate $[OH^-]$ in a solution obtained by adding 0.0100 mol solid NaOH to 1.00 L of 15.0 M NH_3.

178. What mass of $NaOH(s)$ must be added to 1.0 L of 0.050 M NH_3 to ensure that the percent ionization of NH_3 is no greater than 0.0010%? Assume no volume change on addition of NaOH.

179. Consider 1000. mL of a 1.00×10^{-4}-M solution of a certain acid HA that has a K_a value equal to 1.00×10^{-4}. How much water was added or removed (by evaporation) so that a solution remains in which 25.0% of HA is dissociated at equilibrium? Assume that HA is nonvolatile.

180. Calculate the mass of sodium hydroxide that must be added to 1.00 L of 1.00-M $HC_2H_3O_2$ to double the pH of the solution (assume that the added NaOH does not change the volume of the solution).

181. Consider the species PO_4^{3-}, HPO_4^{2-}, and $H_2PO_4^-$. Each ion can act as a base in water. Determine the K_b value for each of these species. Which species is the strongest base?

182. Calculate the pH of a 0.10-M solution of sodium phosphate. (See Exercise 181.)

183. Will 0.10 M solutions of the following salts be acidic, basic, or neutral? See Appendix 5 for K_a values.

   a. ammonium bicarbonate

   b. sodium dihydrogen phosphate

   c. sodium hydrogen phosphate

   d. ammonium dihydrogen phosphate

   e. ammonium formate

184. **a.** The principal equilibrium in a solution of $NaHCO_3$ is

$$HCO_3^-(aq) + HCO_3^-(aq) \rightleftharpoons H_2CO_3(aq) + CO_3^{2-}(aq)$$

   Calculate the value of the equilibrium constant for this reaction.

   b. At equilibrium, what is the relationship between $[H_2CO_3]$ and $[CO_3^{2-}]$?

   c. Using the equilibrium

$$H_2CO_3(aq) \rightleftharpoons 2H^+(aq) + CO_3^{2-}(aq)$$

   derive an expression for the pH of the solution in terms of K_{a_1} and K_{a_2} using the result from part b.

   d. What is the pH of a solution of $NaHCO_3$?

185. A 0.100-g sample of the weak acid HA (molar mass = 100.0 g/mol) is dissolved in 500.0 g water. The freezing point of the resulting solution is $-0.0056°C$. Calculate the value of K_a for this acid. Assume molality equals molarity in this solution.

186. A sample containing 0.0500 mole of $Fe_2(SO_4)_3$ is dissolved in enough water to make 1.00 L of solution. This solution contains hydrated SO_4^{2-} and Fe^{3+} ions. The latter behaves as an acid:

$$Fe(H_2O)_6^{3+}(aq) \rightleftharpoons Fe(H_2O)_5OH^{2+}(aq) + H^+(aq)$$

   a. Calculate the expected osmotic pressure of this solution at 25°C if the above dissociation is negligible.

   b. The actual osmotic pressure of the solution is 6.73 atm at 25°C. Calculate K_a for the dissociation reaction of $Fe(H_2O)_6^{3+}$. (To do this calculation, you must assume that none of the ions go through the semipermeable membrane. Actually, this is not a great assumption for the tiny H^+ ion.)

Integrative Problems

These problems require the integration of multiple concepts to find the solutions.

187. A 2.14-g sample of sodium hypoiodite is dissolved in water to make 1.25 L of solution. The solution pH is 11.32. What is K_b for the hypoiodite ion?

188. Isocyanic acid (HNCO) can be prepared by heating sodium cyanate in the presence of solid oxalic acid according to the equation

$$2NaOCN(s) + H_2C_2O_4(s) \longrightarrow 2HNCO(l) + Na_2C_2O_4(s)$$

Upon isolating pure HNCO(l), an aqueous solution of HNCO can be prepared by dissolving the liquid HNCO in water. What is the pH of a 100.-mL solution of HNCO prepared from the reaction of 10.0 g each of NaOCN and $H_2C_2O_4$, assuming all of the HNCO produced is dissolved in solution? (K_a of HNCO = 1.2×10^{-4}.)

189. A certain acid, HA, has a vapor density of 5.11 g/L when in the gas phase at a temperature of 25°C and a pressure of 1.00 atm. When 1.50 g of this acid is dissolved in enough water to make 100.0 mL of solution, the pH is found to be 1.80. Calculate K_a for the acid.

Marathon Problems

These problems are designed to incorporate several concepts and techniques into one situation.

190. An aqueous solution contains a mixture of 0.0500 M HCOOH ($K_a = 1.77 \times 10^{-4}$) and 0.150 M CH_3CH_2COOH ($K_a = 1.34 \times 10^{-5}$). Calculate the pH of this solution. Because both acids are of comparable strength, the H^+ contribution from both acids must be considered.

191. For the following, mix equal volumes of one solution from Group I with one solution from Group II to achieve the indicated pH. Calculate the pH of each solution.

Group I: 0.20 M NH_4Cl, 0.20 M HCl, 0.20 M $C_6H_5NH_3Cl$, 0.20 M $(C_2H_5)_3NHCl$

Group II: 0.20 M KOI, 0.20 M NaCN, 0.20 M KOCl, 0.20 M $NaNO_2$

a. the solution with the lowest pH

b. the solution with the highest pH

c. the solution with the pH closest to 7.00

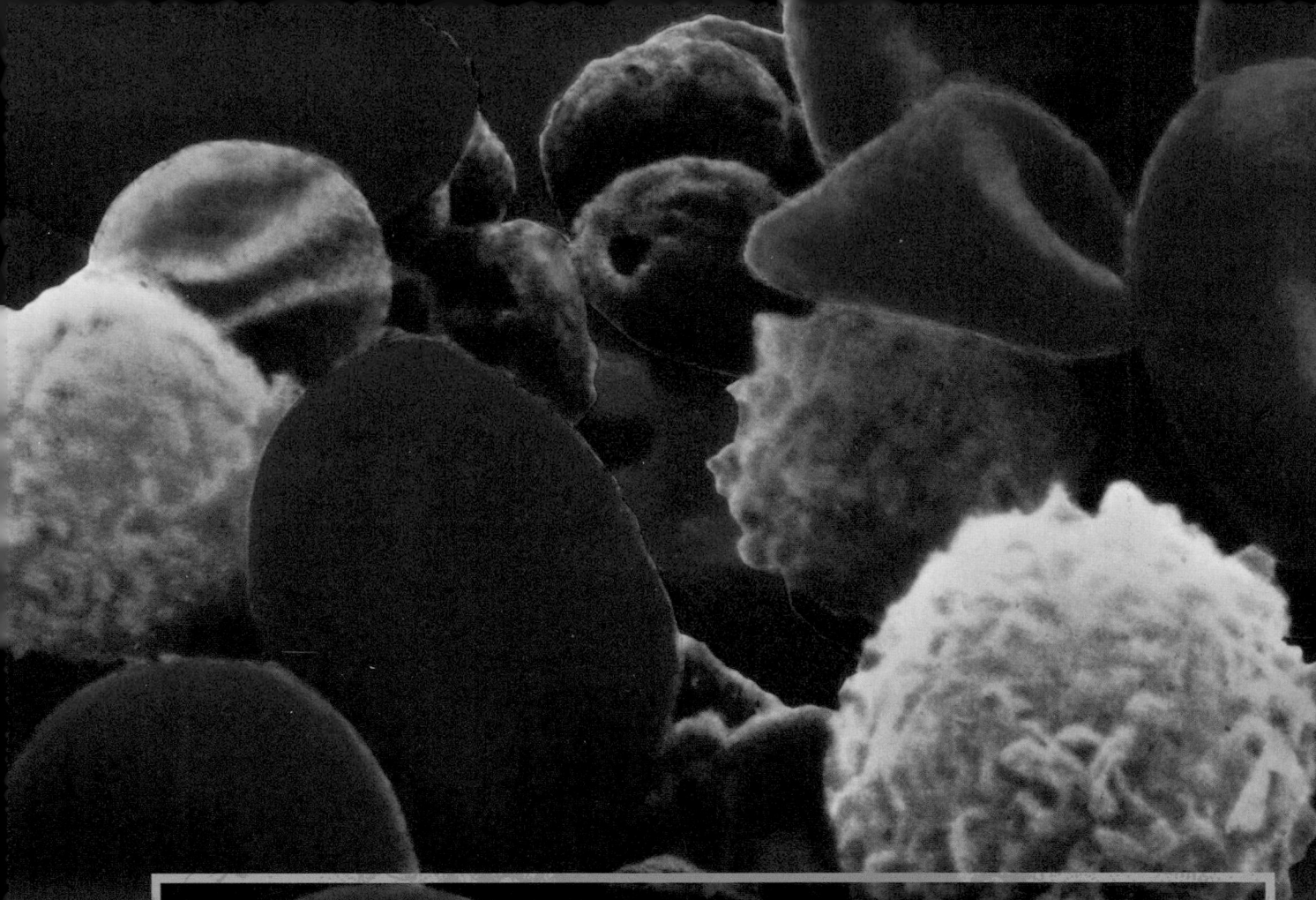

CHAPTER FOURTEEN

Acid–Base Equilibria

Colored Scanning Electron Micrograph (SEM) of human blood, showing red and white cells and platelets. Blood has effective buffers to maintain the pH at a constant value. (National Cancer Institute/Science Source)

Much important chemistry, including almost all the chemistry of the natural world, occurs in aqueous solution. We have already introduced one very significant class of aqueous reactions, those of acids and bases. In this chapter we will explore more applications of acid–base equilibria. In particular, we will examine buffered solutions, which contain components that enable the solution to be resistant to a change in pH. Buffered systems are especially important in living systems, which can survive only in a relatively narrow pH range. For example, although human blood contains many buffering systems, the most important of these consists of a mixture of carbonic acid (~0.0012 M) and bicarbonate ion (~0.024 M). These concentrations produce a pH of 7.4 for normal blood. Because our cells are so sensitive to pH, it is important that this pH value be maintained. So when reactions occur in our bodies, such as the formation of lactic acid ($HC_3H_5O_3$) when our muscles are exerted, the buffering systems must be capable of neutralizing the effects of this acid to maintain the pH at 7.4. We will see in this chapter how a buffered solution can deal with added acid without a significant change in pH.

In this chapter we will also study acid–base titrations to explore how the pH changes when a base is added to an acid and vice versa. This process is important because titrations are often used to determine the amount of acid or base present in an unknown sample. In addition, we will see how indicators can be used to mark the end point of an acid–base titration.

14-1 | Solutions of Acids or Bases Containing a Common Ion

In Chapter 13 we were concerned with calculating the equilibrium concentrations of species (particularly H^+ ions) in solutions containing an acid or a base. In this section we will discuss solutions that contain not only the weak acid HA but also its salt NaA. Although this appears to be a new type of problem, we will see that this case can be handled rather easily using the procedures developed in Chapter 13.

Suppose we have a solution containing the weak acid hydrofluoric acid (HF, $K_a = 7.2 \times 10^{-4}$) and its salt sodium fluoride (NaF). Recall that when a salt dissolves in water, it breaks up completely into its ions—it is a strong electrolyte:

$$NaF(s) \xrightarrow{H_2O(l)} Na^+(aq) + F^-(aq)$$

Since hydrofluoric acid is a weak acid and only slightly dissociated, the major species in the solution are HF, Na^+, F^-, and H_2O. The **common ion** in this solution is F^-, since it is produced by both hydrofluoric acid and sodium fluoride. What effect does the presence of the dissolved sodium fluoride have on the dissociation equilibrium of hydrofluoric acid?

To answer this question, we compare the extent of dissociation of hydrofluoric acid in two different solutions, the first containing 1.0 M HF and the second containing 1.0 M HF and 1.0 M NaF. By Le Châtelier's principle, we would expect the dissociation equilibrium for HF

$$HF(aq) \rightleftharpoons H^+(aq) + F^-(aq)$$

in the second solution to be *driven to the left by the presence of F^- ions from the NaF.* Thus the extent of dissociation of HF will be *less* in the presence of dissolved NaF:

$$HF(aq) \rightleftharpoons H^+(aq) + F^-(aq)$$

Equilibrium shifts away from added component. Fewer H^+ ions present.

Added F^- ions from NaF

ⓘ The common ion effect is an application of Le Châtelier's principle.

The shift in equilibrium position that occurs because of the addition of an ion already involved in the equilibrium reaction is called the **common ion effect**. ◄ ⓘ This effect makes a solution of NaF and HF less acidic than a solution of HF alone.

The common ion effect is quite general. For example, solid NH_4Cl added to a 1.0-M NH_3 solution produces additional ammonium ions:

$$NH_4Cl(s) \xrightarrow{H_2O} NH_4^+(aq) + Cl^-(aq)$$

and this causes the position of the ammonia–water equilibrium to shift to the left:

$$NH_3(aq) + H_2O(l) \rightleftharpoons NH_4^+(aq) + OH^-(aq)$$

This reduces the equilibrium concentration of OH^- ions.

The common ion effect is also important in solutions of polyprotic acids. The production of protons by the first dissociation step greatly inhibits the succeeding dissociation steps, which, of course, also produce protons, the common ion in this case. We will see later in this chapter that the common ion effect is also important in dealing with the solubility of salts.

Equilibrium Calculations

The procedures for finding the pH of a solution containing a weak acid or base plus a common ion are very similar to the procedures, which we covered in Chapter 14, for solutions containing the acids or bases alone. For example, in the case of a weak acid, the only important difference is that the initial concentration of the anion A^- is not zero in a solution that also contains the salt NaA. Example 14-1 illustrates a typical example using the same general approach we developed in Chapter 13.

| INTERACTIVE EXAMPLE 14-1

Acidic Solutions Containing Common Ions

In Section 13-5 we found that the equilibrium concentration of H^+ in a 1.0-M HF solution is 2.7×10^{-2} M, and the percent dissociation of HF is 2.7%. Calculate $[H^+]$ and the percent dissociation of HF in a solution containing 1.0 M HF ($K_a = 7.2 \times 10^{-4}$) and 1.0 M NaF.

Solution

As the aqueous solutions we consider become more complex, it is more important than ever to be systematic and to *focus on the chemistry* occurring in the solution before

Major Species

F⁻

Na⁺

HF

H₂O

thinking about mathematical procedures. The way to do this is *always* to write the major species first and consider the chemical properties of each one.

In a solution containing 1.0 M HF and 1.0 M NaF, the major species are ◄

$$HF, \quad F^-, \quad Na^+, \quad \text{and} \quad H_2O$$

We know that Na^+ ions have neither acidic nor basic properties and that water is a very weak acid (or base). Therefore, the important species are HF and F^-, which participate in the acid dissociation equilibrium that controls $[H^+]$ in this solution. That is, the position of the equilibrium

$$HF(aq) \rightleftharpoons H^+(aq) + F^-(aq)$$

will determine $[H^+]$ in the solution. The equilibrium expression is

$$K_a = \frac{[H^+][F^-]}{[HF]} = 7.2 \times 10^{-4}$$

The important concentrations are shown in the following table.

Initial Concentration (mol/L)		Equilibrium Concentration (mol/L)
$[HF]_0 = 1.0$ (from dissolved HF) $[F^-]_0 = 1.0$ (from dissolved NaF) $[H^+]_0 = 0$ (neglect contribution from H_2O)	$\xrightarrow{\substack{x \text{ mol/L HF} \\ \text{dissociates}}}$	$[HF] = 1.0 - x$ $[F^-] = 1.0 + x$ $[H^+] = x$

Note that $[F^-]_0 = 1.0\ M$ because of the dissolved sodium fluoride and that at equilibrium $[F^-] > 1.0\ M$ because when the acid dissociates it produces F^- as well as H^+. Then

$$K_a = 7.2 \times 10^{-4} = \frac{[H^+][F^-]}{[HF]} = \frac{(x)(1.0 + x)}{1.0 - x} \approx \frac{(x)(1.0)}{1.0}$$

(since x is expected to be small).

Solving for x gives

$$x = \frac{1.0}{1.0}(7.2 \times 10^{-4}) = 7.2 \times 10^{-4}$$

Noting that x is small compared to 1.0, we conclude that this result is acceptable. Thus

$$[H^+] = x = 7.2 \times 10^{-4}\ M \qquad \text{(The pH is 3.14.)}$$

The percent dissociation of HF in this solution is

$$\frac{[H^+]}{[HF]_0} \times 100 = \frac{7.2 \times 10^{-4}\ M}{1.0\ M} \times 100 = 0.072\%$$

Compare these values for $[H^+]$ and the percent dissociation of HF with those for a 1.0-M HF solution, where $[H^+] = 2.7 \times 10^{-2}\ M$ and the percent dissociation is 2.7%. The large difference shows clearly that the presence of the F^- ions from the dissolved NaF greatly inhibits the dissociation of HF. The position of the acid dissociation equilibrium has been shifted to the left by the presence of F^- ions from NaF.

See Exercises 14-23 and 14-24

14-2 | Buffered Solutions

The most important application of acid–base solutions containing a common ion is for buffering. A **buffered solution** is one that *resists a change in its pH* when either hydroxide ions or protons are added. The most important practical example of a buffered solution is our blood, which can absorb the acids and bases produced in biological reactions without changing its pH. A constant pH for blood is vital because cells can survive only in a very narrow pH range. ◄ *i*

A buffered solution may contain a *weak* acid and its salt (for example, HF and NaF) or a *weak* base and its salt (for example, NH_3 and NH_4Cl). By choosing the appropriate components, a solution can be buffered at virtually any pH.

In treating buffered solutions in this chapter, we will start by considering the equilibrium calculations. We will then use these results to show how buffering works. That is, we will answer the question: How does a buffered solution resist changes in pH when an acid or a base is added?

As you do the calculations associated with buffered solutions, keep in mind that these are merely solutions containing weak acids or bases, and the procedures required are the same ones we have already developed. Be sure to use the systematic approach introduced in Chapter 13. ◄ *i*

i The most important buffering system in the blood involves HCO_3^- and H_2CO_3.

i The systematic approach developed in Chapter 13 for weak acids and bases applies to buffered solutions.

INTERACTIVE EXAMPLE 14-2

The pH of a Buffered Solution I

A buffered solution contains 0.50 *M* acetic acid ($HC_2H_3O_2$, $K_a = 1.8 \times 10^{-5}$) and 0.50 *M* sodium acetate ($NaC_2H_3O_2$). Calculate the pH of this solution.

Major Species

 $HC_2H_3O_2$

 $C_2H_3O_2^-$

Na⁺ Na^+

H₂O H_2O

Solution

The major species in the solution are ◄

$$HC_2H_3O_2, \quad Na^+, \quad C_2H_3O_2^-, \quad \text{and} \quad H_2O$$

$\uparrow$ Weak acid $\quad$ $\uparrow$ Neither acid nor base $\quad$ $\uparrow$ Base (conjugate base of $HC_2H_3O_2$) $\quad$ $\uparrow$ Very weak acid or base

Examination of the solution components leads to the conclusion that the acetic acid dissociation equilibrium, which involves both $HC_2H_3O_2$ and $C_2H_3O_2^-$, will control the pH of the solution:

$$HC_2H_3O_2(aq) \rightleftharpoons H^+(aq) + C_2H_3O_2^-(aq)$$

$$K_a = 1.8 \times 10^{-5} = \frac{[H^+][C_2H_3O_2^-]}{[HC_2H_3O_2]}$$

The concentrations are as follows:

Initial Concentration (mol/L)		Equilibrium Concentration (mol/L)
$[HC_2H_3O_2]_0 = 0.50$ $[C_2H_3O_2^-]_0 = 0.50$ $[H^+]_0 \approx 0$	$\xrightarrow[\substack{\text{dissociates} \\ \text{to reach} \\ \text{equilibrium}}]{\substack{x \text{ mol/L of} \\ HC_2H_3O_2}}$	$[HC_2H_3O_2] = 0.50 - x$ $[C_2H_3O_2^-] = 0.50 + x$ $[H^+] = x$

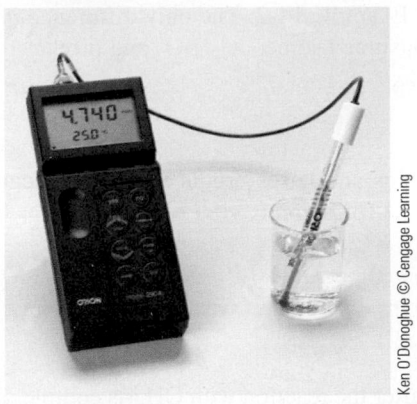

▲ A digital pH meter shows the pH of the buffered solution to be 4.740.

The corresponding ICE table is

	$HC_2H_3O_2(aq)$	$\rightleftharpoons$	$H^+(aq)$	+	$C_2H_3O_2^-(aq)$
Initial	0.50		≈ 0		0.50
Change	$-x$		$+x$		$+x$
Equilibrium	$0.50 - x$		x		$0.50 + x$

Then

$$K_a = 1.8 \times 10^{-5} = \frac{[H^+][C_2H_3O_2^-]}{[HC_2H_3O_2]} = \frac{(x)(0.50 + x)}{0.50 - x} \approx \frac{(x)(0.50)}{0.50}$$

and

$$x \approx 1.8 \times 10^{-5}$$

The approximations are valid (by the 5% rule), so

$$[H^+] = x = 1.8 \times 10^{-5}\,M \qquad \text{and} \qquad pH = 4.74 ◄$$

See Exercises 14-31 and 14-32

INTERACTIVE EXAMPLE 14-3

pH Changes in Buffered Solutions

Calculate the change in pH that occurs when 0.010 mole of solid NaOH is added to 1.0 L of the buffered solution described in Example 14-2. Compare this pH change with that which occurs when 0.010 mole of solid NaOH is added to 1.0 L water.

Solution

Since the added solid NaOH will completely dissociate, the major species in solution *before any reaction occurs* are $HC_2H_3O_2$, Na^+, $C_2H_3O_2^-$, OH^-, and H_2O. ◄ Note that the solution contains a relatively large amount of the very strong base hydroxide ion, which has a great affinity for protons. The best source of protons is the acetic acid, and the reaction that will occur is

$$OH^-(aq) + HC_2H_3O_2(aq) \longrightarrow H_2O(l) + C_2H_3O_2^-(aq)$$

Although acetic acid is a weak acid, the hydroxide ion is such a strong base that the reaction above will *proceed essentially to completion* (until the OH^- ions are consumed).

The best approach to this problem involves two distinct steps: (1) assume that the reaction goes to completion, and carry out the stoichiometric calculations, and then (2) carry out the equilibrium calculations.

Major Species

 $HC_2H_3O_2$

 $C_2H_3O_2^-$

 Na^+

 OH^-

H_2O

1. *The stoichiometry problem.* The stoichiometry for the reaction is shown below.

	$HC_2H_3O_2\,(aq)$	+	$OH^-(aq)$	$\longrightarrow$	$C_2H_3O_2^-\,(aq)$	+	$H_2O(l)$
Before reaction	1.0 L × 0.50 M = 0.50 mol		0.010 mol		1.0 L × 0.50 M = 0.50 mol		
After reaction	0.50 − 0.010 = 0.49 mol		0.010 − 0.010 = 0 mol		0.50 + 0.010 = 0.51 mol		

Major Species

$HC_2H_3O_2$

$C_2H_3O_2^-$

Na^+

H_2O

Note that 0.010 mole of $HC_2H_3O_2$ has been converted to 0.010 mole of $C_2H_3O_2^-$ by the added OH^-.

2. *The equilibrium problem.* After the reaction between OH^- and $HC_2H_3O_2$ is complete, the major species in solution are ◄

$$HC_2H_3O_2, \quad Na^+, \quad C_2H_3O_2^-, \quad \text{and} \quad H_2O$$

The dominant equilibrium involves the dissociation of acetic acid.

This problem is then very similar to that in Example 14-2. The only difference is that the addition of 0.010 mole of OH^- has consumed some $HC_2H_3O_2$ and produced some $C_2H_3O_2^-$, yielding the following ICE table:

	$HC_2H_3O_2(aq)$	$\rightleftharpoons$	$H^+(aq)$	$+$	$C_2H_3O_2^-(aq)$
Initial	0.49		0		0.51
Change	$-x$		$+x$		$+x$
Equilibrium	$0.49 - x$		x		$0.51 + x$

Note that the initial concentrations are defined after the reaction with OH^- is complete but before the system adjusts to equilibrium. Following the usual procedure gives

$$K_a = 1.8 \times 10^{-5} = \frac{[H^+][C_2H_3O_2^-]}{[HC_2H_3O_2]} = \frac{(x)(0.51 + x)}{0.49 - x} \approx \frac{(x)(0.51)}{0.49}$$

and

$$x \approx 1.7 \times 10^{-5}$$

The approximations are valid (by the 5% rule), so

$$[H^+] = x = 1.7 \times 10^{-5} \qquad \text{and} \qquad pH = 4.76$$

The change in pH produced by the addition of 0.01 mole of OH^- to this buffered solution is then

$$\underset{\underset{\text{New solution}}{\uparrow}}{4.76} \quad - \quad \underset{\underset{\text{Original solution}}{\uparrow}}{4.74} \quad = +0.02$$

The pH increased by 0.02 pH units.

Now compare this with what happens when 0.01 mole of solid NaOH is added to 1.0 L water to give 0.01 M NaOH. In this case $[OH^-] = 0.01\ M$ and

$$[H^+] = \frac{K_w}{[OH^-]} = \frac{1.0 \times 10^{-14}}{1.0 \times 10^{-2}} = 1.0 \times 10^{-12}$$

$$pH = 12.00$$

Thus the change in pH is

$$\underset{\underset{\text{New solution}}{\uparrow}}{12.00} \quad - \quad \underset{\underset{\text{Pure water}}{\uparrow}}{7.00} \quad = +5.00$$

The increase is 5.00 pH units. ◄ Note how well the buffered solution resists a change in pH as compared with pure water.

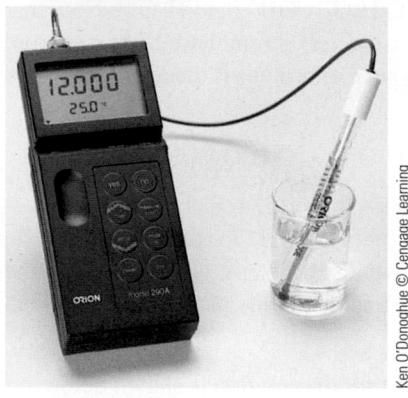

▲ (top) Pure water at pH 7.000. (bottom) When 0.01 mole of NaOH is added to 1.0 L pure water, the pH jumps to 12.000.

See Exercises 14-33 and 14-34

Examples 14-2 and 14-3 represent typical buffer problems that involve all the concepts that you need to know to handle buffered solutions containing weak acids. Pay special attention to the following points:

1. Buffered solutions are simply solutions of weak acids or bases containing a common ion. The pH calculations on buffered solutions require exactly the same procedures introduced in Chapter 13. *This is not a new type of problem.*

2. When a strong acid or base is added to a buffered solution, it is best to deal with the stoichiometry of the resulting reaction first. After the stoichiometric calcula-

tions are completed, then consider the equilibrium calculations. This procedure can be presented as follows:

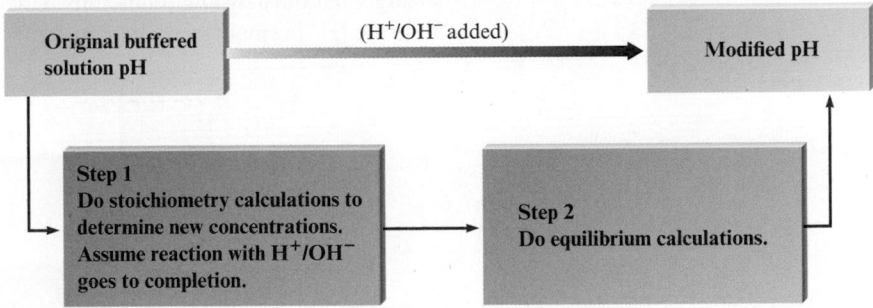

Buffering: How Does It Work?

Examples 14-2 and 14-3 demonstrate the ability of a buffered solution to absorb hydroxide ions without a significant change in pH. *But how does a buffer work?* Suppose a buffered solution contains relatively large quantities of a weak acid HA and its conjugate base A^-. When hydroxide ions are added to the solution, since the weak acid represents the best source of protons, the following reaction occurs:

$$OH^-(aq) + HA(aq) \longrightarrow A^-(aq) + H_2O(l)$$

The net result is that OH^- ions are not allowed to accumulate but are replaced by A^- ions.

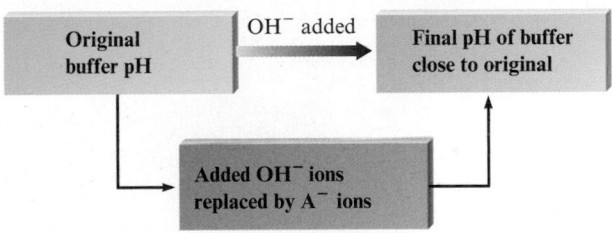

The stability of the pH under these conditions can be understood by examining the equilibrium expression for the dissociation of HA:

$$K_a = \frac{[H^+][A^-]}{[HA]}$$

or, rearranging,

$$[H^+] = K_a\frac{[HA]}{[A^-]}$$

In other words, the *equilibrium concentration of H^+, and thus the pH, is determined by the ratio $[HA]/[A^-]$.* ◄ ⓘ When OH^- ions are added, HA is converted to A^-, and the ratio $[HA]/[A^-]$ decreases. However, *if the amounts of HA and A^- originally present are very large compared with the amount of OH^- added,* the change in the $[HA]/[A^-]$ ratio will be small.

In Examples 14-2 and 14-3,

$$\frac{[HA]}{[A^-]} = \frac{0.50}{0.50} = 1.0 \quad \text{Initially}$$

$$\frac{[HA]}{[A^-]} = \frac{0.49}{0.51} = 0.96 \quad \text{After adding 0.01 mol/L } OH^-$$

The change in the ratio $[HA]/[A^-]$ is very small. Thus the $[H^+]$ and the pH remain essentially constant.

ⓘ In a buffered solution the pH is governed by the ratio $[HA]/[A^-]$.

The essence of buffering, then, is that [HA] and [A⁻] are large compared with the amount of OH⁻ added. Thus when the OH⁻ is added, the concentrations of HA and A⁻ change, but only by small amounts. Under these conditions, the [HA]/[A⁻] ratio and thus the [H⁺] remain virtually constant.

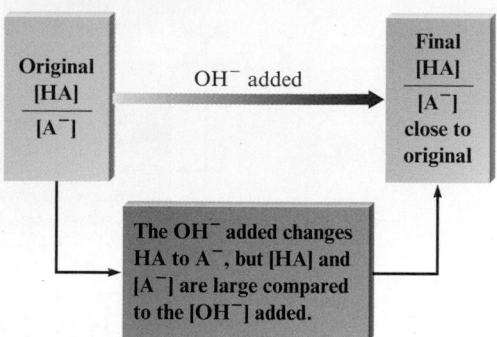

Similar reasoning applies when protons are added to a buffered solution of a weak acid and a salt of its conjugate base. Because the A⁻ ion has a high affinity for H⁺, the added H⁺ ions react with A⁻ to form the weak acid:

$$H^+(aq) + A^-(aq) \longrightarrow HA(aq)$$

and free H⁺ ions do not accumulate. In this case there will be a net change of A⁻ to HA. However, if [A⁻] and [HA] are large compared with the [H⁺] added, little change in the pH will occur.

LET'S REVIEW

$$HA(aq) \rightleftharpoons H^+(aq) + A^-(aq)$$

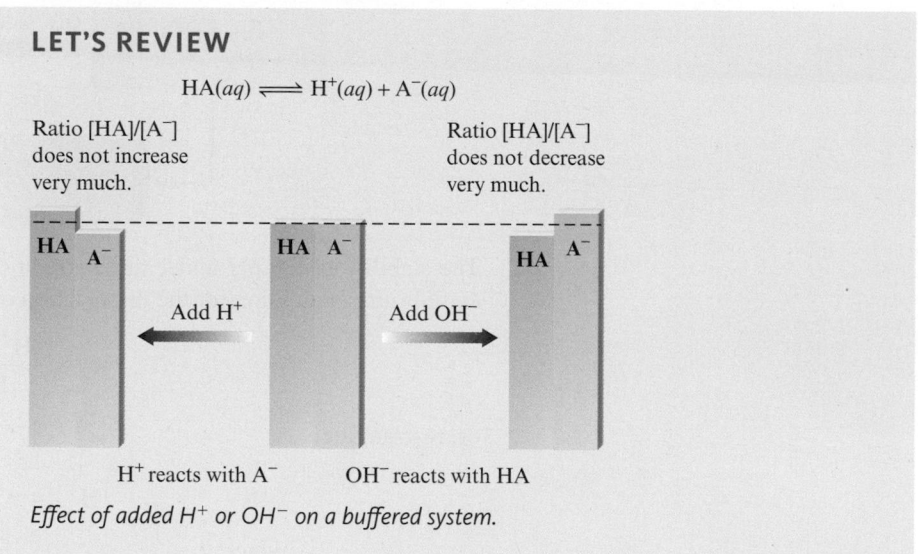

Effect of added H⁺ or OH⁻ on a buffered system.

The form of the acid dissociation equilibrium expression

$$[H^+] = K_a \frac{[HA]}{[A^-]} \tag{14-1}$$

is often useful for calculating [H⁺] in a buffered solution, since [HA] and [A⁻] are known. For example, to calculate [H⁺] in a buffered solution containing 0.10 M HF ($K_a = 7.2 \times 10^{-4}$) and 0.30 M NaF, we simply substitute into Equation (14-1):

$$[H^+] = (7.2 \times 10^{-4}) \frac{0.10}{0.30} = 2.4 \times 10^{-4} \, M$$

Another useful form of Equation (14-1) can be obtained by taking the negative log of both sides:

$$-\log[\text{H}^+] = -\log(K_a) - \log\left(\frac{[\text{HA}]}{[\text{A}^-]}\right)$$

That is,

$$\text{pH} = pK_a - \log\left(\frac{[\text{HA}]}{[\text{A}^-]}\right)$$

or, where inverting the log term reverses the sign:

$$\text{pH} = pK_a + \log\left(\frac{[\text{A}^-]}{[\text{HA}]}\right) = pK_a + \log\left(\frac{[\text{base}]}{[\text{acid}]}\right) \qquad (14\text{-}2)$$

This log form of the expression for K_a is called the **Henderson–Hasselbalch equation** and is useful for calculating the pH of solutions when the ratio $[\text{HA}]/[\text{A}^-]$ is known.

For a particular buffering system (conjugate acid–base pair), all solutions that have the same ratio $[\text{A}^-]/[\text{HA}]$ will have the same pH. For example, a buffered solution containing $5.0\ M$ $\text{HC}_2\text{H}_3\text{O}_2$ and $3.0\ M$ $\text{NaC}_2\text{H}_3\text{O}_2$ will have the same pH as one containing $0.050\ M$ $\text{HC}_2\text{H}_3\text{O}_2$ and $0.030\ M$ $\text{NaC}_2\text{H}_3\text{O}_2$. This can be shown as follows:

System	$[\text{A}^-]/[\text{HA}]$
$5.0\ M$ $\text{HC}_2\text{H}_3\text{O}_2$ and $3.0\ M$ $\text{NaC}_2\text{H}_3\text{O}_2$	$\dfrac{3.0\ M}{5.0\ M} = 0.60$
$0.050\ M$ $\text{HC}_2\text{H}_3\text{O}_2$ and $0.030\ M$ $\text{NaC}_2\text{H}_3\text{O}_2$	$\dfrac{0.030\ M}{0.050\ M} = 0.60$

Therefore,

$$\text{pH} = pK_a + \log\left(\frac{[\text{C}_2\text{H}_3\text{O}_2^-]}{[\text{HC}_2\text{H}_3\text{O}_2]}\right) = 4.74 + \log(0.60) = 4.74 - 0.22 = 4.52$$

Note that in using this equation we have assumed that the equilibrium concentrations of A^- and HA are equal to the initial concentrations. That is, we are assuming the validity of the approximations

$$[\text{A}^-] = [\text{A}^-]_0 + x \approx [\text{A}^-]_0 \quad \text{and} \quad [\text{HA}] = [\text{HA}]_0 - x \approx [\text{HA}]_0$$

where x is the amount of acid that dissociates. Since the initial concentrations of HA and A are relatively large in a buffered solution, this assumption is generally acceptable.

INTERACTIVE EXAMPLE 14-4

The pH of a Buffered Solution II

Calculate the pH of a solution containing $0.75\ M$ lactic acid ($K_a = 1.4 \times 10^{-4}$) and $0.25\ M$ sodium lactate. Lactic acid ($\text{HC}_3\text{H}_5\text{O}_3$) is a common constituent of biological systems. For example, it is found in milk and is present in human muscle tissue during exertion.

Major Species

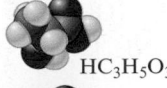
$\text{HC}_3\text{H}_5\text{O}_3$

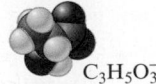
$\text{C}_3\text{H}_5\text{O}_3^-$

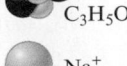
Na^+

H_2O

Solution

The major species in solution are ◄

$$\text{HC}_3\text{H}_5\text{O}_3, \quad \text{Na}^+, \quad \text{C}_3\text{H}_5\text{O}_3^-, \quad \text{and} \quad \text{H}_2\text{O}$$

Since Na^+ has no acid–base properties and H_2O is a weak acid or base, the pH will be controlled by the lactic acid dissociation equilibrium:

$$\text{HC}_3\text{H}_5\text{O}_3(aq) \rightleftharpoons \text{H}^+(aq) + \text{C}_3\text{H}_5\text{O}_3^-(aq)$$

$$K_a = \frac{[\text{H}^+][\text{C}_3\text{H}_5\text{O}_3^-]}{[\text{HC}_3\text{H}_5\text{O}_3]} = 1.4 \times 10^{-4}$$

Since $[HC_3H_5O_3]_0$ and $[C_3H_5O_3^-]_0$ are relatively large,

$$[HC_3H_5O_3] \approx [HC_3H_5O_3]_0 = 0.75\ M$$

and

$$[C_3H_5O_3^-] \approx [C_3H_5O_3^-]_0 = 0.25\ M$$

Thus using the rearranged K_a expression, we have

$$[H^+] = K_a \frac{[HC_3H_5O_3]}{[C_3H_5O_3^-]} = (1.4 \times 10^{-4})\frac{(0.75\ M)}{(0.25\ M)} = 4.2 \times 10^{-4}\ M$$

and

$$pH = -\log(4.2 \times 10^{-4}) = 3.38$$

Alternatively, we could use the Henderson–Hasselbalch equation:

$$pH = pK_a + \log\left(\frac{[C_3H_5O_3^-]}{[HC_3H_5O_3]}\right) = 3.85 + \log\left(\frac{0.25\ M}{0.75\ M}\right) = 3.38$$

See Exercises 14-35 and 14-37

Buffered solutions also can be formed from a weak base and the corresponding conjugate acid. In these solutions, the weak base B reacts with any H^+ added:

$$B + H^+ \longrightarrow BH^+$$

and the conjugate acid BH^+ reacts with any added OH^-:

$$BH^+ + OH^- \longrightarrow B + H_2O$$

The approach needed to perform pH calculations for these systems is virtually identical to that used above. This makes sense because, as is true of all buffered solutions, a weak acid (BH^+) and a weak base (B) are present. A typical case is illustrated in Example 14-5.

INTERACTIVE EXAMPLE 14-5

The pH of a Buffered Solution III

A buffered solution contains $0.25\ M\ NH_3$ ($K_b = 1.8 \times 10^{-5}$) and $0.40\ M\ NH_4Cl$. Calculate the pH of this solution.

Solution

Major Species

Cl^-

NH_4^+

NH_3

H_2O

The major species in solution are ◄

$$NH_3,\ \underbrace{NH_4^+,\ Cl^-}_{\text{From the dissolved } NH_4Cl},\ \text{and}\ H_2O$$

Since Cl^- is such a weak base and water is a weak acid or base, the important equilibrium is

$$NH_3(aq) + H_2O(l) \rightleftharpoons NH_4^+(aq) + OH^-(aq)$$

and

$$K_b = 1.8 \times 10^{-5} = \frac{[NH_4^+][OH^-]}{[NH_3]}$$

The appropriate ICE table is:

	$NH_3(aq)$	+	$H_2O(l)$	$\rightleftharpoons$	$NH_4^+(aq)$	+	$OH^-(aq)$
Initial	0.25		—		0.40		≈ 0
Change	$-x$		—		$+x$		$+x$
Equilibrium	$0.25 - x$		—		$0.40 + x$		x

Then

$$K_b = 1.8 \times 10^{-5} = \frac{[NH_4^+][OH^-]}{[NH_3]} = \frac{(0.40 + x)(x)}{0.25 - x} \approx \frac{(0.40)(x)}{0.25}$$

and

$$x \approx 1.1 \times 10^{-5}$$

The approximations are valid (by the 5% rule), so

$$[OH^-] = x = 1.1 \times 10^{-5}$$
$$pOH = 4.95$$
$$pH = 14.00 - 4.95 = 9.05$$

This case is typical of a buffered solution in that the initial and equilibrium concentrations of buffering materials are essentially the same.

Alternative Solution

There is another way of looking at this problem. Since the solution contains relatively large quantities of *both* NH_4^+ and NH_3, we can use the equilibrium

$$NH_3(aq) + H_2O(l) \rightleftharpoons NH_4^+(aq) + OH^-(aq)$$

to calculate $[OH^-]$ and then calculate $[H^+]$ from K_w as we have just done. Or we can use the dissociation equilibrium for NH_4^+, that is,

$$NH_4^+(aq) \rightleftharpoons NH_3(aq) + H^+(aq)$$

to calculate $[H^+]$ directly. *Either choice will give the same answer,* since the same equilibrium concentrations of NH_3 and NH_4^+ must satisfy both equilibria.

We can obtain the K_a value for NH_4^+ from the given K_b value for NH_3, since $K_a \times K_b = K_w$:

$$K_a = \frac{K_w}{K_b} = \frac{1.0 \times 10^{-14}}{1.8 \times 10^{-5}} - 5.6 \times 10^{-10}$$

Then, using the Henderson–Hasselbalch equation, we have

$$pH = pK_a + \log\left(\frac{[base]}{[acid]}\right)$$

$$- 9.25 + \log\left(\frac{0.25\,M}{0.40\,M}\right) = 9.25 - 0.20 - 9.05$$

See Exercises 14-36 and 14-38

INTERACTIVE EXAMPLE 14-6

Major Species

Cl⁻

H⁺

NH_4^+

NH_3

H_2O

Adding Strong Acid to a Buffered Solution I

Calculate the pH of the solution that results when 0.10 mole of gaseous HCl is added to 1.0 L of the buffered solution from Example 14-5.

Solution

Before any reaction occurs, the solution contains the following major species: ◄

$$NH_3, \quad NH_4^+, \quad \underbrace{Cl^-, \quad H^+,}_{\text{From added HCl}} \quad \text{and} \quad H_2O$$

What reaction can occur? We know that H^+ will not react with Cl^- to form HCl. In contrast to Cl^-, the NH_3 molecule has a great affinity for protons [this is demonstrated by the fact that NH_4^+ is such a weak acid ($K_a = 5.6 \times 10^{-10}$)]. Thus NH_3 will react with H^+ to form NH_4^+:

$$NH_3(aq) + H^+(aq) \longrightarrow NH_4^+(aq)$$

Since this reaction can be assumed to go essentially to completion to form the very weak acid NH_4^+, we will do the stoichiometry calculations before we consider the equilibrium calculations. That is, we will let the reaction run to completion and then consider the equilibrium. ◀ ⓘ

The stoichiometry calculations for this process are shown below.

	NH_3	+	H^+	$\longrightarrow$	NH_4^+
Before reaction	(1.0 L)(0.25 M) = 0.25 mol		0.10 mol ↑ Limiting reactant		(1.0 L)(0.40 M) = 0.40 mol
After reaction	0.25 − 0.10 = 0.15 mol		0		0.40 + 0.10 = 0.50 mol

After the reaction goes to completion, the solution contains the major species ◀

$$NH_3, \quad NH_4^+, \quad Cl^-, \quad and \quad H_2O$$

Major Species

 Cl^-

   NH_4^+

   NH_3

 H_2O

and

$$[NH_3]_0 = \frac{0.15 \text{ mol}}{1.0 \text{ L}} = 0.15 \ M$$

$$[NH_4^+]_0 = \frac{0.50 \text{ mol}}{1.0 \text{ L}} = 0.50 \ M$$

We can use the Henderson–Hasselbalch equation, where

$$[Base] = [NH_3] \approx [NH_3]_0 = 0.15 \ M$$
$$[Acid] = [NH_4^+] \approx [NH_4^+]_0 = 0.50 \ M$$

Then

$$pH = pK_a + \log\left(\frac{[NH_3]}{[NH_4^+]}\right)$$

$$= 9.25 + \log\left(\frac{0.15 \ M}{0.50 \ M}\right) = 9.25 - 0.52 = 8.73$$

Note that the addition of HCl only slightly decreases the pH, as we would expect in a buffered solution.

See Exercise 14-39

See Exercise 14-39

LET'S REVIEW | *Summary of the Most Important Characteristics of Buffered Solutions*

- Buffered solutions contain relatively large concentrations of a weak acid and the corresponding weak base. They can involve a weak acid HA and the conjugate base A^- or a weak base B and the conjugate acid BH^+.

- When H^+ is added to a buffered solution, it reacts essentially to completion with the weak base present:

$$H^+ + A^- \longrightarrow HA \quad or \quad H^+ + B \longrightarrow BH^+$$

(Box continues on following page)

Remember: Think about the chemistry first. Ask yourself if a reaction will occur among the major species.

> **LET'S REVIEW** | *Summary of the Most Important Characteristics of Buffered Solutions (continued)*
>
> ■ When OH^- is added to a buffered solution, it reacts essentially to completion with the weak acid present:
>
> $$OH^- + HA \longrightarrow A^- + H_2O \quad \text{or} \quad OH^- + BH^+ \longrightarrow B + H_2O$$
>
> ■ The pH in the buffered solution is determined by the ratio of the concentrations of the weak acid and weak base. As long as this ratio remains virtually constant, the pH will remain virtually constant. This will be the case as long as the concentrations of the buffering materials (HA and A^- or B and BH^+) are large compared with the amounts of H^+ or OH^- added.

14-3 | Buffering Capacity

The **buffering capacity** of a buffered solution represents the amount of protons or hydroxide ions the buffer can absorb without a significant change in pH. A buffer with a large capacity contains large concentrations of buffering components and so can absorb a relatively large amount of protons or hydroxide ions and show little pH change. *The pH of a buffered solution is determined by the ratio $[A^-]/[HA]$. The capacity of a buffered solution is determined by the magnitudes of $[HA]$ and $[A^-]$.* ◄ ⓘ

ⓘ A buffer with a large capacity contains large concentrations of the buffering components.

INTERACTIVE EXAMPLE 14-7

Adding Strong Acid to a Buffered Solution II

Calculate the change in pH that occurs when 0.010 mole of gaseous HCl is added to 1.0 L of each of the following solutions:

Solution A: 5.00 M $HC_2H_3O_2$ and 5.00 M $NaC_2H_3O_2$

Solution B: 0.050 M $HC_2H_3O_2$ and 0.050 M $NaC_2H_3O_2$

For acetic acid, $K_a = 1.8 \times 10^{-5}$.

Solution

For both solutions the initial pH can be determined from the Henderson–Hasselbalch equation:

$$pH = pK_a + \log\left(\frac{[C_2H_3O_2^-]}{[HC_2H_3O_2]}\right)$$

Major Species

Cl^-

$HC_2H_3O_2$

$C_2H_3O_2^-$

Na^+

H^+

H_2O

In each case, $[C_2H_3O_2^-] = [HC_2H_3O_2]$. Therefore, the initial pH for both A and B is

$$pH = pK_a + \log(1) = pK_a = -\log(1.8 \times 10^{-5}) = 4.74$$

After the addition of HCl to each of these solutions, the major species *before any reaction occurs* are ◄

$$HC_2H_3O_2, \quad Na^+, \quad C_2H_3O_2^-, \quad \underline{H^+, \quad Cl^-}, \quad \text{and} \quad H_2O$$
$$\text{\small From the added HCl}$$

Will any reactions occur among these species? Note that we have a relatively large quantity of H^+, which will readily react with any effective base. We know that Cl^- will not react with H^+ to form HCl in water. However, $C_2H_3O_2^-$ will react with H^+ to form the weak acid $HC_2H_3O_2$:

$$H^+(aq) + C_2H_3O_2^-(aq) \longrightarrow HC_2H_3O_2(aq)$$

Because $HC_2H_3O_2$ is a weak acid, we assume that this reaction runs to completion; the 0.010 mole of added H^+ will convert 0.010 mole of $C_2H_3O_2^-$ to 0.010 mole of $HC_2H_3O_2$.

For solution A (since the solution volume is 1.0 L, the number of moles equals the molarity), the following calculations apply:

	H^+	+	$C_2H_3O_2^-$	$\longrightarrow$	$HC_2H_3O_2$
Before reaction	0.010 M		5.00 M		5.00 M
After reaction	0		4.99 M		5.01 M

The new pH can be obtained by substituting the new concentrations into the Henderson–Hasselbalch equation:

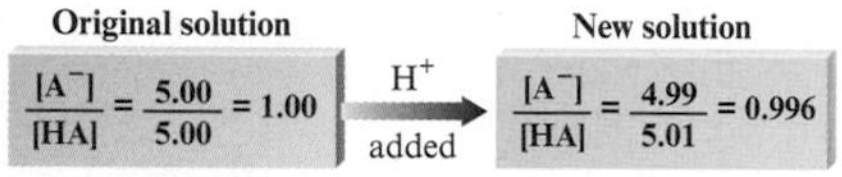

$$pH = pK_a + \log\left(\frac{[C_2H_3O_2^-]}{[HC_2H_3O_2]}\right)$$

$$= 4.74 + \log\left(\frac{4.99}{5.01}\right) = 4.74 - 0.0017 = 4.74$$

There is virtually no change in pH for solution A when 0.010 mole of gaseous HCl is added. ◄

For solution B, the following calculations apply:

	H^+	+	$C_2H_3O_2^-$	$\longrightarrow$	$HC_2H_3O_2$
Before reaction	0.010 M		0.050 M		0.050 M
After reaction	0		0.040 M		0.060 M

The new pH is

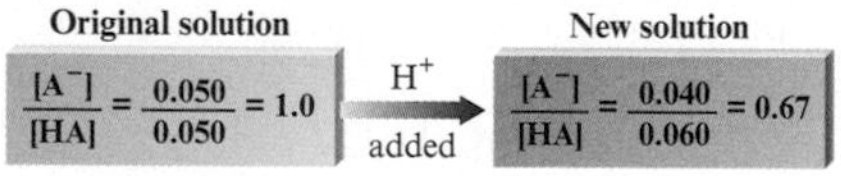

$$pH = 4.74 + \log\left(\frac{0.040}{0.060}\right)$$

$$= 4.74 - 0.18 = 4.56$$

Although the pH change for solution B is small, a change did occur, which is in contrast to solution A. ◄

These results show that solution A, which contains much larger quantities of buffering components, has a much higher buffering capacity than solution B. ◄ ◄

See Exercises 14-39 and 14-40

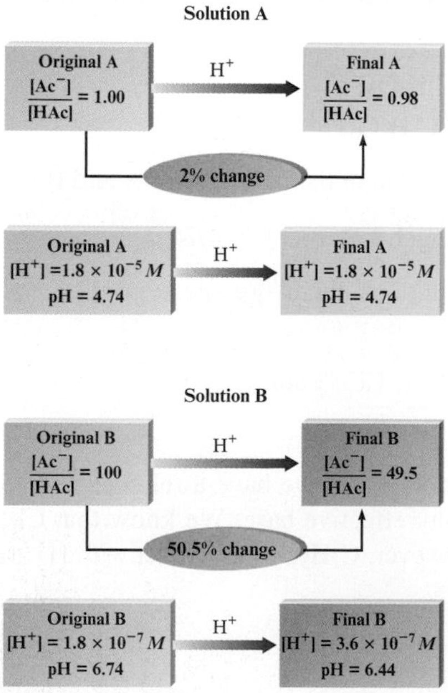

We have seen that the pH of a buffered solution depends on the ratio of the concentrations of buffering components. When this ratio is least affected by added protons or hydroxide ions, the solution is the most resistant to a change in pH. To find the ratio that gives optimal buffering, let's suppose we have a buffered solution containing a large concentration of acetate ion and only a small concentration of acetic acid. Addition of protons to form acetic acid will produce a relatively large *percent* change in the concentration of acetic acid and so will produce a relatively large change in the ratio $[C_2H_3O_2^-]/[HC_2H_3O_2]$ (Table 14-1). Similarly, if hydroxide ions are added to remove some acetic acid, the percent change in the concentration of acetic acid is again large. The same effects are seen if the initial concentration of acetic acid is large and that of acetate ion is small.

Because large changes in the ratio $[A^-]/[HA]$ will produce large changes in pH, we want to avoid this situation for the most effective buffering. This type of reasoning leads us to the general conclusion that optimal buffering occurs when [HA] is equal to $[A^-]$. It is for this condition that the ratio $[A^-]/[HA]$ is most resistant to change when H^+ or OH^- is added to the buffered solution. This means that when choosing the

(margin figures, left column)

Original solution

$$\frac{[A^-]}{[HA]} = \frac{5.00}{5.00} = 1.00$$

H^+ added

New solution

$$\frac{[A^-]}{[HA]} = \frac{4.99}{5.01} = 0.996$$

Original solution

$$\frac{[A^-]}{[HA]} = \frac{0.050}{0.050} = 1.0$$

H^+ added

New solution

$$\frac{[A^-]}{[HA]} = \frac{0.040}{0.060} = 0.67$$

Solution A

Original A
$$\frac{[Ac^-]}{[HAc]} = 1.00$$

H^+

Final A
$$\frac{[Ac^-]}{[HAc]} = 0.98$$

2% change

Original A
$[H^+] = 1.8 \times 10^{-5}\,M$
pH = 4.74

H^+

Final A
$[H^+] = 1.8 \times 10^{-5}\,M$
pH = 4.74

Solution B

Original B
$$\frac{[Ac^-]}{[HAc]} = 100$$

H^+

Final B
$$\frac{[Ac^-]}{[HAc]} = 49.5$$

50.5% change

Original B
$[H^+] = 1.8 \times 10^{-7}\,M$
pH = 6.74

H^+

Final B
$[H^+] = 3.6 \times 10^{-7}\,M$
pH = 6.44

Table 14-1 | Change in $[C_2H_3O_2^-]/[HC_2H_3O_2]$ for Two Solutions When 0.01 Mole of H^+ Is Added to 1.0 L of Each

Solution	$\left(\dfrac{[C_2H_3O_2^-]}{[HC_2H_3O_2]}\right)_{orig}$	$\left(\dfrac{[C_2H_3O_2^-]}{[HC_2H_3O_2]}\right)_{new}$	Change	Percent Change
A	$\dfrac{1.00\ M}{1.00\ M} = 1.00$	$\dfrac{0.99\ M}{1.01\ M} = 0.98$	$1.00 \rightarrow 0.98$	2.00%
B	$\dfrac{1.00\ M}{0.01\ M} = 100$	$\dfrac{0.99\ M}{0.02\ M} = 49.5$	$100 \rightarrow 49.5$	50.5%

buffering components for a specific application, we want $[A^-]/[HA]$ to equal 1. It follows that since

$$pH = pK_a + \log\left(\frac{[A^-]}{[HA]}\right) = pK_a + \log(1) = pK_a$$

the pK_a of the weak acid to be used in the buffer should be as close as possible to the desired pH. For example, suppose we need a buffered solution with a pH of 4.00. The most effective buffering will occur when [HA] is equal to $[A^-]$. From the Henderson–Hasselbalch equation,

$$pH = pK_a + \log\left(\frac{[A^-]}{[HA]}\right)$$

4.00 is wanted

Ratio = 1 for most effective buffer

That is, $4.00 = pK_a + \log(1) = pK_a + 0$ and $pK_a = 4.00$

Thus the best choice of a weak acid is one that has $pK_a = 4.00$ or $K_a = 1.0 \times 10^{-4}$.

Critical Thinking

The text states that "the pK_a for a weak acid to be used in the buffer should be as close as possible to the desired pH." What is the problem with choosing a weak acid whose pK_a is not close to the desired pH when making a buffer?

INTERACTIVE EXAMPLE 14-8

Preparing a Buffer

A chemist needs a solution buffered at pH 4.30 and can choose from the following acids (and their sodium salts):

a. chloroacetic acid ($K_a = 1.35 \times 10^{-3}$)

b. propanoic acid ($K_a = 1.3 \times 10^{-5}$)

c. benzoic acid ($K_a = 6.4 \times 10^{-5}$)

d. hypochlorous acid ($K_a = 3.5 \times 10^{-8}$)

Calculate the ratio $[HA]/[A^-]$ required for each system to yield a pH of 4.30. Which system will work best?

Solution

A pH of 4.30 corresponds to

$$[H^+] = 10^{-4.30} = \text{antilog}(-4.30) = 5.0 \times 10^{-5}\ M$$

Since K_a values rather than pK_a values are given for the various acids, we use Equation (14-1)

$$[H^+] = K_a\frac{[HA]}{[A^-]}$$

rather than the Henderson–Hasselbalch equation. We substitute the required $[H^+]$ and K_a for each acid into Equation (14-1) to calculate the ratio $[HA]/[A^-]$ needed in each case.

Acid	$[H^+] = K_a\dfrac{[HA]}{[A^-]}$	$\dfrac{[HA]}{[A^-]}$
a. Chloroacetic	$5.0 \times 10^{-5} = 1.35 \times 10^{-3}\left(\dfrac{[HA]}{[A^-]}\right)$	3.7×10^{-2}
b. Propanoic	$5.0 \times 10^{-5} = 1.3 \times 10^{-5}\left(\dfrac{[HA]}{[A^-]}\right)$	3.8
c. Benzoic	$5.0 \times 10^{-5} = 6.4 \times 10^{-5}\left(\dfrac{[HA]}{[A^-]}\right)$	0.78
d. Hypochlorous	$5.0 \times 10^{-5} = 3.5 \times 10^{-8}\left(\dfrac{[HA]}{[A^-]}\right)$	1.4×10^{3}

Since $[HA]/[A^-]$ for benzoic acid is closest to 1, the system of benzoic acid and its sodium salt will be the best choice among those given for buffering a solution at pH 4.30. This example demonstrates the principle that the optimal buffering system has a pK_a value close to the desired pH. The pK_a for benzoic acid is 4.19.

See Exercises 14-47 and 14-48

14-4 | Titrations and pH Curves

Science Source

▲ A setup used to conduct the pH titration of an acid or a base.

As we saw in Chapter 6, a titration is commonly used to determine the amount of acid or base in a solution. ◄ This process involves a solution of known concentration (the titrant) delivered from a buret into the unknown solution until the substance being analyzed is just consumed. The stoichiometric (equivalence) point is often signaled by the color change of an indicator. In this section we will discuss the pH changes that occur during an acid–base titration. We will use this information later to show how an appropriate indicator can be chosen for a particular titration.

The progress of an acid–base titration is often monitored by plotting the pH of the solution being analyzed as a function of the amount of titrant added. Such a plot is called a **pH curve** or **titration curve**.

Strong Acid–Strong Base Titrations

The net ionic reaction for a strong acid–strong base titration is

$$H^+(aq) + OH^-(aq) \longrightarrow H_2O(l)$$

To compute $[H^+]$ at a given point in the titration, we must determine the amount of H^+ that remains at that point and divide by the total volume of the solution. Before we proceed, we need to consider a new unit, which is especially convenient for titrations. Since titrations usually involve small quantities (burets are typically graduated in milliliters), the mole is inconveniently large. Therefore, we will use the **millimole** (**mmol**), which, as the prefix indicates, is a thousandth of a mole:

$$1 \text{ mmol} = \frac{1 \text{ mol}}{1000} = 10^{-3} \text{ mol}$$

So far we have defined molarity only in terms of moles per liter. We can now define it in terms of *millimoles per milliliter,* as shown below:

$$\text{Molarity} = \frac{\text{mol solute}}{\text{L solution}} = \frac{\dfrac{\text{mol solute}}{1000}}{\dfrac{\text{L solution}}{1000}} = \frac{\text{mmol solute}}{\text{mL solution}}$$

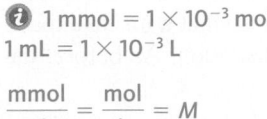

$1 \text{ mmol} = 1 \times 10^{-3} \text{ mol}$
$1 \text{ mL} = 1 \times 10^{-3} \text{ L}$

$$\frac{\text{mmol}}{\text{mL}} = \frac{\text{mol}}{\text{L}} = M$$

A 1.0-*M* solution thus contains 1.0 mole of solute per liter of solution or, *equivalently,* 1.0 mmol of solute per milliliter of solution. ◄ ⓘ Just as we obtain the number of moles of solute from the product of the volume in liters and the molarity, we obtain the number of millimoles of solute from the product of the volume in milliliters and the molarity:

$$\text{Number of mmol} = \text{volume (in mL)} \times \text{molarity}$$

CASE STUDY | *Strong Acid–Strong Base Titration*

We will illustrate the calculations involved in a strong acid–strong base titration by considering the titration of 50.0 mL of 0.200 *M* HNO₃ with 0.100 *M* NaOH. We will calculate the pH of the solution at selected points during the course of the titration, where specific volumes of 0.100 *M* NaOH have been added.

A. No NaOH has been added. ◄

Since HNO_3 is a strong acid (is completely dissociated), the solution contains the major species

$$H^+, \quad NO_3^-, \quad \text{and} \quad H_2O$$

and the pH is determined by the H^+ from the nitric acid. Since 0.200 *M* HNO₃ contains 0.200 *M* H^+,

$$[H^+] = 0.200 \, M \quad \text{and} \quad pH = 0.699$$

B. 10.0 mL of 0.100 *M* NaOH has been added. ◄

In the mixed solution *before any reaction occurs,* the major species are

$$H^+, \quad NO_3^-, \quad Na^+, \quad OH^-, \quad \text{and} \quad H_2O$$

Note that large quantities of both H^+ and OH^- are present. The 1.00 mmol (10.0 mL × 0.100 *M*) of added OH^- will react with 1.00 mmol H^+ to form water:

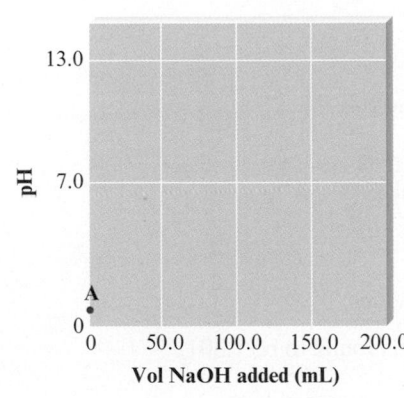

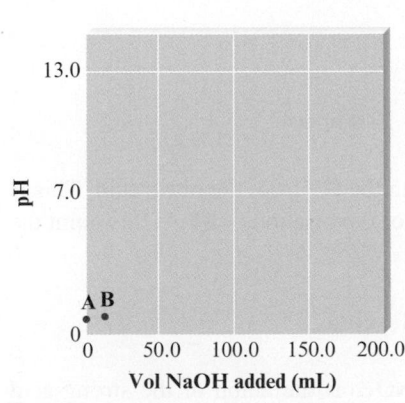

	H^+	+	OH^-	$\longrightarrow$	H_2O
Before reaction	50.0 mL × 0.200 *M* = 10.0 mmol		10.0 mL × 0.100 *M* = 1.00 mmol		
After reaction	10.0 − 1.00 = 9.0 mmol		1.00 − 1.00 = 0		

After the reaction, the solution contains

$$H^+, \quad NO_3^-, \quad Na^+, \quad \text{and} \quad H_2O \text{ (the } OH^- \text{ ions have been consumed)}$$

and the pH will be determined by the H^+ remaining:

The final solution volume is the sum of the original volume of HNO_3 and the volume of added NaOH.

$$[H^+] = \frac{\text{mmol } H^+ \text{ left}}{\text{volume of solution (mL)}} = \frac{9.0 \text{ mmol}}{(50.0 + 10.0) \text{ mL}} = 0.15 \, M$$

Original volume of HNO_3 solution   Volume of NaOH added ◄ ⓘ

$$pH = -\log(0.15) = 0.82$$

C. 20.0 mL (total) of 0.100 *M* NaOH has been added. ◄

We consider this point from the perspective that a total of 20.0 mL NaOH has been added to the *original* solution, rather than that 10.0 mL has been added to the solution from point B. It is best to go back to the original solution each time so that a mistake made at an earlier point does not show up in each succeeding calculation. As before, the added OH^- will react with H^+ to form water:

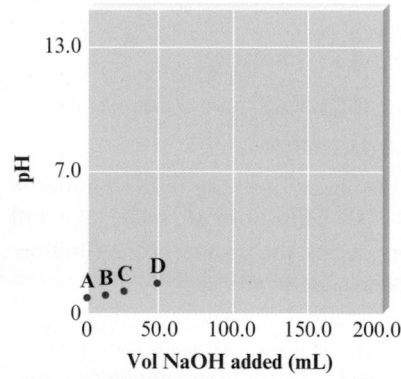

Vol NaOH added (mL)

	H^+	+	OH^-	⟶	H_2O
Before reaction	50.0 mL × 0.200 *M* = 10.0 mmol		20.0 mL × 0.100 *M* = 2.00 mmol		
After reaction	10.0 − 2.00 = 8.00 mmol		2.00 − 2.00 = 0 mmol		

After the reaction

H^+ remaining

$$[H^+] = \frac{8.00 \text{ mmol}}{(50.0 + 20.0) \text{ mL}} = 0.11 \, M$$

$$pH = 0.942$$

D. 50.0 mL (total) of 0.100 *M* NaOH has been added. ◄

Proceeding exactly as for points B and C, the pH is found to be 1.301.

E. 100.0 mL (total) of 0.100 *M* NaOH has been added. ◄

At this point the amount of NaOH that has been added is

$$100.0 \text{ mL} \times 0.100 \, M = 10.0 \text{ mmol}$$

The original amount of nitric acid was

$$50.0 \text{ mL} \times 0.200 \, M = 10.0 \text{ mmol}$$

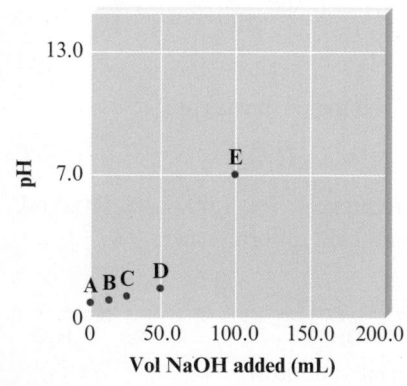

Vol NaOH added (mL)

Enough OH^- has been added to react exactly with the H^+ from the nitric acid. This is the **stoichiometric point**, or **equivalence point**, of the titration. ◄ ⓘ At this point the major species in solution are

$$Na^+, \quad NO_3^-, \quad \text{and} \quad H_2O$$

ⓘ *Equivalence (stoichiometric) point:* The point in the titration where an amount of base has been added to exactly react with all the acid originally present.

Since Na^+ has no acid or base properties and NO_3^- is the anion of the strong acid HNO_3 and is therefore a very weak base, neither NO_3^- nor Na^+ affects the pH, and the solution is neutral (the pH is 7.00).

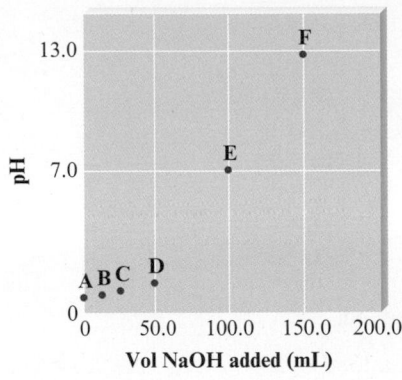

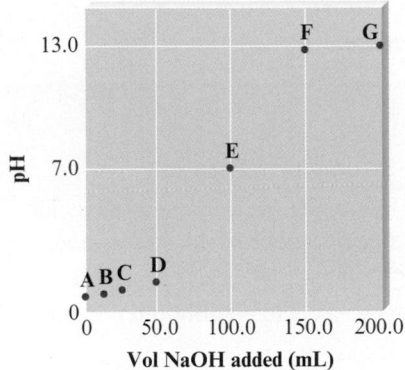

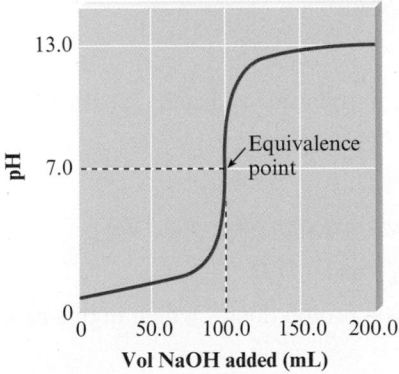

Figure 14-1 | The pH curve for the titration of 50.0 mL of 0.200 M HNO$_3$ with 0.100 M NaOH. Note that the equivalence point occurs at 100.0 mL NaOH added, the point where exactly enough OH$^-$ has been added to react with all the H$^+$ originally present. The pH of 7 at the equivalence point is characteristic of a strong acid–strong base titration.

F. 150.0 mL (total) of 0.100 M NaOH has been added. ◄

The stoichiometric calculations for the titration reaction are as follows:

	H$^+$	+	OH$^-$	$\longrightarrow$	H$_2$O
Before reaction	50.0 mL $\times$ 0.200 M = 10.0 mmol		150.0 mL $\times$ 0.100 M = 15.0 mmol		
After reaction	10.0 $-$ 10.0 = 0 mmol		15.0 $-$ 10.0 = 5.0 mmol ↑ Excess OH$^-$ added		

Now OH$^-$ is *in excess* and will determine the pH:

$$[OH^-] = \frac{\text{mmol OH}^- \text{ in excess}}{\text{volume (mL)}} = \frac{5.0 \text{ mmol}}{(50.0 + 150.0) \text{ mL}} = \frac{5.0 \text{ mmol}}{200.0 \text{ mL}} = 0.025 \, M$$

Since $[H^+][OH^-] = 1.0 \times 10^{-14}$,

$$[H^+] = \frac{1.0 \times 10^{-14}}{2.5 \times 10^{-2}} = 4.0 \times 10^{-13} \, M \qquad \text{and} \qquad pH = 12.40$$

G. 200.0 mL (total) of 0.100 M NaOH has been added. ◄

Proceeding as for point F, the pH is found to be 12.60.

The results of these calculations are summarized by the pH curve shown in Fig. 14-1. Note that the pH changes very gradually until the titration is close to the equivalence point, where a dramatic change occurs. This behavior is due to the fact that early in the titration there is a relatively large amount of H$^+$ in the solution, and the addition of a given amount of OH$^-$ thus produces a small change in pH. However, near the equivalence point [H$^+$] is relatively small, and the addition of a small amount of OH$^-$ produces a large change.

The pH curve in Fig. 14-1, typical of the titration of a strong acid with a strong base, has the following characteristics:

Before the equivalence point, [H$^+$] (and hence the pH) can be calculated by dividing the number of millimoles of H$^+$ remaining by the total volume of the solution in milliliters.

At the equivalence point, the pH is 7.00.

After the equivalence point, [OH$^-$] can be calculated by dividing the number of millimoles of excess OH$^-$ by the total volume of the solution. Then [H$^+$] is obtained from K_w.

The titration of a strong base with a strong acid requires reasoning very similar to that used above, except, of course, that OH$^-$ is in excess before the equivalence point and H$^+$ is in excess after the equivalence point. The pH curve for the titration of 100.0 mL of 0.50 M NaOH with 1.0 M HCl is shown in Fig. 14-2.

Titrations of Weak Acids with Strong Bases

We have seen that since strong acids and strong bases are completely dissociated, the calculations to obtain the pH curves for titrations involving the two are quite straightforward. However, when the acid being titrated is a weak acid, there is a major difference: To calculate [H$^+$] after a certain amount of strong base has been added, we must deal with the weak acid dissociation equilibrium. We have dealt with this same situation earlier in this chapter when we treated buffered solutions. Calculation of the pH curve for a titration of a weak acid with a strong base really amounts to a series of buffer problems. In performing these calculations it is very important to remember that even though the acid is weak, it *reacts essentially to completion* with hydroxide ion, a very strong base.

Figure 14-2 | The pH curve for the titration of 100.0 mL of 0.50 M NaOH with 1.0 M HCl. The equivalence point occurs at 50.00 mL HCl added, since at this point 5.0 mmol H^+ has been added to react with the original 5.0 mmol OH^-.

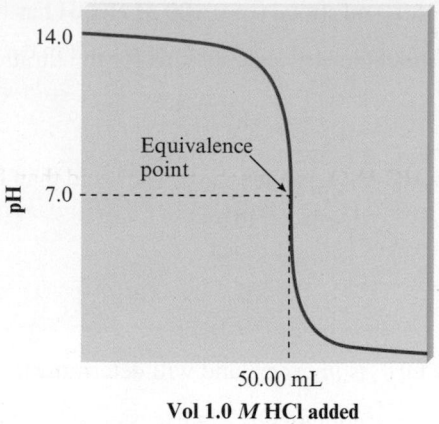

Vol 1.0 M HCl added

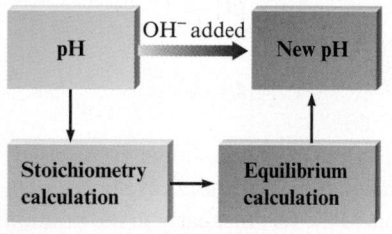

ⓘ Treat the stoichiometry and equilibrium problems separately.

This process always involves a two-step procedure.

PROBLEM-SOLVING STRATEGY

Calculating the pH Curve for a Weak Acid–Strong Base Titration

1. *A stoichiometry problem.* The reaction of hydroxide ion with the weak acid is assumed to run to completion, and the concentrations of the acid *remaining* and the conjugate base *formed* are determined.

2. *An equilibrium problem.* The position of the weak acid equilibrium is determined, and the pH is calculated.

It is *essential* to do these steps *separately.* ◄ ⓘ Note that the procedures necessary to do these calculations have all been used before. ◄

CASE STUDY | *Weak Acid–Strong Base Titration*

As an illustration, we will consider the titration of 50.0 mL of 0.10 M acetic acid ($HC_2H_3O_2$, $K_a = 1.8 \times 10^{-5}$) with 0.10 M NaOH. As before, we will calculate the pH at various points representing volumes of added NaOH.

A. No NaOH has been added.

This is a typical weak acid calculation of the type introduced in Chapter 13. The pH is 2.87. (Check this yourself.)

B. 10.0 mL of 0.10 M NaOH has been added.

The major species in the mixed solution *before any reaction takes place* are

$$HC_2H_3O_2, \quad OH^-, \quad Na^+, \quad and \quad H_2O$$

The strong base OH^- will react with the strongest proton donor, which in this case is $HC_2H_3O_2$.

Stoichiometry Problem

	OH^-	$+$	$HC_2H_3O_2$	$\longrightarrow$	$C_2H_3O_2^-$	$+$	H_2O
Before reaction	10 mL × 0.10 M = 1.0 mmol		50.0 mL × 0.10 M = 5.0 mmol		0 mmol		
After reaction	1.0 − 1.0 = 0 mmol		5.0 − 1.0 = 4.0 mmol		1.0 mmol		
	↑ Limiting reactant				↑ Formed by the reaction		

Equilibrium Problem

We examine the major components left in the solution *after the reaction takes place* to decide on the dominant equilibrium. The major species are

$$HC_2H_3O_2, \quad C_2H_3O_2^-, \quad Na^+, \quad and \quad H_2O$$

Since $HC_2H_3O_2$ is a much stronger acid than H_2O, and since $C_2H_3O_2^-$ is the conjugate base of $HC_2H_3O_2$, the pH will be determined by the position of the acetic acid dissociation equilibrium:

$$HC_2H_3O_2(aq) \rightleftharpoons H^+(aq) + C_2H_3O_2^-(aq)$$

where

$$K_a = \frac{[H^+][C_2H_3O_2^-]}{[HC_2H_3O_2]}$$

We follow the usual steps to complete the equilibrium calculations:

> **The initial concentrations are defined after the reaction with OH⁻ has gone to completion but before any dissociation of $HC_2H_3O_2$ occurs.**

Initial Concentration		Equilibrium Concentration
$[HC_2H_3O_2]_0 = \dfrac{4.0 \text{ mmol}}{(50.0+10.0)\text{ mL}} = \dfrac{4.0}{60.0}$		$[HC_2H_3O_2] = \dfrac{4.0}{60.0} - x$
$[C_2H_3O_2^-]_0 = \dfrac{1.0 \text{ mmol}}{(50.0+10.0)\text{ mL}} = \dfrac{1.0}{60.0}$	$\xrightarrow[\text{dissociates}]{x \text{ mmol/mL} \atop HC_2H_3O_2}$	$[C_2H_3O_2^-] = \dfrac{1.0}{60.0} + x$
$[H^+]_0 \approx 0$		$[H^+] = x$

The appropriate ICE table is

	$HC_2H_3O_2(aq)$	$\rightleftharpoons$	$H^+(aq)$	$+$	$C_2H_3O_2^-(aq)$
Initial	$\dfrac{4.0}{60.0}$		≈ 0		$\dfrac{1.0}{60.0}$
Change	$-x$		$+x$		$+x$
Equilibrium	$\dfrac{4.0}{60.0} - x$		x		$\dfrac{1.0}{60.0} + x$

Therefore,

$$1.8 \times 10^{-5} = K_a = \frac{[H^+][C_2H_3O_2^-]}{[HC_2H_3O_2]} = \frac{x\left(\dfrac{1.0}{60.0} + x\right)}{\dfrac{4.0}{60.0} - x} \approx \frac{x\left(\dfrac{1.0}{60.0}\right)}{\dfrac{4.0}{60.0}} = \left(\dfrac{1.0}{4.0}\right)x$$

> **Note that the approximations made are well within the 5% rule.**

$$x = \left(\dfrac{4.0}{1.0}\right)(1.8 \times 10^{-5}) = 7.2 \times 10^{-5} = [H^+] \quad and \quad pH = 4.14$$

C. 25.0 mL (total) of 0.10 *M* NaOH has been added.

The procedure here is very similar to that used at point B and will only be summarized briefly. The stoichiometry problem is summarized as follows:

Stoichiometry Problem

	OH^-	$+$	$HC_2H_3O_2$	$\longrightarrow$	$C_2H_3O_2^-$	$+$	H_2O
Before reaction	25.0 mL × 0.10 *M* = 2.5 mmol		50.0 mL × 0.10 *M* = 5.0 mmol		0 mmol		
After reaction	2.5 − 2.5 = 0		5.0 − 2.5 = 2.5 mmol		2.5 mmol		

Equilibrium Problem

After the reaction, the major species in solution are

$$HC_2H_3O_2, \quad C_2H_3O_2^-, \quad Na^+, \quad and \quad H_2O$$

The equilibrium that will control the pH is

$$HC_2H_3O_2(aq) \rightleftharpoons H^+(aq) + C_2H_3O_2^-(aq)$$

and the pertinent concentrations are as follows:

Initial Concentration		Equilibrium Concentration
$[HC_2H_3O_2]_0 = \dfrac{2.5 \text{ mmol}}{(50.0 + 25.0) \text{ mL}}$	$\xrightarrow[\text{dissociates}]{x \text{ mmol/mL} \atop HC_2H_3O_2}$	$[HC_2H_3O_2] = \dfrac{2.5}{75.0} - x$
$[C_2H_3O_2^-]_0 = \dfrac{2.5 \text{ mmol}}{(50.0 + 25.0) \text{ mL}}$		$[C_2H_3O_2^-] = \dfrac{2.5}{75.0} + x$
$[H^+]_0 \approx 0$		$[H^+] = x$

The corresponding ICE table is

	$HC_2H_3O_2(aq)$	$\rightleftharpoons$	$H^+(aq)$	$+$	$C_2H_3O_2^-(aq)$
Initial	$\dfrac{2.5}{75.0}$		≈ 0		$\dfrac{2.5}{75.0}$
Change	$-x$		$+x$		$+x$
Equilibrium	$\dfrac{2.5}{75.0} - x$		x		$\dfrac{2.5}{75.0} + x$

Therefore,

$$1.8 \times 10^{-5} = K_a = \frac{[H^+][C_2H_3O_2^-]}{[HC_2H_3O_2]} = \frac{x\left(\dfrac{2.5}{75.0} + x\right)}{\dfrac{2.5}{75.0} - x} \approx \frac{x\left(\dfrac{2.5}{75.0}\right)}{\dfrac{2.5}{75.0}}$$

$$x = 1.8 \times 10^{-5} = [H^+] \quad and \quad pH = 4.74$$

This is a special point in the titration because it is *halfway to the equivalence point.* The original solution, 50.0 mL of 0.10 *M* $HC_2H_3O_2$, contained 5.0 mmol $HC_2H_3O_2$. Thus 5.0 mmol OH^- is required to reach the equivalence point. That is, 50 mL NaOH is required, since

$$(50.0 \text{ mL})(0.10 \text{ } M) = 5.0 \text{ mmol}$$

After 25.0 mL NaOH has been added, half the original $HC_2H_3O_2$ has been converted to $C_2H_3O_2^-$. At this point in the titration $[HC_2H_3O_2]_0$ is equal to $[C_2H_3O_2^-]_0$. ◄ ⓘ We can neglect the effect of dissociation; that is,

$$[HC_2H_3O_2] = [HC_2H_3O_2]_0 - x \approx [HC_2H_3O_2]_0$$
$$[C_2H_3O_2^-] = [C_2H_3O_2^-]_0 + x \approx [C_2H_3O_2^-]_0$$

The expression for K_a at the halfway point is

ⓘ At this point, half the acid has been used up, so

$$[HC_2H_3O_2] = [C_2H_3O_2^-]$$

$$K_a = \frac{[H^+][C_2H_3O_2^-]}{[HC_2H_3O_2]} = \frac{[H^+][C_2H_3O_2^-]_0}{[HC_2H_3O_2]_0} = [H^+]$$

Equal at the
halfway point

Then, *at the halfway point* in the titration,

$$[H^+] = K_a \quad \text{and} \quad pH = pK_a$$

D. 40.0 mL (total) of 0.10 M NaOH has been added.

The procedures required here are the same as those used for points B and C. The pH is 5.35. (Check this yourself.)

E. 50.0 mL (total) of 0.10 M NaOH has been added.

Equilibrium Problem

This is the equivalence point of the titration; 5.0 mmol OH^- has been added, which will just react with the 5.0 mmol $HC_2H_3O_2$ originally present. At this point the solution contains the major species

$$Na^+, \quad C_2H_3O_2^-, \quad \text{and} \quad H_2O$$

Note that the solution contains $C_2H_3O_2^-$, which is a base. Remember that a base wants to combine with a proton, and the only source of protons in this solution is water. Thus the reaction will be

$$C_2H_3O_2^-(aq) + H_2O(l) \rightleftharpoons HC_2H_3O_2(aq) + OH^-(aq)$$

This is a *weak base* reaction characterized by K_b:

$$K_b = \frac{[HC_2H_3O_2][OH^-]}{[C_2H_3O_2^-]} = \frac{K_w}{K_a} = \frac{1.0 \times 10^{-14}}{1.8 \times 10^{-5}} = 5.6 \times 10^{-10}$$

The relevant concentrations are as follows:

Initial Concentration (before any $C_2H_3O_2^-$ reacts with H_2O)		Equilibrium Concentration
$[C_2H_3O_2^-]_0 = \dfrac{5.0 \text{ mmol}}{(50.0 + 50.0) \text{ mL}}$ $= 0.050 \ M$	x mmol/mL $C_2H_3O_2^-$ reacts with H_2O $\longrightarrow$	$[C_2H_3O_2^-] = 0.050 - x$ $[OH^-] = x$ $[HC_2H_3O_2] = x$
$[OH^-]_0 \approx 0$ $[HC_2H_3O_2]_0 = 0$		

The corresponding ICE table is

	$C_2H_3O_2^-(aq)$	$+$	$H_2O(l)$	$\rightleftharpoons$	$HC_2H_3O_2(aq)$	$+$	$OH^-(aq)$
Initial	0.050		—		0		≈ 0
Change	$-x$		—		$+x$		$+x$
Equilibrium	$0.050 - x$		—		x		x

Therefore,

$$5.6 \times 10^{-10} = K_b = \frac{[HC_2H_3O_2][OH^-]}{[C_2H_3O_2^-]} = \frac{(x)(x)}{0.050 - x} \approx \frac{x^2}{0.050}$$

$$x \approx 5.3 \times 10^{-6}$$

The approximation is valid (by the 5% rule), so

$$[OH^-] = 5.3 \times 10^{-6} \ M$$

and

$$[H^+][OH^-] = K_w = 1.0 \times 10^{-14}$$
$$[H^+] = 1.9 \times 10^{-9} \ M$$
$$pH = 8.72$$

This is another important result: *The pH at the equivalence point of a titration of a weak acid with a strong base is always greater than 7.* This is so because the anion of the acid, which remains in solution at the equivalence point, is a base. In contrast, for the titration of a strong acid with a strong base, the pH at the equivalence point is 7.0, because the anion remaining in this case is *not* an effective base. ◀ ⓘ

ⓘ The pH at the equivalence point of a titration of a weak acid with a strong base is always greater than 7.

F. 60.0 mL (total) of 0.10 *M* NaOH has been added.

At this point, excess OH^- has been added. The stoichiometric calculations are as follows:

Stoichiometry Problem

	OH^-	+	$HC_2H_3O_2$	$\longrightarrow$	$C_2H_3O_2^-$	+	H_2O
Before reaction	60.0 mL × 0.10 *M* = 6.0 mmol		50.0 mL × 0.10 *M* = 5.0 mmol		0 mmol		
After reaction	6.0 − 5.0 = 1.0 mmol in excess		5.0 − 5.0 = 0		5.0 mmol		

Equilibrium Problem

After the reaction is complete, the solution contains the major species

$$Na^+, \quad C_2H_3O_2^-, \quad OH^-, \quad and \quad H_2O$$

There are two bases in this solution, OH^- and $C_2H_3O_2^-$. However, $C_2H_3O_2^-$ is a weak base compared with OH^-. Therefore, the amount of OH^- produced by reaction of $C_2H_3O_2^-$ with H_2O will be small compared with the excess OH^- already in solution. You can verify this conclusion by looking at point E, where only 5.3×10^{-6} *M* OH^- was produced by $C_2H_3O_2^-$. The amount in this case will be even smaller, since the excess OH^- will push the K_b equilibrium to the left.

Thus the pH is determined by the excess OH^-:

$$[OH^-] = \frac{\text{mmol of } OH^- \text{ in excess}}{\text{volume (in mL)}} = \frac{1.0 \text{ mmol}}{(50.0 + 60.0) \text{ mL}}$$

$$= 9.1 \times 10^{-3} \, M$$

and

$$[H^+] = \frac{1.0 \times 10^{-14}}{9.1 \times 10^{-3}} = 1.1 \times 10^{-12} \, M$$

$$pH = 11.96$$

G. 75.0 mL (total) of 0.10 *M* NaOH has been added.

The procedure needed here is very similar to that for point F. The pH is 12.30. (Check this yourself.)

The pH curve for this titration is shown in Fig. 14-3. It is important to note the differences between this curve and that in Fig. 14-1. For example, the shapes of the plots are quite different before the equivalence point, although they are very similar after that point. ◀ (The shapes of the strong and weak acid curves are the same after the equivalence points because excess OH^- controls the pH in this region in both cases.) Near the beginning of the titration of the weak acid, the pH increases more rapidly than it does in the strong acid case. It levels off near the halfway point and then increases rapidly again. The leveling off near the halfway point is caused by buffering effects. Earlier in this chapter we saw that optimal buffering occurs when [HA] is equal to [A⁻]. This is exactly the case at the halfway point of the titration. As we can see from the curve, the pH changes least rapidly in this region of the titration.

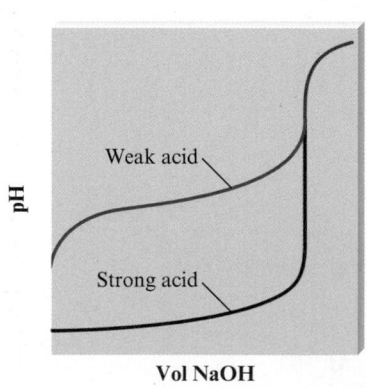

Weak acid

pH

Strong acid

Vol NaOH

Figure 14-3 | The pH curve for the titration of 50.0 mL of 0.100 M $HC_2H_3O_2$ with 0.100 M NaOH. Note that the equivalence point occurs at 50.0 mL NaOH added, where the amount of added OH^- exactly equals the original amount of acid. The pH at the equivalence point is greater than 7.0 because the $C_2H_3O_2^-$ ion present at this point is a base and reacts with water to produce OH^-.

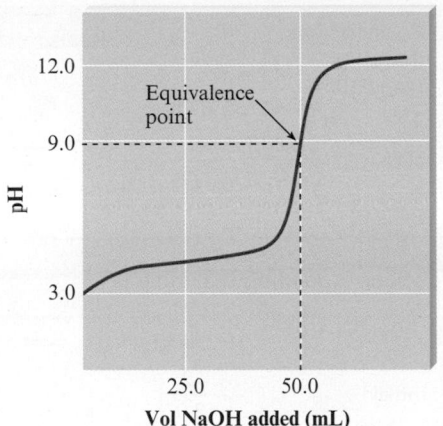

The other notable difference between the curves for strong and weak acids is the value of the pH at the equivalence point. For the titration of a strong acid, the equivalence point occurs at pH 7. For the titration of a weak acid, the pH at the equivalence point is greater than 7 because of the basicity of the conjugate base of the weak acid.

It is important to understand that the equivalence point in an acid–base titration is *defined by the stoichiometry, not by the pH*. The equivalence point occurs when enough titrant has been added to react exactly with all the acid or base being titrated. ◄ ⓘ

ⓘ The equivalence point is defined by the stoichiometry, not by the pH.

INTERACTIVE EXAMPLE 14-9

Titration of a Weak Acid

Hydrogen cyanide gas (HCN), a powerful respiratory inhibitor, is highly toxic. It is a very weak acid ($K_a = 6.2 \times 10^{-10}$) when dissolved in water. If a 50.0-mL sample of 0.100 M HCN is titrated with 0.100 M NaOH, calculate the pH of the solution

a. after 8.00 mL of 0.100 M NaOH has been added.

b. at the halfway point of the titration.

c. at the equivalence point of the titration.

Solution

Stoichiometry Problem

a. After 8.00 mL of 0.100 M NaOH has been added, the following calculations apply:

	HCN	+	OH^-	$\longrightarrow$	CN^-	+	H_2O
Before reaction	50.0 mL × 0.100 M = 5.00 mmol		8.00 mL × 0.100 M = 0.800 mmol		0 mmol		
After reaction	5.00 − 0.800 = 4.20 mmol		0.800 − 0.800 = 0		0.800 mmol		

Equilibrium Problem

Since the solution contains the major species

$$HCN, \quad CN^-, \quad Na^+, \quad and \quad H_2O$$

the position of the acid dissociation equilibrium

$$HCN(aq) \rightleftharpoons H^+(aq) + CN^-(aq)$$

will determine the pH.

	Initial Concentration		Equilibrium Concentration
$[HCN]_0 = \dfrac{4.2 \text{ mmol}}{(50.0 + 8.0) \text{ mL}}$			$[HCN] = \dfrac{4.2}{58.0} - x$
$[CN^-]_0 = \dfrac{0.800 \text{ mmol}}{(50.0 + 8.0) \text{ mL}}$	$\xrightarrow[\text{dissociates}]{\substack{x \text{ mmol/mL} \\ HCN}}$		$[CN^-] = \dfrac{0.80}{58.0} + x$
$[H^+]_0 \approx 0$			$[H^+] = x$

The corresponding ICE table is

	HCN(aq)	$\rightleftharpoons$	H⁺(aq)	+	CN⁻(aq)
Initial	$\dfrac{4.2}{58.0}$		≈ 0		$\dfrac{0.80}{58.0}$
Change	$-x$		$+x$		$+x$
Equilibrium	$\dfrac{4.2}{58.0} - x$		x		$\dfrac{0.80}{58.0} + x$

Substituting the equilibrium concentrations into the expression for K_a gives

The approximations made here are well within the 5% rule.

$$6.2 \times 10^{-10} = K_a = \frac{[H^+][CN^-]}{[HCN]} = \frac{x\left(\dfrac{0.80}{58.0} + x\right)}{\dfrac{4.2}{58.0} - x} \approx \frac{x\left(\dfrac{0.80}{58.0}\right)}{\left(\dfrac{4.2}{58.0}\right)} = x\left(\frac{0.80}{4.2}\right)$$

$$x = 3.3 \times 10^{-9}\, M = [H^+] \quad \text{and} \quad \text{pH} = 8.49$$

b. *At the halfway point of the titration.* The amount of HCN originally present can be obtained from the original volume and molarity:

$$50.0 \text{ mL} \times 0.100\, M = 5.00 \text{ mmol}$$

Thus the halfway point will occur when 2.50 mmol OH⁻ has been added:

$$\text{Volume of NaOH (in mL)} \times 0.100\, M = 2.50 \text{ mmol OH}^-$$

or $\quad\quad\quad\quad$ Volume of NaOH $= 25.0$ mL

As was pointed out previously, at the halfway point [HCN] is equal to [CN⁻] and pH is equal to pK_a. Thus, after 25.0 mL of 0.100 M NaOH has been added,

$$\text{pH} = pK_a = -\log(6.2 \times 10^{-10}) = 9.21$$

c. *At the equivalence point.* The equivalence point will occur when a total of 5.00 mmol OH⁻ has been added. Since the NaOH solution is 0.100 M, the equivalence point occurs when 50.0 mL NaOH has been added. This amount will form 5.00 mmol CN⁻.

Equilibrium Problem

The major species in solution at the equivalence point are

$$CN^-, \quad Na^+, \quad \text{and} \quad H_2O$$

Thus the reaction that will control the pH involves the basic cyanide ion extracting a proton from water:

$$CN^-(aq) + H_2O(l) \rightleftharpoons HCN(aq) + OH^-(aq)$$

and $\quad\quad K_b = \dfrac{K_w}{K_a} = \dfrac{1.0 \times 10^{-14}}{6.2 \times 10^{-10}} = 1.6 \times 10^{-5} = \dfrac{[HCN][OH^-]}{[CN^-]}$

Initial Concentration		Equilibrium Concentration
$[CN^-]_0 = \dfrac{5.00 \text{ mmol}}{(50.0 + 50.0) \text{ mL}}$ $= 5.00 \times 10^{-2} M$	$\xrightarrow[\text{with } H_2O]{x \text{ mmol/mL of } CN^- \text{ reacts}}$	$[CN^-] = (5.00 \times 10^{-2}) - x$
$[HCN]_0 = 0$		$[HCN] = x$
$[OH^-]_0 \approx 0$		$[OH^-] = x$

The corresponding ICE table is

	$CN^-(aq)$	$+$	$H_2O(l)$	$\rightleftharpoons$	$HCN(aq)$	$+$	$OH^-(aq)$
Initial	0.050		—		0		0
Change	$-x$		—		$+x$		$+x$
Equilibrium	$0.050 - x$		—		x		x

Substituting the equilibrium concentrations into the expression for K_b and solving in the usual way gives

$$[OH^-] = x = 8.9 \times 10^{-4}$$

Then, from K_w, we have

$$[H^+] = 1.1 \times 10^{-11} M \quad \text{and} \quad pH = 10.96$$

See Exercises 14-59, 14-61, and 14-62

ⓘ The amount of acid present, not its strength, determines the equivalence point.

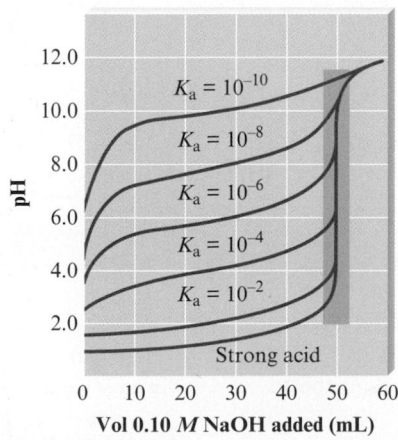

Figure 14-4 | The pH curves for the titrations of 50.0-mL samples of 0.10 M acids with various K_a values with 0.10 M NaOH.

Two important conclusions can be drawn from a comparison of the titration of 50.0 mL of 0.1 M acetic acid covered earlier in this section and that of 50.0 mL of 0.1 M hydrocyanic acid analyzed in Example 14-9. First, the same amount of 0.1 M NaOH is required to reach the equivalence point in both cases. The fact that HCN is a much weaker acid than $HC_2H_3O_2$ has no bearing on the amount of base required. It is the *amount* of acid, not its strength, that determines the equivalence point. ◂ ⓘ Second, the pH value at the equivalence point *is* affected by the acid strength. For the titration of acetic acid, the pH at the equivalence point is 8.72; for the titration of hydrocyanic acid, the pH at the equivalence point is 10.96. This difference occurs because the CN^- ion is a much stronger base than the $C_2H_3O_2^-$ ion. Also, the pH at the halfway point of the titration is much higher for HCN than for $HC_2H_3O_2$, again because of the greater base strength of the CN^- ion (or, equivalently, the smaller acid strength of HCN).

The strength of a weak acid has a significant effect on the shape of its pH curve. Figure 14-4 shows pH curves for 50-mL samples of 0.10 M solutions of various acids titrated with 0.10 M NaOH. Note that the equivalence point occurs in each case when the same volume of 0.10 M NaOH has been added but that the shapes of the curves are dramatically different. The weaker the acid, the greater the pH value at the equivalence point. In particular, note that the vertical region that surrounds the equivalence point becomes shorter as the acid being titrated becomes weaker. We will see in the next section that the choice of an indicator is more limited for such a titration.

Besides being used to analyze for the amount of acid or base in a solution, titrations can be used to determine the values of equilibrium constants, as shown in Example 14-10.

Calculation of K_a

Calculating K_a

A chemist has synthesized a monoprotic weak acid and wants to determine its K_a value. To do so, the chemist dissolves 2.00 mmol of the solid acid in 100.0 mL water and titrates the resulting solution with 0.0500 M NaOH. After 20.0 mL NaOH has been added, the pH is 6.00. What is the K_a value for the acid?

Solution

Stoichiometry Problem

We represent the monoprotic acid as HA. The stoichiometry for the titration reaction is shown below.

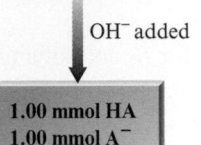

	HA	+	OH$^-$	$\longrightarrow$	A$^-$	+	H$_2$O
Before reaction	2.00 mmol		20.0 mL × 0.0500 M = 1.00 mmol		0 mmol		
After reaction	2.00 − 1.00 = 1.00 mmol		1.00 − 1.00 = 0		1.00 mmol		

Equilibrium Problem ◄

After the reaction the solution contains the major species

$$HA, \quad A^-, \quad Na^+, \quad and \quad H_2O$$

The pH will be determined by the equilibrium

$$HA(aq) \rightleftharpoons H^+(aq) + A^-(aq)$$

for which

$$K_a = \frac{[H^+][A^-]}{[HA]}$$

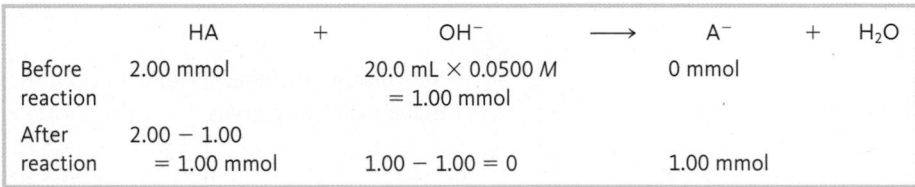

Initial Concentration	Equilibrium Concentration
$[HA]_0 = \dfrac{1.00 \text{ mmol}}{(100.0 + 20.0) \text{ mL}}$ $= 8.33 \times 10^{-3} \, M$	$[HA] = 8.33 \times 10^{-3} - x$
$[A^-]_0 = \dfrac{1.00 \text{ mmol}}{(100.0 + 20.0) \text{ mL}}$ $= 8.33 \times 10^{-3} \, M$	$[A^-] = 8.33 \times 10^{-3} + x$
$[H^+]_0 \approx 0$	$[H^+] = x$

x mmol/mL HA dissociates

The corresponding ICE table is

	HA(aq)	$\rightleftharpoons$	H$^+$(aq)	+	A$^-$(aq)
Initial	8.33×10^{-3}		≈ 0		8.33×10^{-3}
Change	$-x$		$+x$		$+x$
Equilibrium	$8.33 \times 10^{-3} - x$		x		$8.33 \times 10^{-3} + x$

Note that x is known here because the pH at this point is known to be 6.00. Thus

$$x = [H^+] = \text{antilog}(-pH) = 1.0 \times 10^{-6} \, M$$

Substituting the equilibrium concentrations into the expression for K_a allows calculation of the K_a value:

$$K_a = \frac{[H^+][A^-]}{[HA]} = \frac{x(8.33 \times 10^{-3} + x)}{(8.33 \times 10^{-3}) - x}$$

$$= \frac{(1.0 \times 10^{-6})(8.33 \times 10^{-3} + 1.0 \times 10^{-6})}{(8.33 \times 10^{-3}) - (1.0 \times 10^{-6})}$$

$$\approx \frac{(1.0 \times 10^{-6})(8.33 \times 10^{-3})}{8.33 \times 10^{-3}} = 1.0 \times 10^{-6}$$

There is an easier way to think about this problem. The original solution contained 2.00 mmol of HA, and since 20.0 mL of added 0.0500 M NaOH contains 1.0 mmol OH^-, this is the halfway point in the titration (where [HA] is equal to [A^-]). Thus

$$[H^+] = K_a = 1.0 \times 10^{-6}$$

See Exercise 14-67

Titrations of Weak Bases with Strong Acids

Titrations of weak bases with strong acids can be treated using the procedures we introduced previously. As always, you should *think first about the major species in solution* and decide whether a reaction occurs that runs essentially to completion. ◄ ⓘ If such a reaction does occur, let it run to completion and do the stoichiometric calculations. Finally, choose the dominant equilibrium and calculate the pH.

ⓘ Recall from Chapter 13 that the best way to start acid–base problems is to consider the major species present and decide if each acts as an acid, base, both, or neither.

CASE STUDY | *Weak Base–Strong Acid Titration*

The calculations involved for the titration of a weak base with a strong acid will be illustrated by the titration of 100.0 mL of 0.050 M NH_3 with 0.10 M HCl.

A. Before the addition of any HCl.

  1. Major species:

$$NH_3 \quad \text{and} \quad H_2O$$

    NH_3 is a base and will seek a source of protons. In this case H_2O is the only available source.

  2. No reactions occur that go to completion, since NH_3 cannot readily take a proton from H_2O. This is evidenced by the small K_b value for NH_3.

  3. The equilibrium that controls the pH involves the reaction of ammonia with water:

$$NH_3(aq) + H_2O(l) \rightleftharpoons NH_4^+(aq) + OH^-(aq)$$

    Use K_b to calculate [OH^-]. Although NH_3 is a weak base (compared with OH^-), it produces much more OH^- in this reaction than is produced from the autoionization of H_2O.

B. Before the equivalence point.

  1. Major species (before any reaction occurs):

$$NH_3, \quad \underbrace{H^+, \quad Cl^-}_{\substack{\text{From added} \\ \text{HCl}}}, \quad \text{and} \quad H_2O$$

2. The NH_3 will react with H^+ from the added HCl:

$$NH_3(aq) + H^+(aq) \rightleftharpoons NH_4{}^+(aq)$$

This reaction proceeds essentially to completion because the NH_3 readily reacts with a free proton. This case is much different from the previous case, where H_2O was the only source of protons. The stoichiometric calculations are then carried out using the known volume of 0.10 M HCl added.

3. After the reaction of NH_3 with H^+ is run to completion, the solution contains the following major species:

$$NH_3, \quad \underset{\substack{\uparrow \\ \text{Formed in} \\ \text{titration reaction}}}{NH_4{}^+}, \quad Cl^-, \quad \text{and} \quad H_2O$$

Note that the solution contains NH_3 and $NH_4{}^+$, and the equilibria involving these species will determine $[H^+]$. You can use either the dissociation reaction of $NH_4{}^+$

$$NH_4{}^+(aq) \rightleftharpoons NH_3(aq) + H^+(aq)$$

or the reaction of NH_3 with H_2O

$$NH_3(aq) + H_2O(l) \rightleftharpoons NH_4{}^+(aq) + OH^-(aq)$$

C. At the equivalence point.

1. By definition, the equivalence point occurs when all the original NH_3 is converted to $NH_4{}^+$. Thus the major species in solution are

$$NH_4{}^+, \quad Cl^-, \quad \text{and} \quad H_2O$$

2. No reactions occur that go to completion.

3. The dominant equilibrium (the one that controls the $[H^+]$) will be the dissociation of the weak acid $NH_4{}^+$, for which

$$K_a = \frac{K_w}{K_b(\text{for } NH_3)}$$

D. Beyond the equivalence point.

1. Excess HCl has been added, and the major species are

$$H^+, \quad NH_4{}^+, \quad Cl^-, \quad \text{and} \quad H_2O$$

2. No reaction occurs that goes to completion.

3. Although $NH_4{}^+$ will dissociate, it is such a weak acid that $[H^+]$ will be determined simply by the excess H^+:

$$[H^+] = \frac{\text{mmol } H^+ \text{ in excess}}{\text{mL solution}}$$

The results of these calculations are shown in Table 14-2. The pH curve is shown in Fig. 14-5.

Table 14-2 | Summary of Results for the Titration of 100.0 mL 0.050 M NH$_3$ with 0.10 M HCl

Volume of 0.10 M HCl Added (mL)	[NH$_3$]$_0$	[NH$_4^+$]$_0$	[H$^+$]	pH
0	0.05 M	0	1.1×10^{-11} M	10.96
10.0	$\dfrac{4.0 \text{ mmol}}{(100 + 10) \text{ mL}}$	$\dfrac{1.0 \text{ mmol}}{(100 + 10) \text{ mL}}$	1.4×10^{-10} M	9.85
25.0*	$\dfrac{2.5 \text{ mmol}}{(100 + 25) \text{ mL}}$	$\dfrac{2.5 \text{ mmol}}{(100 + 25) \text{ mL}}$	5.6×10^{-10} M	9.25
50.0†	0	$\dfrac{5.0 \text{ mmol}}{(100 + 50) \text{ mL}}$	4.3×10^{-6} M	5.36
60.0‡	0	$\dfrac{5.0 \text{ mmol}}{(100 + 60) \text{ mL}}$	$\dfrac{1.0 \text{ mmol}}{160 \text{ mL}}$ $= 6.2 \times 10^{-3}$ M	2.21

*Halfway point   †Equivalence point   ‡[H$^+$] determined by the 1.0 mmol of excess H$^+$

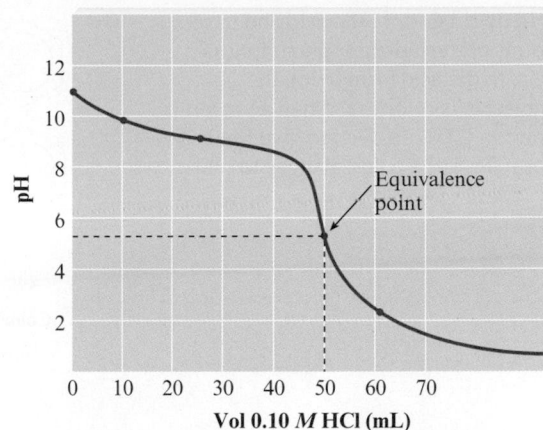

Figure 14-5 | The pH curve for the titration of 100.0 mL of 0.050 M NH$_3$ with 0.10 M HCl. Note the pH at the equivalence point is less than 7, since the solution contains the weak acid NH$_4^+$.

Critical Thinking

You have read about titrations of strong acids with strong bases, weak acids with strong bases, and weak bases with strong acids. What if you titrated a weak acid with a weak base? Sketch a pH curve and defend its shape. Label the equivalence point and discuss the possibilities for the pH value at the equivalence point.

14-5 | Acid–Base Indicators

There are two common methods for determining the equivalence point of an acid–base titration:

1. Use a pH meter (see Fig. 13-7) to monitor the pH and then plot the titration curve. The center of the vertical region of the pH curve indicates the equivalence point (for example, see Figs. 14-1 through 14-5).

2. Use an **acid–base indicator**, which marks the end point of a titration by changing color. Although the *equivalence point of a titration, defined by the stoichiometry, is not necessarily the same as the end point* (where the indicator changes color), careful selection of the indicator will ensure that the error is negligible.

The most common acid–base indicators are complex molecules that are themselves weak acids (represented by HIn). They exhibit one color when the proton is attached to the molecule and a different color when the proton is absent. For example, **phenolphthalein**, a commonly used indicator, is colorless in its HIn form and pink in its In$^-$, or basic, form. ◀ The actual structures of the two forms of phenolphthalein are shown in Fig. 14-6.

To see how molecules such as phenolphthalein function as indicators, consider the following equilibrium for some hypothetical indicator HIn, a weak acid with $K_a = 1.0 \times 10^{-8}$.

$$\text{HIn}(aq) \rightleftharpoons \text{H}^+(aq) + \text{In}^-(aq)$$
$$\text{Red} \qquad\qquad\qquad \text{Blue}$$

$$K_a = \frac{[\text{H}^+][\text{In}^-]}{[\text{HIn}]}$$

▲ The indicator phenolphthalein is colorless in acidic solution and pink in basic solution.

© Cengage Learning

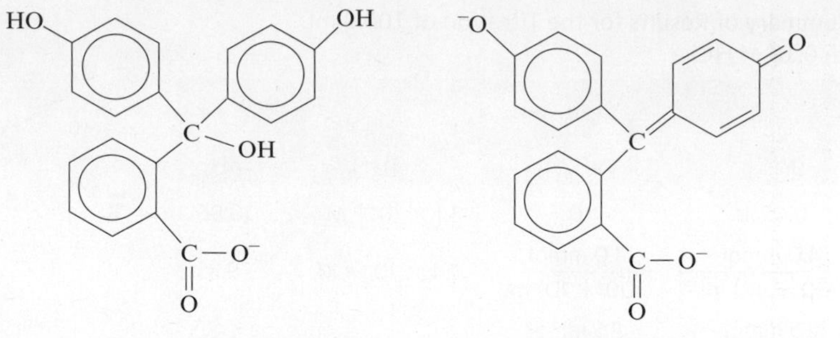

Figure 14-6 | The acid and base forms of the indicator phenolphthalein. In the acid form (HIn), the molecule is colorless. When a proton (plus H_2O) is removed to give the base form (In^-), the color changes to pink.

Colorless acid form, HIn   Pink base form, In⁻

By rearranging, we get

$$\frac{K_a}{[H^+]} = \frac{[In^-]}{[HIn]}$$

Suppose we add a few drops of this indicator to an acidic solution whose pH is 1.0 ($[H^+] = 1.0 \times 10^{-1}$). Then

$$\frac{K_a}{[H^+]} = \frac{1.0 \times 10^{-8}}{1.0 \times 10^{-1}} = 10^{-7} = \frac{1}{10{,}000{,}000} = \frac{[In^-]}{[HIn]}$$

This ratio shows that the predominant form of the indicator is HIn, resulting in a red solution. As OH^- is added to this solution in a titration, $[H^+]$ decreases and the equilibrium shifts to the right, changing HIn to In^-. At some point in a titration, enough of the In^- form will be present in the solution so that a purple tint will be noticeable. That is, a color change from red to reddish purple will occur. ◄ ⓘ

How much In^- must be present for the human eye to detect that the color is different from the original one? For most indicators, about a tenth of the initial form must be converted to the other form before a new color is apparent. We will assume, then, that in the titration of an acid with a base, the color change will occur at a pH where

$$\frac{[In^-]}{[HIn]} = \frac{1}{10}$$

ⓘ The *end point* is defined by the change in color of the indicator. The *equivalence point* is defined by reaction stoichiometry.

INTERACTIVE EXAMPLE 14-11

Indicator Color Change

Bromthymol blue, an indicator with a K_a value of 1.0×10^{-7}, is yellow in its HIn form and blue in its In^- form. Suppose we put a few drops of this indicator in a strongly acidic solution. If the solution is then titrated with NaOH, at what pH will the indicator color change first be visible?

Solution

For bromthymol blue,

$$K_a = 1.0 \times 10^{-7} = \frac{[H^+][In^-]}{[HIn]}$$

We assume that the color change is visible when

$$\frac{[In^-]}{[HIn]} = \frac{1}{10}$$

Figure 14-7 | (a) Yellow acid form of bromthymol blue; (b) a greenish tint is seen when the solution contains 1 part blue and 10 parts yellow; (c) blue basic form.

That is, we assume that we can see the first hint of a greenish tint (yellow plus a little blue) when the solution contains 1 part blue and 10 parts yellow (Fig. 14-7). Thus

$$K_a = 1.0 \times 10^{-7} = \frac{[H^+](1)}{10}$$

$$[H^+] = 1.0 \times 10^{-6} \quad \text{or} \quad pH = 6.00$$

The color change is first visible at pH 6.00.

See Exercises 14-69 through 14-72

The Henderson–Hasselbalch equation is very useful in determining the pH at which an indicator changes color. For example, application of Equation (14-2) to the K_a expression for the general indicator HIn yields

$$pH = pK_a + \log\left(\frac{[In^-]}{[HIn]}\right)$$

where K_a is the dissociation constant for the acid form of the indicator (HIn). Since we assume that the color change is visible when

$$\frac{[In^-]}{[HIn]} = \frac{1}{10}$$

we have the following equation for determining the pH at which the color change occurs:

$$pH = pK_a + \log(\tfrac{1}{10}) = pK_a - 1$$

For bromthymol blue ($K_a = 1 \times 10^{-7}$, or $pK_a = 7$), the pH at the color change is

$$pH = 7 - 1 = 6$$

as we calculated in Example 14-11.

When a basic solution is titrated, the indicator HIn will initially exist as In^- in solution, but as acid is added, more HIn will be formed. In this case the color change will be visible when there is a mixture of 10 parts In^- and 1 part HIn. That is, a color change from blue to blue-green will occur (see Fig. 14-7) due to the presence of some of the yellow HIn molecules. This color change will be first visible when

$$\frac{[In^-]}{[HIn]} = \frac{10}{1}$$

Note that this is the reciprocal of the ratio for the titration of an acid. Substituting this ratio into the Henderson–Hasselbalch equation gives

$$pH = pK_a + \log\left(\tfrac{10}{1}\right) = pK_a + 1$$

For bromthymol blue ($pK_a = 7$), we have a color change at

$$pH = 7 + 1 = 8$$

In summary, when bromthymol blue is used for the titration of an acid, the starting form will be HIn (yellow), and the color change occurs at a pH of about 6. When bromthymol blue is used for the titration of a base, the starting form is In^- (blue), and the color change occurs at a pH of about 8. Thus the useful pH range for bromthymol blue is

$$pK_a(\text{bromthymol blue}) \pm 1 = 7 \pm 1$$

or from 6 to 8. This is a general result. For a typical acid–base indicator with dissociation constant K_a, the color transition occurs over a range of pH values given by $pK_a \pm 1$. The useful pH ranges for several common indicators are shown in Fig. 14-8.

When we choose an indicator for a titration, we want the indicator end point (where the color changes) and the titration equivalence point to be as close as possible. Choosing an indicator is easier if there is a large change in pH near the equivalence point of the titration. The dramatic change in pH near the equivalence point in a strong acid–strong base titration (see Figs. 14-1 and 14-2) produces a sharp end point; that is, the complete color change (from the acid-to-base or base-to-acid colors) usually occurs over one drop of added titrant.

What indicator should we use for the titration of 100.00 mL of 0.100 M HCl with 0.100 M NaOH? We know that the equivalence point occurs at pH 7.00. In the initially acidic solution, the indicator will be predominantly in the HIn form. As OH^- ions are added, the pH increases rather slowly at first (see Fig. 14-1) and then rises rapidly at the equivalence point. This sharp change causes the indicator dissociation equilibrium

$$HIn \rightleftharpoons H^+ + In^-$$

to shift suddenly to the right, producing enough In^- ions to give a color change. Since we are titrating an acid, the indicator is predominantly in the acid form initially. Therefore, the first observable color change will occur at a pH where

$$\frac{[In^-]}{[HIn]} = \frac{1}{10}$$

Thus

$$pH = pK_a + \log\left(\tfrac{1}{10}\right) = pK_a - 1$$

If we want an indicator that changes color at pH 7, we can use this relationship to find the pK_a value for a suitable indicator:

$$pH = 7 = pK_a - 1 \quad \text{or} \quad pK_a = 7 + 1 = 8$$

Thus an indicator with a pK_a value of 8 ($K_a = 1 \times 10^{-8}$) changes color at about pH 7 and is ideal for marking the end point for a strong acid–strong base titration.

How crucial is it for a strong acid–strong base titration that the indicator change color exactly at pH 7? We can answer this question by examining the pH change near the equivalence point of the titration of 100 mL of 0.10 M HCl and 0.10 M NaOH. The data for a few points at or near the equivalence point are shown in Table 14-3.

Table 14-3 | Selected pH Values Near the Equivalence Point in the Titration of 100.0 mL of 0.10 M HCl with 0.10 M NaOH

NaOH Added (mL)	pH
99.99	5.3
100.00	7.0
100.01	8.7

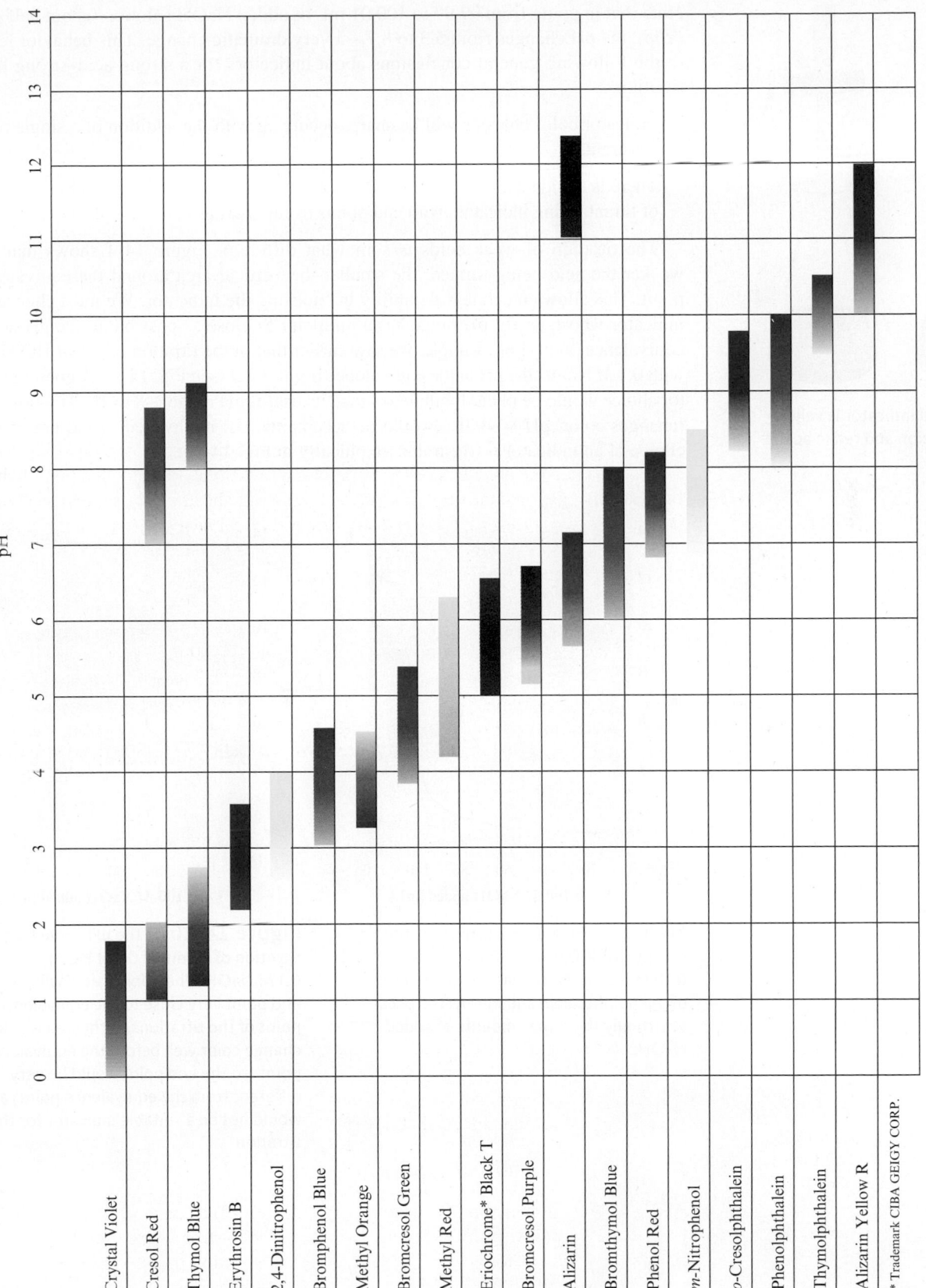

Figure 14-8 | The useful pH ranges for several common indicators. Note that most indicators have a useful range of about two pH units, as predicted by the expression $pK_a \pm 1$.

The pH ranges shown are approximate. Specific transition ranges depend on the indicator solvent chosen.

pH

14
13
12
11
10
9
8
7
6
5
4
3
2
1
0

Crystal Violet
Cresol Red
Thymol Blue
Erythrosin B
2,4-Dinitrophenol
Bromphenol Blue
Methyl Orange
Bromcresol Green
Methyl Red
Eriochrome* Black T
Bromcresol Purple
Alizarin
Bromthymol Blue
Phenol Red
m-Nitrophenol
o-Cresolphthalein
Phenolphthalein
Thymolphthalein
Alizarin Yellow R

* Trademark CIBA GEIGY CORP.

▲ Methyl red indicator is yellow in basic solution and red in acidic solution.

Note that in going from 99.99 to 100.01 mL of added NaOH solution (about half of a drop), the pH changes from 5.3 to 8.7—a very dramatic change. This behavior leads to the following general conclusions about indicators for a strong acid–strong base titration:

Indicator color changes will be sharp, occurring with the addition of a single drop of titrant.

There is a wide choice of suitable indicators. The results will agree within one drop of titrant, using indicators with end points as far apart as pH 5 and pH 9 (Fig. 14-9).

The titration of weak acids is somewhat different. Figure 14-4 shows that the weaker the acid being titrated, the smaller the vertical area around the equivalence point. This allows much less flexibility in choosing the indicator. We must choose an indicator whose useful pH range has a midpoint as close as possible to the pH at the equivalence point. For example, we saw earlier that in the titration of 0.1 M $HC_2H_3O_2$ with 0.1 M NaOH the pH at the equivalence point is 8.7 (see Fig. 14-3). A good indicator choice would be phenolphthalein, since its useful pH range is 8 to 10. Thymol blue (changes color, pH 8–9) also would be acceptable, but methyl red would not. ◄ The choice of an indicator is illustrated graphically in Fig. 14-10.

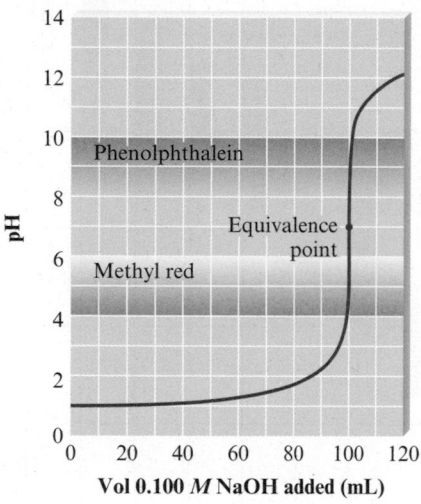

Vol 0.100 M NaOH added (mL)

Figure 14-9 │ The pH curve for the titration of 100.0 mL of 0.10 M HCl with 0.10 M NaOH. Note that the end points of phenolphthalein and methyl red occur at virtually the same amounts of added NaOH.

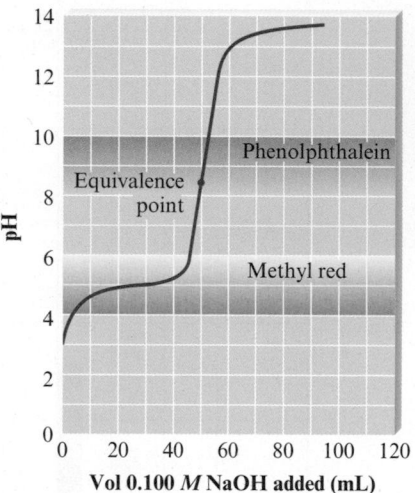

Vol 0.100 M NaOH added (mL)

Figure 14-10 │ The pH curve for the titration of 50 mL of 0.1 M $HC_2H_3O_2$ with 0.1 M NaOH. Phenolphthalein will give an end point very close to the equivalence point of the titration. Methyl red would change color well before the equivalence point (so the end point would be very different from the equivalence point) and would not be a suitable indicator for this titration.

FOR REVIEW

Key Terms

Section 14-1
common ion
common ion effect

Section 14-2
buffered solution
Henderson–Hasselbalch equation

Section 14-3
buffering capacity

Section 14-4
pH curve (titration curve)
millimole (mmol)
equivalence point (stoichiometric point)

Section 14-5
acid–base indicator
phenolphthalein

Buffered Solutions

- Contains a weak acid (HA) and its salt (NaA) or a weak base (B) and its salt (BHCl)
- Resists a change in its pH when H^+ or OH^- is added
- For a buffered solution containing HA and A^-
  - The Henderson–Hasselbalch equation is useful:

$$\text{pH} = \text{p}K_a + \log\left(\frac{[A^-]}{[HA]}\right)$$

  - The capacity of the buffered solution depends on the amounts of HA and A^- present
- The most efficient buffering occurs when the $\frac{[A^-]}{[HA]}$ ratio is close to 1
- Buffering works because the amounts of HA (which reacts with added OH^-) and A^- (which reacts with added H^+) are large enough that the $\frac{[A^-]}{[HA]}$ ratio does not change significantly when strong acids or bases are added

Acid–Base Titrations

- The progress of a titration is represented by plotting the pH of the solution versus the volume of added titrant; the resulting graph is called a pH curve or titration curve
- Strong acid–strong base titrations show a sharp change in pH near the equivalence point
- The shape of the pH curve for a strong base–strong acid titration before the equivalence point is quite different from the shape of the pH curve for a strong base–weak acid titration
  - The strong base–weak acid pH curve shows the effects of buffering before the equivalence point
  - For a strong base–weak acid titration, the pH is greater than 7 at the equivalence point because of the basic properties of A^-
- Indicators are sometimes used to mark the equivalence point of an acid–base titration
  - The end point is where the indicator changes color
  - The goal is to have the end point and the equivalence point be as close as possible

Review Questions

Answers to the Review Questions can be found on the Student Web site (accessible from **www.cengagebrain.com**).

1. What is meant by the presence of a common ion? How does the presence of a common ion affect an equilibrium such as

$$HNO_2(aq) \rightleftharpoons H^+(aq) + NO_2^-(aq)$$

What is an acid–base solution called that contains a common ion?

2. Define a buffer solution. What makes up a buffer solution? How do buffers absorb added H^+ or OH^- with little pH change?

Is it necessary that the concentrations of the weak acid and the weak base in a buffered solution be equal? Explain. What is the pH of a buffer when the weak acid and conjugate base concentrations are equal?

A buffer generally contains a weak acid and its weak conjugate base, or a weak base and its weak conjugate acid, in water. You can solve for the pH by setting up the equilibrium problem using the K_a reaction of the weak acid or the K_b reaction of the conjugate base. Both reactions give the same answer for the pH of the solution. Explain.

A third method that can be used to solve for the pH of a buffer solution is the Henderson–Hasselbalch equation. What is the Henderson–Hasselbalch equation? What assumptions are made when using this equation?

3. One of the most challenging parts of solving acid–base problems is writing out the correct equation. When a strong acid or a strong base is added to solutions, they are great at what they do, and we always react them first. If a strong acid is added to a buffer, what reacts with the H^+ from the strong acid and what are the products? If a strong base is added to a buffer, what reacts with the OH^- from the strong base and what are the products? Problems involving the reaction of a strong acid or strong base are assumed to be stoichiometry problems and not equilibrium problems. What is assumed when a strong acid or strong base reacts to make it a stoichiometry problem?

4. A good buffer generally contains relatively equal concentrations of weak acid and conjugate base. If you wanted to buffer a solution at pH = 4.00 or pH = 10.00, how would you decide which weak acid–conjugate base or weak base–conjugate acid pair to use? The second characteristic of a good buffer is good buffering capacity. What is the *capacity* of a buffer? How do the following buffers differ in capacity? How do they differ in pH?

    0.01 *M* acetic acid/0.01 *M* sodium acetate

    0.1 *M* acetic acid/0.1 *M* sodium acetate

    1.0 *M* acetic acid/1.0 *M* sodium acetate

5. Draw the general titration curve for a strong acid titrated by a strong base. At the various points in the titration, list the major species present before any reaction takes place and the major species present after any reaction takes place. What reaction takes place in a strong acid–strong base titration? How do you calculate the pH at the various points along the curve? What is the pH at the equivalence point for a strong acid–strong base titration? Why?

6. Instead of the titration of a strong acid by a strong base considered in Question 5, consider the titration of a strong base by a strong acid. Compare and contrast a strong acid–strong base titration with a strong base–strong acid titration.

7. Sketch the titration curve for a weak acid titrated by a strong base. When performing calculations concerning weak acid–strong base titrations, the general two-step procedure is to solve a stoichiometry problem first, then to solve an equilibrium problem to determine the pH. What reaction takes place in the stoichiometry part of the problem? What is assumed about this reaction?

At the various points in your titration curve, list the major species present after the strong base (NaOH, for example) reacts to completion with the weak acid, HA. What equilibrium problem would you solve at the various points in your titration curve to calculate the pH? Why is pH > 7.0 at the equivalence point of a weak acid–strong base titration? Does the pH at the halfway point to equivalence have to be less than 7.0? What does the pH at the halfway point equal? Compare and contrast the titration curves for a strong acid–strong base titration and a weak acid–strong base titration.

8. Sketch the titration curve for a weak base titrated by a strong acid. Weak base–strong acid titration problems also follow a two-step procedure. What reaction takes place in the stoichiometry part of the problem? What is assumed about this reaction? At the various points in your titration curve, list the major species present after the strong acid (HNO_3, for example) reacts to completion with the weak base, B. What equilibrium problem would you solve at the various points in your titration curve to calculate the pH? Why is pH < 7.0 at the equivalence point of a weak base–strong acid titration? If pH = 6.0 at the halfway point to equivalence, what is the K_b value for the weak base titrated? Compare and contrast the titration curves for a strong base–strong acid titration and a weak base–strong acid titration.

9. What is an acid–base indicator? Define the equivalence (stoichiometric) point and the end point of a titration. Why should you choose an indicator so that the two points coincide? Do the pH values of the two points have to be within ±0.01 pH unit of each other? Explain.

10. Why does an indicator change from its acid color to its base color over a range of pH values? In general, when do color changes start to occur for indicators? Can the indicator thymol blue contain only a single $-CO_2H$ group and no other acidic or basic functional group? Explain.

Active Learning Questions

These questions are designed to be used by groups of students in class.

1. What are the major species in solution after $NaHSO_4$ is dissolved in water? What happens to the pH of the solution as more $NaHSO_4$ is added? Why? Would the results vary if baking soda ($NaHCO_3$) were used instead?

2. A friend asks the following: "Consider a buffered solution made up of the weak acid HA and its salt NaA. If a strong base like NaOH is added, the HA reacts with the OH^- to form A^-. Thus the amount of acid (HA) is decreased, and the amount of base (A^-) is increased. Analogously, adding HCl to the buffered solution forms more of the acid (HA) by reacting with the base (A^-). Thus how can we claim that a buffered solution resists changes in the pH of the solution?" How would you explain buffering to this friend?

3. Mixing together solutions of acetic acid and sodium hydroxide can make a buffered solution. Explain. How does the amount of each solution added change the effectiveness of the buffer?

4. Could a buffered solution be made by mixing aqueous solutions of HCl and NaOH? Explain. Why isn't a mixture of a strong acid and its conjugate base considered a buffered solution?

5. Sketch two pH curves, one for the titration of a weak acid with a strong base and one for a strong acid with a strong base. How are they similar? How are they different? Account for the similarities and the differences.

6. Sketch a pH curve for the titration of a weak acid (HA) with a strong base (NaOH). List the major species, and explain how you would go about calculating the pH of the solution at various points, including the halfway point and the equivalence point.

7. You have a solution of the weak acid HA and add some HCl to it. What are the major species in the solution? What do you need to know to calculate the pH of the solution, and how would you use this information? How does the pH of the solution of just the HA compare with that of the final mixture? Explain.

8. You have a solution of the weak acid HA and add some of the salt NaA to it. What are the major species in the solution? What do you need to know to calculate the pH of the solution, and how would you use this information? How does the pH of the solution of just the HA compare with that of the final mixture? Explain.

A blue question or exercise number indicates that the answer to that question or exercise appears at the back of this book and a solution appears in the *Solutions Guide*, as found on PowerLecture.

Questions

9. The common ion effect for weak acids is to significantly decrease the dissociation of the acid in water. Explain the common ion effect.

10. Consider a buffer solution where [weak acid] > [conjugate base]. How is the pH of the solution related to the pK_a value of the weak acid? If [conjugate base] > [weak acid], how is pH related to pK_a?

11. A best buffer has about equal quantities of weak acid and conjugate base present as well as having a large concentration of each species present. Explain.

12. Consider the following pH curves for 100.0 mL of two different acids with the same initial concentration each titrated by 0.10 *M* NaOH.

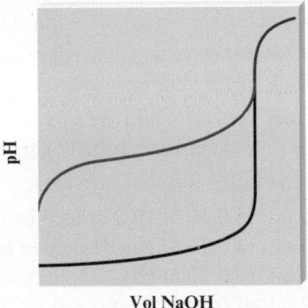

a. Which plot represents a pH curve of a weak acid, and which plot is for a strong acid? How can you tell? Cite three differences between the plots that help you decide.

b. In both cases the pH is relatively constant before the pH changes greatly. Does this mean that at some point in each titration each solution was a buffered solution?

c. True or false? The equivalence point volume for each titration is the same. Explain your answer.

d. True or false? The pH at the equivalence point for each titration is the same. Explain your answer.

13. An acid is titrated with NaOH. The following beakers are illustrations of the contents of the beaker at various times during the titration. These are presented out of order. *Note*: Counter-ions and water molecules have been omitted from the illustrations for clarity.

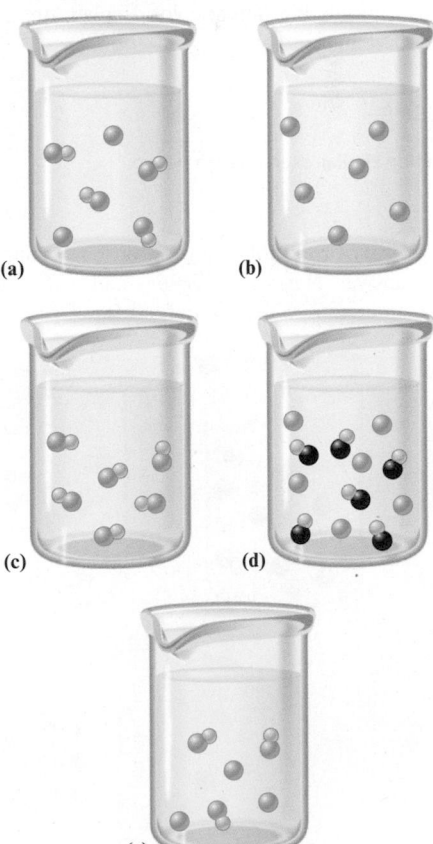

a. Is the acid a weak or strong acid? How can you tell?

b. Arrange the beakers in order of what the contents would look like as the titration progresses.

c. For which beaker would pH = pK_a? Explain your answer.

d. Which beaker represents the equivalence point of the titration? Explain your answer.

e. For which beaker would the K_a value for the acid not be necessary to determine the pH? Explain your answer.

14. Consider the following four titrations.

   i. 100.0 mL of 0.10 M HCl titrated by 0.10 M NaOH

  ii. 100.0 mL of 0.10 M NaOH titrated by 0.10 M HCl

 iii. 100.0 mL of 0.10 M CH_3NH_2 titrated by 0.10 M HCl

 iv. 100.0 mL of 0.10 M HF titrated by 0.10 M NaOH

Rank the titrations in order of:

a. increasing volume of titrant added to reach the equivalence point.

b. increasing pH initially before any titrant has been added.

c. increasing pH at the halfway point in equivalence.

d. increasing pH at the equivalence point.

How would the rankings change if C_5H_5N replaced CH_3NH_2 and if HOC_6H_5 replaced HF?

15. Figure 14-4 shows the pH curves for the titrations of six different acids by NaOH. Make a similar plot for the titration of three different bases by 0.10 M HCl. Assume 50.0 mL of 0.20 M of the bases and assume the three bases are a strong base (KOH), a weak base with $K_b = 1 \times 10^{-5}$, and another weak base with $K_b = 1 \times 10^{-10}$.

16. Acid–base indicators mark the end point of titrations by "magically" turning a different color. Explain the "magic" behind acid–base indicators.

Exercises

In this section similar exercises are paired.

Buffers

17. How many of the following are buffered solutions? Explain your answer. *Note:* Counter-ions and water molecules have been omitted from the illustrations for clarity.

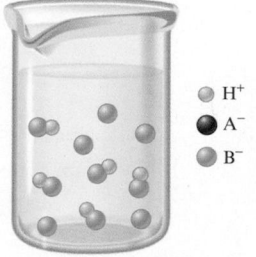

- ○ H^+
- ● A^-
- ○ B^-

18. Which of the following can be classified as buffer solutions?

a. 0.25 M HBr + 0.25 M HOBr

b. 0.15 M $HClO_4$ + 0.20 M RbOH

c. 0.50 M HOCl + 0.35 M KOCl

d. 0.70 M KOH + 0.70 M $HONH_2$

e. 0.85 M H_2NNH_2 + 0.60 M $H_2NNH_3NO_3$

19. A certain buffer is made by dissolving $NaHCO_3$ and Na_2CO_3 in some water. Write equations to show how this buffer neutralizes added H^+ and OH^-.

20. A buffer is prepared by dissolving $HONH_2$ and $HONH_3NO_3$ in some water. Write equations to show how this buffer neutralizes added H^+ and OH^-.

21. Calculate the pH of each of the following solutions.

a. 0.100 M propanoic acid ($HC_3H_5O_2$, $K_a = 1.3 \times 10^{-5}$)

b. 0.100 M sodium propanoate ($NaC_3H_5O_2$)

c. pure H_2O

d. a mixture containing 0.100 M $HC_3H_5O_2$ and 0.100 M $NaC_3H_5O_2$

22. Calculate the pH of each of the following solutions.

a. 0.100 M $HONH_2$ ($K_b = 1.1 \times 10^{-8}$)

b. 0.100 M $HONH_3Cl$

c. pure H_2O

d. a mixture containing 0.100 M $HONH_2$ and 0.100 M $HONH_3Cl$

23. Compare the percent dissociation of the acid in Exercise 21a with the percent dissociation of the acid in Exercise 21d. Explain the large difference in percent dissociation of the acid.

24. Compare the percent ionization of the base in Exercise 22a with the percent ionization of the base in Exercise 22d. Explain any differences.

25. Calculate the pH after 0.020 mole of HCl is added to 1.00 L of each of the four solutions in Exercise 21.

26. Calculate the pH after 0.020 mole of HCl is added to 1.00 L of each of the four solutions in Exercise 22.

27. Calculate the pH after 0.020 mole of NaOH is added to 1.00 L of each of the four solutions in Exercise 21.

28. Calculate the pH after 0.020 mole of NaOH is added to 1.00 L of each of the solutions in Exercise 22.

29. Which of the solutions in Exercise 21 shows the least change in pH upon the addition of acid or base? Explain.

30. Which of the solutions in Exercise 22 is a buffered solution?

31. Calculate the pH of a solution that is 1.00 M HNO_2 and 1.00 M $NaNO_2$.

32. Calculate the pH of a solution that is 0.60 M HF and 1.00 M KF.

33. Calculate the pH after 0.10 mole of NaOH is added to 1.00 L of the solution in Exercise 31, and calculate the pH after 0.20 mole of HCl is added to 1.00 L of the solution in Exercise 31.

34. Calculate the pH after 0.10 mole of NaOH is added to 1.00 L of the solution in Exercise 32, and calculate the pH after 0.20 mole of HCl is added to 1.00 L of the solution in Exercise 32.

35. Calculate the pH of each of the following buffered solutions.

a. 0.10 M acetic acid/0.25 M sodium acetate

b. 0.25 M acetic acid/0.10 M sodium acetate

c. 0.080 M acetic acid/0.20 M sodium acetate

d. 0.20 M acetic acid/0.080 M sodium acetate

36. Calculate the pH of each of the following buffered solutions.

   a. 0.50 M $C_2H_5NH_2$/0.25 M $C_2H_5NH_3Cl$

   b. 0.25 M $C_2H_5NH_2$/0.50 M $C_2H_5NH_3Cl$

   c. 0.50 M $C_2H_5NH_2$/0.50 M $C_2H_5NH_3Cl$

37. Calculate the pH of a buffered solution prepared by dissolving 21.5 g benzoic acid ($HC_7H_5O_2$) and 37.7 g sodium benzoate in 200.0 mL of solution.

38. A buffered solution is made by adding 50.0 g NH_4Cl to 1.00 L of a 0.75-M solution of NH_3. Calculate the pH of the final solution. (Assume no volume change.)

39. Calculate the pH after 0.010 mole of gaseous HCl is added to 250.0 mL of each of the following buffered solutions.

   a. 0.050 M NH_3/0.15 M NH_4Cl

   b. 0.50 M NH_3/1.50 M NH_4Cl

   Do the two original buffered solutions differ in their pH or their capacity? What advantage is there in having a buffer with a greater capacity?

40. An aqueous solution contains dissolved $C_6H_5NH_3Cl$ and $C_6H_5NH_2$. The concentration of $C_6H_5NH_2$ is 0.50 M and pH is 4.20.

   a. Calculate the concentration of $C_6H_5NH_3^+$ in this buffer solution.

   b. Calculate the pH after 4.0 g NaOH(s) is added to 1.0 L of this solution. (Neglect any volume change.)

41. Calculate the mass of sodium acetate that must be added to 500.0 mL of 0.200 M acetic acid to form a pH = 5.00 buffer solution.

42. What volumes of 0.50 M HNO_2 and 0.50 M $NaNO_2$ must be mixed to prepare 1.00 L of a solution buffered at pH = 3.55?

43. Consider a solution that contains both C_5H_5N and $C_5H_5NHNO_3$. Calculate the ratio $[C_5H_5N]/[C_5H_5NH^+]$ if the solution has the following pH values:

   a. pH = 4.50   **c.** pH = 5.23

   b. pH = 5.00   **d.** pH = 5.50

44. Calculate the ratio $[NH_3]/[NH_4^+]$ in ammonia/ammonium chloride buffered solutions with the following pH values:

   a. pH = 9.00   **c.** pH = 10.00

   b. pH = 8.80   **d.** pH = 9.60

45. Carbonate buffers are important in regulating the pH of blood at 7.40. If the carbonic acid concentration in a sample of blood is 0.0012 M, determine the bicarbonate ion concentration required to buffer the pH of blood at pH = 7.40.

$$H_2CO_3(aq) \rightleftharpoons HCO_3^-(aq) + H^+(aq) \qquad K_a = 4.3 \times 10^{-7}$$

46. When a person exercises, muscle contractions produce lactic acid. Moderate increases in lactic acid can be handled by the blood buffers without decreasing the pH of blood. However, excessive amounts of lactic acid can overload the blood buffer system, resulting in a lowering of the blood pH. A condition called *acidosis* is diagnosed if the blood pH falls to 7.35 or lower. Assume the primary blood buffer system is the carbonate buffer system described in Exercise 45. Calculate what happens to the $[H_2CO_3]/[HCO_3^-]$ ratio in blood when the pH decreases from 7.40 to 7.35.

47. Consider the acids in Table 13-2. Which acid would be the best choice for preparing a pH = 7.00 buffer? Explain how to make 1.0 L of this buffer.

48. Consider the bases in Table 13-3. Which base would be the best choice for preparing a pH = 5.00 buffer? Explain how to make 1.0 L of this buffer.

49. Calculate the pH of a solution that is 0.40 M H_2NNH_2 and 0.80 M $H_2NNH_3NO_3$. In order for this buffer to have pH = pK_a, would you add HCl or NaOH? What quantity (moles) of which reagent would you add to 1.0 L of the original buffer so that the resulting solution has pH = pK_a?

50. Calculate the pH of a solution that is 0.20 M HOCl and 0.90 M KOCl. In order for this buffer to have pH = pK_a, would you add HCl or NaOH? What quantity (moles) of which reagent would you add to 1.0 L of the original buffer so that the resulting solution has pH = pK_a?

51. Which of the following mixtures would result in buffered solutions when 1.0 L of each of the two solutions are mixed?

   a. 0.1 M KOH and 0.1 M CH_3NH_3Cl

   b. 0.1 M KOH and 0.2 M CH_3NH_2

   c. 0.2 M KOH and 0.1 M CH_3NH_3Cl

   d. 0.1 M KOH and 0.2 M CH_3NH_3Cl

52. Which of the following mixtures would result in a buffered solution when 1.0 L of each of the two solutions are mixed?

   a. 0.2 M HNO_3 and 0.4 M $NaNO_3$

   b. 0.2 M HNO_3 and 0.4 M HF

   c. 0.2 M HNO_3 and 0.4 M NaF

   d. 0.2 M HNO_3 and 0.4 M NaOH

53. What quantity (moles) of NaOH must be added to 1.0 L of 2.0 M $HC_2H_3O_2$ to produce a solution buffered at each pH?

   a. pH = pK_a **b.** pH = 4.00   **c.** pH = 5.00

54. Calculate the number of moles of HCl(g) that must be added to 1.0 L of 1.0 M $NaC_2H_3O_2$ to produce a solution buffered at each pH.

   a. pH = pK_a **b.** pH = 4.20   **c.** pH = 5.00

Acid–Base Titrations

55. Consider the titration of a generic weak acid HA with a strong base that gives the following titration curve:

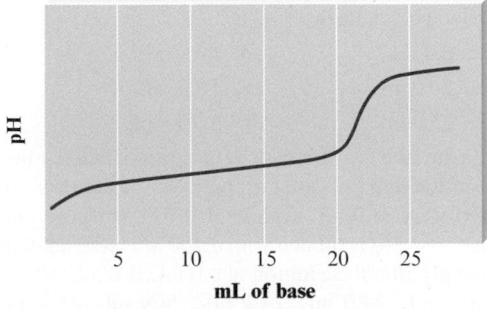

On the curve, indicate the points that correspond to the following:

   a. the stoichiometric (equivalence) point

   b. the region with maximum buffering

   c. pH = pK_a

d. pH depends only on [HA]

e. pH depends only on [A⁻]

f. pH depends only on the amount of excess strong base added

56. Sketch the titration curve for the titration of a generic weak base B with a strong acid. The titration reaction is

$$B + H^+ \rightleftharpoons BH^+$$

On this curve, indicate the points that correspond to the following:

a. the stoichiometric (equivalence) point

b. the region with maximum buffering

c. $pH = pK_a$

d. pH depends only on [B]

e. pH depends only on [BH⁺]

f. pH depends only on the amount of excess strong acid added

57. Consider the titration of 40.0 mL of 0.200 M $HClO_4$ by 0.100 M KOH. Calculate the pH of the resulting solution after the following volumes of KOH have been added.

a. 0.0 mL   d. 80.0 mL

b. 10.0 mL   e. 100.0 mL

c. 40.0 mL

58. Consider the titration of 80.0 mL of 0.100 M $Ba(OH)_2$ by 0.400 M HCl. Calculate the pH of the resulting solution after the following volumes of HCl have been added.

a. 0.0 mL   d. 40.0 mL

b. 20.0 mL   e. 80.0 mL

c. 30.0 mL

59. Consider the titration of 100.0 mL of 0.200 M acetic acid ($K_a = 1.8 \times 10^{-5}$) by 0.100 M KOH. Calculate the pH of the resulting solution after the following volumes of KOH have been added.

a. 0.0 mL   d. 150.0 mL

b. 50.0 mL   e. 200.0 mL

c. 100.0 mL   f. 250.0 mL

60. Consider the titration of 100.0 mL of 0.100 M H_2NNH_2 ($K_b = 3.0 \times 10^{-6}$) by 0.200 M HNO_3. Calculate the pH of the resulting solution after the following volumes of HNO_3 have been added.

a. 0.0 mL   d. 40.0 mL

b. 20.0 mL   e. 50.0 mL

c. 25.0 mL   f. 100.0 mL

61. Lactic acid is a common by-product of cellular respiration and is often said to cause the "burn" associated with strenuous activity. A 25.0-mL sample of 0.100 M lactic acid ($HC_3H_5O_3$, $pK_a = 3.86$) is titrated with 0.100 M NaOH solution. Calculate the pH after the addition of 0.0 mL, 4.0 mL, 8.0 mL, 12.5 mL, 20.0 mL, 24.0 mL, 24.5 mL, 24.9 mL, 25.0 mL, 25.1 mL, 26.0 mL, 28.0 mL, and 30.0 mL of the NaOH. Plot the results of your calculations as pH versus milliliters of NaOH added.

62. Repeat the procedure in Exercise 61, but for the titration of 25.0 mL of 0.100 M propanoic acid ($HC_3H_5O_2$, $K_a = 1.3 \times 10^{-5}$) with 0.100 M NaOH.

63. Repeat the procedure in Exercise 61, but for the titration of 25.0 mL of 0.100 M NH_3 ($K_b = 1.8 \times 10^{-5}$) with 0.100 M HCl.

64. Repeat the procedure in Exercise 61, but for the titration of 25.0 mL of 0.100 M pyridine with 0.100 M hydrochloric acid (K_b for pyridine is 1.7×10^{-9}). Do not calculate the points at 24.9 and 25.1 mL.

65. Calculate the pH at the halfway point and at the equivalence point for each of the following titrations.

a. 100.0 mL of 0.10 M $HC_7H_5O_2$ ($K_a = 6.4 \times 10^{-5}$) titrated by 0.10 M NaOH

b. 100.0 mL of 0.10 M $C_2H_5NH_2$ ($K_b = 5.6 \times 10^{-4}$) titrated by 0.20 M HNO_3

c. 100.0 mL of 0.50 M HCl titrated by 0.25 M NaOH

66. In the titration of 50.0 mL of 1.0 M methylamine, CH_3NH_2 ($K_b = 4.4 \times 10^{-4}$), with 0.50 M HCl, calculate the pH under the following conditions.

a. after 50.0 mL of 0.50 M HCl has been added

b. at the stoichiometric point

67. You have 75.0 mL of 0.10 M HA. After adding 30.0 mL of 0.10 M NaOH, the pH is 5.50. What is the K_a value of HA?

68. A student dissolves 0.0100 mole of an unknown weak base in 100.0 mL water and titrates the solution with 0.100 M HNO_3. After 40.0 mL of 0.100 M HNO_3 was added, the pH of the resulting solution was 8.00. Calculate the K_b value for the weak base.

Indicators

69. Two drops of indicator HIn ($K_a = 1.0 \times 10^{-9}$), where HIn is yellow and In⁻ is blue, are placed in 100.0 mL of 0.10 M HCl.

a. What color is the solution initially?

b. The solution is titrated with 0.10 M NaOH. At what pH will the color change (yellow to greenish yellow) occur?

c. What color will the solution be after 200.0 mL NaOH has been added?

70. Methyl red has the following structure:

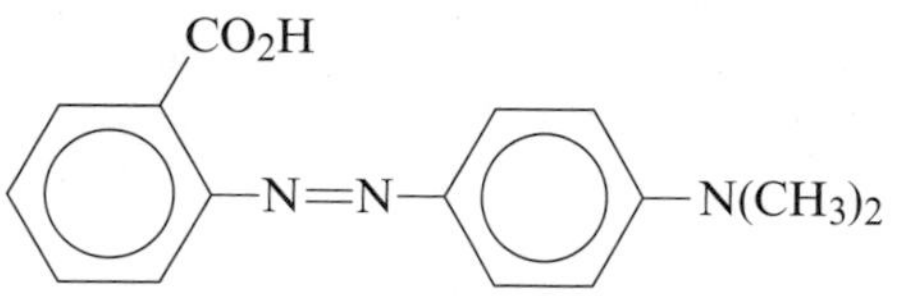

It undergoes a color change from red to yellow as a solution gets more basic. Calculate an approximate pH range for which methyl red is useful. What is the color change and the pH at the color change when a weak acid is titrated with a strong base using methyl red as an indicator? What is the color change and the pH at the color change when a weak base is titrated with a strong acid using methyl red as an indicator? For which of these two types of titrations is methyl red a possible indicator?

71. Potassium hydrogen phthalate, known as KHP (molar mass = 204.22 g/mol), can be obtained in high purity and is used to determine the concentration of solutions of strong bases by the reaction

$$HP^-(aq) + OH^-(aq) \longrightarrow H_2O(l) + P^{2-}(aq)$$

If a typical titration experiment begins with approximately 0.5 g KHP and has a final volume of about 100 mL, what is an appropriate indicator to use? The pK_a for HP^- is 5.51.

72. A certain indicator HIn has a pK_a of 3.00 and a color change becomes visible when 7.00% of the indicator has been converted to In^-. At what pH is this color change visible?

73. Which of the indicators in Fig. 14-8 could be used for the titrations in Exercises 57 and 59?

74. Which of the indicators in Fig. 14-8 could be used for the titrations in Exercises 58 and 60?

75. Which of the indicators in Fig. 14-8 could be used for the titrations in Exercises 61 and 63?

76. Which of the indicators in Fig. 14-8 could be used for the titrations in Exercises 62 and 64?

77. Estimate the pH of a solution in which bromcresol green is blue and thymol blue is yellow. (See Fig. 14-8.)

78. Estimate the pH of a solution in which crystal violet is yellow and methyl orange is red. (See Fig. 14-8.)

79. A solution has a pH of 7.0. What would be the color of the solution if each of the following indicators were added? (See Fig. 14-8.)
  a. thymol blue   **c.** methyl red
  b. bromthymol blue   **d.** crystal violet

80. A solution has a pH of 4.5. What would be the color of the solution if each of the following indicators were added? (See Fig. 14-8.)
  a. methyl orange   **c.** bromcresol green
  b. alizarin   **d.** phenolphthalein

Additional Exercises

81. Derive an equation analogous to the Henderson–Hasselbalch equation but relating pOH and pK_b of a buffered solution composed of a weak base and its conjugate acid, such as NH_3 and NH_4^+.

82. a. Calculate the pH of a buffered solution that is 0.100 M in $C_6H_5CO_2H$ (benzoic acid, $K_a = 6.4 \times 10^{-5}$) and 0.100 M in $C_6H_5CO_2Na$.

  b. Calculate the pH after 20.0% (by moles) of the benzoic acid is converted to benzoate anion by addition of a strong base. Use the dissociation equilibrium

$$C_6H_5CO_2H(aq) \rightleftharpoons C_6H_5CO_2^-(aq) + H^+(aq)$$

  to calculate the pH.

  c. Do the same as in part b, but use the following equilibrium to calculate the pH:

$$C_6H_5CO_2^-(aq) + H_2O(l) \rightleftharpoons C_6H_5CO_2H(aq) + OH^-(aq)$$

  d. Do your answers in parts b and c agree? Explain.

83. Tris(hydroxymethyl)aminomethane, commonly called TRIS or Trizma, is often used as a buffer in biochemical studies. Its buffering range is pH 7 to 9, and K_b is 1.19×10^{-6} for the aqueous reaction

$$\underset{\text{TRIS}}{(HOCH_2)_3CNH_2} + H_2O \rightleftharpoons \underset{\text{TRISH}^+}{(HOCH_2)_3CNH_3^+} + OH^-$$

  a. What is the optimal pH for TRIS buffers?

  b. Calculate the ratio [TRIS]/[TRISH$^+$] at pH = 7.00 and at pH = 9.00.

  c. A buffer is prepared by diluting 50.0 g TRIS base and 65.0 g TRIS hydrochloride (written as TRISHCl) to a total volume of 2.0 L. What is the pH of this buffer? What is the pH after 0.50 mL of 12 M HCl is added to a 200.0-mL portion of the buffer?

84. You make 1.00 L of a buffered solution (pH = 4.00) by mixing acetic acid and sodium acetate. You have 1.00 M solutions of each component of the buffered solution. What volume of each solution do you mix to make such a buffered solution?

85. You have the following reagents on hand:

Solids (pK_a of Acid Form Is Given)	Solutions
Benzoic acid (4.19)	5.0 M HCl
Sodium acetate (4.74)	1.0 M acetic acid (4.74)
Potassium fluoride (3.14)	2.6 M NaOH
Ammonium chloride (9.26)	1.0 M HOCl (7.46)

  What combinations of reagents would you use to prepare buffers at the following pH values?
  a. 3.0   **b.** 4.0   **c.** 5.0   **d.** 7.0   **e.** 9.0

86. Amino acids are the building blocks for all proteins in our bodies. A structure for the amino acid alanine is

$$\underset{\substack{\text{Amino} \\ \text{group}}}{H_2N} - \underset{\underset{H}{|}}{\overset{\overset{CH_3}{|}}{C}} - \underset{\substack{\text{Carboxylic} \\ \text{acid group}}}{\overset{\overset{O}{||}}{C}} - OH$$

  All amino acids have at least two functional groups with acidic or basic properties. In alanine, the carboxylic acid group has $K_a = 4.5 \times 10^{-3}$ and the amino group has $K_b = 7.4 \times 10^{-5}$. Because of the two groups with acidic or basic properties, three different charged ions of alanine are possible when alanine is dissolved in water. Which of these ions would predominate in a solution with $[H^+] = 1.0\ M$? In a solution with $[OH^-] = 1.0\ M$?

87. Phosphate buffers are important in regulating the pH of intracellular fluids at pH values generally between 7.1 and 7.2.

  a. What is the concentration ratio of $H_2PO_4^-$ to HPO_4^{2-} in intracellular fluid at pH = 7.15?

$$H_2PO_4^-(aq) \rightleftharpoons HPO_4^{2-}(aq) + H^+(aq) \qquad K_a = 6.2 \times 10^{-8}$$

  b. Why is a buffer composed of H_3PO_4 and $H_2PO_4^-$ ineffective in buffering the pH of intracellular fluid?

$$H_3PO_4(aq) \rightleftharpoons H_2PO_4^-(aq) + H^+(aq) \qquad K_a = 7.5 \times 10^{-3}$$

88. What quantity (moles) of HCl(g) must be added to 1.0 L of 2.0 M NaOH to achieve a pH of 0.00? (Neglect any volume changes.)

89. Calculate the value of the equilibrium constant for each of the following reactions in aqueous solution.
  a. $HC_2H_3O_2 + OH^- \rightleftharpoons C_2H_3O_2^- + H_2O$
  b. $C_2H_3O_2^- + H^+ \rightleftharpoons HC_2H_3O_2$
  c. $HCl + NaOH \rightleftharpoons NaCl + H_2O$

90. The following plot shows the pH curves for the titrations of various acids by 0.10 M NaOH (all of the acids were 50.0-mL samples of 0.10 M concentration).

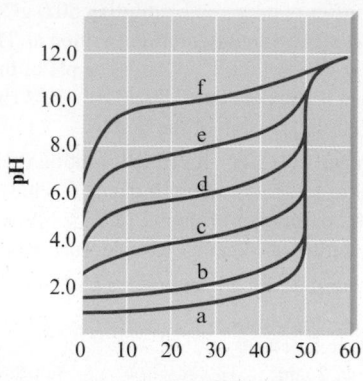

Vol 0.10 M NaOH added (mL)

a. Which pH curve corresponds to the weakest acid?

b. Which pH curve corresponds to the strongest acid? Which point on the pH curve would you examine to see if this acid is a strong acid or a weak acid (assuming you did not know the initial concentration of the acid)?

c. Which pH curve corresponds to an acid with $K_a \approx 1 \times 10^{-6}$?

91. Calculate the volume of 1.50×10^{-2} M NaOH that must be added to 500.0 mL of 0.200 M HCl to give a solution that has pH = 2.15.

92. Repeat the procedure in Exercise 61, but for the titration of 25.0 mL of 0.100 M HNO$_3$ with 0.100 M NaOH.

93. A certain acetic acid solution has pH = 2.68. Calculate the volume of 0.0975 M KOH required to reach the equivalence point in the titration of 25.0 mL of the acetic acid solution.

94. A 0.210-g sample of an acid (molar mass = 192 g/mol) is titrated with 30.5 mL of 0.108 M NaOH to a phenolphthalein end point. Is the acid monoprotic, diprotic, or triprotic?

95. The active ingredient in aspirin is acetylsalicylic acid. A 2.51-g sample of acetylsalicylic acid required 27.36 mL of 0.5106 M NaOH for complete reaction. Addition of 13.68 mL of 0.5106 M HCl to the flask containing the aspirin and the sodium hydroxide produced a mixture with pH = 3.48. Determine the molar mass of acetylsalicylic acid and its K_a value. State any assumptions you must make to reach your answer.

96. One method for determining the purity of aspirin ($C_9H_8O_4$) is to hydrolyze it with NaOH solution and then to titrate the remaining NaOH. The reaction of aspirin with NaOH is as follows:

$$C_9H_8O_4(s) + 2OH^-(aq)$$
Aspirin
$$\xrightarrow[\text{10 min}]{\text{Boil}} C_7H_5O_3^-(aq) + C_2H_3O_2^-(aq) + H_2O(l)$$
Salicylate ion   Acetate ion

A sample of aspirin with a mass of 1.427 g was boiled in 50.00 mL of 0.500 M NaOH. After the solution was cooled, it took 31.92 mL of 0.289 M HCl to titrate the excess NaOH. Calculate the purity of the aspirin. What indicator should be used for this titration? Why?

97. A student intends to titrate a solution of a weak monoprotic acid with a sodium hydroxide solution but reverses the two solutions and places the weak acid solution in the buret. After 23.75 mL of the weak acid solution has been added to 50.0 mL of the 0.100 M NaOH solution, the pH of the resulting solution is 10.50. Calculate the original concentration of the solution of weak acid.

98. A student titrates an unknown weak acid, HA, to a pale pink phenolphthalein end point with 25.0 mL of 0.100 M NaOH. The student then adds 13.0 mL of 0.100 M HCl. The pH of the resulting solution is 4.70. How is the value of pK_a for the unknown acid related to 4.70?

99. A sample of a certain monoprotic weak acid was dissolved in water and titrated with 0.125 M NaOH, requiring 16.00 mL to reach the equivalence point. During the titration, the pH after adding 2.00 mL NaOH was 6.912. Calculate K_a for the weak acid.

ChemWork Problems

These multiconcept problems (and additional ones) are found interactively online with the same type of assistance a student would get from an instructor.

100. Consider 1.0 L of a solution that is 0.85 M HOC$_6$H$_5$ and 0.80 M NaOC$_6$H$_5$. (K_a for HOC$_6$H$_5$ = 1.6×10^{-10}.)

a. Calculate the pH of this solution.

b. Calculate the pH after 0.10 mole of HCl has been added to the original solution. Assume no volume change on addition of HCl.

c. Calculate the pH after 0.20 mole of NaOH has been added to the original buffer solution. Assume no volume change on addition of NaOH.

101. What concentration of NH$_4$Cl is necessary to buffer a 0.52-M NH$_3$ solution at pH = 9.00? (K_b for NH$_3$ = 1.8×10^{-5}.)

102. Consider the following acids and bases:

HCO$_2$H	$K_a = 1.8 \times 10^{-4}$
HOBr	$K_a = 2.0 \times 10^{-9}$
(C$_2$H$_5$)$_2$NH	$K_b = 1.3 \times 10^{-3}$
HONH$_2$	$K_b = 1.1 \times 10^{-8}$

Choose substances from the following list that would be the best choice to prepare a pH = 9.0 buffer solution.

a. HCO$_2$H
b. HOBr
c. KHCO$_2$
d. HONH$_3$NO$_3$

e. (C$_2$H$_5$)$_2$NH
f. (C$_2$H$_5$)$_2$NH$_2$Cl
g. HONH$_2$
h. NaOBr

103. Consider a buffered solution containing CH$_3$NH$_3$Cl and CH$_3$NH$_2$. Which of the following statements concerning this solution is(are) *true*? (K_a for CH$_3$NH$_3^+$ = 2.3×10^{-11}.)

a. A solution consisting of 0.10 M CH$_3$NH$_3$Cl and 0.10 M CH$_3$NH$_2$ would have a higher buffering capacity than one containing 1.0 M CH$_3$NH$_3$Cl and 1.0 M CH$_3$NH$_2$.

b. If [CH$_3$NH$_2$] > [CH$_3$NH$_3^+$], then the pH is larger than the pK_a value.

c. Adding more [CH$_3$NH$_3$Cl] to the initial buffer solution will decrease the pH.

d. If $[CH_3NH_2] < [CH_3NH_3{}^+]$, then pH < 3.36.

e. If $[CH_3NH_2] = [CH_3NH_3{}^+]$, then pH $= 10.64$.

104. Consider the titration of 150.0 mL of 0.100 M HI by 0.250 M NaOH.

   a. Calculate the pH after 20.0 mL of NaOH has been added.

   b. What volume of NaOH must be added so that the pH = 7.00?

105. Consider the titration of 100.0 mL of 0.100 M HCN by 0.100 M KOH at 25°C. (K_a for HCN $= 6.2 \times 10^{-10}$.)

   a. Calculate the pH after 0.0 mL of KOH has been added.

   b. Calculate the pH after 50.0 mL of KOH has been added.

   c. Calculate the pH after 75.0 mL of KOH has been added.

   d. Calculate the pH at the equivalence point.

   e. Calculate the pH after 125 mL of KOH has been added.

106. Consider the titration of 100.0 mL of 0.200 M HONH$_2$ by 0.100 M HCl. (K_b for HONH$_2$ $= 1.1 \times 10^{-8}$.)

   a. Calculate the pH after 0.0 mL of HCl has been added.

   b. Calculate the pH after 25.0 mL of HCl has been added.

   c. Calculate the pH after 70.0 mL of HCl has been added.

   d. Calculate the pH at the equivalence point.

   e. Calculate the pH after 300.0 mL of HCl has been added.

   f. At what volume of HCl added does the pH = 6.04?

107. Consider the following four titrations (i–iv):

   i. 150 mL of 0.2 M NH$_3$ ($K_b = 1.8 \times 10^{-5}$) by 0.2 M HCl

   ii. 150 mL of 0.2 M HCl by 0.2 M NaOH

   iii. 150 mL of 0.2 M HOCl ($K_a = 3.5 \times 10^{-8}$) by 0.2 M NaOH

   iv. 150 mL of 0.2 M HF ($K_a = 7.2 \times 10^{-4}$) by 0.2 M NaOH

   a. Rank the four titrations in order of increasing pH at the halfway point to equivalence (lowest to highest pH).

   b. Rank the four titrations in order of increasing pH at the equivalence point.

   c. Which titration requires the largest volume of titrant (HCl or NaOH) to reach the equivalence point?

Challenge Problems

108. Another way to treat data from a pH titration is to graph the absolute value of the change in pH per change in milliliters added versus milliliters added (ΔpH/ΔmL versus mL added). Make this graph using your results from Exercise 61. What advantage might this method have over the traditional method for treating titration data?

109. A buffer is made using 45.0 mL of 0.750 M HC$_3$H$_5$O$_2$ ($K_a = 1.3 \times 10^{-5}$) and 55.0 mL of 0.700 M NaC$_3$H$_5$O$_2$. What volume of 0.10 M NaOH must be added to change the pH of the original buffer solution by 2.5%?

110. A 0.400-M solution of ammonia was titrated with hydrochloric acid to the equivalence point, where the total volume was 1.50 times the original volume. At what pH does the equivalence point occur?

111. What volume of 0.0100 M NaOH must be added to 1.00 L of 0.0500 M HOCl to achieve a pH of 8.00?

112. Consider a solution formed by mixing 50.0 mL of 0.100 M H$_2$SO$_4$, 30.0 mL of 0.100 M HOCl, 25.0 mL of 0.200 M NaOH, 25.0 mL of 0.100 M Ba(OH)$_2$, and 10.0 mL of 0.150 M KOH. Calculate the pH of this solution.

113. When a diprotic acid, H$_2$A, is titrated with NaOH, the protons on the diprotic acid are generally removed one at a time, resulting in a pH curve that has the following generic shape:

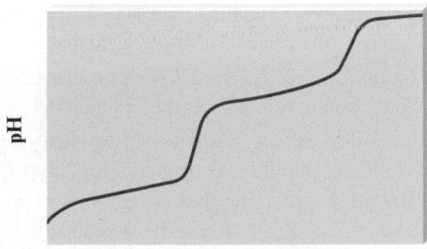

Vol NaOH added

   a. Notice that the plot has essentially two titration curves. If the first equivalence point occurs at 100.0 mL NaOH added, what volume of NaOH added corresponds to the second equivalence point?

   b. For the following volumes of NaOH added, list the major species present after the OH$^-$ reacts completely.

   i. 0 mL NaOH added

   ii. between 0 and 100.0 mL NaOH added

   iii. 100.0 mL NaOH added

   iv. between 100.0 and 200.0 mL NaOH added

   v. 200.0 mL NaOH added

   vi. after 200.0 mL NaOH added

   c. If the pH at 50.0 mL NaOH added is 4.0, and the pH at 150.0 mL NaOH added is 8.0, determine the values K_{a_1} and K_{a_2} for the diprotic acid.

114. Consider the following two acids:

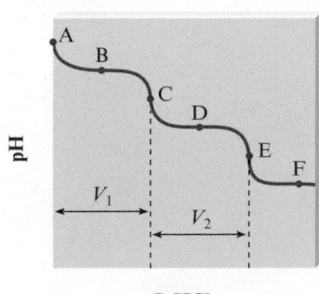

$pK_{a_1} = 2.98$; $pK_{a_2} = 13.40$

Salicylic acid

HO$_2$CCH$_2$CH$_2$CH$_2$CH$_2$CO$_2$H

Adipic acid $pK_{a_1} = 4.41$; $pK_{a_2} = 5.28$

In two separate experiments the pH was measured during the titration of 5.00 mmol of each acid with 0.200 M NaOH. Each experiment showed only one stoichiometric point when the data were plotted. In one experiment the stoichiometric point was at 25.00 mL added NaOH, and in the other experiment the stoichiometric point was at 50.00 mL added NaOH. Explain these results. (See Exercise 113.)

115. The titration of Na$_2$CO$_3$ with HCl has the following qualitative profile:

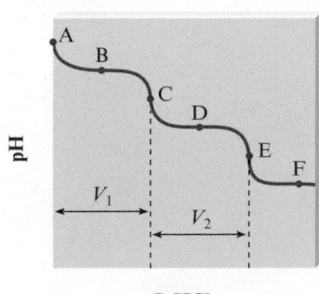

mL HCl

a. Identify the major species in solution at points A–F.

b. Calculate the pH at the halfway points to equivalence, B and D. (*Hint:* Refer to Exercise 113.)

116. Consider the titration curve in Exercise 115 for the titration of Na_2CO_3 with HCl.

a. If a mixture of $NaHCO_3$ and Na_2CO_3 was titrated, what would be the relative sizes of V_1 and V_2?

b. If a mixture of NaOH and Na_2CO_3 was titrated, what would be the relative sizes of V_1 and V_2?

c. A sample contains a mixture of $NaHCO_3$ and Na_2CO_3. When this sample was titrated with 0.100 *M* HCl, it took 18.9 mL to reach the first stoichiometric point and an additional 36.7 mL to reach the second stoichiometric point. What is the composition in mass percent of the sample?

117. A few drops of each of the indicators shown in the accompanying table were placed in separate portions of a 1.0-*M* solution of a weak acid, HX. The results are shown in the last column of the table. What is the approximate pH of the solution containing HX? Calculate the approximate value of K_a for HX.

Indicator	Color of HIn	Color of In$^-$	pK_a of HIn	Color of 1.0 *M* HX
Bromphenol blue	Yellow	Blue	4.0	Blue
Bromcresol purple	Yellow	Purple	6.0	Yellow
Bromcresol green	Yellow	Blue	4.8	Green
Alizarin	Yellow	Red	6.5	Yellow

118. Malonic acid ($HO_2CCH_2CO_2H$) is a diprotic acid. In the titration of malonic acid with NaOH, stoichiometric points occur at pH = 3.9 and 8.8. A 25.00-mL sample of malonic acid of unknown concentration is titrated with 0.0984 *M* NaOH, requiring 31.50 mL of the NaOH solution to reach the phenolphthalein end point. Calculate the concentration of the initial malonic acid solution. (See Exercise 113.)

Integrative Problems

These problems require the integration of multiple concepts to find the solutions.

119. A buffer solution is prepared by mixing 75.0 mL of 0.275 *M* fluorobenzoic acid ($C_7H_5O_2F$) with 55.0 mL of 0.472 *M* sodium fluorobenzoate. The pK_a of this weak acid is 2.90. What is the pH of the buffer solution?

120. A 10.00-g sample of the ionic compound NaA, where A$^-$ is the anion of a weak acid, was dissolved in enough water to make 100.0 mL of solution and was then titrated with 0.100 *M* HCl. After 500.0 mL HCl was added, the pH was 5.00. The experimenter found that 1.00 L of 0.100 *M* HCl was required to reach the stoichiometric point of the titration.

a. What is the molar mass of NaA?

b. Calculate the pH of the solution at the stoichiometric point of the titration.

121. Calculate the pH of a solution prepared by mixing 250. mL of 0.174 *m* aqueous HF (density = 1.10 g/mL) with 38.7 g of an aqueous solution that is 1.50% NaOH by mass (density = 1.02 g/mL). (K_a for HF = 7.2×10^{-4}.)

Marathon Problem

This problem is designed to incorporate several concepts and techniques into one situation.

122. Consider a solution prepared by mixing the following:

50.0 mL of 0.100 *M* Na_3PO_4
100.0 mL of 0.0500 *M* KOH
200.0 mL of 0.0750 *M* HCl
50.0 mL of 0.150 *M* NaCN

Determine the volume of 0.100 *M* HNO_3 that must be added to this mixture to achieve a final pH value of 7.21.

Solubility and Complex Ion Equilibria

Stalactites and stalagmites. These formations are created when carbonate minerals dissolve in groundwater acidified by carbon dioxide and then solidify when the water evaporates. (Jason Patrick Ross/Shutterstock.com)

615

Most of the chemistry of the natural world occurs in aqueous solution. We have already introduced one very significant class of aqueous equilibria, acid–base reactions. In this chapter we will consider more applications of aqueous equilibria, those involving the solubility of salts and those involving the formation of complex ions.

The interplay of acid–base, solubility, and complex ion equilibria is often important in natural processes, such as the weathering of minerals, the uptake of nutrients by plants, and tooth decay. For example, limestone ($CaCO_3$) will dissolve in water made acidic by dissolved carbon dioxide:

$$CO_2(aq) + H_2O(l) \rightleftharpoons H^+(aq) + HCO_3^-(aq)$$
$$H^+(aq) + CaCO_3(s) \rightleftharpoons Ca^{2+}(aq) + HCO_3^-(aq)$$

This two-step process and its reverse account for the formation of limestone caves and the stalactites and stalagmites found therein. In the forward direction of the process, the acidic water (containing carbon dioxide) dissolves the underground limestone deposits, thereby forming a cavern. The reverse process occurs as the water drips from the ceiling of the cave, and the carbon dioxide is lost to the air. This causes solid calcium carbonate to form, producing stalactites on the ceiling and stalagmites where the drops hit the cave floor.

In this chapter we will discuss the formation of solids from an aqueous solution and the resulting equilibria. We will also show how selective precipitation and the formation of complex ions can be used to do qualitative analysis.

15-1 | Solubility Equilibria and the Solubility Product

Solubility is a very important phenomenon. The fact that substances such as sugar and table salt dissolve in water allows us to flavor foods easily. The fact that calcium sulfate is less soluble in hot water than in cold water causes it to coat tubes in boilers, reducing thermal efficiency. Tooth decay involves solubility: When food lodges between the teeth, acids form that dissolve tooth enamel, which contains a mineral called *hydroxyapatite,* $Ca_5(PO_4)_3OH$. Tooth decay can be reduced by treating teeth with fluoride (see Chemical Connections 15-1). Fluoride replaces the hydroxide in hydroxyapatite to produce the corresponding fluorapatite, $Ca_5(PO_4)_3F$, and calcium fluoride, CaF_2, both of which are less soluble in acids than the original enamel. Another important consequence of solubility involves the use of a suspension of barium sulfate to improve the clarity of X rays of the gastrointestinal tract. The very low solubility of barium sulfate, which contains the toxic ion Ba^{2+}, makes ingestion of the compound safe. ◄

In this section we will consider the equilibria associated with solids dissolving to form aqueous solutions. We will assume that when a typical ionic solid dissolves in water, it dissociates completely into separate hydrated cations and anions. ◄ ⓘ For example, calcium fluoride dissolves in water as follows:

$$CaF_2(s) \xrightarrow{H_2O} Ca^{2+}(aq) + 2F^-(aq)$$

When the solid salt is first added to the water, no Ca^{2+} and F^- ions are present. However, as the dissolution proceeds, the concentrations of Ca^{2+} and F^- increase, making

ⓘ For simplicity, we will ignore the effects of ion associations in these solutions.

it more and more likely that these ions will collide and re-form the solid phase. Thus two competing processes are occurring—the dissolution reaction and its reverse:

$$Ca^{2+}(aq) + 2F^-(aq) \longrightarrow CaF_2(s)$$

Ultimately, dynamic equilibrium is reached:

$$CaF_2(s) \rightleftharpoons Ca^{2+}(aq) + 2F^-(aq)$$

At this point no more solid dissolves (the solution is said to be *saturated*).

We can write an equilibrium expression for this process according to the law of mass action:

$$K_{sp} = [Ca^{2+}][F^-]^2$$

where $[Ca^{2+}]$ and $[F^-]$ are expressed in mol/L. The constant K_{sp} is called the **solubility product constant** or simply the **solubility product** for the equilibrium expression.

Since CaF_2 is a pure solid, it is not included in the equilibrium expression. The fact that the amount of excess solid present does not affect the position of the solubility equilibrium might seem strange at first; more solid means more surface area exposed to the solvent, which would seem to result in greater solubility. This is not the case, however. When the ions in solution re-form the solid, they do so on the surface of the solid. Thus doubling the surface area of the solid not only doubles the rate of dissolving, but also doubles the rate of re-formation of the solid. The amount of excess solid present therefore has no effect on the equilibrium position. Similarly, although either increasing the surface area by grinding up the solid or stirring the solution speeds up the attainment of equilibrium, neither procedure changes the amount of solid dissolved at equilibrium. Neither the amount of excess solid nor the size of the particles present will shift the *position* of the solubility equilibrium. ◄ *ⓘ*

It is very important to distinguish between the *solubility* of a given solid and its *solubility product*. The solubility product is an *equilibrium constant* and has only *one* value for a given solid at a given temperature. Solubility, on the other hand, is an *equilibrium position*. In pure water at a specific temperature a given salt has a particular solubility. On the other hand, if a common ion is present in the solution, the solubility varies according to the concentration of the common ion. However, in all cases the product of the ion concentrations must satisfy the K_{sp} expression. ◄ *ⓘ* The K_{sp} values at 25°C for many common ionic solids are listed in Table 15-1. The units are customarily omitted.

Solving solubility equilibria problems requires many of the same procedures we have used to deal with acid–base equilibria, as illustrated in Examples 15-1 and 15-2.

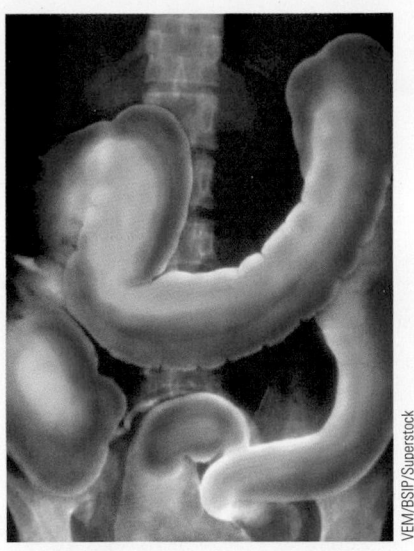

▲ X-ray of barium enema

ⓘ Pure liquids and pure solids are never included in an equilibrium expression.

ⓘ K_{sp} is an equilibrium constant; solubility is an equilibrium position.

INTERACTIVE EXAMPLE 15-1

Calculating K_{sp} from Solubility I

Copper(I) bromide has a measured solubility of 2.0×10^{-4} mol/L at 25°C. Calculate its K_{sp} value.

Solution

In this experiment the solid was placed in contact with water. Thus, before any reaction occurred, the system contained solid CuBr and H_2O. The process that occurs is the dissolving of CuBr to form the separated Cu^+ and Br^- ions:

$$CuBr(s) \rightleftharpoons Cu^+(aq) + Br^-(aq)$$

where

$$K_{sp} = [Cu^+][Br^-]$$

Initially, the solution contains no Cu^+ or Br^-, so the initial concentrations are

$$[Cu^+]_0 = [Br^-]_0 = 0$$

Table 15-1 | K_{sp} Values at 25°C for Common Ionic Solids

Ionic Solid	K_{sp} (at 25°C)	Ionic Solid	K_{sp} (at 25°C)	Ionic Solid	K_{sp} (at 25°C)
Fluorides		Hg_2CrO_4*	2×10^{-9}	$Co(OH)_2$	2.5×10^{-16}
BaF_2	2.4×10^{-5}	$BaCrO_4$	8.5×10^{-11}	$Ni(OH)_2$	1.6×10^{-16}
MgF_2	6.4×10^{-9}	Ag_2CrO_4	9.0×10^{-12}	$Zn(OH)_2$	4.5×10^{-17}
PbF_2	4×10^{-8}	$PbCrO_4$	2×10^{-16}	$Cu(OH)_2$	1.6×10^{-19}
SrF_2	7.9×10^{-10}			$Hg(OH)_2$	3×10^{-26}
CaF_2	4.0×10^{-11}	Carbonates		$Sn(OH)_2$	3×10^{-27}
		$NiCO_3$	1.4×10^{-7}	$Cr(OH)_3$	6.7×10^{-31}
Chlorides		$CaCO_3$	8.7×10^{-9}	$Al(OH)_3$	2×10^{-32}
$PbCl_2$	1.6×10^{-5}	$BaCO_3$	1.6×10^{-9}	$Fe(OH)_3$	4×10^{-38}
$AgCl$	1.6×10^{-10}	$SrCO_3$	7×10^{-10}	$Co(OH)_3$	2.5×10^{-43}
Hg_2Cl_2*	1.1×10^{-18}	$CuCO_3$	2.5×10^{-10}		
		$ZnCO_3$	2×10^{-10}	Sulfides	
Bromides		$MnCO_3$	8.8×10^{-11}	MnS	2.3×10^{-13}
$PbBr_2$	4.6×10^{-6}	$FeCO_3$	2.1×10^{-11}	FeS	3.7×10^{-19}
$AgBr$	5.0×10^{-13}	Ag_2CO_3	8.1×10^{-12}	NiS	3×10^{-21}
Hg_2Br_2*	1.3×10^{-22}	$CdCO_3$	5.2×10^{-12}	CoS	5×10^{-22}
		$PbCO_3$	1.5×10^{-15}	ZnS	2.5×10^{-22}
Iodides		$MgCO_3$	6.8×10^{-6}	SnS	1×10^{-26}
PbI_2	1.4×10^{-8}	Hg_2CO_3*	9.0×10^{-15}	CdS	1.0×10^{-28}
AgI	1.5×10^{-16}			PbS	7×10^{-29}
Hg_2I_2*	4.5×10^{-29}	Hydroxides		CuS	8.5×10^{-45}
		$Ba(OH)_2$	5.0×10^{-3}	Ag_2S	1.6×10^{-49}
Sulfates		$Sr(OH)_2$	3.2×10^{-4}	HgS	1.6×10^{-54}
$CaSO_4$	6.1×10^{-5}	$Ca(OH)_2$	1.3×10^{-6}		
Ag_2SO_4	1.2×10^{-5}	$AgOH$	2.0×10^{-8}	Phosphates	
$SrSO_4$	3.2×10^{-7}	$Mg(OH)_2$	8.9×10^{-12}	Ag_3PO_4	1.8×10^{-18}
$PbSO_4$	1.3×10^{-8}	$Mn(OH)_2$	2×10^{-13}	$Sr_3(PO_4)_2$	1×10^{-31}
$BaSO_4$	1.5×10^{-9}	$Cd(OH)_2$	5.9×10^{-15}	$Ca_3(PO_4)_2$	1.3×10^{-32}
		$Pb(OH)_2$	1.2×10^{-15}	$Ba_3(PO_4)_2$	6×10^{-39}
Chromates		$Fe(OH)_2$	1.8×10^{-15}	$Pb_3(PO_4)_2$	1×10^{-54}
$SrCrO_4$	3.6×10^{-5}				

*Contains Hg_2^{2+} ions. $K = [Hg_2^{2+}][X^-]^2$ for Hg_2X_2 salts, for example.

The equilibrium concentrations can be obtained from the measured solubility of CuBr, which is 2.0×10^{-4} mol/L. This means that 2.0×10^{-4} mole of solid CuBr dissolves per 1.0 L of solution to come to equilibrium with the excess solid. The reaction is

$$CuBr(s) \longrightarrow Cu^+(aq) + Br^-(aq)$$

Thus

2.0×10^{-4} mol/L CuBr(s)

$$\longrightarrow 2.0 \times 10^{-4} \text{ mol/L } Cu^+(aq) + 2.0 \times 10^{-4} \text{ mol/L } Br^-(aq)$$

We can now write the equilibrium concentrations:

$$[Cu^+] = [Cu^+]_0 + \text{ change to reach equilibrium}$$
$$= 0 + 2.0 \times 10^{-4} \text{ mol/L}$$

and

$$[Br^-] = [Br^-]_0 + \text{ change to reach equilibrium}$$
$$= 0 + 2.0 \times 10^{-4} \text{ mol/L}$$

These equilibrium concentrations allow us to calculate the value of K_{sp} for CuBr:

$$K_{sp} = [Cu^+][Br^-] = (2.0 \times 10^{-4} \text{ mol/L})(2.0 \times 10^{-4} \text{ mol/L})$$
$$= 4.0 \times 10^{-8} \text{ mol}^2/L^2 = 4.0 \times 10^{-8}$$

The units for K_{sp} values are usually omitted.

See Exercises 15-21 and 15-22

INTERACTIVE EXAMPLE 15-2

Calculating K_{sp} from Solubility II

Calculate the K_{sp} value for bismuth sulfide (Bi_2S_3), which has a solubility of 1.0×10^{-15} mol/L at 25°C. ◄

Solution

The system initially contains H_2O and solid Bi_2S_3, which dissolves as follows:

$$Bi_2S_3(s) \rightleftharpoons 2Bi^{3+}(aq) + 3S^{2-}(aq)$$

Therefore, $$K_{sp} = [Bi^{3+}]^2[S^{2-}]^3$$

Since no Bi^{3+} and S^{2-} ions were present in solution before the Bi_2S_3 dissolved,

$$[Bi^{3+}]_0 = [S^{2-}]_0 = 0$$

Thus the equilibrium concentrations of these ions will be determined by the amount of salt that dissolves to reach equilibrium, which in this case is 1.0×10^{-15} mol/L. Since each Bi_2S_3 unit contains $2Bi^{3+}$ and $3S^{2-}$ ions: ◄ *(i)*

$$1.0 \times 10^{-15} \text{ mol/L } Bi_2S_3(s)$$
$$\longrightarrow 2(1.0 \times 10^{-15} \text{ mol/L}) Bi^{3+}(aq) + 3(1.0 \times 10^{-15} \text{ mol/L}) S^{2-}(aq)$$

The equilibrium concentrations are ◄ *(i)*

$$[Bi^{3+}] = [Bi^{3+}]_0 + \text{change} = 0 + 2.0 \times 10^{-15} \text{ mol/L}$$
$$[S^{2-}] = [S^{2-}]_0 + \text{change} = 0 + 3.0 \times 10^{-15} \text{ mol/L}$$

Then

$$K_{sp} = [Bi^{3+}]^2[S^{2-}]^3 = (2.0 \times 10^{-15})^2(3.0 \times 10^{-15})^3 = 1.1 \times 10^{-73}$$

See Exercises 15-23 through 15-26

▲ Precipitation of bismuth sulfide.

Photo by Ken O'Donoghue © Cengage Learning

(i) Sulfide is a very basic anion and really exists in water as HS^-. We will not consider this complication.

(i) Solubilities must be expressed in mol/L in K_{sp} calculations.

We have seen that the experimentally determined solubility of an ionic solid can be used to calculate its K_{sp} value.* The reverse is also possible: The solubility of an ionic solid can be calculated if its K_{sp} value is known.

---

*This calculation assumes that all the dissolved solid is present as separated ions. In some cases, such as $CaSO_4$, large numbers of ion pairs exist in solution, so this method yields an incorrect value for K_{sp}.

If dental chemistry continues to progress at the present rate, tooth decay may soon be a thing of the past. Cavities are holes that develop in tooth enamel, which is composed of the mineral hydroxyapatite, $Ca_5(PO_4)_3OH$. Recent research has shown that there is constant dissolving and re-forming of the tooth mineral in the saliva at the tooth's surface. Demineralization (dissolving of tooth enamel) is mainly caused by weak acids in the saliva created by bacteria as they metabolize carbohydrates in food. (The solubility of $Ca_5(PO_4)_3OH$ in acidic saliva should come as no surprise to you if you understand how pH affects the solubility of a salt with basic anions.)

In the first stages of tooth decay, parts of the tooth surface become porous and spongy and develop Swiss-cheese-like holes that, if untreated, eventually turn into cavities (see photo). However, recent results indicate that if the affected tooth is bathed in a solution containing appropriate amounts of Ca^{2+}, PO_4^{3-}, and F^-, it remineralizes. Because the F^- replaces OH^- in the tooth mineral ($Ca_5(PO_4)_3OH$ is changed to $Ca_5(PO_4)_3F$), the remineralized area is more resistant to future decay, since fluoride is a weaker base than hydroxide ion. In addition, it has been shown that the presence of Sr^{2+} in the remineralizing fluid significantly increases resistance to decay.

If these results hold up under further study, the work of dentists will change dramatically. Dentists will be much more

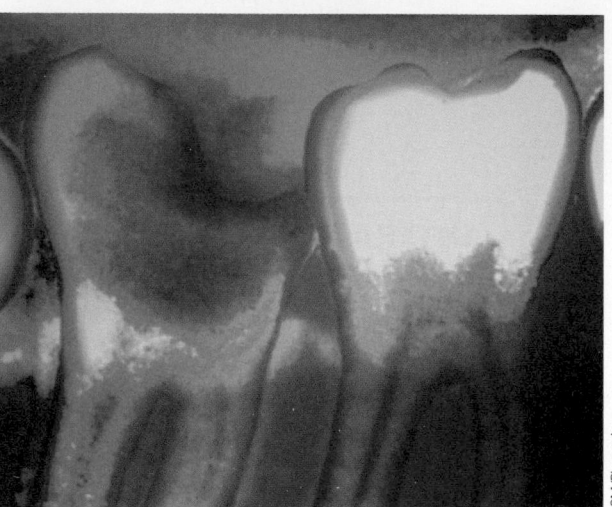

X-ray photo showing decay (dark area) of the molar.

involved in preventing damage to teeth than in repairing damage that has already occurred. One can picture the routine use of a remineralization rinse that will repair problem areas before they become cavities. Dental drills could join leeches as a medical anachronism.

INTERACTIVE EXAMPLE 15-3

Calculating Solubility from K_{sp}

The K_{sp} value for copper(II) iodate, $Cu(IO_3)_2$, is 1.4×10^{-7} at 25°C. Calculate its solubility at 25°C.

Solution

The system initially contains H_2O and solid $Cu(IO_3)_2$, which dissolves according to the following equilibrium:

$$Cu(IO_3)_2(s) \rightleftharpoons Cu^{2+}(aq) + 2IO_3^-(aq)$$

Therefore,

$$K_{sp} = [Cu^{2+}][IO_3^-]^2$$

To find the solubility of $Cu(IO_3)_2$, we must find the equilibrium concentrations of the Cu^{2+} and IO_3^- ions. We do this in the usual way by specifying the initial concentrations (before any solid has dissolved) and then defining the change required to reach equilibrium. Since in this case we do not know the solubility, we will assume that x mol/L of the solid dissolves to reach equilibrium. The 1:2 stoichiometry of the salt means that

$$x \text{ mol/L } Cu(IO_3)_2(s) \longrightarrow x \text{ mol/L } Cu^{2+}(aq) + 2x \text{ mol/L } IO_3^-(aq)$$

The concentrations are as follows:

	Initial Concentration (mol/L) [before any $Cu(IO_3)_2$ dissolves]		Equilibrium Concentration (mol/L)
	$[Cu^{2+}]_0 = 0$ $[IO_3^-]_0 = 0$	$\xrightarrow[\substack{\text{to reach} \\ \text{equilibrium}}]{\substack{x \text{ mol/L} \\ \text{dissolves}}}$	$[Cu^{2+}] = x$ $[IO_3^-] = 2x$

Substituting the equilibrium concentrations into the expression for K_{sp} gives

$$1.4 \times 10^{-7} = K_{sp} = [Cu^{2+}][IO_3^-]^2 = (x)(2x)^2 = 4x^3$$

Then

$$x = \sqrt[3]{3.5 \times 10^{-8}} = 3.3 \times 10^{-3} \text{ mol/L}$$

Thus the solubility of solid $Cu(IO_3)_2$ is 3.3×10^{-3} mol/L.

See Exercises 15-27 and 15-28

Relative Solubilities

A salt's K_{sp} value gives us information about its solubility. However, we must be careful in using K_{sp} values to predict the *relative* solubilities of a group of salts. There are two possible cases:

1. The salts being compared produce the same number of ions. For example, consider

$$AgI(s) \qquad K_{sp} = 1.5 \times 10^{-16}$$
$$CuI(s) \qquad K_{sp} = 5.0 \times 10^{-12}$$
$$CaSO_4(s) \qquad K_{sp} = 6.1 \times 10^{-5}$$

Each of these solids dissolves to produce two ions:

$$\text{Salt} \rightleftharpoons \text{cation} + \text{anion}$$
$$K_{sp} = [\text{cation}][\text{anion}]$$

If x is the solubility in mol/L, then at equilibrium

$$[\text{Cation}] = x$$
$$[\text{Anion}] = x$$
$$K_{sp} = [\text{cation}][\text{anion}] = x^2$$
$$x = \sqrt{K_{sp}} = \text{solubility}$$

Therefore, in this case we can compare the solubilities for these solids by comparing the K_{sp} values:

$$\underset{\substack{\text{Most soluble;} \\ \text{largest } K_{sp}}}{CaSO_4(s)} > CuI(s) > \underset{\substack{\text{Least soluble;} \\ \text{smallest } K_{sp}}}{AgI(s)}$$

2. The salts being compared produce different numbers of ions. For example, consider

$$CuS(s) \qquad K_{sp} = 8.5 \times 10^{-45}$$
$$Ag_2S(s) \qquad K_{sp} = 1.6 \times 10^{-49}$$
$$Bi_2S_3(s) \qquad K_{sp} = 1.1 \times 10^{-73}$$

Table 15-2 | Calculated Solubilities for CuS, Ag$_2$S, and Bi$_2$S$_3$ at 25°C

Salt	K_{sp}	Calculated Solubility (mol/L)
CuS	8.5×10^{-45}	9.2×10^{-23}
Ag$_2$S	1.6×10^{-49}	3.4×10^{-17}
Bi$_2$S$_3$	1.1×10^{-73}	1.0×10^{-15}

Because these salts produce different numbers of ions when they dissolve, the K_{sp} values cannot be compared *directly* to determine relative solubilities. In fact, if we calculate the solubilities (using the procedure in Example 15-3), we obtain the results summarized in Table 15-2. The order of solubilities is

$$\text{Bi}_2\text{S}_3(s) \underset{\text{Most soluble}}{} > \text{Ag}_2\text{S}(s) > \underset{\text{Least soluble}}{\text{CuS}(s)}$$

which is opposite to the order of the K_{sp} values.

Remember that relative solubilities can be predicted by comparing K_{sp} values *only* for salts that produce the same total number of ions.

Common Ion Effect

So far we have considered ionic solids dissolved in pure water. We will now see what happens when the water contains an ion in common with the dissolving salt. For example, consider the solubility of solid silver chromate (Ag$_2$CrO$_4$, $K_{sp} = 9.0 \times 10^{-12}$) in a 0.100-$M$ solution of AgNO$_3$. ◄ Before any Ag$_2$CrO$_4$ dissolves, the solution contains the major species Ag$^+$, NO$_3^-$, and H$_2$O, with solid Ag$_2$CrO$_4$ on the bottom of the container. Since NO$_3^-$ is not found in Ag$_2$CrO$_4$, we can ignore it. The relevant initial concentrations (before any Ag$_2$CrO$_4$ dissolves) are

$$[\text{Ag}^+]_0 = 0.100 \, M \text{ (from the dissolved AgNO}_3\text{)}$$
$$[\text{CrO}_4^{2-}]_0 = 0$$

The system comes to equilibrium as the solid Ag$_2$CrO$_4$ dissolves according to the reaction

$$\text{Ag}_2\text{CrO}_4(s) \rightleftharpoons 2\text{Ag}^+(aq) + \text{CrO}_4^{2-}(aq)$$

for which
$$K_{sp} = [\text{Ag}^+]^2[\text{CrO}_4^{2-}] = 9.0 \times 10^{-12}$$

We assume that x mol/L of Ag$_2$CrO$_4$ dissolves to reach equilibrium, which means that

$$x \text{ mol/L Ag}_2\text{CrO}_4(s) \longrightarrow 2x \text{ mol/L Ag}^+(aq) + x \text{ mol/L CrO}_4^{2-}$$

Now we can specify the equilibrium concentrations in terms of x:

$$[\text{Ag}^+] = [\text{Ag}^+]_0 + \text{change} = 0.100 + 2x$$
$$[\text{CrO}_4^{2-}] = [\text{CrO}_4^{2-}]_0 + \text{change} = 0 + x = x$$

Substituting these concentrations into the expression for K_{sp} gives

$$9.0 \times 10^{-12} = [\text{Ag}^+]^2[\text{CrO}_4^{2-}] = (0.100 + 2x)^2(x)$$

The mathematics required here appear to be complicated, since the multiplication of terms on the right-hand side produces an expression that contains an x^3 term. However, as is usually the case, we can make simplifying assumptions. Since the K_{sp} value

▲ A potassium chromate solution being added to aqueous silver nitrate, forming silver chromate.

Photo by Ken O'Donoghue © Cengage Learning

for Ag_2CrO_4 is small (the position of the equilibrium lies far to the left), x is expected to be small compared with $0.100\ M$. Therefore, $0.100 + 2x \approx 0.100$, which allows simplification of the expression:

$$9.0 \times 10^{-12} = (0.100 + 2x)^2(x) \approx (0.100)^2(x)$$

Then
$$x \approx \frac{9.0 \times 10^{-12}}{(0.100)^2} = 9.0 \times 10^{-10}\ mol/L.$$

Since x is much less than $0.100\ M$, the approximation is valid (by the 5% rule). Thus

Solubility of Ag_2CrO_4 in $0.100\ M\ AgNO_3 = x = 9.0 \times 10^{-10}\ mol/L$

and the equilibrium concentrations are

$$[Ag^+] = 0.100 + 2x = 0.100 + 2(9.0 \times 10^{-10}) = 0.100\ M$$
$$[CrO_4^{2-}] = x = 9.0 \times 10^{-10}\ M$$

Now we compare the solubilities of Ag_2CrO_4 in pure water and in $0.100\ M\ AgNO_3$:

Solubility of Ag_2CrO_4 in pure water $= 1.3 \times 10^{-4}\ mol/L$

Solubility of Ag_2CrO_4 in $0.100\ M\ AgNO_3 = 9.0 \times 10^{-10}\ mol/L$

Note that the solubility of Ag_2CrO_4 is much less in the presence of Ag^+ ions from $AgNO_3$. This is another example of the common ion effect. The solubility of a solid is lowered if the solution already contains ions common to the solid. ◄ ⓘ

ⓘ Note that the common ion effect is completely consistent with Le Châtelier's principle. Le Châtelier's principle tells us that if we had a salt in equilibrium with its ions (for example, AgCl) and added a different salt containing one of the ions in solution (for example, NaCl in this case), equilibrium would shift to the left, thus lowering the solubility (of AgCl in our example). By having the chloride ions in solution before the solid salt (AgCl) is added to water, we get the same effect of lowered solubility.

Critical Thinking

What if all you know about two salts is that the value of K_{sp} for salt A is greater than that of salt B? Why can we not compare relative solubilities of the salts? Use numbers to show how salt A could be more soluble than salt B, and how salt B could be more soluble than salt A.

INTERACTIVE EXAMPLE 15-4

Solubility and Common Ions

Calculate the solubility of solid CaF_2 ($K_{sp} = 4.0 \times 10^{-11}$) in a 0.025-$M$ NaF solution.

Solution

Before any CaF_2 dissolves, the solution contains the major species Na^+, F^-, and H_2O. The solubility equilibrium for CaF_2 is

$$CaF_2(s) \rightleftharpoons Ca^{2+}(aq) + 2F^-(aq)$$

and
$$K_{sp} = 4.0 \times 10^{-11} = [Ca^{2+}][F^-]^2$$

Initial Concentration (mol/L) (before any CaF_2 dissolves)		Equilibrium Concentration (mol/L)
$[Ca^{2+}]_0 = 0$ $[F^-]_0 = 0.025\ M$ ↗ From 0.025 M NaF	$\xrightarrow[\text{to reach equilibrium}]{\substack{x\ mol/L\ CaF_2 \\ \text{dissolves}}}$	$[Ca^{2+}] = x$ $[F^-] = 0.025 + 2x$ ↗ ↗ From NaF From CaF_2

Substituting the equilibrium concentrations into the expression for K_{sp} gives

$$K_{sp} = 4.0 \times 10^{-11} = [Ca^{2+}][F^-]^2 = (x)(0.025 + 2x)^2$$

Assuming that $2x$ is negligible compared with 0.025 (since K_{sp} is small) gives

$$4.0 \times 10^{-11} \approx (x)(0.025)^2$$
$$x \approx 6.4 \times 10^{-8}$$

The approximation is valid (by the 5% rule), and

$$\text{Solubility} = x = 6.4 \times 10^{-8} \text{ mol/L}$$

Thus 6.4×10^{-8} mole of solid CaF_2 dissolves per liter of the 0.025-M NaF solution.

See Exercises 15-37 through 15-42

pH and Solubility

The pH of a solution can greatly affect a salt's solubility. For example, magnesium hydroxide dissolves according to the equilibrium

$$Mg(OH)_2(s) \rightleftharpoons Mg^{2+}(aq) + 2OH^-(aq)$$

Addition of OH^- ions (an increase in pH) will, by the common ion effect, force the equilibrium to the left, decreasing the solubility of $Mg(OH)_2$. On the other hand, an addition of H^+ ions (a decrease in pH) increases the solubility, because OH^- ions are removed from solution by reacting with the added H^+ ions. In response to the lower concentration of OH^-, the equilibrium position moves to the right. This is why a suspension of solid $Mg(OH)_2$, known as *milk of magnesia,* dissolves as required in the stomach to combat excess acidity.

This idea also applies to salts with other types of anions. For example, the solubility of silver phosphate (Ag_3PO_4) is greater in acid than in pure water because the PO_4^{3-} ion is a strong base that reacts with H^+ to form the HPO_4^{2-} ion. The reaction

$$H^+(aq) + PO_4^{3-}(aq) \longrightarrow HPO_4^{2-}(aq)$$

occurs in acidic solution, thus lowering the concentration of PO_4^{3-} and shifting the solubility equilibrium

$$Ag_3PO_4(s) \rightleftharpoons 3Ag^+(aq) + PO_4^{3-}(aq)$$

to the right. This, in turn, increases the solubility of silver phosphate.

Silver chloride (AgCl), however, has the same solubility in acid as in pure water. Why? Since the Cl^- ion is a very weak base (that is, HCl is a very strong acid), no HCl molecules are formed. Thus the addition of H^+ to a solution containing Cl^- does not affect $[Cl^-]$ and has no effect on the solubility of a chloride salt.

The general rule is that if the anion X^- is an effective base—that is, if HX is a weak acid—the salt MX will show increased solubility in an acidic solution. Examples of common anions that are effective bases are OH^-, S^{2-}, CO_3^{2-}, $C_2O_4^{2-}$, and CrO_4^{2-}. Salts containing these anions are much more soluble in an acidic solution than in pure water.

As mentioned at the beginning of this chapter, one practical result of the increased solubility of carbonates in acid is the formation of huge limestone caves such as Mammoth Cave in Kentucky and Carlsbad Caverns in New Mexico. Carbon dioxide dissolved in groundwater makes it acidic, increasing the solubility of calcium carbonate and eventually producing huge caverns. As the carbon dioxide escapes to the air, the pH of the dripping water goes up and the calcium carbonate precipitates, forming stalactites and stalagmites. ◀

▲ Stalactites and stalagmites in Luray Cavern in Virginia.

Adambies/Dreamtime.com

15-2 | Precipitation and Qualitative Analysis

So far we have considered solids dissolving in solutions. Now we will consider the reverse process—the formation of a solid from solution. When solutions are mixed, various reactions can occur. We have already considered acid–base reactions in some detail. In this section we show how to predict whether a precipitate will form when two solutions are mixed. We will use the **ion product**, which is defined just like the expression for K_{sp} for a given solid except that *initial concentrations are used* instead of equilibrium concentrations. For solid CaF_2, the expression for the ion product Q is written

$$Q = [Ca^{2+}]_0[F^-]_0^2$$

If we add a solution containing Ca^{2+} ions to a solution containing F^- ions, a precipitate may or may not form, depending on the concentrations of these ions in the resulting mixed solution. To predict whether precipitation will occur, we consider the relationship between Q and K_{sp}.

Q is used here in a very similar way to the use of the reaction quotient in Chapter 12.

If Q is greater than K_{sp}, precipitation occurs and will continue until the concentrations are reduced to the point that they satisfy K_{sp}.

If Q is less than K_{sp}, no precipitation occurs. ◂ ⓘ

INTERACTIVE EXAMPLE 15-5

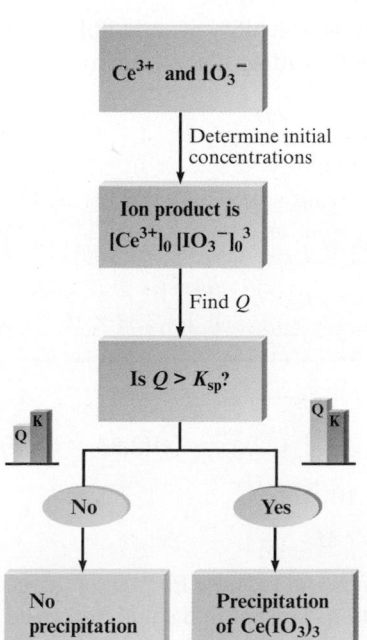

Determining Precipitation Conditions

A solution is prepared by adding 750.0 mL of 4.00×10^{-3} M $Ce(NO_3)_3$ to 300.0 mL of 2.00×10^{-2} M KIO_3. Will $Ce(IO_3)_3$ ($K_{sp} = 1.9 \times 10^{-10}$) precipitate from this solution?

Solution

First, we calculate $[Ce^{3+}]_0$ and $[IO_3^-]_0$ in the mixed solution before any reaction occurs:

$$[Ce^{3+}]_0 = \frac{(750.0 \text{ mL})(4.00 \times 10^{-3} \text{ mmol/mL})}{(750.0 + 300.0) \text{ mL}} = 2.86 \times 10^{-3} \, M$$

$$[IO_3^-]_0 = \frac{(300.0 \text{ mL})(2.00 \times 10^{-2} \text{ mmol/mL})}{(750.0 + 300.0) \text{ mL}} = 5.71 \times 10^{-3} \, M$$

The ion product for $Ce(IO_3)_3$ is

$$Q = [Ce^{3+}]_0[IO_3^-]_0^3 = (2.86 \times 10^{-3})(5.71 \times 10^{-3})^3 = 5.32 \times 10^{-10}$$

Since Q is greater than K_{sp}, $Ce(IO_3)_3$ will precipitate from the mixed solution. ◂

See Exercises 15-49 through 15-52

Sometimes we want to do more than simply predict whether precipitation will occur; we may want to calculate the equilibrium concentrations in the solution after precipitation occurs. For example, let us calculate the equilibrium concentrations of Pb^{2+} and I^- ions in a solution formed by mixing 100.0 mL of 0.0500 M $Pb(NO_3)_2$ and 200.0 mL of 0.100 M NaI. First, we must determine whether solid

PbI_2 ($K_{sp} = 1.4 \times 10^{-8}$) forms when the solutions are mixed. To do so, we need to calculate $[Pb^{2+}]_0$ and $[I^-]_0$ before any reaction occurs:

$$[Pb^{2+}]_0 = \frac{\text{mmol } Pb^{2+}}{\text{mL solution}} = \frac{(100.0 \text{ mL})(0.0500 \text{ mmol/mL})}{300.0 \text{ mL}} = 1.67 \times 10^{-2} \, M$$

$$[I^-]_0 = \frac{\text{mmol } I^-}{\text{mL solution}} = \frac{(200.0 \text{ mL})(0.100 \text{ mmol/mL})}{300.0 \text{ mL}} = 6.67 \times 10^{-2} \, M$$

The ion product for PbI_2 is

$$Q = [Pb^{2+}]_0[I^-]_0^2 = (1.67 \times 10^{-2})(6.67 \times 10^{-2})^2 = 7.43 \times 10^{-5}$$

Since Q is greater than K_{sp}, a precipitate of PbI_2 will form.

Since the K_{sp} for PbI_2 is quite small (1.4×10^{-8}), only very small quantities of Pb^{2+} and I^- can coexist in aqueous solution. In other words, when Pb^{2+} and I^- are mixed, most of these ions will precipitate out as PbI_2. That is, the reaction

$$Pb^{2+}(aq) + 2I^-(aq) \longrightarrow PbI_2(s)$$

(which is the reverse of the dissolution reaction) goes essentially to completion. ◄ ⓘ

If, when two solutions are mixed, a reaction occurs that goes virtually to completion, it is essential to do the stoichiometry calculations before considering the equilibrium calculations. Therefore, in this case we let the system go completely in the direction toward which it tends. Then we will let it adjust back to equilibrium. If we let Pb^{2+} and I^- react to completion, we have the following concentrations:

ⓘ The equilibrium constant for formation of solid PbI_2 is $1/K_{sp}$, or 7×10^7, so this equilibrium lies far to the right.

	Pb^{2+}	+	$2I^-$	$\longrightarrow$	PbI_2
Before reaction	(100.0 mL)(0.0500 M) = 5.00 mmol		(200.0 mL)(0.100 M) = 20.0 mmol		The amount of PbI_2 formed does not influence the equilibrium.
After reaction	0 mmol		20.0 − 2(5.00) = 10.0 mmol ◄ ⓘ		

ⓘ In this reaction, 10.0 mmol I^- is in excess.

Next we must allow the system to adjust to equilibrium. At equilibrium $[Pb^{2+}]$ is not actually zero because the reaction does not go quite to completion. The best way to think about this is that once the PbI_2 is formed, a very small amount redissolves to reach equilibrium. Since I^- is in excess, the PbI_2 is dissolving into a solution that contains 10.0 mmol I^- per 300.0 mL solution, or $3.33 \times 10^{-2} \, M \, I^-$.

We could state this problem as follows: What is the solubility of solid PbI_2 in a $3.33 \times 10^{-2} \, M$-NaI solution? The lead iodide dissolves according to the equation

$$PbI_2(s) \rightleftharpoons Pb^{2+}(aq) + 2I^-(aq)$$

The concentrations are as follows:

Initial Concentration (mol/L)		Equilibrium Concentration (mol/L)
$[Pb^{2+}]_0 = 0$ $[I^-]_0 = 3.33 \times 10^{-2}$	$\xrightarrow[\text{dissolves}]{\begin{array}{c} x \text{ mol/L} \\ PbI_2(s) \end{array}}$	$[Pb^{2+}] = x$ $[I^-] = 3.33 \times 10^{-2} + 2x$

Substituting into the expression for K_{sp} gives

$$K_{sp} = 1.4 \times 10^{-8} = [Pb^{2+}][I^-]^2 = (x)(3.33 \times 10^{-2} + 2x)^2 \approx (x)(3.33 \times 10^{-2})^2$$

Then

$$[Pb^{2+}] = x = 1.3 \times 10^{-5} \, M$$

$$[I^-] = 3.33 \times 10^{-2} \, M$$

Note that $3.33 \times 10^{-2} \gg 2x$, so the approximation is valid. These Pb^{2+} and I^- concentrations thus represent the equilibrium concentrations present in a solution formed by mixing 100.0 mL of 0.0500 M $Pb(NO_3)_2$ and 200.0 mL of 0.100 M NaI.

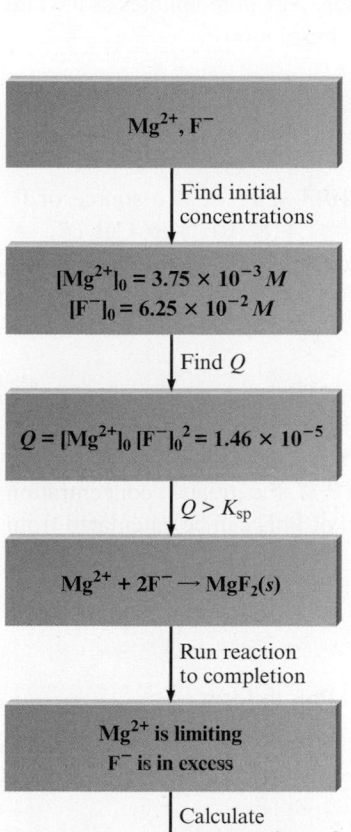

INTERACTIVE EXAMPLE 15-6

Precipitation

A solution is prepared by mixing 150.0 mL of $1.00 \times 10^{-2} \ M$ Mg(NO$_3$)$_2$ and 250.0 mL of $1.00 \times 10^{-1} \ M$ NaF. Calculate the concentrations of Mg^{2+} and F$^-$ at equilibrium with solid MgF$_2$ ($K_{sp} = 6.4 \times 10^{-9}$).

Solution

The first step is to determine whether solid MgF$_2$ forms. To do this, we need to calculate the concentrations of Mg^{2+} and F$^-$ in the mixed solution and find Q:

$$[Mg^{2+}]_0 = \frac{\text{mmol Mg}^{2+}}{\text{mL solution}} = \frac{(150.0 \text{ mL})(1.00 \times 10^{-2} \ M)}{400.0 \text{ mL}} = 3.75 \times 10^{-3} \ M$$

$$[F^-]_0 = \frac{\text{mmol F}^-}{\text{mL solution}} = \frac{(250.0 \text{ mL})(1.00 \times 10^{-1} \ M)}{400.0 \text{ mL}} = 6.25 \times 10^{-2} \ M$$

$$Q = [Mg^{2+}]_0[F^-]_0^2 = (3.75 \times 10^{-3})(6.25 \times 10^{-2})^2 = 1.46 \times 10^{-5}$$

Since Q is greater than K_{sp}, solid MgF$_2$ will form.

The next step is to run the precipitation reaction to completion:

	Mg^{2+}	+	2F$^-$	$\longrightarrow$	MgF$_2$(s)
Before reaction	(150.0)(1.00 $\times$ 10^{-2}) = 1.50 mmol		(250.0)(1.00 $\times$ 10^{-1}) = 25.0 mmol		
After reaction	1.50 − 1.50 = 0		25.0 − 2(1.50) = 22.0 mmol		

Note that excess F$^-$ remains after the precipitation reaction goes to completion. The concentration is

$$[F^-]_{excess} = \frac{22.0 \text{ mmol}}{400.0 \text{ mL}} = 5.50 \times 10^{-2} \ M$$

Although we have assumed that the Mg^{2+} is completely consumed, we know that [Mg^{2+}] will not be zero at equilibrium. We can compute the equilibrium [Mg^{2+}] by letting MgF$_2$ redissolve to satisfy the expression for K_{sp}. How much MgF$_2$ will dissolve in a 5.50×10^{-2}-M NaF solution? We proceed as usual:

$$MgF_2(s) \rightleftharpoons Mg^{2+}(aq) + 2F^-(aq)$$

$$K_{sp} = [Mg^{2+}][F^-]^2 = 6.4 \times 10^{-9}$$

Initial Concentration (mol/L)		Equilibrium Concentration (mol/L)
[Mg^{2+}]$_0$ = 0 [F$^-$]$_0$ = 5.50 $\times$ 10^{-2}	$\xrightarrow[\text{dissolves}]{\begin{array}{c} x \text{ mol/L} \\ \text{MgF}_2(s) \end{array}}$	[Mg^{2+}] = x [F$^-$] = 5.50 $\times$ 10^{-2} + 2x

$$K_{sp} = 6.4 \times 10^{-9} = [Mg^{2+}][F^-]^2$$
$$= (x)(5.50 \times 10^{-2} + 2x)^2 \approx (x)(5.50 \times 10^{-2})^2$$
$$[Mg^{2+}] = x = 2.1 \times 10^{-6} \ M$$
$$[F^-] = 5.50 \times 10^{-2} \ M \ \blacktriangleleft$$

See Exercises 15-53 through 15-56

Selective Precipitation

Mixtures of metal ions in aqueous solution are often separated by **selective precipitation**, that is, by using a reagent whose anion forms a precipitate with only one or a few of the metal ions in the mixture. For example, suppose we have a solution containing both Ba^{2+} and Ag^+ ions. If NaCl is added to the solution, AgCl precipitates as a white solid, but since $BaCl_2$ is soluble, the Ba^{2+} ions remain in solution.

INTERACTIVE EXAMPLE 15-7

Selective Precipitation

A solution contains $1.0 \times 10^{-4}\ M$ Cu^+ and $2.0 \times 10^{-3}\ M$ Pb^{2+}. If a source of I^- is added gradually to this solution, will PbI_2 ($K_{sp} = 1.4 \times 10^{-8}$) or CuI ($K_{sp} = 5.3 \times 10^{-12}$) precipitate first? Specify the concentration of I^- necessary to begin precipitation of each salt.

Solution

For PbI_2, the K_{sp} expression is

$$1.4 \times 10^{-8} = K_{sp} = [Pb^{2+}][I^-]^2$$

Since $[Pb^{2+}]$ in this solution is known to be $2.0 \times 10^{-3}\ M$, the greatest concentration of I^- that can be present without causing precipitation of PbI_2 can be calculated from the K_{sp} expression: ◀ *ⓘ*

ⓘ The approximations made here fall within the 5% rule.

$$1.4 \times 10^{-8} = [Pb^{2+}][I^-]^2 = (2.0 \times 10^{-3})[I^-]^2$$
$$[I^-] = 2.6 \times 10^{-3}\ M$$

Any I^- in excess of this concentration will cause solid PbI_2 to form.

Similarly, for CuI, the K_{sp} expression is

$$5.3 \times 10^{-12} = K_{sp} = [Cu^+][I^-] = (1.0 \times 10^{-4})[I^-]$$

and
$$[I^-] = 5.3 \times 10^{-8}\ M$$

A concentration of I^- in excess of $5.3 \times 10^{-8}\ M$ will cause formation of solid CuI.

As I^- is added to the mixed solution, CuI will precipitate first, since the $[I^-]$ required is less. Therefore, Cu^+ would be separated from Pb^{2+} using this reagent.

See Exercises 15-57 through 15-60

ⓘ We can compare K_{sp} values to find relative solubilities because FeS and MnS produce the same number of ions in solution.

Since metal sulfide salts differ dramatically in their solubilities, the sulfide ion is often used to separate metal ions by selective precipitation. For example, consider a solution containing a mixture of $10^{-3}\ M$ Fe^{2+} and $10^{-3}\ M$ Mn^{2+}. Since FeS ($K_{sp} = 3.7 \times 10^{-19}$) is much less soluble than MnS ($K_{sp} = 2.3 \times 10^{-13}$), careful addition of S^{2-} to the mixture will precipitate Fe^{2+} as FeS, leaving Mn^{2+} in solution. ◀ *ⓘ*

One real advantage of the sulfide ion as a precipitating reagent is that because it is basic, its concentration can be controlled by regulating the pH of the solution. H_2S is a diprotic acid that dissociates in two steps:

$$H_2S \rightleftharpoons H^+ + HS^- \qquad K_{a_1} = 1.0 \times 10^{-7}$$
$$HS^- \rightleftharpoons H^+ + S^{2-} \qquad K_{a_2} \approx 10^{-19}$$

Note from the small K_{a_2} value that S^{2-} ions have a high affinity for protons. In an acidic solution (large $[H^+]$), $[S^{2-}]$ will be relatively small, since under these conditions the dissociation equilibria will lie far to the left. On the other hand, in basic solutions $[S^{2-}]$ will be relatively large, since the very small value of $[H^+]$ will pull both equilibria to the right, producing S^{2-}.

Figure 15-1 | The separation of Cu^{2+} and Hg^{2+} from Ni^{2+} and Mn^{2+} using H_2S. At a low pH, $[S^{2-}]$ is relatively low and only the very insoluble HgS and CuS precipitate. When OH^- is added to lower $[H^+]$, the value of $[S^{2-}]$ increases, and MnS and NiS precipitate.

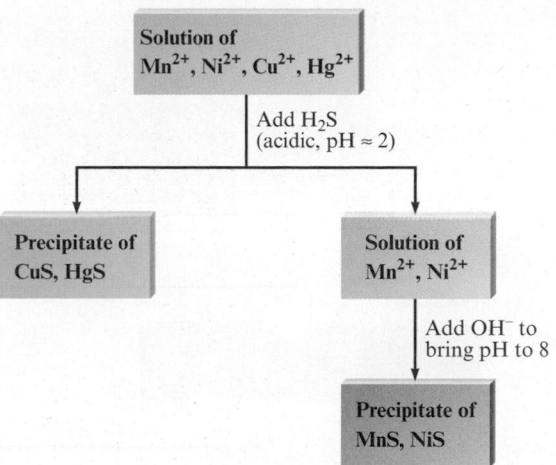

Solution of Mn^{2+}, Ni^{2+}, Cu^{2+}, Hg^{2+}

Add H_2S (acidic, pH ≈ 2)

Precipitate of CuS, HgS

Solution of Mn^{2+}, Ni^{2+}

Add OH^- to bring pH to 8

Precipitate of MnS, NiS

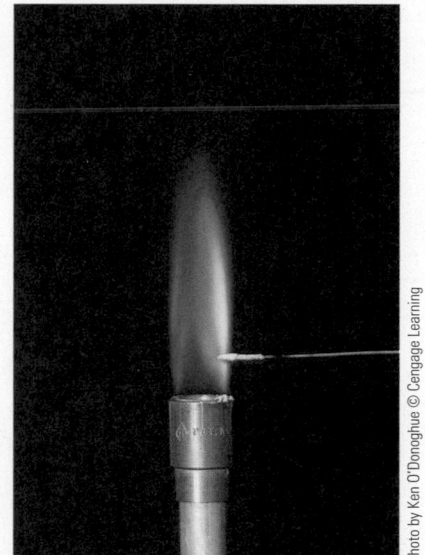

▲ (top) Flame test for potassium. (bottom) Flame test for sodium.

This means that the most insoluble sulfide salts, such as CuS ($K_{sp} = 8.5 \times 10^{-45}$) and HgS ($K_{sp} = 1.6 \times 10^{-54}$), can be precipitated from an acidic solution, leaving the more soluble ones, such as MnS ($K_{sp} = 2.3 \times 10^{-13}$) and NiS ($K_{sp} = 3 \times 10^{-21}$), still dissolved. The manganese and nickel sulfides can then be precipitated by making the solution slightly basic. This procedure is diagrammed in Fig. 15-1.

Qualitative Analysis

The classic scheme for **qualitative analysis** of a mixture containing all the common cations (listed in Fig. 15-2) involves first separating them into five major groups based on solubilities. (These groups are not directly related to the groups of the periodic table.) Each group is then treated further to separate and identify the individual ions. We will be concerned here only with separation of the major groups.

Group I—Insoluble chlorides

When dilute aqueous HCl is added to a solution containing a mixture of the common cations, only Ag^+, Pb^{2+}, and Hg_2^{2+} will precipitate out as insoluble chlorides. All other chlorides are soluble and remain in solution. The Group I precipitate is removed, leaving the other ions in solution for treatment with sulfide ion.

Group II—Sulfides insoluble in acid solution

After the insoluble chlorides are removed, the solution is still acidic, since HCl was added. If H_2S is added to this solution, only the most insoluble sulfides (those of Hg^{2+}, Cd^{2+}, Bi^{3+}, Cu^{2+}, and Sn^{4+}) will precipitate, since $[S^{2-}]$ is relatively low because of the high concentration of H^+. The more soluble sulfides will remain dissolved under these conditions, and the precipitate of the insoluble salt is removed.

Group III—Sulfides insoluble in basic solution

The solution is made basic at this stage, and more H_2S is added. As we saw earlier, a basic solution produces a higher $[S^{2-}]$, which leads to precipitation of the more soluble sulfides. The cations precipitated as sulfides at this stage are Co^{2+}, Zn^{2+}, Mn^{2+}, Ni^{2+}, and Fe^{2+}. If any Cr^{3+} and Al^{3+} ions are present, they also will precipitate, but as insoluble hydroxides (remember the solution is now basic). The precipitate is separated from the solution containing the rest of the ions. ◄

Group IV—Insoluble carbonates

At this point, all the cations have been precipitated except those from Groups 1A and 2A of the periodic table. The Group 2A cations form insoluble carbonates and can be precipitated by the addition of CO_3^{2-}. For example, Ba^{2+}, Ca^{2+}, and Mg^{2+} form solid carbonates and can be removed from the solution.

Figure 15-2 | A schematic diagram of the classic method for separating the common cations by selective precipitation.

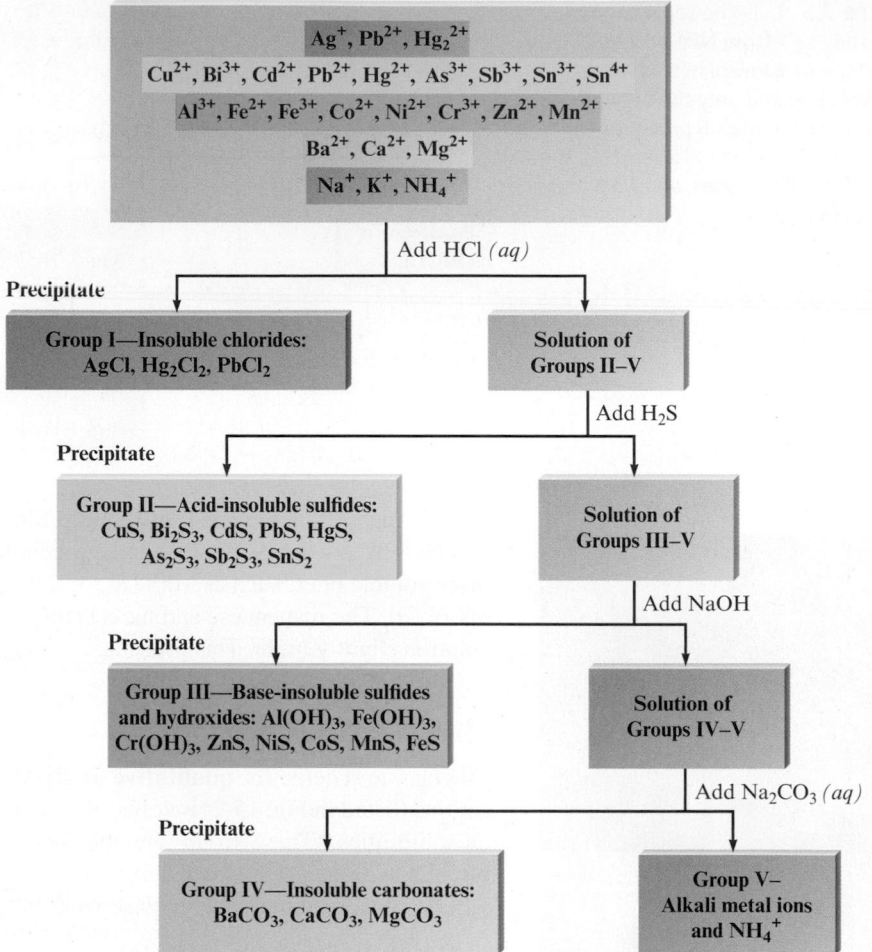

From left to right, cadmium sulfide, chromium(III) hydroxide, aluminum hydroxide, and nickel(II) hydroxide.

ⓘ Note at this point in the analysis that some solutions containing Na$^+$ have been added. Thus the flame test for Na$^+$ must be performed on the original solution.

Group V—Alkali metal and ammonium ions
The only ions remaining in solution at this point are the Group 1A cations and the NH_4^+ ion, all of which form soluble salts with the common anions. The Group 1A cations are usually identified by the characteristic colors they produce when heated in a flame. ◄ These colors are due to the emission spectra of these ions. ◄ ⓘ

The qualitative analysis scheme for cations based on the selective precipitation procedure described above is summarized in Fig. 15-2.

15-3 | Equilibria Involving Complex Ions

$CoCl_4^{2-}$

A **complex ion** is a charged species consisting of a metal ion surrounded by *ligands*. A ligand is simply a Lewis base—a molecule or ion having a lone electron pair that can be donated to an empty orbital on the metal ion to form a covalent bond. Some common ligands are H_2O, NH_3, Cl^-, and CN^-. The number of ligands attached to a metal ion is called the *coordination number*. The most common coordination numbers are 6, for example, in $Co(H_2O)_6^{2+}$ and $Ni(NH_3)_6^{2+}$; 4, for example, in $CoCl_4^{2-}$ and $Cu(NH_3)_4^{2+}$; and 2, for example, in $Ag(NH_3)_2^+$; but others are known. ◄

The properties of complex ions will be discussed in more detail in Chapter 20. For now, we will just look at the equilibria involving these species. Metal ions add ligands one at a time in steps characterized by equilibrium constants called **formation constants** or **stability constants**. For example, when solutions containing Ag^+ ions and NH_3 molecules are mixed, the following reactions take place:

$$Ag^+(aq) + NH_3(aq) \rightleftharpoons Ag(NH_3)^+(aq) \qquad K_1 = 2.1 \times 10^3$$
$$Ag(NH_3)^+(aq) + NH_3(aq) \rightleftharpoons Ag(NH_3)_2^+(aq) \qquad K_2 = 8.2 \times 10^3$$

where K_1 and K_2 are the formation constants for the two steps. In a solution containing Ag^+ and NH_3, all the species NH_3, Ag^+, $Ag(NH_3)^+$, and $Ag(NH_3)_2^+$ exist at equilibrium. Calculating the concentrations of all these components can be complicated. However, usually the total concentration of the ligand is much larger than the total concentration of the metal ion, and approximations can greatly simplify the problems.

For example, consider a solution prepared by mixing 100.0 mL of 2.0 M NH_3 with 100.0 mL of 1.0×10^{-3} M $AgNO_3$. *Before any reaction occurs,* the mixed solution contains the major species Ag^+, NO_3^-, NH_3, and H_2O. What reaction or reactions will occur in this solution? From our discussions of acid–base chemistry, we know that one reaction is

$$NH_3(aq) + H_2O(l) \rightleftharpoons NH_4^+(aq) + OH^-(aq)$$

However, we are interested in the reaction between NH_3 and Ag^+ to form complex ions, and since the position of the preceding equilibrium lies far to the left (K_b for NH_3 is 1.8×10^{-5}), we can neglect the amount of NH_3 used up in the reaction with water. Therefore, before any complex ion formation, the concentrations in the mixed solution are

$$[Ag^+]_0 = \frac{(100.0 \text{ mL})(1.0 \times 10^{-3} M)}{(200.0 \text{ mL})} = 5.0 \times 10^{-4} M$$

Total volume

$$[NH_3]_0 = \frac{(100.0 \text{ mL})(2.0 M)}{(200.0 \text{ mL})} = 1.0 M$$

As mentioned already, the Ag^+ ion reacts with NH_3 in a stepwise fashion to form $AgNH_3^+$ and then $Ag(NH_3)_2^+$:

$$Ag^+(aq) + NH_3(aq) \rightleftharpoons Ag(NH_3)^+(aq) \qquad K_1 = 2.1 \times 10^3$$
$$Ag(NH_3)^+(aq) + NH_3(aq) \rightleftharpoons Ag(NH_3)_2^+(aq) \qquad K_2 = 8.2 \times 10^3$$

▲ A solution containing the blue $CoCl_4^{2-}$ complex ion.

© Cengage Learning

Since both K_1 and K_2 are large, and since there is a large excess of NH_3, *both reactions can be assumed to go essentially to completion.* This is equivalent to writing the net reaction in the solution as follows:

$$Ag^+ + 2NH_3 \longrightarrow Ag(NH_3)_2^+$$

The relevant stoichiometric calculations are as follows:

	Ag^+	$+$	$2NH_3$	$\longrightarrow$	$Ag(NH_3)_2^+$
Before reaction	$5.0 \times 10^{-4}\ M$		$1.0\ M$		0
After reaction	0		$1.0 - 2(5.0 \times 10^{-4}) \approx 1.0\ M$		$5.0 \times 10^{-4}\ M$

↗

Twice as much NH_3 as
Ag^+ is required

Note that in this case we have used molarities when performing the stoichiometry calculations, and we have assumed this reaction to be complete, using all the original Ag^+ to form $Ag(NH_3)_2^+$. In reality, a *very small amount* of the $Ag(NH_3)_2^+$ formed will dissociate to produce small amounts of $Ag(NH_3)^+$ and Ag^+. However, since the amount of $Ag(NH_3)_2^+$ dissociating will be so small, we can safely assume that $[Ag(NH_3)_2^+]$ is $5.0 \times 10^{-4}\ M$ at equilibrium. Also, we know that since so little NH_3 has been consumed, $[NH_3]$ is $1.0\ M$ at equilibrium. We can use these concentrations to calculate $[Ag^+]$ and $[Ag(NH_3)^+]$ using the K_1 and K_2 expressions.

To calculate the equilibrium concentration of $Ag(NH_3)^+$, we use

$$K_2 = 8.2 \times 10^3 = \frac{[Ag(NH_3)_2^+]}{[Ag(NH_3)^+][NH_3]}$$

since $[Ag(NH_3)_2^+]$ and $[NH_3]$ are known. Rearranging and solving for $[Ag(NH_3)^+]$ gives

$$[Ag(NH_3)^+] = \frac{[Ag(NH_3)_2^+]}{K_2[NH_3]} = \frac{5.0 \times 10^{-4}}{(8.2 \times 10^3)(1.0)} = 6.1 \times 10^{-8}\ M$$

Now the equilibrium concentration of Ag^+ can be calculated using K_1:

$$K_1 = 2.1 \times 10^3 = \frac{[Ag(NH_3)^+]}{[Ag^+][NH_3]} = \frac{6.1 \times 10^{-8}}{[Ag^+](1.0)}$$

$$[Ag^+] = \frac{6.1 \times 10^{-8}}{(2.1 \times 10^3)(1.0)} = 2.9 \times 10^{-11}\ M$$

So far we have assumed that $Ag(NH_3)_2^+$ is the dominant silver-containing species in solution. Is this a valid assumption? The calculated concentrations are

$$[Ag(NH_3)_2^+] = 5.0 \times 10^{-4}\ M$$
$$[Ag(NH_3)^+] = 6.1 \times 10^{-8}\ M$$
$$[Ag^+] = 2.9 \times 10^{-11}\ M$$

These values clearly support the conclusion that

ⓘ Essentially all the Ag^+ ions originally present end up in $Ag(NH_3)_2^+$.

$$[Ag(NH_3)_2^+] \gg [Ag(NH_3)^+] \gg [Ag^+]$$

Thus the assumption that $Ag(NH_3)_2^+$ is the dominant Ag^+-containing species is valid, and the calculated concentrations are correct. ◄ ⓘ

This analysis shows that although complex ion equilibria have many species present and look complicated, the calculations are actually quite straightforward, especially if the ligand is present in large excess.

INTERACTIVE EXAMPLE 15-8

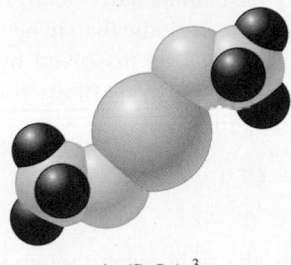

$Ag(S_2O_3)_2{}^{3-}$

Complex Ions

Calculate the concentrations of Ag^+, $Ag(S_2O_3)^-$, and $Ag(S_2O_3)_2{}^{3-}$ in a solution prepared by mixing 150.0 mL of 1.00×10^{-3} M $AgNO_3$ with 200.0 mL of 5.00 M $Na_2S_2O_3$. ◄ The stepwise formation equilibria are

$$Ag^+ + S_2O_3{}^{2-} \rightleftharpoons Ag(S_2O_3)^- \qquad K_1 - 7.4 \times 10^8$$
$$Ag(S_2O_3)^- + S_2O_3{}^{2-} \rightleftharpoons Ag(S_2O_3)_2{}^{3-} \qquad K_2 = 3.9 \times 10^4$$

Solution

The concentrations of the ligand and metal ion in the mixed solution *before any reaction occurs* are

$$[Ag^+]_0 = \frac{(150.0\ \text{mL})(1.00 \times 10^{-3}\ M)}{(150.0\ \text{mL} + 200.0\ \text{mL})} = 4.29 \times 10^{-4}\ M$$

$$[S_2O_3{}^{2-}]_0 = \frac{(200.0\ \text{mL})(5.00\ M)}{(150.0\ \text{mL} + 200.0\ \text{mL})} = 2.86\ M$$

Since $[S_2O_3{}^{2-}]_0 \gg [Ag^+]_0$, and since K_1 and K_2 are large, both formation reactions can be assumed to go to completion, and the net reaction in the solution is as follows:

	Ag^+	$+$	$2S_2O_3{}^{2-}$	$\longrightarrow$	$Ag(S_2O_3)_2{}^{3-}$
Before reaction	$4.29 \times 10^{-4}\ M$		2.86 M		0
After reaction	~0		$2.86 - 2(4.29 \times 10^{-4})$ $\approx 2.86\ M$		$4.29 \times 10^{-4}\ M$

Note that Ag^+ is limiting and that the amount of $S_2O_3{}^{2-}$ consumed is negligible. Also note that since all these species are in the same solution, the molarities can be used to do the stoichiometry problem.

Of course, the concentration of Ag^+ is not zero at equilibrium, and there is some $Ag(S_2O_3)^-$ in the solution. To calculate the concentrations of these species, we must use the K_1 and K_2 expressions. We can calculate the concentration of $Ag(S_2O_3)^-$ from K_2:

$$3.9 \times 10^4 = K_2 = \frac{[Ag(S_2O_3)_2{}^{3-}]}{[Ag(S_2O_3)^-][S_2O_3{}^{2-}]} = \frac{4.29 \times 10^{-4}}{[Ag(S_2O_3)^-](2.86)}$$

$$[Ag(S_2O_3)^-] = 3.8 \times 10^{-9}\ M$$

We can calculate $[Ag^+]$ from K_1:

$$7.4 \times 10^8 = K_1 = \frac{[Ag(S_2O_3)^-]}{[Ag^+][S_2O_3{}^{2-}]} = \frac{3.8 \times 10^{-9}}{[Ag^+](2.86)}$$

$$[Ag^+] = 1.8 \times 10^{-18}\ M$$

These results show that $[Ag(S_2O_3)_2{}^{3-}] \gg [Ag(S_2O_3)^-] \gg [Ag^+]$

Thus the assumption is valid that essentially all the original Ag^+ is converted to $Ag(S_2O_3)_2{}^{3-}$ at equilibrium.

See Exercises 15-67 through 15-70

Complex Ions and Solubility

Often ionic solids that are very nearly water-insoluble must be dissolved somehow in aqueous solutions. For example, when the various qualitative analysis groups are precipitated out, the precipitates must be redissolved to separate the ions within each

▲ (top) Aqueous ammonia is added to silver chloride (white). (bottom) Silver chloride, insoluble in water, dissolves to form $Ag(NH_3)_2^+(aq)$ and $Cl^-(aq)$.

ⓘ When reactions are added, the equilibrium constant for the overall process is the product of the constants for the individual reactions.

group. Consider a solution of cations that contains Ag^+, Pb^{2+}, and Hg_2^{2+}, among others. When dilute aqueous HCl is added to this solution, the Group I ions will form the insoluble chlorides AgCl, $PbCl_2$, and Hg_2Cl_2. Once this mixed precipitate is separated from the solution, it must be redissolved to identify the cations individually. How can this be done? We know that some solids are more soluble in acidic than in neutral solutions. What about chloride salts? For example, can AgCl be dissolved by using a strong acid? The answer is no, because Cl^- ions have virtually no affinity for H^+ ions in aqueous solution. The position of the dissolution equilibrium

$$AgCl(s) \rightleftharpoons Ag^+(aq) + Cl^-(aq)$$

is not affected by the presence of H^+.

How can we pull the dissolution equilibrium to the right, even though Cl^- is an extremely weak base? The key is to lower the concentration of Ag^+ in solution by forming complex ions. For example, Ag^+ reacts with excess NH_3 to form the stable complex ion $Ag(NH_3)_2^+$. As a result, AgCl is quite soluble in concentrated ammonia solutions. ◄ The relevant reactions are

$$AgCl(s) \rightleftharpoons Ag^+(aq) + Cl^-(aq) \qquad K_{sp} = 1.6 \times 10^{-10}$$
$$Ag^+(aq) + NH_3(aq) \rightleftharpoons Ag(NH_3)^+(aq) \qquad K_1 = 2.1 \times 10^3$$
$$Ag(NH_3)^+(aq) + NH_3(aq) \rightleftharpoons Ag(NH_3)_2^+(aq) \qquad K_2 = 8.2 \times 10^3$$

The Ag^+ ion produced by dissolving solid AgCl combines with NH_3 to form $Ag(NH_3)_2^+$, which causes more AgCl to dissolve, until the point at which

$$[Ag^+][Cl^-] = K_{sp} = 1.6 \times 10^{-10}$$

Here $[Ag^+]$ refers only to the Ag^+ ion that is present as a separate species in solution. It is *not* the total silver content of the solution, which is

$$[Ag]_{\text{total dissolved}} = [Ag^+] + [Ag(NH_3)^+] + [Ag(NH_3)_2^+]$$

For reasons discussed in the previous section, virtually all the Ag^+ from the dissolved AgCl ends up in the complex ion $Ag(NH_3)_2^+$. Thus we can represent the dissolving of solid AgCl in excess NH_3 by the equation

$$AgCl(s) + 2NH_3(aq) \rightleftharpoons Ag(NH_3)_2^+(aq) + Cl^-(aq)$$

Since this equation is the *sum of the three stepwise reactions* given above, the equilibrium constant for the reaction is the product of the constants for the three reactions. ◄ ⓘ (Demonstrate this to yourself by multiplying together the three expressions for K_{sp}, K_1, and K_2.) The equilibrium expression is

$$K = \frac{[Ag(NH_3)_2^+][Cl^-]}{[NH_3]^2}$$
$$= K_{sp} \times K_1 \times K_2 = (1.6 \times 10^{-10})(2.1 \times 10^3)(8.2 \times 10^3) = 2.8 \times 10^{-3}$$

Using this expression, we will now calculate the solubility of solid AgCl in a 10.0-M NH_3 solution. If we let x be the solubility (in mol/L) of AgCl in the solution, we can then write the following expressions for the equilibrium concentrations of the pertinent species:

$$[Cl^-] = x \leftarrow$$
$$[Ag(NH_3)_2^+] = x \leftarrow$$

x mol/L of AgCl dissolves to produce x mol/L Cl^- and x mol/L $Ag(NH_3)_2^+$

$$[NH_3] = 10.0 - 2x \leftarrow$$

Formation of x mol/L $Ag(NH_3)_2^+$ requires $2x$ mol/L NH_3, since each complex ion contains two NH_3 ligands

Substituting these concentrations into the equilibrium expression gives

$$K = 2.8 \times 10^{-3} = \frac{[Ag(NH_3)_2^+][Cl^-]}{[NH_3]^2} = \frac{(x)(x)}{(10.0 - 2x)^2} = \frac{x^2}{(10.0 - 2x)^2}$$

No approximations are necessary here. Taking the square root of both sides of the equation gives

$$\sqrt{2.8 \times 10^{-3}} = \frac{x}{10.0 - 2x}$$

$$x = 0.48 \text{ mol/L} = \text{solubility of AgCl}(s) \text{ in } 10.0 \, M \text{ NH}_3$$

Thus the solubility of AgCl in 10.0 M NH$_3$ is much greater than its solubility in pure water, which is

$$\sqrt{K_{sp}} = 1.3 \times 10^{-5} \text{ mol/L}$$

In this chapter we have considered two strategies for dissolving a water-insoluble ionic solid. If the *anion* of the solid is a good base, the solubility is greatly increased by acidifying the solution. In cases where the anion is not sufficiently basic, the ionic solid often can be dissolved in a solution containing a ligand that forms stable complex ions with its *cation*.

Sometimes solids are so insoluble that combinations of reactions are needed to dissolve them. For example, to dissolve the extremely insoluble HgS ($K_{sp} = 10^{-54}$), it is necessary to use a mixture of concentrated HCl and concentrated HNO$_3$, called *aqua regia*. The H$^+$ ions in the aqua regia react with the S^{2-} ions to form H$_2$S, and Cl$^-$ reacts with Hg^{2+} to form various complex ions, including HgCl$_4^{2-}$. In addition, NO$_3^-$ oxidizes S^{2-} to elemental sulfur. These processes lower the concentrations of Hg^{2+} and S^{2-} and thus promote the solubility of HgS.

Since the solubility of many salts increases with temperature, simple heating is sometimes enough to make a salt sufficiently soluble. For example, earlier in this section we considered the mixed chloride precipitates of the Group I ions—PbCl$_2$, AgCl, and Hg$_2$Cl$_2$. The effect of temperature on the solubility of PbCl$_2$ is such that we can precipitate PbCl$_2$ with cold aqueous HCl and then redissolve it by heating the solution to near boiling. The silver and mercury(I) chlorides remain precipitated, since they are not significantly soluble in hot water. However, solid AgCl can be dissolved using aqueous ammonia. The solid Hg$_2$Cl$_2$ reacts with NH$_3$ to form a mixture of elemental mercury and HgNH$_2$Cl:

$$\underset{\text{White}}{\text{Hg}_2\text{Cl}_2(s)} + 2\text{NH}_3(aq) \longrightarrow \underset{\text{Black}}{\text{HgNH}_2\text{Cl}(s)} + \text{Hg}(l) + \text{NH}_4^+(aq) + \text{Cl}^-(aq)$$

The mixed precipitate appears gray. This is an oxidation–reduction reaction in which one mercury(I) ion in Hg$_2$Cl$_2$ is oxidized to Hg^{2+} in HgNH$_2$Cl and the other mercury(I) ion is reduced to Hg, or elemental mercury.

The treatment of the Group I ions is summarized in Fig. 15-3. Note that the presence of Pb^{2+} is confirmed by adding CrO$_4^{2-}$, which forms bright yellow lead(II) chromate (PbCrO$_4$). Also note that H$^+$ added to a solution containing Ag(NH$_3$)$_2^+$ reacts with the NH$_3$ to form NH$_4^+$, destroying the Ag(NH$_3$)$_2^+$ complex. Silver chloride then re-forms:

$$2\text{H}^+(aq) + \text{Ag(NH}_3)_2^+(aq) + \text{Cl}^-(aq) \longrightarrow 2\text{NH}_4^+(aq) + \text{AgCl}(s)$$

Note that the qualitative analysis of cations by selective precipitation involves all the types of reactions we have discussed and represents an excellent application of the principles of chemical equilibrium.

Figure 15-3 | The separation of the Group I ions in the classic scheme of qualitative analysis.

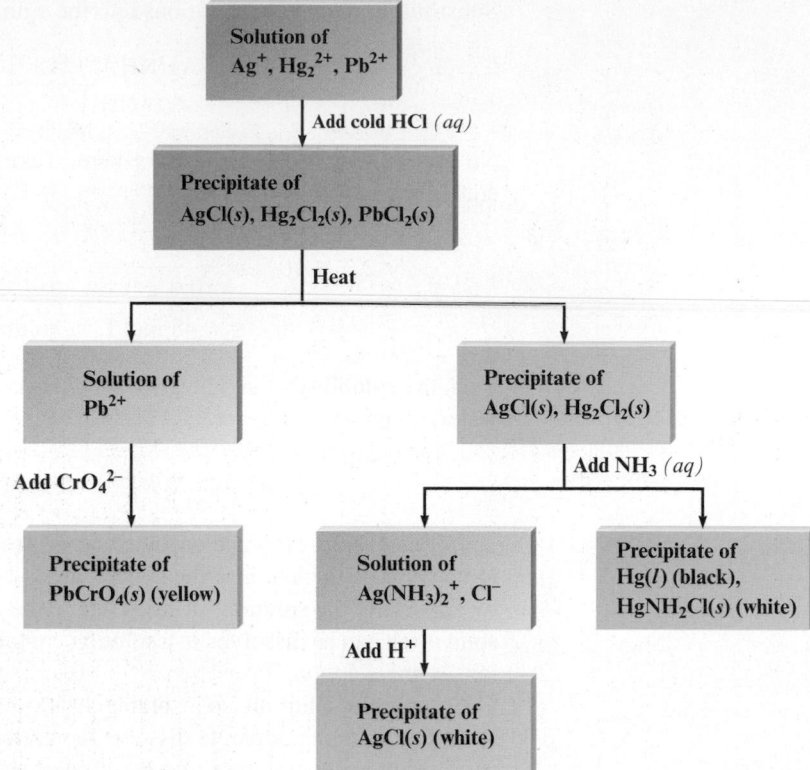

FOR REVIEW

Key Terms

Section 15-1
solubility product constant
  (solubility product)

Section 15-2
ion product
selective precipitation
qualitative analysis

Section 15-3
complex ion
formation (stability) constants

Solids Dissolving in Water

- For a slightly soluble salt, an equilibrium is set up between the excess solid (MX) and the ions in solution:

$$MX(s) \rightleftharpoons M^+(aq) + X^-(aq)$$

- The corresponding constant is called K_{sp}:

$$K_{sp} = [M^+][X^-]$$

  - The solubility of MX(s) is decreased by the presence of another source of either M^+ or X^-; this is called the common ion effect

- Predicting whether precipitation will occur when two solutions are mixed involves calculating Q for the initial concentrations:

  - If $Q > K_{sp}$, precipitation occurs
  - If $Q \leq K_{sp}$, no precipitation occurs

Qualitative Analysis

- A mixture of ions can be separated by selective precipitation
  - The ions are first separated into groups by adding HCl(aq), then H₂S(aq), then NaOH(aq), and finally Na₂CO₃(aq)
  - The ions in the groups are separated and identified by further selective dissolution and precipitation

Complex Ions

- Complex ions consist of a metal ion surrounded by attached ligands
  - A ligand is a Lewis base
  - The number of ligands is called the coordination number, which is commonly 2, 4, or 6
- Complex ion equilibria in solution are described by formation (stability) constants
- The formation of complex ions can be used to selectively dissolve solids in the qualitative analysis scheme

Review Questions Answers to the Review Questions can be found on the Student Web site (accessible from www.cengagebrain.com).

1. To what reaction does the solubility product constant, K_{sp}, refer? Table 15-1 lists K_{sp} values for several ionic solids. For any of these ionic compounds, you should be able to calculate the solubility. What is the solubility of a salt, and what procedures do you follow to calculate the solubility of a salt? How would you calculate the K_{sp} value for a salt given the solubility?

2. Under what circumstances can you compare the relative solubilities of two salts directly by comparing the values of their solubility products? When can relative solubilities not be compared based on K_{sp} values?

3. What is a common ion and how does its presence affect the solubility?

4. List some salts whose solubility increases as the pH becomes more acidic. What is true about the anions in these salts? List some salts whose solubility remains unaffected by the solution pH. What is true about the anions in these salts?

5. What is the difference between the ion product, Q, and the solubility product, K_{sp}? What happens when $Q > K_{sp}$? $Q < K_{sp}$? $Q = K_{sp}$?

6. Mixtures of metal ions in aqueous solution can sometimes be separated by selective precipitation. What is selective precipitation? If a solution contained 0.10 M Mg^{2+}, 0.10 M Ca^{2+}, and 0.10 M Ba^{2+}, how could addition of NaF be used to separate the cations out of solution—that is, what

would precipitate first, then second, then third? How could addition of K_3PO_4 be used to separate out the cations in a solution that is 1.0 M Ag^+, 1.0 M Pb^{2+}, and 1.0 M Sr^{2+}?

7. Figure 15-2 summarizes the classic method for separating a mixture of common cations by selective precipitation. Explain the chemistry involved with each of the four steps in the diagram.

8. What is a complex ion? The stepwise formation constants for the complex ion $Cu(NH_3)_4^{2+}$ are $K_1 \approx 1 \times 10^3$, $K_2 \approx 1 \times 10^4$, $K_3 \approx 1 \times 10^3$, and $K_4 \approx 1 \times 10^3$. Write the reactions that refer to each of these formation constants. Given that the values of the formation constants are large, what can you deduce about the equilibrium concentration of $Cu(NH_3)_4^{2+}$ versus the equilibrium concentration of Cu^{2+}?

9. When 5 M ammonia is added to a solution containing $Cu(OH)_2(s)$, the precipitate will eventually dissolve in solution. Why? If 5 M HNO_3 is then added, the $Cu(OH)_2$ precipitate re-forms. Why? In general, what effect does the ability of a cation to form a complex ion have on the solubility of salts containing that cation?

10. Figure 15-3 outlines the classic scheme for separating a mixture of insoluble chloride salts from one another. Explain the chemistry involved in the various steps of the figure.

Active Learning Questions

These questions are designed to be used by groups of students in class.

1. Which of the following will affect the total amount of solute that can dissolve in a given amount of solvent?
   a. The solution is stirred.
   b. The solute is ground to fine particles before dissolving.
   c. The temperature changes.

2. Devise as many ways as you can to experimentally determine the K_{sp} value of a solid. Explain why each of these would work.

3. You are browsing through the *Handbook of Hypothetical Chemistry* when you come across a solid that is reported to have a K_{sp} value of zero in water at 25°C. What does this mean?

4. A friend tells you: "The constant K_{sp} of a salt is called the solubility product constant and is calculated from the concentrations of ions in the solution. Thus, if salt A dissolves to a

greater extent than salt B, salt A must have a higher K_{sp} than salt B." Do you agree with your friend? Explain.

5. Explain the following phenomenon: You have a test tube with an aqueous solution of silver nitrate as shown in test tube 1 below. A few drops of aqueous sodium chromate solution was added with the end result shown in test tube 2. A few drops of aqueous sodium chloride solution was then added with the end result shown in test tube 3.

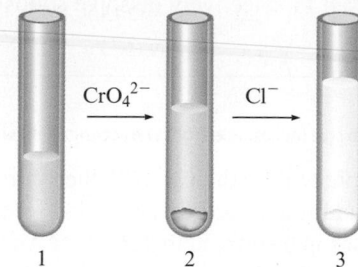

1   2   3

Use the K_{sp} values in the book to support your explanation, and include the balanced equations. Also, list the ions that are present in solution in each test tube.

6. What happens to the K_{sp} value of a solid as the temperature of the solution changes? Consider both increasing and decreasing temperatures, and explain your answer.

7. Which is more likely to dissolve in an acidic solution, silver sulfide or silver chloride? Why?

A blue question or exercise number indicates that the answer to that question or exercise appears at the back of this book and a solution appears in the *Solutions Guide*, as found on PowerLecture.

Questions

8. For which of the following is the K_{sp} value of the ionic compound the largest? The smallest? Explain your answer.

9. $Ag_2S(s)$ has a larger molar solubility than CuS even though Ag_2S has the smaller K_{sp} value. Explain how this is possible.

10. Solubility is an equilibrium position, whereas K_{sp} is an equilibrium constant. Explain the difference.

11. The salts in Table 15-1, with the possible exception of the hydroxide salts, generally have one of the following mathematical relationships between the K_{sp} value and the molar solubility s.

   i. $K_{sp} = s^2$ **iii.** $K_{sp} = 27s^4$
   ii. $K_{sp} = 4s^3$ **iv.** $K_{sp} = 108s^5$

   For each mathematical relationship, give an example of a salt in Table 15-1 that exhibits that relationship.

12. When $Na_3PO_4(aq)$ is added to a solution containing a metal ion and a precipitate forms, the precipitate generally could be one of two possibilities. What are the two possibilities?

13. The common ion effect for ionic solids (salts) is to significantly decrease the solubility of the ionic compound in water. Explain the common ion effect.

14. Sulfide precipitates are generally grouped as sulfides insoluble in acidic solution and sulfides insoluble in basic solution. Explain why there is a difference between the two groups of sulfide precipitates.

15. List some ways one can increase the solubility of a salt in water.

16. The stepwise formation constants for a complex ion usually have values much greater than 1. What is the significance of this?

17. Silver chloride dissolves readily in 2 M NH_3 but is quite insoluble in 2 M NH_4NO_3. Explain.

18. If a solution contains either $Pb^{2+}(aq)$ or $Ag^+(aq)$, how can temperature be manipulated to help identify the ion in solution?

Exercises

In this section similar exercises are paired.

Solubility Equilibria

19. Write balanced equations for the dissolution reactions and the corresponding solubility product expressions for each of the following solids.
   a. $AgC_2H_3O_2$ **b.** $Al(OH)_3$ **c.** $Ca_3(PO_4)_2$

20. Write balanced equations for the dissolution reactions and the corresponding solubility product expressions for each of the following solids.
   a. Ag_2CO_3 **b.** $Ce(IO_3)_3$ **c.** BaF_2

21. Use the following data to calculate the K_{sp} value for each solid.
   a. The solubility of CaC_2O_4 is 4.8×10^{-5} mol/L.
   b. The solubility of BiI_3 is 1.32×10^{-5} mol/L.

22. Use the following data to calculate the K_{sp} value for each solid.
   a. The solubility of $Pb_3(PO_4)_2$ is 6.2×10^{-12} mol/L.
   b. The solubility of Li_2CO_3 is 7.4×10^{-2} mol/L.

23. Approximately 0.14 g nickel(II) hydroxide, $Ni(OH)_2(s)$, dissolves per liter of water at 20°C. Calculate K_{sp} for $Ni(OH)_2(s)$ at this temperature.

24. The solubility of the ionic compound M_2X_3, having a molar mass of 288 g/mol, is 3.60×10^{-7} g/L. Calculate the K_{sp} of the compound.

25. The concentration of Pb^{2+} in a solution saturated with $PbBr_2(s)$ is 2.14×10^{-2} M. Calculate K_{sp} for $PbBr_2$.

26. The concentration of Ag^+ in a solution saturated with $Ag_2C_2O_4(s)$ is 2.2×10^{-4} M. Calculate K_{sp} for $Ag_2C_2O_4$.

27. Calculate the solubility of each of the following compounds in moles per liter. Ignore any acid–base properties.
   a. Ag_3PO_4, $K_{sp} = 1.8 \times 10^{-18}$
   b. $CaCO_3$, $K_{sp} = 8.7 \times 10^{-9}$
   c. Hg_2Cl_2, $K_{sp} = 1.1 \times 10^{-18}$ (Hg_2^{2+} is the cation in solution.)

28. Calculate the solubility of each of the following compounds in moles per liter. Ignore any acid–base properties.

    a. PbI_2, $K_{sp} = 1.4 \times 10^{-8}$

    b. $CdCO_3$, $K_{sp} = 5.2 \times 10^{-12}$

    c. $Sr_3(PO_4)_2$, $K_{sp} = 1 \times 10^{-31}$

29. Cream of tartar, a common ingredient in cooking, is the common name for potassium bitartrate (abbreviated KBT, molar mass − 188.2 g/mol). Historically, KBT was a crystalline solid that formed on the casks of wine barrels during the fermentation process. Calculate the maximum mass of KBT that can dissolve in 250.0 mL of solution to make a saturated solution. The K_{sp} value for KBT is 3.8×10^{-4}.

30. Barium sulfate is a contrast agent for X-ray scans that are most often associated with the gastrointestinal tract. Calculate the mass of $BaSO_4$ that can dissolve in 100.0 mL of solution. The K_{sp} value for $BaSO_4$ is 1.5×10^{-9}.

31. Calculate the molar solubility of $Mg(OH)_2$, $K_{sp} = 8.9 \times 10^{-12}$.

32. Calculate the molar solubility of $Cd(OH)_2$, $K_{sp} = 5.9 \times 10^{-11}$.

33. Calculate the molar solubility of $Al(OH)_3$, $K_{sp} = 2 \times 10^{-32}$.

34. Calculate the molar solubility of $Co(OH)_3$, $K_{sp} = 2.5 \times 10^{-43}$.

35. For each of the following pairs of solids, determine which solid has the smallest molar solubility.

    a. $CaF_2(s)$, $K_{sp} = 4.0 \times 10^{-11}$, or $BaF_2(s)$, $K_{sp} = 2.4 \times 10^{-5}$

    b. $Ca_3(PO_4)_2(s)$, $K_{sp} = 1.3 \times 10^{-32}$, or $FePO_4(s)$, $K_{sp} = 1.0 \times 10^{-22}$

36. For each of the following pairs of solids, determine which solid has the smallest molar solubility.

    a. FeC_2O_4, $K_{sp} = 2.1 \times 10^{-7}$, or $Cu(IO_4)_2$, $K_{sp} = 1.4 \times 10^{-7}$

    b. Ag_2CO_3, $K_{sp} = 8.1 \times 10^{-12}$, or $Mn(OH)_2$, $K_{sp} = 2 \times 10^{-13}$

37. Calculate the solubility (in moles per liter) of $Fe(OH)_3$ ($K_{sp} = 4 \times 10^{-38}$) in each of the following.

    a. water

    b. a solution buffered at pH = 5.0

    c. a solution buffered at pH = 11.0

38. Calculate the solubility of $Co(OH)_2(s)$ ($K_{sp} = 2.5 \times 10^{-16}$) in a buffered solution with a pH of 11.00.

39. The K_{sp} for silver sulfate (Ag_2SO_4) is 1.2×10^{-5}. Calculate the solubility of silver sulfate in each of the following.

    a. water

    b. 0.10 M $AgNO_3$

    c. 0.20 M K_2SO_4

40. The K_{sp} for lead iodide (PbI_2) is 1.4×10^{-8}. Calculate the solubility of lead iodide in each of the following.

    a. water

    b. 0.10 M $Pb(NO_3)_2$

    c. 0.010 M NaI

41. Calculate the solubility of solid $Ca_3(PO_4)_2$ ($K_{sp} = 1.3 \times 10^{-32}$) in a 0.20-M Na_3PO_4 solution.

42. Calculate the solubility of solid $Pb_3(PO_4)_2$ ($K_{sp} = 1 \times 10^{-54}$) in a 0.10-M $Pb(NO_3)_2$ solution.

43. The solubility of $Ce(IO_3)_3$ in a 0.20-M KIO_3 solution is 4.4×10^{-8} mol/L. Calculate K_{sp} for $Ce(IO_3)_3$.

44. The solubility of $Pb(IO_3)_2(s)$ in a 0.10-M KIO_3 solution is 2.6×10^{-11} mol/L. Calculate K_{sp} for $Pb(IO_3)_2(s)$.

45. Which of the substances in Exercises 27 and 28 show increased solubility as the pH of the solution becomes more acidic? Write equations for the reactions that occur to increase the solubility.

46. For which salt in each of the following groups will the solubility depend on pH?

    a. AgF, AgCl, AgBr   **c.** $Sr(NO_3)_2$, $Sr(NO_2)_2$

    b. $Pb(OH)_2$, $PbCl_2$ **d.** $Ni(NO_3)_2$, $Ni(CN)_2$

Precipitation Conditions

47. What mass of ZnS ($K_{sp} = 2.5 \times 10^{-22}$) will dissolve in 300.0 mL of 0.050 M $Zn(NO_3)_2$? Ignore the basic properties of S^{2-}.

48. The concentration of Mg^{2+} in seawater is 0.052 M. At what pH will 99% of the Mg^{2+} be precipitated as the hydroxide salt? [K_{sp} for $Mg(OH)_2 = 8.9 \times 10^{-12}$.]

49. Will a precipitate form when 100.0 mL of 4.0×10^{-4} M $Mg(NO_3)_2$ is added to 100.0 mL of 2.0×10^{-4} M NaOH?

50. A solution contains 1.0×10^{-5} M Ag^+ and 2.0×10^{-6} M CN^-. Will AgCN(s) precipitate? (K_{sp} for AgCN(s) is 2.2×10^{-12}.)

51. A solution is prepared by mixing 100.0 mL of 1.0×10^{-2} M $Pb(NO_3)_2$ and 100.0 mL of 1.0×10^{-3} M NaF. Will $PbF_2(s)$ ($K_{sp} = 4 \times 10^{-8}$) precipitate?

52. A solution contains 2.0×10^{-3} M Ce^{3+} and 1.0×10^{-2} M IO_3^-. Will $Ce(IO_3)_3(s)$ precipitate? [K_{sp} for $Ce(IO_3)_3$ is 3.2×10^{-10}.]

53. Calculate the final concentrations of $K^+(aq)$, $C_2O_4^{2-}(aq)$, $Ba^{2+}(aq)$, and $Br^-(aq)$ in a solution prepared by adding 0.100 L of 0.200 M $K_2C_2O_4$ to 0.150 L of 0.250 M $BaBr_2$. (For BaC_2O_4, $K_{sp} = 2.3 \times 10^{-8}$.)

54. A solution is prepared by mixing 75.0 mL of 0.020 M $BaCl_2$ and 125 mL of 0.040 M K_2SO_4. What are the concentrations of barium and sulfate ions in this solution? Assume only SO_4^{2-} ions (no HSO_4^-) are present.

55. A 50.0-mL sample of 0.00200 M $AgNO_3$ is added to 50.0 mL of 0.0100 M $NaIO_3$. What is the equilibrium concentration of Ag^+ in solution? (K_{sp} for $AgIO_3$ is 3.0×10^{-8}.)

56. A solution is prepared by mixing 50.0 mL of 0.10 M $Pb(NO_3)_2$ with 50.0 mL of 1.0 M KCl. Calculate the concentrations of Pb^{2+} and Cl^- at equilibrium. [K_{sp} for $PbCl_2(s)$ is 1.6×10^{-5}.]

57. A solution contains 1.0×10^{-5} M Na_3PO_4. What is the minimum concentration of $AgNO_3$ that would cause precipitation of solid Ag_3PO_4 ($K_{sp} = 1.8 \times 10^{-18}$)?

58. The K_{sp} of $Al(OH)_3$ is 2×10^{-32}. At what pH will a 0.2-M Al^{3+} solution begin to show precipitation of $Al(OH)_3$?

59. A solution is 1×10^{-4} M in NaF, Na_2S, and Na_3PO_4. What would be the order of precipitation as a source of Pb^{2+} is added gradually to the solution? The relevant K_{sp} values are $K_{sp}(PbF_2) = 4 \times 10^{-8}$, $K_{sp}(PbS) = 7 \times 10^{-29}$, and $K_{sp}[Pb_3(PO_4)_2] = 1 \times 10^{-54}$.

60. A solution contains 0.25 M $Ni(NO_3)_2$ and 0.25 M $Cu(NO_3)_2$. Can the metal ions be separated by slowly adding Na_2CO_3? Assume that for successful separation 99% of the metal ion must be precipitated before the other metal ion begins to precipitate, and assume no volume change on addition of Na_2CO_3.

Complex Ion Equilibria

61. Write equations for the stepwise formation of each of the following complex ions.

   a. $Ni(CN)_4^{2-}$

   b. $V(C_2O_4)_3^{3-}$

62. Write equations for the stepwise formation of each of the following complex ions.

   a. CoF_6^{3-}

   b. $Zn(NH_3)_4^{2+}$

63. In the presence of CN^-, Fe^{3+} forms the complex ion $Fe(CN)_6^{3-}$. The equilibrium concentrations of Fe^{3+} and $Fe(CN)_6^{3-}$ are 8.5×10^{-40} M and 1.5×10^{-3} M, respectively, in a 0.11-M KCN solution. Calculate the value for the overall formation constant of $Fe(CN)_6^{3-}$.

$$Fe^{3+}(aq) + 6CN^-(aq) \rightleftharpoons Fe(CN)_6^{3-}(aq) \qquad K_{overall} = ?$$

64. In the presence of NH_3, Cu^{2+} forms the complex ion $Cu(NH_3)_4^{2+}$. If the equilibrium concentrations of Cu^{2+} and $Cu(NH_3)_4^{2+}$ are 1.8×10^{-17} M and 1.0×10^{-3} M, respectively, in a 1.5-M NH_3 solution, calculate the value for the overall formation constant of $Cu(NH_3)_4^{2+}$.

$$Cu^{2+}(aq) + 4NH_3(aq) \rightleftharpoons Cu(NH_3)_4^{2+}(aq) \qquad K_{overall} = ?$$

65. When aqueous KI is added gradually to mercury(II) nitrate, an orange precipitate forms. Continued addition of KI causes the precipitate to dissolve. Write balanced equations to explain these observations. (*Hint:* Hg^{2+} reacts with I^- to form HgI_4^{2-}.)

66. As sodium chloride solution is added to a solution of silver nitrate, a white precipitate forms. Ammonia is added to the mixture and the precipitate dissolves. When potassium bromide solution is then added, a pale yellow precipitate appears. When a solution of sodium thiosulfate is added, the yellow precipitate dissolves. Finally, potassium iodide is added to the solution and a yellow precipitate forms. Write equations for all the changes mentioned above. What conclusions can you draw concerning the sizes of the K_{sp} values for AgCl, AgBr, and AgI?

67. The overall formation constant for HgI_4^{2-} is 1.0×10^{30}. That is,

$$1.0 \times 10^{30} = \frac{[HgI_4^{2-}]}{[Hg^{2+}][I^-]^4}$$

What is the concentration of Hg^{2+} in 500.0 mL of a solution that was originally 0.010 M Hg^{2+} and 0.78 M I^-? The reaction is

$$Hg^{2+}(aq) + 4I^-(aq) \rightleftharpoons HgI_4^{2-}(aq)$$

68. A solution is prepared by adding 0.10 mole of $Ni(NH_3)_6Cl_2$ to 0.50 L of 3.0 M NH_3. Calculate $[Ni(NH_3)_6^{2+}]$ and $[Ni^{2+}]$ in this solution. $K_{overall}$ for $Ni(NH_3)_6^{2+}$ is 5.5×10^8. That is,

$$5.5 \times 10^8 = \frac{[Ni(NH_3)_6^{2+}]}{[Ni^{2+}][NH_3]^6}$$

for the overall reaction

$$Ni^{2+}(aq) + 6NH_3(aq) \rightleftharpoons Ni(NH_3)_6^{2+}(aq)$$

69. A solution is formed by mixing 50.0 mL of 10.0 M NaX with 50.0 mL of 2.0×10^{-3} M $CuNO_3$. Assume that Cu^+ forms complex ions with X^- as follows:

$$Cu^+(aq) + X^-(aq) \rightleftharpoons CuX(aq) \qquad K_1 = 1.0 \times 10^2$$
$$CuX(aq) + X^-(aq) \rightleftharpoons CuX_2^-(aq) \qquad K_2 = 1.0 \times 10^4$$
$$CuX_2^-(aq) + X^-(aq) \rightleftharpoons CuX_3^{2-}(aq) \qquad K_3 = 1.0 \times 10^3$$

with an overall reaction

$$Cu^+(aq) + 3X^-(aq) \rightleftharpoons CuX_3^{2-}(aq) \qquad K = 1.0 \times 10^9$$

Calculate the following concentrations at equilibrium.

   a. CuX_3^{2-} **b.** CuX_2^- **c.** Cu^+

70. A solution is prepared by mixing 100.0 mL of 1.0×10^{-4} M $Be(NO_3)_2$ and 100.0 mL of 8.0 M NaF.

$$Be^{2+}(aq) + F^-(aq) \rightleftharpoons BeF^+(aq) \qquad K_1 = 7.9 \times 10^4$$
$$BeF^+(aq) + F^-(aq) \rightleftharpoons BeF_2(aq) \qquad K_2 = 5.8 \times 10^3$$
$$BeF_2(aq) + F^-(aq) \rightleftharpoons BeF_3^-(aq) \qquad K_3 = 6.1 \times 10^2$$
$$BeF_3^-(aq) + F^-(aq) \rightleftharpoons BeF_4^{2-}(aq) \qquad K_4 = 2.7 \times 10^1$$

Calculate the equilibrium concentrations of F^-, Be^{2+}, BeF^+, BeF_2, BeF_3^-, and BeF_4^{2-} in this solution.

71. a. Calculate the molar solubility of AgI in pure water. K_{sp} for AgI is 1.5×10^{-16}.

   b. Calculate the molar solubility of AgI in 3.0 M NH_3. The overall formation constant for $Ag(NH_3)_2^+$ is 1.7×10^7.

   c. Compare the calculated solubilities from parts a and b. Explain any differences.

72. Solutions of sodium thiosulfate are used to dissolve unexposed AgBr ($K_{sp} = 5.0 \times 10^{-13}$) in the developing process for black-and-white film. What mass of AgBr can dissolve in 1.00 L of 0.500 M $Na_2S_2O_3$? Ag^+ reacts with $S_2O_3^{2-}$ to form a complex ion:

$$Ag^+(aq) + 2S_2O_3^{2-}(aq) \rightleftharpoons Ag(S_2O_3)_2^{3-}(aq)$$
$$K = 2.9 \times 10^{13}$$

73. K_f for the complex ion $Ag(NH_3)_2^+$ is 1.7×10^7. K_{sp} for AgCl is 1.6×10^{-10}. Calculate the molar solubility of AgCl in 1.0 M NH_3.

74. The copper(I) ion forms a chloride salt that has $K_{sp} = 1.2 \times 10^{-6}$. Copper(I) also forms a complex ion with Cl^-:

$$Cu^+(aq) + 2Cl^-(aq) \rightleftharpoons CuCl_2^-(aq) \qquad K = 8.7 \times 10^4$$

   a. Calculate the solubility of copper(I) chloride in pure water. (Ignore $CuCl_2^-$ formation for part a.)

   b. Calculate the solubility of copper(I) chloride in 0.10 M NaCl.

75. A series of chemicals were added to some $AgNO_3(aq)$. $NaCl(aq)$ was added first to the silver nitrate solution with the end result shown below in test tube 1, $NH_3(aq)$ was then added with the end result shown in test tube 2, and $HNO_3(aq)$ was added last with the end result shown in test tube 3.

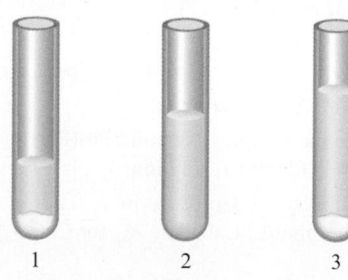

Explain the results shown in each test tube. Include a balanced equation for the reaction(s) taking place.

76. The solubility of copper(II) hydroxide in water can be increased by adding either the base NH_3 or the acid HNO_3. Explain. Would added NH_3 or HNO_3 have the same effect on the solubility of silver acetate or silver chloride? Explain.

Additional Exercises

77. A solution contains 0.018 mole each of I^-, Br^-, and Cl^-. When the solution is mixed with 200. mL of 0.24 M $AgNO_3$, what mass of $AgCl(s)$ precipitates out, and what is $[Ag^+]$? Assume no volume change.

$$AgI: K_{sp} = 1.5 \times 10^{-16}$$
$$AgBr: K_{sp} = 5.0 \times 10^{-13}$$
$$AgCl: K_{sp} = 1.6 \times 10^{-10}$$

78. You have two salts, AgX and AgY, with very similar K_{sp} values. You know that K_a for HX is much greater than K_a for HY. Which salt is more soluble in acidic solution? Explain.

79. Tooth enamel is composed of the mineral hydroxyapatite. The K_{sp} of hydroxyapatite, $Ca_5(PO_4)_3OH$, is 6.8×10^{-37}. Calculate the solubility of hydroxyapatite in pure water in moles per liter. How is the solubility of hydroxyapatite affected by adding acid? When hydroxyapatite is treated with fluoride, the mineral fluorapatite, $Ca_5(PO_4)_3F$, forms. The K_{sp} of this substance is 1×10^{-60}. Calculate the solubility of fluorapatite in water. How do these calculations provide a rationale for the fluoridation of drinking water?

80. The U.S. Public Health Service recommends the fluoridation of water as a means for preventing tooth decay. The recommended concentration is 1 mg F^- per liter. The presence of calcium ions in hard water can precipitate the added fluoride. What is the maximum molarity of calcium ions in hard water if the fluoride concentration is at the USPHS recommended level? (K_{sp} for $CaF_2 = 4.0 \times 10^{-11}$)

81. What mass of $Ca(NO_3)_2$ must be added to 1.0 L of a 1.0-M HF solution to begin precipitation of $CaF_2(s)$? For CaF_2, $K_{sp} = 4.0 \times 10^{-11}$ and K_a for HF = 7.2×10^{-4}. Assume no volume change on addition of $Ca(NO_3)_2(s)$.

82. Calculate the mass of manganese hydroxide present in 1300 mL of a saturated manganese hydroxide solution. For $Mn(OH)_2$, $K_{sp} = 2.0 \times 10^{-13}$.

83. On a hot day, a 200.0-mL sample of a saturated solution of PbI_2 was allowed to evaporate until dry. If 240 mg of solid PbI_2 was collected after evaporation was complete, calculate the K_{sp} value for PbI_2 on this hot day.

84. The active ingredient of Pepto-Bismol is the compound bismuth subsalicylate, which undergoes the following dissociation when added to water:

$$C_7H_5BiO_4(s) + H_2O(l) \rightleftharpoons C_7H_4O_3^{2-}(aq)$$
$$+ Bi^{3+}(aq) + OH^-(aq) \qquad K = ?$$

If the maximum amount of bismuth subsalicylate that reacts by this reaction is 3.2×10^{-19} mol/L, calculate the equilibrium constant for the preceding reaction.

85. Nanotechnology has become an important field, with applications ranging from high-density data storage to the design of "nano machines." One common building block of nanostructured architectures is manganese oxide nanoparticles. The particles can be formed from manganese oxalate nanorods, the formation of which can be described as follows:

$$Mn^{2+}(aq) + C_2O_4^{2-}(aq) \rightleftharpoons MnC_2O_4(aq) \qquad K_1 = 7.9 \times 10^3$$
$$MnC_2O_4(aq) + C_2O_4^{2-}(aq) \rightleftharpoons Mn(C_2O_4)_2^{2-}(aq)$$
$$K_2 = 7.9 \times 10^1$$

Calculate the value for the overall formation constant for $Mn(C_2O_4)_2^{2-}$:

$$K = \frac{[Mn(C_2O_4)_2^{2-}]}{[Mn^{2+}][C_2O_4^{2-}]^2}$$

86. The equilibrium constant for the following reaction is 1.0×10^{23}:

$$Cr^{3+}(aq) + H_2EDTA^{2-}(aq) \rightleftharpoons CrEDTA^-(aq) + 2H^+(aq)$$

$$EDTA^{4-} = \begin{array}{c} ^-O_2C-CH_2 \\ ^-O_2C-CH_2 \end{array} \hspace{-4pt} \diagdown N-CH_2-CH_2-N \diagup \begin{array}{c} CH_2-CO_2^- \\ CH_2-CO_2^- \end{array}$$

Ethylenediaminetetraacetate

EDTA is used as a complexing agent in chemical analysis. Solutions of EDTA, usually containing the disodium salt Na_2H_2EDTA, are used to treat heavy metal poisoning. Calculate $[Cr^{3+}]$ at equilibrium in a solution originally 0.0010 M in Cr^{3+} and 0.050 M in H_2EDTA^{2-} and buffered at pH = 6.00.

87. Calculate the concentration of Pb^{2+} in each of the following.

  a. a saturated solution of $Pb(OH)_2$, $K_{sp} = 1.2 \times 10^{-15}$

  b. a saturated solution of $Pb(OH)_2$ buffered at pH = 13.00

  c. Ethylenediaminetetraacetate ($EDTA^{4-}$) is used as a complexing agent in chemical analysis and has the following structure:

$$\begin{array}{c} ^-O_2C-CH_2 \\ ^-O_2C-CH_2 \end{array} \hspace{-4pt} \diagdown N-CH_2-CH_2-N \diagup \begin{array}{c} CH_2-CO_2^- \\ CH_2-CO_2^- \end{array}$$

Ethylenediaminetetraacetate

Solutions of $EDTA^{4-}$ are used to treat heavy metal poisoning by removing the heavy metal in the form of a soluble complex ion. The reaction of $EDTA^{4-}$ with Pb^{2+} is

$$Pb^{2+}(aq) + EDTA^{4-}(aq) \rightleftharpoons PbEDTA^{2-}(aq)$$
$$K = 1.1 \times 10^{18}$$

Consider a solution with 0.010 mole of $Pb(NO_3)_2$ added to 1.0 L of an aqueous solution buffered at pH = 13.00 and containing 0.050 M Na_4EDTA. Does $Pb(OH)_2$ precipitate from this solution?

88. Will a precipitate of $Cd(OH)_2$ form if 1.0 mL of 1.0 M $Cd(NO_3)_2$ is added to 1.0 L of 5.0 M NH_3?

$$Cd^{2+}(aq) + 4NH_3(aq) \rightleftharpoons Cd(NH_3)_4^{2+}(aq)$$
$$K = 1.0 \times 10^7$$

$$Cd(OH)_2(s) \rightleftharpoons Cd^{2+}(aq) + 2OH^-(aq)$$
$$K_{sp} = 5.9 \times 10^{-15}$$

89. **a.** Using the K_{sp} value for $Cu(OH)_2$ (1.6×10^{-19}) and the overall formation constant for $Cu(NH_3)_4^{2+}$ (1.0×10^{13}), calculate the value for the equilibrium constant for the following reaction:

$$Cu(OH)_2(s) + 4NH_3(aq) \rightleftharpoons Cu(NH_3)_4^{2+}(aq) + 2OH^-(aq)$$

b. Use the value of the equilibrium constant you calculated in part a to calculate the solubility (in mol/L) of $Cu(OH)_2$ in 5.0 M NH_3. In 5.0 M NH_3 the concentration of OH^- is 0.0095 M.

90. Describe how you could separate the ions in each of the following groups by selective precipitation.

a. Ag^+, Mg^{2+}, Cu^{2+} **c.** Pb^{2+}, Bi^{3+}

b. Pb^{2+}, Ca^{2+}, Fe^{2+}

91. The solubility rules outlined in Chapter 6 say that $Ba(OH)_2$, $Sr(OH)_2$, and $Ca(OH)_2$ are marginally soluble hydroxides. Calculate the pH of a saturated solution of each of these marginally soluble hydroxides.

92. In the chapter discussion of precipitate formation, we ran the precipitation reaction to completion and then let some of the precipitate redissolve to get back to equilibrium. To see why, redo Example 15-6, where

Initial Concentration (mol/L)	Equilibrium Concentration (mol/L)
$[Mg^{2+}]_0 = 3.75 \times 10^{-3}$	$[Mg^{2+}] = 3.75 \times 10^{-3} - y$
$[F^-]_0 = 6.25 \times 10^{-2}$	$[F^-] = 6.25 \times 10^{-2} - 2y$
$\xrightarrow[\text{reacts to form } MgF_2]{y \text{ mol/}Mg^{2+}}$	

ChemWork Problems

These multiconcept problems (and additional ones) are found interactively online with the same type of assistance a student would get from an instructor.

93. Assuming that the solubility of $Ca_3(PO_4)_2(s)$ is 1.6×10^{-7} mol/L at 25°C, calculate the K_{sp} for this salt. Ignore any potential reactions of the ions with water.

94. Order the following solids (a–d) from least soluble to most soluble. Ignore any potential reactions of the ions with water.

a. $AgCl$ $K_{sp} = 1.6 \times 10^{-10}$

b. Ag_2S $K_{sp} = 1.6 \times 10^{-49}$

c. CaF_2 $K_{sp} = 4.0 \times 10^{-11}$

d. CuS $K_{sp} = 8.5 \times 10^{-45}$

95. The K_{sp} for $PbI_2(s)$ is 1.4×10^{-8}. Calculate the solubility of $PbI_2(s)$ in 0.048 M NaI.

96. The solubility of $Pb(IO_3)_2(s)$ in a 7.2×10^{-2}-M KIO_3 solution is 6.0×10^{-9} mol/L. Calculate the K_{sp} value for $Pb(IO_3)_2(s)$.

97. A 50.0-mL sample of 0.0413 M $AgNO_3(aq)$ is added to 50.0 mL of 0.100 M $NaIO_3(aq)$. Calculate the $[Ag^+]$ at equilibrium in the resulting solution. [K_{sp} for $AgIO_3(s) = 3.17 \times 10^{-8}$.]

98. The Hg^{2+} ion forms complex ions with I^- as follows:

$$Hg^{2+}(aq) + I^-(aq) \rightleftharpoons HgI^+(aq) \qquad K_1 = 1.0 \times 10^8$$
$$HgI^+(aq) + I^-(aq) \rightleftharpoons HgI_2(aq) \qquad K_2 = 1.0 \times 10^5$$
$$HgI_2(aq) + I^-(aq) \rightleftharpoons HgI_3^-(aq) \qquad K_3 = 1.0 \times 10^9$$
$$HgI_3^-(aq) + I^-(aq) \rightleftharpoons HgI_4^{2-}(aq) \qquad K_4 = 1.0 \times 10^8$$

A solution is prepared by dissolving 0.088 mole of $Hg(NO_3)_2$ and 5.00 moles of NaI in enough water to make 1.0 L of solution.

a. Calculate the equilibrium concentration of $[HgI_4^{2-}]$.

b. Calculate the equilibrium concentration of $[I^-]$.

c. Calculate the equilibrium concentration of $[Hg^{2+}]$.

Challenge Problems

99. The copper(I) ion forms a complex ion with CN^- according to the following equation:

$$Cu^+(aq) + 3CN^-(aq) \rightleftharpoons Cu(CN)_3^{2-}(aq) \qquad K = 1.0 \times 10^{11}$$

a. Calculate the solubility of $CuBr(s)$ ($K_{sp} = 1.0 \times 10^{-5}$) in 1.0 L of 1.0 M $NaCN$.

b. Calculate the concentration of Br^- at equilibrium.

c. Calculate the concentration of CN^- at equilibrium.

100. Consider a solution made by mixing 500.0 mL of 4.0 M NH_3 and 500.0 mL of 0.40 M $AgNO_3$. Ag^+ reacts with NH_3 to form $AgNH_3^+$ and $Ag(NH_3)_2^+$:

$$Ag^+(aq) + NH_3(aq) \rightleftharpoons AgNH_3^+(aq) \qquad K_1 = 2.1 \times 10^3$$
$$AgNH_3^+(aq) + NH_3(aq) \rightleftharpoons Ag(NH_3)_2^+(aq) \qquad K_2 = 8.2 \times 10^3$$

Determine the concentration of all species in solution.

101. **a.** Calculate the molar solubility of $AgBr$ in pure water. K_{sp} for $AgBr$ is 5.0×10^{-13}.

b. Calculate the molar solubility of $AgBr$ in 3.0 M NH_3. The overall formation constant for $Ag(NH_3)_2^+$ is 1.7×10^7, that is,

$$Ag^+(aq) + 2NH_3(aq) \longrightarrow Ag(NH_3)_2^+(aq) \qquad K = 1.7 \times 10^7.$$

c. Compare the calculated solubilities from parts a and b. Explain any differences.

d. What mass of $AgBr$ will dissolve in 250.0 mL of 3.0 M NH_3?

e. What effect does adding HNO_3 have on the solubilities calculated in parts a and b?

102. Calculate the equilibrium concentrations of NH_3, Cu^{2+}, $Cu(NH_3)^{2+}$, $Cu(NH_3)_2^{2+}$, $Cu(NH_3)_3^{2+}$, and $Cu(NH_3)_4^{2+}$ in a solution prepared by mixing 500.0 mL of 3.00 M NH_3 with 500.0 mL of 2.00×10^{-3} M $Cu(NO_3)_2$. The stepwise equilibria are

$$Cu^{2+}(aq) + NH_3(aq) \rightleftharpoons CuNH_3^{2+}(aq)$$
$$K_1 = 1.86 \times 10^4$$
$$CuNH_3^{2+}(aq) + NH_3(aq) \rightleftharpoons Cu(NH_3)_2^{2+}(aq)$$
$$K_2 = 3.88 \times 10^3$$
$$Cu(NH_3)_2^{2+}(aq) + NH_3(aq) \rightleftharpoons Cu(NH_3)_3^{2+}(aq)$$
$$K_3 = 1.00 \times 10^3$$
$$Cu(NH_3)_3^{2+}(aq) + NH_3(aq) \rightleftharpoons Cu(NH_3)_4^{2+}(aq)$$
$$K_4 = 1.55 \times 10^2$$

103. Calculate the solubility of $AgCN(s)$ ($K_{sp} = 2.2 \times 10^{-12}$) in a solution containing $1.0 \ M \ H^+$. (K_a for HCN is 6.2×10^{-10}.)

104. Calcium oxalate (CaC_2O_4) is relatively insoluble in water ($K_{sp} = 2 \times 10^{-9}$). However, calcium oxalate is more soluble in acidic solution. How much more soluble is calcium oxalate in $0.10 \ M \ H^+$ than in pure water? In pure water, ignore the basic properties of $C_2O_4^{2-}$.

105. What is the maximum possible concentration of Ni^{2+} ion in water at 25°C that is saturated with $0.10 \ M \ H_2S$ and maintained at pH 3.0 with HCl?

106. A mixture contains $1.0 \times 10^{-3} \ M \ Cu^{2+}$ and $1.0 \times 10^{-3} \ M \ Mn^{2+}$ and is saturated with $0.10 \ M \ H_2S$. Determine a pH where CuS precipitates but MnS does not precipitate. K_{sp} for $CuS = 8.5 \times 10^{-45}$ and K_{sp} for $MnS = 2.3 \times 10^{-13}$.

107. Sodium tripolyphosphate ($Na_5P_3O_{10}$) is used in many synthetic detergents. Its major effect is to soften the water by complexing Mg^{2+} and Ca^{2+} ions. It also increases the efficiency of surfactants, or wetting agents that lower a liquid's surface tension. The K value for the formation of $MgP_3O_{10}^{3-}$ is 4.0×10^8. The reaction is $Mg^{2+}(aq) + P_3O_{10}^{5-}(aq) \rightleftharpoons MgP_3O_{10}^{3-}(aq)$. Calculate the concentration of Mg^{2+} in a solution that was originally 50. ppm Mg^{2+} (50. mg/L of solution) after 40. g $Na_5P_3O_{10}$ is added to 1.0 L of the solution.

108. You add an excess of solid MX in 250 g water. You measure the freezing point and find it to be $-0.028°C$. What is the K_{sp} of the solid? Assume the density of the solution is $1.0 \ g/cm^3$.

109. **a.** Calculate the molar solubility of SrF_2 in water, ignoring the basic properties of F^-. (For SrF_2, $K_{sp} = 7.9 \times 10^{-10}$.)

     b. Would the measured molar solubility of SrF_2 be greater than or less than the value calculated in part a? Explain.

     c. Calculate the molar solubility of SrF_2 in a solution buffered at pH = 2.00. (K_a for HF is 7.2×10^{-4}.)

Integrative Problems

These problems require the integration of multiple concepts to find the solutions.

110. A solution saturated with a salt of the type M_3X_2 has an osmotic pressure of 2.64×10^{-2} atm at 25°C. Calculate the K_{sp} value for the salt, assuming ideal behavior.

111. Consider 1.0 L of an aqueous solution that contains $0.10 \ M$ sulfuric acid to which 0.30 mole of barium nitrate is added. Assuming no change in volume of the solution, determine the pH, the concentration of barium ions in the final solution, and the mass of solid formed.

112. The K_{sp} for Q, a slightly soluble ionic compound composed of M_2^{2+} and X^- ions, is 4.5×10^{-29}. The electron configuration of M^+ is $[Xe]6s^1 4f^{14} 5d^{10}$. The X^- anion has 54 electrons. What is the molar solubility of Q in a solution of NaX prepared by dissolving 1.98 g NaX in 150. mL solution?

Marathon Problem

This problem is designed to incorporate several concepts and techniques into one situation.

113. Aluminum ions react with the hydroxide ion to form the precipitate $Al(OH)_3(s)$, but can also react to form the soluble complex ion $Al(OH)_4^-$. In terms of solubility, $Al(OH)_3(s)$ will be more soluble in very acidic solutions as well as more soluble in very basic solutions.

     a. Write equations for the reactions that occur to increase the solubility of $Al(OH)_3(s)$ in very acidic solutions and in very basic solutions.

     b. Let's study the pH dependence of the solubility of $Al(OH)_3(s)$ in more detail. Show that the solubility of $Al(OH)_3$, as a function of $[H^+]$, obeys the equation

$$S = [H^+]^3 K_{sp}/K_w^3 + KK_w/[H^+]$$

     where S = solubility = $[Al^{3+}] + [Al(OH)_4^-]$ and K is the equilibrium constant for

$$Al(OH)_3(s) + OH^-(aq) \rightleftharpoons Al(OH)_4^-(aq)$$

     c. The value of K is 40.0 and K_{sp} for $Al(OH)_3$ is 2×10^{-32}. Plot the solubility of $Al(OH)_3$ in the pH range 4–12.

CHAPTER SIXTEEN

Spontaneity, Entropy, and Free Energy

Solid carbon dioxide (dry ice), when placed in water, causes violent bubbling as gaseous CO_2 is released. The "fog" is moisture condensed from the cold air. (Science Source)

The *first law of thermodynamics* is a statement of the law of conservation of energy: Energy can be neither created nor destroyed. In other words, *the energy of the universe is constant*. Although the total energy is constant, the various forms of energy can be interchanged in physical and chemical processes. For example, if you drop a book, some of the initial potential energy of the book is changed to kinetic energy, which is then transferred to the atoms in the air and the floor as random motion. The net effect of this process is to change a given quantity of potential energy to exactly the same quantity of thermal energy. Energy has been converted from one form to another, but the same quantity of energy exists before and after the process.

Now let's consider a chemical example. When methane is burned in excess oxygen, the major reaction is

$$CH_4(g) + 2O_2(g) \longrightarrow CO_2(g) + 2H_2O(g) + \text{energy}$$

This reaction produces a quantity of energy, which is released as heat. This energy flow results from the lowering of the potential energy stored in the bonds of CH_4 and O_2 as they react to form CO_2 and H_2O. This is illustrated in Fig. 16-1. Potential energy has been converted to thermal energy, but the energy content of the universe has remained constant in accordance with the first law of thermodynamics.

The first law of thermodynamics is used mainly for energy bookkeeping, that is, to answer such questions as

How much energy is involved in the change?

Does energy flow into or out of the system?

What form does the energy finally assume?

When accounting for the energy change we need to consider the atoms and molecules in the system and the surroundings. For example, we need to account for the random motions of the atoms in the air and floor as a book drops.

Although the first law of thermodynamics provides the means for accounting for energy, it gives no hint as to why a particular process occurs in a given direction. ◄ 𝒊 This is the main question to be considered in this chapter.

𝒊 The first law of thermodynamics: The energy of the universe is constant.

16-1 | Spontaneous Processes and Entropy

A process is said to be *spontaneous* if it *occurs without outside intervention*. **Spontaneous processes** may be fast or slow. As we will see in this chapter, thermodynamics can tell us the *direction* in which a process will occur but can say nothing about the *speed* of the process. ◄ 𝒊 As we saw in Chapter 11, the rate of a reaction depends on many factors, such as activation energy, temperature, concentration, and catalysts, and we were able to explain these effects using a simple collision model. In describing a chemical reaction, the discipline of chemical kinetics focuses on the pathway between reactants and products; thermodynamics considers only the initial and final

𝒊 *Spontaneous* does not mean fast.

Figure 16-1 | When methane and oxygen react to form carbon dioxide and water, the products have lower potential energy than the reactants. This change in potential energy results in energy flow (heat) to the surroundings.

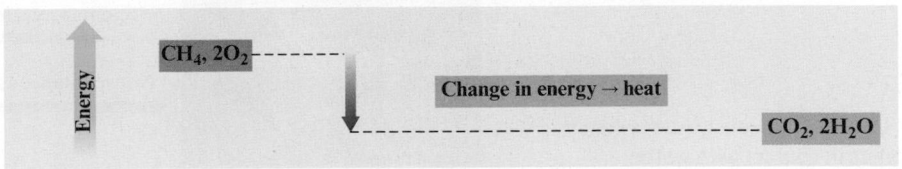

Figure 16-2 | The rate of a reaction depends on the pathway from reactants to products; this is the domain of kinetics. Thermodynamics tells us whether a reaction is spontaneous based only on the properties of the reactants and products. The predictions of thermodynamics do not require knowledge of the pathway between reactants and products.

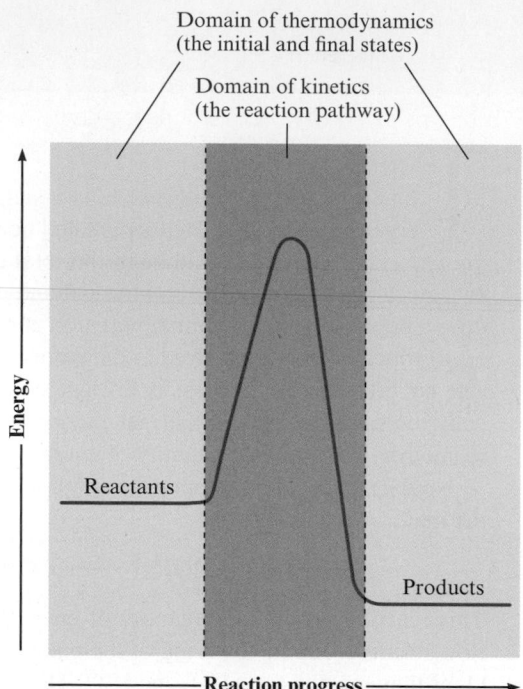

i An alternative way to describe a spontaneous process is to say that it is "thermodynamically favored."

i The process of diamond turning to graphite, while thermodynamically favored, is said to be under kinetic control; that is, it does not proceed at a noticeable rate.

i Rust is an observable spontaneous process. To understand why it is spontaneous, we need to think of the molecules involved in the reaction.

states and does not require knowledge of the pathway between reactants and products (Fig. 16-2). ◀ *i*

In summary, thermodynamics lets us predict whether a process will occur but gives no information about the amount of time required for the process. For example, according to the principles of thermodynamics, a diamond should change spontaneously to graphite. The fact that we do not observe this process does not mean the prediction is wrong; it simply means the process is very slow. Thus we need both thermodynamics and kinetics to describe reactions fully. ◀ *i*

To explore the idea of spontaneity, consider the following physical and chemical processes:

A ball rolls down a hill but never spontaneously rolls back up the hill.

If exposed to air and moisture, steel rusts spontaneously. However, the iron oxide in rust does not spontaneously change back to iron metal and oxygen gas. ◀ ◀ *i*

A gas fills its container uniformly. It never spontaneously collects at one end of the container.

Heat flow always occurs from a hot object to a cooler one. The reverse process never occurs spontaneously.

Wood burns spontaneously in an exothermic reaction to form carbon dioxide and water, but wood is not formed when carbon dioxide and water are heated together.

At temperatures below 0°C, water spontaneously freezes, and at temperatures above 0°C, ice spontaneously melts.

Spontaneous reaction

Nonspontaneous reaction

▲ Iron spontaneously rusts when it comes in contact with water.

What thermodynamic principle will provide an explanation of why, under a given set of conditions, each of these diverse processes occurs in one direction and never in the reverse? In searching for an answer, we could explain the behavior of a ball on a hill in terms of gravity. But what does gravity have to do with the rusting of a nail or the freezing of water? Early developers of thermodynamics thought that exothermicity might be the key—that a process would be spontaneous if it were exothermic. Although this factor does appear to be important, since many spontaneous processes are exothermic, it is not the total answer. For example, the melting of ice, which occurs spontaneously at temperatures greater than 0°C, is an endothermic process.

What common characteristic causes the processes listed above to be spontaneous in one direction only? After many years of observation, scientists have concluded that the characteristic common to all spontaneous processes is an increase in a property called **entropy**, denoted by the symbol S. The driving force for a spontaneous process is an increase in the entropy of the universe.

▲ A disordered pile of playing cards.

What is entropy? Although there is no simple definition that is completely accurate, *entropy can be viewed as a measure of molecular randomness or disorder.* The natural progression of things is from order to disorder, from lower entropy to higher entropy. To illustrate the natural tendency toward disorder, you only have to think about the condition of your room. Your room naturally tends to get messy (disordered), because an ordered room requires everything to be in its place. There are simply many more ways for things to be out of place than for them to be in their places.

As another example, suppose you have a deck of playing cards ordered in some particular way. You throw these cards into the air and pick them all up at random. ◄ Looking at the new sequence of the cards, you would be very surprised to find that it matched the original order. Such an event would be possible, but *very improbable.* There are billions of ways for the deck to be disordered, but only one way to be ordered according to your definition. Thus the chances of picking the cards up out of order are much greater than the chance of picking them up in order. It is natural for disorder to increase.

Entropy is a thermodynamic function that describes the *number of arrangements* (positions and/or energy levels) that are *available to a system* existing in a given state. Entropy is closely associated with probability. The key concept is that the more ways a particular state can be achieved, the greater is the likelihood (probability) of finding that state. In other words, *nature spontaneously proceeds toward the states that have the highest probabilities of existing.* ◄ *(i)* This conclusion is not surprising at all. The difficulty comes in connecting this concept to real-life processes. For example, what does the spontaneous rusting of steel have to do with probability? Understanding the connection between entropy and spontaneity will allow us to answer such questions. We will begin to explore this connection by considering a very simple process, the expansion of an ideal gas into a vacuum, as represented in Fig. 16-3. Why is this process spontaneous? ◄ *(i)* The driving force is probability. Because there are more ways of having the gas evenly spread throughout the container than there are ways for it to be in any other possible state, the gas spontaneously attains the uniform distribution.

(i) Probability refers to likelihood.

(i) Only by considering a gas as a collection of constantly and randomly moving particles can we answer the question of why this process is spontaneous.

To understand this conclusion, we will greatly simplify the system and consider the possible arrangements of only four gas molecules in the two-bulbed container (Fig. 16-4). How many ways can each arrangement (state) be achieved? Arrangements I and V can be achieved in only one way—all the molecules must be in one end. Arrangements II and IV can be achieved in four ways, as shown in Table 16-1. Each configuration that gives a particular arrangement is called a *microstate*. Arrangement I has one microstate, and arrangement II has four microstates. Arrangement III can be achieved in six ways (six microstates), as shown in Table 16-1. *Which arrangement is most likely to occur?* The one that can be achieved in the greatest number of ways. Thus arrangement III is most probable. The relative probabilities of arrangements III, II, and I are 6:4:1. We have discovered an important principle: The probability of occurrence

Figure 16-3 | The expansion of an ideal gas into an evacuated bulb.

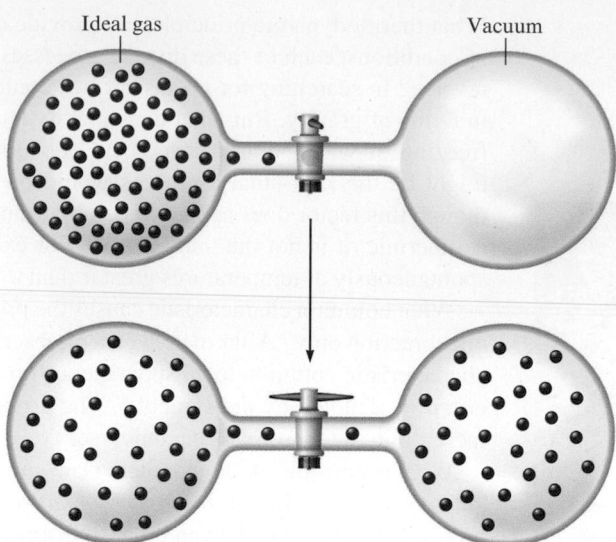

of a particular arrangement (state) depends on the number of ways (microstates) in which that arrangement can be achieved.

The consequences of this principle are dramatic for large numbers of molecules. One gas molecule in the flask in Fig. 16-4 has one chance in two of being in the left bulb. We say that the probability of finding the molecule in the left bulb is $\frac{1}{2}$. For two molecules in the flask, there is one chance in two of finding each molecule in the left bulb, so there is one chance in four ($\frac{1}{2} \times \frac{1}{2} = \frac{1}{4}$) that *both* molecules will be in the left bulb. ◄ 𝒾 As the number of molecules increases, the relative probability of finding all of them in the left

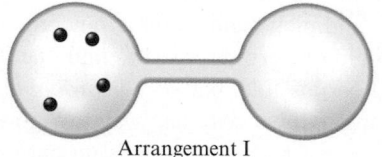

Arrangement I

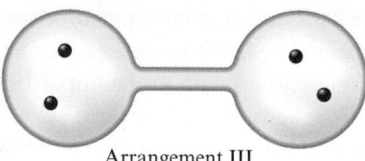

Arrangement II

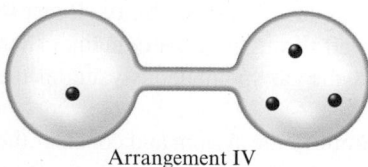

Arrangement III

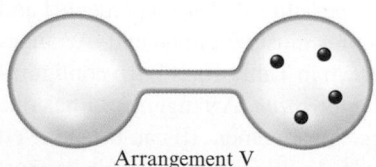

Arrangement IV

Arrangement V

Figure 16-4 | Possible arrangements (states) of four molecules in a two-bulbed flask.

Table 16-1 | The Microstates That Give a Particular Arrangement (State)

Arrangement	Microstates	Number of Microstates
I		1
II		4
III		6
IV		4
V		1

i For two molecules in the flask, there are four possible microstates:

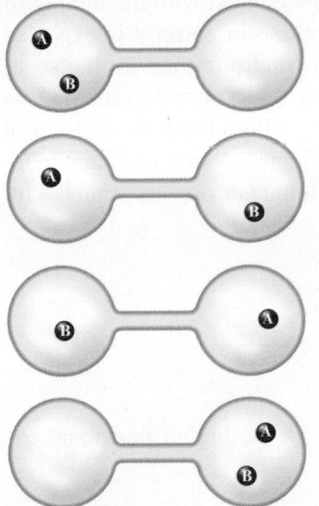

Thus there is one chance in four of finding

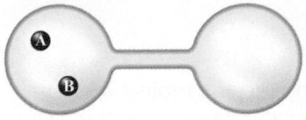

Table 16-2 | Probability of Finding All the Molecules in the Left Bulb as a Function of the Total Number of Molecules

Number of Molecules	Relative Probability of Finding All Molecules in the Left Bulb
1	$\dfrac{1}{2}$
2	$\dfrac{1}{2} \times \dfrac{1}{2} = \dfrac{1}{2^2} = \dfrac{1}{4}$
3	$\dfrac{1}{2} \times \dfrac{1}{2} \times \dfrac{1}{2} = \dfrac{1}{2^3} = \dfrac{1}{8}$
5	$\dfrac{1}{2} \times \dfrac{1}{2} \times \dfrac{1}{2} \times \dfrac{1}{2} \times \dfrac{1}{2} = \dfrac{1}{2^5} = \dfrac{1}{32}$
10	$\dfrac{1}{2^{10}} = \dfrac{1}{1024}$
n	$\dfrac{1}{2^n} = \left(\dfrac{1}{2}\right)^n$
6×10^{23} (1 mole)	$\left(\dfrac{1}{2}\right)^{6 \times 10^{23}} \approx 10^{-(2 \times 10^{23})}$

bulb decreases, as shown in Table 16-2. For 1 mole of gas, the probability of finding all the molecules in the left bulb is so small that this arrangement would "never" occur.

Thus a gas placed in one end of a container will spontaneously expand to fill the entire vessel evenly because, for a large number of gas molecules, there is a huge number of microstates in which equal numbers of molecules are in both ends. On the other hand, the opposite process,

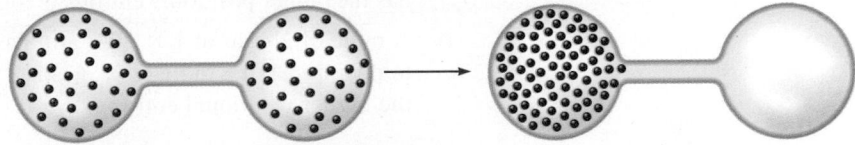

although not impossible, is *highly* improbable, since only one microstate leads to this arrangement. Therefore, this process does not occur spontaneously.

The type of probability we have been considering in this example is called **positional probability** because it depends on the number of configurations in space (positional microstates) that yield a particular state. A gas expands into a vacuum to give a uniform distribution because the expanded state has the highest positional probability, that is, the largest entropy, of the states available to the system.

Positional probability is also illustrated by changes of state. In general, positional entropy increases in going from solid to liquid to gas. A mole of a substance has a much smaller volume in the solid state than it does in the gaseous state. In the solid state, the molecules are close together, with relatively few positions available to them; ◀ *i* in the gaseous state, the molecules are far apart, with many more positions available to them. The liquid state is closer to the solid state than it is to the gaseous state in these terms. We can summarize these comparisons as follows:

i Solids are more ordered than liquids or gases and thus have lower entropy.

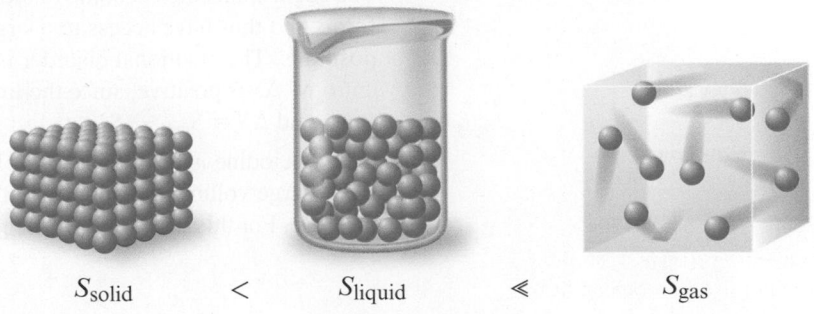

$$S_{\text{solid}} \quad < \quad S_{\text{liquid}} \quad \ll \quad S_{\text{gas}}$$

ⓘ The tendency to mix is due to the increased volume available to the particles of each component of the mixture. For example, when two liquids are mixed, the molecules of each liquid have more available volume and thus more available positions.

Positional entropy is also very important in the formation of solutions. In Chapter 10 we saw that solution formation is favored by the natural tendency for substances to mix. We can now be more precise. The entropy change associated with the mixing of two pure substances is expected to be positive. An increase in entropy is expected because there are many more microstates for the mixed condition than for the separated condition. This effect is due principally to the increased volume available to a given "particle" after mixing occurs. For example, when two liquids are mixed to form a solution, the molecules of each liquid have more available volume and thus more available positions. ◄ ⓘ Therefore, the increase in positional entropy associated with mixing favors the formation of solutions.

INTERACTIVE EXAMPLE 16-1

Positional Entropy

For each of the following pairs, choose the substance with the higher positional entropy (per mole) at a given temperature.

a. Solid CO_2 and gaseous CO_2

b. N_2 gas at 1 atm and N_2 gas at 1.0×10^{-2} atm

Solution

a. Since a mole of gaseous CO_2 has the greater volume by far, the molecules have many more available positions than in a mole of solid CO_2. Thus gaseous CO_2 has the higher positional entropy.

b. A mole of N_2 gas at 1×10^{-2} atm has a volume 100 times that (at a given temperature) of a mole of N_2 gas at 1 atm. Thus N_2 gas at 1×10^{-2} atm has the higher positional entropy.

See Exercise 16-31

INTERACTIVE EXAMPLE 16-2

Predicting Entropy Changes

Predict the sign of the entropy change for each of the following processes.

a. Solid sugar is added to water to form a solution.

b. Iodine vapor condenses on a cold surface to form crystals.

Solution

a. The sugar molecules become randomly dispersed in the water when the solution forms and thus have access to a larger volume and a larger number of possible positions. The positional disorder is increased, and there will be an increase in entropy. ΔS is positive, since the final state has a larger entropy than the initial state, and $\Delta S = S_{final} - S_{initial}$.

b. Gaseous iodine is forming a solid. This process involves a change from a relatively large volume to a much smaller volume, which results in lower positional disorder. For this process ΔS is negative (the entropy decreases).

See Exercise 16-32

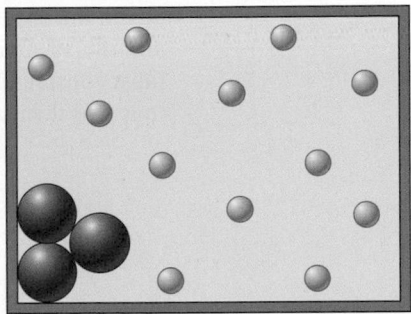

In this text we have emphasized the meaning of the second law of thermodynamics   that the entropy of the universe is always increasing. Although the results of all our experiments support this conclusion, this does not mean that order cannot appear spontaneously in a given part of the universe. The best example of this phenomenon involves the assembly of cells in living organisms. Of course, when a process that creates an ordered system is examined in detail, it is found that other parts of the process involve an increase in disorder such that the sum of all the entropy changes is positive. In fact, scientists are now finding that the search for maximum entropy in one part of a system can be a powerful force for organization in another part of the system.

To understand how entropy can be an organizing force, look at the accompanying figure. In a system containing large and small "balls" as shown in the figure, the small balls can "herd" the large balls into clumps in the corners and near the walls. This clears out the maximum space for the small balls so that they can move more freely, thus maximizing the entropy of the system, as demanded by the second law of thermodynamics.

In essence, the ability to maximize entropy by sorting different-sized objects creates a kind of attractive force, called a *depletion*, or *excluded-volume, force*. These "entropic forces" operate for objects in the size range of approximately 10^{-8} to approximately 10^{-6} m. For entropy-induced ordering to occur, the particles must be constantly jostling each other and must be constantly agitated by solvent molecules, thus making gravity unimportant.

There is increasing evidence that entropic ordering is important in many biological systems. For example, this phenomenon seems to be responsible for the clumping of sickle-cell hemoglobin in the presence of much smaller proteins that act as the "smaller balls." Entropic forces also have been linked to the clustering of DNA in cells without nuclei, and Allen Minton of the National Institutes of Health in Bethesda, Maryland, is studying the role of entropic forces in the binding of proteins to cell membranes.

Entropic ordering also appears in nonbiological settings, especially in the ways polymer molecules clump together. For example, polymers added to paint to improve the flow characteristics of the paint actually caused it to coagulate because of depletion forces.

Thus, as you probably have concluded already, entropy is a complex issue. As entropy drives the universe to its ultimate death of maximum chaos, it provides some order along the way.

16-2 | Entropy and the Second Law of Thermodynamics

ⓘ The total energy of the universe is constant, but the entropy is increasing.

We have seen that processes are spontaneous when they result in an increase in disorder. Nature always moves toward the most probable state available to it. We can state this principle in terms of entropy: In any spontaneous process there is always an increase in the entropy of the universe. ◀ ⓘ This is the **second law of thermodynamics**. Contrast this with the first law of thermodynamics, which tells us that the energy of the universe is constant. Energy is conserved in the universe, but entropy is not. In fact, the second law can be paraphrased as follows: *The entropy of the universe is increasing.*

As in Chapter 7, we find it convenient to divide the universe into a system and the surroundings. Thus we can represent the change in the entropy of the universe as

$$\Delta S_{univ} = \Delta S_{sys} + \Delta S_{surr}$$

where ΔS_{sys} and ΔS_{surr} represent the changes in entropy that occur in the system and surroundings, respectively.

To predict whether a given process will be spontaneous, we must know the sign of ΔS_{univ}. If ΔS_{univ} is positive, the entropy of the universe increases, and the process is spontaneous in the direction written. If ΔS_{univ} is negative, the process is spontaneous in the *opposite* direction. If ΔS_{univ} is zero, the process has no tendency to occur, and the system is at equilibrium. To predict whether a process is spontaneous, we must consider the entropy changes that occur both in the system and in the surroundings and then take their sum.

Critical Thinking

What if ΔS_{univ} was a state function? How would the world be different?

EXAMPLE 16-3   The Second Law

In a living cell, large molecules are assembled from simple ones. Is this process consistent with the second law of thermodynamics?

Solution

To reconcile the operation of an order-producing cell with the second law of thermodynamics, we must remember that ΔS_{univ}, not ΔS_{sys}, must be positive for a process to be spontaneous. A process for which ΔS_{sys} is negative can be spontaneous if the associated ΔS_{surr} is both larger and positive. The operation of a cell is such a process.

See Questions 16-11 and 16-12

16-3 | The Effect of Temperature on Spontaneity

To explore the interplay of ΔS_{sys} and ΔS_{surr} in determining the sign of ΔS_{univ}, we will first discuss the change of state for 1 mole of water from liquid to gas,

$$H_2O(l) \longrightarrow H_2O(g)$$

considering the water to be the system and everything else the surroundings. ◄

What happens to the entropy of water in this process? A mole of liquid water (18 g) has a volume of approximately 18 mL. A mole of gaseous water at 1 atmosphere and 100°C occupies a volume of approximately 31 L. Clearly, there are many more positions available to the water molecules in a volume of 31 L than in 18 mL, and the vaporization of water is favored by this increase in positional probability. That is, for this process the entropy of the system increases; ΔS_{sys} has a positive sign.

What about the entropy change in the surroundings? Although we will not prove it here, entropy changes in the surroundings are determined primarily by the flow of energy into or out of the system as heat. To understand this, suppose an exothermic process transfers 50 J of energy as heat to the surroundings, where it becomes thermal energy, that is, kinetic energy associated with the random motions of atoms. Thus this flow of energy into the surroundings increases the random motions of atoms there and thereby increases the entropy of the surroundings. The sign of ΔS_{surr} is positive. When an endothermic process occurs in the system, it produces the opposite effect. Heat flows from the surroundings to the system, and the random motions of the atoms in the

▲ Boiling water to form steam increases its volume and thus its entropy.

Can Balcioglu/Shutterstock

surroundings decrease, decreasing the entropy of the surroundings. The vaporization of water is an endothermic process. Thus, for this change of state, ΔS_{surr} is negative.

Remember it is the sign of ΔS_{univ} that tells us whether the vaporization of water is spontaneous. We have seen that ΔS_{sys} is positive and favors the process and that ΔS_{surr} is negative and unfavorable. Thus the components of ΔS_{univ} are in opposition. Which one controls the situation? The answer *depends on the temperature*. We know that at a pressure of 1 atmosphere, water changes spontaneously from liquid to gas at all temperatures above 100°C. Below 100°C, the opposite process (condensation) is spontaneous.

Since ΔS_{sys} and ΔS_{surr} are in opposition for the vaporization of water, the temperature must have an effect on the relative importance of these two terms. To understand why this is so, we must discuss in more detail the factors that control the entropy changes in the surroundings. The central idea is that the entropy changes in the surroundings are primarily determined by heat flow. An exothermic process in the system increases the entropy of the surroundings, because the resulting energy flow increases the random motions in the surroundings. This means that exothermicity is an important driving force for spontaneity. In earlier chapters we have seen that a system tends to undergo changes that lower its energy. We now understand the reason for this tendency. When a system at constant temperature moves to a lower energy, the energy it gives up is transferred to the surroundings, leading to an increase in entropy there. ◄ ⓘ

The significance of exothermicity as a driving force *depends on the temperature at which the process occurs*. That is, the magnitude of ΔS_{surr} depends on the temperature at which the heat is transferred. We will not attempt to prove this fact here. Instead, we offer an analogy. Suppose that you have $50 to give away. Giving it to a millionaire would not create much of an impression—a millionaire has money to spare. However, to a poor college student, $50 would represent a significant sum and would be received with considerable joy. The same principle can be applied to energy transfer via the flow of heat. If 50 J of energy is transferred to the surroundings, the impact of that event depends greatly on the temperature. If the temperature of the surroundings is very high, the atoms there are in rapid motion. The 50 J of energy will not make a large percent change in these motions. On the other hand, if 50 J of energy is transferred to the surroundings at a very low temperature, where atomic motion is slow, the energy will cause a large percent change in these motions. The impact of the transfer of a given quantity of energy as heat to or from the surroundings will be greater at lower temperatures. ◄ ⓘ

For our purposes, there are two important characteristics of the entropy changes that occur in the surroundings:

1. *The sign of ΔS_{surr} depends on the direction of the heat flow.* At constant temperature, an exothermic process in the system causes heat to flow into the surroundings, increasing the random motions and thus the entropy of the surroundings. For this case, ΔS_{surr} is positive. The opposite is true for an endothermic process in a system at constant temperature. Note that although the driving force described here really results from the change in entropy, it is often described in terms of energy: Nature tends to seek the lowest possible energy. ◄ ⓘ

2. *The magnitude of ΔS_{surr} depends on the temperature.* The transfer of a given quantity of energy as heat produces a much greater percent change in the randomness of the surroundings at a low temperature than it does at a high temperature. Thus ΔS_{surr} depends directly on the quantity of heat transferred and inversely on temperature. In other words, the tendency for the system to lower its energy becomes a more important driving force at lower temperatures.

ⓘ In an endothermic process, heat flows from the surroundings into the system. In an exothermic process, heat flows into the surroundings from the system.

ⓘ Recall that temperature is a measure of molecular motion. Thus a transfer of energy to the surroundings at a greater temperature will change the molecular motion by a smaller percentage than transferring the same amount of energy to surroundings at a lower temperature.

ⓘ In a process occurring at constant temperature, the tendency for the system to lower its energy results from the positive value of ΔS_{surr}.

$$\begin{array}{c}\text{Driving force}\\\text{provided by}\\\text{the energy flow}\\\text{(heat)}\end{array} = \begin{array}{c}\text{magnitude of the}\\\text{entropy change of}\\\text{the surroundings}\end{array} = \frac{\text{quantity of heat (J)}}{\text{temperature (K)}}$$

These ideas are summarized as follows:

i Exothermic process:
ΔS_{surr} = positive

i Endothermic process:
ΔS_{surr} = negative

Exothermic process:	$\Delta S_{surr} = +\dfrac{\text{quantity of heat (J)}}{\text{temperature (K)}}$ ◂ i
Endothermic process:	$\Delta S_{surr} = -\dfrac{\text{quantity of heat (J)}}{\text{temperature (K)}}$ ◂ i

We can express ΔS_{surr} in terms of the change in enthalpy ΔH for a process occurring at constant pressure, since

$$\text{Heat flow (constant } P) = \text{change in enthalpy} = \Delta H$$

i When no subscript is present, the quantity (for example, ΔH) refers to the system.

Recall that ΔH consists of two parts: a sign and a number. ◂ i The *sign* indicates the direction of flow, where a plus sign means into the system (endothermic) and a minus sign means out of the system (exothermic). The *number* indicates the quantity of energy.

Combining all these concepts produces the following definition of ΔS_{surr} for a reaction that takes place under conditions of constant temperature (in kelvins) and pressure:

$$\Delta S_{surr} = -\frac{\Delta H}{T}$$

i The minus sign changes the point of view from the system to the surroundings.

The minus sign is necessary because the sign of ΔH is determined with respect to the reaction system, and this equation expresses a property of the surroundings. ◂ i This means that if the reaction is exothermic, ΔH has a negative sign, but since heat flows into the surroundings, ΔS_{surr} is positive.

INTERACTIVE EXAMPLE 16-4

▲ The mineral stibnite contains Sb_2S_3.

© Cengage Learning

Determining ΔS_{surr}

In the metallurgy of antimony, the pure metal is recovered via different reactions, depending on the composition of the ore. ◂ For example, iron is used to reduce antimony in sulfide ores:

$$Sb_2S_3(s) + 3Fe(s) \longrightarrow 2Sb(s) + 3FeS(s) \qquad \Delta H = -125 \text{ kJ}$$

Carbon is used as the reducing agent for oxide ores:

$$Sb_4O_6(s) + 6C(s) \longrightarrow 4Sb(s) + 6CO(g) \qquad \Delta H = 778 \text{ kJ}$$

Calculate ΔS_{surr} for each of these reactions at 25°C and 1 atm.

Solution

We use

$$\Delta S_{surr} = -\frac{\Delta H}{T}$$

where

$$T = 25 + 273 = 298 \text{ K}$$

For the sulfide ore reaction,

$$\Delta S_{surr} = -\frac{-125 \text{ kJ}}{298 \text{ K}} = 0.419 \text{ kJ/K} = 419 \text{ J/K}$$

Note that ΔS_{surr} is positive, as it should be, since this reaction is exothermic and heat flow occurs to the surroundings, increasing the randomness of the surroundings.

Table 16-3 | Interplay of ΔS_{sys} and ΔS_{surr} in Determining the Sign of ΔS_{univ}

Signs of Entropy Changes			
ΔS_{sys}	ΔS_{surr}	ΔS_{univ}	Process Spontaneous?
+	+	+	Yes
−	−	−	No (reaction will occur in opposite direction)
+	−	?	Yes, if ΔS_{sys} has a larger magnitude than ΔS_{surr}
−	+	?	Yes, if ΔS_{surr} has a larger magnitude than ΔS_{sys}

For the oxide ore reaction,

$$\Delta S_{surr} = -\frac{778 \text{ kJ}}{298} = -2.61 \text{ kJ/K} = -2.61 \times 10^3 \text{ J/K}$$

In this case ΔS_{surr} is negative because heat flow occurs from the surroundings to the system.

See Exercises 16-33 and 16-34

We have seen that the spontaneity of a process is determined by the entropy change it produces in the universe. We also have seen that ΔS_{univ} has two components, ΔS_{sys} and ΔS_{surr}. If for some process both ΔS_{sys} and ΔS_{surr} are positive, then ΔS_{univ} is positive, and the process is spontaneous. If, on the other hand, both ΔS_{sys} and ΔS_{surr} are negative, the process does not occur in the direction indicated but is spontaneous in the opposite direction. Finally, if ΔS_{sys} and ΔS_{surr} have opposite signs, the spontaneity of the process depends on the sizes of the opposing terms. These cases are summarized in Table 16-3.

We can now understand why spontaneity is often dependent on temperature and thus why water spontaneously freezes below 0°C and melts above 0°C. The term ΔS_{surr} is temperature-dependent. Since

$$\Delta S_{surr} = -\frac{\Delta H}{T}$$

at constant pressure, the value of ΔS_{surr} changes markedly with temperature. The magnitude of ΔS_{surr} will be very small at high temperatures and will increase as the temperature decreases. That is, exothermicity is most important as a driving force at low temperatures.

16-4 | Free Energy

The symbol G for free energy honors Josiah Willard Gibbs (1839–1903), who was professor of mathematical physics at Yale University from 1871 to 1903. He laid the foundations of many areas of thermodynamics, particularly as they apply to chemistry.

So far we have used ΔS_{univ} to predict the spontaneity of a process. However, another thermodynamic function is also related to spontaneity and is especially useful in dealing with the temperature dependence of spontaneity. This function is called the **free energy**, which is symbolized by G and defined by the relationship

$$G = H - TS$$

where H is the enthalpy, T is the Kelvin temperature, and S is the entropy. ◀ *i*

For a process that occurs at constant temperature, the change in free energy (ΔG) is given by the equation

$$\Delta G = \Delta H - T\Delta S$$

Note that all quantities here refer to the system. From this point on we will follow the usual convention that when no subscript is included, the quantity refers to the system.

To see how this equation relates to spontaneity, we divide both sides of the equation by $-T$ to produce

$$-\frac{\Delta G}{T} = -\frac{\Delta H}{T} + \Delta S$$

Remember that at constant temperature and pressure

$$\Delta S_{surr} = -\frac{\Delta H}{T}$$

So we can write

$$-\frac{\Delta G}{T} = -\frac{\Delta H}{T} + \Delta S = \Delta S_{surr} + \Delta S = \Delta S_{univ}$$

We have shown that

$$\Delta S_{univ} = -\frac{\Delta G}{T} \qquad \text{at constant } T \text{ and } P$$

This result is very important. It means that a process carried out at constant temperature and pressure will be spontaneous only if ΔG is negative. A process (at constant T and P) is spontaneous in the direction in which the free energy decreases ($-\Delta G$ means $+\Delta S_{univ}$).

Now we have two functions that can be used to predict spontaneity: the entropy of the universe, which applies to all processes, and free energy, which can be used for processes carried out at constant temperature and pressure. Since so many chemical reactions occur under the latter conditions, free energy is the more useful to chemists.

Let's use the free energy equation to predict the spontaneity of the melting of ice:

$$H_2O(s) \longrightarrow H_2O(l)$$

for which $\Delta H° = 6.03 \times 10^3$ J/mol   and $\Delta S° = 22.1$ J/K · mol

ⓘ The superscript degree symbol (°) indicates all substances are in their standard states.

Results of the calculations of ΔS_{univ} and $\Delta G°$ at $-10°C$, $0°C$, and $10°C$ are shown in Table 16-4. ◄ ⓘ ◄ ⓘ These data predict that the process is spontaneous at $10°C$; that is, ice melts at this temperature because ΔS_{univ} is positive and $\Delta G°$ is negative. The opposite is true at $-10°C$, where water freezes spontaneously.

ⓘ To review the definitions of standard states, see Section 7-5.

Why is this so? The answer lies in the fact that ΔS_{sys} ($\Delta S°$) and ΔS_{surr} oppose each other. The term $\Delta S°$ favors the melting of ice because of the increase in positional entropy, and ΔS_{surr} favors the freezing of water because it is an exothermic process. At temperatures below $0°C$, the change of state occurs in the exothermic direction because ΔS_{surr} is larger in magnitude than ΔS_{sys}. But above $0°C$ the change occurs in the direction in which ΔS_{sys} is favorable, since in this case ΔS_{sys} is larger in magnitude than ΔS_{surr}. At $0°C$ the *opposing tendencies just balance*, and the two states coexist;

Table 16-4 │ Results of the Calculation of ΔS_{univ} and $\Delta G°$ for the Process $H_2O(s) \rightarrow H_2O(l)$ at $-10°C$, $0°C$, and $10°C$*

T (°C)	T (K)	$\Delta H°$ (J/mol)	$\Delta S°$ (J/K · mol)	$\Delta S_{surr} = -\dfrac{\Delta H°}{T}$ (J/K · mol)	$\Delta S_{univ} = \Delta S° + \Delta S_{surr}$ (J/K · mol)	$T\Delta S°$ (J/mol)	$\Delta G° = \Delta H° - T\Delta S°$ (J/mol)
−10	263	6.03×10^3	22.1	−22.9	−0.8	5.81×10^3	$+2.2 \times 10^2$
0	273	6.03×10^3	22.1	−22.1	0	6.03×10^3	0
10	283	6.03×10^3	22.1	−21.3	+0.8	6.25×10^3	-2.2×10^2

*Note that at $10°C$, $\Delta S°$ (ΔS_{sys}) controls, and the process occurs even though it is endothermic. At $-10°C$, the magnitude of ΔS_{surr} is larger than that of $\Delta S°$, so the process is spontaneous in the opposite (exothermic) direction.

Table 16-5 | Various Possible Combinations of ΔH and ΔS for a Process and the Resulting Dependence of Spontaneity on Temperature ◄ ⓘ

Case	Result
ΔS positive, ΔH negative	Spontaneous at all temperatures
ΔS positive, ΔH positive	Spontaneous at high temperatures (where exothermicity is relatively unimportant)
ΔS negative, ΔH negative	Spontaneous at low temperatures (where exothermicity is dominant)
ΔS negative, ΔH positive	Process not spontaneous at *any* temperature (reverse process is spontaneous at *all* temperatures)

ⓘ Note that although ΔH and ΔS are somewhat temperature dependent, it is a good approximation to assume they are constant over a relatively small temperature range.

there is no driving force in either direction. An equilibrium exists between the two states of water. Note that ΔS_{univ} is equal to 0 at 0°C.

We can reach the same conclusions by examining $\Delta G°$. At $-10°$, $\Delta G°$ is positive because the $\Delta H°$ term is larger than the $T\Delta S°$ term. The opposite is true at 10°C. At 0°C, $\Delta H°$ is equal to $T\Delta S°$ and $\Delta G°$ is equal to 0. This means that solid H_2O and liquid H_2O have the same free energy at 0°C ($\Delta G° = G_{\text{liquid}} - G_{\text{solid}}$), and the system is at equilibrium.

We can understand the temperature dependence of spontaneity by examining the behavior of ΔG. For a process occurring at constant temperature and pressure,

$$\Delta G = \Delta H - T\Delta S$$

If ΔH and ΔS favor opposite processes, spontaneity will depend on temperature in such a way that the exothermic direction will be favored at low temperatures. For example, for the process

$$H_2O(s) \longrightarrow H_2O(l)$$

ΔH is positive and ΔS is positive. The natural tendency for this system to lower its energy is in opposition to its natural tendency to increase its positional randomness. At low temperatures, ΔH dominates, and at high temperatures, ΔS dominates. The various possible cases are summarized in Table 16-5.

Critical Thinking

Consider an ideal gas in a container fitted with a frictionless, massless piston. What if weight is added to the top of the piston? We would expect the gas to be compressed at constant temperature. For this to be true, ΔS would be negative (since the gas is compressed) and ΔH would be zero (since the process is at constant temperature). This would make ΔG positive. Does this mean the isothermal compression of the gas is not spontaneous? Defend your answer.

INTERACTIVE EXAMPLE 16-5

Free Energy and Spontaneity

At what temperatures is the following process spontaneous at 1 atm?

$$Br_2(l) \longrightarrow Br_2(g)$$

$$\Delta H° = 31.0\,\text{kJ/mol} \quad \text{and} \quad \Delta S° = 93.0\,\text{J/K} \cdot \text{mol}$$

What is the normal boiling point of liquid Br_2?

Solution

The vaporization process will be spontaneous at all temperatures where $\Delta G°$ is negative. Note that $\Delta S°$ favors the vaporization process because of the increase in positional entropy, and $\Delta H°$ favors the *opposite* process, which is exothermic. These opposite tendencies will exactly balance at the boiling point of liquid Br_2, since at this temperature liquid and gaseous Br_2 are in equilibrium ($\Delta G° = 0$). We can find this temperature by setting $\Delta G° = 0$ in the equation

$$\Delta G° = \Delta H° - T\Delta S°$$
$$0 = \Delta H° - T\Delta S°$$
$$\Delta H° = T\Delta S°$$

Then

$$T = \frac{\Delta H°}{\Delta S°} = \frac{3.10 \times 10^4 \text{ J/mol}}{93.0 \text{ J/K} \cdot \text{mol}} = 333 \text{ K}$$

At temperatures above 333 K, $T\Delta S°$ has a larger magnitude than $\Delta H°$, and $\Delta G°$ (or $\Delta H° - T\Delta S°$) is negative. Above 333 K, the vaporization process is spontaneous; the opposite process occurs spontaneously below this temperature. At 333 K, liquid and gaseous Br_2 coexist in equilibrium. These observations can be summarized as follows (the pressure is 1 atm in each case):

1. $T > 333$ K. The term $\Delta S°$ controls. The increase in entropy when liquid Br_2 is vaporized is dominant.

2. $T < 333$ K. The process is spontaneous in the direction in which it is exothermic. The term $\Delta H°$ controls.

3. $T = 333$ K. The opposing driving forces are just balanced ($\Delta G° = 0$), and the liquid and gaseous phases of bromine coexist. This is the normal boiling point.

See Exercises 16-37 through 16-39

16-5 | Entropy Changes in Chemical Reactions

The second law of thermodynamics tells us that a process will be spontaneous if the entropy of the universe increases when the process occurs. We saw in Section 16-4 that for a process at constant temperature and pressure, we can use the change in free energy of the system to predict the sign of ΔS_{univ} and thus the direction in which it is spontaneous. So far we have applied these ideas only to physical processes, such as changes of state and the formation of solutions. However, the main business of chemistry is studying chemical reactions, and, therefore, we want to apply the second law to reactions.

First, we will consider the entropy changes accompanying chemical reactions that occur under conditions of constant temperature and pressure. As for the other types of processes we have considered, the entropy changes in the *surroundings* are determined by the heat flow that occurs as the reaction takes place. However, the entropy changes in the *system* (the reactants and products of the reaction) are determined by positional probability.

For example, in the ammonia synthesis reaction

$$N_2(g) + 3H_2(g) \longrightarrow 2NH_3(g)$$

four reactant molecules become two product molecules, lowering the number of independent units in the system, which leads to less positional disorder.

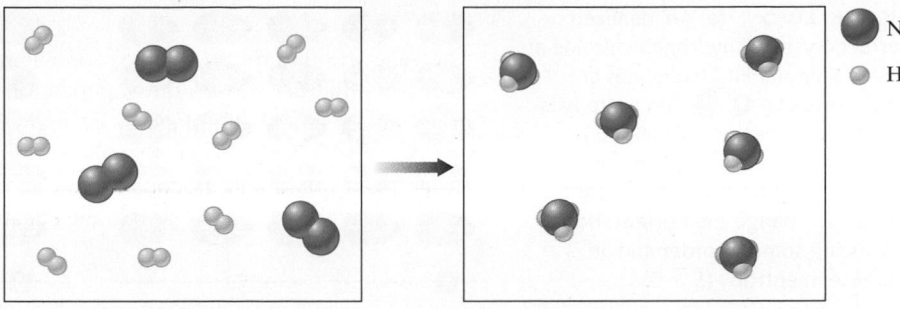

Greater entropy   Less entropy

Fewer molecules mean fewer possible configurations. To help clarify this idea, consider a special container with a million compartments, each large enough to hold a hydrogen molecule. Thus there are a million ways one H_2 molecule can be placed in this container. But suppose we break the H—H bond and place the two independent H atoms in the same container. A little thought will convince you that there are *many* more than a million ways to place the two separate atoms. The number of arrangements possible for the two independent atoms is much greater than the number for the molecule. Thus for the process

$$H_2 \longrightarrow 2H$$

positional entropy increases.

Does positional entropy increase or decrease when the following reaction takes place?

$$4NH_3(g) + 5O_2(g) \longrightarrow 4NO(g) + 6H_2O(g)$$

In this case, 9 gaseous molecules are changed to 10 gaseous molecules, and the positional entropy increases. There are more independent units as products than as reactants. In general, when a reaction involves gaseous molecules, the change in positional entropy is dominated by the relative numbers of molecules of gaseous reactants and products. If the number of molecules of the gaseous products is greater than the number of molecules of the gaseous reactants, positional entropy typically increases, and ΔS will be positive for the reaction. ◄ ⓘ

ⓘ Note that we need to consider the number of molecules and the state of the substance to determine the change in entropy for a process or reaction.

INTERACTIVE EXAMPLE 16-6

Predicting the Sign of $\Delta S°$

Predict the sign of $\Delta S°$ for each of the following reactions.

a. The thermal decomposition of solid calcium carbonate:

$$CaCO_3(s) \longrightarrow CaO(s) + CO_2(g)$$

b. The oxidation of SO_2 in air:

$$2SO_2(g) + O_2(g) \longrightarrow 2SO_3(g)$$

Solution

a. Since in this reaction a gas is produced from a solid reactant, the positional entropy increases, and $\Delta S°$ is positive.

b. Here three molecules of gaseous reactants become two molecules of gaseous products. Since the number of gas molecules decreases, positional entropy decreases, and $\Delta S°$ is negative.

See Exercises 16-41 and 16-42

Figure 16-5 | (a) An idealized perfect crystal of hydrogen chloride at 0 K; the dipolar HCl molecules are represented by 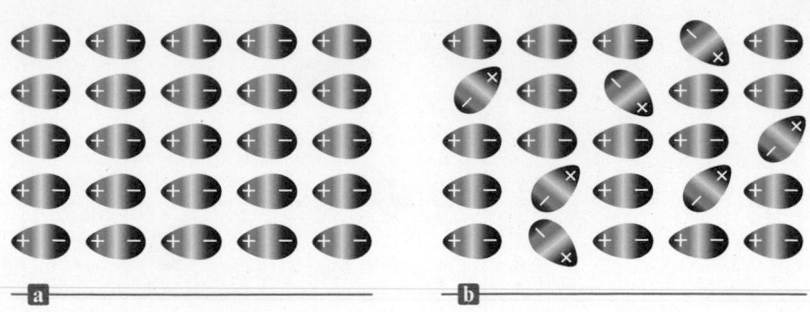. The entropy is zero ($S = 0$) for this crystal at 0 K. (b) As the temperature rises above 0 K, lattice vibrations allow some dipoles to change their orientations, producing some disorder and an increase in entropy ($S > 0$).

In thermodynamics it is the *change* in a certain function that usually is important. The change in enthalpy determines if a reaction is exothermic or endothermic at constant pressure. The change in free energy determines if a process is spontaneous at constant temperature and pressure. It is fortunate that changes in thermodynamic functions are sufficient for most purposes, since absolute values for many thermodynamic characteristics of a system, such as enthalpy or free energy, cannot be determined.

However, we can assign absolute entropy values. Consider a solid at 0 K, where molecular motion virtually ceases. If the substance is a perfect crystal, its internal arrangement is absolutely regular [Fig. 16-5(a)]. There is only *one way* to achieve this perfect order: Every particle must be in its place. For example, with N coins there is only one way to achieve the state of all heads. Thus a perfect crystal represents the lowest possible entropy; that is, *the entropy of a perfect crystal at 0 K is zero*. This is a statement of the **third law of thermodynamics**. ◄ ⓘ

As the temperature of a perfect crystal is increased, the random vibrational motions increase, and disorder increases within the crystal [Fig. 16-5(b)]. Thus the entropy of a substance increases with temperature. Since S is zero for a perfect crystal at 0 K, the entropy value for a substance at a particular temperature can be calculated by knowing the temperature dependence of entropy. (We will not show such calculations here.)

The *standard entropy values* ($S°$) of many common substances at 298 K and 1 atm are listed in Appendix 4. ◄ ⓘ From these values you will see that the entropy of a substance does indeed increase in going from solid to liquid to gas. One especially interesting feature of this table is the very low $S°$ value for diamond. The structure of diamond is highly ordered, with each carbon strongly bound to a tetrahedral arrangement of four other carbon atoms (see Section 9-5). This type of structure allows very little disorder and has a very low entropy, even at 298 K. Graphite has a slightly higher entropy because its layered structure allows for a little more disorder.

Because *entropy is a state function of the system* (it is not pathway dependent), the entropy change for a given chemical reaction can be calculated by taking the difference between the standard entropy values of products and those of the reactants:

$$\Delta S°_{\text{reaction}} = \Sigma n_p S°_{\text{products}} - \Sigma n_r S°_{\text{reactants}}$$

where, as usual, Σ represents the sum of the terms. It is important to note that entropy is an extensive property (it depends on the amount of substance present). This means that the number of moles of a given reactant (n_r) or product (n_p) must be taken into account.

ⓘ A perfect crystal at 0 K is an unattainable ideal, taken as a standard but never actually observed.

ⓘ The standard entropy values represent the increase in entropy that occurs when a substance is heated from 0 K to 298 K at 1 atm pressure.

INTERACTIVE EXAMPLE 16-7

Calculating $\Delta S°$

Calculate $\Delta S°$ at 25°C for the reaction

$$2\text{NiS}(s) + 3\text{O}_2(g) \longrightarrow 2\text{SO}_2(g) + 2\text{NiO}(s)$$

given the following standard entropy values:

Substance	$S°$ (J/K · mol)
$SO_2(g)$	248
$NiO(s)$	38
$O_2(g)$	205
$NiS(s)$	53

Solution

Since

$$\Delta S° = \Sigma n_p S°_{products} - \Sigma n_r S°_{reactants}$$

$$= 2S°_{SO_2(g)} + 2S°_{NiO(s)} - 2S°_{NiS(s)} - 3S°_{O_2(s)}$$

$$= 2 \text{ mol} \left(248 \frac{J}{K \cdot mol} \right) + 2 \text{ mol} \left(38 \frac{J}{K \cdot mol} \right)$$

$$-2 \text{ mol} \left(53 \frac{J}{K \cdot mol} \right) - 3 \text{ mol} \left(205 \frac{J}{K \cdot mol} \right)$$

$$= 496 \text{ J/K} + 76 \text{ J/K} - 106 \text{ J/K} - 615 \text{ J/K}$$

$$= -149 \text{ J/K}$$

We would expect $\Delta S°$ to be negative because the number of gaseous molecules decreases in this reaction.

See Exercise 16-45

INTERACTIVE EXAMPLE 16-8

Calculating $\Delta S°$

Calculate $\Delta S°$ for the reduction of aluminum oxide by hydrogen gas:

$$Al_2O_3(s) + 3H_2(g) \longrightarrow 2Al(s) + 3H_2O(g)$$

Use the following standard entropy values:

Substance	$S°$ (J/K · mol)
$Al_2O_3(s)$	51
$H_2(g)$	131
$Al(s)$	28
$H_2O(g)$	189

Solution

$$\Delta S° = \Sigma n_p S°_{products} - \Sigma n_r S°_{reactants}$$

$$= 2S°_{Al(s)} + 3S°_{H_2O(g)} - 3S°_{H_2(g)} - S°_{Al_2O_3(s)}$$

$$= 2 \text{ mol} \left(28 \frac{J}{K \cdot mol} \right) + 3 \text{ mol} \left(189 \frac{J}{K \cdot mol} \right)$$

$$-3 \text{ mol} \left(131 \frac{J}{K \cdot mol} \right) - 1 \text{ mol} \left(51 \frac{J}{K \cdot mol} \right)$$

$$= 56 \text{ J/K} + 567 \text{ J/K} - 393 \text{ J/K} - 51 \text{ J/K}$$

$$= 179 \text{ J/K}$$

See Exercises 16-46 through 16-48

The reaction considered in Example 16-8 involves 3 moles of hydrogen gas on the reactant side and 3 moles of water vapor on the product side. Would you expect ΔS to be large or small for such a case? We have assumed that ΔS depends on the relative

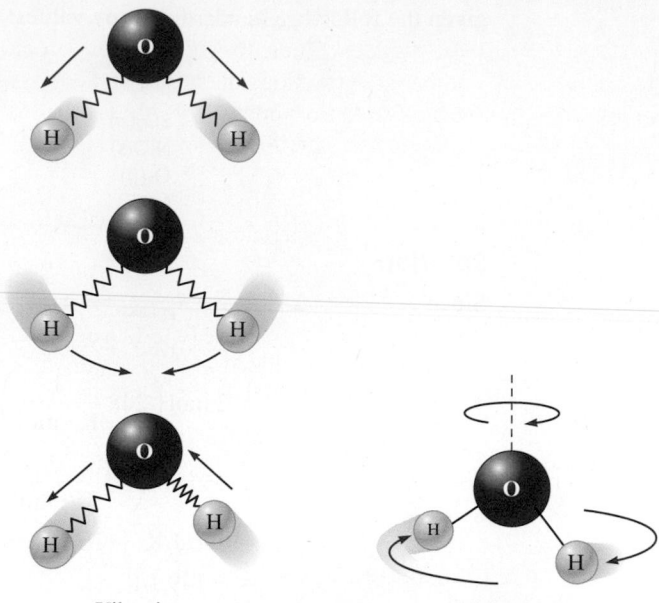

Vibrations   Rotation

Figure 16-6 | The H_2O molecule can vibrate and rotate in several ways, some of which are shown here. This freedom of motion leads to a higher entropy for water than for a substance like hydrogen, a simple diatomic molecule with fewer possible motions.

ⓘ Not only do the numbers of gaseous molecules matter, but so do the structures of the molecules. We need to be able to "see" these structures in order to rank the relative entropies of the substances.

numbers of molecules of gaseous reactants and products. Based on this assumption, ΔS should be near zero for this reaction. However, ΔS is large and positive. Why is this so? The large value for ΔS results from the difference in the entropy values for hydrogen gas and water vapor. The reason for this difference can be traced to the difference in molecular structure. ◄ ⓘ Because it is a nonlinear, triatomic molecule, H_2O has more rotational and vibrational motions (Fig. 16-6) than does the diatomic H_2 molecule. Thus the standard entropy value for $H_2O(g)$ is greater than that for $H_2(g)$. Generally, *the more complex the molecule, the higher the standard entropy value.*

16-6 | Free Energy and Chemical Reactions

For chemical reactions we are often interested in the **standard free energy change** ($\Delta G°$), *the change in free energy that will occur if the reactants in their standard states are converted to the products in their standard states.* For example, for the ammonia synthesis reaction at 25°C,

$$N_2(g) + 3H_2(g) \rightleftharpoons 2NH_3(g) \qquad \Delta G° = -33.3 \text{ kJ} \qquad (16\text{-}1)$$

This $\Delta G°$ value represents the change in free energy when 1 mole of nitrogen gas at 1 atm reacts with 3 moles of hydrogen gas at 1 atm to produce 2 moles of gaseous NH_3 at 1 atm.

It is important to recognize that the standard free energy change for a reaction is not measured directly. For example, we can measure heat flow in a calorimeter to determine $\Delta H°$, but we cannot measure $\Delta G°$ this way. The value of $\Delta G°$ for the ammonia synthesis in Equation (16-1) was *not* obtained by mixing 1 mole of N_2 and 3 moles of H_2 in a flask and measuring the change in free energy as 2 moles of NH_3 formed. For one thing, if we mixed 1 mole of N_2 and 3 moles of H_2 in a flask, the system would go to equilibrium rather than to completion. Also, we have no instrument that measures free energy. However, while we cannot directly measure $\Delta G°$ for a reaction, we can calculate it from other measured quantities, as we will see later in this section.

Why is it useful to know $\Delta G°$ for a reaction? As we will see in more detail later in this chapter, knowing the $\Delta G°$ values for several reactions allows us to compare the relative tendency of these reactions to occur. The more negative the value of $\Delta G°$, the further a reaction will go to the right to reach equilibrium. We must use standard-state free energies to make this comparison because free energy varies with pressure or concentration. Thus, to get an accurate comparison of reaction tendencies, we must compare all reactions under the same pressure or concentration conditions. We will have more to say about the significance of $\Delta G°$ later. ◀ ⓘ

There are several ways to calculate $\Delta G°$. One common method uses the equation

$$\Delta G° = \Delta H° - T\Delta S°$$

which applies to a reaction carried out at constant temperature. For example, for the reaction

$$C(s) + O_2(g) \longrightarrow CO_2(g)$$

the values of $\Delta H°$ and $\Delta S°$ are known to be -393.5 kJ and 3.05 J/K, respectively, and $\Delta G°$ can be calculated at 298 K as follows:

$$\Delta G° = \Delta H° - T\Delta S°$$
$$= -3.935 \times 10^5 \text{ J} - (298 \text{ K})(3.05 \text{ J/K})$$
$$= -3.944 \times 10^5 \text{ J}$$
$$= -394.4 \text{ kJ (per mole of } CO_2)$$

ⓘ The value of $\Delta G°$ tells us nothing about the rate of a reaction, only its eventual equilibrium position.

INTERACTIVE EXAMPLE 16-9

Calculating $\Delta H°$, $\Delta S°$, and $\Delta G°$

Consider the reaction

$$2SO_2(g) + O_2(g) \longrightarrow 2SO_3(g)$$

carried out at 25°C and 1 atm. Calculate $\Delta H°$, $\Delta S°$, and $\Delta G°$ using the following data:

Substance	$H_f°$ (kJ/mol)	$S°$ (J/K · mol)
$SO_2(g)$	−297	248
$SO_3(g)$	−396	257
$O_2(g)$	0	205

Solution

The value of $\Delta H°$ can be calculated from the enthalpies of formation using the equation we discussed in Section 7-5:

$$\Delta H° = \Sigma n_p \Delta H_{f(\text{products})}° - \Sigma n_r \Delta H_{f(\text{reactants})}°$$

Then
$$\Delta H° = 2\Delta H_{f(SO_3(g))}° - 2\Delta H_{f(SO_2(g))}° - \Delta H_{f(O_2(g))}°$$
$$= 2 \text{ mol}(-396 \text{ kJ/mol}) - 2 \text{ mol}(-297 \text{ kJ/mol}) - 0$$
$$= -792 \text{ kJ} + 594 \text{ kJ}$$
$$= -198 \text{ kJ}$$

The value of $\Delta S°$ can be calculated using the standard entropy values and the equation discussed in Section 16-5:

$$\Delta S° = \Sigma n_p S_{\text{products}}° - \Sigma n_r S_{\text{reactants}}°$$

Thus

$$\Delta S° = 2S_{SO_3(g)}° - 2S_{SO_2(g)}° - S_{O_2(g)}°$$
$$= 2 \text{ mol}(257 \text{ J/K} \cdot \text{mol}) - 2 \text{ mol}(248 \text{ J/K} \cdot \text{mol}) - 1 \text{ mol}(205 \text{ J/K} \cdot \text{mol})$$
$$= 514 \text{ J/K} - 496 \text{ J/K} - 205 \text{ J/K}$$
$$= -187 \text{ J/K}$$

We would expect $\Delta S°$ to be negative because three molecules of gaseous reactants give two molecules of gaseous products.

We can see differences in entropy based on the state of matter, the number of molecules, and the nature of the molecules.

For example, positional probability can be illustrated by changes of state. Consider water, for example. As a solid (ice)

it is very organized (low entropy), and as the ice is heated it begins to melt. When ice becomes liquid (water) entropy increases. Heating water even further vaporizes the water to its highest entropy (water vapor).

Solid $\longrightarrow$ Liquid $\longrightarrow$ Gas

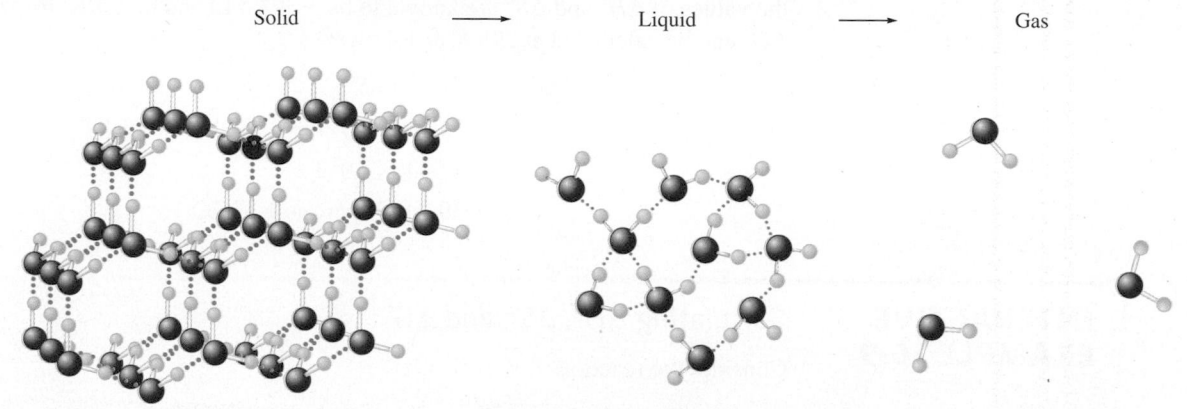

The value of $\Delta G°$ can now be calculated from the equation

$$\Delta G° = \Delta H° - T\Delta S°$$

$$= -198 \text{ kJ} - (298 \text{ K})\left(-187 \frac{\text{J}}{\text{K}}\right)\left(\frac{1 \text{ kJ}}{1000 \text{ J}}\right)$$

$$= -198 \text{ kJ} + 55.7 \text{ kJ} = -142 \text{ kJ}$$

See Exercise 16-53

A second method for calculating ΔG for a reaction takes advantage of the fact that, like enthalpy, *free energy is a state function.* Therefore, we can use procedures for finding ΔG that are similar to those for finding ΔH using Hess's law.

To illustrate this method for calculating the free energy change, we will obtain $\Delta G°$ for the reaction

$$2CO(g) + O_2(g) \longrightarrow 2CO_2(g) \tag{16-2}$$

from the following data:

$$2CH_4(g) + 3O_2(g) \longrightarrow 2CO(g) + 4H_2O(g) \qquad \Delta G° = -1088 \text{ kJ} \tag{16-3}$$
$$CH_4(g) + 2O_2(g) \longrightarrow CO_2(g) + 2H_2O(g) \qquad \Delta G° = -801 \text{ kJ} \tag{16-4}$$

Note that $CO(g)$ is a reactant in Equation (16-2). This means that Equation (16-3) must be reversed, since $CO(g)$ is a product in that reaction as written. When a reaction is reversed, the sign of $\Delta G°$ is also reversed. In Equation (16-4), $CO_2(g)$ is a product, as it is in Equation (16-2), but only one molecule of CO_2 is formed. Thus Equation (16-4)

In a reaction that occurs in the gaseous state the entropy change for the system is determined by positional probability. For example, consider the reaction of hydrogen gas with oxygen gas to form gaseous water.

Greater entropy   Less entropy

The entropy for the system decreases in the reaction as shown because fewer product molecules have fewer possible

microstates. In this case we can simply count the relative numbers of molecules because no phase change is occurring.

However, this simplification can sometimes lead to erroneous results. This is because molecular structure also plays a role in the inherent entropy of a substance. The water molecule, for example, has many rotational and vibrational motions

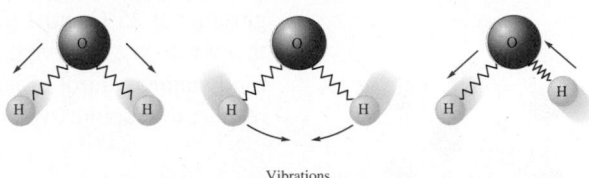

Vibrations

resulting in a higher standard entropy value than that for helium, for example, which is only a single atom.

must be multiplied by 2, which means the $\Delta G°$ value for Equation (16-4) also must be multiplied by 2. Free energy is an extensive property, since it is defined by two extensive properties, H and S.

Reversed Equation (16-3)

$$2CO(g) + 4H_2O(g) \longrightarrow 2CH_4(g) + 3O_2(g) \qquad \Delta G° = -(-1088 \text{ kJ})$$

$2 \times$ Equation (16-4)

$$\underline{2CH_4(g) + 4O_2(g) \longrightarrow 2CO_2(g) + 4H_2O(g) \qquad \Delta G° = 2(-801 \text{ kJ})}$$

$$2CO(g) + O_2(g) \longrightarrow 2CO_2(g) \qquad \begin{aligned} \Delta G° &= -(-1088 \text{ kJ}) \\ &\quad + 2(-801 \text{ kJ}) \\ &= -514 \text{ kJ} \end{aligned}$$

This example shows that the ΔG values for reactions are manipulated in exactly the same way as the ΔH values.

INTERACTIVE EXAMPLE 16-10

Calculating $\Delta G°$

Using the following data (at 25°C)

$$C_{\text{diamond}}(s) + O_2(g) \longrightarrow CO_2(g) \qquad \Delta G° = -397 \text{ kJ} \quad\blacktriangleleft \qquad (16\text{-}5)$$

$$C_{\text{graphite}}(s) + O_2(g) \longrightarrow CO_2(g) \qquad \Delta G° = -394 \text{ kJ} \quad\blacktriangleleft \qquad (16\text{-}6)$$

calculate $\Delta G°$ for the reaction

$$C_{\text{diamond}}(s) \longrightarrow C_{\text{graphite}}(s)$$

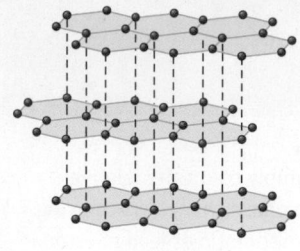

Graphite

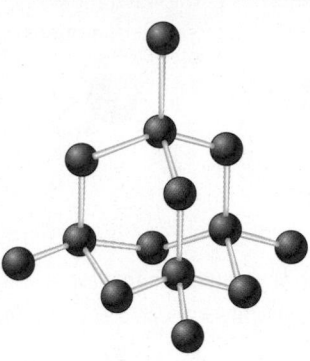

Diamond

Solution

We reverse Equation (16-6) to make graphite a product, as required, and then add the new equation to Equation (16-5):

$$C_{diamond}(s) + O_2(g) \longrightarrow CO_2(g) \qquad \Delta G° = -397 \text{ kJ}$$

Reversed Equation (16-6)

$$\frac{CO_2(g) \longrightarrow C_{graphite}(s) + O_2(g) \qquad \Delta G° = -(-394 \text{ kJ})}{C_{diamond}(s) \longrightarrow C_{graphite}(s) \qquad \quad \Delta G° = -397 \text{ kJ} + 394 \text{ kJ}}$$

$$= -3 \text{ kJ}$$

Since $\Delta G°$ is negative for this process, diamond should spontaneously change to graphite at 25°C and 1 atm. However, the reaction is so slow under these conditions that we do not observe the process. This is another example of kinetic rather than thermodynamic control of a reaction. We can say that diamond is kinetically stable with respect to graphite even though it is thermodynamically unstable.

See Exercises 16-57 and 16-58

In Example 16-10 we saw that the process

$$C_{diamond}(s) \longrightarrow C_{graphite}(s)$$

is spontaneous but very slow at 25°C and 1 atm. The reverse process can be made to occur at high temperatures and pressures. Diamond has a more compact structure and thus a higher density than graphite, so exerting very high pressure causes it to become thermodynamically favored. If high temperatures are also used to make the process fast enough to be feasible, diamonds can be made from graphite. The conditions typically used involve temperatures greater than 1000°C and pressures of about 10^5 atm. About half of all industrial diamonds are made this way.

A third method for calculating the free energy change for a reaction uses standard free energies of formation. The **standard free energy of formation** ($\Delta G_f°$) of a substance is defined as the *change in free energy that accompanies the formation of 1 mole of that substance from its constituent elements with all reactants and products in their standard states.* ◄ *ⓘ* For the formation of glucose ($C_6H_{12}O_6$), the appropriate reaction is

ⓘ The standard state of an element is its most stable state of 25°C and 1 atm.

$$6C(s) + 6H_2(g) + 3O_2(g) \longrightarrow C_6H_{12}O_6(s)$$

The standard free energy associated with this process is called the *free energy of formation of glucose.* Values of the standard free energy of formation are useful in calculating $\Delta G°$ for specific chemical reactions using the equation

$$\Delta G° = \Sigma n_p \Delta G_{f(products)}° - \Sigma n_r \Delta G_{f(reactants)}°$$

ⓘ Calculating $\Delta G°$ is very similar to calculating $\Delta H°$, as shown in Section 7-5.

Values of $\Delta G_f°$ for many common substances are listed in Appendix 4. Note that, analogous to the enthalpy of formation, *the standard free energy of formation of an element in its standard state is zero.* Also note that the number of moles of each reactant (n_r) and product (n_p) must be used when calculating $\Delta G°$ for a reaction. ◄ *ⓘ*

INTERACTIVE EXAMPLE 16-11

Calculating $\Delta G°$

Methanol is a high-octane fuel used in high-performance racing engines. Calculate $\Delta G°$ for the reaction

$$2CH_3OH(g) + 3O_2(g) \longrightarrow 2CO_2(g) + 4H_2O(g)$$

given the following free energies of formation:

Substance	ΔG_f° (kJ/mol)
$CH_3OH(g)$	−163
$O_2(g)$	0
$CO_2(g)$	−394
$H_2O(y)$	−229

Solution

We use the equation

$$\Delta G^\circ = \Sigma n_p \Delta G_{f(products)}^\circ - \Sigma n_r \Delta G_{f(reactants)}^\circ$$

$$= 2\Delta G_{f(CO_2(g))}^\circ + 4\Delta G_{f(H_2O(g))}^\circ - 3\Delta G_{f(O_2(g))}^\circ - 2\Delta G_{f(CH_3OH(g))}^\circ$$

$$= 2\text{ mol}(-394\text{ kJ/mol}) + 4\text{ mol}(-229\text{ kJ/mol}) - 3(0) - 2\text{ mol}(-163\text{ kJ/mol})$$

$$= -1378\text{ kJ}$$

The large magnitude and the negative sign of ΔG° indicate that this reaction is very favorable thermodynamically.

See Exercises 16-59 and 16-60

INTERACTIVE EXAMPLE 16-12

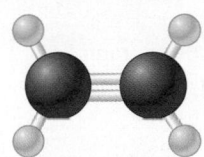

Ethylene

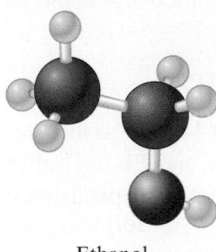

Ethanol

Free Energy and Spontaneity

A chemical engineer wants to determine the feasibility of making ethanol (C_2H_5OH) by reacting water with ethylene (C_2H_4) according to the equation ◀ ◀

$$C_2H_4(g) + H_2O(l) \longrightarrow C_2H_5OH(l)$$

Is this reaction spontaneous under standard conditions?

Solution

To determine the spontaneity of this reaction under standard conditions, we must determine ΔG° for the reaction. We can do this using standard free energies of formation at 25°C from Appendix 4:

$$\Delta G_{f(C_2H_5OH(l))}^\circ = -175\text{ kJ/mol}$$

$$\Delta G_{f(H_2O(l))}^\circ = -237\text{ kJ/mol}$$

$$\Delta G_{f(C_2H_4(g))}^\circ = 68\text{ kJ/mol}$$

Then

$$\Delta G^\circ = \Delta G_{f(C_2H_5OH(l))}^\circ - \Delta G_{f(H_2O(l))}^\circ - \Delta G_{f(C_2H_4(g))}^\circ$$

$$= -175\text{ kJ} - (-237\text{ kJ}) - 68\text{ kJ}$$

$$= -6\text{ kJ}$$

Thus the process is spontaneous under standard conditions at 25°C.

See Exercises 16-61 and 16-62

Although the reaction considered in Example 16-12 is spontaneous, other features of the reaction must be studied to see if the process is feasible. For example, the chemical engineer will need to study the kinetics of the reaction to determine whether it is fast enough to be useful and, if it is not, whether a catalyst can be found to enhance the rate. In doing these studies, the engineer must remember that ΔG° depends on temperature:

$$\Delta G^\circ = \Delta H^\circ - T\Delta S^\circ$$

Thus, if the process must be carried out at high temperatures to be fast enough to be feasible, $\Delta G°$ must be recalculated at that temperature from the $\Delta H°$ and $\Delta S°$ values for the reaction.

16-7 | The Dependence of Free Energy on Pressure

In this chapter we have seen that a system at constant temperature and pressure will proceed spontaneously in the direction that lowers its free energy. This is why reactions proceed until they reach equilibrium. The equilibrium position represents the lowest free energy value available to a particular reaction system. The free energy of a reaction system changes as the reaction proceeds, because free energy is dependent on the pressure of a gas or on the concentration of species in solution. We will deal only with the pressure dependence of the free energy of an ideal gas. The dependence of free energy on concentration can be developed using similar reasoning.

To understand the pressure dependence of free energy, we need to know how pressure affects the thermodynamic functions that comprise free energy, that is, enthalpy and entropy (recall that $G = H - TS$). For an ideal gas, enthalpy is not pressure dependent. However, entropy *does* depend on pressure because of its dependence on volume. Consider 1 mole of an ideal gas at a given temperature. At a volume of 10.0 L, the gas has many more positions available for its molecules than if its volume is 1.0 L. The positional entropy is greater in the larger volume. In summary, at a given temperature for 1 mole of ideal gas

$$S_{\text{large volume}} > S_{\text{small volume}}$$

or, since pressure and volume are inversely related,

$$S_{\text{low pressure}} > S_{\text{high pressure}}$$

We have shown qualitatively that the entropy and therefore the free energy of an ideal gas depend on its pressure. Using a more detailed argument, which we will not consider here, it can be shown that

$$G = G° + RT \ln(P)$$

where $G°$ is the free energy of the gas at a pressure of 1 atm, G is the free energy of the gas at a pressure of P atm, R is the universal gas constant, and T is the Kelvin temperature.

To see how the change in free energy for a reaction depends on pressure, we will consider the ammonia synthesis reaction

$$N_2(g) + 3H_2(g) \longrightarrow 2NH_3(g)$$

In general,
$$\Delta G = \Sigma n_p G_{\text{products}} - \Sigma n_r G_{\text{reactants}}$$

For this reaction
$$\Delta G = 2G_{NH_3} - G_{N_2} - 3G_{H_2}$$

where
$$G_{NH_3} = G°_{NH_3} + RT \ln(P_{NH_3})$$
$$G_{N_2} = G°_{N_2} + RT \ln(P_{N_2})$$
$$G_{H_2} = G°_{H_2} + RT \ln(P_{H_2})$$

Substituting these values into the equation gives

$$\Delta G = 2[G°_{NH_3} + RT \ln(P_{NH_3})] - [G°_{N_2} + RT \ln(P_{N_2})] - 3[G°_{H_2} + RT \ln(P_{H_2})]$$
$$= 2G°_{NH_3} - G°_{N_2} - 3G°_{H_2} + 2RT \ln(P_{NH_3}) - RT \ln(P_{N_2}) - 3RT \ln(P_{H_2})$$
$$= \underbrace{(2G°_{NH_3} - G°_{N_2} - 3G°_{H_2})}_{\Delta G° \text{ reaction}} + RT[2 \ln(P_{NH_3}) - \ln(P_{N_2}) - 3 \ln(P_{H_2})]$$

The first term (in parentheses) is $\Delta G°$ for the reaction. Thus we have

$$\Delta G = \Delta G°_{reaction} + RT[2 \ln(P_{NH_3}) - \ln(P_{N_2}) - 3 \ln(P_{H_2})]$$

and since

$$2 \ln(P_{NH_3}) = \ln(P_{NH_3}^2)$$

$$-\ln(P_{N_2}) = \ln\left(\frac{1}{P_{N_2}}\right)$$

$$-3 \ln(P_{H_2}) = \ln\left(\frac{1}{P_{H_2}^3}\right)$$

the equation becomes

$$\Delta G = \Delta G° + RT \ln\left(\frac{P_{NH_3}^2}{(P_{N_2})(P_{H_2}^3)}\right)$$

But the term

$$\frac{P_{NH_3}^2}{(P_{N_2})(P_{H_2}^3)}$$

is the reaction quotient Q discussed in Section 12-5. Therefore, we have

$$\Delta G = \Delta G° + RT \ln(Q)$$

where Q is the reaction quotient (from the law of mass action), T is the temperature (K), R is the gas law constant and is equal to 8.3145 J/K · mol, $\Delta G°$ is the free energy change for the reaction with all reactants and products at a pressure of 1 atm, and ΔG is the free energy change for the reaction for the specified pressures of reactants and products.

INTERACTIVE EXAMPLE 16-13

Calculating $\Delta G°$

One method for synthesizing methanol (CH_3OH) involves reacting carbon monoxide and hydrogen gases:

$$CO(g) + 2H_2(g) \longrightarrow CH_3OH(l)$$

Calculate ΔG at 25°C for this reaction where carbon monoxide gas at 5.0 atm and hydrogen gas at 3.0 atm are converted to liquid methanol.

Solution

To calculate ΔG for this process, we use the equation

$$\Delta G = \Delta G° + RT \ln(Q)$$

We must first compute $\Delta G°$ from standard free energies of formation (see Appendix 4). Since

$$\Delta G°_{f(CH_3OH(l))} = -166 \text{ kJ}$$
$$\Delta G°_{f(H_2(g))} = 0$$
$$\Delta G°_{f(CO(g))} = -137 \text{ kJ}$$
$$\Delta G° = -166 \text{ kJ} - (-137 \text{ kJ}) - 0 = -29 \text{ kJ} = -2.9 \times 10^4 \text{ J}$$

Note that this is the value of $\Delta G°$ for the reaction of 1 mole of CO with 2 moles of H_2 to produce 1 mole of CH_3OH. We might call this the value of $\Delta G°$ for one "round" of the reaction or for 1 mole of the reaction. Thus the $\Delta G°$ value might better be written as -2.9×10^4 J/mol of reaction, or -2.9×10^4 J/mol rxn.

We can now calculate ΔG using

$$\Delta G° = -2.9 \times 10^4 \text{ J/mol rxn}$$

$$R = 8.3145 \text{ J/K} \cdot \text{mol}$$

$$T = 273 + 25 = 298 \text{ K}$$

$$Q = \frac{1}{(P_{CO})(P_{H_2}{}^2)} = \frac{1}{(5.0)(3.0)^2} = 2.2 \times 10^{-2}$$

> ⓘ Note in this case that ΔG is defined for "1 mole of the reaction," that is, for 1 mole of $CO(g)$ reacting with 2 moles of $H_2(g)$ to form 1 mole of $CH_3OH(l)$. Thus ΔG, $\Delta G°$, and RT $\ln(Q)$ all have units of J/mol of reaction. In this case the units of R are actually J/K · mol of reaction, although they are usually not written this way.

Note that the pure liquid methanol is not included in the calculation of Q. Then

$$\Delta G = \Delta G° + RT \ln(Q)$$
$$= (-2.9 \times 10^4 \text{ J/mol rxn}) + (8.3145 \text{ J/K} \cdot \text{mol rxn})(298 \text{ K}) \ln (2.2 \times 10^{-2})$$
$$= (-2.9 \times 10^4 \text{ J/mol rxn}) - (9.4 \times 10^3 \text{ J/mol rxn}) = -3.8 \times 10^4 \text{ J/mol rxn}$$
$$= -38 \text{ kJ/mol rxn} ◄ ⓘ$$

Note that ΔG is significantly more negative than $\Delta G°$, implying that the reaction is more spontaneous at reactant pressures greater than 1 atm. We might expect this result from Le Châtelier's principle.

See Exercises 16-65 and 16-66

The Meaning of ΔG for a Chemical Reaction

In this section we have learned to calculate ΔG for chemical reactions under various conditions. For example, in Example 16-13 the calculations show that the formation of $CH_3OH(l)$ from $CO(g)$ at 5.0 atm reacting with $H_2(g)$ at 3.0 atm is spontaneous. What does this result mean? Does it mean that if we mixed 1.0 mole of $CO(g)$ and 2.0 moles of $H_2(g)$ together at pressures of 5.0 and 3.0 atm, respectively, that 1.0 mole of $CH_3OH(l)$ would form in the reaction flask? The answer is no. This answer may surprise you in view of what has been said in this section. It is true that 1.0 mole of $CH_3OH(l)$ has a lower free energy than 1.0 mole of $CO(g)$ at 5.0 atm plus 2.0 moles of $H_2(g)$ at 3.0 atm. However, when $CO(g)$ and $H_2(g)$ are mixed under these conditions, there is *an even lower free energy available to this system than 1.0 mole of pure $CH_3OH(l)$*. For reasons we will discuss shortly, *the system can achieve the lowest possible free energy by going to equilibrium, not by going to completion.* At the equilibrium position, some of the $CO(g)$ and $H_2(g)$ will remain in the reaction flask. So even though 1.0 mole of pure $CH_3OH(l)$ is at a lower free energy than 1.0 mole of $CO(g)$ and 2.0 moles of $H_2(g)$ at 5.0 and 3.0 atm, respectively, the reaction system will stop short of forming 1.0 mole of $CH_3OH(l)$. The reaction stops short of completion because the equilibrium mixture of $CH_3OH(l)$, $CO(g)$, and $H_2(g)$ exists at the lowest possible free energy available to the system.

To illustrate this point, we will explore a mechanical example. Consider balls rolling down the two hills shown in Fig. 16-7. Note that in both cases point B has a lower potential energy than point A.

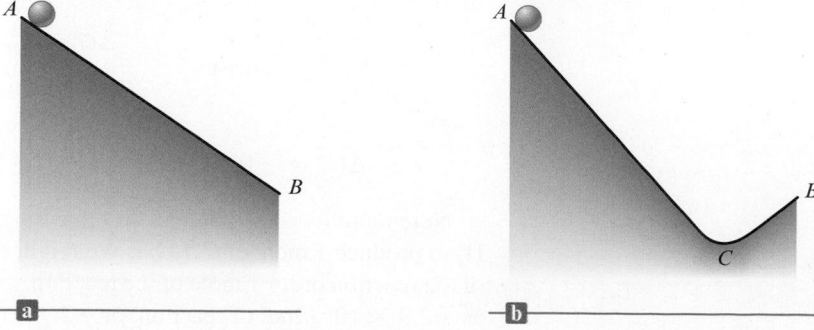

Figure 16-7 | Schematic representations of balls rolling down two types of hills.

In Fig. 16-7(a) the ball will roll to point *B*. This diagram is analogous to a phase change. For example, at 25°C ice will spontaneously change completely to liquid water, because the latter has the lowest free energy. In this case liquid water is the only choice. There is no intermediate mixture of ice and water with lower free energy.

The situation is different for a chemical reaction system, as illustrated in Fig. 16-7(b). In Fig. 16-7(b) the ball will not get to point *B* because there is a lower potential energy at point *C*. Like the ball, a chemical system will seek the *lowest possible* free energy, which, for reasons we will discuss below, is the equilibrium position.

Therefore, although the value of ΔG for a given reaction system tells us whether the products or reactants are favored under a given set of conditions, it does not mean that the system will proceed to pure products (if ΔG is negative) or remain at pure reactants (if ΔG is positive). Instead, the system will spontaneously go to the equilibrium position, the lowest possible free energy available to it. In the next section we will see that the value of $\Delta G°$ for a particular reaction tells us exactly where this position will be.

16-8 | Free Energy and Equilibrium

When the components of a given chemical reaction are mixed, they will proceed, rapidly or slowly depending on the kinetics of the process, to the equilibrium position. In Chapter 12 we defined the equilibrium position as the point at which the forward and reverse reaction rates are equal. In this chapter we look at equilibrium from a thermodynamic point of view, and we find that the **equilibrium point** *occurs at the lowest value of free energy available to the reaction system*. As it turns out, the two definitions give the same equilibrium state, which must be the case for both the kinetic and thermodynamic models to be valid.

To understand the relationship of free energy to equilibrium, let's consider the following simple hypothetical reaction:

$$A(g) \rightleftharpoons B(g)$$

where 1.0 mole of gaseous A is initially placed in a reaction vessel at a pressure of 2.0 atm. The free energies for A and B are diagramed as shown in Fig. 16-8(a). As A reacts to form B, the total free energy of the system changes, yielding the following results:

$$\text{Free energy of A} = G_A = G_A° + RT \ln(P_A)$$
$$\text{Free energy of B} = G_B = G_B° + RT \ln(P_B)$$
$$\text{Total free energy of system} = G = G_A + G_B$$

As A changes to B, G_A will decrease because P_A is decreasing [Fig. 16-8(b)]. In contrast, G_B will increase because P_B is increasing. The reaction will proceed to the right as long as the total free energy of the system decreases (as long as G_B is less than G_A). At some point the pressures of A and B reach the values P_A^e and P_B^e that make G_A equal to G_B. *The system has reached equilibrium* [Fig. 16-8(c)]. Since A at pressure P_A^e and B at pressure P_B^e have the same free energy (G_A equals G_B), ΔG is zero for A at pressure P_A^e changing to B at pressure P_B^e. *The system has reached minimum free energy.* There is no longer any driving force to change A to B or B to A, so the system remains at this position (the pressures of A and B remain constant).

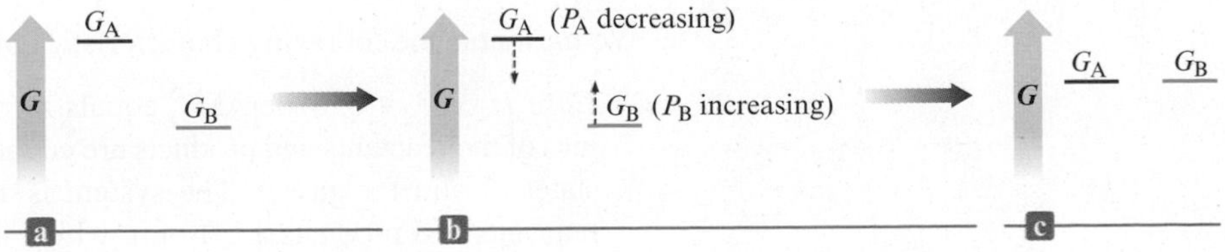

Figure 16-8 | (a) The initial free energies of A and B. (b) As A(*g*) changes to B(*g*), the free energy of A decreases and that of B increases. (c) Eventually, pressures of A and B are achieved such that $G_A = G_B$, the equilibrium position.

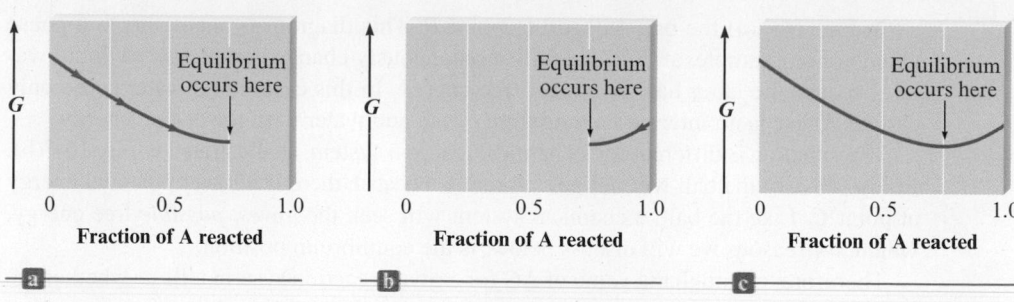

Figure 16-9 | (a) The change in free energy to reach equilibrium, beginning with 1.0 mole of A(g) at P_A = 2.0 atm. (b) The change in free energy to reach equilibrium, beginning with 1.0 mole of B(g) at P_B = 2.0 atm. (c) The free energy profile for A(g) $\rightleftharpoons$ B(g) in a system containing 1.0 mole of gas (A plus B) at P_{TOTAL} = 2.0 atm. Each point on the curve corresponds to the total free energy of the system for a given combination of A and B.

Suppose that for the experiment described above, the plot of free energy versus the mole fraction of A reacted is defined as shown in Fig. 16-9(a). In this experiment, minimum free energy is reached when 75% of A has been changed to B. At this point, the pressure of A is 0.25 times the original pressure, or

$$(0.25)(2.0 \text{ atm}) = 0.50 \text{ atm}$$

The pressure of B is

$$(0.75)(2.0 \text{ atm}) = 1.5 \text{ atm}$$

Since this is the equilibrium position, we can use the equilibrium pressures to calculate a value for K for the reaction in which A is converted to B at this temperature:

$$K = \frac{P_B^e}{P_A^e} = \frac{1.5 \text{ atm}}{0.50 \text{ atm}} = 3.0$$

Exactly the same equilibrium point would be achieved if we placed 1.0 mole of pure B(g) in the flask at a pressure of 2.0 atm. In this case B would change to A until equilibrium ($G_B = G_A$) is reached. This is shown in Fig. 16-9(b).

The overall free energy curve for this system is shown in Fig. 16-9(c). Note that any mixture of A(g) and B(g) containing 1.0 mole of gas (A plus B) at a total pressure of 2.0 atm will react until it reaches the minimum in the curve. ◀ ⓘ

In summary, when substances undergo a chemical reaction, the reaction proceeds to the minimum free energy (equilibrium), which corresponds to the point where

$$G_{\text{products}} = G_{\text{reactants}} \quad \text{or} \quad \Delta G = G_{\text{products}} - G_{\text{reactants}} = 0$$

We can now establish a quantitative relationship between free energy and the value of the equilibrium constant. We have seen that

$$\Delta G = \Delta G° + RT \ln(Q)$$

and at equilibrium ΔG equals 0 and Q equals K.

So

$$\Delta G = 0 = \Delta G° + RT \ln(K)$$

or

$$\Delta G° = -RT \ln(K)$$

We must note the following characteristics of this very important equation.

Case 1: $\Delta G°$ = 0. When $\Delta G°$ equals zero for a particular reaction, the free energies of the reactants and products are equal when all components are in the standard states (1 atm for gases). The system is at equilibrium when the pressures of all reactants and products are 1 atm, which means that K equals 1.

ⓘ For the reaction A(g) $\rightleftharpoons$ B(g), the pressure is constant during the reaction, since the same number of gas molecules is always present.

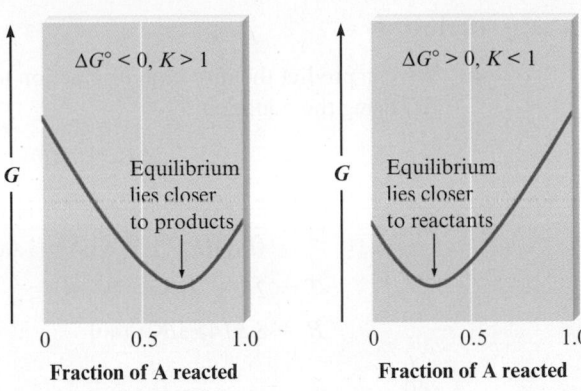

Figure 16-10 | The relationship of $\Delta G°$ for a reaction to its eventual equilibrium position.

Case 2: $\Delta G° < 0$. In this case $\Delta G°$ ($G°_{products} - G°_{reactants}$) is negative, which means that

$$G°_{products} < G°_{reactants}$$

If a flask contains the reactants and products, all at 1 atm, the system will *not* be at equilibrium. Since $G°_{products}$ is less than $G°_{reactants}$, the system will adjust to the right to reach equilibrium. In this case K will be *greater than 1,* since the pressures of the products at equilibrium will be greater than 1 atm and the pressures of the reactants at equilibrium will be less than 1 atm.

Case 3: $\Delta G° > 0$. Since $\Delta G°$ ($G°_{products} - G°_{reactants}$) is positive,

$$G°_{reactants} < G°_{products}$$

If a flask contains the reactants and products, all at 1 atm, the system will *not* be at equilibrium. In this case the system will adjust to the left (toward the reactants, which have a lower free energy) to reach equilibrium. The value of K will be *less than 1,* since at equilibrium the pressures of the reactants will be greater than 1 atm and the pressures of the products will be less than 1 atm.

These results are summarized in Table 16-6 and Figure 16-10. The value of K for a specific reaction can be calculated from the equation

$$\Delta G° = -RT \ln(K)$$

as is shown in Examples 16-14 and 16-15.

Table 16-6 | Qualitative Relationship Between the Change in Standard Free Energy and the Equilibrium Constant for a Given Reaction

$\Delta G°$	K
$\Delta G° = 0$	$K = 1$
$\Delta G° < 0$	$K > 1$
$\Delta G° > 0$	$K < 1$

INTERACTIVE EXAMPLE 16-14

Free Energy and Equilibrium I

Consider the ammonia synthesis reaction

$$N_2(g) + 3H_2(g) \rightleftharpoons 2NH_3(g)$$

where $\Delta G° = -33.3$ kJ per mole of N_2 consumed at 25°C. For each of the following mixtures of reactants and products at 25°C, predict the direction in which the system will shift to reach equilibrium.

a. $P_{NH_3} = 1.00$ atm, $P_{N_2} = 1.47$ atm, $P_{H_2} = 1.00 \times 10^{-2}$ atm

b. $P_{NH_3} = 1.00$ atm, $P_{N_2} = 1.00$ atm, $P_{H_2} = 1.00$ atm

Solution

a. We can predict the direction of reaction to equilibrium by calculating the value of ΔG using the equation

$$\Delta G = \Delta G^\circ + RT \ln(Q)$$

where $\quad Q = \dfrac{P_{NH_3}{}^2}{(P_{N_2})(P_{H_2}{}^3)} = \dfrac{(1.00)^2}{(1.47)(1.00 \times 10^{-2})^3} = 6.80 \times 10^5$

$$T = 25 + 273 = 298 \text{ K}$$

$$R = 8.3145 \text{ J/K} \cdot \text{mol}$$

and

$$\Delta G^\circ = -33.3 \text{ kJ/mol} = -3.33 \times 10^4 \text{ J/mol}$$

> ℹ The units of ΔG, ΔG°, and RT $\ln(Q)$ all refer to the balanced reaction with all amounts expressed in moles. We might say that the units are joules per "mole of reaction," although only the "per mole" is indicated for R (as is customary).

Then

$$\Delta G = (-3.33 \times 10^4 \text{ J/mol}) + (8.3145 \text{ J/K} \cdot \text{mol})(298 \text{ K}) \ln(6.8 \times 10^5)$$
$$= (-3.33 \times 10^4 \text{ J/mol}) + (3.33 \times 10^4 \text{ J/mol}) = 0$$

Since $\Delta G = 0$, the reactants and products have the same free energies at these partial pressures. The system is already at equilibrium, and no shift will occur. ◄ ℹ

b. The partial pressures given here are all 1.00 atm, which means that the system is in the standard state. That is,

$$\Delta G = \Delta G^\circ + RT \ln(Q) = \Delta G^\circ + RT \ln\dfrac{(1.00)^2}{(1.00)(1.00)^3}$$
$$= \Delta G^\circ + RT \ln(1.00) = \Delta G^\circ + 0 = \Delta G^\circ$$

For this reaction at 25°C,

$$\Delta G^\circ = -33.3 \text{ kJ/mol}$$

The negative value for ΔG° means that in their standard states the products have a lower free energy than the reactants. Thus the system will move to the right to reach equilibrium. That is, K is greater than 1.

See Exercise 16-67

INTERACTIVE EXAMPLE 16-15

Free Energy and Equilibrium II

The overall reaction for the corrosion (rusting) of iron by oxygen is ◄

$$4Fe(s) + 3O_2(g) \rightleftharpoons 2Fe_2O_3(s)$$

Using the following data, calculate the equilibrium constant for this reaction at 25°C.

Substance	ΔH_f° (kJ/mol)	S° (J/K · mol)
$Fe_2O_3(s)$	−826	90
$Fe(s)$	0	27
$O_2(g)$	0	205

Solution

To calculate K for this reaction, we will use the equation

$$\Delta G^\circ = -RT \ln(K)$$

Thinkstock/Jupiter Images

▲ Formation of rust on bare steel is a spontaneous process.

We must first calculate $\Delta G°$ from

$$\Delta G° = \Delta H° - T\Delta S°$$

where

$$\Delta H° = 2\Delta H°_{f(Fe_2O_3(s))} - 3\Delta H°_{f(O_2(g))} - 4\Delta H°_{f(Fe(s))}$$
$$= 2 \text{ mol}(-826 \text{ kJ/mol}) - 0 - 0$$
$$= -1652 \text{ kJ} = -1.652 \times 10^6 \text{ J}$$
$$\Delta S° = 2S°_{Fe_2O_3} - 3S°_{O_2} - 4S°_{Fe}$$
$$= 2 \text{ mol}(90 \text{ J/K} \cdot \text{mol}) - 3 \text{ mol}(205 \text{ J/K} \cdot \text{mol}) - 4 \text{ mol}(27 \text{ J/K} \cdot \text{mol})$$
$$= -543 \text{ J/K}$$

and

$$T = 273 + 25 = 298 \text{ K}$$

Then

$$\Delta G° = \Delta H° - T\Delta S° = (-1.652 \times 10^6 \text{ J}) - (298 \text{ K})(-543 \text{ J/K})$$
$$= -1.490 \times 10^6 \text{ J}$$

and

$$\Delta G° = -RT \ln(K) = -1.490 \times 10^6 \text{ J} = -(8.3145 \text{ J/K} \cdot \text{mol})(298 \text{ K}) \ln(K)$$

Thus

$$\ln(K) = \frac{1.490 \times 10^6}{2.48 \times 10^3} = 601$$

and

$$K = e^{601}$$

This is a very large equilibrium constant. The rusting of iron is clearly very favorable from a thermodynamic point of view.

See Exercises 16-69 and 16-70

The Temperature Dependence of K

In Chapter 12 we used Le Châtelier's principle to predict qualitatively how the value of K for a given reaction would change with a change in temperature. Now we can specify the quantitative dependence of the equilibrium constant on temperature from the relationship

$$\Delta G° = -RT \ln(K) = \Delta H° - T\Delta S°$$

We can rearrange this equation to give

$$\ln(K) = -\frac{\Delta H°}{RT} + \frac{\Delta S°}{R} = -\frac{\Delta H°}{R}\left(\frac{1}{T}\right) + \frac{\Delta S°}{R}$$

Note that this is a linear equation of the form $y = mx + b$, where $y = \ln(K)$, $m = -\Delta H°/R$ = slope, $x = 1/T$, and $b = \Delta S°/R$ = intercept. This means that if values of K for a given reaction are determined at various temperatures, a plot of $\ln(K)$ versus $1/T$ will be linear, with slope $-\Delta H°/R$ and intercept $\Delta S°/R$. This result assumes that both $\Delta H°$ and $\Delta S°$ are independent of temperature over the temperature range considered. This assumption is a good approximation over a relatively small temperature range.

16-9 | Free Energy and Work

One of the main reasons we are interested in physical and chemical processes is that we want to use them to do work for us, and we want this work done as efficiently and economically as possible. We have already seen that at constant temperature and pressure, the sign of the change in free energy tells us whether a given process is spontaneous. This is very useful information because it prevents us from wasting effort on a process that has no inherent tendency to occur. Although a thermodynamically favorable chemical reaction may not occur to any appreciable extent because it is too slow, it makes sense in this case to try to find a catalyst to speed up the reaction. On the other hand, if the reaction is prevented from occurring by its thermodynamic characteristics, we would be wasting our time looking for a catalyst.

In addition to its qualitative usefulness (telling us whether a process is spontaneous), the change in free energy is important quantitatively because it can tell us how much work can be done with a given process. The maximum possible useful work obtainable from a process at constant temperature and pressure is equal to the change in free energy:

$$w_{max} = \Delta G$$

This relationship explains why this function is called the *free* energy. Under certain conditions, ΔG for a spontaneous process represents the energy that is *free to do useful work*. ◄ ⓘ On the other hand, for a process that is not spontaneous, the value of ΔG tells us the minimum amount of work that must be *expended* to make the process occur.

Knowing the value of ΔG for a process thus gives us valuable information about how close the process is to 100% efficiency. For example, when gasoline is burned in a car's engine, the work produced is about 20% of the maximum work available.

For reasons we will only briefly introduce in this book, the amount of work we actually obtain from a spontaneous process is *always* less than the maximum possible amount.

To explore this idea more fully, let's consider an electric current flowing through the starter motor of a car. The current is generated from a chemical change in a battery, and we can calculate ΔG for the battery reaction and so determine the energy available to do work. Can we use all this energy to do work? No, because a current flowing through a wire causes frictional heating, and the greater the current, the greater the heat. This heat represents wasted energy—it is not useful for running the starter motor. We can minimize this energy waste by running very low currents through the motor circuit. However, zero current flow would be necessary to eliminate frictional heating entirely, and we cannot derive any work from the motor if no current flows. This represents the difficulty in which nature places us. Using a process to do work requires that some of the energy be wasted, and usually the faster we run the process, the more energy we waste.

Achieving the maximum work available from a spontaneous process can occur only via a hypothetical pathway. Any real pathway wastes energy. If we could discharge the battery infinitely slowly by an infinitesimal current flow, we would achieve the maximum useful work. Also, if we could then recharge the battery using an infinitesimally small current, exactly the same amount of energy would be used to return the battery to its original state. After we cycle the battery in this way, the universe (the system and surroundings) is exactly the same as it was before the cyclic process. This is a **reversible process** (Fig. 16-11).

However, if the battery is discharged to run the starter motor and then recharged using a *finite* current flow, as is the case in reality, *more* work will always be required to recharge the battery than the battery produces as it discharges. This means that even though the battery (the system) has returned to its original state, the surroundings have not, because the surroundings had to furnish a net amount of work as the battery was cycled. The *universe is different* after this cyclic process is performed, and this function is called an **irreversible process**. *All real processes are irreversible.*

In general, after any real cyclic process is carried out in a system, the surroundings have less ability to do work and contain more thermal energy. In any real cyclic process in the system, work is changed to heat in the surroundings, and the entropy of the universe increases. This is another way of stating the second law of thermodynamics.

ⓘ Note that "*PV* work" is not counted as useful work here.

Figure 16-11 | A battery can do work by sending current to a starter motor. The battery can then be recharged by forcing current through it in the opposite direction. If the current flow in both processes is infinitesimally small, $w_1 - w_2$. This is a *reversible process*. But if the current flow is finite, as it would be in any real case, $w_2 > w_1$. This is an *irreversible process* (the *universe is different* after the cyclic process occurs). All real processes are irreversible.

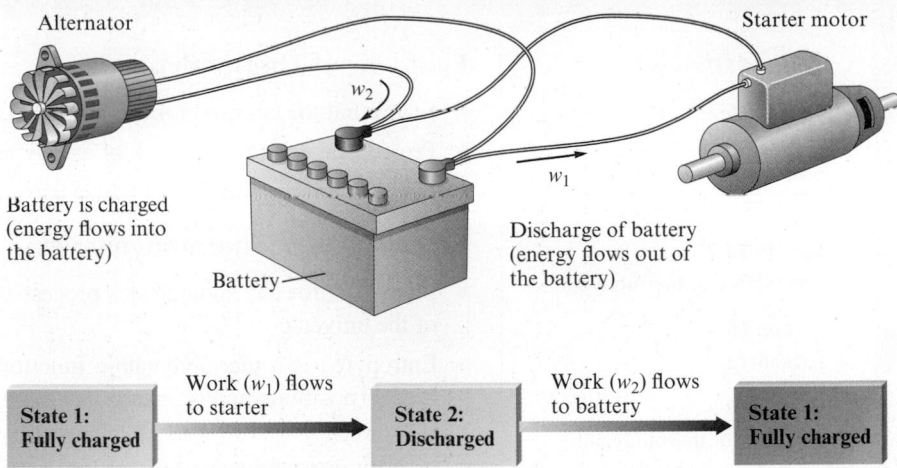

Alternator

w_2

Battery is charged (energy flows into the battery)

Battery

Starter motor

w_1

Discharge of battery (energy flows out of the battery)

| State 1: Fully charged | Work (w_1) flows to starter → | State 2: Discharged | Work (w_2) flows to battery → | State 1: Fully charged |

ⓘ When energy is used to do work, it becomes less organized and less concentrated and thus less useful.

Thus thermodynamics tells us the work potential of a process and then tells us that we can never achieve this potential. In this spirit, thermodynamicist Henry Bent has paraphrased the first two laws of thermodynamics as follows:

First law: You can't win, you can only break even.

Second law: You can't break even.

The ideas we have discussed in this section are applicable to the energy crisis that will probably increase in severity over the next 25 years. The crisis is obviously not one of supply; the first law tells us that the universe contains a constant supply of energy. The problem is the availability of *useful* energy. As we use energy, we degrade its usefulness. ◄ ⓘ For example, when gasoline reacts with oxygen in the combustion reaction, the change in potential energy results in heat flow. Thus the energy concentrated in the bonds of the gasoline and oxygen molecules ends up *spread* over the surroundings as thermal energy, where it is much more difficult to harness for useful work. This is a way in which the entropy of the universe increases: Concentrated energy becomes spread out—more disordered and less useful. Thus the crux of the energy problem is that we are rapidly consuming the concentrated energy found in fossil fuels. It took millions of years to concentrate the sun's energy in these fuels, and we will consume these same fuels in a few hundred years. Thus we must use these energy sources as wisely as possible.

Critical Thinking

What if the first law of thermodynamics was true but the second law was not? How would the world be different?

FOR REVIEW

Key Terms

Section 16-1
spontaneous process
entropy
positional probability

Section 16-2
second law of thermodynamics

Section 16-4
free energy

Section 16-5
third law of thermodynamics

Section 16-6
standard free energy change
standard free energy of formation

Section 16-8
equilibrium point (thermodynamic definition)

Section 16-9
reversible process
irreversible process

First Law of Thermodynamics

- States that the energy of the universe is constant
- Provides a way to keep track of energy as it changes form
- Gives no information about why a particular process occurs in a given direction

Second Law of Thermodynamics

- States that for any spontaneous process there is always an increase in the entropy of the universe
- Entropy(S) is a thermodynamic function that describes the number of arrangements (positions and/or energy levels) available to a system existing in a given state
  - Nature spontaneously proceeds toward states that have the highest probability of occurring
  - Using entropy, thermodynamics can predict the direction in which a process will occur spontaneously:

$$\Delta S_{univ} = \Delta S_{sys} + \Delta S_{surr}$$

  - For a spontaneous process, ΔS_{univ} must be positive
  - For a process at constant temperature and pressure:
    - ΔS_{sys} is dominated by "positional" entropy
    - For a chemical reaction, ΔS_{sys} is dominated by changes in the number of gaseous molecules
    - ΔS_{surr} is determined by heat:

$$\Delta S_{surr} = -\frac{\Delta H}{T}$$

    - ΔS_{surr} is positive for an exothermic process (ΔH is negative)
    - Because ΔS_{surr} depends inversely on T, exothermicity becomes a more important driving force at low temperatures
- Thermodynamics cannot predict the rate at which a system will spontaneously change; the principles of kinetics are necessary to do this

Third Law of Thermodynamics

- States that the entropy of a perfect crystal at 0 K is zero

Free Energy (G)

- Free energy is a state function:

$$G = H - TS$$

- A process occurring at constant temperature and pressure is spontaneous in the direction in which its free energy decreases ($\Delta G < 0$)
- For a reaction the standard free energy change ($\Delta G°$) is the change in free energy that occurs when reactants in their standard states are converted to products in their standard states
- The standard free energy change for a reaction can be determined from the standard free energies of formation (ΔG_f°) of the reactants and products:

$$\Delta G° = \Sigma n_p \Delta G_f^\circ(\text{products}) - \Sigma n_r \Delta G_f^\circ(\text{reactants})$$

- Free energy depends on temperature and pressure:

$$G = G° + RT \ln P$$

- This relationship can be used to derive the relationship between $\Delta G°$ for a reaction and the value of its equilibrium constant K:

$$\Delta G° = -RT \ln K$$

  - For $\Delta G° = 0$, $K = 1$
  - For $\Delta G° < 0$, $K > 1$
  - For $\Delta G° > 0$, $K < 1$

- The maximum possible useful work obtainable from a process at constant temperature and pressure is equal to the change in free energy:

$$w_{max} = \Delta G$$

- In any real process, $w < w_{max}$
- When energy is used to do work in a real process, the energy of the universe remains constant but the usefulness of the energy decreases
  - Concentrated energy is spread out in the surroundings as thermal energy

Review Questions   Answers to the Review Questions can be found on the Student Web site (accessible from www.cengagebrain.com).

1. Define the following:
   a. spontaneous process
   b. entropy
   c. positional probability
   d. system
   e. surroundings
   f. universe

2. What is the second law of thermodynamics? For any process, there are four possible sign combinations for ΔS_{sys} and ΔS_{surr}. Which sign combination(s) always give a spontaneous process? Which sign combination(s) always give a nonspontaneous process? Which sign combination(s) may or may not give a spontaneous process?

3. What determines ΔS_{surr} for a process? To calculate ΔS_{surr} at constant pressure and temperature, we use the following equation: $\Delta S_{surr} = -\Delta H/T$. Why does a minus sign appear in the equation, and why is ΔS_{surr} inversely proportional to temperature?

4. The free energy change, ΔG, for a process at constant temperature and pressure is related to ΔS_{univ} and reflects the spontaneity of the process. How is ΔG related to ΔS_{univ}? When is a process spontaneous? Nonspontaneous? At equilibrium? ΔG is a composite term composed of ΔH, T, and ΔS. What is the ΔG equation? Give the four possible sign combinations for ΔH and ΔS. What temperatures are required for each sign combination to yield a spontaneous process? If ΔG is positive, what does it say about the reverse process? How does the $\Delta G = \Delta H - T\Delta S$ equation reduce when at the melting-point temperature of a solid-to-liquid phase change or at the boiling-point temperature of a liquid-to-gas phase change? What is the sign of ΔG for the solid-to-liquid phase change at temperatures above the freezing point? What is the sign of ΔG for the liquid-to-gas phase change at temperatures below the boiling point?

5. What is the third law of thermodynamics? What are standard entropy values, $S°$, and how are these $S°$ values (listed in Appendix 4) used to calculate $\Delta S°$ for a reaction? How would you use Hess's law to calculate $\Delta S°$ for a reaction? What does the superscript ° indicate?

   Predicting the sign of $\Delta S°$ for a reaction is an important skill to master. For a gas-phase reaction, what do you concentrate on to predict the sign of $\Delta S°$? For a phase change, what do you concentrate on to predict the sign of $\Delta S°$? That is, how are $S°_{solid}$, $S°_{liquid}$, and $S°_{gas}$ related to one another? When a solute dissolves in water, what is usually the sign of $\Delta S°$ for this process?

6. What is the standard free energy change, $\Delta G°$, for a reaction? What is the standard free energy of formation, $\Delta G°_f$, for a substance? How are $\Delta G°_f$ values used to calculate $\Delta G°_{rxn}$? How can you use Hess's law to calculate $\Delta G°_{rxn}$? How can you use $\Delta H°$ and $\Delta S°$ values to calculate $\Delta G°_{rxn}$? Of the functions $\Delta H°$, $\Delta S°$, and $\Delta G°$, which depends most strongly on temperature? When $\Delta G°$ is calculated at temperatures other than 25°C, what assumptions are generally made concerning $\Delta H°$ and $\Delta S°$?

7. If you calculate a value for $\Delta G°$ for a reaction using the values of $\Delta G_f°$ in Appendix 4 and get a negative number, is it correct to say that the reaction is always spontaneous? Why or why not? Free energy changes also depend on concentration. For gases, how is G related to the pressure of the gas? What are standard pressures for gases and standard concentrations for solutes? How do you calculate ΔG for a reaction at nonstandard conditions? The equation to determine ΔG at nonstandard conditions has Q in it: What is Q? A reaction is spontaneous as long as ΔG is negative; that is, reactions always proceed as long as the products have a lower free energy than the reactants. What is so special about equilibrium? Why don't reactions move away from equilibrium?

8. Consider the equation $\Delta G = \Delta G° + RT \ln(Q)$. What is the value of ΔG for a reaction at equilibrium? What does Q equal at equilibrium? At equilibrium, the previous equation reduces to $\Delta G° = -RT \ln(K)$. When $\Delta G° > 0$, what does it indicate about K? When $\Delta G° < 0$, what does it indicate about K? When $\Delta G° = 0$, what does it indicate about K? ΔG predicts spontaneity for a reaction, whereas $\Delta G°$ predicts the equilibrium position. Explain what this statement means. Under what conditions can you use $\Delta G°$ to determine the spontaneity of a reaction?

9. Even if ΔG is negative, the reaction may not occur. Explain the interplay between the thermodynamics and the kinetics of a reaction. High temperatures are favorable to a reaction kinetically but may be unfavorable to a reaction thermodynamically. Explain.

10. Discuss the relationship between w_{max} and the magnitude and sign of the free energy change for a reaction. Also discuss w_{max} for real processes. What is a reversible process?

Active Learning Questions

These questions are designed to be used by groups of students in class.

1. For the process $A(l) \longrightarrow A(g)$, which direction is favored by changes in energy probability? Positional probability? Explain your answers. If you wanted to favor the process as written, would you raise or lower the temperature of the system? Explain.

2. For a liquid, which would you expect to be larger, ΔS_{fusion} or $\Delta S_{evaporation}$? Why?

3. Gas A_2 reacts with gas B_2 to form gas AB at a constant temperature. The bond energy of AB is much greater than that of either reactant. What can be said about the sign of ΔH? ΔS_{surr}? ΔS? Explain how potential energy changes for this process. Explain how random kinetic energy changes during the process.

4. What types of experiments can be carried out to determine whether a reaction is spontaneous? Does spontaneity have any relationship to the final equilibrium position of a reaction? Explain.

5. A friend tells you, "Free energy G and pressure P are related by the equation $G = G° + RT \ln(P)$. Also, G is related to the equilibrium constant K in that when $G_{products} = G_{reactants}$, the system is at equilibrium. Therefore, it must be true that a system is at equilibrium when all the pressures are equal." Do you agree with this friend? Explain.

6. You remember that $\Delta G°$ is related to $RT \ln(K)$ but cannot remember if it's $RT \ln(K)$ or $-RT \ln(K)$. Realizing what $\Delta G°$ and K mean, how can you figure out the correct sign?

7. Predict the sign of ΔS for each of the following and explain.
   a. the evaporation of alcohol
   b. the freezing of water
   c. compressing an ideal gas at constant temperature
   d. dissolving NaCl in water

8. Is ΔS_{surr} favorable or unfavorable for exothermic reactions? Endothermic reactions? Explain.

9. At 1 atm, liquid water is heated above 100°C. For this process, which of the following choices (i–iv) is correct for ΔS_{surr}? ΔS? ΔS_{univ}? Explain each answer.
   i. greater than zero
   ii. less than zero
   iii. equal to zero
   iv. cannot be determined

10. When (if ever) are high temperatures unfavorable to a reaction thermodynamically?

A blue question or exercise number indicates that the answer to that question or exercise appears at the back of this book and a solution appears in the *Solutions Guide*, as found on PowerLecture.

Questions

11. The synthesis of glucose directly from CO_2 and H_2O and the synthesis of proteins directly from amino acids are both non-spontaneous processes under standard conditions. Yet it is necessary for these to occur for life to exist. In light of the second law of thermodynamics, how can life exist?

12. When the environment is contaminated by a toxic or potentially toxic substance (for example, from a chemical spill or the use of insecticides), the substance tends to disperse. How is this consistent with the second law of thermodynamics? In terms of the second law, which requires the least work: cleaning the environment after it has been contaminated or trying to prevent the contamination before it occurs? Explain.

13. Entropy has been described as "time's arrow." Interpret this view of entropy.

14. Human DNA contains almost twice as much information as is needed to code for all the substances produced in the body. Likewise, the digital data sent from *Voyager II* contained one redundant bit out of every two bits of information. The Hubble space telescope transmits three redundant bits for every bit of information. How is entropy related to the transmission of information? What do you think is accomplished by having so many redundant bits of information in both DNA and the space probes?

15. A mixture of hydrogen gas and chlorine gas remains unreacted until it is exposed to ultraviolet light from a burning magnesium strip. Then the following reaction occurs very rapidly:

$$H_2(g) + Cl_2(g) \longrightarrow 2HCl(g)$$

Explain.

16. Consider the following potential energy plots:

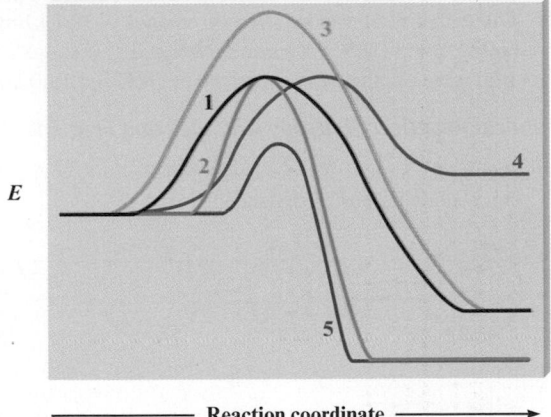

a. Rank the reactions from fastest to slowest and explain your answer. If any reactions have equal rates, explain why.

b. Label the reactions as endothermic or exothermic, and support your answer.

c. Rank the exothermic reactions from greatest to least change in potential energy, and support your answer.

17. ΔS_{surr} is sometimes called the energy disorder term. Explain.

18. Given the following illustration, what can be said about the sign of ΔS for the process of solid NaCl dissolving in water? What can be said about ΔH for this process?

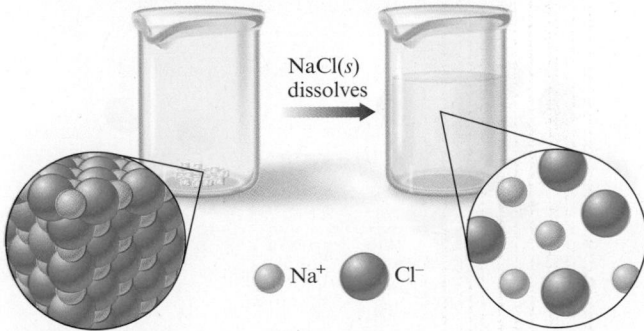

NaCl(s) dissolves

Na$^+$ Cl$^-$

19. The third law of thermodynamics states that the entropy of a perfect crystal at 0 K is zero. In Appendix 4, F$^-$(aq), OH$^-$(aq), and S^{2-}(aq) all have negative standard entropy values. How can $S°$ values be less than zero?

20. The deciding factor on why HF is a weak acid and not a strong acid like the other hydrogen halides is entropy. What occurs when HF dissociates in water as compared to the other hydrogen halides?

21. List three different ways to calculate the standard free energy change, $\Delta G°$, for a reaction at 25°C. How is $\Delta G°$ estimated at temperatures other than 25°C? What assumptions are made?

22. What information can be determined from ΔG for a reaction? Does one get the same information from $\Delta G°$, the standard free energy change? $\Delta G°$ allows determination of the equilibrium constant K for a reaction. How? How can one estimate the value of K at temperatures other than 25°C for a reaction? How can one estimate the temperature where $K = 1$ for a reaction? Do all reactions have a specific temperature where $K = 1$?

23. Monochloroethane (C$_2$H$_5$Cl) can be produced by the direct reaction of ethane gas (C$_2$H$_6$) with chlorine gas or by the reaction of ethylene gas (C$_2$H$_4$) with hydrogen chloride gas. The second reaction gives almost a 100% yield of pure C$_2$H$_5$Cl at a rapid rate without catalysis. The first method requires light as an energy source or the reaction would not occur. Yet $\Delta G°$ for the first reaction is considerably more negative than $\Delta G°$ for the second reaction. Explain how this can be so.

24. At 1500 K, the process

$$I_2(g) \longrightarrow 2I(g)$$
$$\text{10 atm} \qquad \text{10 atm}$$

is not spontaneous. However, the process

$$I_2(g) \longrightarrow 2I(g)$$
$$\text{0.10 atm} \qquad \text{0.10 atm}$$

is spontaneous at 1500 K. Explain.

Exercises

In this section similar exercises are paired.

Spontaneity, Entropy, and the Second Law of Thermodynamics: Free Energy

25. Which of the following processes are spontaneous?
   a. Salt dissolves in H$_2$O.
   b. A clear solution becomes a uniform color after a few drops of dye are added.
   c. Iron rusts.
   d. You clean your bedroom.

26. Which of the following processes are spontaneous?
   a. A house is built.
   b. A satellite is launched into orbit.
   c. A satellite falls back to the earth.
   d. The kitchen gets cluttered.

27. Table 16-1 shows the possible arrangements of four molecules in a two-bulbed flask. What are the possible arrangements if there is one molecule in this two-bulbed flask or two molecules or three molecules? For each, what arrangement is most likely?

28. Consider the following illustration of six molecules of gas in a two-bulbed flask.

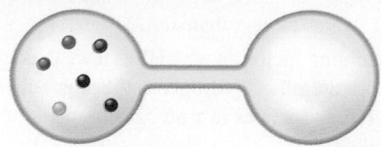

  a. What is the most likely arrangement of molecules? How many microstates are there for this arrangement?

  b. Determine the probability of finding the gas in its most likely arrangement.

29. Consider the following energy levels, each capable of holding two particles:

$$E = 2 \text{ kJ} \underline{\quad}$$

$$E \quad E = 1 \text{ kJ} \underline{\quad}$$

$$E = 0 \quad \underline{XX}$$

Draw all the possible arrangements of the two identical particles (represented by X) in the three energy levels. What total energy is most likely, that is, occurs the greatest number of times? Assume that the particles are indistinguishable from each other.

30. Redo Exercise 29 with two particles A and B, which can be distinguished from each other.

31. Choose the substance with the larger positional probability in each case.

  a. 1 mole of H_2 (at STP) or 1 mole of H_2 (at 100°C, 0.5 atm)

  b. 1 mole of N_2 (at STP) or 1 mole of N_2 (at 100 K, 2.0 atm)

  c. 1 mole of $H_2O(s)$ (at 0°C) or 1 mole of $H_2O(l)$ (at 20°C)

32. Which of the following involve an increase in the entropy of the system?

  a. melting of a solid

  b. sublimation

  c. freezing

  d. mixing

  e. separation

  f. boiling

33. Predict the sign of ΔS_{surr} for the following processes.

  a. $H_2O(l) \longrightarrow H_2O(g)$

  b. $I_2(g) \longrightarrow I_2(s)$

34. Calculate ΔS_{surr} for the following reactions at 25°C and 1 atm.

  a. $C_3H_8(g) + 5O_2(g) \longrightarrow 3CO_2(g) + 4H_2O(l)$
  $$\Delta H° = -2221 \text{ kJ}$$

  b. $2NO_2(g) \longrightarrow 2NO(g) + O_2(g) \qquad \Delta H° = 112 \text{ kJ}$

35. Given the values of ΔH and ΔS, which of the following changes will be spontaneous at constant T and P?

  a. $\Delta H = +25 \text{ kJ}, \Delta S = +5.0 \text{ J/K}, T = 300. \text{ K}$

  b. $\Delta H = +25 \text{ kJ}, \Delta S = +100. \text{ J/K}, T = 300. \text{ K}$

  c. $\Delta H = -10. \text{ kJ}, \Delta S = +5.0 \text{ J/K}, T = 298 \text{ K}$

  d. $\Delta H = -10. \text{ kJ}, \Delta S = -40. \text{ J/K}, T = 200. \text{ K}$

36. At what temperatures will the following processes be spontaneous?

  a. $\Delta H = -18 \text{ kJ}$ and $\Delta S = -60. \text{ J/K}$

  b. $\Delta H = +18 \text{ kJ}$ and $\Delta S = +60. \text{ J/K}$

  c. $\Delta H = +18 \text{ kJ}$ and $\Delta S = -60. \text{ J/K}$

  d. $\Delta H = -18 \text{ kJ}$ and $\Delta S = +60. \text{ J/K}$

37. Ethanethiol (C_2H_5SH; also called ethyl mercaptan) is commonly added to natural gas to provide the "rotten egg" smell of a gas leak. The boiling point of ethanethiol is 35°C and its heat of vaporization is 27.5 kJ/mol. What is the entropy of vaporization for this substance?

38. For mercury, the enthalpy of vaporization is 58.51 kJ/mol and the entropy of vaporization is 92.92 J/K · mol. What is the normal boiling point of mercury?

39. For ammonia (NH_3), the enthalpy of fusion is 5.65 kJ/mol and the entropy of fusion is 28.9 J/K · mol.

  a. Will $NH_3(s)$ spontaneously melt at 200. K?

  b. What is the approximate melting point of ammonia?

40. The enthalpy of vaporization of ethanol is 38.7 kJ/mol at its boiling point (78°C). Determine ΔS_{sys}, ΔS_{surr}, and ΔS_{univ} when 1.00 mole of ethanol is vaporized at 78°C and 1.00 atm.

Chemical Reactions: Entropy Changes and Free Energy

41. Predict the sign of $\Delta S°$ for each of the following changes. Assume all equations are balanced.

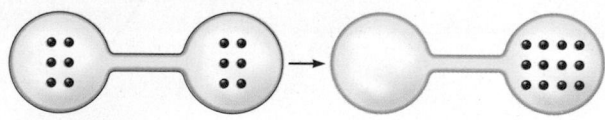

a.

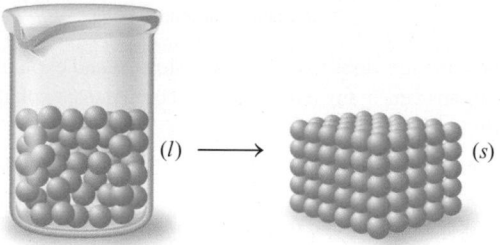

b.

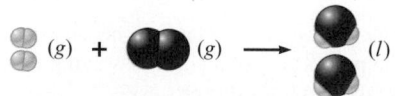

c.

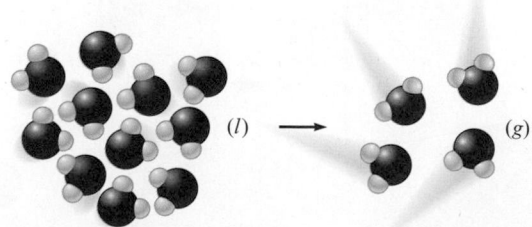

d.

42. Predict the sign of $\Delta S°$ for each of the following changes.

  a. $K(s) + \frac{1}{2}Br_2(g) \longrightarrow KBr(s)$

  b. $N_2(g) + 3H_2(g) \longrightarrow 2NH_3(g)$

  c. $KBr(s) \longrightarrow K^+(aq) + Br^-(aq)$

  d. $KBr(s) \longrightarrow KBr(l)$

43. For each of the following pairs of substances, which substance has the greater value of $S°$?

  a. $C_{graphite}(s)$ or $C_{diamond}(s)$

  b. $C_2H_5OH(l)$ or $C_2H_5OH(g)$

  c. $CO_2(s)$ or $CO_2(g)$

44. For each of the following pairs, which substance has the greater value of S?

  a. N_2O (at 0 K) or He (at 10 K)

  b. $N_2O(g)$ (at 1 atm, 25°C) or He(g) (at 1 atm, 25°C)

  c. $NH_3(s)$ (at 196 K) $\longrightarrow NH_3(l)$ (at 196 K)

45. Predict the sign of $\Delta S°$ and then calculate $\Delta S°$ for each of the following reactions.

  a. $2H_2S(g) + SO_2(g) \longrightarrow 3S_{rhombic}(s) + 2H_2O(g)$

  b. $2SO_3(g) \longrightarrow 2SO_2(g) + O_2(g)$

  c. $Fe_2O_3(s) + 3H_2(g) \longrightarrow 2Fe(s) + 3H_2O(g)$

46. Predict the sign of $\Delta S°$ and then calculate $\Delta S°$ for each of the following reactions.

  a. $H_2(g) + \frac{1}{2}O_2(g) \longrightarrow H_2O(l)$

  b. $2CH_3OH(g) + 3O_2(g) \longrightarrow 2CO_2(g) + 4H_2O(g)$

  c. $HCl(g) \longrightarrow H^+(aq) + Cl^-(aq)$

47. For the reaction

$$C_2H_2(g) + 4F_2(g) \longrightarrow 2CF_4(g) + H_2(g)$$

$\Delta S°$ is equal to -358 J/K. Use this value and data from Appendix 4 to calculate the value of $S°$ for $CF_4(g)$.

48. For the reaction

$$CS_2(g) + 3O_2(g) \longrightarrow CO_2(g) + 2SO_2(g)$$

$\Delta S°$ is equal to -143 J/K. Use this value and data from Appendix 4 to calculate the value of $S°$ for $CS_2(g)$.

49. It is quite common for a solid to change from one structure to another at a temperature below its melting point. For example, sulfur undergoes a phase change from the rhombic crystal structure to the monoclinic crystal form at temperatures above 95°C.

  a. Predict the signs of ΔH and ΔS for the process $S_{rhombic}(s) \longrightarrow S_{monoclinic}(s)$.

  b. Which form of sulfur has the more ordered crystalline structure (has the smaller positional probability)?

50. Two crystalline forms of white phosphorus are known. Both forms contain P_4 molecules, but the molecules are packed together in different ways. The α form is always obtained when the liquid freezes. However, below $-76.9°C$, the α form spontaneously converts to the β form:

$$P_4(s, \alpha) \longrightarrow P_4(s, \beta)$$

  a. Predict the signs of ΔH and ΔS for this process.

  b. Predict which form of phosphorus has the more ordered crystalline structure (has the smaller positional probability).

51. Consider the reaction

$$2O(g) \longrightarrow O_2(g)$$

  a. Predict the signs of ΔH and ΔS.

  b. Would the reaction be more spontaneous at high or low temperatures?

52. Hydrogen cyanide is produced industrially by the following exothermic reaction:

$$2NH_3(g) + 3O_2(g) + 2CH_4(g) \xrightarrow[\text{Pt-Rh}]{100°C} 2HCN(g) + 6H_2O(g)$$

Is the high temperature needed for thermodynamic or kinetic reasons?

53. From data in Appendix 4, calculate $\Delta H°$, $\Delta S°$, and $\Delta G°$ for each of the following reactions at 25°C.

  a. $CH_4(g) + 2O_2(g) \longrightarrow CO_2(g) + 2H_2O(g)$

  b. $6CO_2(g) + 6H_2O(l) \longrightarrow C_6H_{12}O_6(s) + 6O_2(g)$
  Glucose

  c. $P_4O_{10}(s) + 6H_2O(l) \longrightarrow 4H_3PO_4(s)$

  d. $HCl(g) + NH_3(g) \longrightarrow NH_4Cl(s)$

54. The major industrial use of hydrogen is in the production of ammonia by the Haber process:

$$3H_2(g) + N_2(g) \longrightarrow 2NH_3(g)$$

  a. Using data from Appendix 4, calculate $\Delta H°$, $\Delta S°$, and $\Delta G°$ for the Haber process reaction.

  b. Is the reaction spontaneous at standard conditions?

  c. At what temperatures is the reaction spontaneous at standard conditions? Assume $\Delta H°$ and $\Delta S°$ do not depend on temperature.

55. For the reaction at 298 K,

$$2NO_2(g) \rightleftharpoons N_2O_4(g)$$

the values of $\Delta H°$ and $\Delta S°$ are -58.03 kJ and -176.6 J/K, respectively. What is the value of $\Delta G°$ at 298 K? Assuming that $\Delta H°$ and $\Delta S°$ do not depend on temperature, at what temperature is $\Delta G° = 0$? Is $\Delta G°$ negative above or below this temperature?

56. At 100.°C and 1.00 atm, $\Delta H° = 40.6$ kJ/mol for the vaporization of water. Estimate $\Delta G°$ for the vaporization of water at 90.°C and 110.°C. Assume $\Delta H°$ and $\Delta S°$ at 100.°C and 1.00 atm do not depend on temperature.

57. Given the following data:

$$2H_2(g) + C(s) \longrightarrow CH_4(g) \qquad \Delta G° = -51 \text{ kJ}$$
$$2H_2(g) + O_2(g) \longrightarrow 2H_2O(l) \qquad \Delta G° = -474 \text{ kJ}$$
$$C(s) + O_2(g) \longrightarrow CO_2(g) \qquad \Delta G° = -394 \text{ kJ}$$

Calculate $\Delta G°$ for $CH_4(g) + 2O_2(g) \rightarrow CO_2(g) + 2H_2O(l)$.

58. Given the following data:

$$2C_6H_6(l) + 15O_2(g) \longrightarrow 12CO_2(g) + 6H_2O(l)$$
$$\Delta G° = -6399 \text{ kJ}$$

$$C(s) + O_2(g) \longrightarrow CO_2(g) \qquad \Delta G° = -394 \text{ kJ}$$
$$H_2(g) + \frac{1}{2}O_2(g) \longrightarrow H_2O(l) \qquad \Delta G° = -237 \text{ kJ}$$

calculate $\Delta G°$ for the reaction

$$6C(s) + 3H_2(g) \longrightarrow C_6H_6(l)$$

59. For the reaction

$$SF_4(g) + F_2(g) \longrightarrow SF_6(g)$$

the value of $\Delta G°$ is -374 kJ. Use this value and data from Appendix 4 to calculate the value of $\Delta G_f°$ for $SF_4(g)$.

60. The value of $\Delta G°$ for the reaction

$$2C_4H_{10}(g) + 13O_2(g) \longrightarrow 8CO_2(g) + 10H_2O(l)$$

is $-5490.$ kJ. Use this value and data from Appendix 4 to calculate the standard free energy of formation for $C_4H_{10}(g)$.

61. Consider the reaction

$$Fe_2O_3(s) + 3H_2(g) \longrightarrow 2Fe(s) + 3H_2O(g)$$

a. Use $\Delta G_f°$ values in Appendix 4 to calculate $\Delta G°$ for this reaction.

b. Is this reaction spontaneous under standard conditions at 298 K?

c. The value of $\Delta H°$ for this reaction is 100. kJ. At what temperatures is this reaction spontaneous at standard conditions? Assume that $\Delta H°$ and $\Delta S°$ do not depend on temperature.

62. Consider the reaction

$$2POCl_3(g) \longrightarrow 2PCl_3(g) + O_2(g)$$

a. Calculate $\Delta G°$ for this reaction. The $\Delta G_f°$ values for $POCl_3(g)$ and $PCl_3(g)$ are -502 kJ/mol and $-270.$ kJ/mol, respectively.

b. Is this reaction spontaneous under standard conditions at 298 K?

c. The value of $\Delta S°$ for this reaction is 179 J/K · mol. At what temperatures is this reaction spontaneous at standard conditions? Assume that $\Delta H°$ and $\Delta S°$ do not depend on temperature.

63. Using data from Appendix 4, calculate $\Delta H°$, $\Delta S°$, and $\Delta G°$ for the following reactions that produce acetic acid:

$$CH_4(g) + CO_2(g) \longrightarrow CH_3\overset{\displaystyle O}{\overset{\displaystyle \|}{C}}-OH(l)$$

$$CH_3OH(g) + CO(g) \longrightarrow CH_3\overset{\displaystyle O}{\overset{\displaystyle \|}{C}}-OH(l)$$

Which reaction would you choose as a commercial method for producing acetic acid (CH_3CO_2H) at standard conditions? What temperature conditions would you choose for the reaction? Assume $\Delta H°$ and $\Delta S°$ do not depend on temperature.

64. Consider two reactions for the production of ethanol:

$$C_2H_4(g) + H_2O(g) \longrightarrow CH_3CH_2OH(l)$$
$$C_2H_6(g) + H_2O(g) \longrightarrow CH_3CH_2OH(l) + H_2(g)$$

Which would be the more thermodynamically feasible at standard conditions? Why?

Free Energy: Pressure Dependence and Equilibrium

65. Using data from Appendix 4, calculate ΔG for the reaction

$$NO(g) + O_3(g) \longrightarrow NO_2(g) + O_2(g)$$

for these conditions:

$T = 298$ K

$P_{NO} = 1.00 \times 10^{-6}$ atm, $P_{O_3} = 2.00 \times 10^{-6}$ atm

$P_{NO_2} = 1.00 \times 10^{-7}$ atm, $P_{O_2} = 1.00 \times 10^{-3}$ atm

66. Using data from Appendix 4, calculate ΔG for the reaction

$$2H_2S(g) + SO_2(g) \rightleftharpoons 3S_{rhombic}(s) + 2H_2O(g)$$

for the following conditions at 25°C:

$$P_{H_2S} = 1.0 \times 10^{-4} \text{ atm}$$
$$P_{SO_2} = 1.0 \times 10^{-2} \text{ atm}$$
$$P_{H_2O} = 3.0 \times 10^{-2} \text{ atm}$$

67. Consider the reaction

$$2NO_2(g) \rightleftharpoons N_2O_4(g)$$

For each of the following mixtures of reactants and products at 25°C, predict the direction in which the reaction will shift to reach equilibrium.

a. $P_{NO_2} = P_{N_2O_4} = 1.0$ atm

b. $P_{NO_2} = 0.21$ atm, $P_{N_2O_4} = 0.50$ atm

c. $P_{NO_2} = 0.29$ atm, $P_{N_2O_4} = 1.6$ atm

68. Consider the following reaction:

$$N_2(g) + 3H_2(g) \rightleftharpoons 2NH_3(g)$$

Calculate ΔG for this reaction under the following conditions (assume an uncertainty of ± 1 in all quantities):

a. $T = 298$ K, $P_{N_2} = P_{H_2} = 200$ atm, $P_{NH_3} = 50$ atm

b. $T = 298$ K, $P_{N_2} = 200$ atm, $P_{H_2} = 600$ atm, $P_{NH_3} = 200$ atm

69. One of the reactions that destroys ozone in the upper atmosphere is

$$NO(g) + O_3(g) \rightleftharpoons NO_2(g) + O_2(g)$$

Using data from Appendix 4, calculate $\Delta G°$ and K (at 298 K) for this reaction.

70. Hydrogen sulfide can be removed from natural gas by the reaction

$$2H_2S(g) + SO_2(g) \rightleftharpoons 3S(s) + 2H_2O(g)$$

Calculate $\Delta G°$ and K (at 298 K) for this reaction. Would this reaction be favored at a high or low temperature?

71. Consider the following reaction at 25.0°C:

$$2NO_2(g) \rightleftharpoons N_2O_4(g)$$

The values of $\Delta H°$ and $\Delta S°$ are -58.03 kJ/mol and -176.6 J/K · mol, respectively. Calculate the value of K at 25.0°C. Assuming $\Delta H°$ and $\Delta S°$ are temperature independent, estimate the value of K at 100.0°C.

72. The standard free energies of formation and the standard enthalpies of formation at 298 K for difluoroacetylene (C_2F_2) and hexafluorobenzene (C_6F_6) are

	ΔG_f° (kJ/mol)	ΔH_f° (kJ/mol)
$C_2F_2(g)$	191.2	241.3
$C_6F_6(g)$	78.2	132.8

For the following reaction:

$$C_6F_6(g) \rightleftharpoons 3C_2F_2(g)$$

  a. calculate ΔS° at 298 K.

  b. calculate K at 298 K.

  c. estimate K at 3000. K, assuming ΔH° and ΔS° do not depend on temperature.

73. Calculate ΔG° for $H_2O(g) + \frac{1}{2}O_2(g) \rightleftharpoons H_2O_2(g)$ at 600. K, using the following data:

$$H_2(g) + O_2(g) \rightleftharpoons H_2O_2(g) \qquad K = 2.3 \times 10^6 \text{ at 600. K}$$

$$2H_2(g) + O_2(g) \rightleftharpoons 2H_2O(g) \qquad K = 1.8 \times 10^{37} \text{ at 600. K}$$

74. The Ostwald process for the commercial production of nitric acid involves three steps:

$$4NH_3(g) + 5O_2(g) \xrightarrow[825^\circ C]{Pt} 4NO(g) + 6H_2O(g)$$

$$2NO(g) + O_2(g) \longrightarrow 2NO_2(g)$$

$$3NO_2(g) + H_2O(l) \longrightarrow 2HNO_3(l) + NO(g)$$

  a. Calculate ΔH°, ΔS°, ΔG°, and K (at 298 K) for each of the three steps in the Ostwald process (see Appendix 4).

  b. Calculate the equilibrium constant for the first step at 825°C, assuming ΔH° and ΔS° do not depend on temperature.

  c. Is there a thermodynamic reason for the high temperature in the first step, assuming standard conditions?

75. Cells use the hydrolysis of adenosine triphosphate, abbreviated as ATP, as a source of energy. Symbolically, this reaction can be written as

$$ATP(aq) + H_2O(l) \longrightarrow ADP(aq) + H_2PO_4^-(aq)$$

where ADP represents adenosine diphosphate. For this reaction, $\Delta G^\circ = -30.5$ kJ/mol.

  a. Calculate K at 25°C.

  b. If all the free energy from the metabolism of glucose

$$C_6H_{12}O_6(s) + 6O_2(g) \longrightarrow 6CO_2(g) + 6H_2O(l)$$

goes into forming ATP from ADP, how many ATP molecules can be produced for every molecule of glucose?

76. One reaction that occurs in human metabolism is

$$HO_2CCH_2CH_2CHCO_2H(aq) + NH_3(aq) \rightleftharpoons$$

$$| \qquad\qquad NH_2$$

Glutamic acid

$$H_2NCCH_2CH_2CHCO_2H(aq) + H_2O(l)$$

$$\overset{O}{||} \qquad\qquad\qquad | \qquad\qquad\qquad NH_2$$

Glutamine

For this reaction $\Delta G^\circ = 14$ kJ at 25°C.

  a. Calculate K for this reaction at 25°C.

  b. In a living cell this reaction is coupled with the hydrolysis of ATP. (See Exercise 75.) Calculate ΔG° and K at 25°C for the following reaction:

$$\text{Glutamic acid}(aq) + ATP(aq) + NH_3(aq) \rightleftharpoons$$
$$\text{Glutamine}(aq) + ADP(aq) + H_2PO_4^-(aq)$$

77. Consider the following reaction at 800. K:

$$N_2(g) + 3F_2(g) \longrightarrow 2NF_3(g)$$

An equilibrium mixture contains the following partial pressures: $P_{N_2} = 0.021$ atm, $P_{F_2} = 0.063$ atm, $P_{NF_3} = 0.48$ atm. Calculate ΔG° for the reaction at 800. K.

78. Consider the following reaction at 298 K:

$$2SO_2(g) + O_2(g) \longrightarrow 2SO_3(g)$$

An equilibrium mixture contains $O_2(g)$ and $SO_3(g)$ at partial pressures of 0.50 atm and 2.0 atm, respectively. Using data from Appendix 4, determine the equilibrium partial pressure of SO_2 in the mixture. Will this reaction be most favored at a high or a low temperature, assuming standard conditions?

79. Consider the relationship

$$\ln(K) = \frac{-\Delta H^\circ}{RT} + \frac{\Delta S^\circ}{R}$$

The equilibrium constant for some hypothetical process was determined as a function of temperature (Kelvin) with the results plotted below.

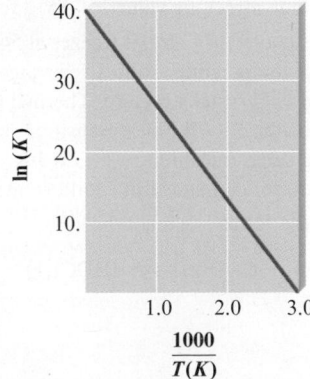

From the plot, determine the values of $\Delta H°$ and $\Delta S°$ for this process. What would be the major difference in the $\ln(K)$ versus $1/T$ plot for an endothermic process as compared to an exothermic process?

80. The equilibrium constant K for the reaction

$$2Cl(g) \rightleftharpoons Cl_2(g)$$

was measured as a function of temperature (Kelvin). A graph of $\ln(K)$ versus $1/T$ for this reaction gives a straight line with a slope of 1.352×10^4 K and a y-intercept of -14.51. Determine the values of $\Delta H°$ and $\Delta S°$ for this reaction. See Exercise 79.

Additional Exercises

81. Using Appendix 4 and the following data, determine $S°$ for $Fe(CO)_5(g)$.

$$Fe(s) + 5CO(g) \longrightarrow Fe(CO)_5(g) \qquad \Delta S° = ?$$
$$Fe(CO)_5(l) \longrightarrow Fe(CO)_5(g) \qquad \Delta S° = 107 \text{ J/K}$$
$$Fe(s) + 5CO(g) \longrightarrow Fe(CO)_5(l) \qquad \Delta S° = -677 \text{ J/K}$$

82. Some water is placed in a coffee-cup calorimeter. When 1.0 g of an ionic solid is added, the temperature of the solution increases from 21.5°C to 24.2°C as the solid dissolves. For the dissolving process, what are the signs for ΔS_{sys}, ΔS_{surr}, and ΔS_{univ}?

83. Consider the following system at equilibrium at 25°C:

$$PCl_3(g) + Cl_2(g) \rightleftharpoons PCl_5(g) \qquad \Delta G° = -92.50 \text{ kJ}$$

What will happen to the ratio of partial pressure of PCl_5 to partial pressure of PCl_3 if the temperature is raised? Explain completely.

84. Calculate the entropy change for the vaporization of liquid methane and liquid hexane using the following data.

	Boiling Point (1 atm)	ΔH_{vap}
Methane	112 K	8.20 kJ/mol
Hexane	342 K	28.9 kJ/mol

Compare the molar volume of gaseous methane at 112 K with that of gaseous hexane at 342 K. How do the differences in molar volume affect the values of ΔS_{vap} for these liquids?

85. As $O_2(l)$ is cooled at 1 atm, it freezes at 54.5 K to form solid I. At a lower temperature, solid I rearranges to solid II, which has a different crystal structure. Thermal measurements show that ΔH for the I $\rightarrow$ II phase transition is -743.1 J/mol, and ΔS for the same transition is -17.0 J/K · mol. At what temperature are solids I and II in equilibrium?

86. Consider the following reaction:

$$H_2O(g) + Cl_2O(g) \rightleftharpoons 2HOCl(g) \qquad K_{298} = 0.090$$

For $Cl_2O(g)$,

$$\Delta G_f° = 97.9 \text{ kJ/mol}$$
$$\Delta H_f° = 80.3 \text{ kJ/mol}$$
$$S° = 266.1 \text{ J/K} \cdot \text{mol}$$

a. Calculate $\Delta G°$ for the reaction using the equation $\Delta G° = -RT \ln(K)$.

b. Use bond energy values (Table 3-3) to estimate $\Delta H°$ for the reaction.

c. Use the results from parts a and b to estimate $\Delta S°$ for the reaction.

d. Estimate $\Delta H_f°$ and $S°$ for $HOCl(g)$.

e. Estimate the value of K at 500. K.

f. Calculate ΔG at 25°C when $P_{H_2O} = 18$ torr, $P_{Cl_2O} = 2.0$ torr, and $P_{HOCl} = 0.10$ torr.

87. Using the following data, calculate the value of K_{sp} for $Ba(NO_3)_2$, one of the least soluble of the common nitrate salts.

Species	$\Delta G_f°$
$Ba^{2+}(aq)$	-561 kJ/mol
$NO_3^-(aq)$	-109 kJ/mol
$Ba(NO_3)_2(s)$	-797 kJ/mol

88. Many biochemical reactions that occur in cells require relatively high concentrations of potassium ion (K^+). The concentration of K^+ in muscle cells is about 0.15 M. The concentration of K^+ in blood plasma is about 0.0050 M. The high internal concentration in cells is maintained by pumping K^+ from the plasma. How much work must be done to transport 1.0 mole of K^+ from the blood to the inside of a muscle cell at 37°C, normal body temperature? When 1.0 mole of K^+ is transferred from blood to the cells, do any other ions have to be transported? Why or why not?

89. Carbon monoxide is toxic because it bonds much more strongly to the iron in hemoglobin (Hgb) than does O_2. Consider the following reactions and approximate standard free energy changes:

$$Hgb + O_2 \longrightarrow HgbO_2 \qquad \Delta G° = -70 \text{ kJ}$$
$$Hgb + CO \longrightarrow HgbCO \qquad \Delta G° = -80 \text{ kJ}$$

Using these data, estimate the equilibrium constant value at 25°C for the following reaction:

$$HgbO_2 + CO \rightleftharpoons HgbCO + O_2$$

90. In the text, the equation

$$\Delta G = \Delta G° + RT \ln(Q)$$

was derived for gaseous reactions where the quantities in Q were expressed in units of pressure. We also can use units of mol/L for the quantities in Q, specifically for aqueous reactions. With this in mind, consider the reaction

$$HF(aq) \rightleftharpoons H^+(aq) + F^-(aq)$$

for which $K_a = 7.2 \times 10^{-4}$ at 25°C. Calculate ΔG for the reaction under the following conditions at 25°C.

a. $[HF] = [H^+] = [F^-] = 1.0\ M$
b. $[HF] = 0.98\ M, [H^+] = [F^-] = 2.7 \times 10^{-2}\ M$
c. $[HF] = [H^+] = [F^-] = 1.0 \times 10^{-5}\ M$
d. $[HF] = [F^-] = 0.27\ M, [H^+] = 7.2 \times 10^{-4}\ M$
e. $[HF] = 0.52\ M, [F^-] = 0.67\ M, [H^+] = 1.0 \times 10^{-3}\ M$

Based on the calculated ΔG values, in what direction will the reaction shift to reach equilibrium for each of the five sets of conditions?

91. Consider the reactions

$$Ni^{2+}(aq) + 6NH_3(aq) \longrightarrow Ni(NH_3)_6{}^{2+}(aq) \quad (1)$$
$$Ni^{2+}(aq) + 3en(aq) \longrightarrow Ni(en)_3{}^{2+}(aq) \quad (2)$$

where

$$en = H_2N-CH_2-CH_2-NH_2$$

The ΔH values for the two reactions are quite similar, yet $K_{\text{reaction 2}} > K_{\text{reaction 1}}$. Explain.

92. Use the equation in Exercise 79 to determine $\Delta H°$ and $\Delta S°$ for the autoionization of water:

$$H_2O(l) \rightleftharpoons H^+(aq) + OH^-(aq)$$

$T(°C)$	K_w
0	1.14×10^{-15}
25	1.00×10^{-14}
35	2.09×10^{-14}
40.	2.92×10^{-14}
50.	5.47×10^{-14}

93. Consider the reaction

$$Fe_2O_3(s) + 3H_2(g) \longrightarrow 2Fe(s) + 3H_2O(g)$$

Assuming $\Delta H°$ and $\Delta S°$ do not depend on temperature, calculate the temperature where $K = 1.00$ for this reaction.

94. Consider the following diagram of free energy (G) versus fraction of A reacted in terms of moles for the reaction $2A(g) \rightarrow B(g)$.

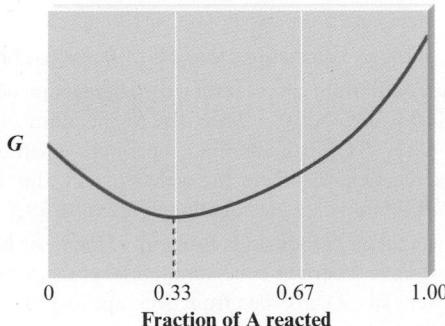

Fraction of A reacted

Before any A has reacted, $P_A = 3.0$ atm and $P_B = 0$. Determine the sign of $\Delta G°$ and the value of K_p for this reaction.

ChemWork Problems

These multiconcept problems (and additional ones) are found interactively online with the same type of assistance a student would get from an instructor.

95. Which of the following reactions (or processes) are expected to have a negative value for $\Delta S°$?
    a. $SiF_6(aq) + H_2(g) \longrightarrow 2HF(g) + SiF_4(g)$
    b. $4Al(s) + 3O_2(g) \longrightarrow 2Al_2O_3(s)$
    c. $CO(g) + Cl_2(g) \longrightarrow COCl_2(g)$
    d. $C_2H_4(g) + H_2O(l) \longrightarrow C_2H_5OH(l)$
    e. $H_2O(s) \longrightarrow H_2O(l)$

96. For rubidium $\Delta H°_{vap} = 69.0$ kJ/mol at 686°C, its boiling point. Calculate $\Delta S°$, q, w, and ΔE for the vaporization of 1.00 mole of rubidium at 686°C and 1.00 atm pressure.

97. Given the thermodynamic data below, calculate ΔS and ΔS_{surr} for the following reaction at 25°C and 1 atm:

$$XeF_6(g) \longrightarrow XeF_4(s) + F_2(g)$$

	$\Delta H°_f$ (kJ/mol)	$S°$ (J/K · mol)
$XeF_6(g)$	-294	300.
$XeF_4(s)$	-251	146
$F_2(g)$	0	203

98. Consider the reaction:

$$H_2S(g) + SO_2(g) \longrightarrow 3S(g) + 2H_2O(l)$$

for which ΔH is -233 kJ and ΔS is -424 J/K.
    a. Calculate the free energy change for the reaction (ΔG) at 393 K.
    b. Assuming ΔH and ΔS do not depend on temperature, at what temperatures is this reaction spontaneous?

99. The following reaction occurs in pure water:

$$H_2O(l) + H_2O(l) \longrightarrow H_3O^+(aq) + OH^-(aq)$$

which is often abbreviated as

$$H_2O(l) \longrightarrow H^+(aq) + OH^-(aq)$$

For this reaction, $\Delta G° = 79.9$ kJ/mol at 25°C. Calculate the value of ΔG for this reaction at 25°C when $[OH^-] = 0.15\ M$ and $[H^+] = 0.71\ M$.

100. Consider the dissociation of a weak acid HA ($K_a = 4.5 \times 10^{-3}$) in water:

$$HA(aq) \rightleftharpoons H^+(aq) + A^-(aq)$$

Calculate $\Delta G°$ for this reaction at 25°C.

101. Consider the reaction:

$$PCl_3(g) + Cl_2(g) \rightleftharpoons PCl_5(g)$$

At 25°C, $\Delta H° = -92.50$ kJ.
Which of the following statements is(are) *true*?
    a. This is an endothermic reaction.
    b. $\Delta S°$ for this reaction is negative.

c. If the temperature is increased, the ratio $\dfrac{PCl_5}{PCl_3}$ will increase.

d. $\Delta G°$ for this reaction has to be negative at all temperatures.

e. When $\Delta G°$ for this reaction is negative, then K_p is greater than 1.00.

102. The equilibrium constant for a certain reaction increases by a factor of 6.67 when the temperature is increased from 300.0 K to 350.0 K. Calculate the standard change in enthalpy ($\Delta H°$) for this reaction (assuming $\Delta H°$ is temperature-independent).

Challenge Problems

103. For the following reactions at constant pressure, predict if $\Delta H > \Delta E$, $\Delta H < \Delta E$, or $\Delta H = \Delta E$.

a. $2HF(g) \rightarrow H_2(g) + F_2(g)$

b. $N_2(g) + 3H_2(g) \rightarrow 2NH_3(g)$

c. $4NH_3(g) + 5O_2(g) \rightarrow 4NO(g) + 6H_2O(g)$

104. The standard enthalpy of formation of $H_2O(l)$ at 298 K is -285.8 kJ/mol. Calculate the change in internal energy for the following process at 298 K and 1 atm:

$$H_2O(l) \rightarrow H_2(g) + \tfrac{1}{2}O_2(g) \qquad \Delta E° = ?$$

(*Hint:* Using the ideal gas equation, derive an expression for work in terms of n, R, and T.)

105. Consider two perfectly insulated vessels. Vessel 1 initially contains an ice cube at 0°C and water at 0°C. Vessel 2 initially contains an ice cube at 0°C and a saltwater solution at 0°C. Consider the process $H_2O(s) \rightarrow H_2O(l)$.

a. Determine the sign of ΔS, ΔS_{surr}, and ΔS_{univ} for the process in vessel 1.

b. Determine the sign of ΔS, ΔS_{surr}, and ΔS_{univ} for the process in vessel 2.

(*Hint:* Think about the effect that a salt has on the freezing point of a solvent.)

106. Liquid water at 25°C is introduced into an evacuated, insulated vessel. Identify the signs of the following thermodynamic functions for the process that occurs: ΔH, ΔS, ΔT_{water}, ΔS_{surr}, ΔS_{univ}.

107. Using data from Appendix 4, calculate $\Delta H°$, $\Delta G°$, and K (at 298 K) for the production of ozone from oxygen:

$$3O_2(g) \rightleftharpoons 2O_3(g)$$

At 30 km above the surface of the earth, the temperature is about 230. K and the partial pressure of oxygen is about 1.0×10^{-3} atm. Estimate the partial pressure of ozone in equilibrium with oxygen at 30 km above the earth's surface. Is it reasonable to assume that the equilibrium between oxygen and ozone is maintained under these conditions? Explain.

108. Entropy can be calculated by a relationship proposed by Ludwig Boltzmann:

$$S = k \ln(W)$$

where $k = 1.38 \times 10^{-23}$ J/K and W is the number of ways a particular state can be obtained. (This equation is engraved on Boltzmann's tombstone.) Calculate S for the five arrangements of particles in Table 16-1.

109. a. Using the free energy profile for a simple one-step reaction, show that at equilibrium $K = k_f/k_r$, where k_f and k_r are the rate constants for the forward and reverse reactions. *Hint:* Use the relationship $\Delta G° = -RT \ln(K)$ and represent k_f and k_r using the Arrhenius equation ($k = Ae^{-E_a/RT}$).

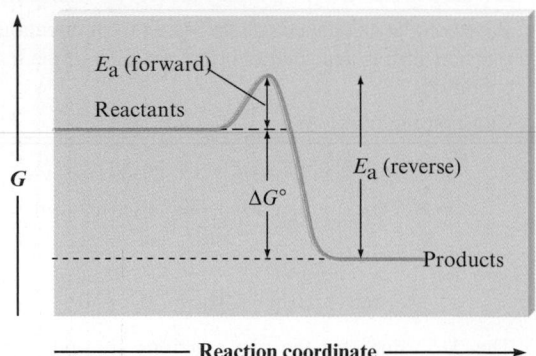

b. Why is the following statement false? "A catalyst can increase the rate of a forward reaction but not the rate of the reverse reaction."

110. Consider the reaction

$$H_2(g) + Br_2(g) \rightleftharpoons 2HBr(g)$$

where $\Delta H° = -103.8$ kJ/mol. In a particular experiment, equal moles of $H_2(g)$ at 1.00 atm and $Br_2(g)$ at 1.00 atm were mixed in a 1.00-L flask at 25°C and allowed to reach equilibrium. Then the molecules of H_2 at equilibrium were counted using a very sensitive technique, and 1.10×10^{13} molecules were found. For this reaction, calculate the values of K, $\Delta G°$, and $\Delta S°$.

111. Consider the system

$$A(g) \longrightarrow B(g)$$

at 25°C.

a. Assuming that $G_A° = 8996$ J/mol and $G_B° = 11{,}718$ J/mol, calculate the value of the equilibrium constant for this reaction.

b. Calculate the equilibrium pressures that result if 1.00 mole of $A(g)$ at 1.00 atm and 1.00 mole of $B(g)$ at 1.00 atm are mixed at 25°C.

c. Show by calculations that $\Delta G = 0$ at equilibrium.

112. The equilibrium constant for a certain reaction decreases from 8.84 to 3.25×10^{-2} when the temperature increases from 25°C to 75°C. Estimate the temperature where $K = 1.00$ for this reaction. Estimate the value of $\Delta S°$ for this reaction. (*Hint:* Manipulate the equation in Exercise 79.)

113. If wet silver carbonate is dried in a stream of hot air, the air must have a certain concentration level of carbon dioxide to prevent silver carbonate from decomposing by the reaction

$$Ag_2CO_3(s) \rightleftharpoons Ag_2O(s) + CO_2(g)$$

$\Delta H°$ for this reaction is 79.14 kJ/mol in the temperature range of 25 to 125°C. Given that the partial pressure of carbon dioxide in equilibrium with pure solid silver carbonate is 6.23×10^{-3} torr at 25°C, calculate the partial pressure of CO_2 necessary to prevent decomposition of Ag_2CO_3 at 110.°C. (*Hint:* Manipulate the equation in Exercise 79.)

114. Carbon tetrachloride (CCl_4) and benzene (C_6H_6) form ideal solutions. Consider an equimolar solution of CCl_4 and C_6H_6 at 25°C. The vapor above the solution is collected and condensed. Using the following data, determine the composition in mole fraction of the condensed vapor.

Substance	ΔG_f°
$C_6H_6(l)$	124.50 kJ/mol
$C_6H_6(g)$	129.66 kJ/mol
$CCl_4(l)$	−65.21 kJ/mol
$CCl_4(g)$	−60.59 kJ/mol

115. Sodium chloride is added to water (at 25°C) until it is saturated. Calculate the Cl^- concentration in such a solution.

Species	ΔG° (kJ/mol)
$NaCl(s)$	−384
$Na^+(aq)$	−262
$Cl^-(aq)$	−131

116. You have a 1.00-L sample of hot water (90.0°C) sitting open in a 25.0°C room. Eventually the water cools to 25.0°C while the temperature of the room remains unchanged. Calculate ΔS_{surr} for this process. Assume the density of water is 1.00 g/cm³ over this temperature range, and the heat capacity of water is constant over this temperature range and equal to 75.4 J/K · mol.

117. Consider a weak acid, HX. If a 0.10-M solution of HX has a pH of 5.83 at 25°C, what is ΔG° for the acid's dissociation reaction at 25°C?

Integrative Problems

These problems require the integration of multiple concepts to find the solutions.

118. Some nonelectrolyte solute (molar mass = 142 g/mol) was dissolved in 150. mL of a solvent (density = 0.879 g/cm³). The elevated boiling point of the solution was 355.4 K. What mass of solute was dissolved in the solvent? For the solvent, the enthalpy of vaporization is 33.90 kJ/mol, the entropy of vaporization is 95.95 J/K · mol, and the boiling-point elevation constant is 2.5 K · kg/mol.

119. For the equilibrium

$$A(g) + 2B(g) \rightleftharpoons C(g)$$

the initial concentrations are [A] = [B] = [C] = 0.100 atm. Once equilibrium has been established, it is found that [C] = 0.040 atm. What is ΔG° for this reaction at 25°C?

120. What is the pH of a 0.125-M solution of the weak base B if $\Delta H^\circ = -28.0$ kJ and $\Delta S^\circ = -175$ J/K for the following equilibrium reaction at 25°C?

$$B(aq) + H_2O(l) \rightleftharpoons BH^+(aq) + OH^-(aq)$$

Marathon Problem

This problem is designed to incorporate several concepts and techniques into one situation.

121. Consider a sample containing 5.00 moles of a monatomic ideal gas that is taken from state A to state B by the following two pathways:

Pathway one: $P_A = 3.00$ atm $\xrightarrow{1}$ $P_C = 3.00$ atm

$V_A = 15.0$ L $V_C = 55.0$ L

$P_B = 6.00$ atm
$\xrightarrow{2}$
$V_B = 20.0$ L

Pathway two: $P_A = 3.00$ atm $\xrightarrow{3}$ $P_D = 6.00$ atm

$V_A = 15.0$ L $V_D = 15.0$ L

$P_B = 6.00$ atm
$\xrightarrow{4}$
$V_B = 20.0$ L

For each step, assume that the external pressure is constant and equals the final pressure of the gas for that step. Calculate q, w, ΔE, and ΔH for each step in kJ, and calculate overall values for each pathway. Explain how the overall values for the two pathways illustrate that ΔE and ΔH are state functions, whereas q and w are path functions. *Hint:* In a more rigorous study of thermochemistry, it can be shown that for an ideal gas:

$$\Delta E = nC_v\Delta T \text{ and}$$
$$\Delta H = nC_p\Delta T$$

where C_v is the molar heat capacity at constant volume and C_p is the molar heat capacity at constant pressure. In addition, for a monatomic ideal gas, $C_v = \frac{3}{2}R$ and $C_p = \frac{5}{2}R$, where $R = 8.3145$ J/K · mol.

122. Impure nickel, refined by smelting sulfide ores in a blast furnace, can be converted into metal from 99.90% to 99.99% purity by the Mond process. The primary reaction involved in the Mond process is

$$Ni(s) + 4CO(g) \rightleftharpoons Ni(CO)_4(g)$$

a. Without referring to Appendix 4, predict the sign of ΔS° for the above reaction. Explain.

b. The spontaneity of the above reaction is temperature-dependent. Predict the sign of ΔS_{surr} for this reaction. Explain.

c. For $Ni(CO)_4(g)$, $\Delta H_f^\circ = -607$ kJ/mol and $S^\circ = 417$ J/K · mol at 298 K. Using these values and data in Appendix 4, calculate ΔH° and ΔS° for the above reaction.

d. Calculate the temperature at which $\Delta G^\circ = 0$ ($K = 1$) for the above reaction, assuming that ΔH° and ΔS° do not depend on temperature.

e. The first step of the Mond process involves equilibrating impure nickel with $CO(g)$ and $Ni(CO)_4(g)$ at about 50°C. The purpose of this step is to convert as much nickel as possible into the gas phase. Calculate the equilibrium constant for the above reaction at 50.°C.

f. In the second step of the Mond process, the gaseous $Ni(CO)_4$ is isolated and heated to 227°C. The purpose of this step is to deposit as much nickel as possible as pure solid (the reverse of the preceding reaction). Calculate the equilibrium constant for the preceding reaction at 227°C.

g. Why is temperature increased for the second step of the Mond process?

h. The Mond process relies on the volatility of $Ni(CO)_4$ for its success. Only pressures and temperatures at which $Ni(CO)_4$ is a gas are useful. A recently developed variation of the Mond process carries out the first step at higher pressures and a temperature of 152°C. Estimate the maximum pressure of $Ni(CO)_4(g)$ that can be attained before the gas will liquefy at 152°C. The boiling point for $Ni(CO)_4$ is 42°C and the enthalpy of vaporization is 29.0 kJ/mol.

[*Hint:* The phase change reaction and the corresponding equilibrium expression are

$$Ni(CO)_4(l) \rightleftharpoons Ni(CO)_4(g) \qquad K = P_{Ni(CO)_4}$$

$Ni(CO)_4(g)$ will liquefy when the pressure of $Ni(CO)_4$ is greater than the K value.]

C H A P T E R   S E V E N T E E N

Electrochemistry

The Hiriko (meaning urban in Basque) is an electric car that folds up to park in a 1.5-meter space. The car was designed at MIT and is produced in Spain by a consortium of small businesses. (© Kika/ZumaPress/Newscom.com)

Electrochemistry constitutes one of the most important interfaces between chemistry and everyday life. Every time you start your car, turn on your calculator, use your smartphone, or listen to a radio at the beach, you are depending on electrochemical reactions. Our society sometimes seems to run almost entirely on batteries. Certainly the advent of small, dependable batteries along with silicon-chip technology has made possible the tiny calculators, portable audio players, and cell phones that we take for granted.

Electrochemistry is important in other less obvious ways. For example, the corrosion of iron, which has tremendous economic implications, is an electrochemical process. In addition, many important industrial materials such as aluminum, chlorine, and sodium hydroxide are prepared by electrolytic processes. In analytical chemistry, electrochemical techniques use electrodes that are specific for a given molecule or ion, such as H^+ (pH meters), F^-, Cl^-, and many others. These increasingly important methods are used to analyze for trace pollutants in natural waters or for the tiny quantities of chemicals in human blood that may signal the development of a specific disease.

Electrochemistry is best defined as *the study of the interchange of chemical and electrical energy*. It is primarily concerned with two processes that involve oxidation–reduction reactions: the generation of an electric current from a spontaneous chemical reaction and, the opposite process, the use of a current to produce chemical change.

17-1 | Balancing Oxidation–Reduction Equations

Oxidation–reduction reactions in aqueous solutions are often complicated, which means that it can be difficult to balance their equations by simple inspection. In this section we will discuss a special technique for balancing the equations of redox reactions that occur in aqueous solutions. It is called the *half-reaction method*.

The Half-Reaction Method for Balancing Oxidation–Reduction Reactions in Aqueous Solutions

For oxidation–reduction reactions that occur in aqueous solution, it is useful to separate the reaction into two **half-reactions**: one involving oxidation and the other involving reduction. For example, consider the unbalanced equation for the oxidation–reduction reaction between cerium(IV) ion and tin(II) ion:

$$Ce^{4+}(aq) + Sn^{2+}(aq) \longrightarrow Ce^{3+}(aq) + Sn^{4+}(aq)$$

This reaction can be separated into a half-reaction involving the substance being *reduced,*

$$Ce^{4+}(aq) \longrightarrow Ce^{3+}(aq)$$

and one involving the substance being *oxidized,*

$$Sn^{2+}(aq) \longrightarrow Sn^{4+}(aq)$$

The general procedure is to balance the equations for the half-reactions separately and then to add them to obtain the overall balanced equation. The half-reaction method for balancing oxidation–reduction equations differs slightly depending on whether the reaction takes place in acidic or basic solution.

PROBLEM-SOLVING STRATEGY

The Half-Reaction Method for Balancing Equations for Oxidation–Reduction Reactions Occurring in Acidic Solution

1. Write separate equations for the oxidation and reduction half-reactions.
2. For each half-reaction,
   a. balance all the elements except hydrogen and oxygen.
   b. balance oxygen using H_2O.
   c. balance hydrogen using H^+.
   d. balance the charge using electrons.
3. If necessary, multiply one or both balanced half-reactions by an integer to equalize the number of electrons transferred in the two half-reactions.
4. Add the half-reactions, and cancel identical species.
5. Check that the elements and charges are balanced.

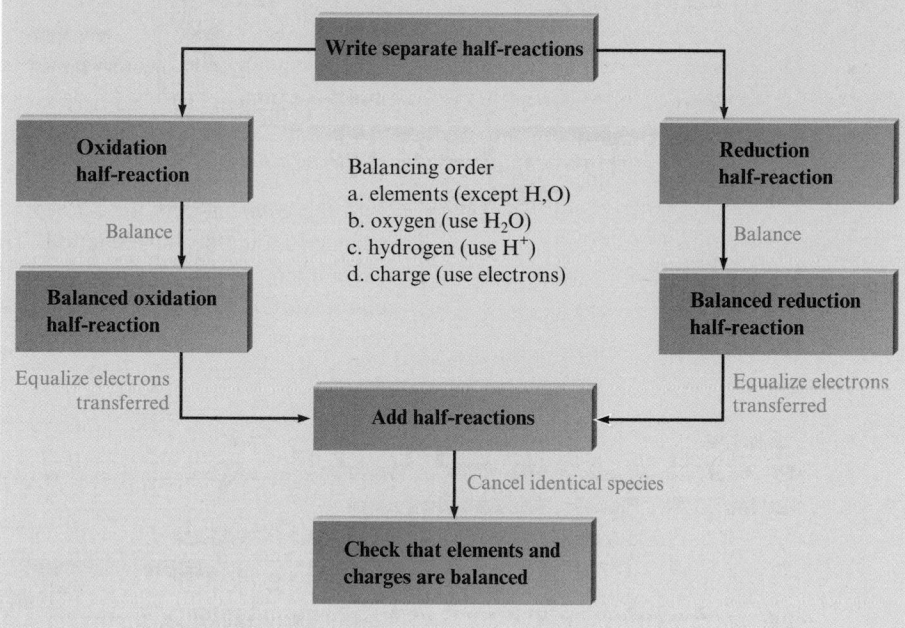

We will illustrate this method by balancing the equation for the reaction between permanganate and iron(II) ions in acidic solution:

$$MnO_4^-(aq) + Fe^{2+}(aq) \xrightarrow{\text{Acid}} Fe^{3+}(aq) + Mn^{2+}(aq)$$

This reaction can be used to analyze iron ore for its iron content.

1. *Identify and write equations for the half-reactions.* The oxidation states for the half-reaction involving the permanganate ion show that manganese is reduced:

$$MnO_4^- \longrightarrow Mn^{2+}$$
$$+7 \quad -2 \text{ (each O)} \quad +2$$

This is the *reduction half-reaction.* The other half-reaction involves the oxidation of iron(II) to iron(III) ion and is the *oxidation half-reaction:*

$$Fe^{2+} \longrightarrow Fe^{3+}$$
$$+2 \qquad\qquad +3$$

2. *Balance each half-reaction.* For the reduction reaction, we have

$$MnO_4^-(aq) \longrightarrow Mn^{2+}(aq)$$

a. The manganese is balanced.
b. We balance oxygen by adding $4H_2O$ to the right side of the equation:

$$MnO_4^-(aq) \longrightarrow Mn^{2+}(aq) + 4H_2O(l)$$

c. Next, we balance hydrogen by adding $8H^+$ to the left side:

$$8H^+(aq) + MnO_4^-(aq) \longrightarrow Mn^{2+}(aq) + 4H_2O(l)$$

d. All the elements have been balanced, but we need to balance the charge using electrons. At this point we have the following overall charges for reactants and products in the reduction half-reaction:

$$\underbrace{8H^+(aq) + MnO_4^-(aq)}_{\underbrace{8+ \quad + \quad 1-}_{7+}} \longrightarrow \underbrace{Mn^{2+}(aq) + 4H_2O(l)}_{\underbrace{2+ \quad + \quad 0}_{2+}}$$

We can equalize the charges by adding five electrons to the left side:

$$\underbrace{5e^- + 8H^+(aq) + MnO_4^-(aq)}_{2+} \longrightarrow \underbrace{Mn^{2+}(aq) + 4H_2O(l)}_{2+}$$

Both the *elements* and the *charges* are now balanced, so this represents the balanced reduction half-reaction. The fact that five electrons appear on the reactant side of the equation makes sense, since five electrons are required to reduce MnO_4^- (Mn has an oxidation state of $+7$) to Mn^{2+} (Mn has an oxidation state of $+2$).

For the oxidation reaction

$$Fe^{2+}(aq) \longrightarrow Fe^{3+}(aq)$$

the elements are balanced, and we must simply balance the charge:

$$\underbrace{Fe^{2+}(aq)}_{2+} \longrightarrow \underbrace{Fe^{3+}(aq)}_{3+}$$

One electron is needed on the right side to give a net $2+$ charge on both sides:

$$\underbrace{Fe^{2+}(aq)}_{2+} \longrightarrow \underbrace{Fe^{3+}(aq) + e^-}_{2+}$$

ⓘ The number of electrons gained in the reduction half-reaction must equal the number of electrons lost in the oxidation half-reaction.

3. *Equalize the electron transfer in the two half-reactions.* ◄ ⓘ Since the reduction half-reaction involves a transfer of five electrons and the oxidation half-reaction involves a transfer of only one electron, the oxidation half-reaction must be multiplied by 5:

$$5Fe^{2+}(aq) \longrightarrow 5Fe^{3+}(aq) + 5e^-$$

4. *Add the half-reactions.* The half-reactions are added to give

$$5e^- + 5Fe^{2+}(aq) + MnO_4^-(aq) + 8H^+(aq) \longrightarrow$$
$$5Fe^{3+}(aq) + Mn^{2+}(aq) + 4H_2O(l) + 5e^-$$

Note that the electrons cancel (as they must) to give the final balanced equation:

$$5Fe^{2+}(aq) + MnO_4^-(aq) + 8H^+(aq) \longrightarrow 5Fe^{3+}(aq) + Mn^{2+}(aq) + 4H_2O(l)$$

5. *Check that elements and charges are balanced.*

Elements balance: 5Fe, 1Mn, 4O, 8H $\longrightarrow$ 5Fe, 1Mn, 4O, 8H

Charges balance: $5(2+) + (1-) + 8(1+) = 17+ \longrightarrow$
$$5(3+) + (2+) + 0 = 17+$$

The equation is balanced.

INTERACTIVE EXAMPLE 17-1

Balancing Oxidation–Reduction Reactions (Acidic)

Potassium dichromate ($K_2Cr_2O_7$) is a bright orange compound that can be reduced to a blue-violet solution of Cr^{3+} ions. ◄ Under certain conditions, $K_2Cr_2O_7$ reacts with ethanol (C_2H_5OH) as follows:

$$H^+(aq) + Cr_2O_7{}^{2-}(aq) + C_2H_5OH(l) \longrightarrow Cr^{3+}(aq) + CO_2(g) + H_2O(l)$$

Balance this equation using the half-reaction method.

Solution

1. The reduction half-reaction is

$$Cr_2O_7{}^{2-}(aq) \longrightarrow Cr^{3+}(aq)$$

Chromium is reduced from an oxidation state of $+6$ in $Cr_2O_7{}^{2-}$ to one of $+3$ in Cr^{3+}.

The oxidation half-reaction is

$$C_2H_5OH(l) \longrightarrow CO_2(g)$$

Carbon is oxidized from an oxidation state of -2 in C_2H_5OH to $+4$ in CO_2.

2. Balancing all elements except hydrogen and oxygen in the first half-reaction, we have

$$Cr_2O_7{}^{2-}(aq) \longrightarrow 2Cr^{3+}(aq)$$

Balancing oxygen using H_2O, we have

$$Cr_2O_7{}^{2-}(aq) \longrightarrow 2Cr^{3+}(aq) + 7H_2O(l)$$

Balancing hydrogen using H^+, we have

$$14H^+(aq) + Cr_2O_7{}^{2-}(aq) \longrightarrow 2Cr^{3+}(aq) + 7H_2O(l)$$

Balancing the charge using electrons, we have

$$6e^- + 14H^+(aq) + Cr_2O_7{}^{2-}(aq) \longrightarrow 2Cr^{3+}(aq) + 7H_2O(l)$$

Next, we turn to the oxidation half-reaction

$$C_2H_5OH(l) \longrightarrow CO_2(g)$$

▲ When potassium dichromate reacts with ethanol, a blue-violet solution containing Cr^{3+} is formed.

© Cengage Learning

Balancing carbon, we have

$$C_2H_5OH(l) \longrightarrow 2CO_2(g)$$

Balancing oxygen using H_2O, we have

$$C_2H_5OH(l) + 3H_2O(l) \longrightarrow 2CO_2(g)$$

Balancing hydrogen using H^+, we have

$$C_2H_5OH(l) + 3H_2O(l) \longrightarrow 2CO_2(g) + 12H^+(aq)$$

We then balance the charge by adding $12e^-$ to the right side:

$$C_2H_5OH(l) + 3H_2O(l) \longrightarrow 2CO_2(g) + 12H^+(aq) + 12e^-$$

3. In the reduction half-reaction, there are 6 electrons on the left-hand side, and there are 12 electrons on the right-hand side of the oxidation half-reaction. Thus we multiply the reduction half-reaction by 2 to give

$$12e^- + 28H^+(aq) + 2Cr_2O_7{}^{2-}(aq) \longrightarrow 4Cr^{3+}(aq) + 14H_2O(l)$$

4. Adding the half-reactions and canceling identical species, we have

Reduction Half-Reaction: $12e^- + 28H^+(aq) + 2Cr_2O_7{}^{2-}(aq) \longrightarrow 4Cr^{3+}(aq) + 14H_2O(l)$

Oxidation Half-Reaction: $C_2H_5OH(l) + 3H_2O(l) \longrightarrow 2CO_2(g) + 12H^+(aq) + 12e^-$

Complete Reaction: $16H^+(aq) + 2Cr_2O_7{}^{2-}(aq) + C_2H_5OH(l) \longrightarrow 4Cr^{3+} + 11H_2O(l) + 2CO_2(g)$

5. Check that elements and charges are balanced.

Elements balance: $22H, 4Cr, 15O, 2C \longrightarrow 22H, 4Cr, 15O, 2C$

Charges balance: $+16 + 2(-2) + 0 = +12 \longrightarrow 4(+3) + 0 + 0 = +12$

See Exercises 17-29 and 17-30

Oxidation–reduction reactions can occur in basic solutions (the reactions involve OH^- ions) as well as in acidic solutions (the reactions involve H^+ ions). The half-reaction method for balancing equations is slightly different for the two cases.

PROBLEM-SOLVING STRATEGY

The Half-Reaction Method for Balancing Equations for Oxidation–Reduction Reactions Occurring in Basic Solution

1. Use the half-reaction method as specified for acidic solutions to obtain the final balanced equation *as if H^+ ions were present.*

2. To both sides of the equation obtained above, add a number of OH^- ions that is equal to the number of H^+ ions. (We want to eliminate H^+ by forming H_2O.)

3. Form H_2O on the side containing both H^+ and OH^- ions, and eliminate the number of H_2O molecules that appear on both sides of the equation.

4. Check that elements and charges are balanced.

(continued)

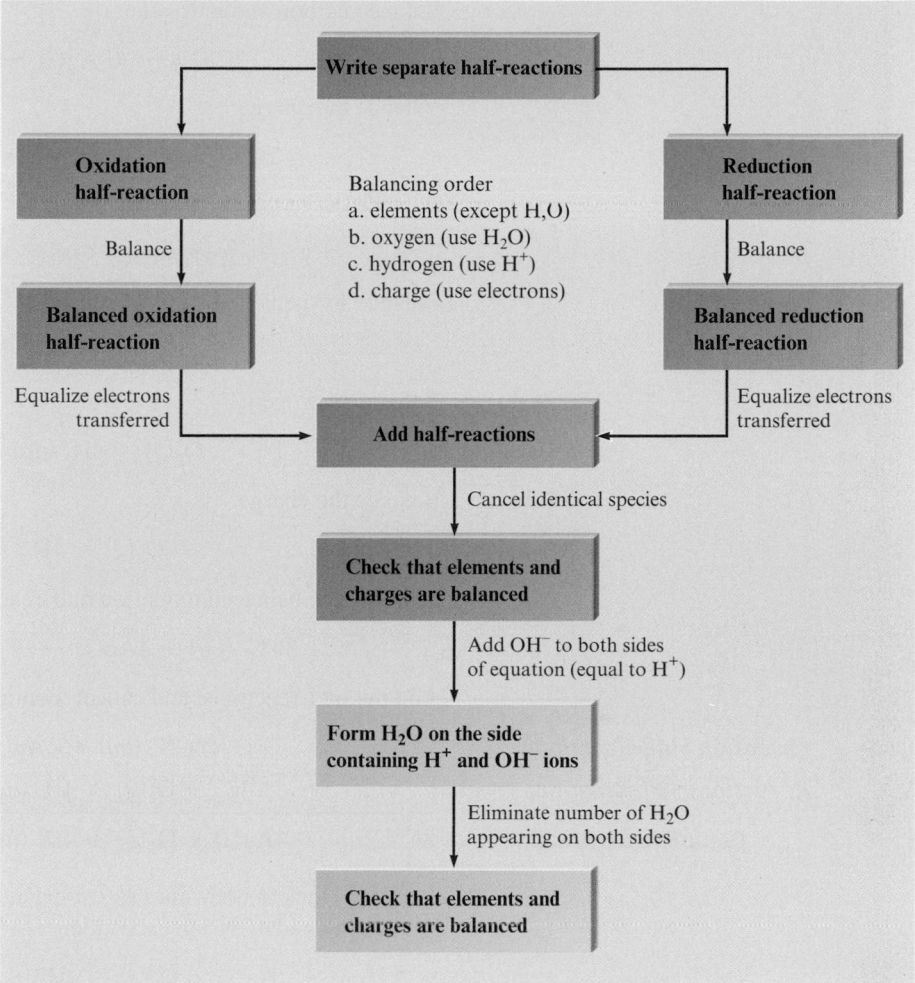

Critical Thinking

When balancing redox reactions occurring in basic solutions, the text instructs you to first use the half-reaction method as specified for acidic solutions. What if you started by adding OH^- first instead of H^+? What potential problem could there be with this approach?

INTERACTIVE EXAMPLE 17-2

Balancing Oxidation–Reduction Reactions (Basic)

Silver is sometimes found in nature as large nuggets; more often it is found mixed with other metals and their ores. An aqueous solution containing cyanide ion is often used to extract the silver using the following reaction that occurs in basic solution:

$$Ag(s) + CN^-(aq) + O_2(g) \xrightarrow{\text{Base}} Ag(CN)_2^-(aq)$$

Balance this equation using the half-reaction method.

Solution

1. Balance the equation as if H^+ ions were present. Balance the oxidation half-reaction:

$$CN^-(aq) + Ag(s) \longrightarrow Ag(CN)_2^-(aq)$$

Balance carbon and nitrogen:

$$2CN^-(aq) + Ag(s) \longrightarrow Ag(CN)_2^-(aq)$$

Balance the charge:

$$2CN^-(aq) + Ag(s) \longrightarrow Ag(CN)_2^-(aq) + e^-$$

Balance the reduction half-reaction:

$$O_2(g) \longrightarrow$$

Balance oxygen:

$$O_2(g) \longrightarrow 2H_2O(l)$$

Balance hydrogen:

$$O_2(g) + 4H^+(aq) \longrightarrow 2H_2O(l)$$

Balance the charge:

$$4e^- + O_2(g) + 4H^+(aq) \longrightarrow 2H_2O(l)$$

Multiply the balanced oxidation half-reaction by 4:

$$8CN^-(aq) + 4Ag(s) \longrightarrow 4Ag(CN)_2^-(aq) + 4e^-$$

Add the half-reactions, and cancel identical species:

Oxidation Half-Reaction: $\quad 8CN^-(aq) + 4Ag(s) \longrightarrow 4Ag(CN)_2^-(aq) + 4e^-$

Reduction Half-Reaction: $\quad 4e^- + O_2(g) + 4H^+(aq) \longrightarrow 2H_2O(l)$

Complete Reaction: $\quad 8CN^-(aq) + 4Ag(s) + O_2(g) + 4H^+(aq) \longrightarrow 4Ag(CN)_2^-(aq) + 2H_2O(l)$

2. Add OH^- ions to both sides of the balanced equation to eliminate the H^+ ions. We need to add $4OH^-$ to each side:

$$8CN^-(aq) + 4Ag(s) + O_2(g) + \underbrace{4H^+(aq) + 4OH^-(aq)}_{4H_2O(l)} \longrightarrow$$

$$4Ag(CN)_2^-(aq) + 2H_2O(l) + 4OH^-(aq)$$

3. Eliminate as many H_2O molecules as possible:

$$8CN^-(aq) + 4Ag(s) + O_2(g) + 2H_2O(l) \longrightarrow 4Ag(CN)_2^-(aq) + 4OH^-(aq)$$

4. Check that elements and charges are balanced.

Elements balance: $\quad$ 8C, 8N, 4Ag, 4O, 4H $\longrightarrow$ 8C, 8N, 4Ag, 4O, 4H

Charges balance: $8(1-) + 0 + 0 + 0 = 8- \longrightarrow 4(1-) + 4(1-) = 8-$

See Exercises 17-31 and 17-32

17-2 | Galvanic Cells

To understand how a redox reaction can be used to generate a current, let's consider the reaction between MnO_4^- and Fe^{2+}:

$$8H^+(aq) + MnO_4^-(aq) + 5Fe^{2+}(aq) \longrightarrow Mn^{2+}(aq) + 5Fe^{3+}(aq) + 4H_2O(l)$$

In this reaction, Fe^{2+} is oxidized and MnO_4^- is reduced; electrons are transferred from Fe^{2+} (the reducing agent) to MnO_4^- (the oxidizing agent).

Figure 17-1 | Schematic of a method to separate the oxidizing and reducing agents of a redox reaction. (The solutions also contain counterions to balance the charge.)

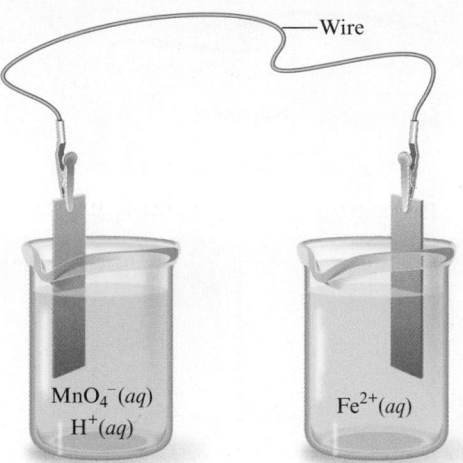

It is useful to break a redox reaction into half-reactions, one involving oxidation and one involving reduction. For the previous reaction, the half-reactions are

$$\text{Reduction:} \quad 8H^+ + MnO_4^- + 5e^- \longrightarrow Mn^{2+} + 4H_2O$$
$$\text{Oxidation:} \quad 5(Fe^{2+} \longrightarrow Fe^{3+} + e^-)$$

The multiplication of the second half-reaction by 5 indicates that this reaction must occur five times for each time the first reaction occurs. The balanced overall reaction is the sum of the half-reactions.

When MnO_4^- and Fe^{2+} are present in the same solution, the electrons are transferred directly when the reactants collide. Under these conditions, no useful work is obtained from the chemical energy involved in the reaction, which instead is released as heat. How can we harness this energy? The key is to separate the oxidizing agent from the reducing agent, thus requiring the electron transfer to occur through a wire. The current produced in the wire by the electron flow can then be directed through a device, such as an electric motor, to provide useful work.

For example, consider the system illustrated in Fig. 17-1. If our reasoning has been correct, electrons should flow through the wire from Fe^{2+} to MnO_4^-. However, when we construct the apparatus as shown, no flow of electrons is apparent. Why? Careful observation shows that when we connect the wires from the two compartments, current flows for an instant and then ceases. The current stops flowing because of charge buildups in the two compartments. If electrons flowed from the right to the left compartment in the apparatus as shown, the left compartment (receiving electrons) would become negatively charged, and the right compartment (losing electrons) would become positively charged. Creating a charge separation of this type requires a large amount of energy. Thus sustained electron flow cannot occur under these conditions.

However, we can solve this problem very simply. The solutions must be connected so that ions can flow to keep the net charge in each compartment zero. This connection might involve a **salt bridge** (a U-tube filled with an electrolyte) or a **porous disk** in a tube connecting the two solutions (Fig. 17-2). Either of these devices allows ions to flow without extensive mixing of the solutions. When we make the provision for ion flow, the circuit is complete. Electrons flow through the wire from reducing agent to oxidizing agent, and ions flow from one compartment to the other to keep the net charge zero.

ⓘ A galvanic cell uses a spontaneous redox reaction to produce a current that can be used to do work.

We now have covered all the essential characteristics of a **galvanic cell**, *a device in which chemical energy is changed to electrical energy.* ◀ ⓘ (The opposite process is called *electrolysis* and will be considered in Section 17-8.)

The reaction in an electrochemical cell occurs at the interface between the electrode and the solution where the electron transfer occurs. The electrode compartment in

Figure 17-2 | Galvanic cells can contain a salt bridge as in (a) or a porous-disk connection as in (b). A salt bridge contains a strong electrolyte held in a Jell-O–like matrix. A porous disk contains tiny passages that allow hindered flow of ions.

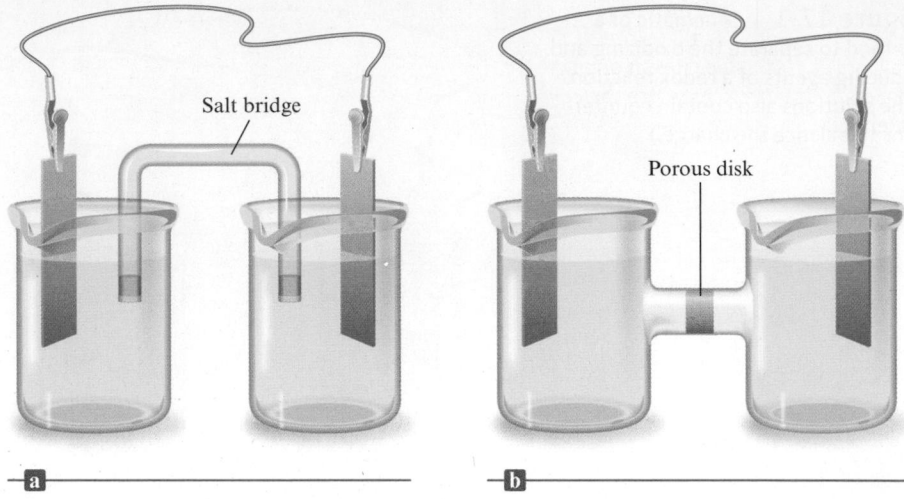

i Oxidation occurs at the anode. Reduction occurs at the cathode.

which *oxidation* occurs is called the **anode**; the electrode compartment in which *reduction* occurs is called the **cathode** (Fig. 17-3). ◄ *i*

Cell Potential

A galvanic cell consists of an oxidizing agent in one compartment that pulls electrons through a wire from a reducing agent in the other compartment. The "pull," or driving force, on the electrons is called the **cell potential** ($\mathscr{E}_{cell}$), or the **electromotive force** (emf) of the cell. The unit of electrical potential is the **volt** (abbreviated V), which is defined as 1 joule of work per coulomb of charge transferred. ◄ *i*

i A volt is 1 joule of work per coulomb of charge transferred: $1 \text{ V} = 1 \text{ J/C}$.

How can we measure the cell potential? One possible instrument is a crude **voltmeter**, which works by drawing current through a known resistance. However, when current flows through a wire, the frictional heating that occurs wastes some of the potentially useful energy of the cell. A traditional voltmeter will therefore measure a potential that is less than the maximum cell potential. The key to determining the maximum potential is to do the measurement under conditions of zero current so that no energy is wasted. Traditionally, this has been accomplished by inserting a variable-voltage device (powered from an external source) in *opposition* to the cell potential. The voltage on this instrument (called a **potentiometer**) is adjusted until no current

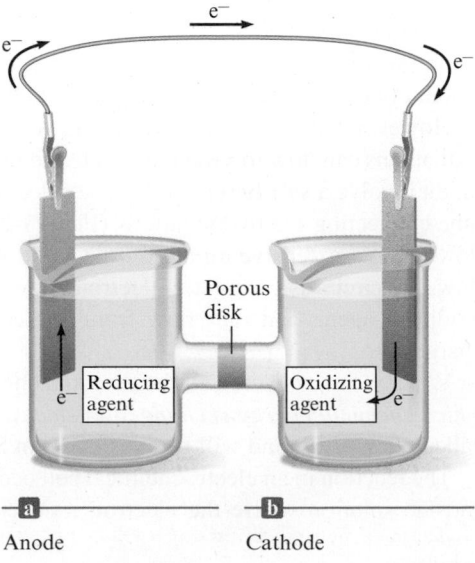

Figure 17-3 | An electrochemical process involves electron transfer at the interface between the electrode and the solution. (a) The species in the solution acting as the reducing agent supplies electrons to the anode. (b) The species in the solution acting as the oxidizing agent receives electrons from the cathode.

Figure 17-4 | Digital voltmeters draw only a negligible current and are convenient to use to measure the cell potential.

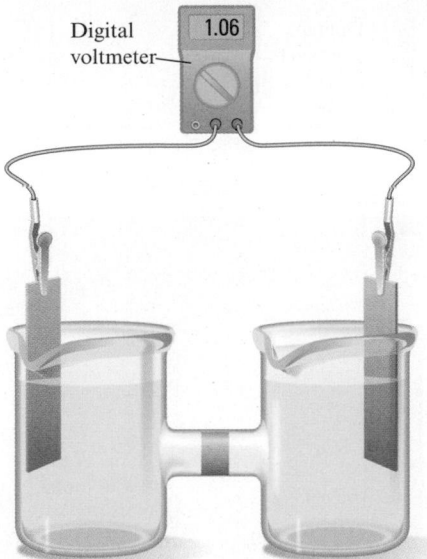

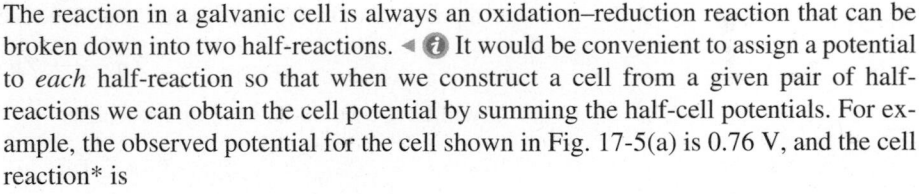

flows in the cell circuit. Under such conditions, the cell potential is equal in magnitude and opposite in sign to the voltage setting of the potentiometer. This value represents the *maximum* cell potential, since no energy is wasted heating the wire. More recently, advances in electronic technology have allowed the design of *digital voltmeters* that draw only a negligible amount of current (Fig. 17-4). Since these instruments are more convenient to use, they have replaced potentiometers in the modern laboratory.

17-3 | Standard Reduction Potentials

ⓘ The name *galvanic cell* honors Luigi Galvani (1737–1798), an Italian scientist generally credited with the discovery of electricity. These cells are sometimes called *voltaic cells* after Alessandro Volta (1745–1827), another Italian, who first constructed cells of this type around 1800.

The reaction in a galvanic cell is always an oxidation–reduction reaction that can be broken down into two half-reactions. ◄ ⓘ It would be convenient to assign a potential to *each* half-reaction so that when we construct a cell from a given pair of half-reactions we can obtain the cell potential by summing the half-cell potentials. For example, the observed potential for the cell shown in Fig. 17-5(a) is 0.76 V, and the cell reaction* is

$$2H^+(aq) + Zn(s) \longrightarrow Zn^{2+}(aq) + H_2(g)$$

For this cell, the anode compartment contains a zinc metal electrode with Zn^{2+} and SO_4^{2-} ions in aqueous solution. The anode reaction is the oxidation half-reaction:

$$Zn \longrightarrow Zn^{2+} + 2e^-$$

The zinc metal, in producing Zn^{2+} ions that go into solution, is giving up electrons, which flow through the wire. For now, we will assume that all cell components are in their standard states, so in this case the solution in the anode compartment will contain $1\ M\ Zn^{2+}$. The cathode reaction of this cell is the reduction half-reaction:

$$2H^+ + 2e^- \longrightarrow H_2$$

The cathode consists of a platinum electrode (used because it is a chemically inert conductor) in contact with $1\ M\ H^+$ ions and bathed by hydrogen gas at 1 atm. Such an electrode, called the **standard hydrogen electrode**, is shown in Fig. 17-5(b).

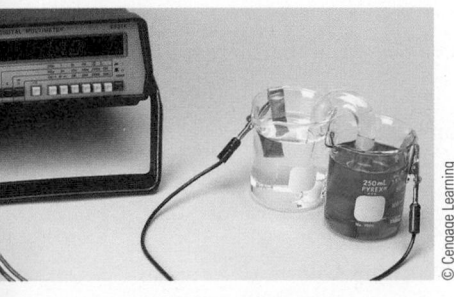

© Cengage Learning

▲ An electrochemical cell with a measured potential of 1.10 V.

*In this text we will follow the convention of indicating the physical states of the reactants and products only in the overall redox reaction. For simplicity, half-reactions will *not* include the physical states.

Figure 17-5 | (a) A galvanic cell involving the reactions $Zn \rightarrow Zn^{2+} + 2e^-$ (at the anode) and $2H^+ + 2e^- \rightarrow H_2$ (at the cathode) has a potential of 0.76 V. (b) The standard hydrogen electrode where $H_2(g)$ at 1 atm is passed over a platinum electrode in contact with $1\ M\ H^+$ ions. This electrode process (assuming ideal behavior) is arbitrarily assigned a value of exactly zero volts.

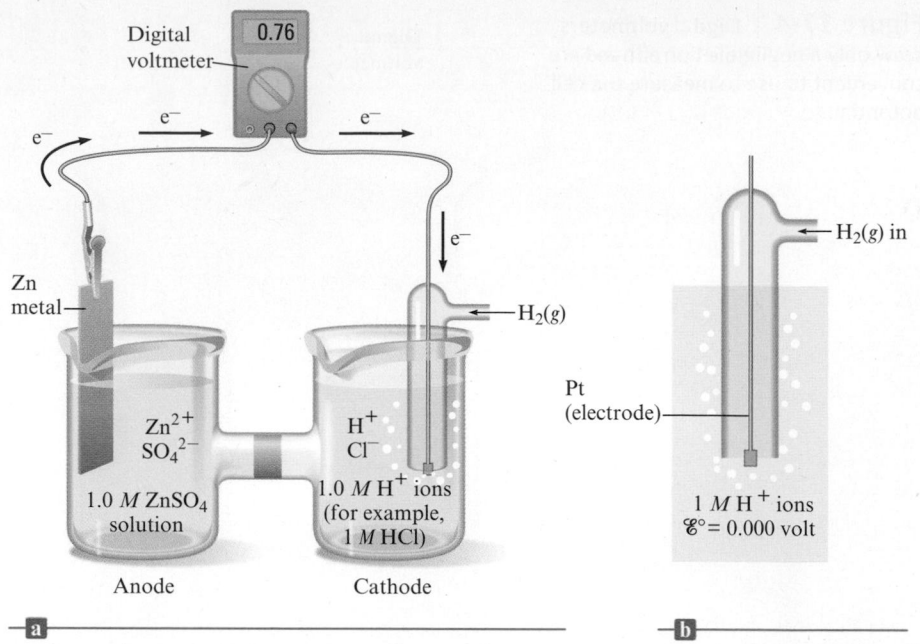

Although we can measure the *total* potential of this cell (0.76 V), there is no way to measure the potentials of the individual electrode processes. Thus, if we want potentials for the half-reactions (half-cells), we must arbitrarily divide the total cell potential. For example, if we assign the reaction

$$2H^+ + 2e^- \longrightarrow H_2$$

where $[H^+] = 1\ M$ and $P_{H_2} = 1$ atm

a potential of exactly zero volts, then the reaction

$$Zn \longrightarrow Zn^{2+} + 2e^-$$

will have a potential of 0.76 V because

$$\mathscr{E}^\circ_{cell} = \mathscr{E}^\circ_{H^+ \rightarrow H_2} + \mathscr{E}^\circ_{Zn \rightarrow Zn^{2+}}$$
$$\uparrow \qquad\qquad \uparrow \qquad\qquad \uparrow$$
$$0.76\ V \qquad 0\ V \qquad\quad 0.76\ V$$

where the superscript ° indicates that *standard states* are used. In fact, by setting the standard potential for the half-reaction $2H^+ + 2e^- \rightarrow H_2$ equal to zero, we can assign values to all other half-reactions.

For example, the measured potential for the cell shown in Fig. 17-6 is 1.10 V. The cell reaction is ◄

$$Zn(s) + Cu^{2+}(aq) \longrightarrow Zn^{2+}(aq) + Cu(s)$$

which can be divided into the half-reactions

Anode: $Zn \longrightarrow Zn^{2+} + 2e^-$
Cathode: $Cu^{2+} + 2e^- \longrightarrow Cu$

Then

$$\mathscr{E}^\circ_{cell} = \mathscr{E}^\circ_{Zn \longrightarrow Zn^{2+}} + \mathscr{E}^\circ_{Cu^{2+} \longrightarrow Cu}$$

Since $\mathscr{E}^\circ_{Zn \rightarrow Zn^{2+}}$ was earlier assigned a value of 0.76 V, the value of $\mathscr{E}^\circ_{Cu^{2+} \rightarrow Cu}$ must be 0.34 V because

$$1.10\ V = 0.76\ V + 0.34\ V$$

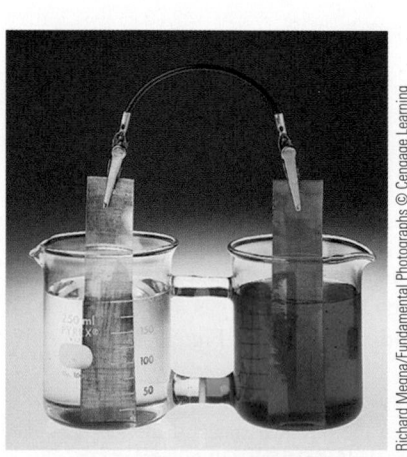

Richard Megna/Fundamental Photographs © Cengage Learning

Figure 17-6 | A galvanic cell involving the half-reactions Zn → Zn²⁺ + 2e⁻ (anode) and Cu²⁺ + 2e⁻ → Cu (cathode), with $\mathscr{E}^\circ_{cell}$ + 1.10 V. ◄ ⓘ

ⓘ Note what is happening at each electrode at the atomic level. At the anode the metal (in this case zinc) is being oxidized to its ion (Zn²⁺) and releasing two electrons; at the cathode the electrons combine with the metal ion (in this case Cu²⁺), which is thereby reduced to its elemental metal (Cu).

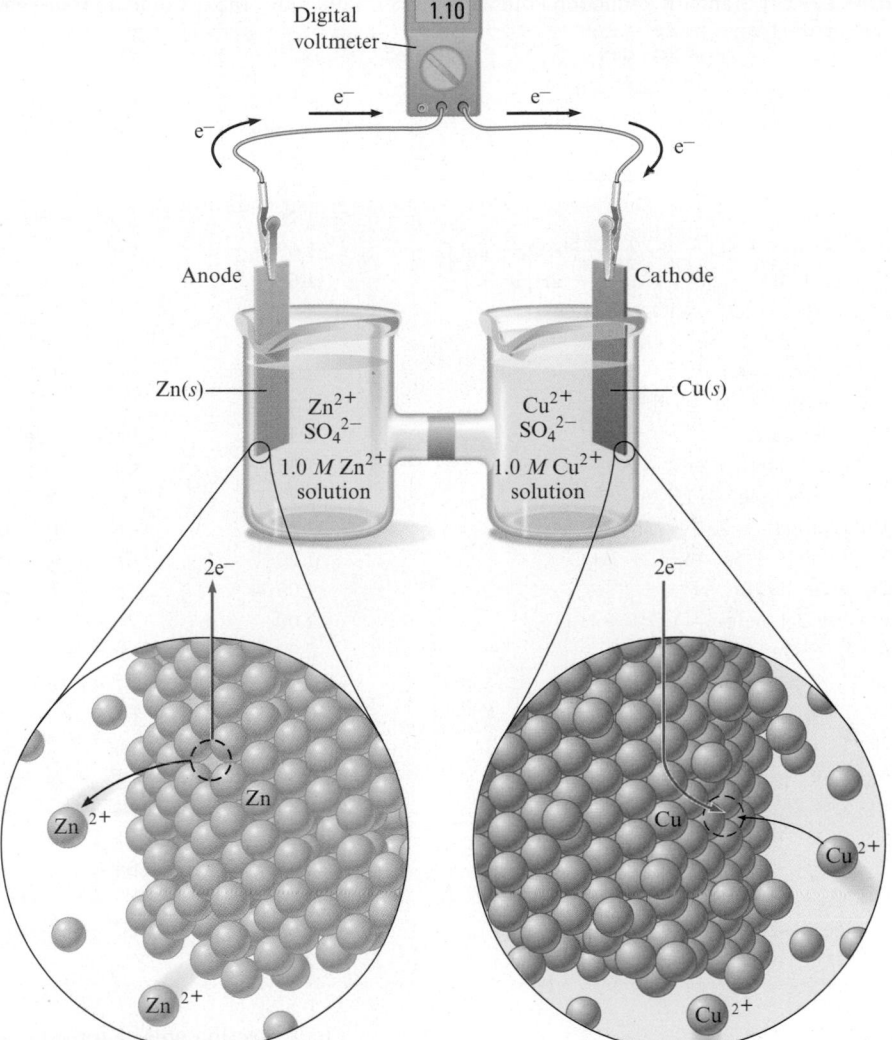

ⓘ The standard hydrogen potential is the reference potential against which all half-reaction potentials are assigned.

ⓘ All half-reactions are given as reduction processes in standard tables.

The scientific community has universally accepted the half-reaction potentials based on the assignment of 0 V to the process $2H^+ + 2e^- \rightarrow H_2$ (under standard conditions where ideal behavior is assumed). ◄ ⓘ However, before we can use these values to calculate cell potentials, we need to understand several essential characteristics of half-cell potentials.

The accepted convention is to give the potentials of half-reactions as *reduction* processes. For example:

$$2H^+ + 2e^- \longrightarrow H_2$$
$$Cu^{2+} + 2e^- \longrightarrow Cu$$
$$Zn^{2+} + 2e^- \longrightarrow Zn$$

The $\mathscr{E}^\circ$ values corresponding to reduction half-reactions with all solutes at 1 *M* and all gases at 1 atm are called **standard reduction potentials**. ◄ ⓘ Standard reduction potentials for the most common half-reactions are given in Table 17-1 and Appendix 5-5.

Combining two half-reactions to obtain a balanced oxidation–reduction reaction often requires two manipulations:

1. One of the reduction half-reactions must be reversed (since redox reactions must involve a substance being oxidized and a substance being reduced). The half-reaction with the largest positive potential will run as written (as a reduction), and

Table 17-1 | Standard Reduction Potentials at 25°C (298 K) for Many Common Half-Reactions

Half-Reaction	$\mathscr{E}°$ (V)	Half-Reaction	$\mathscr{E}°$ (V)
$F_2 + 2e^- \rightarrow 2F^-$	2.87	$O_2 + 2H_2O + 4e^- \rightarrow 4OH^-$	0.40
$Ag^{2+} + e^- \rightarrow Ag^+$	1.99	$Cu^{2+} + 2e^- \rightarrow Cu$	0.34
$Co^{3+} + e^- \rightarrow Co^{2+}$	1.82	$Hg_2Cl_2 + 2e^- \rightarrow 2Hg + 2Cl^-$	0.27
$H_2O_2 + 2H^+ + 2e^- \rightarrow 2H_2O$	1.78	$AgCl + e^- \rightarrow Ag + Cl^-$	0.22
$Ce^{4+} + e^- \rightarrow Ce^{3+}$	1.70	$SO_4^{2-} + 4H^+ + 2e^- \rightarrow H_2SO_3 + H_2O$	0.20
$PbO_2 + 4H^+ + SO_4^{2-} + 2e^- \rightarrow PbSO_4 + 2H_2O$	1.69	$Cu^{2+} + e^- \rightarrow Cu^+$	0.16
$MnO_4^- + 4H^+ + 3e^- \rightarrow MnO_2 + 2H_2O$	1.68	$2H^+ + 2e^- \rightarrow H_2$	0.00
$2e^- + 2H^+ + IO_4^- \rightarrow IO_3^- + H_2O$	1.60	$Fe^{3+} + 3e^- \rightarrow Fe$	−0.036
$MnO_4^- + 8H^+ + 5e^- \rightarrow Mn^{2+} + 4H_2O$	1.51	$Pb^{2+} + 2e^- \rightarrow Pb$	−0.13
$Au^{3+} + 3e^- \rightarrow Au$	1.50	$Sn^{2+} + 2e^- \rightarrow Sn$	−0.14
$PbO_2 + 4H^+ + 2e^- \rightarrow Pb^{2+} + 2H_2O$	1.46	$Ni^{2+} + 2e^- \rightarrow Ni$	−0.23
$Cl_2 + 2e^- \rightarrow 2Cl^-$	1.36	$PbSO_4 + 2e^- \rightarrow Pb + SO_4^{2-}$	−0.35
$Cr_2O_7^{2-} + 14H^+ + 6e^- \rightarrow 2Cr^{3+} + 7H_2O$	1.33	$Cd^{2+} + 2e^- \rightarrow Cd$	−0.40
$O_2 + 4H^+ + 4e^- \rightarrow 2H_2O$	1.23	$Fe^{2+} + 2e^- \rightarrow Fe$	−0.44
$MnO_2 + 4H^+ + 2e^- \rightarrow Mn^{2+} + 2H_2O$	1.21	$Cr^{3+} + e^- \rightarrow Cr^{2+}$	−0.50
$IO_3^- + 6H^+ + 5e^- \rightarrow \frac{1}{2}I_2 + 3H_2O$	1.20	$Cr^{3+} + 3e^- \rightarrow Cr$	−0.73
$Br_2 + 2e^- \rightarrow 2Br^-$	1.09	$Zn^{2+} + 2e^- \rightarrow Zn$	−0.76
$VO_2^+ + 2H^+ + e^- \rightarrow VO^{2+} + H_2O$	1.00	$2H_2O + 2e^- \rightarrow H_2 + 2OH^-$	−0.83
$AuCl_4^- + 3e^- \rightarrow Au + 4Cl^-$	0.99	$Mn^{2+} + 2e^- \rightarrow Mn$	−1.18
$NO_3^- + 4H^+ + 3e^- \rightarrow NO + 2H_2O$	0.96	$Al^{3+} + 3e^- \rightarrow Al$	−1.66
$ClO_2 + e^- \rightarrow ClO_2^-$	0.954	$H_2 + 2e^- \rightarrow 2H^-$	−2.23
$2Hg^{2+} + 2e^- \rightarrow Hg_2^{2+}$	0.91	$Mg^{2+} + 2e^- \rightarrow Mg$	−2.37
$Ag^+ + e^- \rightarrow Ag$	0.80	$La^{3+} + 3e^- \rightarrow La$	−2.37
$Hg_2^{2+} + 2e^- \rightarrow 2Hg$	0.80	$Na^+ + e^- \rightarrow Na$	−2.71
$Fe^{3+} + e^- \rightarrow Fe^{2+}$	0.77	$Ca^{2+} + 2e^- \rightarrow Ca$	−2.76
$O_2 + 2H^+ + 2e^- \rightarrow H_2O_2$	0.68	$Ba^{2+} + 2e^- \rightarrow Ba$	−2.90
$MnO_4^- + e^- \rightarrow MnO_4^{2-}$	0.56	$K^+ + e^- \rightarrow K$	−2.92
$I_2 + 2e^- \rightarrow 2I^-$	0.54	$Li^+ + e^- \rightarrow Li$	−3.05
$Cu^+ + e^- \rightarrow Cu$	0.52		

the other half-reaction will be forced to run in reverse (will be the oxidation reaction). The net potential of the cell will be the *difference* between the two. Since the reduction process occurs at the cathode and the oxidation process occurs at the anode, we can write

$$\mathscr{E}°_{cell} = \mathscr{E}° \text{ (cathode)} - \mathscr{E}° \text{ (anode)}$$

ⓘ When a half-reaction is reversed, the sign of $\mathscr{E}°$ is reversed.

Because subtraction means "change the sign and add," in the examples done here we will change the sign of the oxidation (anode) reaction when we reverse it and add it to the reduction (cathode) reaction. ◄ ⓘ

2. Since the number of electrons lost must equal the number gained, the half-reactions must be multiplied by integers as necessary to achieve the balanced equation. However, the *value of $\mathscr{E}°$ is not changed* when a half-reaction is multiplied by an integer. ◄ ⓘ Since a standard reduction potential is an *intensive property* (it does not depend on how many times the reaction occurs), the potential is *not* multiplied by the integer required to balance the cell reaction.

ⓘ When a half-reaction is multiplied by an integer, $\mathscr{E}°$ remains the same.

Consider a galvanic cell based on the redox reaction

$$Fe^{3+}(aq) + Cu(s) \longrightarrow Cu^{2+}(aq) + Fe^{2+}(aq)$$

The pertinent half-reactions are

$$Fe^{3+} + e^- \longrightarrow Fe^{2+} \qquad \mathscr{E}° = 0.77 \text{ V} \qquad (1)$$

$$Cu^{2+} + 2e^- \longrightarrow Cu \qquad \mathscr{E}° = 0.34 \text{ V} \qquad (2)$$

To balance the cell reaction and calculate the standard cell potential, reaction (2) must be reversed:

$$Cu \longrightarrow Cu^{2+} + 2e^- \qquad -\mathscr{E}° = -0.34 \text{ V}$$

Note the change in sign for the $\mathscr{E}°$ value. Now, since each Cu atom produces two electrons but each Fe^{3+} ion accepts only one electron, reaction (1) must be multiplied by 2:

$$2Fe^{3+} + 2e^- \longrightarrow 2Fe^{2+} \qquad \mathscr{E}° = 0.77 \text{ V}$$

Note that $\mathscr{E}°$ is not changed in this case.

Now we can obtain the balanced cell reaction by summing the appropriately modified half-reactions:

$$2Fe^{3+} + 2e^- \longrightarrow 2Fe^{2+} \qquad\qquad \mathscr{E} \text{ (cathode)} = 0.77 \text{ V}$$
$$Cu \longrightarrow Cu^{2+} + 2e^- \qquad\qquad -\mathscr{E} \text{ (anode)} = -0.34 \text{ V}$$

Cell reaction: $Cu(s) + 2Fe^{3+}(aq) \longrightarrow Cu^{2+}(aq) + 2Fe^{2+}(aq)$ $\qquad \mathscr{E}°_{cell} = \mathscr{E}° \text{ (cathode)} - \mathscr{E}° \text{ (anode)}$
$$= 0.77 \text{ V} - 0.34 \text{ V} = 0.43 \text{ V}$$

Critical Thinking

What if you want to "plate out" copper metal from an aqueous Cu^{2+} solution? Use Table 17-1 to determine several metals you can place in the solution to plate copper metal from the solution. Defend your choices. Why can Zn not be plated out from an aqueous solution of Zn^{2+} using the choices in Table 17-1?

INTERACTIVE EXAMPLE 17-3

Galvanic Cells

a. Consider a galvanic cell based on the reaction

$$Al^{3+}(aq) + Mg(s) \longrightarrow Al(s) + Mg^{2+}(aq)$$

The half-reactions are

$$Al^{3+} + 3e^- \longrightarrow Al \qquad \mathscr{E}° = -1.66 \text{ V} \qquad (1)$$
$$Mg^{2+} + 2e^- \longrightarrow Mg \qquad \mathscr{E}° = -2.37 \text{ V} \qquad (2)$$

Give the balanced cell reaction, and calculate $\mathscr{E}°$ for the cell.

b. A galvanic cell is based on the reaction

$$MnO_4^-(aq) + H^+(aq) + ClO_3^-(aq) \longrightarrow ClO_4^-(aq) + Mn^{2+}(aq) + H_2O(l)$$

The half-reactions are

$$MnO_4^- + 5e^- + 8H^+ \longrightarrow Mn^{2+} + 4H_2O \qquad \mathscr{E}° = 1.51 \text{ V} \qquad (1)$$
$$ClO_4^- + 2H^+ + 2e^- \longrightarrow ClO_3^- + H_2O \qquad \mathscr{E}° = 1.19 \text{ V} \qquad (2)$$

Give the balanced cell reaction, and calculate $\mathscr{E}°$ for the cell.

Solution

a. The half-reaction involving magnesium must be reversed and since this is the oxidation process, it is the anode:

$$Mg \longrightarrow Mg^{2+} + 2e^- \qquad -\mathscr{E}° \text{ (anode)} = -(-2.37 \text{ V}) = 2.37 \text{ V}$$

Also, since the two half-reactions involve different numbers of electrons, they must be multiplied by integers as follows:

$$2(Al^{3+} + 3e^- \longrightarrow Al) \qquad\qquad \mathscr{E}°\text{ (cathode)} = -1.66 \text{ V}$$

$$3(Mg \longrightarrow Mg^{2+} + 2e^-) \qquad\qquad -\mathscr{E}°\text{ (anode)} = 2.37 \text{ V}$$

$$2Al^{3+}(aq) + 3Mg(s) \longrightarrow 2Al(s) + 3Mg^{2+}(aq) \qquad\qquad \mathscr{E}°_{\text{cell}} = \mathscr{E}°\text{ (cathode)} - \mathscr{E}°\text{ (anode)}$$

$$= -1.66 \text{ V} + 2.37 \text{ V} = 0.71 \text{ V}$$

b. Half-reaction (2) must be reversed (it is the anode), and both half-reactions must be multiplied by integers to make the number of electrons equal:

$$2(MnO_4^- + 5e^- + 8H^+ \longrightarrow Mn^{2+} + 4H_2O) \qquad\qquad \mathscr{E}°\text{ (cathode)} = 1.51 \text{ V}$$

$$5(ClO_3^- + H_2O \longrightarrow ClO_4^- + 2H^+ + 2e^-) \qquad\qquad -\mathscr{E}°\text{ (anode)} = -1.19 \text{ V}$$

$$2MnO_4^-(aq) + 6H^+(aq) + 5ClO_3^-(aq) \longrightarrow \qquad\qquad \mathscr{E}°_{\text{cell}} = \mathscr{E}°\text{ (cathode)} - \mathscr{E}°\text{ (anode)}$$

$$2Mn^{2+}(aq) + 3H_2O(l) + 5ClO_4^-(aq) \qquad\qquad = 1.51 \text{ V} - 1.19 \text{ V} = 0.32 \text{ V}$$

See Exercises 17-39 and 17-40

Line Notation

We now will introduce a handy line notation used to describe electrochemical cells. In this notation the anode components are listed on the left and the cathode components are listed on the right, separated by double vertical lines (indicating the salt bridge or porous disk). For example, the line notation for the cell described in Example 17-3(a) is

$$Mg(s)\,|\,Mg^{2+}(aq)\,||\,Al^{3+}(aq)\,|\,Al(s)$$

In this notation a phase difference (boundary) is indicated by a single vertical line. Thus, in this case, vertical lines occur between the solid Mg metal and the Mg^{2+} in aqueous solution and between solid Al and Al^{3+} in aqueous solution. Also note that the substance constituting the anode is listed at the far left and the substance constituting the cathode is listed at the far right.

For the cell described in Example 17-3(b), all the components involved in the oxidation–reduction reaction are ions. Since none of these dissolved ions can serve as an electrode, a nonreacting (inert) conductor must be used. The usual choice is platinum. Thus, for the cell described in Example 17-3(b), the line notation is

$$Pt(s)\,|\,ClO_3^-(aq),\, ClO_4^-(aq),\, H^+(aq)\,||\,H^+(aq),\, MnO_4^-(aq),\, Mn^{2+}(aq)\,|\,Pt(s)$$

Complete Description of a Galvanic Cell

Next we want to consider how to describe a galvanic cell fully, given just its half-reactions. This description will include the cell reaction, the cell potential, and the physical setup of the cell. Let's consider a galvanic cell based on the following half-reactions:

$$Fe^{2+} + 2e^- \longrightarrow Fe \qquad\qquad \mathscr{E}° = -0.44 \text{ V}$$

$$MnO_4^- + 5e^- + 8H^+ \longrightarrow Mn^{2+} + 4H_2O \qquad\qquad \mathscr{E}° = 1.51 \text{ V}$$

In a working galvanic cell, one of these reactions must run in reverse. Which one?

We can answer this question by considering the sign of the potential of a working cell: *A cell will always run spontaneously in the direction that produces a positive cell potential.* ◀ ⓘ Thus, in the present case, it is clear that the half-reaction involving iron must be reversed, since this choice leads to a positive cell potential:

ⓘ A galvanic cell runs spontaneously in the direction that gives a positive value for $\mathscr{E}_{\text{cell}}$.

$$Fe \longrightarrow Fe^{2+} + 2e^- \qquad -\mathscr{E}° = 0.44 \text{ V} \qquad \text{Anode reaction}$$

$$MnO_4^- + 5e^- + 8H^+ \longrightarrow Mn^{2+} + 4H_2O \qquad \mathscr{E}° = 1.51 \text{ V} \qquad \text{Cathode reaction}$$

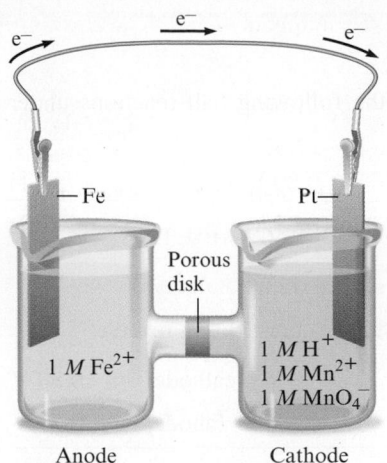

Figure 17-7 | The schematic of a galvanic cell based on the half-reactions:

$$Fe \longrightarrow Fe^{2+} + 2e^-$$
$$MnO_4^- + 5e^- + 8H^+ \longrightarrow Mn^{2+} + 4H_2O$$

where $\quad \mathscr{E}_{cell}^{\circ} = \mathscr{E}^{\circ} \text{(cathode)} - \mathscr{E}^{\circ} \text{(anode)} = 1.51 \text{ V} + 0.44 \text{ V} = 1.95 \text{ V}$

The balanced cell reaction is obtained as follows:

$$5(Fe \longrightarrow Fe^{2+} + 2e^-)$$
$$\underline{2(MnO_4^- + 5e^- + 8H^+ \longrightarrow Mn^{2+} + 4H_2O)}$$
$$2MnO_4^-(aq) + 5Fe(s) + 16H^+(aq) \longrightarrow 5Fe^{2+}(aq) + 2Mn^{2+}(aq) + 8H_2O(l)$$

Now consider the physical setup of the cell, shown schematically in Fig. 17-7. In the left compartment the active components in their standard states are pure metallic iron (Fe) and 1.0 M Fe^{2+}. The anion present depends on the iron salt used. In this compartment the anion does not participate in the reaction but simply balances the charge. The half-reaction that takes place at this electrode is

$$Fe \longrightarrow Fe^{2+} + 2e^-$$

which is an oxidation reaction, so this compartment is the anode. The electrode consists of pure iron metal.

In the right compartment the active components in their standard states are 1.0 M MnO_4^-, 1.0 M H^+, and 1.0 M Mn^{2+}, with appropriate unreacting ions (often called *counterions*) to balance the charge. The half-reaction in this compartment is

$$MnO_4^- + 5e^- + 8H^+ \longrightarrow Mn^{2+} + 4H_2O$$

which is a reduction reaction, so this compartment is the cathode. Since neither MnO_4^- nor Mn^{2+} ions can serve as the electrode, a nonreacting conductor such as platinum must be used.

The next step is to determine the direction of electron flow. In the left compartment the half-reaction involves the oxidation of iron:

$$Fe \longrightarrow Fe^{2+} + 2e^-$$

In the right compartment the half-reaction is the reduction of MnO_4^-:

$$MnO_4^- + 5e^- + 8H^+ \longrightarrow Mn^{2+} + 4H_2O$$

Thus the electrons flow from Fe to MnO_4^- in this cell, or from the anode to the cathode, as is always the case. The line notation for this cell is

$$Fe(s)|Fe^{2+}(aq)||MnO_4^-(aq), Mn^{2+}(aq)|Pt(s)$$

LET'S REVIEW | *Description of a Galvanic Cell*

A complete description of a galvanic cell usually includes four items:

- The cell potential (always positive for a galvanic cell where $\mathscr{E}_{cell}^{\circ} = \mathscr{E}^{\circ}\text{(cathode)} - \mathscr{E}^{\circ}\text{(anode)}$) and the balanced cell reaction.
- The direction of electron flow, obtained by inspecting the half-reactions and using the direction that gives a positive $\mathscr{E}_{cell}$.
- Designation of the anode and cathode.
- The nature of each electrode and the ions present in each compartment. A chemically inert conductor is required if none of the substances participating in the half-reaction is a conducting solid.

EXAMPLE 17-4

Description of a Galvanic Cell

Describe completely the galvanic cell based on the following half-reactions under standard conditions:

$$Ag^+ + e^- \longrightarrow Ag \qquad \mathscr{E}^\circ = 0.80 \text{ V} \qquad (1)$$
$$Fe^{3+} + e^- \longrightarrow Fe^{2+} \qquad \mathscr{E}^\circ = 0.77 \text{ V} \qquad (2)$$

Solution

> Since a positive $\mathscr{E}^\circ_{cell}$ value is required, reaction (2) must run in reverse:

$$Ag^+ + e^- \longrightarrow Ag \qquad\qquad \mathscr{E}^\circ \text{ (cathode)} = 0.80 \text{ V}$$
$$Fe^{2+} \longrightarrow Fe^{3+} + e^- \qquad\qquad -\mathscr{E}^\circ \text{ (anode)} = -0.77 \text{ V}$$

Cell reaction: $Ag^+(aq) + Fe^{2+}(aq) \longrightarrow Fe^{3+}(aq) + Ag(s)$ $\qquad \mathscr{E}^\circ_{cell} = 0.03 \text{ V}$

> Since Ag^+ receives electrons and Fe^{2+} loses electrons in the cell reaction, the electrons will flow from the compartment containing Fe^{2+} to the compartment containing Ag^+.

> Oxidation occurs in the compartment containing Fe^{2+} (electrons flow from Fe^{2+} to Ag^+). Hence this compartment functions as the anode. Reduction occurs in the compartment containing Ag^+, so this compartment functions as the cathode.

> The electrode in the Ag/Ag^+ compartment is silver metal, and an inert conductor, such as platinum, must be used in the Fe^{2+}/Fe^{3+} compartment. Appropriate counterions are assumed to be present. The diagram for this cell is shown in Fig. 17-8. The line notation for this cell is

$$Pt(s)\,|\,Fe^{2+}(aq),\,Fe^{3+}(aq)\,||\,Ag^+(aq)\,|\,Ag(s)$$

See Exercises 17-41 and 17-42

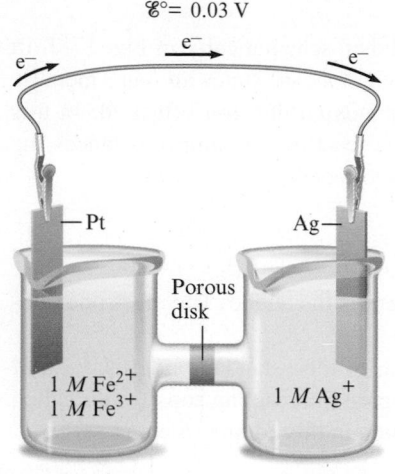

Figure 17-8 | Schematic diagram for the galvanic cell based on the half-reactions:

$$Ag^+ + e^- \longrightarrow Ag$$
$$Fe^{2+} \longrightarrow Fe^{3+} + e^-$$

17-4 | Cell Potential, Electrical Work, and Free Energy

So far we have considered electrochemical cells in a very practical fashion without much theoretical background. The next step will be to explore the relationship between thermodynamics and electrochemistry.

The work that can be accomplished when electrons are transferred through a wire depends on the "push" (the thermodynamic driving force) behind the electrons. This driving force (the emf) is defined in terms of a *potential difference* (in volts) between two points in the circuit. Recall that a volt represents a joule of work per coulomb of charge transferred:

$$\text{emf} = \text{potential difference (V)} = \frac{\text{work (J)}}{\text{charge (C)}}$$

Thus 1 joule of work is produced or required (depending on the direction) when 1 coulomb of charge is transferred between two points in the circuit that differ by a potential of 1 volt.

In this book, *work is viewed from the point of view of the system.* Thus work flowing out of the system is indicated by a minus sign. When a cell produces a current, the cell potential is positive, and the current can be used to do work—to run a motor, for instance. Thus the cell potential $\mathscr{E}$ and the work w have opposite signs:

$$\mathscr{E} = \frac{-w}{q} \begin{array}{l} \leftarrow \text{Work} \\ \leftarrow \text{Charge} \end{array}$$

Therefore,
$$-w = q\mathscr{E}$$

From this equation it can be seen that the maximum work in a cell would be obtained at the maximum cell potential:

$$-w_{max} = q\mathscr{E}_{max} \quad \text{or} \quad w_{max} = -q\mathscr{E}_{max}$$

However, there is a problem. To obtain electrical work, current must flow. When current flows, some energy is inevitably wasted through frictional heating, and the maximum work is not obtained. This reflects the important general principle introduced in Section 16-9: In any real, spontaneous process some energy is always wasted—the actual work realized is always less than the calculated maximum. This is a consequence of the fact that the entropy of the universe must increase in any spontaneous process. Recall from Section 16-9 that the only process from which maximum work could be realized is the hypothetical reversible process. For a galvanic cell this would involve an infinitesimally small current flow and thus an infinite amount of time to do the work. Even though we can never achieve the maximum work through the actual discharge of a galvanic cell, we can measure the maximum potential. There is negligible current flow when a cell potential is measured with a potentiometer or an efficient digital voltmeter. No current flow implies no waste of energy, so the potential measured is the maximum.

Although we can never actually realize the maximum work from a cell reaction, the value for it is still useful in evaluating the efficiency of a real process based on the cell reaction. ◄ ⓘ For example, suppose a certain galvanic cell has a maximum potential (at zero current) of 2.50 V. In a particular experiment 1.33 moles of electrons were passed through this cell at an average actual potential of 2.10 V. The actual work done is

$$w = -q\mathscr{E}$$

where $\mathscr{E}$ represents the actual potential difference at which the current flowed (2.10 V or 2.10 J/C) and q is the quantity of charge in coulombs transferred. The charge on 1 mole of electrons is a constant called the **faraday** (abbreviated F), which has the value *96,485 coulombs of charge per mole of electrons.* ◄ Thus q equals the number of moles of electrons times the charge per mole of electrons:

$$q = nF = 1.33 \text{ mol } e^- \times 96,485 \text{ C/mol } e^-$$

Then, for the preceding experiment, the actual work is

$$w = -q\mathscr{E} = -(1.33 \text{ mol } e^- \times 96,485 \text{ C/mol } e^-)(2.10 \text{ J/C})$$
$$= -2.69 \times 10^5 \text{ J}$$

For the maximum possible work, the calculation is similar, except that the maximum potential is used:

$$w_{max} = -q\mathscr{E}$$
$$= -\left(1.33 \text{ mol } e^- \times 96,485 \frac{C}{\text{mol } e^-}\right)\left(2.50 \frac{J}{C}\right)$$
$$= -3.21 \times 10^5 \text{ J}$$

ⓘ Work is never the maximum possible if any current is flowing.

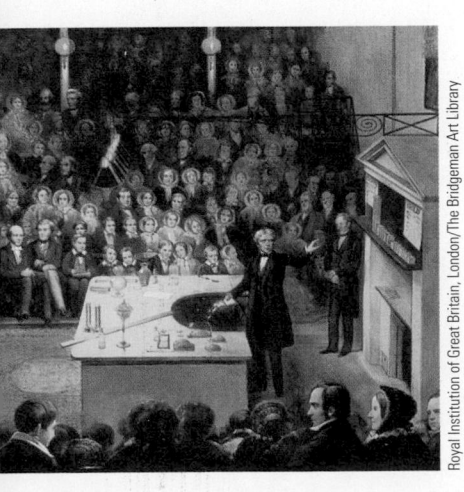

Royal Institution of Great Britain, London/The Bridgeman Art Library

▲ Michael Faraday lecturing at the Royal Institution before Prince Albert and others (1855). The faraday was named in honor of Michael Faraday (1791–1867), an Englishman who may have been the greatest experimental scientist of the nineteenth century. Among his many achievements were the invention of the electric motor and generator and the development of the principles of electrolysis.

Thus, in its actual operation, the efficiency of this cell is

$$\frac{w}{w_{max}} \times 100\% = \frac{-2.69 \times 10^5 \text{ J}}{-3.21 \times 10^5 \text{ J}} \times 100\% = 83.8\%$$

Next we want to relate the potential of a galvanic cell to free energy. In Section 16-9 we saw that for a process carried out at constant temperature and pressure, the change in free energy equals the maximum useful work obtainable from that process:

$$w_{max} = \Delta G$$

For a galvanic cell,

$$w_{max} = -q\mathscr{E}_{max} = \Delta G$$

Since

$$q = nF$$

we have

$$\Delta G = -q\mathscr{E}_{max} = -nF\mathscr{E}_{max}$$

From now on the subscript on $\mathscr{E}_{max}$ will be deleted, with the understanding that any potential given in this book is the maximum potential. Thus

$$\Delta G = -nF\mathscr{E}$$

For standard conditions,

$$\Delta G^\circ = -nF\mathscr{E}^\circ$$

This equation states that the maximum cell potential is directly related to the free energy difference between the reactants and the products in the cell. This relationship is important because it provides an experimental means to obtain ΔG for a reaction. It also confirms that a galvanic cell will run in the direction that gives a positive value for $\mathscr{E}_{cell}$; a positive $\mathscr{E}_{cell}$ value corresponds to a negative ΔG value, which is the condition for spontaneity.

INTERACTIVE EXAMPLE 17-5

Calculating ΔG° for a Cell Reaction

Using the data in Table 17-1, calculate ΔG° for the reaction

$$Cu^{2+}(aq) + Fe(s) \longrightarrow Cu(s) + Fe^{2+}(aq)$$

Is this reaction spontaneous?

Solution

The half-reactions are

$$
\begin{array}{ll}
Cu^{2+} + 2e^- \longrightarrow Cu & \mathscr{E}^\circ \text{ (cathode)} = 0.34 \text{ V} \\
\underline{\phantom{Cu^{2+} + 2e^-} Fe \longrightarrow Fe^{2+} + 2e^-} & \underline{-\mathscr{E}^\circ \text{ (anode)} = 0.44 \text{ V}} \\
Cu^{2+} + Fe \longrightarrow Fe^{2+} + Cu & \mathscr{E}^\circ_{cell} = 0.78 \text{ V}
\end{array}
$$

We can calculate ΔG° from the equation

$$\Delta G^\circ = -nF\mathscr{E}^\circ$$

Since two electrons are transferred per atom in the reaction, 2 moles of electrons are required per mole of reactants and products. Thus $n = 2$ mol e$^-$, $F = 96{,}485$ C/mol e$^-$, and $\mathscr{E}^\circ = 0.78$ V $= 0.78$ J/C. Therefore,

$$\Delta G^\circ = -(2 \text{ mol e}^-)\left(96{,}485 \ \frac{C}{\text{mol e}^-}\right)\left(0.78 \ \frac{J}{C}\right)$$

$$= -1.5 \times 10^5 \text{ J}$$

The process is spontaneous, as indicated by both the negative sign of $\Delta G°$ and the positive sign of $\mathscr{E}°_{cell}$.

This reaction is used industrially to deposit copper metal from solutions resulting from the dissolving of copper ores.

See Exercises 17-49 and 17-50

EXAMPLE 17-6

▲ A gold ring does not dissolve in nitric acid.

Predicting Spontaneity

Using the data from Table 17-1, predict whether 1 M HNO_3 will dissolve gold metal to form a 1-M Au^{3+} solution.

Solution

The half-reaction for HNO_3 acting as an oxidizing agent is

$$NO_3^- + 4H^+ + 3e^- \longrightarrow NO + 2H_2O \qquad \mathscr{E}° \text{ (cathode)} = 0.96 \text{ V}$$

The reaction for the oxidation of solid gold to Au^{3+} ions is

$$Au \longrightarrow Au^{3+} + 3e^- \qquad -\mathscr{E}° \text{ (anode)} = -1.50 \text{ V}$$

The sum of these half-reactions gives the required reaction:

$$Au(s) + NO_3^-(aq) + 4H^+(aq) \longrightarrow Au^{3+}(aq) + NO(g) + 2H_2O(l)$$

and $\mathscr{E}°_{cell} = \mathscr{E}° \text{ (cathode)} - \mathscr{E}° \text{ (anode)} = 0.96 \text{ V} - 1.50 \text{ V} = -0.54 \text{ V}$

Since the $\mathscr{E}°$ value is negative, the process will *not* occur under standard conditions. That is, gold will not dissolve in 1 M HNO_3 to give 1 M Au^{3+}. ◄ In fact, a mixture (1:3 by volume) of concentrated nitric and hydrochloric acids, called *aqua regia,* is required to dissolve gold.

See Exercises 17-59 and 17-60

17-5 | Dependence of Cell Potential on Concentration

So far we have described cells under standard conditions. In this section we consider the dependence of the cell potential on concentration. Under standard conditions (all concentrations 1 M), the cell with the reaction

$$Cu(s) + 2Ce^{4+}(aq) \longrightarrow Cu^{2+}(aq) + 2Ce^{3+}(aq)$$

has a potential of 1.36 V. What will the cell potential be if $[Ce^{4+}]$ is greater than 1.0 M? This question can be answered qualitatively in terms of Le Châtelier's principle. An increase in the concentration of Ce^{4+} will favor the forward reaction and thus increase the driving force on the electrons. The cell potential will increase. On the other hand, an increase in the concentration of a product (Cu^{2+} or Ce^{3+}) will oppose the forward reaction, thus decreasing the cell potential.

These ideas are illustrated in Example 17-7.

INTERACTIVE EXAMPLE 17-7

The Effects of Concentration on $\mathscr{E}$

For the cell reaction

$$2Al(s) + 3Mn^{2+}(aq) \longrightarrow 2Al^{3+}(aq) + 3Mn(s) \qquad \mathscr{E}^{\circ}_{cell} = 0.48 \text{ V}$$

predict whether $\mathscr{E}_{cell}$ is larger or smaller than $\mathscr{E}^{\circ}_{cell}$ for the following cases.

a. $[Al^{3+}] = 2.0 \text{ } M, [Mn^{2+}] = 1.0 \text{ } M$

b. $[Al^{3+}] = 1.0 \text{ } M, [Mn^{2+}] = 3.0 \text{ } M$

Solution

a. A product concentration has been raised above $1.0 \text{ } M$. This will oppose the cell reaction and will cause $\mathscr{E}_{cell}$ to be less than $\mathscr{E}^{\circ}_{cell}$ ($\mathscr{E}_{cell} < 0.48 \text{ V}$).

b. A reactant concentration has been increased above $1.0 \text{ } M$, and $\mathscr{E}_{cell}$ will be greater than $\mathscr{E}^{\circ}_{cell}$ ($\mathscr{E}_{cell} > 0.48 \text{ V}$).

See Exercise 17-67

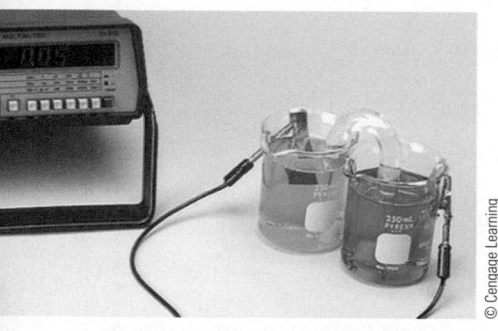

© Cengage Learning

▲ A concentration cell with 1.0 M Cu^{2+} on the right and 0.010 M Cu^{2+} on the left.

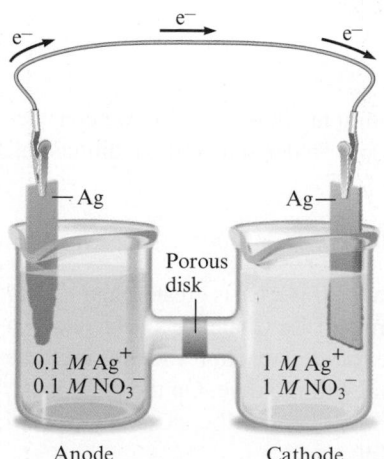

Anode   Cathode

Concentration Cells

Because cell potentials depend on concentration, we can construct galvanic cells where both compartments contain the same components but at different concentrations. For example, in the cell in Fig. 17-9, both compartments contain aqueous $AgNO_3$, but with different molarities. Let's consider the potential of this cell and the direction of electron flow. The half-reaction relevant to both compartments of this cell is

$$Ag^+ + e^- \longrightarrow Ag \qquad \mathscr{E}^{\circ} = 0.80 \text{ V}$$

If the cell had $1 \text{ } M$ Ag^+ in both compartments,

$$\mathscr{E}^{\circ}_{cell} = 0.80 \text{ V} - 0.80 \text{ V} = 0 \text{ V}$$

However, in the cell described here, the concentrations of Ag^+ in the two compartments are $1 \text{ } M$ and $0.1 \text{ } M$. Because the concentrations of Ag^+ are unequal, the half-cell potentials will not be identical, and the cell will exhibit a positive voltage. In which direction will the electrons flow in this cell? The best way to think about this question is to recognize that nature will try to equalize the concentrations of Ag^+ in the two compartments. This can be done by transferring electrons from the compartment containing $0.1 \text{ } M$ Ag^+ to the one containing $1 \text{ } M$ Ag^+ (left to right in Fig. 17-9). This electron transfer will produce more Ag^+ in the left compartment and consume Ag^+ (to form Ag) in the right compartment.

A cell in which both compartments have the same components but at different concentrations is called a **concentration cell**. ◄ The difference in concentration is the only factor that produces a cell potential in this case, and the voltages are typically small.

Figure 17-9 | A concentration cell that contains a silver electrode and aqueous silver nitrate in both compartments. Because the right compartment contains $1 \text{ } M$ Ag^+ and the left compartment contains $0.1 \text{ } M$ Ag^+, there will be a driving force to transfer electrons from left to right. Silver will be deposited on the right electrode, thus lowering the concentration of Ag^+ in the right compartment. In the left compartment the silver electrode dissolves (producing Ag^+ ions) to raise the concentration of Ag^+ in solution.

EXAMPLE 17-8

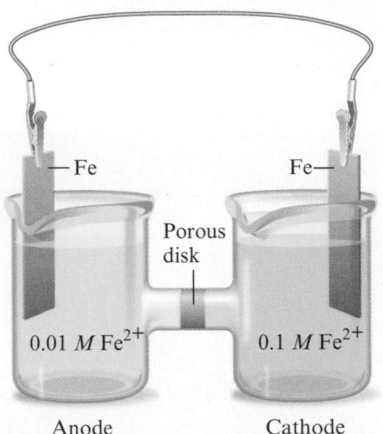

0.01 M Fe^{2+} 0.1 M Fe^{2+}

Anode   Cathode

Figure 17-10 | A concentration cell containing iron electrodes and different concentrations of Fe^{2+} ion in the two compartments.

Nernst was one of the pioneers in the development of electrochemical theory and is generally given credit for first stating the third law of thermodynamics. He won the Nobel Prize in chemistry in 1920.

Concentration Cells

Determine the direction of electron flow, and designate the anode and cathode for the cell represented in Fig. 17-10.

Solution

The concentrations of Fe^{2+} ion in the two compartments can (eventually) be equalized by transferring electrons from the left compartment to the right. This will cause Fe^{2+} to be formed in the left compartment, and iron metal will be deposited (by reducing Fe^{2+} ions to Fe) on the right electrode. Since electron flow is from left to right, oxidation occurs in the left compartment (the anode) and reduction occurs in the right (the cathode).

See Exercise 17-68

The Nernst Equation

The dependence of the cell potential on concentration results directly from the dependence of free energy on concentration. Recall from Chapter 16 that the equation

$$\Delta G = \Delta G^\circ + RT \ln(Q)$$

where Q is the reaction quotient, was used to calculate the effect of concentration on ΔG. Since $\Delta G = -nF\mathscr{E}$ and $\Delta G^\circ = -nF\mathscr{E}^\circ$, the equation becomes

$$-nF\mathscr{E} = -nF\mathscr{E}^\circ + RT \ln(Q)$$

Dividing each side of the equation by $-nF$ gives

$$\mathscr{E} = \mathscr{E}^\circ - \frac{RT}{nF} \ln(Q) \tag{17-1}$$

Equation (17-1), which gives the relationship between the cell potential and the concentrations of the cell components, is commonly called the **Nernst equation**, after the German chemist Walther Hermann Nernst (1864–1941). ◂

The Nernst equation is often given in a form that is valid at 25°C:

$$\mathscr{E} = \mathscr{E}^\circ - \frac{0.0591}{n} \log(Q)$$

Using this relationship, we can calculate the potential of a cell in which some or all of the components are not in their standard states.

For example, $\mathscr{E}^\circ_{cell}$ is 0.48 V for the galvanic cell based on the reaction

$$2Al(s) + 3Mn^{2+}(aq) \longrightarrow 2Al^{3+}(aq) + 3Mn(s)$$

Consider a cell in which

$$[Mn^{2+}] = 0.50 \, M \quad \text{and} \quad [Al^{3+}] = 1.50 \, M$$

The cell potential at 25°C for these concentrations can be calculated using the Nernst equation:

$$\mathscr{E}_{cell} = \mathscr{E}^\circ_{cell} - \frac{0.0591}{n} \log(Q)$$

We know that

$$\mathscr{E}^\circ_{cell} = 0.48 \, V$$

and
$$Q = \frac{[Al^{3+}]^2}{[Mn^{2+}]^3} = \frac{(1.50)^2}{(0.50)^3} = 18$$

Since the half-reactions are

Oxidation: $2Al \longrightarrow 2Al^{3+} + 6e^-$

Reduction: $3Mn^{2+} + 6e^- \longrightarrow 3Mn$

we know that

$$n = 6$$

Thus $$\mathscr{E}_{cell} = 0.48 - \frac{0.0591}{6} \log(18)$$

$$= 0.48 - \frac{0.0591}{6}(1.26) = 0.48 - 0.01 = 0.47 \text{ V}$$

Note that the cell voltage decreases slightly because of the nonstandard concentrations. This change is consistent with the predictions of Le Châtelier's principle (see Example 17-7). In this case, since the reactant concentration is lower than 1.0 M and the product concentration is higher than 1.0 M, $\mathscr{E}_{cell}$ is less than $\mathscr{E}_{cell}^{\circ}$.

The potential calculated from the Nernst equation is the maximum potential before any current flow has occurred. As the cell discharges and current flows from anode to cathode, the concentrations will change, and as a result, $\mathscr{E}_{cell}$ will change. In fact, *the cell will spontaneously discharge until it reaches equilibrium,* at which point

$$Q = K \text{ (the equilibrium constant)} \quad \text{and} \quad \mathscr{E}_{cell} = 0$$

A "dead" battery is one in which the cell reaction has reached equilibrium, and there is no longer any chemical driving force to push electrons through the wire. In other words, *at equilibrium, the components in the two cell compartments have the same free energy,* and $\Delta G = 0$ for the cell reaction at the equilibrium concentrations. The cell no longer has the ability to do work.

Critical Thinking

What if you are told that $\mathscr{E}^{\circ} = 0$ for an electrolytic cell? Does this mean the cell is "dead"? What if $\mathscr{E} = 0$? Explain your answer in each case.

EXAMPLE 17-9   The Nernst Equation

Describe the cell based on the following half-reactions:

$VO_2^+ + 2H^+ + e^- \longrightarrow VO^{2+} + H_2O$	$\mathscr{E}^{\circ} = 1.00 \text{ V}$	(1)
$Zn^{2+} + 2e^- \longrightarrow Zn$	$\mathscr{E}^{\circ} = -0.76 \text{ V}$	(2)

where $T = 25°C$

$[VO_2^+] = 2.0 \ M$

$[H^+] = 0.50 \ M$

$[VO^{2+}] = 1.0 \times 10^{-2} \ M$

$[Zn^{2+}] = 1.0 \times 10^{-1} \ M$

Solution

The balanced cell reaction is obtained by reversing reaction (2) and multiplying reaction (1) by 2:

2 × reaction (1)	$2VO_2^+ + 4H^+ + 2e^- \longrightarrow 2VO^{2+} + 2H_2O$	$\mathscr{E}°$ (cathode) = 1.00 V
Reaction (2) reversed	$Zn \longrightarrow Zn^{2+} + 2e^-$	$-\mathscr{E}°$ (anode) = 0.76 V

Cell reaction: $2VO_2^+(aq) + 4H^+(aq) + Zn(s) \longrightarrow 2VO^{2+}(aq) + 2H_2O(l) + Zn^{2+}(aq)$ $\mathscr{E}°_{cell}$ = 1.76 V

Since the cell contains components at concentrations other than 1 M, we must use the Nernst equation, where $n = 2$ (since two electrons are transferred), to calculate the cell potential. At 25°C we can use the equation

$$\mathscr{E} = \mathscr{E}°_{cell} - \frac{0.0591}{n} \log(Q)$$

$$= 1.76 - \frac{0.0591}{2} \log\left(\frac{[Zn^{2+}][VO^{2+}]^2}{[VO_2^+]^2[H^+]^4}\right)$$

$$= 1.76 - \frac{0.0591}{2} \log\left(\frac{(1.0 \times 10^{-1})(1.0 \times 10^{-2})^2}{(2.0)^2(0.50)^4}\right)$$

$$= 1.76 - \frac{0.0591}{2} \log(4 \times 10^{-5}) = 1.76 + 0.13 = 1.89 \text{ V}$$

The cell diagram is given in Fig. 17-11.

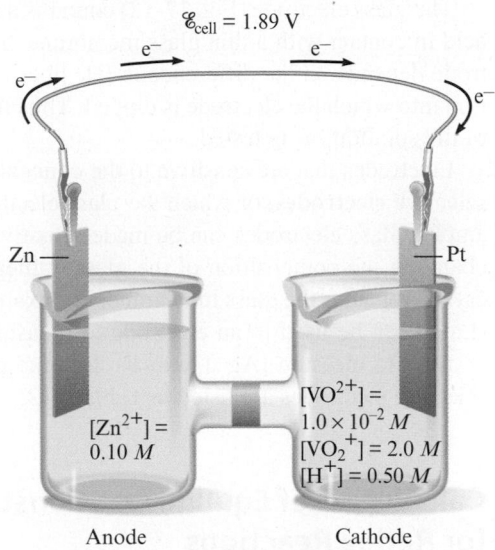

Figure 17-11 | Schematic diagram of the cell described in Example 17-9.

$\mathscr{E}_{cell}$ = 1.89 V

Zn ——   —— Pt

$[Zn^{2+}]$ = 0.10 M

$[VO^{2+}]$ = 1.0×10^{-2} M
$[VO_2^+]$ = 2.0 M
$[H^+]$ = 0.50 M

Anode   Cathode

See Exercises 17-71 through 17-74

Ion-Selective Electrodes

Because the cell potential is sensitive to the concentrations of the reactants and products involved in the cell reaction, measured potentials can be used to determine the concentration of an ion. A pH meter (see Fig. 13-7) is a familiar example of an instrument that measures concentration using an observed potential. The pH meter has three main components: a standard electrode of known potential, a special **glass electrode** that changes potential depending on the concentration of H^+ ions in the solution into which it is dipped, and a potentiometer that measures the potential between the

Figure 17-12 | A glass electrode contains a reference solution of dilute hydrochloric acid in contact with a thin glass membrane in which a silver wire coated with silver chloride has been embedded. When the electrode is dipped into a solution containing H^+ ions, the electrode potential is determined by the difference in $[H^+]$ between the two solutions.

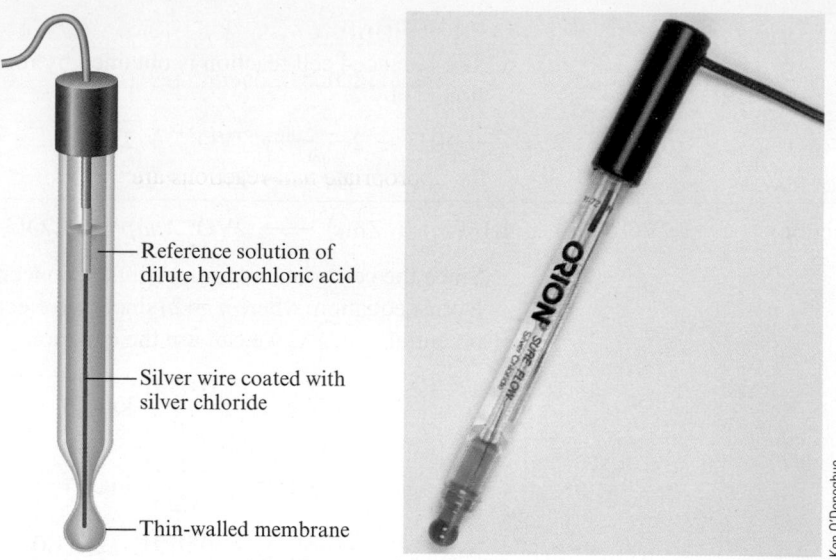

Reference solution of dilute hydrochloric acid

Silver wire coated with silver chloride

Thin-walled membrane

Ken O'Donoghue

Table 17-2 | Some Ions Whose Concentrations Can Be Detected by Ion-Selective Electrodes

Cations	Anions
H^+	Br^-
Cd^{2+}	Cl^-
Ca^{2+}	CN^-
Cu^{2+}	F^-
K^+	NO_3^-
Ag^+	S^{2-}
Na^+	

electrodes. The potentiometer reading is automatically converted electronically to a direct reading of the pH of the solution being tested.

The glass electrode (Fig. 17-12) contains a reference solution of dilute hydrochloric acid in contact with a thin glass membrane. The electrical potential of the glass electrode depends on the difference in $[H^+]$ between the reference solution and the solution into which the electrode is dipped. Thus the electrical potential varies with the pH of the solution being tested.

Electrodes that are sensitive to the concentration of a particular ion are called **ion-selective electrodes**, of which the glass electrode for pH measurement is just one example. Glass electrodes can be made sensitive to such ions as Na^+, K^+, or NH_4^+ by changing the composition of the glass. Other ions can be detected if an appropriate crystal replaces the glass membrane. For example, a crystal of lanthanum(III) fluoride (LaF_3) can be used in an electrode to measure $[F^-]$. Solid silver sulfide (Ag_2S) can be used to measure $[Ag^+]$ and $[S^{2-}]$. Some of the ions that can be detected by ion-selective electrodes are listed in Table 17-2.

Calculation of Equilibrium Constants for Redox Reactions

The quantitative relationship between $\mathscr{E}°$ and $\Delta G°$ allows calculation of equilibrium constants for redox reactions. For a cell at equilibrium,

$$\mathscr{E}_{cell} = 0 \quad \text{and} \quad Q = K$$

Applying these conditions to the Nernst equation valid at 25°C,

$$\mathscr{E} = \mathscr{E}° - \frac{0.0591}{n}\log(Q)$$

gives

$$0 = \mathscr{E}° - \frac{0.0591}{n}\log(K)$$

or

$$\log(K) = \frac{n\mathscr{E}°}{0.0591} \quad \text{at 25°C}$$

INTERACTIVE EXAMPLE 17-10

Equilibrium Constants from Cell Potentials

For the oxidation–reduction reaction

$$S_4O_6^{2-}(aq) + Cr^{2+}(aq) \longrightarrow Cr^{3+}(aq) + S_2O_3^{2-}(aq)$$

the appropriate half-reactions are

$$S_4O_6^{2-} + 2e^- \longrightarrow 2S_2O_3^{2-} \qquad \mathscr{E}° = 0.17 \text{ V} \qquad (1)$$

$$Cr^{3+} + e^- \longrightarrow Cr^{2+} \qquad \mathscr{E}° = -0.50 \text{ V} \qquad (2)$$

Balance the redox reaction, and calculate $\mathscr{E}°$ and K (at 25°C). ◀

Solution

To obtain the balanced reaction, we must reverse reaction (2), multiply it by 2, and add it to reaction (1):

Reaction (1)	$S_4O_6^{2-} + 2e^- \longrightarrow 2S_2O_3^{2-}$	$\mathscr{E}° \text{ (cathode)} = 0.17 \text{ V}$
2 × reaction (2) reversed	$2(Cr^{2+} \longrightarrow Cr^{3+} + e^-)$	$-\mathscr{E}° \text{ (anode)} = -(-0.50) \text{ V}$
Cell reaction:	$2Cr^{2+}(aq) + S_4O_6^{2-}(aq) \longrightarrow 2Cr^{3+}(aq) + 2S_2O_3^{2-}(aq)$	$\mathscr{E}° = 0.67 \text{ V}$

In this reaction, 2 moles of electrons are transferred for every unit of reaction, that is, for every 2 moles of Cr^{2+} reacting with 1 mole of $S_4O_6^{2-}$ to form 2 moles of Cr^{3+} and 2 moles of $S_2O_3^{2-}$. Thus $n = 2$. Then

$$\log(K) = \frac{n\mathscr{E}°}{0.0591} = \frac{2(0.67)}{0.0591} = 22.6$$

The value of K is found by taking the antilog of 22.6:

$$K = 10^{22.6} = 4 \times 10^{22}$$

This very large equilibrium constant is not unusual for a redox reaction.

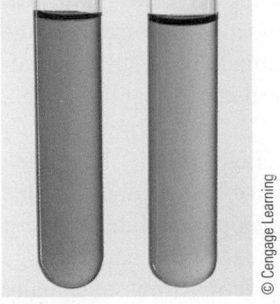

See Exercises 17-75 through 17-78

▲ The blue solution contains Cr^{2+} ions, and the green solution contains Cr^{3+} ions.

17-6 | Batteries

A **battery** is a galvanic cell or, more commonly, a group of galvanic cells connected in series, where the potentials of the individual cells add to give the total battery potential. Batteries are a source of direct current and have become an essential source of portable power in our society. In this section we examine the most common types of batteries. Some new batteries currently being developed are described at the end of the chapter.

Lead Storage Battery

Since about 1915 when self-starters were first used in automobiles, the **lead storage battery** has been a major factor in making the automobile a practical means of transportation. This type of battery can function for several years under temperature extremes from −30°F to 120°F and under incessant punishment from rough roads.

In this battery, lead serves as the anode, and lead coated with lead dioxide serves as the cathode. Both electrodes dip into an electrolyte solution of sulfuric acid. The electrode reactions are

Anode reaction:	$Pb + HSO_4^- \longrightarrow PbSO_4 + H^+ + 2e^-$
Cathode reaction:	$PbO_2 + HSO_4^- + 3H^+ + 2e^- \longrightarrow PbSO_4 + 2H_2O$

Cell reaction: $Pb(s) + PbO_2(s) + 2H^+(aq) + 2HSO_4^-(aq) \longrightarrow 2PbSO_4(s) + 2H_2O(l)$

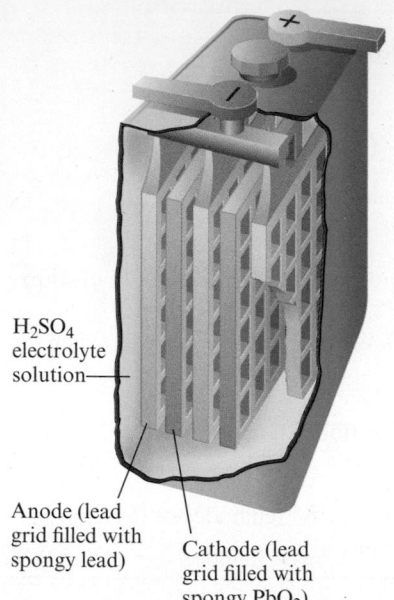

Figure 17-13 | One of the six cells in a 12-V lead storage battery. The anode consists of a lead grid filled with spongy lead, and the cathode is a lead grid filled with lead dioxide. The cell also contains 38% (by mass) sulfuric acid.

H₂SO₄ electrolyte solution—

Anode (lead grid filled with spongy lead)

Cathode (lead grid filled with spongy PbO₂)

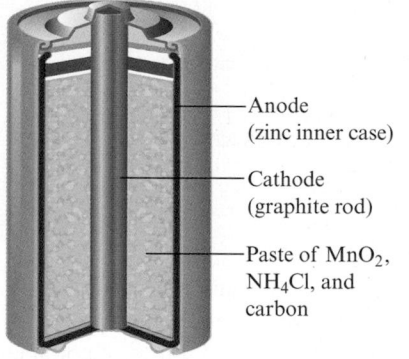

Anode (zinc inner case)

Cathode (graphite rod)

Paste of MnO_2, NH_4Cl, and carbon

Figure 17-14 | A common dry cell battery.

The typical automobile lead storage battery has six cells connected in series. Each cell contains multiple electrodes in the form of grids (Fig. 17-13) and produces approximately 2 V, to give a total battery potential of about 12 V. Note from the cell reaction that sulfuric acid is consumed as the battery discharges. This lowers the density of the electrolyte solution from its initial value of about 1.28 g/cm³ in the fully charged battery. As a result, the condition of the battery can be monitored by measuring the density of the sulfuric acid solution. The solid lead sulfate formed in the cell reaction during discharge adheres to the grid surfaces of the electrodes. The battery is recharged by forcing current through it in the opposite direction to reverse the cell reaction. A car's battery is continuously charged by an alternator driven by the automobile engine.

An automobile with a dead battery can be "jump-started" by connecting its battery to the battery in a running automobile. This process can be dangerous, however, because the resulting flow of current causes electrolysis of water in the dead battery, producing hydrogen and oxygen gases (see Section 17-8 for details). Disconnecting the jumper cables after the disabled car starts causes an arc that can ignite the gaseous mixture. If this happens, the battery may explode, ejecting corrosive sulfuric acid. This problem can be avoided by connecting the ground jumper cable to a part of the engine remote from the battery. Any arc produced when this cable is disconnected will then be harmless.

Traditional types of storage batteries require periodic "topping off" because the water in the electrolyte solution is depleted by the electrolysis that accompanies the charging process. Recent types of batteries have electrodes made of an alloy of calcium and lead that inhibits the electrolysis of water. These batteries can be sealed, since they require no addition of water.

It is rather amazing that in the 100 years in which lead storage batteries have been used, no better system has been found. Although a lead storage battery does provide excellent service, it has a useful lifetime of 3 to 5 years in an automobile. While it might seem that the battery could undergo an indefinite number of discharge/charge cycles, physical damage from road shock and chemical side-reactions eventually cause the battery to fail.

Other Batteries

The calculators, electronic games, digital watches, and portable audio players that are so familiar to us are all powered by small, efficient batteries. The common **dry cell battery** was invented more than 100 years ago by Georges Leclanché (1839–1882), a French chemist. In its *acid version,* the dry cell battery contains a zinc inner case that acts as the anode and a carbon rod in contact with a moist paste of solid MnO_2, solid NH_4Cl, and carbon that acts as the cathode (Fig. 17-14). The half-reactions are complex but can be approximated as follows:

Anode reaction: $Zn \longrightarrow Zn^{2+} + 2e^-$

Cathode reaction: $2NH_4^+ + 2MnO_2 + 2e^- \longrightarrow Mn_2O_3 + 2NH_3 + H_2O$

This cell produces a potential of about 1.5 V.

In the *alkaline version* of the dry cell battery, the solid NH_4Cl is replaced with KOH or $NaOH$. In this case the half-reactions can be approximated as follows:

Anode reaction: $Zn + 2OH^- \longrightarrow ZnO + H_2O + 2e^-$

Cathode reaction: $2MnO_2 + H_2O + 2e^- \longrightarrow Mn_2O_3 + 2OH^-$

The alkaline dry cell lasts longer mainly because the zinc anode corrodes less rapidly under basic conditions than under acidic conditions.

Other types of useful batteries include the *silver cell,* which has a Zn anode and a cathode that uses Ag_2O as the oxidizing agent in a basic environment. *Mercury cells,*

Figure 17-15 | A mercury battery of the type used in calculators.

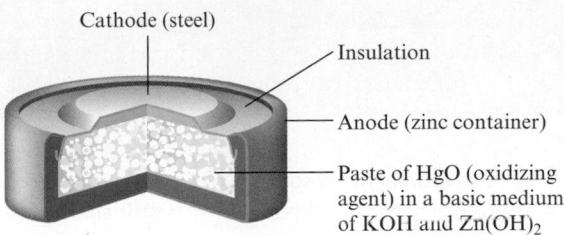

often used in calculators, have a Zn anode and a cathode involving HgO as the oxidizing agent in a basic medium (Fig. 17-15).

An especially important type of battery is the *nickel–cadmium battery,* in which the electrode reactions are

$$\text{Anode reaction:} \qquad Cd + 2OH^- \longrightarrow Cd(OH)_2 + 2e^-$$

$$\text{Cathode reaction:} \quad NiO_2 + 2H_2O + 2e^- \longrightarrow Ni(OH)_2 + 2OH^-$$

As in the lead storage battery, the products adhere to the electrodes. Therefore, a nickel–cadmium battery can be recharged an indefinite number of times.

Lithium-ion batteries involve the migration of Li^+ ions from the cathode to the anode, where they intercalate (enter the interior) as the battery is charged. At the same time, charge-balancing electrons travel to the anode through the external circuit in the charger. On discharge, the opposite process occurs. The cathode of the first successful lithium-ion batteries originally contained $LiCoO_2$ and a lithium-intercalated carbon (LiC_6) anode. More recently manufacturers have included transition metals such as nickel and manganese in the cathode in addition to cobalt. The mixed-metal cathodes have greater charge capacity and power output and shorter recharge times.

Lithium-ion batteries are used in a wide variety of applications, including cell phones, laptop computers, power tools, and even electric drive systems in automobiles and motorcycles. ◄

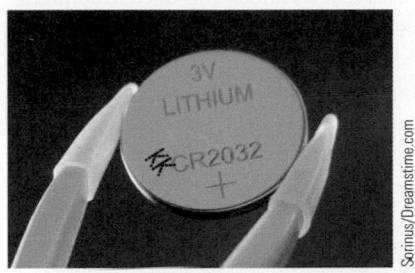

▲ Batteries for electronic watches are, by necessity, very tiny.

Fuel Cells

A **fuel cell** is *a galvanic cell for which the reactants are continuously supplied.* To illustrate the principles of fuel cells, let's consider the exothermic redox reaction of methane with oxygen:

$$CH_4(g) + 2O_2(g) \longrightarrow CO_2(g) + 2H_2O(g) + \text{energy}$$

Usually the energy from this reaction is released as heat to warm homes and to run machines. However, in a fuel cell designed to use this reaction, the energy is used to produce an electric current: The electrons flow from the reducing agent (CH_4) to the oxidizing agent (O_2) through a conductor.

The U.S. space program has supported extensive research to develop fuel cells. The space shuttle uses a fuel cell based on the reaction of hydrogen and oxygen to form water:

$$2H_2(g) + O_2(g) \longrightarrow 2H_2O(l)$$

A schematic of a fuel cell that uses this reaction is shown in Fig. 17-16. The half-reactions are

$$\text{Anode reaction:} \qquad 2H_2 + 4OH^- \longrightarrow 4H_2O + 4e^-$$

$$\text{Cathode reaction:} \quad 4e^- + O_2 + 2H_2O \longrightarrow 4OH^-$$

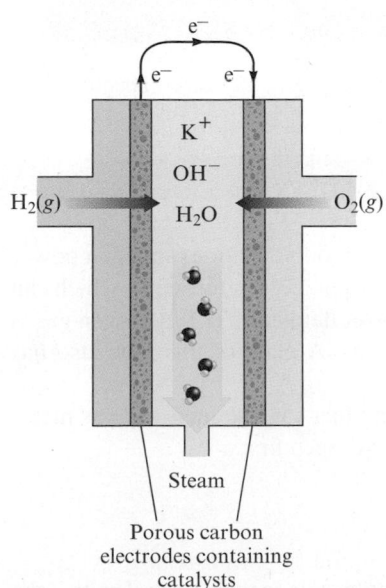

Figure 17-16 | Schematic of the hydrogen–oxygen fuel cell.

A cell of this type weighing about 500 pounds has been designed for space vehicles, but this fuel cell is not practical enough for general use as a source of portable power. However, current research on portable electrochemical power is now proceeding at a rapid pace. In fact, cars powered by fuel cells are now being tested on the streets.

The promise of an energy-efficient, environmentally sound source of electrical power has spurred an intense interest in fuel cells in recent years. Although fuel cells have long been used in the U.S. space program, no practical fuel cell for powering automobiles has been developed. However, we are now on the verge of practical fuel-cell–powered cars. For example, DaimlerChrysler's NECAR 5 was driven across the United States, a 3000-mile trip that took 16 days. NECAR 5 is powered by a H_2/O_2 fuel cell that generates its H_2 from decomposition of methanol (CH_3OH).

General Motors, which has also been experimenting with H_2/O_2 fuel cells, in 2005 introduced the Chevrolet Sequel, which can go from 0 to 60 mph in 10 seconds with a range of 300 miles. The car has composite tanks for storage of 8 kg of liquid hydrogen.

In reality, fuel cells have a long way to go before they can be economically viable in automobiles. The main problem is the membrane that separates the hydrogen electrode from the oxygen electrode. This membrane must prevent H_2 molecules from passing through and still allow ions to pass between the electrodes. Current membranes cost over $3000 for an automobile-size fuel cell. The hope is to reduce this by a factor of 10 in the next few years.

Besides providing power for automobiles, fuel cells are being considered for powering small electronic devices such as cameras, cell phones, and laptop computers. Many of these micro fuel cells currently use methanol as the fuel (reducing agent) rather than H_2. However, these direct-methanol fuel cells are rife with problems. A major difficulty is water management. Water is needed at the anode to react with the methanol and is produced at the cathode. Water is also needed to moisten the electrolyte to promote charge migration.

Although the direct-methanol fuel cell is currently the leader among micro fuel-cell designs, its drawbacks have encouraged the development of other designs. For example, Richard Masel at the University of Illinois at Urbana–Champaign has designed a micro fuel cell that uses formic acid as the fuel. Masel and others are also experimenting with mini hot chambers external to the fuel cell that break down hydrogen-rich fuels into hydrogen gas, which is then fed into the tiny fuel cells.

To replace batteries, fuel cells must be demonstrated to be economically feasible, safe, and dependable. Today, rapid progress is being made to overcome the current problems. A recent estimate indicates that by late in this decade annual sales of the little power plants may reach 200 million units per year. It appears that after years of hype about the virtues of fuel cells, we are finally going to realize their potential.

Part of the fleet of 16 hydrogen fuel cars recently bought by the US Army.

Fuel cells are also finding use as permanent power sources. For example, a power plant built in New York City contains stacks of hydrogen–oxygen fuel cells, which can be rapidly put on-line in response to fluctuating power demands. The hydrogen gas is obtained by decomposing the methane in natural gas. A plant of this type also has been constructed in Tokyo.

In addition, new fuel cells are under development that can use fuels such as methane and diesel directly without having to produce hydrogen first.

17-7 | Corrosion

Corrosion can be viewed as the process of returning metals to their natural state—the ores from which they were originally obtained. Corrosion involves oxidation of the metal. Since corroded metal often loses its structural integrity and attractiveness, this spontaneous process has great economic impact. Approximately one-fifth of the iron and steel produced annually is used to replace rusted metal.

Metals corrode because they oxidize easily. Table 17-1 shows that, with the exception of gold, those metals commonly used for structural and decorative purposes all have standard reduction potentials less positive than that of oxygen gas. ◄ ⓘ When any of these half-reactions is reversed (to show oxidation of the metal) and combined with the reduction half-reaction for oxygen, the result is a positive $\mathscr{E}°$ value. Thus the oxidation of most metals by oxygen is spontaneous (although we cannot tell from the potential how fast it will occur).

In view of the large difference in reduction potentials between oxygen and most metals, it is surprising that the problem of corrosion does not completely prevent the use of metals in air. However, most metals develop a thin oxide coating, which tends to protect their internal atoms against further oxidation. The metal that best demonstrates this phenomenon is aluminum. With a reduction potential of -1.7 V, aluminum should be easily oxidized by O_2. According to the apparent thermodynamics of the reaction, an aluminum airplane could dissolve in a rainstorm. The fact that this very active metal can be used as a structural material is due to the formation of a thin, adherent layer of aluminum oxide (Al_2O_3), more properly represented as $Al_2(OH)_6$, which greatly inhibits further corrosion. The potential of the "passive," oxide-coated aluminum is -0.6 V, a value that causes it to behave much like a noble metal.

Iron also can form a protective oxide coating. This coating is not an infallible shield against corrosion, however; when steel is exposed to oxygen in moist air, the oxide that forms tends to scale off and expose new metal surfaces to corrosion.

The corrosion products of noble metals such as copper and silver are complex and affect the use of these metals as decorative materials. Under normal atmospheric conditions, copper forms an external layer of greenish copper carbonate called *patina*. *Silver tarnish* is silver sulfide (Ag_2S), which in thin layers gives the silver surface a richer appearance. Gold, with a positive standard reduction potential of 1.50 V, significantly larger than that for oxygen (1.23 V), shows no appreciable corrosion in air.

Corrosion of Iron

Since steel is the main structural material for bridges, buildings, and automobiles, controlling its corrosion is extremely important. To do this, we must understand the corrosion mechanism. Instead of being a direct oxidation process as we might expect, the corrosion of iron is an electrochemical reaction, as shown in Fig. 17-17.

Steel has a nonuniform surface because the chemical composition is not completely homogeneous. Also, physical strains leave stress points in the metal. These nonuniformities cause areas where the iron is more easily oxidized (*anodic regions*) than it is at others (*cathodic regions*). In the anodic regions each iron atom gives up two electrons to form the Fe^{2+} ion:

$$Fe \longrightarrow Fe^{2+} + 2e^-$$

ⓘ Some metals, such as copper, gold, silver, and platinum, are relatively difficult to oxidize. These are often called *noble metals*.

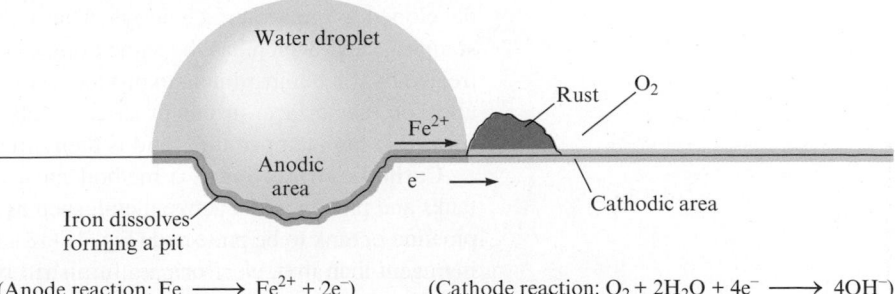

Figure 17-17 | The electrochemical corrosion of iron.

(Anode reaction: $Fe \longrightarrow Fe^{2+} + 2e^-$)   (Cathode reaction: $O_2 + 2H_2O + 4e^- \longrightarrow 4OH^-$)

The electrons that are released flow through the steel, as they do through the wire of a galvanic cell, to a cathodic region, where they react with oxygen:

$$O_2 + 2H_2O + 4e^- \longrightarrow 4OH^-$$

The Fe^{2+} ions formed in the anodic regions travel to the cathodic regions through the moisture on the surface of the steel, just as ions travel through a salt bridge in a galvanic cell. In the cathodic regions Fe^{2+} ions react with oxygen to form rust, which is hydrated iron(III) oxide of variable composition:

$$4Fe^{2+}(aq) + O_2(g) + (4 + 2n)H_2O(l) \longrightarrow 2Fe_2O_3 \cdot nH_2O(s) + 8H^+(aq)$$
Rust

Because of the migration of ions and electrons, rust often forms at sites that are remote from those where the iron dissolved to form pits in the steel. The degree of hydration of the iron oxide affects the color of the rust, which may vary from black to yellow to the familiar reddish brown.

The electrochemical nature of the rusting of iron explains the importance of moisture in the corrosion process. Moisture must be present to act as a kind of salt bridge between anodic and cathodic regions. Steel does not rust in dry air, a fact that explains why cars last much longer in the arid Southwest than in the relatively humid Midwest. Salt also accelerates rusting, a fact all too easily recognized by car owners in the colder parts of the United States, where salt is used on roads to melt snow and ice. The severity of rusting is greatly increased because the dissolved salt on the moist steel surface increases the conductivity of the aqueous solution formed there and thus accelerates the electrochemical corrosion process. Chloride ions also form very stable complex ions with Fe^{3+}, and this factor tends to encourage the dissolving of the iron, again accelerating the corrosion.

Prevention of Corrosion

Prevention of corrosion is an important way of conserving our natural resources of energy and metals. The primary means of protection is the application of a coating, most commonly paint or metal plating, to protect the metal from oxygen and moisture. Chromium and tin are often used to plate steel (see Section 17-9) because they oxidize to form a durable, effective oxide coating. Zinc, also used to coat steel in a process called **galvanizing**, forms a mixed oxide–carbonate coating. Since zinc is a more active metal than iron, as the potentials for the oxidation half-reactions show,

$$Fe \longrightarrow Fe^{2+} + 2e^- \qquad -\mathscr{E}° = 0.44 \text{ V}$$
$$Zn \longrightarrow Zn^{2+} + 2e^- \qquad -\mathscr{E}° = 0.76 \text{ V}$$

any oxidation that occurs dissolves zinc rather than iron. Recall that the reaction with the most positive standard potential has the greatest thermodynamic tendency to occur. Thus zinc acts as a "sacrificial" coating on steel.

Alloying is also used to prevent corrosion. *Stainless steel* contains chromium and nickel, both of which form oxide coatings that change steel's reduction potential to one characteristic of the noble metals. In addition, a new technology is now being developed to create surface alloys. That is, instead of forming a metal alloy such as stainless steel, which has the same composition throughout, a cheaper carbon steel is treated by ion bombardment to produce a thin layer of stainless steel or other desirable alloy on the surface. In this process, a "plasma" or "ion gas" of the alloying ions is formed at high temperatures and is then directed onto the surface of the metal.

Cathodic protection is a method most often used to protect steel in buried fuel tanks and pipelines. An active metal, such as magnesium, is connected by a wire to the pipeline or tank to be protected (Fig. 17-18). Because the magnesium is a better reducing agent than iron, electrons are furnished by the magnesium rather than by the iron, keeping the iron from being oxidized. As oxidation occurs, the magnesium anode

Figure 17-18 | Cathodic protection of an underground pipe.

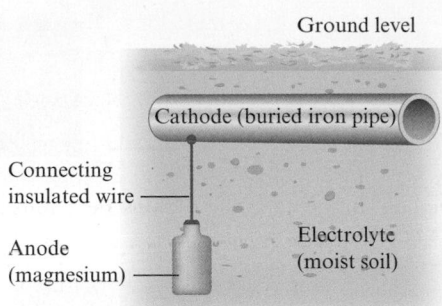

dissolves, and so it must be replaced periodically. Ships' hulls are protected in a similar way by attaching bars of titanium metal to the steel hull. In salt water the titanium acts as the anode and is oxidized instead of the steel hull (the cathode).

17-8 | Electrolysis

A galvanic cell produces current when an oxidation–reduction reaction proceeds spontaneously. A similar apparatus, an **electrolytic cell**, uses electrical energy to produce chemical change. The process of **electrolysis** involves *forcing a current through a cell to produce a chemical change for which the cell potential is negative;* that is, electrical work causes an otherwise nonspontaneous chemical reaction to occur. ◄ ⓘ Electrolysis has great practical importance; for example, charging a battery, producing aluminum metal, and chrome plating an object are all done electrolytically.

ⓘ An electrolytic cell uses electrical energy to produce a chemical change that would otherwise not occur spontaneously.

To illustrate the difference between a galvanic cell and an electrolytic cell, consider the cell shown in Fig. 17-19(a) as it runs spontaneously to produce 1.10 V. In this *galvanic* cell, the reaction at the anode is

$$Zn \longrightarrow Zn^{2+} + 2e$$

whereas at the cathode the reaction is

$$Cu^{2+} + 2e^- \longrightarrow Cu$$

Figure 17-19(b) shows an external power source forcing electrons through the cell in the *opposite* direction to that in (a). This requires an external potential greater than 1.10 V, which must be applied in opposition to the natural cell potential. This device is an *electrolytic cell*. Notice that since electron flow is opposite in the two cases, the anode and cathode are reversed between (a) and (b). Also, ion flow through the salt bridge is opposite in the two cells.

Now we will consider the stoichiometry of electrolytic processes, that is, *how much chemical change occurs with the flow of a given current for a specified time.* Suppose we wish to determine the mass of copper that is plated out when a current of 10.0 amps (an **ampere** [amp], abbreviated A, is *1 coulomb of charge per second*) is passed for 30.0 minutes through a solution containing Cu^{2+}. *Plating* means depositing the neutral metal on the electrode by reducing the metal ions in solution. In this case each Cu^{2+} ion requires two electrons to become an atom of copper metal:

$$Cu^{2+}(aq) + 2e^- \longrightarrow Cu(s)$$

This reduction process will occur at the cathode of the electrolytic cell.

To solve this stoichiometry problem, we need the following steps:

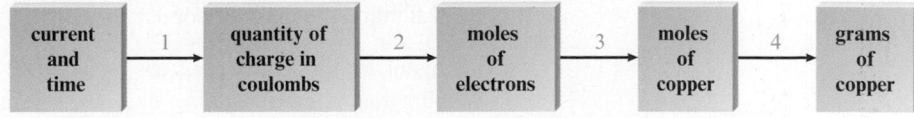

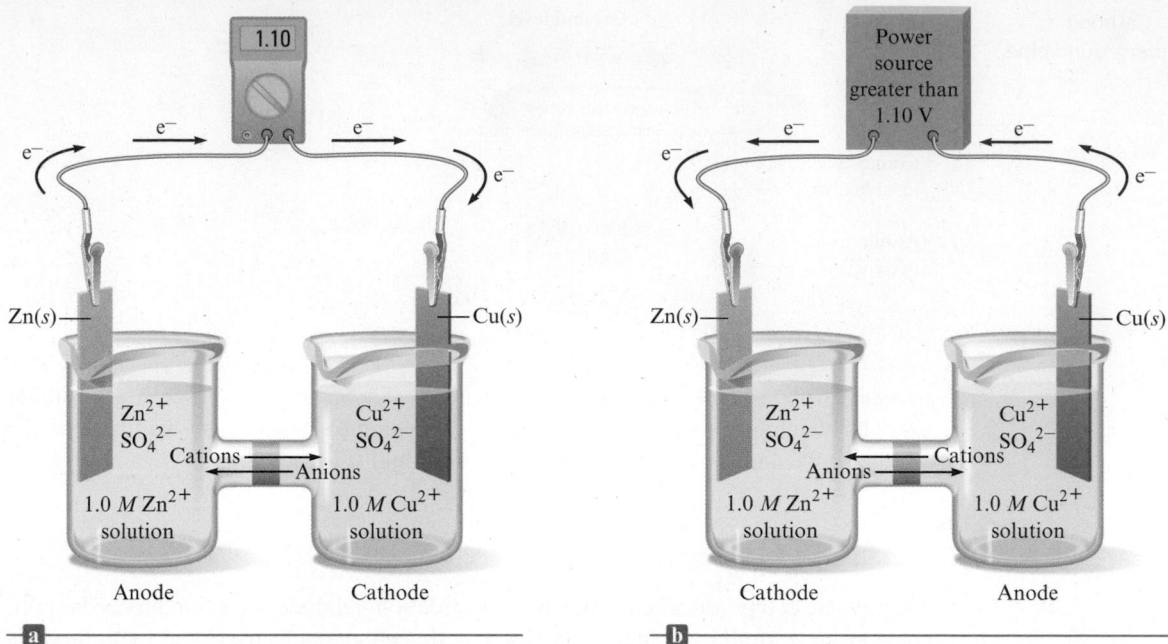

Figure 17-19 | (a) A standard galvanic cell based on the spontaneous reaction

$$Zn + Cu^{2+} \longrightarrow Zn^{2+} + Cu$$

(b) A standard electrolytic cell. A power source forces the opposite reaction

$$Cu + Zn^{2+} \longrightarrow Cu^{2+} + Zn$$

1. Since an amp is a coulomb of charge per second, we multiply the current by the time in seconds to obtain the total coulombs of charge passed into the Cu^{2+} solution at the cathode:

$$\text{Coulombs of charge} = \text{amps} \times \text{seconds} = \frac{C}{s} \times s$$

$$= 10.0 \, \frac{C}{s} \times 30.0 \, \text{min} \times 60.0 \, \frac{s}{\text{min}}$$

$$= 1.80 \times 10^4 \, C$$

2. Since 1 mole of electrons carries a charge of 1 faraday, or 96,485 coulombs, we can calculate the number of moles of electrons required to carry 1.80×10^4 coulombs of charge:

$$1.80 \times 10^4 \, C \times \frac{1 \, \text{mol e}^-}{96{,}485 \, C} = 1.87 \times 10^{-1} \, \text{mol e}^-$$

This means that 0.187 mole of electrons flowed into the Cu^{2+} solution.

3. Each Cu^{2+} ion requires two electrons to become a copper atom. Thus each mole of electrons produces $\frac{1}{2}$ mole of copper metal:

$$1.87 \times 10^{-1} \, \text{mol e}^- \times \frac{1 \, \text{mol Cu}}{2 \, \text{mol e}^-} = 9.35 \times 10^{-2} \, \text{mol Cu}$$

4. We now know the moles of copper metal plated onto the cathode, and we can calculate the mass of copper formed:

$$9.35 \times 10^{-2} \, \text{mol Cu} \times \frac{63.546 \, g}{\text{mol Cu}} = 5.94 \, g \, Cu$$

INTERACTIVE EXAMPLE 17-11

i Example 17-11 describes only the half-cell of interest. There also must be an anode at which oxidation is occurring.

Electroplating

How long must a current of 5.00 A be applied to a solution of Ag^+ to produce 10.5 g silver metal? ◄ *i*

Solution

In this case, we must use the steps given earlier in reverse:

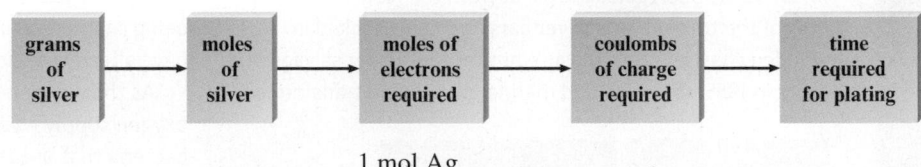

$$10.5 \text{ g Ag} \times \frac{1 \text{ mol Ag}}{107.868 \text{ g Ag}} = 9.73 \times 10^{-2} \text{ mol Ag}$$

Each Ag^+ ion requires one electron to become a silver atom:

$$Ag^+ + e^- \longrightarrow Ag$$

Thus 9.73×10^{-2} mole of electrons is required, and we can calculate the quantity of charge carried by these electrons:

$$9.73 \times 10^{-2} \text{ mol e}^- \times \frac{96{,}485 \text{ C}}{\text{mol e}^-} = 9.39 \times 10^3 \text{ C}$$

The 5.00 A (5.00 C/s) of current must produce 9.39×10^3 C of charge. Thus

$$\left(5.00 \frac{\text{C}}{\text{s}}\right) \times (\text{time, in s}) = 9.39 \times 10^3 \text{ C}$$

$$\text{Time} = \frac{9.39 \times 10^3}{5.00} \text{ s} = 1.88 \times 10^3 \text{ s} = 31.3 \text{ min}$$

See Exercises 17-93 through 17-96

Electrolysis of Water

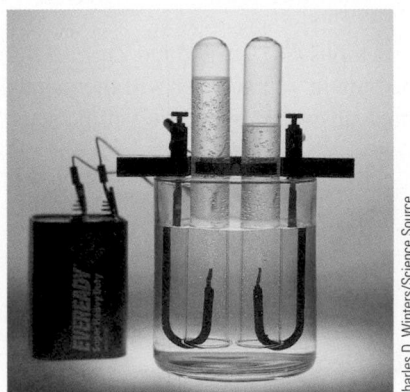

Figure 17-20 | The electrolysis of water produces hydrogen gas at the cathode (on the right) and oxygen gas at the anode (on the left). Note that twice as much hydrogen is produced as oxygen.

Charles D. Winters/Science Source

We have seen that hydrogen and oxygen combine spontaneously to form water and that the accompanying decrease in free energy can be used to run a fuel cell to produce electricity. The reverse process, which is of course nonspontaneous, can be forced by electrolysis:

Anode reaction:	$2H_2O \longrightarrow O_2 + 4H^+ + 4e^-$	$-\mathscr{E}° = -1.23 \text{ V}$
Cathode reaction:	$4H_2O + 4e^- \longrightarrow 2H_2 + 4OH^-$	$\mathscr{E}° = -0.83 \text{ V}$
Net reaction:	$6H_2O \longrightarrow 2H_2 + O_2 + \underbrace{4(H^+ + OH^-)}_{4H_2O}$	$\mathscr{E}° = -2.06 \text{ V}$

or

$$2H_2O \longrightarrow 2H_2 + O_2$$

Note that these potentials assume an anode chamber with 1 *M* H^+ and a cathode chamber with 1 *M* OH^-. In pure water, where $[H^+] = [OH^-] = 10^{-7}$ *M*, the potential for the overall process is -1.23 V.

In practice, however, if platinum electrodes connected to a 6-V battery are dipped into pure water, no reaction is observed because pure water contains so few ions that only a negligible current can flow. However, addition of even a small amount of a soluble salt causes an immediate evolution of bubbles of hydrogen and oxygen, as illustrated in Fig. 17-20.

When the galleon *Atocha* was destroyed on a reef by a hurricane in 1622, it was bound for Spain carrying approximately 47 tons of copper, gold, and silver from the New World. The bulk of the treasure was silver bars and coins packed in wooden chests. When treasure hunter Mel Fisher salvaged the silver in 1985, corrosion and marine growth had transformed the shiny metal into something that looked like coral. Restoring the silver to its original condition required an understanding of the chemical changes that had occurred in 350 years of being submerged in the ocean. Much of this chemistry we have already considered at various places in this text.

As the wooden chests containing the silver decayed, the oxygen supply was depleted, favoring the growth of certain bacteria that use the sulfate ion rather than oxygen as an oxidizing agent to generate energy. As these bacteria consume sulfate ions, they release hydrogen sulfide gas that reacts with silver to form black silver sulfide:

$$2Ag(s) + H_2S(aq) \longrightarrow Ag_2S(s) + H_2(g)$$

Thus, over the years, the surface of the silver became covered with a tightly adhering layer of corrosion, which fortunately protected the silver underneath and thus prevented total conversion of the silver to silver sulfide.

Another change that took place as the wood decomposed was the formation of carbon dioxide. This shifted the equilibrium that is present in the ocean,

$$CO_2(aq) + H_2O(l) \rightleftharpoons HCO_3^-(aq) + H^+(aq)$$

to the right, producing higher concentrations of HCO_3^-. In turn, the HCO_3^- reacted with Ca^{2+} ions present in the seawater to form calcium carbonate:

$$Ca^{2+}(aq) + HCO_3^-(aq) \rightleftharpoons CaCO_3(s) + H^+(aq)$$

Courtesy, Mel Fisher's Treasure

Silver coins and tankards salvaged from the wreck of the Atocha.

Electrolysis of Mixtures of Ions

Suppose a solution in an electrolytic cell contains the ions Cu^{2+}, Ag^+, and Zn^{2+}. If the voltage is initially very low and is gradually turned up, in which order will the metals be plated out onto the cathode? This question can be answered by looking at the standard reduction potentials of these ions:

$$Ag^+ + e^- \longrightarrow Ag \qquad \mathscr{E}° = 0.80 \text{ V}$$
$$Cu^{2+} + 2e^- \longrightarrow Cu \qquad \mathscr{E}° = 0.34 \text{ V}$$
$$Zn^{2+} + 2e^- \longrightarrow Zn \qquad \mathscr{E}° = -0.76 \text{ V}$$

Remember that the more *positive* the $\mathscr{E}°$ value, the more the reaction has a tendency to proceed in the direction indicated. Of the three reactions listed, the reduction of Ag^+ occurs most easily, and the order of oxidizing ability is

$$Ag^+ > Cu^{2+} > Zn^{2+}$$

This means that silver will plate out first as the potential is increased, followed by copper, and finally zinc.

Calcium carbonate is the main component of limestone. Thus, over time, the corroded silver coins and bars became encased in limestone.

Both the limestone formation and the corrosion had to be dealt with. Since $CaCO_3$ contains the basic anion CO_3^{2-}, acid dissolves limestone:

$$2H^+(aq) + CaCO_3(s) \longrightarrow Ca^{2+}(aq) + CO_2(g) + H_2O(l)$$

Soaking the mass of coins in a buffered acidic bath for several hours allowed the individual pieces to be separated, and the black Ag_2S on the surfaces was revealed. An abrasive could not be used to remove this corrosion; it would have destroyed the details of the engraving—a very valuable feature of the coins to a historian or a collector—and it would have washed away some of the silver. Instead, the corrosion reaction was reversed through electrolytic reduction. The coins were connected to the cathode of an electrolytic cell in a dilute sodium hydroxide solution as represented in the figure.

As electrons flow, the Ag^+ ions in the silver sulfide are reduced to silver metal:

$$Ag_2S + 2e^- \longrightarrow Ag + S^{2-}$$

As a by-product, bubbles of hydrogen gas from the reduction of water form on the surface of the coins:

$$2H_2O + 2e^- \longrightarrow H_2(g) + 2OH^-$$

The agitation caused by the bubbles loosens the flakes of metal sulfide and helps clean the coins.

These procedures have made it possible to restore the treasure to very nearly its condition when the *Atocha* sailed many years ago.

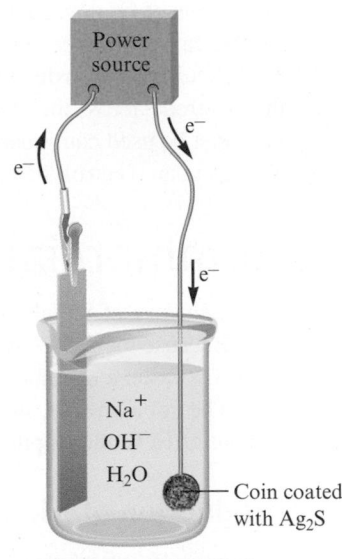

INTERACTIVE EXAMPLE 17-12

Relative Oxidizing Abilities

An acidic solution contains the ions Ce^{4+}, VO_2^+, and Fe^{3+}. Using the $\mathscr{E}°$ values listed in Table 17-1, give the order of oxidizing ability of these species, and predict which one will be reduced at the cathode of an electrolytic cell at the lowest voltage.

Solution

The half-reactions and $\mathscr{E}°$ values are

$$Ce^{4+} + e^- \longrightarrow Ce^{3+} \qquad \mathscr{E}° = 1.70 \text{ V}$$
$$VO_2^+ + 2H^+ + e^- \longrightarrow VO^{2+} + H_2O \qquad \mathscr{E}° = 1.00 \text{ V}$$
$$Fe^{3+} + e^- \longrightarrow Fe^{2+} \qquad \mathscr{E}° = 0.77 \text{ V}$$

The order of oxidizing ability is therefore

$$Ce^{4+} > VO_2^+ > Fe^{3+}$$

The Ce^{4+} ion will be reduced at the lowest voltage in an electrolytic cell.

See Exercise 17-105

The principle described in this section is very useful, but it must be applied with some caution. For example, in the electrolysis of an aqueous solution of sodium chloride, we should be able to use $\mathscr{E}°$ values to predict the products. Of the major species in the solution (Na^+, Cl^-, and H_2O), only Cl^- and H_2O can be readily oxidized. The half-reactions (written as oxidization processes) are

$$2Cl^- \longrightarrow Cl_2 + 2e^- \qquad -\mathscr{E}° = -1.36 \text{ V}$$

$$2H_2O \longrightarrow O_2 + 4H^+ + 4e^- \qquad -\mathscr{E}° = -1.23 \text{ V}$$

Since water has the more positive potential, we would expect to see O_2 produced at the anode because it is easier (thermodynamically) to oxidize H_2O than Cl^-. Actually, this does not happen. As the voltage is increased in the cell, the Cl^- ion is the first to be oxidized. A much higher potential than expected is required to oxidize water. The voltage required in excess of the expected value (called the *overvoltage*) is much greater for the production of O_2 than for Cl_2, which explains why chlorine is produced first.

The causes of overvoltage are very complex. Basically, the phenomenon is caused by difficulties in transferring electrons from the species in the solution to the atoms on the electrode across the electrode–solution interface. Because of this situation, $\mathscr{E}°$ values must be used cautiously in predicting the actual order of oxidation or reduction of species in an electrolytic cell.

17-9 | Commercial Electrolytic Processes

The chemistry of metals is characterized by their ability to donate electrons to form ions. Because metals are typically such good reducing agents, most are found in nature in *ores,* mixtures of ionic compounds often containing oxide, sulfide, and silicate anions. The noble metals, such as gold, silver, and platinum, are more difficult to oxidize and are often found as pure metals.

Production of Aluminum

Aluminum is one of the most abundant elements on the earth, ranking third behind oxygen and silicon. Since aluminum is a very active metal, it is found in nature as its oxide in an ore called *bauxite* (named after Les Baux, France, where it was discovered in 1821). Production of aluminum metal from its ore proved to be more difficult than production of most other metals. In 1782 Lavoisier recognized aluminum to be a metal "whose affinity for oxygen is so strong that it cannot be overcome by any known reducing agent." As a result, pure aluminum metal remained unknown. Finally, in 1854 a process was found for producing metallic aluminum using sodium, but aluminum remained a very expensive rarity. In fact, it is said that Napoleon III served his most honored guests with aluminum forks and spoons, while the others had to settle for gold and silver utensils.

The breakthrough came in 1886 when two men, Charles M. Hall in the United States and Paul Heroult in France, almost simultaneously discovered a practical electrolytic process for producing aluminum (Fig. 17-21). The key factor in the

Figure 17-21 | Charles Martin Hall (1863–1914) was a student at Oberlin College in Ohio when he first became interested in aluminum. One of his professors commented that anyone who could manufacture aluminum cheaply would make a fortune, and Hall decided to give it a try. The 21-year-old Hall worked in a wooden shed near his house with an iron frying pan as a container, a blacksmith's forge as a heat source, and galvanic cells constructed from fruit jars. Using these crude galvanic cells, Hall found that he could produce aluminum by passing a current through a molten Al_2O_3/Na_3AlF_6 mixture. By a strange coincidence, Paul Heroult, a Frenchman who was born and died in the same years as Hall, made the same discovery at about the same time.

Table 17-3 | The Price of
Aluminum over the Past Century

Date	Price of Aluminum ($/lb)*
1855	100,000
1885	100
1890	2
1895	0.50
1970	0.30
1980	0.80
1990	0.74

*Note the precipitous drop in price after
the discovery of the Hall–Heroult process.

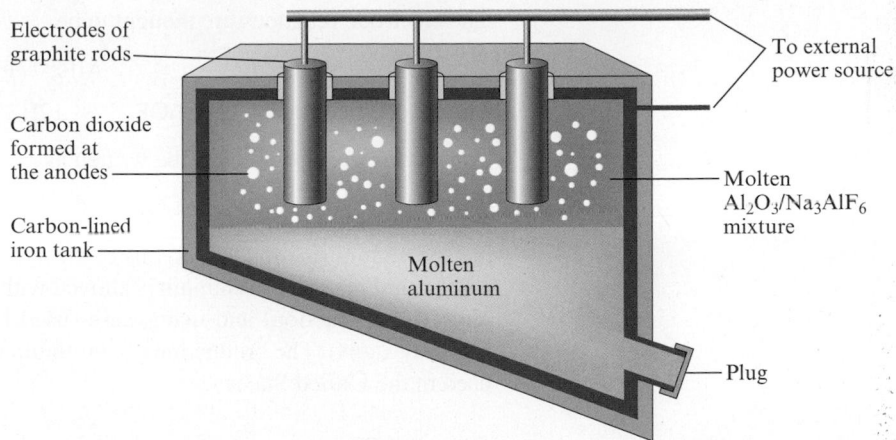

Figure 17-22 | A schematic diagram of an electrolytic cell for producing aluminum by
the Hall–Heroult process. Because molten aluminum is more dense than the mixture of
molten cryolite and alumina, it settles to the bottom of the cell and is drawn off periodically. The graphite electrodes are gradually eaten away and must be replaced from time to
time. The cell operates at a current flow of up to 250,000 A.

Hall–Heroult process is the use of molten cryolite (Na_3AlF_6) as the solvent for the
aluminum oxide.

Electrolysis is possible only if ions can move to the electrodes. A common method
for producing ion mobility is dissolving the substance to be electrolyzed in water. This
is not possible in the case of aluminum because water is more easily reduced than
Al^{3+}, as the following standard reduction potentials show:

$$Al^{3+} + 3e^- \longrightarrow Al \qquad \mathscr{E}° = -1.66 \text{ V}$$
$$2H_2O + 2e^- \longrightarrow H_2 + 2OH^- \qquad \mathscr{E}° = -0.83 \text{ V}$$

Thus aluminum metal cannot be plated out of an aqueous solution of Al^{3+}.

Ion mobility also can be produced by melting the salt. But the melting point of solid
Al_2O_3 is much too high (2050°C) to allow practical electrolysis of the molten oxide. A
mixture of Al_2O_3 and Na_3AlF_6, however, has a melting point of 1000°C, and the resulting molten mixture can be used to obtain aluminum metal electrolytically. Because of
this discovery by Hall and Heroult, the price of aluminum plunged (Table 17-3), and
its use became economically feasible.

Bauxite is not pure aluminum oxide (called *alumina*); it also contains the oxides of
iron, silicon, and titanium, and various silicate materials. To obtain the pure hydrated
alumina ($Al_2O_3 \cdot nH_2O$), the crude bauxite is treated with aqueous sodium hydroxide.
Being amphoteric, alumina dissolves in the basic solution:

$$Al_2O_3(s) + 2OH^-(aq) \longrightarrow 2AlO_2^-(aq) + H_2O(l)$$

The other metal oxides, which are basic, remain as solids. The solution containing the
aluminate ion (AlO_2^-) is separated from the sludge of the other oxides and is acidified
with carbon dioxide gas, causing the hydrated alumina to reprecipitate:

$$2CO_2(g) + 2AlO_2^-(aq) + (n + 1)H_2O(l) \longrightarrow 2HCO_3^-(aq) + Al_2O_3 \cdot nH_2O(s)$$

The purified alumina is then mixed with cryolite and melted, and the aluminum ion
is reduced to aluminum metal in an electrolytic cell of the type shown in Fig. 17-22.
Because the electrolyte solution contains a large number of aluminum-containing ions,
the chemistry is not completely clear. However, the alumina probably reacts with the
cryolite anion as follows:

$$Al_2O_3 + 4AlF_6^{3-} \longrightarrow 3Al_2OF_6^{2-} + 6F^-$$

The electrode reactions are thought to be

Cathode reaction: $$AlF_6^{3-} + 3e^- \longrightarrow Al + 6F^-$$

Anode reaction: $$2Al_2OF_6^{2-} + 12F^- + C \longrightarrow 4AlF_6^{3-} + CO_2 + 4e^-$$

The overall cell reaction can be written as

$$2Al_2O_3 + 3C \longrightarrow 4Al + 3CO_2$$

The aluminum produced in this electrolytic process is 99.5% pure. To be useful as a structural material, aluminum is alloyed with metals such as zinc (used for trailer and aircraft construction) and manganese (used for cooking utensils, storage tanks, and highway signs). The production of aluminum consumes about 5% of all the electricity used in the United States.

Electrorefining of Metals

Purification of metals is another important application of electrolysis. For example, impure copper from the chemical reduction of copper ore is cast into large slabs that serve as the anodes for electrolytic cells. Aqueous copper sulfate is the electrolyte, and thin sheets of ultrapure copper function as the cathodes (Fig. 17-23).

The main reaction at the anode is

$$Cu \longrightarrow Cu^{2+} + 2e^-$$

Other metals such as iron and zinc are also oxidized from the impure anode:

$$Zn \longrightarrow Zn^{2+} + 2e^-$$

$$Fe \longrightarrow Fe^{2+} + 2e^-$$

Noble metal impurities in the anode are not oxidized at the voltage used; they fall to the bottom of the cell to form a sludge, which is processed to remove the valuable silver, gold, and platinum.

The Cu^{2+} ions from the solution are deposited onto the cathode

$$Cu^{2+} + 2e^- \longrightarrow Cu$$

producing copper that is 99.95% pure.

Figure 17-23 | Ultrapure copper sheets that serve as the cathodes are lowered between slabs of impure copper that serve as the anodes into a tank containing an aqueous solution of copper sulfate ($CuSO_4$). It takes about 4 weeks for the anodes to dissolve and for the pure copper to be deposited on the cathodes.

Tom Hollyman/Science Source

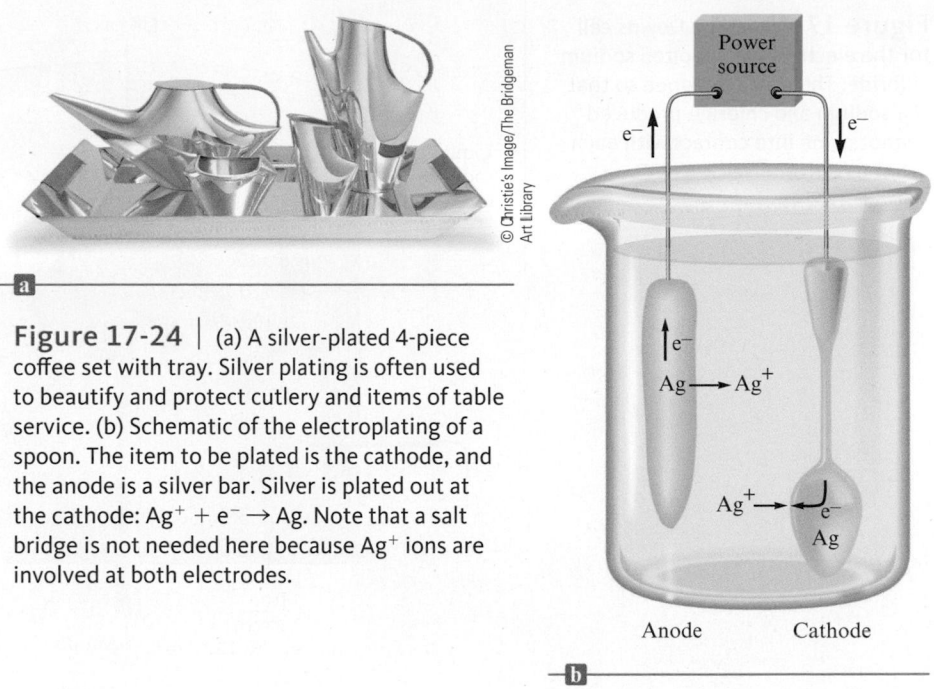

Figure 17-24 | (a) A silver-plated 4-piece coffee set with tray. Silver plating is often used to beautify and protect cutlery and items of table service. (b) Schematic of the electroplating of a spoon. The item to be plated is the cathode, and the anode is a silver bar. Silver is plated out at the cathode: $Ag^+ + e^- \rightarrow Ag$. Note that a salt bridge is not needed here because Ag^+ ions are involved at both electrodes.

Metal Plating

Metals that readily corrode can often be protected by the application of a thin coating of a metal that resists corrosion. Examples are "tin" cans, which are actually steel cans with a thin coating of tin, and chrome-plated steel bumpers for automobiles.

An object can be plated by making it the cathode in a tank containing ions of the plating metal. The silver plating of a spoon is shown schematically in Fig. 17-24(b). In an actual plating process, the solution also contains ligands that form complexes with the silver ion. By lowering the concentration of Ag^+ in this way, a smooth, even coating of silver is obtained.

Electrolysis of Sodium Chloride

Sodium metal is mainly produced by the electrolysis of molten sodium chloride. Because solid NaCl has a rather high melting point (800°C), it is usually mixed with solid $CaCl_2$ to lower the melting point to about (600°C). ◄ *i* The mixture is then electrolyzed in a **Downs cell,** as illustrated in Fig. 17-25, where the reactions are

Anode reaction: $\qquad 2Cl^- \longrightarrow Cl_2 + 2e^-$

Cathode reaction: $\quad Na^+ + e^- \longrightarrow Na$

At the temperatures in the Downs cell, the sodium is liquid and is drained off, then cooled, and cast into blocks. Because it is so reactive, sodium must be stored in an inert solvent, such as mineral oil, to prevent its oxidation.

Electrolysis of aqueous sodium chloride (brine) is an important industrial process for the production of chlorine and sodium hydroxide. In fact, this process is the second largest consumer of electricity in the United States, after the production of aluminum. Sodium is not produced in this process under normal circumstances because H_2O is more easily reduced than Na^+, as the standard reduction potentials show:

$$Na^+ + e^- \longrightarrow Na \qquad\qquad \mathscr{E}° = -2.71 \text{ V}$$

$$2H_2O + 2e^- \longrightarrow H_2 + 2OH^- \qquad \mathscr{E}° = -0.83 \text{ V}$$

Hydrogen, not sodium, is produced at the cathode.

i Addition of a nonvolatile solute lowers the melting point of the solvent, molten NaCl in this case.

Figure 17-25 | The Downs cell for the electrolysis of molten sodium chloride. The cell is designed so that the sodium and chlorine produced cannot come into contact with each other to re-form NaCl.

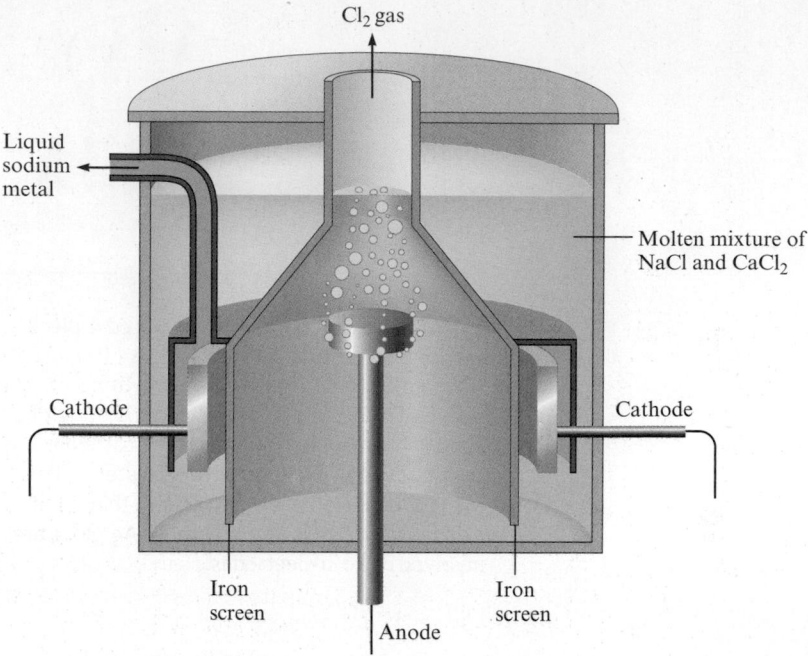

For the reasons we discussed in Section 17-8, chlorine gas is produced at the anode. Thus the electrolysis of brine produces hydrogen and chlorine:

$$\text{Anode reaction:} \qquad 2Cl^- \longrightarrow Cl_2 + 2e^-$$

$$\text{Cathode reaction:} \qquad 2H_2O + 2e^- \longrightarrow H_2 + 2OH^-$$

It leaves a solution containing dissolved NaOH and NaCl.

The contamination of the sodium hydroxide by NaCl can be virtually eliminated using a special **mercury cell** for electrolyzing brine (Fig. 17-26). In this cell, mercury is the conductor at the cathode, and because hydrogen gas has an extremely high overvoltage with a mercury electrode, Na^+ is reduced instead of H_2O. The resulting sodium metal dissolves in the mercury, forming a liquid alloy, which is then pumped to a chamber where the dissolved sodium reacts with water to produce hydrogen:

$$2Na(s) + 2H_2O(l) \longrightarrow 2Na^+(aq) + 2OH^-(aq) + H_2(g)$$

Figure 17-26 | The mercury cell for production of chlorine and sodium hydroxide. The large overvoltage required to produce hydrogen at a mercury electrode means that Na^+ ions are reduced rather than water. The sodium formed dissolves in the liquid mercury and is pumped to a chamber, where it reacts with water.

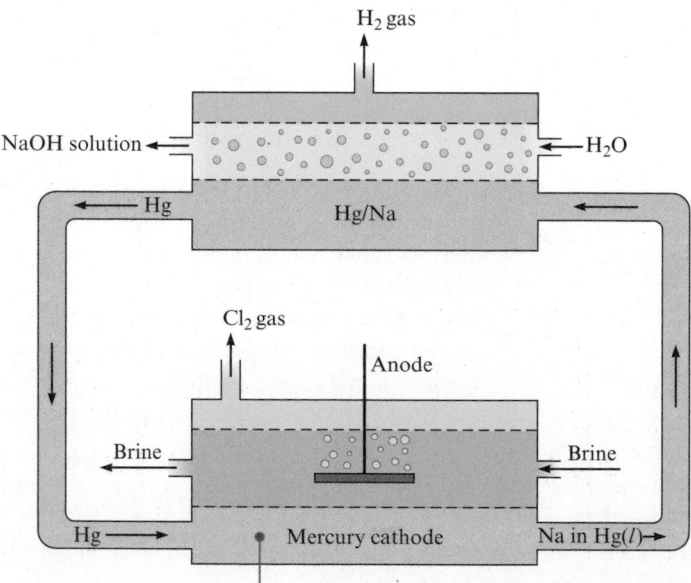

Relatively pure solid NaOH can be recovered from the aqueous solution, and the regenerated mercury is then pumped back to the electrolysis cell. This process, called the **chlor–alkali process**, was the main method for producing chlorine and sodium hydroxide in the United States for many years. However, because of the environmental problems associated with the mercury cell, it has been largely displaced in the chlor–alkali industry by other technologies. In the United States, nearly 75% of the chlor–alkali production is now carried out in diaphragm cells. In a diaphragm cell the cathode and anode are separated by a diaphragm that allows passage of H_2O molecules, Na^+ ions, and, to a limited extent, Cl^- ions. The diaphragm does not allow OH^- ions to pass through it. Thus the H_2 and OH^- formed at the cathode are kept separate from the Cl_2 formed at the anode. The major disadvantage of this process is that the aqueous effluent pumped from the cathode compartment contains a mixture of sodium hydroxide and unreacted sodium chloride, which must be separated if pure sodium hydroxide is a desired product.

In the past 30 years, a new process has been developed in the chlor–alkali industry that uses a membrane to separate the anode and cathode compartments in brine electrolysis cells. The membrane is superior to a diaphragm because the membrane is impermeable to anions. Only cations can flow through the membrane. Because neither Cl^- nor OH^- ions can pass through the membrane separating the anode and cathode compartments, NaCl contamination of the NaOH formed at the cathode does not occur. Although membrane technology is now just becoming prominent in the United States, it is the dominant method for chlor–alkali production in Japan.

FOR REVIEW

Key Terms

electrochemistry

Section 17-1
half-reactions

Section 17-2
salt bridge
porous disk
galvanic cell
anode
cathode
cell potential (electromotive force)
volt
voltmeter
potentiometer

Section 17-3
standard hydrogen electrode
standard reduction potentials

Section 17-4
faraday

Section 17-5
concentration cell
Nernst equation
glass electrode
ion-selective electrode

Electrochemistry

- The study of the interchange of chemical and electrical energy
- Uses oxidation–reduction reactions
- Galvanic cell: chemical energy is transformed into electrical energy by separating the oxidizing and reducing agents and forcing the electrons to travel through a wire
- Electrolytic cell: electrical energy is used to produce a chemical change

Galvanic Cell

- Anode: the electrode where oxidation occurs
- Cathode: the electrode where reduction occurs
- The driving force behind the electron transfer is called the cell potential ($\mathscr{E}_{cell}$)
  - The potential is measured in units of volts (V), defined as a joule of work per coulomb of charge:

$$\mathscr{E}(V) = \frac{-\text{work (J)}}{\text{charge (C)}} = -\frac{w}{q}$$

  - A system of half-reactions, called standard reduction potentials, can be used to calculate the potentials of various cells
    - The half-reaction $2H^+ + 2e^- \longrightarrow H_2$ is arbitrarily assigned a potential of 0 V

Key Terms

Section 17-6
battery
lead storage battery
dry cell battery
fuel cell

Section 17-7
corrosion
galvanizing
cathodic protection

Section 17-8
electrolytic cell
electrolysis
ampere

Section 17-9
Downs cell
mercury cell
chlor–alkali process

Free Energy and Work

- The maximum work that a cell can perform is

$$-w_{\text{max}} = q\mathscr{E}_{\text{max}}$$

  where $\mathscr{E}_{\text{max}}$ represents the cell potential when no current is flowing
- The actual work obtained from a cell is always less than the maximum because energy is lost through frictional heating of the wire when current flows
- For a process carried out at constant temperature and pressure, the change in free energy equals the maximum useful work obtainable from that process:

$$\Delta G = w_{\text{max}} = -q\mathscr{E}_{\text{max}} = -nF\mathscr{E}$$

  where F (faraday) equals 96,485 C and n is the number of moles of electrons transferred in the process

Concentration Cell

- A galvanic cell in which both compartments have the same components but at different concentrations
- The electrons flow in the direction that tends to equalize the concentrations

Nernst Equation

- Shows how the cell potential depends on the concentrations of the cell components:

$$\mathscr{E} = \mathscr{E}_0 - \frac{0.0591}{n}\log Q \qquad \text{at } 25°\text{C}$$

- When a galvanic cell is at equilibrium, $\mathscr{E} = 0$ and $Q = K$

Batteries

- A battery consists of a galvanic cell or group of cells connected in series that serve as a source of direct current.
- Lead storage battery
  - Anode: lead
  - Cathode: lead coated with PbO_2
  - Electrolyte: $H_2SO_4(aq)$
- Dry cell battery
  - Contains a moist paste instead of a liquid electrolyte
  - Anode: usually Zn
  - Cathode: carbon rod in contact with an oxidizing agent (which varies depending on the application)

Fuel Cells

- Galvanic cells in which the reactants are continuously supplied
- The H_2/O_2 fuel cell is based on the reaction between H_2 and O_2 to form water

Corrosion

- Involves the oxidation of metals to form mainly oxides and sulfides
- Some metals, such as aluminum and chromium, form a thin, protective oxide coating that prevents further corrosion

- The corrosion of iron to form rust is an electrochemical process
  - The Fe^{2+} ions formed at anodic areas of the surface migrate through the moisture layer to cathodic regions, where they react with oxygen from the air
  - Iron can be protected from corrosion by coating it with paint or with a thin layer of metal such as chromium, tin, or zinc; by alloying; and by cathodic protection

Electrolysis

- Used to place a thin coating of metal onto steel
- Used to produce pure metals such as aluminum and copper

Review Questions Answers to the Review Questions can be found on the Student Web site (accessible from www.cengagebrain.com).

1. What is a half-reaction? Why must the number of electrons lost in the oxidation half-reaction equal the number of electrons gained in the reduction half-reaction? Summarize briefly the steps in the half-reaction method for balancing redox reactions. What two items must be balanced in a redox reaction (or any reaction)?

2. Galvanic cells harness spontaneous oxidation–reduction reactions to produce work by producing a current. They do so by controlling the flow of electrons from the species oxidized to the species reduced. How is a galvanic cell designed? What is in the cathode compartment? The anode compartment? What purpose do electrodes serve? Which way do electrons always flow in the wire connecting the two electrodes in a galvanic cell? Why is it necessary to use a salt bridge or a porous disk in a galvanic cell? Which way do cations flow in the salt bridge? Which way do the anions flow? What is a cell potential and what is a volt?

3. Table 17-1 lists common half-reactions along with the standard reduction potential associated with each half-reaction. These standard reduction potentials are all relative to some standard. What is the standard (zero point)? If $\mathscr{E}°$ is positive for a half-reaction, what does it mean? If $\mathscr{E}°$ is negative for a half-reaction, what does it mean? Which species in Table 17-1 is most easily reduced? Least easily reduced? The reverse of the half-reactions in Table 17-1 are the oxidation half-reactions. How are standard oxidation potentials determined? In Table 17-1, which species is the best reducing agent? The worst reducing agent?

   To determine the standard cell potential for a redox reaction, the standard reduction potential is added to the standard oxidation potential. What must be true about this sum if the cell is to be spontaneous (produce a galvanic cell)? Standard reduction and oxidation potentials are intensive. What does this mean? Summarize how line notation is used to describe galvanic cells.

4. Consider the equation $\Delta G° = -nF\mathscr{E}°$. What are the four terms in this equation? Why does a minus sign appear in the equation? What does the superscript ° indicate?

5. The Nernst equation allows determination of the cell potential for a galvanic cell at nonstandard conditions. Write out the Nernst equation. What are nonstandard conditions? What do $\mathscr{E}$, $\mathscr{E}°$, n, and Q stand for in the Nernst equation? What does the Nernst equation reduce to when a redox reaction is at equilibrium? What are the signs of $\Delta G°$ and $\mathscr{E}°$ when $K < 1$? When $K > 1$? When $K = 1$? Explain the following statement: $\mathscr{E}$ determines spontaneity, while $\mathscr{E}°$ determines the equilibrium position. Under what conditions can you use $\mathscr{E}°$ to predict spontaneity?

6. What are concentration cells? What is $\mathscr{E}°$ in a concentration cell? What is the driving force for a concentration cell to produce a voltage? Is the higher or the lower ion concentration solution present at the anode? When the anode ion concentration is decreased and/or the cathode ion concentration is increased, both give rise to larger cell potentials. Why? Concentration cells are commonly used to calculate the value of equilibrium constants for various reactions. For example, the silver concentration cell illustrated in Fig. 17-9 can be used to determine the K_{sp} value for $AgCl(s)$. To do so, NaCl is added to the anode compartment until no more precipitate forms. The $[Cl^-]$ in solution is then determined somehow. What happens to $\mathscr{E}_{cell}$ when NaCl is added to the anode compartment? To calculate the K_{sp} value, $[Ag^+]$ must be calculated. Given the value of $\mathscr{E}_{cell}$, how is $[Ag^+]$ determined at the anode?

7. Batteries are galvanic cells. What happens to $\mathscr{E}_{cell}$ as a battery discharges? Does a battery represent a system at equilibrium? Explain. What is $\mathscr{E}_{cell}$ when a battery reaches equilibrium? How are batteries and fuel cells alike? How are they different? The U.S. space program utilizes hydrogen–oxygen fuel

719

cells to produce power for its spacecraft. What is a hydrogen–oxygen fuel cell?

8. Not all spontaneous redox reactions produce wonderful results. Corrosion is an example of a spontaneous redox process that has negative effects. What happens in the corrosion of a metal such as iron? What must be present for the corrosion of iron to take place? How can moisture and salt increase the severity of corrosion? Explain how the following protect metals from corrosion:

   a. paint
   b. durable oxide coatings
   c. galvanizing
   d. sacrificial metal
   e. alloying
   f. cathodic protection

9. What characterizes an electrolytic cell? What is an ampere? When the current applied to an electrolytic cell is multiplied by the time in seconds, what quantity is determined? How is this quantity converted to moles of electrons required? How are moles of electrons required converted to moles of metal plated out? What does plating mean? How do you predict the cathode and the anode half-reactions in an electrolytic cell? Why is the electrolysis of molten salts much easier to predict in terms of what occurs at the anode and cathode than the electrolysis of aqueous dissolved salts? What is overvoltage?

10. Electrolysis has many important industrial applications. What are some of these applications? The electrolysis of molten NaCl is the major process by which sodium metal is produced. However, the electrolysis of aqueous NaCl does not produce sodium metal under normal circumstances. Why? What is purification of a metal by electrolysis?

Active Learning Questions

These questions are designed to be used by groups of students in class.

1. Sketch a galvanic cell, and explain how it works. Look at Figs. 17-1 and 17-2. Explain what is occurring in each container and why the cell in Fig. 17-2 "works" but the one in Fig. 17-1 does not.

2. In making a specific galvanic cell, explain how one decides on the electrodes and the solutions to use in the cell.

3. You want to "plate out" nickel metal from a nickel nitrate solution onto a piece of metal inserted into the solution. Should you use copper or zinc? Explain.

4. A copper penny can be dissolved in nitric acid but not in hydrochloric acid. Using reduction potentials from the book, show why this is so. What are the products of the reaction? Newer pennies contain a mixture of zinc and copper. What happens to the zinc in the penny when the coin is placed in nitric acid? Hydrochloric acid? Support your explanations with data from the book, and include balanced equations for all reactions.

5. Sketch a cell that forms iron metal from iron(II) while changing chromium metal to chromium(III). Calculate the voltage, show the electron flow, label the anode and cathode, and balance the overall cell equation.

6. Which of the following is the best reducing agent: F_2, H_2, Na, Na^+, F^-? Explain. Order as many of these species as possible from the best to the worst oxidizing agent. Why can't you order all of them? From Table 17-1 choose the species that is the best oxidizing agent. Choose the best reducing agent. Explain.

7. You are told that metal A is a better reducing agent than metal B. What, if anything, can be said about A^+ and B^+? Explain.

8. Explain the following relationships: ΔG and w, cell potential and w, cell potential and ΔG, cell potential and Q. Using these relationships, explain how you could make a cell in which both electrodes are the same metal and both solutions contain the same compound, but at different concentrations. Why does such a cell run spontaneously?

9. Explain why cell potentials are not multiplied by the coefficients in the balanced redox equation. (Use the relationship between ΔG and cell potential to do this.)

10. What is the difference between $\mathscr{E}$ and $\mathscr{E}°$? When is $\mathscr{E}$ equal to zero? When is $\mathscr{E}°$ equal to zero? (Consider "regular" galvanic cells as well as concentration cells.)

11. Consider the following galvanic cell:

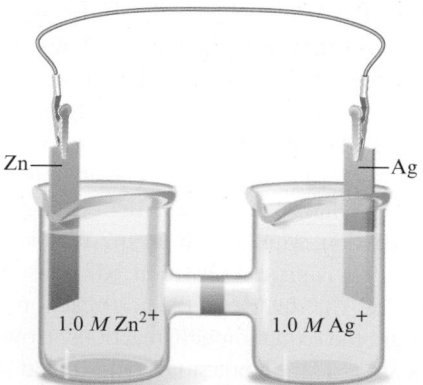

What happens to $\mathscr{E}$ as the concentration of Zn^{2+} is increased? As the concentration of Ag^+ is increased? What happens to $\mathscr{E}°$ in these cases?

12. Look up the reduction potential for Fe^{3+} to Fe^{2+}. Look up the reduction potential for Fe^{2+} to Fe. Finally, look up the reduction potential for Fe^{3+} to Fe. You should notice that adding the reduction potentials for the first two does not give the potential for the third. Why not? Show how you can use the first two potentials to calculate the third potential.

13. If the cell potential is proportional to work and the standard reduction potential for the hydrogen ion is zero, does this mean that the reduction of the hydrogen ion requires no work?

14. Is the following statement true or false? Concentration cells work because standard reduction potentials are dependent on concentration. Explain.

A blue question or exercise number indicates that the answer to that question or exercise appears at the back of this book and a solution appears in the *Solutions Guide*, as found on PowerLecture.

Review of Oxidation–Reduction Reactions

If you have trouble with these exercises, you should review Section 6-9.

15. Define *oxidation* and *reduction* in terms of both change in oxidation number and electron loss or gain.

16. Assign oxidation numbers to all the atoms in each of the following:

a. HNO_3	**e.** $C_6H_{12}O_6$	**i.** $Na_2C_2O_4$
b. $CuCl_2$	**f.** Ag	**j.** CO_2
c. O_2	**g.** $PbSO_4$	**k.** $(NH_4)_2Ce(SO_4)_3$
d. H_2O_2	**h.** PbO_2	**l.** Cr_2O_3

17. Specify which of the following equations represent oxidation–reduction reactions, and indicate the oxidizing agent, the reducing agent, the species being oxidized, and the species being reduced.

a. $CH_4(g) + H_2O(g) \rightarrow CO(g) + 3H_2(g)$

b. $2AgNO_3(aq) + Cu(s) \rightarrow Cu(NO_3)_2(aq) + 2Ag(s)$

c. $Zn(s) + 2HCl(aq) \rightarrow ZnCl_2(aq) + H_2(g)$

d. $2H^+(aq) + 2CrO_4^{2-}(aq) \rightarrow Cr_2O_7^{2-}(aq) + H_2O(l)$

18. The Ostwald process for the commercial production of nitric acid involves the following three steps:

$$4NH_3(g) + 5O_2(g) \longrightarrow 4NO(g) + 6H_2O(g)$$
$$2NO(g) + O_2(g) \longrightarrow 2NO_2(g)$$
$$3NO_2(g) + H_2O(l) \longrightarrow 2HNO_3(aq) + NO(g)$$

a. Which reactions in the Ostwald process are oxidation–reduction reactions?

b. Identify each oxidizing agent and reducing agent.

Questions

19. What is electrochemistry? What are redox reactions? Explain the difference between a galvanic and an electrolytic cell.

20. When balancing equations in Chapter 5, we did not mention that reactions must be charge balanced as well as mass balanced. What do *charge balanced* and *mass balanced* mean? How are redox equations charge balanced?

21. When magnesium metal is added to a beaker of $HCl(aq)$, a gas is produced. Knowing that magnesium is oxidized and that hydrogen is reduced, write the balanced equation for the reaction. How many electrons are transferred in the balanced equation? What quantity of useful work can be obtained when Mg is added directly to the beaker of HCl? How can you harness this reaction to do useful work?

22. How can one construct a galvanic cell from two substances, each having a negative standard reduction potential?

23. The free energy change for a reaction, ΔG, is an extensive property. What is an extensive property? Surprisingly, one can calculate ΔG from the cell potential, $\mathcal{E}$, for the reaction. This is surprising because $\mathcal{E}$ is an intensive property. How can the extensive property ΔG be calculated from the intensive property $\mathcal{E}$?

24. What is wrong with the following statement: The best concentration cell will consist of the substance having the most positive standard reduction potential. What drives a concentration cell to produce a large voltage?

25. When jump-starting a car with a dead battery, the ground jumper should be attached to a remote part of the engine block. Why?

26. In theory, most metals should easily corrode in air. Why? A group of metals called the noble metals are relatively difficult to corrode in air. Some noble metals include gold, platinum, and silver. Reference Table 17-1 to come up with a possible reason why the noble metals are relatively difficult to corrode.

27. Consider the electrolysis of a molten salt of some metal. What information must you know to calculate the mass of metal plated out in the electrolytic cell?

28. Consider the following electrochemical cell:

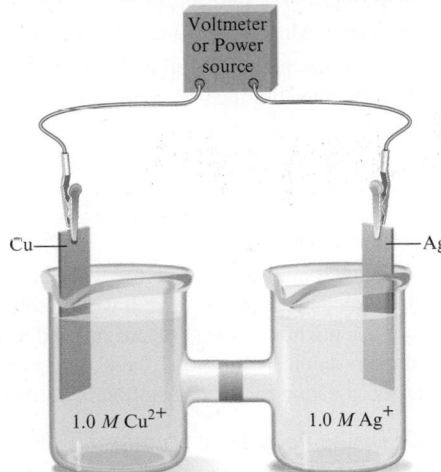

a. If silver metal is a product of the reaction, is the cell a galvanic cell or electrolytic cell? Label the cathode and anode, and describe the direction of the electron flow.

b. If copper metal is a product of the reaction, is the cell a galvanic cell or electrolytic cell? Label the cathode and anode, and describe the direction of the electron flow.

c. If the above cell is a galvanic cell, determine the standard cell potential.

d. If the above cell is an electrolytic cell, determine the minimum external potential that must be applied to cause the reaction to occur.

Exercises

In this section similar exercises are paired.

Balancing Oxidation–Reduction Equations

29. Balance the following oxidation–reduction reactions that occur in acidic solution using the half-reaction method.

a. $I^-(aq) + ClO^-(aq) \rightarrow I_3^-(aq) + Cl^-(aq)$

b. $As_2O_3(s) + NO_3^-(aq) \rightarrow H_3AsO_4(aq) + NO(g)$

c. $Br^-(aq) + MnO_4^-(aq) \rightarrow Br_2(l) + Mn^{2+}(aq)$

d. $CH_3OH(aq) + Cr_2O_7^{2-}(aq) \rightarrow CH_2O(aq) + Cr^{3+}(aq)$

30. Balance the following oxidation–reduction reactions that occur in acidic solution using the half-reaction method.

a. $Cu(s) + NO_3^-(aq) \rightarrow Cu^{2+}(aq) + NO(g)$

b. $Cr_2O_7^{2-}(aq) + Cl^-(aq) \rightarrow Cr^{3+}(aq) + Cl_2(g)$

c. $Pb(s) + PbO_2(s) + H_2SO_4(aq) \rightarrow PbSO_4(s)$

d. $Mn^{2+}(aq) + NaBiO_3(s) \rightarrow Bi^{3+}(aq) + MnO_4^-(aq)$

e. $H_3AsO_4(aq) + Zn(s) \rightarrow AsH_3(g) + Zn^{2+}(aq)$

31. Balance the following oxidation–reduction reactions that occur in basic solution.

a. $Al(s) + MnO_4^-(aq) \rightarrow MnO_2(s) + Al(OH)_4^-(aq)$

b. $Cl_2(g) \rightarrow Cl^-(aq) + OCl^-(aq)$

c. $NO_2^-(aq) + Al(s) \rightarrow NH_3(g) + AlO_2^-(aq)$

32. Balance the following oxidation–reduction reactions that occur in basic solution.

a. $Cr(s) + CrO_4^{2-}(aq) \rightarrow Cr(OH)_3(s)$

b. $MnO_4^-(aq) + S^{2-}(aq) \rightarrow MnS(s) + S(s)$

c. $CN^-(aq) + MnO_4^-(aq) \rightarrow CNO^-(aq) + MnO_2(s)$

33. Chlorine gas was first prepared in 1774 by C. W. Scheele by oxidizing sodium chloride with manganese(IV) oxide. The reaction is

$$NaCl(aq) + H_2SO_4(aq) + MnO_2(s) \longrightarrow$$
$$Na_2SO_4(aq) + MnCl_2(aq) + H_2O(l) + Cl_2(g)$$

Balance this equation.

34. Gold metal will not dissolve in either concentrated nitric acid or concentrated hydrochloric acid. It will dissolve, however, in aqua regia, a mixture of the two concentrated acids. The products of the reaction are the $AuCl_4^-$ ion and gaseous NO. Write a balanced equation for the dissolution of gold in aqua regia.

Galvanic Cells, Cell Potentials, Standard Reduction Potentials, and Free Energy

35. Consider the following galvanic cell:

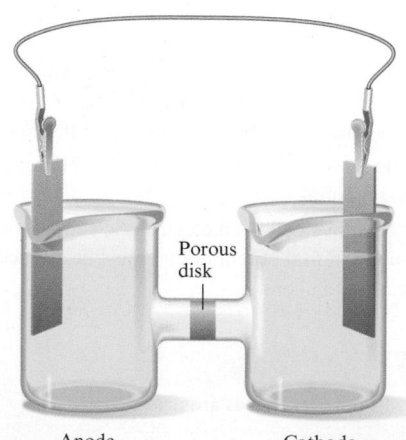

Label the reducing agent and the oxidizing agent, and describe the direction of the electron flow.

36. Consider the following galvanic cell:

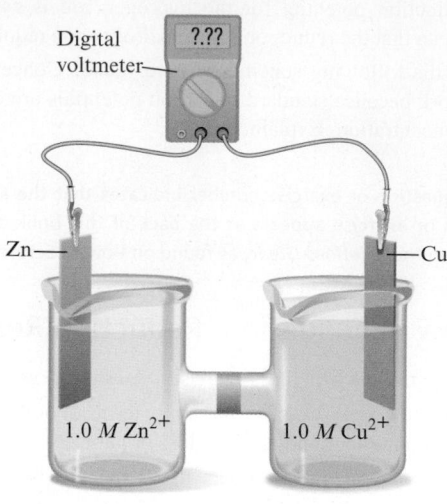

a. Label the reducing agent and the oxidizing agent, and describe the direction of the electron flow.

b. Determine the standard cell potential.

c. Which electrode increases in mass as the reaction proceeds, and which electrode decreases in mass?

37. Sketch the galvanic cells based on the following overall reactions. Show the direction of electron flow, and identify the cathode and anode. Give the overall balanced equation. Assume that all concentrations are 1.0 M and that all partial pressures are 1.0 atm.

a. $Cr^{3+}(aq) + Cl_2(g) \rightleftharpoons Cr_2O_7^{2-}(aq) + Cl^-(aq)$

b. $Cu^{2+}(aq) + Mg(s) \rightleftharpoons Mg^{2+}(aq) + Cu(s)$

38. Sketch the galvanic cells based on the following overall reactions. Show the direction of electron flow, the direction of ion migration through the salt bridge, and identify the cathode and anode. Give the overall balanced equation. Assume that all concentrations are 1.0 M and that all partial pressures are 1.0 atm.

a. $IO_3^-(aq) + Fe^{2+}(aq) \rightleftharpoons Fe^{3+}(aq) + I_2(aq)$

b. $Zn(s) + Ag^+(aq) \rightleftharpoons Zn^{2+}(aq) + Ag(s)$

39. Calculate $\mathscr{E}°$ values for the galvanic cells in Exercise 37.

40. Calculate $\mathscr{E}°$ values for the galvanic cells in Exercise 38.

41. Sketch the galvanic cells based on the following half-reactions. Show the direction of electron flow, show the direction of ion migration through the salt bridge, and identify the cathode and anode. Give the overall balanced equation, and determine $\mathscr{E}°$ for the galvanic cells. Assume that all concentrations are 1.0 M and that all partial pressures are 1.0 atm.

a. $Cl_2 + 2e^- \rightarrow 2Cl^- \quad \mathscr{E}° = 1.36$ V
$Br_2 + 2e^- \rightarrow 2Br^- \quad \mathscr{E}° = 1.09$ V

b. $MnO_4^- + 8H^+ + 5e^- \rightarrow Mn^{2+} + 4H_2O \quad \mathscr{E}° = 1.51$ V
$IO_4^- + 2H^+ + 2e^- \rightarrow IO_3^- + H_2O \quad \mathscr{E}° = 1.60$ V

42. Sketch the galvanic cells based on the following half-reactions. Show the direction of electron flow, show the direction of ion migration through the salt bridge, and identify the cathode and anode. Give the overall balanced equation, and determine $\mathscr{E}°$ for the galvanic cells. Assume that all concentrations are 1.0 M and that all partial pressures are 1.0 atm.

a. $H_2O_2 + 2H^+ + 2e^- \rightarrow 2H_2O$ $\mathscr{E}° = 1.78$ V
   $O_2 + 2H^+ + 2e^- \rightarrow H_2O_2$ $\mathscr{E}° = 0.68$ V

b. $Mn^{2+} + 2e^- \rightarrow Mn$ $\mathscr{E}° = -1.18$ V
   $Fe^{3+} + 3e^- \rightarrow Fe$ $\mathscr{E}° = -0.036$ V

43. Give the standard line notation for each cell in Exercises 37 and 41.

44. Give the standard line notation for each cell in Exercises 38 and 42.

45. Consider the following galvanic cells:

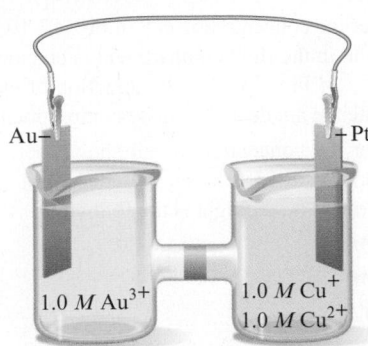

a.

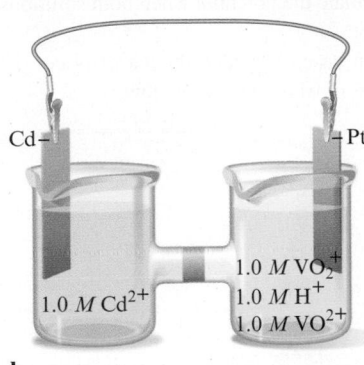

b.

For each galvanic cell, give the balanced cell equation and determine $\mathscr{E}°$. Standard reduction potentials are found in Table 17-1.

46. Give the balanced cell equation and determine $\mathscr{E}°$ for the galvanic cells based on the following half-reactions. Standard reduction potentials are found in Table 17-1.

a. $Cr_2O_7^{2-} + 14H^+ + 6e^- \rightarrow 2Cr^{3+} + 7H_2O$
   $H_2O_2 + 2H^+ + 2e^- \rightarrow 2H_2O$

b. $2H^+ + 2e^- \rightarrow H_2$
   $Al^{3+} + 3e^- \rightarrow Al$

47. Calculate $\mathscr{E}°$ values for the following cells. Which reactions are spontaneous as written (under standard conditions)? Balance the equations. Standard reduction potentials are found in Table 17-1.

a. $MnO_4^-(aq) + I^-(aq) \longrightarrow I_2(aq) + Mn^{2+}(aq)$

b. $MnO_4^-(aq) + F^-(aq) \longrightarrow F_2(g) + Mn^{2+}(aq)$

48. Calculate $\mathscr{E}°$ values for the following cells. Which reactions are spontaneous as written (under standard conditions)? Balance the equations that are not already balanced. Standard reduction potentials are found in Table 17-1.

a. $H_2(g) \longrightarrow H^+(aq) + H^-(aq)$

b. $Au^{3+}(aq) + Ag(s) \longrightarrow Ag^+(aq) + Au(s)$

49. Chlorine dioxide (ClO_2), which is produced by the reaction

$$2NaClO_2(aq) + Cl_2(g) \longrightarrow 2ClO_2(g) + 2NaCl(aq)$$

has been tested as a disinfectant for municipal water treatment. Using data from Table 17-1, calculate $\mathscr{E}°$ and $\Delta G°$ at 25°C for the production of ClO_2.

50. The amount of manganese in steel is determined by changing it to permanganate ion. The steel is first dissolved in nitric acid, producing Mn^{2+} ions. These ions are then oxidized to the deeply colored MnO_4^- ions by periodate ion (IO_4^-) in acid solution.

a. Complete and balance an equation describing each of the above reactions.

b. Calculate $\mathscr{E}°$ and $\Delta G°$ at 25°C for each reaction.

51. Calculate the maximum amount of work that can be obtained from the galvanic cells at standard conditions in Exercise 45.

52. Calculate the maximum amount of work that can be obtained from the galvanic cells at standard conditions in Exercise 46.

53. Estimate $\mathscr{E}°$ for the half-reaction

$$2H_2O + 2e^- \longrightarrow H_2 + 2OH^-$$

given the following values of $\Delta G_f°$:

$$H_2O(l) = -237 \text{ kJ/mol}$$
$$H_2(g) = 0.0$$
$$OH^-(aq) = -157 \text{ kJ/mol}$$
$$e^- = 0.0$$

Compare this value of $\mathscr{E}°$ with the value of $\mathscr{E}°$ given in Table 17-1.

54. The equation $\Delta G° = -nF\mathscr{E}°$ also can be applied to half-reactions. Use standard reduction potentials to estimate $\Delta G_f°$ for $Fe^{2+}(aq)$ and $Fe^{3+}(aq)$. ($\Delta G_f°$ for $e^- = 0$.)

55. Glucose is the major fuel for most living cells. The oxidative breakdown of glucose by our body to produce energy is called *respiration*. The reaction for the complete combustion of glucose is

$$C_6H_{12}O_6(s) + 6O_2(g) \longrightarrow 6CO_2(g) + 6H_2O(l)$$

If this combustion reaction could be harnessed as a fuel cell, calculate the theoretical voltage that could be produced at standard conditions. (*Hint:* Use $\Delta G_f°$ values from Appendix 4.)

56. Direct methanol fuel cells (DMFCs) have shown some promise as a viable option for providing "green" energy to small electrical devices. Calculate $\mathscr{E}°$ for the reaction that takes place in DMFCs:

$$CH_3OH(l) + 3/2 O_2(g) \longrightarrow CO_2(g) + 2H_2O(l)$$

Use values of $\Delta G_f°$ from Appendix 4.

57. Using data from Table 17-1, place the following in order of increasing strength as oxidizing agents (all under standard conditions).

$$Cd^{2+}, \quad IO_3^-, \quad K^+, \quad H_2O, \quad AuCl_4^-, \quad I_2$$

58. Using data from Table 17-1, place the following in order of increasing strength as reducing agents (all under standard conditions).

$$Cu^+, \quad F^-, \quad H^-, \quad H_2O, \quad I_2, \quad K$$

59. Answer the following questions using data from Table 17-1 (all under standard conditions).
    a. Is $H^+(aq)$ capable of oxidizing $Cu(s)$ to $Cu^{2+}(aq)$?
    b. Is $Fe^{3+}(aq)$ capable of oxidizing $I^-(aq)$?
    c. Is $H_2(g)$ capable of reducing $Ag^+(aq)$?

60. Answer the following questions using data from Table 17-1 (all under standard conditions).
    a. Is $H_2(g)$ capable of reducing $Ni^{2+}(aq)$?
    b. Is $Fe^{2+}(aq)$ capable of reducing $VO_2^+(aq)$?
    c. Is $Fe^{2+}(aq)$ capable of reducing $Cr^{3+}(aq)$ to $Cr^{2+}(aq)$?

61. Consider only the species (at standard conditions)

$$Na^+, \quad Cl^-, \quad Ag^+, \quad Ag, \quad Zn^{2+}, \quad Zn, \quad Pb$$

    in answering the following questions. Give reasons for your answers. (Use data from Table 17-1.)
    a. Which is the strongest oxidizing agent?
    b. Which is the strongest reducing agent?
    c. Which species can be oxidized by $SO_4^{2-}(aq)$ in acid?
    d. Which species can be reduced by $Al(s)$?

62. Consider only the species (at standard conditions)

$$Br^-, \quad Br_2, \quad H^+, \quad H_2, \quad La^{3+}, \quad Ca, \quad Cd$$

    in answering the following questions. Give reasons for your answers.
    a. Which is the strongest oxidizing agent?
    b. Which is the strongest reducing agent?
    c. Which species can be oxidized by MnO_4^- in acid?
    d. Which species can be reduced by $Zn(s)$?

63. Use the table of standard reduction potentials (Table 17-1) to pick a reagent that is capable of each of the following oxidations (under standard conditions in acidic solution).
    a. oxidize Br^- to Br_2 but not oxidize Cl^- to Cl_2
    b. oxidize Mn to Mn^{2+} but not oxidize Ni to Ni^{2+}

64. Use the table of standard reduction potentials (Table 17-1) to pick a reagent that is capable of each of the following reductions (under standard conditions in acidic solution).
    a. reduce Cu^{2+} to Cu but not reduce Cu^{2+} to Cu^+
    b. reduce Br_2 to Br^- but not reduce I_2 to I^-

65. Hydrazine is somewhat toxic. Use the half-reactions shown below to explain why household bleach (a highly alkaline solution of sodium hypochlorite) should not be mixed with household ammonia or glass cleansers that contain ammonia.

$$ClO^- + H_2O + 2e^- \longrightarrow 2OH^- + Cl^- \qquad \mathscr{E}° = 0.90 \text{ V}$$
$$N_2H_4 + 2H_2O + 2e^- \longrightarrow 2NH_3 + 2OH^- \qquad \mathscr{E}° = -0.10 \text{ V}$$

66. The compound with the formula TlI_3 is a black solid. Given the following standard reduction potentials,

$$Tl^{3+} + 2e^- \longrightarrow Tl^+ \qquad \mathscr{E}° = 1.25 \text{ V}$$
$$I_3^- + 2e^- \longrightarrow 3I^- \qquad \mathscr{E}° = 0.55 \text{ V}$$

    would you formulate this compound as thallium(III) iodide or thallium(I) triiodide?

The Nernst Equation

67. A galvanic cell is based on the following half-reactions at 25°C:

$$Ag^+ + e^- \longrightarrow Ag$$
$$H_2O_2 + 2H^+ + 2e^- \longrightarrow 2H_2O$$

    Predict whether $\mathscr{E}_{cell}$ is larger or smaller than $\mathscr{E}°_{cell}$ for the following cases.
    a. $[Ag^+] = 1.0 \text{ } M$, $[H_2O_2] = 2.0 \text{ } M$, $[H^+] = 2.0 \text{ } M$
    b. $[Ag^+] = 2.0 \text{ } M$, $[H_2O_2] = 1.0 \text{ } M$, $[H^+] = 1.0 \times 10^{-7} \text{ } M$

68. Consider the concentration cell in Fig. 17-10. If the Fe^{2+} concentration in the right compartment is changed from $0.1 \text{ } M$ to $1 \times 10^{-7} \text{ } M$ Fe^{2+}, predict the direction of electron flow, and designate the anode and cathode compartments.

69. Consider the concentration cell shown below. Calculate the cell potential at 25°C when the concentration of Ag^+ in the compartment on the right is the following.
    a. $1.0 \text{ } M$
    b. $2.0 \text{ } M$
    c. $0.10 \text{ } M$
    d. $4.0 \times 10^{-5} \text{ } M$
    e. Calculate the potential when both solutions are $0.10 \text{ } M$ in Ag^+.

    For each case, also identify the cathode, the anode, and the direction in which electrons flow.

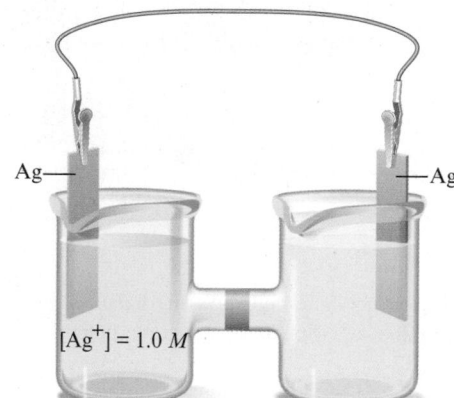

$[Ag^+] = 1.0 \text{ } M$

70. Consider a concentration cell similar to the one shown in Exercise 69, except that both electrodes are made of Ni and in the left-hand compartment $[Ni^{2+}] = 1.0 \text{ } M$. Calculate the cell potential at 25°C when the concentration of Ni^{2+} in the compartment on the right has each of the following values.
    a. $1.0 \text{ } M$
    b. $2.0 \text{ } M$
    c. $0.10 \text{ } M$
    d. $4.0 \times 10^{-5} \text{ } M$
    e. Calculate the potential when both solutions are $2.5 \text{ } M$ in Ni^{2+}.

    For each case, also identify the cathode, anode, and the direction in which electrons flow.

71. The overall reaction in the lead storage battery is

$$Pb(s) + PbO_2(s) + 2H^+(aq) + 2HSO_4^-(aq) \longrightarrow$$
$$2PbSO_4(s) + 2H_2O(l)$$

Calculate $\mathscr{E}$ at 25°C for this battery when $[H_2SO_4] = 4.5\ M$, that is, $[H^+] = [HSO_4^-] = 4.5\ M$. At 25°C, $\mathscr{E}° = 2.04$ V for the lead storage battery.

72. Calculate the pH of the cathode compartment for the following reaction given $\mathscr{E}_{cell} = 3.01$ V when $[Cr^{3+}] = 0.15\ M$, $[Al^{3+}] = 0.30\ M$, and $[Cr_2O_7^{2-}] = 0.55\ M$.

$$2Al(s) + Cr_2O_7^{2-}(aq) + 14H^+(aq) \longrightarrow$$
$$2Al^{3+}(aq) + 2Cr^{3+}(aq) + 7H_2O(l)$$

73. Consider the cell described below:

$$Zn\,|\,Zn^{2+}(1.00\ M)\,||\,Cu^{2+}(1.00\ M)\,|\,Cu$$

Calculate the cell potential after the reaction has operated long enough for the $[Zn^{2+}]$ to have changed by 0.20 mol/L. (Assume $T = 25°C$.)

74. Consider the cell described below:

$$Al\,|\,Al^{3+}(1.00\ M)\,||\,Pb^{2+}(1.00\ M)\,|\,Pb$$

Calculate the cell potential after the reaction has operated long enough for the $[Al^{3+}]$ to have changed by 0.60 mol/L. (Assume $T = 25°C$.)

75. Calculate $\Delta G°$ and K at 25°C for the reactions in Exercises 37 and 41.

76. Calculate $\Delta G°$ and K at 25°C for the reactions in Exercises 38 and 42.

77. Consider the galvanic cell based on the following half-reactions:

$$Zn^{2+} + 2e^- \longrightarrow Zn \quad \mathscr{E}° = -0.76\text{ V}$$
$$Fe^{2+} + 2e^- \longrightarrow Fe \quad \mathscr{E}° = -0.44\text{ V}$$

a. Determine the overall cell reaction and calculate $\mathscr{E}_{cell}°$.
b. Calculate $\Delta G°$ and K for the cell reaction at 25°C.
c. Calculate $\mathscr{E}_{cell}$ at 25°C when $[Zn^{2+}] = 0.10\ M$ and $[Fe^{2+}] = 1.0 \times 10^{-5}\ M$.

78. Consider the galvanic cell based on the following half-reactions:

$$Au^{3+} + 3e^- \longrightarrow Au \quad \mathscr{E}° = 1.50\text{ V}$$
$$Tl^+ + e^- \longrightarrow Tl \quad \mathscr{E}° = -0.34\text{ V}$$

a. Determine the overall cell reaction and calculate $\mathscr{E}_{cell}°$.
b. Calculate $\Delta G°$ and K for the cell reaction at 25°C.
c. Calculate $\mathscr{E}_{cell}$ at 25°C when $[Au^{3+}] = 1.0 \times 10^{-2}\ M$ and $[Tl^+] = 1.0 \times 10^{-4}\ M$.

79. An electrochemical cell consists of a standard hydrogen electrode and a copper metal electrode.

a. What is the potential of the cell at 25°C if the copper electrode is placed in a solution in which $[Cu^{2+}] = 2.5 \times 10^{-4}\ M$?
b. The copper electrode is placed in a solution of unknown $[Cu^{2+}]$. The measured potential at 25°C is 0.195 V. What is $[Cu^{2+}]$? (Assume Cu^{2+} is reduced.)

80. An electrochemical cell consists of a nickel metal electrode immersed in a solution with $[Ni^{2+}] = 1.0\ M$ separated by a porous disk from an aluminum metal electrode.

a. What is the potential of this cell at 25°C if the aluminum electrode is placed in a solution in which $[Al^{3+}] = 7.2 \times 10^{-3}\ M$?
b. When the aluminum electrode is placed in a certain solution in which $[Al^{3+}]$ is unknown, the measured cell potential at 25°C is 1.62 V. Calculate $[Al^{3+}]$ in the unknown solution. (Assume Al is oxidized.)

81. An electrochemical cell consists of a standard hydrogen electrode and a copper metal electrode. If the copper electrode is placed in a solution of 0.10 M NaOH that is saturated with $Cu(OH)_2$, what is the cell potential at 25°C? [For $Cu(OH)_2$, $K_{sp} = 1.6 \times 10^{-19}$.]

82. An electrochemical cell consists of a nickel metal electrode immersed in a solution with $[Ni^{2+}] = 1.0\ M$ separated by a porous disk from an aluminum metal electrode immersed in a solution with $[Al^{3+}] = 1.0\ M$. Sodium hydroxide is added to the aluminum compartment, causing $Al(OH)_3(s)$ to precipitate. After precipitation of $Al(OH)_3$ has ceased, the concentration of OH^- is $1.0 \times 10^{-4}\ M$ and the measured cell potential is 1.82 V. Calculate the K_{sp} value for $Al(OH)_3$.

$$Al(OH)_3(s) \rightleftharpoons Al^{3+}(aq) + 3OH^-(aq) \quad K_{sp} = ?$$

83. Consider a concentration cell that has both electrodes made of some metal M. Solution A in one compartment of the cell contains $1.0\ M\ M^{2+}$. Solution B in the other cell compartment has a volume of 1.00 L. At the beginning of the experiment 0.0100 mole of $M(NO_3)_2$ and 0.0100 mole of Na_2SO_4 are dissolved in solution B (ignore volume changes), where the reaction

$$M^{2+}(aq) + SO_4^{2-}(aq) \rightleftharpoons MSO_4(s)$$

occurs. For this reaction equilibrium is rapidly established, whereupon the cell potential is found to be 0.44 V at 25°C. Assume that the process

$$M^{2+} + 2e^- \longrightarrow M$$

has a standard reduction potential of -0.31 V and that no other redox process occurs in the cell. Calculate the value of K_{sp} for $MSO_4(s)$ at 25°C.

84. You have a concentration cell in which the cathode has a silver electrode with 0.10 M Ag^+. The anode also has a silver electrode with $Ag^+(aq)$, 0.050 M $S_2O_3^{2-}$, and $1.0 \times 10^{-3}\ M$ $Ag(S_2O_3)_2^{3-}$. You read the voltage to be 0.76 V.

a. Calculate the concentration of Ag^+ at the anode.
b. Determine the value of the equilibrium constant for the formation of $Ag(S_2O_3)_2^{3-}$.

$$Ag^+(aq) + 2S_2O_3^{2-}(aq) \rightleftharpoons Ag(S_2O_3)_2^{3-}(aq) \quad K = ?$$

85. Under standard conditions, what reaction occurs, if any, when each of the following operations is performed?

a. Crystals of I_2 are added to a solution of NaCl.
b. Cl_2 gas is bubbled into a solution of NaI.
c. A silver wire is placed in a solution of $CuCl_2$.
d. An acidic solution of $FeSO_4$ is exposed to air.

For the reactions that occur, write a balanced equation and calculate $\mathscr{E}°$, $\Delta G°$, and K at 25°C.

86. A disproportionation reaction involves a substance that acts as both an oxidizing and a reducing agent, producing higher and lower oxidation states of the same element in the products. Which of the following disproportionation reactions are spontaneous under standard conditions? Calculate $\Delta G°$ and K at 25°C for those reactions that are spontaneous under standard conditions.

  a. $2Cu^+(aq) \rightarrow Cu^{2+}(aq) + Cu(s)$

  b. $3Fe^{2+}(aq) \rightarrow 2Fe^{3+}(aq) + Fe(s)$

  c. $HClO_2(aq) \rightarrow ClO_3^-(aq) + HClO(aq)$ (unbalanced)

  Use the half-reactions:

$$ClO_3^- + 3H^+ + 2e^- \longrightarrow HClO_2 + H_2O \qquad \mathscr{E}° = 1.21 \text{ V}$$
$$HClO_2 + 2H^+ + 2e^- \longrightarrow HClO + H_2O \qquad \mathscr{E}° = 1.65 \text{ V}$$

87. Consider the following galvanic cell at 25°C:

$$Pt\,|\,Cr^{2+}(0.30\ M),\ Cr^{3+}(2.0\ M)\,||\,Co^{2+}(0.20\ M)\,|\,Co$$

The overall reaction and equilibrium constant value are

$$2Cr^{2+}(aq) + Co^{2+}(aq) \longrightarrow$$
$$2Cr^{3+}(aq) + Co(s) \quad K = 2.79 \times 10^7$$

Calculate the cell potential, $\mathscr{E}$, for this galvanic cell and ΔG for the cell reaction at these conditions.

88. An electrochemical cell consists of a silver metal electrode immersed in a solution with $[Ag^+] = 1.0\ M$ separated by a porous disk from a copper metal electrode. If the copper electrode is placed in a solution of $5.0\ M$ NH_3 that is also $0.010\ M$ in $Cu(NH_3)_4^{2+}$, what is the cell potential at 25°C?

$$Cu^{2+}(aq) + 4NH_3(aq) \rightleftharpoons Cu(NH_3)_4^{2+}(aq)$$
$$K = 1.0 \times 10^{13}$$

89. Calculate K_{sp} for iron(II) sulfide given the following data:

$$FeS(s) + 2e^- \longrightarrow Fe(s) + S^{2-}(aq) \qquad \mathscr{E}° = -1.01 \text{ V}$$
$$Fe^{2+}(aq) + 2e^- \longrightarrow Fe(s) \qquad \mathscr{E}° = -0.44 \text{ V}$$

90. For the following half-reaction, $\mathscr{E}° = -2.07$ V:

$$AlF_6^{3-}(aq) + 3e^- \longrightarrow Al(s) + 6F^-(aq)$$

Using data from Table 17-1, calculate the equilibrium constant at 25°C for the reaction

$$Al^{3+}(aq) + 6F^-(aq) \rightleftharpoons AlF_6^{3-}(aq) \quad K = ?$$

91. Calculate $\mathscr{E}°$ for the following half-reaction:

$$AgI(s) + e^- \longrightarrow Ag(s) + I^-(aq)$$

(*Hint:* Reference the K_{sp} value for AgI and the standard reduction potential for Ag^+.)

92. The solubility product for $CuI(s)$ is 1.1×10^{-12}. Calculate the value of $\mathscr{E}°$ for the half-reaction

$$CuI(s) + e^- \longrightarrow Cu(s) + I^-(aq)$$

Electrolysis

93. How long will it take to plate out each of the following with a current of 100.0 A?

  a. 1.0 kg Al from aqueous Al^{3+}

  b. 1.0 g Ni from aqueous Ni^{2+}

  c. 5.0 moles of Ag from aqueous Ag^+

94. The electrolysis of BiO^+ produces pure bismuth. How long would it take to produce 10.0 g Bi by the electrolysis of a BiO^+ solution using a current of 25.0 A?

95. What mass of each of the following substances can be produced in 1.0 h with a current of 15 A?

  a. Co from aqueous Co^{2+}

  b. Hf from aqueous Hf^{4+}

  c. I_2 from aqueous KI

  d. Cr from molten CrO_3

96. Aluminum is produced commercially by the electrolysis of Al_2O_3 in the presence of a molten salt. If a plant has a continuous capacity of 1.00 million A, what mass of aluminum can be produced in 2.00 h?

97. An unknown metal M is electrolyzed. It took 74.1 s for a current of 2.00 A to plate out 0.107 g of the metal from a solution containing $M(NO_3)_3$. Identify the metal.

98. Electrolysis of an alkaline earth metal chloride using a current of 5.00 A for 748 s deposits 0.471 g of metal at the cathode. What is the identity of the alkaline earth metal chloride?

99. What volume of F_2 gas, at 25°C and 1.00 atm, is produced when molten KF is electrolyzed by a current of 10.0 A for 2.00 h? What mass of potassium metal is produced? At which electrode does each reaction occur?

100. What volumes of $H_2(g)$ and $O_2(g)$ at STP are produced from the electrolysis of water by a current of 2.50 A in 15.0 min?

101. A single Hall–Heroult cell (as shown in Fig. 17-22) produces about 1 ton of aluminum in 24 h. What current must be used to accomplish this?

102. A factory wants to produce 1.00×10^3 kg barium from the electrolysis of molten barium chloride. What current must be applied for 4.00 h to accomplish this?

103. It took 2.30 min using a current of 2.00 A to plate out all the silver from 0.250 L of a solution containing Ag^+. What was the original concentration of Ag^+ in the solution?

104. A solution containing Pt^{4+} is electrolyzed with a current of 4.00 A. How long will it take to plate out 99% of the platinum in 0.50 L of a 0.010-M solution of Pt^{4+}?

105. A solution at 25°C contains 1.0 M Cd^{2+}, 1.0 M Ag^+, 1.0 M Au^{3+}, and 1.0 M Ni^{2+} in the cathode compartment of an electrolytic cell. Predict the order in which the metals will plate out as the voltage is gradually increased.

106. Consider the following half-reactions:

$$IrCl_6^{3-} + 3e^- \longrightarrow Ir + 6Cl^- \qquad \mathscr{E}° = 0.77 \text{ V}$$
$$PtCl_4^{2-} + 2e^- \longrightarrow Pt + 4Cl^- \qquad \mathscr{E}° = 0.73 \text{ V}$$
$$PdCl_4^{2-} + 2e^- \longrightarrow Pd + 4Cl^- \qquad \mathscr{E}° = 0.62 \text{ V}$$

A hydrochloric acid solution contains platinum, palladium, and iridium as chloro-complex ions. The solution is a constant 1.0 M in chloride ion and 0.020 M in each complex ion. Is it feasible to separate the three metals from this solution by electrolysis? (Assume that 99% of a metal must be plated out before another metal begins to plate out.)

107. In the electrolysis of an aqueous solution of Na_2SO_4, what reactions occur at the anode and the cathode (assuming standard conditions)?

	$\mathscr{E}°$
$S_2O_8^{2-} + 2e^- \longrightarrow 2SO_4^{2-}$	2.01 V
$O_2 + 4H^+ + 4e^- \longrightarrow 2H_2O$	1.23 V
$2H_2O + 2e^- \longrightarrow H_2 + 2OH^-$	-0.83 V
$Na^+ + e^- \longrightarrow Na$	-2.71 V

108. Copper can be plated onto a spoon by placing the spoon in an acidic solution of $CuSO_4(aq)$ and connecting it to a copper strip via a power source as illustrated below:

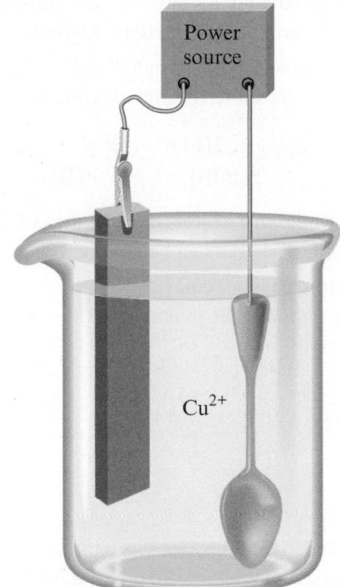

a. Label the anode and cathode, and describe the direction of the electron flow.

b. Write out the chemical equations for the reactions that occur at each electrode.

109. What reactions take place at the cathode and the anode when each of the following is electrolyzed?

a. molten $NiBr_2$ **b.** molten AlF_3 **c.** molten MnI_2

110. What reaction will take place at the cathode and the anode when each of the following is electrolyzed?

a. molten KF   **b.** molten $CuCl_2$ **c.** molten MgI_2

111. What reactions take place at the cathode and the anode when each of the following is electrolyzed? (Assume standard conditions.)

a. 1.0 M $NiBr_2$ solution

b. 1.0 M AlF_3 solution

c. 1.0 M MnI_2 solution

112. What reaction will take place at the cathode and the anode when each of the following is electrolyzed? (Assume standard conditions.)

a. 1.0 M KF solution

b. 1.0 M $CuCl_2$ solution

c. 1.0 M MgI_2 solution

Additional Exercises

113. Gold is produced electrochemically from an aqueous solution of $Au(CN)_2^-$ containing an excess of CN^-. Gold metal and oxygen gas are produced at the electrodes. What amount (moles) of O_2 will be produced during the production of 1.00 mole of gold?

114. The blood alcohol (C_2H_5OH) level can be determined by titrating a sample of blood plasma with an acidic potassium dichromate solution, resulting in the production of $Cr^{3+}(aq)$ and carbon dioxide. The reaction can be monitored because the dichromate ion ($Cr_2O_7^{2-}$) is orange in solution, and the Cr^{3+} ion is green. The unbalanced redox equation is

$$Cr_2O_7^{2-}(aq) + C_2H_5OH(aq) \longrightarrow Cr^{3+}(aq) + CO_2(g)$$

If 31.05 mL of 0.0600 M potassium dichromate solution is required to titrate 30.0 g blood plasma, determine the mass percent of alcohol in the blood.

115. The saturated calomel electrode, abbreviated SCE, is often used as a reference electrode in making electrochemical measurements. The SCE is composed of mercury in contact with a saturated solution of calomel (Hg_2Cl_2). The electrolyte solution is saturated KCl. $\mathscr{E}_{SCE}$ is $+0.242$ V relative to the standard hydrogen electrode. Calculate the potential for each of the following galvanic cells containing a saturated calomel electrode and the given half-cell components at standard conditions. In each case, indicate whether the SCE is the cathode or the anode. Standard reduction potentials are found in Table 17-1.

a. $Cu^{2+} + 2e^- \longrightarrow Cu$ **d.** $Al^{3+} + 3e^- \longrightarrow Al$

b. $Fe^{3+} + e^- \longrightarrow Fe^{2+}$ **e.** $Ni^{2+} + 2e^- \longrightarrow Ni$

c. $AgCl + e^- \longrightarrow Ag + Cl^-$

116. Consider the following half-reactions:

$Pt^{2+} + 2e^- \longrightarrow Pt$	$\mathscr{E}° = 1.188$ V
$PtCl_4^{2-} + 2e^- \longrightarrow Pt + 4Cl^-$	$\mathscr{E}° = 0.755$ V
$NO_3^- + 4H^+ + 3e^- \longrightarrow NO + 2H_2O$	$\mathscr{E}° = 0.96$ V

Explain why platinum metal will dissolve in aqua regia (a mixture of hydrochloric and nitric acids) but not in either concentrated nitric or concentrated hydrochloric acid individually.

117. Consider the standard galvanic cell based on the following half-reactions:

$$Cu^{2+} + 2e^- \longrightarrow Cu$$
$$Ag^+ + e^- \longrightarrow Ag$$

The electrodes in this cell are Ag(s) and Cu(s). Does the cell potential increase, decrease, or remain the same when the following changes occur to the standard cell?

a. $CuSO_4(s)$ is added to the copper half-cell compartment (assume no volume change).

b. $NH_3(aq)$ is added to the copper half-cell compartment. [*Hint:* Cu^{2+} reacts with NH_3 to form $Cu(NH_3)_4^{2+}(aq)$.]

c. NaCl(s) is added to the silver half-cell compartment. [*Hint:* Ag^+ reacts with Cl^- to form AgCl(s).]

d. Water is added to both half-cell compartments until the volume of solution is doubled.

e. The silver electrode is replaced with a platinum electrode.

$$Pt^{2+} + 2e^- \longrightarrow Pt \qquad \mathscr{E}° = 1.19 \text{ V}$$

118. A standard galvanic cell is constructed so that the overall cell reaction is

$$2Al^{3+}(aq) + 3M(s) \longrightarrow 3M^{2+}(aq) + 2Al(s)$$

where M is an unknown metal. If $\Delta G° = -411$ kJ for the overall cell reaction, identify the metal used to construct the standard cell.

119. The black silver sulfide discoloration of silverware can be removed by heating the silver article in a sodium carbonate solution in an aluminum pan. The reaction is

$$3Ag_2S(s) + 2Al(s) \rightleftharpoons 6Ag(s) + 3S^{2-}(aq) + 2Al^{3+}(aq)$$

a. Using data in Appendix 4, calculate $\Delta G°$, K, and $\mathscr{E}°$ for the above reaction at 25°C. [For $Al^{3+}(aq)$, $\Delta G_f° = -480.$ kJ/mol.]

b. Calculate the value of the standard reduction potential for the following half-reaction:

$$2e^- + Ag_2S(s) \longrightarrow 2Ag(s) + S^{2-}(aq)$$

120. In 1973 the wreckage of the Civil War ironclad USS *Monitor* was discovered near Cape Hatteras, North Carolina. [The *Monitor* and the CSS *Virginia* (formerly the USS *Merrimack*) fought the first battle between iron-armored ships.] In 1987 investigations were begun to see if the ship could be salvaged. It was reported in *Time* (June 22, 1987) that scientists were considering adding sacrificial anodes of zinc to the rapidly corroding metal hull of the *Monitor*. Describe how attaching zinc to the hull would protect the *Monitor* from further corrosion.

121. When aluminum foil is placed in hydrochloric acid, nothing happens for the first 30 seconds or so. This is followed by vigorous bubbling and the eventual disappearance of the foil. Explain these observations.

122. Which of the following statements concerning corrosion is(are) *true*? For the false statements, correct them.

a. Corrosion is an example of an electrolytic process.

b. Corrosion of steel involves the reduction of iron coupled with the oxidation of oxygen.

c. Steel rusts more easily in the dry (arid) Southwest states than in the humid Midwest states.

d. Salting roads in the winter has the added benefit of hindering the corrosion of steel.

e. The key to cathodic protection is to connect via a wire a metal more easily oxidized than iron to the steel surface to be protected.

123. A fuel cell designed to react grain alcohol with oxygen has the following net reaction:

$$C_2H_5OH(l) + 3O_2(g) \longrightarrow 2CO_2(g) + 3H_2O(l)$$

The maximum work that 1 mole of alcohol can do is 1.32×10^3 kJ. What is the theoretical maximum voltage this cell can achieve at 25°C?

124. The overall reaction and equilibrium constant value for a hydrogen–oxygen fuel cell at 298 K is

$$2H_2(g) + O_2(g) \longrightarrow 2H_2O(l) \qquad K = 1.28 \times 10^{83}$$

a. Calculate $\mathscr{E}°$ and $\Delta G°$ at 298 K for the fuel cell reaction.

b. Predict the signs of $\Delta H°$ and $\Delta S°$ for the fuel cell reaction.

c. As temperature increases, does the maximum amount of work obtained from the fuel cell reaction increase, decrease, or remain the same? Explain.

125. What is the maximum work that can be obtained from a hydrogen–oxygen fuel cell at standard conditions that produces 1.00 kg water at 25°C? Why do we say that this is the maximum work that can be obtained? What are the advantages and disadvantages in using fuel cells rather than the corresponding combustion reactions to produce electricity?

126. The overall reaction and standard cell potential at 25°C for the rechargeable nickel–cadmium alkaline battery is

$$Cd(s) + NiO_2(s) + 2H_2O(l) \longrightarrow$$
$$Ni(OH)_2(s) + Cd(OH)_2(s) \qquad \mathscr{E}° = 1.10 \text{ V}$$

For every mole of Cd consumed in the cell, what is the maximum useful work that can be obtained at standard conditions?

127. An experimental fuel cell has been designed that uses carbon monoxide as fuel. The overall reaction is

$$2CO(g) + O_2(g) \longrightarrow 2CO_2(g)$$

The two half-cell reactions are

$$CO + O^{2-} \longrightarrow CO_2 + 2e^-$$
$$O_2 + 4e^- \longrightarrow 2O^{2-}$$

The two half-reactions are carried out in separate compartments connected with a solid mixture of CeO_2 and Gd_2O_3. Oxide ions can move through this solid at high temperatures (about 800°C). ΔG for the overall reaction at 800°C under certain concentration conditions is -380 kJ. Calculate the cell potential for this fuel cell at the same temperature and concentration conditions.

128. The ultimate electron acceptor in the respiration process is molecular oxygen. Electron transfer through the respiratory chain takes place through a complex series of oxidation–reduction reactions. Some of the electron transport steps use iron-containing proteins called *cytochromes*. All cytochromes transport electrons by converting the iron in the cytochromes from the +3 to the +2 oxidation state. Consider the following reduction potentials for three different cytochromes used in the transfer process of electrons to oxygen (the potentials have been corrected for pH and for temperature):

$$\text{cytochrome a}(Fe^{3+}) + e^- \longrightarrow \text{cytochrome a}(Fe^{2+})$$
$$\mathscr{E} = 0.385 \text{ V}$$

$$\text{cytochrome b}(Fe^{3+}) + e^- \longrightarrow \text{cytochrome b}(Fe^{2+})$$
$$\mathscr{E} = 0.030 \text{ V}$$

$$\text{cytochrome c}(Fe^{3+}) + e^- \longrightarrow \text{cytochrome c}(Fe^{2+})$$
$$\mathscr{E} = 0.254 \text{ V}$$

In the electron transfer series, electrons are transferred from one cytochrome to another. Using this information, determine the cytochrome order necessary for spontaneous transport of electrons from one cytochrome to another, which eventually will lead to electron transfer to O_2.

129. One of the few industrial-scale processes that produce organic compounds electrochemically is used by the Monsanto Company to produce 1,4-dicyanobutane. The reduction reaction is

$$2CH_2\!\!=\!\!CHCN + 2H^+ + 2e^- \longrightarrow NC\!-\!(CH_2)_4\!-\!CN$$

The $NC-(CH_2)_4-CN$ is then chemically reduced using hydrogen gas to $H_2N-(CH_2)_6-NH_2$, which is used in the production of nylon. What current must be used to produce 150. kg $NC-(CH_2)_4-CN$ per hour?

130. It took 150. s for a current of 1.25 A to plate out 0.109 g of a metal from a solution containing its cations. Show that it is not possible for the cations to have a charge of 1+.

131. It takes 15 kWh (kilowatt-hours) of electrical energy to produce 1.0 kg aluminum metal from aluminum oxide by the Hall–Heroult process. Compare this to the amount of energy necessary to melt 1.0 kg aluminum metal. Why is it economically feasible to recycle aluminum cans? [The enthalpy of fusion for aluminum metal is 10.7 kJ/mol (1 watt = 1 J/s).]

132. In the electrolysis of a sodium chloride solution, what volume of $H_2(g)$ is produced in the same time it takes to produce 257 L $Cl_2(g)$, with both volumes measured at 50.°C and 2.50 atm?

133. An aqueous solution of an unknown salt of ruthenium is electrolyzed by a current of 2.50 A passing for 50.0 min. If 2.618 g Ru is produced at the cathode, what is the charge on the ruthenium ions in solution?

ChemWork Problems

These multiconcept problems (and additional ones) are found interactively online with the same type of assistance a student would get from an instructor.

134. Which of the following statement(s) is/are *true*?
   a. Copper metal can be oxidized by Ag^+ (at standard conditions).
   b. In a galvanic cell the oxidizing agent in the cell reaction is present at the anode.
   c. In a cell using the half reactions $Al^{3+} + 3e^- \longrightarrow Al$ and $Mg^{2+} + 2e^- \longrightarrow Mg$, aluminum functions as the anode.
   d. In a concentration cell electrons always flow from the compartment with the lower ion concentration to the compartment with the higher ion concentration.
   e. In a galvanic cell the negative ions in the salt bridge flow in the same direction as the electrons.

135. Consider a galvanic cell based on the following half-reactions:

	$\mathscr{E}°$ (V)
$La^{3+} + 3e^- \longrightarrow La$	−2.37
$Fe^{2+} + 2e^- \longrightarrow Fe$	−0.44

   a. What is the expected cell potential with all components in their standard states?
   b. What is the oxidizing agent in the overall cell reaction?
   c. What substances make up the anode compartment?

   d. In the standard cell, in which direction do the electrons flow?
   e. How many electrons are transferred per unit of cell reaction?
   f. If this cell is set up at 25°C with $[Fe^{2+}] = 2.00 \times 10^{-4}\ M$ and $[La^{3+}] = 3.00 \times 10^{-3}\ M$, what is the expected cell potential?

136. Consider a galvanic cell based on the following theoretical half-reactions:

	$\mathscr{E}°$ (V)
$M^{4+} + 4e^- \longrightarrow M$	0.66
$N^{3+} + 3e^- \longrightarrow N$	0.39

What is the value of $\Delta G°$ and K for this cell?

137. Consider a galvanic cell based on the following half-reactions:

	$\mathscr{E}°$ (V)
$Au^{3+} + 3e^- \longrightarrow Au$	1.50
$Mg^{2+} + 2e^- \longrightarrow Mg$	−2.37

   a. What is the standard potential for this cell?
   b. A nonstandard cell is set up at 25°C with $[Mg^{2+}] = 1.00 \times 10^{-5}\ M$. The cell potential is observed to be 4.01 V. Calculate $[Au^{3+}]$ in this cell.

138. An electrochemical cell consists of a silver metal electrode immersed in a solution with $[Ag^+] = 1.00\ M$ separated by a porous disk from a compartment with a copper metal electrode immersed in a solution of 10.00 M NH_3 that also contains $2.4 \times 10^{-3}\ M\ Cu(NH_3)_4^{2+}$. The equilibrium between Cu^{2+} and NH_3 is:

$$Cu^{2+}(aq) + 4NH_3(aq) \rightleftharpoons Cu(NH_3)_4^{2+}(aq) \qquad K = 1.0 \times 10^{13}$$

and the two cell half-reactions are:

$$Ag^+ + e^- \longrightarrow Ag \qquad \mathscr{E}° = 0.80\ V$$
$$Cu^{2+} + 2e^- \longrightarrow Cu \qquad \mathscr{E}° = 0.34\ V$$

Assuming Ag^+ is reduced, what is the cell potential at 25°C?

139. An aqueous solution of $PdCl_2$ is electrolyzed for 48.6 seconds, and during this time 0.1064 g of Pd is deposited on the cathode. What is the average current used in the electrolysis?

Challenge Problems

140. Balance the following equations by the half-reaction method.
   a. $Fe(s) + HCl(aq) \longrightarrow HFeCl_4(aq) + H_2(g)$
   b. $IO_3^-(aq) + I^-(aq) \xrightarrow{Acid} I_3^-(aq)$
   c. $Cr(NCS)_6^{4-}(aq) + Ce^{4+}(aq) \xrightarrow{Acid}$
      $Cr^{3+}(aq) + Ce^{3+}(aq) + NO_3^-(aq) + CO_2(g) + SO_4^{2-}(aq)$
   d. $CrI_3(s) + Cl_2(g) \xrightarrow{Base}$
      $CrO_4^{2-}(aq) + IO_4^-(aq) + Cl^-(aq)$
   e. $Fe(CN)_6^{4-}(aq) + Ce^{4+}(aq) \xrightarrow{Base}$
      $Ce(OH)_3(s) + Fe(OH)_3(s) + CO_3^{2-}(aq) + NO_3^-(aq)$

141. Combine the equations

$$\Delta G° = -nF\mathscr{E}° \quad \text{and} \quad \Delta G° = \Delta H° - T\Delta S°$$

to derive an expression for $\mathscr{E}°$ as a function of temperature. Describe how one can graphically determine $\Delta H°$ and $\Delta S°$ from measurements of $\mathscr{E}°$ at different temperatures, assuming that $\Delta H°$ and $\Delta S°$ do not depend on temperature. What property would you look for in designing a reference half-cell that would produce a potential relatively stable with respect to temperature?

142. The overall reaction in the lead storage battery is

$$Pb(s) + PbO_2(s) + 2H^+(aq) + 2HSO_4^-(aq) \longrightarrow$$
$$2PbSO_4(s) + 2H_2O(l)$$

a. For the cell reaction $\Delta H° = -315.9$ kJ and $\Delta S° = 263.5$ J/K. Calculate $\mathscr{E}°$ at $-20.°C$. Assume $\Delta H°$ and $\Delta S°$ do not depend on temperature.

b. Calculate $\mathscr{E}$ at $-20.°C$ when $[HSO_4^-] = [H^+] = 4.5\ M$.

c. Consider your answer to Exercise 71. Why does it seem that batteries fail more often on cold days than on warm days?

143. Consider the following galvanic cell:

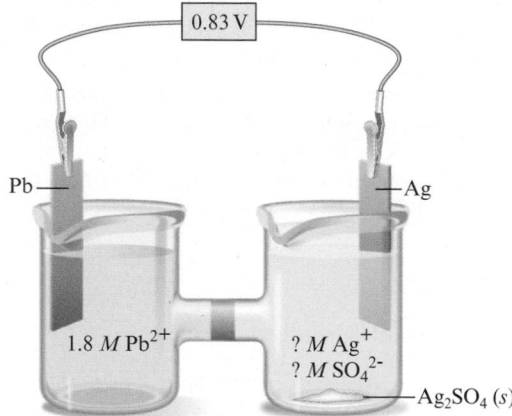

Calculate the K_{sp} value for $Ag_2SO_4(s)$. Note that to obtain silver ions in the right compartment (the cathode compartment), excess solid Ag_2SO_4 was added and some of the salt dissolved.

144. A zinc–copper battery is constructed as follows at 25°C:

$$Zn|Zn^{2+}(0.10\ M)||Cu^{2+}(2.50\ M)|Cu$$

The mass of each electrode is 200. g.

a. Calculate the cell potential when this battery is first connected.

b. Calculate the cell potential after 10.0 A of current has flowed for 10.0 h. (Assume each half-cell contains 1.00 L of solution.)

c. Calculate the mass of each electrode after 10.0 h.

d. How long can this battery deliver a current of 10.0 A before it goes dead?

145. A galvanic cell is based on the following half-reactions:

$$Fe^{2+} + 2e^- \longrightarrow Fe(s) \qquad \mathscr{E}° = -0.440\ V$$
$$2H^+ + 2e^- \longrightarrow H_2(g) \qquad \mathscr{E}° = 0.000\ V$$

where the iron compartment contains an iron electrode and $[Fe^{2+}] = 1.00 \times 10^{-3}\ M$ and the hydrogen compartment contains a platinum electrode, $P_{H_2} = 1.00$ atm, and a weak acid, HA, at an initial concentration of $1.00\ M$. If the observed cell potential is 0.333 V at 25°C, calculate the K_a value for the weak acid HA.

146. Consider a cell based on the following half-reactions:

$$Au^{3+} + 3e^- \longrightarrow Au \qquad \mathscr{E}° = 1.50\ V$$
$$Fe^{3+} + e^- \longrightarrow Fe^{2+} \qquad \mathscr{E}° = 0.77\ V$$

a. Draw this cell under standard conditions, labeling the anode, the cathode, the direction of electron flow, and the concentrations, as appropriate.

b. When enough NaCl(s) is added to the compartment containing gold to make the $[Cl^-] = 0.10\ M$, the cell potential is observed to be 0.31 V. Assume that Au^{3+} is reduced and assume that the reaction in the compartment containing gold is

$$Au^{3+}(aq) + 4Cl^-(aq) \rightleftharpoons AuCl_4^-(aq)$$

Calculate the value of K for this reaction at 25°C.

147. The measurement of pH using a glass electrode obeys the Nernst equation. The typical response of a pH meter at 25.00°C is given by the equation

$$\mathscr{E}_{meas} = \mathscr{E}_{ref} + 0.05916\ pH$$

where $\mathscr{E}_{ref}$ contains the potential of the reference electrode and all other potentials that arise in the cell that are not related to the hydrogen ion concentration. Assume that $\mathscr{E}_{ref} = 0.250$ V and that $\mathscr{E}_{meas} = 0.480$ V.

a. What is the uncertainty in the values of pH and $[H^+]$ if the uncertainty in the measured potential is ± 1 mV (± 0.001 V)?

b. To what precision must the potential be measured for the uncertainty in pH to be ± 0.02 pH unit?

148. You have a concentration cell with Cu electrodes and $[Cu^{2+}] = 1.00\ M$ (right side) and $1.0 \times 10^{-4}\ M$ (left side).

a. Calculate the potential for this cell at 25°C.

b. The Cu^{2+} ion reacts with NH_3 to form $Cu(NH_3)_4^{2+}$ by the following equation:

$$Cu^{2+}(aq) + 4NH_3(aq) \rightleftharpoons Cu(NH_3)_4^{2+}(aq)$$
$$K = 1.0 \times 10^{13}$$

Calculate the new cell potential after enough NH_3 is added to the left cell compartment such that at equilibrium $[NH_3] = 2.0\ M$.

149. A galvanic cell is based on the following half-reactions:

$$Ag^+ + e^- \longrightarrow Ag(s) \qquad \mathscr{E}° = 0.80\ V$$
$$Cu^{2+} + 2e^- \longrightarrow Cu(s) \qquad \mathscr{E}° = 0.34\ V$$

In this cell, the silver compartment contains a silver electrode and excess AgCl(s) ($K_{sp} = 1.6 \times 10^{-10}$), and the copper compartment contains a copper electrode and $[Cu^{2+}] = 2.0\ M$.

a. Calculate the potential for this cell at 25°C.

b. Assuming 1.0 L of 2.0 M Cu^{2+} in the copper compartment, calculate the moles of NH_3 that would have to be added to give a cell potential of 0.52 V at 25°C (assume no volume change on addition of NH_3).

$$Cu^{2+}(aq) + 4NH_3(aq) \rightleftharpoons$$
$$Cu(NH_3)_4^{2+}(aq) \qquad K = 1.0 \times 10^{13}$$

150. Given the following two standard reduction potentials,

$$M^{3+} + 3e^- \longrightarrow M \qquad \mathscr{E}° = -0.10 \text{ V}$$
$$M^{2+} + 2e^- \longrightarrow M \qquad \mathscr{E}° = -0.50 \text{ V}$$

solve for the standard reduction potential of the half-reaction

$$M^{3+} + e^- \longrightarrow M^{2+}$$

(*Hint:* You must use the extensive property $\Delta G°$ to determine the standard reduction potential.)

151. Consider the following galvanic cell:

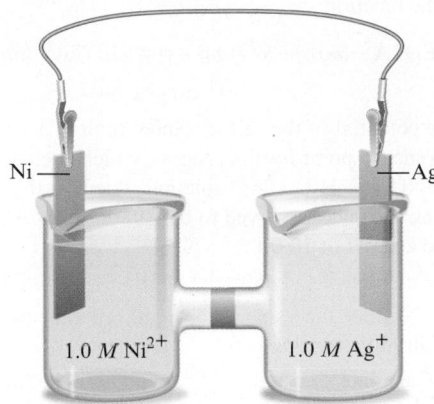

Calculate the concentrations of $Ag^+(aq)$ and $Ni^{2+}(aq)$ once the cell is "dead."

152. A chemist wishes to determine the concentration of CrO_4^{2-} electrochemically. A cell is constructed consisting of a saturated calomel electrode (SCE; see Exercise 115) and a silver wire coated with Ag_2CrO_4. The $\mathscr{E}°$ value for the following half-reaction is 0.446 V relative to the standard hydrogen electrode:

$$Ag_2CrO_4 + 2e^- \longrightarrow 2Ag + CrO_4^{2-}$$

a. Calculate $\mathscr{E}_{cell}$ and ΔG at 25°C for the cell reaction when $[CrO_4^{2-}] = 1.00 \text{ mol/L}$.

b. Write the Nernst equation for the cell. Assume that the SCE concentrations are constant.

c. If the coated silver wire is placed in a solution (at 25°C) in which $[CrO_4^{2-}] = 1.00 \times 10^{-5} M$, what is the expected cell potential?

d. The measured cell potential at 25°C is 0.504 V when the coated wire is dipped into a solution of unknown $[CrO_4^{2-}]$. What is $[CrO_4^{2-}]$ for this solution?

e. Using data from this problem and from Table 17-1, calculate the solubility product (K_{sp}) for Ag_2CrO_4.

153. Consider the following galvanic cell:

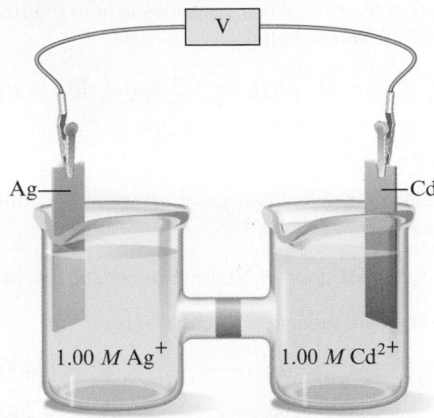

A 15.0-mole sample of NH_3 is added to the Ag compartment (assume 1.00 L of total solution after the addition). The silver ion reacts with ammonia to form complex ions as shown:

$$Ag^+(aq) + NH_3(aq) \rightleftharpoons AgNH_3^+(aq)$$
$$K_1 = 2.1 \times 10^3$$
$$AgNH_3^+(aq) + NH_3(aq) \rightleftharpoons Ag(NH_3)_2^+(aq)$$
$$K_2 = 8.2 \times 10^3$$

Calculate the cell potential after the addition of 15.0 moles of NH_3.

154. When copper reacts with nitric acid, a mixture of $NO(g)$ and $NO_2(g)$ is evolved. The volume ratio of the two product gases depends on the concentration of the nitric acid according to the equilibrium

$$2H^+(aq) + 2NO_3^-(aq) + NO(g) \rightleftharpoons 3NO_2(g) + H_2O(l)$$

Consider the following standard reduction potentials at 25°C:

$$3e^- + 4H^+(aq) + NO_3^-(aq) \longrightarrow NO(g) + 2H_2O(l)$$
$$\mathscr{E}° = 0.957 \text{ V}$$
$$e^- + 2H^+(aq) + NO_3^-(aq) \longrightarrow NO_2(g) + 2H_2O(l)$$
$$\mathscr{E}° = 0.775 \text{ V}$$

a. Calculate the equilibrium constant for the above reaction.

b. What concentration of nitric acid will produce a NO and NO_2 mixture with only 0.20% NO_2 (by moles) at 25°C and 1.00 atm? Assume that no other gases are present and that the change in acid concentration can be neglected.

Integrative Problems

These problems require the integration of multiple concepts to find the solutions.

155. The following standard reduction potentials have been determined for the aqueous chemistry of indium:

$$In^{3+}(aq) + 2e^- \longrightarrow In^+(aq) \qquad \mathscr{E}° = -0.444 \text{ V}$$
$$In^+(aq) + e^- \longrightarrow In(s) \qquad \mathscr{E}° = -0.126 \text{ V}$$

a. What is the equilibrium constant for the disproportionation reaction, where a species is both oxidized and reduced, shown below?

$$3In^+(aq) \longrightarrow 2In(s) + In^{3+}(aq)$$

b. What is ΔG_f° for $In^+(aq)$ if $\Delta G_f^\circ = -97.9$ kJ/mol for $In^{3+}(aq)$?

156. An electrochemical cell is set up using the following unbalanced reaction:

$$M^{a+}(aq) + N(s) \longrightarrow N^{2+}(aq) + M(s)$$

The standard reduction potentials are:

$$M^{a+} + ae^- \longrightarrow M \quad \mathscr{E}^\circ = 0.400 \text{ V}$$
$$N^{2+} + 2e^- \longrightarrow N \quad \mathscr{E}^\circ = 0.240 \text{ V}$$

The cell contains 0.10 M N^{2+} and produces a voltage of 0.180 V. If the concentration of M^{a+} is such that the value of the reaction quotient Q is 9.32×10^{-3}, calculate $[M^{a+}]$. Calculate w_{max} for this electrochemical cell.

157. Three electrochemical cells were connected in series so that the same quantity of electrical current passes through all three cells. In the first cell, 1.15 g chromium metal was deposited from a chromium(III) nitrate solution. In the second cell, 3.15 g osmium was deposited from a solution made of Os^{n+} and nitrate ions. What is the name of the salt? In the third cell, the electrical charge passed through a solution containing X^{2+} ions caused deposition of 2.11 g metallic X. What is the electron configuration of X?

158. A silver concentration cell is set up at 25°C as shown below:

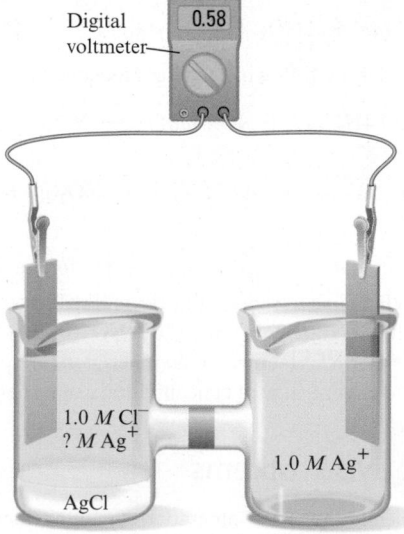

The AgCl(s) is in excess in the left compartment.

a. Label the anode and cathode, and describe the direction of the electron flow.

b. Determine the value of K_{sp} for AgCl at 25°C.

Marathon Problems

These problems are designed to incorporate several concepts and techniques into one situation.

159. A galvanic cell is based on the following half-reactions:

$$Cu^{2+}(aq) + 2e^- \longrightarrow Cu(s) \quad \mathscr{E}^\circ = 0.34 \text{ V}$$
$$V^{2+}(aq) + 2e^- \longrightarrow V(s) \quad \mathscr{E}^\circ = -1.20 \text{ V}$$

In this cell, the copper compartment contains a copper electrode and $[Cu^{2+}] = 1.00$ M, and the vanadium compartment contains a vanadium electrode and V^{2+} at an unknown concentration. The compartment containing the vanadium (1.00 L of solution) was titrated with 0.0800 M H_2EDTA^{2-}, resulting in the reaction

$$H_2EDTA^{2-}(aq) + V^{2+}(aq) \rightleftharpoons VEDTA^{2-}(aq) + 2H^+(aq)$$
$$K = ?$$

The potential of the cell was monitored to determine the stoichiometric point for the process, which occurred at a volume of 500.0 mL H_2EDTA^{2-} solution added. At the stoichiometric point, $\mathscr{E}_{cell}$ was observed to be 1.98 V. The solution was buffered at a pH of 10.00.

a. Calculate $\mathscr{E}_{cell}$ before the titration was carried out.

b. Calculate the value of the equilibrium constant, K, for the titration reaction.

c. Calculate $\mathscr{E}_{cell}$ at the halfway point in the titration.

160. The table below lists the cell potentials for the 10 possible galvanic cells assembled from the metals A, B, C, D, and E, and their respective 1.00 M 2+ ions in solution. Using the data in the table, establish a standard reduction potential table similar to Table 17-1 in the text. Assign a reduction potential of 0.00 V to the half-reaction that falls in the middle of the series. You should get two different tables. Explain why, and discuss what you could do to determine which table is correct.

	A(s) in A²⁺(aq)	B(s) in B²⁺(aq)	C(s) in C²⁺(aq)	D(s) in D²⁺(aq)
E(s) in E²⁺(aq)	0.28 V	0.81 V	0.13 V	1.00 V
D(s) in D²⁺(aq)	0.72 V	0.19 V	1.13 V	—
C(s) in C²⁺(aq)	0.41 V	0.94 V	—	—
B(s) in B²⁺(aq)	0.53 V	—	—	—

CHAPTER EIGHTEEN

The Nucleus: A Chemist's View

A sunset view of a European nuclear power plant. (Tomas Sereda/iStockphoto.com)

Since the chemistry of an atom is determined by the number and arrangement of its electrons, the properties of the nucleus are not of primary importance to chemists. In the simplest view, the nucleus provides the positive charge to bind the electrons in atoms and molecules. However, a quick reading of any daily newspaper will show you that the nucleus and its properties have an important impact on our society. This chapter considers those aspects of the nucleus about which everyone should have some knowledge.

Several aspects of the nucleus are immediately impressive: its very small size, its very large density, and the magnitude of the energy that holds it together. The radius of a typical nucleus appears to be about 10^{-13} cm. This can be compared to the radius of a typical atom, which is on the order of 10^{-8} cm. A visualization will help you appreciate the small size of the nucleus: If the nucleus of the hydrogen atom were the size of a Ping-Pong ball, the electron in the $1s$ orbital would be, on average, 0.5 kilometer (0.3 mile) away. The density of the nucleus is equally impressive—approximately 1.6×10^{14} g/cm^3. A sphere of nuclear material the size of a Ping-Pong ball would have a mass of 2.5 *billion tons!* In addition, the energies involved in nuclear processes are typically millions of times larger than those associated with normal chemical reactions. This fact makes nuclear processes very attractive for feeding the voracious energy appetite of our civilization.

Atomos, the Greek root of the word *atom,* means "indivisible." It was originally believed that the atom was the ultimate indivisible particle of which all matter was composed. However, as we discussed in Chapter 1, Lord Rutherford showed in 1911 that the atom is not homogeneous, but rather has a dense, positively charged center surrounded by electrons. Subsequently, scientists have learned that the nucleus of the atom can be subdivided into particles called **neutrons** and **protons**. In fact, in the past two decades it has become apparent that even the protons and neutrons are composed of smaller particles called *quarks.*

For most purposes, the nucleus can be regarded as a collection of **nucleons** (neutrons and protons), and the internal structures of these particles can be ignored. As we discussed in Chapter 1, the number of protons in a particular nucleus is called the **atomic number** (Z), and the sum of the neutrons and protons is the **mass number** (A). ◄ ⓘ Atoms that have identical atomic numbers but different mass number values are called **isotopes**. ◄ ⓘ However, we usually do not use the singular form *isotope* to refer to a particular member of a group of isotopes. Rather, we use the term *nuclide.* A **nuclide** is a unique atom, represented by the symbol

$$^{A}_{Z}X$$

where X represents the symbol for a particular element. For example, the following nuclides constitute the isotopes of carbon: carbon-12 ($^{12}_{6}$C), carbon-13 ($^{13}_{6}$C), and carbon-14 ($^{14}_{6}$C).

ⓘ The atomic number Z is the number of protons in a nucleus; the mass number A is the sum of protons and neutrons in a nucleus.

ⓘ The term *isotopes* refers to a group of nuclides with the same atomic number. Each individual atom is properly called a *nuclide*, not an isotope.

18-1 | Nuclear Stability and Radioactive Decay

Nuclear stability is the central topic of this chapter and forms the basis for all the important applications related to nuclear processes. Nuclear stability can be considered from both a kinetic and a thermodynamic point of view. **Thermodynamic stability**, as we will use the term here, refers to the potential energy of a particular nucleus as compared with the sum of the potential energies of its component protons and neutrons. We will use the term **kinetic stability** to describe the probability that a nucleus will undergo decomposition to form a different nucleus—a process called **radioactive decay**. We will consider radioactivity in this section.

Many nuclei are radioactive; that is, they decompose, forming another nucleus and producing one or more particles. An example is carbon-14, which decays as follows:

$$^{14}_{6}\text{C} \longrightarrow {}^{14}_{7}\text{N} + {}^{0}_{-1}\text{e}$$

where ${}^{0}_{-1}\text{e}$ represents an electron, which is called a **beta particle**, or **β particle**, in nuclear terminology. This equation is typical of those representing radioactive decay in that both A and Z must be conserved. That is, the Z values must give the same sum on both sides of the equation ($6 = 7 - 1$), as must the A values ($14 = 14 + 0$).

Of the approximately 2000 known nuclides, only 279 are stable with respect to radioactive decay. Tin has the largest number of stable isotopes—10.

It is instructive to examine how the numbers of neutrons and protons in a nucleus are related to its stability with respect to radioactive decay. Figure 18-1 shows a plot of the positions of the stable nuclei as a function of the number of protons (Z) and the number of neutrons ($A - Z$). The stable nuclides are said to reside in the **zone of stability**. ◄ *i*

The following are some important observations concerning radioactive decay:

- All nuclides with 84 or more protons are unstable with respect to radioactive decay.

- Light nuclides are stable when Z equals $A - Z$, that is, when the neutron-to-proton ratio is 1. However, for heavier elements the neutron-to-proton ratio required for stability is greater than 1 and increases with Z.

i Our observations show us which nuclides are stable and which are not stable. An understanding of the structure of the nucleus helps us to explain this.

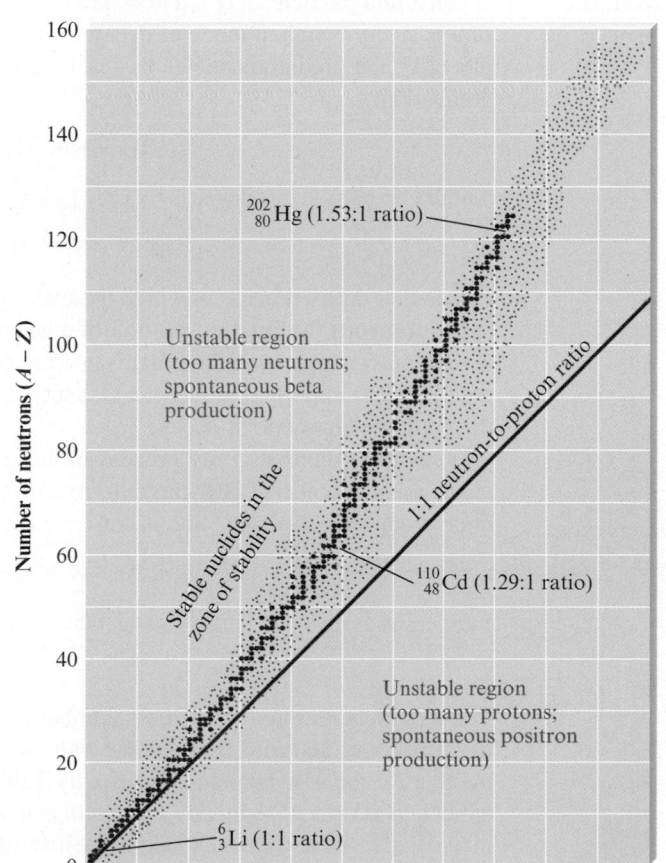

Figure 18-1 | The zone of stability. The red dots indicate the nuclides that *do not* undergo radioactive decay. Note that as the number of protons in a nuclide increases, the neutron-to-proton ratio required for stability also increases.

Table 18-1 | Number of Stable Nuclides Related to Numbers of Protons and Neutrons

Number of Protons	Number of Neutrons	Number of Stable Nuclides	Examples
Even	Even	168	$^{12}_{6}C$, $^{16}_{8}O$
Even	Odd	57	$^{13}_{6}C$, $^{47}_{22}Ti$
Odd	Even	50	$^{19}_{9}F$, $^{23}_{11}Na$
Odd	Odd	4	$^{2}_{1}H$, $^{6}_{3}Li$

Note: Even numbers of protons and neutrons seem to favor stability.

- Certain combinations of protons and neutrons seem to confer special stability. For example, nuclides with even numbers of protons and neutrons are more often stable than those with odd numbers, as shown by the data in Table 18-1.

- There are also certain specific numbers of protons or neutrons that produce especially stable nuclides. These *magic numbers* are 2, 8, 20, 28, 50, 82, and 126. This behavior parallels that for atoms in which certain numbers of electrons (2, 10, 18, 36, 54, and 86) produce special chemical stability (the noble gases).

Types of Radioactive Decay

Radioactive nuclei can undergo decomposition in various ways. These decay processes fall into two categories: those that involve a change in the mass number of the decaying nucleus and those that do not. We will consider the former type of process first.

An **alpha particle**, or **α particle**, is a helium nucleus ($^{4}_{2}He$). **Alpha-particle production** is a very common mode of decay for heavy radioactive nuclides. ◄ ⓘ For example, $^{238}_{92}U$, the predominant (99.3%) isotope of natural uranium, decays by α-particle production:

> ⓘ α-particle production involves a change in *A* for the decaying nucleus; β-particle production has no effect on *A*.

$$^{238}_{92}U \longrightarrow {}^{4}_{2}He + {}^{234}_{90}Th$$

Another α-particle producer is $^{230}_{90}Th$:

$$^{230}_{90}Th \longrightarrow {}^{4}_{2}He + {}^{226}_{88}Ra$$

Another decay process in which the mass number of the decaying nucleus changes is **spontaneous fission**, the splitting of a heavy nuclide into two lighter nuclides with similar mass numbers. Although this process occurs at an extremely slow rate for most nuclides, it is important in some cases, such as for $^{254}_{98}Cf$, where spontaneous fission is the predominant mode of decay.

The most common decay process in which the mass number of the decaying nucleus remains constant is **β-particle production**. For example, the thorium-234 nuclide produces a β particle and is converted to protactinium-234:

$$^{234}_{90}Th \longrightarrow {}^{234}_{91}Pa + {}^{0}_{-1}e$$

Iodine-131 is also a β-particle producer:

$$^{131}_{53}I \longrightarrow {}^{0}_{-1}e + {}^{131}_{54}Xe$$

The β particle is assigned the mass number 0, since its mass is tiny compared with that of a proton or neutron. Because the value of *Z* is -1 for the β particle, the atomic number for the new nuclide is greater by 1 than for the original nuclide. Thus *the net effect of β-particle production is to change a neutron to a proton.* We therefore expect nuclides that lie above the zone of stability (those nuclides whose neutron/proton ratios are too high) to be β-particle producers.

It should be pointed out that although the β particle is an electron, the emitting nucleus does not contain electrons. As we shall see later in this chapter, a given quantity of energy (which is best regarded as a form of matter) can become a particle (another form of matter) under certain circumstances. The unstable nuclide creates an electron as it releases energy in the decay process. The electron thus results from the decay process rather than being present before the decay occurs. Think of this as somewhat like talking: Words are not stored inside us but are formed as we speak. Later in this chapter we will discuss in more detail this very interesting phenomenon where matter in the form of particles and matter in the form of energy can interchange.

A **gamma ray**, or **γ ray**, refers to a high-energy photon. Frequently, γ-ray production accompanies nuclear decays and particle reactions, such as in the α-particle decay of $^{238}_{92}\text{U}$:

$$^{238}_{92}\text{U} \longrightarrow \, ^{4}_{2}\text{He} + \, ^{234}_{90}\text{Th} + 2\,^{0}_{0}\gamma$$

where two γ rays of different energies are produced in addition to the α particle. The emission of γ rays is one way a nucleus with excess energy (in an excited nuclear state) can relax to its ground state.

Positron production occurs for nuclides that are below the zone of stability (those nuclides whose neutron/proton ratios are too small). The positron is a particle with the same mass as the electron but opposite charge. An example of a nuclide that decays by positron production is sodium-22:

$$^{22}_{11}\text{Na} \longrightarrow \, ^{0}_{1}\text{e} + \, ^{22}_{10}\text{Ne}$$

Note that *the net effect is to change a proton to a neutron,* causing the product nuclide to have a higher neutron/proton ratio than the original nuclide.

Besides being oppositely charged, the positron shows an even more fundamental difference from the electron: It is the *antiparticle* of the electron. When a positron collides with an electron, the particulate matter is changed to electromagnetic radiation in the form of high-energy photons:

$$^{0}_{-1}\text{e} + \, ^{0}_{1}\text{e} \longrightarrow 2\,^{0}_{0}\gamma$$

This process, which is characteristic of matter–antimatter collisions, is called *annihilation* and is another example of the interchange of the forms of matter.

Electron capture is a process in which one of the inner-orbital electrons is captured by the nucleus, as illustrated by the process

$$^{201}_{80}\text{Hg} + \underbrace{^{0}_{-1}\text{e}}_{\text{Inner-orbital electron}} \longrightarrow \, ^{201}_{79}\text{Au} + \, ^{0}_{0}\gamma$$

This reaction would have been of great interest to the alchemists, but unfortunately it does not occur at a rate that would make it a practical means for changing mercury to gold. Gamma rays are always produced along with electron capture to release excess energy. The various types of radioactive decay are summarized in Table 18-2.

Critical Thinking

What if a nuclide were to undergo two successive decays such that it became the original nuclide? Which decays could account for this? Provide an example.

Table 18-2 | Various Types of Radioactive Processes Showing the Changes That Take Place in the Nuclides

Process	Change in A	Change in Z	Change in Neutron/Proton Ratio	Example
β-particle (electron) production	0	+1	Decrease	$^{227}_{89}\text{Ac} \longrightarrow \, ^{227}_{90}\text{Th} + \, ^{0}_{-1}\text{e}$
Positron production	0	−1	Increase	$^{13}_{7}\text{N} \longrightarrow \, ^{13}_{6}\text{C} + \, ^{0}_{1}\text{e}$
Electron capture	0	−1	Increase	$^{73}_{33}\text{As} + \, ^{0}_{-1}\text{e} \longrightarrow \, ^{73}_{32}\text{Ge}$
α-particle production	−4	−2	Increase	$^{210}_{84}\text{Po} \longrightarrow \, ^{206}_{82}\text{Pb} + \, ^{4}_{2}\text{He}$
γ-ray production	0	0	—	Excited nucleus $\longrightarrow$ ground-state nucleus + $^{0}_{0}\gamma$
Spontaneous fission	—	—	—	$^{254}_{98}\text{Cf} \longrightarrow$ lighter nuclides + neutrons

INTERACTIVE EXAMPLE 18-1

Nuclear Equations I

Write balanced equations for each of the following processes.

a. $^{11}_{6}C$ produces a positron.

b. $^{214}_{83}Bi$ produces a β particle.

c. $^{237}_{93}Np$ produces an α particle.

Solution

a. We must find the product nuclide represented by $^{A}_{Z}X$ in the following equation:

$$^{11}_{6}C \longrightarrow \underset{\underset{\text{Positron}}{\uparrow}}{^{0}_{1}e} + ^{A}_{Z}X$$

We can find the identity of $^{A}_{Z}X$ by recognizing that the total of the Z and A values must be the same on both sides of the equation. Thus for X, Z must be $6 - 1 = 5$ and A must be $11 - 0 = 11$. Therefore, $^{A}_{Z}X$ is $^{11}_{5}B$. (The fact that Z is 5 tells us that the nuclide is boron.) Thus the balanced equation is

$$^{11}_{6}C \longrightarrow ^{0}_{1}e + ^{11}_{5}B$$

b. Knowing that a β particle is represented by $_{-1}^{0}e$ and that Z and A are conserved, we can write

$$^{214}_{83}Bi \longrightarrow ^{0}_{-1}e + ^{214}_{84}X$$

so $^{A}_{Z}X$ must be $^{214}_{84}Po$.

c. Since an α particle is represented by $^{4}_{2}He$, the balanced equation must be

$$^{237}_{93}Np \longrightarrow ^{4}_{2}He + ^{233}_{91}Pa$$

See Exercises 18-11, 18-14, and 18-15

INTERACTIVE EXAMPLE 18-2

Nuclear Equations II

In each of the following nuclear reactions, supply the missing particle.

a. $^{195}_{79}Au + ? \rightarrow ^{195}_{78}Pt$

b. $^{38}_{19}K \rightarrow ^{38}_{18}Ar + ?$

Solution

a. Since A does not change and Z decreases by 1, the missing particle must be an electron:

$$^{195}_{79}Au + _{-1}^{0}e \longrightarrow ^{195}_{78}Pt$$

This is an example of electron capture.

b. To conserve Z and A, the missing particle must be a positron:

$$^{38}_{19}K \longrightarrow ^{38}_{18}Ar + ^{0}_{1}e$$

Thus potassium-38 decays by positron production.

See Exercises 18-12, 18-13, and 18-16

Figure 18-2 | The decay series from $^{238}_{92}$U to $^{206}_{82}$Pb. Each nuclide in the series except $^{206}_{82}$Pb is radioactive, and the successive transformations (shown by the arrows) continue until $^{238}_{82}$Pb is finally formed. The horizontal red arrows indicate β-particle production (Z increases by 1 and A is unchanged). The diagonal blue arrows signify α-particle production (both A and Z decrease).

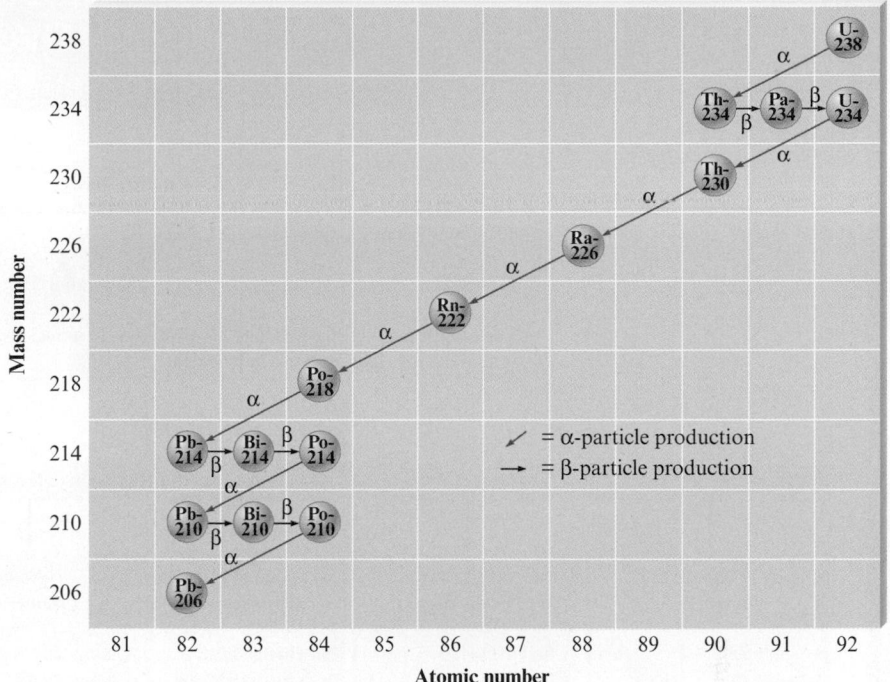

Often a radioactive nucleus cannot reach a stable state through a single decay process. In such a case, a **decay series** occurs until a stable nuclide is formed. A well-known example is the decay series that starts with $^{238}_{92}$U and ends with $^{206}_{82}$Pb, as shown in Fig. 18-2. Similar series exist for $^{235}_{92}$U:

$$^{235}_{92}\text{U} \xrightarrow[\text{decays}]{\text{Series of}} {}^{207}_{82}\text{Pb}$$

and for $^{232}_{90}$Th:

$$^{232}_{90}\text{Th} \xrightarrow[\text{decays}]{\text{Series of}} {}^{208}_{82}\text{Pb}$$

18-2 | The Kinetics of Radioactive Decay

ⓘ Rates of reaction are discussed in Chapter 11.

In a sample containing radioactive nuclides of a given type, each nuclide has a certain probability of undergoing decay. ◄ ⓘ Suppose that a sample of 1000 atoms of a certain nuclide produces 10 decay events per hour. This means that over the span of an hour, 1 out of every 100 nuclides will decay. Given that this probability of decay is characteristic for this type of nuclide, we could predict that a 2000-atom sample would give 20 decay events per hour. Thus, for radioactive nuclides, the **rate of decay**, which is the negative of the change in the number of nuclides per unit time

$$\left(-\frac{\Delta N}{\Delta t}\right)$$

is directly proportional to the number of nuclides N in a given sample:

$$\text{Rate} = -\frac{\Delta N}{\Delta t} \propto N$$

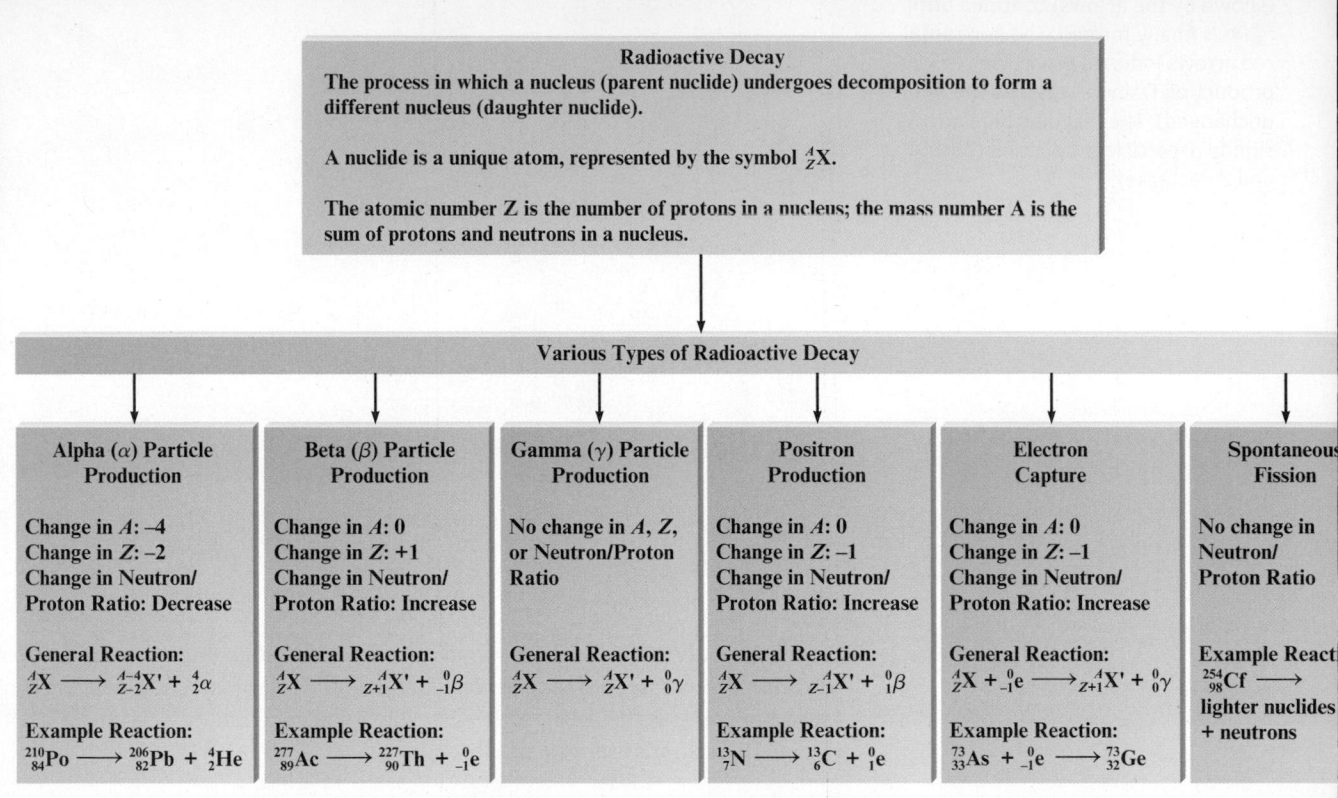

Radioactive Decay

The process in which a nucleus (parent nuclide) undergoes decomposition to form a different nucleus (daughter nuclide).

A nuclide is a unique atom, represented by the symbol $_{Z}^{A}X$.

The atomic number Z is the number of protons in a nucleus; the mass number A is the sum of protons and neutrons in a nucleus.

Various Types of Radioactive Decay

Alpha (α) Particle Production	Beta (β) Particle Production	Gamma (γ) Particle Production	Positron Production	Electron Capture	Spontaneous Fission
Change in A: –4	Change in A: 0	No change in A, Z, or Neutron/Proton Ratio	Change in A: 0	Change in A: 0	No change in Neutron/ Proton Ratio
Change in Z: –2	Change in Z: +1		Change in Z: –1	Change in Z: –1	
Change in Neutron/ Proton Ratio: Decrease	Change in Neutron/ Proton Ratio: Increase		Change in Neutron/ Proton Ratio: Increase	Change in Neutron/ Proton Ratio: Increase	
General Reaction: $_{Z}^{A}X \longrightarrow _{Z-2}^{A-4}X' + _{2}^{4}\alpha$	General Reaction: $_{Z}^{A}X \longrightarrow _{Z+1}^{A}X' + _{-1}^{0}\beta$	General Reaction: $_{Z}^{A}X \longrightarrow _{Z}^{A}X' + _{0}^{0}\gamma$	General Reaction: $_{Z}^{A}X \longrightarrow _{Z-1}^{A}X' + _{1}^{0}\beta$	General Reaction: $_{Z}^{A}X + _{-1}^{0}e \longrightarrow _{Z+1}^{A}X' + _{0}^{0}\gamma$	Example React $_{98}^{254}Cf \longrightarrow$ lighter nuclides + neutrons
Example Reaction: $_{84}^{210}Po \longrightarrow _{82}^{206}Pb + _{2}^{4}He$	Example Reaction: $_{89}^{277}Ac \longrightarrow _{90}^{227}Th + _{-1}^{0}e$		Example Reaction: $_{7}^{13}N \longrightarrow _{6}^{13}C + _{1}^{0}e$	Example Reaction: $_{33}^{73}As + _{-1}^{0}e \longrightarrow _{32}^{73}Ge$	

The negative sign is included because the number of nuclides is decreasing. We now insert a proportionality constant k to give

$$\text{Rate} = -\frac{\Delta N}{\Delta t} = kN$$

This is the rate law for a first-order process, as we saw in Chapter 11. As shown in Section 11-4, the integrated first-order rate law is

$$\ln\left(\frac{N}{N_0}\right) = -kt$$

where N_0 represents the original number of nuclides (at $t = 0$) and N represents the number *remaining* at time t.

Half-Life

The **half-life** ($t_{1/2}$) of a radioactive sample is defined as the time required for the number of nuclides to reach half the original value ($N_0/2$). We can use this definition in connection with the integrated first-order rate law (as we did in Section 11-4) to produce the following expression for $t_{1/2}$:

$$t_{1/2} = \frac{\ln(2)}{k} = \frac{0.693}{k}$$

Thus, if the half-life of a radioactive nuclide is known, the rate constant can be easily calculated, and vice versa.

INTERACTIVE EXAMPLE 18-3

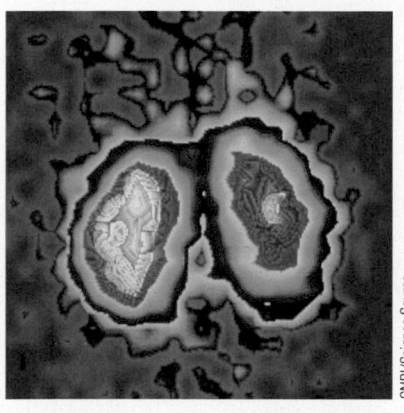

▲ False-color gamma scan of human kidneys showing an infection in the right kidney (purple color). Gamma camera imaging uses the radioisotope technetium-99m

ⓘ The harmful effects of radiation will be discussed in Section 18-7.

Kinetics of Nuclear Decay I

Technetium-99m is used to form pictures of internal organs in the body and is often used to assess heart damage. ◄ The m for this nuclide indicates an excited nuclear state that decays to the ground state by gamma emission. The rate constant for decay of $^{99m}_{43}\text{Tc}$ is known to be 1.16×10^{-1}/h. What is the half-life of this nuclide?

Solution

The half-life can be calculated from the expression

$$t_{1/2} = \frac{0.693}{k} = \frac{0.693}{1.16 \times 10^{-1}/\text{h}}$$
$$= 5.98 \text{ h}$$

Thus it will take 5.98 h for a given sample of technetium-99m to decrease to half the original number of nuclides.

See Exercise 18-25

As we saw in Section 11-4, the half-life for a first-order process is constant. This is shown for the β-particle decay of strontium-90 in Fig. 18-3; it takes 28.9 years for each halving of the amount of $^{90}_{38}\text{Sr}$. Contamination of the environment with $^{90}_{38}\text{Sr}$ poses serious health hazards because of the similar chemistry of strontium and calcium (both are in Group 2A). Strontium-90 in grass and hay is incorporated into cow's milk along with calcium and is then passed on to humans, where it lodges in the bones. Because of its relatively long half-life, it persists for years in humans, causing radiation damage that may lead to cancer. ◄ ⓘ

Figure 18-3 | The decay of a 10.0-g sample of strontium-90 over time. Note that the half-life is a constant 28.9 years.

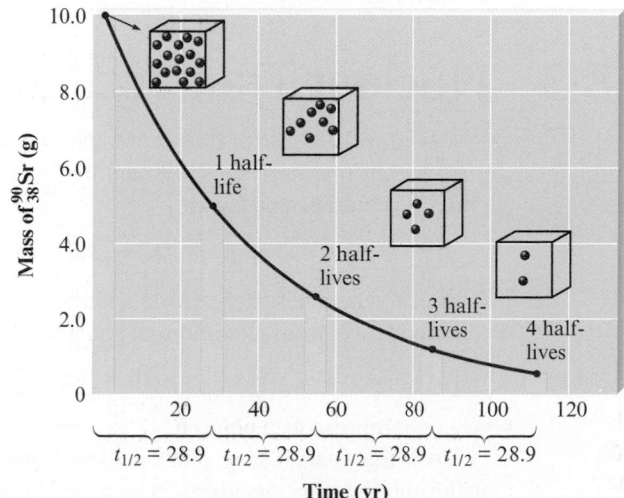

Figure 18-4 | The change in the amount of $^{99}_{42}$Mo with time ($t_{1/2} = 66$ h).

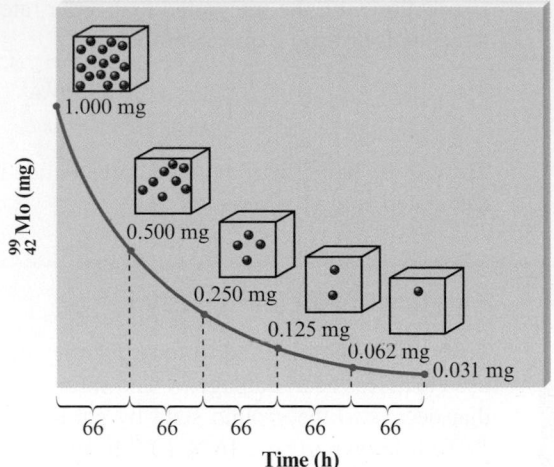

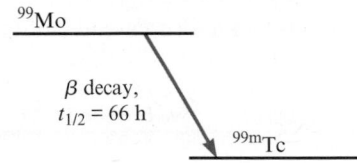

INTERACTIVE EXAMPLE 18-4

Kinetics of Nuclear Decay II

The half-life of molybdenum-99 is 66.0 h. How much of a 1.000-mg sample of $^{99}_{42}$Mo is left after 330 h? ◄

Solution

The easiest way to solve this problem is to recognize that 330 h represents five half-lives for $^{99}_{42}$Mo:

$$330 = 5 \times 66.0$$

We can sketch the change that occurs, as is shown in Fig. 18-4. Thus, after 330 h, 0.031 mg $^{99}_{42}$Mo remains.

See Exercise 18-27

The half-lives of radioactive nuclides vary over a tremendous range. For example, $^{144}_{60}$Nd has a half-life of 2.3×10^{15} years, while $^{214}_{84}$Po has a half-life of 2×10^{-4} second. To give you some perspective on this, the half-lives of the nuclides in the $^{238}_{92}$U decay series are given in Table 18-3.

18-3 | Nuclear Transformations

In 1919 Lord Rutherford observed the first **nuclear transformation**, *the change of one element into another.* He found that by bombarding $^{14}_{7}$N with α particles, the nuclide $^{17}_{8}$O could be produced:

$$^{14}_{7}\text{N} + {}^{4}_{2}\text{He} \longrightarrow {}^{17}_{8}\text{O} + {}^{1}_{1}\text{H}$$

Fourteen years later, Irene Curie and her husband Frederick Joliot observed a similar transformation from aluminum to phosphorus:

$$^{27}_{13}\text{Al} + {}^{4}_{2}\text{He} \longrightarrow {}^{30}_{15}\text{P} + {}^{1}_{0}\text{n}$$

where $^{1}_{0}$n represents a neutron.

Over the years, many other nuclear transformations have been achieved, mostly using **particle accelerators**, which, as the name reveals, are devices used to give

Table 18-3 | The Half-Lives of Nuclides in the $^{238}_{92}$U Decay Series

Nuclide	Particle Produced	Half-Life
Uranium-238 ($^{238}_{92}$U)	α	4.47×10^9 years
Thorium-234 ($^{234}_{90}$Th)	β	24.1 days
Protactinium-234 ($^{234}_{91}$Pa)	β	6.7 hours
Uranium-234 ($^{234}_{92}$U)	α	2.46×10^5 years
Thorium-230 ($^{230}_{90}$Th)	α	7.5×10^4 years
Radium-226 ($^{226}_{88}$Ra)	α	1.60×10^3 years
Radon-222 ($^{222}_{86}$Rn)	α	3.82 days
Polonium-218 ($^{218}_{84}$Po)	α	3.1 minutes
Lead-214 ($^{214}_{82}$Pb)	β	26.8 minutes
Bismuth-214 ($^{214}_{83}$Bi)	β	19.9 minutes
Polonium-214 ($^{214}_{84}$Po)	α	1.6×10^{-4} second
Lead-210 ($^{210}_{82}$Pb)	β	22.2 years
Bismuth-210 ($^{210}_{83}$Bi)	β	5.0 days
Polonium-210 ($^{210}_{84}$Po)	α	138.4 days
Lead-206 ($^{206}_{82}$Pb)	—	Stable

▲ A cyclotron at TRIUMF, Canada's national laboratory of particle and nuclear physics.

particles very high velocities. Because of the electrostatic repulsion between the target nucleus and a positive ion, accelerators are needed when positive ions are used as bombarding particles. The particle, accelerated to a very high velocity, can overcome the repulsion and penetrate the target nucleus, thus effecting the transformation. A schematic diagram of one type of particle accelerator, the **cyclotron**, is shown in Fig. 18-5. The ion is introduced at the center of the cyclotron and is accelerated in an expanding spiral path by use of alternating electric fields in the presence of a

Figure 18-5 | A schematic diagram of a cyclotron. The ion is introduced in the center and is pulled back and forth between the hollow D-shaped electrodes by constant reversals of the electric field. Magnets above and below these electrodes produce a spiral path that expands as the particle velocity increases. When the particle has sufficient speed, it exits the accelerator and is directed at the target nucleus. ◄

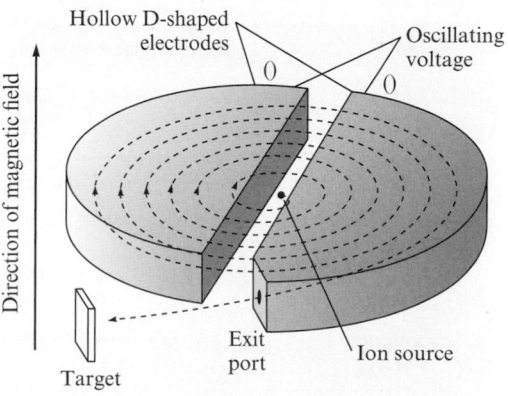

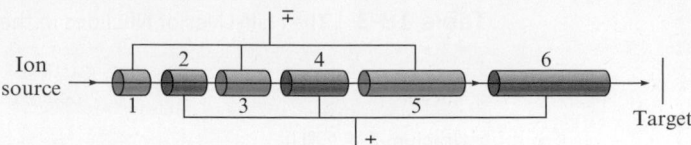

Figure 18-6 | Schematic diagram of a linear accelerator, which uses a changing electric field to accelerate a positive ion along a linear path. As the ion leaves the source, the odd-numbered tubes are negatively charged, and the even-numbered tubes are positively charged. The positive ion is thus attracted into tube 1. As the ion leaves tube 1, the tube polarities are reversed. Now tube 1 is positive, repelling the positive ion, and tube 2 is negative, attracting the positive ion. This process continues, eventually producing high particle velocity.

▲ CERN large Hadron Collider

magnetic field. The **linear accelerator** illustrated in Fig. 18-6 uses changing electric fields to achieve high velocities on a linear pathway.

In addition to positive ions, neutrons are often used as bombarding particles to effect nuclear transformations. Because neutrons are uncharged and thus not repelled electrostatically by a target nucleus, they are readily absorbed by many nuclei, leading to new nuclides. The most common source of neutrons for this purpose is a fission reactor (see Section 18-6).

By using neutron and positive-ion bombardment, scientists have been able to extend the periodic table. ◀ Prior to 1940, the heaviest known element was uranium ($Z = 92$), but in 1940, neptunium ($Z = 93$) was produced by neutron bombardment of $^{238}_{92}U$. The process initially gives $^{239}_{92}U$, which decays to $^{239}_{93}Np$ by β-particle production:

$$^{238}_{92}U + {}^{1}_{0}n \longrightarrow {}^{239}_{92}U \xrightarrow{t_{1/2} = 23 \text{ min}} {}^{238}_{92}Np + {}^{0}_{-1}e$$

In the years since 1940, the elements with atomic numbers greater than 92, called the **transuranium elements,*** have been synthesized. Many of these elements have very short half-lives, as shown in Table 18-4. As a result, only a few atoms of some have ever been formed. This, of course, makes the chemical characterization of these elements extremely difficult.

Table 18-4 | Syntheses of Some of the Transuranium Elements

Element	Neutron Bombardment	Half-Life
Neptunium ($Z = 93$)	$^{238}_{92}U + {}^{1}_{0}n \longrightarrow {}^{239}_{93}Np + {}^{0}_{-1}e$	2.36 days ($^{239}_{93}Np$)
Plutonium ($Z = 94$)	$^{239}_{93}Np \longrightarrow {}^{239}_{94}Pu + {}^{0}_{-1}e$	24,110 years ($^{239}_{94}Pu$)
Americium ($Z = 95$)	$^{239}_{94}Pu + 2{}^{1}_{0}n \longrightarrow {}^{241}_{94}Pu \longrightarrow {}^{241}_{95}Am + {}^{0}_{-1}e$	433 years ($^{241}_{95}Am$)

Element	Positive-Ion Bombardment	Half-Life
Curium ($Z = 96$)	$^{239}_{94}Pu + {}^{4}_{2}He \longrightarrow {}^{242}_{96}Cm + {}^{1}_{0}n$	163 days ($^{242}_{96}Cm$)
Californium ($Z = 98$)	$^{242}_{96}Cm + {}^{4}_{2}He \longrightarrow {}^{245}_{98}Cf + {}^{1}_{0}n$ or $^{238}_{92}U + {}^{12}_{6}C \longrightarrow {}^{246}_{98}Cf + 4{}^{1}_{0}n$	45 minutes ($^{245}_{98}Cf$)
Rutherfordium ($Z = 104$)	$^{249}_{98}Cf + {}^{12}_{6}C \longrightarrow {}^{257}_{104}Rf + 4{}^{1}_{0}n$	
Dubnium ($Z = 105$)	$^{249}_{98}Cf + {}^{15}_{7}N \longrightarrow {}^{260}_{105}Db + 4{}^{1}_{0}n$	
Seaborgium ($Z = 106$)	$^{249}_{98}Cf + {}^{18}_{8}O \longrightarrow {}^{263}_{106}Sg + 4{}^{1}_{0}n$	

*For more information, see G. B. Kauffman, "Beyond uranium," *Chem. Eng. News* (Nov. 19, 1990): 18.

The discovery of element 117 (Uus) provides an excellent illustration of the importance of teamwork in modern scientific activities. Although element 117 was prepared in Dubna, Russia, the target nuclides were prepared at Oak Ridge National Laboratory (ORNL) in Tennessee, and data analysis on the discovery was carried out at Lawrence Livermore Laboratory in California.

The birth of element 117 started at ORNL where a 250-day irradiation experiment produced 22 mg of berkelium-249, which has a 320-day half-life. This process was followed by a 90-day effort to separate and purify the berkelium, after which the berkelium was sent to the Joint Institute for Nuclear Research (JINR) in Dubna, Russia. JINR has an accelerator beam that enabled calcium-48 to be directed at the

berkelium-249 target. This 150-day process resulted in the production of six atoms of element 117 by the following nuclear reactions:

$$^{48}_{20}\text{Ca} + ^{249}_{97}\text{Bk} \rightarrow ^{297}_{117}\text{Uus*} \rightarrow ^{294}_{117}\text{Uus} + 3\,^{1}_{0}\text{n}$$
(1 atom produced)

$$^{48}_{20}\text{Ca} + ^{249}_{97}\text{Bk} \rightarrow ^{297}_{117}\text{Uus*} \rightarrow ^{293}_{117}\text{Uus} + 4\,^{1}_{0}\text{n}$$
(5 atoms produced)

The periodic table shows Uus as the heaviest element in the halogen family. However, because only six atoms of Uus were produced and they existed for only about 0.01 second, no evidence is available at present of the chemical behavior of Uus.

18-4 | Detection and Uses of Radioactivity

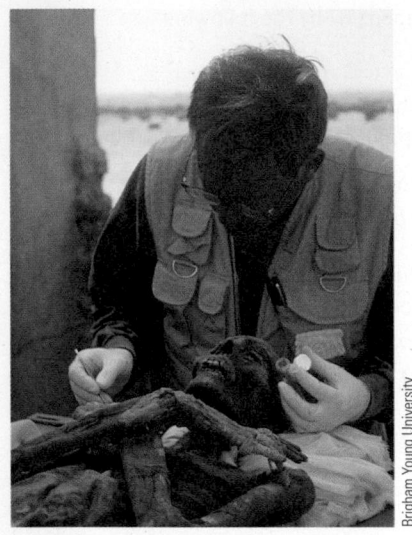

▲ Brigham Young researcher Scott Woodward taking a bone sample for carbon-14 dating at an archaeological site in Egypt.

Brigham Young University

ⓘ Radioactive nuclides are often called *radionuclides*. Carbon dating is based on the radionuclide $^{14}_{6}C$.

Although various instruments measure radioactivity levels, the most familiar of them is the **Geiger–Müller counter**, or **Geiger counter** (Fig. 18-7). ◄ ⓘ This instrument takes advantage of the fact that the high-energy particles from radioactive decay processes produce ions when they travel through matter. The probe of the Geiger counter is filled with argon gas, which can be ionized by a rapidly moving particle. This reaction is demonstrated by the equation:

$$Ar(g) \xrightarrow[\text{particle}]{\text{High-energy}} Ar^+(g) + e^-$$

Normally, a sample of argon gas will not conduct a current when an electrical potential is applied. However, the formation of ions and electrons produced by the passage of the high-energy particle allows a momentary current to flow. Electronic devices detect this current flow, and the number of these events can be counted. Thus the decay rate of the radioactive sample can be determined.

Another instrument often used to detect levels of radioactivity is a **scintillation counter**, which takes advantage of the fact that certain substances, such as zinc sulfide, give off light when they are struck by high-energy radiation. A photocell senses the flashes of light that occur as the radiation strikes and thus measures the number of decay events per unit of time.

Dating by Radioactivity

Archaeologists, geologists, and others involved in reconstructing the ancient history of the earth rely heavily on radioactivity to provide accurate dates for artifacts and rocks. ◄ A method that has been very important for dating ancient articles made from wood or cloth is **radiocarbon dating**, or **carbon-14 dating**, a technique originated in the 1940s by Willard Libby, an American chemist who received a Nobel Prize for his efforts in this field. ◄ ⓘ

Radiocarbon dating is based on the radioactivity of the nuclide $^{14}_{6}C$, which decays via β-particle production:

$$^{14}_{6}C \longrightarrow {}^{0}_{-1}e + {}^{14}_{7}N$$

Carbon-14 is continuously produced in the atmosphere when high-energy neutrons from space collide with nitrogen-14:

$$^{14}_{7}N + {}^{1}_{0}n \longrightarrow {}^{14}_{6}C + {}^{1}_{1}H$$

Thus carbon-14 is continuously produced by this process, and it continuously decomposes through β-particle production. Over the years, the rates for these two processes have become equal, and like a participant in a chemical reaction at equilibrium, the amount of $^{14}_{6}C$ that is present in the atmosphere remains approximately constant.

Figure 18-7 | A schematic representation of a Geiger–Müller counter. The high-energy radioactive particle enters the window and ionizes argon atoms along its path. The resulting ions and electrons produce a momentary current pulse, which is amplified and counted.

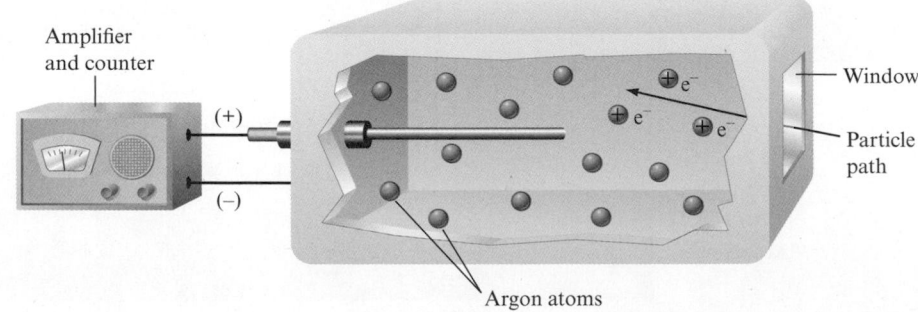

i The $^{14}_{6}C/^{12}_{6}C$ ratio is the basis for carbon-14 dating.

▲ Dr. Thomas Swetnam, a dendro-chronologist at the University of Arizona in Tucson.

Carbon-14 can be used to date wood and cloth artifacts because the $^{14}_{6}C$, along with the other carbon isotopes in the atmosphere, reacts with oxygen to form carbon dioxide. A living plant consumes carbon dioxide in the photosynthesis process and incorporates the carbon, including $^{14}_{6}C$, into its molecules. As long as the plant lives, the $^{14}_{6}C/^{12}_{6}C$ ratio in its molecules remains the same as in the atmosphere because of the continuous uptake of carbon. However, as soon as a tree is cut to make a wooden bowl or a flax plant is harvested to make linen, the $^{14}_{6}C/^{12}_{6}C$ ratio begins to decrease because of the radioactive decay of $^{14}_{6}C$ (the $^{12}_{6}C$ nuclide is stable). Since the half-life of $^{14}_{6}C$ is 5730 years, a wooden bowl found in an archaeological dig showing a $^{14}_{6}C/^{12}_{6}C$ ratio that is half that found in currently living trees is approximately 5730 years old. This reasoning assumes that the current $^{14}_{6}C/^{12}_{6}C$ ratio is the same as that found in ancient times. ◄ *i*

Dendrochronologists, scientists who date trees from annual growth rings, have used data collected from long-lived species of trees, such as bristlecone pines and sequoias, to show that the $^{14}_{6}C$ content of the atmosphere has changed significantly over the ages. ◄ These data have been used to derive correction factors that allow very accurate dates to be determined from the observed $^{14}_{6}C/^{12}_{6}C$ ratio in an artifact, especially for artifacts 10,000 years old or younger. Recent measurements of uranium-to-thorium ratios in ancient coral indicate that dates in the 20,000- to 30,000-year range may have errors as large as 3000 years. As a result, efforts are now being made to recalibrate the $^{14}_{6}C$ dates over this period.

INTERACTIVE EXAMPLE 18-5

^{14}C Dating

The remnants of an ancient fire in a cave in Africa showed a $^{14}_{6}C$ decay rate of 3.1 counts per minute per gram of carbon. Assuming that the decay rate of $^{14}_{6}C$ in freshly cut wood (corrected for changes in the $^{14}_{6}C$ content of the atmosphere) is 13.6 counts per minute per gram of carbon, calculate the age of the remnants. The half-life of $^{14}_{6}C$ is 5730 years.

Solution

The key to solving this problem is to realize that the decay rates given are directly proportional to the number of $^{14}_{6}C$ nuclides present. Radioactive decay follows first-order kinetics:

$$\text{Rate} = kN$$

Thus

$$\frac{3.1 \text{ counts/min} \cdot \text{g}}{13.6 \text{ counts/min} \cdot \text{g}} = \frac{\text{rate at time } t}{\text{rate at time } 0} = \frac{kN}{kN_0}$$

Number of nuclides ↙ present at time t

↖ Number of nuclides present at time 0

$$= \frac{N}{N_0} = 0.23$$

We can now use the integrated first-order rate law:

$$\ln\left(\frac{N}{N_0}\right) = -kt$$

where

$$k = \frac{0.693}{t_{1/2}} = \frac{0.693}{5730 \text{ years}}$$

to solve for t, the time elapsed since the campfire:

$$\ln\left(\frac{N}{N_0}\right) = \ln(0.23) = -\left(\frac{0.693}{5730 \text{ years}}\right)t$$

Solving this equation gives $t = 12{,}000$ years; the campfire in the cave occurred about 12,000 years ago.

See Exercises 18-37 and 18-38

One drawback of radiocarbon dating is that a fairly large piece of the object (from a half to several grams) must be burned to form carbon dioxide, which is then analyzed for radioactivity. Another method for counting $^{14}_{6}C$ nuclides avoids destruction of a significant portion of a valuable artifact. This technique, requiring only about 10^{-3} g, uses a mass spectrometer (see Chapter 5), in which the carbon atoms are ionized and accelerated through a magnetic field that deflects their path. Because of their different masses, the various ions are deflected by different amounts and can be counted separately. This allows a very accurate determination of the $^{14}_{6}C/^{12}_{6}C$ ratio in the sample.

In their attempts to establish the geologic history of the earth, geologists have made extensive use of radioactivity. For example, since $^{238}_{92}U$ decays to the stable $^{206}_{82}Pb$ nuclide, the ratio of $^{206}_{82}Pb$ to $^{238}_{92}U$ in a rock can, under favorable circumstances, be used to estimate the age of the rock. The radioactive nuclide $^{176}_{71}Lu$, which decays to $^{176}_{72}Hf$, has a half-life of 37 billion years (only 186 nuclides out of 10 trillion decay each year!). Thus this nuclide can be used to date very old rocks. With this technique, scientists have estimated that the earth's crust formed 4.3 billion years ago.

Critical Thinking

What if carbon-14 was not continually produced in the atmosphere? How would this affect the effectiveness of radiocarbon dating?

INTERACTIVE EXAMPLE 18-6

Dating by Radioactivity

A rock containing $^{238}_{92}U$ and $^{206}_{82}Pb$ was examined to determine its approximate age. Analysis showed the ratio of $^{206}_{82}Pb$ atoms to $^{238}_{92}U$ atoms to be 0.115. Assuming that no lead was originally present, that all the $^{206}_{82}Pb$ formed over the years has remained in the rock, and that the number of nuclides in intermediate stages of decay between $^{238}_{92}U$ and $^{206}_{82}Pb$ is negligible, calculate the age of the rock. The half-life of $^{238}_{92}U$ is 4.5×10^9 years. ◀ *i*

i Because the half-life of $^{238}_{92}U$ is very long compared with those of the other members of the decay series (see Table 18-3) to reach $^{206}_{82}Pb$, the number of nuclides present in intermediate stages of decay is negligible. That is, once a $^{238}_{92}U$ nuclide starts to decay, it reaches $^{206}_{82}Pb$ relatively fast.

Solution

This problem can be solved using the integrated first-order rate law:

$$\ln\left(\frac{N}{N_0}\right) = -kt = -\left(\frac{0.693}{4.5 \times 10^9 \text{ years}}\right)t$$

where N/N_0 represents the ratio of $^{238}_{92}U$ atoms now found in the rock to the number present when the rock was formed. We are assuming that each $^{206}_{82}Pb$ nuclide present must have come from decay of a $^{238}_{92}U$ atom:

$$^{238}_{92}U \longrightarrow {}^{206}_{82}Pb$$

Thus

$$\boxed{\begin{array}{c}\text{Number of } ^{238}_{92}U \text{ atoms}\\ \text{originally present}\end{array}} = \boxed{\begin{array}{c}\text{number of } ^{206}_{82}Pb \text{ atoms}\\ \text{now present}\end{array}} + \boxed{\begin{array}{c}\text{number of } ^{238}_{92}U \text{ atoms}\\ \text{now present}\end{array}}$$

$$\frac{\text{Atoms of } ^{206}_{82}Pb \text{ now present}}{\text{Atoms of } ^{238}_{92}U \text{ now present}} = 0.115 = \frac{0.115}{1.000} = \frac{115}{1000}$$

Think carefully about what this means. For every 1115 $^{238}_{92}$U atoms originally present in the rock, 115 have been changed to $^{206}_{82}$Pb and 1000 remain as $^{238}_{92}$U. Thus

$$\frac{N}{N_0} = \underbrace{\frac{\overbrace{^{238}_{92}\text{U}}^{\text{Now present}}}{^{206}_{82}\text{Pb} + {}^{238}_{92}\text{U}}}_{^{238}_{92}\text{U originally present}} = \frac{1000}{1115} = 0.8969$$

$$\ln\left(\frac{N}{N_0}\right) = \ln(0.8969) = -\left(\frac{0.693}{4.5 \times 10^9 \text{ years}}\right)t$$

$$t = 7.1 \times 10^8 \text{ years}$$

This is the approximate age of the rock. It was formed sometime in the Cambrian period.

See Exercises 18-39 and 18-40

Medical Applications of Radioactivity

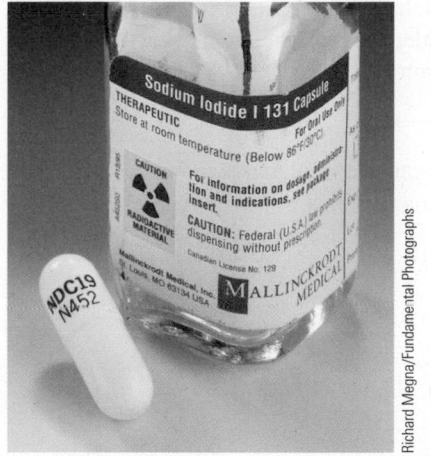

▲ A pellet containing radioactive ^{131}I.

Although the rapid advances of the medical sciences in recent decades are due to many causes, one of the most important has been the discovery and use of **radiotracers**, radioactive nuclides that can be introduced into organisms in food or drugs and whose pathways can be *traced* by monitoring their radioactivity. For example, the incorporation of nuclides such as $^{14}_{6}$C and $^{32}_{15}$P into nutrients has produced important information about metabolic pathways.

Iodine-131 has proved very useful in the diagnosis and treatment of illnesses of the thyroid gland. ◄ Patients drink a solution containing small amounts of Na^{131}I, and the uptake of the iodine by the thyroid gland is monitored with a scanner (Fig. 18-8).

Thallium-201 can be used to assess the damage to the heart muscle in a person who has suffered a heart attack, because thallium is concentrated in healthy muscle tissue. Technetium-99m is also taken up by normal heart tissue and is used for damage assessment in a similar way.

Radiotracers provide sensitive and noninvasive methods for learning about biological systems, for detection of disease, for monitoring the action and effectiveness of drugs, and for early detection of pregnancy, and their usefulness should continue to grow. Some useful radiotracers are listed in Table 18-5.

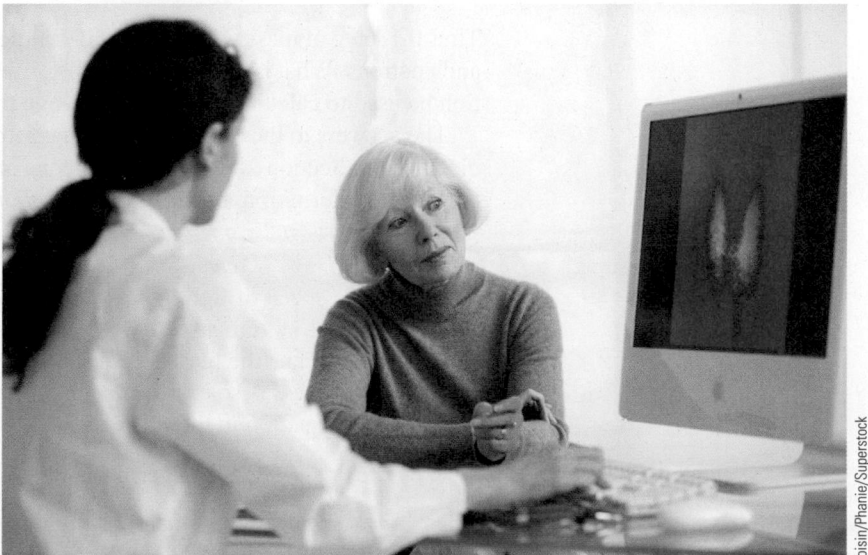

Figure 18-8 | Doctor discussing a thyroid gland scintigram with a patient.

Table 18-5 | Some Radioactive Nuclides, with Half-Lives and Medical Applications as Radiotracers

Nuclide	Half-Life	Area of the Body Studied
^{131}I	8.0 days	Thyroid
^{59}Fe	44.5 days	Red blood cells
^{99}Mo	66 hours	Metabolism
^{32}P	14.3 days	Eyes, liver, tumors
^{51}Cr	27.7 days	Red blood cells
^{87}Sr	2.8 hours	Bones
^{99m}Tc	6.0 hours	Heart, bones, liver, and lungs
^{133}Xe	5.2 days	Lungs
^{24}Na	15.0 hours	Circulatory system

18-5 | Thermodynamic Stability of the Nucleus

We can determine the thermodynamic stability of a nucleus by calculating the change in potential energy that would occur if that nucleus were formed from its constituent protons and neutrons. For example, let's consider the hypothetical process of forming a $^{16}_{8}O$ nucleus from eight neutrons and eight protons:

$$8\,^{1}_{0}n + 8\,^{1}_{1}H \longrightarrow\ ^{16}_{8}O$$

The energy change associated with this process can be calculated by comparing the sum of the masses of eight protons and eight neutrons with that of the oxygen nucleus:

$$\text{Mass of } (8\,^{1}_{0}n + 8\,^{1}_{1}H) = 8(1.67493 \times 10^{-24}\ \text{g}) + 8(1.67262 \times 10^{-24}\ \text{g})$$

$$\uparrow \qquad\qquad\qquad\qquad \uparrow$$
$$\text{Mass of } ^{1}_{0}n \qquad\qquad \text{Mass of } ^{1}_{1}H$$

$$= 2.67804 \times 10^{-23}\ \text{g}$$

$$\text{Mass of } ^{16}_{8}O \text{ nucleus} = 2.65535 \times 10^{-23}\ \text{g}$$

The difference in mass for one nucleus is

$$\text{Mass of } ^{16}_{8}O - \text{mass of } (8\,^{1}_{0}n + 8\,^{1}_{1}H) = -2.269 \times 10^{-25}\ \text{g}$$

The difference in mass for formation of 1 mole of $^{16}_{8}O$ nuclei is therefore

$$(-2.269 \times 10^{-25}\ \text{g/nucleus})(6.022 \times 10^{23}\ \text{nuclei/mol}) = -0.1366\ \text{g/mol}$$

Thus 0.1366 g of mass would be lost if 1 mole of oxygen-16 were formed from protons and neutrons. What is the reason for this difference in mass, and how can this information be used to calculate the energy change that accompanies this process?

The answers to these questions can be found in the work of Albert Einstein. As we discussed in Section 2-2, Einstein's theory of relativity showed that energy should be considered a form of matter. His famous equation

$$E = mc^2$$

where c is the speed of light, gives the relationship between a quantity of energy and its mass. When a system gains or loses energy, it also gains or loses a quantity of mass, given by E/c^2. Thus the mass of a nucleus is less than that of its component nucleons because the process is so exothermic.

Einstein's equation in the form

$$\text{Energy change} = \Delta E = \Delta mc^2$$

where Δm is the change in mass, or the **mass defect**, can be used to calculate ΔE for the hypothetical formation of a nucleus from its component nucleons. ◄ ⓘ

ⓘ The energy changes associated with normal chemical reactions are small enough that the corresponding mass changes are not detectable.

INTERACTIVE EXAMPLE 18-7

Nuclear Binding Energy I

Calculate the change in energy if 1 mole of $^{16}_{8}O$ nuclei was formed from neutrons and protons.

Solution

We have already calculated that 0.1366 g of mass would be lost in the hypothetical process of assembling 1 mole of $^{16}_{8}O$ nuclei from the component nucleons. We can calculate the change in energy for this process from

$$\Delta E = \Delta m c^2$$

where

$$c = 3.00 \times 10^8 \text{ m/s} \quad \text{and} \quad \Delta m = -0.1366 \text{ g/mol} = -1.366 \times 10^{-4} \text{ kg/mol}$$

Thus

$$\Delta E = (-1.366 \times 10^{-4} \text{ kg/mol})(3.00 \times 10^8 \text{ m/s})^2 = -1.23 \times 10^{13} \text{ J/mol}$$

The negative sign for the ΔE value indicates that the process is exothermic. Energy, and thus mass, is lost from the system.

See Exercises 18-41 through 18-43

The energy changes observed for nuclear processes are extremely large compared with those observed for chemical and physical changes. Thus nuclear processes constitute a potentially valuable energy resource.

The thermodynamic stability of a particular nucleus is normally represented as energy released per nucleon. To illustrate how this quantity is obtained, we will continue to consider $^{16}_{8}O$. First, we calculate ΔE per nucleus by dividing the molar value from Example 18-7 by Avogadro's number:

$$\Delta E \text{ per } ^{16}_{8}O \text{ nucleus} = \frac{-1.23 \times 10^{13} \text{ J/mol}}{6.022 \times 10^{23} \text{ nuclei/mol}} = -2.04 \times 10^{-11} \text{ J/nucleus}$$

In terms of a more convenient energy unit, a million electronvolts (MeV), where

$$1 \text{ MeV} = 1.60 \times 10^{-13} \text{ J}$$

$$\Delta E \text{ per } ^{16}_{8}O \text{ nucleus} = (-2.04 \times 10^{-11} \text{ J/nucleus}) \left(\frac{1 \text{ MeV}}{1.60 \times 10^{-13} \text{ J}} \right)$$

$$= -1.28 \times 10^2 \text{ MeV/nucleus}$$

Next, we can calculate the value of ΔE per nucleon by dividing by A, the sum of neutrons and protons:

$$\Delta E \text{ per nucleon for } ^{16}_{8}O = \frac{-1.28 \times 10^2 \text{ MeV/nucleus}}{16 \text{ nucleons/nucleus}}$$

$$= -7.98 \text{ MeV/nucleon}$$

This means that 7.98 MeV of energy per nucleon would be *released* if $^{16}_{8}O$ were formed from neutrons and protons. The energy required to *decompose* this nucleus into its components has the same numeric value but a positive sign (since energy is required). This is called the **binding energy** per nucleon for $^{16}_{8}O$.

The values of the binding energy per nucleon for the various nuclides are shown in Fig. 18-9. Note that the most stable nuclei (those requiring the largest energy per nucleon to decompose the nucleus) occur at the top of the curve. The most stable nucleus known is $^{56}_{26}Fe$, which has a binding energy per nucleon of 8.79 MeV.

Figure 18-9 | The binding energy per nucleon as a function of mass number. The most stable nuclei are at the top of the curve. The most stable nucleus is $^{56}_{26}$Fe.

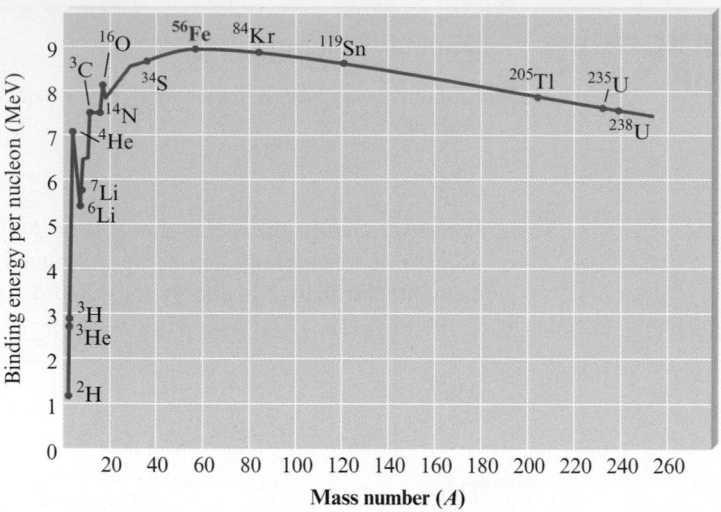

INTERACTIVE EXAMPLE 18-8

Nuclear Binding Energy II

Calculate the binding energy per nucleon for the $^4_2$He nucleus (atomic masses: $^4_2$He = 4.0026 u; $^1_1$H = 1.0078 u).

Solution

First, we must calculate the mass defect (Δm) for $^4_2$He. Since atomic masses (which include the electrons) are given, we must decide how to account for the electron mass: ◀ ⓘ

$$4.0026 = \text{mass of } ^4_2\text{He atom} = \text{mass of } ^4_2\text{He nucleus} + 2m_e$$

Electron mass

$$1.0078 = \text{mass of } ^1_1\text{H atom} = \text{mass of } ^1_1\text{H nucleus} + m_e$$

ⓘ Since atomic masses include the masses of the electrons, to obtain the mass of a given atomic nucleus from its atomic mass, we must subtract the mass of the electrons.

Thus, since a $^4_2$He nucleus is "synthesized" from two protons and two neutrons, we see that

$$\Delta m = \underbrace{(4.0026 - 2m_e)}_{\substack{\text{Mass of } ^4_2\text{He} \\ \text{nucleus}}} - \left[\underbrace{2(1.0078 - m_e)}_{\substack{\text{Mass of } ^1_1\text{H} \\ \text{nucleus (proton)}}} + \underbrace{2(1.0087)}_{\substack{\text{Mass of} \\ \text{neutron}}}\right]$$

$$= 4.0026 - 2m_e - 2(1.0078) + 2m_e - 2(1.0087)$$

$$= 4.0026 - 2(1.0078) - 2(1.0087)$$

$$= -0.0304 \text{ u}$$

Note that in this case the electron mass cancels out in taking the difference. This will always happen in this type of calculation if the atomic masses are used both for the nuclide of interest and for $^1_1$H. Thus 0.0304 of mass is *lost* per $^4_2$He nucleus formed.

The corresponding energy change can be calculated from

$$\Delta E = \Delta mc^2$$

where

$$\Delta m = -0.0304 \frac{\text{u}}{\text{nucleus}} = \left(-0.0304 \frac{\text{u}}{\text{nucleus}}\right)\left(1.66 \times 10^{-27} \frac{\text{kg}}{\text{u}}\right)$$

$$= -5.04 \times 10^{-29} \frac{\text{kg}}{\text{nucleus}}$$

and $$c = 3.00 \times 10^8 \text{ m/s}$$

Thus $$\Delta E = \left(-5.04 \times 10^{-29} \frac{\text{kg}}{\text{nucleus}} \right) \left(3.00 \times 10^8 \frac{\text{m}}{\text{s}} \right)^2$$

$$= -4.54 \times 10^{-12} \text{ J/nucleus}$$

This means that 4.54×10^{-12} J of energy is *released* per nucleus formed and that 4.54×10^{-12} J would be required to decompose the nucleus into the constituent neutrons and protons. Thus the binding energy (BE) per nucleon is

$$\text{BE per nucleon} = \frac{4.54 \times 10^{-12} \text{ J/nucleus}}{4 \text{ nucleons/nucleus}}$$

$$= 1.14 \times 10^{-12} \text{ J/nucleon}$$

$$= \left(1.14 \times 10^{-12} \frac{\text{J}}{\text{nucleon}} \right) \left(\frac{1 \text{ MeV}}{1.60 \times 10^{-13} \text{ J}} \right)$$

$$= 7.13 \text{ MeV/nucleon}$$

See Exercises 18-44 through 18-46

18-6 | Nuclear Fission and Nuclear Fusion

The graph shown in Fig. 18-9 has very important implications for the use of nuclear processes as sources of energy. Recall that energy is released, that is, ΔE is negative, when a process goes from a less stable to a more stable state. The higher a nuclide is on the curve, the more stable it is. This means that two types of nuclear processes will be exothermic (Fig. 18-10):

1. Combining two light nuclei to form a heavier, more stable nucleus. This process is called **fusion**.
2. Splitting a heavy nucleus into two nuclei with smaller mass numbers. This process is called **fission**.

Because of the large binding energies involved in holding the nucleus together, both these processes involve energy changes more than a million times larger than those associated with chemical reactions.

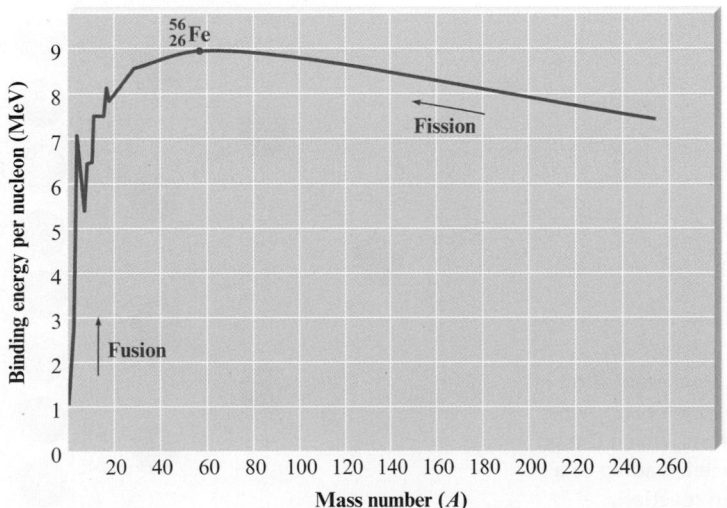

Figure 18-10 | Both fission and fusion produce more stable nuclides and are thus exothermic.

Figure 18-11 | On capturing a neutron, the $^{235}_{92}U$ nucleus undergoes fission to produce two lighter nuclides, free neutrons (typically three), and a large amount of energy.

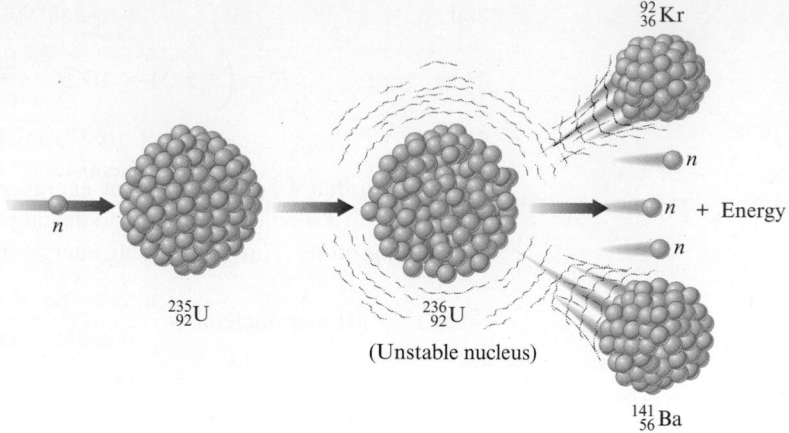

$^{92}_{36}Kr$

n

n + Energy

n

$^{235}_{92}U$

$^{236}_{92}U$

(Unstable nucleus)

$^{141}_{56}Ba$

Nuclear Fission

Nuclear fission was discovered in the late 1930s when $^{235}_{92}U$ nuclides bombarded with neutrons were observed to split into two lighter elements:

$$^1_0n + ^{235}_{92}U \longrightarrow ^{141}_{56}Ba + ^{92}_{36}Kr + 3\,^1_0n$$

This process, shown schematically in Fig. 18-11, releases 3.5×10^{-11} J of energy per event, which translates to 2.1×10^{13} J per mole of $^{235}_{92}U$. Compare this figure with that for the combustion of methane, which releases only 8.0×10^5 J of energy per mole. The fission of $^{235}_{92}U$ produces about 26 million times more energy than the combustion of methane.

The process shown in Fig. 18-11 is only one of the many fission reactions that $^{235}_{92}U$ can undergo. Another is

$$^1_0n + ^{235}_{92}U \longrightarrow ^{137}_{52}Te + ^{97}_{40}Zr + 2\,^1_0n$$

In fact, over 200 different isotopes of 35 different elements have been observed among the fission products of $^{235}_{92}U$.

In addition to the product nuclides, neutrons are produced in the fission reactions of $^{235}_{92}U$. This makes it possible to have a self-sustaining fission process—a **chain reaction** (Fig. 18-12). For the fission process to be self-sustaining, at least one neutron from each fission event must go on to split another nucleus. If, on average, *less than one* neutron causes another fission event, the process dies out and the reaction is said

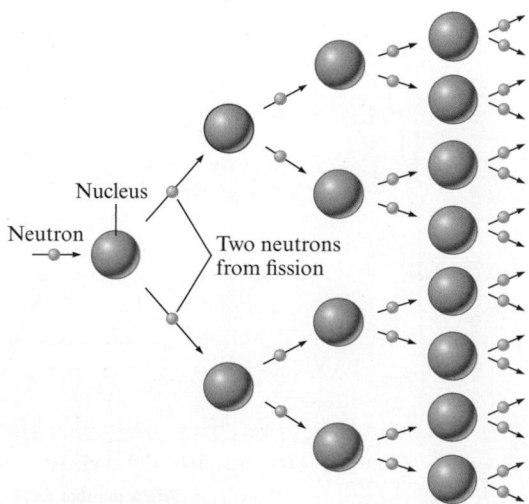

Nucleus

Neutron

Two neutrons
from fission

Figure 18-12 | Representation of a fission process in which each event produces two neutrons, which can go on to split other nuclei, leading to a self-sustaining chain reaction.

Energy—a crucial commodity in today's world—will become even more important as the pace of world development increases. Because the energy content of the universe is constant, the challenge of energy is not its quantity but rather its quality. We must find economical and environmentally friendly ways to change the energy available in the universe to forms useful to humanity. This process always involves tradeoffs.

Currently, about 65% of the world's energy consumption involves combustion of fossil fuels (coal, 39%; natural gas, 19%; oil, 7%). The use of these fuels causes significant pollution and contributes huge amounts of greenhouse gases (mostly CO_2) to the atmosphere.

One of the most abundant sources of energy is the energy that binds the atomic nucleus together. We can derive useful energy by assembling small nuclei (fusion) or splitting large nuclei (fission). Although fusion reactors are being studied, a practical fusion reactor appears to be decades away. By contrast, fission reactors have been used since the 1950s. In fact, the production of electricity via fission reactors is widespread. At present, more than 400 nuclear reactors operate in 31 countries, producing over 355 billion watts of electrical power (see accompanying table). More than 30 reactors are currently under construction, and at least 100 more are in the planning stages. The 103 reactors currently operating in the United States produce almost 100 billion watts of electricity—about 20% of the country's electrical demands.

Forecasts indicate that the United States will need an additional 355 billion watts of generating capacity in the next 20 years. Where will this energy come from? A significant amount will be derived from coal-fired power plants with their inherent environmental problems.

Another potential source of power is solar energy. It should be an excellent pollution-free energy source, but significant technical problems remain to be solved before it sees widespread use. Wind power is also being developed but promises to make only a limited contribution to overall energy use.

The most important power source is nuclear energy. To provide all of the 355 billion watts from nuclear energy would require hundreds of new reactors. However, nuclear power generation is very controversial because of safety, waste disposal, and cost issues. On the other hand, nuclear energy produces no greenhouse gases, has a much lower volume of waste products than the combustion of fossil fuels, has an almost unlimited supply of fuel, and has an excellent safety record. Research is now under way to improve existing reactor designs and to find new types of reactors that will be safer, more efficient, and generate much less waste by finding ways to reprocess the reactor fuel.

Actually, the United States has demonstrated an ability to deal successfully with the long-term storage of nuclear wastes. The Waste Isolation Pilot Plant (WIPP) in New Mexico has been receiving nuclear wastes since 1999 with no accidents in either transporting or storing the wastes. WIPP uses tunnels carved into the salt beds of an ancient ocean. Once a repository room becomes full, the salt will collapse around the waste, encapsulating it forever.

There is no doubt that nuclear energy will be important to the United States and to the world. An excellent source of information about all aspects of nuclear energy use is the book *Power to Save the World—The Truth About Nuclear Energy*, by Gwyneth Cravens (Alfred A. Knopf, New York, 2007). This book is a thorough but very readable treatment of the subject.

Top Ten Countries Producing Electricity by Nuclear Power (in order of total nuclear output)

Country	Percentage of Country's Total Power Production
United States	21.9
France	77.4
Japan	34.0
Germany	30.3
Russia	13.1
Canada	16.0
Ukraine	43.8
United Kingdom	26.0
Sweden	52.4
South Korea	35.8

to be **subcritical**. If *exactly one* neutron from each fission event causes another fission event, the process sustains itself at the same level and is said to be **critical**. If *more than one* neutron from each fission event causes another fission event, the process rapidly escalates and the heat buildup causes a violent explosion. This situation is described as **supercritical**.

To achieve the critical state, a certain mass of fissionable material, called the **critical mass**, is needed. If the sample is too small, too many neutrons escape before they have a chance to cause a fission event, and the process stops. This is illustrated in Fig. 18-13.

Figure 18-13 | If the mass of fissionable material is too small, most of the neutrons escape before causing another fission event, and the process dies out.

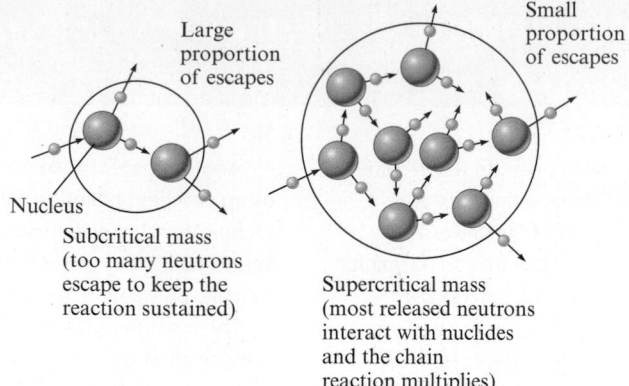

Large proportion of escapes

Small proportion of escapes

Nucleus

Subcritical mass
(too many neutrons escape to keep the reaction sustained)

Supercritical mass
(most released neutrons interact with nuclides and the chain reaction multiplies)

During World War II, an intense research effort called the Manhattan Project was carried out by the United States to build a bomb based on the principles of nuclear fission. This program produced the fission bombs that were used with devastating effects on the cities of Hiroshima and Nagasaki in 1945. Basically, a fission bomb operates by suddenly combining subcritical masses of fissionable material to form a supercritical mass, thereby producing an explosion of incredible intensity.

Nuclear Reactors

Because of the tremendous energies involved, it seemed desirable to develop the fission process as an energy source to produce electricity. To accomplish this, reactors were designed in which controlled fission can occur. The resulting energy is used to heat water to produce steam to run turbine generators, in much the same way that a coal-burning power plant generates energy. A schematic diagram of a nuclear power plant is shown in Fig. 18-14.

In the **reactor core**, shown in Fig. 18-15, uranium that has been enriched to approximately 3% $^{235}_{92}$U (natural uranium contains only 0.7% $^{235}_{92}$U) is housed in cylinders. ◄

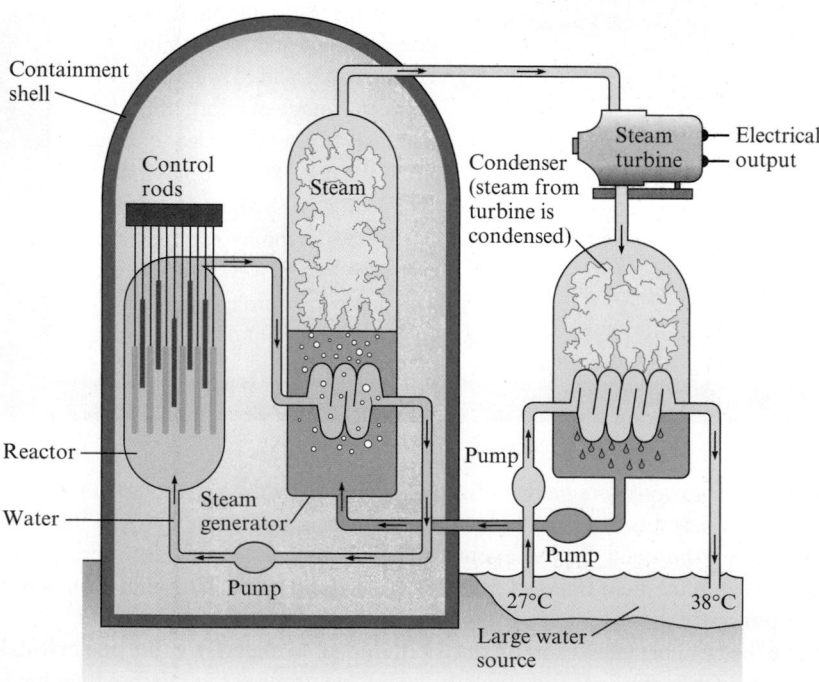

Containment shell

Control rods

Steam

Condenser (steam from turbine is condensed)

Steam turbine

Electrical output

Reactor

Water

Steam generator

Pump

Pump

Pump

27°C   38°C

Large water source

Figure 18-14 | A schematic diagram of a nuclear power plant.

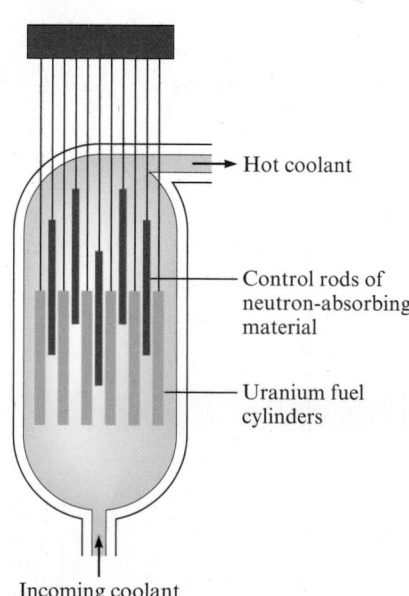

Hot coolant

Control rods of neutron-absorbing material

Uranium fuel cylinders

Incoming coolant

Figure 18-15 | A schematic of a reactor core. The position of the control rods determines the level of energy production by regulating the amount of fission taking place.

▲ Uranium oxide (refined uranium).

A **moderator** surrounds the cylinders to slow down the neutrons so that the uranium fuel can capture them more efficiently. **Control rods**, composed of substances that absorb neutrons, are used to regulate the power level of the reactor. The reactor is designed so that should a malfunction occur, the control rods are automatically inserted into the core to stop the reaction. A liquid (usually water) is circulated through the core to extract the heat generated by the energy of fission; the energy can then be passed on via a heat exchanger to water in the turbine system.

Although the concentration of $^{235}_{92}U$ in the fuel elements is not great enough to allow a supercritical mass to develop in the core, a failure of the cooling system can lead to temperatures high enough to melt the core. As a result, the building housing the core must be designed to contain the core even if meltdown occurs. A great deal of controversy now exists about the efficiency of the safety systems in nuclear power plants. Accidents such as the one at the Three Mile Island facility in Pennsylvania in 1979 and in Chernobyl,* Ukraine, in 1986 have led to questions about the wisdom of continuing to build fission-based power plants.

Breeder Reactors

One potential problem facing the nuclear power industry is the supply of $^{235}_{92}U$. Some scientists have suggested that we have nearly depleted those uranium deposits rich enough in $^{235}_{92}U$ to make production of fissionable fuel economically feasible. Because of this possibility, **breeder reactors** have been developed, in which fissionable fuel is actually produced while the reactor runs. In the breeder reactors now being studied, the major component of natural uranium, nonfissionable $^{238}_{92}U$, is changed to fissionable $^{239}_{94}Pu$. The reaction involves absorption of a neutron, followed by production of two β particles:

$$^{1}_{0}n + {}^{238}_{92}U \longrightarrow {}^{239}_{92}U$$

$$^{239}_{92}U \longrightarrow {}^{239}_{93}Np + {}^{0}_{-1}e$$

$$^{239}_{93}Np \longrightarrow {}^{239}_{94}Pu + {}^{0}_{-1}e$$

As the reactor runs and $^{235}_{92}U$ is split, some of the excess neutrons are absorbed by $^{238}_{92}U$ to produce $^{239}_{94}Pu$. The $^{239}_{94}Pu$ is then separated out and used to fuel another reactor. Such a reactor thus "breeds" nuclear fuel as it operates.

Although breeder reactors are now used in France, the United States is proceeding slowly with their development because of their controversial nature. One problem involves the hazards in handling plutonium, which flames on contact with air and is very toxic.

Fusion

Large quantities of energy are also produced by the fusion of two light nuclei. In fact, stars produce their energy through nuclear fusion. Our sun, which presently consists of 73% hydrogen, 26% helium, and 1% other elements, gives off vast quantities of energy from the fusion of protons to form helium:

$$^{1}_{1}H + {}^{1}_{1}H \longrightarrow {}^{2}_{1}H + {}^{0}_{1}e$$

$$^{1}_{1}H + {}^{2}_{1}H \longrightarrow {}^{3}_{2}He$$

$$^{3}_{2}He + {}^{3}_{2}He \longrightarrow {}^{4}_{2}He + 2 \, {}^{1}_{1}H$$

$$^{3}_{2}He + {}^{1}_{1}H \longrightarrow {}^{4}_{2}He + {}^{0}_{1}e$$

Intense research is under way to develop a feasible fusion process because of the ready availability of many light nuclides (deuterium, $^{2}_{1}H$, in seawater, for example) that

---

*See C. A. Atwood, "Chernobyl—What happened?" *J. Chem. Ed.* 65 (1988): 1037.

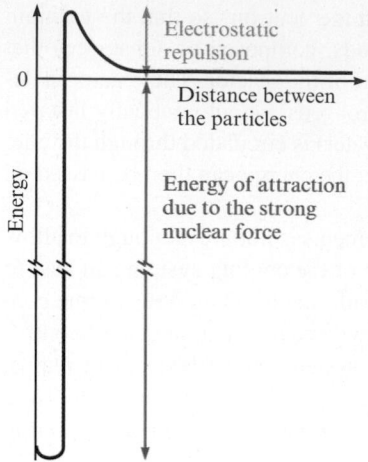

Figure 18-16 | A plot of energy versus the separation distance for two $_1^2$H nuclei. The nuclei must have sufficient velocities to get over the electrostatic repulsion "hill" and get close enough for the nuclear binding forces to become effective, thus "fusing" the particles into a new nucleus and releasing large quantities of energy. The binding force is at least 100 times the electrostatic repulsion.

can serve as fuel in fusion reactors. The major stumbling block is that high temperatures are required to initiate fusion. The forces that bind nucleons together to form a nucleus are effective only at *very small* distances ($\sim 10^{-13}$ cm). Thus, for two protons to bind together and thereby release energy, they must get very close together. But protons, because they are identically charged, repel each other electrostatically. This means that to get two protons (or two deuterons) close enough to bind together (the nuclear binding force is *not* electrostatic), they must be "shot" at each other at speeds high enough to overcome the electrostatic repulsion.

The electrostatic repulsion forces between two $_1^2$H nuclei are so great that a temperature of 4×10^7 K is required to give them velocities large enough to cause them to collide with sufficient energy that the nuclear forces can bind the particles together and thus release the binding energy. This situation is represented in Fig. 18-16.

Currently, scientists are studying two types of systems to produce the extremely high temperatures required: high-powered lasers and heating by electric currents. At present, many technical problems remain to be solved, and it is not clear which method will prove more useful or when fusion might become a practical energy source. However, there is still hope that fusion will be a major energy source sometime in the future.

Critical Thinking

Nuclear fission processes can provide a lot of energy, but they also can be dangerous. What if Congress decided to outlaw all processes that involve fission? How would that change our society?

18-7 | Effects of Radiation

Everyone knows that being hit by a train is very serious. The problem is the energy transfer involved. In fact, any source of energy is potentially harmful to organisms. Energy transferred to cells can break chemical bonds and cause malfunctioning of the cell systems. This fact is behind the concern about the ozone layer in the earth's upper atmosphere, which screens out high-energy ultraviolet radiation from the sun. Radioactive elements, which are sources of high-energy particles, are also potentially hazardous, although the effects are usually quite subtle. The reason for the subtlety of radiation damage is that even though high-energy particles are involved, the quantity of energy actually deposited in tissues *per event* is quite small. However, the resulting damage is no less real, although the effects may not be apparent for years.

Radiation damage to organisms can be classified as somatic or genetic damage. **Somatic damage** is damage to the organism itself, resulting in sickness or death. The effects may appear almost immediately if a massive dose of radiation is received; for smaller doses, damage may appear years later, usually in the form of cancer. **Genetic damage** is damage to the genetic machinery, which produces malfunctions in the offspring of the organism.

The biological effects of a particular source of radiation depend on several factors:

1. *The energy of the radiation.* The higher the energy content of the radiation, the more damage it can cause. Radiation doses are measured in **rads** (which is short for *r*adiation *a*bsorbed *d*ose), where 1 rad corresponds to 10^{-2} J of energy deposited per kilogram of tissue.
2. *The penetrating ability of the radiation.* The particles and rays produced in radioactive processes vary in their abilities to penetrate human tissue: γ rays are highly penetrating, β particles can penetrate approximately 1 cm, and α particles are stopped by the skin.

Table 18-6 | Effects of Short-Term Exposures to Radiation

Dose (rem)	Clinical Effect
0–25	Nondetectable
25–50	Temporary decrease in white blood cell counts
100–200	Strong decrease in white blood cell counts
500	Death of half the exposed population within 30 days after exposure

3. *The ionizing ability of the radiation.* Extraction of electrons from biomolecules to form ions is particularly detrimental to their functions. The ionizing ability of radiation varies dramatically. For example, γ rays penetrate very deeply but cause only occasional ionization. On the other hand, α particles, although not very penetrating, are very effective at causing ionization and produce a dense trail of damage. Thus ingestion of an α-particle producer, such as plutonium, is particularly damaging.

4. *The chemical properties of the radiation source.* When a radioactive nuclide is ingested into the body, its effectiveness in causing damage depends on its residence time. For example, $^{85}_{36}Kr$ and $^{90}_{38}Sr$ are both β-particle producers. However, since krypton is chemically inert, it passes through the body quickly and does not have much time to do damage. Strontium, being chemically similar to calcium, can collect in bones, where it may cause leukemia and bone cancer.

Because of the differences in the behavior of the particles and rays produced by radioactive decay, both the energy dose of the radiation and its effectiveness in causing biological damage must be taken into account. The **rem** (which is short for *r*oentgen *e*quivalent for *m*an) is defined as follows:

$$\text{Number of rems} = (\text{number of rads}) \times \text{RBE}$$

where RBE represents the relative effectiveness of the radiation in causing biological damage.

Table 18-6 shows the physical effects of short-term exposure to various doses of radiation, and Table 18-7 gives the sources and amounts of radiation exposure for a typical person in the United States. Note that natural sources contribute about twice as much as human activities to the total exposure. However, although the nuclear industry contributes only a small percentage of the total exposure, the major controversy associated with nuclear power plants is the *potential* for radiation hazards. These arise mainly from two sources: accidents allowing the release of radioactive materials and improper disposal of the radioactive products in spent fuel elements. The radioactive products of the fission of $^{235}_{92}U$, although only a small percentage of the total products, have half-lives of several hundred years and remain dangerous for a long time. Various schemes have been advanced for the disposal of these wastes. The one that seems to hold the most promise is the incorporation of the wastes into ceramic blocks and the burial of these blocks in geologically stable formations. At present, however, no disposal method has been accepted, and nuclear wastes continue to accumulate in temporary storage facilities.

Even if a satisfactory method for permanent disposal of nuclear wastes is found, there will continue to be concern about the effects of exposure to low levels of radiation. Exposure is inevitable from natural sources such as cosmic rays and radioactive minerals, and many people are also exposed to low levels of radiation from reactors, radioactive tracers, or diagnostic X rays. Currently, we have little reliable information on the long-term effects of low-level exposure to radiation.

Two models of radiation damage, illustrated in Fig. 18-17, have been proposed: the *linear model* and the *threshold model*. The linear model postulates that damage from radiation is proportional to the dose, even at low levels of exposure. Thus any exposure is dangerous. The threshold model, on the other hand, assumes that no significant

Table 18-7 | Typical Radiation Exposures for a Person Living in the United States (1 millirem = 10^{-3} rem)

	Exposure (millirems/year)
Cosmic radiation	50
From the earth	47
From building materials	3
In human tissues	21
Inhalation of air	5
Total from natural sources	126
X-ray diagnosis	50
Radiotherapy	10
Internal diagnosis/therapy	1
Nuclear power industry	0.2
TV tubes, industrial wastes, etc.	2
Radioactive fallout	4
Total from human activities	67
Total	193

Figure 18-17 | The two models for radiation damage. In the linear model, even a small dosage causes a proportional risk. In the threshold model, risk begins only after a certain dosage.

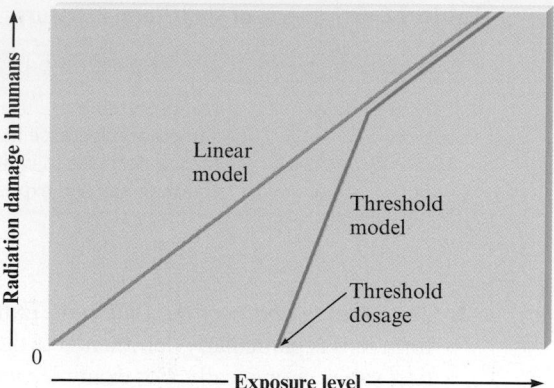

damage occurs below a certain exposure, called the *threshold exposure*. Note that if the linear model is correct, radiation exposure should be limited to a bare minimum (ideally at the natural levels). If the threshold model is correct, a certain level of radiation exposure beyond natural levels can be tolerated. Most scientists feel that since there is little evidence available to evaluate these models, it is safest to assume that the linear hypothesis is correct and to minimize radiation exposure.

FOR REVIEW

Key Terms

neutron
proton
nucleon
atomic number
mass number
isotopes
nuclide

Section 18-1
thermodynamic stability
kinetic stability
radioactive decay
beta (β) particle
zone of stability
alpha (α) particle
α-particle production
spontaneous fission
β-particle production
gamma (γ) ray
positron production
electron capture
decay series

Section 18-2
rate of decay
half-life

Radioactivity

- Certain nuclei decay spontaneously into more stable nuclei
- Types of radioactive decay:
    - α-particle ($_2^4$He) production
    - β-particle ($_{-1}^0$e) production
    - Positron ($_1^0$e) production
    - γ rays are usually produced in a radioactive decay event
- A decay series involves several radioactive decays to finally reach a stable nuclide
- Radioactive decay follows first-order kinetics
    - Half-life of a radioactive sample: the time required for half of the nuclides to decay
- The transuranium elements (those beyond uranium in the periodic table) can be synthesized by particle bombardment of uranium or heavier elements
- Radiocarbon dating uses the $_6^{14}$C/$_6^{12}$C ratio in an object to establish its date of origin

Thermodynamic Stability of a Nucleus

- Compares the mass of a nucleus to the sum of the masses of its component nucleons
- When a system gains or loses energy, it also gains or loses mass as described by the relationship $E = mc^2$
- The difference between the sum of the masses of the component nucleons and the actual mass of a nucleus (called the mass defect) can be used to calculate the nuclear binding energy

Key Terms

Section 18-3
nuclear transformation
particle accelerator
cyclotron
linear accelerator
transuranium elements

Section 18-4
Geiger–Müller counter (Geiger counter)
scintillation counter
radiocarbon dating (carbon-14 dating)
radiotracers

Section 18-5
mass defect
binding energy

Section 18-6
fusion
fission
chain reaction
subcritical reaction
critical reaction
supercritical reaction
critical mass
reactor core
moderator
control rods
breeder reactor

Section 18-7
somatic damage
genetic damage
rad
rem

Nuclear Energy Production

- Fusion: the process of combining two light nuclei to form a heavier, more stable nucleus
- Fission: the process of splitting a heavy nucleus into two lighter, more stable nuclei
  - Current nuclear power reactors use controlled fission to produce energy

Radiation Damage

- Radiation can cause direct (somatic) damage to a living organism or genetic damage to the organism's offspring
- The biological effects of radiation depend on the energy, the penetrating ability, the ionizing ability of the radiation, and the chemical properties of the nuclide producing the radiation

Review Questions

Answers to the Review Questions can be found on the Student Web site (accessible from **www.cengagebrain.com**).

1. Define or illustrate the following terms:
   a. thermodynamic stability
   b. kinetic stability
   c. radioactive decay
   d. beta-particle production
   e. alpha-particle production
   f. positron production
   g. electron capture
   h. gamma-ray emissions

   In radioactive decay processes, *A* and *Z* are conserved. What does this mean?
2. Figure 18-1 illustrates the zone of stability. What is the zone of stability? Stable light nuclides have about equal numbers of neutrons and protons. What happens to the neutron-to-proton ratio for stable nuclides as the number of protons increases? Nuclides that are not already in the zone of stability undergo radioactive processes to get to the zone of stability. If a nuclide has too many neutrons, which process(es) can the nuclide undergo to become more stable? Answer the same question for a nuclide having too many protons.
3. All radioactive decay processes follow first-order kinetics. What does this mean? What happens to the rate of radioactive decay as the number of nuclides is halved? Write the first-order rate law and the integrated first-order rate law. Define the terms in each equation. What is the half-life equa-

tion for radioactive decay processes? How does the half-life depend on how many nuclides are present? Are the half-life and rate constant k directly related or inversely related?

4. What is a nuclear transformation? How do you balance nuclear transformation reactions? Particle accelerators are used to perform nuclear transformations. What is a particle accelerator?

5. What is a Geiger counter, and how does it work? What is a scintillation counter, and how does it work? Radiotracers are used in the medical sciences to learn about metabolic pathways. What are radiotracers? Explain why ^{14}C and ^{32}P radioactive nuclides would be very helpful in learning about metabolic pathways. Why is iodine-131 useful for diagnosis of diseases of the thyroid? How could you use a radioactive nuclide to demonstrate that chemical equilibrium is a dynamic process?

6. Explain the theory behind carbon-14 dating. What assumptions must be made and what problems arise when using carbon-14 dating?

   The decay of uranium-238 to lead-206 is also used to estimate the age of objects. Specifically, $^{206}Pb/^{238}U$ ratios allow dating of rocks. Why is the ^{238}U decay to ^{206}Pb useful for dating rocks but useless for dating objects 10,000 years old or younger? Similarly, why is carbon-14 dating useful for dating objects 10,000 years old or younger but useless for dating rocks?

7. Define *mass defect* and *binding energy*. How do you determine the mass defect for a nuclide? How do you convert the mass defect into the binding energy for a nuclide? Iron-56 has the largest binding energy per nucleon among all known nuclides. Is this good or bad for iron-56? Explain.

8. Define *fission* and *fusion*. How does the energy associated with fission or fusion processes compare to the energy changes associated with chemical reactions? Fusion processes are more likely to occur for lighter elements, whereas fission processes are more likely to occur for heavier elements. Why? (*Hint:* Refer to Fig. 18-10.) The major stumbling block for turning fusion reactions into a feasible source of power is the high temperature required to initiate a fusion reaction. Why are elevated temperatures necessary to initiate fusion reactions but not fission reactions?

9. The fission of uranium-235 is used exclusively in nuclear power plants located in the United States. There are many different fission reactions of uranium-235, but all the fission reactions are self-sustaining chain reactions. Explain. Differentiate between the terms *critical, subcritical,* and *supercritical*. What is the critical mass? How does a nuclear power plant produce electricity? What are the purposes of the moderator and the control rods in a fission reactor? What are some problems associated with nuclear reactors? What are breeder reactors? What are some problems associated with breeder reactors?

10. The biological effects of a particular source of radiation depend on several factors. List some of these factors. Even though ^{85}Kr and ^{90}Sr are both β-particle emitters, the dangers associated with the decay of ^{90}Sr are much greater than those linked to ^{85}Kr. Why? Although γ rays are far more penetrating than α particles, the latter are more likely to cause damage to an organism. Why? Which type of radiation is more effective at promoting the ionization of biomolecules?

A blue question or exercise number indicates that the answer to that question or exercise appears at the back of this book and a solution appears in the *Solutions Guide,* as found on PowerLecture.

Questions

1. When nuclei undergo nuclear transformations, γ rays of characteristic frequencies are observed. How does this fact, along with other information in the chapter on nuclear stability, suggest that a quantum mechanical model may apply to the nucleus?

2. There is a trend in the United States toward using coal-fired power plants to generate electricity rather than building new nuclear fission power plants. Is the use of coal-fired power plants without risk? Make a list of the risks to society from the use of each type of power plant.

3. Which type of radioactive decay has the net effect of changing a neutron into a proton? Which type of decay has the net effect of turning a proton into a neutron?

4. Consider the following graph of binding energy per nucleon as a function of mass number.

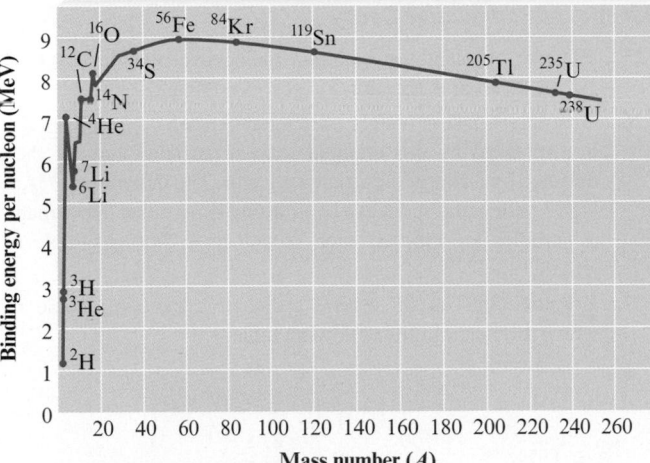

a. What does this graph tell us about the relative half-lives of the nuclides? Explain your answer.

b. Which nuclide shown is the most thermodynamically stable? Which is the least thermodynamically stable?

c. What does this graph tell us about which nuclides undergo fusion and which undergo fission to become more stable? Support your answer.

5. What are transuranium elements and how are they synthesized?

6. Scientists have estimated that the earth's crust was formed 4.3 billion years ago. The radioactive nuclide ^{176}Lu, which decays to ^{176}Hf, was used to estimate this age. The half-life of ^{176}Lu is 37 billion years. How are ratios of ^{176}Lu to ^{176}Hf utilized to date very old rocks?

7. Why are the observed energy changes for nuclear processes so much larger than the energy changes for chemical and physical processes?

8. Natural uranium is mostly nonfissionable ^{238}U; it contains only about 0.7% of fissionable ^{235}U. For uranium to be useful as a nuclear fuel, the relative amount of ^{235}U must be increased to about 3%. This is accomplished through a gas diffusion process. In the diffusion process, natural uranium reacts with fluorine to form a mixture of ^{238}UF$_6(g)$ and ^{235}UF$_6(g)$. The fluoride mixture is then enriched through a multistage diffusion process to produce a 3% ^{235}U nuclear fuel. The diffusion process utilizes Graham's law of effusion (see Chapter 8, Section 8-7). Explain how Graham's law of effusion allows natural uranium to be enriched by the gaseous diffusion process.

9. Much of the research on controlled fusion focuses on the problem of how to contain the reacting material. Magnetic fields appear to be the most promising mode of containment. Why is containment such a problem? Why must one resort to magnetic fields for containment?

10. A recent study concluded that any amount of radiation exposure can cause biological damage. Explain the differences between the two models of radiation damage, the linear model and the threshold model.

Exercises

In this section similar exercises are paired.

Radioactive Decay and Nuclear Transformations

11. Write an equation describing the radioactive decay of each of the following nuclides. (The particle produced is shown in parentheses, except for electron capture, where an electron is a reactant.)

a. $^3_1$H (β)

b. $^8_3$Li (β followed by α)

c. $^7_4$Be (electron capture)

d. $^8_5$B (positron)

12. In each of the following radioactive decay processes, supply the missing particle.

a. ^{60}Co $\rightarrow$ ^{60}Ni + ?

b. ^{97}Tc + ? $\rightarrow$ ^{97}Mo

c. ^{99}Tc $\rightarrow$ ^{99}Ru + ?

d. ^{239}Pu $\rightarrow$ ^{235}U + ?

13. Supply the missing particle, and state the type of decay for each of the following nuclear processes.

a.

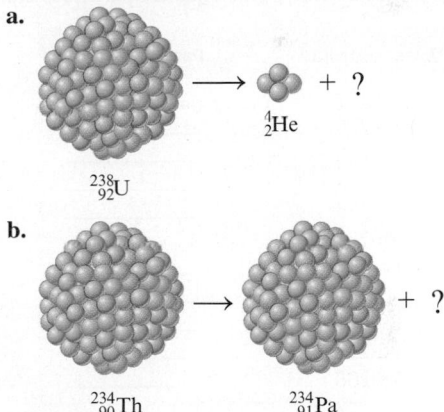

14. Write balanced equations for each of the processes described below.

a. Chromium-51, which targets the spleen and is used as a tracer in studies of red blood cells, decays by electron capture.

b. Iodine-131, used to treat hyperactive thyroid glands, decays by producing a β particle.

c. Phosphorus-32, which accumulates in the liver, decays by β-particle production.

15. Write an equation describing the radioactive decay of each of the following nuclides. (The particle produced is shown in parentheses, except for electron capture, where an electron is a reactant.)

a. ^{68}Ga (electron capture)   **c.** ^{212}Fr (α)

b. ^{62}Cu (positron)   **d.** ^{129}Sb (β)

16. In each of the following radioactive decay processes, supply the missing particle.

a. ^{73}Ga $\rightarrow$ ^{73}Ge + ?   **c.** ^{205}Bi $\rightarrow$ ^{205}Pb + ?

b. ^{192}Pt $\rightarrow$ ^{188}Os + ?   **d.** ^{241}Cm + ? $\rightarrow$ ^{241}Am

17. Uranium-235 undergoes a series of α-particle and β-particle productions to end up as lead-207. How many α particles and β particles are produced in the complete decay series?

18. The radioactive isotope ^{247}Bk decays by a series of α-particle and β-particle productions, taking ^{247}Bk through many transformations to end up as ^{207}Pb. In the complete decay series, how many α particles and β particles are produced?

19. One type of commercial smoke detector contains a minute amount of radioactive americium-241 (^{241}Am), which decays by α-particle production. The α particles ionize molecules in the air, allowing it to conduct an electric current. When smoke particles enter, the conductivity of the air is changed and the alarm buzzes.

    a. Write the equation for the decay of $^{241}_{95}$Am by α-particle production.

    b. The complete decay of ^{241}Am involves successively α, α, β, α, α, β, α, α, α, β, α, and β production. What is the final stable nucleus produced in this decay series?

    c. Identify the 11 intermediate nuclides.

20. Thorium-232 is known to undergo a progressive decay series until it reaches stability at lead-208. For each step of the series indicated in the table below, which nuclear particle is emitted?

Parent Nuclide	Particle Emitted
Th-232	_____
Ra-228	_____
Ac-228	_____
Th-228	_____
Ra-224	_____
Rn-220	_____
Po-216	_____
Pb-212	_____
Bi-212	_____
Po-212	_____
Pb-208	_____

21. There are four stable isotopes of iron with mass numbers 54, 56, 57, and 58. There are also two radioactive isotopes: iron-53 and iron-59. Predict modes of decay for these two isotopes. (See Table 18-2.)

22. The only stable isotope of fluorine is fluorine-19. Predict possible modes of decay for fluorine-21, fluorine-18, and fluorine-17.

23. In 1994 it was proposed (and eventually accepted) that element 106 be named seaborgium, Sg, in honor of Glenn T. Seaborg, discoverer of the transuranium elements.

    a. ^{263}Sg was produced by the bombardment of ^{249}Cf with a beam of ^{18}O nuclei. Complete and balance an equation for this reaction.

    b. ^{263}Sg decays by α emission. What is the other product resulting from the α decay of ^{263}Sg?

24. Many elements have been synthesized by bombarding relatively heavy atoms with high-energy particles in particle accelerators. Complete the following nuclear equations, which have been used to synthesize elements.

    a. $\underline{\hspace{1cm}} + {}^{4}_{2}\text{He} \rightarrow {}^{243}_{97}\text{Bk} + {}^{1}_{0}\text{n}$

    b. $^{238}_{92}\text{U} + {}^{12}_{6}\text{C} \rightarrow \underline{\hspace{1cm}} + 6\,{}^{1}_{0}\text{n}$

    c. $^{249}_{98}\text{Cf} + \underline{\hspace{1cm}} \rightarrow {}^{260}_{105}\text{Db} + 4\,{}^{1}_{0}\text{n}$

    d. $^{249}_{98}\text{Cf} + {}^{10}_{5}\text{B} \rightarrow {}^{257}_{103}\text{Lr} + \underline{\hspace{1cm}}$

Kinetics of Radioactive Decay

25. The rate constant for a certain radioactive nuclide is 1.0×10^{-3} h^{-1}. What is the half-life of this nuclide?

26. Americium-241 is widely used in smoke detectors. The radiation released by this element ionizes particles that are then detected by a charged-particle collector. The half-life of ^{241}Am is 433 years, and it decays by emitting α particles. How many α particles are emitted each second by a 5.00-g sample of ^{241}Am?

27. Krypton consists of several radioactive isotopes, some of which are listed in the following table.

	Half-Life
^{73}Kr	27 s
^{74}Kr	11.5 min
^{76}Kr	14.8 h
^{81}Kr	2.1×10^5 yr

Which of these isotopes is most stable, and which isotope is "hottest"? How long does it take for 87.5% of each isotope to decay?

28. Radioactive copper-64 decays with a half-life of 12.8 days.

    a. What is the value of k in s^{-1}?

    b. A sample contains 28.0 mg ^{64}Cu. How many decay events will be produced in the first second? Assume the atomic mass of ^{64}Cu is 64.0 u.

    c. A chemist obtains a fresh sample of ^{64}Cu and measures its radioactivity. She then determines that to do an experiment, the radioactivity cannot fall below 25% of the initial measured value. How long does she have to do the experiment?

29. A chemist wishing to do an experiment requiring ^{47}Ca^{2+} (half-life = 4.5 days) needs 5.0 μg of the nuclide. What mass of ^{47}CaCO$_3$ must be ordered if it takes 48 h for delivery from the supplier? Assume that the atomic mass of ^{47}Ca is 47.0 u.

30. The curie (Ci) is a commonly used unit for measuring nuclear radioactivity: 1 curie of radiation is equal to 3.7×10^{10} decay events per second (the number of decay events from 1 g radium in 1 s).

    a. What mass of Na$_2$^{38}SO$_4$ has an activity of 10.0 mCi? Sulfur-38 has an atomic mass of 38.0 u and a half-life of 2.87 h.

    b. How long does it take for 99.99% of a sample of sulfur-38 to decay?

31. The first atomic explosion was detonated in the desert north of Alamogordo, New Mexico, on July 16, 1945. What percentage of the strontium-90 ($t_{1/2} = 28.9$ years) originally produced by that explosion still remains as of July 16, 2015?

32. Iodine-131 is used in the diagnosis and treatment of thyroid disease and has a half-life of 8.0 days. If a patient with thyroid disease consumes a sample of Na^{131}I containing 10. μg ^{131}I, how long will it take for the amount of ^{131}I to decrease to 1/100 of the original amount?

33. Technetium-99 has been used as a radiographic agent in bone scans ($^{99}_{43}$Tc is absorbed by bones). If $^{99}_{43}$Tc has a half-life of 6.0 hours, what fraction of an administered dose of 100. μg $^{99}_{43}$Tc remains in a patient's body after 2.0 days?

34. Phosphorus-32 is a commonly used radioactive nuclide in biochemical research, particularly in studies of nucleic acids. The half-life of phosphorus-32 is 14.3 days. What mass of phosphorus-32 is left from an original sample of 175 mg Na$_3^{32}$PO$_4$ after 35.0 days? Assume the atomic mass of ^{32}P is 32.0 u.

35. The bromine-82 nucleus has a half-life of 1.0×10^3 min. If you wanted 1.0 g ^{82}Br and the delivery time was 3.0 days, what mass of NaBr should you order (assuming all of the Br in the NaBr was ^{82}Br)?

36. Fresh rainwater or surface water contains enough tritium ($^3_1$H) to show 5.5 decay events per minute per 100. g water. Tritium has a half-life of 12.3 years. You are asked to check a vintage wine that is claimed to have been produced in 1946. How many decay events per minute should you expect to observe in 100. g of that wine?

37. A living plant contains approximately the same fraction of carbon-14 as in atmospheric carbon dioxide. Assuming that the observed rate of decay of carbon-14 from a living plant is 13.6 counts per minute per gram of carbon, how many counts per minute per gram of carbon will be measured from a 15,000-year-old sample? Will radiocarbon dating work well for small samples of 10 mg or less? (For ^{14}C, $t_{1/2}$ = 5730 years.)

38. Assume a constant ^{14}C/^{12}C ratio of 13.6 counts per minute per gram of living matter. A sample of a petrified tree was found to give 1.2 counts per minute per gram. How old is the tree? (For ^{14}C, $t_{1/2}$ = 5730 years.)

39. A rock contains 0.688 mg ^{206}Pb for every 1.000 mg ^{238}U present. Assuming that no lead was originally present, that all the ^{206}Pb formed over the years has remained in the rock, and that the number of nuclides in intermediate stages of decay between ^{238}U and ^{206}Pb is negligible, calculate the age of the rock. (For ^{238}U, $t_{1/2}$ = 4.5×10^9 years.)

40. The mass ratios of ^{40}Ar to ^{40}K also can be used to date geologic materials. Potassium-40 decays by two processes:

$$^{40}_{19}\text{K} + ^{0}_{-1}\text{e} \longrightarrow ^{40}_{18}\text{Ar} \ (10.7\%) \qquad t_{1/2} = 1.27 \times 10^9 \text{ years}$$

$$^{40}_{19}\text{K} \longrightarrow ^{40}_{20}\text{Ca} + ^{0}_{-1}\text{e} \ (89.3\%)$$

   a. Why are ^{40}Ar/^{40}K ratios used to date materials rather than ^{40}Ca/^{40}K ratios?

   b. What assumptions must be made using this technique?

   c. A sedimentary rock has an ^{40}Ar/^{40}K ratio of 0.95. Calculate the age of the rock.

   d. How will the measured age of a rock compare to the actual age if some ^{40}Ar escaped from the sample?

Energy Changes in Nuclear Reactions

41. The sun radiates 3.9×10^{23} J of energy into space every second. What is the rate at which mass is lost from the sun?

42. The earth receives 1.8×10^{14} kJ/s of solar energy. What mass of solar material is converted to energy over a 24-h period to provide the daily amount of solar energy to the earth? What mass of coal would have to be burned to provide the same amount of energy? (Coal releases 32 kJ of energy per gram when burned.)

43. Many transuranium elements, such as plutonium-232, have very short half-lives. (For ^{232}Pu, the half-life is 36 minutes.) However, some, like protactinium-231 (half-life = 3.34×10^4 years), have relatively long half-lives. Use the masses given in the following table to calculate the change in energy when 1 mole of ^{232}Pu nuclei and 1 mole of ^{231}Pa nuclei are each formed from their respective number of protons and neutrons.

Atom or Particle	Atomic Mass
Neutron	1.67493×10^{-24} g
Proton	1.67262×10^{-24} g
Electron	9.10939×10^{-28} g
^{232}Pu	3.85285×10^{-22} g
^{231}Pa	3.83616×10^{-22} g

(Since the masses of ^{232}Pu and ^{231}Pa are atomic masses, they each include the mass of the electrons present. The mass of the nucleus will be the atomic mass minus the mass of the electrons.)

44. The most stable nucleus in terms of binding energy per nucleon is ^{56}Fe. If the atomic mass of ^{56}Fe is 55.9349 u, calculate the binding energy per nucleon for ^{56}Fe.

45. Calculate the binding energy in J/nucleon for carbon-12 (atomic mass = 12.0000 u) and uranium-235 (atomic mass = 235.0439 u). The atomic mass of $^1_1$H is 1.00782 u and the mass of a neutron is 1.00866 u. The most stable nucleus known is ^{56}Fe (see Exercise 44). Would the binding energy per nucleon for ^{56}Fe be larger or smaller than that of ^{12}C or ^{235}U? Explain.

46. Calculate the binding energy per nucleon for $^2_1$H and $^3_1$H. The atomic masses are $^2_1$H, 2.01410 u; and $^3_1$H, 3.01605 u.

47. The mass defect for a lithium-6 nucleus is −0.03434 g/mol. Calculate the atomic mass of ^{6}Li.

48. The binding energy per nucleon for magnesium-27 is 1.326×10^{-12} J/nucleon. Calculate the atomic mass of ^{27}Mg.

49. Calculate the amount of energy released per gram of hydrogen nuclei reacted for the following reaction. The atomic masses are $^1_1$H, 1.00782 u; $^2_1$H, 2.01410 u; and an electron, 5.4858×10^{-4} u. (*Hint:* Think carefully about how to account for the electron mass.)

$$^1_1\text{H} + ^1_1\text{H} \longrightarrow ^2_1\text{H} + ^0_{+1}\text{e}$$

50. The easiest fusion reaction to initiate is

$$^2_1\text{H} + ^3_1\text{H} \longrightarrow ^4_2\text{He} + ^1_0\text{n}$$

Calculate the energy released per $^4_2$He nucleus produced and per mole of $^4_2$He produced. The atomic masses are $^2_1$H, 2.01410 u; $^3_1$H, 3.01605 u; and $^4_2$He, 4.00260 u. The masses of the electron and neutron are 5.4858×10^{-4} u and 1.00866 u, respectively.

Detection, Uses, and Health Effects of Radiation

51. The typical response of a Geiger–Müller tube is shown below. Explain the shape of this curve.

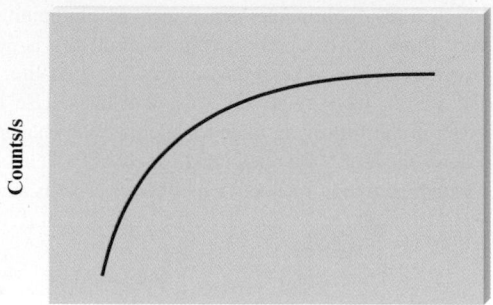

Disintegrations/s from sample

52. When using a Geiger–Müller counter to measure radioactivity, it is necessary to maintain the same geometrical orientation between the sample and the Geiger–Müller tube to compare different measurements. Why?

53. Consider the following reaction to produce methyl acetate:

$$CH_3OH + CH_3\overset{\displaystyle O}{\overset{\displaystyle \|}{C}}OH \longrightarrow CH_3\overset{\displaystyle O}{\overset{\displaystyle \|}{C}}OCH_3 + H_2O$$

Methyl acetate

When this reaction is carried out with CH_3OH containing oxygen-18, the water produced does not contain oxygen-18. Explain.

54. A chemist studied the reaction mechanism for the reaction

$$2NO(g) + O_2(g) \longrightarrow 2NO_2(g)$$

by reacting $N^{16}O$ with $^{18}O_2$. If the reaction mechanism is

$$NO + O_2 \rightleftharpoons NO_3 \text{ (fast equilibrium)}$$
$$NO_3 + NO \longrightarrow 2NO_2 \text{ (slow)}$$

what distribution of ^{18}O would you expect in the NO_2? Assume that N is the central atom in NO_3, assume only $N^{16}O^{18}O_2$ forms, and assume stoichiometric amounts of reactants are combined.

55. Uranium-235 undergoes many different fission reactions. For one such reaction, when ^{235}U is struck with a neutron, ^{144}Ce and ^{90}Sr are produced along with some neutrons and electrons. How many neutrons and β-particles are produced in this fission reaction?

56. Breeder reactors are used to convert the nonfissionable nuclide $^{238}_{92}U$ to a fissionable product. Neutron capture of the $^{238}_{92}U$ is followed by two successive beta decays. What is the final fissionable product?

57. Which do you think would be the greater health hazard: the release of a radioactive nuclide of Sr or a radioactive nuclide of Xe into the environment? Assume the amount of radioactivity is the same in each case. Explain your answer on the basis of the chemical properties of Sr and Xe. Why are the chemical properties of a radioactive substance important in assessing its potential health hazards?

58. Consider the following information:
    i. The layer of dead skin on our bodies is sufficient to protect us from most α-particle radiation.
    ii. Plutonium is an α-particle producer.
    iii. The chemistry of Pu^{4+} is similar to that of Fe^{3+}.
    iv. Pu oxidizes readily to Pu^{4+}.

    Why is plutonium one of the most toxic substances known?

Additional Exercises

59. Predict whether each of the following nuclides is stable or unstable (radioactive). If the nuclide is unstable, predict the type of radioactivity you would expect it to exhibit.
    a. $^{45}_{19}K$ b. $^{56}_{26}Fe$ c. $^{20}_{11}Na$ d. $^{194}_{81}Tl$

60. Each of the following isotopes has been used medically for the purpose indicated. Suggest reasons why the particular element might have been chosen for this purpose.
    a. cobalt-57, for study of the body's use of vitamin B_{12}
    b. calcium-47, for study of bone metabolism
    c. iron-59, for study of red blood cell function

61. The mass percent of carbon in a typical human is 18%, and the mass percent of ^{14}C in natural carbon is $1.6 \times 10^{-10}\%$. Assuming a 180-lb person, how many decay events per second occur in this person due exclusively to the β-particle decay of ^{14}C (for ^{14}C, $t_{1/2} = 5730$ years)?

62. At a flea market, you've found a very interesting painting done in the style of Rembrandt's "dark period" (1642–1672). You suspect that you really do not have a genuine Rembrandt, but you take it to the local university for testing. Living wood shows a carbon-14 activity of 15.3 counts per minute per gram. Your painting showed a carbon-14 activity of 15.1 counts per minute per gram. Could it be a genuine Rembrandt?

63. Define "third-life" in a similar way to "half-life," and determine the "third-life" for a nuclide that has a half-life of 31.4 years.

64. A proposed system for storing nuclear wastes involves storing the radioactive material in caves or deep mine shafts. One of the most toxic nuclides that must be disposed of is plutonium-239, which is produced in breeder reactors and has a half-life of 24,100 years. A suitable storage place must be geologically stable long enough for the activity of plutonium-239 to decrease to 0.1% of its original value. How long is this for plutonium-239?

65. During World War II, tritium (^3H) was a component of fluorescent watch dials and hands. Assume you have such a watch that was made in January 1944. If 17% or more of the original tritium was needed to read the dial in dark places, until what year could you read the time at night? (For ^3H, $t_{1/2} = 12.3$ yr.)

66. A positron and an electron can annihilate each other on colliding, producing energy as photons:

$$_{-1}^{0}e + _{+1}^{0}e \longrightarrow 2\,_0^0\gamma$$

Assuming that both γ rays have the same energy, calculate the wavelength of the electromagnetic radiation produced.

67. A small atomic bomb releases energy equivalent to the detonation of 20,000 tons of TNT; a ton of TNT releases 4×10^9 J of

energy when exploded. Using 2×10^{13} J/mol as the energy released by fission of ^{235}U, approximately what mass of ^{235}U undergoes fission in this atomic bomb?

68. During the research that led to production of the two atomic bombs used against Japan in World War II, different mechanisms for obtaining a supercritical mass of fissionable material were investigated. In one type of bomb, a "gun" shot one piece of fissionable material into a cavity containing another piece of fissionable material. In the second type of bomb, the fissionable material was surrounded with a high explosive that, when detonated, compressed the fissionable material into a smaller volume. Discuss what is meant by critical mass, and explain why the ability to achieve a critical mass is essential to sustaining a nuclear reaction.

69. Using the kinetic molecular theory (see Section 8.6), calculate the root mean square velocity and the average kinetic energy of $^{2}_{1}$H nuclei at a temperature of 4×10^7 K. (See Exercise 50 for the appropriate mass values.)

70. Consider the following reaction, which can take place in particle accelerators:

$$^{1}_{1}H + ^{1}_{0}n \longrightarrow 2\,^{1}_{1}H + ^{1}_{0}n + ^{1}_{-1}H$$

Calculate the energy change for this reaction. Is energy released or absorbed? What is a possible source for this energy?

71. Photosynthesis in plants can be represented by the following overall equation:

$$6CO_2(g) + 6H_2O(l) \xrightarrow{\text{Light}} C_6H_{12}O_6(s) + 6O_2(g)$$

Algae grown in water containing some ^{18}O (in H_2^{18}O) evolve oxygen gas with the same isotopic composition as the oxygen in the water. When algae growing in water containing only ^{16}O were furnished carbon dioxide containing ^{18}O, no ^{18}O was found to be evolved from the oxygen gas produced. What conclusions about photosynthesis can be drawn from these experiments?

72. Strontium-90 and radon-222 both pose serious health risks. ^{90}Sr decays by β-particle production and has a relatively long half-life (28.9 years). Radon-222 decays by α-particle production and has a relatively short half-life (3.82 days). Explain why each decay process poses health risks.

ChemWork Problems

These multiconcept problems (and additional ones) are found interactively online with the same type of assistance a student would get from an instructor.

73. Complete the following table with the nuclear particle that is produced in each nuclear reaction.

Initial Nuclide	Product Nuclide	Particle Produced
$^{239}_{94}$Pu	$^{235}_{92}$U	_____
$^{214}_{82}$Pb	$^{214}_{83}$Bi	_____
$^{60}_{27}$Co	$^{60}_{28}$Ni	_____
$^{99}_{43}$Tc	$^{99}_{44}$Ru	_____
$^{239}_{93}$Np	$^{239}_{94}$Pu	_____

74. A certain radioactive nuclide has a half-life of 3.00 hours.
  a. Calculate the rate constant in s^{-1} for this nuclide.
  b. Calculate the decay rate in decays/s for 1.000 mole of this nuclide.

75. Iodine-131 has a half-life of 8.0 days. How many days will it take for 174 g of ^{131}I to decay to 83 g of ^{131}I?

76. Rubidium-87 decays by β-particle production to strontium-87 with a half-life of 4.7×10^{10} years. What is the age of a rock sample that contains 109.7 μg of ^{87}Rb and 3.1 μg of ^{87}Sr? Assume that no ^{87}Sr was present when the rock was formed. The atomic masses for ^{87}Rb and ^{87}Sr are 86.90919 u and 86.90888 u, respectively.

77. Given the following information:

Mass of proton = 1.00728 u
Mass of neutron = 1.00866 u
Mass of electron = 5.486×10^{-4} u
Speed of light = 2.9979×10^8 m/s

Calculate the nuclear binding energy of $^{24}_{12}$Mg, which has an atomic mass of 23.9850 u.

78. Which of the following statement(s) is(are) *true*?
  a. A radioactive nuclide that decays from 1.00×10^{10} atoms to 2.5×10^9 atoms in 10 minutes has a half-life of 5.0 minutes.
  b. Nuclides with large Z values are observed to be α-particle producers.
  c. As Z increases, nuclides need a greater proton-to-neutron ratio for stability.
  d. Those "light" nuclides that have twice as many neutrons as protons are expected to be stable.

Challenge Problems

79. Naturally occurring uranium is composed mostly of ^{238}U and ^{235}U, with relative abundances of 99.28% and 0.72%, respectively. The half-life for ^{238}U is 4.5×10^9 years, and the half-life for ^{235}U is 7.1×10^8 years. Assuming that the earth was formed 4.5 billion years ago, calculate the relative abundances of the ^{238}U and ^{235}U isotopes when the earth was formed.

80. The curie (Ci) is a commonly used unit for measuring nuclear radioactivity: 1 curie of radiation is equal to 3.7×10^{10} decay events per second (the number of decay events from 1 g radium in 1 s). A 1.7-mL sample of water containing tritium was injected into a 150-lb person. The total activity of radiation injected was 86.5 mCi. After some time to allow the tritium activity to equally distribute throughout the body, a sample of blood plasma containing 2.0 mL water at an activity of 3.6 μCi was removed. From these data, calculate the mass percent of water in this 150-lb person.

81. A 0.10-cm^3 sample of a solution containing a radioactive nuclide (5.0×10^3 counts per minute per milliliter) is injected into a rat. Several minutes later 1.0 cm^3 blood is removed. The blood shows 48 counts per minute of radioactivity. Calculate the volume of blood in the rat. What assumptions must be made in performing this calculation?

82. Zirconium is one of the few metals that retains its structural integrity upon exposure to radiation. The fuel rods in most nuclear reactors therefore are often made of zirconium. Answer the following questions about the redox properties of zirconium based on the half-reaction

$$ZrO_2 \cdot H_2O + H_2O + 4e^- \longrightarrow Zr + 4OH^- \quad \mathscr{E}° = -2.36 \text{ V}$$

a. Is zirconium metal capable of reducing water to form hydrogen gas at standard conditions?

b. Write a balanced equation for the reduction of water by zirconium.

c. Calculate $\mathscr{E}°$, $\Delta G°$, and K for the reduction of water by zirconium metal.

d. The reduction of water by zirconium occurred during the accidents at Three Mile Island in 1979. The hydrogen produced was successfully vented and no chemical explosion occurred. If 1.00×10^3 kg Zr reacts, what mass of H_2 is produced? What volume of H_2 at 1.0 atm and 1000.°C is produced?

e. At Chernobyl in 1986, hydrogen was produced by the reaction of superheated steam with the graphite reactor core:

$$C(s) + H_2O(g) \longrightarrow CO(g) + H_2(g)$$

It was not possible to prevent a chemical explosion at Chernobyl. In light of this, do you think it was a correct decision to vent the hydrogen and other radioactive gases into the atmosphere at Three Mile Island? Explain.

83. In addition to the process described in the text, a second process called the *carbon–nitrogen cycle* occurs in the sun:

$$^1_1H + ^{12}_6C \longrightarrow ^{13}_7N + ^0_0\gamma$$
$$^{13}_7N \longrightarrow ^{13}_6C + ^0_{+1}e$$
$$^1_1H + ^{13}_6C \longrightarrow ^{14}_7N + ^0_0\gamma$$
$$^1_1H + ^{14}_7N \longrightarrow ^{15}_8O + ^0_0\gamma$$
$$^{15}_8O \longrightarrow ^{15}_7N + ^0_{+1}e$$
$$^1_1H + ^{15}_7N \longrightarrow ^{12}_6C + ^4_2He + ^0_0\gamma$$

Overall reaction: $\quad 4\,^1_1H \longrightarrow\,^4_2He + 2\,^0_{+1}e$

a. What is the catalyst in this process?

b. What nucleons are intermediates?

c. How much energy is released per mole of hydrogen nuclei in the overall reaction? (The atomic masses of ^1_1H and ^4_2He are 1.00782 u and 4.00260 u, respectively.)

84. The most significant source of natural radiation is radon-222. ^{222}Rn, a decay product of ^{238}U, is continuously generated in the earth's crust, allowing gaseous Rn to seep into the basements of buildings. Because ^{222}Rn is an α-particle producer with a relatively short half-life of 3.82 days, it can cause biological damage when inhaled.

a. How many α particles and β particles are produced when ^{238}U decays to ^{222}Rn? What nuclei are produced when ^{222}Rn decays?

b. Radon is a noble gas so one would expect it to pass through the body quickly. Why is there a concern over inhaling ^{222}Rn?

c. Another problem associated with ^{222}Rn is that the decay of ^{222}Rn produces a more potent α-particle producer ($t_{1/2} = 3.11$ min) that is a solid. What is the identity of the solid? Give the balanced equation of this species decaying by α-particle production. Why is the solid a more potent α-particle producer?

d. The U.S. Environmental Protection Agency (EPA) recommends that ^{222}Rn levels not exceed 4 pCi per liter of air (1 Ci = 1 curie = 3.7×10^{10} decay events per second; 1 pCi = 1×10^{-12} Ci). Convert 4.0 pCi per liter of air into concentrations units of ^{222}Rn atoms per liter of air and moles of ^{222}Rn per liter of air.

85. To determine the K_{sp} value of Hg_2I_2, a chemist obtained a solid sample of Hg_2I_2 in which some of the iodine is present as radioactive ^{131}I. The count rate of the Hg_2I_2 sample is 5.0×10^{11} counts per minute per mole of I. An excess amount of $Hg_2I_2(s)$ is placed into some water, and the solid is allowed to come to equilibrium with its respective ions. A 150.0-mL sample of the saturated solution is withdrawn and the radioactivity measured at 33 counts per minute. From this information, calculate the K_{sp} value for Hg_2I_2.

$$Hg_2I_2(s) \rightleftharpoons Hg_2^{2+}(aq) + 2I^-(aq) \quad K_{sp} = [Hg_2^{2+}][I^-]^2$$

86. Estimate the temperature needed to achieve the fusion of deuterium to make an α particle. The energy required can be estimated from Coulomb's law [use the form $E = 9.0 \times 10^9 (Q_1Q_2/r)$, using $Q = 1.6 \times 10^{-19}$ C for a proton, and $r = 2 \times 10^{-15}$ m for the helium nucleus; the unit for the proportionality constant in Coloumb's law is $J \cdot m/C^2$].

Integrative Problems

These problems require the integration of multiple concepts to find the solutions.

87. A reported synthesis of the transuranium element bohrium (Bh) involved the bombardment of berkelium-249 with neon-22 to produce bohrium-267. Write a nuclear reaction for this synthesis. The half-life of bohrium-267 is 15.0 seconds. If 199 atoms of bohrium-267 could be synthesized, how much time would elapse before only 11 atoms of bohrium-267 remain? What is the expected electron configuration of elemental bohrium?

88. Radioactive cobalt-60 is used to study defects in vitamin B_{12} absorption because cobalt is the metallic atom at the center of the vitamin B_{12} molecule. The nuclear synthesis of this cobalt isotope involves a three-step process. The overall reaction is iron-58 reacting with two neutrons to produce cobalt-60 along with the emission of another particle. What particle is emitted in this nuclear synthesis? What is the binding energy in J per nucleon for the cobalt-60 nucleus (atomic masses: $^{60}Co = 59.9338$ u; $^1H = 1.00782$ u)? What is the de Broglie wavelength of the emitted particle if it has a velocity equal to $0.90c$, where c is the speed of light?

The Representative Elements

Crystals of bismuth (mitzy/Shutterstock.com)

So far in this book we have covered the major principles and explored the most important models of chemistry. In particular, we have seen that the chemical properties of the elements can be explained very successfully by the quantum mechanical model of the atom. In fact, the most convincing evidence of that model's validity is its ability to relate the observed periodic properties of the elements to the number of valence electrons in their atoms.

We have learned many properties of the elements and their compounds, but we have not discussed extensively the relationship between the chemical properties of a specific element and its position on the periodic table. In this chapter we will explore the chemical similarities and differences among the elements in the several groups of the periodic table and will try to interpret these data using the wave mechanical model of the atom. In the process we will illustrate a great variety of chemical properties and further demonstrate the practical importance of chemistry.

19-1 | A Survey of the Representative Elements

The traditional form of the periodic table is shown in Fig. 19-1. Recall that the **representative elements**, whose chemical properties are determined by the valence-level *s* and *p* electrons, are designated Groups 1A (1) through 8A (18). The **transition metals**, in the center of the table, result from the filling of *d* orbitals. The elements that correspond to the filling of the 4*f* and 5*f* orbitals are listed separately as the **lanthanides** and **actinides**, respectively.

The heavy black line in Fig. 19-1 separates the metals from the nonmetals. Some elements just on either side of this line, such as silicon and germanium, exhibit both metallic and nonmetallic properties. These elements are often called **metalloids**, or **semimetals**. The fundamental chemical difference between metals and nonmetals is that metals tend to lose their valence electrons to form *cations,* which usually have the valence electron configuration of the noble gas from the preceding period. On the other hand, nonmetals tend to gain electrons to form *anions* that exhibit the electron configuration of the noble gas in the same period. Metallic character is observed to increase in going down a given group, which is consistent with the trends in ionization energy, electron affinity, and electronegativity discussed earlier (see Sections 2-12 and 3-2). ◄ *ⓘ*

ⓘ Metallic character increases going down a group in the periodic table.

Atomic Size and Group Anomalies

Although the chemical properties of the members of a group have many similarities, there are also important differences. The most dramatic differences usually occur between the first and second member. For example, hydrogen in Group 1A (1) is a nonmetal, whereas lithium is a very active metal. This extreme difference results primarily from the very large difference in the atomic radii of hydrogen and lithium, as shown in Fig. 19-2. Since the small hydrogen atom has a much greater attraction for electrons than do the larger members of Group 1A (1), it forms covalent bonds with nonmetals. In contrast, the other members of Group 1A (1) lose their valence electrons to nonmetals to form 1+ cations in ionic compounds.

The effect of size is also evident in other groups. For example, the oxides of the metals in Group 2A (2) are all quite basic except for the first member of the series;

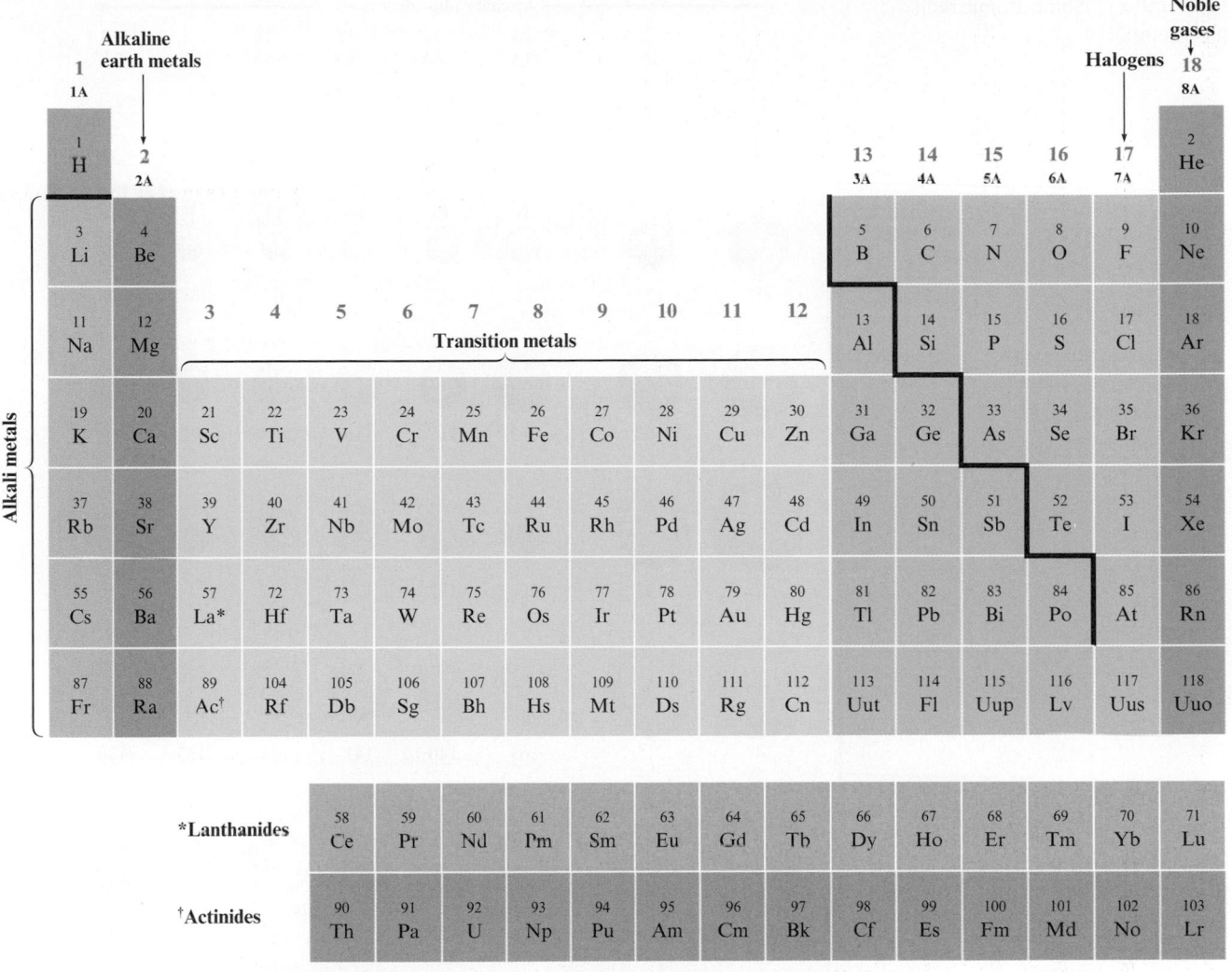

Figure 19-1 | The periodic table. The elements in the A groups are the representative elements. The elements shown in yellow are called transition metals. The heavy black line approximately separates the nonmetals from the metals.

beryllium oxide (BeO) is amphoteric. The basicity of an oxide depends on its ionic character. Ionic oxides contain the O^{2-} ion, which reacts with water to form two OH^- ions. All the oxides of the Group 2A (2) metals are highly ionic except for beryllium oxide, which has considerable covalent character. The small Be^{2+} ion can effectively polarize the electron "cloud" of the O^{2-} ion, thereby producing significant electron sharing. We see the same pattern in Group 3A (13), where only the small boron atom behaves as a nonmetal, or sometimes as a semimetal, whereas aluminum and the other members are active metals.

In Group 4A (14) the effect of size is reflected in the dramatic differences between the chemical properties of carbon and silicon. The chemistry of carbon is dominated by molecules containing chains of C—C bonds, but silicon compounds mainly contain Si—O bonds rather than Si—Si bonds. Silicon does form compounds with chains of Si—Si bonds, but these compounds are much more reactive than the corresponding carbon compounds. The reasons for the difference in reactivity between the carbon and silicon compounds are quite complex but are likely related to the differences in the sizes of the carbon and silicon atoms.

Figure 19-2 | Some atomic radii (in picometers).

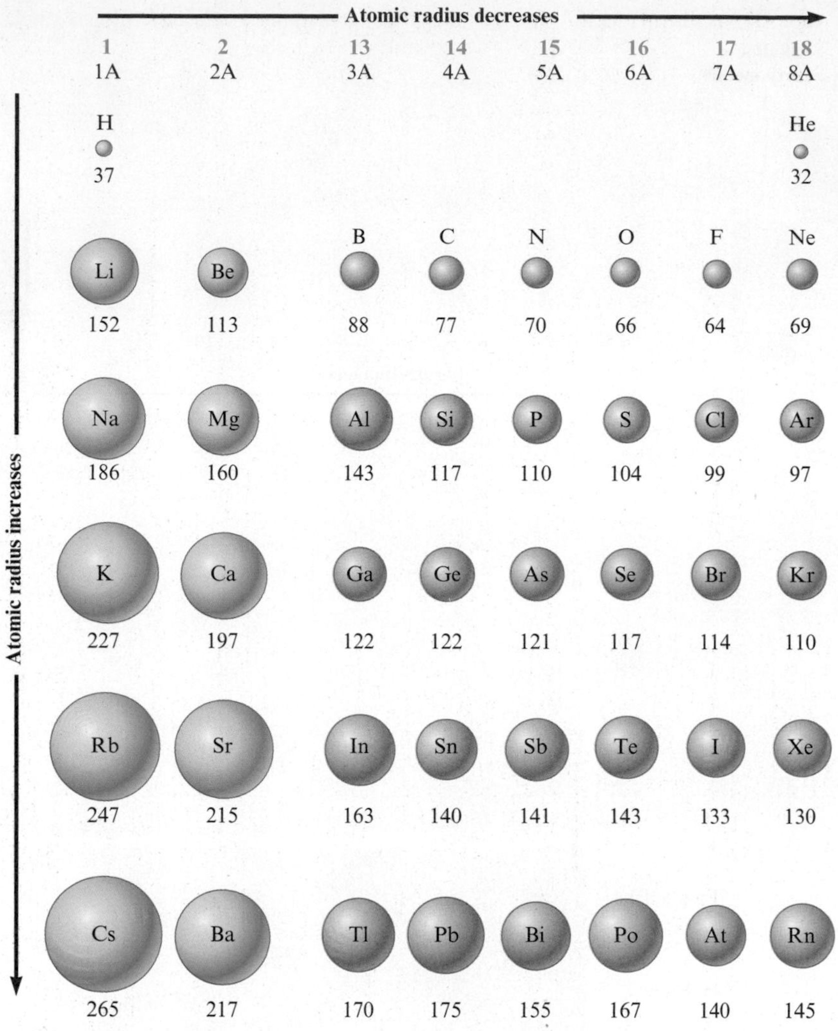

Atomic radius decreases →

	1 1A	2 2A	13 3A	14 4A	15 5A	16 6A	17 7A	18 8A
	H 37							He 32
	Li 152	Be 113	B 88	C 77	N 70	O 66	F 64	Ne 69
	Na 186	Mg 160	Al 143	Si 117	P 110	S 104	Cl 99	Ar 97
	K 227	Ca 197	Ga 122	Ge 122	As 121	Se 117	Br 114	Kr 110
	Rb 247	Sr 215	In 163	Sn 140	Sb 141	Te 143	I 133	Xe 130
	Cs 265	Ba 217	Tl 170	Pb 175	Bi 155	Po 167	At 140	Rn 145

Atomic radius increases ↓

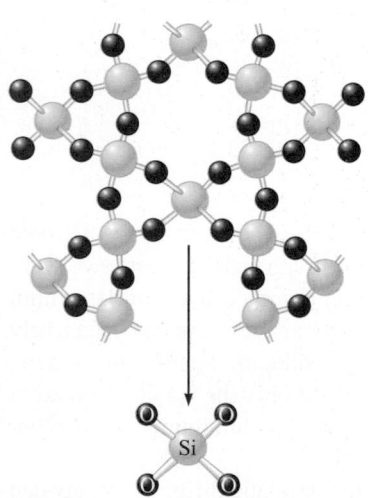

Figure 19-3 | The structure of quartz, which has the empirical formula SiO_2. Note that the structure is based on interlocking SiO_4 tetrahedra (shown below the arrow), in which each oxygen atom is shared by two silicon atoms.

Carbon and silicon also differ markedly in their abilities to form π bonds. As we discussed in Section 4-1, carbon dioxide is composed of discrete CO_2 molecules with the Lewis structure

$$\ddot{\text{O}}=\text{C}=\ddot{\text{O}}$$

where the carbon and oxygen atoms achieve the [Ne] configuration by forming π bonds. In contrast, the structure of silica (empirical formula SiO_2) is based on SiO_4 tetrahedra with Si—O—Si bridges, as shown in Fig. 19-3. The silicon $3p$ valence orbitals do not overlap very effectively with the smaller oxygen $2p$ orbitals to form π bonds; therefore, discrete SiO_2 molecules with the Lewis structure

$$\ddot{\text{O}}=\text{Si}=\ddot{\text{O}}$$

are not stable. Instead, the silicon atoms achieve a noble gas configuration by forming four Si—O single bonds.

The importance of π bonding for the relatively small elements of the second period also explains the different elemental forms of the members of Groups 5A (15) and 6A (16). For example, elemental nitrogen exists as very stable N_2 molecules with the Lewis structure :N≡N:. Elemental phosphorus forms larger aggregates of atoms, the simplest being the tetrahedral P_4 molecules found in white phosphorus (see Fig. 19-18). Like silicon atoms, the relatively large phosphorus atoms do not form strong

π bonds but prefer to achieve a noble gas configuration by forming single bonds to several other phosphorus atoms. In contrast, its very strong π bonds make the N_2 molecule the most stable form of elemental nitrogen. Similarly, in Group 6A (16) the most stable form of elemental oxygen is the O_2 molecule with a double bond. However, the larger sulfur atom forms bigger aggregates, such as the cyclic S_8 molecule (see Fig. 19-22), which contain only single bonds.

The relatively large change in size in going from the first to the second member of a group also has important consequences for the Group 7A (17) elements. For example, fluorine has a smaller electron affinity than chlorine. This violation of the expected trend can be attributed to the fact that the small size of the fluorine $2p$ orbitals causes unusually large electron–electron repulsions. The relative weakness of the bond in the F_2 molecule can be explained in terms of the repulsions among the lone pairs, shown in the Lewis structure:

$$\ddot{\underset{\displaystyle ..}{F}}-\ddot{\underset{\displaystyle ..}{F}}\!:$$

The small size of the fluorine atoms allows close approach of the lone pairs, which leads to much greater repulsions than those found in the Cl_2 molecule with its much larger atoms.

Thus the relatively large increase in atomic radius in going from the first to the second member of a group causes the first element to exhibit properties quite different from the others.

Critical Thinking

Notice that atomic size decreases as atomic number increases across a row, yet atomic size increases as atomic number increases down a column. Why is this?

Abundance and Preparation

Table 19-1 shows the distribution of elements in the earth's crust, oceans, and atmosphere. The major element is, of course, oxygen, which is found in the atmosphere as O_2, in the oceans as H_2O, and in the earth's crust primarily in silicate and carbonate minerals. The second most abundant element, silicon, is found throughout the earth's crust in the silica and silicate minerals that form the basis of most sand, rocks, and soil. ◄ The most abundant metals, aluminum and iron, are found in ores, in which they are combined with nonmetals, most commonly oxygen. One notable fact revealed by Table 19-1 is the small incidence of most transition metals. Since many of these relatively rare elements are assuming increasing importance in our high-technology

Table 19-1 | Distribution (Mass Percent) of the 18 Most Abundant Elements in the Earth's Crust, Oceans, and Atmosphere

Element	Mass Percent	Element	Mass Percent
Oxygen	49.2	Chlorine	0.19
Silicon	25.7	Phosphorus	0.11
Aluminum	7.50	Manganese	0.09
Iron	4.71	Carbon	0.08
Calcium	3.39	Sulfur	0.06
Sodium	2.63	Barium	0.04
Potassium	2.40	Nitrogen	0.03
Magnesium	1.93	Fluorine	0.03
Hydrogen	0.87	All others	0.49
Titanium	0.58		

Table 19-2 | Abundance of Elements in the Human Body

Major Elements	Mass Percent	Trace Elements (in alphabetical order)
Oxygen	65.0	Arsenic
Carbon	18.0	Chromium
Hydrogen	10.0	Cobalt
Nitrogen	3.0	Copper
Calcium	1.4	Fluorine
Phosphorus	1.0	Iodine
Magnesium	0.50	Manganese
Potassium	0.34	Molybdenum
Sulfur	0.26	Nickel
Sodium	0.14	Selenium
Chlorine	0.14	Silicon
Iron	0.004	Vanadium
Zinc	0.003	

society, it is possible that the control of transition metal ores may ultimately have more significance in world politics than will control of petroleum supplies.

The distribution of elements in living materials is very different from that found in the earth's crust. Table 19-2 shows the distribution of elements in the human body. Oxygen, carbon, hydrogen, and nitrogen form the basis for all biologically important molecules. The other elements, even though they are found in relatively small amounts, are often crucial for life. For example, zinc is found in over 150 different biomolecules in the human body.

Only about one-fourth of the elements occur naturally in the free state. Most are found in a combined state. The *process of obtaining a metal from its ore* is called **metallurgy**. Since the metals in ores are found in the form of cations, the chemistry of *metallurgy always involves reduction of the ions to the elemental metal (with an oxidation state of zero).* A variety of reducing agents can be used, but carbon is the usual choice because of its wide availability and relatively low cost. ◄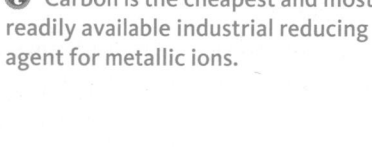

Carbon is the cheapest and most readily available industrial reducing agent for metallic ions.

Electrolysis is often used to reduce the most active metals. In Chapter 17 we considered the electrolytic production of aluminum metal. The alkali metals are also produced by electrolysis, usually of their molten halide salts.

The preparation of nonmetals varies widely. Elemental nitrogen and oxygen are usually obtained from the **liquefaction** of air, which is based on the principle that a gas cools as it expands. After each expansion, part of the cooler gas is compressed, whereas

▲ Sand, such as that found in the massive sand dunes bordering the desert plain near Namib, Namibia, is composed of silicon and oxygen.

Frank Krahmer/Radius Images/Masterfile

the rest is used to carry away the heat of the compression. The compressed gas is then allowed to expand again. This cycle is repeated many times. Eventually, the remaining gas becomes cold enough to form the liquid state. Because liquid nitrogen and liquid oxygen have different boiling points, they can be separated by the distillation of liquid air. Both substances are important industrial chemicals, with nitrogen ranking second in terms of amount manufactured in the United States (approximately 60 billion pounds per year) and oxygen ranking third (over 40 billion pounds per year). Hydrogen can be obtained from the electrolysis of water, but more commonly it is obtained from the decomposition of the methane in natural gas. Sulfur is found underground in its elemental form and is recovered by the Frasch process (see Section 19-12). The halogens are obtained by oxidation of the anions from halide salts (see Section 19-13).

19-2 | The Group 1A (1) Elements

1
1A
H
Li
Na
K
Rb
Cs
Fr

The Group 1A (1) elements with their ns^1 valence electron configurations are all very active metals (they lose their valence electrons very readily), except for hydrogen, which behaves as a nonmetal. ◄ We will discuss the chemistry of hydrogen in the next section. Many of the properties of the **alkali metals** have been given previously (see Section 2-13). The sources and methods of preparation of pure alkali metals are given in Table 19-3. The ionization energies, standard reduction potentials, ionic radii, and melting points for the alkali metals are listed in Table 19-4.

In Section 2-13 we saw that the alkali metals all react vigorously with water to release hydrogen gas:

$$2M(s) + 2H_2O(l) \longrightarrow 2M^+(aq) + 2OH^-(aq) + H_2(g)$$

We will reconsider this process briefly because it illustrates several important concepts. From the ionization energies, we might expect lithium to be the weakest of the alkali metals as a reducing agent in water. However, the standard reduction potentials indicate that it is the strongest. This reversal results mainly from the very large energy of hydration of the small Li^+ ion. Because of its relatively high charge density, the Li^+ ion very effectively attracts water molecules. A large quantity of energy is released in the process, favoring the formation of the Li^+ ion and making lithium a strong reducing agent in aqueous solution. ◄ ⓘ

We also saw in Section 2-13 that lithium, although it is the strongest reducing agent, reacts more slowly with water than sodium or potassium. ◄ From the discussions in Chapters 11 and 16, we know that the *equilibrium position* for a reaction (in this case indicated by the $\mathscr{E}°$ values) is controlled by thermodynamic factors but that the *rate* of a reaction is controlled by kinetic factors. There is *no* direct connection between these factors. Lithium reacts more slowly with water than sodium or potassium because as a solid lithium has a higher melting point than either of the other

ⓘ Several properties of the alkali metals are given in Table 2-8.

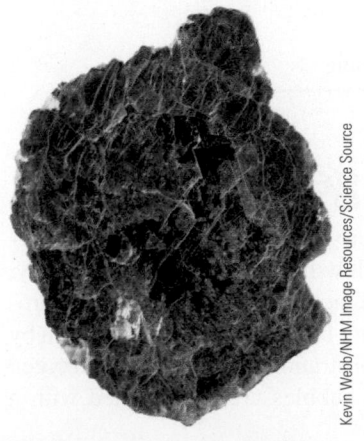
Kevin Webb/NHM Image Resources/Science Source

Figure 19-4 | Lepidolite is composed of mainly lithium, aluminum, silicon, and oxygen, but it also contains significant amounts of rubidium and cesium.

Table 19-3 | Sources and Methods of Preparation of the Pure Alkali Metals

Element	Source	Method of Preparation
Lithium	Silicate minerals such as spodumene, $LiAl(Si_2O_6)$	Electrolysis of molten LiCl
Sodium	NaCl	Electrolysis of molten NaCl
Potassium	KCl	Electrolysis of molten KCl
Rubidium	Impurity in lepidolite, $Li_2(F,OH)_2Al_2(SiO_3)_3$	Reduction of RbOH with Mg and H_2
Cesium	Pollucite ($Cs_4Al_4Si_9O_{26} \cdot H_2O$) and an impurity in lepidolite (Fig. 19-4)	Reduction of CsOH with Mg and H_2

▲ Sodium reacts violently with water.

Table 19-4 | Selected Physical Properties of the Alkali Metals

Element	Ionization Energy (kJ/mol)	Standard Reduction Potential (V) for $M^+ + e^- \rightarrow M$	Radius of M^+ (pm)	Melting Point (°C)
Lithium	520	−3.05	60	180
Sodium	495	−2.71	95	98
Potassium	419	−2.92	133	64
Rubidium	409	−2.99	148	39
Cesium	382	−3.02	169	29

elements. Since lithium does not become molten from the heat of reaction with water as sodium and potassium do, it has a smaller area of contact with the water.

Table 19-5 summarizes some important reactions of the alkali metals.

The alkali metal ions are very important for the proper functioning of biological systems such as nerves and muscles; Na^+ and K^+ ions are present in all body cells and fluids. In human blood plasma, the concentrations are

$$[Na^+] \approx 0.15\ M \qquad \text{and} \qquad [K^+] \approx 0.005\ M$$

In the fluids *inside* the cells, the concentrations are reversed:

$$[Na^+] \approx 0.005\ M \qquad \text{and} \qquad [K^+] \approx 0.16\ M$$

Since the concentrations are so different inside and outside the cells, an elaborate mechanism involving selective ligands is needed to transport Na^+ and K^+ ions through the cell membranes.

Table 19-5 | Selected Reactions of the Alkali Metals

Reaction	Comment
$2M + X_2 \longrightarrow 2MX$	X_2 = any halogen molecule
$4Li + O_2 \longrightarrow 2Li_2O$	Excess oxygen
$2Na + O_2 \longrightarrow Na_2O_2$	
$M + O_2 \longrightarrow MO_2$	M = K, Rb, or Cs
$2M + S \longrightarrow M_2S$	
$6Li + N_2 \longrightarrow 2Li_3N$	Li only
$12M + P_4 \longrightarrow 4M_3P$	
$2M + H_2 \longrightarrow 2MH$	
$2M + 2H_2O \longrightarrow 2MOH + H_2$	
$2M + 2H^+ \longrightarrow 2M^+ + H_2$	Violent reaction!

19-3 | The Chemistry of Hydrogen

Under ordinary conditions of temperature and pressure, hydrogen is a colorless, odorless gas composed of H_2 molecules. Because of its low molar mass and nonpolarity, hydrogen has a very low boiling point (−253°C) and melting point (−260°C). Hydrogen gas is highly flammable; mixtures of air containing from 18% to 60% hydrogen by volume are explosive. In a common lecture demonstration, hydrogen and oxygen gases are bubbled into soapy water. The resulting bubbles are then ignited with a candle on a long stick, producing a loud explosion. ◄

The major industrial source of hydrogen gas is the reaction of methane with water at high temperatures (800–1000°C) and pressures (10–50 atm) in the presence of a metallic catalyst (often nickel):

$$CH_4(g) + H_2O(g) \xrightarrow[\text{Catalyst}]{\text{Heat, pressure}} CO(g) + 3H_2(g)$$

(left) Hydrogen gas being used to blow soap bubbles. (right) As the bubbles float upward, they are lighted by using a candle on a long stick. The orange flame results from the heat of the reaction between hydrogen and oxygen, which excites sodium atoms in the soap bubbles.

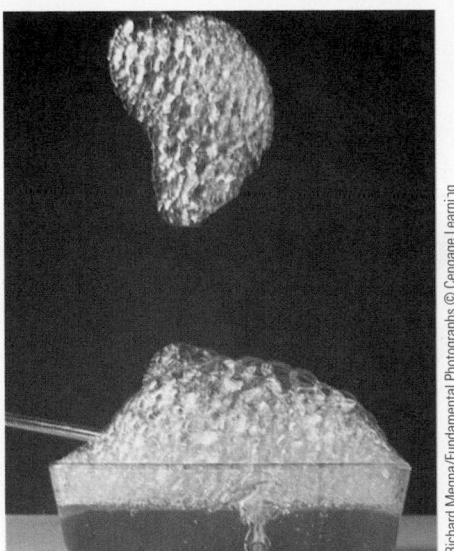

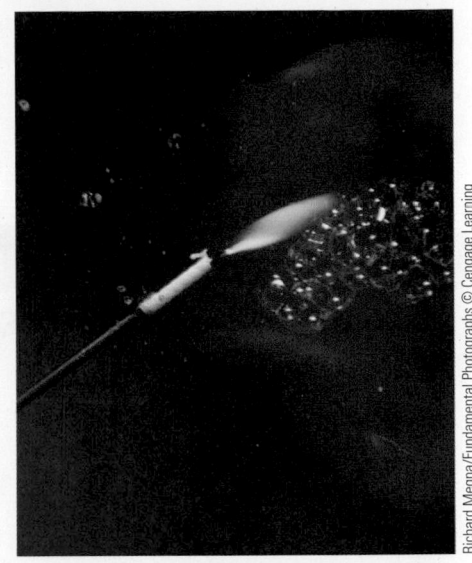

Large quantities of hydrogen are also formed as a by-product of gasoline production, when hydrocarbons with high molecular masses are broken down (or *cracked*) to produce smaller molecules more suitable for use as a motor fuel.

Very pure hydrogen can be produced by the electrolysis of water (see Section 17-8), but this method currently is not economically feasible for large-scale production because of the relatively high cost of electricity.

The major industrial use of hydrogen is in the production of ammonia by the Haber process. Large quantities of hydrogen are also used for hydrogenating unsaturated vegetable oils (those containing carbon–carbon double bonds) to produce solid shortenings that are saturated (containing carbon–carbon single bonds):

$$\underset{\underset{}{}}{\overset{\overset{H\quad H}{|\quad |}}{\sim\sim\sim C = C \sim\sim\sim}} \xrightarrow{\;H_2\;} \underset{\underset{H\quad H}{|\quad |}}{\overset{\overset{H\quad H}{|\quad |}}{\sim\sim\sim C - C \sim\sim\sim}}$$

The catalysis of this process was discussed in Section 11-7.

Chemically, hydrogen behaves as a typical nonmetal, forming covalent compounds with other nonmetals and forming salts with very active metals. Binary compounds containing hydrogen are called **hydrides**, of which there are three classes. The **ionic (or saltlike) hydrides** are formed when hydrogen combines with the most active metals, those from Groups 1A (1) and 2A (2). Examples are LiH and CaH_2, which can best be characterized as containing hydride ions (H^-) and metal cations. Because the presence of two electrons in the small $1s$ orbital produces large electron–electron repulsions and because the nucleus has only a 1+ charge, the hydride ion is a strong reducing agent. For example, when ionic hydrides are placed in water, a violent reaction takes place. This reaction results in the formation of hydrogen gas, as seen in the equation

$$LiH(s) + H_2O(l) \longrightarrow H_2(g) + Li^+(aq) + OH^-(aq)$$

Covalent hydrides are formed when hydrogen combines with other nonmetals. We have encountered many of these compounds already: HCl, CH_4, NH_3, H_2O, and so on. The most important covalent hydride is water. The polarity of the H_2O molecule leads to many of water's unusual properties. Water has a much higher boiling point than is expected from its molar mass. It has a large heat of vaporization and a large heat capacity, both of which make it a very useful coolant. Water has a higher density as a liquid than as a solid because of the open structure of ice, which results from

Figure 19-5 | The structure of ice, showing the hydrogen bonding.

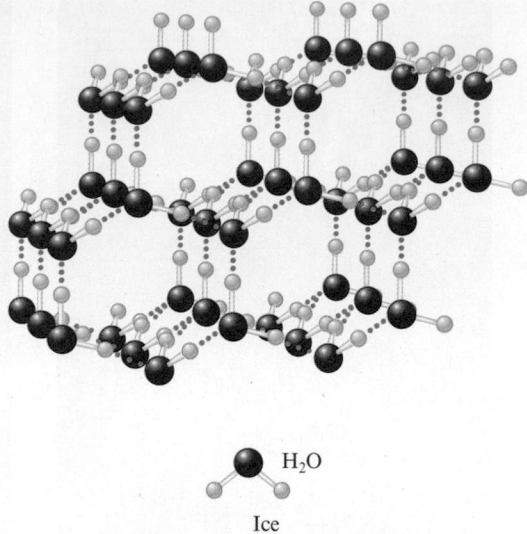

H₂O

Ice

maximizing the hydrogen bonding (Fig. 19-5). Because water is an excellent solvent for ionic and polar substances, it provides an effective medium for life processes. In fact, water is one of the few covalent hydrides that is nontoxic to organisms.

The third class of hydrides is the **metallic**, or **interstitial**, **hydrides**, which are formed when transition metal crystals are treated with hydrogen gas. The hydrogen molecules dissociate at the metal's surface, and the small hydrogen atoms migrate into the crystal structure to occupy holes, or *interstices*. These metal–hydrogen mixtures are more like solid solutions than true compounds. Palladium can absorb about *900 times* its own volume of hydrogen gas. In fact, hydrogen can be purified by placing it under slight pressure in a vessel containing a thin wall of palladium. The hydrogen diffuses into and through the metal wall, leaving the impurities behind.

Although hydrogen can react with transition metals to form compounds such as UH_3 and FeH_6, most of the interstitial hydrides have variable compositions (often called *nonstoichiometric* compositions) with formulas such as $LaH_{2.76}$ and $VH_{0.56}$. The compositions of the nonstoichiometric hydrides vary with the length of exposure of the metal to hydrogen gas.

When interstitial hydrides are heated, much of the absorbed hydrogen is lost as hydrogen gas. Because of this behavior, these materials offer possibilities for storing hydrogen for use as a portable fuel. The internal combustion engines in current automobiles can burn hydrogen gas with little modification, but storage of enough hydrogen to provide an acceptable mileage range remains a problem. One possible solution might be to use a fuel tank containing a porous solid that includes a transition metal. The hydrogen gas could be pumped into the solid to form the interstitial hydride. The hydrogen gas could then be released when the engine requires additional energy. This system is now being tested by several automobile companies.

19-4 | The Group 2A (2) Elements

The Group 2A (2) elements (with the valence electron configuration ns^2) are very reactive, losing their two valence electrons to form ionic compounds that contain M^{2+} cations. ◄ These elements are commonly called the **alkaline earth metals** because of the basicity of their oxides:

$$MO(s) + H_2O(l) \longrightarrow M^{2+}(aq) + 2OH^-(aq)$$

Table 19-6 | Selected Physical Properties, Sources, and Methods of Preparation of the Group 2A (2) Elements

Element	Radius of M^{2+} (pm)	Ionization Energy (kJ/mol)		$\mathscr{E}°$ (V) for $M^{2+} + 2e^- \longrightarrow M$	Source	Method of Preparation
		First	Second			
Beryllium	≈30	900	1760	−1.70	Beryl $(Be_3Al_2Si_6O_{18})$	Electrolysis of molten $BeCl_2$
Magnesium	65	735	1445	−2.37	Magnesite $(MgCO_3)$, dolomite $(MgCO_3 \cdot CaCO_3)$, carnallite $(MgCl_2 \cdot KCl \cdot 6H_2O)$	Electrolysis of molten $MgCl_2$
Calcium	99	590	1146	−2.76	Various minerals containing $CaCO_3$	Electrolysis of molten $CaCl_2$
Strontium	113	549	1064	−2.89	Celestite $(SrSO_4)$, strontianite $(SrCO_3)$	Electrolysis of molten $SrCl_2$
Barium	135	503	965	−2.90	Baryte $(BaSO_4)$, witherite $(BaCO_3)$	Electrolysis of molten $BaCl_2$
Radium	140	509	979	−2.92	Pitchblende (1 g of Ra/7 tons of ore)	Electrolysis of molten $RaCl_2$

2
2A

| Be |
| Mg |
| Ca |
| Sr |
| Ba |
| Ra |

ℹ An amphoteric oxide displays both acidic and basic properties.

▲ Calcium metal reacting with water to form bubbles of hydrogen gas.

Only the amphoteric beryllium oxide (BeO) also shows some acidic properties, such as dissolving in aqueous solutions containing hydroxide ions: ◀ ℹ

$$BeO(s) + 2OH^-(aq) + H_2O(l) \longrightarrow Be(OH)_4^{2-}(aq)$$

The more active alkaline earth metals react with water as the alkali metals do, producing hydrogen gas:

$$M(s) + 2H_2O(l) \longrightarrow M^{2+}(aq) + 2OH^-(aq) + H_2(g)$$

Calcium, strontium, and barium react vigorously at 25°C. ◀ The less easily oxidized beryllium and magnesium show no observable reaction with water at 25°C, although magnesium reacts with boiling water. Table 19-6 summarizes various properties, sources, and methods of preparation of the alkaline earth metals.

The alkaline earth metals have great practical importance. Calcium and magnesium ions are essential for human life. Calcium is found primarily in the structural minerals composing bones and teeth. Magnesium (as the Mg^{2+} ion) plays a vital role in metabolism and in muscle functions. Because magnesium metal has a relatively low density and displays moderate strength, it is a useful structural material, especially if alloyed with aluminum.

Table 19-7 summarizes some important reactions involving the alkaline earth metals.

Table 19-7 | Selected Reactions of the Group 2A (2) Elements

Reaction	Comment
$M + X_2 \longrightarrow MX_2$	X_2 = any halogen molecule
$2M + O_2 \longrightarrow 2MO$	Ba gives BaO_2 as well
$M + S \longrightarrow MS$	
$3M + N_2 \longrightarrow M_3N_2$	High temperatures
$6M + P_4 \longrightarrow 2M_3P_2$	High temperatures
$M + H_2 \longrightarrow MH_2$	M = Ca, Sr, or Ba; high temperatures; Mg at high pressure
$M + 2H_2O \longrightarrow M(OH)_2 + H_2$	M = Ca, Sr, or Ba
$M + 2H^+ \longrightarrow M^{2+} + H_2$	
$Be + 2OH^- + 2H_2O \longrightarrow Be(OH)_4^{2-} + H_2$	

© Cengage Learning

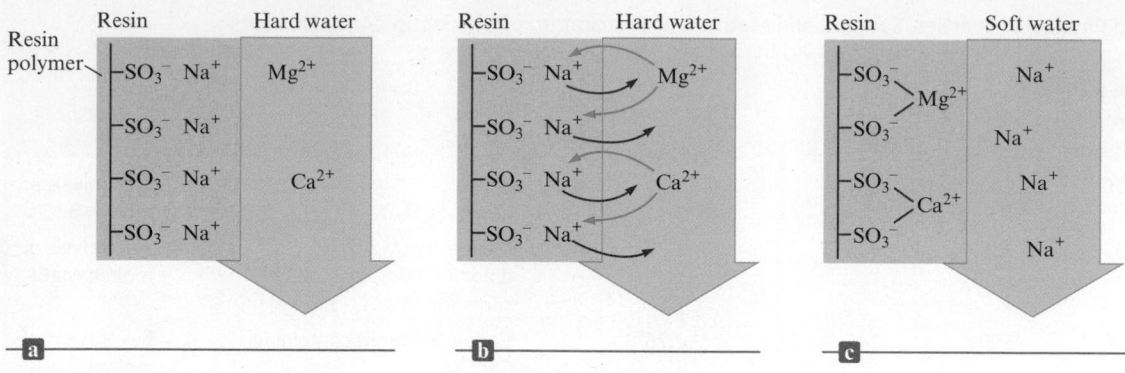

Figure 19-6 | (a) A schematic representation of a typical cation-exchange resin. (b) and (c) When hard water is passed over the cation-exchange resin, the Ca^{2+} and Mg^{2+} bind to the resin.

Relatively large concentrations of Ca^{2+} and Mg^{2+} ions are often found in natural water supplies. These ions in this so-called **hard water** interfere with the action of detergents and form precipitates with soap. The Ca^{2+} ion is often removed by precipitation as $CaCO_3$ in large-scale water softening. In individual homes Ca^{2+}, Mg^{2+}, and other cations are removed by **ion exchange**. An **ion-exchange resin** consists of large molecules (polymers) that have many ionic sites. A cation-exchange resin is represented schematically in Fig. 19-6(a), showing Na^+ ions bound ionically to the SO_3^- groups that are covalently attached to the resin polymer. When hard water is passed over the resin, Ca^{2+} and Mg^{2+} bind to the resin in place of Na^+, which is released into the solution [Fig. 19-6(b)]. Replacing Mg^{2+} and Ca^{2+} by Na^+ [Fig. 19-6(c)] "softens" the water because the sodium salts of soap are soluble.

19-5 | The Group 3A (13) Elements

13
3A
B
Al
Ga
In
Tl

The Group 3A (13) elements (valence electron configuration ns^2np^1) generally show the increase in metallic character in going down the group that is characteristic of the representative elements. ◄ Some physical properties, sources, and methods of preparation of the Group 3A (13) elements are summarized in Table 19-8.

Boron is a typical nonmetal, and most of its compounds are covalent. The most interesting compounds of boron are the covalent hydrides called **boranes**. We might expect BH_3 to be the simplest hydride, since boron has three valence electrons to share with three hydrogen atoms. However, this compound is unstable, and the simplest known member of the series is diborane (B_2H_6), with the structure shown in Fig. 19-7(a). In this

Table 19-8 | Selected Physical Properties, Sources, and Methods of Preparation of the Group 3A (13) Elements

Element	Radius of M^{3+} (pm)	Ionization Energy (kJ/mol)	$\mathscr{E}°$ (V) for $M^{3+} + 3e^- \longrightarrow M$	Source	Method of Preparation
Boron	20	798	—	Kernite, a form of borax ($Na_2B_4O_7 \cdot 4H_2O$)	Reduction by Mg or H_2
Aluminum	50	581	-1.66	Bauxite (Al_2O_3)	Electrolysis of Al_2O_3 in molten Na_3AlF_6
Gallium	62	577	-0.53	Traces in various minerals	Reduction with H_2 or electrolysis
Indium	81	556	-0.34	Traces in various minerals	Reduction with H_2 or electrolysis
Thallium	95	589	0.72	Traces in various minerals	Electrolysis

Figure 19-7 | (a) The structure of B_2H_6 with its two three-center B—H—B bridging bonds and four "normal" B—H bonds. (b) The structure of B_5H_9. There are five "normal" B—H bonds to terminal hydrogens and four three-center bridging bonds around the base.

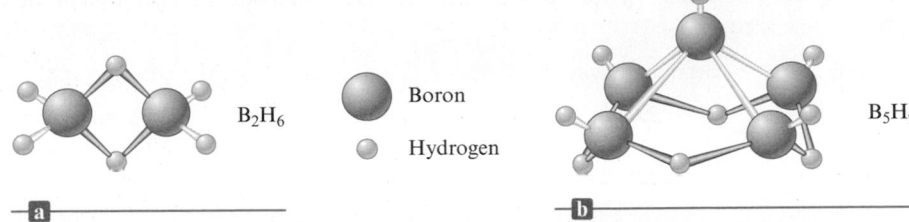

Boron

Hydrogen

B_2H_6

B_5H_9

© Cengage Learning

▲ Gallium melts in the hand.

molecule the terminal B—H bonds are normal covalent bonds, each involving one electron pair. The bridging bonds are three-center bonds similar to those in solid BeH_2. Another interesting borane contains the square pyramidal B_5H_9 molecule [Fig. 19-7(b)], which has four three-center bonds situated around the base of the pyramid. Because the boranes are extremely electron-deficient, they are highly reactive. The boranes react very exothermically with oxygen and were once evaluated as potential fuels for rockets in the U.S. space program.

Aluminum, the most abundant metal on the earth, has metallic physical properties, such as high thermal and electrical conductivities and a lustrous appearance; however, its bonds to nonmetals are significantly covalent. This covalency is responsible for the amphoteric nature of Al_2O_3, which dissolves in acidic or basic solution, and for the acidity of $Al(H_2O)_6^{3+}$ (see Section 13-8):

$$Al(H_2O)_6^{3+}(aq) \rightleftharpoons Al(OH)(H_2O)_5^{2+}(aq) + H^+(aq)$$

One especially interesting property of *gallium* is its unusually low melting point of 29.8°C, which is in contrast to the 660°C melting point of aluminum. ◄ Also, since gallium's boiling point is about 2400°C, it has the largest liquid range of any metal. This makes it useful for thermometers, especially to measure high temperatures. Gallium, like water, expands when it freezes. The chemistry of gallium is quite similar to that of aluminum. For example, Ga_2O_3 is amphoteric.

The chemistry of *indium* is similar to that of aluminum and gallium except that compounds containing the 1+ ion are known, such as InCl and In_2O, in addition to those with the more common 3+ ion. The chemistry of *thallium* is completely metallic.

Table 19-9 summarizes some important reactions of the Group 3A (13) elements.

Table 19-9 | Selected Reactions of the Group 3A (13) Elements

Reaction	Comment
$2M + 3X_2 \longrightarrow 2MX_3$	X_2 = any halogen molecule; Tl gives TlX as well, but no TlI_3
$4M + 3O_2 \longrightarrow 2M_2O_3$	High temperatures; Tl gives Tl_2O as well
$2M + 3S \longrightarrow M_2S_3$	High temperatures; Tl gives Tl_2S as well
$2M + N_2 \longrightarrow 2MN$	M = Al only
$2M + 6H^+ \longrightarrow 2M^{3+} + 3H_2$	M = Al, Ga, or In; Tl gives Tl^+
$2M + 2OH^- + 6H_2O \longrightarrow 2M(OH)_4^- + 3H_2$	M = Al or Ga

14 4A

C
Si
Ge
Sn
Pb

19-6 | The Group 4A (14) Elements

Group 4A (14) (with the valence electron configuration ns^2np^2) contains two of the most important elements on the earth: carbon, the fundamental constituent of the molecules necessary for life, and silicon, which forms the basis of the geological world. ◄ The change from nonmetallic to metallic properties seen in Group 3A (13) is also apparent in going down Group 4A (14) from carbon, a typical nonmetal, to silicon and germanium, usually considered semimetals, to the metals tin and lead. Table 19-10 summarizes some physical properties, sources, and methods of preparation of the elements in this group.

Table 19-10 | Selected Physical Properties, Sources, and Methods of Preparation of the Group 4A (14) Elements

Element	Electronegativity	Melting Point (°C)	Boiling Point (°C)	Source	Method of Preparation
Carbon	2.6	3727 (sublimes)	—	Graphite, diamond, petroleum, coal	—
Silicon	1.9	1410	2355	Silicate minerals, silica	Reduction of K_2SiF_6 with Al, or reduction of SiO_2 with Mg
Germanium	2.0	937	2830	Germinate (mixture of copper, iron, and germanium sulfides)	Reduction of GeO_2 with H_2 or C
Tin	2.0	232	2270	Cassiterite (SnO_2)	Reduction of SnO_2 with C
Lead	2.3	327	1740	Galena (PbS)	Roasting of PbS with O_2 to form PbO_2 and then reduction with C

Table 19-11 | Strengths of C—C, Si—Si, and Si—O Bonds

Bond	Bond Energy (kJ/mol)
C—C	347
Si—Si	340
Si—O	368

ℹ The most commonly proposed basis for an alternative biochemical system is the silicon atom because silicon has many chemical properties similar to those of carbon (which explains its position in the same group). Like carbon, silicon can create molecules that are sufficiently large to carry biological information. This doesn't mean that silicon can be used as an alternative to carbon in life such as ours. Carbon has a much greater ability than silicon to form bonds with diverse types of atoms.

All the Group 4A (14) elements can form four covalent bonds to nonmetals—for example, CH_4, SiF_4, $GeBr_4$, $SnCl_4$, and $PbCl_4$. In each of these tetrahedral molecules, the central atom is described as sp^3 hybridized by the localized electron model.

We have seen that carbon also differs markedly from the other members of Group 4A (14) in its ability to form π bonds. This accounts for the completely different structures and properties of CO_2 and SiO_2. Note from Table 19-11 that C—C bonds and Si—O bonds are stronger than Si—Si bonds. This partly explains why the chemistry of carbon is dominated by C—C bonds, whereas that of silicon is dominated by Si—O bonds.

Carbon occurs in the allotropic forms graphite, diamond, and fullerenes. The most important chemistry of carbon is organic chemistry, which is described in detail in Chapter 21.

Silicon, the second most abundant element in the earth's crust, is a semimetal found widely distributed in silica and silicates (see Section 9-5). About 85% of the earth's crust is composed of these substances. Although silicon is found in some steel and aluminum alloys, its major use is in semiconductors for electronic devices (see Chapter 9). ◄ ℹ

Germanium, a relatively rare element, is a semimetal used mainly in the manufacture of semiconductors for transistors and similar electronic devices.

Tin is a soft, silvery metal that can be rolled into thin sheets (tin foil) and has been used for centuries in various alloys such as bronze (20% Sn and 80% Cu), solder (33% Sn and 67% Pb), and pewter (85% Sn, 7% Cu, 6% Bi, and 2% Sb). Tin exists as three allotropes: *white tin,* stable at normal temperatures; *gray tin,* stable at temperatures below 13.2°C; and *brittle tin,* found at temperatures above 161°C. When tin is exposed to low temperatures, it gradually changes to powdery gray tin and crumbles away; this is known as *tin disease.*

Currently, tin is used mainly as a protective coating for steel, especially for cans used as food containers. The thin layer of tin, applied electrolytically, forms a protective oxide coating that prevents further corrosion.

Lead is easily obtained from its ore, galena (PbS). Because lead melts at such a low temperature, it may have been the first pure metal obtained from its ore. We know that lead was used as early as 3000 B.C. by the Egyptians. It was later used by the Romans to make utensils, glazes on pottery, and even intricate plumbing systems. ◄ The Romans also prepared a sweetener called *sapa* by boiling down grape juice in lead-lined vessels. The sweetness of this syrup was partly caused by the formation of lead(II) acetate (formerly called sugar of lead), a very sweet-tasting compound. The problem with these practices is that lead is very toxic. In fact, the Romans had so much contact with lead that it may have contributed to the demise of their civilization. Analysis of bones from that era shows significant levels of lead.

Ludwig van Beethoven, arguably the greatest composer who ever lived, led a troubled life fraught with sickness, deafness, and personality aberrations. Now we may know the source of these difficulties: lead poisoning. Scientists have recently reached this conclusion through analysis of Beethoven's hair. When Beethoven died in 1827 at age 56, many mourners took samples of the great man's hair. In fact, it was said at the time that he was practically bald by the time he was buried. The hair that was recently analyzed consisted of 582 strands—3 to 6 inches long—bought for the Center of Beethoven Studies for $7300 in 1994 from Sotheby's auction house in London.

According to William Walsh of the Health Research Institute (HRI) in suburban Chicago, Beethoven's hair showed a lead concentration 100 times the normal levels. The scientists concluded that Beethoven's exposure to lead came as an adult, possibly from the mineral water he drank and swam in when he visited spas.

The lead poisoning may well explain Beethoven's volatile temper—the composer was subject to towering rages and sometimes had the look of a wild animal. In rare cases lead poisoning has been known to cause deafness, but the researchers remain unsure if this problem led to Beethoven's hearing loss.

According to Walsh, the scientists at HRI were originally looking for mercury, a common treatment for syphilis in the early nineteenth century, in Beethoven's hair. The absence of mercury supports the consensus of scholars that Beethoven did not have this disease. Not surprisingly, Beethoven himself wanted to know what made him so ill. In a letter to his brothers in 1802, he asked them to have doctors find the cause of his frequent abdominal pain after his death.

Portrait of Ludwig van Beethoven by Josef Karl Stieler ca. 1820

▲ Roman bath

Table 19-12 | Selected Reactions of the Group 4A (14) Elements

Reaction	Comment
$M + 2X_2 \longrightarrow MX_4$	X_2 = any halogen molecule; M = Ge or Sn; Pb gives PbX_2
$M + O_2 \longrightarrow MO_2$	M = Ge or Sn; high temperatures; Pb gives PbO or Pb_3O_4
$M + 2H^+ \longrightarrow M^{2+} + H_2$	M = Sn or Pb

Although lead poisoning has been known since at least the second century B.C., lead continues to be a problem. For example, many children have been exposed to lead by eating chips of lead-based paint. Because of this problem, lead-based paints are no longer used for children's furniture, and many states have banned lead-based paint for interior use. Lead poisoning can also occur when acidic foods and drinks leach the lead from lead-glazed pottery dishes that were improperly fired and when liquor is stored in leaded crystal decanters, producing toxic levels of lead in the drink in a relatively short time. In addition, the widespread use of tetraethyl lead [$(C_2H_5)_4Pb$] as an antiknock agent in gasoline has increased the lead levels in our environment. Concern about the effects of this lead pollution has caused the U.S. government to require the gradual replacement of the lead in gasoline with other antiknock agents. The largest commercial use of lead (over one million tons annually) is for electrodes in the lead storage batteries used in automobiles (see Section 17-6).

Table 19-12 summarizes some important reactions of the Group 4A (14) elements.

19-7 | The Group 5A (15) Elements

15
5A
N
P
As
Sb
Bi

The Group 5A (15) elements (with the valence electron configuration ns^2np^3), which are prepared as shown in Table 19-13, exhibit remarkably varied chemical properties. ◄ As usual, metallic character increases going down the group, as is apparent from the electronegativity values (Table 19-13). Nitrogen and phosphorus are nonmetals that can gain three electrons to form 3− anions in salts with active metals; examples are magnesium nitride (Mg_3N_2) and beryllium phosphide (Be_3P_2). The chemistry of these two important elements is discussed in the next two sections.

Bismuth and *antimony* tend to be metallic, readily losing electrons to form cations. Although these elements have five valence electrons, so much energy is required to remove all five that no ionic compounds containing Bi^{5+} or Sb^{5+} ions are known.

Table 19-13 | Selected Physical Properties, Sources, and Methods of Preparation of the Group 5A (15) Elements

Element	Electronegativity	Source	Method of Preparation
Nitrogen	3.0	Air	Liquefaction of air
Phosphorus	2.2	Phosphate rock [$Ca_3(PO_4)_2$], fluorapatite [$Ca_5(PO_4)_3F$]	$2Ca_3(PO_4)_2 + 6SiO_2 \longrightarrow 6CaSiO_3 + P_4O_{10}$ $P_4O_{10} + 10C \longrightarrow 4P + 10CO$
Arsenic	2.2	Arsenopyrite (Fe_3As_2, FeS)	Heating arsenopyrite in the absence of air
Antimony	2.1	Stibnite (Sb_2S_3)	Roasting Sb_2S_3 in air to form Sb_2O_3 and then reduction with carbon
Bismuth	2.0	Bismite (Bi_2O_3), bismuth glance (Bi_2S_3)	Roasting Bi_2S_3 in air to form Bi_2O_3 and then reduction with carbon

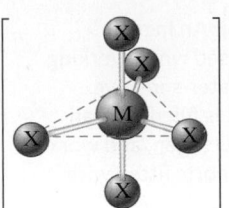

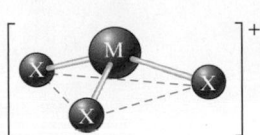

Figure 19-8 | The pyramidal shape of the Group 5A (15) MX$_3$ molecules.

Figure 19-9 | The trigonal bipyramidal shape of the MX$_5$ molecules.

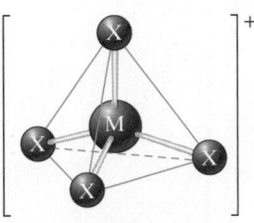

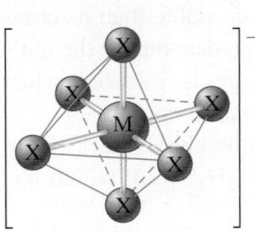

Figure 19-10 | The structures of the tetrahedral MX$_4^+$ and the octahedral MX$_6^-$ ions.

The Group 5A (15) elements can form molecules or ions that involve three, five, or six covalent bonds to the Group 5A (15) atom. Examples involving three single bonds are NH$_3$, PH$_3$, NF$_3$, and AsCl$_3$. Each of these molecules has a lone pair of electrons (and thus can behave as a Lewis base) and a pyramidal shape as predicted by the VSEPR model (see Fig. 19-8).

All the Group 5A (15) elements except nitrogen can form molecules with five covalent bonds (of general formula MX$_5$). Nitrogen cannot form such molecules because of its small size. The MX$_5$ molecules have a trigonal bipyramidal shape (see Fig. 19-9) as predicted by the VSEPR model, and the central atom can be described as dsp^3 hybridized.

Although the MX$_5$ molecules have a trigonal bipyramidal structure in the gas phase, the solids of many of these compounds contain a 1:1 mixture of the ions MX$_4^+$ and MX$_6^-$ (Fig. 19-10). The MX$_4^+$ cation is tetrahedral (the atom represented by M is sp^3 hybridized), and the MX$_6^-$ anion is octahedral (the atom represented by M is d^2sp^3 hybridized). Examples are PCl$_5$ (which in the solid state contains PCl$_4^+$ and PCl$_6^-$) and AsF$_3$Cl$_2$ (which in the solid state contains AsCl$_4^+$ and AsF$_6^-$).

As discussed in Section 19-1, the ability of the Group 5A (15) elements to form π bonds decreases dramatically after nitrogen. This explains why elemental nitrogen exists as N$_2$ molecules containing two π bonds, whereas the other elements in the group exist as larger aggregates containing single bonds. For example, in the gas phase the elements phosphorus, arsenic, and antimony consist of P$_4$, As$_4$, and Sb$_4$ molecules, respectively.

19-8 | The Chemistry of Nitrogen

At the earth's surface virtually all elemental nitrogen exists as the N$_2$ molecule with its very strong triple bond (941 kJ/mol). Because of this large bond strength, the N$_2$ molecule is so unreactive that it can coexist with most other elements under normal conditions without undergoing any appreciable reaction. This property makes nitrogen gas very useful as a medium for experiments involving substances that react with oxygen or water. Such experiments can be done using an inert atmosphere box of the type shown in Fig. 19-11.

The strength of the triple bond in the N$_2$ molecule is important both thermodynamically and kinetically. Thermodynamically, the great stability of the N≡N bond means that most binary compounds containing nitrogen decompose exothermically to the elements, for example:

$$N_2O(g) \longrightarrow N_2(g) + \tfrac{1}{2}O_2(g) \qquad \Delta H° = -82 \text{ kJ}$$

$$NO(g) \longrightarrow \tfrac{1}{2}N_2(g) + \tfrac{1}{2}O_2(g) \qquad \Delta H° = -90 \text{ kJ}$$

$$NO_2(g) \longrightarrow \tfrac{1}{2}N_2(g) + O_2(g) \qquad \Delta H° = -34 \text{ kJ}$$

$$N_2H_4(g) \longrightarrow N_2(g) + 2H_2(g) \qquad \Delta H° = -95 \text{ kJ}$$

$$NH_3(g) \longrightarrow \tfrac{1}{2}N_2(g) + \tfrac{3}{2}H_2(g) \qquad \Delta H° = +46 \text{ kJ}$$

Figure 19-11 | An inert atmosphere box used when working with oxygen- or water-sensitive materials. The box is filled with an inert gas such as nitrogen, and work is done through the ports fitted with large rubber gloves.

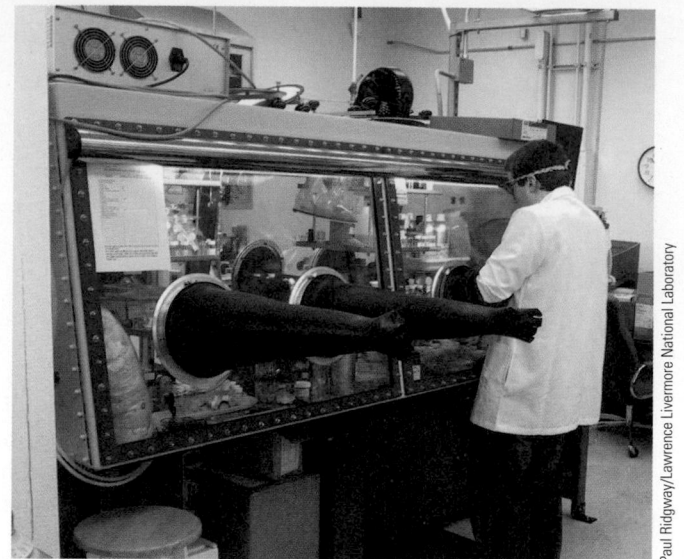

Paul Ridgway/Lawrence Livermore National Laboratory

Of these compounds, only ammonia is thermodynamically more stable than its component elements. That is, only for ammonia is energy required to decompose the molecule to its elements. For the remaining molecules, energy is released when decomposition to the elements occurs, as a result of the great stability of N_2.

The importance of the thermodynamic stability of N_2 can be clearly seen in the power of nitrogen-based explosives, such as nitroglycerin ($C_3H_5N_3O_9$), which has the following skeletal structure:

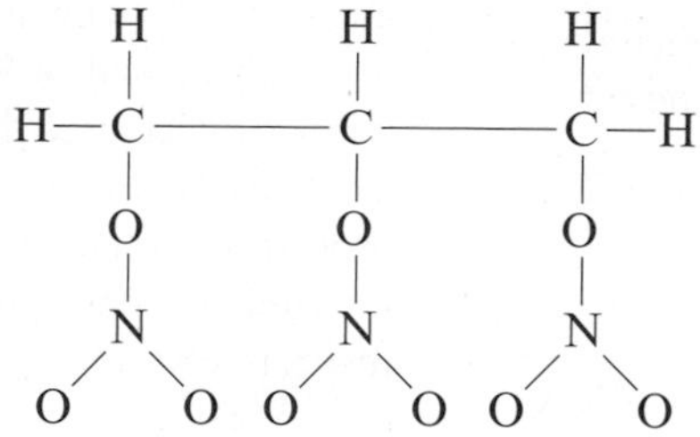

$$
\begin{array}{ccccc}
& H & & H & & H \\
& | & & | & & | \\
H-C & & C & & C-H \\
& | & & | & & | \\
& O & & O & & O \\
& | & & | & & | \\
& N & & N & & N \\
O & & O \; O & & O \; O & & O
\end{array}
$$

When ignited or subjected to sudden impact, nitroglycerin decomposes very rapidly and exothermically:

$$4C_3H_5N_3O_9(l) \longrightarrow 6N_2(g) + 12CO_2(g) + 10H_2O(g) + O_2(g) + energy$$

An explosion occurs; that is, large volumes of gas are produced in a fast, highly exothermic reaction. Note that 4 moles of liquid nitroglycerin produce 29 (6 + 12 + 10 + 1) moles of gaseous products. This alone produces a large increase in volume. However, also note that the products, which include N_2, are very stable molecules with strong bonds. Their formation is therefore accompanied by the release of large quantities of energy as heat, which increases the gaseous volume. The hot, rapidly expanding gases produce a pressure surge and damaging shock wave.

Most high explosives are organic compounds that, like nitroglycerin, contain nitro (—NO_2) groups and produce nitrogen and other gases as products. Another example is *trinitrotoluene* (TNT), a solid at normal temperatures, which decomposes as follows: ◄

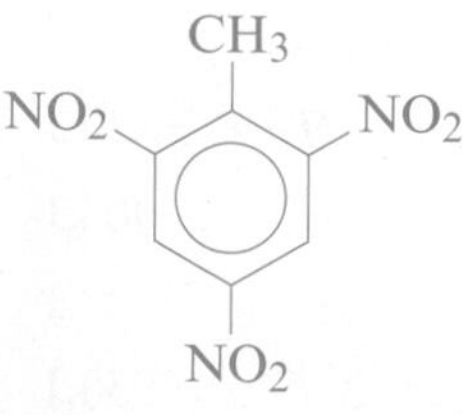

$$2C_7H_5N_3O_6(s) \longrightarrow 12CO(g) + 5H_2(g) + 3N_2(g) + 2C(s) + energy$$

Note that 2 moles of solid TNT produce 20 moles of gaseous products plus energy.

The effect of bond strength on the kinetics of reactions involving the N_2 molecule is illustrated by the synthesis of ammonia from nitrogen and hydrogen, a reaction we

NO_2 ⬡ NO_2 with CH₃ at top and NO₂ at bottom

TNT

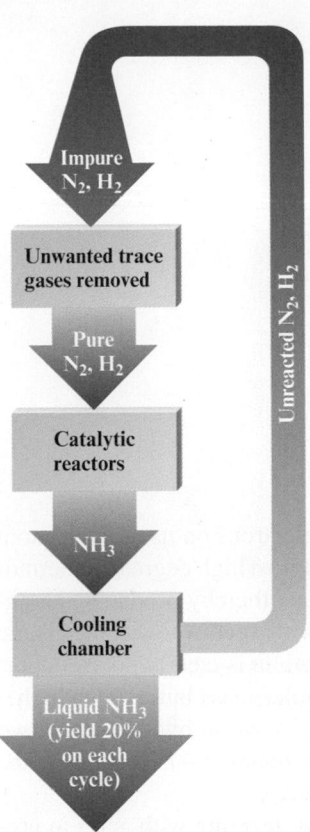

Figure 19-12 | A schematic diagram of the Haber process for the manufacture of ammonia.

▲ Nodules on the roots of pea plants contain nitrogen-fixing bacteria.

have discussed many times before. Because a large quantity of energy is required to disrupt the N≡N bond, the ammonia synthesis reaction occurs at a negligible rate at room temperature, even though the equilibrium constant is very large ($K \approx 10^8$). Of course, the most direct way to increase the rate of a reaction is to raise the temperature. However, since this reaction is very exothermic,

$$N_2(g) + 3H_2(g) \longrightarrow 2NH_3(g) \qquad \Delta H° = -92 \text{ kJ}$$

the value of K decreases significantly with a temperature increase (at 500°C, $K \approx 10^{-2}$).

Obviously, the kinetics and the thermodynamics of this reaction are in opposition. A compromise must be reached, involving high pressure to force the equilibrium to the right and high temperature to produce a reasonable rate. The **Haber process** for manufacturing ammonia represents such a compromise (Fig. 19-12). The process is carried out at a pressure of about 250 atm and a temperature of approximately 400°C. Even higher temperatures would be required if a catalyst consisting of a solid iron oxide mixed with small amounts of potassium oxide and aluminum oxide were not used to facilitate the reaction.

Nitrogen is essential to living systems. The problem with nitrogen is not one of supply—we are surrounded by it—but rather one of changing it from the inert N_2 molecule to a form usable by plants and animals. The process of transforming N_2 to other nitrogen-containing compounds is called **nitrogen fixation**. The Haber process is one example of nitrogen fixation. The ammonia produced can be applied to the soil as a fertilizer, since plants can readily use the nitrogen in ammonia to make the nitrogen-containing biomolecules essential for their growth.

Nitrogen fixation also results from the high-temperature combustion process in automobile engines. The nitrogen in the air is drawn into the engine and reacts at a significant rate with oxygen to form nitric oxide (NO), which further reacts with oxygen from the air to form nitrogen dioxide (NO_2). This nitrogen dioxide, which contributes to photochemical smog in many urban areas (see Section 11-7), reacts with moisture in the air and eventually reaches the soil to form nitrate salts, which are plant nutrients.

Nitrogen fixation also occurs naturally. For example, lightning provides the energy to disrupt N_2 and O_2 molecules in the air, producing highly reactive nitrogen and oxygen atoms. These atoms in turn attack other N_2 and O_2 molecules to form nitrogen oxides that eventually become nitrates. Although lightning has traditionally been credited with forming about 10% of the total fixed nitrogen, recent studies indicate that lightning may account for as much as half of the fixed nitrogen available on the earth. Another natural nitrogen fixation process involves bacteria that reside in the root nodules of plants such as beans, peas, and alfalfa. These **nitrogen-fixing bacteria** readily allow the conversion of nitrogen to ammonia and to other nitrogen-containing compounds useful to plants. The efficiency of these bacteria is intriguing: They produce ammonia at soil temperatures and 1 atm pressure, whereas the Haber process requires severe conditions of 400°C and 250 atm. For obvious reasons, researchers are studying these bacteria intensively. ◄

When plants and animals die and decompose, the elements they consist of are returned to the environment. In the case of nitrogen, the return of the element to the atmosphere as nitrogen gas, called **denitrification**, is carried out by bacteria that change nitrates to nitrogen. The complex **nitrogen cycle** is summarized in Fig. 19-13. It has been estimated that as much as 10 million tons more nitrogen per year is currently being fixed by natural and human processes than is being returned to the atmosphere. This fixed nitrogen is accumulating in soil, lakes, rivers, and oceans, where it promotes the growth of algae and other undesirable organisms.

Nitrogen Hydrides

By far the most important hydride of nitrogen is **ammonia**. A toxic, colorless gas with a pungent odor, ammonia is manufactured in huge quantities (approximately 40 billion pounds per year), mainly for use in fertilizers.

Figure 19-13 | The nitrogen cycle. To be used by plants and animals, nitrogen must be converted from N_2 to nitrogen-containing compounds, such as nitrates, ammonia, and proteins. The nitrogen is returned to the atmosphere by natural decay processes.

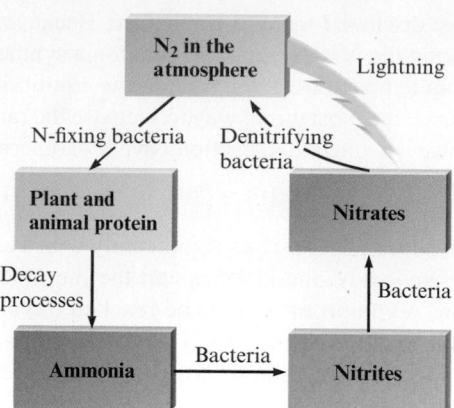

The pyramidal ammonia molecule has a lone pair of electrons on its nitrogen atom (see Fig. 19-8) and polar N—H bonds. This structure leads to a high degree of intermolecular interaction by hydrogen bonding in the liquid state, thereby producing an unusually high boiling point ($-33.4°C$) for a substance with such a low molar mass. Note, however, that the hydrogen bonding in liquid ammonia is clearly not as important as that in liquid water, which has about the same molar mass but a much higher boiling point. The water molecule has two polar bonds involving hydrogen and two lone pairs—the right combination for optimum hydrogen bonding—in contrast to the one lone pair and three polar bonds of the ammonia molecule.

As we saw in Chapter 13, ammonia behaves as a base, reacting with acids to produce ammonium salts. For example,

$$NH_3(g) + HCl(g) \longrightarrow NH_4Cl(s)$$

A second nitrogen hydride of major importance is **hydrazine** (N_2H_4). The Lewis structure of hydrazine

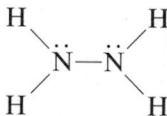

indicates that each nitrogen atom should be sp^3 hybridized with bond angles close to 109.5° (the tetrahedral angle), since the nitrogen atom is surrounded by four electron pairs. The observed structure with bond angles of 112° (Fig. 19-14) agrees reasonably well with these predictions. Hydrazine, a colorless liquid with an ammoniacal odor, freezes at 2°C and boils at 113.5°C. This boiling point is quite high for a compound with a molar mass of 32; this suggests that considerable hydrogen bonding occurs among the polar hydrazine molecules.

Hydrazine is a powerful reducing agent and has been widely used as a rocket propellant. For example, its reaction with oxygen is highly exothermic:

$$N_2H_4(l) + O_2(g) \longrightarrow N_2(g) + 2H_2O(g) \qquad \Delta H° = -622 \text{ kJ}$$

Since hydrazine also reacts vigorously with the halogens, fluorine is often used instead of oxygen as the oxidizer in rocket engines. Substituted hydrazines, where one or more of the hydrogen atoms are replaced by other groups, are also useful rocket fuels. For example, monomethylhydrazine,

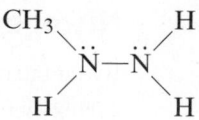

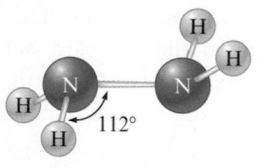

Figure 19-14 | The molecular structure of hydrazine (N_2H_4). This arrangement minimizes the repulsion between the lone pairs on the nitrogen atoms by placing them on opposite sides.

is used with the oxidizer dinitrogen tetroxide (N_2O_4) to power the U.S. space shuttle orbiter. The reaction is

$$5N_2O_4(l) + 4N_2H_3(CH_3)(l) \longrightarrow 12H_2O(g) + 9N_2(g) + 4CO_2(g)$$

Because of the large number of gaseous molecules produced and the exothermic nature of this reaction, a very high thrust per mass of fuel is achieved. The reaction is also self-starting—it begins immediately when the fuels are mixed—which is a useful property for rocket engines that must be started and stopped frequently.

The use of hydrazine as a rocket propellant is a rather specialized application. The main industrial use of hydrazine is as a "blowing" agent in the manufacture of plastics. ◄ Hydrazine decomposes to form nitrogen gas, which causes foaming in the liquid plastic and results in a porous texture. Another major use of hydrazine is in the production of agricultural pesticides. Of the many hundreds of hydrazine derivatives (substituted hydrazines) that have been tested, 40 are used as fungicides, herbicides, insecticides, or plant growth regulators.

Nitrogen Oxides

Nitrogen forms a series of oxides in which its oxidation state ranges from $+1$ to $+5$, as shown in Table 19-14.

Dinitrogen monoxide (N_2O), more commonly called *nitrous oxide* or laughing gas, has an inebriating effect and has been used as a mild anesthetic by dentists. Because of its high solubility in fats, nitrous oxide is widely used as a propellant in aerosol cans of whipped cream. It is dissolved in the liquid inside the can at high pressure and forms bubbles that produce foaming as the liquid is released from the can. A significant amount of N_2O exists in the atmosphere, mostly produced by soil microorganisms, and

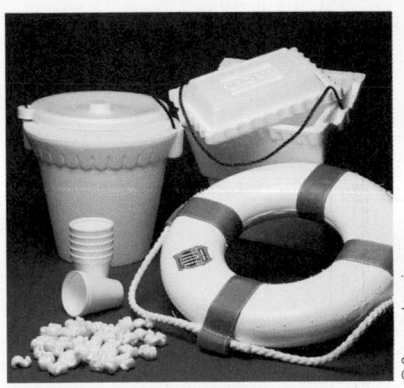

▲ Blowing agents—such as hydrazine, which forms nitrogen gas on decomposition—are used to produce porous plastics like these polystyrene products.

Table 19-14 │ Some Common Nitrogen Compounds

Oxidation State of Nitrogen	Compound	Formula	Lewis Structure*
−3	Ammonia	NH_3	H—N̈—H with H below
−2	Hydrazine	N_2H_4	H—N̈—N̈—H with H below each N
−1	Hydroxylamine	NH_2OH	H—N̈—Ö—H with H below N
0	Nitrogen	N_2	:N≡N:
+1	Dinitrogen monoxide (nitrous oxide)	N_2O	:N̈=N=Ö:
+2	Nitrogen monoxide (nitric oxide)	NO	:N̈=Ö:
+3	Dinitrogen trioxide	N_2O_3	N—N̈=Ö: with two O on N
+4	Nitrogen dioxide	NO_2	:Ö—N̈=Ö:
+5	Nitric acid	HNO_3	:Ö—N—Ö—H with O below

*In some cases additional resonance structures are needed to fully describe the electron distribution.

Figure 19-15 | The molecular orbital energy-level diagram for nitric oxide (NO). The bond order is 2.5, or $(8 - 3)/2$.

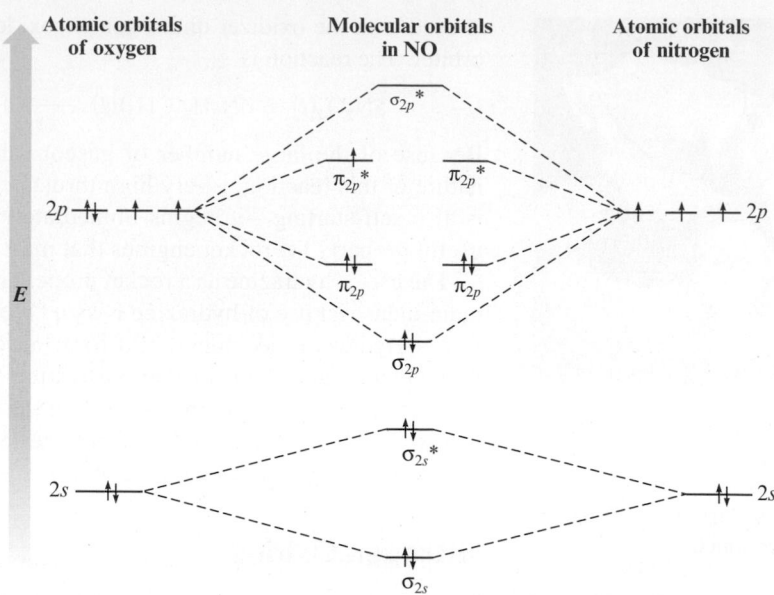

© Cengage Learning

▲ A copper penny reacts with nitric acid to produce NO gas, which is immediately oxidized in air to give reddish brown NO_2.

Table 19-15 | Comparison of the Bond Lengths and Bond Energies for Nitric Oxide and the Nitrosyl Ion

	NO	NO$^+$
Bond length (Å)	1.15	1.09
Bond energy (kJ/mol)	630	1020
Bond order (predicted by MO model)	2.5	3

its concentration appears to be gradually increasing. Because it can strongly absorb infrared radiation, nitrous oxide plays a small but probably significant role in controlling the earth's temperature in the same way that atmospheric carbon dioxide and water vapor do (see the discussion of the greenhouse effect in Section 7-6). Some scientists fear that the rapid decrease of tropical rain forests resulting from the development of countries such as Brazil will significantly affect the rate of production of N_2O by soil organisms and thus will have important effects on the earth's temperature.

Nitrogen monoxide (NO), commonly called *nitric oxide,* has been found to be an important regulator in biological systems. Nitric oxide is a colorless gas under normal conditions and can be produced in the laboratory by reacting 6 *M* nitric acid with copper metal:

$$8H^+(aq) + 2NO_3^-(aq) + 3Cu(s) \longrightarrow 3Cu^{2+}(aq) + 4H_2O(l) + 2NO(g)$$

When this reaction is carried out in the air, the nitric oxide is immediately oxidized by O_2 to reddish brown nitrogen dioxide (NO_2). ◄

Since the NO molecule has an odd number of electrons, it is most conveniently described in terms of the molecular orbital model. The molecular orbital energy-level diagram is shown in Fig. 19-15. Note that the NO molecule should be paramagnetic and have a bond order of 2.5, predictions that are supported by experimental observations. Since the NO molecule has one high-energy electron, it is not surprising that it can be rather easily oxidized to form NO$^+$, the *nitrosyl ion.* Because an antibonding electron is removed in going from NO to NO$^+$, the resulting ion should have a stronger bond (the predicted bond order is 3) than the molecule. This is borne out by experiment. The bond lengths and bond energies for nitric oxide and the nitrosyl ion are shown in Table 19-15.

Nitric oxide is thermodynamically unstable and decomposes to nitrous oxide and nitrogen dioxide:

$$3NO(g) \longrightarrow N_2O(g) + NO_2(g)$$

Nitrogen dioxide (NO_2), which is also an odd-electron molecule, has a V-shaped structure. The reddish brown, paramagnetic NO_2 molecule readily dimerizes to form dinitrogen tetroxide,

$$2NO_2(g) \rightleftharpoons N_2O_4(g)$$

Nitrous oxide (N$_2$O), more properly called dinitrogen monoxide, is a compound with many interesting uses. It was discovered in 1772 by Joseph Priestley (who is also given credit for discovering oxygen gas), and its intoxicating effects were noted almost immediately. In 1798, the 20-year-old Humphry Davy became director of the Pneumatic Institute, which was set up to investigate the medical effects of various gases. Davy tested the effects of N$_2$O on himself, reporting that after inhaling 16 quarts of the gas in 7 minutes, he became "absolutely intoxicated."

Over the next century "laughing gas," as nitrous oxide became known, was developed as an anesthetic, particularly for dental procedures. Nitrous oxide is still used as an anesthetic, although it has been primarily replaced by more modern drugs.

One major use of nitrous oxide today is as the propellant in cans of "instant" whipped cream. The high solubility of N$_2$O in the whipped cream mixture makes it an excellent candidate for pressurizing the cans of whipping cream.

Another current use of nitrous oxide is to produce "instant horsepower" for street racers. Because the reaction of N$_2$O with O$_2$ to form NO actually absorbs heat, this reaction has a cooling effect when placed in the fuel mixture in an automobile engine. This cooling effect lowers combustion temperatures, thus allowing the fuel–air mixture to be significantly more dense (the density of a gas is inversely proportional to temperature). The effect can produce a burst of additional power in excess of 200 horsepower. Because engines are not designed to run steadily at such high power levels, the nitrous oxide is injected from a tank when extra power is desired.

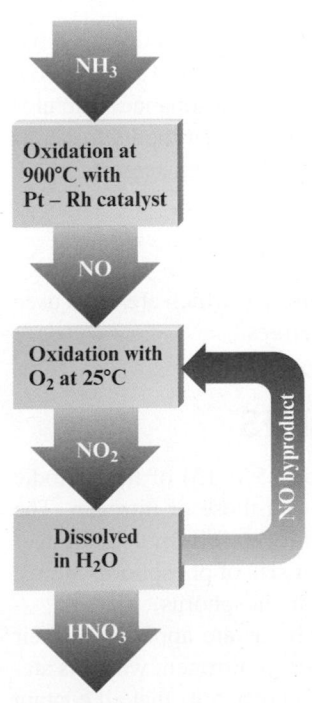

Figure 19-16 | The Ostwald process.

which is diamagnetic and colorless. The value of the equilibrium constant is approximately 1 for this process at 55°C, and since the dimerization is exothermic, K decreases as the temperature increases.

The least common of the nitrogen oxides are *dinitrogen trioxide* (N$_2$O$_3$), a blue liquid that readily dissociates to gaseous nitric oxide and nitrogen dioxide, and *dinitrogen pentoxide* (N$_2$O$_5$), which under normal conditions is a solid that is best viewed as a mixture of NO$_2^+$ and NO$_3^-$ ions. Although N$_2$O$_5$ molecules can exist in the gas phase, they readily dissociate to nitrogen dioxide and oxygen:

$$2N_2O_5(g) \rightleftharpoons 4NO_2(g) + O_2(g)$$

This reaction follows first-order kinetics, as discussed in Section 11-4.

Oxyacids of Nitrogen

Nitric acid is an important industrial chemical (approximately 8 million tons produced annually) used in the manufacture of many products, such as nitrogen-based explosives and ammonium nitrate for use as a fertilizer.

Nitric acid is produced commercially by the oxidation of ammonia in the **Ostwald process** (Fig. 19-16). In the first step of this process, ammonia is oxidized to nitric oxide:

$$4NH_3(g) + 5O_2(g) \longrightarrow 4NO(g) + 6H_2O(g) \qquad \Delta H° = -905 \text{ kJ}$$

Although this reaction is highly exothermic, it is very slow at 25°C. A side reaction occurs between nitric oxide and ammonia:

$$4NH_3(g) + 6NO(g) \longrightarrow 5N_2(g) + 6H_2O(g)$$

which is particularly undesirable because it traps the nitrogen in the very unreactive N$_2$ molecules. The desired reaction can be accelerated and the effects of the competing reaction can be minimized if the ammonia oxidation is carried out by using a catalyst

Figure 19-17 | (a) The molecular structure of HNO_3. (b) The resonance structures of HNO_3.

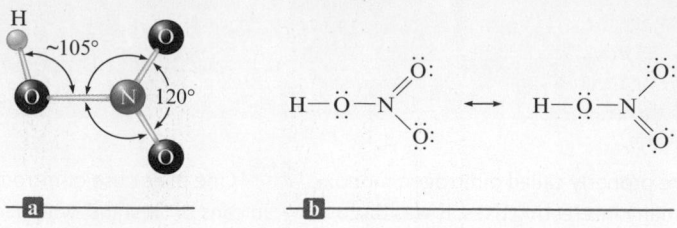

of a platinum–rhodium alloy heated to 900°C. Under these conditions, there is a 97% conversion of the ammonia to nitric oxide.

In the second step, nitric oxide is reacted with oxygen to produce nitrogen dioxide:

$$2NO(g) + O_2(g) \longrightarrow 2NO_2(g) \qquad \Delta H° = -113 \text{ kJ}$$

This oxidation reaction has a rate that *decreases* with increasing temperature. Because of this very unusual behavior, the reaction is carried out at approximately 25°C and is kept at this temperature by cooling with water.

The third step in the Ostwald process is the absorption of nitrogen dioxide by water:

$$3NO_2(g) + H_2O(l) \longrightarrow 2HNO_3(aq) + NO(g) \qquad \Delta H° = -139 \text{ kJ}$$

The gaseous NO produced in the reaction is recycled so that it can be oxidized to NO_2. The aqueous nitric acid from this process is about 50% HNO_3 by mass, which can be increased to 68% by distillation to remove some of the water. The maximum concentration attainable by this method is 68% because nitric acid and water form an *azeotrope* at this concentration. ◄ 🛈 The solution can be further concentrated to 95% HNO_3 by treatment with concentrated sulfuric acid, which strongly absorbs water; H_2SO_4 is often used as a *dehydrating (water-removing) agent*.

🛈 An *azeotrope* is a solution that, like a pure liquid, distills at a constant temperature without a change in composition.

Nitric acid is a colorless, fuming liquid (bp = 83°C) with a pungent odor; it decomposes in sunlight by the following reaction:

$$4HNO_3(l) \xrightarrow{h\nu} 4NO_2(g) + 2H_2O(l) + O_2(g)$$

As a result, nitric acid turns yellow as it ages because of the dissolved nitrogen dioxide. The common laboratory reagent called *concentrated nitric acid* is 15.9 M HNO_3 (70.4% HNO_3 by mass) and is a very strong oxidizing agent. The resonance structures and molecular structure of HNO_3 are shown in Fig. 19-17. Note that the hydrogen is bound to an oxygen atom rather than to nitrogen as the formula might suggest.

Nitrous acid (HNO_2) is a weak acid,

$$HNO_2(aq) \rightleftharpoons H^+(aq) + NO_2^-(aq) \qquad K_a = 4.0 \times 10^{-4}$$

that forms pale yellow nitrite (NO_2^-) salts. In contrast to nitrates, which are often used as explosives, nitrites are quite stable even at high temperatures.

19-9 | The Chemistry of Phosphorus

Although phosphorus lies directly below nitrogen in Group 5A (15) of the periodic table, its chemical properties are significantly different from those of nitrogen. The differences arise mainly from four factors: nitrogen's ability to form much stronger π bonds, the greater electronegativity of nitrogen, the larger size of phosphorus atoms, and the potential availability of empty valence d orbitals on phosphorus.

The chemical differences between nitrogen and phosphorus are apparent in their elemental forms. In contrast to the diatomic form of elemental nitrogen, which is stabilized by strong π bonds, there are several solid forms of phosphorus that all contain aggregates of atoms. *White phosphorus,* which contains discrete tetrahedral P_4 molecules [Fig. 19-18(a)], is very reactive; it bursts into flames on contact with air (it is said

Figure 19-18 | (a) The P$_4$ molecule found in white phosphorus. (b) The crystalline network structure of black phosphorus. (c) The chain structure of red phosphorus.

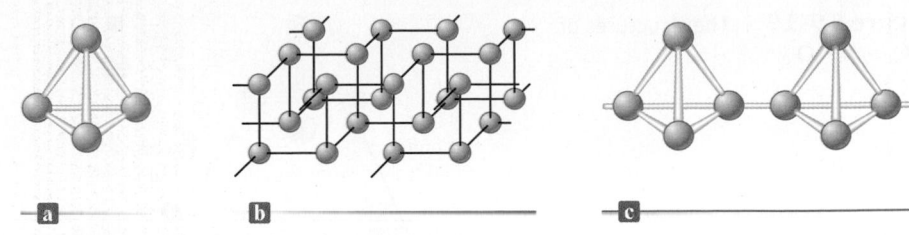

Charles D. Winters

▲ White phosphorus reacts vigorously with the oxygen in air and must be stored under water. Red phosphorus is stable in air.

to be *pyrophoric*). ◄ Consequently, white phosphorus is commonly stored under water. White phosphorus is quite toxic; the P$_4$ molecules are very damaging to tissue, particularly the cartilage and bones of the nose and jaw. The much less reactive forms, called *black phosphorus* and *red phosphorus,* are network solids (see Section 9-5). Black phosphorus has a regular crystalline structure [Fig. 19-18(b)], but red phosphorus is amorphous and is thought to consist of chains of P$_4$ units [Fig. 19-18(c)]. Red phosphorus can be obtained by heating white phosphorus in the absence of air at 1 atm. Black phosphorus is obtained from either white or red phosphorus by heating at high pressures.

Even though phosphorus has a lower electronegativity than nitrogen, it will form phosphides (ionic substances containing the P^{3-} anion) such as Na$_3$P and Ca$_3$P$_2$. Phosphide salts react vigorously with water to produce *phosphine* (PH$_3$), a toxic, colorless gas:

$$2Na_3P(s) + 6H_2O(l) \longrightarrow 2PH_3(g) + 6Na^+(aq) + 6OH^-(aq)$$

Phosphine is analogous to ammonia, although it is a much weaker base ($K_b \approx 10^{-26}$) and is much less soluble in water.

Phosphine has the Lewis structure

$$H-\overset{\displaystyle ..}{P}-H$$
$$|$$
$$H$$

and a pyramidal molecular structure, as we would predict from the VSEPR model. However, it has bond angles of 94° rather than 107°, as found in the ammonia molecule. The reasons for this are complex; therefore, we will simply regard phosphine as an exception to the simple version of the VSEPR model that we use.

Phosphorus Oxides and Oxyacids

ⓘ The terminal oxygens are the nonbridging oxygen atoms.

Phosphorus reacts with oxygen to form oxides in which its oxidation states are +5 and +3. The oxide P$_4$O$_6$ is formed when elemental phosphorus is burned in a limited supply of oxygen, and P$_4$O$_{10}$ is produced when the oxygen is in excess. Picture these oxides (shown in Fig. 19-19) as being constructed by adding oxygen atoms to the fundamental P$_4$ structure. The intermediate states, P$_4$O$_7$, P$_4$O$_8$, and P$_4$O$_9$, which contain one, two, and three terminal oxygen atoms, respectively, are also known. ◄ ⓘ

Tetraphosphorus decoxide (P$_4$O$_{10}$), which was formerly represented as P$_2$O$_5$ and called phosphorus pentoxide, has a great affinity for water and thus is a powerful dehydrating agent. For example, it can be used to convert HNO$_3$ and H$_2$SO$_4$ to their parent oxides, N$_2$O$_5$ and SO$_3$, respectively.

When tetraphosphorus decoxide dissolves in water, **phosphoric acid** (H$_3$PO$_4$), also called **orthophosphoric acid**, is produced:

$$P_4O_{10}(s) + 6H_2O(l) \longrightarrow 4H_3PO_4(aq)$$

Pure phosphoric acid is a white solid that melts at 42°C. Aqueous phosphoric acid is a much weaker acid ($K_{a_1} \approx 10^{-2}$) than nitric acid or sulfuric acid and is a poor oxidizing agent.

Figure 19-19 | The structures of P_4O_6 and P_4O_{10}.

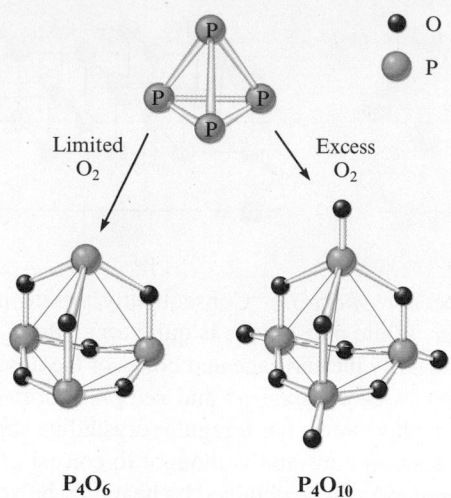

Limited O_2

Excess O_2

O

P

P_4O_6 P_4O_{10}

When the oxide P_4O_6 is placed in water, **phosphorous acid** (H_3PO_3) is formed [Fig. 19-20(a)]. Although the formula suggests a triprotic acid, phosphorous acid is a *diprotic* acid. The hydrogen atom bonded directly to the phosphorus atom is not acidic in aqueous solution; only those hydrogen atoms bonded to the oxygen atoms in H_3PO_3 can be released as protons.

A third oxyacid of phosphorus is *hypophosphorous acid* (H_3PO_2) [Fig. 19-20(b)], which is a monoprotic acid.

Phosphorus in Fertilizers

Phosphorus is essential for plant growth. Although most soil contains large amounts of phosphorus, it is often present in insoluble minerals, making it inaccessible to the plants. Soluble phosphate fertilizers are manufactured by treating phosphate rock with sulfuric acid to make **superphosphate of lime**, a mixture of $CaSO_4 \cdot 2H_2O$ and $Ca(H_2PO_4)_2 \cdot H_2O$. If phosphate rock is treated with phosphoric acid, $Ca(H_2PO_4)_2$, known as *triple phosphate,* is produced. The reaction of ammonia with phosphoric acid gives *ammonium dihydrogen phosphate* ($NH_4H_2PO_4$), a very efficient fertilizer that furnishes both phosphorus and nitrogen.

Figure 19-20 | (a) The structure of phosphorous acid (H_3PO_3). (b) The structure of hypophosphorous acid (H_3PO_2).

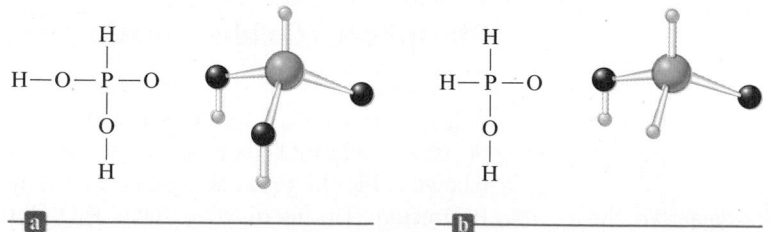

a   b

19-10 | The Group 6A (16) Elements

| 16 |
| 6A |
| O |
| S |
| Se |
| Te |
| Po |

Although in Group 6A (16) (Table 19-16) there is the usual tendency for metallic properties to increase going down the group, none of the Group 6A (16) elements behaves as a typical metal. ◀ The most common chemical behavior of a Group 6A (16) element involves reacting with a metal to achieve a noble gas electron configuration by adding two electrons to become a 2− anion in ionic compounds. In fact, for most metals, the oxides and sulfides constitute the most common minerals.

The Group 6A (16) elements can form covalent bonds with other nonmetals. For example, they combine with hydrogen to form a series of covalent hydrides of the general formula H_2X. Those members of the group that have valence d orbitals

Table 19-16 | Selected Physical Properties, Sources, and Methods of Preparation of the Group 6A (16) Elements

Element	Electronegativity	Radius of X^{2-} (pm)	Source	Method of Preparation
Oxygen	3.4	140	Air	Distillation from liquid air
Sulfur	2.6	184	Sulfur deposits	Melted with hot water and pumped to the surface
Selenium	2.6	198	Impurity in sulfide ores	Reduction of H_2SeO_4 with SO_2
Tellurium	2.1	221	Nagyagite (mixed sulfide and telluride)	Reduction of ore with SO_2
Polonium	2.0	230	Pitchblende	

▲ Walnuts contain trace amounts of selenium.

available (all except oxygen) commonly form molecules in which they are surrounded by more than eight electrons. Examples are SF_4, SF_6, TeI_4, and $SeBr_4$.

The two heaviest members of Group 6A (16) can lose electrons to form cations. Although they do not lose all six valence electrons because of the high energies that would be required, tellurium and polonium appear to exhibit some chemistry involving their 4+ cations. However, the chemistry of these Group 6A (16) cations is much more limited than that of the Group 5A (15) elements bismuth and antimony.

In recent years there has been a growing interest in the chemistry of selenium, an element found throughout the environment in trace amounts. Selenium's toxicity has long been known, but some medical studies have shown an *inverse* relationship between the incidence of cancer and the selenium levels in soil. It has been suggested that the greater dietary intake of selenium by people living in areas with relatively high selenium levels somehow furnishes protection from cancer. ◄ These studies are only preliminary, but selenium is definitely known to be physiologically important (it is involved in the activity of vitamin E and certain enzymes). Selenium (as well as tellurium) is also a semiconductor and therefore finds some application in the electronics industry.

Polonium was discovered in 1898 by Marie and Pierre Curie in their search for the sources of radioactivity in pitchblende. Polonium has 27 isotopes and is highly toxic and very radioactive. It has been suggested that the isotope ^{210}Po, a natural contaminant of tobacco and an α-particle producer (see Section 18-1), might be at least partly responsible for the incidence of cancer in smokers.

19-11 | The Chemistry of Oxygen

It is hard to overstate the importance of oxygen, the most abundant element in and near the earth's crust. ◄ Oxygen is present in the atmosphere as oxygen gas and ozone; in soil and rocks in oxide, silicate, and carbonate minerals; in the oceans in water; and in our bodies in water and a myriad of other molecules. In addition, most of the energy we need to live and run our civilization comes from the exothermic reactions of oxygen with carbon-containing molecules.

The most common elemental form of oxygen (O_2) constitutes 21% of the volume of the earth's atmosphere. Since nitrogen has a lower boiling point than oxygen, nitrogen can be boiled away from liquid air, leaving oxygen and small amounts of argon, another component of air. Liquid oxygen is a pale blue liquid that freezes at $-219°C$ and boils at $-183°C$. The paramagnetism of the O_2 molecule can be demonstrated by pouring liquid oxygen between the poles of a strong magnet, where it "sticks" until it boils away (see Fig. 4-51). The paramagnetism of the O_2 molecule can be accounted for by the molecular orbital model (Fig. 4-50), which also explains its bond strength.

▲ A U.S. Navy test pilot in his F/A-18F Super Hornet wearing an oxygen mask.

The other form of elemental oxygen is **ozone** (O_3), a molecule that can be represented by the resonance structures

$$\ddot{O} \diagup \diagdown \ddot{O} \longleftrightarrow \ddot{O} \diagup \diagdown \ddot{O}$$

The bond angle in the O_3 molecule is 117°, in reasonable agreement with the prediction of the VSEPR model (three effective pairs require a trigonal planar arrangement). That the bond angle is slightly less than 120° can be explained by concluding that more space is required for the lone pair than for the bonding pairs.

Ozone can be prepared by passing an electrical discharge through pure oxygen gas. The electrical energy disrupts the bonds in some O_2 molecules, thereby producing oxygen atoms, which react with other O_2 molecules to form O_3. Ozone is much less stable than oxygen at 25°C and 1 atm. For example, $K \approx 10^{-56}$ for the equilibrium

$$3O_2(g) \rightleftharpoons 2O_3(g)$$

A pale blue, highly toxic gas, ozone is a much more powerful oxidizing agent than oxygen. The strong oxidizing power of ozone makes it useful for killing bacteria in swimming pools, hot tubs, and aquariums. It is also increasingly being used in municipal water treatment and for washing produce after it comes out of the fields. One of the main advantages of using ozone for water purification is that it does not leave potentially toxic residues behind. On the other hand, chlorine, which is widely used for water purification, leaves residues of chloro compounds, such as chloroform ($CHCl_3$), which may cause cancer after long-term exposure. Although ozone effectively kills the bacteria in water, one problem with **ozonolysis** is that the water supply is not protected against recontamination, since virtually no ozone remains after the initial treatment. In contrast, for chlorination, significant residual chlorine remains after treatment.

The oxidizing ability of ozone can be detrimental, especially when it is present in the pollution from automobile exhausts (see Section 8-10).

Ozone exists naturally in the upper atmosphere of the earth. ◀ ⓘ The *ozone layer* is especially important because it absorbs ultraviolet light and thus acts as a screen to prevent this radiation, which can cause skin cancer, from penetrating to the earth's surface. When an ozone molecule absorbs this energy, it splits into an oxygen molecule and an oxygen atom:

$$O_3 \xrightarrow{h\nu} O_2 + O$$

If the oxygen molecule and atom collide, they will not stay together as ozone unless a "third body," such as a nitrogen molecule, is present to help absorb the energy released by bond formation. The third body absorbs this energy as kinetic energy; its temperature is increased. Therefore, the energy originally absorbed as ultraviolet radiation is eventually changed to thermal energy. Thus the ozone prevents the harmful high-energy ultraviolet light from reaching the earth.

ⓘ Scientists have become concerned that Freons and nitrogen dioxide are promoting the destruction of the ozone layer (see Section 11-7).

19-12 | The Chemistry of Sulfur

Sulfur is found in nature both in large deposits of the free element and in widely distributed ores, such as galena (PbS), cinnabar (HgS), pyrite (FeS_2), gypsum ($CaSO_4 \cdot 2H_2O$), epsomite ($MgSO_4 \cdot 7H_2O$), and glauberite ($Na_2SO_4 \cdot CaSO_4$). ◀

About 60% of the sulfur produced in the United States comes from the underground deposits of elemental sulfur found in Texas and Louisiana. This sulfur is recovered by using the **Frasch process** developed by Herman Frasch in the 1890s. ◀ Superheated water is pumped into the deposit to melt the sulfur (mp = 113°C), which is then forced to the surface by air pressure (Fig. 19-21). The remaining 40% of sulfur produced in

Figure 19-21 | The Frasch method for recovering sulfur from underground deposits.

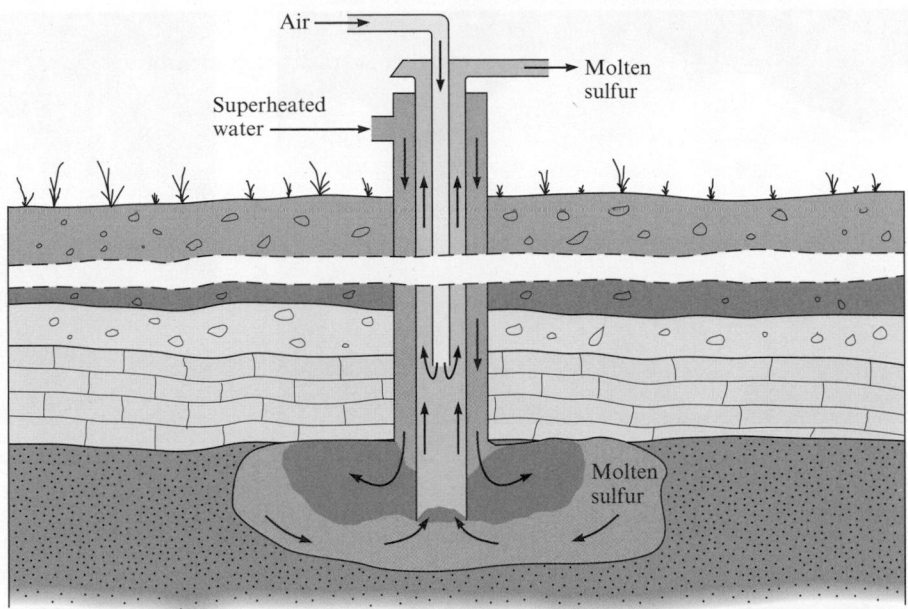

▲ The mineral cinnabar

▲ Sulfur obtained from underground deposits by the Frasch process.

the United States either is a by-product of the purification of fossil fuels before combustion to prevent pollution or comes from the sulfur dioxide (SO_2) scrubbed from the exhaust gases when sulfur-containing fuels are burned.

In contrast to oxygen, elemental sulfur exists as S_2 molecules only in the gas phase at high temperatures. Because sulfur atoms form much stronger σ bonds than π bonds, S_2 is less stable at 25°C than larger aggregates such as S_6 and S_8 rings and S_n chains (Fig. 19-22). The most stable form of sulfur at 25°C and 1 atm is called *rhombic sulfur* [Fig. 19-23(a)], which contains stacked S_8 rings. If rhombic sulfur is melted and heated to 120°C, it forms *monoclinic sulfur* as it slowly cools [Fig. 19-23(b)]. The monoclinic form also contains S_8 rings, but the rings are stacked differently than in rhombic sulfur.

Sulfur Oxides

From its position below oxygen in the periodic table, we might expect the simplest stable oxide of sulfur to have the formula SO. However, *sulfur monoxide,* which can be produced in small amounts when gaseous sulfur dioxide (SO_2) is subjected to an electrical discharge, is very unstable. The difference in the stabilities of the O_2 and SO molecules probably reflects the much stronger π bonding between oxygen atoms than between a sulfur and an oxygen atom.

Sulfur burns in air with a bright blue flame to give *sulfur dioxide* (SO_2), a colorless gas with a pungent odor, which condenses to a liquid at −10°C and 1 atm. Sulfur dioxide is a V-shaped molecule, which is a very effective antibacterial agent often used to preserve stored fruit.

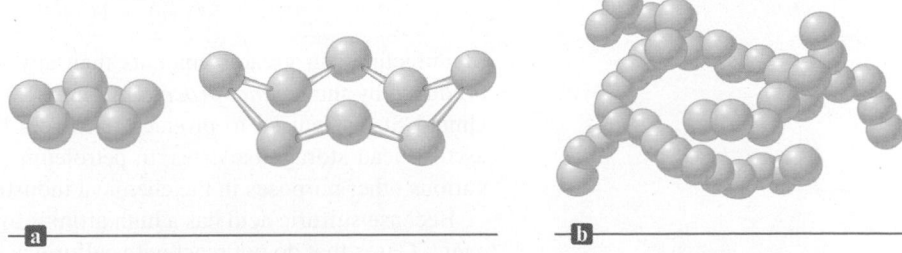

Figure 19-22 | (a) The S_8 molecule. (b) Chains of sulfur atoms in viscous liquid sulfur. The chains may contain as many as 10,000 atoms.

Figure 19-23 | (a) Crystals of rhombic sulfur. (b) Crystals of monoclinic sulfur.

Sulfur dioxide reacts with oxygen to produce *sulfur trioxide* (SO_3):

$$2SO_2(g) + O_2(g) \longrightarrow 2SO_3(g)$$

However, this reaction is very slow in the absence of a catalyst. One of the mysteries during early research on air pollution was how the sulfur dioxide produced from the combustion of sulfur-containing fuels is so rapidly converted to sulfur trioxide in the atmosphere. It is now known that dust and other particles can act as heterogeneous catalysts for this process (see Section 11-7).

Oxyacids of Sulfur

Sulfur dioxide dissolves in water to form an acidic solution. The reaction is often represented as

$$SO_2(g) + H_2O(l) \longrightarrow H_2SO_3(aq)$$

where H_2SO_3 is called *sulfurous acid*. However, very little H_2SO_3 actually exists in the solution. The major form of sulfur dioxide in water is SO_2, and the acid dissociation equilibria are best represented as

$$SO_2(aq) + H_2O(l) \rightleftharpoons H^+(aq) + HSO_3^-(aq) \qquad K_{a_1} = 1.5 \times 10^{-2}$$
$$HSO_3^-(aq) \rightleftharpoons H^+(aq) + SO_3^{2-}(aq) \qquad K_{a_2} = 1.0 \times 10^{-7}$$

This situation is analogous to the behavior of carbon dioxide in water (see Section 13-7). Although H_2SO_3 cannot be isolated, salts of SO_3^{2-} (*sulfites*) and HSO_3^- (*hydrogen sulfites*) are well known.

Sulfur trioxide reacts violently with water to produce the diprotic acid **sulfuric acid**:

$$SO_3(g) + H_2O(l) \longrightarrow H_2SO_4(aq)$$

Manufactured in greater amounts than any other chemical, sulfuric acid is usually produced by the *contact process*. About 60% of the sulfuric acid manufactured in the United States is used to produce fertilizers from phosphate rock. The other 40% is used in lead storage batteries, in petroleum refining, in steel manufacturing, and for various other purposes in the chemical industry.

Because sulfuric acid has a high affinity for water, it is often used as a dehydrating agent. Gases that do not react with sulfuric acid, such as oxygen, nitrogen, and carbon dioxide, are often dried by bubbling them through concentrated solutions of the acid. Sulfuric acid is such a powerful dehydrating agent that it will remove hydrogen and

Figure 19-24 | The reaction of H_2SO_4 with sucrose (left) to produce a blackened column of carbon (right).

Ken O'Donoghue © Cengage Learning

oxygen from a substance in a 2:1 ratio even when the substance contains no molecular water. For example, concentrated sulfuric acid reacts vigorously with common table sugar (sucrose), leaving a charred mass of carbon (Fig. 19-24):

$$C_{12}H_{22}O_{11}(s) + 11H_2SO_4(conc) \longrightarrow 12C(s) + 11H_2SO_4 \cdot H_2O(l)$$
$$\text{Sucrose}$$

19-13 | The Group 7A (17) Elements

17
7A

| F |
| Cl |
| Br |
| I |
| At |

In our coverage of the representative elements we have progressed from the groups of metallic elements (Groups 1A (1) and 2A (2)), through groups in which the lighter members are nonmetals and the heavier members are metals (Groups 3A (13), 4A (14), and 5A (15)), to a group containing all nonmetals (Group 6A (16)—although some might prefer to call polonium a metal). The Group 7A (17) elements, the **halogens** (with the valence electron configuration ns^2np^5), are all nonmetals whose properties generally vary smoothly going down the group. ◄ ◄ The only notable exceptions are the unexpectedly low value for the electron affinity of fluorine and the unexpectedly small bond energy of the F_2 molecule (see Section 19-1). Table 19-17 summarizes the trends in some physical properties of the halogens.

Because of their high reactivities, the halogens are not found as free elements in nature. Instead, they are found as halide ions (X^-) in various minerals and in seawater (Table 19-18).

Although astatine is a member of Group 7A (17), its chemistry is of no practical importance because all its known isotopes are radioactive. The longest-lived isotope, ^{210}At, has a half-life of only 8.3 hours.

The halogens, particularly fluorine, have very high electronegativity values (see Table 19-17). They tend to form polar covalent bonds with other nonmetals and ionic bonds with metals in their lower oxidation states. When a metal ion is in a higher

© Cengage Learning

▲ Samples of chlorine gas, liquid bromine, and solid iodine.

Table 19-17 | Trends in Selected Physical Properties of the Group 7A (17) Elements

Element	Electronegativity	Radius of X^- (pm)	$\mathcal{E}°$ (V) for $X_2 + 2e \rightarrow 2X^-$	Bond Energy of X_2 (kJ/mol)
Fluorine	4.0	136	2.87	154
Chlorine	3.2	181	1.36	239
Bromine	3.0	195	1.09	193
Iodine	2.7	216	0.54	149
Astatine	2.2	—	—	—

Table 19-18 | Some Physical Properties, Sources, and Methods of Preparation of the Group 7A (17) Elements

Element	Color and State	Percentage of Earth's Crust	Melting Point (°C)	Boiling Point (°C)	Source	Method of Preparation
Fluorine	Pale yellow gas	0.07	−220	−188	Fluorospar (CaF_2), cryolite (Na_3AlF_6), fluorapatite [$Ca_5(PO_4)_3F$]	Electrolysis of molten KHF_2
Chlorine	Yellow-green gas	0.14	−101	−34	Rock salt (NaCl), halite (NaCl), sylvite (KCl)	Electrolysis of aqueous NaCl
Bromine	Red-brown liquid	2.5×10^{-4}	−7.3	59	Seawater, brine wells	Oxidation of Br^- by Cl_2
Iodine	Violet-black solid	3×10^{-5}	113	184	Seaweed, brine wells	Oxidation of I^- by electrolysis or MnO_2

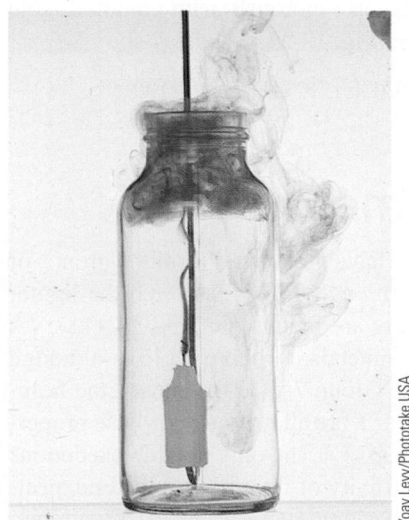

▲ A candle burning in an atmosphere of $Cl_2(g)$. The exothermic reaction, which involves breaking C—C and C—H bonds in the wax and forming C—Cl bonds in their places, produces enough heat to make the gases in the region incandescent (a flame results).

oxidation state, such as +3 or +4, the metal–halogen bonds are polar and covalent. For example, $TiCl_4$ and $SnCl_4$ are both covalent compounds that are liquids under normal conditions.

Hydrogen Halides ◄

The hydrogen halides can be prepared by a reaction of the elements

$$H_2(g) + X_2(g) \longrightarrow 2HX(g)$$

This reaction occurs with explosive vigor when fluorine and hydrogen are mixed. On the other hand, hydrogen and chlorine can coexist with little apparent reaction for relatively long periods in the dark. However, ultraviolet light causes an explosively fast reaction, and this is the basis of a popular lecture demonstration, the "hydrogen–chlorine cannon." Bromine and iodine also react with hydrogen, but more slowly.

Some physical properties of the hydrogen halides are listed in Table 19-19. Note the very high boiling point for hydrogen fluoride, which results from extensive hydrogen bonding among the very polar HF molecules (Fig. 19-25). Fluoride ion has such a high affinity for protons that in concentrated aqueous solutions of hydrogen fluoride, the ion $[F---H---F]^-$ exists, in which an H^+ ion is centered between two F^- ions.

When dissolved in water, the hydrogen halides behave as acids, and all except hydrogen fluoride are completely dissociated. Because water is a much stronger base than the Cl^-, Br^-, or I^- ion, the acid strengths of HCl, HBr, and HI cannot be differentiated in water. However, in a less basic solvent, such as glacial (pure) acetic acid, the acids show different strengths:

$$\text{H—I} > \text{H—Br} > \text{H—Cl} \gg \text{H—F}$$

Strongest   Weakest
acid   acid

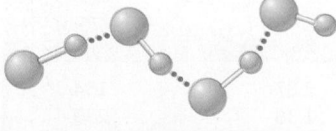

Figure 19-25 | The hydrogen bonding among HF molecules in liquid hydrogen fluoride.

Table 19-19 | Some Physical Properties of the Hydrogen Halides

HX	Melting Point (°C)	Boiling Point (°C)	H—X Bond Energy (kJ/mol)
HF	−83	20	565
HCl	−114	−85	427
HBr	−87	−67	363
HI	−51	−35	295

To see why hydrogen fluoride is the only weak acid in water among the HX molecules, let's consider the dissociation equilibrium,

$$HX(aq) \rightleftharpoons H^+(aq) + X^-(aq) \qquad \text{where} \qquad K_a = \frac{[H^+][X^-]}{[HX]}$$

from a thermodynamic point of view. Recall that acid strength is reflected by the magnitude of K_a—a small K_a value means a weak acid. Also recall that the value of an equilibrium constant is related to the standard free energy change for the reaction,

$$\Delta G° = -RT \ln(K)$$

As $\Delta G°$ becomes more negative, K becomes larger; a *decrease* in free energy favors a given reaction. As we saw in Chapter 16, free energy depends on enthalpy, entropy, and temperature. For a process at constant temperature,

$$\Delta G° = \Delta H° - T\Delta S°$$

Thus, to explain the various acid strengths of the hydrogen halides, we must focus on the factors that determine $\Delta H°$ and $\Delta S°$ for the acid dissociation reaction.

What energy terms are important in determining $\Delta H°$ for the dissociation of HX in water? (Keep in mind that large, positive contributions to the value of $\Delta H°$ will tend to make $\Delta G°$ more highly positive, K_a smaller, and the acid weaker.) One important factor is certainly the H—X bond strength. Note from Table 19-19 that the H—F bond is much stronger than the other H—X bonds. This factor tends to make HF a weaker acid than the others.

Another important contribution to $\Delta H°$ is the enthalpy of hydration (see Section 10-2) of X^- (Table 19-20). As we would expect, the smallest of the halide ions, F^-, has the most negative value—its hydration is the most exothermic. This term favors the dissociation of HF into its ions more so than it does for the other HX molecules.

So far we have two conflicting factors: The large HF bond energy tends to make HF a weaker acid than the other hydrogen halides, but the enthalpy of hydration favors the dissociation of HF more than that of the others. When we compare data for HF and HCl, the difference in bond energy (138 kJ/mol) is slightly smaller than the difference in the enthalpies of hydration for the anions (144 kJ/mol). If these were the *only* important factors, HF should be a stronger acid than HCl because the large enthalpy of hydration of F^- more than compensates for the large HF bond strength.

As it turns out, the *deciding factor appears to be entropy.* Note from Table 19-20 that the entropy of hydration for F^- is much more negative than the entropy of hydration for the other halides because of the high degree of ordering that occurs as the water molecules associate with the small F^- ion. ◀ ⓘ Remember that a negative change in entropy is unfavorable. Thus, although the enthalpy of hydration favors dissociation of HF, the *entropy* of hydration strongly opposes it. ◀ ⓘ

When all these factors are taken into account, $\Delta G°$ for the dissociation of HF in water is positive; that is, K_a is small. In contrast, $\Delta G°$ for dissociation of the other HX molecules in water is negative (K_a is large). This example illustrates the complexity of the processes that occur in aqueous solutions and the importance of entropy effects in that medium.

In practical terms, **hydrochloric acid** is the most important of the **hydrohalic acids,** the aqueous solutions of the hydrogen halides. About 3 million tons of hydrochloric acid are produced annually for use in cleaning steel before galvanizing and in the manufacture of many other chemicals.

Hydrofluoric acid is used to etch glass by reacting with the silica in glass to form the volatile gas SiF_4:

$$SiO_2(s) + 4HF(aq) \longrightarrow SiF_4(g) + 2H_2O(l)$$

Table 19-20 | The Enthalpies and Entropies of Hydration for the Halide Ions

$X^-(g) \xrightarrow{H_2O} X^-(aq)$		
X^-	$\Delta H°$ (kJ/mol)	$\Delta S°$ (J/K mol)
F^-	−510	−159
Cl^-	−366	−96
Br^-	−334	−81
I^-	−291	−64

ⓘ Hydration becomes more exothermic as the charge density of an ion increases. Thus, for ions of a given charge, the smallest is most strongly hydrated.

ⓘ When H_2O molecules cluster around an ion, an ordering effect occurs; thus $\Delta S°_{hyd}$ is negative.

Table 19-21 | The Known Oxyacids of the Halogens

Oxidation State of Halogen	Fluorine	Chlorine	Bromine	Iodine*	General Name of Acids	General Name of Salts
+1	HOF†	HOCl	HOBr	HOI	Hypohalous acid	Hypohalites, MOX
+3	‡	HOClO	‡	‡	Halous acid	Halites, MXO_2
+5	‡	$HOClO_2$	$HOBrO_2$	$HOIO_2$	Halic acid	Halates, MXO_3
+7	‡	$HOClO_3$	$HOBrO_3$	$HOIO_3$	Perhalic acid	Perhalates, MXO_4

*Iodine also forms $H_4I_2O_9$ (mesodiperiodic acid) and H_5IO_6 (paraperiodic acid).
†HOF oxidation state is best represented as -1.
‡Compound is unknown.

Oxyacids and Oxyanions

All the halogens except fluorine combine with various numbers of oxygen atoms to form a series of oxyacids, as shown in Table 19-21. The strengths of these acids vary in direct proportion to the number of oxygen atoms attached to the halogen, with the acid strength increasing as more oxygens are added.

The only member of the chlorine series that has been obtained in the pure state is *perchloric acid* ($HOClO_3$), a strong acid and a powerful oxidizing agent. Because perchloric acid reacts explosively with many organic materials, it must be handled with great caution. The other oxyacids of chlorine are known only in solution, although salts containing their anions are well known (Fig. 19-26).

Hypochlorous acid (HOCl) is formed when chlorine gas is dissolved in cold water:

$$Cl_2(aq) + H_2O(l) \rightleftharpoons HOCl(aq) + H^+(aq) + Cl^-(aq)$$

Note that in this reaction chlorine is both oxidized (from 0 in Cl_2 to $+1$ in HOCl) and reduced (from 0 in Cl_2 to -1 in Cl^-). Such a reaction, *in which a given element is both oxidized and reduced,* is called a **disproportionation reaction**. Hypochlorous acid and its salts are strong oxidizing agents; their solutions are widely used as household bleaches and disinfectants.

Chlorate salts, such as $KClO_3$, are also strong oxidizing agents and are used as weed killers and as oxidizers in fireworks (see Chapter 2) and explosives.

Fluorine forms only one oxyacid, hypofluorous acid (HOF), but it forms at least two oxides. When fluorine gas is bubbled into a dilute solution of sodium hydroxide, the compound *oxygen difluoride* (OF_2) is formed: ◄ ⓘ

$$4F_2(g) + 3H_2O(l) \longrightarrow 6HF(aq) + OF_2(g) + O_2(g)$$

ⓘ The name for OF_2 is oxygen difluoride rather than difluorine oxide because fluorine has a higher electronegativity than oxygen and thus is named as the anion.

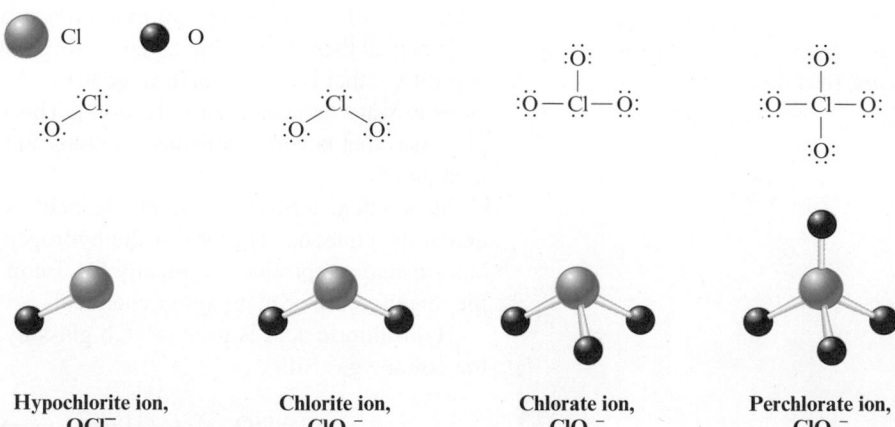

Figure 19-26 | The structures of the oxychloro anions.

Hypochlorite ion, OCl^-

Chlorite ion, ClO_2^-

Chlorate ion, ClO_3^-

Perchlorate ion, ClO_4^-

Oxygen difluoride is a pale yellow gas (bp $= -145°C$) that is a strong oxidizing agent. The oxide *dioxygen difluoride* (O_2F_2) is an orange solid that can be prepared by an electric discharge in an equimolar mixture of fluorine and oxygen gases:

$$F_2(g) + O_2(g) \xrightarrow[\text{discharge}]{\text{Electric}} O_2F_2(s)$$

19-14 | The Group 8A (18) Elements

18
8A
He
Ne
Ar
Kr
Xe
Rn

The Group 8A (18) elements, the **noble gases**, are characterized by filled s and p valence orbitals (electron configurations of $2s^2$ for helium and ns^2np^6 for the others). ◄ Because of their completed valence shells, these elements are very unreactive. In fact, no noble gas compounds were known prior to 1962. Selected properties of the Group 8A (18) elements are summarized in Table 19-22.

Helium was identified by its characteristic emission spectrum as a component of the sun before it was found on the earth. The major sources of helium on the earth are natural gas deposits, where helium was formed from the α-particle decay of radioactive elements. The α particle is a helium nucleus that can easily pick up electrons from the environment to form a helium atom. Although helium forms no compounds, it is an important substance that is used as a coolant, as a pressurizing gas for rocket fuels, as a diluent in the gases used for deep-sea diving and spaceship atmospheres, and as the gas in lighter-than-air airships (blimps).

Like helium, *neon* forms no compounds, but it is a very useful element. ◄ For example, neon is widely used in luminescent lighting (neon signs). *Argon,* which recently has been shown to form chemical bonds under special circumstances, is used to provide the noncorrosive atmosphere in incandescent light bulbs, which prolongs the life of the tungsten filament.

Krypton and *xenon* have been observed to form many stable chemical compounds. The first of these was prepared in 1962 by Neil Bartlett (1932-2008), an English chemist who made an ionic compound that he thought had the formula $XePtF_6$. Subsequent studies indicated that the compound might be better represented as $XeFPtF_6$ and contains the XeF^+ and PtF_6^- ions.

Less than a year after Bartlett's report, a group at Argonne National Laboratory near Chicago prepared xenon tetrafluoride by reacting xenon and fluorine gases in a nickel reaction vessel at $400°C$ and 6 atm:

$$Xe(g) + 2F_2(g) \longrightarrow XeF_4(s)$$

Xenon tetrafluoride forms stable colorless crystals. Two other xenon fluorides, XeF_2 and XeF_6, were synthesized by the group at Argonne, and a highly explosive xenon oxide (XeO_3) was also found. The xenon fluorides react with water to form hydrogen fluoride and oxycompounds. For example:

$$XeF_6(s) + 3H_2O(l) \longrightarrow XeO_3(aq) + 6HF(aq)$$
$$XeF_6(s) + H_2O(l) \longrightarrow XeOF_4(aq) + 2HF(aq)$$

▲ Neon artist Chris Bracey puts the finishing touches to the artwork "Heart of the World"

Matthew Lloyd/Getty Images

Table 19-22 | Selected Properties of Group 8A (18) Elements

Element	Melting Point (°C)	Boiling Point (°C)	Atmospheric Abundance (% by volume)	Examples of Compounds
Helium	-270	-269	5×10^{-4}	None
Neon	-249	-246	1×10^{-3}	None
Argon	-189	-186	9×10^{-1}	HArF
Krypton	-157	-153	1×10^{-4}	KrF_2
Xenon	-112	-107	9×10^{-6}	XeF_4, XeO_3, XeF_6

Figure 19-27 | The structures of several known xenon compounds.

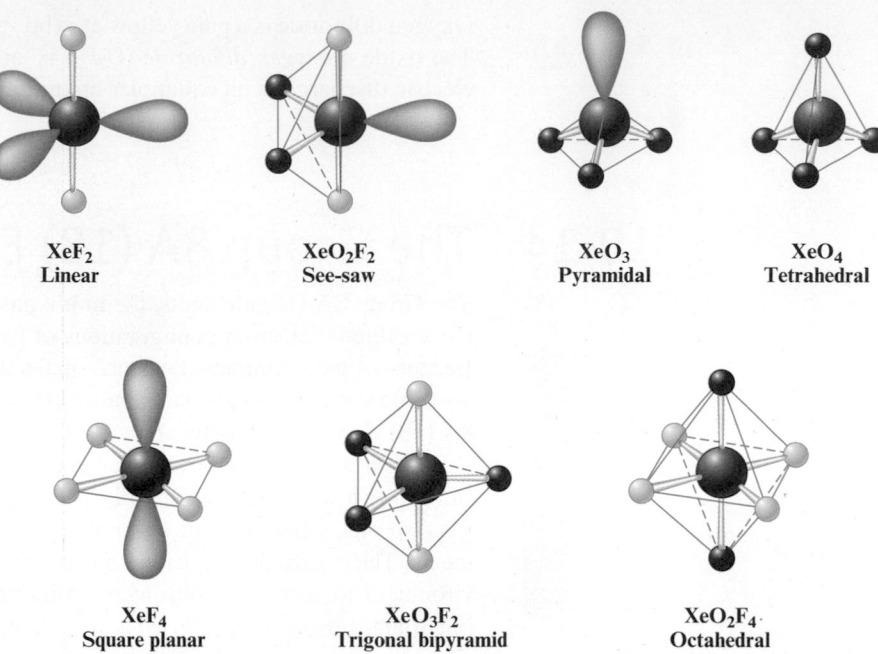

XeF$_2$
Linear

XeO$_2$F$_2$
See-saw

XeO$_3$
Pyramidal

XeO$_4$
Tetrahedral

XeF$_4$
Square planar

XeO$_3$F$_2$
Trigonal bipyramid

XeO$_2$F$_4$
Octahedral

In the past 35 years, other xenon compounds have been prepared. Examples are XeO$_4$ (explosive), XeOF$_4$, XeOF$_2$, and XeO$_3$F$_2$. These compounds contain discrete molecules with covalent bonds between the xenon atom and the other atoms. A few compounds of krypton, such as KrF$_2$ and KrF$_4$, have also been observed. The structures of several known xenon compounds are shown in Fig. 19-27. Radon also has been shown to form compounds similar to those of xenon and krypton.

FOR REVIEW

Key Terms

Section 19-1
representative elements
transition metals
lanthanides
actinides
metalloids (semimetals)
metallurgy
liquefaction

Section 19-2
alkali metals

Section 19-3
hydride
ionic (saltlike) hydride
covalent hydride
metallic (interstitial) hydride

Representative Elements

- Chemical properties are determined by their *s* and *p* valence-electron configurations
- Metallic character increases going down the group
- The properties of the first element in a group usually differ most from the properties of the other elements in the group due to a significant difference in size
  - In Group 1A (1), hydrogen is a nonmetal and the other members of the group are active metals
  - The first member of a group forms the strongest π bonds, causing nitrogen and oxygen to exist as N$_2$ and O$_2$ molecules

Elemental Abundances on the Earth

- Oxygen is the most abundant element, followed by silicon
- The most abundant metals are aluminum and iron, which are found as ores

Key Terms

Section 19-4
alkaline earth metals
hard water
ion exchange
ion-exchange resin

Section 19-5
boranes

Section 19-8
Haber process
nitrogen fixation
nitrogen-fixing bacteria
denitrification
nitrogen cycle
ammonia
hydrazine
nitric acid
Ostwald process

Section 19-9
phosphoric (orthophosphoric) acid
phosphorous acid
superphosphate of lime

Section 19-11
ozone
ozonolysis

Section 19-12
Frasch process
sulfuric acid

Section 19-13
halogens
hydrochloric acid
hydrohalic acids
disproportionation reaction

Section 19-14
noble gases

Group 1A (1) Elements (Alkali Metals)

- Have valence configuration ns^1
- Except for hydrogen, readily lose one electron to form M^+ ions in their compounds with nonmetals
- React vigorously with water to form M^+ and OH^- ions and hydrogen gas
- Form a series of oxides of the types M_2O (oxide), M_2O_2 (peroxide), and MO_2 (superoxide)
    - Not all metals form all types of oxide compounds
- Hydrogen forms covalent compounds with nonmetals
- With very active metals, hydrogen forms hydrides that contain the H^- ion

Group 2A (2) Elements (Alkaline Earth Metals)

- Have valence configuration ns^2
- React less violently with water than alkali metals
- The heavier alkaline earth metals form nitrides and hydrides
- Hard water contains Ca^{2+} and Mg^{2+} ions
    - Form precipitates with soap
    - Usually removed by ion-exchange resins that replace the Ca^{2+} and Mg^{2+} ions with Na^+

Group 3A (13) Elements

- Have valence configuration ns^2np^1
- Show increasing metallic character going down the group
- Boron is a nonmetal that forms many types of covalent compounds, including boranes, which are highly electron-deficient and thus are very reactive
- The metals aluminum, gallium, and indium show some covalent tendencies

Group 4A (14) Elements

- Have valence configuration ns^2np^2
- Lighter members are nonmetals; heavier members are metals
    - All group members can form covalent bonds to nonmetals
- Carbon forms a huge variety of compounds, most of which are classified as organic compounds

Group 5A (15) Elements

- Elements show a wide variety of chemical properties
    - Nitrogen and phosphorus are nonmetals
    - Antimony and bismuth tend to be metallic, although no ionic compounds containing Sb^{5+} and Bi^{5+} are known; the compounds containing Sb(V) and Bi(V) are molecular rather than ionic
    - All group members except N form molecules with five covalent bonds
    - The ability to form π bonds decreases dramatically after N
- Chemistry of nitrogen
    - Most nitrogen-containing compounds decompose exothermically, forming the very stable N_2 molecule, which explains the power of nitrogen-based explosives

- The nitrogen cycle, which consists of a series of steps, shows how nitrogen is cycled in the natural environment
- Nitrogen fixation changes the N_2 in air into compounds useful to plants
  - The Haber process is a synthetic method of nitrogen fixation
  - In the natural world, nitrogen fixation occurs through nitrogen-fixing bacteria in the root nodules of certain plants and through lightning in the atmosphere
- Ammonia is the most important hydride of nitrogen
  - Contains pyramidal NH_3 molecules
  - Widely used as a fertilizer
- Hydrazine (N_2H_4) is a powerful reducing agent
- Nitrogen forms a series of oxides including N_2O, NO, NO_2, and N_2O_5
- Nitric acid (HNO_3) is a very important strong acid manufactured by the Ostwald process
- Chemistry of phosphorus
  - Exists in three elemental forms: white (contains P_4 molecules), red, and black
  - Phosphine (PH_3) has bond angles close to 90 degrees
  - Phosphorus forms oxides including P_4O_6 and P_4O_{10} (which dissolves in water to form phosphoric acid, H_3PO_4)

Group 6A (16) Elements

- Metallic character increases going down the group but no element behaves as a typical metal
- The lighter members tend to gain two electrons to form X^{2-} ions in compounds with metals
- Chemistry of oxygen
  - Elemental forms are O_2 and O_3
  - Oxygen forms a wide variety of oxides
  - O_2 and especially O_3 are powerful oxidizing agents
- Chemistry of sulfur
  - The elemental forms are called rhombic and monoclinic sulfur, both of which contain S_8 molecules
  - The most important oxides are SO_2 (which forms H_2SO_3 in water) and SO_3 (which forms H_2SO_4 in water)
  - Sulfur forms a wide variety of compounds in which it shows the oxidation states $+6$, $+4$, $+2$, 0, and -2

Group 7A (17) Elements (Halogens)

- All nonmetals
- Form hydrides of the type HX that behave as strong acids in water except for HF, which is a weak acid
- The oxyacids of the halogens become stronger as more oxygen atoms are present
- The interhalogens contain two or more different halogens

Group 8A (18) Elements (Noble Gases)

- All elements are monatomic gases and are generally very unreactive
- The heavier elements form compounds with electronegative elements such as fluorine and oxygen

1. What are the two most abundant elements by mass in the earth's crust, oceans, and atmosphere? Does this make sense? Why? What are the four most abundant elements by mass in the human body? Does this make sense? Why?

2. What evidence supports putting hydrogen in Group 1A (1) of the periodic table? In some periodic tables hydrogen is listed separately from any of the groups. In what ways is hydrogen unlike a typical Group 1A (1) element? What is the valence electron configuration for the alkali metals? List some common properties of alkali metals. How are the pure metals prepared? Predict the formulas of the compounds formed when an alkali metal reacts with F_2, S, P_4, H_2, and H_2O.

3. What is the valence electron configuration for alkaline earth metals? List some common properties of alkaline earth metals. How are alkaline earth metals prepared? Predict the formulas of the compounds formed when an alkaline earth metal reacts with F_2, O_2, S, N_2, H_2, and H_2O.

4. What is the valence electron configuration for the Group 3A (13) elements? How does metallic character change as one goes down this group? How are boron and aluminum different? Predict the formulas of the compounds formed when aluminum reacts with F_2, O_2, S, and N_2.

5. What is the valence electron configuration for Group 4A (14) elements? Group 4A (14) contains two of the most important elements on the earth. What are they, and why are they so important? How does metallic character change as one goes down Group 4A (14)? What are the three allotropic forms of carbon? List some properties of germanium, tin, and lead. Predict the formulas of the compounds formed when Ge reacts with F_2 and O_2.

6. What is the valence electron configuration for Group 5A (15) elements? Metallic character increases when going down a group. Give some examples illustrating how Bi and Sb have metallic characteristics not associated with N, P, and As. Elemental nitrogen exists as N_2, whereas in the gas phase the elements phosphorus, arsenic, and antimony consist of P_4, As_4, and Sb_4 molecules, respectively. Give a possible reason for this difference between N_2 and the other Group 5A (15) elements. White phosphorus is much more reactive than black or red phosphorus. Explain.

7. Table 19-14 lists some common nitrogen compounds having oxidation states ranging from -3 to $+5$. Rationalize this spread in oxidation states. For each substance listed in Table 19-14, list some of its special properties. Ammonia forms hydrogen-bonding intermolecular forces resulting in an unusually high boiling point for a substance with the small size of NH_3. Can hydrazine (N_2H_4) also form hydrogen-bonding interactions? How is phosphine's (PH_3) structure different from that of ammonia?

8. What is the valence electron configuration of Group 6A (16) elements? What are some property differences between oxygen and polonium? What are the Lewis structures for the two allotropic forms of oxygen? What is the molecular structure and the bond angle in ozone? The most stable form of solid sulfur is the rhombic form; however, a solid form called monoclinic sulfur can also form. What is the difference between rhombic and monoclinic sulfur? Explain why O_2 is much more stable than S_2 or SO. When $SO_2(g)$ or $SO_3(g)$ reacts with water, an acidic solution forms. Explain. What are the molecular structures and bond angles in SO_2 and SO_3? H_2SO_4 is a powerful dehydrating agent: What does this mean?

9. What is the valence electron configuration of the halogens? Why do the boiling points and melting points of the halogens increase steadily from F_2 to I_2? Give two reasons why F_2 is the most reactive of the halogens. Explain why the boiling point of HF is much higher than the boiling points of HCl, HBr, and HI. In nature, the halogens are generally found as halide ions in various minerals and seawater. What is a halide ion, and why are halide salts so stable? The oxidation states of the halogens vary from -1 to $+7$. Identify compounds of chlorine that have -1, $+1$, $+3$, $+5$, and $+7$ oxidation states.

10. What special property of the noble gases makes them unreactive? The boiling points and melting points of the noble gases increase steadily from He to Xe. Explain. The noble gases were among the last elements discovered; their existence was not predicted by Mendeleev when he published his first periodic table. Explain. In chemistry textbooks written before 1962, the noble gases were referred to as the inert gases. Why do we no longer use this term? For the structures of the xenon compounds in Fig. 19-27, give the bond angles exhibited and the hybridization of the central atom in each compound.

A blue question or exercise number indicates that the answer to that question or exercise appears at the back of this book and a solution appears in the *Solutions Guide*, as found on PowerLecture.

Questions

1. Although the earth was formed from the same interstellar material as the sun, there is little elemental hydrogen (H_2) in the earth's atmosphere. Explain.

2. List two major industrial uses of hydrogen.

3. How do the acidities of the aqueous solutions of the alkaline earth metal ions (M^{2+}) change in going down the group?

4. Diagonal relationships in the periodic table exist as well as the vertical relationships. For example, Be and Al are similar in some of their properties as are B and Si. Rationalize why these diagonal relationships hold for properties such as size, ionization energy, and electron affinity.

5. Atomic size seems to play an important role in explaining some of the differences between the first element in a group and the subsequent group elements. Explain.

6. Silicon carbide (SiC) is an extremely hard substance. Propose a structure for SiC.

7. In most compounds, the solid phase is denser than the liquid phase. Why isn't this true for water?

8. What is nitrogen fixation? Give some examples of nitrogen fixation.

9. All the Group 1A (1) and 2A (2) metals are produced by electrolysis of molten salts. Why?

10. Why are the tin(IV) halides more volatile than the tin(II) halides?

Exercises

In this section similar exercises are paired.

Group 1A (1) Elements

11. Hydrogen is produced commercially by the reaction of methane with steam:

$$CH_4(g) + H_2O(g) \rightleftharpoons CO(g) + 3H_2(g)$$

   a. Calculate $\Delta H°$ and $\Delta S°$ for this reaction (use the data in Appendix 4).

   b. What temperatures will favor product formation at standard conditions? Assume $\Delta H°$ and $\Delta S°$ do not depend on temperature.

12. The major industrial use of hydrogen is in the production of ammonia by the Haber process:

$$3H_2(g) + N_2(g) \longrightarrow 2NH_3(g)$$

   a. Using data from Appendix 4, calculate $\Delta H°$, $\Delta S°$, and $\Delta G°$ for the Haber process reaction.

   b. Is the reaction spontaneous at standard conditions?

   c. At what temperatures is the reaction spontaneous at standard conditions? Assume $\Delta H°$ and $\Delta S°$ do not depend on temperature.

13. Write balanced equations describing the reaction of lithium metal with each of the following: O_2, S, Cl_2, P_4, H_2, H_2O, and HCl.

14. The electrolysis of aqueous sodium chloride (brine) is an important industrial process for the production of chlorine and sodium hydroxide. In fact, this process is the second largest consumer of electricity in the United States, after the production of aluminum. Write a balanced equation for the electrolysis of aqueous sodium chloride (hydrogen gas is also produced).

15. Refer to Table 19-5 and give examples of the three types of alkali metal oxides that form. How do they differ?

16. Label the following hydrides as ionic, covalent, or interstitial and support your answer. *Note:* The light blue atoms are hydrogen atoms.

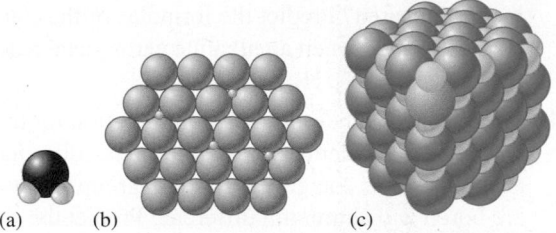

(a)   (b)   (c)

17. Many lithium salts are hygroscopic (absorb water), but the corresponding salts of the other alkali metals are not. Why are lithium salts different from the others?

18. What will be the atomic number of the next alkali metal to be discovered? How would you expect the physical properties of the next alkali metal to compare with the properties of the other alkali metals summarized in Table 19-4?

Group 2A (2) Elements

19. One harmful effect of acid rain is the deterioration of structures and statues made of marble or limestone, both of which are essentially calcium carbonate. The reaction of calcium carbonate with sulfuric acid yields carbon dioxide, water, and calcium sulfate. Because calcium sulfate is marginally soluble in water, part of the object is washed away by the rain. Write a balanced chemical equation for the reaction of sulfuric acid with calcium carbonate.

20. Write balanced equations describing the reaction of Sr with each of the following: O_2, S, Cl_2, P_4, H_2, H_2O, and HCl.

21. What mass of barium is produced when molten $BaCl_2$ is electrolyzed by a current of 2.50×10^5 A for 6.00 h?

22. How long will it take to produce 1.00×10^3 kg of magnesium metal by the electrolysis of molten magnesium chloride using a current of 5.00×10^4 A?

23. Beryllium shows some covalent characteristics in some of its compounds, unlike the other alkaline earth compounds. Give a possible explanation for this phenomenon.

24. What ions are found in hard water? What happens when water is "softened"?

Group 3A (13) Elements

25. Consider element 113. What is the expected electron configuration for element 113? What oxidation states would be exhibited by element 113 in its compounds?

26. Thallium and indium form +1 and +3 oxidation states when in compounds. Predict the formulas of the possible compounds between thallium and oxygen and between indium and chlorine. Name the compounds.

27. Boron hydrides were once evaluated for possible use as rocket fuels. Complete and balance the following equation for the combustion of diborane.

$$B_2H_6(g) + O_2(g) \longrightarrow B(OH)_3(s)$$

28. Elemental boron is produced by reduction of boron oxide with magnesium to give boron and magnesium oxide. Write a balanced equation for this reaction.

29. Write equations describing the reactions of Ga with each of the following: F_2, O_2, S, and HCl.

30. Write a balanced equation describing the reaction of aluminum metal with concentrated aqueous sodium hydroxide.

31. Al_2O_3 is amphoteric. What does this mean?

32. What are three-centered bonds?

Group 4A (14) Elements

33. Discuss the importance of the C—C and Si—Si bond strengths and of π bonding to the properties of carbon and silicon.

34. Besides the central atom, what are the differences between CO_2 and SiO_2?

35. The following illustration shows the orbitals used to form the bonds in carbon dioxide.

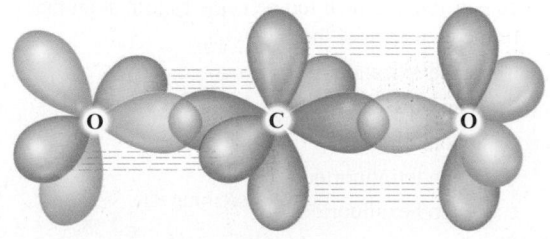

Each color represents a different orbital. Label each orbital, draw the Lewis structure for carbon dioxide, and explain how the localized electron model describes the bonding in CO_2.

36. In addition to CO_2, two additional stable oxides of carbon form. The space-filling models for CO_2 and the other two stable oxides are:

What are the formulas for the two additional stable oxides of carbon? Explain the bonding in each of these two forms using the localized electron model.

37. Silicon is produced for the chemical and electronics industries by the following reactions. Give the balanced equation for each reaction.

a. $SiO_2(s) + C(s) \longrightarrow Si(s) + CO(g)$

b. Silicon tetrachloride is reacted with very pure magnesium, producing silicon and magnesium chloride.

c. $Na_2SiF_6(s) + Na(s) \longrightarrow Si(s) + NaF(s)$

38. Write equations describing the reactions of Sn with each of the following: Cl_2, O_2, and HCl.

39. The compound Pb_3O_4 (red lead) contains a mixture of lead(II) and lead(IV) oxidation states. What is the mole ratio of lead(II) to lead(IV) in Pb_3O_4?

40. Tin forms compounds in the +2 and +4 oxidation states. Therefore, when tin reacts with fluorine, two products are possible. Write balanced equations for the production of the two tin halide compounds and name them.

Group 5A (15) Elements

41. The oxyanion of nitrogen in which it has the highest oxidation state is the nitrate ion (NO_3^-). The corresponding oxyanion of phosphorus is PO_4^{3-}. The NO_4^{3-} ion is known but not very stable. The PO_3^- ion is not known. Account for these differences in terms of the bonding in the four anions.

42. In each of the following pairs of substances, one is stable and known, and the other is unstable. For each pair, choose the stable substance, and explain why the other is unstable.

a. NF_5 or PF_5

b. AsF_5 or AsI_5

c. NF_3 or NBr_3

43. Write balanced equations for the reactions described in Table 19-13 for the production of Bi and Sb.

44. Arsenic reacts with oxygen to form oxides that react with water in a manner analogous to that of the phosphorus oxides. Write balanced chemical equations describing the reaction of arsenic with oxygen and the reaction of the resulting oxide with water.

45. The Group 5A (15) elements can form molecules or ions that involve three, five, or six covalent bonds; NH_3, $AsCl_5$, and PF_6^- are examples. Draw the Lewis structure for each of these substances, and predict the molecular structure and hybridization for each. Why doesn't NF_5 or NCl_6^- form?

46. Compare the Lewis structures with the molecular orbital view of the bonding in NO, NO^+, and NO^-. Account for any discrepancies between the two models.

47. Many oxides of nitrogen have positive values for the standard free energy of formation. Using NO as an example, explain why this is the case.

48. Using data from Appendix 4 calculate ΔH°, ΔS°, and ΔG° for the reaction

$$N_2(g) + O_2(g) \longrightarrow 2NO(g)$$

Why does NO form in an automobile engine but then does not readily decompose back to N_2 and O_2 in the atmosphere?

49. In many natural waters, nitrogen and phosphorus are the least abundant nutrients available for plant life. Some waters that become polluted from agricultural runoff or municipal sewage become infested with algae. The algae flourish, and fish life dies off as a result. Describe how these events are chemically related.

50. Phosphate buffers are important in regulating the pH of intracellular fluids. If the concentration ratio of $H_2PO_4^-/HPO_4^{2-}$ in a sample of intracellular fluid is 1.1:1, what is the pH of this sample of intracellular fluid?

$$H_2PO_4^-(aq) \rightleftharpoons HPO_4^{2-}(aq) + H^+(aq) \quad K_a = 6.2 \times 10^{-8}$$

51. Phosphoric acid (H_3PO_4) is a triprotic acid, phosphorous acid (H_3PO_3) is a diprotic acid, and hypophosphorous acid (H_3PO_2) is a monoprotic acid. Explain this phenomenon.

52. Trisodium phosphate (TSP) is an effective grease remover. Like many cleaners, TSP acts as a base in water. Write a balanced equation to account for this basic behavior.

Group 6A (16) Elements

53. Use bond energies to estimate the maximum wavelength of light that will cause the reaction

$$O_3 \xrightarrow{h\nu} O_2 + O$$

54. The xerographic (dry writing) process was invented in 1938 by C. Carlson. In xerography, an image is produced on a photoconductor by exposing it to light. Selenium is commonly used, since its conductivity increases three orders of magnitude upon exposure to light in the range from 400 to 500 nm. What color light should be used to cause selenium to become conductive? (See Figure 2-2.)

55. Write a balanced equation describing the reduction of H_2SeO_4 by SO_2 to produce selenium.

56. Complete and balance each of the following reactions.
    a. the reaction between sulfur dioxide gas and oxygen gas
    b. the reaction between sulfur trioxide gas and water
    c. the reaction between concentrated sulfuric acid and sucrose ($C_{12}H_{22}O_{11}$)

57. Ozone is desirable in the upper atmosphere but undesirable in the lower atmosphere. A dictionary states that ozone has the scent of a spring thunderstorm. How can these seemingly conflicting statements be reconciled in terms of the chemical properties of ozone?

58. Ozone is a possible replacement for chlorine in municipal water purification. Unlike chlorine, virtually no ozone remains after treatment. This has good and bad consequences. Explain.

59. How can the paramagnetism of O_2 be explained using the molecular orbital model?

60. Describe the bonding in SO_2 and SO_3 using the localized electron model (hybrid orbital theory). How would the molecular orbital model describe the π bonding in these two compounds?

Group 7A (17) Elements

61. Write the Lewis structure for O_2F_2. Predict the bond angles and hybridization of the two central oxygen atoms. Assign oxidation states and formal charges to the atoms in O_2F_2. The compound O_2F_2 is a vigorous and potent oxidizing and fluorinating agent. Are oxidation states or formal charges more useful in accounting for these properties of O_2F_2?

62. Give the Lewis structure, molecular structure, and hybridization of the oxygen atom for OF_2. Would you expect OF_2 to be a strong oxidizing agent like O_2F_2 discussed in Exercise 61?

63. Fluorine reacts with sulfur to form several different covalent compounds. Three of these compounds are SF_2, SF_4, and SF_6. Draw the Lewis structures for these compounds, and predict the molecular structures (including bond angles). Would you expect OF_4 to be a stable compound?

64. Predict some possible compounds that could form between chlorine and selenium. (*Hint:* See Exercise 63.)

65. How does the oxyacid strength of the halogens vary as the number of oxygens in the formula increases?

66. Explain why HF is a weak acid, whereas HCl, HBr, and HI are all strong acids.

Group 8A (18) Elements

67. The xenon halides and oxides are isoelectronic with many other compounds and ions containing halogens. Give a molecule or ion in which iodine is the central atom that is isoelectronic with each of the following.
    a. xenon tetroxide
    b. xenon trioxide
    c. xenon difluoride
    d. xenon tetrafluoride
    e. xenon hexafluoride

68. For each of the following, write the Lewis structure(s), predict the molecular structure (including bond angles), and give the expected hybridization of the central atom.
    a. KrF_2 c. XeO_2F_2
    b. KrF_4 d. XeO_2F_4

69. Although He is the second most abundant element in the universe, it is very rare on the earth. Why?

70. The noble gas with the largest atmospheric abundance is argon. Using the data in Table 19-22, calculate the mass of argon at 25°C and 1.0 atm in a room 10.0 m × 10.0 m × 10.0 m. How many Ar atoms are in this room? How many Ar atoms do you inhale in one breath (approximately 2 L) of air at 25°C and 1.0 atm? Argon gas is inert, so it poses no serious health risks. However, if significant amounts of radon are inhaled into the lungs, lung cancer is a possible result. Explain the health risk differences between argon gas and radon gas.

71. There is evidence that radon reacts with fluorine to form compounds similar to those formed by xenon and fluorine. Predict the formulas of these RnF_x compounds. Why is the chemistry of radon difficult to study?

72. For the RnF_x compounds you predicted in Exercise 71, give the molecular structure (including bond angles).

Additional Exercises

73. Hydrazine (N_2H_4) is used as a fuel in liquid-fueled rockets. When hydrazine reacts with oxygen gas, nitrogen gas and water vapor are produced. Write a balanced equation and use bond energies from Table 3-3 to estimate ΔH for this reaction.

74. The inert-pair effect is sometimes used to explain the tendency of heavier members of Group 3A (13) to exhibit +1 and +3 oxidation states. What does the inert-pair effect reference? [*Hint:* Consider the valence electron configuration for Group 3A (13) elements.]

75. How could you determine experimentally whether the compound Ga_2Cl_4 contains two gallium(II) ions or one gallium(I) and one gallium(III) ion? (*Hint:* Consider the electron configurations of the three possible ions.)

76. The resistivity (a measure of electrical resistance) of graphite is $(0.4 \text{ to } 5.0) \times 10^{-4}$ ohm · cm in the basal plane. (The basal plane is the plane of the six-membered rings of carbon atoms.) The resistivity is 0.2 to 1.0 ohm · cm along the axis perpendicular to the plane. The resistivity of diamond is 10^{14} to 10^{16} ohm · cm and is independent of direction. How can you account for this behavior in terms of the structures of graphite and diamond?

77. Slaked lime, $Ca(OH)_2$, is used to soften hard water by removing calcium ions from hard water through the reaction

$$Ca(OH)_2(aq) + Ca^{2+}(aq) + 2HCO_3^-(aq) \rightarrow$$
$$2CaCO_3(s) + 2H_2O(l)$$

Although $CaCO_3(s)$ is considered insoluble, some of it does dissolve in aqueous solutions. Calculate the molar solubility of $CaCO_3$ in water ($K_{sp} = 8.7 \times 10^{-9}$).

78. EDTA is used as a complexing agent in chemical analysis. Solutions of EDTA, usually containing the disodium salt Na_2H_2EDTA, are also used to treat heavy metal poisoning. The equilibrium constant for the following reaction is 6.7×10^{21}:

$$Pb^{2+}(aq) = + H_2EDTA^{2-}(aq) \rightleftharpoons PbEDTA^{2-}(aq) + 2H^+(aq)$$

$$EDTA^{4-} = \begin{array}{c} ^-O_2C-CH_2 \\ \\ ^-O_2C-CH_2 \end{array} N-CH_2-CH_2-N \begin{array}{c} CH_2-CO_2^- \\ \\ CH_2-CO_2^- \end{array}$$

Ethylenediaminetetraacetate

Calculate $[Pb^{2+}]$ at equilibrium in a solution originally 0.0050 *M* in Pb^{2+}, 0.075 *M* in H_2EDTA^{2-}, and buffered at pH = 7.00.

79. Photogray lenses contain small embedded crystals of solid silver chloride. Silver chloride is light-sensitive because of the reaction

$$AgCl(s) \xrightarrow{h\nu} Ag(s) + Cl$$

Small particles of metallic silver cause the lenses to darken. In the lenses this process is reversible. When the light is removed, the reverse reaction occurs. However, when pure white silver chloride is exposed to sunlight it darkens; the reverse reaction does not occur in the dark.

a. How do you explain this difference?

b. Photogray lenses do become permanently dark in time. How do you account for this?

80. Draw Lewis structures for the $AsCl_4^+$ and $AsCl_6^-$ ions. What type of reaction (acid–base, oxidation–reduction, or the like) is the following?

$$2AsCl_5(g) \longrightarrow AsCl_4AsCl_6(s)$$

81. Provide a reasonable estimate for the number of atoms in a 150-lb adult human. Use the information given in Table 19-2.

82. In large doses, selenium is toxic. However, in moderate intake, selenium is a physiologically important element. How is selenium physiologically important?

83. In the 1950s and 1960s, several nations conducted tests of nuclear warheads in the atmosphere. It was customary, following each test, to monitor the concentration of strontium-90 (a radioactive isotope of strontium) in milk. Why would strontium-90 tend to accumulate in milk?

84. What is a disproportionation reaction? Use the following reduction potentials

$$ClO_3^- + 3H^+ + 2e^- \longrightarrow HClO_2 + H_2O \qquad \mathscr{E}° = 1.21 \text{ V}$$
$$HClO_2 + 2H^+ + 2e^- \longrightarrow HClO + H_2O \qquad \mathscr{E}° = 1.65 \text{ V}$$

to predict whether $HClO_2$ will disproportionate.

85. Sulfur forms a wide variety of compounds in which it has +6, +4, +2, 0, and −2 oxidation states. Give examples of sulfur compounds having each of these oxidation states.

86. Halogens form a variety of covalent compounds with each other. For example, chlorine and fluorine form the compounds ClF, ClF_3, and ClF_5. Predict the molecular structure (including bond angles) for each of these three compounds. Would you expect FCl_3 to be a stable compound? Explain.

ChemWork Problems

These multiconcept problems (and additional ones) are found interactively online with the same type of assistance a student would get from an instructor.

87. Hydrogen gas is being considered as a fuel for automobiles. There are many chemical means for producing hydrogen gas from water. One of these reactions is

$$C(s) + H_2O(g) \longrightarrow CO(g) + H_2(g)$$

In this case the form of carbon used is graphite.
   a. Calculate $\Delta H°$ and $\Delta S°$ for this reaction using data from Appendix 4.
   b. At what temperature is $\Delta G° =$ zero for this reaction? Assume $\Delta H°$ and $\Delta S°$ do not depend on temperature.

88. Molten $CaCl_2$ is electrolyzed for 8.00 h to produce $Ca(s)$ and $Cl_2(g)$.
   a. What current is needed to produce 5.52 kg of calcium metal?
   b. If 5.52 kg calcium metal is produced, what mass (in kg) of Cl_2 is produced?

89. Calculate the solubility of $Mg(OH)_2$ ($K_{sp} = 8.9 \times 10^{-12}$) in an aqueous solution buffered at pH = 9.42.

90. Which of the following statement(s) is(are) *true*?
   a. The alkali metals are found in the earth's crust in the form of pure elements.
   b. Gallium has one of the highest melting points known for metals.
   c. When calcium metal reacts with water, one of the products is $H_2(g)$.
   d. When $AlCl_3$ is dissolved in water, it produces an acidic solution.
   e. Lithium reacts in the presence of excess oxygen gas to form lithium superoxide.

91. What is the hybridization of the underlined nitrogen atom in each of the following molecules or ions?
   a. $\underline{N}O^+$
   b. N_2O_3 ($O_2N\underline{N}O$)
   c. $\underline{N}O_2^-$
   d. $\underline{N}_2$

92. Nitrous oxide (N_2O) can be produced by thermal decomposition of ammonium nitrate:

$$NH_4NO_3(s) \xrightarrow{\text{heat}} N_2O(g) + 2H_2O(l)$$

What volume of $N_2O(g)$ collected over water at a total pressure of 94.0 kPa and 22°C can be produced from thermal decomposition of 8.68 g NH_4NO_3? The vapor pressure of water at 22°C is 21 torr.

93. What is the hybridization of the central atom in each of the following molecules?
   a. SF_6
   b. ClF_3
   c. $GeCl_4$
   d. XeF_4

94. What is the molecular structure for each of the following molecules or ions?
   a. OCl_2
   b. ClO_4^-
   c. ICl_5
   d. PF_6^-

95. The atmosphere contains $9.0 \times 10^{-6}\%$ Xe by volume at 1.0 atm and 25°C.
   a. Calculate the mass of Xe in a room 7.26 m by 8.80 m by 5.67 m.
   b. A typical person takes in about 2 L of air during a breath. How many Xe atoms are inhaled in each breath?

96. Which of following statement(s) is/are *true*?
   a. Phosphoric acid is a stronger acid than nitric acid.
   b. The noble gas with the lowest boiling point is helium.
   c. Sulfur is found as the free element in the earth's crust.
   d. One of the atoms in Teflon is fluorine.
   e. The P_4 molecule has a square planar structure.

Challenge Problems

97. Suppose 10.00 g of an alkaline earth metal reacts with 10.0 L water to produce 6.10 L hydrogen gas at 1.00 atm and 25°C. Identify the metal and determine the pH of the solution.

98. From the information on the temperature stability of white and gray tin given in this chapter, which form would you expect to have the more ordered structure (have the smaller positional probability)?

99. Lead forms compounds in the +2 and +4 oxidation states. All lead(II) halides are known (and are known to be ionic). Only PbF_4 and $PbCl_4$ are known among the possible lead(IV) halides. Presumably lead(IV) oxidizes bromide and iodide ions, producing the lead(II) halide and the free halogen:

$$PbX_4 \longrightarrow PbX_2 + X_2$$

Suppose 25.00 g of a lead(IV) halide reacts to form 16.12 g of a lead(II) halide and the free halogen. Identify the halogen.

100. Many structures of phosphorus-containing compounds are drawn with some P=O bonds. These bonds are not the typical π bonds we've considered, which involve the overlap of two p orbitals. Instead, they result from the overlap of a d orbital on the phosphorus atom with a p orbital on oxygen. This type of π bonding is sometimes used as an explanation for why H_3PO_3 has the first structure below rather than the second:

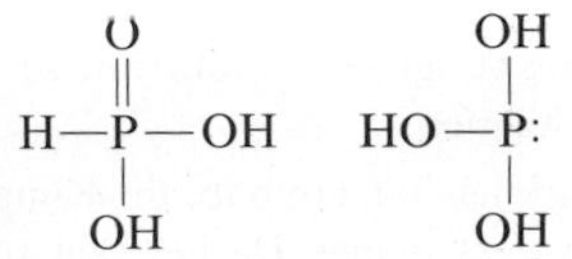

O   OH

‖ |

H—P—OH   HO—P:

| |

OH   OH

Draw a picture showing how a d orbital and a p orbital overlap to form a π bond.

101. Use bond energies (see Table 3-3) to show that the preferred products for the decomposition of N_2O_3 are NO_2 and NO rather than O_2 and N_2O. (The N—O single bond energy is 201 kJ/mol.) (*Hint:* Consider the reaction kinetics.)

102. A proposed two-step mechanism for the destruction of ozone in the upper atmosphere is

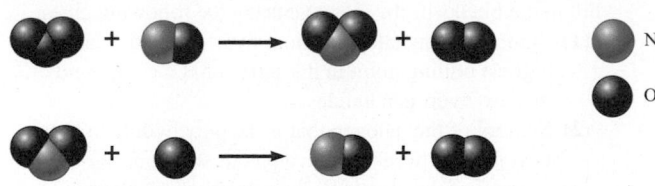

- N
- O

a. What is the overall balanced equation for the ozone destruction reaction?

b. Which species is a catalyst?

c. Which species is an intermediate?

d. What is the rate law derived from this mechanism if the first step in the mechanism is slow and the second step is fast?

e. One of the concerns about the use of Freons is that they will migrate to the upper atmosphere, where chlorine atoms can be generated by the reaction

$$CCl_2F_2 \xrightarrow{h\nu} CF_2Cl + Cl$$
Freon-12

Chlorine atoms also can act as a catalyst for the destruction of ozone. The first step of a proposed mechanism for chlorine-catalyzed ozone destruction is

$$Cl(g) + O_3(g) \longrightarrow ClO(g) + O_2(g) \quad \text{Slow}$$

Assuming a two-step mechanism, propose the second step in the mechanism and give the overall balanced equation.

103. You travel to a distant, cold planet where the ammonia flows like water. In fact, the inhabitants of this planet use ammonia (an abundant liquid on their planet) much as earthlings use water. Ammonia is also similar to water in that it is amphoteric and undergoes autoionization. The K value for the auto-ionization of ammonia is 1.8×10^{-12} at the standard temperature of the planet. What is the pH of ammonia at this temperature?

104. Nitrogen gas reacts with hydrogen gas to form ammonia gas (NH_3). Consider the following illustration representing the original reaction mixture in a 15.0 L container (the numbers of each molecule shown are relative numbers):

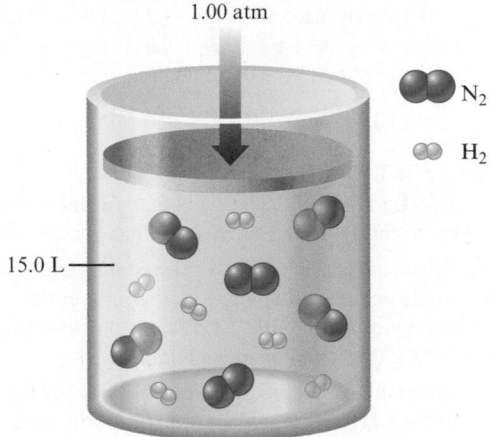

- N_2
- H_2

Assume this reaction mixture goes to completion. The piston apparatus allows the container volume to change in order to keep the pressure constant at 1.00 atm. Assume ideal behavior and constant temperature.

a. What is the partial pressure of ammonia in the container when the reaction is complete?

b. What is the mole fraction of ammonia in the container when the reaction is complete?

c. What is the volume of the container when the reaction is complete?

105. A cylinder fitted with a movable piston initially contains 2.00 moles of $O_2(g)$ and an unknown amount of $SO_2(g)$. The oxygen is known to be in excess. The density of the mixture is 0.8000 g/L at some T and P. After the reaction has gone to completion, forming $SO_3(g)$, the density of the resulting gaseous mixture is 0.8471 g/L at the same T and P. Calculate the mass of SO_3 formed in the reaction.

Integrative Problems

These problems require the integration of multiple concepts to find the solutions.

106. The heaviest member of the alkaline earth metals is radium (Ra), a naturally radioactive element discovered by Pierre and Marie Curie in 1898. Radium was initially isolated from the uranium ore pitchblende, in which it is present as approximately 1.0 g per 7.0 metric tons of pitchblende. How many atoms of radium can be isolated from 1.75×10^8 g pitchblende (1 metric ton = 1000 kg)? One of the early uses of radium was as an additive to paint so that watch dials coated with this paint would glow in the dark. The longest-lived isotope of radium has a half-life of 1.60×10^3 years. If an antique watch, manufactured in 1925, contains 15.0 mg radium, how many atoms of radium will remain in 2025?

107. Indium(III) phosphide is a semiconducting material that has been frequently used in lasers, light-emitting diodes (LED), and fiber-optic devices. This material can be synthesized at 900. K according to the following reaction:

$$In(CH_3)_3(g) + PH_3(g) \longrightarrow InP(s) + 3CH_4(g)$$

a. If 2.56 L $In(CH_3)_3$ at 2.00 atm is allowed to react with 1.38 L PH_3 at 3.00 atm, what mass of $InP(s)$ will be produced assuming the reaction has an 87% yield?

b. When an electric current is passed through an optoelectronic device containing InP, the light emitted has an energy of 2.03×10^{-19} J. What is the wavelength of this light and is it visible to the human eye?

c. The semiconducting properties of InP can be altered by doping. If a small number of phosphorus atoms are replaced by atoms with an electron configuration of $[Kr]5s^24d^{10}5p^4$, is this n-type or p-type doping?

108. Although nitrogen trifluoride (NF_3) is a thermally stable compound, nitrogen triiodide (NI_3) is known to be a highly explosive material. NI_3 can be synthesized according to the equation

$$BN(s) + 3IF(g) \longrightarrow BF_3(g) + NI_3(g)$$

a. What is the enthalpy of formation for $NI_3(s)$ given the enthalpy of reaction (-307 kJ) and the enthalpies of formation for $BN(s)$ (-254 kJ/mol), $IF(g)$ (-96 kJ/mol), and $BF_3(g)$ (-1136 kJ/mol)?

b. It is reported that when the synthesis of NI_3 is conducted using 4 moles of IF for every 1 mole of BN, one of the by-products isolated is $[IF_2]^+[BF_4]^-$. What are the molecular geometries of the species in this by-product? What are the hybridizations of the central atoms in each species in the by-product?

109. While selenic acid has the formula H_2SeO_4 and thus is directly related to sulfuric acid, telluric acid is best visualized as H_6TeO_6 or $Te(OH)_6$.

a. What is the oxidation state of tellurium in $Te(OH)_6$?

b. Despite its structural differences with sulfuric and selenic acid, telluric acid is a diprotic acid with $pK_{a_1} = 7.68$ and $pK_{a_2} = 11.29$. Telluric acid can be prepared by hydrolysis of tellurium hexafluoride according to the equation

$$TeF_6(g) + 6H_2O(l) \longrightarrow Te(OH)_6(aq) + 6HF(aq)$$

Tellurium hexafluoride can be prepared by the reaction of elemental tellurium with fluorine gas:

$$Te(s) + 3F_2(g) \longrightarrow TeF_6(g)$$

If a cubic block of tellurium (density = 6.240 g/cm³) measuring 0.545 cm on edge is allowed to react with 2.34 L fluorine gas at 1.06 atm and 25°C, what is the pH of a solution of $Te(OH)_6$ formed by dissolving the isolated $TeF_6(g)$ in 115 mL solution? Assume 100% yield in all reactions.

Marathon Problems

These problems are designed to incorporate several concepts and techniques into one situation.

110. Captain Kirk has set a trap for the Klingons who are threatening an innocent planet. He has sent small groups of fighter rockets to sites that are invisible to Klingon radar and put a decoy in the open. He calls this the "fishhook" strategy. Mr. Spock has sent a coded message to the chemists on the fighters to tell the ships what to do next. The outline of the message is

___ ___   ___   ___ ___ ___
(1)   (2)   (3)   (4)   (5)   (6)
___   ___ ___   ___ ___ ___ ___ ___
(7)   (8)   (9)   (10) (11) (12) (10) (11)

Fill in the blanks of the message using the following clues.

(1) Symbol of the halogen whose hydride has the second highest boiling point in the series of HX compounds that are hydrogen halides.

(2) Symbol of the halogen that is the only hydrogen halide, HX, that is a weak acid in aqueous solution.

(3) Symbol of the element whose existence on the sun was known before its existence on the earth was discovered.

(4) The Group 5A (15) element in Table 19-13 that should have the most metallic character.

(5) Symbol of the Group 6A (16) element that, like selenium, is a semiconductor.

(6) Symbol for the element known in rhombic and monoclinic forms.

(7) Symbol for the element that exists as diatomic molecules in a yellow-green gas when not combined with another element.

(8) Symbol for the most abundant element in and near the earth's crust.

(9) Symbol for the element that seems to give some protection against cancer when a diet rich in this element is consumed.

(10) Symbol for the smallest noble gas that forms compounds with fluorine having the general formula AF_2 and AF_4 (reverse the symbol and split the letters as shown).

(11) Symbol for the toxic element that, like phosphorus and antimony, forms tetrameric molecules when uncombined with other elements (split the letters of the symbol as shown).

(12) Symbol for the element that occurs as an inert component of air but is a very prominent part of fertilizers and explosives.

111. Use the symbols of the elements described in the following clues to fill in the blanks that spell out the name of a famous American scientist. Although this scientist was better known as a physicist than as a chemist, the Philadelphia institute that bears his name does include a biochemistry research facility.

___ ___ ___   ___ ___ ___ ___
(1)   (2)   (3)   (4)   (5) (6)   (7)

(1) The oxide of this alkaline earth metal is amphoteric.
(2) The element that makes up approximately 3.0% by mass of the human body.
(3) The element having a $7s^1$ valence electron configuration.
(4) This element is the alkali metal with the least negative standard reduction potential. Write its symbol in reverse order.
(5) The alkali metal whose ion is more concentrated in intra-cellular fluids as compared with blood plasma.
(6) This is the only alkali metal that reacts directly with nitrogen to make a binary compound with formula M_3N.
(7) This element is the first in Group 3A (13) for which the +1 oxidation state is exhibited in stable compounds. Use only the second letter of its symbol.

CHAPTER TWENTY

Transition Metals and Coordination Chemistry

The chrome plated archer on the hood of a classic Pierce-Arrow automobile. (Larry Klees)

ransition metals have many uses in our society. Iron is used for steel; copper for electrical wiring and water pipes; titanium for paint; silver for photographic paper; manganese, chromium, vanadium, and cobalt as additives to steel; platinum for industrial and automotive catalysts; and so on.

One indication of the importance of transition metals is the great concern shown by the U.S. government for continuing the supply of these elements. In recent years the United States has been a net importer of about 60 "strategic and critical" minerals, including cobalt, manganese, platinum, palladium, and chromium. All these metals play a vital role in the U.S. economy and defense, and approximately 90% of the required amounts must be imported (Table 20-1).

In addition to being important in industry, transition metal ions play a vital role in living organisms. For example, complexes of iron provide for the transport and storage of oxygen, molybdenum and iron compounds are catalysts in nitrogen fixation, zinc is found in more than 150 biomolecules in humans, copper and iron play a crucial role in the respiratory cycle, and cobalt is found in essential biomolecules such as vitamin B_{12}.

In this chapter we will explore the general properties of transition metals, paying particular attention to the bonding, structure, and properties of the complex ions of these metals.

20-1 | The Transition Metals: A Survey

General Properties

One striking characteristic of the representative elements is that their chemistry changes markedly across a given period as the number of valence electrons changes. The chemical similarities occur mainly within the vertical groups. In contrast, *the transition metals show great similarities within a given period as well as within a given vertical group.* This difference occurs because the last electrons added for transition metals are inner electrons: *d* electrons for the *d*-block transition metals and *f* electrons for the lanthanides and actinides. These inner *d* and *f* electrons cannot participate as easily in bonding as can the valence *s* and *p* electrons. Thus the chemistry of the transition elements is not affected as greatly by the gradual change in the number of electrons as is the chemistry of the representative elements.

Table 20-1 | Some Transition Metals Important to the U.S. Economy and Defense

Metal	Uses	Percentage Imported
Chromium	Stainless steel (especially for parts exposed to corrosive gases and high temperatures)	~91%
Cobalt	High-temperature alloys in jet engines, magnets, catalysts, drill bits	~93%
Manganese	Steelmaking	~97%
Platinum and palladium	Catalysts	~87%

Figure 20-1 | The position of the transition elements on the periodic table. The *d*-block elements correspond to filling the 3*d*, 4*d*, 5*d*, or 6*d* orbitals. The inner transition metals correspond to filling the 4*f* (lanthanides) or 5*f* (actinides) orbitals.

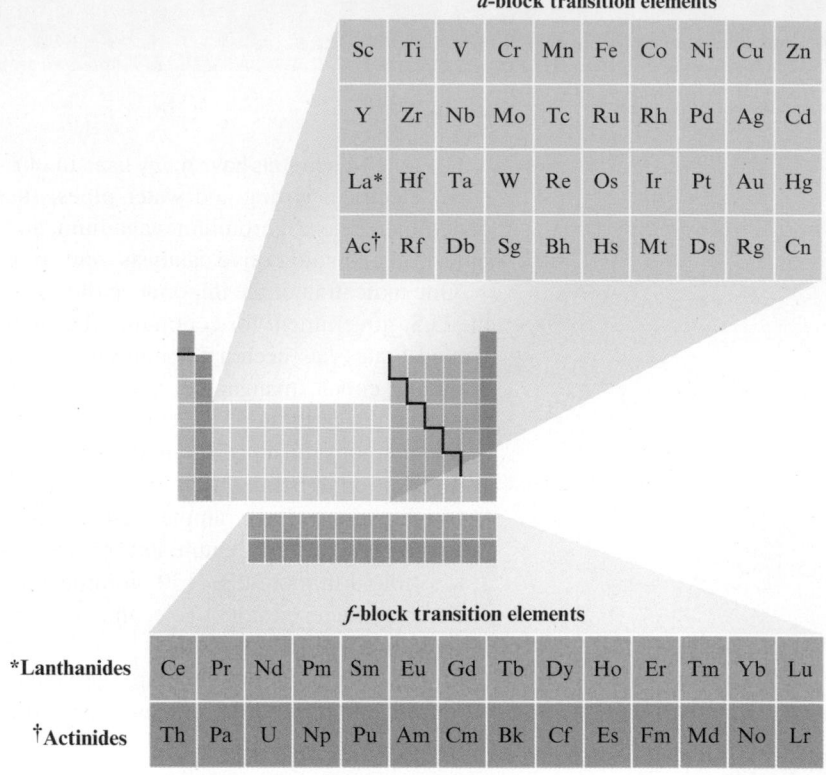

d-block transition elements

Sc	Ti	V	Cr	Mn	Fe	Co	Ni	Cu	Zn
Y	Zr	Nb	Mo	Tc	Ru	Rh	Pd	Ag	Cd
La*	Hf	Ta	W	Re	Os	Ir	Pt	Au	Hg
Ac†	Rf	Db	Sg	Bh	Hs	Mt	Ds	Rg	Cn

f-block transition elements

*Lanthanides	Ce	Pr	Nd	Pm	Sm	Eu	Gd	Tb	Dy	Ho	Er	Tm	Yb	Lu
†Actinides	Th	Pa	U	Np	Pu	Am	Cm	Bk	Cf	Es	Fm	Md	No	Lr

Group designations are traditionally given on the periodic table for the *d*-block transition metals (Fig. 20-1). However, these designations do not relate as directly to the chemical behavior of these elements as they do for the representative elements (the A groups), so we will not use them.

As a class, the transition metals behave as typical metals, possessing metallic luster and relatively high electrical and thermal conductivities. Silver is the best conductor of heat and electric current. ◄ However, copper is a close second, which explains copper's wide use in the electrical systems of homes and factories.

Despite their many similarities, the transition metals do vary considerably in certain properties. For example, tungsten has a melting point of 3400°C and is used for filaments in light bulbs; mercury is a liquid at 25°C. Some transition metals such as iron and titanium are hard and strong and make very useful structural materials; others such as copper, gold, and silver are relatively soft. The chemical reactivity of the transition metals also varies significantly. Some react readily with oxygen to form oxides. Of these metals, some, such as chromium, nickel, and cobalt, form oxides that adhere tightly to the metallic surface, protecting the metal from further oxidation. Others, such as iron, form oxides that scale off, constantly exposing new metal to the corrosion process. On the other hand, the noble metals—primarily gold, silver, platinum, and palladium—do not readily form oxides.

In forming ionic compounds with nonmetals, the transition metals exhibit several typical characteristics:

More than one oxidation state is often found. For example, iron combines with chlorine to form $FeCl_2$ and $FeCl_3$.

The cations are often **complex ions**, species where *the transition metal ion is surrounded by a certain number of ligands* (molecules or ions that behave as Lewis

▲ Sport trophies are often made from silver.

bases). For example, the compound $[Co(NH_3)_6]Cl_3$ contains the $Co(NH_3)_6^{3+}$ cation and Cl^- anions.

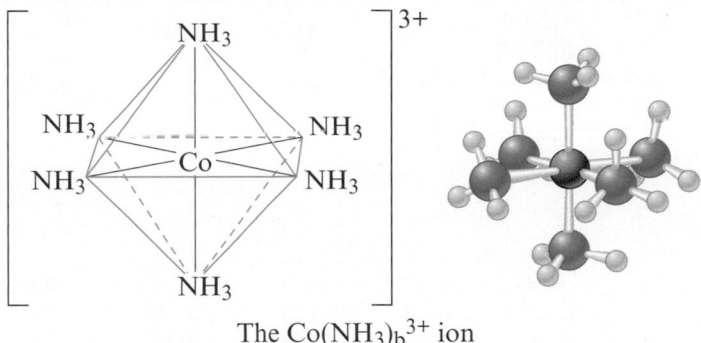

The $Co(NH_3)_b^{3+}$ ion

Most compounds are colored, because the transition metal ion in the complex ion can absorb visible light of specific wavelengths. ◄

Many compounds are paramagnetic (they contain unpaired electrons).

In this chapter we will concentrate on the **first-row transition metals** (scandium through zinc) because they are representative of the other transition series and because they have great practical significance. Some important properties of these elements are summarized in Table 20-2 and are discussed in the next section.

Electron Configurations

The electron configurations of the first-row transition metals were discussed in Section 2-11. The $3d$ orbitals begin to fill after the $4s$ orbital is complete, that is, after calcium ($[Ar]4s^2$). The first transition metal, *scandium*, has one electron in the $3d$ orbitals; the

▲ (clockwise from upper left) Calcite stalactites colored by traces of iron. Quartz is often colored by the presence of transition metals such as Mn, Fe, and Ni. Wulfenite contains $PbMoO_4$. Rhodochrosite is a mineral containing $MnCO_3$.

Table 20-2 | Selected Properties of the First-Row Transition Metals

	Scandium	Titanium	Vanadium	Chromium	Manganese	Iron	Cobalt	Nickel	Copper	Zinc
Atomic number	21	22	23	24	25	26	27	28	29	30
Electron configuration*	$4s^23d^1$	$4s^23d^2$	$4s^23d^3$	$4s^13d^5$	$4s^23d^5$	$4s^23d^6$	$4s^23d^7$	$4s^23d^8$	$4s^13d^{10}$	$4s^23d^{10}$
Atomic radius (pm)	162	147	134	130	135	126	125	124	128	138
Ionization energies (eV/atom)										
First	6.54	6.82	6.74	6.77	7.44	7.87	7.86	7.64	7.73	9.39
Second	12.80	13.58	14.65	16.50	15.64	16.18	17.06	18.17	20.29	17.96
Third	24.76	27.49	29.31	30.96	33.67	30.65	33.50	35.17	36.83	39.72
Reduction potential† (V)	−2.08	−1.63	−1.2	−0.91	−1.18	−0.44	−0.28	−0.23	+0.34	−0.76
Common oxidation states	+3	+2,+3, +4	+2,+3, +4,+5	+2,+3, +6	+2,+3, +4,+7	+2,+3	+2,+3	+2	+1,+2	+2
Melting point (°C)	1397	1672	1710	1900	1244	1530	1495	1455	1083	419
Density (g/cm³)	2.99	4.49	5.96	7.20	7.43	7.86	8.9	8.90	8.92	7.14
Electrical conductivity‡	—	2	3	10	2	17	24	24	97	27

*Each atom has an argon inner-core configuration.
†For the reduction process $M^{2+} + 2e^- \rightarrow M$ (except for scandium, where the ion is Sc^{3+}).
‡Compared with an arbitrarily assigned value of 100 for silver.

second, *titanium,* has two; and the third, *vanadium,* has three. We would expect *chromium,* the fourth transition metal, to have the electron configuration $[Ar]4s^23d^4$. However, the actual configuration is $[Ar]4s^13d^5$, which shows a half-filled 4s orbital and a half-filled set of 3d orbitals (one electron in each of the five 3d orbitals). ◀ ⓘ It is tempting to say that the configuration results because half-filled "shells" are especially stable. Although there are some reasons to think that this explanation might be valid, it is an oversimplification. For instance, tungsten, which is in the same vertical group as chromium, has the configuration $[Xe]6s^24f^{14}5d^4$, where half-filled s and d shells are not found. There are several similar cases.

ⓘ Chromium has the electron configuration $[Ar]4s^13d^5$.

Basically, the chromium configuration occurs because the energies of the 3d and 4s orbitals are very similar for the first-row transition elements. We saw in Section 2-11 that when electrons are placed in a set of degenerate orbitals, they first occupy each orbital singly to minimize electron repulsions. ◀ ⓘ Since the 4s and 3d orbitals are virtually degenerate in the chromium atom, we would expect the configuration

ⓘ A set of orbitals with the same energy is said to be *degenerate.*

$$4s\ \uparrow \qquad 3d\ \uparrow\uparrow\uparrow\uparrow\uparrow$$

rather than

$$4s\ \uparrow\downarrow \qquad 3d\ \uparrow\uparrow\uparrow\uparrow\underline{\ \ }$$

since the second arrangement has greater electron–electron repulsions and thus a higher energy.

The only other unexpected configuration among the first-row transition metals is that of copper, which is $[Ar]4s^13d^{10}$ rather than the expected $[Ar]4s^23d^9$. ◀ ⓘ

ⓘ Copper has the electron configuration $[Ar]4s^13d^{10}$.

In contrast to the neutral transition metals, where the 3d and 4s orbitals have very similar energies, the *energy of the 3d orbitals in transition metal ions is significantly less than that of the 4s orbital.* ◀ ⓘ This means that the electrons remaining after the ion is formed occupy the 3d orbitals, since they are lower in energy. *First-row*

ⓘ In transition metal *ions,* the 3d orbitals are lower in energy than the 4s orbitals.

Figure 20-2 | Plots of the first (red dots) and third (blue dots) ionization energies for the first-row transition metals.

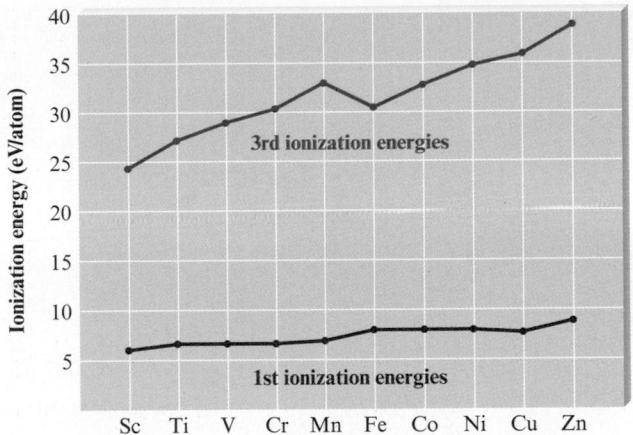

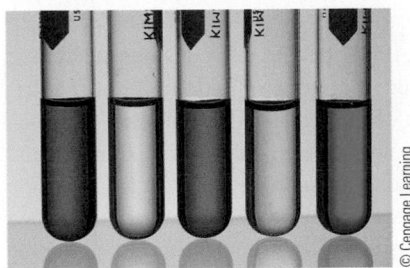

▲ (from left to right) Aqueous solutions containing the metal ions Co^{2+}, Mn^{2+}, Cr^{3+}, Fe^{3+}, and Ni^{2+}.

transition metal ions do not have 4s electrons. For example, manganese has the configuration $[Ar]4s^2 3d^5$, while that of Mn^{2+} is $[Ar]3d^5$. The neutral titanium atom has the configuration $[Ar]4s^2 3d^2$, while that of Ti^{3+} is $[Ar]3d^1$.

Oxidation States and Ionization Energies

The transition metals can form a variety of ions by losing one or more electrons. The common oxidation states of these elements are shown in Table 20-2. Note that for the first five metals the maximum possible oxidation state corresponds to the loss of all the 4s and 3d electrons. For example, the maximum oxidation state of chromium ($[Ar]4s^1 3d^5$) is +6. Toward the right end of the period, the maximum oxidation states are not observed; in fact, the 2+ ions are the most common. ◄ The higher oxidation states are not seen for these metals because the 3d orbitals become lower in energy as the nuclear charge increases, and the electrons become increasingly difficult to remove. From Table 20-2 we see that ionization energy increases gradually going from left to right across the period. However, the third ionization energy (when an electron is removed from a 3d orbital) increases faster than the first ionization energy, clear evidence of the significant decrease in the energy of the 3d orbitals going across the period (Fig. 20-2).

Standard Reduction Potentials

When a metal acts as a *reducing agent*, the half-reaction is

$$M \longrightarrow M^{n+} + ne^-$$

This is the reverse of the conventional listing for half-reactions in tables. Thus, to rank the transition metals in order of reducing ability, it is most convenient to reverse the reactions and the signs given in Table 20-2. The metal with the most positive potential is then the best reducing agent. The transition metals are listed in order of reducing ability in Table 20-3.

Since $\mathscr{E}°$ is zero for the process

$$2H^+ + 2e^- \longrightarrow H_2$$

all the metals except copper can reduce H^+ ions to hydrogen gas in 1 *M* aqueous solutions of strong acid:

$$M(s) + 2H^+(aq) \longrightarrow H_2(g) + M^{2+}(aq)$$

As Table 20-3 shows, the reducing abilities of the first-row transition metals generally decrease going from left to right across the period. Only chromium and zinc do not follow this trend.

Table 20-3 | Relative Reducing Abilities of the First-Row Transition Metals in Aqueous Solution

Reaction	Potential (V)
$Sc \rightarrow Sc^{3+} + 3e^-$	2.08
$Ti \rightarrow Ti^{2+} + 2e^-$	1.63
$V \rightarrow V^{2+} + 2e^-$	1.2
$Mn \rightarrow Mn^{2+} + 2e^-$	1.18
$Cr \rightarrow Cr^{2+} + 2e^-$	0.91
$Zn \rightarrow Zn^{2+} + 2e^-$	0.76
$Fe \rightarrow Fe^{2+} + 2e^-$	0.44
$Co \rightarrow Co^{2+} + 2e^-$	0.28
$Ni \rightarrow Ni^{2+} + 2e^-$	0.23
$Cu \rightarrow Cu^{2+} + 2e^-$	−0.34

Reducing ability ↑

Figure 20-3 | Atomic radii of the 3d, 4d, and 5d transition series.

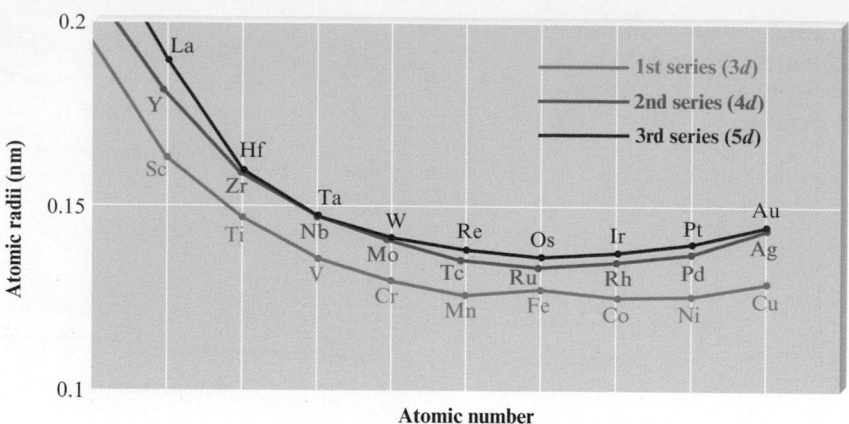

The 4d and 5d Transition Series

In comparing the 3d, 4d, and 5d transition series, it is instructive to consider the atomic radii of these elements (Fig. 20-3). Note that there is a general, although not regular, decrease in size going from left to right for each of the series. Also note that although there is a significant increase in radius in going from the 3d to the 4d metals, the 4d and 5d metals are remarkably similar in size. This latter phenomenon is the result of the **lanthanide contraction**. In the **lanthanide series**, consisting of the elements between lanthanum and hafnium (see Fig. 20-1), electrons are filling the 4f orbitals. Since the 4f orbitals are buried in the interior of these atoms, the additional electrons do not add to the atomic size. In fact, the increasing nuclear charge (remember that a proton is added to the nucleus for each electron) causes the radii of the lanthanide elements to decrease significantly going from left to right. This lanthanide contraction just offsets the normal increase in size due to going from one principal quantum level to another. Thus the 5d elements, instead of being significantly larger than the 4d elements, are almost identical to them in size. This leads to a great similarity in the chemistry of the 4d and 5d elements in a given vertical group. For example, the chemical properties of hafnium and zirconium are remarkably similar, and they always occur together in nature. Their separation, which is probably more difficult than the separation of any other pair of elements, often requires fractional distillation of their compounds.

In general, the differences between the 4d and 5d elements in a group increase gradually going from left to right. For example, niobium and tantalum are also quite similar, but less so than zirconium and hafnium.

Although generally less well known than the 3d elements, the 4d and 5d transition metals have certain very useful properties. For example, zirconium and zirconium oxide (ZrO_2) have great resistance to high temperatures and are used, along with niobium and molybdenum alloys, for space vehicle parts that are exposed to high temperatures during re-entry into the earth's atmosphere. Niobium and molybdenum are also important alloying materials for certain types of steel. ◄ ⓘ Tantalum, which has a high resistance to attack by body fluids, is often used for replacement of bones. ◄ The *platinum group metals*—ruthenium, osmium, rhodium, iridium, palladium, and platinum—are all quite similar and are widely used as catalysts for many types of industrial processes.

ⓘ Niobium was originally called *columbium* and is still occasionally referred to by that name.

20-2 | The First-Row Transition Metals

We have seen that the transition metals are similar in many ways but also show important differences. We will now explore some of the specific properties of each of the 3d transition metals.

Scandium is a rare element that exists in compounds mainly in the +3 oxidation state—for example, in $ScCl_3$, Sc_2O_3, and $Sc_2(SO_4)_3$. The chemistry of scandium

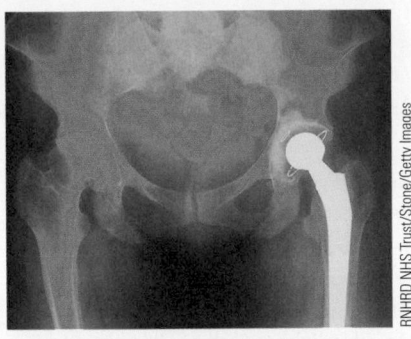

▲ An X ray of a patient who has had a hip replacement. The normal hip joint is on the left; the hip joint constructed from the transition metal tantalum is on the right.

▲ Ti(H₂O)₆³⁺ is purple in solution.

Table 20-4 | Oxidation States and Species for Vanadium in Aqueous Solution

Oxidation State of Vanadium	Species in Aqueous Solution
+5	VO_2^+ (yellow)
+4	VO^{2+} (blue)
+3	$V^{3+}(aq)$ (blue-green)
+2	$V^{2+}(aq)$ (violet)

strongly resembles that of the lanthanides, with most of its compounds being colorless and diamagnetic. This is not surprising; as we will see in Section 20-6, the color and magnetism of transition metal compounds usually arise from the d electrons on the metal ion, and Sc^{3+} has no d electrons. Scandium metal, which can be prepared by electrolysis of molten $ScCl_3$, is not widely used because of its rarity, but it is found in some electronic devices, such as high-intensity lamps.

Titanium is widely distributed in the earth's crust (0.6% by mass). Because of its relatively low density and high strength, titanium is an excellent structural material, especially in jet engines, where light weight and stability at high temperatures are required. Nearly 5000 kg of titanium alloys is used in each engine of a Boeing 747 jetliner. In addition, the resistance of titanium to chemical attack makes it a useful material for pipes, pumps, and reaction vessels in the chemical industry.

The most familiar compound of titanium is no doubt responsible for the white color of this paper. Titanium dioxide, or more correctly, *titanium(IV) oxide* (TiO_2), is a highly opaque substance used as the white pigment in paper, paint, linoleum, plastics, synthetic fibers, whitewall tires, and cosmetics (sunscreens, for example). More than one million tons are used annually in these and other products. Titanium(IV) oxide is widely dispersed in nature, but the main ores are rutile (impure TiO_2) and ilmenite ($FeTiO_3$). Rutile is processed by treatment with chlorine to form volatile $TiCl_4$, which is separated from the impurities and burned to form TiO_2:

$$TiCl_4(g) + O_2(g) \longrightarrow TiO_2(s) + 2Cl_2(g)$$

Ilmenite is treated with sulfuric acid to form a soluble sulfate:

$$FeTiO_3(s) + 2H_2SO_4(aq) \longrightarrow Fe^{2+}(aq) + TiO^{2+}(aq) + 2SO_4^{2-}(aq) + 2H_2O(l)$$

When this aqueous mixture is allowed to stand, under vacuum, solid $FeSO_4 \cdot 7H_2O$ forms first and is removed. The mixture is then heated, and the insoluble titanium(IV) oxide hydrate ($TiO_2 \cdot H_2O$) forms. The water of hydration is driven off by heating to form pure TiO_2:

$$TiO_2 \cdot H_2O(s) \xrightarrow{\text{Heat}} TiO_2(s) + H_2O(g)$$

In its compounds, titanium is most often found in the +4 oxidation state. Examples are TiO_2 and $TiCl_4$, the latter a colorless liquid (bp = 137°C) that fumes in moist air to produce TiO_2:

$$TiCl_4(l) + 2H_2O(l) \longrightarrow TiO_2(s) + 4HCl(g)$$

Titanium(III) compounds can be produced by reduction of the +4 state. In aqueous solution, Ti^{3+} exists as the purple $Ti(H_2O)_6^{3+}$ ion, which is slowly oxidized to titanium(IV) by air. ◄ Titanium(II) is not stable in aqueous solution but does exist in the solid state in compounds such as TiO and the dihalides of general formula TiX_2.

Vanadium is widely spread throughout the earth's crust (0.02% by mass). It is used mostly in alloys with other metals such as iron (80% of vanadium is used in steel) and titanium. Vanadium(V) oxide (V_2O_5) is used as an industrial catalyst in the production of materials such as sulfuric acid.

Pure vanadium can be obtained from the electrolytic reduction of fused salts, such as VCl_2, to produce a metal similar to titanium that is steel gray, hard, and corrosion resistant. Often the pure element is not required for alloying. For example, *ferrovanadium,* produced by the reduction of a mixture of V_2O_5 and Fe_2O_3 with aluminum, is added to iron to form *vanadium steel,* a hard steel used for engine parts and axles.

The principal oxidation state of vanadium is +5, found in compounds such as the orange V_2O_5 (mp = 650°C) and the colorless VF_5 (mp = 19.5°C). The oxidation states from +5 to +2 all exist in aqueous solution (Table 20-4). The higher oxidation states, +5 and +4, do not exist as hydrated ions of the type $V^{n+}(aq)$ because the highly charged ion causes the attached water molecules to be very acidic. The H^+ ions are lost to give the oxycations VO_2^+ and VO^{2+}. The hydrated V^{3+} and V^{2+} ions are easily oxidized and thus can function as reducing agents in aqueous solution.

Titanium dioxide, more properly called titanium(IV) oxide, is a very important material. Approximately 1.5 million tons of the substance are produced each year in the United States for use as a pigment in paper and paints and as a component of sunscreens.

In recent years, however, scientists have found a new use for TiO_2. When surfaces are coated with titanium dioxide, they become resistant to dirt and bacteria. For example, the Pilkington Glass Company is now making glass coated with TiO_2 that cleans itself. All the glass needs is sun and rain to keep itself clean. The self-cleaning action arises from two effects. First, the coating of TiO_2 acts as a catalyst in the presence of ultraviolet (UV) light to break down carbon-based pollutants to carbon dioxide and water. Second, because TiO_2 reduces surface tension, rainwater "sheets" instead of forming droplets on the glass, thereby washing away the grime on the surface of the glass. Although this self-cleaning glass is bad news for window washers, it could save millions of dollars in maintenance costs for owners of commercial buildings.

Because the TiO_2-treated glass requires UV light for its action, it does not work well for interior surfaces where UV light is present only in small amounts. However, a team of Japanese researchers has found that if the TiO_2 coating is doped with nitrogen atoms, it will catalyze the breakdown of dirt in the presence of visible light as well as UV light. Studies also show that this N-doped TiO_2 surface coating kills many types of bacteria in the presence of visible or ultraviolet light. This discovery could lead to products such as self-sterilizing bathroom tiles, counters, and toilets. In addition, because the TiO_2 on the surface of glass has such a strong attraction for water molecules (greatly lowering the surface tension), water does not bead up to form droplets. Just as this effect produces sheeting action on exterior glass, so it prevents interior windows and mirrors from "fogging up."

Titanium dioxide, a cheap and plentiful material, may prove to be worth its weight in gold as a surface coating.

Although *chromium* is relatively rare, it is a very important industrial material. The chief ore of chromium is chromite ($FeCr_2O_4$), which can be reduced by carbon to give *ferrochrome*,

$$FeCr_2O_4(s) + 4C(s) \longrightarrow \underbrace{Fe(s) + 2Cr(s)}_{\text{Ferrochrome}} + 4CO(g)$$

which can be added directly to iron in the steelmaking process. Chromium metal, which is often used to plate steel, is hard and brittle and maintains a bright surface by developing a tough invisible oxide coating.

Chromium commonly forms compounds in which it has the oxidation state $+2$, $+3$, or $+6$, as shown in Table 20-5. The Cr^{2+} (chromous) ion is a powerful reducing agent in aqueous solution. In fact, traces of O_2 in other gases can be removed by bubbling through a Cr^{2+} solution:

$$4Cr^{2+}(aq) + O_2(g) + 4H^+(aq) \longrightarrow 4Cr^{3+}(aq) + 2H_2O(l)$$

The chromium(VI) species are excellent oxidizing agents, especially in acidic solution, where chromium(VI) as the dichromate ion ($Cr_2O_7^{2-}$) is reduced to the Cr^{3+} ion:

$$Cr_2O_7^{2-}(aq) + 14H^+(aq) + 6e^- \longrightarrow 2Cr^{3+}(aq) + 7H_2O(l) \qquad \mathscr{E}° = 1.33 \text{ V}$$

The oxidizing ability of the dichromate ion is strongly pH-dependent, increasing as $[H^+]$ increases, as predicted by Le Châtelier's principle. In basic solution, chromium(VI) exists as the chromate ion, a much less powerful oxidizing agent:

$$CrO_4^{2-}(aq) + 4H_2O(l) + 3e^- \longrightarrow Cr(OH)_3(s) + 5OH^-(aq) \qquad \mathscr{E}° = -0.13 \text{ V}$$

The structures of the $Cr_2O_7^{2-}$ and CrO_4^{2-} ions are shown in Fig. 20-4.

Table 20-5 | Typical Chromium Compounds

Oxidation State of Chromium	Examples of Compounds (X = halogen)
+2	CrX_2
+3	CrX_3
	Cr_2O_3 (green)
	$Cr(OH)_3$ (blue-green)
+6	$K_2Cr_2O_7$ (orange)
	Na_2CrO_4 (yellow)
	CrO_3 (red)

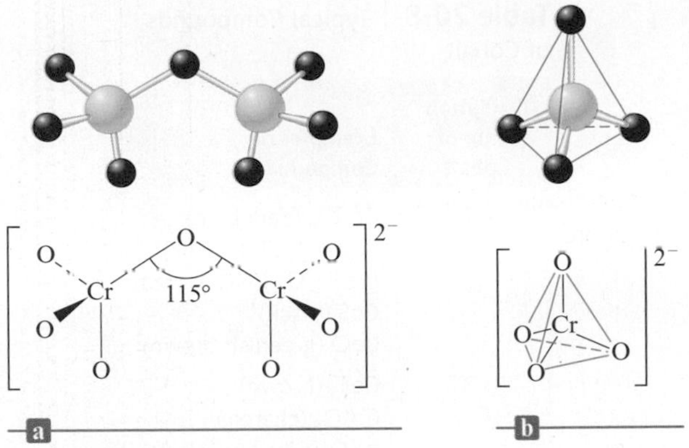

Figure 20-4 | The structures of the chromium(VI) anions: (a) $Cr_2O_7^{2-}$, which exists in acidic solution, and (b) CrO_4^{2-}, which exists in basic solution.

Red chromium(VI) oxide (CrO_3) dissolves in water to give a strongly acidic, red-orange solution:

$$2CrO_3(s) + H_2O(l) \longrightarrow 2H^+(aq) + Cr_2O_7^{2-}(aq)$$

It is possible to precipitate bright orange dichromate salts, such as $K_2Cr_2O_7$, from these solutions. When made basic, the solution turns yellow, and chromate salts such as Na_2CrO_4 can be obtained. A mixture of chromium(VI) oxide and concentrated sulfuric acid, commonly called *cleaning solution,* is a powerful oxidizing medium that can remove organic materials from analytical glassware, yielding a very clean surface.

Manganese is relatively abundant (0.1% of the earth's crust), although no significant sources are found in the United States. The most common use of manganese is in the production of an especially hard steel used for rock crushers, bank vaults, and armor plate. One interesting source of manganese is from *manganese nodules* found on the ocean floor. These roughly spherical "rocks" contain mixtures of manganese and iron oxides as well as smaller amounts of other metals such as cobalt, nickel, and copper. Apparently, the nodules were formed at least partly by the action of marine organisms. Because of the abundance of these nodules, there is much interest in developing economical methods for their recovery and processing.

Manganese can exist in all oxidation states from $+2$ to $+7$, although $+2$ and $+7$ are the most common. Manganese(II) forms an extensive series of salts with all the common anions. In aqueous solution Mn^{2+} forms $Mn(H_2O)_6^{2+}$, which has a light pink color. Manganese(VII) is found in the intensely purple permanganate ion (MnO_4^-). Widely used as an analytical reagent in acidic solution, the MnO_4^- ion behaves as a strong oxidizing agent, with the manganese becoming Mn^{2+}:

$$MnO_4^-(aq) + 8H^+(aq) + 5e^- \longrightarrow Mn^{2+}(aq) + 4H_2O(l) \qquad \mathscr{E}° = 1.51 \text{ V}$$

Several typical compounds of manganese are shown in Table 20-6.

Iron is the most abundant heavy metal (4.7% of the earth's crust) and the most important to our civilization. It is a white, lustrous, not particularly hard metal that is very reactive toward oxidizing agents. For example, in moist air it is rapidly oxidized by oxygen to form rust, a mixture of iron oxides.

The chemistry of iron involves mainly its $+2$ and $+3$ oxidation states. Typical compounds are listed in Table 20-7. In aqueous solutions iron(II) salts are generally light green because of the presence of $Fe(H_2O)_6^{2+}$. Although the $Fe(H_2O)_6^{3+}$ ion is colorless, aqueous solutions of iron(III) salts are usually yellow to brown in color due to the presence of $Fe(OH)(H_2O)_5^{2+}$, which results from the acidity of $Fe(H_2O)_6^{3+}$ ($K_a = 6 \times 10^{-3}$):

$$Fe(H_2O)_6^{3+}(aq) \rightleftharpoons Fe(OH)(H_2O)_5^{2+}(aq) + H^+(aq)$$

Although *cobalt* is relatively rare, it is found in ores such as smaltite ($CoAs_2$) and cobaltite ($CoAsS$) in large enough concentrations to make its production economically

Table 20-6 | Some Compounds of Manganese in Its Most Common Oxidation States

Oxidation State of Manganese	Examples of Compounds
+2	Mn(OH)₂ (pink)
	MnS (salmon)
	MnSO₄ (reddish)
	MnCl₂ (pink)
+4	MnO₂ (dark brown)
+7	KMnO₄ (purple)

Table 20-7 | Typical Compounds of Iron

Oxidation State of Iron	Examples of Compounds
+2	FeO (black)
	FeS (brownish black)
	FeSO₄ · 7H₂O (green)
	K₄Fe(CN)₆ (yellow)
+3	FeCl₃ (brownish black)
	Fe₂O₃ (reddish brown)
	K₃Fe(CN)₆ (red)
	Fe(SCN)₃ (red)
+2, +3 (mixture)	Fe₃O₄ (black)
	KFe[Fe(CN)₆] (deep blue, "Prussian blue")

Table 20-8 | Typical Compounds of Cobalt

Oxidation State of Cobalt	Examples of Compounds
+2	$CoSO_4$ (dark blue)
	$[Co(H_2O)_6]Cl_2$ (pink)
	$[Co(H_2O)_6](NO_3)_2$ (red)
	CoS (black)
	CoO (greenish brown)
+3	CoF_3 (brown)
	Co_2O_3 (charcoal)
	$K_3[Co(CN)_6]$ (yellow)
	$[Co(NH_3)_6]Cl_3$ (yellow)

Table 20-9 | Typical Compounds of Nickel

Oxidation State of Nickel	Examples of Compounds
+2	$NiCl_2$ (yellow)
	$[Ni(H_2O)_6]Cl_2$ (green)
	NiO (greenish black)
	NiS (black)
	$[Ni(H_2O)_6]SO_4$ (green)
	$[Ni(NH_3)_6](NO_3)_2$ (blue)

© Cengage Learning

▲ An aqueous solution containing the Ni^{2+} ion.

ⓘ Copper roofs and bronze statues, such as the Statue of Liberty, turn green in air because $Cu_3(OH)_4SO_4$ and $Cu_4(OH)_6SO_4$ form.

feasible. Cobalt is a hard, bluish white metal mainly used in alloys such as stainless steel and stellite, an alloy of iron, copper, and tungsten that is used in surgical instruments.

The chemistry of cobalt involves mainly its +2 and +3 oxidation states, although compounds containing cobalt in the 0, +1, or +4 oxidation state are known. Aqueous solutions of cobalt(II) salts contain the $Co(H_2O)_6^{2+}$ ion, which has a characteristic rose color. Cobalt forms a wide variety of coordination compounds, many of which will be discussed in later sections of this chapter. Some typical cobalt compounds are listed in Table 20-8.

Nickel, which ranks 24th in elemental abundance in the earth's crust, is found in ores, where it is combined mainly with arsenic, antimony, and sulfur. Nickel metal, a silvery white substance with high electrical and thermal conductivities, is quite resistant to corrosion and is often used for plating more active metals. Nickel is also widely used in the production of alloys such as steel.

Nickel in compounds is almost exclusively in the +2 oxidation state. Aqueous solutions of nickel(II) salts contain the $Ni(H_2O)_6^{2+}$ ion, which has a characteristic emerald green color. ◄ Coordination compounds of nickel(II) will be discussed later in this chapter. Some typical nickel compounds are listed in Table 20-9.

Copper, widely distributed in nature in ores containing sulfides, arsenides, chlorides, and carbonates, is valued for its high electrical conductivity and its resistance to corrosion. ◄ ⓘ It is widely used for plumbing, and 50% of all copper produced annually is used for electrical applications. Copper is a major constituent in several well-known alloys (Table 20-10).

Although copper is not highly reactive (it will not reduce H^+ to H_2, for example), the reddish metal does slowly corrode in air, producing the characteristic green *patina* consisting of basic copper sulfate

$$3Cu(s) + 2H_2O(l) + SO_2(g) + 2O_2(g) \longrightarrow Cu_3(OH)_4SO_4(s)$$

Basic copper sulfate

and other similar compounds.

Table 20-10 | Alloys Containing Copper

Alloy	Composition (% by mass)
Brass	Cu (20–97), Zn (2–80), Sn (0–14), Pb (0–12), Mn (0–25)
Bronze	Cu (50–98), Sn (0–35), Zn (0–29), Pb (0–50), P (0–3)
Sterling silver	Cu (7.5), Ag (92.5)
Gold (18-karat)	Cu (5–15), Au (75), Ag (10–20)
Gold (14-karat)	Cu (12–28), Au (58), Ag (4–30)

Table 20-11 | Typical Compounds of Copper

Oxidation State of Copper	Examples of Compounds
+1	Cu_2O (red)
	Cu_2S (black)
	CuCl (white)
+2	CuO (black)
	$CuSO_4 \cdot 5H_2O$ (blue)
	$CuCl_2 \cdot 2H_2O$ (green)
	$[Cu(H_2O)_6](NO_3)_2$ (blue)

The chemistry of copper principally involves the +2 oxidation state, but many compounds containing copper(I) are also known. Aqueous solutions of copper(II) salts are a characteristic bright blue color due to the presence of the $Cu(H_2O)_6^{2+}$ ion. Table 20-11 lists some typical copper compounds.

Although trace amounts of copper are essential for life, copper in large amounts is quite toxic; copper salts are used to kill bacteria, fungi, and algae. For example, paints containing copper are used on ship hulls to prevent fouling by marine organisms.

Widely dispersed in the earth's crust, *zinc* is refined mainly from sphalerite (ZnS), which often occurs with galena (PbS). Zinc is a white, lustrous, very active metal that behaves as an excellent reducing agent and tarnishes rapidly. About 90% of the zinc produced is used for galvanizing steel. Zinc forms colorless salts in the +2 oxidation state.

20-3 | Coordination Compounds

Transition metal ions characteristically form coordination compounds, which are usually colored and often paramagnetic. A **coordination compound** typically consists of a *complex ion,* a transition metal ion with its attached ligands (see Section 14-8), and **counterions,** anions or cations as needed to produce a compound with no net charge. The substance $[Co(NH_3)_5Cl]Cl_2$ is a typical coordination compound. The brackets indicate the composition of the complex ion, in this case $Co(NH_3)_5Cl^{2+}$, and the two Cl^- counterions are shown outside the brackets. Note that in this compound one Cl^- acts as a ligand along with the five NH_3 molecules. In the solid state this compound consists of the large $Co(NH_3)_5Cl^{2+}$ cations and twice as many Cl^- anions, all packed together as efficiently as possible. When dissolved in water, the solid behaves like any ionic solid; the cations and anions are assumed to separate and move about independently:

$$[Co(NH_3)_5Cl]Cl_2(s) \xrightarrow{H_2O} Co(NH_3)_5Cl^{2+}(aq) + 2Cl^-(aq)$$

Coordination compounds have been known since about 1700, but their true nature was not understood until the 1890s when a young Swiss chemist named Alfred Werner (1866–1919) proposed that transition metal ions have two types of valence (combining ability). One type of valence, which Werner called the *secondary valence,* refers to the ability of a metal ion to bind to Lewis bases (ligands) to form complex ions. The other type, the *primary valence,* refers to the ability of the metal ion to form ionic bonds with oppositely charged ions. Thus Werner explained that the compound, originally written as $CoCl_3 \cdot 5NH_3$, was really $[Co(NH_3)_5Cl]Cl_2$, where the Co^{3+} ion has a primary valence of 3, satisfied by the three Cl^- ions, and a secondary valence of 6, satisfied by the six ligands (five NH_3 and one Cl^-). We now call the primary valence the **oxidation state** and the secondary valence the **coordination number**, which reflects the number of bonds formed between the metal ion and the ligands in the complex ion.

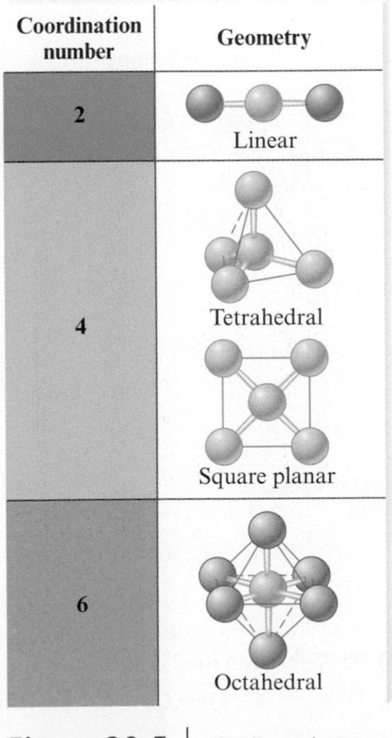

Coordination Number

The number of bonds formed by metal ions to ligands in complex ions varies from two to eight depending on the size, charge, and electron configuration of the transition metal ion. As shown in Table 20-12, 6 is the most common coordination number, followed closely by 4, with a few metal ions showing a coordination number of 2. Many metal ions show more than one coordination number, and there is really no simple way to predict what the coordination number will be in a particular case. The typical geometries for the various common coordination numbers are shown in Fig. 20-5. Note that six ligands produce an octahedral arrangement around the metal ion. Four ligands can form either a tetrahedral or a square planar arrangement, and two ligands give a linear structure.

Coordination number	Geometry
2	Linear
4	Tetrahedral
	Square planar
6	Octahedral

Figure 20-5 | The ligand arrangements for coordination numbers 2, 4, and 6.

Table 20-12 | Typical Coordination Numbers for Some Common Metal Ions

M	Coordination Numbers	M^{2+}	Coordination Numbers	M^{3+}	Coordination Numbers
Cu^+	2, 4	Mn^{2+}	4, 6	Sc^{3+}	6
Ag^+	2	Fe^{2+}	6	Cr^{3+}	6
Au^+	2, 4	Co^{2+}	4, 6	Co^{3+}	6
		Ni^{2+}	4, 6		
		Cu^{2+}	4, 6	Au^{3+}	4
		Zn^{2+}	4, 6		

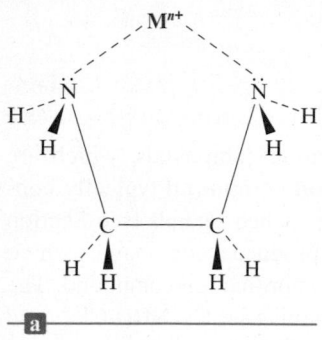

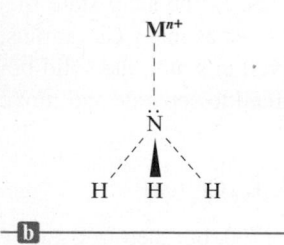

Figure 20-6 | (a) The bidentate ligand ethylenediamine can bond to the metal ion through the lone pair on each nitrogen atom, thus forming two coordinate covalent bonds. (b) Ammonia is a monodentate ligand.

Ligands

A **ligand** is a *neutral molecule or ion having a lone electron pair that can be used to form a bond to a metal ion*. The formation of a metal–ligand bond therefore can be described as the interaction between a Lewis base (the ligand) and a Lewis acid (the metal ion). The resulting bond is often called a **coordinate covalent bond**.

A *ligand that can form one bond to a metal ion* is called a **monodentate ligand**, or a **unidentate ligand** (from root words meaning "one tooth"). Examples of unidentate ligands are shown in Table 20-13.

Some ligands have more than one atom with a lone electron pair that can be used to bond to a metal ion. Such ligands are said to be **chelating ligands**, or **chelates** (from the Greek word *chele* for "claw"). A ligand that can form two bonds to a metal ion is called a **bidentate ligand**. A very common bidentate ligand is ethylenediamine (abbreviated en), which is shown coordinating to a metal ion in Fig. 20-6(a). Note the

Table 20-13 | Some Common Ligands

Type	Examples
Unidentate/ monodentate	H_2O CN^- SCN^- (thiocyanate) X^- (halides) NH_3 NO_2^- (nitrite) OH^-
Bidentate	Oxalate / Ethylenediamine (en)
Polydentate	Diethylenetriamine (dien) $H_2\ddot{N}$—$(CH_2)_2$—$\ddot{N}H$—$(CH_2)_2$—$\ddot{N}H_2$ Three coordinating atoms Ethylenediaminetetraacetate (EDTA) Six coordinating atoms

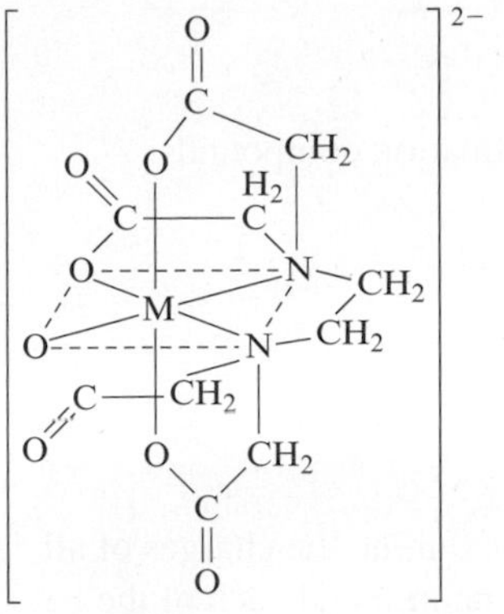

Figure 20-7 | The coordination of EDTA with a 2+ metal ion.

Table 20-14 | Names of Some Common Unidentate Ligands

Neutral Molecules	
Aqua	H_2O
Ammine	NH_3
Methylamine	CH_3NH_2
Carbonyl	CO
Nitrosyl	NO

Anions	
Fluoro	F^-
Chloro	Cl^-
Bromo	Br^-
Iodo	I^-
Hydroxo	OH^-
Cyano	CN^-

Table 20-15 | Latin Names Used for Some Metal Ions in Anionic Complex Ions

Metal	Name in an Anionic Complex
Iron	Ferrate
Copper	Cuprate
Lead	Plumbate
Silver	Argentate
Gold	Aurate
Tin	Stannate

relationship between this ligand and the unidentate ligand ammonia [Fig. 20-6(b)]. Oxalate, another typical bidentate ligand, is shown in Table 20-13.

Ligands that can form more than two bonds to a metal ion are called *polydentate ligands*. Some ligands can form as many as six bonds to a metal ion. The best-known example is ethylenediaminetetraacetate (abbreviated EDTA), which is shown in Table 20-13. This ligand virtually surrounds the metal ion (Fig. 20-7), coordinating through six atoms (a *hexadentate ligand*). As might be expected from the large number of coordination sites, EDTA forms very stable complex ions with most metal ions and is used as a "scavenger" to remove toxic heavy metals such as lead from the human body. It is also used as a reagent to analyze solutions for their metal ion content. EDTA is found in countless consumer products, such as soda, beer, salad dressings, bar soaps, and most cleaners. In these products EDTA ties up trace metal ions that would otherwise catalyze decomposition and produce unwanted precipitates.

Some even more complicated ligands are found in biological systems, where metal ions play crucial roles in catalyzing reactions, transferring electrons, and transporting and storing oxygen. A discussion of these complex ligands will follow in Section 20-7.

Nomenclature

In Werner's lifetime, no system was used to name coordination compounds. Names of the compounds were commonly based on colors and names of discoverers. As the field expanded and more coordination compounds were identified, an orderly system of nomenclature became necessary. A simplified version of this system is summarized by the following rules.

Rules for Naming Coordination Compounds

- As with any ionic compound, *the cation is named before the anion.*
- In naming a complex ion, *the ligands are named before the metal ion.*
- In naming ligands, *an o is added to the root name of an anion.* For example, the halides as ligands are called *fluoro, chloro, bromo,* and *iodo;* hydroxide is *hydroxo;* cyanide is *cyano;* and so on. *For a neutral ligand, the name of the molecule is used,* with the exception of H_2O, NH_3, CO, and NO, as illustrated in Table 20-14.
- *The prefixes mono-, di-, tri-, tetra-, penta-,* and *hexa-* are used to denote the number of simple ligands. The prefixes *bis-, tris-, tetrakis-,* and so on are also used, especially for more complicated ligands or ones that already contain *di-, tri-,* and so on.
- *The oxidation state of the central metal ion is designated by a Roman numeral in parentheses.*
- *When more than one type of ligand is present, they are named alphabetically.** Prefixes do not affect the order.
- *If the complex ion has a negative charge, the suffix -ate is added to the name of the metal. Sometimes the Latin name is used to identify the metal* (Table 20-15).

The application of these rules is shown in Example 20-1.

---

*In an older system the negatively charged ligands were named first, then neutral ligands, with positively charged ligands named last. We will follow the newer convention in this text.

INTERACTIVE EXAMPLE 20-1

Naming Coordination Compounds I

Give the systematic name for each of the following coordination compounds.

a. $[Co(NH_3)_5Cl]Cl_2$ ◄

b. $K_3Fe(CN)_6$ ◄

c. $[Fe(en)_2(NO_2)_2]_2SO_4$

Solution

a. To determine the oxidation state of the metal ion, we examine the charges of all ligands and counterions. The ammonia molecules are neutral and each of the chloride ions has a $1-$ charge, so the cobalt ion must have a $3+$ charge to produce a neutral compound. Thus cobalt has the oxidation state $+3$, and we use cobalt(III) in the name.

The ligands include one Cl^- ion and five NH_3 molecules. The chloride ion is designated as *chloro,* and each ammonia molecule is designated as *ammine.* The prefix *penta-* indicates that there are five NH_3 ligands present. The name of the complex cation is therefore pentaamminechlorocobalt(III). Note that the ligands are named alphabetically, disregarding the prefix. Since the counterions are chloride ions, the compound is named as a chloride salt:

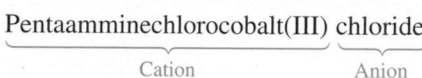

Pentaamminechlorocobalt(III) chloride

        Cation   Anion

b. First, we determine the oxidation state of the iron by considering the other charged species. The compound contains three K^+ ions and six CN^- ions. Therefore, the iron must carry a charge of $3+$, giving a total of six positive charges to balance the six negative charges. The complex ion present is thus $Fe(CN)_6^{3-}$. The cyanide ligands are each designated *cyano,* and the prefix *hexa-* indicates that six are present. Since the complex ion is an anion, we use the Latin name *ferrate.* The oxidation state is indicated by (III) at the end of the name. The anion name is therefore hexacyanoferrate(III). The cations are K^+ ions, which are simply named potassium. Putting this together gives the name

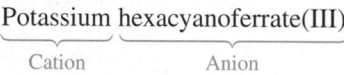

Potassium hexacyanoferrate(III)

       Cation   Anion

(The common name of this compound is potassium ferricyanide.)

c. We first determine the oxidation state of the iron by looking at the other charged species: four NO_2^- ions and one SO_4^{2-} ion. The ethylenediamine is neutral. Thus the two iron ions must carry a total positive charge of 6 to balance the six negative charges. This means that each iron has a $+3$ oxidation state and is designated as iron(III).

Since the name ethylenediamine already contains *di,* we use *bis-* instead of *di-* to indicate the two en ligands. The name for NO_2^- as a ligand is *nitro,* and the prefix *di-* indicates the presence of two NO_2^- ligands. Since the anion is sulfate, the compound's name is

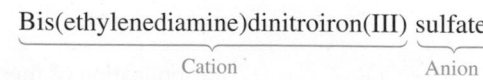

Bis(ethylenediamine)dinitroiron(III) sulfate

          Cation   Anion

Because the complex ion is a cation, the Latin name for iron is not used.

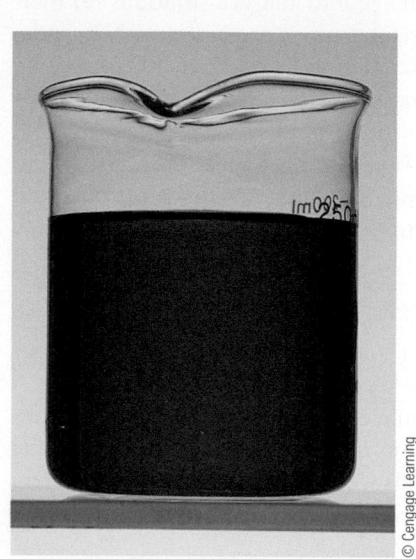

▲ An aqueous solution of $[Co(NH_3)_5Cl]Cl_2$.

▲ Solid $K_3Fe(CN)_6$.

See Exercises 20-34 through 20-36

INTERACTIVE EXAMPLE 20-2

Naming Coordination Compounds II

Given the following systematic names, give the formula of each coordination compound.

a. Triamminebromoplatinum(II) chloride

b. Potassium hexafluorocobaltate(III)

Solution

a. *Triammine* signifies three ammonia ligands, and *bromo* indicates one bromide ion as a ligand. The oxidation state of platinum is $+2$, as indicated by the Roman numeral II. Thus the complex ion is $[Pt(NH_3)_3Br]^+$. One chloride ion is needed to balance the $1+$ charge of this cation. The formula of the compound is $[Pt(NH_3)_3Br]Cl$. Note that brackets enclose the complex ion.

b. The complex ion contains six fluoride ligands attached to a Co^{3+} ion to give $CoF_6{}^{3-}$. Note that the *-ate* ending indicates that the complex ion is an anion. The cations are K^+ ions, and three are required to balance the $3-$ charge on the complex ion. Thus the formula is $K_3[CoF_6]$.

See Exercises 20-37 and 20-38

20-4 | Isomerism

When two or more species have the same formula but different properties, they are said to be **isomers**. Although isomers contain exactly the same types and numbers of atoms, the arrangements of the atoms differ, and this leads to different properties. We will consider two main types of isomerism: **structural isomerism**, where the isomers contain the same atoms but one or more bonds differ, and **stereoisomerism**, where all the bonds in the isomers are the same but the spatial arrangements of the atoms are different. Each of these classes also has subclasses (Fig. 20-8), which we will now consider.

Structural Isomerism

The first type of structural isomerism we will consider is **coordination isomerism**, in which the composition of the complex ion varies. For example, $[Cr(NH_3)_5SO_4]Br$ and $[Cr(NH_3)_5Br]SO_4$ are coordination isomers. In the first case, $SO_4{}^{2-}$ is coordinated to Cr^{3+}, and Br^- is the counterion; in the second case, the roles of these ions are reversed.

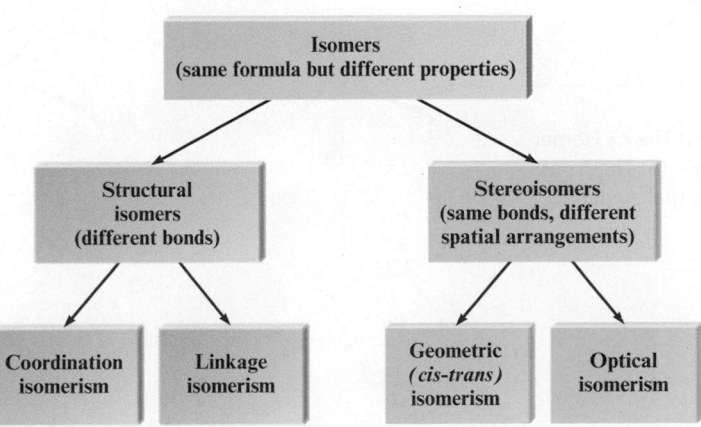

Figure 20-8 | Some classes of isomers.

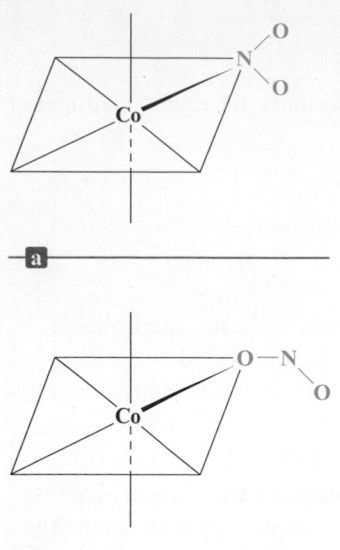

Figure 20-9 │ As a ligand, NO_2^- can bond to a metal ion (a) through a lone pair on the nitrogen atom or (b) through a lone pair on one of the oxygen atoms.

Another example of coordination isomerism is the $[Co(en)_3][Cr(ox)_3]$ and $[Cr(en)_3]$ $[Co(ox)_3]$ pair, where ox represents the oxalate ion, a bidentate ligand shown in Table 20-13.

In a second type of structural isomerism, **linkage isomerism**, the composition of the complex ion is the same, but the point of attachment of at least one of the ligands differs. Two ligands that can attach to metal ions in different ways are thiocyanate (SCN^-), which can bond through lone electron pairs on the nitrogen or the sulfur atom, and the nitrite ion (NO_2^-), which can bond through lone electron pairs on the nitrogen or the oxygen atom. For example, the following two compounds are linkage isomers:

$$[Co(NH_3)_4(NO_2)Cl]Cl$$
Tetraamminechloronitrocobalt(III) chloride
(yellow)

$$[Co(NH_3)_4(ONO)Cl]Cl$$
Tetraamminechloronitritocobalt(III) chloride
(red)

In the first case, the NO_2^- ligand is called *nitro* and is attached to Co^{3+} through the nitrogen atom; in the second case, the NO_2^- ligand is called *nitrito* and is attached to Co^{3+} through an oxygen atom (Fig. 20-9).

Stereoisomerism

Stereoisomers have the same bonds but different spatial arrangements of the atoms. One type, **geometrical isomerism**, or *cis–trans* isomerism, occurs when atoms or groups of atoms can assume different positions around a rigid ring or bond. An important example is the compound $Pt(NH_3)_2Cl_2$, which has a square planar structure. The two possible arrangements of the ligands are shown in Fig. 20-10. In the **trans isomer**, the ammonia molecules are across (*trans*) from each other. In the **cis isomer**, the ammonia molecules are next (*cis*) to each other.

Geometrical isomerism also occurs in octahedral complex ions. For example, the compound $[Co(NH_3)_4Cl_2]Cl$ has *cis* and *trans* isomers (Fig. 20-11).

A second type of stereoisomerism is called **optical isomerism** because the isomers have opposite effects on plane-polarized light. When light is emitted from a source such as a glowing filament, the oscillating electric fields of the photons in the beam are oriented randomly, as shown in Fig. 20-12. If this light is passed through a polarizer, only the photons with electric fields oscillating in a single plane remain, constituting *plane-polarized light*.

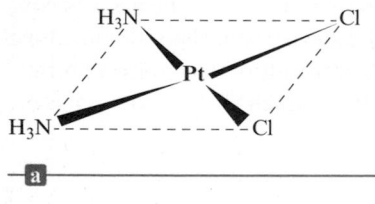

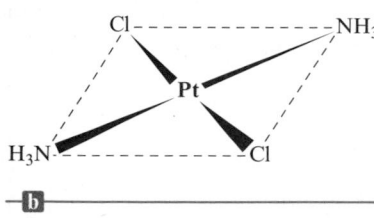

Figure 20-10 │ (a) The *cis* isomer of $Pt(NH_3)_2Cl_2$ (yellow). (b) The *trans* isomer of $Pt(NH_3)_2Cl_2$ (pale yellow).

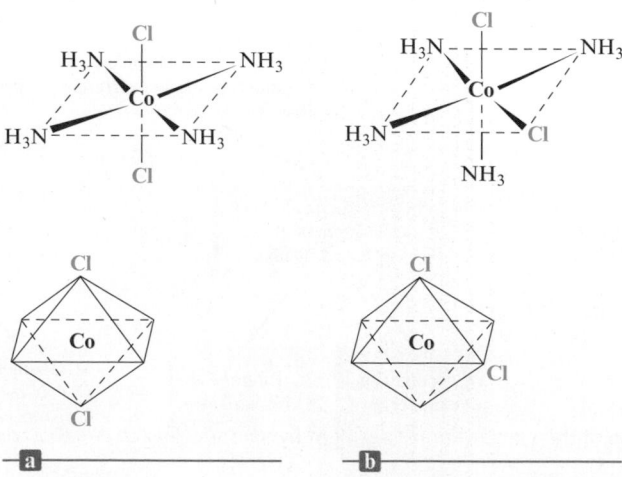

Figure 20-11 │
(a) The *trans* isomer of $[Co(NH_3)_4Cl_2]^+$. The chloride ligands are directly across from each other. (b) The *cis* isomer of $[Co(NH_3)_4Cl_2]^+$. The chloride ligands in this case share an edge of the octahedron. Because of their different structures, the *trans* isomer of $[Co(NH_3)_4Cl_2]Cl$ is green and the *cis* isomer is violet.

Figure 20-12 | Unpolarized light consists of waves vibrating in many different planes (indicated by the arrows). The polarizing filter blocks all waves except those vibrating in a given plane.

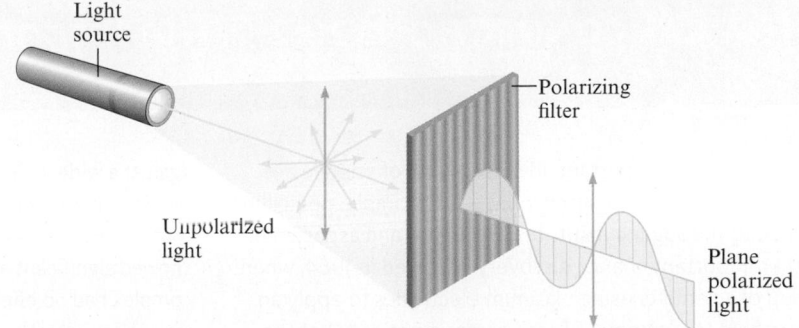

In 1815, a French physicist, Jean Biot (1774–1862), showed that certain crystals could rotate the plane of polarization of light. Later it was found that solutions of certain compounds could do the same thing (Fig. 20-13). Louis Pasteur (1822–1895) was the first to understand this behavior. In 1848 he noted that solid sodium ammonium tartrate ($NaNH_4C_4H_4O_4$) existed as a mixture of two types of crystals, which he painstakingly separated with tweezers. Separate solutions of these two types of crystals rotated plane-polarized light in exactly opposite directions. This led to a connection between optical activity and molecular structure.

We now realize that optical activity is exhibited by molecules that have *nonsuperimposable mirror images.* Your hands are nonsuperimposable mirror images (Fig. 20-15). The two hands are related like an object and its mirror image; one hand cannot be turned to make it identical to the other. Many molecules show this same feature, such as the complex ion $[Co(en)_3]^{3+}$ shown in Fig. 20-16. Objects that have nonsuperimposable mirror images are said to be **chiral** (from the Greek word *cheir*, meaning "hand").

The isomers of $[Co(en)_3]^{3+}$ (Fig. 20-17) are nonsuperimposable mirror images called **enantiomers**, which rotate plane-polarized light in opposite directions and are thus optical isomers. The isomer that rotates the plane of light to the right (when viewed down the beam of oncoming light) is said to be *dextrorotatory,* designated by *d.* The isomer that rotates the plane of light to the left is *levorotatory (l).* An equal mixture of the *d* and *l* forms in solution, called a *racemic mixture,* does not rotate the plane of the polarized light at all because the two opposite effects cancel each other.

Geometrical isomers are not necessarily optical isomers. For instance, the *trans* isomer of $[Co(en)_2Cl_2]^+$ shown in Fig. 20-17 is identical to its mirror image. Since this

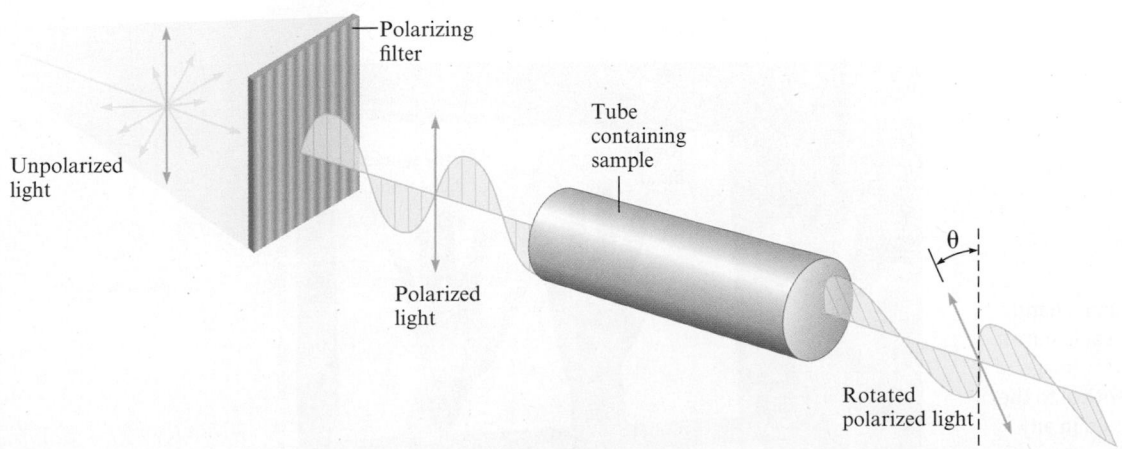

Figure 20-13 | The rotation of the plane of polarized light by an optically active substance. The angle of rotation is called theta (θ).

CHEMICAL CONNECTIONS 20-2 | *The Importance of Being* cis

Some of the most important advancements of science are the results of accidental discoveries—for example, penicillin, Teflon, and the sugar substitutes cyclamate and aspartame. Another important chance discovery occurred in 1964, when a group of scientists using platinum electrodes to apply an electric field to a colony of *E. coli* bacteria noticed that the bacteria failed to divide but continued to grow, forming long, fibrous cells. Further study revealed that cell division was inhibited by small concentrations of *cis*-$Pt(NH_3)_2Cl_2$ and *cis*-$Pt(NH_3)_2Cl_4$ formed electrolytically in the solution.

Cancerous cells multiply very rapidly because cell division is uncontrolled. Thus these and similar platinum complexes were evaluated as *antitumor agents,* which inhibit the division of cancer cells. The results showed that *cis*-$Pt(NH_3)_2Cl_2$ was active against a wide variety of tumors, including testicular and ovarian tumors, which are very resistant to treatment by more traditional methods. However, although the *cis* complex showed significant antitumor activity, the corresponding *trans* complex had no effect on tumors. This shows the importance of isomerism in biological systems. When drugs are synthesized, great care must be taken to obtain the correct isomer.

Although *cis*-$Pt(NH_3)_2Cl_2$ has proven to be a valuable drug, unfortunately it has some troublesome side effects, the most serious being kidney damage. As a result, the search continues for even more effective antitumor agents. Promising candidates are shown in Fig. 20-14. Note that they are all *cis* complexes.

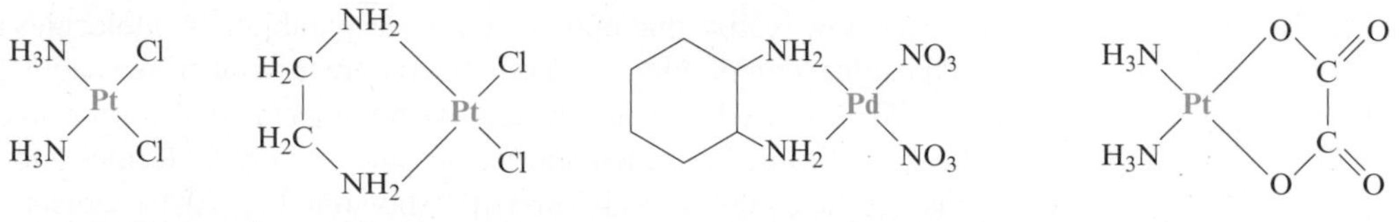

Figure 20-14 | Some *cis* complexes of platinum and palladium that show significant antitumor activity. It is thought that the *cis* complexes work by losing two adjacent ligands and forming coordinate covalent bonds to adjacent bases on a DNA molecule.

isomer is superimposable on its mirror image, it does not exhibit optical isomerism and is not chiral. On the other hand, *cis*-$[Co(en)_2Cl_2]^+$ is *not* superimposable on its mirror image; a pair of enantiomers exists for this complex ion (the *cis* isomer is chiral).

Most important biomolecules are chiral, and their reactions are highly structure dependent. For example, a drug can have a particular effect because its molecules can

Left hand
Right hand
Mirror image of right hand
Martin Bough/Fundamental Photographs © Cengage Learning

Figure 20-15 | A human hand exhibits a nonsuperimposable mirror image. Note that the mirror image of the right hand (while identical to the left hand) cannot be turned in any way to make it identical to (superimposable on) the actual right hand.

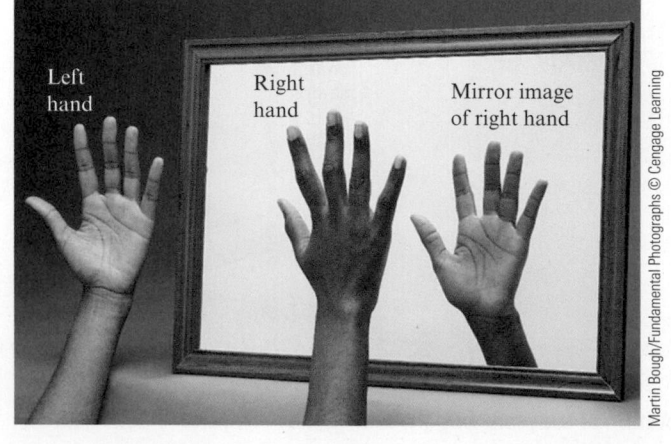

Figure 20-16 | Isomers I and II of Co(en)$_3$$^{3+}$ are mirror images (the image of I is identical to II) that cannot be superimposed. That is, there is no way that I can be turned in space so that it is the same as II.

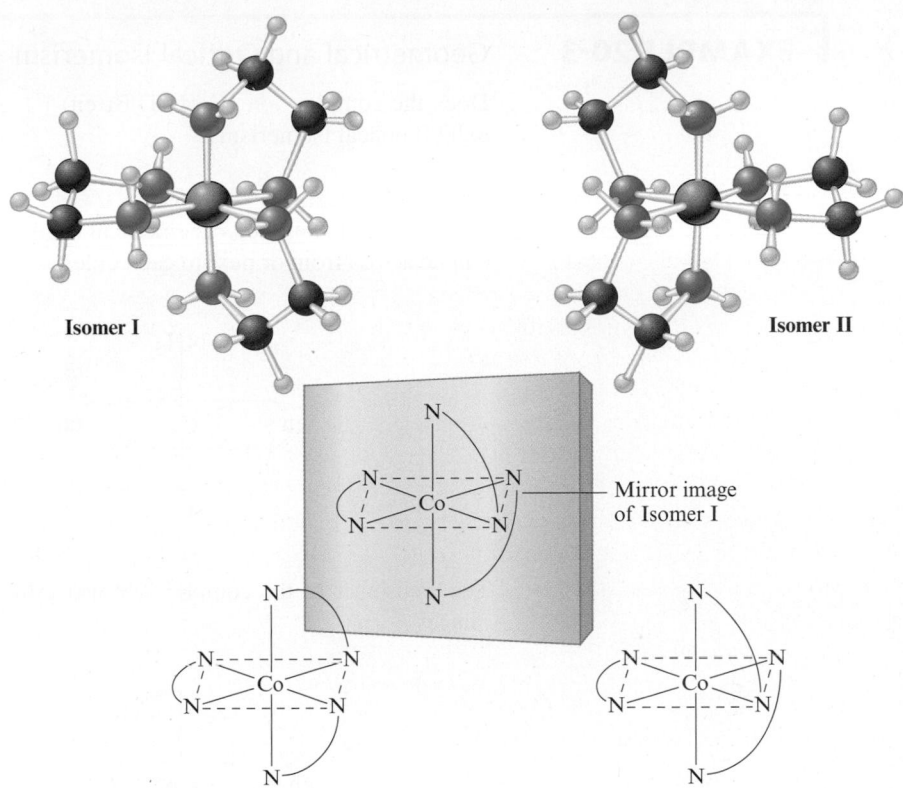

bind to chiral molecules in the body. To bind correctly, however, the correct optical isomer of the drug must be administered. Just as the right hand of one person requires the right hand of another to perform a handshake, a given isomer in the body requires a specific isomer of the drug to bind together. Because of this, the syntheses of drugs, which are usually very complicated molecules, must be carried out in a way that produces the correct "handedness," a requirement that greatly adds to the synthetic difficulties.

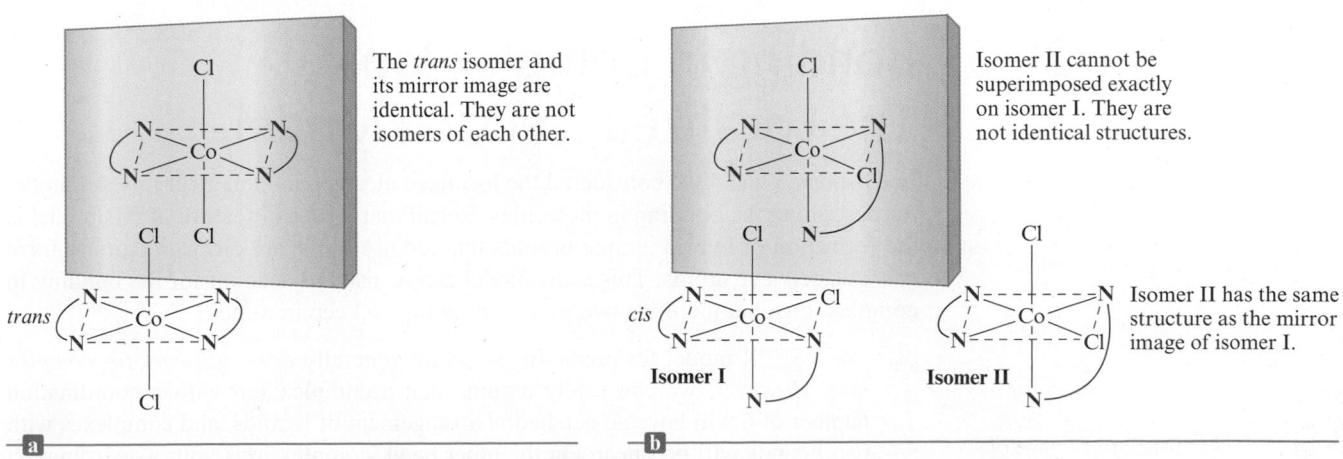

Figure 20-17 | (a) The *trans* isomer of Co(en)$_2$Cl$_2$$^+$ and its mirror image are identical (superimposable). (b) The *cis* isomer of Co(en)$_2$Cl$_2$$^+$ and its mirror image are not superimposable and are thus a pair of optical isomers.

EXAMPLE 20-3   Geometrical and Optical Isomerism

Does the complex ion $[Co(NH_3)Br(en)_2]^{2+}$ exhibit geometrical isomerism? Does it exhibit optical isomerism?

Solution

The complex ion exhibits geometrical isomerism because the ethylenediamine ligands can be across from or next to each other:

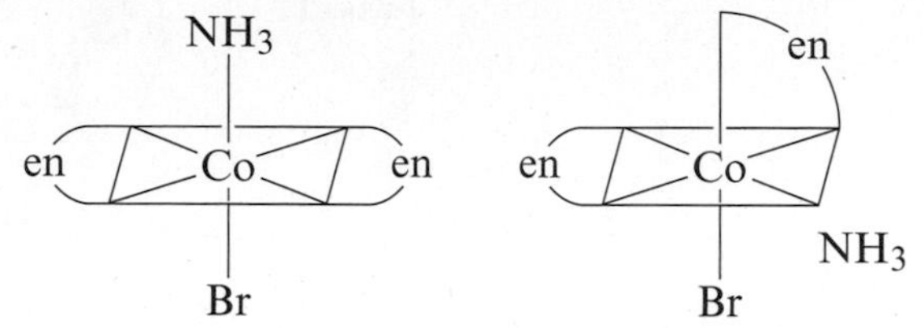

The *cis* isomer of the complex ion also exhibits optical isomerism because its mirror images

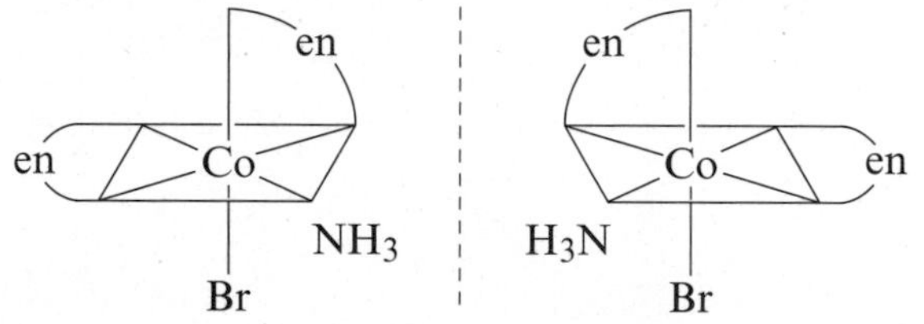

cannot be turned in any way to make them superimposable. Thus these mirror-image isomers of the *cis* complex are shown to be enantiomers that will rotate plane-polarized light in opposite directions.

See Exercises 20-47 and 20-48

20-5 | Bonding in Complex Ions: The Localized Electron Model

In Chapters 3 and 4 we considered the localized electron model, a very useful model for describing the bonding in molecules. Recall that a central feature of this model is the formation of hybrid atomic orbitals that are used to share electron pairs to form σ bonds between atoms. This same model can be used to account for the bonding in complex ions, but there are two important points to keep in mind:

1. The VSEPR model for predicting structure generally *does not work for complex ions*. However, we can safely assume that a complex ion with a coordination number of 6 will have an octahedral arrangement of ligands, and complexes with two ligands will be linear. On the other hand, complex ions with a coordination number of 4 can be either tetrahedral or square planar, and there is no completely reliable way to predict which will occur in a particular case.

Figure 20-18 | A set of six d^2sp^3 hybrid orbitals on Co^{3+} can accept an electron pair from each of six NH_3 ligands to form the $Co(NH_3)_6^{3+}$ ion.

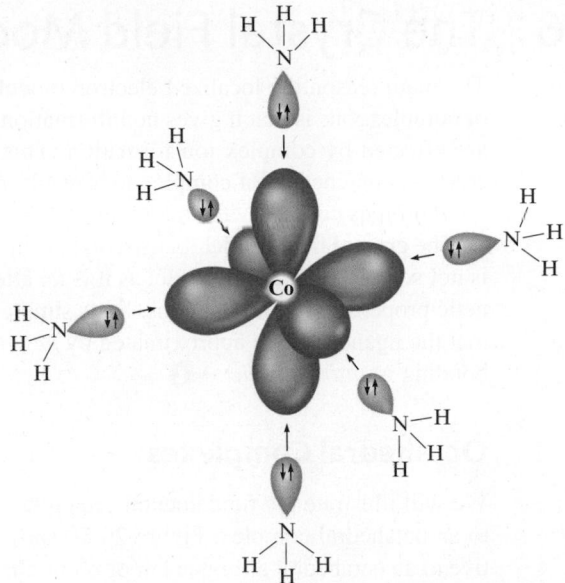

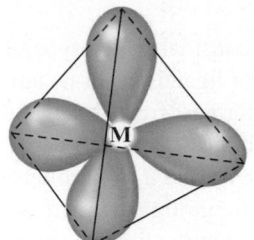

Tetrahedral ligand arrangement; sp^3 hybridization

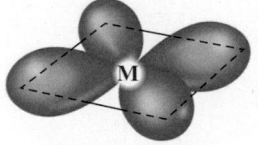

Square planar ligand arrangement; dsp^2 hybridization

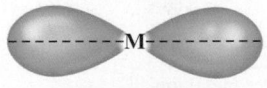

Linear ligand arrangement; sp hybridization

Figure 20-19 | The hybrid orbitals required for tetrahedral, square planar, and linear complex ions. The metal ion hybrid orbitals are empty, and the metal ion bonds to the ligands by accepting lone pairs.

2. The interaction between a metal ion and a ligand can be viewed as a Lewis acid–base reaction with the ligand donating a lone pair of electrons to an *empty* orbital of the metal ion to form a coordinate covalent bond:

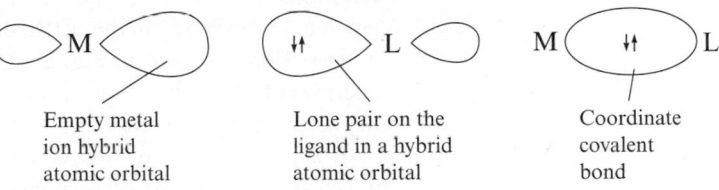

| Empty metal ion hybrid atomic orbital | Lone pair on the ligand in a hybrid atomic orbital | Coordinate covalent bond |

The hybrid orbitals used by the metal ion depend on the number and arrangement of the ligands. For example, accommodating the lone pairs from the six ammonia molecules in the octahedral $Co(NH_3)_6^{3+}$ ion requires a set of six empty hybrid atomic orbitals in an octahedral arrangement. As we discussed in Section 4-3, an octahedral set of orbitals is formed by the hybridization of two d, one s, and three p orbitals to give a set of six d^2sp^3 orbitals (Fig. 20-18).

The hybrid orbitals required on a metal ion in a four-coordinate complex depend on whether the structure is tetrahedral or square planar. For a tetrahedral arrangement of ligands, an sp^3 hybrid set is required (Fig. 20-19). For example, in the tetrahedral $CoCl_4^{2-}$ ion, the Co^{2+} can be described as sp^3 hybridized. A square planar arrangement of ligands requires a dsp^2 hybrid orbital set on the metal ion (Fig. 20-19). For example, in the square planar $Ni(CN)_4^{2-}$ ion, the Ni^{2+} is described as dsp^2 hybridized.

A linear complex requires two hybrid orbitals 180 degrees from each other. This arrangement is given by an sp hybrid set (see Fig. 20-19). Thus, in the linear $Ag(NH_3)_2^{+}$ ion, the Ag^+ can be described as sp hybridized.

Although the localized electron model can account in a general way for metal–ligand bonds, it is rarely used today because it cannot readily account for important properties of complex ions, such as magnetism and color. Thus we will not pursue the model any further.

20-6 | The Crystal Field Model

ⓘ Recall that models in science are always simplifications, and different models are used to answer different questions. The localized electron model is an attempt to answer the question, "Why would a metal ion bond with a ligand?," while the crystal field model is an attempt to answer questions about the colors and magnetic properties of complex ions.

The main reason the localized electron model cannot fully account for the properties of complex ions is that it gives no information about how the energies of the d orbitals are affected by complex ion formation. This is critical because, as we will see, the color and magnetism of complex ions result from changes in the energies of the metal ion d orbitals caused by the metal–ligand interactions.

The **crystal field model** focuses on the energies of the d orbitals. In fact, this model is not so much a bonding model as it is an attempt to account for the colors and magnetic properties of complex ions. In its simplest form, the crystal field model assumes that the ligands can be approximated by *negative point charges* and that metal–ligand bonding is *entirely ionic.* ◄ ⓘ

Octahedral Complexes

We will illustrate the fundamental principles of the crystal field model by applying it to an octahedral complex. Figure 20-20 shows the orientation of the $3d$ orbitals relative to an octahedral arrangement of point-charge ligands. The important thing to note is that two of the orbitals, d_{z^2} and $d_{x^2-y^2}$, point their lobes *directly at* the point-charge ligands and three of the orbitals, d_{xz}, d_{yz}, and d_{xy}, point their lobes *between* the point charges.

To understand the effect of this difference, we need to consider which type of orbital is lower in energy. Because the negative point-charge ligands repel negatively charged electrons, the electrons will first fill the d orbitals farthest from the ligands to minimize repulsions. In other words, the d_{xz}, d_{yz}, and d_{xy} orbitals (known as the t_{2g} set) are at a *lower energy* in the octahedral complex than are the d_{z^2} and $d_{x^2-y^2}$ orbitals (the e_g set). This is shown in Fig. 20-21. The negative point-charge ligands increase the energies of all the d orbitals. However, the orbitals that point at the ligands are raised in energy more than those that point between the ligands.

It is this **splitting of the 3d orbital energies** (symbolized by Δ) that explains the color and magnetism of complex ions of the first-row transition metal ions. For

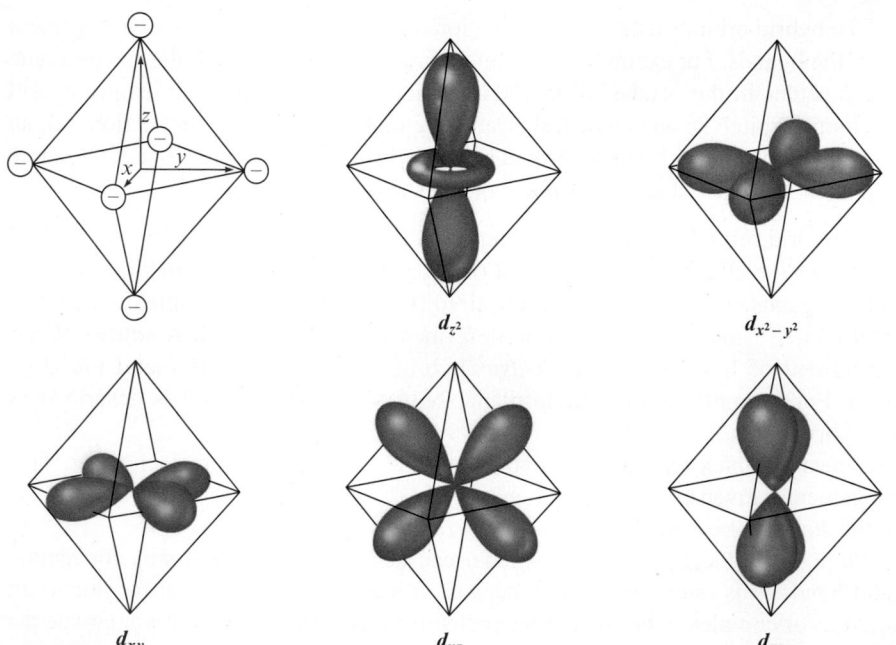

d_{z^2} $d_{x^2-y^2}$

d_{xy} d_{yz} d_{xz}

Figure 20-20 | An octahedral arrangement of point-charge ligands and the orientation of the 3d orbitals.

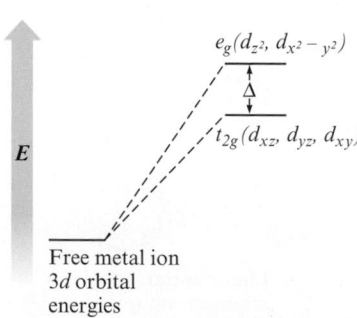

E

$e_g(d_{z^2}, d_{x^2-y^2})$

Δ

$t_{2g}(d_{xz}, d_{yz}, d_{xy})$

Free metal ion
3d orbital
energies

Figure 20-21 | The energies of the 3d orbitals for a metal ion in an octahedral complex. The 3d orbitals are degenerate (all have the same energy) in the free metal ion. In the octahedral complex the orbitals are split into two sets as shown. The difference in energy between the two sets is designated as Δ (delta).

Figure 20-22 | Possible electron arrangements in the split 3d orbitals in an octahedral complex of Co^{3+} (electron configuration 3d^6). (a) In a strong field (large Δ value), the electrons fill the t_{2g} set first, giving a diamagnetic complex. (b) In a weak field (small Δ value), the electrons occupy all five orbitals before any pairing occurs.

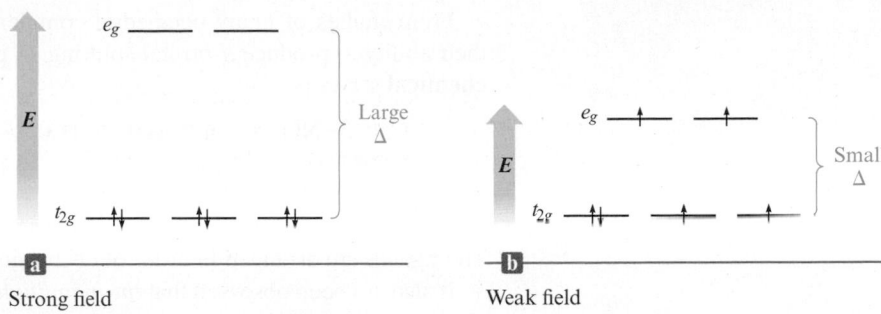

example, in an octahedral complex of Co^{3+} (a metal ion with six 3d electrons), there are two possible ways to place the electrons in the split 3d orbitals (Fig. 20-22). If the splitting produced by the ligands is very large, a situation called the **strong-field case**, the electrons will pair in the lower-energy t_{2g} orbitals. This gives a *diamagnetic* complex in which all the electrons are paired. On the other hand, if the splitting is small (the **weak-field case**), the electrons will occupy all five orbitals before pairing occurs. In this case the complex has four unpaired electrons and is *paramagnetic*.

The crystal field model allows us to account for the differences in the magnetic properties of Co(NH$_3$)$_6$$^{3+}$ and CoF$_6$$^{3-}$. The Co(NH$_3$)$_6$$^{3+}$ ion is known to be diamagnetic and thus corresponds to the strong-field case, also called the **low-spin case**, since it yields the *minimum* number of unpaired electrons. In contrast, the CoF$_6$$^{3-}$ ion, which is known to have four unpaired electrons, corresponds to the weak-field case, also known as the **high-spin case**, since it gives the *maximum* number of unpaired electrons.

Critical Thinking

What if you are told the number of unpaired electrons for a coordinate covalent ion and are asked to tell if the ligand produced a strong or weak field? Give an example of a coordinate covalent ion for which you could decide if it produced a strong or weak field and one for which you couldn't, and explain your answers.

INTERACTIVE EXAMPLE 20-4

Crystal Field Model I

The Fe(CN)$_6$$^{3-}$ ion is known to have one unpaired electron. Does the CN$^-$ ligand produce a strong or weak field?

Solution

Since the ligand is CN$^-$ and the overall complex ion charge is 3$-$, the metal ion must be Fe^{3+}, which has a 3d^5 electron configuration. The two possible arrangements of the five electrons in the d orbitals split by the octahedrally arranged ligands are

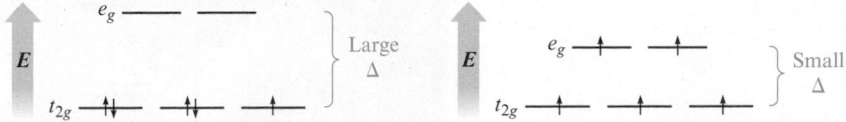

The strong-field case gives one unpaired electron, which agrees with the experimental observation. The CN$^-$ ion is a strong-field ligand toward the Fe^{3+} ion.

See Exercises 20-53 and 20-54

From studies of many octahedral complexes, we can arrange ligands in order of their ability to produce d-orbital splitting. A partial listing of ligands in this **spectrochemical series** is

$$CN^- > NO_2^- > en > NH_3 > H_2O > OH^- > F^- > Cl^- > Br^- > I^-$$

Strong-field
ligands
(large Δ)

Weak-field
ligands
(small Δ)

The ligands are arranged in order of decreasing Δ values toward a given metal ion.

It also has been observed that *the magnitude of Δ for a given ligand increases as the charge on the metal ion increases.* For example, NH_3 is a weak-field ligand toward Co^{2+} but acts as a strong-field ligand toward Co^{3+}. This makes sense; as the metal ion charge increases, the ligands will be drawn closer to the metal ion because of the increased charge density. As the ligands move closer, they cause greater splitting of the d orbitals and produce a larger Δ value.

INTERACTIVE EXAMPLE 20-5

Crystal Field Model II

Predict the number of unpaired electrons in the complex ion $Cr(CN)_6^{4-}$.

Solution

The net charge of $4-$ means that the metal ion present must be Cr^{2+} $(-6 + 2 = -4)$, which has a $3d^4$ electron configuration. Since CN^- is a strong-field ligand (see the spectrochemical series), the correct crystal field diagram for $Cr(CN)_6^{4-}$ is

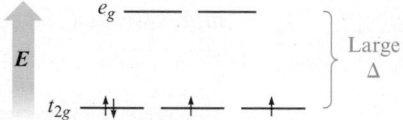

The complex ion will have two unpaired electrons. Note that the CN^- ligand produces such a large splitting that all four electrons will occupy the t_{2g} set even though two of the electrons must be paired in the same orbital.

See Exercises 20-55 and 20-56

We have seen how the crystal field model can account for the magnetic properties of octahedral complexes. The same model also can explain the colors of these complex ions. For example, $Ti(H_2O)_6^{3+}$, an octahedral complex of Ti^{3+}, which has a $3d^1$ electron configuration, is violet because it absorbs light in the middle of the visible region of the spectrum (Fig. 20-23). When a substance absorbs certain wavelengths of light in the visible region, the color of the substance is determined by the wavelengths of visible light that remain. We say that the substance exhibits the color *complementary* to those absorbed. The $Ti(H_2O)_6^{3+}$ ion is violet because it absorbs light in the

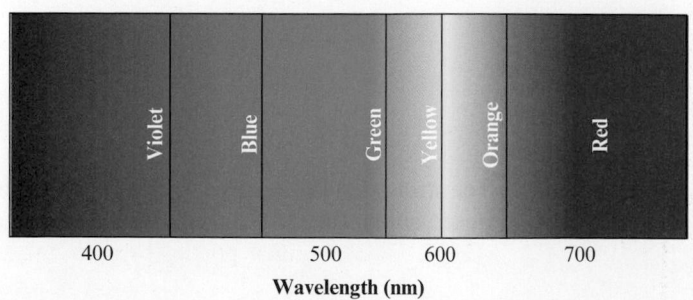

Figure 20-23 | The visible spectrum.

Wavelength (nm)

The beautiful, pure color of gems, so valued by cultures everywhere, arises from trace transition metal ion impurities in minerals that would otherwise be colorless. For example, the stunning red of a ruby, the most valuable of all gemstones, is caused by Cr^{3+} ions, which replace about 1% of the Al^{3+} ions in the mineral corundum, which is a form of aluminum oxide (Al_2O_3) that is nearly as hard as diamond. In the corundum structure the Cr^{3+} ions are surrounded by six oxide ions at the vertices of an octahedron. This leads to the characteristic octahedral splitting of chromium's $3d$ orbitals, such that the Cr^{3+} ions absorb strongly in the blue-violet and yellow-green regions of the visible spectrum but transmit red light to give the characteristic ruby color. (On the other hand, if some of the Al^{3+} ions in the corundum are replaced by a mixture of Fe^{2+}, Fe^{3+}, and Ti^{4+} ions, the gem is a sapphire with its brilliant blue color; or if some of the Al^{3+} ions are replaced by Fe^{3+} ions, the stone is a yellow topaz.)

Emeralds are derived from the mineral beryl, a beryllium aluminum silicate (empirical formula $3BeO \cdot Al_2O_3 \cdot 6SiO_2$). When some of the Al^{3+} ions in beryl are replaced by Cr^{3+} ions, the characteristic green color of emerald results. In this environment the splitting of the Cr^{3+} $3d$ orbitals causes it to strongly absorb yellow and blue-violet light and to transmit green light.

A gem closely related to ruby and emerald is alexandrite, named after Alexander II of Russia. This gem is based on the mineral chrysoberyl, a beryllium aluminate with the empirical formula $BeO \cdot Al_2O_3$ in which approximately 1% of the Al^{3+} ions are replaced by Cr^{3+} ions. In the chrysoberyl environment Cr^{3+} absorbs strongly in the yellow region of the spectrum. Alexandrite has the interesting property of changing colors depending on the light source. When the first alexandrite stone was discovered deep in a mine in the Russian Ural Mountains in 1831, it appeared to be a deep red color in the firelight of the miners' lamps. However, when the stone was brought to the surface, its color was blue. This seemingly magical color change occurs because the firelight of a miner's helmet is rich in the yellow and red wavelengths of the visible spectrum but does not contain much blue. Absorption of the yellow by the stone produces a reddish color. However, daylight has much more intensity in the blue region than firelight. Thus the extra blue in the light transmitted by the stone gives it bluish color in daylight.

Once the structure of a natural gem is known, it is usually not very difficult to make the gem artificially. For example, rubies and sapphires are made on a large scale by fusing $Al(OH)_3$ with the appropriate transition metal salts at approximately 1200°C to make the "doped" corundum. With these techniques gems of astonishing size can be manufactured: Rubies as large as 10 lb and sapphires up to 100 lb have been synthesized. Smaller synthetic stones produced for jewelry are virtually identical to the corresponding natural stones, and it takes great skill for a gemologist to tell the difference.

Alexandrite changes color depending on the light source. The image on the left is in daylight and the one on the right is illuminated by incandescent lighting.

yellow-green region, thus letting red light and blue light pass, which gives the observed violet color. This is shown schematically in Fig. 20-24. Table 20-16 shows the general relationship between the wavelengths of visible light absorbed and the approximate color observed.

The reason that the $Ti(H_2O)_6^{3+}$ ion absorbs a specific wavelength of visible light can be traced to the transfer of the lone d electron between the split d orbitals, as shown in Fig. 20-25. A given photon of light can be absorbed by a molecule only if the

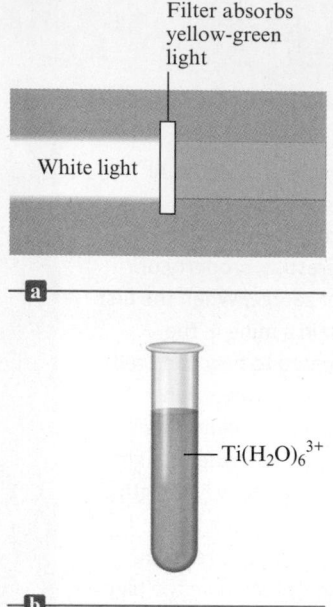

Filter absorbs
yellow-green
light

White light

a

Ti(H₂O)₆³⁺

b

Figure 20-24 | (a) When white light shines on a filter that absorbs in the yellow-green region, the emerging light is violet. (b) Because the complex ion $Ti(H_2O)_6^{3+}$ absorbs yellow-green light, a solution of it is violet.

Table 20-16 | Approximate Relationship of Wavelength of Visible Light Absorbed to Color Observed

Absorbed Wavelength in nm (Color)	Observed Color
400 (violet)	Greenish yellow
450 (blue)	Yellow
490 (blue-green)	Red
570 (yellow-green)	Violet
580 (yellow)	Dark blue
600 (orange)	Blue
650 (red)	Green

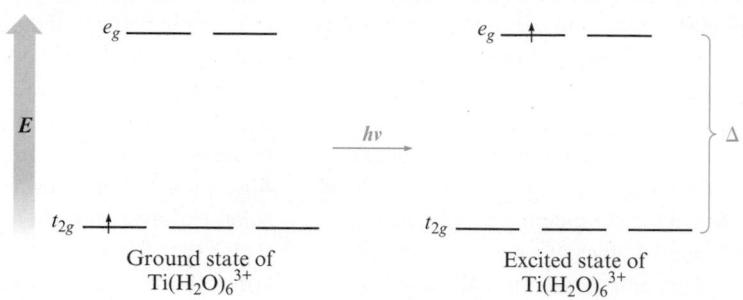

Ground state of
$Ti(H_2O)_6^{3+}$

Excited state of
$Ti(H_2O)_6^{3+}$

Figure 20-25 | The complex ion $Ti(H_2O)_6^{3+}$ can absorb visible light in the yellow-green region to transfer the lone d electron from the t_{2g} to the e_g set.

wavelength of the light provides exactly the energy needed by the molecule. In other words, the wavelength absorbed is determined by the relationship

$$\Delta E = \frac{hc}{\lambda}$$

where ΔE represents the energy spacing in the molecule (we have used simply Δ in this chapter) and λ represents the wavelength of light needed. Because the d-orbital splitting in most octahedral complexes corresponds to the energies of photons in the visible region, octahedral complex ions are usually colored.

Since the ligands coordinated to a given metal ion determine the size of the d-orbital splitting, the color changes as the ligands are changed. ◄ This occurs because a change in Δ means a change in the wavelength of light needed to transfer electrons between the t_{2g} and e_g orbitals. Several octahedral complexes of Cr^{3+} and their colors are listed in Table 20-17.

Table 20-17 | Several Octahedral Complexes of Cr³⁺ and Their Colors

Isomer	Color
[Cr(H₂O)₆]Cl₃	Violet
[Cr(H₂O)₅Cl]Cl₂	Blue-green
[Cr(H₂O)₄Cl₂]Cl	Green
[Cr(NH₃)₆]Cl₃	Yellow
[Cr(NH₃)₅Cl]Cl₂	Purple
[Cr(NH₃)₄Cl₂]Cl	Violet

Other Coordination Geometries

Using the same principles developed for octahedral complexes, we will now consider complexes with other geometries. For example, Fig. 20-26 shows a tetrahedral arrangement of point charges in relation to the $3d$ orbitals of a metal ion. There are two important facts to note:

1. None of the $3d$ orbitals "point at the ligands" in the tetrahedral arrangement, as the $d_{x^2-y^2}$ and d_{z^2} orbitals do in the octahedral case. Thus the tetrahedrally arranged ligands do not differentiate the d orbitals as much in the tetrahedral case as in the octahedral case. That is, the difference in energy between the split d orbitals will be significantly less in tetrahedral complexes. Although we will not derive it here, the tetrahedral splitting is $\frac{4}{9}$ that of the octahedral splitting for a given ligand and metal ion:

$$\Delta_{tet} = \tfrac{4}{9}\Delta_{oct}$$

▲ Solutions of [Cr(NH₃)₆]Cl₃ (left) and [Cr(NH₃)₅Cl]Cl₂ (right).

Ken O'Donoghue © Cengage Learning

Figure 20-26 | (a) Tetrahedral and octahedral arrangements of ligands shown inscribed in cubes. Note that in the two types of arrangements, the point charges occupy opposite parts of the cube; the octahedral point charges are at the centers of the cube faces, and the tetrahedral point charges occupy opposite corners of the cube. (b) The orientations of the $3d$ orbitals relative to the tetrahedral set of point charges.

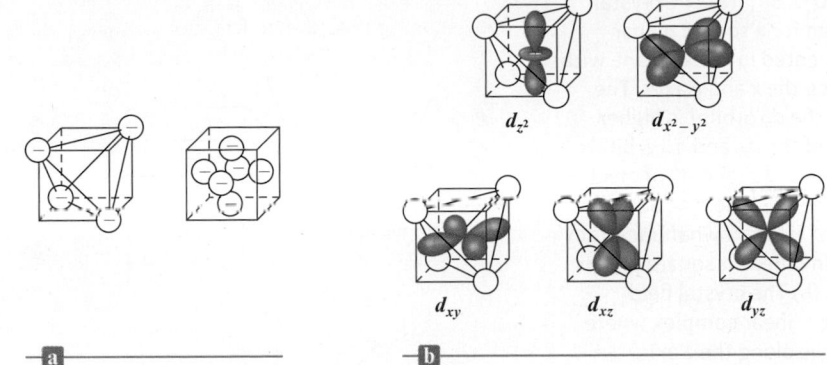

2. Although not exactly pointing at the ligands, the d_{xy}, d_{xz}, and d_{yz} orbitals are closer to the point charges than are the d_{z^2} and $d_{x^2-y^2}$ orbitals. This means that the tetrahedral d-orbital splitting will be opposite to that for the octahedral arrangement. The two arrangements are contrasted in Fig. 20-27. Because the d-orbital splitting is relatively small for the tetrahedral case, the weak-field case (high-spin case) *always* applies. There are no known ligands powerful enough to produce the strong-field case in a tetrahedral complex.

INTERACTIVE EXAMPLE 20-6

Crystal Field Model III

Give the crystal field diagram for the tetrahedral complex ion $CoCl_4^{2-}$.

Solution

The complex ion contains Co^{2+}, which has a $3d^7$ electron configuration. The splitting of the d orbitals will be small, since this is a tetrahedral complex, giving the high-spin case with three unpaired electrons.

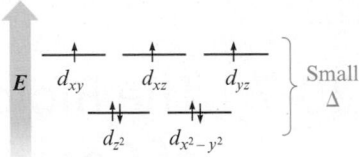

See Exercises 20-61 and 20-63

The crystal field model also applies to square planar and linear complexes. The crystal field diagrams for these cases are shown in Fig. 20-28. The ranking of orbitals in these diagrams can be explained by considering the relative orientations of the point charges and the orbitals. The diagram in Fig. 20-27 for the octahedral arrangement can be used to obtain these orientations. We can obtain the square planar complex by

Figure 20-27 | The crystal field diagrams for octahedral and tetrahedral complexes. The relative energies of the sets of d orbitals are reversed. For a given type of ligand, the splitting is much larger for the octahedral complex ($\Delta_{oct} > \Delta_{tet}$) because in this arrangement the d_{z^2} and $d_{x^2-y^2}$ orbitals point their lobes directly at the point charges and are thus relatively high in energy.

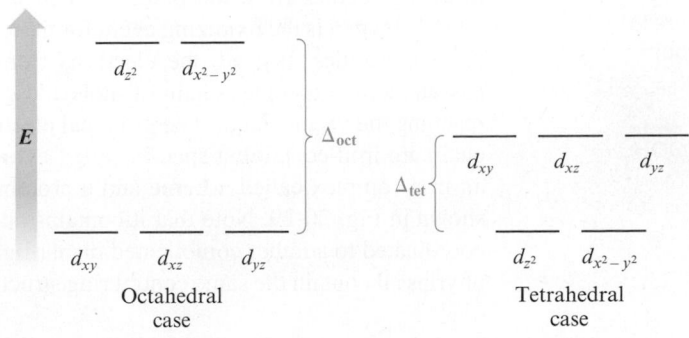

Figure 20-28 | (a) The crystal field diagram for a square planar complex oriented in the xy plane with ligands along the x and y axes. The position of the d_{z^2} orbital is higher than those of the d_{xz} and d_{yz} orbitals because of the "doughnut" of electron density in the xy plane. The actual position of d_{z^2} is somewhat uncertain and varies in different square planar complexes. (b) The crystal field diagram for a linear complex where the ligands lie along the z axis.

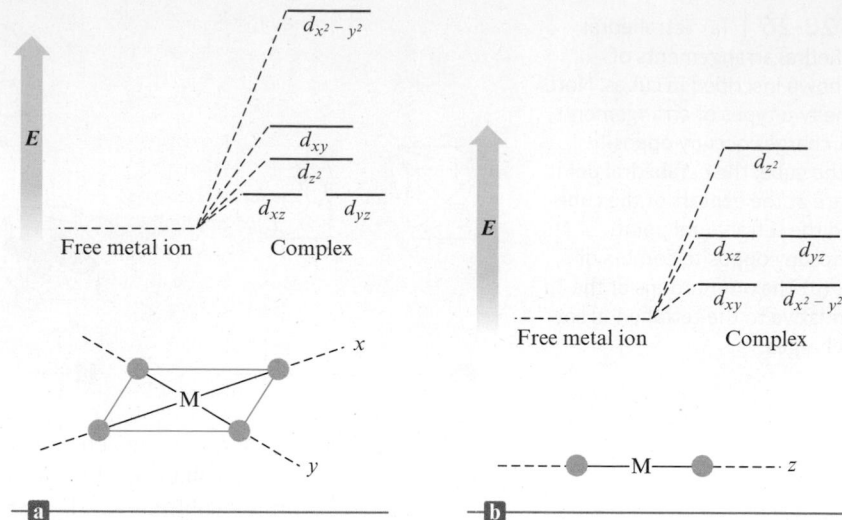

removing the two point charges along the z axis. This will greatly lower the energy of d_{z^2}, leaving only $d_{x^2-y^2}$, which points at the four remaining ligands as the highest-energy orbital. We can obtain the linear complex from the octahedral arrangement by leaving the two ligands along the z axis and removing the four in the xy plane. This means that only the d_{z^2} points at the ligands and is highest in energy.

Critical Thinking

Figure 20-28(a) shows a crystal field diagram for a square planar complex oriented in the xy plane. What if you oriented the complex in the xz plane? Sketch the crystal field diagram and contrast it with Fig. 20-28(a).

20-7 | The Biological Importance of Coordination Complexes

The ability of metal ions to coordinate with and release ligands and to easily undergo oxidation and reduction makes them ideal for use in biological systems. For example, metal ion complexes are used in humans for the transport and storage of oxygen, as electron-transfer agents, as catalysts, and as drugs. Most of the first-row transition metals are essential for human health, as summarized in Table 20-18. We will concentrate on iron's role in biological systems, since several of its coordination complexes have been studied extensively.

Iron plays a central role in almost all living cells. In mammals, the principal source of energy comes from the oxidation of carbohydrates, proteins, and fats. ◄ *ⓘ* Although oxygen is the oxidizing agent for these processes, it does not react directly with these molecules. Instead, the electrons from the breakdown of these nutrients are passed along a complex chain of molecules, called the *respiratory chain,* eventually reaching the O_2 molecule. The principal electron-transfer molecules in the respiratory chain are iron-containing species called **cytochromes**, consisting of two main parts: an iron complex called a **heme** and a protein. The structure of the heme complex is shown in Fig. 20-29. Note that it contains an iron ion (it can be either Fe^{2+} or Fe^{3+}) coordinated to a rather complicated planar ligand called a **porphyrin**. As a class, porphyrins all contain the same central ring structure but have different substituent groups

ⓘ A protein is a large molecule assembled from amino acids, which have the general structure in which R represents various groups.

$$H_2N - \underset{\underset{H}{|}}{\overset{\overset{R}{|}}{C}} - COOH$$

Table 20-18 | The First-Row Transition Metals and Their Biological Significance

First-Row Transition Metal	Biological Function(s)
Scandium	None known.
Titanium	None known.
Vanadium	None known in humans.
Chromium	Assists Insulin in the control of blood sugar; may also be involved in the control of cholesterol.
Manganese	Necessary for a number of enzymatic reactions.
Iron	Component of hemoglobin and myoglobin; involved in the electron-transport chain.
Cobalt	Component of vitamin B_{12}, which is essential for the metabolism of carbohydrates, fats, and proteins.
Nickel	Component of the enzymes urease and hydrogenase.
Copper	Component of several enzymes; assists in iron storage; involved in the production of color pigments of hair, skin, and eyes.
Zinc	Component of insulin and many enzymes.

at the edges of the rings. The various porphyrin molecules act as tetradentate ligands for many metal ions, including iron, cobalt, and magnesium. In fact, *chlorophyll,* a substance essential to the process of photosynthesis, is a magnesium–porphyrin complex of the type shown in Fig. 20-30.

In addition to participating in the transfer of electrons from nutrients to oxygen, iron plays a principal role in the transport and storage of oxygen in mammalian blood and tissues. Oxygen is stored in a molecule called **myoglobin**, which consists of a heme complex and a protein in a structure very similar to that of the cytochromes. In myoglobin, the Fe^{2+} ion is coordinated to four nitrogen atoms of the porphyrin ring

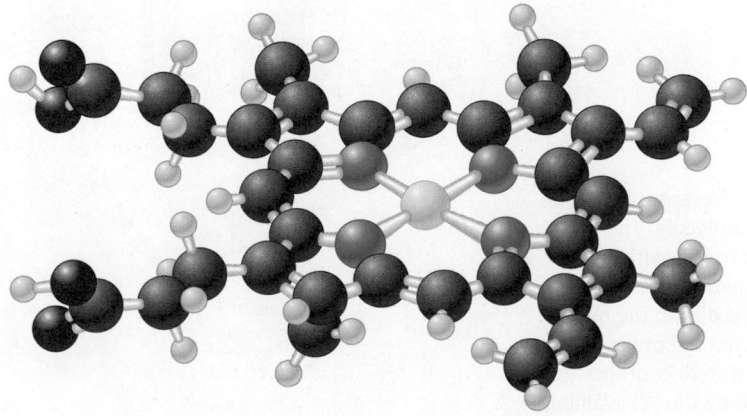

Figure 20-29 | The heme complex, in which an Fe^{2+} ion is coordinated to four nitrogen atoms of a planar porphyrin ligand.

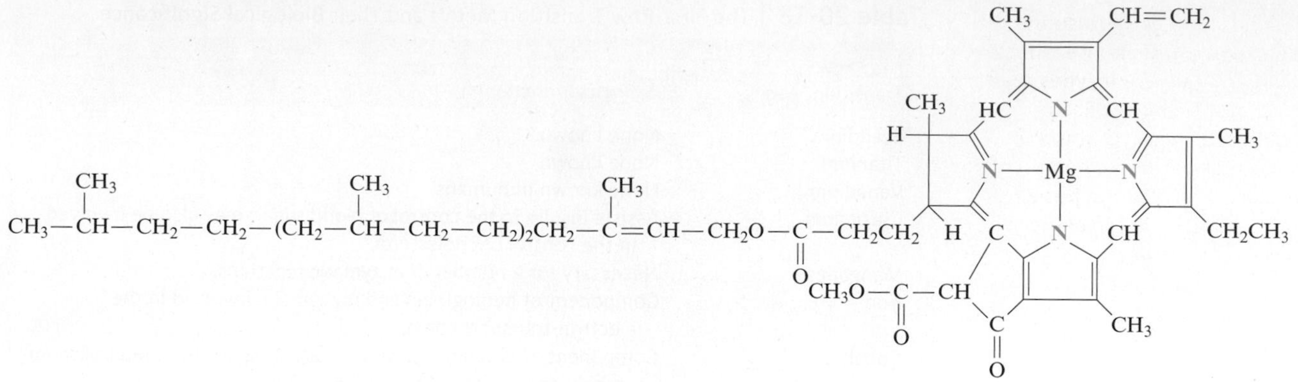

Figure 20-30 | Chlorophyll is a porphyrin complex of Mg^{2+}. There are two similar forms of chlorophyll, one of which is shown here.

and to a nitrogen atom of the protein chain, as shown in Fig. 20-31. Since Fe^{2+} is normally six-coordinate, this leaves one position open for attachment of an O_2 molecule.

One especially interesting feature of myoglobin is that it involves an O_2 molecule attaching directly to Fe^{2+}. However, if gaseous O_2 is bubbled into an aqueous solution containing heme, the Fe^{2+} is immediately oxidized to Fe^{3+}. This oxidation of the Fe^{2+} in heme does not happen in myoglobin. This fact is of crucial importance because Fe^{3+} does not form a coordinate covalent bond with O_2, and myoglobin would not function if the bound Fe^{2+} could be oxidized. Since the Fe^{2+} in the "bare" heme complex can be oxidized, it must be the protein that somehow prevents the oxidation. How? Based on much research, the answer seems to be that the oxidation of Fe^{2+} to Fe^{3+} involves an oxygen bridge between two iron ions (the circles indicate the ligands):

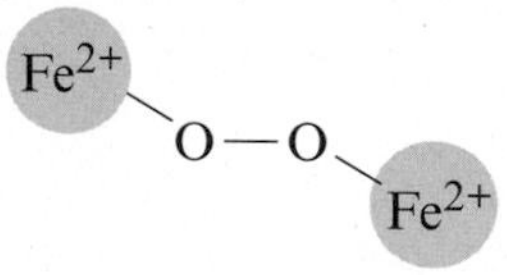

$$Fe^{2+} \quad O-O \quad Fe^{2+}$$

The bulky protein around the heme group in myoglobin prevents two molecules from getting close enough to form the oxygen bridge, and so oxidation of the Fe^{2+} is prevented.

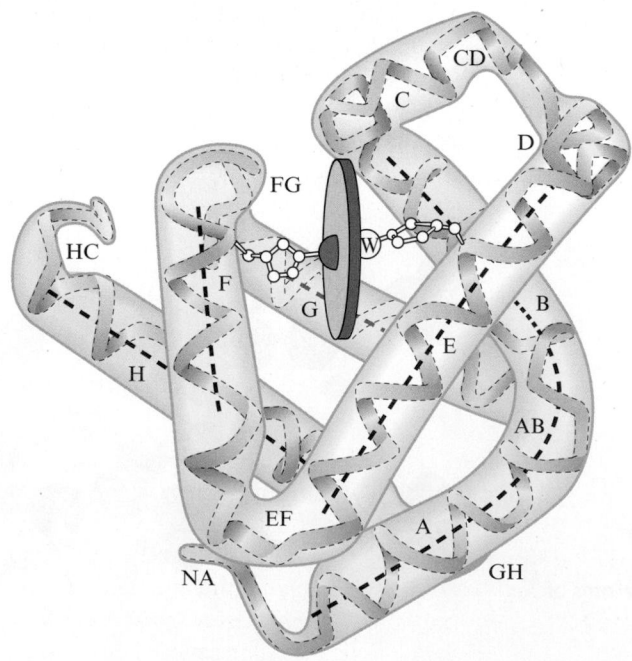

Figure 20-31 | A representation of the myoglobin molecule. The Fe^{2+} ion is coordinated to four nitrogen atoms in the porphyrin of the heme (represented by the disk in the figure) and on nitrogen from the protein chain. This leaves a sixth coordination position (indicated by the W) available for an oxygen molecule.

Figure 20-32 | A representation of the hemoglobin structure. There are two slightly different types of protein chains (α and β). Each hemoglobin has two α chains and two β chains, each with a heme complex near the center. Thus each hemoglobin molecule can complex with four O_2 molecules.

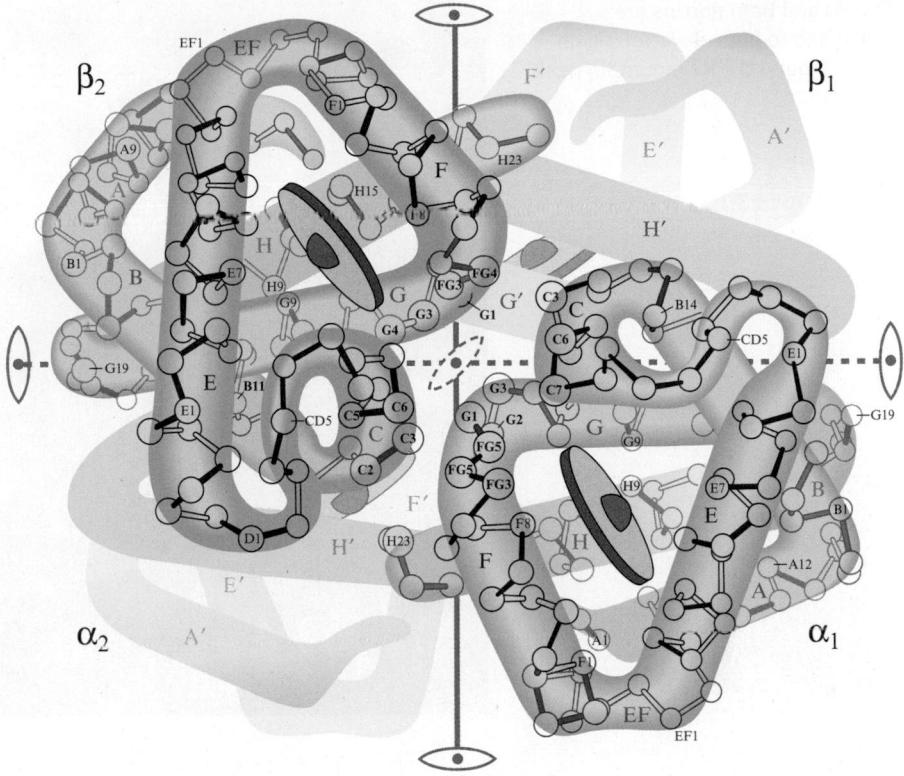

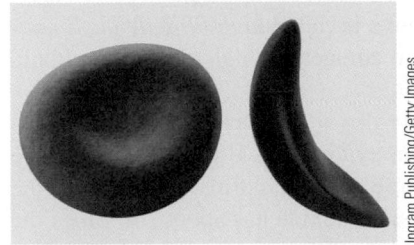

Figure 20-33 | Normal vs. sickle blood cells

The transport of O_2 in the blood is carried out by **hemoglobin**, a molecule consisting of four myoglobin-like units, as shown in Fig. 20-32. Each hemoglobin can therefore bind four O_2 molecules to form a bright red diamagnetic complex. The diamagnetism occurs because oxygen is a strong-field ligand toward Fe^{2+}, which has a $3d^6$ electron configuration. When the oxygen molecule is released, water molecules occupy the sixth coordination position around each Fe^{2+}, giving a bluish paramagnetic complex (H_2O is a weak-field ligand toward Fe^{2+}) that gives venous blood its characteristic bluish tint.

Hemoglobin dramatically demonstrates how sensitive the function of a biomolecule is to its structure. In certain people, in the synthesis of the proteins needed for hemoglobin, an improper amino acid is inserted into the protein in two places. This may not seem very serious, since there are several hundred amino acids present. However, because the incorrect amino acid has a nonpolar substituent instead of the polar one found on the proper amino acid, the hemoglobin drastically changes its shape. The red blood cells are then sickle-shaped rather than disk-shaped, as shown in Fig. 20-33. The misshapen cells can aggregate, causing clogging of tiny capillaries. This condition, known as *sickle cell anemia,* is the subject of intense research.

Our knowledge of the workings of hemoglobin allows us to understand the effects of high altitudes on humans. The reaction between hemoglobin and oxygen can be represented by the following equilibrium:

$$Hb(aq) + 4O_2(g) \rightleftharpoons Hb(O_2)_4(aq)$$
Hemoglobin   Oxyhemoglobin

At high altitudes, where the oxygen content of the air is lower, the position of this equilibrium will shift to the left, according to Le Châtelier's principle. Because less oxyhemoglobin is formed, fatigue, dizziness, and even a serious illness called *high-altitude sickness* can result. One way to combat this problem is to use supplemental oxygen, as most high-altitude mountain climbers do. However, this is impractical for people who live at high elevations. In fact, the human body adapts to the lower oxygen concentrations by making more hemoglobin, causing the equilibrium to shift back to

▲ Sherpa and Balti porters are acclimatized to high elevations such as those around the K2 mountain peak in Pakistan.

Gopal Chitrakar/Reuters

the right. Someone moving from Chicago to Boulder, Colorado (5300 feet), would notice the effects of the new altitude for a couple of weeks, but as the hemoglobin level increased, the effects would disappear. This change is called *high-altitude acclimatization,* which explains why athletes who want to compete at high elevations should practice there for several weeks prior to the event. ◄

Our understanding of the biological role of iron also allows us to explain the toxicities of substances such as carbon monoxide and the cyanide ion. Both CO and CN^- are very good ligands toward iron and so can interfere with the normal workings of the iron complexes in the body. For example, carbon monoxide has about 200 times the affinity for the Fe^{2+} in hemoglobin as oxygen does. The resulting stable complex, **carboxyhemoglobin**, prevents the normal uptake of O_2, thus depriving the body of needed oxygen. Asphyxiation can result if enough carbon monoxide is present in the air. The mechanism for the toxicity of the cyanide ion is somewhat different. Cyanide coordinates strongly to cytochrome oxidase, an iron-containing cytochrome enzyme that catalyzes the oxidation–reduction reactions of certain cytochromes. The coordinated cyanide thus prevents the electron-transfer process and rapid death results. Because of its behavior, cyanide is called a *respiratory inhibitor.*

20-8 | Metallurgy and Iron and Steel Production

In the preceding section we saw the importance of iron in biological systems. Of course, iron is also very important in many other ways in our world. In this section we will discuss the isolation of metals from their natural sources and the formulation of metals into useful materials, with special emphasis on the role of iron.

Metals are very important for structural applications, electrical wires, cooking utensils, tools, decorative items, and many other purposes. However, because the main chemical characteristic of a metal is its ability to give up electrons, almost all metals in nature are found in ores, combined with nonmetals such as oxygen, sulfur, and the

▲ Pouring molten metal in a steel mill.

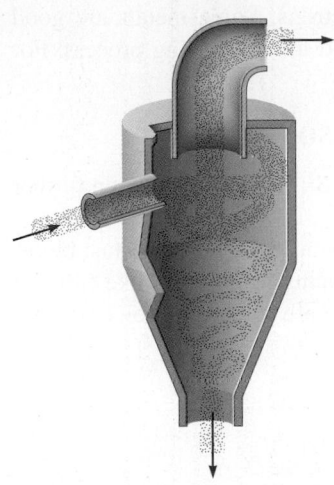

Figure 20-34 | A schematic diagram of a cyclone separator. The ore is pulverized and blown into the separator. The more dense mineral particles are thrown toward the walls by centrifugal force and fall down the funnel. The lighter particles (gangue) tend to stay closer to the center and are drawn out through the top by the stream of air.

halogens. To recover and use these metals, we must separate them from their ores and reduce the metal ions. Then, because most metals are unsuitable for use in the pure state, we must form alloys that have the desired properties. The process of separating a metal from its ore and preparing it for use is known as **metallurgy**. ◄ The steps in this process are typically

1. Mining
2. Pretreatment of the ore
3. Reduction to the free metal
4. Purification of the metal (refining)
5. Alloying

An ore can be viewed as a mixture containing **minerals** (relatively pure metal compounds) and **gangue** (sand, clay, and rock). Some typical minerals are listed in Table 20-19. Although silicate minerals are the most common in the earth's crust, they are typically very hard and difficult to process, making metal extraction relatively expensive. Therefore, other ores are used when available.

After mining, an ore must be treated to remove the gangue and to concentrate the mineral. The ore is first pulverized and then processed in a variety of devices, including cyclone separators (Fig. 20-34), inclined vibrating tables, and flotation tanks.

In the **flotation process**, the crushed ore is fed into a tank containing a water–oil–detergent mixture. Because of the difference in the surface characteristics of the mineral particles and the silicate rock particles, the oil wets the mineral particles. A stream of air blown through the mixture causes tiny bubbles to form on the oil-covered pieces, which then float to the surface, where they can be skimmed off.

After the mineral has been concentrated, it is often chemically altered in preparation for the reduction step. For example, nonoxide minerals are often converted to oxides before reduction. Carbonates and hydroxides can be converted by simple heating:

$$CaCO_3(s) \xrightarrow{\text{Heat}} CaO(s) + CO_2(g)$$

$$Mg(OH)_2(s) \xrightarrow{\text{Heat}} MgO(s) + H_2O(g)$$

Table 20-19 | Common Minerals Found in Ores

Anion	Examples
None (free metal)	Au, Ag, Pt, Pd, Rh, Ir, Ru
Oxide	Fe_2O_3 (hematite)
	Fe_3O_4 (magnetite)
	Al_2O_3 (bauxite)
	SnO_2 (cassiterite)
Sulfide	PbS (galena)
	ZnS (sphalerite)
	FeS_2 (pyrite)
	HgS (cinnabar)
	Cu_2S (chalcocite)
Chloride	NaCl (rock salt)
	KCl (sylvite)
	$KCl \cdot MgCl_2$ (carnalite)
Carbonate	$FeCO_3$ (siderite)
	$CaCO_3$ (limestone)
	$MgCO_3$ (magnesite)
	$MgCO_3 \cdot CaCO_3$ (dolomite)
Sulfate	$CaSO_4 \cdot 2H_2O$ (gypsum)
	$BaSO_4$ (barite)
Silicate	$Be_3Al_2Si_6O_{18}$ (beryl)
	$Al_2(Si_2O_8)(OH)_4$ (kaolinite)
	$LiAl(SiO_3)_2$ (spodumene)

Figure 20-35 | A schematic representation of zone refining.

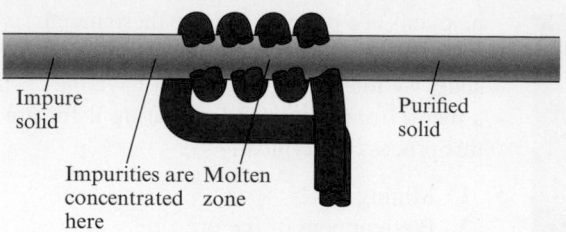

Sulfide minerals can be converted to oxides by heating in air at temperatures below their melting points, a process called **roasting**:

$$2ZnS(s) + 3O_2(g) \xrightarrow{\text{Heat}} 2ZnO(s) + 2SO_2(g)$$

As we have seen earlier, sulfur dioxide causes severe problems if released into the atmosphere, and modern roasting operations collect this gas and use it in the manufacture of sulfuric acid.

The method chosen to reduce the metal ion to the free metal, a process called **smelting**, depends on the affinity of the metal ion for electrons. Some metals are good enough oxidizing agents that the free metal is produced in the roasting process. For example, the roasting reaction for cinnabar is

$$HgS(s) + O_2(g) \xrightarrow{\text{Heat}} Hg(l) + SO_2(g)$$

where the Hg^{2+} is reduced by electrons donated by the S^{2-} ion, which is then further oxidized by O_2 to form SO_2.

The roasting of a more active metal produces the metal oxide, which must be reduced to obtain the free metal. The most common reducing agents are coke (impure carbon), carbon monoxide, and hydrogen. The following are some common examples of the reduction process:

$$Fe_2O_3(s) + 3CO(g) \xrightarrow{\text{Heat}} 2Fe(l) + 3CO_2(g)$$

$$WO_3(s) + 3H_2(g) \xrightarrow{\text{Heat}} W(l) + 3H_2O(g)$$

$$ZnO(s) + C(s) \xrightarrow{\text{Heat}} Zn(l) + CO(g)$$

The most active metals, such as aluminum and the alkali metals, must be reduced electrolytically, usually from molten salts (see Section 17-9).

The metal obtained in the reduction step is invariably impure and must be refined. The methods of refining include electrolytic refining (see Section 17-9), oxidation of impurities (as for iron, see below), and distillation of low-boiling metals such as mercury and zinc. One process used when highly pure metals are needed is **zone refining**. In this process a bar of the impure metal travels through a heater (Fig. 20-35), which causes melting and recrystallizing of the metal as the bar cools. Purification of the metal occurs because as the crystal re-forms, the metal ions are likely to fit much better in the crystal lattice than are the atoms of impurities. Thus the impurities tend to be excluded and carried to the end of the bar. Several repetitions of this process give a very pure metal bar.

Hydrometallurgy

The metallurgical processes we have considered so far are usually called **pyrometallurgy** (*pyro* means "at high temperatures"). These traditional methods require large quantities of energy and have two other serious problems: atmospheric pollution (mainly by sulfur dioxide) and relatively high costs that make treatment of low-grade ores economically unfeasible.

In the past hundred years, a different process, **hydrometallurgy** (*hydro* means "water"), has been used to extract metals from ores by use of aqueous chemical solutions, a process called **leaching**. The first two uses of hydrometallurgy were for the extraction of gold from low-grade ores and for the production of aluminum oxide, or alumina, from bauxite, an aluminum-bearing ore.

Gold is sometimes found in ores in the elemental state, but it usually occurs in relatively small concentrations. A process called **cyanidation** treats the crushed ore with an aqueous cyanide solution in the presence of air to dissolve the gold by forming the complex ion $Au(CN)_2^-$:

$$4Au(s) + 8CN^-(aq) + O_2(g) + 2H_2O(l) \longrightarrow 4Au(CN)_2^-(aq) + 4OH^-(aq)$$

Pure gold is then recovered by reaction of the solution of $Au(CN)_2^-$ with zinc powder to reduce Au^+ to Au:

$$2Au(CN)_2^-(aq) + Zn(s) \longrightarrow 2Au(s) + Zn(CN)_4^{2-}(aq)$$

The extraction of alumina from bauxite (the Bayer process) leaches the ore with sodium hydroxide at high temperatures and pressures to dissolve the amphoteric aluminum oxide:

$$Al_2O_3(s) + 2OH^-(aq) \longrightarrow 2AlO_2^-(aq) + H_2O(l)$$

This process leaves behind solid impurities such as SiO_2, Fe_2O_3, and TiO_2, which are not appreciably soluble in basic solution. After the solid impurities are removed, the pH of the solution is lowered, causing the pure aluminum oxide to re-form. It is then electrolyzed to produce aluminum metal (see Section 17-9).

As illustrated by these processes, hydrometallurgy involves two distinct steps: *selective leaching* of a given metal ion from the ore and recovery of the metal ion from the solution by *selective precipitation* as an ionic compound.

The leaching agent can simply be water if the metal-containing compound is a water-soluble chloride or sulfate. However, most commonly, the metal is present in a water-insoluble substance that must somehow be dissolved. The leaching agents used in such cases are usually aqueous solutions containing acids, bases, oxidizing agents, salts, or some combination of these. Often the dissolving process involves the formation of complex ions. For example, when an ore containing water-insoluble lead sulfate is treated with an aqueous sodium chloride solution, the soluble complex ion $PbCl_4^{2-}$ is formed:

$$PbSO_4(s) + 4Na^+(aq) + 4Cl^-(aq) \longrightarrow 4Na^+(aq) + PbCl_4^{2-}(aq) + SO_4^{2-}(aq)$$

Formation of a complex ion also occurs in the cyanidation process for the recovery of gold. However, since the gold is present in the ore as particles of metal, it must first be oxidized by oxygen to produce Au^+, which then reacts with CN^- to form the soluble $Au(CN)_2^-$ species. Thus, in this case, the leaching process involves a combination of oxidation and complexation.

Sometimes just oxidation is used. For example, insoluble zinc sulfide can be converted to soluble zinc sulfate by pulverizing the ore and suspending it in water to form a slurry through which oxygen is bubbled:

$$ZnS(s) + 2O_2(g) \longrightarrow Zn^{2+}(aq) + SO_4^{2-}(aq)$$

One advantage of hydrometallurgy over the traditional processes is that sometimes the leaching agent can be pumped directly into the ore deposits in the earth. For example, aqueous sodium carbonate (Na_2CO_3) can be injected into uranium-bearing ores to form water-soluble complex carbonate ions.

Recovering the metal ions from the leaching solution involves forming an insoluble solid containing the metal ion to be recovered. This step may involve addition of an anion to form an insoluble salt, reduction to the solid metal, or a combination of reduction and precipitation of a salt. Examples of these processes are shown in Table 20-20. Because of its suitability for treating low-grade ores economically and without significant pollution, hydrometallurgy is becoming more popular for recovering many important metals such as copper, nickel, zinc, and uranium.

Table 20-20 | Examples of Methods for Recovery of Metal Ions from Leaching Solutions

Method		Examples
Precipitation of a salt		$Cu^{2+}(aq) + S^{2-}(aq) \longrightarrow CuS(s)$
		$Cu^+(aq) + HCN(aq) \longrightarrow CuCN(s) + H^+(aq)$
	Chemical	$Au^+(aq) + Fe^{2+}(aq) \longrightarrow Au(s) + Fe^{3+}(aq)$
Reduction		$Cu^{2+}(aq) + Fe(s) \longrightarrow Cu(s) + Fe^{2+}(aq)$
		$Ni^{2+}(aq) + H_2(g) \longrightarrow Ni(s) + 2H^+(aq)$
	Electrolytic	$Cu^{2+}(aq) + 2e^- \longrightarrow Cu(s)$
		$Al^{3+}(aq) + 3e^- \longrightarrow Al(s)$
Reduction plus precipitation		$2Cu^{2+}(aq) + 2Cl^-(aq) + H_2SO_3(aq) + H_2O(l) \longrightarrow$
		$ 2CuCl(s) + 3H^+(aq) + HSO_4^-(aq)$

The Metallurgy of Iron

Iron is present in the earth's crust in many types of minerals. *Iron pyrite* (FeS_2) is widely distributed but is not suitable for production of metallic iron and steel because it is almost impossible to remove the last traces of sulfur. The presence of sulfur makes the resulting steel too brittle to be useful. *Siderite* ($FeCO_3$) is a valuable iron mineral that can be converted to iron oxide by heating. The iron oxide minerals are *hematite* (Fe_2O_3), the more abundant, and *magnetite* (Fe_3O_4, really $FeO \cdot Fe_2O_3$). *Taconite* ores contain iron oxides mixed with silicates and are more difficult to process than the others. However, taconite ores are being increasingly used as the more desirable ores are consumed.

To concentrate the iron in iron ores, advantage is taken of the natural magnetism of Fe_3O_4 (hence its name, *magnetite*). The Fe_3O_4 particles can be separated from the gangue by magnets. The ores that are not magnetic are often converted to Fe_3O_4; hematite is partially reduced to magnetite, while siderite is first converted to FeO thermally, then oxidized to Fe_2O_3, and then reduced to Fe_3O_4:

$$FeCO_3(s) \xrightarrow{\text{Heat}} FeO(s) + CO_2(g)$$

$$4FeO(s) + O_2(g) \longrightarrow 2Fe_2O_3(s)$$

$$3Fe_2O_3(s) + C(s) \longrightarrow 2Fe_3O_4(s) + CO(g)$$

Sometimes the nonmagnetic ores are concentrated by flotation processes.

The most commonly used reduction process for iron takes place in the **blast furnace** (Fig. 20-36). The raw materials required are concentrated iron ore, coke, and limestone (which serves as a *flux* to trap impurities). The furnace, which is approximately 25 feet in diameter, is charged from the top with a mixture of iron ore, coke, and limestone. A very strong blast (~350 mi/h) of hot air is injected at the bottom, where the oxygen reacts with the carbon in the coke to form carbon monoxide, the reducing agent for the iron. The temperature of the charge increases as it travels down the furnace, with reduction of the iron to iron metal occurring in steps:

$$3Fe_2O_3 + CO \longrightarrow 2Fe_3O_4 + CO_2$$

$$Fe_3O_4 + CO \longrightarrow 3FeO + CO_2$$

$$FeO + CO \longrightarrow Fe + CO_2$$

Iron can reduce carbon dioxide,

$$Fe + CO_2 \longrightarrow FeO + CO$$

so complete reduction of the iron occurs only if the carbon dioxide is destroyed by adding excess coke:

$$CO_2 + C \longrightarrow 2CO$$

Figure 20-36 | The blast furnace used in the production of iron.

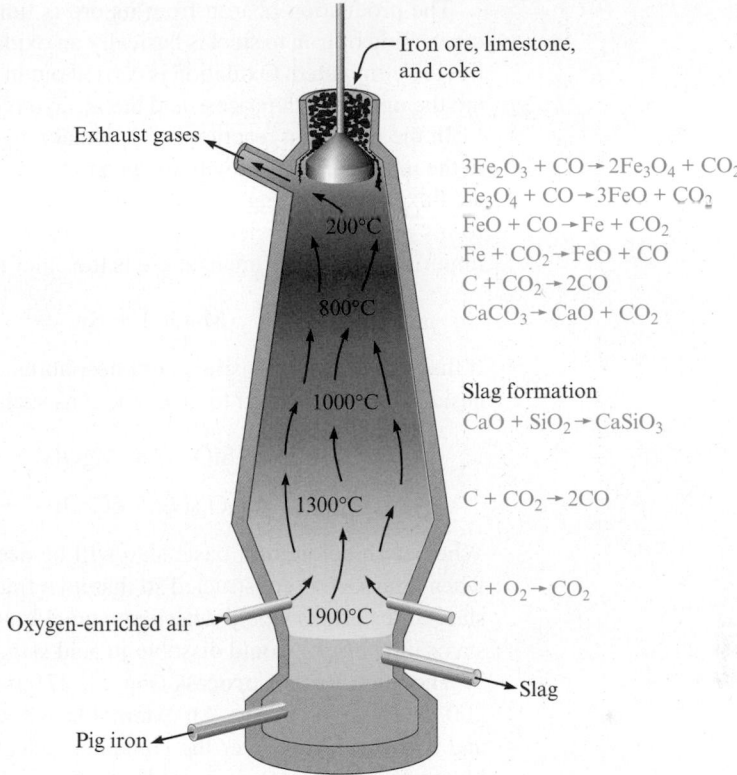

The limestone ($CaCO_3$) in the charge loses carbon dioxide, or *calcines,* in the hot furnace and combines with silica and other impurities to form **slag**, which is mostly molten calcium silicate, $CaSiO_3$,

$$CaO + SiO_2 \longrightarrow CaSiO_3$$

and alumina (Al_2O_3). The slag floats on the molten iron and is skimmed off. The gas that escapes from the top of the furnace contains carbon monoxide, which is combined with air to form carbon dioxide. The energy released in this exothermic reaction is collected in a heat exchanger and used in heating the furnace.

The iron collected from the blast furnace, called **pig iron**, is quite impure. It contains ~90% iron, ~5% carbon, ~2% manganese, ~1% silicon, ~0.3% phosphorus, and ~0.04% sulfur (from impurities in the coke). The production of 1 ton of pig iron requires approximately 1.7 tons of iron ore, 0.5 ton of coke, 0.25 ton of limestone, and 2 tons of air.

Iron oxide also can be reduced in a **direct reduction furnace**, which operates at much lower temperatures (1300–2000°F) than a blast furnace and produces a solid "sponge iron" rather than molten iron. Because of the milder reaction conditions, the direct reduction furnace requires a higher grade of iron ore (with fewer impurities) than that used in a blast furnace. The iron from the direct reduction furnace is called *DRI (directly reduced iron)* and contains ~95% iron, with the balance mainly silica and alumina.

Production of Steel

Steel is an alloy and can be classified as **carbon steel**, which contains up to about 1.5% carbon, or **alloy steel**, which contains carbon plus other metals such as Cr, Co, Mn, and Mo. The wide range of mechanical properties associated with steel is determined by its chemical composition and by the heat treatment of the final product.

The production of iron from its ore is fundamentally a reduction process, but the conversion of iron to steel is basically an oxidation process in which unwanted impurities are eliminated. Oxidation is carried out in various ways, but the two most common are the open hearth process and the basic oxygen process.

In the oxidation reactions of steelmaking, the manganese, phosphorus, and silicon in the impure iron react with oxygen to form oxides, which in turn react with appropriate fluxes to form slag. Sulfur enters the slag primarily as sulfides, and excess carbon forms carbon monoxide or carbon dioxide. The flux chosen depends on the major impurities present. If manganese is the chief impurity, an acidic flux of silica is used:

$$MnO(s) + SiO_2(s) \xrightarrow{\text{Heat}} MnSiO_3(l)$$

If the main impurity is silicon or phosphorus, a basic flux, usually lime (CaO) or magnesia (MgO), is needed to give reactions such as

$$SiO_2(s) + MgO(s) \xrightarrow{\text{Heat}} MgSiO_3(l)$$

$$P_4O_{10}(s) + 6CaO(s) \xrightarrow{\text{Heat}} 2Ca_3(PO_4)_2(l)$$

Whether an acidic or a basic slag will be needed is a factor that must be considered when a furnace is constructed so that its refractory linings will be compatible with the slag. Silica bricks would deteriorate quickly in the presence of basic slag, and magnesia or lime bricks would dissolve in acid slag.

The **open hearth process** (Fig. 20-37) uses a dishlike container that holds 100 to 200 tons of molten iron. An external heat source is required to keep the iron molten, and a concave roof over the container reflects heat back toward the iron surface. A blast of air or oxygen is passed over the surface of the iron to react with impurities. Silicon and manganese are oxidized first and enter the slag, followed by oxidation of carbon to carbon monoxide, which causes agitation and foaming of the molten bath. The exothermic oxidation of carbon raises the temperature of the bath, causing the limestone flux to calcine:

$$CaCO_3 \xrightarrow{\text{Heat}} CaO + CO_2$$

The resulting lime floats to the top of the molten mixture (an event called the *lime boil*), where it combines with phosphates, sulfates, and silicates. Next comes the refining process, which involves continued oxidation of carbon and other impurities. Because the melting point increases as the carbon content decreases, the bath temperatures must be increased during this phase of the operation. If the carbon content falls below that desired in the final product, coke or pig iron may be added.

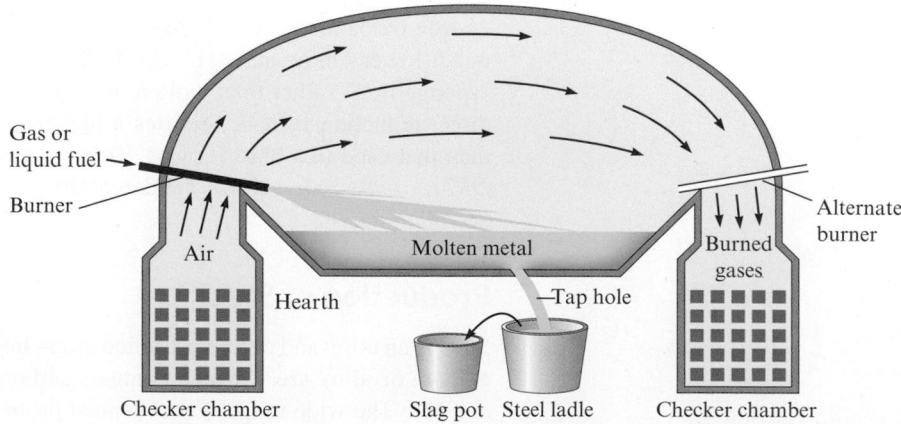

Figure 20-37 │ A schematic diagram of the open hearth process for steelmaking. The checker chambers contain bricks that absorb heat from gases passing over the molten charge. The flow of air and gases is reversed periodically.

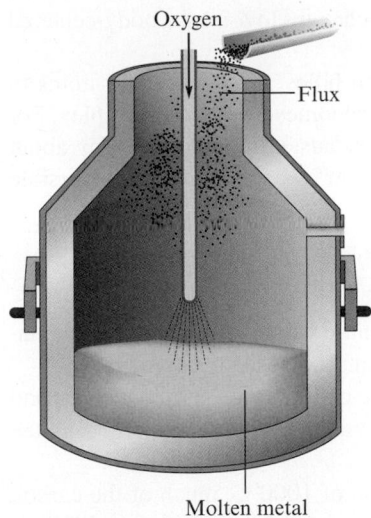

Oxygen

Flux

Molten metal

Figure 20-38 | The basic oxygen process for steelmaking.

The final composition of the steel is "fine-tuned" after the charge is poured. For example, aluminum is sometimes added at this stage to remove trace amounts of oxygen via the reaction

$$4Al + 3O_2 \longrightarrow 2Al_2O_3$$

Alloying metals such as vanadium, chromium, titanium, manganese, and nickel are also added to give the steel the properties needed for a specific application.

The processing of a batch of steel by the open hearth process is quite slow, taking up to 8 hours. The **basic oxygen process** is much faster. Molten pig iron and scrap iron are placed in a barrel-shaped container (Fig. 20-38) that can hold as much as 300 tons of material. A high-pressure blast of oxygen is directed at the surface of the molten iron, oxidizing impurities in a manner very similar to that used in the open hearth process. Fluxes are added after the oxygen blow begins. One advantage of this process is that the exothermic oxidation reactions proceed so rapidly that they produce enough heat to raise the temperature nearly to the boiling point of iron without an external heat source. Also, at these high temperatures only about an hour is needed to complete the oxidation processes.

The **electric arc method**, which was once used only for small batches of specialty steels, is being utilized more and more in the steel industry. In this process an electric arc between carbon electrodes is used to melt the charge. This means that no fuel-borne impurities are added to the steel, since no fuel is needed. Also, higher temperatures are possible than in the open hearth or basic oxygen processes, and this leads to more effective removal of sulfur and phosphorus impurities. Oxygen is added in this process so that the oxide impurities in the steel can be controlled effectively.

Heat Treatment of Steel

One way of producing the desired physical properties in steel is by controlling the chemical composition (Table 20-21). Another method for tailoring the properties of steel involves heat treatment. Pure iron exists in two different crystalline forms, depending on the temperature. At any temperature below 912°C, iron has a body-centered cubic structure and is called *α-iron*. Between 912°C and 1394°C, iron has a face-centered cubic

Table 20-21 | Percent Composition and Uses of Various Types of Steel

Type of Steel	% Carbon	% Manganese	% Phosphorus	% Sulfur	% Silicon	% Nickel	% Chromium	% Other	Uses
Plain carbon	≤1.35	≤1.65	≤0.04	≤0.05	≤0.60	—	—	—	Sheet steel, tools
High-strength (low-alloy)	≤0.25	≤1.65	≤0.04	≤0.05	0.15–0.9	0.4–1	0.3–1.3	Cu (0.2–0.6) Sb (0.01–0.08) V (0.01–0.08)	Transportation equipment, structural beams
Alloy	≤1.00	≤3.50	≤0.04	≤0.05	0.15–2.0	0.25–10.0	0.25–4.0	Mo (0.08–4.0) V (0–0.2) W (0–18) Co (0–5)	Automobile and aircraft engine parts
Stainless	0.03–1.2	1.0–10	0.04–0.06	≤0.03	1–3	1–22	4.0–27	—	Engine parts, steam turbine parts, kitchen utensils
Silicon	—	—	—	—	0.5–5.0	—	—	—	Electric motors and transformers

structure called *austentite,* or γ-*iron.* At 1394°C, iron changes to δ-*iron,* a body-centered cubic structure identical to α-iron.

When iron is alloyed with carbon, which fits into holes among the iron atoms to form the interstitial alloy *carbon steel,* the situation becomes even more complex. For example, the temperature at which α-iron changes to austentite is lowered by about 200°C. Also, at high temperatures iron and carbon react by an endothermic reversible reaction to form an iron carbide called *cementite:*

$$3Fe + C + energy \rightleftharpoons Fe_3C$$
$$\text{(Heat)} \qquad \text{Cementite}$$

By Le Châtelier's principle, we can predict that cementite will become more stable relative to iron and carbon as the temperature is increased. This is the observed result.

Thus steel is really a mixture of iron metal in one of its crystal forms, carbon, and cementite. The proportions of these components are very important in determining the physical properties of steel.

When steel is heated to temperatures in the region of 1000°C, much of the carbon is converted to cementite. If the steel is then allowed to cool slowly, the equilibrium shown above shifts to the left, and small crystals of carbon precipitate, giving a steel that is relatively ductile. If the cooling is very rapid, the equilibrium does not have time to adjust. The cementite is trapped, and the steel has a high cementite content, making it quite brittle. The proportions of carbon crystals and cementite can be "fine-tuned" to give the desired properties by heating to intermediate temperatures followed by rapid cooling, a process called **tempering**. The rate of heating and cooling determines not only the amounts of cementite present but also the size of its crystals and the form of crystalline iron present.

FOR REVIEW

Key Terms

Section 20-1
complex ion
first-row transition metals
lanthanide contraction
lanthanide series

Section 20-3
coordination compound
counterion
oxidation state
coordination number
ligand
coordinate covalent bond
monodentate (unidentate) ligand
chelating ligand (chelate)
bidentate ligand

Section 20-4
isomers
structural isomerism
stereoisomerism
coordination isomerism
linkage isomerism

First-Row Transition Metals (Scandium–Zinc)

- All have one or more electrons in the $4s$ orbital and various numbers of $3d$ electrons
- All exhibit metallic properties
    - A particular element often shows more than one oxidation state in its compounds
- Most compounds are colored, and many are paramagnetic
- Most commonly form coordination compounds containing a complex ion involving ligands (Lewis bases) attached to a central transition metal ion
    - The number of attached ligands (called the coordination number) can vary from 2 to 8, with 4 and 6 being most common
- Many transition metal ions have major biological importance in molecules such as enzymes and those that transport and store oxygen
    - Chelating ligands form more than one bond to the transition metal ion

Isomerism

- Isomers: two or more compounds with the same formula but different properties
    - Coordination isomerism: the composition of the coordination sphere varies
    - Linkage isomerism: the point of attachment of one or more ligands varies
    - Stereoisomerism: isomers have identical bonds but different spatial arrangements
        - Geometric isomerism: ligands assume different relative positions in the coordination sphere; examples are *cis* and *trans* isomers
        - Optical isomerism: molecules with nonsuperimposable mirror images rotate plane-polarized light in opposite directions

Key Terms

geometrical (*cis–trans*)
   isomerism
trans isomer
cis isomer
optical isomerism
chiral
enantiomers

Section 20-6

crystal field model
d-orbital splitting
strong-field (low-spin) case
weak-field (high-spin) case
spectrochemical series

Section 20-7

cytochromes
heme
porphyrin
myoglobin
hemoglobin
carboxyhemoglobin

Section 20-8

metallurgy
minerals
gangue
flotation process
roasting
smelting
zone refining
pyrometallurgy
hydrometallurgy
leaching
cyanidation
blast furnace
slag
pig iron
direct reduction furnace
carbon steel
alloy steel
open hearth process
basic oxygen process
electric arc method
tempering

Spectral and Magnetic Properties

- Usually explained in terms of the crystal field model
- Model assumes the ligands are point charges that split the energies of the $3d$ orbitals
- Color and magnetism are explained in terms of how the $3d$ electrons occupy the split $3d$ energy levels
  - Strong-field case: relatively large orbital splitting
  - Weak-field case: relatively small orbital splitting

Metallurgy

- The processes connected with separating a metal from its ore
  - The minerals in ores are often converted to oxides (roasting) before being reduced to the metal (smelting)
- The metallurgy of iron: most common method for reduction uses a blast furnace; process involves iron ore, coke, and limestone
  - Impure product (~90% iron) is called pig iron
- Steel is manufactured by oxidizing the impurities in pig iron

Review Questions Answers to the Review Questions can be found on the Student Web site (accessible from **www.cengagebrain.com**).

1. What two first-row transition metals have unexpected electron configurations? A statement in the text says that first-row transition metal ions do not have $4s$ electrons. Why not? Why do transition metal ions often have several oxidation states, whereas representative metals generally have only one?

2. Define each of the following terms:
   a. coordination compound
   b. complex ion
   c. counterions
   d. coordination number
   e. ligand

f. chelate

g. bidentate

How would transition metal ions be classified using the Lewis definition of acids and bases? What must a ligand have to bond to a metal? What do we mean when we say that a bond is a "coordinate covalent bond"?

3. When a metal ion has a coordination number of 2, 4, or 6, what are the observed geometries and associated bond angles? For each of the following, give the correct formulas for the complex ions.

   a. linear Ag^+ complex ions having CN^- ligands

   b. tetrahedral Cu^+ complex ions having H_2O ligands

   c. tetrahedral Mn^{2+} complex ions having oxalate ligands

   d. square planar Pt^{2+} complex ions having NH_3 ligands

   e. octahedral Fe^{3+} complex ions having EDTA ligands

   f. octahedral Co^{2+} complex ions having Cl^- ligands

   g. octahedral Cr^{3+} complex ions having ethylenediamine ligands

   What is the electron configuration for the metal ion in each of the complex ions in a–g?

4. What is wrong with the following formula–name combinations? Give the correct names for each.

   a. $[Cu(NH_3)_4]Cl_2$
   copperammine chloride

   b. $[Ni(en)_2]SO_4$
   bis(ethylenediamine)nickel(IV) sulfate

   c. $K[Cr(H_2O)_2Cl_4]$
   potassium tetrachlorodiaquachromium(III)

   d. $Na_4[Co(CN)_4C_2O_4]$
   tetrasodium tetracyanooxalatocobaltate(II)

5. Define each of the following and give examples of each.

   a. isomer

   b. structural isomer

   c. stereoisomer

   d. coordination isomer

   e. linkage isomer

   f. geometrical isomer

   g. optical isomer

   Consider the *cis* and *trans* forms of the octahedral complex $Cr(en)_2Cl_2$. Are both of these isomers optically active? Explain.

   Another way to determine whether a substance is optically active is to look for a plane of symmetry in the molecule. If a substance has a plane of

symmetry, then it will not exhibit optical activity (the mirror image will be superimposable). Show the plane of symmetry in the *trans* isomer and prove to yourself that the *cis* isomer does not have a plane of symmetry.

6. What is the major focus of the crystal field model? Why are the *d* orbitals split into two sets for an octahedral complex? What are the two sets of orbitals?

   Define each of the following.

   a. weak-field ligand

   b. strong-field ligand

   c. low-spin complex

   d. high-spin complex

   Why is $Co(NH_3)_6^{3+}$ diamagnetic whereas CoF_6^{3-} is paramagnetic? Some octahedral complex ions have the same *d*-orbital splitting diagrams whether they are high-spin or low-spin. For which of the following is this true?

   a. V^{3+}

   b. Ni^{2+}

   c. Ru^{2+}

7. The crystal field model predicts magnetic properties of complex ions and explains the colors of these complex ions. How? Solutions of $[Cr(NH_3)_6]Cl_3$ are yellow, but $Cr(NH_3)_6^{3+}$ does not absorb yellow light. Why? What color light is absorbed by $Cr(NH_3)_6^{3+}$? What is the spectrochemical series, and how can the study of light absorbed by various complex ions be used to develop this series? Would you expect $Co(NH_3)_6^{2+}$ to absorb light of a longer or shorter wavelength than $Co(NH_3)_6^{3+}$? Explain.

8. Why do tetrahedral complex ions have a different crystal field diagram than octahedral complex ions? What is the tetrahedral crystal field diagram? Why are virtually all tetrahedral complex ions "high spin"?

   Explain the crystal field diagram for square planar complex ions and for linear complex ions.

9. Review Table 20-18, which lists some important biological functions associated with different first-row transition metals. The transport of O_2 in the blood is carried out by hemoglobin. Briefly explain how hemoglobin transports O_2 in the blood.

10. Define and give an example of each of the following.

   a. roasting   **d.** leaching

   b. smelting   **e.** gangue

   c. flotation

   What are the advantages and disadvantages of hydrometallurgy? Describe the process by which a metal is purified by zone refining.

Active Learning Questions

These questions are designed to be used by groups of students in class.

1. You isolate a compound with the formula $PtCl_4 \cdot 2KCl$. From electrical conductance tests of an aqueous solution of the compound, you find that three ions per formula unit are present, and you also notice that addition of $AgNO_3$ does not cause a precipitate. Give the formula for this compound that shows the complex ion present. Explain your findings. Name this compound.

2. Both $Ni(NH_3)_4^{2+}$ and $Ni(SCN)_4^{2-}$ have four ligands. The first is paramagnetic, and the second is diamagnetic. Are the complex ions tetrahedral or square planar? Explain.

3. Which is more likely to be paramagnetic, $Fe(CN)_6^{4-}$ or $Fe(H_2O)_6^{2+}$? Explain.

4. A metal ion in a high-spin octahedral complex has two more unpaired electrons than the same ion does in a low-spin octahedral complex. Name some possible metal ions for which this would be true.

A blue question or exercise number indicates that the answer to that question or exercise appears at the back of this book and a solution appears in the *Solutions Guide*, as found on PowerLecture.

Questions

5. What is the lanthanide contraction? How does the lanthanide contraction affect the properties of the $4d$ and $5d$ transition metals?

6. Four different octahedral chromium coordination compounds exist that all have the same oxidation state for chromium and have H_2O and Cl^- as the ligands and counterions. When 1 mole of each of the four compounds is dissolved in water, how many moles of silver chloride will precipitate upon addition of excess $AgNO_3$?

7. Figure 20-17 shows that the *cis* isomer of $Co(en)_2Cl_2^+$ is optically active while the *trans* isomer is not optically active. Is the same true for $Co(NH_3)_4Cl_2^+$? Explain.

8. A certain first-row transition metal ion forms many different colored solutions. When four coordination compounds of this metal, each having the same coordination number, are dissolved in water, the colors of the solutions are red, yellow, green, and blue. Further experiments reveal that two of the complex ions are paramagnetic with four unpaired electrons and the other two are diamagnetic. What can be deduced from this information about the four coordination compounds?

9. Oxalic acid is often used to remove rust stains. What properties of oxalic acid allow it to do this?

10. For the following crystal field diagrams, label each as low spin, high spin, or cannot tell. Explain your answers.

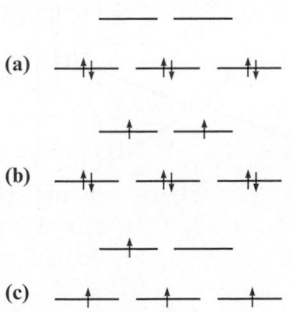

11. $CoCl_4^{2-}$ forms a tetrahedral complex ion and $Co(CN)_6^{3-}$ forms an octahedral complex ion. What is wrong about the following statements concerning each complex ion and the d orbital splitting diagrams?

   a. $CoCl_4^{2-}$ is an example of a strong-field case having two unpaired electrons.

   b. Because CN^- is a weak-field ligand, $Co(CN)_6^{3-}$ will be a low-spin case having four unpaired electrons.

12. The following statements discuss some coordination compounds. For each coordination compound, give the complex ion and the counterions, the electron configuration of the transition metal, and the geometry of the complex ion.

   a. $CoCl_2 \cdot 6H_2O$ is a compound used in novelty devices that predict rain.

   b. During the developing process of black-and-white film, silver bromide is removed from photographic film by the fixer. The major component of the fixer is sodium thiosulfate. The equation for the reaction is:

   $$AgBr(s) + 2Na_2S_2O_3(aq) \longrightarrow$$
   $$Na_3[Ag(S_2O_3)_2](aq) + NaBr(aq)$$

   c. In the production of printed circuit boards for the electronics industry, a thin layer of copper is laminated onto an insulating plastic board. Next, a circuit pattern made of a chemically resistant polymer is printed on the board. The unwanted copper is removed by chemical etching, and the protective polymer is finally removed by solvents. One etching reaction is:

   $$Cu(NH_3)_4Cl_2(aq) + 4NH_3(aq) + Cu(s) \longrightarrow$$
   $$2Cu(NH_3)_4Cl(aq)$$

   Assume these copper complex ions have tetrahedral geometry.

13. When concentrated hydrochloric acid is added to a red solution containing the $Co(H_2O)_6^{2+}$ complex ion, the solution turns blue as the tetrahedral $CoCl_4^{2-}$ complex ion forms. Explain this color change.

14. Tetrahedral complexes of Co^{2+} are quite common. Use a d-orbital splitting diagram to rationalize the stability of Co^{2+} tetrahedral complex ions.

15. Which of the following ligands are capable of linkage isomerism? Explain your answer.

   $$SCN^-, N_3^-, NO_2^-, NH_2CH_2CH_2NH_2, OCN^-, I^-$$

16. Compounds of copper(II) are generally colored, but compounds of copper(I) are not. Explain. Would you expect $Cd(NH_3)_4Cl_2$ to be colored? Explain.

17. Compounds of Sc^{3+} are not colored, but those of Ti^{3+} and V^{3+} are. Why?

18. Almost all metals in nature are found as ionic compounds in ores instead of being in the pure state. Why? What must be done to a sample of ore to obtain a metal substance that has desirable properties?

19. What causes high-altitude sickness, and what is high-altitude acclimatization?

20. Why are CN^- and CO toxic to humans?

Exercises

In this section similar exercises are paired.

Transition Metals and Coordination Compounds

21. Write electron configurations for the following metals.
    a. Ni   **b.** Cd   **c.** Zr   **d.** Os

22. Write electron configurations for the following ions.
    a. Ni^{2+} **c.** Zr^{3+} and Zr^{4+}
    b. Cd^{2+} **d.** Os^{2+} and Os^{3+}

23. Write electron configurations for each of the following.
    a. Ti, Ti^{2+}, Ti^{4+} **b.** Re, Re^{2+}, Re^{3+} **c.** Ir, Ir^{2+}, Ir^{3+}

24. Write electron configurations for each of the following.
    a. Cr, Cr^{2+}, Cr^{3+} **b.** Cu, Cu^+, Cu^{2+} **c.** V, V^{2+}, V^{3+}

25. What is the electron configuration for the transition metal ion in each of the following compounds?
    a. $K_3[Fe(CN)_6]$
    b. $[Ag(NH_3)_2]Cl$
    c. $[Ni(H_2O)_6]Br_2$
    d. $[Cr(H_2O)_4(NO_2)_2]I$

26. What is the electron configuration for the transition metal ion(s) in each of the following compounds?
    a. $(NH_4)_2[Fe(H_2O)_2Cl_4]$
    b. $[Co(NH_3)_2(NH_2CH_2CH_2NH_2)_2]I_2$
    c. $Na_2[TaF_7]$
    d. $[Pt(NH_3)_4I_2][PtI_4]$
    Pt forms $+2$ and $+4$ oxidation states in compounds.

27. Molybdenum is obtained as a by-product of copper mining or is mined directly (primary deposits are in the Rocky Mountains in Colorado). In both cases it is obtained as MoS_2, which is then converted to MoO_3. The MoO_3 can be used directly in the production of stainless steel for high-speed tools (which accounts for about 85% of the molybdenum used). Molybdenum can be purified by dissolving MoO_3 in aqueous ammonia and crystallizing ammonium molybdate. Depending on conditions, either $(NH_4)_2Mo_2O_7$ or $(NH_4)_6Mo_7O_{24} \cdot 4H_2O$ is obtained.
    a. Give names for MoS_2 and MoO_3.
    b. What is the oxidation state of Mo in each of the compounds mentioned above?

28. Titanium dioxide, the most widely used white pigment, occurs naturally but is often colored by the presence of impurities. The chloride process is often used in purifying rutile, a mineral form of titanium dioxide.
    a. Show that the unit cell for rutile, shown below, conforms to the formula TiO_2. (*Hint:* Recall the discussion in Sections 9-4 and 9-7.)

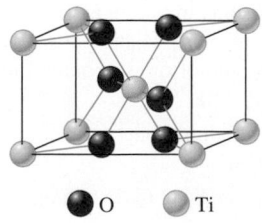

    ● O   ○ Ti

    b. The reactions for the chloride process are
    $$2TiO_2(s) + 3C(s) + 4Cl_2(g)$$
    $$\xrightarrow{950°C} 2TiCl_4(g) + CO_2(g) + 2CO(g)$$
    $$TiCl_4(g) + O_2(g) \xrightarrow{1000–1400°C} TiO_2(s) + 2Cl_2(g)$$

    Assign oxidation states to the elements in both reactions. Which elements are being reduced, and which are being oxidized? Identify the oxidizing agent and the reducing agent in each reaction.

29. When 6 *M* ammonia is added gradually to aqueous copper(II) nitrate, a white precipitate forms. The precipitate dissolves as more 6 *M* ammonia is added. Write balanced equations to explain these observations. [*Hint:* Cu^{2+} reacts with NH_3 to form $Cu(NH_3)_4^{2+}$.]

30. When an aqueous solution of KCN is added to a solution containing Ni^{2+} ions, a precipitate forms, which redissolves on addition of more KCN solution. Write reactions describing what happens in this solution. [*Hint:* CN^- is a Brønsted–Lowry base $(K_b \approx 10^{-5})$ and a Lewis base.]

31. Consider aqueous solutions of the following coordination compounds: $Co(NH_3)_6I_3$, $Pt(NH_3)_4I_4$, Na_2PtI_6, and $Cr(NH_3)_4I_3$. If aqueous $AgNO_3$ is added to separate beakers containing solutions of each coordination compound, how many moles of AgI will precipitate per mole of transition metal present? Assume that each transition metal ion forms an octahedral complex.

32. A coordination compound of cobalt(III) contains four ammonia molecules, one sulfate ion, and one chloride ion. Addition of aqueous $BaCl_2$ solution to an aqueous solution of the compound gives no precipitate. Addition of aqueous $AgNO_3$ to an aqueous solution of the compound produces a white precipitate. Propose a structure for this coordination compound.

33. Name the following complex ions:
    a.
    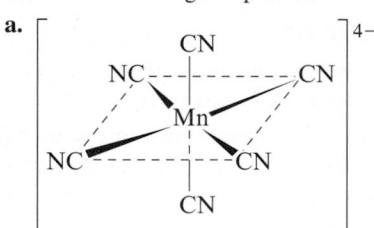
    b.
    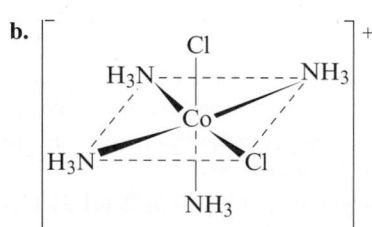
    c.
    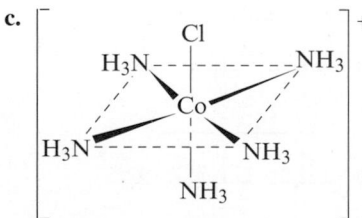

34. Name the following complex ions.

a. $Ru(NH_3)_5Cl^{2+}$

b. $Fe(CN)_6^{4-}$

c. $Mn(NH_2CH_2CH_2NH_2)_3^{2+}$

d. $Co(NH_3)_5NO_2^{2+}$

e. $Ni(CN)_4^{2-}$

f. $Cr(NH_3)_4Cl_2^+$

g. $Fe(C_2O_4)_3^{3-}$

h. $Co(SCN)_2(H_2O)_4^+$

35. Name the following coordination compounds.

a. $[Co(NH_3)_6]Cl_2$

b. $[Co(H_2O)_6]I_3$

c. $K_2[PtCl_4]$

d. $K_4[PtCl_6]$

e. $[Co(NH_3)_5Cl]Cl_2$

f. $[Co(NH_3)_3(NO_2)_3]$

36. Name the following coordination compounds.

a. $[Cr(H_2O)_5Br]Br_2$

b. $Na_3[Co(CN)_6]$

c. $[Fe(NH_2CH_2CH_2NH_2)_2(NO_2)_2]Cl$

d. $[Pt(NH_3)_4I_2][PtI_4]$

37. Give formulas for the following.

a. potassium tetrachlorocobaltate(II)

b. aquatricarbonylplatinum(II) bromide

c. sodium dicyanobis(oxalato)ferrate(III)

d. triamminechloroethylenediaminechromium(III) iodide

38. Give formulas for the following complex ions.

a. tetrachloroferrate(III) ion

b. pentaammineaquaruthenium(III) ion

c. tetracarbonyldihydroxochromium(III) ion

d. amminetrichloroplatinate(II) ion

39. Draw geometrical isomers of each of the following complex ions.

a. $Co(C_2O_4)_2(H_2O)_2^-$

b. $Pt(NH_3)_4I_2^{2+}$

c. $Ir(NH_3)_3Cl_3$

d. $Cr(en)(NH_3)_2I_2^+$

40. Draw structures of each of the following.

a. *cis*-dichloroethylenediamineplatinum(II)

b. *trans*-dichlorobis(ethylenediamine)cobalt(II)

c. *cis*-tetraamminechloronitrocobalt(III) ion

d. *trans*-tetraamminechloronitritocobalt(III) ion

e. *trans*-diaquabis(ethylenediamine)copper(II) ion

41. The carbonate ion (CO_3^{2-}) can act as either a monodentate or a bidentate ligand. Draw a picture of CO_3^{2-} coordinating to a metal ion as a monodentate and as a bidentate ligand. The carbonate ion can also act as a bridge between two metal ions. Draw a picture of a CO_3^{2-} ion bridging between two metal ions.

42. Amino acids can act as ligands toward transition metal ions. The simplest amino acid is glycine ($NH_2CH_2CO_2H$). Draw a structure of the glycinate anion ($NH_2CH_2CO_2^-$) acting as a bidentate ligand. Draw the structural isomers of the square planar complex $Cu(NH_2CH_2CO_2)_2$.

43. How many bonds could each of the following chelating ligands form with a metal ion?

a. acetylacetone (acacH), a common ligand in organometallic catalysts:

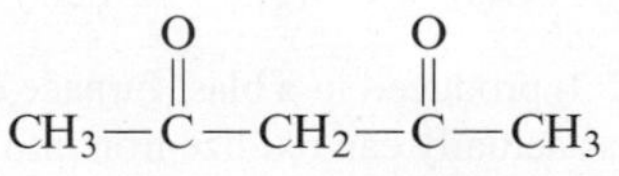

b. diethylenetriamine, used in a variety of industrial processes:

$$NH_2-CH_2-CH_2-NH-CH_2-CH_2-NH_2$$

c. salen, a common ligand for chiral organometallic catalysts:

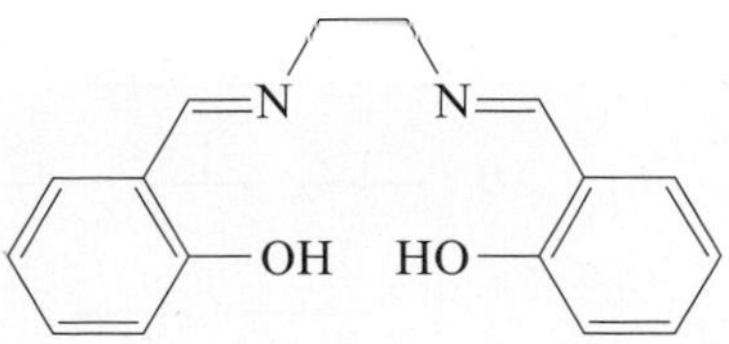

d. porphine, often used in supermolecular chemistry as well as catalysis; biologically, porphine is the basis for many different types of porphyrin-containing proteins, including heme proteins:

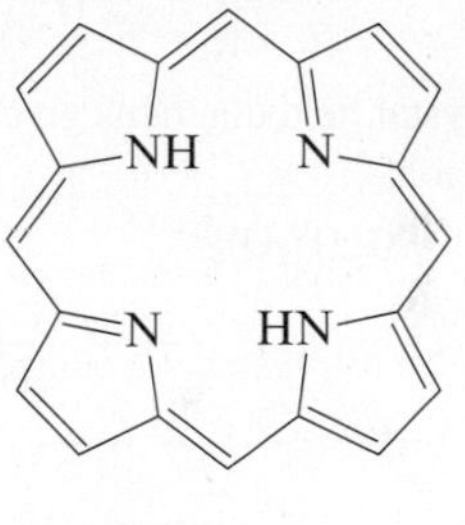

44. BAL is a chelating agent used in treating heavy metal poisoning. It acts as a bidentate ligand. What type of linkage isomers are possible when BAL coordinates to a metal ion?

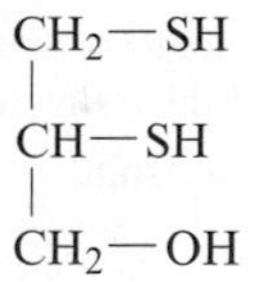

BAL

45. Draw all geometrical and linkage isomers of $Co(NH_3)_4(NO_2)_2$.

46. Draw all geometrical and linkage isomers of square planar $Pt(NH_3)_2(SCN)_2$.

47. Acetylacetone, abbreviated acacH, is a bidentate ligand. It loses a proton and coordinates as acac⁻, as shown below, where M is a transition metal:

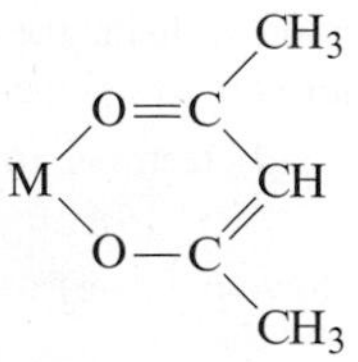

Which of the following complexes are optically active: *cis*-$Cr(acac)_2(H_2O)_2$, *trans*-$Cr(acac)_2(H_2O)_2$, and $Cr(acac)_3$?

48. Draw all geometrical isomers of $Pt(CN)_2Br_2(H_2O)_2$. Which of these isomers has an optical isomer? Draw the various optical isomers.

Bonding, Color, and Magnetism in Coordination Compounds

49. Match the crystal field diagrams given below with the following complex ions.

$Cr(NH_3)_5Cl^{2+}$ $Co(NH_3)_4Br_2^+$ $Fe(H_2O)_6^{3+}$
                (assume strong field)   (assume weak field)

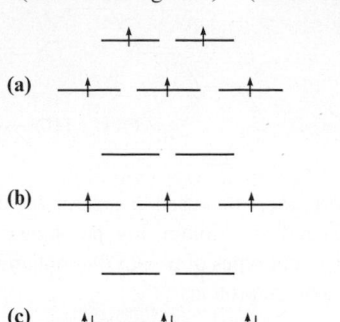

50. Match the crystal field diagrams given below with the following complex ions.

$Fe(CN)_6^{3-}$ $Mn(H_2O)_6^{2+}$

51. Draw the *d*-orbital splitting diagrams for the octahedral complex ions of each of the following.

a. Fe^{2+} (high and low spin)

b. Fe^{3+} (high spin)

c. Ni^{2+}

52. Draw the *d*-orbital splitting diagrams for the octahedral complex ions of each of the following.

a. Zn^{2+}

b. Co^{2+} (high and low spin)

c. Ti^{3+}

53. The CrF_6^{4-} ion is known to have four unpaired electrons. Does the F^- ligand produce a strong or weak field?

54. The $Co(NH_3)_6^{3+}$ ion is diamagnetic, but $Fe(H_2O)_6^{2+}$ is paramagnetic. Explain.

55. How many unpaired electrons are in the following complex ions?

a. $Ru(NH_3)_6^{2+}$ (low-spin case)

b. $Ni(H_2O)_6^{2+}$

c. $V(en)_3^{3+}$

56. The complex ion $Fe(CN)_6^{3-}$ is paramagnetic with one unpaired electron. The complex ion $Fe(SCN)_6^{3-}$ has five unpaired electrons. Where does SCN^- lie in the spectrochemical series relative to CN^-?

57. Rank the following complex ions in order of increasing wavelength of light absorbed.

$Co(H_2O)_6^{3+}, Co(CN)_6^{3-}, CoI_6^{3-}, Co(en)_3^{3+}$

58. The complex ion $Cu(H_2O)_6^{2+}$ has an absorption maximum at around 800 nm. When four ammonias replace water, $Cu(NH_3)_4(H_2O)_2^{2+}$, the absorption maximum shifts to around 600 nm. What do these results signify in terms of the relative field splittings of NH_3 and H_2O? Explain.

59. The following test tubes each contain a different chromium complex ion.

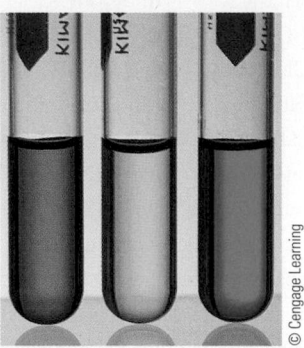

© Cengage Learning

For each complex ion, predict the predominant color of light absorbed. If the complex ions are $Cr(NH_3)_6^{3+}$, $Cr(H_2O)_6^{3+}$, and $Cr(H_2O)_4Cl_2^+$, what is the identity of the complex ion in each test tube? (*Hint:* Reference the spectrochemical series.)

60. Consider the complex ions $Co(NH_3)_6^{3+}$, $Co(CN)_6^{3-}$, and CoF_6^{3-}. The wavelengths of absorbed electromagnetic radiation for these compounds (in no specific order) are 770 nm, 440 nm, and 290 nm. Match the complex ion to the wavelength of absorbed electromagnetic radiation.

61. The wavelength of absorbed electromagnetic radiation for $CoBr_4^{2-}$ is 3.4×10^{-6} m. Will the complex ion $CoBr_6^{4-}$ absorb electromagnetic radiation having a wavelength longer or shorter than 3.4×10^{-6} m? Explain.

62. The complex ion $NiCl_4^{2-}$ has two unpaired electrons, whereas $Ni(CN)_4^{2-}$ is diamagnetic. Propose structures for these two complex ions.

63. How many unpaired electrons are present in the tetrahedral ion $FeCl_4^-$?

64. The complex ion $PdCl_4^{2-}$ is diamagnetic. Propose a structure for $PdCl_4^{2-}$.

Metallurgy

65. A blast furnace is used to reduce iron oxides to elemental iron. The reducing agent for this reduction process is carbon monoxide.

a. Given the following data:

$$Fe_2O_3(s) + 3CO(g) \longrightarrow 2Fe(s) + 3CO_2(g) \qquad \Delta H° = -23 \text{ kJ}$$
$$3Fe_2O_3(s) + CO(g) \longrightarrow 2Fe_3O_4(s) + CO_2(g) \qquad \Delta H° = -39 \text{ kJ}$$
$$Fe_3O_4(s) + CO(g) \longrightarrow 3FeO(s) + CO_2(g) \qquad \Delta H° = 18 \text{ kJ}$$

determine $\Delta H°$ for the reaction

$$FeO(s) + CO(g) \longrightarrow Fe(s) + CO_2(g)$$

b. The CO_2 produced in a blast furnace during the reduction process actually can oxidize iron into FeO. To eliminate this reaction, excess coke is added to convert CO_2 into CO by the reaction

$$CO_2(g) + C(s) \longrightarrow 2CO(g)$$

Using data from Appendix 4, determine $\Delta H°$ and $\Delta S°$ for this reaction. Assuming $\Delta H°$ and $\Delta S°$ do not depend on temperature, at what temperature is the conversion reaction of CO_2 into CO spontaneous at standard conditions?

66. Use the data in Appendix 4 for the following.

   a. Calculate $\Delta H°$ and $\Delta S°$ for the reaction

   $$3Fe_2O_3(s) + CO(g) \longrightarrow 2Fe_3O_4(s) + CO_2(g)$$

   that occurs in a blast furnace.

   b. Assume that $\Delta H°$ and $\Delta S°$ are independent of temperature. Calculate $\Delta G°$ at 800.°C for this reaction.

67. Iron is present in the earth's crust in many types of minerals. The iron oxide minerals are hematite (Fe_2O_3) and magnetite (Fe_3O_4). What is the oxidation state of iron in each mineral? The iron ions in magnetite are a mixture of Fe^{2+} and Fe^{3+} ions. What is the ratio of Fe^{3+} to Fe^{2+} ions in magnetite? The formula for magnetite is often written as $FeO \cdot Fe_2O_3$. Does this make sense? Explain.

68. What roles do kinetics and thermodynamics play in the effect that the following reaction has on the properties of steel?

   $$3Fe + C \rightleftharpoons Fe_3C$$

69. Silver is sometimes found in nature as large nuggets; more often it is found mixed with other metals and their ores. Cyanide ion is often used to extract the silver by the following reaction that occurs in basic solution:

   $$Ag(s) + CN^-(aq) + O_2(g) \xrightarrow{\text{Basic}} Ag(CN)_2^-(aq)$$

   Balance this equation by using the half-reaction method.

70. One of the classic methods for the determination of the manganese content in steel involves converting all the manganese to the deeply colored permanganate ion and then measuring the absorption of light. The steel is first dissolved in nitric acid, producing the manganese(II) ion and nitrogen dioxide gas. This solution is then reacted with an acidic solution containing periodate ion; the products are the permanganate and iodate ions. Write balanced chemical equations for both of these steps.

Additional Exercises

71. Acetylacetone (see Exercise 43, part a), abbreviated acacH, is a bidentate ligand. It loses a proton and coordinates as acac⁻, as shown below:

Acetylacetone reacts with an ethanol solution containing a salt of europium to give a compound that is 40.1% C and 4.71% H by mass. Combustion of 0.286 g of the compound gives 0.112 g Eu_2O_3. Assuming the compound contains only C, H, O, and Eu, determine the formula of the compound formed from the reaction of acetylacetone and the europium salt. (Assume that the compound contains one europium ion.)

72. The compound cisplatin, $Pt(NH_3)_2Cl_2$, has been studied extensively as an antitumor agent. The reaction for the synthesis of cisplatin is:

   $$K_2PtCl_4(aq) + 2NH_3(aq) \longrightarrow Pt(NH_3)_2Cl_2(s) + 2KCl(aq)$$

   Write the electron configuration for platinum ion in cisplatin. Most d^8 transition metal ions exhibit square planar geometry. With this and the name in mind, draw the structure of cisplatin.

73. Use standard reduction potentials to calculate $\mathscr{E}°$, $\Delta G°$, and K (at 298 K) for the reaction that is used in production of gold:

   $$2Au(CN)_2^-(aq) + Zn(s) \longrightarrow 2Au(s) + Zn(CN)_4^{2-}(aq)$$

   The relevant half-reactions are

   $$Au(CN)_2^- + e^- \longrightarrow Au + 2CN^- \qquad \mathscr{E}° = -0.60 \text{ V}$$
   $$Zn(CN)_4^{2-} + 2e^- \longrightarrow Zn + 4CN^- \qquad \mathscr{E}° = -1.26 \text{ V}$$

74. Until the discoveries of Alfred Werner, it was thought that carbon had to be present in a compound for it to be optically active. Werner prepared the following compound containing OH^- ions as bridging groups and separated the optical isomers.

   a. Draw structures of the two optically active isomers of this compound.

   b. What are the oxidation states of the cobalt ions?

   c. How many unpaired electrons are present if the complex is the low-spin case?

75. Draw all the geometrical isomers of $Cr(en)(NH_3)_2BrCl^+$. Which of these isomers also have an optical isomer? Draw the various isomers.

76. A compound related to acetylacetone is 1,1,1-trifluoroacetyl-acetone (abbreviated Htfa):

   Htfa forms complexes in a manner similar to acetylacetone. (See Exercise 47.) Both Be^{2+} and Cu^{2+} form complexes with tfa⁻ having the formula $M(tfa)_2$. Two isomers are formed for each metal complex.

   a. The Be^{2+} complexes are tetrahedral. Draw the two isomers of $Be(tfa)_2$. What type of isomerism is exhibited by $Be(tfa)_2$?

   b. The Cu^{2+} complexes are square planar. Draw the two isomers of $Cu(tfa)_2$. What type of isomerism is exhibited by $Cu(tfa)_2$?

77. Would it be better to use octahedral Ni^{2+} complexes or octahedral Cr^{2+} complexes to determine whether a given ligand is a strong-field or weak-field ligand by measuring the number of unpaired electrons? How else could the relative ligand field strengths be determined?

78. Name the following coordination compounds.

   a. $Na_4[Ni(C_2O_4)_3]$

   b. $K_2[CoCl_4]$

   c. $[Cu(NH_3)_4]SO_4$

   d. $[Co(en)_2(SCN)Cl]Cl$

79. Give formulas for the following.

   a. hexakis(pyridine)cobalt(III) chloride

   b. pentaammineiodochromium(III) iodide

   c. tris(ethylenediamine)nickel(II) bromide

   d. potassium tetracyanonickelate(II)

   e. tetraamminedichloroplatinum(IV) tetrachloroplatinate(II)

80. The complex ion $Ru(phen)_3^{2+}$ has been used as a probe for the structure of DNA. (Phen is a bidentate ligand.)

   a. What type of isomerism is found in $Ru(phen)_3^{2+}$?

   b. $Ru(phen)_3^{2+}$ is diamagnetic (as are all complex ions of Ru^{2+}). Draw the crystal field diagram for the d orbitals in this complex ion.

Phen = 1,10-phenanthroline =

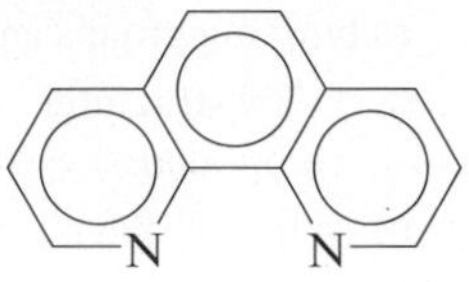

81. Carbon monoxide is toxic because it binds more strongly to iron in hemoglobin (Hb) than does O_2. Consider the following reactions and approximate standard free energy changes:

$$Hb + O_2 \longrightarrow HbO_2 \qquad \Delta G° = -70 \text{ kJ}$$
$$Hb + CO \longrightarrow HbCO \qquad \Delta G° = -80 \text{ kJ}$$

Using these data, estimate the equilibrium constant value at 25°C for the following reaction:

$$HbO_2 + CO \rightleftharpoons HbCO + O_2$$

82. For the process

$$Co(NH_3)_5Cl^{2+} + Cl^- \longrightarrow Co(NH_3)_4Cl_2^+ + NH_3$$

what would be the expected ratio of *cis* to *trans* isomers in the product?

ChemWork Problems

These multiconcept problems (and additional ones) are found interactively online with the same type of assistance a student would get from an instructor.

83. In which of the following is(are) the electron configuration(s) correct for the species indicated?

   a. Cu $[Ar]4s^23d^9$

   b. Fe^{3+} $[Ar]3d^5$

   c. Co $[Ar]4s^23d^7$

   d. La $[Ar]6s^24f^1$

   e. Pt^{2+} $[Xe]4f^{14}5d^8$

84. Which of the following molecules exhibit(s) optical isomerism?

   a. *cis*-$Pt(NH_3)_2Cl_2$

   b. *trans*-$Ni(en)_2Br_2$ (en is ethylenediamine)

   c. *cis*-$Ni(en)_2Br_2$ (en is ethylenediamine)

   d.

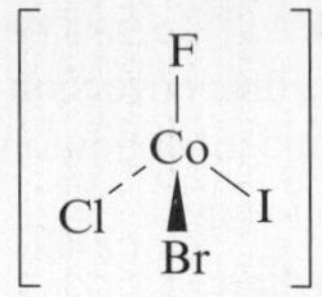

$$\left[\begin{array}{c} F \\ | \\ Cl\!-\!Co\!-\!I \\ | \\ Br \end{array} \right]^{2-}$$

85. Which of the following ions is(are) expected to form colored octahedral aqueous complex ions?

   a. Zn^{2+}

   b. Cu^{2+}

   c. Mn^{3+}

   d. Ti^{4+}

86. The following table indicates the number of unpaired electrons in the crystal field diagrams for some complexes. Complete the table by classifying each species as weak field, strong field, or insufficient information.

Species	Unpaired Electrons	Classification
$Fe(CNS)_6^{4-}$	0	_____
$CoCl_4^{2-}$	3	_____
$Fe(H_2O)_6^{3+}$	5	_____
$Fe(CN)_6^{4-}$	0	_____

87. Which of the following crystal field diagram(s) is(are) correct for the complex given?

   a. $Zn(NH_3)_4^{2+}$ (tetrahedral)

   b. $Mn(CN)_6^{3-}$ (strong field)

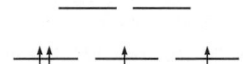

   c. $Ni(CN)_4^{2-}$ (square planar, diamagnetic)

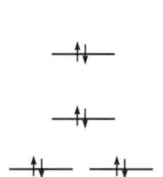

88. Which of the following statement(s) is(are) *true*?

   a. The coordination number of a metal ion in an octahedral complex ion is 8.

   b. All tetrahedral complex ions are low-spin.

   c. The formula for triaquatriamminechromium(III) sulfate is $[Cr(H_2O)_3(NH_3)_3]_2(SO_4)_3$.

   d. The electron configuration of Hf^{2+} is $[Xe]4f^{12}6s^2$.

   e. Hemoglobin contains Fe^{3+}.

Challenge Problems

89. Consider the following complex ion, where A and B represent ligands.

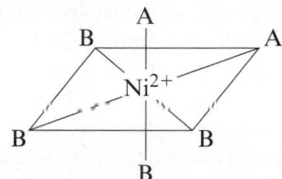

The complex is known to be diamagnetic. Do A and B produce very similar or very different crystal fields? Explain.

90. Consider the pseudo-octahedral complex ion of Cr^{3+}, where A and B represent ligands.

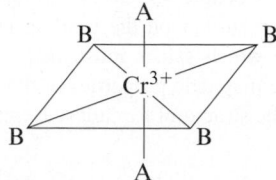

Ligand A produces a stronger crystal field than ligand B. Draw an appropriate crystal field diagram for this complex ion (assume the A ligands are on the z-axis).

91. Consider the following data:

$$Co^{3+} + e^- \longrightarrow Co^{2+} \qquad \mathscr{E}° = 1.82 \text{ V}$$
$$Co^{2+} + 3en \longrightarrow Co(en)_3^{2+} \qquad K = 1.5 \times 10^{12}$$
$$Co^{3+} + 3en \longrightarrow Co(en)_3^{3+} \qquad K = 2.0 \times 10^{47}$$

where en = ethylenediamine.

a. Calculate $\mathscr{E}°$ for the half-reaction

$$Co(en)_3^{3+} + e^- \longrightarrow Co(en)_3^{2+}$$

b. Based on your answer to part a, which is the stronger oxidizing agent, Co^{3+} or $Co(en)_3^{3+}$?

c. Use the crystal field model to rationalize the result in part b.

92. Henry Taube, 1983 Nobel Prize winner in chemistry, has studied the mechanisms of the oxidation–reduction reactions of transition metal complexes. In one experiment he and his students studied the following reaction:

$$Cr(H_2O)_6^{2+}(aq) + Co(NH_3)_5Cl^{2+}(aq)$$
$$\longrightarrow Cr(III) \text{ complexes} + Co(II) \text{ complexes}$$

Chromium(III) and cobalt(III) complexes are substitutionally inert (no exchange of ligands) under conditions of the experiment. Chromium(II) and cobalt(II) complexes can exchange ligands very rapidly. One of the products of the reaction is $Cr(H_2O)_5Cl^{2+}$. Is this consistent with the reaction proceeding through formation of $(H_2O)_5Cr—Cl—Co(NH_3)_5$ as an intermediate? Explain.

93. Chelating ligands often form more stable complex ions than the corresponding monodentate ligands with the same donor atoms. For example,

$$Ni^{2+}(aq) + 6NH_3(aq) \rightleftharpoons Ni(NH_3)_6^{2+}(aq) \qquad K = 3.2 \times 10^8$$
$$Ni^{2+}(aq) + 3en(aq) \rightleftharpoons Ni(en)_3^{2+}(aq) \qquad K = 1.6 \times 10^{18}$$
$$Ni^{2+}(aq) + penten(aq) \rightleftharpoons Ni(penten)^{2+}(aq) \qquad K = 2.0 \times 10^{19}$$

where en is ethylenediamine and penten is

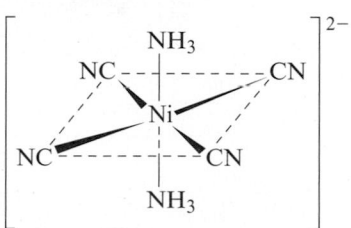

This increased stability is called the *chelate effect*. Based on bond energies, would you expect the enthalpy changes for the above reactions to be very different? What is the order (from least favorable to most favorable) of the entropy changes for the above reactions? How do the values of the formation constants correlate with $\Delta S°$? How can this be used to explain the chelate effect?

94. Qualitatively draw the crystal field splitting of the d orbitals in a trigonal planar complex ion. (Let the z axis be perpendicular to the plane of the complex.)

95. Qualitatively draw the crystal field splitting for a trigonal bipyramidal complex ion. (Let the z axis be perpendicular to the trigonal plane.)

96. Sketch a d-orbital energy diagram for the following.

a. a linear complex ion with ligands on the x axis

b. a linear complex ion with ligands on the y axis

97. Sketch and explain the most likely crystal field diagram for the following complex ion:

$$\left[\begin{array}{c} NH_3 \\ NC\!\!-\!\!-\!\!-\!\!-\!\!-CN \\ Ni \\ NC\!\!-\!\!-\!\!-\!\!-\!\!-CN \\ NH_3 \end{array} \right]^{2-}$$

Note: The CN^- ligand produces a *much* stronger crystal field than NH_3. Assume the NH_3 ligands lie on the z axis.

Integrative Problems

These problems require the integration of multiple concepts to find the solutions.

98. The ferrate ion, FeO_4^{2-}, is such a powerful oxidizing agent that in acidic solution, aqueous ammonia is reduced to elemental nitrogen along with the formation of the iron(III) ion.

a. What is the oxidation state of iron in FeO_4^{2-}, and what is the electron configuration of iron in this polyatomic ion?

b. If 25.0 mL of a 0.243 M FeO_4^{2-} solution is allowed to react with 55.0 mL of 1.45 M aqueous ammonia, what volume of nitrogen gas can form at 25°C and 1.50 atm?

99. Ammonia and potassium iodide solutions are added to an aqueous solution of $Cr(NO_3)_3$. A solid is isolated (compound A), and the following data are collected:

i. When 0.105 g of compound A was strongly heated in excess O_2, 0.0203 g CrO_3 was formed.

ii. In a second experiment it took 32.93 mL of 0.100 M HCl to titrate completely the NH_3 present in 0.341 g compound A.

iii. Compound A was found to contain 73.53% iodine by mass.

iv. The freezing point of water was lowered by 0.64°C when 0.601 g compound A was dissolved in 10.00 g H_2O ($K_f = 1.86°C \cdot$ kg/mol).

What is the formula of the compound? What is the structure of the complex ion present? (*Hints:* Cr^{3+} is expected to be six-coordinate, with NH_3 and possibly I^- as ligands. The I^- ions will be the counterions if needed.)

100. **a.** In the absorption spectrum of the complex ion $Cr(NCS)_6^{3-}$, there is a band corresponding to the absorption of a photon of light with an energy of 1.75×10^4 cm^{-1}. Given 1 $cm^{-1} = 1.986 \times 10^{-23}$ J, what is the wavelength of this photon?

  b. The $Cr-N-C$ bond angle in $Cr(NCS)_6^{3-}$ is predicted to be 180°. What is the hybridization of the N atom in the NCS^- ligand when a Lewis acid–base reaction occurs between Cr^{3+} and NCS^- that would give a 180° $Cr-N-C$ bond angle? $Cr(NCS)_6^{3-}$ undergoes substitution by ethylenediamine (en) according to the equation

$$Cr(NCS)_6^{3-} + 2en \longrightarrow Cr(NCS)_2(en)_2^+ + 4NCS^-$$

Does $Cr(NCS)_2(en)_2^+$ exhibit geometric isomerism? Does $Cr(NCS)_2(en)_2^+$ exhibit optical isomerism?

Marathon Problem

This problem is designed to incorporate several concepts and techniques into one situation.

101. There are three salts that contain complex ions of chromium and have the molecular formula $CrCl_3 \cdot 6H_2O$. Treating 0.27 g of the first salt with a strong dehydrating agent resulted in a mass loss of 0.036 g. Treating 270 mg of the second salt with the same dehydrating agent resulted in a mass loss of 18 mg. The third salt did not lose any mass when treated with the same dehydrating agent. Addition of excess aqueous silver nitrate to 100.0-mL portions of 0.100 *M* solutions of each salt resulted in the formation of different masses of silver chloride; one solution yielded 1430 mg AgCl; another, 2870 mg AgCl; the third, 4300 mg AgCl. Two of the salts are green and one is violet.

Suggest probable structural formulas for these salts, defending your answer on the basis of the preceding observations. State which salt is most likely to be violet. Would a study of the magnetic properties of the salts be helpful in determining the structural formulas? Explain.

CHAPTER TWENTY-ONE

Organic and Biological Molecules

This web is formed from a natural polymer spun by spiders. ("Moved On" by Julie Falk)

Two Group 4A elements, carbon and silicon, form the basis of most natural substances. Silicon, with its great affinity for oxygen, forms chains and rings containing Si—O—Si bridges to produce the silica and silicates that form the basis for most rocks, sands, and soils. What silicon is to the geological world, carbon is to the biological world. Carbon has the unusual ability of bonding strongly to itself to form long chains or rings of carbon atoms. In addition, carbon forms strong bonds to other nonmetals such as hydrogen, nitrogen, oxygen, sulfur, and the halogens. Because of these bonding properties, there are a myriad of carbon compounds; several million are now known, and the number continues to grow rapidly. Among these many compounds are the **biomolecules**, those responsible for maintaining and reproducing life.

The study of carbon-containing compounds and their properties is called **organic chemistry**. Although a few compounds involving carbon, such as its oxides and carbonates, are considered to be inorganic substances, the vast majority are organic compounds that typically contain chains or rings of carbon atoms.

Originally, the distinction between inorganic and organic substances was based on whether a compound was produced by living systems. For example, until the early nineteenth century it was believed that organic compounds had some sort of "life force" and could be synthesized only by living organisms. This view was dispelled in 1828 when the German chemist Friedrich Wöhler (1800–1882) prepared urea from the inorganic salt ammonium cyanate by simple heating:

$$NH_4OCN \xrightarrow{\text{Heat}} H_2N-\underset{\underset{O}{\|}}{C}-NH_2$$

Ammonium cyanate   Urea

Urea is a component of urine, so it is clearly an organic material; yet here was clear evidence that it could be produced in the laboratory as well as by living things.

Organic chemistry plays a vital role in our quest to understand living systems. Beyond that, the synthetic fibers, plastics, artificial sweeteners, and drugs that are such an accepted part of modern life are products of industrial organic chemistry. In addition, the energy on which we rely so heavily to power our civilization is based mostly on the organic materials found in coal and petroleum.

Because organic chemistry is such a vast subject, we can provide only a brief introduction to it in this book. We will begin with the simplest class of organic compounds, the hydrocarbons, and then show how most other organic compounds can be considered to be derivatives of hydrocarbons.

21-1 | Alkanes: Saturated Hydrocarbons

As the name indicates, **hydrocarbons** are compounds composed of carbon and hydrogen. Those compounds whose carbon–carbon bonds are all single bonds are said to be **saturated**, because each carbon is bound to four atoms, the maximum number. Hydrocarbons containing carbon–carbon multiple bonds are described as being

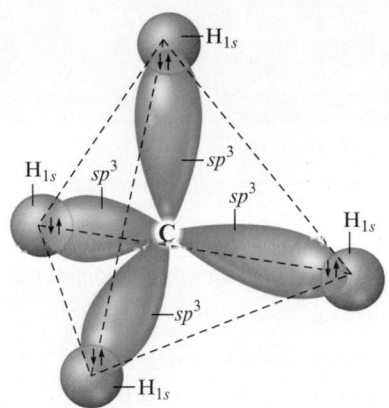

Figure 21-1 | The C—H bonds in methane.

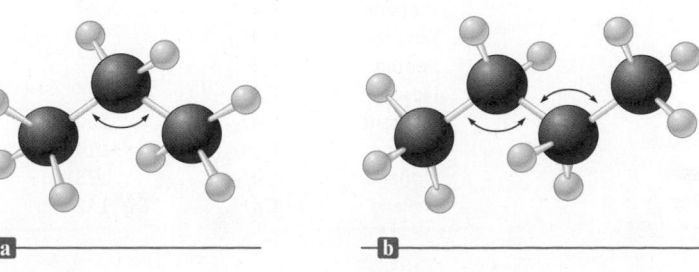

Figure 21-2 | (a) The Lewis structure of ethane (C_2H_6). (b) The molecular structure of ethane represented by space-filling and ball-and-stick models.

unsaturated, since the carbon atoms involved in a multiple bond can react with additional atoms, as shown by the *addition* of hydrogen to ethylene:

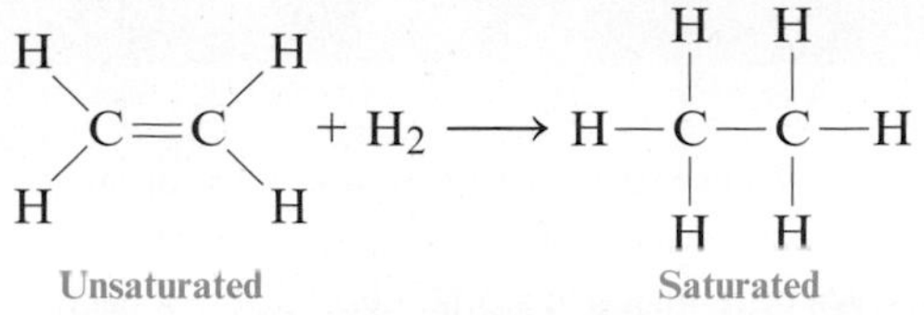

$$\underset{\text{Unsaturated}}{\underset{H}{\overset{H}{>}}C=C\underset{H}{\overset{H}{<}}} + H_2 \longrightarrow \underset{\text{Saturated}}{H-\underset{\underset{H}{|}}{\overset{\overset{H}{|}}{C}}-\underset{\underset{H}{|}}{\overset{\overset{H}{|}}{C}}-H}$$

Note that each carbon in ethylene is bonded to three atoms (one carbon and two hydrogens) but that each can bond to one additional atom if one bond of the carbon–carbon double bond is broken.

The simplest member of the saturated hydrocarbons, which are also called the **alkanes**, is *methane* (CH_4). As discussed in Section 4-3, methane has a tetrahedral structure and can be described in terms of a carbon atom using an sp^3 hybrid set of orbitals to bond to the four hydrogen atoms (Fig. 21-1). The next alkane, the one containing two carbon atoms, is *ethane* (C_2H_6), as shown in Fig. 21-2. Each carbon in ethane is surrounded by four atoms and thus adopts a tetrahedral arrangement and sp^3 hybridization, as predicted by the localized electron model.

The next two members of the series are *propane* (C_3H_8) and *butane* (C_4H_{10}), shown in Fig. 21-3. Again, each carbon is bonded to four atoms and is described as sp^3 hybridized.

Alkanes in which the carbon atoms form long "strings" or chains are called **normal**, **straight-chain**, or **unbranched hydrocarbons**. As can be seen from Fig. 21-3, the chains in normal alkanes are not really straight but zig-zag, since the tetrahedral C—C—C angle is 109.5°. The normal alkanes can be represented by the structure

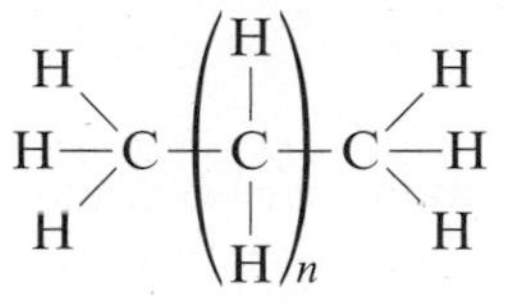

$$H-\underset{\underset{H}{|}}{\overset{\overset{H}{|}}{C}}\left(\underset{\underset{H}{|}}{\overset{\overset{H}{|}}{C}}\right)_{\!n}\!\!-\underset{\underset{H}{|}}{\overset{\overset{H}{|}}{C}}-H$$

where n is an integer. Note that each member is obtained from the previous one by inserting a *methylene* (CH_2) group. We can condense the structural formulas by omitting some of the C—H bonds. For example, the general formula for normal alkanes shown above can be condensed to

$$CH_3-(CH_2)_n-CH_3$$

The first ten normal alkanes and some of their properties are listed in Table 21-1. Note that all alkanes can be represented by the general formula C_nH_{2n+2}. For example, nonane, which has nine carbon atoms, is represented by $C_9H_{(2\times9)+2}$, or C_9H_{20}. Also note from Table 21-1 that the melting points and boiling points increase as the molar masses increase, as we would expect.

Figure 21-3 | The structures of (a) propane ($CH_3CH_2CH_3$) and (b) butane ($CH_3CH_2CH_2CH_3$). Each angle shown in red is 109.5°.

In this chapter we will use various ways to represent organic molecules. As with any model, each has its own limitations and variations. We have included a brief description of each of these models to help you as you study this chapter.

Space-Filling and Ball-and-Stick

These models represent the three-dimensional structures of organic molecules. The space-filling model is especially good at showing the relative sizes of the atoms in the molecule and their spatial relationships along with the overall shape of the molecule. For example, propane, C_3H_8, can be seen as

Ball-and-stick models are most often used in organic molecule "kits" and are helpful in showing how atoms or groups of atoms can rotate.

Structural Formulas

While not generally showing the three-dimensional nature of the molecules, structural formulas often show the bonds (as indicated by lines). These are essentially Lewis structures for organic molecules except they do not generally include lone pairs of electrons. For example, the structural formula for ethanol is

$$H-\overset{\overset{\displaystyle H}{|}}{\underset{\underset{\displaystyle H}{|}}{C}}-\overset{\overset{\displaystyle H}{|}}{\underset{\underset{\displaystyle H}{|}}{C}}-OH.$$

For simplicity, we sometimes condense the formula so that atoms connected to a central atom are grouped. For example, we write the condensed structural formula for ethanol as CH_3CH_2OH. This helps differentiate it from dimethyl ether, which has the same molecular formula, but is better represented as

$$H-\overset{\overset{\displaystyle H}{|}}{\underset{\underset{\displaystyle H}{|}}{C}}-O-\overset{\overset{\displaystyle H}{|}}{\underset{\underset{\displaystyle H}{|}}{C}}-H \quad \text{or} \quad CH_3OCH_3.$$

Structural formulas are simplified even more using what is called bond line notation. In this model, the lines represent the chemical bonds and the vertices represent implicit carbon atoms. In this text we will often use this notation for cyclic molecules. For example, cyclohexene can be written as

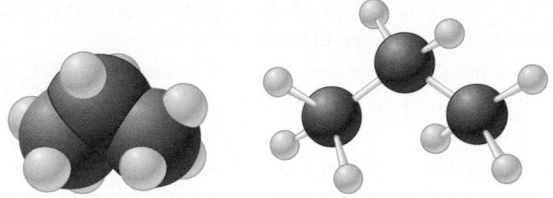

which has a condensed structural formula of ⬡ using bond line notation.

Table 21-1 | Selected Properties of the First Ten Normal Alkanes

Name	Formula	Molar Mass	Melting Point (°C)	Boiling Point (°C)	Number of Structural Isomers
Methane	CH_4	16	−182	−162	1
Ethane	C_2H_6	30	−183	−89	1
Propane	C_3H_8	44	−187	−42	1
Butane	C_4H_{10}	58	−138	0	2
Pentane	C_5H_{12}	72	−130	36	3
Hexane	C_6H_{14}	86	−95	68	5
Heptane	C_7H_{16}	100	−91	98	9
Octane	C_8H_{18}	114	−57	126	18
Nonane	C_9H_{20}	128	−54	151	35
Decane	$C_{10}H_{22}$	142	−30	174	75

Figure 21-4 | (a) Normal butane (abbreviated *n*-butane). (b) The branched isomer of butane (called isobutane).

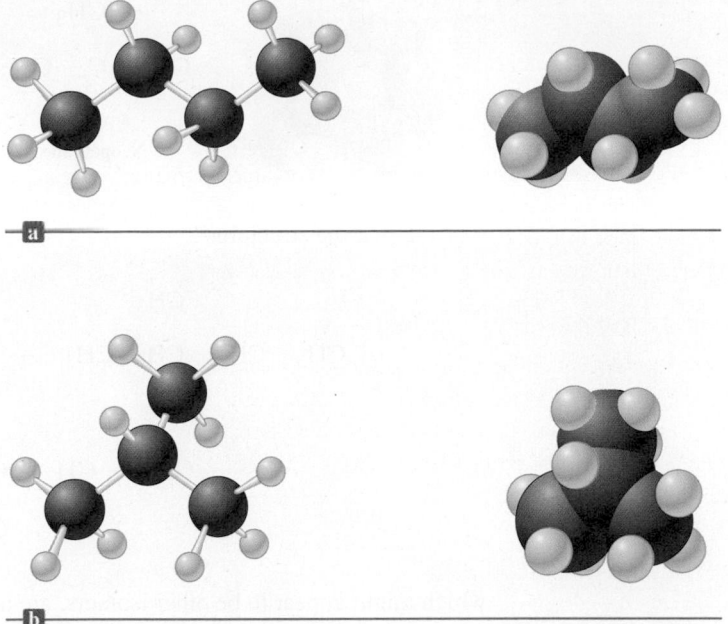

Isomerism in Alkanes

Butane and all succeeding members of the alkanes exhibit **structural isomerism**. Recall from Section 20-4 that structural isomerism occurs when two molecules have the same atoms but different bonds. For example, butane can exist as a straight-chain molecule (normal butane, or *n*-butane) or with a branched-chain structure (called isobutane), as shown in Fig. 21-4. Because of their different structures, these molecules exhibit different properties. For example, the boiling point of *n*-butane is −0.5°C, whereas that of isobutane is −12°C.

EXAMPLE 21-1   Structural Isomerism

Draw the isomers of pentane.

Solution

Pentane (C_5H_{12}) has the following isomeric structures:

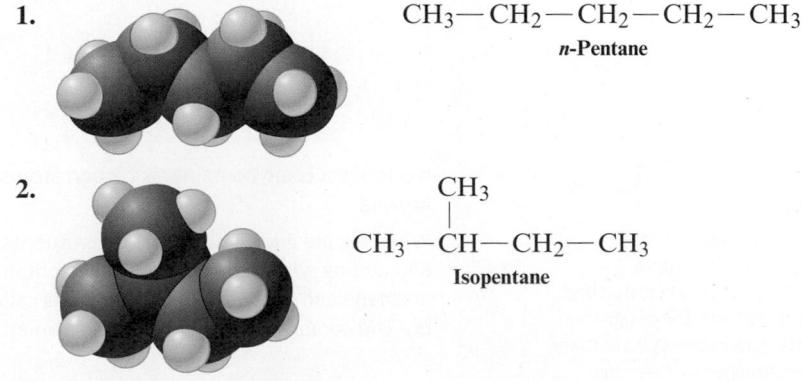

1.

$$CH_3 — CH_2 — CH_2 — CH_2 — CH_3$$

n-**Pentane**

2.

$$CH_3 — \underset{\underset{CH_3}{|}}{CH} — CH_2 — CH_3$$

Isopentane

3.

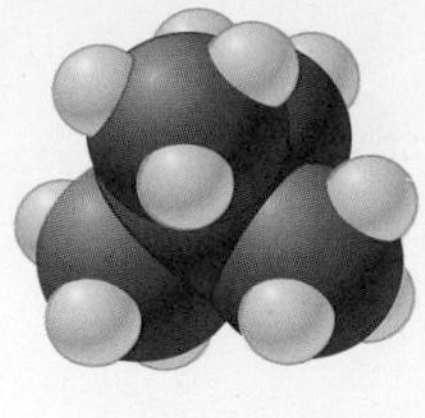

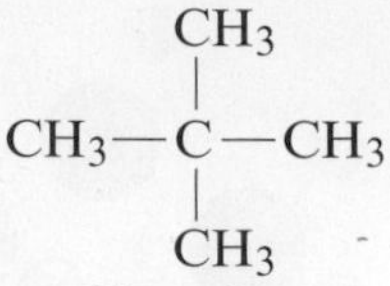

$$CH_3-\underset{\underset{\displaystyle CH_3}{|}}{\overset{\overset{\displaystyle CH_3}{|}}{C}}-CH_3$$

Neopentane

Note that the structures

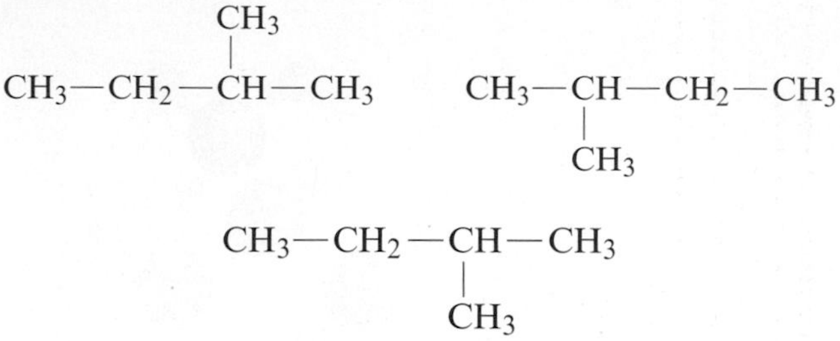

$$CH_3-CH_2-\underset{\underset{\displaystyle CH_3}{|}}{CH}-CH_3 \qquad CH_3-\underset{\underset{\displaystyle CH_3}{|}}{CH}-CH_2-CH_3$$

$$CH_3-CH_2-\underset{\underset{\displaystyle CH_3}{|}}{CH}-CH_3$$

which might appear to be other isomers, are actually identical to structure 2.

See Exercise 21-13

Table 21-2 | The Most Common Alkyl Substituents and Their Names

Structure*	Name†
—CH₃	Methyl
—CH₂CH₃	Ethyl
—CH₂CH₂CH₃	Propyl
CH₃CHCH₃	Isopropyl
—CH₂CH₂CH₂CH₃	Butyl
CH₃CHCH₂CH₃	sec-Butyl
—CH₂—C(H)(CH₃)—CH₃	Isobutyl
—C(CH₃)(CH₃)—CH₃	tert-Butyl

*The bond with one end open shows the point of attachment of the substituent to the carbon chain.

†For the butyl groups, *sec-* indicates attachment to the chain through a secondary carbon, a carbon atom attached to *two* other carbon atoms. The designation *tert-* signifies attachment through a tertiary carbon, a carbon attached to *three* other carbon atoms.

Nomenclature

Because there are literally millions of organic compounds, it would be impossible to remember common names for all of them. We must have a systematic method for naming them. The following rules are used in naming alkanes.

Rules for Naming Alkanes

1. The names of the alkanes beyond butane are obtained by adding the suffix *-ane* to the Greek root for the number of carbon atoms (*pent-* for five, *hex-* for six, and so on). For a branched hydrocarbon, the longest continuous chain of carbon atoms determines the root name for the hydrocarbon. For example, in the alkane

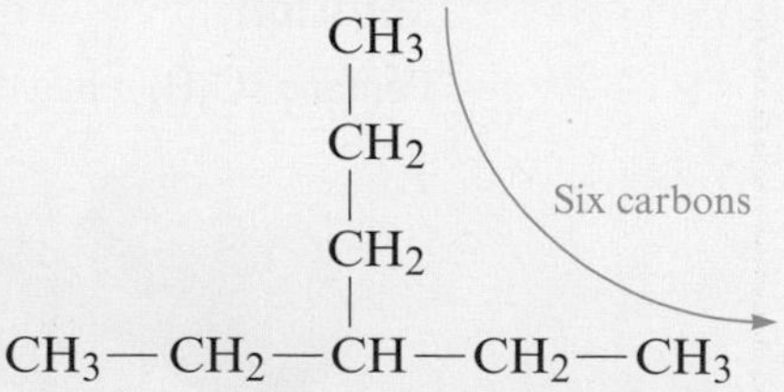

$$\begin{array}{c} CH_3 \\ | \\ CH_2 \\ | \\ CH_2 \\ | \\ CH_3-CH_2-CH-CH_2-CH_3 \end{array}$$

Six carbons

the longest chain contains six carbon atoms, and this compound is named as a hexane.

2. When alkane groups appear as substituents, they are named by dropping the *-ane* and adding *-yl*. For example, —CH₃ is obtained by removing a hydrogen from methane and is called *methyl*, —C₂H₅ is called *ethyl*, —C₃H₇ is called *propyl*, and so on. The compound above is therefore an ethylhexane (Table 21-2).

continued

3. The positions of substituent groups are specified by numbering the longest chain of carbon atoms sequentially, starting at the end closest to the branching. For example, the compound

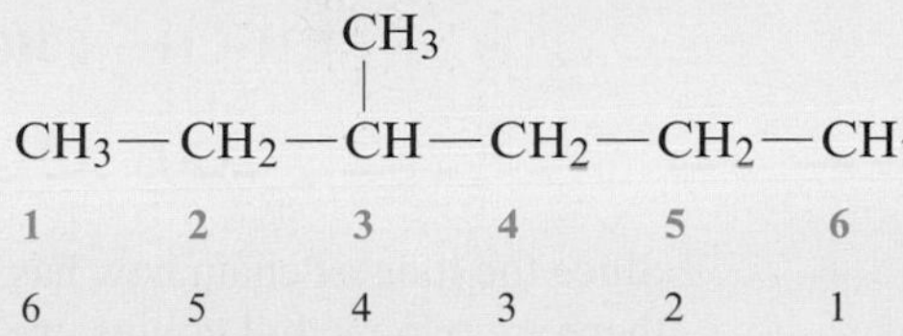

$$CH_3$$
$$|$$
$$CH_3-CH_2-CH-CH_2-CH_2-CH_3$$

| 1 | 2 | 3 | 4 | 5 | 6 | Correct numbering |
| 6 | 5 | 4 | 3 | 2 | 1 | Incorrect numbering |

is called 3-methylhexane. Note that the top set of numbers is correct since the left end of the molecule is closest to the branching, and this gives the smallest number for the position of the substituent. Also, note that a hyphen is written between the number and the substituent name.

4. The location and name of each substituent are followed by the root alkane name. The substituents are listed in alphabetical order, and the prefixes *di-*, *tri-*, and so on, are used to indicate multiple, identical substituents.

EXAMPLE 21-2   Isomerism and Nomenclature

Draw the structural isomers for the alkane C_6H_{14} and give the systematic name for each one.

Solution

We will proceed systematically, starting with the longest chain and then rearranging the carbons to form the shorter, branched chains.

1. $CH_3CH_2CH_2CH_2CH_2CH_3$ Hexane

Note that although a structure such as

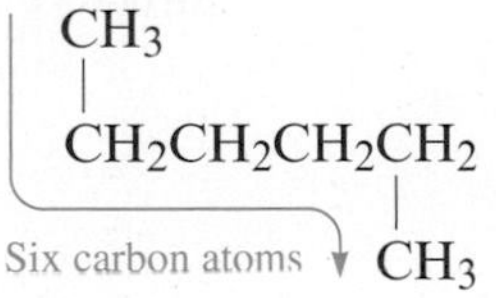

$$CH_3$$
$$|$$
$$CH_2CH_2CH_2CH_2$$
$$|$$
Six carbon atoms $\downarrow$ CH_3

may look different it is still hexane, since the longest carbon chain has six atoms.

2. We now take one carbon out of the chain and make it a methyl substituent.

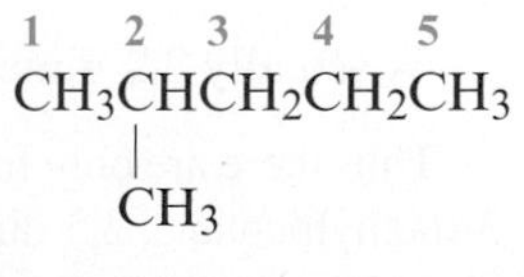

1 2 3 4 5
$$CH_3CHCH_2CH_2CH_3$$ 2-Methylpentane
$$|$$
$$CH_3$$

Since the longest chain consists of five carbons, this is a substituted pentane: 2-methylpentane. The 2 indicates the position of the methyl group on the chain. Note that if we numbered the chain from the right end, the methyl group would be on carbon 4. Because we want the smallest possible number, the numbering shown is correct.

3. The methyl substituent can also be on carbon 3 to give

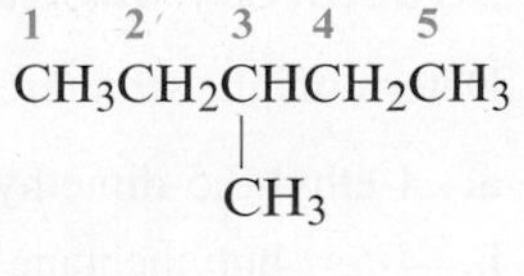

1 2 3 4 5
$$CH_3CH_2CHCH_2CH_3$$ **3-Methylpentane**
$$|$$
$$CH_3$$

Note that we have now exhausted all possibilities for placing a single methyl group on pentane.

4. Next, we can take two carbons out of the original six-member chain:

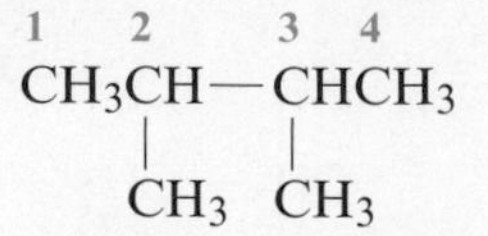

$$\overset{1}{C}H_3\overset{2}{C}H—\overset{3}{C}H\overset{4}{C}H_3 \qquad \textbf{2,3-Dimethylbutane}$$
$$\qquad\ \ |\qquad |$$
$$\qquad\ \ CH_3\ \ CH_3$$

Since the longest chain now has four carbons, the root name is butane. Since there are two methyl groups, we use the prefix *di-*. The numbers denote that the two methyl groups are positioned on the second and third carbons in the butane chain. Note that when two or more numbers are used, they are separated by a comma.

5. The two methyl groups can also be attached to the same carbon atom as shown here:

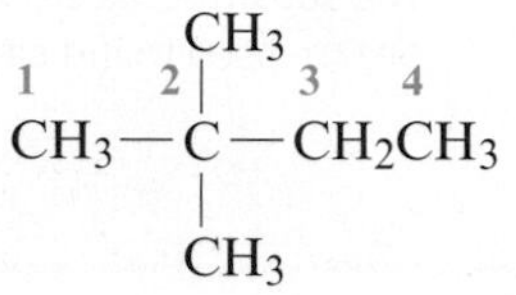

$$\qquad\qquad CH_3$$
$$\overset{1}{C}H_3—\overset{2}{\underset{|}{C}}—\overset{3}{C}H_2\overset{4}{C}H_3 \qquad \textbf{2,2-Dimethylbutane}$$
$$\qquad\ \ |$$
$$\qquad\ \ CH_3$$

We might also try ethyl-substituted butanes, such as

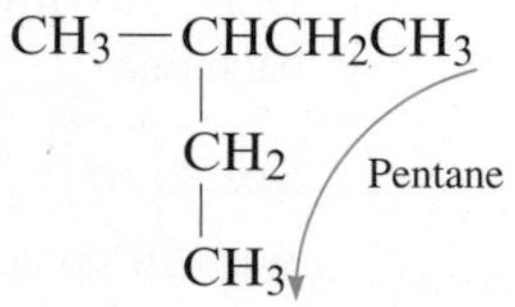

$$CH_3—CHCH_2CH_3$$
$$\qquad\quad |$$
$$\qquad\quad CH_2 \quad \text{Pentane}$$
$$\qquad\quad |$$
$$\qquad\quad CH_3$$

However, note that this is instead a pentane (3-methylpentane), since the longest chain has five carbon atoms. Thus it is not a new isomer. Trying to reduce the chain to three atoms provides no further isomers either. For example, the structure

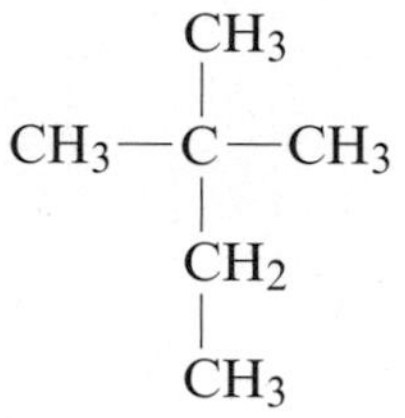

$$\qquad\qquad CH_3$$
$$\qquad\qquad |$$
$$CH_3—C—CH_3$$
$$\qquad\qquad |$$
$$\qquad\qquad CH_2$$
$$\qquad\qquad |$$
$$\qquad\qquad CH_3$$

is actually 2,2-dimethylbutane.

Thus there are only five distinct structural isomers of C_6H_{14}: hexane, 2-methylpentane, 3-methylpentane, 2,3-dimethylbutane, and 2,2-dimethylbutane.

See Exercises 21-15 and 21-16

INTERACTIVE EXAMPLE 21-3

Structures from Names

Determine the structure for each of the following compounds.

a. 4-ethyl-3,5-dimethylnonane

b. 4-*tert*-butylheptane

Solution

a. The root name *nonane* signifies a nine-carbon chain. Thus we have

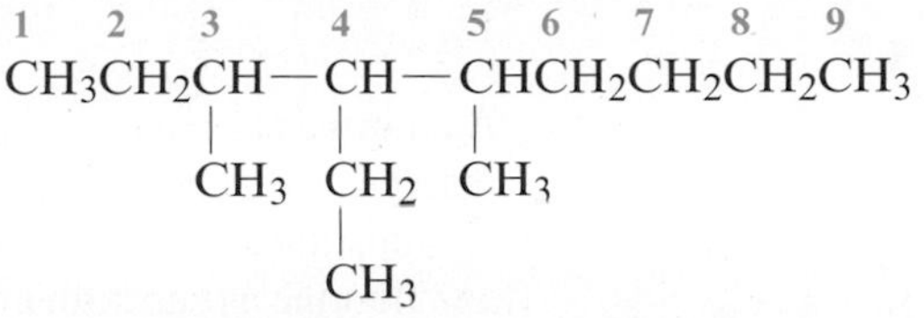

$$\underset{1}{CH_3}\underset{2}{CH_2}\underset{3}{CH}-\underset{4}{CH}-\underset{5}{CH}\underset{6}{CH_2}\underset{7}{CH_2}\underset{8}{CH_2}\underset{9}{CH_3}$$
$$\qquad\qquad\; |\qquad |\qquad |$$
$$\qquad\qquad CH_3\;\; CH_2\;\; CH_3$$
$$\qquad\qquad\qquad\; |$$
$$\qquad\qquad\qquad CH_3$$

b. Heptane signifies a seven-carbon chain, and the *tert*-butyl group is

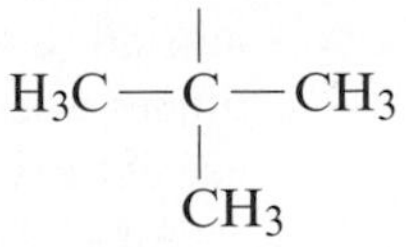

$$\qquad\qquad\;\; |$$
$$H_3C-C-CH_3$$
$$\qquad\;\; |$$
$$\qquad CH_3$$

Thus we have

$$\underset{1}{CH_3}\underset{2}{CH_2}\underset{3}{CH_2}\underset{4}{CH}\underset{5}{CH_2}\underset{6}{CH_2}\underset{7}{CH_3}$$
$$\qquad\qquad\;\; |$$
$$\qquad H_3C-C-CH_3$$
$$\qquad\qquad\;\; |$$
$$\qquad\qquad CH_3$$

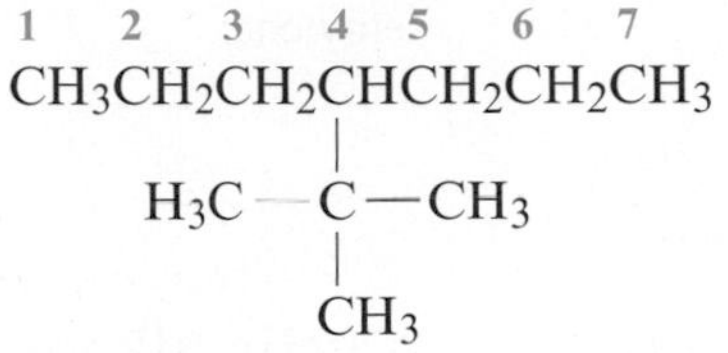

See Exercises 21-19 and 21-20

Reactions of Alkanes

Because they are saturated compounds and because the C—C and C—H bonds are relatively strong, the alkanes are fairly unreactive. For example, at 25°C they do not react with acids, bases, or strong oxidizing agents. This chemical inertness makes them valuable as lubricating materials and as the backbone for structural materials such as plastics.

At a sufficiently high temperature alkanes do react vigorously and exothermically with oxygen, and these **combustion reactions** are the basis for their widespread use as fuels. ◄ For example, the reaction of butane with oxygen is

$$2C_4H_{10}(g) + 13O_2(g) \longrightarrow 8CO_2(g) + 10H_2O(g)$$

The alkanes can also undergo **substitution reactions**, primarily where halogen atoms replace hydrogen atoms. For example, methane can be successively chlorinated as follows: ◄ *i*

$$CH_4 + Cl_2 \xrightarrow{h\nu} \underset{\text{Chloromethane}}{CH_3Cl} + HCl$$

$$CH_3Cl + Cl_2 \xrightarrow{h\nu} \underset{\text{Dichloromethane}}{CH_2Cl_2} + HCl$$

$$CH_2Cl_2 + Cl_2 \xrightarrow{h\nu} \underset{\substack{\text{Trichloromethane}\\\text{(chloroform)}}}{CHCl_3} + HCl$$

$$CHCl_3 + Cl_2 \xrightarrow{h\nu} \underset{\substack{\text{Tetrachloromethane}\\\text{(carbon tetrachloride)}}}{CCl_4} + HCl$$

Note that the products of the last two reactions have two names; the systematic name is given first, followed by the common name in parentheses. (This format will be used

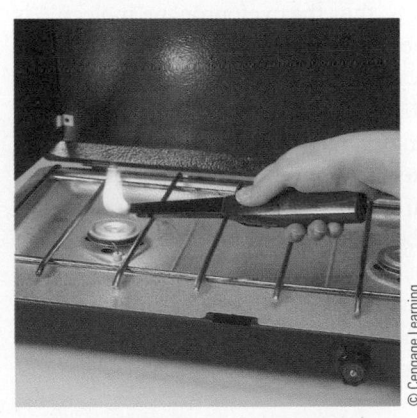

▲ A butane lighter used for camping.

© Cengage Learning

i The *hv* above the arrow represents ultraviolet light.

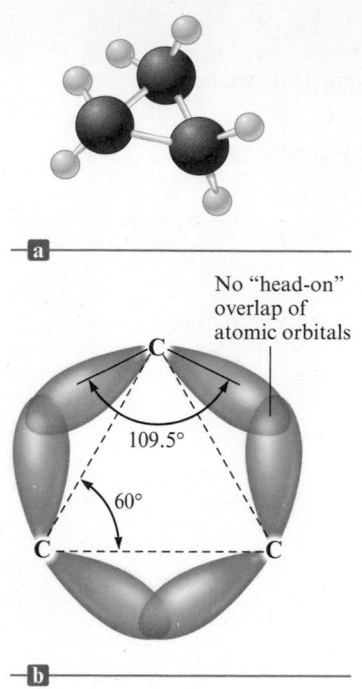

No "head-on"
overlap of
atomic orbitals

Figure 21-5 | (a) The molecular structure of cyclopropane (C_3H_6). (b) The overlap of the sp^3 orbitals that form the C—C bonds in cyclopropane.

throughout this chapter for compounds that have common names.) Also, note that ultraviolet light (hv) furnishes the energy to break the Cl—Cl bond to produce chlorine atoms:

$$Cl_2 \longrightarrow Cl \cdot + Cl \cdot$$

A chlorine atom has an unpaired electron, as indicated by the dot, which makes it very reactive and able to attack the C—H bond.

Substituted methanes with the general formula CF_xCl_{4-x} containing both chlorine and fluorine as substituents are called chlorofluorocarbons (CFCs) and are also known as *Freons*. These substances are very unreactive and have been extensively used as coolant fluids in refrigerators and air conditioners. Unfortunately, their chemical inertness allows Freons to remain in the atmosphere so long that they eventually reach altitudes where they are a threat to the protective ozone layer (see Section 11-7), and the use of these compounds is being rapidly phased out.

Alkanes can also undergo **dehydrogenation reactions** in which hydrogen atoms are removed and the product is an unsaturated hydrocarbon. For example, in the presence of chromium(III) oxide at high temperatures, ethane can be dehydrogenated, yielding ethylene:

$$CH_3CH_3 \xrightarrow[500°C]{Cr_2O_3} CH_2{=}CH_2 + H_2$$
Ethylene

Cyclic Alkanes

Besides forming chains, carbon atoms also form rings. The simplest of the **cyclic alkanes** (general formula C_nH_{2n}) is cyclopropane (C_3H_6), shown in Fig. 21-5(a). Since the carbon atoms in cyclopropane form an equilateral triangle with 60° bond angles, their sp^3 hybrid orbitals do not overlap head-on as in normal alkanes [Fig. 21-5(b)]. This results in unusually weak, or *strained,* C—C bonds; thus the cyclopropane molecule is much more reactive than straight-chain propane. The carbon atoms in cyclobutane (C_4H_8) form a square with 88° bond angles, and cyclobutane is also quite reactive.

The next two members of the series, cyclopentane (C_5H_{10}) and cyclohexane (C_6H_{12}), are quite stable, because their rings have bond angles very close to the tetrahedral angles, which allows the sp^3 hybrid orbitals on adjacent carbon atoms to overlap head-on and form normal C—C bonds, which are quite strong. To attain the tetrahedral angles, the cyclohexane ring must "pucker"—that is, become nonplanar. Cyclohexane can exist in two forms, the *chair* and the *boat* forms, as shown in Fig. 21-6. The two hydrogen atoms above the ring in the boat form are quite close to each other, and the resulting repulsion between these atoms causes the chair form to be preferred. At 25°C more than 99% of cyclohexane exists in the chair form.

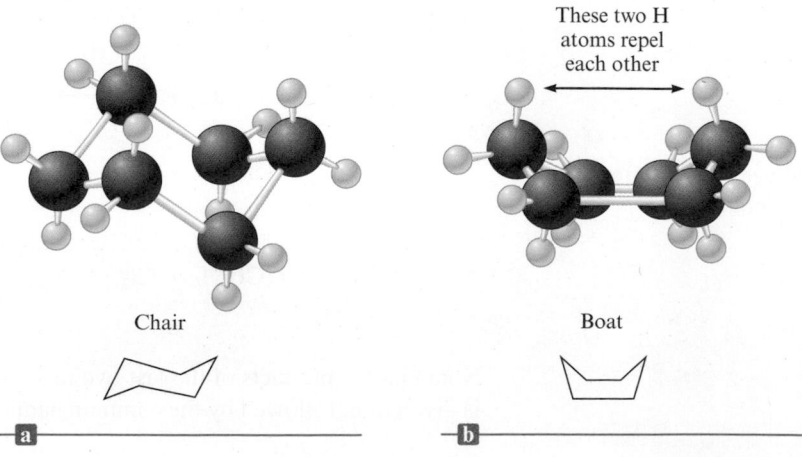

These two H atoms repel each other

Chair

Boat

Figure 21-6 | The (a) chair and (b) boat forms of cyclohexane.

For simplicity, the cyclic alkanes are often represented by the following structures:

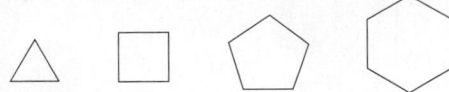

Thus the structure CH$_3$ represents methylcyclopropane.

The nomenclature for cycloalkanes follows the same rules as for the other alkanes except that the root name is preceded by the prefix *cyclo-*. The ring is numbered to yield the smallest substituent numbers possible.

INTERACTIVE EXAMPLE 21-4

Naming Cyclic Alkanes

Name each of the following cyclic alkanes.

a. 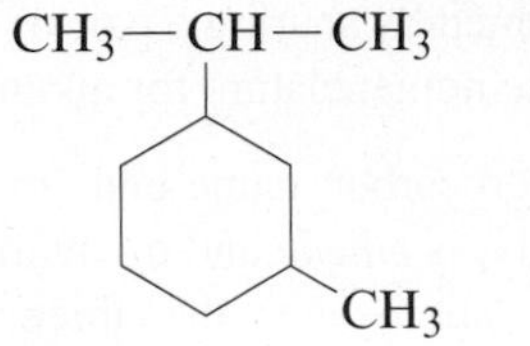 CH$_3$—CH—CH$_3$

b. 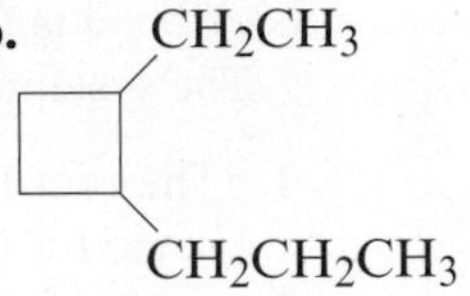 CH$_2$CH$_3$

CH$_2$CH$_2$CH$_3$

CH$_3$

Solution

a. The six-carbon cyclohexane ring is numbered as follows:

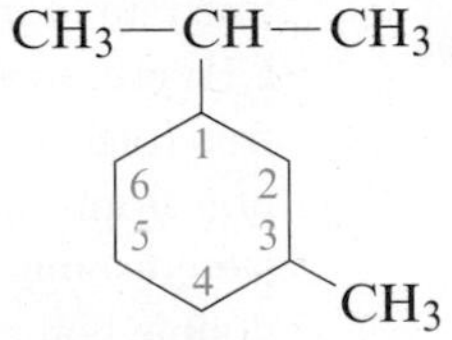

CH$_3$—CH—CH$_3$

CH$_3$

There is an isopropyl group at carbon 1 and a methyl group at carbon 3. The name is 1-isopropyl-3-methylcyclohexane, since the alkyl groups are named in alphabetical order.

b. This is a cyclobutane ring, which is numbered as follows:

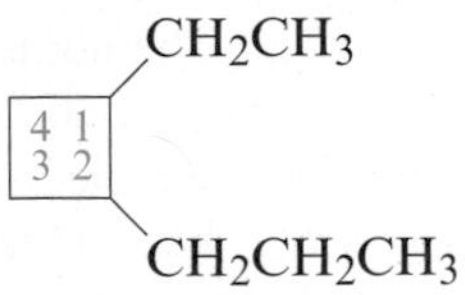

CH$_2$CH$_3$

CH$_2$CH$_2$CH$_3$

The name is 1-ethyl-2-propylcyclobutane.

See Exercise 21-22

21-2 | Alkenes and Alkynes

Multiple carbon–carbon bonds result when hydrogen atoms are removed from alkanes. Hydrocarbons that contain at least one carbon–carbon double bond are called **alkenes** and have the general formula C$_n$H$_{2n}$. The simplest alkene (C$_2$H$_4$), commonly known as *ethylene,* has the Lewis structure

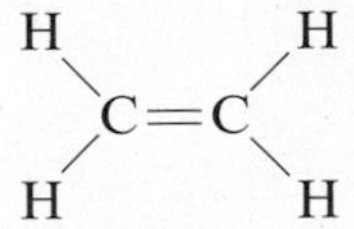

$$\underset{H}{\overset{H}{\diagdown}}C=C\underset{H}{\overset{H}{\diagup}}$$

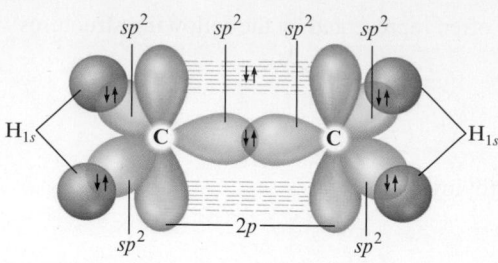

Figure 21-7 | The bonding in ethylene.

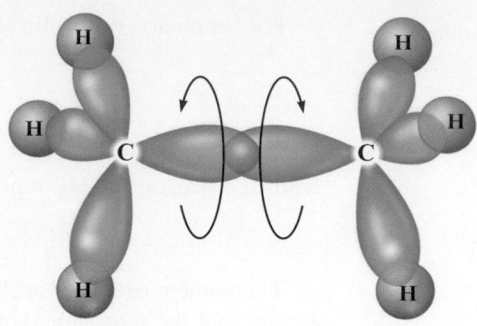

Figure 21-8 | The bonding in ethane.

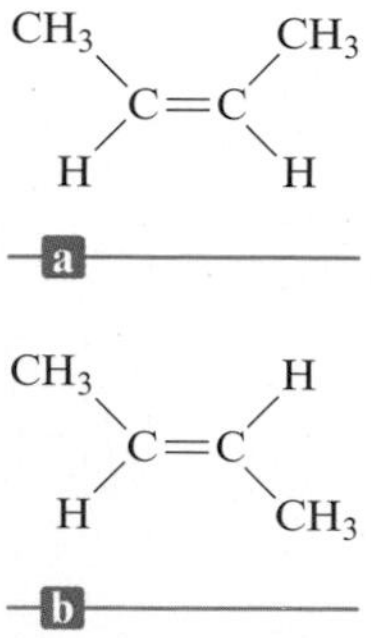

CH₃   CH₃

C=C

H   H

a

CH₃   H

C=C

H   CH₃

b

Figure 21-9 | The two stereo-isomers of 2-butene: (a) *cis*-2-butene and (b) *trans*-2-butene.

i For cyclic alkenes, number through the double bond toward the substituent.

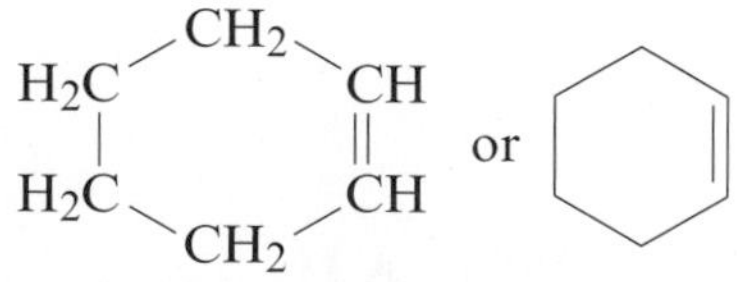

As discussed in Section 4-3, each carbon in ethylene can be described as sp^2 hybrid-ized. The C—C σ bond is formed by sharing an electron pair between sp^2 orbitals, and the π bond is formed by sharing a pair of electrons between p orbitals (Fig. 21-7).

The systematic nomenclature for alkenes is quite similar to that for alkanes.

1. The root hydrocarbon name ends in *-ene* rather than *-ane*. Thus the systematic name for C_2H_4 is *ethene* and the name for C_3H_6 is *propene*.
2. In alkenes containing more than three carbon atoms, the location of the double bond is indicated by the lowest-numbered carbon atom involved in the bond. Thus $CH_2=CHCH_2CH_3$ is called 1-butene, and $CH_3CH=CHCH_3$ is called 2-butene.

Note from Fig. 21-7 that the p orbitals on the two carbon atoms in ethylene must be lined up (parallel) to allow formation of the π bond. This prevents rotation of the two CH_2 groups relative to each other at ordinary temperatures, in contrast to alkanes, where free rotation is possible (Fig. 21-8). The restricted rotation around doubly bonded carbon atoms means that alkenes exhibit ***cis–trans* isomerism**. For example, there are two stereoisomers of 2-butene (Fig. 21-9). Identical substituents on the same side of the double bond are designated *cis* and those on opposite sides are labeled *trans*.

Alkynes are unsaturated hydrocarbons containing at least one triple carbon–carbon bond. The simplest alkyne is C_2H_2 (commonly called *acetylene*), which has the systematic name *ethyne*. ◄ As discussed in Section 4-3, the triple bond in acetylene can be described as one σ bond between two sp hybrid orbitals on the two carbon atoms and two π bonds involving two $2p$ orbitals on each carbon atom (Fig. 21-10).

The nomenclature for alkynes involves the use of *-yne* as a suffix to replace the *-ane* of the parent alkane. Thus the molecule $CH_3CH_2C≡CCH_3$ has the name 2-pentyne.

Like alkanes, unsaturated hydrocarbons can exist as ringed structures, for example, ◄ *i*

H₂C—CH₂—CH

                              ‖   or

H₂C—CH₂—CH

Cyclohexene

            CH₃

            CH   CH₃

H₂C   CH₂   or

HC=CH

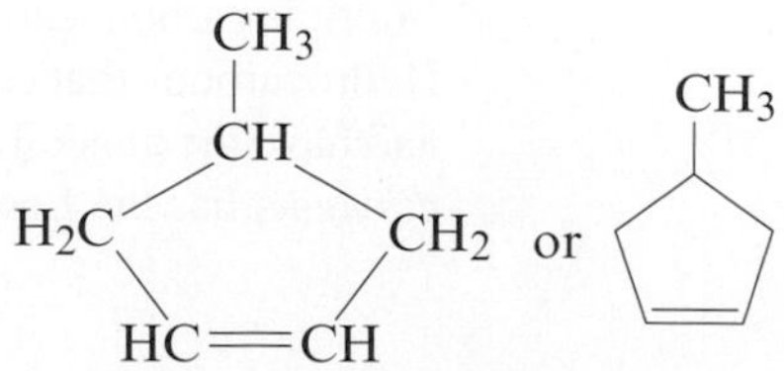

4-Methyl-cyclopentene

Figure 21-10 | The bonding in acetylene.

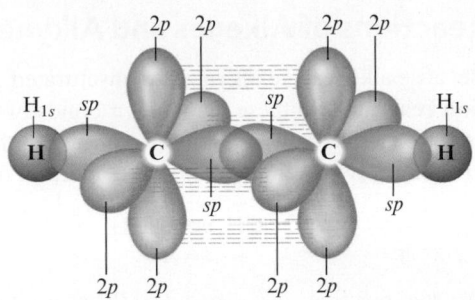

INTERACTIVE EXAMPLE 21-5

Naming Alkenes and Alkynes

Name each of the following molecules.

▲ A worker using an oxyacetylene torch.

Bob Bartee/Alamy

a.

$$\begin{array}{ccc} & H & CH_3 \\ & \diagdown & \diagup \\ & C = C \\ & \diagup & \diagdown \\ CH_3CH_2CH & & H \\ & | & \\ & CH_3 & \end{array}$$

b. $CH_3CH_2C{\equiv}CCHCH_2CH_3$
$\qquad\qquad\qquad\quad |$
$\qquad\qquad\qquad\ CH_2$
$\qquad\qquad\qquad\quad |$
$\qquad\qquad\qquad\ CH_3$

Solution

a. The longest chain, which contains six carbon atoms, is numbered as follows:

$$\begin{array}{ccc} & \overset{3}{H} & \overset{1}{CH_3} \\ & \diagdown & \diagup \\ & \overset{3}{C} = \overset{2}{C} \\ \underset{6}{CH_3}\underset{5}{CH_2}\underset{4}{CH} & & H \\ & | & \\ & CH_3 & \end{array}$$

Thus the hydrocarbon is a 2-hexene. Since the hydrogen atoms are located on opposite sides of the double bond, this molecule corresponds to the *trans* isomer. The name is *trans*-4-methyl-2-hexene.

b. The longest chain, consisting of seven carbon atoms, is numbered as shown (giving the triple bond the lowest possible number):

$$\overset{1}{CH_3}\overset{2}{CH_2}\overset{3}{C}{\equiv}\overset{4}{C}\overset{5}{C}H\overset{6}{CH_2}\overset{7}{CH_3}$$
$$| $$
$$CH_2$$
$$| $$
$$CH_3$$

The hydrocarbon is a 3-heptyne. The full name is 5-ethyl-3-heptyne, where the position of the triple bond is indicated by the lower-numbered carbon atom involved in this bond.

See Exercises 21-25, 21-26, and 21-44

Reactions of Alkenes and Alkynes

Because alkenes and alkynes are unsaturated, their most important reactions are **addition reactions**. In these reactions π bonds, which are weaker than the C—C σ bonds, are broken, and new σ bonds are formed to the atoms being added. For example, **hydrogenation reactions** involve the addition of hydrogen atoms:

$$CH_2{=}CHCH_3 + H_2 \xrightarrow{\text{Catalyst}} CH_3CH_2CH_3$$
$$\underset{\text{1-Propene}}{} \qquad\qquad \underset{\text{Propane}}{}$$

For this reaction to proceed rapidly at normal temperatures, a catalyst of platinum, palladium, or nickel is used. The catalyst serves to help break the relatively strong H—H bond, as was discussed in Section 11-7. Hydrogenation of alkenes is an important industrial process, particularly in the manufacture of solid shortenings where unsaturated fats (fats containing double bonds), which are generally liquid, are converted to solid saturated fats.

Halogenation of unsaturated hydrocarbons involves addition of halogen atoms. For example,

$$CH_2{=}CHCH_2CH_2CH_3 + Br_2 \longrightarrow CH_2BrCHBrCH_2CH_2CH_3$$
$$\underset{\text{1-Pentene}}{} \qquad\qquad\qquad \underset{\text{1,2-Dibromopentane}}{}$$

Another important reaction involving certain unsaturated hydrocarbons is **polymerization**, a process in which many small molecules are joined together to form a large molecule. Polymerization will be discussed in Section 21-5.

21-3 | Aromatic Hydrocarbons

A special class of cyclic unsaturated hydrocarbons is known as the **aromatic hydrocarbons**. The simplest of these is benzene (C_6H_6), which has a planar ring structure, as shown in Fig. 21-11(a). In the localized electron model of the bonding in benzene, resonance structures of the type shown in Fig. 21-11(b) are used to account for the known equivalence of all the carbon–carbon bonds. But as we discussed in Section 4-7, the best description of the benzene molecule assumes that sp^2 hybrid orbitals on each carbon are used to form the C—C and C—H σ bonds, while the remaining $2p$ orbital on each carbon is used to form π molecular orbitals. The delocalization of these π electrons is usually indicated by a circle inside the ring [Fig. 21-11(c)].

The delocalization of the π electrons makes the benzene ring behave quite differently from a typical unsaturated hydrocarbon. As we have seen previously, unsaturated hydrocarbons generally undergo rapid addition reactions. However, benzene does not. Instead, it undergoes substitution reactions in which *hydrogen atoms are replaced by other atoms*. For example,

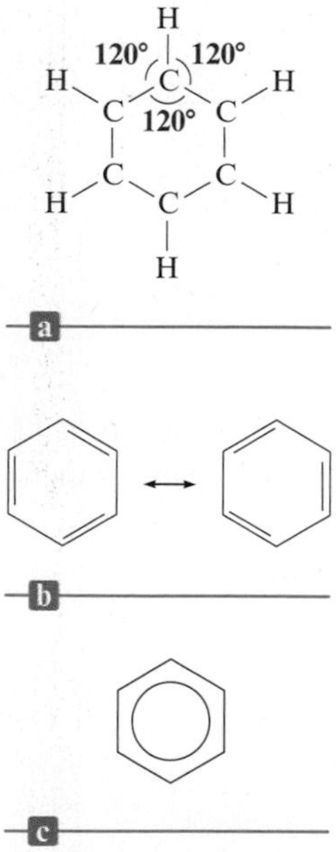

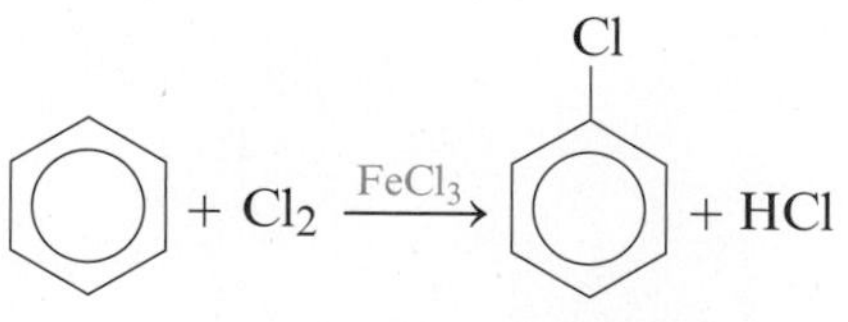

Figure 21-11 | (a) The structure of benzene, a planar ring system in which all bond angles are 120°. (b) Two of the resonance structures of benzene. (c) The usual representation of benzene. The circle represents the electrons in the delocalized π system. All C—C bonds in benzene are equivalent.

Chlorobenzene

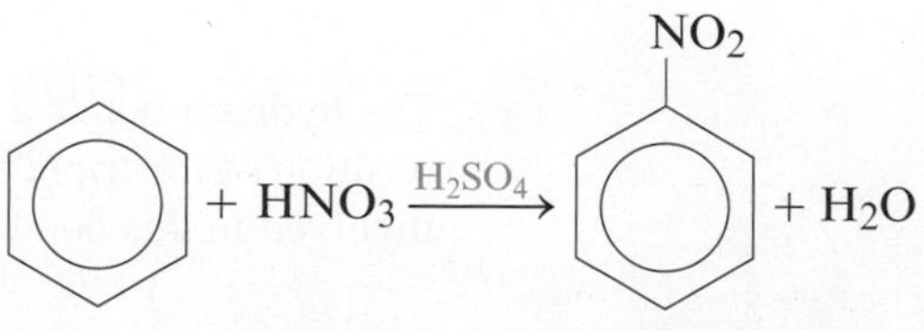

Nitrobenzene

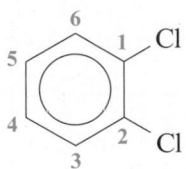

Toluene

In each case the substance shown over the arrow is needed to catalyze these substitution reactions.

Substitution reactions are characteristic of saturated hydrocarbons, and addition reactions are characteristic of unsaturated ones. The fact that benzene reacts more like a saturated hydrocarbon indicates the great stability of the delocalized π electron system.

The nomenclature of benzene derivatives is similar to the nomenclature for saturated ring systems. If there is more than one substituent present, numbers are used to indicate substituent positions. For example, the compound

is named 1,2-dichlorobenzene. Another nomenclature system uses the prefix *ortho-* (*o-*) for two adjacent substituents, *meta-* (*m-*) for two substituents with one carbon between them, and *para-* (*p-*) for two substituents opposite each other. When benzene is used as a substituent, it is called the **phenyl group**. Examples of some aromatic compounds are shown in Fig. 21-12.

Benzene is the simplest aromatic molecule. More complex aromatic systems can be viewed as consisting of a number of "fused" benzene rings. Some examples are given in Table 21-3. ◄

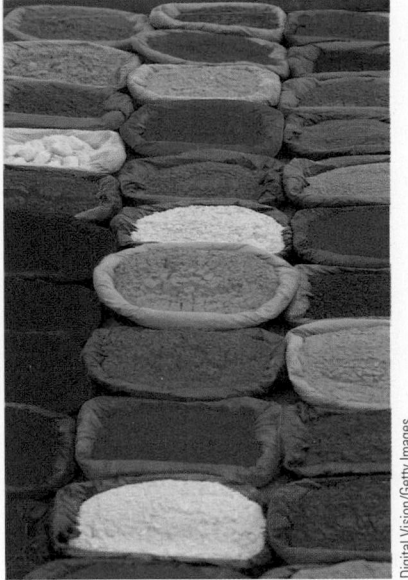

▲ Compounds containing aromatic rings are often used in dyes, such as these for sale in a market in Nepal.

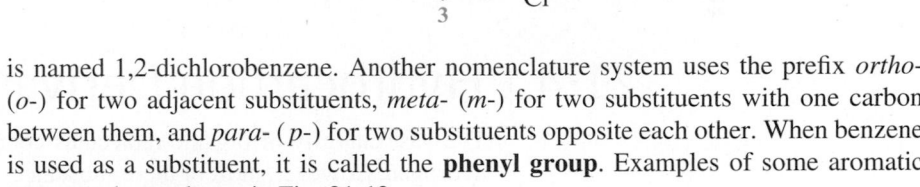

1,2-Dibromobenzene
(*o*-dibromobenzene)

1,3-Dibromobenzene
(*m*-dibromobenzene)

1,4-Dibromobenzene
(*p*-dibromobenzene)

Methylbenzene
(toluene)

3-Bromonitrobenzene
(*m*-bromonitrobenzene)

3-Chlorotoluene
(*m*-chlorotoluene)

Phenyl group

4-Chloro-2-phenylhexane

Figure 21-12 | Some selected substituted benzenes and their names. Common names are given in parentheses.

Table 21-3 | More Complex Aromatic Systems

Structural Formula	Name	Use of Effect
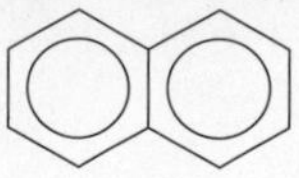	Naphthalene	Formerly used in mothballs
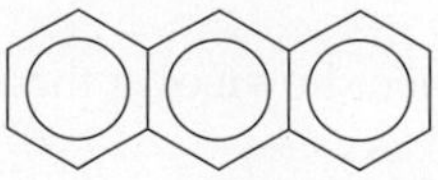	Anthracene	Dyes
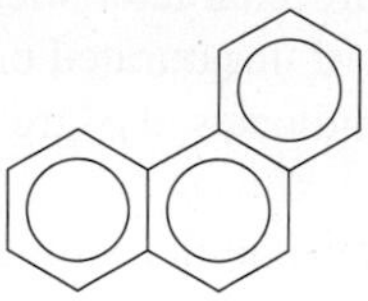	Phenanthrene	Dyes, explosives, and synthesis of drugs
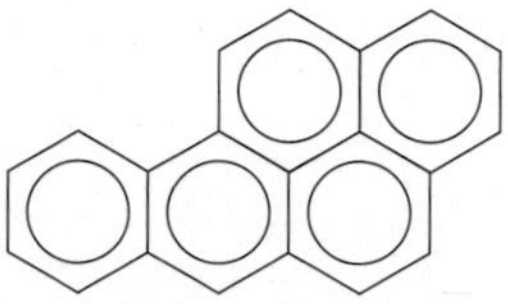	3,4-Benzpyrene	Active carcinogen found in smoke and smog

21-4 | Hydrocarbon Derivatives

The vast majority of organic molecules contain elements in addition to carbon and hydrogen. However, most of these substances can be viewed as **hydrocarbon derivatives**, molecules that are fundamentally hydrocarbons but that have additional atoms or groups of atoms called **functional groups**. The common functional groups are listed in Table 21-4. Because each functional group exhibits characteristic chemistry, we will consider the groups separately.

Alcohols

Alcohols are characterized by the presence of the hydroxyl group (—OH). Some common alcohols are listed in Table 21-5. The systematic name for an alcohol is obtained by replacing the final -*e* of the parent hydrocarbon with -*ol*. The position of the —OH group is specified by a number (where necessary) chosen so that it is the smallest of the substituent numbers. Alcohols are classified according to the number of hydrocarbon fragments bonded to the carbon where the —OH group is attached (see below), where R, R′, and R″ (which may be the same or different) represent hydrocarbon fragments.

$$R—CH_2OH \qquad \begin{matrix} R \\ \diagdown \\ CHOH \\ \diagup \\ R' \end{matrix} \qquad \begin{matrix} R \\ | \\ R'—COH \\ | \\ R'' \end{matrix}$$

Primary alcohol   *Secondary* alcohol   *Tertiary* alcohol
(one R group)   (two R groups)   (three R groups)

Alcohols usually have much higher boiling points than might be expected from their molar masses. For example, both methanol and ethane have a molar mass of 30, but the boiling point for methanol is 65°C while that for ethane is −89°C. This difference can be understood if we consider the types of intermolecular attractions that

Table 21-4 | The Common Functional Groups

Class	Functional Group	General Formula*	Example
Halohydrocarbons	—X (F, Cl, Br, I)	R—X	CH_3I Iodomethane (methyl iodide)
Alcohols	—OH	R—OH	CH_3OH Methanol (methyl alcohol)
Ethers	—O—	R—O—R′	CH_3OCH_3 Dimethyl ether
Aldehydes	$\overset{\displaystyle O}{\overset{\displaystyle \|}{-C}}-H$	$\overset{\displaystyle O}{\overset{\displaystyle \|}{R-C}}-H$	CH_2O Methanal (formaldehyde)
Ketones	$\overset{\displaystyle O}{\overset{\displaystyle \|}{-C}}-$	$\overset{\displaystyle O}{\overset{\displaystyle \|}{R-C}}-R′$	CH_3COCH_3 Propanone (dimethyl ketone or acetone)
Carboxylic acids	$\overset{\displaystyle O}{\overset{\displaystyle \|}{-C}}-OH$	$\overset{\displaystyle O}{\overset{\displaystyle \|}{R-C}}-OH$	CH_3COOH Ethanoic acid (acetic acid)
Esters	$\overset{\displaystyle O}{\overset{\displaystyle \|}{-C}}-O-$	$\overset{\displaystyle O}{\overset{\displaystyle \|}{R-C}}-O-R′$	$CH_3COOCH_2CH_3$ Ethyl acetate
Amines	$-NH_2$	$R-NH_2$	CH_3NH_2 Aminomethane (methylamine)

*R and R′ represent hydrocarbon fragments.

occur in these liquids. Ethane molecules are nonpolar and exhibit only weak London dispersion interactions. However, the polar —OH group of methanol produces extensive hydrogen bonding similar to that found in water (see Section 9-1), which results in the relatively high boiling point.

Although there are many important alcohols, the simplest ones, methanol and ethanol, have the greatest commercial value. Methanol, also known as *wood alcohol* because it was formerly obtained by heating wood in the absence of air, is prepared industrially (approximately 4 million tons annually in the United States) by the hydrogenation of carbon monoxide:

$$CO + 2H_2 \xrightarrow[\text{ZnO/Cr}_2\text{O}_3]{400°C} CH_3OH$$

▲ A winemaker draws off a glass of wine in a modern wine cellar.

Table 21-5 | Some Common Alcohols

Formula	Systematic Name	Common Name
CH_3OH	Methanol	Methyl alcohol
CH_3CH_2OH	Ethanol	Ethyl alcohol
$CH_3CH_2CH_2OH$	1-Propanol	*n*-Propyl alcohol
$CH_3\underset{\displaystyle \underset{OH}{\|}}{C}HCH_3$	2-Propanol	Isopropyl alcohol

Methanol is used as a starting material for the synthesis of acetic acid and for many types of adhesives, fibers, and plastics. It is also used (and such use may increase) as a motor fuel. Methanol is highly toxic to humans and can lead to blindness and death if ingested.

Ethanol is the alcohol found in beverages such as beer, wine, and whiskey; it is produced by the fermentation of glucose in corn, barley, grapes, and so on: ◄

$$C_6H_{12}O_6 \xrightarrow{\text{Yeast}} 2CH_3CH_2OH + 2CO_2$$
Glucose   Ethanol

The reaction is catalyzed by the enzymes found in yeast. This reaction can proceed only until the alcohol content reaches about 13% (the percentage found in most wines), at which point the yeast can no longer survive. Beverages with higher alcohol content are made by distilling the fermentation mixture.

Ethanol, like methanol, can be burned in the internal combustion engines of automobiles and is now commonly added to gasoline to form gasohol. ◄ It is also used in industry as a solvent and for the preparation of acetic acid. One common method for the production of ethanol in the chemical industry is by reaction of water with ethylene:

$$CH_2{=}CH_2 + H_2O \xrightarrow[\text{Catalyst}]{\text{Acid}} CH_3CH_2OH$$

Many polyhydroxyl (more than one —OH group) alcohols are known, the most important being *1,2-ethanediol* (ethylene glycol),

$$\begin{array}{c} H_2C-OH \\ | \\ H_2C-OH \end{array}$$

a toxic substance that is the major constituent of most automobile antifreeze solutions.

The simplest aromatic compound with an attached —OH group is

which is commonly called **phenol**. While this compound looks like an alcohol, its properties are very different from alcohols. Most of the 1 million tons of phenol produced annually in the United States is used to make polymers for adhesives and plastics.

▲ An E85 fuel pump.

AP Photo/Seth Perlman

INTERACTIVE EXAMPLE 21-6

Naming and Classifying Alcohols

For each of the following alcohols, give the systematic name and specify whether the alcohol is primary, secondary, or tertiary.

a. $\underset{\text{OH}}{\text{CH}_3\text{CHCH}_2\text{CH}_3}$

b. $\text{ClCH}_2\text{CH}_2\text{CH}_2\text{OH}$

c. $\underset{\text{OH}}{\overset{\text{CH}_3}{\text{CH}_3\text{CCH}_2\text{CH}_2\text{CH}_2\text{CH}_2\text{Br}}}$

Solution

a. The chain is numbered as follows:

$$\overset{1}{C}H_3\overset{2}{C}H\overset{3}{C}H_2\overset{4}{C}H_3$$
$$\underset{OH}{|}$$

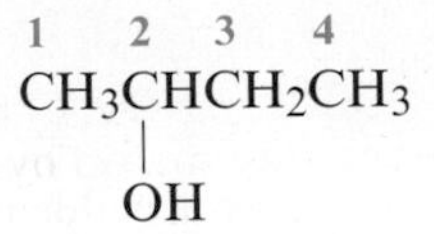

The compound is called 2-butanol, since the —OH group is located at the number 2 position of a four-carbon chain. Note that the carbon to which the —OH is attached also has —CH_3 and —CH_2CH_3 groups attached:

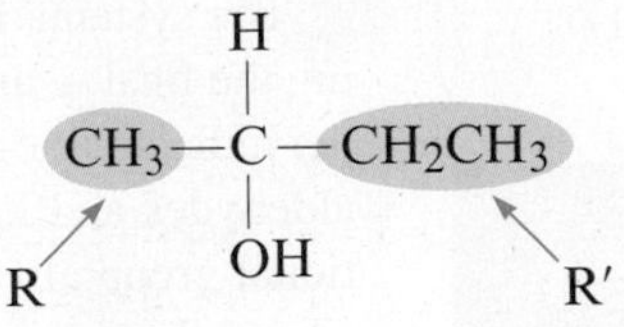

$$CH_3 \overset{\overset{H}{|}}{\underset{\underset{OH}{|}}{C}} CH_2CH_3$$

R   R′

Therefore, this is a *secondary* alcohol.

b. The chain is numbered as follows:

$$Cl-\overset{3}{C}H_2-\overset{2}{C}H_2-\overset{1}{C}H_2-OH$$

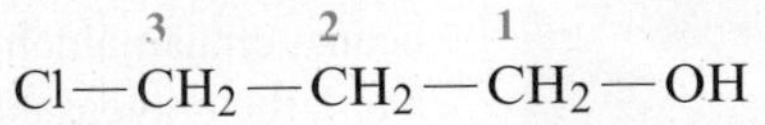

The name is 3-chloro-1-propanol. This is a *primary* alcohol:

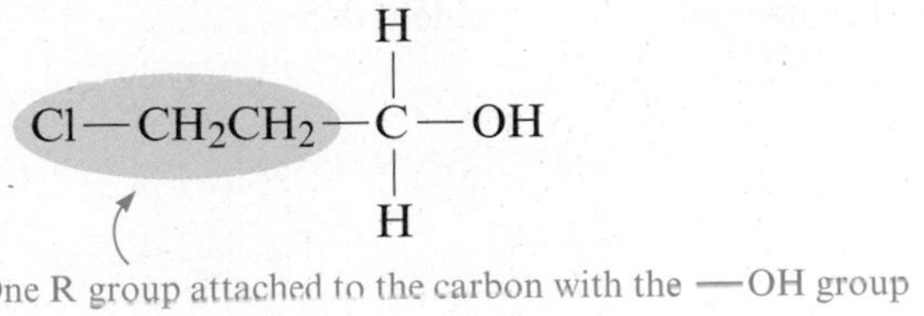

$$Cl-CH_2CH_2 \overset{\overset{H}{|}}{\underset{\underset{H}{|}}{C}} OH$$

One R group attached to the carbon with the —OH group

c. The chain is numbered as follows:

$$\overset{1}{C}H_3 \overset{\overset{CH_3}{|}}{\underset{\underset{OH}{|}}{\overset{2}{C}}} \overset{3}{C}H_2 - \overset{4}{C}H_2 - \overset{5}{C}H_2 - \overset{6}{C}H_2Br$$

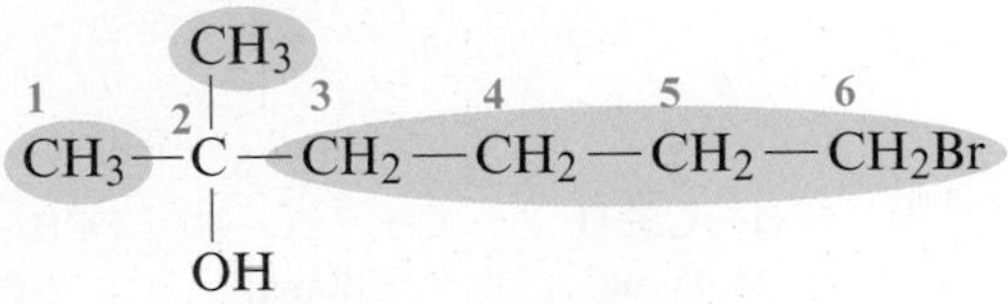

The name is 6-bromo-2-methyl-2-hexanol. This is a *tertiary* alcohol since the carbon where the —OH is attached also has three other R groups attached.

See Exercise 21-51

Aldehydes and Ketones

Aldehydes and ketones contain the **carbonyl group**,

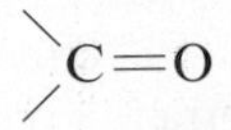

$$\overset{\diagdown}{\underset{\diagup}{C}} = O$$

In **ketones** this group is bonded to two carbon atoms, as in acetone,

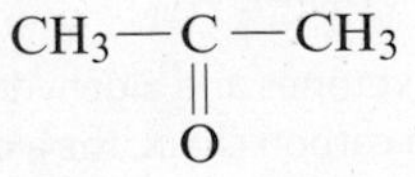

$$CH_3 - \overset{\overset{}{C}}{\underset{\underset{O}{\|}}{}} - CH_3$$

In **aldehydes** the carbonyl group is bonded to at least one hydrogen atom, as in formaldehyde,

$$H-\overset{\displaystyle ||}{\underset{O}{C}}-H$$

or acetaldehyde,

$$CH_3-\overset{\displaystyle ||}{\underset{O}{C}}-H$$

The systematic name for an aldehyde is obtained from the parent alkane by removing the final -e and adding -al. For ketones the final -e is replaced by -one, and a number indicates the position of the carbonyl group where necessary. Examples of common aldehydes and ketones are shown in Fig. 21-13. Note that since the aldehyde functional group always occurs at the end of the carbon chain, the aldehyde carbon is assigned the number 1 when substituent positions are listed in the name.

Ketones often have useful solvent properties (acetone is found in nail polish remover, for example) and are frequently used in industry for this purpose. Aldehydes typically have strong odors. Vanillin is responsible for the pleasant odor in vanilla beans; cinnamaldehyde produces the characteristic odor of cinnamon. ◄ On the other hand, the unpleasant odor in rancid butter arises from the presence of butyraldehyde.

Aldehydes and ketones are most often produced commercially by the oxidation of alcohols. For example, oxidation of a *primary* alcohol yields the corresponding aldehyde:

▲ Cinnamaldehyde produces the characteristic odor of cinnamon.

$$CH_3CH_2OH \xrightarrow{\text{Oxidation}} CH_3\overset{\displaystyle O}{\underset{H}{C}}$$

Figure 21-13 | Some common ketones and aldehydes. Note that since the aldehyde functional group always appears at the end of a carbon chain, the aldehyde carbon is assigned the number 1 when the compound is named.

Oxidation of a *secondary* alcohol results in a ketone:

$$CH_3CHCH_3 \xrightarrow{\text{Oxidation}} CH_3CCH_3$$
$$\underset{OH}{|} \qquad\qquad \underset{O}{\|}$$

Carboxylic Acids and Esters

Carboxylic acids are characterized by the presence of the **carboxyl group**

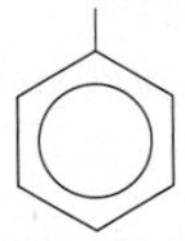

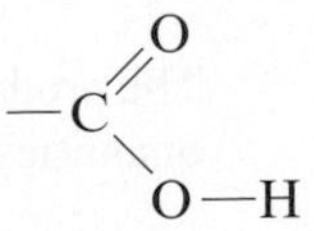

$$-C\overset{\displaystyle O}{\underset{\displaystyle O-H}{\Big\|}}$$

that gives an acid of the general formula RCOOH. Typically, these molecules are weak acids in aqueous solution (see Section 13.5). Organic acids are named from the parent alkane by dropping the final *-e* and adding *-oic*. Thus CH_3COOH, commonly called acetic acid, has the systematic name ethanoic acid, since the parent alkane is ethane. Other examples of carboxylic acids are shown in Fig. 21-14.

Many carboxylic acids are synthesized by oxidizing primary alcohols with a strong oxidizing agent. For example, ethanol can be oxidized to acetic acid by using potassium permanganate:

$$CH_3CH_2OH \xrightarrow{\text{KMnO}_4(aq)} CH_3COOH$$

A carboxylic acid reacts with an alcohol to form an **ester** and a water molecule. For example, the reaction of acetic acid with ethanol produces ethyl acetate and water:

$$\underset{}{CH_3C}\overset{\displaystyle O}{\overset{\|}{-}}OH \quad H-OCH_2CH_3 \longrightarrow CH_3C\overset{\displaystyle O}{\overset{\|}{-}}OCH_2CH_3 + H_2O$$

React to
form water

Esters often have a sweet, fruity odor that is in contrast to the often pungent odors of the parent carboxylic acids. For example, the odor of bananas is caused by *n*-amyl acetate,

$$CH_3C\overset{\displaystyle O}{\underset{\displaystyle OCH_2CH_2CH_2CH_2CH_3}{\diagup}}$$

and that of oranges is caused by *n*-octyl acetate,

$$CH_3C-OC_8H_{17}$$
$$\underset{O}{\|}$$

The systematic name for an ester is formed by changing the *-oic* ending of the parent acid to *-oate*. The parent alcohol chain is named first with a *-yl* ending. For example, the systematic name for *n*-octyl acetate is *n*-octylethanoate (from ethanoic acid).

$$CH_3CH_2CH_2COOH$$
Butanoic acid

$$\underset{\text{Benzoic acid}}{\overset{\displaystyle COOH}{\bigcirc}}$$

Benzoic acid

$$CH_3CHCH_2CH_2COOH$$
$$\underset{Br}{|}$$
4-Bromopentanoic acid

$$Cl-\underset{\underset{Cl}{|}}{\overset{\overset{Cl}{|}}{C}}-COOH$$

**Trichloroethanoic acid
(trichloroacetic acid)**

Figure 21-14 | Some carboxylic acids.

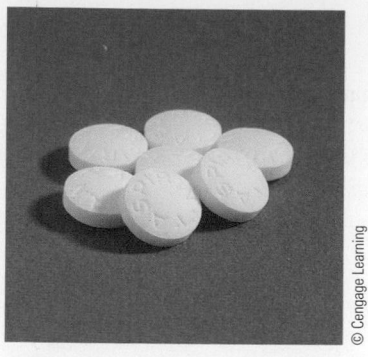

▲ Aspirin tablets.

▲ Computer-generated space-filling model of acetylsalicylic acid (aspirin).

A very important ester is formed from the reaction of salicylic acid and acetic acid:

Salicylic acid   **Acetic acid**   **Acetylsalicylic acid**

The product is acetylsalicylic acid, commonly known as *aspirin,* which is used in huge quantities as an analgesic (painkiller). ◄ ◄

Amines

Amines are probably best viewed as derivatives of ammonia in which one or more N—H bonds are replaced by N—C bonds. The resulting amines are classified as *primary* if one N—C bond is present, *secondary* if two N—C bonds are present, and *tertiary* if all three N—H bonds in NH_3 have been replaced by N—C bonds (Fig. 21-15). Examples of some common amines are given in Table 21-6.

Common names are often used for simple amines; the systematic nomenclature for more complex molecules uses the name *amino-* for the —NH$_2$ functional group. For example, the molecule

$$CH_3CHCH_2CH_3$$
$$|$$
$$NH_2$$

is named 2-aminobutane.

Many amines have unpleasant "fishlike" odors. For example, the odors associated with decaying animal and human tissues are caused by amines such as putrescine ($H_2NCH_2CH_2CH_2NH_2$) and cadaverine ($H_2NCH_2CH_2CH_2CH_2CH_2NH_2$).

Aromatic amines are used primarily to make dyes. Since many of them are carcinogenic, they must be handled with great care.

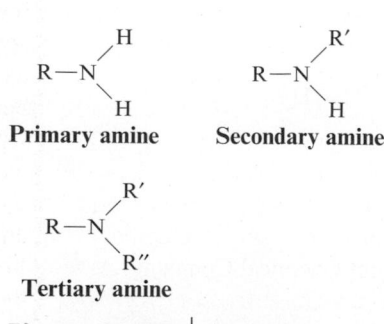

Primary amine   **Secondary amine**

Tertiary amine

Figure 21-15 | The general formulas for primary, secondary, and tertiary amines. R, R′, and R″ represent carbon-containing substituents.

Table 21-6 │ Some Common Amines

Formula	Common Name	Type
CH_3NH_2	Methylamine	Primary
$CH_3CH_2NH_2$	Ethylamine	Primary
$(CH_3)_2NH$	Dimethylamine	Secondary
$(CH_3)_3N$	Trimethylamine	Tertiary
NH$_2$	Aniline	Primary
H N	Diphenylamine	Secondary

21-5 | Polymers

Polymers are large, usually chainlike molecules that are built from small molecules called *monomers*. Polymers form the basis for synthetic fibers, rubbers, and plastics and have played a leading role in the revolution that has been brought about in daily life by chemistry. It has been estimated that about 50% of the industrial chemists in the United States work in some area of polymer chemistry, a fact that illustrates just how important polymers are to our economy and standard of living.

The Development and Properties of Polymers

The development of the polymer industry provides a striking example of the importance of serendipity in the progress of science. Many discoveries in polymer chemistry arose from accidental observations that scientists followed up.

The age of plastics might be traced to a day in 1846 when Christian Schoenbein, a chemistry professor at the University of Basel in Switzerland, spilled a flask containing nitric and sulfuric acids. In his hurry to clean up the spill, he grabbed his wife's cotton apron, which he then rinsed out and hung up in front of a hot stove to dry. Instead of drying, the apron flared and burned.

Very interested in this event, Schoenbein repeated the reaction under more controlled conditions and found that the new material, which he correctly concluded to be nitrated cellulose, had some surprising properties. As he had experienced, the nitrated cellulose is extremely flammable and, under certain circumstances, highly explosive. In addition, he found that it could be molded at moderate temperatures to give objects that were, upon cooling, tough but elastic. Predictably, the explosive nature of the substance was initially of more interest than its other properties, and cellulose nitrate rapidly became the basis for smokeless gunpowder. Although Schoenbein's discovery cannot be described as a truly synthetic polymer (because he simply found a way to modify the natural polymer cellulose), it formed the basis for a large number of industries that grew up to produce photographic films, artificial fibers, and molded objects of all types.

The first synthetic polymers were produced as by-products of various organic reactions and were regarded as unwanted contaminants. Thus the first preparations of many of the polymers now regarded as essential to our modern lifestyle were thrown away in disgust. One chemist who refused to be defeated by the "tarry" products obtained when he reacted phenol with formaldehyde was the Belgian-American chemist Leo H. Baekeland (1863–1944). Baekeland's work resulted in the first completely synthetic plastic (called Bakelite), a substance that when molded to a certain shape under high pressure and temperature cannot be softened again or dissolved. ◄ Bakelite is a **thermoset polymer**. In contrast, cellulose nitrate is a **thermoplastic polymer**; that is, it can be remelted after it has been molded.

The discovery of Bakelite in 1907 spawned a large plastics industry, producing telephones, billiard balls, and insulators for electrical devices. During the early days of polymer chemistry, there was a great deal of controversy over the nature of these materials. Although the German chemist Hermann Staudinger speculated in 1920 that polymers were very large molecules held together by strong chemical bonds, most chemists of the time assumed that these materials were much like colloids, in which small molecules are aggregated into large units by forces weaker than chemical bonds.

One chemist who contributed greatly to the understanding of polymers as giant molecules was Wallace H. Carothers of the DuPont Chemical Company. Among his accomplishments was the preparation of nylon. The nylon story further illustrates the importance of serendipity in scientific research. When nylon is first prepared, the resulting product is a sticky material with little structural integrity. Because of this, it was initially put aside as having no apparently useful characteristics. However, Julian Hill, a chemist in the Carothers research group, one day put a small ball of this nylon

▲ A radio from the 1930s made of Bakelite.

Phil Nelson

Wallace H. Carothers, a brilliant organic chemist who was principally responsible for the development of nylon and the first synthetic rubber (Neoprene), was born in 1896 in Burlington, Iowa. As a youth, Carothers was fascinated by tools and mechanical devices and spent many hours experimenting. In 1915 he entered Tarkio College in Missouri. Carothers so excelled in chemistry that even before his graduation, he was made a chemistry instructor.

Carothers eventually moved to the University of Illinois at Urbana–Champaign, where he was appointed to the faculty when he completed his Ph.D. in organic chemistry in 1924. He moved to Harvard University in 1926, and then to DuPont in 1928 to participate in a new program in fundamental research. At DuPont, Carothers headed the organic chemistry division, and during his ten years there played a prominent role in laying the foundations of polymer chemistry.

By the age of 33, Carothers had become a world-famous chemist whose advice was sought by almost everyone working in polymers. He was the first industrial chemist to be elected to the prestigious National Academy of Sciences.

Carothers was an avid reader of poetry and a lover of classical music. Unfortunately, he also suffered from severe bouts of depression that finally led to his suicide in 1937 in a Philadelphia hotel room, where he drank a cyanide solution. He was 41 years old. Despite the brevity of his career, Carothers

Wallace H. Carothers.

Courtesy, Dupont Company

was truly one of the finest American chemists of all time. His great intellect, his love of chemistry, and his insistence on perfection produced his special genius.

▲ Nylon netting magnified 62 times.

Ron Boardman/Frank Lane Picture Agency/Corbis

on the end of a stirring rod and drew it away from the remaining sticky mass, forming a string. He noticed the silky appearance and strength of this thread and realized that nylon could be drawn into useful fibers. ◄

The reason for this behavior of nylon is now understood. When nylon is first formed, the individual polymer chains are oriented randomly, like cooked spaghetti, and the substance is highly amorphous. However, when drawn out into a thread, the chains tend to line up (the nylon becomes more crystalline), which leads to increased hydrogen bonding between adjacent chains. This increase in crystallinity, along with the resulting increase in hydrogen-bonding interactions, leads to strong fibers and thus to a highly useful material. Commercially, nylon is produced by forcing the raw material through a *spinneret,* a plate containing small holes, which forces the polymer chains to line up.

Another property that adds strength to polymers is **crosslinking**, the existence of covalent bonds between adjacent chains. The structure of Bakelite is highly crosslinked, which accounts for the strength and toughness of this polymer. Another example of crosslinking occurs in the manufacture of rubber. Raw natural rubber consists of chains of the type

$$\text{---CH}_2\text{---CH}_2\text{---CH}{=}\text{C---CH}_2\text{---CH}_2\text{---CH}{=}\text{C---CH}_2\text{---}$$
$$\qquad\qquad\qquad\quad\underset{\text{CH}_3}{|}\qquad\qquad\qquad\qquad\qquad\underset{\text{CH}_3}{|}$$

One of the most amazing characteristics of insects is their ability to cling to almost any surface, whether it is vertical or upside down. However, the walls of the *Nepenthes* pitcher plant are so slippery that any insect that lands on those walls slips to its death in the digestive juices at the bottom of the "pitcher." One can envision this type of slippery surface being useful in human activities, such as in pipes for handling biomedical fluids and fuels. Slippery surfaces also could be useful in repelling ice inside freezers or on ship hulls used in polar regions.

Joanna Aizenberg and her colleagues at Harvard University have designed a synthetic system that mimics the slippery

A pitcher plant.

Decha Thapanya/Shutterstock.com

pitcher plant surface. Their omniphobic (repels virtually all liquids) surface, which they call SLIPS, is prepared from a porous network of Teflon nanofibers that is infused with a special oil-and-water fluid. It is the layer of fluid on the surface of the Teflon "sponge" that makes SLIPS so slippery. Another advantage of SLIPS is its self-healing property. If any damage occurs to the surface, more fluid flows from the interior to heal cracks or dents. The surface can be maintained indefinitely because a reservoir of fluid sustains the fluid.

SLIPS is a great example of how useful it can be to mimic nature's problem-solving skills.

ⓘ Charles Goodyear tried for many years to change natural rubber into a useful product. In 1839 he accidentally dropped some rubber containing sulfur on a hot stove. Noting that the rubber did not melt as expected, Goodyear pursued this lead and developed vulcanization.

and is a soft, sticky material unsuitable for tires. However, in 1839 Charles Goodyear (1800–1860), an American chemist, accidentally found that if sulfur is added to rubber and the resulting mixture is heated (a process called **vulcanization**), the resulting rubber is still elastic (reversibly stretchable) but is much stronger. ◄ This change in character occurs because sulfur atoms become bonded between carbon atoms on different chains. These sulfur atoms form bridges between the polymer chains, thus linking the chains together. ◄ *ⓘ*

Types of Polymers

The simplest and one of the best-known polymers is *polyethylene*, which is constructed from ethylene monomers:

$$n\mathrm{CH_2{=}CH_2} \xrightarrow{\text{Catalyst}} \left(\begin{array}{cc} \overset{\displaystyle H}{\underset{\displaystyle H}{|}} & \overset{\displaystyle H}{\underset{\displaystyle H}{|}} \\ \mathrm{C} & \mathrm{C} \end{array}\right)_n$$

where n represents a large number (usually several thousand). Polyethylene is a tough, flexible plastic used for piping, bottles, electrical insulation, packaging films, garbage bags, and many other purposes. Its properties can be varied by using substituted ethylene monomers. For example, when tetrafluoroethylene is the monomer, the polymer Teflon is obtained:

$$n\left(\begin{array}{c} \mathrm{F} \quad\quad \mathrm{F} \\ \diagdown\; \mathrm{C}{=}\mathrm{C} \;\diagup \\ \mathrm{F} \quad\quad \mathrm{F} \end{array}\right) \longrightarrow \left(\begin{array}{cc} \overset{\displaystyle F}{\underset{\displaystyle F}{|}} & \overset{\displaystyle F}{\underset{\displaystyle F}{|}} \\ \mathrm{C} & \mathrm{C} \end{array}\right)_n$$

Tetrafluoroethylene   **Teflon**

Ashiga/Shutterstock.com

▲ Crosslinking gives the rubber in these tires strength and toughness.

The discovery of Teflon, a very important substituted polyethylene, is another illustration of the role of chance in chemical research. In 1938 a DuPont chemist named

Roy Plunkett was studying the chemistry of gaseous tetrafluoroethylene. He synthesized about 100 pounds of the chemical and stored it in steel cylinders. When one of the cylinders failed to produce tetrafluoroethylene gas when the valve was opened, the cylinder was cut open to reveal a white powder. This powder turned out to be a polymer of tetrafluoroethylene, which was eventually developed into Teflon. Because of the resistance of the strong C—F bonds to chemical attack, Teflon is an inert, tough, and nonflammable material widely used for electrical insulation, nonstick coatings on cooking utensils, and bearings for low-temperature applications.

Other polyethylene-type polymers are made from monomers containing chloro, methyl, cyano, and phenyl substituents, as summarized in Table 21-7. In each case the double carbon–carbon bond in the substituted ethylene monomer becomes a single bond in the polymer. The different substituents lead to a wide variety of properties.

The polyethylene polymers illustrate one of the major types of polymerization reactions, called **addition polymerization**, in which the monomers simply "add together" to produce the polymer. No other products are formed. The polymerization process is initiated by a **free radical** (a species with an unpaired electron) such as the hydroxyl

Table 21-7 | Some Common Synthetic Polymers and Their Monomers and Applications

Monomer		Polymer		
Name	Formula	Name	Formula	Uses
Ethylene	$H_2C{=}CH_2$	Polyethylene	$-(CH_2-CH_2)_n-$	Plastic piping, bottles, electrical insulation, toys
Propylene	$H_2C{=}C$ with H above and CH_3 below	Polypropylene	$-(CH-CH_2-CH-CH_2)_n-$ with CH_3 groups below	Film for packaging, carpets, lab wares, toys
Vinyl chloride	$H_2C{=}C$ with H above and Cl below	Polyvinyl chloride (PVC)	$-(CH_2-CH)_n-$ with Cl below	Piping, siding, floor tile, clothing, toys
Acrylonitrile	$H_2C{=}C$ with H above and CN below	Polyacrylonitrile (PAN)	$-(CH_2-CH)_n-$ with CN below	Carpets, fabrics
Tetrafluoro-ethylene	$F_2C{=}CF_2$	Teflon	$-(CF_2-CF_2)_n-$	Cooking utensils, electrical insulation, bearings
Styrene	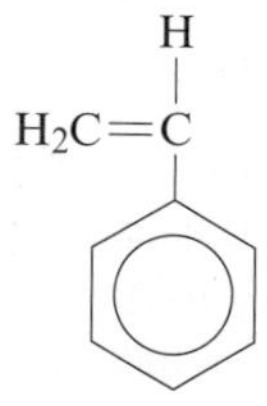 $H_2C{=}C$ with H above and phenyl ring below	Polystyrene	 $-(CH_2CH)_n-$ with phenyl ring below	Containers, thermal insulation, toys
Butadiene	$H_2C{=}C-C{=}CH_2$ with H, H above	Polybutadiene	$-(CH_2CH{=}CHCH_2)_n-$	Tire tread, coating resin
Butadiene and styrene	(See above.)	Styrene-butadiene rubber	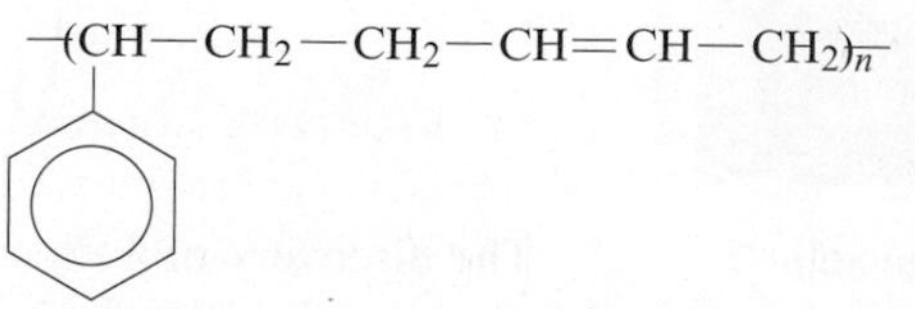 $-(CH-CH_2-CH_2-CH{=}CH-CH_2)_n-$ with phenyl ring below	Synthetic rubber

radical (HO·). The free radical attacks and breaks the π bond of an ethylene molecule to form a new free radical,

which is then available to attack another ethylene molecule:

Repetition of this process thousands of times creates a long-chain polymer. Termination of the growth of the chain occurs when *two radicals* react to form a bond, a process that consumes two radicals without producing any others.

Another common type of polymerization is **condensation polymerization**, in which a small molecule, such as water, is formed for each extension of the polymer chain. The most familiar polymer produced by condensation is *nylon*. Nylon is a **copolymer**, since two different types of monomers combine to form the chain; a **homopolymer** is the result of polymerizing a single type of monomer. One common form of nylon is produced when hexamethylenediamine and adipic acid react by splitting out a water molecule to form a C—N bond:

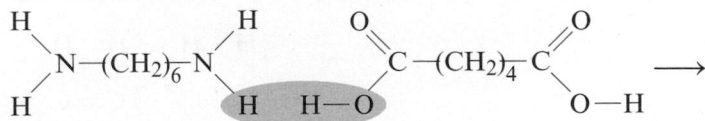

Hexamethylenediamine   **Adipic acid**

The molecule formed, called a **dimer** (two monomers joined), can undergo further condensation reactions since it has an amino group at one end and a carboxyl group at the other. Thus both ends are free to react with another monomer. Repetition of this process leads to a long chain of the type

which is the basic structure of nylon. The reaction to form nylon occurs quite readily and is often used as a lecture demonstration (Fig. 21-16). The properties of nylon can be varied by changing the number of carbon atoms in the chain of the acid or amine monomer.

Figure 21-16 | The reaction to form nylon can be carried out at the interface of two immiscible liquid layers in a beaker. The bottom layer contains adipoyl chloride,

$$Cl-\underset{\underset{O}{\|}}{C}-(CH_2)_4-\underset{\underset{O}{\|}}{C}-Cl$$

dissolved in CCl_4, and the top layer contains hexamethylenediamine,

$$H_2N-(CH_2)_6-NH_2$$

dissolved in water. A molecule of HCl is formed as each C—N bond forms.

More than 1 million tons of nylon is produced annually in the United States for use in clothing, carpets, rope, and so on. Many other types of condensation polymers are also produced. For example, Dacron is a copolymer formed from the condensation reaction of ethylene glycol (a dialcohol) and *p*-terephthalic acid (a dicarboxylic acid):

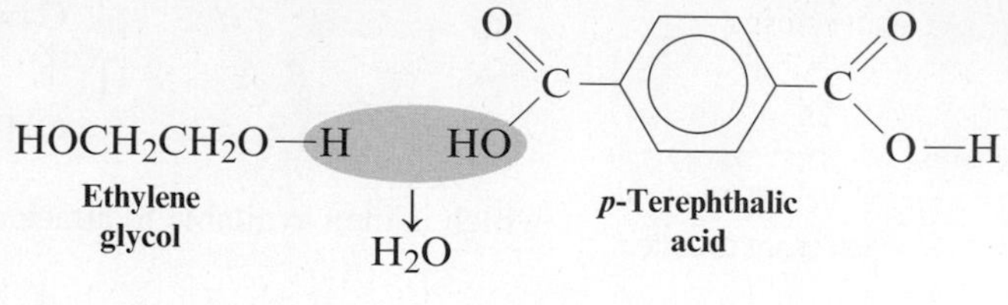

The repeating unit of Dacron is

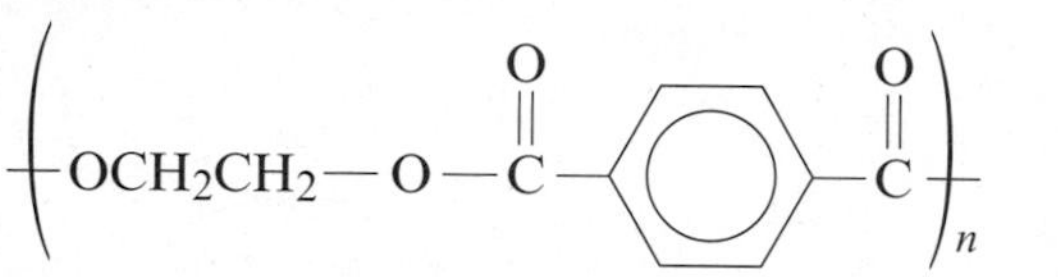

Note that this polymerization involves a carboxylic acid and an alcohol forming an ester group:

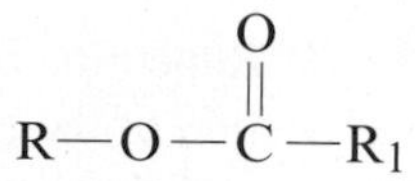

Thus Dacron is called a **polyester**. By itself or blended with cotton, Dacron is widely used in fibers for the manufacture of clothing.

Polymers Based on Ethylene

A large section of the polymer industry involves the production of macromolecules from ethylene or substituted ethylenes. As discussed previously, ethylene molecules polymerize by addition after the double bond has been broken by some initiator:

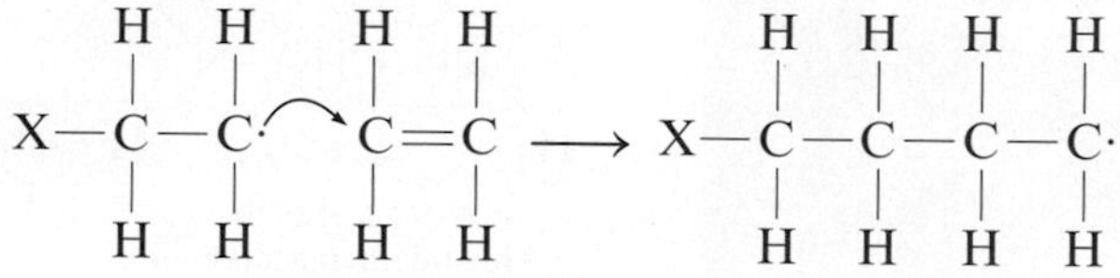

This process continues by adding new ethylene molecules to eventually give polyethylene, a thermoplastic material.

There are two forms of polyethylene: low-density polyethylene (LDPE) and high-density polyethylene (HDPE). The chains in LDPE contain many branches and thus do not pack as tightly as those in HDPE, which consist of mostly straight-chain molecules.

Traditionally, LDPE has been manufactured under conditions of high pressure ($\approx$20,000 psi) and high temperature (500°C). ◀ ⓘ These severe reaction conditions require specially designed equipment, and for safety reasons the reaction usually has been run behind a reinforced concrete barrier. More recently, lower reaction pressures and temperatures have become possible through the use of catalysts. One catalytic system using triethylaluminum, $Al(C_2H_5)_3$, and titanium(IV) chloride was developed by Karl Ziegler in Germany and Giulio Natta in Italy. Although this catalyst is very efficient, it catches fire on contact with air and must be handled very carefully. A safer catalytic system was developed at Phillips Petroleum Company. It uses a chromium(III) oxide (Cr_2O_3) and aluminosilicate catalyst and has mainly taken over in the United States. The product of the catalyzed reaction is highly linear (unbranched) and is often called *linear low-density polyethylene*. It is very similar to HDPE.

ⓘ psi is the abbreviation for pounds per square inch: 15 psi $\approx$ 1 atm.

Figure 21-17 | A major use of HDPE is for blow-molded objects such as bottles for soft drinks, shampoos, bleaches, and so on. (a) A tube composed of HDPE is inserted into the mold (die). (b) The die closes, sealing the bottom of the tube. (c) Compressed air is forced into the warm HDPE tube, which then expands to take the shape of the die. (d) The molded bottle is removed from the die.

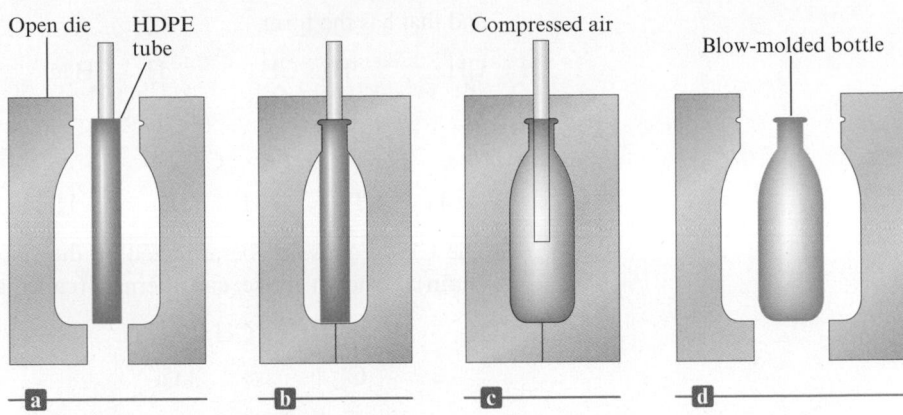

ⓘ Molecular weight (not molar mass) is the common terminology in the polymer industry.

The major use of LDPE is in the manufacture of the tough, transparent film that is used in packaging so many consumer goods. Two-thirds of the approximately 10 billion pounds of LDPE produced annually in the United States are used for this purpose. The major use of HDPE is for blow-molded products, such as bottles for consumer products (Fig. 21-17).

The useful properties of polyethylene are due primarily to its high molecular weight (molar mass). ◄ *ⓘ* Although the strengths of the interactions between specific points on the nonpolar chains are quite small, the chains are so long that these small attractions accumulate to a very significant value, so that the chains stick together very tenaciously. There is also a great deal of physical tangling of the lengthy chains. The combination of these interactions gives the polymer strength and toughness. However, a material like polyethylene can be melted and formed into a new shape (thermoplastic behavior), because in the melted state the molecules can readily flow past one another.

Since a high molecular weight gives a polymer useful properties, one might think that the goal would be to produce polymers with chains as long as possible. However, this is not the case—polymers become much more difficult to process as the molecular weights increase. Most industrial operations require that the polymer flow through pipes as it is processed. But as the chain lengths increase, viscosity also increases. In practice, the upper limit of a polymer's molecular weight is set by the flow requirements of the manufacturing process. Thus the final product often reflects a compromise between the optimal properties for the application and those needed for ease of processing.

Although many polymer properties are greatly influenced by molecular weight, some other important properties are not. For example, chain length does not affect a polymer's resistance to chemical attack. Physical properties such as color, refractive index, hardness, density, and electrical conductivity are also not greatly influenced by molecular weight.

We have already seen that one way of altering the strength of a polymeric material is to vary the chain length. Another method for modifying polymer behavior involves varying the substituents. For example, if we use a monomer of the type

$$\begin{array}{c} H \\ \diagdown \\ \end{array} C = C \begin{array}{c} H \\ \diagup \\ \diagdown X \end{array}$$

the properties of the resulting polymer depend on the identity of X. The simplest example is polypropylene, whose monomer is

$$\begin{array}{c} H \\ \diagdown \\ H \diagup \end{array} C = C \begin{array}{c} H \\ \diagup \\ \diagdown CH_3 \end{array}$$

and that has the form

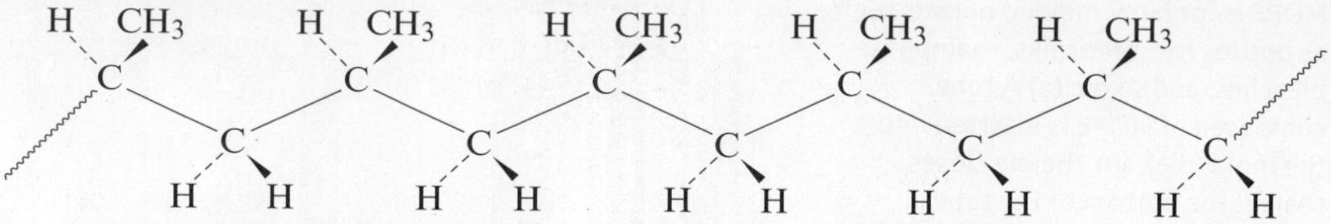

The CH_3 groups can be arranged on the same side of the chain (called an **isotactic chain**) as shown above, can alternate (called a **syndiotactic chain**) as shown below,

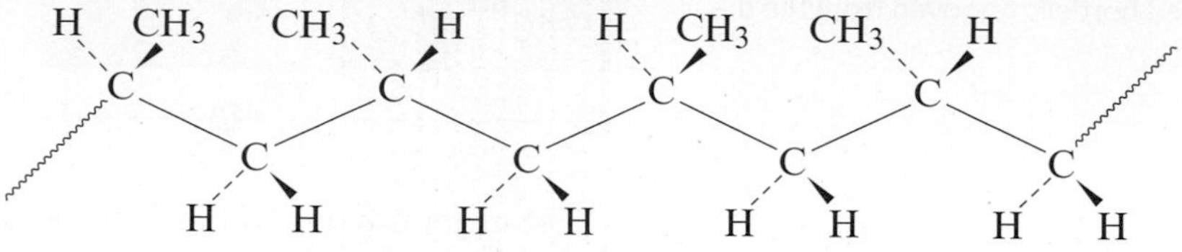

or can be randomly distributed (called an **atactic chain**).

The chain arrangement has a significant effect on the polymer's properties. Most polypropylene is made using the Ziegler-Natta catalyst, $Al(C_2H_5)_3 \cdot TiCl_4$, which produces highly isotactic chains that pack together quite closely. As a result, polypropylene is more crystalline, and therefore stronger and harder, than polyethylene. The major uses of polypropylene are for molded parts (40%), fibers (35%), and packaging films (10%). Polypropylene fibers are especially useful for athletic wear because they do not absorb water from perspiration, as cotton does. Rather, the moisture is drawn away from the skin to the surface of the polypropylene garment, where it can evaporate. The annual U.S. production of polypropylene is about 7 billion pounds.

Another related polymer, **polystyrene**, is constructed from the monomer styrene,

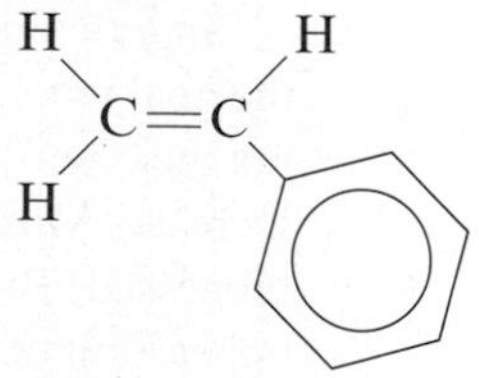

Pure polystyrene is too brittle for many uses, so most polystyrene-based polymers are actually *copolymers* of styrene and butadiene,

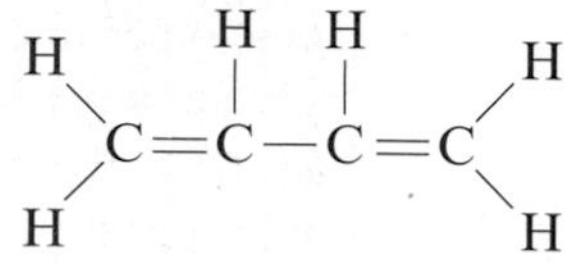

thus incorporating bits of butadiene rubber into the polystyrene matrix. The resulting polymer is very tough and is often used as a substitute for wood in furniture.

Another polystyrene-based product is acrylonitrile-butadiene-styrene (ABS), a tough, hard, and chemically resistant plastic used for pipes and for items such as radio housings, telephone cases, and golf club heads, for which shock resistance is an essential property. Originally, ABS was produced by copolymerization of the three monomers:

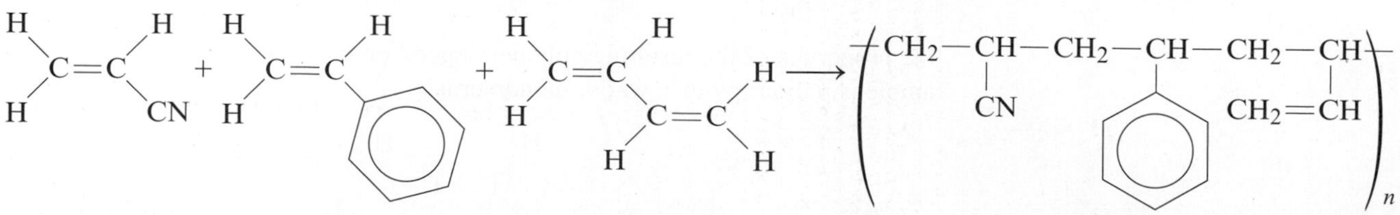

Acrylonitrile   **Styrene**   **Butadiene**

▲ PVC pipe is widely used in industry.

It is now prepared by a special process called *grafting,* in which butadiene is polymerized first, and then the cyanide and phenyl substituents are added chemically.

Another high-volume polymer, **polyvinyl chloride (PVC)**, is constructed from the monomer vinyl chloride, ◄

$$\underset{H}{\overset{H}{>}}C=C\underset{Cl}{\overset{H}{<}}$$

21-6 | Natural Polymers

Proteins

▲ The protein in muscles enables them to contract.

We have seen that many useful synthetic materials are polymers. Thus it should not be surprising that a great many natural materials are also polymers: starch, hair, silicate chains in soil and rocks, silk and cotton fibers, and the cellulose in woody plants, to name only a few.

In this section we will consider a class of natural polymers, the **proteins**, which make up about 15% of our bodies and have molecular weights (molar masses) that range from about 6000 to over 1,000,000 grams per mole. Proteins perform many functions in the human body. **Fibrous proteins** provide structural integrity and strength for many types of tissue and are the main components of muscle, hair, and cartilage. ◄ Other proteins, usually called **globular proteins** because of their roughly spherical shape, are the "worker" molecules of the body. These proteins transport and store oxygen and nutrients, act as catalysts for the thousands of reactions that make life possible, fight invasion by foreign objects, participate in the body's many regulatory systems, and transport electrons in the complex process of metabolizing nutrients.

The building blocks of all proteins are the **α-amino acids**, where R may represent H, CH_3, or a more complex substituent. ◄ ⓘ These molecules are called α-amino acids because the amino group ($-NH_2$) is always attached to the α-carbon, the one next to the carboxyl group ($-CO_2H$). The 20 amino acids most commonly found in proteins are shown in Fig. 21-18.

Note from Fig. 21-18 that the amino acids are grouped into polar and nonpolar classes, determined by the R groups, or **side chains**. ◄ ⓘ Nonpolar side chains contain mostly carbon and hydrogen atoms, whereas polar side chains contain large numbers of nitrogen and oxygen atoms. This difference is important, because polar side chains are *hydrophilic* (water-loving), but nonpolar side chains are *hydrophobic* (water-fearing), and this characteristic greatly affects the three-dimensional structure of the resulting protein.

The protein polymer is built by condensation reactions between amino acids. For example,

ⓘ α-Carbon

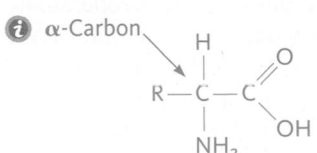

ⓘ At the pH in biological fluids, the amino acids shown in Fig. 21-18 exist in a different form, with the proton of the —COOH group transferred to the —NH$_2$ group. For example, glycine would be in the form $H_3^+NCH_2COO^-$.

The product shown before is called a **dipeptide**. This name is used because the structure

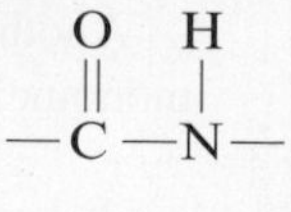

$$
\begin{array}{cc}
\text{O} & \text{H} \\
\| & | \\
-\text{C}- & \text{N}-
\end{array}
$$

is called a **peptide linkage** by biochemists. (The same grouping is called an amide by organic chemists.) Additional condensation reactions lengthen the chain to produce a **polypeptide**, eventually yielding a protein.

You can imagine that with 20 amino acids, which can be assembled in any order, there is essentially an infinite variety possible in the construction of proteins. This flexibility allows an organism to tailor proteins for the many types of functions that must be carried out.

The order, or sequence, of amino acids in the protein chain is called the **primary structure**, conveniently indicated by using three-letter codes for the amino acids (see Fig. 21-18), where it is understood that the terminal carboxyl group is on the right and the terminal amino group is on the left. For example, one possible sequence for a tripeptide containing the amino acids lysine, alanine, and leucine is

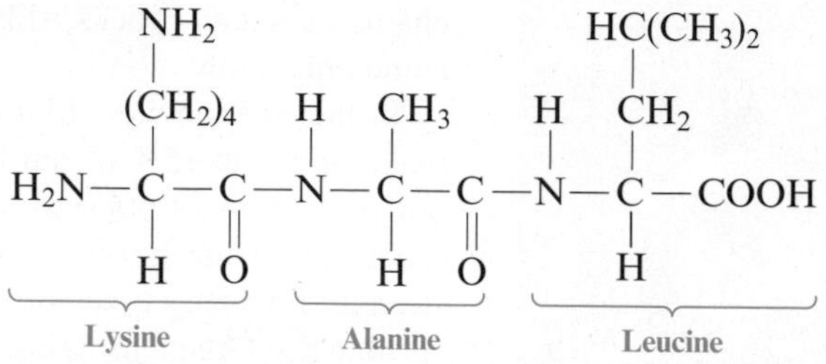

which is represented in the shorthand notation by

lys-ala-leu

Note from Example 21-7 that there are six sequences possible for a polypeptide with three given amino acids. There are three possibilities for the first amino acid (any one of the three given amino acids), there are two possibilities for the second amino acid (one has already been accounted for), but there is only one possibility left for the third amino acid. Thus the number of sequences is $3 \times 2 \times 1 = 6$. The product $3 \times 2 \times 1$ is often written 3! (and is called 3 factorial). Similar reasoning shows that for a polypeptide with four amino acids, there are 4!, or $4 \times 3 \times 2 \times 1 = 24$, possible sequences.

INTERACTIVE EXAMPLE 21-7

Tripeptide Sequences

Write the sequences of all possible tripeptides composed of the amino acids tyrosine, histidine, and cysteine.

Solution

There are six possible sequences:

tyr-his-cys	his-tyr-cys	cys-tyr-his
tyr-cys-his	his-cys-tyr	cys-his-tyr

See Exercise 21-89

Nonpolar R groups

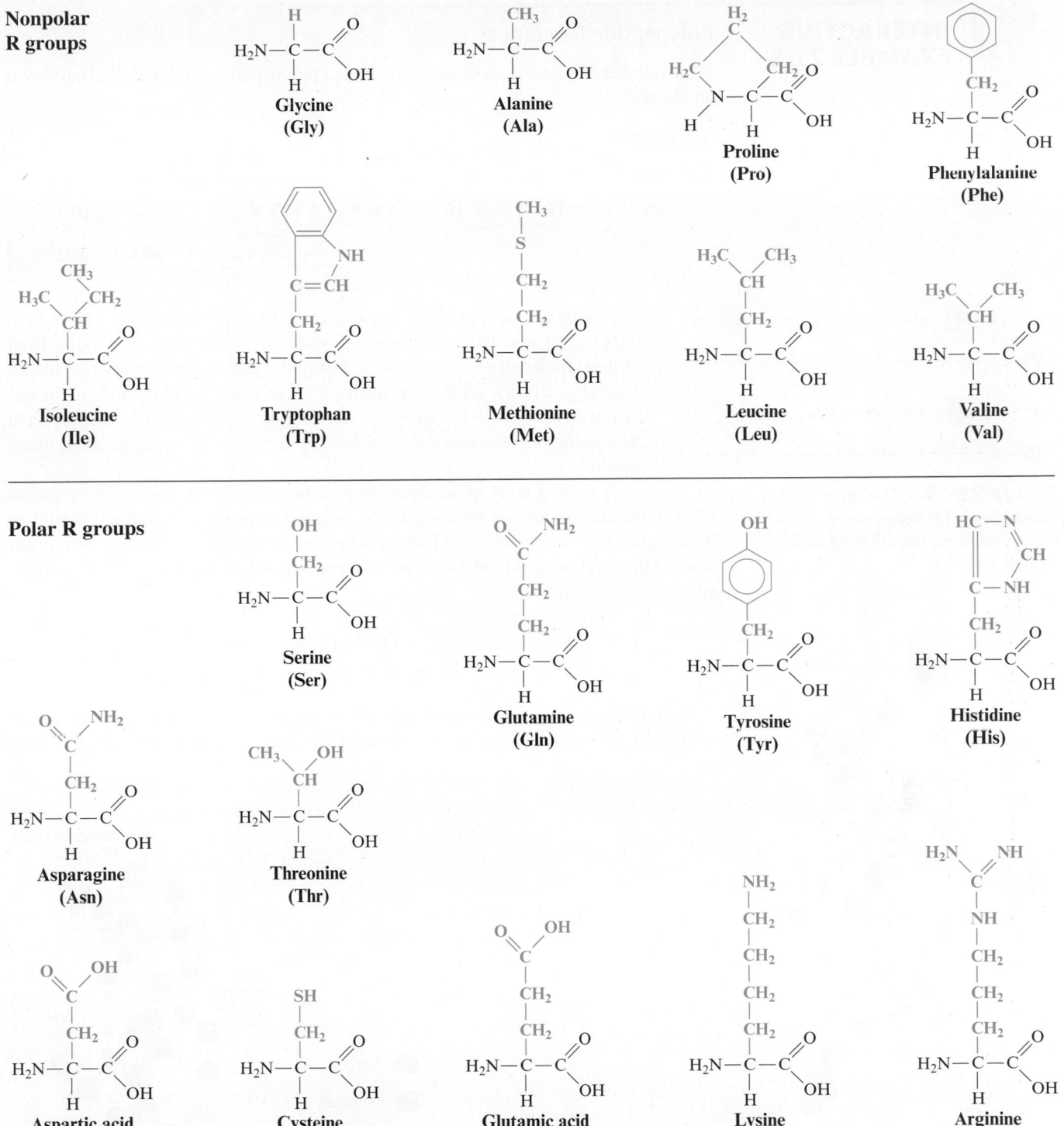

Polar R groups

Figure 21-18 | The 20 α-amino acids found in most proteins. The R group is shown in color.

Polypeptide Sequences

What number of possible sequences exists for a polypeptide composed of 20 different amino acids?

Solution

The answer is 20!, or

$$20 \times 19 \times 18 \times 17 \times 16 \times \cdots \times 5 \times 4 \times 3 \times 2 \times 1 = 2.43 \times 10^{18}$$

See Exercise 21-90

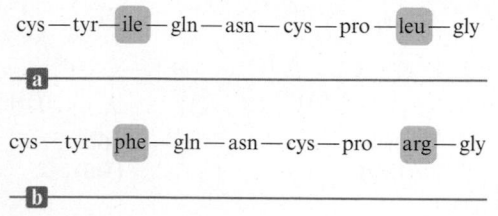

cys — tyr — ile — gln — asn — cys — pro — leu — gly

a

cys — tyr — phe — gln — asn — cys — pro — arg — gly

b

Figure 21-19 | The amino acid sequences in (a) oxytocin and (b) vasopressin. The differing amino acids are boxed.

A striking example of the importance of the primary structure of polypeptides can be seen in the differences between *oxytocin* and *vasopressin*. Both of these molecules are nine-unit polypeptides that differ by only two amino acids (Fig. 21-19), yet they perform completely different functions in the human body. Oxytocin is a hormone that triggers contraction of the uterus and milk secretion. Vasopressin raises blood pressure levels and regulates kidney function.

A second level of structure in proteins, beyond the sequence of amino acids, is the arrangement of the chain of the long molecule. The **secondary structure** is determined to a large extent by hydrogen bonding between lone pairs on an oxygen atom in the carbonyl group of an amino acid and a hydrogen atom attached to a nitrogen of another amino acid:

$$\text{C}=\overset{..}{\text{O}}: \text{---} \text{H} \text{---} \text{N}$$
$$\quad\quad \delta^- \quad\quad \delta^+$$

Such interactions can occur *within* the chain coils to form a spiral structure called an **α-helix**, as shown in Figs. 21-20 and 21-21. This type of secondary structure gives

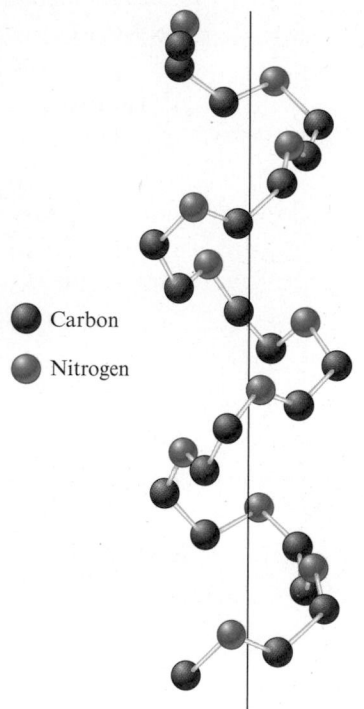

● Carbon
● Nitrogen

Figure 21-20 | Hydrogen bonding within a protein chain causes it to form a stable helical structure called the α-helix. Only the main atoms in the helical backbone are shown here. The hydrogen bonds are not shown.

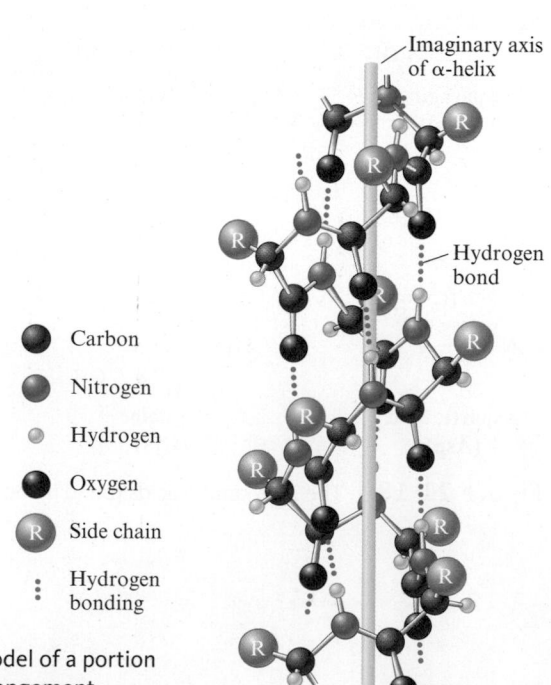

Imaginary axis of α-helix

Hydrogen bond

● Carbon
● Nitrogen
○ Hydrogen
● Oxygen
Ⓡ Side chain
⋮ Hydrogen bonding

Figure 21-21 | Ball-and-stick model of a portion of a protein chain in the α-helical arrangement, showing the hydrogen-bonding interactions.

Figure 21-22 | When hydrogen bonding occurs between protein chains rather than within them, a stable structure (the pleated sheet) results. This structure contains many protein chains and is found in natural fibers, such as silk, and in muscles.

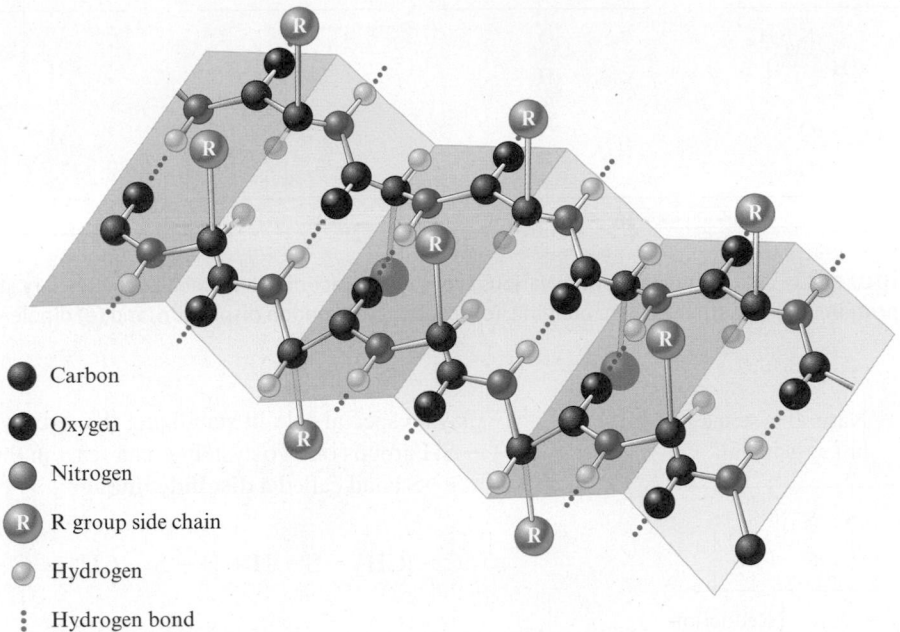

- ● Carbon
- ● Oxygen
- ● Nitrogen
- Ⓡ R group side chain
- ○ Hydrogen
- ⋮ Hydrogen bond

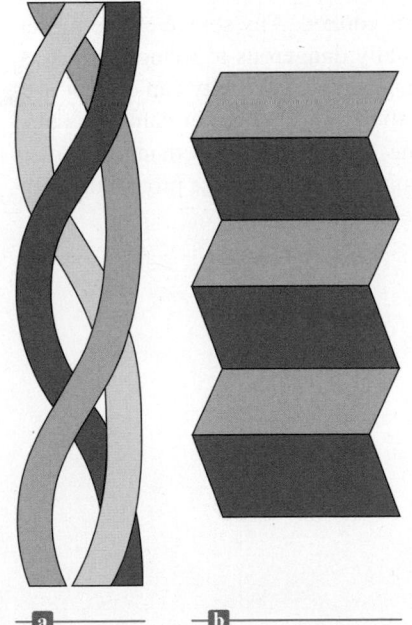

Figure 21-23 | (a) Collagen, a protein found in tendons, consists of three protein chains (each with a helical structure) twisted together to form a superhelix. The result is a long, relatively narrow protein. (b) The pleated-sheet arrangement of many proteins bound together to form the elongated protein found in silk fibers.

the protein elasticity (springiness) and is found in the fibrous proteins in wool, hair, and tendons. Hydrogen bonding can also occur *between different* protein chains, joining them together in an arrangement called a **pleated sheet**, as shown in Fig. 21-22. Silk contains this arrangement of proteins, making its fibers flexible yet very strong and resistant to stretching. The pleated sheet is also found in muscle fibers. The hydrogen bonds in the α-helical protein are called *intrachain* (within a given protein chain), and those in the pleated sheet are said to be *interchain* (between protein chains).

As you might imagine, a molecule as large as a protein has a great deal of flexibility and can assume a variety of overall shapes. The specific shape that a protein assumes depends on its function. For long, thin structures, such as hair, wool and silk fibers, and tendons, an elongated shape is required. This may involve an α-helical secondary structure, as found in the protein α-keratin in hair and wool or in the collagen found in tendons [Fig. 21-23(a)], or it may involve a pleated-sheet secondary structure, as found in silk [Fig. 21-23(b)]. Many of the proteins in the body having nonstructural functions are globular, such as myoglobin (see Fig. 20-31). Note that the secondary structure of myoglobin is basically α-helical. However, in the areas where the chain bends to give the protein its compact globular structure, the α-helix breaks down to give a secondary configuration known as the **random-coil arrangement**.

The overall shape of the protein, long and narrow or globular, is called its **tertiary structure** and is maintained by several different types of interactions: hydrogen bonding, dipole–dipole interactions, ionic bonds, covalent bonds, and London dispersion forces between nonpolar groups. These bonds, which represent all the bonding types discussed in this text, are summarized in Fig. 21-24.

The amino acid *cysteine*

$$HS-CH_2-\overset{\overset{\displaystyle H}{|}}{C}-\overset{\overset{\displaystyle O}{||}}{C}-OH$$
$$\underset{\overset{\displaystyle |}{\underset{H \quad H}{N}}}{}$$

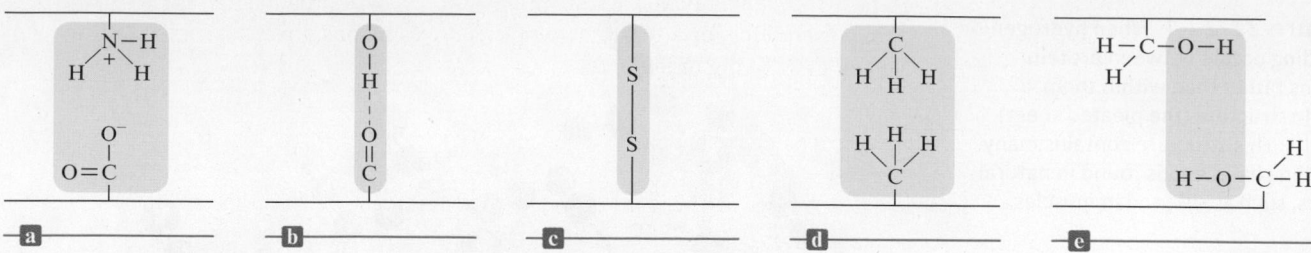

Figure 21-24 | Summary of the various types of interactions that stabilize the tertiary structure of a protein: (a) ionic, (b) hydrogen bonding, (c) covalent, (d) London dispersion, and (e) dipole–dipole.

Natural cysteine linkages in hair

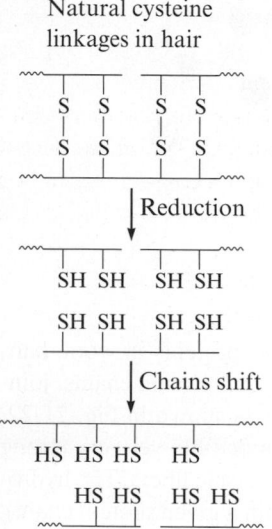

Hair set in curlers alters tertiary structures

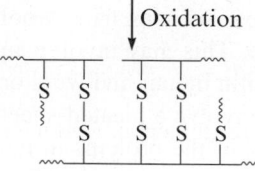

New cysteine linkages in waved hair

Figure 21-25 | The permanent waving of hair.

plays a special role in stabilizing the tertiary structure of many proteins because the —SH groups on two cysteines can react in the presence of an oxidizing agent to form a S—S bond called a **disulfide linkage**:

$$C-CH_2-S-H + H-S-CH_2-C \longrightarrow C-CH_2-\boxed{S-S}-CH_2-C$$

A practical application of the chemistry of disulfide bonds is permanent waving of hair, as summarized in Fig. 21-25. The S—S linkages in the protein of hair are broken by treatment with a reducing agent. The hair is then set in curlers to change the tertiary protein structure to the desired shape. Then treatment with an oxidizing agent causes new S—S bonds to form, which allow the hair protein to retain the new structure.

The three-dimensional structure of a protein is crucial to its function. The process of breaking down this structure is called **denaturation** (Fig. 21-26). For example, the denaturation of egg proteins occurs when an egg is cooked. Any source of energy can cause denaturation of proteins and is thus potentially dangerous to living organisms. For example, ultraviolet and X-ray radiation or nuclear radioactivity can disrupt protein structure, which may lead to cancer or genetic damage. Protein damage is also caused by chemicals like benzene, trichloroethane, and 1,2-dibromoethane. The metals lead and mercury, which have a very high affinity for sulfur, cause protein denaturation by disrupting disulfide bonds between protein chains.

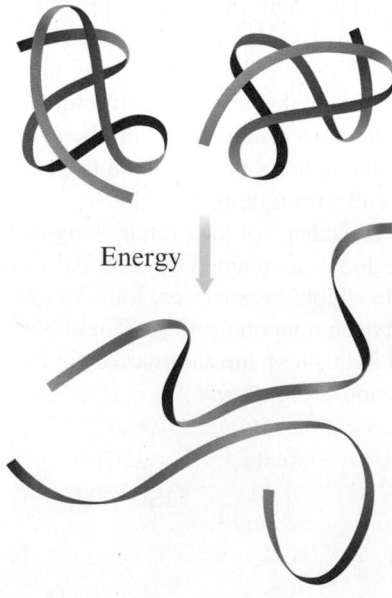

Energy

Figure 21-26 | A schematic representation of the thermal denaturation of a protein.

The tremendous flexibility in the various levels of protein structure allows the tailoring of proteins for a wide range of specific functions. Proteins are the "workhorse" molecules of living organisms.

> **Critical Thinking**
> What if you contracted a disease that prevents all hydrogen bonding in proteins? Could you live with such a condition?

Carbohydrates

Carbohydrates form another class of biologically important molecules. They serve as a food source for most organisms and as a structural material for plants. Because many carbohydrates have the empirical formula CH_2O, it was originally believed that these substances were hydrates of carbon, thus accounting for the name.

Most important carbohydrates, such as starch and cellulose, are polymers composed of monomers called **monosaccharides**, or **simple sugars**. The monosaccharides are polyhydroxy ketones and aldehydes. The most important contain five carbon atoms (**pentoses**) or six carbon atoms (**hexoses**). One important hexose is *fructose*, a sugar found in honey and fruit. Its structure is

$$
\begin{array}{c}
CH_2OH \\
| \\
C = O \\
| \\
HO - {}^*C - H \\
| \\
H - {}^*C - OH \\
| \\
H - {}^*C - OH \\
| \\
CH_2OH
\end{array}
$$

Fructose

where the asterisks indicate chiral carbon atoms. In Section 20-4 we saw that molecules with nonsuperimposable mirror images exhibit optical isomerism. A carbon atom with four *different* groups bonded to it in a tetrahedral arrangement *always* has a nonsuperimposable mirror image (Fig. 21-27), which gives rise to a pair of optical isomers. For example, the simplest sugar, glyceraldehyde,

$$
\begin{array}{c}
H \quad\quad O \\
\diagdown\,/ \\
C \\
| \\
H - {}^*C - OH \\
| \\
CH_2OH
\end{array}
$$

which has one chiral carbon, has two optical isomers, as shown in Fig. 21-28.

In fructose each of the three chiral carbon atoms satisfies the requirement of being surrounded by four different groups. This leads to a total of 2^3, or 8, isomers that differ in their ability to rotate polarized light. The particular isomer whose structure is shown in Table 21-8 is called D-fructose. Generally, monosaccharides have one isomer that is more common in nature than the others. The most important pentoses and hexoses are shown in Table 21-8.

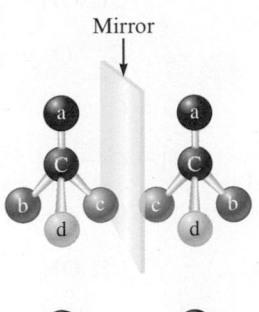

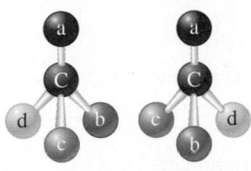

Figure 21-27 | When a tetrahedral carbon atom has four different substituents, there is no way that its mirror image can be superimposed. The lower two forms show other possible orientations of the molecule. Compare these with the mirror image and note that they cannot be superimposed.

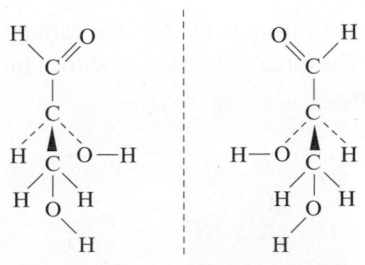

Figure 21-28 | The mirror image optical isomers of glyceraldehyde. Note that these mirror images cannot be superimposed.

General Name of Sugar	Number of Carbon Atoms
Triose	3
Tetrose	4
Pentose	5
Hexose	6
Heptose	7
Octose	8
Nonose	9

Table 21-8 | Some Important Monosaccharides

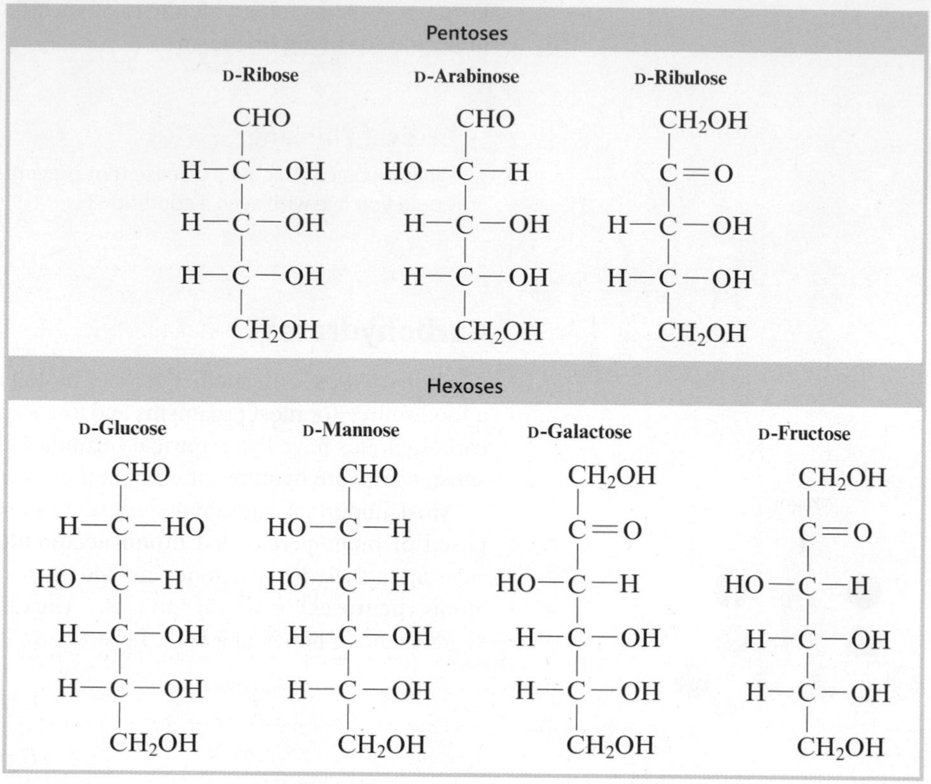

Pentoses

D-Ribose, D-Arabinose, D-Ribulose

Hexoses

D-Glucose, D-Mannose, D-Galactose, D-Fructose

INTERACTIVE EXAMPLE 21-9

Chiral Carbons in Carbohydrates

Determine the number of chiral carbon atoms in the following pentose:

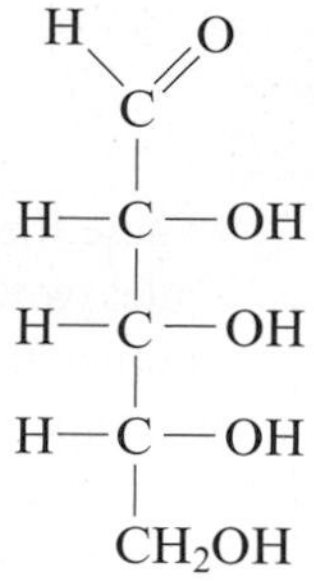

Solution

We must look for carbon atoms that have four different substituents. The top carbon has only three substituents and thus cannot be chiral. The three carbon atoms shown in blue each have four different groups attached to them:

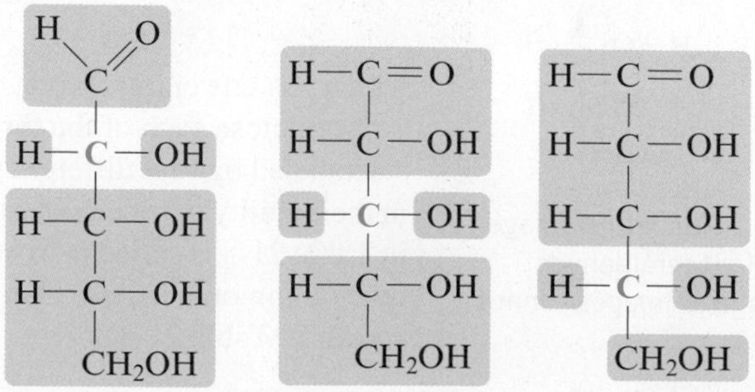

Since the fifth carbon atom has only three types of substituents (it has two hydrogen atoms), it is not chiral.

Thus the three chiral carbon atoms in this pentose are those shown in blue:

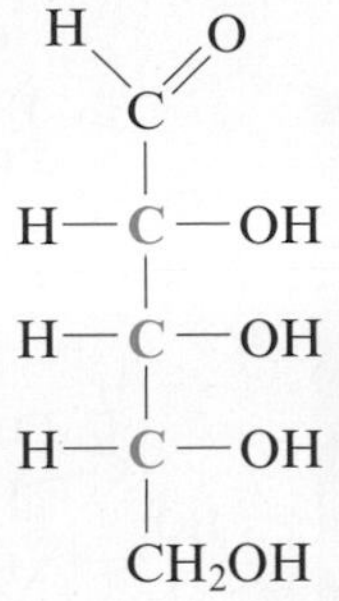

Note that D-ribose and D-arabinose, shown in Table 21-8, are two of the eight isomers of this pentose.

See Exercises 21-96 and 21-101 through 21-104

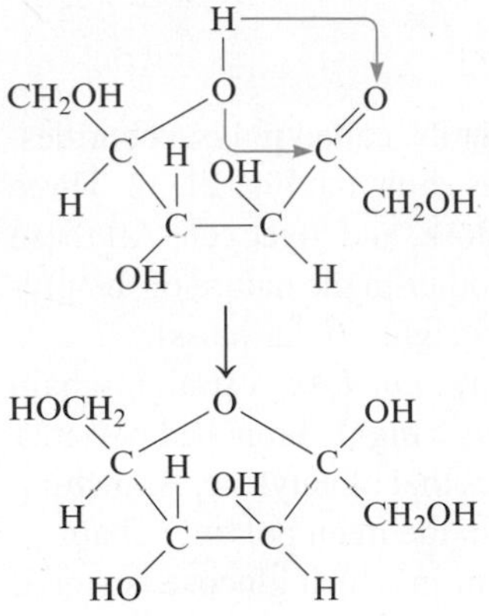

Although we have so far represented the monosaccharides as straight-chain molecules, they usually cyclize, or form a ring structure, in aqueous solution. Figure 21-29 shows this reaction for fructose. Note that a new bond is formed between the oxygen of the terminal hydroxyl group and the carbon of the ketone group. In the cyclic form fructose is a five-membered ring containing a C—O—C bond. The same type of reaction can occur between a hydroxyl group and an aldehyde group, as shown for D-glucose in Fig. 21-30. In this case a six-membered ring is formed.

More complex carbohydrates are formed by combining monosaccharides. For example, **sucrose**, common table sugar, is a **disaccharide** formed from glucose and fructose by elimination of water to form a C—O—C bond between the rings, which is called a **glycoside linkage** (Fig. 21-31). When sucrose is consumed in food, the above reaction is reversed. ◄ An enzyme in saliva catalyzes the breakdown of this disaccharide.

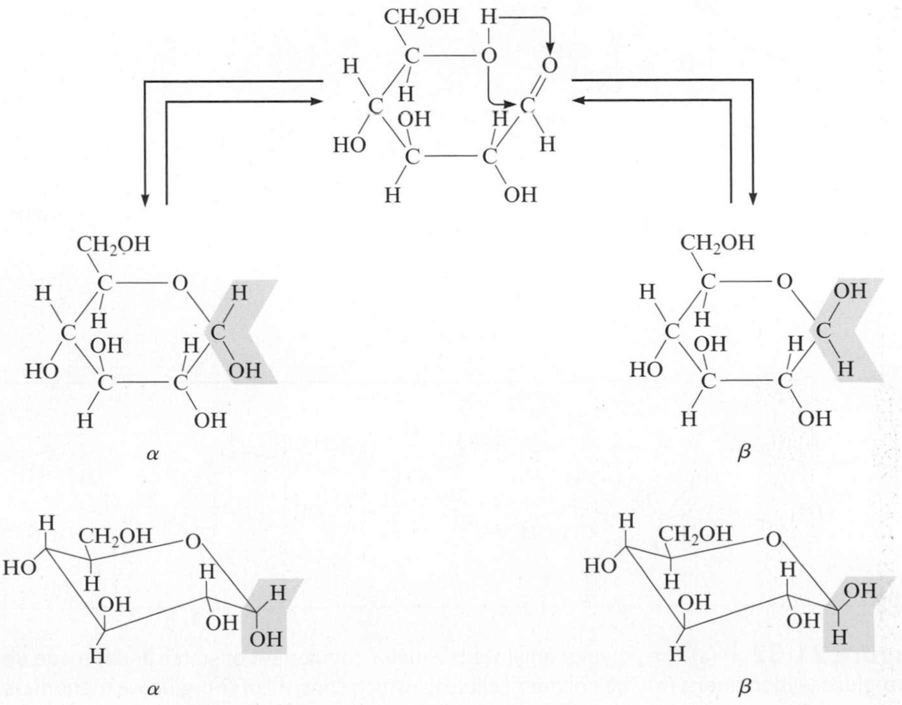

Figure 21-29 | The cyclization of D-fructose.

Figure 21-30 | The cyclization of glucose. Two different rings are possible; they differ in the orientation of the hydroxy group and hydrogen on one carbon, as indicated. The two forms are designated α and β and are shown here in two representations.

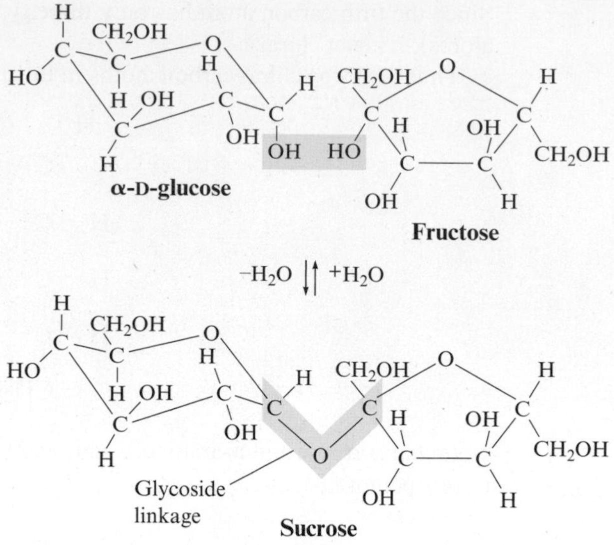

Figure 21-31 | Sucrose is a disaccharide formed from α-D-glucose and fructose.

α-D-glucose

Fructose

$-H_2O \updownarrow +H_2O$

Glycoside linkage

Sucrose

▲ Bowl of sugar cubes.

© Cengage Learning

Large polymers consisting of many monosaccharide units, called polysaccharides, can form when each ring forms two glycoside linkages, as shown in Fig. 21-32. Three of the most important of these polymers are starch, cellulose, and glycogen. All these substances are polymers of glucose, differing from each other in the nature of the glycoside linkage, the amount of branching, and molecular weight (molar mass).

Starch, a polymer of α-D-glucose, consists of two parts: *amylose,* a straight-chain polymer of α-glucose [see Fig. 21-32(a)], and *amylopectin,* a highly branched polymer of α-glucose with a molecular weight that is 10 to 20 times that of amylose. Branching occurs when a third glycoside linkage attaches a branch to the main polymer chain.

Starch, the carbohydrate reservoir in plants, is the form in which glucose is stored by the plant for later use as cellular fuel. Glucose is stored in this high-molecular-weight form because it results in less stress on the plant's internal structure by osmotic

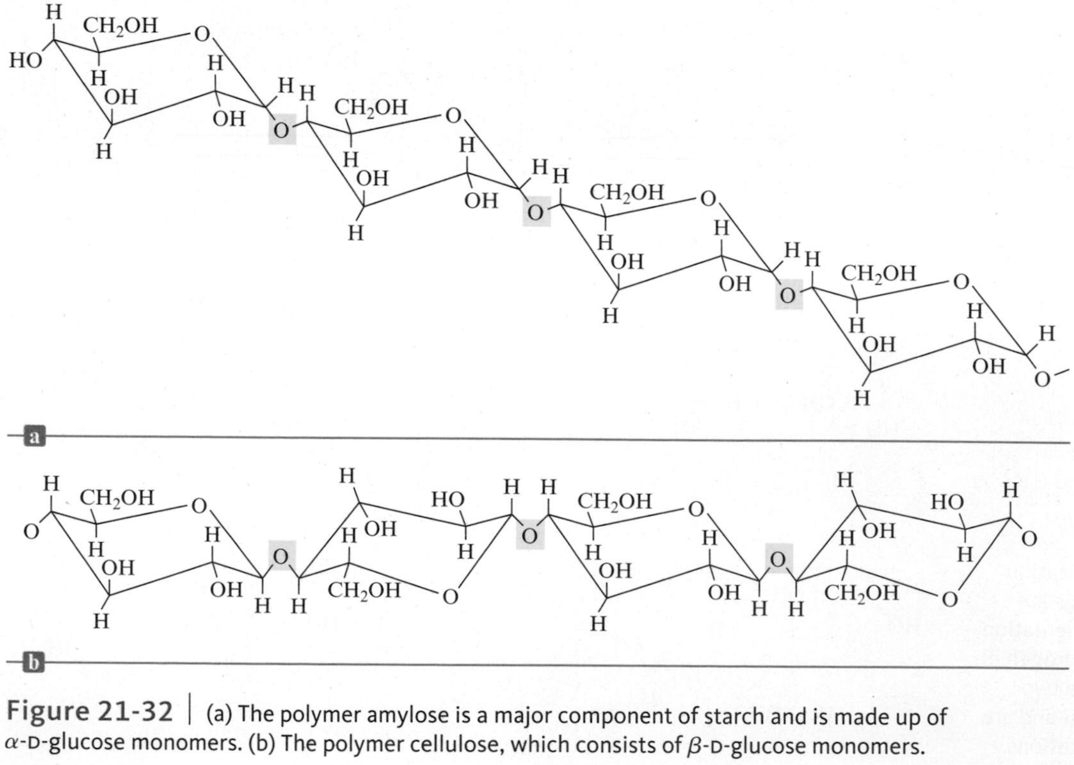

a

b

Figure 21-32 | (a) The polymer amylose is a major component of starch and is made up of α-D-glucose monomers. (b) The polymer cellulose, which consists of β-D-glucose monomers.

pressure. Recall from Section 10-6 that it is the concentration of solute molecules (or ions) that determines the osmotic pressure. Combining the individual glucose molecules into one large chain keeps the concentration of solute molecules relatively low, minimizing the osmotic pressure.

Cellulose, the major structural component of woody plants and natural fibers (such as cotton), is a polymer of β-D-glucose and has the structure shown in Fig. 21-32(b). Note that the β-glycoside linkages in cellulose give the glucose rings a different relative orientation than is found in starch. Although this difference may seem minor, it has very important consequences. The human digestive system contains α-glycosidases, enzymes that can catalyze breakage of the α-glycoside bonds in starch. These enzymes are not effective on the β-glycoside bonds of cellulose, presumably because the different structure results in a poor fit between the enzyme's active site and the carbohydrate. The enzymes necessary to cleave β-glycoside linkages, the β-glycosidases, are found in bacteria that exist in the digestive tracts of termites, cows, deer, and many other animals. Thus, unlike humans, these animals can derive nutrition from cellulose.

Glycogen, the main carbohydrate reservoir in animals, has a structure similar to that of amylopectin but with more branching. It is this branching that is thought to facilitate the rapid breakdown of glycogen into glucose when energy is required.

Nucleic Acids

Life is possible only because each cell, when it divides, can transmit the vital information about how it works to the next generation. It has been known for a long time that this process involves the chromosomes in the nucleus of the cell. Only since 1953, however, have scientists understood the molecular basis of this intriguing cellular "talent."

The substance that stores and transmits the genetic information is a polymer called **deoxyribonucleic acid (DNA)**, a huge molecule with a molecular weight as high as several billion grams per mole. ◄ Together with other similar nucleic acids called the **ribonucleic acids (RNA)**, DNA is also responsible for the synthesis of the various proteins needed by the cell to carry out its life functions. The RNA molecules, which are found in the cytoplasm outside the nucleus, are much smaller than DNA polymers, with molecular weights of only 20,000 to 40,000 grams per mole.

The monomers of the nucleic acids, called **nucleotides**, are composed of three distinct parts:

1. a *five-carbon sugar,* deoxyribose in DNA and ribose in RNA (Fig. 21-33)
2. a *nitrogen-containing organic base* of the type shown in Fig. 21-34
3. a *phosphoric acid molecule* (H_3PO_4)

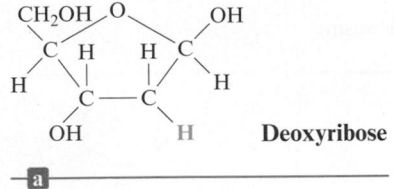

Figure 21-33 | The structure of the pentoses (a) deoxyribose and (b) ribose. Deoxyribose is the sugar molecule present in DNA; ribose is found in RNA.

Alfred Pasieka/Science Source

▲ A computer image of the base pairs of DNA. The blue lines represent the sugar–phosphate backbone and the colored bars represent the hydrogen bonding between the base pairs.

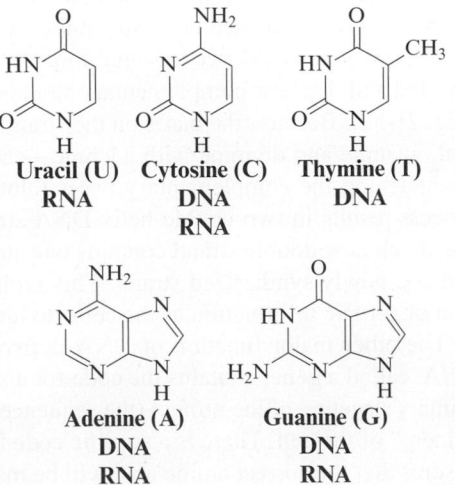

Uracil (U) Cytosine (C) Thymine (T)
RNA   DNA   DNA
           RNA

Adenine (A)   Guanine (G)
DNA   DNA
RNA   RNA

Figure 21-34 | The organic bases found in DNA and RNA.

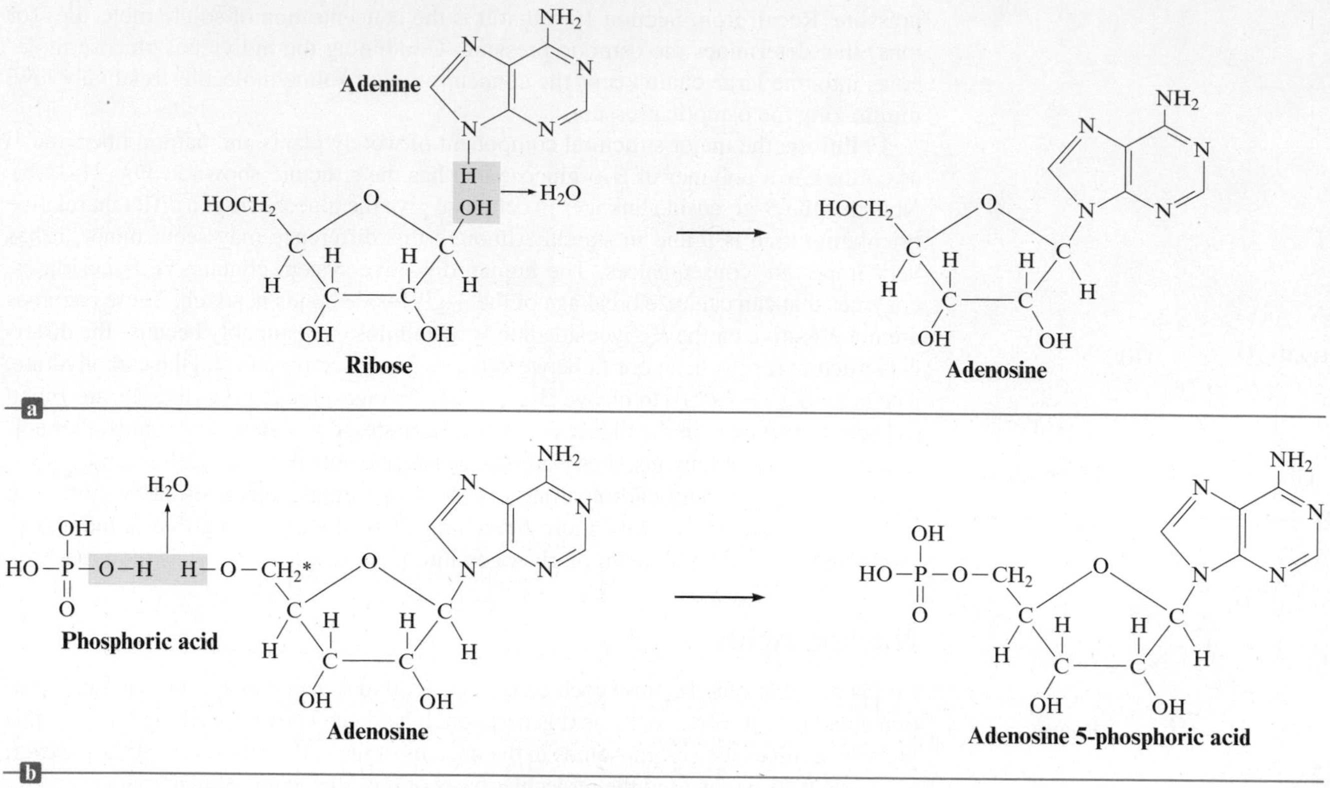

Figure 21-35 | (a) Adenosine is formed by the reaction of adenine with ribose. (b) The reaction of phosphoric acid with adenosine to form the ester adenosine 5-phosphoric acid, a nucleotide. (At biological pH, the phosphoric acid would not be fully protonated as is shown here.)

The base and the sugar combine as shown in Fig. 21-35(a) to form a unit that in turn reacts with phosphoric acid to create the nucleotide, which is an ester [Fig. 21-35(b)]. The nucleotides become connected through condensation reactions that eliminate water to give a polymer of the type represented in Fig. 21-36; such a polymer can contain a *billion* units.

The key to DNA's functioning is its *double-helical structure with complementary bases on the two strands*. The bases form hydrogen bonds to each other, as shown in Fig. 21-37. Note that the structures of cytosine and guanine make them perfect partners for hydrogen bonding, and they are *always* found as pairs on the two strands of DNA. Thymine and adenine form similar hydrogen-bonding pairs.

There is much evidence to suggest that the two strands of DNA unwind during cell division and that new complementary strands are constructed on the unraveled strands (Fig. 21-38). Because the bases on the strands always pair in the same way—cytosine with guanine and thymine with adenine—each unraveled strand serves as a template for attaching the complementary bases (along with the rest of the nucleotide). This process results in two double-helix DNA structures that are identical to the original one. Each new double strand contains one strand from the original DNA double helix and one newly synthesized strand. This replication of DNA allows for the transmission of genetic information as the cells divide.

The other major function of DNA is **protein synthesis**. A given segment of the DNA, called a **gene**, contains the code for a specific protein. These codes transmit the primary structure of the protein (the sequence of amino acids) to the construction "machinery" of the cell. There is a specific code for each amino acid in the protein, which ensures that the correct amino acid will be inserted as the protein chain grows. A code consists of a set of three bases called a **codon**.

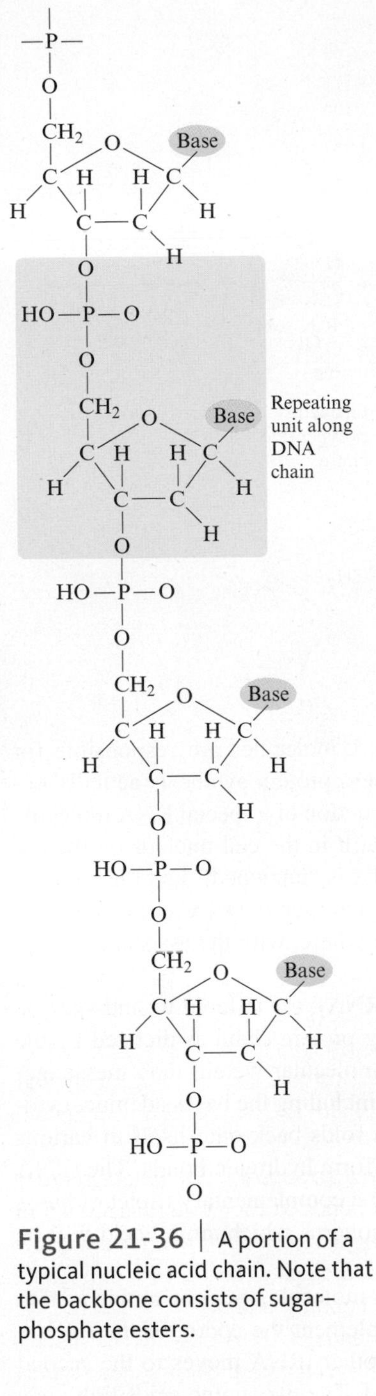

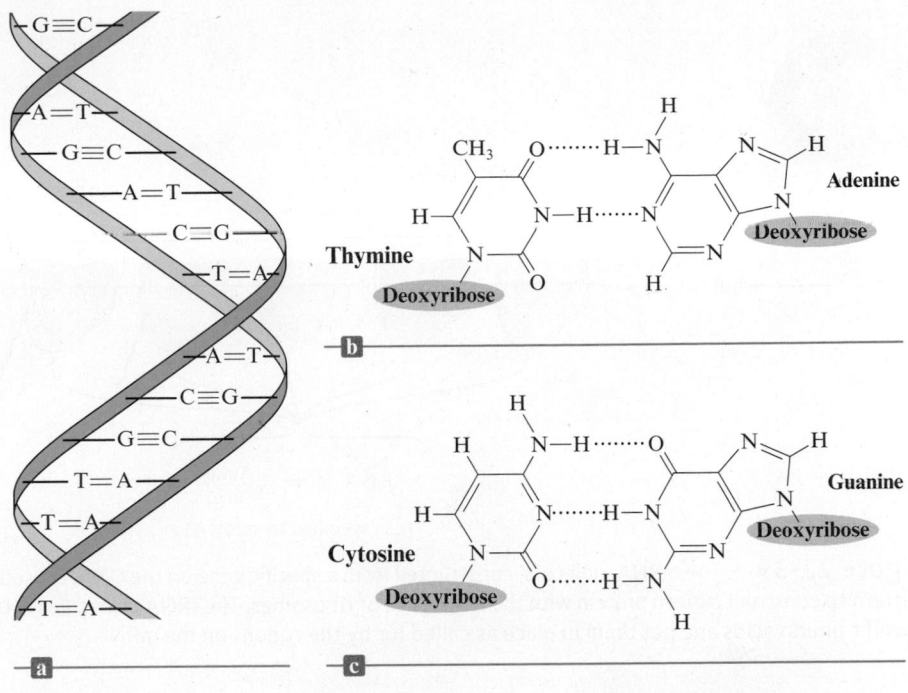

Figure 21-36 | A portion of a typical nucleic acid chain. Note that the backbone consists of sugar–phosphate esters.

Figure 21-37 | (a) The DNA double helix contains two sugar–phosphate backbones, with the bases from the two strands hydrogen-bonded to each other. The complementarity of the (b) thymine-adenine and (c) cytosine-guanine pairs.

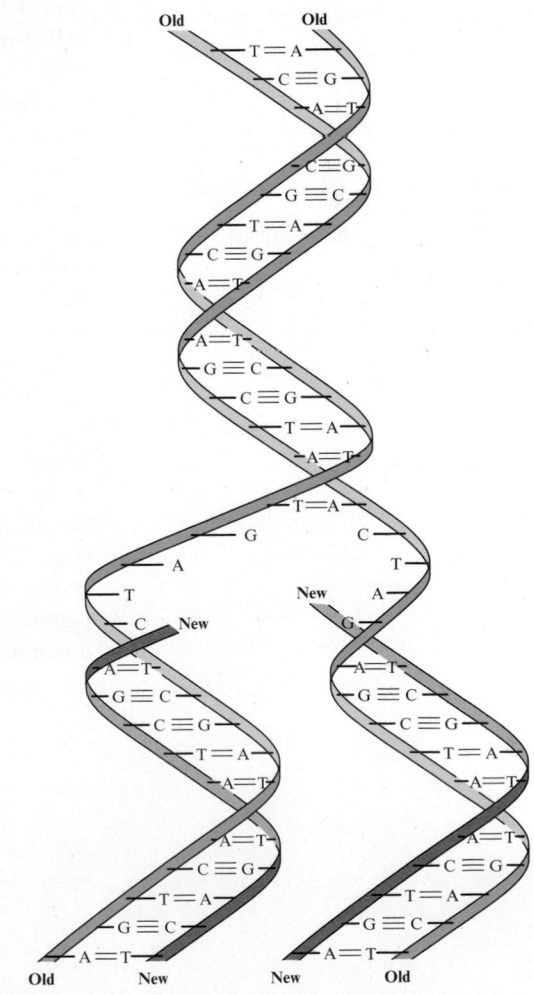

Figure 21-38 | During cell division the original DNA double helix unwinds and new complementary strands are constructed on each original strand.

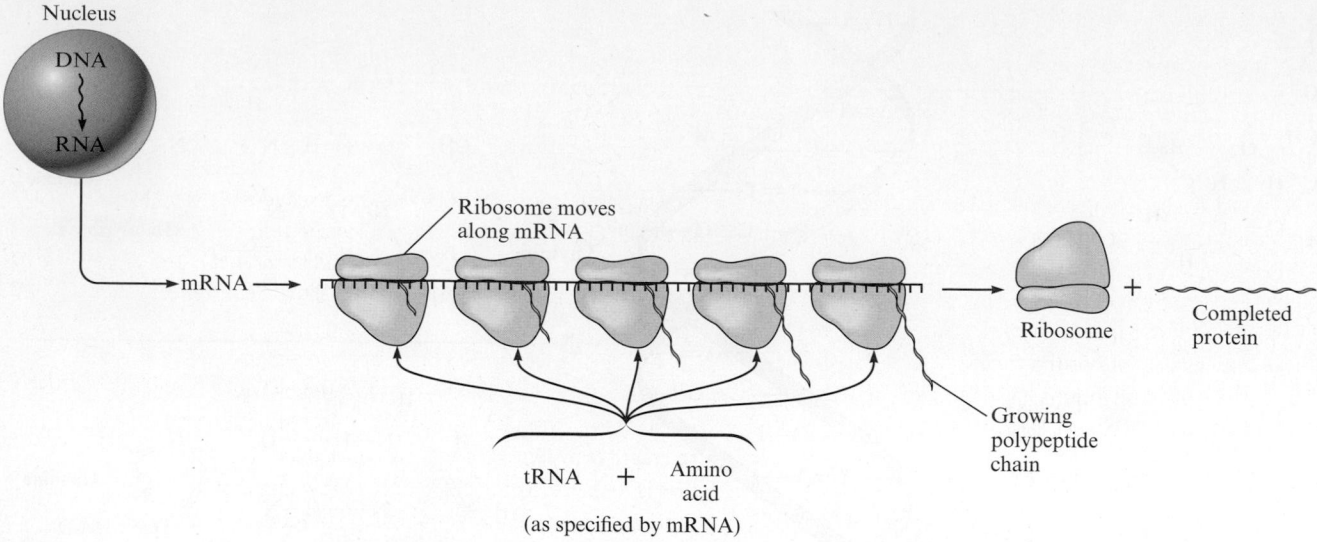

Figure 21-39 | The mRNA molecule, constructed from a specific gene on the DNA, is used as the pattern to construct a given protein with the assistance of ribosomes. The tRNA molecules attach to specific amino acids and put them in place as called for by the codons on the mRNA.

DNA stores the genetic information, while RNA molecules are responsible for transmitting this information to the ribosomes, where protein synthesis actually occurs. This complex process involves, first, the construction of a special RNA molecule called **messenger RNA (mRNA)**. The mRNA is built in the cell nucleus on the appropriate section of DNA (the gene); the double helix is "unzipped," and the complementarity of the bases is used in a process similar to that used in DNA replication. The mRNA then migrates into the cytoplasm of the cell where, with the assistance of the ribosomes, the protein is synthesized.

Small RNA fragments, called **transfer RNA (tRNA)**, are tailored to find specific amino acids and then to attach them to the growing protein chain as dictated by the codons in the mRNA. Transfer RNA has a lower molecular weight than messenger RNA. It consists of a chain of 75 to 80 nucleotides, including the bases adenine, cytosine, guanine, and uracil, among others. The chain folds back onto itself in various places as the complementary bases along the chain form hydrogen bonds. The tRNA decodes the genetic message from the mRNA, using a complementary triplet of bases called an **anticodon**. The nature of the anticodon governs which amino acid will be brought to the protein under construction.

The protein is built in several steps. First, a tRNA molecule brings an amino acid to the mRNA [the anticodon of the tRNA must complement the codon of the mRNA (Fig. 21-39)]. Once this amino acid is in place, another tRNA moves to the second codon site of the mRNA with its specific amino acid. The two amino acids link via a peptide bond, and the tRNA on the first codon breaks away. The process is repeated down the chain, always matching the tRNA anticodon with the mRNA codon.

Among today's best-selling cosmetics are self-tanning lotions. Many light-skinned people want to look like they have just spent a vacation in the Caribbean, but they recognize the dangers of too much sun—it causes premature aging and may lead to skin cancer. Chemistry has come to the rescue in the form of lotions that produce an authentic-looking tan. All of these lotions have the same active ingredient: dihydroxyacetone (DHA). DHA, which has the structure

$$H-O-\overset{\overset{\displaystyle H}{|}}{\underset{\underset{\displaystyle H}{|}}{C}}-\overset{\overset{\displaystyle O}{\|}}{C}-\overset{\overset{\displaystyle H}{|}}{\underset{\underset{\displaystyle H}{|}}{C}}-OH$$

is a nontoxic, simple sugar that occurs as an intermediate in carbohydrate metabolism in higher-order plants and animals. The DHA used in self-tanners is prepared by bacterial fermentation of glycerine,

$$H-O-\overset{\overset{\displaystyle H}{|}}{\underset{\underset{\displaystyle H}{|}}{C}}-\overset{\overset{\displaystyle \overset{\displaystyle H}{|}}{O}}{\underset{\underset{\displaystyle H}{|}}{C}}-\overset{\overset{\displaystyle H}{|}}{\underset{\underset{\displaystyle H}{|}}{C}}-O-H$$

The tanning effects of DHA were discovered by accident in the 1950s at Children's Hospital at the University of Cincinnati, where DHA was being used to treat children with glycogen storage disease. When the DHA was accidentally spilled on the skin, it produced brown spots.

The mechanism of the browning process involves the Maillard reaction, which was discovered by Louis-Camille Maillard in 1912. In this process amino acids react with sugars to create brown or golden brown products. The same reaction is responsible for much of the browning that occurs during the manufacture and storage of foods. It is also the reason that beer is golden brown.

The browning of skin occurs in the stratum corneum—the outermost, dead layer—where the DHA reacts with free amino ($-NH_2$) groups of the proteins found there.

DHA is present in most tanning lotions at concentrations between 2% and 5%, although some products designed to give a deeper tan are more concentrated. Because the lotions themselves turn brown above pH 7, the tanning lotions are buffered at pH 5.

Thanks to these new products, tanning is now both safe and easy.

Ingredients: Water, Aloe Vera Gel, Glycerin, Propylene Glycol, Polysorbate 20, Fragrance, Tocopheryl Acetate (Vitamin E Acetate), Imidazolidinyl Urea, Dihydroxyacetone.

© 1995, 1998 Distributed by Schering-Plough HealthCare Products, Inc., Memphis, TN 38151, USA. All rights Reserved. Made in USA.

www.coppertone.com

21621-C

Self-tanning products and a close-up of a label showing the contents.

FOR REVIEW

Key Terms

biomolecule
organic chemistry

Section 21-1
hydrocarbons
saturated
unsaturated
alkanes
normal (straight-chain or
    unbranched) hydrocarbons
structural isomerism
combustion reaction
substitution reaction
dehydrogenation reaction
cyclic alkanes

Section 21-2
alkenes
cis–trans isomerism
alkynes
addition reaction
hydrogenation reaction
halogenation
polymerization

Section 21-3
aromatic hydrocarbons
phenyl group

Section 21-4
hydrocarbon derivatives
functional group
alcohols
phenol
carbonyl group
ketones
aldehydes
carboxylic acids
carboxyl group
ester
amines

Section 21-5
polymers
thermoset polymer
thermoplastic polymer
crosslinking
vulcanization
addition polymerization
free radical
condensation polymerization
copolymer
homopolymer

Hydrocarbons

- Compounds composed of mostly carbon and hydrogen atoms that typically contain chains or rings of carbon atoms
- Alkanes
  - Contain compounds with only C—C single bonds
  - Can be represented by the formula C_nH_{2n+2}
  - Are said to be saturated because each carbon present is bonded to the maximum number of atoms (4)
  - The carbon atoms are described as being sp^3 hybridized
  - Their structural isomerism involves the formation of branched chains
  - React with O_2 to form CO_2 and H_2O (called a combustion reaction)
  - Undergo substitution reactions
- Alkenes
  - Contain one or more C=C double bonds
  - Simplest alkene is C_2H_4 (ethylene), which is described as containing sp^2 hybridized carbon atoms
  - Restricted rotation about the C=C bonds in alkenes can lead to *cis–trans* isomerism
  - Undergo addition reactions
- Alkynes
  - Contain one or more C≡C triple bonds
  - Simplest example is C_2H_2 (acetylene), described as containing sp-hybridized carbon atoms
  - Undergo addition reactions
- Aromatic hydrocarbons
  - Contain rings of carbon atoms with delocalized π electrons
  - Undergo substitution reactions rather than addition reactions

Hydrocarbon Derivatives

- Contain one or more functional groups
- Alcohols: contain the —OH group
- Aldehydes: contain a $\overset{\displaystyle}{\underset{\text{H}}{>}}$C=O group
- Ketones: contain the $>$C=O group
- Carboxylic acids: contain the —C$\overset{\text{O}}{\underset{\text{OH}}{}}$ group

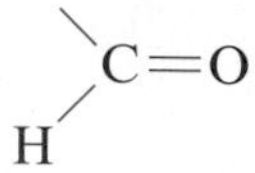

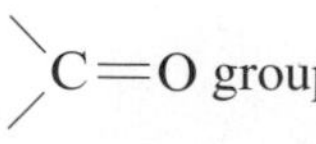

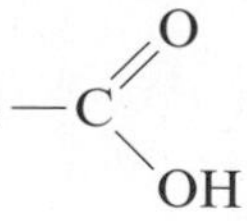

Polymers

- Large molecules formed from many small molecules (called monomers)
  - Addition polymerization: monomers add together by a free radical mechanism
  - Condensation polymerization: monomers connect by splitting out a small molecule, such as water

Key Terms

dimer
polyester
isotactic chain
syndiotactic chain
atactic chain
polystyrene
polyvinyl chloride (PVC)

Section 21-6
proteins
fibrous proteins
globular proteins
α-amino acids
side chains
dipeptide
peptide linkage
polypeptide
primary structure
secondary structure
α-helix
pleated sheet
random-coil arrangement
tertiary structure
disulfide linkage
denaturation
carbohydrates
monosaccharides (simple sugars)
pentoses
hexoses
sucrose
disaccharide
glycoside linkage
starch
cellulose
glycogen
deoxyribonucleic acid (DNA)
ribonucleic acid (RNA)
nucleotides
protein synthesis
gene
codon
messenger RNA (mRNA)
transfer RNA (tRNA)
anticodon

Proteins

- A class of natural polymers with molar masses ranging from 600 to 1,000,000
- Fibrous proteins form the structural basis of muscle, hair, and cartilage
- Globular proteins perform many biologic functions, including transport and storage of oxygen, catalysis of biological reactions, and regulation of biological systems
- Building blocks of proteins (monomers) are α-amino acids, which connect by a condensation reaction to form a peptide linkage
- Protein structure
  - Primary: the order of amino acids in the chain
  - Secondary: the arrangement of the protein chain
    - α-helix
    - pleated sheet
  - Tertiary structure: the overall shape of the protein

Carbohydrates

- Contain carbon, hydrogen, and oxygen
- Serve as food sources for most organisms
- Monosaccharides are most commonly five-carbon and six-carbon polyhydroxy ketones and aldehydes
  - Monosaccharides combine to form more complex carbohydrates, such as sucrose, starch, and cellulose

Genetic Processes

- When a cell divides, the genetic information is transmitted via deoxyribonucleic acid (DNA), which has a double helical structure
  - During cell division, the double helix unravels and a new polymer forms along each strand of the original DNA
  - The genetic code is carried by organic bases that hydrogen-bond to each other in specific pairs in the interior of the DNA double helix

---

Review Questions Answers to the Review Questions can be found on the Student Web site (accessible from **www.cengagebrain.com**).

1. What is a hydrocarbon? What is the difference between a saturated hydrocarbon and an unsaturated hydrocarbon? Distinguish between normal and branched hydrocarbons. What is an alkane? What is a cyclic alkane? What are the two general formulas for alkanes? What is the hybridization of carbon atoms in alkanes? What are the bond angles in alkanes? Why are cyclopropane and cyclobutane so reactive?

The normal (unbranched) hydrocarbons are often referred to as *straight-chain hydrocarbons*. What does this name refer to? Does it mean that the carbon atoms in a straight-chain hydrocarbon really have a linear arrangement? Explain. In the shorthand notation for cyclic alkanes, the hydrogens are usually omitted. How do you determine the number of hydrogens bonded to each carbon in a ring structure?

2. What is an alkene? What is an alkyne? What are the general formulas for alkenes and alkynes, assuming one multiple bond in each? What are the bond angles in alkenes and alkynes? Describe the bonding in alkenes and alkynes using C_2H_4 and C_2H_2 as your examples. Why is there restricted rotation in alkenes and alkynes? Is the general formula for a cyclic alkene C_nH_{2n}? If not, what is the general formula, assuming one multiple bond?

3. What are aromatic hydrocarbons? Benzene exhibits resonance. Explain. What are the bond angles in benzene? Give a detailed description of the bonding in benzene. The π electrons in benzene are delocalized, while the π electrons in simple alkenes and alkynes are localized. Explain the difference.

4. Summarize the nomenclature rules for alkanes, alkenes, alkynes, and aromatic compounds. Correct the following false statements regarding nomenclature of hydrocarbons.

   a. The root name for a hydrocarbon is based on the shortest continuous chain of carbon atoms.

   b. The suffix used to name all hydrocarbons is *-ane*.

   c. Substituent groups are numbered so as to give the largest numbers possible.

   d. No number is required to indicate the positions of double or triple bonds in alkenes and alkynes.

   e. Substituent groups get the lowest number possible in alkenes and alkynes.

   f. The *ortho-* term in aromatic hydrocarbons indicates the presence of two substituent groups bonded to carbon-1 and carbon-3 in benzene.

5. What functional group distinguishes each of the following hydrocarbon derivatives?

   a. halohydrocarbons

   b. alcohols

   c. ethers

   d. aldehydes

   e. ketones

   f. carboxylic acids

   g. esters

   h. amines

   Give examples of each functional group. What prefix or suffix is used to name each functional group? What are the bond angles in each? Describe the bonding in each functional group. What is the difference between a primary, secondary, and

tertiary alcohol? For the functional groups in a–h, when is a number required to indicate the position of the functional group? Carboxylic acids are often written as RCOOH. What does —COOH indicate and what does R indicate? Aldehydes are sometimes written as RCHO. What does —CHO indicate?

6. Distinguish between isomerism and resonance. Distinguish between structural and geometric isomerism. When writing the various structural isomers, the most difficult task is identifying which are different isomers and which are identical to a previously written structure—that is, which are compounds that differ only by the rotation of a carbon single bond. How do you distinguish between structural isomers and those that are identical?

   Alkenes and cycloalkanes are structural isomers of each other. Give an example of each using C_4H_8. Another common feature of alkenes and cycloalkanes is that both have restricted rotation about one or more bonds in the compound, so both can exhibit *cis–trans* isomerism. What is required for an alkene or cycloalkane to exhibit *cis–trans* isomerism? Explain the difference between *cis* and *trans* isomers.

   Alcohols and ethers are structural isomers of each other, as are aldehydes and ketones. Give an example of each to illustrate. Which functional group in Table 21-4 can be structural isomers of carboxylic acids?

   What is optical isomerism? What do you look for to determine whether an organic compound exhibits optical isomerism? 1-Bromo-1-chloroethane is optically active whereas 1-bromo-2-chloroethane is not optically active. Explain.

7. What type of intermolecular forces do hydrocarbons exhibit? Explain why the boiling point of *n*-heptane is greater than that of *n*-butane. A general rule for a group of hydrocarbon isomers is that as the amount of branching increases, the boiling point decreases. Explain why this would be true.

   The functional groups listed in Table 21-4 all exhibit London dispersion forces, but they also usually exhibit additional dipole–dipole forces. Explain why this is the case for each functional group. Although alcohols and ethers are structural isomers of each other, alcohols always boil at significantly higher temperatures than similar-size ethers. Explain. What would you expect when comparing the boiling points of similar-size carboxylic acids to esters? $CH_3CH_2CH_3$, CH_3CH_2OH, CH_3CHO, and HCOOH all have about the same molar mass, but they boil at very different temperatures. Why? Place these compounds in order by increasing boiling point.

8. Distinguish between substitution and addition reactions. Give an example of each type of reaction. Alkanes and aromatics are fairly stable compounds. To make them react, a special catalyst must be present. What catalyst must be present when reacting Cl_2 with an alkane or with benzene? Adding Cl_2 to an alkene or alkyne does not require a special catalyst. Why are alkenes and alkynes more reactive than alkanes and aromatic compounds? All organic compounds can be combusted. What is the other reactant in a combustion reaction, and what are the products, assuming the organic compound contains only C, H, and perhaps O?

The following are some other organic reactions covered in Section 21-4. Give an example to illustrate each type of reaction.

a. Adding H_2O to an alkene (in the presence of H^+) yields an alcohol.

b. Primary alcohols are oxidized to aldehydes, which can be further oxidized to carboxylic acids.

c. Secondary alcohols are oxidized to ketones.

d. Reacting an alcohol with a carboxylic acid (in the presence of H^+) produces an ester.

9. Define and give an example of each of the following.

a. addition polymer

b. condensation polymer

c. copolymer

d. homopolymer

e. polyester

f. polyamide

Distinguish between a thermoset polymer and a thermoplastic polymer. How do the physical properties of polymers depend on chain length and extent of chain branching? Explain how crosslinking agents are used to change the physical properties of polymers. Isotactic polypropylene makes stronger fibers than atactic polypropylene. Explain. In which polymer, polyethylene or polyvinyl chloride, would you expect to find the stronger intermolecular forces (assuming the average chain lengths are equal)?

10. Give the general formula for an amino acid. Some amino acids are labeled hydrophilic and some are labeled hydrophobic. What do these terms refer to? Aqueous solutions of amino acids are buffered solutions. Explain. Most of the amino acids in Fig. 21-18 are optically active. Explain. What is a peptide bond? Show how glycine, serine, and alanine react to form a tripeptide. What is a protein, and what are the monomers in proteins? Distinguish between the primary, secondary, and tertiary structures of a protein. Give examples of the types of forces that maintain each type of structure. Describe how denaturation affects the function of a protein.

11. What are carbohydrates, and what are the monomers in carbohydrates? The monosaccharides in Table 21-8 are all optically active. Explain. What is a *disaccharide?* Which monosaccharide units make up the disaccharide sucrose? What do you call the bond that forms between the monosaccharide units? What forces are responsible for the solubility of starch in water? What is the difference between starch, cellulose, and glycogen?

12. Describe the structural differences between DNA and RNA. The monomers in nucleic acids are called nucleotides. What are the three parts of a nucleotide? The compounds adenine, guanine, cytosine, and thymine are called the nucleic acid bases. What structural features in these compounds make them bases? DNA exhibits a double-helical structure. Explain. Describe how the complementary base pairing between the two individual strands of DNA forms the overall double-helical structure. How is complementary base pairing involved in the replication of the DNA molecule during cell division? Describe how protein synthesis occurs. What is a codon, and what is a gene? The deletion of a single base from a DNA molecule can constitute a fatal mutation, whereas substitution of one base for another is often not as serious a mutation. Explain.

A blue question or exercise number indicates that the answer to that question or exercise appears at the back of this book and a solution appears in the *Solutions Guide,* as found on PowerLecture.

Questions

1. A confused student was doing an isomer problem and listed the following six names as different structural isomers of C_7H_{16}.

a. 1-sec-butylpropane
b. 4-methylhexane
c. 2-ethylpentane
d. 1-ethyl-1-methylbutane
e. 3-methylhexane
f. 4-ethylpentane

How many different structural isomers are actually present in these six names?

2. For the following formulas, what types of isomerism could be exhibited? For each formula, give an example that illustrates

the specific type of isomerism. The types of isomerism are structural, geometric, and optical.

  a. C_6H_{12} **b.** $C_5H_{12}O$ **c.** $C_6H_4Br_2$

3. What is wrong with the following names? Give the correct name for each compound.

  a. 2-ethylpropane

  b. 5-iodo-5,6-dimethylhexane

  c. *cis*-4-methyl-3-pentene

  d. 2-bromo-3-butanol

4. The following organic compounds cannot exist. Why?

  a. 2-chloro-2-butyne

  b. 2-methyl-2-propanone

  c. 1,1-dimethylbenzene

  d. 2-pentanal

  e. 3-hexanoic acid

  f. 5,5-dibromo-1-cyclobutanol

5. If you had a group of hydrocarbons, what structural features would you look at to rank the hydrocarbons in order of increasing boiling point?

6. Which of the functional groups in Table 21-4 can exhibit hydrogen bonding intermolecular forces? Can CH_2CF_2 exhibit hydrogen bonding? Explain.

7. A polypeptide is also called a polyamide. Nylon is also an example of a polyamide. What is a polyamide? Consider a polyhydrocarbon, a polyester, and a polyamide. Assuming average chain lengths are equal, which polymer would you expect to make the strongest fibers and which polymer would you expect to make the weakest fibers? Explain.

8. Give an example reaction that would yield the following products. Name the organic reactant and product in each reaction.

  a. alkane

  b. monohalogenated alkane

  c. dihalogenated alkane

  d. tetrahalogenated alkane

  e. monohalogenated benzene

  f. alkene

9. Give an example reaction that would yield the following products as major organic products. See Exercises 21-62 and 21-65 for some hints. For oxidation reactions, just write oxidation over the arrow and don't worry about the actual reagent.

  a. primary alcohol   **e.** ketone

  b. secondary alcohol   **f.** carboxylic acid

  c. tertiary alcohol   **g.** ester

  d. aldehyde

10. What is polystyrene? The following processes result in a stronger polystyrene polymer. Explain why in each case.

  a. addition of catalyst to form syndiotactic polystyrene

  b. addition of 1,3-butadiene and sulfur

  c. producing long chains of polystyrene

  d. addition of a catalyst to make linear polystyrene

11. Answer the following questions regarding the formation of polymers.

  a. What structural features must be present in a monomer in order to form a homopolymer polyester?

  b. What structural features must be present in the monomers in order to form a copolymer polyamide? (*Hint:* Nylon is an example of a polyamide. When the monomers link together to form nylon, an amide functional group results from each linkage.)

  c. What structural features must be present in a monomer that can form both an addition polymer and a condensation polymer?

12. In Section 21-6, three important classes of biologically important natural polymers are discussed. What are the three classes, what are the monomers used to form the polymers, and why are they biologically important?

Exercises

In this section similar exercises are paired.

Hydrocarbons

13. Draw the five structural isomers of hexane (C_6H_{14}).

14. Name the structural isomers in Exercise 13.

15. Draw all the structural isomers for C_8H_{18} that have the following root name (longest carbon chain). Name the structural isomers.

  a. heptane   **b.** butane

16. Draw all the structural isomers for C_8H_{18} that have the following root name (longest carbon chain). Name the structural isomers.

  a. hexane   **b.** pentane

17. Draw a structural formula for each of the following compounds.

  a. 2-methylpropane

  b. 2-methylbutane

  c. 2-methylpentane

  d. 2-methylhexane

18. Draw a structural formula for each of the following compounds.

  a. 2,2-dimethylheptane   **c.** 3,3-dimethylheptane

  b. 2,3-dimethylheptane   **d.** 2,4-dimethylheptane

19. Draw the structural formula for each of the following.

  a. 3-isobutylhexane

  b. 2,2,4-trimethylpentane, also called *isooctane*. This substance is the reference (100 level) for octane ratings.

  c. 2-*tert*-butylpentane

  d. The names given in parts a and c are incorrect. Give the correct names for these hydrocarbons.

20. Draw the structure for 4-ethyl-2,3-diisopropylpentane. This name is incorrect. Give the correct systematic name.

21. Name each of the following.

  a.

$$CH_3-\underset{\underset{CH_3}{|}}{\overset{\overset{CH_3}{|}}{C}}-CH_2-\underset{\underset{CH_3}{|}}{CH}-CH_2-CH_3$$

  b.

$$\underset{\underset{CH_3}{|}}{CH_2}-CH_2-CH_2-\underset{\underset{CH_3}{|}}{CH}-CH_2-CH_2-\underset{\underset{CH_3}{|}}{CH_2}$$

c.

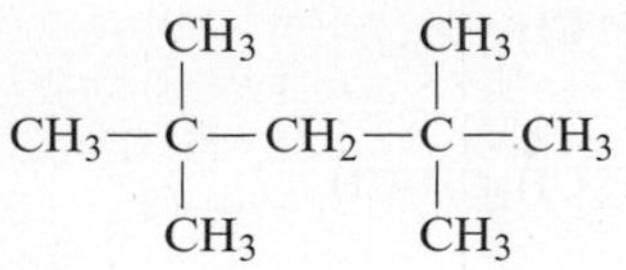

$$CH_3-\underset{\underset{CH_3}{|}}{\overset{\overset{CH_3}{|}}{C}}-CH_2-\underset{\underset{CH_3}{|}}{\overset{\overset{CH_3}{|}}{C}}-CH_3$$

d.

$$CH_3-\underset{\underset{CH_2-CH_3}{|}}{\overset{\overset{CH_2-CH_3}{|}}{C}}-CH_2-CH_2-CH_2-CH_2-CH_3$$

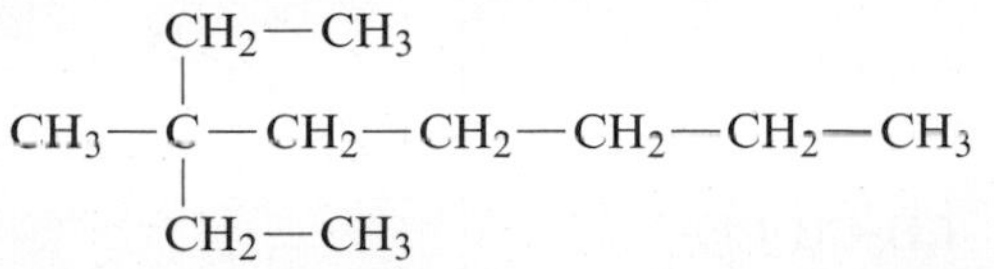

22. Name each of the following cyclic alkanes, and indicate the formula of the compound.

a. ☐—CHCH₃ with CH₃ below

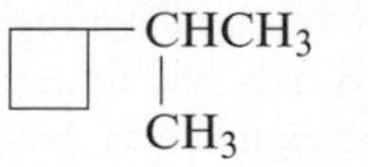

b. cyclopentane with CH₃, CCH₃, CH₃ substituents

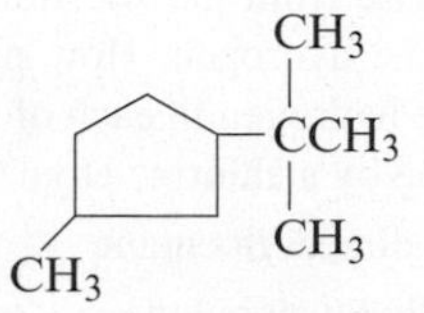

c. CH₃ CH₂CH₂CH₃ cyclohexane with —CH₃

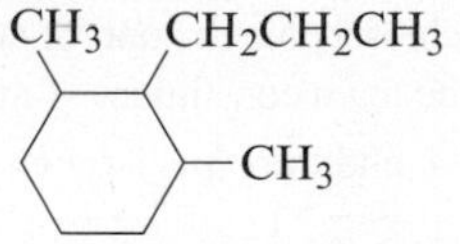

23. Give two examples of saturated hydrocarbons. How many other atoms are bonded to each carbon in a saturated hydrocarbon?

24. Draw the structures for two examples of *unsaturated* hydrocarbons. What structural feature makes a hydrocarbon unsaturated?

25. Name each of the following alkenes.

a. $CH_2{=}CH-CH_2-CH_3$

b.

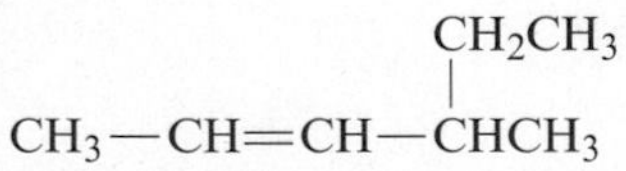

$$CH_3-CH{=}CH-\underset{\underset{CH_3}{|}}{\overset{\overset{CH_2CH_3}{|}}{CH}}$$

c.

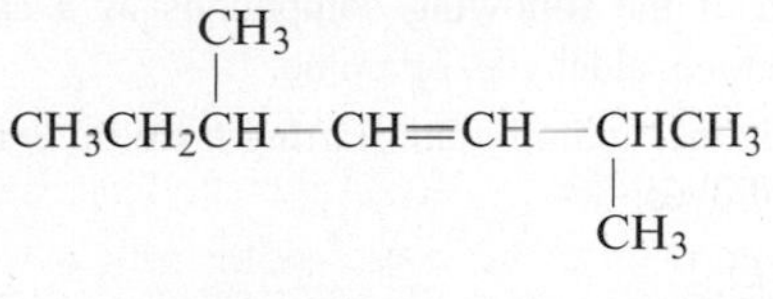

$$CH_3CH_2\underset{\underset{}{\overset{\overset{CH_3}{|}}{CH}}}{}-CH{=}CH-\underset{\underset{CH_3}{|}}{CHCH_3}$$

26. Name each of the following alkenes or alkynes.

a.

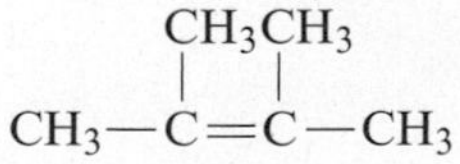

$$CH_3-\underset{\underset{CH_3}{|}}{C}{=}\underset{\overset{CH_3}{|}}{C}-CH_3$$

b.

$$\underset{}{C}{\equiv}C-\underset{\underset{}{\overset{\overset{CH_3}{|}}{CH}}}{}-CH_2-CH_3 \quad (CH_3 \text{ above first C})$$

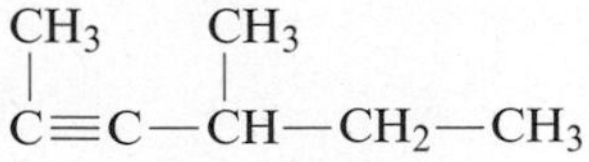

c.

$$CH_2{=}C-CH-CH_3 \text{ with } CH_3 \text{ and } CH_2-CH_3 \text{ below}$$

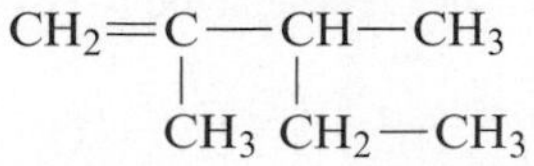

27. Give the structure for each of the following.

a. 3-hexene
b. 2,4-heptadiene
c. 2-methyl-3-octene

28. Give the structure for each of the following.

a. 4-methyl-1-pentyne
b. 2,3,3-trimethyl-1-hexene
c. 3-ethyl-4-decene

29. Give the structure of each of the following aromatic hydrocarbons.

a. *o*-ethyltoluene
b. *p*-di-*tert*-butylbenzene
c. *m*-diethylbenzene
d. 1-phenyl-2-butene

30. Cumene is the starting material for the industrial production of acetone and phenol. The structure of cumene is

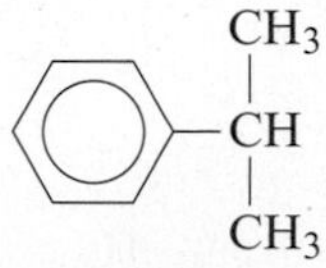

benzene ring—CH with two CH₃ groups

Give the systematic name for cumene.

31. Name each of the following.

a.

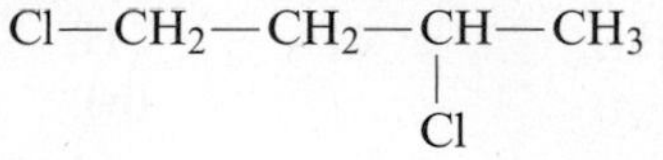

$$Cl-CH_2-CH_2-\underset{\underset{Cl}{|}}{CH}-CH_3$$

b. $CH_3CH_2CH_2CCl_3$

c.

$$\underset{CH_3}{\overset{CH_3}{>}}CCl-CH-CH-CH_3 \text{ with } Cl \text{ and } CH_2-CH_3 \text{ below}$$

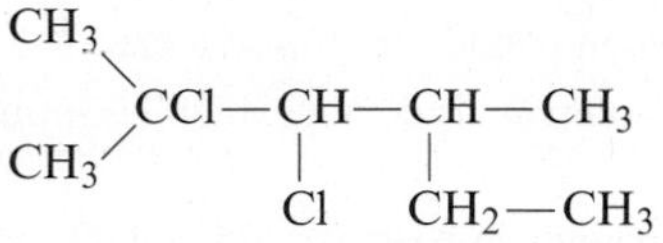

d. CH_2FCH_2F

32. Name each of the following compounds.

a.

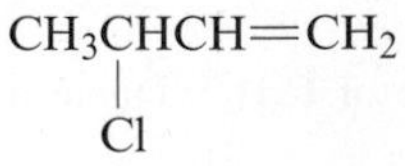

$$CH_3CHCH{=}CH_2 \text{ with } Cl \text{ below}$$

b. CH₃ cyclopentene with CH₂CH₃

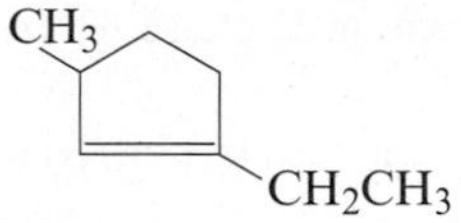

c. Cl cyclopentene with —CH₂CH₂CH₃

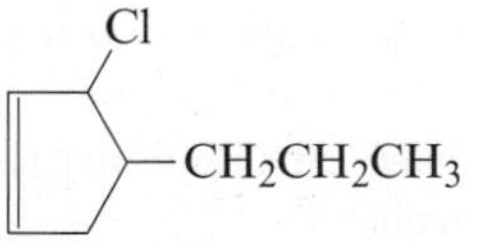

d. cyclohexane with CH₃, CH₃, CH₃ substituents

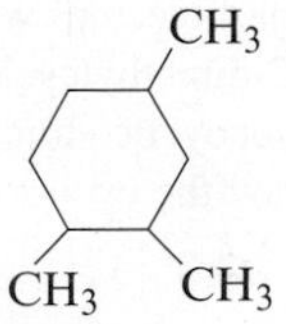

e. CH₃ Br benzene

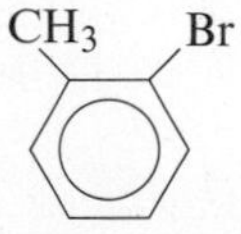

f. CH₃ Br cyclohexane

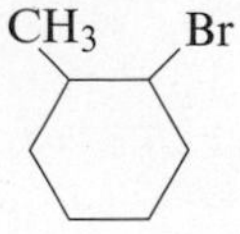

g. 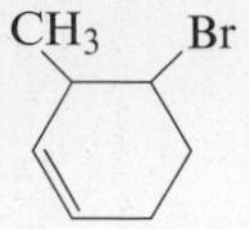 CH₃   Br

Isomerism

33. There is only one compound that is named 1,2-dichloroethane, but there are two distinct compounds that can be named 1,2-dichloroethene. Why?

34. Consider the following four structures.

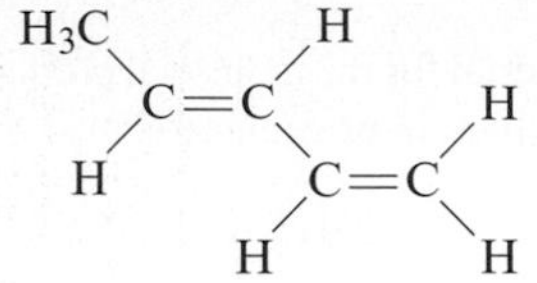

(i)

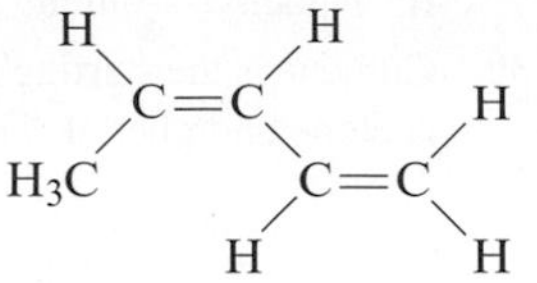

(ii)

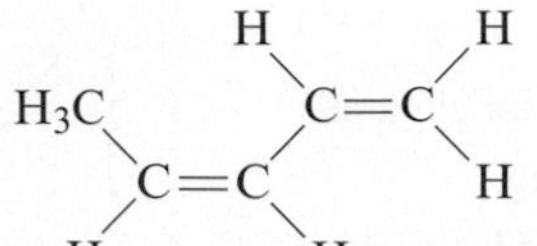

(iii)

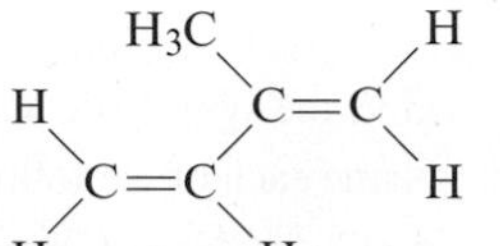

(iv)

a. Which of these compounds would have the same physical properties (melting point, boiling point, density, and so on)?

b. Which of these compounds are *trans* isomers?

c. Which of these compounds do not exhibit *cis–trans* isomerism?

35. Which of the compounds in Exercises 25 and 27 exhibit *cis–trans* isomerism?

36. Which of the compounds in Exercises 26 and 28 exhibit *cis–trans* isomerism?

37. Draw all the structural isomers of C_5H_{10}. Ignore any cyclic isomers.

38. Which of the structural isomers in Exercise 37 exhibit *cis–trans* isomerism?

39. Draw all the structural and geometrical (*cis–trans*) isomers of C_3H_5Cl.

40. Draw all the structural and geometrical (*cis–trans*) isomers of bromochloropropene.

41. Draw all structural and geometrical (*cis–trans*) isomers of C_4H_7F. Ignore any cyclic isomers.

42. *Cis–trans* isomerism is also possible in molecules with rings. Draw the *cis* and *trans* isomers of 1,2-dimethylcyclohexane. In Exercise 41, you drew all of the noncyclic structural and geometric isomers of C_4H_7F. Now draw the cyclic structural and geometric isomers of C_4H_7F.

43. Draw the following.

a. *cis*-2-hexene

b. *trans*-2-butene

c. *cis*-2,3-dichloro-2-pentene

44. Name the following compounds.

a. 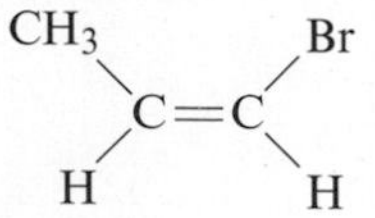 CH₃   Br
       C=C
     H   H

b.   CH₃   CH₂CH₃
            C=C
     CH₃CH₂   CH₂CH₂CH₃

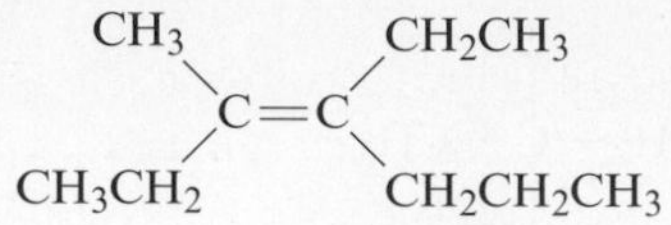

c.   I
          |
     CH₃CHCH₂   H
                  C=C
     CH₃CH₂CH₂   I

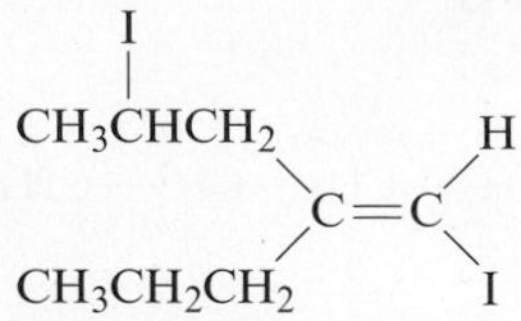

45. If one hydrogen in a hydrocarbon is replaced by a halogen atom, the number of isomers that exist for the substituted compound depends on the number of types of hydrogen in the original hydrocarbon. Thus there is only one form of chloroethane (all hydrogens in ethane are equivalent), but there are two isomers of propane that arise from the substitution of a methyl hydrogen or a methylene hydrogen. How many isomers can be obtained when one hydrogen in each of the compounds named below is replaced by a chlorine atom?

a. *n*-pentane   **c.** 2,4-dimethylpentane

b. 2-methylbutane   **d.** methylcyclobutane

46. There are three isomers of dichlorobenzene, one of which has now replaced naphthalene as the main constituent of mothballs.

a. Identify the *ortho,* the *meta,* and the *para* isomers of dichlorobenzene.

b. Predict the number of isomers for trichlorobenzene.

c. It turns out that the presence of one chlorine atom on a benzene ring will cause the next substituent to add *ortho* or *para* to the first chlorine atom on the benzene ring. What does this tell you about the synthesis of *m*-dichlorobenzene?

d. Which of the isomers of trichlorobenzene will be the hardest to prepare?

Functional Groups

47. Identify each of the following compounds as a carboxylic acid, ester, ketone, aldehyde, or amine.

a. Anthraquinone, an important starting material in the manufacture of dyes:

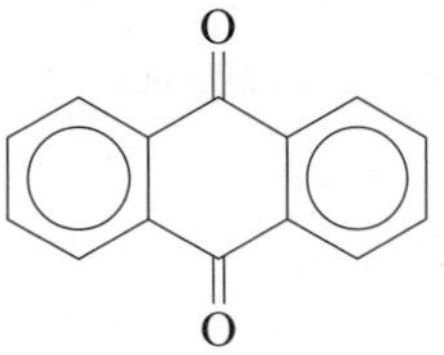

b.

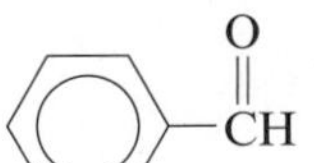

c. 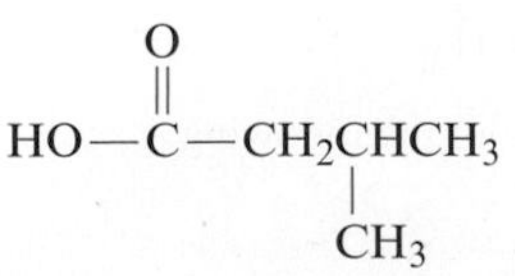   O
          ‖
     HO—C—CH₂CHCH₃
                |
               CH₃

d.

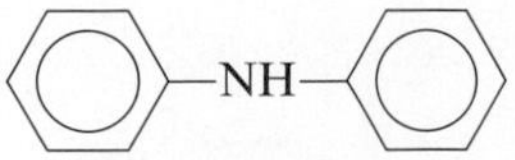

48. Identify the functional groups present in the following compounds.

a.

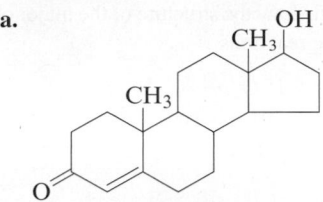

Testosterone

b. 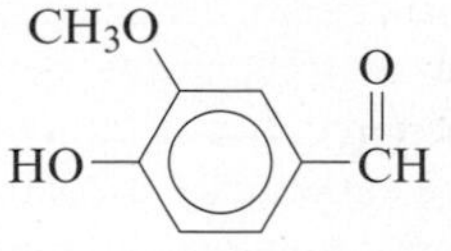 CH₃O

HO— ⬡ —CH with O double bond

Vanillin

c.

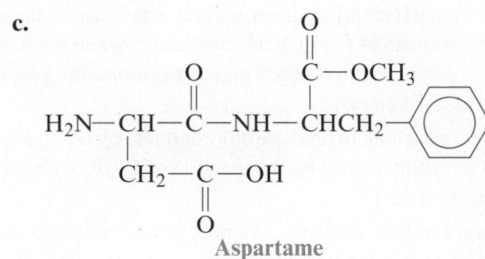

Aspartame

49. Mimosine is a natural product found in large quantities in the seeds and foliage of some legume plants and has been shown to cause inhibition of hair growth and hair loss in mice.

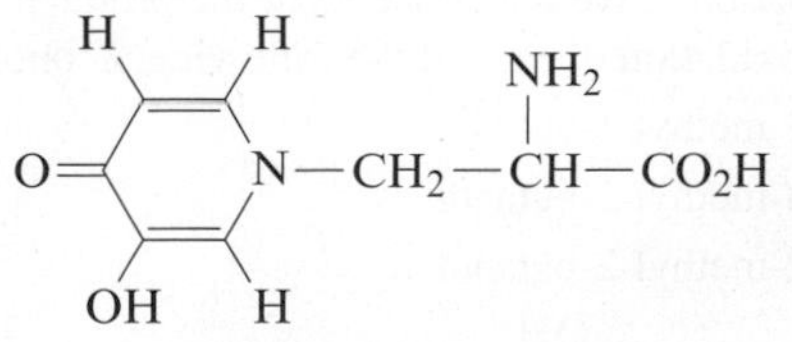

$$\text{Mimosine, } C_8H_{10}N_2O_4$$

a. What functional groups are present in mimosine?

b. Give the hybridization of the eight carbon atoms in mimosine.

c. How many σ and π bonds are found in mimosine?

50. Minoxidil ($C_9H_{15}N_5O$) is a compound produced by Pharmacia Company that has been approved as a treatment of some types of male pattern baldness.

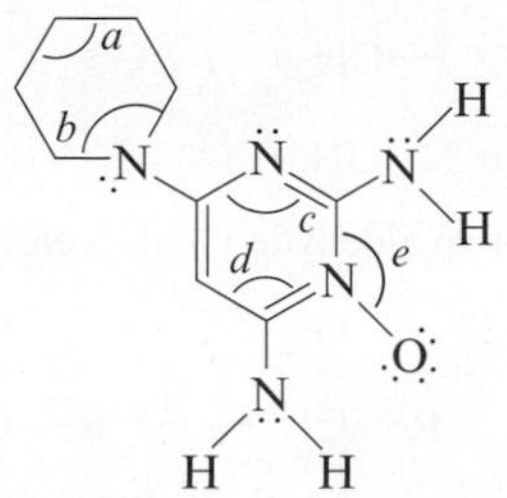

a. Would minoxidil be more soluble in acidic or basic aqueous solution? Explain.

b. Give the hybridization of the five nitrogen atoms in minoxidil.

c. Give the hybridization of each of the nine carbon atoms in minoxidil.

d. Give approximate values of the bond angles marked *a, b, c, d,* and *e.*

e. Including all the hydrogen atoms, how many σ bonds exist in minoxidil?

f. How many π bonds exist in minoxidil?

51. For each of the following alcohols, give the systematic name and specify whether the alcohol is primary, secondary, or tertiary.

a. 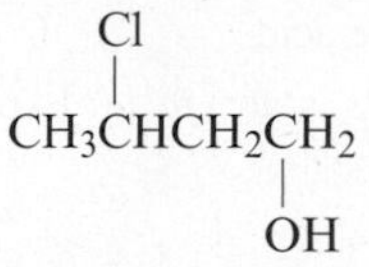   Cl
         |
CH₃CHCH₂CH₂
            |
            OH

b. 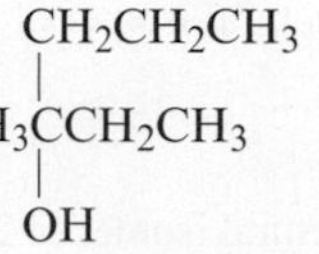   CH₂CH₂CH₃
         |
CH₃CCH₂CH₃
   |
   OH

c. 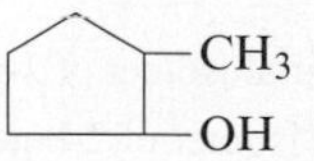   ⬠—CH₃
        —OH

52. Draw structural formulas for each of the following alcohols. Indicate whether the alcohol is primary, secondary, or tertiary.

a. 1-butanol   **c.** 2-methyl-1-butanol
b. 2-butanol   **d.** 2-methyl-2-butanol

53. Name all the alcohols that have the formula $C_5H_{12}O$. How many ethers have the formula $C_5H_{12}O$?

54. Name all the aldehydes and ketones that have the formula $C_5H_{10}O$.

55. Name the following compounds.

a. 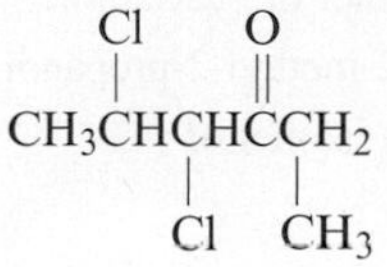   Cl   O
         | ||
CH₃CHCHCCH₂
         | |
         Cl   CH₃

b. 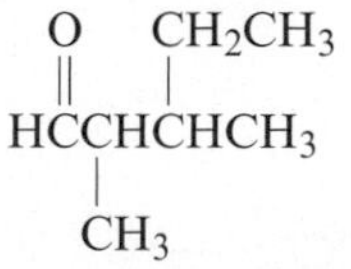   O   CH₂CH₃
        || |
HCCHCHCH₃
            |
            CH₃

c. 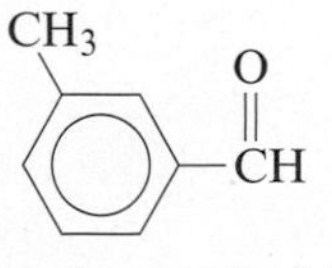 CH₃
       ⬡—CH with O double bond

56. Draw the structural formula for each of the following.

a. formaldehyde (methanal)
b. 4-heptanone
c. 3-chlorobutanal
d. 5,5-dimethyl-2-hexanone

57. Name the following compounds.

a.

Cl—⬡—C—OH with O double bond

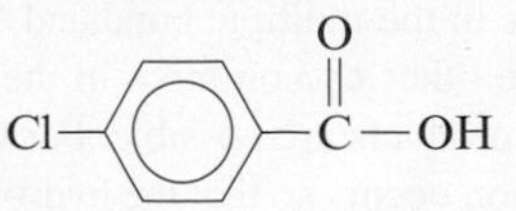

b.

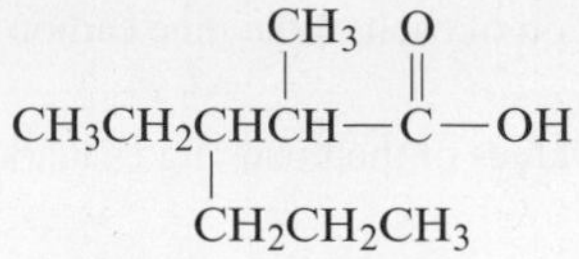

$$CH_3CH_2CHCH-\overset{\overset{\displaystyle O}{\|}}{C}-OH$$

with CH_3 on the CH and $CH_2CH_2CH_3$ below

c. HCOOH

58. Draw a structural formula for each of the following.

  a. 3-methylpentanoic acid

  b. ethyl methanoate

  c. methyl benzoate

  d. 3-chloro-2,4-dimethylhexanoic acid

59. Which of the following statements is(are) *false*? Explain why the statement(s) is(are) *false*.

  a. 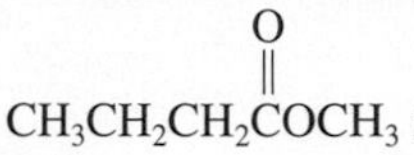 $CH_3CH_2CH_2\overset{\overset{\displaystyle O}{\|}}{C}OCH_3$ is a structural isomer of pentanoic acid.

  b. 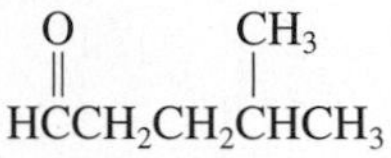 $HCCH_2CH_2CHCH_3$ (with O double bonded to first C and CH_3 on fourth C) is a structural isomer of 2-methyl-3-pentanone.

  c. $CH_3CH_2OCH_2CH_2CH_3$ is a structural isomer of 2-pentanol.

  d. 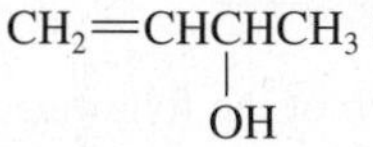 $CH_2{=}CHCHCH_3$ with OH below is a structural isomer of 2-butenal.

  e. Trimethylamine is a structural isomer of $CH_3CH_2CH_2NH_2$.

60. Draw the isomer(s) specified. There may be more than one possible isomer for each part.

  a. a cyclic compound that is an isomer of *trans*-2-butene

  b. an ester that is an isomer of propanoic acid

  c. a ketone that is an isomer of butanal

  d. a secondary amine that is an isomer of butylamine

  e. a tertiary amine that is an isomer of butylamine

  f. an ether that is an isomer of 2-methyl-2-propanol

  g. a secondary alcohol that is an isomer of 2-methyl-2-propanol

Reactions of Organic Compounds

61. Complete the following reactions.

  a. $CH_3CH{=}CHCH_3 + H_2 \xrightarrow{Pt}$

  b. 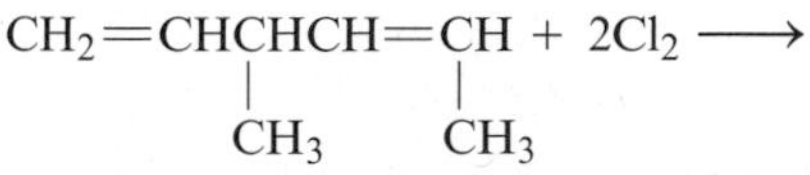 $CH_2{=}CHCHCH{=}CH + 2Cl_2 \longrightarrow$ (with CH_3 and CH_3 substituents)

  c. 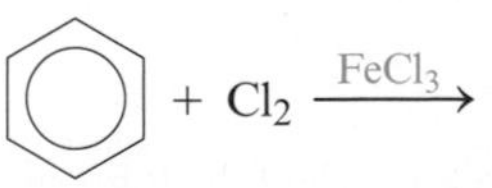 benzene ring $+ Cl_2 \xrightarrow{FeCl_3}$

  d. 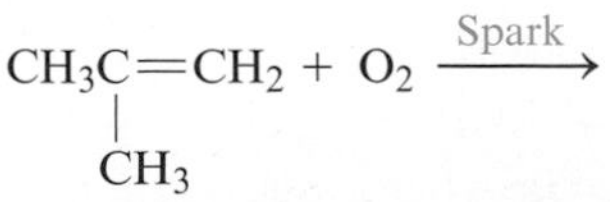 $CH_3C{=}CH_2 + O_2 \xrightarrow{Spark}$ (with CH_3 substituent)

62. Reagents such as HCl, HBr, and HOH (H_2O) can add across carbon–carbon double and triple bonds, with H forming a bond to one of the carbon atoms in the multiple bond and Cl, Br, or OH forming a bond to the other carbon atom in the multiple bond. In some cases, two products are possible. For the major organic product, the addition occurs so that the hydrogen atom

in the reagent attaches to the carbon atom in the multiple bond that already has the greater number of hydrogen atoms bonded to it. With this rule in mind, draw the structure of the major product in each of the following reactions.

  a. $CH_3CH_2CH{=}CH_2 + H_2O \xrightarrow{H^+}$

  b. $CH_3CH_2CH{=}CH_2 + HBr \longrightarrow$

  c. $CH_3CH_2C{\equiv}CH + 2HBr \longrightarrow$

  d. 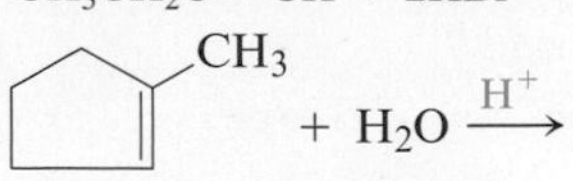 cyclohexene with CH_3 substituent $+ H_2O \xrightarrow{H^+}$

  e. 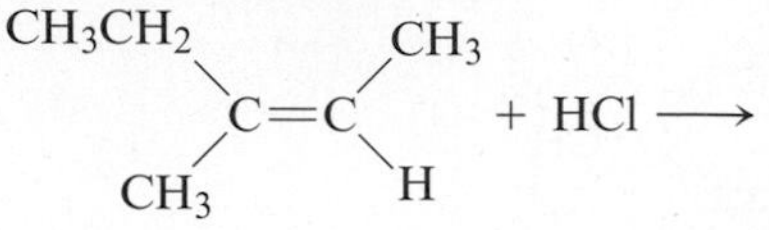 $\underset{CH_3}{\overset{CH_3CH_2}{C}}{=}\underset{H}{\overset{CH_3}{C}} + HCl \longrightarrow$

63. When toluene ($C_6H_5CH_3$) reacts with chlorine gas in the presence of iron(III) catalyst, the product is a mixture of the *ortho* and *para* isomers of $C_6H_4ClCH_3$. However, when the reaction is light-catalyzed with no Fe^{3+} catalyst present, the product is $C_6H_5CH_2Cl$. Explain.

64. Why is it preferable to produce chloroethane by the reaction of $HCl(g)$ with ethene than by the reaction of $Cl_2(g)$ with ethane? (See Exercise 62.)

65. Using appropriate reactants, alcohols can be oxidized into aldehydes, ketones, and/or carboxylic acids. Primary alcohols can be oxidized into aldehydes, which can then be oxidized into carboxylic acids. Secondary alcohols can be oxidized into ketones, while tertiary alcohols do not undergo this type of oxidation. Give the structure of the product(s) resulting from the oxidation of each of the following alcohols.

  a. 3-methyl-1-butanol

  b. 3-methyl-2-butanol

  c. 2-methyl-2-butanol

  d. 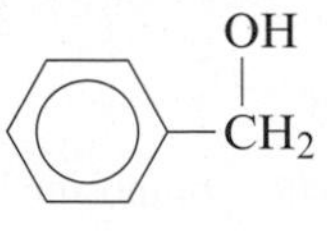 benzene ring with $\overset{OH}{\underset{}{CH_2}}$

  e. 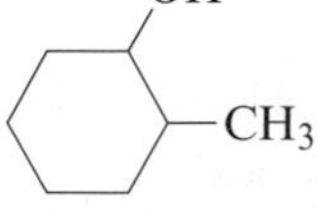 cyclohexane with OH and CH_3

  f. 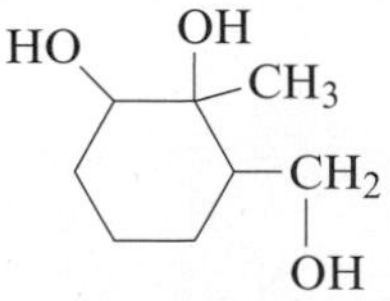 cyclohexane with HO, OH, CH_3, CH_2, OH substituents

66. Oxidation of an aldehyde yields a carboxylic acid:

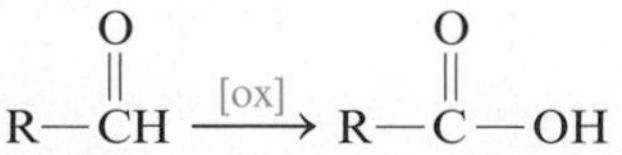

$$R-\overset{\overset{\displaystyle O}{\|}}{C}H \xrightarrow{[ox]} R-\overset{\overset{\displaystyle O}{\|}}{C}-OH$$

Draw the structures for the products of the following oxidation reactions.

  a. propanal $\xrightarrow{[ox]}$

  b. 2,3-dimethylpentanal $\xrightarrow{[ox]}$

  c. 3-ethylbenzaldehyde $\xrightarrow{[ox]}$

67. How would you synthesize each of the following?

    a. 1,2-dibromopropane from propene

    b. acetone (2-propanone) from an alcohol

    c. *tert*-butyl alcohol (2-methyl-2-propanol) from an alkene (See Exercise 62.)

    d. propanoic acid from an alcohol

68. What tests could you perform to distinguish between the following pairs of compounds?

    a. $CH_3CH_2CH_2CH_3$, CH_2=$CHCH_2CH_3$

    b.

    $CH_3CH_2CH_2COOH$, $CH_3CH_2\overset{\overset{\displaystyle O}{\|}}{C}CH_3$

    c.

    $CH_3CH_2CH_2OH$, $CH_3\overset{\overset{\displaystyle O}{\|}}{C}CH_3$

    d. $CH_3CH_2NH_2$, CH_3OCH_3

69. How would you synthesize the following esters?

    a. *n*-octylacetate

    b.

    $CH_3CH_2CH_2CH_2CH_2CH_2O-\overset{\overset{\displaystyle O}{\|}}{C}CH_2CH_3$

70. Complete the following reactions.

    a. $CH_3CO_2H + CH_3OH \rightarrow$

    b. $CH_3CH_2CH_2OH + HCOOH \rightarrow$

Polymers

71. Kel-F is a polymer with the structure

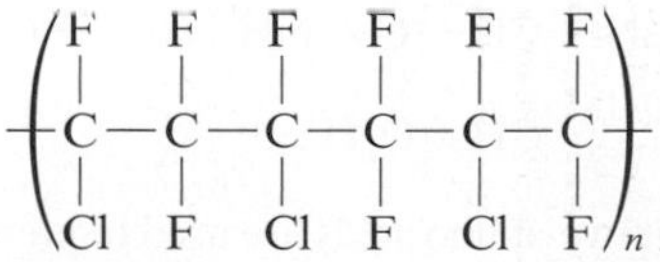

    What is the monomer for Kel-F?

72. What monomer(s) must be used to produce the following polymers?

    a.

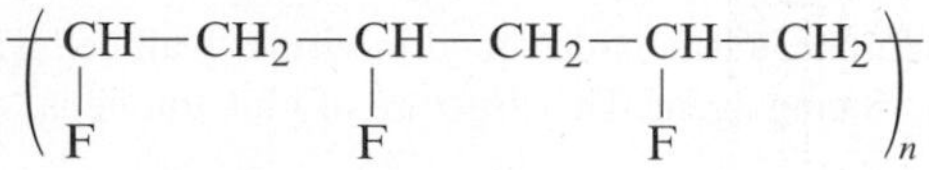

    b.

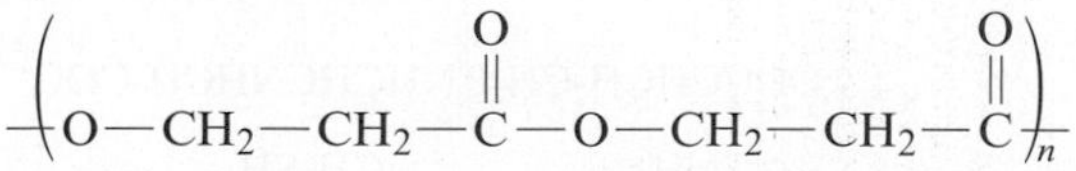

    c.

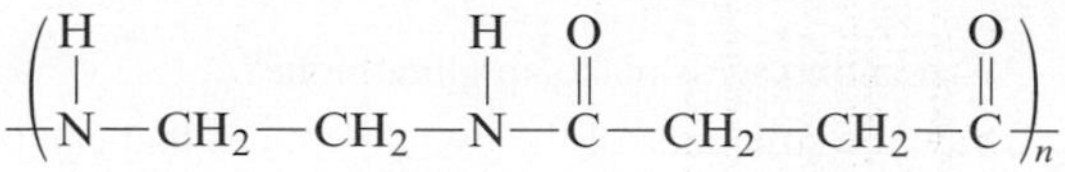

    d.

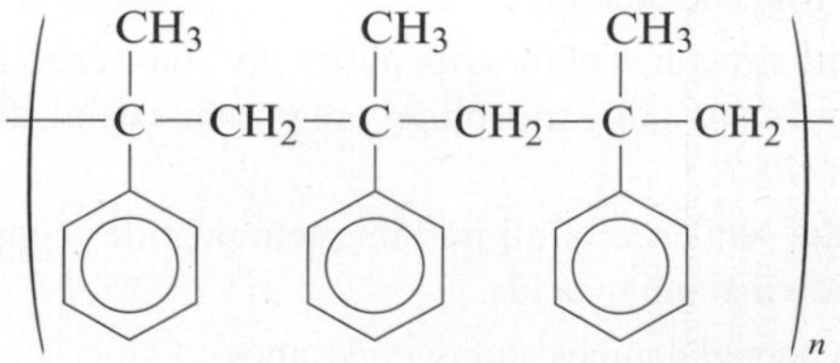

    e.

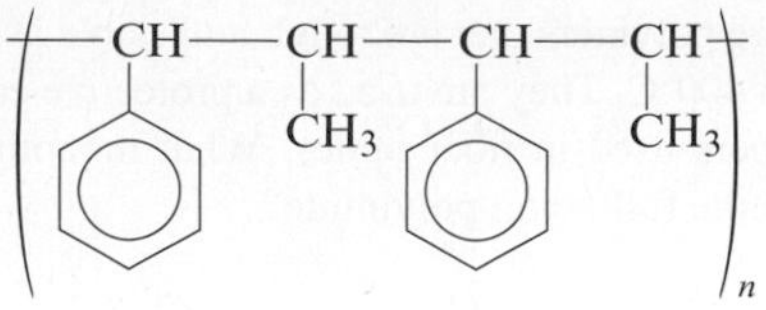

    f.

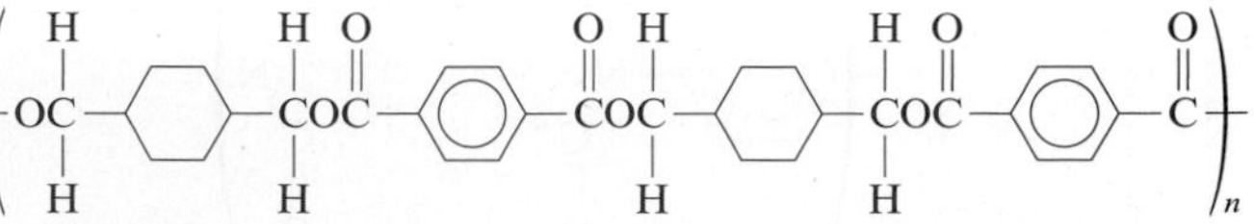

    (This polymer is Kodel, used to make fibers of stain-resistant carpeting.)

    Classify these polymers as condensation or addition polymers. Which are copolymers?

73. "Super glue" contains methyl cyanoacrylate,

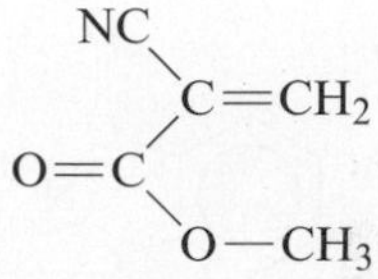

    which readily polymerizes upon exposure to traces of water or alcohols on the surfaces to be bonded together. The polymer provides a strong bond between the two surfaces. Draw the structure of the polymer formed by methyl cyanoacrylate.

74. Isoprene is the repeating unit in natural rubber. The structure of isoprene is

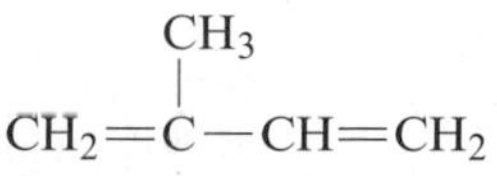

    a. Give a systematic name for isoprene.

    b. When isoprene is polymerized, two polymers of the form

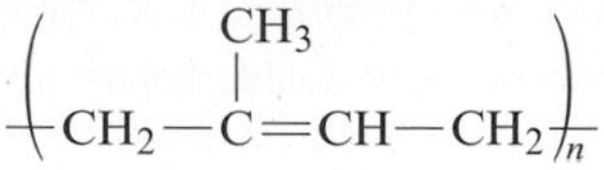

    are possible. In natural rubber, the *cis* configuration is found. The polymer with the *trans* configuration about the double bond is called gutta percha and was once used in the manufacture of golf balls. Draw the structure of natural rubber and gutta percha showing three repeating units and the configuration about the carbon–carbon double bonds.

75. Kevlar, used in bulletproof vests, is made by the condensation copolymerization of the monomers

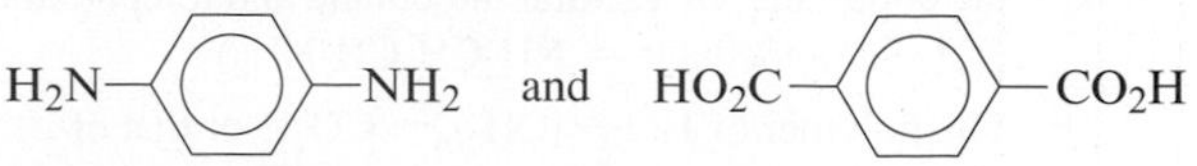

    Draw the structure of a portion of the Kevlar chain.

76. The polyester formed from lactic acid,

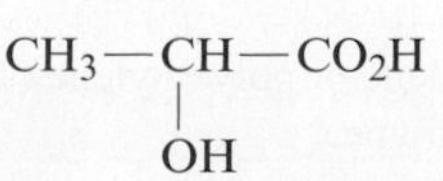

    is used for tissue implants and surgical sutures that will dissolve in the body. Draw the structure of a portion of this polymer.

77. Polyimides are polymers that are tough and stable at temperatures of up to 400°C. They are used as a protective coating on the quartz fibers used in fiber optics. What monomers were used to make the following polyimide?

78. The Amoco Chemical Company has successfully raced a car with a plastic engine. Many of the engine parts, including piston skirts, connecting rods, and valve-train components, were made of a polymer called *Torlon:*

What monomers are used to make this polymer?

79. Polystyrene can be made more rigid by copolymerizing styrene with divinylbenzene:

How does the divinylbenzene make the copolymer more rigid?

80. Polyesters containing double bonds are often crosslinked by reacting the polymer with styrene.

a. Draw the structure of the copolymer of

$$HO-CH_2CH_2-OH \quad and \quad HO_2C-CH=CH-CO_2H$$

b. Draw the structure of the crosslinked polymer (after the polyester has been reacted with styrene).

81. Which of the following polymers would be stronger or more rigid? Explain your choices.

a. The copolymer of ethylene glycol and terephthalic acid or the copolymer of 1,2-diaminoethane and terephthalic acid (1,2-diaminoethane = $NH_2CH_2CH_2NH_2$)

b. The polymer of $HO-(CH_2)_6-CO_2H$ or that of

c. Polyacetylene or polyethylene (The monomer in polyacetylene is ethyne.)

82. Poly(lauryl methacrylate) is used as an additive in motor oils to counter the loss of viscosity at high temperature. The structure is

The long hydrocarbon chain of poly(lauryl methacrylate) makes the polymer soluble in oil (a mixture of hydrocarbons with mostly 12 or more carbon atoms). At low temperatures the polymer is coiled into balls. At higher temperatures the balls uncoil and the polymer exists as long chains. Explain how this helps control the viscosity of oil.

Natural Polymers

83. Which of the amino acids in Fig. 21-18 contain the following functional groups in their R group?

a. alcohol   **c.** amine

b. carboxylic acid   **d.** amide

84. When pure crystalline amino acids are heated, decomposition generally occurs before the solid melts. Account for this observation. (*Hint:* Crystalline amino acids exist as $H_3\overset{+}{N}CRHCOO^-$, called *zwitterions.*)

85. Aspartame, the artificial sweetener marketed under the name NutraSweet, is a methyl ester of a dipeptide:

a. What two amino acids are used to prepare aspartame?

b. There is concern that methanol may be produced by the decomposition of aspartame. From what portion of the molecule can methanol be produced? Write an equation for this reaction.

86. Glutathione, a tripeptide found in virtually all cells, functions as a reducing agent. The structure of glutathione is

What amino acids make up glutathione?

87. Draw the structures of the two dipeptides that can be formed from serine and alanine.

88. Draw the structures of the tripeptides gly–ala–ser and ser–ala–gly. How many other tripeptides are possible using these three amino acids?

89. Write the sequence of all possible tetrapeptides composed of the following amino acids.

a. two phenylalanines and two glycines

b. two phenylalanines, glycine, and alanine

90. How many different pentapeptides can be formed using five different amino acids?

91. Give an example of amino acids that could give rise to the interactions pictured in Fig. 21-24 that maintain the tertiary structures of proteins.

92. What types of interactions can occur between the side chains of the following amino acids that would help maintain the tertiary structure of a protein?

  a. cysteine and cysteine
  b. glutamine and serine
  c. glutamic acid and lysine
  d. proline and leucine

93. Oxygen is carried from the lungs to tissues by the protein hemoglobin in red blood cells. Sickle cell anemia is a disease resulting from abnormal hemoglobin molecules in which a valine is substituted for a single glutamic acid in normal hemoglobin. How might this substitution affect the structure of hemoglobin?

94. Over 100 different kinds of mutant hemoglobin molecules have been detected in humans. Unlike sickle cell anemia (see Exercise 93), not all of these mutations arc as serious. In one nonlethal mutation, glutamine substitutes for a single glutamic acid in normal hemoglobin. Rationalize why this substitution is nonlethal.

95. Draw cyclic structures for D-ribose and D-mannose.

96. Indicate the chiral carbon atoms found in the monosaccharides D-ribose and D-mannose.

97. In addition to using *numerical* prefixes in the general names of sugars to indicate how many carbon atoms are present, we often use the prefixes *keto-* and *aldo-* to indicate whether the sugar is a ketone or an aldehyde. For example, the monosaccharide fructose is frequently called a *ketohexose* to emphasize that it contains six carbons as well as the ketone functional group. For each of the monosaccharides shown in Table 21-8 classify the sugars as aldohexoses, aldopentoses, ketohexoses, or ketopentoses.

98. Glucose can occur in three forms: two cyclic forms and one open-chain structure. In aqueous solution, only a tiny fraction of the glucose is in the open-chain form. Yet tests for the presence of glucose depend on reaction with the aldehyde group, which is found only in the open-chain form. Explain why these tests work.

99. What are the structural differences between α- and β-glucose? These two cyclic forms of glucose are the building blocks to form two different polymers. Explain.

100. Cows can digest cellulose, but humans can't. Why not?

101. Which of the amino acids in Fig. 21-18 contain more than one chiral carbon atom? Draw the structures of these amino acids and indicate all chiral carbon atoms.

102. Why is glycine not optically active?

103. Which of the noncyclic isomers of bromochloropropene are optically active?

104. How many chiral carbon atoms does the following structure have?

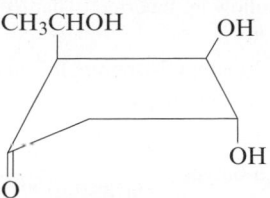

105. Part of a certain DNA sequence is G–G–T–C–T–A–T–A–C. What is the complementary sequence?

106. The codons (words) in DNA (that specify which amino acid should be at a particular point in a protein) are three bases long. How many such three-letter words can be made from the four bases adenine, cytosine, guanine, and thymine?

107. Which base will hydrogen-bond with uracil within an RNA molecule? Draw the structure of this base pair.

108. Tautomers are molecules that differ in the position of a hydrogen atom. A tautomeric form of thymine has the structure

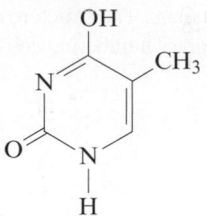

If this tautomeric form, rather than the stable form of thymine, were present in a strand of DNA during replication, what would be the result?

109. The base sequences in mRNA that code for certain amino acids are

Glu: GAA, GAG
Val: GUU, GUC, GUA, GUG
Met: AUG
Trp: UGG
Phe: UUU, UUC
Asp: GAU, GAC

These sequences are complementary to the sequences in DNA.

  a. Give the corresponding sequences in DNA for the amino acids listed above.
  b. Give a DNA sequence that would code for the peptide trp–glu–phe–met.
  c. How many different DNA sequences can code for the tetrapeptide in part b?
  d. What is the peptide that is produced from the DNA sequence T–A–C–C–T–G–A–A–G?
  e. What other DNA sequences would yield the same tripeptide as in part d?

110. The change of a single base in the DNA sequence for normal hemoglobin can encode for the abnormal hemoglobin giving rise to sickle cell anemia. Which base in the codon for glu in DNA is replaced to give the codon(s) for val? (See Exercises 93 and 109.)

Additional Exercises

111. Draw the following incorrectly named compounds and name them correctly.

  a. 2-ethyl-3-methyl-5-isopropylhexane

  b. 2-ethyl-4-*tert*-butylpentane

  c. 3-methyl-4-isopropylpentane

  d. 2-ethyl-3-butyne

112. In the presence of light, chlorine can substitute for one (or more) of the hydrogens in an alkane. For the following reactions, draw the possible monochlorination products.

  a. 2,2-dimethylpropane + Cl_2 $\xrightarrow{h\nu}$

  b. 1,3-dimethylcyclobutane + Cl_2 $\xrightarrow{h\nu}$

  c. 2,3-dimethylbutane + Cl_2 $\xrightarrow{h\nu}$

113. Polychlorinated dibenzo-*p*-dioxins (PCDDs) are highly toxic substances that are present in trace amounts as by-products of some chemical manufacturing processes. They have been implicated in a number of environmental incidents—for example, the chemical contamination at Love Canal and the herbicide spraying in Vietnam. The structure of dibenzo-*p*-dioxin, along with the customary numbering convention, is

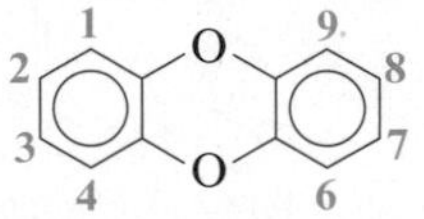

The most toxic PCDD is 2,3,7,8-tetrachloro-dibenzo-*p*-dioxin. Draw the structure of this compound. Also draw the structures of two other isomers containing four chlorine atoms.

114. Consider the following five compounds.

  a. $CH_3CH_2CH_2CH_2CH_3$

  b.
$$CH_3CH_2CH_2CH_2\overset{\displaystyle OH}{\underset{\displaystyle |}{}}$$

  c. $CH_3CH_2CH_2CH_2CH_2CH_3$

  d.
$$CH_3CH_2CH_2\overset{\displaystyle O}{\overset{\displaystyle \|}{C}}H$$

  e.
$$CH_3\overset{\displaystyle CH_3}{\underset{\displaystyle CH_3}{\overset{\displaystyle |}{\underset{\displaystyle |}{C}}}}CH_3$$

The boiling points of these five compounds are 9.5°C, 36°C, 69°C, 76°C, and 117°C. Which compound boils at 36°C? Explain.

115. The two isomers having the formula C_2H_6O boil at −23°C and 78.5°C. Draw the structure of the isomer that boils at −23°C and of the isomer that boils at 78.5°C.

116. Ignoring ring compounds, which isomer of $C_2H_4O_2$ should boil at the lowest temperature?

117. Explain why methyl alcohol is soluble in water in all proportions, while stearyl alcohol [$CH_3(CH_2)_{16}OH$] is a waxy solid that is not soluble in water.

118. Is octanoic acid more soluble in 1 *M* HCl, 1 *M* NaOH, or pure water? Explain. Drugs such as morphine ($C_{17}H_{19}NO_3$) are often treated with strong acids. The most commonly used form of morphine is morphine hydrochloride ($C_{17}H_{20}ClNO_3$). Why is morphine treated in this way? (*Hint:* Morphine is an amine.)

119. Consider the compounds butanoic acid, pentanal, *n*-hexane, and 1-pentanol. The boiling points of these compounds (in no specific order) are 69°C, 103°C, 137°C, and 164°C. Match the boiling points to the correct compound.

120. A compound containing only carbon and hydrogen is 85.63% C by mass. Reaction of this compound with H_2O produces a secondary alcohol as the major product and a primary alcohol as the minor product. (See Exercise 62.) If the molar mass of the hydrocarbon is between 50 and 60 g/mol, name the compound.

121. Three different organic compounds have the formula C_3H_8O. Only two of these isomers react with $KMnO_4$ (a strong oxidizing agent). What are the names of the products when these isomers react with excess $KMnO_4$?

122. Consider the following polymer:

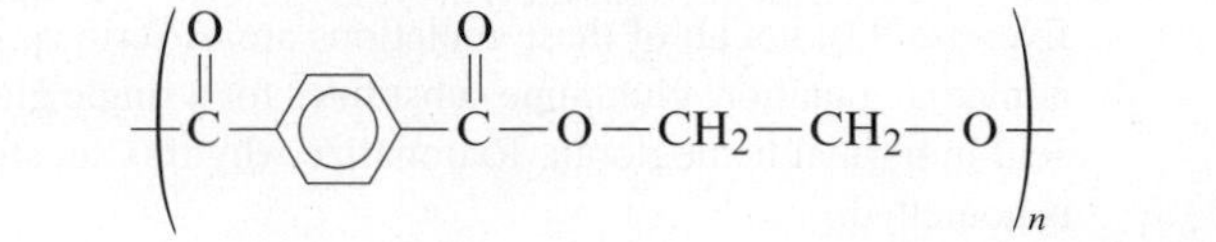

Is this polymer a homopolymer or a copolymer, and is it formed by addition polymerization or condensation polymerization? What is (are) the monomer(s) for this polymer?

123. Nylon is named according to the number of C atoms between the N atoms in the chain. Nylon-46 has 4 C atoms, then 6 C atoms, and this pattern repeats. Nylon-6 always has 6 carbon atoms in a row. Speculate as to why nylon-46 is stronger than nylon-6. (*Hint:* Consider the strengths of interchain forces.)

124. The polymer nitrile is a copolymer made from acrylonitrile and butadiene; it is used to make automotive hoses and gaskets. Draw the structure of nitrile. (*Hint:* See Table 21-7.)

125. *Polyaramid* is a term applied to polyamides containing aromatic groups. These polymers were originally made for use as tire cords but have since found many other uses.

  a. Kevlar is used in bulletproof vests and many high-strength composites. The structure of Kevlar is

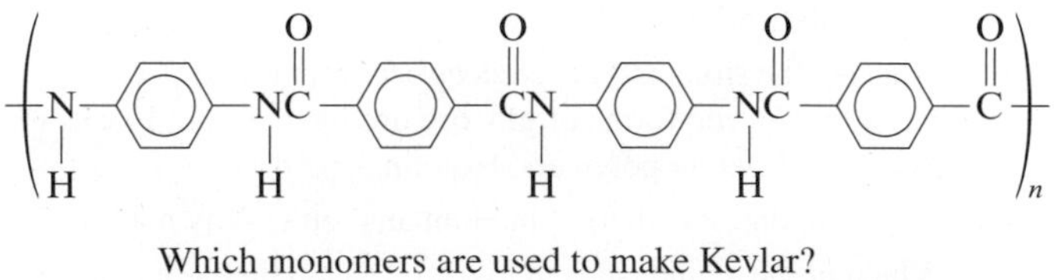

Which monomers are used to make Kevlar?

  b. Nomex is a polyaramid used in fire-resistant clothing. It is a copolymer of

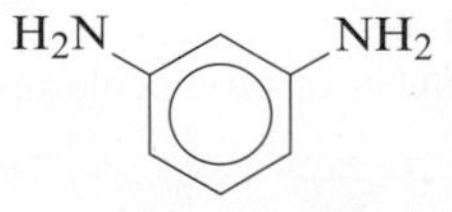

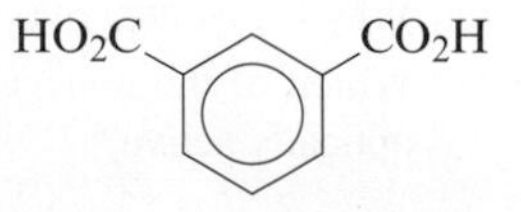

Draw the structure of the Nomex polymer. How do Kevlar and Nomex differ in their structures?

126. When acrylic polymers are burned, toxic fumes are produced. For example, in many airplane fires, more passenger deaths have been caused by breathing toxic fumes than by the fire it-self. Using polyacrylonitrile as an example, what would you expect to be one of the most toxic, gaseous combustion prod-ucts created in the reaction?

127. Ethylene oxide,

$$CH_2 \text{—} CH_2$$
$$\diagdown O \diagup$$

is an important industrial chemical. Although most ethers are unreactive, ethylene oxide is quite reactive. It resembles C_2H_4 in its reactions in that addition reactions occur across the C—O bond in ethylene oxide.

 a. Why is ethylene oxide so reactive? (*Hint:* Consider the bond angles in ethylene oxide as compared with those predicted by the VSEPR model.)

 b. Ethylene oxide undergoes addition polymerization, form-ing a polymer used in many applications requiring a non-ionic surfactant. Draw the structure of this polymer.

128. Another way of producing highly crosslinked polyesters is to use glycerol. Alkyd resins are a polymer of this type. The poly-mer forms very tough coatings when baked onto a surface and is used in paints for automobiles and large appliances. Draw the structure of the polymer formed from the condensation of

$$CH_2\text{—}CH\text{—}CH_2 \text{ and}$$
$$\quad| \qquad | \qquad |$$
$$OH \quad OH \quad OH$$
Glycerol

Phthalic acid (with CO_2H and CO_2H on benzene ring)

Explain how crosslinking occurs in this polymer.

129. Monosodium glutamate (MSG) is commonly used as a flavor-ing in foods. Draw the structure of MSG.

130. a. Use bond energies (see Table 3-3) to estimate ΔH for the reaction of two molecules of glycine to form a peptide linkage.

 b. Would you predict ΔS to favor the formation of peptide linkages between two molecules of glycine?

 c. Would you predict the formation of proteins to be a spon-taneous process?

131. The reaction to form a phosphate–ester linkage between two nucleotides can be approximated as follows:

$$Sugar\text{—}O\text{—}P\text{—}OH + HO\text{—}CH_2\text{—}sugar \longrightarrow$$

$$Sugar\text{—}O\text{—}P\text{—}O\text{—}CH_2\text{—}sugar + H_2O$$

Would you predict the formation of a dinucleotide from two nucleotides to be a spontaneous process?

132. Considering your answers to Exercises 130 and 131, how can you justify the existence of proteins and nucleic acids in light of the second law of thermodynamics?

133. All amino acids have at least two functional groups with acidic or basic properties. In alanine, the carboxylic acid group has $K_a = 4.5 \times 10^{-3}$ and the amino group has $K_b = 7.4 \times 10^{-5}$. Three ions of alanine are possible when alanine is dissolved in water. Which of these ions would predominate in a solution with $[H^+] = 1.0\ M$? In a solution with $[OH^-] = 1.0\ M$?

134. The average molar mass of one base pair of nucleotides in DNA is approximately 600 g/mol. The spacing between suc-cessive base pairs is about 0.34 nm, and a complete turn in the helical structure of DNA occurs about every 3.4 nm. If a DNA molecule has a molar mass of 4.5×10^9 g/mol, approximately how many complete turns exist in the DNA α-helix structure?

135. When heat is added to proteins, the hydrogen bonding in the secondary structure is disrupted. What are the algebraic signs of ΔH and ΔS for the denaturation process?

136. In glycine, the carboxylic acid group has $K_a = 4.3 \times 10^{-3}$ and the amino group has $K_b = 6.0 \times 10^{-5}$. Use these equilibrium constant values to calculate the equilibrium constants for the following.

 a. $^+H_3NCH_2CO_2^- + H_2O \rightleftharpoons H_2NCH_2CO_2^- + H_3O^+$

 b. $H_2NCH_2CO_2^- + H_2O \rightleftharpoons H_2NCH_2CO_2H + OH^-$

 c. $^+H_3NCH_2CO_2H \rightleftharpoons 2H^+ + H_2NCH_2CO_2^-$

ChemWork Problems

These multiconcept problems (and additional ones) are found interac-tively online with the same type of assistance a student would get from an instructor.

137. Name each of the following alkanes.

 a. $CH_3\text{—}CH_2\text{—}CH_2\text{—}CH_2\text{—}CH_3$

 b. $CH_3\text{—}CH\text{—}CH\text{—}CH_2\text{—}CH$ with CH_3 groups and $CH_2\text{—}CH_3$ branches

 c. $CH_3\text{—}CH_2\text{—}CH_2\text{—}CH\text{—}CH\text{—}CH_2\text{—}CH_2\text{—}CH_3$ with $CH\text{—}CH_3 / CH_3$ and $CH_3\text{—}CH_2$ branches

138. Name each of the following alkanes.

 a. $CH_3\text{—}C\text{—}CH\text{—}CH_2\text{—}CH_3$ with Br, Br, CH_3

 b. $CH_3\text{—}CH\text{—}CH\text{—}CH$ with CH_3, I, $CH_2\text{—}CH_3$, $CH_2\text{—}CH_3$

 c. $CH_3\text{—}CH\text{—}CH_2\text{—}CH\text{—}CH_2\text{—}CH_2\text{—}CH_2\text{—}CH_2\text{—}CH_3$ with CH_3 and F

139. Name each of the following cyclic alkanes.

a. (cyclopropane with I substituent)

b. (cyclohexane with CH₃, CH₃, Cl, CH₃ substituents)

c. Cl (cyclopentane with CH₂CH₃ substituent)

d. (cyclopentene with CH₃ and CH₂CH₃ substituents)

140. Name each of the following alkenes and alkynes.

a. $CH_3CH_2-\underset{\underset{CH_3}{|}}{C}=CH_2$

b. $CH_2=\underset{\underset{CH_3}{|}}{C}-CH_2-\underset{\underset{CH_3}{|}}{C}=CH_2$

c. $CH_3CH_2CH-CH=CH-CH_2\underset{\underset{CH_3}{|}}{CH}$
      $\underset{CH_2CH_3}{|}$ $\underset{CH_3}{|}$

d. $CH_3CH_2CH_2CH_2\underset{\underset{Br}{|}}{CH}-C\equiv CH$

e. $CH_3-\underset{\underset{CH_2CH_2}{|}}{\overset{\overset{CH_3}{|}}{C}}-C\equiv C-\underset{\underset{CH_3}{|}}{\overset{\overset{CH_3}{|}}{CH}}$
      $\underset{Cl}{|}$

f. $HC\equiv C-\underset{\underset{CH_3}{|}}{CH}-\underset{\underset{CH_2CH_2CH_2CH_3}{|}}{\overset{\overset{CH_2CH_3}{|}}{CH}}$

141. **a.** Name each of the following alcohols.

$HO-CH_2CH_2CH_2CH_2CH_3$

$HO-\underset{\underset{CH_3}{|}}{\overset{\overset{CH_3}{|}}{C}}-CH_2-\underset{\underset{CH_3}{|}}{\overset{\overset{OH}{|}}{CH}}$

b. Name each of the following alcohols, including the stereochemistry if *cis–trans* isomers are possible.

(cyclohexane with two OH groups)

$CH_2=CH-\underset{\underset{OH}{|}}{CH}-\underset{\underset{CH_3}{|}}{\overset{\overset{CH_3}{|}}{CH}}$

142. Name each of the following organic compounds.

a. $CH_3CH_2CH_2CH_2CH_2CH_2\overset{\overset{O}{\|}}{C}H$

b. $CH_3\underset{\underset{CH_3}{|}}{CH}CH_2CHCH_2\overset{\overset{O}{\|}}{C}H$
      $\overset{\overset{CH_3CH_2}{|}}{}$

c. $CH_3CH_2\overset{\overset{O}{\|}}{C}CH_2CH_2CH_3$

d. $CH_3CH_2\overset{\overset{O}{\|}}{C}\underset{\underset{CH_3}{|}}{CH}-\underset{\underset{CH_3}{|}}{\overset{\overset{CH_3}{|}}{CH}}$

e. $CH_3CH_2\underset{\underset{CH_3}{|}}{CH}CHCH_2\overset{\overset{O}{\|}}{C}OH$
      $\overset{\overset{CH_3}{|}}{}$

143. Esterification reactions are carried out in the presence of a strong acid such as H_2SO_4. A carboxylic acid is warmed with an alcohol, and an ester and water are formed. You may have made a fruity-smelling ester in the lab when studying organic functional groups. Name the carboxylic acid that is necessary to complete the following esterification reaction.

$? \quad + \quad HO-\underset{\underset{CH_3}{|}}{\overset{\overset{CH_3}{|}}{C}}-CH_3 \xrightarrow{\text{conc. acid}} CH_3-\underset{\underset{CH_3}{|}}{CH}-\overset{\overset{O}{\|}}{C}-O-\underset{\underset{CH_3}{|}}{\overset{\overset{CH_3}{|}}{C}}-CH_3$
          $\overset{\overset{Cl}{|}}{}$

$+ H_2O$

144. Rank these organic compounds in terms of increasing water solubility (from least water soluble to most water soluble).

$CH_3CH_2CH_2\overset{\overset{O}{\|}}{C}OH$

A

$CH_3CH_2CH=\underset{\underset{CH_3}{|}}{\overset{\overset{CH_3}{|}}{C}}$

B

$CH_3CH_2CH_2\overset{\overset{O}{\|}}{C}CH_3$

C

Challenge Problems

145. The isoelectric point of an amino acid is the pH at which the molecule has no net charge. For glycine, that point would be the pH at which virtually all glycine molecules are in the form $^+H_3NCH_2CO_2^-$. This form of glycine is amphoteric since it can act as both an acid and a base. If we assume that the principal equilibrium at the isoelectric point has the best acid reacting with the best base present, then the reaction is

$$2\,^+H_3NCH_2CO_2^- \rightleftharpoons H_2NCH_2CO_2^- + \,^+H_3NCH_2CO_2H \quad \text{(i)}$$

Assuming this reaction is the principal equilibrium, then the following relationship must hold true:

$$[H_2NCH_2CO_2^-] = [^+H_3NCH_2CO_2H] \quad \text{(ii)}$$

Use this result and your answer to part c of Exercise 136 to calculate the pH at which equation (ii) is true. This pH will be the isoelectric point of glycine.

146. In 1994 chemists at Texas A & M University reported the synthesis of a non-naturally occurring amino acid (*C & E News*, April 18, 1994, pp. 26–27):

a. To which naturally occurring amino acid is this compound most similar?

b. A tetrapeptide, phe–met–arg–phe—NH₂, is synthesized in the brains of rats addicted to morphine and heroin. (The

$$-NH_2 \text{ indicates that the peptide ends in } -\overset{\overset{\displaystyle O}{\|}}{C}-NH_2$$

instead of —CO₂H.) The TAMU scientists synthesized a similar tetrapeptide, with the synthetic amino acid above replacing one of the original amino acids. Draw a structure for the tetrapeptide containing the synthetic amino acid.

c. Indicate the chiral carbon atoms in the synthetic amino acid.

147. The structure of tartaric acid is

a. Is the form of tartaric acid pictured below optically active? Explain.

(*Note:* The dashed lines show groups behind the plane of the page. The wedges show groups in front of the plane.)

b. Draw the optically active forms of tartaric acid.

148. Consider a sample of a hydrocarbon at 0.959 atm and 298 K. Upon combusting the entire sample in oxygen, you collect a mixture of gaseous carbon dioxide and water vapor at 1.51 atm and 375 K. This mixture has a density of 1.391 g/L and occu-

pies a volume four times as large as that of the pure hydrocarbon. Determine the molecular formula of the hydrocarbon and name it.

149. Mycomycin, a naturally occurring antibiotic produced by the fungus *Nocardia acidophilus,* has the molecular formula $C_{13}H_{10}O_2$ and the systematic name 3,5,7,8-tridecatetraene-10,12-diynoic acid. Draw the structure of mycomycin.

150. Sorbic acid is used to prevent mold and fungus growth in some food products, especially cheeses. The systematic name for sorbic acid is 2,4-hexadienoic acid. Draw structures for the four geometrical isomers of sorbic acid.

151. Consider the following reactions. For parts b–d, see Exercise 62.

a. When C_5H_{12} is reacted with $Cl_2(g)$ in the presence of ultraviolet light, four different monochlorination products form. What is the structure of C_5H_{12} in this reaction?

b. When C_4H_8 is reacted with H_2O, a tertiary alcohol is produced as the major product. What is the structure of C_4H_8 in this reaction?

c. When C_7H_{12} is reacted with HCl, 1-chloro-1-methylcyclohexane is produced as the major product. What are the two possible structures for C_7H_{12} in this reaction?

d. When a hydrocarbon is reacted with H_2O and the major product of this reaction is then oxidized, acetone (2-propanone) is produced. What is the structure of the hydrocarbon in this reaction?

e. When $C_5H_{12}O$ is oxidized, a carboxylic acid is produced. What are the possible structures for $C_5H_{12}O$ in this reaction?

152. Polycarbonates are a class of thermoplastic polymers that are used in the plastic lenses of eyeglasses and in the shells of bicycle helmets. A polycarbonate is made from the reaction of bisphenol A (BPA) with phosgene ($COCl_2$):

$$\xrightarrow{\text{Catalyst}} \text{polycarbonate} + 2n\text{HCl}$$

Phenol (C_6H_5OH) is used to terminate the polymer (stop its growth).

a. Draw the structure of the polycarbonate chain formed from the above reaction.

b. Is this reaction a condensation or an addition polymerization?

153. A urethane linkage occurs when an alcohol adds across the carbon–nitrogen double bond in an isocyanate:

Polyurethanes (formed from the copolymerization of a diol with a diisocyanate) are used in foamed insulation and a variety of other construction materials. What is the structure of the polyurethane formed by the following reaction?

$$HOCH_2CH_2OH + O=C=N \overset{}{\underset{}{\bigcirc}} N=C=O \longrightarrow$$

154. ABS plastic is a tough, hard plastic used in applications requiring shock resistance. The polymer consists of three monomer units: acrylonitrile (C_3H_3N), butadiene (C_4H_6), and styrene (C_8H_8).

   a. Draw two repeating units of ABS plastic assuming that the three monomer units react in a 1:1:1 mole ratio and react in the same order as the monomers listed above.

   b. A sample of ABS plastic contains 8.80% N by mass. It took 0.605 g Br_2 to react completely with a 1.20-g sample of ABS plastic. What is the percent by mass of acrylonitrile, butadiene, and styrene in this polymer sample?

   c. ABS plastic does not react in a 1:1:1 mole ratio among the three monomer units. Using the results from part b, determine the relative numbers of the monomer units in this sample of ABS plastic.

155. Stretch a rubber band while holding it gently to your lips. Then slowly let it relax while still in contact with your lips.

   a. What happens to the temperature of the rubber band on stretching?

   b. Is the stretching an exothermic or endothermic process?

   c. Explain the above result in terms of intermolecular forces.

   d. What is the sign of ΔS and ΔG for stretching the rubber band?

   e. Give the molecular explanation for the sign of ΔS for stretching.

156. Alcohols are very useful starting materials for the production of many different compounds. The following conversions, starting with 1-butanol, can be carried out in two or more steps. Show the steps (reactants/catalysts) you would follow to carry out the conversions, drawing the formula for the organic product in each step. For each step, a major product must be produced. (See Exercise 62.) (*Hint:* In the presence of H^+, an alcohol is converted into an alkene and water. This is the exact reverse of the reaction of adding water to an alkene to form an alcohol.)

   a. 1-butanol $\longrightarrow$ butane

   b. 1-butanol $\longrightarrow$ 2-butanone

157. A chemical "breathalyzer" test works because ethanol in the breath is oxidized by the dichromate ion (orange) to form acetic acid and chromium(III) ion (green). The balanced reaction is

$$3C_2H_5OH(aq) + 2Cr_2O_7^{2-}(aq) + 2H^+(aq) \longrightarrow$$
$$3HC_2H_3O_2(aq) + 4Cr^{3+}(aq) + 11H_2O(l)$$

You analyze a breathalyzer test in which 4.2 mg $K_2Cr_2O_7$ was reduced. Assuming the volume of the breath was 0.500 L at 30.°C and 750. mm Hg, what was the mole percent alcohol of the breath?

158. Estradiol is a female hormone with the following structure:

How many chiral carbon atoms are in estradiol?

Integrative Problems

These problems require the integration of multiple concepts to find the solutions.

159. An organometallic compound is one containing at least one metal–carbon bond. An example of an organometallic species is $(CH_3CH_2)MBr$, which contains a metal–ethyl bond.

   a. If M^{2+} has the electron configuration $[Ar]3d^{10}$, what is the percent by mass of M in $(CH_3CH_2)MBr$?

   b. A reaction involving $(CH_3CH_2)MBr$ is the conversion of a ketone to an alcohol as illustrated here:

      How does the hybridization of the starred carbon atom change, if at all, in going from reactants to products?

   c. What is the systematic name of the product? (*Hint:* In this shorthand notation, all the C—H bonds have been eliminated and the lines represent C—C bonds, unless shown differently. As is typical of most organic compounds, each carbon atom has four bonds to it and the oxygen atoms have only two bonds.)

160. Helicenes are extended fused polyaromatic hydrocarbons that have a helical or screw-shaped structure.

   a. A 0.1450-g sample of solid helicene is combusted in air to give 0.5063 g CO_2. What is the empirical formula of this helicene?

   b. If a 0.0938-g sample of this helicene is dissolved in 12.5 g of solvent to give a 0.0175 M solution, what is the molecular formula of this helicene?

   c. What is the balanced reaction for the combustion of this helicene?

Marathon Problems

These problems are designed to incorporate several concepts and techniques into one situation.

161. For each of the following, fill in the blank with the correct response. All of these fill-in-the-blank problems pertain to material covered in the sections on alkanes, alkenes and alkynes, aromatic hydrocarbons, and hydrocarbon derivatives.

   a. The first "organic" compound to be synthesized in the laboratory, rather than being isolated from nature, was _____, which was prepared from _____.

b. An organic compound whose carbon–carbon bonds are all single bonds is said to be _____.

c. The general orientation of the four pairs of electrons around the carbon atoms in alkanes is _____.

d. Alkanes in which the carbon atoms form a single unbranched chain are said to be _____ alkanes.

e. Structural isomerism occurs when two molecules have the same number of each type of atom but exhibit different arrangements of the _____ between those atoms.

f. The systematic names of all saturated hydrocarbons have the ending _____ added to a root name that indicates the number of carbon atoms in the molecule.

g. For a branched hydrocarbon, the root name for the hydrocarbon comes from the number of carbon atoms in the _____ continuous chain in the molecule.

h. The positions of substituents along the hydrocarbon framework of a molecule are indicated by the _____ of the carbon atom to which the substituents are attached.

i. The major use of alkanes has been in _____ reactions, as a source of heat and light.

j. With very reactive agents, such as the halogen elements, alkanes undergo _____ reactions, whereby a new atom replaces one or more hydrogen atoms of the alkane.

k. Alkenes and alkynes are characterized by their ability to undergo rapid, complete _____ reactions, by which other atoms attach themselves to the carbon atoms of the double or triple bond.

l. Unsaturated fats may be converted to saturated fats by the process of _____.

m. Benzene is the parent member of the group of hydrocarbons called _____ hydrocarbons.

n. An atom or group of atoms that imparts new and characteristic properties to an organic molecule is called a _____ group.

o. A _____ alcohol is one in which there is only one hydrocarbon group attached to the carbon atom holding the hydroxyl group.

p. The simplest alcohol, methanol, is prepared industrially by the hydrogenation of _____.

q. Ethanol is commonly prepared by the _____ of certain sugars by yeast.

r. Both aldehydes and ketones contain the _____ group, but they differ in where this group occurs along the hydrocarbon chain.

s. Aldehydes and ketones can be prepared by _____ of the corresponding alcohol.

t. Organic acids, which contain the _____ group, are typically weak acids.

u. The typically sweet-smelling compounds called _____ result from the condensation reaction of an organic acid with an _____.

162. Choose one of the following terms to match the description given in statements (1)–(17). All of the following pertain to proteins or carbohydrates.

a. aldohexose g. disaccharides m. ketohexoses
b. saliva h. disulfide n. oxytocin
c. cellulose i. globular o. pleated sheet
d. CH_2O j. glycogen p. polypeptide
e. cysteine k. glycoside linkage q. primary structure
f. denaturation l. hydrophobic

(1) polymer consisting of many amino acids
(2) linkage that forms between two cysteine species
(3) peptide hormone that triggers milk secretion
(4) proteins with roughly spherical shape
(5) sequence of amino acids in a protein
(6) silk protein secondary structure
(7) water-repelling amino acid side chain
(8) amino acid responsible for permanent wave in hair
(9) breakdown of a protein's tertiary and/or secondary structure
(10) animal polymer of glucose
(11) —C—O—C— bond between rings in disaccharide sugars
(12) empirical formula leading to the name carbohydrate
(13) where enzymes catalyzing the breakdown of glycoside linkages are found
(14) six-carbon ketone sugars
(15) structural component of plants, polymer of glucose
(16) sugars consisting of two monomer units
(17) six-carbon aldehyde sugars

163. For each of the following, fill in the blank with the correct response(s). All of the following pertain to nucleic acids.

a. The substance in the nucleus of the cell that stores and transmits genetic information is DNA, which stands for _____.

b. The basic repeating monomer units of DNA and RNA are called _____.

c. The pentose deoxyribose is found in DNA, whereas _____ is found in RNA.

d. The basic linkage in DNA or RNA between the sugar molecule and phosphoric acid is a phosphate _____ linkage.

e. The bases on opposite strands of DNA are said to be _____ to each other, which means the bases fit together specifically by hydrogen bonding to one another.

f. In a strand of normal DNA, the base _____ is always found paired with the base adenine, whereas _____ is always found paired with cytosine.

g. A given segment of the DNA molecule, which contains the molecular coding for a specific protein to be synthesized, is referred to as a _____.

h. During protein synthesis, _____ RNA molecules attach to and transport specific amino acids to the appropriate position on the pattern provided by _____ RNA molecules.

i. The codes specified by _____ are responsible for assembling the correct primary structure of proteins.

Appendixes

| Mathematical Procedures

A1-1 | Exponential Notation

The numbers characteristic of scientific measurements are often very large or very small; thus it is convenient to express them using powers of 10. For example, the number 1,300,000 can be expressed as 1.3×10^6, which means multiply 1.3 by 10 six times, or

$$1.3 \times 10^6 = 1.3 \times \underbrace{10 \times 10 \times 10 \times 10 \times 10 \times 10}_{10^6 = 1 \text{ million}}$$

Note that each multiplication by 10 moves the decimal point one place to the right:

$$1.3 \times 10 = 13.$$
$$13 \times 10 = 130.$$
$$130 \times 10 = 1300.$$
$$\vdots$$

Thus the easiest way to interpret the notation 1.3×10^6 is that it means move the decimal point in 1.3 to the right six times:

$$1.3 \times 10^6 = 1\underset{1\ 2\ 3\ 4\ 5\ 6}{3\ 0\ 0\ 0\ 0\ 0} = 1,300,000$$

Using this notation, the number 1985 can be expressed as 1.985×10^3. Note that the usual convention is to write the number that appears before the power of 10 as a number between 1 and 10. To end up with the number 1.985, which is between 1 and 10, we had to move the decimal point three places to the left. To compensate for that, we must multiply by 10^3, which says that to get the intended number we start with 1.985 and move the decimal point three places to the right; that is:

$$1.985 \times 10^3 = 1\underset{1\ 2\ 3}{9\ 8\ 5.}$$

Some other examples are given below.

Number	Exponential Notation
5.6	5.6×10^0 or 5.6×1
39	3.9×10^1
943	9.43×10^2
1126	1.126×10^3

So far we have considered numbers greater than 1. How do we represent a number such as 0.0034 in exponential notation? We start with a number between 1 and 10 and *divide* by the appropriate power of 10:

$$0.0034 = \frac{3.4}{10 \times 10 \times 10} = \frac{3.4}{10^3} = 3.4 \times 10^{-3}$$

Division by 10 moves the decimal point one place to the *left*. Thus the number

$$0.\underset{7\ \ 6\ \ 5\ \ 4\ \ 3\ \ 2\ \ 1}{0\ 0\ 0\ 0\ 0\ 0\ 1}\,4$$

can be written as 1.4×10^{-7}.

To summarize, we can write any number in the form

$$N \times 10^{\pm n}$$

where N is between 1 and 10 and the exponent n is an integer. If the sign preceding n is positive, it means the decimal point in N should be moved n places to the right. If a negative sign precedes n, the decimal point in N should be moved n places to the left.

Multiplication and Division

When two numbers expressed in exponential notation are multiplied, the initial numbers are multiplied and the exponents of 10 are added:

$$(M \times 10^m)(N \times 10^n) = (MN) \times 10^{m+n}$$

For example (to two significant figures, as required),

$$(3.2 \times 10^4)(2.8 \times 10^3) = 9.0 \times 10^7$$

When the numbers are multiplied, if a result greater than 10 is obtained for the initial number, the decimal point is moved one place to the left and the exponent of 10 is increased by 1:

$$(5.8 \times 10^2)(4.3 \times 10^8) = 24.9 \times 10^{10}$$
$$= 2.49 \times 10^{11}$$
$$= 2.5 \ \ \times 10^{11} \quad \text{(two significant figures)}$$

Division of two numbers expressed in exponential notation involves normal division of the initial numbers and *subtraction* of the exponent of the divisor from that of the dividend. For example,

$$\frac{4.8 \times 10^8}{\underbrace{2.1 \times 10^3}_{\text{Divisor}}} = \frac{4.8}{2.1} \times 10^{(8-3)} = 2.3 \times 10^5$$

If the initial number resulting from the division is less than 1, the decimal point is moved one place to the right and the exponent of 10 is decreased by 1. For example,

$$\frac{6.4 \times 10^3}{8.3 \times 10^5} = \frac{6.4}{8.3} \times 10^{(3-5)} = 0.77 \times 10^{-2}$$
$$= 7.7 \times 10^{-3}$$

Addition and Subtraction

To add or subtract numbers expressed in exponential notation, *the exponents of the numbers must be the same.* For example, to add 1.31×10^5 and 4.2×10^4, we must rewrite one number so that the exponents of both are the same. The number 1.31×10^5 can be written 13.1×10^4, since moving the decimal point one place to the right can be compensated for by decreasing the exponent by 1. Now we can add the numbers:

$$13.1 \times 10^4$$
$$+ \ 4.2 \times 10^4$$
$$17.3 \times 10^4$$

In correct exponential notation the result is expressed as 1.73×10^5.

To perform addition or subtraction with numbers expressed in exponential notation, only the initial numbers are added or subtracted. The exponent of the result is the same as those of the numbers being added or subtracted. To subtract 1.8×10^2 from 8.99×10^3, we write

$$8.99 \times 10^3$$
$$-0.18 \times 10^3$$
$$8.81 \times 10^3$$

Powers and Roots

When a number expressed in exponential notation is taken to some power, the initial number is taken to the appropriate power and the exponent of 10 is multiplied by that power:

$$(N \times 10^n)^m = N^m \times 10^{m \cdot n}$$

For example,*

$$(7.5 \times 10^2)^3 = 7.5^3 \times 10^{3 \cdot 2}$$
$$= 422 \times 10^6$$
$$= 4.22 \times 10^8$$
$$= 4.2 \ \times 10^8 \quad \text{(two significant figures)}$$

When a root is taken of a number expressed in exponential notation, the root of the initial number is taken and the exponent of 10 is divided by the number representing the root:

$$\sqrt{N \times 10^n} = (n \times 10^n)^{1/2} = \sqrt{N} \times 10^{n/2}$$

For example,

$$(2.9 \times 10^6)^{1/2} = \sqrt{2.9} \times 10^{6/2}$$
$$= 1.7 \times 10^3$$

Because the exponent of the result must be an integer, we may sometimes have to change the form of the number so that the power divided by the root equals an integer. For example,

$$\sqrt{1.9 \times 10^3} = (1.9 \times 10^3)^{1/2} = (0.19 \times 10^4)^{1/2}$$
$$= \sqrt{0.19} \times 10^2$$
$$= 0.44 \times 10^2$$
$$= 4.4 \times 10^1$$

In this case, we moved the decimal point one place to the left and increased the exponent from 3 to 4 to make $n/2$ an integer.

The same procedure is followed for roots other than square roots. For example,

$$\sqrt[3]{6.9 \times 10^5} = (6.9 \times 10^5)^{1/3} = (0.69 \times 10^6)^{1/3}$$
$$= \sqrt[3]{0.69} \times 10^2$$
$$= 0.88 \times 10^2$$
$$= 8.8 \times 10^1$$

and

$$\sqrt[3]{4.6 \times 10^{10}} = (4.6 \times 10^{10})^{1/3} = (46 \times 10^9)^{1/3}$$
$$= \sqrt[3]{46} \times 10^3$$
$$= 3.6 \times 10^3$$

---

*Refer to the instruction booklet for your calculator for directions concerning how to take roots and powers of numbers.

A1-2 | Logarithms

A logarithm is an exponent. Any number N can be expressed as follows:

$$N = 10^x$$

For example,

$$1000 = 10^3$$
$$100 = 10^2$$
$$10 = 10^1$$
$$1 = 10^0$$

The common, or base 10, logarithm of a number is the power to which 10 must be taken to yield the number. Thus, since $1000 = 10^3$,

$$\log 1000 = 3$$

Similarly,

$$\log 100 = 2$$
$$\log 10 = 1$$
$$\log 1 = 0$$

For a number between 10 and 100, the required exponent of 10 will be between 1 and 2. For example, $65 = 10^{1.8129}$; that is, $\log 65 = 1.8129$. For a number between 100 and 1000, the exponent of 10 will be between 2 and 3. For example, $650 = 10^{2.8129}$ and $\log 650 = 2.8129$.

A number N greater than 0 and less than 1 can be expressed as follows:

$$N = 10^{-x} = \frac{1}{10^x}$$

For example,

$$0.001 = \frac{1}{1000} = \frac{1}{10^3} = 10^{-3}$$

$$0.01 = \frac{1}{100} = \frac{1}{10^2} = 10^{-2}$$

$$0.1 = \frac{1}{10} = \frac{1}{10^1} = 10^{-1}$$

Thus

$$\log 0.001 = -3$$
$$\log 0.01 = -2$$
$$\log 0.1 = -1$$

Although common logs are often tabulated, the most convenient method for obtaining such logs is to use an electronic calculator. On most calculators the number is first entered and then the log key is punched. The log of the number then appears in the display.* Some examples are given below. You should reproduce these results on your calculator to be sure that you can find common logs correctly.

Number	Common Log
36	1.56
1849	3.2669
0.156	−0.807
1.68×10^{-5}	−4.775

*Refer to the instruction booklet for your calculator for the exact sequence to obtain logarithms.

Note that the number of digits after the decimal point in a common log is equal to the number of significant figures in the original number.

Since logs are simply exponents, they are manipulated according to the rules for exponents. For example, if $A = 10^x$ and $B = 10^y$, then their product is

$$A \cdot B = 10^x \cdot 10^y = 10^{x+y}$$

and

$$\log AB = x + y = \log A + \log B$$

For division, we have

$$\frac{A}{B} = \frac{10^x}{10^y} = 10^{x-y}$$

and

$$\log \frac{A}{B} = x - y = \log A - \log B$$

For a number raised to a power, we have

$$A^n = (10^x)^n = 10^{nx}$$

and

$$\log A^n = nx = n \log A$$

It follows that

$$\log \frac{1}{A^n} = \log A^{-n} = -n \log A$$

or, for $n = 1$,

$$\log \frac{1}{A} = -\log A$$

When a common log is given, to find the number it represents, we must carry out the process of exponentiation. For example, if the log is 2.673, then $N = 10^{2.673}$. The process of exponentiation is also called taking the antilog, or the inverse logarithm. This operation is usually carried out on calculators in one of two ways. The majority of calculators require that the log be entered first and then the keys (INV) and (LOG) pressed in succession. For example, to find $N = 10^{2.673}$ we enter 2.673 and then press (INV) and (LOG). The number 471 will be displayed; that is, $N = 471$. Some calculators have a (10^x) key. In that case, the log is entered first and then the (10^x) key is pressed. Again, the number 471 will be displayed.

Natural logarithms, another type of logarithm, are based on the number 2.7183, which is referred to as e. In this case, a number is represented as $N = e^x = 2.7183^x$. For example,

$$N = 7.15 = e^x$$

$$\ln 7.15 = x = 1.967$$

To find the natural log of a number using a calculator, the number is entered and then the (ln) key is pressed. Use the following examples to check your technique for finding natural logs with your calculator:

Number (e^x)	Natural Log (x)
784	6.664
1.61×10^3	7.384
1.00×10^{-7}	−16.118
1.00	0

If a natural logarithm is given, to find the number it represents, exponentiation to the base e (2.7183) must be carried out. With many calculators this is done using a key marked $\boxed{e^x}$ (the natural log is entered, with the correct sign, and then the $\boxed{e^x}$ key is pressed). The other common method for exponentiation to base e is to enter the natural log and then press the $\boxed{\text{INV}}$ and $\boxed{\text{ln}}$ keys in succession. The following examples will help you check your technique:

ln N(x)	N(e^x)
3.256	25.9
−5.169	5.69×10^{-3}
13.112	4.95×10^{5}

Since natural logarithms are simply exponents, they are also manipulated according to the mathematical rules for exponents given earlier for common logs.

A1-3 | Graphing Functions

In interpreting the results of a scientific experiment, it is often useful to make a graph. If possible, the function to be graphed should be in a form that gives a straight line. The equation for a straight line (a *linear equation*) can be represented by the general form

$$y = mx + b$$

where y is the *dependent variable*, x is the *independent variable*, m is the *slope*, and b is the *intercept* with the y axis.

To illustrate the characteristics of a linear equation, the function $y = 3x + 4$ is plotted in Fig. A-1. For this equation $m = 3$ and $b = 4$. Note that the y intercept occurs when $x = 0$. In this case the intercept is 4, as can be seen from the equation ($b = 4$).

The slope of a straight line is defined as the ratio of the rate of change in y to that in x:

$$m = \text{slope} = \frac{\Delta y}{\Delta x}$$

For the equation $y = 3x + 4$, y changes three times as fast as x (since x has a coefficient of 3). Thus the slope in this case is 3. This can be verified from the graph. For the triangle shown in Fig. A-1,

$$\Delta y = 34 - 10 = 24 \quad \text{and} \quad \Delta x = 10 - 2 = 8$$

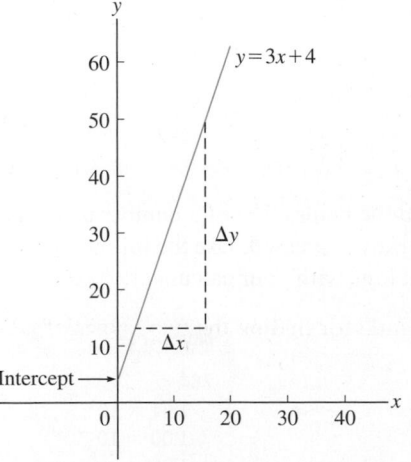

Figure A-1 | Graph of the linear equation $y = 3x + 4$.

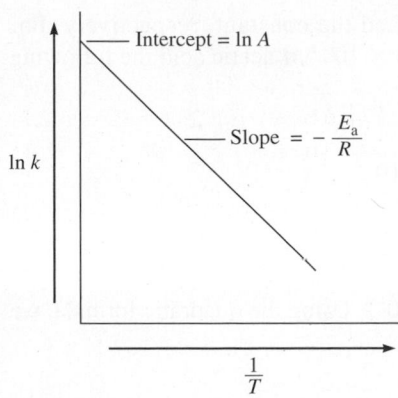

Figure A-2 | Graph of ln k versus $1/T$.

Table A-1 | Some Useful Linear Equations in Standard Form

Equation ($y = mx + b$)	What Is Plotted (y vs. x)	Slope (m)	Intercept (b)	Section in Text
$[A] = -kt + [A]_0$	$[A]$ vs. t	$-k$	$[A]_0$	1-4
$\ln[A] = -kt + \ln[A]_0$	$\ln[A]$ vs. t	$-k$	$\ln[A]_0$	1-4
$\dfrac{1}{[A]} = kt + \dfrac{1}{[A]_0}$	$\dfrac{1}{[A]}$ vs. t	k	$\dfrac{1}{[A]_0}$	1-4
$\ln P_{vap} = -\dfrac{\Delta H_{vap}}{R}\left(\dfrac{1}{T}\right) + C$	$\ln P_{vap}$ vs. $\dfrac{1}{T}$	$\dfrac{-\Delta H_{vap}}{R}$	C	9-9

Thus

$$\text{Slope} = \frac{\Delta y}{\Delta x} = \frac{24}{8} = 3$$

The preceding example illustrates a general method for obtaining the slope of a line from the graph of that line. Simply draw a triangle with one side parallel to the y axis and the other parallel to the x axis as shown in Fig. A-1. Then determine the lengths of the sides to give Δy and Δx, respectively, and compute the ratio $\Delta y/\Delta x$.

Sometimes an equation that is not in standard form can be changed to the form $y = mx + b$ by rearrangement or mathematical manipulation. An example is the equation $k = Ae^{-E_a/RT}$ described in Section 11-6, where A, E_a, and R are constants; k is the dependent variable; and $1/T$ is the independent variable. This equation can be changed to standard form by taking the natural logarithm of both sides,

$$\ln k = \ln Ae^{-E_a/RT} = \ln A + \ln e^{-E_a/RT} = \ln A - \frac{E_a}{RT}$$

noting that the log of a product is equal to the sum of the logs of the individual terms and that the natural log of $e^{-E_a/RT}$ is simply the exponent $-E_a/RT$. Thus, in standard form, the equation $k = Ae^{-E_a/RT}$ is written

$$\underset{y}{\ln k} = \underset{m}{-\frac{E_a}{R}}\underset{x}{\left(\frac{1}{T}\right)} + \underset{b}{\ln A}$$

A plot of ln k versus $1/T$ (see Fig. A-2) gives a straight line with slope $-E_a/R$ and intercept ln A.

Other linear equations that are useful in the study of chemistry are listed in standard form in Table A-1.

A1-4 | Solving Quadratic Equations

A *quadratic equation*, a polynomial in which the highest power of x is 2, can be written as

$$ax^2 + bx + c = 0$$

One method for finding the two values of x that satisfy a quadratic equation is to use the *quadratic formula*:

$$x = \frac{-b \pm \sqrt{b^2 - 4ac}}{2a}$$

where a, b, and c are the coefficients of x^2, x, and the constant, respectively. For example, in determining $[H^+]$ in a solution of $1.0 \times 10^{-4}\,M$ acetic acid the following expression arises:

$$1.8 \times 10^{-5} = \frac{x^2}{1.0 \times 10^{-4} - x}$$

which yields

$$x^2 + (1.8 \times 10^{-5})x - 1.8 \times 10^{-9} = 0$$

where $a = 1$, $b = 1.8 \times 10^{-5}$, and $c = -1.8 \times 10^{-9}$. Using the quadratic formula, we have

$$x = \frac{-b \pm \sqrt{b^2 - 4ac}}{2a}$$

$$= \frac{-1.8 \times 10^{-5} \pm \sqrt{3.24 \times 10^{-10} - (4)(1)(-1.8 \times 10^{-9})}}{2(1)}$$

$$= \frac{-1.8 \times 10^{-5} \pm \sqrt{3.24 \times 10^{-10} + 7.2 \times 10^{-9}}}{2}$$

$$= \frac{-1.8 \times 10^{-5} \pm \sqrt{7.5 \times 10^{-9}}}{2}$$

$$= \frac{-1.8 \times 10^{-5} \pm 8.7 \times 10^{-5}}{2}$$

Thus

$$x = \frac{6.9 \times 10^{-5}}{2} = 3.5 \times 10^{-5}$$

and

$$x = \frac{-10.5 \times 10^{-5}}{2} = -5.2 \times 10^{-5}$$

Note that there are two roots, as there always will be, for a polynomial in x^2. In this case x represents a concentration of H^+ (see Section 13-3). Thus the positive root is the one that solves the problem, since a concentration cannot be a negative number.

A second method for solving quadratic equations is by *successive approximations,* a systematic method of trial and error. A value of x is guessed and substituted into the equation everywhere x (or x^2) appears, except for one place. For example, for the equation

$$x^2 + (1.8 \times 10^{-5})x - 1.8 \times 10^{-9} = 0$$

we might guess $x = 2 \times 10^{-5}$. Substituting that value into the equation gives

$$x^2 + (1.8 \times 10^{-5})(2 \times 10^{-5}) - 1.8 \times 10^{-9} = 0$$

or

$$x^2 = 1.8 \times 10^{-9} - 3.6 \times 10^{-10} = 1.4 \times 10^{-9}$$

Thus

$$x = 3.7 \times 10^{-5}$$

Note that the guessed value of x (2×10^{-5}) is not the same as the value of x that is calculated (3.7×10^{-5}) after inserting the estimated value. This means that $x = 2 \times 10^{-5}$ is not the correct solution, and we must try another guess. We take the calculated value (3.7×10^{-5}) as our next guess:

$$x^2 + (1.8 \times 10^{-5})(3.7 \times 10^{-5}) - 1.8 \times 10^{-9} = 0$$
$$x^2 = 1.8 \times 10^{-9} - 6.7 \times 10^{-10} = 1.1 \times 10^{-9}$$

Thus

$$x = 3.3 \times 10^{-5}$$

Now we compare the two values of x again:

Guessed: $\quad x = 3.7 \times 10^{-5}$

Calculated: $x = 3.3 \times 10^{-5}$

These values are closer but not close enough. Next we try 3.3×10^{-5} as our guess:

$$x^2 + (1.8 \times 10^{-5})(3.3 \times 10^{-5}) - 1.8 \times 10^{-9} = 0$$
$$x^2 = 1.8 \times 10^{-9} - 5.9 \times 10^{-10} = 1.2 \times 10^{-9}$$

Thus

$$x = 3.5 \times 10^{-5}$$

Again we compare:

Guessed: $\quad x = 3.3 \times 10^{-5}$

Calculated: $x = 3.5 \times 10^{-5}$

Next we guess $x = 3.5 \times 10^{-5}$ to give

$$x^2 + (1.8 \times 10^{-5})(3.5 \times 10^{-5}) - 1.8 \times 10^{-9} = 0$$
$$x^2 = 1.8 \times 10^{-9} - 6.3 \times 10^{-10} = 1.2 \times 10^{-9}$$

Thus

$$x = 3.5 \times 10^{-5}$$

Now the guessed value and the calculated value are the same; we have found the correct solution. Note that this agrees with one of the roots found with the quadratic formula in the first method.

To further illustrate the method of successive approximations, we will solve Example 13-17 using this procedure. In solving for $[H^+]$ for 0.010 M H_2SO_4, we obtain the following expression:

$$1.2 \times 10^{-2} = \frac{x(0.010 + x)}{0.010 - x}$$

which can be rearranged to give

$$x = (1.2 \times 10^{-2})\left(\frac{0.010 - x}{0.010 + x}\right)$$

We will guess a value for x, substitute it into the right side of the equation, and then calculate a value for x. In guessing a value for x, we know it must be less than 0.010, since a larger value will make the calculated value for x negative and the guessed and calculated values will never match. We start by guessing $x = 0.005$.

The results of the successive approximations are shown in the following table:

Trial	Guessed Value for x	Calculated Value for x
1	0.0050	0.0040
2	0.0040	0.0051
3	0.00450	0.00455
4	0.00452	0.00453

Note that the first guess was close to the actual value and that there was oscillation between 0.004 and 0.005 for the guessed and calculated values. For trial 3, an average of these values was used as the guess, and this led rapidly to the correct value (0.0045 to the correct number of significant figures). Also, note that it is useful to carry extra digits until the correct value is obtained. That value can then be rounded off to the correct number of significant figures.

The method of successive approximations is especially useful for solving polynomials containing x to a power of 3 or higher. The procedure is the same as for quadratic equations: Substitute a guessed value for x into the equation for every x term but one, and then solve for x. Continue this process until the guessed and calculated values agree.

A1-5 | Uncertainties in Measurements

Like all the physical sciences, chemistry is based on the results of measurements. Every measurement has an inherent uncertainty, so if we are to use the results of measurements to reach conclusions, we must be able to estimate the sizes of these uncertainties.

For example, the specification for a commercial 500-mg acetaminophen (the active painkiller in Tylenol) tablet is that each batch of tablets must contain 450 to 550 mg of acetaminophen per tablet. Suppose that chemical analysis gave the following results for a batch of acetaminophen tablets: 428 mg, 479 mg, 442 mg, and 435 mg. How can we use these results to decide if the batch of tablets meets the specification? Although the details of how to draw such conclusions from measured data are beyond the scope of this text, we will consider some aspects of how this is done. We will focus here on the types of experimental uncertainty, the expression of experimental results, and a simplified method for estimating experimental uncertainty when several types of measurement contribute to the final result.

Types of Experimental Error

There are two types of experimental uncertainty (error). A variety of names are applied to these types of errors:

$$\text{Precision} \longleftrightarrow \text{random error} \equiv \text{indeterminate error}$$
$$\text{Accuracy} \longleftrightarrow \text{systematic error} \equiv \text{determinate error}$$

The difference between the two types of error is well illustrated by the attempts to hit a target shown in Fig. R-2 in the Review Chapter.

Random error is associated with every measurement. To obtain the last significant figure for any measurement, we must always make an estimate. For example, we interpolate between the marks on a meter stick, a buret, or a balance. The precision of replicate measurements (repeated measurements of the same type) reflects the size of the random errors. Precision refers to the reproducibility of replicate measurements.

The accuracy of a measurement refers to how close it is to the true value. An inaccurate result occurs as a result of some flaw (systematic error) in the measurement: the presence of an interfering substance, incorrect calibration of an instrument, operator error, and so on. The goal of chemical analysis is to eliminate systematic error, but random errors can only be minimized. In practice, an experiment is almost always done to find an unknown value (the true value is not known—someone is trying to obtain that value by doing the experiment). In this case the precision of several replicate determinations is used to assess the accuracy of the result. The results of the replicate experiments are expressed as an average (which we assume is close to the true value) with an error limit that gives some indication of how close the average value may be to the true value. The error limit represents the uncertainty of the experimental result.

Expression of Experimental Results

If we perform several measurements, such as for the analysis for acetaminophen in painkiller tablets, the results should express two things: the average of the measurements and the size of the uncertainty.

There are two common ways of expressing an average: the mean and the median. The mean ($\bar{x}$) is the arithmetic average of the results, or

$$\text{Mean} = \bar{x} = \sum_{i=1}^{n} \frac{x_i}{n} = \frac{x_1 + x_2 + \cdots + x_n}{n}$$

where Σ means take the sum of the values. The mean is equal to the sum of all the measurements divided by the number of measurements. For the acetaminophen results given previously, the mean is

$$\bar{x} = \frac{428 + 479 + 442 + 435}{4} = 446 \text{ mg}$$

The median is the value that lies in the middle among the results. Half the measurements are above the median and half are below the median. For results of 465 mg, 485 mg, and 492 mg, the median is 485 mg. When there is an even number of results, the median is the average of the two middle results. For the acetaminophen results, the median is

$$\frac{442 + 435}{2} = 438 \text{ mg}$$

There are several advantages to using the median. If a small number of measurements is made, one value can greatly affect the mean. Consider the results for the analysis of acetaminophen: 428 mg, 479 mg, 442 mg, and 435 mg. The mean is 446 mg, which is larger than three of the four weights. The median is 438 mg, which lies near the three values that are relatively close to one another.

In addition to expressing an average value for a series of results, we must express the uncertainty. This usually means expressing either the precision of the measurements or the observed range of the measurements. The range of a series of measurements is defined by the smallest value and the largest value. For the analytical results on the acetaminophen tablets, the range is from 428 mg to 479 mg. Using this range, we can express the results by saying that the true value lies between 428 mg and 479 mg. That is, we can express the amount of acetaminophen in a typical tablet as 446 ± 33 mg, where the error limit is chosen to give the observed range (approximately).

The most common way to specify precision is by the standard deviation, s, which for a small number of measurements is given by the formula

$$s = \left[\frac{\sum_{i=1}^{n} (x_i - \bar{x})^2}{n - 1} \right]^{1/2}$$

where x_i is an individual result, $\bar{x}$ is the average (either mean or median), and n is the total number of measurements. For the acetaminophen example, we have

$$s = \left[\frac{(428 - 446)^2 + (479 - 446)^2 + (442 - 446)^2 + (435 - 446)^2}{4 - 1} \right]^{1/2} = 23$$

Thus we can say the amount of acetaminophen in the typical tablet in the batch of tablets is 446 mg with a sample standard deviation of 23 mg. Statistically this means that any additional measurement has a 68% probability (68 chances out of 100) of being between 423 mg ($446 - 23$) and 469 mg ($446 + 23$). Thus the standard deviation is a measure of the precision of a given type of determination.

The standard deviation gives us a means of describing the precision of a given type of determination using a series of replicate results. However, it is also useful to be able to estimate the precision of a procedure that involves several measurements by combining the precisions of the individual steps. That is, we want to answer the following question: How do the uncertainties propagate when we combine the results of several different types of measurements? There are many ways to deal with the propagation of uncertainty. We will discuss only one simple method here.

A Simplified Method for Estimating Experimental Uncertainty

To illustrate this method, we will consider the determination of the density of an irregularly shaped solid. In this determination we make three measurements. First, we measure the mass of the object on a balance. Next, we must obtain the volume of the solid. The easiest method for doing this is to partially fill a graduated cylinder with a liquid and record the volume. Then we add the solid and record the volume again. The difference in the measured volumes is the volume of the solid. We can then calculate the density of the solid from the equation

$$D = \frac{M}{V_2 - V_1}$$

where M is the mass of the solid, V_1 is the initial volume of liquid in the graduated cylinder, and V_2 is the volume of liquid plus solid. Suppose we get the following results:

$$M = 23.06 \text{ g}$$
$$V_1 = 10.4 \text{ mL}$$
$$V_2 = 13.5 \text{ mL}$$

The calculated density is

$$\frac{23.06 \text{ g}}{13.5 \text{ mL} - 10.4 \text{ mL}} = 7.44 \text{ g/mL}$$

Now suppose that the precision of the balance used is ± 0.02 g and that the volume measurements are precise to ± 0.05 mL. How do we estimate the uncertainty of the density? We can do this by assuming a worst case. That is, we assume the largest uncertainties in all measurements, and see what combinations of measurements will give the largest and smallest possible results (the greatest range). Since the density is the mass divided by the volume, the largest value of the density will be that obtained using the largest possible mass and the smallest possible volume:

Largest possible mass = 23.06 + .02

$$D_{max} = \frac{23.08}{13.45 - 10.45} = 7.69 \text{ g/mL}$$

Smallest possible V_2 Largest possible V_1

The smallest value of the density is

Smallest possible mass

$$D_{min} = \frac{23.04}{13.35 - 10.35} = 7.20 \text{ g/mL}$$

Largest possible V_2 Smallest possible V_1

Thus the calculated range is from 7.20 to 7.69 and the average of these values is 7.44. The error limit is the number that gives the high and low range values when added and subtracted from the average. Therefore, we can express the density as 7.44 $\pm$ 0.25 g/mL, which is the average value plus or minus the quantity that gives the range calculated by assuming the largest uncertainties.

Analysis of the propagation of uncertainties is useful in drawing qualitative conclusions from the analysis of measurements. For example, suppose that we obtained the preceding results for the density of an unknown alloy and we want to know if it is one of the following alloys:

Alloy A: D = 7.58 g/mL

Alloy B: D = 7.42 g/mL

Alloy C: D = 8.56 g/mL

We can safely conclude that the alloy is not C. But the values of the densities for alloys A and B are both within the inherent uncertainty of our method. To distinguish between A and B, we need to improve the precision of our determination: The obvious choice is to improve the precision of the volume measurement.

The worst-case method is very useful in estimating uncertainties when the results of several measurements are combined to calculate a result. We assume the maximum uncertainty in each measurement and calculate the minimum and maximum possible result. These extreme values describe the range and thus the error limit.

APPENDIX 2 | # The Quantitative Kinetic Molecular Model

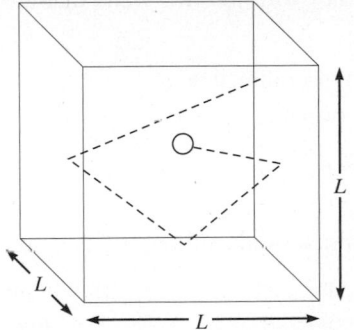

Figure A-3 | An ideal gas particle in a cube whose sides are of length L. The particle collides elastically with the walls in a random, straight-line motion.

We have seen that the kinetic molecular model successfully accounts for the properties of an ideal gas. This appendix will show in some detail how the postulates of the kinetic molecular model lead to an equation corresponding to the experimentally obtained ideal gas equation.

Recall that the particles of an ideal gas are assumed to be volumeless, to have no attraction for each other, and to produce pressure on their container by colliding with the container walls.

Suppose there are n moles of an ideal gas in a cubical container with sides each of length L. Assume each gas particle has a mass m and that it is in rapid, random, straight-line motion colliding with the walls, as shown in Fig. A-3. The collisions will be assumed to be *elastic*—no loss of kinetic energy occurs. We want to compute the force on the walls from the colliding gas particles and then, since pressure is force per unit area, to obtain an expression for the pressure of the gas.

Before we can derive the expression for the pressure of a gas, we must first discuss some characteristics of velocity. Each particle in the gas has a particular velocity u that can be divided into components u_x, u_y, and u_z, as shown in Fig. A-4. First, using u_x and u_y and the Pythagorean theorem, we can obtain u_{xy} as shown in Fig. A-4(c):

$$u_{xy}^2 = u_x^2 + u_y^2$$

Hypotenuse of   Sides of
right triangle   right triangle

Then, constructing another triangle as shown in Fig. A-4(c), we find

$$u^2 = u_{xy}^2 + u_z^2$$

or

$$u^2 = u_x^2 + u_y^2 + u_z^2$$

Now let's consider how an "average" gas particle moves. For example, how often does this particle strike the two walls of the box that are perpendicular to the x axis? It is important to realize that only the x component of the velocity affects the particle's impacts on these two walls, as shown in Fig. A-5(a). The larger the x component of the velocity, the faster the particle travels between these two walls, and the more impacts per unit time it will make on these walls. Remember, the pressure of the gas is due to these collisions with the walls.

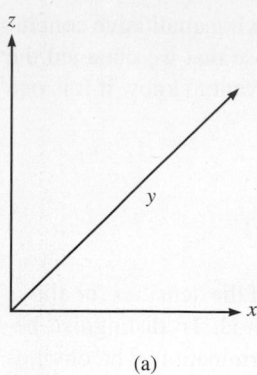

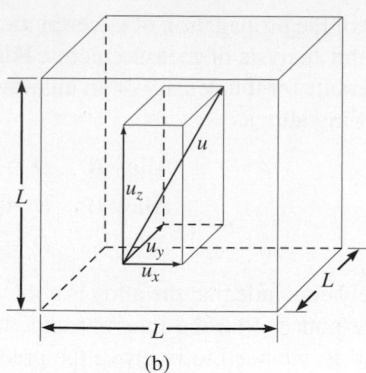

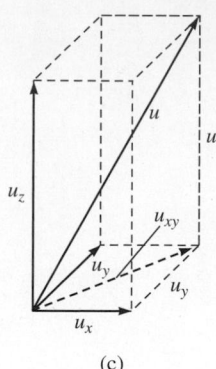

Figure A-4 | (a) The Cartesian coordinate axes.

(b) The velocity u of any gas particle can be broken down into three mutually perpendicular components: u_x, u_y, and u_z. This can be represented as a rectangular solid with sides u_x, u_y, and u_z and body diagonal u.

(c) In the xy plane,

$$u_x^2 + u_y^2 = u_{xy}^2$$

by the Pythagorean theorem. Since u_{xy} and u_x are also perpendicular,

$$u^2 = u_{xy}^2 + u_z^2 = u_x^2 + u_y^2 + u_z^2$$

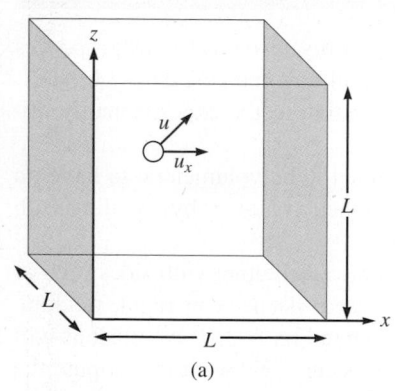

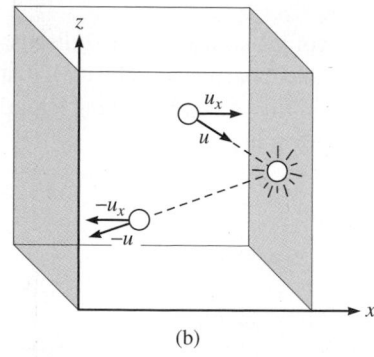

Figure A-5 |

(a) Only the x component of the gas particle's velocity affects the frequency of impacts on the shaded walls, the walls that are perpendicular to the x axis.

(b) For an elastic collision, there is an exact reversal of the x component of the velocity and of the total velocity. The change in momentum (final − initial) is then

$$-mu_x - mu_x = -2mu_x$$

The collision frequency (collisions per unit of time) with the two walls that are perpendicular to the x axis is given by

$$(\text{Collision frequency})_x = \frac{\text{velocity in the } x \text{ direction}}{\text{distance between the walls}}$$

$$= \frac{u_x}{L}$$

Next, what is the force of a collision? Force is defined as mass times acceleration (change in velocity per unit of time):

$$F = ma = m\left(\frac{\Delta u}{\Delta t}\right)$$

where F represents force, a represents acceleration, Δu represents a change in velocity, and Δt represents a given length of time.

Since we assume that the particle has constant mass, we can write

$$F = \frac{m\Delta u}{\Delta t} = \frac{\Delta(mu)}{\Delta t}$$

The quantity mu is the momentum of the particle (momentum is the product of mass and velocity), and the expression $F = \Delta(mu)/\Delta t$ implies that force is the change in momentum per unit of time. When a particle hits a wall perpendicular to the x axis, as shown in Fig. A-5(b), an elastic collision results in an *exact reversal* of the x component of velocity. That is, the *sign*, or direction, of u_x reverses when the particle collides with one of the walls perpendicular to the x axis. Thus the final momentum is the *negative*, or opposite, of the initial momentum. Remember that an elastic collision means that there is no change in the *magnitude* of the velocity. The change in momentum in the x direction is then

$$\text{Change in momentum} = \Delta(mu_x) = \text{final momentum} - \text{initial momentum}$$

$$= \underset{\substack{\nearrow \\ \text{Final} \\ \text{momentum} \\ \text{in } x \text{ direction}}}{-mu_x} \underset{\substack{\nwarrow \\ \text{Initial} \\ \text{momentum} \\ \text{in } x \text{ direction}}}{- mu_x}$$

$$= -2mu_x$$

But we are interested in the force the gas particle exerts on the walls of the box. Since we know that every action produces an equal but opposite reaction, the change in momentum with respect to the wall on impact is $-(-2mu_x)$, or $2mu_x$.

Recall that since force is the change in momentum per unit of time,

$$\text{Force}_x = \frac{\Delta(mu_x)}{\Delta t}$$

for the walls perpendicular to the x axis.

This expression can be obtained by multiplying the change in momentum per impact by the number of impacts per unit of time:

$$\text{Force}_x = (2mu_x)\left(\frac{u_x}{L}\right) = \text{change in momentum per unit of time}$$

Change in momentum per impact   Impacts per unit of time

That is,

$$\text{Force}_x = \frac{2mu_x^2}{L}$$

So far we have considered only the two walls of the box perpendicular to the x axis. We can assume that the force on the two walls perpendicular to the y axis is given by

$$\text{Force}_y = \frac{2mu_y^2}{L}$$

and that on the two walls perpendicular to the z axis by

$$\text{Force}_z = \frac{2mu_z^2}{L}$$

Since we have shown that

$$u^2 = u_x^2 + u_y^2 + u_z^2$$

the total force on the box is

$$\text{Force}_{\text{TOTAL}} = \text{force}_x + \text{force}_y + \text{force}_z$$

$$= \frac{2mu_x^2}{L} + \frac{2mu_y^2}{L} + \frac{2mu_z^2}{L}$$

$$= \frac{2m}{L}(u_x^2 + u_y^2 + u_z^2) = \frac{2m}{L}(u^2)$$

Now since we want the average force, we use the average of the square of the velocity $(\overline{u^2})$ to obtain

$$\overline{\text{Force}}_{\text{TOTAL}} = \frac{2m}{L}(\overline{u^2})$$

Next, we need to compute the pressure (force per unit of area)

$$\text{Pressure due to "average" particle} = \frac{\overline{\text{force}}_{\text{TOTAL}}}{\text{area}_{\text{TOTAL}}}$$

$$= \frac{\dfrac{2m\overline{u^2}}{L}}{6L^2} = \frac{m\overline{u^2}}{3L^3}$$

The 6 sides of the cube   Area of each side

Since the volume V of the cube is equal to L^3, we can write

$$\text{Pressure} = P = \frac{m\overline{u^2}}{3V}$$

So far we have considered the pressure on the walls due to a single, "average" particle. Of course, we want the pressure due to the entire gas sample. The number of particles in a given gas sample can be expressed as follows:

$$\text{Number of gas particles} = nN_A$$

where n is the number of moles and N_A is Avogadro's number.

The total pressure on the box due to n moles of a gas is therefore

$$P = nN_A \frac{m\overline{u^2}}{3V}$$

Next we want to express the pressure in terms of the kinetic energy of the gas molecules. Kinetic energy (the energy due to motion) is given by $\frac{1}{2}mu^2$, where m is the mass and u is the velocity. Since we are using the average of the velocity squared $(\overline{u^2})$, and since $m\overline{u^2} = 2(\frac{1}{2}m\overline{u^2})$, we have

$$P = \left(\frac{2}{3}\right)\frac{nN_A(\frac{1}{2}m\overline{u^2})}{V}$$

or

$$\frac{PV}{n} = \left(\frac{2}{3}\right)N_A(\tfrac{1}{2}m\overline{u^2})$$

Thus, based on the postulates of the kinetic molecular model, we have been able to derive an equation that has the same form as the ideal gas equation,

$$\frac{PV}{n} = RT$$

This agreement between experiment and theory supports the validity of the assumptions made in the kinetic molecular model about the behavior of gas particles, at least for the limiting case of an ideal gas.

APPENDIX 3   Spectral Analysis

Although volumetric and gravimetric analyses are still very commonly used, spectroscopy is the technique most often used for modern chemical analysis. *Spectroscopy* is the study of electromagnetic radiation emitted or absorbed by a given chemical species. Since the quantity of radiation absorbed or emitted can be related to the quantity of the absorbing or emitting species present, this technique can be used for quantitative analysis. There are many spectroscopic techniques, as electromagnetic radiation spans a wide range of energies to include X rays; ultraviolet, infrared, and visible light; and microwaves, to name a few of its familiar forms. We will consider here only one procedure, which is based on the absorption of visible light.

If a liquid is colored, it is because some component of the liquid absorbs visible light. In a solution the greater the concentration of the light-absorbing substance, the more light absorbed, and the more intense the color of the solution.

The quantity of light absorbed by a substance can be measured by a *spectrophotometer,* shown schematically in Fig. A-6. This instrument consists of a source that emits all wavelengths of light in the visible region (wavelengths of ~400 to 700 nm); a monochromator, which selects a given wavelength of light; a sample holder for the

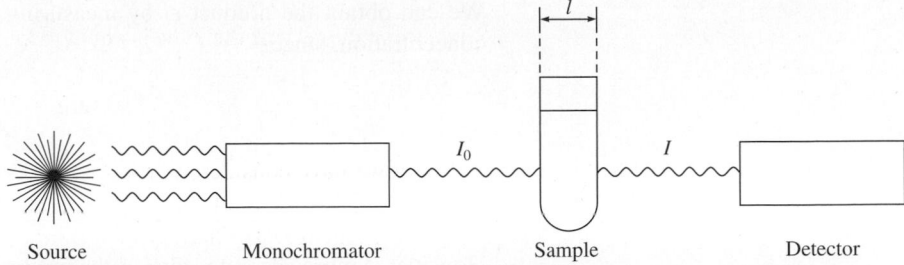

Figure A-6 | A schematic diagram of a simple spectrophotometer. The source emits all wavelengths of visible light, which are dispersed using a prism or grating and then focused, one wavelength at a time, onto the sample. The detector compares the intensity of the incident light (I_0) to the intensity of the light after it has passed through the sample (I).

solution being measured; and a detector, which compares the intensity of incident light I_0 to the intensity of light after it has passed through the sample I. The ratio I/I_0, called the *transmittance,* is a measure of the fraction of light that passes through the sample. The amount of light absorbed is given by the *absorbance A,* where

$$A = -\log \frac{I}{I_0}$$

The absorbance can be expressed by the *Beer–Lambert law:*

$$A = \epsilon lc$$

where ϵ is the molar absorptivity or the molar extinction coefficient (in L/mol · cm), l is the distance the light travels through the solution (in cm), and c is the concentration of the absorbing species (in mol/L). The Beer–Lambert law is the basis for using spectroscopy in quantitative analysis. If ϵ and l are known, measuring A for a solution allows us to calculate the concentration of the absorbing species in the solution.

Suppose we have a pink solution containing an unknown concentration of $Co^{2+}(aq)$ ions. A sample of this solution is placed in a spectrophotometer, and the absorbance is measured at a wavelength where ϵ for $Co^{2+}(aq)$ is known to be 12 L/mol · cm. The absorbance A is found to be 0.60. The width of the sample tube is 1.0 cm. We want to determine the concentration of $Co^{2+}(aq)$ in the solution. This problem can be solved by a straightforward application of the Beer–Lambert law,

$$A = \epsilon lc$$

where

$$A = 0.60$$

$$\epsilon = \frac{12\ \text{L}}{\text{mol} \cdot \text{cm}}$$

$$l = \text{light path} = 1.0\ \text{cm}$$

Solving for the concentration gives

$$c = \frac{A}{\epsilon l} = \frac{0.60}{\left(12\ \dfrac{\text{L}}{\text{mol} \cdot \text{cm}}\right)(1.0\ \text{cm})} = 5.0 \times 10^{-2}\ \text{mol/L}$$

To obtain the unknown concentration of an absorbing species from the measured absorbance, we must know the product ϵl, since

$$c = \frac{A}{\epsilon l}$$

We can obtain the product ϵl by measuring the absorbance of a solution of *known* concentration, since

Measured using a
↙ spectrophotometer

$$\epsilon l = \frac{A}{c}$$

↖ Known from making up
the solution

However, a more accurate value of the product ϵl can be obtained by plotting A versus c for a series of solutions. Note that the equation $A = \epsilon l c$ gives a straight line with slope ϵl when A is plotted against c.

For example, consider the following typical spectroscopic analysis. A sample of steel from a bicycle frame is to be analyzed to determine its manganese content. The procedure involves weighing out a sample of the steel, dissolving it in strong acid, treating the resulting solution with a very strong oxidizing agent to convert all the manganese to permanganate ion (MnO_4^-), and then using spectroscopy to determine the concentration of the intensely purple MnO_4^- ions in the solution. To do this, however, the value of ϵl for MnO_4^- must be determined at an appropriate wavelength. The absorbance values for four solutions with known MnO_4^- concentrations were measured to give the following data:

Solution	Concentration of MnO_4^- (mol/L)	Absorbance
1	7.00×10^{-5}	0.175
2	1.00×10^{-4}	0.250
3	2.00×10^{-4}	0.500
4	3.50×10^{-4}	0.875

A plot of absorbance versus concentration for the solutions of known concentration is shown in Fig. A-7. The slope of this line (change in A/change in c) is 2.48×10^3 L/mol. This quantity represents the product ϵl.

A sample of the steel weighing 0.1523 g was dissolved and the unknown amount of manganese was converted to MnO_4^- ions. Water was then added to give a solution with a final volume of 100.0 mL. A portion of this solution was placed in a spectrophotometer, and its absorbance was found to be 0.780. Using these data, we want to calculate the percent manganese in the steel. The MnO_4^- ions from the manganese in the

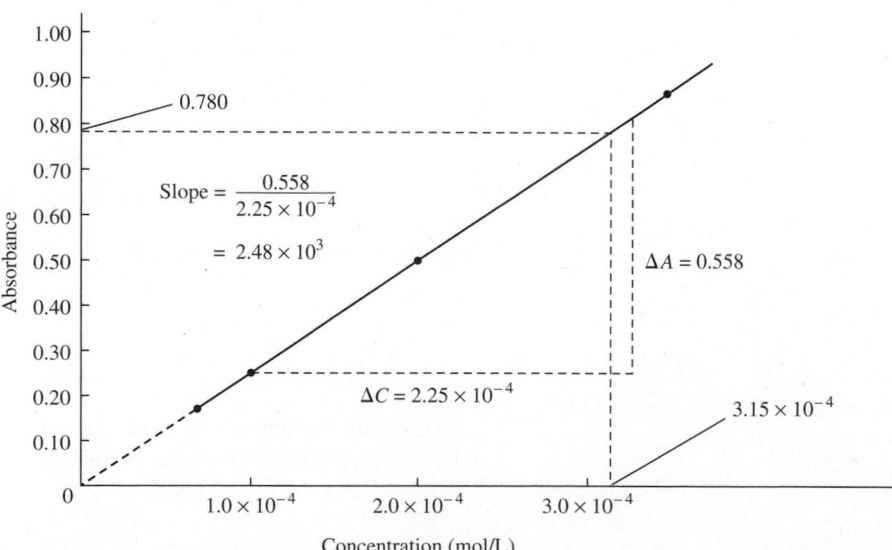

Figure A-7 | A plot of absorbance versus concentration of MnO_4^- in a series of solutions of known concentration.

dissolved steel sample show an absorbance of 0.780. Using the Beer–Lambert law, we calculate the concentration of MnO_4^- in this solution:

$$c = \frac{A}{\epsilon l} = \frac{0.780}{2.48 \times 10^3 \text{ L/mol}} \times 3.15 \times 10^{-4} \text{ mol/L}$$

There is a more direct way for finding c. Using a graph such as that in Fig. A-7 (often called a *Beer's law plot*), we can read the concentration that corresponds to $A = 0.780$. This interpolation is shown by dashed lines on the graph. By this method, $c = 3.15 \times 10^{-4}$ mol/L, which agrees with the value obtained above.

Recall that the original 0.1523-g steel sample was dissolved, the manganese was converted to permanganate, and the volume was adjusted to 100.0 mL. We now know that $[MnO_4^-]$ in that solution is 3.15×10^{-4} M. Using this concentration, we can calculate the total number of moles of MnO_4^- in that solution:

$$\text{mol of } MnO_4^- = 100.0 \text{ mL} \times \frac{1 \text{ L}}{1000 \text{ mL}} \times 3.15 \times 10^{-4} \frac{\text{mol}}{\text{L}}$$

$$= 3.15 \times 10^{-5} \text{ mol}$$

Since each mole of manganese in the original steel sample yields a mole of MnO_4^-, that is,

$$1 \text{ mol of Mn} \xrightarrow{\text{Oxidation}} 1 \text{ mol of } MnO_4^-$$

the original steel sample must have contained 3.15×10^{-5} mol of manganese. The mass of manganese present in the sample is

$$3.15 \times 10^{-5} \text{ mol of Mn} \times \frac{54.938 \text{ g of Mn}}{1 \text{ mol of Mn}} = 1.73 \times 10^{-3} \text{ g of Mn}$$

Since the steel sample weighed 0.1523 g, the present manganese in the steel is

$$\frac{1.73 \times 10^{-3} \text{ g of Mn}}{1.523 \times 10^{-1} \text{ g of sample}} \times 100 = 1.14\%$$

This example illustrates a typical use of spectroscopy in quantitative analysis. The steps commonly involved are as follows:

1. Preparation of a calibration plot (a Beer's law plot) by measuring the absorbance values of a series of solutions with known concentrations.
2. Measurement of the absorbance of the solution of unknown concentration.
3. Use of the calibration plot to determine the unknown concentration.

APPENDIX 4   Selected Thermodynamic Data

Note: All values are assumed precise to at least ± 1.

Substance and State	ΔH°_f (kJ/mol)	ΔG°_f (kJ/mol)	S° (J/K · mol)
Aluminum			
Al(s)	0	0	28
Al₂O₃(s)	−1676	−1582	51
Al(OH)₃(s)	−1277		
AlCl₃(s)	−704	−629	111
Barium			
Ba(s)	0	0	67
BaCO₃(s)	−1219	−1139	112

Substance and State	ΔH°_f (kJ/mol)	ΔG°_f (kJ/mol)	S° (J/K · mol)
Barium (*continued*)			
BaO(s)	−582	−552	70
Ba(OH)₂(s)	−946		
BaSO₄(s)	−1465	−1353	132
Beryllium			
Be(s)	0	0	10
BeO(s)	−599	−569	14
Be(OH)₂(s)	−904	−815	47

(continued)

Substance and State	ΔH°_f (kJ/mol)	ΔG°_f (kJ/mol)	S° (J/K · mol)
Bromine			
$Br_2(l)$	0	0	152
$Br_2(g)$	31	3	245
$Br_2(aq)$	−3	4	130
$Br^-(aq)$	−121	−104	82
$HBr(g)$	−36	−53	199
Cadmium			
$Cd(s)$	0	0	52
$CdO(s)$	−258	−228	55
$Cd(OH)_2(s)$	−561	−474	96
$CdS(s)$	−162	−156	65
$CdSO_4(s)$	−935	−823	123
Calcium			
$Ca(s)$	0	0	41
$CaC_2(s)$	−63	−68	70
$CaCO_3(s)$	−1207	−1129	93
$CaO(s)$	−635	−604	40
$Ca(OH)_2(s)$	−987	−899	83
$Ca_3(PO_4)_2(s)$	−4126	−3890	241
$CaSO_4(s)$	−1433	−1320	107
$CaSiO_3(s)$	−1630	−1550	84
Carbon			
$C(s)$ (graphite)	0	0	6
$C(s)$ (diamond)	2	3	2
$CO(g)$	−110.5	−137	198
$CO_2(g)$	−393.5	−394	214
$CH_4(g)$	−75	−51	186
$CH_3OH(g)$	−201	−163	240
$CH_3OH(l)$	−239	−166	127
$H_2CO(g)$	−116	−110	219
$HCOOH(g)$	−363	−351	249
$HCN(g)$	135.1	125	202
$C_2H_2(g)$	227	209	201
$C_2H_4(g)$	52	68	219
$CH_3CHO(g)$	−166	−129	250
$C_2H_5OH(l)$	−278	−175	161
$C_2H_6(g)$	−84.7	−32.9	229.5
$C_3H_6(g)$	20.9	62.7	266.9
$C_3H_8(g)$	−104	−24	270
$C_2H_4O(g)$ (ethylene oxide)	−53	−13	242
$CH_2{=}CHCN(g)$	185.0	195.4	274
$CH_3COOH(l)$	−484	−389	160
$C_6H_{12}O_6(s)$	−1275	−911	212
CCl_4	−135	−65	216
Chlorine			
$Cl_2(g)$	0	0	223
$Cl_2(aq)$	−23	7	121
$Cl^-(aq)$	−167	−131	57
$HCl(g)$	−92	−95	187
Chromium			
$Cr(s)$	0	0	24
$Cr_2O_3(s)$	−1128	−1047	81
$CrO_3(s)$	−579	−502	72
Copper	*		
$Cu(s)$	0	0	33
$CuCO_3(s)$	−595	−518	88
$Cu_2O(s)$	−170	−148	93
$CuO(s)$	−156	−128	43

Substance and State	ΔH°_f (kJ/mol)	ΔG°_f (kJ/mol)	S° (J/K · mol)
Copper (*continued*)			
$Cu(OH)_2(s)$	−450	−372	108
$CuS(s)$	−49	−49	67
Fluorine			
$F_2(g)$	0	0	203
$F^-(aq)$	−333	−279	−14
$HF(g)$	−271	−273	174
Hydrogen			
$H_2(g)$	0	0	131
$H(g)$	217	203	115
$H^+(aq)$	0	0	0
$OH^-(aq)$	−230	−157	−11
$H_2O(l)$	−286	−237	70
$H_2O(g)$	−242	−229	189
Iodine			
$I_2(s)$	0	0	116
$I_2(g)$	62	19	261
$I_2(aq)$	23	16	137
$I^-(aq)$	−55	−52	106
Iron			
$Fe(s)$	0	0	27
$Fe_3C(s)$	21	15	108
$Fe_{0.95}O(s)$ (wustite)	−264	−240	59
FeO	−272	−255	61
$Fe_3O_4(s)$ (magnetite)	−1117	−1013	146
$Fe_2O_3(s)$ (hematite)	−826	−740	90
$FeS(s)$	−95	−97	67
$FeS_2(s)$	−178	−166	53
$FeSO_4(s)$	−929	−825	121
Lead			
$Pb(s)$	0	0	65
$PbO_2(s)$	−277	−217	69
$PbS(s)$	−100	−99	91
$PbSO_4(s)$	−920	−813	149
Magnesium			
$Mg(s)$	0	0	33
$MgCO_3(s)$	−1113	−1029	66
$MgO(s)$	−602	−569	27
$Mg(OH)_2(s)$	−925	−834	64
Manganese			
$Mn(s)$	0	0	32
$MnO(s)$	−385	−363	60
$Mn_3O_4(s)$	−1387	−1280	149
$Mn_2O_3(s)$	−971	−893	110
$MnO_2(s)$	−521	−466	53
$MnO_4^-(aq)$	−543	−449	190
Mercury			
$Hg(l)$	0	0	76
$Hg_2Cl_2(s)$	−265	−211	196
$HgCl_2(s)$	−230	−184	144
$HgO(s)$	−90	−59	70
$HgS(s)$	−58	−49	78
Nickel			
$Ni(s)$	0	0	30
$NiCl_2(s)$	−316	−272	107
$NiO(s)$	−241	−213	38
$Ni(OH)_2(s)$	−538	−453	79
$NiS(s)$	−93	−90	53

Substance and State	ΔH°_f (kJ/mol)	ΔG°_f (kJ/mol)	S° (J/K · mol)
Nitrogen			
$N_2(g)$	0	0	192
$NH_3(g)$	−46	−17	193
$NH_3(aq)$	−80	−27	111
$NH_4^+(aq)$	−132	−79	113
$NO(g)$	90	87	211
$NO_2(g)$	34	52	240
$N_2O(g)$	82	104	220
$N_2O_4(g)$	10	98	304
$N_2O_4(l)$	−20	97	209
$N_2O_5(s)$	−42	134	178
$N_2H_4(l)$	51	149	121
$N_2H_3CH_3(l)$	54	180	166
$HNO_3(aq)$	−207	−111	146
$HNO_3(l)$	−174	−81	156
$NH_4ClO_4(s)$	−295	−89	186
$NH_4Cl(s)$	−314	−203	96
Oxygen			
$O_2(g)$	0	0	205
$O(g)$	249	232	161
$O_3(g)$	143	163	239
Phosphorus			
$P(s)$ (white)	0	0	41
$P(s)$ (red)	−18	−12	23
$P(s)$ (black)	−39	−33	23
$P_4(g)$	59	24	280
$PF_5(g)$	−1578	−1509	296
$PH_3(g)$	5	13	210
$H_3PO_4(s)$	−1279	−1119	110
$H_3PO_4(l)$	−1267	—	—
$H_3PO_4(aq)$	−1288	−1143	158
$P_4O_{10}(s)$	−2984	−2698	229
Potassium			
$K(s)$	0	0	64
$KCl(s)$	−436	−408	83
$KClO_3(s)$	−391	−290	143
$KClO_4(s)$	−433	−304	151
$K_2O(s)$	−361	−322	98
$K_2O_2(s)$	−496	−430	113
$KO_2(s)$	−283	−238	117
$KOH(s)$	−425	−379	79
$KOH(aq)$	−481	−440	9.20
Silicon			
$SiO_2(s)$ (quartz)	−911	−856	42
$SiCl_4(l)$	−687	−620	240
Silver			
$Ag(s)$	0	0	43
$Ag^+(aq)$	105	77	73
$AgBr(s)$	−100	−97	107
$AgCN(s)$	146	164	84
$AgCl(s)$	−127	−110	96
$Ag_2CrO_4(s)$	−712	−622	217
$AgI(s)$	−62	−66	115
$Ag_2O(s)$	−31	−11	122
$Ag_2S(s)$	−32	−40	146

Substance and State	ΔH°_f (kJ/mol)	ΔG°_f (kJ/mol)	S° (J/K · mol)
Sodium			
$Na(s)$	0	0	51
$Na^+(aq)$	−240	−262	59
$NaBr(s)$	−360	−347	84
$Na_2CO_3(s)$	−1131	−1048	136
$NaHCO_3(s)$	−948	−852	102
$NaCl(s)$	−411	−384	72
$NaH(s)$	−56	−33	40
$NaI(s)$	−288	−282	91
$NaNO_2(s)$	−359		
$NaNO_3(s)$	−467	−366	116
$Na_2O(s)$	−416	−377	73
$Na_2O_2(s)$	−515	−451	95
$NaOH(s)$	−427	−381	64
$NaOH(aq)$	−470	−419	50
Sulfur			
$S(s)$ (rhombic)	0	0	32
$S(s)$ (monoclinic)	0.3	0.1	33
$S^{2-}(aq)$	33	86	−15
$S_8(g)$	102	50	431
$SF_6(g)$	−1209	−1105	292
$H_2S(g)$	−21	−34	206
$SO_2(g)$	−297	−300	248
$SO_3(g)$	−396	−371	257
$SO_4^{2-}(aq)$	−909	−745	20
$H_2SO_4(l)$	−814	−690	157
$H_2SO_4(aq)$	−909	−745	20
Tin			
$Sn(s)$ (white)	0	0	52
$Sn(s)$ (gray)	−2	0.1	44
$SnO(s)$	−285	−257	56
$SnO_2(s)$	−581	−520	52
$Sn(OH)_2(s)$	−561	−492	155
Titanium			
$TiCl_4(g)$	−763	−727	355
$TiO_2(s)$	−945	−890	50
Uranium			
$U(s)$	0	0	50
$UF_6(s)$	−2137	−2008	228
$UF_6(g)$	−2113	−2029	380
$UO_2(s)$	−1084	−1029	78
$U_3O_8(s)$	−3575	−3393	282
$UO_3(s)$	−1230	−1150	99
Xenon			
$Xe(g)$	0	0	170
$XeF_2(g)$	−108	−48	254
$XeF_4(s)$	−251	−121	146
$XeF_6(g)$	−294		
$XeO_3(s)$	402		
Zinc			
$Zn(s)$	0	0	42
$ZnO(s)$	−348	−318	44
$Zn(OH)_2(s)$	−642		
$ZnS(s)$ (wurtzite)	−193		
$ZnS(s)$ (zinc blende)	−206	−201	58
$ZnSO_4(s)$	−983	−874	120

APPENDIX 5 | Equilibrium Constants and Reduction Potentials

A5-1 | Values of K_a for Some Common Monoprotic Acids

Name	Formula	Value of K_a
Hydrogen sulfate ion	HSO_4^-	1.2×10^{-2}
Chlorous acid	$HClO_2$	1.2×10^{-2}
Monochloracetic acid	$HC_2H_2ClO_2$	1.35×10^{-3}
Hydrofluoric acid	HF	7.2×10^{-4}
Nitrous acid	HNO_2	4.0×10^{-4}
Formic acid	HCO_2H	1.8×10^{-4}
Lactic acid	$HC_3H_5O_3$	1.38×10^{-4}
Benzoic acid	$HC_7H_5O_2$	6.4×10^{-5}
Acetic acid	$HC_2H_3O_2$	1.8×10^{-5}
Hydrated aluminum(III) ion	$[Al(H_2O)_6]^{3+}$	1.4×10^{-5}
Propanoic acid	$HC_3H_5O_2$	1.3×10^{-5}
Hypochlorous acid	$HOCl$	3.5×10^{-8}
Hypobromous acid	$HOBr$	2×10^{-9}
Hydrocyanic acid	HCN	6.2×10^{-10}
Boric acid	H_3BO_3	5.8×10^{-10}
Ammonium ion	NH_4^+	5.6×10^{-10}
Phenol	HOC_6H_5	1.6×10^{-10}
Hypoiodous acid	HOI	2×10^{-11}

A5-2 | Stepwise Dissociation Constants for Several Common Polyprotic Acids

Name	Formula	K_{a_1}	K_{a_2}	K_{a_3}
Phosphoric acid	H_3PO_4	7.5×10^{-3}	6.2×10^{-8}	4.8×10^{-13}
Arsenic acid	H_3AsO_4	5×10^{-3}	8×10^{-8}	6×10^{-10}
Carbonic acid	H_2CO_3	4.3×10^{-7}	5.6×10^{-11}	
Sulfuric acid	H_2SO_4	Large	1.2×10^{-2}	
Sulfurous acid	H_2SO_3	1.5×10^{-2}	1.0×10^{-7}	
Hydrosulfuric acid	H_2S	1.0×10^{-7}	$\sim 10^{-19}$	
Oxalic acid	$H_2C_2O_4$	6.5×10^{-2}	6.1×10^{-5}	
Ascorbic acid (vitamin C)	$H_2C_6H_6O_6$	7.9×10^{-5}	1.6×10^{-12}	
Citric acid	$H_3C_6H_5O_7$	8.4×10^{-4}	1.8×10^{-5}	4.0×10^{-6}

A5-3 | Values of K_b for Some Common Weak Bases

Name	Conjugate Formula	Acid	K_b
Ammonia	NH_3	NH_4^+	1.8×10^{-5}
Methylamine	CH_3NH_2	$CH_3NH_3^+$	4.38×10^{-4}
Ethylamine	$C_2H_5NH_2$	$C_2H_5NH_3^+$	5.6×10^{-4}
Diethylamine	$(C_2H_5)_2NH$	$(C_2H_5)_2NH_2^+$	1.3×10^{-3}
Triethylamine	$(C_2H_5)_3N$	$(C_2H_5)_3NH^+$	4.0×10^{-4}
Hydroxylamine	$HONH_2$	$HONH_3^+$	1.1×10^{-8}
Hydrazine	H_2NNH_2	$H_2NNH_3^+$	3.0×10^{-6}
Aniline	$C_6H_5NH_2$	$C_6H_5NH_3^+$	3.8×10^{-10}
Pyridine	C_5H_5N	$C_5H_5NH^+$	1.7×10^{-9}

A5-4 | K_{sp} Values at 25°C for Common Ionic Solids

Ionic Solid	K_{sp} (at 25°C)	Ionic Solid	K_{sp} (at 25°C)	Ionic Solid	K_{sp} (at 25°C)
Fluorides		Hg_2CrO_4*	2×10^{-9}	$Ni(OH)_2$	1.6×10^{-16}
BaF_2	2.4×10^{-5}	$BaCrO_4$	8.5×10^{-11}	$Zn(OH)_2$	4.5×10^{-17}
MgF_2	6.4×10^{-9}	Ag_2CrO_4	9.0×10^{-12}	$Cu(OH)_2$	1.6×10^{-19}
PbF_2	4×10^{-8}	$PbCrO_4$	2×10^{-16}	$Hg(OH)_2$	3×10^{-26}
SrF_2	7.9×10^{-10}			$Sn(OH)_2$	3×10^{-27}
CaF_2	4.0×10^{-11}	**Carbonates**		$Cr(OH)_3$	6.7×10^{-31}
		$NiCO_3$	1.4×10^{-7}	$Al(OH)_3$	2×10^{-32}
		$CaCO_3$	8.7×10^{-9}	$Fe(OH)_3$	4×10^{-38}
Chlorides		$BaCO_3$	1.6×10^{-9}	$Co(OH)_3$	2.5×10^{-43}
$PbCl_2$	1.6×10^{-5}	$SrCO_3$	7×10^{-10}		
$AgCl$	1.6×10^{-10}	$CuCO_3$	2.5×10^{-10}	**Sulfides**	
Hg_2Cl_2*	1.1×10^{-18}	$ZnCO_3$	2×10^{-10}	MnS	2.3×10^{-13}
		$MnCO_3$	8.8×10^{-11}	FeS	3.7×10^{-19}
Bromides		$FeCO_3$	2.1×10^{-11}	NiS	3×10^{-21}
$PbBr_2$	4.6×10^{-6}	Ag_2CO_3	8.1×10^{-12}	CoS	5×10^{-22}
$AgBr$	5.0×10^{-13}	$CdCO_3$	5.2×10^{-12}	ZnS	2.5×10^{-22}
Hg_2Br_2*	1.3×10^{-22}	$PbCO_3$	1.5×10^{-15}	SnS	1×10^{-26}
		$MgCO_3$	1×10^{-5}	CdS	1.0×10^{-28}
Iodides		Hg_2CO_3*	9.0×10^{-15}	PbS	7×10^{-29}
PbI_2	1.4×10^{-8}			CuS	8.5×10^{-45}
AgI	1.5×10^{-16}	**Hydroxides**		Ag_2S	1.6×10^{-49}
Hg_2I_2*	4.5×10^{-29}	$Ba(OH)_2$	5.0×10^{-3}	HgS	1.6×10^{-54}
		$Sr(OH)_2$	3.2×10^{-4}		
Sulfates		$Ca(OH)_2$	1.3×10^{-6}	**Phosphates**	
$CaSO_4$	6.1×10^{-5}	$AgOH$	2.0×10^{-8}	Ag_3PO_4	1.8×10^{-18}
Ag_2SO_4	1.2×10^{-5}	$Mg(OH)_2$	8.9×10^{-12}	$Sr_3(PO_4)_2$	1×10^{-31}
$SrSO_4$	3.2×10^{-7}	$Mn(OH)_2$	2×10^{-13}	$Ca_3(PO_4)_2$	1.3×10^{-32}
$PbSO_4$	1.3×10^{-8}	$Cd(OH)_2$	5.9×10^{-15}	$Ba_3(PO_4)_2$	6×10^{-39}
$BaSO_4$	1.5×10^{-9}	$Pb(OH)_2$	1.2×10^{-15}	$Pb_3(PO_4)_2$	1×10^{-54}
		$Fe(OH)_2$	1.8×10^{-15}		
Chromates		$Co(OH)_2$	2.5×10^{-16}		
$SrCrO_4$	3.6×10^{-5}				

*Contains Hg_2^{2+} ions. $K_{sp} = [Hg_2^{2+}][X^-]^2$ for Hg_2X_2 salts.

A5-5 | Standard Reduction Potentials at 25°C (298 K) for Many Common Half-Reactions

Half-Reaction	$\mathscr{E}°$ (V)	Half-Reaction	$\mathscr{E}°$ (V)
$F_2 + 2e^- \rightarrow 2F^-$	2.87	$O_2 + 2H_2O + 4e^- \rightarrow 4OH^-$	0.40
$Ag^{2+} + e^- \rightarrow Ag^+$	1.99	$Cu^{2+} + 2e^- \rightarrow Cu$	0.34
$Co^{3+} + e^- \rightarrow Co^{2+}$	1.82	$Hg_2Cl_2 + 2e^- \rightarrow 2Hg + 2Cl^-$	0.34
$H_2O_2 + 2H^+ + 2e^- \rightarrow 2H_2O$	1.78	$AgCl + e^- \rightarrow Ag + Cl^-$	0.22
$Ce^{4+} + e^- \rightarrow Ce^{3+}$	1.70	$SO_4^{2-} + 4H^+ + 2e^- \rightarrow H_2SO_3 + H_2O$	0.20
$PbO_2 + 4H^+ + SO_4^{2-} + 2e^- \rightarrow PbSO_4 + 2H_2O$	1.69	$Cu^{2+} + e^- \rightarrow Cu^+$	0.16
$MnO_4^- + 4H^+ + 3e^- \rightarrow MnO_2 + 2H_2O$	1.68	$2H^+ + 2e^- \rightarrow H_2$	0.00
$2e^- + 2H^+ + IO_4^- \rightarrow IO_3^- + H_2O$	1.60	$Fe^{3+} + 3e^- \rightarrow Fe$	-0.036
$MnO_4^- + 8H^+ + 5e^- \rightarrow Mn^{2+} + 4H_2O$	1.51	$Pb^{2+} + 2e^- \rightarrow Pb$	-0.13
$Au^{3+} + 3e^- \rightarrow Au$	1.50	$Sn^{2+} + 2e^- \rightarrow Sn$	-0.14
$PbO_2 + 4H^+ + 2e^- \rightarrow Pb^{2+} + 2H_2O$	1.46	$Ni^{2+} + 2e^- \rightarrow Ni$	-0.23
$Cl_2 + 2e^- \rightarrow 2Cl^-$	1.36	$PbSO_4 + 2e^- \rightarrow Pb + SO_4^{2-}$	-0.35
$Cr_2O_7^{2-} + 14H^+ + 6e^- \rightarrow 2Cr^{3+} + 7H_2O$	1.33	$Cd^{2+} + 2e^- \rightarrow Cd$	-0.40
$O_2 + 4H^+ + 4e^- \rightarrow 2H_2O$	1.23	$Cr^{3+} + e^- \rightarrow Cr^{2+}$	-0.41
$MnO_2 + 4H^+ + 2e^- \rightarrow Mn^{2+} + 2H_2O$	1.21	$Fe^{2+} + 2e^- \rightarrow Fe$	-0.44
$IO_3^- + 6H^+ + 5e^- \rightarrow \frac{1}{2}I_2 + 3H_2O$	1.20	$Cr^{3+} + 3e^- \rightarrow Cr$	-0.73
$Br_2 + 2e^- \rightarrow 2Br^-$	1.09	$Zn^{2+} + 2e^- \rightarrow Zn$	-0.76
$VO_2^+ + 2H^+ + e^- \rightarrow VO^{2+} + H_2O$	1.00	$2H_2O + 2e^- \rightarrow H_2 + 2OH^-$	-0.83
$AuCl_4^- + 3e^- \rightarrow Au + 4Cl^-$	0.99	$Mn^{2+} + 2e^- \rightarrow Mn$	-1.18
$NO_3^- + 4H^+ + 3e^- \rightarrow NO + 2H_2O$	0.96	$Al^{3+} + 3e^- \rightarrow Al$	-1.66
$ClO_2 + e^- \rightarrow ClO_2^-$	0.954	$H_2 + 2e^- \rightarrow 2H^-$	-2.23
$2Hg^{2+} + 2e^- \rightarrow Hg_2^{2+}$	0.91	$Mg^{2+} + 2e^- \rightarrow Mg$	-2.37
$Ag^+ + e^- \rightarrow Ag$	0.80	$La^{3+} + 3e^- \rightarrow La$	-2.37
$Hg_2^{2+} + 2e^- \rightarrow 2Hg$	0.80	$Na^+ + e^- \rightarrow Na$	-2.71
$Fe^{3+} + e^- \rightarrow Fe^{2+}$	0.77	$Ca^{2+} + 2e^- \rightarrow Ca$	-2.76
$O_2 + 2H^+ + 2e^- \rightarrow H_2O_2$	0.68	$Ba^{2+} + 2e^- \rightarrow Ba$	-2.90
$MnO_4^- + e^- \rightarrow MnO_4^{2-}$	0.56	$K^+ + e^- \rightarrow K$	-2.92
$I_2 + 2e^- \rightarrow 2I^-$	0.54	$Li^+ + e^- \rightarrow Li$	-3.05
$Cu^+ + e^- \rightarrow Cu$	0.52		

APPENDIX 6 | SI Units and Conversion Factors

Length

SI unit: meter (m)

1 meter = 1.0936 yards

1 centimeter = 0.39370 inch

1 inch = 2.54 centimeters (exactly)

1 kilometer = 0.62137 mile

1 mile = 5280 feet

= 1.6093 kilometers

1 angstrom = 10^{-10} meter

= 100 picometers

Mass

SI unit: kilogram (kg)

1 kilogram = 1000 grams

= 2.2046 pounds

1 pound = 453.59 grams

= 0.45359 kilogram

= 16 ounces

1 ton = 2000 pounds

= 907.185 kilograms

1 metric ton = 1000 kilograms

= 2204.6 pounds

1 atomic mass unit = 1.66056×10^{-27} kilograms

Volume

SI unit: cubic meter (m³)

1 liter = 10^{-3} m³

= 1 dm³

= 1.0567 quarts

1 gallon = 4 quarts

= 8 pints

= 3.7854 liters

1 quart = 32 fluid ounces

= 0.94633 liter

Temperature

SI unit: kelvin (K)

0 K = −273.15°C

= −459.67°F

K = °C + 273.15

$°C = \dfrac{5}{9}(°F - 32)$

$°F = \dfrac{9}{5}(°C) + 32$

Energy

SI unit: joule (J)

1 joule = 1 kg · m²/s²

= 0.23901 calorie

= 9.4781×10^{-4} btu

(British thermal unit)

1 calorie = 4.184 joules

= 3.965×10^{-3} btu

1 btu = 1055.06 joules

= 252.2 calories

Pressure

SI unit: pascal (Pa)

1 pascal = 1 N/m²

= 1 kg/m · s²

1 atmosphere = 101.325 kilopascals

= 760 torr (mmHg)

= 14.70 pounds per square inch

1 bar = 10^5 pascals

Glossary

Accuracy the agreement of a particular value with the true value. (R-2)

Acid a substance that produces hydrogen ions in solution; a proton donor. (6)

Acid–base indicator a substance that marks the end point of an acid–base titration by changing color. (14-5)

Acid dissociation constant (K_a) the equilibrium constant for a reaction in which a proton is removed from an acid by H_2O to form the conjugate base and H_3O^+. (13-1)

Acid rain a result of air pollution by sulfur dioxide. (8-10)

Acidic oxide a covalent oxide that dissolves in water to give an acidic solution. (13-10)

Actinide series a group of 14 elements following actinium in the periodic table, in which the $5f$ orbitals are being filled. (2-11; 19-1)

Activated complex (transition state) the arrangement of atoms found at the top of the potential energy barrier as a reaction proceeds from reactants to products. (11-6)

Activation energy the threshold energy that must be overcome to produce a chemical reaction. (11-6)

Addition polymerization a type of polymerization in which the monomers simply add together to form the polymer, with no other products. (21-5)

Addition reaction a reaction in which atoms add to a carbon–carbon multiple bond. (21-2)

Adsorption the collection of one substance on the surface of another. (11-7)

Air pollution contamination of the atmosphere, mainly by the gaseous products of transportation and production of electricity. (8-10)

Alcohol an organic compound in which the hydroxyl group is a substituent on a hydrocarbon. (21-4)

Aldehyde an organic compound containing the carbonyl group bonded to at least one hydrogen atom. (21-4)

Alkali metal a Group 1A metal. (2-13; 19-2)

Alkaline earth metal a Group 2A metal. (2-13; 19-4)

Alkane a saturated hydrocarbon with the general formula C_nH_{2n+2}. (21-1)

Alkene an unsaturated hydrocarbon containing a carbon–carbon double bond. The general formula is C_nH_{2n}. (21-2)

Alkyne an unsaturated hydrocarbon containing a triple carbon–carbon bond. The general formula is C_nH_{2n-2}. (21-2)

Alloy a substance that contains a mixture of elements and has metallic properties. (9-4)

Alloy steel a form of steel containing carbon plus other metals such as chromium, cobalt, manganese, and molybdenum. (20-8)

Alpha (α) particle a helium nucleus. (18-1)

Alpha-particle production a common mode of decay for radioactive nuclides in which the mass number changes. (18-1)

Amine an organic base derived from ammonia in which one or more of the hydrogen atoms are replaced by organic groups. (13-6; 21-4)

α-Amino acid an organic acid in which an amino group and an R group are attached to the carbon atom next to the carboxyl group. (21-6)

Amorphous solid a solid with considerable disorder in its structure. (9-3)

Ampere the unit of electric current equal to one coulomb of charge per second. (17-8)

Amphoteric substance a substance that can behave either as an acid or as a base. (13-2)

Angular momentum quantum number (ℓ) the quantum number relating to the shape of an atomic orbital, which can assume any integral value from 0 to $n-1$ for each value of n. (2-6)

Anion a negative ion. (1-7)

Anode the electrode in a galvanic cell at which oxidation occurs. (17-2)

Antibonding molecular orbital an orbital higher in energy than the atomic orbitals of which it is composed. (4-4)

Aqueous solution a solution in which water is the dissolving medium or solvent. (6)

Aromatic hydrocarbon one of a special class of cyclic unsaturated hydrocarbons, the simplest of which is benzene. (21-3)

Arrhenius concept a concept postulating that acids produce hydrogen ions in aqueous solution, while bases produce hydroxide ions. (13-1)

Arrhenius equation the equation representing the rate constant as $k = Ae^{-E_a/RT}$, where A represents the product of the collision frequency and the steric factor, and $e^{-E_a/RT}$ is the fraction of collisions with sufficient energy to produce a reaction. (11-6)

Atactic chain a polymer chain in which the substituent groups such as CH_3 are randomly distributed along the chain. (21-5)

Atmosphere the mixture of gases that surrounds the earth's surface. (8-10)

Atomic number the number of protons in the nucleus of an atom. (1-7; 17)

Atomic radius half the distance between the nuclei in a molecule consisting of identical atoms. (2-12)

Atomic solid a solid that contains atoms at the lattice points. (9-3)

Atomic weight the weighted average mass of the atoms in a naturally occurring element. (1-5)

Aufbau principle the principle stating that as protons are added one by one to the nucleus to build up the elements, electrons are similarly added to hydrogen-like orbitals. (2-11)

Autoionization the transfer of a proton from one molecule to another of the same substance. (13-2)

Avogadro's law equal volumes of gases at the same temperature and pressure contain the same number of particles. (8-2)

Avogadro's number the number of atoms in exactly 12 grams of pure ^{12}C, equal to 6.022×10^{23}. (5-4)

Ball-and-stick model a molecular model that distorts the sizes of atoms but shows bond relationships clearly. (3-4)

Band model a molecular model for metals in which the electrons are assumed to travel around the metal crystal in molecular orbitals formed from the valence atomic orbitals of the metal atoms. (9-4)

Barometer a device for measuring atmospheric pressure. (8-1)

Base a substance that produces hydroxide ions in aqueous solution; a proton acceptor. (6-8)

Basic oxide an ionic oxide that dissolves in water to produce a basic solution. (13-10)

Basic oxygen process a process for producing steel by oxidizing and removing the impurities in iron using a high-pressure blast of oxygen. (20-8)

Battery a group of galvanic cells connected in series. (17-6)

Beta (β) particle an electron produced in radioactive decay. (18-1)

Beta-particle production a decay process for radioactive nuclides in which the mass number remains constant and the atomic number changes. The net effect is to change a neutron to a proton. (18-1)

Bidentate ligand a ligand that can form two bonds to a metal ion. (20-3)

Bimolecular step a reaction involving the collision of two molecules. (11-5)

Binary compound a two-element compound. (3-12)

Binding energy (nuclear) the energy required to decompose a nucleus into its component nucleons. (18-5)

Biomolecule a molecule responsible for maintaining and/or reproducing life. (21)

Blast furnace a furnace in which iron oxide is reduced to iron metal by using a very strong blast of hot air to produce carbon monoxide from coke, and then using this gas as a reducing agent for the iron. (20-8)

Bond energy the energy required to break a given chemical bond. (4-1)

Bond length the distance between the nuclei of the two atoms connected by a bond; the distance where the total energy of a diatomic molecule is minimal. (4-1)

Bond order the difference between the number of bonding electrons and the number of antibonding electrons, divided by two. It is an index of bond strength. (4-4)

Bonding molecular orbital an orbital lower in energy than the atomic orbitals of which it is composed. (4-4)

Bonding pair an electron pair found in the space between two atoms. (3-8)

Borane a covalent hydride of boron. (19-5)

Boyle's law the volume of a given sample of gas at constant temperature varies inversely with the pressure. (8-2)

Breeder reactor a nuclear reactor in which fissionable fuel is produced while the reactor runs. (7-3)

Brønsted–Lowry model a model proposing that an acid is a proton donor, and a base is a proton acceptor. (13-1)

Buffered solution a solution that resists a change in its pH when either hydroxide ions or protons are added. (14-2)

Buffering capacity the ability of a buffered solution to absorb protons or hydroxide ions without a significant change in pH; determined by the magnitudes of [HA] and [A$^-$] in the solution. (14-3)

Calorimetry the science of measuring heat flow. (7-3)

Capillary action the spontaneous rising of a liquid in a narrow tube. (9-2)

Carbohydrate a polyhydroxyl ketone, a polyhydroxyl aldehyde, or a polymer composed of these. (21-6)

Carbon steel an alloy of iron containing up to about 1.5% carbon. (20-8)

Carboxyhemoglobin a stable complex of hemoglobin and carbon monoxide that prevents normal oxygen uptake in the blood. (20-7)

Carboxyl group the —COOH group in an organic acid. (13-2; 21-4)

Carboxylic acid an organic compound containing the carboxyl group; an acid with the general formula RCOOH. (21-4)

Catalyst a substance that speeds up a reaction without being consumed. (11-7)

Cathode the electrode in a galvanic cell at which reduction occurs. (17-2)

Cathode rays the "rays" emanating from the negative electrode (cathode) in a partially evacuated tube; a stream of electrons. (1-6)

Cathodic protection a method in which an active metal, such as magnesium, is connected to steel to protect it from corrosion. (17-6)

Cation a positive ion. (1-7)

Cell potential (electromotive force) the driving force in a galvanic cell that pulls electrons from the reducing agent in one compartment to the oxidizing agent in the other. (17-2)

Ceramic a nonmetallic material made from clay and hardened by firing at high temperature; it contains minute silicate crystals suspended in a glassy cement. (9-5)

Chain reaction (nuclear) a self-sustaining fission process caused by the production of neutrons that proceed to split other nuclei. (18-6)

Charles's law the volume of a given sample of gas at constant pressure is directly proportional to the temperature in kelvins. (8-2)

Chelating ligand (chelate) a ligand having more than one atom with a lone pair that can be used to bond to a metal ion. (20-3)

Chemical bond the force or, more accurately, the energy that holds two atoms together in a compound. (3-4)

Chemical change the change of substances into other substances through a reorganization of the atoms; a chemical reaction. (R-8)

Chemical equation a representation of a chemical reaction showing the relative numbers of reactant and product molecules. (5-8)

Chemical equilibrium a dynamic reaction system in which the concentrations of all reactants and products remain constant as a function of time. (12)

Chemical formula the representation of a molecule in which the symbols for the elements are used to indicate the types of atoms present and subscripts are used to show the relative numbers of atoms. (3-4)

Chemical kinetics the area of chemistry that concerns reaction rates. (11)

Chemical stoichiometry the calculation of the quantities of material consumed and produced in chemical reactions. (9)

Chirality the quality of having nonsuperimposable mirror images. (20-4)

Chlor–alkali process the process for producing chlorine and sodium hydroxide by electrolyzing brine in a mercury cell. (17-9)

Chromatography the general name for a series of methods for separating mixtures by employing a system with a mobile phase and a stationary phase. (R-8)

Coagulation the destruction of a colloid by causing particles to aggregate and settle out. (10-8)

Codons organic bases in sets of three that form the genetic code. (21-6)

Colligative properties properties of a solution that depend only on the number, and not on the identity, of the solute particles. (10-5)

Collision model a model based on the idea that molecules must collide to react; used to account for the observed characteristics of reaction rates. (11-7)

Colloid (colloidal dispersion) a suspension of particles in a dispersing medium. (10-8)

Combustion reaction the vigorous and exothermic reaction that takes place between certain substances, particularly organic compounds, and oxygen. (21-1)

Common ion effect the shift in an equilibrium position caused by the addition or presence of an ion involved in the equilibrium reaction. (14-1)

Complete ionic equation an equation that shows all substances that are strong electrolytes as ions. (6-6)

Complex ion a charged species consisting of a metal ion surrounded by ligands. (15-3; 20-1)

Compound a substance with constant composition that can be broken down into elements by chemical processes. (R-8)

Concentration cell a galvanic cell in which both compartments contain the same components, but at different concentrations. (17-5)

Condensation the process by which vapor molecules re-form a liquid. (9-9)

Condensation polymerization a type of polymerization in which the formation of a small molecule, such as water, accompanies the extension of the polymer chain. (21-5)

Condensation reaction a reaction in which two molecules are joined, accompanied by the elimination of a water molecule. (19-3)

Condensed states of matter liquids and solids. (9-1)

Conjugate acid the species formed when a proton is added to a base. (13-1)

Conjugate acid–base pair two species related to each other by the donating and accepting of a single proton. (13-1)

Conjugate base what remains of an acid molecule after a proton is lost. (13-1)

Continuous spectrum a spectrum that exhibits all the wavelengths of visible light. (2-3)

Control rods rods in a nuclear reactor composed of substances that absorb neutrons. These rods regulate the power level of the reactor. (18-6)

Coordinate covalent bond a metal–ligand bond resulting from the interaction of a Lewis base (the ligand) and a Lewis acid (the metal ion). (20-3)

Coordination compound a compound composed of a complex ion and counter ions sufficient to give no net charge. (20-3)

Coordination isomerism isomerism in a coordination compound in which the composition of the coordination sphere of the metal ion varies. (20-4)

Coordination number the number of bonds formed between the metal ion and the ligands in a complex ion. (20-3)

Copolymer a polymer formed from the polymerization of more than one type of monomer. (21-5)

Core electron an inner electron in an atom; one not in the outermost (valence) principal quantum level. (2-11)

Corrosion the process by which metals are oxidized in the atmosphere. (17-7)

Coulomb's law $E = 2.31 \times 10^{-19} \left(\dfrac{Q_1 Q_2}{r} \right)$, where E is the energy of interaction between a pair of ions, expressed in joules; r is the distance between the ion centers in nm; and Q_1 and Q_2 are the numerical ion charges. (3-1)

Counterions anions or cations that balance the charge on the complex ion in a coordination compound. (20-3)

Covalent bonding a type of bonding in which electrons are shared by atoms. (3-4; 3-1)

Critical mass the mass of fissionable material required to produce a self-sustaining chain reaction. (18-6)

Critical point the point on a phase diagram at which the temperature and pressure have their critical values; the end point of the liquid–vapor line. (9-10)

Critical pressure the minimum pressure required to produce liquefaction of a substance at the critical temperature. (9-10)

Critical reaction (nuclear) a reaction in which exactly one neutron from each fission event causes another fission event, thus sustaining the chain reaction. (18-6)

Critical temperature the temperature above which vapor cannot be liquefied no matter what pressure is applied. (9-10)

Crosslinking the existence of bonds between adjacent chains in a polymer, thus adding strength to the material. (21-5)

Crystal field model a model used to explain the magnetism and colors of coordination complexes through the splitting of the d orbital energies. (20-6)

Crystalline solid a solid with a regular arrangement of its components. (9-3)

Cubic closest packed (ccp) structure a solid modeled by the closest packing of spheres with an *abcabc* arrangement of layers; the unit cell is face-centered cubic. (9-4)

Cyanidation a process in which crushed gold ore is treated with an aqueous cyanide solution in the presence of air to dissolve the gold. Pure gold is recovered by reduction of the ion to the metal. (20-8)

Cyclotron a type of particle accelerator in which an ion introduced at the center is accelerated in an expanding spiral path by the use of alternating electrical fields in the presence of a magnetic field. (18-3)

Cytochromes a series of iron-containing species composed of heme and a protein. Cytochromes are the principal electron-transfer molecules in the respiratory chain. (20-7)

Dalton's law of partial pressures for a mixture of gases in a container, the total pressure exerted is the sum of the pressures that each gas would exert if it were alone. (8-5)

Degenerate orbitals a group of orbitals with the same energy. (2-7)

Dehydrogenation reaction a reaction in which two hydrogen atoms are removed from adjacent carbons of a saturated hydrocarbon, giving an unsaturated hydrocarbon. (21-1)

Denaturation the breaking down of the three-dimensional structure of a protein resulting in the loss of its function. (21-6)

Denitrification the return of nitrogen from decomposed matter to the atmosphere by bacteria that change nitrates to nitrogen gas. (19-2)

Density a property of matter representing the mass per unit volume. (R-7)

Deoxyribonucleic acid (DNA) a huge nucleotide polymer having a double-helical structure with complementary bases on the two strands. Its major functions are protein synthesis and the storage and transport of genetic information. (21-6)

Desalination the removal of dissolved salts from an aqueous solution. (10-6)

Dialysis a phenomenon in which a semipermeable membrane allows transfer of both solvent molecules and small solute molecules and ions. (10-6)

Diamagnetism a type of magnetism, associated with paired electrons, that causes a substance to be repelled from the inducing magnetic field. (4-5)

Differential rate law an expression that gives the rate of a reaction as a function of concentrations; often called the rate law. (11-2)

Diffraction the scattering of light from a regular array of points or lines, producing constructive and destructive interference. (2-2)

Diffusion the mixing of gases. (8-7)

Dilution the process of adding solvent to lower the concentration of solute in a solution. (6-3)

Dimer a molecule formed by the joining of two identical monomers. (21-5)

Dipole–dipole attraction the attractive force resulting when polar molecules line up so that the positive and negative ends are close to each other. (9-1)

Dipole moment a property of a molecule whose charge distribution can be represented by a center of positive charge and a center of negative charge. (4-2)

Direct reduction furnace a furnace in which iron oxide is reduced to iron metal using milder reaction conditions than in a blast furnace. (20-8)

Disaccharide a sugar formed from two monosaccharides joined by a glycoside linkage. (21-6)

Disproportionation reaction a reaction in which a given element is both oxidized and reduced. (19-7)

Distillation a method for separating the components of a liquid mixture that depends on differences in the ease of vaporization of the components. (R-8)

Disulfide linkage an S—S bond that stabilizes the tertiary structure of many proteins. (21-6)

Double bond a bond in which two pairs of electrons are shared by two atoms. (3-7)

Downs cell a cell used for electrolyzing molten sodium chloride. (17-9)

Dry cell battery a common battery used in calculators, watches, radios, and portable audio players. (17-6)

Dual nature of light the statement that light exhibits both wave and particulate properties. (2-2)

Effusion the passage of a gas through a tiny orifice into an evacuated chamber. (8-7)

Electrical conductivity the ability to conduct an electric current. (6-2)

Electrochemistry the study of the interchange of chemical and electrical energy. (17)

Electrolysis a process that involves forcing a current through a cell to cause a nonspontaneous chemical reaction to occur. (17-8)

Electrolyte a material that dissolves in water to give a solution that conducts an electric current. (6-2)

Electrolytic cell a cell that uses electrical energy to produce a chemical change that would otherwise not occur spontaneously. (17-8)

Electromagnetic radiation radiant energy that exhibits wavelike behavior and travels through space at the speed of light in a vacuum. (2-1)

Electron a negatively charged particle that moves around the nucleus of an atom. (1-6)

Electron affinity the energy change associated with the addition of an electron to a gaseous atom. (2-12)

Electron capture a process in which one of the inner-orbital electrons in an atom is captured by the nucleus. (18-1)

Electron spin quantum number a quantum number representing one of the two possible values for the electron spin; either $+\frac{1}{2}$ or $-\frac{1}{2}$. (2-8)

Electronegativity the tendency of an atom in a molecule to attract shared electrons to itself. (3-2)

Element a substance that cannot be decomposed into simpler substances by chemical or physical means. (R-8)

Elementary step a reaction whose rate law can be written from its molecularity. (11-5)

$E = mc^2$ Einstein's equation proposing that energy has mass; E is energy, m is mass, and c is the speed of light. (2-2)

Empirical formula the simplest whole-number ratio of atoms in a compound. (5-6)

Enantiomers isomers that are nonsuperimposable mirror images of each other. (20-4)

End point the point in a titration at which the indicator changes color. (6-8)

Endothermic refers to a reaction where energy (as heat) flows into the system. (7-1)

Energy the capacity to do work or to cause heat flow. (7-1)

Enthalpy a property of a system equal to $E + PV$, where E is the internal energy of the system, P is the pressure of the system, and V is the volume of the system. At constant pressure the change in enthalpy equals the energy flow as heat. (7-2)

Enthalpy (heat) of fusion the enthalpy change that occurs to melt a solid at its melting point. (9-9)

Entropy a thermodynamic function that measures randomness or disorder. (16-1)

Enzyme a large molecule, usually a protein, that catalyzes biological reactions. (11-7)

Equilibrium constant the value obtained when equilibrium concentrations of the chemical species are substituted in the equilibrium expression. (12-2)

Equilibrium expression the expression (from the law of mass action) obtained by multiplying the product concentrations and dividing by the multiplied reactant concentrations, with each concentration raised to a power represented by the coefficient in the balanced equation. (12-2)

Equilibrium point (thermodynamic definition) the position where the free energy of a reaction system has its lowest possible value. (16-8)

Equilibrium position a particular set of equilibrium concentrations. (12-2)

Equivalence point (stoichiometric point) the point in a titration when enough titrant has been added to react exactly with the substance in solution being titrated. (6-9; 14-4)

Ester an organic compound produced by the reaction between a carboxylic acid and an alcohol. (21-4)

Exothermic refers to a reaction where energy (as heat) flows out of the system. (7-1)

Exponential notation expresses a number as $N \times 10^M$, a convenient method for representing a very large or very small number and for easily indicating the number of significant figures. (R-3)

Faraday a constant representing the charge on one mole of electrons; 96,485 coulombs. (17-4)

Filtration a method for separating the components of a mixture containing a solid and a liquid. (R-8)

First law of thermodynamics the energy of the universe is constant; same as the law of conservation of energy. (7-1)

Fission the process of using a neutron to split a heavy nucleus into two nuclei with smaller mass numbers. (18-6)

Flotation process a method of separating the mineral particles in an ore from the gangue that depends on the greater wettability of the mineral pieces. (20-8)

Formal charge the charge assigned to an atom in a molecule or polyatomic ion derived from a specific set of rules. (3-11)

Formation constant (stability constant) the equilibrium constant for each step of the formation of a complex ion by the addition of an individual ligand to a metal ion or complex ion in aqueous solution. (15-3)

Formula equation an equation representing a reaction in solution showing the reactants and products in undissociated form, whether they are strong or weak electrolytes. (6-6)

Fossil fuel coal, petroleum, or natural gas; consists of carbon-based molecules derived from decomposition of once-living organisms. (7-6)

Frasch process the recovery of sulfur from underground deposits by melting it with hot water and forcing it to the surface by air pressure. (19-6)

Free energy a thermodynamic function equal to the enthalpy (H) minus the product of the entropy (S) and the Kelvin temperature (T); $G = H - TS$. Under certain conditions the change in free energy for a process is equal to the maximum useful work. (16-4)

Free radical a species with an unpaired electron. (21-5)

Frequency the number of waves (cycles) per second that pass a given point in space. (2-1)

Fuel cell a galvanic cell for which the reactants are continuously supplied. (17-6)

Functional group an atom or group of atoms in hydrocarbon derivatives that contains elements in addition to carbon and hydrogen. (21-4)

Fusion the process of combining two light nuclei to form a heavier, more stable nucleus. (18-6)

Galvanic cell a device in which chemical energy from a spontaneous redox reaction is changed to electrical energy that can be used to do work. (17-2)

Galvanizing a process in which steel is coated with zinc to prevent corrosion. (17-7)

Gamma (γ) ray a high-energy photon. (18-1)

Gangue the impurities (such as clay or sand) in an ore. (20-8)

Geiger–Müller counter (Geiger counter) an instrument that measures the rate of radioactive decay based on the ions and electrons produced as a radioactive particle passes through a gas-filled chamber. (18-4)

Gene a given segment of the DNA molecule that contains the code for a specific protein. (21-6)

Geometrical (*cis–trans*) isomerism isomerism in which atoms or groups of atoms can assume different positions around a rigid ring or bond. (20-4; 21-2)

Glass an amorphous solid obtained when silica is mixed with other compounds, heated above its melting point, and then cooled rapidly. (9-5)

Glass electrode an electrode for measuring pH from the potential difference that develops when it is dipped into an aqueous solution containing H^+ ions. (17-5)

Glycoside linkage a C—O—C bond formed between the rings of two cyclic monosaccharides by the elimination of water. (21-6)

Graham's law of effusion the rate of effusion of a gas is inversely proportional to the square root of the mass of its particles. (8-7)

Greenhouse effect a warming effect exerted by the earth's atmosphere (particularly CO_2 and H_2O) due to thermal energy retained by absorption of infrared radiation. (7-6)

Ground state the lowest possible energy state of an atom or molecule. (2-4)

Group (of the periodic table) a vertical column of elements having the same valence electron configuration and showing similar properties. (2-13)

Haber process the manufacture of ammonia from nitrogen and hydrogen, carried out at high pressure and high temperature with the aid of a catalyst. (5-10; 19-2)

Half-life (of a radioactive sample) the time required for the number of nuclides in a radioactive sample to reach half of the original value. (18-2)

Half-life (of a reactant) the time required for a reactant to reach half of its original concentration. (11-4)

Half-reactions the two parts of an oxidation–reduction reaction, one representing oxidation, the other reduction. (6-10; 16-1)

Halogen a Group 7A element. (2-13; 19-7)

Halogenation the addition of halogen atoms to unsaturated hydrocarbons. (21-2)

Hard water water from natural sources that contains relatively large concentrations of calcium and magnesium ions. (19-4)

Heat energy transferred between two objects due to a temperature difference between them. (7-1)

Heat capacity the amount of energy required to raise the temperature of an object by one degree Celsius. (7-3)

Heat of fusion the enthalpy change that occurs to melt a solid at its melting point. (9-9)

Heat of hydration the enthalpy change associated with placing gaseous molecules or ions in water; the sum of the energy needed to expand the solvent and the energy released from the solvent–solute interactions. (10-2)

Heat of solution the enthalpy change associated with dissolving a solute in a solvent; the sum of the energies needed to expand both solvent and solute in a solution and the energy released from the solvent–solute interactions. (10-2)

Heat of vaporization the energy required to vaporize one mole of a liquid at a pressure of one atmosphere. (9-9)

Heating curve a plot of temperature versus time for a substance where energy is added at a constant rate. (9-9)

Heisenberg uncertainty principle a principle stating that there is a fundamental limitation to how precisely both the position and momentum of a particle can be known at a given time. (2-5)

Heme an iron complex. (20-7)

Hemoglobin a biomolecule composed of four myoglobin-like units (proteins plus heme) that can bind and transport four oxygen molecules in the blood. (20-7)

Henderson–Hasselbalch equation an equation giving the relationship between the pH of an acid–base system and the concentrations of base and acid: $\mathrm{pH} = \mathrm{p}K_\mathrm{a} + \log\left(\dfrac{[\text{base}]}{[\text{acid}]}\right)$. (14-2)

Henry's law the amount of a gas dissolved in a solution is directly proportional to the pressure of the gas above the solution. (10-3)

Hess's law in going from a particular set of reactants to a particular set of products, the enthalpy change is the same whether the reaction takes place in one step or in a series of steps; in summary, enthalpy is a state function. (7-4)

Heterogeneous equilibrium an equilibrium involving reactants and/or products in more than one phase. (12-4)

Hexagonal closest packed (hcp) structure a structure composed of closest packed spheres with an *ababab* arrangement of layers; the unit cell is hexagonal. (9-4)

Homogeneous equilibrium an equilibrium system where all reactants and products are in the same phase. (12-4)

Homopolymer a polymer formed from the polymerization of only one type of monomer. (21-5)

Hund's rule the lowest energy configuration for an atom is the one having the maximum number of unpaired electrons allowed by the Pauli exclusion principle in a particular set of degenerate orbitals, with all unpaired electrons having parallel spins. (2-11)

Hybrid orbitals a set of atomic orbitals adopted by an atom in a molecule different from those of the atom in the free state. (4-3)

Hybridization a mixing of the native orbitals on a given atom to form special atomic orbitals for bonding. (4-3)

Hydration the interaction between solute particles and water molecules. (6-1)

Hydride a binary compound containing hydrogen. The hydride ion, H^-, exists in ionic hydrides. The three classes of hydrides are covalent, interstitial, and ionic. (19-3)

Hydrocarbon a compound composed of carbon and hydrogen. (21-1)

Hydrocarbon derivative an organic molecule that contains one or more elements in addition to carbon and hydrogen. (21-4)

Hydrogen bonding unusually strong dipole–dipole attractions that occur among molecules in which hydrogen is bonded to a highly electronegative atom. (9-1)

Hydrogenation reaction a reaction in which hydrogen is added, with a catalyst present, to a carbon–carbon multiple bond. (21-2)

Hydrohalic acid an aqueous solution of a hydrogen halide. (19-7)

Hydrometallurgy a process for extracting metals from ores by use of aqueous chemical solutions. Two steps are involved: selective leaching and selective precipitation. (20-8)

Hydronium ion the H_3O^+ ion; a hydrated proton. (13-1)

Hypothesis one or more assumptions put forth to explain the observed behavior of nature. (1-2)

Ideal gas law an equation of state for a gas, where the state of the gas is its condition at a given time; expressed by $PV = nRT$, where P = pressure, V = volume, n = moles of the gas, R = the universal gas constant, and T = absolute temperature. This equation expresses behavior approached by real gases at high T and low P. (8-3)

Ideal solution a solution whose vapor pressure is directly proportional to the mole fraction of solvent present. (10-4)

Indicator a chemical that changes color and is used to mark the end point of a titration. (6-8; 14-5)

Integrated rate law an expression that shows the concentration of a reactant as a function of time. (11-2)

Interhalogen compound a compound formed by the reaction of one halogen with another. (19-7)

Intermediate a species that is neither a reactant nor a product but that is formed and consumed in the reaction sequence. (11-5)

Intermolecular forces relatively weak interactions that occur between molecules. (9-1)

Internal energy a property of a system that can be changed by a flow of work, heat, or both; $\Delta E = q + w$, where ΔE is the change in the internal energy of the system, q is heat, and w is work. (7-1)

Ion an atom or a group of atoms that has a net positive or negative charge. (1-7)

Ion exchange (water softening) the process in which an ion-exchange resin removes unwanted ions (for example, Ca^{2+} and Mg^{2+}) and replaces them with Na^+ ions, which do not interfere with soap and detergent action. (19-4)

Ion pairing a phenomenon occurring in solution when oppositely charged ions aggregate and behave as a single particle. (10-7)

Ion-product (dissociation) constant (K_w) the equilibrium constant for the auto-ionization of water; $K_\mathrm{w} = [H^+][OH^-]$. At 25°C, $K_\mathrm{w} = 1.0 \times 10^{-14}$. (13-2)

Ion-selective electrode an electrode sensitive to the concentration of a particular ion in solution. (17-5)

Ionic bonding the electrostatic attraction between oppositely charged ions. (3-1; 3-4)

Ionic compound (binary) a compound that results when a metal reacts with a nonmetal to form a cation and an anion. (3-1)

Ionic solid a solid containing cations and anions that dissolves in water to give a solution containing the separated ions, which are mobile and thus free to conduct an electric current. (3-4; 9-3)

Irreversible process any real process. When a system undergoes the changes State 1 → State 2 → State 1 by any real pathway, the universe is different than before the cyclic process took place in the system. (16-9)

Isoelectronic ions ions containing the same number of electrons. (3-3)

Isomers species with the same formula but different properties. (20-4)

Isotactic chain a polymer chain in which the substituent groups such as CH_3 are all arranged on the same side of the chain. (21-5)

Isotonic solutions solutions having identical osmotic pressures. (10-6)

Isotopes atoms of the same element (the same number of protons) with different numbers of neutrons. They have identical atomic numbers but different mass numbers. (1-7; 17)

Ketone an organic compound containing the carbonyl group bonded

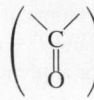

to two carbon atoms. (21-4)

Kinetic energy $(\frac{1}{2}mv^2)$ energy due to the motion of an object; dependent on the mass of the object and the square of its velocity. (7-1)

Kinetic molecular theory (KMT) a model that assumes that an ideal gas is composed of tiny particles (molecules) in constant motion. (8-6)

Lanthanide contraction the decrease in the atomic radii of the lanthanide series elements, going from left to right in the periodic table. (20-1)

Lanthanide series a group of 14 elements following lanthanum in the periodic table, in which the 4f orbitals are being filled. (2-11; 19-1; 20-1)

Lattice a three-dimensional system of points designating the positions of the centers of the components of a solid (atoms, ions, or molecules). (9-3)

Lattice energy the energy change occurring when separated gaseous ions are packed together to form an ionic solid. (3-4)

Law of conservation of energy energy can be converted from one form to another but can be neither created nor destroyed. (7-1)

Law of conservation of mass mass is neither created nor destroyed. (1-2; 1-4)

Law of definite proportion a given compound always contains exactly the same proportion of elements by mass. (1-4)

Law of mass action a general description of the equilibrium condition; it defines the equilibrium constant expression. (12-2)

Law of multiple proportions a law stating that when two elements form a series of compounds, the ratios of the masses of the second element that combine with one gram of the first element can always be reduced to small whole numbers. (1-4)

Leaching the extraction of metals from ores using aqueous chemical solutions. (20-8)

Lead storage battery a battery (used in cars) in which the anode is lead, the cathode is lead coated with lead dioxide, and the electrolyte is a sulfuric acid solution. (17-6)

Le Châtelier's principle if a change is imposed on a system at equilibrium, the position of the equilibrium will shift in a direction that tends to reduce the effect of that change. (12-7)

Lewis acid an electron-pair acceptor. (13-11)

Lewis base an electron-pair donor. (13-11)

Lewis structure a diagram of a molecule showing how the valence electrons are arranged among the atoms in the molecule. (3-9)

Ligand a neutral molecule or ion having a lone pair of electrons that can be used to form a bond to a metal ion; a Lewis base. (20-3)

Lime–soda process a water-softening method in which lime and soda ash are added to water to remove calcium and magnesium ions by precipitation. (13-6)

Limiting reactant (limiting reagent) the reactant that is completely consumed when a reaction is run to completion. (5-11)

Line spectrum a spectrum showing only certain discrete wavelengths. (2-3)

Linear accelerator a type of particle accelerator in which a changing electrical field is used to accelerate a positive ion along a linear path. (18-3)

Linkage isomerism isomerism involving a complex ion where the ligands are all the same but the point of attachment of at least one of the ligands differs. (20-4)

Liquefaction the transformation of a gas into a liquid. (19-1)

Localized electron (LE) model a model that assumes that a molecule is composed of atoms that are bound together by sharing pairs of electrons using the atomic orbitals of the bound atoms. (3-8)

London dispersion forces the forces, existing among noble gas atoms and nonpolar molecules, that involve an accidental dipole that induces a momentary dipole in a neighbor. (9-1)

Lone pair an electron pair that is localized on a given atom; an electron pair not involved in bonding. (3-8)

Magnetic quantum number (m_ℓ) the quantum number relating to the orientation of an orbital in space relative to the other orbitals with the same ℓ quantum number. It can have integral values between ℓ and $-\ell$, including zero. (2-6)

Main-group (representative) elements elements in the groups labeled 1A, 2A, 3A, 4A, 5A, 6A, 7A, and 8A in the periodic table. The group number gives the sum of the valence s and p electrons. (2-11; 17-1)

Major species the components present in relatively large amounts in a solution. (13-3)

Manometer a device for measuring the pressure of a gas in a container. (8-1)

Mass the quantity of matter in an object. (R-1)

Mass defect the change in mass occurring when a nucleus is formed from its component nucleons. (18-5)

Mass number the total number of protons and neutrons in the atomic nucleus of an atom. (1-7; 17)

Mass percent the percent by mass of a component of a mixture (10-1) or of a given element in a compound. (5-6)

Mass spectrometer an instrument used to determine the relative masses of atoms by the deflection of their ions on a magnetic field. (5-2)

Matter the material of the universe. (R-8)

Messenger RNA (mRNA) a special RNA molecule built in the cell nucleus that migrates into the cytoplasm and participates in protein synthesis. (21-6)

Metal an element that gives up electrons relatively easily and is lustrous, malleable, and a good conductor of heat and electricity. (2-13)

Metalloids (semimetals) elements along the division line in the periodic table between metals and nonmetals. These elements exhibit both metallic and nonmetallic properties. (2-13; 19-1)

Metallurgy the process of separating a metal from its ore and preparing it for use. (19-1; 20-8)

Millimeters of mercury (mm Hg) a unit of pressure, also called a torr, 760 mm Hg = 760 torr = 101,325 Pa = 1 standard atmosphere. (8-1)

Mineral a relatively pure compound as found in nature. (20-8)

Model (theory) a set of assumptions put forth to explain the observed behavior of matter. The models of chemistry usually involve assumptions about the behavior of individual atoms or molecules. (1-2)

Moderator a substance used in a nuclear reactor to slow down the neutrons. (18-6)

Molal boiling-point elevation constant a constant characteristic of a particular solvent that gives the change in boiling point as a function of solution molality; used in molecular weight determinations. (10-5)

Molal freezing-point depression constant a constant characteristic of a particular solvent that gives the change in freezing point as a function of the solution molality; used in molecular weight determinations (10-5)

Molality (m) the number of moles of solute per kilogram of solvent in a solution. (10-1)

Molar heat capacity the energy required to raise the temperature of one mole of a substance by one degree Celsius. (7-3)

Molar mass the mass in grams of one mole of molecules or formula units of a substance; also called *molecular weight.* (5-5)

Molar volume the volume of one mole of an ideal gas; equal to 22.42 liters at STP. (8-4)

Molarity (M) moles of solute per volume of solution in liters. (6-3; 10-1)

Mole (mol) the number equal to the number of carbon atoms in exactly 12 grams of pure ^{12}C: Avogadro's number. One mole represents 6.022×10^{23} units. (5-4)

Mole fraction (x) the ratio of the number of moles of a given component in a mixture to the total number of moles in the mixture. (8-5; 10-1)

Mole ratio (stoichiometry) the ratio of moles of one substance to moles of another substance in a balanced chemical equation. (5-9)

Molecular formula the exact formula of a molecule, giving the types of atoms and the number of each type. (5-6)

Molecular orbital (MO) model a model that regards a molecule as a collection of nuclei and electrons, where the electrons are assumed to occupy orbitals much as they do in atoms, but having the orbitals extend over the entire molecule. In this model the electrons are assumed to be delocalized rather than always located between a given pair of atoms. (4-4; 9-4)

Molecular orientations (kinetics) orientations of molecules during collisions, some of which can lead to reaction while others cannot. (11-6)

Molecular solid a solid composed of neutral molecules at the lattice points. (9-3)

Molecular structure the three-dimensional arrangement of atoms in a molecule. (4-1)

Molecularity the number of species that must collide to produce the reaction represented by an elementary step in a reaction mechanism. (11-5)

Molecule a bonded collection of two or more atoms of the same or different elements. (3-4)

Monodentate (unidentate) ligand a ligand that can form one bond to a metal ion. (20-3)

Monoprotic acid an acid with one acidic proton. (13-2)

Monosaccharide (simple sugar) a polyhydroxy ketone or aldehyde containing from three to nine carbon atoms. (21-6)

Myoglobin an oxygen-storing biomolecule consisting of a heme complex and a proton. (20-7)

Natural law a statement that expresses generally observed behavior. (1-2)

Nernst equation an equation relating the potential of an electrochemical cell to the concentrations of the cell components:

$$\mathscr{E} = \mathscr{E}° - \frac{0.0591}{n} \log(Q) \text{ at } 25°C \qquad (17\text{-}5)$$

Net ionic equation an equation for a reaction in solution, where strong electrolytes are written as ions, showing only those components that are directly involved in the chemical change. (6-6)

Network solid an atomic solid containing strong directional covalent bonds. (9-5)

Neutralization reaction an acid–base reaction. (6-8)

Neutron a particle in the atomic nucleus with mass virtually equal to the proton's but with no charge. (1-7; 18)

Nitrogen cycle the conversion of N_2 to nitrogen-containing compounds, followed by the return of nitrogen gas to the atmosphere by natural decay processes. (19-2)

Nitrogen fixation the process of transforming N_2 to nitrogen-containing compounds useful to plants. (19-2)

Nitrogen-fixing bacteria bacteria in the root nodules of plants that can convert atmospheric nitrogen to ammonia and other nitrogen-containing compounds useful to plants. (19-2)

Noble gas a Group 8A element. (2-13; 19-8)

Node an area of an orbital having zero electron probability. (2-7)

Nonelectrolyte a substance that, when dissolved in water, gives a nonconducting solution. (6-2)

Nonmetal an element not exhibiting metallic characteristics. Chemically, a typical nonmetal accepts electrons from a metal. (2-13)

Normal boiling point the temperature at which the vapor pressure of a liquid is exactly one atmosphere. (9-9)

Normal melting point the temperature at which the solid and liquid states have the same vapor pressure under conditions where the total pressure on the system is one atmosphere. (9-9)

Normality (N) the number of equivalents of a substance dissolved in a liter of solution. (10-1)

Nuclear atom an atom having a dense center of positive charge (the nucleus) with electrons moving around the outside. (1-6)

Nuclear transformation the change of one element into another. (18-3)

Nucleon a particle in an atomic nucleus, either a neutron or a proton. (18)

Nucleotide a monomer of the nucleic acids composed of a five-carbon sugar, a nitrogen-containing base, and phosphoric acid. (21-6)

Nucleus the small, dense center of positive charge in an atom. (1-6)

Nuclide the general term applied to each unique atom; represented by $^{A}_{Z}X$, where X is the symbol for a particular element. (18)

Octet rule the observation that atoms of nonmetals tend to form the most stable molecules when they are surrounded by eight electrons (to fill their valence orbitals). (3-9)

Open hearth process a process for producing steel by oxidizing and removing the impurities in molten iron using external heat and a blast of air or oxygen. (20-8)

Optical isomerism isomerism in which the isomers have opposite effects on plane-polarized light. (20-4)

Orbital a specific wave function for an electron in an atom. The square of this function gives the probability distribution for the electron. (2-5)

***d*-Orbital splitting** a splitting of the *d* orbitals of the metal ion in a complex such that the orbitals pointing at the ligands have higher energies than those pointing between the ligands. (20-6)

Order (of reactant) the positive or negative exponent, determined by experiment, of the reactant concentration in a rate law. (11-2)

Organic acid an acid with a carbon-atom backbone; often contains the carboxyl group. (13-2)

Organic chemistry the study of carbon-containing compounds (typically chains of carbon atoms) and their properties. (21)

Osmosis the flow of solvent into a solution through a semipermeable membrane. (10-6)

Osmotic pressure (π) the pressure that must be applied to a solution to stop osmosis; $\pi = MRT$. (10-6)

Ostwald process a commercial process for producing nitric acid by the oxidation of ammonia. (19-2)

Oxidation an increase in oxidation state (a loss of electrons). (6-9; 16-1)

Oxidation–reduction (redox) reaction a reaction in which one or more electrons are transferred. (6-9; 16-1)

Oxidation states a concept that provides a way to keep track of electrons in oxidation–reduction reactions according to certain rules. (6-9; 20-3)

Oxidizing agent (electron acceptor) a reactant that accepts electrons from another reactant. (6-9; 16-1)

Oxyacid an acid in which the acidic proton is attached to an oxygen atom. (13-2)

Ozone O_3, the form of elemental oxygen in addition to the much more common O_2. (19-5)

Paramagnetism a type of induced magnetism, associated with unpaired electrons, that causes a substance to be attracted into the inducing magnetic field. (4-5)

Partial pressures the independent pressures exerted by different gases in a mixture. (8-5)

Particle accelerator a device used to accelerate nuclear particles to very high speeds. (18-3)

Pascal the SI unit of pressure; equal to newtons per meter squared. (8-1)

Pauli exclusion principle in a given atom no two electrons can have the same set of four quantum numbers. (2-8)

Peptide linkage the bond resulting from the condensation reaction between amino acids; represented by:

$$\begin{matrix} & O & & H & \\ & \| & & | & \\ -&C&-&N&- \end{matrix} \qquad (21\text{-}6)$$

Percent dissociation the ratio of the amount of a substance that is dissociated at equilibrium to the initial concentration of the substance in a solution, multiplied by 100. (13-5)

Percent yield the actual yield of a product as a percentage of the theoretical yield. (5-11)

Periodic table a chart showing all the elements arranged in columns with similar chemical properties. (2-13)

pH curve (titration curve) a plot showing the pH of a solution being analyzed as a function of the amount of titrant added. (14-4)

pH scale a log scale based on 10 and equal to $-\log[H^+]$; a convenient way to represent solution acidity. (13-3)

Phase diagram a convenient way of representing the phases of a substance in a closed system as a function of temperature and pressure. (9-10)

Phenyl group the benzene molecule minus one hydrogen atom. (21-3)

Photochemical smog air pollution produced by the action of light on oxygen, nitrogen oxides, and unburned fuel from auto exhaust to form ozone and other pollutants. (8-10)

Photon a quantum of electromagnetic radiation. (2-2)

Physical change a change in the form of a substance, but not in its chemical composition; chemical bonds are not broken in a physical change. (R-8)

Pi (π) bond a covalent bond in which parallel *p* orbitals share an electron pair occupying the space above and below the line joining the atoms. (4-3)

Planck's constant the constant relating the change in energy for a system to the frequency of the electromagnetic radiation absorbed or emitted; equal to 6.626×10^{-34} J · s. (2-2)

Polar covalent bond a covalent bond in which the electrons are not shared equally because one atom attracts them more strongly than the other. (4-1)

Polar molecule a molecule that has a permanent dipole moment. (6-1)

Polyatomic ion an ion containing a number of atoms. (3-4)

Polyelectronic atom an atom with more than one electron. (2-9)

Polymer a large, usually chainlike molecule built from many small molecules (monomers). (21-5)

Polymerization a process in which many small molecules (monomers) are joined together to form a large molecule. (21-2)

Polypeptide a polymer formed from amino acids joined together by peptide linkages. (21-6)

Polyprotic acid an acid with more than one acidic proton. It dissociates in a stepwise manner, one proton at a time. (13-7)

Porous disk a disk in a tube connecting two different solutions in a galvanic cell that allows ion flow without extensive mixing of the solutions. (17-2)

Porphyrin a planar ligand with a central ring structure and various substituent groups at the edges of the ring. (20-7)

Positional probability a type of probability that depends on the number of arrangements in space that yield a particular state. (16-1)

Positron production a mode of nuclear decay in which a particle is formed having the same mass as an electron but opposite charge. The net effect is to change a proton to a neutron. (18-1)

Potential energy energy due to position or composition. (7-1)

Precipitation reaction a reaction in which an insoluble substance forms and separates from the solution. (6-5)

Precision the degree of agreement among several measurements of the same quantity; the reproducibility of a measurement. (R-2)

Primary structure (of a protein) the order (sequence) of amino acids in the protein chain. (21-6)

Principal quantum number (n) the quantum number relating to the size and energy of an orbital; it can have any positive integer value. (2-6)

Probability distribution the square of the wave function indicating the probability of finding an electron at a particular point in space. (2-5)

Product a substance resulting from a chemical reaction. It is shown to the right of the arrow in a chemical equation. (5-8)

Protein a natural high-molecular-weight polymer formed by condensation reactions between amino acids. (21-6)

Proton a positively charged particle in an atomic nucleus. (1-7; 18)

Pure substance a substance with constant composition. (R-8)

Pyrometallurgy recovery of a metal from its ore by treatment at high temperatures. (20-8)

Quantization the concept that energy can occur only in discrete units called *quanta*. (2-2)

Rad a unit of radiation dosage corresponding to 10^{-2} J of energy deposited per kilogram of tissue (from *r*adiation *a*bsorbed *d*ose). (18-7)

Radioactive decay (radioactivity) the spontaneous decomposition of a nucleus to form a different nucleus. (1-6; 18-1)

Radiocarbon dating (carbon-14 dating) a method for dating ancient wood or cloth based on the rate of radioactive decay of the nuclide $^{14}_6C$. (18-4)

Radiotracer a radioactive nuclide, introduced into an organism for diagnostic purposes, whose pathway can be traced by monitoring its radioactivity. (18-4)

Random error an error that has an equal probability of being high or low. (R-2)

Raoult's law the vapor pressure of a solution is directly proportional to the mole fraction of solvent present. (10-4)

Rate constant the proportionality constant in the relationship between reaction rate and reactant concentrations. (11-2)

Rate of decay the change in the number of radioactive nuclides in a sample per unit time. (18-2)

Rate-determining step the slowest step in a reaction mechanism, the one determining the overall rate. (11-5)

Rate law (differential rate law) an expression that shows how the rate of reaction depends on the concentration of reactants. (11-2)

Reactant a starting substance in a chemical reaction. It appears to the left of the arrow in a chemical equation. (5-8)

Reaction mechanism the series of elementary steps involved in a chemical reaction. (11-5)

Reaction quotient (Q) a quotient obtained by applying the law of mass action to initial concentrations rather than to equilibrium concentrations. (12-5)

Reaction rate the change in concentration of a reactant or product per unit time. (11-1)

Reactor core the part of a nuclear reactor where the fission reaction takes place. (18-6)

Reducing agent (electron donor) a reactant that donates electrons to another substance to reduce the oxidation state of one of its atoms. (6-9; 16-1)

Reduction a decrease in oxidation state (a gain of electrons). (6-9; 16-1)

Rem a unit of radiation dosage that accounts for both the energy of the dose and its effectiveness in causing biological damage (from *r*oentgen *e*quivalent for *m*an). (18-7)

Resonance a condition occurring when more than one valid Lewis structure can be written for a particular molecule. The actual electronic structure is not represented by any one of the Lewis structures but by the average of all of them. (3-11)

Reverse osmosis the process occurring when the external pressure on a solution causes a net flow of solvent through a semipermeable membrane from the solution to the solvent. (10-6)

Reversible process a cyclic process carried out by a hypothetical pathway, which leaves the universe exactly the same as it was before the process. No real process is reversible. (16-9)

Ribonucleic acid (RNA) a nucleotide polymer that transmits the genetic information stored in DNA to the ribosomes for protein synthesis. (21-6)

Roasting a process of converting sulfide minerals to oxides by heating in air at temperatures below their melting points. (20-8)

Root mean square velocity the square root of the average of the squares of the individual velocities of gas particles. (8-6)

Salt an ionic compound. (13-8)

Salt bridge a U-tube containing an electrolyte that connects the two compartments of a galvanic cell, allowing ion flow without extensive mixing of the different solutions. (17-1)

Scientific method the process of studying natural phenomena, involving observations, forming laws and theories, and testing of theories by experimentation. (1-2)

Scintillation counter an instrument that measures radioactive decay by sensing the flashes of light produced in a substance by the radiation. (18-4)

Second law of thermodynamics in any spontaneous process, there is always an increase in the entropy of the universe. (16-2)

Secondary structure (of a protein) the three-dimensional structure of the protein chain (for example, α-helix, random coil, or pleated sheet). (21-6)

Selective precipitation a method of separating metal ions from an aqueous mixture by using a reagent whose anion forms a precipitate with only one or a few of the ions in the mixture. (6-7; 15-2)

Semiconductor a substance conducting only a slight electric current at room temperature, but showing increased conductivity at higher temperatures. (9-5)

Semipermeable membrane a membrane that allows solvent but not solute molecules to pass through. (10-6)

SI system International System of units based on the metric system and units derived from the metric system. (R-1)

Side chain (of amino acid) the hydrocarbon group on an amino acid represented by H, CH_3, or a more complex substituent. (21-6)

Sigma (σ) bond a covalent bond in which the electron pair is shared in an area centered on a line running between the atoms. (4-3)

Significant figures the certain digits and the first uncertain digit of a measurement. (R-2)

Silica the fundamental silicon–oxygen compound, which has the empirical formula SiO_2, and forms the basis of quartz and certain types of sand. (9-5)

Silicates salts that contain metal cations and polyatomic silicon–oxygen anions that are usually polymeric. (9-5)

Single bond a bond in which one pair of electrons is shared by two atoms. (3-7)

Smelting a metallurgical process that involves reducing metal ions to the free metal. (20-8)

Solubility the amount of a substance that dissolves in a given volume of solvent at a given temperature. (6-2)

Solubility product constant the constant for the equilibrium expression representing the dissolving of an ionic solid in water. (15-1)

Solute a substance dissolved in a liquid to form a solution. (6-2; 10-1)

Solution a homogeneous mixture. (R-8)

Solvent the dissolving medium in a solution. (6-2)

Somatic damage radioactive damage to an organism resulting in its sickness or death. (18-7)

Space-filling model a model of a molecule showing the relative sizes of the atoms and their relative orientations. (3-4)

Specific heat capacity the energy required to raise the temperature of one gram of a substance by one degree Celsius. (7-3)

Spectator ions ions present in solution that do not participate directly in a reaction. (6-6)

Spectrochemical series a listing of ligands in order based on their ability to produce *d*-orbital splitting. (20-6)

Spontaneous fission the spontaneous splitting of a heavy nuclide into two lighter nuclides. (18-1)

Spontaneous process a process that occurs without outside intervention. (16-1)

Standard atmosphere a unit of pressure equal to 760 mm Hg. (8-1)

Standard enthalpy of formation the enthalpy change that accompanies the formation of one mole of a compound at 25°C from its elements, with all substances in their standard states at that temperature. (7-5)

Standard free energy change the change in free energy that will occur for one unit of reaction if the reactants in their standard states are converted to products in their standard states. (16-6)

Standard free energy of formation the change in free energy that accompanies the formation of one mole of a substance from its constituent elements with all reactants and products in their standard states. (16-6)

Standard hydrogen electrode a platinum conductor in contact with 1 M H^+ ions and bathed by hydrogen gas at one atmosphere. (17-3)

Standard reduction potential the potential of a half-reaction under standard state conditions, as measured against the potential of the standard hydrogen electrode. (17-3)

Standard solution a solution whose concentration is accurately known. (6-3)

Standard state a reference state for a specific substance defined according to a set of conventional definitions. (7-5)

Standard temperature and pressure (STP) the condition 0°C and 1 atmosphere of pressure. (8-4)

Standing wave a stationary wave as on a string of a musical instrument; in the wave mechanical model, the electron in the hydrogen atom is considered to be a standing wave. (2-5)

State function (property) a property that is independent of the pathway. (7-1)

States (of matter) the three different forms in which matter can exist; solid, liquid, and gas. (R-8)

Stereoisomerism isomerism in which all the bonds in the isomers are the same but the spatial arrangements of the atoms are different. (20-4)

Steric factor the factor (always less than 1) that reflects the fraction of collisions with orientations that can produce a chemical reaction. (11-6)

Stoichiometric quantities quantities of reactants mixed in exactly the correct amounts so that all are used up at the same time. (5-10)

Strong acid an acid that completely dissociates to produce an H^+ ion and the conjugate base. (6-2; 13-2)

Strong base a metal hydroxide salt that completely dissociates into its ions in water. (6-2; 13-6)

Strong electrolyte a material that, when dissolved in water, gives a solution that conducts an electric current very efficiently. (6-2)

Structural formula the representation of a molecule in which the relative positions of the atoms are shown and the bonds are indicated by lines. (3-4)

Structural isomerism isomerism in which the isomers contain the same atoms but one or more bonds differ. (20-4; 21-1)

Subcritical reaction (nuclear) a reaction in which less than one neutron causes another fission event and the process dies out. (18-6)

Sublimation the process by which a substance goes directly from the solid to the gaseous state without passing through the liquid state. (9-9)

Subshell a set of orbitals with a given azimuthal quantum number. (2-6)

Substitution reaction (hydrocarbons) a reaction in which an atom, usually a halogen, replaces a hydrogen atom in a hydrocarbon. (21-1)

Supercooling the process of cooling a liquid below its freezing point without its changing to a solid. (9-9)

Supercritical reaction (nuclear) a reaction in which more than one neutron from each fission event causes another fission event. The process rapidly escalates to a violent explosion. (18-6)

Superheating the process of heating a liquid above its boiling point without its boiling. (9-9)

Superoxide a compound containing the O_2^- anion. (18-2)

Surface tension the resistance of a liquid to an increase in its surface area. (9-2)

Surroundings everything in the universe surrounding a thermodynamic system. (7-1)

Syndiotactic chain a polymer chain in which the substituent groups such as CH_3 are arranged on alternate sides of the chain. (21-5)

System (thermodynamic) that part of the universe on which attention is to be focused. (7-1)

Systematic error an error that always occurs in the same direction. (R-2)

Tempering a process in steel production that fine-tunes the proportions of carbon crystals and cementite by heating to intermediate temperatures followed by rapid cooling. (20-8)

Termolecular step a reaction involving the simultaneous collision of three molecules. (11-5)

Tertiary structure (of a protein) the overall shape of a protein, long and narrow or globular, maintained by different types of intramolecular interactions. (21-6)

Theoretical yield the maximum amount of a given product that can be formed when the limiting reactant is completely consumed. (5-11)

Theory a set of assumptions put forth to explain some aspect of the observed behavior of matter. (1-2)

Thermal pollution the oxygen-depleting effect on lakes and rivers of using water for industrial cooling and returning it to its natural source at a higher temperature. (10-3)

Thermodynamic stability (nuclear) the potential energy of a particular nucleus as compared to the sum of the potential energies of its component protons and neutrons. (18-1)

Thermodynamics the study of energy and its interconversions. (7-1)

Thermoplastic polymer a substance that when molded to a certain shape under appropriate conditions can later be remelted. (21-5)

Thermoset polymer a substance that when molded to a certain shape under pressure and high temperatures cannot be softened again or dissolved. (21-5)

Third law of thermodynamics the entropy of a perfect crystal at 0 K is zero. (16-5)

Titration a technique in which one solution is used to analyze another. (6-8)

Torr another name for millimeter of mercury (mm Hg). (8-1)

Transfer RNA (tRNA) a small RNA fragment that finds specific amino acids and attaches them to the protein chain as dictated by the codons in mRNA. (21-6)

Transition metals several series of elements in which inner orbitals (d or f orbitals) are being filled. (2-11; 19-1)

Transuranium elements the elements beyond uranium that are made artificially by particle bombardment. (18-3)

Triple bond a bond in which three pairs of electrons are shared by two atoms. (3-7)

Triple point the point on a phase diagram at which all three states of a substance are present. (9-10)

Tyndall effect the scattering of light by particles in a suspension. (10-8)

Uncertainty (in measurement) the characteristic that any measurement involves estimates and cannot be exactly reproduced. (R-2)

Unimolecular step a reaction step involving only one molecule. (11-5)

Unit cell the smallest repeating unit of a lattice. (9-3)

Unit factor method an equivalence statement between units used for converting from one unit to another. (R-5)

Universal gas constant the combined proportionality constant in the ideal gas law; 0.08206 L · atm/K · mol or 8.3145 J/K · mol. (8-3)

Valence electrons the electrons in the outermost principal quantum level of an atom. (2-11)

Valence shell electron-pair repulsion (VSEPR) model a model whose main postulate is that the structure around a given atom in a molecule is determined principally by minimizing electron-pair repulsions. (4-1)

Van der Waals equation a mathematical expression for describing the behavior of real gases. (8-8)

van't Hoff factor the ratio of moles of particles in solution to moles of solute dissolved. (10-7)

Vapor pressure the pressure of the vapor over a liquid at equilibrium. (9-9)

Vaporization (evaporization) the change in state that occurs when a liquid evaporates to form a gas. (9-9)

Viscosity the resistance of a liquid to flow. (9-2)

Volt the unit of electrical potential defined as one joule of work per coulomb of charge transferred. (17-2)

Voltmeter an instrument that measures cell potential by drawing electric current through a known resistance. (17-2)

Volumetric analysis a process involving titration of one solution with another. (6-8)

Vulcanization a process in which sulfur is added to rubber and the mixture is heated, causing crosslinking of the polymer chains and thus adding strength to the rubber. (21-5)

Wave function a function of the coordinates of an electron's position in three-dimensional space that describes the properties of the electron. (2-5)

Wave mechanical model a model for the hydrogen atom in which the electron is assumed to behave as a standing wave. (2-5)

Wavelength the distance between two consecutive peaks or troughs in a wave. (2-1)

Weak acid an acid that dissociates only slightly in aqueous solution. (6-2; 13-2)

Weak base a base that reacts with water to produce hydroxide ions to only a slight extent in aqueous solution. (6-2; 13-6)

Weak electrolyte a material that, when dissolved in water, gives a solution that conducts only a small electric current. (6-2)

Weight the force exerted on an object by gravity. (R-1)

Work force acting over a distance. (7-1)

X-ray diffraction a technique for establishing the structure of crystalline solids by directing X rays of a single wavelength at a crystal and obtaining a diffraction pattern from which interatomic spaces can be determined. (9-3)

Zone of nuclear stability the area encompassing the stable nuclides on a plot of their positions as a function of the number of protons and the number of neutrons in the nucleus. (18-1)

Zone refining a metallurgical process for obtaining a highly pure metal that depends on continuously melting the impure material and recrystallizing the pure metal. (20-8)

Answers to Selected Exercises

The answers listed here are from the *Complete Solutions Guide*, in which rounding is carried out at each intermediate step in a calculation in order to show the correct number of significant figures for that step. Therefore, an answer given here may differ in the last digit from the result obtained by carrying extra digits throughout the entire calculation and rounding at the end (the procedure you should follow).

Chapter Review

1. A random error has equal probability of being too high or too low. This type of error occurs when estimating the value of the last digit of a measurement. A systematic error is one that always occurs in the same direction, either too high or too low. For example, this type of error would occur if the balance you were using weighed all objects 0.20 g too high, that is, if the balance wasn't calibrated correctly. A random error is an indeterminate error, whereas a systematic error is a determinate error. **3.** The results, average = $14.91 \pm 0.03\%$, are precise (are close to each other) but are not accurate (are not close to the true value). **5.** Significant figures are the digits we associate with a number. They contain all of the certain digits and the first uncertain digit (the first estimated digit). What follows is one thousand indicated to varying numbers of significant figures: 1000 or 1×10^3 (1 S.F.); 1.0×10^3 (2 S.F.); 1.00×10^3 (3 S.F.); 1000. or 1.000×10^3 (4 S.F.). To perform the calculation, the addition/subtraction significant rule is applied to $1.5 - 1.0$. The result of this is the one significant figure answer of 0.5. Next, the multiplication/division rule is applied to 0.5/0.50. A one significant number divided by a two significant number yields an answer with one significant figure (answer = 1). **7.** In a subtraction, the result gets smaller, but the uncertainties add. If the two numbers are very close together, the uncertainty may be larger than the result. For example, let's assume we want to take the difference of the following two measured quantities, $999,999 \pm 2$ and $999,996 \pm 2$. The difference is 3 ± 4. Because of the uncertainty, subtracting two similar numbers is poor practice. **9.** The slope of the T_F vs. T_C plot is 1.8 (= 9/5) and the y-intercept is 32°F. The slope of T_C vs. T_K plot is 1 and the y-intercept is −273°C. **11. a.** coffee; saltwater; the air we breathe (N_2 + O_2 + others); brass (Cu + Zn) **b.** book; human being; tree; desk **c.** sodium chloride (NaCl); water (H_2O); glucose ($C_6H_{12}O_6$); carbon dioxide (CO_2) **d.** nitrogen (N_2); oxygen (O_2); copper (Cu); zinc (Zn) **e.** boiling water; freezing water; melting a popsicle; dry ice subliming **f.** Electrolysis of molten sodium chloride to produce sodium and chlorine gas; the explosive reaction between oxygen and hydrogen to produce water; photosynthesis, which converts H_2O and CO_2 into $C_6H_{12}O_6$ and O_2; the combustion of gasoline in our cars to produce CO_2 and H_2O **13.** The law of conservation of energy states that energy can be converted from one form to another but can neither be destroyed nor created; the energy content of the universe is constant. The two categories of energy are kinetic energy and potential energy. Kinetic energy is energy due to motion, and potential energy is stored energy due to position. For example, the energy stored in the bonds between atoms in a molecule is potential energy. **15. a.** exact; **b.** inexact; **c.** exact; **d.** inexact **17. a.** 3; **b.** 4; **c.** 4; **d.** 1; **e.** 7; **f.** 1; **g.** 3; **h.** 3 **19. a.** 3.42×10^{-4}; **b.** 1.034×10^4; **c.** 1.7992×10^1; **d.** 3.37×10^5 **21.** 2.85 + 0.280 = 3.13 mL; the graduated cylinder on the left limits the precision of the total volume; it is the least precise measuring device between the two graduated cylinders. **23. a.** 641.0; **b.** 1.327; **c.** 77.34; **d.** 3215; **e.** 0.420 **25. a.** 188.1; **b.** 12; **c.** 4×10^{-7}; **d.** 6.3×10^{-26}; **e.** 4.9; Uncertainty appears in the first decimal place. The average of several numbers can be only as precise as the least precise number. Averages can be exceptions to the significant figure

rules. **f.** 0.22 **27. a.** 84.3 mm; **b.** 2.41 m; **c.** 2.945×10^{-5} cm; **d.** 14.45 km; **e.** 2.353×10^5 mm; **f.** 0.9033 μm **29. a.** 8 lb and 9.9 oz; $20\frac{1}{4}$ in; **b.** 4.0×10^4 km, 4.0×10^7 m; **c.** 1.2×10^{-2} m³, 12 L, 730 in³, 0.42 ft³ **31. a.** 4.00×10^2 rods; 10.0 furlongs; 2.01×10^3 m; 2.01 km; **b.** 8390.0 rods; 209.75 furlongs; 42,195 m; 42.195 km **33. a.** 0.373 kg, 0.822 lb; **b.** 31.1 g, 156 carats; **c.** 19.3 cm³ **35.** 24 capsules **37.** 2.95×10^9 knots; 3.36×10^9 mi/h **39.** To the proper number of significant figures, the car is traveling at 40. mi/h, which would not violate the speed limit. **41.** \$1.47/lb **43.** 7×10^5 kg **45. a.** −273°C, 0 K; **b.** −40.°C, 233 K; **c.** 20.°C, 293 K; **d.** 4×10^7 °C, 4×10^7 K **47. a.** 312.4 K; 102.6°F; **b.** 248 K; −13°F; **c.** 0 K; −459°F; **d.** 1074 K; 1470°F **49.** 160.°C = 320.°F **51. a.** $T_C = T_A \times \dfrac{8°C}{5°A} - 45°C$; **b.** $T_F = T_A \times \dfrac{72°F}{25°A} - 49°F$; **c.** 75°C = 75°A; **d.** 93°C; 2.0×10^2 °F; **e.** 56°A **53.** It will float (density = 0.80 g/cm³). **55.** 1×10^6 g/cm³ **57. a.** 0.28 cm³; **b.** 49 carats **59.** 3.8 g/cm³ **61. a.** Both are the same mass; **b.** 1.0 mL mercury; **c.** Both are the same mass; **d.** 1.0 L benzene **63. a.** 1.0 kg feather; **b.** 100 g water; **c.** same **65.** 2.77 cm **67. a.** Picture iv represents a gaseous compound. Pictures ii and iii also contain a gaseous compound but have a gaseous element present. **b.** Picture vi represents a mixture of two gaseous elements. **c.** Picture v represents a solid element. **d.** Pictures ii and iii both represent a mixture of a gaseous element and a gaseous compound. **69.** Homogeneous: having visibly indistinguishable parts; heterogeneous: having visibly distinguishable parts; **a.** heterogeneous; **b.** homogeneous (hopefully); **c.** homogeneous; **d.** homogeneous (hopefully); **e.** heterogeneous; **f.** heterogeneous **71. a.** pure; **b.** mixture; **c.** mixture; **d.** pure; **e.** mixture (copper and zinc); **f.** pure; **g.** mixture; **h.** mixture; **i.** mixture. Iron and uranium are elements. Water is a compound. **73.** Compound **75. a.** physical; **b.** chemical; **c.** physical; **d.** chemical **77.** 9×10^{13} dollars per person **79.** 0.010 kg **81.** 1.0×10^5 bags **83.** 3.0×10^{17} m **85.** 56.56°C **87. a.** Volume × density = mass; the orange block is more dense. Since mass (orange) > mass (blue) and volume (orange) < volume (blue), the density of the orange block must be greater to account for the larger mass of the orange block. **b.** Which block is more dense cannot be determined. Since mass (orange) > mass (blue) and volume (orange) > volume (blue), the density of the orange block may or may not be larger than the blue block. If the blue block is more dense, then its density cannot be so large that the mass of the smaller blue block becomes larger than the orange block mass. **c.** The blue block is more dense. Since mass (blue) = mass (orange) and volume (blue) < volume (orange), the density of the blue block must be larger to equate the masses. **d.** The blue block is more dense. Since mass (blue) > mass (orange) and the volumes are equal, the density of the blue block must be larger to give the blue block the larger mass. **89.** 8.5 ± 0.5 g/cm³ **91.** 7.0% Cu. **93.** At point II, kinetic energy = 78 J and potential energy = 118 J. **95.** 3.648×10^4 cable lengths, 6.671×10^6 m, 3648 nautical miles. **97.** Phosphorus would be a liquid. **99. a.** False; sugar is generally considered to be the pure compound sucrose, $C_{12}H_{22}O_{11}$. **b.** False; elements and compounds are pure substances. **c.** True; air is a mixture of mostly nitrogen and oxygen gases. **d.** False; gasoline has many additives, so it is a mixture. **e.** True; compounds are broken down to elements by chemical change.

Chapter 1

11. A law summarizes what happens, e.g., law of conservation of mass in a chemical reaction or the ideal gas law, PV = nRT. A theory (model) is an attempt to explain why something happens. **13.** The fundamental steps are

(1) making observations; (2) formulating hypotheses; (3) performing experiments to test the hypotheses. The key to the scientific method is performing experiments to test hypotheses. If after the test of time the hypotheses seem to account satisfactorily for some aspect of natural behavior, then the set of tested hypotheses turns into a theory (model). However, scientists continue to perform experiments to refine or replace existing theories. Hence, science is a dynamic or active process, not a static one.   **15.** Law of conservation of mass: mass is neither created nor destroyed. The mass before a chemical reaction always equals the mass after a chemical reaction. Law of definite proportion: a given compound always contains exactly the same proportion of elements by mass. Water is always 1 g H for every 8 g oxygen. Law of multiple proportions: When two elements form a series of compounds, the ratios of the mass of the second element that combine with one gram of the first element can always be reduced to small whole numbers. For CO_2 and CO, discussed in Section 1.4, the mass ratios of oxygen that react with 1 g of carbon in each compound are in a 2:1 ratio.   **17.** Natural niacin and commercially produced niacin have the exact same formula of $C_6H_5NO_2$. Therefore, both sources produce niacin having an identical nutritional value. There may be other compounds present in natural niacin that would increase the nutritional value, but the nutritional value due to just niacin is identical to the commercially produced niacin.   **19.** J. J. Thomson's study of cathode-ray tubes led him to postulate the existence of negatively charged particles, which we now call electrons. Ernest Rutherford and his alpha bombardment of metal foil experiments led him to postulate the nuclear atom—an atom with a tiny dense center of positive charge (the nucleus) with electrons moving about the nucleus at relatively large distances away; the distance is so large that an atom is mostly empty space.   **21.** The number and arrangement of electrons in an atom determine how the atom will react with other atoms. The electrons determine the chemical properties of an atom. The number of neutrons determines the isotope identity.   **23.** 2; the ratio increases   **25. a.** The composition of a substance depends on the number of atoms of each element making up the compound (depends on the formula of the compound) and not on the composition of the mixture from which it was formed.   **b.** Avogadro's hypothesis implies that volume ratios are equal to molecule ratios at constant temperature and pressure. $H_2(g) + Cl_2(g) \rightarrow 2\ HCl(g)$; from the balanced equation, the volume of HCl produced will be twice the volume of H_2 (or Cl_2) reacted.   **27.** 299 g   **29.** All the masses of hydrogen in these three compounds can be expressed as simple whole-number ratios. The g H/g N in hydrazine, ammonia, and hydrogen azide are in the ratios 6:9:1.   **31.** For CO and CO_2, it is easiest to concentrate on the mass of oxygen that combines with 1 g of carbon. From the formulas, CO_2 has twice the mass of oxygen that combines per gram of carbon as does CO. For CO_2 and C_3O_2, it is easiest to concentrate on the mass of carbon that combines with 1 g of oxygen. From the formulas, C_3O_2 has three times the mass of carbon that combines per gram of oxygen as does CO_2. As expected, the mass ratios are whole numbers as predicted by the law of multiple proportions.   **33.** Mass is conserved in a chemical reaction because atoms are conserved. Chemical reactions involve the reorganization of atoms, so formulas change in a chemical reaction, but the number and types of atoms do not change. Because the atoms do not change in a chemical reaction, mass must not change.   **35.** O, 7.94; Na, 22.8; Mg, 11.9.   **37.** Using $r = 5 \times 10^{-14}$ cm, $d_{nucleus} = 3 \times 10^{15}$ g/cm³; using $r = 1 \times 10^{-8}$ cm, $d_{atom} = 0.4$ g/cm³   **39.** 37   **41. a.** $^{23}_{11}$Na **b.** $^{19}_{9}$F **c.** $^{16}_{8}$O   **43. a.** 35 p, 44 n, 35 e; **b.** 35 p, 46 n, 35 e; **c.** 94 p, 145 n, 94 e; **d.** 55 p, 78 n, 55 e; **e.** 1 p, 2 n, 1 e; **f.** 26 p, 30 n, 26 e   **45. a.** $^{17}_{8}$O; **b.** $^{37}_{17}$Cl; **c.** $^{60}_{27}$Co; **d.** $^{57}_{26}$Fe; **e.** $^{131}_{53}$I; **f.** $^{7}_{3}$Li   **47. a.** 56 protons, 54 electrons; **b.** 30 protons, 28 electrons; **c.** 7 protons, 10 electrons; **d.** 37 protons, 36 electrons; **e.** 27 protons, 24 electrons; **f.** 52 protons, 54 electrons; **g.** 35 protons, 36 electrons   **49.** $^{151}_{63}$Eu³⁺; $^{118}_{50}$Sn²⁺   **51.** $^{238}_{92}$U, 92 protons, 146 neutrons, 92 electrons, neutral charge (0); $^{40}_{20}$Ca²⁺, 18 electrons; $^{51}_{23}$V³⁺, 3+; $^{89}_{39}$Y, 39 protons, 50 neutrons, 39 electrons, neutral charge (0); $^{79}_{35}$Br⁻, 1−; $^{31}_{15}$P³⁻, 18 electrons   **53.** 26 protons, 27 neutrons, 24 electrons   **55. a.** P, 15 p, 16 n; **b.** I, 53 p, 74 n; **c.** K, 19 p, 20 n; **d.** Yb, 70 p, 103 n   **57.** radium; 142 neutrons   **59.** 22.8 g   **61.** $^{4}_{2}$He, 2 protons, 2 neutrons; $^{20}_{10}$Ne, 10 protons, 10 neutrons; $^{48}_{22}$Ti, 22 protons, 26 neutrons; $^{190}_{76}$Os, 76 protons, 114 neutrons; $^{50}_{27}$Co, 27 protons, 23 neutrons   **63. a.** True; **b.** False; this was J. J. Thomson.

c. False; a proton is about 1800 times more massive than an electron. **d.** The nucleus contains the protons and the neutrons.   **65. a.** The compounds are isomers of each other. Isomers are compounds with the same formula but the atoms are attached differently, resulting in different properties. **b.** When wood burns, most of the solid material is converted to gases, which escape. **c.** Atoms are not indivisible particles. Atoms are composed of electrons, neutrons, and protons. **d.** The two hydride samples contain different isotopes of either hydrogen and/or lithium. Isotopes may have different masses but have similar chemical properties.   **67.** C:H = 8:18 or 4:9   **69.** The relative mass of Y is 1.14 times that of X. Thus, if X has an atomic mass of 100, then Y will have an atomic mass of 114.   **71.** CO_2

Chapter 2

19. The equations relating the terms are $v\lambda = c$, $E = hv$, and $E = hc/\lambda$. From the equations, wavelength and frequency are inversely related, photon energy and frequency are directly related, and photon energy and wavelength are inversely related. The unit of 1 joule (J) = 1 kg m²/s². This is why you must change mass units to kg when using the de Broglie equation.   **21.** The *photoelectric effect* refers to the phenomenon in which electrons are emitted from the surface of a metal when light strikes it. The light must have a certain minimum frequency (energy) in order to remove electrons from the surface of a metal. Light having a frequency below the minimum results in no electrons being emitted, whereas light at or higher than the minimum frequency does cause electrons to be emitted. For light having a frequency higher than the minimum frequency, the excess energy is transferred into kinetic energy for the emitted electron. Albert Einstein explained the photoelectric effect by applying quantum theory.   **23.** Example 2-3 calculates the de Broglie wavelength of a ball and of an electron. The ball has a wavelength on the order of 10^{-34} m. This is incredibly short and, as far as the wave-particle duality is concerned, the wave properties of large objects are insignificant. The electron, with its tiny mass, also has a short wavelength: on the order of 10^{-10} m. However, this wavelength is significant as it is on the same order as the spacing between atoms in a typical crystal. For very tiny objects like electrons, the wave properties are important. The wave properties must be considered, along with the particle properties, when hypothesizing about the electron motion in an atom.   **25.** The Bohr model was an important step in the development of the current quantum mechanical model of the atom. The idea that electrons can occupy only certain, allowed energy levels is illustrated nicely (and relatively easily). We talk about the Bohr model to present the idea of quantized energy levels.   **27.** When the p and d orbital functions are evaluated at various points in space, the results sometimes have positive values and sometimes have negative values. The term *phase* is often associated with the + and − signs. For example, a sine wave has alternating positive and negative phases. This is analogous to the positive and negative values (phases) in the p and d orbitals.   **29.** If one more electron is added to a half-filled subshell, electron–electron repulsions will increase because two electrons must now occupy the same atomic orbital. This may slightly decrease the stability of the atom.   **31.** The valence electrons are strongly attracted to the nucleus for elements with large ionization energies. One would expect these species to readily accept another electron and have very negative electron affinities. The noble gases are an exception; they have a large ionization energy but a positive electron affinity. Noble gases have a filled valence shell. The added electron must go into a higher n value atomic orbital, which would have a significantly higher energy. This is unfavorable.   **33.** For hydrogen, all orbitals with the same value of n have the same energy. For polyatomic atoms/ions, the energy of the orbitals also depends on ℓ. Because there are more nondegenerate energy levels for polyatomic atoms/ions as compared with hydrogen, there are many more possible electronic transitions, resulting in more complicated line spectra.   **35.** Yes, the maximum number of unpaired electrons in any configuration corresponds to a minimum in electron–electron repulsions.   **37.** Ionization energy applies to the removal of the electron from the atom in the gas phase. The work function applies to the removal of an electron from the solid element.   **39.** 3.84×10^{14} s⁻¹   **41.** 3.0×10^{10} s⁻¹, 2.0×10^{-23} J/photon   **43.** from 9.4×10^{14} s⁻¹ to 1.1×10^{15} s⁻¹   **45.** Wave a has the longer wavelength (4.0×10^{-4} m).

Wave b has the higher frequency (1.5×10^{12} s^{-1}) and larger photon energy (9.9×10^{-22} J). Since both of these waves represent electromagnetic radiation, they both should travel at the same speed, c, the speed of light. Both waves represent infrared radiation. **47.** 1.50×10^{23} atoms **49.** 427.7 nm **51.** 276 nm **53. a.** 2.4×10^{-11} m; **b.** 3.4×10^{-34} m **55.** 1.6×10^{-27} kg **57. a.** 656.7 nm (visible); **b.** 486.4 nm (visible); **c.** 121.6 nm (ultraviolet) **59.**

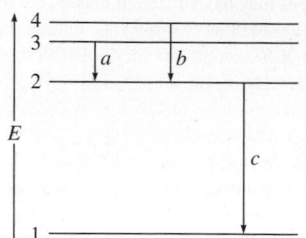

61. 7462 nm; 93.79 nm **63.** $n = 1 \rightarrow n = 5$, $\lambda = 95.00$ nm; $n = 2 \rightarrow n = 6$, $\lambda = 410.4$ nm; visible light has sufficient energy for the $n = 2 \rightarrow n = 6$ transition but does not have sufficient energy for the $n = 1 \rightarrow n = 5$ transition. **65.** $n = 1$, 91.20 nm; $n = 2$, 364.8 nm **67.** $n = 2$ **69. a.** 5.79×10^{-4} m; **b.** 3.64×10^{-33} m. The uncertainty in position of the electron is much larger than the size of an atom, whereas the uncertainty for the ball is insignificant as compared to the size of a ball. **71.** $n = 1, 2, 3, \ldots$; $\ell = 0, 1, 2, \ldots (n-1)$; $m_\ell = -\ell, \ldots, -2, -1, 0, 1, 2, \ldots, +\ell$. **73. b.** For $\ell = 3$, m_ℓ can range from -3 to $+3$; thus $+4$ is not allowed. **c.** n cannot equal zero. **d.** ℓ cannot be a negative number. **75.** ψ^2 gives the probability of finding the electron at that point. **77.** He: $1s^2$; Ne: $1s^22s^22p^6$; Ar: $1s^22s^22p^63s^23p^6$; each peak in the diagram corresponds to a subshell with different values of n. Corresponding subshells are closer to the nucleus for heavier elements because of the increased nuclear charge. **79.** 3; 1; 5; 25; 16 **81. a.** 32; **b.** 8; **c.** 25; **d.** 10; **e.** 6

83. a.

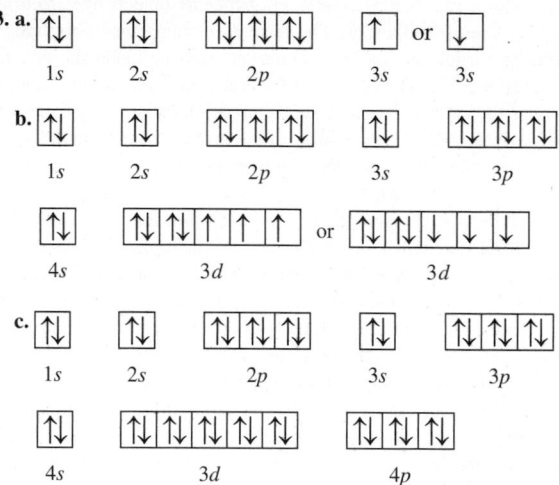

Na has 1 unpaired electron, Co has 3 unpaired electrons, and Kr has 0 unpaired electrons.

85. Si: $1s^22s^22p^63s^23p^2$ or [Ne]$3s^23p^2$; Ga: $1s^22s^22p^63s^23p^64s^23d^{10}4p^1$ or [Ar]$4s^23d^{10}4p^1$; As: [Ar]$4s^23d^{10}4p^3$; Ge: [Ar]$4s^23d^{10}4p^2$; Al: [Ne]$3s^23p^1$; Cd: [Kr]$5s^24d^{10}$; S: [Ne]$3s^23p^4$; Se: [Ar]$4s^23d^{10}4p^4$ **87.** Sc: $1s^22s^22p^63s^23p^6$ $4s^23d^1$; Fe: $1s^22s^22p^63s^23p^64s^23d^6$; P: $1s^22s^22p^63s^23p^3$; Cs: $1s^22s^22p^63s^23p^6$ $4s^23d^{10}4p^65s^24d^{10}5p^66s^1$; Eu: $1s^22s^22p^63s^23p^64s^23d^{10}4p^65s^24d^{10}5p^66s^24f^65d^1$ (Actual: [Xe]$6s^24f^7$); Pt: $1s^22s^22p^63s^23p^64s^23d^{10}4p^65s^24d^{10}5p^66s^24f^{14}5d^8$ (Actual: [Xe]$6s^14f^{14}5d^9$); Xe: $1s^22s^22p^63s^23p^64s^23d^{10}4p^65s^24d^{10}5p^6$; Br: $1s^22s^2$ $2p^63s^23p^64s^23d^{10}4p^5$ **89.** O: $1s^22s^22p^4$; C: $1s^22s^22p^2$; H: $1s^1$; N: $1s^22s^22p^3$; Ca: [Ar]$4s^2$; P: [Ne]$3s^23p^3$; Mg: [Ne]$3s^2$; K: [Ar]$4s^1$ **91. a.** I: [Kr]$5s^24d^{10}5p^5$; **b.** element 120: [Rn]$7s^25f^{14}6d^{10}7p^68s^2$; **c.** Rn: [Xe]$6s^24f^{14}5d^{10}6p^6$; **d.** Cr: [Ar]$4s^13d^5$ **93. a.** S; **b.** N; **c.** Se$^-$ **95. a.** 18; **b.** 30; **c.** 8; **d.** 40

97. B: $1s^22s^22p^1$

	n	ℓ	m_ℓ	m_s
$1s$	1	0	0	$+\frac{1}{2}$
$1s$	1	0	0	$-\frac{1}{2}$
$2s$	2	0	0	$+\frac{1}{2}$
$2s$	2	0	0	$-\frac{1}{2}$
$2p$	2	1	-1	$+\frac{1}{2}$

N: $1s^22s^22p^3$

	n	ℓ	m_ℓ	m_s
$1s$	1	0	0	$+\frac{1}{2}$
$1s$	1	0	0	$-\frac{1}{2}$
$2s$	2	0	0	$+\frac{1}{2}$
$2s$	2	0	0	$-\frac{1}{2}$
$2p$	2	1	-1	$+\frac{1}{2}$
$2p$	2	1	0	$+\frac{1}{2}$
$2p$	2	1	$+1$	$+\frac{1}{2}$

For boron, there are six possibilities for the $2p$ electron. For nitrogen, all the $2p$ electrons could have $m_s = -\frac{1}{2}$. **99.** 1A: 1, ns^1, Li, $2s^1$; 2A: 2, ns^2, Ra, $7s^2$; 3A: 3, ns^2np^1, Ga, $4s^24p^1$; 4A: 4, ns^2np^2, Si, $3s^23p^2$; 5A: 5, ns^2np^3, Sb, $5s^25p^3$; 6A: 6, ns^2np^4, Po, $6s^26p^4$; 7A: 7, ns^2np^5, 117, $7s^27p^5$; 8A: 8, ns^2np^6, Ne, $2s^22p^6$ **101.** none; an excited state; energy released **103.** C, O, Si, S, Ti, Ni, Ge, Se **105.** Li (1 unpaired electron), N (3 unpaired electrons), Ni (2 unpaired electrons), and Te (2 unpaired electrons) are all expected to be paramagnetic because they have unpaired electrons. **107. a.** S < Se < Te; **b.** Br < Ni < K; **c.** F < Si < Ba **109. a.** Te < Se < S; **b.** K < Ni < Br; **c.** Ba < Si < F **111. a.** He; **b.** Cl; **c.** element 116; **d.** Si; **e.** Na$^+$ **113. a.** [Rn]$7s^25f^{14}6d^4$; **b.** W **115.** Se is an exception to the general ionization trend. There are extra electron–electron repulsions in Se because two electrons are in the same $4p$ orbital, resulting in a lower ionization energy than expected. **117. a.** As we remove succeeding electrons, the electron being removed is closer to the nucleus, and there are fewer electrons left repelling it. The remaining electrons are more strongly attracted to the nucleus, and it takes more energy to remove these electrons. **b.** Al: $1s^22s^22p^63s^23p^1$; for I_4, we begin removing an electron with $n = 2$. For I_3, we remove an electron with $n = 3$ (the last valence electron). In going from $n = 3$ to $n = 2$, there is a big jump in ionization energy because the $n = 2$ electrons are much closer to the nucleus on average than the $n = 3$ electrons. Since $n = 2$ electrons are closer to the nucleus, they are held more tightly and require a much larger amount of energy to remove compared to the $n = 3$ electrons. In general, valence electrons are much easier to remove than inner-core electrons. **119. a.** C, Br; **b.** N, Ar; **c.** C, Br **121.** Al (-44), Si (-120), P (-74), S (-200.4), Cl (-348.7); Based on the increasing nuclear charge, we would expect the electron affinity (EA) values to become more negative as we go from left to right in the period. Phosphorus is out of line. The equation for the EA of P is

$$P(g) + e^- \rightarrow P^-(g)$$
$$[\text{Ne}]3s^23p^3 \quad °[\text{Ne}]3s^23p^4$$

The additional electron in P$^-$ will have to go into an orbital that already has one electron. There will be greater repulsions between the paired electrons in P$^-$, causing the EA of P to be less favorable than predicted based solely on attractions to the nucleus. **123. a.** Se < S; **b.** I < Br < F < Cl **125.** Electron–electron repulsions are much greater in O$^-$ than in S$^-$ because the electron goes into a smaller $2p$ orbital versus the larger $3p$ orbital in sulfur. This results in a more favorable (more negative) EA for sulfur. **127. a.** Se$^{3+}(g) \rightarrow$ Se$^{4+}(g) + e^-$; **b.** S$^-(g) + e^- \rightarrow$ S$^{2-}(g)$; **c.** Fe$^{3+}(g) + e^- \rightarrow$ Fe$^{2+}(g)$; **d.** Mg$(g) \rightarrow$ Mg$^+(g) + e^-$ **129.** 6.582×10^{14} s^{-1}; 4.361×10^{-19} J **131.** Yes; the ionization energy general trend decreases down a group, and the atomic radius trend increases down a group. The data in Table 2-8 confirm both of these general trends. **133. a.** $6\text{Li}(s) + \text{N}_2(g) \rightarrow 2\text{Li}_3\text{N}(s)$; **b.** $2\text{Rb}(s) + \text{S}(s) \rightarrow \text{Rb}_2\text{S}(s)$ **135.** No; lithium metal is very reactive. It will react somewhat violently with water, making it completely unsuitable for human consumption. Lithium has a low first ionization energy, so it is more likely that the lithium prescribed will be in the form of a soluble lithium salt (a soluble ionic compound with Li$^+$ as the cation). **137.** 200 s **139.** $\lambda = 4.104 \times 10^{-5}$ cm so violet light is emitted. **141. a.** true for H only; **b.** true for all atoms; **c.** true for all atoms **143.** smallest size: O; largest size: K; smallest 1st ionization energy: K; largest 1st ionization energy: N   **145.** Valence electrons are easier to remove than inner-core electrons. The large difference in energy between I_2 and I_3 indicates that this element has two valence electrons. The element is most likely an alkaline earth metal because alkaline earth metals have two valence electrons. **147.** None of the noble gases and no subatomic particles had been discovered when Mendeleev published his periodic table. There was no element out of place in terms of reactivity, and there was no reason to predict an entire family of missing elements. Mendeleev ordered his table by mass; he had no way of knowing there were gaps in atomic numbers (they hadn't been discovered yet). **149.** 386 nm **151.** 4053 nm; 95.00 nm **153.** $2f$ and $2d_{xy}$ orbitals are forbidden energy levels (they do not exist), so zero electrons can occupy these. The $3p$ and $4p$ sets each contain three degenerate orbitals, so each set can hold 6 electrons, and the $5d_{yz}$ represents a singular energy level that can hold 2 electrons. **155. a.** False; an electron in a $2s$ orbital would be, on average, closer to the nucleus than an electron in a $3s$ orbital. **b.** True; the Bohr model is fundamentally incorrect. **c.** True; **d.** False; these two terms come from different atomic structure models. **e.** True; when $n = 3$, ℓ can equal 0 (s orbital), 1 (p orbitals), and 2 (d orbitals). **157.** More favorable electron affinity: K, Br, and Se (follows general electron affinity trend). Higher ionization energy: K, Br, and Se (follows general ionization energy trend). Larger size: Cs, Te, and Ge (follows general atomic radius trend). **159.** Mg : $1s^2 2s^2 2p^6 3s^2$: 0.734 MJ/mol : 160 pm

S : $1s^2 2s^2 2p^6 3s^2 3p^4$: 0.999 MJ/mol : 104 pm

Ca : $1s^2 2s^2 2p^6 3s^2 3p^6 4s^2$: 0.590 MJ/mol : 197 pm

161. a. line A, $n = 6$ to $n = 3$; line B, $n = 5$ to $n = 3$; **b.** 121.6 nm **163.** X = C, $m = 5$; C^{5+} is the ion. **165.** For $r = a_0$ and $\theta = 0°$, $\psi^2 = 2.46 \times 10^{28}$. For $r = a_0$ and $\theta = 90°$, $\psi^2 = 0$. As expected, the xy plane is a node for the $2p_z$ atomic orbital.

167. a.

1							2
3							4
5	6	7	8	9	10	11	12
13	14	15	16	17	18	19	20

b. 2, 4, 12, and 20; **c.** There are many possibilities. One example of each formula is XY = 1 + 11, XY_2 = 6 + 11, X_2Y = 1 + 10, XY_3 = 7 + 11, and X_2Y_3 = 7 + 10; **d.** 6; **e.** 0; **f.** 18 **169.** The ratios for Mg, Si, P, Cl, and Ar are about the same. However, the ratios for Na, Al, and S are higher. For Na, the second IE is extremely high because the electron is taken from $n = 2$ (the first electron is taken from $n = 3$). For Al, the first electron requires a bit less energy than expected by the trend due to the fact it is a $3p$ electron. For S, the first electron requires a bit less energy than expected by the trend due to electrons being paired in one of the p orbitals. **171. a.** 43.33; **b.** Silver is element 47, so $Z = 47$ for silver. Our calculated Z_{eff} value is less than 47. Electrons in other orbitals can penetrate the $1s$ orbital. Thus a $1s$ electron can be slightly shielded from the nucleus by these penetrating electrons, giving a Z_{eff} close to but less than Z. **173. a.** 265 nm; **b.** Yes, 259-nm EMR will eject an electron; **c.** Cu (copper)

Chapter 3

17. In H_2 and HF, the bonding is covalent in nature, with an electron pair being shared between the atoms. In H_2, the two atoms are identical, so the sharing is equal; in HF, the two atoms are different, with different electronegativities, so the sharing is unequal, and as a result, the bond is polar covalent. Both are in marked contrast to the situation in NaF: NaF is an ionic compound—an electron has been completely transferred from sodium to fluorine—producing separate ions. **19.** In Cl_2 the bonding is pure covalent, with the bonding electrons shared equally between the two chlorine atoms. In HCl, there is also a shared pair of bonding electrons, but the shared pair is drawn more closely to the chlorine atom. This is called a polar covalent bond as opposed to the pure covalent bond in Cl_2. **21. a.** This represents ionic bonding. Ionic bonding is the electrostatic attraction between anions and cations. **b.** This represents covalent bonding where electrons are shared between two atoms. This could be the space-filling model for H_2O or SF_2 or

NO_2, etc. **23.** Electronegativity increases left to right across the periodic table and decreases from top to bottom. Hydrogen has an electronegativity value between B and C in the second row, and identical to P in the third row. Going further down the periodic table, H has an electronegativity value between As and Se (row 4) and identical to Te (row 5). It is important to know where hydrogen fits into the electronegativity trend, especially for rows 2 and 3. If you know where H fits into the trend, then you can predict bond dipole directions for nonmetals bonded to hydrogen. **25.** Two other factors that must be considered are the ionization energy needed to produce more positively charged ions and the electron affinity needed to produce more negatively charged ions. The favorable lattice energy more than compensates for the unfavorable ionization energy of the metal and for the unfavorable electron affinity of the nonmetal as long as electrons are added to or removed from the valence shell. Once the valence shell is full, the ionization energy required to remove another electron is extremely unfavorable; the same is true for electron affinity when an electron is added to a higher n shell. These two quantities are so unfavorable after the valence shell is complete that they overshadow the favorable lattice energy, and the higher charged ionic compounds do not form. **27.** Statements a and c are true. For statement b, SF_4 has five electron pairs around the sulfur in the best Lewis structure; it is an exception to the octet rule. Because OF_4 has the same number of valence electrons as SF_4, OF_4 would also have to be an exception to the octet rule. However, row 2 elements such as O never have more than 8 electrons around them, so OF_4 does not exist. For statement d, two resonance structures can be drawn for ozone:

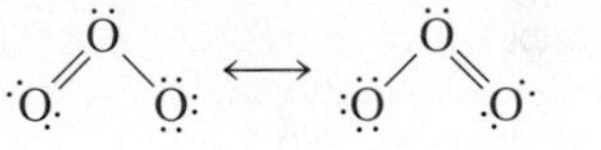

When resonance structures can be drawn, the actual bond lengths and strengths are all equal to each other. Even though each Lewis structure implies the two O—O bonds are different, this is not the case; both of the O—O bonds are equivalent. When resonance structures can be drawn, you can think of the bonding as an average of all of the resonance structures. **29.** Na_3AsO_4, sodium arsenate; H_3AsO_4, arsenic acid; $Mg_3(SbO_4)_2$, magnesium antimonate **31. a.** C < N < O; **b.** Se < S < Cl; **c.** Sn < Ge < Si; **d.** Tl < Ge < S **33. a.** Ge—F; **b.** P—Cl; **c.** S—F; **d.** Ti—Cl **35.** Order of electronegativity from Fig. 3-4: **a.** C(2.5) < N (3.0) < O (3.5), same; **b.** Se (2.4) < S (2.5) < Cl (3.0), same; **c.** Si = Ge = Sn (1.8), different; **d.** Tl (1.8) = Ge (1.8) < S (2.5), different. Most polar bonds using actual electronegativity values: **a.** Si—F and Ge—F equal polarity (Ge—F predicted); **b.** P—Cl (same as predicted); **c.** S—F (same as predicted); **d.** Ti—Cl (same as predicted) **37.** Incorrect: b, d, e; **b.** $^{\delta-}$Cl—I$^{\delta+}$; **d.** nonpolar bond so no partial charges; **e.** $^{\delta-}$O—P$^{\delta+}$ **39. a.** ionic; **b.** covalent; **c.** polar covalent; **d.** ionic; **e.** polar covalent; **f.** covalent **41.** F—H > O—H > N—H > C—H > P—H **43. a.** Lose 2 e^- to form Ra^{2+}. **b.** Lose 3 e^- to form In^{3+}. **c.** Gain 3 e^- to form P^{3-}. **d.** Gain 2 e^- to form Te^{2-}. **e.** Gain 1 e^- to form Br^-. **f.** Lose 1 e^- to form Rb^+. **45.** Al^{3+}: $[He]2s^2 2p^6$; Ba^{2+}: $[Kr]5s^2 4d^{10} 5p^6$; Se^{2-}: $[Ar]4s^2 3d^{10} 4p^6$; I^-: $[Kr]5s^2 4d^{10} 5p^6$ **47. a.** Li^+ and N^{3-} are the expected ions. The formula of the compound would be Li_3N (lithium nitride). **b.** Ga^{3+} and O^{2-}; Ga_2O_3, gallium(III) oxide or gallium oxide; **c.** Rb^+ and Cl^-; RbCl, rubidium chloride; **d.** Ba^{2+} and S^{2-}; BaS, barium sulfide **49. a.** Mg^{2+} and Al^{3+}: $1s^2 2s^2 2p^6$; K^+: $1s^2 2s^2 2p^6 3s^2 3p^6$; **b.** N^{3-}, O^{2-}, and F^-: $1s^2 2s^2 2p^6$; Te^{2-}: $[Kr]5s^2 4d^{10} 5p^6$ **51. a.** Sc^{3+}; **b.** Te^{2-}; **c.** Ce^{4+}, Ti^{4+}; **d.** Ba^{2+} **53. a.** F^-, O^{2-}, or N^{3-}; **b.** Cl^-, S^{2-}, or P^{3-}; **c.** F^-, O^{2-}, or N^{3-}; **d.** Br^-, Se^{2-}, or As^{3-} **55.** Mg^{2+}, Na^+, F^-, O^{2-}, and N^{3-} are all isoelectronic with neon. $Mg^{2+} < Na^+ < F^- < O^{2-} < N^{3-}$ **57. a.** Cu > Cu^+ > Cu^{2+}; **b.** Pt^{2+} > Pd^{2+} > Ni^{2+}; **c.** O^{2-} > O^- > O; **d.** La^{3+} > Eu^{3+} > Gd^{3+} > Yb^{3+}; **e.** Te^{2-} > I^- > Cs^+ > Ba^{2+} > La^{3+} **59. a.** NaCl, Na^+ smaller than K^+; **b.** LiF, F^- smaller than Cl^-; **c.** MgO, O^{2-} greater charge than OH^-; **d.** $Fe(OH)_3$, Fe^{3+} greater charge than Fe^{2+}; **e.** Na_2O, O^{2-} greater charge than Cl^-; **f.** MgO, Mg^{2+} smaller than Ba^{2+}, and O^{2-} smaller than S^{2-}. **61.** −411 kJ/mol **63.** The lattice energy for $Mg^{2+}O^{2-}$ will be much more negative than for Mg^+O^-. **65.** 181 kJ/mol **67.** Ca^{2+} has greater charge than Na^+, and Se^{2-} is smaller than Te^{2-}. Charge differences affect lattice energy values more than size differences, and we expect the trend from most negative to least negative to be:

$$CaSe > CaTe > Na_2Se > Na_2Te$$
$$(-2862) \quad (-2721) \quad (-2130) \quad (-2095)$$

69. a. −183 kJ; **b.** −109 kJ **71.** −42 kJ **73.** −1282 kJ **75.** −1228 kJ **77.** 485 kJ/mol

79. a. :F—F: **b.** Ö=Ö **c.** :C≡O:

d.
```
     H
     |
H — C — H
     |
     H
```
e.
```
H — N̈ — H
     |
     H
```
f. H—Ö—H **g.** H—F̈:

81. a.
```
  :Cl:
   |
:Cl—C—Cl:
   |
  :Cl:
```
b.
```
:Cl—N̈—Cl:
     |
    :Cl:
```
c. :Cl—S̈e—Cl: **d.** :Ï—Cl:

83. H—Be—H

```
     H
     |
     B
    / \
   H   H
```

85.

PF₅
```
   :F:   :F:
    \    |
:F̈ ⟍ P — F:
    /    |
   :F̈   :F:
```

SF₄
```
      :F:
  :F̈   |
    \   S:
  :F̈ / |
      :F:
```

ClF₃
```
   :F:
    |
:Cl—F:
    |
   :F:
```

Br₃⁻ [:B̈r—B̈r—B̈r:]⁻

Row 3 and heavier nonmetals can have more than 8 electrons. Row 3 and heavier elements have empty *d* orbitals that are close in energy to the valence *s* and *p* orbitals. These empty *d* orbitals can accept extra electrons if needed.

87. a. NO₂⁻ [Ö=N—Ö:]⁻ ⟷ [:Ö—N=Ö]⁻

NO₃⁻
```
    [   :O:        ]⁻
    [    ‖         ]
    [    N         ]  ⟷
    [  ⁄   ⁀       ]
    [ :Ö:   :Ö:    ]
```
```
[   :O:    ]⁻        [   :Ö:    ]⁻
[    ‖     ]          [    |     ]
[    N     ]   ⟷      [    N     ]
[  ⁄  ⁀    ]          [  ⁄  ⁀    ]
[ :Ö:  :Ö:]          [ :Ö:  :Ö: ]
```

N₂O₄

b. OCN⁻ [:Ö—C≡N:]⁻ ⟷ [Ö=C=N̈]⁻ ⟷ [:O=C—N̈:]⁻

SCN⁻ [:S̈—C≡N:]⁻ ⟷ [S̈=C=N̈]⁻ ⟷ [:S≡C—N̈:]⁻

N₃⁻ [:N̈—N≡N:]⁻ ⟷ [N̈=N=N̈]⁻ ⟷ [:N≡N—N̈:]⁻

89.

91. With resonance, all carbon–carbon bonds are equivalent (we indicate this with a circle in the ring), giving three different structures:

Localized double bonds give four unique structures.

93.
```
     H                          H
     |                          |
H — C — N=C=Ö:   ⟷    H — C — N≡C—Ö:   ⟷
     |                          |
     H                          H
```
```
     H
     |
H — C — N̈—C≡O:
     |
     H
```

95. Longest → shortest C—O bond: $CH_3OH > CO_3^{2-} > CO_2 > CO$; weakest → strongest C—O bond: $CH_3OH < CO_3^{2-} < CO_2 < CO$

97.
```
      F +1                    :F: 0
      ‖                        |
      B −1                     B  0
    ⁄   ⁀                    ⁄   ⁀
 0 :F   F: 0            0 :F   :F: 0
```

The first Lewis structure obeys the octet rule but has a +1 formal charge on the most electronegative element there is, fluorine, and a negative formal charge on a much less electronegative element, boron. This is just the opposite of what we expect: negative formal charge on F and positive formal charge on B. The other Lewis structure does not obey the octet rule for B but has a zero formal charge on each element in BF₃. Since structures generally want to minimize formal charge, BF₃ with only single bonds is best from a formal charge point of view. **99.** a–f and h all have similar Lewis structures:

```
        :Ÿ:
         |
   :Ÿ — X — Ÿ:
         |
        :Ÿ:
```

g. ClO_3^-

$$\left[:\ddot{O}-\underset{\displaystyle :\ddot{O}:}{\overset{\displaystyle }{Cl}}-\ddot{O}: \right]$$

Formal charges: **a.** +1; **b.** +2; **c.** +3; **d.** +1; **e.** +2; **f.** +4; **g.** +2; **h.** +1

101.

$$H-\underset{\displaystyle :O:}{\overset{\displaystyle H}{C}}=\underset{\displaystyle }{\overset{\displaystyle C\equiv N:}{C}}-\underset{\displaystyle }{\overset{\displaystyle }{C}}-\ddot{O}-\underset{\displaystyle H}{\overset{\displaystyle H}{C}}-H$$

103. OCN^-:

$$\left[\ddot{O}=C=\ddot{N}: \right]^- \longleftrightarrow \left[:\ddot{O}-C\equiv N: \right]^- \longleftrightarrow \left[:O\equiv C-\ddot{N}: \right]^-$$

Formal charge   0   0   −1   −1   0   0   +1   0   −2

Only the first two resonance structures should be important. The third places a positive formal charge on the most electronegative atom in the ion and a −2 formal charge on N.

CNO^-:

$$\left[\ddot{C}=N=\ddot{O}: \right]^- \longleftrightarrow \left[:C\equiv N-\ddot{O}: \right]^- \longleftrightarrow \left[:\ddot{C}-N\equiv O: \right]^-$$

Formal charge   −2   +1   0   −1   +1   −1   −3   +1   +1

All the resonance structures for fulminate (CNO^-) involve greater formal charges than in cyanate (OCN^-), making fulminate more reactive (less stable). **105. a.** sodium bromide; **b.** rubidium oxide; **c.** calcium sulfide; **d.** aluminum iodide; **e.** SrF_2; **f.** Al_2Se_3; **g.** K_3N; **h.** Mg_3P_2 **107. a.** cesium fluoride; **b.** lithium nitride; **c.** silver sulfide (Silver only forms stable 1+ ions in compounds, so no Roman numerals are needed.) **d.** manganese(IV) oxide; **e.** titanium(IV) oxide; **f.** strontium phosphide **109. a.** barium sulfite; **b.** sodium nitrite; **c.** potassium permanganate; **d.** potassium dichromate; **111. a.** dinitrogen tetroxide; **b.** iodine trichloride; **c.** sulfur dioxide; **d.** diphosphorus pentasulfide **113. a.** copper(I) iodide; **b.** copper(II) iodide; **c.** cobalt(II) iodide; **d.** sodium carbonate; **e.** sodium hydrogen carbonate or sodium bicarbonate; **f.** tetrasulfur tetranitride; **g.** sulfur tetrafluoride; **h.** sodium hypochlorite; **i.** barium chromate; **j.** ammonium nitrate **115.** SeO_4^{2-}: selenate; SeO_3^{2-}: selenite; TeO_4^{2-}: tellurate; TeO_3^{2-}: tellurite **117. a.** SF_2; **b.** SF_6; **c.** NaH_2PO_4; **d.** Li_3N; **e.** $Cr_2(CO_3)_3$; **f.** SnF_2; **g.** $NH_4C_2H_3O_2$; **h.** NH_4HSO_4; **i.** $Co(NO_3)_3$; **j.** Hg_2Cl_2; mercury(I) exists as Hg_2^{2+} ions. **k.** $KClO_3$; **l.** NaH **119. a.** Na_2O; **b.** Na_2O_2; **c.** KCN; **d.** $Cu(NO_3)_2$; **e.** $SeBr_4$; **f.** HIO_2; **g.** PbS_2; **h.** CuCl; **i.** GaAs (we would predict the stable ions to be Ga^{3+} and As^{3-}); **j.** CdSe (cadmium only forms 2+ charged ions in compounds); **k.** ZnS (zinc only forms 2+ charged ions in compounds); **l.** HNO_2; **m.** P_2O_5 **121. a.** nitric acid, HNO_3; **b.** perchloric acid, $HClO_4$; **c.** acetic acid, $HC_2H_3O_2$; **d.** sulfuric acid, H_2SO_4; **e.** phosphoric acid, H_3PO_4 **123. a.** radius: $N^+ < N < N^-$; I.E.: $N^- < N < N^+$; **b.** radius: $Cl^+ < Cl < Se < Se^-$; I.E.: $Se^- < Se < Cl < Cl^+$; **c.** radius: $Sr^{2+} < Rb^+ < Br^-$; I.E.: $Br^- < Rb^+ < Sr^{2+}$; **125. a.** 1549 kJ; **b.** 1390. kJ **c.** 1312 kJ; **d.** 1599 kJ **127. a.** NaBr: In $NaBr_2$, the sodium ion would have a +2 charge assuming each bromine has a −1 charge. Sodium doesn't form stable Na^{2+} compounds. **b.** ClO_4^-: ClO_4 has 31 valence electrons so it is impossible to satisfy the octet rule for all atoms in ClO_4. The extra electron from the −1 charge in ClO_4^- allows for complete octets for all atoms. **c.** XeO_4: We can't draw a Lewis structure that obeys the octet rule for SO_4 (30 electrons), unlike XeO_4 (32 electrons). **d.** SeF_4: Both compounds require the central atom to expand its octet. O is too small and doesn't have low-energy d orbitals to expand its octet (which is true for all row 2 elements).
129. The general structure of the trihalide ions is:

$$\left[:\ddot{X}-\ddot{X}-\ddot{X}: \right]^-$$

Bromine and iodine are large enough and have low-energy, empty d orbitals to accommodate the expanded octet. Fluorine is small, and its valence shell contains only $2s$ and $2p$ orbitals (four orbitals) and cannot expand its octet. The lowest-energy d orbitals in F are $3d$; they are too high in energy compared with $2s$ and $2p$ to be used in bonding. **131. a.** lead(II) acetate; **b.** copper(II) sulfate; **c.** calcium oxide; **d.** magnesium sulfate; **e.** magnesium hydroxide; **f.** calcium sulfate; **g.** dinitrogen monoxide or nitrous oxide (common name) **133.** Because this is a relatively small number of neutrons, the number of protons will be very close to the number of neutrons present. The heavier elements have significantly more neutrons than protons in their nuclei. Because this element forms anions, it is a nonmetal and will be a halogen because halogens form stable 1− charged ions in ionic compounds. From the halogens listed, chlorine, fits the data. The two isotopes are ^{35}Cl and ^{37}Cl, and the number of electrons in the 1− ion is 18.
135. $:\ddot{Cl}-\ddot{S}-\ddot{S}-\ddot{Cl}:$
137. carbon tetrabromide, CBr_4; cobalt(II) phosphate, $Co_3(PO_4)_2$; magnesium chloride, $MgCl_2$; nickel(II) acetate, $Ni(C_2H_3O_2)_2$; calcium nitrate, $Ca(NO_3)_2$ **139.** K will lose 1 e^- to form K^+. Cs will lose 1 e^- to form Cs^+. Br will gain 1 e^- to form Br^-. Sulfur will gain 2 e^- to form S^{2-}. Se will gain 2 e^- to form Se^{2-}. **141. a.** nonpolar covalent; **b.** ionic; **c.** ionic; **d.** polar covalent; **e.** polar covalent; **f.** polar covalent; **g.** polar covalent; **h.** ionic **143. a.** $O^{2-} > O^- > O$; **b.** $Fe^{2+} > Ni^{2+} > Zn^{2+}$; **c.** $Cl^- > K^+ > Ca^{2+}$ **145.** −303 kJ **147. a.** -5×10^{-18} J; **b.** -6×10^{-18} J **149.** −562 kJ **151.** See Fig. 3-8 to see the data supporting MgO as an ionic compound. Note that the lattice energy is large enough to overcome all of the other processes (removing 2 electrons from Mg, and so on). The bond energy for O_2 (247 kJ/mol) and electron affinity (737 kJ/mol) are the same when making CO. However, the energy needed to ionize carbon to form a C^{2+} ion must be too large. Fig. 2-35 shows that the first ionization energy for carbon is about 350 kJ/mol greater than the first IE for magnesium. If all other numbers were equal, the overall energy change would be ~250 kJ/mol (see Fig. 3-8). It is not unreasonable to assume that the second ionization energy for carbon is more than 250 kJ/mol greater than the second ionization energy for magnesium. This would result in a positive ΔE value for the formation of CO as an ionic compound. One wouldn't expect CO to be ionic if the energetics are unfavorable. **153.** reaction i: −2636 kJ; reaction ii: −3471 kJ; reaction iii: −3543 kJ; reaction iii yields the most energy per kg (−8085 kJ/kg)

155.

$$H-\underset{\displaystyle \underset{\displaystyle H}{\overset{\displaystyle |}{C}}-H}{\overset{\displaystyle O}{\overset{\displaystyle \|}{C}}}-\underset{\displaystyle }{\overset{\displaystyle H}{\ddot{N}}}-\underset{\displaystyle H}{\overset{\displaystyle H}{C}}-H \quad\longleftrightarrow\quad H-\underset{\displaystyle \underset{\displaystyle H}{\overset{\displaystyle |}{C}}-H}{\overset{\displaystyle :\ddot{O}:}{C}}=\underset{\displaystyle }{\overset{\displaystyle H}{N}}-\underset{\displaystyle H}{\overset{\displaystyle H}{C}}-H$$

157.

159. Cr < As < P < Cl

Chapter 4

9. Only statement c is true. The bond dipoles in CF_4 and KrF_4 are arranged in a manner that they all cancel each other out, making them nonpolar molecules (CF_4 has a tetrahedral molecular structure, whereas KrF_4 has a square planar molecular structure). In SeF_4, the bond dipoles in this see-saw molecule do not cancel each other out, so SeF_4 is polar. For statement a, all the molecules have either a trigonal planar geometry or a trigonal bipyramid geometry, both of which have 120° bond angles. However, $XeCl_2$ has three lone pairs and two bonded chlorine atoms around it. $XeCl_2$ has a linear molecular structure with a 180° bond angle. With three lone pairs, we no longer have a 120° bond angle in $XeCl_2$. For statement b, SO_2 has a V-shaped molecular structure with a bond angle of about 120°. CS_2 is linear with a 180° bond angle, and SCl_2 is V-shaped but with an approximate 109.5° bond angle. The three compounds do not have the same bond angle. For statement d, central atoms adopt a geometry to minimize electron repulsions, not maximize them.   **11.** In hybrid orbital theory, some or all of the valence atomic orbitals of the central atom in a molecule are mixed together to form hybrid orbitals; these hybrid orbitals point to where the bonded atoms and lone pairs are oriented. The σ bonds are formed from the hybrid orbitals overlapping head to head with an appropriate orbital on the bonded atom. The π bonds in hybrid orbital theory are formed from unhybridized p atomic orbitals. The p orbitals overlap side to side to form the π bond where the π electrons occupy the space above and below a line joining the atoms (the internuclear axis). Assuming the z-axis is the internuclear axis, then the p_z atomic orbital will always be hybridized whether the hybridization is sp, sp^2, sp^3, dsp^3 or d^2sp^3. For sp hybridization, the p_x and p_y atomic orbitals are unhybridized; they are used to form two π bonds to the bonded atom(s). For sp^2 hybridization, either the p_x or p_y atomic orbital is hybridized (along with the s and p_z orbitals); the other p orbital is used to form a π bond to a bonded atom. For sp^3 hybridization, the s and all of the p orbitals are hybridized; no unhybridized p atomic orbitals are present, so typical π bonds do not form with sp^3 hybridization. For dsp^3 and d^2sp^3 hybridization, we just mix in one or two d orbitals into the hybridization process. Which specific d orbitals are used is not important to our discussion.   **13. a.** If this structure is a trigonal planar arrangement of electron pairs, the central atom is sp^2 hybridized. If this structure is a tetrahedral arrangement of electron pairs, the central atom is sp^3 hybridized; **b.** dsp^3; **c.** sp^3; **d.** dsp^3; **e.** sp^3 **15.** We use d orbitals when we have to; i.e., we use d orbitals when the central atom on a molecule has more than eight electrons around it. The d orbitals are necessary to accommodate the electrons over eight. Row 2 elements never have more than eight electrons around them so they never hybridize d orbitals. We rationalize this by saying there are no d orbitals close in energy to the valence $2s$ and $2p$ orbitals ($2d$ orbitals are forbidden energy levels). However, for row 3 and heavier elements, there are $3d$, $4d$, $5d$, etc. orbitals that will be close in energy to the valence s and p orbitals. It is row 3 and heavier nonmetals that hybridize d orbitals when they have to. For sulfur, the valence electrons are in $3s$ and $3p$ orbitals. Therefore, $3d$ orbitals are closest in energy and are available for hybridization. Arsenic would hybridize $4d$ orbitals to go with the valence $4s$ and $4p$ orbitals while iodine would hybridize $5d$ orbitals since the valence electrons are in $n = 5$.   **17.** Bonding and antibonding molecular orbitals are both solutions to the quantum mechanical treatment of the molecule. Bonding orbitals form when in-phase orbitals combine to give constructive interference. This results in enhanced electron probability located between the two nuclei. The end result is that a bonding MO is lower in energy than the atomic orbitals of which it is composed. Antibonding orbitals form when out-of-phase orbitals combine. The mismatched phases produce destructive interference leading to a node in the electron probability between the two nuclei. With electron distribution pushed to the outside, the energy of an antibonding orbital is higher than the energy of the atomic orbitals of which it is composed.   **19.** The localized electron model does not deal effectively with molecules containing unpaired electrons. We can draw all of the possible resonance structures for NO but still not have a good feel for whether the bond in NO is weaker or stronger than the bond in NO^-. MO theory can handle odd electron species without any modifications. In addition, hybrid orbital theory does not predict that NO^- is paramagnetic. The MO theory correctly makes this prediction.   **21. a.** trigonal planar; 120°; **b.** V-shaped; ~120°   **23.** For Exercise 81: **a.** tetrahedral, 109.5°; **b.** trigonal pyramid, <109.5°; **c.** V-shaped, <109.5°; **d.** linear, no bond angle in diatomic molecules. For Exercise 87: **a.** NO_2^-: V-shaped, ~120°; NO_3^-: trigonal planar, 120°; N_2O_4: trigonal planar about both N atoms, 120°; **b.** all are linear, 180°   **25.** Br_3^-: linear; ClF_3: T-shaped; SF_4: see-saw   **27. a.** linear; 180°; **b.** T-shaped; ~90°; **c.** see-saw; ~90° and ~120°; **d.** trigonal bipyramid; 90° and 120°   **29. a.** Has a permanent dipole; **b.** Has no permanent dipole; **c.** Has no permanent dipole; **d.** Has a permanent dipole; **e.** Has no permanent dipole; **f.** Has no permanent dipole.   **31.** SeO_2, PCl_3, and SCl_2 are polar molecules (have net dipole moments).

33. a. OCl_2 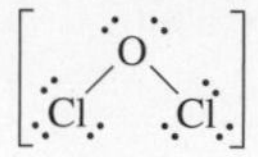   V-shaped, polar

KrF_2 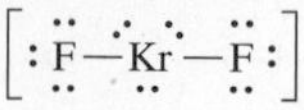   Linear, nonpolar

BeH_2 H—Be—H   Linear, nonpolar

SO_2 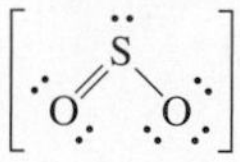   V-shaped, polar

(one other resonance structure possible)

b. SO_3 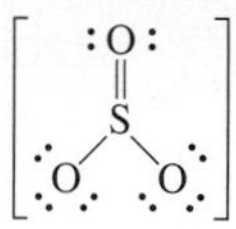   Trigonal planar, nonpolar

(two other resonance structures possible)

NF_3 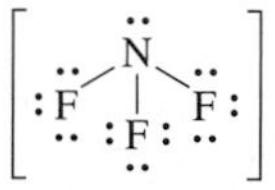   Trigonal pyramid, polar

IF_3 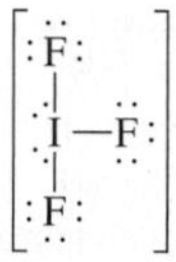   T-shaped, polar

c. CF_4 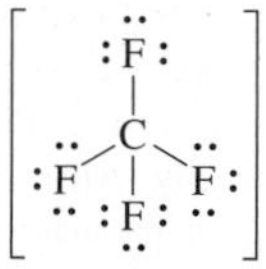   Tetrahedral, nonpolar

SeF_4 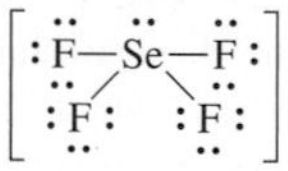   See-saw, polar

KrF_4 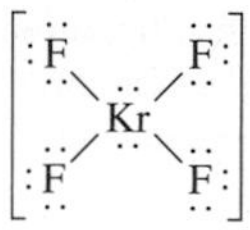   Square planar, nonpolar

d. IF_5 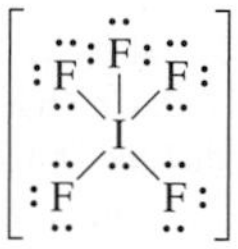   Square pyramid, polar

AsF_5 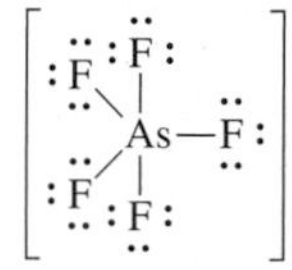   Trigonal bipyramid, nonpolar

35. Element E is a halogen (F, Cl, Br, or I); trigonal pyramid; <109.5°   **37.** The polar bonds are symmetrically arranged about the central atoms, and all the individual bond dipoles cancel to give no net dipole moment for each molecule, i.e., the molecules are nonpolar.

39. H_2O:

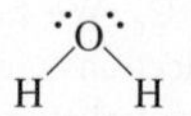

H_2O has a tetrahedral arrangement of the electron pairs about the O atom that requires sp^3 hybridization. Two of the sp^3 hybrid orbitals are used to form bonds to the two hydrogen atoms, and the other two sp^3 hybrid orbitals hold the two lone pairs on oxygen.

41. H_2CO:

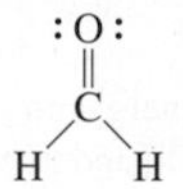

The central carbon atom has a trigonal planar arrangement of the electron pairs that requires sp^2 hybridization. Two of the sp^2 hybrid orbitals are used to form the two bonds to hydrogen. The other sp^2 hybrid orbital forms the σ bond to oxygen. The unchanged (unhybridized) p orbital on carbon is used to form the π bond between carbon and oxygen.

43. Ethane:

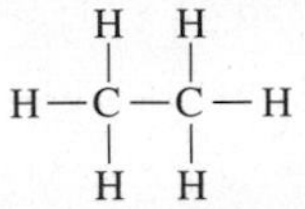

The carbon atoms are sp^3 hybridized. The six C—H bonds are formed from the sp^3 hybrid orbitals on C with the $1s$ atomic orbitals from the hydrogen atoms. The carbon–carbon bond is formed from an sp^3 hybrid orbital on each C atom.

Ethanol:

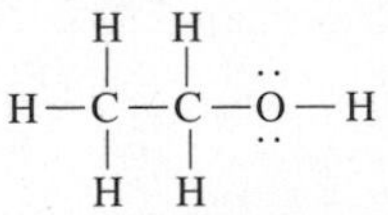

The two C atoms and the O atom are all sp^3 hybridized. All bonds are formed from these sp^3 hybrid orbitals. The C—H and O—H bonds form from sp^3 hybrid orbitals and the $1s$ atomic orbitals from the hydrogen atom. The C—C and C—O bonds are formed from sp^3 hybrid orbitals on each atom. **45.** For Exercise 81: All are sp^3 hybridized. For Exercise 87: **a.** NO_2^-, sp^2; NO_3^-, sp^2; N_2O_4, both N atoms are sp^2 hybridized; **b.** All are sp hybridized. **47.** The molecules in Exercise 21 all exhibit sp^2 hybridization about the central atom; the molecules in Exercise 22 all exhibit sp^3 hybridization about the central atom. **49. a.** tetrahedral, 109.5°, sp^3, nonpolar

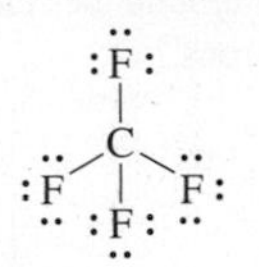

b. trigonal pyramid, <109.5°, sp^3, polar

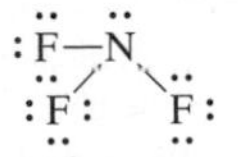

c. V-shaped, <109.5°, sp^3, polar

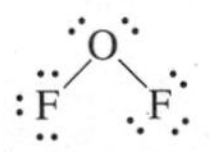

d. trigonal planar, 120°, sp^2, nonpolar

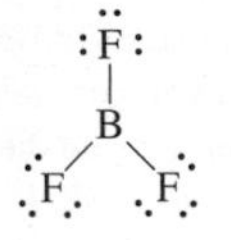

e. linear, 180°, sp, nonpolar

H—Be—H

f. a ≈120°, see-saw, dsp^3, polar
b ≈90°

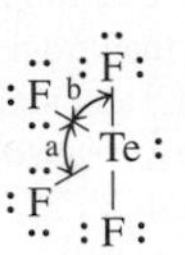

g. a = 90°, trigonal bipyramid, dsp^3, nonpolar
b = 120°

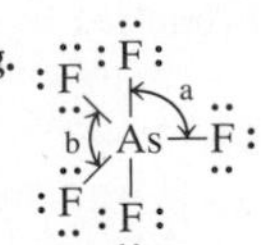

h. :F̈—Kr̈—F̈: linear, 180°, dsp^3, nonpolar

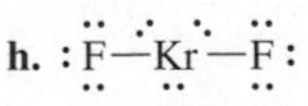

i. square planar, 90°, d^2sp^3, nonpolar

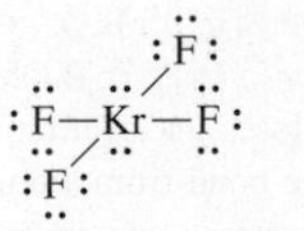

j. octahedral, 90°, d^2sp^3, nonpolar

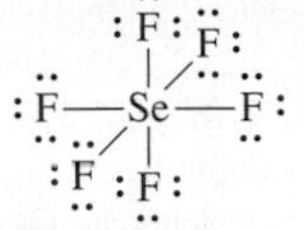

k. **l.**

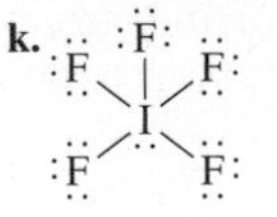

square pyramid, ≈90°,   T-shaped, ≈90°,
d^2sp^3, polar dsp^3, polar

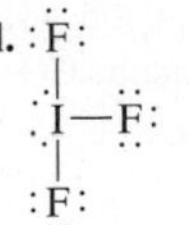

51. The π bond forces all six atoms into the same plane. **53. a.** There are 33 σ and 9 π bonds. **b.** All C atoms are sp^2 hybridized since all have a trigonal planar arrangement of the electrons. **55.** Biacetyl:

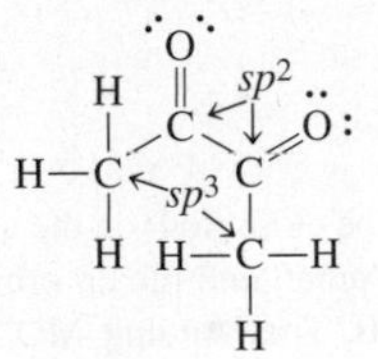

All CCO angles are 120°. The six atoms are not forced to lie in the same plane. 11 σ and 2 π bonds; Acetoin:

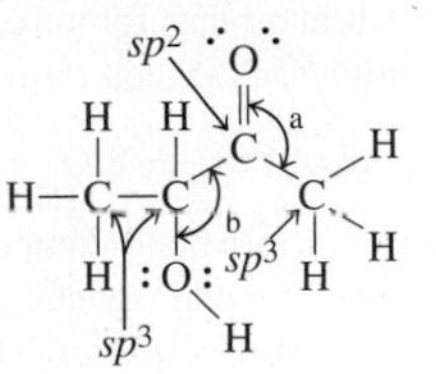

angle a = 120°, angle b = 109.5°, 13 σ bonds and 1 π bond

57. a.

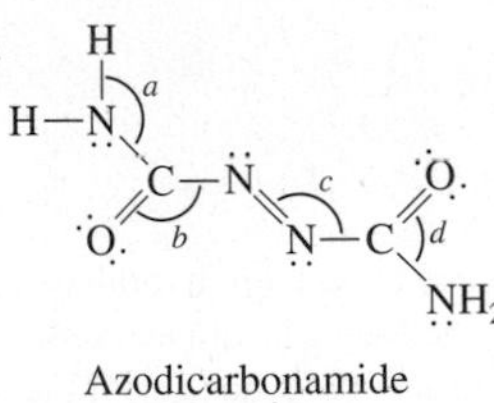

Azodicarbonamide 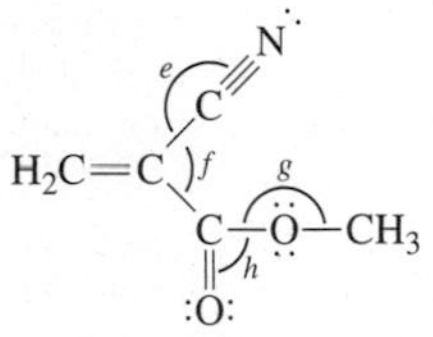   Methyl cyanoacrylate

Note: NH_2, CH_2, and CH_3 are shorthand for N or C atoms singly bonded to hydrogen atoms. **b.** In azodicarbonamide, the two carbon atoms are sp^2 hybridized. The two nitrogen atoms with hydrogens attached are sp^3 hybridized and the other two nitrogens are sp^2 hybridized. In methyl cyanoacrylate, the CH_3 carbon is sp^3 hybridized, the carbon with the triple bond is sp hybridized, and the other three carbons are sp^2 hybridized. **c.** Azodicarbonamide contains 3 π bonds and methyl cyanoacrylate contains 4 π bonds. **d.** a: ≈109.5°, b: 120°, c: ≈120°, d: 120°, e: 180°, f: 120°, g: ≈109.5°, h: 120° **59.** To complete the Lewis structure, add lone pairs to complete octets for each atom. **a.** 6; **b.** 4; **c.** The center N in —N=N=N group; **d.** 33 σ; **e.** 5 π; **f.** 180°; **g.** ≈109.5°; **h.** sp^3 **61. a.** The bonding molecular orbital is on the right, and the antibonding molecular orbital is on the left. The bonding MO has the greatest electron probability between the nuclei, while the antibonding MO has greatest electron probability on either side of the nuclei. **b.** The bonding MO is lower in energy. Because the electrons have greatest probability of being between the two nuclei, these electrons are attracted to two different nuclei, resulting in lower energy. **63. a.** H_2^+, H_2, H_2^-; **b.** He_2^{2+} and He_2^+ **65. a.** $(\sigma_{2s})^2$;

B.O. = 1; diamagnetic (0 unpaired electrons); **b.** $(\sigma_{2s})^2(\sigma_{2s}*)^2(\pi_{2p})^4$; B.O. = 2; diamagnetic (0 unpaired electrons); **c.** $(\sigma_{3s})^2(\sigma_{3s}*)^2(\sigma_{3p})^2(\pi_{3p})^4(\pi_{3p}*)^2$; B.O. = 2; paramagnetic (2 unpaired electrons)   **67.** When O_2 loses an electron, it comes from an antibonding orbital, which strengthens the bond from a bond order of 2 to a bond order of 2.5. When N_2 loses an electron, it comes from a bonding orbital, which changes the bond order from 3 to 2.5 (the bond weakens).   **69.** CO: $(\sigma_{2s})^2(\sigma_{2s}*)^2(\pi_{2p})^4(\sigma_{2p})^2$, B.O. = $(8 - 2)/2 = 3$; 0 unpaired electrons; O_2: $(\sigma_{2s})^2(\sigma_{2s}*)^2(\sigma_{2p})^2(\pi_{2p})^4(\pi_{2p}*)^2$, B.O. = $(8 - 4)/2 = 2$; 2 unpaired electrons; The most obvious differences are that CO has a larger bond order than O_2 (3 vs. 2) and that CO is diamagnetic, whereas O_2 is paramagnetic.   **71. a.** $(\sigma_{2s})^2(\sigma_{2s}*)^2(\pi_{2p})^4(\sigma_{2p})^2$; B.O. = 3; diamagnetic; **b.** $(\sigma_{2s})^2(\sigma_{2s}*)^2(\pi_{2p})^4(\sigma_{2p})^1$; B.O. = 2.5; paramagnetic; **c.** $(\sigma_{2s})^2(\sigma_{2s}*)^2(\pi_{2p})^4$; B.O. = 2; diamagnetic; bond length: CO < CO$^+$ < CO^{2+}; bond energy: CO^{2+} < CO$^+$ < CO   **73.** H_2; B_2; C_2^{2-}
75.

77. a. The electrons would be closer to F on the average. The F atom is more electronegative than the H atom, and the $2p$ orbital of F is lower in energy than the $1s$ orbital of H; **b.** The bonding MO would have more fluorine $2p$ character because it is closer in energy to the fluorine $2p$ orbital; **c.** The antibonding MO would place more electron density closer to H and would have a greater contribution from the higher-energy hydrogen $1s$ atomic orbital.   **79.** $[:C \equiv C:]^{2-}$ sp hybrid orbitals form the σ bond and the two unhybridized p atomic orbitals from each carbon form the two π bonds.

$$\text{MO: } (\sigma_{2s})^2(\sigma_{2s}*)^2(\pi_{2p})^4(\sigma_{2p})^2, \text{ B.O.} = (8 - 2)/2 = 3$$

Both give the same picture, a triple bond composed of a σ bond and two π bonds. Both predict the ion to be diamagnetic. Lewis structures deal well with diamagnetic (all electrons paired) species. The Lewis model cannot really predict magnetic properties.   **81.** O_3 and NO_2^- have identical Lewis structures, so we need to discuss only one of them. The Lewis structures for O_3 are

Localized electron model: The central oxygen atom is sp^2 hybridized, which is used to form the two σ bonds and hold the lone pair of electrons. An unchanged (unhybridized) p atomic orbital forms the π bond with the neighboring oxygen atoms. The actual structure of O_3 is an average of the two resonance structures. Molecular orbital model: There are two localized σ bonds and a π bond that is delocalized over the entire surface of the molecule. The delocalized π bond results from overlap of a p atomic orbital on each oxygen atom in O_3.
83. a. Trigonal pyramid; sp^3

b. Tetrahedral; sp^3

c. Square pyramid; d^2sp^3

d. T-shaped; dsp^3

e. Trigonal bipyramid; dsp^3

85. a. No, some atoms are attached differently; **b.** Structure 1: All N = sp^3, all C = sp^2; structure 2: All C and N = sp^2; **c.** The first structure with the carbon–oxygen double bonds is slightly more stable.   **87. a.** Both have one or more 180° bond angles; both are made up entirely of Xe and Cl; both have the individual bond dipoles arranged so they cancel each other (both are nonpolar); both have lone pairs on the central Xe atom; both have a central Xe atom that has more than 8 electrons around it. **b.** All have lone pairs on the central atom; all have a net dipole moment (all are polar).   **89.** 267 kJ/mol; this amount of energy must be supplied to break the π bond.   **91.** Yes, each structure has the same number of effective pairs around the central atom. (We count a multiple bond as a single group of electrons.)

93.

The lone pair of electrons around Te exerts a stronger repulsion than the bonding pairs. The stronger repulsion pushes the four square planar F atoms away from the lone pair, reducing the bond angles between the axial F atom and the square planar F atoms.   **95. a.** The NNO structure is correct. From the Lewis structures we would predict both NNO and NON to be linear, but NON would be nonpolar. NNO is polar.

The central N is sp hybridized. We can probably ignore the third resonance structure on the basis of formal charge. **c.** sp hybrid orbitals on the center N overlap with atomic orbitals (or hybrid orbitals) on the other two atoms to form two σ bonds. The remaining p orbitals on the center N overlap with p orbitals on the other N to form two π bonds.   **97.** F_2: $(\sigma_{2s})^2(\sigma_{2s}*)^2(\sigma_{2p})^2(\pi_{2p})^4(\pi_{2p}*)^4$; F_2 should have a lower ionization energy than F. The electron removed from F_2 is in a $\pi_{2p}*$ antibonding molecular orbital, which is higher in energy than the $2p$ atomic orbitals from which the electron in atomic fluorine is removed. Since the electron removed from F_2 is higher in energy than the electron removed from F, it should be easier to remove an electron from F_2 than from F.   **99.** π molecular orbital   **101.** CO_2: carbon dioxide, linear; NH_3: ammonia, trigonal pyramid; SO_3: sulfur trioxide; trigonal planar; H_2O: dihydrogen oxide (water), V-shaped or bent; ClO_4^-: perchlorate ion; tetrahedral

103. SO_2 (18 e$^-$)

V-shaped, $\approx 120°$, polar, sp^2 hybridized

PCl_3 (26 e$^-$)

trigonal pyramid, < 109.5°, polar, sp^3 hybridized

NNO (16 e$^-$)

+2 others

linear, 180°, polar, sp hybridized

COS (16 e⁻)   S=C=O   linear, 180°,
+2 others   polar, sp hybridized

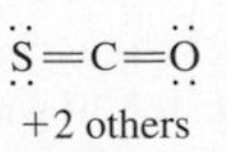

PF₃ (26 e⁻) 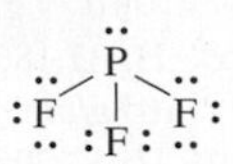   trigonal pyramid, < 109.5°,
polar, sp^3 hybridized

All of these compounds are polar. In each compound, the bond dipoles do not cancel each other out, leading to a polar compound. SO₂ is the only compound exhibiting ≈120° bond angles. PCl₃ and PF₃ both have a tetrahedral arrangement of electron pairs, so these are the compounds that have central atoms that are sp^3 hybridized. NNO and COS both have linear molecular structure.

105.   F₃ClO   F₂ClO₂⁺

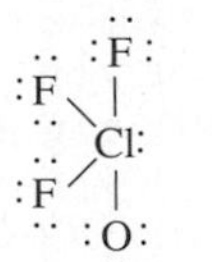

see-saw, dsp^3 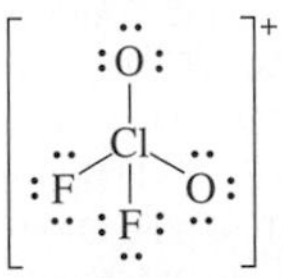   tetrahedral, sp^3

F₃ClO₂

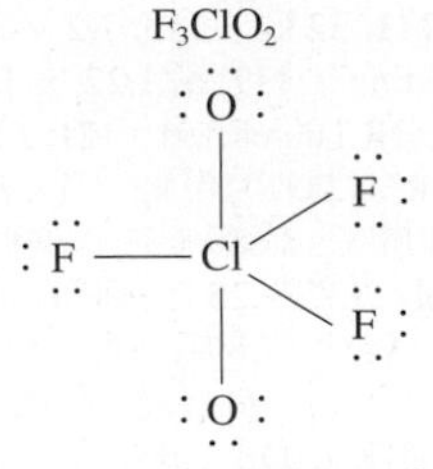

trigonal bipyramid, dsp^3

107. A Lewis structure for this compound is:

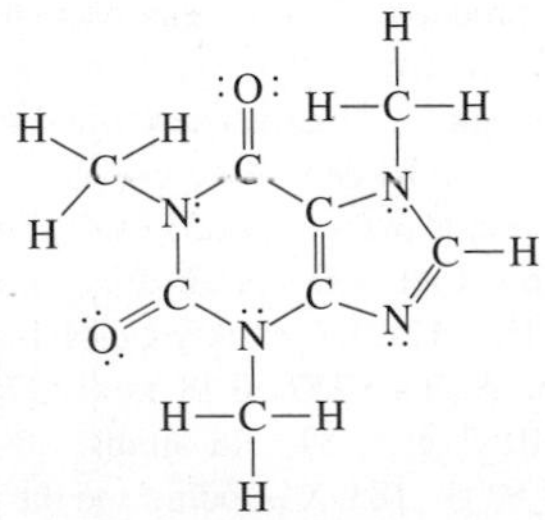

All of the carbon atoms in the rings have a trigonal planar arrangement of electron pairs, so these 5 carbon atoms are predicted to have 120° bond angles. The other 3 carbon atoms have a tetrahedral arrangement of electron pairs with predicted 109.5° bond angles. The three nitrogen atoms that are bonded to a —CH₃ group all have a tetrahedral arrangement of electron pairs with predicted 109.5° bond angles. The remaining nitrogen in the 5 atom ring is predicted to have a 120° bond angle due to the trigonal planar arrangement of electron pairs about the central nitrogen atom. There are 8 lone pairs of electrons and the molecule has 4 double bonds.
109. N₂: $(\sigma_{2s})^2(\sigma_{2s}{}^*)^2(\pi_{2p})^4(\sigma_{2p})^2$; N₂⁺: $(\sigma_{2s})^2(\sigma_{2s}{}^*)^2(\pi_{2p})^4(\sigma_{2p})^1$; N₂⁻: $(\sigma_{2s})^2(\sigma_{2s}{}^*)^2(\pi_{2p})^4(\sigma_{2p})^2(\pi_{2p}{}^*)^1$
111. As the halogen atoms get larger, it becomes more difficult to fit three halogen atoms around the small nitrogen atom and the NX₃ molecule becomes less stable.
113.

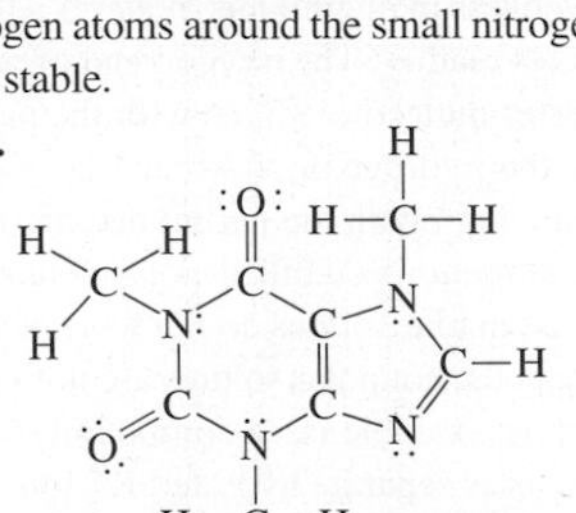

6 C and N atoms are sp^2 hybridized;
6 C and N atoms are sp^3 hybridized;
0 C and N atoms are sp hybridized;
25 σ bonds and 4 π bonds
115. a. NCN²⁻:

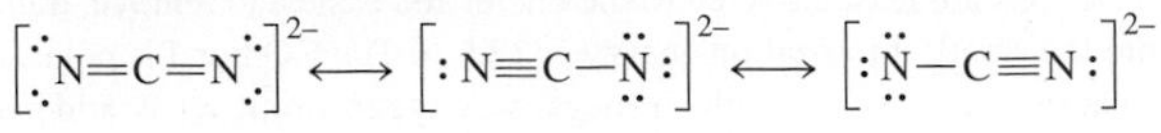

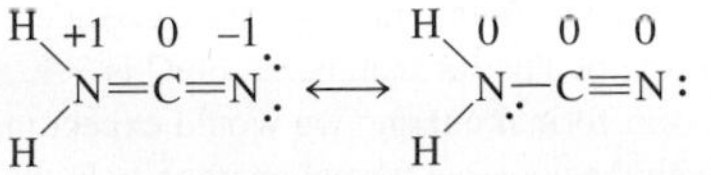

Favored by formal charge

Dicyandiamide:

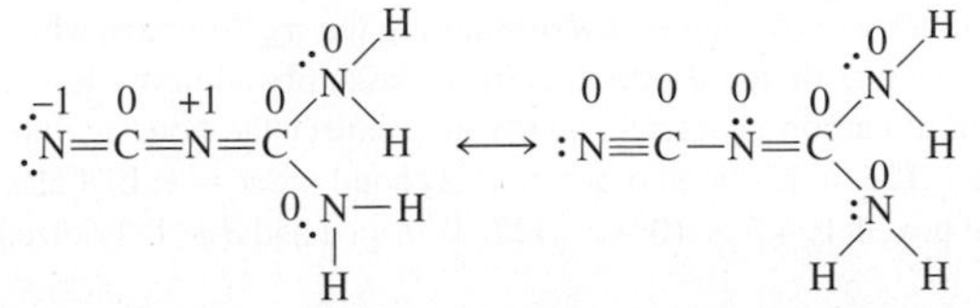

Favored by formal charge

Melamine:

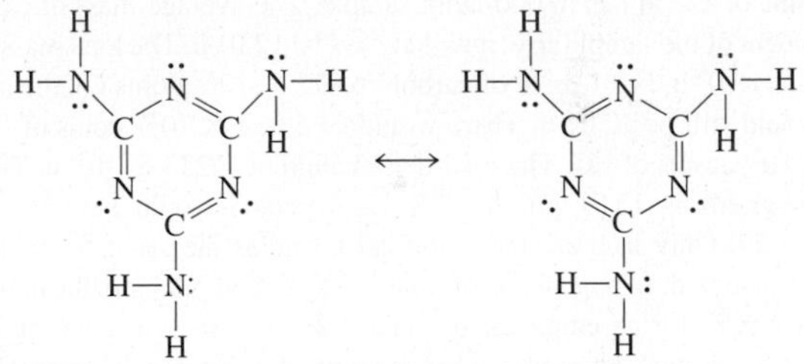

b. NCN²⁻: C is sp hybridized. Each resonance structure predicts a different hybridization for the N atoms. For the remaining compounds, we will predict hybrids for the favored resonance structures only.

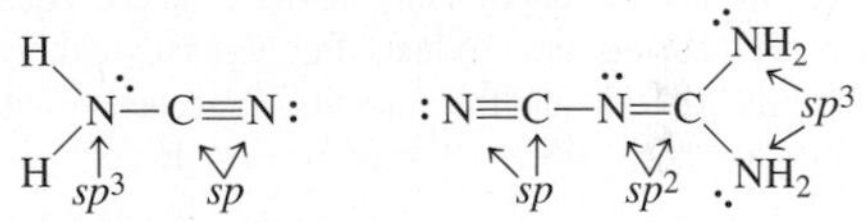

Melamine: N in NH₂ groups are all sp^3 hybridized. Atoms in ring are all sp^2 hybridized; **c.** NCN²⁻: 2 σ and 2 π bonds; H₂NCN: 4 σ and 2 π bonds; dicyandiamide: 9 σ and 3 π bonds; melamine: 15 σ and 3 π bonds; **d.** The π system forces the ring to be planar just as the benzene ring is planar. **e.** The structure

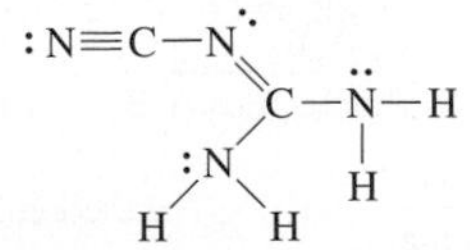

best agrees with experiment because it has three different CN bonds. This structure is also favored on the basis of formal charge.   **117.** Both reactions apparently involve only the breaking of the N—Cl bond. However, in the reaction ONCl → NO + Cl some energy is released in forming the stronger NO bond, lowering the value of ΔH. Therefore, the apparent N—Cl bond energy is artificially low for this reaction. The first reaction involves only the breaking of the N—Cl bond.   **119.** The ground state MO electron configuration for He₂ is $(\sigma_{1s})^2(\sigma_{1s}{}^*)^2$ giving a bond order of 0. Therefore, He₂ molecules are not predicted to be stable (and are not stable) in the lowest-energy ground state. However, in a high-energy environment, the electron(s) from the antibonding orbitals in He₂ can be promoted into higher-energy bonding orbitals, thus giving a nonzero bond order and a "reason" to form. For example, a possible excited-state MO electron configuration for He₂ would be $(\sigma_{1s})^2(\sigma_{1s}{}^*)^1(\sigma_{2s})^1$, giving a bond order of (3 − 1)/2 = 1. Thus excited

He$_2$ molecules can form, but they spontaneously break apart as the electron(s) fall back to the ground state, where the bond order equals zero.   **121.** The species with the smallest ionization energy has the highest energy electron. O$_2$, N$_2{}^{2-}$, N$_2{}^-$, and O$_2{}^+$ all have at least one electron in the high-energy $\pi_{2p}*$ orbitals. Because N$_2{}^{2-}$ has the smallest ratio of protons to electrons, the $\pi_{2p}*$ electrons are least attracted to the nuclei and easiest to remove, translating into the smallest ionization energy.   **123. a.** The CO bond is polar, with the negative end at the more electronegative oxygen atom. We would expect metal cations to be attracted to, and to bond to, the oxygen end of CO on the basis of electronegativity. **b.** The formal charge on C is -1, and the formal charge on O is $+1$. From formal charge, we would expect metal cations to bond to the carbon (with the negative formal charge). **c.** In molecular orbital theory, only orbitals with proper symmetry overlap to form bonding orbitals. The metals that form bonds to CO are usually transition metals, all of which have outer electrons in the d orbitals. The only molecular orbitals of CO that have proper symmetry to overlap with d orbitals are the $\pi_{2p}*$ orbitals, whose shape is similar to that of the d orbitals. Since the antibonding molecular orbitals have more carbon character, one would expect the bond to form through carbon.   **125. a.** Li$_2$ bond order $= 1$; B$_2$ bond order $= 1$; **b.** 4 electrons must be removed; **c.** 4.5×10^5 kJ   **127.** T-shaped and dsp^3 hybridized

Chapter 5

23. From the relative abundances, there would be 9889 atoms of ^{12}C and 111 atoms of ^{13}C in the 10,000-atom sample. The average mass of carbon is independent of the sample size; it will always be 12.01 u. The total mass would be 1.201×10^5 u. For 1 mole of carbon (6.022×10^{23} atoms C), the average mass would still be 12.01 u. There would be 5.955×10^{23} atoms of ^{12}C and 6.68×10^{21} atoms of ^{13}C. The total mass would be 7.233×10^{24} u. The total mass in grams is 12.01 g/mol.   **25.** Each person would have 90 trillion dollars.   **27.** Only in b are the empirical formulas the same for both compounds illustrated. In b, general formulas of X$_2$Y$_4$ and XY$_2$ are illustrated, and both have XY$_2$ for an empirical formula.   **29.** The mass percent of a compound is a constant no matter what amount of substance is present. Compounds always have constant composition.   **31.** c   **33.** The theoretical yield is the stoichiometric amount of product that should form if the limiting reactant is completely consumed and the reaction has 100% yield.   **35.** The information needed is mostly the coefficients in the balanced equation and the molar masses of the reactants and products. For percent yield, we would need the actual yield of the reaction and the amounts of reactants used.
a. mass of CB produced $= 1.00 \times 10^4$ molecules A$_2$B$_2$ $\times$

$$\frac{1 \text{ mol A}_2\text{B}_2}{6.022 \times 10^{23} \text{ molecules A}_2\text{B}_2} \times \frac{2 \text{ mol CB}}{1 \text{ mol A}_2\text{B}_2} \times \frac{\text{molar mass of CB}}{\text{mol CB}}$$

b. atoms of A produced $= 1.00 \times 10^4$ molecules A$_2$B$_2$ $\times \dfrac{2 \text{ atoms A}}{1 \text{ molecule A}_2\text{B}_2}$

c. mol of C reacted $= 1.00 \times 10^4$ molecules A$_2$B$_2$ $\times$

$$\frac{1 \text{ mol A}_2\text{B}_2}{6.022 \times 10^{23} \text{ molecules A}_2\text{B}_2} \times \frac{2 \text{ mol C}}{1 \text{ mol A}_2\text{B}_2}$$

d. % yield $= \dfrac{\text{actual mass}}{\text{theoretical mass}} \times 100$; The theoretical mass of CB produced was calculated in part a. If the actual mass of CB produced is given, then the percent yield can be determined for the reaction using the percent yield equation.   **37.** 207.2 u, Pb   **39.** 185 u   **41.** 48% ^{151}Eu and 52% ^{153}Eu   **43.** There are three peaks in the mass spectrum, each 2 mass units apart. This is consistent with two isotopes, differing in mass by two mass units.   **45.** 4.64 $\times 10^{-20}$ g Fe   **47.** 1.00×10^{22} atoms C   **49.** Al$_2$O$_3$, 101.96 g/mol; Na$_3$AlF$_6$, 209.95 g/mol   **51. a.** 17.03 g/mol; **b.** 32.05 g/mol; **c.** 252.08 g/mol   **53. a.** 0.0587 mol NH$_3$; **b.** 0.0312 mol N$_2$H$_4$; **c.** 3.97×10^{-3} mol (NH$_4$)$_2$Cr$_2$O$_7$ **55. a.** 85.2 g NH$_3$; **b.** 160. g N$_2$H$_4$; **c.** 1260 g (NH$_4$)$_2$Cr$_2$O$_7$ **57. a.** 70.1 g N; **b.** 140. g N; **c.** 140. g N   **59. a.** 3.54×10^{22} molecules NH$_3$; **b.** 1.88×10^{22} molecules N$_2$H$_4$; **c.** 2.39×10^{21} formula units (NH$_4$)$_2$Cr$_2$O$_7$ **61. a.** 3.54×10^{22} atoms N; **b.** 3.76×10^{22} atoms N; **c.** 4.78×10^{21} atoms N   **63.** 2.77×10^{19} molecules; 3.26 mg Cl   **65. a.** 0.9393 mol; **b.** 2.17×10^{-4} mol; **c.** 2.5×10^{-8} mol   **67. a.** 4.01×10^{22} atoms N; **b.** 5.97×10^{22} atoms N;

c. 3.67×10^{22} atoms N; **d.** 6.54×10^{22} atoms N   **69.** 176.12 g/mol; 2.839×10^{-3} mol; 1.710×10^{21} molecules   **71. a.** 165.39 g/mol; **b.** 3.023 mol; **c.** 3.3 g; **d.** 5.5×10^{22} atoms; **e.** 1.6 g; **f.** 1.373×10^{-19} g   **73. a.** 50.00% C, 5.595% H, 44.41% O; **b.** 55.80% C, 7.025% H, 37.18% O; **c.** 67.90% C, 5.699% H, 26.40% N   **75.** NO   **77.** 6.54×10^4 g/mol   **79. a.** 39.99% C, 6.713% H, 53.30% O; **b.** 40.00% C, 6.714% H, 53.29% O; **c.** 40.00% C, 6.714% H, 53.29% O (all the same except for rounding differences)   **81. a.** NO$_2$; **b.** CH$_2$; **c.** P$_2$O$_5$; **d.** CH$_2$O   **83.** C$_3$H$_6$O$_2$ **85.** compound I: HgO; compound II: Hg$_2$O   **87.** SN; S$_4$N$_4$ **89.** C$_3$H$_5$Cl; C$_6$H$_{10}$Cl$_2$ **91.** CH$_2$O   **93.** C$_3$H$_4$, C$_9$H$_{12}$ **95. a.** C$_6$H$_{12}$O$_6$(s) $+$ 6O$_2$(g) $\rightarrow$ 6CO$_2$(g) $+$ 6H$_2$O(g); **b.** Fe$_2$S$_3$(s) $+$ 6HCl(g) $\rightarrow$ 2FeCl$_3$(s) $+$ 3H$_2$S(g); **c.** CS$_2$(l) $+$ 2NH$_3$(g) $\rightarrow$ H$_2$S(g) $+$ NH$_4$SCN(s)   **97.** 2H$_2$O$_2$(aq) $\xrightarrow{\text{MnO}_2}$ 2H$_2$O(l) $+$ O$_2$(g)   **99. a.** 3Ca(OH)$_2$(aq) $+$ 2H$_3$PO$_4$(aq) $\rightarrow$ 6H$_2$O(l) $+$ Ca$_3$(PO$_4$)$_2$(s); **b.** Al(OH)$_3$(s) $+$ 3HCl(aq) $\rightarrow$ AlCl$_3$(aq) $+$ 3H$_2$O(l); **c.** 2AgNO$_3$(aq) $+$ H$_2$SO$_4$(aq) $\rightarrow$ Ag$_2$SO$_4$(s) $+$ 2HNO$_3$(aq)   **101. a.** 2C$_6$H$_6$(l) $+$ 15O$_2$(g) $\rightarrow$ 12CO$_2$(g) $+$ 6H$_2$O(g); **b.** 2C$_4$H$_{10}$(g) $+$ 13O$_2$(g) $\rightarrow$ 8CO$_2$(g) $+$ 10H$_2$O(g); **c.** C$_{12}$H$_{22}$O$_{11}$(s) $+$ 12O$_2$(g) $\rightarrow$ 12CO$_2$(g) $+$ 11H$_2$O(g); **d.** 4Fe(s) $+$ 3O$_2$(g) $\rightarrow$ 2Fe$_2$O$_3$(s); **e.** 4FeO(s) $+$ O$_2$(g) $\rightarrow$ 2Fe$_2$O$_3$(s)   **103. a.** SiO$_2$(s) $+$ 2C(s) $\rightarrow$ Si(s) $+$ 2CO(g); **b.** SiCl$_4$(l) $+$ 2Mg(s) $\rightarrow$ Si(s) $+$ 2MgCl$_2$(s); **c.** Na$_2$SiF$_6$(s) $+$ 4Na(s) $\rightarrow$ Si(s) $+$ 6NaF(s)   **105.** 7.26 g Al; 21.5 g Fe$_2$O$_3$; 13.7 g Al$_2$O$_3$ **107.** 4.355 kg   **109. a.** 0.076 g; **b.** 0.052 g   **111.** 32 kg   **113.** 7.2×10^4 g   **115.** 2NO(g) $+$ O$_2$(g) $\rightarrow$ 2NO$_2$(g); NO is limiting.   **117. a.** 1.22×10^3 g; 284 g H$_2$ unreacted   **119.** 0.301 g; 3.6×10^{-2} g HCl unreacted   **121.** 2.81×10^6 g HCN; 5.63×10^6 g H$_2$O   **123.** 82.9%   **125.** 1.20×10^3 kg $=$ 1.20 metric tons   **127.** ^{12}C^{1}H$_2{}^{16}$O   **129.** 6   **131.** 71.40% C, 8.689% H, 5.648% F, 14.26% O   **133.** C$_8$H$_{11}$O$_3$N   **135.** 1360 g/mol   **137.** 9.25×10^{22} H atoms   **139.** 4.30×10^{-2} mol; 2.50 g   **141.** 5   **143.** 42.8%   **145.** 81.1 g   **147.** 86.2%   **149.** C$_{20}$H$_{30}$O   **151.** 14.3% X, 85.7% Z   **153.** 1.242 mol Mg$_3$(PO$_4$)$_2$; 1.846 mol Ca(NO$_3$)$_2$; 0.7291 mol K$_2$CrO$_4$; 3.761 mol N$_2$O$_5$ **155.** C$_7$H$_8$O   **157. a.** False; this is what is done when equations are balanced. **b.** False; the coefficients give molecule ratios as well as mole ratios between reactants and products. **c.** False; the reactants are on the left, with the products on the right. **d.** True; **e.** True; in order for mass to be conserved, there must be the same number of atoms as well as the same type of atoms on both sides.   **159.** 51.7 g Na$_2$SO$_3$, 7.39 g H$_2$O   **161.** I: NH$_3$; II: N$_2$H$_4$; III: HN$_3$. If we set the atomic mass of H equal to 1.008, the atomic mass for nitrogen is 14.01.   **163.** C$_7$H$_5$N$_3$O$_6$ **165.** 5.7 g FeO and 22.4 g Fe$_2$O$_3$ **167.** 40.08%   **169.** N$_4$H$_6$ **171.** 1.8×10^6 g Cu(NH$_3$)$_4$Cl$_2$; 5.9×10^5 g NH$_3$ **173.** 207 u, Pb   **175.** Al$_2$Se$_3$ **177.** 0.48 mol   **179. a.** 113 Fe atoms; **b.** mass $= 9.071 \times 10^{-20}$ g; **c.** 540 Ru atoms   **181.** M $=$ Y, X $=$ Cl, yttrium(III) chloride; 1.84 g   **183.** X is iodine and the molecular structure of IF$_5$ is square pyramid.

Chapter 6

13. a. Polarity is a term applied to covalent compounds. Polar covalent compounds have an unequal sharing of electrons in bonds that results in an unequal charge distribution in the overall molecule. Polar molecules have a partial negative end and a partial positive end. These are not full charges as in ionic compounds, but are charges much less in magnitude. Water is a polar molecule and dissolves other polar solutes readily. The oxygen end of water (the partial negative end of the polar water molecule) aligns with the partial positive end of the polar solute while the hydrogens of water (the partial positive end of the polar water molecule) align with the partial negative end of the solute. These opposite charged attractions stabilize polar solutes in water. This process is called hydration. Nonpolar solutes do not have permanent partial negative and partial positive ends; nonpolar solutes are not stabilized in water and do not dissolve. **b.** KF is a soluble ionic compound so it is a strong electrolyte. KF(aq) actually exists as separate hydrated K$^+$ ions and hydrated F$^-$ ions in solution: C$_6$H$_{12}$O$_6$ is a polar covalent molecule that is a nonelectrolyte. C$_6$H$_{12}$O$_6$ is hydrated as described in part a. **c.** RbCl is a soluble ionic compound so it exists as separate hydrated Rb^+ ions and hydrated Cl^- ions in solution. AgCl is an insoluble ionic compound so the ions stay together in solution and fall to the bottom of the container as a precipitate. **d.** HNO_3 is a strong acid and exists as separate hydrated H^+ ions and hydrated NO_3^- ions in solution. CO is a polar covalent molecule and is hydrated as explained in part a. **15.** Water is a polar solvent and will dissolve other polar covalent solutes. F_2 is a nonpolar compound and will not be soluble in water. Both CF_4 and SF_2 have polar bonds, but only SF_2 is polar. The individual bond dipoles cancel each other in the tetrahedral CF_4 molecule, so CF_4 is nonpolar and will not be soluble in water. However, the individual bond dipoles do not cancel each other out in the bent SF_2 molecule, and this molecule is polar. Of the three molecules, the polar SF_2 molecule will be the most soluble in water. **17.** Only statement b is true. A concentrated solution can also contain a nonelectrolyte dissolved in water, e.g., concentrated sugar water. Acids are either strong or weak electrolytes. Some ionic compounds are not soluble in water, so they are not labeled as a specific type of electrolyte. **19.** Bromides: NaBr, KBr, and NH_4Br (and others) would be soluble and AgBr, $PbBr_2$, and Hg_2Br_2 would be insoluble. Sulfates: Na_2SO_4, K_2SO_4, and $(NH_4)_2SO_4$ (and others) would be soluble and $BaSO_4$, $CaSO_4$, and $PbSO_4$ (or Hg_2SO_4) would be insoluble. Hydroxides: NaOH, KOH, and $Ca(OH)_2$ (and others) would be soluble and $Al(OH)_3$, $Fe(OH)_3$, and $Cu(OH)_2$ (and others) would be insoluble. Phosphates: Na_3PO_4, K_3PO_4, and $(NH_4)_3PO_4$ (and others) would be soluble and Ag_3PO_4, $Ca_3(PO_4)_2$, and $FePO_4$ (and others) would be insoluble. Lead: $PbCl_2$, $PbBr_2$, PbI_2, $Pb(OH)_2$, $PbSO_4$, and PbS (and others) would be insoluble. $Pb(NO_3)_2$ would be a soluble Pb^{2+} salt. **21.** The Brønsted–Lowry definitions are best for our purposes. An acid is a proton donor and a base is a proton acceptor. A proton is an H^+ ion. Neutral hydrogen has 1 electron and 1 proton, so an H^+ ion is just a proton. An acid–base reaction is the transfer of an H^+ ion (a proton) from an acid to a base. **23. a.** The species reduced is the element that gains electrons. The reducing agent causes reduction to occur by itself being oxidized. The reducing agent is generally listed as the entire formula of the compound/ion that contains the element oxidized. **b.** The species oxidized is the element that loses electrons. The oxidizing agent causes oxidation to occur by itself being reduced. The oxidizing agent is generally listed as the entire formula of the compound/ion that contains the element reduced. **c.** For simple binary ionic compounds, the actual charges on the ions are the oxidation states. For covalent compounds, nonzero oxidation states are imaginary charges the elements would have if they were held together by ionic bonds (assuming the bond is between two different nonmetals). Nonzero oxidation states for elements in covalent compounds are not actual charges. Oxidation states for covalent compounds are a bookkeeping method to keep track of electrons in a reaction.

25.

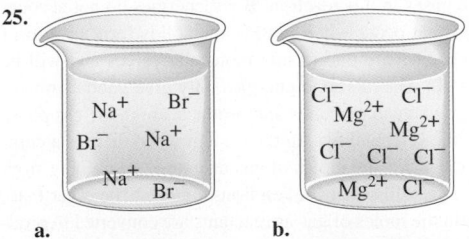

a.   **b.**

c. For answers c–i, we will describe what should be in each solution. For c, the drawing should have three times as many NO_3^- anions as Al^{3+} cations. **d.** The drawing should have twice as many NH_4^+ cations as SO_4^{2-} anions. **e.** The drawing should have equal numbers of Na^+ cations and OH^- anions. **f.** The drawing should have equal numbers of Fe^{2+} cations and SO_4^{2-} anions. **g.** The drawing should have equal numbers of K^+ cations and MnO_4^- anions. **h.** The drawing should have equal numbers of H^+ cations and ClO_4^- anions. **i.** The drawing should have equal numbers of NH_4^+ cations and $C_2H_3O_2^-$ anions. **27.** $CaCl_2(s) \rightarrow Ca^{2+}(aq) + 2Cl^-(aq)$ **29. a.** 0.2677 M; **b.** 1.255×10^{-3} M; **c.** 8.065×10^{-3} M **31. a.** $M_{Ca^{2+}} = 1.00$ M, $M_{NO_3^-} = 2.00$ M; **b.** $M_{Na^+} = 4.0$ M, $M_{SO_4^{2-}} = 2.0$ M; **c.** $M_{NH_4^+} = M_{Cl^-} = 0.187$ M; **d.** $M_{K^+} = 0.0564$ M, $M_{PO_4^{3-}} = 0.0188$ M **33.** 100.0 mL of 0.30 M $AlCl_3$ **35.** 4.00 g **37. a.** Place 20.0 g NaOH in a 2-L volumetric flask; add water to dissolve the NaOH and fill to the mark. **b.** Add 500. mL of the 1.00 M NaOH

stock solution to a 2-L volumetric flask; fill to the mark with water. **c.** As in a, instead using 38.8 g K_2CrO_4. **d.** As in b, instead using 114 mL of 1.75 M K_2CrO_4 stock solution. **39.** $M_{NH_4^+} = 0.272$ M, $M_{SO_4^{2-}} = 0.136$ M **41.** 4.5 M **43.** 5.94×10^{-8} M **45.** Aluminum nitrate, magnesium chloride, and rubidium sulfate are soluble. **47. a.** No precipitate forms; **b.** $Al(OH)_3(s)$; **c.** $CaSO_4(s)$; **d.** NiS(s) **49. a.** No reaction occurs because all possible products are soluble salts. **b.** $2Al(NO_3)_3(aq) + 3Ba(OH)_2(aq) \rightarrow 2Al(OH)_3(s) + 3Ba(NO_3)_2(aq)$; $2Al^{3+}(aq) + 6NO_3^-(aq) + 3Ba^{2+}(aq) + 6OH^-(aq) \rightarrow 2Al(OH)_3(s) + 3Ba^{2+}(aq) + 6NO_3^-(aq)$, $Al^{3+}(aq) + 3OH^-(aq) \rightarrow Al(OH)_3(s)$; **c.** $CaCl_2(aq) + Na_2SO_4(aq) \rightarrow CaSO_4(s) + 2NaCl(aq)$; $Ca^{2+}(aq) + 2Cl^-(aq) + 2Na^+(aq) + SO_4^{2-}(aq) \rightarrow CaSO_4(s) + 2Na^+(aq) + 2Cl^-(aq)$; $Ca^{2+}(aq) + SO_4^{2-}(aq) \rightarrow CaSO_4(s)$; **d.** $K_2S(aq) + Ni(NO_3)_2(aq) \rightarrow 2KNO_3(aq) + NiS(s)$; $2K^+(aq) + S^{2-}(aq) + Ni^{2+}(aq) + 2NO_3^-(aq) \rightarrow 2K^+(aq) + 2NO_3^-(aq) + NiS(s)$; $Ni^{2+}(aq) + S^{2-}(aq) \rightarrow NiS(s)$ **51. a.** $CuSO_4(aq) + Na_2S(aq) \rightarrow CuS(s) + Na_2SO_4(aq)$; $Cu^{2+}(aq) + S^{2-}(aq) \rightarrow CuS(s)$; the gray spheres are the Na^+ spectator ions and the blue-green spheres are the SO_4^{2-} spectator ions; **b.** $CoCl_2(aq) + 2NaOH(aq) \rightarrow Co(OH)_2(s) + 2NaCl(aq)$; $Co^{2+}(aq) + 2OH^-(aq) \rightarrow Co(OH)_2(s)$; the gray spheres are the Na^+ spectator ions and the green spheres are the Cl^- spectator ions; **c.** $AgNO_3(aq) + KI(aq) \rightarrow AgI(s) + KNO_3(aq)$; $Ag^+(aq) + I^-(aq) \rightarrow AgI(s)$; the red spheres are the K^+ spectator ions and the blue spheres are the NO_3^- spectator ions **53. a.** $Ba^{2+}(aq) + SO_4^{2-}(aq) \rightarrow BaSO_4(s)$; **b.** $Pb^{2+}(aq) + 2Cl^-(aq) \rightarrow PbCl_2(s)$; **c.** no reaction; **d.** no reaction; **e.** $Cu^{2+}(aq) + 2OH^-(aq) \rightarrow Cu(OH)_2(s)$ **55.** Ca^{2+}, Sr^{2+}, or Ba^{2+} could all be present. **57.** 0.607 g **59.** 0.520 g $Al(OH)_3$ **61.** 2.82 g **63. a.** $2KOH(aq) + Mg(NO_3)_2(aq) \rightarrow Mg(OH)_2(s) + 2KNO_3(aq)$; **b.** magnesium hydroxide; **c.** 0.583 g; **d.** 0 M OH^-, 5.00×10^{-2} M Mg^{2+}, 0.100 M K^+, 0.200 M NO_3^- **65.** 23 u; Na **67. a.** $2HClO_4(aq) + Mg(OH)_2(s) \rightarrow Mg(ClO_4)_2(aq) + 2H_2O(l)$; $2H^+(aq) + 2ClO_4^-(aq) + Mg(OH)_2(s) \rightarrow Mg^{2+}(aq) + 2ClO_4^-(aq) + 2H_2O(l)$; $2H^+(aq) + Mg(OH)_2(s) \rightarrow Mg^{2+}(aq) + 2H_2O(l)$; **b.** $HCN(aq) + NaOH(aq) \rightarrow NaCN(aq) + H_2O(l)$; $HCN(aq) + Na^+(aq) + OH^-(aq) \rightarrow Na^+(aq) + CN^-(aq) + H_2O(l)$; $HCN(aq) + OH^-(aq) \rightarrow H_2O(l) + CN^-(aq)$; **c.** $HCl(aq) + NaOH(aq) \rightarrow NaCl(aq) + H_2O(l)$; $H^+(aq) + Cl^-(aq) + Na^+(aq) + OH^-(aq) \rightarrow Na^+(aq) + Cl^-(aq) + H_2O(l)$; $H^+(aq) + OH^-(aq) \rightarrow H_2O(l)$ **69. a.** $KOH(aq) + HNO_3(aq) \rightarrow H_2O(l) + KNO_3(aq)$; **b.** $Ba(OH)_2(aq) + 2HCl(aq) \rightarrow 2H_2O(l) + BaCl_2(aq)$; **c.** $3HClO_4(aq) + Fe(OH)_3(s) \rightarrow 3H_2O(l) + Fe(ClO_4)_3(aq)$; **d.** $AgOH(s) + HBr(aq) \rightarrow AgBr(s) + H_2O(l)$; **e.** $Sr(OH)_2(aq) + 2HI(aq) \rightarrow 2H_2O(l) + SrI_2(aq)$ **71. a.** 100. mL; **b.** 66.7 mL; **c.** 50.0 mL **73.** 2.0×10^{-2} M excess OH^- **75.** 0.102 M **77.** 43.8 mL **79.** 0.4178 g **81. a.** K, +1; O, −2; Mn, +7; **b.** Ni, +4; O, −2; **c.** Na, +1; Fe, +2; O, −2; H, +1 **d.** H, +1; O, −2; N, −3; P, +5; **e.** P, +3; O, −2; **f.** O, −2; Fe, $+\frac{8}{3}$; **g.** O, −2; F, −1; Xe, +6; **h.** S, +4; F, −1; **i.** C, +2; O, −2; **j.** C, 0; H, +1; O, −2 **83. a.** −3; **b.** −3; **c.** −2; **d.** +2; **e.** +1; **f.** +4; **g.** +3; **h.** +5; **i.** 0

85.

Redox?	Oxidizing Agent	Reducing Agent	Substance Oxidized	Substance Reduced
a. Yes	Ag^+	Cu	Cu	Ag^+
b. No	—	—	—	—
c. No	—	—	—	—
d. Yes	$SiCl_4$	Mg	Mg	$SiCl_4$ (Si)
e. No	—	—	—	—

In b, c, and e, no oxidation numbers change from reactants to products. **87.** Each sodium atom will lose one electron, and each fluorine atom will gain one electron. Two sodium atoms are required to react with one fluorine molecule in order to equalize the electrons lost with the electrons gained. $2Na(s) + F_2(g) \rightarrow 2NaF(s)$ **89. a.** $2C_2H_6(g) + 7O_2(g) \rightarrow 4CO_2(g) + 6H_2O(g)$; **b.** $Mg(s) + 2HCl(aq) \rightarrow Mg^{2+}(aq) + 2Cl^-(aq) + H_2(g)$; **c.** $2Co^{3+}(aq) + Ni(s) \rightarrow 2Co^{2+}(aq) + Ni^{2+}(aq)$; **d.** $Zn(s) + H_2SO_4(aq) \rightarrow ZnSO_4(aq) + H_2(g)$ **91.** We should weigh out between 4.24 and 4.32 g of KIO_3. We should weigh it to the nearest milligram or 0.1 mg. Dissolve the KIO_3 in water, and dilute to the mark in a 1-L volumetric flask. This will produce a solution whose concentration is within the limits and is known to at least the fourth decimal place. **93.** increase by 0.019 M **95. a.** $AgNO_3$, $Pb(NO_3)_2$, and $Hg_2(NO_3)_2$ would form precipitates with the Cl^- ion; $Ag^+(aq) + Cl^-(aq) \rightarrow$

$AgCl(s)$; $Pb^{2+}(aq) + 2Cl^-(aq) \rightarrow PbCl_2(s)$; $Hg_2^{2+}(aq) + 2Cl^-(aq) \rightarrow Hg_2Cl_2(s)$; **b.** Na_2SO_4, Na_2CO_3, and Na_3PO_4 would form precipitates with the Ca^{2+} ion; $Ca^{2+}(aq) + SO_4^{2-}(aq) \rightarrow CaSO_4(s)$; $Ca^{2+}(aq) + CO_3^{2-}(aq) \rightarrow CaCO_3(s)$; $3Ca^{2+}(aq) + 2PO_4^{3-}(aq) \rightarrow Ca_3(PO_4)_2(s)$; **c.** NaOH, Na_2S, and Na_2CO_3 would form precipitates with the Fe^{3+} ion; $Fe^{3+}(aq) + 3OH^-(aq) \rightarrow Fe(OH)_3(s)$; $2Fe^{3+}(aq) + 3S^{2-}(aq) \rightarrow Fe_2S_3(s)$; $2Fe^{3+}(aq) + 3CO_3^{2-}(aq) \rightarrow Fe_2(CO_3)_3(s)$; **d.** $BaCl_2$, $Pb(NO_3)_2$, and $Ca(NO_3)_2$ would form precipitates with the SO_4^{2-} ion; $Ba^{2+}(aq) + SO_4^{2-}(aq) \rightarrow BaSO_4(s)$; $Pb^{2+}(aq) + SO_4^{2-}(aq) \rightarrow PbSO_4(s)$; $Ca^{2+}(aq) + SO_4^{2-}(aq) \rightarrow CaSO_4(s)$; **e.** Na_2SO_4, NaCl, and NaI would form precipitates with the Hg_2^{2+} ion; $Hg_2^{2+}(aq) + SO_4^{2-}(aq) \rightarrow Hg_2SO_4(s)$; $Hg_2^{2+}(aq) + 2Cl^-(aq) \rightarrow Hg_2Cl_2(s)$; $Hg_2^{2+}(aq) + 2I^-(aq) \rightarrow Hg_2I_2(s)$; **f.** NaBr, Na_2CrO_4, and Na_3PO_4 would form precipitates with the Ag^+ ion; $Ag^+(aq) + Br^-(aq) \rightarrow AgBr(s)$; $2Ag^+(aq) + CrO_4^{2-}(aq) \rightarrow Ag_2CrO_4(s)$; $3Ag^+(aq) + PO_4^{3-}(aq) \rightarrow Ag_3PO_4(s)$ **97.** Ba **99.** 1.465% **101.** 2.00 *M* **103.** 180.2 g/mol **105.** $C_6H_8O_6$ **107.** 4.05% **109.** 63.74% **111.** Chromium is reduced and carbon is oxidized. Twelve (12) electrons are transferred in the balanced reaction. **113.** 2.979×10^{-3} *M* **115.** 747 mL **117.** 0.052 *M* Cl^- in excess after reaction **119.** 0.328 *M* **121.** $MgSO_4$: Sulfur has a +6 oxidation state. $PbSO_4$: Lead has a +2 oxidation state. O_2: O has an oxidation state of zero. Ag: Ag has an oxidation state of zero. $CuCl_2$: Copper has a +2 oxidation state. **123. a.** 2.5×10^{-8} *M*; **b.** 8.4×10^{-9} *M*; **c.** 1.33×10^{-4} *M*; **d.** 2.8×10^{-7} *M* **125. a.** 24.8% Co, 29.7% Cl, 5.09% H, 40.4% O; **b.** $CoCl_2 \cdot 6H_2O$; **c.** $CoCl_2 \cdot 6H_2O(aq) + 2AgNO_3(aq) \rightarrow 2AgCl(s) + Co(NO_3)_2(aq) + 6H_2O(l)$, $CoCl_2 \cdot 6H_2O(aq) + 2NaOH(aq) \rightarrow Co(OH)_2(s) + 2NaCl(aq) + 6H_2O(l)$, $4Co(OH)_2(s) + O_2(g) \rightarrow 2Co_2O_3(s) + 4H_2O(l)$ **127.** 14.6 g Zn and 14.4 g Ag **129.** 0.123 g SO_4^{2-}, 60.0% SO_4^{2-}; 61% K_2SO_4 and 39% Na_2SO_4 **131.** 4.90 *M* **133.** Y, 2.06 mL/min; Z, 4.20 mL/min **135.** 57.6 mL **137.** 4.7×10^{-2} *M* **139.** Citric acid has three acidic hydrogens per citric acid molecule. **141.** 0.07849 ± 0.00016 *M* or 0.0785 ± 0.0002 *M* **143.** $3(NH_4)_2CrO_4(aq) + 2Cr(NO_3)_3(aq) \rightarrow 6NH_4NO_3(aq) + Cr_2(CrO_4)_3(s)$; 7.34 g **145.** X = Se; H_2Se is hydroselenic acid; 0.252 g

Chapter 7

11. Path-dependent functions for a trip from Chicago to Denver are those quantities that depend on the route taken. One can fly directly from Chicago to Denver or one could fly from Chicago to Atlanta to Los Angeles and then to Denver. Some path-dependent quantities are miles traveled, fuel consumption of the airplane, time traveling, airplane snacks eaten, etc. State functions are path-independent; they only depend on the initial and final states. Some state functions for an airplane trip from Chicago to Denver would be longitude change, latitude change, elevation change, and overall time zone change. **13.** Both *q* and *w* are negative. **15. a.** 446 kJ released; **b.** 446 kJ released

17.

$$H_2O(l) + \tfrac{1}{2}CO_2(g) \rightarrow \tfrac{1}{2}CH_4(g) + O_2(g) \qquad \Delta H_1 = -\tfrac{1}{2}(-891 \text{ kJ})$$
$$\tfrac{1}{2}CH_4(g) + O_2(g) \rightarrow \tfrac{1}{2}CO_2(g) + H_2O(g) \qquad \Delta H_2 = \tfrac{1}{2}(-803 \text{ kJ})$$
$$H_2O(l) \rightarrow H_2O(g) \qquad \Delta H = \Delta H_1 + \Delta H_2 = 44 \text{ kJ}$$

19. The zero points for ΔH_f° values are elements in their standard state. All substances are measured in relationship to this zero point. **21.** No matter how insulated your thermos bottle, some heat will always escape into the surroundings. If the temperature of the thermos bottle (the surroundings) is high, less heat initially will escape from the coffee (the system); as a result your coffee will stay hotter for a longer period of time. **23.** Fossil fuels contain carbon; the incomplete combustion of fossil fuels produces $CO(g)$ instead of $CO_2(g)$. This occurs when the amount of oxygen reacting is not sufficient to convert all of the carbon in fossil fuels to CO_2. Carbon monoxide is a poisonous gas to humans. **25.** 150 J **27. a.** Potential energy is energy due to position. Initially, ball A has a higher potential energy than ball B because the position of ball A is higher than that of ball B. In the final position, ball B has the higher position, so ball B has the higher potential energy. **b.** As ball A rolled down the hill, some of the potential energy lost by A was converted to random motion of the components of the hill (frictional heating). The remainder of the lost potential energy was added to B to

initially increase its kinetic energy and then to increase its potential energy. **29.** 16 kJ **31. a.** 41 kJ; **b.** 35 kJ; **c.** 47 kJ; **d.** part a only **33.** 375 J heat transferred to the system **35.** −13.2 kJ **37.** 11.0 L **39.** $q = 30.9$ kJ, $w = -12.4$ kJ, $\Delta E = 18.5$ kJ **41.** This is an endothermic reaction, so heat must be absorbed to convert reactants into products. The high-temperature environment of internal combustion engines provides the heat. **43. a.** endothermic; **b.** exothermic; **c.** exothermic; **d.** endothermic **45. a.** 1650 kJ released; **b.** 826 kJ released; **c.** 7.39 kJ released; **d.** 34.4 kJ released **47.** 4400 g C_3H_8 **49.** $H_2O(l)$; 2.30×10^3 J; $Hg(l)$; 140°C **51.** $Al(s)$ **53.** 311 K **55.** 23.7°C **57.** 0.25 J/°C · g **59.** −66 kJ/mol **61.** 170 J/g; 20. kJ/mol **63.** 39.2°C **65. a.** 31.5 kJ/°C; **b.** -1.10×10^3 kJ/mol **67.** −220.8 kJ **69.** 1268 kJ; No, because this reaction is very endothermic, it would not be a practical way of making ammonia due to the high energy costs. **71.** −233 kJ **73.** −713 kJ **75.** The enthalpy change for the formation of one mole of a compound from its elements, with all substances in their standard states. $Na(s) + \tfrac{1}{2}Cl_2(g) \rightarrow NaCl(s)$; $H_2(g) + \tfrac{1}{2}O_2(g) \rightarrow H_2O(l)$; $6C_{graphite}(s) + 6H_2(g) + 3O_2(g) \rightarrow C_6H_{12}O_6(s)$; $Pb(s) + S_{rhombic}(s) + 2O_2(g) \rightarrow PbSO_4(s)$ **77. a.** −940. kJ; **b.** −265 kJ; **c.** −176 kJ **79. a.** −908 kJ, −112 kJ, −140. kJ; **b.** $12NH_3(g) + 21O_2(g) \rightarrow 8HNO_3(aq) + 4NO(g) + 14H_2O(g)$, exothermic **81.** −2677 kJ **83.** −169 kJ/mol **85.** 132 kJ **87.** −29.67 kJ/g **89.** For $C_3H_8(g)$, −50.37 kJ/g vs. −47.7 kJ/g for octane. Because of the low boiling point of propane, there are extra safety hazards associated with storing the propane in high-pressure compressed gas tanks. **91.** 4900 g **93. a.** $2SO_2(g) + O_2(g) \rightarrow 2SO_3(g)$; $w > 0$; **b.** $COCl_2(g) \rightarrow CO(g) + Cl_2(g)$; $w < 0$; **c.** $N_2(g) + O_2(g) \rightarrow 2NO(g)$; $w = 0$; Compare the sum of the coefficients of all the product gases in the balanced equation to the sum of the coefficients of all the reactant gases. When a balanced reaction has more moles of product gases than moles of reactant gases, then the reaction will expand in volume (ΔV is positive) and the system does work on the surroundings ($w < 0$). When a balanced reaction has a decrease in the moles of gas from reactants to products, then the reaction will contract in volume (ΔV is negative) and the surroundings does compression work on the system ($w > 0$). When there is no change in the moles of gas from reactants to products, then $\Delta V = 0$ and $w = 0$. **95. a.** $C_{12}H_{22}O_{11}(s) + 12 O_2(g) \rightarrow 12 CO_2(g) + 11 H_2O(l)$; **b.** −5630 kJ/mol **97.** 25 J **99.** 4.2 kJ heat released **101.** The calculated ΔH value will be less positive (smaller) than it should be. **103.** 25.91°C **105.** $\tfrac{1}{2}C + F \rightarrow A + B + D$; 47.0 kJ **107. a.** 632 kJ; **b.** $C_2H_2(g)$ **109.** 28 g **111. a.** −361 kJ; **b.** −199 kJ; **c.** −227 kJ; **d.** −112 kJ **113.** $CH_3OH(g) + CO(g) \rightarrow CH_3COOH(l)$ Using Appendix 4 standard enthalpies of formation, $\Delta H^\circ = -173$ kJ. From Exercise 72 of Chapter 3, $\Delta H = -20.$ kJ when using bond energies. For this reaction, bond energies give a poor estimate for ΔH. The major reason for the large discrepancy is that not all species are gases in this reaction. Bond energies do not account for the energy changes that occur when liquids and solids form instead of gases. These energy changes are due to intermolecular forces and will be discussed in Chapter 9. In general, bond energies only give good estimates for the enthalpy change when all reactants and products are in the gas phase **115.** Work is done by the surroundings on the system when there is a compression. This will occur when the moles of gas decrease when going from reactants to products. This will occur for reactions a and c. The other reactions have an increase in the moles of gas as reactants are converted to products. In these reactions, the system does PV work on the surroundings. **117.** 3.97×10^3 kJ heat released **119.** 51.5°C **121.** An element in its standard state has a standard enthalpy of formation equal to zero. At 25°C and 1 atm, chlorine is found as $Cl_2(g)$ and hydrogen is found as $H_2(g)$. So these two elements (a and b) have enthalpies of formation equal to zero. The other two choices (c and d) do not have the elements in their standard state. The standard state for nitrogen is $N_2(g)$ and the standard state for chlorine is $Cl_2(g)$. **123.** When a liquid is converted into a gas, there is an increase in volume. The 2.5 kJ/mol quantity can be considered as the work done by the vaporization process in pushing back the atmosphere. **125.** 32 m^2 **127.** $D_{calc} = 389$ kJ/mol as compared with 391 kJ/mol in the table. There is good agreement. **129.** 628 kJ/mol **131.** 56.9 kJ **133.** 3.3 cm **135. a.** 146 kJ; **b.** $\Delta H = 407$ kJ; **c.** $\Delta H = 1117$ kJ; **d.** $\Delta H = 1524$ kJ

Chapter 8

21. higher than; 13.6 times taller; When the pressure of the column of liquid standing on the surface of the liquid is equal to the pressure exerted by air on the rest of the surface of the liquid, then the height of the column of liquid is a measure of atmospheric pressure. Because water is 13.6 times less dense than mercury, the column of water must be 13.6 times longer than that of mercury to match the force exerted by the columns of liquid standing on the surface. **23.** The P versus $1/V$ plot is incorrect. The plot should be linear with positive slope and a y-intercept of zero. $PV = k$ so $P = k\,(1/V)$, which is in the form of the straight-line equation $y = mx + b$. **25.** $d =$ (molar mass)P/RT; density is directly proportional to the molar mass of the gas. Helium, which has the smallest molar mass of all the noble gases, will have the smallest density. **27.** At STP ($T = 273.2$ K and $P = 1.000$ atm), the volume of 1.000 mole of gas is:

$$V = \frac{nRT}{P} = \frac{1.000\ \text{mol} \times \dfrac{0.08206\ \text{L atm}}{\text{K mol}} \times 273.2\ \text{K}}{1.000\ \text{atm}} = 22.42\ \text{L}$$

The volume of 1.000 mole of any gas is 22.42 L, assuming the gas behaves ideally. Therefore, the molar volume of He(g) and $N_2(g)$ at STP both equal 22.42 L/mol. If the temperature increases to 25.0°C (298.2 K), the molar volume of a gas will be larger than 22.42 L/mol because volume and temperature are directly related at constant pressure. If 1.000 mole of a gas is collected over water at a total pressure of 1.000 atm, the partial pressure of the collected gas will be less than 1.000 atm because water vapor is present ($P_{\text{total}} = P_{\text{gas}} + P_{\text{H}_2\text{O}}$). At some partial pressure below 1.000 atm, the molar volume of a gas will be larger the 22.42 L/mol because volume and pressure are inversely related at constant temperature. **29.** No; At any nonzero Kelvin temperature, there is a distribution of kinetic energies. Similarly, there is a distribution of velocities at any nonzero Kelvin temperature. **31.** $2NH_3(g) \rightarrow N_2(g) + 3H_2(g)$; At constant P and T, volume is directly proportional to the moles of gas present. In the reaction, the moles of gas double as reactants are converted to products, so the volume of the container should double. At constant V and T, P is directly proportional to the moles of gas present. As the moles of gas double, the pressure will double. The partial pressure of N_2 will be 1/2 the initial pressure of NH_3 and the partial pressure of H_2 will be 3/2 the initial pressure of NH_3. The partial pressure of H_2 will be three times the partial pressure of N_2. **33.** CO_2 **35.** From the plot, Ne appears to behave most ideally. **37. a.** 3.6×10^3 mm Hg; **b.** 3.6×10^3 torr; **c.** 4.9×10^5 Pa; **d.** 71 psi **39.** 65 torr, 8.7×10^3 Pa, 8.6×10^{-2} atm **41. a.** 642 torr, 0.845 atm; 8.56×10^4 Pa; **b.** 975 torr; 1.28 atm; 1.30×10^5 Pa; **c.** 517 torr; 850. torr **43.** The balloon will burst. **45.** 0.89 mol **47. a.** 14.0 L; **b.** 4.72×10^{-2} mol; **c.** 678 K; **d.** 133 atm **49.** 88.9 g He; 44.8 g H_2 **51.** 0.0449 mol **53. a.** 69.6 K; **b.** 32.3 atm **55.** 1.27 mol **57.** $P_B = 2P_A$ **59. a.** 12.8 atm; **b.** 161 K; **c.** 620. K **61.** 5.1×10^4 torr **63.** The volume of the balloon increases from 1.00 L to 2.82 L, so the change in volume is 1.82 L. **65.** 3.21 g Al **67.** 135 g NaN_3 **69.** 1.5×10^7 g Fe, 2.6×10^7 g 98% H_2SO_4 **71.** 33.2 g **73.** 2.47 mol H_2O **75. a.** $2CH_4(g) + 2NH_3(g) + 3O_2(g) \rightarrow 2HCN(g) + 6H_2O(g)$; **b.** 13.3 L **77.** Cl_2 **79.** 12.6 g/L **81.** $P_{\text{He}} = 0.50$ atm, $P_{\text{Ne}} = 0.30$ atm, $P_{\text{Ar}} = 0.20$ atm **83.** 1.1 atm, $P_{\text{CO}_2} = 1.1$ atm, $P_{\text{TOTAL}} = 2.1$ atm **85.** $P_{\text{H}_2} = 317$ torr, $P_{\text{N}_2} = 50.7$ torr, $P_{\text{TOTAL}} = 368$ torr **87.** $P_{\text{He}} = 50.0$ torr; $P_{\text{Ne}} = 76.0$ torr; $P_{\text{Ar}} = 90.0$ torr; $P_{\text{TOTAL}} = 216.0$ torr **89. a.** $\chi_{\text{CH}_4} = 0.412$, $\chi_{\text{O}_2} = 0.588$; **b.** 0.161 mol; **c.** 1.06 g CH_4, 3.03 g O_2 **91.** 0.990 atm; 0.625 g **93.** 18.0% **95.** 46.5% **97.** $P_{\text{TOTAL}} = 6.0$ atm; $P_{\text{H}_2} = 1.5$ atm; $P_{\text{N}_2} = 4.5$ atm **99.** $P_{\text{N}_2} = 0.98$ atm; $P_{\text{TOTAL}} = 2.9$ atm **101.** Both $CH_4(g)$ and $N_2(g)$ have the same average kinetic energy at the various temperatures. 273 K, 5.65×10^{-21} J/molecule; 546 K, 1.13×10^{-20} J/molecule **103.** CH_4: 652 m/s (273 K); 921 m/s (546 K); N_2: 493 m/s (273 K); 697 m/s (546 K) **105.** The number of gas particles is constant, so at constant moles of gas, either a temperature change or a pressure change results in the smaller volume. If the temperature is constant, an increase in the external pressure would cause the volume to decrease. Gases are mostly empty space, so gases are easily compressible. If the pressure is constant, a decrease in temperature would cause the volume to decrease. As the temperature is lowered, the gas particles move with a slower average velocity and

don't collide with the container walls as frequently and as forcefully. As a result, the internal pressure decreases. In order to keep the pressure constant, the volume of the container must decrease in order to increase the gas particle collisions per unit area.

107.

	Average KE	Average Velocity	Wall-Collision Frequency
a.	Increase	Increase	Increase
b.	Decrease	Decrease	Decrease
c.	Same	Same	Increase
d.	Same	Same	Increase

109. a. All the same; **b.** Flask C **111.** CF_2Cl_2 **113.** The relative rates of effusion of $^{12}C^{16}O$, $^{12}C^{17}O$, and $^{12}C^{18}O$ are 1.04, 1.02, and 1.00. Advantage: CO_2 isn't as toxic as CO. Disadvantage: Can get a mixture of oxygen isotopes in CO_2, so some species would effuse at about the same rate. **115. a.** 12.24 atm; **b.** 12.13 atm; **c.** The ideal gas law is high by 0.91%. **117.** 5.2×10^{-6} atm; 1.3×10^{14} atoms He/cm^3 **119.** $S(s) + O_2(g) \rightarrow SO_2(g)$; $2SO_2(g) + O_2(g) \rightarrow 2SO_3(g)$; $SO_3(g) + H_2O(l) \rightarrow H_2SO_4(aq)$ **121. a.** 0.19 torr; **b.** 6.6×10^{21} molecules CO/m^3; **c.** 6.6×10^{15} molecules CO/cm^3

123. a.

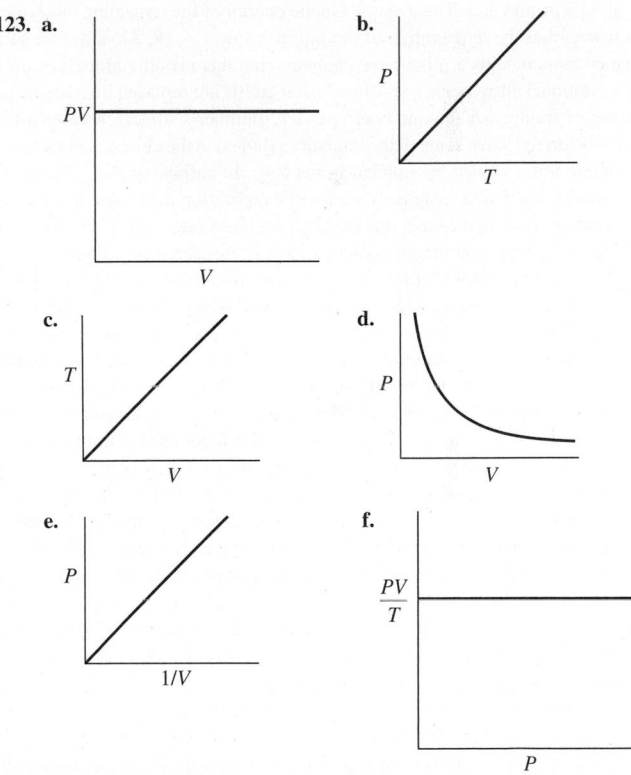

125. 0.772 atm · L; in Example 8-3, 1.0 mole of gas was present at 0°C. The moles of gas and/or the temperature must have been different for Boyle's data. **127.** $P_{\text{He}} = 582$ torr, $P_{\text{Xe}} = 18$ torr **129.** 13.3% **131.** 1490 **133.** 24 torr **135.** 4.1×10^6 L air; 7.42×10^5 L H_2 **137.** C_2H_3N; C_2H_3N **139.** $C_{12}H_{21}NO$; $C_{24}H_{42}N_2O_2$ **141.** 2.3×10^9 kJ of heat released **143.** Processes b, c, and d all will result in a doubling of the pressure. Process b doubles the pressure because the absolute temperature is doubled (from 200. K to 400. K). Process c has the effect of halving the volume, which would double the pressure by Boyle's law. Process d doubles the pressure because the moles of gas are doubled (28 g N_2 is 1 mol of N_2 and 32 g O_2 is 1 mol of O_2). Process a won't double the pressure because 28 g O_2 is less than 1 mol of gas, and process e won't double the temperature since the absolute temperature is not doubled (goes from 303 K to 333 K). **145.** $\Delta V = 997 - 855 = 142$ L **147.** 51.4 g **149.** All have the same average kinetic energy. The average velocity order is $O_2 < N_2 <$ He. **151.** 13.4% CaO.

86.6% BaO   **153.** C_2H_6 **155. a.** 8.7×10^3 L air/min; **b.** $\chi_{CO} = 0.0017$, $\chi_{CO_2} = 0.032$, $\chi_{O_2} = 0.13$, $\chi_{N_2} = 0.77$, $\chi_{H_2O} = 0.067$ **157. a.** 1.01×10^4 g; **b.** 6.65×10^4 g; **c.** 8.7×10^3 g   **159. a.** Due to air's larger average molar mass, a given volume of air at a given set of conditions has a higher density than helium. We need to heat the air to greater than 25°C to lower the air density (by driving air out of the hot-air balloon) until the density is the same as that for helium (at 25°C and 1.00 atm). **b.** 2150 K   **161.** C_3H_8 is the most likely identity.   **163.** $w = -2.94$ kJ, $\Delta E = 27.8$ kJ   **165.** 0.023 mol   **167.** 4.8 g/L; UF_3 will effuse 1.02 times faster.

Chapter 9

13. Intermolecular forces are the forces between molecules that hold the substances together in the solid and liquid phases. Hydrogen bonding is a specific type of intermolecular force. In this figure, the dotted lines represent the hydrogen bonding interactions that hold individual H_2O molecules together in the solid and liquid phases (answer a is correct). The solid lines represent the O—H covalent bonds.   **15.** Atoms have an approximately spherical shape. It is impossible to pack spheres together without some empty space between the spheres.   **17.** Evaporation takes place when some molecules at the surface of a liquid have enough energy to break the intermolecular forces holding them in the liquid phase. When a liquid evaporates, the molecules that escape have high kinetic energies. The average kinetic energy of the remaining molecules is lower; thus the temperature of the liquid is lower.   **19.** An alloy is a substance that contains a mixture of elements and has metallic properties. In a substitutional alloy, some of the host metal atoms are replaced by other metal atoms of similar size (e.g., in brass, pewter, plumber's solder). An interstitial alloy is formed when some of the interstices (holes) in the closest packed metal structure are occupied by smaller atoms (e.g., in carbon steels).   **21. a.** As intermolecular forces increase, the rate of evaporation decreases. **b.** increase T: increase rate; **c.** increase surface area: increase rate   **23.** $C_2H_5OH(l) \rightarrow C_2H_5OH(g)$ is an endothermic process. Heat is absorbed when liquid ethanol vaporizes; the internal heat from the body provides this heat, which results in the cooling of the body.   **25.** Sublimation will occur, allowing water to escape as $H_2O(g)$.   **27.** The strength of intermolecular forces determines relative boiling points. The types of intermolecular forces for covalent compounds are London dispersion forces, dipole forces, and hydrogen bonding. Because the three compounds are assumed to have similar molar mass and shape, the strength of the London dispersion forces will be about equal among the three compounds. One of the compounds will be nonpolar so it only has London dispersion forces. The other two compounds will be polar so they have additional dipole forces and will boil at a higher temperature than the nonpolar compound. One of the polar compounds will have an H covalently bonded to either N, O, or F. This gives rise to the strongest type of covalent intermolecular force, hydrogen bonding. This compound exhibiting hydrogen bonding will have the highest boiling point while the polar compound with no hydrogen bonding will boil at an intermediate temperature.   **29. a.** Both CO_2 and H_2O are molecular solids. Both have an ordered array of the individual molecules, with the molecular units occupying the lattice points. A difference within each solid lattice is the strength of the intermolecular forces. CO_2 is nonpolar and only exhibits London dispersion forces. H_2O exhibits the relatively strong hydrogen bonding interactions. The difference in strength is evidenced by the solid phase change that occurs at 1 atm. CO_2 sublimes at a relatively low temperature of $-78°C$. In sublimation, all of the intermolecular forces are broken. However, H_2O doesn't have a solid phase change until 0°C, and in this phase change from ice to water, only a fraction of the intermolecular forces are broken. The higher temperature and the fact that only a portion of the intermolecular forces are broken are attributed to the strength of the intermolecular forces in H_2O as compared to CO_2. Related to the intermolecular forces are the relative densities of the solid and liquid phases for these two compounds. $CO_2(s)$ is denser than $CO_2(l)$ while $H_2O(s)$ is less dense than $H_2O(l)$. For $CO_2(s)$ and for most solids, the molecules pack together as close as possible, which is why solids are usually more dense than the liquid phase. Water is an exception to this. Water molecules are particularly well suited to

interact with each other because each molecule has two polar O—H bonds and two lone pairs on each oxygen. This can lead to the association of four hydrogen atoms with each oxygen: two by covalent bonds and two by dipoles. To keep this symmetric arrangement (which maximizes the hydrogen bonding interactions), the $H_2O(s)$ molecules occupy positions that create empty space in the lattice. This translates into a smaller density for $H_2O(s)$ (less mass per unit volume). **b.** Both NaCl and CsCl are ionic compounds with the anions at the lattice points of the unit cell and the cations occupying the empty spaces created by the anions (called holes). In NaCl, the Cl^- anions occupy the lattice points of a face-centered unit cell with the Na^+ cations occupying the octahedral holes. Octahedral holes are the empty spaces created by six Cl^- ions. CsCl has the Cl^- ions at the lattice points of a simple cubic unit cell with the Cs^+ cations occupying the middle of the cube.   **31.** If TiO_2 conducts electricity as a liquid, then it would be ionic.   **33.** In the $\ln(P_{vap})$ versus $1/T$ plot, the slope of the straight line is equal to $-\Delta H_{vap}/R$. Because ΔH_{vap} is always positive, the slope of the line will always be negative.   **35. a.** LD (London dispersion); **b.** dipole, LD; **c.** hydrogen bonding, LD; **d.** ionic; **e.** LD; **f.** dipole, LD; **g.** ionic   **37. a.** OCS; **b.** SeO_2; **c.** $H_2NCH_2CH_2NH_2$; **d.** H_2CO; **e.** CH_3OH **39. a.** Neopentane is more compact than n-pentane. There is less surface-area contact among neopentane molecules. This leads to weaker London dispersion forces and a lower boiling point. **b.** HF is capable of hydrogen bonding; HCl is not. **c.** LiCl is ionic, and HCl is a molecular solid with only dipole forces and London dispersion forces. Ionic forces are much stronger than the forces for molecular solids. **d.** n-Hexane is a larger molecule, so it has stronger London dispersion forces.   **41. a.** HBr has dipole forces in addition to LD forces; **b.** NaCl, stronger ionic forces; **c.** I_2, larger molecule so stronger LD forces; **d.** N_2, smallest nonpolar compound present, has weakest LD forces; **e.** CH_4, smallest nonpolar compound present, has weakest LD forces; **f.** HF, can form relatively strong hydrogen bonding interactions, unlike the other compounds; **g.** $CH_3CH_2CH_2OH$, unlike others, has relatively strong hydrogen bonding.   **43.** H_2O is attracted to glass while Hg is not.   **45.** The structure of H_2O_2 produces greater hydrogen bonding than water.   **47.** 313 pm   **49.** 0.704 Å   **51.** 1.54 g/cm³   **53.** 174 pm; 11.6 g/cm³   **55.** Ag   **57.** edge, 328 pm; radius, 142 pm   **59.** face-centered cubic unit cell   **61.** For a cubic closest packed structure, 74.06% of the volume of each unit cell is occupied by atoms; in a simple cubic unit cell structure, 52.36% is occupied. The cubic closest packed structure provides the more efficient means for packing atoms.   **63.** Doping silicon with phosphorus produces an n-type semiconductor. The phosphorus adds electrons at energies near the conduction band of silicon. Electrons do not need as much energy to move from filled to unfilled energy levels so conduction increases. Doping silicon with gallium produces a p-type semiconductor. Because gallium has fewer valence electrons than silicon, holes (unfilled energy levels) at energies in the previously filled molecular orbitals are created, which induces greater electron movement (greater conductivity).   **65.** p-type   **67.** 5.0×10^2 nm   **69.** NaCl: $4Na^+$, $4Cl^-$; CsCl: $1Cs^+$, $1Cl^-$; ZnS: $4Zn^{2+}$, $4S^{2-}$; TiO_2: $2Ti^{4+}$, $4O^{2-}$ **71.** CoF_2 **73.** $ZnAl_2S_4$ **75.** MF_2 **77.** NaCl structure; $r_{O^{2-}} = 1.49 \times 10^{-8}$ cm; $r_{Mg^{2+}} = 6.15 \times 10^{-9}$ cm   **79.** The calculated distance between ion centers is 358 pm. Ionic radii give a distance of 350. pm. The distance calculated from the density is 8 pm (2.3%) greater than that calculated from the table of ionic radii.   **81. a.** CO_2: molecular; **b.** SiO_2: network; **c.** Si: atomic, network; **d.** CH_4: molecular; **e.** Ru: atomic, metallic; **f.** I_2: molecular; **g.** KBr: ionic; **h.** H_2O: molecular; **i.** NaOH: ionic; **j.** U: atomic, metallic; **k.** $CaCO_3$: ionic; **l.** PH_3: molecular   **83. a.** The unit cell consists of Ni at the cube corners and Ti at the body center or Ti at the cube corners and Ni at the body center. **b.** NiTi; **c.** Both have a coordination number of 8.   **85.** $CaTiO_3$; each structure has six oxygens around each Ti.   **87. a.** $YBa_2Cu_3O_9$; **b.** The structure of this superconductor material is based on the second perovskite structure. The $YBa_2Cu_3O_9$ structure is three of these cubic perovskite unit cells stacked on top of each other. The oxygens are in the same places, Cu takes the place of Ti, two Ca are replaced by two Ba, and one Ca is replaced by Y. **c.** $YBa_2Cu_3O_7$ **89.** Li, 158 kJ/mol; Mg, 139 kJ/mol. Bonding is stronger in Li.   **91.** 89°C   **93.** 77°C

95.

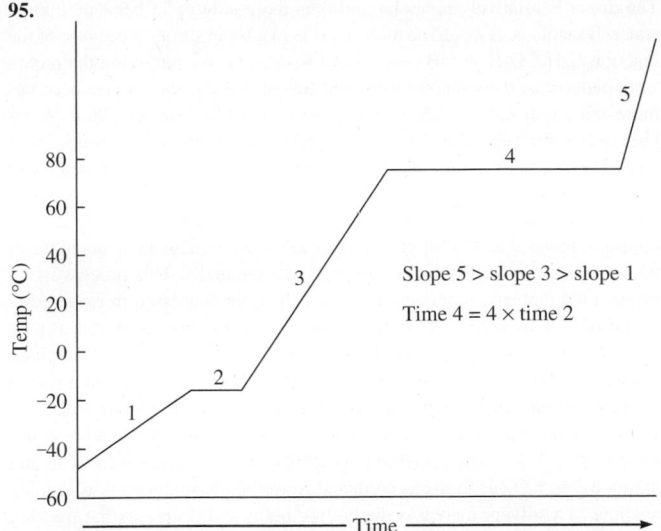

Slope 5 > slope 3 > slope 1

Time 4 = 4 × time 2

97. a. Much more energy is required to break the intermolecular forces when going from a liquid to a gas than is required to go from a solid to a liquid. Hence ΔH_{vap} is much larger than ΔH_{fus}; **b.** 113 J; **c.** 4220 J; **d.** 4220 J released   **99.** 1680 kJ   **101.** 1490 g   **103.** A: solid; B: liquid; C: vapor; D: solid + vapor; E: solid + liquid + vapor (triple point); F: liquid + vapor; G: liquid + vapor (critical point); H: vapor; the first dashed line (at the lower temperature) is the normal freezing point, and the second dashed line is the normal boiling point. The solid phase is denser.   **105. a.** two; **b.** higher pressure triple point: graphite, diamond, and liquid; lower pressure triple point: graphite, liquid and vapor; **c.** It is converted to diamond (the more dense solid form); **d.** Diamond is more dense, which is why graphite can be converted to diamond by applying pressure.   **107.** Because the density of the liquid phase is greater than the density of the solid phase, the slope of the solid/liquid boundary line is negative (as in H_2O). With a negative slope, the melting points increase with a decrease in pressure so the normal melting point of X should be greater than 225°C.   **109.** As the physical properties indicate, the intermolecular forces are slightly stronger in D_2O than in H_2O.   **111.** A: CH_4; B: SiH_4; C: NH_3 **113.** XeF_2 **115.** B_2H_6, molecular; SiO_2, network; CsI, ionic; W, metallic   **117.** 57.8 torr   **119.** 4.65 kg/h   **121. a.** dipole, LD (SF_4 is a polar compound); **b.** LD only (CO_2 is a nonpolar compound); **c.** H-bonding, LD (H-bonding due to the O—H bond); **d.** H-bonding, LD; **e.** dipole, LD (ICl_5 is a polar compound); **f.** LD only (XeF_4 is a nonpolar compound.)   **123. a.** False; the ionic forces in LiF are much stronger than the molecular intermolecular forces found in H_2S. **b.** True; HF is capable of H-bonding, HBr is not. **c.** True; the larger Cl_2 molecule will have the stronger London dispersion forces. **d.** True; HCl is a polar compound while CCl_4 is a nonpolar compound. **e.** False; the ionic forces in MgO are much stronger than the molecular intermolecular forces found in CH_3CH_2OH.   **125.** Each manganese ion has a +3 charge.   **127. a.** Kr: group 8A; **b.** SO_2: molecular; **c.** Ni: metallic; **d.** SiO_2: network; **e.** NH_3: molecular; **f.** Pt: metallic   **129.** 207 torr   **131.** $\Delta E = 27.86$ kJ/mol; $\Delta H = 30.79$ kJ/mol   **133.** 46.7 kJ/mol; 90.%

135.

$$\begin{array}{cccccc} & H & H & & H & H \\ & | & | & \;\;\ddots & & | & \;\;\ddots & | \\ H-&C-&C-&O-H & \text{and} & H-C-&O-&C-H \\ & | & | & & & | & \;\;\ddots & | \\ & H & H & & H & & H \end{array}$$

Liquid (H-bonding)   Gas (no H-bonding)

The first structure with the —OH bond is capable of forming hydrogen bonding; the other structure is not. Therefore, the liquid (which has the stronger intermolecular forces) is the first structure, and the gas is the second structure.   **137.** The solids with high melting points (NaCl, $MgCl_2$, NaF, MgF_2, AlF_3) are all ionic solids. $SiCl_4$, SiF_4, Cl_2, F_2, PF_5, and SF_6 are nonpolar covalent molecules with LD forces. PCl_3 and SCl_2 are polar molecules with LD and dipole forces. In these 8 molecular substances the intermolecular forces are weak and the melting points low. $AlCl_3$ is intermediate. The melting point

indicates there are stronger forces present than in the nonmetal halides, but not as strong as for an ionic solid. $AlCl_3$ illustrates a gradual transition from ionic to covalent bonding; from an ionic solid to discrete molecules.   **139.** $TiO_{1.182}$ or $Ti_{0.8462}O$; 63.7% Ti^{2+}, 36.3% Ti^{3+} (a 1.75:1 ion ratio)   **141.** 6.58 g/cm³   **143.**

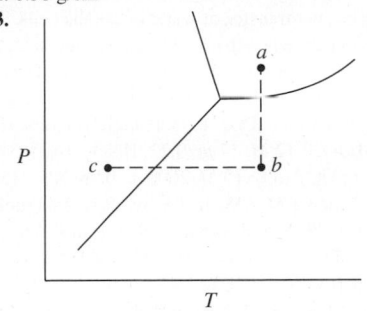

As P is lowered, we go from a to b on the phase diagram. The water boils. The boiling of water is endothermic and the water is cooled ($b \rightarrow c$), forming some ice. If the pump is left on, the ice will sublime until none is left. This is the basis of freeze drying.

145. The volume of the hole is $\frac{4}{3}\pi[(\sqrt{3}-1)r]^3$ **147.** CdS; n-type

149. 253 torr; 6.38×10^{22} atoms

Chapter 10

11. 9.74 M **13.** 160 mL   **15.** 4.5 M **17.** As the temperature increases, the gas molecules will have a greater average kinetic energy. A greater fraction of the gas molecules in solution will have kinetic energy greater than the attractive forces between the gas molecules and the solvent molecules. More gas molecules will escape to the vapor phase, and the solubility of the gas will decrease.   **19.** The levels of the liquids in each beaker will become constant when the concentration of solute is the same in both beakers. Because the solute is less volatile, the beaker on the right will have a larger volume when the concentrations become equal. Water will initially condense in this beaker in a larger amount than solute is evaporating, while the net change occurring initially in the other beaker is for water to evaporate in a larger amount than solute is condensing. Eventually the rate that solute and H_2O leave and return to each beaker will become equal when the concentrations become equal.   **21.** No. For an ideal solution, $\Delta H_{soln} = 0$.   **23.** Normality is the number of equivalents per liter of solution. For an acid or a base, an equivalent is the mass of acid or base that can furnish 1 mole of protons (if an acid) or accept 1 mole of protons (if a base). A proton is an H^+ ion. Molarity is defined as the moles of solute per liter of solution. When the number of equivalents equals the number of moles of solute, then normality = molarity. This is true for acids that only have one acidic proton in them and for bases that accept only one proton per formula unit. Examples of acids where equivalents = moles solute are HCl, HNO_3, HF, and $HC_2H_3O_2$. Examples of bases where equivalents = moles solute are NaOH, KOH, and NH_3. When equivalents ≠ moles solute, then normality ≠ molarity. This is true for acids that donate more than one proton (H_2SO_4, H_3PO_4, H_2CO_3, etc.) and for bases that react with more than one proton per formula unit [$Ca(OH)_2$, $Ba(OH)_2$, $Sr(OH)_2$, etc.].   **25.** Only statement b is true. A substance freezes when the vapor pressures of the liquid and solid phases are the same. When a solute is added to water, the vapor pressure of the solution at 0°C is less than the vapor pressure of the solid; the net result is for any ice present to convert to liquid in order to try to equalize the vapor pressures (which never can occur at 0°C). A lower temperature is needed to equalize the vapor pressures of water and ice, hence the freezing point is depressed. For statement a, the vapor pressure of a solution is directly related to the mole fraction of solvent (not solute) by Raoult's law. For statement c, colligative properties depend on the number of solute particles present and not on the identity of the solute. For statement d, the boiling point of water is increased because the sugar solute decreases the vapor pressure of the water; a higher temperature is required for the vapor pressure of the solution to equal the external pressure so boiling can occur.   **27.** Isotonic solutions are those that have identical osmotic pressures. Crenation and hemolysis refer to phenomena that occur when red blood cells are bathed in solutions having a mismatch in osmotic pressure between the inside and the outside of the cell. When red blood cells are in a solution having a higher osmotic pressure than that of the cells, the cells shrivel as there is a net transfer of water out of the cells. This is called crenation. Hemolysis occurs when the red blood cells are bathed in a solution having lower osmotic pressure than that inside the cell. Here, the cells rupture as there is a net transfer of water to inside the red blood cells.   **29.** 1.06 g/mL; 0.0180 mole fraction H_3PO_4, 0.9820 mole fraction H_2O; 0.981 mol/L; 1.02 mol/kg   **31.** HCl: 12 M, 17 m, 0.23; HNO_3: 16 M, 37 m, 0.39; H_2SO_4: 18 M, 200 m, 0.76; $HC_2H_3O_2$: 17 M, 2000 m, 0.96; NH_3: 15 M, 23 m, 0.29   **33.** 35%; 0.39; 7.3 m; 3.1 M **35.** 10.1% by mass; 2.45 mol/kg   **37.** 23.9%; 1.6 m, 0.028, 4.11 N **39.** $NaI(s) \rightarrow Na^+(aq) + I^-(aq)$ $\Delta H_{soln} = -8$ kJ/mol   **41.** The attraction of water molecules for Al^{3+} and OH^- cannot overcome the larger lattice energy of $Al(OH)_3$.   **43. a.** CCl_4; **b.** H_2O; **c.** H_2O; **d.** CCl_4; **e.** H_2O; **f.** H_2O; **g.** CCl_4 **45. a.** NH_3; **b.** CH_3CN; **c.** CH_3COOH **47.** As the length of the hydrocarbon chain increases, the solubility decreases because the nonpolar hydrocarbon chain interacts poorly with the polar water molecules.   **49.** 1.04×10^{-3} mol/L · atm; 1.14×10^{-3} mol/L   **51.** 50.0 torr   **53.** 0.918   **55.** 23.2 torr at 25°C; 70.0 torr at 45°C   **57. a.** 290 torr; **b.** 0.69   **59.** $\chi_{methanol} = \chi_{propanol} = 0.500$ **61.** solution c   **63. a.** These two molecules are named acetone (CH_3COCH_3) and water. As discussed in Section 10-4 on nonideal solutions, acetone–water solutions exhibit negative deviations from Raoult's law. Acetone and water have the ability to hydrogen bond with each other, which gives the solution stronger intermolecular forces compared to the pure states of both solute and solvent. In the pure state, acetone cannot hydrogen bond with itself, so the middle diagram illustrating negative deviations from Raoult's law is the correct choice for acetone–water solutions. **b.** These two molecules are named ethanol (CH_3CH_2OH) and water. Ethanol–water solutions show positive deviations from Raoult's law. Both substances can hydrogen bond in the pure state, and they can continue this in solution. However, the solute–solvent interactions are weaker for ethanol–water solutions due to the significant nonpolar part of ethanol (CH_3—CH_2 is the nonpolar part of ethanol). This nonpolar part of ethanol weakens the intermolecular forces in solution, so the first diagram illustrating positive deviations from Raoult's law is the correct choice for ethanol–water solutions. **c.** These two molecules are named heptane (C_7H_{16}) and hexane (C_6H_{14}). Heptane and hexane are very similar nonpolar substances; both are composed entirely of nonpolar C—C bonds and relatively nonpolar C—H bonds, and both have a similar size and shape. Solutions of heptane and hexane should be ideal, so the third diagram illustrating no deviation from Raoult's law is the correct choice for heptane–hexane solutions. **d.** These two molecules are named heptane (C_7H_{16}) and water. The interactions between the nonpolar heptane molecules and the polar water molecules will certainly be weaker in solution compared to the pure solvent and pure solute interactions. This results in positive deviations from Raoult's law (the first diagram).   **65.** 101.5°C   **67.** 14.8 g $C_3H_8O_3$ **69.** $T_f = -29.9$°C, $T_b = 108.2$°C   **71.** 6.6×10^{-2} mol/kg; 590 g/mol (610 g/mol if no rounding of numbers)   **73. a.** $\Delta T = 2.0 \times 10^{-5}$°C, $\pi = 0.20$ torr; **b.** Osmotic pressure is better for determining the molar mass of large molecules. A temperature change of 10^{-5}°C is very difficult to measure. A change in height of a column of mercury by 0.2 mm is not as hard to measure precisely.   **75.** 2.51×10^5 g/mol   **77.** Dissolve 210 g sucrose in some water and dilute to 1.0 L in a volumetric flask. To get 0.62 ± 0.01 mol/L, we need 212 ± 3 g sucrose.   **79. a.** 0.010 m Na_3PO_4 and 0.020 m KCl; **b.** 0.020 m HF; **c.** 0.020 m $CaBr_2$ **81. a.** $T_f = -13$°C; $T_b = 103.5$°C; **b.** $T_f = -4.7$°C; $T_b = 101.3$°C   **83.** 1.67   **85. a.** $T_f = -0.28$°C; $T_b = 100.077$°C; **b.** $T_f = -0.37$°C; $T_b = 100.10$°C   **87.** 2.63 (0.0225 m), 2.60 (0.0910 m), 2.57 (0.278 m); $i_{average} = 2.60$ **89. a.** yes; **b.** no   **91.** Benzoic acid is capable of hydrogen bonding, but a significant part of benzoic acid is the nonpolar benzene ring. In benzene, a hydrogen bonded dimer forms.

The dimer is relatively nonpolar and thus more soluble in benzene than in water. Benzoic acid would be more soluble in a basic solution because of the reaction $C_6H_5CO_2H + OH^- \rightarrow C_6H_5CO_2^- + H_2O$. By removing the proton from benzoic acid, an anion forms, and like all anions, the species becomes more soluble in water.   **93. a.** 26.6 kJ/mol; **b.** -657 kJ/mol   **95. a.** Water boils when the vapor pressure equals the pressure above the water. In an open pan, $P_{atm} \approx 1.0$ atm. In a pressure cooker, $P_{inside} > 1.0$ atm and water boils at a higher temperature. The higher the cooking temperature, the faster the cooking time. **b.** Salt dissolves in water, forming a solution with a melting point lower than that of pure water ($\Delta T_f = K_f m$). This happens in water on the surface of ice. If it is not too cold, the ice melts. This process won't occur if the ambient temperature is lower than the depressed freezing point of the salt solution. **c.** When water freezes from a solution, it freezes as pure water, leaving behind a more concentrated salt solution. **d.** On the CO_2 phase diagram, the triple point is above 1 atm and $CO_2(g)$ is the stable phase at 1 atm and room temperature. $CO_2(l)$ can't exist at normal atmospheric pressures, which explains why dry ice sublimes rather than boils. In a fire extinguisher, $P > 1$ atm and $CO_2(l)$ can exist. When CO_2 is released from the fire extinguisher, $CO_2(g)$ forms as predicted from the phase diagram. **e.** Adding a solute to a solvent increases the boiling point and decreases the freezing point of the solvent. Thus, the solvent is a liquid over a wider range of temperatures when a solute is dissolved.   **97.** 0.600   **99.** $P_{ideal} = 188.6$ torr; $\chi_{acetone} = 0.512$, $\chi_{methanol} = 0.488$; the actual vapor pressure of the solution is smaller than the ideal vapor pressure, so this solution exhibits a negative deviation from Raoult's law. This occurs when solute–solvent attractions are stronger than for the pure substances.   **101.** 776 g/mol   **103.** $C_2H_4O_3$; 151 g/mol (exp.); 152.10 g/mol (calc.); $C_4H_8O_6$ **105.** 1.97% NaCl   **107. a.** 100.77°C; **b.** 23.1 mm Hg; **c.** Assume an ideal solution; assume no ions form ($i = 1$); assume the solute is nonvolatile.   **109.** 639 g/mol; 33.7 torr   **111.** 3 kJ/mol   **113.** 0.7316   **115.** 100.25°C   **117.** 43% $NaNO_3$ and 57% $Mg(NO_3)_2$ by mass

119. 30.% A: $^\circ\chi_A = \dfrac{0.30y}{0.70x + 0.30y}$, $\chi_B = 1 - \chi_A$;

50.% A: $^\circ\chi_A = \dfrac{y}{x + y}$, $\chi_B = 1 - \dfrac{y}{x + y}$;

80.% A: $^\circ\chi_A = \dfrac{0.80y}{0.20x + 0.80y}$, $\chi_B = 1 - \chi_A$;

30.% A: $^\circ\chi_A^{\ V} = \dfrac{0.30x}{0.30x + 0.70y}$, $\chi_B^{\ V} = 1 - \dfrac{0.30x}{0.30x + 0.70y}$;

50.% A: $^\circ\chi_A^{\ V} = \dfrac{x}{x + y}$, $\chi_B^{\ V} = 1 - \chi_A^{\ V}$;

80.% A: $^\circ\chi_A^{\ V} = \dfrac{0.80x}{0.80x + 0.20y}$, $\chi_B^{\ V} = 1 - \chi_A^{\ V}$

121. 72.5% sucrose and 27.5% NaCl by mass; 0.313   **123.** 0.050   **125.** 44% naphthalene, 56% anthracene   **127.** -0.20°C, 100.056°C   **129. a.** 46 L; **b.** No; a reverse osmosis system that applies 8.0 atm can only purify water with solute concentrations less than 0.32 mol/L. Salt water has a solute concentration of 2(0.60 M) = 1.2 M ions. The solute concentration of salt water is much too high for this reverse osmosis unit to work.   **131.** $i = 3.00$; $CdCl_2$

Chapter 11

11. In a unimolecular reaction, a single reactant molecule decomposes to products. In a bimolecular reaction, two molecules collide to give products. The probability of the simultaneous collision of three molecules with enough energy and orientation is very small, making termolecular steps very unlikely.   **13.** All of these choices would affect the rate of the reaction, but only b and c affect the rate by affecting the value of the rate constant k. The value of the rate constant is dependent on temperature. It also depends on the activation energy. A catalyst will change the value of k because the activation energy changes. Increasing the concentration (partial pressure) of either O_2 or NO does not affect the value of k, but it does increase the rate of the reaction because both concentrations appear in the rate law.   **15.** The average rate decreases with time because the reverse reaction occurs more frequently as the concentration of products increases. Initially, with no products present, the rate of the forward reaction is at its fastest; but as time goes on, the rate gets slower and slower since products are converting back into reactants. The instantaneous rate will also decrease with time. The only rate that is constant is the initial rate. This is the instantaneous rate taken at $t \approx 0$. At this time, the amount of products is insignificant and the rate of the reaction only depends on the rate of the forward reaction. **17.** When the rate doubles as the concentration quadruples, the order is 1/2. For a reactant that has an order of -1, the rate will decrease by a factor of 1/2 when the concentrations are doubled. **19.** Two reasons are: **1)** the collision must involve enough energy to produce the reaction; i.e., the collision energy must equal or exceed the activation energy. **2)** the relative orientation of the reactants must allow formation of any new bonds necessary to produce products. **21.** Enzymes are very efficient catalysts. As is true for all catalysts, enzymes speed up a reaction by providing an alternative pathway for reactants to convert to products. This alternative pathway has a smaller activation energy and, hence, a faster rate. Also true is that catalysts are not used up in the overall chemical reaction. Once an enzyme comes in contact with the correct reagent, the chemical reaction quickly occurs, and the enzyme is then free to catalyze another reaction. Because of the efficiency of the reaction step, only a relatively small amount of enzyme is needed to catalyze a specific reaction, no matter how complex the reaction. **23.** P_4: 6.0×10^{-4} mol/L · s; H_2: 3.6×10^{-3} mol/L · s **25. a.** average rate of decomposition of $H_2O_2 = 2.31 \times 10^{-5}$ mol/L · s, rate of production of $O_2 = 1.16 \times 10^{-5}$ mol/L · s; **b.** average rate of decomposition of $H_2O_2 = 1.16 \times 10^{-5}$ mol/L · s, rate of production of $O_2 = 5.80 \times 10^{-6}$ mol/L · s **27. a.** mol/L · s; **b.** mol/L · s; **c.** s^{-1}; **d.** L/mol · s; **e.** L^2/mol^2 · s **29. a.** rate = $k[NO]^2[Cl_2]$; **b.** 1.8×10^2 L^2/mol^2 · min **31. a.** rate = $k[NOCl]^2$; **b.** 6.6×10^{-29} cm^3/molecules · s; **c.** 4.0×10^{-8} L/mol · s **33. a.** rate = $k[I^-][OCl^-]$; **b.** 3.7 L/mol · s; **c.** 0.083 mol/L · s **35. a.** first order in Hb and first order in CO; **b.** rate = $k[Hb][CO]$; **c.** 0.280 L/μmol · s; **d.** 2.26 μmol/L · s **37.** rate = $k[H_2O_2]$; $\ln[H_2O_2] = -kt + \ln[H_2O_2]_0$; $k = 8.3 \times 10^{-4}$ s^{-1}; 0.037 M **39.** rate = $k[NO_2]^2$; $\dfrac{1}{[NO_2]} = kt + \dfrac{1}{[NO_2]_0}$; $k = 2.08 \times 10^{-4}$ L/mol · s; 0.131 M **41. a.** rate = k; $[C_2H_5OH] = -kt + [C_2H_5OH]_0$; because slope = $-k$, $k = 4.00 \times 10^{-5}$ mol/L · s; **b.** 156 s; **c.** 313 s **43.** rate = $k[C_4H_6]^2$; $\dfrac{1}{[C_4H_6]} = kt + \dfrac{1}{[C_4H_6]_0}$; $k = 1.4 \times 10^{-2}$ L/mol · s **45.** second order; 0.1 M **47. a.** $[A] = -kt + [A]_0$; **b.** 1.0×10^{-2} s; **c.** 2.5×10^{-4} M **49.** 9.2×10^{-3} s^{-1}; 75 s **51.** 150. s **53.** 12.5 s **55. a.** 1.1×10^{-2} M; **b.** 0.025 M **57.** 1.6 L/mol · s **59. a.** rate = $k[CH_3NC]$; **b.** rate = $k[O_3][NO]$; **c.** rate = $k[O_3]$; **d.** rate = $k[O_3][O]$ **61.** Rate = $k[C_4H_9Br]$; $C_4H_9Br + 2H_2O \rightarrow C_4H_9OH + Br^- + H_3O^+$; the intermediates are $C_4H_9^+$ and $C_4H_9OH_2^+$.

63.

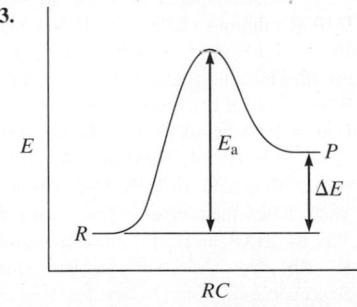

65. 341 kJ/mol **67.** The graph of $\ln(k)$ versus $1/T$ is linear with slope = $-E_a/R = -1.2 \times 10^4$ K; $E_a = 1.0 \times 10^2$ kJ/mol **69.** 9.5×10^{-5} L/mol · s **71.** 51°C **73.** $H_3O^+(aq) + OH^-(aq) \rightarrow 2H_2O(l)$ should have the faster rate. H_3O^+ and OH^- will be electrostatically attracted to each other; Ce^{4+} and Hg_2^{2+} will repel each other (so E_a is much larger). **75. a.** NO; **b.** NO$_2$; **c.** 2.3 **77.** CH_2D—CH_2D should be the product. If the mechanism is possible, then the reaction must be $C_2H_4 + D_2 \rightarrow CH_2DCH_2D$. If we got this product, then we could conclude that this is a possible mechanism. If we got some other product, e.g., CH_3CHD_2, then we would conclude that the mechanism is wrong. Even though this mechanism correctly predicts the products of the reaction, we cannot say conclusively that this is the correct mechanism; we might be able to conceive of other mechanisms that would give the same product as our proposed one. **79.** The rate depends on the number of

reactant molecules adsorbed on the surface of the catalyst. This quantity is proportional to the concentration of reactant. However, when all the catalyst surface sites are occupied, the rate becomes independent of the concentration of reactant. **81.** 215°C **83. a.** 20 min; **b.** 30 min; **c.** 15 min **85.** 427 s **87.** 1.0×10^2 kJ/mol **89.** 6.58×10^{-6} mol/L · s **91. a.** 25 kJ/mol; **b.** 12 s;

c.

T	Interval	$54 - 2$(Intervals)
21.0°C	16.3 s	21°C
27.8°C	13.0 s	28°C
30.0°C	12 s	30.°C

This rule of thumb gives excellent agreement to two significant figures. **93. a.** 115 L^3/mol^3 · s; **b.** 87.0 s; **c.** [A] = 1.27×10^{-5} M, [B] = 1.00 M **95.** The reaction is first order with respect to both A and B; rate = $k[A][B]$; $k = 1.6$ L/mol · s **97.** 35.4 s **99. a.** False; from the half-life expression for a zero-order reaction in Table 11-6 of the text, there is a direct relationship between concentration and the half-life value. As the reaction proceeds for a zero-order reaction, the half-life value will decrease because concentration decreases. **b.** True **c.** False; from the half-life expression for a first-order reaction in Table 11-6, the half-life does not depend on concentration. **d.** True; the half-life for a second-order reaction increases with time because there is an inverse relationship between concentration and the half-life value. **101.** 53.9 kJ/mol; the chirping rate will be about 45 chirps per minute at 7.5°C. **103.** rate = $\dfrac{k[I^-][OCl^-]}{[OH^-]}$; $k = 6.0 \times 10^1$ s^{-1} **105. a.** first order with respect to both reactants; **b.** rate = $k[NO][O_3]$; **c.** $k' = 1.8$ s^{-1}; $k'' = 3.6$ s^{-1}; **d.** $k = 1.8 \times 10^{-14}$ cm^3/molecules · s **107. a.** For a three-step reaction with the first step limiting, the energy-level diagram could be

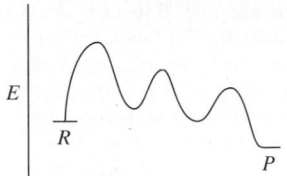

Note that the heights of the second and third humps must be lower than the first-step activation energy. However, the height of the third hump could be higher than the second hump. One cannot determine this absolutely from the information in the problem.

b.

$F_2 \rightarrow 2F$	slow
$F + H_2 \rightarrow HF + H$	fast
$H + F \rightarrow HF$	fast

$$F_2 + H_2 \rightarrow 2HF$$

c. F_2 was the limiting reactant. **109. a.** [B] $\gg$ [A] so that [B] can be considered constant over the experiments. (This gives us a pseudo-order rate law equation.) **b.** Rate = $k[A]^2[B]$, $k = 0.050$ L^2/mol^2 · s **111.** Rate = $k[A][B]^2$, $k = 1.4 \times 10^{-2}$ L^2/mol^2 · s **113.** 2.20×10^{-5} s^{-1}; 5.99×10^{21} molecules **115.** 1.3×10^{-5} s^{-1}; 112 torr

Chapter 12

11. No, equilibrium is a dynamic process. Both the forward and reverse reactions are occurring at equilibrium, just at equal rates. Thus the forward and reverse reactions will distribute ^{14}C atoms between CO and CO_2. **13.** A large value for K ($K \gg 1$) indicates there are relatively large concentrations of product gases and/or solutes as compared with the concentrations of reactant gases and/or solutes at equilibrium. A reaction with a large K value is a good source of products. **15.** 4 molecules H_2O, 2 molecules CO, 4 molecules H_2, and 4 molecules CO_2 are present at equilibrium. **17.** K and K_p are equilibrium constants as determined by the law of mass action. For K, the units used for concentrations are mol/L, and for K_p, partial pressures in units of atm are used (generally). Q is called the reaction quotient. Q has the exact same form as K or K_p, but instead of equilibrium concentrations, initial concentrations are used to calculate the Q value. Q is of use when it is compared to the K value.

When $Q = K$ (or when $Q_p = K_p$), the reaction is at equilibrium. When $Q \neq K$, the reaction is not at equilibrium and one can determine what has to be the net change for the system to get to equilibrium.   **19.** We always try to make good assumptions that simplify the math. In some problems, we can set up the problem so that the net change, x, that must occur to reach equilibrium is a small number. This comes in handy when you have expressions like $0.12 - x$ or $0.727 + 2x$. When x is small, we assume that it makes little difference when subtracted from or added to some relatively big number. When this is true, $0.12 - x \approx 0.12$ and $0.727 + 2x \approx 0.727$. If the assumption holds by the 5% rule, then the assumption is assumed valid. The 5% rule refers to x (or $2x$ or $3x$, etc.) that was assumed small compared to some number. If x (or $2x$ or $3x$, etc.) is less than 5% of the number the assumption was made against, then the assumption will be assumed valid. If the 5% rule fails to work, one can generally use a math procedure called the method of successive approximations to solve the quadratic or cubic equation.

21. a. $K = \dfrac{[NO]^2}{[N_2][O_2]}$;   **b.** $K = \dfrac{[NO_2]^2}{[N_2O_4]}$;   **c.** $K = \dfrac{[SiCl_4][H_2]^2}{[SiH_4][Cl_2]^2}$;

d. $K = \dfrac{[PCl_3]^2[Br_2]^3}{[PBr_3]^2[Cl_2]^3}$ **23. a.** 0.11; **b.** 77; **c.** 8.8; **d.** 1.7×10^{-4} **25.** 4.0×10^6

27. 1.7×10^{-5} **29.** 6.3×10^{-13} **31.** 4.6×10^3

33. a. $K = \dfrac{[H_2O]}{[NH_3]^2[CO_2]}$, $K_p = \dfrac{P_{H_2O}}{P_{NH_3}^2 \times P_{CO_2}}$;   **b.** $K = [N_2][Br_2]^3$,

$K_p = P_{N_2} \times P_{Br_2}^3$; **c.** $K = [O_2]^3$, $K_p = P_{O_2}^3$; **d.** $K = \dfrac{[H_2O]}{[H_2]}$, $K_p = \dfrac{P_{H_2O}}{P_{H_2}}$

35. only reaction d   **37.** 8.0×10^9 **39. a.** not at equilibrium; $Q > K$, shift left; **b.** at equilibrium; **c.** not at equilibrium; $Q > K$, shift left   **41. a.** decrease; **b.** no change; **c.** no change; **d.** increase   **43.** 0.16 mol   **45.** 3.4   **47.** 0.056   **49.** $[N_2]_0 = 10.0\ M$, $[H_2]_0 = 11.0\ M$ **51.** $[SO_3] = [NO] = 1.06\ M$; $[SO_2] = [NO_2] = 0.54\ M$ **53.** 7.8×10^{-2} atm   **55.** $P_{SO_2} = 0.38$ atm; $P_{O_2} = 0.44$ atm; $P_{SO_3} = 0.12$ atm   **57. a.** $[NO] = 0.032\ M$, $[Cl_2] = 0.016\ M$, $[NOCl] = 1.0\ M$; **b.** $[NO] = [NOCl] = 1.0\ M$, $[Cl_2] = 1.6 \times 10^{-5}\ M$; **c.** $[NO] = 8.0 \times 10^{-3}\ M$, $[Cl_2] = 1.0\ M$, $[NOCl] = 2.0\ M$ **59.** $[CO_2] = 0.39\ M$, $[CO] = 8.6 \times 10^{-3}\ M$, $[O_2] = 4.3 \times 10^{-3}\ M$ **61.** 0.27 atm   **63. a.** no effect; **b.** shifts left; **c.** shifts right   **65. a.** right; **b.** right; **c.** no effect; **d.** left; **e.** no effect   **67. a.** left; **b.** right; **c.** left; **d.** no effect; **e.** no effect; **f.** right   **69.** increase   **71.** 2.6×10^{81} **73.** 6.74×10^{-6} **75. a.** 0.379 atm; **b.** 0.786   **77.** $[Fe^{3+}] = 2 \times 10^{-4}\ M$, $[SCN^-] = 0.08\ M$, $[FeSCN^{2+}] = 0.020\ M$ **79.** 1.43×10^{-2} atm   **81.** pink   **83.** Added OH^- reacts with H^+ to produce H_2O. As H^+ is removed, the reaction shifts right to produce more H^+ and CrO_4^{2-}. Because more CrO_4^{2-} is produced, the solution turns yellow.   **85.** $9.0 \times 10^{-3}\ M$ **87.** 0.50 **89.** $3.0 \times 10^{-6}\ M$ **91.** $2.1 \times 10^{-4}\ M$ **93.** 0.81

95. a.

	$2\ NOCl(g)$	$2\ NO(g)$	$Cl_2(g)$
Initial	$2.6\ M$	0	0
Change	$-2x$	$+2x$	$+x$
Equilibrium	$2.6 - 2x$	$2x$	x

b. $[NO] = 0.060\ M$; $[Cl_2] = 0.030\ M$; $[NOCl] = 2.5\ M$

97. Add $N_2(g)$: reaction shifts right to reestablish equilibrium. $[N_2]$ will increase overall, $[H_2]$ will decrease, and $[NH_3]$ will increase. Note that only some of the added N_2 will react to get back to equilibrium. Therefore, $[N_2]$ will show a net increase overall.

Remove $H_2(g)$: reaction shifts left to reestablish equilibrium. $[N_2]$ will increase, $[H_2]$ will decrease overall, and $[NH_3]$ will decrease.

Add $NH_3(g)$: reaction will shift left to reestablish equilibrium. $[N_2]$ and $[H_2]$ will increase, while $[NH_3]$ will also increase overall.

Add $Ne(g)$ at constant volume: no change for $[N_2]$, $[H_2]$, or $[NH_3]$.

Increase temperature (add heat): because this is an exothermic reaction, heat is a product in this reaction. As temperature is increased, the reaction will shift left to reestablish equilibrium. $[N_2]$ and $[H_2]$ will increase, while $[NH_3]$ will decrease.

Decrease volume: because there are more moles of gaseous reactants compared to gaseous products, a decrease in volume will cause this reaction to shift right to reestablish equilibrium. $[N_2]$ and $[H_2]$ will decrease, while $[NH_3]$ will increase.

Add a catalyst: no change for $[N_2]$, $[H_2]$, or $[NH_3]$.

99. $P_{CO} = 0.58$ atm, $P_{CO_2} = 1.65$ atm   **101.** $[NOCl] = 2.0\ M$, $[NO] = 0.050\ M$, $[Cl_2] = 0.025\ M$ **103.** 2.1×10^{-3} atm   **105.** $P_{NO_2} = 0.704$ atm, $P_{N_2O_4} = 0.12$ atm   **107. a.** Assuming mol/L units for concentrations, $K = 8.16 \times 10^{47}$; **b.** 33.9 L   **109.** 0.63   **111.** 0.240 atm   **113. a.** 2.33×10^{-4}; **b.** The argon gas will increase the volume of the container. This is because the container is a constant-pressure system, and if the number of moles increases at constant T and P, the volume must increase. An increase in volume will dilute the concentrations of all gaseous reactants and gaseous products. Because there are more moles of product gases than reactant gases (3 moles vs. 2 moles), the dilution will decrease the numerator of K more than the denominator will decrease. This causes $Q < K$ and the reaction shifts right to get back to equilibrium. Because temperature was unchanged, the value of K will not change. K is a constant as long as temperature is constant.   **115.** 192 g; 1.3 atm   **117.** 33

Chapter 13

19. as an acid: $HCO_3^-(aq) + H_2O(l) \rightleftharpoons CO_3^{2-}(aq) + H_3O^+(aq)$; as a base: $HCO_3^-(aq) + H_2O(l) \rightleftharpoons H_2CO_3(aq) + OH^-(aq)$; as an acid: $H_2PO_4^-(aq) + H_2O(l) \rightleftharpoons HPO_4^{2-}(aq) + H_3O^+(aq)$; as a base: $H_2PO_4^-(aq) + H_2O(l) \rightleftharpoons H_3PO_4(aq) + OH^-(aq)$.   **21.** b, c, and d   **23.** 10.78 (4 significant figures); 6.78 (3 significant figures); 0.78 (2 significant figures); a pH value is a logarithm. The numbers to the left of the decimal place identify the power of 10 to which $[H^+]$ is expressed in scientific notation—for example, 10^{-11}, 10^{-7}, 10^{-1}. The number of decimal places in a pH value identifies the number of significant figures in $[H^+]$. In all three pH values, the $[H^+]$ should be expressed only to two significant figures since these pH values have only two decimal places.

25.

NH_3	+	NH_3	$\rightleftharpoons$	NH_2^-	+	NH_4^+
Acid		Base		Conjugate Base		Conjugate Acid

One of the NH_3 molecules acts as a base and accepts a proton to form NH_4^+. The other NH_3 molecule acts as an acid and donates a proton to form NH_2^-. NH_4^+ is the conjugate acid of the NH_3 base. In the reverse reaction, NH_4^+ donates a proton. NH_2^- is the conjugate base of the NH_3 acid. In the reverse reaction, NH_2^- accepts a proton. Conjugate acid–base pairs differ only by an H^+ in the formula.   **27. a.** These would be 0.10 M solutions of strong acids like HCl, HBr, HI, HNO$_3$, H$_2$SO$_4$, or HClO$_4$. **b.** These are salts of the conjugate acids of the bases in Table 13-3. These conjugate acids are all weak acids. Three examples would be 0.10 M solutions of NH$_4$Cl, CH$_3$NH$_3$NO$_3$, and C$_2$H$_5$NH$_3$Br. Note that the anions used to form these salts are conjugate bases of strong acids; this is because they have no acidic or basic properties in water (with the exception of HSO$_4^-$, which has weak acid properties). **c.** These would be 0.10 M solutions of strong bases like LiOH, NaOH, KOH, RbOH, CsOH, Ca(OH)$_2$, Sr(OH)$_2$, and Ba(OH)$_2$. **d.** These are salts of the conjugate bases of the neutrally charged weak acids in Table 13-2. The conjugate bases of weak acids are weak bases themselves. Three examples would be 0.10 M solutions of NaClO$_2$, KC$_2$H$_3$O$_2$, and CaF$_2$. The cations used to form these salts are Li$^+$, Na$^+$, K$^+$, Rb$^+$, Cs$^+$, Ca^{2+}, Sr^{2+}, and Ba^{2+} since these cations have no acidic or basic properties in water. Notice that these are the cations of the strong bases that you should memorize. **e.** There are two ways to make a neutral salt. The easiest way is to combine a conjugate base of a strong acid (except for HSO$_4^-$) with one of the cations from the strong bases. These ions have no acidic/basic properties in water so salts of these ions are neutral. Three examples would be 0.10 M solutions of NaCl, KNO$_3$, and SrI$_2$. Another type of strong electrolyte that can produce neutral solutions are salts that contain an ion with weak acid properties combined with an ion of opposite charge having weak base properties. If the K_a for the weak acid ion is equal to the K_b for the weak base ion, then the salt will produce a neutral solution. The most common example of this type of salt is ammonium acetate, NH$_4$C$_2$H$_3$O$_2$. For this salt, K_a for NH$_4^+$ = K_b for C$_2$H$_3$O$_2^-$ = 5.6×10^{-10}. This salt, at any concentration, produces a neutral solution.

29. a. $H_2O(l) + H_2O(l) \rightleftharpoons H_3O^+(aq) + OH^-(aq)$ or
$H_2O(l) \rightleftharpoons H^+(aq) + OH^-(aq)$ $K = K_w = [H^+][OH^-]$
b. $HF(aq) + H_2O(l) \rightleftharpoons F^-(aq) + H_3O^+(aq)$ or

$$HF(aq) \rightleftharpoons H^+(aq) + F^-(aq) \qquad K = K_a = \frac{[H^+][F^-]}{[HF]}$$

c. $C_5H_5N(aq) + H_2O(l) \rightleftharpoons C_5H_5NH^+(aq) + OH^-(aq)$

$$K = K_b - \frac{[C_5H_5NH^+][OH^-]}{[C_5H_5N]}$$

31. a. This expression holds true for solutions of strong acids having a concentration greater than 1.0×10^{-6} M. For example, 0.10 M HCl, 7.8 M HNO$_3$, and 3.6×10^{-4} M HClO$_4$ are solutions where this expression holds true. **b.** This expression holds true for solutions of weak acids where the two normal assumptions hold. The two assumptions are that the contribution of H$^+$ from water is negligible and that the acid is less than 5% dissociated in water (from the assumption that x is small compared to some number). This expression will generally hold true for solutions of weak acids having a K_a value less than 1×10^{-4}, as long as there is a significant amount of weak acid present. Three example solutions are 1.5 M HC$_2$H$_3$O$_2$, 0.10 M HOCl, and 0.72 M HCN. **c.** This expression holds true for strong bases that donate 2 OH$^-$ ions per formula unit. As long as the concentration of the base is above 5×10^{-7} M, this expression will hold true. Three examples are 5.0×10^{-3} M Ca(OH)$_2$, 2.1×10^{-4} M Sr(OH)$_2$, and 9.1×10^{-5} M Ba(OH)$_2$. **d.** This expression holds true for solutions of weak bases where the two normal assumptions hold. The assumptions are that the OH$^-$ contribution from water is negligible and that the base is less than 5% ionized in water. For the 5% rule to hold, you generally need bases with $K_b < 1 \times 10^{-4}$ and concentrations of weak base greater than 0.10 M. Three examples are 0.10 M NH$_3$, 0.54 M C$_6$H$_5$NH$_2$, and 1.1 M C$_5$H$_5$N. **33.** One reason HF is a weak acid is that the H—F bond is unusually strong and thus, is difficult to break. This contributes to the reluctance of the HF molecules to dissociate in water. **35. a.** $HClO_4(aq) + H_2O(l) \rightarrow H_3O^+(aq) + ClO_4^-(aq)$ or $HClO_4(aq) \rightarrow H^+(aq) + ClO_4^-(aq)$; water is commonly omitted from K_a reactions. **b.** $CH_3CH_2CO_2H(aq) \rightleftharpoons H^+(aq) + CH_3CH_2CO_2^-(aq)$; **c.** $NH_4^+(aq) \rightleftharpoons H^+(aq) + NH_3(aq)$ **37. a.** H$_2$O, base; H$_2$CO$_3$, acid; H$_3$O$^+$, conjugate acid; HCO$_3^-$, conjugate base; **b.** C$_5$H$_5$NH$^+$, acid; H$_2$O, base; C$_5$H$_5$N, conjugate base; H$_3$O$^+$, conjugate acid; **c.** HCO$_3^-$, base; C$_5$H$_5$NH$^+$, acid; H$_2$CO$_3$, conjugate acid; C$_5$H$_5$N, conjugate base **39. a.** HClO$_4$, strong acid; **b.** HOCl, weak acid; **c.** H$_2$SO$_4$, strong acid; **d.** H$_2$SO$_3$, weak acid **41.** HClO$_4$ > HClO$_2$ > NH$_4^+$ > H$_2$O **43. a.** HCl; **b.** HNO$_2$; **c.** HCN since it has a larger K_a value. **45. a.** 1.0×10^{-7} M, neutral; **b.** 12 M, basic; **c.** 8.3×10^{-16} M, acidic; **d.** 1.9×10^{-10} M, acidic **47. a.** endothermic; **b.** $[H^+] = [OH^-] = 2.34 \times 10^{-7}$ M **49.** [45] **a.** pH − pOH = 7.00; **b.** pH − 15.08, pOH = −1.08; **c.** pH = −1.08, pOH = 15.08; **d.** pH = 4.27, pOH = 9.73 [46] **a.** pH = 14.18, pOH = −0.18; **b.** pH = −0.44, pOH = 14.44; **c.** pH = pOH = 7.00; **d.** pH = 10.86, pOH = 3.14 **51. a.** pH = 6.88, pOH = 7.12, $[H^+] = 1.3 \times 10^{-7}$ M, $[OH^-] = 7.6 \times 10^{-8}$ M, acidic; **b.** pH = 0.92, pOH = 13.08, $[H^+] = 0.12$ M, $[OH^-] = 8.4 \times 10^{-14}$ M, acidic; **c.** pH = 10.89, pOH = 3.11, $[H^+] = 1.3 \times 10^{-11}$ M, $[OH^-] = 7.8 \times 10^{-4}$ M, basic; **d.** pH = pOH = 7.00, $[H^+] = [OH^-] = 1.0 \times 10^{-7}$ M, neutral **53.** pOH = 11.9, $[H^+] = 8 \times 10^{-3}$ M, $[OH^-] = 1 \times 10^{-12}$ M, acidic **55. a.** H$^+$, ClO$_4^-$, H$_2$O; 0.602; **b.** H$^+$, NO$_3^-$, H$_2$O; 0.602 **57. a.** 1.00; **b.** −0.70; **c.** 7.00 **59.** 3.2×10^{-3} M **61.** Add 4.2 mL of 12 M HCl to water with mixing; add enough water to bring the solution volume to 1600 mL. **63. a.** HNO$_2$ and H$_2$O, 2.00; **b.** HC$_2$H$_3$O$_2$ and H$_2$O, 2.68 **65.** $[H^+] = [F^-] = 3.5 \times 10^{-3}$ M, $[OH^-] = 2.9 \times 10^{-12}$ M, [HF] = 0.017 M, 2.46 **67.** [HC$_3$H$_5$O$_2$] = 0.099 M, [C$_3$H$_5$O$_2^-$] = [H$^+$] = 1.1×10^{-3} M, $[OH^-] = 9.1 \times 10^{-12}$, pH = 2.96, 1.1% dissociated. **69.** 1.96 **71.** 1.57; 5.9×10^{-9} M **73. a.** 0.60%; **b.** 1.9%; **c.** 5.8%; **d.** Dilution shifts equilibrium to the side with the greater number of particles (% dissociation increases). **e.** [H$^+$] also depends on initial concentration of weak acid. **75.** 1.4×10^{-4} **77.** 0.16 **79.** 0.024 M **81.** 6×10^{-3}

83. a. $NH_3(aq) + H_2O(l) \rightleftharpoons NH_4^+(aq) + OH^-(aq)$ $K_b = \dfrac{[NH_4^+][OH^-]}{[NH_3]}$;

b. $C_5H_5N(aq) + H_2O(l) \rightleftharpoons C_5H_5NH^+(aq) + OH^-(aq)$

$$K_b = \frac{[C_5H_5NH^+][OH^-]}{[C_5H_5N]}$$

85. NH$_3$ > C$_5$H$_5$N > H$_2$O > NO$_3^-$ **87. a.** C$_6$H$_5$NH$_2$; **b.** C$_6$H$_5$NH$_2$; **c.** OH$^-$; **d.** CH$_3$NH$_2$ **89. a.** 13.00; **b.** 7.00; **c.** 14.30   **91. a.** K$^+$, OH$^-$, and H$_2$O, 0.015 M, 12.18; **b.** Ba^{2+}, OH$^-$, and H$_2$O, 0.030 M, 12.48   **93.** 0.16 g   **95.** NH$_3$ and H$_2$O, 1.6×10^{-3} M, 11.20   **97. a.** $[OH^-] = 8.9 \times 10^{-3}$ M, $[H^+] = 1.1 \times 10^{-12}$ M, 11.96; **b.** $[OH^-] = 4.7 \times 10^{-5}$ M, $[H^+] = 2.1 \times 10^{-10}$ M, 9.68   **99.** 12.00   **101. a.** 1.3%; **b.** 4.2%; **c.** 6.4%   **103.** 1.0×10^{-9}

105. $H_2SO_3(aq) \rightleftharpoons HSO_3^-(aq) + H^+(aq)$ K_{a_1} reaction
$HSO_3^-(aq) \rightleftharpoons SO_3^{2-}(aq) + H^+(aq)$ K_{a_2} reaction

107. 3.00   **109.** 4.00; 1.0×10^{-19} M   **111.** −0.30   **113.** HCl > NH$_4$Cl > KNO$_3$ > KCN > KOH   **115.** OCl$^-$ **117. a.** $[OH^-] = [H^+] = 1.0 \times 10^{-7}$ M, pH = 7.00; **b.** $[OH^-] = 2.4 \times 10^{-5}$ M, $[H^+] = 4.2 \times 10^{-10}$ M, pH = 9.38   **119. a.** 5.82; **b.** 10.95   **121.** $[HN_3] = [OH^-] = 2.3 \times 10^{-6}$ M, $[Na^+] = 0.010$ M, $[N_3^-] = 0.010$ M, $[H^+] = 4.3 \times 10^{-9}$ M   **123.** NaF   **125.** 3.66   **127.** 3.08   **129. a.** neutral; **b.** basic; $NO_2^- + H_2O \rightleftharpoons HNO_2 + OH^-$; **c.** acidic; $C_5H_5NH^+ \rightleftharpoons C_5H_5N + H^+$; **d.** acidic because NH$_4^+$ is a stronger acid than NO$_2^-$ is a base; $NH_4^+ \rightleftharpoons NH_3 + H^+$; $NO_2^- + H_2O \rightleftharpoons HNO_2 + OH^-$; **e.** basic; $OCl^- + H_2O \rightleftharpoons HOCl + OH^-$; **f.** basic because OCl$^-$ is a stronger base than NH$_4^+$ is an acid; $OCl^- + H_2O \rightleftharpoons HOCl + OH^-$, $NH_4^+ \rightleftharpoons NH_3 + H^+$ **131. a.** HIO$_3$ < HBrO$_3$; as the electronegativity of the central atom increases, acid strength increases. **b.** HNO$_2$ < HNO$_3$; as the number of oxygen atoms attached to the central atom increases, acid strength increases. **c.** HOI < HOCl; same reasoning as in part a. **d.** H$_3$PO$_3$ < H$_3$PO$_4$; same reasoning as in part b. **133. a.** H$_2$O < H$_2$S < H$_2$Se; acid strength increases as bond energy decreases. **b.** CH$_3$CO$_2$H < FCH$_2$CO$_2$H < F$_2$CHCO$_2$H < F$_3$CCO$_2$H; as the electronegativity of the neighboring atoms increases, acid strength increases. **c.** NH$_4^+$ < HONH$_3^+$; same reasoning as in part b. **d.** NH$_4^+$ < PH$_4^+$; same reasoning as in part a. **135. a.** basic; $CaO(s) + H_2O(l) \rightarrow Ca(OH)_2(aq)$; **b.** acidic; $SO_2(g) + H_2O(l) \rightarrow H_2SO_3(aq)$; **c.** acidic; $Cl_2O(g) + H_2O(l) \rightarrow 2HOCl(aq)$ **137. a.** B(OH)$_3$, acid; H$_2$O, base; **b.** Ag$^+$, acid; NH$_3$, base; **c.** BF$_3$, acid; F$^-$, base **139.** $Al(OH)_3(s) + 3H^+(aq) \rightarrow Al^{3+}(aq) + 3H_2O(l)$; $Al(OH)_3(s) + OH^-(aq) \rightarrow Al(OH)_4^-(aq)$ **141.** Fe^{3+}; because it is smaller with a greater positive charge, Fe^{3+} will be more strongly attracted to a lone pair of electrons from a Lewis base. **143.** 990 mL H$_2$O   **145.** NH$_4$Cl   **147. a.** Hb(O$_2$)$_4$ in lungs, HbH$_4^{4+}$ in cells; **b.** Decreasing [CO$_2$] will decrease [H$^+$], favoring Hb(O$_2$)$_4$ formation. Breathing into a bag raises [CO$_2$]. **c.** NaHCO$_3$ lowers the acidity from accumulated CO$_2$.   **149.** 4.2×10^{-2} M   **151. a.** 2.62; **b.** 2.4%; **c.** 8.48   **153. a.** HA is a weak acid. Most of the acid is present as HA molecules; only one set of H$^+$ and A$^-$ ions is present. In a strong acid, all of the acid would be dissociated into H$^+$ and A$^-$ ions. **b.** 2.2×10^{-3} **155.** 9.2×10^{-7} **157. a.** 1.66; **b.** Fe^{2+} ions will produce a less acidic solution (higher pH) due to the lower charge on Fe^{2+} as compared with Fe^{3+}. As the charge on a metal ion increases, acid strength of the hydrated ion increases.   **159.** acidic; $HSO_4^- \rightleftharpoons SO_4^{2-} + H^+$; 1.54   **161. a.** H$_2SO_3$; **b.** HClO$_3$; **c.** H$_3PO_3$; NaOH and KOH are ionic compounds composed of either Na$^+$ or K$^+$ cations and OH$^-$ anions. When soluble ionic compounds dissolve in water, they form the ions from which they are formed. The acids in this problem are all covalent compounds. When these acids dissolve in water, the covalent bond between oxygen and hydrogen breaks to form H$^+$ ions.   **163.** 0.0070 M HNO$_3$: $[H^+] = 0.0070$ M H$^+$; pH = 2.15; pOH = 11.85; $[OH^-] = 1.4 \times 10^{-12}$ M; 3.0 M KOH: $[H^+] = 3.3 \times 10^{-15}$ M; pH = 14.48; pOH = −0.48; $[OH^-] = 3.0$ M   **165. a.** C$_2$H$_5$NH$_2$ is a weak base, so the major species are C$_2$H$_5$NH$_2$ and H$_2$O. **b.** 12.28   **167.** NaCl: neutral salt (pH = 7.0); Na$^+$ and Cl$^-$ have no acidic or basic properties. RbOCl: basic; OCl$^-$ is the conjugate base of the weak acid HOCl, so it is a weak base. Rb$^+$ is a neutral species. KI: neutral salt (pH = 7.0); K$^+$ and I$^-$ have no acidic or basic properties. Ba(ClO$_4$)$_2$: neutral (pH = 7.0); Ba^{2+} and ClO$_4^-$ have no acidic or basic properties. NH$_4$NO$_3$: acidic; NH$_4^+$ is the conjugate acid of the weak base NH$_3$, so it is a weak acid. NO$_3^-$ is a neutral species.   **169.** AlCl$_3$ < CsClO$_4$ < NaF < NaCN < KOH   **171.** 7.20.   **173.** 4540 mL   **175.** 4.17

177. 0.022 M **179.** 917 mL of water evaporated **181.** PO_4^{3-}, $K_b = 0.021$; HPO_4^{2-}, $K_b = 1.6 \times 10^{-7}$; $H_2PO_4^-$, $K_b = 1.3 \times 10^{-12}$; from the K_b values, PO_4^{3-} is the strongest base. **183. a.** basic; **b.** acidic; **c.** basic; **d.** acidic; **e.** acidic **185.** 1.0×10^{-3} **187.** 5.4×10^{-4} **189.** 2.5×10^{-3}

Chapter 14

9. When an acid dissociates, ions are produced. A common ion is when one of the product ions in a particular equilibrium is added from an outside source. For a weak acid dissociating to its conjugate base and H^+, the common ion would be the conjugate base; this would be added by dissolving a soluble salt of the conjugate base into the acid solution. The presence of the conjugate base from an outside source shifts the equilibrium to the left so less acid dissociates. **11.** The more weak acid and conjugate base present, the more H^+ and/or OH^- that can be absorbed by the buffer without significant pH change. When the concentrations of weak acid and conjugate base are equal (so that pH = pK_a), the buffer system is equally efficient at absorbing either H^+ or OH^-. If the buffer is overloaded with weak acid or with conjugate base, then the buffer is not equally efficient at absorbing either H^+ or OH^-. **13. a.** Let's call the acid HA, which is a weak acid. When HA is present in the beakers, it exists in the undissociated form, making it a weak acid. A strong acid would exist as separate H^+ and A^- ions. **b.** beaker c → beaker a → beaker e → beaker b → beaker d **c.** pH = pK_a when a buffer solution is present that has equal concentrations of the weak acid and conjugate base. This is beaker e. **d.** The equivalence point is when just enough OH^- has been added to exactly react with all of the acid present initially. This is beaker b. **e.** Past the equivalence point, the pH is dictated by the concentration of excess OH^- added from the strong base. We can ignore the amount of hydroxide added by the weak conjugate base that is also present. This is beaker d. **15.** The three key points to emphasize in your sketch are the initial pH, the pH at the halfway point to equivalence, and the pH at the equivalence point. For the two weak bases titrated, pH = pK_a at the halfway point to equivalence (50.0 mL HCl added) because [weak base] = [conjugate acid] at this point. For the initial pH, the strong base has the highest pH (most basic), while the weakest base has the lowest pH (least basic). At the equivalence point, the strong base titration has pH = 7.0. The weak bases titrated have acidic pHs at the equivalence point because the conjugate acids of the weak bases titrated are the major species present. The weakest base has the strongest conjugate acid so its pH will be lowest (most acidic) at the equivalence point.

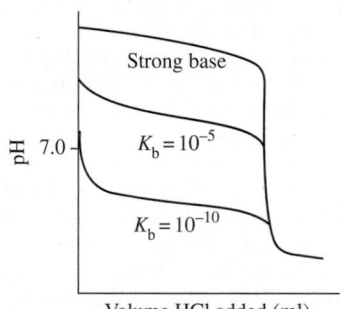

17. Only the third beaker represents a buffer solution. A weak acid and its conjugate base both must be present in large quantities in order to have a buffer solution. This is only the case in the third beaker. The first beaker represents a beaker full of strong acid that is 100% dissociated. The second beaker represents a weak acid solution. In a weak acid solution, only a small fraction of the acid is dissociated. In this representation, 1/10 of the weak acid has dissociated. The only B^- present in this beaker is from the dissociation of the weak acid. A buffer solution has B^- added from another source. **19.** When strong acid or strong base is added to a sodium bicarbonate/sodium carbonate buffer mixture, the strong acid/base is neutralized. The reaction goes to completion resulting in the strong acid/base being replaced with a weak acid/base. This results in a new buffer solution. The reactions are $H^+(aq) + CO_3^{2-}(aq) \rightarrow HCO_3^-(aq)$; $OH^-(aq) + HCO_3^-(aq) \rightarrow CO_3^{2-}(aq)$

$+ H_2O(l)$ **21. a.** 2.96; **b.** 8.94; **c.** 7.00; **d.** 4.89 **23.** 1.1% vs. 1.3×10^{-2}% dissociated; the presence of $C_3H_5O_2^-$ in solution 21d greatly inhibits the dissociation of $HC_3H_5O_2$. This is called the *common ion effect*. **25. a.** 1.70; **b.** 5.49; **c.** 1.70; **d.** 4.71 **27. a.** 4.29; **b.** 12.30; **c.** 12.30; **d.** 5.07 **29.** solution d; solution d is a buffer solution that resists pH changes. **31.** 3.40 **33.** 3.48; 3.22 **35. a.** 5.14; **b.** 4.34; **c.** 5.14; **d.** 4.34 **37.** 4.37 **39. a.** 7.97; **b.** 8.73; both solutions have an initial pH = 8.77. The two solutions differ in their buffer capacity. Solution b with the larger concentrations has the greater capacity to resist pH change. **41.** 15 g **43. a.** 0.19; **b.** 0.59; **c.** 1.0; **d.** 1.9 **45.** 1.3×10^{-2} M **47.** HOCl; there are many possibilities. One possibility is a solution with [HOCl] = 1.0 M and [NaOCl] = 0.35 M. **49.** 8.18; add 0.20 mole NaOH **51.** solution d **53. a.** 1.0 mole; **b.** 0.30 mole; **c.** 1.3 moles **55. a.** ~22 mL base added; **b.** buffer region is from ~1 mL to ~21 mL base added. The maximum buffering region would be from ~5 mL to ~17 mL base added with the halfway point to equivalence (~11 mL) as the best buffer point. **c.** ~11 mL base added; **d.** 0 mL base added; **e.** ~22 mL base added (the stoichiometric point); **f.** any point after the stoichiometric point (volume base added > ~22 mL) **57. a.** 0.699; **b.** 0.854; **c.** 1.301; **d.** 7.00; **e.** 12.15 **59. a.** 2.72; **b.** 4.26; **c.** 4.74; **d.** 5.22; **e.** 8.79; **f.** 12.15

61.

Volume (mL)	pH
0.0	2.43
4.0	3.14
8.0	3.53
12.5	3.86
20.0	4.46
24.0	5.24
24.5	5.6
24.9	6.3
25.0	8.28
25.1	10.3
26.0	11.30
28.0	11.75
30.0	11.96

See *Solutions Guide* for pH plot.

63.

Volume (mL)	pH
0.0	11.11
4.0	9.97
8.0	9.58
12.5	9.25
20.0	8.65
24.0	7.87
24.5	7.6
24.9	6.9
25.0	5.28
25.1	3.7
26.0	2.71
28.0	2.24
30.0	2.04

See *Solutions Guide* for pH plot.

65. a. 4.19, 8.45; **b.** 10.74; 5.96; **c.** 0.89, 7.00 **67.** 2.1×10^{-6} **69. a.** yellow; **b.** 8.0; **c.** blue **71.** phenolphthalein **73.** Phenol red is one possible indicator for the titration in Exercise 57. Phenolphthalein is one possible indicator for the titration in Exercise 59. **75.** Phenolphthalein is one possible indicator for Exercise 61. Bromcresol green is one possible indicator for Exercise 63. **77.** The pH is between 5 and 8. **79. a.** yellow; **b.** green; **c.** yellow; **d.** blue **81.** $pOH = pK_b + \log\dfrac{[acid]}{[base]}$ **83. a.** The optimal buffer pH is about 8.1; **b.** 0.083 at pH = 7.00; 8.3 at pH = 9.00; **c.** 8.08; 7.95 **85. a.** potassium fluoride + HCl; **b.** benzoic acid + NaOH; **c.** acetic acid + sodium acetate; **d.** HOCl + NaOH; **e.** ammonium chloride + NaOH **87. a.** 1.1 ≈ 1; **b.** A best buffer has approximately equal concentrations of weak acid and conjugate base, so pH ≈ pK_a for a best buffer. The pK_a value for an $H_3PO_4/H_2PO_4^-$ buffer is $-\log(7.5 \times 10^{-3}) = 2.12$. A pH of 7.15 is too high for an $H_3PO_4/H_2PO_4^-$ buffer to be effective. At this high of a pH, there would be so little H_3PO_4 present that we could hardly consider it a buffer; this solution would not be effective in resisting pH changes, especially when a strong base is added. **89. a.** 1.8×10^9; **b.** 5.6×10^4; **c.** 1.0×10^{14} **91.** 4.4 L **93.** 65 mL **95.** 180. g/mol; 3.3×10^{-4}; assumed acetylsalicylic acid is a weak monoprotic acid. **97.** 0.210 M **99.** 1.74×10^{-8} **101.** 0.92 M **103. a.** False; the buffer with the largest concentrations has the highest buffer capacity. **b.** True; when the base component of the buffer has a larger concentration than the acid component, then pH > pK_a. **c.** True; as more of the acid component of the buffer is added, the pH will become more acidic (pH will decrease). **d.** False; for this buffer, $pK_a = -\log(2.3 \times 10^{-11}) = 10.64$. Here, we have more of the acid component of the buffer than the base component. When this occurs, pH < pK_a. So we can say that for this situation, pH < 10.64, not that pH < 3.36. **e.** True; when [weak acid] = [conjugate base] in a buffer, pH = pK_a. **105. a.** 5.10; **b.** 9.21; **c.** 9.69; **d.** 10.95; **e.** 12.04 **107. a.** ii < iv < iii < i; **b.** i < ii < iv < iii; **c.** All require the same volume of titrant to reach the equivalence point. **109.** 49 mL **111.** 3.9 L **113. a.** 200.0 mL; **b. i.** H_2A, H_2O; **ii.** H_2A, HA^-, H_2O, Na^+; **iii.** HA^-, H_2O, Na^+; **iv.** HA^-, A^{2-}, H_2O, Na^+; **v.** A^{2-}, H_2O, Na^+; **vi.** A^{2-}, H_2O, Na^+, OH^-; **c.** $K_{a_1} = 1 \times 10^{-4}$; $K_{a_2} = 1 \times 10^{-8}$ **115. a.** The major species present at the various points after H^+ reacts completely are A: CO_3^{2-}, H_2O, Na^+; B: CO_3^{2-}, HCO_3^-, H_2O, Cl^-, Na^+; C: HCO_3^-, H_2O, Cl^-, Na^+; D: HCO_3^-, CO_2 (H_2CO_3), H_2O, Cl^-, Na^+; E: CO_2 (H_2CO_3), H_2O, Cl^-, Na^+; F: H^+, CO_2 (H_2CO_3), H_2O, Cl^-, Na^+; **b.** B: pH = 10.25; D: pH = 6.37 **117.** pH ≈ 5.0; $K_a ≈ 1 \times 10^{-10}$ **119.** 3.00 **121.** 2.78

Chapter 15

9. K_{sp} values can only be compared to determine relative solubilities when the salts produce the same number of ions. Here, Ag_2S and CuS do not produce the same number of ions when they dissolve, so each has a different mathematical relationship between the K_{sp} value and the molar solubility. To determine which salt has the larger molar solubility, you must do the actual calculations and compare the two molar solubility values. **11. i.** This is the result when you have a salt that breaks up into two ions. Examples of these salts would be AgCl, $SrSO_4$, $BaCrO_4$, and $ZnCO_3$. **ii.** This is the result when you have a salt that breaks up into three ions, either two cations and one anion or one cation and two anions. Some examples are SrF_2, Hg_2I_2, and Ag_2SO_4. **iii.** This is the result when you have a salt that breaks up into four ions, either three cations and one anion (Ag_3PO_4) or one cation and three anions (ignoring the hydroxides, there are no examples of this type of salt in Table 15-1). **iv.** This is the result when you have a salt that breaks up into five ions, either three cations and two anions [$Sr_3(PO_4)_2$] or two cations and three anions (no examples of this type of salt are in Table 15-1). **13.** For the K_{sp} reaction of a salt dissolving into its respective ions, the common ion would be if one of the ions in the salt was added from an outside source. When a common ion is present, the K_{sp} equilibrium shifts to the left resulting in less of the salt dissolving into its ions. **15.** Some people would automatically think that an increase in temperature would increase the solubility of a salt. This is not always the case because some salts show a decrease in solubility as temperature increases. The two major methods used to increase solubility of a salt both involve removing one of the ions in the salt by reaction. If the salt has an ion with basic properties, adding H^+ will increase the solubility of the salt because the added H^+ will react with the basic ion, thus removing it from solution. More salt dissolves in order to make up for the lost ion. Some examples of salts with basic ions are AgF, $CaCO_3$, and $Al(OH)_3$. The other way to remove an ion is to form a complex ion. For example, the Ag^+ ion in silver salts forms the complex ion $Ag(NH_3)_2^+$ as ammonia is added. Silver salts increase their solubility as NH_3 is added because the Ag^+ ion is removed through complex ion formation. **17.** In 2.0 M NH_3, the soluble complex ion $Ag(NH_3)_2^+$ forms, which increases the solubility of AgCl(s). The reaction is $AgCl(s) + 2NH_3 \rightleftharpoons Ag(NH_3)_2 + Cl^-$. In 2.0 M NH_4NO_3, NH_3 is only formed by the dissociation of the weak acid NH_4^+. There is not enough NH_3 produced by this reaction to dissolve AgCl(s) by the formation of the complex ion. **19. a.** $AgC_2H_3O_2(s) \rightleftharpoons Ag^+(aq) + C_2H_3O_2^-(aq)$; $K_{sp} = [Ag^+][C_2H_3O_2^-]$; **b.** $Al(OH)_3(s) \rightleftharpoons Al^{3+}(aq) + 3OH^-(aq)$; $K_{sp} = [Al^{3+}][OH^-]^3$; **c.** $Ca_3(PO_4)_2(s) \rightleftharpoons 3Ca^{2+}(aq) + 2PO_4^{3-}(aq)$; $K_{sp} =$

$[Ca^{2+}]^3[PO_4^{3-}]^2$ **21. a.** 2.3×10^{-9}; **b.** 8.20×10^{-19} **23.** 1.4×10^{-8} **25.** 3.92×10^{-5} **27. a.** 1.6×10^{-5} mol/L; **b.** 9.3×10^{-5} mol/L; **c.** 6.5×10^{-7} mol/L **29.** 0.89 g **31.** 1.3×10^{-4} mol/L **33.** 2×10^{-11} mol/L **35. a.** CaF_2; **b.** $FePO_4$ **37. a.** 4×10^{-17} mol/L; **b.** 4×10^{-11} mol/L; **c.** 4×10^{-29} mol/L **39. a.** 1.4×10^{-2} mol/L; **b.** 1.2×10^{-3} mol/L; **c.** 3.9×10^{-3} mol/L **41.** 2.3×10^{-11} mol/L **43.** 3.5×10^{-10} **45.** If the anion in the salt can act as a base in water, then the solubility of the salt will increase as the solution becomes more acidic. Added H^+ will react with the base, forming the conjugate acid. As the basic anion is removed, more of the salt will dissolve to replenish the basic anion. The salts with basic anions are Ag_3PO_4, $CaCO_3$, $CdCO_3$, and $Sr_3(PO_4)_2$. Hg_2Cl_2 and PbI_2 do not have any pH dependence because Cl^- and I^- are terrible bases (the conjugate bases of strong acids).

$$Ag_3PO_4(s) + H^+(aq) \longrightarrow 3Ag^+(aq) + HPO_4^{2-}(aq) \xrightarrow{\text{excess } H^+}$$
$$3Ag^+(aq) + H_3PO_4(aq)$$

$$CaCO_3(s) + H^+(aq) \longrightarrow Ca^{2+}(aq) + HCO_3^-(aq) \xrightarrow{\text{excess } H^+}$$
$$Ca^{2+}(aq) + H_2CO_3(aq) \, [H_2O(l) + CO_2(g)]$$

$$CdCO_3(s) + H^+(aq) \longrightarrow Cd^{2+}(aq) + HCO_3^-(aq) \xrightarrow{\text{excess } H^+}$$
$$Cd^{2+}(aq) + H_2CO_3(aq) \, [H_2O(l) + CO_2(g)]$$

$$Sr_3(PO_4)_2(s) + 2H^+(aq) \longrightarrow 3Sr^{2+}(aq) + 2HPO_4^{2-}(aq) \xrightarrow{\text{excess } H^+}$$
$$3Sr^{2+}(aq) + 2H_3PO_4(aq)$$

47. 1.5×10^{-19} g **49.** No precipitate forms. **51.** $PbF_2(s)$ will not form. **53.** $[K^+] = 0.160\,M$, $[C_2O_4^{2-}] = 3.3 \times 10^{-7}\,M$, $[Ba^{2+}] = 0.0700\,M$, $[Br^-] = 0.300\,M$ **55.** 7.5×10^{-6} mol/L **57.** $[AgNO_3] > 5.6 \times 10^{-5}\,M$ **59.** $PbS(s)$ will form first, followed by $Pb_3(PO_4)_2(s)$, and $PbF_2(s)$ will form last.

61. a.
$$Ni^{2+} + CN^- \rightleftharpoons NiCN^+$$
$$NiCN^+ + CN^- \rightleftharpoons Ni(CN)_2$$
$$Ni(CN)_2 + CN^- \rightleftharpoons Ni(CN)_3^-$$
$$\underline{Ni(CN)_3^- + CN^- \rightleftharpoons Ni(CN)_4^{2-}}$$
$$Ni^{2+} + 4CN^- \rightleftharpoons Ni(CN)_4^{2-}$$

b.
$$V^{3+} + C_2O_4^{2-} \rightleftharpoons VC_2O_4^+$$
$$VC_2O_4^+ + C_2O_4^{2-} \rightleftharpoons V(C_2O_4)_2^-$$
$$\underline{V(C_2O_4)_2^- + C_2O_4^{2-} \rightleftharpoons V(C_2O_4)_3^{3-}}$$
$$V^{3+} + 3C_2O_4^{2-} \rightleftharpoons V(C_2O_4)_3^{3-}$$

63. 1.0×10^{42} **65.** $Hg^{2+}(aq) + 2I^-(aq) \rightleftharpoons HgI_2(s)$ (orange precipitate); $HgI_2(s) + 2I^-(aq) \rightleftharpoons HgI_4^{2-}(aq)$ (soluble complex ion) **67.** $3.3 \times 10^{-32}\,M$ **69. a.** $1.0 \times 10^{-3}\,M$; **b.** $2.0 \times 10^{-7}\,M$; **c.** $8.0 \times 10^{-15}\,M$ **71. a.** 1.2×10^{-8} mol/L; **b.** 1.5×10^{-4} mol/L; **c.** The presence of NH_3 increases the solubility of AgI. Added NH_3 removes Ag^+ from solution by forming the complex ion $Ag(NH_3)_2^+$. As Ag^+ is removed, more AgI will dissolve to replenish the Ag^+ concentration. **73.** 4.7×10^{-2} mol/L **75.** Test tube 1: added Cl^- reacts with Ag^+ to form the silver chloride precipitate. The net ionic equation is $Ag^+(aq) + Cl^-(aq) \rightarrow AgCl(s)$. Test tube 2: added NH_3 reacts with Ag^+ ions to form the soluble complex ion $Ag(NH_3)_2^+$. As this complex ion forms, Ag^+ is removed from solution, which causes AgCl(s) to dissolve. When enough NH_3 is added, then all of the silver chloride precipitate will dissolve. The equation is $AgCl(s) + 2NH_3(aq) \rightarrow Ag(NH_3)_2^+(aq) + Cl^-(aq)$. Test tube 3: added H^+ reacts with the weak base NH_3 to form NH_4^+. As NH_3 is removed, Ag^+ ions are released to solution, which can then react with Cl^- to re-form AgCl(s). The equations are $Ag(NH_3)_2^+(aq) + 2H^+(aq) \rightarrow Ag^+(aq) + 2NH_4^+(aq)$ and $Ag^+(aq) + Cl^-(aq) \rightarrow AgCl(s)$. **77.** 1.7 g AgCl; 5×10^{-9} mol/L **79.** 2.7×10^{-5} mol/L; the solubility of hydroxyapatite will increase as a solution gets more acidic, since both phosphate and hydroxide can react with H^+. 6×10^{-8} mol/L; the hydroxyapatite in the tooth enamel is converted to the less soluble fluorapatite by fluoride-treated water. The less soluble fluorapatite will then be more difficult to dissolve, making teeth less susceptible to decay. **81.** Addition of more than 9.0×10^{-6} g $Ca(NO_3)_2$ should start precipitation of $CaF_2(s)$. **83.** 7.0×10^{-8} **85.** 6.2×10^5

87. a. 6.7×10^{-6} mol/L; **b.** 1.2×10^{-13} mol/L; **c.** $Pb(OH)_2(s)$ will not form since $Q < K_{sp}$. **89. a.** 1.6×10^{-6}; **b.** 0.056 mol/L **91.** $Ba(OH)_2$, pH = 13.34; $Sr(OH)_2$, pH = 12.93; $Ca(OH)_2$, pH = 12.15 **93.** 1.1×10^{-32} **95.** 6.1×10^{-6} mol/L; **97.** 1.08×10^{-6} mol/L **99. a.** 0.33 mol/L; **b.** 0.33 M; **c.** 4.8×10^{-3} M **101. a.** 7.1×10^{-7} mol/L; **b.** 8.7×10^{-3} mol/L; **c.** The presence of NH_3 increases the solubility of AgBr. Added NH_3 removes Ag^+ from solution by forming the complex ion, $Ag(NH_3)_2^+$. As Ag^+ is removed, more AgBr(s) will dissolve to replenish the Ag^+ concentration. **d.** 0.41 g AgBr; **e.** Added HNO_3 will have no effect on the AgBr(s) solubility in pure water. Neither H^+ nor NO_3^- react with Ag^+ or Br^- ions. Br^- is the conjugate base of the strong acid HBr, so it is a terrible base. However, added HNO_3 will reduce the solubility of AgBr(s) in the ammonia solution. NH_3 is a weak base ($K_b = 1.8 \times 10^{-5}$). Added H^+ will react with NH_3 to form NH_4^+. As NH_3 is removed, a smaller amount of the $Ag(NH_3)_2^+$ complex ion will form, resulting in a smaller amount of AgBr(s) that will dissolve. **103.** 5.7×10^{-2} mol/L **105.** 3 M **107.** 4.8×10^{-11} M **109. a.** 5.8×10^{-4} mol/L; **b.** greater; F^- is a weak base ($K_b = 1.4 \times 10^{-11}$), so some of the F^- is removed by reaction with water. As F^- is removed, more SrF_2 will dissolve; **c.** 3.5×10^{-3} mol/L **111.** pH = 0.70; 0.20 M Ba^{2+}; 23 g $BaSO_4(s)$

Chapter 16

11. Living organisms need an external source of energy to carry out these processes. For all processes combined, ΔS_{univ} must be greater than zero (the 2nd law). **13.** As any process occurs, ΔS_{univ} will increase; ΔS_{univ} cannot decrease. Time also goes in one direction, just as ΔS_{univ} goes in one direction. **15.** This reaction is kinetically slow but thermodynamically favorable ($\Delta G < 0$). Thermodynamics only tells us if a reaction can occur. To answer the question will the reaction occur, one also needs to consider the kinetics (speed of reaction). The ultraviolet light provides the activation energy for this slow reaction to occur. **17.** $\Delta S_{surr} = -\Delta H/T$; heat flow ($\Delta H$) into or out of the system dictates ΔS_{surr}. If heat flows into the surroundings, the random motions of the surroundings increase, and the entropy of the surroundings increases. The opposite is true when heat flows from the surroundings into the system (an endothermic reaction). Although the driving force described here really results from the change in entropy of the surroundings, it is often described in terms of energy. Nature tends to seek the lowest possible energy. **19.** Note that these substances are not in the solid state but are in the aqueous state; water molecules are also present. There is an apparent increase in ordering when these ions are placed in water. The hydrating water molecules must be in a highly ordered state when surrounding these anions. **21.** One can determine $\Delta S°$ and $\Delta H°$ for the reaction using the standard entropies and standard enthalpies of formation in Appendix 4, then use the equation $\Delta G° = \Delta H° - T\Delta S°$. One can also use the standard free energies of formation in Appendix 4. And finally, one can use Hess's law to calculate $\Delta G°$. Here, reactions having known $\Delta G°$ values are manipulated to determine $\Delta G°$ for a different reaction. For temperatures other than 25°C, $\Delta G°$ is estimated using the $\Delta G° = \Delta H° - T\Delta S°$ equation. The assumptions made are that the $\Delta H°$ and $\Delta S°$ values determined from Appendix 4 data are temperature independent. We use the same $\Delta H°$ and $\Delta S°$ values as determined when $T = 25°C$, then plug in the new temperature in kelvins into the equation to estimate $\Delta G°$ at the new temperature. **23.** The light source for the first reaction is necessary for kinetic reasons. The first reaction is just too slow to occur unless a light source is available. The kinetics of a reaction are independent of the thermodynamics of a reaction. Even though the first reaction is more favorable thermodynamically (assuming standard conditions), it is unfavorable for kinetic reasons. The second reaction has a negative $\Delta G°$ value and is a fast reaction, so the second reaction occurs very quickly. When considering if a reaction will occur, thermodynamics and kinetics must both be considered. **25.** a, b, c **27.** Possible arrangements for one molecule:

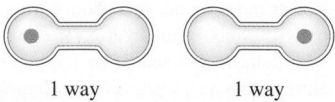

1 way   1 way

Both are equally probable.

Possible arrangements for two molecules:

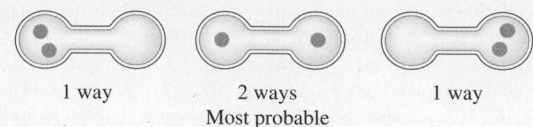

1 way   2 ways   1 way
Most probable

Possible arrangement for three molecules:

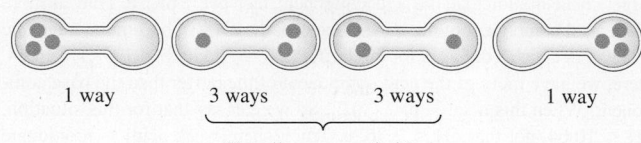

1 way   3 ways   3 ways   1 way

Equally most probable

29. We draw all of the possible arrangements of the two particles in the three levels.

2 kJ	—	—	x	—	x	xx
1 kJ	—	x	—	xx	x	—
0 kJ	xx	x	x	—	—	—
Total E =	0 kJ	1 kJ	2 kJ	2 kJ	3 kJ	4 kJ

The most likely total energy is 2 kJ. **31. a.** H_2 at 100°C and 0.5 atm; **b.** N_2 at STP; **c.** $H_2O(l)$ **33. a.** negative; **b.** positive **35.** Spontaneous ($\Delta G < 0$) for b, c, d **37.** 89.3 J/K · mol **39. a.** yes ($\Delta G < 0$); **b.** 196 K **41. a.** negative; **b.** negative; **c.** negative; **d.** positive **43. a.** $C_{graphite}(s)$; **b.** $C_2H_5OH(g)$; **c.** $CO_2(g)$ **45. a.** negative, -186 J/K; **b.** positive, 187 J/K; **c.** hard to predict since $\Delta n = 0$; 138 J/K **47.** 262 J/K · mol **49. a.** ΔH and ΔS are both positive; **b.** $S_{rhombic}$ **51. a.** ΔH and ΔS are both negative; **b.** low temperatures **53. a.** $\Delta H° = -803$ kJ, $\Delta S° = -4$ J/K, $\Delta G° = -802$ kJ; **b.** $\Delta H° = 2802$ kJ, $\Delta S° = -262$ J/K, $\Delta G° = 2880.$ kJ; **c.** $\Delta H° = -416$ kJ, $\Delta S° = -209$ J/K, $\Delta G° = -354$ kJ; **d.** $\Delta H° = -176$ kJ, $\Delta S° = -284$ J/K, $\Delta G° = -91$ kJ **55.** -5.40 kJ; 328.6 K; $\Delta G°$ is negative below 328.6 K. **57.** -817 kJ **59.** -731 kJ/mol **61. a.** 53 kJ; **b.** No, the reaction is not spontaneous at standard concentrations and 298 K. **c.** $T > 630$ K **63.** $CH_4(g) + CO_2(g) \rightarrow CH_3CO_2H(l)$, $\Delta H° = -16$ kJ, $\Delta S° = -240.$ J/K, $\Delta G° = 56$ kJ; $CH_3OH(g) + CO(g) \rightarrow CH_3CO_2H(l)$, $\Delta H° = -173$ kJ, $\Delta S° = -278$ J/K, $\Delta G° = -90.$ kJ; the second reaction is preferred. It should be run at temperatures below 622 K. **65.** -188 kJ **67. a.** shifts right; **b.** no shift since the reaction is at equilibrium; **c.** shifts left **69.** -198 kJ; 5.07×10^{34} **71.** 8.72; 0.0789 **73.** 140 kJ **75. a.** 2.22×10^5; **b.** 94.3 **77.** -71 kJ/mol **79.** $\Delta H° = 1.1 \times 10^5$ J/mol; $\Delta S° = 330$ J/K · mol; the major difference in the plot is the slope of the line. An endothermic process has a negative slope for the ln(K) versus $1/T$ plot, whereas an exothermic process has a positive slope. **81.** 447 J/K · mol **83.** decreases; ΔS will be negative since 2 moles of gaseous reactants form 1 mole of gaseous product. For ΔG to be negative, ΔH must be negative (exothermic). For exothermic reactions, K decreases as T increases, so the ratio of the partial pressure of PCl_5 to the partial pressure of PCl_3 will decrease. **85.** 43.7 K **87.** 7.0×10^{-4} **89.** 60 **91.** ΔS is more favorable (less negative) for reaction 2 than for reaction 1, resulting in $K_2 > K_1$. In reaction 1, seven particles in solution form one particle. In reaction 2, four particles form one, which results in a smaller decrease in positional probability than for reaction 1. **93.** 725 K **95.** A negative value for $\Delta S°$ indicates a process that has a decrease in positional probability. Processes b, c, and d all are expected to have negative values for $\Delta S°$. **97.** $\Delta S° = 49$ J/K; $\Delta S_{surr} = -140$ J/K **99.** 74.4 kJ/mol **101. a.** False; $\Delta H°$ is negative, so this is an exothermic reaction. **b.** True; since the number of gaseous molecules decreases in this reaction, positional probability will decrease. **c.** False; heat is a product in this exothermic reaction. As heat is added (temperature increases), this reaction will shift left by Le Châtelier's principle. This results in a decrease in the value of K. At equilibrium at the higher temperature, reactant concentrations will increase and the product concentration will decrease. Therefore, the ratio of $[PCl_5]/[PCl_3]$ will decrease with an increase in temperature. **d.** False; when the signs of $\Delta H°$ and $\Delta S°$ both are negative, the reaction will only be spontaneous at temperatures below some value (at low temperatures) where the favorable $\Delta H°$ term dominates. **e.** True; from

$\Delta G° = -RT \ln K$, when $\Delta G° < 0$, K must be greater than 1. **103. a.** $\Delta H = \Delta E$; **b.** $\Delta H < \Delta E$; **c.** $\Delta H > \Delta E$. **105. a.** Vessel 1: At 0°C, this system is at equilibrium, so $\Delta S_{univ} = 0$ and $\Delta S = \Delta S_{surr}$. Because the vessel is perfectly insulated, $q = 0$ so $\Delta S_{surr} = 0 = \Delta S_{sys}$. **b.** Vessel 2: The presence of salt in water lowers the freezing point of water to a temperature below 0°C. In vessel 2 the conversion of ice into water will be spontaneous at 0°C, so $\Delta S_{univ} > 0$. Because the vessel is perfectly insulated, $\Delta S_{surr} = 0$. Therefore, ΔS_{sys} must be positive ($\Delta S > 0$) in order for ΔS_{univ} to be positive. **107.** $\Delta H° = 286$ kJ; $\Delta G° = 326$ kJ; $K = 7.22 \times 10^{-58}$; $P_{O_3} = 3.3 \times 10^{-41}$ atm; this partial pressure represents one molecule of ozone per 9.5×10^{17} L of air. Equilibrium is probably not maintained under the conditions because the concentration of ozone is not large enough to maintain equilibrium
109. a.

$$k_f = A \exp\left(\frac{-E_a}{RT}\right) \text{ and } k_r = A \exp\left(\frac{-(E_a - \Delta G°)}{RT}\right),$$

$$\frac{k_f}{k_r} = \exp\left(\frac{-E_a}{RT} + \frac{(E_a - \Delta G°)}{RT}\right) = \exp\left(\frac{-\Delta G°}{RT}\right)$$

From $\Delta G° = -RT \ln K$, K also equals the same expression:

$$K = \exp\left(\frac{-\Delta G°}{RT}\right), \text{ so } K = \frac{k_f}{k_r}.$$

b. A catalyst increases the value of the rate of the rate constant (increases rate) by lowering the activation energy. For the equilibrium constant K to remain constant, both k_f and k_r must increase by the same factor. Therefore, a catalyst must increase the rate of both the forward and the reverse reactions. **111. a.** 0.333; **b.** $P_A = 1.50$ atm; $P_B = 0.50$ atm; **c.** $\Delta G = \Delta G° + RT \ln(P_B/P_A) = 2722$ J $- 2722$ J $= 0$ **113.** greater than 7.5 torr **115.** 6 M **117.** 61 kJ/mol **119.** -4.1 kJ/mol

Chapter 17

15. Oxidation: increase in oxidation number, loss of electrons; reduction: decrease in oxidation number, gain of electrons **17.** Reactions a, b, and c are oxidation–reduction reactions.

Oxidizing Agent	Reducing Agent	Substance Oxidized	Substance Reduced
a. H_2O	CH_4	$CH_4(C)$	$H_2O(H)$
b. $AgNO_3$	Cu	Cu	$AgNO_3(Ag)$
c. HCl	Zn	Zn	HCl(H)

19. Electrochemistry is the study of the interchange of chemical and electrical energy. A redox (oxidation–reduction) reaction is a reaction in which one or more electrons are transferred. In a galvanic cell, a spontaneous redox reaction occurs which produces an electric current. In an electrolytic cell, electricity is used to force a nonspontaneous redox reaction to occur. **21.** Magnesium is an alkaline earth metal; Mg will oxidize to Mg^{2+}. The oxidation state of hydrogen in HCl is $+1$. To be reduced, the oxidation state of H must decrease. The obvious choice for the hydrogen product is $H_2(g)$ where hydrogen has a zero oxidation state. The balanced reaction is: $Mg(s) + 2HCl(aq) \rightarrow MgCl_2(aq) + H_2(g)$. Mg goes from the 0 to the $+2$ oxidation state by losing two electrons. Each H atom goes from the $+1$ to the 0 oxidation state by gaining one electron. Since there are two H atoms in the balanced equation, then a total of two electrons are gained by the H atoms. Hence, two electrons are transferred in the balanced reaction. When the electrons are transferred directly from Mg to H^+, no work is obtained. In order to harness this reaction to do useful work, we must control the flow of electrons through a wire. This is accomplished by making a galvanic cell, which separates the reduction reaction from the oxidation reaction in order to control the flow of electrons through a wire to produce a voltage. **23.** An extensive property is one that depends on the amount of substance. The free energy change for a reaction depends on whether 1 mole of product is produced or 2 moles of product is produced or 1 million moles of product is produced. This is not the case for cell potentials, which do not depend on the amount of substance. The equation that relates ΔG to E is $\Delta G = -nFE$. It is the n term that converts the intensive property E into the extensive property ΔG. n is the number of moles of electrons transferred in the balanced reaction that ΔG is associated

with. **25.** A potential hazard when jump-starting a car is that the electrolysis of $H_2O(l)$ can occur. When $H_2O(l)$ is electrolyzed, the products are the explosive gas mixture of $H_2(g)$ and $O_2(g)$. A spark produced during jump-starting a car could ignite any $H_2(g)$ and $O_2(g)$ produced. Grounding the jumper cable far from the battery minimizes the risk of a spark nearby the battery where $H_2(g)$ and $O_2(g)$ could be collecting. **27.** You need to know the identity of the metal so you know which molar mass to use. You need to know the oxidation state of metal ion in the salt so the moles of electrons transferred can be determined. And finally, you need to know the amount of current and the time the current was passed through the electrolytic cell. If you know these four quantities, then the mass of metal plated out can be calculated. **29. a.** $3I^-(aq) + 2H^+(aq) + ClO^-(aq) \rightarrow I_3^-(aq) + Cl^-(aq) + H_2O(l)$; **b.** $7H_2O(l) + 4H^+(aq) + 3As_2O_3(s) + 4NO_3^-(aq) \rightarrow 4NO(g) + 6H_3AsO_4(aq)$; **c.** $16H^+(aq) + 2MnO_4^-(aq) + 10Br^-(aq) \rightarrow 5Br_2(l) + 2Mn^{2+}(aq) + 8H_2O(l)$; **d.** $8H^+(aq) + 3CH_3OH(aq) + Cr_2O_7^{2-}(aq) \rightarrow 2Cr^{3+}(aq) + 3CH_2O(aq) + 7H_2O(l)$ **31. a.** $2H_2O(l) + Al(s) + MnO_4^-(aq) \rightarrow Al(OH)_4^-(aq) + MnO_2(s)$; **b.** $2OH^-(aq) + Cl_2(g) \rightarrow Cl^-(aq) + OCl^-(aq) + H_2O(l)$; **c.** $OH^-(aq) + H_2O(l) + NO_2^-(aq) + 2Al(s) \rightarrow NH_3(g) + 2AlO_2^-(aq)$ **33.** $4NaCl(aq) + 2H_2SO_4(aq) + MnO_2(s) \rightarrow 2Na_2SO_4(aq) + MnCl_2(aq) + Cl_2(g) + 2H_2O(l)$ **35.** The reducing agent causes reduction to occur; it does this by containing the species that is oxidized. Oxidation occurs at the anode, so the reducing agent will be in the anode compartment. The oxidizing agent causes oxidation to occur; it does this by containing the species that is reduced. Reduction occurs at the cathode, so the oxidizing agent will be in the cathode compartment. Electron flow is always from the anode compartment to the cathode compartment. **37.** See Fig. 17-3 of the text for a typical galvanic cell. The anode compartment contains the oxidation half-reaction compounds/ions, and the cathode compartment contains the reduction half-reaction compounds/ions. The electrons flow from the anode to the cathode. For each of the following answers, all solutes are 1.0 M and all gases are at 1.0 atm. **a.** $7H_2O(l) + 2Cr^{3+}(aq) + 3Cl_2(g) \rightarrow Cr_2O_7^{2-}(aq) + 6Cl^-(aq) + 14H^+(aq)$; cathode: Pt electrode; Cl_2 bubbled into solution, Cl^- in solution; anode: Pt electrode; Cr^{3+}, H^+, and $Cr_2O_7^{2-}$ in solution; **b.** $Cu^{2+}(aq) + Mg(s) \rightarrow Cu(s) + Mg^{2+}(aq)$; cathode: Cu electrode; Cu^{2+} in solution; anode: Mg electrode; Mg^{2+} in solution **39. a.** 0.03 V; **b.** 2.71 V **41.** See Exercise 37 for a description of a galvanic cell. For each of the following answers, all solutes are 1.0 M and all gases are at 1.0 atm. In the salt bridge, cations flow to the cathode and anions flow to the anode. **a.** $Cl_2(g) + 2Br^-(aq) \rightarrow Br_2(aq) + 2Cl^-(aq)$, $\mathscr{E}° = 0.27$ V; cathode: Pt electrode; $Cl_2(g)$ bubbled in, Cl^- in solution; anode: Pt electrode; Br_2 and Br^- in solution; **b.** $3H_2O(l) + 5IO_4^-(aq) + 2Mn^{2+}(aq) \rightarrow 5IO_3^-(aq) + 2MnO_4^-(aq) + 6H^+(aq)$, $\mathscr{E}° = 0.09$ V; cathode: Pt electrode; IO_4^-, IO_3^-, and H_2SO_4 (as a source of H^+) in solution; anode: Pt electrode; Mn^{2+}, MnO_4^-, and H_2SO_4 in solution **43.** 37a. $Pt|Cr^{3+}$ (1.0 M), H^+ (1.0 M), $Cr_2O_7^{2-}$ (1.0 M)$||Cl_2$ (1.0 atm)$|Cl^-$ (1.0 M)$|Pt$; 37b. $Mg|Mg^{2+}$ (1.0 M)$||Cu^{2+}$ (1.0 M)$|Cu$; 41a. $Pt|Br^-$ (1.0 M), Br_2 (1.0 M)$||Cl_2$ (1.0 atm)$|Cl^-$ (1.0 M)$|Pt$; 41b. $Pt|Mn^{2+}$ (1.0 M), MnO_4^- (1.0 M), H^+ (1.0 M)$||IO_4^-$ (1.0 M), H^+ (1.0 M), IO_3^- (1.0 M) $|Pt$ **45. a.** $Au^{3+}(aq) + 3Cu^+(aq) \rightarrow 3Cu^{2+}(aq) + Au(s)$, $\mathscr{E}° = 1.34$ V; **b.** $2VO_2^+(aq) + 4H^+(aq) + Cd(s) \rightarrow Cd^{2+}(aq) + 2VO^{2+}(aq) + 2H_2O(l)$, $\mathscr{E}° = 1.40$ V **47. a.** $16H^+ + 2MnO_4^- + 10I^- \rightarrow 5I_2 + 2Mn^{2+} + 8H_2O$, $\mathscr{E}_{cell} = 0.97$ V, spontaneous; **b.** $16H^+ + 2MnO_4^- + 10F^- \rightarrow 5F_2 + 2Mn^{2+} + 8H_2O$, $\mathscr{E}_{cell} = -1.36$ V, not spontaneous **49.** $\mathscr{E}° = 0.41$ V, $\Delta G° = -79$ kJ **51.** 45a. -388 kJ; 45b. $-270.$ kJ **53.** -0.829 V; the two values agree to two significant figures. **55.** 1.24 V **57.** $K^+ < H_2O < Cd^{2+} < I_2 < AuCl_4^- < IO_3^-$ **59. a.** no; **b.** yes; **c.** yes; **61. a.** Of the species available, Ag^+ would be the best oxidizing agent because it has the largest $\mathscr{E}°$ value. **b.** Of the species available, Zn would be the best reducing agent because it has the largest $-\mathscr{E}°$ value. **c.** $SO_4^{2-}(aq)$ can oxidize Pb(s) and Zn(s) at standard conditions. When $SO_4^{2-}(aq)$ is coupled with these reagents, $\mathscr{E}_{cell}$ is positive. **d.** Al(s) can reduce $Ag^+(aq)$ and $Zn^{2+}(aq)$ at standard conditions because $\mathscr{E}_{cell} > 0$. **63. a.** $Cr_2O_7^{2-}$, O_2, MnO_2, IO_3^-; **b.** $PbSO_4$, Cd^{2+}, Fe^{2+}, Cr^{3+}, Zn^{2+}, H_2O **65.** $ClO^-(aq) + 2NH_3(aq) \rightarrow Cl^-(aq) + N_2H_4(aq) + H_2O(l)$, $\mathscr{E}_{cell} = 1.00$ V; because $\mathscr{E}_{cell}$ is positive for this reaction, at standard conditions ClO^- can spontaneously oxidize NH_3 to the somewhat toxic N_2H_4. **67. a.** larger; **b.** smaller **69.** Electron flow is always from the anode to the cathode. For the cells with a nonzero cell potential, we will identify the cathode, which means the other compartment is the anode. **a.** 0; **b.** 0.018 V; compartment with $[Ag^+] = 2.0$ M is cathode; **c.** 0.059 V; compartment with $[Ag^+] = 1.0$ M is cathode; **d.** 0.26 V; compartment with $[Ag^+] = 1.0$ M is cathode; **e.** 0   **71.** 2.12 V   **73.** 1.09 V   **75.** [37]. **a.** $\Delta G^\circ = -20$ kJ; 1×10^3; **b.** $\Delta G^\circ = -523$ kJ; 5.12×10^{91}; [41]. **a.** $\Delta G^\circ = -52$ kJ; 1.4×10^9; **b.** $\Delta G^\circ = -90$ kJ; 2×10^{15} **77. a.** $Fe^{2+}(aq) + Zn(s) \rightarrow Zn^{2+}(aq) + Fe(s)$ $\mathscr{E}_{cell} = 0.32$ V; **b.** -62 kJ; 6.8×10^{10}; **c.** 0.20 V   **79. a.** 0.23 V; **b.** 1.2×10^{-5} M **81.** 0.16 V, copper is oxidized.   **83.** 1.7×10^{-30} **85. a.** no reaction; **b.** $Cl_2(g) + 2I^-(aq) \rightarrow I_2(s) + 2Cl^-(aq)$, $\mathscr{E}_{cell} = 0.82$ V; $\Delta G^\circ = -160$ kJ; $K = 5.6 \times 10^{27}$; **c.** no reaction; **d.** $4Fe^{2+}(aq) + 4H^+(aq) + O_2(g) \rightarrow 4Fe^{3+}(aq) + 2H_2O(l)$, $\mathscr{E}_{cell} = 0.46$ V; $\Delta G^\circ = -180$ kJ; $K = 1.3 \times 10^{31}$;   **87.** 0.151 V; -29.1 kJ   **89.** 5.1×10^{-20} **91.** -0.14 V   **93. a.** 30. hours; **b.** 33 s; **c.** 1.3 hours   **95. a.** 16 g; **b.** 25 g; **c.** 71 g; **d.** 4.9 g   **97.** Bi   **99.** 9.12 L F_2 (anode), 29.2 g K (cathode)   **101.** 1×10^5 A   **103.** 1.14×10^{-2} M **105.** Au followed by Ag followed by Ni followed by Cd   **107.** Cathode: $2H_2O + 2e^- \rightarrow H_2(g) + 2OH^-$; anode: $2H_2O \rightarrow O_2(g) + 4H^+ + 4e^-$ **109. a.** cathode: $Ni^{2+} + 2e^- \rightarrow Ni$; anode: $2Br^- \rightarrow Br_2 + 2e^-$; **b.** cathode: $Al^{3+} + 3e^- \rightarrow Al$; anode: $2F^- \rightarrow F_2 + 2e^-$; **c.** cathode: $Mn^{2+} + 2e^- \rightarrow Mn$; anode: $2I^- \rightarrow I_2 + 2e^-$ **111. a.** cathode: $Ni^{2+} + 2e^- \rightarrow Ni$; anode: $2Br^- \rightarrow Br_2 + 2e^-$; **b.** cathode: $2H_2O + 2e^- \rightarrow H_2 + 2OH^-$; anode: $2H_2O \rightarrow O_2 + 4H^+ + 4e^-$; **c.** cathode: $2H_2O + 2e^- \rightarrow H_2 + 2OH^-$; anode: $2I^- \rightarrow I_2 + 2e^-$ **113.** 0.250 mol   **115. a.** 0.10 V, SCE is anode; **b.** 0.53 V, SCE is anode; **c.** 0.02 V, SCE is cathode; **d.** 1.90 V, SCE is cathode; **e.** 0.47 V, SCE is cathode   **117. a.** decrease; **b.** increase; **c.** decrease; **d.** decrease; **e.** same   **119. a.** $\Delta G^\circ = -582$ kJ; $K = 3.45 \times 10^{102}$; $\mathscr{E}^\circ = 1.01$ V; **b.** -0.65 V   **121.** Aluminum has the ability to form a durable oxide coating over its surface. Once the HCl dissolves this oxide coating, Al is exposed to H^+ and is easily oxidized to Al^{3+}. Thus, the Al foil disappears after the oxide coating is dissolved.   **123.** 1.14 V   **125.** $w_{max} = -13,200$ kJ; the work done can be no larger than the free energy change. If the process were reversible all of the free energy released would go into work, but this does not occur in any real process. Fuel cells are more efficient in converting chemical energy to electrical energy; they are also less massive. Major disadvantage: They are expensive.   **127.** 0.98 V   **129.** 7.44×10^4 A   **131.** To produce 1.0 kg Al by the Hall-Heroult process requires 5.4×10^4 kJ of energy. To melt 1.0 kg Al requires 4.0×10^2 kJ of energy. It is feasible to recycle Al by melting the metal because, in theory, it takes less than 1% of the energy required to produce the same amount of Al by the Hall-Heroult process.   **133.** $+3$ **135. a.** 1.93 V; **b.** Fe^{2+}; **c.** The anode would be composed of an La electrode and 1.0 M La^{3+}. **d.** The electrons flow from the La/La^{3+} compartment to the Fe^{2+}/Fe (from the anode to the cathode). **e.** Six electrons are transferred per unit of cell reaction. **f.** 1.87 V   **137. a.** 3.87 V; **b.** 0.40 M **139.** 3.97 A   **141.** $\mathscr{E}^\circ = \dfrac{T\Delta S^\circ}{nF} - \dfrac{\Delta H^\circ}{nF}$; if we graph $\mathscr{E}^\circ$ versus T, we should get a straight line ($y = mx + b$). The slope of the line is equal to $\Delta S^\circ/nF$ and the y-intercept is equal to $-\Delta H^\circ/nF$. $\mathscr{E}^\circ$ will have little temperature dependence for cell reactions with ΔS° close to zero.   **143.** 9.8×10^{-6} **145.** 2.39×10^{-7} **147. a.** ± 0.02 pH units; $\pm 6 \times 10^{-6}$ M H^+; **b.** ± 0.001 V   **149. a.** 0.16 V; **b.** 8.6 mol   **151.** $[Ag^+] = 4.6 \times 10^{-18}$ M; $[Ni^{2+}] = 1.5$ M **153.** 0.64 V   **155. a.** 5.77×10^{10}; **b.** -12.2 kJ/mol   **157.** Osmium(IV) nitrate; $[Ar]4s^1 3d^{10}$

Chapter 18

1. The characteristic frequencies of energies emitted in a nuclear reaction suggest that discrete energy levels exist in the nucleus. The extra stability of certain numbers of nucleons and the predominance of nuclei with even numbers of nucleons suggest that the nuclear structure might be described by using quantum numbers.   **3.** β-particle production has the net effect of turning a neutron into a proton. Radioactive nuclei having too many neutrons typically undergo β-particle decay. Positron production has the net effect of turning a proton into a neutron. Nuclei having too many protons typically undergo positron decay.   **5.** The transuranium elements are the elements having more protons than uranium. They are synthesized by bombarding heavier nuclei with neutrons and positive ions in a particle accelerator.   **7.** $\Delta E = \Delta mc^2$; The key difference is the mass change when going from reactants to

products. In chemical reactions, the mass change is indiscernible. In nuclear processes, the mass change is discernible. It is the conversion of this discernible mass change into energy that results in the huge energies associated with nuclear processes.   **9.** The temperatures of fusion reactions are so high that all physical containers would be destroyed. At these high temperatures, most of the electrons are stripped from the atoms. A plasma of gaseous ions is formed that can be controlled by magnetic fields.   **11. a.** $^3_1H \rightarrow {}^3_2He + {}^0_{-1}e$; **b.** $^8_3Li \rightarrow {}^8_4Be + {}^0_{-1}e$, $^8_4Be \rightarrow 2\,{}^4_2He$; overall reaction: $^8_3Li \rightarrow 2\,{}^4_2He + {}^0_{-1}e$; **c.** $^7_4Be + {}^0_{-1}e \rightarrow {}^7_3Li$; **d.** $^8_5B \rightarrow {}^8_4Be + {}^0_{+1}e$ **13. a.** $^{234}_{90}Th$; this is α-particle production. **b.** $^0_{-1}e$; this is β-particle production.   **15. a.** $^{68}_{31}Ga + {}^0_{-1}e \rightarrow {}^{68}_{30}Zn$; **b.** $^{62}_{29}Cu \rightarrow {}^0_{+1}e + {}^{62}_{28}Ni$; **c.** $^{212}_{87}Fr \rightarrow {}^4_2He + {}^{208}_{85}At$; **d.** $^{129}_{51}Sb \rightarrow {}^0_{-1}e + {}^{129}_{52}Te$ **17.** 7 α particles; 4 β particles   **19. a.** $^{241}_{95}Am \rightarrow {}^4_2He + {}^{237}_{93}Np$; **b.** $^{209}_{83}Bi$; **c.** The intermediate radionuclides are $^{237}_{93}Np$, $^{233}_{91}Pa$, $^{233}_{92}U$, $^{229}_{90}Th$, $^{225}_{88}Ra$, $^{225}_{89}Ac$, $^{221}_{87}Fr$, $^{217}_{85}At$, $^{213}_{83}Bi$, $^{213}_{84}Po$, and $^{209}_{82}Pb$.   **21.** $^{53}_{26}Fe$ has too many protons. It will undergo positron production, electron capture, and/or alpha-particle production. $^{59}_{26}Fe$ has too many neutrons and will undergo beta-particle production. (See Table 18-2 of the text.)   **23. a.** $^{249}_{98}Cf + {}^{18}_8O \rightarrow {}^{263}_{106}Sg + 4\,{}^1_0n$; **b.** $^{259}_{104}Rf$ **25.** 690 hours   **27.** ^{81}Kr is most stable since it has the longest half-life. ^{73}Kr is "hottest" since it decays very rapidly due to its very short half-life. ^{73}Kr, 81s; ^{74}Kr, 34.5 min; ^{76}Kr, 44.4 h; ^{81}Kr, 6.3×10^5 yr   **29.** 15 μg $^{47}CaCO_3$ should be ordered at a minimum.   **31.** 18.6%   **33.** The fraction that remains is 0.0041, or 0.41%.   **35.** 26 g   **37.** 2.3 counts per minute per gram of C. No; for a 10.-mg C sample, it would take roughly 40 min to see a single disintegration. This is too long to wait, and the background radiation would probably be much greater than the ^{14}C activity. Thus ^{14}C dating is not practical for very small samples.   **39.** 3.8×10^9 yr   **41.** 4.3×10^6 kg/s   **43.** ^{232}Pu, -1.715×10^{14} J/mol; ^{231}Pa, -1.714×10^{14} J/mol   **45.** ^{12}C: 1.230×10^{-12} J/nucleon; ^{235}U: 1.2154×10^{-12} J/nucleon; since ^{56}Fe is the most stable known nucleus, the binding energy per nucleon for ^{56}Fe would be larger than that of ^{12}C or ^{235}U. (See Fig. 18-9 of the text.)   **47.** 6.01513 u   **49.** -2.0×10^{10} J/g of hydrogen nuclei   **51.** The Geiger–Müller tube has a certain response time. After the gas in the tube ionizes to produce a "count," some time must elapse for the gas to return to an electrically neutral state. The response of the tube levels off because, at high activities, radioactive particles are entering the tube faster than the tube can respond to them.   **53.** Water is produced in this reaction by removing an OH group from one substance and an H from the other substance. There are two ways to do this:

i.

$$\underset{\text{CH}_3\text{C}}{\overset{\text{O}}{\|}}\!\!\!-\!\!\boxed{\text{OH} + \text{H}}\!\!-\!\!^{18}\text{OCH}_3 \longrightarrow \underset{\text{CH}_3\text{C}}{\overset{\text{O}}{\|}}\!\!\!-\!\!^{18}\text{OCH}_3 + \text{HO}\!-\!\text{H}$$

ii.

$$\underset{\text{CH}_3\text{CO}}{\overset{\text{O}}{\|}}\!\!\!-\!\!\boxed{\text{H} + \text{H}\,{}^{18}\text{O}}\!\!-\!\!\text{CH}_3 \longrightarrow \underset{\text{CH}_3\text{CO}}{\overset{\text{O}}{\|}}\!\!\!-\!\!\text{CH}_3 + \text{H}\!-\!{}^{18}\text{OH}$$

Because the water produced is not radioactive, methyl acetate forms by the first reaction, where all of the oxygen-18 ends up in methyl acetate.   **55.** 2 neutrons; 4 β particles   **57.** Strontium. Xe is chemically unreactive and not readily incorporated into the body. Sr can be easily oxidized to Sr^{2+}. Strontium is in the same family as calcium and could be absorbed and concentrated in the body in a fashion similar to Ca^{2+}. The chemical properties determine where radioactive material may be concentrated in the body or how easily it may be excreted.   **59. a.** unstable; beta production; **b.** stable; **c.** unstable; positron production or electron capture; **d.** unstable, positron production, electron capture, or alpha production.   **61.** 3800 decays/s   **63.** The third-life will be the time required for the number of nuclides to reach one-third of the original value ($N_0/3$). The third-life of this nuclide is 49.8 years.   **65.** 1975   **67.** 900 g ^{235}U **69.** 7×10^5 m/s; 8×10^{-16} J/nuclei;   **71.** All evolved $O_2(g)$ comes from water.   **73.** The equations for the nuclear reactions are:

$^{239}_{94}Pu \rightarrow {}^{235}_{92}U + {}^4_2He$; $^{214}_{82}Pb \rightarrow {}^{214}_{83}Bi + {}^0_{-1}e$; $^{60}_{27}Co \rightarrow {}^{60}_{28}Ni + {}^0_{-1}e$;

$^{99}_{43}Tc \rightarrow {}^{99}_{44}Ru + {}^0_{-1}e$; $^{239}_{93}Np \rightarrow {}^{239}_{94}Pu + {}^0_{-1}e$

75. 8.5 days   **77.** 3.177×10^{-11} J   **79.** 77% ^{238}U and 23% ^{235}U **81.** Assuming that (1) the radionuclide is long lived enough that no significant decay occurs during the time of the experiment, and (2) the total activity is uniformly distributed only in the rat's blood, $V = 10.$ mL.   **83. a.** $^{12}_6C$; **b.** ^{13}N, ^{13}C, ^{14}N, ^{15}O, and ^{15}N; **c.** -5.950×10^{11} J/mol ^1H **85.** 4.3×10^{-29} **87.** $^{249}_{97}Bk + ^{22}_{10}Ne \rightarrow ^{267}_{107}Bh + 4\ ^1_0n$; 62.7 s; $[Rn]7s^25f^{14}6d^5$

Chapter 19

1. The gravity of the earth cannot keep the light H_2 molecules in the atmosphere.   **3.** The acidity decreases. Solutions of Be^{2+} are acidic, while solutions of the other M^{2+} ions are neutral.   **5.** For Groups 1A–3A, the small size of H (as compared to Li), Be (as compared to Mg), and B (as compared to Al) seems to be the reason why these elements have nonmetallic properties, while others in Groups 1A–3A are strictly metallic. The small size of H, Be, and B also causes these species to polarize the electron cloud in nonmetals, thus forcing a sharing of electrons when bonding occurs. For Groups 4A–6A, a major difference between the first and second members of a group is the ability to form π bonds. The smaller elements form stable π bonds, while the larger elements do not exhibit good overlap between parallel p orbitals and, in turn, do not form strong π bonds. For Group 7A, the small size of F as compared to Cl is used to explain the low electron affinity of F and the weakness of the F—F bond.   **7.** In order to maximize hydrogen bonding interactions in the solid phase, ice is forced into an open structure. This open structure is why $H_2O(s)$ is less dense than $H_2O(l)$.   **9.** Group 1A and 2A metals are all easily oxidized. They must be produced in the absence of materials (H_2O, O_2) that are capable of oxidizing them.   **11. a.** $\Delta H° = 207$ kJ, $\Delta S° = 216$ J/K; **b.** $T > 958$ K   **13.** $4Li(s) + O_2(g) \rightarrow 2Li_2O(s)$; $2Li(s) + S(s) \rightarrow Li_2S(s)$; $2Li(s) + Cl_2(g) \rightarrow 2LiCl(s)$; $12Li(s) + P_4(s) \rightarrow 4Li_3P(s)$; $2Li(s) + H_2(g) \rightarrow 2LiH(s)$; $2Li(s) + 2H_2O(l) \rightarrow 2LiOH(aq) + H_2(g)$; $2Li(s) + 2HCl(aq) \rightarrow 2LiCl(aq) + H_2(g)$ **15.** When lithium reacts with excess oxygen, Li_2O forms, which is composed of Li^+ and O_2^- ions. This is called an oxide salt. When sodium reacts with oxygen, Na_2O_2 forms, which is composed of Na^+ and O_2^{2-} ions. This is called a peroxide salt. When potassium (or rubidium or cesium) reacts with oxygen, KO_2 forms, which is composed of K^+ and O_2^- ions. For your information, this is called a superoxide salt. So the three types of alkali metal oxides that can form differ in the oxygen anion part of the formula (O^{2-} vs. O_2^{2-} vs. O_2^-). Each of these anions have unique bonding arrangements and oxidation states.   **17.** The small size of the Li^+ cation results in a much greater attraction to water. The attraction to water is not so great for the other alkali metal ions. Thus lithium salts tend to absorb water.   **19.** $CaCO_3(s) + H_2SO_4(aq) \rightarrow CaSO_4(aq) + H_2O(l) + CO_2(g)$ **21.** 3.84×10^6 g Ba   **23.** Beryllium has a small size and a large electronegativity as compared with the other alkaline earth metals. The electronegativity of Be is so high that it does not readily give up electrons to nonmetals, as is the case for the other alkaline earth metals. Instead, Be has significant covalent character in its bonds; it prefers to share valence electrons rather than give them up to form ionic bonds.   **25.** element 113: $[Rn]7s^25f^{14}6d^{10}7p^1$; element 113 would fall below Tl in the periodic table. We would expect element 113, like Tl, to form $+1$ and $+3$ oxidation states in its compounds.   **27.** $B_2H_6(g) + 3O_2(g) \rightarrow 2B(OH)_3(s)$ **29.** $2Ga(s) + 3F_2(g) \rightarrow 2GaF_3(s)$; $4Ga(s) + 3O_2(g) \rightarrow 2Ga_2O_3(s)$; $2Ga(s) + 3S(s) \rightarrow Ga_2S_3(s)$; $2Ga(s) + 6HCl(aq) \rightarrow 2GaCl_3(aq) + 3H_2(g)$ **31.** An amphoteric substance is one that can behave as either an acid or a base. Al_2O_3 dissolves in both acidic and basic solutions. The reactions are $Al_2O_3(s) + 6H^+(aq) \rightarrow 2Al^{3+}(aq) + 3H_2O(l)$ and $Al_2O_3(s) + 2OH^-(aq) + 3H_2O(l) \rightarrow 2Al(OH)_4^-(aq)$.   **33.** Compounds containing Si—Si single and multiple bonds are rare, unlike compounds of carbon. The bond strengths of the Si—Si and C—C single bonds are similar. The difference in bonding properties must be for other reasons. One reason is that silicon does not form strong pi bonds, unlike carbon. Another reason is that silicon forms particularly strong sigma bonds to oxygen, resulting in compounds with Si—O bonds instead of Si—Si bonds.   **35.** O=C=O The darker green orbitals about carbon are sp hybrid orbitals. The lighter green orbitals about each oxygen are sp^2 hybrid orbitals, and the gold orbitals about all of the atoms are unhybridized p atomic orbitals. In each double bond in CO_2, one σ bond and one π bond exist. The two carbon–oxygen σ bonds are formed from overlap of sp hybrid orbitals from carbon with an sp^2 hybrid orbital from each oxygen. The two carbon–oxygen π bonds are formed from side-to-side overlap of the unhybridized p atomic orbitals from carbon with an unhybridized p atomic orbital from each oxygen, as illustrated in the figure.   **37. a.** $SiO_2(s) + 2C(s) \rightarrow Si(s) + 2CO(g)$; **b.** $SiCl_4(l) + 2Mg(s) \rightarrow Si(s) + 2MgCl_2(s)$; **c.** $Na_2SiF_6(s) + 4Na(s) \rightarrow Si(s) + 6NaF(s)$ **39.** 2:1   **41.** Nitrogen's small size does not provide room for all four oxygen atoms, making NO_4^{3-} unstable. Phosphorus is larger so PO_4^{3-} is more stable. To form NO_3^-, a pi bond must form. Phosphorus doesn't form strong pi bonds as readily as nitrogen.   **43.** $2Bi_2S_3(s) + 9O_2(g) \rightarrow 2Bi_2O_3(s) + 6SO_2(g)$; $2Bi_2O_3(s) + 3C(s) \rightarrow 4Bi(s) + 3CO_2(g)$; $2Sb_2S_3(s) + 9O_2(g) \rightarrow 2Sb_2O_3(s) + 6SO_2(g)$; $2Sb_2O_3(s) + 3C(s) \rightarrow 4Sb(s) + 3CO_2(g)$

45.

Trigonal bipyramid; dsp^3

Trigonal bipyramid; dsp^3

Octahedral; d^2sp^3

Nitrogen does not have low-energy d orbitals it can use to expand its octet. Both NF_5 and NCl_6^- would require nitrogen to have more than 8 valence electrons around it; this never happens.   **47.** $\frac{1}{2}N_2(g) + \frac{1}{2}O_2(g) \rightarrow NO(g)$; $\Delta G° = \Delta G°_{f(NO)}$; NO (and some other oxides of nitrogen) have weaker bonds as compared with the triple bond of N_2 and the double bond of O_2. Because of this, NO (and some other oxides of nitrogen) has a higher (positive) standard free energy of formation as compared to the relatively stable N_2 and O_2 molecules.   **49.** The pollution provides nitrogen and phosphorus nutrients so the algae can grow. The algae consume dissolved oxygen, causing fish to die.   **51.** The acidic hydrogens in the oxyacids of phosphorus all are bonded to oxygen. The hydrogens bonded directly to phosphorus are not acidic. H_3PO_4 has three oxygen-bonded hydrogens, and it is a triprotic acid. H_3PO_3 has only two of the hydrogens bonded to oxygen, and it is a diprotic acid. The third oxyacid of phosphorus, H_3PO_2, has only one of the hydrogens bonded to an oxygen; it is a monoprotic acid.   **53.** 821 nm   **55.** $H_2SeO_4(aq) + 3SO_2(g) \rightarrow Se(s) + 3SO_3(g) + H_2O(l)$ **57.** In the upper atmosphere, O_3 acts as a filter for ultraviolet (UV) radiation:

$$O_3 \xrightarrow{h\nu} O_2 + O$$

O_3 is also a powerful oxidizing agent. It irritates the lungs and eyes, and at high concentration, it is toxic. The smell of a "spring thunderstorm" is O_3 formed during lightning discharges. Toxic materials don't necessarily smell bad. For example, HCN smells like almonds.   **59.** The M.O. electron configuration of O_2 has two unpaired electrons in the degenerate pi antibonding (π^*_{2p}) orbitals. A substance with unpaired electrons is paramagnetic.   **61.** From the following Lewis structure, each oxygen atom has a tetrahedral arrangement of electron pairs. Therefore, bond angles are $\approx 109.5°$, and each O is sp^3 hybridized.

| Formal charge: | 0 | 0 | 0 | 0 |
| Oxidation state: | -1 | $+1$ | $+1$ | -1 |

Oxidation states are more useful. We are forced to assign $+1$ as the oxidation state for oxygen. Oxygen is very electronegative, and $+1$ is not a stable oxidation state for this element.

63.

V-shaped; <109.5°

See-saw; ≈120°, ≈90°   Octahedral; 90°

OF$_4$ would not form. This compound would require oxygen to have more than 8 valence electrons around it. This never occurs for oxygen; oxygen does not have low-energy d orbitals it can use to expand its octet.   **65.** The oxyacid strength increases as the number of oxygens in the formula increases. Therefore, the order of the oxyacids from weakest to strongest acid is HOCl < HClO$_2$ < HClO$_3$ < HClO$_4$.   **67. a.** IO$_4^-$; **b.** IO$_3^-$; **c.** IF$_2^-$; **d.** IF$_4^-$; **e.** IF$_6^-$ **69.** Helium is unreactive and doesn't combine with any other elements. It is a very light gas and would easily escape the earth's gravitational pull as the planet was formed.   **71.** One would expect RnF$_2$, RnF$_4$, and maybe RnF$_6$ to form in fashion similar to XeF$_2$, XeF$_4$, and XeF$_6$. The chemistry of radon is difficult to study because radon isotopes are all radioactive. The hazards of dealing with radioactive materials are immense.   **73.** N$_2$H$_4$(l) + O$_2$(g) → N$_2$(g) + 2H$_2$O(g); $\Delta H = -590.$ kJ   **75.** If the compound contained Ga(II), it would be paramagnetic and if the compound contained Ga(I) and Ga(III), it would be diamagnetic. Paramagnetic compounds have an apparent greater mass in a magnetic field.   **77.** 9.3 × 10^{-5} mol/L   **79. a.** AgCl(s) $\xrightarrow{h\nu}$ Ag(s) + Cl; the reactive chlorine atom is trapped in the crystal. When light is removed, Cl reacts with silver atoms to re-form AgCl; i.e., the reverse reaction occurs. In pure AgCl, the Cl atoms escape, making the reverse reaction impossible. **b.** Over time, chlorine is lost and the dark silver metal is permanent.   **81.** 6.5 × 10^{27} atoms **83.** Strontium and calcium are both alkaline earth metals, so both have similar chemical properties. Since milk is a good source of calcium, strontium could replace some calcium in milk without much difficulty.   **85.** +6 oxidation state: SO$_4^{2-}$, SO$_3$, SF$_6$; +4 oxidation state: SO$_3^{2-}$, SO$_2$, SF$_4$; +2 oxidation state: SCl$_2$; 0 oxidation state: S$_8$ and all other elemental forms of sulfur; −2 oxidation state: H$_2$S, Na$_2$S   **87. a.** $\Delta H° = 132$ kJ; $\Delta S° = 134$ J/K; **b.** 985 K   **89.** 0.013 mol/L   **91. a.** sp; **b.** sp^2; **c.** sp^2; **d.** sp **93. a.** S is d^2sp^3 hybridized; **b.** Cl is dsp^3 hybridized; **c.** Ge is sp^3 hybridized; **d.** Xe is d^2sp^3 hybridized.   **95. a.** 0.17 g; **b.** 5 × 10^{15} atoms   **97.** Ca; 12.698 **99.** I   **101.** For the reaction

the activation energy must in some way involve the breaking of a nitrogen–nitrogen single bond. For the reaction

at some point nitrogen–oxygen bonds must be broken. N—N single bonds (160 kJ/mol) are weaker than N—O single bonds (201 kJ/mol). In addition, resonance structures indicate that there is more double-bond character in the N—O bonds than in the N—N bond. Thus NO$_2$ and NO are preferred by kinetics because of the lower activation energy.   **103.** 5.89   **105.** 20. g **107. a.** 7.1 g; **b.** 979 nm; this electromagnetic radiation is not visible to humans; it is in the infrared region of the electromagnetic radiation spectrum; **c.** n-type   **109. a.** +6; **b.** 4.42

Chapter 20

5. The lanthanide elements are located just before the 5d transition metals. The lanthanide contraction is the steady decrease in the atomic radii of the lanthanide elements when going from left to right across the periodic table. As a result of the lanthanide contraction, the sizes of the 4d and 5d elements are very similar. This leads to a greater similarity in the chemistry of the 4d and 5d elements in a given vertical group.   **7.** No; both the *trans* and the *cis* forms of Co(NH$_3$)$_4$Cl$_2^+$ have mirror images that are superimposable. For the *cis* form, the mirror image only needs a 90° rotation to produce the original structure. Hence, neither the *trans* nor the *cis* form is optically active. **9.** Fe$_2$O$_3$(s) + 6H$_2$C$_2$O$_4$(aq) → 2Fe(C$_2$O$_4$)$_3^{3-}$(aq) + 3H$_2$O(l) + 6H$^+$(aq); the oxalate anion forms a soluble complex ion with iron in rust (Fe$_2$O$_3$), which allows rust stains to be removed.

11. a.

small Δ

CoCl$_4^{2-}$ is an example of a weak-field case having three unpaired electrons.

b.

large Δ

CN$^-$ is a strong-field ligand so Co(CN)$_6^{3-}$ will be a low-spin case having zero unpaired electrons.

13. From Table 20-16, the red octahedral Co(H$_2$O)$_6^{2+}$ complex ion absorbs blue-green light ($\lambda \approx 490$ nm), whereas the blue tetrahedral CoCl$_4^{2-}$ complex ion absorbs orange light ($\lambda \approx 600$ nm). Because tetrahedral complexes have a d-orbital splitting much less than octahedral complexes, one would expect the tetrahedral complex to have a smaller energy difference between split d orbitals. This translates into longer-wavelength light absorbed ($E = hc/\lambda$) for tetrahedral complex ions compared to octahedral complex ions. Information from Table 20-16 confirms this.   **15.** SCN$^-$, NO$_2^-$, and OCN$^-$ can form linkage isomers; all are able to bond to the metal ion in two different ways.   **17.** Sc^{3+} has no electrons in d orbitals. Ti^{3+} and V^{3+} have d electrons present. The color of transition metal complexes results from electron transfer between split d orbitals. If no d electrons are present, no electron transfer can occur, and the compounds are not colored.   **19.** At high altitudes, the oxygen content of air is lower, so less oxyhemoglobin is formed which diminishes the transport of oxygen in the blood. A serious illness called high-altitude sickness can result from the decrease of O$_2$ in the blood. High-altitude acclimatization is the phenomenon that occurs in the human body in response to the lower amounts of oxyhemoglobin in the blood. This response is to produce more hemoglobin, and, hence, increase the oxyhemoglobin in the blood. High-altitude acclimatization takes several weeks to take hold for people moving from lower altitudes to higher altitudes.   **21. a.** Ni: [Ar]4$s^2$3d^8; **b.** Cd: [Kr]5$s^2$4d^{10}; **c.** Zr: [Kr]5$s^2$4d^2; **d.** Os: [Xe]6$s^2$4f^{14}5d^6 **23. a.** Ti: [Ar]4$s^2$3d^2; Ti^{2+}: [Ar]3d^2; Ti^{4+}: [Ne]3$s^2$3p^6 or [Ar]; **b.** Re: [Xe]6$s^2$4f^{14}5d^5; Re^{2+}: [Xe]4f^{14}5d^5; Re^{3+}: [Xe]4f^{14}5d^4; **c.** Ir: [Xe]6$s^2$4f^{14}5d^7; Ir^{2+}: [Xe]4f^{14}5d^7; Ir^{3+}: [Xe]4f^{14}5d^6 **25. a.** Fe^{3+}: [Ar]3d^5; **b.** Ag$^+$: [Kr]4d^{10}; **c.** Ni^{2+}: [Ar]3d^8; **d.** Cr^{3+}: [Ar]3d^3 **27. a.** molybdenum(IV) sulfide, molybdenum(VI) oxide; **b.** MoS$_2$, +4; MoO$_3$, +6; (NH$_4$)$_2$Mo$_2$O$_7$, +6; (NH$_4$)$_6$Mo$_7$O$_{24}$ · 4H$_2$O, +6   **29.** NH$_3$ is a weak base that produces OH$^-$ ions in solution. The white precipitate is Cu(OH)$_2$(s). Cu^{2+}(aq) + 2OH$^-$(aq) → Cu(OH)$_2$(s); with excess NH$_3$ present, Cu^{2+} forms a soluble complex ion, Cu(NH$_3$)$_4^{2+}$. Cu(OH)$_2$(s) + 4NH$_3$(aq) → Cu(NH$_3$)$_4^{2+}$(aq) + 2OH$^-$(aq)   **31.** [Co(NH$_3$)$_6$]I$_3$: 3 moles of AgI; [Pt(NH$_3$)$_4$]I$_2$: 2 moles of AgI; Na$_2$[PtI$_6$]: 0 moles of AgI; [Cr(NH$_3$)$_4$I$_2$]I: 1 mole of AgI **33. a.** hexacyanomanganate(II) ion; **b.** *cis*-tetraamminedichlorocobalt(III) ion; **c.** pentaamminechlorocobalt(II) ion   **35. a.** hexaamminecobalt(II) chloride; **b.** hexaaquacobalt(III) iodide; **c.** potassium tetrachloroplatinate(II); **d.** potassium hexachloroplatinate(II); **e.** pentaamminechlorocobalt(III) chloride; **f.** triamminetrinitrocobalt(III)   **37. a.** K$_2$[CoCl$_4$]; **b.** [Pt(H$_2$O)(CO)$_3$]Br$_2$; **c.** Na$_3$[Fe(CN)$_2$(C$_2$O$_4$)$_2$]; **d.** [Cr(NH$_3$)$_3$Cl(H$_2$NCH$_2$CH$_2$NH$_2$)]I$_2$

39. a.

cis   trans

b.

cis   trans

c.

d.

en = N⌒N=NH₂CH₂CH₂NH₂

41. monodentate   bidentate   bridging

43. a. 2; **b.** 3; **c.** 4; **d.** 4

45.

47. Cr(acac)₃ and cis-Cr(acac)₂(H₂O)₂ are optically active.   **49.** With five electrons each in a different orbital, diagram (a) is for the weak-field Fe(H₂O)₆³⁺ complex ion. With three electrons, diagram (b) is for the Cr(NH₃)₅Cl²⁺ complex ion. With six electrons all paired up, diagram (c) is for the strong-field Co(NH₃)₄Br₂⁺ complex ion.

51. a. Fe²⁺

High spin   Low spin

b. Fe³⁺   **c.** Ni²⁺

High spin

53. weak field   **55. a.** 0; **b.** 2; **c.** 2   **57.** Co(CN)₆³⁻ < Co(en)₃³⁺ < Co(H₂O)₆³⁺ < CoI₆³⁻   **59.** The violet complex ion absorbs yellow-green light ($\lambda \approx 570$ nm), the yellow complex ion absorbs blue light ($\lambda \approx 450$ nm),

and the green complex ion absorbs red light ($\lambda \approx 650$ nm). The violet complex ion is Cr(H₂O)₆³⁺, the yellow complex ion is Cr(NH₃)₆³⁺, and the green complex ion is Cr(H₂O)₄Cl₂⁺.   **61.** CoBr₄²⁻ is a tetrahedral complex ion, while CoBr₆⁴⁻ is an octahedral complex ion. Since tetrahedral d-orbital splitting is less than one-half the octahedral d-orbital splitting, the octahedral complex ion (CoBr₆⁴⁻) will absorb higher-energy light, which will have a shorter wavelength than 3.4×10^{-6} m ($E = hc/\lambda$).   **63.** 5   **65. a.** -11 kJ; **b.** $\Delta H° = 172.5$ kJ; $\Delta S° = 176$ J/K; $T > 980$. K   **67.** Fe₂O₃: iron has a $+3$ oxidation state; Fe₃O₄: iron has a $+8/3$ oxidation state. The three iron ions in Fe₃O₄ must have a total charge of $+8$. The only combination that works is to have two Fe³⁺ ions and one Fe²⁺ ion per formula unit. This makes sense from the other formula for magnetite, FeO · Fe₂O₃. FeO has an Fe²⁺ ion and Fe₂O₃ has two Fe³⁺ ions.   **69.** 8CN⁻(aq) + 4Ag(s) + O₂(g) + 2H₂O(l) → 4Ag(CN)₂⁻(aq) + 4OH⁻(aq)   **71.** The molecular formula is EuC₁₅H₂₁O₆. Because each acac⁻ is C₅H₇O₂⁻, an abbreviated molecular formula is Eu(acac)₃.   **73.** 0.66 V; -130 kJ; 2.2×10^{22} **75.** There are four geometrical isomers (labeled i–iv). Isomers iii and iv are optically active, and the nonsuperimposable mirror images are shown.

i.

ii.

iii.

optically active   mirror   mirror image of iii
(nonsuperimposable)

iv.

optically active   mirror   mirror image of iv
(nonsuperimposable)

77. Octahedral Cr²⁺ complexes should be used. Cr²⁺: [Ar]3d^4; high-spin (weak-field) Cr²⁺ complexes have four unpaired electrons and low-spin (strong-field) Cr²⁺ complexes have two unpaired electrons. Ni²⁺: [Ar]3d^8; octahedral Ni²⁺ complexes will always have two unpaired electrons, whether high or low spin. Therefore, Ni²⁺ complexes cannot be used to distinguish weak- from strong-field ligands by examining magnetic properties. Alternatively, the ligand field strengths can be measured using visible spectra. Either Cr²⁺ or Ni²⁺ complexes can be used for this method.   **79. a.** [Co(C₅H₅N)₆]Cl₃; **b.** [Cr(NH₃)₅I]I₂; **c.** [Ni(NH₂CH₂CH₂NH₂)₃]Br₂; **d.** K₂[Ni(CN)₄]; **e.** [Pt(NH₃)₄Cl₂][PtCl₄]   **81.** 60   **83. a.** [Ar]4$s^1$3d^{10} is correct for Cu; **b.** correct; **c.** correct; **d.** [Xe]6$s^2$5d^1 is correct for La; **e.** correct   **85.** Zn²⁺: [Ar]3d^{10}; Cu²⁺: [Ar]3d^9; Mn³⁺: [Ar]3d^4; Ti⁴⁺: [Ne]3$s^2$3p^6; color is a result of the electron transfer between split d orbitals. This cannot occur for the filled d orbitals in Zn²⁺. This also cannot occur for Ti⁴⁺, which has no d electrons. So Zn²⁺ and Ti⁴⁺ compounds/ions will not be colored. Cu²⁺ and Mn³⁺ do have d orbitals that are partially filled, so we would expect compounds/ions of Cu²⁺ and Mn³⁺ to be colored.   **87. a.** Zn²⁺ has a 3d^{10} electron configuration. This diagram is for a d^{10} octahedral complex ion, not a d^{10} tetrahedral complex as is given. So this diagram is incorrect. **b.** This diagram is correct. In this complex ion, Mn has a $+3$ oxidation state and an [Ar]3d^4 electron configuration. The strong field diagram for a d^4 ion is shown. **c.** This diagram is correct. In this complex ion, Ni²⁺ is present, which has a $3d^8$ electron configuration. The diagram shown is correct for a diamagnetic square planar complex.   **89.** $Ni^{2+} = d^8$; if ligands A and B produced very similar crystal fields, the $cis\text{-}[NiA_2B_4]^{2+}$ complex ion would give the following octahedral crystal field diagram for a d^8 ion:

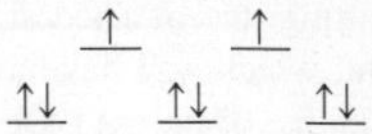

$$\underline{\uparrow} \qquad \underline{\uparrow}$$
$$\underline{\uparrow\downarrow} \qquad \underline{\uparrow\downarrow} \qquad \underline{\uparrow\downarrow}$$

This is paramagnetic.

Because it is given that the complex ion is diamagnetic, the A and B ligands must produce different crystal fields, giving a unique d-orbital splitting diagram that would result in a diamagnetic species.   **91. a.** -0.26 V; **b.** From standard reduction potentials, Co^{3+} ($\mathscr{E}° = 1.82$ V) is a much stronger oxidizing agent than $Co(en)_3^{3+}$ ($\mathscr{E}° = -0.26$ V); **c.** In aqueous solution, Co^{3+} forms the hydrated transition metal complex, $Co(H_2O)_6^{3+}$. In both complexes, $Co(H_2O)_6^{3+}$ and $Co(en)_3^{3+}$, cobalt exists as Co^{3+}, which has 6 d electrons. If we assume a strong-field case for each complex ion, then the d-orbital splitting diagram for each has the six electrons paired in the lower-energy t_{2g} orbitals. When each complex ion gains an electron, the electron enters the higher-energy e_g orbitals. Since en is a stronger-field ligand than H_2O, then the d-orbital splitting is larger for $Co(en)_3^{3+}$, and it takes more energy to add an electron to $Co(en)_3^{3+}$ than to $Co(H_2O)_6^{3+}$. Therefore, it is more favorable for $Co(H_2O)_6^{3+}$ to gain an electron than for $Co(en)_3^{3+}$ to gain an electron.   **93.** No, in all three cases six bonds are formed between Ni^{2+} and nitrogen. So ΔH values should be similar. $\Delta S°$ for formation of the complex ion is most negative for 6 NH_3 molecules reacting with a metal ion (7 independent species become 1). For penten reacting with a metal ion, 2 independent species become 1, so $\Delta S°$ is the least negative. Thus the chelate effect occurs because the more bonds a chelating agent can form to the metal, the less unfavorable $\Delta S°$ becomes for the formation of the complex ion, and the larger the formation constant.

95.
$$\underline{\quad} \qquad d_{z^2}$$
$$\underline{\quad} \quad \underline{\quad} \qquad d_{x^2-y^2}, d_{xy}$$
$$\underline{\quad} \quad \underline{\quad} \qquad d_{xz}, d_{yz}$$

97. The coordinate system for the complex ion is shown below. From the coordinate system, the CN^- ligands are in a square planar arrangement. Since CN^- produces a much stronger crystal field, the diagram will most resemble that of a square planar complex:

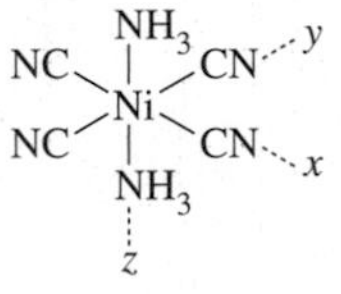

$$
\begin{aligned}
d_{x^2-y^2} &\quad \underline{\quad} \\
d_{z^2} &\quad \underline{\uparrow\downarrow} \\
d_{xy} &\quad \underline{\uparrow\downarrow} \\
d_{xz} &\quad \underline{\uparrow\downarrow} \;\; \underline{\uparrow\downarrow} \;\; d_{yz}
\end{aligned}
$$

With the NH_3 ligands on the z axis, we will assume that the d_{z^2} orbital is destabilized more than the d_{xy} orbital. This may or may not be the case.
99. $[Cr(NH_3)_5I]I_2$; octahedral

Chapter 21

1. a. 1-sec-butylpropane

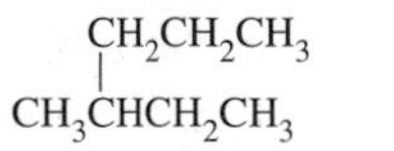

$$CH_3CHCH_2CH_3$$
$$\quad\; | $$
$$\quad CH_2CH_2CH_3$$

3-methylhexane is correct.

b. 4-methylhexane

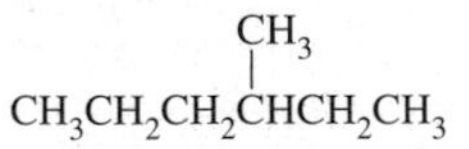

$$CH_3CH_2CH_2CHCH_2CH_3$$
$$\qquad\qquad\qquad | $$
$$\qquad\qquad\qquad CH_3$$

3-methylhexane is correct.

c. 2-ethylpentane

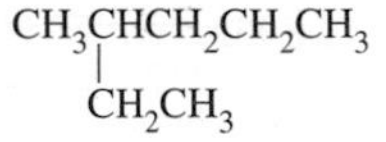

$$CH_3CHCH_2CH_2CH_3$$
$$\quad\; | $$
$$\quad CH_2CH_3$$

3-methylhexane is correct.

d. 1-ethyl-1-methylbutane

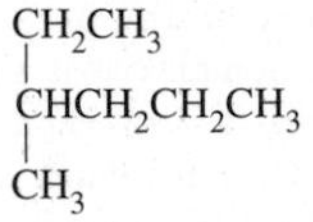

$$CH_2CH_3$$
$$| $$
$$CHCH_2CH_2CH_3$$
$$| $$
$$CH_3$$

3-methylhexane is correct.

e. 3-methylhexane

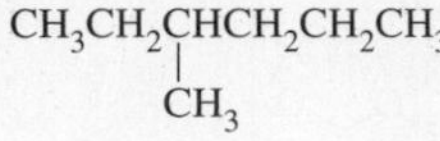

$$CH_3CH_2CHCH_2CH_2CH_3$$
$$\qquad\quad | $$
$$\qquad\quad CH_3$$

f. 4-ethylpentane

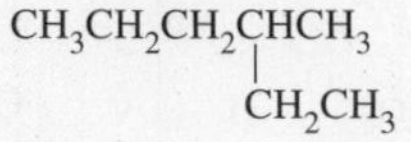

$$CH_3CH_2CH_2CHCH_3$$
$$\qquad\qquad | $$
$$\qquad\qquad CH_2CH_3$$

3-methylhexane is correct.

All six of these are the same compound. They only differ from each other by rotations about one or more carbon–carbon single bonds. Only one isomer of C_7H_{16} is present in all of these names: 3-methylhexane.

3. a.
$$CH_3CHCH_3$$
$$\quad\; | $$
$$\quad CH_2CH_3$$

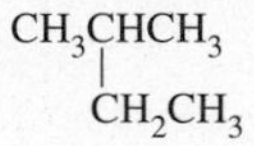

The longest chain is 4 carbons long. The correct name is 2-methylbutane.

b.
$$\qquad\qquad\qquad I \;\; CH_3$$
$$\qquad\qquad\qquad | \qquad | $$
$$CH_3CH_2CH_2CH_2C\!-\!CH_2$$
$$\qquad\qquad\qquad\qquad | $$
$$\qquad\qquad\qquad\qquad CH_3$$

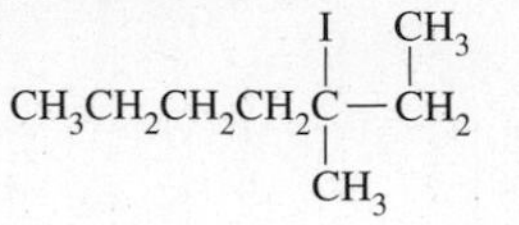

The longest chain is 7 carbons long and we would start the numbering system at the other end for lowest possible numbers. The correct name is 3-iodo-3-methylheptane.

c.
$$\qquad\qquad\qquad CH_3$$
$$\qquad\qquad\qquad | $$
$$CH_3CH_2CH\!=\!C\!-\!CH_3$$

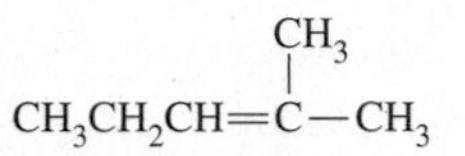

This compound cannot exhibit $cis-trans$ isomerism since one of the double-bonded carbons has the same two groups (CH_3) attached. The numbering system should also start at the other end to give the double bond the lowest possible number. 2-methyl-2-pentene is correct.

d.
$$\quad\; Br\; OH$$
$$\quad\; | \quad | $$
$$CH_3CHCHCH_3$$

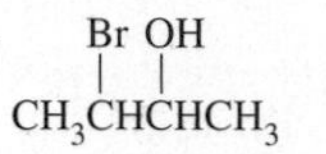

The OH functional group gets the lowest number. 3-bromo-2-butanol is correct.

5. Hydrocarbons are nonpolar substances exhibiting only London dispersion forces. Size and shape are the two most important structural features relating to the strength of London dispersion forces. For size, the bigger the molecule (the larger the molar mass), the stronger the London dispersion forces and the higher the boiling point. For shape, the more branching present in a compound, the weaker the London dispersion forces and the lower the boiling point.   **7.** The amide functional group is:

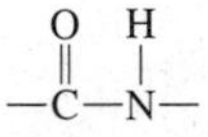

$$\quad O \quad H$$
$$\quad \| \quad\; | $$
$$-C\!-\!N-$$

When the amine end of one amino acid reacts with the carboxylic acid end of another amino acid, the two amino acids link together by forming an amide functional group. A polypeptide has many amino acids linked together, with each linkage made by the formation of an amide functional group. Because all linkages result in the presence of the amide functional group, the resulting polymer is called a polyamide. For nylon, the monomers also link together by forming the amide functional group (the amine end of one monomer reacts with a carboxylic acid end of another monomer to give the amide functional group linkage). Hence nylon is also a polyamide. The correct order of strength is polyhydrocarbon < polyester < polyamide. The difference in strength is related to the types of intermolecular forces present. All of these polymers have London dispersion forces. However, polyhydrocarbons only have London dispersion forces. The polar ester group in polyesters and the polar amide group in polyamides give rise to additional dipole forces. The polyamide has the ability to form relatively strong hydrogen bonding interactions, hence why it would form the strongest fibers.

9. a.
$$CH_2\!=\!CH_2 + H_2O \xrightarrow{H^+} \begin{array}{c} OH \;\; H \\ | \quad\; | \\ CH_2\!-\!CH_2 \end{array} \quad 1° \text{ alcohol}$$

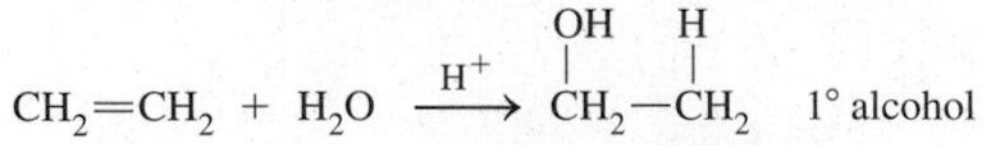

b. 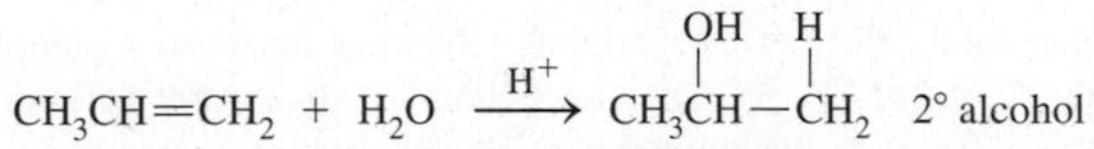
$$CH_3CH\!=\!CH_2 + H_2O \xrightarrow{H^+} \begin{array}{c} OH \;\; H \\ | \quad\; | \\ CH_3CH\!-\!CH_2 \end{array} \quad 2° \text{ alcohol}$$
$$\qquad\qquad\qquad\qquad\qquad\qquad\qquad \text{major product}$$

c.

$$CH_3C{=}CH_2 + H_2O \xrightarrow{H^+} CH_3\overset{\underset{|}{OH}}{C}{-}\overset{\underset{|}{H}}{CH_2} \quad 3° \text{ alcohol}$$
$$\underset{CH_3}{|} \qquad\qquad\qquad \underset{CH_3}{|}$$

major product

d.

$$CH_3CH_2OH \xrightarrow{\text{oxidation}} CH_3\overset{O}{\overset{\|}{C}}H \quad \text{aldehyde}$$

e.

$$CH_3\overset{\underset{|}{OH}}{C}HCH_3 \xrightarrow{\text{oxidation}} CH_3\overset{O}{\overset{\|}{C}}CH_3 \quad \text{ketone}$$

f.

$$CH_3CH_3CH_3OH \xrightarrow{\text{oxidation}} CH_3CH_2\overset{O}{\overset{\|}{C}}{-}OH \quad \text{carboxylic acid}$$

or

$$CH_3CH_2\overset{O}{\overset{\|}{C}}H \xrightarrow{\text{oxidation}} CH_3CH_2\overset{O}{\overset{\|}{C}}{-}OH$$

g.

$$CH_3OH + HO\overset{O}{\overset{\|}{C}}CH_3 \longrightarrow CH_3{-}O{-}\overset{O}{\overset{\|}{C}}CH_3 + H_2O \quad \text{ester}$$

11. a. A polyester forms when an alcohol functional group reacts with a carboxylic acid functional group. The monomer for a homopolymer polyester must have an alcohol functional group and a carboxylic acid functional group present in the structure. **b.** A polyamide forms when an amine functional group reacts with a carboxylic acid functional group. For a copolymer polyamide, one monomer would have at least two amine functional groups present and the other monomer would have at least two carboxylic acid functional groups present. For polymerization to occur, each monomer must have two reactive functional groups present. **c.** To form an addition polymer, a carbon–carbon double bond must be present. To form a polyester, the monomer would need the alcohol and carboxylic acid functional groups present. To form a polyamide, the monomer would need the amine and carboxylic acid functional groups present. The two possibilities are for the monomer to have a carbon–carbon double bond, an alcohol functional group, and a carboxylic acid functional group all present, or to have a carbon–carbon double bond, an amine functional group, and a carboxylic acid functional group present.

13. $CH_3{-}CH_2{-}CH_2{-}CH_2{-}CH_2{-}CH_3$

$$CH_3{-}\overset{\underset{|}{CH_3}}{C}H{-}CH_2{-}CH_2{-}CH_3$$

$$CH_3{-}CH_2{-}\overset{\underset{|}{CH_3}}{C}H{-}CH_2{-}CH_3$$

$$CH_3{-}\overset{\overset{CH_3}{|}}{\underset{\underset{CH_3}{|}}{C}}{-}CH_2{-}CH_3$$

$$CH_3{-}\overset{\underset{|}{CH_3}}{C}H{-}\overset{\underset{|}{CH_3}}{C}H{-}CH_3$$

15. a.

$$CH_3\overset{\underset{|}{CH_3}}{C}HCH_2CH_2CH_2CH_3$$
2-methylheptane

$$CH_3CH_2\overset{\underset{|}{CH_3}}{C}HCH_2CH_2CH_3$$
3-methylheptane

$$CH_3CH_2CH_2\overset{\underset{|}{CH_3}}{C}HCH_2CH_2CH_3$$
4-methylheptane

b.

$$CH_3{-}\overset{\overset{CH_3}{|}}{\underset{\underset{CH_3}{|}}{C}}{-}\overset{\overset{CH_3}{|}}{\underset{\underset{CH_3}{|}}{C}}{-}CH_3$$
2,2,3,3-tetramethylbutane

17. a.
$$CH_3\overset{\underset{|}{CH_3}}{C}HCH_3$$

b.
$$CH_3\overset{\underset{|}{CH_3}}{C}HCH_2CH_3$$

c.
$$CH_3\overset{\underset{|}{CH_3}}{C}HCH_2CH_2CH_3$$

d.
$$CH_3\overset{\underset{|}{CH_3}}{C}HCH_2CH_2CH_2CH_3$$

19. a. $CH_3CH_2{-}CH{-}CH_2CH_2CH_3$
$$\underset{\underset{\underset{CH_3}{|}}{CH{-}CH_3}}{\overset{|}{CH_2}}$$

b.
$$CH_3{-}\overset{\overset{CH_3}{|}}{\underset{\underset{CH_3}{|}}{C}}{-}CH_2{-}\overset{\underset{CH_3}{|}}{C}H{-}CH_3$$

c. $CH_3{-}CH{-}CH_2CH_2CH_3$
$$\overset{\overset{CH_3}{|}}{\underset{\underset{CH_3}{|}}{C}}{-}CH_3$$

d. 4-ethyl-2-methylheptane; 2,2,3-trimethylhexane

21. a. 2,2,4-trimethylhexane; **b.** 5-methylnonane; **c.** 2,2,4,4-tetramethylpentane; **d.** 3-ethyl-3-methyloctane

23. $CH_3{-}CH_2{-}CH_2{-}CH_3$;

$$\begin{array}{c} H\ \ H \\ | \ \ | \\ H{-}C{-}C{-}H \\ | \ \ | \\ H{-}C{-}C{-}H \\ | \ \ | \\ H\ \ H \end{array}$$

Each carbon is bonded to four other atoms.

25. a. 1-butene; **b.** 4-methyl-2-hexene; **c.** 2,5-dimethyl-3-heptene

27. a. $CH_3CH_2CH{=}CHCH_2CH_3$; **b.** $CH_3CH{=}CHCH{=}CHCH_2CH_3$;

c. $CH_3\underset{\underset{CH_3}{|}}{C}HCH{=}CHCH_2CH_2CH_2CH_3$

29. a.

b.

c.

d.

31. a. 1,3-dichlorobutane; **b.** 1,1,1-trichlorobutane; **c.** 2,3-dichloro-2,4-dimethylhexane; **d.** 1,2-difluoroethane **33.** $CH_2Cl{-}CH_2Cl$, 1,2-dichloroethane: There is free rotation about the C—C single bond that doesn't lead to different compounds. $CHCl{=}CHCl$, 1,2-dichloroethene: There is no rotation about the C=C double bond. This creates the *cis* and *trans* isomers, which are different compounds. **35.** [25], compounds b and c; [27], all compounds

37. $CH_2=CHCH_2CH_2CH_3$ $CH_3CH=CHCH_2CH_3$

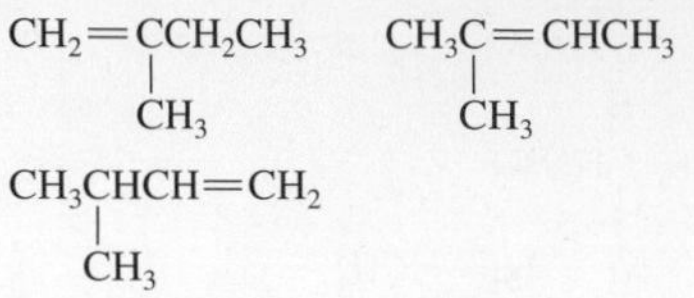

$CH_2=CCH_2CH_3$ $CH_3C=CHCH_3$
with CH_3 substituents

$CH_3CHCH=CH_2$
with CH_3 substituent

39.

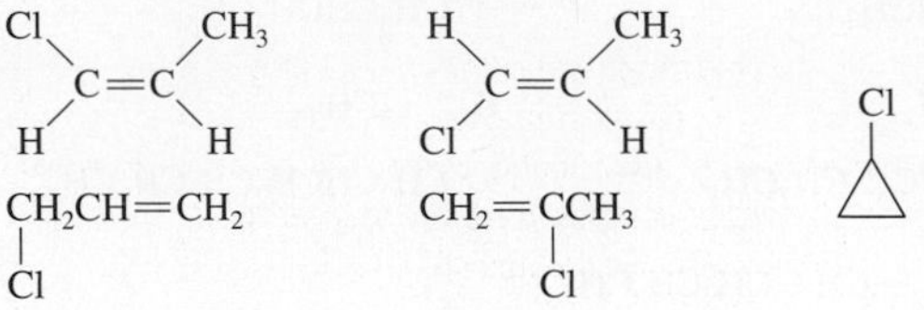

41.

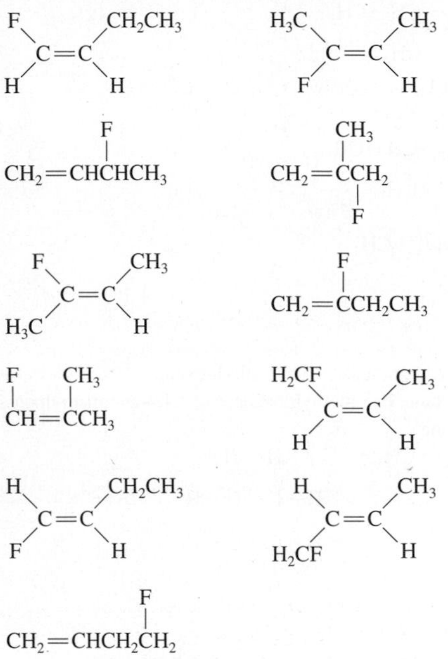

43. 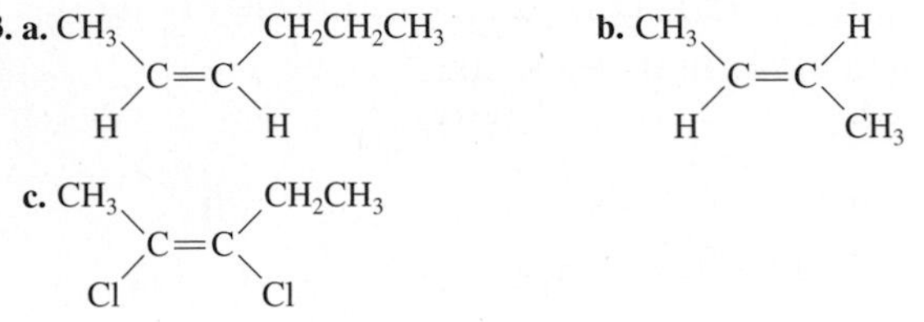 a.   b.

c.

45. a. 3 monochloro isomers of *n*-pentane; b. 4 monochloro isomers of 2-methylbutane; c. 3 monochloro isomers of 2,4-dimethylpentane; d. 4 monochloro isomers of methylcyclobutane   47. a. ketone; b. aldehyde; c. carboxylic acid; d. amine

49. a.

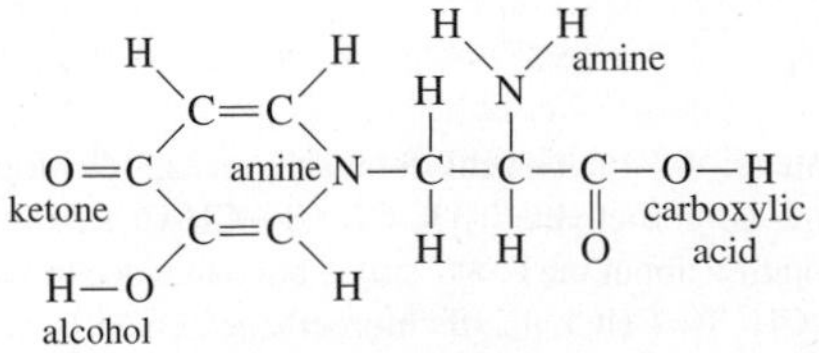

b. 5 carbons in ring and the carbon in —CO_2H: sp^2; the other two carbons: sp^3; c. 24 sigma bonds, 4 pi bonds   51. a. 3-chloro-1-butanol, primary; b. 3-methyl-3-hexanol, tertiary; c. 2-methylcyclopentanol, secondary   53. 1-pentanol; 2-pentanol; 3-pentanol; 2-methyl-1-butanol; 2-methyl-2-butanol; 3-methyl-2-butanol; 3-methyl-1-butanol; 2,2-dimethyl-1-propanol; 6 ethers   55. a. 4,5-dichloro-3-hexanone; b. 2,3-dimethylpentanal; c. 3-methylbenzaldehyde or *m*-methylbenzaldehyde   57. a. 4-chlorobenzoic

acid or *p*-chlorobenzoic acid; b. 3-ethyl-2-methylhexanoic acid; c. methanoic acid (common name = formic acid)   59. Only statement d is false.

2-butenol:

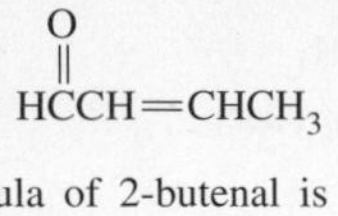

The formula of 2-butenal is C_4H_6O, while the alcohol has a formula of C_4H_8O.

61. a. $CH_3CH_2CH_2CH_3$

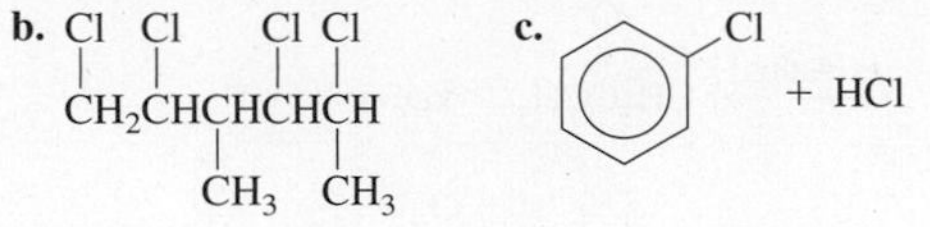

b.   c. + HCl

d. $C_4H_8(g) + 6O_2(g) \longrightarrow 4CO_2(g) + 4H_2O(g)$

63. For the iron-catalyzed reaction, one of the *ortho* or *para* hydrogens in benzene is replaced by chlorine. When an iron catalyst is not present, the benzene hydrogens are unreactive. To substitute for an alkane hydrogen, light must be present. For toluene, the light-catalyzed reaction substitutes a chlorine for a hydrogen in the methyl group attached to the benzene ring.

65. a.

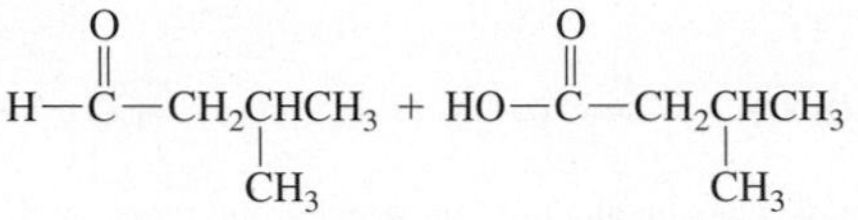

b.

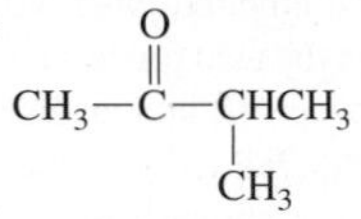

c. No reaction

d.

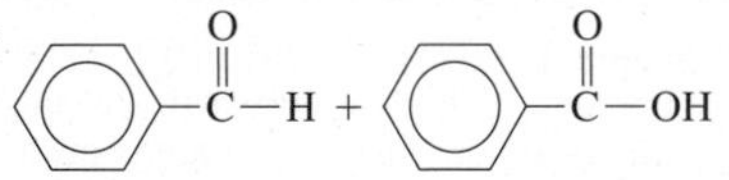

e.

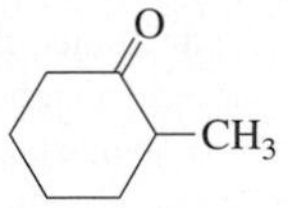

f.

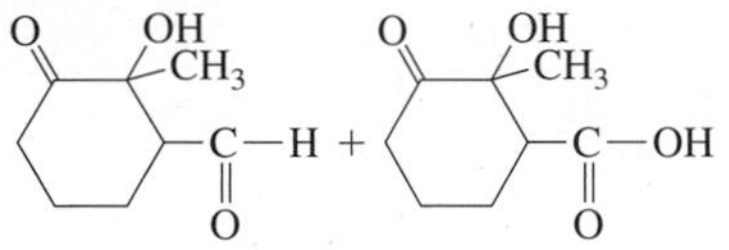

67. a. $CH_3CH=CH_2 + Br_2 \rightarrow CH_3CHBrCH_2Br$;

b.

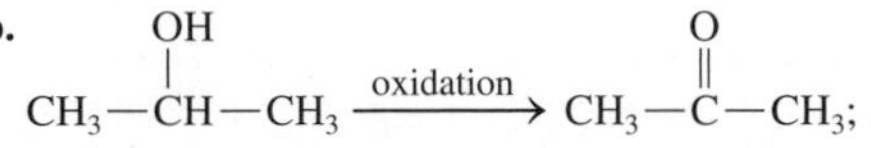

c.

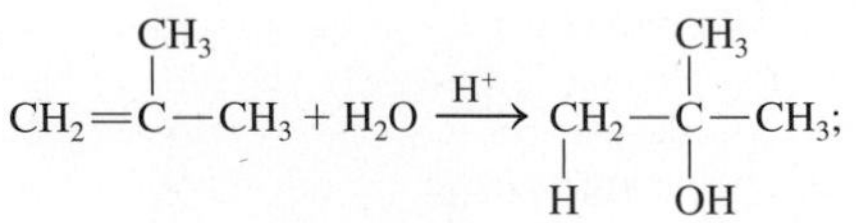

d.

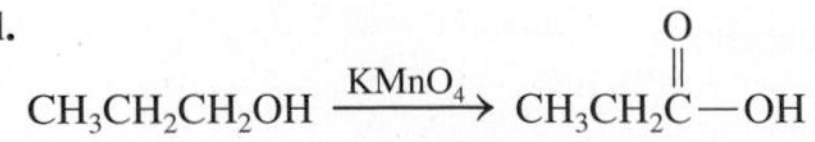

69. a.

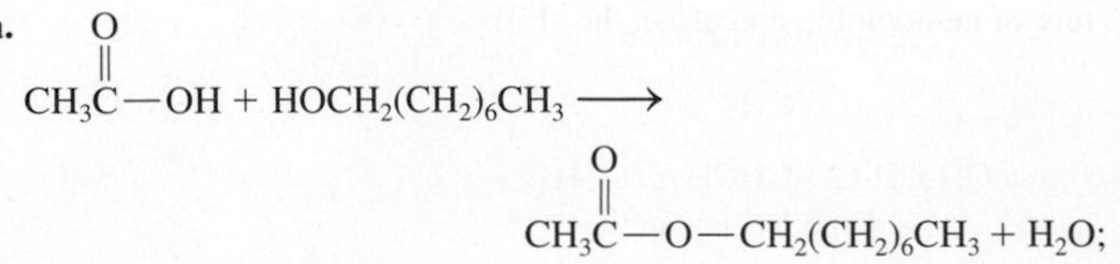

b.

$$CH_3CH_2\overset{O}{\overset{\|}{C}}-OH + HOCH_2(CH_2)_4CH_3 \longrightarrow$$

$$CH_3CH_2\overset{O}{\overset{\|}{C}}-O-CH_2(CH_2)_4CH_3 + H_2O$$

71. $CFCl{=}CF_2$

73.

75.

77.

79. Divinylbenzene has two reactive double bonds that are used during formation of the polymer. The key is for the double bonds to insert themselves into two different polymer chains. When this occurs, the two chains are bonded together (are crosslinked). The chains cannot move past each other because of the crosslinks making the polymer more rigid. **81. a.** The polymer from 1,2-diaminoethane and terephthalic acid is stronger because of the possibility of hydrogen bonding between chains. **b.** The polymer of

is more rigid because the chains are stiffer due to the rigid benzene rings in the chains. **c.** Polyacetylene is $nHC{\equiv}CH \rightarrow -(CH{=}CH-)_n$. Polyacetylene is more rigid because the double bonds in the chains make the chains stiffer. **83. a.** serine; tyrosine; threonine; **b.** aspartic acid; glutamic acid; **c.** histidine; lysine; arginine; tryptophan; **d.** glutamine; asparagine **85. a.** aspartic acid and phenylalanine; **b.** Aspartame contains the methyl ester of phenylalanine. This ester can hydrolyze to form methanol, $RCO_2CH_3 + H_2O \rightleftharpoons RCO_2H + CH_3OH$.

87.

ser–ala   ala–ser

89. a. Six tetrapeptides are possible. From NH_2 to CO_2H end: phe–phe–gly–gly, gly–gly–phe–phe, gly–phe–phe–gly, phe–gly–gly–phe, phe–gly–phe–gly, gly–phe–gly–phe; **b.** Twelve tetrapeptides are possible. From NH_2 to CO_2H end: phe–phe–gly–ala, phe–phe–ala–gly, phe–gly–phe–ala, phe–gly–ala–phe, phe–ala–phe–gly, phe–ala–gly–phe, gly–phe–phe–ala, gly–phe–ala–phe, gly–ala–phe–phe, ala–phe–phe–gly, ala–phe–gly–phe, ala–gly–phe–phe **91.** Ionic: his, lys, or arg with asp or glu; hydrogen bonding: ser, glu, tyr, his, arg, asn, thr, asp, gln, or lys with any amino acid; covalent: cys with cys; London dispersion: all amino acids with nonpolar R groups (gly, ala, pro, phe, ile, trp, met, leu, val); dipole–dipole: tyr, thr, and ser with each other **93.** Glutamic acid has a polar R group and valine has a nonpolar R group. The change in polarity of the R groups could affect the tertiary structure of hemoglobin and affect the ability of hemoglobin to bond to oxygen.

95.

D-Ribose   D-Mannose

97. aldohexose: glucose, mannose, galactose; aldopentose: ribose, arabinose; ketohexose: fructose; ketopentose; ribulose **99.** They differ in the orientation of a hydroxy group on a particular carbon. Starch is composed from α-D-glucose, and cellulose is composed of β-D-glucose.
101. The chiral carbons are marked with asterisks.

Isoleucine   Threonine

103.

is optically active. The chiral carbon is marked with an asterisk.
105. C–C–A–G–A–T–A–T–G **107.** Uracil will H-bond to adenine.

Uracil   Adenine

109. a. glu: CTT, CTC; val: CAA, CAG, CAT, CAC; met: TAC; trp: ACC; phe: AAA, AAG; asp: CTA, CTG; **b.** ACC–CTT–AAA–TAC or ACC–CTC–AAA–TAC or ACC–CTT–AAG–TAC or ACC–CTC–AAG–TAC; **c.** four (see answer to part b); **d.** met–asp–phe; **e.** TAC–CTA–AAG; TAC–CTA–AAA; TAC–CTG–AAA **111. a.** 2,3,5,6-tetramethyloctane; **b.** 2,2,3,5-tetramethylheptane; **c.** 2,3,4-trimethylhexane; **d.** 3-methyl-1-pentyne

113.

There are many possibilities for isomers. Any structure with four chlorines replacing four hydrogens in any of the numbered positions would be an isomer; i.e., 1,2,3,4-tetrachloro-dibenzo-p-dioxin is a possible isomer.
115. $-23°C$: $CH_3{-}O{-}CH_3$; $78.5°C$: $CH_3{-}CH_2{-}OH$ **117.** Alcohols consist of two parts, the polar OH group and the nonpolar hydrocarbon chain attached to the OH group. As the length of the nonpolar hydrocarbon chain increases, the solubility of the alcohol decreases. In methyl alcohol (methanol), the polar OH group can override the effect of the nonpolar CH_3 group, and methyl alcohol is soluble in water. In stearyl alcohol, the molecule consists mostly of the long nonpolar hydrocarbon chain, so it is insoluble in water. **119.** n-hexane, 69°C; pentanal, 103°C; 1-pentanol, 137°C; butanoic acid, 164°C. **121.** 2-propanone; propanoic acid **123.** In nylon, hydrogen-bonding interactions occur due to the presence of N—H bonds in the polymer. For a given polymer chain length, there are more N—H groups in nylon-46 as compared to nylon-6. Hence, nylon-46 forms a stronger polymer compared to nylon-6 due to the increased hydrogen-bonding interactions.

125. a.

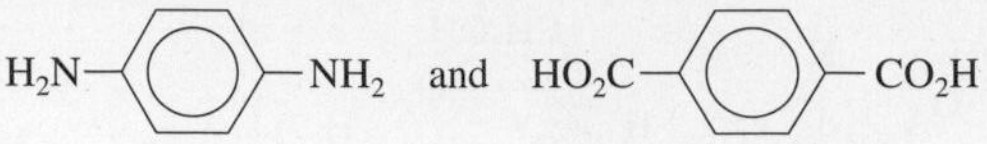

H_2N—⟨ ⟩—NH_2 and HO_2C—⟨ ⟩—CO_2H

b. Repeating unit:

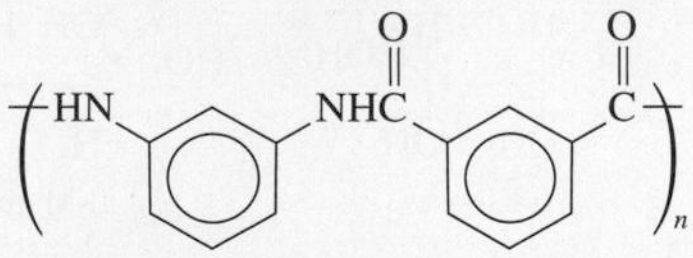

The two polymers differ in the substitution pattern on the benzene rings. The Kevlar chain is straighter, and there is more efficient hydrogen bonding between Kevlar chains than between Nomex chains.   **127. a.** The bond angles in the ring are about 60°. VSEPR predicts bond angles close to 109°. The bonding electrons are much closer together than they prefer, resulting in strong electron–electron repulsions. Thus ethylene oxide is unstable (reactive). **b.** The ring opens up during polymerization and the monomers link together through the formation of O—C bonds.

$+O-CH_2CH_2-O-CH_2CH_2-O-CH_2CH_2+_n$

129. 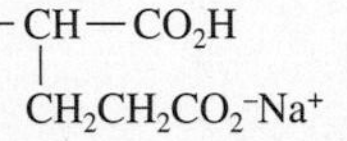 $H_2N-CH-CO_2H$ or $H_2N-CH-CO_2^-Na^+$
        | |
     $CH_2CH_2CO_2^-Na^+$ $CH_2CH_2CO_2H$

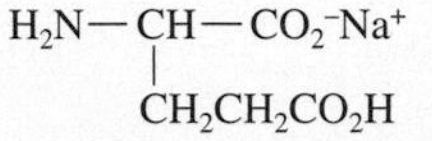

The first structure is MSG, which is impossible for you to predict.   **131.** $\Delta G = \Delta H - T\Delta S$; for the reaction, we break a P—O and O—H bond and form a P—O and O—H bond, so $\Delta H \approx 0$ based on bond dissociation energies. ΔS for this process is negative (unfavorable) because positional probability decreases. Thus, $\Delta G > 0$ due to the unfavorable ΔS term, and the reaction is not expected to be spontaneous.

133.

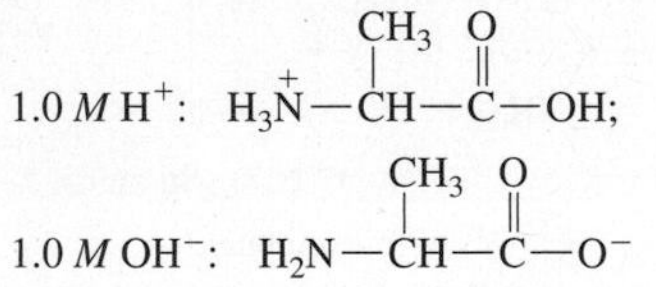

1.0 M H$^+$: $H_3\overset{+}{N}-CH-\overset{O}{\overset{\|}{C}}-OH$; with CH_3 on CH

1.0 M OH$^-$: $H_2N-CH-\overset{O}{\overset{\|}{C}}-O^-$ with CH_3 on CH

135. Both ΔH and ΔS are positive values.   **137. a.** pentane or *n*-pentane; **b.** 3-ethyl-2,5-dimethylhexane; **c.** 4-ethyl-5-isopropyloctane   **139. a.** iodocycloproane; **b.** 2-chloro-1,3,5-trimethylcyclohexane; **c.** 1-chloro-2-ethylcyclopentane; **d.** 1-ethyl-2-methylcyclopentene   **141. a.** 1-pentanol, 2-methyl-2,4-pentadiol; **b.** *trans*-1,2-cyclohexadiol, 4-methyl-1-penten-3-ol (does not exhibit *cis–trans* isomerism)   **143.** 2-chloropropanoic acid   **145.** 6.07   **147. a.** No; the mirror image is superimposable.

b.

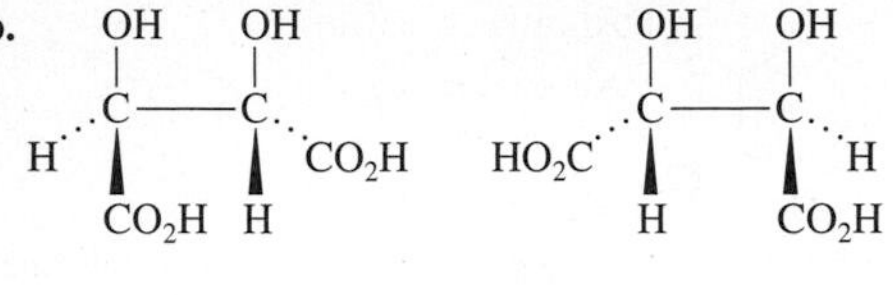

149.

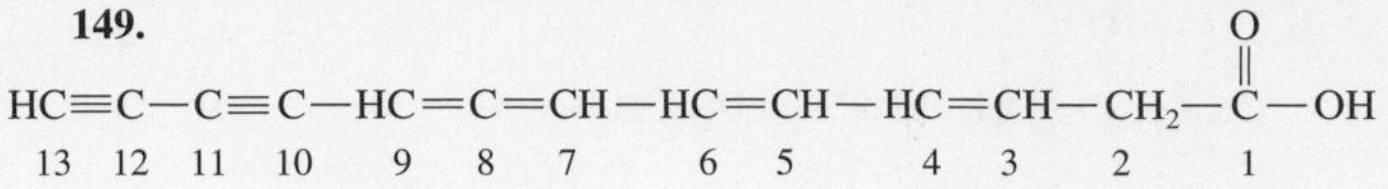

$HC{\equiv}C-C{\equiv}C-HC{=}C{=}CH-HC{=}CH-HC{=}CH-CH_2-\overset{O}{\overset{\|}{C}}-OH$

13 12 11 10   9   8   7   6   5   4   3   2   1

151. a. 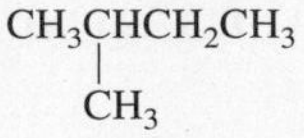 $CH_3CHCH_2CH_3$
              |
             CH_3

b. 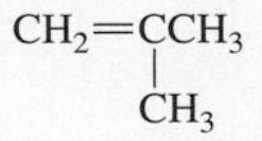 $CH_2{=}CCH_3$
              |
             CH_3

c.

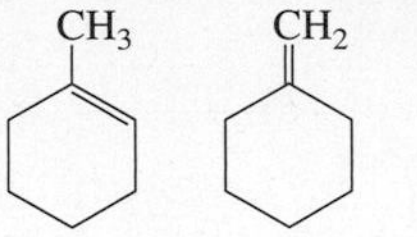

d. $CH_2{=}CHCH_3$

e. 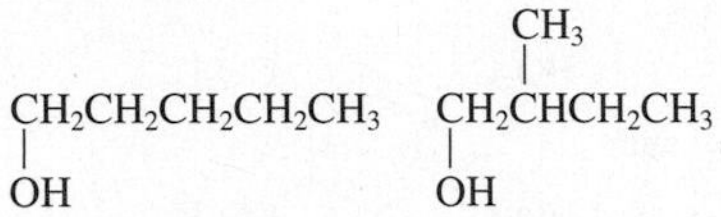 $CH_2CH_2CH_2CH_2CH_3$ $CH_3CHCH_2CH_3$ (with CH_3 and OH)
         |
        OH

$CH_3CHCH_2CH_2$ (with OH) CH_2-C-CH_3 (with OH and CH_3)

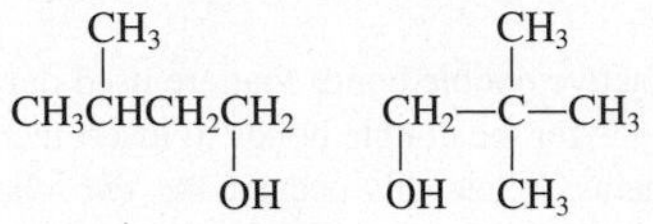

153.

$+OCH_2CH_2OCNH$—⟨ ⟩—$NHCOCH_2CH_2OCNH$—⟨ ⟩—$NHC+_n$ (with C=O groups)

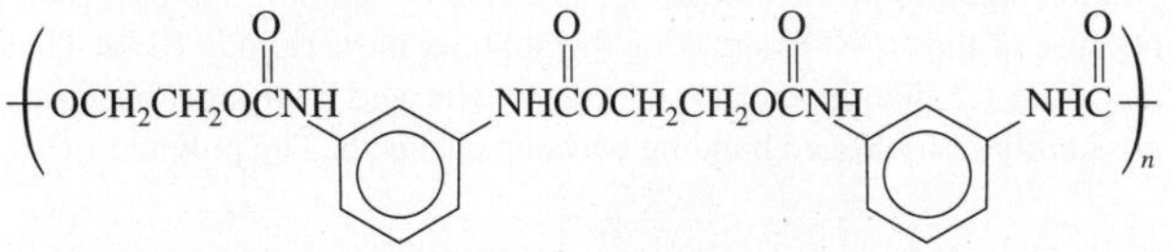

155. a. The temperature of the rubber band increases when it is stretched; **b.** exothermic (heat is released); **c.** As the chains are stretched, they line up more closely resulting in stronger London dispersion forces between the chains. Heat is released as the strength of the intermolecular forces increases. **d.** ΔG is positive and ΔS is negative; **e.** The structure of the stretched polymer chains is more ordered (has a smaller positional probability) than in unstretched rubber. Therefore, entropy decreases as the rubber band is stretched.   **157.** 0.11%   **159. a.** 37.50%; **b.** The hybridization changes from sp^2 to sp^3; **c.** 3,4-dimethyl-3-hexanol

Index